ORGANIC SYNTHESIS

· · · · · · · · · · · ·

THIRD EDITION

MICHAEL B. SMITH
UNIVERSITY OF CONNECTICUT

Wavefunction, Inc.

ORGANIC SYNTHESIS, Third EDITION

Some ancillaries, including electronic and print components, may not be available to customers outside the United States.

This book is printed on acid-free paper.

1 2 3 4 5 6 7 8 9 0 KGP/KGP 0 9 8 7 6 5 4 3 2 1

ISBN-13: 978-1-890661-40-3
ISBN-10: 1-890661-40-6

Lead production supervisor: Pamela Ohsan

Cover designer: Pamela Ohsan

Printer: R R Donnelley

Cover image: The photo on the front cover shows the Atomium in Brussels, Belgium. The photo was taken by Michael Smith.

Library of Congress Cataloging-in-Publication Data

Smith, Michael B., 1946 Oct. 17

Organic synthesis / Michael B. Smith. — 3rd ed.
    p.   cm.

Includes bibliographical references and index.

    ISBN 978-1-890661-40-3 (acid-free paper)
    1. Organic compounds—Synthesis. I. Title.

About the Author

· ·

PROFESSOR MICHAEL B. SMITH was born in Detroit, Michigan in 1946 and moved to Madison Heights, Virginia in 1957, where he attended high school. He received an A.A. from Ferrum College in 1967 and a B.S. in chemistry from Virginia Polytechnic Institute in 1969. After working for three years at the Newport News Shipbuilding and Dry Dock Co. in Newport News VA as an analytical chemist, he entered graduate school at Purdue University. He received a Ph.D. in Organic chemistry in 1977, under the auspices of Professor Joe Wolinsky. Professor Smith spent one year as a faculty research associate at the Arizona State University with Professor G. Robert Pettit, working on the isolation of cytotoxic principles from plants and sponges. He spent a second year of postdoctoral work with Professor Sidney M. Hecht at the Massachusetts Institute of Technology, working on the synthesis of bleomycin A2. Professor Smith moved to the University of Connecticut in 1979, where he is currently professor of chemistry. In 1986 he spent a sabbatical leave in the laboratories of Professor Leon Ghosez, at the Université Catholique de Louvain in Louvain-la-Neuve, Belgium, as a visiting professor.

Professor Smith's research interests focus on (1) the synthesis and structure-proof of bioactive lipids obtained from the human dental pathogen *Porphyromonas gingivalis*. The initial work will be extended to study the structure-biological activity relationships of these compounds. (2) A systematic study of the influence of conductivity-enhancing dopants on conducting polymers, and a green-chemistry study of chemical reactions mediated by conducting polymers in a neutral, two-phase system. (3) The development of chemical probes that targets and allows fluorescent imaging of cancerous hypoxic tumors.

In addition to this research, Professor Smith is the author of the 5th and 6th editions of *March's Advanced Organic Chemistry* and editor of the *Compendium of Organic Synthetic Methods*, Volumes 6-12. He is the author of *Organic Chemistry: Two Semesters*, in it's second edition, which is an outline of undergraduate organic chemistry to be used as a study guide for the first organic course. He has authored a research monograph entitled *Synthesis of Non-alpha Amino Acids*. He is also the author of an undergraduate textbook in organic chemistry entitled *Organic Chemistry. An Acid-Base Approach.*

Table of Contents

Chapter 1   Retrosynthesis, Stereochemistry, and Conformations   1

   1.1.   Introduction   0001
   1.2.   The disconnection protocol   0005
   1.3.   Bond proximity and implications for chemical reactions   0011
   1.4.   Stereochemistry   0012
   1.5.   Conformations   0035
   1.6.   Conclusion   0071
   Homework   0072

Chapter 2   Acids, Bases and Functional Group Exchange Reactions   77

   2.1.   Introduction   0077
   2.2.   Brønsted-Lowry acids and bases   0081
   2.3.   Lewis acids   0093
   2.4   Hard-Soft acid-base theory   0096
   2.5.   Acyl addition, substitution and conjugate addition   0106
   2.6.   Substitution reactions   0112
   2.7.   Characteristics of reactions involving nucleophiles   0137
   2.8.   Substitution by halogen   0144
   2.9.   Elimination reactions   0153
   2.10. Addition reactions   0177
   2.11. Functional group manipulation by rearrangement   0188
   2.12. Aromatic substitution   0193
   2.13. Conclusion   0210
   Homework   0211

Chapter 3   Oxidation   219

   3.1.   Introduction   0219
   3.2.   Alcohols to carbonyls (CH–OH $\rightarrow$ C=O)   0226
   3.3.   Formation of phenols and quinones   0261
   3.4.   Conversion of alkenes to epoxides   0268
   3.5.   Conversion of alkenes to diols (C=C $\rightarrow$ CHOH–CHOH)   0291
   3.6.   Baeyer-Villiger oxidation RCOR' $\rightarrow$ RCO_2R')   0307
   3.7.   Oxidative bond cleavage (C=C $\rightarrow$ C=O + O=C)   0313
   3.8.   Oxidation of alkyl or alkenyl fragments (CH $\rightarrow$ C=O or C–OH)   0324
   3.9.   Oxidation of sulfur, selenium, and nitrogen   0334
   3.10. Conclusion   0340
   Homework   0340

Chapter 4   Reduction   **347**

    4.1.   Introduction   0347
    4.2.   Reduction with complex metal hydrides   0348
    4.3.   Alkoxyaluminate reagents   0363
    4.4.   Reductions with borohydride   0369
    4.5.   Alkoxy- and alkylborohydrides   0377
    4.6.   Borane, aluminum hydride, and derivatives   0385
    4.7.   Stereoselectivity in reductions   0391
    4.8.   Catalytic hydrogenation   0422
    4.9.   Dissolving metal reductions   0451
    4.10. Nonmetallic reducing agents   0472
    4.11. Conclusion   0483
    Homework   0484

Chapter 5   Hydroboration   **491**

    5.1.   Introduction   0491
    5.2.   Preparation of alkyl and alkenyl boranes   0491
    5.3.   Synthetic transformations   0505
    5.4.   Formation of oxygen-containing functional groups 0514
    5.5.   Amines and sulfides via hydroboration   0534
    5.6.   Conclusion   0537
    Homework   0537

Chapter 6   Stereocontrol and Ring Formation   **541**

    6.1.   Introduction   0541
    6.2.   Stereocontrol in acyclic systems   0541
    6.3.   Stereocontrol in cyclic systems   0551
    6.4.   Neighboring group effects and chelation effects   0560
    6.5.   Acyclic stereocontrol via cyclic precursors   0563
    6.6.   Ring-forming reactions   0564
    6.7.   Conclusion   0582
    Homework   0582

Chapter 7   Protecting Groups   **587**

    7.1.   Introduction   0587
    7.2.   When are protecting groups needed?   0589
    7.3.   Protecting groups for alcohols, carbonyls, and amines   0593
    7.4.   Conclusion   0619
    Homework   0620

Chapter 8   C^d Disconnect Products: Nucleophilic Species That Form Carbon-Carbon Bonds   **623**

8.1. Introduction   0623
8.2. Cyanide   0623
8.3. Alkyne anions (R–C≡C:⁻)   0629
8.4   Grignard reagents (C–Mg)   0636
8.5. Organolithium reagents (C–Li)   0670
8.6. Sulfur stabilized carbanions and umpolung   0696
8.7   Organocopper reagents (C–Cu)   0713
8.8. Ylids   0729
8.9. Other organometallic carbanionic compounds   0758
8.10. Allylic tin, alkyltitanium, and allylic silane complexes   0766
8.11. Phenolic carbanions   0771
8.12. Conclusion   0773
Homework   0774

Chapter 9   C^d Disconnect Products: Nucleophilic Species That Form Carbon-Carbon Bonds: Enolate Anions   **781**

9.1. Introduction   0781
9.2. Formation of enolate anions   0781
9.3. Reactions of enolate anions with electrophiles   0803
9.4. Enolate condensation reactions   0816
9.5. Stereoselective enolate reactions   0850
9.6   Enamines   0873
9.7. Michael addition and related reactions   0877
9.8. Enolate reactions of α-halo carbonyl derivatives   0883
9.9. Conclusion   0888
Homework   0889

Chapter 10   Synthetic Strategies   **897**

10.1. Introduction   0897
10.2. Target selection   0898
10.3. Retrosynthesis   0907
10.4. Synthetic strategies   0914
10.5   The strategic bond approach   0920
10.6. Strategic bonds in rings   0934
10.7. Selected synthetic strategies: pancratistatin   0942
10.8. Biomimetic approach to retrosynthesis   0948
10.9. The chiral template approach   0954
10.10. Computer generated strategies   0961
10.11. Degradation techniques as a tool for retrosynthesis   0973
10.12. Combinatorial chemistry   0978
10.13. Conclusion   0988
Homework   0989

Chapter 11 Pericyclic Carbon-Carbon Bond Forming Reactions: Multiple Bond Disconnections   999

    11.1. Introduction   0999
    11.2. Frontier molecular orbital theory   0999
    11.3. Allowed and forbidden reactions   1010
    11.4. [4 + 2]-cycloadditions   1013
    11.5. Inverse electron demand and the retro Diels-Alder   1032
    11.6. Rate enhancement in Diels-Alder reactions   1037
    11.7. Heteroatom Diels-Alder reactions   1049
    11.8. Intramolecular Diels-Alder reactions   1059
    11.9. Enantioselective Diels-Alder reactions   1066
    11.10. [2+2]-Cycloaddition reactions   1076
    11.11. Electrocyclic reactions   1097
    11.12. [3+2]-Cycloaddition reactions   1101
    11.13. Sigmatropic rearrangements   1116
    11.14. The ene reaction   1143
    11.15. Conclusion   1152
    Homework   1153

Chapter 12 Cᵃ Disconnect Products: Electrophilic Carbon-Carbon Bond Forming Reactions   1161

    12.1. Introduction   1161
    12.2. Carbocations   1161
    12.3. Carbon-carbon bond forming reactions of carbocations   1182
    12.4. Friedel-Crafts reactions   1192
    12.5. Friedel-Crafts reactions: formation of heteroatom-containing derivatives   1210
    12.6. π-Allyl palladium complexes   1225
    12.7. Named palladium coupling reactions   1236
    12.8. π-Allyl nickel complexes   1245
    12.9. Electrophilic iron complexes   1248
    12.10. Conclusion   1250
    Homework   1251

Chapter 13 Carbon Radical Disconnect Products: Formation of Carbon-Carbon Bonds Via Radicals and Carbenes   1257

    13.1. Introduction   1257
    13.2. Structure of radicals   1258
    13.3. Formation of radicals by thermolysis   1259
    13.4. Photochemical formation of radicals   1262
    13.5. Reactions of free radicals   1267
    13.6. Intermolecular radical reactions   1278
    13.7. Intramolecular radical reactions (radical cyclization)   1283
    13.8. Metal-induced radical reactions   1296
    13.9. Carbenes and carbenoids   1311
    13.10. Metathesis reactions   1334
    13.11. Pauson-Khand reaction   1342

      13.12. Conclusion   1345

      Homework   1346

Chapter 14. Student Syntheses: The First Synthetic Problem   **1353**

      14.1. Introduction   1353

      14.2. Total synthesis of securamine C   1355

      14.3. Total synthesis of variecolol   1364

Disconnection Index   **1377**

General Index   **1383**

Preface to the 3rd edition

The new edition of *Organic Synthesis* has been revised and rewritten from front to back. I want to thank all who used the book in its first and second editions. The book has been out of print for several years, but the collaboration of Warren Hehre and Wavefunction, Inc. made the third edition possible. It is the same graduate level textbook past users are familiar with, with two major exceptions. First, the book has been revised and updated. Second, molecular modeling problems are included in a manner that is not obtrusive to the theme of understanding reactions and synthesis. More than 60 molecular modeling problems are incorporated into various discussions, spread throughout 11 of the 13 reactions-synthesis oriented chapters. These are intended for the **SpartanModel** program, a copy of which is included in the purchase price of this book. **SpartanModel** will allow the reader to manipulate each model and, in most cases, change or create model compounds of interest to the reader. It is our belief that the selected molecular modeling problems will offer new insights into certain aspects of chemical reactivity, conformational analysis, and stereoselectivity.

Updated examples are used throughout the new edition when possible, and new material is added that make this edition reflect current synthetic methodology. The text has been modified in countless places to improve readability and pedagogy. This new edition contains references taken from more than 6100 journal articles, books, and monographs. Of these references, more than 950 are new to this edition, all taken from the literature after 2002. More than 600 updated or new reactions have been added. There are several entirely new sections that discuss topics missing from the 2nd edition. These include S_N2 type reactions with epoxides; the Burgess Reagent; functional group rearrangements (Beckmann, Schmidt, Curtius, Hofmann, Lossen); Oxidation of allylic carbon with ruthenium compounds; A comparison of LUMO-mapping with the Cram model and Felkin-Anh models in chapter 4; Electrocyclic reactions; [2.3]-sigmatropic rearrangement (Wittig rearrangement); consolidation of C–C bond forming reactions of carbocations and nucleophiles into a new section.

Homework in each chapter has been extensively revised. There are more than 800 homework problems, and more than 300 of the homework problems are new. Most of the homework problems do not contain leading references for the answers. The answers to all problems from chapters 1-9, and 11-13 are available in an on-line *Student Solutions Manual* for this book. As in previous edition, a few leading references are provided for the synthesis problems in chapter 10. Although answers are given for homework that relates to all other chapters, in chapter 10 most problems do not have answers. The student is encouraged to discuss any synthetic problem with their instructor.

With the exception of scanned figures, all drawings in this book were prepared using ChemDraw, provided by CambridgeSoft, and all 3D graphics are rendered with Spartan, provided by Wavefunction, Inc. I thank both organizations for providing the software that made this project possible.

I express my gratitude to all of those who were kind enough to go through the first and second editions and supply me with comments, corrections, and suggestions.

For this new edition, special thanks and gratitude are given to Warren Hehre. Not only did he design the molecular modeling problems, but provided the solutions to the problems and the accompanying software. Warren also helped me think about certain aspects of organic synthesis in a different way because of the modeling, and I believe this has greatly improved the book and the approaches presented in the book. Special thanks are also given to Ms. Pamela Ohsan, who converted the entire book into publishable form. Without her extraordinary efforts, this third edition would not be possible.

Finally, I thank my students, who have provided the inspiration over the years for this book. They have also been my best sounding board, allowing me to test new ideas and organize the text as it now appears. I thank my friends and colleagues who have provided countless suggestions and encouragement over the years, particularly Spencer Knapp (Rutgers), George Majetich (Georgia), Frederick Luzzio (Louisville), and Phil Garner (Washington State). You have all helped more than you can possibly know, and I am most grateful.

A special thanks to my wife Sarah and son Steven, whose patience and understanding made the work possible.

If there are errors, corrections, and suggestions, please let me know by Email or normal post. Any errors will be posted at http://books.wavefun.com/organicsysnthesis3rd.

Thank you again. I hope this new edition is useful to you in your studies.

Michael B. Smith
Storrs, Connecticut
April, 2010

University of Connecticut
Department of Chemistry
55 N. Eagleville Road
Storrs, Connecticut 06269-3060

Email: michael.smith@uconn.edu
Fax: (860) 486-2981
Homepage: http://orgchem.chem.uconn.edu/home/mbs-home.html

Preface to the 1st edition. Why I wrote this book!

A reactions oriented course is a staple of most graduate organic programs, and synthesis is taught either as a part of that course or as a special topic. Ideally, the incoming student is an organic major, who has a good working knowledge of basic reactions, stereochemistry and conformational principles. In fact, however, many (often most) of the students in a first year graduate level organic course have deficiencies in their undergraduate work, are not organic majors and are not synthetically inclined. Does one simply tell the student to "go away and read about it," giving a list of references, or does one take class time to fill in the deficits? The first option works well for highly motivated students with a good background, less well for those with a modest background. In many cases, the students spend so much time catching up that it is difficult to focus them on the cutting edge material we all want to teach. If one exercises the second option of filling in all the deficits, one never gets to the cutting edge material. This is especially punishing to the outstanding students and to the organic majors. a compromise would provide the student with a reliable and readily available source for background material that could be used as needed. The instructor could then feel comfortable that the proper foundations have been laid and push on to more interesting areas of organic chemistry.

Unfortunately, such a source of background material either is lacking altogether or consists of several books and dozens of review articles. I believe my teaching experience at UConn as just described is rather typical, with a mix of non-organic majors, outstanding and well-motivated students, and many students with weak backgrounds who have the potential to go on to useful and productive careers if time is taken to help them. Over the years I have assigned what books were available in an attempt to address these problems, but found that "graduate level textbooks" left much to be desired. I assembled a large reading list and mountains of handouts and spent half of my life making up problems that would give my students a reasonable chance at practicing the principles we were discussing. I came to the conclusion that a single textbook was needed that would give me the flexibility I craved to present the course I wanted to teach, but yet would give the students the background they needed to succeed. As I tried different things in the classroom, I solicited the opinions of the graduate students who took the course and tried to develop an approach that worked for them and allowed me to present the information I wanted. The result is this book. I hope that it is readable, provides background information, and also provides the research oriented information that is important for graduate organic students. I also hope it will be of benefit to instructors who face the same challenges I do. I hope this book will be a useful tool to the synthetic community and to graduate level education.

From talks with many people I know that courses for which this book is targeted can be for either one or two semesters. The course can focus only on functional groups, only on making carbon-carbon bonds, on some combination of both (like my course), or only on synthesis.

I have tried to organize the book in such a way that one is not a slave to its organization. Every chapter is internally cross-referenced. If the course is to focus upon making carbon-carbon bonds, for example, there are unavoidable references to oxidation reagents, reducing agents, stereochemical principles, etc. When such a reaction or principle appears, the section and chapter where it is discussed elsewhere in the book is given "in line" so the student can easily find it. It is impossible to write each chapter so it will stand alone, but the chapters are reasonably independent in their presentations. I have organized the book so that functional groups are discussed in the first few chapters and carbon-carbon bond formation reactions are discussed in later chapters, making it easier to use the one book for two different courses or for a combination course. The middle chapters are used for review and to help the student make the transitions from functional group manipulations to applying reactions and principles and thence to actually building molecules. I believe that a course devoted to making carbon-carbon bonds could begin with chapter 8, knowing that all pertinent peripheral material is in the book and readily available to the student. The ultimate goal of the book is to cut down on the mountains of handouts, provide homework to give the student proper practice, give many literature citations to tell the student exactly where to find more information, and allow the instructor to devote time to their particular focus.

This book obviously encompasses a wide range of organic chemistry. Is there a theme? Should there be? The beautiful and elegant total syntheses of interesting and important molecules published by synthetic organic chemists inspired me to become an organic chemist, and I believe that synthesis focuses attention on the problems of organic chemistry in a unique way. To solve a synthetic problem, all elements of organic chemistry must be brought to bear: reactions, mechanism, stereochemistry, conformational control, and strategy. Synthesis therefore brings a perspective on all aspects of organic chemistry and provides a theme for understanding it. The theme of this book is therefore the presentation of reactions in the context of organic synthesis. Wherever possible, examples of a given reaction, process, or strategy are taken from a published total synthesis. The disconnection approach is presented in the first chapter, and as each new functional group transform and carbon-carbon bond forming reaction is discussed, the retrosynthetic analysis (the disconnect products for that reaction) is given. An entire chapter (chapter 10) is devoted to synthetic strategies, and chapter 14 provides examples of first year students' first syntheses. I believe that this theme is a reasonable and useful device for presenting advanced organic chemistry.

The text is fully referenced to facilitate further study, and (where feasible) the principal researcher who did the work is mentioned by name, so the student can follow that person's work in the literature and gain even more insight into a given area. As far as it is known to me, the pioneering work of the great chemists of the past has been referenced. Many of the "named reactions" are no longer referenced in journals, but when they are first mentioned in this book, the original references are given. I believe the early work should not be lost to a new generation of students.

In many cases I have used 3-D drawings to help illustrate stereochemical arguments for a given process. I give the structure of each reagent cited in the text, where that reagent is mentioned, so a beginning student does not have to stop and figure it out. This is probably unnecessary for many students, but it is there if needed.

This is a reaction oriented book, but an attempt is made to give brief mechanistic discussions when appropriate. In addition, some physical organic chemistry is included to try to answer the obvious if unasked questions: why does that alkyl group move, why does that bond break, why is that steric interaction greater than the other one, or why is that reaction diastereoselective?

Most of all, a student needs to practice. Chapters 1-13 have end-of chapter problems that range from those requiring simple answers based on statements within the text to complex problems taken from research literature. In a large number of cases literature citations are provided so answers can be found.

The first part of the book (chapters 1-4) is a review of functional group transforms and basic principles: retrosynthesis, stereochemistry, and conformations. Basic organic reactions are covered, including substitution reactions, addition reactions, elimination reactions, acid/base chemistry, oxidation and reduction. The first two chapters are very loosely organized along the lines of an undergraduate book for presenting the functional group reactions (basic principles, substitution, elimination, addition, acyl addition, aromatic chemistry). Chapter 1 begins with the disconnection approach. I have found that this focuses the students' attention on which reactions they can actually apply and instantly shows them why it is important to have a larger arsenal of reactions to solve a synthetic problem. This has been better than any other device I have tried and that is why it is placed first. Most of the students I see come into our program deficient in their understanding of stereochemistry and conformational control, and so those topics are presented next. Some of this information is remedial material and where unneeded can be skipped, but it is there for those who need it (even if they will not admit that they do). Chapter 2 presents a mini-review of undergraduate organic chemistry reactions and also introduces some modern reactions and applications. Chapter 3 is on oxidation and chapter 4 is on reduction. Each chapter covers areas that are woefully under-emphasized in undergraduate textbooks.

Chapter 5 covers hydroboration, an area that is discussed in several books and reviews. I thought it useful to combine this material into a tightly focused presentation which (1) introduces several novel functional group transforms that appear nowhere else and (2) gives a useful review of many topics introduced in chapters 1-4. Chapter 6 reviews the basic principles that chemists use to control a reaction rather than be controlled by it. It shows the techniques chemists use to "fix" the stereochemistry, if possible, when the reaction does not do what it is supposed to. It shows how stereochemical principles guide a synthesis. An alternative would be to separate stereochemistry into a chapter that discusses all stereochemical principles. However, the theme is synthesis, and stereochemical considerations are as important a part of a synthesis as the reagents being chosen. For that reason, stereochemistry is presented with the reactions in each chapter. Chapter 6 simply ties together the basic principles. This chapter also includes the basics of ring-forming reactions. Chapter 7 completes the first part of the book and gives a brief overview of what protecting groups are and when to use them.

The last half of the book focuses on making carbon-carbon bonds. It is organized fundamentally by the disconnection approach. In Chapter 1, breaking a carbon-carbon bond generated a disconnect product that was labeled as C^d (a nucleophilic species), C^a (an electrophilic species), or $C^{radical}$ (a radical intermediate). In some cases, multiple bonds were disconnected, and many of these disconnections involved pericyclic reactions to reassemble the target. the nucleophilic regents that are equivalent to C^d disconnect products are covered in Chapters 8 and 9, with the very important enolate anion chemistry separated into Chapter 9. Chapter 10 presents various synthetic strategies that a student may apply to a given synthetic problem. This information needs to be introduced as soon as possible, but until the student "knows some chemistry", it cannot really be applied. Placement of synthetic strategies after functional group transforms and nucleophilic methods for making carbon-carbon bonds is a reasonable compromise. Chapter 11 introduces the important Diels-Alder cyclization, as well as dipolar cycloadditions and sigmatropic rearrangements that are critically important to synthesis. Chapter 12 explores electrophilic carbons (C^a), including organometallics that generally react with nucleophilic species. Chapter 13 introduces radical and carbene chemistry. Chapter 14 is included to give the student a taste of a first time student proposal and some of the common mistakes. The point is not to reiterate the chemistry but to show how strategic shortsightedness, poor drawings, and deficiencies in overall presentation can influence how the proposal is viewed. It is mainly intended to show some common mistakes and also some good things to do in presenting a synthesis. It is not meant to supersede the detailed discussions of how and why a completed elegant synthesis is done but to assist the first-time student in preparing a proposal.

The goal of this work is to produce a graduate level textbook, and it does not assume that a student should already know the information, *before* the course. I hope that it will be useful to students and to the synthetic community. Every effort has been made to keep the manuscript error-free. Where there are errors, I take full responsibility and encourage those who find them to contact me directly, at the address given below, with corrections. Suggestions for improving the text, including additions and general comments about the book are also welcome. My goal is to incorporate such changes in future editions of this work. If anyone wishes to contribute homework problems to future editions, please send them to me and I will, of course, give full credit for any I use.

I must begin my "thank yous" with the graduate students at UConn, who inspired this work and worked with me through several years to develop the pedagogy of the text. I must also thank Dr. Chris Lipinski and Dr. David Burnett of Pfizer Central Research (Groton, CT) who organized a reactions/synthesis course for their research assistants. This allowed me to test this book upon an "outside" and highly trained audience. I am indebted to them for their suggestions and their help.

There are many other people to thank. Professor Janet Carlson (Macalester College) reviewed a primeval version of this book and made many useful comments. Professors Al Sneden and Suzanne Ruder (Virginia Commonwealth University) classroom tested an early version of this text and both made many comments and suggestions that assisted me in putting together the final form of this book. Of the early reviewers of this book, I would particularly like to thank Professor Brad Mundy (Colby College) and Professor Marye Anne Fox (University of Texas, Austin), who made insightful and highly useful suggestions that were important for shaping the focus of the book.

Along the way, many people have helped me with portions or sections of the book. Professor Barry Sharpless (Scripps) reviewed the oxidation chapter and also provided many useful insights into his asymmetric epoxidation procedures. Dr. Peter Wuts (Upjohn) was kind enough to review the protecting group chapter (chapter 7) and helped me focus it in the proper way. Professor Ken Houk (UCLA), Professor Stephen Hanessian (Université de Montréal), Professor Larry Weiler (U. of British Columbia), Professor James Hendrickson (Brandeis), Professor Tomas Hudlicky (U. Florida), and Professor Michael Taschner (U. of Akron) reviewed portions of work that applied to their areas of research and I am grateful for their help.

Several people provided original copies of figures or useful reprints or comments. These include Professor Dieter Seebach (ETH), Professor Paul Williard (Brown), Professor E.J. Corey (Harvard), Dr. Frank Urban (Pfizer Central Research), Professor Rene Barone (Université de Marseilles), and Professor Wilhelm Meier (Essen).

Two professors reviewed portions of the final manuscript and not only pointed out errors but made enormously helpful suggestions that were important for completing the book: Professor Fred Ziegler (Yale) and Professor Douglass Taber (U. of Delaware). I thank both of them very much.

There were many other people who reviewed portions of the book and their reviews were very important in shaping my own perception of the book, what was needed and what needed to be changed. These include: Professor Winfield M. Baldwin, Jr. (U. of Georgia), Professor Albert W. Burgstahler (U. of Kansas), Professor George B. Clemens (Bowling Green State University), Professor Ishan Erden (San Francisco State University), Professor Raymond C. Fort, Jr. (U. of Maine), Professor John F. Helling (U. of Florida), Professor R. Daniel Little (U. of California), Professor Gary W. Morrow (U. of Dayton), Professor Michael Rathke (Michigan State University), Professor Bryan W. Roberts (U. of Pennsylvania), Professor James E. Van Verth (Canisius College), Professor Frederick G. West (U. of Utah), and Professor Kang Zhao (New York University). I thank all of them.

I must also thank the many people who have indulged me at meetings, at Gordon conferences, and as visitors to UConn and who discussed their thoughts, needs, and wants in graduate level education. These discussions helped shape the way I put the book together.

Finally, but by no means last in my thoughts, I am indebted to Professors Joe Wolinsky and Jim Brewster of Purdue University. Their dedication and skill taught me how to teach. Thank you!

I particularly want to thank my wife Sarah and son Steven. They endured the many days and nights of my being in the library and the endless hours on the computer with patience and understanding. My family provided the love, the help, and the fulfillment required for me to keep going and helped me to put this project into its proper perspective. They helped me in ways that are too numerous to mention. I thank them and I dedicate this work to them.

Michael B. Smith

Introducing SpartanModel

· ·

SpartanModel is an virtual model kit, designed to provide students of organic chemistry information about molecular structure, stability and properties. In the simplest of terms, *SpartanModel* is the 21st century equivalent of "plastic models" used by students of previous generations. Both provide the means to move from the two-dimensional drawings of molecules to accurate three-dimensional portrayals. However, *SpartanModel* offers a number of significant advantages over plastic models.

The first advantage is that *SpartanModel* overcomes the fact that a plastic model kit contains only a limited number of "parts", perhaps ten or twenty "carbon atoms" and a much smaller number of nitrogen and oxygen atoms. Therefore, only relatively small molecules can be constructed. More importantly, a shortage of parts means that a molecule needs to be disassembled before another molecule can be assembled. This makes it impossible, or in the best circumstances, unnecessarily difficult, to compare the structures of different molecules. *SpartanModel* is unbounded, and molecules with dozens or even hundreds of atoms can be accommodated. Comparisons between different molecules can easily be made.

A related shortcoming of plastic model kits is that they are able to show off just a single aspect of molecular structure, most commonly, the connections (bonds) between atoms. Plastic models that depict overall size and shape are available, but need to be purchased and used separately. On the other hand, models made with *SpartanModel* may be portrayed either to emphasize bonding or to convey information about a molecule's overall size and shape. A further disadvantage is that plastic models "show" but do not "tell" us about important aspects of molecular structure, for example, about the volume that a molecule requires or its surface area. *SpartanModel* provides both a visual image as well as numerical values for these quantities. Of even greater practical value, *SpartanModel* assigns and displays R/S chirality, both for simple molecules where the rules are relatively easy to apply as well as for complex molecules where even an "expert" would be challenged.

Neither *SpartanModel* nor a plastic model kit is able to build proteins. However, *SpartanModel* connects seamlessly to the on-line *Protein Data Bank*[1] (PDB), providing access to ~60,000 experimental protein crystal structures. A PDB entry is automatically retrieved given its identifier, and displayed as a "ribbon model" (eliminating atomic details so as to emphasize the backbone structure). The model may be manipulated enabling detailed visual inspection.

The second advantage is that the structures obtained by *SpartanModel* are not based on the fixed dimensions of the "parts" as they are with plastic models, but rather result from application of

---

1. H.M. Berman, K. Henrick, H. Nakamura, *Nat. Struct. Bio.*, **10** (12), 980 (2003).

quantum mechanics. This means that *SpartanModel* is actually a predictive tool, not merely one following empirical rules. It can be used to explore chemistry.

The third advantage is that the information provide by *SpartanModel* is not restricted to molecular structure as it is with plastic models. Energies, atomic charges and dipole moments, molecular orbitals and orbital energies and electrostatic potential maps may be obtained for any molecule. In addition, heats of formation and infrared spectra for approximately 6000 molecules obtained from high-quality quantum chemical calculations (beyond those provided in *SpartanModel*) are available from a database included with and accessed from *SpartanModel*. The availability of energies (as well as selected heats of formation) allows the most stable isomer to be identified and to say whether a reaction is *exothermic* or *endothermic*. Calculated energies may also be employed to assign the lowest-energy conformation of a molecule and to examine the likelihood that other conformers will be present. The infrared database provides realistic spectra and allows association of individual features in the spectrum with the motions of specific groups of atoms.

In short, plastic models are severely restricted, in terms of the complexity of what can be built, the accuracy of the presentations and in what information they are able to provide.

Some may argue that plastic models are "tried and tested", and that an electronic model kit is unfamiliar or intimidating. We suspect that the vast majority of students will have the opposite perspective. After all, today's students have grown up with computers and expect to use them during their college education. In the final analysis, the choice is not between plastic and computer-based models, but whether or not models have something to offer in a chemist's education. We think that they do.

GETTING STARTED WITH SpartanModel

The easiest way to learn how to get started with *SpartanModel* is to spend an hour to complete the set of tutorials that have been provided. They cover both the use of the program and the interpretation of the quantities that result. Start with **Basic Operations**, which shows how to manipulate molecules, query molecular properties, display molecular orbitals and electrostatic potential maps and draw infrared spectra. Building molecules and performing quantum chemical calculations is not illustrated, but deferred to **Acrylonitrile**. This tutorial should be completed next. The remainder of the tutorials can be completed in any order. **Camphor** and **Androsterone** illustrate construction of successively more complex organic molecules. The first of these provides another example of displaying an infrared spectrum, and both tutorials illustrate the identification and assignment of chiral (R/S) centers. **Acetic Acid Dimer** shows how a molecular complex may be assembled and its binding energy calculated. **1,3-Butadiene** and *trans*-**Cyclooctene** show how energy comparisons among molecules are made. The first involves different conformers of the same molecule and shows how the conformation of a molecule can be changed. The second involves different stereoisomers. **2-Methylpropene** and **Comparing Acid Strengths** illustrate comparisons of molecular orbitals and electrostatic potential maps for different molecules. Finally, **Hemoglobin** shows how to access the Protein Databank (PDB). A part of this tutorial requires that the user be connected to the internet.

PROBLEMS KEYED TO ORGANIC SYNTHESIS

A set of ~60 *problems*, accessible under **Problems** in the **Welcome** screen, has been keyed to the 3rd edition of *Organic Synthesis*. Many of the problems are made up of text (html) files only, opening up a "blank screen" in *SpartanModel*. However, some of the problems include materials that cannot be generated with *SpartanModel* and have been prepared using *Spartan*. These include problems involving transition states and those using LUMO maps and local ionization potential maps. These materials are "read only" and while the models may be examined and manipulated and measurements taken, they may not be altered.

The instructor is free to make additional tutorials and problems using *Spartan*, and to add these to the existing collections. Instructions are provided under *Adding Tutorials and Problems* under the **Help** menu.

TECHNICAL OVERVIEW OF SpartanModel

SpartanModel may be viewed in terms of its components: a molecule builder including a molecular mechanics based scheme for preliminary structure refinement, a real-time quantum chemical engine and two databases.

SpartanModel's builder uses atomic fragments (for example, sp, sp^2 and sp^3 carbon fragments), functional groups (for example, amide and carbonyl groups) and rings (for example, cyclohexane and benzene). Some molecules can be made in just one or two steps ("mouse clicks"), while most others require fewer than ten steps. For example, few than 20 steps are required to build the steroid androsterone, the most complicated molecule provided in a tutorial that accompanies *SpartanModel*. Once constructed, molecules can be displayed as to depict bonding (as with most types of plastic models) or overall size and shape (so-called space-filling or CPK models). Associated with the builder is a simple "molecular mechanics" procedure to provide a refined geometry as well as measurement tools for bond distances and angles, volumes, surface areas and polar surface areas (of space-filling models) and for assignment of R/S chirality.

The quantum chemical engine provided in *SpartanModel* may be used to obtain the geometries and properties of the vast majority of molecules encountered in elementary organic chemistry. The desire for open-endedness (any "reasonable size" molecule may be calculated) together with practical concerns, requires use of a very simple quantum chemical model. The procedure used in *SpartanModel* involves two quantum chemical steps and is preceded by a molecular mechanics[2] step to ensure a reasonable starting geometry. The first quantum chemical step is calculation of geometry using the PM3[3] semi-empirical model and the second step is calculation of the energy and wavefunction at this geometry using the Hartree-Fock 3-21G[4] model. The resulting wavefunction is used for calculation of the dipole moment and atomic charges and (if requested) graphical displays of the molecular orbitals and electrostatic potential map. PM3 geometry calculations and 3-21G energy calculations for molecules comprising up to 30-40 heavy (non-hydrogen) atoms are likely to require less than one minute on a present day Windows or Macintosh computer.

The quantum chemical calculations in *SpartanModel* properly account for geometry and provide a sound basis for graphical displays of molecular orbitals and electrostatic potential maps. They also provide a qualitatively accurate account of the energies of most types of chemical reactions as well as conformational energy differences. Heats of formation for ~6000 molecules obtained from the T1[5] thermochemical recipe and included in a database can be used to supplement calculated energies where higher accuracy may be necessary. T1 has been shown to reproduce experimental heats of formation with an rms error of ~8 kJ/mol.

The second database contains infrared spectra for ~6000 molecules obtained from the EDF2[6]/ 6-31G* density functional model, adjusted to account for known systematic errors and for finite temperature. The resulting spectra are visually and quantitatively very similar to observed infrared spectra. Vibrational modes associated with individual lines in the spectrum may be "animated".

2. T.A. Halgren, J. Computational Chem., 17, 490 (1996).
3. J.J.P. Stewart, *J. Computational Chem.*, **10**, 209 (1989).
4. J.S. Binkley, J.A. Pople and W.J. Hehre, *J. Am. Chem. Soc.*, **102**, 939 (1980); W.J. Pietro, M.M. Francl, W.J. Hehre, D.J. DeFrees, J.A. Pople and J.S. Binkley, *ibid.*, **104**, 5039, (1982). For a review and assessment of techniques and computational methods, see: *A Guide to Molecular Mechanics and Quantum Chemical Calculations*, Wavefunction, Inc., Irvine, CA, USA (2003).
5. W.S. Ohlinger, P.E. Klunzinger, B.J. Deppmeier, W.J. Hehre, *J. Phys. Chem. A.*, **113** 10, 2165 (2009).
6. R.D. Adamson, P.M.W. Gill, and J.A. Pople, *Chem. Phys. Lett.*, **284**, 6, (1998).

THIS BOOK IS DEDICATED TO:

· · · · · · · · · · · ·

MY WIFE SARAH

AND

MY SON STEVEN

Common Abbreviations

Other, less common abbreviations are given in the text when the term is used.

Ac	Acetyl	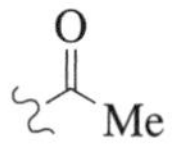
Acac	Acetylacetonate	
AIBN	*azo-bis*-isobutyronitrile	
All	Allyl	
Am	Amyl	$-CH_2(CH_2)_3CH_3$
aq	Aqueous	
Ax	Axial	
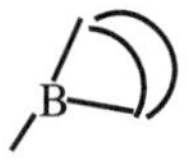	9-Borabicyclo[3.3.1]nonylboryl	
9-BBN	9-Borabicyclo[3.3.1]nonane	
BINAP	*2R,3S*-2,2'-*bis*-(diphenylphosphino)-1,1'-binapthyl	
Bn	Benzyl	$-CH_2Ph$
Boc	*t*-Butoxycarbonyl	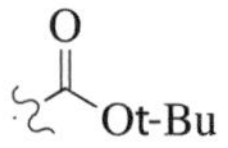
Bpy (Bipy)	2,2'-Bipyridyl	
Bu	*n*-Butyl	$-CH_2CH_2CH_2CH_3$
Bz	Benzoyl	
CAM	Carboxamidomethyl	
CAN	Ceric ammonium nitrate	$(NH_4)_2Ce(NO_3)_6$
c-	Cyclo-	
cat	Catalytic	
Cbz	Carbobenzyloxy	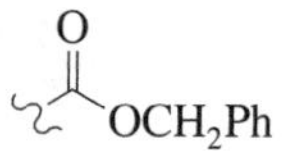
chap	Chapter(s)	
Chirald	2S,3R-(+)-4-dimethylamino-1,2-diphenyl-3-methylbutan-2-ol	

CIP	Cahn–Ingold-Prelog	
COD	1,5-Cyclooctadienyl	
COT	1,3,5-cyclooctatrienyl	
Cp	Cyclopentadienyl	
CSA	Camphorsulfonic acid	
CTAB	Cetyltrimethylammonium bromide	$C_{16}H_{33}NMe_3^+\ Br^-$
Cy (c-C_6H_{11})	Cyclohexyl	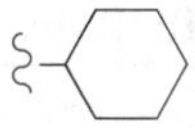
°C	Temperature in Degrees Centigrade	
2D	Two-dimensional	
3D	Three-dimensional	
DABCO	1,4-Diazabicyclo[2.2.2]octane	
d	Day(s)	
dba	Dibenzylidene acetone	
DBE	1,2-Dibromoethane	$BrCH_2CH_2Br$
DBN	1,5-Diazabicyclo[4.3.0]non-5-ene	
DBU	1,8-Diazabicyclo[5.4.0]undec-7-ene	
DCC	1,3-Dicyclohexylcarbodiimide	c-C_6H_{11}-N=C=N-c-C_6H_{11}
DCE	1,2-Dichloroethane	$ClCH_2CH_2Cl$
DDQ	2,3-Dichloro-5,6-dicyano-1,4-benzoquinone	
% de	% Diasteromeric excess	
DEA	Diethylamine	$HN(CH_2CH_3)_2$
DEAD	Diethylazodicarboxylate	EtO_2C-N=NCO_2Et
DET	Diethyl tartrate	
DHP	Dihydropyran	
DIBAL-H	Diisobutylaluminum hydride	$(Me_2CHCH_2)_2AlH$
Diphos (**dppe**)	1,2-bis-(Diphenylphosphino)ethane	$Ph_2PCH_2CH_2PPh_2$
Diphos-4 (**dppb**)	1,4-bis-(Diphenylphosphino)butane	$Ph_2P(CH_2)_4PPh_2$
DIPT	Diisopropyl tartrate	
DMAP	4-Dimethylaminopyridine	
DME	Dimethoxyethane	$MeOCH_2CH_2OMe$
DMF	N,N'-Dimethylformamide	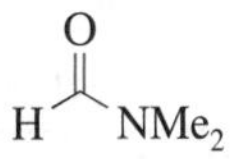
DMS	Dimethyl sulfide	

DMSO	Dimethyl sulfoxide	
dppb	1,4-*bis*-(Diphenylphosphino)butane	$Ph_2P(CH_2)_4PPh_2$
dppe	1,2-*bis*-(Diphenylphosphino)ethane	$Ph_2PCH_2CH_2PPh_2$
dppf	*bis*-(Diphenylphosphino)ferrocene	
dppp	1,3-*bis*-(Diphenylphosphino)propane	$Ph_2P(CH_2)_3PPh_2$
dvb	Divinylbenzene	
e$^-$	Electrolysis	
EA	Electron affinity	
% ee	% Enantiomeric excess	
EE	1-Ethoxyethoxy	EtO(Me)CH-
Et	Ethyl	-CH_2CH_3
EDA	Ethylenediamine	$H_2NCH_2CH_2NH_2$
EDTA	Ethylenediaminetetraacetic acid	
Equiv	Equivalent(s)	
ESR	Electron Spin Resonance Spectroscopy	
FMN	Flavin mononucleotide	
FMO	Frontier Molecular Orbital	
fod	*tris*-(6,6,7,7,8,8,8)-Heptafluoro-2,2-dimethyl-3,5-octanedionate	
Fp	Cyclopentadienyl-*bis*-carbonyl iron	
FVP	Flash Vacuum Pyrolysis	
GC	Gas chromatography	
gl	Glacial	
FVP	Flash Vacuum Pyrolysis	
GC	Gas chromatography	
gl	Glacial	
h	Hour (hours)	
*h*v	Irradiation with light	
1,5-HD	1,5-Hexadienyl	
HMPA	Hexamethylphosphoramide	$(Me_2N)_3P=O$
HMPT	Hexamethylphosphorus triamide	$(Me_2N)_3P$
^{1}H NMR	Proton Nuclear Magnetic Resonance Spectroscopy	
HOMO	Highest occupied molecular orbital	
HPLC	High performance liquid chromatography	
HSAB	Hard/Soft Acid/Base	

IP	Ionization potential	
i-Pr	Isopropyl	-CH(Me)$_2$
IR	Infrared spectroscopy	
LICA (LIPCA)	Lithium cyclohexylisopropylamide	
LDA	Lithium diisopropylamide	LiN(*i*-Pr)$_2$
LHMDS	Lithium hexamethyl disilazide	LiN(SiMe$_3$)$_2$
LTMP	Lithium 2,2,6,6-tetramethylpiperidide	
LUMO	Lowest unoccupied molecular orbital	
MCPBA	*meta*-Chloroperoxybenzoic acid	
Me	Methyl	-CH$_3$ or Me
MEM	β–Methoxyethoxymethyl	MeOCH$_2$CH$_2$OCH$_2$-
Mes	Mesityl	2,4,6-tri-Me-C$_6$H$_2$
min	minutes	
MOM	Methoxymethyl	MeOCH$_2$-
Ms	Methanesulfonyl	MeSO$_2$-
MS	Molecular Sieves (3Å or 4Å)	
MTM	Methylthiomethyl	
MVK	Methyl vinyl ketone	MeSCH$_2$-
NAD	Nicotinamide adenine dinucleotide	
NADP	Sodium triphosphopyridine nucleotide	
Napth	Napthyl (C$_{10}$H$_8$)	
NBD	Norbornadiene	
NBS	*N*-Bromosuccinimide	
NCS	*N*-Chlorosuccinimide	
NIS	*N*-Iodosuccinimide	
Ni(R)	Raney nickel	
NMO	*N*-Methylmorpholine *N*-oxide	
Nu (Nuc)	Nucleophile	
OBs	O-Benzenesulfonate	
Oxone	2 KHSO$_5$•KHSO$_4$•K$_2$SO$_4$	
(P) or  ⬤	Polymeric backbone	
PCC	Pyridinium chlorochromate	
PDC	Pyridinium dichromate	
PEG	Polyethylene glycol	

Ph	Phenyl	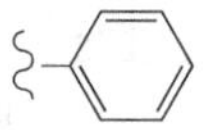
PhH	Benzene	
PhMe	Toluene	
Phth	Phthaloyl	
Pip	Piperidino	
PPA	Polyphosphoric acid	
PPTS	*para*-Toluenesulfonic acid	
Pr	*n*-Propyl	-CH$_2$CH$_2$CH$_3$
Py	Pyridine	
Quant	Quantitative yield	
Red-Al	[(MeOCH$_2$CH$_2$O)$_2$AlH$_2$]Na	
RT	Room temperature	
sBu	*sec*-Butyl	CH$_3$CH$_2$CH(CH$_3$)
sBuLi	*sec*-Butyllithium	CH$_3$CH$_2$CH(Li)CH$_3$
s	seconds	
sec	sections(s)	
SEM	2-(trimethylsilyl)ethoxymethyl	
SET	Single electron transfer	
Siamyl	*sec*-Isoamyl	(CH$_3$)$_2$CHCH(CH$_3$)-
(Sia)$_2$BH	Disiamylborane	
TASF	*tris*-(Diethylamino)sulfonium difluorotrimethyl silicate	
TBAF	Tetrabutylammonium fluoride	*n*-Bu$_4$N$^+$ F$^-$
TBDMS	*t*-Butyldimethylsilyl	*t*-BuMe$_2$Si
TBHP (*t*-BuOOH)	*t*-Butylhydroperoxide	Me$_3$COOH
t-Bu	*tert*-Butyl	-CMe$_3$
TEBA	Triethylbenzylammonium	Bn(Et$_3$)$_3$N$^+$
TEMPO	Tetramethylpiperidinyloxy free radical	
TFA	Trifluoroacetic acid	CF$_3$COOH
TFAA	Trifluoroacetic anhydride	(CF$_3$CO)$_2$O
Tf (OTf)	Triflate	-SO$_2$CF$_3$ (-OSO$_2$CF$_3$)

ThexBH$_2$	Thexylborane (*tert*-hexylborane)	
THF	Tetrahydrofuran	
THP	Tetrahydropyran	
TMEDA	Tetramethylethylenediamine	Me$_2$NCH$_2$CH$_2$NMe$_2$
TMS	Trimethylsilyl	-Si(CH$_3$)$_3$
TMP	2,2,6,6-Tetramethylpiperidine	
Tol	Tolyl	4-(Me)C$_6$H$_4$
TPAP	Tetrapropylperruthenate	
Tr	Trityl	-CPh$_3$
TRIS	Triisopropylphenylsulfonyl	
Ts(Tos)	Tosyl = *p*-Toluenesulfonyl	4-(Me)C$_6$H$_4$SO$_2$
UV	Ultraviolet spectroscopy	
X$_c$	Chiral auxiliary	

chapter 1
Retrosynthesis, Stereochemistry, and Conformations

1.1. INTRODUCTION

Where does one begin a book that will introduce and discuss hundreds of chemical reactions? To synthesize a complex organic molecule many reactions must be used, and the strategy used for that synthesis must consider not only the type of reaction but also the mechanism of that reaction. We begin with a brief introduction to synthetic planning. A full discussion of strategies for total synthesis will be introduced in chapter 10, but surveying the fundamental approach can help one understand how reactions are categorized.

The total synthesis of complex natural products usually demands a thorough knowledge of reactions that form carbon-carbon bonds as well as those that change one functional group into another. Examination of many syntheses of both large and small molecules, reveals that building up a carbon skeleton by carbon-carbon bond forming reactions is rarely done successfully unless all aspects of chemical reactivity, functional group interactions, conformations, and stereochemistry are well understood. The largest number of actual chemical reactions that appear in a synthesis do not make carbon-carbon bonds but rather manipulate functional groups.

Changing one functional group into another is defined as a **functional group interchange** (FGI). Simple examples are the loss of H and Br from 2-bromo-2-methylpentane (**1**) to form 2-methyl-2-pentene (**2**), or oxidizing the alcohol unit in 2-pentanol (**3**) to the carbonyl unit in 2-pentanone (**4**). Contrast these reactions with a reaction that brings reactive fragments together to form a new bond between two carbon atoms, such as the condensation of two molecules of butanal (**5**) under basic conditions to give **6**, which is known as the **aldol condensation**. The aldol condensation forms a carbon-carbon bond but before such a reaction can occur one usually must incorporate or change key functional groups. This observation is an important reason that more functional group exchange reactions are typically required, relative to carbon-carbon bond forming reactions. Several different functional groups may also be structural units of the molecule being synthesized.

Nowadays, the relationship of two molecules in a synthesis is commonly shown using a device known as a **transform**, defined by Corey and co-workers[1] as: "the exact reverse of a synthetic reaction to a target structure". The **target structure** is the final molecule one is attempting to prepare. The synthetic transformation that converts butanal (**5**) to hydroxy-aldehyde **6** via the **aldol condensation** (see sec. 9.4.A) is an example. The transform for this synthetic step is, therefore, **6** ⇒ **5**. Inspection of **6** and **5** reveals that mentally breaking the highlighted bond (bond *a*) in **6** (represented by the squiggly line) leads to disconnect fragments **5** and **7** and in this process bond *a* is said to be **disconnected**. An elementary disconnection approach quickly becomes an integral part of how one thinks about molecules. The focus here, however, is on how to put molecules together. How does the disconnection approach assist us in this endeavor? Understanding why bond *a* is important comes from a thorough knowledge of the chemical properties of compound **6**. When we disconnect bond *a*, we eventually want to make that bond by a chemical reaction. To understand the molecular characteristics of **6** that led us to disconnect bond *a*, we must understand the chemical reactions required to form that bond.

Pentanoic acid (**8**) is a simple target that illustrates the approach. Disconnection of the carbon-carbon bond marked *a* leads to 1-butanol (**9**) as the precursor (the starting material). Analysis of the

targeted carboxylic acid as well as the product alcohol shows a change in oxidation state, and a one-carbon extension of the carbon chain. Of the four carbon-carbon bonds in target **8**, bond *a* must be used to attach the carbonyl to the starting material **9**. Disconnection of bond *a* leads to the simplified structure **10** and the CO₂H (carboxyl) fragment, which is not a real molecule. These two structures are termed **disconnect products** but **10** contains an unspecified X group, that must be a reactive functional group and could be hydroxyl (as in **9**). In addition, the carboxyl fragment shown does not exist and a **synthetic equivalent** of the disconnect product that is a real molecule must be found. In other words, to complete any reaction a real molecule or reagent must be used who reaction generates the disconnect fragment directly. Alternatively, a surrogate can be used to give a product that can be converted to the fragment. For example, in the disconnection C–NH₂ to C and NH₂, ammonia reacts with an alkyl halide to give an amine, but there are problems with over-alkylation. The NH₃ is the direct equivalent

---

1. Corey, E.J.; Cheng, X. *The Logic of Chemical Synthesis*, Wiley-Interscience, NY, *1989*.

of NH_2. To overcome such problems, the phthalimide anion reacts with an alkyl halide to give the phthalimide, and subsequent reaction with hydrazine liberates the amine. Phthalimide is a surrogate for NH_2, used to circumvent problems with the ammonia reaction. In the case of **8**, we require a COOH surrogate, since COOH is not a real molecule. It is known that carbon dioxide (CO_2) reacts with Grignard reagents (**10**, where X = MgBr) to give a carboxylic acid after hydrolysis of the resulting carboxylate salt in a second step. Therefore, the disconnection shown is a viable process since we know a chemical reaction to make that bond. Working backward in this manner is termed **retrosynthetic analysis** or **retrosynthesis**, defined by Corey as "a problem-solving technique for transforming the structure of a synthetic target molecule to a sequence of progressively simple materials along a pathway which ultimately leads to a simple or commercially available starting material for chemical synthesis".[2]

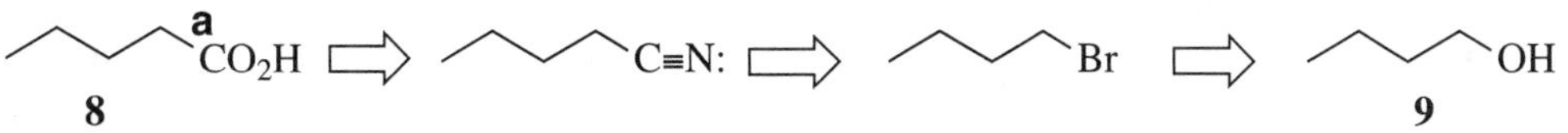

Based on our retrosynthetic analysis, one solution to this synthetic problem is to use the reaction of cyanide ion with primary halides via second order nucleophilic substitution (S_N2, see sec. 2.6.A), to generate a nitrile. The cyano unit is readily hydrolyzed to an acid. Is there an alternative? The answer is yes via a Grignard reaction with carbon dioxide (sec. 8.4.C.iv), but this strategy requires that alcohol **9** be converted to an alkyl halide and then to the corresponding Grignard reagent. Subsequent reaction with carbon dioxide and hydrolysis will give **8**. Formation of an alkyl halide from an alcohol is a functional group interchange (FGI) reaction (see sec. 2.8.A), as is conversion of the halide to a nitrile and the nitrile to an acid. The retrosynthesis effectively describes the reactions necessary for the real synthesis starting from **8**, but reagents must be added to complete the synthesis. Determining the key bond for disconnection in the target led to the conclusion that the synthesis required a carbon-carbon bond forming reaction as well as FGI reactions. Thinking about the synthesis and disconnections led us to analyze and understand the reactions that must be used.

It is important to point out that rarely does one take a retrosynthetic analysis and use the exact reverse track with simple reagents to synthesize the target. For mono-functional molecules this approach often works, but for molecules with multiple functionality, particularly complex natural products, the idea of doing a retrosynthetic analysis and simply providing reagents to convert the starting material to the target is very naïve. There are usually steps that simply do not work using available reagents or those suggested by literature precedent, and the reactions may give poor yields or the wrong stereochemistry. There are unanticipated interactions of functional groups and unexpected requirements for protecting groups. In short, the approach shown here is a beginning, intended to get you to think about how to pull molecules apart, what reactions may be appropriate to put them together again, and to think about strategies

---

2. Reference 1, p 6.

for synthesis. These ideas are elaborated in chapter 10. When we ask how **8** is prepared from **9**, we are forced to examine how many reactions that form carbon-carbon bonds we actually know how to use, as well as those reactions that prepare and interchange functional groups. In other words, it forces us to review our understanding of Organic chemistry. Other important concepts in Organic chemistry must be brought to bear, including stereochemistry and conformational theory.

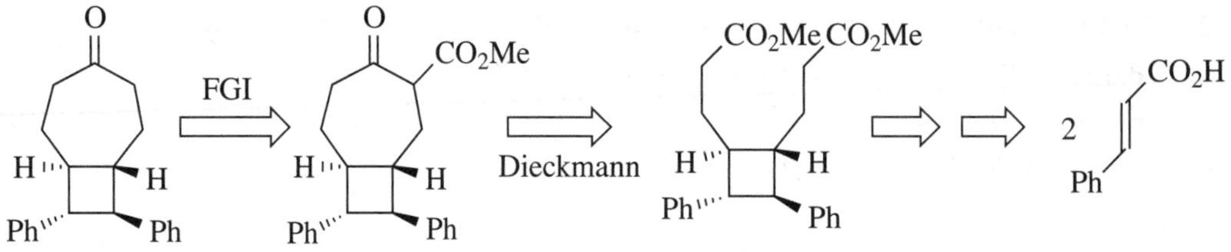

The importance of stereochemistry is illustrated by the disconnection sequence **11** ⇒ **12** ⇒ **13** ⇒ **14**, described by Corey.[3] The second transform involves a **Dieckmann condensation reaction** (see sec. 9.4.B.ii), and several steps are required to prepare **13** from 14. Focus attention on the stereochemical relationship of the various groups and of the ring juncture. It is insufficient to consider methods that simply make a bond, because we must be able to form the bond with control of the relative and absolute stereochemistry.

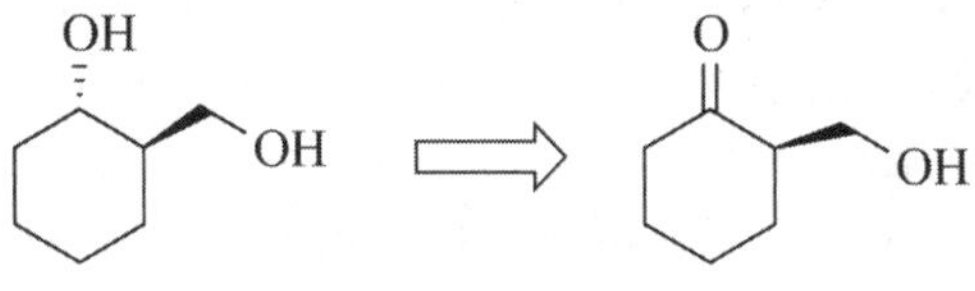

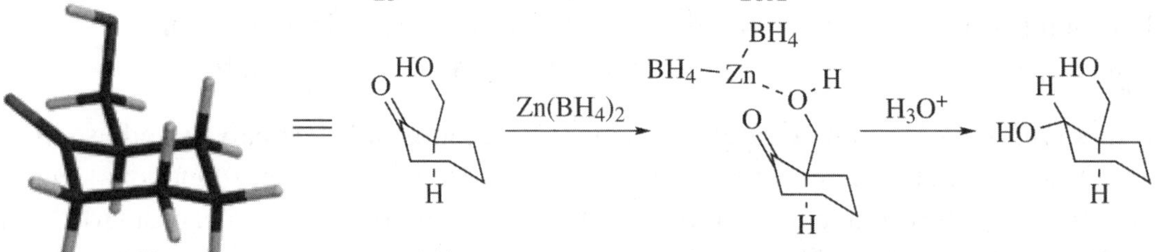

Another important analysis of organic reactions can be found in the transform **15** ⇒ **16A**. Why does this transform lead to the relative stereochemistry (trans) shown in **15**? The proper spatial relationship of the functional groups in the target must be known prior to making the choice for a chemical reaction. This relationship can be difficult to see using the two-dimensional (2D) structures shown. A different model of the molecule such as a 3D-model will usually

---

3. Reference 1, p 13.

give a better understanding of stereochemical and conformational relationships. In this case, representation **16B** (produced by the computer program Spartan06™)[4] provides much more detail than **16A**, and shows that the hydroxyl group is positioned on one side of the carbonyl. If a chelating metal such as zinc is used with the reducing step (zinc borohydride, see sec. 4.4.B), coordination with the hydroxyl (**17**) may deliver hydride primarily from the same face as the CH_2OH unit, leading to the observed trans stereochemistry in **18**. In this example, the **conformation** (3D shape) of the molecule provides a model that helps us understand the stereochemical outcome of the reaction.

Three important factors will be reviewed in this chapter: (*1*) disconnection and retrosynthesis, (*2*) stereochemistry, (*3*) conformational analysis. This review will not to formally introduce synthesis (synthetic theory will be presented in chapter 10),[5] but focus attention on concepts that help us understand reactions.

1.2. THE DISCONNECTION PROTOCOL

When retrosynthetic theory is introduced, it is easy to become embroiled in making the "correct" disconnection rather than focusing on the chemical reactions and concepts required to form the disconnected bond in the synthesis. In reality, the disconnection is dictated by the ability to form the bond in the context of stereochemistry and selectivity, not the other way around. The synthesis of organic molecules dates to the nineteenth century, but the work of Perkin, Robinson and others in the early twentieth century demonstrated the importance of synthetic planning.[1] In the 1940s and 1950s, Woodward, Robinson, Eschenmoser, Stork and others clearly showed how molecules could be synthesized in a logical and elegant manner. In the 1960s, Corey identified the rationale behind his syntheses, and such logical synthetic plans (termed retrosynthetic analyses) are now a common feature of the synthetic literature.[2] The disconnection approach is used to teach synthesis, and Warren has several books that describe this approach in great detail.[6] Several different strategies for the synthesis of organic molecules are available (see chapter 10) and all are useful. When these strategies are applied to a first synthesis in an introductory course, the most critical issue raised after a disconnection is what to do with the disconnect products. All disconnection approaches assign priorities to bonds in a molecule and disconnect those bonds with the highest priorities, as with Corey's

4.   Spartan™ is a trademark of Wavefunction Inc., Irvine, CA, Spartan'06.
5.   (a) Corey, E.J.; Wipke, W.T. *Science*, *1969*, *166*, 178; (b) Corey, E.J.; Long, A.K.; Rubinstein, S.D. *Ibid* *1985*, *228*, 408; (c) Corey, E.J.; Howe, W.J.; Pensak, D.A. *J. Am. Chem. Soc. 1974*, *96*, 7724; (d) Corey, E.J. Quart. Rev. Chem. Soc. *1971*, *25*, 455; (e) Corey, E.J.; Wipke, W.T.; Cramer III, R.D.; Howe, W.J. *J. Am. Chem. Soc. 1972*, *94*, 421; (f) Corey, E.J.; Jorgensen, W.L. Ibid *1976*, *98*, 189.
6.   (a) Warren, S. *Organic Synthesis: The Disconnection Approach*, John Wiley, Chichester *1982*; (b) Warren S. *Workbook for Organic Synthesis: The Disconnection Approach*, John Wiley, Chichester *1982*; (c) Warren, S. *Designing Organic Synthesis: A Programmed Introduction to the Synthon Approach*, Wiley, Chichester *1978*. Also see (d) Wyatt, P.; Warren, *S. Organic Synthesis, Strategy and Control*, Wiley, New Jersey, *2007*.

strategic bond analysis (see sec. 10.5).[1,5] Smith described a simple method where the priorities are based only on the relative ability to chemically form the bond broken in the disconnection,[7] based on known reactions.

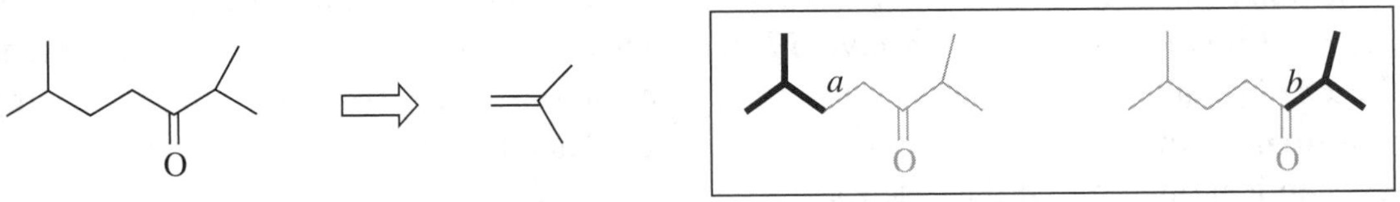

| 19A | 20 | 19B | 19C |

In a typical introductory Organic chemistry course, there are two fundamental types of synthesis problems. In the first, both the starting material and the target are specified. In the second, only the final target is given and the synthetic chemist must deduce the starting material. This latter case usually poses a more difficult problem. We can illustrate the fundamentals of a retrosynthetic analysis with the first type of problem, and will delay discussion of the second type until Chapter 10. In the disconnection **19** ⇒ **20**, 2-butene is the designated starting material, which means that the target (**19**) is to be synthesized from 2-methylpropene (**20**). The four carbons of **20** must be "located" in **19** since this will define the carbon atoms that must be disconnected for the synthesis. The presence of two methyl groups in **20** limits the possibilities, and there are two different locations where the four carbons in isobutene may be found (see **19B** and **19C**). If we choose the pattern shown in **19B**, this dictates that bond *a* must be disconnected. If we choose the pattern in **19C**, then bond *b* must be disconnected. Disconnection of bond *a* in **19B** leads to two fragments (disconnect products), **21** and **22**. Similarly, disconnection of bond *b* in **19C** leads to disconnect products **23** and **24**. Both disconnections must be considered, but structures **21-24** are not real molecules so we cannot evaluate the relative merit of each disconnection using these fragments. Real molecules and real reactions are required. Therefore, two assumptions will be made: (*1*) The key carbon-carbon bonds will be formed by a small subset of reactions and (*2*) the bonds will be made by reactions involving polarized or ionic intermediates.

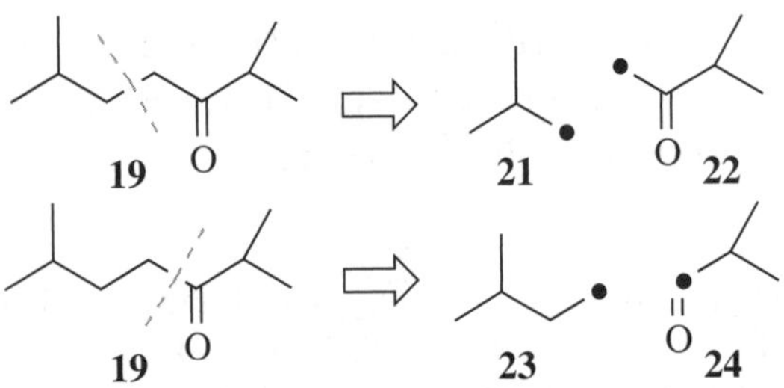

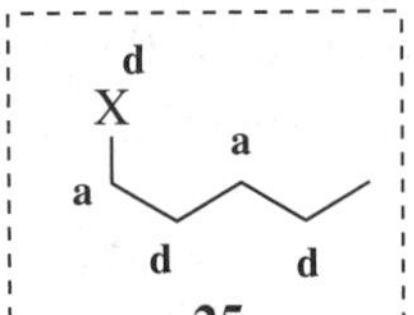

The first assumption is based on the carbon-carbon bond-forming reactions shown in Table 1.1,[7] which are usually presented in a typical sophomore Organic chemistry course. The second

7.   Smith, M.B. *J. Chem. Educ.* **1990**, 67, 848.

assumption is based on the observation that all reactions in the table except entry 10 (the **Diels-Alder reaction**, see sec. 11.4.A) involve highly polarized or ionic intermediates. With this assumption in hand, it is reasonable to assume that the disconnections generated from **19** will lead to ionic or polarized intermediates, and we can form the carbon-carbon bonds with one of the reactions found in Table 1.1. These assumptions allow us to convert each disconnect product (**21-24**) into a polarized fragment.

Table 1.1. Carbon-Carbon Bond-Forming Reactions

1. Cyanide	$R\text{–}X \; + \; {}^{-}C{\equiv}N \longrightarrow R\text{–}C{\equiv}N$
2. Acetylides (Alkyne anions)	$R\text{–}X \; + \; {}^{-}C{\equiv}C\text{–}R' \longrightarrow R\text{–}C{\equiv}C\text{–}R'$
3. Organometallics (not Cu or enolate anions)	(a) $RM \; + \; R'\text{–}X \longrightarrow R\text{–}R'$
$M = MgX, Li, ZnX,.....$	(b) $RM \; + \; R^1\text{C(O)}R^2 \longrightarrow$ tertiary alcohol
	(c) $RM \; + \;$ epoxide$(R^1) \longrightarrow$ β-hydroxy product
4. Organocuprates	(a) $R'\text{–}X \; + \; R_2CuLi \longrightarrow R\text{–}R'$
	(b) $R'M \; + \;$ enone$(R) \longrightarrow$ 1,4-addition product
5. Enolate anion alkylation	enolate$(R) \; + \; R'\text{–}X \longrightarrow$ α-alkyl ketone
6. Enolate anion condensation	(a) enolate$(R) \; + \; R^1\text{C(O)}R^2 \longrightarrow$ aldol product
	(b) enolate$(R) \; + \; R'\text{C(O)}X \longrightarrow$ 1,3-diketone
7. Friedel-Crafts acylation	$Ar\text{–}H \; + \; R'\text{C(O)}X \longrightarrow R'\text{C(O)}Ar$
8. Friedel-Crafts alkylation	$Ar\text{–}H \; + \; R\text{–}X \longrightarrow Ar\text{–}X$
9. Wittig reaction	$R^1\text{C(O)}R^2 \; + \;$ ylide$(R^3,R^4,PPh_3) \longrightarrow$ alkene
10. Diels-Alder cyclization	diene $\; + \;$ dienophile $\longrightarrow$ cyclohexene

The concept of nucleophilic and electrophilic atoms in ionic and polarized intermediates is well known. Polarized bond notation such as $C^{\delta+}\text{–}Br^{\delta-}$ and $C^{\delta-}\text{–}Li^{\delta+}$ is commonly used in describing the reactivity of such bonds. The polarizability of various atoms and molecules is also reflected in hard and soft acid and base (HSAB) theory (see sec. 2.4).[8] Seebach used structure **25** to formalize a bond polarization model.[9] The sites marked *d* in **25** represent *donor* sites or *nucleophilic* atoms. The sites marked *a* are *acceptor* sites and correspond to

8. Ho, T-L. *Hard and Soft Acids and Bases Principle in Organic Chemistry*, Academic Press, New York, *1971*, pp 1-3 and 27-34.
9. Seebach, D. *Angew. Chem. Int. Ed.* *1979*, *18*, 239.

electrophilic atoms. Bond polarization induced by the heteroatom extends down the carbon chain, due to the usual inductive effects that are a combination of through-space and through-bond effects.[10] The electrophilic carbon adjacent to X (C1 for example) is designated C^a (an acceptor atom) since proximity to the $\delta-$ electronegative atom (X) induces the opposite polarity. Similarly, C2 is a donor atom (C^d), but less polarized than X (this carbon is further away from the electrons that induce the bond polarization), and C3 is a weak acceptor atom. As a practical matter, the effect is negligible beyond C4 and will be ignored.

If this protocol is applied to disconnect fragments **21** → **24**, either carbon at the point of disconnection can be assigned as a donor (d) or an acceptor (a) giving four different possibilities. Disconnect products **21** and **22**, for example, become **26** or **27** and **28** or **29**, respectively. Fragments **26** → **29** are not real molecules, and another step is required before the best donor-acceptor pair can be chosen. Each fragment must be correlated with a synthetic equivalent. Table 1.2[7] provides a list of common synthetic equivalents, leading to the definition of **synthetic equivalent** as a molecular fragment that is equivalent to a real molecule by virtue of its chemical reactivity.[11] A C^d site on an unfunctionalized carbon, for example, is equivalent to the $C^{\delta-}$ of a Grignard reagent (see sec. 8.4) and a C^a site on an unfunctionalized carbon is equivalent to the electrophilic carbon on an alkyl halide. Each synthetic equivalent in Table 1.2 is based on an ionic or highly polarized nucleophilic substitution or nucleophilic acyl addition reaction. Using Table 1.2, fragment **26** is the equivalent of a Grignard reagent, **30** (an organocuprate such as $[Me_2CHCH_2]_2CuLi$ can also be used as in sec. 8.7.A). The equivalent of its partner (fragment **27**) is α-chloroketone **31**. Similarly, the synthetic equivalent of **28** is bromide, **32** (another halide can be used) and the equivalent of **29** is an enolate anion, **33** (see sec. 9.2, 9.3). In the actual synthesis, **30** will react with **31** and **32** will react with **33**, and both reactions will produce **19**. Bromide **32** is derived from the 2-methylpropene starting material, but the functional group must be modified. To accomplish this, the chemical relationship between the C-C-Br and C=C must be known. In other words, what reagents are required to transform an alkyl a halide into an alkene, and vice-versa. Figure 1.1[7] provides a functional group reaction diagram to interconvert one functional group into another by known chemical pathways. Since

| 26 | 30 | 27 | 31 | and | 28 | 32 | 29 | 33 |

10. (a) Baker, F.W.; Parish, R.C.; Stock, L.M. *J. Am. Chem. Soc.* **1967**, *89*, 5677; (b) Golden, R.; Stock, L.M. *Ibid* **1966**, *88*, 5928; (c) Holtz, H.D.; Stock, L.M. *Ibid* **1964**, *86*, 5188; (d) Branch, G.E.K.; Calvin, M. *The Theory of Organic Chemistry*, Prentice-Hall: New York, **1941**, chapter 6; (e) Ehrenson, S. *Progr. Phys. Org. Chem.* **1964**, *2*, 195; (f) Roberts, J.D.; Carboni, C.A. *J. Am. Chem. Soc.* **1955**, *77*, 5554; (g) Clark, J.; Perrin, D.D. *Quart. Rev. Chem. Soc.* **1964**, *18*, 295.
11. See Reference 1, p 30.

Table 1.2.   **Common Synthetic Equivalents for Disconnect Products**

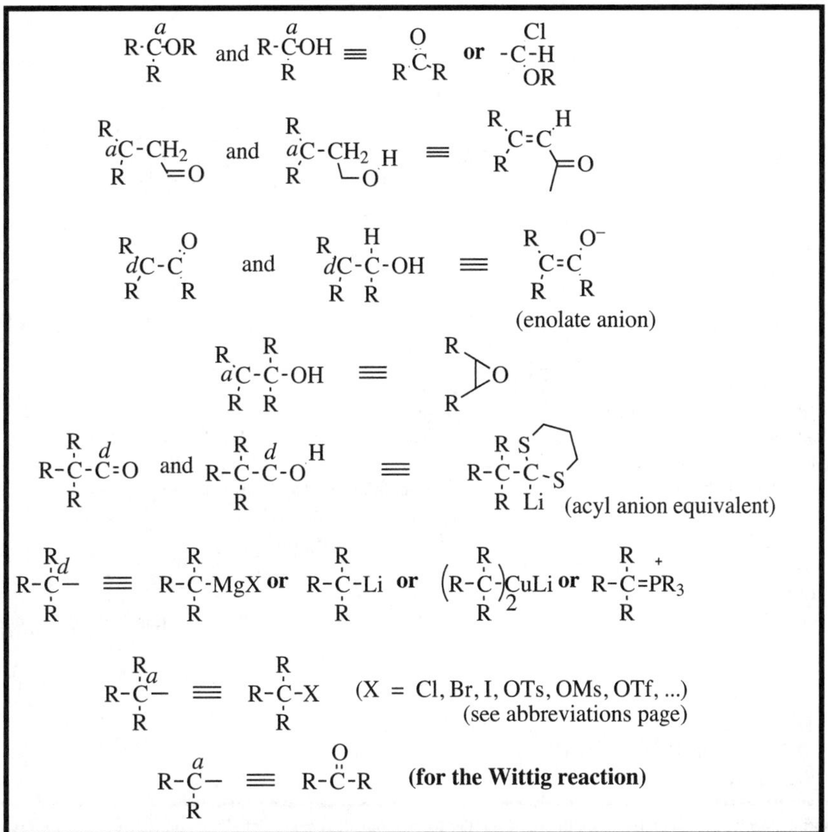

$$R\text{-}\overset{a}{\underset{R}{C}}OR \quad \text{and} \quad R\text{-}\overset{a}{\underset{R}{C}}OH \quad \equiv \quad R\overset{O}{\underset{}{C}}R \quad \text{or} \quad \overset{Cl}{\underset{OR}{-C-H}}$$

$$\overset{R}{\underset{R}{a}C}\text{-}CH_2 \quad \text{and} \quad \overset{R}{\underset{R}{a}C}\text{-}CH_2\text{-}H \quad \equiv \quad \overset{R}{\underset{R'}{C}}=\overset{H}{\underset{}{C}} \Bigg\rangle = O$$

$$\overset{R}{\underset{R'}{d}C}\text{-}\overset{O}{\underset{R}{C}} \quad \text{and} \quad \overset{R}{\underset{R'}{d}C}\text{-}\overset{H}{\underset{R}{C}}\text{-}OH \quad \equiv \quad \overset{R}{\underset{R'}{C}}=\overset{O^-}{\underset{R}{C}}$$

(enolate anion)

$$\overset{R}{\underset{R'}{a}C}\text{-}\overset{R}{\underset{R'}{C}}\text{-}OH \quad \equiv \quad \overset{R}{\underset{R'}{\triangleright}}\!O$$

$$R\text{-}\overset{R}{\underset{R}{C}}\text{-}\overset{d}{C}{=}O \quad \text{and} \quad R\text{-}\overset{R}{\underset{R}{C}}\text{-}\overset{d}{C}\text{-}O\text{-}H \quad \equiv \quad R\text{-}\overset{R}{\underset{R}{C}}\text{-}\overset{S}{\underset{Li}{C}} \quad \text{(acyl anion equivalent)}$$

$$R\text{-}\overset{R}{\underset{R}{\overset{d}{C}}}- \quad \equiv \quad R\text{-}\overset{R}{\underset{R}{C}}\text{-}MgX \; \text{or} \; R\text{-}\overset{R}{\underset{R}{C}}\text{-}Li \; \text{or} \; \Big(R\text{-}\overset{R}{\underset{R}{C}}\Big)_2 CuLi \; \text{or} \; R\text{-}\overset{R}{\underset{R}{C}}{=}PR_3^+$$

$$R\text{-}\overset{R}{\underset{R}{\overset{a}{C}}}- \quad \equiv \quad R\text{-}\overset{R}{\underset{R}{C}}\text{-}X \qquad (X = Cl, Br, I, OTs, OMs, OTf, ...)$$

(see abbreviations page)

$$R\text{-}\overset{R}{\underset{R}{\overset{a}{C}}}- \quad \equiv \quad R\text{-}\overset{O}{\underset{}{C}}\text{-}R \qquad \textbf{(for the Wittig reaction)}$$

Reprinted with permission from Smith, M.B. and the Journal of Chemical Education, Vol. 67, **1990**, 848-856. Copyright 1990, Divison of Chemical Education, Inc.

an alkene and a halide are chemically related by the reactions indicated, either could be used in the disconnect product since one can easily be converted to the other at the appropriate time. Similar analysis of disconnect fragments **23** and **24** may lead to different reactive fragments, giving a different synthesis.

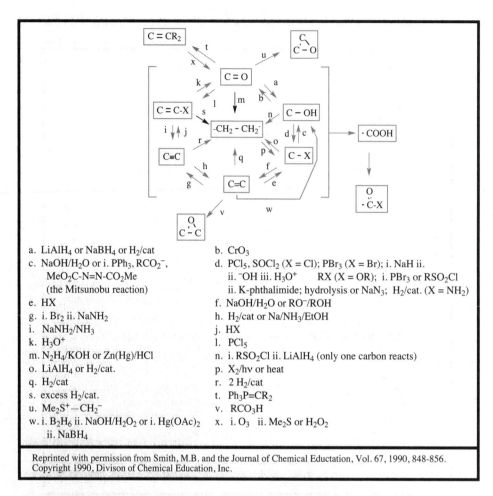

a. LiAlH₄ or NaBH₄ or H₂/cat

$a.$ $LiAlH_4$ or $NaBH_4$ or H_2/cat
$b.$ CrO_3
$c.$ $NaOH/H_2O$ or i. $PPh_3, RCO_2^-,$ $MeO_2C-N=N-CO_2Me$ (the Mitsunobu reaction)
$d.$ $PCl_5, SOCl_2$ (X = Cl); PBr_3 (X = Br); i. NaH ii. ii. ^-OH iii. H_3O^+ RX (X = OR); i. PBr_3 or RSO_2Cl ii. K-phthalimide; hydrolysis or NaN_3; $H_2/cat.$ (X = NH_2)
$e.$ HX
$f.$ $NaOH/H_2O$ or RO^-/ROH
$g.$ i. Br_2 ii. $NaNH_2$
$h.$ H_2/cat or $Na/NH_3/EtOH$
$i.$ $NaNH_2/NH_3$
$j.$ HX
$k.$ H_3O^+
$l.$ PCl_5
$m.$ N_2H_4/KOH or $Zn(Hg)/HCl$
$n.$ i. RSO_2Cl ii. $LiAlH_4$ (only one carbon reacts)
$o.$ $LiAlH_4$ or $H_2/cat.$
$p.$ X_2/hv or heat
$q.$ H_2/cat
$r.$ 2 H_2/cat
$s.$ excess $H_2/cat.$
$t.$ $Ph_3P=CR_2$
$u.$ $Me_2S^+-CH_2^-$
$v.$ RCO_3H
$w.$ i. B_2H_6 ii. $NaOH/H_2O_2$ or i. $Hg(OAc)_2$ ii. $NaBH_4$
$x.$ i. O_3 ii. Me_2S or H_2O_2

Figure. 1.1. Functional Group Exchange Reaction Wheel. A Visual Reminder of the Interconvertibility of Functional Groups.

In most undergraduate Organic chemistry textbooks, the functional group approach to introducing chemistry means that groups with related chemical properties may be presented in different semesters. Although alkenes and carbonyl both react as a base in the presence of an acid, these groups are presented at different places and at different times during a typical course. The approach can make the relationship of one functional group to another difficult to see. Figure 1.1[7] provides a visual reminder of these relationships, which are often essential for completion of a total synthesis. In the context of this chapter, it illustrates chemical relationships that can be used to understand functional group interchange reactions. In the example at hand, the relationship between a bromide and an alkene is apparent from the table (C-C-Br ⇒ C=C). It is also possible to write a functional group exchange based on the alkene as C=C ⇒ C-C-Br.

It is important to emphasize that Figure 1.1 is *not* intended as a device to memorize specific reactions, but rather to introduce the idea of using various functional groups in a synthesis by recognizing their synthetic relationships. Synthetic sequences for *both* disconnections *a* and *b* are shown. The best route is probably involves conversion of **20** to 1-bromomethyl-propane and reaction with enolate **33** (see sec. 9.2, 9.3.A), which is derived from 3-methyl-2-butanone. This synthesis is considered to be better than the second one shown (based on disconnection *b*) since it involves simpler reagents and is shorter (fewer chemical steps). The synthetic chemist must decide which is best, however, based upon his or her own experience and goals. The best route is a subjective judgment, although it usually makes more sense to follow a short and simple route rather than a long and complex one.

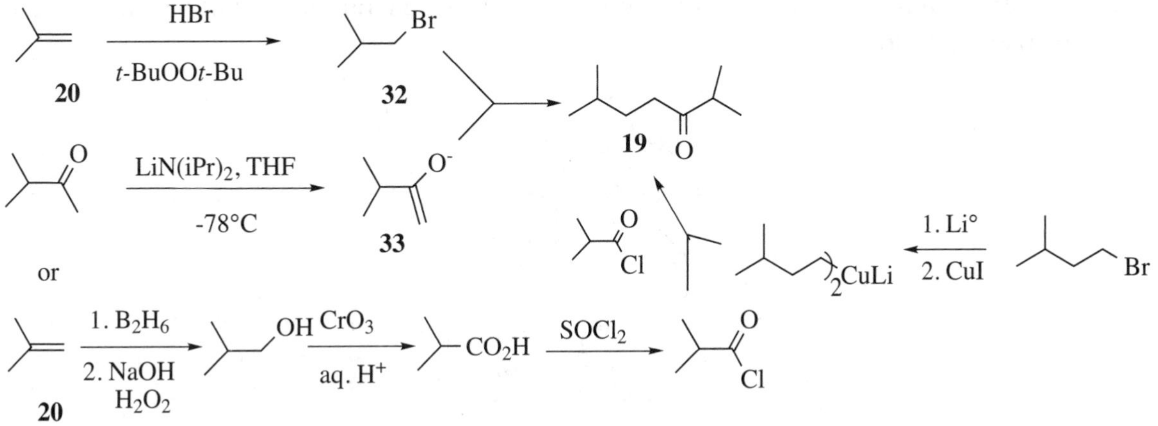

1.3.   BOND PROXIMITY AND IMPLICATIONS FOR CHEMICAL REACTIONS

How is the synthetic analysis just presented related to a study of chemical reactions? In all of these disconnections, we converted the disconnect products to actual compounds so we could predict which chemical reactions might work. In all cases, the disconnected bonds were either the one directly attached to the functional group (C-X) or the one next to it (C-C-X). We can use these observations to understand what reaction types are important for making C-C bonds.

When a functional group (X) is attached to a carbon atom, the bond polarity is important to about the third bond, as indicated in **25**. If we make the *assumption* that most reactions involve highly polarized species, then the three bonds just mentioned become very important. The three important bonds are the C-X bond (called the α bond), the adjacent one (the β bond), and the γ bond, as seen in **34**. If we disconnect the α bond, the synthetic step to form that bond requires direct attack of an X group (where X is an heteroatom) on a suitably functionalized carbon. This process is illustrated by **35**, where a nucleophilic X group attacks the electrophilic carbon. A simple example is the reaction of azide ion (N_3^-) with 1-iodopropane to give 1-azidopropane. If we disconnect the β bond, we obtain **36**, where the natural bond polarization suggests a reaction in which a nucleophilic carbon species attacks the electropositive carbon of the C-X unit. If we assume the functional group is C=O, addition of a Grignard reagent to an aldehyde fits this description. Finally, disconnection of bond γ leads to **37** (also see Chapter 10), and conjugate addition of an organocuprate to a α,β-unsaturated ketone fits this type of disconnection (where X is C=O).

The point of this section is to show how disconnection of molecules that contain polarized functional groups c an be based on natural bond polarization characteristics. This analysis is based on a variety of common reactions, which is important not only for planning a synthesis but for understanding how chemical reactions work. This general theme will be elaborated in many chapters to follow.

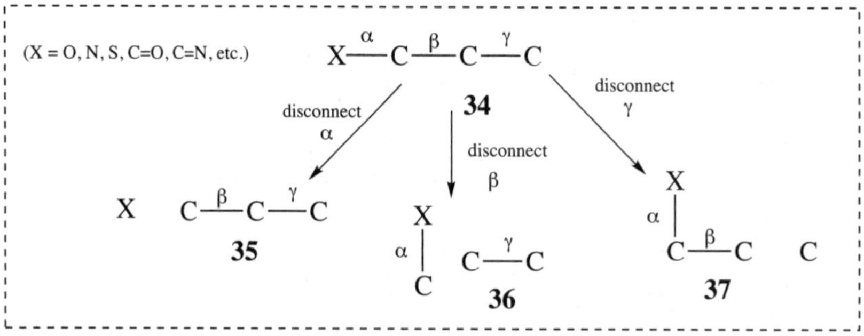

1.4.   STEREOCHEMISTRY

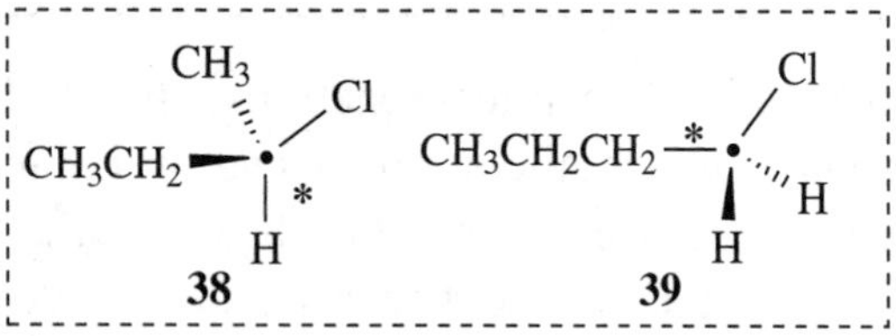

The concept of chirality and absolute configuration is introduced early in undergraduate Organic chemistry courses. For that reason, this section is intended as a review and it is assumed the reader has some familiarity with the concepts. This section will discuss stereoisomers, particularly enantiomers and diastereomers. We will begin with chlorides **38** and **39**, which have the same empirical formula but are clearly different molecules. They are isomers: two or more molecules that have the same empirical formula. They are different molecules! The term isomer does not adequately characterize the relationship between **38** and **39**, since they are of the same chemical type. The term used to define the relationship between **39** and **38** is **regioisomer**: two or more molecules with the same empirical formula, but with a different attachment of the atoms (*different connectivity*).

Another type of isomerism occurs when two molecules have the same empirical formula and also have the same conductivity of atoms but are different molecules. They differ only in the relative spatial positions of the atoms, and are called **stereoisomers**.[12] The following sections will discuss different types of stereoisomers and their characteristic properties.

1.4.A. Absolute Configuration in Chiral Nonracemic Molecules

When an atom is bound to four different atoms or groups in a tetrahedral arrangement, that atom is said to be **stereogenic** or chiral. The second carbon (C*) in 2-chlorobutane (**38**) is a stereogenic center. The C* in **39** is not stereogenic, since that carbon has two identical atoms (H) attached to it. Why is identification of C* in **38** as a stereogenic center important? The answer appears when the mirror image of **38** is drawn (see **40**), and the Cl of **38** is reflected into the Cl in **40**. In addition, hydrogen reflects to hydrogen, methyl to methyl, and ethyl to ethyl, but

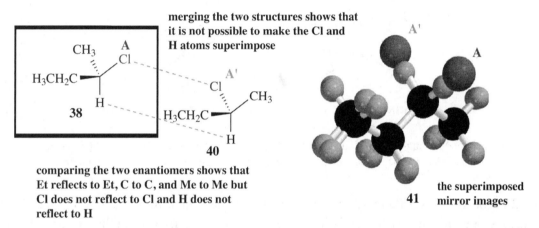

merging the two structures shows that it is not possible to make the Cl and H atoms superimpose

comparing the two enantiomers shows that Et reflects to Et, C to C, and Me to Me but Cl does not reflect to Cl and H does not reflect to H

the superimposed mirror images

this process leads to two different molecules. 2-Chlorobutanes **38** and **40** are *not superimposable* by any rotation of bonds, or positioning of the molecules, and they are *different molecules.* More precisely, the molecules have no symmetry (they are asymmetric)[13] as can be seen in **41**, which is an attempt to superimpose **38** and **40**. It is clear that while the Me and Et groups superimpose in **41**, the Cl and H atoms do not. Confirm this for yourself with any manual or electronic model, by making both enantiomers and trying to make all groups superimpose. They do not! 2-Chlorobutanes **38** and **40** are different molecules, but they are clearly isomers, and they are closely related since they are mirror images. They are not regioisomers because the connectivity of atoms is the same. These two molecules are **stereoisomers** (molecules that differ in their spatial arrangement of atoms but have the same point of attachment).[12] 2-Chlorobutanes **38** and **40** are further distinguished as a special type of stereoisomer called **enantiomers**, that is, stereoisomers that are nonsuperimposable mirror images. If a structure

---

12. For a general discussion of stereoisomers, see Eliel, E.L.; Wilen, S.H.; Mander, L.N. *Stereochemistry of OrganicCompounds,* Wiley-Interscience, New York, *1994,* pp 49-70.

13. For a general discussion of symmetry, symmetry operators and their importance in organic chemistry, see Ref. 12, pp 71-99.

and its mirror image are superimposable by rotation or any motion other than bond making and breaking, then they are identical (a single molecule and not enantiomers; **39** is an example). Make a model of **39** and its mirror image and demonstrate that they are superimposable. When molecules contain more than one stereogenic center, diastereomers (a new type of stereoisomer) will be present, but this will be discussed in sec. 1.4.B.

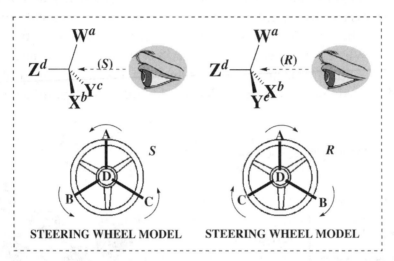

Figure 1.2. The Steering-Wheel Model. In Part, Rreprinted with permission from Cahn, R.S. and the Journal of Chemical Eductation, Vol. 41, **1964**, 116-125. Copyright 1964, Divison of Chemical Education, Inc.

Since 2-chlorobutanes **38** and **40** are different molecules, each must have a unique identifier that allows one enantiomer to be distinguished from the other. The method used is the (R/S) system, which employs a set of priority rules developed by Cahn, Ingold, and Prelog.[14] In this system, each atom attached to the stereogenic center is assigned a priority (***a*** to ***d***, where ***a*** is the highest priority atom and ***d*** is the lowest priority atom). Using the **steering wheel model** (see Figure 1.2) the viewer positions the molecule in such a way that one can sight down the C*-Z^d bond, *with Z^d projected away from the viewer*, as shown. An imaginary line $a{\rightarrow}b{\rightarrow}c$ is drawn and inspected to see if that line of rotation around the wheel follows a clockwise (right) or a counterclockwise (left) path. If the $a{\rightarrow}b{\rightarrow}c$ sequence is clockwise the enantiomer is assigned the (R) configuration. Conversely, if it is counterclockwise, the enantiomer is assigned the (S) configuration. When applying this model, the molecule *must* be rotated so that d is Z^d prior to sighting down the C*-Z^d bond. Several different molecular representations are shown below, along with the correct orientation for determining the (R/S) configuration.

14. (a) Prelog, V.; Helmchen, G. *Angew. Chem. Int. Ed.* **1982**, *21*, 567; (b) Cahn, R.S. *J. Chem. Educ.* **1964**, *41*, 116 (see p 508); (c) Cahn, R.S.; Ingold, C.; Prelog, V. *Angew. Chem. Int. Ed.* **1966**, *5*, 385; (d) Cahn, R.S.; Ingold, C.K. *J. Chem. Soc. (London)* **1951**, 612; (e) Cahn, R.S.; Ingold, C.K.; Prelog, V. *Experientia* **1956**, *12*, 81.

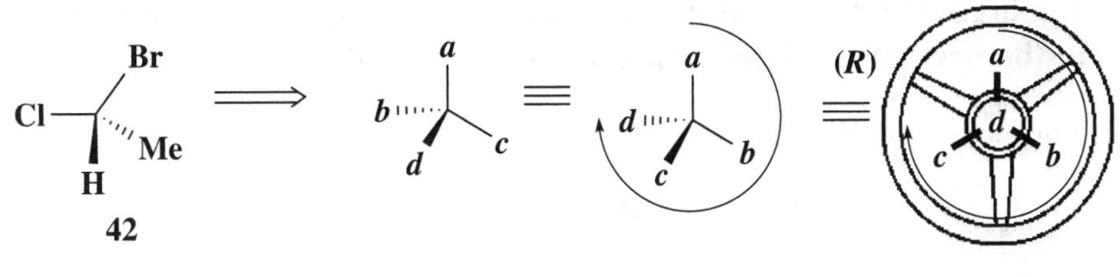

42

Determining the relative priority (*a* to *d*) is accomplished by a series of rules just mentioned, the so-called **Cahn-Ingold-Prelog (or CIP) selection rules**.[14,15,16]

1. **Atoms with a higher atomic number precede atoms of lower atomic number**

    $$O \; > \; N \; > \; C \; > \; H$$

2. **Isotopes with a higher mass number precede isotopes of lower mass number**

    This rule applies to isotopes such as tritium, deuterium and protium ($^3H > {}^2H > {}^1H$). These two rules are illustrated first by 1-bromo-1-chloroethane (**42**), where Br > Cl > Me > H and the *a*→*b*→*c* rotation is clockwise, giving an (*R*) configuration. Similarly, in 1-deuterioethanol (**43**) the priority is O > C > D > H and the *a*→*b*→*c* sequence is also clockwise for an (*R*) configuration. Note the use of a **Fischer projection** for **43**, where the horizontal lines project out of the paper towards you, and the vertical line projects behind the paper. Fischer projections are not used as much nowadays.

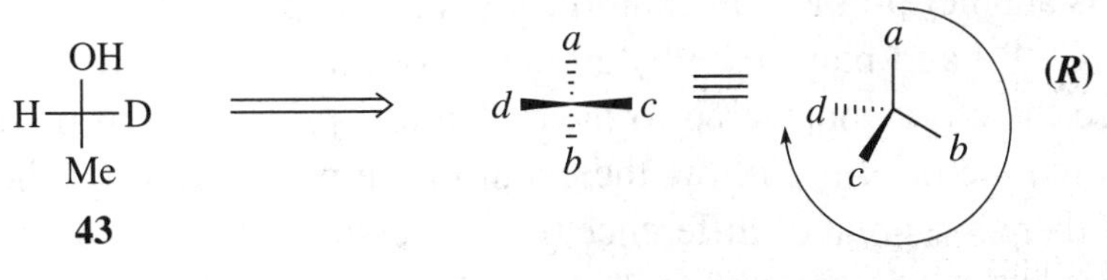

43

A problem arises when two of the atoms (not the groups) attached to C* are the same. In 4-methylhex-5-en-1,3-diol (**44**), the priorities of the atoms attached to C* are O > C ≈ C > H. Rules 1 and 2 do not allow us to distinguish the two carbon atoms. A new rule is required to solve this problem.

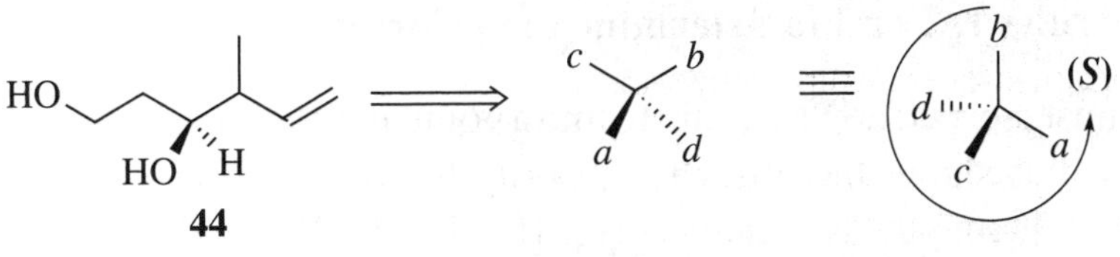

44

---

15. Reference 12, pp 101-112.
16. IUPAC Commission on Nomenclature of Organic Chemistry *Pure Appl. Chem.* **1974**, *45*, 13.

3. **When the highest priority atoms are identical, compare the number of priority atoms at the first point of difference to distinguish them.**

Generic priorities by this rule are:

$$C^{CCC} > C^{CCH} > C^{CHH} > C^{HHH}$$

and

$$C^{OOO} > C^{OOC} > C^{OCC} > C^{CCC}$$

Examination of **44** reveals that the two carbon atoms attached to C* are C^{CHH} and C^{CCH}. The secondary carbon has a higher priority than the primary carbon. The absolute configuration of **44** is (*S*). Another example is the ketal, **45**. The priority analysis is [C*-O-C^{CHH} = *b*], [C*-O-C^{CCH} = *a*] and [C*-C^{CHH}-C^{HHH} = *d*], [C*-C^{CHH}-C^{CHH} = *c*] so the absolute configuration of **45** is (*R*).

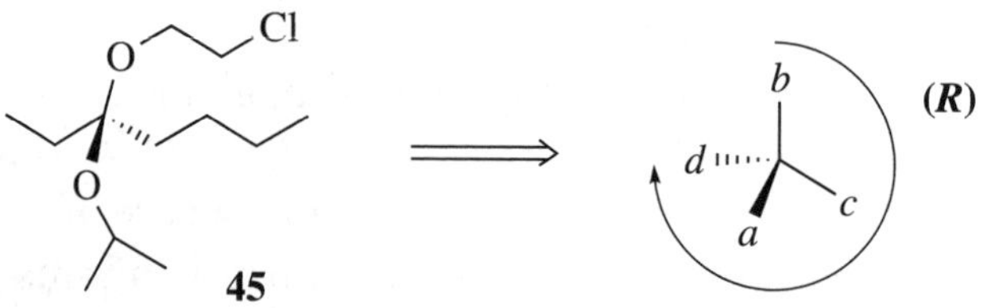

There is another problem in example **45**. As one proceeds down the carbon chain, the ethyl (C^{CHH}) and butyl (C^{CHH}) groups have the same atoms attached directly to C*, and are therefore considered to have identical priorities (no point of difference). To determine the priority, follow the chain to the next carbon of the ethyl and the butyl where there is a point of difference (C^{HHH} versus C^{CHH}). Similarly, the oxygen chains show no point of difference with oxygen attached to C* in both cases (O^{CHH}), but a point of difference was reached at the carbon attached to each oxygen atoms (O-C^{CHH} versus O-C^{CCH}), and this analysis can be formalized as a general rule.

4. **When the first point of difference contains two or more identical atoms, proceed atom by atom down the highest priority chain to the next point of difference and apply rules 1, 2 or 3 to determine the priority.**

One must proceed down a chain to find a point of difference in 3-hydroxy-1-aminohexane (**46**), which shows the following priority for atoms attached to C*: O (*a*) > C ~ C > H (*d*). Comparing the two chains gives [C*-C^{CHH}-C^{NHH} = *b*] and [C*-C^{CHH}-CH^{CHH} = *c*].

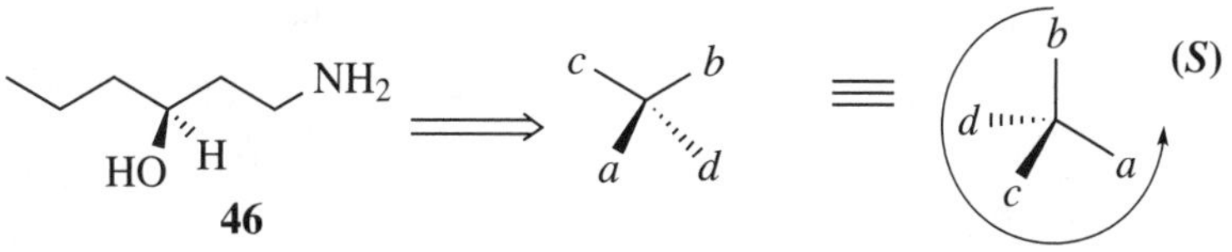

It is necessary to go to the second atom to find a point of difference that will distinguish these groups, and **46** has an (*S*) configuration. Similarly, 1-bromo-7-methyl-7-tetradecanol (**47**) shows the hydroxyl O and the methyl group to be (*a*) and (*d*), respectively, but to determine (*b*) and (*c*) requires analysis to the sixth carbon in each chain. This priority establishes the (*S*) configuration.

$$O(a) > C \approx C \approx C \qquad C^*-C^{HHH} = (d)$$

$$C^*-C^{CHH}-C^{CHH}-C^{CHH}-C^{CHH}-C^{CHH}-C^{BrHH} = (b)$$

$$C^*-C^{CHH}-C^{CHH}-C^{CHH}-C^{CHH}-C^{CHH}-C^{CHH} = (c)$$

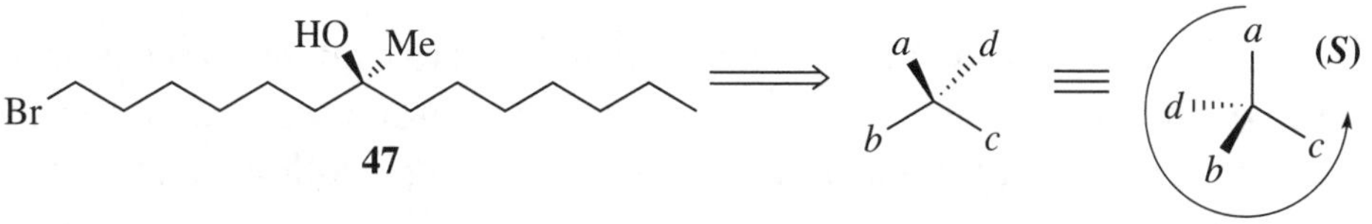

In **48**, the methyl and hydrogen are assigned the priorities (*c*) and (*d*). To determine the *a* and *b* priorities we must focus on the two arms that contain OCH. The highest priority chain is along C-O-C (the oxygen chain), but both groups are O^C-C^{HHH}. They are identical and we cannot use them to establish the priority. In such a case, the lower priority carbon chain must be used (in this case, the isopropyl and ethyl groups rather than the methoxy groups), giving $C^*-C^{OCH}-C^{CCH}$ for (*a*) and $C^*-C^{OCH}-C^{CCH}$ for (*b*) and the molecule has the (*S*) configuration.

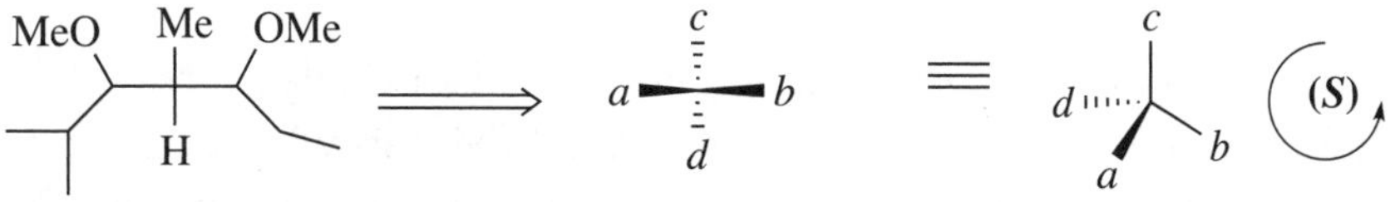

A common misconception for rules 3 and 4 arises from an incorrect interpretation of rules 1 and 2. In **49**, methyl and hydrogen are the (*c*) and (*d*) groups. An analysis by rule 3 predicts $C^{CCH} > C^{CHH}$. In the C^{CHH} chain (-CH$_2$CH$_2$CH$_2$OH), however, there is an oxygen. The CH(Me)CH$_2$CH$_2$CH$_2$CH$_3$ (C^{CCH}) chain does not contain a heteroatom. *The rules dictate that the priority be determined at the first point of difference, not the second.* The presence of the oxygen is, therefore, irrelevant and the C^{CCH} chain will have priority over the C^{CHH} chain. The absolute configuration of **49** is (*S*). Cyclic compounds are treated by the same rules, but each arm of the ring is viewed as a different group and then compared atom-by-atom. 3-Chlorocycloheptanone (**50**), for example, has Cl(*a*) > C ≈ C > H(*d*). The carbonyl chain is C*-C^{CHH}-$C^{OO°C}$ (*b*), in contrast with the carbon arm, C*-C^{CHH}-C^{CHH} (*c*). The configuration is (*R*).

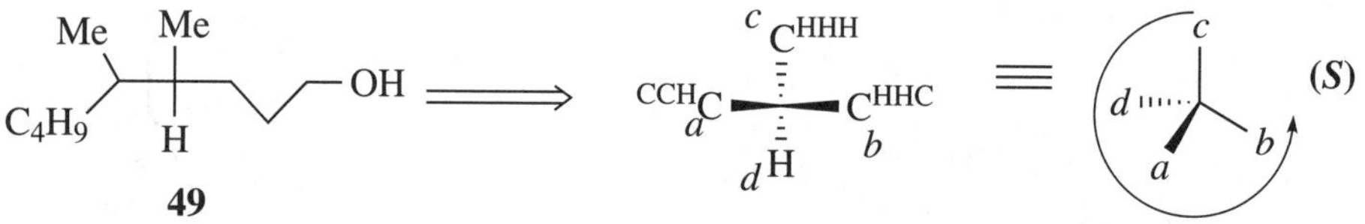

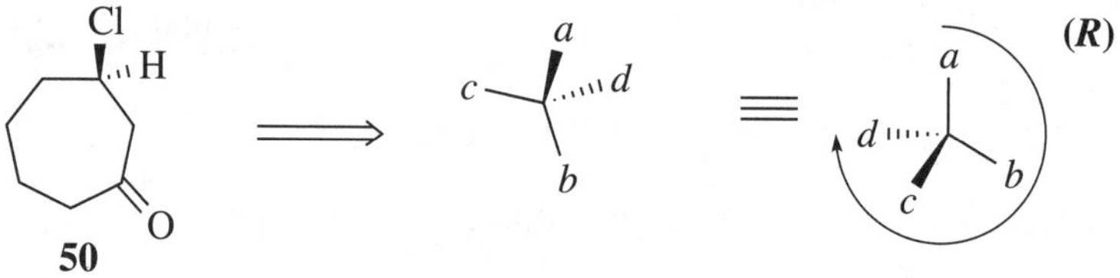

In **51**, the methyl and hydrogen are again (*c*) and (*d*), but the rings pose a problem, one arm of the ring points to chlorine and one to hydroxyl. Since the secondary hydroxy carbon is closest to C*, the bottom arm of each ring is followed rather than the top arm. By this route, the left ring is highest priority:

(C*-C^{CCH}-C^{COH}-C^{CHH}-C^{ClCH} vs. C*-C^{CCH}-C^{COH}-C^{CHH}-C^{CHH}), and the absolute configuration is (*S*).

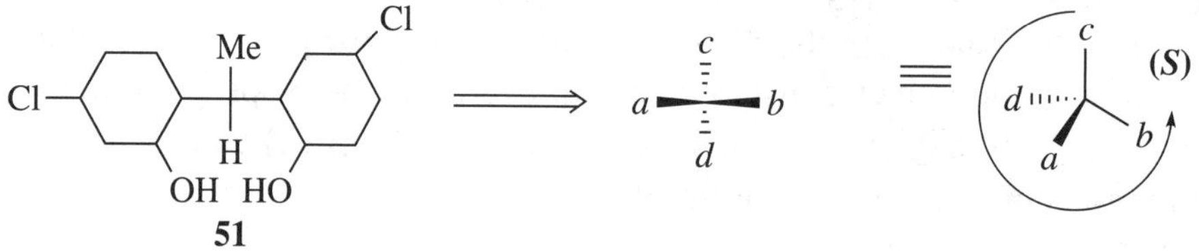

Cycloheptanone (**50**) contained a carbonyl, and the analysis treated that carbon as C^{OOC}. Where did the second oxygen come from? Multiply bonded atoms such as this are found in alkenes, alkynes, carbonyls, and nitriles, and they require yet another rule.

5. **If an atom is attached to another by a multiple bond (double or triple) both atoms are considered to be duplicated. The duplicated atom ($X°$) is considered to have a valence of zero and has a lower priority than a real atom (X), if that is the only point of difference.**

This rule will convert the following common functional groups into their phantom counterparts for consideration in the priority scheme.

$$C^*-CH=CH_2 \equiv C^*-C^{HCC^\circ}-C^{C^\circ HH}$$

$$C^*-C\equiv C\text{-Me} \equiv C^*-C^{CC^\circ C^\circ}-C^{CC^\circ C^\circ}-C^{HHH}$$

$$C^*-C\equiv N \equiv C^*-C^{NN^\circ N^\circ}-N^{CC^\circ C^\circ}$$

$$C^*-C(=O)-C \equiv C^*-C^{OO^\circ C}\_OCC^{\circ\bullet\bullet}$$

$$C^*-CO_2H \equiv C^*-C^{OO^\circ O}-O^{CC^{\circ\bullet\bullet}}$$

Generally, both X° and X are simply treated as real atoms. If the only choice is between a duplicated atom X° and a real atom (X), however, the duplicated atom has a lower priority (in this case, all other rules have failed to establish the priorities). With the *tert*-butyl fragment, for example, proceeding down the chain to the first point of difference leads to *tert*-butyl with a lower priority than the alkynyl fragment:

$$C^*-C\equiv C\text{-Me vs. } C^*-CMe_3 \qquad\qquad C^*-C^{CC^\circ C}-C^{CC^\circ C} > C^*-C^{CCC}-C^{HHH}$$

The C and C° are taken as equal in this analysis and rule 4 determines the priority. We compare $C^{CC^\circ C}$ with C^{HHH}, giving the alkyl fragment a higher priority. Another example of this rule is the analysis of 6-methyl-hept-1-en-4-ol (**52**), which shows O(*a*) > C ≈ C > H(*d*). The alkenyl arm is $C^*-C^{CHH}-C^{CHC^\circ}$ and the alkyl arm is $C^*-C^{CHH}-C^{CCH}$. In this case the rules are unable to distinguish the priority of these two fragments. One proceeds to the next atom ($C^{C^\circ HH}$), which is compared with C^{HHH} ($=CH_2$ vs. CH_3). The former atom is higher in priority and becomes group (*b*). The absolute configuration of **52** is (*S*).

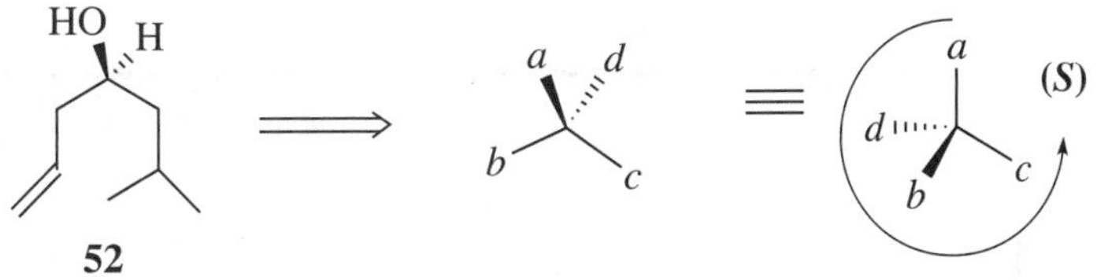

52

The International Union of Pure and Applied Chemistry (IUPAC) Commission assembled a list of common substituents, sorted by increasing order of sequence rule preference (number 76, iodo, is the highest priority and number 1, hydrogen, is the lowest priority in that table).[16] With this table two or more substituents can be evaluated in order to determine their relative priority using the CIP selection rules.[17] Higher numbers have a higher priority. Iodo (76) has a higher priority than bromo (75) and hydrogen (1) has the lowest priority. Similarly, allyl (10) has a higher priority than propyl (4), acetylenyl (21, C≡CH) has a higher priority than *tert*-butyl (19), and hydroxyl (57) has a higher priority than amino (43).

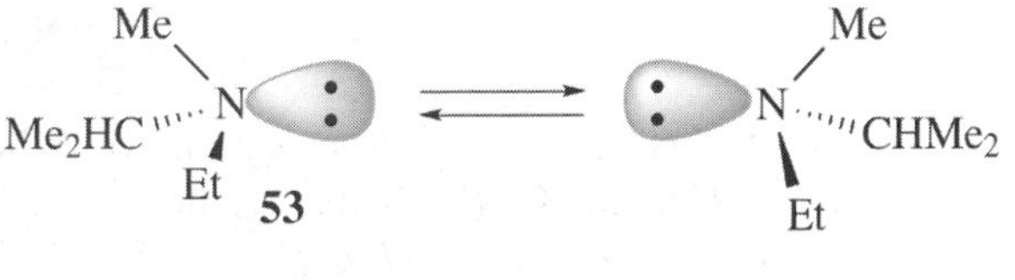

53

17. Cahn, R.S. *J. Chem. Educ.* **1964**, *41*, 116.

In principle, a nitrogen atom can be stereogenic if there are three different alkyl groups on nitrogen (the fourth group is the lone pair of electrons on nitrogen). The nitrogen may be a stereogenic center, but rapid inversion at nitrogen leads to the mirror image (see **53**) being present at the same time. Because of this facile racemization such compounds are not optically active unless this **fluxional inversion** can be inhibited. Approximately 2×10^{11} inversions occur each second for ammonia.[18] The energy barrier for this inversion is somewhat higher in amines (see Table 1.3)[20] due the presence of alkyl groups on nitrogen that are bulkier than the hydrogens in ammonia. Nonetheless, alkyl amines undergo rapid inversion.[19] The magnitude of the energy barrier to inversion in amines is determined by the inter-group bond angle (α, C1-N-C2), and the corresponding bending force constant.[20] Table 1.3[20] shows values of the parameter (α) for ammonia, trimethylamine, phosphine, and trimethylphosphine, along with an estimate for the energy barrier to inversion. Older estimates of these energy barriers are also included.[21] The inversion at nitrogen in alkyl amines cannot be stopped at the reaction temperatures usually employed in organic chemical reactions. When inversion is impossible due to structural features of a molecule such as those found in bicyclic and polycyclic amines, the nitrogen is stereogenic and enantiomers are observed.

Table 1.3.   Energy Barrier to Inversion of Amines and Phosphines

Compound	α	$E_{inversion}$ (kcal mol^{-1})	$E_{inversion}$ (kJ mol^{-1})
NH_3	106.77°	5.58[a]	23.2[a]
		11[b]	46.0[b]
PH_3	93.3°	27[a]	112.9[a]
		47[b]	196.5[b]
NMe_3	109.0°	7.46[a]	31.2[a]
		15[b]	62.7[b]
PMe_3	100°	20.4[a]	85.3[a]
		57[b]	283.3[b]

[a] Reference 20   [b] Reference 21

18. Smith, M.B.; March, J. *March's Advanced Organic Chemistry, 6th ed.*, Wiley, New York, **2007**, pp 142-144.
19. (a) Mislow, K. *Pure Appl. Chem.* **1968**, *25*, 549; (b) Rauk, A.; Allen, L.C.; Mislow, K. *Angew. Chem. Int. Ed.* **1970**, *9*, 400; (c) Lambert, J.B. *Top. Stereochem.* **1971**, *6*, 19.
20. Koeppl, G.W.; Sagatys, D.S.; Krishnamurthy, G.S.; Miller, S.I. *J. Am. Chem. Soc.* **1967**, *89*, 3396.
21. Kincaid, J.F.; Henriques, Jr., F.C. *J. Am. Chem. Soc.* **1940**, *62*, 1474.

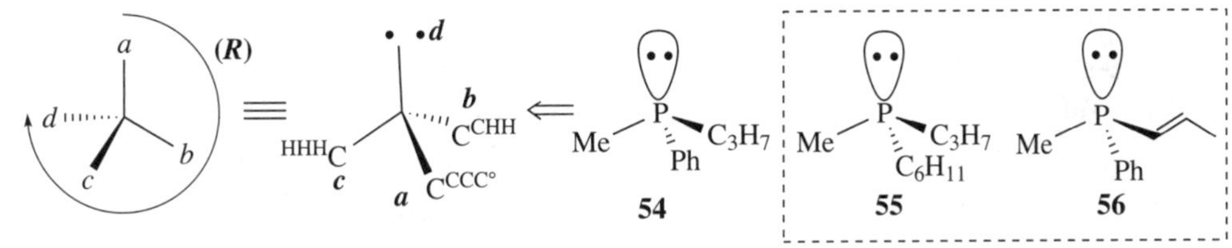

The energy barrier for inversion is low for second row elements (C, O, N), and rapid inversion occurs. With elements in the third row (such as P and S), however, inversion is slow at ambient temperature and those molecules may exist as enantiomers. Methylphenylphosphine (**54**) is configurationally stable at 25°C, although it rapidly inverts at 130°C.[22b] At 130°C, this barrier was measured to be 30.7 kcal mol^{-1} (128.4 kJ mol^{-1}),[22a] for a rate of inversion of 3.34 x 10^5 s^{-1}.[22] At 130°C, phosphine **55** showed a rate of inversion of 0.043 x 10^5 s^{-1} and the rate for **56** was 1.44 x 10^5 s^{-1} [E_a = 32.2 kcal mol^{-1} (134.6 kJ mol^{-1})].[23]

If a nitrogen atom is a stereogenic center, it must be assigned a configuration by the CIP selection rules. An example is the bridgehead nitrogen in (–)-castoramine (**57**),[24] which is incapable of inversion at nitrogen because of its rigid bicyclic structure.[25] Castoramine is also drawn such that the *trans*-orientation of the electron pair is indicated, which is preferred to the *cis*- orientation by ≈ 2.4 kcal mol^{-1} (10.0 kJ mol^{-1}).[25] This barrier effectively locks the molecule into the trans conformation. In this case, the electron pair must be considered a group but the first five rules do not allow it to be assigned a priority. A sixth rule is necessary.

6. **Lone electron pairs receive an atomic number of zero and are assigned the lowest priority. The duplicated C, O and N atoms from rule 5 have a higher priority rather than an electron pair.**

For (–) castoramine (**57**), rule 6 sets the priority such that the stereogenic nitrogen center is (*R*). Just as a nitrogen atom can be stereogenic, a phosphorus atom in a phosphine such as **54** can be considered chiral at temperatures up to about 130°C, as noted in Table 1.3. The lone pair electrons have the lowest priority by rule 6, and the absolute configuration is (*R*).

The six rules for determining the priority of groups on stereogenic centers can be applied to any molecule. Natural products are particularly interesting, since they are

22. (a) Horner, L.; Winkler, H.; Rapp, A.; Mentrup, A.; Hoffmann, H.; Beck, P. *Tetrahedron Lett.* **1961**, 161; (b) Horner, L.; Winkler, H. *Ibid* **1964**, 461.
23. Baechler, R.D.; Mislow, K. *J. Am. Chem. Soc.* **1970**, *92*, 3090.
24. LaLonde, R.T.; Muhammad, N.; Wong, C.F.; Sturiale, E.R. *J. Org. Chem,* **1980**, *45*, 3664.
25. (a) Aaron H.S.; Ferguson, C.P *Tetrahedron Lett.* **1968**, 6191; (b) Aaron H.S. *Chem. Ind. (London)* **1965**, 1338.

the targets of many synthetic endeavors. As noted above, the absolute and relative configuration of such molecules must be known if they become synthetic targets, and this must be factored into the retrosynthetic plan. The *Amaryllidaceae* alkaloid crinine (**58**) has four stereogenic centers, including the ring-fused nitrogen atom. The absolute stereochemistry for each stereogenic center, as the molecule is drawn, is indicated where

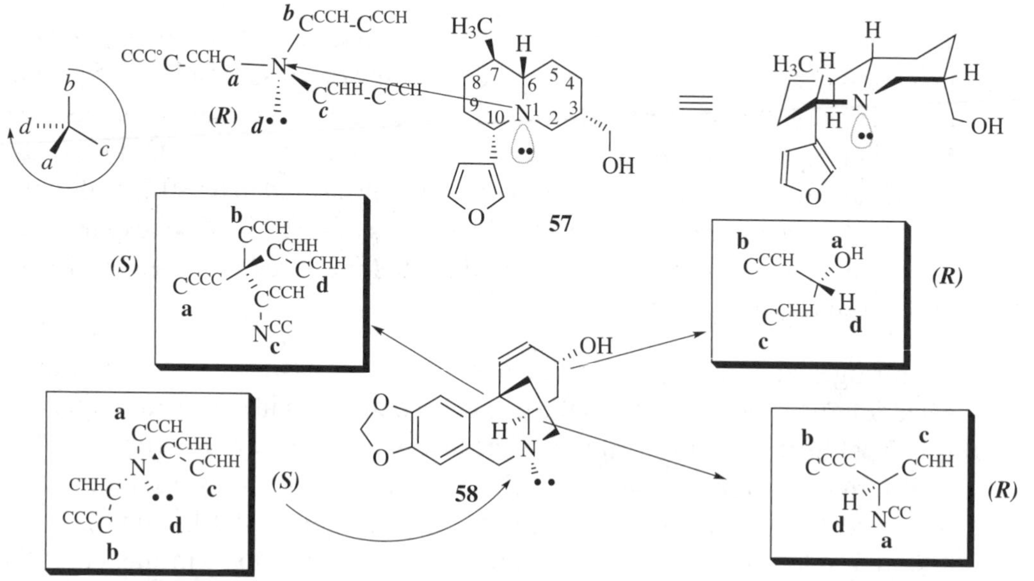

each is treated as if it were an individual and isolated atom. In other words, focus on the carbon bearing the OH group, treat it as an individual tetrahedral carbon with four attached groups (independent of the fact they are all part of one molecule) and then assign priorities. Once done, that carbon atom is determined to have an absolute configuration of (*R*). This fundamental approach is used for all the stereogenic centers, and it can be applied to any molecule, regardless of the complexity. In some cases assigning priorities can be tedious and occasionally confusing, but there are rules to cover all contingencies.

The rules given above allow (*R*) or (*S*) configuration to be assigned for relatively simple molecules, and both **57** and **58** are included as a relatively simple molecule, but application to complex molecules can be problematic. Compound **59**, for example, is non-trivial, and assigning all stereogenic centers by this manual method is at best time consuming, at worst difficult and prone to error. Modern computational methods, with algorithms to asses the absolute configuration, make such assignments rapid and rather easy. One simply draws the structure with the correct stereochemical relationships in the model, and the program makes the correct assignment. Indeed, the computer-generated assignments are shown in **59**. Draw structures **60** and **61**, and assign the absolute configuration using Spartan, and compare with the results that are given. Try it using the rules just discussed as well! Compound **60** is the oligocyclopropane FR-900848,[26] and **61** is the macrolide toxin pectenotoxin 2.[27] A similar analysis of the absolute configuration is shown for all stereogenic centers in ciguatoxin, **62**.[28]

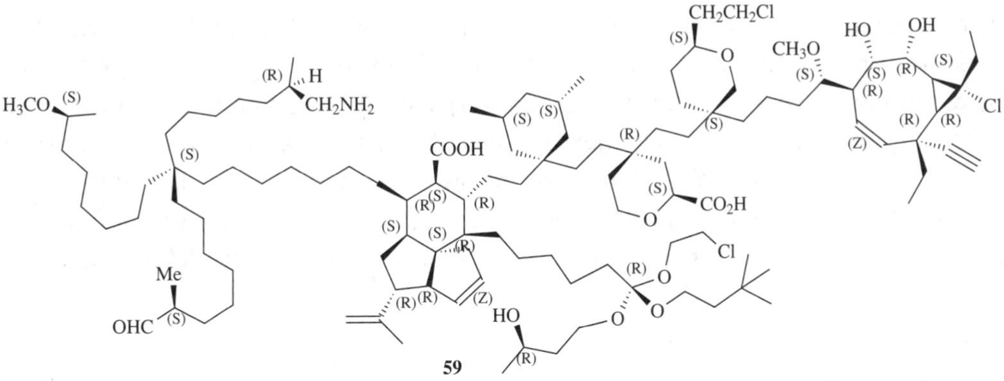

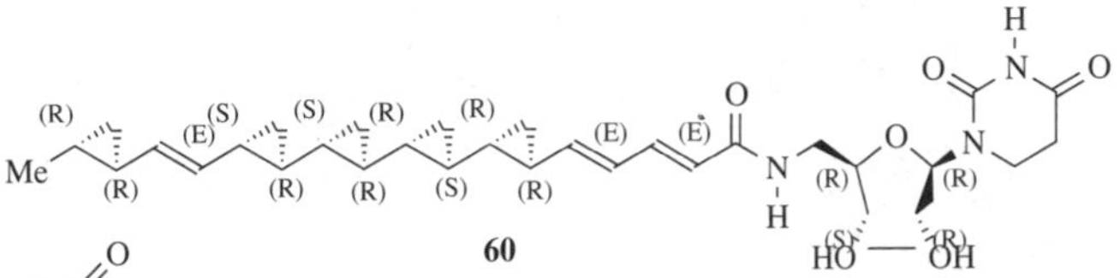

26. Yoshida, M.; Ezaki, M.; Hashimoto, M.; Yamashita, M.; Shigematsu, N.; Okuhara, M.; Kohsaka, M.; Horikoshi, K. *J. Antibiot.* **1990**, *43*, 748.

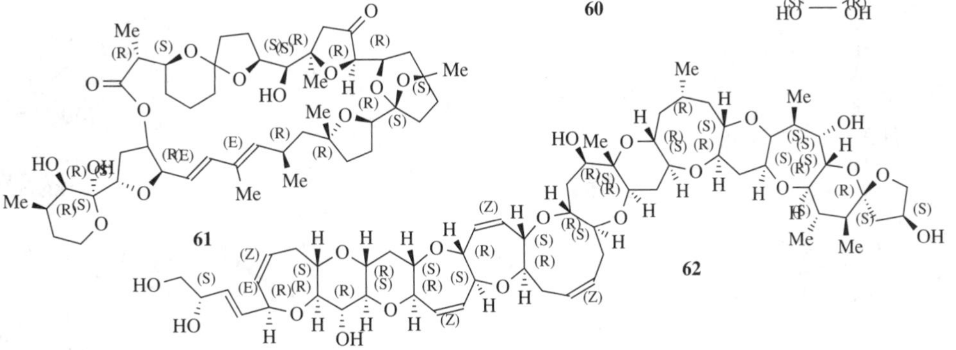

27. (a) Yasumoto, T.; Murata, M.; Oshima, Y.; Sano, G. K.; Matsumoto, J. *Tetrahedron* **1985**, *41*, 1019; (b) Murata, M.; Sano, M.; Iwashita, T.; Naoki, H.; Yasumoto, T. *Agric. Biol. Chem.* **1986**, *50*, 2693; (c) Sasaki, K.; Wright, J.L.C.; Yasumoto, T. *J. Org. Chem.* **1998**, *63*, 2475; (d) Suzuki, T.; Beuzenberg, V.; Mackenzie, L.; Quilliam, M.A. *J. Chromatogr. A* **2003**, *992*, 141; (e) Miles, C.O.; Wilkins, A.L.; Samdal, I.A.; Sandvik, M.; Petersen, D.; Quilliam, M.A.; Naustvoll, L.J.; Rundberget, T.; Torgesen, T.; Hovgaard, P.; Jensen, D.J.; Cooney, J.M. *Chem. Res. Toxicol.* **2004**, *17*, 1423; (f) Halim, R.; Brimble, M.A.; Merten, J. *Org. Lett.* **2005**, *7*, 265.

28. (a) Scheuer, P.J.; Takahashi, W.; Tsutsumi, J.; Yoshida, T. *Science* **1967**, *155*, 1267; (b) Murata, M.; Legurand, A.M.; Ishibashi, Y.; Fukui, M.; Yasumoto, T. *J. Am. Chem. Soc.* **1990**, *112*, 4380; (c) Satake, M.; Morohashi, A.; Oguri, H.; Oishi, T.; Hirama, M.; Harada, N.; Yasumoto, T. *J. Am. Chem. Soc.* **1997**, *119*, 11325; (d) Hamajima, A.; Isobe, M. *Org. Lett.* **2006**, *8*, 1205.

1.4.B. Diastereomers

When there is more than one stereogenic center, the maximum number of possible stereoisomers is predicted by the 2^n **rule** (for n stereogenic centers there is a maximum of 2^n stereoisomers). Therefore, two stereogenic centers in a molecule such as **63** lead to four possible stereoisomers with the configurations C1(R)-C2(R), C1(S)-C2(S), C1(R)-C2(S) and C1(S)-C$_2$(R). Bromohydrin **63** has the configuration (*SS*), and its enantiomeric mirror image (**64**) is (*RR*). Similarly, the (*RS*) compound (**65**) and its mirror image (**66**) with the (*SR*) configuration are enantiomers. Comparison of **63** and **65** shows they are also stereoisomers, but they are *not* mirror images *nor* are they superimposable. This type of stereoisomer is given the name **diastereomer**: stereoisomers with two or more stereogenic centers that are not superimposable and not mirror images. By this definition, **64** and **66** are also diastereomers, as are **64** and **65** or **63** and **66**.

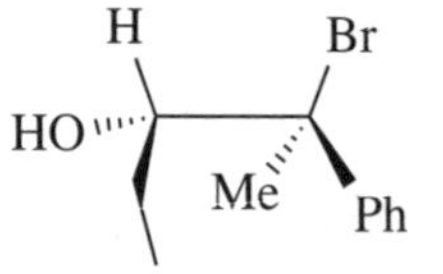

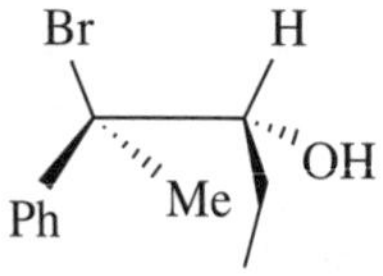

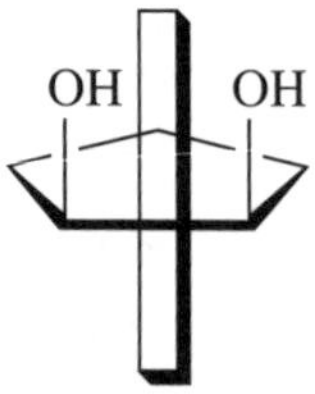

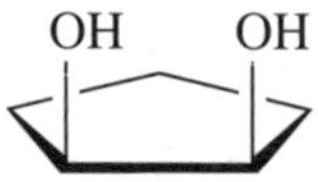

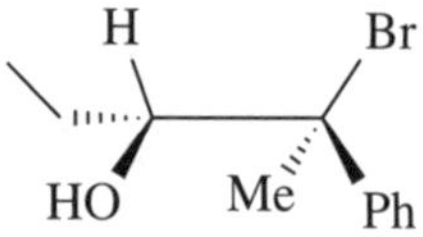

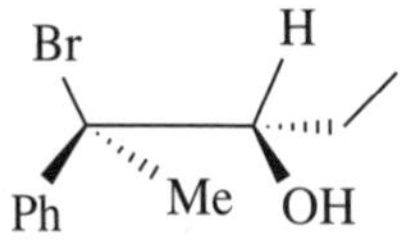

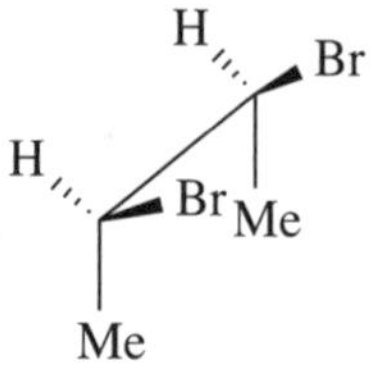

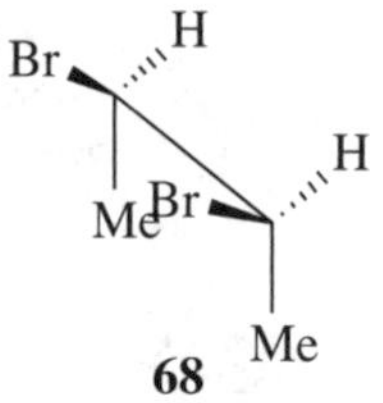

There are cases when a plane of symmetry bisects a diastereomer, as in *cis*-1,2-cyclopentanediol (**67**) and (2R,3S)-dibromobutane **68**. The mirror image of (2R,3S)-dibromobutane is (2S,3R)-dibromobutane, the same compound. In effect, one half of the molecule is the mirror image of the other half as a result of the symmetry in the molecule, which leads to *fewer* stereoisomers than the maximum predicted by the 2^n rule. Structures **67** and **68** are superimposable upon their mirror images so *they are not enantiomers but are the same compound* and are called **meso compounds**. The cis-diol **67** is a meso compound but its diastereomer, the trans-diol, exists as enantiomers **69** and **70**. There are a total of three stereoisomers: isomers **67** and **69** or **67** and **70** are diastereomers, **69** and **70** are enantiomers and **67** is a meso compound. Similarly, **67** is a meso compound, but **71** [(2S,3S)-dibromobutane] and **72** [(2R,3R)-dibromobutane]

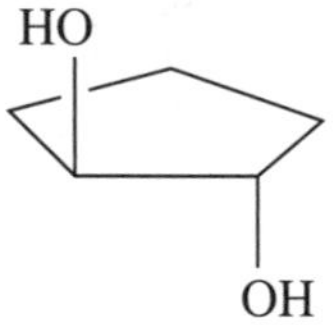

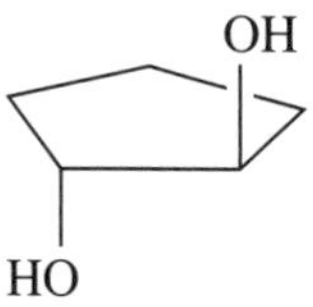

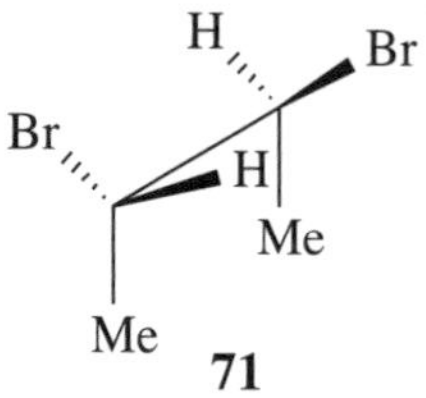

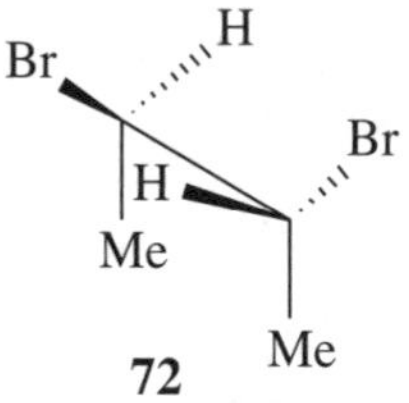

are enantiomers giving the three stereoisomers. Note that changing the configuration of one stereogenic center ($2R \rightarrow 2S$) cannot be done by rotation but only by making and breaking bonds. Dibromides **68** and **71** are diastereomers, as are **68** and **72**.

In chapters to come we will see that many different types of organic reactions produce diastereomers. These diastereomeric pairs show rotation about the carbon-carbon bond (see sec. 1.5.A), which precludes the configurational rigidity necessary for using the cis- and trans-isomer designations. Separate nomenclature systems have been developed based on the relationship of the groups on the stereogenic centers. One system for distinguishing diastereomers labels them threo- and erythro. Unfortunately, there is more than one definition for threo- and erythro. Winstein and co-workers gave the following definition of these diastereomers: in a compound with two asymmetric carbons that has two common ligands and a third that differs, the isomers that would be meso if the third ligand were identical are erythro diastereomers.[29] An alternative definition is: if two asymmetric carbons have only one ligand in common, then the other four ligands are paired in the same commonsense way and isomers that would have equal pairs eclipsed in any conformation are erythro.[30] Eliel, Mislow and their co-workers[31] defined erythro and threo in terms of Fischer projections. The aldol products (see sec. 9.4.A) **73** and **74**, and the ester-aldehyde condensation products **75** and **76**[32] are shown with the erythro and threo notation.

To alleviate the confusion of the various uses and perceptions of the erythro/threo notation,[33] Masamune proposed the terms syn and anti to describe the relative stereochemistry of diastereomers.[34] In this notation, adjacent groups on the same side of an extended (zigzag) structure are syn, and those on opposite sides are anti. This model is illustrated with **77-79**.[34,35] An extended structure is used as the basis of the model.

The clearest way to show differences is to use the absolute configuration (*R*) or (*S*) nomenclature for each stereogenic center in both enantiomers or in enantiomerically pure diastereomers. Throughout this book, the correct configuration will be cited or the syn- and anti- terminology will be used.

---

29. Winstein, S.; Lucas, H.J. *J. Am. Chem. Soc.* **1939**, *61*, 1576, 2845.
30. (a) Lucas H.J.; Schlatter, M.J.; Jones, R.C. *J. Am. Chem. Soc.* **1941**, *63*, 22; (b) Cram, D.J. *Ibid* **1952**, *74*, 2149; (c) Curtin, D.Y.; Kellom, D.B. *Ibid* **1953**, *75*, 6011; (d) House, H.O. *Ibid* **1955**, *77*, 5083.
31. (a) Eliel, E.L.; Wilen, S.H.; Mander, L.N. *Stereochemistry of Organic Compounds*, Wiley-Interscience, New York, **1994**; (b) Eliel, E.L. *Stereochemistry of Carbon Compounds*, McGraw-Hill, New York, **1962**; (c) Mislow, K. *Introduction to Stereochemistry*, W.A. Benjamin, New York, **1965**.
32. Meyers, A.I.; Reider, P.J. *J. Am. Chem. Soc.* **1979**, *101*, 2501.
33. Seebach, D.; Prelog, V. *Angew. Chem. Int. Ed.* **1982**, *21*, 654.
34. Masamune, S.; Ali, Sk.A.; Snitman, D.L.; Garvey, D.S. *Angew. Chem. Int. Ed.* **1980**, *19*, 557.
35. Heathcock, C.H. in *Asymmetric Synthesis, Vol. 3*, Morrison, J.D. (Ed.), Academic Press, New York, **1984**.

1.4.C. Chiral Molecules without a Stereogenic Center (Molecules Containing a Chiral Axis)

A few classes of organic molecules have a chiral axis although they do *not* have a stereogenic center.[36] The mirror image of such a molecule is not superimposable, *which means it is possible to have enantiomers without the presence of a stereogenic center.* Four important classes of compounds that exhibit this property are biaryls such as **80**, alkylidene cyclohexanes (**81**), substituted allenes such as **82** and substituted spiranes such as 3,9-diphenylspiro[5.5]tridecane (**83**).[17,36] Chiral biaryls are important chiral catalysts and are used in many reactions

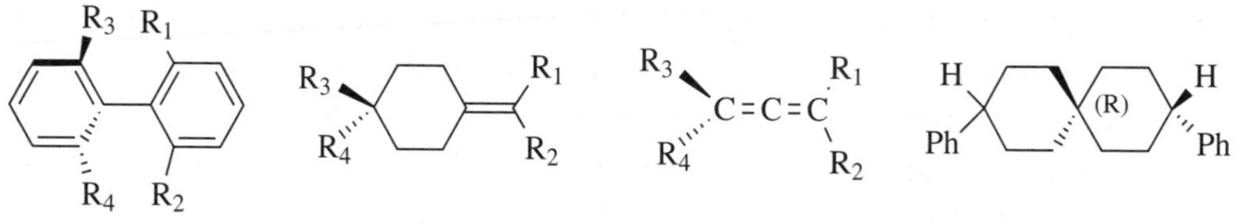

(see sec. 4.8.A, 4.9.G, and sec. 11.9). Obviously, it is important to determine the absolute configuration of both the chiral reactants and the asymmetric products resulting from their use. Allenes are common partners in pericyclic reactions (see secs. 11.4, 11.10 and 11.11) and they are chiral partners in some of these reactions. Alkylidene cyclohexanes are produced by phosphorus ylids upon reaction with cyclohexanone derivatives (see sec. 8.8.A) and the potential for creating asymmetric products is a key consideration in planning a synthesis of such compounds. However, no chiral atom is present and the CIP rules just described do not directly apply.

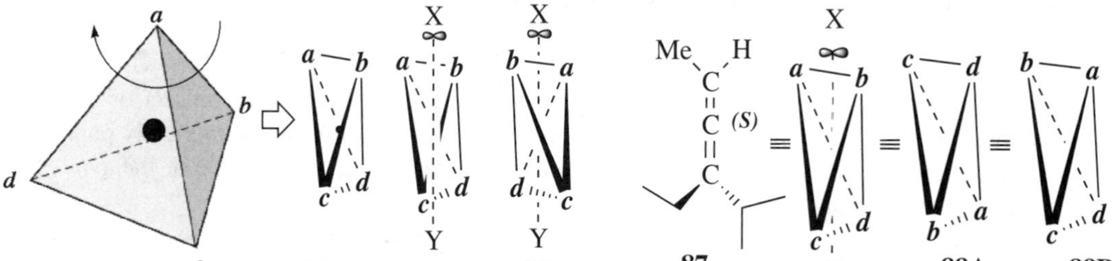

Chiral molecules that have no stereogenic center are evaluated by recognizing the presence of an extended tetrahedron (**85A**).[37] A normal tetrahedral atom is shown by **84**, and if the bond lengths for atoms $a \rightarrow d$ were distorted to give an extended tetrahedron **85A** would be obtained. Rather than a chiral atom, **85A** contains a **chiral axis**, which can be used to assign priorities (see X---Y in **85B**). This model requires that **85B** not be interconvertible with its mirror image

---

36. For a general discussion of this concept and molecules that exhibit this property, see Reference 12, pp 1119-1190.
37. Reference 12, pp 1119-1122.

86 (i.e., rotation about the chiral axis X---Y must not interconvert **85** and **86**).

Allene **87** has a chiral axis since there are four different groups at each corner of the extended tetrahedron. These groups can be assigned priorities $a \rightarrow d$ (**88A**) by the usual rules, and a first glance suggests **88B** (obtained by rotating **88A** so the $a \rightarrow b \rightarrow c$ sequence is in front and the d-atom is to the rear) as the structure to be used for determining the absolute configuration. This is ***incorrect***. There is nothing here to suggest the appropriate angle from which to view the extended tetrahedron. There is no C*-C^d axis from which to view the molecule. The CIP rules were modified to accommodate an extended tetrahedron.[17] (*1*) The top edge of the extended tetrahedron is prioritized a,b and the bottom edge of the extended tetrahedron is also prioritized a,b. (*2*) Near groups precede far groups when viewed from the top of the extended tetrahedron. With these rules, **87** is converted to **89** where the top a,b pair is associated with 1,2 ($a = 1$, $b = 2$). Similarly, the bottom a,b pair is assigned 3,4 ($a = 3$, $b = 4$) since top has priority over bottom. The model is rotated to put (4) to the rear and follows the order $1 \rightarrow 2 \rightarrow 3$ that gives a counterclockwise pathway and an (*S*) configuration. This new model can also be used with allene **90**, where only two different groups are present. The methyl-hydrogen priorities are shown in **86** and converted to the $1 \rightarrow 4$ priority scheme, which places (4) to the rear, giving an (*R*) configuration.[17] Molecules such as this can also be evaluated *a priori* using the **Lowe-Brewster rules**.[38] Eliel et al.[39] point out that the configuration of allenes and alkylidenecycloalkanes can be predicted by these rules in most case, but the model fails for many spirans.

Chiral axes occur for cyclic molecules containing an exocyclic alkylidene moiety such as **91**, which has two different groups at C4 of the cyclohexyl system and two different groups attached to the π bond. The C^{ooo} > H priority for CO_2H and H on the alkene is straightforward, and this constitutes the top of the extended tetrahedron. The cyclohexyl arms are in the plane of the π bond, but the methyl and hydrogen at C4 of the cyclohexane ring are in a different plane that constitutes the bottom of the extended tetrahedron. Since the methyl carbon has priority over the hydrogen the extended tetrahedron is **92A**, which leads to **92B** and the (*R*) configuration.[17]

38. (a) Lowe, G. *Chem. Commun.* **1965**, 411; (b) Brewster, J.H. *Top. Stereochem.* **1967**, *2*, 1; (c) Reference 12, pp 1129-1132.
39. Reference 12, p 1091.

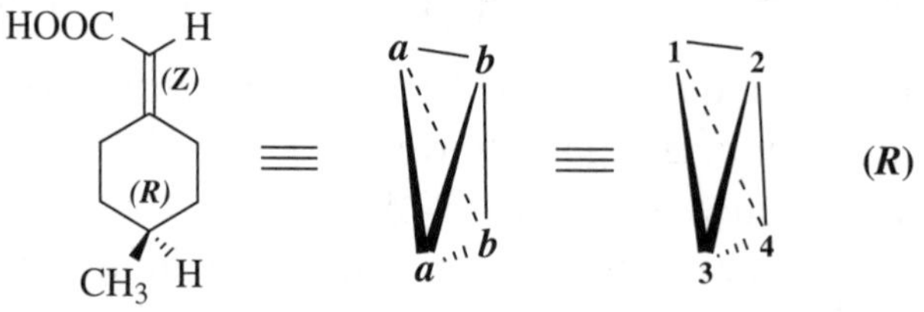

Biaryls such as **93A** can be analyzed with the extended tetrahedron model. The top aromatic ring is prioritized as *a,b* for the 2,6 substituents, as is the bottom ring (see **94A**). The near-far rule leads to **94B** and an (*R*) configuration.[19] A reasonable question asks which aryl ring is on top and which is on the bottom of the extended tetrahedron. Do the rules accommodate these two orientations? Structure **93B** is identical to **93A** except the former has been rotated by 180°. Analysis of both structures leads to an (*R*) configuration. Prioritizing top and bottom *separately* accounts for rotating the molecule in this manner, but the top must be assigned **before** prioritizing the molecule and must **not** be changed after the process has begun. An example of a fused aromatic system used in synthesis is the reducing agent BINAL-H (an abbreviation for **95**, see sec. 4.8.A.).[40] The extended tetrahedron reveals an (*R*)-. Note that binding the aluminum atom into a ring, as shown in **95**, effectively locks the aromatic rings into a single configuration and does not allow racemization unless the C-C-C-C-Al-O- ring is disrupted.

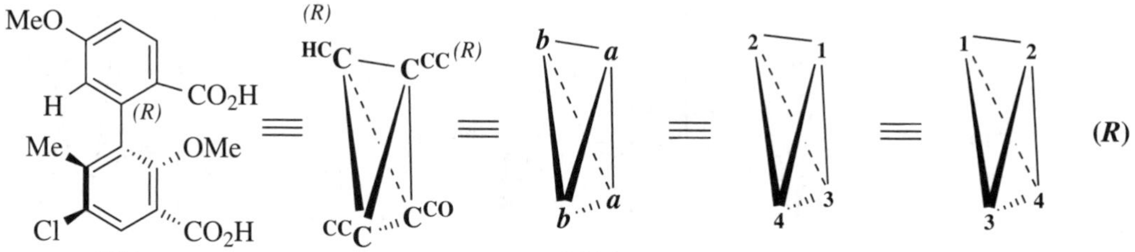

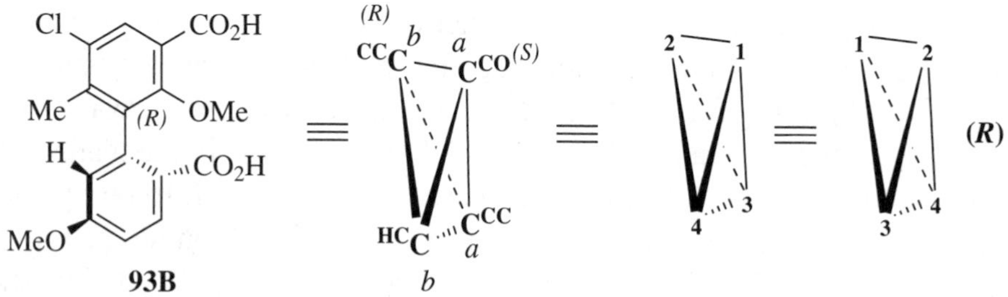

40. (a) Noyori, R.; Tomino, I.; Tanimoto, Y.; Nishizawa, M. *J. Am. Chem. Soc.* **1984**, *106,* 6709; (b) Noyori, R.; Tomino, I.; Yamada, M.; Nishizawa, M. *Ibid* **1984**, *106,* 6717.

95

1.4.D. (*E/Z*) Isomers

Alkenes can exist as stereoisomers. If asked to draw the structure of 3-hexene, two different molecules can be drawn (**96** and **97**). They are not superimposable, nor are they mirror images. Rotation is not possible around the C=C moiety, so the ethyl groups and the hydrogens have different spatial relationships. Those groups cannot be interconverted by rotation. Although no stereogenic center is present, **96** and **97** are stereoisomers and are formally considered to be diastereomers. In this particular case, the C=C unit has identical groups (ethyl) attached and the ethyl groups can be identified as being on the same side of the C=C or on opposite sides. When an alkene contains identical groups and those groups are on the same side, it is called a cis-alkene. When the two identical groups are on opposite sides, it is called a trans-alkene. Alkene **96** is *cis*-3-hexene and **97** is *trans*-3-hexene. These terms are part of the nomenclature as shown.

The terms cis and trans are applied in a straightforward manner for simple alkenes such as *cis*- and *trans*-3-hexene or *cis*- and *trans*-1,2-dibromoethene. For 3-(1-bromo-1-methyl-ethyl)-5-hydroxy-4-(1-methylethyl)-hex-3-enoic acid (**98**), however, the cis-trans nomenclature does not apply. Which of the groups are used to determine sidedness? A more general nomenclature system is required, the (*E/Z*) system.[17,41] The CIP selection rules are used for this system, to assign priorities to each carbon of the double bond. In **99**, *a* and *a'* have the higher priorities and *b* and *b'* have the lower. Since *a* and *a'* are on the same side of the double bond, they are given the designation (*Z*) (from zusammen = together). Similarly, *a* and *a'* are on opposite sides in **100** and this arrangement is given the designation (*E*) (from entgegen = opposite). With this system, the hydroxyl-bearing carbon of C4 in **98** is assigned the highest priority (C^{CHO} vs. C^{CCH} for the isopropyl carbon). Analysis of C3 shows the bromine-bearing carbon to be the highest priority (C^{CCBr} vs. C^{CHH}). This is analogous to **100**, and the name of **98** is (*E*)-(1-bromo-1-methylethyl)- 5-hydroxy-4-(1-methylethyl)-hex-3-enoic acid. This nomenclature is generally applicable to all alkenes that do not have identical groups on one of the alkenyl carbons. 3-Hexene **96** is (*Z*)-3-hexene and **97** is (*E*)-3-hexene. The (*E/Z*) system will be used extensively throughout this book, but the cis/trans designations will be used for simple molecules and to describe relative stereochemistry. For example, the carbonyl group and the hydroxyl-bearing moiety in **98** are cis to each other.

---

41. Reference 12, pp 541-543.

96   97   98

6 OH 2 O
5   1 OH
   3
  4   Br

$a C^{CHO} > C^{CCH}_b$

OH
5
6
4

99
Z

a   a'
b   b'

a   b'
b   a'

100
E

2 O
1 OH
4   3
Br

$a' C^{CCBr} > C^{CHH}_{b'}$

1.4.E. Prochiral Centers

There are molecules that do not possess a stereogenic center but generate a product with a stereogenic center after a chemical reaction. In terms of planning a reaction, it is important that we try to predict whether the product will have an (R)- or an (S)-configuration. To make this prediction, we must know from which face of the molecule the reagent will approach during the reaction. If it approaches from one face the (R)-enantiomer is generated; if it approaches from the opposite face the product is the (S)-enantiomer.

It is obvious that ketone **101** (2-butanone), and alkene **104** do not contain a stereogenic center. If **101** reacts with a Grignard reagent such as phenylmagnesium bromide (PhMgBr), however, the product is a racemic alcohol with enantiomers **102** and **103** (see sec. 8.4.C). Reaction of alkene **104** with a borane such as 9-BBN (9-borabicyclo[3.3.1]nonane) followed by oxidation gives a racemic alcohol, that is, enantiomers **105** and **106** (see sec. 5.4.A). In both cases, the reaction has produced a product that contains a stereogenic carbon. The ketone and alkene are described as **prochiral**,[42] and a working definition of **prochirality** was provided by Hanson:[43] "If a chiral assembly is obtained when a point ligand in a finite non-chiral assembly of point ligands is replaced by a new point ligand, the original assembly is prochiral". A ligand is simply a group attached to the prochiral atom (*a, b, c, d*). A point ligand for a prochiral center has two characteristics: "(*1*) any two point ligands may be identical or non-identical, and (*2*) two point ligands may not occupy the same position in space."[43] Addition of a point ligand to a prochiral center creates a new chiral center.[43] If the ligands are assigned a priority *a-b-c-d* by the CIP selection rules, configurations for each face of the prochiral center can be obtained. This rule is illustrated by conversion of the prochiral atom in **107** to either **108** or **109**. In this example, **109** represents a (S) chiral center and is formed by replacement of a^l with *b*. Similarly, replacing a^2 generates a (R) chiral center **108**. The a^l and a^2 atoms are described by the terms ***pro-S*** and ***pro-R***, respectively. A *pro-R* center is a point ligand in a prochiral molecule whose replacement leads to an (R) center. Analogously, a *pro-S* center is a point ligand in a prochiral molecule whose replacement leads to an (S) center. In **107**, ligand a^2 is

42. Reference 12, pp 465-488.
43. Hanson, K.R. *J. Am. Chem. Soc.* **1966**, *88*, 2731.

pro-R and a^1 is *pro-S*, and an example is D-glyceraldehyde **110**. The methylene group adjacent to the chiral center is prochiral, and replacement of H_b with a group X gives a (*S*) center in **111** (assuming X has a priority of *c* with H = *d*). Replacement of H_a gives a (*R*) center (in **112**) and that hydrogen is *pro-R*.

101 **102** **103** **104** **105** **106**

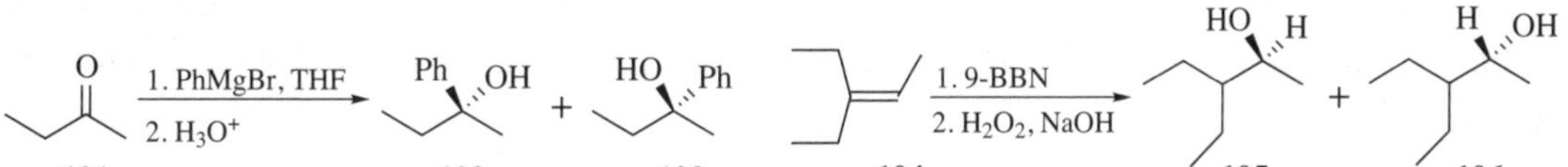

Hanson[43] described rules that accommodate most situations observed with prochiral centers. The reader is referred to this work for specific examples that do not yield to a simple analysis. The intent of this section is to familiarize the reader with nomenclature and the uses of *pro-R* and *pro-S* sites in reactions.

107 **108** **109**

110 **111** **112**

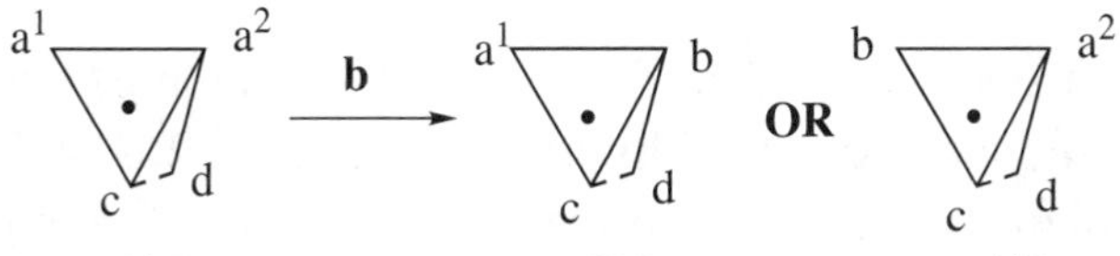

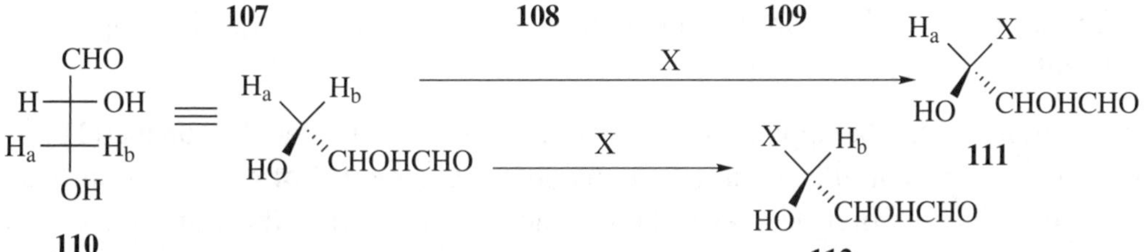

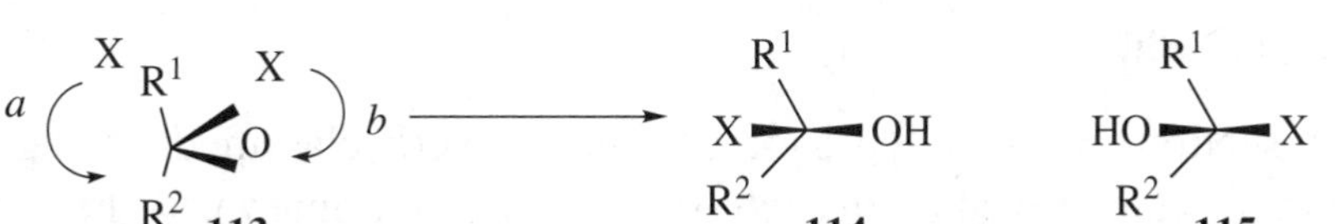

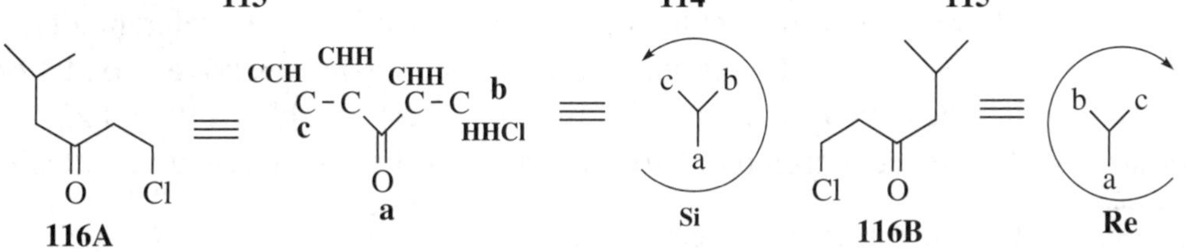

As noted above, synthetically important prochiral centers are the carbonyl of an unsymmetrical ketone or aldehyde and the double bond of an alkene. These functional groups do not contain a *pro-R* or *pro-S* group but it is clear that delivery of a fourth point ligand from one face or the other will lead to an (*R*) or (*S*) stereogenic center, as in conversion of **113** to **114** and/or **115**. If the carbonyl group is oriented as in ketone **116A**, priorities can be

113 **114** **115**

116A **116B**

assigned to the three atoms connected to the prochiral atom based on the CIP rules. For **116A**, the $a \rightarrow b \rightarrow c$ priority is counterclockwise and is analogous to (*S*). It is not really an (*S*) configuration, of course, since it is not a stereogenic center but this (*S*) sequence is termed *si*, from the Latin *sinister*.[43] When the incoming group approaches the π-bond from this face (with the orientation shown in **116A**) it is called the ***si*** face. If the molecule is given the opposite orientation, as in **116B**, the priority sequence $a \rightarrow b \rightarrow c$ is clockwise or (*R*) and this face is termed

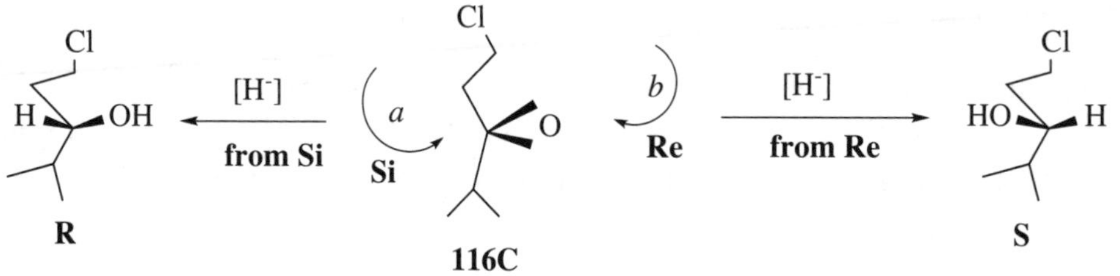

116C

re, from the Latin *rectus*. That face is the ***re*** face.[43] In this example, face ***a*** (see **116C**) is the *si* face and face ***b*** is the *re* face. Attack of hydride (see secs. 4.2.B, 4.7) from face ***a*** (the *si* face) leads to the (*R*) alcohol and attack from face ***b*** (the *re* face) leads to the (*S*) alcohol. The configuration of the final product depends on the priority of the new group added to the prochiral center.

Similar terminology can be applied to alkenes, as with (*Z*)-1-bromo-1-propene (**117**). In this case, there are two prochiral centers to be considered (C1 and C2 of the C=C bond), and *re* or *si* is assigned to each carbon. For **117**, the top faces of both C1 and C2 are *re* (*re-re*) and the bottom faces are *si* (*si-si*)

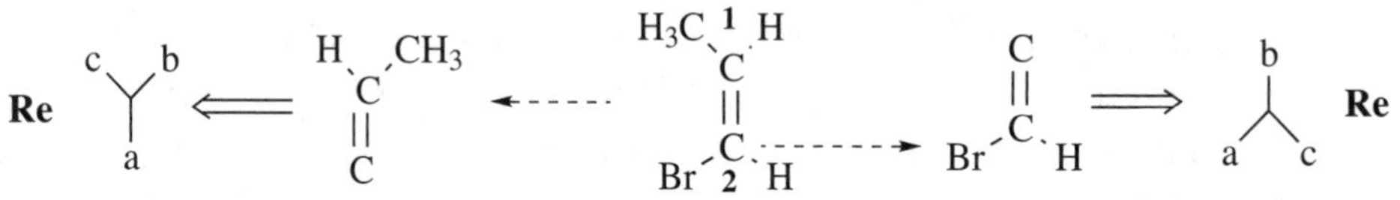

117

1.4.F. Definitions of Selectivity

In the preceding sections various types of molecules were classified as regioisomers or stereoisomers (further categorized as diastereomers and enantiomers). When there are two different functional groups in a molecule, a given reagent may react preferentially with one rather than the other. Such a reaction is sometimes termed **chemoselective**. Oxidation of **118** with manganese dioxide (MnO_2, see sec. 3.2.F.iii) gave a 50% yield of **119** and 25% of **120**.[44] Manganese dioxide showed a preference for oxidation of the secondary allylic alcohol at the

44. Hlubucek, J.R.; Hora, J.; Russell, S.W.; Toube, T.P.; Weedon, B.C.L. *J. Chem. Soc. Perkin Trans. 1* **1974**, 848.

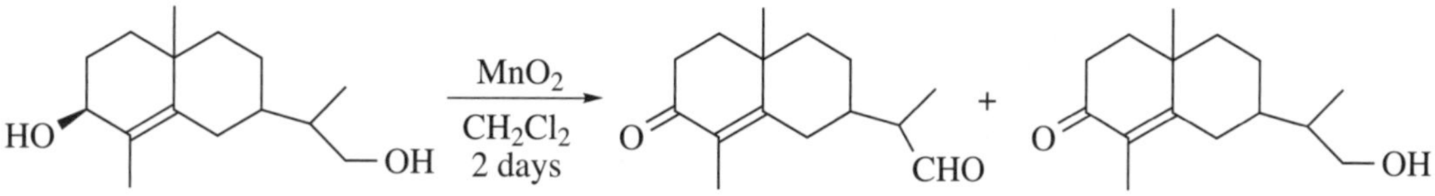

expense of the primary alcohol. The reagent selected one over the other, which leads to the term **chemoselective** (of two or more reactive functional groups, one reacted preferentially to give the major product in the mixture). When both products are formed but one is formed in greater proportions the term selective applies. Contrast this reaction with Gribble's reduction of ketoaldehyde **121** with tetrabutylammonium triacetoxyborohydride [$Bu_4N^+ BH(OAc)_3^-$, where Ac = acetate and Bu = butyl, see sec. 4.5.A], to give **122** in 88%.[45] If the result had been different, and the aldehyde carbonyl was reduced and a portion of the ketone carbonyl was also reduced, the reaction would be chemoselective rather than chemospecific. If there had been 100% reduction of the aldehyde but 0% of the ketone, the reaction would be termed **chemospecific**. Trost defined these terms: "of two or more reactive functional groups, only one reacts (specific), or one predominates (selective)". Reduction of alkenyl ketone **123** gave alcohol **124** and only the carbonyl reacts with sodium borohydride (NaBH$_4$). Since 0% of the alkene reacted, the reaction is chemospecific.

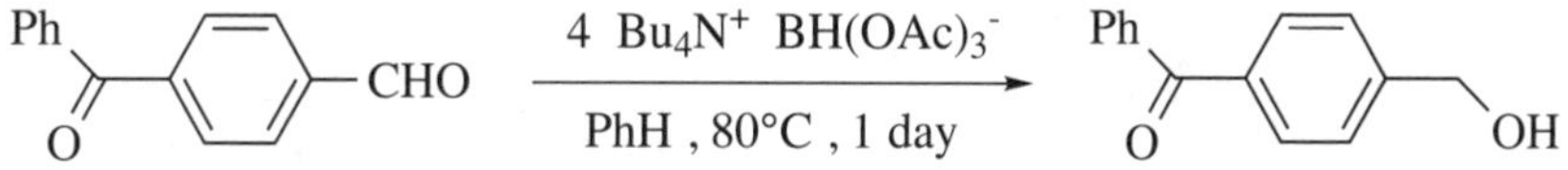

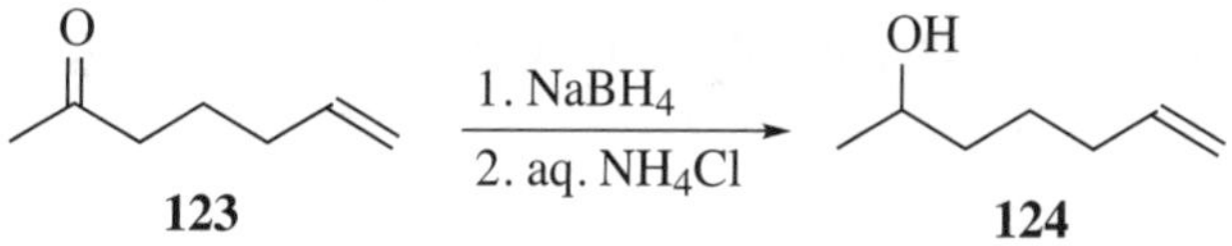

As noted above, the use of the terms selective and specific for giving a preponderance of a given product or only that product, respectively, can be applied for all reactions involving stereochemistry. If a reaction can produce two or more regioisomers, it is regioselective or regiospecific. The second order (E2) elimination (see sec. 2.9.A) of the racemic bromide **125** gave a mixture of **126** and **127** with **126** being the major product, so the reaction is regioselective. Elimination of the enantiopure (*S,S*)-diastereomer (**128**), however, gave *only* the (*Z*)-alkene **129** with none of the (*E*)-isomer and none of the alkene formed by removal the β-hydrogen on the methyl. The reaction is regiospecific and also stereospecific. This result was confirmed by reaction of the (*R,S*)-diastereomer **130** under E2 conditions, which gave only **131**. If **128** and **130** gave an unequal mixture of **129** and **131** the reaction would be stereoselective. One product is formed, so the reaction is stereospecific.

45. Nutaitis, C.F.; Gribble, G.W. *Tetrahedron Lett.* **1983**, *24*, 4287.

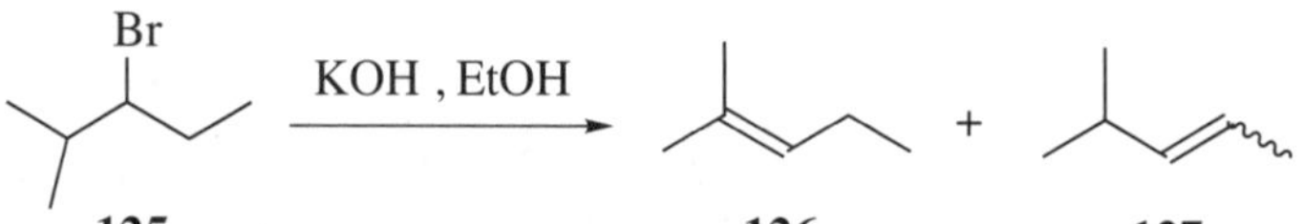

Similar terminology is applied to formation of diastereomers. March and Smith[46] gave the example of maleic acid (**132**) and fumaric acid (**134**) and their reactions with bromine. On addition of bromine to **132**, a racemic mixture of **133** was formed. Addition to **134** gave only *meso*-**135**. In both cases the reaction was diastereospecific. If **132** gave predominantly **133** with only a trace of the diastereomer (**135**), the reaction would be diastereoselective rather than diastereospecific.

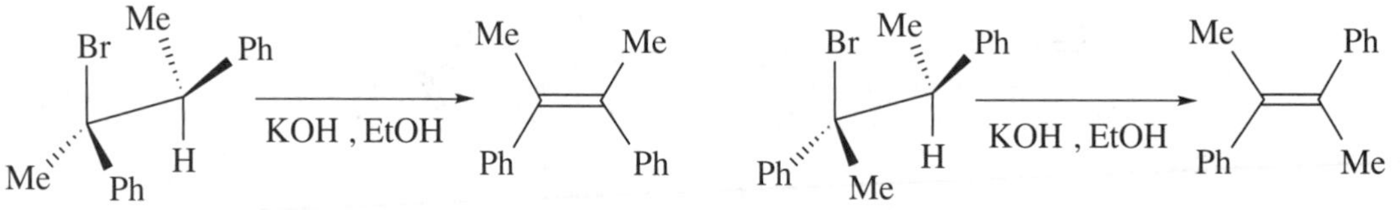

The term stereoselectivity is applied to reaction products. If a reaction produces at least one substance that is not a stereoisomer of the major product, that reaction cannot be stereospecific but at most stereoselective.[47] If stereoisomeric starting materials react to give a single stereoisomeric product, the reaction is **stereospecific**, but if

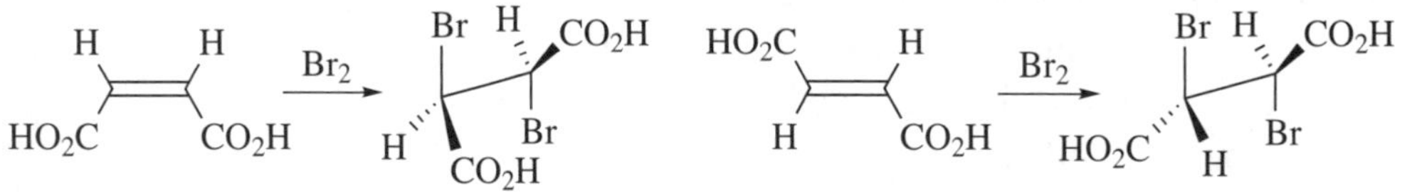

another stereoisomer is also produced (giving a mixture of products) the reaction is **stereoselective**. These terms apply to all types of stereoisomers, including enantiomers (**enantioselective** and **enantiospecific**), diastereomers (**diastereoselective** and **diastereospecific**) and regioisomers (**regioselective** and **regiospecific**). Reduction of keto ester **136** with zinc borohydride [$Zn(BH_4)_2$, see sec. 4.4.B and 4.7.B] gave 98% reduction with a >99:1 preference for **137** over **138**. The reaction produced two diastereomers, and it is **diastereoselective**.[47] Since traces of **138** are produced, the reduction cannot be diastereospecific.

46. Reference 18, pp 1002.
47. Oishi, T. in *New Synthetic Methodology and Functionally Interesting Compounds*, Yoshida, Z. (Ed.), Kodansha/Elsevier (Tokyo/Amsterdam), *1986*, pp 81-98.

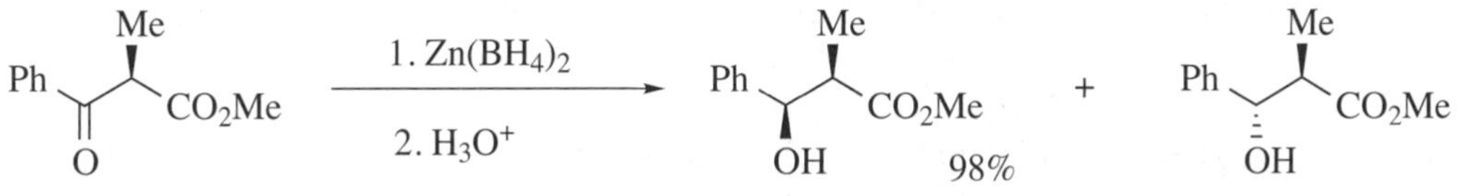

136   **137**   **138**

The final definition concerns formation of enantiomers, and the terms enantioselective and enantiospecific are used. If a reaction produces an unequal mixture of enantiomers it is enantioselective. If it generates only one enantiomer of two possibilities it is **enantiospecific**. The baker's yeast reduction (see sec. 4.10.F) of **139** gave **140** with >99% ee (S).[48] (Here **% ee** means percent of **enantiomeric excess**.) A 0% ee means a 50:50 mixture (racemic mixture), 50% ee means a 75:25 mixture and 90% ee means a 95:5 mixture. The predominance of the (S) enantiomer makes this reaction highly enantioselective.

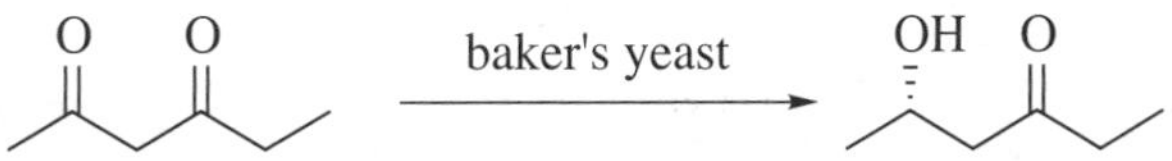

139   **140**

The selectivity terms introduced in this section will be used throughout the book for reactions that generate stereoisomers. In addition to % ee, % de (diastereomeric excess, defined in the same way as ee but for diastereomers), or % dr (diastereomeric ratio) will be used throughout.

1.5. CONFORMATIONS

Organic molecules can largely be categorized as having tetrahedral, trigonal or digonal (linear) geometry around each sp^3, sp^2, or sp hybridized carbon atom, respectively. When organic molecules absorb energy from the environment they can partially dissipate excess energy by molecular vibrations, including rotation or twisting about all carbon-carbon single bonds. Generally, the more energy absorbed, the more facile will be the rotation. The three dimensional nature of carbon compounds, and rotation around carbon-carbon bonds leads to different spatial orientations of molecules called **rotamers**. Different rotamers are not different structures, but different orientations of the same molecule. Analysis of the spatial relationships of the groups can give information about the interactions of the groups as they rotate around the carbon-carbon bond. Based on these ideas, the shape that a molecule assumes is determined by understanding the rotamer population for all bonds, which leads to the overall shape (conformation) of the molecule. This shape is generally taken as the lowest energy **conformation** of that molecule.. Most molecules have more than one conformation, but one or at least a small subset will constitute the low energy conformation(s). This information is

---

48. Bolte, J.; Gourcy, J.-G.; Veschambre, H. *Tetrahedron Lett. 1986, 27*, 565.

used to determine how reagents will approach the molecule, and even the stereochemistry of certain reactions.

When we draw the structure of an acyclic molecule we usually draw the lowest energy rotamer, so we draw it with a regular shape rather than as an amorphous shape. For cyclic and polycyclic molecules the various interactions of the groups will lead to at least one and often several energy minima that are drawn to represent the shape of that molecule. In chemical reactions, the shape of the molecule will influence the way an incoming reagent interacts with it and this can have a significant effect on both reactivity and stereochemical induction, especially with the more rigid cyclic and polycyclic molecules. This section will discuss the fundamental conformational preferences of simple acyclic and cyclic molecules. These principles will be applied throughout the book for discussions of chemoselectivity and stereoselectivity.

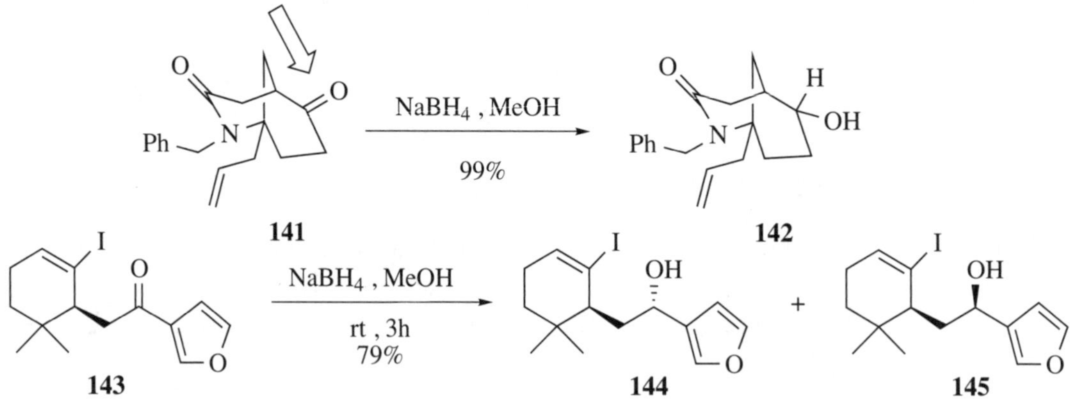

An example of the link between conformation and stereochemistry is illustrated with bicyclic ketone **141**, which assumes a rigid structure so that a simple drawing is sufficient to conclude that one face is more encumbered than the other. It is, therefore, possible to anticipate the direction of hydride attack (see sec. 4.4.A) to be from the more open face of the molecule, from the direction of the arrow in **141** as it is drawn to give equatorial alcohol **142** as the only isomer isolated in 99% yield. This reaction was take from the total synthesis of the microbial immunosuppresive agent FR901483, by Weinreb and co-workers.[49] Acyclic molecules are problematic, however, since they do not assume a rigid conformation. They are quite flexible and all but the simplest molecules may exist in a variety of different shapes or conformations. Indeed, the shape of such molecules depends on the torsional angles involving the various single bonds. Generally, it is not possible to say which carbonyl face is likely to be the less crowded in a flexible ketone, and what the stereochemistry of a reaction product will be. The second example using ketone **143** is far more typical, where there are several possible conformations. Each has a different three-dimensional structure with different hindered or exposed regions. When **143**, taken from Jan and Liu's synthesis of (+)-ricciocarpin A, was

---

49. Kropf, J.E.; Meigh, I.C.; Bebbington, M.W.P.; Weinreb, S.M. *J. Org. Chem.* **2006**, *71*, 2046.

reduced with NaBH4 (sec. 4.4.A), a 1:1.7 mixture of diastereomeric alcohols **144** and **145** was formed.[50] Even this relatively simple molecule shows enough conformational flexibility that a mixture of diastereomers is formed. In sec. 4.7.B, the Cram model and the Felkin-Ahn model will be discussed in order to predict selectivity in acyclic systems. However, both models assume that one key conformation is present in order to make the prediction. Identification of the "relevant" conformer or set of conformers is a necessary first step to anticipate product selectivity, and a conformational analysis is required.

1.5.A. Conformations of Acyclic Molecules

1.5.A.1. Conformations of Simple Alkanes

Different conformers result from rotation about single bonds and also from restricted rotations in rings. The number of different conformers depends on the number of single bonds and on the number and size of the flexible rings. A simple rule of thumb is that each single bond multiplies the number of possible conformers by three. Thus, a molecule with one single bond has three conformers, a molecule with two single bonds has nine conformers, and so forth. It is more difficult to elaborate conformers for flexible rings. Three-membered rings are conformationally rigid and four and (for the most part) five-membered rings can be considered to be rigid. Six-membered rings comprising sp^3 centers typically exhibit two "chair" conformers, although higher-energy "twist-boat" conformers are also possible. Seven-membered and larger rings generally posses an even grater number of conformers. In the final analysis, even a reasonably sized organic molecule can exhibit hundreds to thousands of possible conformations. Thus, identifying the relevant shape cannot be accomplished by simply looking at a single drawing or by simple manipulation of a model.

This section will introduce and illustrate computer-based conformational analysis tools. Properly used, these tools can furnish reliable predictions as to what shapes flexible molecules adopt and, in so doing provide insight into how they are likely to react. We start by introducing important conformational preferences established for organic molecules with only a single degree of conformational freedom. Our treatment relies heavily on data from quantum chemical calculations. This allows more detailed analysis than would otherwise be possible.

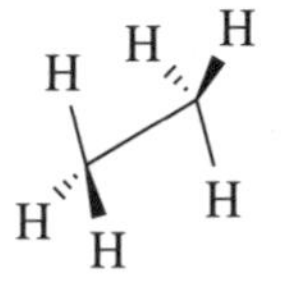

146 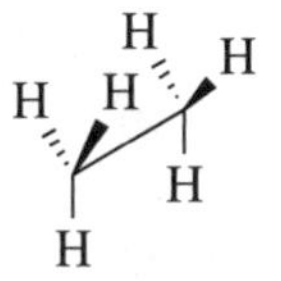   147 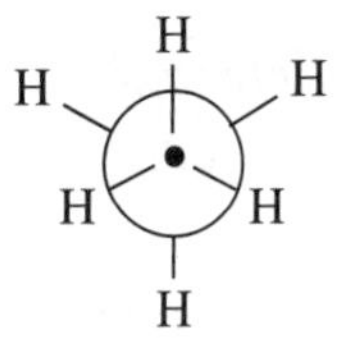   148 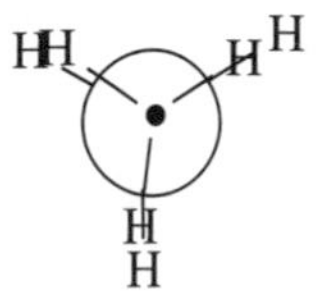   149

50. Jan, N.-W.; Liu, H.-J. *Org. Lett.* **2006**, *8*, 151.

Ethane is a simple example of an acyclic molecule capable of rotation about two carbon atoms. If one could freeze this motion at different positions, the result would be different rotamers. Inspection of rotamer **146** and its **Newman projection** (the dot and circle model used in **148**)[8] reveals that the C–H bonds and the hydrogen atoms attached to each carbon are as far removed from each other as possible. Rotamer **147** (its Newman projection is **149**) sharply contrasts with **146**, since the C–H bonds and the hydrogen atoms attached to each carbon are as close together as possible. The traditional view is that non-bonded interaction between the two eclipsing hydrogens atoms and the electronic repulsion of the C-H bonds destabilizes this rotamer, making it higher in energy. Indeed, a review categorizes the destabilizing interactions as Pauli exchange steric repulsion, hyperconjugation, and relaxation energy changes.[51] In addition, skeletal relaxation effects play a role.[51] An analysis of ethane shows that the energy depends on the angle of torsion about the carbon-carbon bond. Complete (360°) rotation about the C–C bond (see Fig. 1.3) leads to three identical staggered structures (**146A**) that are energy minima and three identical eclipsed structures (**148A**) that are energy maxima. The difference in energy between the staggered and eclipsed structures is referred to as the barrier to rotation. At ~12 kJ/mol,[52] it is small enough that bond rotation will be very fast at normal temperatures.[53] This means that any measured properties for ethane actually represent an average of properties of the three (identical) conformers. The three (identical) eclipsed structures **148A** do not contribute to the properties of ethane because they do not exist in the sense of contributing to the population.

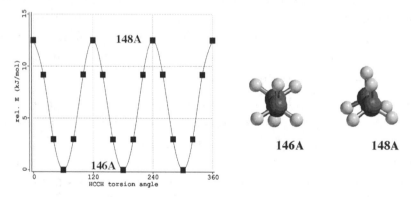

Figure 1.3. Energy barrier for ethane over a rotation of 360°

Propane (**150**), with two identical carbon-carbon bonds, gives rise to nine identical energy minima and nine identical energy maxima. As with ethane, only the minima contribute to the

51. Goodman, L.; Pophristic, V.; Weinhold, F. *Acc. Chem. Res.* **1999**, *32*, 983.
52. Measured at 2.9 kcal mol[-1] (12.1 kJ mol[-1]):. (a) Gunstone, F.D. *Guidebook to Stereochemistry*, Longman, London, *1975*, p 56; (b) Reference 12, p 599; (c) Reference 29b, p 125 (d) Kagan, H. *Organic Stereochemistry*, Halsted Press/Wiley, New York, *1979*, p 50; (e) Hine, J. *Physical Organic Chemistry, 2nd Ed.*, McGraw-Hill, New York, *1962*, p 36.
53. The relative energies of conformers will be expressed in units of kJ mol[-1].

properties of the molecule. 2-Methylpropane (**151**) has three identical C–C bonds that give rise to 27 identical sets of energy minima and maxima, and 2,2-dimethylpropane (**152**) has four identical C–C bonds that give rise to 81 identical sets of energy minima and maxima. *Energy plots for all four molecules are identical, except for different rotational barriers.*

Butane has a new structural feature that must be addressed. The C1-C2 and the C3-C4 bonds are identical in terms of their substitution pattern, but the C2–C3 bond is different with each carbon having one methyl and two hydrogen atoms: H_3C—CH_2-CH_2—CH_3 versus H_3C-CH_2—CH_2-CH_3 (see **153A**). Rotation can occur about both C1–C2 and C3–C4, as seen by comparing **153A** and **153B**. Rotation of the central carbon-carbon bond in *n*-butane leads to three energy minima and three energy maxima, as indicated in Fig. 1.4. Two of the minima

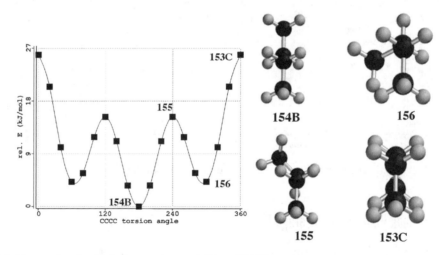

Figure 1.4. Energy barrier for butane over a rotation of 360°

minima (so-called *gauche* conformers) are identical, while the third minimum (the so-called *anti* conformer) is different. Likewise, two of the energy maxima (connecting *gauche* and *anti* minima) are the same, while the third maximum (connecting the two *gauche* minima) is different. The magnitude of the barriers connecting *anti* and *gauche* conformers depends on the direction of rotation (*anti* to *gauche* or *gauche* to *anti*), the difference being the same as the energy difference between *anti* and *gauche* *n*-butane conformers. The maximum barrier to rotation is estimated to be 4.5-4.9 kcal mol[-1] (19-21 kJ mol[-1]),[54] which represents the methyl-

54. (a) Reference 52c; (b) Reference 52a, p 57; (c) Reference 12, p 602; (d) Reference 31b, p 126.

methyl interaction. The other energy barrier represents each Me-H eclipsing interaction and is estimated to be 3.6 kcal mol^{-1} (15 kJ mol^{-1}).[55] The energy estimated for each *gauche* conformation is 0.96 kcal mol^{-1} (4 kJ mol^{-1}),[52b] an indication that the methyl groups are close but do not eclipse.

It is possible to categorize butane as a dimethyl substituted ethane. Energy profiles for rotation about the central carbon-carbon bond for a molecule such as 1,2-dichloroethane other disubstituted ethanes are qualitatively similar, in that each shows two *gauche* minima and one *anti* minimum (see Fig. 1.5). The relative energies of *gauche* and *anti* conformers and rotation barriers depend on the substituents. *Anti* conformers are usually preferred over *gauche* arrangements, for example, for 1,2-dichloroethane and for 1-chloropropane (**A** and **B** in Fig. 1.5, respectively). One reason for this preference is a desire to minimize unfavorable non-bonded contacts between the substituents. Another factor in some cases is electrostatics, the desire to align bond dipoles as to minimize the total dipole moment, for example, to align the $^{\delta+}$C-F$^{\delta-}$ bond dipoles in difluoroethane such that they subtract rather than add.

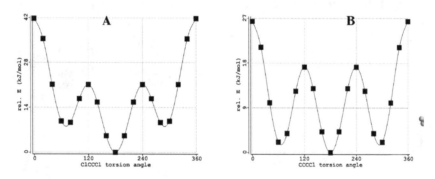

Figure 1.5. Energy barrier for substituted ethanes over a rotation of 360°

1.5.A.2. The Boltzmann Distribution: Average Properties of Flexible Molecules

To obtain the properties of any sample of *n*-butane, it is necessary to average the properties of the *anti* and *gauche* conformers. More generally, to find the average value of a property (**A**) of a flexible molecule, it is necessary to sum over all possible different conformers, taking into account the value of the property for that conformer (**a**) as well as both the number of times that the conformer appears (**n**) and its Boltzmann weight (**W**).

$$\mathbf{A} = \sum \mathbf{a}_i \, \mathbf{n}_i \, \mathbf{W}_i$$

The Boltzmann weight depends on the energy of the conformer relative to the energy of the lowest energy conformer (ΔE), and on the temperature (**T**): $\mathbf{W_i} = \exp(-\Delta E_i/kT)/ \sum \exp(-\Delta E_i/kT)$ where k is the Boltzmann constant. An energy difference of 4 kJ mol^{-1} leads to a Boltzmann

55. (a) Reference 52a, p 57; (b) Reference 52c, pp 48-53.

weight of ~0.1 (10%) at room temperature, an energy difference of 1.9 kcal mol^{-1} (8 kJ mol^{-1}) to a weight of ~0.05 (5%), and a difference of 12 kJ mol^{-1} to a weight of ~0.01 (1%). Only rarely will more a few of the possible conformers have Boltzmann weights in excess of 1% and contribute significantly to the equilibrium. For molecules like ethane and propane where all conformers are the same, the average is independent of temperature. For *n*-butane, where the *anti* and the two equivalent *gauche* conformers contribute, the average depends on temperature. At very low temperatures, the average will be dominated by the lowest energy (*anti*) conformer, but it will limit to an equal weighting of both conformers as the temperature increases.

Note that measurements of some quantities, such as the NMR spectrum, yield averages over possible conformations while others such as the IR spectrum provide information about individual conformers. In the latter case, low Boltzmann weights will almost always preclude actually "seeing" any but the few lowest-energy conformers. Whether a particular measurement yields an average or discrete quantities depends on the time scale of the underlying physical process. Relaxation of magnetic spin (the basis of NMR) is typically slower than conformational equilibrium at normal temperatures, while molecular vibration (the basis of infrared spectroscopy) is much faster. The time scale of the experimental measurement must be established before interpreting the result.

Energy barriers for various molecules are quantified in terms of enthalpy (H°), and calculation of the energy of a given rotamer allows an estimate the relative population of that rotamer (see Boltzmann distribution above). The relative energy of the conformations can be correlated with the relative percentage of each rotamer. An enthalpy difference of 1 kcal mol^{-1} (4.186 kJ mol^{-1}) between the *anti* rotamer and the next most populous rotamer corresponds to the presence of about 72% of the anti-isomer at room temperature.[56] For butane, **154** is the all-anti conformation. The zigzag or extended look to conformation **154** is typically drawn to represent straight-chain alkanes. *n*-Tridecane (**157**), for example, is drawn as the all anti conformation, the extended conformation. Virtually all acyclic hydrocarbon chains are *assumed* to exist primarily in this conformation. This assumption is, of course, incorrect and conformer **157** probably does not exist to a significant extent in a real distribution. The previous discussion establishes that if butane exists in several conformations, the more flexible tridecane will exist in many more and we can use a Boltzmann distribution to estimate the contribution of different conformations. When drawing molecules, we usually draw straight-chain alkanes or alkyl fragments in this fully extended form. In reality, the conformational picture is much more complex and it is unlikely that molecules really look this way. This is a convention that is useful for drawing, but not for an analysis of chemical properties.

56. (a) Reference 31b, pp 131-133.

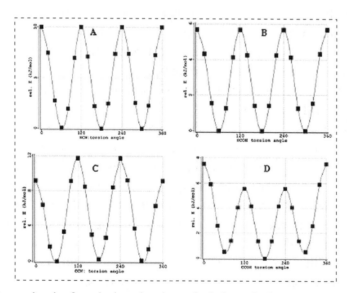

154   157

1.5.A.3. Heteroatom Substituents

Thus far, discussion has concerned rotation about carbon-carbon bonds. The same principles apply to rotation about carbon-heteroatom bonds, particularly carbon-nitrogen and carbon-oxygen bonds. Consider methylamine and methanol. Different examples are methylamine and methanol. The nitrogen in organic amines is sp³ hybridized. Three of the hybrids form single bonds to other atoms while the fourth hybrid is thought of as a non-bonded pair of electrons (a lone pair). Similarly, the oxygen in alcohols or ethers is sp³ hybridized. In both cases, two of the hybrids are employed to bond to other atoms, leaving two lone pairs. This implies that both sp³ nitrogen and oxygen centers are roughly tetrahedral, consistent with the fact that nitrogen centers in amines are pyramidal and oxygen centers in alcohols and ethers are bent. Energy plots for methylamine (**A**) and methanol (**B**) are similar to that for ethane as seen in Fig. 1.6. All show three identical sets of energy minima and maxima. Rotational barriers are somewhat smaller than that in ethane: ~8 kJ mol⁻¹ in methylamine and ~4 kJ mol⁻¹ in methanol.

Figure 1.6. Energy barrier for substituted amines, alcohols, and ethers over a rotation of 360°

Substitution of a hydrogen on nitrogen in methylamine to give a secondary amine or on oxygen in methanol to give an ether, does not change the fundamental nature of the energy curves seen for the parent compounds, although it does change the details (energy barriers). On the other hand, substitution on carbon in both methylamine (to give ethylamine) and methanol (to give ethanol) leads to energy curves which, like that for *n*-butane, show two different minima (**C** for ethylamine in Fig. 1.6 and **D** for ethanol). Note that *anti-gauche* energy differences for both molecules are very small.

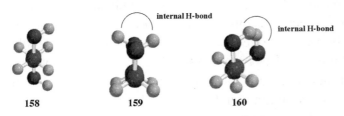

158   159   160

Groups capable of hydrogen bonding provide a stabilizing interaction that can compensate for the destabilizing interaction expected by steric repulsion. One can draw both an anti conformation (**158**) and a *syn*-conformation (**159**) for 1,2-ethanediol (ethylene glycol), and the hydroxyl group is capable of hydrogen bonding. Rotating the molecule to the *gauche* conformation **160** relieves the eclipsing interaction but maintains the hydrogen bonding, making **160** the expected low-energy conformation *if no hydrogen-bonding solvent is present*. If the solvent is unable to hydrogen bond with ethylene glycol, the molecule will hydrogen bond with itself as shown. If the solvent can hydrogen bond, intermolecular hydrogen bonding will compete with intramolecular hydrogen bonding, and intermolecular hydrogen bonding will win on the basis of statistics (entropy).

Another way to stabilize an eclipsed or *gauche* conformation is to coordinate a heteroatom substituent with a metal ion (chelation). Oishi and co-worker's reduction of **161** with zinc borohydride proceeds via a chelated species, **162**.[57] Chelation of zinc to the hydroxyl and carbonyl groups effectively immobilizes the reactive components into a single conformation in the transition state required for reaction, as shown in **162**. This fixed conformation sets the position of the methyl and hydrogen at the α-carbon, which leads to facial bias, and the hydride is delivered from the less hindered face over the hydrogen in **162** (see secs. 4.4.B and 4.7.B). Since transition metal salts usually behave as Lewis acids, the presence of a heteroatom that functions as a Lewis base (O, S, N, or P, see sec. 2.3) will lead to chelation. The most favored acyclic conformation is usually a *gauche* or analogous rotamer.

1.5.A.4. Heteroatom-Heteroatom Bonds

Many molecules have heteroatom-heteroatom bonds, such as N-N in hydrazines or O-O in peroxides. The energy profile for hydrazine (**163** in Fig. 1.7) is quite different from those of previous systems, containing two identical minima corresponding to arrangements in which the nitrogen lone pairs are perpendicular, and two different maxima. The higher maximum is ~44 kJ mol⁻¹ above the minima and corresponds to an arrangement in which both NH bonds eclipse

57. Nakata, T.; Tanaka, T.; Oishi, T. *Tetrahedron Lett.* **1983**, *26*, 2653.

and the two lone pairs eclipse. The other maximum is only 13 kJ mol^{-1} above the minima and is very broad. It corresponds to a range of conformers centered around a structure in which the two lone pairs are *anti* to each other. Unfavorable lone pair-lone pair interactions are much larger than bond-bond or bond-lone-pair interactions even if the lone pairs point away from each other. The best way to minimize them is to keep the lone pairs perpendicular. Note that the preference due to lone pair interactions is much larger than preferences previously noted due to bond-bond and bond-lone pair interactions. Hydrogen peroxide provides another example of the consequences of interaction of lone pairs. The curve (**164** in Fig. 1.7) shows a pair of identical minima with torsional angles around 120° and 240°. The curve shows two different maxima. The higher is ~40 kJ mol^{-1} above the minima and corresponds to a *syn* conformer (HOOH angle = 0°), while the lower is ~5 kJ mol^{-1} above the minima and corresponds to an *anti* conformer (HOOH angle = 180°). In summary, hydrazine and hydrogen peroxide, like ethane, methylamine and methanol, are both described in terms of a single structure.

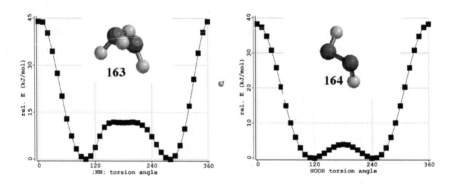

Figure 1.7. Energy barrier for substituted amines, alcohols, and ethers over a rotation of 360°

1.5.A.5. Bonds Connecting sp³ and sp² Hybrids: Propene and 1-Butene

The planar sp^2 carbon atoms in alkenes are relatively easy to distinguish, but what effect does the C=C unit have on the adjacent sp^3–sp^2 bond? A plot of energy versus the C=C-C-H torsional angle in propene (**A** in Fig. 1.8) shows three identical minima and three identical maxima, just like that for ethane. Even the rotation barrier is similar (~8 kJ mol^{-1} in propene versus ~12 kJ mol^{-1} in ethane). However, the methyl group in propene sees a different environment than the methyl group in ethane. If it staggers the vinylic C-H bond, then it must eclipse the C=C unit and vice versa. The minima in propene correspond to arrangements in which one of the methyl C-H bonds eclipse the carbon-carbon double bond but two methyl carbons stagger the alkene C-H bond. The plot for the central carbon-carbon single bond in 1-butene (**B** in Fig. 1.8) closely resembles the corresponding curve for *n*-butane. It shows three minima, two that are identical corresponding to arrangements in which a (methylene) C-H bond eclipses the carbon-carbon double bond, and the third only ~3 kJ mol^{-1} higher in energy in which the C–C single bond eclipses the double bond. There is a general rule that single bonds eclipse double bonds.

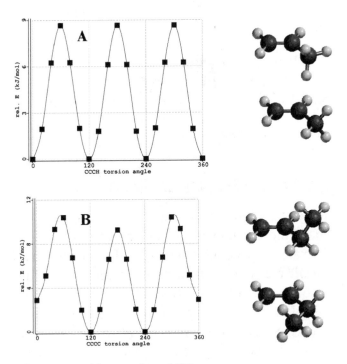

Figure 1.8. Rotation of propene and butene about 360°

Much larger conformational preferences involving single and double bonds can occur. The most conspicuous are the C-O bonds in carboxylic acids and carboxylic acid esters, between sp^2 (carbonyl group) and sp^3 (oxygen) centers. Energy curves for CO bond rotation in acetic acid and in methyl acetate (**A** and **B**, respectively, in Fig. 1.9) both show two minima corresponding to *syn* (O=CCH and O=CCC = 0°) and *anti* (O=CCH and O=CCC = 180°) arrangements, and a pair of identical maxima in between. The *syn* conformer is preferred by nearly 60 kJ mol^{-1} for both molecules. In both conformers, C=O and O-H (O-Me) bonds eclipse but the C=O and C–O bond dipoles subtract in the *syn*-conformer and they add in the *anti*-conformer (see **A** and **B**).

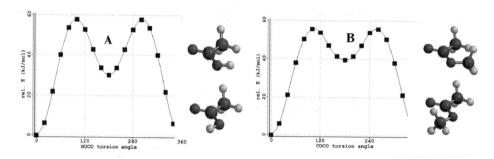

Figure 1.9. Rotation of propene and butene about 360°

1.5.A.6. Bonds Connecting sp² Hybrids: 1,3-Butadiene and Styrene

A final case is presented that focuses on the carbon-carbon single bond connecting two sp^2 hybridized atoms, such as those found in conjugated dienes or styrene derivatives. The conventional wisdom is that the two double bonds need to be coplanar in order to maximize "conjugation". While *trans* planar 1,3-butadiene is an energy minimum, the corresponding *cis* planar conformer is not (see **A** in Fig. 1.10). There is an energy minimum nearby (CCCC torsion angle ~40°) that is ~12 kJ mol⁻¹ higher than a cis form (an energy maximum), and ~12 kJ mol⁻¹ higher in energy than the *trans* form. The energy barrier to rotation about C_2-C_3 has been shown to be 5 kcal mol⁻¹ (21 kJ mol⁻¹).[58] The two rotamers mentioned arise by rotation about the C_2-C_3 single bond: **165** with the two π bonds cis- to each other and **166** with the π-bonds trans. Rotamer **165** is called the s-cis conformation, and **166** is the s-trans-conformation. When the diene is substituted, (*E,E*)-, (*Z,E*)- and (*E,Z*)- dienes are possible. The s-cis conformation of (*Z,Z*)-3,5-hexadiene (**167**) is even less favorable than the s-trans conformation (**168**), when compared with 1,3-butadiene. In **167**, the Me-Me interaction is quite apparent, and although the s-trans conformer (**168**) has two methyl-hydrogen interactions, it is lower in energy than the Me-Me interaction in **167**.

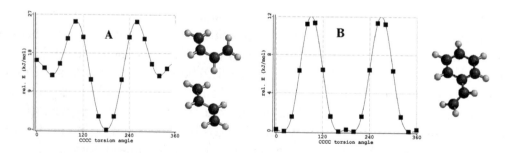

Figure 1.10. Rotation of propene and butene about 360°

---

58. Reference 52c, p 54.

Similar expectations apply to styrene (see **B** in Fig. 1.10) where there is only one unique conformer. Here the vinyl and phenyl groups are nearly coplanar and the energy barrier through planar styrene is tiny. With the advent of coupling reactions such as the Heck reaction (sec. 12.7.A), such compounds have gained greater importance.

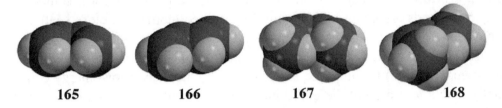

165   **166**   **167**   **168**

1.5.B. Conformations of Cyclic Molecules

Steric and electronic interactions influence the rotation about bonds within acyclic molecules, leading to a relatively small population of low-energy rotamers. The cyclic nature of these molecules imposes additional energy barriers to conformational mobility relative to acyclic molecules. Although the C–C bonds in cyclic compounds are not capable of complete rotation, they undergo what is known as pseudorotation, largely by twisting and bending around the various carbon-carbon bonds in the ring. Depending on the flexibility of the ring, which is related to the size of the ring, many conformations are possible. As with acyclic compounds, there are usually a small number of low energy conformation that dominate the population.

An important difference that is observed in cyclic molecules is deformation of the bond angles from 109°28' as the molecule attempts to rotate around the C-C bonds, which can be severe when the ring is small. Cyclopropane, for example, has a C-C-C bond angle of ~ 60° (see **169**) which induces great strain in the molecule. This type of strain is called **Baeyer** or **angle strain**, formally defined as the increase in energy of cyclic compounds that arises from the deformation of the optimum valence angle of 109°28' for sp^3 carbon or 120° for sp^2 carbon. The higher ground state energy for cyclopropane is the result of such strain. Formation of three-membered rings is common, but the higher ground state energy is important in the reactions of cyclopropane as well as those that form cyclopropanes.[58] Deformation of the σ bonds leads to significant p character (the hybridization is Csp$^{2.3}$), and many ring-opening reactions of cyclopropane mimic the chemistry of alkenes. The increased p-character is reflected in the diminished electron density of the C–C bonds in cyclopropane, which is slightly displaced from linearity between the nuclei.[59] The four-carbon cyclic compound is cyclobutane (see **170**) with bond angles of ~90°, and in planar cyclopentane (**171A**) with bond angles of ~108°. Cyclobutane has significant Baeyer strain, less than in cyclopropane but more than in cyclopentane.

59. (a) Bernett, W.A. *J. Chem. Educ.* **1967**, *44*, 17; (b) de Meijere, A. *Angew. Chem. Int. Ed.* **1979**, *18*, 809; (c) Reference 12, pp 676-678; (d) Reference 31b, pp 204-306 and 124-179.

Cyclopropane (**169**), planar cyclobutane (**170A**) and planar cyclopentane (**171A**) exhibit a second major type of strain, which is the same as that seen in acyclic molecules. All C-H bonds in these planar molecules are eclipsed, with severe nonbonded interactions of the eclipsed hydrogens and electronic repulsion of the eclipsed bonds. This type of strain is called **Pitzer** or **bond opposition strain** and is defined as the increase in energy for a compound arising when adjacent bonds are eclipsed, bringing the attached atoms into close spatial proximity. Cyclic molecules generally do not exist in the planar form due to the Pitzer strain. Although rotation of 360° about the C-C bonds is not possible, twisting and bending is possible via partial rotation (pseudorotation) to minimize Pitzer strain.[59] One or two low-energy conformations are usually drawn to represent the entire molecule, although as the ring becomes larger and the flexibility increases more low energy conformations are possible. For cyclobutane, a bent or puckered conformation (**170B**) is taken as the low energy conformation,[60] which minimizes Baeyer strain relative to the planar form (**170A**).[61] For cyclopentane, the bent conformation assumes an envelope shape as in **171B**.[61] Although Baeyer strain is increased in the envelope conformation (**171B**) relative to the planar form (**171A**),[62] the decrease in Pitzer strain more than compensates and **171B** is the low-energy conformation of cyclopentane. Analysis of planar cyclohexane (**172A**) reveals extensive Pitzer strain and the bond angles would be 120°, introducing Baeyer strain. Pseudorotation leads to the familiar chair conformation (**172B**) as the low energy conformation. Chair cyclohexane has virtually no Baeyer strain with bond angles of 109°28'. Cycloheptane has similar bond angles[63] but is more flexible, and there are several low energy conformations, although only **173A** is shown. Cycloheptane also has a boatform (**173B**) that is very close in energy to a chair form **173A**.[64] The lowest energy form of cycloheptane, however, is the twist-chair conformation, **173C**.[65]

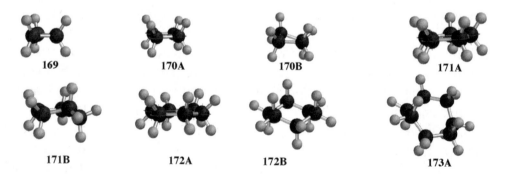

169   170A   170B   171A

171B   172A   172B   173A

60. (a) Reference 12, p 676; (b) Reference 31b, p 248.
61. (a) Dunitz, J.D.; Schomaker, V. *J. Chem. Phys.* **1952**, *20*, 1703; (b) Rathjens, Jr., G.W.; Freeman, N.K.; Gwinn, W.D.; Pitzer, K.S. *J. Am. Chem. Soc.* **1953**, *75*, 5634.
62. Kilpatrick, J.E.; Pitzer, K.S.; Spitzer, R. *J. Am. Chem. Soc.* **1947**, *69*, 2483.
63. (a) Reference 12, pp 686-689; (b) Reference 29b, p 252.
64. Allinger, N.L. *J. Am. Chem. Soc.* **1959**, *81*, 5727.
65. Wiberg, K.B. *J. Org. Chem.* **2003**, *68*, 9322.

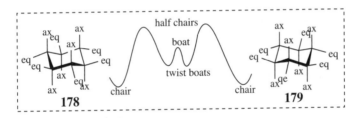

173B   173C   174A   175   176   177

As mentioned, cyclohexane is known to favor a "chair" structure in which all six carbons are equivalent but the hydrogens divide into two sets of six *equatorial* hydrogens and six *axial* hydrogens, marked in **178**. However, the fact that only one proton resonance is seen in its room temperature NMR spectrum suggests a rapid conformation change in which *equatorial* and *axial* hydrogens exchange. An energy profile for the process (Fig.1.11) shows the two identical chair structures as the starting and ending points, respectively, marked **178** and **179**. There is a boat conformation (**175**), as well as two identical higher-energy "twist-boat" minima (**176**). In **175**, there is a cross ring (**transannular**) interaction of the 1,4-hydrogens (referred to as flagpole hydrogens). The curve contains three energy maxima. Two are identical "half chairs" (**177**) and connect the chair a twist boat forms, while the boat (**175**) connects the two twist-boat forms.

Figure. 1.11. Conformational mobility of cyclohexane.

Chair cyclohexane is ~28 kJ mol^{-1} lower in energy than the twist-boat (or just twist) conformation, suggesting that the latter will have little influence on the properties of cyclohexane. It is not possible to attach a simple label to the geometrical coordinate in Fig. 1.11 responsible for this process. However, interconversion of chair cyclohexane into the twist-boat form can be viewed as a restricted rotation about one of the ring bonds (see **180** to **181**). Correspondingly, the interconversion of the twist-boat intermediate into the other chair form can be viewed as rotation about the opposite ring bond.

The molecular conformations induced by relief of Baeyer strain and Pitzer strain are reflected in the energy required for intramolecular cyclization reactions to form each ring. As two reactive ends of the acyclic fragment come together, the strain inherent to the ring product becomes important (see sec. 6.6.B.i). In other words, the acyclic precursor assumes the conformation of the ring that is being formed in the transition state of the reaction. Strain in this transition state is important for the formation of the cyclic product. Figure 1.12(a) shows the relative reactivity for formation of lactones from ω-hydroxy acids of ring size C3 to C17.[66] There is a reactivity maximum at C5 and reactivity minima at C3 and C8.[66] Figure 1.12(b) shows the enthalpy (ΔH) of several cyclic alkanes, with a minimum at C6 and a maximum at C9.[67] There are two other important features of Figure 1.12 (b). Rings of C8-C13 members are significantly higher in energy than the small rings (C3 → C7) or the large rings (≥ C14). These medium size rings are extremely difficult to form using intramolecular cyclization.[66] The other feature of Figure 1.12(b) is the higher energy required to form rings with an odd number of atoms when compared to the energy required to form rings with an even number of atoms. Compare cyclohexane and cycloheptane to see the difference. Cycloheptane has an odd carbon that is not easily accommodated by the low-energy chair (all *gauche*) form, and its presence leads to an increase in Pitzer strain.[67,66] Similar effects are seen for all odd carbon rings.

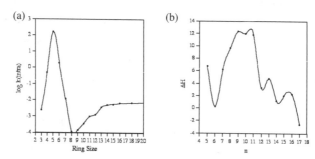

Figure 1.12. (a) Reactivity Profile for Lactone Formation. (b) Enthalpy of Cycloalkanes [CnH2n]. [Reprinted with permission from Illuminati, G.; Mandolini, L. Acc. Chem. Res. 1981, 14, 95. Copyright © 1981 American Chemical Society.]

As the ring size increases to eight members and higher, increased flexibility leads to an increasing number of lower energy conformations. Indeed, an inspection of cyclooctane reveals there are several conformations, including **174B** (crown), **174C** (boat-chair), a boat-boat conformation (**174D**)[68,64] as well as twist-boat-chair and twist chair-chair conformations.[69] These latter conformations are higher in energy than the crown (**174B**).[70] Wiberg used calculations to show that the boat-chair is the lowest energy conformer, with the twist boat-chair, twist chair-chair and the crown as the next higher energy conformations.[64] The boat-boat conformation is relatively high in energy.[64] A crown conformation is observed for cyclodecane, **182A**,[67b,71]

66. Illuminati, G.; Mandolini, L. *Acc. Chem. Res.* **1981**, 14, 95. Also see Ref. 12, p 680.
67. Also see (a) Dunitz, J.D.; Prelog, V.P *Angew. Chem.* **1960**, 72, 896; (b) Prelog, V. in *Perspectives in Organic Chemistry*, Todd, A.R. (Ed.), Interscience, New York, **1956**, p 96.
68. Bellis, H.E.; Slowinski, E.J. *Spectrochim. Acta* **1959**, 15, 1103;
69. Pauncz, R.; Ginsburg, D. *Tetrahedron* **1960**, 9, 40.
70. (a) Reference 12, pp 765-766; (b) Reference 31b, p 253; (c) Allinger, N.L.; Hu, S. *J. Am. Chem. Soc.* **1961**, 83, 1664.
71. Prelog, V. *J. Chem. Soc.* 1950, 420.

174B   **174C**   **174D**   **182A**

which conforms to the same extended chair conformation found in diamond (see Fig. 1.13). Crown-cyclodecane[64] fits on this diamond lattice[68] and it is usually taken as an important low energy conformation. The odd-membered ring in cyclononane shows a twist, also found in cycloheptane when compared to chair cyclohexane[60] and a slight increase in Pitzer strain. As seen in Figure 1.5, the twist in crown-like cyclononane makes it unable to superimpose on the diamond lattice. Even large rings such as cyclooctadecane can show this chair-like or crown-like conformation in a low energy form.

cyclodecane   cycloundecane

Figure 1.13. Cyclodecane and Cycloundecane on a Diamond Lattice.

Another type of strain that is found in large ring compounds must be introduced To better see this interaction, models of cyclopentane-cyclooctadecane are shown looking down on the top of each ring, exposing a cavity in the center of each ring. If the conformation of a ring brings atoms in close proximity, *within the cavity of the ring*, this is known as a **transannular interaction**.[66] Hydrogen atoms do not intrude into the internal cavities of envelope cyclopentane (**171C**) or chair cyclohexane (see **172C**), so there are no transannular interactions. In cycloheptane (see **173D**), two of the hydrogen atoms can move into positions that are slightly within the cavity,

171C   **172C**   **173D**

contributing to modest transannular strain and partially accounting for the higher energy of cycloheptane. In crown-cyclooctane (**174E**), pseudorotation moves hydrogen atoms that are on opposite sides of the ring into the cavity, contributing to a significant transannular interaction, and a significant increase in the inherent energy for that ring. There is also significant transannular strain in cyclononane (**183**) and crown-cyclodecane (**182B**). As the ring size increases, the cavity becomes larger, and the net energy decreases after reaching a maximum at cyclodecane. Although the potential for a transannular interaction is quite high in cyclooctadecane (**184**), the large cavity and increased flexibility allow the transannular hydrogens to sweep past each other upon pseudorotation. The cavity inside the ring is large enough to accommodate these atoms without generating a large energy gradient.

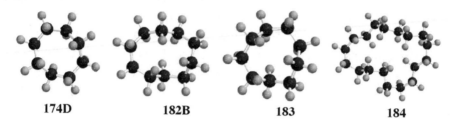

As shown above, the higher energy of medium size rings results from unfavorable dihedral angles.[71] This type of strain has been termed ***I-strain*** (internal strain).[72] Formally, *I strain* involves the change in strain when going from a tetrahedral to a trigonal carbon or vice versa (as in changing an alcohol to a ketone). For medium-size rings, tetrahedral bond angles bring the transannular substituents into close proximity. Conversion to a trigonal planar carbon relieves this interaction somewhat and lowers the energy of the system.[73] An example is the oxidation of cyclooctanol (**185**) to cyclooctanone (**186**, see sec. 3.2.A). The reverse process (the reduction, see secs. 4.2.B, 4.8.C, 4.9.B: **186** ® **185**) is more difficult due to the increased *I* strain as well as increased transannular strain in the alcohol.[74]

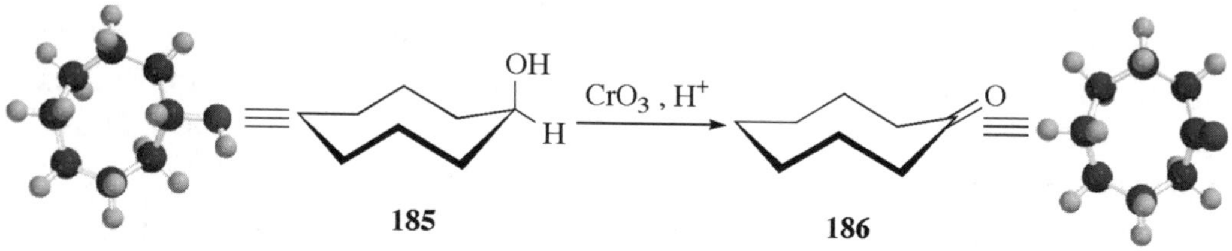

The presence of a heteroatom, a C=C unit, or a carbonyl influences the conformation of medium-size rings, but the changes are often subtle. The calculated low-energy conformation of cyclooctane (**187**) is compared with those of *cis*-cyclooctene (**188**), the ether oxocane (**189**) and cyclooctanone (**190**). Although **188** is somewhat flattened, the conformations of the other three eight-membered rings are rather similar.[75]

72. (a) Brown, H.C.; Fletcher, R.S.; Johannesen, R.B. *J. Am. Chem. Soc.* **1951**, *73*, 212; (b) Brown, H.C.; Borkowski, M. *Ibid* **1952**, *74*, 1894.
73. Prelog, V. *Bull. Chim. Soc. Fr.* **1960**, 1433.
74. Reference 31b, pp 266-269.
75. Pawar, D.M.; Moody, E.M.; Noe, E.A. *J. Org. Chem.* **1999**, *64*, 4586.

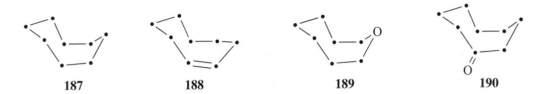

187   **188**   **189**   **190**

Estimating the conformations of large ring compounds (macrocycles) is more complicated, and more difficult than with compounds discussed in the section 1.4. Since this class of compounds includes macrolide antibiotics,[76a] important commercial products such as muscone and also crown ethers,[76b] there is a great deal of interest in macrocycles.[76] Both the chemical and physical properties of macrocycles depend on the conformations of the large ring.[77] An analysis of conformations of small and medium ring compounds was reported that used an analysis of the sign of torsional angles,[78] and when combined with the use of dihedral maps.[79] Weiler[80] produced a **polar map analysis** of macrocycles that yielded conformational information. This method has also been applied to the conformational analysis of macrocyclic ethers.[81] Another way to represent this conformation is based on Dale's numerical system describing the number of bonds found between corner atoms.[82c] Dale's system is sometimes called **wedge notation**.[83] Unfortunately, calculations can be difficult, and the method does not seem to be widely used. The system is based on the idea that even-membered rings can exist in four quadrangular (four-cornered system) conformations (see **191**). (*1*) All four sides contain an odd number of bonds (*a-d* = odd). (*2*) Two adjacent sides are odd with the others even (*ab* odd, *cd* even; or *bc* odd, *ad* even). (*3*) Two adjacent sides are odd with the others even (*ac* odd, *bd* even; or bd odd, ac even). (*4*) All sides contain an even number of bonds (*a→d* = even). For odd membered rings, the best conformation arises when there are three (a triangular system) or five corners (a quinguangular system). When there are three corners all three sets can be odd, or two can be even with one odd and one side must have one or more convex faces. For five corners, all can be odd, two adjacent or next to adjacent can be even with the other three odd, or one can be odd and four even, so there must be one or more concave faces.

76. (a) Paterson, I.; Mansuri, M.M. *Tetrahedron* **1985**, *41*, 3568; (b) Hayward, R.C. Chem. Soc. Rev. 1983, 12, 285; (c) Still, W.C.; Galynker, I. *Tetrahedron* **1981**, *37*, 3981 and references cited therein.

77. (a) Dale, J. *Angew. Chem. Int. Ed.* **1966**, *5*, 1000; (b) Idem *Top Stereochem.* **1976**, *9*, 199; (c) Idem *Acta Chem. Scand.* **1973**, *27*, 1115.

78. (a) Bucourt, R. *Top Stereochem.* **1974**, *8*, 159; (b) DeClerq, P.J. *Tetrahedron* **1984**, *40*, 3729; (c) Idem *Ibid* **1981**, *37*, 4277; (d) Toromanoff, E. *Ibid* **1980**, *36*, 2809.

79. Ogura, H.; Furuhata, K.; Harada, Y.; Iitaka, Y. *J. Am. Chem. Soc.* **1978**, *100*, 6733.

80. (a) Ounsworth, J.P.; Weiler, L. *J. Chem. Educ.* **1987**, *64*, 568; (b) Keller, T.H.; Neeland, E.G.; Rettig, S.; Trotter, J.; Weiler, L. *J. Am. Chem. Soc.* **1988**, *110*, 7858.

81. Clyne, D.S.; Weiler, L. *Tetrahedron* **2000**, *56*, 1281.

82. (a) Dale, J. *Acta Chem. Scand.* **1973**, *27*, 1115, 1130; (b) Dale, J. *Acta Chem. Scand.* **1973**, *27*, 1149; (c) Björnstad, S.L.; Borgen, G.; Dale, J.; Gaupset, G. *Acta Chem. Scand. Ser. B* **1975**, *B29*, 320.

83. Reference 12, pp 763-764, 766-769.

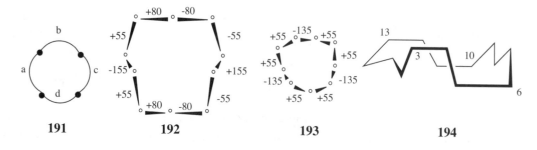

| 191 | 192 | 193 | 194 |

The actual notation is illustrated for the diamond form of cyclodecane, **192** (see Fig. 1.13). The dihedral angles were measured from a Dreiding Model™ (this can also be done with appropriate computer software), and the corners determined for a given conformation. For **192**, those angles for the four corners are +55/+80 and -55/-80. This method takes the number of bonds on each side, starting with the short side and gives the lowest possible combination of numbers. For **192**, there are 2, 3, 2 and 3 bonds, respectively, and this is [2323]-decane. Nonane has an odd number of atoms, and has three corners. Conformation **193** represents nonane and it has three corners with angles determined to be +55/+55, leading to the notation [333]nonane. For cyclotetradecane, the lowest energy conformation is **194**, the [3434] conformation. When substituents are attached to large rings, they tend to occupy only exterior positions to minimize the large transannular interactions.[84] A fully substituted atom of a macrocycle will usually occupy a corner position since this is the only position where it does not cause severe transannular strain.[82] For both substituted and unsubstituted macrocycles, the conformation of the ring will impose restraints on the torsional angles available to each unit of carbons (four carbons define a torsional angle). The use of polar maps of torsional angles may be of value for determining the conformation of many macrocyclic compounds. Fourteen-membered 3-keto lactones, for example, have been synthesized and their conformations analyzed using this technique.[85] This ability to predict conformational bias is important to the stereochemical outcome of chemical reactions involving macrocyclic rings.

1.5.C. Estimating Conformational Populations

The preceding discussion focused on a low-energy conformation of several ring compounds, but in fact there may be several conformations for each ring that lie close together in energy. To obtain a complete picture several conformations of equal, close, or higher energy must be considered. The conformational mobility of cyclohexane includes, for example, the two low-energy chair conformations, six degenerate twist-boat conformations, and six degenerate

84. Dale, J. *Stereochemistry and Conformational Analysis*, Verlag-Chemie, New York, *1978*.
85. Neeland, E.G.; Ounsworth, J.P.; Sims, R.J.; Weiler, L. *J. Org. Chem.* **1994**, *59*, 7383.

boat conformations.[86,87] If all the energies for the various conformations are known, it should be possible to estimate the percent of each. Since the chair conformation usually constitutes the majority of this conformer population, a typical assumption is usually be made to ignore the other conformations for an initial estimation. Building cyclohexane using Spartan and minimizing the structure will generate a chair cyclohexane. Eliel et al.[88] estimated that the energy difference between the boat and chair forms is about 4 kcal mol^{-1} (16 kJ mol^{-1}), which means "only one molecule in a thousand will be in the boat form" at 298 K (25°C). Note that the boat form exists in certain bridged cyclohexane molecules such as bicyclo[2.2.2]-octane (**195**) which is all boat, or it may be forced on the ring by bulky substituents (e.g., *trans*-1,2-di-*tert*-butylcyclohexane) which contains about 12% of the boat **196A**.[88,89]

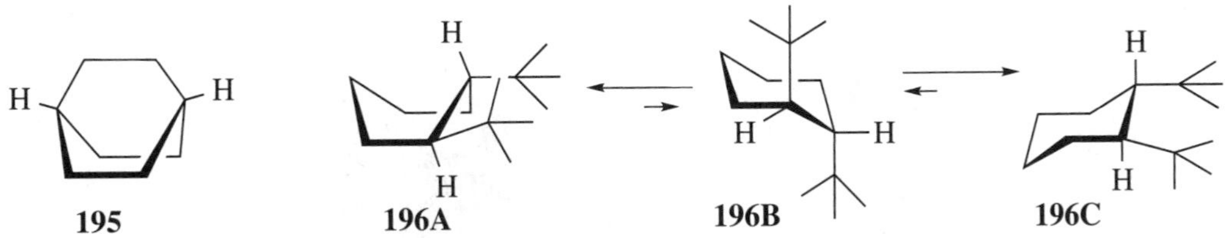

The discussions of cyclohexane given above have made it clear that the chair conformations are lower in energy. The boat and two chair conformations are shown for **196** (trans-1,2-di-*tert*-butylcyclohexane). It is clear that there are significant steric interactions in the boat (**191A**) because the two *tert*-butyl groups are in close proximity, whereas in the diaxial chair (**196B**) those groups are on opposite sides of the ring. Interestingly, the interaction of the *tert*-butyl groups is somewhat relieved in the twist-boat conformation, which exists alongside the more stable **196A**,[88] but the structures shown indicate that **196C** (the *tert*-butyl groups are both equatorial) is lower in energy than **196B** (the *tert*-butyl groups are both axial). Why is **196B** higher in energy than **196C**, and why is the low energy form the boat conformation **196A** or the twist-boat conformation mentioned? The answer reveals two major sources of strain in chair cyclohexanes, which have both axial and equatorial bonds (see **178** and **179**). In cyclohexane, this interaction of the axial hydrogen atoms on either side of the ring is minimal because hydrogen atoms are relatively small as seen in the space filling model **172D**. Replacing the hydrogen atoms in the axial positions with substituents leads to competition for the space above and below the ring, giving rise to a transannular interaction (*A* **strain**, or *A*^{1,3}-**strain**, a **1,3-diaxial interaction**, see Fig. 1.14). When one hydrogen is replaced with a methyl group (methylcyclohexane, **197**), the interaction between methyl and hydrogen is

86. (a) Lowry, T.H.; Richardson, K.S. *Mechanism and Theory in Organic Chemistry, 3rd Ed.,* Harper and Row, New York, *1987*, p 141; (b) Squillacote, M.; Sheriden, R.S.; Chapman, O.L.; Anet, F.A.L. *J. Am. Chem. Soc. 1975*, *97*, 3244; (c) Hirsch, J.A. *Concepts in Theoretical Organic Chemistry*, Allyn and Bacon, Boston, *1974*, pp 249-252.
87. Barton, D.H.R. *Experientia 1950*, *6*, 316.
88. van de Graaf, B.; Baas, J.M.A.; Wepster, B.M. *Recueil Trav. Chim. Pays-Bas 1978*, *97*, 286.
89. Allinger, N.L.; Freiberg, L.A. *J. Am. Chem. Soc. 1960*, *82*, 2393.

greater, making the transannular interaction much higher in energy relative to the hydrogen-hydrogen interaction of cyclohexane. ***This is $A^{1,3}$ strain.*** In 1,3-dimethylcyclohexane (**198**) and in 1,3,5-trimethylcyclohexane (**199**) $A^{1,3}$-strain increases significantly as the steric interactions increase. This $A^{1,3}$-strain destabilizes the conformations with the axial substituents, shifting the equilibrium to the lower energy chair form that has the substituents in the equatorial position. Returning to **196B**, it is clear that each axial *tert*-butyl group will interact with the two axial hydrogen atoms on its side of the ring, leading to significant $A^{1,3}$-strain) estimated to be about 12 kcal mol^{-1} (50.2 kJ mol^{-1}) or 6 kcal/*tert*-butyl group.[90] There is no question that the equilibrium for the two chair conformations will shift to favor the chair with the two *tert*-butyl groups in the equatorial position, effectively eliminating the $A^{1,3}$-strain.

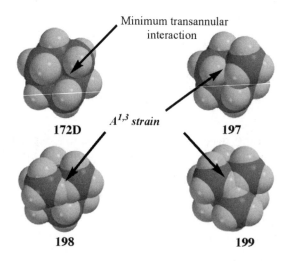

Figure 1.14. $A^{1,3}$ Strain in Cyclohexane, Methylcyclohexanes.

Closer inspection of **196C** reveals that $A^{1,3}$-strain has been eliminated but the 1,2-diequatorial *tert*-butyl groups are in close proximity, producing a new type of strain called **G strain**. The energy of the interaction (the **G value**)[91] is 2.5 kcal mol^{-1} (10.47 kJ mol^{-1}), per *tert*-butyl group (see Fig. 1.15). This type of steric interaction is analogous to the interactions found in the *gauche* butane conformation, and is often referred to as a *gauche* interaction. If we take cyclohexane as a model the trans-diequatorial hydrogens in **172E** show little interaction, but replacing them with methyl (see *trans*-1,2-dimethylcyclohexane, **200**) causes significant G-strain. When the two *tert*-butyl groups in **196** are trans-diequatorial (see **196C**), the interaction is very large and destabilizes that conformation. If both chair conformations have high energy interactions, the molecule will distort to minimizes these interaction, and **196** exists largely in either the boat or the twist-boat.

90. Winstein, S.; Holness, N.J. *J. Am. Chem. Soc.* **1955**, *77*, 5562.
91. Corey, E.J.; Feiner, N.F. *J. Org. Chem.* **1980**, *45*, 765.

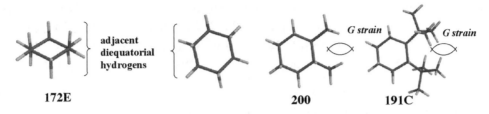

Figure 1.15. *G* Strain in 1,2-Dimethylcyclohexane and 1,2-Di-*tert*-butylcyclohexane.

Why are these physical chemistry concepts being discussed in a reactions-synthesis book? It will be apparent throughout this book that the conformation of a molecule has a critical effect on the reactivity and stereochemistry of many reactions. Strain energies that influence the conformation of a molecule can be quantified in some cases, and the percentage of each conformation can be calculated. With such information we can make more reasonable predictions of reactivity and stereochemistry. The following protocols are presented in an attempt to reinforce the idea that conformational analysis is a critical part of synthetic analysis, and to introduce the most basic approaches to conformational analysis.

We will focus on cyclohexane derivatives to illustrate the fundamental idea. When a reagent approaches a cyclohexane ring, the conformation of the ring will influence how the reagent interacts with any functional group on that ring. The populations of the two chair conformations can influence the relative rate and the stereochemical outcome of a given reaction. The relative populations of both chair conformations in cyclohexane derivatives can be predicted with some accuracy by calculating the *A* and/or *G* values. For a monosubstituted cyclohexane, two chair forms are possible (**201** and **202**) and they are in a dynamic equilibrium. If X ≠ H, the diaxial steric interaction (*A* strain) in **201** will be larger than in **202** and a higher percentage of **202** is expected. This non-bonded interaction can be measured in terms of $F°$, the conformational free energy.[92] Since there are two conformations of differing energies, this difference is represented by $(\Delta F°)$,[86a] and it is possible to relate this energy term to the free energy of the system by: $\Delta G° = \Delta H° - T\Delta S°$. For a given temperature, the entropy term will be small relative to $\Delta F°$ (taken to be $\Delta H°$ hereafter) and a second assumption can be made, that $\Delta S° = 0$, leading to $\Delta G° \approx \Delta H°$.

The $\Delta S°$ term is not zero, but is usually measured in calories (J) and the $\Delta H°$ term in kilocalories (KJ), so ignoring $\Delta S°$ will introduce only a small error into the calculation. Lowry and

92. (a) Eliel, E.L. *J. Chem. Educ.* **1960**, *37*, 126; (b) Reference 12, pp 694-698; (c) Reference 31b, pp 234-239; (d) Reference 86a, pp 138-140; (e) Hammett, L.P. *J. Am. Chem. Soc.* **1937**, *59*, 96; (f) Hammett, L.P. *Physical Organic Chemistry, 2nd Ed.*, McGraw-Hill, New York, **1970**, p 347ff; (g) Jaffé, H.H. *Chem. Rev.* **1953**, *53*, 191; (h) Hansch, C.; Leo, A.J. *Substituent Constants for Correlation Analysis in Chemistry and Biology*, Wiley, New York, **1979**; (i) Reference 12, p 654.

Richardson's[93] list of $\Delta H°$ and $S°_{298}$ values for a variety of specific bonds demonstrates this difference. In the C- Me bond, for example, $\Delta H°$ is -10.08 kcal mol^{-1} (42.2 kJ mol^{-1}) and $S°_{298}$ is 30.4 cal mol^{-1} (127.2 J mol^{-1}). Similarly, $\Delta H°$ for cyclopropane is listed as 27.6 kcal mol^{-1} (115.5 kJ mol^{-1}) and $S°_{298}$ is 32.1 cal mol^{-1} (134.3 J mol^{-1}).[93]

The $\Delta H°$ term is related to $\Delta G°$, the free energy, which is related to the equilibrium constant by the expression: $\Delta G° = -2.303$ RT log K_{eq}. The conformational equilibrium between **201** and **202** is represented by K_{eq}, which can be calculated from the percentage of the two conformers: $K_{eq} = \%$ **202** $/ \%$ **201**, where $\%$ **201** $+ \%$ **202** $= 100\%$. Note that a general discussion of the Curtin-Hammett principle[92c-f] (product composition is related to the relative concentrations of the conformers)[92i] may be of value in any discussion of this equilibrium reaction.

For this process to work, some estimation of $F°$ (hereafter called $H°$) must be made. In connection with the development of the interactive synthesis computer program *Logic and Heuristics Applied to Synthetic Analysis* (LHASA, sec. 10.4.B),[1,5] Corey developed a protocol for determining the conformational energies of cyclohexane derivatives,[91] and assembled a list of H values for various substituent interactions in Table 1.4.[91] This table is related to a similar one, first assembled by Hirsch and co-workers.[94,95] The energy for a given conformation is taken to be the sums of all monoaxial interactions (A values), any 1,2-diequatorial interactions (G values) and any multiple diaxial interactions (defined by Corey as U values[86] and found in the two structures labeled). Simple doubling or tripling of the appropriate A value did not give correct results in this latter case and more accurate U-values were required. Note that the A and U values are identical in many cases. They usually differ if the substituent contains unsaturation.

The protocol for calculating the percentage of each conformation is illustrated by the two chair conformations of 2-chloro-3,4-dimethylcyclohexanol, **203** and **204**. Using Table 1.4, the pertinent interactions are U_{OR}, the two U_{CH_2R} and U_{Cl} in **203** and the two G_{CH_2R}, G_{OR} and G_{Cl} in **204**.

$$\Delta H° = H_{product} - H_{reactant} = H_{204} - H_{203} = (G_{CH_2R} + G_{Cl} + G_{CH_2R} + G_{OR}) -$$
$$(U_{OR} + U_{CH_2R} + U_{CH_2R} + U_{Cl}) = (0.4 + 0.5 + 0.4 + 0.2) - (0.8 + 1.8 + 1.8 + 0.4) =$$
$$(1.5) - (4.8) = -\textbf{3.3} \text{ kcal mol}^{-1} (-13.8 \text{ kJ mol}^{-1}).$$

93. Reference 86a, pp 164-165.
94. Hirsch, J.A. in *Topics in Stereochemistry, Vol. 1,* Allinger, N.L.; Eliel, E.L. (Eds.), Wiley-Interscience, New York, *1967*, pp 199-222.
95. (a) Reference 31b, pp 236-237; (b) Barrett, J.W.; Linstead, R.P. *J. Chem. Soc. 1936*, 611. For a list of energy values for several functional groups, see Reference 12, pp 696-697.

Table 1.4. Corey's A, G, and U Values for Calculating Conformational Energies of Cyclohexane Derivatives.

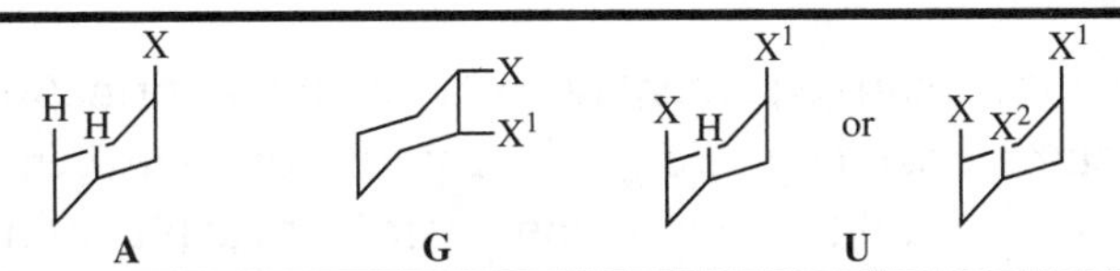

Group	A (kcal mol^{-1})[a]		G (kcal mol^{-1})[a]		U (kcal mol^{-1})[a]	
H	0		0		0	
F	0.2	(0.84)	0		0	
Cl	0.4	(1.67)	0.5	(2.09)	0.4	(1.67)
Br	0.4	(1.67)	0.8	(3.35)	0.4	(1.67)
I	0.4	(1.67)	1.0	(4.19)	0.4	(1.67)
PR$_3$	1.6	(6.7)	1.6	(6.7)	1.6	(6.7)
SR	0.8	(3.35)	0.5	(2.09)	0.8	(3.35)
S(O)R	1.9	(7.95)	2.7	(11.30)	1.9	(7.95)
S(O$_2$)R	2.5	(10.47)	3.5	(14.65)	2.5	(10.47)
OR	0.8	(3.35)	0.2	(0.84)	0.8	(3.35)
NH$_3^+$	2.0	(8.37)	0.5	(2.09)	2.0	(8.37)
NR$_3^+$	2.1	(8.79)	0.5	(2.09)	2.1	(8.79)
NHR	1.3	(5.44)	0.3	(1.26)	1.3	(5.44)
N=	0.5	(2.09)	0.1	(0.42)	0.5	(2.09)
N≡	0.2	(0.84)	0.1	(0.42)	1.2	(5.02)
NO$_2$	1.1	(4.61)	0.3	(1.26)	0.5	(2.09)
C≡	0.2	(0.84)	0		1.2	(5.02)
aryl	3.0	(12.56)	1.2	(5.02)	0.5	(2.09)
CO$_2^-$	2.0	(8.37)	0.5	(2.09)	2.0	(8.37)
CHO	0.8	(3.35)	0.3	(1.26)	0.8	(3.35)
C=	1.3	(5.44)	0.2	(0.84)	0.9	(3.77)
CR$_3$	6.0	(25.11)	2.5	(10.47)	6.0	(25.11)
CHR$_2$	2.1	(8.79)	0.8	(3.35)	2.1	(8.79)
CH$_2$R	1.8	(7.54)	0.4	91.67)	1.8	(7.54)

[a] The value in parentheses is the energy in kJ mol^{-1}

[Reprinted with permission from Corey, E.J.; Feiner, N.F. *J. Org. Chem.* **1980**, *45*, 765. Copyright © *1980* American Chemical Society.]

If $\Delta H° \approx \Delta G°$, then *assume* $\Delta G° = -3.3 = -2.303\ RT \log K_{eq}$. At 25°C, $-2.303\ RT = -1.364$. Therefore,

$$\Delta G° = -3.3 = -1.364 \log K_{eq} \text{ sp } \log K_{eq} = -3.3/-1.364 = 2.42$$
$$\text{and } K_{eq} = 10^{2.42} = 262.6.$$

Since %**203** + %**204** = 100%, where **204** = 1 - **203**, then K_{eq} = 204/203 = 1-204/203 and K_{eq} (**203**) = 1 - **203**);

so, 262.6 (**203**) = 1 - (**203**), and 262.6 (**203**) + (**203**) = 1 or 262.6 + 1 (**203**) = 1, which leads to

$$(\textbf{203})\ (263.6) = 1 \text{ and } (\textbf{203}) = 1263.6 = 0.0038$$

Therefore, there is 0.38% of **203** and 99.62% of **204**. *Remember that two major assumptions were made for this calculation*: (*1*) all conformations except **203** and **204** were ignored, and (*2*) ΔS° was assumed to be zero.

As first shown by Barton,[87] the relative percentage of a given conformation directly influences the rate of a reaction. As summarized by Eliel,[92a] the equilibrium between **205** and **206** is given by the equilibrium constant (K), and the rate of conversion of each conformation to product is given by k_{205} and k_{206}, as developed by Winstein and Eliel.[92a] The overall rate will be Rate = k [C], where k is the observed specific rate and [C] is the stoichiometric concentration of the cyclohexane.[92a] Rate = k_{205} [**205**] + k_{206} [**206**] for this system, and the value k is related to the equilibrium constant as shown in the following:

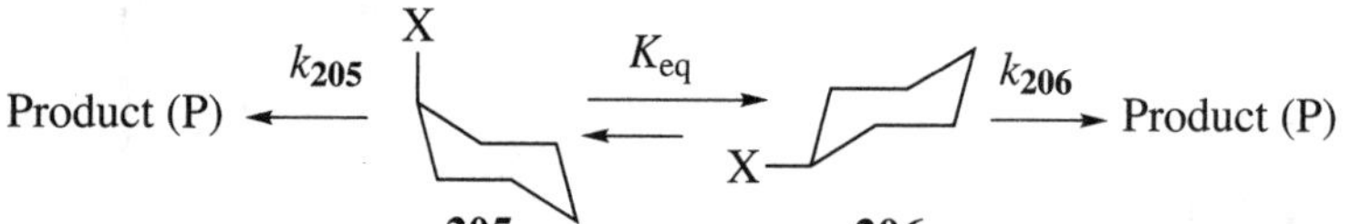

$$K_{eq} = \frac{k_{205} - k}{k - k_{206}}$$

and

$$k = \frac{k_{206} + k_{205}}{K + 1}$$

As mentioned above, the percentage of a given conformation is an essential factor in determining the rate and viability of a given reaction and these calculations allow the synthetic chemist to estimate conformational populations. Just as the minor chair conformation influences the rate, so other conformations (boat, twist-boat, etc.) can also exert an influence, depending on their relative populations.

MOLECULAR MODELING: Conformers of Cyclohexane

It is well known that cyclohexane adopts a chair conformation. Actually, the molecule exists in both chair and twist-boat conformers. The chair structure is the lower-energy conformer (quantum chemical calculations suggest that the twist-boat conformer is ~30 kJ/mol higher in energy), and the two interconvert via a low-energy half-chair transition state. The large energy difference and the small barrier to interconversion mean that detecting (let alone isolating) twist-boat cyclohexane is nearly impossible. Quantum chemical calculations show that a temperature in excess of 1500 K would be needed to achieve 10% of an equilibrium mixture of these conformers.

Interconversion of equivalent chair conformers of cyclohexane may be thought of as proceeding in two stages that "mirror" each other. The first is interconversion of one chair conformer to the twist-boat conformer via a half-chair transition state, and the second is interconversion of the twist-boat conformer to the second chair conformer via a second (equivalent) half-chair transition state. A plot of the energy profile for the process appears on screen. You can step through the individual structures that make up the sequence or "animate" it using the **Step** and **Play** keys, respectively at the bottom left of the screen.

In addition to the values in Table 1.4 used for calculations of substituted cyclohexane derivatives,[85] Corey developed formulas for several other cyclic molecules. Incorporation of two halogens with a 1,2- or 1,4-diaxial relationship on a cyclohexane ring led to a higher proportion of the diaxial conformer[96] due to the more favorable electrostatic interactions of the lone pair electrons.[97] Corey's models in this case were **207** and **208**, and the equation used to calculate the conformational energy is E = 1/2 ($A_x + A_y$). Introduction of a π-bond or a heteroatom in the cyclohexane ring changes the energy relationships, since the presence of a π bond flattened the ring. The usual *U* and G values were used for monosubstituted or disubstituted derivatives such as **209** or **210**, respectively. A carbonyl or an aldehyde unit led to a diminished value for *A*, and a spirocyclic or fused ring structure also diminished *A*. Introduction of an oxygen in the ring (to form a pyran) led to a diminished value for *A*.

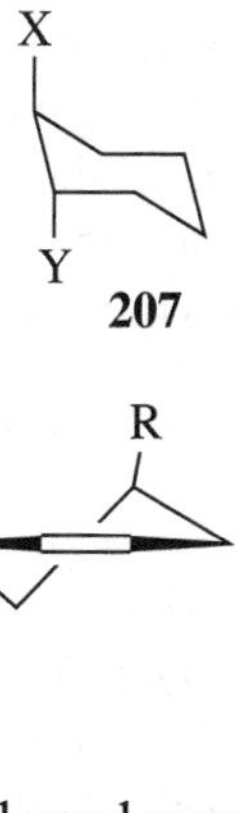

Corey and co-workers[91] also developed parameters to evaluate boat conformations, which are often important contributors to the conformational population of cyclohexane derivatives (see above). The energies for boat conformations with one or more substituents in the flagpole

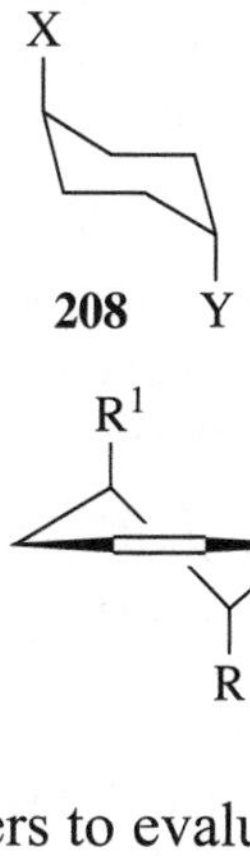

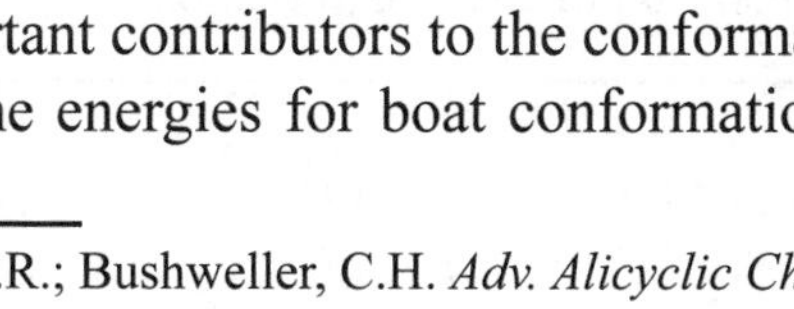

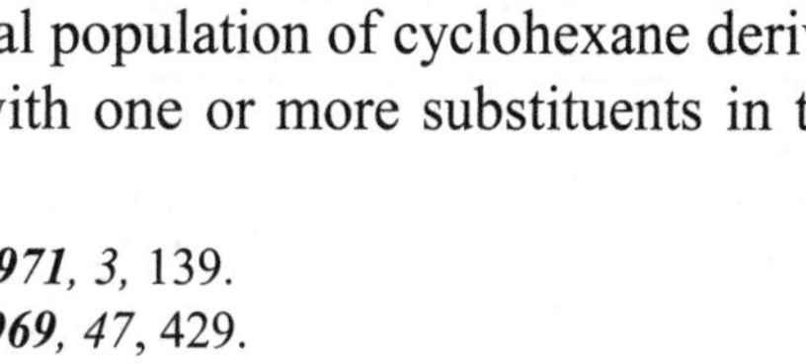

96. Jensen, F.R.; Bushweller, C.H. *Adv. Alicyclic Chem.* **1971**, *3*, 139.
97. Wood, G.; Woo, E.P.; Miskow, M.H. *Can. J. Chem.* **1969**, *47*, 429.

positions use a new term (b): $E = (b^1 + b^2) A^R$ for **211**, and $E = (b^1 + b^2) (U^R + U^{R'})$ for **212**), derived from the bowsprit-flagpole interactions (R-H in **211** or R^1-R in **212**). Introduction of a π bond, a cyclopropane or a cyclobutane ring flattens the boat, leading to increased destabilization. Fused five-membered rings and larger fused rings do not flatten the ring of interest.

These structural types emphasize that the presence of π bonds and heteroatoms will influence the overall conformation. "Flattening" of a ring is often an important structural feature of natural products and other attractive synthetic targets. The π bond in a cyclohexene moiety flattens the ring in **213A** and its other low-energy conformer, **213B**. These are actually half-chair conformations analogous to **177**.[98] The isopropyl group (*i*-Pr) is less stable in the pseudo-axial position and more stable in the pseudo-equatorial position.[99] The *i*-Pr↔Br interaction is largely alleviated in **213B**. This effect is compatible with **213B** as the major conformation and higher relative energy for **213A**.

MOLECULAR MODELING: *Axial/Equatorial* **Preferences in Disubstituted Cyclohexanes**

It is well known that substituents on cyclohexane generally (but not always) prefer to be *equatorial*. Where there are two different substituents and only one can be *equatorial*, for example, in a *cis*-1,4-disubstituted cyclohexane, the substituents will "compete" for the favored position. Build the conformer of cis-1-bromo-4-*tert*-butylcyclohexane with the Br *axial* and the *tert*-butyl group *equatorial* and obtain its energy. Next, *click* on > to the right of the tab at the top of the screen and select **Continue** to bring up a fresh builder palette. Build the conformer in which the Br is *equatorial* and the *tert*-butyl group is *axial* and determine its energy. Which substituent, bromine or *tert*-butyl, "wins" the competition and assumes the *equatorial* position? Is the higher energy conformer likely to be observed in a room-temperature equilibrium distribution? Hint: a 95:5 ratio of major to minor conformers corresponds to an energy difference of around 8 kJ/mol.

A similar competition exists for disubstituted cyclohexenes, for example for cis-3-bromo-6-*tert*-butyl-1-cyclohexene. Obtain energies for the two possible conformers (bromine *axial* and *tert*-butyl *equatorial* and *vice versa*). Which substituent, bromine or *tert*-butyl, assumes the *equatorial* position? Is the difference in energy between the conformers, smaller, larger or about the same as for *cis*-1-bromo-4-*tert*-butylcyclohexane? If smaller or larger, examine the geometries of the two cyclohexane conformers and two cyclohexene conformers to provide an explanation.

---

98. (a) Böeseken, J.; de Rijck van der Gracht, W.J.F. *Rec. Trav. Chim.* **1937**, *56*, 1203; (b) Reference 31b, p 239.
99. (a) Reference 31b, p 240; (b) Dauben, W.G.; Pitzer, K.S. in *Conformational Analysis in Steric Effects in Organic Chemistry*, Newman, M.S. (Ed.), Wiley, New York, **1956**, pp 38-39.

Similar conformational effects arise in other functionalized systems, including reactive intermediates. Conversion of cyclohexanone to its enolate anion (**214A**, see sec. 9.2) flattens the ring (see **214B**), and leads to a cyclohexene-like conformation. Note the conformational similarity of **214A** to cyclohexene derivative **213A**. The dimethylamino enamine (see sec. 9.6) of cyclohexanone (**215A**) also shows this half-chair conformation (see **215B**). In these cases, the lone pair electrons (on oxygen in **214** and on nitrogen in **215**) overlap with the π bond.

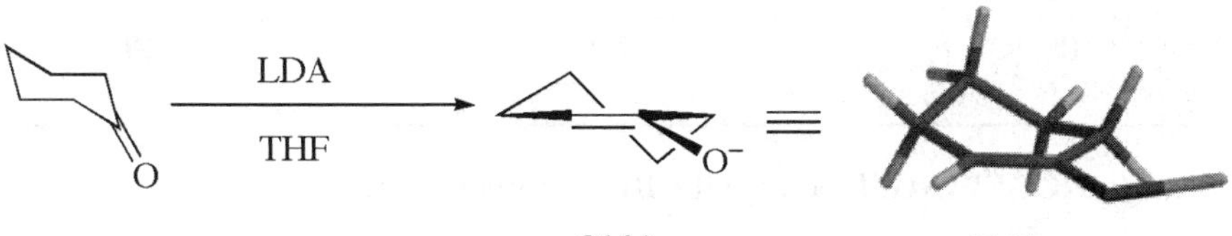

214A   **214B**

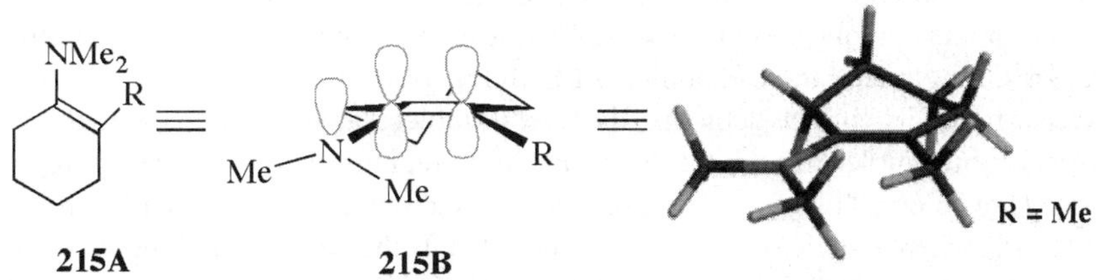

215A   **215B**   R = Me

When compared to chair cyclohexane, the presence of the trigonal planar carbonyl in cyclohexanone removes the 1,3-diaxial hydrogen atom interactions and, as seen in the models for Corey's calculations, partially flattens the ring. The conversion of alcohol **216** to ketone **217** illustrates this point. Although **216** is an equilibrating mixture with an axial or equatorial hydroxyl group, upon oxidation to the ketone the flattening effect of the carbonyl is easy to see. Energetically, the R↔OH interaction (*U* value) in **216** is removed upon oxidation to **217**. In the ketone, the carbonyl effectively eclipses the α-equatorial hydrogens, further lowering the conformational energy.[100] Ketone **217** exists as an equilibrating conformational mixture favoring the R group in an equatorial position.

Exocyclic methylene compounds such as **218** will exhibit the same flattening effect observed with the carbonyl group. An interesting effect is seen in the conversion of **218** to **219** or **220**, where *A* strain is increased in the products, relative to the methylene group. Further, the greater *A* strain of methyl (1.8 kcal mol^{-1} in Table 1.4) versus 0.4 kcal mol^{-1} for Br (7.5 and 1.7 kJ mol^{-1} respectively) suggests that **219** may predominate after addition of HBr to the alkene (see sec. 2.10.A). This analysis ignores the relative trajectories of approach of the reagents (see sec. 6.6.A) and other steric interactions, however.

---

100.   (a) Allinger, N.L.; Blatter, H.M. *J. Am. Chem. Soc.* **1961**, *83*, 994; (b) Allinger, N.L.; Freiberg, L.A. *Ibid* **1962**, *84*, 2201.

MOLECULAR MODELING: Electrophilic Bromination of Methylenecyclohexane

Electrophilic bromination of 4-*tert*-butylmethylenecyclohexane can lead to either **219** or **220**. The mechanism presumably involves addition of "H$^+$" to the external methylene carbon, follow by attack of Br$^-$ on the resulting carbocation. Examine a LUMO map for the intermediate carbocation. This colors an electron density surface depicting the overall molecular size and shape with the value of the lowest-unoccupied molecular orbital. By convention, regions on the surface where the LUMO is concentrated are colored "blue" while those where it is absent are colored "red". Bromide would be expected to add where the LUMO is most concentrated. Is there a clear preference for addition to one face of the carbocation over the other? If there is, which product would be favored, **219** or **220**?

In addition to the endocyclic double bond usually found in an enolate anion such as **214**, enolate anions can be formed that have an exocyclic double bond (also see sec. 9.5.D). Relief of strain will presumably assist formation of the enolate anion, and may influence the final stereochemistry of groups in the product when the enolate anion reacts with an a suitable reagent (see sec. 9.3.A, 9.4).[101] This effect can also be seen in the equilibration of the cis-ester **221** to the trans-ester **223** via the planar enolate (**221**).[98] This equilibration (accomplished by treating **221** with a base followed by a proton source) favors **223** due to reduction of the *A* strain of the axial carboethoxy group in [A_{COOR} = 1.20 kcal mol^{-1} (4.99 kJ mol^{-1})].

Six-membered rings containing oxygen (pyran derivatives based on **224**) are an integral part of carbohydrates such as α-D-glucopyranose (**225**). Less substituted pyrans show similar effects. The presence of the α-alkoxy group in **229** leads to a significant effect also seen in α-alkoxy pyrans, the **anomeric effect**.[102] An alkyl group on cyclohexane or pyran usually prefers the equatorial position due to increased *A* strain. When an alkoxy group is attached to pyran, however, it prefers the axial position. This effect is probably due to dipolar interactions of the oxygen lone pairs (see **226A** and **226B**).[103] The anomeric effect is evident in glucose, where α-D-glucopyranose (**225**) accounts for a significant portion of the conformational equilibrium.

101.   Reference 52a, p 67.
102.   (a) Lemieux, R.U. *Pure Appl. Chem.* **1971**, *27*, 527; (b) Angyal, S.J. *Angew. Chem. Int. Ed.* **1969**, *8*, 157; (c) Kirby, A.J. *The Anomeric Effect and Related Stereoelectronic Effects at Oxygen*, Springer-Verlag, New York, *1983*.
103.   David, S.; Eisenstein, O.; Salem, L.; Hehre, W.J.; Hoffmann, R. *J. Am. Chem. Soc.* **1973**, *95*, 3806.

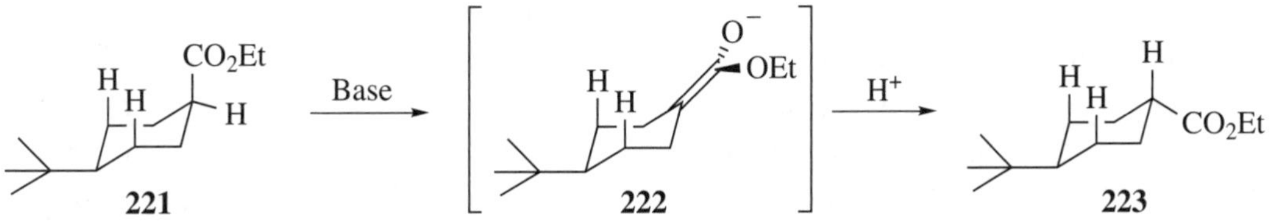

The influence of substituents on conformational stability is also evident in amines such as *N*-ethyl piperidine. The chair conformation can be drawn with the ethyl group axial (N lone pair equatorial in **227A**) or equatorial (N lone pair axial as in **227B**). The greater *A* value for ethyl than for the lone pair suggests a predominance of **227B** in this equilibrium mixture. This preference for the larger group being in an equatorial position is analogous to the effects seen in pyrans and cyclohexanes.

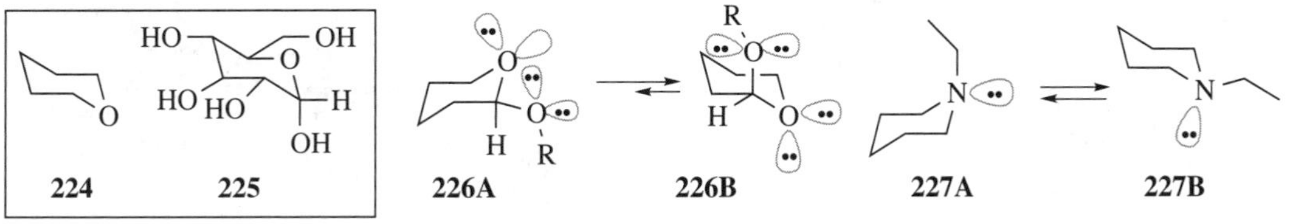

MOLECULAR MODELING: Conformational Mobility of Tetrahydropyran Rings

Compounds **A-D** were isolated from *Aplysia kurodai* by Kakisawa and co-workers (*J. Org. Chem.* **1987**, *52*, 4597-4600). Proton NMR was used to assign the structures of **C** and **D**, but led to ambiguous results for **A** and **B**. (An X-ray crystal structure was required to verify the structure of **B**, which was then used to elucidate the structure of **A**). The authors interpreted the NMR results to indicate that **A** and **B** are conformationally mobile whereas **C** and **D** are rigid.

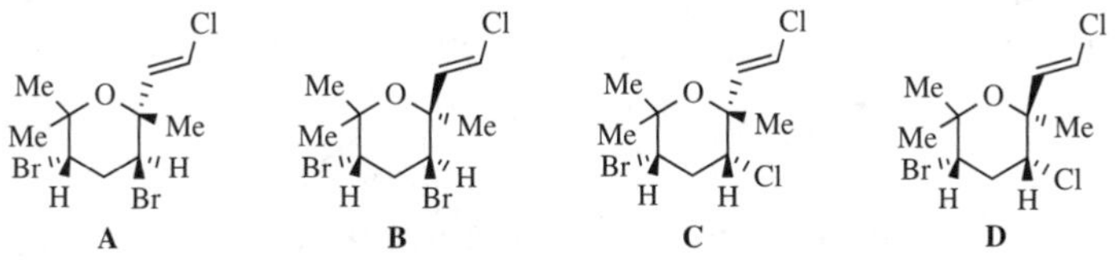

While rotation about single bonds is normally too rapid to be "frozen out" in an NMR experiment, chair-chair interconversion in cyclohexane (and presumably tetrahydropyran) derivatives is much slower and can be stopped at low temperatures. One after the other, build and obtain the energy for the two different chair conformers or either **A** or **B** (or both) and of **C** or **D** (or both). Note that you need to consider the possible conformations about the bond connecting the ring to the vinyl group, choosing the lowest energy for each. Hint: Try the vinyl group "up" and the vinyl group "down" to begin, determine the energy of each and then try one or two other vinyl conformations. For the second and following calculations, *click* on > to the right of the tab at the top of the screen and select ***Continue*** to bring up a fresh builder palette. Are the energies of the two chairs for **A** (or **B**) similar or significantly different? Are the energies of the two chairs for **C** (or **D**) similar or significantly different? If (**A** (or **B**) and **C** (or **D**) exhibit different behavior, suggest why. Are your results in line with the conclusion from the NMR experiments, that A and B are conformationally mobile while C and D are not?

The presence of bulky substituents, or substituents capable of chelation or hydrogen bonding can alter conformational populations,[104,105] as first seen with ethylene glycol. In **228A**, there are two effects that influence the conformation.[106] Bulky substituents lead to large *A* and *G* strain in both chair conformations. In addition, internal hydrogen bonding of the glutarimide moieties helps stabilize the twist-boat conformation shown in **228B**, which is the lowest energy conformation of this molecule. The 3D model shows the twist in the cyclohexane moiety as well as the proximity of the two diketopiperazine units. The twist-boat conformation is also observed in the absence of internal hydrogen bonding. As confirmed by X-ray crystallography, *cis,trans,trans*-1,2,4,4-tetraisopropylcyclohexane exists as a twist-boat.[107]

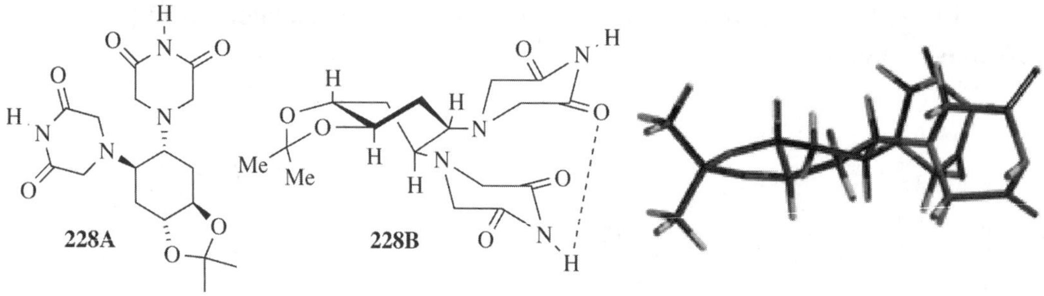

As mentioned in an earlier example, the reduction of cyclooctanone (**186**) shows how conformations can influence reactivity. Treatment of **186** with lithium aluminum hydride (LiAlH$_4$, see sec. 4.2.B) generates the alcohol (**185**). The alcohol will have greater transannular strain than the ketone. The reduction may be somewhat sluggish relative to similar reduction of cyclohexanone. The oxidation, however, converts **185** to **186** using chromium trioxide (see sec. 3.2.A) and leads to diminished strain as the newly introduced carbonyl flattens the ring slightly. As the product is somewhat more stable than the starting material, the oxidation is more facile. Similarly, formation of the enolate (**186** → **229**) with lithium diisopropylamide (LDA, see sec. 9.2.B) should be facile due to relief of transannular strain for formation of the planar enolate moiety. A subsequent alkylation step (**229** → **230**, see sec. 9.3.A) may be sluggish, however, since the methyl group in **230** may introduce a bit more strain in the product (remember there are several other low energy conformations for the eight-membered ring) than in the original unsubstituted cyclooctanone (**186**).

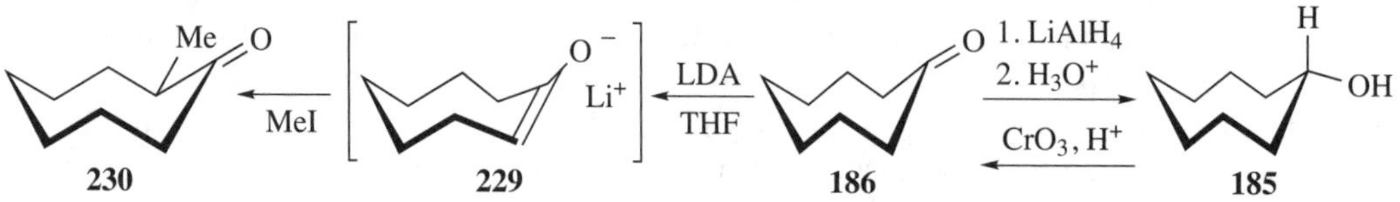

104. Reference 18, p 200.
105. (a) Stolow, R.D. *J. Am. Chem. Soc.* **1961**, *83*, 2592; (b) Stolow, R.D.; McDonagh, P.M.; Bonaventura, M.M. *Ibid* **1964**, *86*, 2165.
106. Witiak, D.T.; Wei, Y. *J. Org. Chem.* **1991**, *56*, 5408.
107. Columbus, I.; Cohen, S.; Biali, S.E. *J. Am. Chem. Soc.* **1994**, *116*, 10306.

MOLECULAR MODELING: Thermodynamic Control of Product Distributions

Where reactions have very low energy transition states, thermodynamics is likely to control product distribution. Reactions that involve protonation followed by deprotonation are good examples. In the gas phase and in non-interacting solvents, both protonation and deprotonation occur with little or no barrier, meaning that final product distributions are dictated only by relative product energies. In the presence of aqueous acid, the hydroxy acid shown below does not give rise to its lactone (lactone A) but rather to lactone B. This reaction occurs via protonation of the hydroxyl unit of the acid, followed by displacement of the hydronium ion by the carboxyl unit. The carbon in lactone B proximal to the lactone oxygen atom has undergone epimerization from (*S*) to (*R*). This transformation requires that the initially formed oxonium ion is displaced by water, with inversion, to form a new oxonium ion, and subsequent reaction with the carboxyl group leads to lactone-B. Why should the epimerization reaction be preferred to direct lactone formation?

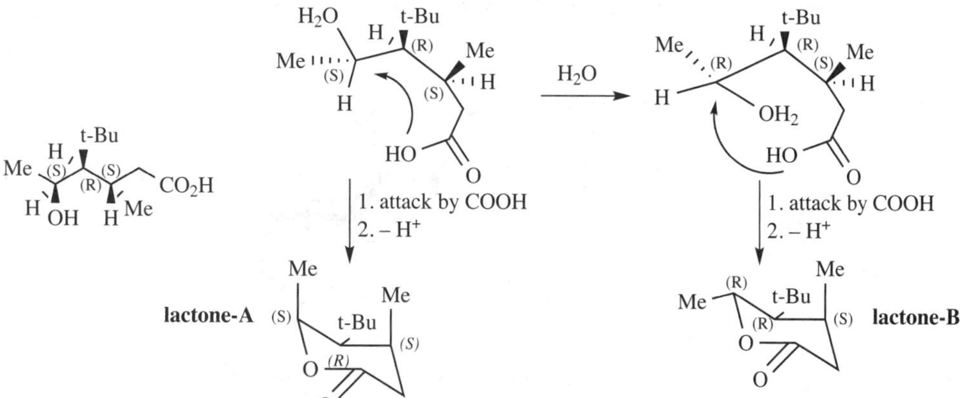

Assuming that the reaction is under thermodynamic control, the lactone that is formed will be the more stable lactone, irrespective of how it was formed. Build lactone-A and obtain its energy. *Click* on > to the right of the tab at the top of the screen and select ***Continue*** to bring up a fresh builder palette. Build lactone-B and obtain the energy. Which lactone is lower in energy? Suggest a reason for this preference. Does the information you have obtained suggest a reason for the observation that epimerization followed by formation of-B occurs preferentially to direct formation of lactone-A? Elaborate.

1.5.D. Conformations in Polycyclic Molecules

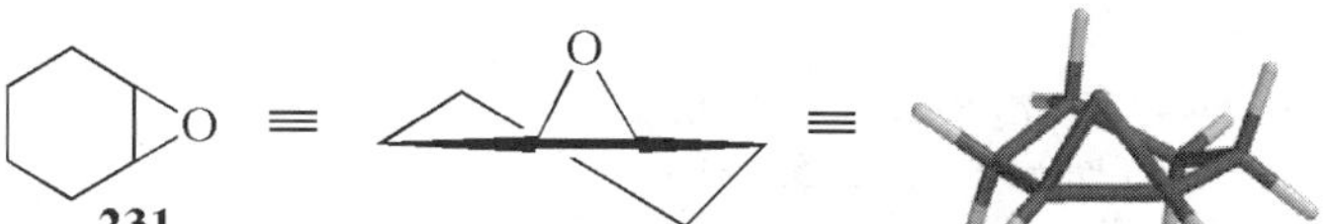

231

The conformational constraints in monocyclic molecules are magnified in bi-, tri-, and polycyclic molecules. A simple case is cyclohexene oxide (**231**) where the planar three-membered ring flattens the ring. Similar effects are seen in cyclohexene, and also occur when cyclopropane is fused to a ring. Bicyclo[3.1.0]hexanes also show this flattening effect. An example is carane (**232**), which exists primarily in the conformation shown (note that the geminal dimethyl groups are perpendicular to the plane of the cyclopropane ring). The cis-isomer (**233**) is expected to show a significant Me-Me interaction, which is not present in the trans-isomer, **232** (carane).

The conformational constraints peculiar to [m.n.0]alkanes (n ≠ 1) can be illustrated by comparing the cis- and trans- isomers of bicyclo[3.3.0]octane (**234**), hydrindane (**235**), as well as **236** (decalin). Each five- and six-membered ring tends to assume an envelope conformation or a chair conformation respectively in their lowest energy form. *cis*-Bicyclo[3.3.0]octane (**237**) shows a bent conformation with the two envelope shapes relieving some of the cross ring interactions.[108,92] *trans*-Bicyclo[3.3.0]octane (**238**) shows a more extended or open structure, and there is some distortion of the five-membered ring. Compound **238** is ~ 6 kcal mol^{-1} (25.1 kJ mol^{-1}) higher in energy than **237**.[109] Both cis- and trans- isomers are easily prepared, however.[110]

Hydrindanes have the five-membered ring in the envelope conformation with the six-membered ring in a chair. The *cis*-indane (**239**) shows a 1,3-diaxial interaction with substituents, and there is a *gauche*-like interaction. In the trans-isomer (**240**), the 1,3-diaxial interaction is diminished, and the extended conformation in **240** is slightly lower in energy (about 1 kcal mol^{-1} or 4.19 kJ mol^{-1}) than **239**.[111] Similar analysis of *cis*-decalin (**241**) and *trans*-decalin (**242**) leads to the same bent and extended forms, with both six-membered rings in a chair conformation.[112] In **241**, one ring can assume more of a boat like structure in some conformations. There is a 1,3-diaxial interaction in substituted **241**, and the overall energy of **241** is lower due to decreased nonbonded interactions.[113] Eliel describes the conformations of several other bicyclic and polycyclic compounds.[114]

108. See reference 12, p 776.
109. (a) Reference 12, pp 774-775; (b) Reference 31b, p 274.
110. Owen, L.N.; Peto, A.G. *J. Chem. Soc, 1955*, 2383.
111. (a) Reference 12, pp 776-779; (b) Reference 31b,p275; (c) Browne, C.C.; Rossini, F.D. *J. Phys. Chem.* *1960*, 64, 927.
112. Turner, R.B. *J. Am. Chem. Soc.* *1954*, 74, 2118.
113. (a) Reference 31b, p 279; (b) Hückel, W. *Annalen, 1925, 441,* 1; (c) Hückel, W.; Friedrich, H. *Ibid 1926*, *451,* 132.
114. Reference 12, pp 780-793.

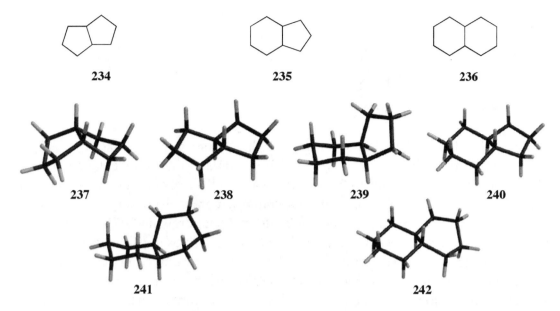

234   **235**   **236**

237   **238**   **239**   **240**

241   **242**

Incorporation of a π bond from an alkene or aryl moiety into a bicyclic system leads to significant flattening of the ring. Examples are **243A** (9-decalene) and the benzene-containing derivatives tetrahydronaphthalene (**244A**) and 1*H*-dihydroindene (**245A**). The planar alkene moiety in **243A** (see **243B**) forces the four allylic carbons to be coplanar and each ring more or less behaves as if it were cyclohexene. The planar benzene ring in **244A** (see **244B**) imposes similar conformational constraints, and the carbons of the nonaromatic ring assume a conformation similar to cyclohexene. In **245A** (see **245B**), the nonplanar cyclopentane ring behaves more or less like cyclopentene. The presence of three or more fused rings is rather common in nature. The natural product hirsutene (**246A**)[115] also shows the conformational bias imposed by three fused five-membered rings (see model **246B**).

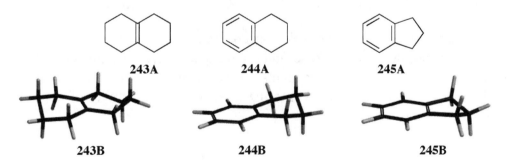

243A   **244A**   **245A**

243B   **244B**   **245B**

---

115. For leading references relating to the structure and synthesis of this molecule, see (a) Hua, D.H.; Venkataraman, S.; Ostrander, R.A.; Sinai, G.-Z.; McCann, P.J.; Coulter, M.J.; Xu, M.R. *J. Org. Chem.* ***1988**, 53,* 507; (b) Curran, D.P.; Raciewicz, D.M. *Tetrahedron **1985**, 41,* 3943.

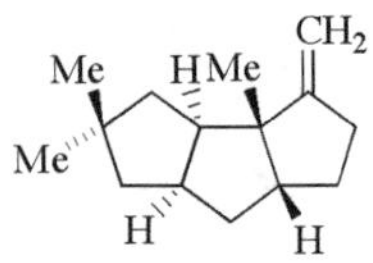

246A 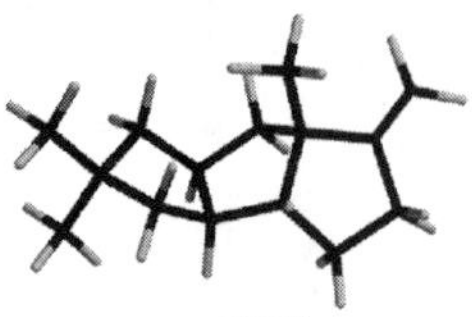 246B 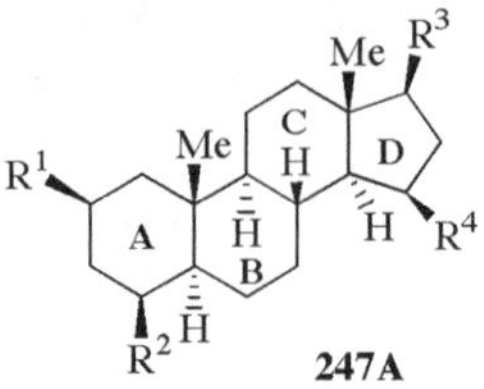 247A 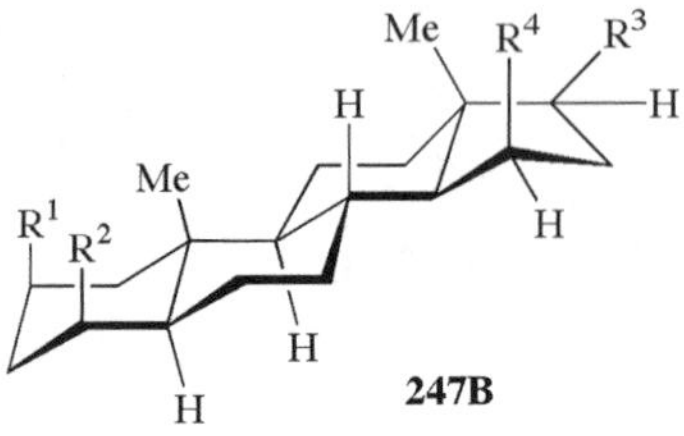 247B

A last example of this phenomenon is seen in the steroid nucleus (**247A**). Each of the cyclohexane moieties will exist in a chair conformation and the cyclopentane will be in an envelope conformation, as seen in **247B**. If a carbonyl or alkene moiety is included, the six-membered ring will be flattened and its conformation will be very similar to that of cyclohexanone or cyclohexene. As seen in conformational drawing **247B**, each trans-fused cyclohexane ring resembles *trans*-decalin and the four fused rings impart great conformational rigidity to the molecule. Only the terminal cyclohexyl and cyclopentyl rings have some mobility. The interaction R^1-R^2-Me of the cyclohexane ring is important, as are the R^3-Me and R-R^4 interaction of the cyclopentane ring. Examples of the conformational variations in steroids are seen by examination of cholesterol (**248A**) and β-estradiol (**249A**).[116] When the A ring is aromatic, as in estradiol (**249**), the A/B ring system is flattened, inducing significant distortion in the C ring as well (see **249B** relative to **248B**). Changes in conformation for polycyclic systems can often be predicted using the simple five- and six-ring models discussed above.

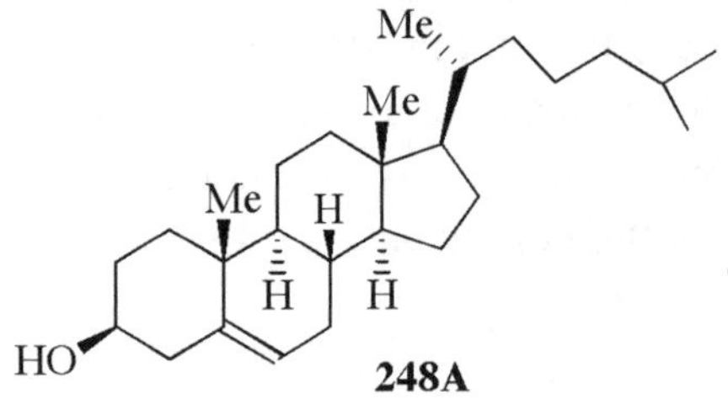

248A 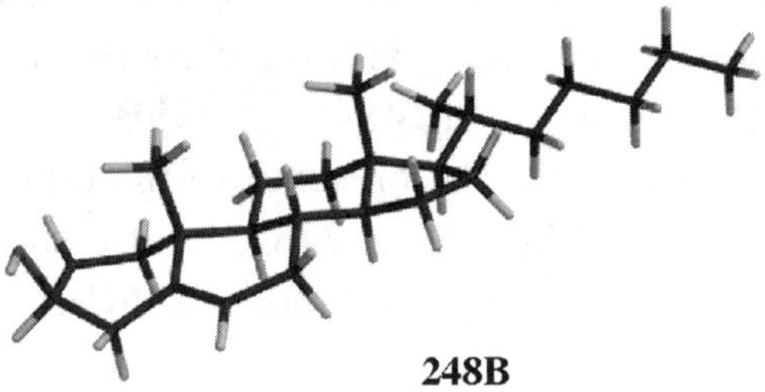 248B

It is reasonable to assume that the simple concepts introduced for monocyclic rings such as cyclopentane, cyclohexane, cyclohexene, and cyclohexanone can be applied with little change to complex molecules containing those units. Throughout this book the analysis of simple rings will be used to understand the conformational bias and reactivity of larger and more difficult synthetic targets.

---

116. Fieser, L.F.; Fieser, M. *Steroids,* Van Nostrand Reinhold, New York, *1959*.

1.6. CONCLUSION

This chapter has reviewed the most fundamental concepts of stereochemistry and conformational analysis, with some applications to more complex problems. For students who have had a good undergraduate Organic chemistry course, it is intended as a review since these concepts are integral to the understanding of organic reactions. The concept of disconnection and retrosynthetic analysis is a prelude to a discussion of synthetic strategies in Chapter 10. The disconnection method and pertinent transforms will, however, be presented in virtually every chapter as new reactions are introduced.

The introduction to organic reactions will be continued in the functional group exchange reactions in Chapter 2, which will review the reactions introduced in a typical first Organic chemistry course. Some of the concepts will be expanded and updated to include more synthetically useful reactions. Oxidation and reduction will be presented in separate chapters to focus attention on the vast number of reactions that fall under these categories.

HOMEWORK

1. For each of the following molecules calculate the percentage for both chair conformations using the values in Table 1.4. Calculate using the temperatures 150°C for (a), 35°C for (c) and 25°C for (b) and (c).

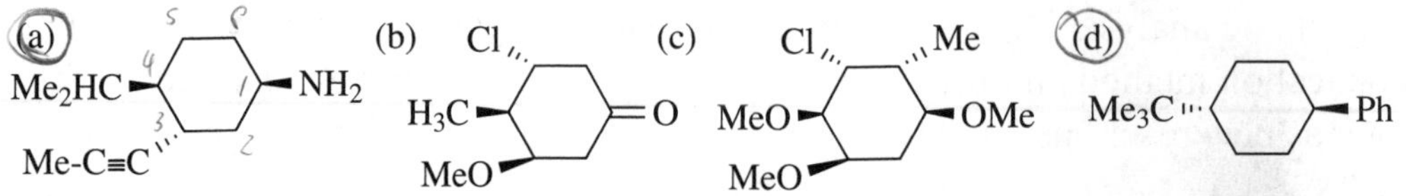

2. Determine the absolute configuration (*R* or *S*) for every stereogenic center in the following molecules:

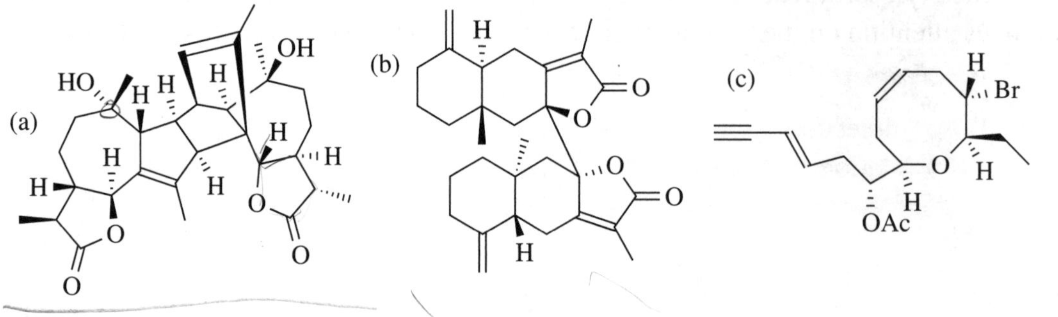

3. Determine the absolute configuration for every stereogenic center in the following molecules:

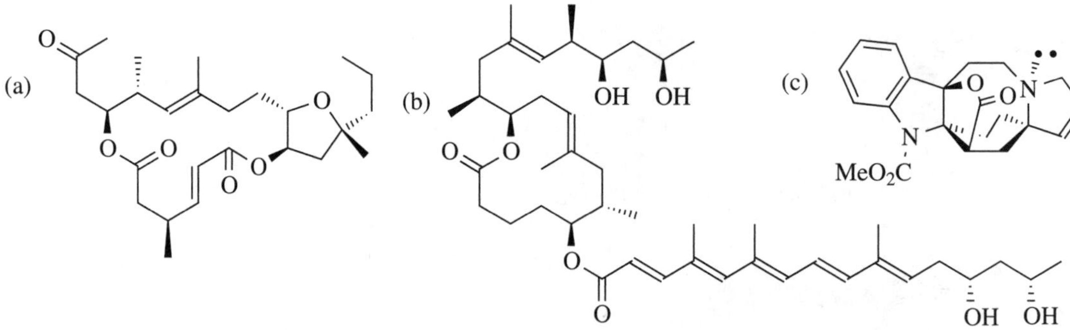

4. For disconnect fragments **23** and **24** in Section 1.2, convert each to a real molecule using synthetic equivalents in Table 1.2. Briefly discuss the retrosynthetic and synthetic sequences and show the complete retrosynthetic analysis based on this disconnection.

5. Using the reaction wheel (Fig. 1.1) give reasonable syntheses, including reagents and all intermediates (no mechanisms).

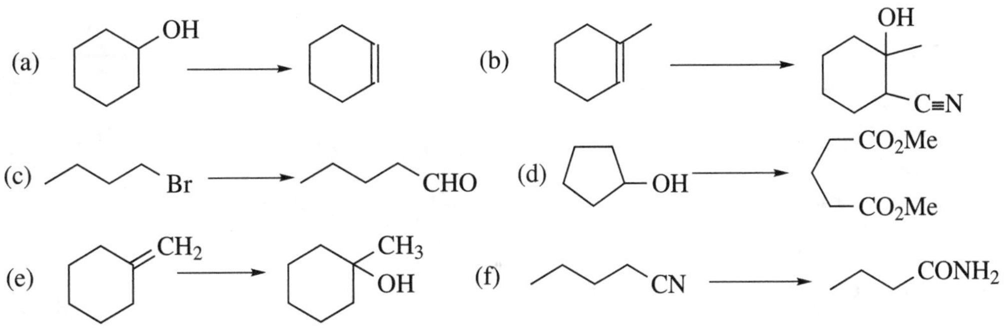

(a) [cyclohexanol with OH] → [cyclohexene]

(b) [methylcyclohexene] → [structure with OH, C≡N, CO₂Me]

(c) [chain with Br] → [chain with CHO]

(d) [cyclopentane with OH] → [chain with CO₂Me, CO₂Me]

(e) [cyclohexane with =CH₂] → [cyclohexane with CH₃, OH]

(f) [chain with CN] → [chain with CONH₂]

6. Use the *Compendium of Organic Synthetic Methods* (Vol. 11) to give three *different* reactions (with literature references and reactions) for each of the following:

   (a) acids from nitriles
   (b) aldehydes from nitriles
   (c) amides from halides
   (d) amines from nitriles
   (e) ethers from halides
   (f) halides from amines
   (g) ketones from olefins
   (h) alkenes from aldehydes

7. Draw molecule **A** in what should be the low-energy conformation. Briefly discuss the conformation for the highlighted rings in molecule A. Also discuss the conformational preference (axial or equatorial) for all both methyl groups as well as the hydroxyl.

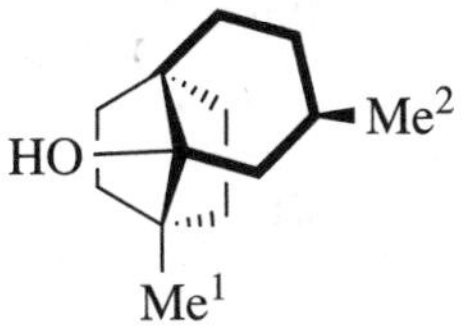

[structure of molecule A with HO, Me², Me¹ labels] **A**

8. Use a 3D computer drawing program, draw molecules (a) and (b) in 3D form, from the indicated perspective (see the example below).

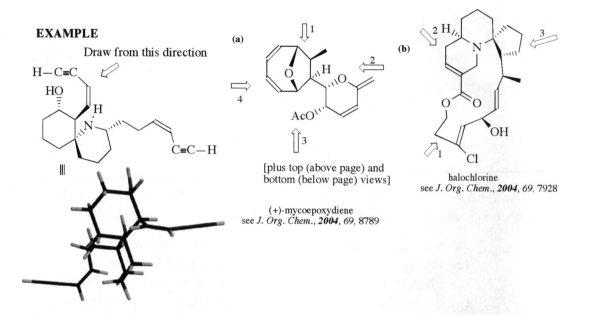

EXAMPLE

Draw from this direction

(+)-mycoepoxydiene
see *J. Org. Chem.*, **2004**, *69*, 8789

[plus top (above page) and
bottom (below page) views]

halochlorine
see *J. Org. Chem.*, **2004**, *69*, 7928

(9.) Determine the absolute configuration (R or S) for each chiral axis in the following molecules:

(a)   (b)   (c)   (d)   (e)

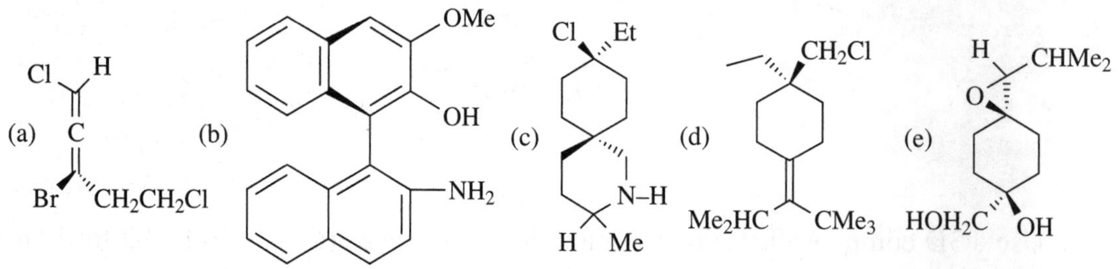

10. Determine the absolute configuration for each chiral center in the following molecules.:

(a)   (b)   (c)   (d)

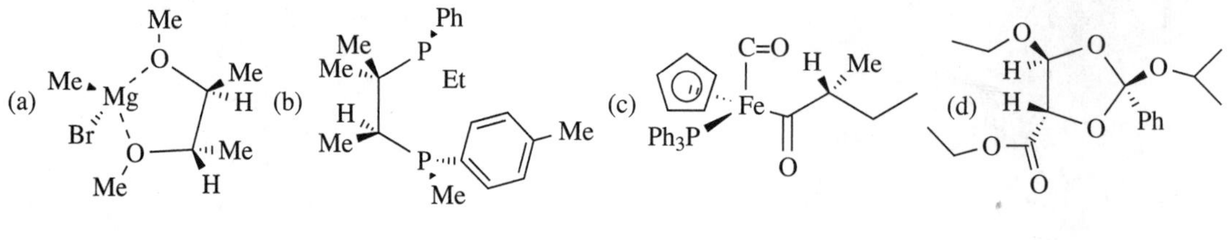

11. The conversion of the hydroxy-acid shown to the right, to the corresponding lactone, is rather slow. When pushed under aqueous acid conditions, epimerization at the carbon bearing the hydroxyl can occur. Draw the lactone and explain the observations noted in this question.

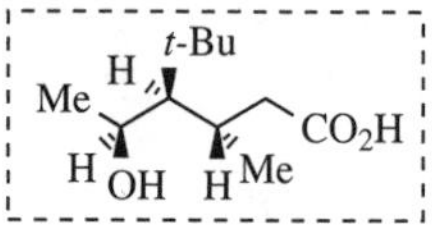

12. Explain why *cis*-1,3-cyclohexanediol exists mainly in the diaxial conformation in aqueous ethanol solution, when by Corey's ΔG calculations (see Table 1.4), the diequatorial conformation should be lower in energy.

13. Determine the correct *re/si* label for each prochiral atom in the following molecules.

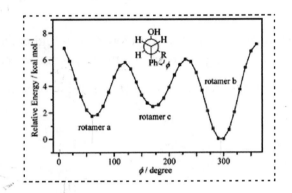

(a)   (b) 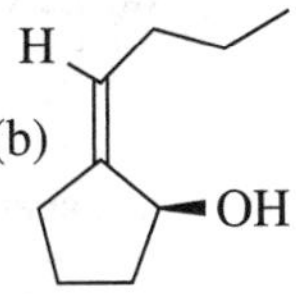   (c)

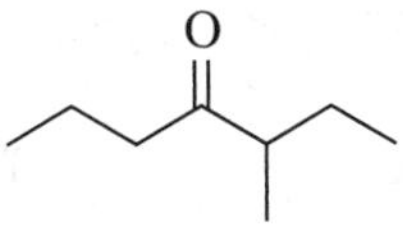

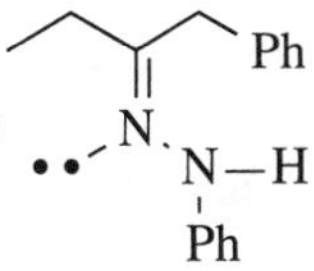

14. Draw the three most significant anti-rotamers of 1-phenyl-1-ethanol (R = Me in the accompanying diagram), and correlate them with the rotamers marked a-c on the energy curve.

15. (a) Draw the (2*S*, 3*R*,4*R*) and the (2*S*, 3*S*, 4*S*) derivative of molecule **A**. (b) Draw both the erythro/threo and the syn/anti forms of molecule **B**.

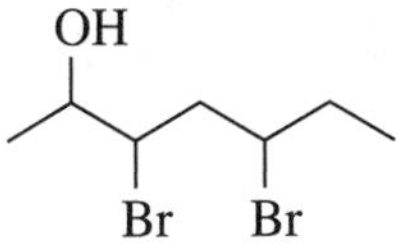

OH

Br   Br   **A**

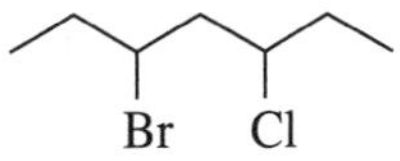

Br   Cl   **B**

16. Classify each of the following reactions as stereospecific/stereoselective and/or regiospecific/regioselective.

(a)

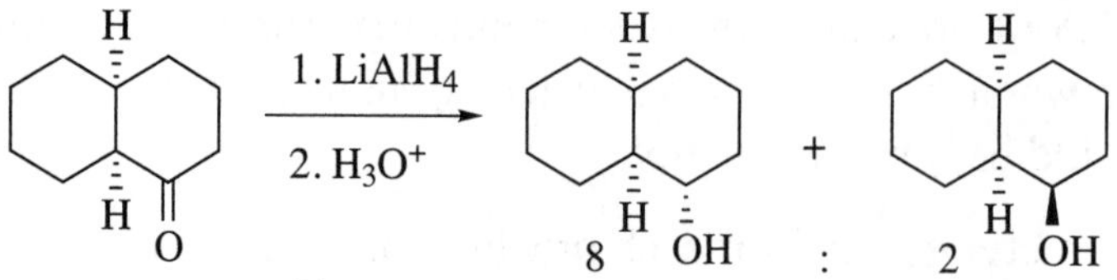

1. LiAlH₄

2. H₃O⁺

8 : 2

(b)

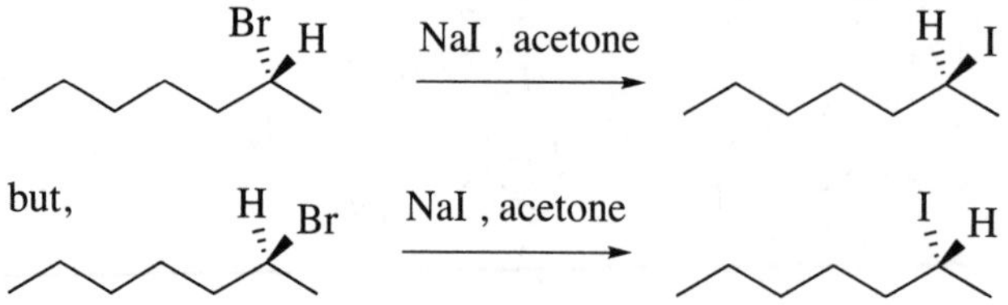

Br H   NaI , acetone

but,   H Br   NaI , acetone   I H

(c)

1. LiN(i-Pr)₂

2. MeI

90   :   10

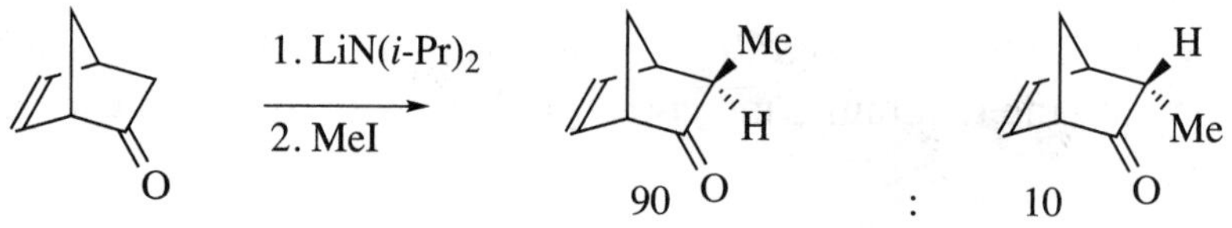

17. Explain why the ground state energy of cyclodecane is higher than that of cyclooctadecane. Why is cyclopentadecane higher in energy than cyclohexadecane?

18. Draw the boat conformation of *cis-,trans-,trans*-1,2,3,4-tetraisopropylcyclohexane.

chapter 2

Acids, Bases and Functional Group Exchange Reactions

2.1. INTRODUCTION

In most syntheses, the main focus is usually on construction of the molecule using carbon-carbon bond forming reactions. Most of the actual chemical reactions in a synthesis, however, are those that incorporate or change functional groups. Such reactions are known as functional group exchange reactions, and this chapter will review major reaction types involved in functional group exchanges. For all practical purposes, this discussion also constitutes a review of a typical undergraduate organic course.

Figure 1.1 showed that many functional group interchange reactions are chemically related. These transformations usually proceed via ionic intermediates and/or polarized C-X bonds, or they involve π bonds. A short list of typical reagents was provided in that Figure, but many other reagents are usually available to effect a given transformation. Functional group exchange reactions generally have two functions in syntheses. Functional groups are (*1*) a structural feature of the final target, and (*2*) a necessary feature incorporated into a molecule that must be manipulated (transformed) to alter or form a carbon-carbon bond. Functional group interchange reactions are, therefore, very important in synthetic organic chemistry.

Many different types of reactions are used in functional group interchanges. We can categorize reactions according to the type of transformation, which is a subset of functional group interchanges. This categorization can help us understand when and how to use a particular reaction. Grouping reactions by similar or related mechanisms is very common.

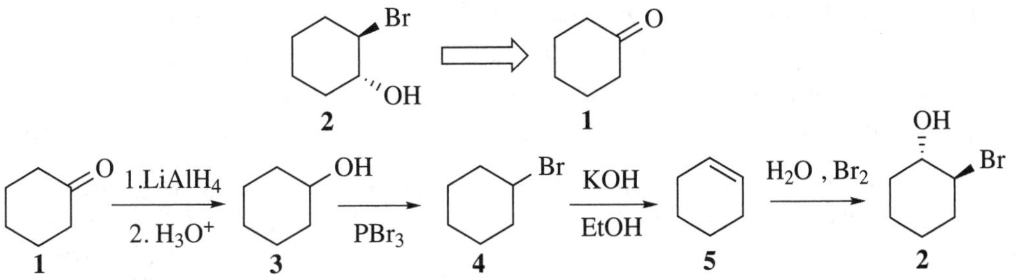

The importance of functional group exchange reactions is illustrated by a simple synthetic problem, in which bromohydrin **2** is prepared from cyclohexanone (**1**). This particular synthesis requires only functional group transformations. The synthetic relationship between **1** and **2** is represented in the diagram as **2** ⇒ **1**, which is meant to show that the bromohydrin can be formed from the ketone by an as yet unknown number of synthetic steps. Corey and co-workers[1,2] called this representation *a transform*, and used it to define retrosynthetic relationships (sec. 1.2). If we analyze the **2** ⇒**1** relationship, it is clear that Br and OH must be incorporated on adjacent carbons, and one possibility is an addition reaction to an alkene. We must somehow convert the ketone to an alkene. One approach would involve reduction of the carbonyl to an alcohol, which could be converted to a leaving group that would allow an elimination reaction to give the alkene. This analysis is based on the synthetic conversion of **1** to **2**, and the reaction types suggested by this analysis leads to a choice of the reagents that are required for each transformation. Initial reduction of the ketone, with a suitable reducing agent such as $LiAlH_4$ (sec. 4.2.B) gives alcohol **3**. Subsequent reaction with phosphorus tribromide (sec. 2.8.A) gives bromide **4**, in what is known as a substitution reaction. Base induced elimination gives alkene **5**, which illustrates both elimination and an acid-base reaction. The final reaction with bromine and water (sec. 2.10.C) gives **2,** and illustrates an addition reaction. This sequence used five major reaction types: acid-base, reduction, substitution, elimination, and addition. Te reaction types used for functional group exchanges are collected into seven traditional categories:

1. Acid - base reactions $HA + B \longrightarrow HB + A$
2. Substitution reactions $RX + A \longrightarrow RA + X$
3. Elimination reactions $X\text{-}C\text{-}C\text{-}Y \longrightarrow C{=}C + X\text{-}Y$
4. Addition reactions $C{=}C + X\text{-}X \longrightarrow X\text{-}C\text{-}C\text{-}X$
5. Nucleophilic acyl addition $C{=}O \longrightarrow Nuc\text{-}C\text{-}O\text{-}E$
6. Oxidation reactions   An example is $H\text{-}C\text{-}O\text{-}H \longrightarrow C{=}O$
7. Reduction reactions   An example is $C{=}C \longrightarrow H\text{-}C\text{-}C\text{-}H$

Most functional group exchanges and transformations fall into one of these seven categories, which it is useful for classification of reagents, as in Table 2.1.[3] Many reactions can be placed into more than one of these simple categories. Elimination reactions, for example, involve an initial removal of hydrogen by a base (an acid-base reaction). Addition reactions are two-step reactions that involve a substitution. Oxidation and reduction reactions constitute such a large proportion of functional group exchanges that they will be discussed in separate chapters (chaps. 3 and 4, respectively), to properly examine their synthetic applications. This chapter

1. Corey, E.J.; Cheng, X. *The Logic of Chemical Synthesis,* Wiley Interscience, New York, *1989.*
2. (a) Corey, E.J.; Wipke, W.T. *Science, 1969, 166,* 178; (b) Corey, E.J.; Long, A.K.; Rubinstein, S.D. *Ibid 1985, 228,* 408; (c) Corey, E.J.; Howe, W.J.; Pensak, D.A. *J. Am. Chem. Soc. 1974, 96,* 7724; (d) Corey, E.J. *Quart. Rev. Chem. Soc. 1971, 25,* 455; (e) Corey, E.J.; Wipke, W.T.; Cramer, III, R.D.; Howe, W.J. *J. Am. Chem. Soc. 1972, 94,* 421; (f) Corey, E.J.; Jorgensen, W.L. *Ibid 1976, 98,* 189.
3. Many examples of these types of reactions are found in Larock, R.C. *Comprehensive Organic Transformations,* 2nd Ed. Wiley-VCH, New York, *1999.*

will begin with the first five reaction types to discuss their application to chemical reactions, synthesis, and functional group transformations. The ultimate goal is to establish a firm base upon which to introduce both the reactions required for making important carbon-carbon bonds, and also the techniques employed by the modern synthetic chemist.

Substitution (S_N1 or S_N2) is a reaction where one functional group attached to n aliphatic carbon is replaced by another called a **nucleophile** (nucleophilic aliphatic substitution). Nucleophiles can also form a new bond to an acyl carbon (nucleophilic acyl addition). Both of these **retrosynthetic transforms** are represented by the C-Nuc species, where Nuc = CN^-, AcO^-, RO^-, N_3^-, CO_2^-, etc. and X = Cl, Br, I, OAc, OSO_2R, etc., for S_N2 reactions. In nucleophilic acyl substitutions, the nucleophile is usually a carbon, nitrogen or oxygen species.

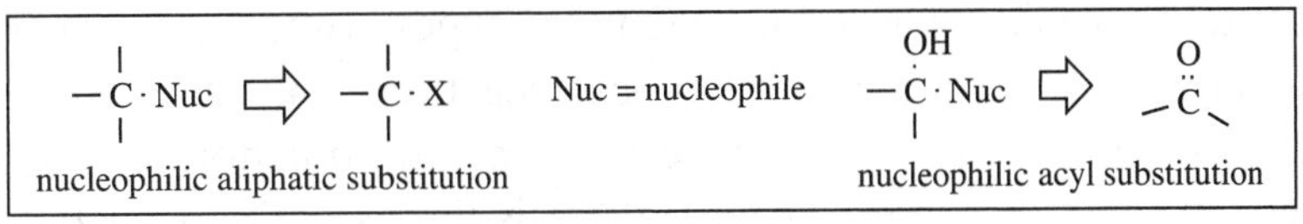

Table 2.1. Correlation of Common Reagents with General Reaction Types

Class of Reagent (Reaction Type)	Reagents
Acids (acid-base; addition)	HCl, HBr, HF, HI, H_2SO_4, HNO_3, H_3PO_4, RCO_2H, RSO_3H, HOH, ROH, AlX_3, FeX_3, ZnX_2, SnX_4
Bases (acid-base; elimination)	NaOH, KOH, LiOH, $Ca(OH)_2$, NaH, KH, RO–M+ (t-BuO, MeO, EtO; M = Na^+, K^+), RLi, RMgX, NH_3, R_3N, $NaNH_2$, $LiNH_2$, $NaNR_2$ or $LiNR_2$ (HNR_2 + Na or Li), HOH, ROH, ROR
Nucleophiles (substitution; nucleophilic acyl substitution)	MX (X=Cl^-, Br^-, I^-, CN^-), HO^-, RO^-, NH_2^-, NR_2^-, $RC\equiv X^-$, $R(CO)$, HS^-, RS^-; M=Na^+, K^+, Li^+, RLi, RMgX, NH_3, NR_3, HOH, ROH, R_2CuLi
Oxidizing agents (oxidation; addition; acid-base)	CrO_3, $K_2Cr_2O_7$, H_2O, ROOR, $HCrO_4$, $KMnO_4$, Cu/heat, MnO_2, SeO_2, HIO_4, O_3, NaOX(NaOH/X_2), NBS, NCS, RCO_3H, HONO ($NaNO_2$/HX), HNO_3, H_2SO_4, Ag^+, Cu_2^+
Reducing agents (reduction; addition)	Sn/HCl, Fe/HCl, H_2 + catalyst, (Pd, Pt, Ni(R), Ru), $NaBR_4$ (R=H, alkyl, OR), $LiAlR_4$(=H, alkyl, OR)

Reagents that are classified as bases can be used with both Lewis and Brønsted-Lowry acids, and one application is the elimination of alkyl halides (E2 reactions; sec. 2.9.A) in the presence of a base. Elimination involves conversion of a saturated moiety containing a leaving group to a molecule with a multiple bond. An example of this latter transform is:

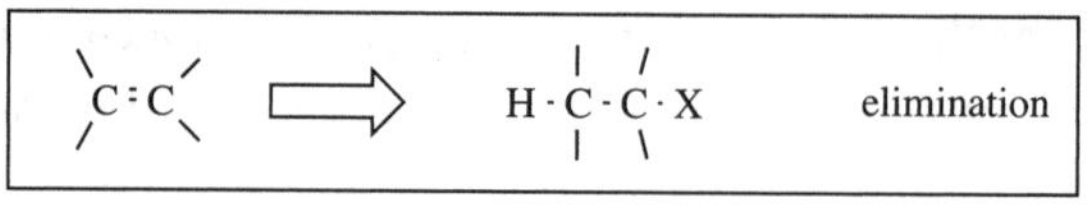

where the leaving group (X) is OH, halogen, OTs, and so on. The E2 reaction is discussed in section 2.9.A. Note that another class of elimination reactions includes polar species such as sulfoxides, amine oxides, and selenoxides, which undergo thermal syn elimination (sec. 2.9.C).

Acid-base reactions are among the most general that are known. They are used to generate reactive intermediates that can be part of other reaction types. Both Lewis and Brønsted-Lowry bases donate two electrons to an acid. Acid-base reactions are integral to many reactions, although they are not always easy to describe by a specific transform since they may be an adjunct to the desired transformation. An example is the use of conjugate bases of acids as nucleophiles in substitution reactions. Typical reactions that generate nucleophiles are HCN → CN⁻ and HN₃ → N₃⁻. Conjugate bases of weak acids are usually required to initiate E2 type reactions, and typical examples are ethoxide and amide bases from the reactions EtOH ® EtO⁻ and HNEt₂ → LiNEt₂. Many reagents that add to alkenes are acids (HCl, HOBr, etc.), and the reaction mechanism for these reactions begins with the alkene reacting as a base, donating two electrons to the acid (H⁺). Addition reactions involve the transformation of sp hybridized carbons to sp² or to sp³, or the transformation of sp² hybridized carbon atoms to sp³. Addition reactions that add strong protonic acids HX, or diatomic bromine, chlorine, etc. to alkenes, are common in synthesis. The retrosynthetic transform can be generalized as:

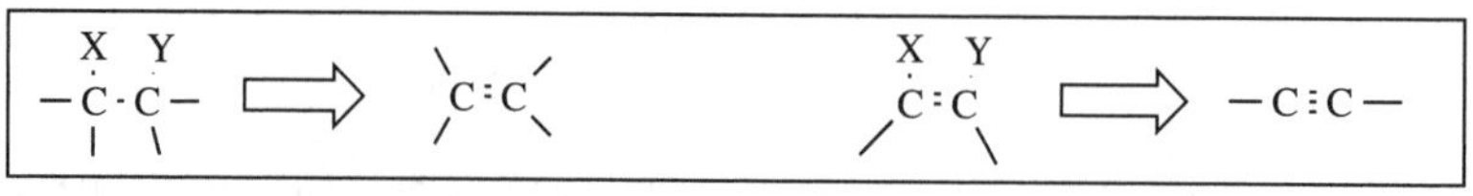

As each of these reaction types is discussed in later sections, it will be apparent that virtually all substrates that are functionally transformed are polarized or contain π bonds. Pericyclic reactions will be presented in Chapter 11, and are an important exception. When analyzing a substrate for a functional group transform, the following criteria are usually important.

1. Examine the degree and direction of polarization of a bond. Where will the electrons go?

2. How many bonds away will the polarization be effective?

3. What is the attraction of a given reagent for the polarized bond?

4. Does the polarized bond contain more than one leaving group.

5. If there is a π bond, how efficiently can it donate electrons.

6. If the π bond is polarized, will the δ+ center undergo substitution or will the δ- pole act as an electron donor.

7. Does the reactive center possess special strain or steric hindrance that will modify the reactivity?

8. Is the reactive center conjugated?

Perhaps most importantly,

9. Is there more than one reactive center in the molecule and can more than one reaction occur competitively?

2.2. BRØNSTED-LOWRY ACIDS AND BASES

Many organic molecules react via an initial acid-base reaction that is formally classified by a Brønsted-Lowry definition. Remember that a Brønsted-Lowry acid is a proton donor and a Brønsted-Lowry base is a proton acceptor. The key to this definition is an understanding that there is a proton transfer and change the focus to understand that a base removes the proton from the acid by donating two electrons to the proton. We are therefore looking for proton donors and acceptors, and specifically for molecules that donate electrons to a protonic acid to form a new bond to that hydrogen atom. Understanding the properties of acid-base equilibria is an aid to understanding the mechanistic details of many different reactions.

2.2.A. Acidity in Organic Molecules

The reaction of HA and B to give BH^+ and A^- is the general expression written for acid-base reactions. The acid (HA) reacts with a base (B:) to give the conjugate acid HB^+ (CA) and a conjugate base, A^- (CB). The hydrogen atom is transferred to B: as a proton (H^+). Since the product HB^+ is also an acid and A^- is a base, they react to establish an equilibrium. The position of the equilibrium is given by the equilibrium constant K, where K_a is used for acid-base reactions. A small value of K_a represents little ionization of HA, which means that the equilibrium is shifted to the left and this shift is associated with decrease in the acidity of HA. In other words, if the equilibrium lies to the left, HA did not react with the base (B) to any great extent and it is considered to be a weaker acid. If K_a is large, HA reacted with B to produce BH^+ and A^-, so the equilibrium is shifted to the right and HA is considered to be a stronger acid. The position of the equilibrium is dependent on many factors, including the solvent. A more convenient way to express the relative strength of an acid is pK_a. A large K_a is equivalent to a small pK_a so as the pK_a decreases, the acid strength increases. Note that our definition of a strong acid (large K_a) implies that the reaction of HA and B is more facile than the reverse reaction of BH^+ and A^-. Conversely, a weak acid (small K_a) implies that the reaction of BH^+ and A^- is more facile than that of HA and B.

$$HA + B \xrightleftharpoons{K_a} BH^+ + A^-$$

$$K_z = \frac{[HB^+]\,[A^-]}{[HA]\,[B:]} \quad \text{where } pK_a = -\log k_a \text{ and } K = 10^{-pK_a}$$

Many organic compounds such as carboxylic acids, phenols and 1,3-dicarbonyl compounds are acids, but they are typically much weaker than the familiar mineral acids HCl or H_2SO_4. Likewise, many organic compounds (particularly amines and phosphines) are bases, but such bases are generally weaker than alkali bases such as NaOH. Amphoteric compounds such as alcohols are common. Alcohols are weak acids in the presence of a strong base, but they are weak bases in the presence of a strong acid. This chemical reactivity can be exploited in many chemical transformations. In the reaction of 3-pentanol (**6**) and H^+ (HCl), the initial product is oxonium ion (**7**). Relating this reaction to the generic acid-base reaction shown above [HA] is the acid catalyst (HCl), [B] is the oxygen atom of 3-pentanol, the chloride ion is [A$^-$] (the gegenion of the acid catalyst), and [HB$^+$] is $Et_2CH\text{-}OH_2^+$ (**7**). An acid catalyst that is strong enough to react with the alcohol, which is a weak base, initiates the reaction. In other words, the acid catalyst must be a stronger acid than the OH unit of the alcohol. In this particular reaction **7** is formed and reacts with the nucleophilic chloride ion at carbon, with loss of the leaving group water to give 3-chloropentane as the final product.

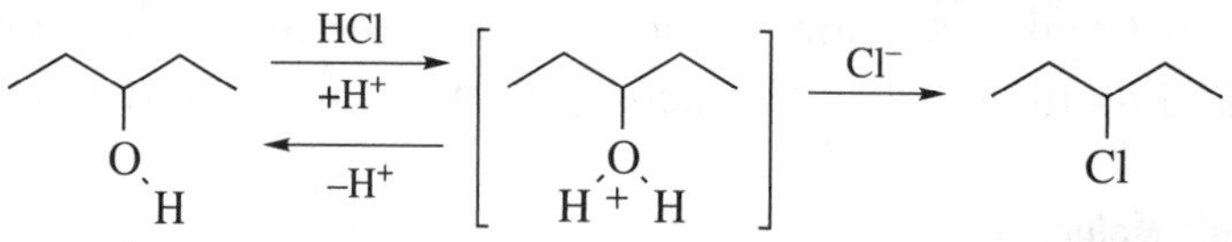

Ethers are weak bases. Although they are commonly used as solvents, diethyl ether and THF are weak bases in the presence of a strong acid and THF is a stronger base than diethyl ether. The relative base strength can be seen upon examination of the experimentally determined acidity of the conjugate acid derived from diethyl ether [Et_2OH^+], which has a $pK_a = -3.12$.[4] This acid is stronger than the protonated form of THF [$C_4H_8OH^+$], which has a $pK_a = -2.08$.[4] In reactions where the ether behaves as a Lewis base, as in Grignard reactions (sec. 8.4.A) or hydroboration (sec. 5.2.A), understanding the relative basicity is important.

The hydrogen on the α-carbon of a ketone such as 2-butanone is an important weak acid in organic chemistry. When 2-butanone (A) reacts with sodium ethoxide (B), the conjugate base is enolate anion **8** (CB), and ethanol is the conjugate acid (CA). Ethanol (pK_a about 17) is a stronger acid than the ketone (pK_a about 19-20) and will react with the basic enolate to regenerate the ketone and shift the equilibrium back to the left. The net result is an equilibrium

---

4.   (a) Arnett, E.M.; Wu, C.Y. *J. Am. Chem. Soc.* **1960**, *82*, 4999; (b) Arnett, E.M.; Wu, C.Y. *Ibid* **1962**, *84*, 1680, 1684; (c) Deno, N.C.; Turner, J.O. *J. Org. Chem.* **1966**, *31*, 1969.

solution that contains enolate **8**, ethanol, 2-butanone, and NaOEt (sec. 9.2.E).[5] It is important to note that in any acid-base reaction, HA does not function as a Brønsted-Lowry acid unless a sufficiently strong base (B) is present to remove and accept the acidic proton. If the base under consideration is too weak, the equilibrium is shifted almost exclusively to the left (i.e., we call HA a weak acid).

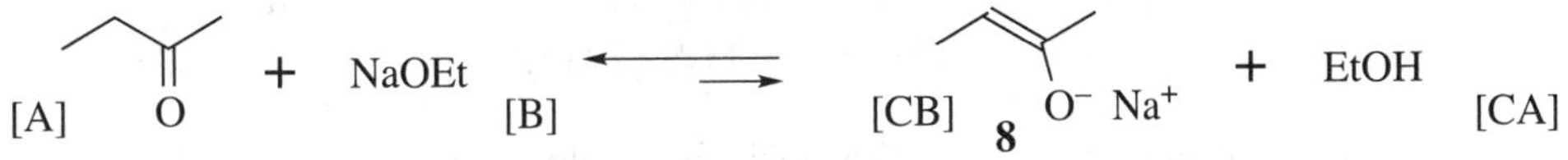

When comparing two acids, several factors can be used to evaluate relative acid strength. In an acid-base reaction where acids are on both sides of the equilibrium (A and CA from above), this information can be useful to estimate the position of the equilibrium, which is essentially an estimation of K_a.

1. **In general, a strong acid generates a weak conjugate base and, conversely, a weak acid generates a strong conjugate base.**

    Comparison of HCl and NH_3 indicates that HCl is a strong acid that generates the weak base, Cl^-:

    $$HCl \rightarrow H^+ + Cl^-$$

    Conversely, ammonia is a weak acid that generates the strong base, NH_2^-.

    $$NH_3 \rightarrow H^+ + Cl^-$$

A base will usually deprotonate an acid that has a pK_a lower than its own conjugated acid, and the extent to which the reaction produces products if the main criterion for labeling a base as strong or weak.

2. **Across the periodic table, acidity of an X-H species generally increases from left to right.[6]**

    $$\text{weak acid } CH_4 < NH_3 < H_2O < HF \text{ strong acid}$$
    $$pK_a \quad (\approx 50) \quad (\approx 31) \quad (15.7) \quad (3.17)$$
    $$\text{strong conj. base } CH_3^- > NH_2^- > HO^- > F^- \text{ weak conj. base}$$

5. (a) House, H.O. *Modern Synthetic Reactions*, 2nd Ed. W.A. Benjamin, *1972*, pp 452-595; (b) House, H.O.; Phillips, W.V.; Sayer, T.S.B.; Yau, C.C. *J. Org. Chem. 1978, 43*, 700; (c) Etheredge, S.J. *Ibid 1966, 31*, 1990; (d) House, H.O.; Czuba, L.J.; Gall, M.; Olmstead, H.D. *Ibid 1969, 34*, 2324; (e) House, H.O.; Gall, M.; Almstead, H.D. *Ibid 1971, 36*, 2361.

6. Lowry, T.H.; Richardson, K.S. *Mechanism and Theory in Organic Chemistry*, 3rd Ed. Harper and Row, New York, *1987*, pp 297-298.

A consequence of this trend in acidity is that basicity increases from right to left across the periodic table.

3. Acidity increases going down the periodic table, despite a decrease in electronegativity.

$$HF\ (3.17)\ <\ HCl\ (-7)\ <\ HBr\ (-9)\ <\ HI\ (-10)^7\ and,$$
$$H_2O\ (15.74)\ <\ H_2S\ (7.00)^6$$

Based on rule 3, basicity increases going up the periodic table. Fluoride is smaller and more electronegative than iodide, has a greater attraction for protons, and forms a stronger covalent bond. These factors make it more difficult to lose a proton from HF than from HI, hence, the equilibrium is shifted to the left relative to HI, and HF is a weaker acid. Clearly, the increased size of the iodine atom leads to a greater H-I bond distance and, due to the resultant decrease in internuclear electron density, a weaker bond. The acid strength increases, since the weaker I-H bond is more easily ionized, which ignores solvation effects, however, as well as product stability, which are critical to this analysis (see below). However, *we must look at the both the starting materials and the products to determine the position of the equilibrium, and thereby the relative acidity.* The iodide ion is larger than the fluoride ion and the charge is dispersed to a great extent, leading to greater stability. The larger ions are also more easily solvated, again leading to greater stability. Greater stability of one product relative to the other means the equilibrium is shifted further to the right for HI-iodide, and this leads to HI being a stronger acid.

These three concepts allow a quick inspection of a simple acid (HX) for its relative acid strength. Experimental analysis of the acids commonly encountered in organic reactions requires knowledge of at least three factors: (*1*) electronic effects, (*2*) resonance effects, and (*3*) solvent effects.

The presence of an electron-donating or an electron-withdrawing heteroatom (or group) on a carbon induces bond polarization through adjacent carbon atoms in a chain, as illustrated by **9** (first introduced in section 1.3).[8] Electron-donating groups release electrons via inductive or field effects, leading to the so-called +*I* effect. Conversely, an electron-withdrawing group removes electron density, the so-called –*I* effect. With respect to an acid, the +*I* effect lowers the acid strength whereas the –*I* effect raises it. Electronic effects are described by the inductive model[9,10] and the field effect model.[9,11]

---

7. Smith, M.B.; March, J. March's *Advanced Organic Chemistry, 6th ed.*; Wiley, New York, *2007*, pp 384, 375.
8. (a) Seebach, D. Angew. *Chem. Int. Ed. 1979, 18*, 239; (b) Cram, D.J.; Cram, J.M. *Chem. Forsch. 1972, 31,* 1 [*Chem. Abstr. 77*:163690d *1972*].
9. (a) Baker, F.W.; Parish, R.C.; Stock, L.M. *J. Am. Chem. Soc. 1967, 89,* 5677, (b) Golden, R.; Stock, L.M. *Ibid 1966 88*, 5928, (c) Holtz, H.D.; Stock, L.M. *Ibid 1964, 86*, 5188.
10. Branch, G.E.K; Calvin, M., *The Theory of Organic Chemistry*, Prentice Hall, New York, *1941*, Chapter 6.
11. (a) Kirkwood, J.G.; Westheimer, F.H. *J. Chem. Phys. 1938, 6,* 506; (b) Westheimer, F.H.; Kirkwood, J.G. *Ibid 1938, 6,* 513.

The inductive model assumes that substituent effects are propagated by the successive polarization of the bonds between the substituent and the reaction site (as in **9**). This effect is transmitted through the σ bond network (σ inductive effect) as well as the π-bond network (π inductive effect).[10] The field effect model assumes that the polar effect originates in bond dipole moments, and is propagated according to the classical laws of electrostatics. The appropriate description of this effect is the **Kirkwood-Westheimer model**,[11] in which the molecule is treated as a cavity of low dielectric constant submerged in a solvent continuum.[9a]

Attempts to distinguish inductive and field effects have been frustrated by the lack of a molecule that provides an unambiguous answer.[9b] Such a molecule would require that the low dielectric cavity of the field effect model[12] be occupied by the chemical bonds important to the inductive effect model.[10] For example, o-chlorophenylpropionic acid is weaker than expected by the inductive model.[12] If a small attenuation factor is adopted for through-bond transmission of the polar effect, both models predict similar results.[13]

Stock reported the pK_a of two systems, **10**[9c] and **11**,[9b] that allow the field effect to be studied, and deter-mined that the dominant term associated with the change in pK_a was the positive charge, transmitted through the cavity.[14a] This finding is in general agreement with the Kirkwood-Westheimer model,[11] which suggests that the field effect is the dominant term.[9] Ammonium salts have an –I effect,[15] which lowers pK_a (strengthens the acid) relative to acetic acid.[16] Due to the localized character of the σ electrons, the inductive effect in this system diminishes with increasing distance between separation of substituent and reaction center. The change in pK_a is actually due to a combination of inductive, steric and hyperconjugative influences. Chloroacetic acid ($pK_a = 2.87$)[17] can be compared with propionic acid ($pK_a = 4.76$), since there is no change in the number of α hydrogens, no steric or hyperconjugative effect, and the inductive effect is the only significant contribution to pK_a.[14] The –I effect of chlorine induces a significant reduction in pK_a (stronger acid). Electron releasing groups with a +I effect will

12. (a) Roberts, J.D.; Carboni, R.A. *J. Am. Chem. Soc.* **1955**, 77, 5554; (b) also see Wells, P.R.; Adcock, W. *Australian J. Chem.* **1965**, 18, 1365.
13. Ehrenson, S. *Progr. Phys. Org. Chem.* **1964**, 2, 195.
14. (a) Dewar, M.J.S.; Grisdale, P.J. *J. Am. Chem. Soc.* **1962**, 84, 3539, 3548; (b) Murrell, J.N.; Kettle, S.F.A.; Tedder, J.M. *Valence Theory*, John Wiley, Londo, **1965**, Chapter 16.
15. Brown, H.C.; McDaniel, D.H.; Hafliger, O. *Determination of Organic Structures by Physical Methods*, Ed. Braude, E.A.; Nachod, F.C., Academic Press, New York, **1955**, p 569.
16. Reference 15, see Chapter 14 therein.
17. Bolton, P.D.; Hepler, L.G. *Quart. Rev. Chem. Soc.* **1971**, 25, 521.

have the opposite effect and lead to a decrease in acidity.[14]

The fundamental concept of inductive and field effects introduced for carboxylic acids can be applied to weaker acids such as the α-hydrogen of ketones. Reaction of **12** with the strong base lithium diisopropyl amide (LDA) in Zoretic and co-worker's synthesis of megaphone[18] illustrates this premise. Both H_a and H_b in **12** are acidic, but H_a has a lower pK_a by virtue of its proximity to the carbonyl (secs. 9.1, 9.2) and is removed preferentially. The conjugating effect of the π bond makes H_b acidic (it is a *vinylogous* acid), but that acidity is diminished due to the distance from the carbonyl. Formation of the enolate anion of **12** was followed by alkylation (secs. 2.6.A, 9.3.A) to provide **13** in good yield.

It is of paramount important to remember that acid-base reactions are equilibrium reactions, where the position of the equilibrium (K_a) is determined by the interactions of the acid-base pair A and B and the acid-base pair on the other side of the reaction (CA and CB). One cannot discuss the acidity of a molecule without considering the products formed by the reaction. For example resonance delocalization in the final product makes that product more stable relative to a competing reaction or even a different reaction where the product is not resonance stabilized. The result is that the resonance-stabilized product helps to shift the reaction equilibrium to the right, towards that product. Resonance stabilizes the product due to the existence of low energy molecular orbitals that lead to p electron delocalization, transmitted over unsaturated systems. Electron delocalization diminishes the net charge, leading to additional stabilization. This effect is clear when acetic acid and methanol are compared as acids. Acetic acid (pK_a, 4.76) owes its greater acidity, in part, to the resonance stabilized acetate anion (**14**), whereas methanol (pK_a, 18) generates the methoxide anion (**15**). The charge is *localized* on the oxygen atom in methoxide, but is *delocalized* over three atoms in the acetate anion. Delocalization of charge lowers the net energy of the system, making it more stable (and thereby, less reactive). The increased stability of the conjugate base of acetic acid (the acetate anion) shifts the equilibrium towards the products to a much greater extent than is possible in the methanol reaction (which generates the methoxide anion). A shift to a higher concentration of products is reflected by a larger K_a, which means a lower pK_a for acetic acid relative to methanol.

18. Zoretic, P.A.; Bhakta, C.; Khan, R.H. *Tetrahedron Lett.* **1983**, *24*, 1125.

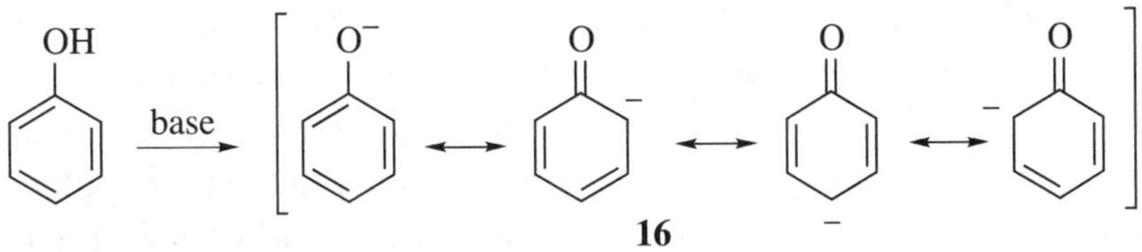

16

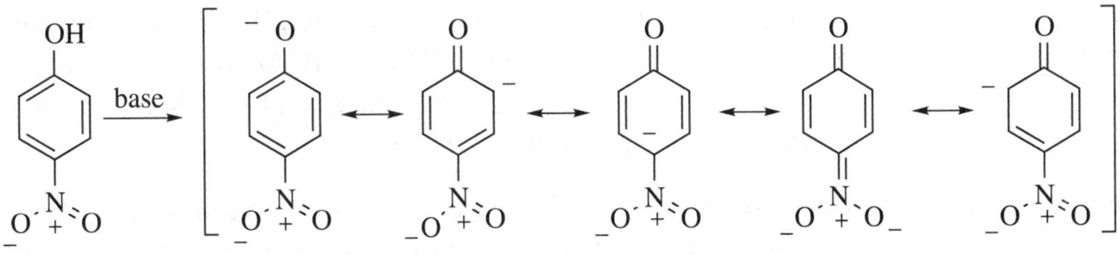

17

Resonance effects can also be transmitted via the π bonds of aromatic rings. A comparison of phenol (pK_a, 10.0), 4-nitrophenol (pK_a, 7.2) and 2,6-dinitrophenol (pK_a, 3.6) shows that the presence of a nitro group enhances acidity, in part by delocalizing electron density in the conjugated phenoxide base. Removal of the acidic proton of phenol leads to the resonance-stabilized anion **16**. The presence of a nitro group at C4 leads to delocalization of the charge on the nitro group (see **17**) and increased resonance stability. A second nitro group at C2 or C6 (as in 2,6-dinitrophenol) will lead to greater resonance stability and enhanced acidity of that phenol. In other words stabilization of the anion by nitro substituents is reflected in the pK_a values.

Table 2.2. Influence of Solvents on the Acidity of Common Protonic Acids

	pK				
	in HOH	in MeOH	in DMF	in DMSO	in MeCN
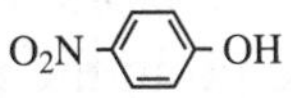 O_2N—⟨⟩—OH	7.15	11.2	10.9	9.9	7.0
MeCO$_2$H	4.76	9.6	11.1	11.4	22.3
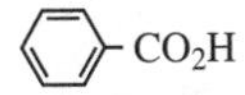 ⟨⟩—CO$_2$H	4.20	9.1	10.2	10.0	20.7

[Reprinted with permission from Clare, B.W.; Cook, D.; Ko, E.C.F.; Mac, Y.C.; Parker, A.J., *J. Am. Chem. Soc., 1966, 88*, 1911. Copyright © 1966 American Chemical Society.]

The acidity of a compound changes with the polarity of the solvent, and is different in protic versus aprotic solvents. Table 2.2,[19] compares the pK's of several acids in various solvents.[20] Most reactions take place in solution rather than in the gas phase,[20] and the factors that account for the enormous differences between solution and gas-phase chemistry are operationally

---

19. (a) Clare, B.W.; Cook, D.; Ko, E.C.F.; Mac, Y.C.; Parker, A.J. *J. Am. Chem. Soc. 1966, 88*, 1911; (b) Kolthoff, I.M.; Chantooni, Jr., M.K.; Bhowmik, S. *Ibid 1968, 90*, 23.
20. Arnett, E.M. *Acc. Chem. Res. 1973, 6*, 404.

defined as **solvation**.[19] Protic solvents facilitate charge separation, and solvation of both the conjugate acid and conjugate base can shift the equilibrium to the right (the compound is more acidic). Aprotic solvents effectively solvate only the cationic product but this can influence the equilibrium and the acids in Table 2.2 show enhanced acidity in the more polar DMSO, relative to DMF. Solvating ability is not the only factor however. Acidity is diminished in methanol due to decreased hydrogen-bonding, relative to the acidity in water. Basicity of the solvent is an important factor. A comparison of aprotic solvents (DMF, DMSO, and acetonitrile) shows that the more basic the solvent, the greater the apparent acidity of HA. This effect is due, in part, to disruption of the internal hydrogen bonding of HA.[21]

In later chapters (particularly chapters 8 and 9), we will examine carbon acids (C-H) that generate carbanions as the conjugate base. The solvent plays a critical role in the relative pK_a of carbon acids (15.8 for phenylacetylene in diethyl ether, 26.5 in DMSO for example).[22] In this particular example, the difference in pK_a is explained by the extent of ion-pairing or ion aggregation. The equilibrium for the deprotonation of phenylacetylene is also shifted to the right when the solvent is changed from cyclohexylamine ($pK_a = 20.5$) to diethyl ether due to increased stability of the ion pair (for the potassium anion, $PhC\equiv C^- K^+$) or the ion aggregate $(PhC\equiv C^- K^+)_n$.

It is apparent that acidity in organic molecules is influenced by several factors. Among these, inductive effects, solvent effects, and effects that influence the stability of both the acid and the conjugate base are important. The base is obviously important since the strength of an acid is directly dependent on the strength of the base.

2.2.B Basicity in Organic Molecules[23]

Since the base is part of the acid-base equilibrium, the relative strength of a base is determined by the position of the equilibrium and is influenced by the same factors discussed for acids. For common bases, base strength *increases* going to the left and up in the periodic table. An example is

$$F^- > Cl^- > Br^- > I^-$$

Base strength is usually expressed by the pK_a of its *conjugate acid, BH⁺* according to the equation

$$BH^+ \rightleftharpoons B: + H^+ \qquad K_a = \frac{[B:]\ [H^+]}{BH^+}$$

21. Grunwald, E.; Ralph, E.K. *Acc. Chem. Res.* **1971**, *4*, 107.
22. Bordwell, F.G.; Matthews, W.S. *J. Am. Chem. Soc.* **1974**, *96*, 1214.
23. Clark, J.; Perrin, D.D. *Quart. Rev. Chem. Soc.* **1964**, *18*, 295.

Table 2.3. +*I*, +*R*, -*I* and -*R* Effects in Bases

Effect		Examples
+*I*	(base strengthening)	RO^-, R_2N^-, alkyl
–*I*	(base weakening)	SO_2R, $NH_3^{\neq}$, CF_3, NO_2, CN, Cl, Br CO_2H, CO_2R, I, COR, SR, Ph, NR_3^+
+*R*	(base strengthening)	F, Cl, Br, I, OH, OR, NH_2, NR_2 NHCOR, O^-, HN^-, Me, alkyl
–*R*	(base weakening)	NO_2, CN^-, CO_2H, CO_2R, $CONH_2$, Ph, COR, SO_2R

The inductive and field effects lead to +*I* (electron donating groups, base strengthening) and -*I* effects (electron withdrawing groups, base weakening). Examination of amines shows that electron releasing groups intensify the electron density around nitrogen, and to a first approximation, this enhances basicity. Electron-withdrawing groups should have the opposite effect. Resonance effects (+*R*) also apply, as described earlier for acids. Resonance effects are also known as **mesomeric** effects (+*R* is +*M*; -*R* is -*M*), as described in the discussion of aromatic substitution in section 2.11. Electron donating groups can be characterized as having +*R* effects, and electron-withdrawing groups as having -*R* effects. When comparing 4-nitroaniline and 4-methylaniline, the methyl group is electron releasing (+*R*) but the nitro group is electron withdrawing (-*R*). The +*R* effect of the methyl is transmitted through the aromatic ring, intensifying the electron density on nitrogen and increasing the basicity.

The *I* and *R* effects may operate in opposite directions. An example is a slight resonance effect of a group in the meta position of an aromatic nucleus, where the inductive effect is rather large. In the para position, the resonance effect is also large. Table 2.3[24] correlates several substituents with *I* and *R* effects.

There are two categories of steric hindrance around the basic atom that influence relative base strength.

1. **Primary steric effect.** Steric impedance to protonation occurs when bulky groups surround the basic center. When this results in greater strain in the cationic conjugate acid produced upon protonation than in the neutral molecule, the base is weaker.[25]

2. **Secondary steric effect**. There is hindrance to solvation and it is base weakening.

The primary steric effect is credited with the greater basicity of aniline relative to 2,6-

24. (a) Trotman-Dickenson, A.F. *J. Chem. Soc.* ***1949***, 1293, (b) Arnett, E.M.; Jones III, F.M.; Taagepera, M., Henderson, W.G., Beauchamp, J.L.; Holtz, D.; Taft, R.W. *J. Am. Chem. Soc.* ***1972***, *94*, 4724.
25. Taft, Jr., R.W *Steric Effects in Organic Chemistry*, John Wiley, New York, ***1956***, Chapter 13.

dimethylaniline, due to greater steric inhibition for reaction with a proton. The secondary effect is probably responsible for the difference in basicity between *N,N*-diethylaniline (pK_a, = 6.65) and *N,N*-dimethylaniline (pK_a, = 5.8). The two ethyl groups inhibit solvation of the protonated conjugated acid,[16] which is the ammonium salt. Solvation varies with the available space around the basic center, and should be less for pyridine than for aniline. The diminished basicity of tertiary amines relative to secondary amines in solution has been attributed to this effect.[24a]

It has been known for many years that the proton acceptor ability of amines in solution [expressed as pK_a of the conjugate acid (BH$^+$)] takes the order:[19] NH_3 < R_3N < RNH_2 ≈ R_2NH. This anomalous order is understood in terms of two opposed influences, one base strengthening and the other base weakening. The base strengthening effect is the electron releasing inductive effects of the alkyl groups on nitrogen. This effect should make tertiary amines more basic when compared to secondary or primary amines. As described immediately above, the other major influence is solvation. Solvation is greatest for the ammonium salt produced from primary amines, leading to greater stabilization and enhanced basicity. Using high pressure mass spectrometry (there are no solvation effects), Munson demonstrated the order of gas-phase basicity is:[26] NH_3 < RNH_2 < R_2NH < R_3N.

It was assumed that the inversion of base strength in solution was primarily a solvation phenomenon. Indeed, this effect is largely the result of hydrogen bonding in an aqueous solvent when comparing ammonia ($NH_3 \rightarrow NH_4^+$) with four sites for hydrogen-bonding in the ammonium salt versus a tertiary amine ($R_3N \rightarrow R_3NH^+$), with only one site for hydrogen bonding in the ammonium salt. The larger the number of alkyl groups on nitrogen leads to fewer protons in the ammonium salt, and diminished hydrogen bonding. This structural feature works in opposition to the electron-releasing effect of the increasing number of alkyl groups and dispersal of charge by the inductive effect.[24]

Clearly, a discussion of relative acidity or basicity must specify the solvent. The actual order of basicity for amines in solvents, however, is a combination of inductive and solvation effects. Steric hindrance around the electron pair on nitrogen also plays an important role (see above). When all of these factors are accounted for, secondary amines tend to be the most basic, as shown above.

The pK of several common amines used in synthesis have been determined in THF at 25°C.[27] 4-*N,N*-Dimethylaminopyridine (DMAP, **18**), proton sponge (**19**), quinuclidine, (**20**), DBU (1,8-diazabicyclo-[5.4.0]-undec-7-ene, **21**),[28b] and diazabicyclo-[2.2.2]octane (DABCO, **22**) are shown, along with triethylamine, tributylamine, and tribenzylamine for comparison.[27]

26. Munson, M.S.B. *J. Am. Chem. Soc.* **1965**, *87*, 2332.
27. Streitwieser, A.; Kim, Y.-J. *J. Am. Chem. Soc.* **2000**, *122*, 11783 and references cited therein.
28. For synthetic examples, see (a) Achab, S.; Das, B.C. *J. Chem. Soc. Chem. Commun.* **1983**, 391; (b) Oediger, H.; Möller, Fr. *Angew. Chem. Int. Ed.* **1967**, *6*, 76.

The pK values for these bases are shown in THF, DMSO, and acetonitrile in Table 2.4,[27] determined from proton-transfer ion pairs of the type BH$^+$ with acidic indicator hydrocarbons. The values obtained in THF differ substantially from the ionic pK_a values for NH$^+$ in DMSO or acetonitrile. Streitwieser and co-workers[27] state that "at the present time amines cannot be placed quantitatively on any of the ion pair acidity scales currently in use for neutral acids in THF". The realization that it may be necessary to use different bases in different solvents, despite the fact that all are considered to be good bases in organic reactions, makes it clear why a quantitative comparison of bases is useful.

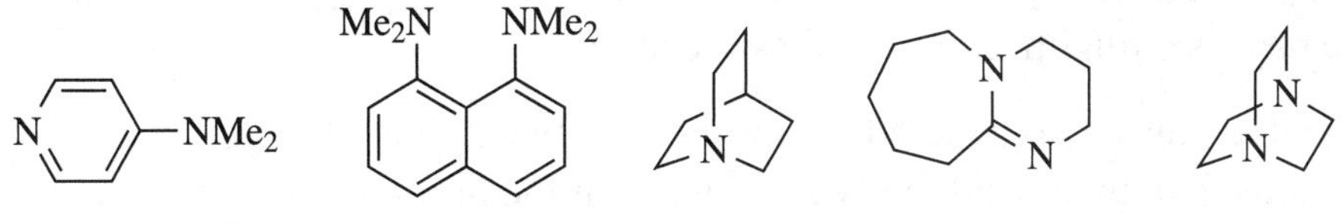

Many different types of bases are used in organic chemistry. Common inorganic bases include the anions of weak bases such as water and ammonia (hydroxide and amide: $^-$OH and $^-$NH$_2$). Similar deprotonation of alcohols leads to alkoxide bases (RO$^-$). Methoxide (MeO$^-$), ethoxide (EtO$^-$), and *tert*-butoxide (Me$_3$C-O$^-$) are very common, and are commonly used in the alcohol solvents from which they were made (methoxide in methanol, *tert*-butoxide in *tert*-butanol, etc.). Bases typically used to deprotonate water (pK_a = 15.8) or alcohols (pK_a ~ 18) are sodium hydride (NaH), potassium hydride (KH), sodium and potassium hydroxide [(NaOH, KOH), sodium metal [Na(0)], and sodium or potassium amide (NaNH$_2$ and KNH$_2$).

Table 2.4. pK values of several common amines in various solvents.

Amine	pK$_a$ (THF)	pK$_a$ (DMSO)	pK$_a$ (MeCN)
18	0.61		18.18
19	2.15	7.47	17.28
20	0.15	9.8	19.51
21	–3.78		23.9
22	0.80	8.93	18.29
NEt$_3$	2.11	9.0	18.70
N(CH$_2$Ph)$_3$	2,41	3.65	12.85
NBu$_3$	3.40		18.09

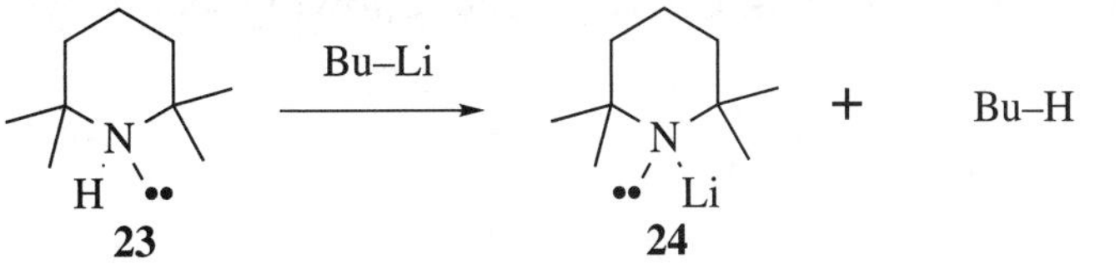

Amines are acids only in the presence of very strong bases that generate a conjugate acid that is significantly weaker than the amine. Amines are acids in the presence of Grignard reagents (R–MgX) and organolithium reagents (R–Li, secs. 8.4.G.iii and 8.5.G), which give amide bases ($R_2N^-M^+$) along with an alkane as the very weak conjugate acid (R-H). In a typical acid-base reaction, butyllithium is a strong base and diethylamine is the acid that gives lithium diethylamide (Et_2NLi) and the conjugate base and butane as the conjugate acid. Similarly, diisopropylamine gives lithium diisopropylamide (*i*-Pr_2NLi, often called LDA) and butane, and 2,2,6,6-tetramethylpiperidine (**23**) gives lithium 2,2,6,6-tetramethylpiperidide (**24**) and butane.[29] Since butane has a pK_a of >40, and the amines have pK_a values around 25, formation of the amide base is straightforward in these reactions.

The parent amine can be considered as amphoteric but it can function as a base only in the presence of acids that are considerably stronger than the ammonium salt of the base. In other words, triethylamine reacts with HCl to form triethylammonium chloride, where HCl is a much stronger acid than the ammonium salt, pushing the equilibrium in that direction. Triethylamine is a very weak base in the presence of a weak acid such as water, where the equilibrium favors the amine rather than the ammonium salt. Tertiary amines such as triethylamine (NEt_3) or pyridine (C_5H_5N, Py) are most commonly used as bases when an amine base is required, and these amines listed in Table 2.4. Imidazole (**25**) is an amine base used in systems that require a base that is weaker than triethylamine or pyridine. The stronger amide bases discussed above are used for deprotonation of weak acids (ketones, alkynes, esters, sulfides, etc., see Chaps. 8 and 9). The amide bases are often called non-nucleophilic bases due to their poor reactivity as nucleophiles in S_N2 (secs. 9.2.B, 2.7.C) and in acyl addition reactions (sec. 2.5). Their diminished nucleophilic strength results from the bulky alkyl substituents about the basic nitrogen which inhibits approach to carbon, but does not encumber approach to hydrogen. Two other non-nucleophilic bases are 1,5-diazabicyclo[4.3.0]-non-5-ene (DBN, **26**)[28b] and DBU (**21** from above),[28a] in which the bridgehead nitrogen is sterically encumbered. When a very powerful base is required for deprotonation (pK_a of the acid is > 25), organometallic bases such as an organolithium reagent are used most often (sec. 8.5.G). Commercially available reagents such as butyllithium (BuLi), methyllithium (MeLi), and phenyllithium (PhLi) are commonly used in synthesis (see sec. 8.5.H).

25   **26**

The acid-base concepts described in this section focus on bases that react with protons, the classical Brønsted-Lowry definition of a base. There are other types of electron deficient or electron-donating molecules that do not involve transfer of protons. Such compounds are called Lewis acids or Lewis bases, respectively. The following section will discuss Lewis acids.

---

29. Olofson, R.A.; Dougherty, C.M. *J. Am. Chem. Soc.* **1973**, *95*, 581.

2.3. LEWIS ACIDS

There are two working definitions of acidity and basicity. A **Lewis acid** is an electron pair acceptor and a **Brønsted-Lowry acid** is a proton donor. For basicity, a **Lewis base** is an electron pair donor and a **Brønsted-Lowry base** is a proton acceptor. Section 2.2.B focused primarily on the Brønsted-Lowry definitions and reactions of protonic acids (H^+). This contrasts with a Lewis acid, which is an atom other than hydrogen that accepts two electrons from an electron donating species (the Lewis base). Group 13 atoms such as boron and aluminum are electron deficient since they can form only three covalent bonds and remain neutral. They can satisfy the octet rule by accepting an electron pair from a Lewis base, forming an adduct (a charged complex called an **ate complex**), as shown below. Such complexes will be of particular importance in hydride reductions (sec. 4.2.B) and in hydroboration reactions (see sec. 5.3, 5.4).

$$\underset{acid}{MX_n} \ + \ \underset{base}{B} \ \longrightarrow \ \underset{adduct}{MX_n^- \ B^+}$$

Transition metal salts are usually characterized by the presence of a metal atom that is capable of assuming multiple valences, and in many cases they can function as a Lewis acid. The relative order of Lewis acidity is not always straightforward since the electrophilic metal can have different valences and ligands (the groups or atoms attached to the metal). A number of general statements can be made, however.[30]

Lewis acids usually take the form MX_n, where X is the ligand (a halogen atom, an amine or phosphine, CO, cyclopentadienyl, an alkene, etc.). The metal is M and n in X_n is the normal valency of M. The acid MX_n exhibits acidic properties in an equilibrium reaction with the base B (see above). Formation of a conjugate acid-base adduct ($MX^- B^+$, the *ate* complex mentioned above) is characteristic of Lewis acid-base reactions. If chelation effects and double bond formation between M and any given atom of the base are excluded, three general statements can be made concerning acid strength.[30]

1. In MX_n ($n < 4$) acidity arises from the central atom's requirement for completion of an outer electron octet by accepting one or more pairs of electrons from the base. Acidity is diminished when two electron pairs are required. There is a smaller energy gain upon receipt of the first pair and an accumulation of negative charge on M if two pairs are received. Therefore, Group 13 acids are more acidic than transition metal acids: $BF_3 > AlCl_3 > FeCl_3$

2. The acidity of M will decrease within any group with increasing atomic volume (effectively, with increasing atomic number) owing to the weaker attraction between

30.   (a) Satchell, D.P.N.; Satchell, R.S. *Chem. Rev.* **1969**, *69*, 251, (b) Idem *Quart. Rev. Chem. Soc.* **1971**, *25*, 171.

nuclear charge and incoming electron pairs (**Fajans' rules**).[30a]   The result of these effects leads to the order: $BX_3 > AlX_3 > GaX_3 > InX_3$

3. In general, the energies of different atomic orbitals lie closer together with increasing atomic number, because orbital contractions arising from the electronegativity of the nucleus tend to decrease with increasing atomic number. As a result, there is more effective overlap of hybridized orbitals. The availability of *d* orbitals will be easier (especially with *d* outer orbitals) and more effective, the heavier the element. This leads to a decrease in acidity, as in the series (B → Al → Ga → In) above.

These three statements are taken directly from the review by Satchell and Satchell,[30] and can be used to estimate the strength of Lewis acids according their Group in the Periodic Table.[30]

a. **Group 11** (Cu, Ag, Au) and **12** (Zn, Cd, Hg) usually give covalent species MX_2: ($ZnCl_2$, CdI_2, etc.); M has four electrons in the outer group and needs two additional pairs. By rule 1 these should have only moderate acidity and by rule 2: $ZnCl_2 > CdCl_2 > HgCl_2$

b. **Group 13** (B, Al, In, Tl) and **3** (Sc, Y, La, Ac) by rule 1 tend to form the most acidic compounds (MX_3). The metal (M) requires only one electron pair to complete the octet (BF_3, $AlCl_3$, $GaBr_3$, etc.). Rule 2 is then applied.

   $BF_3 > AlCl_3 > GaBr_3 > InCl_3 > TlCl_3$ and, $ScCl_3 > YCl_3 > LaCl_3$.

c. **Group 14** (C, Si, Ge, Sn, Pb) and **4** (Ti, Zr, Hf) form covalent species such as MX_4. There are no acidic properties when M = C, and when M = Si only weak acidic properties are observed.

d. **Group 15** (N, P, As, Sb, Bi) and **5** (V, Nb, Ta) usually take the form MX_3 or MX_5. When M is nitrogen, there are no acidic properties, indeed these are bases. When M is phosphorus the molecule is a weak base, but the octet can expand to give weakly acid properties in some reactions. In general, MX_5 compounds are more acidic than Groups 15 and 5 MX_3 compounds due to the expanded octet and the utilization of d orbitals. Notethat both SbX_5 and NbX_5 are powerful acids.

e. **Group 16** (O, S, Se, Te, Po) and **6** (Cr, Mo, W) have the normal covalent character of MX_2. There are close parallels with Groups 14 and 4. Both OX_2 and SX_2 have virtually no acidity and TeX_2 is very weak. Species arising from abnormal valences, such as TeX_4, can exhibit enhanced acidity since only one electron pair is then required to complete the outer shell of 12.

f. **Group 17** (F, Cl, Br, I, At) and **7** (Mn, Tc, Re) resemble 14 and 4 and 16 and 6 in that the lower members are only weakly acidic. The higher members can expand the octet when attached to electron-withdrawing substituents, and have enhanced acidity (I_3^- and Ph-ICl^-).[31]

With these six general statements as a guide, it is possible to order the more common Lewis acids by decreasing acid strength: BX_3 > AlX_3 > FeX_3 > GaX_3 > SbX_5 > InX_3 > SnX_4 > AsX_5 > SbX_3 > ZrX_4

The relative strength of Lewis acids is important to many major classes of organic reactions utilized in synthesis. In sections 11.6.A and 12.4, Lewis acids will be used in connection with several important carbon-carbon bond forming reactions. These include **Friedel-Crafts** type reactions[32] and **Diels-Alder reactions**[33,34,35] (sec. 11.4.A), as well as other pericyclic reactions (chap. 11). Changing the Lewis acids can have dramatic effects on the rate of reaction, regio- and stereoselectivity, and product distribution in both the Diels Alder and Friedel-Crafts reactions.[32] If the Lewis acid is too reactive, degradation of products or starting materials can be a serious problem. For these reasons, several Lewis acids covering a wide range of acid strengths should be available for a synthesis. Aluminum chloride is a very reactive and unselective catalyst, reacting with virtually all functional groups that have Lewis base properties.[34,36] Zinc chloride ($ZnCl_2$), however, is a mild and selective catalyst in reactions where halides or alcohols are required to react selectively with an olefinic double bond.[32,37] Tin tetrachloride ($SnCl_4$) is a very mild catalyst that can be used for acylation of reactive aromatic nuclei such as thiophene. Catalysis of reactions involving thiophene with $AlCl_3$ is too vigorous and extensive decomposition accompanies the process.[32,38] Tin tetrachloride has also been applied to Friedel-Crafts reactions of active aromatics[32,39] which cannot tolerate vigorous reaction conditions, as with furan derivatives for example.[32,40] These few examples clearly demonstrate that a good working knowledge of the relative acid strength of Lewis acids is essential for understanding many important reactions in synthesis.

31. Andrews, L.J.; Keefer, R.M. *J. Am. Chem. Soc.* **1952**, *74*, 4500.
32. Olah, G.A., *Friedel Crafts and Related Reactions*, Vol. I, Interscience Publishers, **1963**, New York, pp 201-366.
33. (a) Inukai, T.; Kasai, M. *J. Org. Chem.* **1965**, *30*, 3567, (b) Inukai, T.; Kojima, T. *Ibid* **1967**, *32*, 869, 872.
34. For references dealing with coordination of the Lewis acid to ester dienophiles, see (a) Lappert, M.F. *J. Chem. Soc.* **1961**, 817, (b) Lappert, M.F. *Ibid* **1962**, 542.
35. (a) Houk, K.N.; Strozier, R.W. *J. Am. Chem. Soc.* **1973**, *95*, 4094, (b) Alston, P.V.; Ottenbrite, R.M. *J. Org. Chem.* **1975**, *40*, 1111, (c) Ansell, M.F.; Nash, B.W.; Wilson, D.A. *J. Chem. Soc.* **1963**, 3012.
36. Thomas, C.A. *Anhydrous Aluminum Chloride in Organic Chemistry*, Reinhold Publishers, New York, **1941**.
37. Nenitzescu, C.D.; Isa˘cescu, D.A. *Berichte* **1933**, *66*, 1100.
38. Goldfarb, I. J. *Russ. Phys. Chem. Soc.* **1930**, *62*, 1073 [*Chem. Abstr. 25*: 2719, **1931**].
39. Stadnikoff, G.; Banyschewa, A. *Berichte* **1928**, *61*, 1996.
40. Gilman, H.; Calloway, N.O. *J. Am. Chem. Soc.* **1933**, *55*, 4197.

2.4. HARD-SOFT ACID-BASE THEORY[41,42]

2.4.A. Hard and Soft Acids and Bases

There is an alternative view of acid-base reactions that relies on the Lewis acid-base definition, but is used to classify a wide range of organic reaction types. Cations are classified as Lewis acids, anions are Lewis bases and salts are viewed as acid-base complexes. When one writes the normal acid-base equation, certain conventions are observed.

$$A' + A–B \longrightarrow A'–B + A$$

implies that A' is a stronger acid than A, and

$$B' + A–B \longrightarrow A–B' + B$$

implies that B' is a stronger base than B. For

$$A + B \longrightarrow A–B$$

where the equilibrium constant is defined by

$$\log K = [S_A] [S_B]$$

where the S terms represent the relative strengths, which are functions of the environment and the temperature. There are solvent effects that make it difficult to establish a universal order of acid or base strength[43] due to variations in the parameter S as well as other factors.

For example, if the acid (A) is chromic acid there is a different basicity order than if A is cuprous ion. Usually, the simple equation shown above is inadequate and requires additional terms

$$\log K = [S_A] [S_B] + [s_A] [s_B] \qquad\qquad 2.1$$

where s is a parameter for the strength of each acid and base, and is called the softness parameter.

The **Edward's equation**[44] (Eqn. 2.2) can be of value for determining the relative strength of Lewis acids and is written as

$$\log \frac{K}{K_0} = \alpha\, E_n + \beta\, H \qquad\qquad 2.2$$

---

41. (a) Pearson, R.G. *Hard and Soft Acids and Bases*, Dowden, Hutchinson and Roe Inc., Stroudsburg, PA, *1973*; (b) Pearson, R.G. *J. Am. Chem. Soc. 1963, 85*, 3533; (c) Pearson, R.G.; Songstad, J. *Ibid 1967, 89*, 1827; (d) *Idem J. Org. Chem. 1967, 32*, 2899.
42. (a) Pearson, R.G. *J. Chem. Educ. 1968, 45*, 581; (b) *Idem Ibid 1968, 45*, 643.
43. (a) Lewis, G.N. *Valence and the Structure of Atoms and Molecules*, The Chemical Catalog Co., New York, *1923*; (b) Lewis, G.N. *J. Franklin Inst. 1938, 226*, 293.
44. Edwards, J.O. *J. Am. Chem. Soc. 1954, 76*, 1540.

where H is a proton basicity factor: $H = 2.74 + pK_a$ for water, at 25°C. The term E_n is a redox factor and is equal to $E° + 2.60$ [$E_n = E° + 2.60$), where E° is the standard oxidation potential for the base, defined by the reaction:

$$B^- \longrightarrow B^+ + 2e^-$$

There are obviously many acids, but the proton (H^+) and an alkyl mercuric ion (RHg^+) have roughly opposite affinities for binding bases. The values of the parameters for the Edward's equation for H^+ are $\alpha = 0.000$ and $\beta = 2.000$ and for $MeHg^+$ the values are $\alpha = 5.786$ and $\beta = -0.031$.[42,45] The reactions of these two acids with Lewis bases define the relative strengths of the Lewis bases, which gives rise to two observations. (1) Bases in which the donor atom is O, N, or F prefer to coordinate to the proton. (2) Bases in which the donor atom is P, S, I, Br, Cl, or C prefer to coordinate to mercury.

Table 2.5. Classification of hard and soft acids [a]

Hard acids	Borderline cases	Soft acids
H^+, Li^+, Na^+, K^+, Mg^{2+}, Ca^{2+}, La^{3+}, Ce^{4+}, Gd^{3+}, Ti^{4+}, Cr^{3+}, $Cr^{\wedge+}$, MoO^{3+}, Mn^{2+}, Mn^{7+}, Fe^{3+}, Co^{3+}, BF_3, BCl_3, $B(OR)_3$, Al^{3+}, $AlMe_3$, $AlCl_3$, AlH_3, CO_2, RCO^+, Sn^{4+}, Me_2Sn^{2+}, RPO_2^+, $ROPO_2^+$, SO_3, RSO_2^{+}, $ROSO_2^+$, HX (hydrogen bonding molecules), metal atoms, bulk metals	Fe^{2+}, Co^{2+}, Ni^{2+}, Cu^{2+}, Zn^{2+}, Rh^{3+}, Ir^{3+}, Ru^{3+}, Os^{2+}, BMe_2. R_2C^+, $C_6H_5^+$, Sn^{2+}, Pb^{2+}, NO^+, Sb^{3+}, Bi^{3+}, SO_2	Pd^{2+}, Pt^{2+}, Pt^{4+}, Cu^+, Ag^+, Au^+, Cd^{2+}, Hg^+, Hg^{2+}, $MeHg^+$, BH_3, Tl^+, $TlMe_3$, CH_2, carbenes, $\neq$-acceptors, (chloronil, quinones, tetracyanoethylene, etc.), HO^+, RO^+, RS^+, RSe^+, Br_2, I_2, O, Cl, Br, I, N., RO, RO_2, M°

[a] In all cases, the molecules are listed in descending order (the first entry in each column is the hardest or softest).

Reprinted with permission from Pearson, R.G. and the Journal of Chemical Eductation, Vol. 45, **1868**, 581-587. Copyright 1968, Divison of Chemical Education, Inc.

In Group 1, the donor atoms are highly electronegative, have a low polarizability, and are hard to oxidize. These are called **HARD BASES** (valence electrons are bound tightly). In group 2, the donor atoms are of low electronegativity, are highly polarizable and are easy to oxidize.[41c] These are called **SOFT BASES** (valence electrons are loosely bound).[41c] Recognition of these categories forms the basis of the HSAB principle, initially developed by Pearson[41] and also described by Ho.[45]

In separate reactions with hard bases, there are two distinct classes of acids called **HARD ACIDS** and **SOFT ACIDS**. For **HARD ACIDS**,[41c] acceptor atoms are small in size, have a high positive charge, and do not contain unshared pairs of electrons in their valence shell. They are of low polarizability and of high electronegativity.   For **SOFT ACIDS**,[41c] acceptor atoms are large in size, have a low positive charge, and contain unshared *p* or *d* electrons in the

45. Ho, T.-L. *Hard and Soft Acids and Bases Principles in Organic Chemistry*, Academic Press, New York, *1977*.

valence shell. They are highly polarizable and of low electronegativity.

In order to rank these acids according to their relative strength, the criteria of Schwarzenbach[46a] and Ahrland and co-workers[46b] were used. Hard acids form complexes with the following relative stability:[45]

$$
\begin{array}{ccccccc}
N & \gg & P & > & As & > & Sb \\
O & \gg & S & > & Se & > & Te \\
F & > & Cl & > & Br & > & I
\end{array}
$$

Similarly, soft acids react in the following manner to produce stable complexes:[45]

$$
\begin{array}{ccccccc}
N & \ll & P & > & As & > & Sb \\
O & \ll & S & < & Se & < & Te \\
F & < & Cl & < & Br & < & I
\end{array}
$$

A list of commonly used hard and soft acids is shown in Table 2.5. [42a,47] Lewis acids with large α and β values tend to be soft, and Lewis acids with small α and β values are hard. Soft acids form stable complexes with bases that are highly polarizable and good reducing agents (not necessarily good bases toward the proton). Hard acids usually form stable complexes with bases that are good bases toward the proton. A comparison of the ability of Lewis acids to bind donating atoms of Lewis bases is conveniently made by examining the order of the donor atoms as a function of increasing electronegativity:

$$
As, P \; < \; C, Se, S, I \; < \; Br \; < \; N, Cl \; < \; O \; < \; F
$$

where the atoms on the left form more stable complexes with soft Lewis acids, and those on the right form more stable complexes with hard Lewis acids.

In general, hard acids prefer to bind to hard bases and soft acids prefer to bind to soft bases. Several other general comments can be made.

1. Heavier and less electronegative members of a series are usually softer.

$$
R_3P \; > \; R_3N
$$
$$
\textbf{(softer)} \quad I^- \; > \; Br^- \; > \; Cl^- \; > \; F^- \quad \textbf{(harder)}
$$

2. The more p character, the softer the acid.

$$
\textbf{(softer)} \quad sp^3C \; > \; sp^2C \; > \; spC \quad \textbf{(harder)}
$$

3. The more electron releasing groups on the moiety, the harder the acid.

$$
\textbf{(harder)} \quad (CH_3)_3C^{\oplus} \; > \; (CH_3)_2HC^{\oplus} \; > \; (CH_3)H_2C^{\oplus} \; > \; H_3C^{\oplus} \quad \textbf{(softer)}
$$

46. (a) Schwarzenbach, G. *Experientia Suppl.* **1956**, *5*, 162 [*Chem. Abstr. 51*:7293a **1957**]; (b) Ahrland, S.; Chatt, J.; Davies, N.R. *Quart. Rev. (London)* **1958**, *12*, 265.
47. Reference 45, p 5.

Metals require a minimum of a half-filled outer d shell to be good acceptors,[48] emphasizing the importance of d electrons for metal ions. A comparison of the series Ca→Zn suggests that their ions should become harder due to increasing nuclear charge. One also expects an increase in electronegativity, and therefore an increase in hardness. In fact, the increase in d electrons leads to increasing softness in this series.

The softness of a base is defined by the equilibrium established in the reaction of $MeHg^+$ and BH^+:

$$MeHg^+_{(aq)} + BH^+_{(aq)} \xrightleftharpoons{K} MeHgB^+_{(aq)} + H^+_{(aq)}$$

If K is much greater than one, the base is soft. Conversely, if K is ≤ 1, the base is hard. Table 2.6 shows common hard and soft bases.[47,42a,49] The nature of the outer groups on the acceptor atom is important. The hard acid BF_3 possesses hard fluoride ions and readily adds to hard bases. This contrasts with BH_3, which is a soft acid, where soft hydride ions readily add to soft anions. Soft bases tend to group together on a given central atom as do hard ligands. There is a mutual stabilizing effect called **symbiosis**.[50]

Table 2.6. Classification of Hard and Soft Bases[a]

Hard acids	Borderline cases	Soft acids
NH_3, RNH_2, N_2H_4, H_2O, ^-OH, O^{2-}, ROH, RO^-, R_2O, AcO^-, CO_3^{2-}, NO_3^-, PO_4^{3-}, SO_4^{2-}, ClO_4^-, F^- (Cl^-)	$C_6H_5NH_2$, C_5H_5N, N_3^-, N_2, NO_2^-, SO_2^{2-}, Br^-, R_2S, RSH, RS^-, $S_2O_3^{2-}$, I^-	H^-, R^-, $CH_2{=}CH_2$, C_6H_6, CN^-, RNC, CO, SCN^-, R_3P, $(RO)_3P$

a In all cases, the molecules are listed in descending order (the first entry in each column is the hardest or softest).

Reprinted with permission from Pearson, R.G. and the Journal of Chemical Education, Vol. 45, **1868**, 581-587. Copyright 1968, Division of Chemical Education, Inc.

In the case under consideration, the hard fluoride ligands form a complex that makes boron essentially 'B+' and hard. The soft hydride donates negative charge to the central boron by via a covalent bond, or by simple polarization, making the boron more neutral and soft. It appears that the actual charge on the central atom, rather than the formal charge, determines the degree of softness. Pearson pointed out "while the HSAB principle has proved useful in many ways,[41a] it has been justifiably criticized because of the lack of a precise definition of hardness and the inability to assign numbers to this property".[51] The hardness parameter (η) is now defined in terms of electronegativity as in Equation 2.3[52]

48. Ahrland S. *Structure and Bonding, 1966, 1,* 207.
49. Reference 45, p 6.
50. Jorgensen, C.K. *Structure and Bonding (Berlin) 1966, 1,* 234.
51. Pearson, R.G. *J. Chem. Educ. 1987, 64,* 561.
52. Parr, R.G.; Pearson, R.G. *J. Am. Chem. Soc. 1983, 105,* 7512.

$$\eta \ = \ \frac{1}{2} \ \frac{\partial^2 E}{\partial N^2} \ = \ \frac{1}{2} \ \frac{\partial M}{\partial N} \qquad\qquad 2.3$$

where ∂E and ∂N are obtained from the Mulliken electronegativity (χ_M).[53] Pearson used an operational definition of hardness (Equation 2.4).

$$\eta \ = \ \frac{I \ - \ A}{2} \qquad\qquad 2.4$$

In this definition, I is the ionization potential and A is the electron affinity. It was noted that a hard molecule will have a large value of η. For a reaction A + B $\rightarrow$ A:B, the following expression can be used to decide which is the acid, A or B:[51]

$$(I_A \ - \ A_B) \ - \ (I_B \ - \ A_A) \ = \ 2 \ (X°_A \ - \ X°_B) \qquad\qquad 2.5$$

If this quantity is positive, transferring an electron from B to A requires less energy than from A to B. The A group is then a Lewis acid. The direction of electron flow can be determined by the difference in absolute electronegativities, whose magnitude is a measure of the driving force for the transfer.[51]

In most cases, there will be electron transfer in both directions. Equation 2.6 gives the total energy when the amount of electron transfer in both directions is near unity (~ 1).

$$(I_A \ - \ A_B) \ + \ (I_B \ - \ I_A) \ = \ 2 \ (\eta_A \ - \ \eta_B) \qquad\qquad 2.6$$

If both A and B are hard, then η_A and η_B will be large, and transfer of electrons in *both* directions will be disfavored. If A and B are soft, however, η_A and η_B will be small, favoring electron transfer.[51] Table 2.7[51] provides typical values for electronegativity ($X°$) and for hardness (η) for several cations, anions and neutrals.

From Table 2.7, it is apparent that Al^{3+} ($\eta = 45.8$ eV) is a hard Lewis acid, whereas Ag^+ ($\eta = 6.9$ eV) is a soft Lewis acid. To illustrate neutral molecules, hydrogen (H_2) is harder than iodine (I_2), and chloromethane ($\eta = 7.5$ eV) is harder than iodomethane ($\eta = 4.7$ eV). Trimethylamine ($\eta = 6.3$) is a harder base than trimethylphosphine ($\eta = 5.9$).

53. Mulliken, R.S. *J. Chem. Phys.* **1934**, 2, 782.

Table 2.7. Electronegativity and Hardness Parameters

Ion/Molecule	X° (eV)	η (eV)	Ion/Molecule	X° (eV)	η (eV)
Na^+	26.2	21.1	Fe^{2+}	23.4	7.3
Ag^+	14.6	6.9	Zn^{2+}	28.8	10.8
I^+	14.8	4.3	Pd^{2+}	26.2	6.8
Mn^{2+}	24.4	9.3	Hg^{2+}	26.5	7.7
Ca^{2+}	31.6	19.7	Al^{3+}	74.2	45.8
NEUTRALS					
BF_3	7.8	7.8	C_4H_5N	4.4	6.0
SO_3	7.2	5.5		4.4	6.1
N_2	7.0	8.6	H_2S	4.3	6.4
Cl_2	7.0	4.6	C_6H_6	4.0	5.2
SO_2	6.7	5.6	PH_3	4.0	6.0
H_2	6.7	8.7	H_2O	3.1	9.5
O_2	6.3	5.9	MeCl	3.8	7.5
CO	6.1	7.9	NH_3	2.9	7.9
I_2	6.0	3.4	PMe_3	2.8	5.9
Pt	5.6	3.5	Me_2S	2.7	6.0
MeI	4.9	4.7	Me_2O	2.0	8.0
HCl	4.7	8.0	NMe_3	1.5	6.3
F^-	3.40	7.0	Br^-	3.36	4.2
HO^-	1.83	5.7	I^-	3.06	3.7
H_2N^-	0.74	5.3	H^-	0.74	6.8
H_3C^-	0.08	4.9	MeS^-	1.9	3.1
Cl^-	3.62	4.7	$t\text{-}Bu^-$	–0.3	3.6
HS^-	2.30	4.1	$C_6H_5^-$	1.1	4.1
H_2P^-	1.25	4.3	NO_2^-	3.1	3.9

[Reprinted with permission from Parr, R.G.; Pearson, R.G., *J. Am. Chem. Soc.*, *1983*, *105*, 7512. Copyright © *1983* American Chemical Society.]

2.4.B. HSAB and Molecular Orbital Theory

Frontier molecular orbital (FMO) theory (sec. 11.2) correlates molecular orbital energies and orbital coefficients with the relative reactivity of molecules. In a molecule, the orbital containing the highest energy bonding electrons is called the highest occupied molecular orbital (HOMO) and the lowest lying orbital that does not contain electrons but can receive them is the lowest unoccupied molecular orbital (LUMO). Both HOMO and LUMO are shown in Figure 2.1.[51] The difference in energy between these orbitals determines the reactivity of many types of reactions. Experimentally, the energy of the HOMO is the ionization potential (*I*) and the energy of the LUMO is the electron affinity (*A*), leading to the equations:

$$I = -\varepsilon_{HOMO} \quad \text{and} \quad A = -\varepsilon_{LUMO}$$

In Figure 2.1, the electronegativity (X°) is taken as the midpoint (average) between the HOMO and LUMO energies.[51] If the HOMO is –10 eV and the LUMO is +1 eV, then the midpoint is

5.5 eV and the value of X° is 1 eV –5.5 eV = –4.5 eV. The parameter η will then be 5.5 eV (see Fig. 2.1), which allows one to define a hard molecule or ion as having a large energy gap between the HOMO and LUMO ($\Delta_{HOMO\text{-}LUMO}$).[51] A soft molecule or ion will have a small $\Delta_{HOMO\text{-}LUMO}$.[51] Soft acids and bases have properties that lead to a high energy HOMO and a low energy LUMO (a small energy gap). Hard acids and bases have the opposite properties: a low energy HOMO and a high-energy LUMO (a large energy gap). This generalization can be applied to cations, anions and radicals as well as neutral molecules. Radicals contain a singly occupied molecular orbital (SOMO) as shown in Figure 2.2.[51] The electron affinity term in this case arises from adding a second electron to the SOMO and X° is the average energy of the two electrons in that orbital. The quantity *I–A* is the average value of the interaction energy of the two electrons.[54]

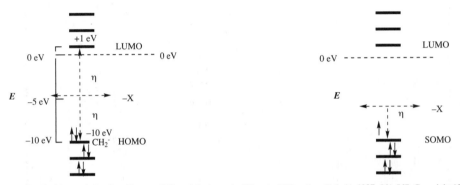

Reprinted with permission from Pearson, R.G. and the Journal of Chemical Education, Vol. 64, **1987**, 561-567. Copyright 1987, Division of Chemical Education, Inc.

Figure 2.1. Correlation of HOMO and LUMO with Electronegativity and Hardness

Figure 2.2. Energy Diagram for Radicals

The HOMO-LUMO energy gap is usually the lowest energy electron absorption band. Soft acids and bases tend to absorb light closer to the visible region than do hard acids and bases. Alkenes are chromophoric and it is typical for them to have high-energy HOMOs and low energy LUMOs. In general, unsaturation increases softness. It is also apparent that soft molecules are more reactive than hard molecules (due to a smaller energy gap *I-A*) in unimolecular and bimolecular reactions.

These principles finally allowed Pearson to define hardness (η) as follows:[51] a hard molecule resists changes in its electron charge cloud, both the total amount of charge and the charge distribution in space. A soft molecule has an easily changed electron distribution.

2.4.C. Applications of HSAB Theory

HSAB theory has been applied to many chemical reactions not usually classified as acid-base

54. Klopman, G. *J. Am. Chem. Soc.* **1964**, *86*, 1463.

reactions. Care should be exercised in nontrivial cases, but this approach may be of value in predicting the likelihood of a number of reactions in a synthetic plan. Several examples are illustrated below.[51]

2.4.C.i. Substitution (also see sec. 2.6). Saville showed that nucleophilic displacement of alkyl halides can be characterized by HSAB theory.[55] In general, the reaction works best when the R group and nucleophile partner are matched. The two major categories are

$$Nu = nucleophile$$

When R–C has a greater differential in hardness and softness, the reaction proceeds best. Table 2.7 shows that chloromethane is harder than iodomethane. Therefore, a hard base such as methoxide will show a larger energy difference for iodomethane, and should react faster.

2.4.C.ii. E2 versus S_N2 (see secs. 2.9.A and 2.6.A). Ethoxide and the malonate anion are about equal in basicity,[56] but ethoxide is a much harder base. Reaction of 2-bromopropane with ethoxide gives primarily elimination, but similar reaction with the malonate anion gives the substitution product **27**. In general, hard bases give elimination and soft bases give substitution, but the gegenion has a profound influence on the course of the reaction. Reaction of 1-bromo-2-chloroethane with **28** gave only elimination (to bromoethene or chloroethene), leaving behind the neutral substrate **29**. However, reaction with chloromagnesium salt **30** gave nucleophilic displacement with formation of **31** after a second displacement by the enamine (sec. 9.6) formed in the first substitution.[57]

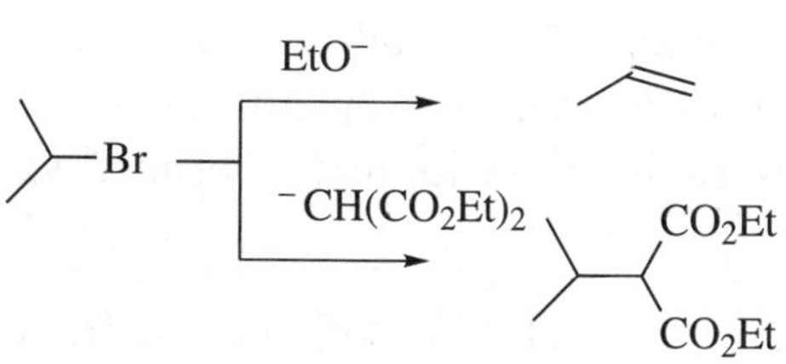

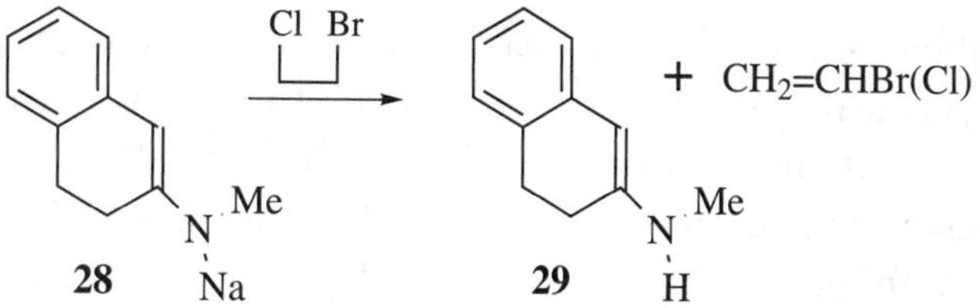

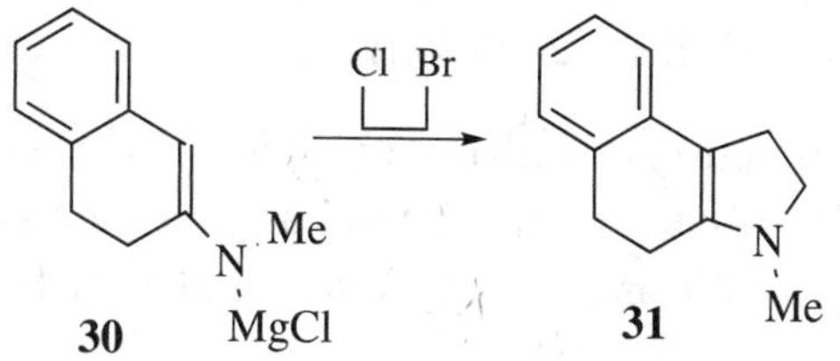

55. Saville, B. *Angew. Chem. Int. Ed.* **1967**, *6*, 928.
56. Pearson, R.G.; *J. Am. Chem. Soc.* **1949**, *71*, 2212.
57. (a) Evans, D.A.; Bryan, C.A.; Wahl, G.M. *J. Org. Chem.* **1970**, *35*, 4122; (b) Reference 44, pp 56-57.

2.4.C.iii. Epoxidation. Peroxyacids such as **32** (sec. 3.4.C) contain an unusual soft-hard combination that leads to attack by an alkene (a soft base) on the soft oxygen of the peroxyacid.

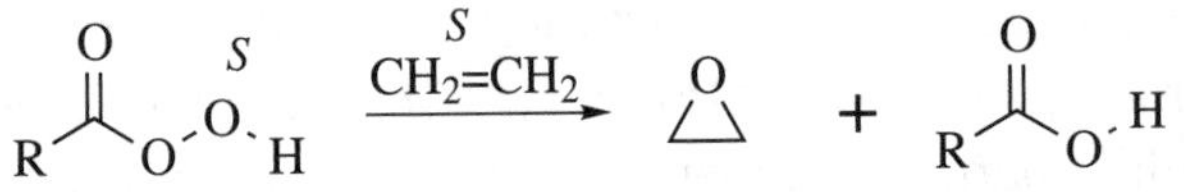

2.4.C.iv. Carbonyls. The carbon of a carbonyl is a hard acceptor and the oxygen is a hard donor.

$$\text{Hard acceptor} \implies C{=}O \impliedby \text{Hard donor}$$

An example is **33**, which reacts with the soft thiophenoxide (PhS$^-$) at the tosylate center (C-OTs) to give **34**.[58] Reaction of **33** with the harder cyanide ion, however, gives **35** and subsequent ring closure, where the alkoxide unit displaces tosylate to generate oxetane **36**.[58] Thiophenoxide is a soft base that reacts with the soft acid (CH$_2$OTs),

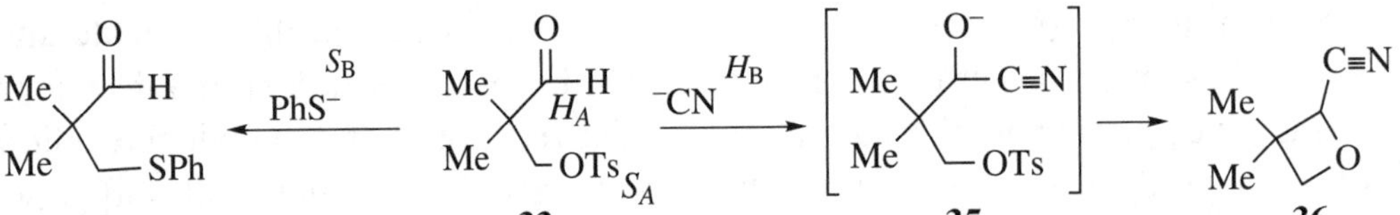

whereas cyanide is a hard base that reacts with the hard acid (the carbonyl). Reductive hydrogenolysis of α-bromoketones (**37**) by some metal borohydrides (not NaBH$_4$, see sec. 4.4.A) is another example. Transfer of the soft base (hydride) leads to loss of the halide (a soft base). The reaction works best when the metal (M$^+$) is a soft acid.[59] The sodium counterion in sodium borohydride is a hard acid and does not work well in this reaction. A soft acid such as copper or lead, however, will facilitate the reduction.[60,61] Reduction with metal borohydrides will be discussed in section 4.4.B.

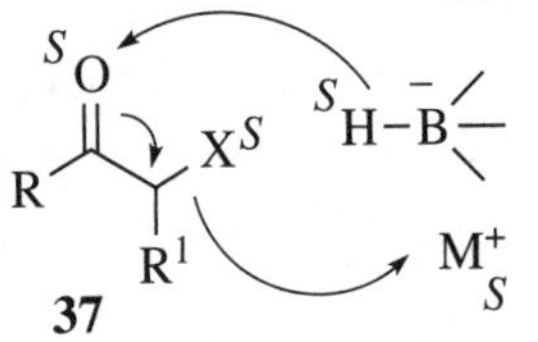

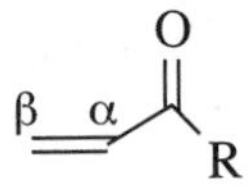

Conjugate reduction of α,β-unsaturated ketones (sec. 4.2-4.5) can also be explained by HSAB theory.[62] The β-carbon of a conjugated ketone (**38**) is softer than the carbonyl. A covalent M-H reagent such as hydride is soft and LiAlH$_4$ reduction of cyclohexenone gave 22% of

58. (a) Nerdel, F.; Weyerstahl, P.; Lucas, K. *Tetrahedron Lett.* **1968**, 5751; (b) Reference 44, p 86.
59. Reference 45, p 92.
60. Goto, T.; Kishi, Y. *Tetrahedron Lett.* **1961**, 513.
61. Goto, T.; Kishi, Y. *Nippon Kagaku Zasshi* **1962**, *83*, 1135 [*Chem. Abstr. 58:* 14059a, **1963**].
62. Bottin, J.; Eisenstein, O.; Minot, C.; Anh, N.T. *Tetrahedron Lett.* **1972**, 3015.

the conjugate reduction product. Conversely, lithium trimethoxyaluminum hydride (sec. 4.3) suppresses the soft conjugate addition since the alkoxy groups are hard and only 5% of conjugate addition was observed with this reagent.[63] The presence of an alkyl group increases the softness of the borohydride. Sodium borohydride gives mainly 1,2-addition but potassium tri-*sec*-butylborohydride (K-Selectride, sec. 4.5.C) gives mainly 1,4-reduction.[64] The B-H bond is more covalent than Al-H, and borohydrides are softer than aluminum hydrides, and this accounts for the observation that BH_4^- is relatively inert to the hard acid H^+. Softer cations such as Na^+ favor 1,4-reduction and it is known that $NaBH_4$ gives more 1,4-reduction than $LiBH_4$.[65]

The reaction of enones with organometallics also follows this pattern. Dialkylcuprates (with a soft copper) are soft and give primarily 1,4-addition. When organolithium reagents (with a hard Li atom) react with enones, the major product is 1,2-addition. The hardness of common organometallics follows the order:[66]

$$(\textbf{harder}) \quad RLi > RMgX > RCu > R_2CuLi \quad (\textbf{softer})$$

The alkyl portion also plays an important role. The hardness order of alkyl fragments is:[66]

$$(\textbf{harder}) \quad Me > Et > i\text{-Pr} > t\text{-Bu} \quad (\textbf{softer})$$

This reactivity order suggests that *tert*-butylmagnesium halides are softer and will give more conjugate addition than methylmagnesium halides.[67,66] When the alkyl fragment of an organocuprate is hard (such as lithium trialkynyl cuprates) the reagent adds to the carbonyl via 1,2-addition.[68]

39

2.4.C.v. Enolate Anions. Enolate anions such as **39** are bidentate nucleophiles (the carbanion and alkoxy units can both donate electrons) and bidentate bases. Oxygen is a harder base than carbon and reaction with hard acids tends to give more *O*-alkylation. The hardness of alkyl halides is $RCl > RBr > RI$, and more C alkylation is expected with R-I, but more O alkylation with R-Cl.[42] The product distribution varies as the polarity of the solvent is changed, however.[42] In general, acylation and silylation of enolates (classified as hard acids) give more *O*-substituted product. The ease of C-X bond cleavage parallels the softness of X. In general, the hard Si atoms tend to form strong bonds with oxygenated compounds.

63. Durand, J.; Anh, N.T.; Huet, J. *Tetrahedron Lett.* **1974**, 2397.
64. Ganem, B. *J. Org. Chem.* **1975**, *40*, 146.
65. Wheeler, J.W.; Chung, R.H. *J. Org. Chem.* **1969**, *34*, 1149.
66. Reference 45, p 94.
67. Damiano, J.C.; Diara, A. *C.R. Hebd. Seances Acad. Sci. Ser. C* **1973**, *276*, 441.
68. Palmisano, G.; Pellegata, R. *J. Chem. Soc. Chem. Commun.* **1975**, 892.

2.4.C.vi. Hydroboration

$$\text{acceptor} \implies \text{B-H} \impliedby \text{soft base (H}^-)$$

The reaction between alkenes and boranes forms the basis of hydroboration (see Chap. 5). Borane (B_2H_6, formally known as diborane) is a soft Lewis acid and will complex with soft bases such as alkenes. When BH_3

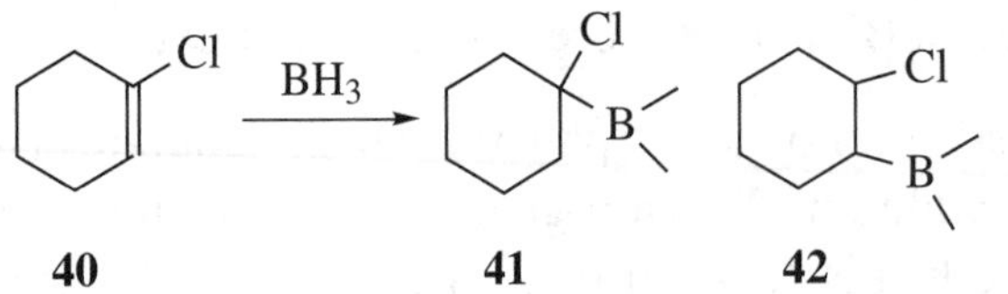

40 **41** **42** 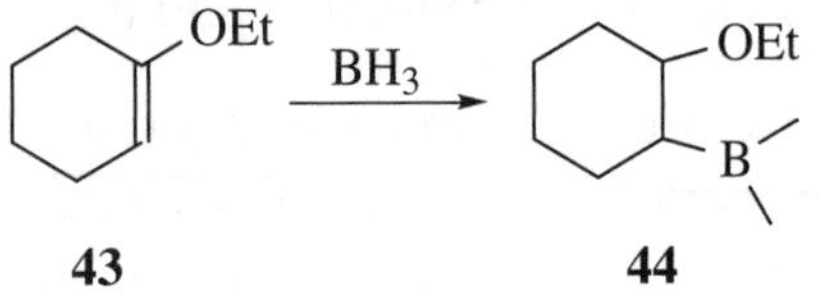 **43** **44**

(the monomeric form of diborane) reacted with a vinyl chloride, the boron was attached primarily to the α-carbon (**40** gave 60% of **41** and 40% of **42**).[69] Reaction with enol ether **43**, however, gave only **44**.[69] Both chlorine and ethoxy are hard bases and influence hardness at the *ipso*-carbon.[70] Ethoxy is harder and accounts for the predominance of **44**. The chlorine substituent leads to resonance stabilization that swamps the increase in hardness. The increased regioselectivity observed in reactions of chloroborane, for example ($ClBH_2$, sec. 5.2.A), is attributed to the heightened hardness of the boron atom,[71] induced by the chlorine.

2.5. ACYL ADDITION, SUBSTITUTION AND CONJUGATE ADDITION

In general, nucleophiles react with ketones and aldehydes at the acyl carbon (the carbonyl carbon), breaking the π bond and forming a new bond to carbon. This class of organic reactions is called nucleophilic acyl addition. When the carbonyl contains a leaving group, acyl addition is followed by loss of that group to give a substitution product, and the overall process is called acyl substitution. Conjugate carbonyl compounds have a C=C unit attached to the carbonyl, and the π-bond extends the reactivity: $\overset{\delta+}{C}=\overset{\delta-}{C}-\overset{\delta+}{C}=\overset{\delta-}{O}$. In other words, the nucleophile can react with the C=C unit as well as the C=O unit. This extension of reactivity by conjugation to a π-bond is called **vinylogy**. A nucleophile can, therefore, attack the C=C unit in the conjugated system in what is called conjugate addition or Michael addition. All of these types of reactions will be discussed.

2.5.A. Nucleophilic Attack

The discussion of Lewis acids made it clear that electrophilic centers other than a proton can

69. (a) Brown, H.C.; Sharp, R.L.*J. Am. Chem. Soc.* **1968**, *90*, 2915; (b) Pasto, D.J.; Snyder, R. *J. Org. Chem.* **1966**, *31*, 2773.
70. Reference 45, p 156.
71. Reference 45, p 157.

react with an electron-donating species. Bond polarization, induced by the presence of more electronegative atoms, will polarize atoms that are connected to the electronegative atom by covalent bonds. If the bond polarization leads to a δ+ dipole on an atom, that atom can react with reagents that donate electrons. A simple example is the reaction of carbonyls (C=O) with the nucleophilic alkyne anion. The carbonyl carbon has a positive dipole ($^{\delta+}C$) because of the electronegative oxygen atom, and that carbon can accept an electron pair (it is termed an **electrophile** or electron loving) from the negatively charged carbon of MeC≡C:⁻ to give **45**. The halogen-bearing carbon of an alkyl halide also has a positive dipole that makes it electrophilic, and attack by the carbanionic carbon will lead to displacement of the iodide (to form **46**) in a S_N2 reaction, which will be discussed in sec. 2.6.A. Formation of both **45** and **46** are reasonable extensions of Lewis acid-Lewis base reactions, but the electrophilic center is a carbon atom in both cases. Although the attacking species is formally a Lewis base, it is convenient to designate it as a **nucleophile** (nucleus loving) to emphasize its attack on carbon rather than other electrophilic atoms. A **nucleophile** is therefore defined as a species able to donate two electrons to carbon, forming a new bond to that carbon. As we will see in section 2.9.A., if a nucleophile is a relatively strong base as well as a nucleophile, elimination can compete with substitution.

The reactivity of a nucleophile reacts with an electrophile depends on both the strength of the nucleophile (its ability to donate electrons to carbon) and the nature of the carbonyl substrate. An order of nucleophilic strength for reaction of different nucleophiles *with a carbonyl* is[72]

$$C{=}N{-}O^- \;>\; EtO^- \;>\; MeO^- \;>\; HO^- \;>\; N_3^- \;>\; F^- \;>\; HOH \;>\; Br^- \;>\; I^-$$

It is important to mention that nucleophilic strength for a given nucleophile changes with the electrophile. In other words, the nucleophilic alkyne anion from above will exhibit different nucleophilic strength with a ketone than with an alkyl halide. A complete discussion of nucleophilic strength will be delayed until sec. 2.7.A, in connection with reactions of nucleophiles and alkyl halides.[73] This section will focus on nucleophilic reactions of carbonyl compounds. There are three fundamental types of reactions: (*1*) acyl addition, (*2*) acyl substitution, and (*3*) conjugate addition when the carbonyl is conjugated to a π bond. As mentioned above, inductive effects make the β-carbon of the π bond electrophilic (see **47**) and subject to nucleophilic attack.

72. Koskikallio, J. *Acta Chem. Scand.* **1969**, *23*, 1477, 1490.
73. Harris, J.C.; Kurz, J.L. *J. Am. Chem. Soc.* **1970**, *92*, 349.

2.5.B. Acyl Addition to Carbonyls[74]

Nucleophilic addition to a carbonyl can be reversible if the attacking nucleophilic species becomes a good leaving group in the product (leaving groups are discussed in sec. 2.7.E). Water and hydroxide, for example, add reversibly. When water adds to cyclohexanone, the product is **48**, but H_2O is a good leaving group in the presence of the alkoxide unit, making the reaction reversible as shown. As a practical matter, such reversibility occurs when weak nucleophiles react with aldehydes and ketones. The process becomes effectively irreversible if the nucleophile becomes a poor leaving group in the product. When $Na^+-C{\equiv}CMe$ reacts with a carbonyl a carbon-carbon bond is formed which is a very poor leaving group for the reverse reaction, making the reverse reaction difficult. For this reason, cyclohexanone reacts with the sodium salt of 1-propyne to give alkoxide **49**. Acetylides such as $^-C{\equiv}CMe$ add to carbonyls and the $C-C{\equiv}C$ unit is a very poor leaving group so the addition is essentially irreversible. A hydrolysis step generates the final alcohol product, **50**. Clearly, this analysis is overly simplistic and the nucleophilic strength of the incoming reagent as well as the groups attached to the acyl carbon play a major role. This analysis does illustrate the important role played by the nucleophile and the nature of the product that results from acyl addition.

As with water, ethanol or other alcohols are weak nucleophiles and acyl addition to cyclohexanone is reversible. One method to improve the reaction is to protonate the acyl unit of cyclohexanone to give an oxygen-stabilized carbocation by adding an acid catalyst. Protonation of cyclohexanone, for example, gives resonance stabilized cation **51** (sec. 2.6.B.i and 12.2) that reacts with ethanol to give **52**. Loss of a proton regenerates the acid catalysts and forms hemi-acetal **53**. All steps are reversible, but in the presence of the acid catalyst and ethanol solvent (i.e., a large excess of ethanol), the equilibrium favors formation of the hemi-ketal. Subsequent protonation of OH, loss of water, addition of a second molecule of ethanol, and loss of a proton gives the diethyl ketal of cyclohexanone (1,1-diethoxycyclohexane, **54**). Aldehydes react with alcohols to give acetals. Ketals and acetals are common protecting groups for the carbonyl group (sec. 7.3.B.i). When thiols are used in this reaction, dithioketals or dithioacetals are formed. Primary amines react to give imines and secondary amines react to give enamines (sec. 9.6). Although acyl addition is usually reversible, addition of excess reagent (such as ethanol) or removal of a product (water) will drive the equilibrium from ketone to ketal (or another addition product). Conversely, if the ketal is treated with an acid catalyst in the presence of excess water (aqueous acid for example), the equilibrium will shift

---

74. For examples see reference 3, pp 1125-1188.

to favor the ketone. In other words, the ketal will be hydrolyzed to the ketone.

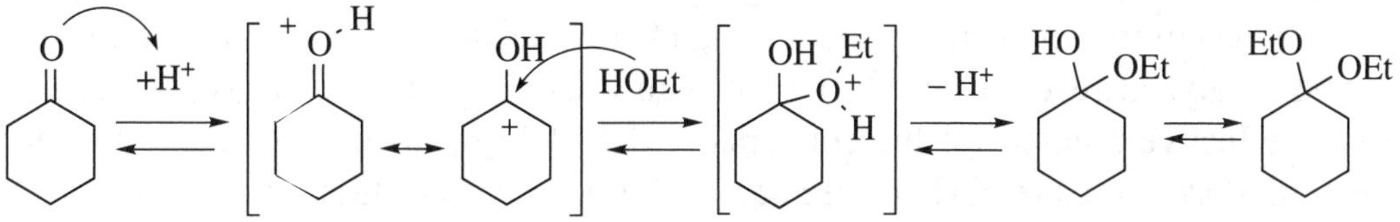

51   **52**   **53**   **54**

In general, addition of carbon nucleophiles to a carbonyl is irreversible (or the equilibrium greatly favors the product) and addition of heteroatom nucleophiles is reversible. Even when the reaction equilibrates, the equilibrium may lie to the side of the products. Heteroatom nucleophiles often require an acid catalyst in order to establish an equilibrium that leads to good yields of acyl addition products. All of these reactions are discussed in any undergraduate Organic chemistry book so we will keep the focus on addition of strong nucleophiles.

2.5.C. Acyl Substitution of Carbonyl Compounds[75]

Derivatives of carboxylic acids such as acid chlorides and esters have an interesting structural feature when compared to a ketone or an aldehyde. The carbonyl carbon of these groups bear a chlorine or an OR group, respectively. When a nucleophile adds to the carbonyl the resulting alkoxide intermediate, called a **tetrahedral intermediate** bears the Cl or OR units, which are putative leaving groups. If the newly formed bond is rather strong, such as a C–C bond, both the C–Cl or C–OR bonds are relatively easy to break. Regeneration of the C=O unit is accompanied by loss of Cl⁻ or RO⁻, which are rather stable ionic entities, and the overall result of this process is a substitution reaction where the nucleophile replaces chlorine on the carbonyl. This reaction with carboxylic acid derivatives (acid chlorides, esters, anhydrides, amides) is called **acyl substitution**. Many types of nucleophiles can be used, including hydroxide, alkoxides or amines in addition to carbon nucleophiles (see chap. 8). Acyl addition of hydroxide ion to benzoyl chloride, for example, gives **55** as the initial product. It is a reactive intermediate and loss of the chloride ion leads to benzoic acid. Under these basic conditions, the initially formed acid is converted to the carboxylate anion, requiring a neutralization step to recover the acid.

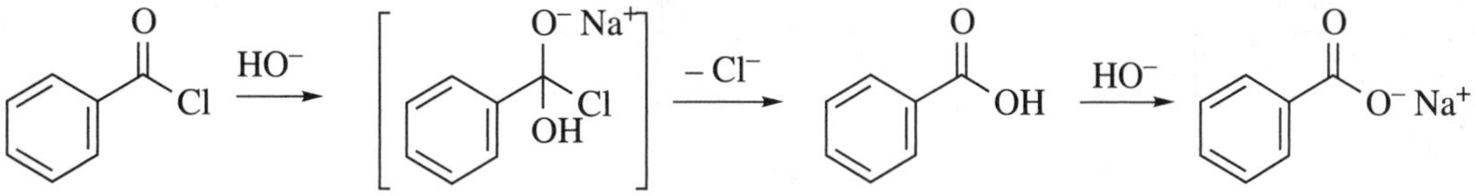

55

With ketones and aldehydes, addition of an anionic nucleophile leads to a species such as **48**

---

75. For examples see reference 3, pp 1717-1808, 1929-1994.

or **49**, where there are strong C–C or C–H bonds, and the H or C units are not considered to be leaving groups. The presence of a leaving group in acid derivatives leads to this fundamental difference in reactivity between acyl addition and acyl substitution. This reaction is the basis of the acid catalyzed reaction of an alcohol with a carboxylic acid to give an ester. Note that a very effective method for the conversion of a carboxylic acid to a methyl ester is the reaction with diazomethane (CH_2N_2, see sec. 13.9.C). Acyl substitution also forms the basis for conversion of a carboxylic acid to an acid chloride with thionyl chloride, for conversion of an ester to an amide by reaction with an amine, and the acid or base hydrolysis of esters, acid chlorides, anhydrides, or amides to give carboxylic acids. Acyl addition can be reversible when the leaving group ability of the acyl derivative is close to that of the new nucleophile, as in the transesterification reaction of ethyl butanoate and methanol. The acid-catalyzed reaction of ethyl butanoate and methanol generates a tetrahedral intermediate bearing both OMe and OEt, and the equilibrium is driven to the methyl ester only if a large excess of methanol is present. Recognizing the mechanistic details of this process allows one to control the reaction, making a process like transesterification useful in synthesis. In a synthesis of (+)-aldosterone by Shea and co-workers,[76] methyl ester **56** was transesterified[77] with the chiral dimethyl-*tert*-butylsilyl-protected (*S,S*)-hydrobenzoin **57** to give ester **58** in 75% yield and >99% ee. Formation of **58** allowed Shea to proceed with an asymmetric synthesis.

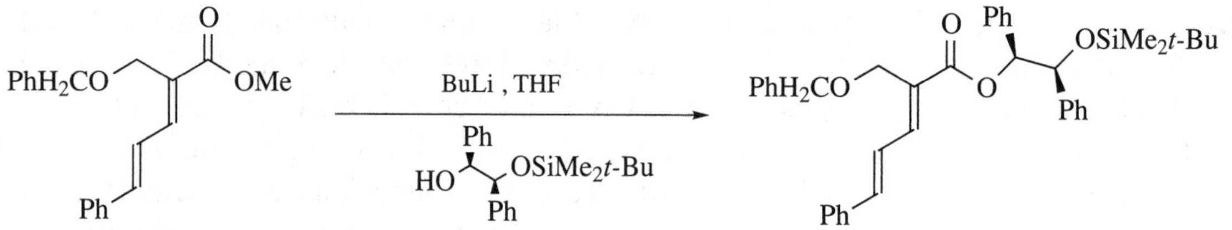

Nucleophilic addition of Grignard reagents to carboxylic acid derivatives is another example where acyl addition gives a product that can react with the reagent.[78] Initial addition to the carbonyl of methyl butanoate gives the tetrahedral intermediate **59**, which loses methoxide to give the ketone, 1-phenyl-1-butanone (**60**). Phenylmagnesium bromide reacts with the ketone faster than it does with the ester, however, to give **61**. Hydrolysis generates tertiary alcohol **62** (sec. 8.4.C.i).

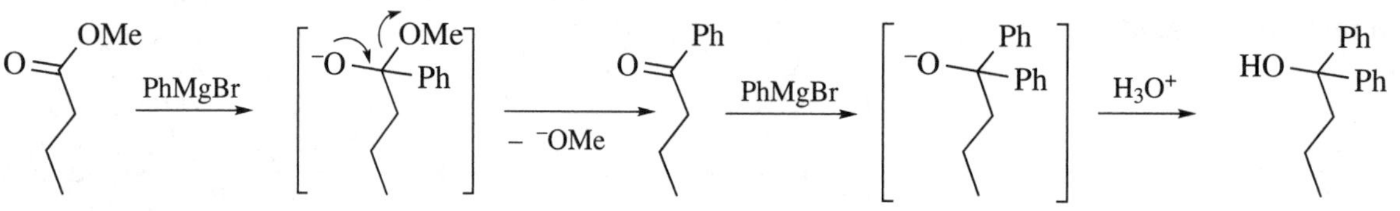

76. Bear, B.R.; Parnes, J.S.; Shea, K.J. *Org. Lett.* **2003**, *5*, 1613.
77. Meth-Cohn, O. *J. Chem. Soc. Chem. Commun.* **1986**, 695.
78. Kharasch, M.S.; Reinmuth, O. *Grignard Reactions of Nonmetallic Substances*, Prentice-Hall, Inc., Englewood Cliffs, New Jersey, *1954*, pp 549-766, 846-869.

Esters, anhydrides, and amides give similar products via acyl substitution, followed by acyl addition to the initially formed product. The best leaving group is chloride followed closely by carboxylate (acid chlorides and anhydrides, respectively) and then the alkoxy group from an ester. Amides have a $^-NR_2$ leaving group and are the least reactive. Therefore, acid chlorides and anhydrides should react with Grignard reagents to give a higher percentage of alcohol, whereas an ester might give a mixture of ketone and alcohol products. For the most part this is correct, but the product distribution depends on the reactivity of the Grignard reagent, the nature of the acid derivative, the solvent and temperature of the reaction. A useful approximation correlates the reactivity of the carbonyl moiety with the leaving group propensity of the ligand attached to carbon. This leads to the usual order of reactivity to nucleophilic acyl substitution as:

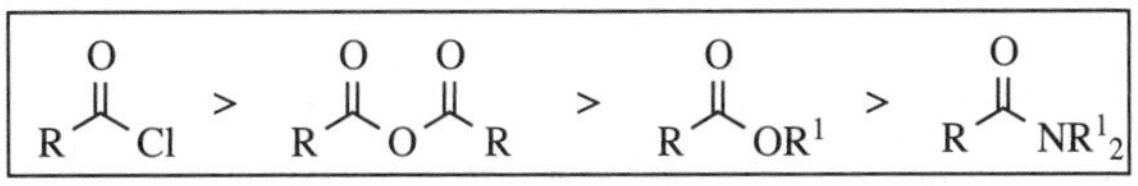

If a large excess of the nucleophile (such as excess Grignard) is used, most reactions are driven to give the tertiary alcohol as the major (often the only) product. In later chapters, the problem of reactivity of the ketone product with the organometallic reagents will be circumvented by use of a dialkyl cadmium reagent (sec. 8.4.C.ii).[79] The reaction of acid chlorides with organocuprates also gives the ketone in good yield (sec. 8.7). In some cases, functionalized amides can give good yields of the acyl addition product. A N-Methoxy-N-methyl amide such as **63** is referred to as a **Weinreb amide**.[80] In a synthesis of pumiliotoxins 209F and 251D,[81] Jamison and Woodin reacted **63** with methylmagnesium bromide, and obtained an 94% yield of ketone **64** (note that the allylic carbamate unit is referred to as N-Alloc). Grignard reactions will be discussed further in section 8.4.C.ii.

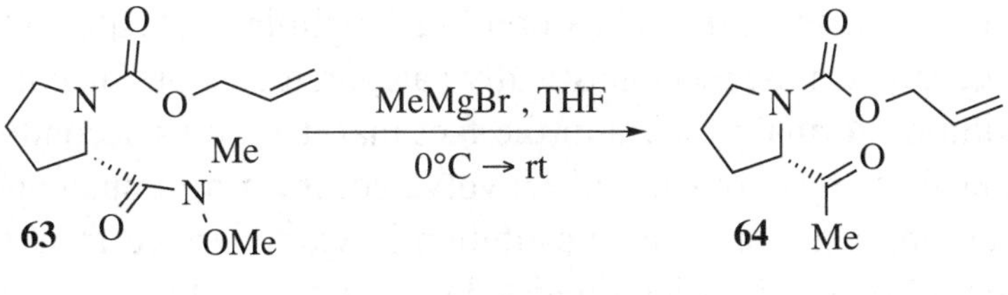

2.5.D. Conjugate Addition[82]

Conjugated ketones, aldehydes, esters and related molecules have a carbonyl unit connected directly to the π bond of an alkene (as in **65**). Inductive effects make the terminal carbon of the alkene electrophilic, and subject to attack by a nucleophile in what is formally a vinylogous

79. (a) Jones, P.R.; Desio, P.J. *Chem. Rev.* **1978**, *78*, 491; (b) Cason, J.; Fessenden, R. *J. Org. Chem.* **1960**, *25*, 477.
80. (a) Nahm, S.; Weinreb, S.M. *Tetrahedron Lett.* **1981**, *22*, 3815; (b) Mundy, B.P.; Ellerd, M.G.; Favaloro Jr., F.G. *Name Reactions and Reagents in Organic Synthesis, 2nd Ed.* Wiley-Interscience, New Jersey, **2005**, p. 866.
81. Woodin, K.S.; Jamison, T.F. *J. Org. Chem.* **2007**, *72*, 7451.
82. For examples see reference 3, pp 1567-1616, 1809-1842.

acyl addition. This type of reaction is referred to as **conjugate addition** or **Michael addition**[83] (sec. 9.7.A) and it is facilitated by formation of a resonance-stabilized product, the enolate anion (**66**). Hydrolysis gives the final addition product, ketone **67**. Conjugate addition competes with normal nucleophilic attack at the acyl carbon (acyl addition), and the major product is usually determined by steric hindrance at the carbonyl or at the alkene carbon, and by the nature of the nucleophile. When the R group in **65** is hindered, attack at the acyl carbon can be difficult, leading to conjugate addition. When R is small, such as the H in conjugate aldehydes, normal 1,2-addition to the carbonyl (acyl addition) is usually preferred. Acyl addition is usually preferred with conjugated aldehydes with nucleophiles such as alkyne anions and Grignard reagents, but organocuprates give conjugate addition with conjugated aldehydes and with most α,β-unsaturated carbonyl derivatives (sec. 8.7.A.vi).[84,85]

Conjugate addition will be a key feature of the **Robinson annulation**[86,83] in sec. 9.7.C, in which a ketone enolate reacts with a conjugated ketone to give a cyclohexenone derivative. Michael additions are common in synthesis and we will discuss examples in chapter 8 and 9.

2.6. SUBSTITUTION REACTIONS

Replacement of one atom or functional group for another is referred to as substitution. The two major types of substitution previously discussed involve reaction at an sp^3 carbon (nucleophilic aliphatic substitution) or at sp^2 carbonyl carbon (nucleophilic acyl substitution). Substitution reactions at an sp^3 carbon can be mechanistically categorized as bimolecular or unimolecular. Bimolecular substitution formally refers to the fact that it follows second-order kinetics, but as a practical matter these reactions usually involve collision of a nucleophile with a carbon bearing a leaving group. Unimolecular substitution is again derived from the rate expression because the reaction follows first-order kinetics, but ionization of a substrate bearing a leaving group to form a carbocation intermediate is the rate-determining step, and this is followed by reaction with a nucleophile. The former reaction is labeled S_N2 and the latter is labeled S_N1.

83. Bergmann, D.; Ginsburg, D.; Pappo, R. *Org. Reactions, 1959, 10,* 179.
84. (a) Posner, G.H. *An Introduction to Synthesis Using Organocopper Reagents*, Wiley-Interscience, New York, *1980*, (b) Mandeville, W.H.; Whitesides, G.M. *J. Org. Chem. 1974, 39,* 400, (c) House, H.O.; Wilkins, J.M. *Ibid 1978, 43,* 2443, (d) Whitesides, G.M.; San Filippo, Jr. J.; Casey, Jr., C.P.; Panek, E.J. *J. Am. Chem. Soc. 1967, 89,* 5302.
85. (a) Davis, R.; Untch, K.G., *J. Org. Chem. 1979, 44,* 3755; (b) Stork, G.; Isobe, M. *J. Am. Chem. Soc. 1975, 97,* 6260, 4745.
86. duFeu, E.C.; McQuillin, F.J.; Robinson, R. *J. Chem. Soc. 1937,* 53.

Both S_N2 and S_N1 reactions are extremely useful for the introduction of heteroatom and carbon functional groups into a carbon skeleton.

2.6.A. Aliphatic Bimolecular Nucleophilic Substitution

When a nucleophile collides with a sp^3 carbon that is part of a polarized bond bearing a leaving group (an electrophilic carbon), it leads to the transform:

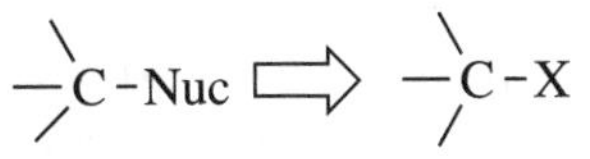

$$\overset{\backslash}{\underset{/}{C}}\text{-Nuc} \implies \overset{\backslash}{\underset{/}{C}}\text{-X}$$

and the experimental reaction required to accomplish this transform is

$$\text{Nuc: } + \overset{\backslash}{\underset{/}{C}}\text{-X} \longrightarrow \text{Nuc-}\overset{/}{\underset{\backslash}{C}} + \text{X:}^-$$

where X is a suitable leaving group and the reaction is called **nucleophilic aliphatic substitution**.[87]

2.6.A.i. The S_N2 Reaction. In reactions where the substrate (see **68**) has a leaving group (X) connected to a sp^3 carbon, displacement by a nucleophile leads to a product in which X has been replaced by the nucleophile to give **70**. This process can be described as a bimolecular (it follows second order rate equation) substitution that involves a nucleophile and the mechanistic descriptor is S_N2. Reactions labeled as S_N2 may be observed in several different functional group exchange reactions and the substrate and the reagent can be either simple or complex. All S_N2 reactions can be represented by the simple reaction sequence: **68 → 70**, which proceeds by a pentacoordinate transition state (**69**).[154b] This scheme arises from many experimental observations in a wide variety of S_N2 type reactions and no intermediate has ever been detected. The primary features of the reaction have been determined by many years of experiments and include (*1*) it follows second order kinetics, (*2*) inversion of configuration at the stereogenic center in chiral molecules and (*3*) the dependence of the reaction rate on the structure of the halide.

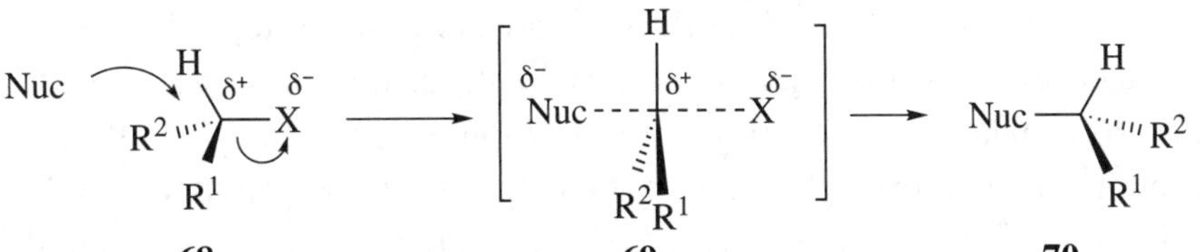

Approach of the nucleophile from the side of the molecule bearing the group X is unfavorable

---

87. For examples see reference 3, pp 611-618 (halogenation of hydrocarbons), 667-676 (interconversion of halides), 779-784 (amines from alkyl halides), 889-910 (ethers by alkylation of alcohols), 970-972 (alcohol by halide substitution), and pp 1717-1808 (alkylation and substitution of nitriles and acid derivatives).

due to electrostatic and steric repulsion. Approach from the side opposite X (backside or anti attack) will minimize these repulsive forces and is energetically preferred. To initiate the reaction, the nucleophile collides with the δ^+ carbon, and transfers two electrons to that carbon. These two electrons are used to form a new bond, and the C-X bond is broken with both of those electrons being transferred to the more electronegative heteroatom leaving group, completing the substitution process.

The experimentally observed characteristics of many of these substitution reactions led to this mechanism, which requires a five-coordinate transition state such as **69**. The transition state structure is taken to be the logical midpoint of the reaction, where the nucleophile-carbon bond is partially formed and the C-X bond is partially broken. Transition state **69** is consistent with the following observations.

1. Backside attack dictates that inversion of configuration must occur in the final product, relative to the starting material. Inversion can be detected when the starting halide possesses a stereogenic carbon bearing a leaving group, as in the conversion of **68** → **69** (R^1, R^2 = H, alkyl, aryl).

2. Since the reaction requires a collision of the nucleophile and the halide,[154a] the rate of reaction is proportional to the concentrations of both reactants and the nature of the solvent. An S_N2 reaction is termed bimolecular because it is described by the rate expression: (Rate = k [N] [halide]). The rate of reaction is also dependent upon the strength of the nucleophile (see above).[154] Protic solvents slow the reaction, making aprotic solvents the preferred reaction medium.

3. The rate of reaction is fast with MeBr and slower with secondary halides. Tertiary halides such as *tert*-butyl bromide, $Me_3C–Br$, do not undergo substitution via this mechanism at any reasonable rate.

The structure of the halide is critically important for a S_N2 reaction, as shown in Table 2.8.[88] This table gives the rate for reaction for the reaction of alkyl bromides with potassium iodide. As the steric bulk around the carbon being attacked by iodide increases, the rate of reaction slows. This observation correlates directly with an increase in activation energy required to achieve the pentacoordinate transition state, **69**. With tertiary halides, steric hindrance in the transition state (**69**) makes the rate of an S_N2 reaction so slow that for all practical purposes there is no reaction. The neopentyl system, which appears to be primary, is in fact so sterically hindered (**69**, R^1 = H, R^2 = CMe_3) that the rate of an S_N2 reaction is slower than that of a tertiary halide. Methyl halides easily attain the transition state (**69**, R^1 = R^2 = H) and an S_N2 reaction is very facile. In allyl and benzyl halides, the π-bond participates in removal of the leaving group in the transition state, and both halides react faster than bromomethane.[88]

---

88. (a) Streitwieser, Jr., A. *Chem. Rev.* **1956**, *56*, 571, (b) Streitwieser, Jr., A. *Solvolytic Displacement Reactions*, McGraw-Hill, New York, **1962**, p.13, (c) Reference 155, p.343.

Table 2.8. Reaction Rates of Alkyl Halides with Potassium Iodide

R–Br + K–I $\longrightarrow$ R–I + K–Br	
R–X	**Relative Rate**
MeBr	1
EtBr	2.2×10^{-2}
Me_2CHBr	8.3×10^{-4}
Me_3CBr	5.5×10^{-5}
Me_3CCH_2Br	3.3×10^{-7}
$CH_2=CHCH_2Br$	2.3
$PhCH_2Br$	4.0

[Reprinted with permission from Streitwieser, Jr., A. *Chem. Rev.* *1956*, *56*, 571. Copyright © *1956* American Chemical Society.]

MOLECULAR MODELING: Electrostatic Potential Maps and the Strengths of Nucleophiles

In addition to steric requirements, the effectiveness of a molecule to act as a nucleophile should depend on the extent to which negative charge is localized. Specifically, a molecule in which the charge is delocalized would be expected to be a poorer nucleophile than one where the charge is concentrated on a single atom. One way to rank related molecules is to compare electrostatic potential maps. The electrostatic potential is the energy felt by a positive charge as a function of location in a molecule. The more negative the electrostatic potential, the better able a molecule is to be an effective nucleophile. An electrostatic potential map colors the electron density surface for the molecule, which conveys its overall size and shape, with the value of the potential. By convention, regions on the surface that are colored red indicate areas of high negative potential (attraction of the positive charge) while regions that are colored blue indicate areas of positive potential (repulsion of the positive charge).

One after another, build the hydroxy, methoxy, *tert*-butoxy, trifluoromethoxy, trichloromethoxy and phenoxy anions and obtain their structures and electrostatic potential maps. For each anion following the first, *click* on > to the right of the (rightmost) tab at the top of the screen and select **Continue**. This will bring up a fresh builder palette while keeping this text on screen. Display the maps for all five anions simultaneously on screen. To do this, *check* the box to the left of each of the five tabs at the top of the screen. Manipulate each image in turn to best see the oxygen. It is perhaps easier to do this with transparent surfaces. *Click* on an image and select ***Transparent*** from the **Style** menu at the bottom of the screen. When you are done, return to solid surfaces.

Because these are negative ions, the test charge is never repelled and the maps contain no "blue regions". Focus your attention on the most "red" region (associated with the oxygen) and rank the anions in order of decreasing negative charge. You can measure the potential for a region of a map by clicking on the region. The value will be presented at the bottom of the screen. Which anion should be the "best" nucleophile? Which should be the poorest? Justify your observations.

Optional: Examine other "oxygen" nucleophiles to see if you can identify one that should be stronger than the best of those above.

MOLECULAR MODELING: Local Ionization Potential Maps and the Strengths of Nucleophiles

The previous problem examined the ability of electrostatic potential maps to account for the relative strengths of oxyanions as nucleophiles. The hypothesis was that the more negative the potential at oxygen, the more nucleophilic the anion. An alternative but closely-related measure of nucleophile strength is provided by local ionization potential maps. What is depicted here is not the energy between a positive test charge and the nuclei and electrons of a molecule as a function of location, but rather the energy required to eject ("ionize") an electron from a particular location on the surface. Neither the electrostatic potential nor the local ionization potential can actually be measured, although the energy required to eject an electron from a molecule can be measured (it is the ionization potential).

Examine local ionization potential maps for hydroxy, methoxy, *tert*-butoxy, trifluoromethoxy, trichloromethoxy and phenoxy anions (left to right, top to bottom on screen). (*SpartanModel* cannot calculate local ionization potential maps and these results were obtained from *Spartan*.) By convention, regions on the surface that are colored red indicate areas where while regions that are colored blue indicate regions where ionization is difficult. You are looking to compare regions of low ionization potential ("red" regions). The maps are initially displayed as transparent models to allow you to see the underlying structures. Once you are oriented, switch to solid models to facilitate comparison of the maps. *Click* on one of the maps and select *Solid* from the **Style** menu at the bottom of the screen.

Which anion should be the "best" nucleophile? Which should be the poorest? It might help you to measure the value of the local potential for the appropriate region of a map. *Click* on this region and the value will be presented at the bottom of the screen. Is the ordering of nucleophile strengths the same as that found from electrostatic potential maps?

MOLECULAR MODELING: Solvent Effects on S_N2 Reactions

Quantum chemical calculations refer strictly to isolated molecules, that is, molecules in the gas phase. While it is not yet practical to carry out calculations with explicit solvent, it is possible to empirically correct the gas-phase calculations to account (at least qualitatively) for the effects of solvent. Caution is required as the solvation models are primitive, taking into account only "bulk" effects and ignoring individual interactions, most important, individual hydrogen bonds.

Consider, for example, the role of the solvent on the equilibrium between chloride ion and methyl iodide on one side and methyl chloride and iodide ion on the other. In solution, the reaction as written is known to be *exothermic,* that is, chloride displaces iodide. Iodide is said to be the better *leaving group*.

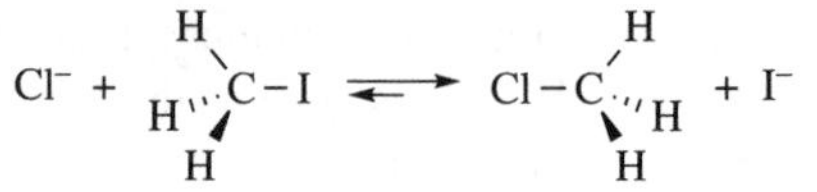

$$Cl^- + \ _{H^{\prime\prime\prime}}\!\!\overset{\displaystyle H}{\underset{\displaystyle H}{}}\!\!C\!-\!I \ \rightleftharpoons \ Cl\!-\!\overset{\displaystyle H}{\underset{\displaystyle H}{C}}_{\prime\prime\prime}H + I^-$$

Examine energy profiles for S_N2 reaction of chloride and methyl iodide obtained from quantum chemical calculations. This starts from "separated" chloride anion and methyl iodide, proceeds through the transition state involving five coordinate carbon, and goes onto methyl chloride and iodide anion. The black curve corresponds to the reaction in the gas, and the red curve corresponds

MOLECULAR MODELING: Solvent Effects on S_N2 Reactions, con't.

to reaction in "water", that is, data on the isolated molecule have been empirically corrected to account for the effects of the solvent. Do the solvent-corrected calculations reproduce the known direction of the equilibrium, that is, is the reaction *exothermic* in solvent? Is the reaction *endothermic*, *exothermic* or thermoneutral in the gas phase?

The electrostatic potential map for one step in the overall reaction sequence is shown on screen. This indicates what a point positive charge "feels" at different regions on the accessible surface of molecule. If the charge is attracted this suggests that the region on the surface is locally "negative". If it is repelled, this suggests that the region is locally positive. By convention, "negatively-charged" regions on the surface are colored "red", "positively-charged" regions are colored "blue" and neutral regions are colored "green". Because this is a negative ion, there are no blue regions. The map is shown as a transparent model, which means that you can also see the underlying structure (Cl on the left and I on the right). You can switch to a solid model by *clicking* on the map and selecting ***Solid*** from the **Style** menu at the bottom of the screen. This will allow you to better assess the transfer of charge as the reaction proceeds.

To "visualize" the transfer of negative charge from chloride to iodide, step through the sequence of electrostatic potential maps using the **Step** keys at the bottom left of the screen, or "animate" the sequence using the **Play** key. Note that this data refers to the gas phase as the solvent corrections discussed above apply only to energies. For which side of the reaction is the negative charge more extreme? Given this result, are the reactants or products likely to be more stabilized by the solvent? Elaborate. Is this consistent with any change in energy between isolated molecule and solvated calculations that you previously observed?

The S_N2 reaction is a powerful method for incorporating functional groups into a carbon skeleton, but the mechanism of the reaction imposes several important synthetic restrictions. The poor reactivity of sterically hindered halides dictates that primary or secondary substrates be used, and inversion of configuration makes the reaction suitable for manipulating stereochemistry. A typical example is the S_N2 reaction of cyanide ion with the primary tosylate, formed by reaction of alcohol **71** with tosyl chloride, to give nitrile **72** in 80% overall yield, taken from Panek and Zhang's synthesis of herboxidiene/GEX 1A.[89]

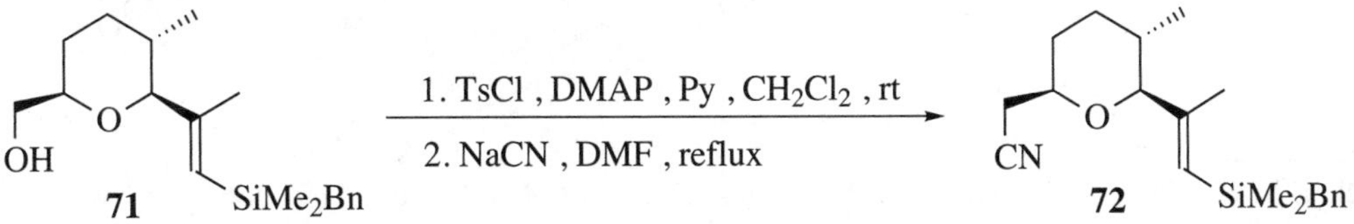

OH
O
SiMe2Bn
1. TsCl , DMAP , Py , CH2Cl2 , rt
2. NaCN , DMF , reflux
CN
O
SiMe2Bn

71 → **72**

An S_N2 reaction using the nucleophilic azide anion is another good way to introduce nitrogen into a molecule, with control of stereochemistry and usually without the complications often accompanying the use of amine nucleophiles (elimination, salt formation, reversibility). An example is the conversion of one hydroxyl unit in diethyl *L*-tartrate **73** to azide **74** in 74% yield,

89. Zhang, Y.; Panek, J.S. *Org. Lett.* ***2007***, *9*, 3141.

with clean inversion of configuration, in Overman and co-worker's synthesis of (–)-sarain A.[90] Note that the polar aprotic solvent DMSO was used to maximize the yield of S_N2 products and suppress side reactions such as elimination (secs. 2.7, 2.9.A). Alkyl azides can be explosive, however, so care should be exercised when azide in used. Typical S_N2 reactions involve functional group transforms such as $Br \rightarrow OMe$, $OTs \rightarrow N_3 \rightarrow NH_2$, and $Cl \rightarrow I$, showing the range and power of this reaction in synthesis. Many other types of nucleophilic reactions are possible, including reactions of enolate anions with alkyl halides (sec. 9.3.A), which proceed via an S_N2 pathway.[91] A classical reaction that involves a S_N2 reaction is the **Williamson ether synthesis**,[92] illustrated by the conversion of the OH unit in **75** to the alkoxide, allowing S_N2 displacement of one bromide in 1,2-dibromodecane to give ether **76** in 65% yield in Thelakkat and co-worker's synthesis of a donor acceptor dyad (donor-bridge acceptor), used for the study of energy- and electron transfer processes in organic semiconductors.[93] Intramolecular Williamson ether syntheses are possible. In a synthesis of galanthamine, Tu and co-workers showed that treatment of **77** with the base DBU converted the –O-SiMe$_2t$-Bu group to the alkoxide (RO^-), which displaced the bromide to form the benzofuran unit in **78**, in 90% yield.[94] The use of microwaves in conjunction with phase transfer catalysis is also quite effective in these reactions.[95]

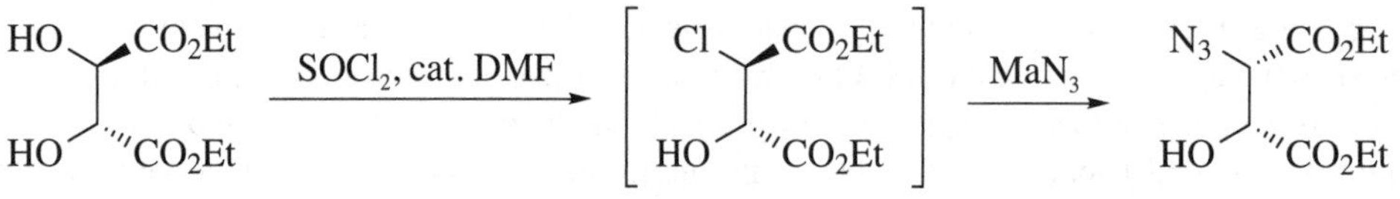

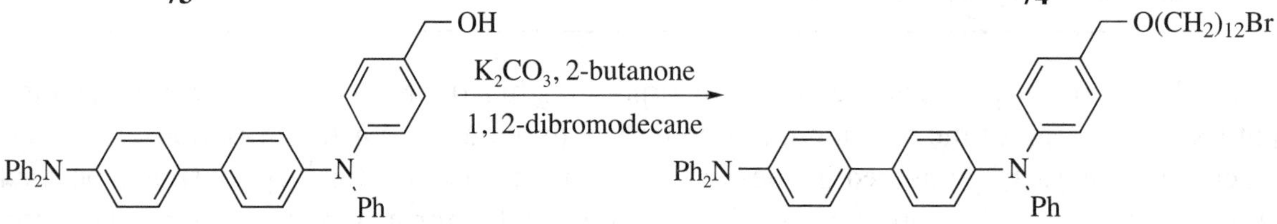

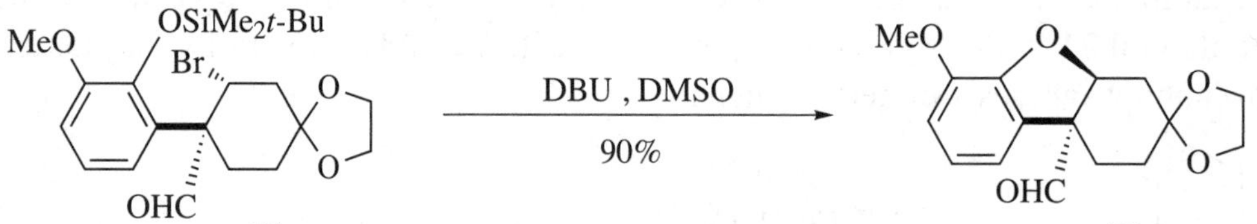

90. Becker, M.H.; Chua, P.; Downham, R.; Douglas, C.J.; Garg, N.K.; Hiebert, S.; Jaroch, S.; Matsuoka, R.T.; Middleton, J.A.; Ng, F.W.; Overman, L E. J. *Am. Chem. Soc.* **2007**, *129*, 11987..

91. For a synthetic example, see Marshall, J.A.; Ellison, R.H. *J. Am. Chem. Soc.* **1976**, *98*, 4312.

92. (a) Williamson, A.W. *J. Chem. Soc.* **1851**, *4*, 229; (b) Dermer, O.C. *Chem. Rev.* **1934**, *14*, 385 (see p 409); (c) *The Merck Index, 14th Ed.* Merck & Co., Inc., Whitehouse Station, New Jersey, **2006**, p ONR-101.

93. Bauer, P.; Wietasch, H.; Lindner, S.; Thelakkat, M. *Chem. Mater.* **2007**, *19*, 88.

94. Hu, X.-D.; Tu, Y. Q.; Zhang, E.; Gao, S.; Wang, S.; Wang, A.; Fan, C.-A.; Wang, M. *Org. Lett.* **2006**, *8*, 1823.

95. For a review of microwave activation in phase transfer catalysis see Deshayes, S.; Liagre, M.; Loupy, A.; Luche, J.-L.; Petit, A. *Tetrahedron* **1999**, *55*, 10851.

The reactions involving **71**, **73**, and **75** used three different types of nucleophiles, cyanide, an azide, and an alkoxide. The ability to evaluate the relative strength and effectiveness of different nucleophiles is obviously as important for S_N2 reactions as it was for nucleophilic acyl substitution reactions. A general list of nucleophiles in order of their ability to form bonds to sp^3 carbon in a S_N2 reaction is:

$$RS^- > ArS^- > S_2O_3^- > (H_2N)_2C{=}S > I^- > CN^- > SCN^- > HO^- > N_3^- > Br^- >$$
$$ArO^- > Cl^- > Pyridine > AcO^- > HOH$$

where the dominant factor is assumed to be polarizability.[101]

Nucleophilic strength depends on two factors, the nature of the substrate (s) and the strength of the nucleophile, measured by its nucleophilicity (n). Nucleophilic strength for a given species is given by the **Swain-Scott equation**[96a] [$\log (k/k_O) = (s)(n)$], where $n = 0$ for water at 25°C, $s = 2.00$ for bromomethane,[97a] and n is a reagent parameter measuring nucleophilic power in a particular system. The parameter s is the susceptibility of the substrate (here an halide) to being attacked (it measures the power of a substrate to discriminate between various nucleophiles). An electrophlicity index has been established based on electronegativity divided by chemical hardness.[96b,c] The important qualitative message of this equation is that nucleophilic strength varies with the substrate. Table 2.9 gives a partial list of nucleophiles,[97] ordered by use of the Swain-Scott equation[96] for their reaction with bromomethane. This list constitutes a general order of nucleophilic strength in S_N2 reactions. It is known that the nucleophilicity order differs for S_N2 reactions when compared with acyl addition to a carbonyl (secs. 2.6.A,B and 2.5), and tends to follow more closely the order established for basicity [the carbonyl carbon is a harder acid than saturated carbon (sec. 2.4.C)].

There are obviously several parameters contributing to nucleophilic strength, and a few general statements can be made. *1.* A nucleophile with a negative charge is stronger than its conjugate acid. $NH_2^- > NH_3$ and $HO^- > HOH$. *2.* In the same row, nucleophilicity parallels basicity, and $R_3C^- > R_2N^- > RO^- > F^-$.

Rules 1 and 2 predict the following order for nucleophilic strength:

$$NH_2^- > RO^- > HO^- > R_2NH > ArO^{-*} > NH_3 > C_5H_5N > F^- > HOH > ClO_4^{-**}$$
* less nucleophilic due to resonance delocalization   ** charge spread out due to resonance, less nucleophilic

---

96. (a) Swain, C.G.; Scott, C.B. *J. Am. Chem. Soc.* **1953**, *75*, 141; (b) Parr, R.G.; Szentpály, L.V.; Liu, S. *J. Am. Chem. Soc.* **1999**, *121*, 1922; (c) Maynard, A.T.; Huang, M.; Rice, W.G.; Covel, D.G. *Proc. Natl. Acad. Sci. USA* **1998**, *95*, 11578.
97. Edwards, J.O.; Pearson, R.G. *J. Am. Chem. Soc.* **1961**, *84*, 16.

Table 2.9. Nucleophilic Strength with Bromoethane as Ordered by the Swain-Scott Equation

Nucleophile	n	Nucleophile	n	Nucleophile	n
$S_2O_3^{2-}$	6.35	SO_3^{2-}	5.67	^-CN	5.13
PO_4^-	5.06	I^-	4.93	^-SCN	4.80
CO_3^{2-}	4.36	HO^-	4.23	^-OPh	4.16
Br^-	4.02	N_3^-	3.92	$^-OPhp-Cl$	3.86
NO_2^-	3.71	HPO_3^-	3.52	$B_2O_7^{2-}$	3.45
$p-NO_2PhO^-$	3.08	Cl^-	2.99	AcO^-	2.76
HCO_2^-	2.62	HSO_3^-	2.3	$H_2PO_4^-$	2.2
F^-	1.88	SO_4^{2-}	2.1	HNO_3^-	2.1
HSO_4^-	2.1	ClO_3^-	2.1	ClO_4^-	2.1
HOH	0.01				

[Reprinted with permission from Edwards, J.O.; Pearson, R.G. *J. Am. Chem. Soc.* **1961**, *84*, 16. Copyright ©*1961* American Chemical Society.]

Nucleophilic strength for a given substituent can be measured in terms of the rate of the S_N2 reaction or the rate in reactions with carbonyl derivatives.[98] The relative rates of several nucleophiles determined by reaction with iodomethane are shown in Table 2.10.[99] Electronic effects are important as illustrated by methoxide, where the electron releasing methyl group makes methoxide more nucleophilic relative to hydroxide. The rate of the S_N2 reaction of sodium hydroxide with iodomethane is 1.3×10^{-4} $M^{-1}s^{-1}$ whereas the rate with sodium methoxide with iodomethane is 2.51×10^2 $M^{-1}s^{-1}$.[99]

Table 2.10. Relative Rates of Reaction of Nucleophiles with Iodomethane

Nucleophile	k (x10^{-3} M^{-1} s^{-1})
MeO^-	0.251
F^-	0.00005
$NHEt_2$	2.2
PhO^-	0.073
NH_3	0.041
pyridine	0.022

Reprinted with permission from Pearson, R.G.; Sobel, H.; Songstad, J. *J. Am. Chem. Soc.*, **1968**, *90*, 319. Copyright *1968* American Chemical Society.

A species is less nucleophilic if there is resonance delocalization of the electron density. An example is the phenoxide anion, where the electrons on oxygen are delocalized away from oxygen by the adjacent phenyl ring, meaning that oxygen cannot donate electrons as effectively. Phenoxide is less nucleophilic than an alkoxide that is incapable of resonance, such as methoxide. Nucleophilic reactions are usually under kinetic control whereas acid-base reactions are under thermodynamic control. Under kinetic control, the most nucleophilic

98. (a) Hudson, R.F.; Green, M. *J. Chem. Soc.* **1962**, 1055, (b) Bender, M.L.; Glasson, W.A. *J. Am. Chem. Soc.* **1959**, *81*, 1590, (c) Jencks, W.P.; Gilchrist, M. *Ibid* **1965**, *90*, 2622.
99. Pearson, R.G.; Sobel, H.; Songstad, J. *J. Am. Chem. Soc.* **1968**, *90*, 319.

species will react faster in substitution reactions.

3. Going down the periodic table, nucleophilicity increases but basicity decreases [I⁻ > Br⁻ > Cl⁻ > F⁻ (this ordering is solvent dependent), and (RS⁻ > RO⁻)].

Comparing the rate of reaction of MeI with PhSe⁻ [7000 x 10^3 $M^{-1}s^{-1}$], PhS⁻ [1070 x 10^3 $M^{-1}s^{-1}$] and PhO⁻ [0.073 x 10^3 $M^{-1}s^{-1}$] shows the trend mentioned in (*3*).[99] Likewise, trialkylphosphines are more nucleophilic than trialkylamines ($R_3P > R_3N$). The rate of reaction of Et_3P with iodomethane is 1.29 x 10^3 $M^{-1}s^{-1}$ whereas the rate for Et_3N is 0.595 x 10^3 $M^{-1}s^{-1}$.[99] These latter reactions are particularly useful in organic synthesis. Phenoxide is commonly used in S_N2 reactions for the preparation of phenyl (or aryl) ethers, a structural unit found in many naturally occurring molecules. Thiophenol is used to generate phenyl sulfides via the thiophenoxide anion, and phenyl selenides can be formed from the phenyl selenide anion. Oxidation of the sulfide or selenide generates the corresponding sulfoxide (*S*-O) or selenoxide (*Se*-O), which are subject to thermal syn elimination to form the corresponding alkene[100] (secs. 2.9.C.v, 2.9.C.vi). These transformations are used to introduce an alkenyl unit into the molecule. Thioalkyl groups are also used to enhance the acidity of the adjacent hydrogen and the chemistry of sulfur-stabilized carbanions is important for reactions that generate carbon–carbon bonds, including many of the Umpolung reagents (sec. 8.6.B).

The reaction of trialkylphosphines, especially triphenylphosphine, with alkyl halides is particularly useful since the resultant phosphonium salts are easily converted to a phosphonium ylid on treatment with a suitable base (sec. 8.8.A). Ylids are, of course, the reactive species in the well-known **Wittig olefination** reaction,[101] which will be discussed in section 8.8.A.i. A related S_N2 process involves reaction of a trialkylphosphite with an alkyl halide, the **Arbuzov reaction** (sometimes called the **Michaelis-Arbuzov reaction**).[102] Triethylphosphite (**79**) reacts with iodomethane to give the phosphonium salt, **80**. Heating generates the monoalkyl phosphonic ester (**81**). This type of phosphonic ester can be converted to an ylid and used in the well-known **Horner-Wadsworth-Emmons olefination**[103] (sec. 8.8.A.iii).

100. DePuy, C.H.; King, R.W. *Chem. Rev.* **1960**, *60*, 431.
101. (a) Wittig, G.; Schöllkopf, U. *Berichte* **1954**, *87*, 1318; (b) Wittig, G.; Haag, W. *Berichte* **1955**, *88*, 1654; (c) Trippett, S. *Quart. Rev. Chem. Soc. (London)* **1963**, *17*, 406; (d) Wittig, G. *Acc. Chem. Res.* **1974**, *7*, 6.
102. (a) Michaelis, A.; Kaehne, R. *Berichte* **1898**, *31*, 1048; (b) Arbuzow, A.E. *J. Russ. Phys. Chem. Soc.* **1906**, *38*, 687; (c) Idem *Chem. Zentr.*, **1906**, *II*, 1639; (d) Arbuzow, B.A. *Pure Appl. Chem.* **1964**, *9*, 307; (e) Kosolapoff, G.M. *Org. React.* **1951**, *6*, 273 (see pp 276-277).
103. (a) Wadsworth, W.S.; Emmons, W.D. *J. Am. Chem. Soc.* **1961**, *83*, 1733; (b) Boutagy, R.; Thomas, Jr., R. *Chem. Rev.* **1974**, *74*, 87.

The analogous but slower reaction with amines can be used to form ammonium salts. When the ammonium salt has a hydrogen atom on the carbon β- to the nitrogen atom, these salts are useful in the **Hoffmann elimination** sequence (sec. 2.9.C.i).[104] Triphenylamine reacts with iodomethane to give methyltriphenylammonium iodide ($MePh_3N^+I^-$) and alkyl ammonium salts such as this can be used as phase-transfer catalysts.[105] Phase transfer ammonium salts often contain large alkyl groups (dodecyl, hexadecyl, and octadecyl), and sometimes small groups. Tetrabutylammonium chloride is a useful phase-transfer reagent.

Substitution by amines or phosphines with an alkyl halide is clearly an S_N2 type process. Normally, the presence of water in the reaction medium suppresses the rate of S_N2 reactions because they usually involve the reaction of a charged nucleophile and a neutral substrate. An exception can occur, however, if both starting materials are neutral but the products are charged, as in the reaction of amines with alkyl halides. This reaction generates transition state **82** with a δ+ nitrogen and a δ– halogen. A protic solvent will separate the developing charge in this transition state, ultimately leading to the solvent separated ions $^+NRH_2R'$ and X^-. Solvation and ion separation therefore accelerates the S_N2 process in this preparation of ammonium salts. Amine alkylation of this type finds its way into synthesis. An interesting example is taken from Loh and Lee's synthesis of (–)-epibatidine,[106] in which the intramolecular displacement of one bromide in **83** gave an 85% yield of **84**. Polyalkylation is sometimes a problem when amines or ammonia react with alkyl halides, and amine surrogates are often used, including phthalimide. In a synthesis of a mitochondrial complex 1 inhibitor,[107] Radeke and co-workers prepared phthalimide **86** from benzylic bromide **85**, by initial coupling with phthalimide in 86% yield followed by three steps to convert the aryl iodide to the hydroxy-butyl unit. Final treatment with hydrazine in refluxing butanol liberated the amino group in **87** in 84% yield, with phthalhydrazide (**88**) as a by-product.

82

104.   (a) Hofmann, A.W. *Annalen* **1851**, *78*, 253; (b) Hofmann, A.W. *Annalen* **1851**, *79*, 11; (c) Brown, H.C.; Moritani, I. *J. Am. Chem. Soc.* **1956**, *78*, 2203; (d) Cope, A.C.; Trumbull, E.R. *Org. Reactions*, **1960**, *11*, 317; (e) Hofmann, A.W. *Berichte* **1881**, *14*, 659; (f) Wall, E.N.; McKenna, J. *J. Chem. Soc. B*, **1970**, 318; (g) Wall, E.N.; McKenna, J. *J. Chem. Soc. C* **1970**, 188; (h) Bach, R.D.; Bair, K.W.; Andrezejewski, D. *J. Am. Chem. Soc.* **1972**, *94*, 8608; (i) Archer, D.A.; Booth, H. *J. Chem. Soc.* **1963**, 322.

105.   (a) Jones, R.A. *Aldrichimica Acta* **1976**, *9*, 35; (b) Starks, C.M. *J. Am. Chem. Soc.* **1971**, *93*, 195; (c) Starks, C.M.; Owens, R.M. *J. Am. Chem. Soc.* **1973**, *95*, 3613; (d) Herriott, A.W.; Picker, D. *J. Am. Chem. Soc.* **1975**, *97*, 2345.

106.   Lee, C.-L.K., Loh, T.-P. *Org. Lett.* **2005**, *7*, 2965.

107.   Radeke, H.; Hanson, K.; Yalamanchili, P.; Hayes, M.; Zhang, Z.-Q.; Azure, M.; Yu, M.; Guaraldi, M.; Kagan, M.; Robinson, S.; Casebier, D. *J. Med. Chem.* **2007**, *50*, 4304.

4. Greater steric encumbrance diminishes nucleophilicity. Comparing reactions with pyridine and 2,6-dimethylpyridine (2,6-lutidine) illustrates this point. The rate of reaction of pyridine with methyl iodide is 0.022×10^3 M^{-1}s^{-1}. The two methyl groups proximal to the nitrogen atom in 2,6-lutidine sterically encumber approach of nitrogen to the carbon of methyl iodide, and the rate declines to 0.00042×10^3 M^{-1}s^{-1}, about 50 times slower.[99] As pyridine approaches the electrophilic carbon, the ortho hydrogens interact with the alkyl groups on that carbon, but this represents a minimal steric interaction when compared to the ortho-methyl groups in 2,6-lutidine Steric hindrance makes lutidine less nucleophilic than pyridine.

A key feature of S$_N$2 reactions is inversion of configuration at the electrophilic center (see **73 → 74** and **83 → 84** above). The S$_N$2 reaction is an effective method for inversion of the stereocenter with insertion of a different functional group. Alcohols are convenient synthetic intermediates, but the OH moiety is a poor leaving group (sec. 2.7.E). Conversion of the OH unit into an ester, sulfonate ester or phosphonate derivative often allows an S$_N$2 type displacement to proceed, with inversion of that stereogenic center.

2.6.A.ii. The Mitsunobu Reaction. If one sets a goal of inverting the stereochemistry of 2(S)-pentanol to form 2(R)-pentanol, there is a problem. Hydroxide will not displace OH, and if the OH unit is converted to a halide or sulfonate ester, treatment with hydroxide can lead to elimination as a major competitive reaction. The use of nucleophile such as acetate would minimize elimination, but acetate is a poor nucleophile for substitution reaction. There are many times when a nucleophile is too weak for facile S$_N$2 displacement, and substitution is also difficult if the substrate contains a poor leaving group. A successful and highly useful solution to this problem is the **Mitsunobu reaction**.[108] This procedure is typically used to invert the

108. (a) Mitsunobu, O. *Synthesis* **1981**, 1; (b) The Merck Index, 14th Ed. Merck & Co., Inc., Whitehouse Station, New Jersey, **2006**, p ONR-62.

stereochemistry of an alcohol by converting a poor leaving group to an excellent leaving group for subsequent reaction with a nucleophile. A synthetic example shows the conversion of (*R*)-alcohol **89** to (*S*)-alcohol **90** in 51% overall yield using diethyl azodicarboxylate (DEAD, **91**), *p*-nitrobenzoic acid and triphenylphosphine, followed by saponification of the resulting ester with LiOH in aqueous methanol in a synthesis of (+)-ricciocarpin A by Liu and Jan.[109] As initially formulated, the process involved reaction of **91** with triphenylphosphine (Ph₃P) to form **92**. Disopropyl azodiacarboxylate (DIAD) is commonly used in this reaction rather than DEAD. Dipolar ion **92** reacts with HX, which is present in the initial reaction, to give phosphonium salt **93**. An alcohol is then added and the subsequent reaction generates alkoxyphosphonium salt **95** and diimide **94**. Displacement of triphenylphosphine oxide (a neutral leaving group) by X

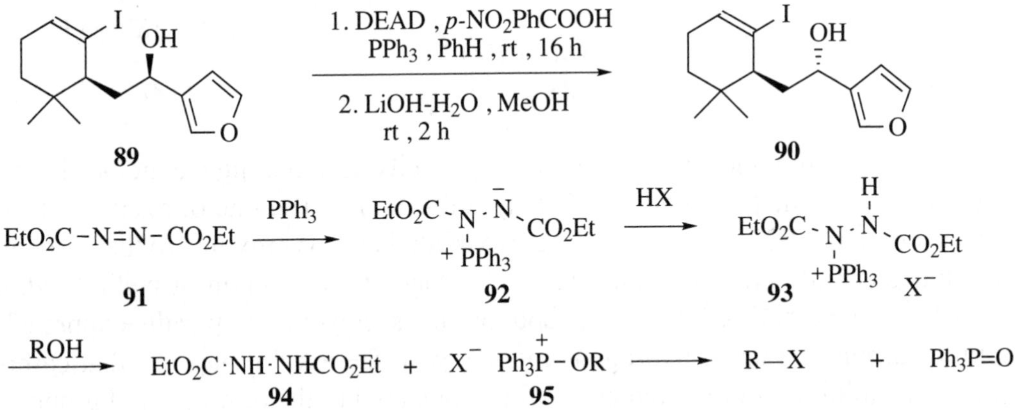

gives the substitution product, R-X. The X moiety in HX can be a variety of groups, including carboxylate, azide, imido, and R (an active methylene compound such as malonate or an α-cyanoester).[108] As suggested by the range of functional groups cited, there are many variations. In one, taken from DeBrabender and co-worker's synthesis of gustastatin,[110] selective etherification of one phenolic OH unit in **96** gave benzyl ether **97** in 73% yield. Note the use of DIAD in this reaction. Lepore and He have shown that ether formation via the Mitsunobu reaction is facilitated with sonication.[111] Ghosh and Moon reported an example of what may be the most common application of the Mitsunobu reaction in an enantioselective synthesis of (+)-jasplakinolide,[112] where the alcohol unit in **98** was converted to the *p*-nitrobenzoate ester with clean inversion of configuration, and then hydrolyzed to the alcohol in **99** in 78% yield. Recent work to improve this reaction has focused on several aspects of the reaction,[113] including the use of alternative reagents, phase-switching modifications of phosphine, azodicarboxylate,

109.   Jan, N.-W.; Liu, H.-J. *Org. Lett* **2006**, *8*, 151.
110.   Garcia-Fortanet, J.; Debergh, J.R.; De Brabander, J.K. *Org. Lett.* **2005**, *7*, 685.
111.   Lepore, S.D.; He, Y. *J. Org. Chem.* **2003**, *68*, 8261.
112.   Ghosh, A.K.; Moon, D.K. *Org. Lett.* **2007**, *9*, 2425.
113.   Dembinski, R. *Eur. J. Org. Chem.* **2004**, 2763.

and nucleophilic components are emphasized and also the separation of products from by-products[114] using fluorous compounds.

Other transformations are possible. The reaction of succinimide with 2-propanol under Mitsunobu conditions gave *N*-isopropyl succinimide in 73% yield.[115] Hydroxyl can also be replaced with other functional groups using this method. Mitsunobu coupling can be used to insert functional groups that might be difficult to prepare under other conditions. As just discussed, reaction of an alcohol and an acid,[116] generates an ester. Azides can be conveniently prepared, as illustrated in Banwell and co-worker's synthesis of (+)-brunsvigine,[117] in which (*S*)-alcohol **100** was converted to (*R*)-azide **101** in 93% yield using Mitsunobu conditions. Note that the reagent diphenylphosphoryl azide (DPPA) was used, since it is often more efficient than the KN_3 or NaN_3 reagents.[118] Alcohols react with phthalimide under Mitsunobu conditions to give the N-substituted phthalimide, which can be unmasked to the primary amine by reaction with hydrazine, as noted above.[119] This latter reaction of phthalimide with alkyl halides followed by treatment with hydrazine is a variation of a classical reaction called the **Gabriel synthesis**.[120] The use of hydrazine in this manner is called the **Ing-Manske modification**[121] of the Gabriel synthesis. The use of Mitsunobu conditions simply expands the reaction to include alcohols as starting materials. The same transformation has been accomplished using $TsNHCO_2t$-Bu (abbreviated TsNHBoc) to incorporate a nitrogen moiety.[122]

114. Dandapani, S.; Curran, D.P. *Chem. Eur. J.* **2004**, *10*, 3131.
115. Reference 108, p 5.
116. For an example, see Keck, G.E.; McHardy, S.F.; Murry, J.A. *J. Org. Chem.* **1999**, *64*, 4465.
117. Banwell, M.G.; Kokas, O.J.; Willis, A.C. *Org. Lett.* **2007**, *9*, 3503.
118. See Lal, B.; Pramanik, B.N.; Manhas, M.S.; Bose, A.K. *Tetrahedron Lett.* **1977**, *18*, 1977.
119. For examples, see (a) Brosius, A.D.; Overman, L.E.; Schwink, L. *J. Am. Chem. Soc.* **1999**, *121*, 700; (b) Mulzer, J.; Scharp, M. *Synthesis* **1993**, 615.
120. (a) Gabriel, S. *Berichte* **1887**, *20*, 2224; (b) Gibson, M.S.; Bradshaw, R.W. *Angew. Chem. Int. Ed.* **1968**, *7*, 919; (c) Mundy, B.P.; Ellerd, M.G.; Fabaloro Jr., F.G. *Name Reactions and Reagents in Organic Synthesis 2nd Ed.* Wiley, New York, **2005**, pp 264; (d) The Merck Index, 14th Ed. Merck & Co., Inc., Whitehouse Station, New Jersey, **2006**, p ONR-36.
121. Ing, H.R.; Manske, R.H.F. *J. Chem. Soc.* **1926**, 2348.
122. Trost, B.M.; Oslob, J.D. *J. Am. Chem. Soc.* **1999**, *121*, 3057.

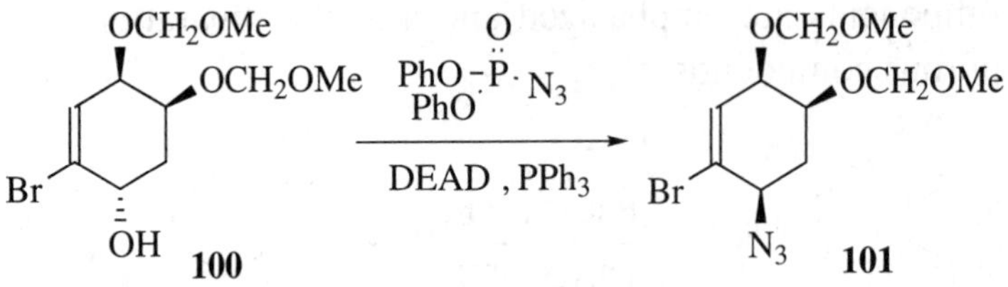

Another modification of the Mitsunobu reaction illustrates two features that can be useful in synthesis. When **102** was treated with triphenylphosphine and DEAD, cyclization occurred to give benzoxepin **103** in 87% yield.[123] This example shows that Mitsunobu conditions can be used for intramolecular reactions, and also that an alcohol can be coupled with another OH unit. Yamaguchi and co-workers used this transformation in a synthesis of radulanins.[123]

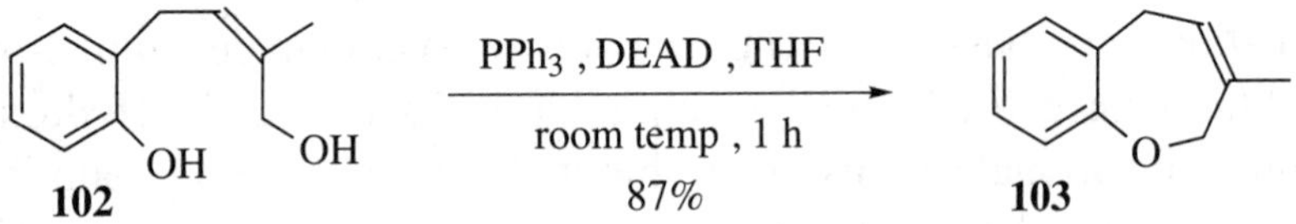

The Mitsunobu reaction is a clever and highly useful exploitation of the S_N2 reaction that allows one to control the stereochemistry of a given stereogenic center (see Chap. 6). It also allows one to correct the stereochemistry if an alcohol is produced that has the opposite stereochemistry from that desired (sec. 6.2.B).

2.6.A.iii. The S_N2' Reaction. Normal S_N2 reactions were illustrated in the previous sections. In general, allylic halides react faster with nucleophiles than do simple alkyl halides (see above) because the π bond can participate in expulsion of the leaving group. If the leaving group bearing

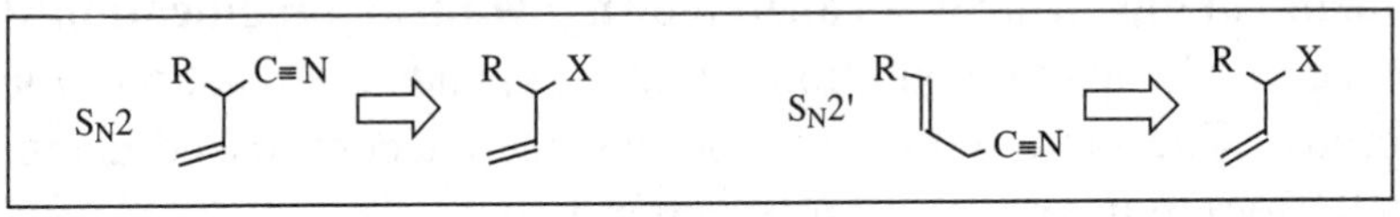

carbon becomes too sterically hindered in the allylic system, however, the normal S_N2 path is inhibited and an alternative mode of attack occurs, at the end of the π system. Such attack is termed an **S_N2' reaction**[124] (nucleophilic bimolecular substitution with allylic rearrangement). The retrosynthetic analysis of each reaction, using cyanide as the nucleophile, offers a useful comparison, especially for regiochemical differences in the product distribution. The two pathways often compete and can lead to annoying mixtures of regioisomers, although the reaction may be synthetically useful. Formation of piperidine derivative **105** via the S_N2' displacement of the allylic 3,5-dichlorobenzoate (**104**) is an excellent example of this process.[125]

123. Yamaguchi, S.; Furihata, K.; Miyazawa, M.; Yokoyama, H.; Hirai, Y. *Tetrahedron Lett.* **2000**, *41*, 4787.
124. Magid, R.M. *Tetrahedron* **1980**, *36*, 1901.
125. (a) Stork, G.; White, W.N. *J. Am. Chem. Soc.* **1953**, *75*, 4119; (b) Stork, G.; White, W.N. *Ibid* **1956**, *78*, 4609; (c) Stork, G.; Kreft III, A.F. *Ibid* **1977**, *99*, 3850, 8373.

The nucleophile attacks the π bond since the *O*-benzyl carbon is sterically encumbered. The π bond assists the removal of the leaving group in an S_N2' reaction.

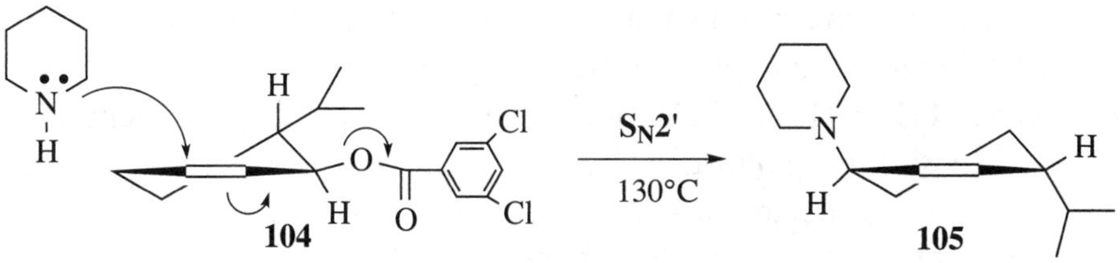

The S_N2' reaction proceeds with a high degree of stereochemical control, as in the nucleophilic displacement **104 → 105**. Bordwell and co-workers described the S_N2' reaction with amines as second order, but invoked a tight ion pair to explain the process.[126] A tight ion pair will have significant ionic character, but the ions are not solvent separated. Smith and coworker's work with cyclopropylcarbinyl halides supports this mechanistic rationale. Reaction of piperidine with the *tert*-butyl derivative (**106**)[127] gave the expected **107** in 88% yield but also 8% of **109**. Formation of this latter product is not compatible with a simple S_N2 displacement due to the great steric hindrance imposed by the neopentyl-like structure, but the tight ion pair mechanism explains the presence of **108** very well. When it can be applied as the major process, stereochemical and regiochemical control makes the S_N2' reaction a useful tool in synthetic planning. With other allylic substrates, the possibility of this reaction as a competitive side reaction must be acknowledged as a process that can diminish the yield of the S_N2 product.

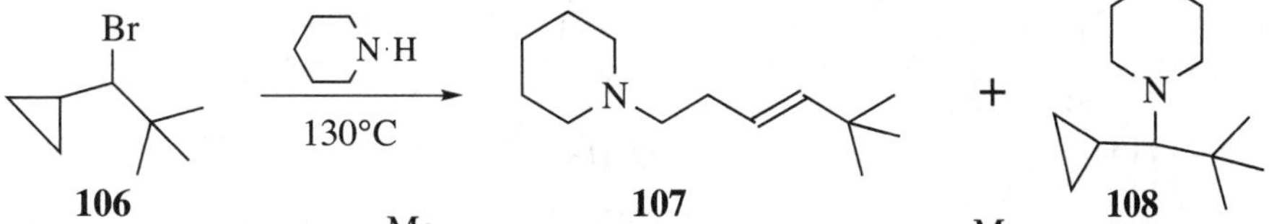

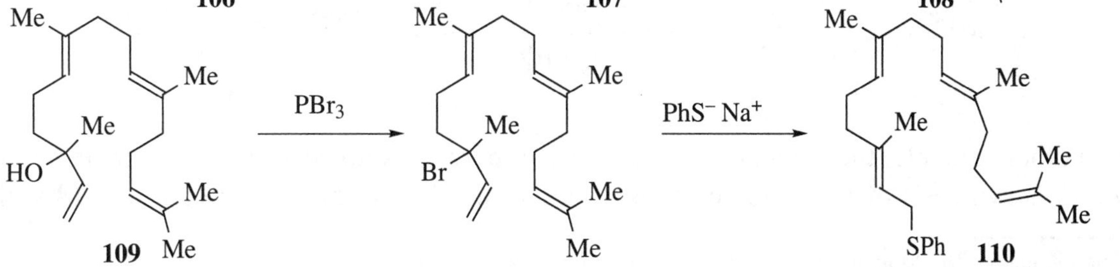

A synthetic example of an S_N2' reaction is the conversion of **109** to a tertiary bromide with phosphorus tribromide. Subsequent reaction with the sodium salt of thiophenol gave an S_N2'

126. (a) Bordwell, F.G.; Pagani, G.A. *J. Am. Chem. Soc.* **1975**, *97*, 118; (b) Bordwell, F.G.; Mecca, T.G. *Ibid* **1975**, *97*, 123, 127; (c) Bordwell, F.G.; Wiley, P.F.; Mecca, T.G. *Ibid* **1975**, *97*, 132.
127. (a) Smith, M.B.; Hrubiec, R.T.; Zezza, C.A. *J. Org. Chem.* **1985**, *50*, 4815; (b) Hrubiec, R.T.; Smith, M.B. *Tetrahedron Lett.* **1983**, *24*, 5031.

displacement to give **110** in Itô and co-worker's synthesis of neocembrene.[128]

Organocuprates (sec. 8.7.A) react via an S_N2' type process in some cases.[129] The organocuprate reaction may actually proceed by copper stabilized radical intermediates, rather than a formal S_N2' mechanism.[130] Magnesium cuprates also undergo S_N2' reactions, as in the conversion of tertiary allylic acetate **111** to **112**, in 72% yield.[131] Note that the displacement proceeded with good diastereoselectivity for the anti-diastereomer **112**. An S_N2' like reaction has also been observed in reactions of organocuprates with cyclopropylcarbinyl halides, where the strained cyclopropane ring behaves similarly to the π bond of an alkene.[132]

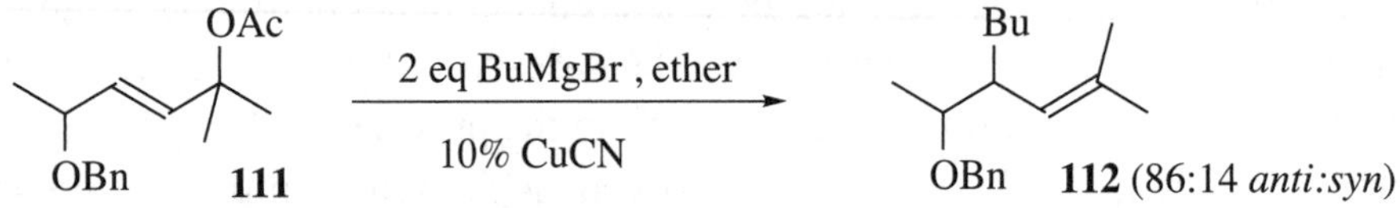

2.6.A.iv. S_N2 **Type Reactions with Epoxides.** Epoxides are highly reactive, and many nucleophiles can open the three-membered ring to give a substituted or functionalized alcohol. Nucleophilic opening on an epoxide in aprotic solvents is best described as an S_N2 reaction. The nucleophile attacks the less sterically hindered carbon atom in **113** to give alkoxide **114**, and subsequent protonation gives alcohol **115**. In Feldman and Saunders' synthesis[133] of (–)-agelastatin A, epoxide **116** was treated with sodium azide, and the resulting alkoxide was protonated with aqueous ammonium chloride to give azido alcohol **117**.

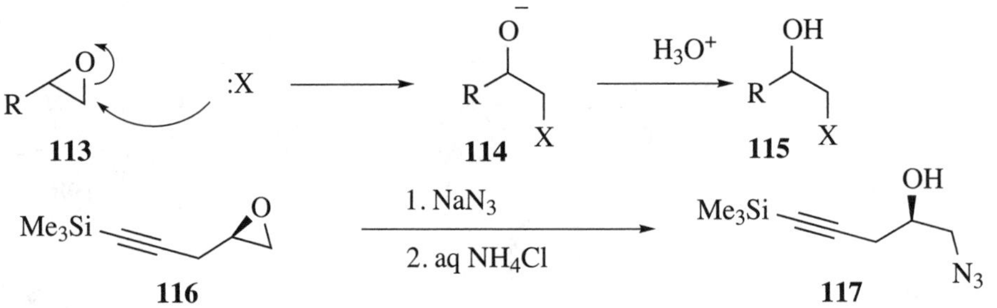

In another example, the nucleophile attacked the more substituted site. In Baskaran and co-worker's synthesis of indolizidine 167B,[134] initial reaction of the epoxide unit in **118** with a

128. Kodama, M.; Matsuki, Y.; Itô, S. *Tetrahedron Lett.* **1975**, 3065.
129. For older examples, see (a) Kreft, A. *Tetrahedron Lett.* **1977**, 1035; (b) Trost, B.M.; Tanigawa, Y. *J. Am. Chem. Soc.* **1979**, *101*, 4413.
130. (a) Johnson, C.R.; Dutra, G.A. *J. Am. Chem. Soc.* **1973**, *95*, 7777, 7783; (b) Ashby, E.C.; DePriest, R.N.; Tuncay, A.; Srivastava, S. *Tetrahedron Lett.* **1982**, *23*, 5251.
131. Belelie, J.L.; Chong, J.M. *J. Org. Chem.* **2002**, *67*, 3000.
132. (a) Hrubiec, R.T.; Smith, M.B. *Tetrahedron* **1984**, *40*, 1457; (b) *Idem J. Org. Chem.* **1984**, *49*, 385; (c) Posner, G.H.; Ting, J.-S.; Lentz, C.M. *Tetrahedron* **1976**, *32*, 2281.
133. Feldman, K.S.; Saunders, J.C. *J. Am. Chem. Soc.* **2002**, *124*, 9060.
134. Reddy, P.G.; Varghese, B.; Baskaran, S. *Org. Lett.* **2003**, *5*, 583.

Lewis acid (BF_3) generated an oxonium ion, and reduction of the azide gave a primary amine, and attack of the ion paired system at the more hindered site gave **119** in 50% overall yield.

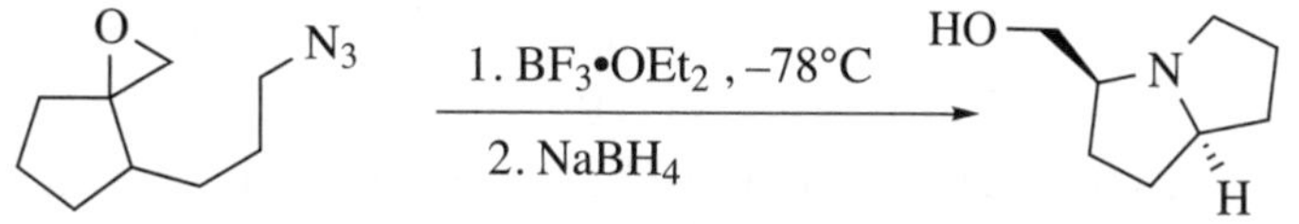

2.6.B. UNIMOLECULAR NUCLEOPHILIC SUBSTITUTION

Unimolecular substitution (ionization followed by substitution, S_N1) is generally less useful in synthesis, but may be the best specific reaction to give a particular functional group exchange. In Chapter 12 we will see many examples where cationic reactions are useful. Unimolecular substitution occurs as a side reaction in aqueous media, and the extent of ionization can influence the yield, stereochemistry and regiochemistry of the final product.

2.6.B.i. Carbocations (Carbenium ions). A carbocation is an electron-deficient reaction intermediate bearing an sp^2 hybridized carbon that is, in effect, an empty p orbital. Such species are now called carbenium ions, but since we are focusing on the ionic nature of intermediates the older term will be used throughout. Carbocations are key electrophilic intermediates in many ionization reactions, as will be discussed in Chapter 12, and react with many suitable nucleophiles. From a mechanistic standpoint, ionization of an alkyl halide or a sulfonate ester is a unimolecular process since the rate-determining step is not collision of two molecules but rather slow ionization of the C–Y bond with loss of the leaving group Y. Ionization of an alkyl halide, for example, leads to cation **92** and Y^-. Carbocation **120** will react with the best available nucleophile, which can be Y^- from the ionization process, the solvent or an added nucleophile. The overall substitution reaction is a two-step process (ionization followed by substitution) that is termed S_N1 (a unimolecular nucleophilic substitution).[135] Electrostatic attraction between the positive carbon of the carbocation and the (–) or (δ–) site of the nucleophile makes the second reaction (coupling) much faster than the initial ionization reaction. For ionization, the term leaving group is a bit of a misnomer since it does not leave but is pulled off, usually by the solvent by solvation and separation of the resulting ions.

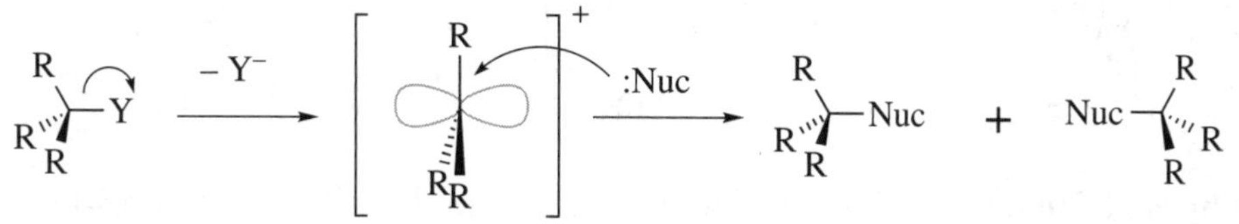

135. (a) Bateman, L.C.; Hughes, E.D.; Ingold, C.K. *J. Chem. Soc.* **1940**, 960, 1011, (b) Harris, J.M. *Prog. Phys. Org. Chem.* **1974**, *11*, 89.

The central carbon of carbocation **120** is sp² hybridized and trigonal planar and, as mentioned above, is conveniently viewed as being a trisubstituted carbon atom with an empty p orbital. The relative stability of carbocations is briefly discussed below and again in section 12.2.A in connection with forming carbon-carbon bonds via carbocation intermediates. Since the carbocation has an sp² carbon, **120** has a trigonal planar shape and an attacking nucleophile can approach from either face. This model predicts a racemized substitution product, but this simple model is misleading because stereochemical retention or inversion is often observed in S$_N$1 reactions. If the initial ionization generates a tight ion pair (represented by **121**), then the leaving group (Y) provides some steric hindrance to the attacking nucleophile and the inverted substitution product (**122**) may predominate, although it is usually not the exclusive product. If the leaving group (Y) can chelate to or otherwise coordinate with the nucleophile, then an ion pair such as **123** may result, and delivery of the nucleophile will be from the same face as the leaving group. Such ion-pairing will lead to a product with retention of configuration (see **124**). Depending on the solvent, a S$_N$1 reaction may have **120**, **121** or **123** as an intermediate,

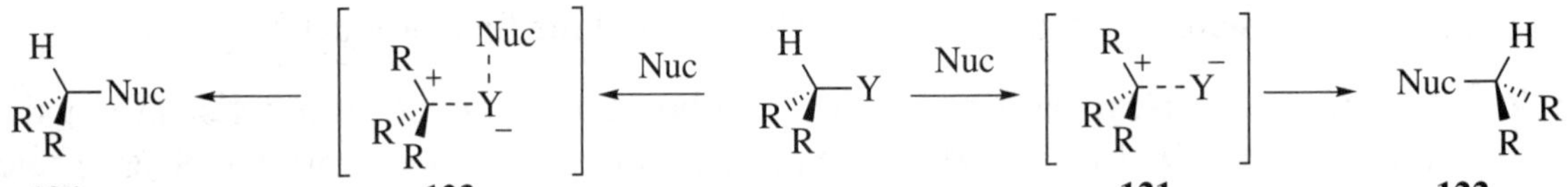

but in general the reaction proceeds with significant if not complete racemization at the electrophilic center in acyclic and simple monocyclic compounds. Polycyclic molecules can be an exception, however, since the conformational constraints of the ring system may allow only one face to react with an attacking nucleophile. Adamantyl alcohol (**125**, for example, gives carbocation **126** upon treatment with aqueous HBr. The rigid skeletal structure of **126** limits approach of the nucleophilic bromide to the top of the molecule as shown since one side of the cationic carbon is sterically blocked. As a result, ionization of **125** leads to exclusive formation of **127**.[136] In general, cationic reactions of this type can be highly stereoselective in substitution reactions if there is steric bias for one face over the other. In the absence of such steric bias, these reactions often show poor to moderate stereoselectivity.

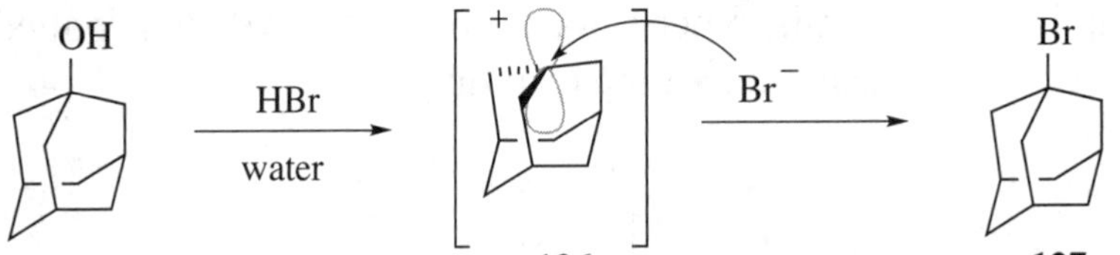

2.6.B.ii. The S$_N$1 Reaction. The mechanistic rationale described above results from many years of experimental observations about the S$_N$1 reaction (statements 1-6 below).[135] In general,

136. Dubowchik, G.M.; Padilla, L.; Edinger, K.; Firestone, R.A. *J. Org. Chem.* **1996**, *61*, 4676.

the S_N1 process occurs in those cases where water is present as a solvent or co-solvent, there is a good leaving group, and the substrate is tertiary or secondary. A S_N1 reaction can occur in any polar protic solvent such as methanol, ethanol, or acetic acid, but ionization is much slower in these solvents relative to water because they are not as efficient for the separation of ions (sec. 2.6.B). If ionization is relatively slow, faster reactions (S_N2 or elimination processes) often predominate. In the absence of competing reactions, however, ionization reactions are quite common, and solvolysis of halides in alcohol solvents is quiet common. It is convenient to *assume* that S_N1 reactions occur only in aqueous media, and use this assumption to predict product distributions for a given set of reaction conditions. *As with any assumption, this one fails to accurately predict the products in many cases, but it provides a good working model to begin analyzing reactions.* The ultimate test, of course, is the actual experimental result. The S_N1 process is characterized by the following:

1. The rate of reaction with tertiary halides that is much faster than with primary halides,[137] which is shown in Table 2.11 for reaction of bromoalkanes with potassium iodide in aqueous media.[137]

2. The rate equation can be described using only the concentration of the halide and not that of the nucleophile. The rate determining (slow) step is ionization of the C-Br bond and the fast second step does not greatly influence the overall rate. The rate expression is Rate $= k$ [RX] for this unimolecular reaction.[129]

3. Reaction with chiral halides proceeds with significant racemization rather than with inversion (see above).

4. A cationic intermediate (**120**) is formed rather than the transition state noted for an S_N2 reaction (see **69**).

5. The reaction often proceeds with rearrangement of the carbon skeleton of the organic substrate.

6. Unless water is a solvent or co-solvent, the initial ionization reaction is rather slow and may not be competitive with other possible reactions.

Table 2.11. Reaction of Alkyl Halides with Potassium Iodide in Aqueous Media

R–Br	Relative Rate
MeBr	1.0
EtBr	1.0
Me$_2$CHBr	11.6
Me$_3$CBr	1.2×10^6

---

137. Reference 88b, p 43.

A carbocation is the product of ionization, and its stability is critical to the overall rate of the S_N1 process.[138] A general ranking of carbocations by their relative stability is[138]

$$R_3C^+ > H_2C=CHCH_2^+ \approx PhCH_2^+ > R_2CH^+ > RCH_2^+ > CH_3^+$$

This ranking is based on the electron-releasing inductive properties of the carbon groups attached to the positive center. The three alkyl groups on the tertiary cation release electrons to partially diminish the point charge of the cation, leading to greater stability of the tertiary cation relative to the primary, which has only one electron releasing group. Both allyl and benzyl cations disperse the charge by resonance and are more stable than primary alkyl or secondary alkyl cations. In general, tertiary alkyl cations are probably more stable than a primary allylic cation, although a secondary or tertiary allylic cation will be more stable than a tertiary alkyl. Cation stability is also discussed in section 12.2.A.

With cations ranked according to their relative stability, we then assume that given a choice the more stable cation will be formed in a reaction because it is lower in energy. The addition of HCl to methylenecyclobutane (**128**) illustrates this point, where the final product is 1-chloro-1-methylcyclobutane (**131**), used by Fitjer and Mandelt in a synthesis of cuparene.[139] The π bond reacts as a base, donating a pair of electrons to the acid (H⁺ = HCl). When the new C–H bond is formed, a carbocation is generated at the other carbon of the π bond, and there are two possible cations, **129** and **130**. The electron-releasing effects of the alkyl groups make the tertiary cation (**129**) more stable than the primary cation (**130**). The reaction will show a preference for **129**, which then traps the nucleophilic chloride (generated by cleavage of HCl after transfer of the proton to the alkene) to give the major product, **131**. The stability of the intermediate cation will largely determine the course of unimolecular reactions.

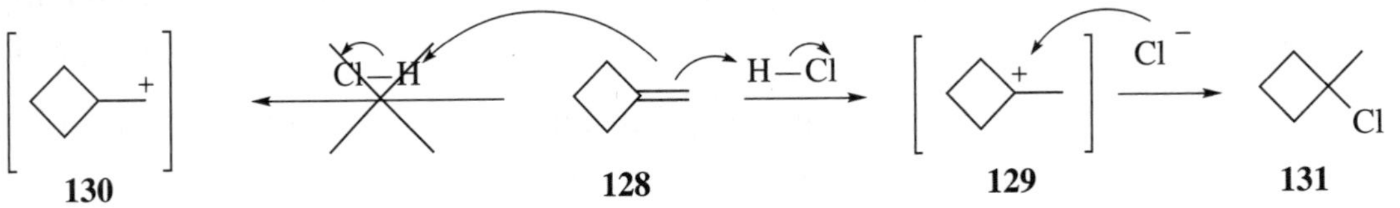

As mentioned, secondary allylic and secondary benzylic cations are more stable than secondary alkyl cations. This stability is due to participation of the π bond in the ionization process and also by stabilization of the cationic intermediate by resonance. The resonance delocalization in both the allyl cation (**132**) and the benzylic cation (**133**) diminishes the net charge at the cationic carbon, making them more stable and easier to form. Although they are more stable, they are highly reactive intermediates and easily react with nucleophiles in a S_N1 reaction. See Figure 12.12.1, sec. 12.2.B, for a more detailed analysis of the relative stability of various

138. (a) Olah, G.A.; Olah, J.A. *Carbonium Ions, Vol. 2*, Olah, G.A.; Schleyer, P.v.R., Ed. Wiley, New York, *1969*, pp 715-782, (b) Richey, J.M. *The Chemistry of Alkenes*, Vol. 2, Zabicky, J., Ed. Interscience, New York, *1970*, p 44.
139. Mandelt, K.; Fitjer, L. *Synthesis 1998*, 1523.

types of carbocations.

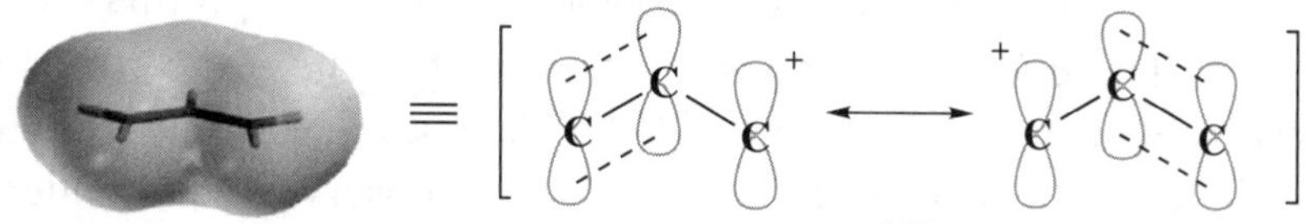

132

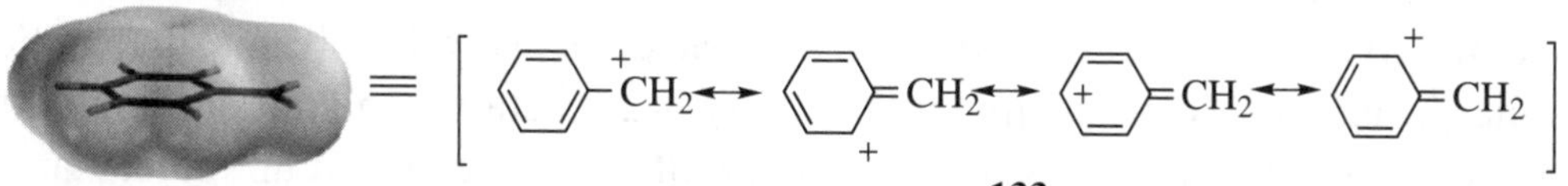

133

When heteroatoms are attached to a positive carbon, donation of electrons (back donation) to the electron deficient center leads to greater stability, often by resonance delocalization. Two common examples are oxygen-stabilized cations such as **134**, and sulfur-stabilized cations such as **135**. In both cases, back donation from the heteroatom leads to resonance stabilized cations, as shown. Both **134** and **135** are considered to be more stable than a simple tertiary carbocation, but the relative stability really depends on the nature of the groups of the sp^2 carbon atom. Direct protonation of a carbonyl is the most common method for generating oxo-stabilized cations such as **134**. Carbonyl protonation leads to the formation of enamines and imines from ketones and aldehydes by reaction with amines,[140] although the key intermediate in both cases is an ammonium ion that functions as the active proton-transfer species (the actual acid catalyst). Another method for generating this cation is by addition of acid to a vinyl ether. Reaction of **136**, for example, gives **137** rather than the tertiary cation **138**, an indication of the greater stability of that oxygen-stabilized cation. Similar results are observed upon protonation of thiocarbonyls or thioenol ethers. As noted above, sec. 2.2.B will provide a more detailed discussion of the stability of heteroatom-stabilized carbocations.

134   **135**

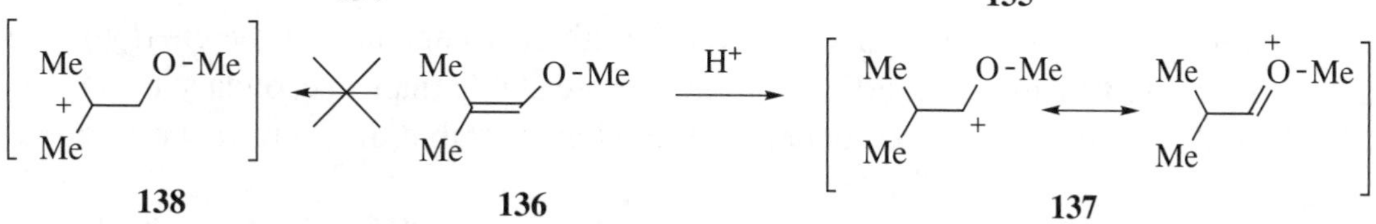

138   **136**   **137**

Relief of steric strain influences the rate of S_N1 reactions, although the effect is relatively small. When R is large, the planar carbocation has less steric hindrance than the tetrahedral precursor,

140. (a) Hayne, L.W. in *Enamines: Synthesis Structure and Reactions*, Cook, A.G. (Ed.), Marcel Dekker, New York, *1969*, pp. 55-100; (b) Herr, M.E.; Heyl, F.W. *J. Am. Chem. Soc.* *1953*, *75*, 5927; (c) *Idem Ibid* *1952*, *74*, 3627; (d) Heyl, F.W.; Herr, M.E. *Ibid* *1953*, *75*, 1918.

accelerating the ionization process. For example, EtMe$_2$CCl is ionized to the corresponding carbocation 1.7 times faster than Me$_3$CCl, Et$_2$MeCCl is ionized 2.6 times faster, and Et$_3$CCl 3.0 times faster.[141g] These ionization rates illustrate that as the steric crowding in the halide increases, the rate of ionization is accelerated.[141g] This type of crowding, called **back strain** by Brown,[141d] contributes to the increased rate of ionization observed with hindered halides.[142d]

In general, S$_N$2 reactions are faster than S$_N$1 for primary and secondary halides, and tertiary halides do not give S$_N$2 reactions but do give S$_N$1 reactions in protic solvents. In reactions with nucleophiles, primary halides commonly give only S$_N$2 reactions, and tertiary halides undergo S$_N$1 processes. Secondary halides give a mixture of bimolecular and unimolecular products in aqueous solvents, and the relative amounts of each type of product are dependent upon the nucleophile, the substrate, and (of course) the ionizing power of the solvent.

2.6.B.iii. Cationic Rearrangements. Reactions that generate cationic intermediates have a potential liability that makes the choice of an S$_N$1 reaction unsuitable for many syntheses. Carbocations can undergo skeletal rearrangement to give a thermodynamically more stable cation, *prior* to trapping the nucleophile. This means that $k_{rearrangement} > k_{trapping}$ for this type of reaction. If a σ bond is adjacent to the cationic center and parallel to the p-orbital, as illustrated in **139**, electron density is donated from the σ-bond. As the electrons in the bond distort toward the positive center, the C-H σ bond weakens and the adjacent orbitals begin to share electron density, as illustrated in **140**. As the electrons in the σ bond shift, the atom attached to carbon (H in the case of **139** but it can also be another carbon group) is transferred to the adjacent carbon via transition state **140**. As this occurs, a new carbocation is formed as well as a new C-H bond (see **141**) to complete the molecular rearrangement (a 1,2-hydride shift). If we examine **139** and **141**, the product of the rearrangement is a more stable tertiary cation, generated from a

139   140   141

less stable secondary cation. The driving force for this rearrangement is the energy gained in forming a more stable cation. A tertiary cation is more stable than a secondary by 11-15 kcal mol^{-1} (46.0-62.7 kJ mol^{-1}) and a secondary cation is more stable than a primary by about the

141.   (a) Fry, J.L.; Engler, E.M.; Schleyer, P.v.R. *J. Am. Chem. Soc.* **1972**, *94*, 4628; (b) Schleyer, P.v.R.; Nicholas, R.D. *Ibid* **1961**, *83*, 182; (c) Tanida, H.; Tsushima, T. *Ibid* **1970**, *92*, 3397; (d) Brown, H.C. *J. Chem. Soc.* **1956**, 1248; (e) Brown, H.C.; Borkowski, M. *J. Am. Chem. Soc.* **1952**, *74*, 1894. (f) Brown, H.C.; Fletcher, R.S. *J. Am. Chem. Soc.* **1949**, *71*, 1845; (g) Harris, J.M.; Wamser, C.C. *Fundamentals of Organic Reaction Mechanisms*, Wiley, New York, **1976**, p118.

142.   (a) Fittig, R. *Ann.* **1860**, *114*, 54; (b) Collins, C.J. *Quart. Rev.* **1960**, *14*, 357; (c) Mundy, B.P.; Otzenberger, R.D. *J. Chem. Educ.* **1971**, *48*, 431; (d) Büchi, G.; MacLeod, Jr., W.D.; Padilla, O.J. *J. Am. Chem. Soc.* **1964**, *86*, 4438.

same amount.[143] The substituent can only be transferred (a 1,2-hydride shift for **139 → 141**) if the electrons in the σ bond are parallel to the p orbital of the cation.[144] As mentioned above, alkyl groups can move (called 1,2-alkyl shifts) and aryl groups can be transferred (called 1,2-aryl shifts). The energy barrier to transfer of hydrogen is lower than that for alkyl or aryl shifts, but the energy barrier for a 1,2-methyl shift is estimated to be < 5 kcal mol⁻¹ (20.9 kJ mol⁻¹).[145] For the degenerate rearrangement of **142A** to **142B**, the energy barrier for a 1,2-hydrogen shift was shown to be 3.4 kcal mol⁻¹ (14.2 kJ mol⁻¹).[146] Although the difference in energy between a hydrogen shift and a methyl shift is small, hydrogen will migrate preferentially in virtually all cases. The cationic rearrangements mentioned (1,2-hydride shifts, 1,2-alkyl shifts and 1,2-aryl shifts) all occur with great facility,[138] to form the more stable carbocation.

| **142A** | **142B** | **143** | **144** |

The rearrangement just described is not limited to migration of a hydrogen atom. Indeed, many alkyl or even aryl groups (see sec. 12.2.C) can move, as illustrated by the rearrangement of **143** to **144**. In **142**, hydrogen was shifted in preference to a bulkier ethyl group, but in **143** one of the adjacent carbon atoms has three hydrogen atoms and the other carbon has three attached ethyl groups. In this case, migration of a hydrogen from the methyl group would lead to a less stable primary carbocation, and does not occur. To form a more stable tertiary carbocation, one of the ethyl groups must migrate and the 1,2-ethyl shift occurs to give **144**. In general, the smaller substituent migrates preferentially, as in hydrogen versus ethyl in **142**. Rearrangements such as those just described must occur faster than the S_N1-type coupling reaction[147] between the nucleophile and the carbocation. The chemistry of carbocations, including rearrangements, is discussed in more detail in section 12.2.B.

A synthetically useful example of this process is the conversion of **145** to **148**, which involves a 1,2-alkyl shift, and was part of Hwu and Wetzel's synthesis of (−)-solavetivone.[148] The alkyl fragment is actually part of the bicyclic ring system, one arm of the bicyclo[4.4.0]decane ring system. Reaction of the OH unit with the Lewis acid resulted in formation of tertiary cation **146**. A 1,2-alkyl shift gave **147**, where the new cation was stabilized by the well-known

143. (a) Lossing, F.D.; Semeluk, G.P. *Can. J. Chem.* **1970**, *48*, 955; (b) Radom, L.; Pople, J.A.; Schleyer, P.v.R. *J. Am. Chem. Soc.* **1972**, *94*, 5935; (c) Arnett, E.M.; Petro, C. *Ibid* **1978**, *100*, 5408.

144. Reference 6, pp 427-434.

145. Olah, G.A.; Lukas, J. *J. Am. Chem. Soc.* **1967**, *89*, 4739.

146. (a) Saunders, M.; Cline, G.W. *J. Am. Chem. Soc.* **1990**, *112*, 3955; (b) Saunders, M.; Kates, M.R. *Ibid* **1978**, *100*, 7082.

147. Reference 138a, pp 766-778.

148. Hwu, J.R.; Wetzel, J.M. *J. Org. Chem.* **1992**, *57*, 922.

β-effect of adjacent silicon of the trimethylsilyl group.[149] Loss of the trimethylsilyl group from **147** gave the final product, spiran **148**.

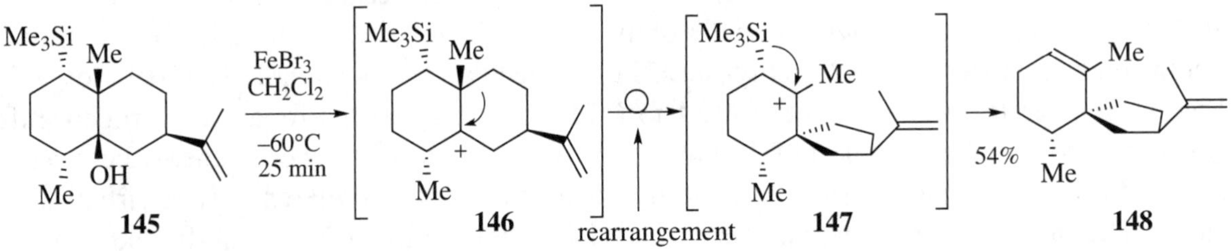

A second synthetic example illustrates several useful features of cation rearrangements. When **149** (OPiv is a pivaloyl ester) was treated with *p*-toluenesulfonic acid in methanol, protonation occurred at the ether oxygen to generate oxonium ion **150**. Fragmentation of the adjacent C-O bond generated resonance stabilized acyl cation **151**, a more stable cation. There are two reactive hydroxyl groups in **151**, but the acyl cation in **151** can react with one hydroxyl unit to generate a six-membered ring or the other to generate a five-membered ring. The latter mode of attack was favored, and **152** was isolated in 90% yield from this reaction as a part of Danishefsky and co-worker's synthesis of elutherobin.[150]

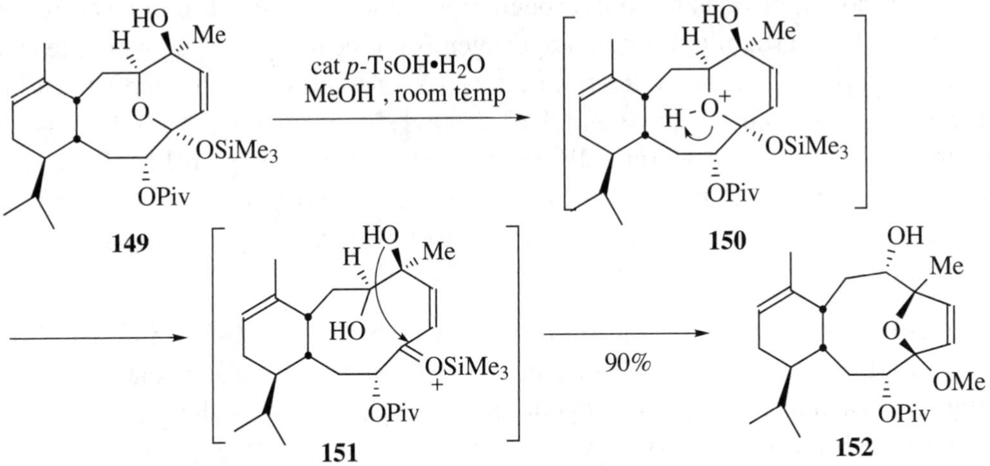

Marshall and co-worker's transformation of the [4.1.0]-bicycloheptane unit in **153** to the

---

149. See (a) Davidson, A.H.; Fleming, I.; Grayson, J.I.; Pearce, A.; Snowden, R.L.; Warren, S. *J. Chem. Soc. Perkin Trans. 1* ***1977***, 550; (b) Fleming, I.; Paterson, I.; Pearce, A. *Ibid* ***1981***, 256; (c) Fleming, I.; Patel, S. *Tetrahedron Lett.* ***1981***, *22*, 2321; (d) Fleming, I.; Michael, J.P. *J. Chem. Soc. Chem. Commun.* ***1978***, 245; (e) Roush, W.R.; D'Ambra, T.E. *J. Am. Chem. Soc.* ***1983***, *105*, 1058.
150. Chen, X.-T.; Bhattacharya, S.K.; Zhou, B.; Gutteridge, C.E.; Pettus, T.R.R.; Danishefsky, S.J. *J. Am. Chem. Soc.* ***1999***, *121*, 6563.

cycloheptene unit in **155**[151] is an example of a reaction that raises mechanistic questions. After protonation of the alcohol unit in **153**, loss of water generated carbocation **154**. A 1,2-alkyl shift to give the cyclopropane unit and trapping of carboxylate would give lactone **155**. Alternatively, nucleophilic attack by the carboxyl oxygen on the cyclopropane ring would open the ring with concomitant expulsion of the leaving group (water) to give **155** without invoking the presence of cation **154**. This latter mechanistic pathway is a synchronous process rather than a stepwise ionic process, and cannot be ruled out based on the formation of the observed product. A tight ion pair in which the ions are not solvent separated offers another reasonable pathway, where the carboxylate anion would react with the cationic center as described above. Additional experimental information about this transformation is required to determine the precise mechanistic pathway. Note that cyclopropylcarbinyl cations such as **154** rapidly open to homoallylic derivatives or cyclobutane compounds,[152] and the cyclopropane ring probably assists in expulsion of the leaving group (in this case an $-OH_2^+$ moiety).

The presence of a free carbocation or the operation of a synchronous mechanism is an important mechanistic consideration when planning any sequence involving cations. Divergent mechanisms can have different influences on the product distribution and stereochemistry of a reaction, and it is important to understand those differences. A detailed kinetic study would presumably determine if a reaction was concerted, stepwise, or a mixture of both. Alternatively, mechanistic studies can be found by examining the literature, or by experimental methods if a synthetic outcome is at all in doubt. Indeed, many synthetic organic chemists regularly publish mechanistically oriented papers. This information will usually solve questions problems with a planned transformation.

2.7. CHARACTERISTICS OF REACTIONS INVOLVING NUCLEOPHILES

It is apparent from section 2.5 that many factors influence the course of nucleophilic reactions, including the nature of the nucleophile, its electron donating capability, the solvent, the substrate

---

151. (a) Marshall, J.A.; Tuller, F.N.; Ellison, R.H. *Synth. Commun.* **1973**, *3*, 465. (b) Marshall, J.A.; Ellison, R.H. *J. Org. Chem.* **1975**, *40*, 2070; (c) Marshall, J.A.; Ellison, R.H. *J. Am. Chem. Soc.* **1976**, *98*, 4312.
152. (a) Julia, M.; Noël, Y.; Guégan, R. *Bull. Chim. Soc. France*, **1968**, 3742; (b) Julia, M.; Noël, Y. Ibid **1968**, 3749, 3756; (c) Brady, S.F.; Ilton, M.A.; Johnson, W.S. *J. Am. Chem. Soc.* **1968**, *90*, 2882; (d) Poulter, C.D.; Winstein, S. *Ibid* **1970**, *92*, 4282; (e) Poulter, C.D.; Friedrich, E.C.; Winstein, S. *J. Am. Chem. Soc.* **1970**, *92*, 4274.

and the nature of the leaving group. When substitution (sec. 2.6.A,B) and elimination (sec. 2.9.A,B) in an aliphatic substrate are discussed, these reactions compete if the nucleophiles used are also bases (MeO⁻ for example). The factors mentioned above influence the extent of this competition. It would be very useful to have a list of parameters for such situations that allow one to make predictions. This section will focus on several factors that influence both nucleophilic and elimination reactions. Analysis of these factors lead to key assumptions that allows one to predict the major product in many cases.

2.7.A. The Solvent

A solvent should solubilize all reactants and also absorb excess heat that may be liberated by the reaction. A very important property of a solvent is polarity, which largely determines its ability to solvate and separate ions (solvation). Solvation is an important factor in most nucleophilic reactions, and is very important in acid-base reactions as seen in sec. 2.2. A good measure of the ability to separate ions is the dielectric constant.[153,154] Table 2.12[153,154] shows the dielectric constants for several common solvents, where a high dielectric constant means the solution can conduct a current, which is associated with an increase in ion formation (the solvent facilitates ionization).

Table 2.12[154] Dielectric Constants and Relative Polarity of Common Organic Solvents

Protic Solvent	Dielectric Constant		Aprotic Solvent	Dielectric Constant
HOH	78.5	**POLAR**	$H(C=O)NH_2$	109.0
DOD	78.25		DMSO	46.68
EtOH (80% aq)	67		DMF	36.71
HCO_2H	58.5		MeCN	37.5
MeOH	32.7		$MeNO_2$	35.87
EtOH	24.55		HMPA	30
n-BuOH	17.8		acetone	20.7
NH_3	16.9		Ac_2O	20.7
t-BuOH	12.47		pyridine	12.4
phenol	9.78		*o*-dichlorobenzene	9.93
$MeNH_2$	9.5		CH_2Cl_2	9.08
$MeCO_2H$	6.15		THF	7.58
			ethyl acrylate	6.02
			chlorobenzene	5.71
			ether	4.34
			CS_2	2.64
			NEt_3	2.42
			benzene	2.28
			CCl_4	2.24
			dioxane	2.21
		NONPOLAR	hexane	1.88

[Reprinted with permission from Lowry, T.H.; Richardson, K.S. *Mechanism and Theory in Organic Chemistry, 3rd ed.*, Harper and Row, New York, *1987*, pp. 177-181. Copyright © 1987 by Harper Collins.]

---

153. (a) Parker, A.J. *Quart. Rev. Chem. Soc.* **1962**, *16*, 163, (b) Parker, A.J. *Chem. Rev.* **1969**, *69*, 1.

154. (a) Reference 6, pp 177-181; (b) Cowdrey, W.A.; Hughes, E.D.; Ingold, C.K.; Masterman, S.; Scott, A.D. *J. Chem. Soc.* **1937**, 1252. For a discussion of solvation energy, see (c) van Bochove, M.A.; Bickelhaupt. F.M. *Eur. J. Org. Chem.* **2008**, 649.

Table 2.12 is further divided into two classes of solvents, those that contain an acidic proton (X–H, **protic**) and those that do not possess an acidic proton (**aprotic**). Within each class, the higher the dielectric constant is associated with the more polar solvent. The essential difference between protic and aprotic solvents is the ability of protic solvents to solvate *both* cations and anions, whereas aprotic solvents efficiently solvate *only* cations. Ionization is favored only when both ions are solvated, allowing them to be separated. The O and the H of the polarized OH unit in the protic solvents water or alcohols are not sterically encumbered, which allows the δ^-O end to solvate cations and the δ^+H end to solvate anions (via hydrogen bonding). The negative dipole of aprotic solvents can solvate cations, but the positive dipole is sterically hindered and anion solvation is poor.

If the solvent is water or contains water, the bimolecular (collision) processes between a neutral substrate and a charged nucleophile (such as nucleophilic acyl addition reactions and nucleophilic displacement with alkyl halides) are generally slow due to solvation effects. Water is an excellent solvent for the solvation and separation of ions, however, so unimolecular processes (which involve ionization to carbocations, see sec. 2.6.B.i) may be competitive. If the solvent is not water, ionization is much slower. Ionization certainly occurs in alcohols or solvents like acetic acid, but the process is so slow that other reactions may occur faster. Based on this observation, we can *assume* that ionization (unimolecular reactions) will be competitive in water, but not in other solvents, leading to the *assumption* that bimolecular reactions should dominate in solvents other than water. This statement is clearly an *assumption*, and it is not entirely correct since ionization can occur in ethanol, acetic acid, and so on, but the assumption is remarkably accurate in many simple reactions and it allows one to begin making predictions about nucleophilic reactions. Even if those predictions are shown experimentally to be incorrect, one begins to understand the factors that are important to each reaction. In those cases where the assumption fails and the products are different from those predicted, the analysis will lead to a better understanding of the overall process. Remember, however, that the various devices used in this section to predict the major product are *assumptions*.

The solvation effect of water is based on its ability to surround the nucleophilic and electrophilic centers (by hydrogen bonding to the electrophilic H or the nucleophilic O: HO–H----X$^-$ and H_2O----X$^+$), making collision more difficult in a bimolecular process. For ionization to compete with bimolecular processes, the counterion should be a weak nucleophile and/or a weak base in most cases. If the counterion is too basic, it can induce elimination that can be faster than ionization. If the counterion is too nucleophilic, bimolecular substitution processes will be faster.

In an ionization reaction involving water, the δ^+C and the X$^{\delta^-}$ groups coordinate with the oxygen and proton of water, respectively. Such coordination "pulls" the C–X bond, thus increasing the bond polarity, allowing greater solvation by the water. Solvation of the polarized atoms eventually leads to separation of the developing charges, which accelerates the ionization process and stabilizes the ion products. The net result is ionization of the C–X bond into a carbocation and the X⁻ anion, both solvated by water. In an aprotic solvent such as diethyl ether, solvation of the cation is possible but the anion is not solvated very well. In practical terms, the solvent does *not* help to pull the leaving group away in the rate determining step so ionization is much slower and generally not competitive with bimolecular processes. Even in protic solvents such as ethanol, the dielectric is so small that charge separation is inefficient, again slowing the rate determining ionization step in a unimolecular reaction.

In nucleophilic substitution reactions of alkyl halides, a pentacoordinate transition state (see **69**) is formed at the midpoint of this bimolecular process. No intermediate has ever been observed. If water is the solvent, separating charges will hinder formation of the transition state, which will slow both bimolecular substitution and elimination reactions. A large number of nucleophiles and bases are ionic, and they will be solvated by water. The solvated nucleophile must penetrate its own solvent sheath as well as any solvation sphere associated with the electrophile to reach the carbon for collision in a bimolecular process, which effectively reduces its nucleophilicity. Nucleophilic strength is, therefore, dependent on the solvent. Aprotic solvents solvate cations but the nucleophile is not well solvated. Therefore, the nucleophile has virtually no solvent sheath and can approach the electrophilic center with greater ease, effectively increasing the nucleophilicity. Using an aprotic solvent therefore allows the maximum rate possible for the bimolecular process. In all cases, the more polar the solvent, protic or aprotic, the more pronounced will be the solvation effect. Solvation will influence the height of reaction barriers, often due to changes in the solvation energy of the transition state.[154c] "Variations in ΔG_{Esolv} (transition state) are larger than those in ΔG_{Esolv}(reactants)."[154c]

2.7.B The Substrate

Bimolecular processes in general are dependent on the nature of the substrate (the molecule containing the electrophilic center). Bimolecular processes generally dominate reactions of primary alkyl halides. The low lying transition state available for bimolecular substitution (secs. 2.6.A,B) is responsible for this observation, combined with the high energy required to form a primary cation (sec. 2.6.B.i). Substitution reactions tend to be faster than elimination for primary halides, even in protic solvents. Exceptions occur when a very hindered base (or nucleophile) such as DBN or DBU is used (**26** and **21**, respectively), since approach of those reagents to the carbon bearing the leaving group is difficult.

In reactions of tertiary halides, unimolecular processes dominate in protic solvents (especially water and aqueous solvents) where substitution is usually faster than elimination. A carbocation is formed in the ionization process and substitution involves nucleophilic attack directly at a positive carbon, whereas elimination involves removal of an acidic hydrogen two bonds removed from the positive carbon. Clearly, a nucleophilic species is strongly attracted to the most positive center and that should lead to the major product. In aprotic solvents, bimolecular substitution is not observed for tertiary halides due to the high energy required to form the pentacoordinate transition state (sec. 2.6.A.i). Under conditions that favor bimolecular reactions, and in the presence of a suitable base, elimination is the dominant process.

For secondary halides in aqueous solvents, unimolecular and bimolecular processes compete, and the result is usually a mixture of products. With a strong base and a protic solvent other than water, bimolecular elimination is usually faster than substitution, although this is only an assumption and accurate predictions can be difficult with secondary substrates. In polar aprotic solvents, bimolecular processes are usually faster. If a strong base is present and a protic solvent is used, bimolecular elimination is usually preferred to bimolecular substitution, but this is another assumption. These assumptions can be summarized by saying that elimination is preferred in protic solvents and substitution in aprotic solvents, for secondary halides and sulfonate esters. If ethanol is used as a solvent and sodium ethoxide is a nucleophilic base, Table 2.13 shows the competition between bimolecular substitution (S_N2) and bimolecular elimination (E2) for a series of alkyl bromides. The preference for E2 reactions of secondary and tertiary halides in this protic solvent is clearly shown.

Table 2.13.[155] Relative Rates of Substitution vs. Elimination for Alkyl Bromides

Halide	k (S_N2)	k (E2)	% Alkene
EtBr	172	2.6	0.9
n-PrBr	54.7	5.3	8.9
i-PrBr	0.058	0.237	80.3
t-BuBr	0.1	4.17	97

[Reprinted with permission from Hine, *J. Physical Organic Chemistry*, 2nd ed., McGraw-Hill, New York, *1962*, p. 203. Copyright *1962*, McGraw-Hill.]

2.7.C. The Nucleophile-Base

In a bimolecular process, the stronger the nucleophile, the greater the attraction for $^{\delta+}C$, which is favored in aprotic solvents. When the nucleophile is also a base, the rate of elimination will increase relative to substitution as base strength increases. In general, high nucleophilicity and/or high basicity leads to a competition between S_N2 and E2. If the attacking species is a strong nucleophile but a weak base, the reaction will proceed almost entirely by S_N2 when primary and secondary substrates are involved, but tertiary substrates show elimination or no reaction at all depending on the strength of the base. With low nucleophilicity, high basicity species, E2 dominates for tertiary and secondary substrates. Even primary halides can undergo elimination with a nonnucleophilic base such as DBN (**26**) or DBU (**21**).

For unimolecular reactions, unimolecular substitution (S_N1, sec. 2.6.B.ii) is favored with strong, non-basic nucleophiles in aqueous media. Unimolecular elimination (E1, sec. 2.9.B) processes are favored with bases of weak to moderate strength that are also poor nucleophiles. Such bases

---

155.   Hine, J. *Physical Organic Chemistry*, 2nd Ed. McGraw-Hill, New York, *1962*, p 203.

include the hydrogen sulfate anion and the perchlorate anion, which are resonance stabilized and very poor nucleophiles. In general, the S_N1 reaction is faster than the E1 reaction since the cationic carbon is most strongly attracted to the electron rich center, but this assumption is complicated by the fact that the S_N1 product is often unstable and/or difficult to isolate. The hydrogen sulfate or perchlorate S_N1 products are classical examples of unstable substitution products, and in such a case, the E1 process may lead to the major product..

2.7.D. Concentration of the Reacting Species

For bimolecular processes, the rate of reaction is given by the expression: Rate $= k$ [nucleophile or base] [RX], where k is the rate constant. Increasing the concentration of the nucleophile, halide, or base leads to an increase in rate. Decreasing the concentration will decrease the rate. To ensure a rapid and complete reaction, an excess of base or nucleophile is often used.

For unimolecular processes, the rate of the reaction is given by Rate $= k_1$ [RX] $+ k_2$ [nucleophile or base], because this is a two-step process, ionization followed by substitution. The two steps (two reactions) are (1) ionization of the substrate to form a cation (rate $= k_1$ [RX]) and (2) reaction of the cation with the nucleophile or base to produce the product (rate $= k_2$ [R$^+$] [nucleophile or base]). For the unimolecular reaction, the overall rate is actually shown to be Rate $= k$ [RX], where the observed rate constant (k) is consistent with the first process (k_1) being very slow and the second (k_2) being so fast that changes in k_2 have little or no effect. Ionization is indeed the slow step and trapping the resulting carbocation with a nucleophile is very fast, so ionization is the rate-determining step in the reaction. An increase or decrease in the concentration of nucleophile or base, therefore, has only a minor effect on the overall rate of the reaction in most cases. Unimolecular reactions are commonly done in an aqueous solvent mixture (aq THF, aq acetone, etc.) with as a high a concentration of substrate and nucleophile or base as is possible in the aqueous medium.

2.7.E. The Leaving Group

A leaving group (X) is the atom or group of atoms displaced in a substitution or elimination reaction. A good leaving group should have a relatively weak and polarized C-X bond. After departure, X should be a very stable ion, or a stable neutral molecule such as water. If this ion or molecule can be effectively solvated, its leaving group ability is enhanced. In general, gegenions derived from strong acids are good leaving groups and those derived from weak acids are poor leaving groups. Typical good leaving groups that generate ions after their departure are Br, I, OTs, OMs, and Cl and typical poor leaving groups are OAc, OH and NH_2, although acetate is a common leaving group for many reactions and it is significantly better than hydroxyl or amino. What this really says is that more vigorous conditions are required for OAc to leave than would be required for Br to leave. In general, this statement is true. Although ⁻OH and ⁻NH_2 are poor leaving groups, alcohols can be protonated to generate oxonium ions, where water is a good leaving group. Similarly, amines can be protonated to form an ammonium ion, and ammonia can be a good leaving group.

Halides are probably the most common leaving groups. A general order of leaving group ability for halides is I > Br > Cl >>> F. Large atoms such as iodide form a C–I unit, which has a longer bond than C–F so one expects iodide to be a better leaving group than fluoride. The longer bond is generally weaker due to less electron density between the nuclei, and easier to break (it is actually displaced by the nucleophile). After the halide departs, the larger iodide is better solvated by the solvent, diminishing the net charge of the ion and making it more stable, which assists the overall reaction. The role of iodide as a leaving group is complicated by the fact that it is a better nucleophile than fluoride (sec. 2.6.A.i). If an excess of iodide is used in a nucleophilic displacement reaction with a chiral allylic iodide, the iodide functions as both a leaving group and a nucleophile, and iodide displaces iodide with successive S_N2 inversions leading to racemic product. The less basic the group, the more easily it departs. Table 2.14[156] provides a limited rate comparison of leaving group ability in a specific S_N2 reaction.

Table 2.14. Leaving group ability in the reaction of haloethanes and ethoxide:

Et–X + EtO⁻ ⟶ Et–O–Et + X⁻	
X	**Relative Rate (k_X/k_{Br})**
Cl	0.0024
Br	2.0
I	2.9
OTs	3.6

[Reprinted with permission from Lowry, T.H.; Richardson, K.S. *Mechanism and Theory in Organic Chemistry, 3rd ed.*, Harper and Row, New York, *1987*, pp. 374. Copyright © 1987 by Harper Collins.]

The problems of nucleophilic leaving groups such as iodide (see above) could be alleviated if the leaving group were not nucleophilic. Tosylate [–OSO$_2$–C$_6$H$_4$(4–Me)] and mesylate (–OSO$_2$–CH$_3$) groups are derived from sulfonic acids and they are good leaving groups. However, the resonance-stabilized anion (RSO_3^-) has a diffuse charge and is a poor nucleophile. Indeed, sulfonate esters possess several qualities of a good leaving group: (*1*) resonance stabilization of the departed ion, (*2*) relief of steric strain upon leaving due to the bulk of the group, and (*3*) the relative weakness of the C-O (sulfonate) bond

Groups involving second-row elements are usually better leaving groups than comparable first row element groups. Charged species (which generate neutrals after departure) are better leaving groups than neutral leaving groups (which generate charged species after departure).

These preferences are illustrated by $RS^- > RO^-$ and $R_2O^+ >>>>> RO^-$.

In general, good leaving groups are large, of low nucleophilicity, more stable after departure,

---

156.   Reference 6, p 374.

and of low electronegativity. High polarizability with large atoms such as iodide or sulfur is usually associated with weaker bonds and better leaving group ability. Poor leaving groups are small, very nucleophilic, and highly electronegative.

2.8. SUBSTITUTION BY HALOGEN

Halogen can be introduced into a molecule is several ways, including substitution of one halogen atom for another. An important method involves replacement of the OH unit of an alcohol with halogen. The hydrogen atom of an alkyl fragment, particularly an allylic or benzylic fragment can be replaced with halogen by a free-radical mechanism. Both of these transformations will be examined in this section.

2.8.A. Halogenation of Alcohols

The reaction of a tertiary alcohol with HBr or HCl to give a halide via a carbocation intermediate is a general reaction in which alcohols react with mineral acids (HCl, HBr, and HI) to give the corresponding alkyl halide.[157] We can explain the process by a simple example in which 1-methyl-1-cyclohexanol (**156**) reacts with HBr to initially give oxonium ion **157**, which loses H_2O to produce tertiary methylcyclohexyl cation **158**. This cation traps the gegenion of the acid (Br$^-$ in this case) to give **159**.

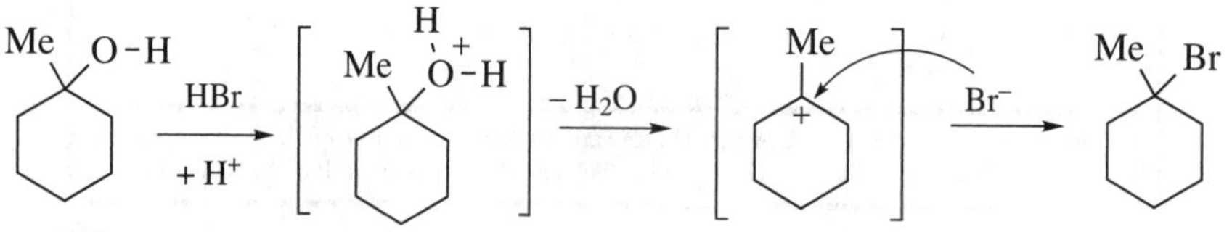

Reaction of 1-butanol with HBr also gives the expected primary halide but the reaction follows an S_N2 mechanism, with bromide displacing $^+OH_2$ in the sequence **160** → **161** → **162**. The high energy barrier for ionization of **161** to the relatively unstable primary cation, coupled with the low energy demands to the S_N2 transition state lead to this preference. For secondary alcohols, both mechanisms can operate simultaneously and competitively. In aqueous media, the cationic mechanism usually dominates for secondary substrates, but if no solvent is used (neat) or if the reaction occurs in aprotic solvents the S_N2 mechanism dominates. The mechanism obviously depends on the specific substrate and the reagents used, as well as on the reaction conditions. As a functional group exchange, however, reaction of primary, secondary or tertiary alcohols with HBr or HCl is an excellent method for the preparation of alkyl bromides or chlorides.

---

157. For examples see Ref. 3, pp 689-696.

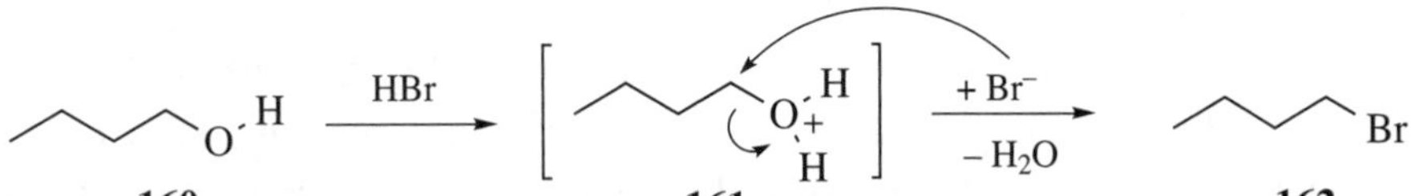

Several halogenating reagents other than HBr, HCl, or HI have been discovered that convert alcohols to halides,[157] usually sulfur or phosphorus halides or oxyhalides. Two of the most common are thionyl chloride ($SOCl_2$) and thionyl bromide ($SOBr_2$). Thionyl chloride reacts with an alcohol to form an alkyl chlorosulfite (**163**).

Subsequent loss of $ClSO_2^-$ generates a cation that dissociates to a tight ion pair (represented by **164**), and chloride is transferred to carbon intramolecularly with retention of configuration.[158] Chlorine is unable to approach the rear of the cation and the ion pair does not dissociate sufficiently to allow a S_N1-type reaction. In this case, 2(R)-butanol was converted to 2(R)-chlorobutane. The mechanistic designator for this process is (Substitution Nucleophilic Internal), S_Ni, and requires that the leaving group possess an atom capable of being transferred to the carbon substrate.[159] When primary alcohols are treated with thionyl chloride, the chloride is usually formed in excellent yield. An example, taken from Chai and co-worker's synthesis of hybocarpone,[160] reacted benzylic alcohol **165** with thionyl chloride to give a 96% yield of benzylic chloride **166**. Subsequent reaction with KCN in DMSO gave **167** in 99% yield, in a classical S_N2 reaction (see 2.6.A.i).

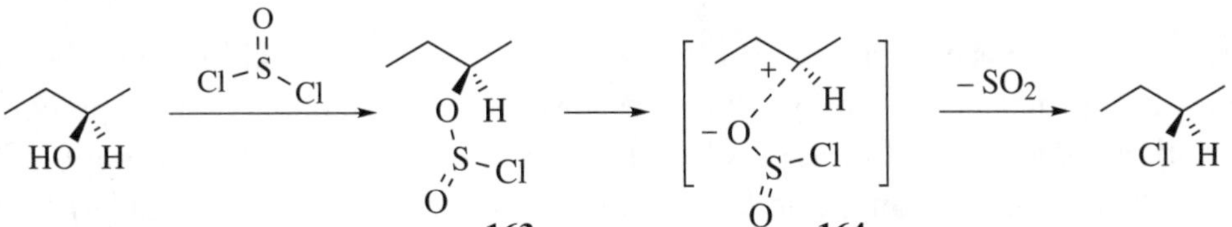

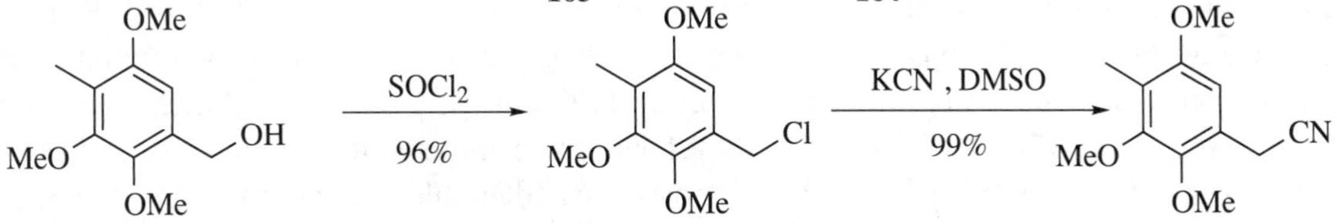

If a basic amine such as pyridine is added to this reaction mixture, the stereochemistry of the final product is different. Initial reaction with alcohol **168** gives **169**, but pyridine (or another basic tertiary amine) reacts with the HCl by-product to give pyridinium hydrochloride (see

158. (a) Lee, C.C.; Clayton, J.W.; Lee, D.G.; Finlayson, A.J. *Tetrahedron* **1962**, *18*, 1395; (b) Lee, C.C.; Finlayson, A.J. *Can. J. Chem.* **1961**, *39*, 260.
159. (a) Reference 7, p 420; (b) Lewis, E.S.; Boozer, C.E. *J. Am. Chem. Soc.* **1952**, *74*, 308.
160. Chai, C.L.L.; Elix, J.A.; Moore, F.K.E. *J. Org. Chem.* **2006**, *71*, 992.

170). The nucleophilic chloride ion is now in solution rather than being lost as HCl, and it displaces the chlorosulfite unit in an S_N2 reaction that gives alkyl chloride **171** with inversion of configuration. Reaction of an alcohol with thionyl chloride and pyridine is called the **Darzens' procedure**.[161] A modification that improves the stereoselectivity premixes the alcohol with thionyl chloride and pyridine, and the resulting sulfinate is treated with thionyl bromide.[162] Thionyl bromide converts alcohols to the bromide,[163] but the extent of inversion or retention is not as clean as with thionyl chloride. Thionyl chloride is a common reagent for the R–OH → R–Cl transformation. An acid by-product is produced, however, and sensitive systems (particularly allylic alcohols) may be subject to acid catalyzed rearrangements and side reactions.

Phosphorus halides give the ROH → RX transformation for primary, secondary, and tertiary alcohols. The most common reagents are phosphorus trichloride (PCl_3), phosphorus pentachloride (PCl_5), phosphorus oxychloride (phosphoryl chloride, $POCl_3$), phosphorus tribromide (PBr_3), and phosphorus pentabromide (PBr_5). The chloride reagents give less rearrangement in reactions with alcohols than is observed with HBr or HI. Some rearrangement is commonly observed with secondary alcohols, even with the phosphorus reagents.

Phosphorus trichloride is commonly used for the preparation of acid chlorides[164] but less often for the conversion of alcohols to chlorides because other reagents are more practical. This contrasts with phosphorus tribromide, which is a common reagent for conversion of alcohols to bromides.[165] A typical example is taken from a synthesis of the C(1)–C(19) fragment of tetrafibricin by Roush and Lira,[166] in which the reaction of allylic alcohol **172** with PBr_3 and pyridine gave bromide **173** in good yield. An alternative and convenient method for converting alcohol to bromides uses a mixture of carbon tetrabromide and triphenylphosphine. An example is the conversion of **174** to **175** in 96% yield, taken from Jamison and Woodin's synthesis of pumiliotoxins 209F and 251D.[167] Triphenylphosphine and *N*-chlorosuccinimide (NCS, see sec. 2.8.B) converts allylic alcohols to allylic chlorides.[168] Conversion of an alcohol to a mesylate

161.   Darzens, G. *Compt. Rend.* **1911**, *152*, 1601.
162.   Frazer, M.J.; Gerrard, W.; Machell, G.; Shepherd, B.D. *Chem. Ind.* **1954**, 931.
163.   Elderfield, R.C.; Kremer, C.B.; Kupchan, S.M.; Birstein, O.; Cortes, G. *J. Am. Chem. Soc.* **1947**, *69*, 1258.
164.   Allen, C.F.H.; Barker, W.E. *Org. Synth. Coll. Vol. 2* **1943**, 156.
165.   Fieser, L.F.; *Fieser, M. Reagents for Organic Synthesis Vol. 1*, Wiley, New York, **1967**, p 873.
166.   Lira, R.; Roush, W.R. *Org. Lett.* **2007**, *9*, 533.
167.   Woodin, K.S.; Jamison, T.F. *J. Org. Chem.* **2007**, *72*, 7451.
168.   For an example from a synthesis see Lan, J.; Li, J.; Liu, Z.; Li, Y.; Chan, A.S.C. *Tetrahedron: Asymmetry* **1999**, *10*, 1877.

followed by treatment with LiBr is a mild method for preparing bromides,[169] and treatment with LiCl gives the chloride.[170] A mixture of carbon tetrachloride and hexamethylphosphorus triamide (HMPT) has been used to convert alcohols to chlorides.[171]

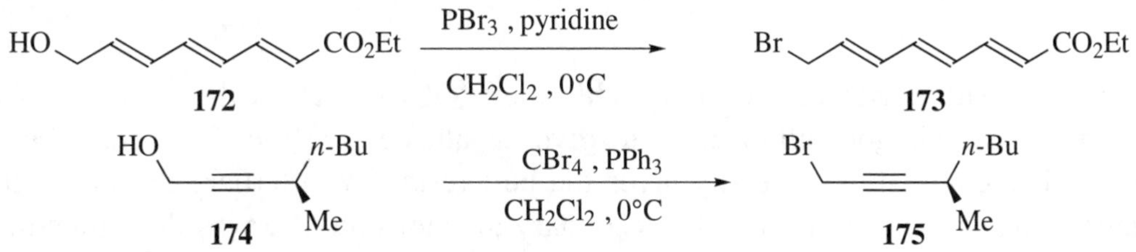

The bromination method mentioned above using triphenylphosphine and either CBr_4 or elemental bromine is a particularly mild method for this transformation. Cyclopropylcarbinyl alcohols are acid sensitive substrates, for example, usually giving ring-opening products or cyclobutane derivatives on treatment with acid (see above).[159,172] When 176 was treated with PPh_3 and bromine, however, a 79% yield of 177 was obtained.[173,174] It was essential that the bromide-triphenylphosphine-alcohol complex be heated in order to generate this bromide, since mixing the reagents without heating gave only starting material.[173]

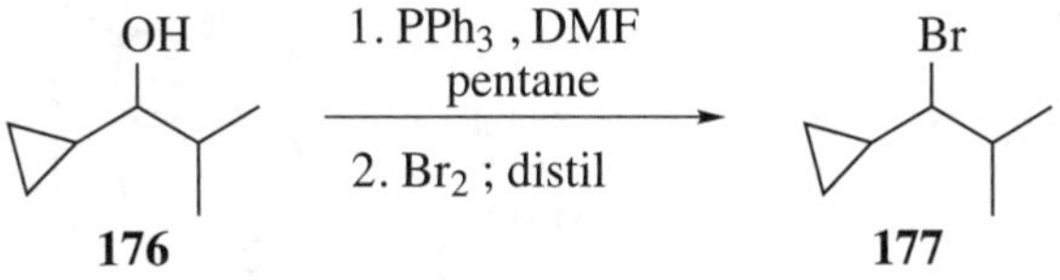

Phosphorus pentachloride is commonly used for the preparation of acid chlorides,[175] along with thionyl chloride.[176] Bromination of aromatic rings has been reported using $POBr_3$ in DMF, in the conversion of 178 to 179 in DeShong and McElroy's synthesis of the CD-ring of streptonigrin.[177]

169. For examples from syntheses, see (a) Clive, D.L.J.; Hisaindee, *J. Org. Chem.* **2000**, *65*, 4923; (b) Huang, A.X.; Xiong, Z.; Corey, E.J. *J. Am. Chem. Soc.* **1999**, *121*, 9999.
170. Smith III, A.B.; Wan, Z. *J. Org. Chem.* **2000**, *65*, 3738.
171. For an example taken from a synthesis of (+)-australine, see White, J.D.; Hrnciar, P. *J. Org. Chem.* **2000**, *65*, 9129.
172. Descoins, C.; Samain, D. *Tetrahedron Lett.* **1976**, 745.
173. Hrubiec, R.T.; Smith, M.B. *J. Org. Chem.* **1984**, *49*, 431.
174. (a) Kirmse, W.; Kapps, M.; Hager, R.B. *Chem. Ber.,* **1966**, *99*, 2855; (b) Wiley, G.A.; Hershkowitz, R.L.; Rein, B.M.; Chung, B.C. *J. Am. Chem. Soc.* **1964**, *86*, 964.
175. Reference 165, p 866.
176. For an example from a synthesis see Mori, K.; Tashiro, T.; Sano, S. *Tetrahedron Lett.* **2000**, *41*, 5243.
177. McElroy, W.T.; DeShong, P. *Tetrahedron* **2006**, *62*, 7155. For the use of PBr_5 in this type of reaction, see Kaslow, C.E.; Marsh, M.M. *J. Org. Chem.* **1947**, *12*, 456.

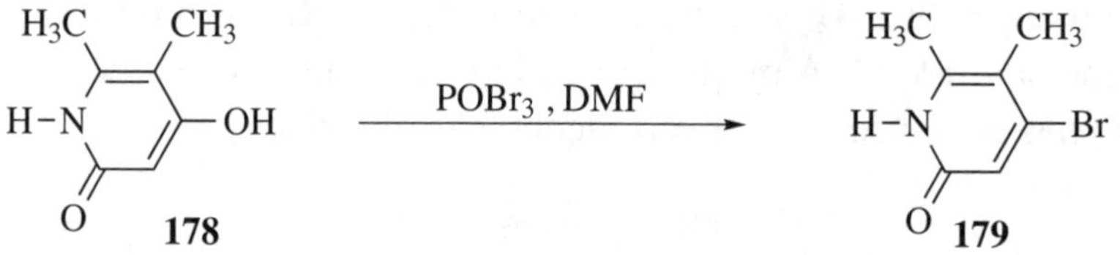

In combination with pyridine, phosphoryl chloride ($POCl_3$) reacts with alcohols to form an intermediate that undergoes elimination to give an alkene (see sec. 2.9.A for elimination reactions). The elimination generally gives the best results with tertiary alcohols, although secondary alcohols work as well. The secondary alcohol unit **180** was dehydrated to give **181** by treatment with $POCl_3$ and pyridine, albeit in only 26% yield in Basabe and co-workers enantioselective synthesis of luffolide.[178] A mixture of thionyl chloride and pyridine can also lead to elimination.[179] Substitution by chloride ion sometimes competes with elimination when secondary alcohols are substrates. Treatment of alcohol **182** with $POCl_3$ and pyridine, for example, gave chloride **183** rather than the conjugated ene-yne.[180] As with other halogenating agents, primary alcohols react to give mainly the substitution product, the alkyl halide. Phosphorus pentabromide can be used with secondary alcohols in place of PBr_3 since much less rearrangement occurs.[175]

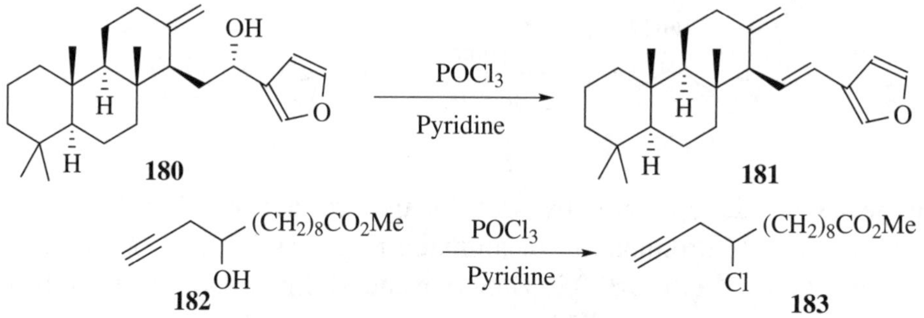

Although most phosphorus chlorides and bromides are commercially available, it is sometimes necessary to prepare the appropriate reagent as it is needed. The phosphorus iodides have poor shelf lives (they are unstable and decompose under mild conditions) and are usually prepared *in situ*, or immediately prior to use by reaction of red phosphorus with iodine. Using P and I_2 is a common method for the conversion of aliphatic alcohols to aliphatic iodides. An example is the conversion of cetyl alcohol (**184**) to cetyl iodide (**185**) in 85% yield.[181] Another popular method is illustrated by treatment of **186** with triphenylphosphine, iodine, and imidazole. In

178. Basabe, P.; Delgado, S.; Marcos, I.S.; Diez, D.; Diego, A.; De Roman, M.; Urones, J.G. *J. Org. Chem.* **2005**, *70*, 9480

179. For an example taken from a synthesis of (–)-cacospongionolide F, see Demeke, D.; Forsyth, C.J. *Org. Lett.* **2003**, *5*, 991.

180. Truscheit, E.; Eiter, K. *Liebigs Ann. Chem.* **1962**, *658*, 65.

181. (a) Reference 165, p 862; (b) Hartman, W.W.; Byers, J.R.; Dickey, J.B. *Org. Synth. Coll. Vol. 2* **1943**, 322.

this example, taken from Marshall and Schaaf's synthesis of leptofuranin D,[182] the primary alcohol unit was converted to iodide **187** in 94% yield. Phosphorus bromides can also be prepared from phosphorus and bromine. Reaction of 1,2,3-propanetriol with red phosphorus and bromine, for example, gave 1,3-dibromo-2-propanol.[177]

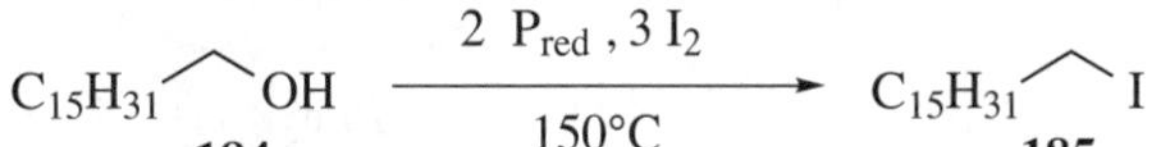

The **Finkelstein reaction** is highly useful for converting various halides into an iodide[183] in which an alkyl chloride, bromide, mesylate, or tosylate is treated with NaI or KI to produce alkyl iodides via an S_N2 reaction. In a synthesis of mosin B,[184] Tanaka, Monden, and co-workers converted tosylate **188** to iodide **189** in 79% yield with sodium iodide in acetone. This reaction is required when a better leaving group is required for a given transformation. Since alkyl iodides are known to decompose upon long standing, the Finkelstein reaction allows the iodide to be generated *in situ* and used immediately.

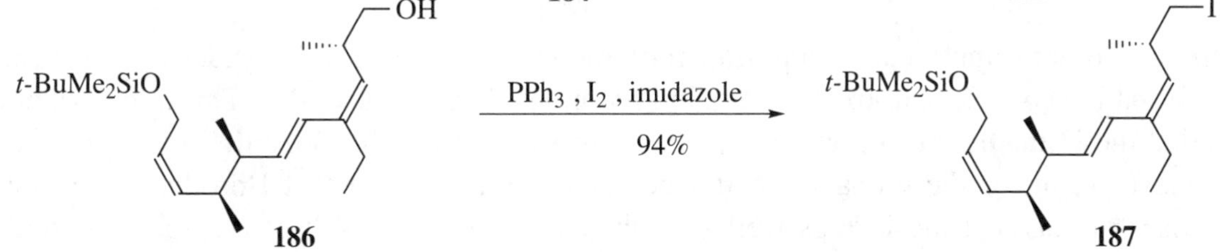

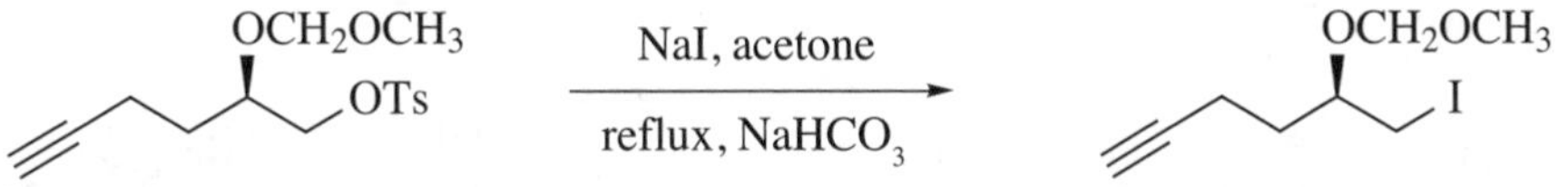

When allylic alcohols are subjected to acidic conditions, formation of the allyl cation can lead to regioselectivity problems as well as polyalkylation or S_N2' reaction by-products. Most of the halogenating reagents described can give deleterious side reactions. Such problems can be circumvented or alleviated by using a different type of halogenating reagent. Magid showed that allylic alcohols such as **190** are converted to allylic chloride **191** by a mixture of triphenylphosphine (PPh₃) and hexachloroacetone that gives chlorotriphenylphosphonium

182. Marshall, J.A.; Schaaf, G.M. *J. Org. Chem.* **2003**, *68*, 7428. Alcohols are converted to iodides using $(PhO)_3P^+CH_3I^-$ in DMF: see Su, Q.; Dakin, L.A.; Panek, J.S. *J. Org. Chem.* **2007**, *72*, 2 in a synthesis of leucascandrolide A.

183. (a) Finkelstein, H. *Berichte* **1910**, *43*, 1528; (b) Ingold, C.K. *Structure and Mechanism in Organic Chemistry*, 2nd Ed. Cornell Univ. Press, London, **1969**, p 435; (c) The Merck Index, 14th Ed. Merck & Co., Inc., Whitehouse Station, New Jersey, **2006**, p ONR-31.

184. Maezaki, N.; Kojima, N.; Sakamoto, A.; Tominaga, H.; Iwata, C.; Tanaka, T.; Monden, M.; Damdinsuren, B.; Nakamori, S. *Chem. Eur. J.* **2003**, *9*, 390.

chloride (Cl–PPh$_3^+$Cl$^-$),[185] the chlorinating species. The reaction of triphenylphosphine with bromine or carbon tetrabromide (CBr$_4$) similarly converts an alcohol into the corresponding bromide, as we have seen above.

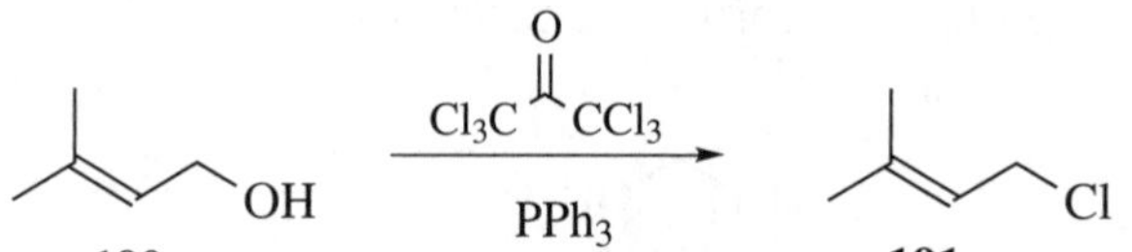

There are other functional group transformations that give alkyl halides. A classical transformation converts carboxylates to alkyl bromides (C-CO$_2^-$ ® C-Br). This transformation is called the **Hunsdiecker reaction**[186] (it has also been called the **Borodin reaction**),[187] and it is most useful for the preparation of secondary halides. Note that Borodin was a noted composer of classical music[188] as well as a chemist. The silver salt of a carboxylic acid is heated with bromine to give the bromide via decarboxylation. An example is conversion of the silver salt of cyclohexane carboxylic acid (**192**) to bromocyclohexane.[189] In a different process, phosphorus and bromine react with acids in the **Hell-Volhard-Zelinsky reaction**.[190] An example is the conversion of cyclopropanecarboxylic acid to the α-bromo acid bromide **193**, which was reacted with 4-aminopyridine **194** to give **195** in Storey and Ladwa's synthesis of 5-azaoxindoles.[190e]

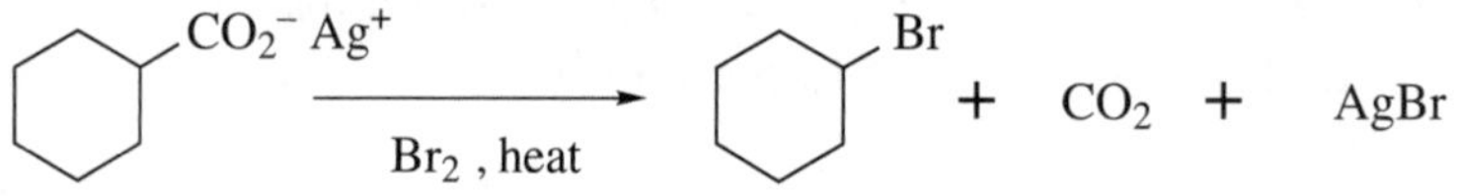

185. (a) Magid, R.M.; Fruchey, O.S.; Johnson, W.L.; Allen, T.G. *J. Org. Chem.* **1979**, *44*, 359; (b) Magid, R.M.; Fruchey, O.S. *J. Am. Chem. Soc.* **1977**, *99*, 8368.
186. (a) Hunsdiecker, H.; Hunsdiecker, Cl. *Berichte* **1942**, *75*, 291; (b) Johnson, R.G.; Ingham, R.K. *Chem. Rev.* **1956**, *56*, 219; (c) Wilson, C.V. *Org. React.* **1957**, *9*, 332 (see p 341); (d) The Merck Index, 14th Ed. Merck & Co., Inc., Whitehouse Station, New Jersey, **2006**, p ONR-47. For a microwavae-promoted version see Bazin, M.-A.; El Kihel, L.; Lancelot, J.-C.; Rault, S. *Tetrahedron Lett.* **2007**, *48*, 4347.
187. Borodine, A. *Annalen* **1861**, *119*, 121.
188. See the biography by (a) Abraham, G.; Seroff, V.I. *The Mighty Five* (1948); See also, (b) Zetlin, M.O. *The Five* (tr. 1959).
189. (a) Eliel, E.L.; Haber, R.G. *J. Org. Chem.* **1959**, *24*, 143; (b) Marvell, E.N.; Sexton, H., *Ibid* **1964**, *29*, 2919.
190. (a) Hell, C. *Berichte* **1881**, *14*, 891; (b) Volhard, J. *Annalen* **1887**, *242*, 141; (c) Zelinsky, N. *Berichte* **1887**, *20*, 2026; (d) Watson, H.B. *Chem. Rev.* **1930**, *7*, 180; (e) The Merck Index, 14th Ed. Merck & Co., Inc., Whitehouse Station, New Jersey, **2006**, p ONR-42; (e) Storey, J.M.D.; Ladwa, M.M. *Tetrahedron Lett.* **2006**, *47*, 381.

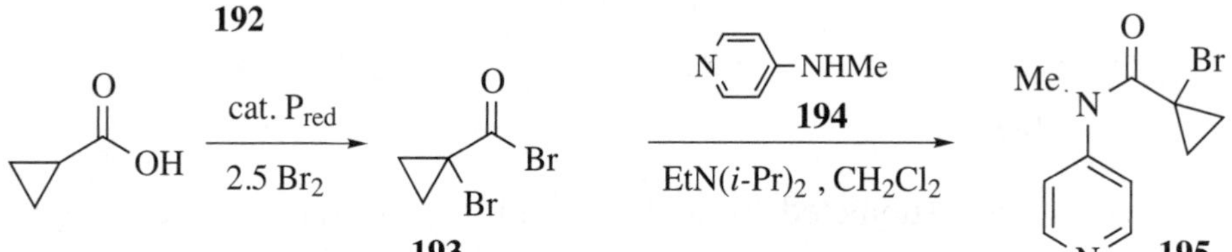

2.8.B. ALLYLIC AND BENZYLIC HALOGENATION

Bromine or chlorine can be introduced into a molecule by processes that generate radical intermediates.[191] The reaction of bromine or chlorine with allylic and benzylic compounds, in the presence of heat, light or radical initiators (secs. 13.3, 13.4) leads to resonance stabilized free radicals that react with additional bromine or chlorine to give the corresponding halide. Addition of diatomic bromine to **196** (in the presence benzoyl peroxide and photochemical initiation) gave benzylic bromide **197** in high yield in a synthesis of indole alkaloids.[192] This reaction is just another example of the well-known free-radical substitution reaction of alkanes with chorine or bromine in the presence of light (*hv*) or heat. The resonance stability of an allylic or a benzylic radical is a significant contributor to the benzylic or allylic selectivity. In non-allylic or benzylic systems, radical chlorination of alkanes by chlorine is unselective, converting virtually all C–H bonds in the molecule to C–Cl. Bromination, on the other hand, is more selective in such systems, showing a great propensity to react with tertiary positions (tertiary C–H) to give tertiary bromides, as in the conversion of 2-methylpentane to **198**. Chlorination or bromination of allylic and benzylic sites is usually a highly useful synthetic transformation.

Halogenation of allylic and benzylic C–H moieties is more convenient using the readily available *N*-bromosuccinimide (NBS, **199**) or *N*-chlorosuccinimide (NCS, **200**) with radical initiators, heat and/or light,[193] in what is called the **Wohl-Ziegler reaction**.[193f,g,h,194] Reactions with NBS are commonly carried out in refluxing carbon tetrachloride, with addition of a catalytic amount of radical initiator [such as *azobis*-isobutyronitrile (AIBN) or benzoyl peroxide, sec. 13.3], or with an incandescent lamp held close to the reaction. Heating AIBN leads to homolytic cleavage and formation of a cyano radical ($Me_2C^{\bullet}$–CN), which initiates the radical chain process (sec. 13.3). This chain process generates the allylic (or benzylic) radical that reacts with bromine (or additional NBS) to give the allylic or benzylic bromide product and another radical chain carrier.[193] Reaction with NCS proceeds by a similar mechanism. A synthetic example, taken

191. For examples see Ref. 3, pp 611-615.
192. Söderberg, B.C.; Chisnell, A.C.; O'Neil, S.N.; Shriver, J.A. *J. Org. Chem.* **1999**, *64*, 9731.
193. (a) Goldfinger, P.; Gosselain, P.A.; Martin, R.H. *Nature (London)* **1951**, *168*, 30; (b) McGrath, B.P.; Tedder, J.M. *Proc. Chem. Soc.* **1961**, 80; (c) Pearson, R.E.; Martin, J.C. *J. Am. Chem. Soc.* **1963**, *85*, 354; (d) Russell, G.G.; DeBoer, C.; Desmond, K.M. *Ibid* **1963**, *85*, 365; (e) Walling, C.; Rieger, A.L.; Tanner, D.D. *Ibid* **1963**, *85*, 3129; (f) Wohl, A. *Berichte* **1919**, *52*, 51; (g) Ziegler, K. Spāth, A.; Schaaf, E.; Schumann, W.; Winkelmann, E. *Annalen* **1942**, *551*, 80; (h) Djerassi, C. *Chem. Rev.* **1948**, *43*, 271.
194. The Merck Index, 14th Ed. Merck & Co., Inc., Whitehouse Station, New Jersey, **2006**, p ONR-102.

from Goosen and Melzer's synthesis of valsartan,[195] in which the benzylic methyl group in **201** was converted to the bromomethyl unit in **202** by reaction with **199** and AIBN.

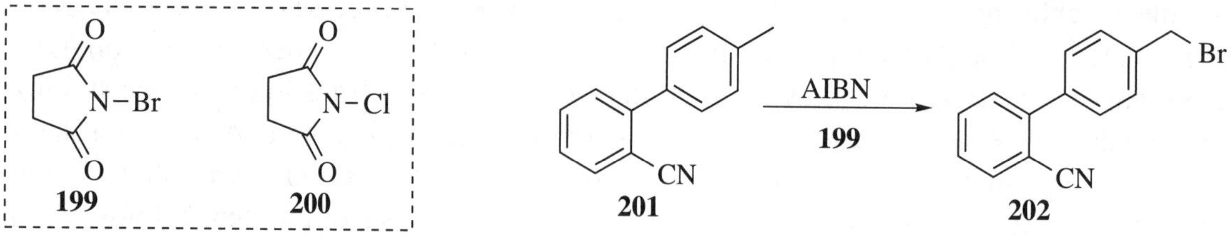

Ketones that have an enolizable hydrogen can be halogenated at the α position (the carbon adjacent to the carbonyl) with bromine, chlorine, NBS, or NCS.[196] The reaction probably proceeds via addition of X_2 to the enol form of the carbonyl (secs. 9.2.A, 9.8.A). Elimination of HX from the addition product generates a new enol, which tautomerizes to the α-haloketone.[196] Reaction of cyclohexanone with bromine, for example, gives 2-bromocyclohexanone (**204**) and reaction with NCS (**200**) gives 2-chlorocyclohexanone (**203**).

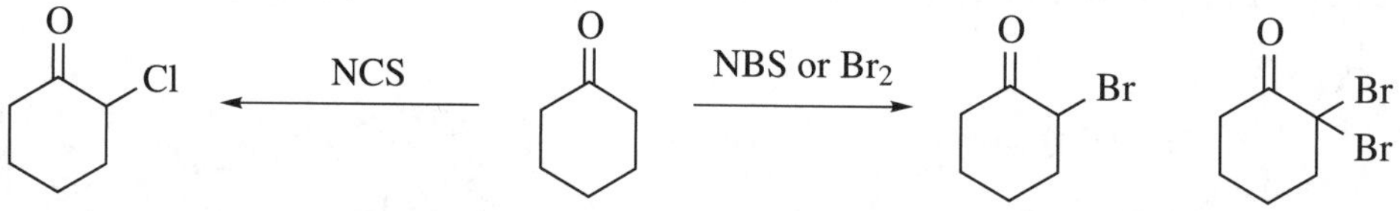

Halogenation of carbonyl compounds is often accompanied by secondary reactions. The initially formed α-halo carbonyl can react with another equivalent of halogen to give the α,α-dihalocarbonyl. In the bromination of cyclohexanone, 2,2,-dibromocyclohexanone (**205**) was also produced. Usually the mono-halogenated product predominates with ketones, but this depends on the substrate and the reaction conditions. Halogenation of lactams using this method often gives a α,α-dihalide. 2-Pyrrolidinone, for example, reacts with NCS or PCl_5, to give a mixture of 3-chloro- and 3,3-dichloro-2-pyrrolidinone.[197] When phosphorus halides are used the dichloride is usually the major product, but the monochloride can be formed in reasonable yield using NCS.

In unsymmetrical ketones, halogenation usually occurs at the more highly substituted α position, but regioselectivity can be a problem. Heating an unsymmetrical ketone with *tert*-butyl bromide and DMSO at 65°C is a mild and selective solution to the problem,[198] usually giving the more substituted bromide. Another problem in halogenation reactions of carbonyls is elimination (note that $POCl_3$ is produced as a side product when PCl_5 is used). When

195. Goosen, L.L.; Melzer, B. *J. Org. Chem.* **2007**, *72*, 7473.
196. Catch, J.R.; Hey, D.H.; Jones, E.R.H.; Wilson, W. *J. Chem. Soc.* **1948**, 276.
197. Elberling, J.A.; Nagasawa, H.T. *J. Heterocyclic Chem.* **1972**, *9*, 411.
198. Armani, E.; Dossena, A.; Marchelli, R.; Casnati, G. *Tetrahedron* **1984**, *40*, 2035.

cyclopentanone reacted with PCl₅, for example, 1-chlorocyclopentene was the major product, via elimination, but 1,1-dichlorocyclopentane was also formed by addition of chloride ion to the α-chloro-carbocation.[199] Similarly, cyclohexanone gave 1-chlorocyclohexene (**206**).[200]

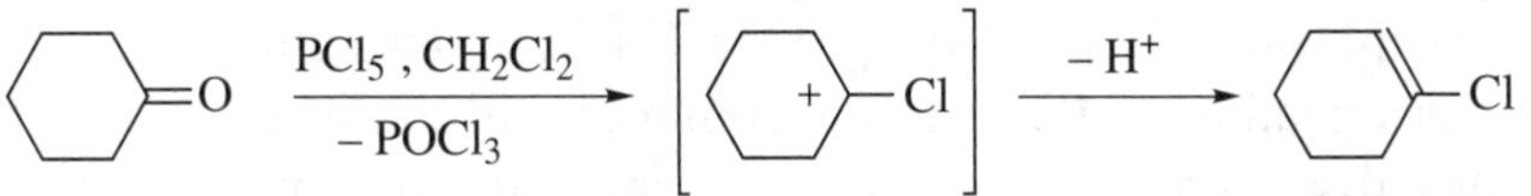

Incorporation of halogen into a molecule is important because halides are often precursors for substitution or elimination reactions. The ability to convert alcohols to halides is therefore an important functional group transformation. The transform is

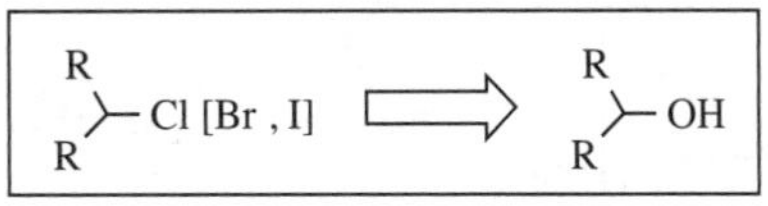

2.9. ELIMINATION REACTIONS

There are several types of reactions in which a halogen, a sulfonate ester, or another functional group is lost from a molecule, along with hydrogen, or sometimes, functional group, to generate a carbon-carbon double bond. Alkenes and alkynes are formed by this process. This section will examine methods for the formation of such molecules via elimination reactions.[201] A monograph describes several different preparations of alkenes.[202]

2.9.A. Bimolecular 1,2-Elimination

Elimination is a major class of functional group exchange reactions in which a H atom and a leaving group are removed from a molecule, resulting in a π bond. However, removal of the hydrogen in this reaction relies on acid-base chemistry. Mechanistically, elimination is characterized by removal of a proton by a base, followed by expulsion of an atom or group (the leaving group, X) to form the π-bond in what appears to be a synchronous process. The reaction eliminates the elements of H and X, which leads to the name. We have discussed the characteristic features of nucleophiles and their reactions. In many cases, the nucleophile used in a substitution reaction is also a good Brønsted-Lowry base. Although there is strong attraction between the δ⁺ dipole on the carbon bearing the leaving group and the base, collision with this carbon is unproductive in tertiary substrates due to the high energy of activation for the S_N2 transition state **69**. The bond polarization induced by the halide extends down the carbon

199.   Charpentier-Morize, M.; Sansoulet, J. *Bull. Chim. Soc. Fr.* **1977**, 331.
200.   Baldwin, J.E.; Burrell, R.C. *J. Org. Chem.* **1999**, *64*, 3567.
201.   For examples see reference 3, pp 251-263, 569-580.
202.   Williams, J.M.J. (Ed.) *Preparation of Alkenes*, Oxford Press, Oxford, **1996**.

chain to the β-hydrogen, which has a δ⁺ dipole. Collision of the base with this β-hydrogen is an acid-base reaction that removes the hydrogen as indicated in **207**. As the base forms a bond to the β-H, negative charge develops on the β-carbon, and migration of the electron density toward the δ⁺ dipole of the carbon bearing X generates a new π-bond (in alkene product **209**), with concomitant expulsion of the leaving group. This reaction follows second-order kinetics, and is given the designation of **E2 reaction** (bimolecular elimination). The E2 reaction is synchronous rather than stepwise,[203] and all experimental observations can be explained by a transition state such as **208**. Note, however, that a "unified rule for elimination" has been proposed that predicts regioselectivity for a wide range of substrates.[204] This rule appears to apply best when ion-pairing occurs, and is not as useful in cases where there is no ion paring.

The elimination sequence begins with an acid-base reaction, where the acid is a hydrogen atom attached to the β-carbon relative to the leaving group (the β-hydrogen, see **208**). Bond polarization resulting from the presence of electronegative nature of X makes that hydrogen δ⁺ so it functions as a weak acid. It has been determined that as the added base reacts with the β-hydrogen, the favored rotamer for the molecule will have the leaving group and the β-hydrogen in an anti relationship (i.e. 180° apart).[203] Electronic repulsion between the developing charge (electron pair) on the β-carbon and the leaving group is minimized in an anti- transition state. This requirement for an anti-relationship makes the stereochemistry of the alkene product predictable. In **208**, the relative position of the alkyl groups in the anti transition state (note that R¹ and R³ are on the same side) is retained in the alkene (**209**) as the π bond is formed. When halide **210A**, drawn as its Fischer projection, is treated with a base such as *tert*-butoxide, only the (*Z*)-alkene (**211**) is formed via E2 elimination. Halide **210A** is the (2*R*,3*S*) diastereomer of 2-bromo-3-phenylbutane, and the rotamer that has the bromine leaving group and the β-hydrogen in an anti conformation is **210B**. The relative positions of phenyl and methyl are syn in **210B**, and after elimination of the elements of H (to *tert*-butoxide) and Br as the bromide ion (*Z*)-**211** is formed. Similar reaction with the (2*R*,3*R*)-diastereomer of 2-bromo-3-phenyl-butane gives only (*E*)-2-phenyl-2-butene.

203.   (a) Saunders, Jr., W.H.; Cockerill, A.F. *Mechanisms of Elimination Reactions*, Wiley-Interscience, New York, *1973*; (b) Bordwell, F.G. *Acc. Chem. Res. 1972*, *5*, 374; (c) Dhar, M.L.; Hughes, E.D.; Ingold, C.K.; Masterman, S. *J. Chem. Soc. 1948*, 2055, 2058, 2065; (d) Sneen, R.A. *Acc. Chem. Res. 1973*, *6*, 46, (e) Bordwell, F. G. *Ibid 1970*, *3*, 281.
204.   Gevorkyan, A.A.; Arakelyan, A.S.; Cockerill, A.F. *Tetrahedron 1997, 53*, 7947.

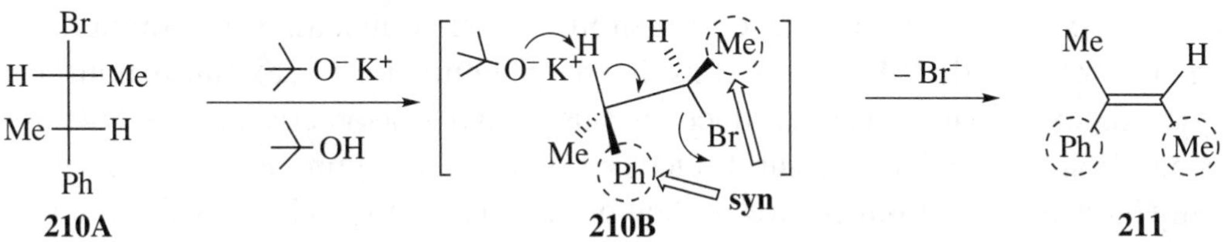

Br
H—Me
Me—H
Ph
210A

$\nearrow$O⁻ K⁺

$\nearrow$OH

[structure with O⁻ K⁺, H, H, Me, Me, Br, Ph, **syn**]
210B

− Br⁻

Me   H
(Ph)   (Me)
211

Using the **Hammond postulate** (a transition state will be structurally and energetically similar to the species nearest to it on the reaction path), the transition state for this reaction is product-like (a late transition state), so that factors stabilizing the alkene product will have a strong influence on the transition state. Alkyl groups that are attached to an sp^2 carbon release electrons to the π bond and enhance the strength of the π bond. A more highly substituted alkene is therefore more stable, and a relative order of alkene stability is:[205]

When there are two different β-hydrogen atoms, there is the possibility of forming two different alkene products (regioisomers). Removal of hydrogen in **212A** will lead to a double bond between C2-C3 [(Z)-3-ethyl-2-hexene], but removal of hydrogen from **212B** will lead to a double bond between C1-C2 (3-ethyl-1-hexene). These two alkenes are regioisomers. The transition state for **182A** leads to a more highly substituted alkene (3-ethyl-2-hexene) than does the transition state for **212B**, which leads to 3-ethyl-1-hexene. A trisubstituted alkene is thermodynamically more stable than a monosubstituted alkene. Since the Hammond postulate assumes a product-like transition state, **212A** leads to the more stable product and is favored over **212B**. Acid-base reactions are equilibrium processes and under thermodynamic control, so formation of the more stable product is favored. Formation of the more stable alkene is characteristic of E2 reactions, and is usually termed **Saytzeff (Zaitsev) elimination**.[206] Formation of the more substituted alkene and this term is also associated with E1 reactions.

The E2 transition state leads to strict conformational restraints with cyclic structures.[207]

$R_2C=CR_2$ > $R_2C=CHR$ > $RCH=CHR$ > $R_2C=CH_2$ > $RCH=CH_2$ > $H_2C=CH_2$

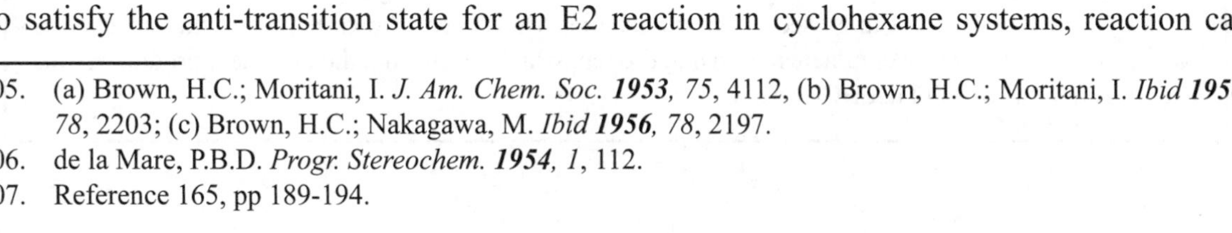

Et   H
C_3H_7 Me

[transition state with O⁻ K⁺, H, H, CH₃, Et, Br, C_3H_7]
212A

[transition state with O⁻ K⁺, Et, H, H, C_3H_7, Br, H, H]
212B

H   H
C_3H_7(Et)HC   H

To satisfy the anti-transition state for an E2 reaction in cyclohexane systems, reaction can

205. (a) Brown, H.C.; Moritani, I. *J. Am. Chem. Soc.* **1953**, *75*, 4112, (b) Brown, H.C.; Moritani, I. *Ibid* **1956**, *78*, 2203; (c) Brown, H.C.; Nakagawa, M. *Ibid* **1956**, *78*, 2197.
206. de la Mare, P.B.D. *Progr. Stereochem.* **1954**, *1*, 112.
207. Reference 165, pp 189-194.

occur only if the β-hydrogen and the leaving group are trans-diaxial. Both equilibrating chair conformations **213A** and **213B** must be considered for elimination of bromine in trans-1-bromo-3-methylcyclohexane when a base is added, because two regioisomeric alkenes **214** and/or **215** are possible. Both alkenes have a disubstituted C=C unit and are approximately equal in stability. Based only on stability of products, we predict formation of both products in close to a 1:1 ratio.

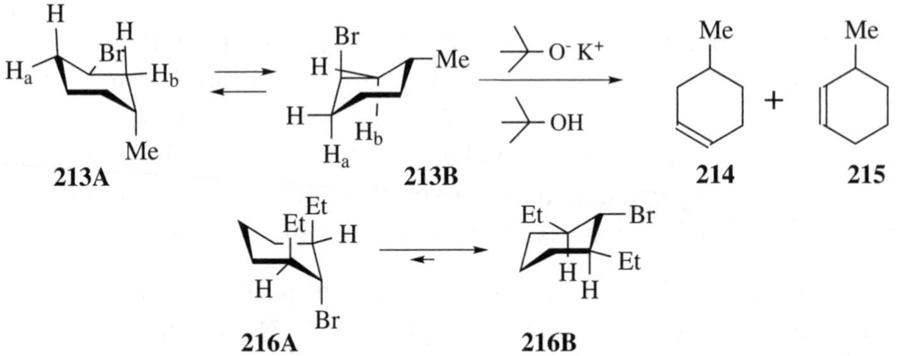

213A   **213B**   **214**   +   **215**

216A   **216B**

Taking a closer look at this system, the bromine is equatorial in **213A**, and there are no β-hydrogens that can attain an anti conformation relative to the bromine. An E2 elimination cannot occur from **213A**. In **213B**, however, both H_a and H_b are anti to the axial bromine and elimination leads to **214** and **215**, respectively. Bromide **216** poses a different problem in that there are two β-hydrogens. There are two chair conformations (**216A** and **216B**), but **216B** is the thermodynamically more stable due to decreased *A* strain (sec. 1.5.D). In **216A** the two ethyl groups are axial and both β-hydrogens are equatorial and, therefore, not susceptible to base induced elimination. In **216B** both β-hydrogens are axial, but the bromine is equatorial, and again E2 elimination is not possible. We therefore predict that treatment of **216** with base will *not* lead to an E2 reaction. The conformational characteristics of a cyclohexane, and possibly other ring systems, must be evaluated carefully because they may have a significant influence on reactivity in an E2 reaction.

MOLECULAR MODELING: Stereochemistry of Elimination of HX

Elimination of HX from a haloalkane or halocycloalkane typically follows an E2 mechanism that requires the leaving group to be *anti* to the β-hydrogen. The E2 reaction is an acid-base process in which the added base reacts with a slightly acidic proton on the carbon atom β to the leaving group, removing that proton. As the proton is removed, the developing charge migrates towards the α-carbon atom, leading to displacement of the leaving group and formation of a carbon-carbon double bond. Where there is more than one (different) β-hydrogen atom, the relatively acidity of the β-hydrogen atoms is an important issue in E2 reactions. As can be seen from the labeling experiment shown below, in the presence of base, 2-norbornyl chloride eliminates the *syn* hydrogen, which means that the dominant product results in elimination of DCl and not HCl.

MOLECULAR MODELING: Stereochemistry of Elimination of HX, con't.

The accepted mechanistic explanation is that elimination probably involves initial deprotonation of 2-norbornyl chloride by alkoxide, followed by loss of chloride. In this case, the question is which "proton" (*syn* or *anti* to chlorine) is more susceptible to attack by alkoxide.

Examine a LUMO map for 2-norbornyl chloride. It is shown as a transparent model allowing you to see the underlying structure. Once you are oriented, you can switch to a solid model by *clicking* on the map and selecting **Solid** from the **Style** menu at the bottom of the screen. A LUMO map colors an electron density surface that gives the overall molecular size and shape (where a reagent can approach), with the (absolute) value of the lowest-unoccupied molecular orbital (LUMO). By convention, "blue" regions indicate large LUMO concentration and are likely to undergo attack by a nucleophile, while "red" regions indicate small LUMO concentrations. (Discussion of the use of LUMO maps to describe selectivity in nucleophilic additions to carbonyl compounds is provided in Chapter 4, section 4.7.B). Is the *exo* hydrogen (*syn* to Cl) more or less blue than the *endo* hydrogen (*anti* to Cl)? Which is more likely to be abstracted by the base? Is your result in agreement with what is actually observed? Is any other hydrogen "blue", that is, susceptible to nucleophilic attack? If so, which hydrogen and what is the product that should result? Is there any experimental evidence for this?

The leaving group in an E2 reaction can be a chloride, a bromide or an iodide, and also sulfonate esters such as a mesylate ($-OSO_2Me$), a tosylate ($-OSO_2p-Me-C_6H_4$), or a triflate ($-OSO_2CF_3$). The previously mentioned non-nucleophilic base DBN (**26**)[208a] can be used to induce an E2 type reaction with primary halides. In a synthetic example taken from Alvarez-Manzaneda and co-worker's synthesis of (+)-austrodoral, DBU (**21**) reacted with acetate **217** to give conjugated ketone **218** in 95% yield.[209] It does not remove H_b with elimination of acetate (an E2 process) nor H_c to generate the ester enolate (sec. 9.2). The nitrogen atom in the base DBN is sterically hindered, and a poor nucleophile. This means that the normal nucleophilic acyl addition to a carbonyl group in an ester or a ketone is very slow and not competitive with the elimination process. The two non-nucleophilic bases give moderate to good yields of alkene in E2 reactions with alkyl halides, and little or no substitution products. Even primary halides, which give mostly S_N2 reactions with nucleophilic bases, usually give an alkene in good yield. Non-conjugate alkenes can be formed, as in a study toward the synthesis of stenine by Padwa

208. For synthetic examples, see (a) Achab, S.; Das, B.C. *J. Chem. Soc. Chem. Commun.* **1983**, 391; (b) Oediger, H.; Möller, Fr. *Angew. Chem. Int. Ed.* **1967**, *6*, 76.
209. Alvarez-Manzaneda, E.; Chahboun, R.; Barranco, I.; Cabrera, E.; Alvarez, E.; Lara, A.; Alvarez-Manzaneda, R.; Hmamouchi, M.; Es-Samti, H. *Tetrahedron* **2007**, *63*, 11943.

and Ginn[210] in which initial conversion of the alcohol unit in **219** to a mesylate followed by heating with DBU gave **220** in 64% overall yield.

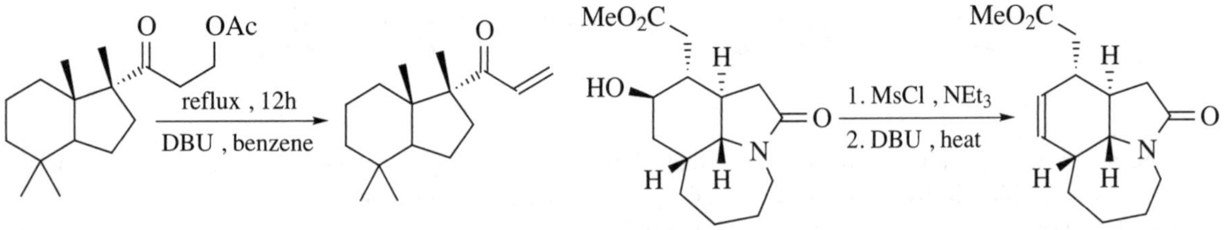

The use of DBN and DBU in E2 reactions is somewhat selective, and the most common bases used are hydroxide (in water) or alkoxides in alcohol solvent (sodium methoxide in methanol, sodium ethoxide in ethanol, potassium *tert*-butoxide in *tert*-butanol).[203] Amide bases such as sodium amide or lithium diethylamide can be used in ammonia or amine solvents. In these latter cases, the bases are also good nucleophiles, but in reactions with tertiary halides the elimination process dominates. With secondary halides, substitution can compete and with primary halides, S_N2 may be the major reaction, but this depends on the nature of the base/nucleophile. Apart from DBN and DBU, sterically hindered amines such as 2,2,6,6-tetramethylpiperidine have been used and the weaker base imidazole is important for elimination reactions with sensitive substrates. As with **219** → **220**, sulfonate esters are good leaving groups commonly used in elimination reactions.

It is clear that a variety of substrates with different leaving groups can be converted to alkenes via elimination. β-Elimination of sulfones is also known.[211] A general transform for the E2 reaction is

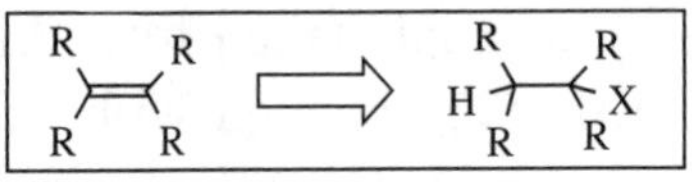

where X = halogen, OH, OR, OSO$_2$R

Note that vicinal dibromides can be treated with base in such a way that *both* bromine atoms are lost to generate an alkene. Many different reagents have been used for this transformation.[212] Recently, indium metal was shown to be an effective reagent.[213]

The E2 reaction can be applied to the synthesis of alkynes. When a vinyl halide is treated with a strong base, loss of HX by what is essentially an E2 process leads to formation of a triple

210. Padwa, A.; Ginn, J.D. *J. Org. Chem.* **2005**, *70*, 5197.
211. For a review of desulfonylation reactions see Nájera, C.; Yus, M. *Tetrahedron* **1999**, *55*, 10547.
212. (a) Butcher, T.S.; Detty, M.R. *J. Org. Chem,.* **1998**, *63*, 177; (b) Malanga, C.; Mannuccki,. S.; Lardicci, L. *Tetrahedron* **1998**, *54*, 1021; (c) Khurana, J.M.; Maikap, G.C. *J. O rg. Chem.* **1991**, *56*, 2582; (d) Savoia, E.; Tagliavini, C.; Trombini, C.; Umani-Ronchi, A. *J. Org. Chem.* **1982**, *47*, 876; (e) Allred, E.L.; Beck, B.R.; Voorhees, K.J. *J. Org. Chem.* **1974**, *39*, 1426.
213. Ranu, B.C. *Eur. J. Org. Chem.* **2000**, 2347 (see pp 2348-2349).

bond. In a simple example, taken from Mori and co-worker's synthesis of the sphingosine derivative sulfobacin B,[214] 12-methyltridec-1-ene was treated with bromine in dichloromethane to give dibromide **221**. When this was treated with potassium *tert*-butoxide in petroleum ether, in the presence of 18-crown-6, initial elimination gave vinyl bromide **222**. Subsequent reaction with the base initiated an E2 reaction that gave alkyne **223** in 72% yield for both chemical steps. Note that this method of alkyne formation from vicinal bromides can lead to isomerization of the alkyne unit to give internal alkynes rather than terminal alkynes. Higher reaction temperatures and a large excess of base should be avoided.

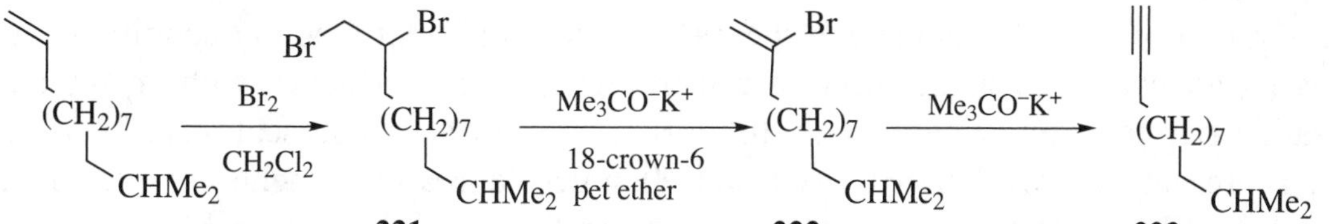

An alternative method for preparing vinyl bromides can be incorporated into a synthesis of alkynes when combined with an olefination reaction (see sec. 8.8.A). In Ramachandran and co-worker's[215] synthesis of (–)-dictyostatin, aldehyde **224** was treated with carbon tetrabromide, triphenylphosphine and triethylamine to give vinyl dibromide **225** initially. Subsequent treatment with butyllithium led to elimination and an alkyne anion, and workup gave alkyne **226** in 83% yield. This sequence is commonly known as the **Corey-Fuchs procedure** (see chapter 8).[216] Note that this sequence is compatible with the presence of reactive functionality such as a tosylate leaving group. In Martin and co-workers' synthesis of (+)-8-*epi*-xanthatin,[217] aldehyde **227** was treated with carbon tetrabromide and triphenylphosphine to give vinyl dibromide (**228**).

Subsequent reaction with *n*-butyllithium and then triisopropylsilyl triflate (see chap. 7, sect. **7.3.A.ii**) gave the TIPS alkyne **229** in 70% overall yield. It is clear that extending the E2 reaction to prepare alkynes is quite useful as well as versatile.

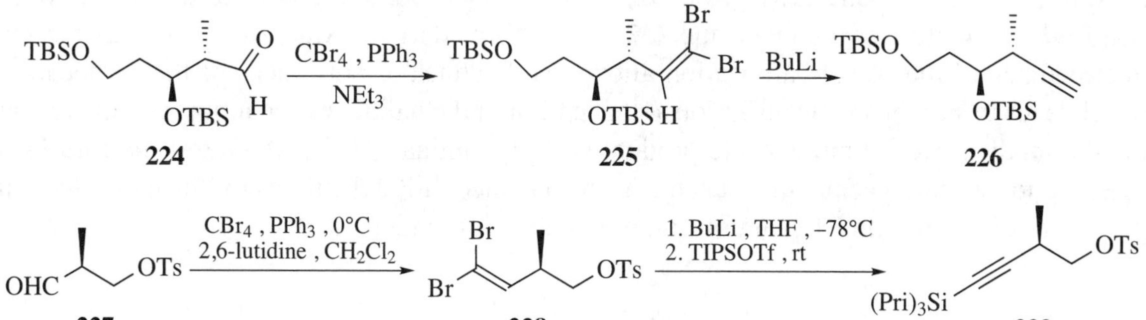

214. Takikawa, H.; Nozawa, D.; Kayo, A.; Muto, S.-e.; Mori, K. *J. Chem. Soc. Perkin Trans. 1* **1999**, 2467.
215. Ramachandran, P.V.; Srivastava, A.; Hazra, D. *Org. Lett.* **2007**, *9*, 157.
216. Corey, E.J.; Fuchs, P.L. *Tetrahedron Lett.* **1972**, *36*, 3769.
217. Kummer, D.A.; Brenneman, J.B.; Martin, S.F. *Org. Lett* **2005**, *7*, 4621.

The E2 disconnections to prepare alkynes are

2.9.B. UNIMOLECULAR 1,2-ELIMINATION

The E2 reaction follows second-order kinetics, and the rate of reaction is proportional to the concentration of both the halide and the base. Under different reaction conditions, usually involving aqueous media, the reaction can follow first-order kinetics with an initial, slow ionization step to form a carbocation. In most cases (*1*) water is present, (*2*) a good leaving group is present, (*3*) a weak base is used, and (*4*) molecular rearrangements are often observed. As with the S_N1 reaction in sec 2.6.D, the rate of the reaction is given by the expression: Rate = k_1 [RX] + k_2 [nucleophile or base]. The slow, and rate determining step is ionization, followed by a fast removal of the β-hydrogen from the carbocation intermediate. Therefore, in this mechanism, there is no dependence on the concentration of the base and the observed overall rate is dependent only on the halide, leading to Rate = k [RX]. A simple example based on the generic mechanism of ionization to a carbocation and removal of a proton involves heating bromide **230** in aqueous media, in which the water assists in breaking the C–Br bond to generate a solvent separated ion pair (**231**). The resulting carbocation and bromide ions are solvated by water. The β-hydrogen of the carbocation (H_a in **231**) is adjacent to a full positive charge, and polarized to a greater extent than the β-hydrogen in **230**. Note that **231** may not be completely solvent separated, but we are assuming this is the case for simplicity. A base is required to remove H_a, but H_a in **231** is a stronger acid than H_a in **208** so a weaker base can be used for deprotonation. In this case, the water present in the reaction medium will suffice, and the electron pair from the C–H_a bond migrates toward the positive carbon to form a new π bond in the alkene product, **232**. Rotation about the C–C bond in **231** prior to removal of H_a will lead to a mixture of rotamers, and **232** will be formed as a mixture of (*E/Z*)-isomers. The intermediacy of the planar carbocation allows facile attack at both faces of the carbocation, which leads to significant racemization at that carbon if the halide was enantiopure. In general, the thermodynamically more stable product will predominate (called **Saytzeff elimination**). The reaction is unimolecular in mechanism, and a nucleophile substitutes for the halide leaving group, so it is termed an E1 **reaction** (unimolecular elimination).[218,219]

218. (a) Seib, R.C.; Shiner, Jr., V.J.; Sendijarevi'c, V.; Humski, K. *J. Am. Chem. Soc.* **1978**, *100*, 8133, (b) McLennon, D.J. *Quart. Rev. Chem. Soc.* **1967**, *21*, 490.
219. Willstätter, R.; Waser, E. *Berichte* **1911**, *44*, 3423.

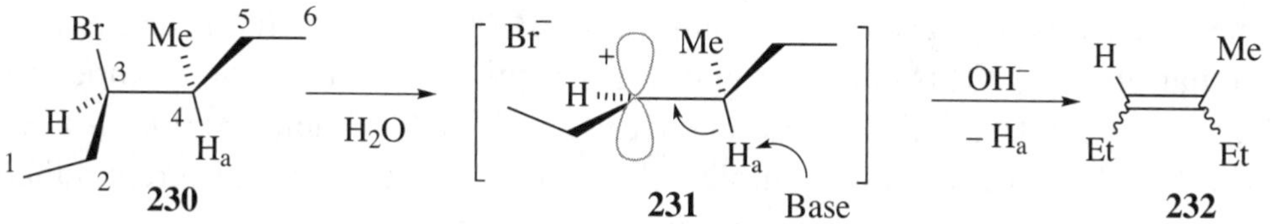

The S_N1 reaction usually competes with the E1 process in aqueous media since most bases used in this reaction are nucleophilic. Only when the gegenion is a very weak nucleophile will the E1 process dominate and produce substantial amounts of alkene product. Formation of an alkene when alcohols react with concentrated sulfuric acid or perchloric acid is an example, as when **233** was treated with concentrated sulfuric acid to give alkene **234**.[219] Non-oxidizing acids can be used, as in the treatment of **235** with *p*-toluenesulfonic acid in refluxing benzene to give an 86% yield of *E*-**236** in Shiina and co-worker's synthesis of tamoxifen.[220] It is not always clear if the reaction proceeds by an E1 or an E2 mechanism since the oxonium ion derived from **235** can undergo an E2 reaction in the presence of sodium sulfate. Such mechanistic blurring is common, and often additives such as sodium sulfate are used to drive the reaction to the elimination product. If a carbocation intermediate is present, rearrangements are possible. Although sometimes useful, the E1 reaction is more commonly observed as a side reaction in S_N1 processes, or in E2 reactions in aqueous or alcohol media. If there is a mixture of E1 and E2 in a reaction, there is the possibility of E1 reactions when carbocation intermediates are generated in protic solvents.

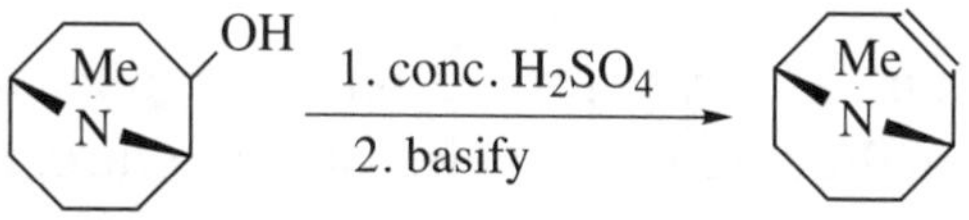

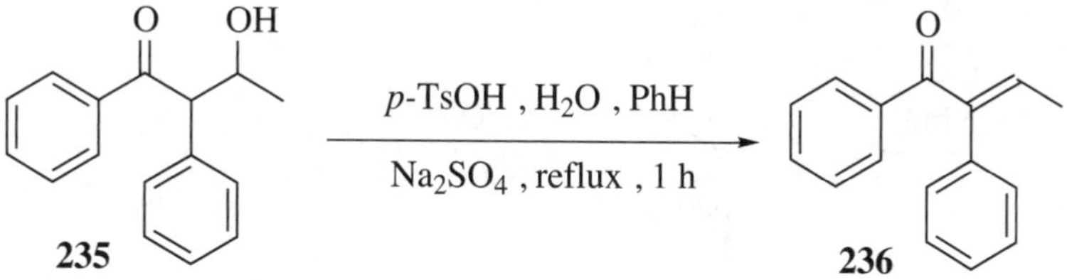

2.9.C. Syn-Elimination Reactions

Both E2 and E1 reactions require a base and are intermolecular processes. For E2 reactions the hydrogen removed by the base must have an anti relationship to the leaving group. Another type of elimination process is possible if the basic atom is part of, or tethered to the substrate, but the β-hydrogen can only be removed if it eclipses the basic atom in a syn conformation.

220. Shiinaa, I.; Sanoa, Y.; Nakata, K.; Suzuki, M.; Yokoyama, T.; Sasaki, A.; Orikasa, T.; Miyamoto, T.; Ikekita, M.; Nagahara, Y.; Hasome, Y. *Bioorganic & Med. Chem.* **2007**, *15*, 7599.

If the tethered base is also a leaving group, removal of the syn β-hydrogen is possible (an intramolecular process). This elimination process requires higher reaction temperatures since the requisite syn conformation is higher in energy (sec. 1.5.A). Such reactions are termed **syn elimination**.[104] There are several examples of this reaction involving different tethered bases that can also serve as leaving groups, and several are named reactions.

2.9.C.i. Hofmann Elimination. In a syn elimination, experimental results show that the less stable alkene is formed, and that the reaction is under kinetic control rather than thermodynamic control. The thermodynamically controlled E2 reactions require an anti-transition state. The kinetically controlled syn-elimination, however, requires a syn-transition state in which the β-hydrogen and the leaving group assume an eclipsed conformation. When there is more than one possibility, the lowest energy syn-eclipsed rotamer will be important for syn-eliminations. A classical example of this process is the **Hofmann elimination**,[108,104] illustrated by conversion of 2-bromo-3-methylbutane (**237**) to 3-methyl-1-butene (**240**). The bromide is first converted to ammonium bromide **238** by an S_N2 reaction with trimethylamine. Ammonium salt **238** does not possess a basic atom or group (the bromide counterion is a weak basic). Bromide must be replaced with a stronger base, such as hydroxide before elimination can occur. Reaction with silver oxide (Ag_2O) in the presence of 1 equivalent of water (water cannot be the solvent or a co-solvent) gives the corresponding trimethylalkylammonium hydroxide (**239**). If water is present in excess, the ionic base would be solvent separated from the ionic ammonium moiety and an E2 reaction would result. In the absence of an ion separating solvent, however, the basic hydroxide is tethered to the ammonium ion by ionic bonding. Hydroxide can react with the β-hydrogen only when that C–H bond eclipses the C–NMe$_3^+$ ⁻OH bond. When hydroxide reacts with this hydrogen, water is formed along with a π bond via expulsion of trimethylamine. In some cases, there is evidence that a nitrogen ylid intermediate is generated,[221] (sec. 8.8.C) but direct removal of hydrogen by hydroxide accounts for the product in most cases.[222]

It is possible to use the standard E2 model to evaluate the selectivity of the Hofmann elimination, when protic solvents such as water or ethanol are employed. In such a case, solvent separated ammonium and hydroxide ions are expected with an anti-transition state for removal of a β-hydrogen. First, note that in **239** there are two acidic β-hydrogens, H_a and H_b. The *anti-*

---

221. (a) Cope, A.C; Mehta, A.S. *J. Am. Chem. Soc.* **1963**, *85*, 1949, (b) Baldwin, M.A.; Banthorpe, D.V.; Loudon, A.G.; Waller, F.D. *J. Chem. Soc. B*, **1967**, 509; (c) Ingold, C.K. *Proc. Chem. Soc.* **1962**, 265; (d) Banthorpe, D.V.; Hughes, E.D.; Ingold, C.K. *J. Chem. Soc.* **1960**, 4054.
222. (a) Julian, P.L.; Meyer, E.W.; Printy, H.C. *J. Am. Chem. Soc.* **1948**, *70*, 887; (b) Banwell, M.G.; Austin, K.A.B.; Willis, A.C. *Tetrahedron* **2007**, *63*, 6388.

conformations for removal of H_a and also H_b in **239** are considered. The *anti*-conformation **241** leads to the less substituted alkene 3-methyl-1-butene (**240**) via removal of H_a. The *anti*-conformation **242** leads to the more substituted alkene 2-methyl-2-butene via removal of H_b. In **242** there is a *gauche*-interaction (see circled groups) that raises the energy of the resulting transition state relative to that from **241**, where this interaction is less. Conformation **241** is lower in energy and its transition state gives the major product, the less substituted alkene 1-pentene.

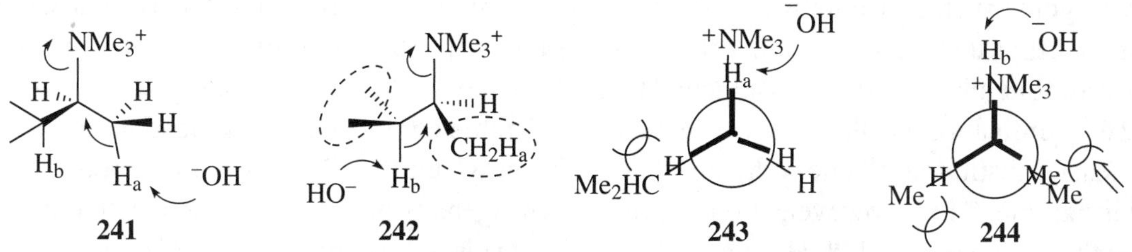

An alternative analysis of the Hofmann elimination will focus on a *syn*-conformation and intramolecular deprotonation via a *N*-ylid (see sec. 8.8.C). If the solvent is aprotic rather than or if there is no solvent at all (neat), one expects the ammonium hydroxide moiety will exist as a tight ion pair (no solvent separation). The only facile mechanism for removal of a β-hydrogen is, therefore, an intramolecular process that demands a *syn*-transition state. We return to **239** and the problem of two acidic β-hydrogens, H_a and H_b. Removal of either hydrogen via an intramolecular process requires a syn conformation, so **243** and **244** are the two important eclipsed conformations that must be considered. The methyl-methyl and methyl-hydrogen interactions in **244** (see ⟸) are significantly higher in energy than the hydrogen-hydrogen and isopropyl-hydrogen interactions that arise in **243** (see sec. 1.5.A). Removal of H_a from the lower energy **243** will lead to the favored lower energy transition state and the observed major product **240**. A synthetic transformation taken from Banwell and co-worker's synthesis of (+)-hirsutic acid uses the Hofmann elimination and also illustrates an alternative approach to this reaction. Reaction of iodomethane with the dimethylamine unit in **245** leads to the trimethylammonium salt. The usual Hofmann elimination sequence, with syn-elimination, gave the vinyl derivative, **246**.[222b] This variation is important since using an amine as a Hofmann precursor is often more convenient than relying on an S_N2 reaction of an amine and a halide to produce the requisite ammonium salt (sec. 2.6.A).

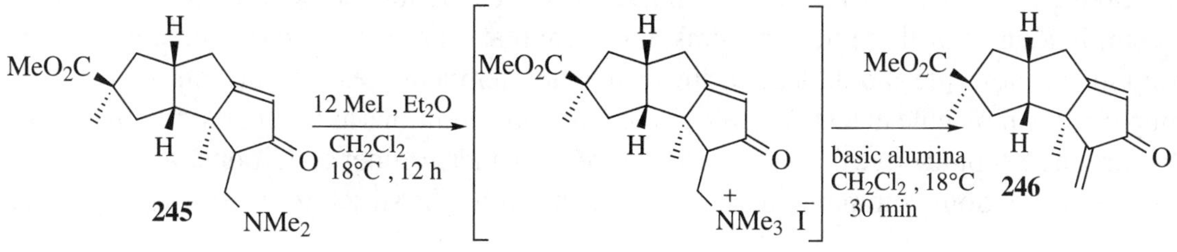

Note that solvent effects are quite important in this reaction. A solvent that promotes complete solvent separation of the ammonium ion and the hydroxide ion will favor E2 and *anti*-elimination, whereas a solvent that favors a tight ion pair will favor syn-elimination. In water, one anticipates extensive ion separation, but in ethanol or methanol, solvent separation is less efficient and the elimination may proceed by more than one mechanistic pathway.

Vital to the success of the Hofmann elimination is the use of an amine that does not possess β-hydrogens, such as trimethylamine or triphenylamine. Triethylamine has β-hydrogens on all three ethyl units and reaction with 3-iodo-2-methylpentane will give ammonium salt **247** where all four alkyl substituents attached to nitrogen have a β-hydrogen that can be removed by a base. Elimination via the most energetically favorable conformation will lead to the smallest and least substituted alkene, ethylene (from the ethyl units).[221a,b] The other product is the tertiary amine, **248**. However, if one of the β-hydrogens is more acidic, such as that adjacent to the thiophene ring in **249**, Hofmann elimination leads to formation of a single alkene, **250**.[223] This example also illustrates opening a bicyclic ring system to give a 10-membered ring, which is difficult to form using cyclization techniques (see sec. 6.6.B).

Syn-elimination can be extended to systems other than those required for Hofmann elimination, but a common feature is the presence of a basic atom connected to the molecule that is also a leaving group. The basic end of the molecule in the new system should be negatively charged or at least negatively polarized. As the dipole moment of the basic end of the molecule increases, the attraction for the δ^+ proton should increase, leading to greater facility of the syn elimination via transfer of hydrogen to the basic atom. Typically, this means that the reaction should occur at a lower reaction temperature. If the tethered base is a negatively charged ion, such as hydroxide in the Hofmann elimination, ion pairing with the cation is required if the syn-elimination is to occur and this places stringent solvent requirements on the system (see above). The leaving group ability of the base will influence the reaction temperature and facility of the syn elimination. For these new systems, the combination of a good base and a good leaving group in the molecule should lead to lower reaction temperatures. If one can build a leaving group into the molecule that is unstable to the reaction conditions as it departs, fragmentation to other products should drive the reaction to completion under milder conditions. A number of syn elimination reactions commonly used in synthesis exhibit the structural features just presented.

223. Berkes, D.; Netchitaïlo, P.; Morel, J.; Decroix, B. *Synth. Commun.* **1998**, *28*, 949.

2.9.C.ii. Amine Oxide Pyrolysis (Cope Elimination). Using the Hofmann elimination as a prototype, modification of the leaving group can facilitate syn-elimination. Specifically, the hydroxide ion (the base) and ammonium ion (the leaving group) pair can be replaced with a basic atom or group that is part of the leaving group. Cope found that incorporation of an amine *N*-oxide ($R_3N–O$) satisfied this new criterion (see **251**). The negative oxygen dipole of an amine oxide function as a base, and removal of the β-hydrogen via a syn conformation leads to cleavage of the C–N bond and loss of the resulting neutral leaving group, *N,N*-dimethylhydroxylamine. The base is part of the molecule and *not* a separate molecule, and the E2 elimination is not competitive. Such reactions generally require higher reaction temperatures relative to an E2 process. Heating amine oxides to produce alkenes is known as **Cope elimination**,[224,108d,104] and a typical reaction temperature is 120°C, significantly lower than that required for the Hofmann elimination as shown for the conversion of **251** to **252**.[108d] A Cope elimination requires the synthesis of *N*-oxides, and these are usually formed by oxidation of the corresponding amine with hydrogen peroxide or *m*-chloroperoxybenzoic acid[225] (sec. 3.9.B). The reaction is not limited to dimethylamines, although the temperatures required for elimination will vary with the structure of the amine oxide. In a synthesis of (–)-cuparene,[226] Grainger and Patel oxidized amine **253** to *N*-oxide **254** in 83% yield with *m*-chloroperoxybenzoic acid. Subsequent heating to 200°C in DMSO, in the presence of microwave irradiation for one minute, gave a 72% yield of **255**. In the absence of microwave irradiation, heating **254** to 60°C in THF for 8 hours gave a 40% yield of **255**. Note that alkenyl-*N*-methylhydroxylamines can undergo a reverse Cope elimination to give pyrrolidine- or piperidine *N*-oxides.[227]

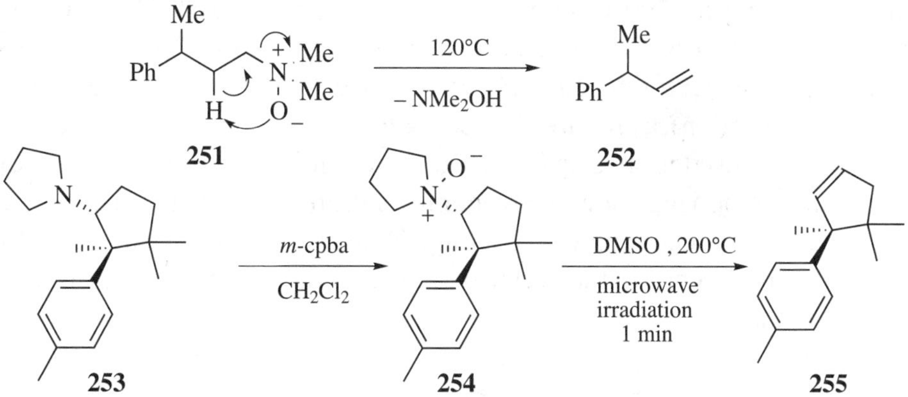

224. (a) Cope, A.C.; Foster, T.T.; Towle, P.H. *J. Am. Chem. Soc.* **1949**, *71*, 3929; (b) Cope, A.C.; Pike, R.A.; Spencer, C.F. *Ibid* **1953**, *75*, 3212; (c) The Merck Index, 14th Ed. Merck & Co., Inc., Whitehouse Station, New Jersey, **2006**, p ONR-19.
225. (a) Cope, A.C.; Ciganek, E. *Org. Syn. Coll. Vol. 4* **1963**, 612, (b) Craig, J.C.; Purushothaman, K.K. *J. Org. Chem.* **1970**, *35*, 1721.
226. Grainger, R.S.; Patel, A. *Chem. Commun.* **2003**, 1072.
227. Ciganek, E.; Read, Jr., J.M.; Calabrese, J.C. *J. Org. Chem.* **1995**, *60*, 5795.

2.9.C.iii. Ester Pyrolysis.[228,104] Other functional groups can be incorporated into a molecule that will facilitate syn-elimination via intramolecular removal of a β-hydrogen. A simple example is an acetoxy group (–OCOMe), where the carbonyl oxygen of the ester functions as a tethered base. Clearly, this oxygen is a weak base and acetic acid from acetate is a poorer leaving group relative to an amine from an ammonium salt or a *N,N*-dialkylhydroxylamine from an amine oxide. Indeed, syn elimination of acetates (with loss of acetic acid) requires higher reaction temperatures. On the other hand, acetates are readily available by reaction of an alcohol with acetic anhydride or acetyl chloride and pyridine (or another base). For **256**, heating to 450°C was required for elimination to the alkene product (**257**) and in this example the deuterium was retained.[228b,229] Similar pyrolysis of the enantiomer (**258**) gave alkene **259** in which the deuterium was lost. This finding is completely consistent with elimination of hydrogen or deuterium from the β-carbon by the acetoxy group from the lowest energy syn-rotamer.[228c,104]

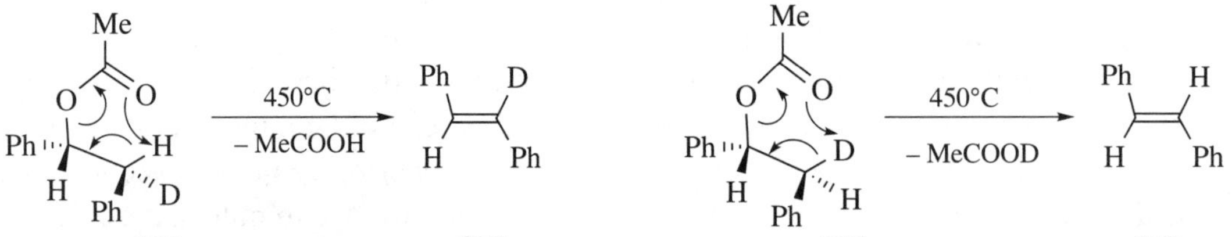

The importance of the syn relationship of a β-hydrogen and the leaving group is seen in hydrindanes **260** and **261**, which both produced **262** upon pyrolysis. The syn-derivative (**260**) required temperatures at least 200°C higher than the anti-acetate (**261**).[228b] Asynthetic example taken from Basabe and co-worker's synthesis of (+)-lagerstronolide from (+)-sclareol,[230] illustrates two key points with this reaction. When diacetate **264** was heated on silica gel (a variation of the basic reaction), a mixture of three alkenes were formed in 90% yield, which were difficult to separate and used in subsequent reactions as a mixture. It is likely that **264** was the favored product and isomerized under the reaction conditions. Note that such isomerization of the exo-methylene group is common with heating. Note also that the acetate unit at the tertiary carbon eliminated in preference to the acetate at the secondary carbon, which is also typical in this type of syn elimination.

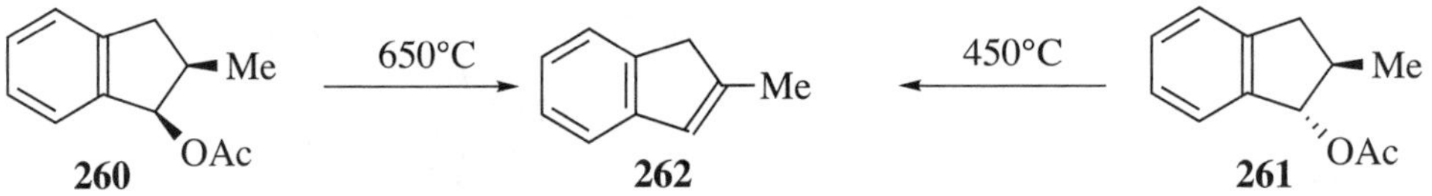

228. (a) Alexander, E.R.; Mudrak, A. *J. Am. Chem. Soc.* **1950**, *72*, 1810; (b) Idem *Ibid* **1950**, *72*, 3194; (c) Curtin, C.Y.; Kellom, D.B. *Ibid* **1953**, *75*, 6011.
229. Jung, M.E.; Lyster, M.A. *J. Org. Chem.* **1977**, *42*, 3761.
230. Basabe, P.; Bodero, O.; Marcos, I.S.; Diez, D.; de Román, M.; Blanco, A.; Urones, J.G. *Tetrahedron* **2007**, *63*, 11838.

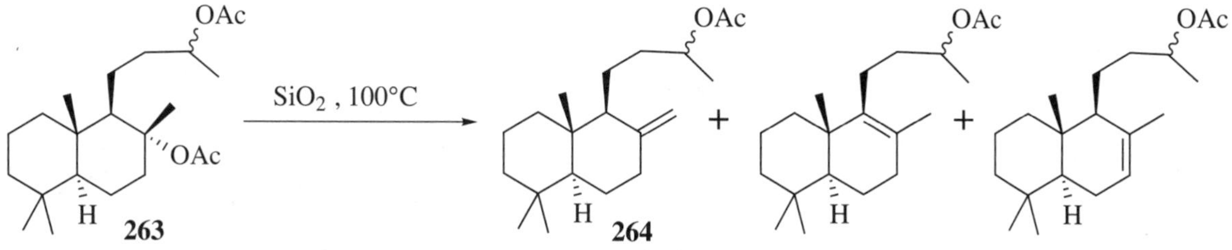

2.9.C.iv. Xanthate Ester Pyrolysis.[231,104,228a,b] The high temperatures required for acetate pyrolysis greatly limits the synthetic utility of this reaction. Improved leaving group ability and a greater basicity of the tethered base would improve the reaction if departure of the leaving group generated a product that was more stable or if that product decomposed after its departure. In either case the elimination would be more facile. The sulfur analog of a carbonate is called a xanthate, O–(C=S)–SR, and satisfies the goals for improved syn-elimination. The xanthatae group is a better leaving group than acetate, but sulfur is a poorer base than oxygen, although it is a better nucleophile.[232,93,101,102] Elimination occurred at significantly lower temperatures relative to an acetate (~ 200°C), and the neutral molecule **268** was lost as a byproduct. This product fragmented to COS and methanethiol, making the elimination essentially irreversible. In a specific example, the thiocarbonyl group of xanthate **265** fragmented to **266** and **267**, with the less substituted **266** being the major isomer. This reaction is one step of the **Chugaev elimination**[233,231] used for conversion of alcohols to alkenes.[231] The xanthate ester was prepared from the corresponding alcohol by reaction with carbon disulfide to give a thioanion (R–OCS$_2^-$), followed by reaction with iodomethane to give **265**. A synthetic example is taken from a synthesis of (–)-kainic acid by Ogasawara and co-workers,[234] in which alcohol **269** was converted to xanthate ester **270** in 92% yield. Subsequent heating in refluxing diphenyl ether gave **271**, but under these conditions an ene reaction (sec. 11.13) led to formation of **272** in 72% yield. Since we have not yet discussed the ene reaction, the key carbon atoms are numbered in **271** and **272** so that one can follow the transformation.

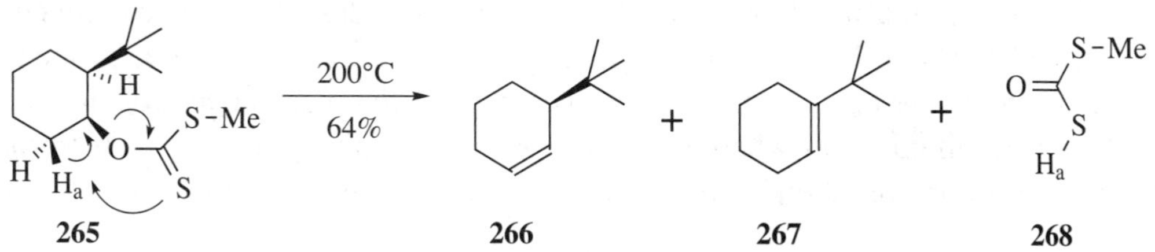

231. Nace, H.R. *Org. React.* **1962**, *12*, 57.
232. (a) Wells, P.R. *Chem. Rev.* **1963**, *63*, 171, (b) Koskikallio, J. *Acta Chem. Scand.* **1969**, *23*, 1477, 1490.
233. (a) Tschugaeff, L. *Berichte* **1899**, *32*, 3332; (b) DePuy, C.H.; King, R.W. *Chem. Rev.* **1960**, *60*, 441 (see p 444); (c) The Merck Index, 14th Ed. Merck & Co., Inc., Whitehouse Station, New Jersey, **2006**, p ONR-17.
234. Nakagawa, H.; Sugahara, T.; Ogasawara, K. *Org. Lett.* **2000**, *2*, 3181.

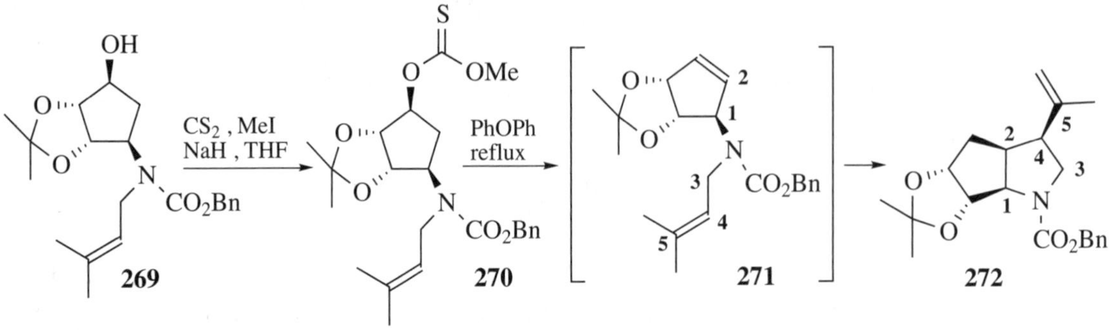

There are alternative elimination pathways for xanthate esters. Epoxy-alcohol **273** was converted to xanthate ester **274** in quantitative yield using standard conditions. The elimination step used triethylsilane (Et_3SiH), a radical initiator such as AIBN, and treatment with tetrabutylammonium fluoride (TBAF). Under these conditions, opening of the epoxide ring was accompanied by elimination to give **275** in 72% yield in Ogasawara and co-worker's synthesis of (+)-frontalin.[235] This elimination is related to the **Barton-McCombie reaction** involving reduction of xanthates under radical conditions (see sec. 4.9.H).

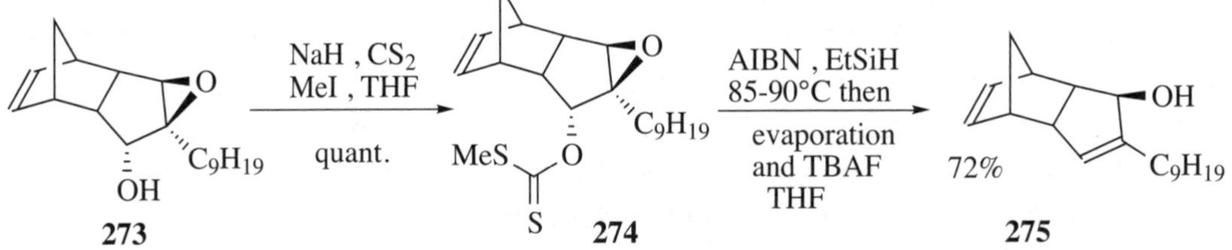

2.9.C.v. Sulfoxide Pyrolysis.[236] A very useful change in the syn elimination precursor incorporates the polarized oxygen of a sulfoxide. If a sulfide is oxidized with peroxide or with sodium *meta*-periodate ($NaIO_4$),[237] the resulting sulfoxide has a highly polarized S–O bond (sec. 3.9.A). The negatively polarized oxygen of sulfoxides (*S*-oxides) can function as a tethered base with the O–S–R moiety serving as a leaving group. If the β-hydrogen and the S–O moiety assume an eclipsed conformation, the basic oxygen removes the hydrogen to generate PhS–OH (benzenesulfenic acid where sulfur is bonded to a phenyl group). This product is unstable to the reaction conditions and decomposes, facilitating the elimination. An example

235. Kanada, R.M.; Tankguchi, T.; Ogawawara, K. *Tetrahedron Lett.* **2000**, *41*, 3631.
236. (a) Grieco, P.A.; Reap, J.J. *Tetrahedron Lett.* **1974**, 1097; (b) Montanari, F. *Int. J. Sulfur Chem. C,* **1971**, *6*, 137 [*Chem. Abstr.* 76:24243b **1972**]; (c) Trost, B.M.; Salzmann, T.N. *J. Org. Chem.* **1975**, *40*, 148; (d) Kingsbury, C.A.; Cram, D.J. *J. Am. Chem. Soc.* **1960**, *82*, 1810.
237. (a) Grieco, P.A.; Miyashita, M. *J. Org. Chem.* **1974**, *39*, 120; (b) Mitchell, R.H. *J. Chem. Soc. Chem. Commun.* **1974**, 990; (c) Sharpless, K.B.; Young, M.W.; Lauer, R.L. *J. Am. Chem. Soc.* **1973**, *95*, 2697; (d) Sharpless, K.B.; Young, M.W.; Lauer, R.L. *Tetrahedron Lett.* **1973**, 1979; (e) Jones, D.N.; Mundy, D.; Whitehouse, R.D. *J. Chem. Soc. Chem. Commun.* **1970**, 86; (f) Reich, H.J.; Renga, J.M.; Reich, I.L. *J. Org. Chem.* **1974**, *39*, 2133, (g) Sharpless, K.B.; Young, M.W. *Ibid* **1975**, *40*, 947.

is taken from a synthesis of precursors of the agalacto (exo) fragment of the quartromicins by Roush and Qi,[238] in which the sulfide unit in **276** was oxidized to the corresponding sulfoxide (**276**) with *m*-chloroperoxybenzoic acid (sect. 3.9.A.i). Subsequent heating in toluene gave **278** in good yield.

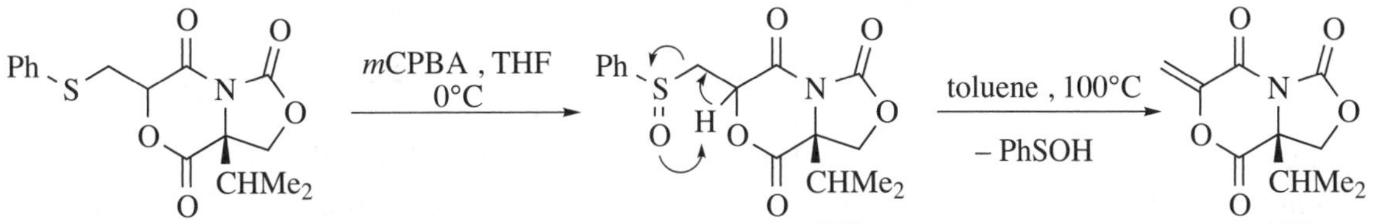

In addition to phenyl sulfoxides, methyl sulfoxides (–S(O)Me) can also be prepared, and they undergo syn-elimination when heated, but the temperature required for pyrolysis is ≈ 150°C. When R is phenyl, elimination can often be observed by heating to only 80°C, although toluene at 100°C was used for **277**.

2.9.C.vi. Selenoxide Pyrolysis.[238] Just as sulfides are oxidized to sulfoxides, selenides (R-Se-R) can be oxidized to selenoxides. Analogous to a sulfoxide, heating a selenoxide leads to thermal syn-elimination to give the less substituted alkene. The increased polarity of the Se-O bond of the selenoxide, relative to the S-O bond of the sulfoxide, and the loss of the unstable R-Se-OH leads to even lower temperatures for thermal syn-elimination (typically 0-25°C). Elimination of PhSeOH from **279** gave the exocyclic-methylene derivative (**280**) as the major product, rather than the endocyclic alkene.[250a] This elimination occurred at 0°C, and selenoxide elimination is a particularly attractive method for the generation of sensitive alkenes such as the less stable exocyclic methylene lactone (**280**). In this particular case, the steric constraints for isomerization found in the bicyclic selenoxide **280** are also important. Note that *selenium compounds are **toxic*** and great care should be exercised in the handling and disposal of all materials.

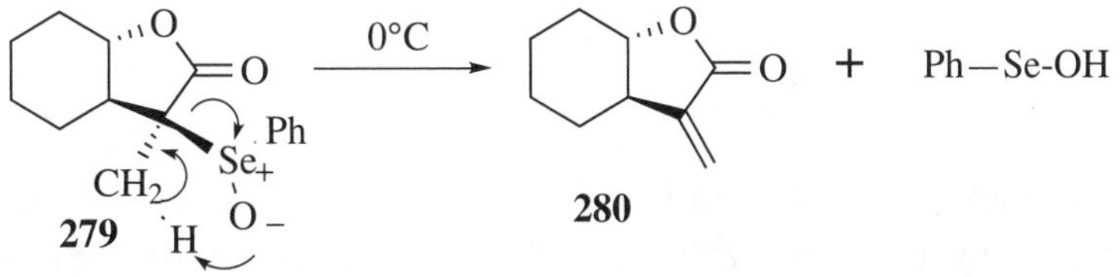

There are many synthetic examples that use this technique. In a synthesis of (–)-aristolochene by Pedro and co-workers,[239] selenide **281** was treated with aqueous hydrogen peroxide at 0°C to give the selenoxide. Simply warming to ambient temperature induced elimination to give alkene **282** in 69% yield, despite the presence the sensitive epoxide functional group.

---

238.   Qi, J.; Roush, W.R. *Org. Lett.* **2006**, *8*, 2795.
239.   Blay, G.; Cardona, L.; Collado, A.M.; Garcia, B.; Pedro, J.R. *J. Org. Chem.* **2006**, *71*, 4929.

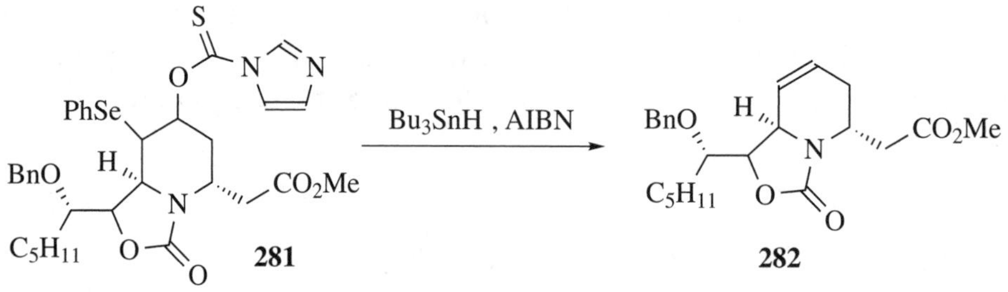

Selenides can undergo another type of elimination reaction, when the selenium unit is attached to a carbon adjacent to a thiocarbamate group as in **281**. Radical induced elimination with AIBN in the presence of a hydrogen atom donating species such as tributyltin hydride (see sec. 13.3) gave the elimination product **282** in 82% yield, which was used in Comins and Kuethe' synthesis of (+)-cannabisativine.[240] Note that **281** was generated *in situ* from an alcohol precursor and the thiocarbamate formation-elimination steps were done in a one-pot reaction.

Factors other than bond polarization and leaving group ability influence the ease of thermolysis, illustrated by xanthate ester pyrolysis. DePuy and King state "the much greater ease of xanthate pyrolysis as compared to the pyrolysis of esters must be due to an added driving force as a consequence of conversion, in the transition state, of the system –O–CS into the system O=C–S–. Examination of the relative bond energies shows that conversion should be exothermic to the extent of nearly 20 kcal mol^{-1} (83.7 kJ mol^{-1})".[104] Bond polarization is, however, a convenient handle for predicting the relative ease of syn elimination in many cases, although it accounts for only part of the difference in actual reactivity and product formation. It is also known that *n*-alkyl aryl selenides that have an ortho-substituent on the aromatic ring are superior reagents for oxidation and syn-elimination when compared to those compounds having a para-substituent.[241] Structural effects therefore play a role in the reaction.

The retrosynthetic transform for syn elimination is that shown for the anti elimination reactions except that the less substituted alkene is the target:

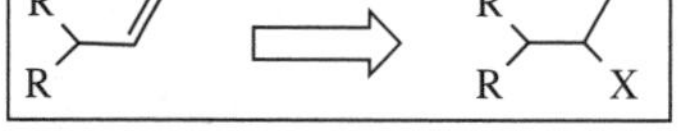

2.9.C.vii.   Burgess Reagent.[242]   Another reagent, (carboxysulfamoyl)triethylammonium hydroxide inner salt methyl ester (**283**, the **Burgess reagent**) follows a syn-elimination mechanism, and is quite useful for introducing an alkene unit into a substrate. Reaction of

240.   Kuethe, J.T.; Comins, D.L. *Org. Lett.* **2000**, *2*, 855.
241.   Sayama, S.; Onami, T. *Tetrahedron Lett.* **2000**, *41*, 5557.
242.   (a) Burgess, E.M.; Penton Jr., H.R.; Taylor, E.A. *J. Org. Chem.* **1973**, *38*, 26; (b) Atkins Jr., G.M.; Burgess, E.M. *J. Am. Chem. Soc.* **1968**, *90*, 4744; (c) Crabbé, P.; León, C. *J. Org. Chem.* **1970**, *35*, 2594. For a chiral Burgess reagent, see (d) Leisch, H.; Saxon, R.; Sullivan, B.; Hudlicky, T. *Synlett* **2006**, 445; (e) Sullivan, B.; Gilmet, J.; Leisch, H.; Hudlicky, T. *J. Nat. Prod.* **2008**, *71*, 346.

an alcohol with **283** leads to **284**, and heating initiates the elimination to give the alkene. The Burgess reagent does not appear to be as selective for syn elimination as the preceding reagents. In general, as stated directly by Burgess, there is a cis stereochemical constraint and Saytzeff elimination is observed in aliphatic cases.[242] When **283** was heated with **285** in acetonitrile, for example, a 1:1 mixture of the exocyclic alkene (**286**) and endocyclic alkene (**287**) was formed in 80% yield.[242] Rearrangement is observed in some cases. Nonetheless, the Burgess reagent can be quite valuable in fine organic synthesis. In a synthesis of momilactone A by Deslongchamps and Germain,[243] alcohol **288** was heated with **283** in refluxing toluene, to give a 40-50% yield of momilactone A (**289**).

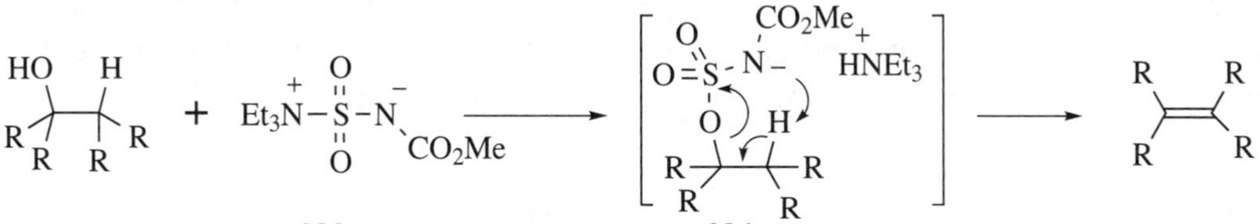

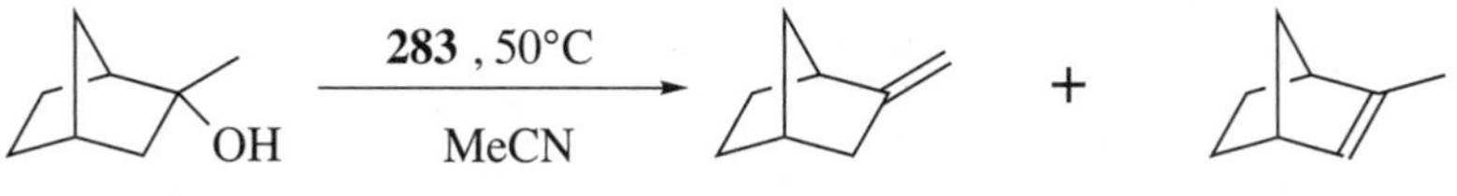

2.9.D. 1,3-Elimination (Decarboxylation)

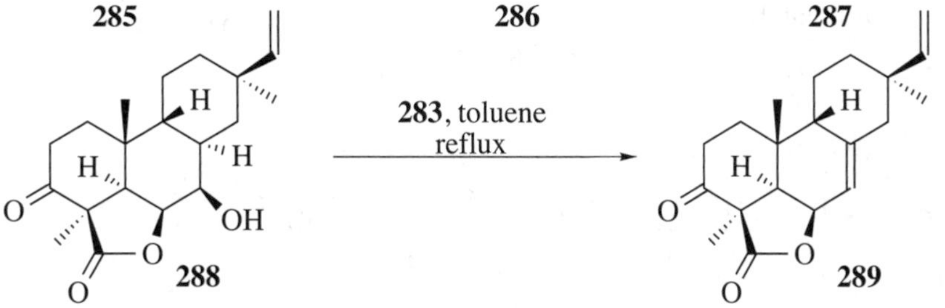

Sections 2.9.A-C focused on 1,2-elimination reactions. The syn elimination reactions just discussed showed that intramolecular removal of an acidic hydrogen leads to loss of a leaving group and formation of a π-bond. Elimination can also occur when an acidic hydrogen and a leaving group are separated by three atoms, in what is termed 1,3-elimination. When a carboxyl carbon is β to a π-bond (a carbonyl, alkene or aryl) the acidic hydrogen of the acid can be transferred to the oxygen of the carbonyl (or to the sp² carbon of alkenes or aryls) via a six-center transition state. This requires a syn-orientation of the carbonyl oxygen and the carboxyl hydrogen, as was observed in syn 1,2-elimination. This thermal process results in cleavage of the bond connecting the carboxyl carbon, loss of carbon dioxide and formation of a new π bond. This 1,3-elimination process is called **decarboxylation** and is a common reaction of β-keto-acids, β-carboxyl esters, and 1,3-diacids (malonic acid derivatives). A simple example

243. Germain, J.; Deslongchamps, P. *J. Org. Chem.* **2002**, *67*, 5269.

is the elimination of carbon dioxide from 3-oxopentanoic acid (**290**) to give an enol, which tautomerizes to 2-butanone. Structure **290** is drawn in a way that emphasizes the transfer of hydrogen to the oxygen of the ketone moiety occurs via a six-center transition state. The **malonic acid synthesis** (sec. 9.3.A) generates substituted malonic esters that can be saponified to diacids such as **292**. In one example, treatment of diethyl malonate with sodium ethoxide and then 6-bromo-(2*E*)-hexene gave a 72% yield of **291**.[244] Saponification with aqueous KOH gave a 99% yield of **292**, which upon heating to 160°C for 5 h gave a 94% yield of **294**, in Clive and Hisaindee's synthesis of brevioxime.[244] The initially formed product of decarboxylation is an enol (in this case **293**) and this enol form of the acid tautomerized to the isolated product, **294**. Decarboxylation begins with an internal acid-base reaction, where the acid is the O–H unit of the carboxylic acid, and the base is the oxygen of the carbonyl β to the acid moiety. Decarboxylation is assisted by loss of a neutral leaving group carbon dioxide.

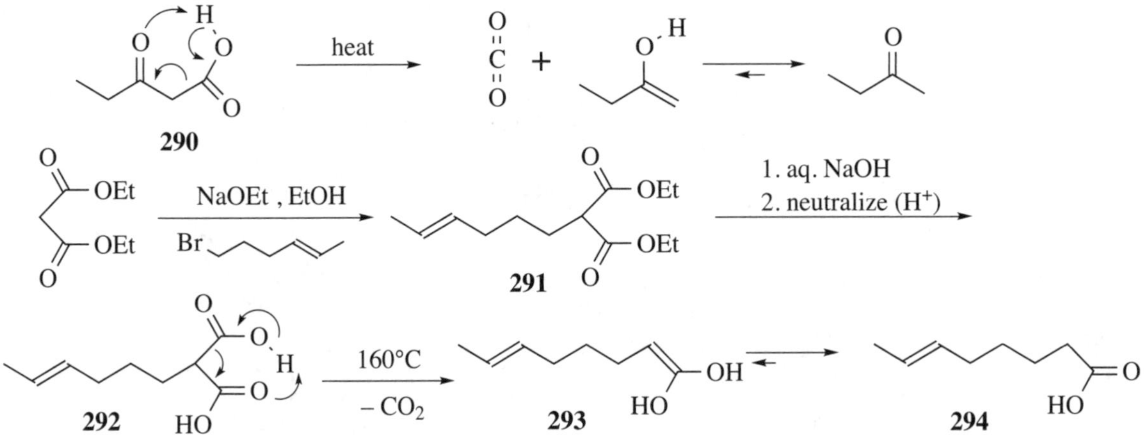

Krapcho and co-workers developed a mild decarboxylation procedure of esters.[245] In a synthesis of deacetoxyalcyonin acetate by Molander and co-workers,[246] β-keto methyl ester **295** (drawn in its enol form) was heated with LiCl in aqueous DMSO to give decarboxylation and a good yield of **296**. This mild procedure is referred to as **Krapcho decarboxylation**.[247] An alternative procedure has been reported. In a synthesis of (–)-gilbertine by Blechert and Jiricek,[248] heating **297** with DABCO, in toluene and water, gave **298** in 66% yield.

244. Clive, D.J.; Hisaindee, S. *J. Org. Chem.* **2000**, *65*, 4923.
245. (a) Krapcho, A.P.; Weimaster, J.F.; Eldridge, J.M.; Jahngen, Jr., E.G.E.; Lovey, A.J.; Stephens, W.P. *J. Org. Chem.* **1978**, *43*, 138; (b) Krapcho, A.P. *Synthesis* **1982**, 805, 893.
246. Molander, G.A.; St. Jean,Jr., D.J.; Haas, J. *J. Am. Chem. Soc.* **2004**, *126*, 1642.
247. (a) The Merck Index, 14th Ed. Merck & Co., Inc., Whitehouse Station, New Jersey, **2006**, p ONR-53; (b) Mundy, B.P.; Ellerd, M.G.; Favaloro Jr., F.G. *Name Reactions and Reagents in Organic Synthesis, 2nd Ed.* Wiley-Interscience, New Jersey, **2005**, pp. 380-381.
248. Jiricek, J.; Blechert, S. *J. Am. Chem. Soc.* **2004**, *126*, 3534.

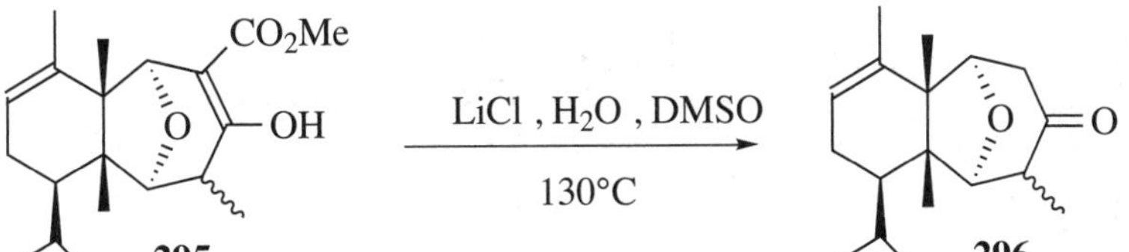

Decarboxylation is also observed with conjugated carboxylic acids such as **299**. In conjugated acids, the C=C unit attacks the acidic proton of the COOH unit via a six-centered transition state leading to decarboxylation. The alkene unit can be even further removed from the acid, as in Smith and co-worker's synthesis of zampanolide[249] in which **299** was heated to 130°C in quinoline for 3 hours. The proximal C=C unit functioned as the internal base, and decarboxylation led to a 79% yield of **300**. In general, the π bond of an alkene is a much weaker base than the oxygen of a carbonyl so the hydrogen transfer is less efficient. This fact usually means that higher reaction temperatures are required for the decarboxylation, and additives such as quinoline are often required.

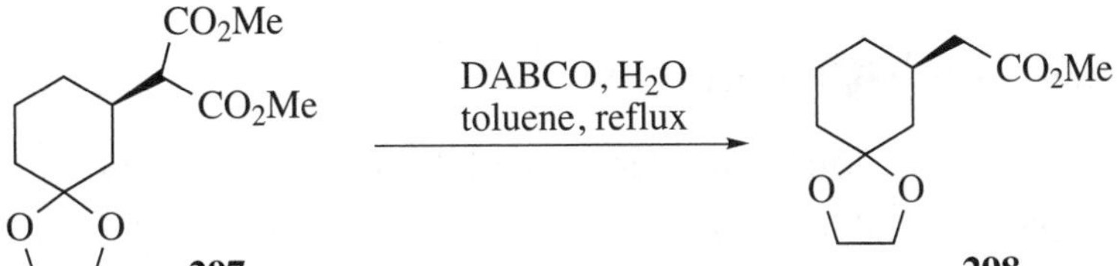

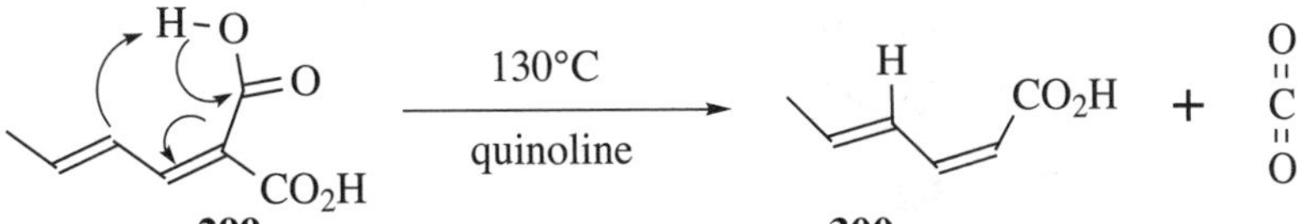

Decarboxylation of conjugated acids proceeds with bond migration, as shown in the transformation of **299**→**300**. Such bond migration must be accounted for in any synthetic plan. Another source of π bonds are aromatic rings. Transfer of the acidic hydrogen of the acid to the π-bond of an aryl carbon is more difficult than with simple alkenes. First, the π bond is a weaker base due to resonance delocalization and second, accepting the hydrogen requires that the aromatic nature of the benzene ring be disrupted. For both reasons higher energy is required for the reaction, which means higher reaction temperatures for thermolysis of **301** relative to benzylic systems such as **303**. The decarboxylation proceeds by the usual six-center transition state to give benzylidene intermediate **302**, which is unstable to the thermal conditions of the reaction and quickly aromatizes to **303**.

---

249. Smith III, A.B.; Safonov, I.G.; Corbett, R.M. *J. Am. Chem. Soc.* **2002**, *124* 11102.

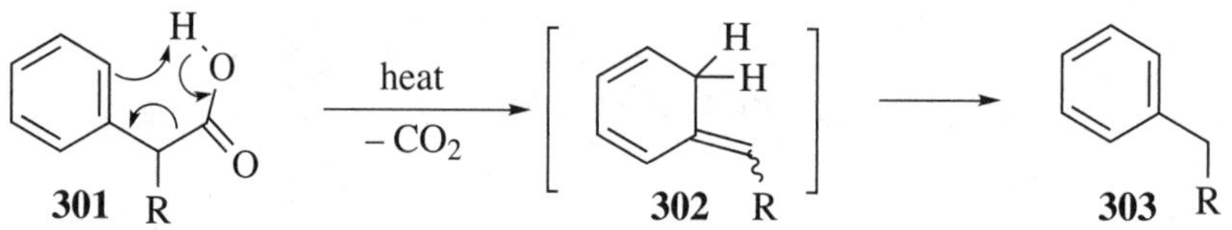

2.9.E. 1,3-Elimination (Grob Fragmentation)

Both 1,2-elimination reactions and decarboxylation reactions require attack at an acidic hydrogen, which is accompanied by concomitant bond cleavage that releases a leaving group and forms a π bond. Other types of elimination reactions are possible if the electron-donating species and the leaving group are properly positioned. If the basic atom and the leaving group are separated by at least three carbons and the bonds that are made and broken assume an anti relationship, elimination is possible. Prelog first observed this reaction in work that solved the structure of quinine and other Cinchona alkaloids. This work involved degradation studies that included treatment of **304** with KOH and silver nitrate. Silver ion reacted with the bromide to give a carbocation (**305**).

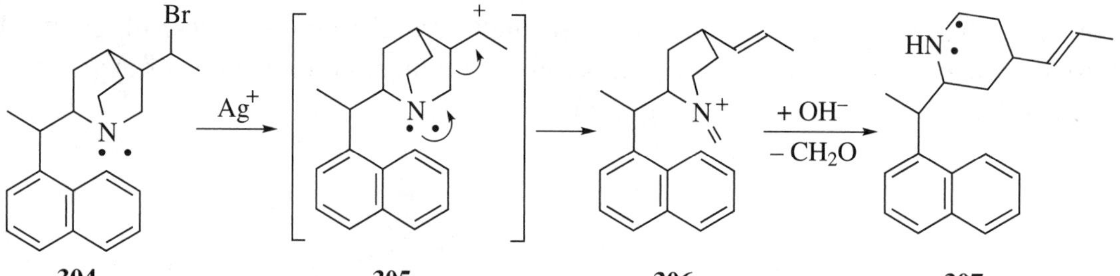

Subsequently, the nitrogen lone pair induced a 1,3-elimination ring-opening reaction, as shown, to give iminium salt **306**.[250] In the presence of aqueous hydroxide in the reaction medium, addition to the iminium salt led to loss of formaldehyde and formation of the final product, **307**. Transfer of the nitrogen lone electron pair toward the leaving group (or the carbocation center) required the N–C bond and the C–C$^+$ bond to have an anti relationship. This anti relationship is taken from Prelog's proposed synchronous mechanism for the 1,3-elimination, and can be generalized to say that "both the C_α–X bond and the orbital of the nitrogen lone pair be antiperiplanar to the C_β-C_γ bond".[251,252] Grob, et al. later expanded this 1,3-elimination to include amino halides and amino sulfonates.[253] The synchronous nature of the reaction was confirmed by Grob and is illustrated by the elimination of the tosylate group in **308** and **310**.[252] Elimination of the 5β-isomer (**308**) gave the (*E*)-isomer (**309**), whereas elimination of

250. (a) Prelog, V.; Zalán, E. *Helv. Chim. Acta* **1944**, *27*, 535; (b) Prelog, V.; Häfliger, O. *Ibid* **1950**, *33*, 2021.
251. Klyne, W.; Prelog, V. *Experientia* **1960**, *16*, 521.
252. Grob, C.A. *Angew. Chem. Int. Ed.* **1969**, *8*, 535 and references cited therein.
253. Grob, C.A.; Kiefer, H.R.; Lutz, H.J.; Wilkens, H.J. *Helv. Chim. Acta* **1967**, *50*, 416.

the 5α-isomer gave the (*Z*)-isomer (**311**), along with **311**. The stereochemistry of the products is best explained if the pertinent orbitals are antiparallel in the transition state leading to the elimination. The reaction rates for both of these substrates were rather high, but the 5α-isomer (**310**) reacted much slower and gave a mixture of the substitution product (**311**) in addition to the elimination product (**312**). This latter result led to a more general conclusion that the usual S_N2 reaction expected with a tosylate and a nucleophile such as hydroxide will occur unless the rate of the synchronous elimination is fast. Although the reaction was discovered by Prelog, Grob's contributions to this reaction led to its bearing his name, the **Grob fragmentation**.[254]

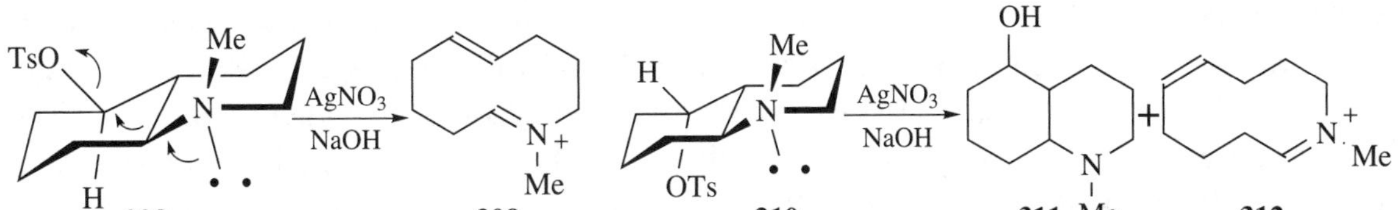

Both ionization and elimination processes are possible in these systems, as illustrated by generic structure **313**. Compound **313** gives the elimination products (iminium ion **314** and the alkene) with a rate constant k_f and the ionization product **315** with a rate of k_i. This latter pathway leads to substitution. The synchronous fragmentation can occur only if the k_f is comparable with or greater than the rate of ionization k_i for the γ-amino carbon ($k_f \geq k_i$).[253] If the stereochemical requirements are satisfied, k_f is the dominant term. This increase in reactivity due to synchronous fragmentation is referred to as the **frangomeric effect**, and expressed by $f = k_f/k_i$. This effect is not limited to amines or carbocations derived from alkyl halides and silver salts. Suitable derivatives of 1,3-diols also undergo this fragmentation.[255] A synthetic example is taken from Paquette and co-worker's synthesis of jatrophatrione,[256] in which mesyl-alcohol **316** was treated with potassium *tert*-butoxide, and fragmentation gave alkoxide **317** (hydrogen atoms removed in the 3D model for clarity). As shown, the mesylate leaving group is properly oriented relative to the alkoxide for a Grob fragmentation leading to **318**, in 98% yield.

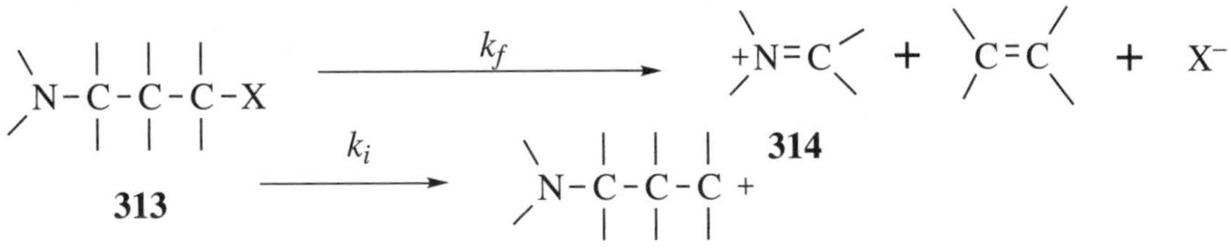

254. (a) The Merck Index, 14th Ed. Merck & Co., Inc., Whitehouse Station, New Jersey, *2006*, p ONR-39; (b) Mundy, B.P.; Ellerd, M.G.; Favaloro Jr., F.G. *Name Reactions and Reagents in Organic Synthesis, 2nd Ed.* Wiley-Interscience, New Jersey, *2005*, pp. 288-289.
255. Zimmerman, H.E.; English, Jr., J. *J. Am. Chem. Soc.* *1954*, *76*, 2285, 2291, 2294.
256. Paquette, L.A.; Yang, J.; Long, Y.O. *J. Am. Chem. Soc.* *2002*, *124*, 6542.

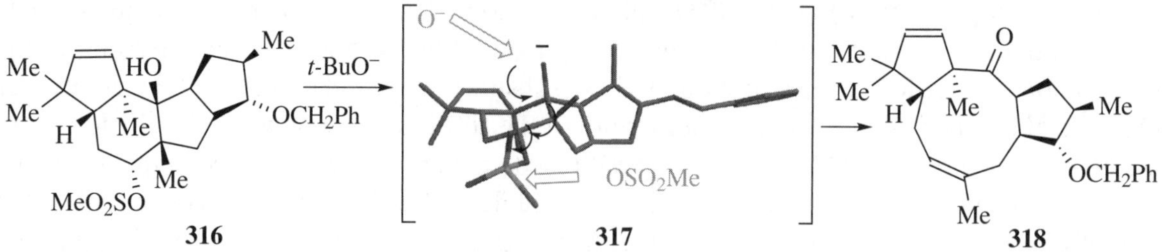

The alkoxide required to initiate a Grob fragmentation can be generated in several ways. Direct reaction of an alcohol with a suitable base to form the alkoxide is one way, as seen in **317**. An alternative route is carbanion addition to a carbonyl followed by a Grob fragmentation, as was used in Magnus and Buddhsukh's synthesis of hinesol.[257] Another example is taken from Baran and co-workers synthesis of vinigrol in which **319** was treated with potassium hexamethyldisilazide to give alkoxide **320**. The mesylate leving group is posiinted to allow the Grob fragmention, whcihn gave **321** in 93% yield[258] A radical-induced Grob-type fragmentation has also been reported in a synthesis of terpenoids.[259]

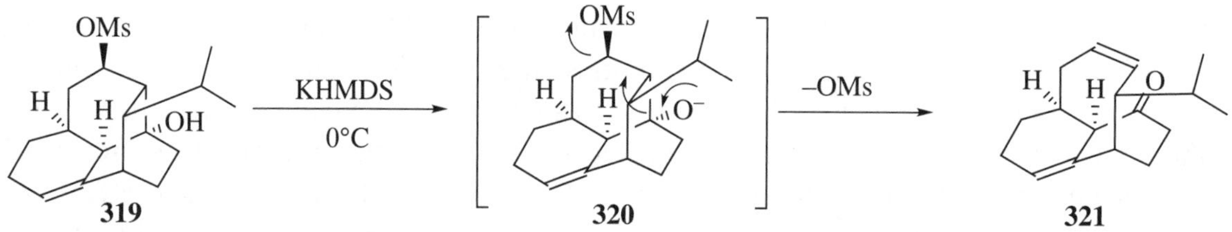

2.9.F. Other 1,3-Elimination Reactions

A number of 1,3-elimination processes other than those described above have been observed. A brief introduction of a few typical reactions will suffice. Treatment of diketone **322** with hydroxide initially gave the expected addition product **323**, but ring opening generated an enolate anion (**324**). Formation of the resonance stabilized enolate anion (sec. 9.2) was the driving force of this fragmentation. Internal proton transfer gave **325**, and hydrolysis led to the observed product, keto-acid **326**.[260]

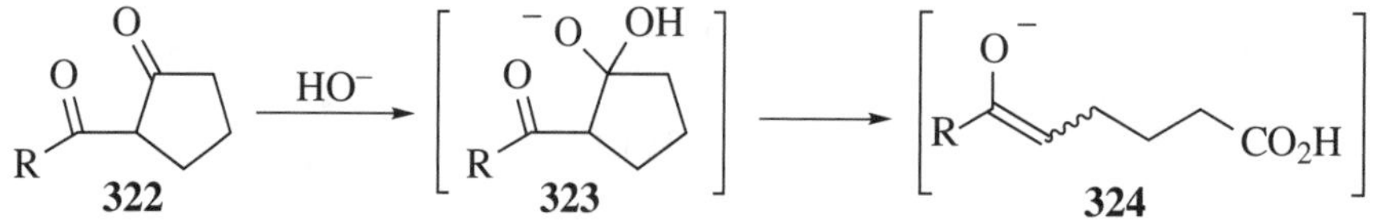

257. (a) Buddhsukh, D.; Magnus, P. *J. Chem. Soc. Chem. Commun.* **1975**, 952; (b) Chass, D.A.; Buddhsukh, D.; Magnus, P.D. *J. Org. Chem.* **1978**, *43*, 1750.
258. Maimone, T.J.; Voica, A.-F.; Baran, P.S. *Angew. Chemie. Int. Ed.* **2008**, *47*, 3054.
259. Lange, G.L.; Gottardo, C.; Merica, A. *J. Org. Chem.* **1999**, *64*, 6738.
260. Stetter, H. *Angew. Chem.* **1955**, *67*, 769.

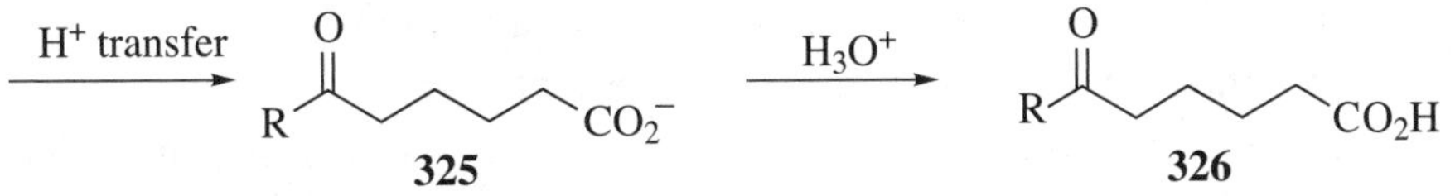

Elimination of carbon dioxide (CO_2) has been observed from β-halo carboxylates. When β-bromo acrylic acid **327** was treated with aqueous hydroxide, for example, 1,3-elimination of bromide from the carboxylate anion **328** gave alkene **329** along with carbon dioxide.[261] In some cases, homoallylic alcohols can eliminate an aldehyde moiety. An example is the pyrolysis of **330** at 500°C to give methylenecyclohexane (**331**) along with the aldehyde heptanal (**332**). This fragmentation occurs via a six-center transition state,[262] although it could be also be described as a *retro*-ene reaction (see sec. 11.13).

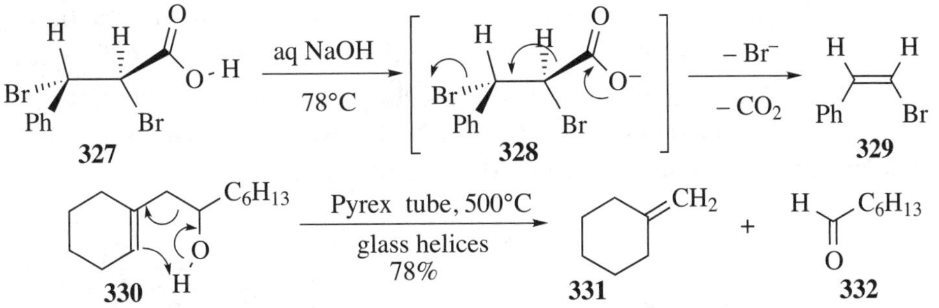

This section has described several synthetically important 1,2- and 1,3-elimination processes. In both cases substitution can be a competitive reaction, but proper choice of reaction conditions will maximize formation of the elimination products. Several interesting disconnections that are possible with substrates capable of undergoing 1,2- and 1,3-eliminations are shown.

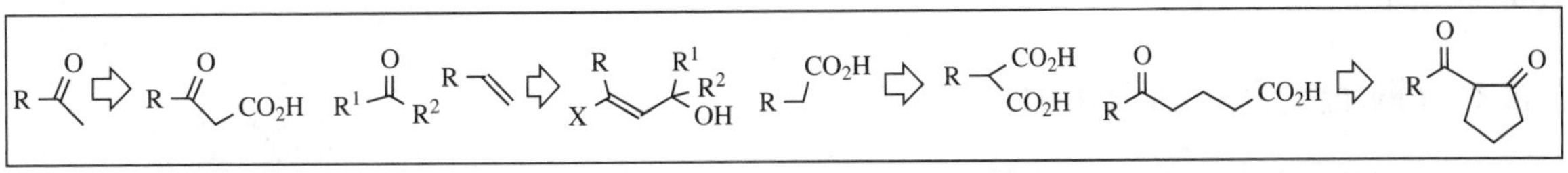

2.10. ADDITION REACTIONS

Acids and bases have figured prominently in all of the reactions described in this chapter. We have previously explored the concept that an alkene can function as a base. In this section, the π bond of an alkene or an alkyne will be used as a base with protonic acids, Lewis acids, and with other electrophilic atoms. When an alkene donates an electron pair (acts as a base) to a proton, the π bond breaks and forms a new C–H bond to one carbon of the old C=C unit. The

261. (a) Grovenstein, Jr., E.; Lee, D.E. *J. Am. Chem. Soc.* **1953**, *75*, 2639; (b) Cristol, S.J.; Norris, W.P. *Ibid* **1953** *75*, 2645.
262. Arnold, R.T.; Smolinsky, G. *J. Am. Chem. Soc.* **1959**, *81*, 6443.

other carbon of the C=C unit becomes a carbocation, and is subject to subsequent substitution or elimination reactions. If the alkene donates the electrons to a Lewis acid, the resulting complex will lead to other products. Recognizing the basicity of alkenes and alkynes will help explain the reactions discussed in this section.

The elimination reactions presented in Sec. 2.9 are important in synthesis, and usually involve conversion of an alcohol or alkyl halide to an alkene or alkyne. Many functional group interchange reactions involve interconversion functional groups, as in the conversion of an alkene or alkyne to an alkyl halide or an alkyl halide to an alkene or alkyne. The reaction in which a new atom or group is transferred to each carbon of the π bond forms two new sp^3 bonds as the π bond is broken, and the overall process is called an **addition reaction** since the atoms re "added" to the π-bond.[263] A general reaction for addition of a reagent to an alkene is the conversion of **333** to **334**. This addition process encompasses many reaction types, but for synthetic planning, it is convenient to classify addition reactions into four mechanistic categories. (*1*) Those involving free and metal stabilized ions as intermediates. (*2*) Those involving symmetrically bridged ions as intermediates. (*3*) Addition via a four-center transition state. (*4*) Nucleophilic acyl addition, discussed previously in sec. 2.5.

Nucleophilic acyl addition was discussed in sec. 2.5. The main examples of four-center transition state additions are hydroboration, which will be discussed in chapter 5, and hydride reduction of carbonyls, which will be discussed in section 4.2-4.5. This section will therefore focus on categories 1 and 2.

2.10.A. π-Bonds as Brønsted-Lowry Bases

The π-bond of an alkene can react as a Brønsted-Lowry base in the presence of a protonic acid such as HCl or HBr, where an electron pair is donated to the acid (H$^+$) to form a new C–H bond. The reaction only occurs with relatively strong mineral acids, so the alkene is considered to be a weak base. In a reaction with HBr and cyclohexene, the π-bond is broken as the new Br–C bond is formed and a carbocation is generated on the adjacent carbon atom of the C=C unit. The reaction of cyclohexene with HBr therefore generates secondary cation **335**. The cationic center in **335** is subsequently attacked by the nucleophilic gegenion (Br$^-$ from HBr) to produce bromocyclohexane. The latter portion of this sequence is analogous to the second step (coupling) of an S$_N$1 reaction. It is usually assumed that the initial reaction will form of a solvent separated carbocation intermediate, but this depends on the solvent. A tight ion pair intermediate can react in the substitution step to give the same product. The net result of this

---

263. For examples see reference 3, pp 629-666, 912-915.

cationic reaction is ***addition*** of H and Br across the π bond. The carbocation center is planar, allowing the nucleophilic bromide to approach from either face, so the addition reaction give a racemic product.

335

Alkenes are rather weak bases and will react only with strong acids. This limits the reaction to strong acids HX (such as HCl, HBr, H_2SO_4, HNO_3, etc.), which react with alkenes to form an alkyl halide, sulfate, nitrate, and so on. Alkyl sulfates and nitrates are often unstable, and undergo elimination reactions or decompose. In many cases, this instability (including elimination and ionization) means those products cannot be formed at all. When $HClO_2$ or HBF_4 are used, ClO_4^- and BF_4^- are very weak nucleophiles and react slowly with the cation intermediate. In such a case, another nucleophile can be added that will react preferentially with the cation. If water is present (it may come from an aqueous solvent), reaction with the carbocation intermediate will give an alcohol product. If a more nucleophilic reagent such as KCN is added to the reaction, a nitrile can be formed. From a synthetic viewpoint, this option greatly expands the utility of the reaction, allowing introduction of a nucleophilic unit (Y) that is not a part of the initial acid (HX). Since a carbocation is formed during the addition process, it is subject to the rearrangement problems inherent to such species (sec. 2.6.B.iii). This process is under thermodynamic control and tends to give the more stable ion, and the more stable final product. Before the mechanistic nature of this reaction was understood it was observed that reactions of alkenes and acids always produced the more substituted product, which is the so-called **Markovnikov addition**,[264,265] of simple alkenes. In other words, addition of HCl, HBr, or HI to substituted alkenes will lead to an alkyl halide where the nucleophile is attached to the more substituted carbon of the π-bond. The basis of this regioselectivity is, of course, formation of the more stable tertiary carbocation **326**, which is attacked by bromide ion to give 1-bromo-1-methylcyclohexane (**337**). This is the normal course of all reactions with acid and an alkene if the intermediate is a cation. Another example was shown in section 2.6.B.ii, where **128** reacted with HCl to give a tertiary carbocation rather than a primary carbocation.

336   **337**

In the reaction of unfunctionalized alkenes with acids such as HX, the key to predicting the regiochemistry is to identify the more stable cation (or radical) intermediate. The presence

---

264.   Markovnikoff, V. *Compt. Rend.* **1875**, *81*, 668.
265.   Isenberg, N.; Grdinic, M. *J. Chem. Educ.* **1969**, *46*, 601.

of a functional group or a heteroatom substituent can led to a different outcome. HBr reacted with the conjugated π bond of methyl acrylate to give methyl 3-bromopropanoate (**339**) as the final product, an anti-Markovnikov addition.[266] This result can be predicted by comparing the possible cation intermediates. Addition of H^+ to the π bond generates cation **338** as the precursor to **339**. Addition to give the alternative cation (**340**), although more substituted would place the positive charge adjacent to a δ^+ carbon of the carbonyl. Such an intermediate is destabilized and is not formed under these conditions. The greater stability of **338** relative to **340** predicts the preference for the observed product **339**.

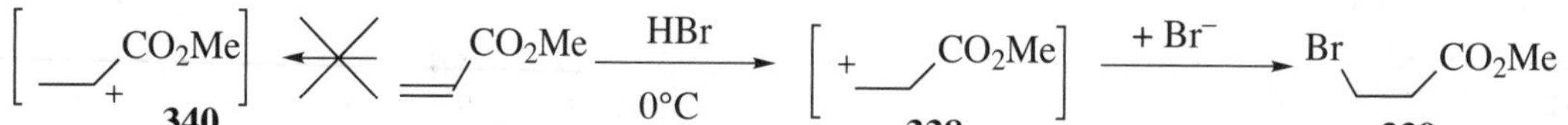

In all cases the carbocation mechanism for addition of acids such as HX to an alkene gives the more stable carbocation, and the substitution product will result from trapping that carbocation with the nucleophile. Many of these intermediate cations are, of course, subject to rearrangement as seen with S_N1 reactions in sec. 2.6.B.ii.

The disconnection for carbocation addition with strong acids HX is:

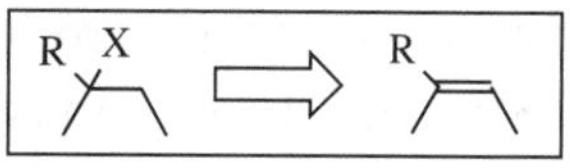

If we want to prepare an anti-Markovnikov halide from an alkene, we must change the mechanism of the reaction so there is no intermediate carbocation. The new route presented below is more-or-less specific for the reaction of HBr with alkenes. Reaction of methylcyclohexene with HBr, in the presence of the radical initiator *azobis*-isobutyronitrile (AIBN, sec. 13.3), gave the less substituted bromide. This regiochemical outcome immediately showed that the reaction cannot proceed by a cation intermediate, and mechanistic work has shown that it proceeds by a radical intermediate. Alternatively, reaction of an alkene with HBr in the presence of a peroxide leads to the same outcome, also via a radical intermediate (secs. 13.3-13.7). Formation of a radical from AIBN or a peroxide is followed by reaction with HBr (before reaction with the alkene) to generate a bromine radical (Br·). This highly reactive intermediate is attacked by the π-bond of methylcyclohexene to give the more stable carbon radical (**341**).[265] This mechanism demands that the bromine atom is added to the C=C unit before the hydrogen atom (contrast with the carbocation reaction where the acidic hydrogen atom added first to form **336**). In the example at hand, methylcyclohexene reacts with the bromide radical to generate the more stable tertiary radical (**341**) with Br on the secondary carbon.

---

266.   Mozingo, R.; Patterson, L.A. *Org. Synth. Coll. Vol. 3* **1955**, 576.

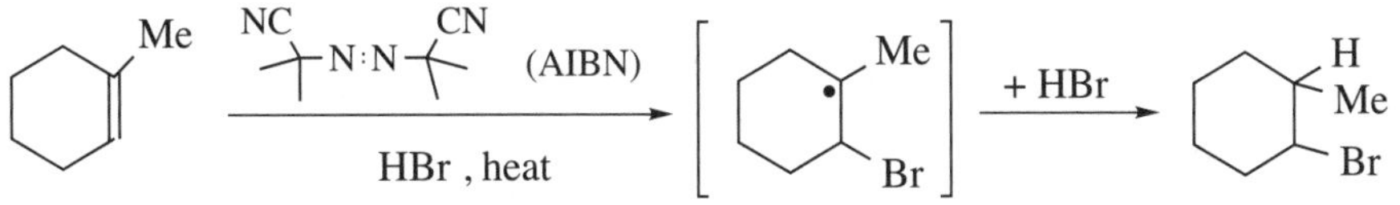

341 **342**

The tertiary radical is formed preferentially rather than the less stable secondary radical where the Br would be on the tertiary carbon. Formation of the more stable radical can be rationalized by essentially the same arguments used for stating that a tertiary cation is more stable than a secondary cation. In the final step of this mechanism, carbon radical intermediate **341** removes a hydrogen atom from HBr to give the final product (1-bromo-2-methylcyclohexane, **342**), and also generate a new bromine radical in a chain propagation step. The net result is addition of H and Br, but the bromine resides on the less substituted position of the double bond. This regiochemical outcome is referred to as **anti-Markovnikov addition**.[265] This particular transformation is limited to HBr since HCl and HI do not show such selectivity. Radical addition of HBr to alkenes requires initiation by exposure to light in the presence of a peroxide or another radical initiator such as AIBN (see sec. 13.3).[264] Despite these limitations, the reaction can be quite useful in synthesis. Photolysis of **343** with HBr, for example, gave a 73% yield of primary bromide **344** in Coates and co-worker's synthesis of cameroonanol-7α-ol.[267]

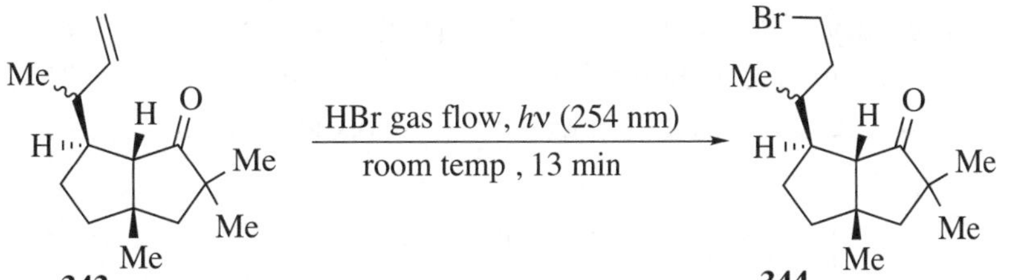

343 **344**

2.10.B π-Bonds as Lewis Bases: Oxymercuration

As mentioned in previous sections, carbocation intermediates are subject to rearrangement to a more stable ion (sec. 2.6.B.iii). Rearrangement can be minimized or sometimes prevented altogether by attaching a heteroatom to the electrophilic carbon. An example is the oxygen-stabilized cation (**345**) generated by reaction of a ketone with an acid catalyst. The electrons on the heteroatom are donated to the positive center leading to resonance stabilization (this is called back donation). Such a cation is typically more stable than the non-oxygen stabilized secondary or tertiary cations that would result after a rearrangement. In this case, rearrangement does not occur. The presence of some Lewis acids adjacent to the carbocation, such as the mercuric ion, stabilizes the cation via back-donation from the metal to the positive center.

---

267.   Davis, C.E.; Duffy, B.C.; Coates, R.M. *J. Org. Chem.* **2003**, *68*, 6935.

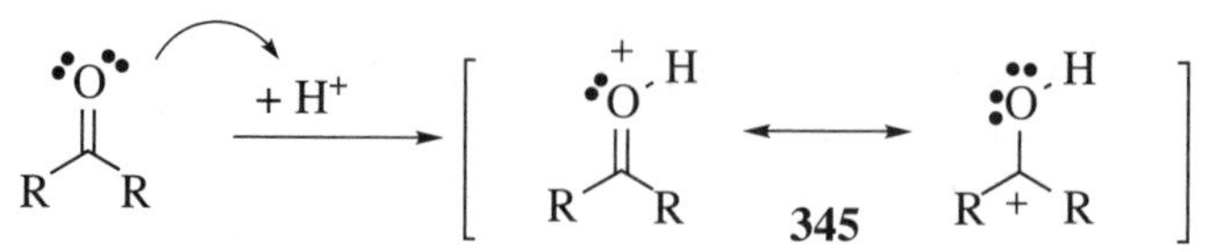

In the case of mercury, a bridge-like intermediate such as that depicted in **347** represents the back-donation from the metal atom. The reaction of norbornene (**346**) and mercuric acetate [Hg(OAc)$_2$] generated **347**, which reacted with water to give **348**. This cation does not undergo the normally facile norbornyl rearrangement (sec. 12.2.B), indicating the stability imparted by back-donation from the adjacent mercury. Indeed, mercury-stabilized cations are remarkably free of the rearrangements that plague reactions where the cation is generated by other means.[268] Despite the stabilization, the cation readily reacts with a suitable nucleophile. In the reaction of **347**, water is the reaction medium and also serves as a nucleophile, leading to mercuric alcohol **348**. The mercury is removed from **348** by reduction with sodium borohydride to give the final alcohol product, **349** via hydrogenolysis (cleavage of the C-Hg bond, see sec. 4.4.A and 4.8.E).[269] This overall process is known as **oxymercuration-demercuration**.[270] This reaction usually proceeds with high regioselectivity for the product, with the hydroxyl at the more substituted position (the norbornyl structure also leads to formation of >99% exo alcohol in **349**) as observed in reactions of any other cation. In a synthesis of guaiane by Deprés and co-workers,[271] **350** was treated with mercury (II) bistrifluoroacetate (trifluoroacetate ion is a weak nucleophile and less prone to competitive substitution reactions) to give alcohol **351** after reduction with sodium borohydride. The reaction can be stereoselective, but it is not stereospecific.

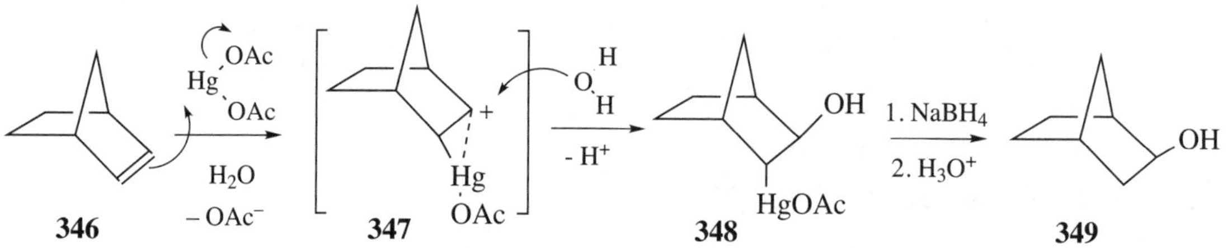

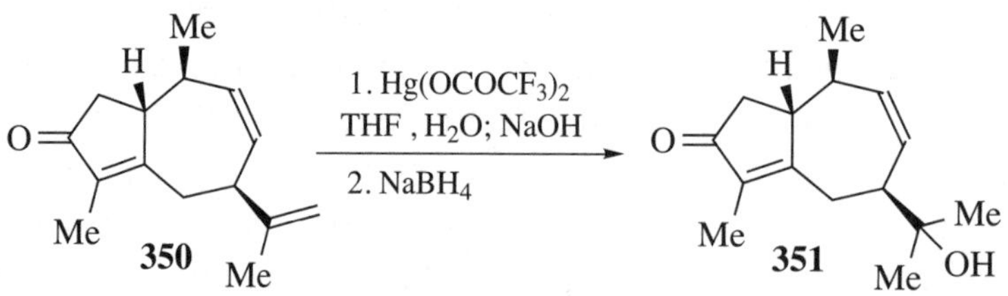

268. (a) Brown, H.C.; Lynch, G.J. *J. Org. Chem.* **1981**, *46*, 930; (b) Brown, H.C.; Geoghegan, Jr., P.J.; Kurek, J.T. *Ibid* **1981**, *46*, 3810; (c) Thies, R.W.; Boop, J.L.; Schiedler, M.; Zimmerman, D.C.; LaPage, T.H. *Ibid* **1983**, *48*, 2021.
269. Brown, H.C.; Kawakami, J.H.; Ikegami, S. *J. Am. Chem. Soc.* **1967**, *89*, 1525.
270. (a) Kitching, W. *Organomet. React.* **1972**, *3*, 319 [*Chem. Abstr.* 77:87189 **1972**], (b) *Idem. Organomet. Chem. Rev.* **1968**, *3*, 61 [*Chem. Abstr.* 69:18335x **1968**].
271. Coquerel, Y.; Greene, A.E.; Deprés, J.-P. *Org. Lett.* **2003**, *5*, 4453.

The use of an alcohol rather than water as the nucleophilic solvent will lead to formation of an ether, and is a simple variation of oxymercuration.[269,272] In Donaldson and Greer's synthesis of the C3-C15 segment of phorboxazole, pyrone **352** was treated with mercuric acetate in methanol.[273] Subsequent reduction with sodium cyanoborohydride (sec. 4.5.D) gave a 69% yield of **353**. The intramolecular variation of this reaction is quite useful, as illustrated by Hiemstra and co-worker's conversion of **354** to **355** in 88% yield, as part of a synthesis of ent-gelsedine.[274] When an alkyne is the reactive substrate, oxymercuration leads to a ketone. In a synthesis of (+)-leucascandrolide A by Paterson and Tudge,[275] alkyne **356** reacted with mercuric acetate and *p*-toluenesulfonic acid in wet THF to give an 86% yield of methyl ketone **357**. Oxymercuration of terminal alkynes is a valuable route to methyl ketones.

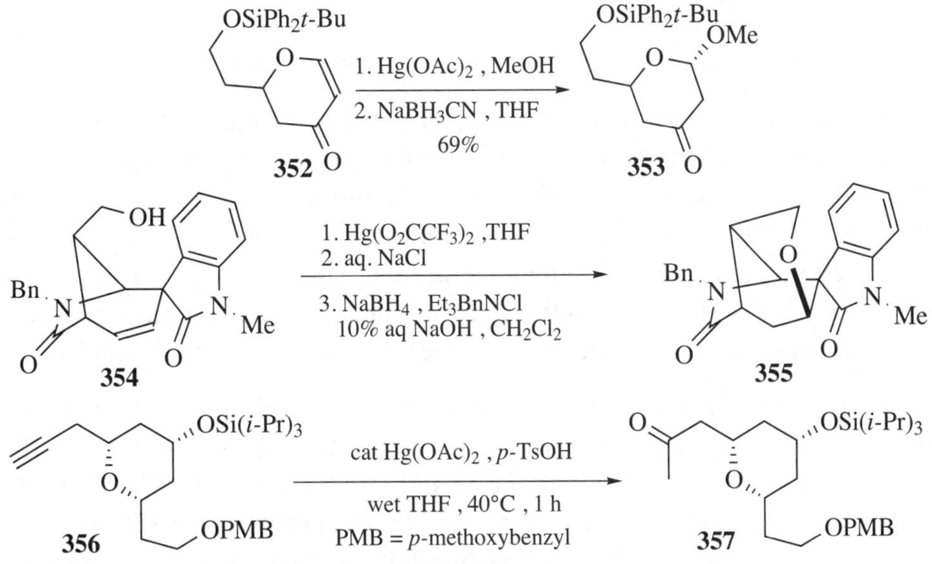

The functional group transform for this reaction type is shown.

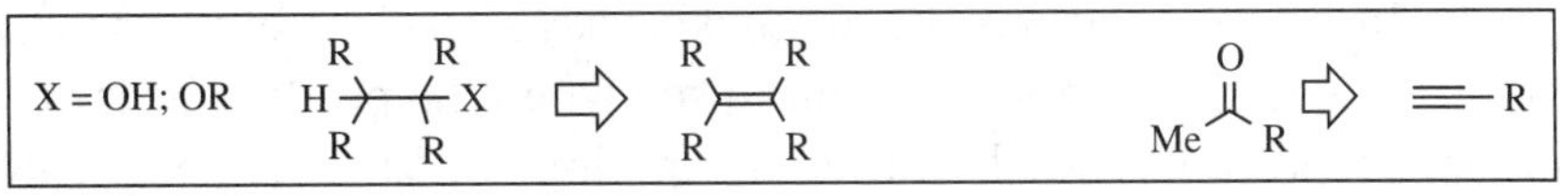

2.10.C. π-Bonds as Lewis Bases: Reaction With Dihalogens and Related Reagents

The previous section discussed the reaction of alkenes as Lewis basses with mercuric compounds, which are Lewis acids. Alkenes also react with other Lewis acids, including

272. Coxon, T.M.; Hartshorn, M.P.; Mitchell, J.W.; Richards, R.E. *Chem. Ind., **1968**,* 652.
273. Greer, P.B.; Donaldson, W.A. *Tetrahedron Lett.* **2000**, *41,* 3801.
274. van Henegouwen, W.G.B.; Fieseler, R.M.; Rutjes, F.P.J.T.; Hiemstra, H. *J. Org. Chem.* **2000**, *65,* 8317.
275. Paterson, I.; Tudge, M. *Angew. Chem. Int. Ed.* **2003**, *42,* 343.

dihalogens. Symmetrical reagents such as bromine or chlorine are not normally polarized but the halogens are *polarizable*, and when brought into proximity with a π bond a dipole is induced in the X-X bond [C=C→X⁺X⁻]. This type of structure is sometimes called a charge transfer complex. The polarized halogen reacts as a Lewis acid in the presence of an alkene where electron donation from the π-bond to X⁺ leads to displacement of X⁻ and formation of a new C–X bond. This reaction should generate a positive center adjacent to the newly formed C–X bond, but the X group contains unshared electrons that are back donated to the developing positive center, forming a symmetrically bridged (three membered ring) intermediate. The three-membered ring intermediate is called a **halonium ion** (chloronium, bromonium or iodonium). This experimental observation suggests that the three-membered ring intermediate is more stable than the carbocation. Molecules that are not polarizable, such as diatomic hydrogen, do not react in this way. Indeed, dihydrogen requires the presence of a transition metal to first break the H–H bond before it can react with a π-bond.

Evidence suggests that formation of a three-membered ring intermediate is *not* a stepwise process in aprotic solvents. Subsequent S_N2 type attack by the X⁻ gegenion (i.e., backside attack) gives the final addition product, a *vicinal* dihalide. Therefore, reaction with diatomic bromine gives a dibromide, diatomic chlorine gives a dichloride, and diatomic iodine gives a diiodide. Reaction of a diatomic halogen with an alkene in this manner gives the trans (anti) dihalide. Anti-addition of X⁻ (backside attack) at the carbon of the halonium ion intermediate, which is a rigid ion, gives facial selectivity during attack of the nucleophilic X⁻ species and a trans product.

Addition of halogens (X₂) to alkenes is diastereospecific. To explain the specificity for the trans product, initial addition of dihalogen leads to the three-membered halonium ion. Subsequent attack by X⁻ will occur to minimize electronic and steric repulsion. Therefore, addition of diatomic chlorine to *cis*-3-hexene leads to chloronium ion intermediate **358**, and displacement by chloride ion gives the product **359**. Similarly, addition of diatomic bromine to cyclohexene leads to **360** as an intermediate, and anti-attack by bromide ion gives **361**. These two simple examples are typical for the reaction of diatomic halogen and alkenes. The trans relationship easy to see in **361**. In reactions of acyclic alkenes such as *cis*-3-hexene, the trans addition of the chlorine atoms to give **359** is more difficult to see when the product is drawn in the typical *zig-zag* drawing that represents the all-anti conformation. In a stereochemically pure alkene such as *cis*-3-hexene, the positions of the ethyl groups and the chlorine atom are fixed. Once the three-membered ring is formed, rotation is also impossible, so the positions of those groups and atoms remain fixed. Subsequent anti-attack by the chloride ion fixes all stereocenters in the final product, and the overall reaction is diastereospecific. The trans addition of the chlorine atoms is clear in the dichloride product **359**, where the relative positions of all groups and atoms is retained from the starting alkene, but rotation about that C–C bond is possible, and the molecule is redrawn in the extended conformation marked *syn* (sec. 1.4.B).[276] *Note, however, that the stereochemical relationship of the two chlorine atoms refers to anti addition although it*

---

276. Masamune, S.; Ali, Sk.A.; Snitman, D.L.; Garvey, D.S. *Angew. Chem. Int. Ed.* **1980**, *19*, 557.

is drawn in the syn conformation. In Boeckman and co-worker's synthesis of (+)-tetronolide,[277] vinyl ether **362** reacted with bromine in dichloromethane at –78°C gave **363** in near quantitative yield. There are practical problems when using liquid bromine, and alternative brominating reagents are now available, including tribromide derivatives. Tribromide reagents that have been used for brominating alkenes, and also benzene derivatives, include tetrabutylammonium tribromide,[278] 1,8-diazabicyclo-[5.4.0]-undec-7-ene hydrogen tribromide [DBU–H•Br$_3$][279] (see **21**), cetyltrimethylammonium tribromide,[280] and pyridinium bromide perbromide (Py–H•Br$_3$).[281] In a synthesis of latrunculin B,[282] Fürstner and co-workers treated alkene **364** with 4-dimethylaminopyridinium bromide perbromide (pyridinium tribromide, **365**), and obtained an 87% yield of **366**.

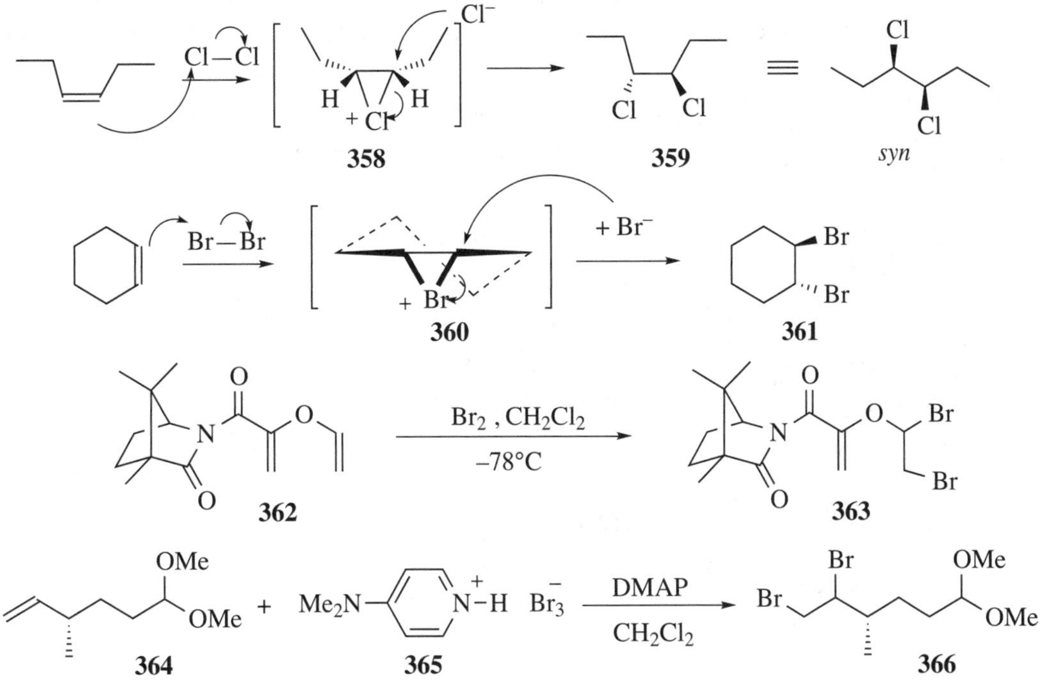

The diastereoselectivity that arises from the intermediacy of a halonium ion has been exploited in a new type of reaction that produces lactones. When an alkene contains a carboxylic acid unit

277. Boeckman, Jr., R.K.; Shao, P.; Wrobleski, S.T.; Boehmler, D.J.; Heintzelman, G.R.; Barbosa, A.J. *J. Am. Chem. Soc.* **2006**, *128*, 10572.
278. (a) Buckels, R.E.; Popov, A.I.; Zelezny, F.; Smith, R.J. *J. Am. Chem. Soc.* **1951**, *73*, 4525; (b) Chaudhuri, M.K.; Khan, A.T.; Patel, B.K.; Dey, D.; Kharmawophlang, W.; Lakshmipradha, T.R.; Mandal, G.C. *Tetrahedron Lett.* **1998**, *39*, 8163.
279. Muathen, H.A. *J. Org. Chem.* **1992**, *57*, 2740.
280. Kajigaeshi, S.; Kakinami, T.; Tokiyama, H.; Hirakawa, T.; Okamoto, T. *Chem. Lett.* **1987**, 627.
281. Fieser, L. F.; *Fieser, M. Reagents for Organic Synthesis,* Wiley: New York, *1967*, Vol. 1, p 967.
282. Fürstner, A.; De Souza, D.; Parra-Rapado, L.; Jensen, J.T. *Angew. Chem. Int. Ed.* **2003**, *42*, 5358.

elsewhere in the molecule (such as **367**) reaction with diatomic iodine and sodium bicarbonate gives an iodonium ion such as **368**, which is opened by the carboxylate anion to give an iodo-lactone (**369**),[16] diastereoselectively. This reaction is commonly referred to as **iodolactonization**,[283] and it can be applied to acyclic systems as well.[284] In a synthesis of (−)-cinatrin B by Rizzacasa and co-workers,[285] alkene-acid **370** formed iodonium-carboxylate **371** during the course of the reaction with iodine and sodium bicarbonate, leading to a 97% yield of **372** (88% ds). A variation of this cyclization reaction uses an ester. In a synthesis of (+)-epoxyquinol A,[286] Hayashi and co-workers treated ester **373** with iodine, and obtained an 81% yield of **374** (>99% ee after crystallization). Several approaches have been developed for asymmetric iodolactonization reactions.[287] Both bromolactonization and chlorolactonization reactions are known. The use of liquid bromine or gaseous chlorine can be problematic, but as mentioned above, pyridinium bromide perbromide in THF, can be used to generate bromolactones.[288]

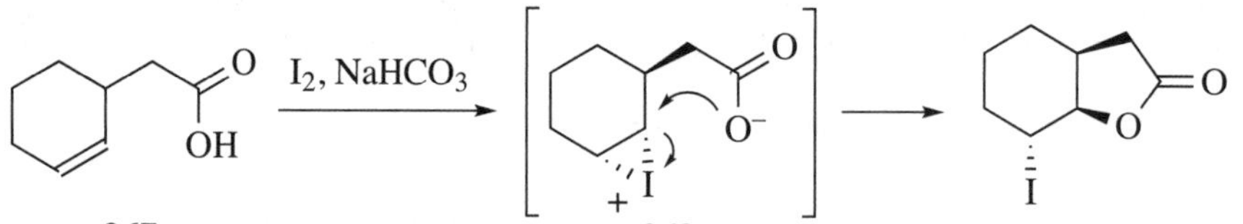

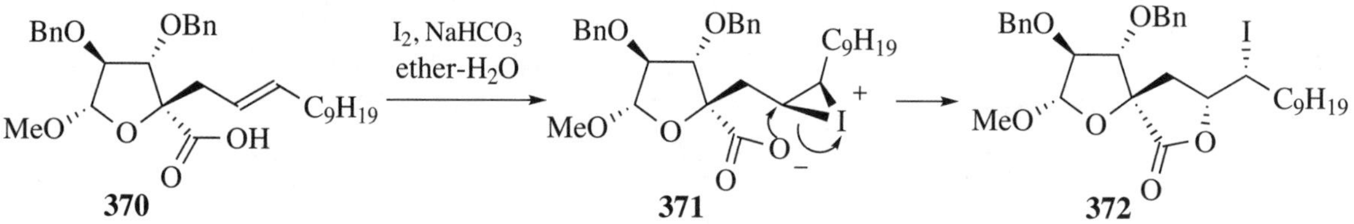

283. (a) Klein, J. *J. Am. Chem. Soc.* **1959**, *81*, 3611; (b) van Tamelen, E.E.; Shamma, M. *Ibid* **1954**, *76*, 2315; (c) House, H.O.; Carlson, R.G.; Babad, H. *J. Org. Chem.* **1963**, *28*, 3359; (d) Corey, E.J.; Albonico, S.M.; Koelliker, V.; Schaaf, T.K.; Varma, R.K. *J. Am. Chem. Soc.* **1971**, *93*, 1491.
284. (a) Cardillo, G.; Orena, M. *Tetrahedron* **1990**, *46*, 3321; (b) Robin, S.; Rousseau, G. *Tetrahedron* **1998**, *54*, 13681.
285. Cuzzupe, A.N.; Di Florio, R.; Rizzacasa, M.A. *J. Org. Chem.* **2002**, *67*, 4392.
286. Shoji, M.; Yamaguchi, J.; Kakeya, H.; Osada, H.; Hayashi, Y. *Angew. Chem. Int. Ed.* **2002**, *41*, 3192.
287. (a) Blot, V.; Reboul, V.; Metzner, P. *J. Org. Chem.* **2004**, *69*, 1196; (b) Moon, H.-S.; Eisenberg, S.W.E.; Wilson, M.E.; Schore, N.E.; Kurth, M.J. *J. Org. Chem.* **1994**, *59*, 6504; (c) Haas, J.; Piguel, S.; Wirth, T. *Org. Lett.* **2002**, *4*, 297; (d) Wang, M.; Gao, L.X.; Mai, W.P.; Xia, A.X.; Wang, F.; Zhang, S.B. *J. Org. Chem.* **2004**, *69*, 2874.
288. For an example taken from a synthesis see Trost, B.M.; Haffner, C.D.; Jebaratnam, D.J.; Krische, M.J.; Thomas, A.P. *J. Am. Chem. Soc.* **1999**, *121*, 6183.

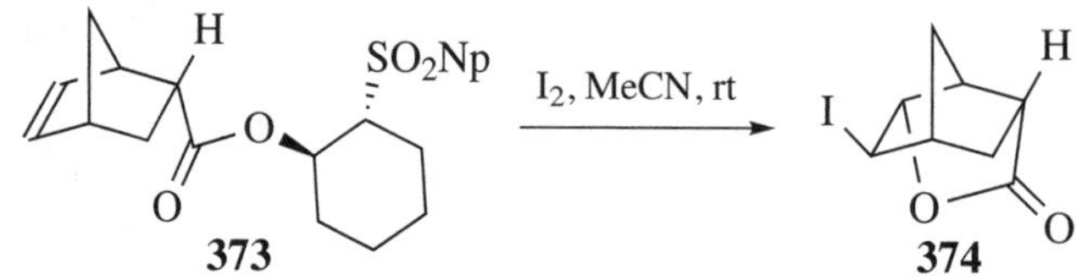

Iodolactamization is possible via a related process, but problems relating to the bidentate nature of amide anion require modification of conditions, as illustrated by formation of the lactone as the major product in a previous study.[287a] Knapp and Rodriques described a synthesis in which amide **375** was first converted to the corresponding triflate, and subsequent reaction with iodine gave iodo-lactam **376** in 68% yield.[289] When a chiral oxazolidine auxiliary was attached to the nitrogen, iodolactamization occurred with excellent enantioselectivity.[290] A related cyclization of *N*-sulfonyl-amino-alkenes and NBS gave the bromolactam,[291] and a dichloro-*N,N*-bis(allyl) amide was converted to a dichlorolactam with $FeCl_2$.[292]

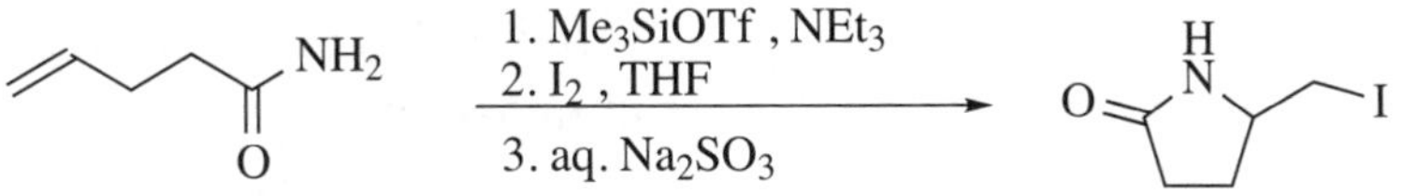

Another addition involves the reaction of unfunctionalized alkenes with aqueous solutions of bromine or chlorine. Halogenation can proceed by formation of a cation under aqueous conditions, but the course of the reaction is complicated by the fact that dissolution of chlorine or bromine in water results in formation of a hypohalous acid (XOH, X = Cl, Br), represented by the reaction:

$$X_2 \quad + \quad HOH \quad \longrightarrow \quad X\text{-}OH \; = \; X^+ \, {}^-OH$$

When pure XOH (ClOH or BrOH) is used in an aprotic medium, the X^+ portion of this reagent is electrophilic and reacts with an alkene to generate the same halonium ion observed with bromine and chlorine additions. The ${}^-$OH gegenion is the nucleophilic partner and opens the halonium ion to produce a halohydrin, analogous to the reactions of bromine or chlorine. It is more typical for these reagents to be generated in water, however, and the initially formed halohydrin will exist primarily as the ring-opened cation or an ion pair, and the hydroxide nucleophile generally attacks the cationic center to give the more substituted alcohol.[293] Indeed, with relatively simple systems, reactions of alkenes with aqueous chlorine or bromine usually give the halohydrin with the hydroxyl at the *more* substituted position.[283,293] Reaction of methylcyclohexene with HOCl in water, for example, gives the usual Markovnikov product, chlorohydrin (**377**).

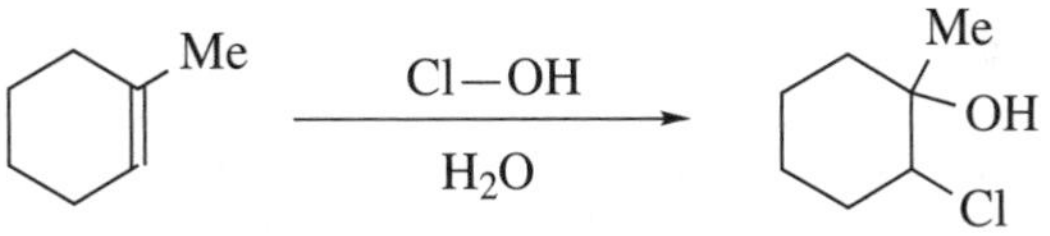

289. Knapp, S.; Rodriques, K.E.; Levorse, A.T.; Ornaf, R.M. *Tetrahedron Lett.* **1985**, *26*, 1803.
290. Shen, M.; Li, C. *J. Org. Chem.* **2004**, *69*, 7906.
291. Tamaru, Y.; Kawamura, S.; Tanaka, K.; Yoshida, Z. *Tetrahedron Lett.* **1984**, *25*, 1063.
292. Tseng, C.K.; Teach, E.G.; Simons, R.W. *Synth. Commun.* **1984**, *14*, 1027.
293. Boguslavskaya, L.S. *Russ. Chem. Rev.* **1972**, *41*, 740.

Reaction of NBS or NCS with a minimal amount of H_2O in DMSO is a convenient method for the preparation of HOBr or HOCl, respectively.[294] Subsequent reaction with an alkene leads to a chlorohydrin or a bromohydrin. The reaction is usually selective for the trans product, and this suggests the presence of the halonium ion intermediate. This reaction is one that can be mechanistically ambiguous in some cases since the outcome depends on the solvent, the nature of the substrate, and the concentration of the reagents. A synthetic illustration of this reaction is addition of hypobromous acid (formed in aqueous solution with NBS) to **378**, taken from Elango and Yan's synthesis of (+)-narciclasine.[295] This reaction generated a 1:1 mixture of the two regioisomers, **379** and **380**, each having the trans-geometry shown. Although some halohydrins are formed with good regioselectivity, this example shows that one should anticipate a mixture in the absence of regiocontrolling factors.

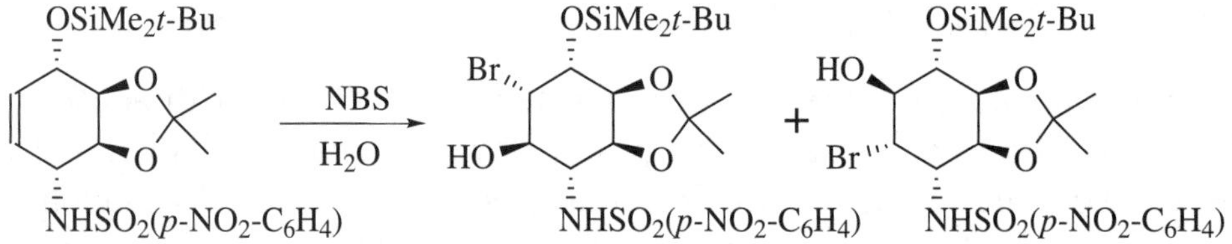

These reactions give the useful transform:

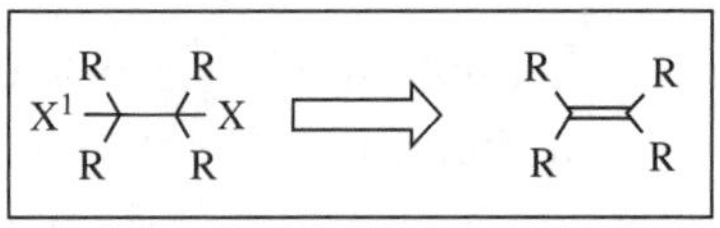

$X = X^1 = Br, Cl, I$ or $X = Br, Cl, I; X^1 = OH$

2.11. FUNCTIONAL GROUP MANIPULATION BY REARRANGEMENT

There are several common functional group manipulations that involve skeletal rearrangement. This section will discuss a few examples that correlate with Named reactions, both to illustrate the principle and to show the details of these specific transformations.

2.11.A. Beckmann Rearrangement

Oximes (**381**) are readily obtained by reaction of aldehydes or ketones with hydroxylamine. When oximes are treated with sulfuric acid, PCl_5 and related reagents, rearrangement occurs to

294. (a) Dalton, D.R.; Hendrickson, J.B.; Jones, D. *Chem. Commun.* **1966**, 591; (b) Dalton, D.R.; Dutta, V.P., *J. Chem. Soc. B* **1971**, 85; (c) Sisti, A.J. *J. Org. Chem.* **1970**, *35*, 2670.
295. Elango, S.; Yan, T.-H. *J. Org. Chem.* **2002**, *67*, 6954

give substituted amides in what is called the **Beckmann rearrangement**.[296] Initial protonation of the hydroxyl group to give **382** is followed by rearrangement, with displacement of water to give **383**. Subsequent reaction with water and loss of a proton yields the amide.[297] The group that migrates is generally the one anti to the hydroxyl, and this is often used as a method of determining the configuration of the oxime. However, the syn group may migrate in some oximes, especially where R and R' are both alkyl. In most cases however, it is likely that the oxime undergoes isomerization under the reaction conditions *before* migration takes place.[298] It is possible to get mixtures of the two different amides from oximes derives from unsymmetrical ketones. Note that hydrogen seldom *migrates*, and most alkyl groups are compatible with the reaction. In alkyl aryl ketones, the aryl group generally migrates preferentially. The oximes of cyclic ketones such as **384** give ring enlargement under these conditions, with formation of a lactam (**385**).[299] Beckmann rearrangements have also been carried out photochemically.[300]

Esters of oximes also undergo the Beckmann rearrangement with many organic and inorganic acids. A side reaction with many substrates is the formation of nitriles. Under the correct conditions, cycloalkane carboxylic acids rearrange to form lactams. Azacyclododecan-2-one, for example, was prepared from cycloundecan-2-carboxylic acid by reaction with nitrosyl sulfuric acid in chlorosulfonic acid.[301] Another useful variation of the Beckmann rearrangement is the direct treatment of a ketone with hydroxylamine-*O*-sulfonic acid in the presence of an acid such as formic acid, or with sulfuric acid in a second step. Initial formation of the oxime-

296. For reviews, see (a) Gawley, R.E. *Org. React.* **1988**, *35*, 1; McCarty, C.G. in Patai *The Chemistry of the Carbon-Nitrogen Double Bond*, Wiley: NY, **1970**, pp. 408-439; (b) The Merck Index, 14th Ed. Merck & Co., Inc., Whitehouse Station, New Jersey, **2006**, p ONR-7; (c) Mundy, B.P.; Ellerd, M.G.; Favaloro Jr., F.G. *Name Reactions and Reagents in Organic Synthesis, 2nd Ed.* Wiley-Interscience, New Jersey, **2005**, pp. 80-81. Also see (d) Nguyen, M.T.; Raspoet, G.; Vanquickenborne, L.G. *J. Am. Chem. Soc.* **1997**, *119*, 2552.
297. Donaruma, L.G.; Heldt, W.Z. *Org. React.* **1960**, *11*, 1, see p. 5.
298. Lansbury, P.T.; Mancuso, N.R. *Tetrahedron Lett.* **1965**, 2445.
299. Vinnik, M.I.; Zarakhani, N.G. *Russ. Chem. Rev.* **1967**, *36*, 51. .
300. (a) Izawa, H.; de Mayo, P.; Tabata, T. *Can. J. Chem.* **1969**, *47*, 51; (b) Cunningham, M.; Ng Lim, L.S.; Just, T. *Can. J. Chem.* **1971**, *49*, 2891; (c) Suginome, H.; Yagihashi, F. *J. Chem. Soc. Perkin Trans. 1* **1977**, 2488.
301. (a) Imperial Chemical Industries LTD Belgian Patent 616 544, Oct 17 1962 [*Chem. Abstr.* **1963**, 59:452]; (b) Nagasawa, H.T.; Elberling, J.A.; Fraser, P.S.; Mizuno, N.S. *J. Med. Chem.* **1971**, *14*, 501.

O-sulfonic acid (**387**) is followed by *in situ* formation of an oxime, which rearranges in the presence of the acid.[302]   In the example shown, 2-pentylcyclohexanone (**386**) reacted with hydroxylamine-*O*-sulfonic acid and formic acid to give **387**, and rearrangament *in situ* gave a 71% yield of 7-pentyl-hexahydroazepin-2-one, **388**.[303]

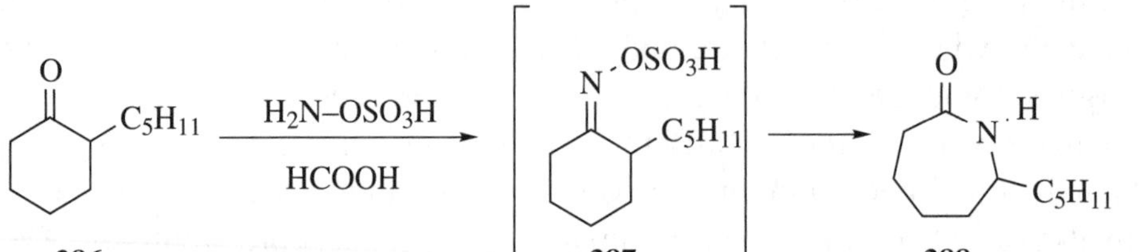

A synthetic example, illustrating another variation in the Beckmann rearrangement, is taken from the Denis and Agouridas synthesis of the ketolide analog of 6-*O*-methylazithromycin.[304] Studies related to the synthesis of the targeted compound generated oxime **389**, which was treated with tosyl chloride to give the *O*-tosyl oxime. Subsequent Beckmann rearrangement gave lactam **390** in 70% yield. It is clear from this example that variations of the Beckmann rearrangement that do not use strong acid are compatible with a vast array of functionality.

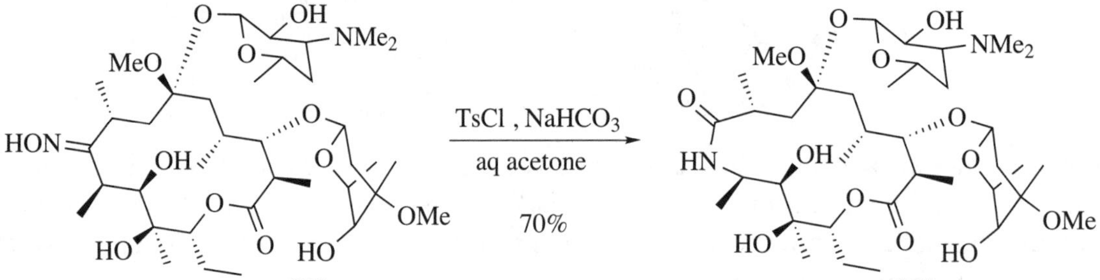

2.11.B. Schmidt Rearrangement

The addition of hydrazoic acid to carboxylic acids, aldehydes and ketones, and alcohols and alkenes all give amides, and all go by the name **Schmidt reaction**.[305]   The reaction with carboxylic acids is probably the most common, involving initial formation of an acyl azide (**391**), which rearrangements with loss of nitrogen to give a transient isocyanate (**392**).

302.   (a) Smith, P.A.S. *J. Am. Chem. Soc.* **1948,** *70,* 323; (b) Sanford, J.K.; Blair, F.T.;. Arroya, J.; Sherk, E.W. *J. Am. Chem. Soc.* **1945,** *67,* 1941.

303.   Duhamel, P.; Kotera, M.; Monteil, T.; Marabout, B.; Davoust, D. *J. Org. Chem.* **1989,** *54,* 4419.

304.   Denis, A.; Agouridas, C. *Bioorg. Med. Chem. Lett.* **1998**, *8*, 2427.

305.   (a) Banthorpe, D.V. in Patai, S. *The Chemistry of the Azido Group*, Wiley, NY, 1971, pp. 405-434; (b) *The Merck Index, 14th Ed.* Merck & Co., Inc., Whitehouse Station, New Jersey, **2006**, p ONR-83; (c) Mundy, B.P.; Ellerd, M.G.; Favaloro Jr., F.G. *Name Reactions and Reagents in Organic Synthesis, 2nd Ed.* Wiley-Interscience, New Jersey, **2005**, pp. 574-575.

Subsequent reaction with water gives **393**, which decarboxylates to give the amine product.[306] Sulfuric acid and many Lewis acids can be used as catalysts in this reaction. Under normal reaction conditions, the isocyanate is not isolated. When hydrazoic acid reacts with a ketone, the product of the rearrangement is an amide (**394**).[307]

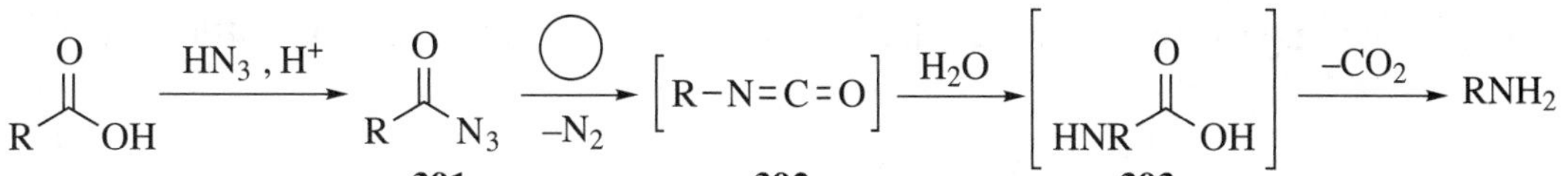

Dialkyl ketones and cyclic ketones tend to react faster than alkyl aryl ketones, and these faster than diaryl ketones. Cyclic ketones give lactams,[308] and a simple example is the reaction of cyclohexanone with HBr and sodium azide. A 63% yield of hexahydroazepine-2-one (**395**) was obtained in water, and a 66% yield was obtained in acetic acid (HBr, $NaNO_2$).[309] A variation of the Schmidt reaction treated the ketone with hydrazoic acid with another acid and also obtained a lactam. Under these conditions 2-isopropylcyclopentanone reacted with HN_3/H_2SO_4 at 3-7°C to give a 63% yield of 6-isopropyl-2-piperidone, **396**.[310]

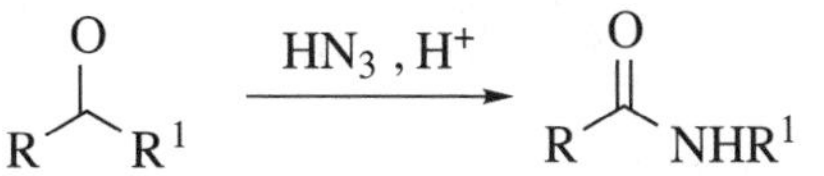

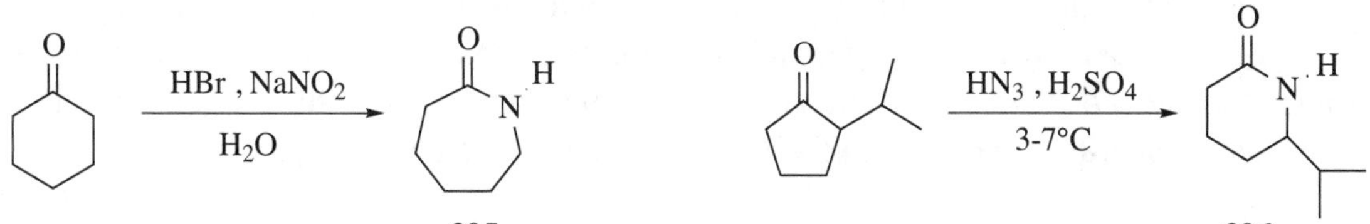

The reaction is rarely applied to aldehydes, where the product is usually a nitrile. Nitriles are observed as by-products with ketones as well. Regioselectivity can be problem when unsymmetrical cyclic ketones are converted to the lactam, but the reaction often proceeds with good regioselectivity. When 2-methyl-cyclohexanone reacted with sodium azide in the presence of polyphosphoric acid,[311] For example, the two lactam products were 7-methylhexahydroazepin-2-one (**397**) in 54% yield and **398** in 10% yield.[312] A useful variation

306. Koldobskii, G.I.; Ostrovskii, V.A.; Gidaspov, B.V. *Russ. Chem. Rev.* **1978**, *47*, 1084.
307. Koldobskii, G.I.; Tereschenko, G.F.; Gerasimova, E.S.; Bagal, L.I. *Russ. Chem. Rev.* **1971**, *40*, 835; (b) Beckwith, A.L.J. in Zabicky, J. *The Chemistry of Amides*, Wiley, NY, **1970**, pp. 137-145.
308. Krow, G.R. *Tetrahedron* **1981**, *37*, 1283.
309. Smith, P.A.S. *J. Am. Chem. Soc.* **1948**, *70*, 320-323.
310. Shechter, H.; Kirk, J.C. *J. Am. Chem. Soc.* **1951**, *73*, 3087.
311. Conley, R.T. *J. Org. Chem.* **1958**, *23*, 1330.
312. Overberger, C.G.; Parker, G.M. *J. Polym. Sci., A-***1968**, *6*, 513.

of the Schmidt reaction involves reaction of a silyl enol ether of a cyclic ketone with TMSN$_3$,[313] followed by photolysis of the product with UV light to give a lactam.

A synthetic example illustrates an intramolecular variation. In a synthesis of (+)-aspidospermidine by Aubé and co-workers, ketone **399** (containing the alkyl azido substituent shown) was treated with TiCl$_4$,[314] leading to the intramolecular Schmidt reaction that gave lactam **400** in 82% yield.

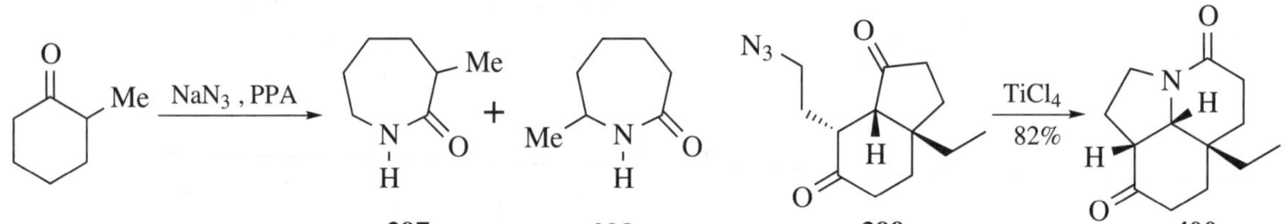

2.11.C. Related Rearrangements

An intermediate in the Schmidt rearrangement was an acyl azide **391**, and the intermediate isocyanate **392** was not isolated under those conditions. Without the aqueous conditions, the thermal rearrangement of acyl azides to isocyanates is known as the **Curtius rearrangement**.[315] Subsequent reaction with water, alcohols or amines lead to amines, carbamates, or acylureas.[316] In general, acyl azides are prepared by treatment of acylhydrazines (hydrazides) with nitrous acid. Lewis acids can catalyze the Curtius rearrangement, but they are not required. In a synthesis of (+)-yatakemycin by Boger and co-workers,[317] acid **401** was treated with diphenyl phosphoryl azide (DPPA) to give the acyl azide, **402**. Under the aqueous conditions, the product was amine **403**, via reaction of water with the intermediate isocyanate.

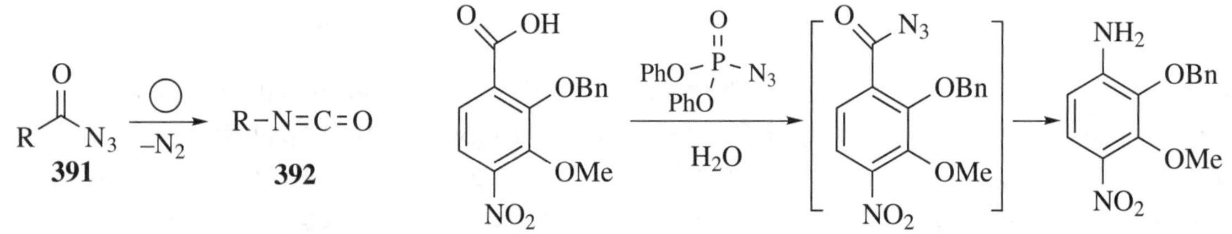

---

313. Evans, P.A.; Modi, D.P. *J. Org. Chem.* **1995**, *60*, 6662.
314. Iyengar, R.; Schildknegt, K.; Morton, M.; Aube, J. *J. Org. Chem.* **2005**, *70*, 10645.
315. (a) Banthorpe, D.V. in Patai, S. *The Chemistry of the Azido Group*; Wiley, NY, **1971**, pp. 397-405; (b) The MerckIndex, 14th Ed. Merck & Co., Inc., Whitehouse Station, New Jersey, **2006**, p ONR-21 (c) Mundy, B.P.; Ellerd, M.G.; Favaloro Jr., F.G. *Name Reactions and Reagents in Organic Synthesis, 2nd Ed.* Wiley-Interscience, New Jersey, **2005**, pp. 192-193.
316. (a) Pfister, J.R.; Wyman, W.E. *Synthesis* **1983**, 38. See also (b) Capson, T.L.; Poulter, C.D. *Tetrahedron Lett.* **1984**, *25*, 3515.
317. Tichenor, M.S.; Kastrinsky, D.B.; Boger, D.L. *J. Am. Chem. Soc.* **2004**, *126*, 8396.

When an amide is treated with sodium hypobromite (NaOBr; NaOH and Br$_2$) an isocyanate is formed,[318] and subsequent hydrolysis liberates an amine with one less carbon that the starting amide, in what is known as the **Hofmann rearrangement**.[319] Ureas and acylureas are sometimes formed in this reaction. *N*-Acyl derivatives of hydroxamic acids[320] give isocyanates when treated with base, or upon heating, in the **Lossen rearrangement**.[321] Similarly, aromatic acyl halides are converted to amines when treated with hydroxylamine-*O*-sulfonic acid.[322]

These reactions give the retrosynthetically useful transforms:

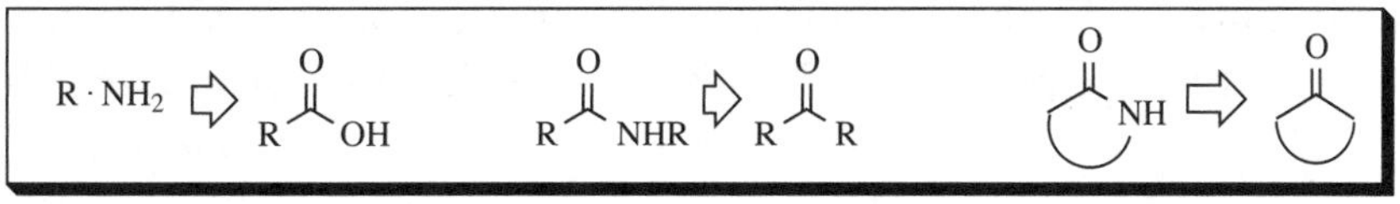

2.12. AROMATIC SUBSTITUTION

In many ways, the principles of substitution, elimination, and addition converge in aromatic systems in what is generically called aromatic substitution.[323] Addition to electrophilic centers, substitution of carbocations, nucleophilic displacement, and elimination of leaving groups are all mechanistic features of various aromatic substitution reactions.

2.12.A. Defining Aromatic Substitution

The substitution reactions discussed in previous sections occurred at a sp^3 hybridized carbon, and direct substitution at a sp^2 carbon was not observed. Addition reactions, however, occurred at sp^2 hybridized carbons in the presence of acids. Substitution reactions are possible, but the conditions are much different. If we mix benzene with Cl–Cl at ambient temperature and pressure, for example, there is no reaction. For a substitution reaction to occur benzene must act as a base, but benzene does *not* react with HCl or HBr when they are mixed together so it is clear that it is too weak a base for even these strong acids. Benzene does not react with the polarizable Cl–Cl or Br–Br, indicating that it is too weak a Lewis base as well. Benzene is resonance stabilized, so these results are quite reasonable. The poor reactivity can be changed by reaction of a formal cation (X$^+$) with benzene, where benzene is now a strong enough

318.   Sy, A.O.; Raksis, J.W. *Tetrahedron Lett.* **1980**, *21*, 2223.
319.   (a) Wallis, E.S.; Lane, J.F. *Org. React.* **1946**, *3*, 267-306; (b) *The Merck Index, 14th Ed.* Merck & Co., Inc., Whitehouse Station, New Jersey, **2006**, p ONR-46; (c) Mundy, B.P.; Ellerd, M.G.; Favaloro Jr., F.G. *Name Reactions and Reagents in Organic Synthesis, 2nd Ed.* Wiley-Interscience, New Jersey, **2005**, pp. 326-327.
320.   Bauer, L.; Exner, O. *Angew. Chem. Int. Ed.* **1974**, *13*, 376.
321.   Salomon, C.J.; Breuer, E. *J. Org. Chem.* **1997**, *62*, 3858.
322.   Wallace, R.G.; Barker, J.M.; Wood, M.L. *Synthesis* **1990**, 1143.
323.   For examples see Ref. 3, pp 129-134, 619-629, 703-705, 759-761, 911-912.

Lewis base to react, forming a new C–X bond as the aromatic ring is disrupted. This process is generalized for the reaction of benzene with an electrophilic species (X⁺), where the cation (**404**) formed by the reaction is resonance stabilized. Loss of a proton to form **405**, in what is essentially an E1 process, is very rapid because a stable aromatic ring will be regenerated. The E1 analogy is reasonable when it recognized that the proton on the carbon adjacent to the cationic center is acidic. O'Ferrall and co-workers have estimated the pK_a of this proton to be about –24.3.[324] The intermediate cation **404A** is sometimes referred to as a benzenonium ion[325] but old terminology called it a **Wheland intermediate**, especially when drawn as the delocalized cation **404B**.[326] In this reaction, the electrophilic species (X⁺) replaced an aromatic hydrogen via an electrophilic process, and is termed **electrophilic aromatic substitution**.[327]

Nucleophilic attack on unsubstituted benzene rings is difficult since an electron-rich species must react with an electron-rich substrate. The reaction is possible if a good leaving group is attached to the aromatic ring, but reaction conditions are usually harsh. The reaction of an aryl halide (**406**) with a nucleophile (Y⁻) produces a resonance stabilized anion (**407**) as an intermediate, but the transition state for the formation of **407** is relatively high in energy and requires high temperatures, high pressures, and long reaction times. The presence of groups on the aromatic substrate that can disperse the charge by resonance, such as nitro groups, stabilize the carbanionic intermediate and the reaction proceeds under much milder reaction conditions. Elimination of X⁻ from **407** gives the final substitution product, **408**. The overall reaction is called **nucleophilic aromatic substitution**.

An alternative mechanism leads to nucleophilic aromatic substitution. When an unactivated aryl halide (**405**) is treated with a strong base, removal of the ortho hydrogen and subsequent loss of the halide ion (X⁻) gives a highly reactive intermediate called a benzyne (**409**). Benzyne will be attacked by nucleophiles in a reaction that opens the π bond that is not part of the aromatic cloud, to generate a carbanion (**410**). Protonation completes the sequence to give the

324. McCormack, A.C.; McDonnell, C.M.; More O'Ferrall, R.A.; O'Donoghue, A.C.; Rao, S.N. *J. Am. Chem. Soc.* **2002**, *124*, 8575.
325. Stock, M.M. *Aromatic Substitution Reactions,* Prentice-Hall: Englewood Cliffs, NJ, **1968**, p 21.
326. Wheland, G.W. *J. Am. Chem. Soc.* **1942**, *64*, 900.
327. Taylor, R. *Electrophilic Aromatic Substitution*, Wiley, Chichester, **1990**.

aromatic substitution product, **411**. The chemistry of benzyne, and arynes in general, is more than 100 years old,[328] and these reactive intermediates remain quite important in a variety of synthetic transformations.[329]

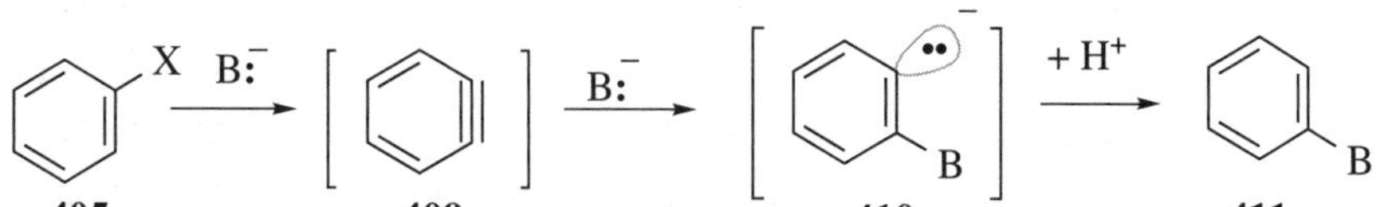

Section 2.11.B will deal with the most fundamental aromatic substitution reactions. This subject is usually presented in detail in undergraduate textbooks. The treatment varies greatly with the text, however, and leading references are usually absent. This discussion is intended only as a brief review.

2.12.B. Electrophilic Aromatic Substitution

The π bonds in benzene are poor electron donors because, in part, such a reaction would disrupt the stable aromatic sextet. In other words, benzene is a weak base and normally will not react with an electrophilic species such as HCl, as mentioned above. Although benzene can form a weak charge transfer complex with chlorine (see **412**), mixing benzene and chlorine does *not* give a substitution reaction. When a strong Lewis acid such as AlCl₃ is added, however, a charge transfer complex such as **413** is formed *between the Lewis acid and chlorine rather than with benzene*. In this complex, the chlorine-chlorine bond is weakened sufficiently so that it can be attacked by benzene, transferring one chlorine to benzene (with disruption of the aromatic system) and transfer of the other chlorine to AlCl₃ (see **414**). For all practical purposes, **413** can be treated as Cl^+ in its reaction with benzene. The stronger the Lewis acid (see sec. 2.3), the faster the halogen and benzene react. The result is formation of a Wheland intermediate, which loses a proton to give chlorobenzene.[330] It is important to note that the charge-transfer complex mechanism may not be valid for chlorination of alkyl benzenes and is solvent dependent.[330a]

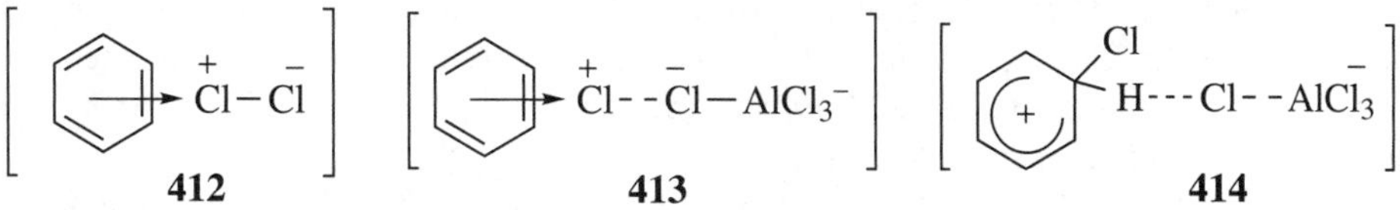

Two charge-transfer complexes are formed during the course of this reaction.[331] The first will be **413** and the second (**414**) is generated during loss of the proton in the aromatization

328. Stoermer, R.; Kahlert, B. *Ber. Dtsch. Chem. Ges.* **1902**, *35*, 1633.
329. Wenk, H.H.; Winkler, M.; Sander, W. *Angew. Chem. Int. Ed.* **2003**, *42*, 502.
330. (a) de la Mare, P.B.D. *Acc. Chem. Res.* **1974**, *7*, 361 and references cited therein; (b) Le Page, L.; Jungers, J.C. *Bull. Soc. Chim. Fr.* **1960**, 525; (c) Olah, G.A.; Kuhn, S.J.; Hardie, B.A. *J. Am. Chem. Soc.* **1964**, *86*, 1055.
331. Reference 325, p 27.

process. In the chlorination reaction, the rate-determining step (the slowest) is conversion of the charge transfer complex to the Wheland complex (**413** → complex).[327] Formation of **413**, conversion of the Wheland complex to **414,** and elimination of the proton are all fast processes. Bromine reacts in a manner identical to chlorine,[332,330b] producing Br^+ as a reactive intermediate (in the charge-transfer complex). Reaction of bromine, benzene, and aluminum chloride gives bromobenzene. Typical Lewis acid catalysts for this reaction are iodine, $AlCl_3$, $SbCl_3$, PCl_3, PCl_5, and $SnCl_4$.[333] When benzene is treated with a mixture of nitric and sulfuric acids, nitrobenzene is the product,[334] via a nitronium ion (NO_2^+).[335] Benzene also reacts with sulfuric acid [or sulfuric acid saturated with sulfur trioxide (SO_3), which is known as fuming sulfuric] to give benzenesulfonic acid.[336] Benzenesulfonic acid is also formed by reaction of benzene with chlorosulfonic acid ($ClSO_3H$)[337] or fluorosulfonic acid (FSO_3H).[338]

When substituents are present on the aromatic ring, these substitution reactions can proceed with great selectivity. Majetich and co-workers, for example, showed that aniline reacted with HBr/DMSO to give a 76% yield of 4-bromoaniline, in a remarkably selective reaction.[339] Bromination of **415** gave a 75% yield of **416**[340] in a synthesis of lateriflorone by Theodorakis and co-workers. Note that the highly activated ring in **415** was brominated without the use of a Lewis acid. Such reactivity is common when strong activating groups such as OH and OR are present on the aromatic ring. In a synthesis of 11,12-demethoxypauciflorine,[341] Magnus and co-workers treated **417** with fuming nitric acid, and obtained an 82% yield of **418**.

332. Reference 327, pp 377-380.
333. (a) Price, C.C. *J. Am. Chem. Soc.* **1936**, *58*, 2101; (b) *Idem Chem. Rev.* **1941**, *29*, 37; (c) Price, C.C.; Arntzen, C.E. *J. Am. Chem. Soc.* **1938**, *60*, 2835.
334. (a) Hoggett, J.G.; Moodie, R.B.; Penton, J.R.; Schofield, K. *Nitration and Aromatic Reactivity*, Cambridge Univ. Press, Cambridge, *1971*; (b) Schofield, K. *Aromatic Nitration*, Cambridge Univ. Press, Cambridge, *1980*.
335. (a) Westheimer, F.H.; Kharasch, M.S. *J. Am. Chem. Soc.* **1946**, *68*, 1871; (b) Martinsen, H. *Z. Phys. Chem.* **1905**, *50*, 385; (c) Idem *Ibid* **1907**, *59*, 605; (d) Marziano, N.C.; Sampoli, M.; Pinna, F.; Passerini, A. *J. Chem. Soc. Perkin Trans. 2* **1984**, 1163.
336. (a) Brand, J.C.D.; Jarvie, A.W.P.; Horning, W.C. *J. Chem Soc.* **1959**, 3844; (b) Cerfontain, H.; Sixma, F.L.J.; Vollbracht, L. *Recl. Trav. Chim. Pays-Bas* **1963**, *82*, 659; (c) Cerfontain, H. *Ibid* **1961**, *80*, 296; (d) *Idem Ibid* **1965**, *84*, 551
337. (a) Harding, L. *J. Chem. Soc.* **1921**, 1261; (b) Levina, L.I.; Patrakova, S.N.; Patruskev, D.A. *J. Gen. Chem. USSR* **1958**, *28*, 2427.
338. Gillespie, R.J. *Acc. Chem. Res.* **1968**, *1*, 202.
339. Majetich, G.; Hicks, R.; Reister, S. *J. Org. Chem.* **1997**, *62*, 4321.
340. Tisdale, E.J.; Li, H.; Vong, B.G.; Kim, S.H.; Theodorakis, E.A. *Org. Lett.* **2003**, *5*, 1491.
341. Magnus, P.; Gazzard, L.; Hobson, L.; Payne, A.H.; Rainey, T.J.; Westlund, N.; Lynch, V. *Tetrahedron* **2002**, *58*, 3423.

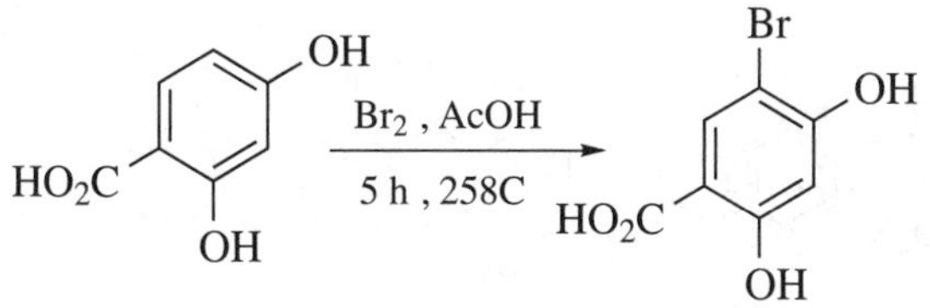

415   **416** 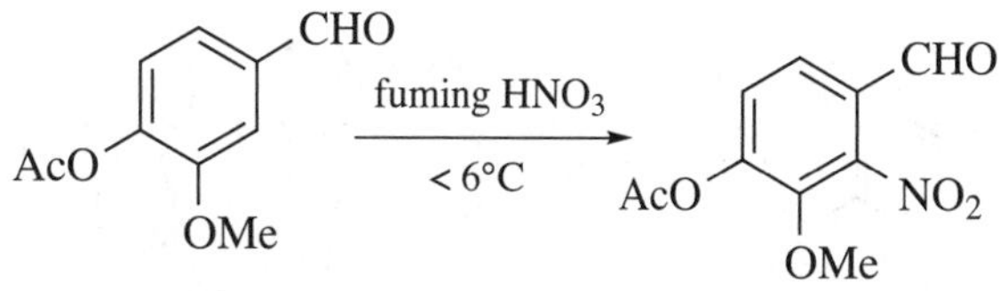   **417**   **418**

Electrophilic aromatic substitution occurs readily many aromatic derivatives. Naphthalene, for example, reacts with a mixture of bromine and aluminum chloride to give 99% of 1-bromonapthalene (**419**) with only 1% of 2-bromonapthalene (**420**). Substitution at the 1-position of naphthalene is greatly preferred,[342] and this can be explained by examining the reactive intermediate. Reaction of naphthalene and Br$^+$ at the C1-position generates the resonance stabilized Wheland type intermediate, **421**. There are a total of seven resonance structures in **421**, and an important feature is the presence of a fully aromatic (intact benzene ring) structure in four of those resonance intermediates. Similar analysis of the intermediates for attack at C2 reveals there are six structures, but only two of them have fully aromatic rings. This simple analysis predicts **421** to be more stable than the intermediate resulting from attack at C2, consistent with the experimentally observed major product **419** in a 99:1 ratio (**419/420**).

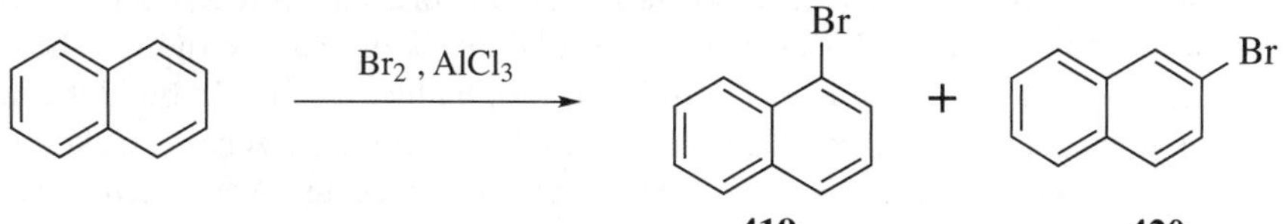

419   **420**

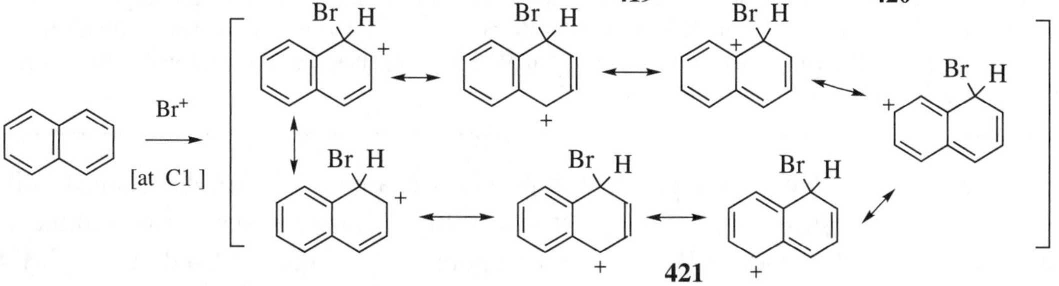

421

In aromatic substrates where reaction with X$^+$ generates several different cationic intermediates that are close in energy, little selectivity is expected and a mixture of products is predicted. An illustration is the nitration reaction of phenanthrene, which gave five products: **422** (6%), **423** (23%), **424** (7%), **425** (27%) and **426** (37%).[343] Although substitution at C9 is preferred, a complex mixture is obtained that may be difficult to separate. Substitution at several positions is a common occurrence for substitution reactions of polycyclic aromatic derivatives, although the product distribution is usually influenced by the nature of the catalyst and by the reaction conditions that are used.

---

342. Reference 325, p 76.
343. Reference 325, p 78.

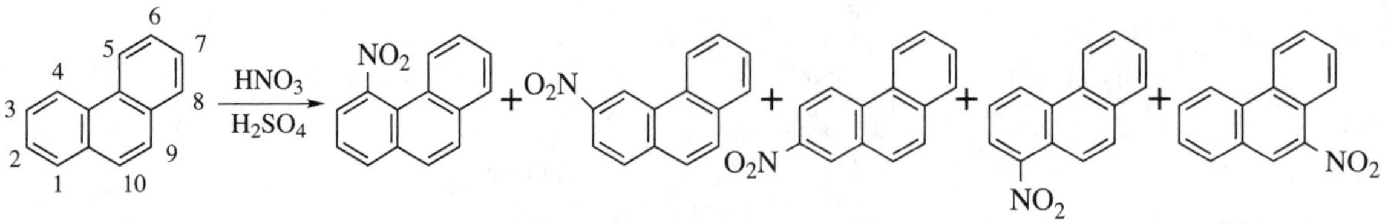

422 (c4) **423** (c3) **424** (c2) **425** (c1) **426** (c9)

MOLECULAR MODELING: Reactive Intermediates Anticipate Selectivity in Nitration of Phenanthrene

According to the **Hammond Postulate**, the geometry of the reactants in an *exothermic* reaction should resemble the geometry of the transition state. This suggests that the relative energies of the intermediates will reflect the relative energies of the transition states connecting them to products, and will anticipate the kinetic product distribution. Consider the electrophilic nitration of phenanthrene. Each of five different products (**422-426**), follows from a different reactive ("Wheland") intermediate. Of the five products, **423**, **425** and **426** account for 87% of the total.

Build the Wheland intermediate that leads to **422** and obtain its energy. Start from **Naphthalene** under the **Rings** menu, and build the third ring from a pair of sp^2 carbons, a carbocation center and an sp^3 carbon. Add **Nitro** (**Groups** menu) to the sp^3 carbon. In a similar way, obtain energies for the intermediates leading to **423-425**. Before you start each, *click* on > at the right of the tab(s) at the top of the screen and select **Continue** to bring up a fresh builder palette. To build the intermediate leading to **426**, start from **Benzene** (**Rings** menu), add a second benzene to make biphenyl and connect the two rings with an sp^3 carbon and a carbocation center. Add **Nitro** (**Groups** menu) to the sp^3 carbon. Rank the five Wheland intermediates in order of increasing energy. Are the intermediates corresponding to **423**, **425** and **426** the three with the lowest energy? Are all five intermediates sufficiently close in energy (within 5-20 kJ/mol) such that it is reasonable to expect that all five products will actually be observed?

Despite the many products produced in the previous example, electrophilic substitution of aromatic molecules can proceed with great selectivity in many cases. An example is the nitration of benzo[*c*]fluorene (**427**), which gave a good yield of the 5-nitro derivative (**428**).[344] Nitration of 2-hydroxyazulene (**429**) gave a mixture of 1-nitroazulen-2-one (**430**) and 6-nitro-2-hydroxyazulene (**431**).[345]

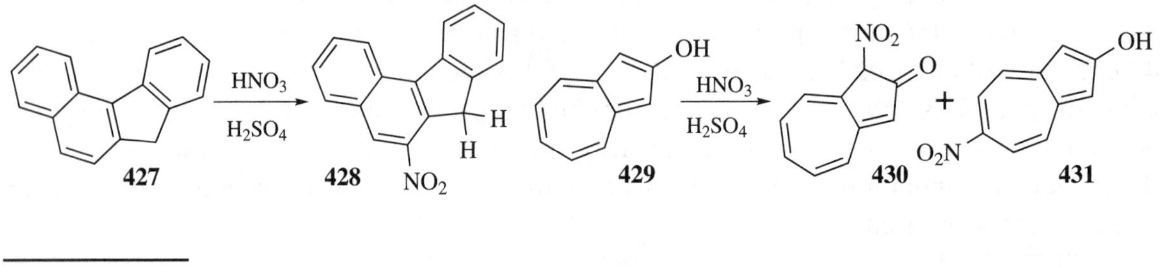

---

344. Bolton, R. *J. Chem. Soc. (S)* **1977**, 149.
345. Nozoe, T.; Asao, T.; Oda, M. *Bull. Chem. Soc. Jpn.* **1974**, *47*, 681.

Heterocyclic aromatic derivatives also undergo electrophilic aromatic substitution reactions. Pyrrole, for example, reacts with nitric acid in acetic anhydride, at 0°C, to give 50% of 2-nitropyrrole and 15% of 3-nitro-pyrrole. Pyrrole is more reactive in these reactions when compared to benzene, and the mild reaction conditions described are sufficient. One caution is that the products may decompose with prolonged exposure to a powerful Lewis acid. The preference for substitution at C2 in pyrrole is seen in other five-membered ring heterocycles such as furan and thiophene. This preference can be explained by the relative stability of the cationic intermediates, just as in substitution reactions with benzene derivatives. Inspection of the cation resulting from attack at C2 (**432**) shows that the lone pair electrons on the heteroatom participates in resonance delocalization, giving a total of three resonance structures. Since attack at C3 (see **433**) will lead to only two structures, attack at C2 is preferred. If there are substituents at C2 and C5, however, attack will occur at C3. Pyrrole, with the more basic nitrogen atom, is better able to stabilize the cationic intermediate than furan or thiophene and is more reactive.

Pyridine, unlike five-membered ring heterocycles, undergoes substitution with difficulty and at the C3 position. This reaction illustrated by its reaction with mercuric sulfate (HgSO$_4$) and H$_2$SO$_4$ (at 220°C) to give pyridine 3-sulfonic acid. The poor reactivity relative to pyrrole is clear by simply examining the vigorous conditions that are required for the reaction. The preference for C3 substitution in six-membered ring heterocycles is again explained by generation of a more stable cationic intermediate.

Other heterocyclic ring systems react with the reagents in this section, including bicyclic and tricyclic compounds. In some cases, one aromatic ring contains a heteroatom while the other does not, as in quinoline or isoquinoline. The aromatic ring in quinoline (**434**) that does not contain nitrogen undergoes electrophilic aromatic substitution preferentially. It is known that pyridine is less reactive than benzene, so it is reasonable that the pyridine portion of quinoline should also be less reactive than the benzene portion. The two products of this reaction (HNO$_3$, H$_2$SO$_4$, 0°C, 3 min.) are 8-nitroquinoline (**435**) and 5-nitroquinoline (**436**, the nitrogen is 1), formed in roughly equal amounts. The preference for attack at C5 and C8 is explained by the usual stability arguments for the cationic intermediates.

The products arising from nitration of quinoline clearly show that one ring reacted faster than the other, although the isomer distribution depends on reaction conditions. The deactivated pyridine ring in **434** lowers the rate of electrophilic substitution relative to the other aromatic ring. This general phenomenon can be applied to monocyclic heterocycles, polycyclic heterocycles, polycyclic aromatic hydrocarbons, and benzene derivatives.

The presence of electron-withdrawing or electron-releasing groups on an aromatic ring will deactivate or activate that ring to electrophilic aromatic substitution, which suggests that the primary effect is inductive. It is also true that the inductive effect will be transmitted to or from the aromatic ring at the carbon bearing the substituent, the *ipso* carbon. Reaction of X^+ with a benzene ring containing an electron-rich substituent leads to a Wheland intermediate that has a negative charge or a δ- dipole attached to the ipso carbon. The expected electron releasing effect will stabilize the cationic intermediate. If we focus only on inductive effects that occur in the reactive intermediates, the effect can be used as a convenient method for predicting reactivity. If the substituent releases electrons (X has a negative charge or a δ- dipole, as in **437**) that substituent is said to have a positive inductive effect (+I) and should stabilize the cation intermediate. A negative inductive effect (–*I*) occurs when the substituent at the ipso carbon is electron withdrawing (Z has a positive charge or δ+ dipole, as in **438**), and the cation intermediate is destabilized making its formation more difficult.

$$437 \qquad\qquad 438 \qquad\qquad\qquad\qquad 439$$

A second factor is important in electrophilic aromatic substitution reactions. If the group at the ipso carbon is capable of dispersing the positive charge by resonance delocalization, that intermediate will be more stable. This resonance effect is often called a mesomeric effect, and labeled +*M* or –*M*. The OR group in **439** exhibits a +M effect since the electrons on oxygen can be donated to the positive charge at the ipso carbon, which results in transferring the charge to oxygen, as shown, by resonance. It is reasonable to assume that a substituent at the ipso carbon with unshared electrons (such as oxygen, sulfur, nitrogen, halogen) will have a +M effect, which will stabilize the cationic intermediate of electrophilic aromatic substitution. Conversely, a group bearing a positive charge or a partially positive dipole will not be able to delocalize a positive charge, and indeed, the two like charges on adjacent atoms will destabilize that intermediate by electronic repulsion, as in **440**. An intermediate such as **439** should be easier to form and since cation formation is the rate-determining step (see above), the substitution should be faster. Anisole, for example, is very reactive in electrophilic aromatic substitution, and in many cases a Lewis acid is not required for the reaction. Toluene, on the other hand, has a +*I* effect (see **441**) but no mesomeric effect. The substitution should be faster than for benzene (with no activating substituents) but slower than for anisole with both a +*I* and a +*M* effect (see **442**).

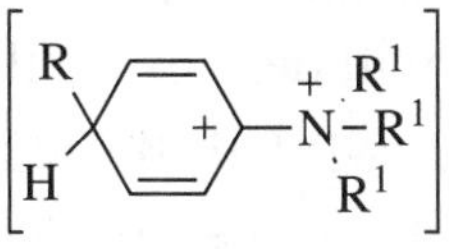

440 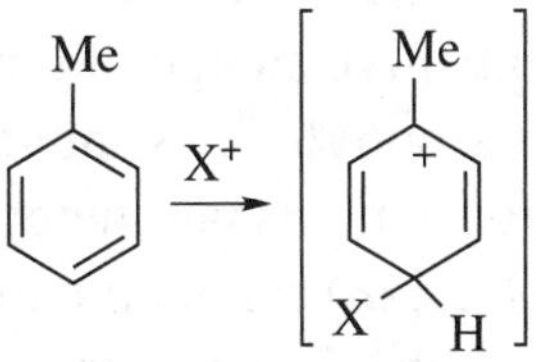 **441** 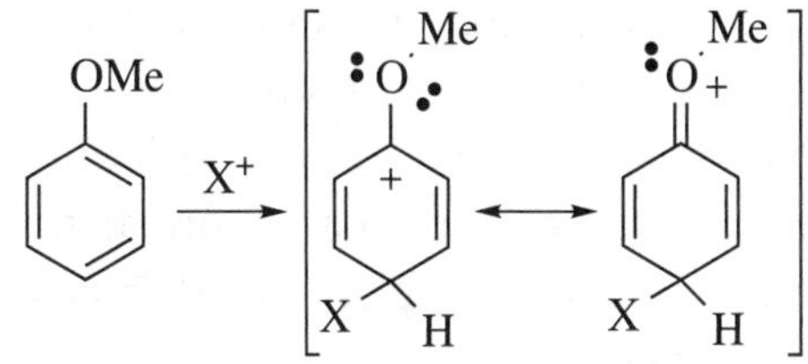 **442**

It is important to note that the +*M* effect observed with anisole occurs only with ortho and para substitution. In those cases the positive charge in the intermediate can be transferred to the ipso carbon and thereby to the oxygen. The presence of an extra resonance form is indicative of increased stability for those intermediates because delocalization occurs over more atoms. This greater stability means that the cation is formed with greater ease and electrophilic aromatic substitution is faster. Attack at the meta-position generates a cation that cannot transfer the charge to the ipso carbon and no +*M* effect is available. For substituents such as OR, therefore, ortho/para substitution should be faster than meta substitution, and bromination of anisole will give, primarily, 2-bromoanisole (**443**) and 4-bromoanisole (**444**) in roughly equal amounts.

Aromatic compounds that have substituents with a +*I* effect or a +*I* and a +*M* effect show an increase in the rate of electrophilic aromatic substitution relative to benzene and are said to be activating substituents.[346] Since ortho and para products predominate with these activating groups, activators are also said to be ortho/para directors. A preference for either the ortho- or the para product is common, particularly in aromatic compounds that have several substituents. When 2-methylanisole was treated with bromine in chloroform, for example, a 94% yield of **445** was obtained as part of Vyvan Looper's synthesis of heliannuol D.[347] Note that the presence of two activating groups allowed the reaction to proceed without adding a Lewis acid, which is typical.

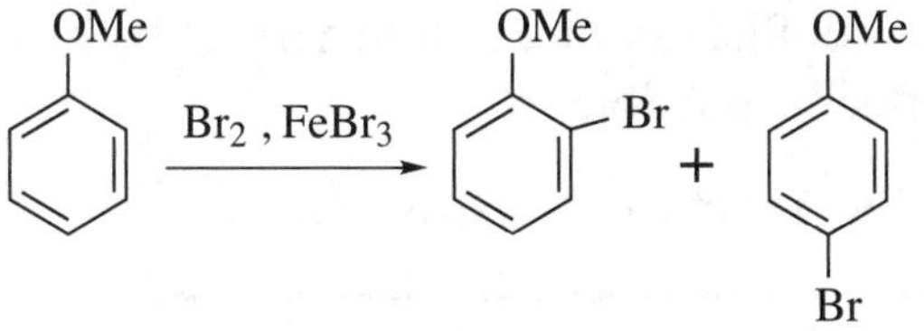

443   **444** 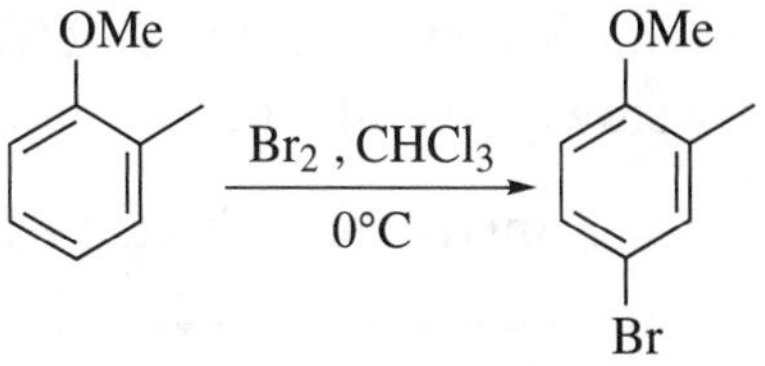   **445**

Aromatic compounds that have substituents with a -*I* effect or a +*I* effect, and a +*M* effect show a decrease in the rate of electrophilic aromatic substitution, relative to benzene, and that substituent is said to be deactivating.[346] This occurs when the substituent at the ipso carbon has a positive charge or a δ+ dipole (SR_2, NO_2, CO_2R, SO_3R, NR_3, C=O or C≡N). Reaction with the aromatic ring at the ortho or para position generates an intermediate that places a positive charge adjacent to the electrophilic center. An acetyl group is deactivating and the key

---

346. Reference 327, pp 41-48, 51.
347. Vyvyan, J.R.; Looper, R.E. *Tetrahedron Lett.* **2000**, *41*, 1151.

resonance structures for ortho, para and meta attack by X^+ are shown in **446**, **447**, and **448**. In **446** and **447** a distinct $-I$ effect is seen, as the positive charge of the intermediate resides on the ipso carbon, adjacent to the $\delta+$ dipole of the carbonyl. This interaction will destabilize these intermediates, making them more difficult to form in this rate-determining step and slowing the overall reaction. Attack in the meta position (see **448**) cannot put the positive charge at the *ipso* carbon. The two electrophilic centers are relatively close together and this is destabilizing, but less destabilizing than having the charges on adjacent atoms. Intermediate **448** is, therefore, *less destabilized* than **446** or **447** and will lead to the major product in this particular reaction, meta-substitution. Groups such as acetyl, nitro, and so on are deactivating and meta-directors. This analysis leads to the correct observation that nitration of nitrobenzene gives primarily 1,3-dinitrobenzene. Analysis of benzene derivatives that contain many types of substituents led to Table 2.15,[348] which shows activating and deactivating groups, classified by their I and M effects. The table also shows if the substituent is an ortho/para or a meta director.

Halogenated benzenes show a deactivating effect similar to that for the deactivating groups listed, but give primarily ortho and para substitution products as in the formation of 1,2-dichlorobenzene and 1,4-dichlorobenzene from the reaction of chlorobenzene with Cl_2 and $AlCl_3$. This is explained by the fact that chlorine has a $-I$ effect (making it deactivating) but a $+M$ effect, making it an ortho/para director. This mesomeric effect is seen in **449** for para substitution. When the I and M effects are in conflict, as in chlorobenzene, the reaction is generally slower than if a pure $+I$ and/or $+M$ effect is possible.[346]

Table 2.15. Substituent Effects in Electrophilic Aromatic Substitution

Substituent Type	Effect	Directing Effect	Example
$+I, +M$	Activating	ortho/para	alkyl groups, O^-
$+I, -M$	Activating	ortho/para	$SiMe_3$, CO_2^-
$-I, +M$	Activating	ortho/para	NMe_2, $NHCO_2Me$, OMe, Ph
$-I, +M$	Deactivting	ortho/para	Cl, Br, $CH=CHNO_2$
$-I$	Deactivting	meta	Me_3N^+
$-I, -M$	Deactivting	meta	NO_2, $C\equiv N$, COR

[Reprinted with permission from Taylor, R. *Electrophilic Aromatic Substitution*, Wiley, Chichester, *1990*, p 51. Copyright © *1990* by John Wiley and Sons, Inc.]

---

348.   Reference 327, p 51.

MOLECULAR MODELING: Local Ionization Potential Maps and Electrophilic Reactivity/Selectivity

The degree to which substituents affect the electron distribution of a benzene ring to which they are attached may be accessed by a **local ionization potential map.** This colors an electron density surface that indicates overall molecular size and shape (where a reagent can approach), with the value of the energy required to expel an electron from a specific location on the surface. By convention, regions where the local ionization potential is large (electron ejection is difficult) are colored "blue", while regions where it is small are colored "red". Thus, it is possible to differentiate regions on the accessible surface that are best able to attack (donate electrons) to an electrophilic center (colored red) from those that are less likely to donate electrons and more likely to be attacked (accept electron density) in a reaction (colored blue).

Examine local ionization maps for benzene (acting as the standard), styrene, acetophenone, anisole, chlorobenzene and trimethylsilylbenzene (top left to right and bottom left to right). ***SpartanModel*** cannot calculate local ionization potential maps and these results were obtained using ***Spartan***. The maps are initially presented as transparent models to allow you to see the underlying structures. Change to solid models to allow easier comparison. *Click* on one of the maps and select ***Solid*** from the **Style** menu at the bottom of the screen. Which substituents lower the local ionization potential (on the benzene ring) and hence would be expected to lead to an increase in electrophilic reactivity? For these substituents, is reaction at the *ortho* and *para* positions favored or disfavored over reaction at the *meta* positions? Which substituents raise the local ionization potential, leading to a decrease in electrophilic reactivity? For these substituents, is reaction at the *ortho* and *para* positions favored or disfavored over reaction at the *meta* positions? Are the effects at the *ortho* positions smaller, larger, or about the same as at the *para* position? Are the conclusions from the local ionization potential maps consistent with arguments based on "resonance theory"? Elaborate.

Other factors influence the rate of electrophilic aromatic substitution, including steric hindrance and strain.[349] When a substituent (X in **450**) is physically large, ortho substitution is inhibited by steric hindrance in the transition state which, in principle, makes it more difficult for that intermediate to form. This effect is observed when X is an alkyl group (see Friedel-Crafts alkylation, sec. 12.4.B). Many of the groups encountered in this section have heteroatoms, and the steric effect is often less important if the X group can coordinate with the incoming group. This type of coordination (or chelation) effect is represented in **451** and the result is an increase in the amount of the ortho product. With the heteroatom substituents discussed in this section, this model is more important, and the increase in ortho products at the expense of para products is called the **ortho effect**. Steric hindrance does play a significant role in substitution reactions of aromatic derivatives such as **452**, where the 1,3-relationship of the substituents makes approach along path *c* sterically difficult. Paths *a* and *b* are accessible so substitution of **452** will lead to products resulting from attack along paths *a* and *b*.

349. Reference 327, p 52.

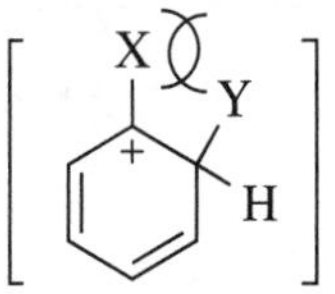

[450] 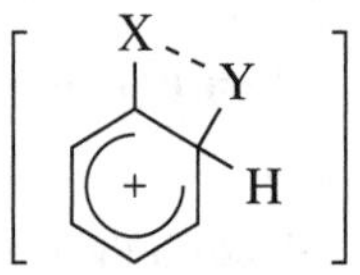 [451] 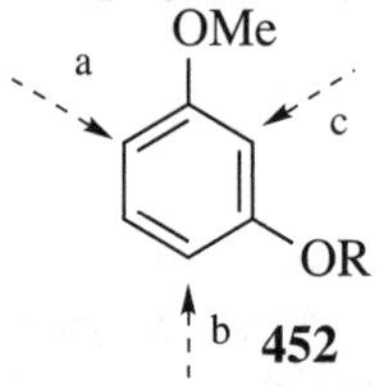 [452]

Deformation of an aromatic ring out of planarity leads to strain. 4,5-Dimethylphenanthrene (**453**) is an example, where the two methyl groups interact and the net result of this interaction is that the molecule is more reactive.[350] Strain also occurs in molecules such as β-hydroxyindane (**454**), which is brominated at the γ-position to give **455**.[350] The strain imparted by the five-membered ring makes the α-position less susceptible to electrophilic aromatic substitution. Formation of resonance intermediate **456** demands a shortened bond for the sp^2 hybridized carbon, and this increases the strain at that position. When this reaction occurs in a six-membered ring such as β-hydroxytetralin (**457**), however, the strain in the intermediate is greatly diminished and bromination gives primarily the α-product, **458**. The explanation for this selectivity is known as the **Mills-Nixon effect**.[351]

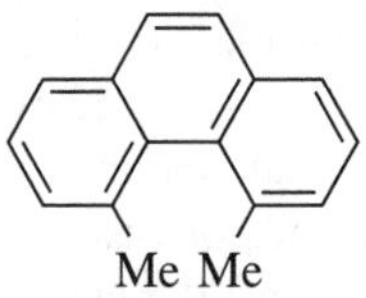

Me Me
453 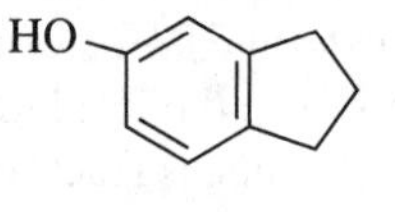   **454** 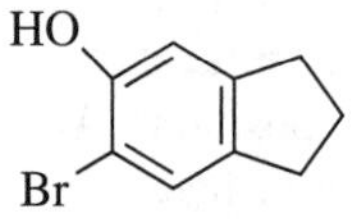   **455**

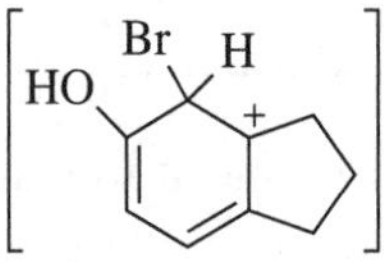

456 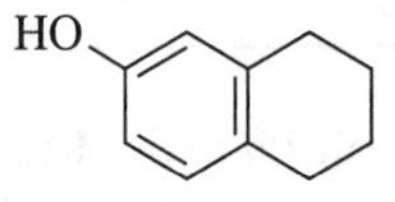   **457** 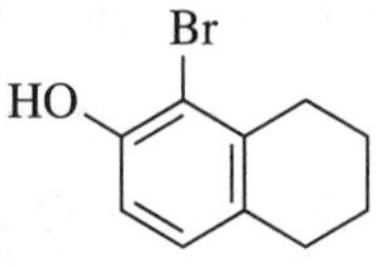   **458**

All of these effects are observed when comparing the rates of various electrophilic aromatic substitution reactions. Activating substituents increase the rate of reaction relative to benzene. The rate of reaction for the nitration of anisole, for example, was 9.7×10^6 times faster than nitration of benzene. The reaction of anisole with nitric and sulfuric acids, gave 44% of o-nitroanisole, 56% of p-nitroanisole and < 1% of m-nitro-anisole.[352] This contrasts with reactions involving deactivating substituents, where selectivity for the meta product is usually very good. Nitration of nitrobenzene, for example, gave 1,3-dinitrobenzene in 94% yield, with only 6% of the ortho product and < 1% of the para product.[352]

---

350. Reference 327, p 53.
351. Mills, W.H.; Nixon, I.G. *J. Chem. Soc.* **1930**, 2510.
352. Reference 325, p 43.

Nitration of aromatic rings has been reported in ionic liquids such as [bmpy][N(Tf)$_2$] (**459**).[353] In this variation, an acyl nitrate formed from HNO$_3$-Ac$_2$O reacts with toluene to give a 93% yield of *ortho-* and *para-*nitrotoluene (*o/p* = 1.3) at room temperature in dichloromethane, in one hour.

2.12.C. Nucleophilic Aromatic Substitution

An aromatic ring is electron rich and nucleophilic attack on that ring is expected to be difficult under normal reaction conditions. Nucleophilic aromatic substitution[354] does occur, however, and is synthetically useful with a variety of substituents and nucleophiles. A variation of this aromatic substitution involves the reaction of aryl halides with cuprous salts, but this will be discussed in sec. 8.2.C. When dinitrochlorobenzene was treated with sodium hydroxide in aqueous media, or neat, no reaction occurred. When that mixture was heated to ~ 100°C, however, a substitution reaction occurred that converted 2,4-dinitrochlorobenzene into 2,4-dinitrophenol.[355] Indeed, this is a commercial method for the production of phenol derivatives, despite the harsh reaction conditions. Direct substitution must occur by attack of the nucleophilic hydroxide at the *ipso* carbon, generating the resonance stabilized carbanion intermediate **460**. The chloride is a leaving group, expelled by the negative charge (two electrons) in **460** and assisted by rearomatization of the ring to give the phenol derivative. The rate-determining step is the initial collision to form **460**, and factors that make this collision more favorable will increase the rate of the overall substitution process. The *I* and *M* effects noted with electrophilic aromatic substitution are operative, but their effect will be the exact opposite for nucleophilic attack on **460** since the charges are different. An electron withdrawing group at the ortho or para positions relatively to the Cl, for example, will delocalize the negative charge and stabilize that intermediate. Conversely, an electron-releasing group at those positions will destabilize

the intermediate. Inspection of resonance contributor **461**, which arises by nucleophilic attack of hydroxide on 4-chloroanisole, shows the negative charge on the carbon adjacent to the OMe moiety. The electron rich oxygen will repel the negative charge and destabilize **461**, making nucleophilic substitution of 4-chloroanisole more difficult than for chlorobenzene. Resonance

353. Lancaster, N.L.; Llopis-Mestre, V. *Chem. Commun.* **2003**, 2812.
354. Rossi, R.A.; de Rossi, R.H. *Aromatic Substitution by the SRN1 Mechanism*, American Chemical Society, Washington, DC, **1983**.
355. Reference 325, p 134.

contributor **462**, arising from reaction of 4-chloronitrobenzene and hydroxide, has a negative charge on the carbon adjacent to the positive nitrogen of the nitro group. The charge can be delocalized out of the ring and onto the nitro oxygen, leading to an additional resonance structure and a more stable intermediate. Therefore, 4-chloronitrobezene should react faster than chlorobenzene. Stabilization of the anionic intermediate by a nitro group will be most effective at the ortho and para positions and much less effective when the substituents are in the meta position, all relative to the *ipso* carbon.

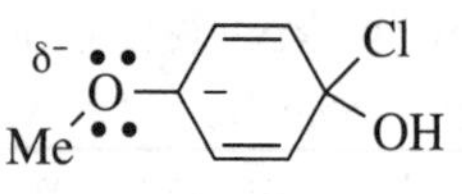

461 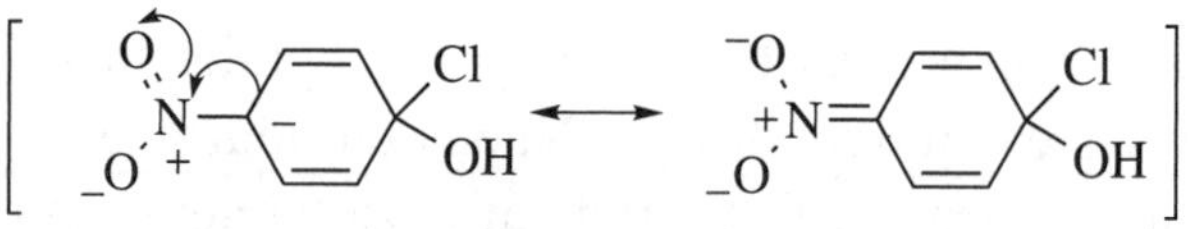   **462**

Table 2.16. Influence of Substituents on Nucleophilic Aromatic Substitution.

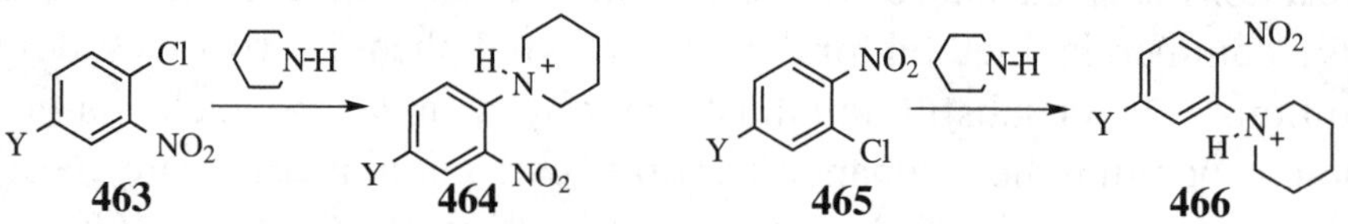

Y	k/k_B for 463	k/k_B for 465
NO$_2$	41000	
CN	6000	5
CO$_2$Et	920	5
Br	10	35
Cl	6	32
OMe	0.025	4
H	1	1
Me	0.2	0.9

[Leon M. Stock, *AROMATIC SUBSTITUTION REACTIONS*, ©*1968*, pp. 87.
Reprinted by permission of Prentice Hall, Englewood Cliffs, New Jersey.

For nucleophilic aromatic substitution, electron-withdrawing groups stabilize the reactive intermediates and enhance the rate of the reaction. These effects are shown in Table 2.16,[356] which compares the relative rates of nucleophilic substitution of chlorobenzene derivatives with piperidine. Electron-donating substituents destabilize the reactive intermediate and diminish the rate of the reaction, which is dramatically illustrated by the reaction of 2,4,6-trinitrochlorobenzene with dilute hydroxide, which gave trinitrophenol upon simple warming and did not require high temperatures or pressures. In Table 2.16, rate enhancement is greatest when the substituent (Y) is in the para position, although some rate enhancement is observed for the meta substituent as well. Compounds with a para substituent in **463**, relative to the chlorine leaving group, stabilized the intermediate that led to formation of **464**, and there was a large rate effect for all electron withdrawing substituents. The derivative with a meta

---

356.   Reference 325, p 87.

substituent relative to the chlorine leaving group (**465**), generated **466** and showed a smaller rate effect. Electron withdrawing groups exerted the largest influence (the reference compound was Y = H [k_H = 1] in all cases.

2.12.D. Benzyne Derivatives

When aryl halides are treated with powerful bases such as potassium amide (KNH$_2$) or *tert*-butyllithium (sec. 8.5.B), a hydrogen atom may be removed from the aromatic ring in an acid-base reaction to give an aryl carbanion. The presence of the halogen in aromatic compounds such as **467** makes the ortho hydrogens more acidic. Removal of the ortho hydrogen generates a carbanion (**468**), but the negative charge is generated perpendicular to the plane of the aromatic π-cloud (see **468B**). The charge resides on the carbon next to the halogen, which is a good leaving group. Elimination of the halide forms a new π bond that is perpendicular to the π-cloud of the benzene ring in a highly reactive species (**469** and see **409**) called a benzyne.

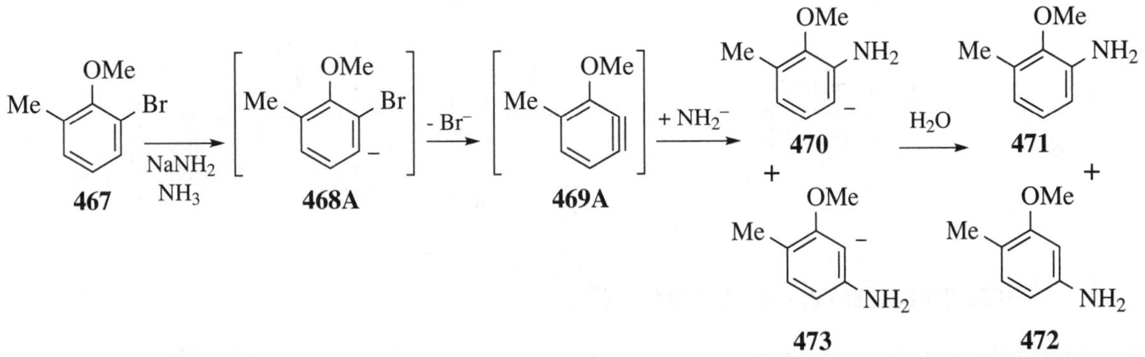

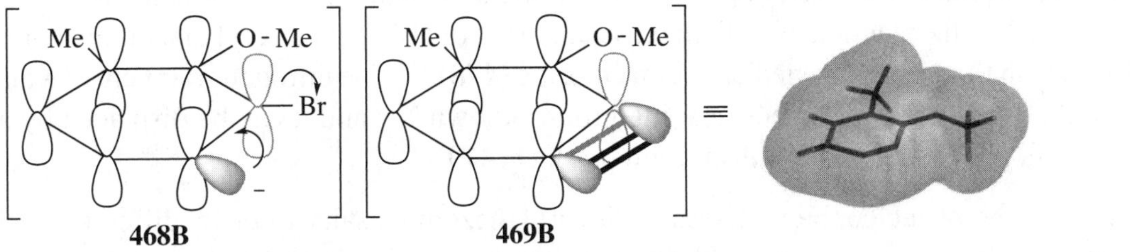

In a specific example, **467** was treated with sodium amide and deprotonation gave **468A**, from which the bromide was expelled to give the benzyne intermediate, **469A**.[357] When **468A** is redrawn as **468B**, it is clear that the carbanionic center is perpendicular to the aromatic π cloud and parallel to the C-Br bond. Expulsion of this adjacent leaving group forms the new π bond of the benzyne intermediate (**468B** → **469B**). This π-bond is subject to nucleophilic attack, and the LUMO map (see sec. 2.9.A) shown for **469B** indicates a blue color for the two

357. (a) Roberts, J.D.; Simmons, Jr., H.E.; Carlsmith, L.A.; Vaughn, C.W. *J. Am. Chem. Soc.* **1953**, *75*, 3290; (b) Roberts, J.D.; Semenow, D.A.; Simmons, Jr., H.E.; Carlsmith, L.A. *Ibid* **1956**, *78*, 601; (c) Roberts, J.D.; Vaughn, C.W.; Carlsmith, L.A.; Semenow, D.A. *Ibid* **1956**, *78*, 611.

benzyne carbon atoms (see **Spartan file C2-469B** under the **Chapter 2 directory**), suggesting that they are susceptible to attack by a nucleophile. The more intense blue at C3 relative to the methoxy group suggests that the 3-amino product will predominate. Indeed, the nucleophilic amide reacts with **469** to produce a mixture of carbanions **470** and **473**, the result of attack at both carbons of the double bond. Subsequent reaction with water generates the neutral aniline derivatives **471** and **472**. One substitution product (**473**) has the new amino substituent adjacent to the carbon that bore the bromide leaving group, and this product is said to result from **cine substitution**.

Benzyne formation and their reactions are not limited to simple benzene derivatives. Heterocyclic molecules can also be used for benzene reactions. One important use of benzyne derivatives is in the Diels-Alder reaction (see sec. 11.4.A). When 3-chloro-2-methoxypyridine (**474**) was treated with LDA at -78°C, for example, lithium-hydrogen exchange occurred to give **475**. Loss of LiBr led to benzyne **476**, which reacted with furan to give the Diels-Alder product (**477**) in 74% yield.[358]

2.12.E. Substitution Reactions of Aryl Diazonium Salts

When a primary aromatic amine such as aniline is treated with nitrous acid (HONO), usually generated by the reaction of sodium nitrite and HCl or H_2SO_4,[359] an aryl diazonium compound is formed, in this case benzenediazonium chloride (**478**).[360] The nitrogen fragment of a diazonium ion ($Ar-N_2^+$) is one of the best leaving groups known[360a,b] and it can be displaced by a variety of nucleophiles to give the substitution product (**479**).[361]

Many different nucleophiles can react with aryl diazonium salts, often by different mechanisms that range from S_N1 type processes to those involving radicals.[362] Heating **478** in aqueous acid generates phenol.[363] This reaction is also illustrated by the conversion of 3-nitroaniline (**480**)

358. Connon, S.J.; Hegarty, A.F. *J. Chem. Soc. Perkin Trans. 1* **2000**, 1245.
359. (a) Ridd, J.H. *Quart. Rev. Chem. Soc.* **1961**, *15*, 418; (b) Hegarty, A.F. *The Chemistry of Diazonium and Diazo Groups*, Part 2, Patai, S. (Ed.), Wiley, New York, **1978**, pp 511-591.
360. (a) Collins, C.J. *Acc. Chem. Res.* **1971**, *4*, 315; (b) Friedman, L. in *Carbonium Ions, Vol. II*, Olah, G A.; Schleyer, P.v.R. (Eds.), Wiley, New York, **1970**, p 655; (c) Wistar, R.; BartLett. P.D. *J. Am. Chem. Soc.* **1941**, *63*, 413.
361. Reference 325, pp 92-100.
362. Wulfman, D.S. in *The Chemistry of Diazonium and Diazo Groups, Part 1*, Patai, S. (Ed.), Wiley, New York, **1978**, pp 286-297.
363. Horning, D.E.; Ross, D.A.; Muchowski, J.M. *Can. J. Chem.* **1973**, *51*, 2347.

to the diazonium salt (**481**), which was followed by heating in aqueous sulfuric acid to give 3-nitrophenol (**482**).[364]

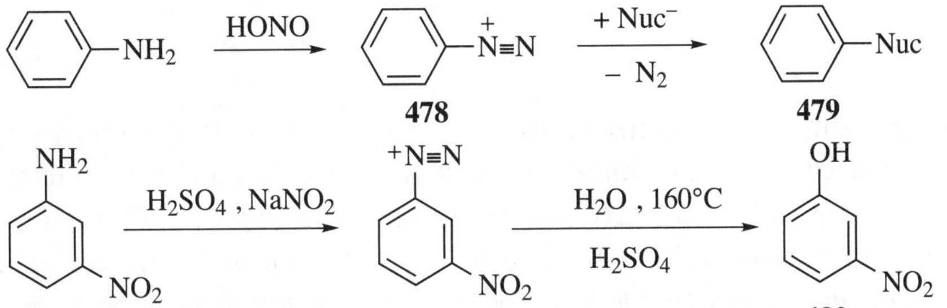

Treatment of diazonium salts with cuprous, Cu(I), salts generates aryl halides. When **478** reacts with CuCl (cuprous chloride) or CuBr (cuprous bromide), the products are chlorobenzene or bromobenzene via what is probably a radical reaction.[365] This conversion is known as the **Sandmeyer reaction**.[366] The use of copper powder rather than cuprous salts for this transformation is often called the **Gattermann reaction**.[367,366b,c] Aryl iodides are also produced from diazonium salts by reaction with potassium iodide (KI), but the actual reactive species may be I_3^-.[368,369] In a synthesis of dichroanal B, Banerjee and co-workers treated aniline **483** with sodium nitrite in aqueous HBr to form the diazonium salt, and subsequent heating with CuBr gave an 88% yield of aryl bromide **484**.[370] Aryl nitriles are generated under Sandmeyer conditions using cuprous cyanide (CuCN), as in the conversion of **485** to benzonitrile derivative **487**, via diazonium chloride **486**.

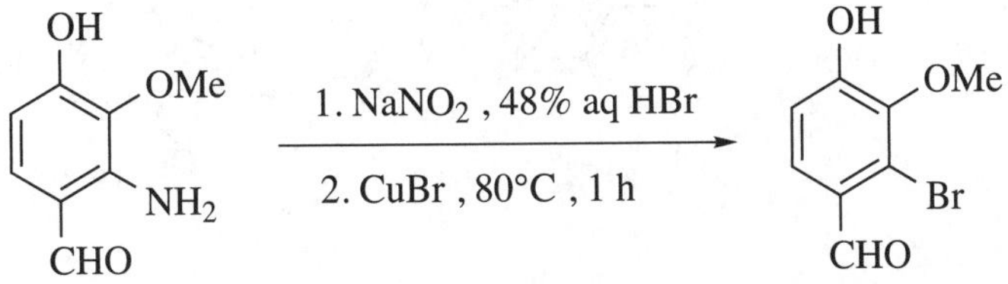

364. Reference 325, p 95.
365. (a) Dickerman, S.C.; DeSouza, D.J.; Jacobson, N. *J. Org. Chem.* **1969**, *34*, 710; (b) Kochi, J.K. *J. Am. Chem. Soc.* **1957**, *79*, 2942.
366. (a) Sandmeyer, T. *Berichte* **1884**, *17*, 1633, 2650; (b) Mowry, D.T. *Chem. Rev.* **1948**, *42*, 189 (see p 213); (c) Hodgson, H.H. *Ibid* **1947**, *40*, 251; (d) The Merck Index, 14th Ed. Merck & Co., Inc., Whitehouse Station, New Jersey, **2006**, p ONR-83; (e) Mundy, B.P.; Ellerd, M.G.; Favaloro Jr., F.G. *Name Reactions and Reagents in Organic Synthesis, 2nd Ed.* Wiley-Interscience, New Jersey, **2005**, pp. 288-289.
367. Gatterman, L. *Berichte* **1890**, *23*, 1218.
368. Carey, J.G.; Millar, I.T. *Chem. Ind. (London)* **1960**, 97.
369. Reference 325, p 96.
370. Banerjee, M.; Mukhopadhyay, R.; Achari, B.; Banerjee, A.Kr. *Org. Lett.* **2003**, *5*, 3931.

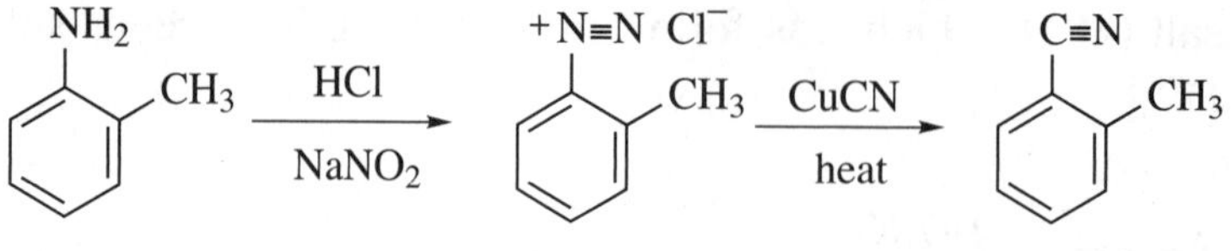

485   486   487

Nitrogen can be completely removed from a diazonium salt by reduction with H_3PO_2 (hypophosphorus acid).[371] This technique is often used in conjunction with using the electron withdrawing properties of a nitro group to position another group. Reduction to an amine, diazotization and then reduction resulted in loss of nitrogen and removal of nitrogen from the molecule. The amine moiety itself can be used to position groups, and this is very useful since the amine unit is electron releasing. Diazotization and reduction removes nitrogen as before. The deamination process is illustrated by the conversion of aryl diamine **488** to the bis(diazonium) salt **489**, and subsequent reduction to the biaryl **490**.[372] In basic solutions, of an activated aromatic substrate, diazonium salts decompose and couple to give biaryls (Ar–Ar).[373] Reaction of benzenediazonium ion (**385**) and toluene gave a 67% yield of **491**, along with 19% of **492** and 14% of **493**.[373] This result suggests the coupling is rather unselective. Selectivity is observed only when the substituent is part of the aryl diazonium ion, as in the conversion of **494** to **495**.[373]

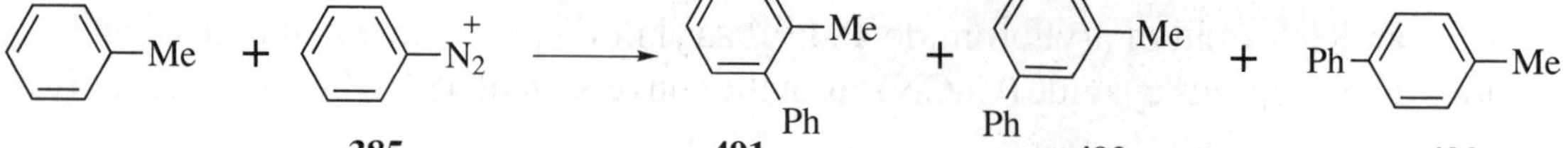

385   491   492   493

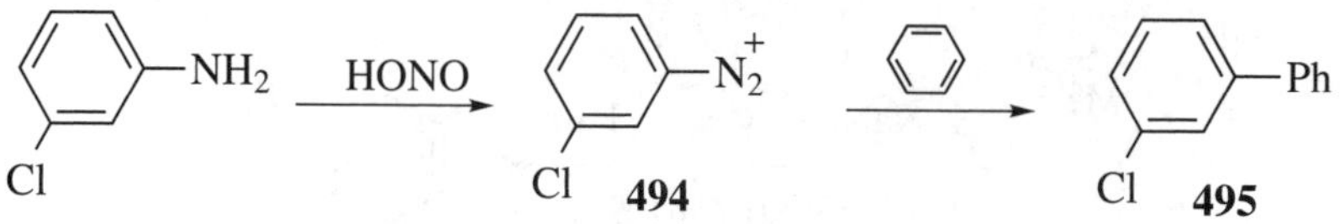

494   495

2.13. CONCLUSION

This chapter has attempted to show that a few reaction types dominate much of the chemistry used by organic chemists on a daily basis. It is also meant as a review of the fundamental functional group transformations encountered in a typical undergraduate organic chemistry course. Chapters 3 and 4 will discuss the place of oxidation and reduction reactions in organic chemical transformations, which will be followed by hydroboration in Chapter 5 and then a discussion of protecting groups in chapter seven. Chapter 6 will discuss methods for

371. Kornblum, N. *Org. React.* **1944**, *2*, 262.
372. Reference 325, p 97.
373. Reference 325, p 99.

controlling stereochemistry and regiochemistry and then macrocyclization reactions. Some of the important concepts in Chapter 6 will be introduced in Chapters 3-5, and Chapter 6 will attempt to tie those concepts together.

HOMEWORK

1. From the rate data given for compound **A** (in the table below), it is clear that the presence of a methoxy group significantly enhances the rate of solvolysis of **A**. Offer a reasonable mechanistic explanation of this solvolysis reaction that is consistent with the observed rate data.

    Assume that differences in the rate due to changing the leaving group is insignificant for this question.

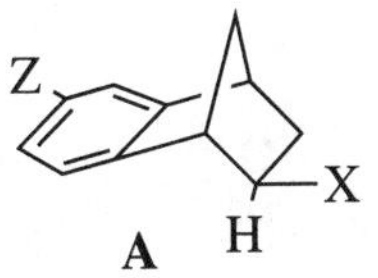

Z	X	Relative Rate
H	OTs	1
MeO	Cl	130
Me	OTs	4.96
Cl	Cl	0.15

2. Reaction of 3-pentanol with potassium iodide in aqueous ethanol gave no reaction under several different reaction conditions. If one adds a catalytic amount of *para*-toluenesulfonic acid, however, the reaction proceeds to give 3-iodopentane. Explain these observations.

3. Explain why the obvious S_N2 conditions shown in the following reaction give only elimination and no substitution with this secondary bromide.

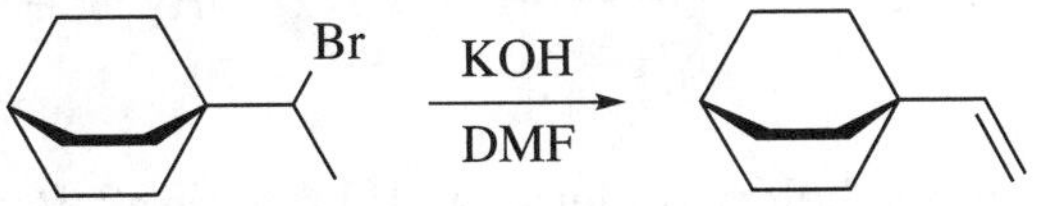

4. Explain each of the following trends.

    (a) Order of basicity.

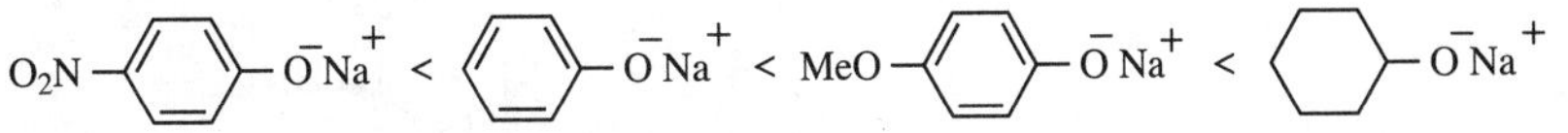

    (b) Order of acidity. $HCO_2H > CH_3CO_2H$
    CH_3CO_2H in water $> CH_3CO_2H$ in THF

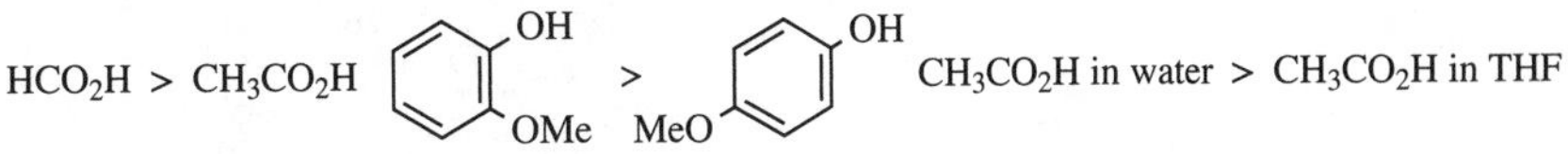

5. Give the major product for each of the following transformations involving conversion of a carboxylic acid or a derivative into another acid derivative.

(a) 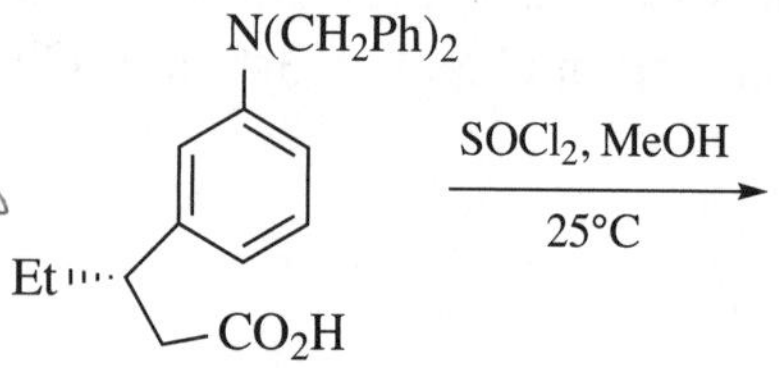 [structure with N(CH₂Ph)₂, Et, CO₂H] → SOCl₂, MeOH / 25°C

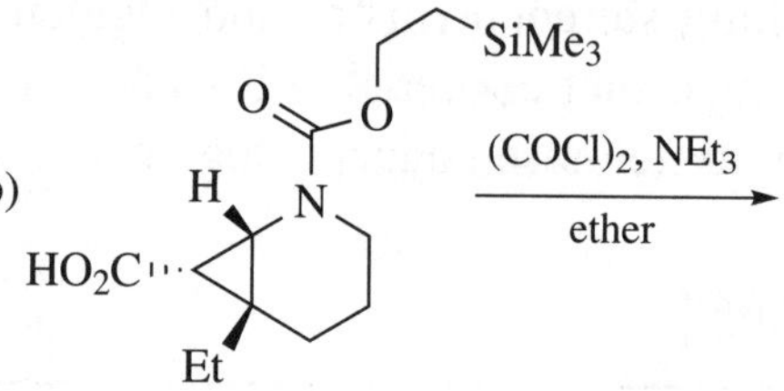

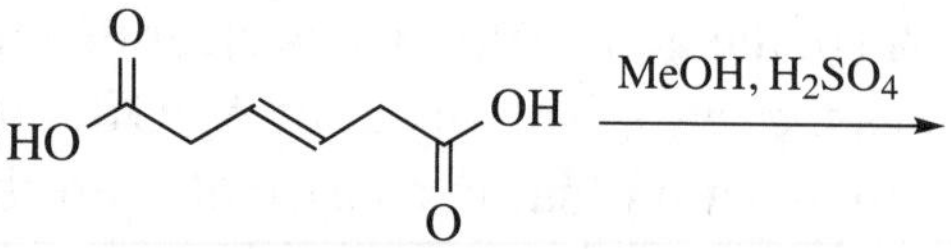

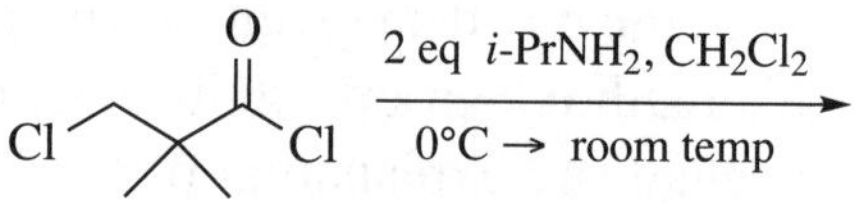

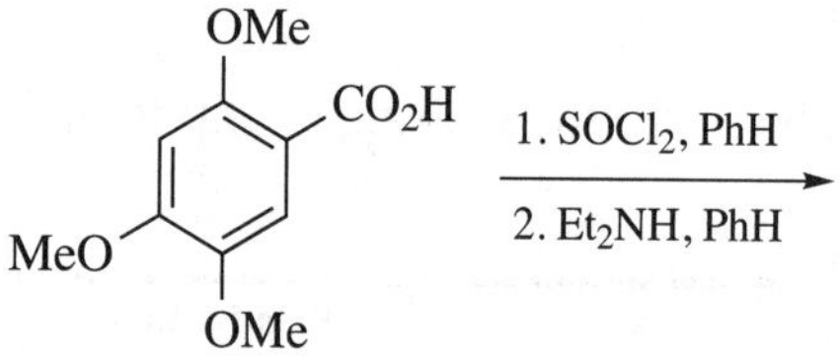

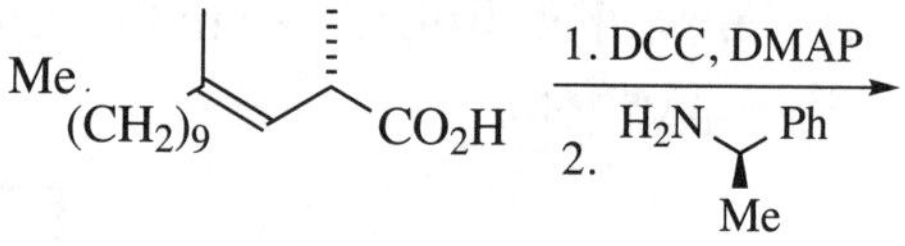

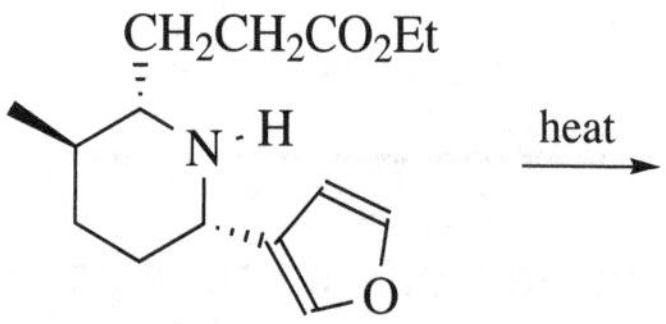

(b) [bicyclic structure with carbamate O=, O, SiMe₃, H, N, HO₂C, Et] → (COCl)₂, NEt₃ / ether

(c) HO₂C...transfer... OH → MeOH, H₂SO₄

(d) [Cl, O, Cl structure] → 2 eq i-PrNH₂, CH₂Cl₂ / 0°C → room temp

(e) [aromatic structure with OMe, CO₂H, MeO, OMe] → 1. SOCl₂, PhH / 2. Et₂NH, PhH

(f) Me, (CH₂)₉, CO₂H → 1. DCC, DMAP / 2. H₂N, Ph, Me

(g) [CH₂CH₂CO₂Et, Me, N-H, furan structure] → heat

(h) [polycyclic structure with O, MeO, O, O] → NH₃, MeOH/THF / 70°C, 2d

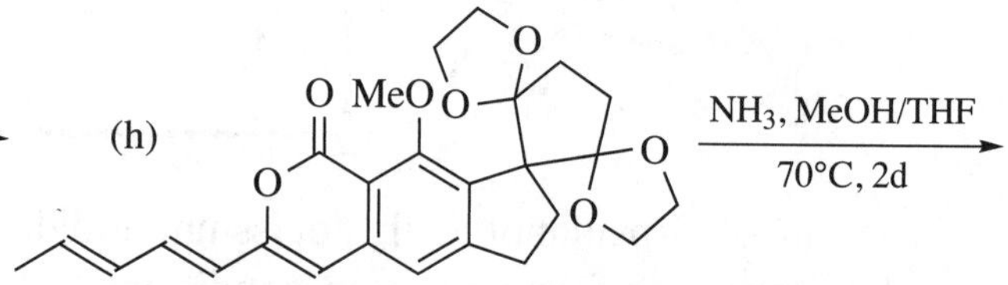

6. Explain why the ¹⁴C label is found at both C1 and C2 in the aniline product despite the fact that the label is localized on C1 in the chlorobenzene precursor. The * indicates the position of the ¹⁴C label.

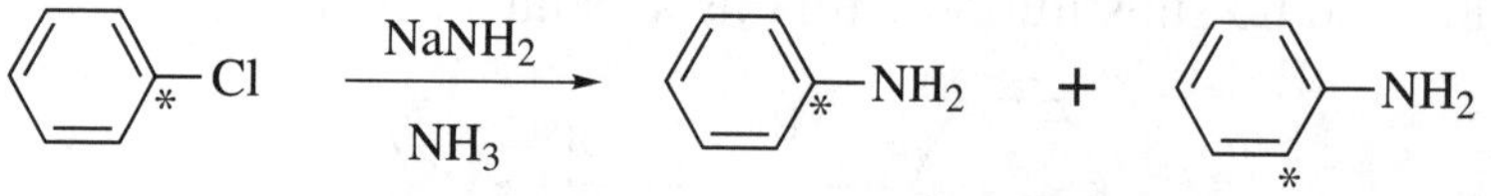

[benzene-Cl, * label] → NaNH₂ / NH₃ → [benzene-NH₂, * label] + [benzene-NH₂, * label]

7. Does the bromide shown below undergo an E2 reaction? If so, draw the major product(s). If not, why not?

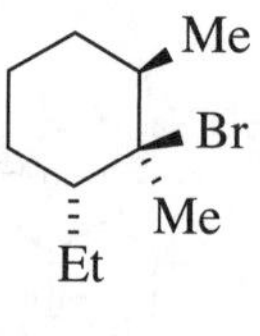

[cyclohexane structure with Me, Br, Me, Et]

8. Offer a mechanistic explanation for the following transformation.

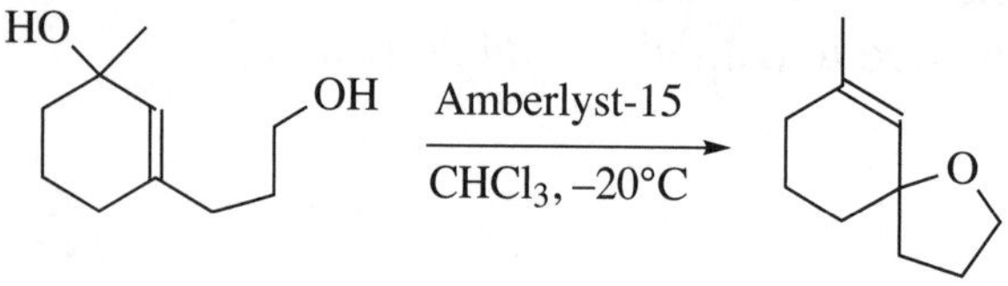

[HO, Me, OH cyclohexene structure] → Amberlyst-15 / CHCl₃, −20°C → [spirocyclic ether structure, O]   Amberlyst is an acidic resin

9. The following reaction gives a mixture of the alcohol and the ether, in the proportions given later in this question. Only 0.03 eq of 1-octene were used. In each case, a 1:1 ratio of water to alcohol was used.

$$C_6H_{17}CH=CH_2 \quad \xrightarrow[\text{2. NaBH}_4]{\substack{\text{1. Hg(OAc)}_2,\ \text{H}_2\text{O/ROH} \\ \text{sodium dodecyl sulfate}}} \quad \underset{\overset{|}{OH}}{C_6H_{17}CHCH_3} \;+\; \underset{\overset{|}{OR}}{C_6H_{17}CHCH_3}$$

Reaction with 1.0 equivalent of 1-octanol (to 0.03 eq of 1-octene) gave a 98:2 mixture of alcohol/ether in 50% aq THF. The use of 10 equivalents of 1-octanol in 0.3 M THF in water gave a 52:48 mixture of alcohol/ether. Explain what the 52:48 mixture tells you about the relative nucleophilicity of water and 1-octanol. Explain why increasing the amount of 1-octanol leads to more ether product.

10. Two products are formed in the following reaction, A and B. (a) Offer a reasonable explanation of how B could arise in this reaction. (b) Would the use of the 4-chlorophenyl derivative, rather than the 4-methoxy derivative help or hurt formation of B? Why or why not?

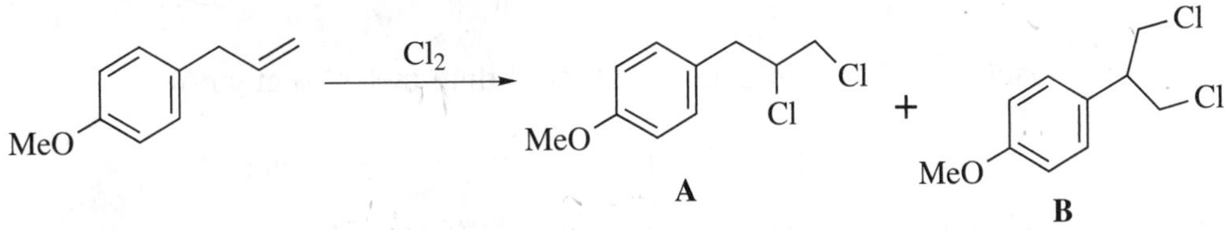

11. Offer a mechanistic explanation for the following transformation.

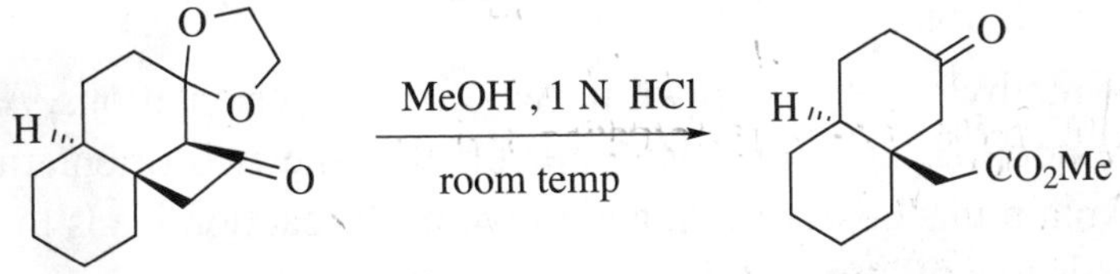

12. Treatment of alcohol A leads to B in 70% yield. Explain this product in terms of a mechanism and comment on why the C=C unit is generated between C2-C3 rather than between C2-C1 [hint, see sec. 6.3.B].

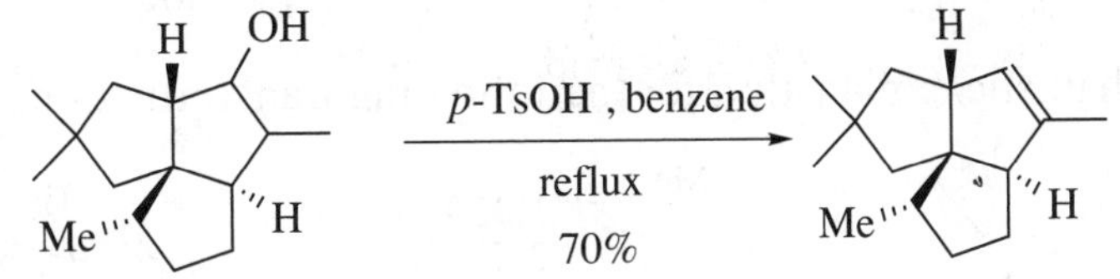

13. Suggest a mechanism for the following transformation.

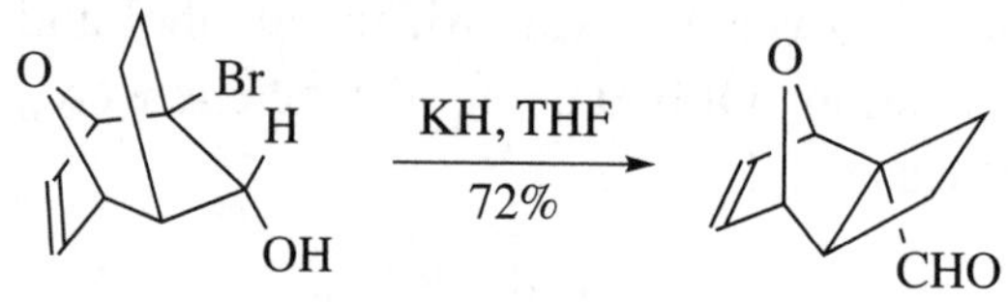

14. Suggest a reasonable mechanism for this reaction.

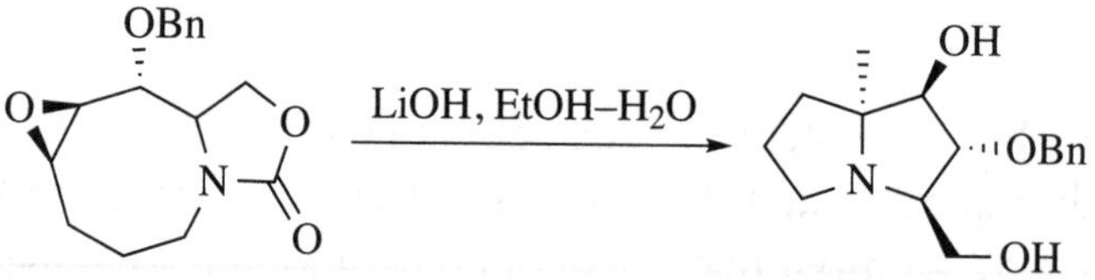

15. Explain each of the following.

(a) Heating 2-N,N,N-triethylammonium)butane gives ethene rather than 1-butene.

(b) The reaction of the epoxide shown gives the S_N2' product rather than the S_N2 product. Draw the S_N2' product, and explain why it is formed preferentially.

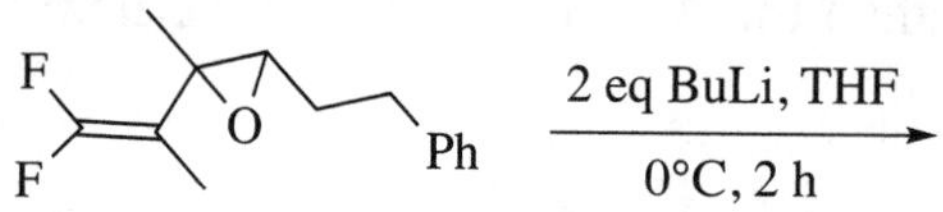

(c) This reaction is *faster* in aqueous ethanol than in 100% ethanol:

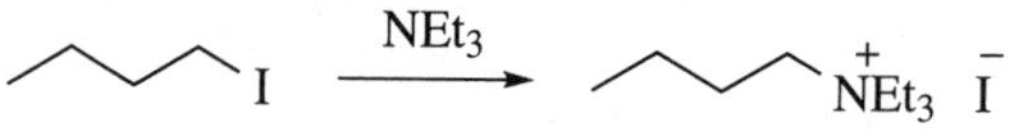

(d) $H_3N^+CH_2CO_2H$ is a stronger acid than $H_3N^+(CH_2)_4CO_2H$.

(e) 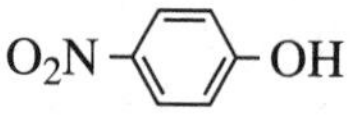 is a stronger acid than

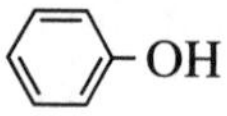

16. 3-Bromo-4-methylhexane is drawn in two different orientations, A and B. Using an E2 reaction, remove H_a and convince yourself that the two orientations lead to the same alkene. Explain this observation, and why an E2 reaction leads to one alkene but an E1 reaction leads to a mixture of (E) and (Z) isomers.

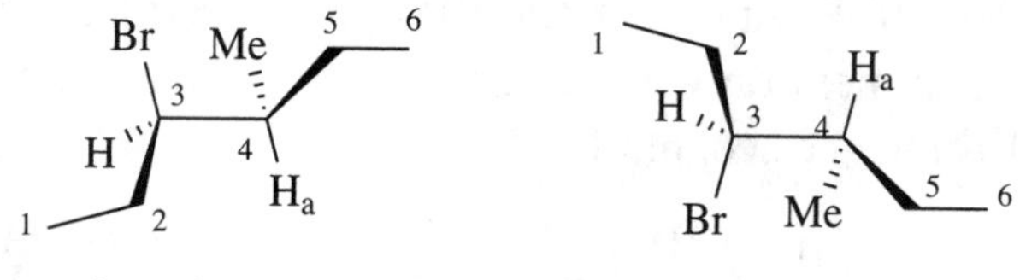

17. Offer a mechanistic explanation for the following transformation.

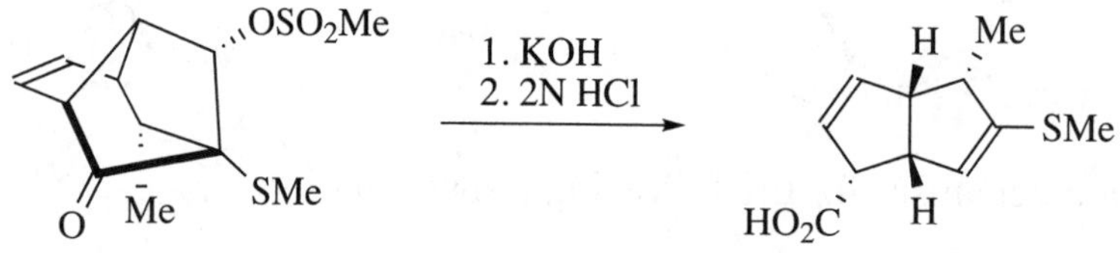

18. Offer a mechanistic explanation for the following transformation.

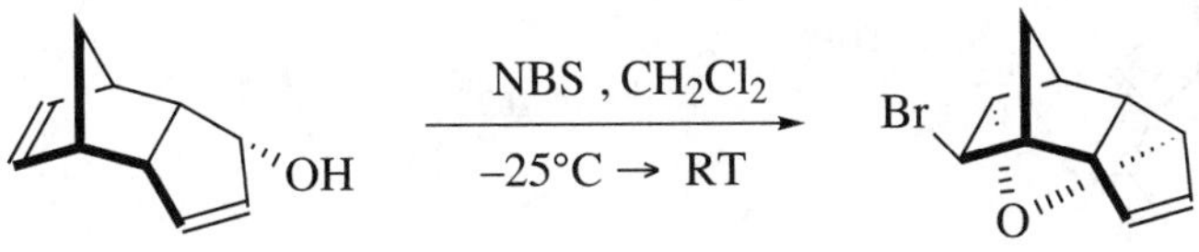

OH
NBS , CH_2Cl_2
–25°C → RT
Br
O

19. Offer a mechanistic explanation for the following transformation.

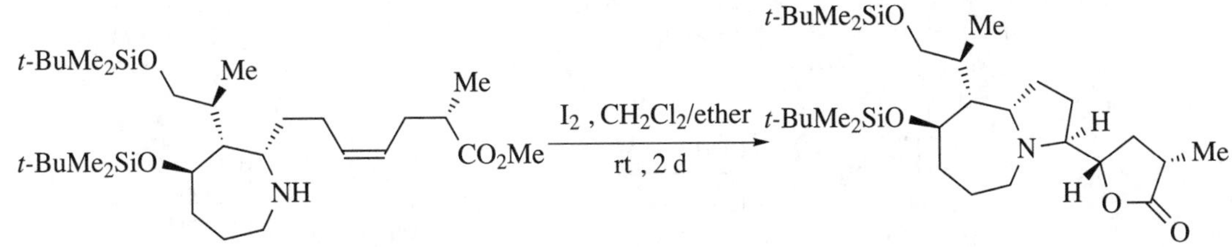

t-BuMe₂SiO
Me
t-BuMe₂SiO
NH
Me
CO₂Me
I₂ , CH₂Cl₂/ether
rt , 2 d
t-BuMe₂SiO
Me
t-BuMe₂SiO
N
H
H
O
O
Me

20. For each of the following give the major product, with all appropriate stereochemistry. If there is no reaction, indicate by N.R.

(a)
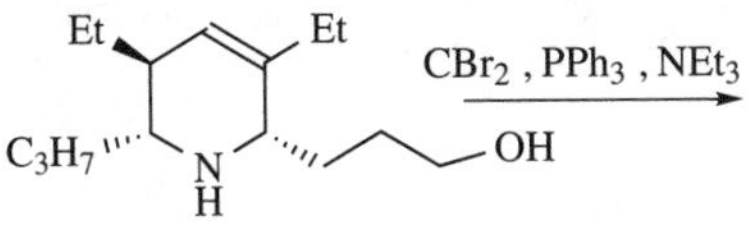

Et
Et
C₃H₇
N
H
OH
CBr₂ , PPh₃ , NEt₃

(b)
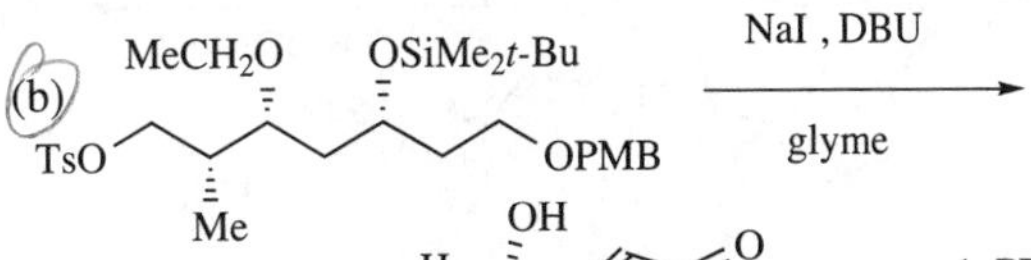

MeCH₂O
OSiMe₂t-Bu
TsO
Me
OPMB
NaI , DBU
glyme

(c)
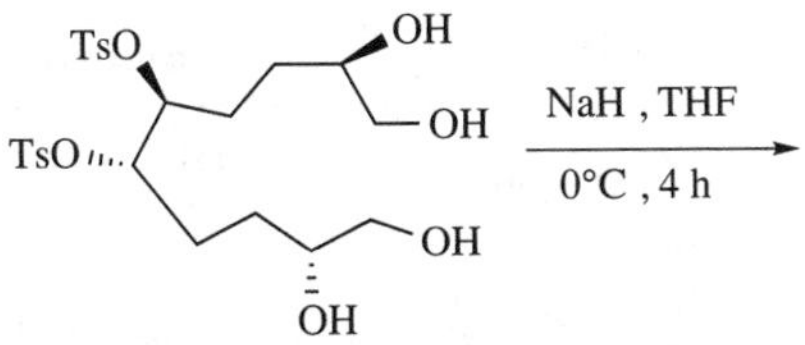

TsO
TsO
OH
OH
OH
OH
NaH , THF
0°C , 4 h

(d)
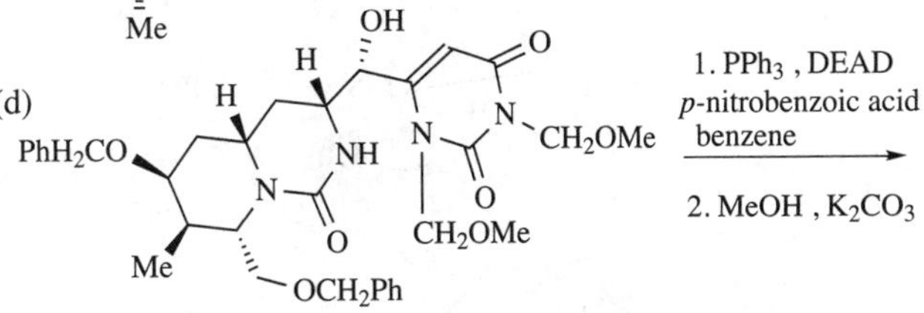

PhH₂CO
H
H
OH
O
NH
N
N
CH₂OMe
N
O
O
CH₂OMe
Me
OCH₂Ph
1. PPh₃ , DEAD
p-nitrobenzoic acid
benzene
2. MeOH , K₂CO₃

(e)
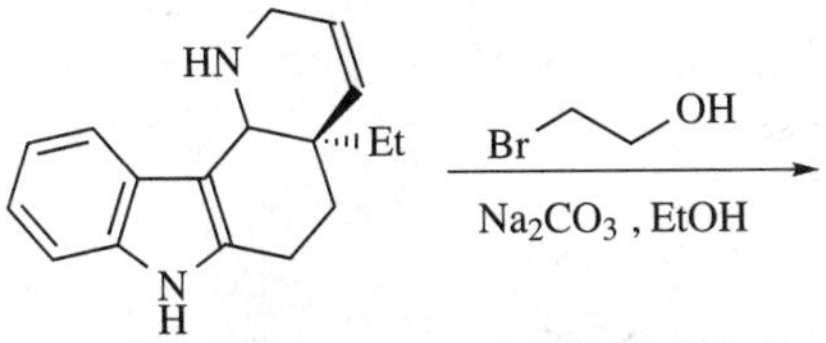

HN
Et
N
H
Br
OH
Na₂CO₃ , EtOH

(f)
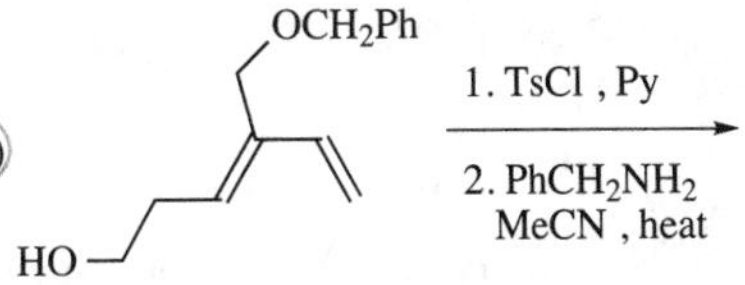

OCH₂Ph
HO
1. TsCl , Py
2. PhCH₂NH₂
MeCN , heat

(g)
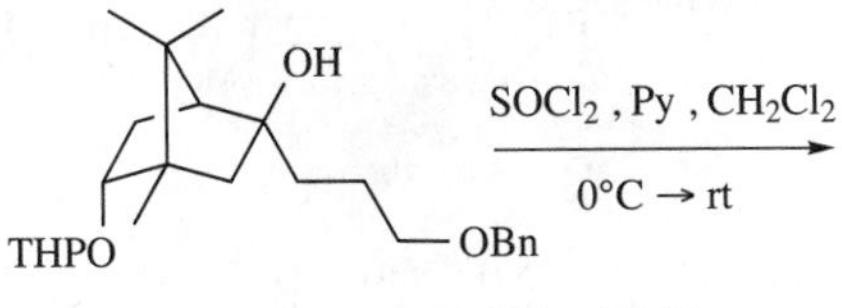

OH
THPO
OBn
SOCl₂ , Py , CH₂Cl₂
0°C → rt

(h)
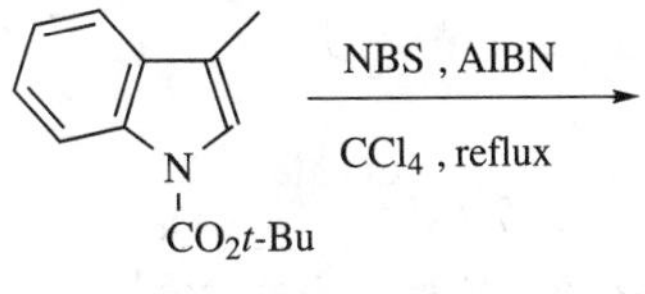

N
CO₂t-Bu
NBS , AIBN
CCl₄ , reflux

(i)
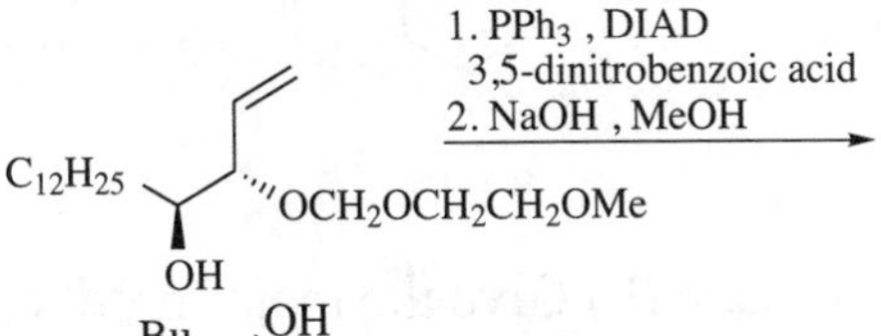

C₁₂H₂₅
OH
OCH₂OCH₂CH₂OMe
1. PPh₃ , DIAD
3,5-dinitrobenzoic acid
2. NaOH , MeOH

(j)
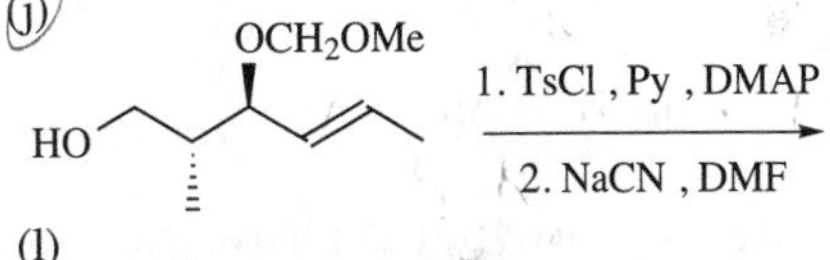

OCH₂OMe
HO
1. TsCl , Py , DMAP
2. NaCN , DMF

(k)
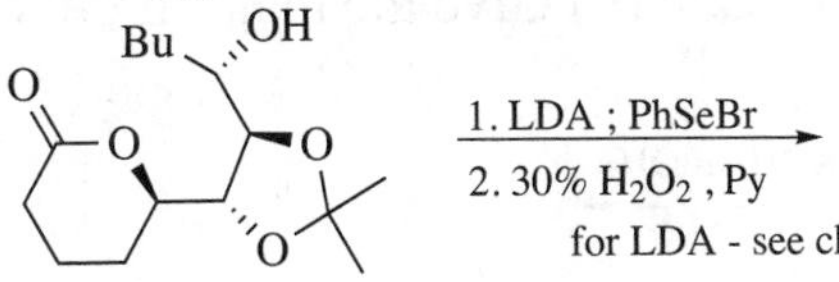

Bu
OH
O
O
O
O
1. LDA ; PhSeBr
2. 30% H₂O₂ , Py

for LDA - see chap 9, Sec 9.2

(l)
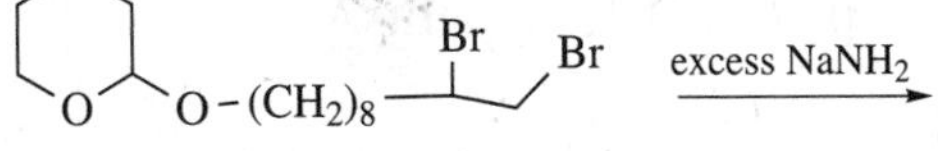

O
O-(CH₂)₈
Br
Br
excess NaNH₂

(m) 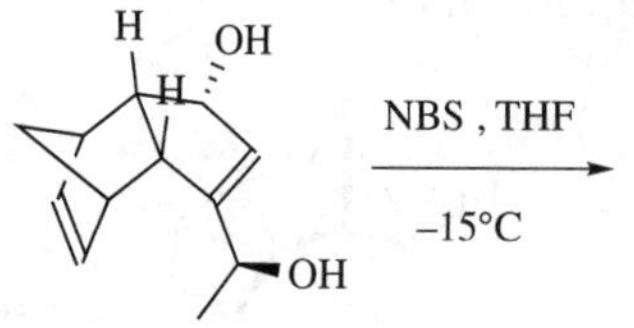 NBS , THF
−15°C
(n)

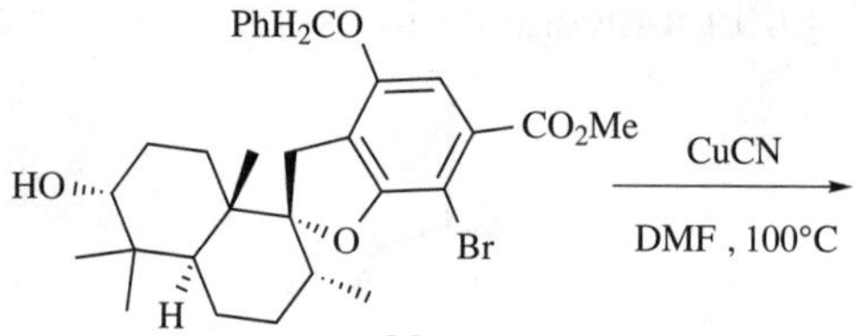

(n) CuCN
DMF , 100°C

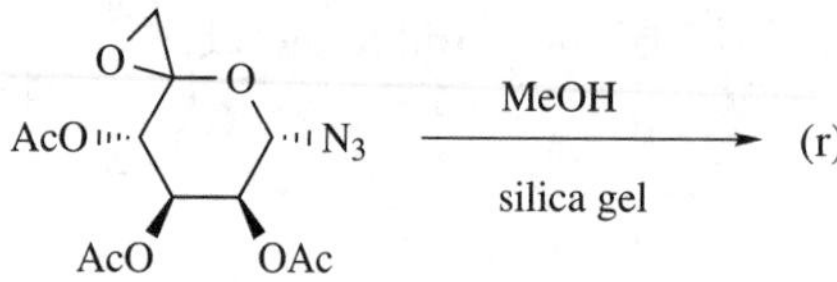

(o) 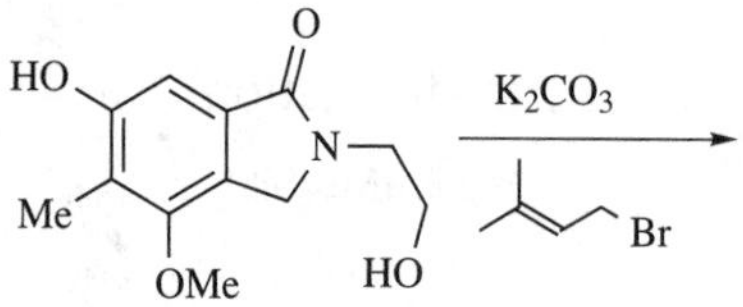 K$_2$CO$_3$
(p)

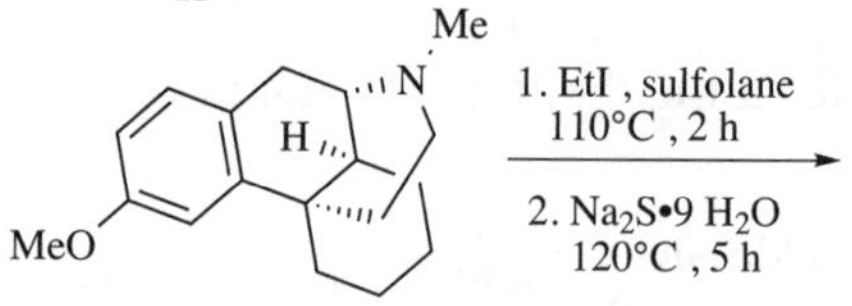

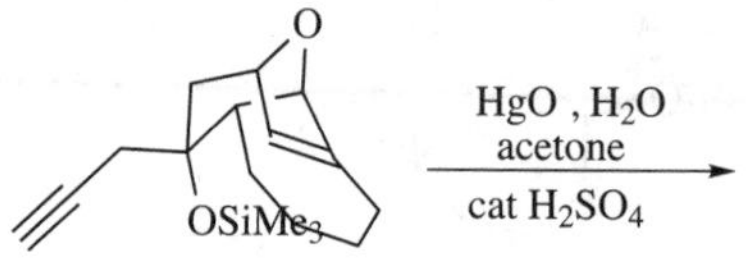

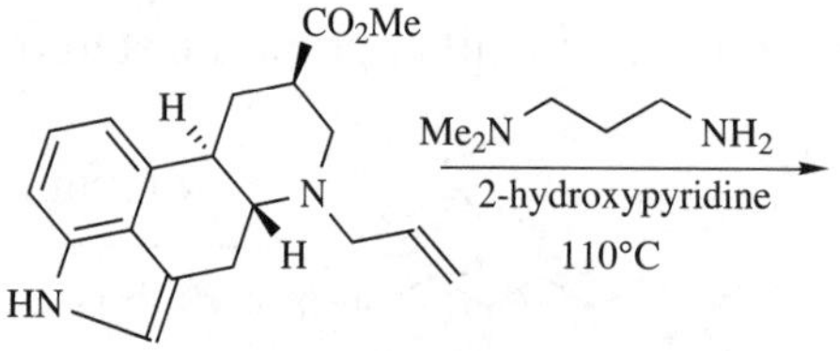

(p) 1. EtI , sulfolane
110°C , 2 h
2. Na$_2$S•9 H$_2$O
120°C , 5 h

(q) MeOH
silica gel
(r)

(r) HgO , H$_2$O
acetone
cat H$_2$SO$_4$

(s) Me$_2$N$\diagdown\diagup$NH$_2$
2-hydroxypyridine
110°C
(t)

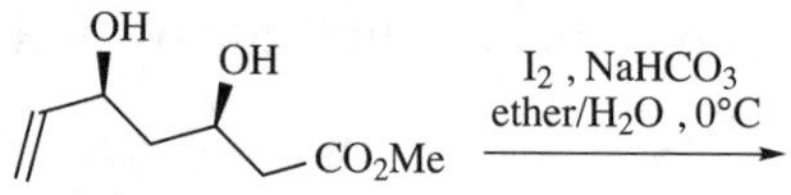

(t) I$_2$, NaHCO$_3$
ether/H$_2$O , 0°C

(u) 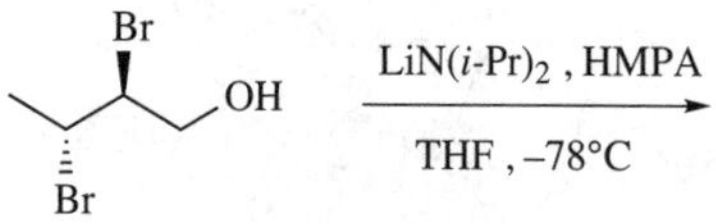 LiN(i-Pr)$_2$, HMPA
THF , −78°C
(v)

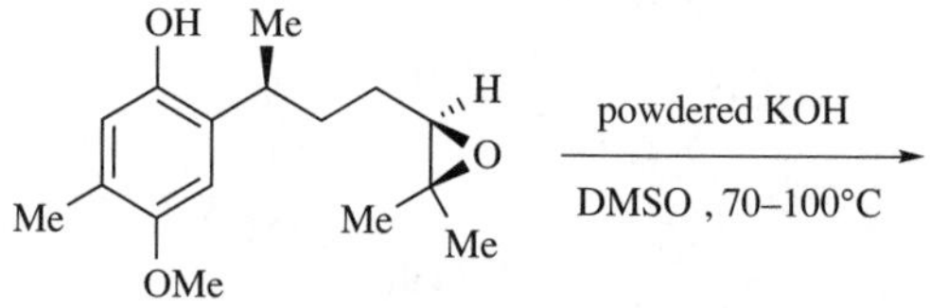

(v) powdered KOH
DMSO , 70–100°C

(w) 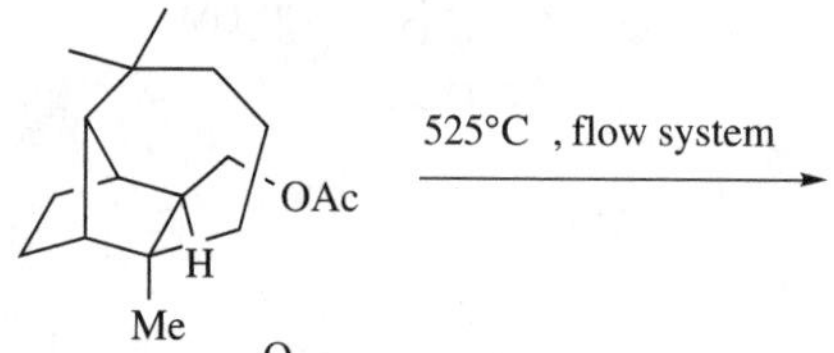 525°C , flow system
(x)

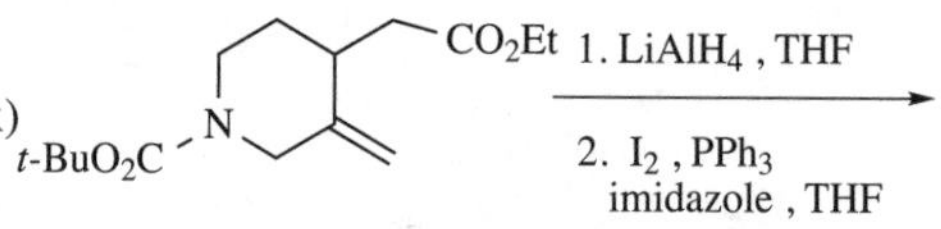

(x) 1. LiAlH$_4$, THF
2. I$_2$, PPh$_3$
imidazole , THF

(y) 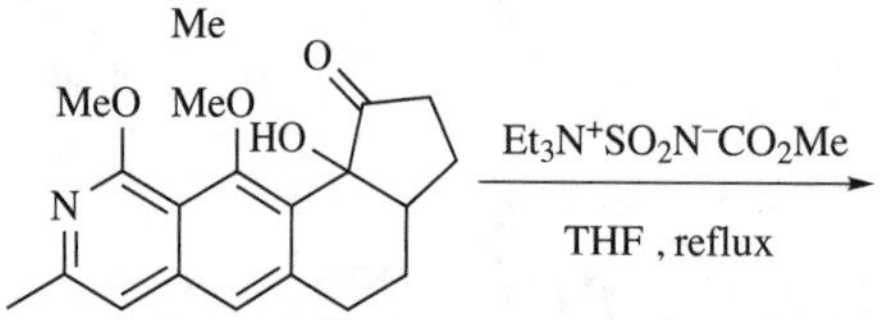 Et$_3$N$^+$SO$_2$N$^-$CO$_2$Me
THF , reflux
(z)

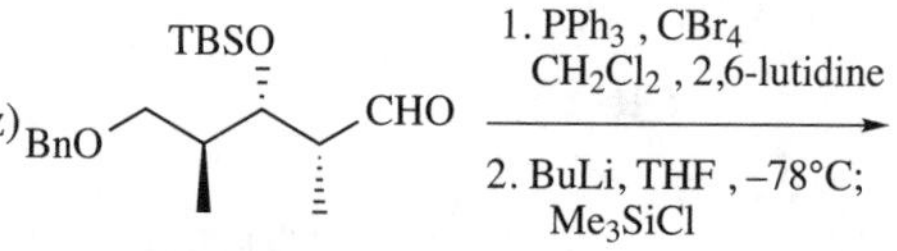

(z) 1. PPh$_3$, CBr$_4$
CH$_2$Cl$_2$, 2,6-lutidine
2. BuLi, THF , −78°C;
Me$_3$SiCl

(aa) 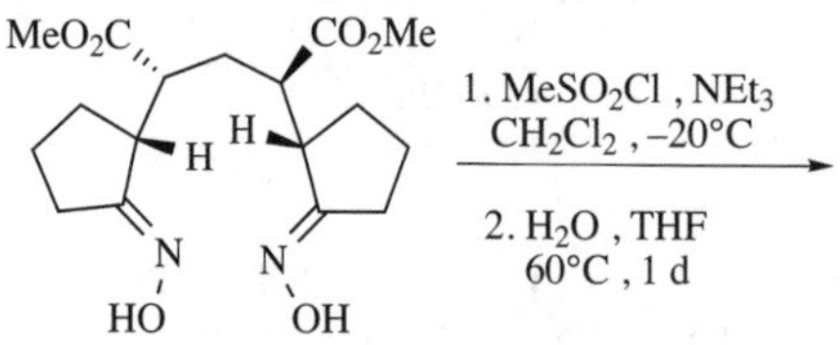 1. MeSO$_2$Cl , NEt$_3$
CH$_2$Cl$_2$, −20°C
2. H$_2$O , THF
60°C , 1 d

(bb) 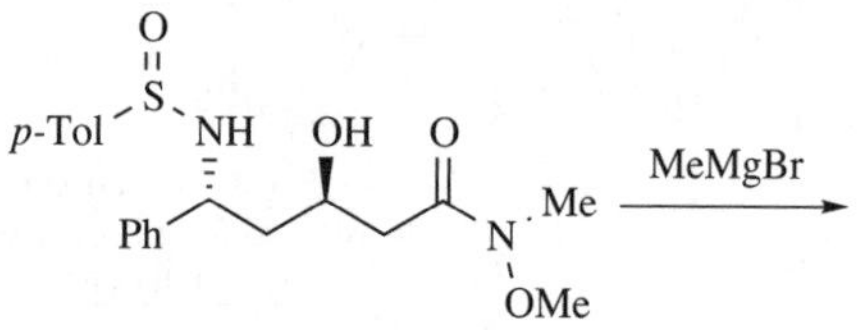 MeMgBr

21. For the reaction shown, (a) Draw the transition state (b) Give the major product (c) What type of reaction is this?

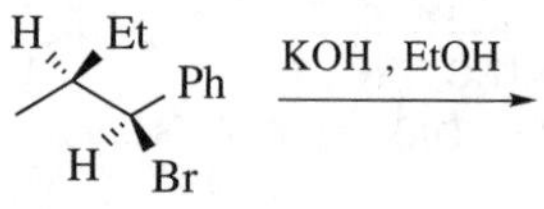

KOH , EtOH

22. In each case, predict the major product and explain your answer.

(a) 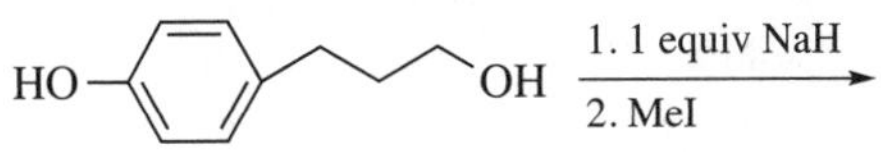 HO—⟨benzene ring⟩—CH₂CH₂CH₂OH   1. 1 equiv NaH / 2. MeI

(b) 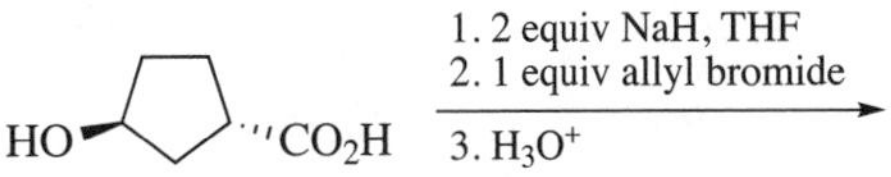 HO—⟨cyclopentane⟩···CO₂H   1. 2 equiv NaH, THF / 2. 1 equiv allyl bromide / 3. H₃O⁺

(c)  AcO~~~~OSO₂Me   1 equic NaCN / DMF

(d) 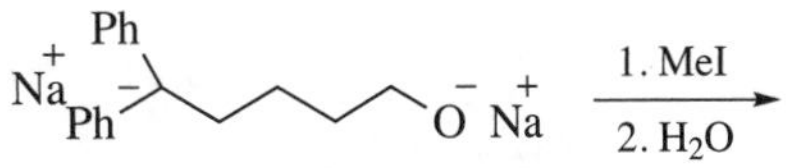 Na⁺ Ph–⟨=⟩–Ph~~~O⁻ Na⁺   1. MeI / 2. H₂O

(e) 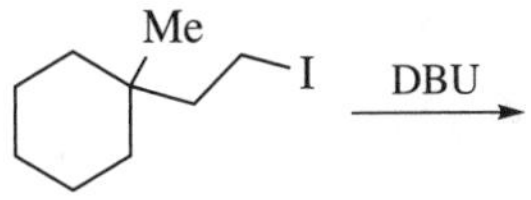 ⟨cyclohexane with Me and CH₂CH₂I⟩   DBU

(f) 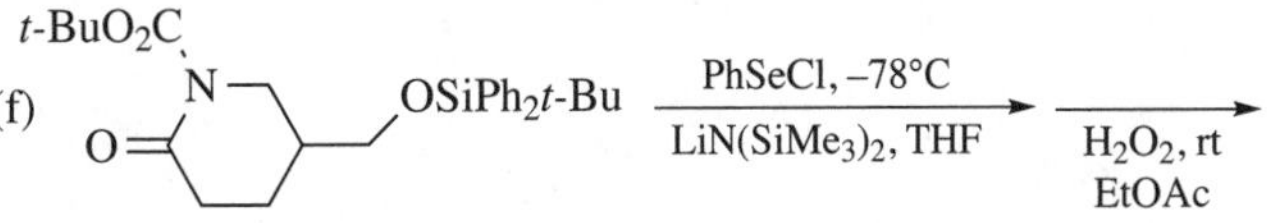 t-BuO₂C–N⟨piperidine, O=⟩–CH₂OSiPh₂t-Bu   PhSeCl, –78°C / LiN(SiMe₃)₂, THF   H₂O₂, rt / EtOAc

23. For the following reaction, predict the product and show the correct stereochemistry. Use Newman projections to indicate how you chose your product and to justify your answer.

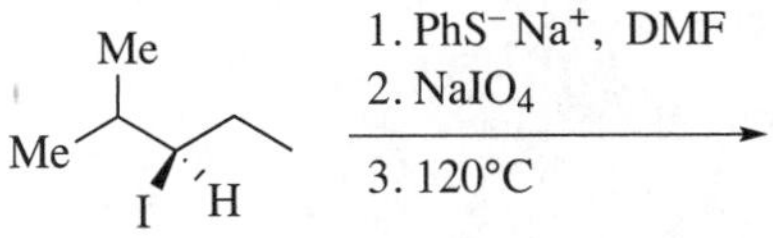

Me—CH(Me)—CH(I)(H)—CH₂CH₃   1. PhS⁻ Na⁺, DMF / 2. NaIO₄ / 3. 120°C

24. For each of the following, show the major product. Explain your choice.

(a) 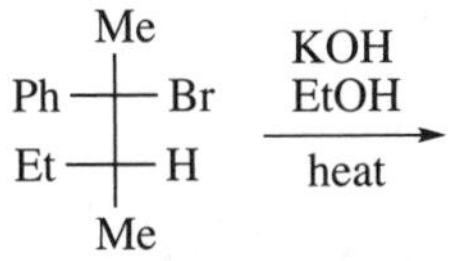 Ph—C(Me)(Br)—C(Et)(H)—Me   KOH EtOH heat

(b) 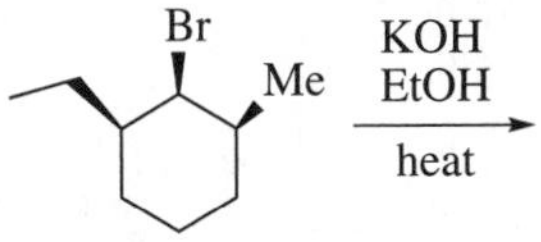 ⟨cyclohexane with Br and Me, ethyl⟩   KOH EtOH heat

(c) ⟨cyclohexane with Me, Br, ethyl⟩   KOH EtOH heat

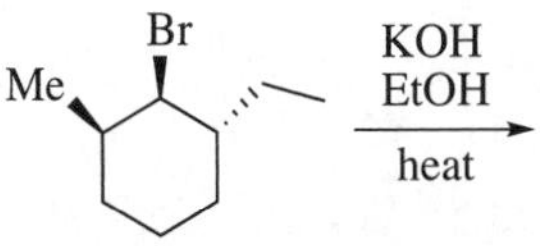

25. For each of the following provide a suitable synthesis. Show all intermediate products and all reagents.

(a)  ⟨cyclohexane⟩–CHO → ⟨cyclohexane⟩–C≡CMe

(b) 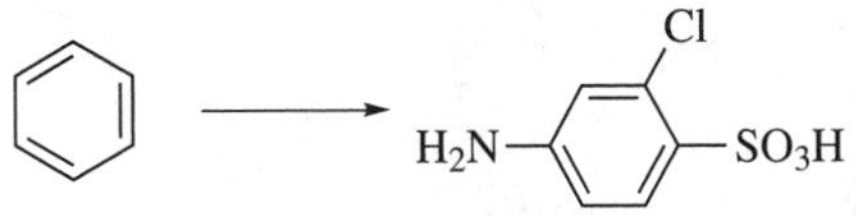 ⟨benzene⟩ → H₂N—⟨benzene, Cl⟩—SO₃H

(c) ⟨benzene⟩ → Cl—⟨benzene⟩—OH

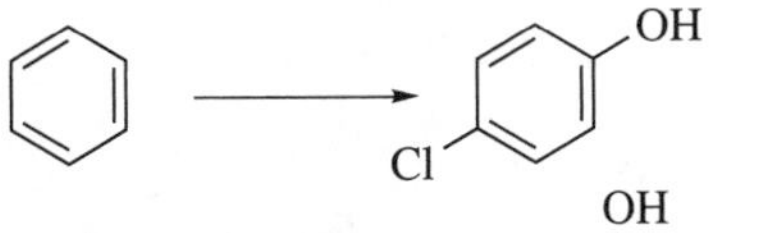

(d) 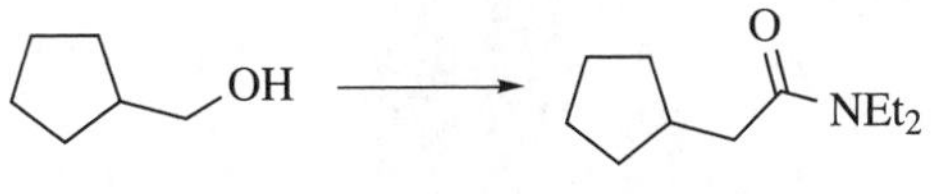 ⟨cyclopentane⟩–CH₂OH → ⟨cyclopentane⟩–CH₂C(=O)NEt₂

(e) 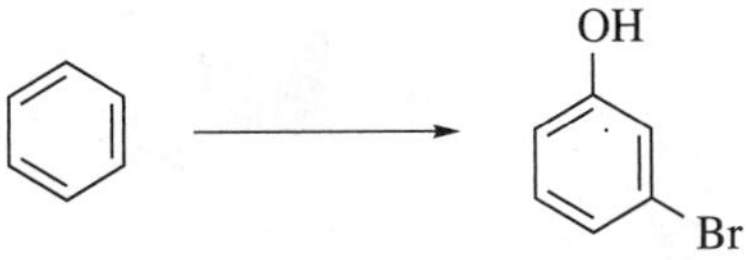 ⟨benzene⟩ → ⟨benzene with OH and Br⟩

(f) 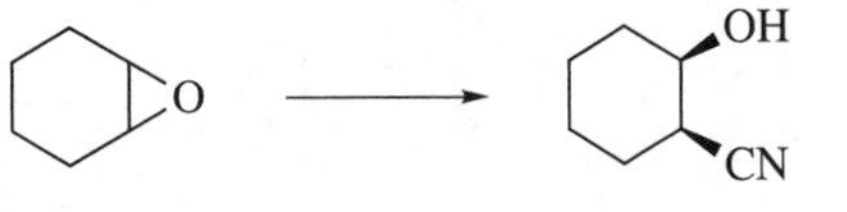 ⟨epoxide on cyclohexane⟩ → ⟨cyclohexane with OH and CN⟩

(g) 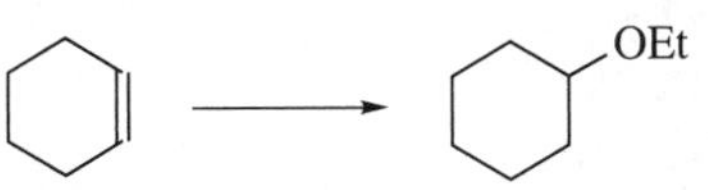 ⟨cyclohexene⟩ → ⟨cyclohexane⟩–OEt

(h) 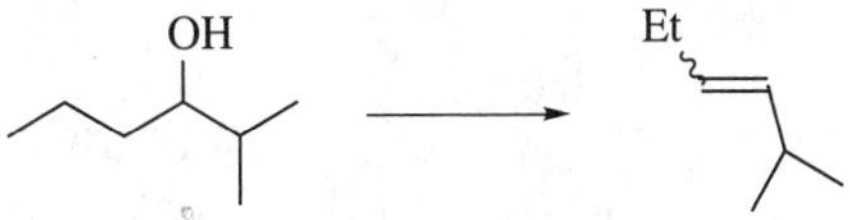 ⟨CH₃CH₂CH₂CH(OH)CH(CH₃)CH₃⟩ → ⟨alkene with Et⟩

(i) 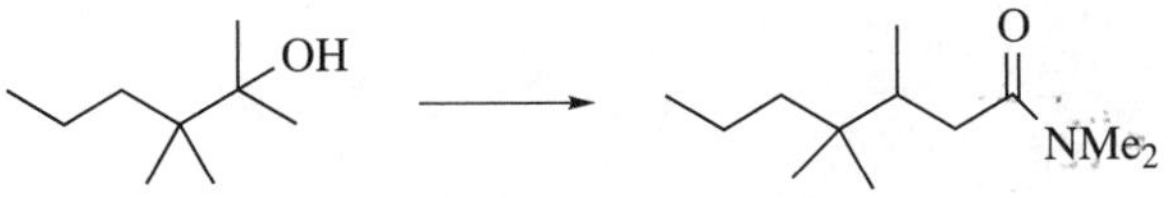 ~~~C(Me)₂—C(Me)₂—CH₂OH → ~~~C(Me)₂—CH(CH₃)CH₂C(=O)NMe₂

chapter 3

Oxidation

3.1. INTRODUCTION

Oxidation in organic chemistry has historically been defined in terms of reactivity. A broad definition by Sheldon and Kochi states "oxidation in Organic chemistry refers to either (*1*) the elimination of hydrogens, as in the sequential dehydrogenation of ethane, or (*2*) the replacement of a hydrogen atom bonded to carbon with another more electronegative element such as oxygen in the following series of oxidative transformations of methane:"[1]

$$CH_4 \longrightarrow CH_3OH \longrightarrow CH_2O \longrightarrow HCO_2H \longrightarrow CO_2$$

Oxidation can also be defined as reaction of an element with oxygen,[2] analogous to the second criterion stated above. A more general definition, applied most often in inorganic chemistry, involves the loss of one or more electrons from an atom or group.[2] Soloveichik and Krakauer[3] listed five criteria that could be used to identify an oxidation process for a given organic reaction, and arrange compounds by their ability to be oxidized.

1. **Pauling's electronegativity scal**e.[4]

2. The oxidation number of the parent substance of each homologous series under consideration.

3. The ratio of bond moment to bond length, μ/l (net charge).[5]

4. Hammett's sigma function as revised by Taft (Taft's σ^*).[6]

5. Chemical reactions of the substance involved, particularly hydrolysis, which does not provide fundamental changes in oxidation stage if certain conventions are maintained.[3]

---

1. Sheldon, R.A.; Kochi, J.K. *Metal-Catalyzed Oxidations of Organic Compounds,* Academic Press, New York, *1981*, p 6.
2. Yalman, R.G. *J. Chem. Educ.* *1959, 36,* 215.
3. Soloveichik, S.; Krakauer, H. *J. Chem. Educ., 1966, 43,* 532.
4. Pauling, L. *Nature of the Chemical Bond,* 3rd ed., Cornell U. Press., Ithaca, New York, *1940*, pp 221-231.
5. (a) Del Re, G. *Electronic Aspects of Biochemistry,* Pullman, B. (Ed.), Academic Press, New York, *1964*, pp 221-231; (b) Smyth, C.P. *Dielectric Behavior and Structure*, McGraw-Hill, New York, *1955*, pp 244-245, 247.
6. (a) Taft, Jr., R.W. *J. Am. Chem. Soc. 1952, 74,* 2729; (b) Taft, Jr., R.W. *Ibid. 1958, 75,*4231; (c) Taft, Jr., R.W. *Steric Effects in Organic Chemistry*, Chapter 13, Newman, M.S. (Ed.), Wiley, New York, *1956*, pp 556-675.

These criteria allow the oxidation state of a molecule to be determined, which is useful for categorizing a reaction as an oxidation or a reduction. Soloverchik and Krakauer listed another set of rules for determining the oxidation state of a molecule:[3]

1. The oxidation state of carbon in hydrocarbons of the methane series is zero

2. The oxidation state is increased by one for each bond with a hydrogen atom or an alkyl group replaced by a bond with a more strongly electron attracting atom or group ($-NH_2$, SH, halogen, CO_2, etc.) for every removal of a pair of hydrogen atoms from two adjacent carbon atoms.

3. The group on a carbon is defined by its substituent carbons bearing the same substituents belonging to the same group. Groups within an oxidation state are arranged in order of increasing induced electron deficiency of the function bearing carbon. ($CH_3X >$ $RCH_2X > R_2CHX > R_3CX$).

With these rules, one can determine if a starting material and a product are in a higher or a lower oxidation state. There are several structural types, and an oxidation is defined as conversion of a group from a lower stage to a higher stage (I → II, II → III, etc.).[3] Examples are R_3CNH_2 → $R_2C=NH$ (I → II); $RCONH_2$ → $(H_2N)_2C=O$ (III → IV); and, $AcCR_3$ → $R_2C=O$ → RCl_3 → Cl_3CCO_2H (I → II → III → IV).[3]

A loss of electrons is associated with the change in oxidation state of an atom, and electron loss can be determined by a device known as **oxidation number**. Holleran and Jespersen[7] listed several rules to determine the oxidation number or each atom within a molecule. The change in the numbers between starting material and products shows the loss or gain of electrons and the change in oxidation state.

1. The sum of the oxidation numbers of all the atoms in a molecule or ion must equal the total charge (0 for a neutral molecule).

2. Metallic atoms of groups 1 and 2 have oxidation numbers of +1 and +2, respectively.

3. H and F are assigned oxidation numbers of +1 and –1, respectively, when part of a molecule.

4. Oxygen is assigned an oxidation number of –2 in its compounds except for peroxide, where it is –1, and for OF_2, where it is +2.

5. Atoms of group 7 are assigned oxidation numbers of –1 in their binary compounds.

6. In binary compounds, atoms from groups 16 and 15 are assigned oxidation numbers of –2 and –3, respectively.

7. If two rules conflict, the rule that occurs first takes precedence.

8. If a conflict occurs within the same group, the lighter element takes the charge (I is +1

---

7. Holleran, E.M.; Jespersen, N.D. *J. Chem. Educ.* **1980**, *57*, 670.

in ICl).

Oxidation numbers can be assigned to any atom, as shown in the following examples:

H_2O_2 (1 for O) BrO_3 (+5 for Br) OF_2 (+2 for O) LiO_2 (−12 for O)

If an element is not included explicitly in one of the rules its oxidation number is found by difference, as in the case of the nitrite ion. The symbol Σ below refers to the net charge for the molecule, as determined by the usual formal charge calculations. For the N in NO_2^-, NO_2^- (each O = −2, Σ = −1, therefore, N = +5).

A simplified version of this approach was used by Hendrickson, Cram and Hammond[8] and later by Pine[9] in their undergraduate organic textbooks. The rules shown above were applied only to those atoms most commonly encountered in simple organic molecules: H, O, C, N, and the halogens. Hendrickson, Cram, Hammond and Pine used the following convention for assigning oxidation numbers.[8,9]

−1   for hydrogen

0   for carbon (bonds to same atom)

+1   for a covalent bond to a heteroatom (usually oxygen or halogen)

The oxidation number should be determined only for those atoms that are modified or changed as a result of the reaction. If the sums of the oxidation numbers of the various atoms in the starting material and in the product are compared, the change in oxidation number will indicate an oxidation or a reduction.

An example is the oxidation of 1-propanol to propanal, which is subsequently oxidized to propanoic acid in a second step. In 1-propanol, the carbon bearing the hydroxyl group has an oxidation number of −1 (C is 0, each H is −1, and O is +1). The oxidation number of carbon in propanal is +1 (C is 0, H is −1, O is +1, and O, from the second bond to the carbonyl, is +1). The change in oxidation number is −1 → +1, for a net **loss** of two electrons, and an oxidation. Remember that as the net charge becomes more positive, this is associated with loss of electrons, which have a negative charge. Similarly, the carbon of interest in propanoic acid has an oxidation number of +3 (C is 0, the three bonds to oxygen total +3). Conversion of propanal to propanoic acid involves the *loss* of two additional electrons (+1 → +3), and is an oxidation. Each of these two electron oxidations is a one-stage oxidation, but a direct conversion of 1-propanol to propanoic acid would be a four electron, two-stage oxidation. Only the carbon bearing the oxygen atoms was examined in this example since the other carbons are not changed during the course of the reaction.

8.   Hendrickson, J.B.; Cram, D.J.; Hammond, G.S. *Organic Chemistry*, 3rd Ed., McGraw-Hill, New York, *1970*.
9.   Pine, S.H. *Organic Chemistry*, 5th Ed., McGraw-Hill, New York, *1987*.

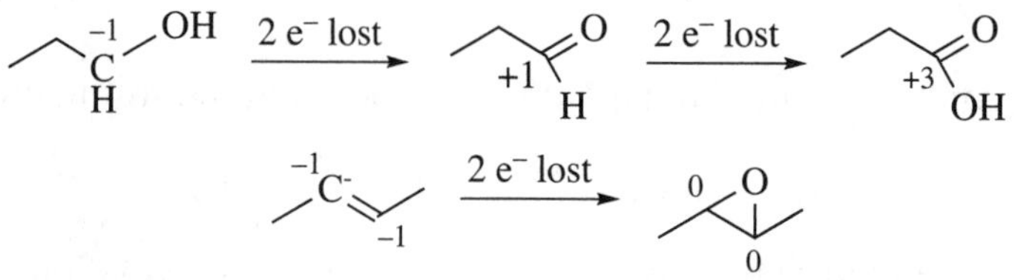

The oxidation of 2-butene to the corresponding epoxide (sec. 3.4) involves modification of two carbons. The oxidation number of each carbon comprising the π bond is –1 (C,C is 0, and H is –1), and the oxidation number of each epoxy carbon in the product is 0 (C,C is 0, O is +1 and H is –1), for a change of one electron (–1 → 0) for *each* of those two carbons. The *net* change for each molecule is loss of two electrons for a one-stage oxidation.

If there is a choice of more than one reagent for a particular oxidation reaction, the relative strength of those reagents is important. It is also important to determine whether a given reagent will induce a one or two-stage (two electron or four electron) oxidation (1-propanol to propanal or 1-propanol to propanoic acid, for example). Oxidation and reduction are linked together, since an oxidation of one species is accompanied by reduction of another. A useful measure of the propensity of a compound or a reagent for oxidation-reduction is the reduction potential. The reduction potential is obtained from polarographic (electrolysis) experiments that involve transfer of electrons between electrodes.[10] "An electrode process is a heterogeneous transfer of electrons between electrodes and an organic molecule".[11] The potential of the electrode controls the availability of electrons in the medium. The tendency toward electron loss or electron gain in the reduction system is given by the reaction

$$ M \quad \rightleftharpoons \quad M^{2+} \quad + \quad 2\,e^- $$

Reduction potential can be measured as an electrical driving force, expressed as the electrode potential.[11] As written, this equation represents a loss of two electrons from M and is an oxidation. If written as the reverse reaction, it is a reduction. Both oxidized and reduced species must be present in this equilibrium, and changes in potential occur as the equilibrium shifts. It is important to note that the electron transfer process shown above does not always correlate with the functional group transformation associated with oxidation of organic substrates. It does allow comparisons of the relative strengths of various oxidizing agents, however. For selected reactions involving electrochemistry of organic substrates see section 4.9.J.

If the electrode potential in an electrolysis reaction is made more positive, oxidation (with loss of electrons) will compensate for this positive change in potential.[12] A more negative potential will be compensated for by addition of electrons to the oxidized form. The equilibrium potential of a reversible system of this type can, therefore, be given by the equation

10. Ross, S.D.; Finkelstein, M.; Rudd, E.J. *Anodic Oxidations*, Academic Press, New York, *1975*.
11. Reference 10, p 7.
12. Allen, M.J. *Organic Electrode Processes*, Reinhold Pub. Corp. New York, *1958*, pp 2.

$$E = E_o - \frac{RT}{nF} \ln \frac{[Ox]}{[Red]} \qquad\qquad 3.1$$

where E_O is the standard reduction potential and F is **Faraday's constant** (96,400 Coulombs per equivalent). Ox and Red refer to the oxidized and reduced forms, respectively. Electrode potential is always expressed relative to an arbitrary standard, and the zero electrode potential is usually taken as that of "the reversible equilibrium between hydrogen gas at one standard atmosphere pressure and hydrogen ions at unit activity" (the normal reversible hydrogen electrode potential).[13] For the reaction,

$$M \rightleftharpoons M^{2+} + e^-$$

the metal (M) is assumed to be coupled with a metal having less tendency to ionize, generating a negative potential in the electrolytic cell. The electrode potential (ε) is proportional to the free energy (ΔG) of the reaction by

$$\Delta G = -z\,F\,E \text{ for the cell} \qquad\qquad 3.2$$

$$\Delta G = -z\,F\,\varepsilon \text{ for a single electrode} \qquad\qquad 3.3$$

where E is the electromotive force, F is the Faraday constant and z is the number of electrons involved in the overall cell reaction. If ε is negative, ΔG is positive and the reaction is *not* spontaneous. Using the **European convention**,[14,15] ε° for a metal is taken as negative when the polarity of the metal in that cell is negative. Therefore, the expression $\Delta G = -[-z\,F\,\varepsilon]$ will be used for the reaction. An example is

$$Na \longrightarrow Na^+ + e^-$$

where ε° is negative, ΔG is negative and ionization (a one-electron oxidation) is a spontaneous reaction.

The oxidation potential represents the ability of a metal atom (M) to be ionized to an ion (M^+) with loss of an electron. For the oxidation of organic molecules, transition metal compounds containing chromium, manganese, ruthenium, selenium, silver, or cerium are often used. The oxidation potential can be a useful method for examining the oxidizing power of these reagents.

Permanganate (MnO_4^-) is a powerful oxidizing agent used in many organic oxidations, but several other oxidation states areavailable to manganese, including manganate (MnO_4^{2-})and hypomanganate (MnO_4^{3-}). Their reactivity toward reductants (the organic substrate) varies

13. Reference 10, p 9.
14. Reference 10, p 10.
15. Ives, D.J.G.; Janz, G.J. *Reference Electrodes: Theory and Practice*, Academic Press, New York, *1961*.

inversely with the charge on the oxidant,[16a] and the *relative order of oxidizing power* is

$$MnO_4^- > MnO_4^{2-} > MnO_4^{3-}.$$

The reduction potentials for these species are known to be[16a]

$$MnO_4^- + e^- \longrightarrow MnO_4^{2-} \qquad (\varepsilon = +0.558 \text{ V})$$

$$MnO_4^{2-} + e^- \longrightarrow MnO_4^{3-} \qquad (\varepsilon = +0.285 \text{ V})$$

These reactions are shown as the standard reduction potential (see below) since reduction potential reflects the propensity of a metal or metal salt to be reduced, with concomitant oxidation of an organic substrate. The larger reduction potential for permanganate is correlated with its stronger oxidizing ability. The point of this example is that the oxidizing agents used in this chapter are not always metals but are often metal derivatives. Their oxidizing power varies with the structural features of each derivative, and it is important to understand the oxidizing power of each when they are used for the oxidation of an organic substrate. Oxidizing power varies with the reaction medium. Permanganate, for example, shows different reduction potentials in acidic and basic media:[16a]

$$MnO_4^- + 3\,e^- + 4\,H^+ \longrightarrow MnO_2 + 2\,H_2O \quad (\varepsilon = +1.679 \text{ V})$$

$$MnO_4^- + 3\,e^- + 2\,H_2O \longrightarrow MnO_2 + 4\,OH^- \quad (\varepsilon = +0.59 \text{ V})$$

The standard reduction potential for many common metal salts and other compounds are known,[17] and the pH of the media plays an important role in the oxidizing power of a given reagent. Acidic, neutral, and basic media will be employed in the following sections for the oxidation of different substrates and a variety of oxidizing agents will be used. The choice of reagent and conditions will influence the overall reaction, and determine if the oxidation is a two electron (one-stage) or a four-electron (two-stage) oxidation.

In most compilations of reduction potentials, the reaction is shown as the reduction potential $(M^+ + e^- \rightarrow M)$. The listed potential is also the oxidation potential for the reverse of the reaction shown. In general, the larger the reduction potential, the greater will be the oxidizing power of that reagent. Ozone, hydrogen peroxide, oxygen, lead [Pb(IV)], silver (Ag$^+$), chromium [Cr(VI)], permanganate (MnO$_4^-$), and periodic acid are strong oxidizing agents and will figure prominently in this chapter.

Reduction potential can also be correlated with ionization potential (*I*), which is defined by

---

16. (a) Stewart, R. in *Oxidation in Organic Chemistry, Part A*, Wiberg, K. (Ed.), Academic Press, New York, *1965*, p 11; (b) Hudlicky´, M. *Oxidations in Organic Chemistry*, American Chemical Society Monograph 186, Amer. Chem. Soc. Washington D.C., *1990*.
17. Handbook of Chemistry and Physics, 87th ed., CRC/Taylor and Francis, Boca Raton, Fla., *2006*, pp 8-20-8-32.

$$M_{gas} \xrightarrow{I} M^+ \quad + \quad e^- \qquad\qquad 3.4^{18}$$

The primary difference between I and the reduction potential (ε) is that I is defined for a gas-phase reaction, whereas ε is measured in solution. The solvent has a great influence on the course of the oxidation. It is difficult to directly correlate ε and I, but the latter can be used for a general comparison of the ability of a molecule or reagent to be oxidized or reduced. In general, larger ionization potentials are correlated with greater oxidizing power. In Table 3.1, osmium tetroxide ($I = 12.97$ eV)[19] is a stronger oxidizing agent than ruthenium tetroxide ($I = 12.33$ eV)[19] by this criterion. It would be useful to know the reduction potentials of the organic substrates that are being oxidized, but these are difficult to obtain since the reduction potential will be different for each type of oxidation. For this reason, comparison of one-electron oxidations (the usual first half-wave potential reported for most organic molecules) can be misleading. The ionization potential is a useful alternative, but suffers from the same problem since it represents loss of one electron to form the ion. Table 3.1[19] lists the ionization potential for common organic substrates that can be oxidized (many will be discussed in this chapter), as well as the ionization potential for a few selected oxidizing agents. There are different types of oxidations for the same substrate. Taking an alkene as an example, the reader is cautioned to remember that the ability of an alkene to be oxidized will depend on whether the reaction is an epoxidation, dihydroxylation, halogenation, or an oxidative cleavage. With that caution in mind, the data in Table 3.1 suggest that primary alcohols are more easily oxidized than secondary alcohols (the lower value of I represents a substrate that is more easily oxidized), and alkenes are more easily oxidized than alkynes. The ionization potentials also suggest that sulfides and tertiary amines are easily oxidized.

Table 3.1.   Ionization Potentials of Common Oxidizable Organic Substrates.

Molecule	I (eV)	Molecule	I (eV)	Molecule	I (eV)
MeC≡CH	10.36	$MeCO_2Me$	10.27	MeCHO	10.20
benzene	9.24	Me–S–Me	8.68	EtCHO	9.98
NMe_3	8.82	Cl_2	11.48	acetone	9.69
O_2	12.06	I_2	9.28	cyclohexanone	9.14
H_2O_2	11.00	OsO_4	12.97	MeCOOH	10.36
EtOH	10.49	2-butene	9.13	PMe_3	8.60
1-propanol	10.10	$EtNH_2$	8.86	Et–S–Et	8.43
2-propanol	10.15	Et_2NH	8.01	Br_2	10.54
phenol	8.51	O_3	12.30	RuO_4	12.33

[Reprinted with permission from *Handbook of Chemistry and Physics, 87th ed.*, CRC Taylor and Francis, Boca Raton, Fla., *2006*, pp 10-203-10-223. Copyright CRC Taylor & Francis, Boca Raton, FL.]

18. Reference 10, p 9.
19. Handbook of Chemistry and Physics, 87th ed., CRC/Taylor and Francis, Boca Raton, Fla., *2006*, pp 10-203-10-223.

In the following sections, the emphasis will be on oxidation reactions that transform one functional group into another. Important transformations will be the oxidation of alcohols to carbonyls, oxidation of alkenes to diols or epoxides, and oxidative cleavage of alkenes to carbonyl derivatives. A few methods will be discussed for the oxidation of alkyl fragments, such as oxidation of a methyl group to hydroxymethyl, an aldehyde, or a carboxylic acid. The oxidation of sulfur and nitrogen compounds will also be briefly discussed. As mentioned, any attempt to classify organic reactions by the oxidizing power of the reagent is difficult since the classification depends on both reaction and reagent. Similarly, classification by organic substrate is difficult since there are oxidations involving transfers of different numbers of electrons, giving different products. The choice of classification by functional group transform is intended to convey a general sense of the use of oxidation in organic synthesis and is specifically correlated with the disconnection-transform theme of this book. The reader is referred to Hudlicky 'sexcellent monograph[16b] for a thorough synthetic discussion of oxidation reactions.

3.2. ALCOHOLS TO CARBONYLS (CH-OH → C=O)

One of the most important oxidations in organic chemistry converts alcohols to aldehydes, ketones, or carboxylic acid derivatives. Many oxidizing agents can be used, but the particular product formed depends on the structure of the alcohol as well as the reagent.[20] Primary alcohols are initially oxidized to aldehydes, which can be isolated, but that aldehyde can be oxidized further to a carboxylic acid. Each of these oxidations is a two-electron process. In many cases, the aldehyde is not isolated but converted directly to the carboxylic acid. Secondary alcohols can be oxidized to ketones via a two electron process, but oxidative cleavage (this would be a four-electron oxidation) can occur if the oxidation conditions are too harsh. The discussion of alcohol oxidation must focus on the relative strength of the reagent, the conditions under which it is used, and structural variations of the alcohol substrate. This section will begin with the most common alcohol oxidizing reagents, which are based on Cr(VI). A discussion of other reagents, including the use of DMSO reagents, the Dess-Martin periodinane reagent (sec. 3.D), tetrapropylammonium perruthenate (TPAP, $Pr_4N^+RuO_4^-$), and others will follow.

3.2.A. Oxidation with Chromium (VI)

Chromium (VI) is a strong oxidizing agent. In sulfuric acid, the reaction of Cr(VI) has a reduction potential of 1.10 V,[17] but several chromium species may be present:

$$Cr^{6+} + 3\,e^- \rightarrow Cr^{3+} \quad (2N\ H_2SO_4) \qquad\qquad 1.10\ V$$

$$Cr_2O_7^{2-} + 14\ H^+ + 6\,e^- \rightarrow 2\ Cr^{3+} + 7\ H_2O \qquad\qquad 1.33\ V$$

---

20. Larock, R.C. *Comprehensive Organic Transformations,* 2nd ed., Wiley-VCH, New York, *1999*, pp 1234-1256

$$HCrO_4^- + 7\,H^+ + 3\,e^- \rightarrow Cr^{3+} + 4\,H_2O \qquad\qquad 1.195\ V$$

3.2.A.i. Chromium Reagents. Oxidation of an alcohol is accompanied by reduction of the chromium [Cr (VI) $\rightarrow$ Cr(III)]. As implied by the reduction potentials of the three chromium species shown, there are many Cr(VI) reagents.[21,22] The inorganic reagent chromium trioxide is commonly used, which is a polymeric species usually written as $(CrO_3)_n$. In aqueous media, chromium trioxide exists in equilibrium with several other Cr(VI) species, including H_2CrO_4, $HCrO_4^-$, CrO_4^{2-}, HCr_2O_7, $Cr_2O_7^{2-}$, $H_2Cr_2O_7$, and $HCr_2O_7^-$.[23] Of the Cr(VI) species listed above, dichromate is the strongest. Oxidation of alcohols requires that both the alcohol and the chromium oxidant are soluble or partly soluble in the medium, so most oxidations are carried out in aqueous media. The use of water means that several Cr(VI) species are present, however. At high dilution, the excess of water shifts the overall equilibrium towards dichromate.[23,24] At high concentrations (less water), polymeric chromium trioxide (polychromates) and chromic acid are the predominate species.[24] Dichromate is generated by "complex formation between the two acid chromate ions to give a dichromate anion".[25] Formation of this complex and its relative concentration are dependent on the pK_a of the acid. In dilute solution the concentration of Cr(VI) is independent of the acid, but only if the acidity is below the apparent pK_a of chromic acid.[25] Modification of the acid will increase the pK_a of the complex as the electron-withdrawing power of the acid decreases. The electrons on the oxygen atoms attached to chromium are less available for protonation,[25] leading to a larger dissociation constant,[26] which suggests that the position of the equilibrium depends on the acid. Indeed, adding a different acid (HA) to the mixture will influence the position of the overall equilibrium according to the reaction:[27]

$$HCrO_4^- + 2\,H^+ + A^- \longrightarrow HCrO_3A + H_2O$$

If the acid is more effective at withdrawing electrons from the complex ($HCrO_3A$), the pK_a for the reaction increases. There is, therefore, a correlation with the strength of the mineral acid and the $HCrO_3A$ species:[25,26]

$$H_3PO_4 < HCl < H_2SO_4 < HClO_4$$

21. Wiberg, K.B. in chapter 2 of *Oxidation in Organic Chemistry, Part A*, Wiberg, K.B. (Ed.), Academic Press, New York, *1965*,pp 69-70.
22. (a) Byström, A.; Wilhelmi, K.A. *Acta Chem. Scand.* *1950, 4,* 1131; (b) Hanic, F.; Stempelóv, D. *Chem. Zvestii* *1960, 14,* 165 [*Chem. Abstr. 54*: 20402a *1960*].
23. (a) Reference 21a, pp 71; (b) Neuss, J.D.; Rieman, W. *J. Am. Chem. Soc.* *1934, 56,* 2238; (c) Tong, J.Y.P.; King, E.L. *Ibid.* *1953, 75,* 6180; (d) Davies, W.G.; Prue, J.E. *Trans. Faraday Soc.* *1955, 51,* 1045; (e) Howard, J.R.; Nair, V.S.K.; Nancollas, G.H. *Ibid.* *1958, 54,* 1034; (f) Schwarzenbach, G.; Meier, J. *J. Inorg. Nuc. Chem.* *1958, 8,* 302; (g) Bailey, N.; Carrington, A.; Lott, K.A.K.; Symons, M.C.R. *J. Chem. Soc.* *1960,* 290; (h) Sasaki, Y. *Acta Chem. Scand.* *1962, 16,* 719.
24. Freedman, M.L. *J. Am. Chem. Soc.* *1958, 80,* 2072.
25. Lee, D.G.; Stewart, R *J. Am. Chem. Soc.* *1964, 86,* 3051.
26. Reference 16a, p 72.
27. (a) Boyd, R.H. *J. Am. Chem. Soc.* *1961, 83,* 4288; (b) Boyd, R.H. *J. Phys. Chem,* *1963, 67,* 737.

The nature of the oxidizing species is obviously important for determining the mechanism of oxidation of an alcohol to a carbonyl with chromium (VI). As mentioned above, in aqueous solutions CrO_3 generates CrO_4^{2-}, $HCrO_3^-$ and $HCr_2O_7^{2-}$ as the primary oxidizing species, along with other chromium species.[28] At pH $\geq$ 1, CrO_3 is almost completely ionized.[28] Since organic molecules are being oxidized with chromium reagents, the focus tends to be on the organic molecule (the alcohol in this case) and Wiberg presented the following scheme to represent the oxidation of a secondary alcohol to a ketone:[29]

R_2CHOH	+	Cr(VI)	$\longrightarrow$	$R_2C{=}O$	+	Cr(IV)	
R_2CHOH	+	Cr(IV)	$\longrightarrow$	$R_2\overset{\bullet}{C}OH$	+	Cr(III)	
$R_2\overset{\bullet}{C}OH$	+	Cr(VI)	$\longrightarrow$	$R_2C{=}O$	+	Cr(V)	
R_2CHOH	+	Cr(V)	$\longrightarrow$	$R_2C{=}O$	+	Cr(III)	
R_2CHOH	+	Cr(VI)	$\longrightarrow$	3 $R_2C{=}O$	+	2	Cr(III)

[Reprinted with permission from Wiberg, K.B.; Mukerjee, S.K. *J. Am. Chem. Soc.* **1974**, *96*, 1884. Copyright © 1974 American Chemical Society.]

This mechanism shows that Cr(VI), Cr(IV) and Cr(V) species are involved in the overall oxidation process. Clearly, this is not a balanced equation and the Cr species are undefined. The Cr(VI) species are usually $HCrO_4^-$ and CrO_3, the Cr(V) is usually $HCrO_4^-$ and the Cr(IV) is usually $HCrO_3^-$.[29] The oxidation involves both water (H_2O) and an acid catalyst (H^+), and a balanced equation for the oxidation of 2-propanol is[28]

$$3 \underset{}{\overset{OH}{\wedge}} + 2\,HCrO_4^- + 8\,H^+ \longrightarrow 3 \underset{}{\overset{O}{\wedge}} + 2\,Cr^{+3} + 8\,H_2O$$

There is strong evidence that the oxidation begins by formation of a chromate ester,[30] which collapses to the carbonyl compound and a Cr(IV) species in the rate-determining step.[28,31,32] The driving force of the reaction is reduction of Cr(VI) to Cr(IV).[33] Westheimer determined that the rate term for the oxidation of 2-propanol was[32] $v = k_a\,[HCrO_4^-]\,[R_2CHOH]\,[H^+] + k_b[HCrO_4^-]\,R_2CHOH]\,[H^+]^2$. If sulfuric acid is used as the acid in an aqueous medium

28. (a) Westheimer, F.H.; Nicolaides, W. *J. Am. Chem. Soc.* **1949**, *71*, 25; (b) Westheimer, F.H. *Chem. Rev.* **1949**, *45*, 419 (see p 427).
29. (a) Wiberg, K.B.; Mukerjee, S.K. *J. Am. Chem. Soc.* **1974**, *96*, 1884; (b) Rahman, M.; Roček, J. *Ibid.* **1971**, *93*, 5462, 5455.
30. (a) Reference 21, p 161 and references cited therein; (b) Kläning, U. *Acta Chem. Scand.* **1957**, *11*, 1313 and *Ibid.* **1958**, *12*, 576.
31. Brownell, R.; Leo, A.; Chang, Y.W.; Westheimer, F.H. *J. Am. Chem. Soc.* **1960**, *82*, 406.
32. (a) Westheimer, F.H.; Novich, A. *J. Chem. Phys.* **1943**, *11*, 506; (b) Reference 21, see p 159 and reference 102 cited therein.
33. Reference 21, p 167.

containing chromium trioxide and 2-propanol, the initial chromate ester is probably **1**, which arises from reaction of $HCrO_4^-$ and H_2SO_4 (HA in the above equation).[25] In water containing no external acid, the chromate ester probably takes the form *i*-$PrOCrO_3H$.[34]

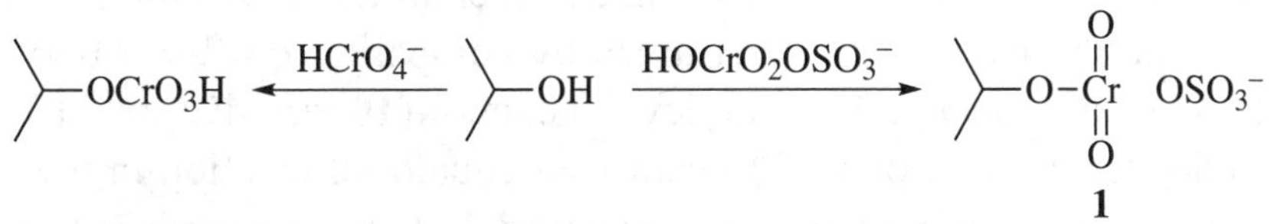

1

Decomposition of chromate ester *i*-$PrOCrO_3H$ involves removal of the proton attached to the oxygen-bearing carbon. Westheimer initially proposed that water behaves as a base to remove this hydrogen,[31,32,35c] although later work showed that addition of pyridine to the reaction (a stronger base than water) did *not* enhance the rate of oxidation.[35] At least two mechanisms have been proposed for this step. Westheimer proposed an intermolecular attack by water as mentioned (see **2**), but later retracted this proposal.[35c,36] Kwart and Francis[34] proposed an intramolecular reaction for removal of the proton (see **3**), by an oxygen on the chromate ester. Roček discounted both mechanisms for several reasons, including the important observation that protonation should *not* lead to rate enhancement with a species such as **3**, but in fact it does. Roček proposed the mechanism shown for **4**,[35b] in which oxygen atoms from the chromic acid removed hydrogens both from carbon and from oxygen. Roček later proposed[37] formation of a coordination complex such as **5** (shown exactly as reported in the paper), presumably formed from Cr(IV) in a similar way to **5**, that decomposed to a radical species (**6**). Further oxidation of **6** generated the products.

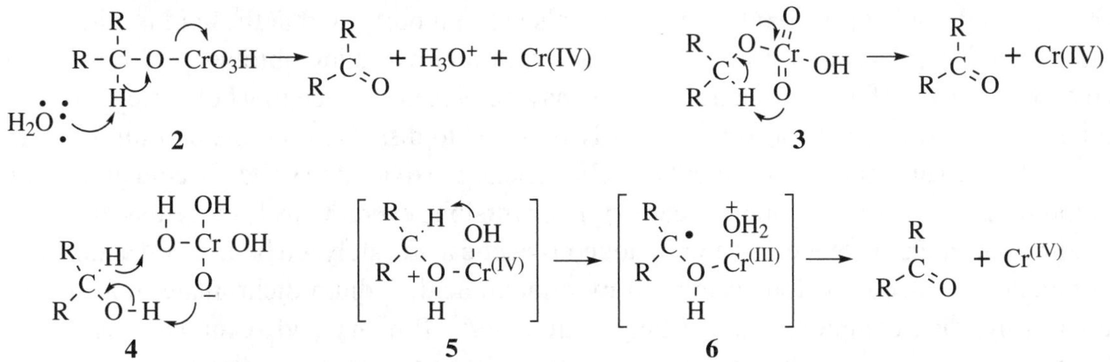

There is a steric component to this reaction. As the steric bulk around the carbon bearing the OH group increases, removing that hydrogen becomes increasingly difficult and the overall

34. Reference 21, p 162.
35. (a) Roček, J.; Krupicka, J. *Chem & Ind. (London)* **1957**, 1668; (b) Roček, J. *Collect. Czech. Chem. Commun.* **1960**, *25,* 1052; (c) Westheimer, F.H.; Chang, Y.W. *J. Phys. Chem.* **1959**, *63,* 438.
36. Kwart, H.; Francis, P.S. *J. Am. Chem. Soc.* **1959**, *81,* 2116.
37. Rahman, M.; Roček, J. *J. Am. Chem. Soc.* **1971**, *93,* 5455.

rate of oxidation will be diminished.[38] The relative rate of oxidation of endo-borneol (see **7**) compared to exo-borneol (see **8**), for example, is 25.0:49.1.[39a] This difference is correlated with the more facile removal of H_a from chromate ester complex **7** (where H_a is on the exo face) than from complex **8** (where is H_a is on the more hindered endo face). Another example of steric hindrance can be seen in cyclohexyl derivatives by comparing *cis-2-tert*-butylcyclohexanol (**9**, relative rate 50.6) and *trans-2-tert*-butylcyclohexanol (**10**, relative rate 10.7).[39a] In both **9** and **10**, the bulky *tert*-butyl group will occupy an equatorial position in the lowest energy conformation (sec. 1.5.D). Removal of the equatorial H_a from the chromate ester of **9** is inhibited by a large steric interaction with the adjacent *tert*-butyl group (*G* strain, sec. 1.5.D). This rate difference is also observed with modified chromium reagents such as pyridinium chlorochromate (sec. 3.2.B.ii).[39b]

From this discussion, it is apparent that oxidation of alcohols with Cr(VI) depends on several factors. The solvent and any acidic additives (or acidic solvents) will influence the rate of the oxidation, as will the structure of the alcohol. It is therefore not surprising that there are many variations in reaction conditions for Cr(VI) oxidations of alcohols. As the substrate and the oxidizing requirements of the chromium are changed, the reaction medium will be changed. Reaction conditions for the oxidation of alcohols are to modify so that the acid is added to the medium. Indeed, chromium trioxide will have different oxidizing abilities in different acids (see above) since different structural entities may be present. Since most organic compounds are insoluble in water, a co-solvent is usually required to dissolve both the chromium reagent and the alcohol substrate. This solvent must be resistant to oxidation so acetic acid or acetone is commonly used. For the alcohol → carbonyl conversion several Cr(VI) reagents can be used, including chromium trioxide in water or aqueous acetic acid catalyzed by mineral acid, sodium dichromate in aqueous acetone catalyzed by mineral acid, sodium dichromate in acetic acid, the CrO_3•pyridine complex, and *tert*-butyl chromate.[40] Primary and secondary alcohols are oxidized to the aldehyde or ketone, respectively. Aldehydes may be oxidized to the carboxylic acid under some conditions.

3.2.A.ii. Chromium Trioxide in Water or Aqueous Acetic Acid. Low molecular weight alcohols are usually sufficiently soluble in water that no co-solvent is required for oxidation

---

38. Reference 21, p 165.
39. (a) Reference 21, p 166; (b) Suggs, J.W., Ph.D. Thesis Harvard Univ., *1976*, Chapter III (see *Dissertation Abstracts On-Line*).
40. Reference 21, p 142.

with chromium reagents. Oxidation of 2-propanol in water gives excellent yields of acetone, although the miscibility of acetone in water makes its isolation difficult. The presence of other oxidizable groups on the organic substrate such as alkenes, sulfides, phenolic, and amines can lead to side reactions in the oxidation of water soluble alcohols, significantly lowering the yield of carbonyl products. Sulfides are oxidized to sulfoxides, amines to hydroxylamines, and phenols to quinones, for example. Phenylalkyl carbinols are subject to oxidative cleavage with chromium trioxide in aqueous acid, as observed in the oxidation shown.[41] When the R group is isopropyl (*i*-Pr), only 6% cleavage was observed, but changing R to *tert*-butyl led to 60% cleavage, presumably due to greater facility for generating a tertiary cation in the latter case.[41] Cleavage of this type is greatly suppressed by addition of manganous ion. Small amounts of cleavage products can be observed with hindered carbinols, but this is a problem only when a stable carbocation can be produced.[42]

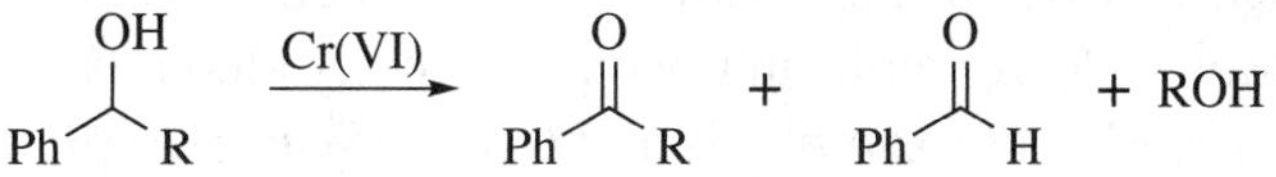

Oxidation of primary alcohols leads to aldehydes in moderate to good yield. Aldehydes are relatively easy to oxidize to the corresponding carboxylic acid, however. In most cases, the oxidation can be stopped at the aldehyde, but small amounts of the acid are a common byproduct. Heating and long reaction times lead to increased amounts of the acid, and in some cases the aldehyde is the minor product. Where feasible, removal of the aldehyde as it forms will minimize side reactions. When the reaction is pushed to give the carboxylic acid, there are at least two reasonable mechanistic rationales for this conversion.[43] Both mechanisms involve formation of a chromate ester. Removal of the α-hydrogen (analogous to the alcohol to aldehyde conversion) either by an external base (as in **11**), or intramolecularly (as in **12**),

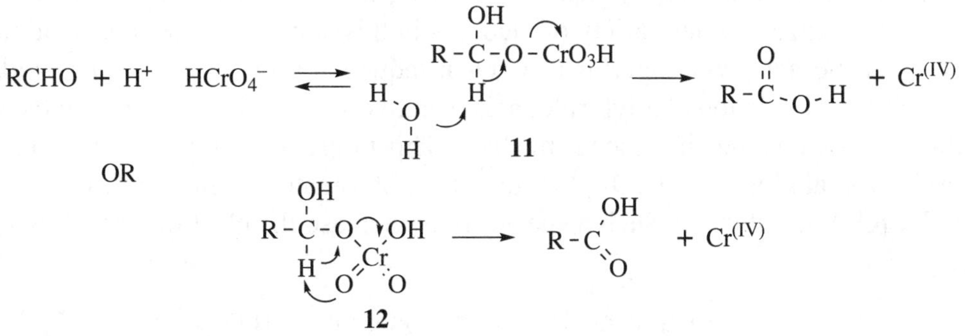

generates the carboxylic acid. This oxidation may also occur by initial formation of a hydrate,

---

41. (a) Reference 21, p 143 and reference 4 therein; (b) Hampton, J.; Leo, A.; Westheimer, F.H. *J. Am. Chem. Soc.* **1956**, *78*, 306.
42. Lee, D.G. in *Oxidation*, Vol 1, Augustine, R.L (Ed.), Marcel Dekker, New York, **1969**, p 58.
43. Reference 21, p 174.

followed by removal of the α-hydrogen from **13**. This mechanism gives the acid moiety, although a radical process that generates **14** is also possible.[43]

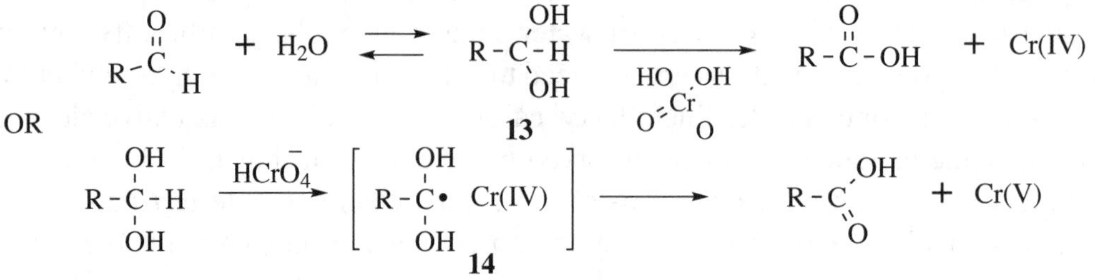

With some primary alcohols, it is possible to establish an equilibrium between the initially formed aldehyde and the precursor alcohol, in which unreacted alcohol is converted to an hemiacetal (**15**), which can be further oxidized to an ester.[44] This transformation is often observed in the oxidation of lower molecular weight primary alcohols. Oxidation of 1-butanol to butyl butyrate with Cr(VI), for example, is an *Organic Syntheses* preparation.[45]

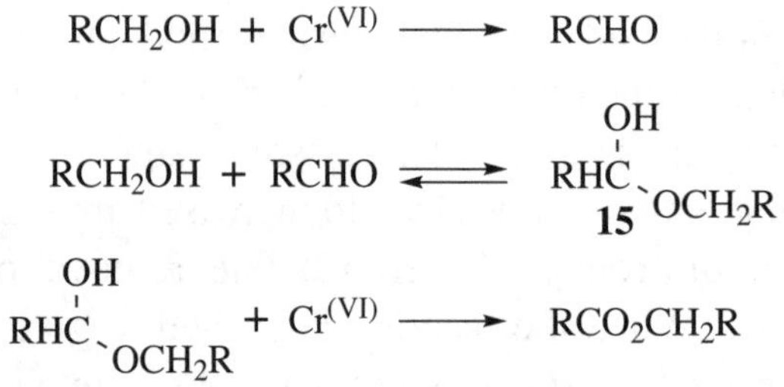

Some of the problems that accompany oxidation in water can be alleviated or moderated by using an organic co-solvent. Oxidation in aqueous acetic acid will suppress the oxidative cleavage to some extent, as well as greatly improving the solubility of the organic substrate. It is possible to oxidize a wider variety of alcohols in this solvent. Bowman found that CrO_3 in anhydrous acetic acid was superior to CrO_3 in aqueous acetic acid for the oxidation of saturated alcohols. Oxidation of allylic alcohols in anhydrous acetic acid gave yields that were comparable to those obtained in aqueous media.[46] This improvement may be due to formation of an acetyl chromate ion ($AcOCrO_3^-$), analogous to chromate complexes formed with other acids (see $HCrO_3A$ see above). Such a species may increase the electron accepting power of chromium.[25,47]

$$H_2CrO_4 + AcOH \rightleftharpoons AcOCrO_3^- + H_2O$$

44. (a) Reference 21, p 143; (b) Mosher, W.A.; Preiss, D.M. *J. Am. Chem. Soc.* **1953**, *75*, 5605.
45. Robertson, G.R. *Org. Synth. Coll. Vol. 1* **1941**, 138.
46. (a) Reference 21, p 152; (b) Bowman, M.I.; Moore, C.E.; Deutsch, H.R.; Hartman, J.L. *Trans. Kentucky Acad. Sci.* **1953**, *14*, 33 [*Chem. Abstr.* 48:1250b **1954**].
47. (a) Cohen, M.; Westheimer, F.H. *J. Am. Chem. Soc.* **1952**, *74*, 4387; (b) Symons, M.C.R. *J. Chem. Soc.* **1963**, 4331.

3.2.A.iii. Chromium Trioxide in Acetone: Jones Oxidation. Organic solvents other than acetic acid can be used with aqueous chromium trioxide. Jones used acetone as a co-solvent in a dilute sulfuric acid solution, and found that oxidation of alkynyl carbinols was improved when compared to other procedures known at that time. Secondary alcohols are oxidized to ketones, and primary alcohols can be oxidized to either an aldehyde or a carboxylic acid. This chromium trioxide/acetone/sulfuric acid reagent is often referred to as the **Jones reagent**, and oxidation of alcohols with this reagent is called **Jones oxidation**.[48] Jones oxidation is especially useful for molecules that contain alkenyl or alkynyl groups.[49] Oxidation of alcohols is usually faster in acetone than in acetic acid, and using a large excess of acetone protects the ketone product from further oxidation.[50] An example is the oxidation of **16** to give ketone **17** in 67% yield, taken from a synthesis of dysidiolide by Forsyth and Demeke.[51] Note that in this oxidation, the secondary alcohol was oxidized to the ketone and the primary alcohol was simultaneously oxidized to the carboxylic acid. Direct conversion of a primary alcohol to an acid is also seen in the oxidation of **18** to **19**, in Crimmins and Vanier's

synthesis of the microbial metabolite designated SCH 351448.[52] Many primary alcohols are oxidized to aldehydes with this procedure,[53] but the reaction conditions are strongly acidic and the final product is sometimes a carboxylic acid. However, there are many examples in which Jones oxidation has been used in the presence of reactive functional groups. If the starting material is an aldehyde rather than an alcohol, Jones oxidation gives the carboxylic acid. An example is seen in Hu and Panek's synthesis of (–)-motuporin, in which aldehyde (**20**) was

48. The Merck Index, 14th ed., Merck & Co., Inc., Whitehouse Station, New Jersey, *2006*, p ONR-49.
49. (a) Heilbron I.; Jones, E.R.H.; Sondheimer, F. *J. Chem. Soc. 1949*, 604; (b) Haynes, L.J.; Heilbron I.; Jones, E.R.H.; Sondheimer, F. *Ibid. 1947*, 1583; (c) Heilbron I.M.; Jones, E.R.H.; Sondheimer, J. *Ibid. 1947*, 1586.
50. Reference 21, p 145.
51. Demeke, D.; Forsyth, C.J. *Tetrahedron 2002, 58,* 6531.
52. Crimmins, M. .; Vanier, G.S. *Org. Lett. 2006, 8,* 2887.
53. (a) Hurd, C.D.; Meinert, R.N. *Org. Synth. Coll. Vol. 2, 1943*, 541; (b) Fossek, W. *Monatsh 1881, 2,* 614 and 1883, 4, 660; (c) Bouveault, L.; Rousset, L. *Bull. Soc. Chim. Fr. 1894, 11,* 300; (d) Jacobson, M. *J. Am. Chem. Soc. 1950, 72,* 1489; (e) Sauer, J. *Org. Synth. Coll. Vol. 4 1963*, 813.

oxidized to the acid, **21** in 88% yield.[54] Jones oxidation is compatible with complex molecules that contain a variety of functional groups such as alkenes, tertiary alcohols, esters, ketones, oxetanes and amides.[55]

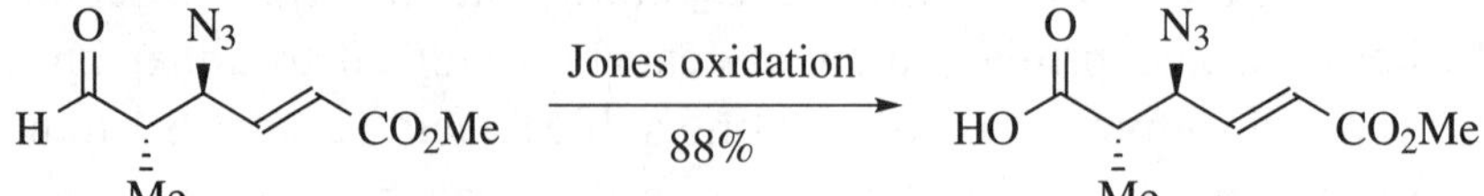

3.2.B. Modified Chromium (VI) Oxidants

Apart from changing the solvent, additives can be included in the reaction medium to create a new oxidizing reagent *in situ*. The fundamental Cr(VI) reagent can also be chemically modified prior to being used as an oxidizing agent. This section will examine both of these approaches.[57a]

3.2.B.i. Chromium Trioxide-Pyridine. In 1948, Sisler and co-workers isolated and characterized a stable complex from the reaction of chromium trioxide and pyridine.[56b] Sisler did not use this reagent for the oxidation of organic molecules, but Sarett and co-workers recognized its utility in the synthesis of steroids. In this connection, alcohol **22** was oxidized to **24** in 89% yield.[57] The reagent, which probably has the trigonal bipyramidal structure shown in **23**, proved useful for the general oxidation of primary and secondary alcohols even in the presence of double bonds and thioethers. The oxidation usually requires pyridine as a solvent and was referred to as **Sarett oxidation**[58] for many years. Care must be exercised in preparing the reagent.

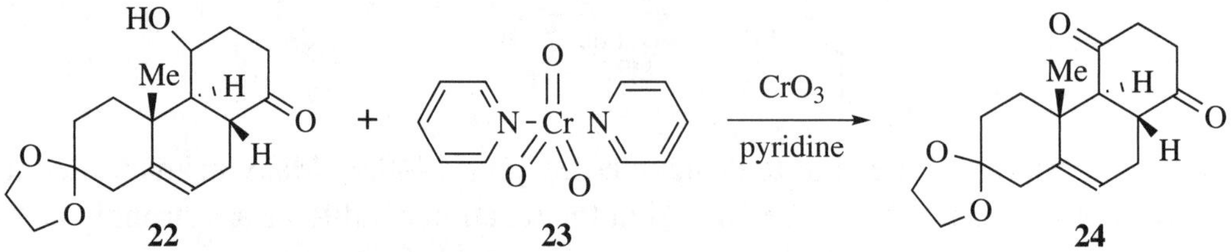

The solution *"must always be prepared by cautious addition of chromium trioxide to pyridine which has been carefully purified by distillation from potassium permanganate."*[59] *Reversing this order of addition may cause the mixture to ignite spontaneously.*[60]

54. Hu, T.; Panek, J.S. *J. Am. Chem. Soc.* **2002**, *124*, 11368.
55. Magri, N.F.; Kingston, D.G.I. *J. Org. Chem.* **1986**, *51*, 797.
56. (a) For a review of alcohol oxidation using oxochromium(VI)-amine reagents, see Luzzio, F.A. *Org. React.* **1998**, *53*, 1; (b) Sisler, H.H.; Bush, J.D.; Accountius, O.E. *J. Am. Chem. Soc.* **1948**, *70*, 3827.
57. Poos, G.I.; Arth. G.E.; Beyler, R.E.; Sarett, L.H. *J. Am. Chem. Soc.* **1953**, *75*, 422.
58. The Merck Index, 14th ed., Merck & Co., Inc., Whitehouse Station, New Jersey, **2006**, p ONR-83.
59. Reference 42, p 60.
60. Holum, J.R. *J. Org. Chem.* **1961**, *26*, 4814.

A second problem with Sarett oxidation is difficulty in isolating the products from a pyridine solution. An advantage of the technique, as mentioned above is that alkenes, ketals, sulfides, and tetrahydropyranyl ethers are oxidized much slower than alcohols and rarely give competitive side reactions.[61] Oxidation of secondary alcohols is usually facile, but oxidation of primary aliphatic alcohols often gives low yields of the aldehyde.[62] Benzylic and allylic alcohols give good yields, however. An example is taken from Kelly and co-worker's synthesis of louisianin C,[63] in which benzylic alcohol **25** was treated with chromium trioxide in pyridine at room temperature to give a 75% yield of louisianin C, **26**.

A modification introduced by Collins et al.[64] was applied to the oxidation of alcohols, and has come to be known as **Collins oxidation**. This modification was developed to deal with the problem of poor yields in the oxidation of primary alcohols to aldehydes, and to improve the isolation of the carbonyl products. The Sisler-Sarett reagent formed by reaction of chromium trioxide and pyridine was first removed from the pyridine solvent and added to dichloromethane, and this mixture was then treated with the alcohol. The oxidation typically required a 5:1 or 6:1 ratio of complex/alcohol, and reaction occurred at ambient temperatures.[64] Cyclohexanol was oxidized to cyclohexanone in 98% yield by this method, and 1-heptanol was oxidized to heptanal in 93% yield. The yield of aldehyde products from primary alcohols was significantly better, but a large excess of the reagent was required and the isolation problems persisted. Nonetheless, Collins oxidation became a mainstay of organic synthesis. An example is taken from the synthesis of hybocarpone, by Nicolaou and Gray,[65] in which oxidation of **27** gave ketone **28** in 86% yield. Allylic alcohols are easily oxidized to the corresponding conjugated ketone or aldehyde with this reagent, and Urones and co-workers used this type of oxidation in a synthesis of (–)-hyrtiosal.[66] This reagent can be used with groups that are more sensitive to hydrolysis or to over-oxidation. These include lactols, as illustrated by Ando and co-worker's conversion of lactol **29** to lactone **30**.[67] Ratcliffe and co-workers further improved this procedure by preparing dichloromethane solutions of the chromium trioxide-pyridine complex directly.[68]

61. Reference 21, pp 154-158.
62. Korytnyk, W.; Kris, E.J.; Singh, R.P. *J. Org. Chem.* **1964**, *29*, 574.
63. Beierle, J.M.; Osimboni, E.B.; Metallinos, C.; Zhao, Y.; Kelly, T.R. *J. Org. Chem.* **2003**, *68*, 4970.
64. Collins, J.C.; Hess, W.W.; Frank, F.J. *Tetrahedron Lett.* **1968**, 3363.
65. Nicolaou, K.C.; Gray, D.L.F. *J. Am. Chem. Soc.* **2004**, *126*, 607.
66. Basabe, P.; Diego, A.; Díez, D.; Marcos, I.S.; Urones, J.G. *Synlett* **2000**, 1807.
67. (a) Ando, M.; Akahane, A.; Takase, K. *Bull. Chem. Soc. Jpn.* **1978**, *51*, 283; (b) Ando, M.; Tajima, K.; Takase, K. *Chem. Lett.* **1978**, 617.
68. Ratcliffe, R.; Rodehorst, R. *J. Org. Chem.* **1970**, *35*, 4000.

The procedure typically used today is addition of chromium trioxide to dichloromethane and pyridine, followed by addition of the alcohol, but it is still referred to as Collins oxidation.

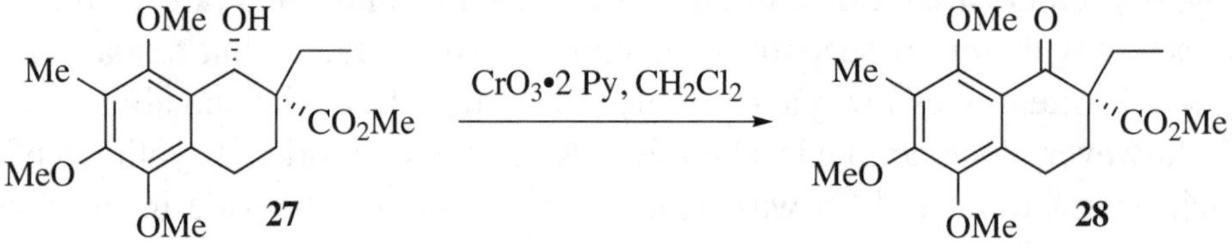

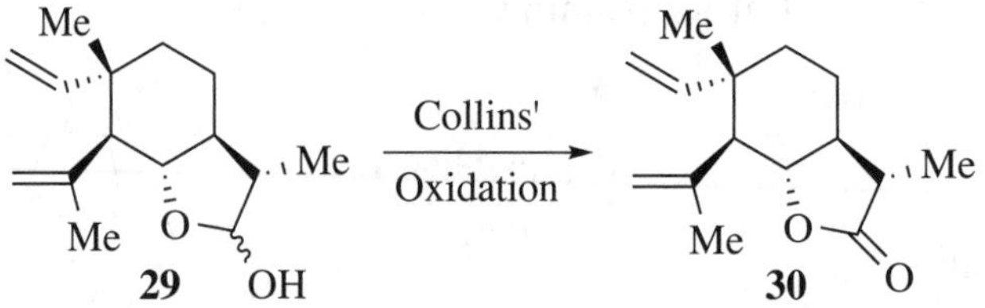

3.2.B.ii. Pyridinium Chlorochromate. The need for improved selectivity in the oxidation of primary alcohols, and greater ease for isolation of products prompted further research into the nature of Cr(VI) reagents. Corey and Suggs found that addition of pyridine to a solution of chromium trioxide in aqueous HCl led to crystallization of a solid reagent (**31**, pyridinium chlorochromate or **PCC**).[69] In dichloromethane, this reagent was superior for the conversion of primary alcohols to aldehydes, but less efficient than the Collins oxidation when applied to allylic alcohols.[70] Oxidation of 1-heptanol with PCC in dichloromethane gave 78% of heptanal. A synthetic example is taken from Corey and Hong's synthesis of arborone, in which **32** was oxidized by PCC to the ketone (**33**) in 82% yield.[71] As stated by Corey, PCC is an effective oxidant in dichloromethane although aqueous chlorochromate species are not very effective oxidants.[70] Oxidation of secondary alcohols to ketones is straightforward.

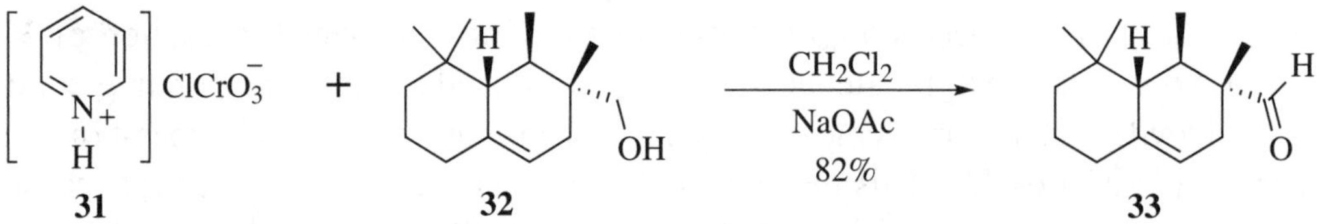

A drawback to the use of pyridinium chlorochromate is its mild acidity (note the use of sodium acetate to buffer the oxidation of **32** to **33**). It is inferior to Collins oxidation when the oxidation products or the starting alcohol is acid sensitive. This acidity can be used to synthetic advantage, however, and Corey described its use to promote oxidative cationic cyclizations.[72] Reaction of citronellol with PCC in dichloromethane was accompanied by cationic cyclization reaction

69. Corey, E.J.; Suggs, J.W. *Tetrahedron Lett.* **1975**, 2647.
70. (a) Banerji, K.K. *Bull Chem. Soc. Jpn.* **1978**, *51*, 2732; (b) Banerji, K.K. *J. Chem. Soc. Perkin Trans. 2* **1978**, 639.
71. Hong, S.; Corey, E.J. *J. Am. Chem. Soc.* **2006**, *128*, 1346.
72. Corey, E.J.; Boger, D.L. *Tetrahedron Lett.* **1978**, 2461.

to give (−)-pulegone.[73a,69]   Oxidative transpositions can also occur.73b   In studies toward a synthesis of guanacastepene A,[74] Mehta and Umarye oxidized of a mixture of **34** and **35** with PCC to give a 90% yield of the rearranged ketone **36**.   Oxidations using PCC on alumina[75] have also been reported, presumably to minimize acid-catalyzed rearrangement.   The PCC reagent has been used for the conversion of oximes to ketones,[76] and for the direct oxidation of tetrahydropyranyl protected alcohols to aldehydes.[77]

3.2.B.iii. Pyridinium Dichromate.   Problems caused by the acidity of PCC can be largely eliminated, by using the more neutral reagent pyridinium dichromate (**PDC**) **37**.   Although first used by Coates and Corrigan[78] it was not exploited synthetically until Corey prepared the reagent by addition of pyridine to neutral chromium trioxide solutions and used it for the oxidation of alcohols to aldehydes, ketones, and acids.[79]   The reagent is not acidic and the neutral conditions required for the oxidation are superior for the oxidation of allylic alcohols. In dichloromethane (non-aqueous workup) the oxidation is similar to that of PCC.   Addition of catalytic amounts of pyridinium trifluoroacetate in dichloromethane significantly increases the rate of oxidation.   Allylic alcohols are oxidized faster than aliphatic alcohols, making PDC the reagent of choice for this transformation.   Cyclohexenol, for example, is oxidized 10 times faster than cyclohexanol with PDC in dichloromethane at 25°C.[79]   Primary aliphatic alcohols are also oxidized under very mild conditions.[80]   This oxidation occurs under essentially neutral conditions, as shown by the reaction of **38** with PDC to give aldehyde **39** in 92% yield, despite the presence of the acid sensitive enol ether.[79]   A synthetic example is taken from Nemoto and co-worker's synthesis of macrosphelide B,[81] in which the allylic secondary alcohol unit in **40** was oxidized to conjugated ketone **41**, in 82% yield using PDC in dichloromethane.

73. (a) Corey, E.J.; Ensley, H.E.; Suggs, J.W. *J. Org. Chem.* **1976**, *41,* 380; (b) Dauben, W.G.; Michno, D.M. *J. Org. Chem.* **1977**, *42,* 682. Also see, (c) Corey, E.J.; Ha, D.C. *Tetrahedron Lett.* **1988**, *29,* 3171; (d) Waddell, T.G.; Carter, A.D.; Miller, T.J.; Pagni, R.M. *J. Org. Chem.* **1992**, *57,* 381; (e) Luzzio, F.A.; Guziec, Jr., F.S. *Org. Prep. Proceed. Int.* **1988**, *20,* 533; (f) Schlecht, M.F.; Kim, H.-J. *J. Org. Chem.* **1989**, *54,* 583.
74. Mehta, G.; Jayant D. Umarye, J.B. *Org. Lett.* **2002**, *4,* 1063.
75. For an example taken from a synthesis see Moreno-Dorado, F.J.; Guerra, F.M.; Aladro, F.J.; Bustamante, J.M.; Jorge, Z.D.; Massanet, G.M. *J. Nat. Prod.* **2000**, *63,* 934.
76. Maloney, J.R.; Lyle, R.E.; Saavedra, J.E.; Lyle, G.G *Synthesis* **1978**, 212.
77. Sonnet, P.E. *Org. Prep. Proceed. Int.* **1978**, *10,* 91.
78. Coates, W.M.; Corrigan, J.R. *Chem. Ind.* **1969**, 1594.
79. Corey, E.J.; Schmidt, G. *Tetrahedron Lett.* **1979**, 399.
80. Boeckman, Jr., R.K.; Perni, R.B. *J. Org. Chem.* **1986**, *51,* 5486.
81. Kawaguchi, T.; Funamori, N.; Matsuya, Y.; Nemoto, H. *J. Org. Chem.* **2004**, *69,* 505.

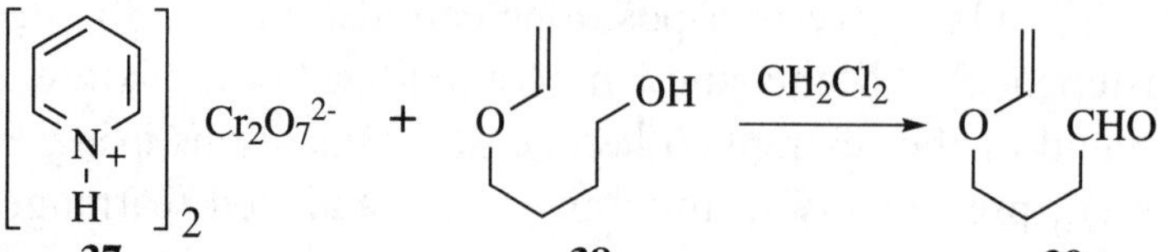

There can be a significant difference in product formation in the oxidation of aliphatic primary alcohols versus that of allylic primary alcohols, when the solvent is DMF. Cyclohexenol is oxidized to cyclohexenone in 86% yield with PDC in DMF at 0°C.[79] Similarly, geraniol was oxidized to geranial in 92% yield. However, treatment of non-conjugated primary alcohols with PDC in DMF (ambient temperature) often gave carboxylic acids rather than an aldehyde. For example, in Cossy and co-worker's synthesis of zoapatanol,[82] oxidation of **42** with PDC in DMF at 25°C gave hydroxy-acid **43**, which cyclized to lactone **44** in greater than 85% yield. Aldehydes can also be oxidized to the carboxylic acid, as in Williams and Jain's synthesis of (+)-negamycin,[83] where the aldehyde unit in **45** was converted to carboxylic acid **46** in 97% yield.

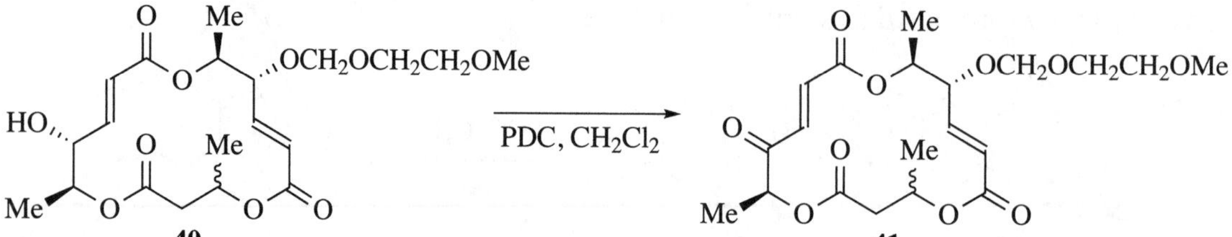

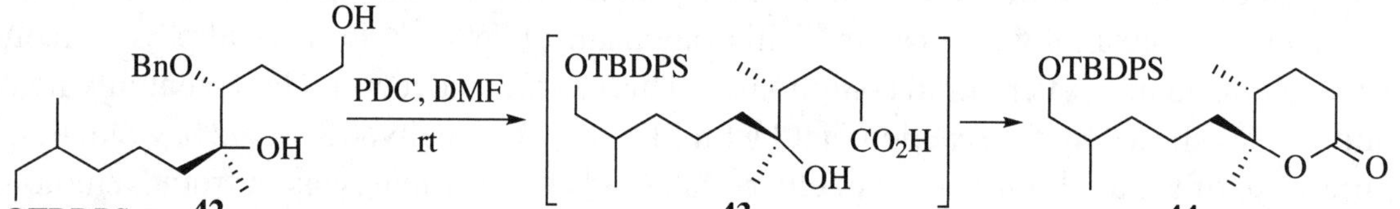

3.2.B.iv. Structurally Modified Chromium Reagents. The literature is full of chromium trioxide complexes developed by adding amine or phosphine bases, usually with the goal of varying the specificity of the oxidation.

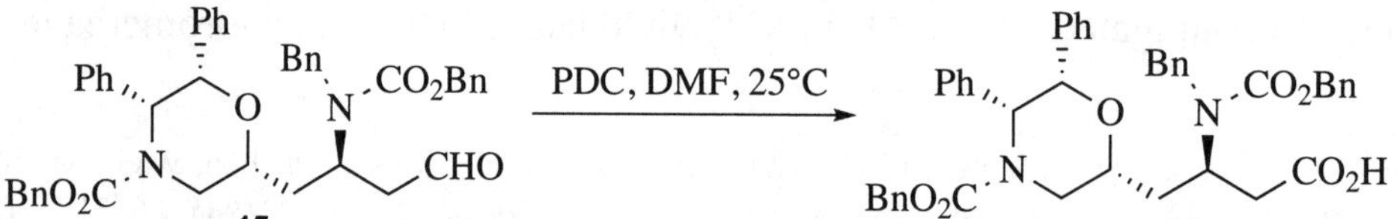

Alternatively, stabilized chromate esters have been prepared and found to be effective oxidizing agents. If the oxidizing power of these modified reagents can be generalized at all, it may be said

82. Taillier, C.; Gille, B.; Bellosta, V.; Cossy, J. *J. Org. Chem.* **2005,** *70*, 2097.
83. Jain, R.P.; Williams, R.M. *J. Org. Chem.* **2002,** *67*, 6361.

they behave similarly to PCC or PDC. Some examples are **47**,[84] **48**,[85] **49**,[86] and **50**,[87] and **51**.[88] The reasons for developing modified reagents can vary with the application. In many instances, they were developed to alleviate problems associated with isolation of products, or the use of large excess of the oxidizing reagent (as with Sarett and Collins oxidations). In other cases, the need for decreased acidity, for improved solubility in low-boiling aprotic organic solvents, or for increasing (or decreasing) the relative oxidizing power of the chromium reagent provided the driving force for developing the new reagent. For **47** and **49**, chromyl chloride (CrO_2Cl_2) was the chromium precursor rather than chromium trioxide. Another modification used amine bases other than pyridine, giving the PDC analog **48** and the PCC analog, **50**. Using a phosphine as the base, the bis(phosphine) 1,1-bis(triphenyl)phosphinomethane generated PDC analog **51**. Oxidation with modified chromium reagents is similar to PCC and PDC oxidations. Reagent **49** was developed by Corey and co-workers, and is of particular interest in that it can be used as a *catalytic* oxidizing agent (only 2% of **49** was required for the oxidation of cyclooctanone from cyclooctanol) in the presence of peroxyacetic acid (in dichloromethane).[86]

The functional group transforms for the oxidations in this section are

3.2.C. Dimethyl Sulfoxide and Related Oxidations[89]

Section 3.2.B.iv discussed the oxidation of alcohols using derivatives of the transition metal Cr(VI), which often required acidic conditions. Many alcohols, especially those bearing sensitive functional groups, require neutral oxidizing conditions. The development of PDC and related reagents helped to alleviate this problem, but PCC is an acidic reagent. There are cases where PDC or PCC give poor yields, especially with hindered alcohols. The key to Cr(VI) oxidations was formation of a chromate ester and removal of the acidic α-hydrogen (as

84. Sharpless, K.B.; Akashi, K. *J. Am. Chem. Soc.* **1975**, *97*, 5927.
85. López, C.; González, A.; Cossío, F.P.; Palomo, C. *Synth. Commun.* **1985**, *15*, 1197.
86. Corey, E.J.; Barrette, E.P.; Magriotis, P.A. *Tetrahedron Lett.* **1985**, *26*, 5855.
87. (a) Davis, H.B.; Sheets, R.M.; Brannfors, J.M.; Paudler, W.W.; Gard, G.L. *Heterocycles* **1983**, *20*, 2029; (b) Firouzabadi, H.; Iranpoor, N.; Kiaeezadeh, F.; Toofan, J. *Tetrahedron* **1986**, *42*, 719.
88. Cristau, H.-J.; Torreilles, E.; Morand, P.; Christol, H. *Tetrahedron Lett.* **1986**, *27*, 1775.
89. Moffatt, J.G. in *Oxidation*, Vol 2, Augustine, R.L.; Trecker, D.J. (Eds.), Marcel Dekker Inc., New York, **1971**, pp. 1-64.

in **52**). In one sense, the chromate ester behaved as a leaving group and removal of the hydrogen that is β to the chromium makes the reaction *conceptually* analogous to an E2 reaction (sec. 2.9.A). The problem with this analogy is, of course, that there is no direct evidence for removal of the hydrogen in **52** by an external base.[34,35] Removal of the hydrogen via an intramolecular process is also viable, as in **53**, and this has a close analogy to the syn-elimination discussed in section 2.9.C. Reagents other than Cr(VI) can be used to oxidize alcohols. It has been shown by a number of workers that DMSO-based reagents form a sulfoxonium intermediate, where the β-hydrogen can be removed, and dimethyl sulfide (Me$_2$S, abbreviated DMS) functions as the leaving group. This overall process leads to oxidation of alcohols and there are many variations of this reaction.

Dimethyl sulfoxide functions as both a solvent and a reactant for a variety of alcohol substrates. The nucleophilic oxygen can react with electrophilic centers to form a sulfoxonium salt, which gives ketones or aldehydes under neutral conditions. Kornblum et al. first observed this process, but with halides rather than with alcohols. He reported that reaction of α-halo ketones with DMSO at elevated temperatures gave good yields of the corresponding glyoxal

(an α-keto-aldehyde).[90] By this procedure, phenylacetyl bromide (**54**) was oxidized to **55** in 71% yield[90] in what has been called the **Kornblum aldehyde synthesis**.[91] The contact times were usually very short, and if the glyoxal could be removed from the reaction medium by distillation as it was formed the reaction was very efficient. It was often difficult to isolate high boiling glyoxals from DMSO, however. This transformation is not limited to α-halo ketones. Primary[92] and secondary[93] alkyl iodides or tosylates can be converted to aldehydes or ketones, although they are much less reactive than α-halo ketones. Reaction of 1-bromooctane with DMSO and NaHCO$_3$ at 100°C, for example, gave octanal in 74% yield after 5 minutes.[92] This outcome can be compared with the conversion of an alkyl tosylate to a ketone, which required temperatures of 150-170°C and the presence of an acid receptor such as sodium bicarbonate. Elimination is a serious side reaction, and attempts to oxidize secondary α-halo ketones often give poor yields.[94] The mechanism probably involves nucleophilic displacement of halides by

90. Kornblum, N.; Powers, J.W.; Anderson, G.J.; Jones, W.J.; Larson, H.O.; Levand, O.; Weaver, W.M. *J. Am. Chem. Soc.* **1957**, *79*, 6562.

91. Mundy, B.P.; Ellerd, M.G.; Favaloro Jr., F.G. *Name Reactions and Reagents in Organic Synthesis, 2nd ed.*, Wiley-Interscience, New Jersey, **2005**, pp. 376-377.

92. (a) Kornblum, N.; Jones, W.J.; Anderson, G.J. *J. Am. Chem. Soc.* **1959**, *81*, 4113; (b) Nace, H.R.; Monagle, J.J. *J. Org. Chem.* **1959**, *24*, 1792.

93. Baizer, M.M. *J. Org. Chem.* **1960**, *25*, 670.

94. (a) Jones, D.N.; Saeed, M.A. *J. Chem. Soc.* **1963**, 4657; (b) Iacona, R.N.; Rowland, A.T.; Nace, H.R. *J. Org. Chem.* **1964**, *29*, 3495.

DMSO to form an alkoxysulfoxonium salt, **56**. Removal of the α-proton, probably by DMSO, leads to the carbonyl product and DMS (which functions as the leaving group).[95]

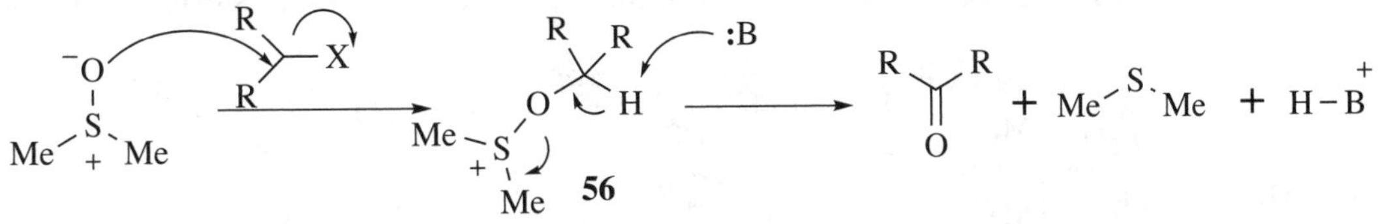

A similar type of oxidation was observed when alcohols reacted with DMSO. Traynelis and Hergenrother showed that benzylic and allylic alcohols were converted to the corresponding aldehyde in high yield by refluxing in DMSO, with air bubbling through the medium.[96] Air was the oxidant, DMSO the solvent and cinnamyl alcohol was oxidized in this manner to cinnamaldehyde in 90% yield.[96] Despite this example, air or oxygen is not required. A mixture of DMSO and several co-reagents react with alcohols to form a complex that leads to formation of aldehydes and ketones. There are several variations of this fundamental reaction involving DMSO. In each case, a key reagent is added to activate the DMSO-alcohol oxidation reaction. Several variations will be presented.

3.2.C.i. Swern Oxidation. This variation is quite common in synthesis and was discovered by Swern, who found that DMSO can be activated for the oxidation of alcohols by addition of trifluoroacetic anhydride.[97] The reaction is usually done in dichloromethane at temperatures below –30°C.

> *In the absence of a moderating solvent, admixture of DMSO and (CF₃CO)₂O proceeds*
> *explosively.*

The initially formed trifluoroacetyl derivative (**57**) is stable below –°C but at higher temperatures undergoes a **Pummerer rearrangement** to give **58**.[98] At lower temperatures, the initially formed sulfoxonium intermediate **57** reacts with an alcohol to form **58**, and then loses DMS to form the ketone or aldehyde. This mechanism is common for virtually all of the oxidations presented in this section, but the reagent used to generate **56** varies. An important and interesting feature of Swern oxidation is the low sensitivity to steric encumbrance. Oxidation of the moderately hindered alcohol 2,4-dimethyl-3-pentanol gave 2,4-dimethyl-3-pentanone in 86% yield.[97] Oxidation of a primary alcohol to the corresponding aldehyde is facile. Amines are commonly added to facilitate decomposition of the initially formed complex. Oxidation of 1-decanol gave a

95. Hunsberger, I.M.; Tien, J.M. *Chem. & Ind.* *1959*, 88.
96. Traynelis, V.J.; Hergenrother, W.L. *J. Am. Chem. Soc.* *1964, 86,* 298.
97. Mancuso, A.J.; Swern, D. *Synthesis 1981,* 165.
98. (a) Sharma, A.K.; Swern, D. *Tetrahedron Lett.* *1974,* 1503; (b) Sharma, A.K.; Ku, T.; Dawson, A.D.; Swern, D. *J. Org. Chem.* *1975, 40,* 2758; (c) Pummerer, R. *Berichte 1910, 43,* 1401.

56% yield of decanal when triethylamine was added. Addition of diisopropylamine rather than triethylamine increased the yield to 81%.[99] Swern oxidation can be used in highly functionalized molecules, and as mentioned it tolerates alcohols that are somewhat sterically hindered.

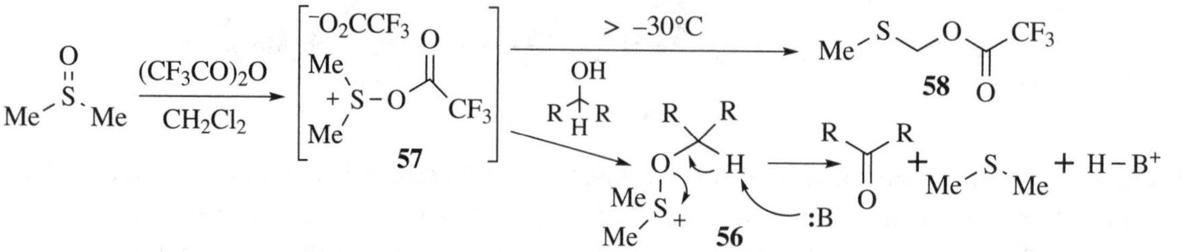

Swern also found that oxalyl chloride activates DMSO for the oxidation of alcohols.

Oxalyl chloride and DMSO react violently and exothermically at ambient temperatures.

The resulting reagent is superior to the DMSO-trifluoroacetic anhydride reagent.[100] The reaction probably proceeds via intermediate complex **59**, which is unstable above the preferred reaction temperature of -60°C.

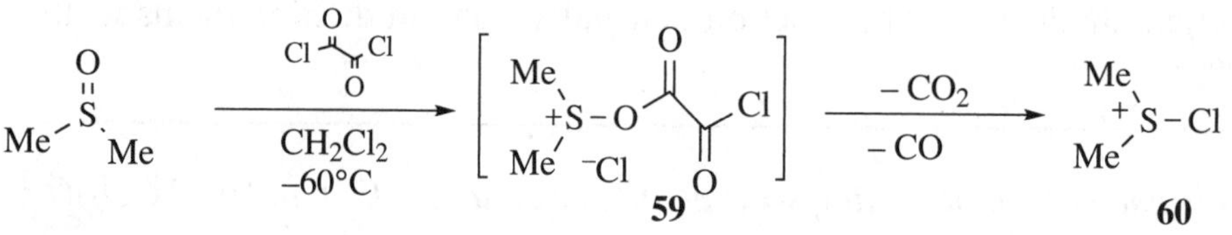

The reaction proceeds with loss of both carbon dioxide and carbon monoxide to give chlorosulfonium salt **60**.[97] When **60** is generated in the presence of an alcohol, **56** is formed and deprotonation gives the carbonyl and dimethyl sulfide. Cyclododecanol was oxidized to cyclododecanone in 97% yield with this reagent, and hex-2-en-1-ol was oxidized to the conjugated aldehyde 2-hexenal in good yield.[97] Oxidation with DMSO and oxalyl chloride is often called **Swern oxidation**,[101] although this term can also be used with the DMSO-trifluoroacetic anhydride oxidation. Note that an odorless Swern oxidation protocol has been developed using dodecyl methyl sulfide.[102]

A typical synthetic use involving oxalyl chloride and DMSO is the conversion of the diol unit in **61** to keto-aldehyde **62** in 74% yield, taken from Ito, Iguchi, and co-worker's synthesis

99. Huang, S.L.; Omura, K.; Swern, D. *Synthesis* **1978**, 297.
100. (a) Omura, K.; Swern, D. *Tetrahedron* **1978**, *34*, 1651; (b) Mancuso, A.J.; Huang, S.-L.; Swern, D. *J. Org. Chem.* **1978**, *43*, 2480.
101. The Merck Index, 14th ed., Merck & Co., Inc., Whitehouse Station, New Jersey, **2006**, p ONR-92.
102. Ohsugi, S-i.; Nishide, K.; Oono, K.; Okuyama, K.; Fudesaka, M.; Kodama, S.; Node, M. *Tetrahedron* **2003**, *59*, 8393.

of (+)-tricycloclavulone.[103] In a synthesis of phorboxazole A by Williams and co-workers,[104] trifluoroacetic anhydride and DMSO was used to oxidize the primary alcohol unit in **63** to aldehyde **64**, in >90% yield. In many synthetic sequences, one will find the term Swern oxidation or simply Swern over a reaction arrow rather than giving the reagents. This is quite common, forcing the reader to recognize certain named reactions.

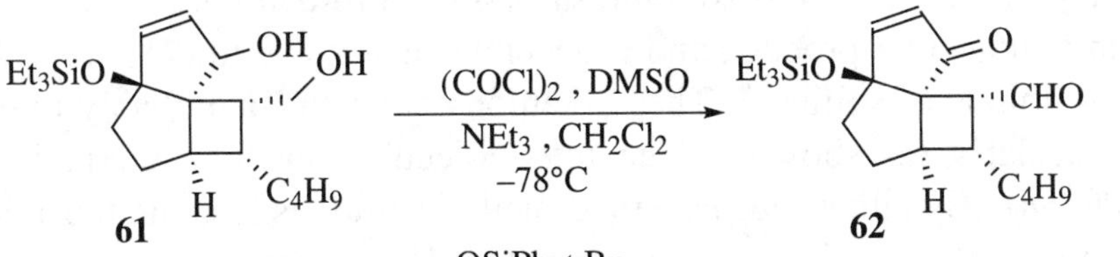

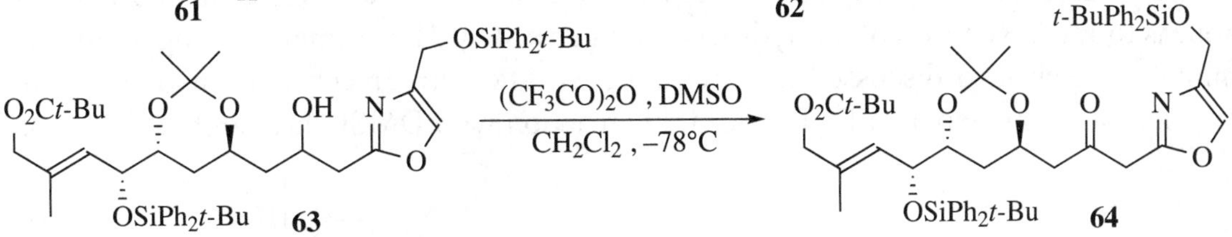

The functional group transform for DMSO-type oxidations is

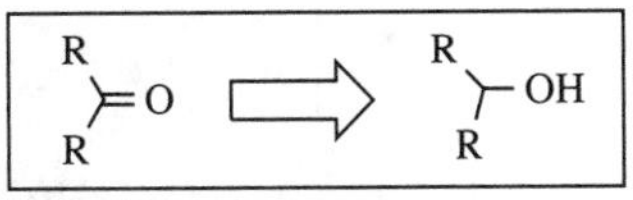

3.2.C.ii. Moffatt Oxidation. Another useful DMSO based oxidation predates the Swern oxidation, and uses a mixture of DMSO and dicyclohexylcarbodiimide (DCC, **66**) in the presence of an acid catalyst to generate an intermediate such as **56**.[105] This reaction is known as **Moffatt oxidation**,[105] and an example is the conversion of **65** to **67** (71% yield), used in Smith and co-worker's synthesis of (−)-penitrem D.[106]

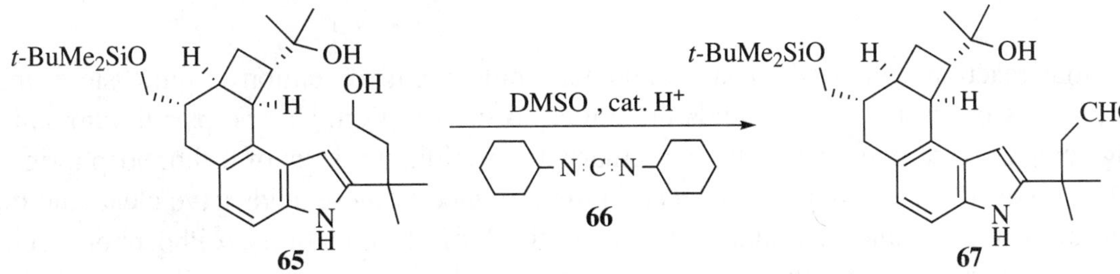

103. Ito, H.; Hasegawa, M.; Takenaka, Y.; Kobayashi, T.; Iguchi, K. *J. Am. Chem. Soc. 2004, 126*, 4520.
104. Williams, D.R.; Kiryanov, A.A.; Emde, U.; Clark, M.P.; Berliner, M.A.; Reeves, J.T. *Angew. Chem. Int. Ed. 2003, 42*, 1258.
105. Pfitzner, K.E.; Moffatt, J.G. *J. Am. Chem. Soc. 1965, 87*, 5661.
106. Smith III, A.B.; Kanoh, N.; Ishiyama, H.; Minakawa, N.; Rainier, J.D.; Hartz, R.A.; Cho,Y.S.; Cui, H.; Moser, W.H. *J. Am. Chem. Soc. 2003, 125*, 8228.

Moffatt provided a mechanistic rationale for this oxidation,[107,105] where the driving force of the reaction is formation and separation of the highly insoluble dicyclohexylurea (**70**). The initial reaction of DMSO and DCC (**66**) forms sulfoxonium intermediate **68**, which now contains a urea leaving group (see **69**). Alkoxysulfonium intermediate **69** is clearly related to Swern intermediate **57**. When the alcohol attacks the electrophilic sulfur atom, dicyclohexylurea (**70**) is displaced to generate the requisite sulfoxonium salt **56**. Deprotonation of the sulfoxonium salt is probably an intermolecular process, and it generates a sulfur ylid (**71**; sec. 8.8.B) that is stabilized by the *d*-orbitals of sulfur.[108] The carbanion center in **71** probably removes the α-hydrogen intramolecularly (as shown). The intermolecular reaction alternative is also possible (analogous to **56**). By either mechanism, dimethyl sulfide is lost and the aldehyde or ketone is formed. Removal of the hydrogen as suggested by **71** is formally analogous to the syn-elimination mechanism discussed in section 2.9.C. An alternative mechanism involving complex formation of the alcohol and DCC prior to its reaction with DMSO has not been observed.[109]

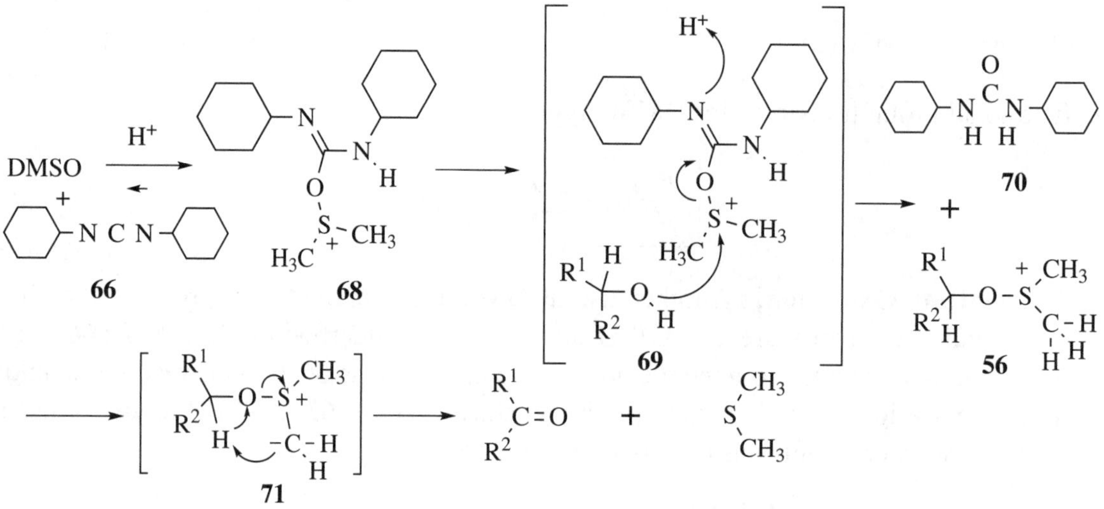

The initial reaction with DMSO is usually fast, but overall oxidation is quite slow unless mineral acids or strong organic acids are added to the reaction.[110] The pyridinium salts of strong acids are also good catalysts, particularly the pyridinium salts of orthophosphoric acid and trifluoroacetic acid. Reagents formed in the presence of these acids gave clean and rapid oxidation of primary and secondary alcohols in the Moffatt oxidation. *o*-Phosphoric acid,[111] dichloroacetic acid, and pyridinium trifluoroacetate are the most common additives used in this oxidation.[112] Moffatt oxidation has several problems associated with it. One is formation of the

107. Reference 89, p 12.
108. (a) Cilento, G. *Chem. Rev.* **1960**, *60*, 147; (b) Johnson, C.R.; Phillips, W.G. *J. Am. Chem. Soc.* **1969**, *91*, 682.
109. Reference 89, pp 13-15.
110. Reference 89, p 6.
111. Pfitzner, K.E.; Moffatt, J.G. *J. Am. Chem. Soc.* **1963**, *85*, 3027.
112. Reference 89, p 7.

urea product **70**, which can be very difficult to separate from the other products. Treatment with oxalic acid is the most common method employed for the removal of the last vestiges of the dicyclohexylurea by-product.[113] Oxidation of homoallylic alcohols is sometimes accompanied by isomerization of the double bond into conjugation with the carbonyl group. This can be minimized or prevented by addition of pyridinium trifluoroacetate. A threefold excess of DCC and an excess of DMSO are usually required for Moffatt oxidation, and these must be removed from the product. The DMSO need not be the solvent, however, and addition of a co-solvent such as ethyl acetate often leads to better results. Despite the drawbacks, Moffatt oxidation is useful in many synthetic applications. Tertiary alcohols are, of course, resistant to Moffatt oxidation but dehydration has been observed in such systems,[114] to give the alkene and regenerate DMSO.[114] Some steroidal allylic alcohols dehydrate to a heteroannular diene under these conditions.[115]

3.2.C.iii. Other DMSO Oxidations. Albright and Goldman developed a reagent using DMSO and acetic anhydride that formed an active sulfoxonium complex.[116] In the initial work, yohimbine was oxidized to the ketone (yohimbinone) in 85% yield, at ambient temperatures in 24 h.[116] As with DCC, DMSO initially reacts with acetic anhydride to form sulfoxonium salt **72**, which then reacts with the alcohol. Acetate is the leaving group attached to sulfur, and reaction with an alcohol generates **56**. The acetate anion serves as the base in this case to

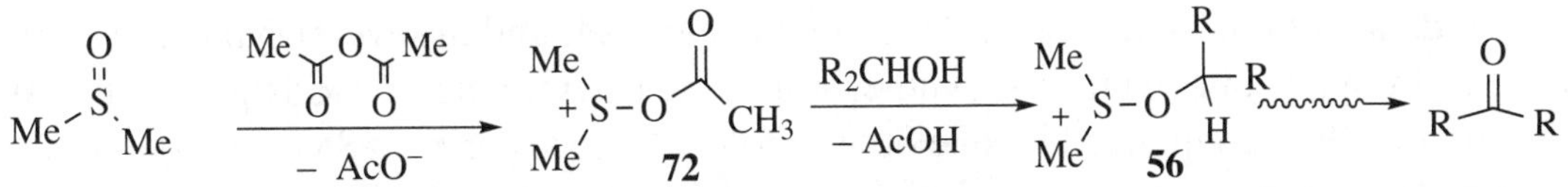

remove the α-proton and generate the ketone or aldehyde. A major advantage of this method is the production of water soluble by-products, and it is especially effective with hindered alcohols. An example, taken from Jeon, Kim and co-worker's synthesis of valienamine,[117] is the oxidation of the secondary alcohol **73** to give a 94% yield of **74** in the presence of the dithioacetal unit. With unhindered alcohols the products are often acetate esters and thiomethoxymethyl esters.

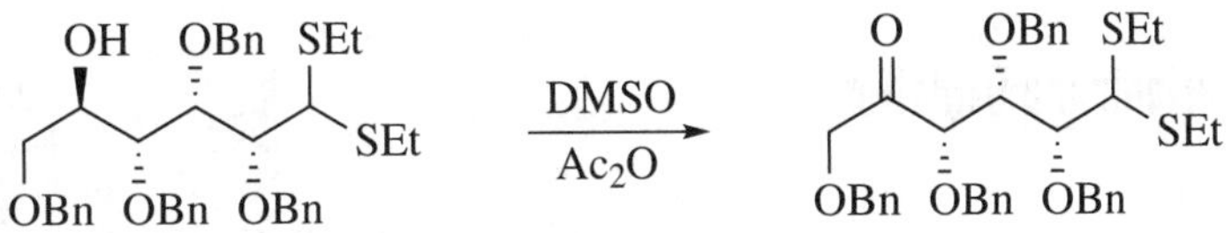

113. Reference 89, p 44.
114. Reference 89, p 38.
115. Reference 89, p 39.
116. Albright, J.D.; Goldman, L. *J. Am. Chem. Soc.* **1967**, *89*, 2416.
117. Chang, Y-K.; Lee, B.-Y.; Kim, D.J.; Lee, G.S.; Jeon, H.B.; Kim, K.S. *J. Org. Chem.* **2005**, *70*, 3299.

Onodera et al. developed a reagent for the oxidation of alcohols using DMSO and phosphorus pentoxide. This method is very efficient for the oxidation the secondary alcohols units found in carbohydrates.[118] The alcohol moiety in **75**, for example, was converted to ketone **76** in 65% yield.[118] The initial intermediate was probably the sulfoxonium derivative (**77**), which was attacked by the alcohol. Deprotonation of the alkoxy-sulfoxonium ion in the usual manner gave the ketone. Virtually all examples of oxidation with this reagent involved carbohydrates.

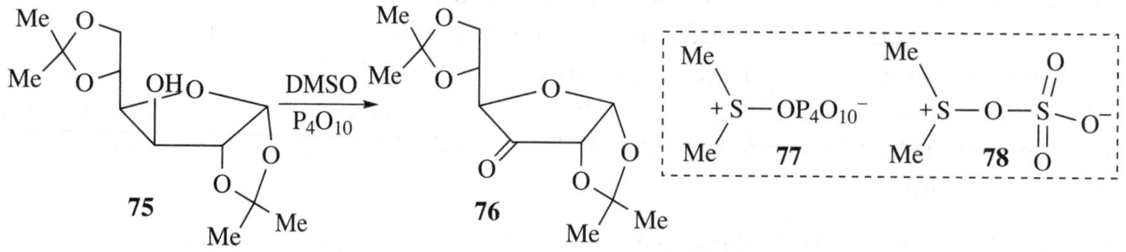

The complex formed by pyridine-sulfur trioxide was shown to activate DMSO for oxidation, in the presence of a base such as triethylamine.[119] The presumed intermediate is **78**, which reacts with the alcohol to give the usual complex **56**, and deprotonation by triethylamine gives the ketone or aldehyde. Alcohols with different stereochemical and conformational properties can exhibit great differences in their rate of oxidation, which leads to differences in the selectivity for the oxidation of related alcohols. This modification is found in several syntheses of natural products. In Wipf and Spencer's synthesis of tuberostemonine,[120] oxidation of the allylic alcohol unit in **79** to conjugated ketone **80** was found to be superior (>88% yield) to oxidation with tetrapropylammonium perruthenate (see 3.2.F.i), when done on large scale.

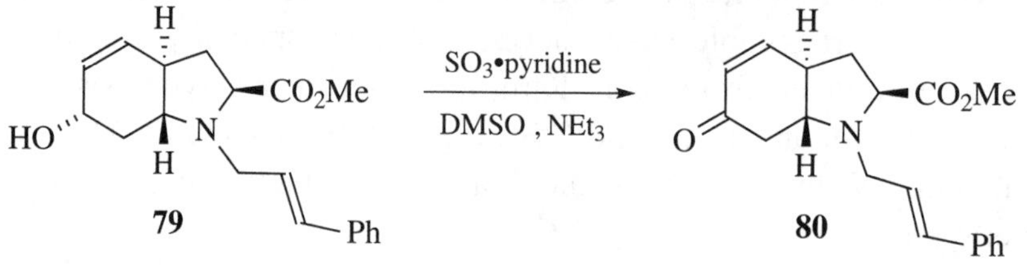

3.2.D. Dess-Martin Periodinane Oxidation

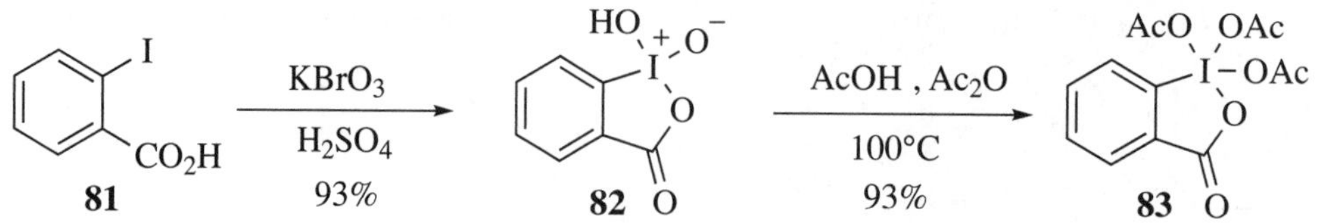

---

118. Onodera, K.; Hirano, S.; Kashimura, N. *J. Am. Chem. Soc.* **1965**, *87*, 4651.
119. Parikh, J.R.; von E. Doering, W. *J. Am. Chem. Soc.* **1967**, *89*, 5505.
120. Wipf, P.; Spencer, S.R. *J. Am. Chem. Soc.* **2005**, *127*, 225

A mild oxidizing reagents has been developed that contains a hypervalent iodine.[121] Dess and Martin showed that 2-iodobenzoic acid (**81**) reacted with KBrO$_3$ in sulfuric acid to give a 93% yield of **82**. Subsequent heating to 100°C with acetic anhydride and acetic acid gave **83** in 93% yield, the so-called **Dess-Martin periodinane** [1,1,1-tris(acetyloxy)-1,1-dihydro-1,2-beniodoxo-3-9(1*H*)-one].[122]

> The Dess-Martin reagent can be shock sensitive under some conditions, and it explodes above 200°C.[123] An improved procedure for its preparation is available.[124]

Alcohols are readily oxidized to ketones or aldehydes, as in the oxidation of cyclohexanol to cyclohexanone in 90% yield, in dichloromethane at 25°C.[122] In this particular reaction, an intermediate was trapped and its structure shown to be **84**.[122] The reagent appears to have an indefinite shelf-life in a sealed container, but hydrolysis occurs upon long-term exposure to atmospheric moisture. Hypervalent iodine reagents are well-known oxidizing agents,[125] and oxidation of the α-position of methyl ketones to give the corresponding α-hydroxyketone is an important application.[126] An inert atmosphere is not required, but addition of catalytic amount of trifluoroacetic acid accelerates the oxidation. The catalyst is not required, however. It has also been shown that a Dess-Martin intermediate can be generated in aqueous media and used for oxidation.[127] Dess and Martin prepared several other periodinane reagents that also oxidizes alcohols,[128] as well as a chiral hypervalent organoiodinane derived from a chiral binaphthyl.[129]

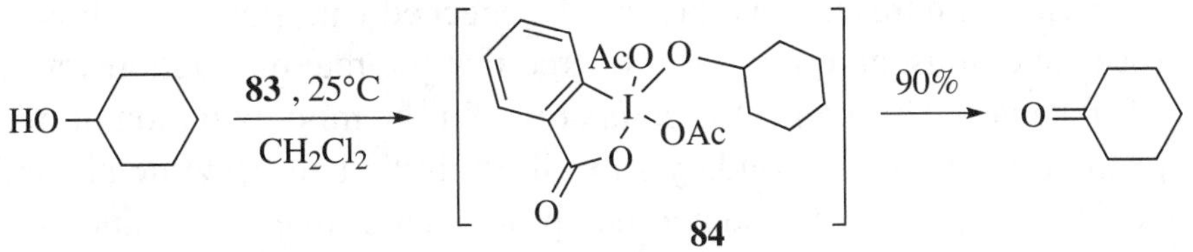

121. For reactions of hypervalent iodine reagents in the synthesis of heterocyclic compounds, see Prakash, O.; Singh, S.P. *Aldrichimica Acta* **1994**, *27*, 15.
122. (a) Dess, D.B.; Martin, J.C. *J. Org. Chem.* **1983**, *48*, 4155. (b) See The Merck Index, 14th ed., Merck & Co., Inc., Whitehouse Station, New Jersey, **2006**, p ONR-23.
123. Plumb, J.B.; Harper, D.J. *Chem. Eng. News* **1990**, *68* (No. 29, July 16), p 3.
124. (a) Ireland, R.E.; Liu, L.J. *J. Org. Chem.* **1993**, *58*, 2899. Also see, (b) Frigerio, M.; Santagostino, M.; Sputore, S. *J. Org. Chem.* **1999**, *64*, 4537.
125. Moriarty, R.M.; Prakash, O. *Acc. Chem. Res.* **1986**, *19*, 244.
126. (a) Moriarty, R.M.; John, L.S.; Du, P.C. *J. Chem. Soc. Chem. Commun.* **1981**, 641; (b) Moriarty, R.M.; Gupta, S.; Hu, H.; Berenschot, D.R.; White, K.B. *J. Am. Chem. Soc.* **1981**, *103*, 686; (c) Moriarty, R.M.; Hu, H.; Gupta, S.C. *Tetrahedron Lett.* **1981**, *22*, 1283.
127. Meyer, S.D.; Schreiber, S.L. *J. Org. Chem.* **1994**, *59*, 7549.
128. Dess, D.B.; Martin, J.C. *J. Am. Chem. Soc.* **1991**, *113*, 7277.
129. Ochiai, M.; Takaoka, Y.; Masaki, Y.; Nagao, Y.; Shiro, M. *J. Am. Chem. Soc.* **1990**, *112*, 5677.

There are many synthetic applications in the recent literature. The oxidation of a primary alcohol unit in a substrate that can tolerate only a narrow range of reaction conditions, or is part of a highly functionalized molecule is common.[130] In Paquette and Chang's synthesis of the polyol domain of amphidinol 3,[131] the primary alcohol unit in **85** was oxidized to an aldehyde in **86**, in quantitative yield (BOM = benzyloxymethyl ether). Secondary alcohols are also oxidized under mild conditions.

3.2.E. Oxidation with TEMPO and Oxammonium Salts

Oxammonium salts, such as **87** (**Bobbitt's reagent**),[132] are useful oxidizing agents for the selective oxidation of alcohols to aldehydes or ketones. Such salts can be generated catalytically from a nitroxide in the presence of a secondary oxidation procedure, either chemical or electrochemical,[133] or with two equivalents of acid and two equivalents of a nitroxide.[134] When **87** was mixed with acetylenic alcohol **88** in dichloromethane, aldehyde **89** was isolated in 93% yield.[135] The reaction is readily monitored as the initial yellow slurry changes to a white slurry and the presence of unreacted oxidant can be checked with starch.[135] It is not necessary to use anhydrous conditions, and it was discovered that the rate of reaction was enhanced by the presence of silica gel. This reagent is compatible for the mild oxidation of many alcohols, including aliphatic primary and secondary as well as allylic and benzylic alcohols, although nitrogen-containing compounds are sometimes problematic unless pyridine is added to the reaction. Note that the stable free radical TEMPO (**91**) also oxidizes alcohols in the presence of an acid catalyst. An example is the conversion of colletodiol (**90**) to grahamimycin A (**92**), in 76% yield in a synthesis reported by O'Doherty and Hunter.[136]

130. Evans, D.A.; Sheppard, G.S. *J. Org. Chem.* **1990**, *55*, 5192.
131. Paquette, L.A.; Chang, S.-K. *Org. Lett.* **2005**, *7*, 3111.
132. (a) Bobbitt, J.M.; Flores, M.C.L. *Heterocycles* **1988**, *27*, 509. Also see, (b) Golubev, V.A.; Zhdanov, R.I.; Gida, V.M.; Rozantsev, É.G. *Bull. Acad. Sci. USSR, Chem. Sci.* **1971**, *20*, 768; (c) Golubev, V.A.; Miklyush, R.V. *Russ. J. Org. Chem.* **1972**, *8*, 1376.
133. (a) de Nooy, A.E.J.; Besemer, A.C.; van Bekkum, H. *Synthesis* **1996**, 1153; (b) Anelli, P.L.; Montanari, F.; Quici, S. *Org. Synth. Coll. Vol. VIII* **1993**, p 367.
134. Ma, Z.; Bobbitt, J.M. *J. Org. Chem.* **1991**, *56*, 6110.
135. Bobbitt, J.M. *J. Org. Chem.* **1998**, *63*, 9367.
136. Hunter, T.J.; O'Doherty, G.A. *Org. Lett,* **2002**, *4*, 4447.

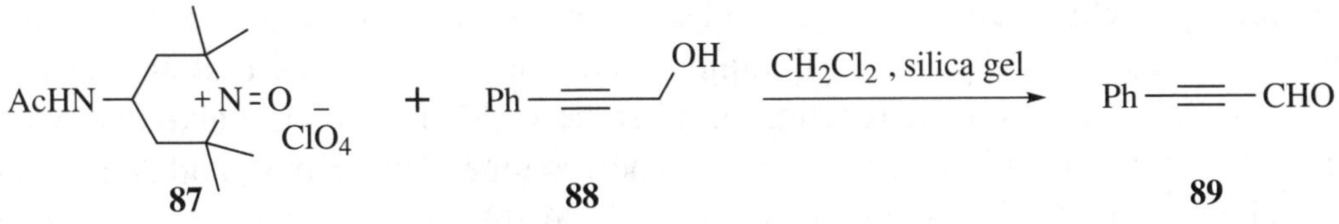

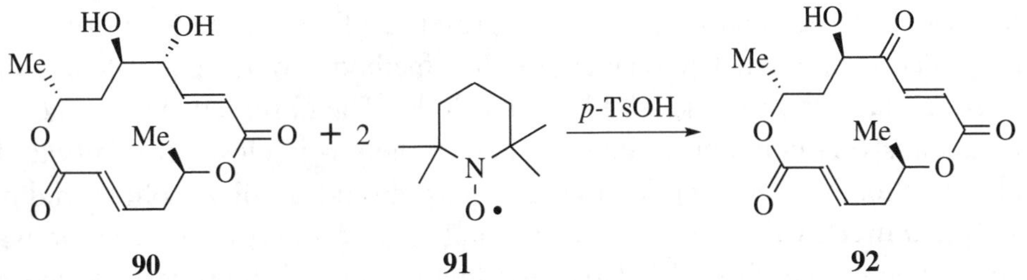

3.2.F. Alternative Metal Compounds As Useful Oxidizing Agents

The oxidation of alcohols is not limited to chromium derivatives, DMSO derivatives, or hypervalent iodine compounds. There are several metal-based reagents that are very effective, particularly with sensitive functionality. This section will examine several of the more important reagents.

3.2.F.i. Tetrapropylammonium Perruthenate-Ley Oxidation. Ruthenium derivatives have been used for oxidation reactions. A particularly useful one is tetrapropylammonium perruthenate (TPAP), introduced by Ley for the oxidation of alcohols and now called **Ley oxidation.**[137] It is a very mild oxidizing agent that is compatible with many sensitive functional groups. In a synthesis of (−)-cylindricine C, for example, Hsung and co-workers oxidized the secondary alcohol unit in **93** with TPAP to give an excellent yield of ketone **94**, without disturbing the α-chloro group.[138]

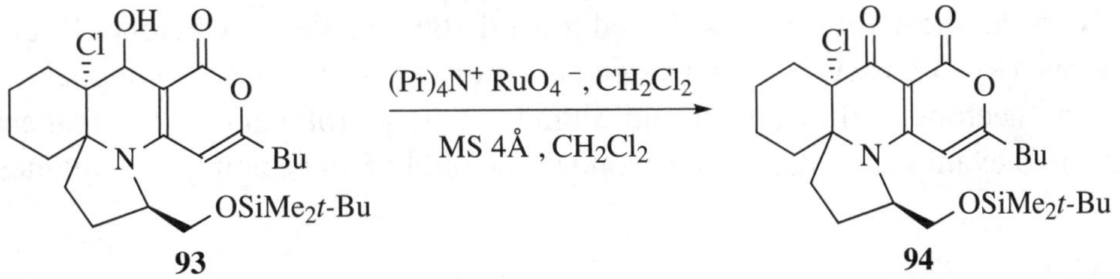

3.2.F.ii. Oppenauer Oxidation. In oxidation reactions involving both Cr(VI) and DMSO reagents, the alcohol is converted to a complex and the α-hydrogen of an alcohol removed as an acid. A classical and alternative method for the oxidation of alcohols focuses on the reversible

137. (a) Griffith, W.P.; Ley, S.V.; Whitcombe, G.P.; White, A.D. *J. Chem. Soc. Chem. Commun.* **1987**, 1625; (b) Griffith, W.P.; Ley, S.V. *Aldrichimica Acta* **1990**, *23*, 13-19; (c) Ley, S.V.; Norman, J.; Griffith, W.P.; Marsden, S.P. *Synthesis* **1994**, 639.
138. Swidorski, J.J.; Wang, J.; Hsung, R.P. *Org. Lett.* **2006**, *8*, 777.

reaction between ketones and metal alkoxides, which is especially effective when the metal is aluminum[139] Reversibility in the aluminum alkoxide reaction was first demonstrated by Verley[140] and Ponndorf[141] for the reaction of a ketone with an aluminum alkoxide, which led to formation of a new aluminum alkoxide and a new ketone. Oppenauer oxidized unsaturated steroidal alcohols using aluminum triisopropoxide [Al(Oi-Pr)$_3$] in acetone, in 1937.[142] The acetone acts as a hydrogen acceptor, and the presence of excess acetone drives the reaction toward the oxidation product. Oppenauer used this method to oxidize the alcohol unit of Δ^5-3-hydroxy steroids such as **95** to the Δ^4-3 ketone (**96**).[142] The observed shift of the double bond into conjugation is very common when the alkene moiety is in close proximity to the reactive center. This reaction has been applied to many steroids and simple alcohols, and has come to be called **Oppenauer oxidation**.[139,143] An uncatalyzed version of this reaction was reported to proceed in supercritical fluids.[144] Note that this reaction is the reverse of the **Meerwein-Ponndorf-Verley reduction**[145] (see section 4.9.I).

The reaction proceeds via an aluminum coordination complex (**97**), with transfer of a hydrogen atom from one alkoxy carbon to another to give **100**.[139,146] Since **97** is a Lewis acid it reacts with the available Lewis base (the carbonyl oxygen of the ketone) to form an ate complex (**98**), and this is followed by transfer of H$_a$ from the alkoxide carbon of **98** to the alkoxy unit from the ketone carbon (giving **99**) via a six-center transition state. Transfer of H$_a$ between **98** and **99** is a reversible process. Just as **97** and **98** are in equilibrium, so the new *ate* complex **99** is in equilibrium with a new alkoxide (**100**) and a new ketone. If an excess of the original ketone (RCOR$_1$) is added, the equilibrium is shifted toward **100**, and the new ketone (R$_2$COR$_3$). The original ketone (RCOR$_1$) acts as a **hydrogen acceptor** in this process. If the goal is oxidation of an alcohol, acetone is the solvent and aluminum triisopropoxide is the reagent. In the previously cited example, oxidation of secondary alcohol **95** in acetone gave ketone **96**.

139. Djerassi, C. *Org. React.* **1951**, *6*, 207.
140. Verley, A. *Bull. Chim. Soc. Fr.* **1925**, *37*, 537.
141. Ponndorf, W. *Angew. Chem.* **1926**, *39*, 138.
142. Oppenauer, R.V. *Rec. Trav. Chim.* **1937**, *56*, 137.
143. (a) The Merck Index, 14th ed., Merck & Co., Inc., Whitehouse Station, New Jersey, **2006**, p ONR-68; (b) de Graauw, C.F.; Peters, J.A.; van Bekkum, H.; Huskens, J. *Synthesis* **1994**, 1007.
144. Sominsky, L.; Rozental, E.; Gottlieb, H.; Gedanken, A.; Hoz, S. *J. Org. Chem.* **2004**, *69*, 1492.
145. The Merck Index, 14th ed., Merck & Co., Inc., Whitehouse Station, New Jersey, **2006**, p ONR-59.
146. (a) Woodward, R.B.; Wendler, N.L.; Brutschy, F.J. *J. Am. Chem. Soc.* **1945**, *67*, 1425; (b) Lutz, R.E.; Gillespie, Jr., J.S. *Ibid.* **1950**, *72*, 344; (c) Von E. Doering, W.; Young, R.W. *Ibid;* **1950**, *72*, 631.

97 **98** **99** **100**

Oppenauer oxidation of saturated alcohols is often sluggish, but this problem can be overcome by changing the structure of the metal alcoholate, the hydrogen acceptor or by changing the solvent. Aluminum tri-*tert*-butoxide, triisopropoxide, or tri-*n*-propoxide are the most commonly used alkoxides, but aluminum triphenoxide gave better yields with saturated hydroxy steroids.[147] Acetone (often with benzene as a co-solvent) is the most commonly used hydrogen acceptor, although cyclohexanone in a solution of toluene or xylene is also used. The latter conditions allow higher reaction temperatures and shorter reaction times.[97] Ideally, the ketone that serves as the hydrogen acceptor should have a high reduction potential (sec. 3.1), but a large excess of the ketone can compensate for a low reduction potential, as with acetone.[148] Benzophenone is used occasionally because it is resistant to condensation reactions (Chaps. 8 and 9) that can compete.

Nonsteroidal alcohols can be oxidized by this method,[149,150] but Oppenauer oxidation of primary alcohols tends to give a low yield of the aldehyde. Although oxidation of benzyl alcohol gave about 60% of benzaldehyde, geraniol (**101**) gave only 35% of geranial (**102**).[151] Schinz and co-workers found that aldehydes are effective hydrogen acceptors when used for the oxidation of primary alcohols to aldehydes. If the boiling point of that aldehyde was ≥50°C *higher* than the expected product, the aldehyde product could be distilled as it was formed.[152,151b]

Two common side reactions were mentioned previously. One is migration of a nonconjugated double bond into conjugation, and the other is condensation of the aldehyde product with the

147. (a) Reference 139, p 226; (b) Reich, H.; Reichstein, T. *Arch. Inter. Pharmacodynamie* **1941**, *65*, 415; [*Chem. Abstr. 35:* 55262 **1941**].
148. (a) Adkins, H.; Cox, F.W. *J. Am. Chem. Soc.* **1938**, *60*, 1151; (b) Cox, F.W.; Adkins, H. *Ibid.* **1939**, *61*, 3364; (c) Baker, R.H.; Adkins, H. *Ibid.* **1940**, *62*, 3305; (d) Adkins, H.; Elofson, R.M.; Rossow, A.G.; Robinson, C.C. *Ibid.* **1949**, *71*, 3622.
149. (a) Reference 139, p 211; (b) Ungnade, H.E.; Ludutsky, A. *J. Am. Chem. Soc.* **1947**, *69*, 2629.
150. (a) Reference 139, p 219; (b) Adams, R.; Hamlin, Jr., K.E. *J. Am. Chem. Soc.* **1942**, *64*, 2597.
151. (a) Yamashita, M.; Matsumura, R. *J. Chem. Soc. Jpn.* **1943**, *64*, 506 [*Chem. Abstr. 41:* 3753g **1947**]; (b) Reference 139, p 222.
152. (a) Schinz, H.; Lauchenauer, A.; Jeger, O.; Rüegg, R. *Helv. Chim. Acta* **1948**, *31*, 2235; (b) Rüegg, R.; Jeger, O. *Ibid.* **1948**, *31*, 1753; (c) Lauchenauer, A.; Schinz, H. *Ibid.* **1949**, *32*, 1265.

carbonyl hydrogen acceptor (this is an acid catalyzed **aldol condensation**, sec. 9.4.A). In recent years, metal catalysts other than aluminum have diminished the occurrence of these side reactions, although other aluminum catalysts have also been developed. Maruoka, for example, reported that the modified aluminum catalyst **103** was highly effective for Oppenauer oxidation of alcohols under mild conditions.[153] An example of an alternative metal catalyst is the cyclopentadienyl zirconium reagent (Cp_2ZrH_2) developed by Ishii et al.[154] that was shown to be an effective catalyst in the Oppenauer oxidation using a 1:1 ratio of alcohol to hydrogen acceptor. The use of this catalyst led to excellent yields of aldehydes from primary alcohols, but heating in an autoclave without solvent was required despite being catalytic in the zirconium reagent. Geraniol (**101**) was oxidized to geranial (**102**) in 92% yield by this method, using benzophenone as the hydrogen acceptor. Magnesium-catalyzed Oppenauer oxidations have been observed in Grignard alkylation reactions with aldehydes (sec. 8.4.C), where direct oxidation of the magnesium alkoxide generated during the reaction prior to hydrolysis is possible.[155]

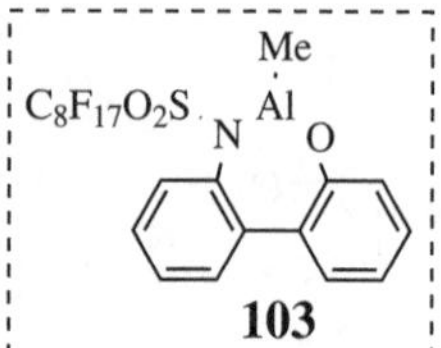

3.2.F.iii. Oxidation with Manganese Dioxide.

Although manganese dioxide (MnO_2) is a common form of Mn(IV), it is rather insoluble in most organic solvents. Manganese dioxide is the usual end-product of permanganate oxidations in basic solution (sec. 3.5.A). The reduction potential of MnO_2 is 1.208 volts for the following reaction.[17]

$$MnO_2 + 4\,H^+ + 2\,e^- \rightarrow Mn^{2+} + 2\,H_2O \qquad\qquad 1.208\ V$$

Manganese dioxide is capable of oxidizing alcohols to ketones or aldehydes, and the reaction proceeds via a radical intermediate (see below), producing MnO (which is Mn^{2+}) as the by-product. Manganese dioxide oxidizes primary and secondary alcohols to the aldehyde or ketone, respectively, in neutral media.[156] Ball, et al discovered this reaction, by precipitating manganese dioxide and then converting vitamin A (**104**) to retinal (**105**) in 80% yield.[157]

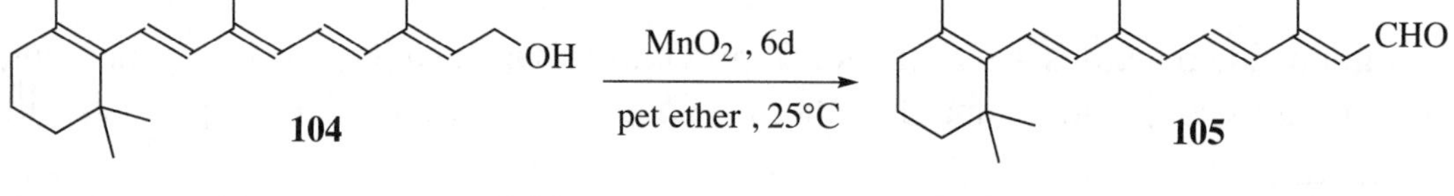

There are several ways to prepare this reagent, and its oxidizing power is strongly influenced by the method of preparation.[156] One of the more common methods involves precipitation of MnO_2 from a warm aqueous solution of $MnSO_4$ and $KMnO_4$ at a particular pH. The precipitated

153. Ooi, T.; Otsuka, H.; Miuraa, T.; Ichikawa, H.; Maruoka, K. *Org. Lett.* **2002**, *4*, 2669.
154. Ishii, Y.; Nakano, T.; Inada, A.; Kishigami, Y.; Sakurai, K.; Ogawa, M. *J. Org. Chem.* **1986**, *51*, 240.
155. Byrne, B.; Karras, M. *Tetrahedron Lett.* **1987**, *28*, 769.
156. (a) Fatiadi, A.J. *Synthesis* **1976**, 65, 133; (b) Evans, R.M. *Quart. Rev. Chem. Soc. (London)* **1959**, *13*, 61.
157. Ball, S.; Goodwin, T.W.; Morton, R.A. *Biochem. J.* **1948**, *42*, 516.

reagent is then activated by heating to 100-200°C or higher[158] for several hours. The solvent used for the oxidation is important, since the reaction proceeds by coordination of both substrate and reagent to the MnO_2. The solvent can influence the degree of adsorption and desorption of the alcohol on the manganese dioxide. If a primary or secondary alcohol is used as the solvent, competition for adsorption sites with the alcohol will diminish the yield of oxidation products. Many other solvents can been used, including alkanes (petroleum ether, cyclopentane), chlorinated hydrocarbons (CH_2Cl_2, $CHCl_3$), ethers (diethyl ether, THF), acetone, acetonitrile, DMSO, DMF, and pyridine. Reaction times range from a few minutes for allylic alcohols to several hours for saturated alcohols. The oxidation of **104** required six days, an indication that activating the manganese dioxide prior to addition of organic substrate is important.

Both Goldman[159] and Henbest et al.[160] proposed a radical intermediate in oxidations using manganese dioxide. Goldman's mechanism is shown for reaction with benzyl alcohol, and it involves adsorption of the alcohol on manganese dioxide (see **106**), followed by formation of a coordination complex (**107**). Coordination in this manner allows electron transfer, which generates a radical (**108**) in a process that is accompanied by reduction of Mn(IV) to Mn(III). A second electron transfer generates the carbonyl product adsorbed on $Mn(OH)_2$ as in **109**. The product is desorbed, with loss of water, to complete the oxidation. Hall and Story proposed an alternative ionic mechanism for oxidation, which invoked formation of a manganate ester.[161]

$$
\begin{array}{ccccccccc}
\underset{\textbf{106}}{\underset{\text{(Adsorption)}}{\overset{MnO_2}{\rightleftharpoons}}} & & \textbf{107} & \rightleftharpoons & \textbf{108} & \rightarrow & \textbf{109} & \rightarrow & PhCHO + MnO/H_2O \\
& & \text{(Coordination)} & & & & \text{(Desorption)}
\end{array}
$$

[Reprinted with permission from Goldman, I.M. *J. Org. Chem.* **1969**, *34*, 3289. Copyright © 1969 American Chemical Society.]

Manganese dioxide is probably most useful for the oxidation of allylic and benzylic alcohols. In a synthesis of sordaricin,[162] Mander and Thomson oxidized the allylic alcohol unit in **110** to conjugated aldehyde **111** in >86% yield using manganese dioxide. Oxidation of alcohols by manganese dioxide is influenced by the steric encumbrance around the carbon bearing the hydroxyl moiety. In general, there is no significant difference in the rate of oxidation of cis- and trans- isomeric alkenes, and there appears to be little or no cis/trans isomerization during the oxidation. Reaction of *cis*- or *trans*-2-methyl-2-penten-1-ol with MnO_2 led to either the cis- or

158. Pratt, E.F.; Van de Castle, J.F. *J. Org. Chem.* **1961**, *26*, 2973.
159. Goldman, I.M. *J. Org. Chem.* **1969**, *34*, 3289.
160. Henbest, H.B.; Stratford, M.J.W. *Chem. Ind. (London)* **1961**, 1170.
161. Hall, T.K.; Story, P.R. *J. Am. Chem. Soc.* **1967**, *89*, 6759.
162. Mander, L.N.; Thomson, R.J. *J. Org. Chem.* **2005**, *70*, 1654

the trans-aldehyde in 64% or 83% yield, respectively.[163] Boehm et al. noted, however, that some cis-trans isomerization occurred in similar oxidation of pentadienols and pentenynols.[164]

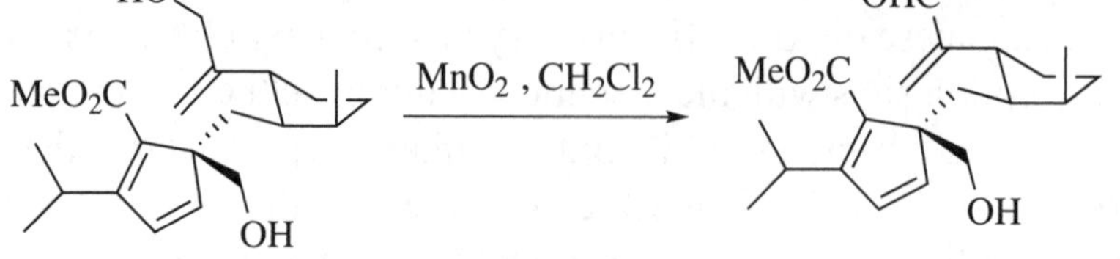

110   **111**

Manganese dioxide oxidizes allylic and benzylic alcohols faster than primary saturated alcohols, but primary and secondary allylic alcohols react at about the same rate.[165] Oxidation of a secondary benzylic alcohol is faster than a primary saturated alcohol. The secondary benzylic alcohol group in **112** was oxidized to give aryl ketone **113** (94% yield), for example, rather than the primary aliphatic hydroxyl.[166]

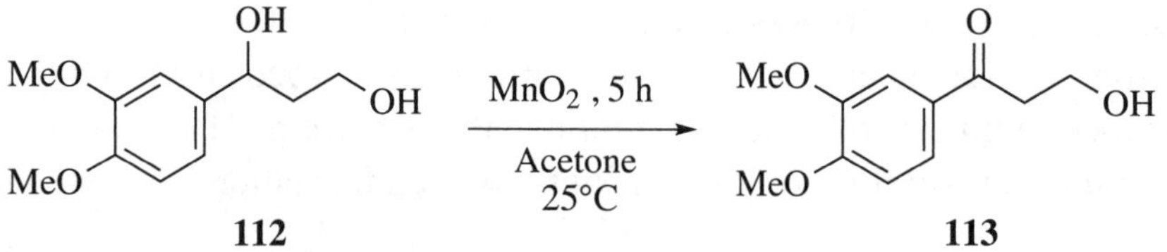

112   **113**

Although the oxidation of allylic and benzylic alcohols is faster, saturated alcohols do react with manganese dioxide. Their oxidation requires a neutral medium, freshly prepared and activated manganese dioxide, the proper solvent and long reaction times. Simple examples are the oxidation of 2-propanol to acetone and 2-methyl-1-propanol to 2-methyl-propanal, both in 50% yield.[167] The conditions are not too vigorous, however, and complex molecules are easily oxidized with manganese dioxide. Even saturated secondary alcohols, which are relatively unreactive,[167] can be oxidized if no allylic or benzylic alcohols are present elsewhere in the molecule.[168] There is little difference in rate between secondary allylic alcohols and primary saturated alcohols, but steric and/or conjugative effects may have an influence on the reaction. Primary allylic and secondary allylic alcohols are often oxidized with essentially the same rate of reaction, leading to poor selectivity.[169] If a large excess of MnO_2 is used, it is possible to

163. Chan, K.C.; Jewel, R.A.; Nutting, W.H.; Rapoport, H. *J. Org. Chem.* **1968**, *33*, 3382.
164. (a) Boehm, E.E.; Thaller, V.; Whiting, M.C. *J. Chem. Soc.* **1963**, 2535; (b) Boehm, E.E.; Whiting, M.C. *Ibid.* **1963**, 2541.
165. For a synthetic example involving primary allylic vs. primary saturated alcohols, see Graham, S.H.; Jonas, D.A. *J. Chem. Soc. C* **1969**, 188.
166. (a) Adler, E.; Becker, H.D. *Acta Chem. Scand.* **1961**, *15*, 849; (b) for a related reaction, see Freudenberg, K.; Lehmann, B. *Chem. Ber.* **1960**, *93*, 1354.
167. Barakat, M.Z.; Abdel-Wahab, M.F.; El-Sadr, M.M. *J. Chem. Soc.* **1956**, 4685.
168. Wada, K.; Enomoto, Y.; Matsui, K.; Munakata, K. *Tetrahedron Lett.* **1968**, 4673.
169. Kienzle, F.; Mayer, H.; Minder, R.E.; Thommen, W. *Helv. Chim. Acta* **1978**, *61*, 2616.

oxidize more than one alcoholic position without affecting other labile functionality.[170]

Allylic alcohols can react with MnO$_2$ under certain conditions to give the conjugated acid or ester (in alcoholic solvents). In an example taken from Weinreb and co-worker's synthesis of cylindrospermopsin, allylic alcohol **114** was treated with MnO$_2$ in the presence of NaCN and methanol to give a 62% yield of methyl ester **115**.[171]

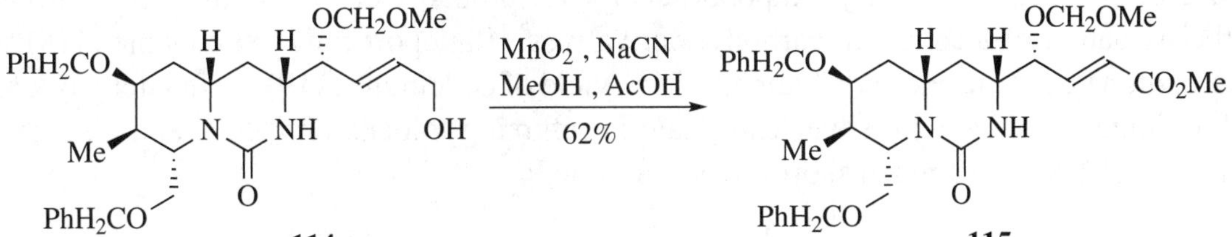

The purpose of cyanide in this oxidation is to convert the conjugated aldehyde to a conjugated ester. In a related reaction using HCN, Corey and co-workers[172] showed that reaction of a conjugated aldehyde with HCN/CN$^-$ generates a cyanohydrin such as **116**. Subsequent oxidation with MnO$_2$ gave cyanoketone, **117**, which reacted with the alcoholic solvent (methanol) to give the corresponding methyl ester.

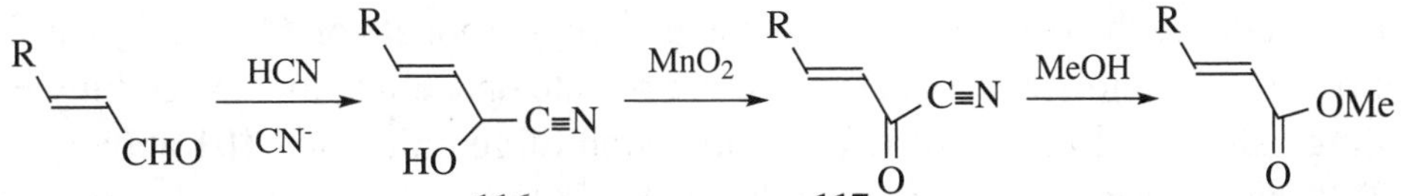

The functional group transforms available from reactions of MnO$_2$ are

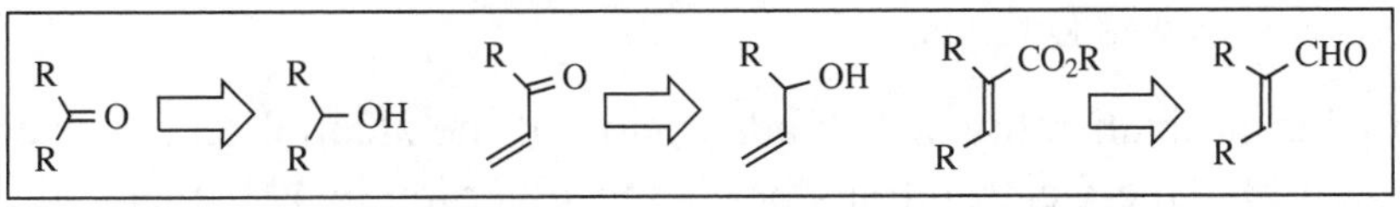

3.2.G. Oxidations with Silver (Silver Carbonate and Silver Oxide)

Silver ion is an excellent oxidizing agent. The reduction potentials for several silver reagents rank them among the more potent of reagents. The most common reagents are silver carbonate (Ag$_2$CO$_3$, reduction potential = 0.4769 V), silver(I) oxide (Ag$_2$O, reduction potential = 0.342 V), and silver(II) oxide (AgO, reduction potential = 0.599 V), where the reduction potentials are taken from the following reactions:[17]

170. (a) Meinwald, J.; Opheim, K.; Eisner, T. *Tetrahedron Lett.* **1973**, 281; (b) Miller, C.H.; Katzenellenbogen, J.A.; Bowlus, S.B. *Ibid.* **1973**, 285.
171. Heintzelman, G.R.; Fang, W.-K.; Keen, S.P.; Walalce, G.A.; Weinreb, S.M. *J. Am. Chem. Soc.* **2002**, *124*, 3939.
172. Corey, E.J.; Gilman, N.W.; Ganem, B.E. *J. Am. Chem. Soc.* **1968**, *90*, 5616.

$$Ag_2CO_3 + 2\,e^- \rightarrow 2\,Ag + CO_3^{2-} \qquad\qquad 0.4769\ V$$

$$Ag_2O + H_2O + 2\,e^- \rightarrow 2\,Ag + 2\,OH^- \qquad\qquad 0.342\ V$$

$$2\,AgO + H_2O + 2\,e^- \rightarrow Ag_2O + 2\,OH^- \qquad\qquad 0.599\ V$$

3.2.G.i. Silver Carbonate. Silver carbonate (Ag_2CO_3) is not a powerful oxidizing agent but it is useful in Organic chemistry. Rapoport et al. were probably the first to use silver carbonate for the oxidation of alcohols to carbonyl derivatives. Rapoport refluxed codeine (**118**) with silver carbonate in benzene and obtained a 75% yield of codeinone (**119**).[173] In later work King et al. oxidized codeine with silver carbonate in refluxing toluene or xylene and obtained an 85% yield of **119** with a much shorter reaction time.[174]

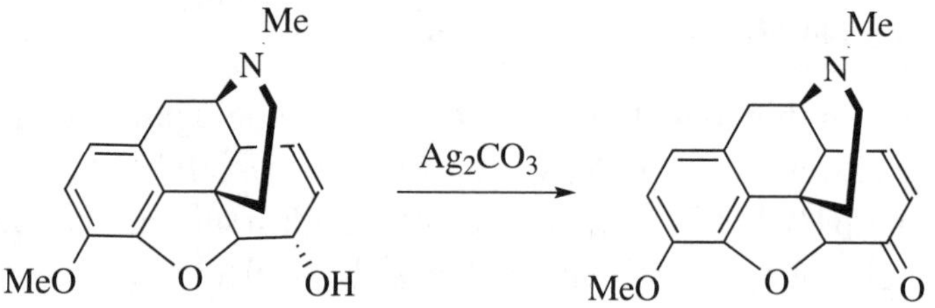

Fétizon, et al. modified the reagent by condensing silver carbonate on Celite. It was found that reaction of primary alcohols with excess silver carbonate on Celite gave excellent yields of the corresponding aldehyde, illustrated by the conversion of geraniol (see **101** above) to geranial (**102**) in 97% yield, using four equivalents of Ag_2CO_3 on Celite, in refluxing benzene.[175] This modified reagent is now known as **Fétizon's reagent**.[176] Saturated primary alcohols are converted to aldehydes in excellent yields and, as seen with Rapoport's work, secondary alcohols are also oxidized to ketones.

Fétizon proposed a mechanism by which adsorption of the alcohol on silver carbonate gave a species such as **120**. One-electron transfer via the silver ion in **121** liberated the protonated carbonyl and carbonic acid (see **122**), which decomposed to carbon dioxide and water.[177] This mechanism involved four distinct processes:[178]

173. Rapoport, H.; Reist, H.N. *J. Am. Chem. Soc.* **1955**, *77*, 490.
174. King, W.; Penprase, W.G.; Kloetzel, M.C. *J. Org. Chem.* **1961**, *26*, 3558.
175. Fétizon, M.; Golfier, M. *C.R. Acad. Sci. Ser. C* **1968**, *267*, 900.
176. Mundy, B.P.; Ellerd, M.G.; Favaloro Jr., F.G. *Name Reactions and Reagents in Organic Synthesis, 2nd ed.*, Wiley-Interscience, New Jersey, **2005**, p. 778.
177. Fétizon, M.; Golfier, M.; Morgues, P. *Tetrahedron Lett.* **1972**, 4445.
178. Kakis, F.J.; Fétizon, M.; Douchkine, N.; Golfier, M.; Mourgues, P.; Prange, T. *J. Org. Chem.* **1974**, *39*, 523.

120   **121**   **122**

1. Reversible adsorption of alcohol on the surface of the oxidizing medium with the electron of oxygen forming a coordinated covalent bond with the silver ions.

2. Oxidation of the C-H bond so the HCOH group is *coplanar* and *perpendicular* to the silver carbonate/Celite surface.

3. A concerted irreversible homolytic shift of electrons to generate reduced silver atoms, hydrogen ions, and a protonated carbonyl.

4. Collapse to products with carbonate ion acting as a hydrogen ion acceptor, generating carbonic acid, which decomposes to carbon dioxide and water.

Polar solvents inhibit the reaction, presumably by interfering with the adsorption process as noted in the mechanism proposed for manganese dioxide oxidations. Oxidation of 1-heptanol to heptanal with Fétizon's reagent was quantitative when the solvent was 35% hexanes. When benzene was used as a solvent, the yield of heptanal was 90% but the yield was < 1% in ethyl acetate, methyl ethyl ketone, or acetonitrile.[178] Since the oxidation is a heterogeneous reaction requiring adsorption of the alcohol substrate, as the surface area of the reagent increases (increased by precipitation on Celite) the rate of oxidation increases. An optimum ratio is reached beyond which increasing the silver carbonate/Celite ratio slows the oxidation.[178]

The reaction does not have carbocation intermediates and even optically pure 1,2,2-triphenylethanol, which is known to have a strong tendency toward rearrangement,[179] undergoes oxidation without racemization or rearrangement.[178] The oxidation of hindered alcohols suffers because of steric hindrance, due to the requirement that the alcohol must be adsorbed on the surface of the reagent. All groups undergoing a bond change must be coplanar in the transition state and perpendicular to the plane of the solid surface. Any combination of steric and conformational factors that inhibit this transition state will slow the reaction. 5α-Androstan-6α-ol, for example, is extremely unreactive with silver carbonate. The diminished rate is attributed to the difficulty in adsorption to the silver, which is hindered by the C19 methyl and the C4 methylene group.[178] This contrasts sharply with 5α-androstan-6β-ol, which is not sterically impeded and undergoes rapid oxidation. The accessibility of the hydrogen attached to the carbon bearing the hydroxyl group is also important. *endo*-2-Norbornanol is

179.   (a) Collins, C.J.; Bonner, W.A.; Lester, C.T. *J. Am. Chem. Soc.* **1959**, *81*, 466; (b) Bonner, W.A.; Collins, C.J. *Ibid.* **1953**, *75*, 5372, 5379; (c) Collins, C.J.; Bonner, W.A. *Ibid.* **1955**, *77*, 92, 99, 6725; (d) Bonner, W.A.; Collins, C.J. *Ibid.* **1956**, *78*, 5587.

oxidized about 50 times faster than *exo*-2-norbornanol, for example.[180]

It is possible to produce a controlled over-oxidation reaction with silver carbonate. Fétizon discovered that 1,3- and 1,4-diols are smoothly oxidized to butyrolactone and valerolactone derivatives in good yield.[181] Oxidation of a diol can be accomplished in the presence of a variety of functional groups, even another alcohol, and a primary alcohol is oxidized faster than a secondary alcohol.[182]a An example of lactone formation is the conversion of 1,4-diol **123** to lactone **124** in Inoue and co-worker's synthesis of an unnatural enantiomer of merrilactone A.[182b]

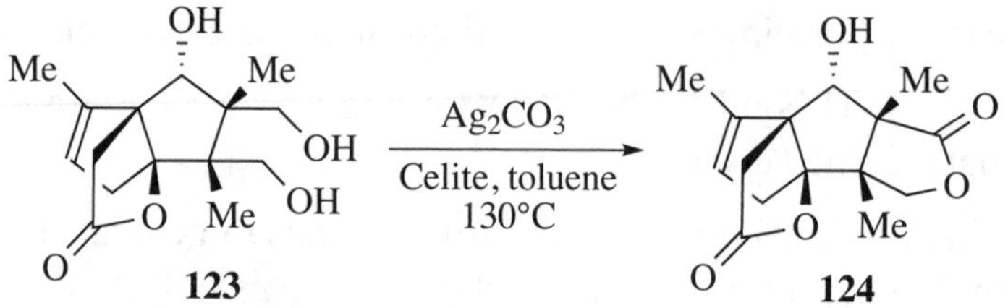

The functional group transforms for silver carbonate are

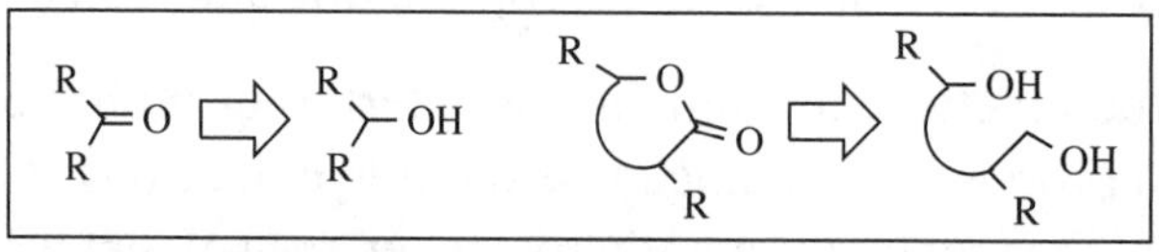

3.2.G.ii. Silver(I) Oxide. Silver(I) oxide (Ag_2O) is the weakest of the silver oxidizing agents introduced in this section. It is used primarily for the selective oxidation of aldehydes to carboxylic acids. One might expect that the oxidation of an aldehyde to a carboxylic acid is facile, but there are cases where the substrate bearing the aldehyde moiety is also susceptible to oxidation. In such cases, a mild and selective oxidant is required and Ag(I) oxide is often the reagent of choice. Silver(I) oxide was used by Rizzacasa and co-workers in the oxidation of **125** to carboxylic acid **126** in a synthesis of (–)-cinatrin B.[183] Oxidation of the alcohol unit

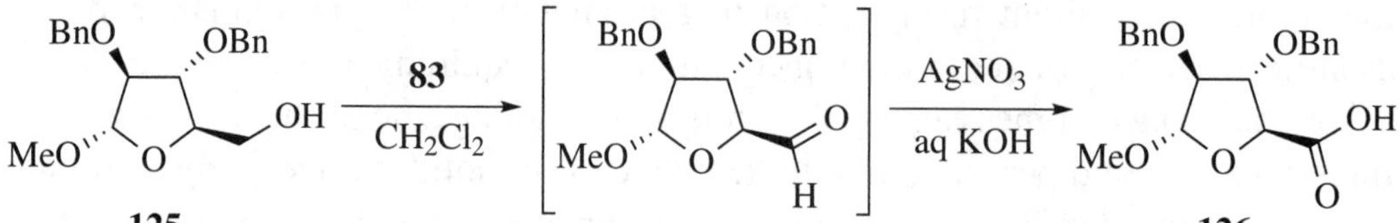

180. Reference 178, see citation 17 therein, Eckert-Maksic, M.; Tusek, L.; Sunko, D.E. *Croat. Chem. Acta* **1971**, *43*, 79. [*Chem. Abstr. 75*: 4999d **1971**].

181. (a) Fétizon, M.; Golfiere, M.; Louis, J.-M. *J. Chem. Soc. Chem. Commu* **1969**, 1118, 1102; (b) Fétizon, M.; Golfier, M.; Louis, J.-M. *Tetrahedron* **1975**, *31*, 171.

182. (a) Chandrasekhar, M.; Chandra, K.L.; Singh, V.K. *J. Org. Chem.* **2003**, *68*, 4039 for a synthesis of (+)-boronolide; (b) Inoue, M.; Lee, N.; Kasuya, S.; Sato, T.; Hirama, M.; Moriyama, M.; Fukuyama, Y *.J. Org. Chem.* **2007**, *72*, 3065.

183. Cuzzupe, A.N.; Di Florio, R.; Rizzacasa, M.A. *J. Org. Chem.* **2002**, *67*, 4392.

in **125** with the **Dess-Martin reagent** (see sec. 3.2.D) gave an aldehyde that was oxidized *in situ* with silver oxide. The silver oxide reagent was prepared *in situ* by mixing a solution of silver nitrate (AgNO$_3$) and excess KOH.[184,185] In a synthesis of strigol, Pepperman examined the oxidation of **127** to keto aldehyde **128**,[186] with the goal of converting **128** to **129**. Several oxidizing agents were used (see Table 3.2),[186] all converting the secondary alcohol unit in **127** to the ketone. Further oxidation to acid **129** was a problem because of extensive decomposition that lowered the yield. However, oxidation of **128** followed by treatment with silver (I) oxide gave **129** in good yield with few by-products. Despite the good yield of **129** eventually obtained in Table 3.2, oxidation of an α,β-unsaturated aldehyde usually gives a poor yield of the acid. Oxidation of **128** had previously given a maximum of 40% yield of **129**.[187] Frank[188] found that oxidation of β-cyclocitral gave only 23% conversion to β-cyclogeranic acid with silver oxide. In **128**, the extended π configuration seemed to facilitate oxidation under the conditions used and it behaves differently than conjugated aldehydes. The oxidation of saturated aldehydes is straightforward.[189] Silver oxide can also oxidize primary alcohols directly to the acid (e.g., 1-dodecanol to dodecanoic acid).[190] Cleavage to the diacid accompanies oxidation of 1,2-diols (sec. 3.7.C), as in the conversion of 1,2-cyclohexanediol to adipic acid in 89% yield.[190]

Table 3.2. Oxidation of 128 to Keto Acid 129

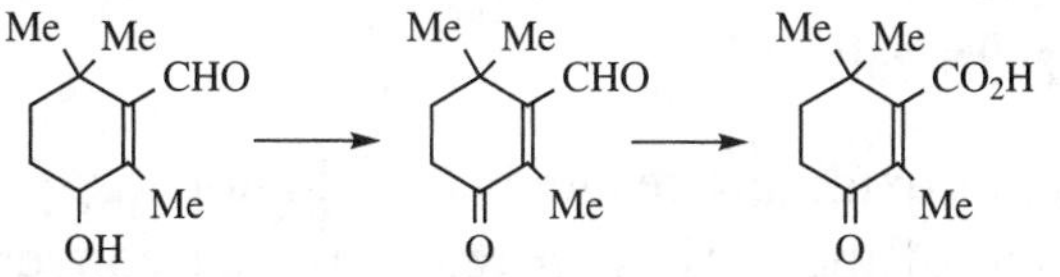

Reagent	% 129[a]
KMnO$_4$•H$_2$SO$_4$	13.5
air (bubbled)	37
air (thin film)	46
Ag$_2$O	72
MnO$_2$•NaCN	80 (as Me ester)
Jones	55

[a] Based on keto-aldehyde **128**

[Reprinted with permission from Pepperman, A.B. *J. Org. Chem.* **1981**, *46*, 5039. Copyright © **1981** American Chemical Society.]

The functional group transforms available to silver (I) oxide are

184. Campaigne, E.; LeSuer, W.M. *Org. Synth. Coll. Vol. 4* **1963**, 919.
185. Pearl, I.A. *Org. Synth. Coll. Vol. 4* **1963**, 972.
186. Pepperman, A.B. *J. Org. Chem.* **1981**, *46*, 5039.
187. Thomason, S.C.; Kubler, D.G. *J. Chem. Educ.* **1968**, *45*, 546.
188. Frank, A.W. *J. Heterocyclic Chem.* **1981**, *18*, 549.
189. For a synthetic example, see Kuroda, C.; Theramongkol, P.; Engebrecht, J.R.; White, J.D. *J. Org. Chem.* **1986**, *51*, 956.
190. Kubias, J. *Collect. Czech. Chem. Commun.* **1966**, *31*, 1666.

3.2.G.iii. Silver(II) Oxide. Silver(II) oxide (AgO) is the most powerful of the three oxidants presented in this section. Although it is not widely used, Ag(II) oxide converts aldehydes and alcohols to carboxylic acids or ketones.[191] The reagent can be prepared by oxidation of silver nitrate with potassium peroxydisulfate ($K_2S_2O_8$) in an alcoholic solvent.[192] Benzyl alcohol was oxidized to benzaldehyde in aqueous solutions of AgO,[192a] and in neutral or alkaline solution further oxidation gave benzoic acid in 57% yield.[192b] Corey found that a mixture of AgO and sodium cyanide converted aldehydes such as **130** to the acid (**131**) in 97% yield, without isomerization of the double bond into conjugation.[172] Oxidation with manganese dioxide and sodium cyanide in acetic acid gave the ethyl ester.[172] Corey observed that cinnamaldehyde gave little or no oxidation, but oxidation of **128** with this reagent gave 80% of **129**.

3.2.H. Other Transition Metal Oxidants

This section contains several oxidizing reagents that do not easily fall into a single category. The intent is to show some of the variety in reagents that convert alcohols to ketones and aldehydes. These reagents are not as widely used as the reagents presented above. The main purpose is to show that alternative reagents (sometimes they are exotic) are currently available or in development that may solve a difficult oxidation problem. The reader is encouraged to conduct a thorough search of the literature since several alternative reagents will probably be found that may solve a given synthetic problem. Most of these alternative reagents use transition metal derivatives.

It is known that potassium permanganate can over-oxidize primary alcohols to the corresponding carboxylic acid, and secondary alcohols can undergo oxidative cleavage. This oxidizing power can be moderated by complexing permanganate ion with another cationic species. Trimethylcetylammonium permanganate,[193] bis(bipyridylcopper) permanganate (bpy_2CuMnO_4)[194] and *n*-butyl permanganate[195] have all been used in the oxidation of alcohols.

191.   (a) Syper, L. *Tetrahedron Lett.* **1967**, 4193; (b) Clarke, T.G.; Hampson, N.A.; Lee, J.B.; Morley, J.R.; Scanlon, B. *Ibid.* **1968**, 5685.
192.   (a) Hammer, R.N.; Klemberg, J. *Inorg. Synth.* **1953**, *4*, 12; (b) Fieser, L.; Fieser, L. *Reagents For Organic Synthesis,* Vol. 2, Wiley, New York, **1969**, p 369.
193.   Rathore, R.; Bhushan, V.; Chandrasekaran, S. *Chem. Lett.* **1984**, 2131.
194.   Firouzabadi, H.; Sardarian, A.R.; Naderi, M.; Vessal, B. *Tetrahedron* **1984**, *40*, 5001.
195.   Firouzabadi, H.; Mostafavipoor, Z. *Bull. Chem. Soc. Jpn.* **1983**, *56*, 914.

Molybdenum reagents have been used for oxidations, and there are many variations in the useful ligands, including: $(NH_4)_6Mo_7O_{24} \cdot 2\ H_2O$,[196,197] $Mo(CO)_6 \cdot Py\text{-cetyl}$,[198,199] $Mo(N_2)_2(dpe)_2$,[200] and $BnNMe_3^+ OMoBr_4^-$.[201] Many of these reagents have been shown to be selective for the oxidation of secondary alcohols, as in the oxidation of 1,10-undecanediol (**132**) to keto alcohol, **133**.[196]

Cerium(IV) has been used extensively,[202] and the two most common reagents are probably ceric ammonium sulfate $[Ce(SO_4)_2 \cdot 2(NH_4)_2SO_4 \cdot 2\ H_2O]$ and ceric ammonium nitrate $[Ce(NH_4)_2(NO_3)_6]$.[203] Firouzabadi et al. have used modified cerium reagents such as $Ce(OH)_3O_2H$[204] and $[(NO_3)_3Ce]_3H_2IO_6$[205] for the oxidation of primary alcohols, especially benzylic and allylic alcohols.

3.3. FORMATION OF PHENOLS AND QUINONES

In addition to the oxidation of aliphatic hydroxy compounds, aromatic derivatives that contain hydroxyl groups (phenol derivatives) can also be oxidized.[206] The course of the oxidation disrupts the aromatic ring to give quinones, which are important components in a variety of natural products.[207] It is also possible to convert aromatic hydrocarbons to phenols by oxidation. Quinones are obtained by oxidation of phenols and a phenol can be very loosely viewed as an enol. Inclusion of this chemistry immediately after discussing the oxidation of alcohols to ketones, aldehydes, or acids is done with the goal of providing some continuity in studying the oxidation of hydroxyl compounds.

196. Trost, B.M.; Masuyama, Y. *Tetrahedron Lett.* **1984**, *25*, 173.
197. Sur, B.; Adak, M.M.; Pathak, T.; Hazra, B.; Banerjee, A. *Synthesis* **1985**, 652.
198. Yamawaki, K.; Yoshida, T.; Suda, T.; Ishii, Y.; Ogawa, M. *Synthesis* **1986**, 59.
199. Yamawaki, K.; Yoshida, T.; Nishihara, H.; Ishii, Y.; Ogawa, M. *Synth. Commun.* **1986**, *16*, 537.
200. Tatsumi, T.; Hashimoto, K.; Tominaga, H.; Mizuta, Y.; Hata, K.; Hidai, M.; Uchida, Y. *J. Organomet. Chem.* **1983**, *252*, 105.
201. Masuyama, Y.; Takahashi, M.; Kurusu, Y. *Tetrahedron Lett.* **1984**, *25*, 4417.
202. Richardson, W.H. in *Oxidation in Organic Chemistry*, Part A, Wiberg, K.B. (Ed.), Academic Press, New York, **1965**, pp 244-277
203. Kanemoto, S.; Tomioka, H.; Oshima, K.; Nozaki, H. *Bull. Chem. Soc. Jpn.* **1986**, *59*, 105.
204. Firouzabadi, H.; Iranpoor, N. *Synth. Commun.* **1984**, *14*, 875.
205. Firouzabadi, H.; Iranpoor, N.; Hajipoor, G.; Toofan, J. *Synth. Commun.* **1984**, *14*, 1033.
206. Reference 20, p 1234
207. For a review of the synthesis of quinones, particularly those useful in synthesis, see Owton, W.M. *J. Chem. Soc. Perkin Trans. 1* **1999**, 2409.

3.3.A. Quinones

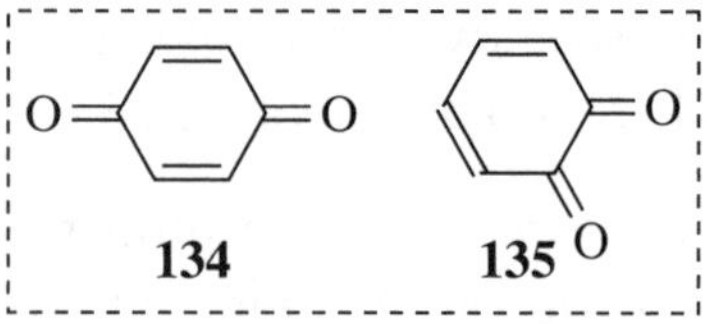

134   **135**

The quinone structure is incorporated in many naturally occurring compounds, generically called *quinoids*. The most common are the 1,4-quinones (*p*-quinones, **134**) and the 1,2-quinones (*o*-quinones, **135**).[208] Quinoids can be prepared by many methods, including oxidation of non-quinoid precursors, cyclization methods, condensation methods, and annulation methods.[209] This section will focus only on oxidation reactions, which constitute the only completely general methodology.[209]

Phenolic derivatives are readily oxidized to quinoids with **Fremy's salt** (potassium nitrosodisulphonate $[(KO_3S)_2NO\bullet]$.[209,210] Fremy[211] prepared this reagent in 1845 as the disodium salt, but Raschig prepared the potassium salt, which was more effective and remains the reagent of choice.[212] Acidic solutions are very unstable, and the salt also decomposes in aqueous solutions at pH greater than 10. The most common solvents are aqueous alcohol and aqueous acetone, and the reaction is usually buffered with phosphate or acetate. The reaction is rapid and may be monitored by disappearance of the purple color of the radical. Teuber and Rau proposed a simple mechanism for the oxidation of phenol.[213] Electron transfer from Fremy's salt to phenol generates a radical species (**136**), and this reacts with excess Fremy's salt to give an addition product, **137**. Loss of the hydrogen α to the oxygen moiety in **137** leads to formation of the quinone (**134**) and potassium aminodisulfonate $[HN(SO_3K)_2]$.

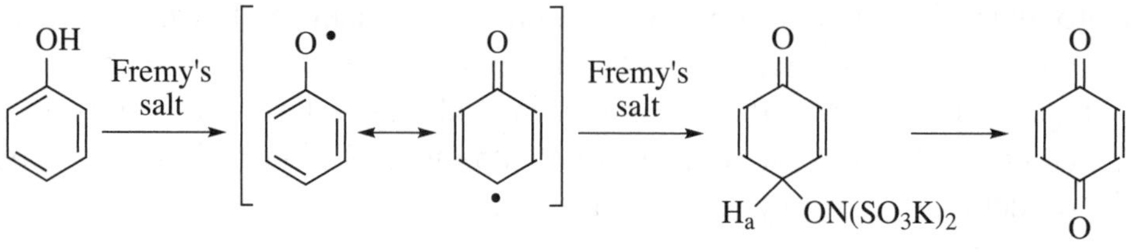

136   **137**   **134**

Fremy's salt <u>sometimes</u> undergoes <u>violent</u>, <u>spontaneous decomposition</u> but this appears to be <u>due to the presence of catalytic amount of nitrite ion</u>.

---

208. *The Chemistry of the Quinoid Compounds*, Parts 1 and 2, Patai, S. (Ed.), Wiley, New York, *1974*.
209. Thomson, R.H. in *The Chemistry of the Quinoid Compounds*, Part 1, Patai, S. (Ed.), Wiley, New York, *1974*, pp 111-161.
210. Zimmer, H.; Lankin, D.C.; Horgan, S.W. *Chem. Rev. 1971*, *71*, 229.
211. Fremy, E. *Ann. Chim. Phys. 1845*, *15*, 408.
212. Raschig, F. *Schwefel und Stickstoff-Studien*, Verlag-Chemie, Leipzig-Berlin, *1924* (from citation 2 in reference 210).
213. (a) Reference 209, p 112; (b) Teuber, H.J.; Rau, W. *Chem. Ber. 1953*, *86*, 1036.

The steric demands on the reaction sometimes lead to formation of the less hindered *p*-quinone as the major product, with low yields of the ortho-quinones. The yield of *o*-quinone can be low even when the para- position is blocked. When both the ortho- and para-positions are not sterically hindered, the *p*-quinoid is usually the major product. When X = H in **138**, oxidation follows *path a* to give the 1,4-quinone. Formation of **139** is consistent with para-attack, and loss of the X group will give **134**. When X in **138** is alkoxy (X = OR) or alkyl (X = R), however, *path b* is preferred. Since the para-position is blocked the ortho- intermediate (**140**) is formed, and loss of the bis(disulfonate) in the usual manner (see above) gives the 1,2-quinone **141**. When X = Cl in **139**, the chlorine is usually lost with formation of the 1,4-quinone.[209]

Two important factors that influence the oxidation. One is electronic stabilization of the incipient phenoxy radical, and the other is the steric requirement connected with formation of the cyclohexadiene intermediate. The presence of electron-withdrawing groups can cause the oxidation to fail, or at least lead to a poorer yield.[209,210,214]

Oxidation of phenols with Fremy's salt can be applied to many phenolic derivatives, as in the synthesis of verapliquinone A by Moody and co-workers[215] in which oxidation of **142** gave 1,4-quinone **143** in 65% yield. Oxidation of phenols with Fremy's salt requires a buffered medium, since the salt is stable in only a limited pH range, as previously mentioned. Many *o*-quinones are unstable to acidic conditions, and may give other products if the pH exceeds 7. For example, oxidation of **144** gave **145** in 85% yield under standard conditions but in acid solution that product dimerizes completely.[216] When the phenyl ring of the phenol contains a *p*-chlorine[209,217] or a *tert*-butyl group,[218] these groups can be eliminated from the intermediate, generating a quinone.[217,218] Elimination of the amino moiety to give a quinoid can accompany oxidation of aniline derivatives by Fremy's salt. Oxidation of 3-ethoxy-5-methylaniline, for example, gave 3-ethoxy-5-methyl-1,4-benzoquinone in 93% yield.[219]

214. Maruyama, K.; Otsuki, T. *Bull. Chem. Soc. Jpn.* **1971**, *44*, 2873.
215. Davis, C.J.; Hurst, T.E.; Jacob, A.M.; Moody, C.J. *J. Org. Chem.* **2005**, *70*, 4414.
216. Teuber, H.J.; Thaler, G. *Chem. Ber.* **1958**, *91*, 2253.
217. Teuber, H.J.; Thaler, G. *Chem. Ber.* **1959**, *92*, 667.
218. Magnusson, R. *Acta Chem. Scand.* **1964**, *18*, 759.
219. Teuber, H.J.; Hasselbach, M. *Chem. Ber.* **1959**, *92*, 674.

142 143

144 145

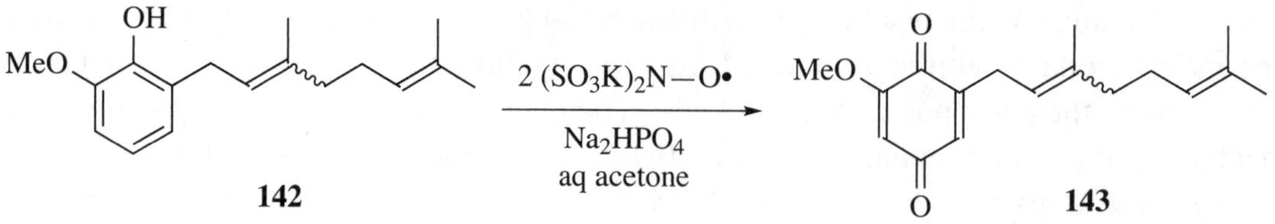

Table 3.3. Oxidation of 2,6-Xylenol (146)

146 147 148 149

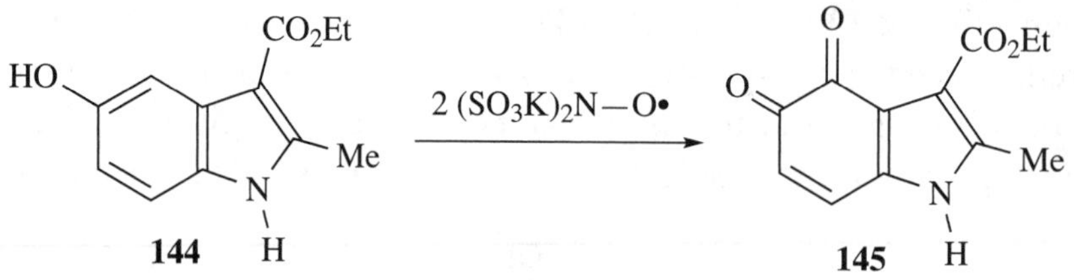

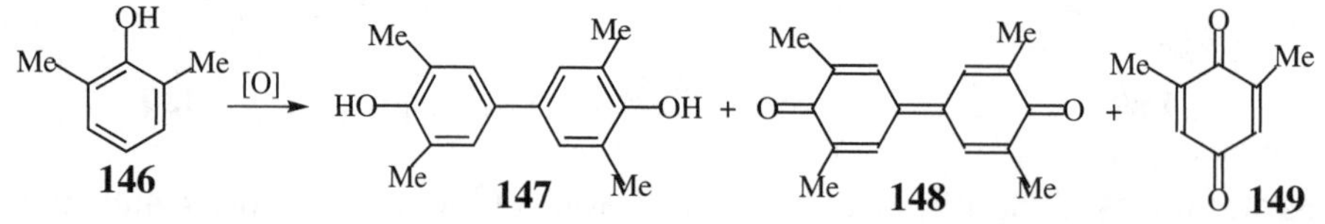

Reagent	% 147	% 148	% 149	% Polymer
$AgNO_3/K_2S_2O_8$	27	51	5	5
$CuSO_4/Na_2S_2O_8$	1	47	27	15
Cu(III) ClO_4		30		45
$CuSO_4$/NaOCl		17	26	60
$CuSO_4$/air stream		3	9	20
$FeCl_3$		54	14	15
$K_3Fe(CN)_6$/NaOH		52		25
MnO_2/4M H_2SO_4			16	50
MnO_2		53		4
$CuSO_4/H_2O_2$		33	9	40
$FeSO_4/H_2O_2$		22	3	55
$TiCl_3/H_2O_2$	9	20	51	5
$TiCl_4/H_2O_2$		50	38	10
SeO_2/H_2O_2	4			
$NaIO_4$		0.3	9	
NaClO	13	08		75
$NO(SO_3K)_2$			**74**	

Many other oxidizing agents convert phenols to quinones, including Cr(VI), but in general most give lower yields of quinone and higher percentages of dimerized products. Chromic acid is an effective oxidant for the conversion of phenols to quinones only when a para- substituent such as halogen, hydroxyl, or amino is present. The yields are often poor,[220] but not always.[221] Anodic oxidation of phenols to quinones[222] can be very efficient. Bacon and Izzat described the

220. Reference 209, p 117.
221. Conant, J.B.; Fieser, L.F. *J. Am. Chem. Soc.* ***1923**, 45*, 2194.
222. Reference 209, p 119.

oxidation of 2,6-xylenol (146) with a variety of oxidizing agents.[223] In most cases, oxidation gave not only the expected quinone (149) but also dimerized products such as the coupled phenol (147) and the coupled quinone (148) as well as polymeric material.[223] Table 3.3[223] shows the product distribution obtained with different oxidizing agents. Clearly, Fremy's salt is the superior reagent in this list for preparation of quinone 149. It is also clear that the coupled quinone predominates with most oxidizing agents, and that polymerization is a problem. Table 3.3 shows the most common reagents used for the oxidation of phenols, including Fremy's salt, ferric ion, copper(II), titanium(III), and titanium(IV), as well as manganese dioxide,[224] selenium dioxide, and sodium periodate.

Note that catechols (1,2-dihyroxybenzenes) are readily oxidized to *o*-quinones,[225] but the products are often sensitive to any electrophilic or nucleophilic species in the reaction medium. Catechol itself gives 145. Dimerization is as much a problem with catechols as with monophenols (see Table 3.3).[226] The conversion of catechol to 145 used silver carbonate, and note that silver salts are the classical oxidation reagent for such transformations.[227] Other reagent have been used to oxidize catechol derivatives, including ceric sulfate, lead tetraacetate, DDQ (2,3-dichloro-5,6-dicyano-1,4-benzoquinone),[228] iodate, and periodate.[229]

There are many syntheses of molecules containing a quinone unit, and in most cases the quinone is oxidized late in the synthesis from a resorcinol or catechol derivative. In one example, taken from a synthesis of bisbolene by Vyvyan and co-workers[230] oxidation of 150 with ceric ammonium nitrate gave an 83% yield of quinone 151. As mentioned above, quinones can be formed from alkoxy derivatives of resorcinol or catechol, as well as from the parent hydroxyl compounds. In a synthesis of shikonin,[231] Couladouros and co-workers treated the *O*-silyl protected derivative 152 with ceric ammonium nitrate (CAN) and obtained a 92% yield of quinone 153.

223. Bacon, R.G.R.; Izzat, A.R. *J. Chem. Soc. C, 1966,* 791.
224. Fatiadi, A.J. *Synthesis 1976,* 133.
225. References 209 and 210.
226. (a) Harley-Mason, J.; Laird, A.H. *J. Chem. Soc. 1958,* 1718; (b) Horner, L.; Dürkheimer, W. *Chem. Ber. 1958, 91,* 2532.
227. (a) Balogh, V.; Fétizon, M.; Golfier, M. *J. Org. Chem. 1971, 36,* 1339; (b) Cason, J. *Org. React. 1948, 4,* 305.
228. Boldt, P. *Chem. Ber. 1966, 99,* 2322.
229. Reference 209, pp 124-125.
230. Vyvyan, J.R.; Loitz, C.; Looper, R.E.; Mattingly, C.S.; Peterson, E.A.; Staben, S.T. *J. Org. Chem. 2004, 69,* 2461.
231. Couladouros, E.A.; Strongilos, A.T.; Papageorgiou, V.P.; Plyta, Z.F. *Chem. Eur. J., 2002, 8,* 1795.

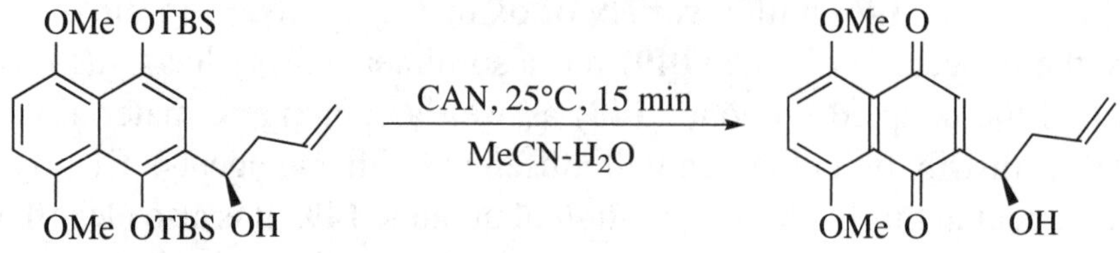

Oxidation of phenol leads to several interesting functional group transforms, including

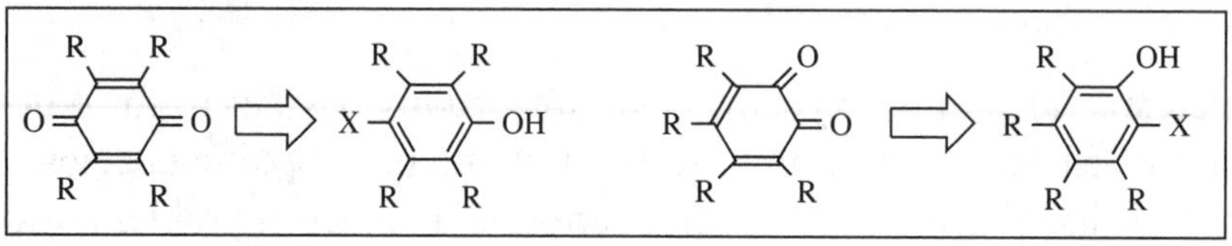

3.3.B. Phenols

As discussed in section 2.11.E, phenols are conveniently prepared by hydrolysis of diazonium salts. There are also several oxidative methods that convert aromatic derivatives to phenols.[232] Direct hydroxylation can be accomplished by using free radical reagents such as a mixture of ferrous ion (Fe^{2+}) and H_2O_2 (**Fenton's reagent**).[233] The yields are usually poor, in the 5-20% range, and formation of biphenyls (a coupling product) predominates. Oxidation of benzene, for example, gave a low yield of phenol but a relatively large yield of biphenyl. In general, Fenton's reagent is not very useful for the preparation of phenols. Udenfriend et al introduced a modification of this procedure as a model for the biogenic hydroxylation of tyramine, using an oxygen-ferrous ion-ascorbic acid system in the presence of ethylenediaminetetraacetic acid (EDTA) (**Udenfriend's reagent**).[234,233a] This reagent gave good yields of ortho- and para-phenolic derivatives from phenyl acetamide. An electrolytic version of this oxidation used anodic oxidation in the presence of Undenfriend's reagent and oxygen to convert tyramine to a mixture of hydroxytyramines and dihydroxytyramine (DOPA).[235]

Alkyl benzenes react with hydrogen peroxide, Lewis acid catalysts (such as boron trifluoride),[236] or strong protonic acids such as trifluoroperoxyacetic acid[237,233a] to give the corresponding phenol. Chambers et al. used trifluoroperoxyacetic acid to oxidize mesitylene (1,3,5-trimethylbenzene)

232. Reference 20, pp 977-978.
233. (a) Haines, A.H. *Methods For the Oxidation of Organic Compounds*, Academic Press, London, *1985*, p 173; (b) Smith, J.R.L.; Norman, R.O.C. *J. Chem. Soc. 1963*, 2897.
234. Udenfriend, S.; Clark, C.T.; Axelrod, J.; Brodie, B.B. *J. Biol. Chem. 1954, 208,* 731.
235. (a) Blanchard, M.; Bouchoule, C.; Djaneye-Boundjou, G.; Canesson, P. *Tetrahedron Lett. 1988, 29,* 2177; (b) Maissant, J.M.; Bouchoule, C.; Canesson, P.; Blanchard, M. *J. Mol. Catal., 1983, 18,* 189; (c) Maissant, J.M.; Bouchoule, C.; Blanchard, M. *Ibid. 1982, 14,* 333.
236. Hart, H.; Buehler, C.A. *J. Org. Chem. 1964, 29,* 2397.
237. Chambers, R.D.; Goggin, P.; Musgrave, W.K.R. *J. Chem. Soc. 1959*, 1804.

to 2,4,6-trimethylphenol in high yield based on consumed mesitylene (only a 17% conversion based on the peroxyacid).[237] Similarly, McClure and Williams[238] reported the oxidation of anisole to phenolic products in low yield (ortho- derivative in 27% yield and para- derivative in 7% yield), with a conversion of only 44%. Hart and co-workers found that the yield of 2,4,6-trimethylphenol from 1,3,5-trimethylbenzene was greatly improved (88% yield with good overall conversion) by using a mixture of boron trifluoride (BF_3) and trifluoroperoxyacetic acid (in dichloromethane at $-7°C$).[236] This modification is probably the best method for peroxyacid oxidation of aromatic hydrocarbons to the corresponding phenol. Aluminum chloride is also an effective catalyst in this transformation, giving primarily a mixture of ortho- and para-hydroxy derivatives.[239]

Phenol can be hydroxylated with potassium persulfate ($K_2S_2O_8$) in alkaline media (the **Elbs persulfate oxidation**).[240] In 1893 Elbs used ammonium persulfate to oxidize 2-nitrophenol to nitroquinol.[241] Later workers used the potassium salt,[242] which is now the reagent of choice, in order to obtain synthetically useful yields of para- hydroxy phenols. The initial oxidation product is the persulfate (**154**), which is hydrolyzed to hydroquinone (**155**) as shown.[243] The ortho-product (a pyrocatechol derivative) is formed if the para-position is blocked, but the reaction is slower and the yields are in the 30-50% range.[244] The presence of electron-withdrawing groups on the aromatic ring improves the yield, but electron-releasing groups such as methoxy (OMe) can also be tolerated.[245]

An alternative to direct hydroxylation is the thermal or photochemical decomposition of diacyl peroxides. Aryl derivatives react to give acyloxylation.[246] p-Xylene was converted to **156** in the presence of cupric chloride,[247] although the yield was only 24%. The benzoate ester in **156**

238. McClure, J.D.; Williams, P.H. *J. Org. Chem.* **1962**, *27*, 627.
239. Kurz, M.E.; Johnson, G.J. *J. Org. Chem.* **1971**, *36*, 3184.
240. (a) Reference 233a, p 174; (b) Sethna, S.M. *Chem. Rev.* **1951**, *49*, 91; (c) Behrman, E.J. *Beilstein J. Org. Chem.* **2006**, *2*:22 (an on-line journal); (d) The Merck Index, 14th ed., Merck & Co., Inc., Whitehouse Station, New Jersey, **2006**, p ONR-27.
241. Elbs, K. *J. Prakt. Chem.* **1893**, *48*, 179.
242. Chemische Fabrik auf Aktien vorm E. Schering: German Patents 81,068, 81,297 and 81,298: see Friedlander, Fortschritte der Teerfarbenfabrikation 1894-1897, 4th Part, pp 126, 127, 121 respectively, taken from citation 12 in reference 240b.
243. (a) Reference 233a, pp 180-181; (b) Baker, W.; Brown, N.C. *J. Chem. Soc.* **1948**, 2303.
244. Reference 233a, p 180.
245. Baker, W.; Savage, R.I. *J. Chem. Soc.* **1938**, 1602.
246. (a) Reference 233a, pp 177-178; (b) Rawlinson, D.J.; Sosnovsky, G. *Synthesis* **1972**, 1.
247. Reid, C.G.; Kovacic, P. *J. Org. Chem.* **1969**, *34*, 3308.

was easily hydrolyzed to the phenolic derivative. Lead tetraacetate [$Pb(OAc)_4$, abbreviated LTA, see sec. 3.7.C.i] gives a similar reaction, but this reagent gives the *O*-acetyl derivative from aryl derivatives. Polycyclic aromatics and aryl derivatives[248],[249] containing electron-releasing groups give the best yields of oxidation products.

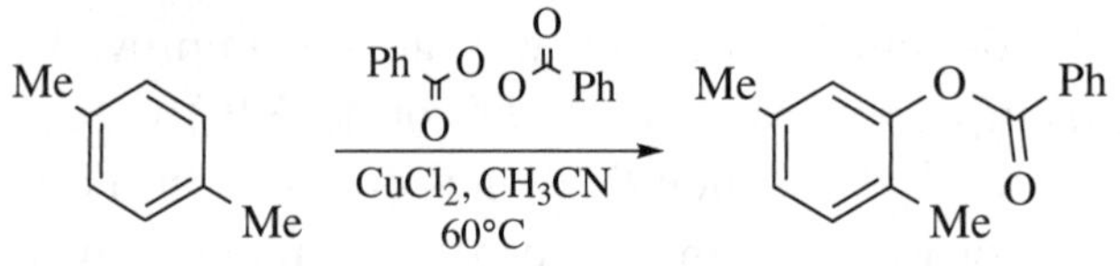

156

The functional group transform for this reaction is

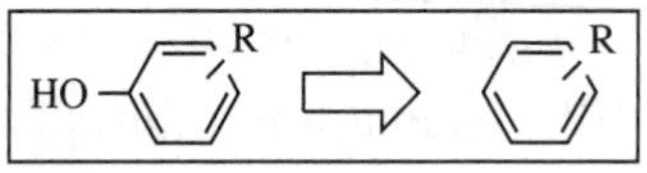

3.4. CONVERSION OF ALKENES TO EPOXIDES

The previous sections focused attention on the oxidation of a C–OH moiety to a C=O (carbonyl) group. Other functional groups are subject to oxidation, including the π bond of alkenes. There are several different oxidative functional group transformations that involve alkenes, including incorporation of one oxygen (epoxidation), two oxygens (dihydroxylation), or 1 oxygen and 1 nitrogen (aminohydroxylation), and finally, oxidative cleavage (usually to carbonyl derivatives). Sections 3.4.A-D will discuss various methods for the conversion of alkenes to epoxides (oxiranes)[250] and the other functional group transformations will be discussed in succeeding sections.

3.4.A. Epoxides from Halohydrins

A simple method for producing epoxides uses the reaction of alkenes and hypohalous acids (generated by the reactions $Cl_2 + H_2O \rightarrow HOCl$ or $Br_2 + H_2O \rightarrow HOBr$) to give the trans- (or anti-) halohydrin as the major product. An example is the conversion of cyclohexene to 2-bromocyclohexanol (sec. 2.10.C). Subsequent treatment with a base such as sodium hydride generates the alkoxide, and displacement of the adjacent bromine via an S_N2 process (an *intramolecular* **Williamson ether synthesis**, sec. 2.6.A.i) gives cyclohexene oxide. The S_N2 nature of this base-induced cyclization restricts the reaction to halohydrins with a primary or secondary halide, and the halogen must be trans- to the hydroxyl group for epoxide formation. The halohydrin is often generated with aqueous NBS or NCS (see sec. 2.10.C). In a scalable synthesis of the potent fungicide 1-cytosinyl-*N*-malayamycin A by Loiseleur, Hanessian and

248. Fieser, L.F.; Clapp, R.C.; Daudt, W.H. *J. Am. Chem. Soc.* **1947**, *64*, 2052.
249. Ritcher, H.J.; Dressler, R.L. *J. Org. Chem.* **1962**, *27*, 4066.
250. Reference 20, pp 915-927.

co-workers,[251] treatment of **157** with aqueous NBS gave bromohydrin **158** (one regioisomer is shown). Subsequent reaction of this mixture with NaOH led to formation of the epoxide unit in **159**. A protected form of the halohydrin can be used in this reaction, with the requisite alkoxide unit unmasked under controlled conditions. In Marino and co-worker's synthesis of (–)-macrolactin A,[252] deprotection of the *O*-silyl group (Sec 7.3.A.i) in **160** with cesium fluoride generated the alkoxide intermediate (**161**) *in situ*, and intramolecular displacement of the chloride on the terminal carbon atom gave epoxide **162** in 86% yield. It should be noted that this epoxide-forming reaction is not restricted to halide leaving groups, and sulfonate esters such as mesylate, tosylate, or triflate can be used. 1,2-Diols can be converted to a mono-sulfonate ester, and subsequent reaction with potassium hydroxide leads to an alkoxide that displaces the leaving group to give the epoxide.[253]

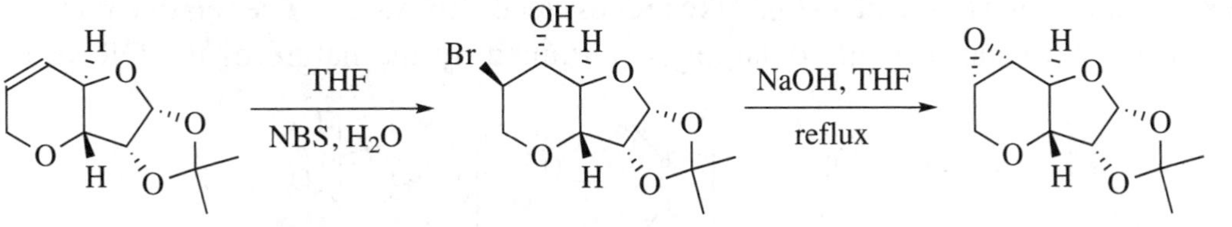

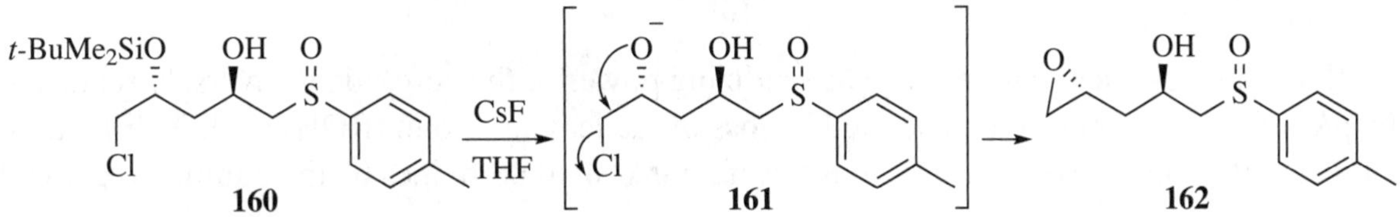

The functional group transform is simply:

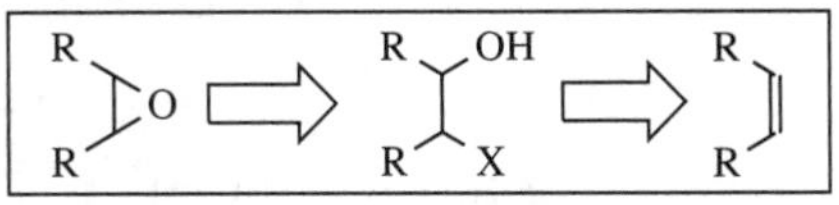

3.4.B. Peroxide Induced Epoxidations

The reaction of peroxides with alkenes is a common method for the preparation of epoxides (oxiranes), but the nature of the peroxide is very important to the success of the oxidation. Peroxides are a source of electrophilic oxygen when they react with the nucleophilic π bond of an alkene. Hydrogen peroxide (H_2O_2) is a powerful oxidizing agent, with a reduction potential of 1.77 V[17] for the following reaction:

251. Loiseleur, O.; Schneider, H.; Huang, G.; Machaalani, R.; Selles, P.; Crowley, P.; Hanessian, S. *Org. Process Res. Dev.* **2006**, *10*, 518,
252. Marino, J.P.; McClure, M.S.; Holub, D.P.; Comasseto, J.V.; Tucci, F.C. *J. Am. Chem. Soc.* **2002**, *124*, 1664.
253. For an example of epoxide formation from a bromohydrin, taken from a synthesis of (–)-coriolin, see Mizuno, H.; Domon, K.; Masuya, K.; Tanino, K.; Kuwajima, I. *J. Org. Chem.* **1999**, *64*, 2648.

$$H_2O_2 + 2e^- + 2H^+ \rightarrow 2H_2O \qquad\qquad 1.77\text{ V}$$

While most peroxide reactions involve homolytic cleavage of the O–O bond, generating free radicals (sec. 13.3), the reaction of H_2O_2 and its monosubstituted derivatives with an alkene can proceed via either a concerted or ionic mechanism.[254] Three categories of peroxides are used for epoxidation, hydrogen peroxide, alkyl hydroperoxides (**163**) and peroxyacids (**164**). The alkene can be viewed as a Lewis base in an initial reaction with the peroxide ROOH forming a coordination complex **165** that generates an oxygen with a positive dipole. A heterolytic cleavage transfers that oxygen to the alkene, and subsequent proton transfer liberates the byproduct (water from hydrogen peroxide, an alcohol from an alkyl hydroperoxide, or a carboxylic acid from a peroxy acid). This sequence involves backside attack at oxygen in an S_N2-like process (sec. 2.6.A.i).[255] The relative rate of oxidation for various peroxide reagents is largely determined by the nature of the OR group in **165**,

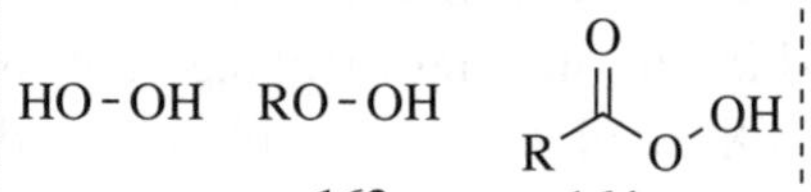

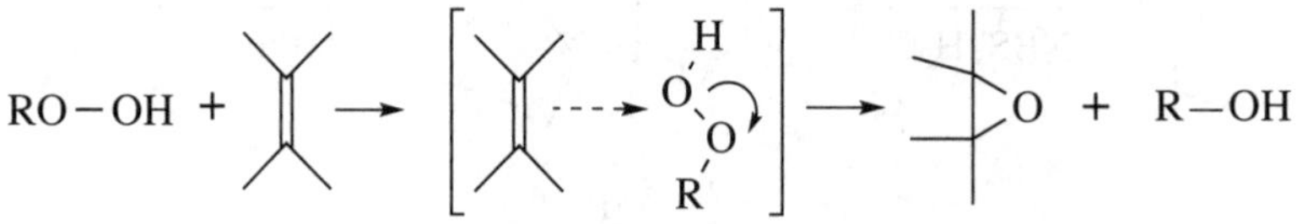

which behaves as a leaving group. The oxidizing power of the peroxide is inversely related to the pK_a of the conjugate acid generated by loss of the leaving group (ROH).[256] It is difficult to compare different types of reagents, but a useful ranking of peroxides by their ability to convert alkenes to epoxides is[256,255]

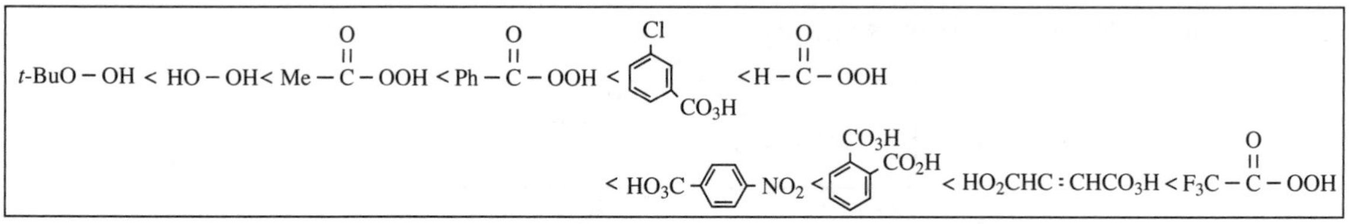

Therefore, the strength of the peroxide reagent correlates more or less with its reactivity with alkenes. In the following sections, the emphasis will be on the synthetic utility and generality of peroxide induced reactions.

3.4.B.i. Hydrogen Peroxide. The reduction potential of oxygen[17] is given by the following reactions:

$$O_2 + 2H^+ + 2e^- \rightarrow H_2O_2 \qquad\qquad 0.682\text{ V}$$
$$O_2 + 4H^+ + 4e^- \rightarrow 2H_2O \qquad\qquad 1.229\text{ V}$$

254. Lewis, S.N. in *Oxidation*, Vol. 1, Augustine, R.L. (Ed.), Marcel Dekker, New York, *1969*, p 214.
255. Reference 254, p 215.
256. Roberts, J.D.; Caserio, M.C. *Basic Principles of Organic Chemistry*, W.A. Benjamin, New York, *1965*, p 301.

Hydrogen peroxide is used in water or water miscible solvents, but epoxidation of alkenes is slow unless metal ion catalysts such as vanadium pentoxide,[257] tungsten trioxide,[258] or molybdenum trioxide[259] are used. In allylic alcohols, oxidation of the hydroxyl moiety is usually faster than oxidation of the alkene as in the oxidation of crotyl alcohol with hydrogen peroxide to give crotonaldehyde. The H_2O_2 oxidation of alkenes is not used unless the alkene is soluble in aqueous media since other peroxide reagents are more effective. Many of the comments presented in section 3.4.B.ii also apply to hydrogen peroxide.

There is an exception to the poor reactivity exhibited by hydrogen peroxide. When **166** was treated with 30% hydrogen peroxide in the presence of benzonitrile and potassium hydrogen carbonate, epoxide **167** was formed in 86% yield,[260] in what is known as the **Payne epoxidation**.[261] The reagents used generate peroxybenzimidic acid *in situ*. This transformation was taken from Smith and co-worker's synthesis of (+)-calyculin A.[260] *The Payne epoxidation is known to give a diastereofacial preference that is opposite to epoxidations with peroxyacids.* In this particular example, the epoxide **167** was a 3:1 (α:β) mixture of isomers at the epoxy carbon.

3.4.B.ii. Alkyl Hydroperoxides. Alkyl hydroperoxides are prepared by reaction of tertiary alcohols or alkenes with hydrogen peroxide, in the presence of sulfuric acid.[262] Alkyl hydroperoxides[263] are also prepared by the autoxidation of alkanes with oxygen, which is also a potent oxidizing agent. The relatively stable reagent *tert*-butyl hydroperoxide (*t*-BuOOH, TBHP), for example was prepared in 75% yield (8% conversion) by reaction of isobutane with oxygen.[264] In this case, the reaction was carried out in the liquid phase with di-*tert*-

257. Kaman, A.J. British Patent 837 464, 1957, [*Chem. Abstr. 55*: 566f, *1961*].
258. (a) Payne, G.B.; Williams, P.H. *J. Org. Chem.* **1959**, *24*, 54; (b) Sergeev, P.G.; Bukreeva, L.M. *Zh. Obshch. Khim.*, **1958**, *28*, 101 (*Chem. Abstr. 52*: 12758, *1958*); (c) Raciszewski, Z. *J. Am. Chem. Soc.* **1960**, *82*, 1267; (d) Stevens, H.C.; Kaman, A.J. *Ibid.* **1965**, *87*, 734.
259. Carlson, G.J.; Skinner, J.R.; Smith, C.W.; Wilcoxen, Jr., C.H. U.S. Patent 2 833 787, 1958 [*Chem. Abstr. 52*: 16367d, *1958*].
260. Smith III, A.B.; Friestad, G.K.; Barbosa, J.; Bertounesque, E.; Duan, J.J.-W.; Hull, K.G.; Iwashima, M.; Qiu, Y.; Spoors, G.; Salvatore, B.A. *J. Am. Chem. Soc.* **1999**, *121*, 10478.
261. Payne, G.B. *Tetrahedron* **1962**, *18*, 763.
262. (a) Milas, N.A.; Surgenor, D.M. *J. Am. Chem. Soc.* **1946**, *68*, 205; (b) Ross, H.; Huttel, R. *Chem. Ber.* **1956**, *89*, 2641.
263. Reference 1, pp 17-32, 315-339 and 340-349.
264. (a) Mayo, F.R. *Acc. Chem. Res.* **1968**, *1*, 193; (b) Winkler, D.E.; Hearne, G.W. *Ind. Eng. Chem.* **1961**, *53*, 655; (c) Bell, E.R.; Dickey, F.H.; Raley, J.H.; Rust, F.F.; Vaughan, W.E. *Ibid.* **1949**, *41*, 2597.

butyl peroxide as an initiator (see sec. 13.3). Most tertiary alkanes react[265] and probably involves a radical intermediate that reacts with oxygen[265] and then with the alkane to give the hydroperoxide and regenerate the radical carrier (sec. 13.3).

Table 3.4. Alkene Epoxidation by Transition Metal Catalyzed Reaction with Alkyl Hydroperoxides[268]

$$RO-OH \quad + \quad \overset{R}{\underset{R}{[}} \quad \overset{M}{\longrightarrow} \quad O \overset{R}{\underset{R}{<}}$$

R	Alkene	Solvent	M	% M	Conditions	% Epoxide
H	cyclohexene	dioxane	MoO_3	10	50°C, 10h	38
t-Bu	cyclohexene	dioxane	MoO_3	12	80°C, 8h	99
t-Bu	cyclohexene	neat	$MoO_2(acac)_2$	2	70°C, 1h	86
cumyl	cyclohexene	PhEt	V-naphth.	0.03	90°C, 1h	97
sec-Bu	cyclopentene	neat	V-octanoate	1	46°C, 6h	98
t-Bu	1-octene	EtOAc	$Na_2MoO_4/Na_4(PMo_{12}O_{40})$	8	135°C, 2h	89
cumyl	1-octene	cumene	MoO_3	19	100°C, 6h	84
tert-amyl	1-propene	neat	Mo-naphth.	2	130°C, 2h	78
cumyl	2-methyl-2-pentene	neat	$Na_2MoO_4/Na_4(PMo_{12}O_{40})$	1.5	110°C, 0.75h	97

[Reprinted with permission of Hiatt, R., In *Oxidation, Vol. 2*, by Augustine, R.L.; Trecher, D.J. (eds.), Marcel Dekker, Inc., New York, *1971*, p. 124 by courtesy of Marcel Dekker Inc.]

Tertiary alkylhydroperoxides are the most common reagents used to oxidize alkenes, since primary or secondary alkylhydroperoxides are susceptible to rearrangement and decomposition. Alkylhydroperoxides are relatively soluble in organic solvents, are more stable and are easier to handle than hydrogen peroxide.[266] Both TBHP and cumyl hydroperoxide are commercially available and widely used. As with H_2O_2, the reaction of alkenes with hydroperoxides usually requires transition metal catalysts in order to form an epoxide.[267] Several common catalyst systems are given in Table 3.4.[268] Epoxidation with hydroperoxide gives racemic products in most cases. An important exception is the **Sharpless asymmetric epoxidation**, which generates chiral epoxides and will be discussed in sec. 3.4.D.i.

More highly substituted alkenes are epoxidized faster than less substituted alkenes.[269] Reaction of **168** with TBHP and molybdenum hexacarbonyl selectively gave oxidation of the trisubstituted alkene, leading to **169**.[270] Electron deficient alkenes are less reactive, as illustrated by the

265. (a) Pritzkow, W.; Gröbe, K.H. *Chem. Ber. 1960, 93*, 2156; (b) Ester, W.; Sommer, A. U.S. Patent 3,259,661, British Patent 874,603 *1961* [*Chem. Abstr. 57*: 1190e, *1962*].
266. Reference 254, p 216.
267. Hiatt, R. in *Oxidation*, Vol. 2; Augustine, R.L.; Trecher, D.J. (Eds.), Marcel Dekker, New York, *1971*, pp 117-123.
268. Reference 267, p 124.
269. Reference 267, p 126.
270. (a) Tolstikov, G.A.; Yur'ev, V.P.; Dzhemilev, U.M. *Russ. Chem. Rev. 1975, 44*, 319; (b) Sheng, M.N.; Zajacek, J.G. *J. Org. Chem. 1970, 35*, 1839.

epoxidation of geranial (**102**), which was epoxidized selectively at the unconjugated alkene moiety to give **170**.[271]   A neighboring group effect is possible when a coordinating group is present, as in allylic alcohols. Epoxidation of allylic alcohol **171**, in Tadano and co-worker's synthesis of (+)-cheimonophyllon E,[272] gave a 78% yield of a 5.6:1 mixture of epoxides **172:173**.

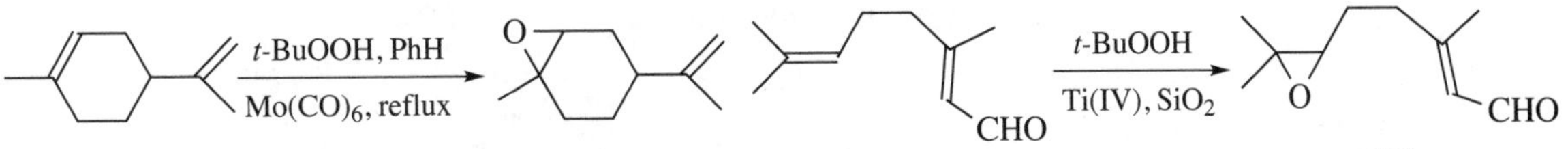

Such reactions involve initial coordination of the metal to both the allylic alcohol and the hydroperoxide, followed by displacement at the peroxy group by the alkene moiety.[273] Although free radicals are not formally involved in the epoxidation, small amounts of radical decomposition products are sometimes observed.[274] Sharpless proposed a mechanism for the vanadium-catalyzed epoxidation[275a] that explains the stereoselectivity and provides insight into metal-catalyzed epoxidations. A vanadium(V) species such as **174** reacts with the allylic alcohol and TBHP to give **175** via displacement of two alkoxy ligands. Coordination of the second oxygen of the peroxide generates a metal peroxide, **176**. The rate-determining step of this process is the epoxidation (**176** → **177**), and the stereochemistry of the epoxide is determined at this point in the mechanism. Displacement of the epoxide probably proceeds via **178**, to regenerate the vanadium catalyst.

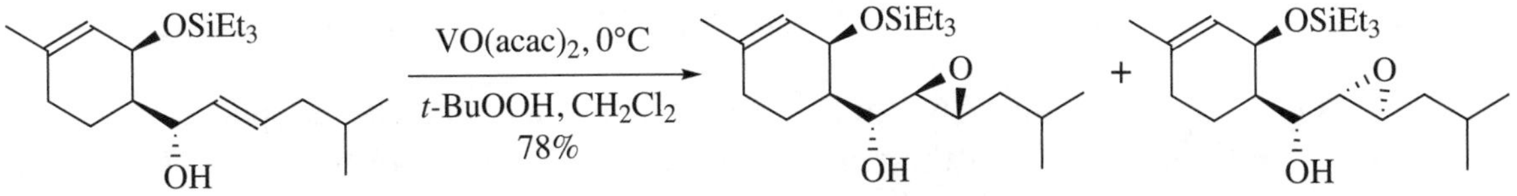

271. (a) Sheldon, R.A. *J. Mol. Catal., 1980, 7,* 107; (b) Reference 1, p 278.
272. Takao, K.; Tsujita, T.; Hara, M.; Tadano, K. *J. Org. Chem. 2002, 67,* 6690.
273. (a) Gould, E.S.; Rado, M. *J. Catal. 1969, 13,* 238; (b) Masuo, F.; Kato, S. *Yuki Gosei Kagaku Kyokai Shi 1968, 26,* 367 [*Chem. Abstr. 69:* 18476u, *1968*].
274. Sheng, N.M.; Zajacek, J.C. *Advances in Chemistry Series No. 76,* American Chemical Society, Washington D.C., *1968,* and see citation 30 in reference 267, p 135.
275. (a) Sharpless, K.B.; Verhoeven, T.R. *Aldrichimica Acta 1979, 12,* 63; (b) Sharpless, K.B.; Michaelson, R.C. *J. Am. Chem. Soc. 1973, 95,* 6136.

In **176**, the alkene π bond will attack the peroxide moiety from the rear along the axis of the O–O bond.[275a] In order to achieve backside attack, the allylic portion of the complex must deform as in **179**, giving an **O–C–C=C** bond angle of ~ 50°.

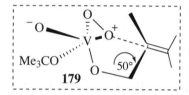

This model predicts the diastereoselectivity for the epoxidation of acyclic allylic alcohols such as **180**, which gave a mixture of **181** and **182** in a ratio of 71:29.[275a] Examination of the transition states for the epoxidation of **180** (**183** and **184**) show a destabilizing interaction between methyl and hydrogen as compared to a hydrogen-hydrogen interaction. In **184**, a severe steric interaction exists between the alkenyl methyl and the methyl on the adjacent carbon, as well as the oxygen-vanadium complex. In **183**, the lower energy destabilizing interaction of the hydrogen and oxygen-vanadium complex suggests is lower in energy. If **183** is lower in energy, **181** is predicted to be the major product. The energy difference between **183** and **184** is not great, and significant amounts of **182** are observed.

The stereochemical consequences of this neighboring group effect were apparent in the transformation **171 → 172** above. The stereoselectivity is a bit easier to see when a cyclic alkene is involved. In a synthesis of phytuberin by Suárez and co-workers, allylic alcohol **185** was treated with VO(acac)$_2$ and t-BuOOH in benzene, and an 85% yield of **186** was obtained.[276] The hydroxyl unit directed the epoxidation to the top face of the bicyclic system, as shown. Note the oxidation of the sulfide to the corresponding sulfoxide (see Sec 3.9.A.i) during the epoxidation.

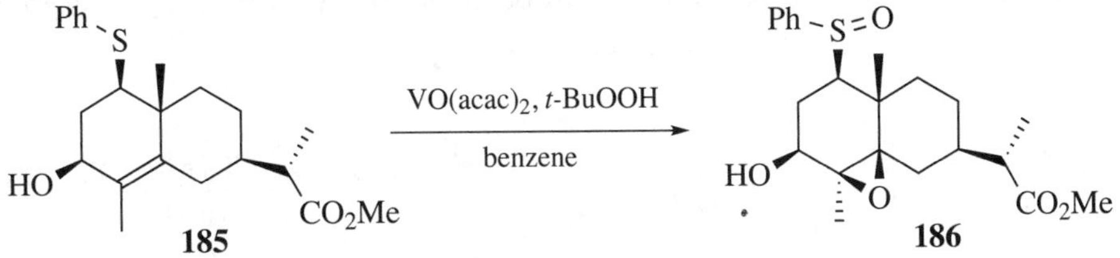

In basic media, peroxides react as acids and deprotonation gives a peroxide-anion that reacts with conjugated carbonyl derivatives via 1,4-addition (conjugate addition). Hydrogen peroxide reacts with base to give the hydroperoxide anion (HOO$^-$), which adds to α,β-unsaturated ketones and aldehydes[277,278] or esters.[279] In a synthesis of (−)-phyllostine by Okamura and co-workers,[280] the reaction of conjugated ketone **187** with basic hydrogen peroxide converted it to the corresponding epoxide (**189**) via an intermediate such as **188**, in which the enolate anion displaced the hydroxyl group to yield the epoxide. Note that the conjugate addition proceeded from the less hindered bottom face of the molecule.

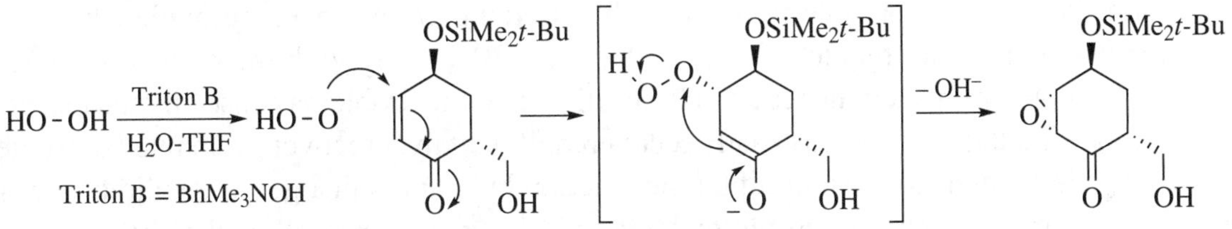

For α,β-unsaturated aldehydes and ketones, the presence of a single carbonyl group is sufficient to activate the double bond to conjugate addition with hydroperoxide anions. This reactivity does not extend to α,β-unsaturated esters, however, and epoxidation does not occur unless two ester groups are present [such as is found with alkylidene malonate derivatives, CH$_2$=C(CO$_2$R)$_2$].[281] Peroxide anion epoxidation is usually stereoselective but rarely stereospecific.[282]

Alkyl hydroperoxides react with a suitable base to give a hydroperoxide anion (ROO$^-$), directly analogous to the reactions of hydrogen peroxide. The reaction with conjugated carbonyl

276. Prangé, T.; Rodriguez, M.S.; Suárez, E. *J. Org. Chem.* **2003**, *68*, 4422. For a different example where the steric bias of the ring system does not cloud the issue of neighboring group assistance, see Kita, Y.; Furukawa, A.; Futamura, J.; Higuchi, K.; Ueda, K.; Fujioka, H. *Tetrahedron Lett.* **2000**, *41*, 2133.
277. Nishigawa, M.; Grieco, P.A.; Burke, S.D.; Metz, W. *J. Chem. Soc. Chem. Commun.* **1978**, 76.
278. Ohfune, Y.; Grieco, P.A.; Wang, C.-L.; Majetich, G. *J. Am. Chem. Soc.* **1978**, *100*, 5946.
279. Payne, G.B.; Williams, P.H. *J. Org. Chem.* **1959**, *24*, 284.
280. Okamura, H.; Shimizu, H.; Yamashita, N.; Iwagawa, T.; Nakatani, M. *Tetrahedron* **2003**, *59*, 10159.
281. Payne, G.B. *J. Org. Chem.* **1959**, *24*, 2048.
282. House, H.O.; Ro, R.S. *J. Am. Chem. Soc.* **1958**, *80*, 2428.

derivatives is virtually the same for both HOO⁻ and ROO⁻, except for reactions with α,β-unsaturated nitriles.[283] Alkaline hydrogen peroxide converts such nitriles to epoxy amides,[284] as in the reaction of acrylonitrile to give **191**, via initial formation of **190**.[284] The related reaction of **192** with *tert*-butylhydroperoxide in basic solution, however, gave good yields of the epoxy-nitrile (**193**) rather than the amide.[284]

3.4.B.iii. Juliá-Colonna Epoxidation.

A useful innovation involving the hydroperoxide anion epoxidation of conjugated carbonyl compounds uses a polyamino acid such as poly-*L*-alanine or poly-*L*-leucine in what is known as the **Juliá-Colonna epoxidation**.[285] Originally, the reaction was done in a water-immiscible solvent such as toluene, in which the substrate was treated with an aqueous NaOH–H_2O_2 solution containing the polyamino acid. In general, harsh reaction conditions were required, including long reaction times and recovering the product from the gel-like catalyst as well as the catalyst was often difficult.[286] These problems were diminished by using a biphasic system, that used a urea-hydrogen peroxide complex as the peroxide donor in an organic solvent containing DBU, along with the polyamino acid.[287] This modification diminished the reaction time, alleviated the problems caused by the use of an aqueous base, and the product was easily recovered.[286] Catalyst recovery remained a problem however.[286] Gellar and Roberts modified the reaction by using silica gel as a solid carrier for the poly-*L*-leucine, and performed the reaction in THF.[286] An example is the epoxidation of **194** to **195** in 93% yield, an improvement in both yield and reaction time when compared to the same reaction without the use of silica gel.[286]

The functional group transforms for this section are

---

283. Reference 254, p 232.
284. Payne, G.B. *Tetrahedron* **1962**, *18*, 763.
285. (a) Banfi, S.; Colonna, S.; Molinari, H.; Juliá, S.; Guixer, J. *Tetrahedron* **1984**, *40*, 5207. For recent reviews of this reaction, see (b) Pu, L. *Tetrahedron: Asymmetry* **1998**, *9*, 1457; (c) Ebrahim, S.; Wills, M. *Tetrahedron; Asymmetry* **1997**, *8*, 3163.
286. Geller, T.; Roberts, S.M. *J. Chem. Soc. Perkin Trans. 1* **1999**, 1397.
287. (a) Bentley, P.A.; Bergeron S.; Cappi, M.W.; Hibbs, D.E.; Hursthouse, M.B.; Nugent, T.C.; Pulido, R.; Roberts, S.M.; Wu, L.E. *Chem. Commun.* **1997**, 739; (b) Adger, B.M.; Barkley, J.V.; Bergeron S.; Cappi, M.W.; Flowerdew, B.E.; Jackson, M.P.; McCague, R.; Nugent, T.C.; Roberts, S.M. *J. Chem. Soc. Perkin Trans. 1* **1997**, 3501; (c) Savizky, R.M.; Suzuki, N.; Bové, J.L. *Tetrahedron: Asymmetry* **1998**, *9*, 3967.

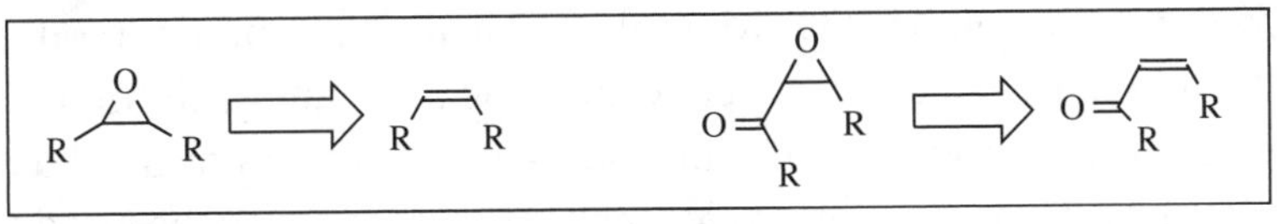

3.4.C. Organic Peroxyacids

Peroxyacids are the most common reagents for the epoxidation of alkenes, which has been called the **Prilezhaev (Prileschajew) reaction**.[288,289] This reaction does not require transition metal catalysis, and gives good yields of the epoxide. Peroxyacids such as **164** are prepared by reaction of carboxylic acids with hydrogen peroxide. An equilibrium is established during the reaction that favors the peroxyacid, although several alternative methods are available.[288a] Strong acids such as formic or trifluoroacetic acid usually generate useful equilibrium concentrations of the peroxyacids upon reaction with hydrogen peroxide (the equilibrium is shifted to the right, forming **164**). Most other alkyl and aryl acids, however, require catalytic amounts of strong mineral acids or p-toluenesulfonic acid to give the corresponding peroxyacid.[290] Several peroxyacids have been commercially available (peroxyformic, peroxybenzoic, trifluoroperoxyacetic and m-chloroperoxybenzoic [mcpba, **196**]), although the purity of these reagents can be as low as 45-60% depending of the purity of the peroxide used in their manufacture. Peroxyacids are significantly less acidic than their carboxylic acid counterparts.[291] The major byproduct of the alkene epoxidation reaction is the carboxylic acid precursor to the peroxyacid (see below). If necessary, this acid can be neutralized during an epoxidation reaction by addition of sodium acetate (or another buffering agent).

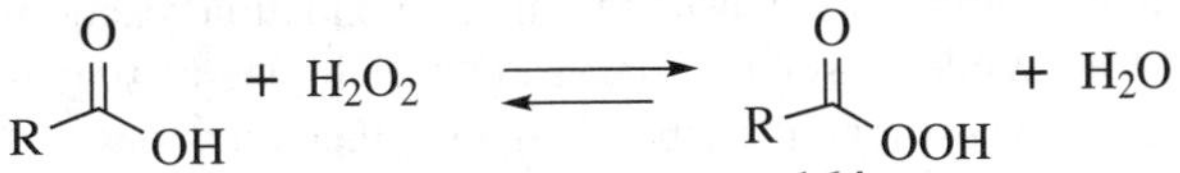

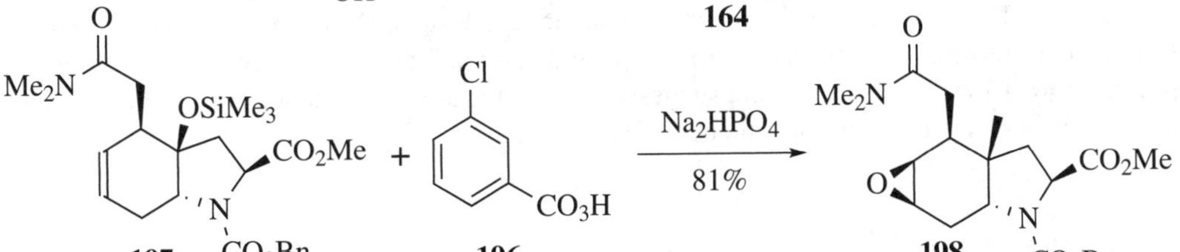

Trifluoroperoxyacetic acid (CF_3CO_3H) gives the strong organic acid trifluoroacetic acid (CF_3CO_2H) as a byproduct. This strong acid can catalyze rearrangement or other reactions of the epoxide product. Such secondary reactions are suppressed by addition of a buffer,

288. (a) Prileschajew, N. *Berichte* **1909**, *42*, 4811; (b) The Merck Index, 14th ed., Merck & Co., Inc., Whitehouse Station, New Jersey, **2006**, p ONR-76.
289. (a) Swern, D. *Chem. Rev.* **1949**, *45*, 1 (see p 16); (b) Swern, D. *Org. React.* **1953**, *7*, 378.
290. Reference 254, pp 215-220.
291. White, R.W.; Emmons, W.D. *Tetrahedron* **1962**, *17*, 31.

which is usually the acid salt ($RCO_2^- M^+$) of the acid by-product (sodium acetate with peroxyacetic acid, sodium benzoate with peroxybenzoic acid, and sodium trifluoroacetate with trifluoroperoxyacetic acid). Both sodium and potassium acetate are used as generic buffers, however, and sodium or potassium hydrogen phosphates can also be used. An example of a buffered reaction is the conversion of **197** to **198** in 81% yield, in Wipf and Li's synthetic studies for a synthesis of the tuberstemonone ring system.[292]

In one example of alkene epoxidation taken from a synthesis of (R)-7-hydroxycarvone from (S)-α-pinene by McIntosh and co-workers,[293] **199** was converted to epoxide **200** with mcpba in dichloromethane buffered with sodium bicarbonate. Although mcpba is one of the more popular epoxidation reagents, shortages of concentrated hydrogen peroxide limit the availaibity of high purity mcpba, although less pure reagent is usually available as mentioned above. Development of alternative reagents has therefore been an ongoing process. The peroxyacid magnesium monoperphthalate (abbreviated MMPP,[294] see the parent peroxy acid **216** below) was prepared by Brougham et al. and is an efficient reagent for the conversion of alkenes to epoxides in aqueous media.[295]

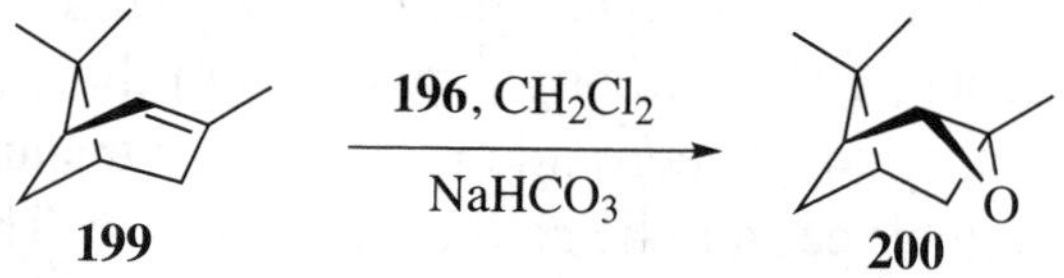

MOLECULAR MODELING: Facial Selectivity in Epoxidation

Facial selectivity is often an important consideration in epoxidation reactions, for example, in the conversion of **199** to **200**. While in some cases, a 2D sketch clearly suggests that one face of a molecule bond is more accessible than the other, in many other situations the 2D drawing may not be sufficient. The, mcpba epoxidation of the macrocyclic lactone shown below is one such case. In principle, substituents may shield both "top" and "bottom" faces of the double bond (see also problem 9p). However, the 3D model suggests that the ring atoms shield one face of the C=C unit, but the extent of shielding by substituents is not clear.

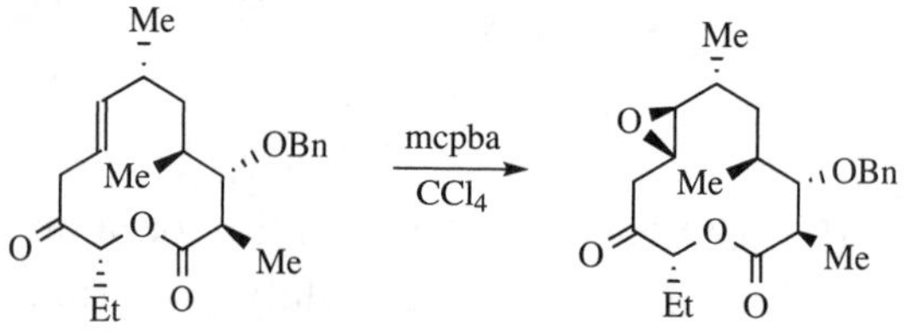

292. Wipf, P.; Li, W. *J. Org. Chem.* **1999**, *64*, 4576.
293. Lakshmi, R.; Bateman, T.D.; McIntosh, M.C. *J. Org. Chem.* **2005**, *70*, 5313.
294. For reactions of MMPP and urea hydrogen peroxide, see Heaney, H. *Aldrichimica Acta* **1993**, *26*, 35.
295. For an example, see Brougham, P.; Cooper, M.S.; Cummerson, D.A.; Heaney, H.; Thompson, N. *Synthesis* **1987**, 1015.

> **MOLECULAR MODELING: Facial Selectivity in Epoxidation, con't.**
>
> Examine the 3D model provided on screen. (This corresponds to the lowest-energy conformer resulting from examination of all conformers using molecular mechanics.) You might find it helpful to switch to a space-filling model to get a better idea of the difference between the two faces (select ***Space Filling*** from the **Model** menu at the top of the screen). Before you do so, orient the molecule as in the sketch, placing the trans-double bond to the top-left and the lactone carbonyl to the bottom. Is one face more shielded that the other? If so, which diastereomeric epoxide is likely to be observed?

A major advantage of using peroxyacids is the rapid epoxidation of unfunctionalized alkenes, which are normally unreactive with hydrogen peroxide or alkyl hydroperoxides without the presence of transition metal catalysts.[296] The reaction probably proceeds via a concerted mechanism as proposed by Bartlett,[297] in which the alkene and the oxygen of the OH unit in **164** coordinate and the epoxide is formed with loss of the carboxylic acid The transition state for this reaction has been analyzed using computational methods[298] and it is usually represented by **201**. The geometry of the transition state dictates a syn delivery of oxygen, without changing the geometry of the alkene. Other mechanisms have been proposed as well.[299]

164 **201**

202

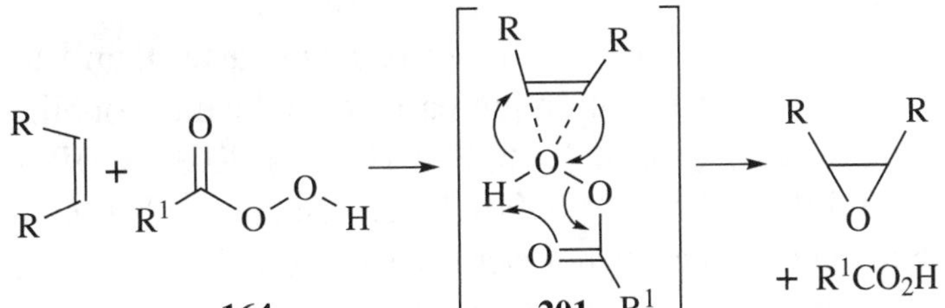

As first noted with hydroperoxides, neighboring group effects are also important in reactions of peroxyacids. Reexamination of the epoxidation of **180** using mcpba rather than TBHP reveals that the peroxyacid gave a 95:5 mixture of **181** and **182**. Using mcpba rather than TBHP led to a slight increase in diastereoselectivity for the reaction, although this depends on the substitution pattern of the alkene. Sharpless and co-workers proposed an intermediate based on

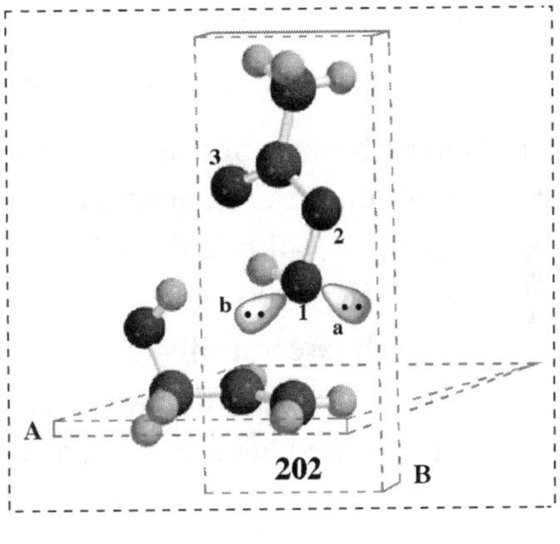

296. Reference 254, p 223.
297. Bartlett, P.D. *Rec. Chem. Progr.* ***1950**, 11*, 47.
298. Freccero, M.; Gandolfi, R.; Sarzi-Amadè; Rastelli, A. *J. Org. Chem.* ***2000**, 65*, 8948.
299. (a) Kwart, H.; Hoffman, D.M. *J. Org. Chem.* ***1966**, 31*, 419; (b) Hanzlik, R.P.; Shearer, G.O. *J. Am. Chem. Soc.* ***1975**, 97*, 5231; (c) Dryuk, V.G. *Tetrahedron* ***1976**, 32*, 2855; (d) Bach, R.D.; Canepa, C.; Winter, J.E.; Blanchette, P.E. *J. Org. Chem.* ***1997**, 62*, 5191.

202 (a generic model for any peroxyacid, although peroxyacetic acid is used here),[275] in which one plane (defined by the peroxyacid molecule in plane B) is oriented at an angle of ≈ 60° to the plane of the π bond (plane A). One of the non-bonding electron pairs on oxygen (electron pair a) lies in plane B and is properly situated to begin bonding with the alkene carbon. The other lone pair electrons (electron pair b) is positioned so it can hydrogen bond with the allylic hydroxyl group, as in **203** or in **204**. The selectivity of epoxidations involving peroxy acids was previously explained by hydrogen bonding to O^2 or O^3 in **202**.[300] If backside displacement of the peroxy bond is required, hydrogen bonding to O^2 or O^3 is not possible. A hydrogen bond to O^1, however, will have a dihedral angle for the **O–C–C=C** bond from ~ 50-130°, with 120° predicted as optimal (shown in **203** and **204**).[275a] In **203**, the destabilizing methyl-hydrogen interaction is less than the methyl-alkenylmethyl interaction in **204**. The peroxy moiety will interact with the alkenylmethyl moiety in both cases. The greater destabilizing interaction in **204** leads to a predominance of **181** (from **203**) with only small amounts of **182** (from **204**).

Alkenes with more substituents are electron rich, and react with peroxyacids faster than less substituted alkenes. In general, the relative rate of epoxidation increases with the nucleophilic character of the alkene[301] ($CH_2=CH_2$ has a relative rate of 1, $RCH=CH_2 = 24$, $RCH=CHR = 500$, $R_2C=CH_2 = 500$, $R_2C=CHR = 6500$, and $R_2C=CR_2 = >>6500$).[302] Tri- and tetrasubstituted double bonds are rapidly epoxidized with peroxy acids, but peroxybenzoic acids are more effective than peroxyalkanoic acids. If a molecule contains two alkene moieties, the more highly substituted alkene is epoxidized faster. 1,2-Dimethyl-1,4-cyclohexadiene, for example, gave primarily the tetrasubstituted epoxide upon reaction with mcpba.[303] This preference is seen in Sarpong and Simmons' synthesis of salviasperanol,[304] where the tetrasubstituted diene in **205** was epoxidized in preference to the disubstituted C=C unit to give **206**. In this case, the trisubstituted alkene allylic to the OH is epoxidized in preference to the non-allylic trisubstituted alkene, presumably via a neighboring group effect of the hydroxyl (Sec. 6.4).

300. Chamberlain, P.; Roberts, M.L.; Whitham, G.H. *J. Chem. Soc. B* **1970**, 1374.
301. Swern, D. *J. Am. Chem. Soc.* **1947**, *69*, 1692.
302. Reference 254, p 225.
303. Paquette, L.A.; Barrett, J.H. *Org. Syn. Coll. Vol. 5*, **1973**, 467.
304. Simmons, E.M.; Sarpong, R. *Org. Lett.* **2006**, *8*, 2883.

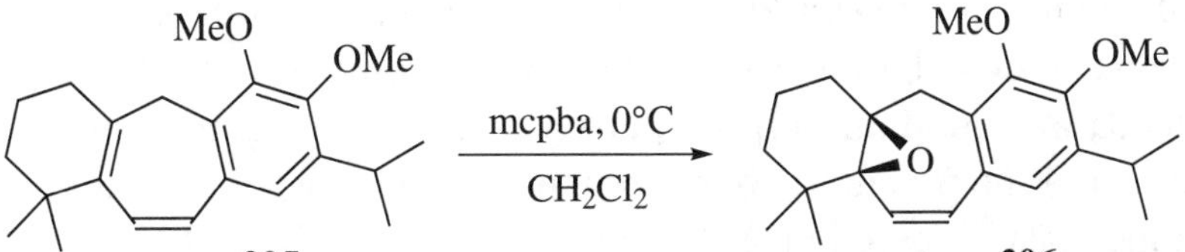

205 → **206**

mcpba, 0°C
CH₂Cl₂

MOLECULAR MODELING: Selectivity in Epoxidation Reactions

Both positional and stereoselectivity in epoxidation reactions can be anticipated using local ionization potential maps. This colors an electron density surface which gives the overall molecular size and shape (where a reagent can approach) with the value of the local ionization potential (the energy required to expel an electron from a given location on the surface). By convention, the surface is colored "red" for small values of the local ionization potential (loosely-bound electrons) and "blue" for large values of the local ionization potential (tightly bound electron). Red regions are nucleophilic and are more likely to donate electrons to an electrophile such as the electrophilic oxygen of a peroxyacid.

Consider for example epoxidation of diene shown below.

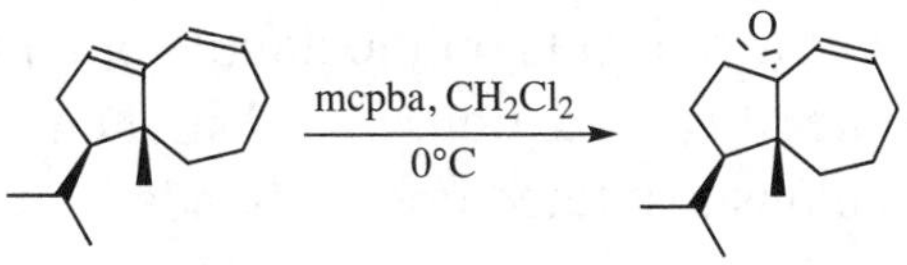

mcpba, CH₂Cl₂
0°C

While this reaction can give rise to one of four products depending on which face of which double bond is most susceptible, the authors characterize only a single product derived from epoxidation of the double bond on the five-membered ring *anti* to both alkyl substituents. (see homework problem 9f). They further note while that epoxidation at the five-member ring is clean, other (unspecified) products are observed.

Examine the local ionization potential map for this diene. It is depicted as a transparent model, allowing the underlying structures to be seen. You can switch to a solid model, allowing the maps to be seen more easily, by *clicking* on the map and then selecting **Solid** for the **Style** menu at the bottom of the screen. For which double bond, that belonging to the five or seven member ring, is the local ionization potential smaller? For which face on this ring, *syn* or *anti* to the two alkyl groups, is the local ionization potential smaller? *Click* on different regions on the map and the value of the local ionization potential will be displayed at the bottom of the screen. Does addition of the peroxyacid to the preferred face of the preferred double bond yield the observed epoxidation product? According to the local ionization potential map, are any other sites on the diene likely to be competitive? If so, identify the product(s) that would result.

Optional: Is the observed epoxidation product also the thermodynamic product? Obtain energies for all four products. Is this reaction likely to be thermodynamically controlled?

Monosubstituted alkenes are generally less reactive and epoxidation may require the use of a strong peroxyacid such as trifluoroperoxyacetic acid, although mcpba is often effective.[305] For most alkenes mcpba is a good all-purpose peroxyacid. Even highly hindered alkenes can be epoxidized.[306] If there is no possibility of hydrogen bonding, the peroxyacid usually approaches the alkene from the least hindered face,[307,308] and such reactions are highly diastereoselective. Selectivity depends on the peroxyacid used as well as the reaction conditions. Epoxidation of **207** with mcpba in THF, for example, gave an 84% yield of epoxide **208** with only 4% of **209** in an example taken from the Azuma et al. synthesis of sphingosine.[309] When done in benzene, the reaction was complete in only 4 hours but the ratio of **208/209** was 82:18. Although **208** was the major product in all cases, diastereoselectivity varied with solvent and reaction time and in some cases the stereoselectivity was much lower.

The presence of electronegative substituents on the double bond lowers the nucleophilicity of the alkene and decreases the rate of epoxidation.[310,311] Vinyl chlorides,[312] vinyl acetates[313] and vinyl ethers[314] react slower than unsubstituted double bonds.[315] Glycidic esters (epoxy esters) such as **210**, for example, can be formed from conjugated esters such as ethyl 2-butenoate using mcpba[316] or trifluoroperoxyacetic acid, with a disodium phosphate or a sodium or potassium hydrogen phosphate buffer.[317] The epoxides resulting from epoxidation of these substrates are often labile, however, and can rearrange to an α-halo or α-carboalkoxy ketone in unbuffered reactions.[318]

305. Reference 254, p 226.
306. Abruscato, G.J.; Tidwell, T.T. *J. Org. Chem.* **1972**, *37*, 4151.
307. (a) Christenson, P.A.; Willis, B.J. *J. Org. Chem.* **1979**, *44*, 2012; (b) Hickinbottom, W.J.; Wood, D.G.M. *J. Chem. Soc.* **1953**, 1906.
308. Swindell, C.S.; Britcher, S.F. *J. Org. Chem.* **1986**, *51*, 793.
309. Azuma, H.; Tamagaki, S.; Ogino, K. *J. Org. Chem.* **2000**, *65*, 3538.
310. Ogata, Y.; Tabushi, I. *J. Am. Chem. Soc.* **1961**, *83*, 3440.
311. White, R.W.; Emmons, W.D. *Tetrahedron* **1962**, *17*, 31.
312. Mousseron M.; Jacquier, R. *Bull. Chim. Soc. Fr.* **1950**, 698.
313. Shine, H.J.; Hunt, G.E. *J. Am. Chem. Soc.* **1958**, *80*, 2434.
314. Stevens, C.L.; Tazuma, J. *J. Am. Chem. Soc.* **1954**, *76*, 715.
315. Moffett, R.B.; Slomp, Jr., G. *J. Am. Chem. Soc.* **1954**, *76*, 3678.
316. Schwartz, N.N.; Blumbergs, J.H. *J. Org. Chem.* **1964**, *29*, 1976.
317. Emmons, W.D.; Pagano, A.S. *J. Am. Chem. Soc.* **1955**, *77*, 89.
318. (a) McDonald, R.N.; Schwab, P.A. *J. Am. Chem. Soc.* **1963**, *85*, 820, 4004; (b) Stevens, C.L.; Dykstra, C.J. *Ibid.* **1953**, *75*, 5975; (c) Gardner, P.D. *Ibid.* **1956**, *78*, 3421.

Conjugation diminishes the reactivity of an alkene in epoxidation reactions.[319] Conjugated esters, for example, are less reactive than simple alkenes due to the electron-withdrawing effect of the carbonyl. Isolated (unconjugated) alkenes are usually epoxidized before conjugated alkenes as seen in the reaction of **211** to give **212**, taken from the synthesis of (+)-salvadione-A by Majetich and co-workers.[320]

The neighboring group effect observed in the reaction of allylic alcohols and hydroperoxides is observed with peroxyacids, as it was in the epoxidation reactions discussed in previous sections. The so-called **Henbest rule** states that allylic alcohols exert a strong directing influence upon epoxidation when peroxyacids are used.[321] ***However, this effect does not operate in medium sized rings, and rings with significant flexibility tend to give the trans-product.***[322] An example involving a cyclopentenol derivative is shown in a synthesis of (+)-streptazolin by Miller and Li.[323] Epoxidation of **213** with mcpba and sodium bicarbonate led to a 92% yield of **214** with

epoxidation occurring on the same face as the ring hydroxyl group. Functional groups other than hydroxyl are capable of a neighboring group effect. The relative ability of a group adjacent to a double bond to coordinate with a peroxy acid in such a way that there is syn delivery of the epoxide oxygen is[324]

$$OH > CO_2H > CO_2R > OCOR$$

Occasionally, the proximity of the alkene to another reactive group can lead to a competing reaction. In the case of **215**, the hydroxyl is too far away to show a significant neighboring group effect and epoxidation with perphthalic acid (**216**) proceeded from the exo-face.[302,325] The trifluoroacetic acid byproduct of the epoxidation protonated the exo-epoxide (see **217**), however, allowing the hydroxyl unit on the other side of the molecule to react as shown to give alcohol-ether **218** as the final product.[302,325] This type of secondary reaction can be prevented

319. Hiyama, T.; Kanakura, A.; Yamamoto, H.; Nozaki, H. *Tetrahedron Lett.* **1978**, 3051.
320. Majetich, G.; Wang, Y.; Li, Y.; Vohs, J.K.; Robinson, G.H. *Org. Lett.* **2003**, *5*, 3847.
321. Henbest, H.B.; Wilson, R.A. *J. Chem. Soc.* **1957**, 1958.
322. Itoh, T.; Jitsukawa, K.; Kaneda, K.; Teranishi, S. *J. Am. Chem. Soc.* **1979**, *101*, 159.
323. Li, F.; Miller, M.J. *J. Org. Chem.* **2006**, *71*, 5221.
324. Reference 233a, p 101.
325. Rosowsky, A. in *Heterocyclic Compounds, Part 1*, Weissberger, A. (Ed.), Wiley Interscience, New York, ***1964***, pp 1-523. For another example, see Tinsley, S.W. *J. Org. Chem.* **1957**, *24*, 1197.

by protection of the OH unit (sec. 7.3.A) or, in some cases, buffering will minimize acid-catalyzed secondary processes.

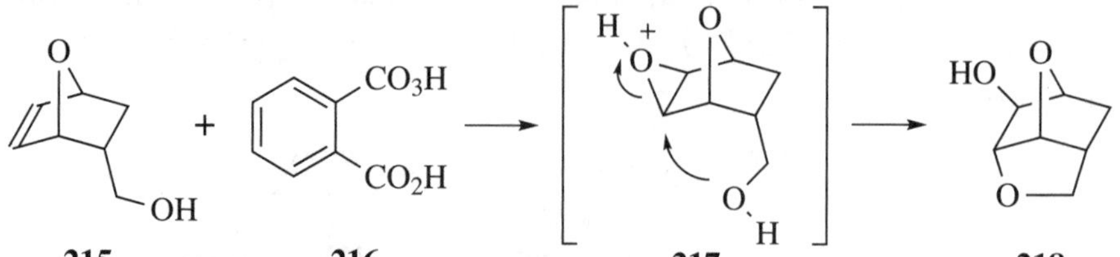

215   216   217   218

The functional group transforms are

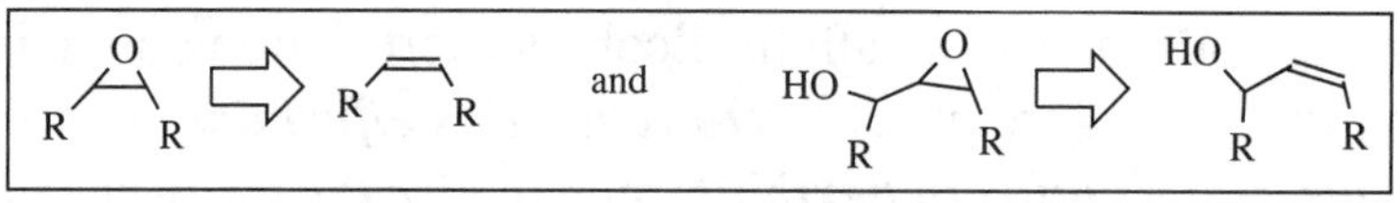

3.4.D.   Asymmetric Epoxidation Reactions

The previous section described the metal catalyzed epoxidation of allylic alcohols by alkyl hydroperoxides, and **202** was proposed as a model to predict the diastereoselectivity of these reactions.   In the cases presented, the epoxidation reaction was diastereoselective but not enantioselective (sec. 1.4.F), generating racemic epoxides.   To achieve asymmetric induction one must control both the orientation of the alkene relative to the peroxide and also the face of the substrate from which the electrophilic oxygen is delivered.   Such control can be accomplished by providing a chiral ligand that will coordinate to the metal catalyst, which is also the peroxide and the alkene unit.   There are two major asymmetric epoxidation reactions currently used in synthesis.   One can be applied only to allylic alcohols and is the prototype for asymmetric induction in these systems.   The other is a procedure that can be applied to simple alkenes.   Both use a metal-catalyzed epoxidation that employs alkyl hydroperoxides.

3.4.D.i.   Sharpless Asymmetric Epoxidation.[326]   The epoxidation reactions described in previous sections generated racemic products even when they were highly diastereoselective. The reagents presented simply did not provide the facial bias required for asymmetric induction. That changed in 1980, when Sharpless reported that epoxidation of allylic alcohols with *tert*-butyl hydroperoxide, catalyzed by tetraisopropoxy titanium (IV) [Ti(O*i*-Pr)$_4$], gave the epoxy alcohol with high enantioselectivity when (+)- or (–)-diethyl tartrate was added to the reaction medium.[327,328,275]   Reaction of geraniol (**101**) under these conditions, in the presence of L-(+)-diethyl tartrate (DET), gave **219** in 77% yield and 95% ee for the (2*S*,3*S*) stereoisomer.[327]   Note

326.   The Merck Index, 14th ed., Merck & Co., Inc., Whitehouse Station, New Jersey, *2006*, p ONR-85.
327.   Katsuki, T.; Sharpless, K.B. *J. Am. Chem. Soc.* *1980, 102,* 5974.
328.   (a) Rossiter, B.E.; Katsuki, T.; Sharpless, K.B. *J. Am. Chem. Soc.* *1981, 103,* 464; (b) Martin, V.S.; Woodard, S.S.; Katsuki, T.; Yamada, Y.; Ikeda, M.; Sharpless, K.B. *Ibid.* *1981, 103,* 6237.

that the natural isomer of diethyl tartrate [L-(+)-] is readily available as is the *unnatural* tartrate [(D-(−)-], although the latter is significantly more expensive. When the epoxy alcohol product is water soluble, optical activity and yield were diminished.[327]

When the previously cited transition structure **202** is applied to the asymmetric epoxidation of allylic alcohols, it must be modified to include binding of the peroxide, the allylic alcohol and also the chiral tartrate. The metal in this reagent is titanium rather than vanadium, and tetraisopropoxy titanium was found to react with two equivalents of diethyl tartrate to form a species such as **220**, where OR = Oi-Pr and CO$_2$R = CO$_2$Et.[329,330] The tartrate can bind to titanium from either the bottom or the top face.[330] The nature of the OR and E groups will determine the facial preference and, thereby, the selectivity of the epoxidation. When **220** reacted with TBHP and an allylic alcohol, two equivalents of 2-propanol were displaced to generate **221**. The peroxide linkage and the allylic alcohol in **221** are bound to the metal and backside attack of the alkene moiety along the axis of the O–O bond will lead to the epoxide. The titanium serves as a template for the reaction and the presence of the chiral tartrate makes **221** a chiral molecule. If the ester groups shown as E in **221** were not present, the allylic alcohol could bind to titanium on both the top and the bottom. When the metal binds tartrate (the source of the E groups in **220** and **221**), a pocket is created that allows the allylic alcohol and the peroxide to coordinate. The face opposite this pocket is too sterically hindered for efficient binding, so binding occurs preferentially from one face and the usual backside attack leads to a single enantiomer. The extent of this enantioselectivity is determined by the extent to which binding of the allylic alcohol and peroxide is limited to a single face. Sharpless showed that this process proceeds by a series of reversible (equilibrium) steps for binding alcohol, alkoxy groups, and peroxide to titanium.[330]

[Reprinted with permission from Burns, C.J.; Martin, C.A.; Sharpless, K.B. *J. Org. Chem.* **1989**, *54*, 2826. Copyright © 1989 American Chemical Society.]

329. (a) Burns, C.J.; Martin, C.A.; Sharpless, K.B. *J. Org. Chem.* **1989**, *54*, 2826; (b) Carlier, P.R.; Sharpless, K.B. *Ibid.* **1989**, *54*, 4016.
330. Johnson, R.A.; Sharpless, K.B. *Comprehensive Organic Synthesis Volume 7*, Chapter 3.2, *Asymmetric Epoxidations*, Pergamon Press, Oxford, **1990**.

Sharpless provided an empirical model that enables one to predict the stereochemistry of the epoxide formed in this reaction. When using the natural L-(+)- (2R, 3R) tartrate, the allylic alcohol should be oriented as shown for **222**, with the hydroxyl group in the plane of the double bond and to the right.[328c] With this orientation, the (+)-tartrate (in this case diisopropyl tartrate, DIPT) delivers oxygen (remember it is the alkene that attacks the peroxide) from the face bearing the hydroxyl (the *re-si* face) to give primarily the anti-product (**223**) with trace amounts of the syn-isomer **224**.[328c] If the model in **222** is drawn again as **225** and **226** for the reaction that forms **223** and **224**, the source of this selectivity can be determined. In **225**, the CO$_2$*i*-Pr↔H interaction is small compared to the cyclohexyl↔CO$_2$*i*-Pr interaction in **218** (see **217B** and **218B**; hydrogen atoms removed for clarity). The preference for **225** leads to approach of peroxide oxygen from the si face of the alkene. In **225**, the (*S*)-chiral center of the allylic alcohol gives a better fit with the stereogenic center of the tartrate (the chirality is matched) than does the (*R*)-stereogenic center of the alcohol in **226**. Such matching of stereogenic centers is referred to as **consonance** (as in **225**) whereas the interaction of mismatched stereogenic centers (in **226**) is called **dissonance**.

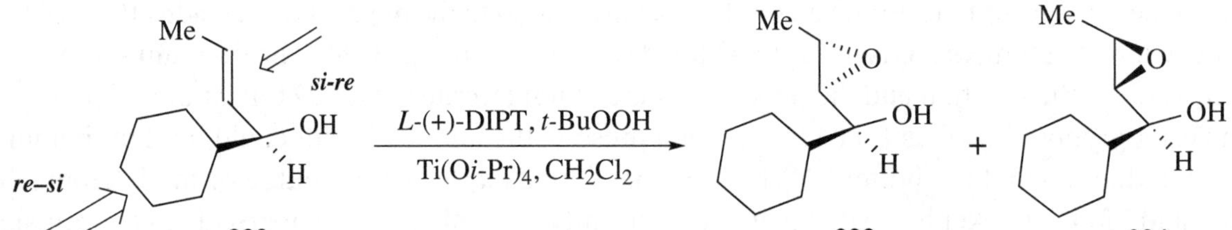

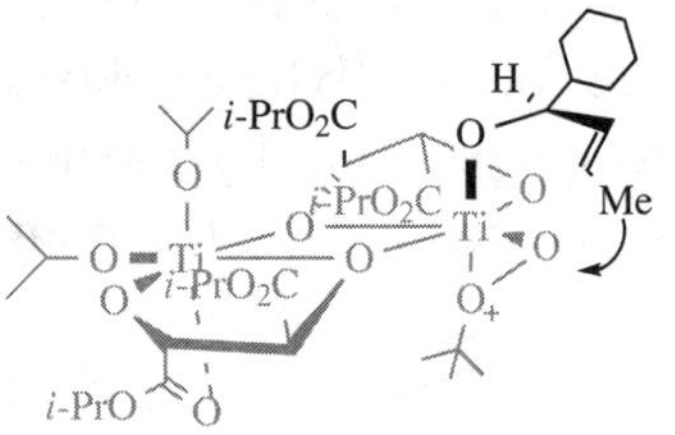

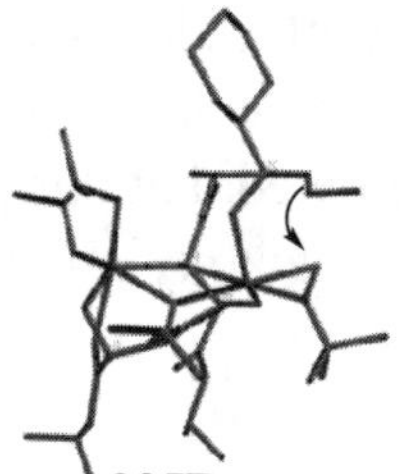

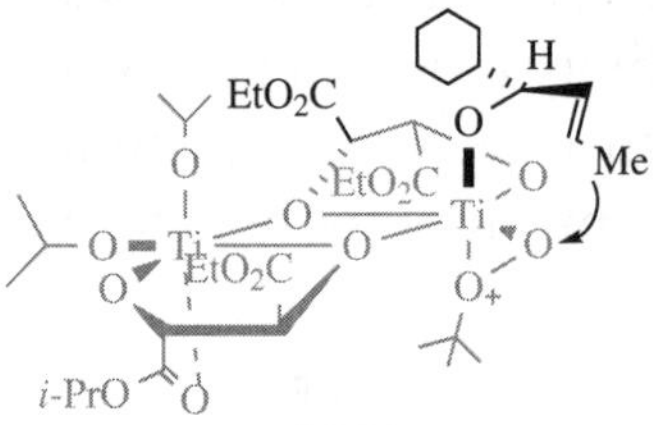

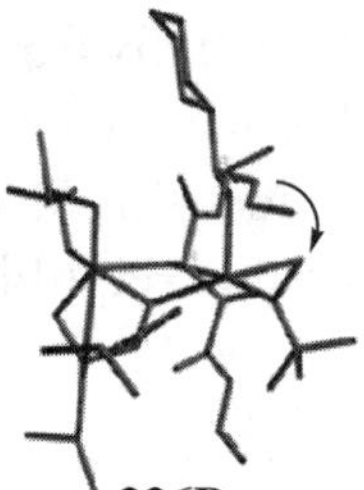

This model is formally illustrated in Figure 3.1[330] for DET, which shows the propensity for (+)-DET to give the anti-product from the *re-si* face and for (−)-DET to give the syn-product from the *si-re* face. Figure 3.1 illustrates consonance when the chirality of natural (+)-DET leads to faster reaction from the *si* face to give the anti-product. Conversely, the chirality of (−)-DET is mismatched to the *si* face making that reaction slower, although it is matched to the *re* face and that epoxidation is faster. Allylic alcohol **222**, for example, reacts 104 times faster with (L)-(+)-DIPT than the enantiomer (**229**), due to the consonance with the (*S*)-enantiomer (and the corresponding dissonance of this tartrate with the (*R*)-enantiomer). This difference in reactivity led to **kinetic resolution** (one enantiomer reacts faster to give a predominance of that epoxide) of the allylic alcohols.[328c] In general, the additive (L)-(+)-DIPT reacts best with allylic alcohols of the type represented by generic structure **228**, where (*D*)-(–)-DIPT would be the choice for **227**. The (*R*)-enantiomer of alcohol **222** is **229**, and it reacts with (*L*)-(+)-DIPT to give a mixture of diastereomers (**230** and **231**), with the syn-isomer (**231**) predominating (38:62).

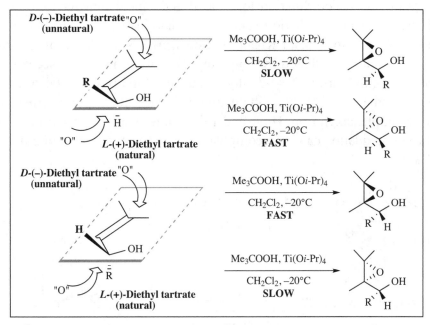

Figure 3.1. Facial selectivity on allylic alcohols for (+) and (–)-diethyl tartrate. [Reprinted with permission from Johnson, R.A.; Sharpless, K.B. *Comprehensive Organic Synthesis*, Volume 7, Chapter 3.2, *Asymmetric Epoxidations*, Pergamon Press, Oxford, *1990*. Copyright 1990, with permission from Elsevier Science.]

The rate for reaction of one enantiomeric allylic alcohol increases in comparison with the other, as the size of the alkyl group on tartrate increases (Me → Et → *i*-Pr).[330] Inspection of **230/231** reveals that changing the tartrate ester from methyl to ethyl and then to isopropyl increases the steric preference for one enantiomer (see above). The CO_2R^1↔cyclohexyl interaction is greater when R^1 is isopropyl than when R^1 is methyl, which will lead to faster binding of **222** than of **229** and faster epoxidation. Figure 3.1 also illustrates the source of kinetic resolution in the matched and mismatched interaction of (+)- or (–)-DET with a chiral allylic alcohol. This methodology can be applied in a variety of synthetic manipulations.[331,332]

Sharpless asymmetric epoxidation is a reaction where an achiral precursor is converted to a chiral substrate with high enantioselectivity. This discovery led to the application of the basic principles of ligand coordination to a metal and ligand transfer to a variety of other reactions. There are many synthetic examples of Sharpless asymmetric epoxidation. One is the conversion of **232** to **233**, in Knapp and co-worker's synthesis of the liposidomycin diazepanone nucleoside,[333] mediated by (–)-diethyl D-tartrate (DET). A second example uses the enantiomer of diethyl tartrate [(+)-diethyl L-tartrate (DET)] and is taken from Crimmins and Caussanel's synthesis of FD-891, where allylic alcohol **234** was converted to **235** in 72% yield.[334] A last example, taken from Hoye and Ye's synthesis of (+)-parviflorin,[335] shows that a double-Sharpless epoxidation can be accomplished, as in the conversion of **236** to **237** in 87% yield in ~ 97% ee.

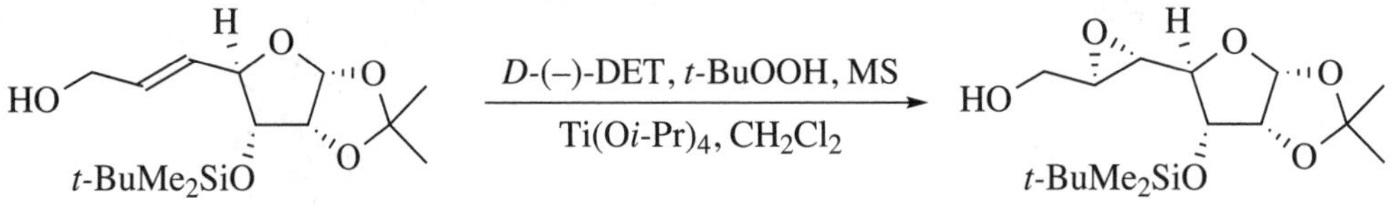

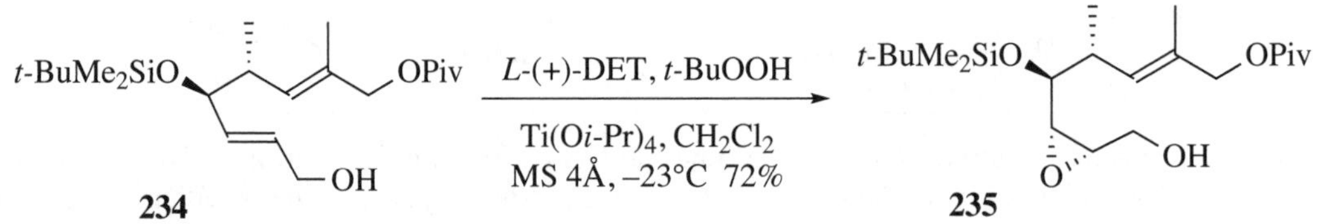

3.4.D.ii. Jacobsen-Katsuki Asymmetric Epoxidation. The Sharpless asymmetric epoxidation only works with allylic alcohols, as illustrated by the conversion of **232** to **233**. By changing the catalyst it is now possible to convert simple alkenes to the corresponding epoxide with high asymmetric induction. In independent work, Katsuki and Jacobsen showed that asymmetric epoxidation occurs using manganese-salen catalysts in the presence of *tert*-butylhydroperoxide,

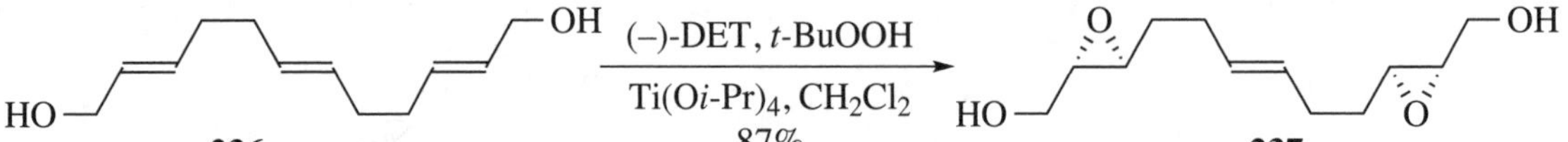

331. (a) Sharpless, K.B.; Behrens, C.H.; Katsuki, T.; Lee, A.W.M.; Martin, V.S.; Takatani, M.; Viti, S.M.; Walker, F.J.; Woodard, S.S. *Pure Appl. Chem.* **1983**, *55*, 589; (b) Reference 265; (c) Behrens, C.H.; Sharpless, K.B. *Aldrichimica Acta* **1983**, *16*, 67.
332. Katsuki, T.; Lee, A.W.M.; Ma, P.; Martin, V.S.; Masamune, S.; Sharpless, K.B.; Tuddenham, D.; Walker, F.J. *J. Org. Chem.* **1982**, *47*, 1373.
333. Knapp, S.; Morriello, G.J.; Doss, G.A. *Org. Lett.* **2002**, *4*, 603.
334. Crimmins, M.T.; Caussanel, F. *J. Am. Chem. Soc.* **2006**, *128*, 3128.
335. Hoye, T.R.; Ye, Z. *J. Am. Chem. Soc.* **1996**, *118*, 1801.

where salen = bis(salicylidene)ethylenediamine. A typical catalyst is **238** where R^1 can be an aryl group or cycloalkyl, bromine, trialkylsilyloxy and other groups. The manganese-salen catalyst **239** is called the **Jacobsen catalyst**, [N,N'-bis(3,5-di-*tert*-butylsalicylidene)-1,2-cyclohexane-diamine]manganese(III)] chloride. Metals other than manganese have also been used. It has been established that the order of mixing of the components is mechanistically important.[336] Although Mn(IV)-salen complexes are formed in the absence of alkenes, they play a minor role in common epoxidation reactions.[336]

Simple alkenes are oxidized with high asymmetric induction (the so- called the **Jacobsen-Katsuki reaction**).[337] In a typical transformation, styrene was converted to styrene oxide (**240**) in 89% yield and 86% ee using 2-8 mol% of **238** (R^1 = Ph, R^2 = OSi(*i*-Pr)$_3$, with the combination of mcpba and *N*-methylmorpholine *N*-oxide (NMO) as the oxidizing agent.[337k] Jacobsen used this procedure in a synthesis of leucotrine A$_4$ (LTA$_4$).[337d] In that synthesis, **241** was oxidized with sodium hypochlorite (NaOCl) and 20% 4-phenylpyridine *N*-oxide at pH 11.3, using 4% of the **239** catalyst). Epoxide **242** was obtained in 62% yield (82% ee and a trans/cis ratio of 8:1).[337d] Katsuki found that the degree of enantioselection in some Mn-salen complexes depends on the conformation of the ligand and the nucleophilicity of the substrate,[338] which is explained by enthalpy and entropy factors.

336.   Adam, W.; Mock-Knoblauch, C.; Saha-Möller, C.R.; Herderich, M. *J. Am. Chem. Soc.* **2000**, *122*, 9685.
337.   (a) Hatayama, A.; Hosoya, N.; Irie, R.; Ito, Y.; Katsuki, T. *Synlett* **1992**, 407; (b) Yamada, T.; Imagawa, K.; Nagata, T.; Mukaiyama, T. *Chem. Lett.* **1992**, 2231; (c) Schwenkreis, T.; Berkessel, A. *Tetrahedron Lett.* **1993**, *34*, 4785; (d) Chang, S.; Lee, N.H.; Jacobsen, E.N. *J. Org. Chem.* **1993**, *58*, 6939; (e) Sasaki, H.; Irie, R.; Katsuki, T. *Synlett* **1994**, 356; (f) Chang, S.; Heid, R.M.; Jacobsen, E.N. *Tetrahedron Lett.* **1994**, *35*, 669; (g) Hosoya, N.; Hatayama, A.; Irie, R.; Sasaki, H.; Katsuki, T. *Tetrahedron* **1994**, *50*, 4311; (h) Brandes, B.D.; Jacobsen, E.N. *J. Org. Chem.* **1994**, *59*, 4378; (i) Sasaki, H.; Irie, R.; Hamada, T.; Suzuki, K.; Katsuki, T. *Tetrahedron* **1994**, *50*, 11827; (j) Brandes, B.D.; Jacobsen, E.N. *Tetrahedron Lett.* **1995**, *36*, 5123; (k) Palucki, M.; McCormick, G.J.; Jacobsen, E.N. *Tetrahedron Lett.* **1995**, *36*, 5457; (l) The Merck Index, 14th ed., Merck & Co., Inc., Whitehouse Station, New Jersey, **2006** p ONR-48; (m) Mundy, B.P.; Ellerd, M.G.; Favaloro Jr., F.G. *Name Reactions and Reagents in Organic Synthesis, 2nd ed.*, Wiley-Interscience, New Jersey, **2005**, pp. 344-345.
338.   Nishida, T.; Miyafuji, A.; Ito, Y.N.; Katsuki, T. *Tetrahedron Lett.* **2000**, *41*, 7053.

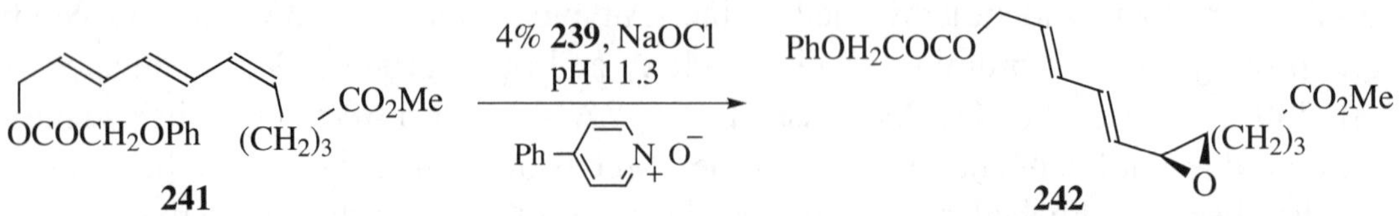

This epoxidation reaction is useful for disubstituted alkenes and also for trisubstituted alkenes.[337h] Epoxidation of trisubstituted alkenes proceeds with absolute asymmetric induction that was opposite to that observed with cis- or trans-disubstituted alkenes.[337h] Jacobsen suggested a general, skewed side-on approach transition state model[339] for the epoxidation that proceeds by a stepwise radical mechanism.[339b] Song et al. introduced a variation in which catalyst **239** is used in an ionic liquid consisting of 1,3-dialkylimidazolium cations with a PF_6^- counterion,[340] which allows the manganese-salen catalyst to be easily recycled. The mechanism of this reaction has been discussed elsewhere.[341] As mentioned above, radical intermediates have been postulated.[342] There is no doubt that this reaction will increase in importance in syntheses. Note that manganese-salen catalysts have been used in epoxide ring-opening reactions.[343]

Salen catalysts are increasingly used for kinetic resolution. The reaction of a racemic epoxide with cobalt salen catalysts in aqueous media leads to selective hydrolysis of one enantiomer, giving a diol and one enantiopure epoxide. An example of this **Jacobsen hydrolytic kinetic resolution** is the conversion of **243** to 47% of enantiopure diol **245** and 49% of enantiopure epoxide **246**, using 0.01 equivalents of salen catalyst **244**[344] in Smith and Kim's synthesis of (−)-indolizidine 223AB.

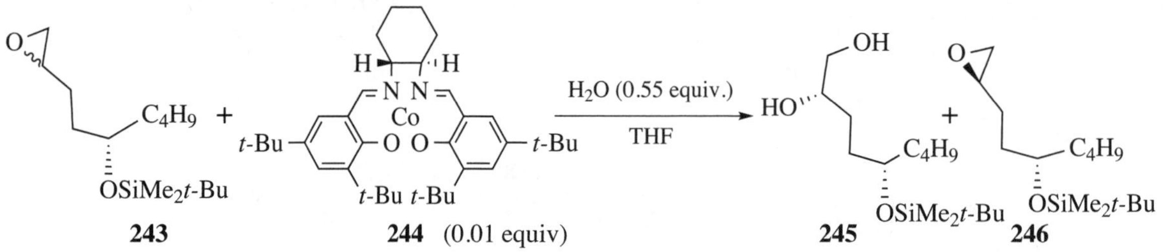

3.4.D.iii.   Epoxidation with Dioxiranes. There is growing interest in non-metal based epoxidations.[345] Dialkyl dioxiranes[346] such as **248** and **251** are versatile oxidizing agents

---

339. (a) Groves, J.T.; Myers, R.S. *J. Am. Chem. Soc.* **1983**, *105*, 5791; (b) Jacobsen, E.N.; Deng, L.; Furukawa, Y.; Martínez, L.E. *Tetrahedron* **1994**, *50*, 4323

340. Song, C.E.; Roh, E.J. *Chem. Commun.* **2000**, 837.

341. See Linker, T. *Angew. Chem,. Int. Ed.* **1997**, *36*, 2060 for a discussion of the mechanism.

342. Cavallo, L.; Jacobsen, H. *Angew. Chem. Int. Ed.* **2000**, *39*, 589.

343. Jacobsen, E.N. *Acc. Chem. Res.* **2000**, *33*, 421.

344. Smith III, A.B.; Kim, D.-S. *J. Org. Chem.* **2006**, *71*, 2547.

345. For a recent review, see Dalko, P.; Moisan, L. *Angew. Chem. Int. Ed.*. **2001**, *40*, 3726.

346. (a)Adam,W.;Curci,R.;Edwards,J.O.*Acc.Chem.Res.***1989**,*22*,205;(b)Murray,R.W.*Chem.Rev.***1989**,*89*,1187; (c)Curci,R.;Dinoi,A.;Rubino,M.F.*Pure.Appl.Chem.***1995**,*67*,811;(d)Denmark,S.E.;Wu,Z.*Synlett***1999**,847.

that can be generated from an oxidant and a ketone. The most commonly used oxidant in this epoxidation is potassium peroxomonosulfate ($KHSO_5$) generated from Oxone® ($2KHSO_5 \cdot KHSO_4 \cdot K_2SO_4$). When the dioxirane is generated in the presence of an alkene, an epoxide is formed. Dimethyldioxirane (**248**) is the parent of dialkyl dioxiranes used for the epoxidation of alkenes.[347] In a synthesis of merrilactone A,[348] Inoue and co-workers converted **247** to epoxide **249** in >95% yield (dr > 13:1) using **248**. If done in the presence of a chiral ketone, a chiral epoxide is formed.[349] In 1984, Curci was the first to report this chiral epoxidation using (+)-isopinocamphone (**250**) as a catalyst to convert (*E*)-β-methylstyrene to epoxide **252**, but in only about 12% ee (1*R*,2*R*).[350]

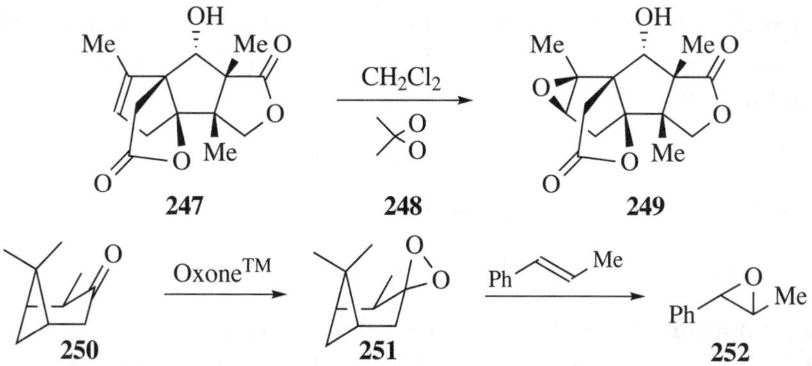

There are many examples that use Oxone® in conjunction with chiral ketones other than **250** for the enantioselective epoxidation of alkenes, and with high %ee. Yang and co-workers used a C_2-symmetric biaryl ketone that gave highly enantioselective oxidations.[351] Denmark and Matsuhashi employed chiral fluoro ketones with good chiral efficiency for asymmetric epoxidation.[352] This area of research is clearly growing in importance as an alternative to the titanium-based and manganese-salen-based oxidations.

3.5. CONVERSION OF ALKENES TO DIOLS (C=C → CHOH-CHOH)

Several reagents react with alkenes to generate 1,2-diols (dihydroxylation).[353] 1,2-Diols and

347. (a) Frohn, M.; Wang, Z.-X.; Shi, Y. *J Org. Chem.* **1998**, *63*, 6425; (b) Baumstark, A.L.; Harden Jr., D.B., *J. Org. Chem.* **1993**, *58*, 7615; (c) Baumstark, A.L.; Vasquez, P.C. *J. Org. Chem.* **1988**, *53*, 3437; (d) Murray, R.W.; Jeyaraman, R. *J. Org. Chem.* **1985**, *50*, 2847.
348. Inoue, M.; Lee, N.; Kasuya, S.; Sato, T.; Hirama, M.; Moriyama, M.; Fukuyama, Y. *J. Org. Chem.* **2007**, *72*, 3065.
349. (a) Frohn, M.; Shi, Y. *Synthesis* **2000**, 1979; (b) Reference 346d.
350. (a) Curci, R.; D'Accolti, L.; Fiorentino, M.; Rosa, A. *Tetrahedron Lett.* **1995**, *36*, 5831; (b) Curci, R.; Fiorentino, M.; Serio, M.R. *J. Chem. Soc. Chem. Commun.* **1984**, 155.
351. (a) Yang, D.; Yip, Y.-C.; Tang, M.W.; Wong, M.-K.; Zheng, J.-H.; Cheung, K.-K.; *J. Am. Chem. Soc.* **1996**, *118*, 491; (b) Yang, D.; Wang, X.-C.; Wong, M.-K.; Yip, Y.-C.; Tang, M.-W. *J. Am. Chem. Soc.* **1996**, *118*, 11311.
352. Denmark, S.E.; Matsuhashi, H. *J. Org. Chem.* **2002**, *67*, 3479.
353. Reference 20, pp 996-1003.

polyols containing several hydroxyl moieties are important structural features of many natural products. 1,2-Diols are usually formed by the syn-addition of manganese or osmium reagents to an alkene. Mechanistically, these are syn- dipolar additions and will be discussed in section 11.11, but a cursory introduction will be given here. Potassium permanganate and osmium tetroxide are the most common reagents used to prepare 1,2-diols, and both reagents react with alkenes in this manner. An alternative route to 1,2-diols is available using silver salts of carboxylic acids will conclude this section.

3.5.A. Dihydroxylation with Potassium Permanganate

Permanganate (MnO_4^-) is a powerful oxidizing agent in basic, acidic and neutral media. In acidic solution, the reduction potential is 1.679 V or 1.491 V, but it is 0.588 V in basic solution according to the following reactions.[17]

$$MnO_4^- + 4\,H^+ + 3\,e^- \rightarrow MnO_2 + 2\,H_2O \qquad\qquad 1.679\ V$$

$$MnO_4^- + 8\,H^+ + 5\,e^- \rightarrow Mn^{2+} + 4\,H_2O \qquad\qquad 1.491\ V$$

$$MnO_4^- + 2\,H_2O + 3\,e^- \rightarrow MnO_2 + 4\,HO^- \qquad\qquad 0.588\ V$$

In this section, permanganate [principally potassium permanganate ($KMnO_4$)] will be used to oxidize alkenes to 1,2-diols. Permanganate (MnO_4^-, **253**) is a 1,3-dipolar molecule (δ^+ O and O^-). An alkene will donate electrons to the electrophilic oxygen and form a C–O bond, and as the π-bond (O=Mn) breaks positive character develops on the other carbon. Subsequent attack by the negative oxygen of the dipolar molecule forms a ring (see **254**). It is likely that this is a concerted process, controlled by the frontier orbital energies of the alkene and the dipole (see sec. 11.2, 11.11.A). Formation of cyclic manganate ester with a cis geometry for the two C-O bond in **254**[354] is accompanied by reduction of Mn(VII) to Mn(VI).[354] Aqueous hydroxide leads to decomposition of the manganate ester **254** to generate a 1,2-diol (**256**), via stepwise cleavage of the Mn-O bonds (see **255**), and the overall reaction is an oxidation of the alkene. The hydroxide ion attacks manganese in **254**, rather than carbon, to form **255**, so the stereochemistry in **254** is retained. Subsequent reaction with water and additional hydroxide liberates the cis- diol and the Mn(V) byproduct, which disproportionates to Mn(VII) [MnO_4^-] and Mn(IV) [usually MnO_2]. The cis geometry of the oxygen atoms is fixed at the time of the dipolar addition, so the timing for C–O bond formation must be close if not concerted (sec. 11.11.A).[355] Reaction of hydroxide at manganese rather than carbon leads to retention of the cis- stereochemistry.

---

354.   (a) Stewart, R. in *Oxidation in Organic Chemistry*, Wiberg, K.B. (Ed.), Academic Press, New York, *1965*, chapter 1; (b) Wiberg, K.B.; Saegebarth, K.A. *J. Am. Chem. Soc. 1957, 79,* 2822; (c) Wagner, G. *J. Russ. Phys. Chem. Soc. 1895, 27,* 219; (d) Böeseken, J. *Rec. Trav. Chim. 1928, 47,* 683.

355.   For a discussion that establishes the [3+2]-mechanism, see Houk, K.N.; Strassner, T. *J. Org. Chem. 1999, 64,* 800.

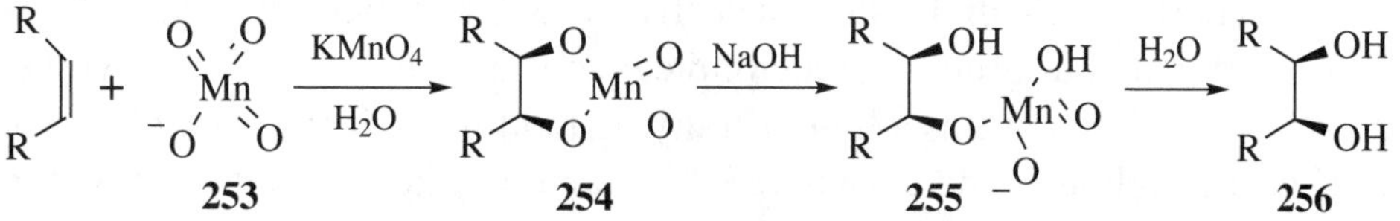

Permanganate reacts with an alkene in an exothermic process.[356] This oxidation with $KMnO_4$ is compatible with several solvents (aqueous EtOH, acetone, *t*-BuOH) under acidic, basic or neutral conditions.[357] Permanganate oxidation to the diol usually requires dilute conditions, since vigorous conditions (75°C, 0.2 M KOH) can lead to extensive carbon-carbon bond cleavage[358] (sec. 3.7). In neutral media (and very low hydroxide concentration) large quantities of α-hydroxy ketone (ketols such as **258**) can be formed.[359] Under these conditions the alkene reacts with permanganate to generate **255**, with Mn^{6+} as the byproduct. Aqueous hydrolysis generates **257** before hydroxide can attack manganese a second time, leading to the ketol and MnO_2. If strongly basic conditions are used to decompose **255**, the diol is the major product. Henbest showed that the presence of electron-withdrawing groups diminish the rate of diol formation and promoted the formation of diketones and cleavage products.[360]

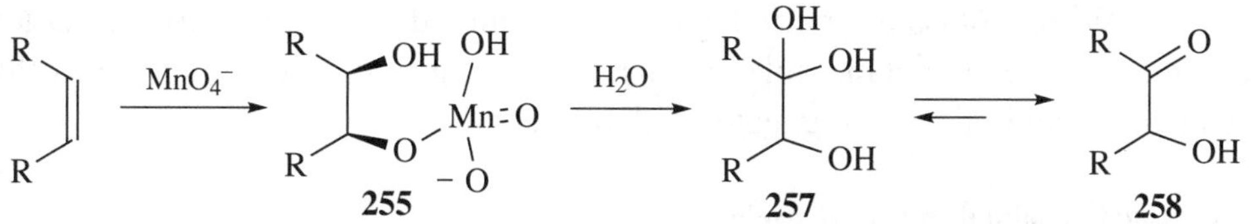

Swern showed that the permanganate oxidation of oleic acid (**259**) was pH dependent. Earlier, Lapworth and Mottram oxidized oleic acid to 9,10-dihydroxy stearic acid (**260**)[361] in essentially quantitative yield by what might be considered standard conditions for hydroxylation: 0.1% oleic acid, no more than 1% $KMnO_4$, short reaction time (5 min), slight excess of alkali and a reaction temperature of 0-10°C.[361] Swern found that reaction of **259** with potassium permanganate and sodium hydroxide without pH control gave 45% of a mixture of 9-keto-10-hydroxystearic acid (**261**) and 10-keto-9-hydroxystearic acid (**262**), along with 30% of **260**.[362] If the pH of the reaction medium was controlled to pH 9-9.5 (by addition of sulfuric acid to the hydroxide solution), a 45% yield of a mixture of **261** and **262** was produced along with 20% of **260**.

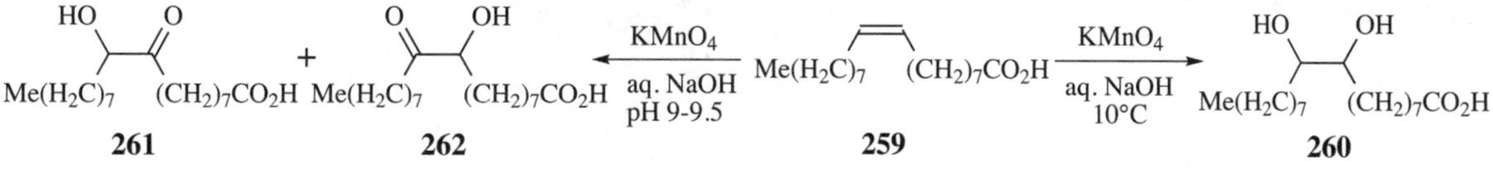

356. Reference 42, p 6.
357. Reference 42, p 7.
358. Hill, J.W.; McEwen, W.L. *Org. Synth. Coll. Vol. 2* **1943**, 53.
359. (a) Reference 16a, pp 42; (b) Snyder, C.H. *J. Chem. Educ.* **1966**, *43*, 141.
360. Henbest, H.B.; Jackson, W.R.; Robb, B.C.G. *J. Chem. Soc. B* **1966**, 803.
361. Lapworth, A.; Mottram, E.N. *J. Chem. Soc.* **1925**, 1628.
362. Coleman, J.E.; Ricciuti, C.; Swern, D. *J. Am. Chem. Soc.* **1956**, *78*, 5342.

The ability to generate cis- diols stereoselectively is important in synthesis, where the ability to position functional groups with complete control is essential. In a synthesis of 21-fluoroprogestin-16,17-dioxolane as a high-affinity ligand for PET imaging of the progesterone receptor by Katzenellenbogen and co-workers,[363] reaction of keto-acetate 263 with potassium permanganate in acetone, gave diol 264 in 85% yield. Note that dihydroxylation occurred exclusively from the bottom face of the molecule.

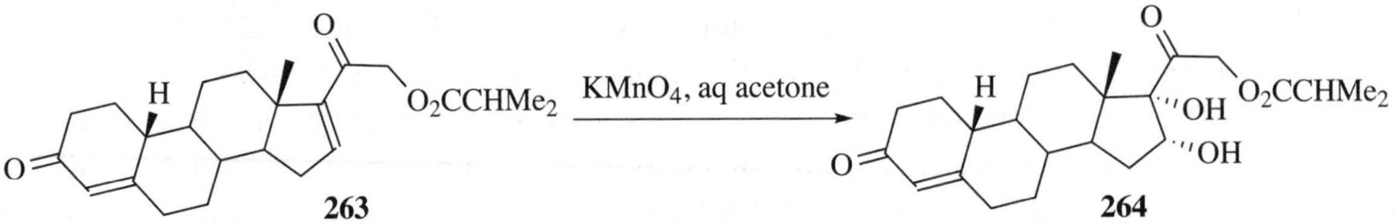

Potassium permanganate oxidation in the presence of phase transfer catalysts permits non-aqueous solvents such as dichloromethane to be used, which generally leads to an increase in yield of the oxidation products.[364] A solution of dichloromethane and aqueous sodium hydroxide (+ alkene) uses a phase transfer agent such as benzyltriethylammonium chloride to solubilize the reactants.[365] When potassium permanganate was added to this mixture, oxidation of *cis*-cyclooctene to *cis*-1,2-cyclooctanediol was observed in significantly higher yields than could be obtained under standard permanganate oxidation conditions.[366]

3.5.B. Dihydroxylation with Osmium Tetroxide

Osmium tetroxide (OsO_4) had a relatively high reduction potential (0.85 V) in acidic media, with OsO_4 being reduced to osmium metal.[17]

$$OsO_4 + 8\,H^+ + 8\,e^- \rightarrow Os + 4\,H_2O \qquad\qquad 0.85\ V$$

For applications to organic chemistry, osmium tetroxide is most often used in neutral, aqueous media and the primary use is for the conversion of alkenes to 1,2-diols.

3.5.B.i. Dihydroxylation of Alkenes. Criegee found that osmium tetroxide (**265**) is reduced by alkenes to give the cis-diol, via an intermediate osmium complex that must be decomposed.[367]

363. Vijaykumar, D.; Mao, W.; Kirschbaum, K.S.; Katzenellenbogen, J.A. *J. Org. Chem.* **2002**, *67*, 4904.
364. (a) Okimoto, T.; Swern, D. *J. Am. Oil Chem. Soc.* **1977**, *54*, 867A; (b) Foglia, T.A.; Barr, P.A.; (c) Malloy, A.J. *Ibid.* **1977**, *54*, 858A.
365. Reference 233a, p 84.
366. Weber, W.P.; Shepherd, J.P. *Tetrahedron Lett.* **1972**, 4907.
367. (a) Criegee, R.; Marchand, B.; Wannowius, H. *Annalen* **1942**, *550*, 99; (b) Criegee, R. *Ibid.* **1936**, *522*, 75.

> *Osmium tetroxide vapors are poisonous and result in damage to the respiratory tract and temporary damage to the eyes.*[368] *Use OsO$_4$ powder only in a well-ventilated hood with* <u>*extreme*</u> *caution.*

In the absence of tertiary amines, osmium tetroxide reacts with alkenes via 1,3-dipolar addition to generate a monomeric Os(VI) ester such as **266**,[369] where L is a **ligand** (a solvent molecule or an added substrate such as pyridine). A reducing agent is usually added to drive the reaction to the cis-diol by reducing Os(VI) to osmium metal, preventing reformation of the ester.[368] When the reaction is done in the presence of a base such a pyridine, an ester complex like **267** is formed by coordination with two equivalents of the ligand (pyridine). Complexes such as **267** can sometimes be isolated.[369] In most osmylation reactions of alkenes, the osmium ester is not isolated but decomposed with a reducing agent, producing the cis-1,2-diol. A variety of reagents can be used to decompose the osmium ester, including bisulfite ion in pyridine,[370] sulfite, or bisulfite in aqueous ethanol,[371,372,367] mannitol in alkaline solution,[366] or potassium chlorate (KClO$_3$) under acidic conditions.[366]

Sharpless et al proposed an alternative mechanism.[373] Octahedral osmate esters (generated directly by [3+2]-cycloaddition) are stable and isolable, but there would be severe angle strain in the initial tetrahedral metallocycle due to the long (ca. 2.2 A) osmium-oxygen bonds.[374] A four-membered intermediate should be substantially less strained since the long Os–C and Os–O bonds would have the effect of relieving the angle strain in the four-membered ring.[373] Sharpless et al. proposed that hydroxylation proceeds by a four-membered organoosmium intermediate, where coordination of a ligand (e.g., pyridine) would produce the octahedral complex and trigger reductive insertion of the Os–C bond into an oxo group yielding the osmium(V1) ester. This intermediate could react reversibly with additional ligand to give

368. Reference 42, p 11.
369. (a) Reference 233a, pp 75-76; (b) Schröder., M. *Chem. Rev.* **1980**, *80*, 187.
370. Baran, J.S. *J. Org. Chem.* **1960**, *25*, 257.
371. Barton, D.H.R.; Eland, D. *J. Chem. Soc.* **1956**, 2085.
372. Vyas, D.M.; Hay, G.W. *Can. J. Chem.* **1975**, *53*, 1362.
373. Sharpless, K.B.; Teranishi, A.Y.; Bäckvall, J.E. *J. Am. Chem. Soc.* **1977**, *99*, 3120.
374. (a) Conn, J.F.; Kim, J.J.; Suddath, F.L.; Blattmann, P.; Rich, A. *J. Am. Chem. Soc.* **1974**, *96*, 7152; (b) Collin, R.; Jones, J.; Griffith, W.P. *J. Chem. Soc. Dalton Trans.*, **1974**, 1094; (c) Collin, R.; Griffith, W.P.; Phillips, F.L.;Skapski, A.C. *Biochim. Biophys. Acta* **1973**, *320*, 745; (d) *Ibid.* **1974**, *354*, 152.

266. Sharpless pointed out,[373] however, that objections to the [3+2]-mechanism are removed if one invokes prior coordination of the nucleophile (the ligand, L) to OsO_4, where [3+2]-cycloaddition could proceed directly to the five coordinate ester.

An osmate complex is formed in the reaction, which can be decomposed in aqueous or alcoholic solution, but the hydrolysis is reversible. The relatively bulky osmium complex is sterically demanding and osmylation generally involves reaction at the less hindered face of an alkene. In a synthesis of himandrine by Mander and co-workers,[375] catalytic dihydroxylation (see below) of **268** failed, but oxidation with a stoichiometric quantity of osmium tetroxide gave the isomeric triol **269** in about 75% yield. Dihydroxylation occurred from the less sterically hindered top (exo) face of **268**.

When stoichiometric amounts of osmium tetroxide are required, applications are often limited to small-scale reactions due to the toxicity and expense.[376] Procedures have been developed where osmium tetroxide is used as a catalyst,[377] greatly expanding the utility of this reagent. Effective catalysts in this reaction are metal chlorates, hydrogen peroxide, t-BuOOH and amine N-oxides. Tertiary amine N-oxides such as NMO (N-methylmorpholine N-oxide) are effective catalysts developed by Van Rheenan et al.[378] These reagents are both a catalyst for the reaction and also as agents to decompose the osmylate ester. Osmylations with NMO can be accomplished with about 1 mol % of OsO_4 at ambient temperatures. This procedure is superior to other syn-hydroxylation procedures since only small amounts of osmium tetroxide are used.[379]

Relatively unsubstituted alkenes react quickly, but the oxidation failed with several highly substituted alkenes.[380] Note that a trisubstituted alkene was successfully oxidized using this reagent.[381] The use of trimethylamine N-oxide often gives better results than NMO in reactions with hindered alkenes.[382] The reaction is usually run in aqueous acetone, THF or

375. O'Connor, P.D.; Mander, L.N.; McLachlan, M.M.W. *Org. Lett.* **2004**, *6*, 703.
376. Reference 233a, p 75.
377. (a) Reference 233a, pp 77-84; (b) Reference 42, pp 13-14.
378. (a) Van Rheenen, V.; Kelly, R.C.; Cha, D.Y. *Tetrahedron Lett.* **1976**, 1973; (b) Van Rheenen, V.; Cha, D.Y.; Hartley, W.M. *Org. Synth. Coll. Vol. 6* **1988**, 342.
379. Reference 233a, p 81.
380. Akashi, K.; Palermo, R.E.; Sharpless, K.B. *J. Org. Chem.* **1978**, *43*, 2063.
381. (a) Larsen, S.D.; Monti, S.A. *J. Am. Chem. Soc.* **1977**, *99*, 8015; (b) Schneider, W.P.; McIntosh, A.V. U.S. Patent, 2 769 824 **1956** [*Chem. Abstr. 51:* 8822, *1957*].
382. Ray, R.; Matteson, D.S. *Tetrahedron Lett.* **1980**, *21*, 449.

tert-butanol as a one or two-phase reaction. Riera and co-workers used NMO in a synthesis of deoxymannojirimycin, to convert **268** to **269** in 96% yield (12/1 dr).[383] Alkenes bearing electron-withdrawing groups react more slowly with OsO_4.[360,384]

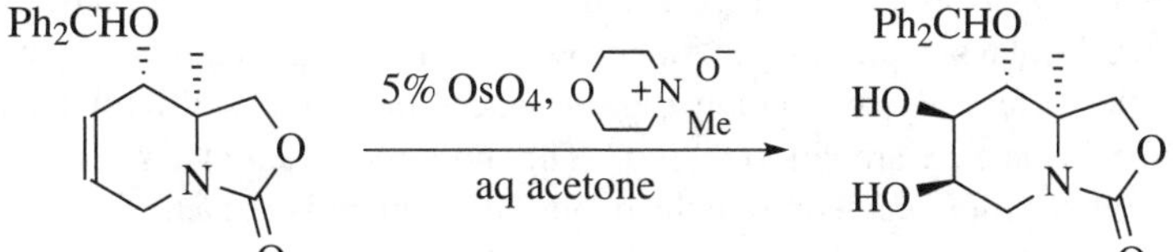

As with permanganate oxidations, α-hydroxy ketones can be formed as side products. In some cases, structural features make the osmium complex relatively unstable. In an aqueous medium it can react with water to give a hydroxy-hydrate, which is then converted to an α-keto alcohol. Sharpless et al. developed a procedure that used *tert*-butyl hydroperoxide with a catalytic amount of osmium tetroxide,[385] in the presence of tetraethylammonium hydroxide ($Et_4N^+ OH^-$). The procedure gave improved yields of the cis-diol and a little α-hydroxyketone, as shown in the conversion of oct-(4*E*)-ene to a mixture of **270** and **271** in 73% yield. This method is more reliable for oxidation of tri- and tetrasubstituted alkenes than the Van Rheenan procedure. The reaction was not suitable for base sensitive alkenes, but later work showed that changing the solvent to acetone allowed the use of tetraethylammonium acetate (Et_4NOAc)[386] for the hydroxylation of sensitive alkenes such as ethyl crotonate.

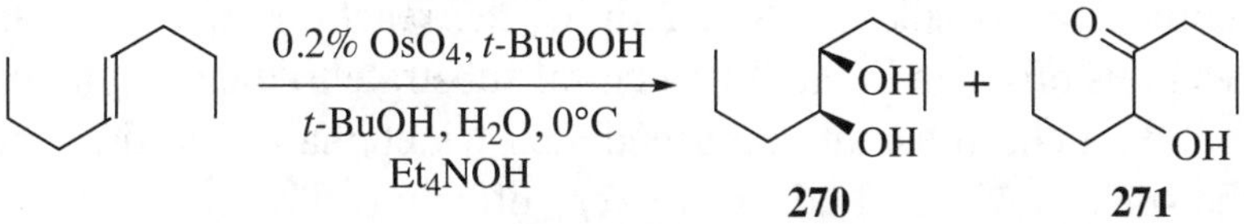

MOLECULAR MODELING: Oxidation of a Conjugated Diene

A simple 3D model (or even a 2D sketch) is sufficient to conclude that osmium tetroxide oxidation of the conjugated diene shown below will occur from the ring face away from the bulky OSiMe₂*t*-Bu group. (The model corresponds to the lowest-energy conformer resulting from examination of all conformers using molecular mechanics.) However, such a model is not likely to reveal that the double bond adjacent to the bulky group and not the one bonded to the ester will be oxidized.

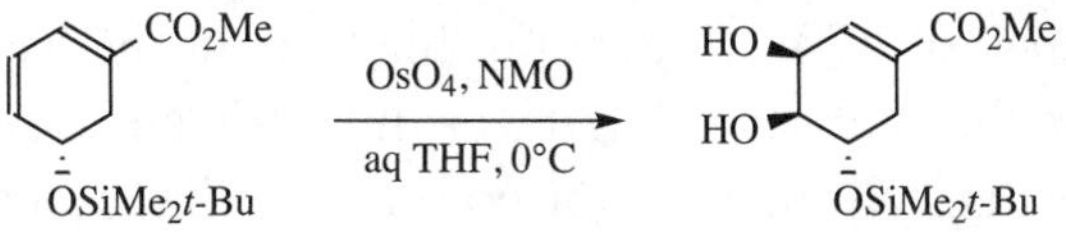

---

383. Martín, R.; Murruzzu, C.; Pericàs, M.A.; Riera, A. *J. Org. Chem.* **2005**, *70*, 2325.
384. Ferretti, A.; Tesi, G. *J. Chem. Soc.* **1965**, 5203.
385. Sharpless, K.B.; Akashi, K. *J. Am. Chem. Soc.* **1976**, *98*, 1986.
386. Akashi, K.; Palermo, R.E.; Sharpless, K.B. *J. Org. Chem.* **1978**, *43*, 2063.

MOLECULAR MODELING: Oxidation of a Conjugated Diene, con't.

Assignment of which double bond is likely to be oxidized as well as the stereochemistry of attack can be accomplished using a local ionization potential map. This colors an electron density surface that indicates overall molecular size and shape with the value of the energy required to remove an electron from a region on the surface. By convention, regions where ionization is difficult are colored "blue", while regions where it is easy are colored "red". The more red a region associated with a carbon-carbon double bond, the more nucleophilic the bond. The more red the face of a particular bond, the more likely the face will approach an electrophilic reagent. Thus, a local ionization potential map makes it possible to identify regions that are nucleophilic and will donate electrons to an electron acceptor (colored red) from those most likely to accept electrons from a donor (colored blue).

Examine the local ionization potential map for this diene. It is depicted as a transparent model, allowing the underlying structures to be seen. You can switch to a solid model, allowing the maps to be seen more easily, by *clicking* on the map and then selecting *Solid* for the **Style** menu at the bottom of the screen.) For which double bond is the local ionization potential smaller? You can get the value of the local ionization on a particular region of the surface by *clicking* on that region. It will be displayed at the bottom of the screen. For which face, *syn* or *anti* to the OSiMe$_2$t-Bu, is the local ionization potential smaller? Does this lead to the observed oxidation product? According to the local ionization potential map, are any other sites on the diene likely to be competitive? If so, identify the product(s) that would result.

3.5.B.ii. Sharpless Asymmetric Dihydroxylation. Although osmylation is an attractive method for the conversion of alkenes to 1,2-diols, the reaction generates racemic products. Sharpless[387] solved this problem by adding a chiral substrate to the osmylation reagents, with the goal of generating a chiral osmate intermediate. Cinchona alkaloids, especially esters of dihydroquinidines such as **272** and **273** were very effective. The % ee of the diol product is good-to-excellent with a wide range of alkenes in what is called the **Sharpless asymmetric dihydroxylation**.[388] The chiral alkaloid forms a complex with the osmium and the alkene, directing attack from the *opposite* enantioface of the alkene.[387] The cis-addition imposed by the cyclic osmium intermediate is very important in reactions of acyclic alkenes, because the two hydroxyl moieties in the product aree syn-, making the reaction highly diastereoselective. This selectivity is also dependent on the geometry of the alkene precursor, and high diastereoselectivity is obtained only with regiochemically pure alkenes. This selectivity is illustrated by the reaction of *trans*-stilbene with OsO$_4$ and NMO, which gave the diol **274** with 99% ee in the presence of **272**. In general, the %ee in the asymmetric dihydroxylation procedure were lower than those observed in the asymmetric epoxidation reaction (sec. 3.4.D). Ionic liquids such as 1,3-dialkylimidazolium compounds have been used as a medium for Sharpless asymmetric dihydroxylation, using a recoverable and reusable osmium-ligand.[389]

387. (a) Jacobsen, E.N.; Markó, I.; Mungall, W.S.; Schröder, G.; Sharpless, K.B. *J. Am. Chem. Soc. 1988, 110,* 1968; (b) Hentges, S.G.; Sharpless, K.B. *Ibid. 1980, 102,* 4263.
388. The Merck Index, 14th ed., Merck & Co., Inc., Whitehouse Station, New Jersey, *2006*, p ONR-85.
389. Branco, L.C.; Afonso, C.A.M. *J. Org. Chem. 2004, 69,* 4381.

The source of enantioselectivity can be discerned from the structure of the osmium complex.[390] An X-ray crystal structure of the OsO₄ complex of (dimethylcarbamoyl) dihydroquinidine (**275**) has been published.[391] Using **275** as a guide, the osmium complex of stilbene can be

275

[Reprinted with permission from Svendsen, J.S.; Markó, I.; Jacobsen, E.N.; Pulla Rao, Ch.; Bott, S.; Sharpless, K.B. *J. Org. Chem.* **1989**, *54*, 2263. Copyright © 1989 American Chemical Society.] Structure drawn using Spartan software by Wavefunction, Inc. Hydrogen atoms omitted for clarity.

represented as **276** (the *p*-methoxy ester was used in this case), which may not be the active complex, but it illustrates how selectivity arises in the reaction. Stilbene coordinates with osmium in a manner that minimizes any interaction of the phenyl groups with the azabicyclo[2.2.2] octane moiety and/or with the methylquinoline ring. As seen in **276**, O1 is significantly more hindered than O2. The phenyl rings of the stilbene molecule are oriented so that those rings are away from O2 and the azabicyclooctane ring. If stilbene binds in this way, the threo (syn) isomer (**274**) will be the major diol product. The chirality of the Cinchona alkaloid is important since it provides facial selectivity for complexation of the alkene and control of orientation via the steric interactions described above. Figure 3.2[330] is a graphic similar to Figure 3.1 (which showed a model for Sharpless asymmetric epoxidation) to illustrate the facial. If the alkene is oriented as in Figure 3.2, using the naturally-occuring dihydroquinidine ester forces delivery of the hydroxyls from the top face. Conversely, dihydroquinoline esters deliver hydroxyls from the bottom face. Highest % ee values (70-90%) are obtained with trans-1,2-disubstituted

390. For an *ab initio* approach to this mechanism, see Veldkamp, A.; Frenking, G. *J. Am. Chem. Soc.* **1994**, *116*, 4937.
391. Svendsen, J.S.; Markó, I.; Jacobsen, E.N.; Pulla Rao, Ch.; Bott, S.; Sharpless, K.B. *J. Org. Chem.* **1989**, *54*, 2263.

alkenes,[330] and the lowest are obtained with cis-1,2-disubstituted alkenes.[294] Trisubstituted and monosubstituted alkenes show greater variation in asymmetric induction (30-80% ee and 25-60% ee, respectively).[330]

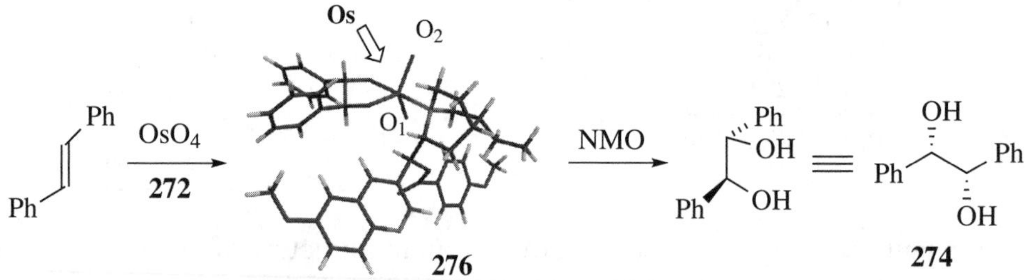

New catalysts are available for this reaction that improve the efficiency and ease of use. Two phthalazine derivatives,[392] (DHQD)$_2$PHAL (**277**) and (DHQ)$_2$PHAL (**278**) have been used in conjunction with an osmium reagent, resulting in two commercial catalysts, AD-mix-β[393] (using **277**) and AD-mix-α[393] (using **278**). The abbreviations shown here refer to the reagents used to make each catalyst. Catalyst **277** is prepared from dihydroquinidine (DHQD) and 1,4-dichlorophthalazine (PHAL), and **278** is prepared from dihydroquinine (DHQ) and PHAL. The actual oxidizing medium labeled AD-mix α or β- uses **278** or **277**, respectively, mixed with potassium osmate [K$_2$OsO$_2$(OH)$_4$], powdered K$_3$Fe(CN)$_6$, and powdered K$_2$CO$_3$ in an aqueous solvent mixture.[392] These mixes have become standard catalysts for dihydroxylation. Sharpless introduced a model, shown in Figure 3.3,[392] to predict the enantioselectivity of each catalyst. With the R$_L$, R$_M$, and R$_S$ substituents of the alkene arranged as shown, the AD-mix-β catalyst gives dihydroxylation from the top, whereas the AD-mix-α catalyst gives dihydroxylation from the bottom, as shown.

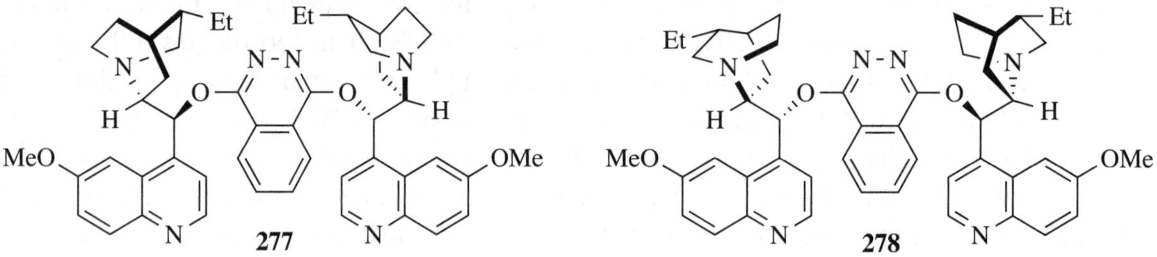

392. Sharpless, K.B.; Amberg, W.; Bennani, Y.L.; Crispino, G.A.; Hartung, J.; Jeong, K.-S.; Kwong, H.-L.; Morikawa, K.; Wang, Z.-M.; Xu, D.; Zhang, X.-L. *J. Org. Chem.* **1992**, *57*, 2768.
393. A trademark of the Aldrich Chemical Co., Inc.

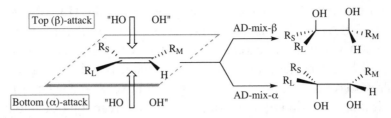

Figure 3.3. Facial selectivity in asymmetric dihydroxylation using AD-mix-α and AD-mix-β. [Reprinted with permission from Sharpless, K.B.; Amberg, W.; Bennani, Y.L.; Crispino, G.A.; Hartung, J.; Jeong, K.-S.; Kwong, H.-L.; Morikawa, K.; Wang, Z.-M.; Xu, D.; Zhang, X.-L. *J. Org. Chem.* **1992**, *57*, 2768. Copyright © 1992 American Chemical Society.]

In a synthetic example, McDonald and Fei used AD-mix-β[393] to convert **279** to **280** in 89% yield and and enantiomeric ratio of 13:1, in a synthesis of altromycin aglycone.[394] Similarly, Hecht and co-workers reacted AD-mix-α[393] with **281**, as part of synthesis of (+)-myristinin A[395] to give an 84% yield of diol **282**.

3.5.C. The Prévost Reaction

An alternative route to 1,2-diols reacts alkenes with a mixture of silver carboxylate and iodine rather than with a peroxide. Prévost found that silver benzoate and iodide converted styrene to 1,2-dibenzoate **283**, which could be saponified to the 1,2-diol.[396a,358] The transformation is

---

394. Fei, Z.B.; McDonald, F.E. *Org. Lett.* **2005,** *7*, 3617.
395. Maloney, D.J.; Deng, J.-Z.; Starck, S.R.; Gao, Z.; Hecht, S.M. *J. Am. Chem. Soc.* **2005,** *127*, 4140.
396. (a) Prévost, C. *Compt. Rend,* **1933,** *196*, 1129; (b) Wilson, C.V. *Org. React.,* **1957,** *9*, 332 (see p 350).

called the **Prévost reaction**.[396,397] This reaction is related to the **Hunsdiecker reaction**[396b,398] (described in sec. 2.8.A). Birkenbach et al. reported a similar reaction of alkenes and silver carboxylates in which cyclohexene was treated with silver acetate and iodide to give 2-iodocyclohexanol.[399] Brunel had earlier reported a related reaction using iodine and mercuric acetate.[400] The Prévost reaction is illustrated using cyclohexene, and water is not involved in the transformation. Initial formation of an iodonium salt from reaction of cyclohexene and iodine is followed by reaction with silver benzoate to give **284a**.[396] Attack by a second molecule of benzoate gives the trans diacetate, **287a**, presumably via a cyclic species such as **285a**, which is opened by benzoate to give the trans diester. Saponification generates the trans-diol (**288**). The Prévost reaction therefore generates trans diols, in contrast to permanganate and osmium hydroxylation reactions, which give cis diols. Alternatively, water can attack **285** to give **286**, which in turn is attacked by the carboxylate anion to give **287**. An example is the conversion of **289** to **290** in 94% yield.[401]

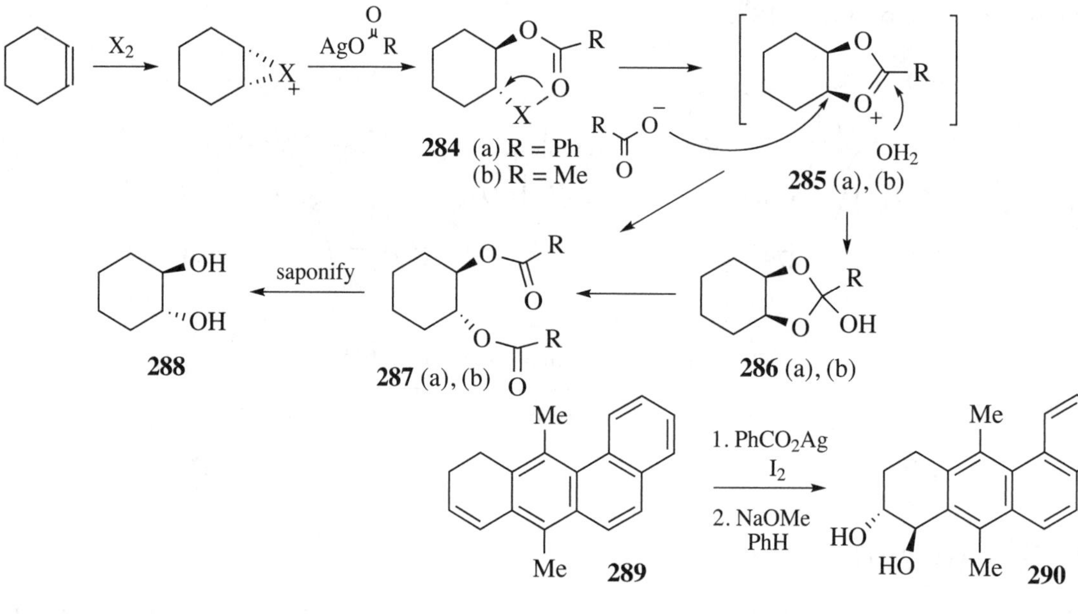

As mentioned, the Prévost reaction proceeds via initial formation of an iodonium salt that is opened by the silver carboxylate. Winstein studied reactions of silver acetate and bromine with alkenes, and showed that initial formation of the bromonium ion was followed by

397. The Merck Index, 14th ed., Merck & Co., Inc., Whitehouse Station, New Jersey, *2006*, p ONR-76.
398. The Merck Index, 14th ed., Merck & Co., Inc., Whitehouse Station, New Jersey, *2006*, p ONR-47.
399. Birkenbach, L.; Goubeau, J.; Berninger, E. *Berichte, 1932, 65,* 1339.
400. Brunel, L. *Bull. Soc. Chim. Fr. 1905, 33,* 382.
401. (a) Harvey, R.G.; Fu, P.P.; Cortez, C.; Pataki, J. *Tetrahedron Lett. 1977,* 3533; (b) Mundy, B.P.; Ellerd, M.G.; Favaloro Jr., F.G. *Name Reactions and Reagents in Organic Synthesis, 2nd Ed.*, Wiley, New York, *2005*, pp 338-339.

formation of a bromoacetate (**284b**), which is analogous to **284a**.[402] In aqueous media, attack by the neighboring acetate generates an oxonium salt (**285b**) that reacts with water to form **286b**. Reaction with additional silver acetate gives the final product, a trans 1,2-diester **287b**. The presence of water in the reaction apparently changed its course from an intermolecular displacement of **285**, which gave the trans-product, to attack by water with ring opening, which preserved the cis stereochemistry. This is called the **Woodward modification**[403] of the Prévost reaction and gives the cis-diol rather than the trans-diol. Woodward used this modification in the conversion of **291** to the cis acetate, which was saponified to give diol **292** in 65% overall yield.[403]

An alternative route to trans diols is available by hydrolysis of epoxides. Reaction of cyclohexene with mcpba gives cyclohexene oxide and hydrolysis (acid or base) gives *trans*-1,2-cyclohexanediol.

The functional group transforms for all of the hydroxylation reactions in this section are

3.5.D. Dihydroxylation of Aromatic Rings

Enzymatic transformations of organic molecules are an increasingly important source of chiral, nonracemic products. Transformations in previous sections focused on oxidation of the π bonds in alkenes, but these reagents generally do not work with aromatic rings. It is possible to convert one π-bond of an aromatic ring into racemic or chiral, nonracemic cyclohexadiene 1,2-diols using enzymes associated with the organism *Pseudomonas putida*.[404,405] Early work by Gibson et al. established that racemic **293** could be formed in good to moderate yield by

402. Winstein, S.; Buckles, R.E. *J. Am. Chem. Soc.* **1942**, *64*, 2787, 2780.
403. (a) Woodward, R.B.; Brutcher, Jr., F.V. *J. Am. Chem. Soc.* **1958**, *80*, 209; (b) The Merck Index, 14th ed., Merck & Co., Inc., Whitehouse Station, New Jersey, **2006**, p ONR-103. Brimble found that iodo-acetates can be isolated in some cases, as found in (c) Brimble, M.A.; Nairn, M.R. *J. Org. Chem.* **1996**, *61*, 4801.
404. (a) Brown, S.M. in *Organic Synthesis: Theory and Practice*, T. Hudlicky (Ed.), JAI Press, Greenich, CT., **1993**, Vol. 2, p 113; (b) Carless, H.A.J. *Tetrahedron Asymmetry* **1992**, *3*, 795; (c) Widdowson, D.A.; Ribbons, D.A.; Thomas, S.D. *Janssenchimica Acta* **1990**, *8*, 3.
405. For recent reviews, see (a) Hudlicky, T.; Gonzalez, D.; Gibson, D.T. *Aldrichimica Acta* **1999**, *32*, 35; (b) Hudlicky, T.; Abboud, K.A.; Entwistle, D.A.; Fan, R.; Maurya, R.; Thorpe, A.J.; Bolonick, J.; Myers, B. *Synthesis* **1996**, 897. Also see, Reference 20, p 978.

exposure of benzene to *P. putida*,[406] based on earlier work by Marr and Stone[407] and by Walker and Wiltshire.[408] In even earlier work, Ballard et al. had postulated **293** as an intermediate in the oxygen assisted oxidation of benzene derivatives in the presence of *P. putida*.[409]

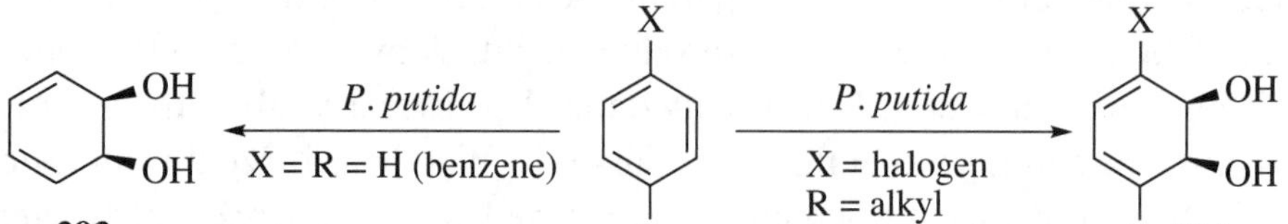

Gibson showed that this transformation is not limited to benzene. The reaction can be generalized to give products such as **294**. In specific cases chlorobenzene gave **295**,[410] bromobenzene gave **296**, 4-chlorotoluene gave **297**[411] and toluene gave **298**,[411] all with excellent asymmetric induction since the aromatic ring contains a substituent and introduction of the hydroxyl groups generates a chiral molecule. In one sense, this oxidation belongs in section 2.11, which described various transformations of benzene derivatives. Since this reaction formally involves the conversion of a π bond in a benzene ring to a cis diol, however, it is placed in this section for the sake of continuity.

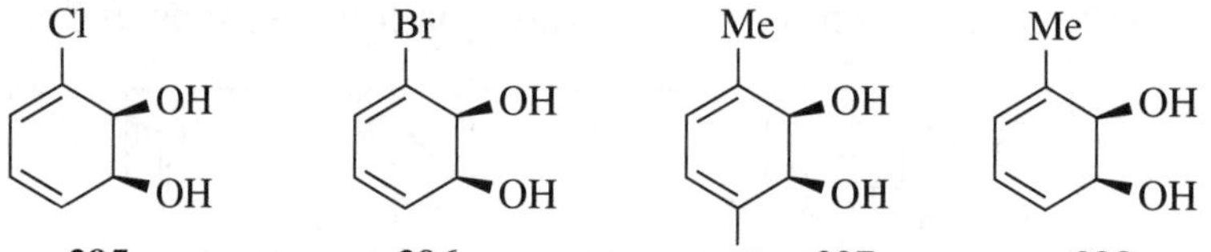

Recently, this transformation has been used in a variety of interesting syntheses, particularly by Hudlicky and by Ley. Ley showed that **293** could be converted to conduritol F (**299**) in four steps (25% overall yield).[412] Carless and Hensley also prepared conduritol A and D using similar methodology.[413] In other work, Ley et al. converted **295** to myoinositol-1,4,5-triphosphate (IP$_3$)[414] and **293** to pinitol, an antidiabetic agent.[415] Hudlicky et al. prepared (+)-pinitol in an enantiodivergent synthesis that used the chiral derivative **296** as a starting material.[416] This

406. Gibson, D.T.; Koch, J.R.; Kallio, R.E. *Biochemistry* **1968**, *7*, 2653.
407. Marr, E.K.; Stone, R.W. *J. Bacteriol.* **1961**, *81*, 425.
408. Walker, N.; Wiltdshire, G.H. *J. Gen. Microbiol.* **1953**, *8*, 273.
409. Ballard, D.G.H.; Courtis, A.; Shirley, I.M.; Taylor, S.C. *J. Chem. Soc. Chem. Commun.* **1983**, 954.
410. Gibson, D.T.; Koch, J.R.; Schuld, C.L.; Kallio, R.E. *Biochemistry* **1968**, *7*, 3795.
411. Gibson, D.T.; Hensley, M.; Yoshioka, H.; Mabry, T.J. *Biochemistry* **1970**, *9*, 1626.
412. Ley, S.V.; Redgrave, A.J. *Synlett* **1990**, 393.
413. Carless, H.A.J.; Oak, O.Z. *Tetrahedron Lett.* **1989**, *30*, 1719.
414. Ley, S.V.; Sternfeld, F. *Tetrahedron Lett.* **1988**, *29*, 5305.
415. Ley, S.V.; Sternfeld, F.; Taylor, S. *Tetrahedron Lett.* **1987**, *28*, 225.
416. Hudlicky, T.; Price, J.D.; Rulin, F.; Tsunoda, T. *J. Am. Chem. Soc.* **1990**, *112*, 9439.

latter example illustrates the true power of this method, since chiral nonracemic synthetic targets can be produced from readily available starting materials.

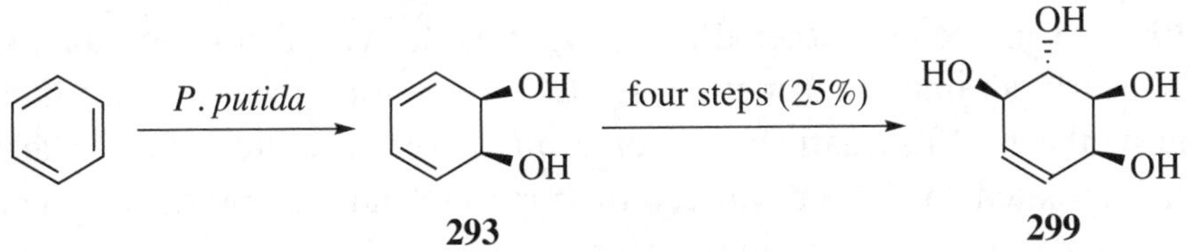

A wide range of chiral nonracemic targets can be prepared using aromatic rings as precursors to diols. Hudlicky prepared **295** in >99.9% ee,[417] and used this diol to synthesize 2,3-isopropylidene (*L*)-ribose γ-lactone (**300**) in four steps (18% overall yield).[417,418] *O*-Linked diinositols have also been prepared, using **285** as the chiral building block.[419] Diols such as **295** and **296** are useful in a variety of syntheses. Diol **296**[420] was used as the diene partner in a Diels-Alder reaction (sec. 11.4) for the synthesis of amino cyclitols. After oxidation of styrene with *P. putida*, the resulting diol was used to synthesize zeylena via a Diels-Alder reaction.[421] The methyl derivative (**297**) was used for the preparation of terpene and prostanoid synthons.[422] Johnson and Penning used **295** as a precursor for the synthesis of prostaglandin PGE$_2$, resolving the starting material to provide the asymmetric precursor.[423]

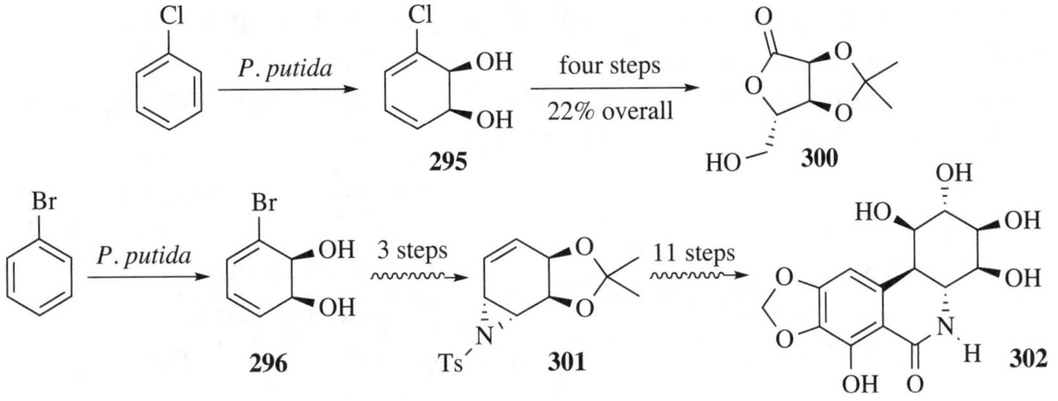

The synthesis of nitrogen-containing natural products is also possible via this route. Hudlicky and co-workers prepared the bicyclic alkaloid (+)-kifunensine, for example, from **295**.[424] More complex molecules can also be prepared, as illustrated by the conversion of **296** into

417. Hudlicky, T.; Price, J.D. *Synlett* **1990**, 159.
418. Hudlicky, T.; Luna, H.; Price, J.D.; Rulin, F. *Tetrahedron Lett.* **1989**, *30*, 4053.
419. Paul, B.J.; Willis, J.; Martinot, T.A.; Ghiviriga, I.; Abboud, K.A. Hudlicky, T. *J. Am. Chem. Soc.* **2002**, *124*, 10416.
420. Hudlicky, T.; Olivo, H.F. *Tetrahedron Lett.* **1991**, *32*, 6077.
421. Hudlicky, T.; Seoane, G.; Pettus, T. *J. Org. Chem.* **1989**, *54*, 4239.
422. Hudlicky, T.; Luna, H.; Barbieri, G.; Kwart, L.D. *J. Am. Chem. Soc.* **1988**, *110*, 4735.
423. Johnson, C.R.; Penning, T.D. *J. Am. Chem. Soc.* **1986**, *108*, 5655.
424. Rouden, J.; Hudlicky, T. *J. Chem. Soc. Perkin Trans. 1* **1993**, 1095.

(+)-pancratistatin (**302**)[425] via **301**. Banwell et al. reported an asymmetric synthesis of (−)-patchoulenone from **298** using this *P. putida*-mediated approach.[426]

The use of **295-298** as well as other diol derivatives derived from functionalized aromatic compounds is clearly an important method for the preparation of molecules that can be used in asymmetric synthesis. Enzymatic oxidation is an important source of diol starting materials that can be manipulated to form a variety of important natural products. This interesting oxidation leads to the functional group transform:

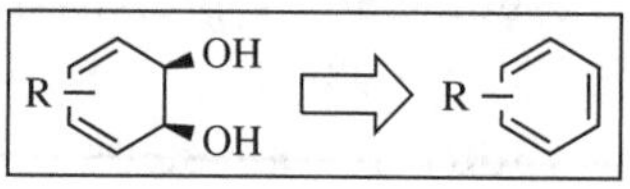

3.5.E. Aminohydroxylation

Asymmetric dihydroxylation of alkenes can be modified to synthesize amino alcohols,[427] in what is known as **Sharpless asymmetric aminohydroxylation** (or **oxyamination**).[428] When conjugated alkenes were treated with nitrogen-containing reagents in the presence of (DHQ)$_2$PHAL (**278**), using what is effectively the AD-mix-α,[393] chiral, nonracemic amino alcohols were formed. The nitrogen sources include TsNClNa,[428] MsNClNa,[429] CbzNClNa,[430] and BocBNClNa.[431] In a typical example treatment of cinnamate **303** with **378**, 4% K$_2$OsO$_2$(OH)$_4$ and MsNClNa in aqueous acetonitrile gave a 65% yield of **304** in 94% ee.[427a] Boger and co-workers used this procedure in a synthesis of ramoplanin A2,[432] where **305** was converted to protected amino-alcohol **306** in 71% yield (99% ee).

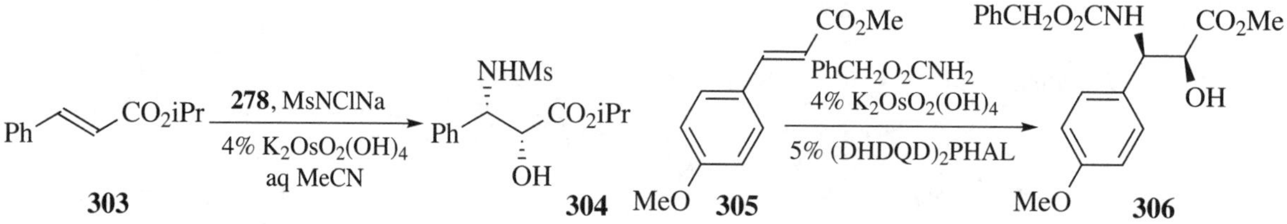

425. Tian, X.; Hudlicky, T.; Königsberger, K. *J. Am. Chem. Soc.* **1995**, *117*, 3643.
426. Banwell, M.; McLeod, M. *Chem. Commun.* **1998**, 1851.
427. (a) O'Brien, P. *Angew. Chem. Int. Ed.* **1999**, *38*, 326; (b) Bodkin, J.A.; McLeod, M.D. *J. Chem. Soc. Perkin Trans. 1* **2002**, 2733.
428. (a) Li, K.B.; Chang, H.-T.; Sharpless, K.B. *Angew. Chem. Int. Ed.* **1996**, *35*, 451; (b) The Merck Index, 14th ed., Merck & Co., Inc., Whitehouse Station, New Jersey, **2006**, p ONR-86.
429. Rudolph, J.; Sennhenn, P.C.; Vlaar, C.P.; Sharpless, K.B. *Angew. Chem. Int. Ed.* **1996**, *35*, 2810.
430. Li, G.; Angert, H.H.; Sharpless, K.B. *Angew. Chem. Int. Ed. Engl.* **1996**, *35*, 2813.
431. (a) Reddy, K.L.; Sharpless, K.B. *J. Am. Chem. Soc,* **1998**, *120*, 1207; (b) O'Brien, P.; Osborne, S.A.; Parker, D.D. *J. Chem. Soc. Perkin Trans 1,* **1998**, 2519.
432. (a) Jiang, W.; Wanner, J.; Lee, R.J.; Bounaud, P.-Y.; Boger, D.L. *J. Am. Chem. Soc.* **2003**, *125*, 1877; (b) Boger, D. L.; Lee, R. J.; Bounaud, P.-Y.; Meier, P. *J. Org. Chem.* **2000**, *65*, 6770.

3.6. BAEYER-VILLIGER OXIDATION (RCOR' → RCO₂R')

3.6.A. Peroxide Induced Baeyer-Villiger Oxidation

We have just seen that the π bond of an alkene reacts with peroxides. The π bond of a carbonyl group can also react with peroxides, but the reaction is accompanied by a rearrangement. In the most common application, ketones react with peroxyacids to give esters via a C → O rearrangement known as the **Baeyer-Villiger reaction**.[433] Both acid and base catalyzed mechanisms have been proposed. The acid catalysis mechanism[434] proceeds via **307**,[435] and base catalyzed reaction with hydrogen peroxide[436] have proposed **308** as the key intermediate.[435] A solid-state variation of this reaction has been reported in which peroxy acids supported on hexagonal mesoporous silica are used to induce the Baeyer-Villiger reaction.[437]

In all cases, an alkyl group migrates to the electrophilic oxygen with concomitant loss of the acyl or hydroxyl leaving group. Since hydroxyl is a poorer leaving group than acyl, the base-catalyzed reaction is "inferior to the peroxyacid route."[438] With peroxyacids, the conversion of alkenes to epoxides can compete with the Baeyer-Villiger reaction. Ketones react with peroxyacids more slowly than do alkenes, usually requiring long reaction times, strong acid catalysts and highly reactive peroxyacids.[439] As stated by Lewis, however, "this rate difference is not large with isolated carbon-carbon double bonds and Baeyer-Villiger oxidation can be competitive with epoxidation."[440] Lewis illustrated this difference in reactivity in the oxidation of **309** with peroxyacetic acid, which gave a 44% yield of lactone **310** and no epoxide. Changing to hydrogen peroxide (benzonitrile was the solvent) led to epoxide **311** in 54% yield rather than the lactone. When an alkene and a carbonyl are present in the same molecule, the presence of an acid catalyst leads to preferential attack at the carbonyl. Selective epoxidation is best accomplished

433. (a) House, H.O. *Modern Synthetic Reactions*, 2nd Ed., Benjamin, Menlo Park, CA, *1972*, pp 306-307, 321-328; (b) Reference 254, p 237; (c) The Merck Index, 14th ed., Merck & Co., Inc., Whitehouse Station, New Jersey, *2006*, p ONR-4. Also see, Reference 20, p 1845.
434. Hawthorne, M.F.; Emmons, W.D. *J. Am. Chem. Soc.* **1958**, *80*, 6398.
435. Reference 254, p 237.
436. House, H.O.; Wasson, R.L. *J. Org. Chem.* **1957**, *22,* 1157.
437. Lambert, A.; Elings, J.A.; Macquarrie, D.J.; Carr, G.; Clark, J.H. *Synlett* **2000**, 1052.
438. Reference 254, p 238.
439. Reference 433a, p 321.
440. (a) Reference 254, p 241; (b) Payne, G.B. *Tetrahedron* **1962**, *18*, 763.

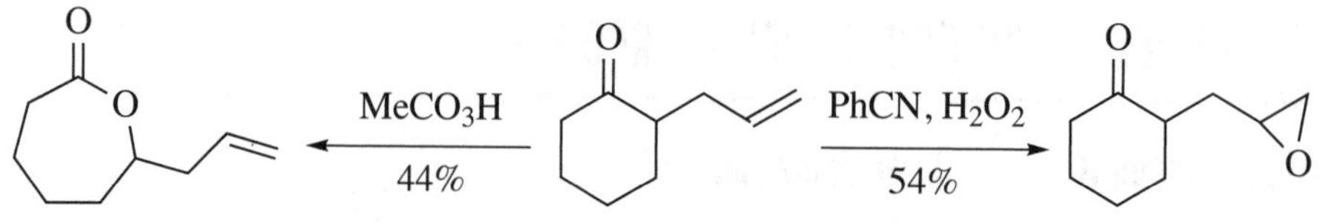

310 **309** **311**

in a reaction medium that does not contain acid. The use of mcpba in an inert solvent at low temperature usually gives selective epoxidation.[370] Silverstein and co-workers exploited this selectivity for the conversion of **312** to **313** in a synthesis of α-multistriatin.[441] Conjugated ketones react with mcpba to give mixtures of products, and steric hindrance can have a significant influence on the course of the reaction. Oxidation of 3-hydroxycholesten-4-en-6-ones, for example, gave mixtures of an epoxide and a lactone where proportions varied with the stereochemistry of the hydroxy group (α or β).[442] Glotter and co-workers concluded that for this system, the Baeyer-Villiger reaction was faster than epoxidation.[442]

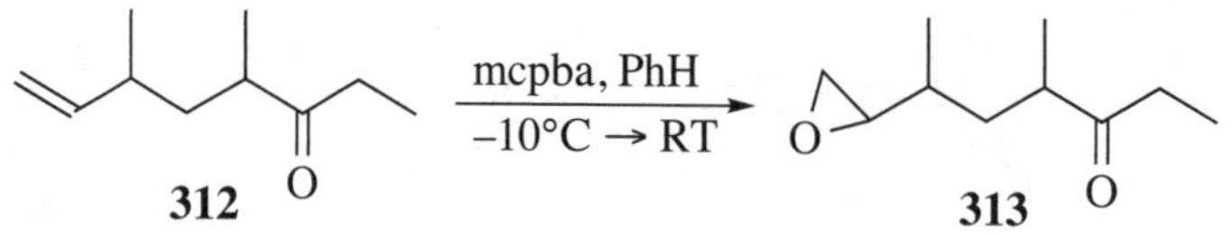

312 **313**

Although most peroxyacids can be used in the Baeyer-Villiger reaction, trifluoroperoxyacetic acid[443] is one of the more efficient reagents for this rearrangement. Such reactions are usually buffered with disodium phosphate (Na_2HPO_4)[444] since the by-product of this reaction is the carboxylic acid precursor of the peroxyacid (sec. 3.4.C). Generation of a strong acid such as trifluoroacetic acid (from trifluoroperoxyacetic acid) may cause problems in an unbuffered reaction. In the generalized reaction of **314**, the final product is an ester (**315**) and CF_3COOH. The presence of both compounds in solution raises the possibility of transesterification reactions. Transesterification of **315** would give a trifluoroacetyl ester (**316**) and acid **317**.[443] Addition of a buffer to the initial reaction minimizes transesterification and other acid catalyzed reactions. Typical buffers include disodium phosphate or the salt of the original acid ($RCO_2^-Na^+$ for example). In the case of **315**, $CF_3CO_2^-M^+$ or even sodium acetate might be a suitable as a buffer.

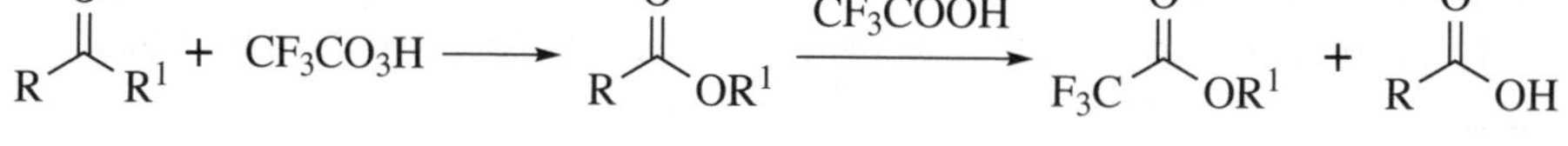

314 **315** **316** **317**

A useful synthetic example of the Baeyer-Villiger oxidation taken from Liebeskind and Zhang's synthesis of (–)-Bao Gong Teng A, in which methyl ketone **318** was converted to ester **319**

441.   Pearce, G.T.; Gore, W.E.; Silverstein. R.M. *J. Org. Chem.* **1976**, *41*, 2797.
442.   Mendelovici, M.; Glotter, E. *J. Chem. Soc. Perkin Trans. 1* **1992**, 1735.
443.   Reference 433a, p 322.
444.   For an example, see Emmons, W.D.; Lucas, G.B. *J. Am. Chem. Soc.* **1965**, *77*, 2287.

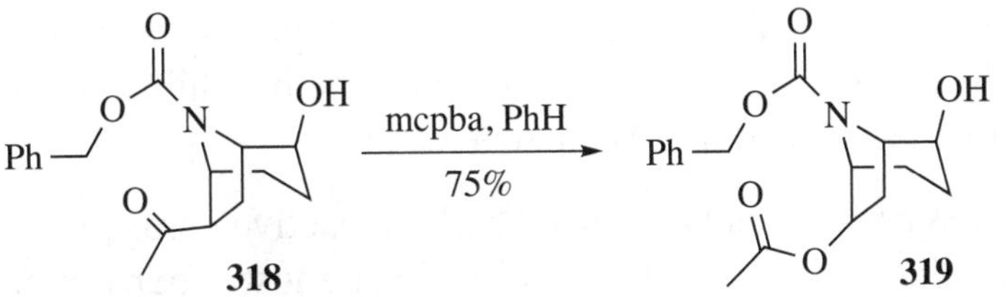

318 → **319**

in 75% yield using mcpba.[445] Note that the larger group migrated in preference to the methyl group, which is characteristic of this reaction. When an unsymmetrical ketone such as **314** is treated with a peroxyacid, two esters are possible (**315** and/or **320**). Different groups (R and R¹ in **314**) will show different migratory aptitudes in the Baeyer-Villiger reaction. Both Lewis[446],[435] and House[447] reported the migratory aptitude of alkyl groups (R) in **314** to be

> *tert*-alkyl > cyclohexyl ≈ 2° alkyl ≈ benzyl ≈ phenyl >
> vinylic[448],[446] > 1° alkyl > cyclopropyl > methyl

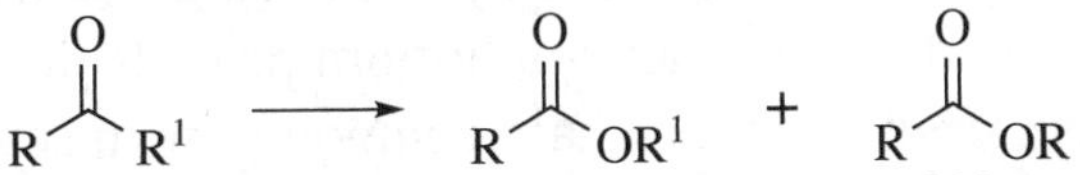

314 → **315** + **320**

This migratory preference is clear in Fukuyama and co-worker's oxidation of **321** to **322** in >80% yield, taken from a synthesis of (+)-vinblastine.[449] There may be some correlation between migratory ability and the ability of that group to stabilize a δ+ charge, but as stated by House "this property, arising from the electron distribution in the migratory group, cannot be the only factor determining migratory aptitude".[450] Electron-releasing groups on aromatic derivatives accelerate the rearrangement and electron-withdrawing groups slow it. Aryl groups generally migrate in preference to alkyl groups.[451] Although the reaction is usually selective, other products are formed and they can sometimes be difficult to separate.

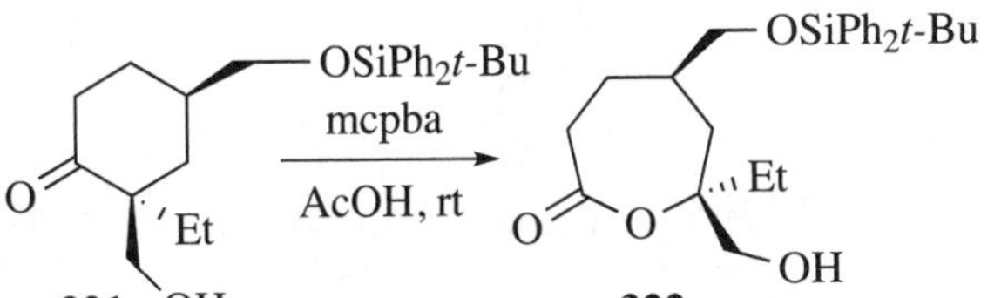

321 → **322** 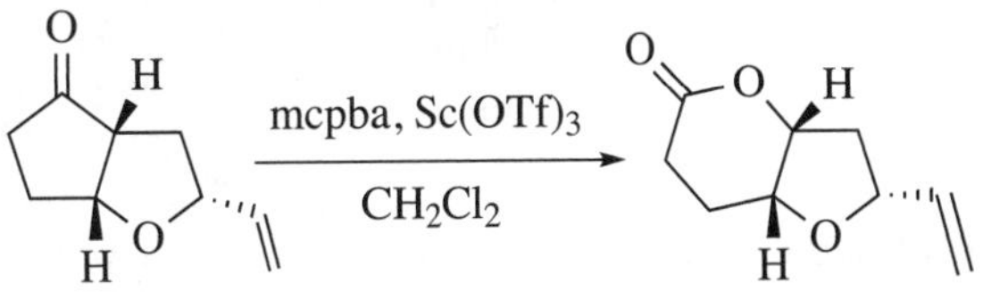   **323** → **324**

445.   Zhang, Y.; Liebeskind, L.S. *J. Am. Chem. Soc.* **2006**, *128*, 465.
446.   (a)Reference254,p241;(b)Hawthorne,M.F.;Emmons,W.D.;McCallum,K.S.*J.Am.Chem.Soc.***1958**,*80*,6393.
447.   Reference 433a, p 324.
448.   Walton, H.M. *J. Org. Chem.* **1957**, *22*, 1161.
449.   Yokowhima, S.; Ueda, T.; Kobayashi, S.; Sato, A.; Kuboyama, T.; Tokuyama, H.; Fukuyama, T. *J. Am. Chem. Soc.* **2002**, *124*, 2137.
450.   Reference 433a, p 326.
451.   (a) Reference 254, pp 241-242; (b) Friess, S.L.; Farnham, N. *J. Am. Chem. Soc.* **1950**, *72*, 5518.

Scandium triflate has been used in conjunction with mcpba to facilitate the Baeyer-Villiger rearrangement.[452] In a synthesis of (±)-*trans*-kumausyne by Philips and Chandler,[453] bicyclic ketone **323** was converted to lactone **324** in 86% yield.

Simple dialkyl ketones are not easily oxidized by less reactive peroxyacids and usually require the more powerful trifluoroperoxyacetic acid.[454] Since most groups migrate in preference to methyl, "the Baeyer-Villiger reaction of methyl ketones serves as a stereospecific source of optically active alcohols by way of the acetate"[455] as seen in the conversion of **325** to **326** in 85% yield.[456] This example also illustrates that the stereochemical integrity of the migratory group is retained, which appears to be a general phenomenon.[457] Saponification of the ester product from the Baeyer-Villiger reaction is easily accomplished, leading to the alcohol and the interesting functional group transform (C–C=O → C–OH) for acyclic ketones. Cyclic ketones (such as **309** and **321**) are more reactive than their acyclic counterparts and oxidation with a variety of peroxyacids leads to good yield of the corresponding lactone, as in the conversion of cyclopentanone to δ-valerolactone in 81% yield[458] (also see **321** → **322**). Hydrolysis of the lactone leads to a ω-hydroxy acid but the facility of the ring opening depends on the size of the lactone ring. Aqueous hydrolysis is an equilibrium process that favors the closed ring in five-membered ring systems, but the open chain compound with seven-membered and higher lactones. The peroxyacid reaction[459] should be buffered to prevent unwanted acid-catalyzed ring opening in such cases. Six-membered ring lactones are intermediate in stability between those just cited, and hydrolysis will usually give a mixture of the lactone and the open-chain acid. The migratory aptitude of the alkyl moieties in mono- and disubstituted cyclic ketones is the same as that observed with acyclic ketones.

Other carbonyl derivatives react with peroxyacids. 1,2-Diketones, *o*-quinones, and α-keto-

452. Kotsuki, H.; Arimura, K.; Araki, T.; Shinohara, T. *Synlett* **1999**, 462.
453. Chandler, C.L.; Philips, A.J. *Org. Lett.* **2005**, *7*, 3493.
454. Reference 254, p 239.
455. Reference 254, p 242.
456. Baggiolini, E.G.; Iacobelli, J.A.; Hennessy, B.M.; Batcho, A.D.; Sereno, J.F.; Uskokovic´, M.R. *J. Org. Chem.* **1986**, *51*, 3098
457. (a) Reference 254, p 237; (b) Mislow, K.; Brenner, J. *J. Am. Chem. Soc.* **1953**, *75*, 2318.
458. (a) Reference 433a, p 323; (b) Sager, W.F.; Duckworth, A. *J. Am. Chem. Soc.* **1955**, *77*, 188.
459. (a) Friess, S.L. *J. Am. Chem. Soc.* **1949**, *71*, 2571; (b) Friess, S.L.; Frankenburg, R.E. *Ibid.* **1952**, *74*, 2679; (c) Mateos, J.L.; Menchaca, H. *J. Org. Chem.* **1964**, *29*, 2026.

esters react via cleavage of the bond between the two carbonyls.[460,451] Similarly, 1,3-dicarbonyl compounds react with peroxyacids to give 2-hydroxy-1,3-dicarbonyls (enediols),[461] but 2,2-disubstituted-1,3-dicarbonyl compounds are resistant to such oxidations.[462] Ethyl acetoacetate is an example of the first type, and it gives a glycidic ester upon treatment with a peroxyacid.[461,462] Aldehydes are easily oxidized, and reaction with peroxyacids or hydrogen peroxide usually gives a carboxylic acid (the transformation is $RCHO \rightarrow RCO_2H$) with little or no rearrangement. Exceptions are electron-rich aldehydes such as functionalized benzaldehyde derivatives. In these cases, formation of the formate ester is facile and is called the **Dakin reaction**.[463] This transformation is compatible with a variety of other functionality in the substrate. In Joullie and co-worker's synthesis of ustiloxin D,[464] for example, aryl aldehyde **327** (Boc = -CO2t-Bu) was treated with 30% hydrogen peroxide and a diselenide to give formate ester **328** via the Dakin reaction. Hydrolysis was accomplished with KOH in aqueous methanol to give phenol **329** in a yield of 73% for both steps.

3.6.B. Enzymatic Baeyer-Villiger Oxidation

Enzymatic conversion of ketones to esters is very common in microbial degradations.[465] Several monooxygenases are now available for this reaction, and a correlation of enantioselectivity with a variety of substrates has been published.[466] A typical transformation that can be synthetically important is the enzymatic Baeyer-Villiger oxidation reported by Walsh, which gave lactone **330** from cyclohexanone[467] using a purified cyclohexanone oxygenase enzyme.[468] Walsh and co-workers studied the mechanism of this reaction, and also showed that this enzyme converted

460. Karrer, P.; Haab, F. *Helv. Chim. Acta* **1949**, *32,* 950.
461. (a) Reference 254, p 243; (b) Karrer, P.; Kebrle, J.; Thakkar, R.M. *Helv. Chim. Acta* **1950**, *33*, 1711.
462. Hassall, C.H. *Org. React.* **1957**, *9*, 73.
463. (a) Dakin, H.D. *Am. Chem. J.* **1909**, *42*, 477; (b) Lee, J.D.; Uff, B.C. *Quart. Rev.* **1967**, *21*, 429 (see p 454); (c) Dakin, H.D. *Org. Synth. Coll. Vol. 1* **1941**, 149; (d) The Merck Index, 14th ed., Merck & Co., Inc., Whitehouse Station, New Jersey, **2006**, p ONR-21.
464. Cao, B.; Park, H.; Joullie, M.M. *J. Am. Chem. Soc.* **2002**, *124*, 520.
465. (a) Sih, C.J.; Rosazza, J.P. in *Applications of Biochemical Systems in Organic Chemistry*, Jones, J.B.; Sih, C.J.; Perlman, D. (Eds.), Wiley, New York, **1976**, pII, pp 100-102; (b) Fonken, G.S.; Johnson, R.A. *Chemical Oxidations with Microorganisms*, Marcel-Dekker, New York, **1972**, pp 157-164; (c) Britton, L.N.; Brand, J.M.; Markovetz, A.J. *Biochim Biophys. Acta* **1974**, *369*, 45; (d) Cripps, R.E. *Biochem. J.*, **1975**, *152*, 233.
466. Kyte, B.G.; Rouvière, P.; Cheng, Q.; Stewart, J.D. *J. Org. Chem.* **2004**, *69*, 12.
467. Ryerson, C.C.; Ballou, D.P.; Walsh, C. *Biochemistry* **1982**, *21*, 2644.
468. Donoghue, N.A.; Norris, D.B.; Trudgill, P.W. *Eur. J. Biochem.* **1976**, *63*, 175.

phenylacetaldehyde to phenylacetic acid (65% yield) accompanied by only a 12% yield of benzyl formate ($PhCH_2OCHO$).[469] The rearrangement induced by this enzyme proceeds with completely stereochemical retention as shown by Schwab,[470] analogous to the Baeyer-Villiger rearrangement that used *m*-chloroperoxybenzoic acid.

FAD = flavin adenine dinucleotide
NADPH = reduced nicotinamide adenine dinucleotide phosphate

330

331 **332**

333 **334**

Taschner and co-workers showed that cyclohexanone oxygenase, obtained from *Acinetobacter* NCIB 9871,[470b] converted 4-methylcyclohexanone to the corresponding γ-methyl caprolactone in 80% yield with >98% ee.[471a] In another example, lactone **332** was prepared from bicyclic ketone **331** in 80% yield and 97% ee.[470b] The enzymatic reaction appears to proceed in a predictable manner using the peroxide induced reaction as a model, although the enantioselectivity of the final product is influenced by the concentration of the substrate.[472] Taschner et al. found that **333** was converted to **334** in 80% yield, and in >98% ee.[471] This rearrangement is clearly related to similar rearrangements that use mcpba.[473] Other enzymatic systems have been found that give a Baeyer-Villiger reaction, including the reaction of myristic acid and hydrogen peroxide catalyzed by *Candida anatarctica* lipase.[474] Changing the enzymatic system can sometimes control the absolute stereochemistry of the lactone product. Oxidation of **335** with *Acinetobacter sp.* NCIB 9872 gave lactone **336**,[475] but oxidation with *P. putida* NCIB 10007 gave

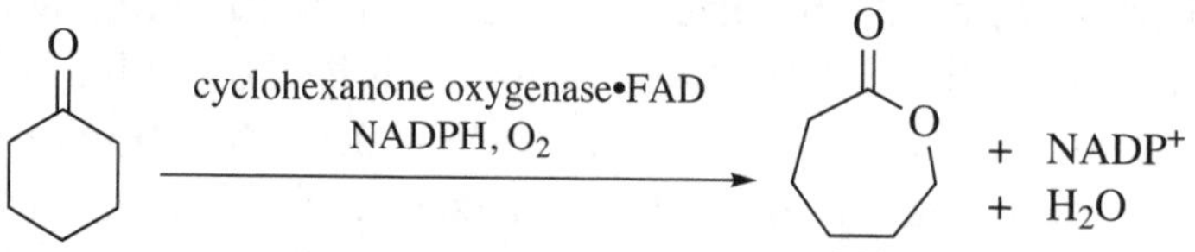

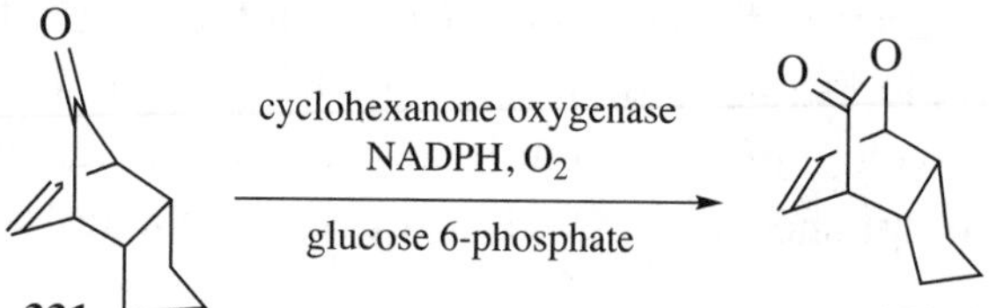

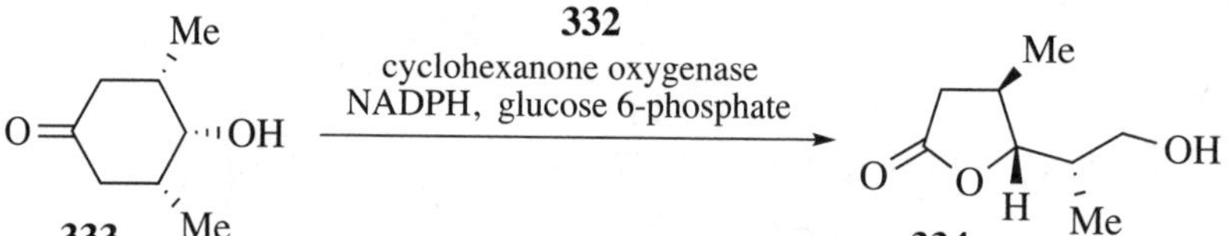

469. Branchand, B.P.; Walsh, C.T. *J. Am. Chem. Soc,* ***1985****, 107,* 2153.
470. (a) Schwab, J.M. *J. Am. Chem. Soc.* ***1981****, 103*, 1876; (b) Schwab, J.M.; Li, W.; Thomas, L.P. *J. Am. Chem. Soc.* ***1983****, 105*, 4800.
471. (a) Taschner, M.J.; Black, D.J. *J. Am. Chem. Soc.* ***1988****, 110,* 6892; (b) Taschner, M.J.; Peddada, L. *J. Chem. Soc. Chem. Commun.* ***1992****,* 1384; (c) Taschner, M.J.; Chen, Q.-Z., *Bioorg. Med. Chem. Lett.* ***1991****, 1*, 535.
472. Zambianchi, F.; Pasta, P.; Ottolina, G.; Carrea, G.; Colonna, S.; Gaggero, N.; Ward, J.M. *Tetrahedron: Asymmetry* ***2000****, 11*, 3653.
473. Ouazzani-Chahdi, J.; Buisson, D.; Azerad, R. *Tetrahedron Lett.* ***1987****, 28*, 1109.
474. Lemoult, S.C.; Richardson, P.F.; Roberts, S.M. *J. Chem. Soc. Perkin Trans 1* ***1995****,* 89.
475. Carnell, A.J.; Roberts, S.M.; Sik, V.; Willets, A.J. *J. Chem. Soc. Perkin Trans. 1* ***1991****,* 2385.

lactone **337**.[476] Stereochemical control is a key feature of the enzymatic Bayer-Villiger reaction.

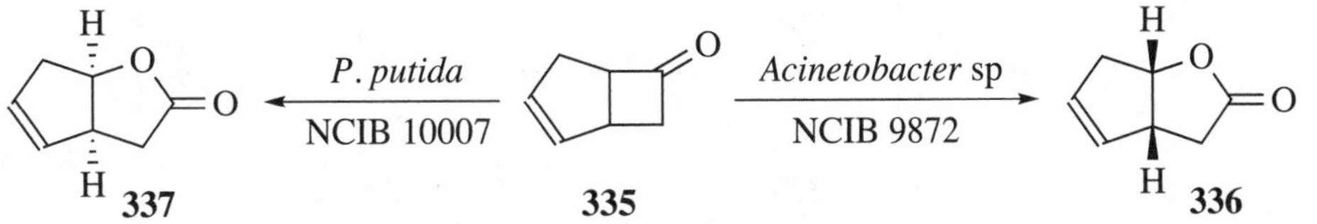

The functional group transforms available from the Baeyer-Villiger reaction are

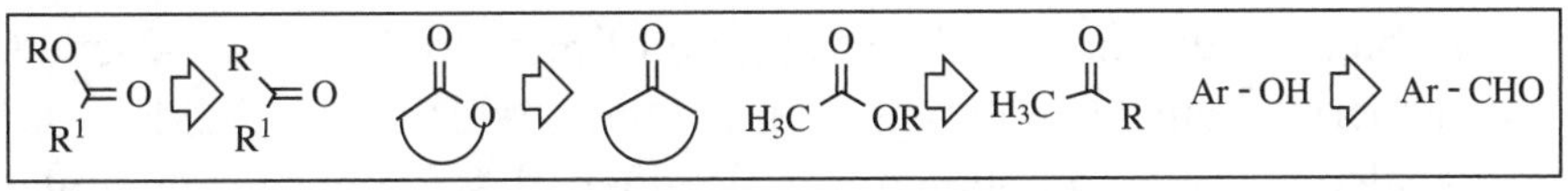

3.7. OXIDATIVE BOND CLEAVAGE (C=C ⟶ C=O + O=C)

3.7.A. Cleavage of Alkenes by Transition Metals

Alkenes can be oxidized to epoxides or 1,2-diols by various reagents.[477] If more powerful oxidizing agents are used, however, the C=C moiety can be cleaved into two carbonyl derivatives. This section will discuss the various reagents that effect this transformation.

3.7.A.i. Potassium Permanganate. In Section 3.5.A, treatment of alkenes with dilute solutions of potassium permanganate was shown to give a one-stage oxidation to 1,2-diols in excellent yield. More concentrated solutions and higher temperatures lead to cleavage of the alkene to a diacid via oxidation of an intermediate diol (**338**),usually in acidic media[478] The terms hot and concentrated are ambiguous and are used for concentrations that range from a >10% solution of permanganate to 10 M solutions, and temperatures that range from ambient to > 100°C. An important modification that increases the synthetic utility of the reaction mixes periodic acid (HIO_4)[479] with a catalytic amount of permanganate (called the **Lemieux-von Rudloff reagent**).[480] The stoichiometric oxidant HIO_4 oxidizes the manganate product back to permanganate, allowing oxidative cleavage of alkenes under catalytic conditions. A solution of potassium permanganate in benzene with the well-known potassium complexing agent dicyclohexyl-18-crown-6 is very effective for cleaving alkenes to acid products,[481] as illustrated

476. Grogan, G.; Roberts, S.M.; Willets, A.J. *J. Chem. Soc. Chem. Commun.* *1993*, 699. For the use of this lactone in a synthesis, see Velázquez, F.; Olivo, H.F. *Org. Lett.* *2000*, 2, 1931.
477. Reference 20, pp 1650-1652.
478. (a) Wolfe, S.; Ingold, C.K.; Lemieux, R.U. *J. Am. Chem. Soc.* *1981*, 103, 938; (b) Wolfe, S.; Ingold, C.K. *Ibid.* *1981*, 103, 940.
479. Smith, M.B.; March, J. *March's Advanced Organic Chemistry, 6th ed.*, Wiley, New York, *2007*, p. 1743.
480. (a) Lemieux, R.U.; von Rudloff, E. *Can. J. Chem.* *1955*, 33, 1701, 1710; (b) von Rudloff, E. *Ibid.* *1955*, 33, 1714; (c) von Rudloff, E. *Ibid.* *1956*, 34, 1413; (d) von Rudloff, E. *Ibid.* *1965*, 43, 1784.
481. (a) Sam, D.J.; Simmons, H.E. *J. Am. Chem. Soc.* *1972*, 94, 4024; (b) Lee, D.G.; Chang, V.S. *J. Org. Chem.* *1978*, 43, 1532.

by the cleavage of cyclohexene to adipic acid.

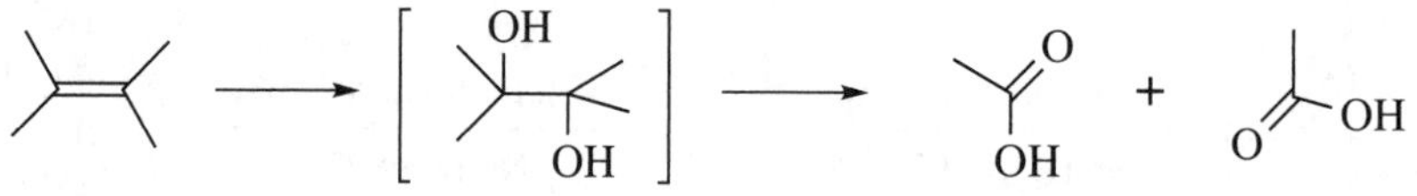

338

Oxidation of disubstituted alkenes bearing one or more C=C-H units leads to aldehydes, but the aldehyde is further oxidized to a carboxylic acid. Although isolation of the aldehyde can be difficult, it is possible. An example is the oxidation of bicyclo[3.3.1]hept-2-ene (norbornene, **339**) with aqueous permanganate and magnesium sulfate, which gave dialdehyde **340** in up to 66% yield.[482] It is also possible to obtain aromatic aldehydes from aryl substituted alkenes by permanganate oxidation in aqueous media, but good yields are obtained only when the aromatic ring has ortho- or para-electron withdrawing substituents.[483,484] The presence of conjugating groups on the alkene increases the rate of oxidation, but bulky groups on the alkene greatly reduce the yield of aldehyde products. It is also possible to oxidize alkynes with permanganate. Midland showed that **341** was oxidized to 2-acetoxyheptanoic acid (**342**) in 92% yield using an excess of permanganate in acetic acid.[485]

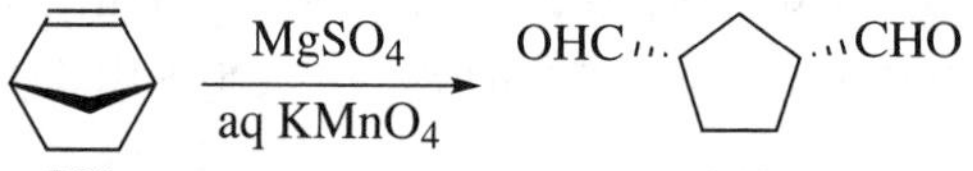

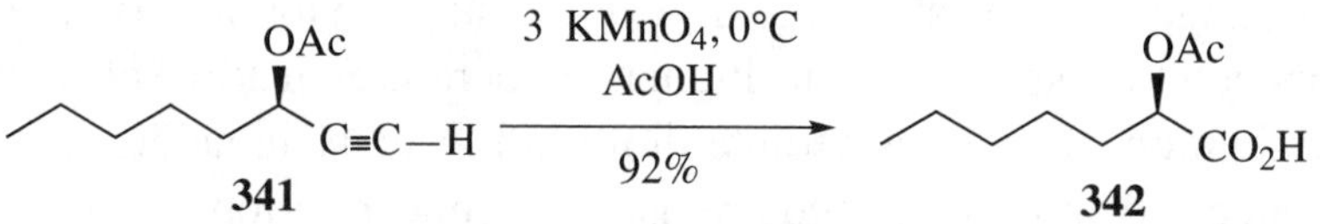

Potassium permanganate is a very powerful reagent and can oxidize alkyl fragments (sec. 3.8) as well as alkenes. Straight-chain alkyl groups attached to aromatic rings are oxidized with concentrated permanganate to give benzoic acid derivatives, as in the oxidation of 2-chlorotoluene to 2-chlorobenzoic acid.[486] Similarly, 2-methylpyridine was oxidized to pyridine 2-carboxylic acid.[487]

3.7.A.ii. Osmium Reagents. When alkenes are oxidized with a mixture of OsO_4 and $NaIO_4$ (the **Lemieux-Johnson reagent**),[488] oxidative cleavage leads to aldehydes and ketone products. This

482. (a) Reference 233a, p 125; (b) Wiberg, K.B.; Saegebarth, K.A. *J. Am. Chem. Soc.* **1957**, *79*, 2822.
483. Reference 233a, pp 124.
484. Viski, P.; Szeverényi, Z.; Simándi, L. *J. Org. Chem.* **1986**, *51*, 3213.
485. (a) Midland, M.M.; Lee, P.E. *J. Org. Chem.* **1981**, *46*, 3933; (b) Krapcho, A.P.; Larson, J.R.; Eldridge, J.M. *Ibid.* **1977**, *42*, 3749.
486. Clarke, H.T.; Taylor, E.R. *Org. Synth. Coll. Vol. 2* **1943**, 135.
487. Singer, A.W.; McElvain, S.M. *Org. Synth. Coll. Vol. 3* **1955**, 740.
488. (a) Reference 233a, p 122; (b) Pappo, R.; Allen, Jr., D.S.; Lemieux, R.U.; Johnson, W.S. *J. Org. Chem.* **1956**, *21*, 478.

technique requires only 1-5 mol% of osmium tetroxide, which is important given the expense and toxicity of osmium compounds. Nemoto and co-workers used this reagent to cleave the vinyl unit in **343** to the aldehyde moiety in **344** as part of the synthesis of macrosphelide A.[489] *N*-Methylmorpholine *N*-oxide is sometimes added to the reaction. Paquette et al. used this variation in a synthesis of (–)-polycavernoside A.[490]

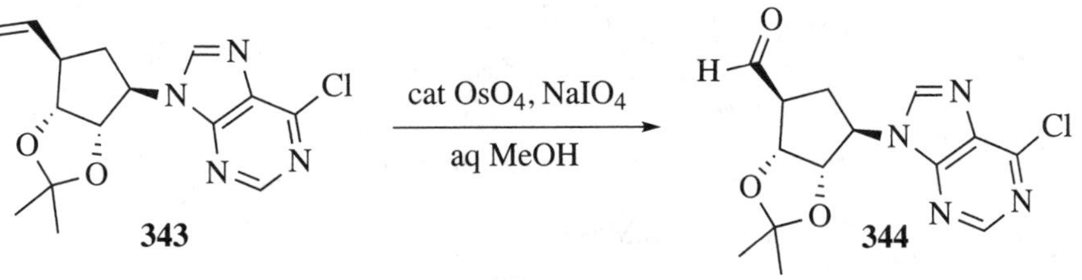

3.7.A.iii. Ruthenium Reagents.
Ruthenium compounds are powerful oxidizing agents that are capable of cleaving alkenes. Ruthenium tetroxide (RuO_4) showed a reduction potential of 0.59 V in the following reaction:[17]

$$RuO_4 + e^- \rightarrow RuO_4^{2-} \qquad\qquad 0.59 \text{ V}$$

Ruthenium tetroxide cleaves alkenes to give ketones, aldehydes, or carboxylic acids.[491] Halogenated solvents such as dichloromethane are commonly used, since ether, benzene and pyridine *react violently* with ruthenium tetroxide.[492] The yield of cleavage products is variable and sometimes low (cyclohexene, for example, gave adipaldehyde in only 10% yield).[493,491] Yields of the oxidative cleavage products were greatly improved by addition of sodium periodate ($NaIO_4$) to the reaction.[494] A disadvantage of the $RuO_4/NaIO_4$ reagent is that alkenyl carbons bearing a hydrogen atom are usually oxidized to the acid rather than the aldehyde.[495] Other ruthenium salts also give excellent yields of cleavage products, particularly when mixed with sodium periodate. In its initial formulation, Sharpless used RuO_2 and $NaIO_4$ for the oxidation but he also showed that ruthenium trichloride ($RuCl_3$) and sodium periodate give selective oxidative cleavage to the acid in good yield.[496] It is relatively easy to oxidize tetrasubstituted alkenes, as in Mehta's synthesis of precapnelladiene using a mixture of RuO_2 and $NaIO_4$ to oxidize a bicyclic alkene to a cyclooctadione derivative.[497] This reagent is useful for more complex targets. In a semisynthesis of heterocyclic analogs of squamocin, for example, Lewin

489. Kawaguchi, T.; Funamori, N.; Matsuya, Y.; Nemoto, H. *J. Org. Chem.* **2004**, *69*, 505.
490. Paquette, L.A.; Barriault, L.; Pissarnitski, D. *J. Am. Chem. Soc.* **1999**, *121*, 4542.
491. Reference 233a, p 128.
492. Berkowitz, L.M.; Rylander, P.N. *J. Am. Chem. Soc.* **1958**, *80*, 6682.
493. (a) Djerassi, C.; Engle, R.R. *J. Am. Chem. Soc.* **1953**, *75*, 3838; (b) see also, Munavalli, S.; Ourisson, G. *Bull. Chim. Soc. Fr.* **1964**, 729 for use in a synthesis of longifolene.
494. (a) Reference 224a, p 129; (b) Pappo, R.; Becker, A. *Bull. Res. Counc. Isr. Sect. A* **1956**, *5A*, 300; (c) Sarel, S.; Yanuka, Y. *J. Org. Chem.* **1959**, *24*, 2018.
495. Stork, G.; Meisels, A.; Davies, J.E. *J. Am. Chem. Soc.* **1963**, *85*, 3419.
496. Carlsen, P.H.J.; Katsuki, T.; Martin, V.S.; Sharpless, K.B. *J. Org. Chem.* **1981**, *46*, 3936.
497. Mehta, G.; Murthy, A.N. *J. Org. Chem.* **1987**, *52*, 2875.

and co-workers treated OSiMe$_2$t-Bu-protected (sec. 7.3.A.1) squamocin (**345**) with RuCl$_3$ and NaIO$_4$, and obtained keto-ester **346** in 86% yield.[498]

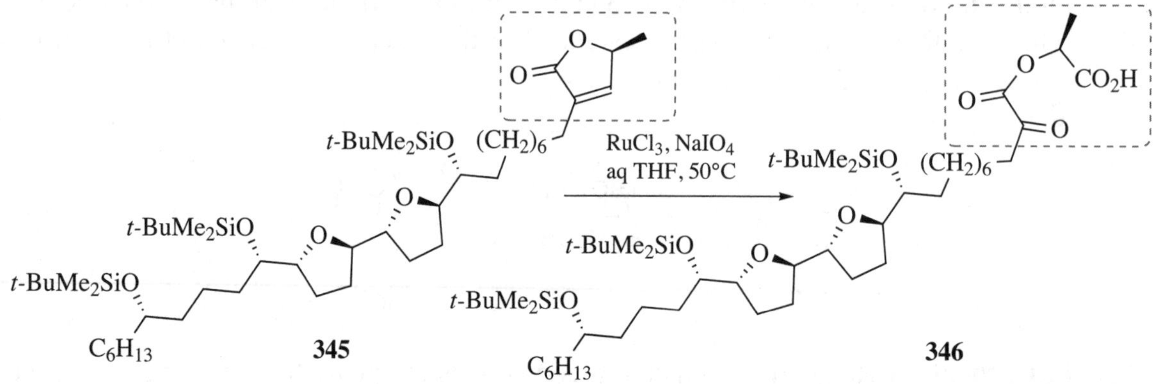

A key ingredient in the RuCl$_3$/NaIO$_4$ oxidation appears to be the choice of solvent, and acetonitrile is frequently used. Alkenes containing a C=C–H unit, such as 1-nonene, are cleaved to acids (nonanoic acid, 89% yield).[496] Alcohols (see **347**) can be oxidized to the acid (in this case, **348** was formed without disturbing the adjacent epoxide) and 1,2-diols are cleaved to the acid in high yield.[496] The oxidations are not accompanied by racemization or rearrangement, which is important when the site of oxidation is adjacent to a stereogenic center (see **345** or **347**).[496] In some cases, secondary reactions can accompany the oxidation. In Silverstein's synthesis of optically pure grandisol,[499] **349** was opened to the diacid (**350**) and ketone **351** was formed after oxidative decarboxylation under the reaction conditions.

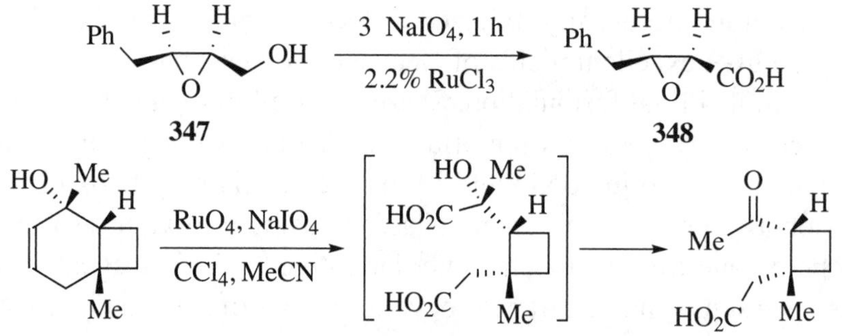

3.7.A.iv. Chromium Reagents.

It is clear from previous discussions in this chapter that Cr(VI) is a powerful oxidant. Indeed, Cr(VI) can cleave alkenes to the corresponding carbonyl compound. It is often difficult to limit the extent of oxidation, however, and the products include "epoxides, ketols, and acids and ketones arising from cleavage or rearrangement reactions".[500] Chromium

498. Duval, R.; Lewin, G.; Hocquemiller, R. *Bioorg. Med., Chem.* **2003**, *11*, 3439.
499. Webster, F.X.; Silverstein, R.M. *J. Org. Chem.* **1986**, *51*, 5226.
500. Reference 233a, p 148.

oxidation of 1,1-diphenylalkenes, however, is the basis of the classical **Barbier-Wieland degradation**, which removes one carbon from the starting molecule.[501] Wieland used this method for the conversion of desoxycholic acid (**352**) to nordesoxycholic acid (**353**), in which two carbons are cleaved (one at a time) from the carboxylic acid chain by sequential Barbier-Wieland degradations.[502] The first sequence with **352** (esterification, reaction with phenylmagnesium bromide and then oxidative cleavage) gave **353**, and the second sequence gave **359**.

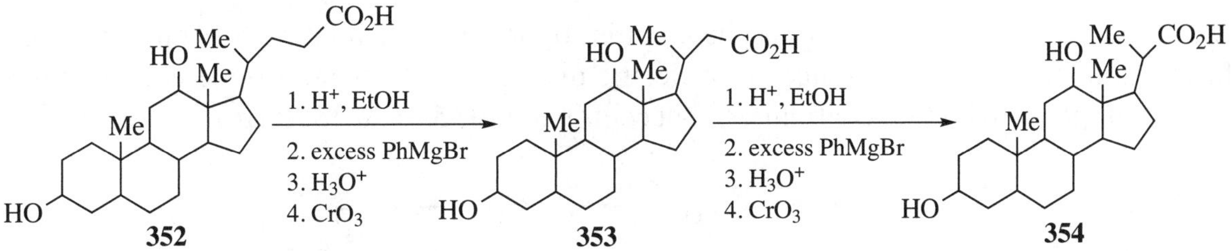

The **Gallagher-Hollander degradation**[503] is a related oxidative cleavage sequence that removes two carbons from the original molecule. The original acid is converted to a diazoketone (see sec. 13.9.B.iii for other applications of diazoketones) and hydrolysis generates a methyl ketone. Bromination of the methyl ketone is followed by elimination (induced by collidine), which generates a conjugated ketone. Oxidative cleavage with chromium trioxide completes the sequence, giving a new acid with a net loss of two carbons.[503] A modification of the Barbier-Wieland degradation allows removal of three carbons and is called the **Miescher degradation**.[504] Initial reaction converts the acid to an ester and reaction with phenylmagnesium bromide generates the diphenylalkylidene intermediate that was important to the Barbier-Wieland sequence. Subsequent treatment with NBS (sec. 2.8.B) gives an allylic bromide, and elimination with methanolic potassium carbonate (K_2CO_3) leads to a conjugated diene. Oxidative cleavage with chromium trioxide cleaves three carbons from the original molecule. These techniques shorten a long chain carboxylic acid by one, two or three carbons in a convenient manner.

The functional group transforms for oxidative cleavage with metals are

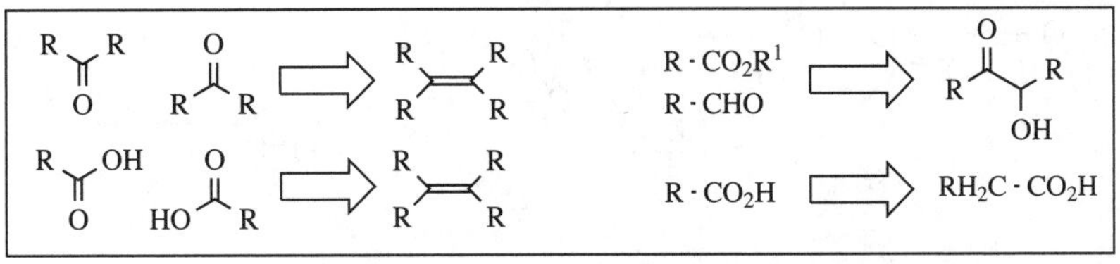

501. (a) Wieland, H. *Berichte* **1912**, *45*, 484; (b) Barbier, P.; Locquin, R. *Compt. Rend.* **1913**, *156*, 1443; (c) The Merck Index, 14th ed., Merck & Co., Inc., Whitehouse Station, New Jersey, **2006**, p ONR-5.

502. Riegel, B.; Moffatt, R.B.; McIntosh, A.V. *Org. Synth. Coll. Vol. 3* **1955**, 237, 234.

503. Hollander, V.P.; Gallagher, T.F. *J. Biol. Chem.* **1946**, *162*, 549.

504. (a) Meystre, C.; Frey, H.; Wettstein, A.; Miescher, K. *Helv. Chim. Acta* **1944**, *27*, 1815; (b) Rodd, E.H. *Chemistry of Carbon Compounds, 2*, **1962**, pp 117, 796, 903; (c) The Merck Index, 14th ed., Merck & Co., Inc., Whitehouse Station, New Jersey, **2006**, p ONR-61.

3.7.B. Cleavage of Alkenes with Ozone

Ozone (O_3) is a powerful oxidizing agent. The reduction potential is high in both acidic and basic media, is 2.07 and 1.24 V, respectively for the following reactions.[17]

$$O_3 + 2\,H^+ + 2\,e^- \rightarrow O_2 + H_2O \qquad\qquad 2.07\ \text{V}$$

$$O_3 + H_2O + 2\,e^- \rightarrow O_2 + 2\,HO^- \qquad\qquad 1.24\ \text{V}$$

Ozone occurs naturally in the atmosphere but in too low a concentration for convenient use. In the laboratory, ozone is produced by electric discharge in oxygen. Ozone is a resonance stabilized species with four contributors, but contributors **355A** and **355B** are most useful[505] for

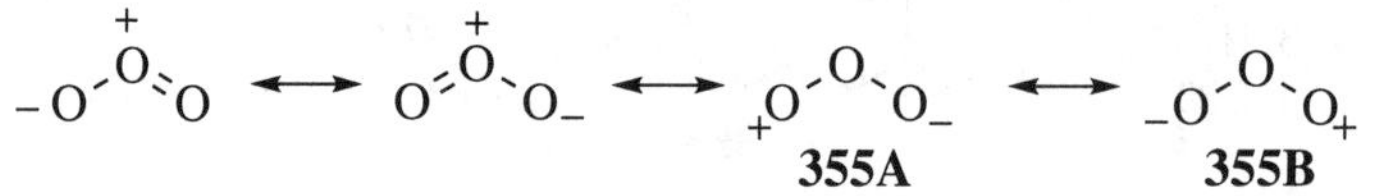

understanding reactions with alkenes. Alkenes react via a 1,3-dipolar addition[506] mechanism (discussed in sec. 11.11.A) to give a 1,2,3-trioxolane product (**356**). This trioxolane is thermally unstable, even at low temperatures, and rapidly rearranges to a 1,2,4-trioxolane, better known as an ozonide (**359**).[505,506] The accepted mechanism of this rearrangement was proposed by Criegee,[507b] in which cleavage of an O-O bond in trioxolane **356** gives a ketone (or aldehyde) and a carbonyl oxide (**357**). These two products do not drift apart, however, but react as shown to give the ozonide, **359**. Other reaction pathways are possible with the carbonyl oxide. When the carbonyl partner is an aldehyde, ozonide formation is preferred in virtually all cases. When the partner is a ketone, another pathway is sometimes seen, in which **357** dimerizes to a tetraoxone (**358**) or polymerizes to a polyperoxide.[508] Criegee showed that alkoxyhydroperoxides (**360**) were formed when methanol was used as a solvent, and carboalkoxyhydroperoxides (**361**) was observed when acetic acid was used as a solvent.[509]

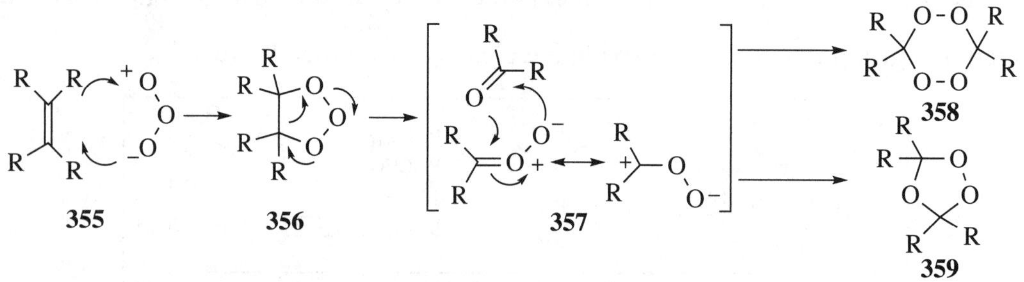

505. (a) Belew, J.S. *Oxidation*, Vol. 1, Augustine, R.L. (Ed.), Marcel Dekker, New York, *1969*, p 262; (b) Meinwald, J. *Chem. Ber.* *1955*, *88*, 1889; (c) Wibaut, J.P.; Sixma, F.L.J.; Kampschmidt, L.W.F.; Boer, H. *Rec. Trav. Chim.* *1950*, *69*, 1355.
506. Criegee, R. *Angew. Chem. Int. Ed.* *1975*, *14*, 745.
507. (a) Reference 505a, p 263 and reference 7 therein; (b) Creigee, R. *Record Chem. Progr.* *1957*, 18, 111.
508. (a) Reference 505a, pp 262-267; (b) Reference 233a, p 120.
509. Reference 505a, p 265.

R O-OH / R ... Me ←(AcOH)— R + O / R O⁻ (357) —(MeOH)→ R O-OH / R OMe (360)

361   **357**   **360**

Several studies have examined cis/trans isomerism in the ozonide[510] but this section will focus only on products derived *from* the ozonide (aldehydes, ketones, and acids). The geometry of the alkyl groups on the ozonide will not be discussed except to say that cis-alkenes give both cis- and trans-ozonides, with the cis generally predominating. Similarly, trans-alkenes give higher yields of the trans-ozonide. In many cases, however, the cis/trans ratio is close to 1:1.[511] In general, electron rich alkenes are oxidized faster than electron-poor alkenes.

An ozonide such as **362** can be decomposed by four methods: (*1*) reduction, (*2*) hydrolysis, (*3*) oxidation, or (*4*) thermolysis.[512] If the ozonide carbons bear a hydrogen (CHR^3 in **362**), reductive decomposition usually gives the corresponding aldehyde whereas an acidic or oxidative workup will yield the acid (by oxidation of the aldehyde in most cases).[513] When that carbon bears two carbon groups (CR^{1}R$_2$ in **362**), the product is a ketone by either oxidative or reductive methods. It is likely that reduction of the peroxide linkage in **362** first yields a bridged hemiacetal (**363**),[514] which fragments to the carbonyl derivatives with loss of water. Hydrolysis of the ozonide apparently occurs at a O-C-O linkage and the peroxide linkage (C-O-O-C) to give the peroxy hemiacetal (structurally related to **360** but with an OH rather than OMe).[514]

$$R^1R^2C{=}O + O{=}CR^3(OH) \xleftarrow{\text{oxidation}} \textbf{362} \xrightarrow{\text{reduction}} [\textbf{363}] \rightarrow R^1R^2C{=}O + O{=}CR^3H$$

362   **363**

Reduction of the ozonide with lithium aluminum hydride (sec. 4.2.C.ii) or sodium borohydride (sec. 4.4.A) generates alcohol products[515] as expected, by further reducing the initially formed carbonyl products. Note that one equivalent of LiAlH$_4$ is required per equivalent of ozonide, and the temperature is typically below –10°C.[516] To prevent over-reduction, the most common reagents that decompose an ozonide to produce aldehyde and ketone products are zinc/acetic acid, dimethyl sulfide (DMS) or triphenylphosphine.

A typical example using ozonolysis in synthesis is the cleavage of the alkene unit in **364** to

510. Reference 505a, pp 267-272 and references cited therein.
511. (a) Reference 505a, p 268; (b) Murray, R.W.; Youssefyeh, R.D.; Story, P.R. *J. Am. Chem. Soc.* **1967**, *89*, 2429.
512. Reference 505a, p 298.
513. Reference 20, pp 1632-1633.
514. Reference 505a, p 299.
515. (a) Reference 233a, p 121; (b) Reference 505a, p 299.
516. (a) Reference 505a, p 301; (b) Greenwood, F.L. *J. Org. Chem.* **1955**, *20*, 803.

give aldehyde **365** with loss of formaldehyde, taken from Fukuyama and co-worker's synthesis of (–)-kainic acid.[517] Ozonolysis of bicyclic alkenes can be used to generate relative large, difunctionalized compound. In a synthesis of deacetoxyalcyonin acetate,[518] Molander and co-workers subjected **366** to ozonolysis and a reductive workup with dimethyl sulfide gave diketone **367**.

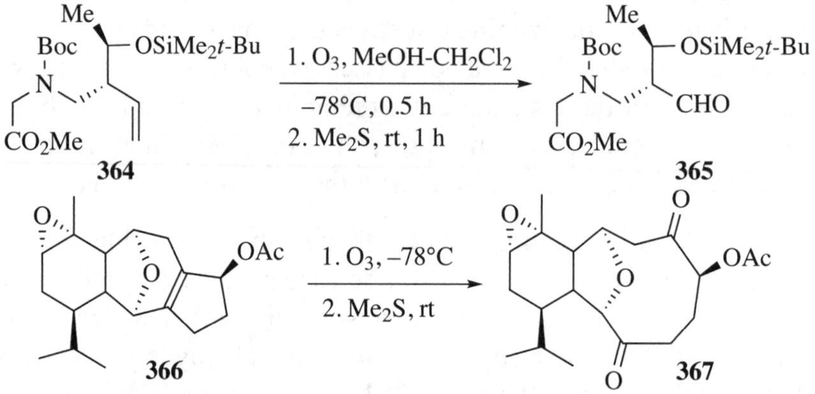

It is possible to trap an aldehyde product of ozonolysis as an acetal using what is called the **Schreiber protocol.**[519] In Fürstner's synthesis of (R)-(+)-lasiodiplodin, cycloheptene was ozonized in the presence of methanol at –78°C,[520] which was followed by neutralization with bicarbonate and dimethyl sulfide to give aldehyde-acetal **368**. This method is quite useful for the chemical differentiation of the two aldehyde units that result from simple ozonolysis of the cyclic alkene.

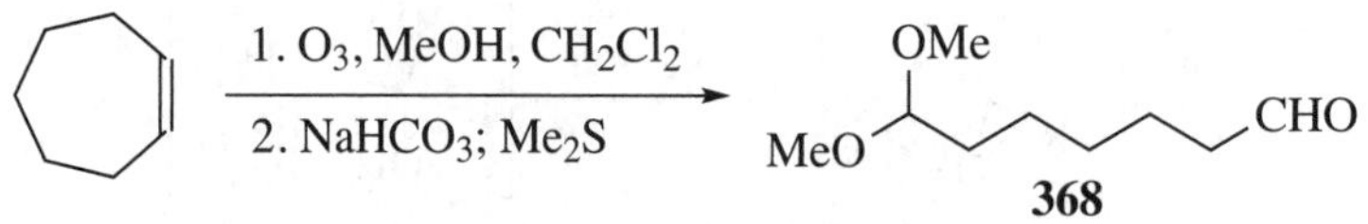

An oxidative workup with ozonides such as **362** uses reagents such as hydrogen peroxide, peroxy acids, silver oxide, chromic acid or permanganate, and produces acids. The conversion of cyclohexene to adipic acid by treatment with (1) O_3 and (2) H_2O_2 is a simple example.[521] As noted above, when the ozonide is disubstituted (two carbon groups on the initial carbon of the alkene), the product is a ketone and either oxidative and reductive workups can be used. If that carbon bears an hydrogen, the oxidative workup will produce an acid and a reductive workup of the aldehyde, as discussed above.

Sterically hindered alkenes react with ozone approximately as fast as unsubstituted alkenes. Ozonolysis of 2,2,5,5-tetramethyl-3-(1,1-dimethylethyl)-3-hexene, for example, gave

517.   Torihata, M.; Nakahata, T.; Kuwahara, S. *Org. Lett.* **2007**, *9*, 2557.
518.   Molander, G.A.; St. Jean Jr., D.J.; Haas, J. *J. Am. Chem. Soc.* **2004**, *126*, 1642.
519.   Schreiber, S.L.; Clause, R.E.; Reagan, J. *Tetrahedron Lett.* **1982**, *23*, 3867.
520.   Fürstner, A.; Thiel, O.R.; Kindler, N.; Bartkowska, B. *J. Org. Chem.* **2000**, *65*, 7990.
521.   Bailey, P.S. *J. Org. Chem.* **1957**, *22*, 1548.

pivalaldehyde and di-*tert*-butyl ketone using trimethyl phosphite as a reducing agent.[522] Steric hindrance is a factor in some cases, however, as in the ozonolysis of **369**, which gave an epoxide (**370**) in 70% yield rather than oxidative cleavage.[523a] This is what Bailey and Lane called partial cleavage, and it is sometimes observed with highly hindered alkenes.[523b] The product distribution depends upon the solvent that is used.[523,524,525]

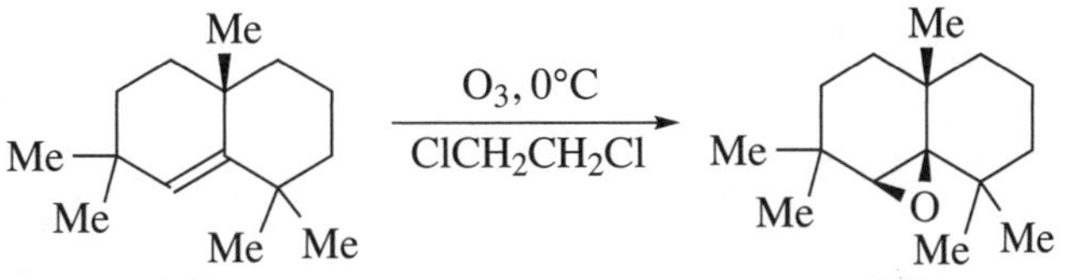

An ozonide can rearrange prior to cleavage if a reactive functional group is adjacent to the carbon bearing the ozonide oxygen. Cargill and Wright used this abnormal reaction in a synthesis of grandisol,[526] where ozonolysis of **371** initially formed the expected ozonide (**372**). Fragmentation of **371**, initiated by the adjacent hydroxyl group, gave keto acid **373** as shown.

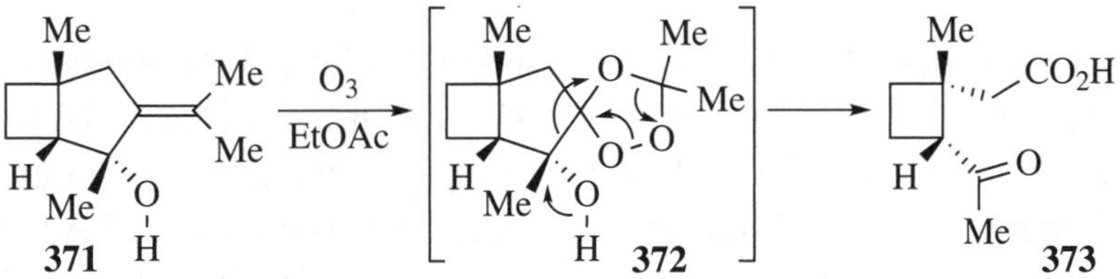

Some aromatic compounds can also react with ozone, although most simple aromatic rings such as benzene derivatives and naphthalene do not. Anthracene reacted with ozone to give anthraquinone in 73% yield after an oxidative workup.[527] Indeed, quinone formation is a more typical example of the reaction of ozone with aromatic compounds (for other oxidative cleavage reactions see sec. 3.8 and for oxidation to phenols and quinones see sec. 3.3). Only a reactive π bond, such as the 9,10 bond of phenanthrene, will oxidatively cleave. Oxidation of phenanthrene (**374**) gave dialdehyde **375** upon treatment with ozone in methanol followed by a workup with potassium iodide in acetic acid.[528]

522. Abruscato, G.J.; Tidwell, T.T. *J. Org. Chem.* **1972**, *37*, 4151.

523. (a) Hochstetler, A.R. *J. Org. Chem.* **1975**, *40*, 1536; (b) Bailey, P.S.; Lane, A.G. *J. Am. Chem. Soc.* **1967**, *89*, 4473.

524. Fuson, R.C.; Armstrong, M.D.; Wallace, W.E.; Kneisley, J.W. *J. Am. Chem. Soc.* **1944**, *66*, 1274.

525. (a) Criegee, R. *Product of Ozonization of Some Olefins*, *Advances in Chemistry Series*, No. 21, American Chem. Soc. Washington, D.C.; **1959**, p 133. See reference 523 and citation 3 therein; (b) Bailey, P.S.; Ward, J.W.; Hornish, R.E. *J. Am. Chem. Soc.* **1971**, *93*, 3552.

526. Cargill, R.L.; Wright, B.W. *J. Org. Chem.* **1975**, *40*, 120.

527. Bailey, P.S.; Kolsaker, P.; Sinha, B.; Ashton, J.B.; Dobinson, F.; Batterbee, J.E. *J. Org. Chem.* **1964**, *29*, 1400.

528. (a) Reference 433a, p 358; (b) Bailey, P.S.; Erickson, R.E. *Org. Synth. Coll Vol. 5* **1973**, 489.

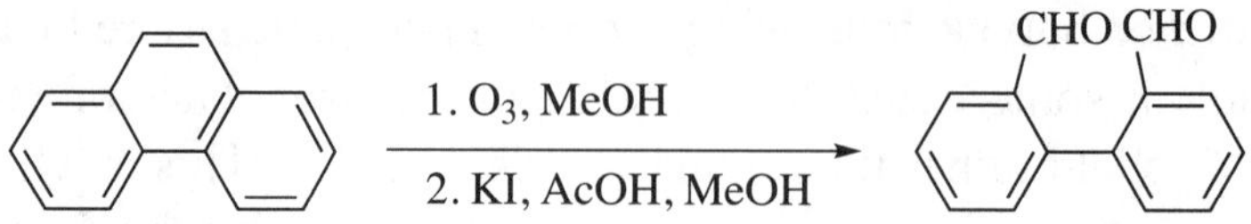

374 → **375**

1. O_3, MeOH
2. KI, AcOH, MeOH

The functional group transform for ozonolysis of alkenes and all oxidative cleavage reactions of alkenes can be summarized by

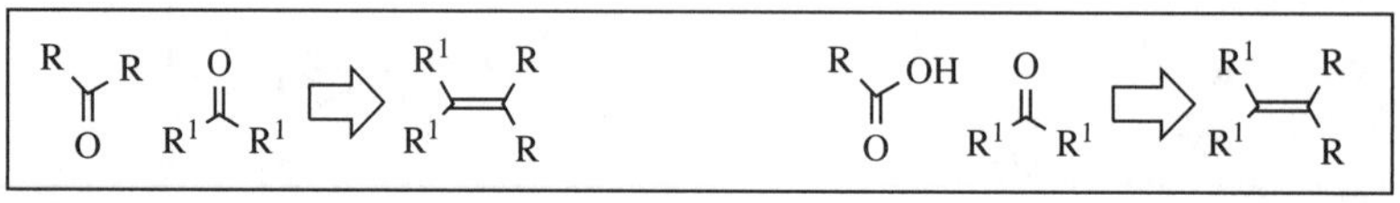

3.7.C. Cleavage of 1,2-Diols

Sections 3.7.A and 3.7.B showed that alkenes are oxidatively cleaved by transition metal oxidants. With certain reagents, 1,2-diols can also be oxidatively cleaved to aldehydes or ketones.[529]

3.7.C.i. Lead Tetraacetate. The reduction potentials for lead compounds were not included in Table 3.1, but lead tetraacetate [Pb(OAc)$_4$, LTA] is a common reagent for oxidizing organic compounds. The reduction potential for LTA in perchloric acid has been reported to be 1.6 V,[530] making it one of the more powerful oxidants examined in this chapter. LTA is used in solvents such as acetic acid or benzene primarily for the oxidative cleavage of 1,2-diols, α-hydroxy ketones, 1,2-diketones or α-hydroxy acids.[531] Cleavage of diols probably proceeds via a cyclic intermediate such as **377**, as observed in the conversion of **376** to **378**.[532] The diol displaces the acetate ligands, allowing fragmentation to formaldehyde, **378** and lead (II) acetate, Pb(OAc)$_2$. Formation of complex **377** is structurally similar to complexes that were observed in reactions of alkenes with permanganate and osmium tetroxide (sec. 3.5). syn-Diols are oxidized faster than *anti*-diols, because cyclic intermediates **377** are formed faster with the former.[533]

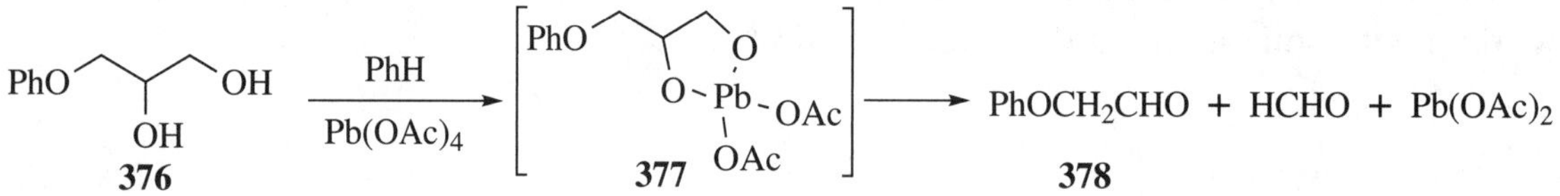

Similar results are observed in the oxidative cleavage of diols with periodic acid. A synthetic application is the oxidative cleavage of the diol in **379** to give formaldehyde and aldehyde

---

529. Reference 20, p 1650.
530. Criegee, R. in *Oxidation in Organic Chemistry, Part A*, Wiberg, K.B. (Ed.), Academic Press, *1965*, p 281.
531. Reference 433a, pp 359-387.
532. (a) Reference 433a, p 360; (b) Speer, R.J.; Mahler, H.R. *J. Am. Chem. Soc.* *1949*, *71*, 1133.
533. Reference 433a, p 361.

380, taken from Trost and Brennan's synthesis of horsfiline.[534] Unfunctionalized alkenes are unaffected by the reagent.[535] α-Hydroxy ketones are cleaved with LTA, as in the oxidation of ketol **381** to give **382** taken from Henrick and Anderson's synthesis of gossyplure.[536] Note that the alkene was not oxidized by LTA.

Functionalized alkenes are oxidized with LTA and a variety of interesting transformations are possible.[537] Ethyl vinyl ether, a typical vinyl ether, gave a diacetate (2-ethoxy-1,2-ethanediol diacetate) upon treatment with LTA in acetic acid.[538] Vinyl esters can also give α-acetoxy ketones.[539] Oxidation of silyl enol ethers leads to α-acetoxy ketones, illustrated by the conversion of **383** to **384** in good yield.[540] Ketones and aldehydes exist in an enol form in part, and this enol may be oxidatively cleaved with LTA to give an α-acetoxy ketone.[541] The yields of the α-acetoxylation product derived from methyl ketones can be improved by the addition of boron trifluoride, which acts as a catalyst.[542] Unsymmetrical dialkyl ketones react selectively at the less substituted position, presumably via the kinetic enol (sec. 9.1, 9.2.E). Oxidation of 2-butanone, for example, gave exclusively 1-acetoxy-2-butanone.[543] Note that oxidation of α-epoxy alcohols give α-acetoxy ketones,[544] and enamines undergo cleavage by LTA to form an α-acetoxy ketone but the yields are often low.[545]

534. Trost, B.M.; Brennan, M.K. *Org. Lett.* **2006**, *8*, 2027.
535. For a synthetic example, see Cernigliaro, G.J.; Kocienski, P.J. *J. Org. Chem.* **1977**, *42*, 3622.
536. Anderson, R.J.; Henrick, C.A. *J. Am. Chem. Soc.* **1975**, *97*, 4327.
537. Butler, R.N. in *Synthetic Reagents Vol. 3*, Pizey, J.S. (Ed.), Ellis Horwood Ltd., Chichester, **1977**, p 277-419.
538. (a) Criegee, R.; Dimroth, P.; Noll, K.; Simon, R.; Weis, C. *Chem. Ber.* **1957**, *90*, 1070; (b) Levas, M. *Ann. Chim. Paris* **1952**, *7*, 697 [*Chem. Abstr. 48*: 1243a, **1954**].
539. Johnson, W.S.; Gastambide, B.; Pappo, R. *J. Am. Chem. Soc.* **1957**, *79*, 1991.
540. Rubottom, J.M.; Gruber, J.M.; Kincaid, K. *Synth. Commun.* **1976**, *6*, 59; (b) Rubottom, G.M.; Marrero, R.; Gruber, J.M. *Tetrahedron* **1983**, *39*, 861.
541. (a) Ichikawaki, K.; Yamaguchi, Y. *J. Chem. Soc. Jpn.* **1952**, *73*, 415; (b) Fuson, R.C.; Maynert, E.W.; Tau, T.-L.; Trumbell, E.R.; Wassmundt, F.W. *J. Am. Chem. Soc.* **1957**, *79*, 1938.
542. Henbest, H.B.; Jones, D.N.; Slater, G.P. *J. Chem. Soc.* **1961**, 4472.
543. Moon, S.; Bohm, H. *J. Org. Chem.* **1972**, *37*, 4338.
544. Brocksom, T.J.; Ferreira, J.T.B.; Braga, A.L. *J. Chem. Res. Synop.* **1981**, 334 [*Chem. Abstr. 96*: 85342t, **1982**].
545. (a) Corbani, F.; Rindone, B.; Scolastico, C. *Tetrahedron Lett.* **1972**, 2597; (b) Corbani, F.; Rindone, B.; Scolastico, C. *Tetrahedron* **1973**, *29*, 3253 and references cited therein.

3.7.C.ii. Sodium Periodate. In section 3.7.A.ii, a mixture of osmium tetroxide and periodic acid oxidatively cleaved alkenes. Sodium periodate can cleave diols directly, as in the Carreira and Lercher synthesis of strychnofoline,[546] were treatment of **385** with NaIO$_4$ gave aldehyde **386** in >90% yield. Sodium periodate in dichloromethane, in the presence of sodium bicarbonate, has also been used.[547]

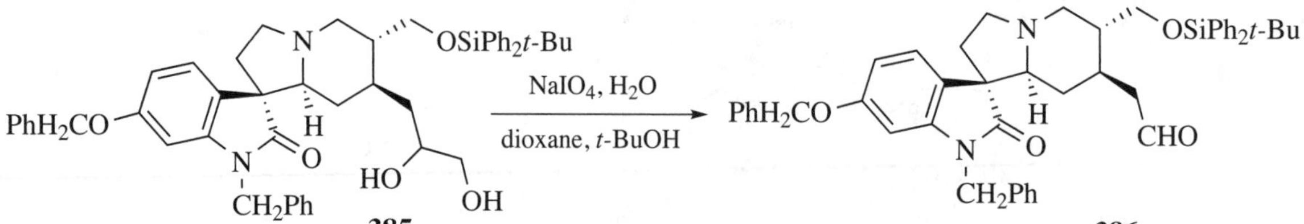

The disconnections for this section are

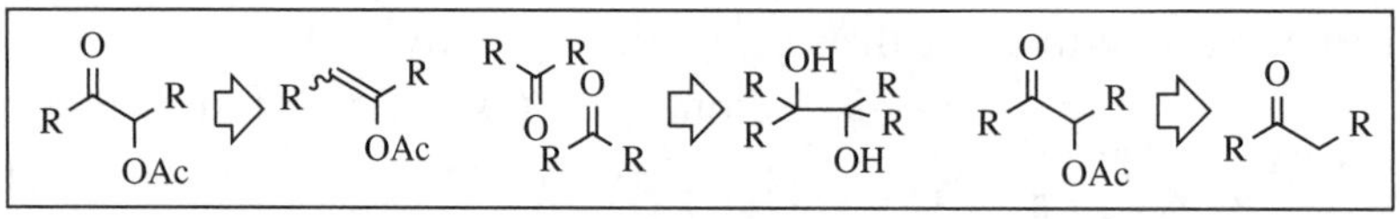

3.8. OXIDATION OF ALKYL OR ALKENYL FRAGMENTS (CH → C=O OR C–OH)

When relatively powerful oxidizing agents are employed with *alkane* fragments that are activated by the presence of a π bond (allylic, benzylic, α-carbonyl compounds), that fragment can be oxidized to an alcohol, aldehyde or ketone, or it can be oxidatively cleaved.[548] The most common reagents for these transformations are selenium dioxide (SeO$_2$), chromium trioxide (CrO$_3$), chromyl chloride (CrO$_2$Cl$_2$) or LTA. Oxidation with ruthenium compounds has been reported. Cleavage of silanes of the type R-SiMe$_2$Ph can be included in this section, using methodology that converts the silyl unit to an alcohol (R–OH). It has also been shown that *t*-BuOOH and CuI oxidizes the allylic carbon of alkenes to conjugated ketones.[549]

3.8.A   Selenium Dioxide

Selenium dioxide is a common reagent used to selectively oxidize an alkyl fragment, and it is effective for converting allylic or benzylic C–H fragments to the corresponding allylic

546.   Lercher, A.; Carreira, E.M. *J. Am. Chem. Soc.* **2002**, *124*, 14826.

547.   For an example, see Alcaide, B.; Almendros, P.; Salgado, N.R. *J. Org. Chem.* **2000**, *65*, 3310.

548.   For oxidations of this type, see Reference 20, pp 975-976.

549.   (a) Salvador, J.A.R.; e Melo, M.L.Sa.; Campos Neves, A.S. *Tetrahedron Lett.* **1997**, *38*, 119. For an example taken from the synthesis of nudenoic acid, see (b) Ho, T.-L.; Su, C.-Y. *J. Org. Chem.* **2000**, *65*, 3566.

alcohol.[550] In most applications, carbonyl derivatives are oxidized to 1,2-dicarbonyl compounds and allylic hydrocarbons to alcohols or esters.[551]

The reaction of SeO_2 and allylic molecules proceeds via an initial **ene reaction** (see the arrows in the example and sec. 11.13) to give **387**. A 2,3-sigmatropic shift (see the arrows in the example and sec. 11.12.A) affords **388**, which is hydrolyzed to allylic alcohol **389**.[552]

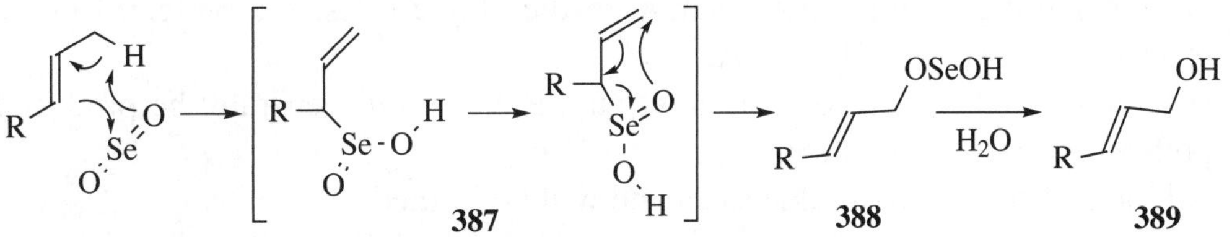

It is important to note that the reaction shows an overall transformation that appears to be a direct conversion of C–H to C–OH, but in fact, the reaction is accompanied by two migrations of the π bond, as shown. Synthetic applications include oxidation of allylic positions to allylic alcohols, as in Danishefsky and co-worker's synthesis of steroid NGA0187,[553] where **390** was converted to **391**. Notice the high selectivity of this reaction for oxidation at the less substituted carbon and in preference to the allylic methyl group side chain, for the diastereomer shown. In the presence of acetic anhydride, selenium dioxide oxidations convert the resulting allylic alcohol product to the corresponding acetate.[554]

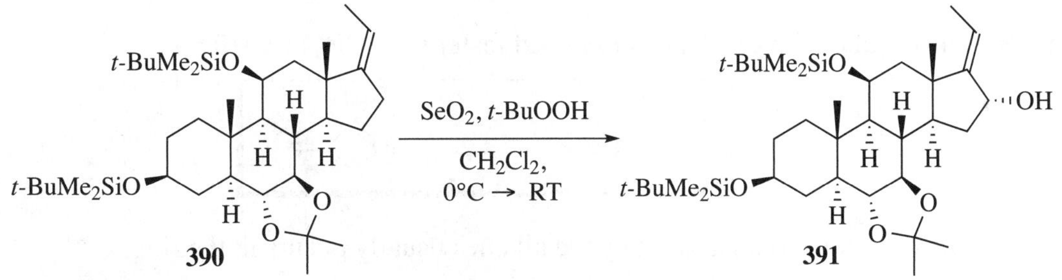

Guillemonat assembled a set of rules that predict the oxidation products for different alkenes.[555] Trachtenberg[556] expanded the discussion of these rules to include general statements concerning oxidation with selenium dioxide and these are listed below.

550. (a) Trachtenberg, E.N. in *Oxidation*, Vol. 1, Augustine, R.L. (Ed.); Marcel-Dekker, New York, *1969*, pp 119-187; (b) Waitkins, G.R.; Clark, C.W. *Chem. Rev. 1945, 36*, 235; (c) Reference 233a, p 25.
551. Rabjohn, N. *Org. React. 1976, 24*, pp 261-426.
552. Reference 233a, p 25
553. Hua, Z.; Carcache, D.A.; Tian, Y.; Li, Y.-M.; Danishefsky, S.J. *J. Org. Chem. 2005, 70*, 9849.
554. Campbell, W.P.; Harris, G.C. *J. Am. Chem. Soc. 1941, 63*, 2721.
555. (a) Guillemonat, A. *Ann. Chim. (Paris) 1939, 11*, 143; (b) Reference 552, p 267.
556. Reference 550a, pp 130-136.

1. Oxidation always occurs on the disubstituted side of a double bond if there is a non-bridged alkenyl hydrogen available there.
2. Oxidation of 1-alkylcyclohexenes occurs in the ring rather than in the side chain.
3. Oxidation never occurs at bridgehead positions in bicyclic systems falling within the limits of **Bredt's rule** (sec. 6.3.B).
4. All other things being equal relative to the above rules, the preferred order of reactivity is $CH_2 > CH_3 > CH$.
5. If the preferred allylic position is tertiary, a diene will generally be produced in preference to a tertiary alcohol.
6. Allylic rearrangement products can and will be formed.
7. Skeletal rearrangement can occur if the preferred allylic position is adjacent to a quaternary carbon or a cyclopropyl ring.

Rabjohn summarized these rules with pertinent examples.[551]

a. In trisubstituted alkenes, oxidation occurs on the more highly substituted side of the double bond.

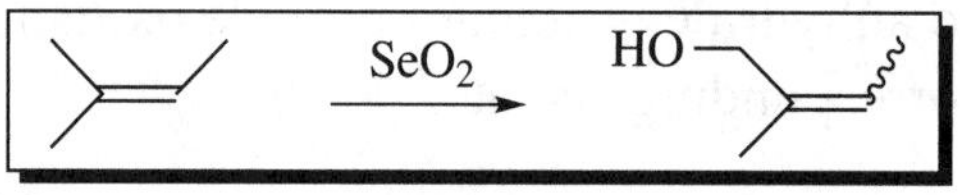

b. By rule (4) above, methylene is oxidized faster than CH_3 or CHR_2:

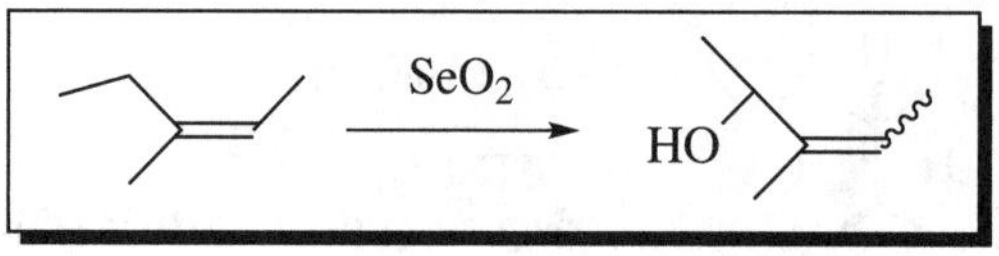

c. By rule (2) above, oxidation of cyclic alkenes usually occurs in the ring.

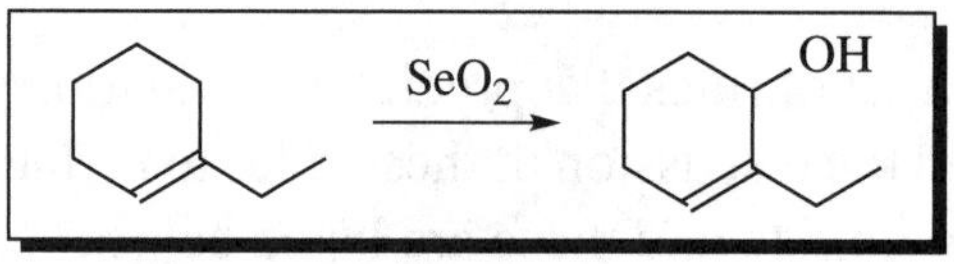

d. Disubstituted alkenes are oxidized in the α-position and methylene is oxidized faster than methyl or methine (rule 4) and example (b).

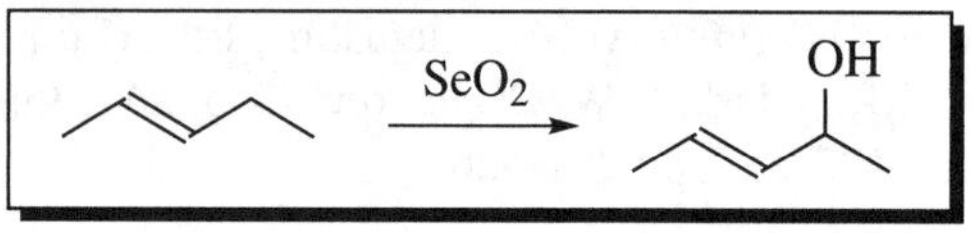

If the alkene is flanked by two methylene groups, both are oxidized.

e. With terminal monosubstituted alkenes the problem of allylic rearrangement is marked.

1-Butene gives a mixture of 2-buten-1-ol and 3-buten-3-ol.[557] Terminal disubstituted alkenes have at least two allylic positions, as do terminal disubstituted alkenes and both are subject to oxidation. Oxidation of the unsymmetrical 4-methylcyclohexene, for example, gives different products as the reaction conditions are changed (**392-394**).

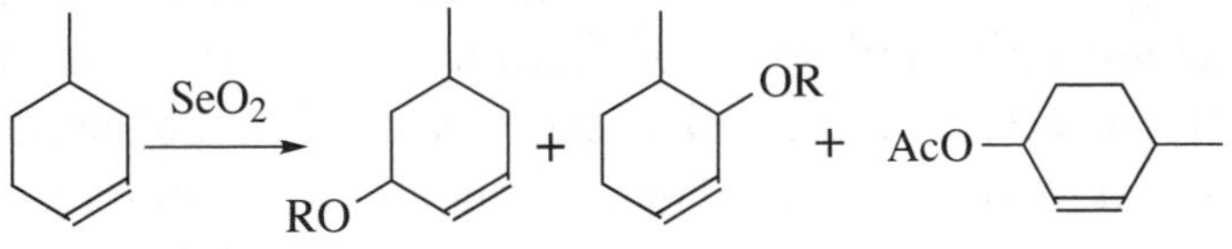

392   **393**   **394**

Mouseron and Jacquier showed that **392** (R=H) was the major product when 4-methylcyclohexene was oxidized with selenium dioxide in acetic acid.[558] When Guillemonat changed the solvent to a mixture of acetic acid and acetic anhydride, **393** (R = Ac) was the major product with trace amounts of **394**. The combined yield of **392-394**, however, was only 19%.[555] In many cases, ketone products are obtained, as in the Trost and Tang synthesis of (–)-codeine[559] where **395** was converted to **396** in 58% yield.

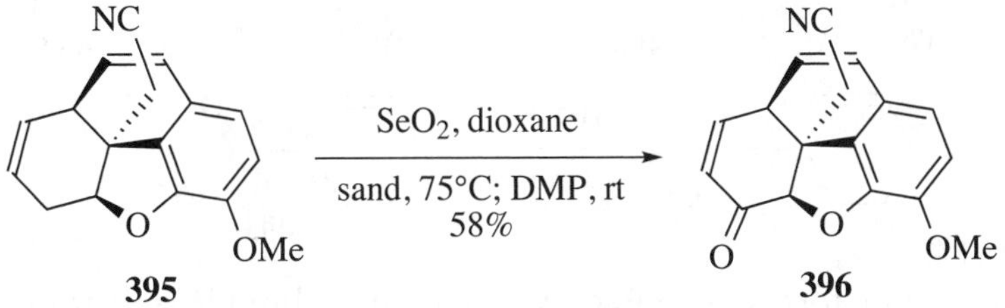

395   **396**

When an electron-withdrawing group is attached to a double bond, the reaction with SeO_2 is slow relative to an unconjugated double bond. Benzylic positions react much slower than allylic positions, as seen in the conversion of the allylic site in **397** to alcohol **398**, in Mori and Tanada's synthesis of (–)-subersic acid.[560]

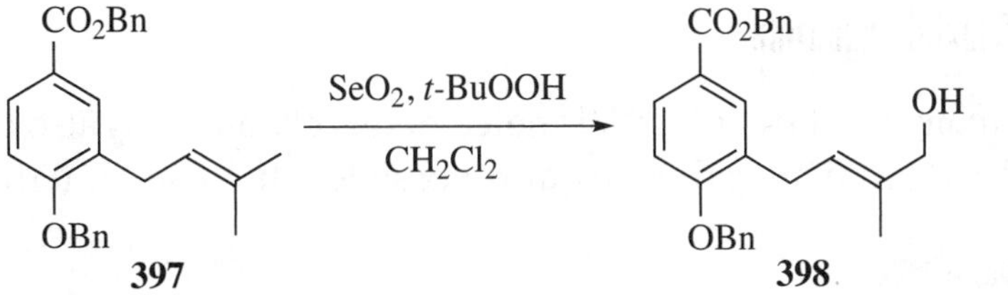

397   **398**

Oxidation of aldehydes and ketones to the corresponding 1,2-dicarbony is another major application of selenium dioxide.[561] This conversion proceeds by reaction of selenium dioxide

557. Reference 550a, p 136.
558. Mousseron M.; Jacquier, R. *Bull. Chim. Soc. Fr.* **1952**, 467.
559. Trost, B.M.; Tang, W. *J. Am. Chem. Soc.* **2002**, *124*, 14542.
560. Tanada, Y.; Mori, K. *Eur. J. Org. Chem.* **2003**, 848.
561. Reference 550a, pp 153-174.

with the enol form of the carbonyl (**399**) to give the selenous enol ester, **400**. Oxidative rearrangement to **401** is followed by loss of selenium and water to give the dicarbonyl (here, diketone **402**).[562] Cyclohexanone was converted to cyclohexane-1,2-dione in 60% yield with SeO$_2$,[563] and similar oxidation of aldehydes led to glyoxals in good yield. Methyl ketones can be oxidized to glyoxals, as in the conversion of acetophenone to **403** in 72% yield.[564] In unsymmetrical ketones oxidation occurs preferentially at the less sterically hindered position,[565] which is probably related to the relative stability of the enol assumed to be the reactive intermediate in the formation of **403**. The hydrogen on the carbon β to the oxoseleno moiety can be removed in some cases, forming an aldehyde by an E2 type elimination.[566] Aldehyde formation is an important reaction only when the α-hydrogen is sterically hindered[567] and is especially important in reactions with ketones.[568] In addition to oxidizing an allylic CH unit or positions α to a carbonyl, selenium dioxide also oxidizes benzylic positions to a carbonyl derivative, without oxidizing the aromatic ring.

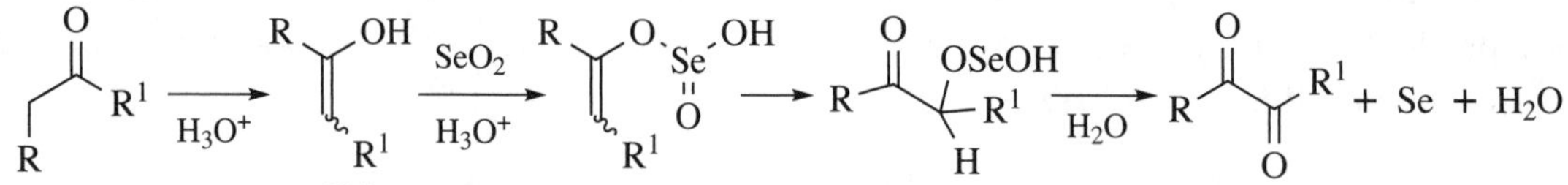

The functional group transforms available from selenium dioxide oxidations are

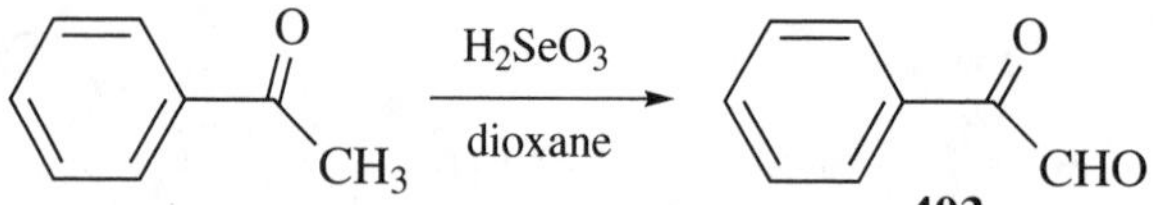

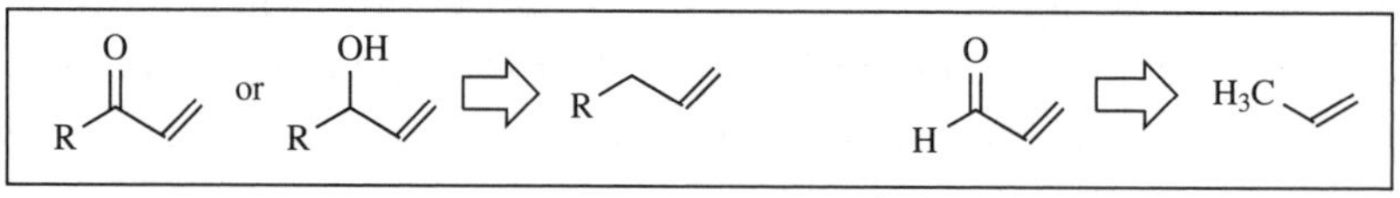

3.8.B. Chromium Trioxide Oxidation

In addition to the many reactions of Cr(VI) noted in this chapter, it is also capable of oxidizing the C-H bond of certain hydrocarbons is also possible. In general, tertiary C-H is cleaved

562. Reference 550a, p 163.
563. (a) Riley, H.L.; Morley, J.F.; Friend, N.A.C. *J. Chem. Soc.* **1932**, 1875; (b) Hach, C.C.; Banks, C.V.; Diehl, H. *Org. Synth. Coll. Vol. 4* **1963**, 229.
564. (a) Riley, H.A.; Gray, A.R. *Org. Synth. Coll. Vol. 2* **1943**, 509; (b) Fieser, L.F.; Fieser, M. *Reagents for Organic Synthesis Vol. 1;* Wiley, New York, **1967**, p 993.
565. Reference 550a, p 158.
566. Reference 550a, pp 166-174.
567. (a) Reference 551, p 273; (b) Meystre, Ch.; Frey, H.; Voser, W.; Wettstein, A. *Helv. Chim. Acta* **1956**, *39*, 734; (c) Szpilfogel, S.; Posthumus, T.; DeWinter, M.; Van Dorp, D.A. *Rec. Trav. Chim.* **1956**, *75*, 475.
568. Reference 551, pp 273-275.

before secondary and secondary before primary. The initial product of this oxidation is usually an alcohol, which can be eliminated to give an alkene or be further oxidized to a ketone or acid. It is also possible to oxidize ketones to acids by oxidative cleavage. In the reactions just mentioned, it is difficult to control overoxidation and this method has little synthetic value, although methyl ketones can be oxidized to the acid $[R_2CH(C=O)Me \rightarrow R_2C\text{-}CO_2H]$, with CrO_3. Chromium oxide can also oxidize benzylic positions, as in the **Étard reaction** (see sec. 3.8.E), or allylic positions. In a synthesis of the shark repellent pavoninin-4,[569] Williams and co-workers oxidized the allylic position in **404** with chromium trioxide in the presence of 3,5-dimethylpyrazole to give a 62% yield of conjugated ketone **405**.

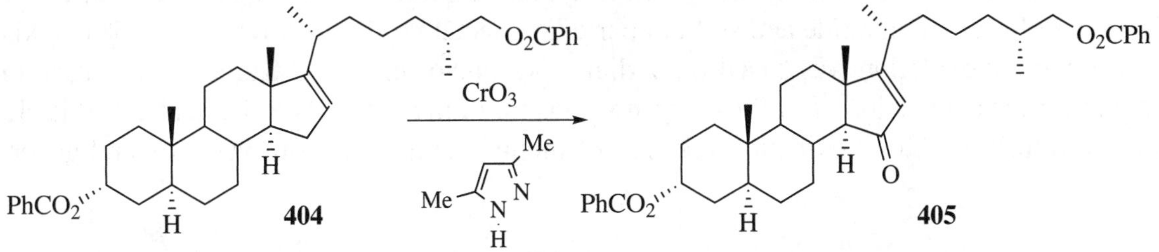

Another hydrocarbon reaction is somewhat dated, the classical **Kuhn-Roth oxidation**.[570] This method allows one to determine the number of methyl groups present in a molecule. Chromium trioxide in concentrated sulfuric acid oxidatively cleaves methyl groups to acetic acid, and titration with standardized base gives to the number of equivalents of acid formed, which corresponds to the number of methyl groups in the original molecule. Kuhn-Roth oxidation of 2-methylpyridine, for example gave about one equivalent of acetic acid.[571b] and oxidation of fatty acids or alcohols (with chain lengths up to 20 carbons)[571] gave one or more equivalents of acetic acid. Modification of the oxidation step allows determination of the number of methyl groups in fatty acids of > 20 carbons.[572] For fatty acids of > 20 carbon atoms, poor solubility of the organic compound led to incomplete oxidation and too few equivalents of acetic acid were observed. Prior dissolution of the acid or alcohol in sulfuric acid, followed by addition of chromium trioxide, gave improved yields and reduced decomposition. Kirste and Stenhagen[572] showed that **406** released about two equivalents of acetic acid. Ginger[571] observed that 1.65 equivalents of acetic acid were formed from α-methylstearic acid (**407**), in what is probably a typical experimental result for a Kuhn-Roth oxidation.

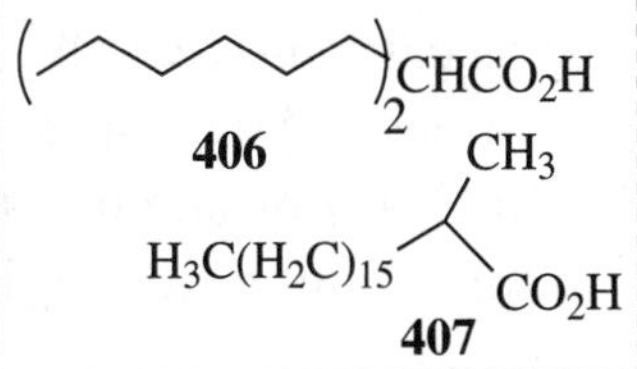

569. Williams, J.R.; Gong, H.; Hoff, N.; Olubodun, O.I. *J. Org. Chem.* **2005**, *70,* 10732.
570. (a) Kühn, R.; Roth, H. *Berichte,* **1933**, *66*, 1274; (b) Franck, B.; Knoke, J. *Ibid.* **1962**, *95*, 579.
571. Ginger, L.G. *J. Biol. Chem.* **1944**, *156*, 453.
572. Kirsten, W.; Stenhagen, E. *Acta Chem. Scand.* **1952**, *6*, 682.

3.8.C. Oxidation with Ruthenium Compounds

Ruthenium salts in combination with peroxide, particularly alkyl hydroperoxides, oxidize methylene units to the corresponding carbonyl. The reaction is most effective for allylic methylene units and methylene units adjacent to heteroatoms. Ruthenium (III) chloride and *tert*-butyl hydroperoxide is an effective reagent for this transformation.[573] In a synthesis of the anti-asthma agent IPL576,092, Shen and Burgoyne converted **408** to **409** in 56% yield.[574] Those authors accomplished the same transformation using a mixture of chromium trioxide (see above), 3,5-dimethylpyrazole in dichloromethane at -20°C. As suggested above, the oxidation is not limited to the formation of ketones. When 3-hydroxyproline derivative **410** was treated with a mixture of ruthenium oxide and sodium periodate, lactam **411** was formed in >95% yield.[575] This particular oxidation was based on earlier work of Yoshifuji and Kamane, who used this reagent for the conversion of cyclic α-amino acids to α-amino dicarboxylic acids.[576] It is clear that the ruthenium-based oxidation reaction of methylene units is both versatile and general.

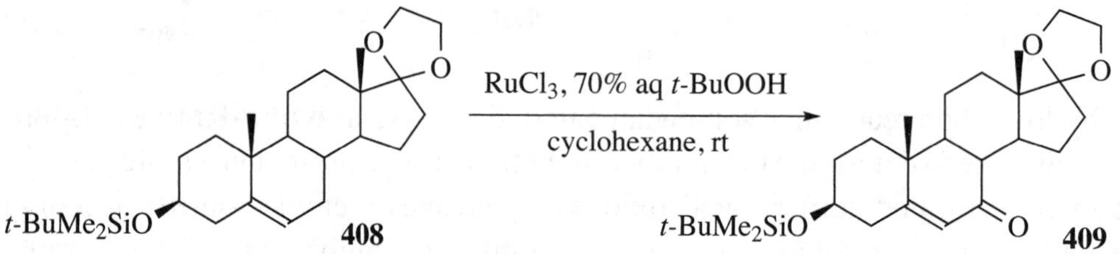

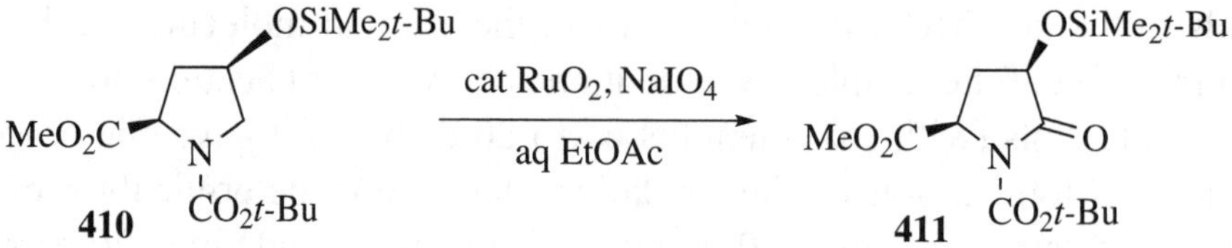

3.8.D. Tamao-Fleming Oxidation

Just as carbon-based alkyl groups can be oxidized, a silane unit can be oxidized under certain conditions. Alkylsilanes can be converted to a hydroxy unit, but either an aryl group[577] (R–SiR'$_2$Ar) or another silyl group[578] (RSi–SiR'$_3$) must be attached to silicon. In an early version of this reaction, Fleming used a two-step process to transform the silane unit to an

573. Miller, R.A.; Li, W.; Humphrey, J.L. *Tetrahedron Lett.* **1996**, *37*, 3429.
574. Shen, Y.; Burgyone, D.L. *J. Org. Chem.* **2002**, *67*, 3908.
575. Zhang, X.; Schmitt, A.C.; Jiang, W. *Tetrahedron Lett,* **2001**, *42,* 5335.
576. (a) Yoshifuji, S.; Kamane, M. *Chem. Pharm. Bull.* **1995**, *43*, 1617; (b) Kamane, M.; Yoshifuji, S. *Tetrahedron Lett.* **1992**, *33*, 8103; (c) Tanaka, K.; Sawanishi, H. *Tetrahedron: Asymmetry* **1998**, *9*, 71.
577. (a) Kumada, M.; Tamao, K.; Yoshida, J.I. *J. Organomet. Chem.* **1982**, *239*, 115; (b) Tamao, K.; Kakui, T.; Akita, M.; Iwahara, T.; Kanatani, R.; Yoshida, J.; Kumada, M. *Tetrahedron* **1983**, *39*, 983; (c) Fleming, I.; Henning, R.; Plaut, H. *J. Chem. Soc. Chem. Commun.* **1984**, 29.
578. Suginome, M.; Matsunaga, S.; Ito, Y. *Synlett* **1995**, 941.

alcohol unit. Treatment with mercuric acetate and peroxyacetic acid, followed by reduction with lithium aluminum hydride (see sec. 4.2.A for reductions with $LiAlH_4$) gave the alcohol. Malacria and co-workers used this approach to convert **412** to alcohol **413** in 78% yield, in a synthesis of deoxymannojirimycin.[579] Note that a reduction step was not required.

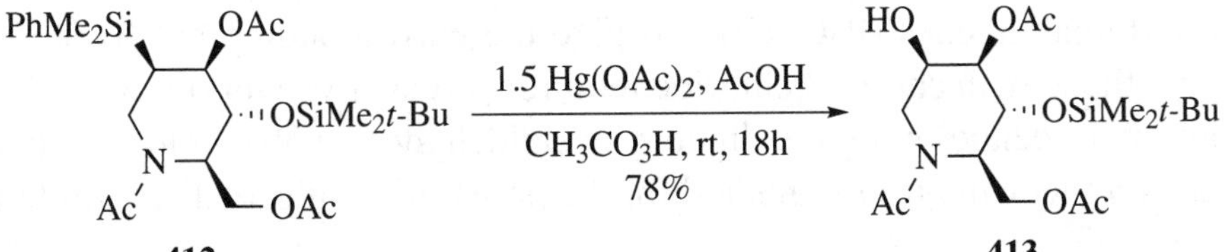

There were several cases when this original procedure failed, but it was discovered that if the silane units are treated with a fluorinating agent such as TBAF or CsF, the Ar or SiR'_3 unit is replaced with F.[580] Subsequent treatment with hydrogen peroxide or a peroxy acid gave the alcohol. This sequence is often called the **Tamao-Fleming oxidation**.[577] In synthetic studies for (+)-pramanicin, Barrett and co-workers found the original procedure failed but the Tamao-Fleming procedure using mcpba-KHF_2 converted **414** to **415** in 70% yield.[581] One of the more attractive features of this oxidation is that the absolute stereochemistry of the C–Si bond in **414** is retained in the C–O bond of **415**. In other words, the reaction proceeds with retention of stereochemistry of the silyl group. Another variation uses KF and hydrogen peroxide for the oxidation, a reaction used by Overman and co-workers for one step of a synthesis of (+)-aloperine.[582]

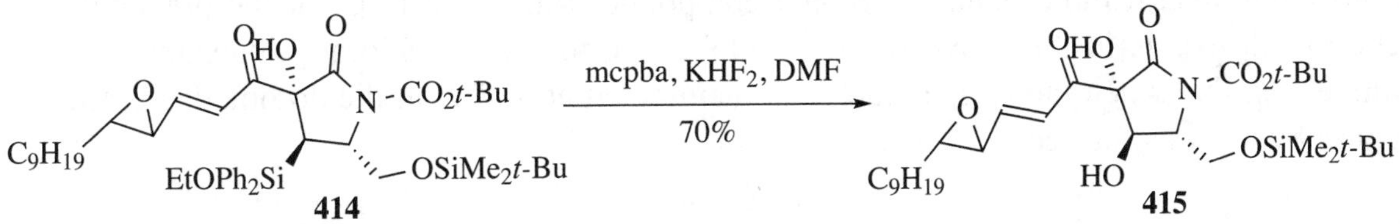

Incorporation of the silyl unit can be done in several ways,[583] including conjugate addition of silylcuprates to conjugated carbonyl compounds (see sec. 8.7.A.vi.). A brief discussion of silane carbanions will be presented in section 8.10.

The functional group transforms available from selenium dioxide oxidations are

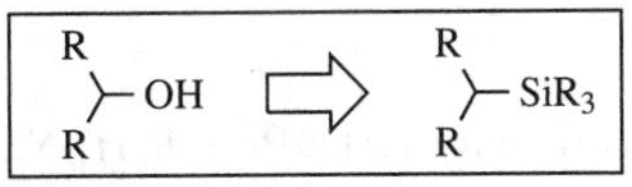

579. Boglio, C.; Stahlke, S.; Thorimbert, S.; Malacria, M. *Org. Lett* **2005**, *7*, 4851.
580. For the protodesilylation step see (a) Häbich, D.; Effenberger, F. *Synthesis* **1979**, 841. For the peroxyacid reaction see (b) Buncel. E.; Davies, A.G. *J. Chem. Soc.* **1958**, 1550.
581. Barrett, A.G.M.; Head, J.; Smith, M.L.; Stock, N.S.; White, A.J.P.; Williams, D.J. *J. Org. Chem.* **1999**, *64*, 6005.
582. Brosius, A.D.; Overman, L.E.; Schwink, L. *J. Am. Chem. Soc.* **1999**, *121*, 700.
583. For examples see (a) Matsumoto, Y.; Hayashi, T.; Ito, Y. *Tetrahedron* **1994**, *50*, 335; Uozumi, Y.; Kitayama, K.; (b) Hayashi, T.; Yanagi, K.; Fukuyo, E. *Bull. Chem. Soc. Jpn.* **1995**, *68*, 713.

3.8.E. Alkene Oxidation (C=C ⟶ C-C=O)

Tansition metal derivatives can convert alkenes into a carbonyl derivative with rearrangement, but *without* bond cleavage.[584] This section will briefly discuss several of these reagents and their reactions. A common one is chromyl chloride (CrO_2Cl_2), which reacts with 1,1-disubstituted terminal alkenes [$R_2C=CH_2$] to give the corresponding aldehyde.[585] Oxidation of 1,1-diphenylethene with chromyl chloride required a reductive workup with zinc (to reduce the chromium intermediate) but gave diphenylacetaldehyde in 63% yield.[586] This interesting conversion is a synthetically useful method for the preparation of α,α-disubstituted aldehydes.

The **Étard reaction** (see sec. 3.8.B)[587] used this reagent for the oxidation of methylated aryl derivatives to the corresponding aryl aldehyde. Tillotson and Houston found that the Étard reaction is catalyzed by small amounts of alkene, added to or present in the reaction medium.[588] The reaction involves addition of chromyl chloride to a carbon disulfide or carbon tetrachloride solution of the arene. A dark brown, insoluble and *explosive* intermediate usually precipitates. Dilute sulfurous acid is added to decompose the precipitate to the aldehyde. Toluene is converted to benzaldehyde and ethylbenzene was oxidized to phenylacetaldehyde with this reagent.

Chromyl chloride can also produce ketones, but formation of a α-chloro ketone is common, as in the conversion of cyclododecene to α-chlorocyclododecanone (**416**) in 79% yield.[589] The chloride moiety in **416** was reduced with zinc and acetic acid in a subsequent reaction to give cyclododecanone in 95% yield. Alkenes react with chromyl chloride to give a trans-chlorohydrin rather than a carbonyl compound (see sec. 2.10.C for the preparation of chlorohydrins), as in the conversion of cyclohexene to *trans*-2-chlorocyclohexanol.[590] It is interesting that when the temperature was maintained at −78°C in dichloromethane, the cis-chlorohydrin was formed.[591]

416

Lead tetraacetate initiates oxidation with terminal alkenes in the presence of acid, to give an

---

584. Reference 233a, pp 132-142.
585. (a) Reference 233a, pp 140, 321; (b) Freeman, F.; DuBois, R.H.; Yamachika, N.J. *Tetrahedron* **1969**, *25*, 3441.
586. (a) Reference 233a, p 140; (b) Freeman, F.; DuBois, R.H.; McLaughlin, T.G. *Org. Synth. Coll. Vol. 6* **1988**, 1028.
587. (a) Étard, M.A. *Compt. Rend.* **1880**, *90*, 534; (b) Ferguson, L.N. *Chem. Rev.* **1946**, *38*, 227 (see p 237); (c) The Merck Index, 14th ed., Merck & Co., Inc., Whitehouse Station, New Jersey, **2006**, p ONR-29.
588. Tillotson, A.; Houston, B. *J. Am. Chem. Soc.* **1951** 73, 221.
589. Sharpless, K.B.; Teranishi, A.Y. *J. Org. Chem.* **1973**, *38*, 185.
590. Cristol, S.J.; Eilar, K.R. *J. Am. Chem. Soc.* **1950**, *72*, 4353.
591. (a) Sharpless, K.B.; Teranishi, A.Y.; Bäckvall, J.E. *J. Am. Chem. Soc.* **1977**, *99*, 3120; (b) Fieser, M. *Reagents for Organic Synthesis Vol. 8* Wiley, New York, **1980**, p 112.

aldehyde by selective oxidation of the terminal carbon.[592] An example is the conversion of styrene to phenylacetaldehyde in 98% yield.[593] Palladium chloride ($PdCl_2$) reacts with terminal alkenes, in the presence of oxygen and copper salts, to give a methyl ketone (this reaction is called the **Wacker process** which is discussed in sec. 12.6.A), and is more useful than the LTA oxidation. Oxidation of terminal alkenes with LTA leads to the aldehyde, whereas oxidation with $PdCl_2$ leads to the methyl ketone. The $PdCl_2$ oxidation of a terminal alkene to a methyl ketone is illustrated by conversion of **417** to **418**, in >80% yield, in Leighton and co-worker's synthesis of dolabelide D.[594]

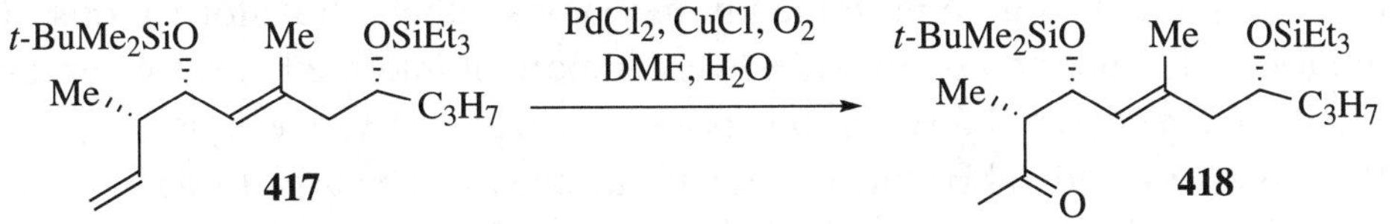

Mercuric salts induce oxidative rearrangement of cyclic alkenes to give cycloalkane carboxaldehydes.[595]

Cyclohexene, for example, gave cyclopentane carboxaldehyde (**419**) in 53% yield.[596,592] Acyclic alkenes such as 2-butene can also be oxidized with mercuric sulfate and sulfuric acid, giving 2-butanone in this case.[597] Thallium nitrate [$Tl(NO_3)_3$] is an important reagent for the oxidative rearrangement of cyclic alkenes.[598] The reaction is not restricted to carbocyclic compounds, but can also be applied to heterocyclic compounds as shown by the oxidation of 3,4-dihydro-2H-pyran (**420**) to the dimethyl acetal (**421**) in 65% yield.[599]

The functional group transforms presented in this section are

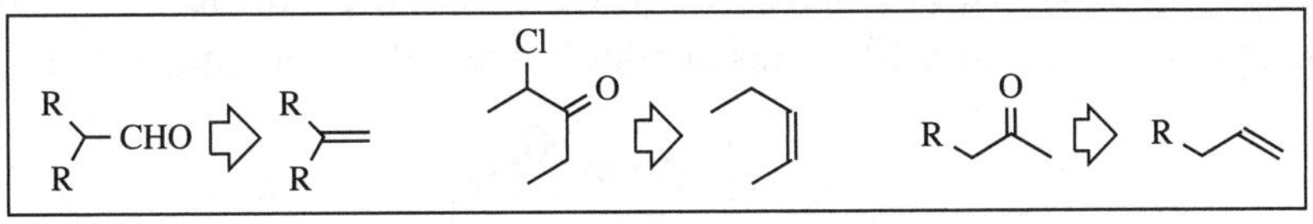

592. Reference 233a, p 136.
593. Lethbridge, A.; Norman, R.O.C.; Thomas, C.B. *J. Chem. Soc. Perkin Trans. 1* **1973**, 35.
594. Park, P.K.; O'Malley, S.J.; Schmidt, D.R.; Leighton, J L. *J. Am. Chem. Soc.* **2006**, *128*, 2796.
595. (a) Reference 233a, pp 135-136; (b) Arzoumanian, H.; Metzger, J. *Synthesis* **1971**, 527.
596. (a) Grummitt, O.; Liska, J.; Greull, G. *Org. Synth. Coll. Vol. 5* **1973**, 320; (b) English, Jr., J.E.; Gregory, J.D.; Trowbridge II, J.R. *J. Am. Chem. Soc.* **1951**, *73*, 615.
597. Charavel, B.; Metzger, J. *Bull. Chim. Soc. Fr.* **1968**, 4102.
598. Reference 233a, pp 137-139.
599. McKillop, A.; Hunt, J.D.; Kienzle, F.; Bigham, E.; Taylor, E.C. *J. Am. Chem. Soc.* **1973**, *95*, 3635.

3.9. OXIDATION OF SULFUR, SELENIUM, AND NITROGEN

It is possible to oxidize atoms other than carbon. Sulfur, selenium and nitrogen are easily oxidized by many of the oxidizing agents presented in this chapter. Several oxidation stages are available, leading to an interesting array of functional groups. Sulfur, for example, can be oxidized from a sulfide to a sulfoxide, which is turn can be oxidized to a sulfone. Sulfonic acids and sulfonate esters represent even higher oxidation states for sulfur. Selenium behaves similarly to sulfur. Nitrogen can be oxidized to hydroxylamines, nitrones, N-oxides or nitro compounds. These groups figure prominently in many synthetic transformations. In section 2.9.C, N-oxides, sulfoxides and selenoxides were important intermediates for syn-elimination reactions. In this chapter (sec. 3.5.B), N-oxides were used to catalyze osmylation reactions with alkenes. In sections 8.6 and 8.8.B, sulfoxides and sulfones will be used to stabilize carbanions, which react to form carbon-carbon bonds. It is therefore fitting that this chapter conclude with a brief survey of the methods for oxidizing these important synthetic intermediates.

3.9.A. Oxidation of Sulfur Compounds

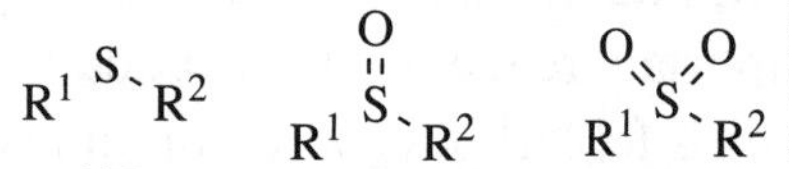

3.9.A.i. Oxidation of Sulfides to Sulfoxides. Sulfides (thioethers, **422**) are readily oxidized to sulfoxides (**423**) by a variety of oxidizing agents, but sodium periodate ($NaIO_4$) is a convenient reagent. Further oxidation of sulfoxides to sulfones (**424**) requires more vigorous conditions but is straightforward.[600] Sulfides can be oxidized directly to the corresponding sulfoxide with peroxyacids, hydrogen peroxide or alkyl hydroperoxides. The overall mechanism for oxidations with peroxy acids and alkyl hydroperoxides is shown.[600] In the reaction with peroxyacids, the sulfide attacks the O–O bond along an S_N2 trajectory (see **425**). The O–O bond is broken, and the hydrogen is transferred to the carbonyl group, liberating a carboxylic acid and the sulfoxide.[599] This type of reaction with alkyl hydroperoxides usually requires acid catalysis. When the sulfide attacks the O-O bond of a hydroperoxide, hydrogen transfer liberates an alcohol and the acid catalyst, as shown in **426**.[600]

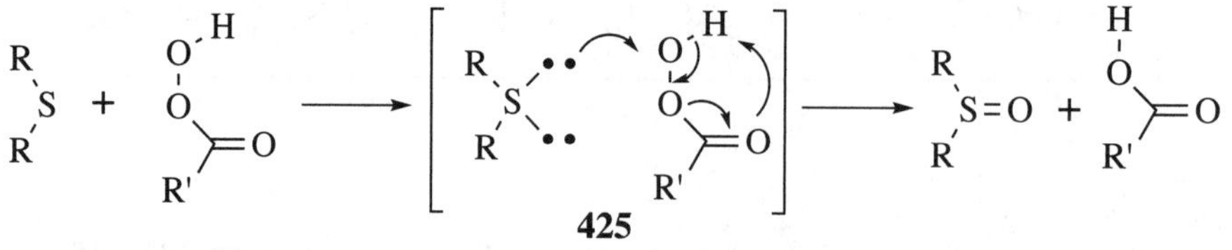

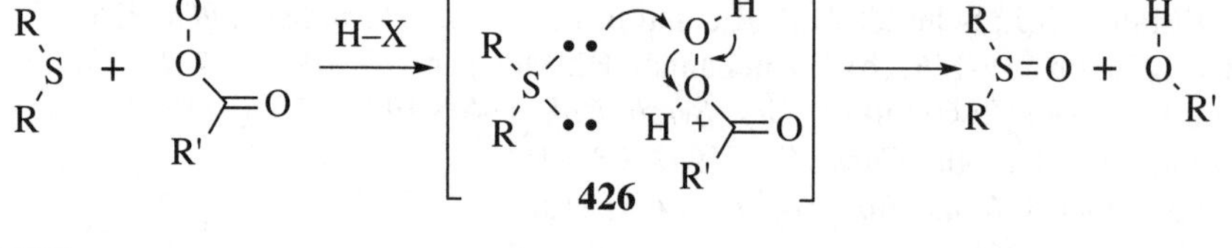

---

600.   Reference 254, pp 244-247.

Sulfides react faster than alkenes with hydrogen peroxide and alkyl hydroperoxides. For this reason, transition metal catalysts are rarely necessary, but these reactions are acid catalyzed and first order in both sulfide and peroxide. The acid (HX) can be as weak as alcohol or water but the "effectiveness (of the oxidation) is determined by the pK_a of the acid."[601] Sulfides also react faster than ketones with peroxides (see the Baeyer-Villiger reaction, sec. 3.6). Formation of the sulfone in these reactions is straightforward, but this over-oxidation usually requires more vigorous reaction conditions. It is usually easy to isolate the sulfoxide from oxidation of a sulfide. Direct conversion of a sulfide to a sulfone requires excess peroxide and vigorous reaction conditions (heating, long reaction times, more concentrated peroxide).

Sodium meta-periodate ($NaIO_4$) is a common oxidizing agent. Leonard and Johnson used sodium periodate to oxidize methylphenyl sulfide to the sulfoxide in near-quantitative yield.[602] In that report, many methods were summarized for oxidizing sulfides, including nitric acid, hydrogen peroxide, chromic acid, ozone, peroxyacids, manganese dioxide, and selenium dioxide.[603] Sykes and Todd were apparently the first to use sodium periodate with sulfides, oxidizing benzylpenicillin methyl ester (**427**) to the corresponding sulfoxide (**428**).[604] Sodium periodate is often the reagent of choice for the oxidation of sulfides to sulfoxides, but mcpba can also be used as in the oxidation of sulfide **429** to sulfide **430**.[605]

3.9.A.ii. Preparation of Chiral Sulfoxides. The sulfur atom in a sulfoxide has four different groups (R, R^1, O, and the lone pair electrons). There is virtually no inversion at sulfur (in contrast to nitrogen) so the sulfur can be a stereogenic center under these circumstances, which raises two points when using unsymmetrical sulfoxides. When a stereogenic center is present

601. Reference 254, p 245.
602. Leonard, N.J.; Johnson, C.R. *J. Org. Chem.* **1962**, *27*, 282.
603. (a) Märcker, C. *Annalen* **1865**, *136*, 75; (b) Reid, E.E. *Organic Chemistry of Bivalent Sulfur, Vol. II*, Chemical Pub. Co. Inc., New York, **1960**, pp 64-66; (c) Bordwell, F.G.; Boutan, P.J. *J. Am. Chem. Soc.* **1957**, *79*, 717; (d) Gazdar, M.; Smiles, S. *J. Chem. Soc.* **1908**, *93*, 1833; (e) Hünig, S.; Boes, O. *Annalen* **1953**, *579*, 23; (f) Peak, D.A.; Watkins, Y.I. *J. Chem. Soc.* **1950**, 445; (g) Edwards, D.; Stenlake, J.B. *J. Chem. Soc.* **1954**, 3272; (h) Horner, L.; Schaefer, H.; Ludwig, W. *Chem. Ber.* **1958**, *91*, 75; (i) Overberger, C.G.; Cummins, R.W. *J. Am. Chem. Soc.* **1953**, *75*, 4250; (j) Barnard, D. *J. Chem. Soc.* **1956**, 489; (k) Mel'nikov, N.N. *Uspekhi Khim.* **1936**, *5*, 443 [*Chem. Abstr. 30*: 51809 **1936**]; (l) Ford-Moore, A.H. *J. Chem. Soc.* **1949**, 2126.
604. (a) Sykes, P.; Todd, A.R., Committee on Penicillin Synthesis Report 526,677; (b) *The Chemistry of Penicillin*, Clarke, H.T.; Johnson, J.R.; Robinson, R.(Eds), Princeton Univ. Press, Princeton, N.J., **1949**, pp 156, 927, 946, 1008.
605. Jiang, B. *Chem. Commun.* **1996**, 861.

in a given molecule, incorporation of a sulfoxide moiety can lead to formation of diastereomers that can complicate separation and identification. The second point is that methods have been developed to resolve enantiomeric sulfoxides, or produce one enantioselectively, and use this material as a chiral auxiliary or as a chiral template (sec. 10.9).

Cram and Kingsbury reported an example that illustrates the oxidation of sulfides to sulfoxides with formation of diastereomers.[606] Oxidation of erythro- or threo-1-alkylthio-1,2-diphenylpropane with peroxybenzoic acid gave syn- and anti- sulfides. The erythro-sulfide (anti-sulfide **431**) gave two erythro- (anti) sulfoxides, **432** and **433**.[606] Similarly, the threo- sulfide (syn-sulfide **434**) gave two threo-(syn-sulfides, **435** and **436**. Both stereogenic carbons retain

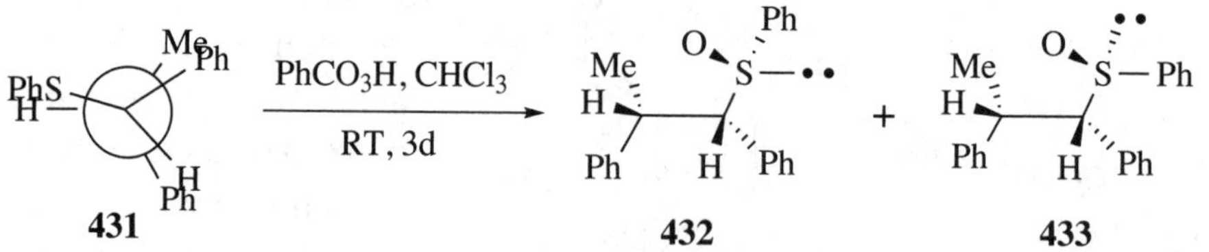

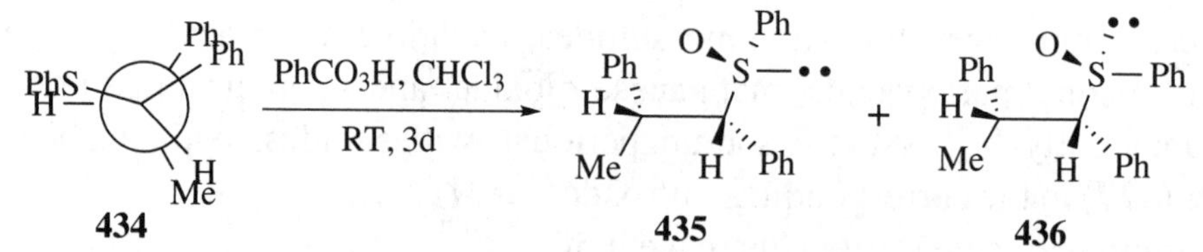

their absolute configuration during the oxidation, but the oxygen is delivered to different faces of sulfur to generate the observed diastereomers. In general, the achiral A-S-B sulfide (**438**) is converted to the chiral sulfoxide **439**. When (+)-peroxycamphoric acid (**437**) was used, the configuration of the predominant optical isomer was the same as in **439** if A was bulkier than B. Later work by Folli et al. confirmed this preference.[607] Approach of sulfur to the peroxyacid occurs in the direction of the lone pair of electrons on sulfur (an S_N2 like trajectory toward the O-O bond of the peroxide; *vide supra*) and can lead to either the (*R*) or the (*S*) enantiomer. As the difference between A and B became larger, or the chiral peroxyacid was made bulkier, greater selectivity was observed. The nature of the solvent influenced the observed optical yields (% ee), which were in the range of 1.2-4.5% ee.[607] Oxidation of phenylmethyl sulfide with (+)-monoperoxycamphoric acid gave the chiral, nonracemic sulfoxide with a 1.2% ee in acetonitrile, and similar oxidation gave 1.7% ee in 2-propanol, 3.8% ee in chloroform, 4.5% ee in ether, 3.6% ee in benzene, and 3.1 % ee in dioxane.[608,607]

606. Kingsbury, C.A.; Cram, D.J. *J. Am. Chem. Soc.* **1960**, *82*, 1810.
607. Folli, U.; Iarossi, D.; Montanari, F.; Torre, G. *J. Chem. Soc. C* **1968**, 1317.
608. (a) Mislow, K.; Green, M.M.; Raban, M. *J. Am. Chem. Soc.* **1965**, *87*, 2761; (b) Mislow, K.; Green, M.M.; Laur, P.; Melillo, J.T.; Simmons, T.; Ternay, Jr., A.L. *Ibid.* **1965**, *87*, 1958.

437   **438**   **439**

Many methods have been reported that generate chiral sulfoxides with high %ee.[609] Kagan and Duñach[610] as well as DiFuria et al.,[611] used titanium tetraisopropoxide [Ti(OiPr)$_4$] and optically active diethyl tartrate (DET) to catalyze the reaction, analogous to the **Sharpless asymmetric epoxidation** discussed in section 3.4.D.i. DiFuria and co-workers oxidized benzylmethyl sulfide (**440**) to **441** in dichloromethane, in 46% ee.[611] Kagan, Duñach, and co-workers oxidized methylphenyl sulfide to the corresponding (*S*)-sulfoxide in water with 89% ee under similar conditions.[610] Katsuki and Saito used a titanium-salen catalyst to obtain chiral sulfoxides with up to 96% ee.[612]

440   **441**

Davis et al. used a chiral oxaziridine for the asymmetric oxidation of sulfides to sulfoxides.[613] Oxidation of isopropyl-*p*-tolyl sulfide (**442**) with oxaziridine **443**, for example, gave 60.3% ee (S) of **444** (at –78°C in chloroform).[613] The absolute configuration of the sulfide is determined by approach of the sulfide to the oxaziridine oxygen, as illustrated by **445**. Steric factors appear to be the primarily reason for the chiral recognition.[614] In this model, attack by sulfur minimizes the R$_L$ and R$_S$ interactions with the oxaziridine moiety (R$_S$ ↔ SO$_2$R^1 and R$_L$ ↔ Ar or R$_S$ ↔ Ar and R$_L$ ↔ SO$_2$R^1). When R^1 in **445** is (+)-10-camphor, the resulting oxaziridine derivative reacted with sulfide to give the (+)-(*R*)-sulfoxide whereas the adduct derived from (–)-camphor gave the (–)-(*S*)-sulfoxide.[580] Davis reported several other chiral oxaziridine reagents that gave good to excellent asymmetric induction for the oxidation of sulfides to sulfoxides. These regents are a mild and excellent alternative to oxidation using chiral peroxyacids.

609. Smith, M.B. *Compendium of Organic Synthetic Methods, Vol. 10,* Wiley, NY, *2002*, pp 231-236; Vol. 11, *2003*, pp 474-476.

610. (a) Pitchen, P.; Duñach, E.; Deshmukh, M.M.; Kagan, H.B. *J. Am. Chem. Soc. 1984, 106,* 8188; (b) Pitchen, P.; Kagan, H.B. *Tetrahedron Lett. 1984, 25,* 1049.

611. DiFuria, F.; Modena, G.; Seraglia, R. *Synthesis 1984,* 325.

612. Saito, B.; Katsuki, T. *Tetrahedron Lett. 2001, 42,* 3873.

613. Davis, F.A.; McCauley, Jr., J.P.; Chattopadhyay, S.; Harakal, M.E.; Towson, J.C.; Watson, W.H.; Tavanaiepour, I. *J. Am. Chem. Soc. 1987, 109,* 3370.

614. (a) Davis, F.A.; Jenkins, Jr., R.H.; Awad, S.B.; Stringer, O.D.; Watson, W.H.; Galloy, J. *J. Am. Chem. Soc. 1982, 104,* 5412; (b) Davis, F.A.; McCauley, Jr., J.P.; Harakal, M.E. *J. Org. Chem. 1989, 49,* 1465.

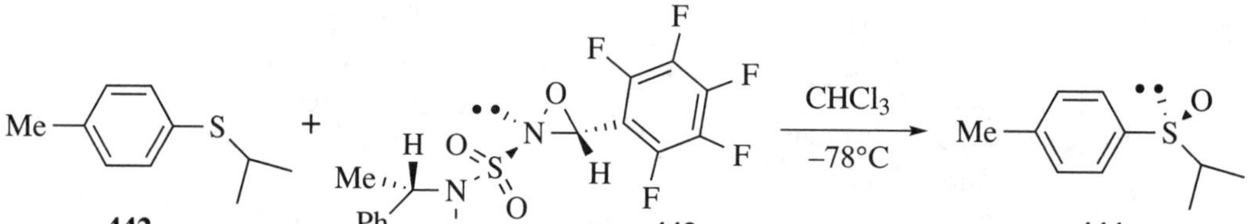

3.9.A.iii. Oxidation of Sulfides or Sulfoxides to Sulfones and Oxidation of Selenium Compounds.

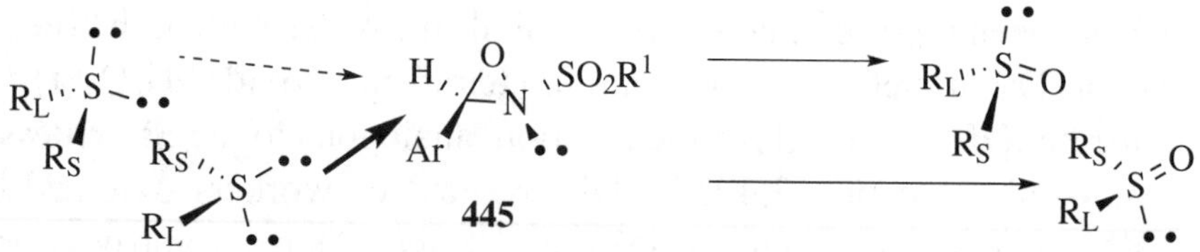

The mechanism for the oxidation of sulfoxides to sulfones is probably similar to that for oxidation of sulfides,[615] but sulfoxides are less reactive than sulfides and the reaction slower.[616] Sulfoxides are oxidized to sulfones using the same oxidants that converted the sulfide to a sulfoxide, but longer reaction times and/or more vigorous conditions are usually required. Direct oxidation of a sulfide to a sulfone is also easily accomplished in many cases. Oxidation of sulfide **446** with mcpba gave sulfone **447** directly without oxidation of the sensitive pyrrole unit.[617] Note that the sulfur in a sulfone is achiral since there are two oxygen atoms.

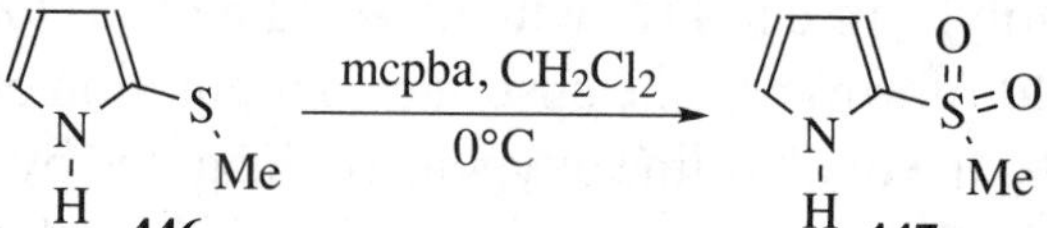

The selenium analogs of the sulfide, sulfoxide and sulfone are known, and the generic structures are **448**, **449**, and **450**, respectively. The oxidation of these compounds is virtually identical to those observed with sulfur. The main drawback to their use is the toxicity of the selenium compounds. Selenoxides are relatively unstable and decompose at ambient temperatures via syn-elimination. Selenoxide elimination is more facile than the analogous sulfoxide elimination (0°C versus 120°C), and is often preferred in organic synthesis (secs. 2.9.C.v, 2.9.C.vi). The thermal instability of selenoxides is also reflected in selenones, which are labile and rarely used.

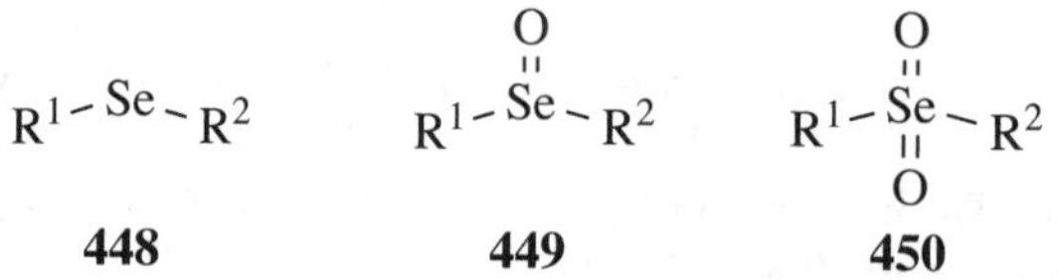

615. Reference 254, p 246.
616. Boeseken, J.; Arrias, E. *Rec. Trav. Chim.* **1935**, *54*, 711.
617. Miranda, L.D.; Cruz-Almanza, R.; Alvazez-García, A.; Muchowski, J.M. *Tetrahedron Lett.* **2000**, *41*, 3035.

3.9.B. Oxidation of Amines

The oxidation of nitrogen compounds is also common in organic chemistry. Aromatic nitro compounds are very important commercial compounds and their preparation was discussed in section 2.11.B. The reader is referred to the references found here and to specific applications for the preparation of oxides of nitrogen. Two categories of nitrogen compounds that are subject to oxidation are amines and amides. Amides "do not react well with hydrogen peroxide or peroxyacids, but most other nitrogen compounds react readily."[618] Oxidation of primary amines gives the nitroso (**452**) and the nitro compound (**453**) via the hydroxylamine derivative (**451**).[619] Ozonolysis of phosphinimines also produces nitro compounds.[620] Secondary amines are also oxidized via the hydroxylamine (**454**) but eliminate to give the nitrone (**455**).

$$R-NH_2 \longrightarrow \underset{\textbf{451}}{R-NHOH} \longrightarrow \underset{\textbf{452}}{R-N=O} \longrightarrow \underset{\textbf{453}}{R-NO_2}$$

$$RCH_2-\underset{H}{\overset{R}{N}} \longrightarrow RCH_2-\underset{\underset{\textbf{454}}{OH}}{\overset{R}{N}} \longrightarrow RCH=\underset{\underset{\textbf{455}}{O-}}{\overset{R}{N+}}$$

As mentioned above, nitro compounds are obviously of great importance in organic chemistry and aryl nitro compounds are an important source of aniline derivatives (secs. 4.2.C.v, 4.8.D). Both amine oxides and nitrones have been synthetically exploited. Alkyl nitroso derivatives, however, usually cannot be isolated since they decompose in solution, although the aromatic derivatives are more stable in solution and can be used in synthesis (sec. 2.11.E). Treatment of a primary amine with excess peroxyacid is a useful preparative route to alkyl nitro compounds.[621] Yields are highest for tertiary alkyl primary amines, then secondary followed by primary alkyl. Peroxyacid oxidation of oximes also provides a route to alkyl nitro compounds.[622] This method is convenient for preparing aromatic nitro compounds as in the oxidation of 2,6-dichloroaniline to 2,6-dichloronitrobenzene (**456**).[623] Nitrones are 1,3-dipoles and have been used in 1,3-dipolar cycloaddition reactions (sec. 11.11.D).

Tertiary amines give *N*-oxides directly and in high yield. Katritzky[624] described the chemistry of heterocyclic *N*-oxides. Thermal elimination of N-oxides is the basis of the classical Cope reaction. (sec. 2.9.C.ii).[625]

618. Reference 254, pp 248-253.
619. Reference 254, p 250.
620. Corey, E.J. ; Samuelsson, B.; Luzzio, F.A. *J. Am. Chem. Soc.* **1984**, *106*, 3682.
621. Emmons, W.D. *J. Am. Chem. Soc.* **1957**, *79*, 5528.
622. Emmons, W.D.; Pagano, A.S. *J. Am. Chem. Soc.* **1955**, *77*, 4557.
623. (a) Reference 433a, p 332; (b) Pagano, A.S.; Emmons, W.D. *Org. Synth. Coll. Vol. 5* **1973**, 367.
624. Katritzky, A.R. *Quart. Rev.* **1956**, *10*, 395.
625. Cope, A.C.; Trumbull, E.R. *Org. React.* **1966**, *11*, 317.

3.10. CONCLUSION

This chapter focused on synthetically useful oxidation reactions for a variety of functional group transformations. This treatment is not comprehensive, but illustrates the various oxidation methods and reagents. These techniques will resurface in many syntheses and as part of other methodology in succeeding chapters. In Chapter 4, the reverse process, reduction, will be presented, emphasizing synthetically useful reactions and reagents.

HOMEWORK

1. The following reaction was reported in the synthesis of niphatoxin B. Give the product and suggest a mech-anistic rationale for this reaction.

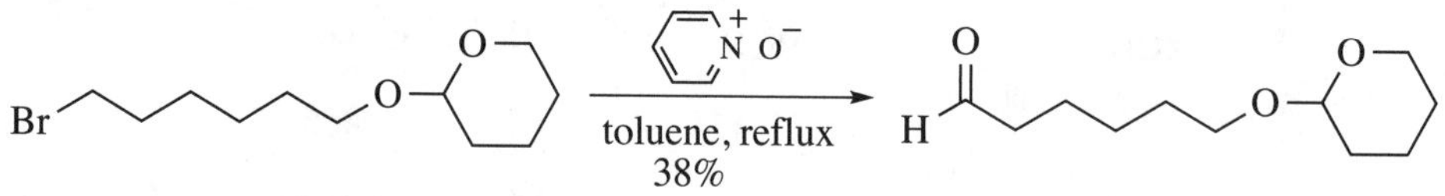

2. The product of this reaction was expected to be a diol-acid, and the infrared (IR) showed an OH peak and a strong carbonyl, but there was no apparent CO_2H absorption. Likewise, the ^1H NMR showed there was no CO_2H group. Show the actual product and explain why the anticipated diol-acid was not isolated.

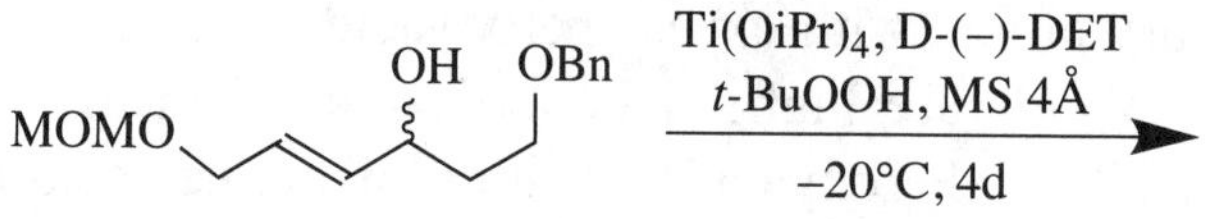

3. Explain each of the following:

    (a) The following reaction was reported in Paquette and Geng's synthesis of ceratopicanol. Explain the product as well as the stereochemistry of the alcohol unit.

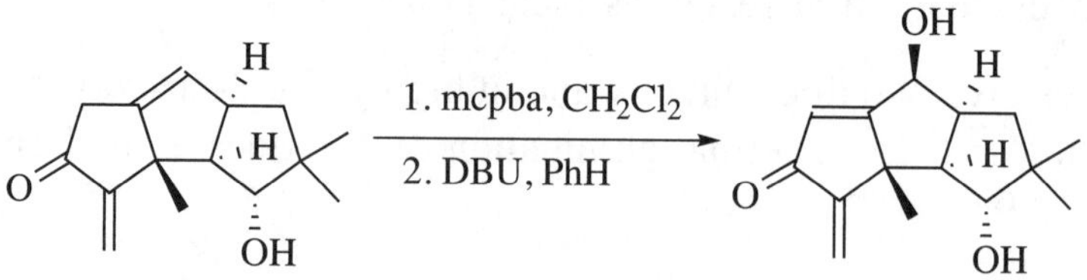

    (b) Explain the regioselectivity of this reaction.

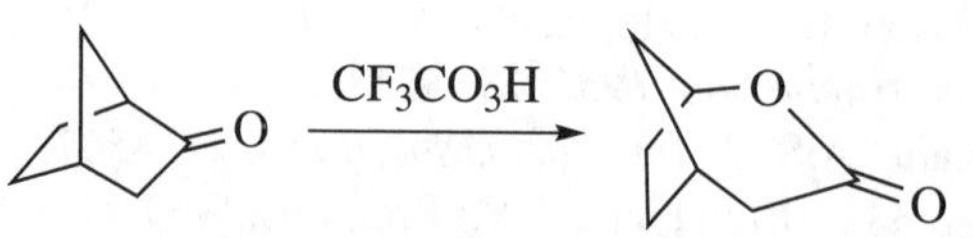

    (c) Explain why p-quinone is the major product.

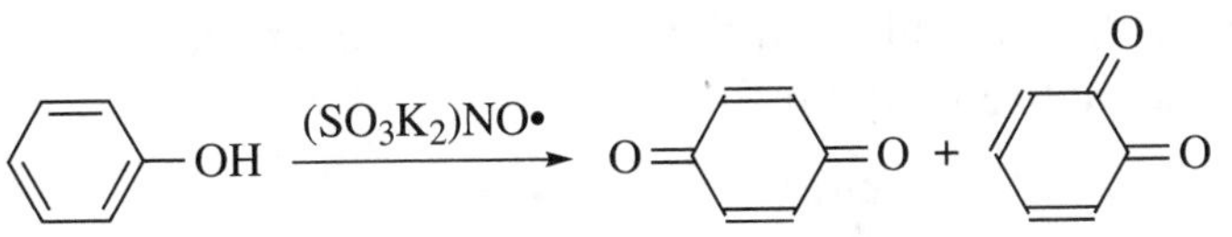

OH
(SO3K2)NO•
O
O
+
O
O

(d) Explain why the nonconjugated alkene reacts preferentially.

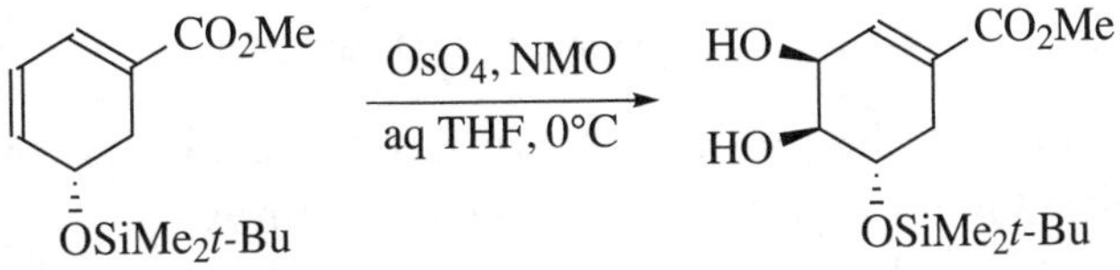

CO2Me
OSiMe2t-Bu
OsO4, NMO
aq THF, 0°C
HO
HO
CO2Me
OSiMe2t-Bu

4. Give the major product of each reaction and explain the regiochemistry and/or stereochemistry where applicable.

(a)

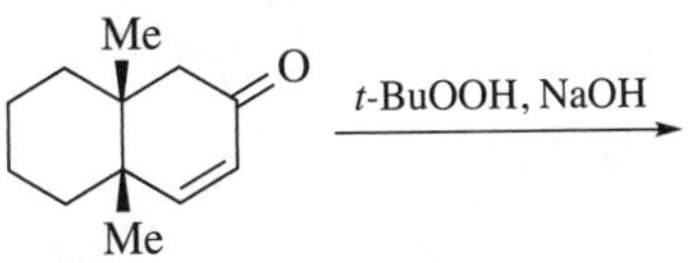

Me
Me
O
t-BuOOH, NaOH

(b)

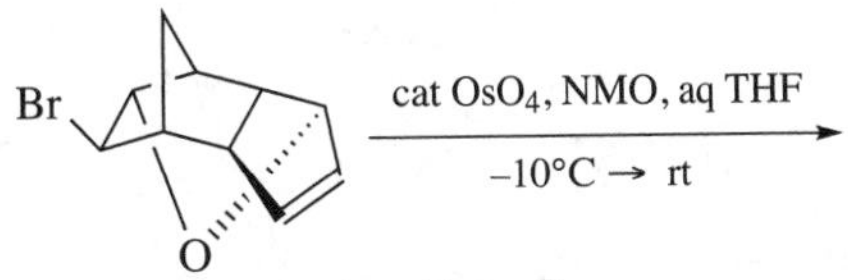

Br
O
cat OsO4, NMO, aq THF
–10°C → rt

(c)

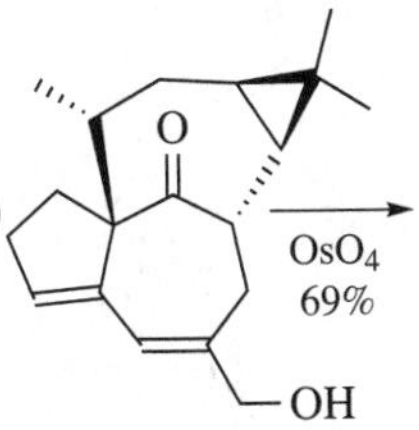

O
OH
OsO4
69%

(d)

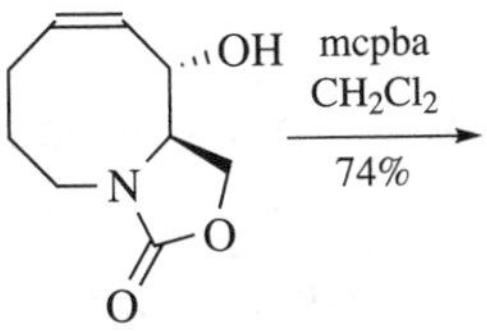

OH
N
O
O
mcpba
CH2Cl2
74%

(e)

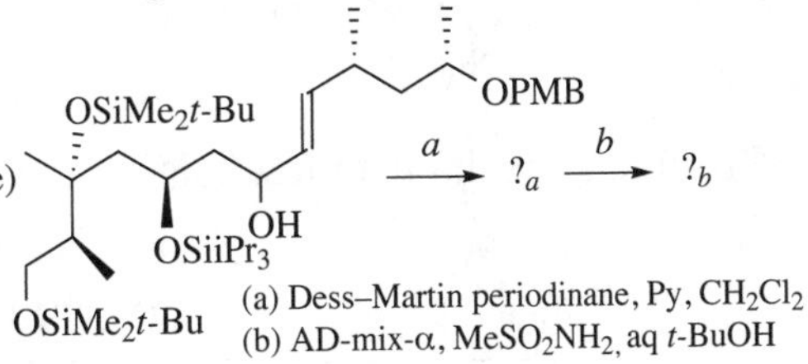

OSiMe2t-Bu
OPMB
OH
OSiiPr3
OSiMe2t-Bu
a
?a
b
?b
(a) Dess–Martin periodinane, Py, CH2Cl2
(b) AD-mix-α, MeSO2NH2, aq t-BuOH

5. Predict the major product. Draw an appropriate transition state for each reaction.

(a)

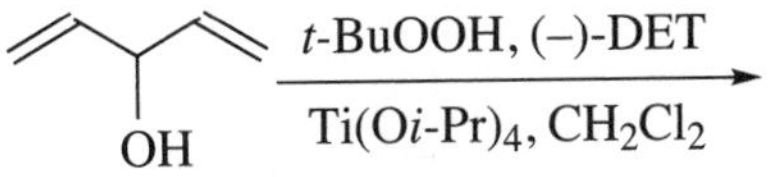

OH
t-BuOOH, (–)-DET
Ti(Oi-Pr)4, CH2Cl2

(b)

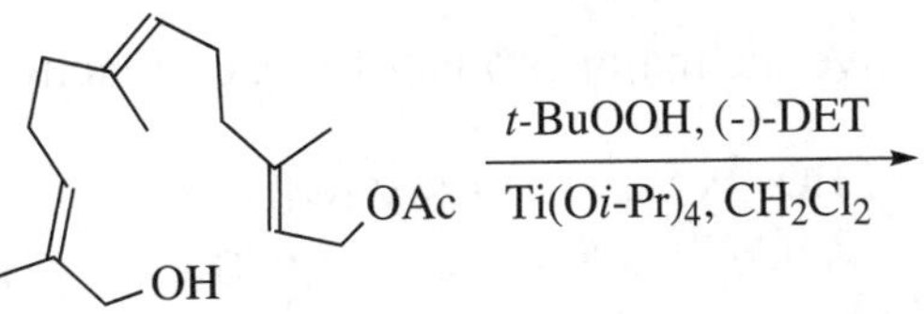

OAc
OH
t-BuOOH, (-)-DET
Ti(Oi-Pr)4, CH2Cl2

(c)

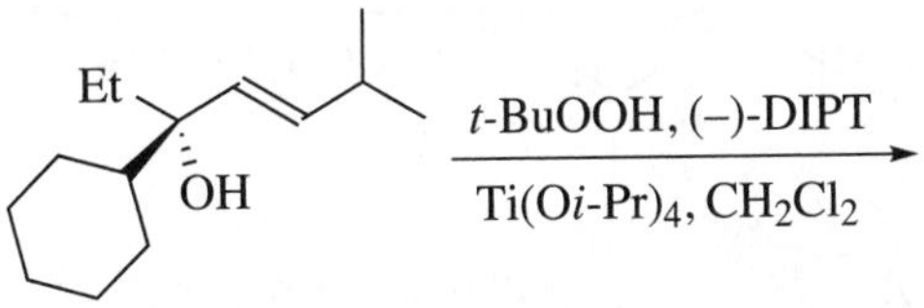

Et
OH
t-BuOOH, (–)-DIPT
Ti(Oi-Pr)4, CH2Cl2

6. (a) Identify each product (?) in the following sequence, taken from Taber's synthesis of (–)-fumagillin.

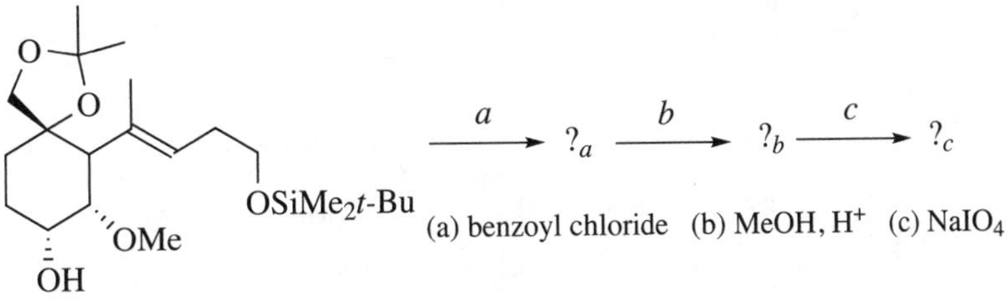

O
O
OMe
OH
OSiMe2t-Bu
a
?a
b
?b
c
?c
(a) benzoyl chloride (b) MeOH, H+ (c) NaIO4

(b) Identify $?_a$, $?_b$, and $?_c$ in the following sequence, taken from Smith's synthesis of (+)-thiazinotrienomycin E.

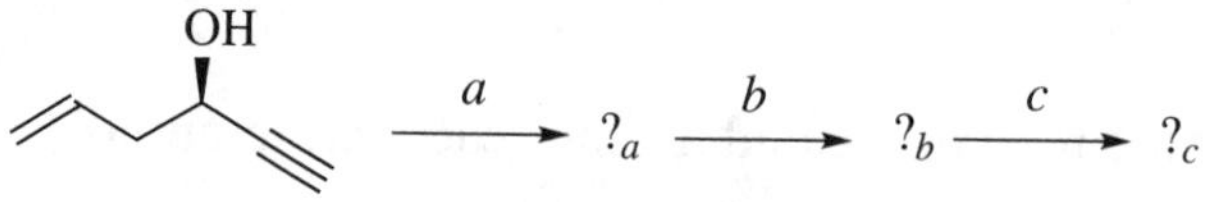

(a) BuLi, ether-DMSO; MeI   (b) O_3; PPh_3 (c) PDC, DMF

7. Briefly explain how the product shown is formed in this Swern oxidation, taken from Feldman and Saunders synthesis of (−)-agelastatin A.

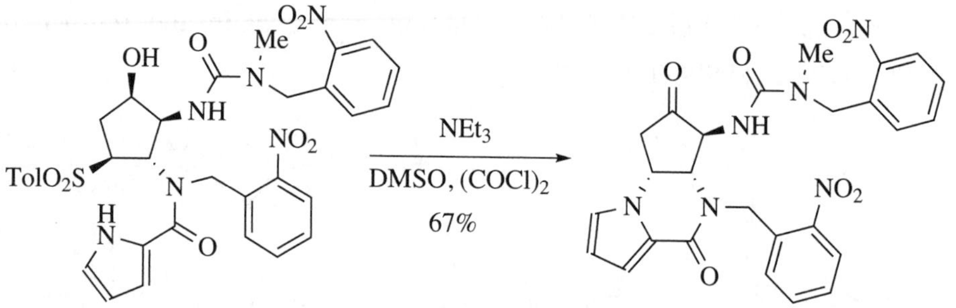

NEt$_3$

DMSO, (COCl)$_2$

67%

8. Provide a mechanism to explain the following transformation.

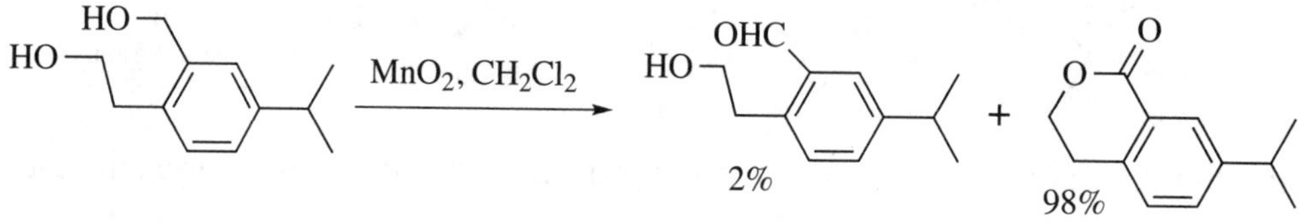

MnO$_2$, CH$_2$Cl$_2$

2%

+

98%

9. Give the major product for each reaction.

(a)

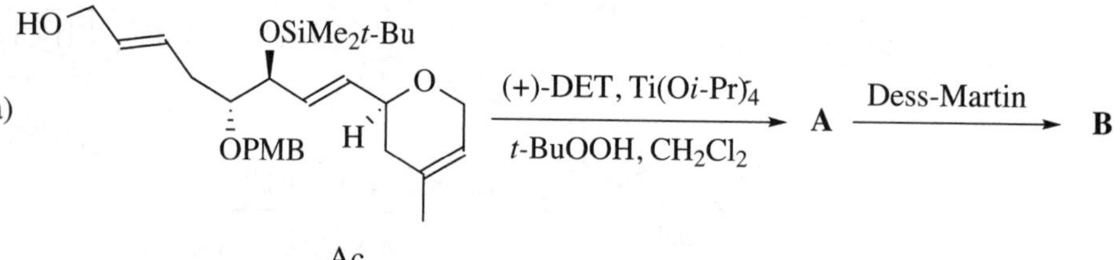

(+)-DET, Ti(Oi-Pr)$_4$

t-BuOOH, CH$_2$Cl$_2$

A

Dess-Martin

B

(b)

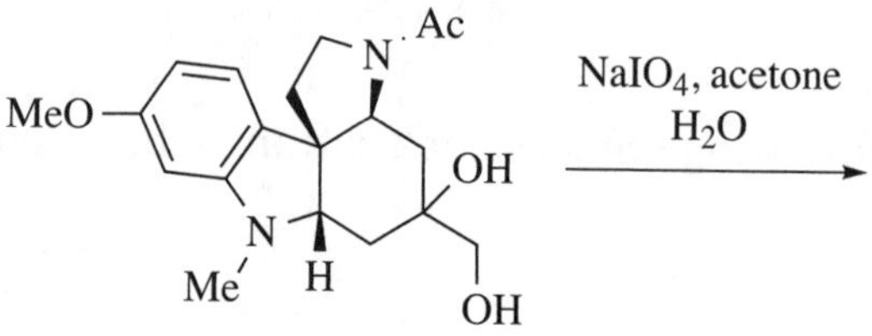

NaIO$_4$, acetone

H$_2$O

(c)

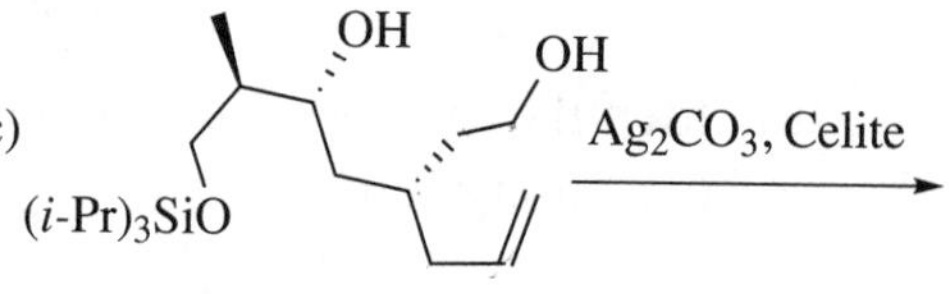

(i-Pr)$_3$SiO

Ag$_2$CO$_3$, Celite

(d)

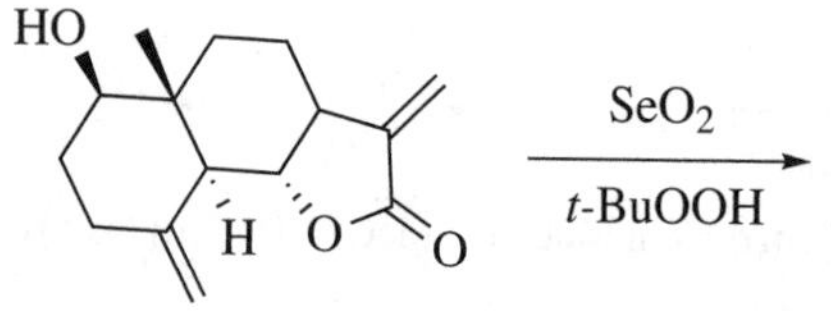

SeO$_2$

t-BuOOH

(e)

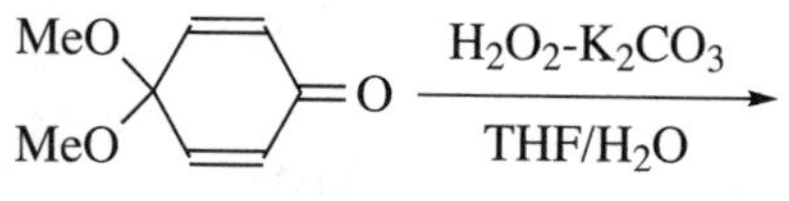

H$_2$O$_2$-K$_2$CO$_3$

THF/H$_2$O

(f) → mcpba, CH$_2$Cl$_2$ / 0°C

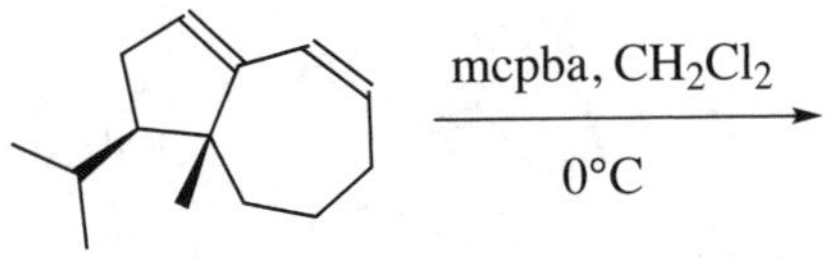

(g)
t-BuMe$_2$SiO
HO ... OPMB
→ SO$_3$•pyridine, 0°C / DMSO, NEt$_3$ / PMB – p-methoxybenzoyl

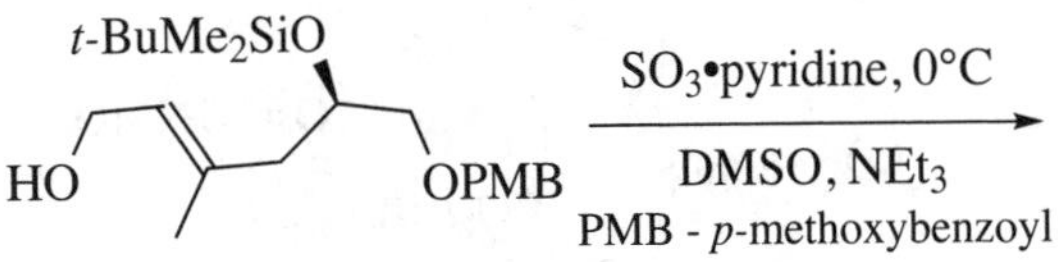

(h)
SiMe$_3$
NH
OH
NHCO$_2$t-Bu
NH
O
H

$\left[\text{N H} \right]_2^+ \text{Cr}_2\text{O}_7^{2-}$ / CH$_2$Cl$_2$

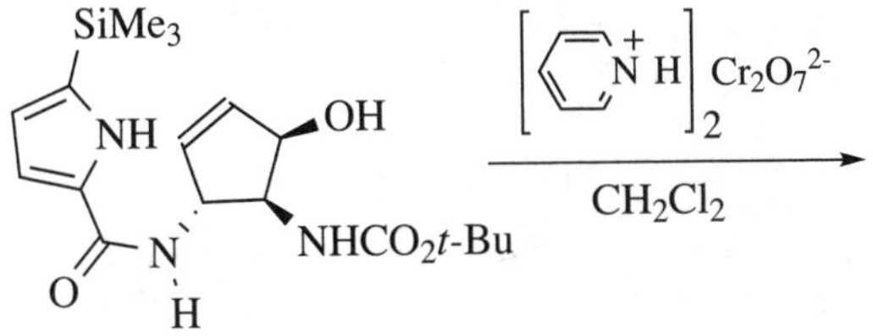

(i) → 1. O$_3$, CH$_2$Cl$_2$, –78°C / 2. Me$_2$S →

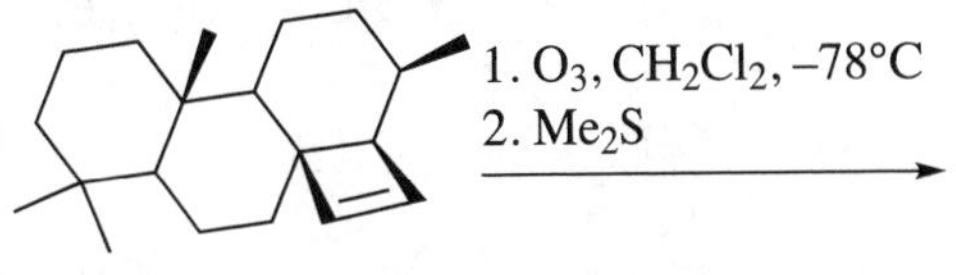

(j) Cl → P. putida 39D →

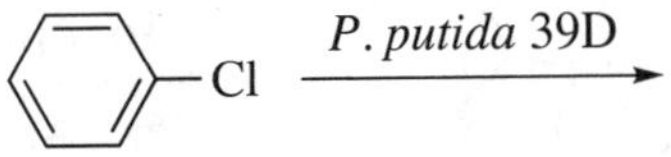

(k)
Me OH OH
H,
O
O
Me Me
→ DMSO, DCC / NEt$_3$, reflux

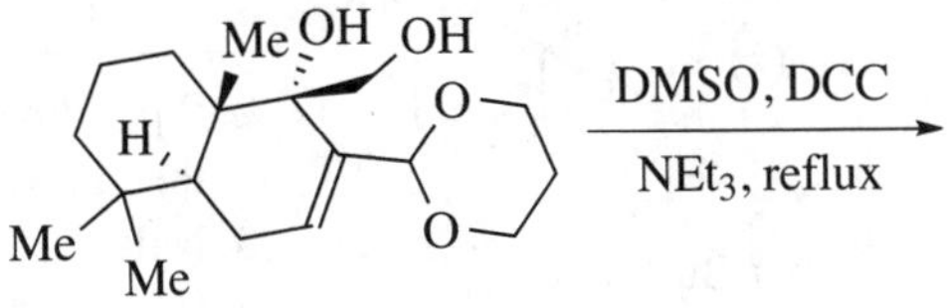

(l) → MnO$_2$, PhH →
OH
O OH
Et

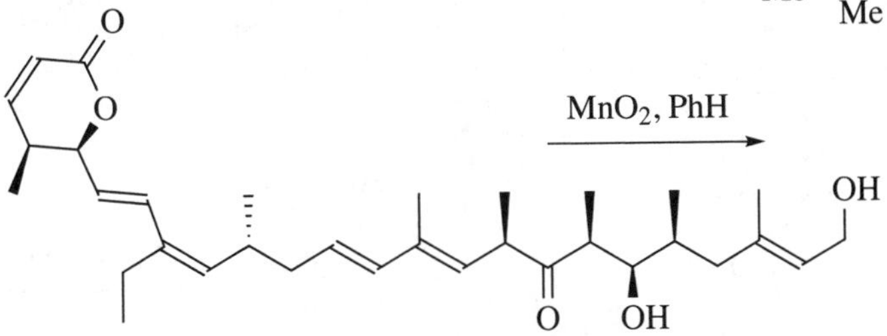

(m) 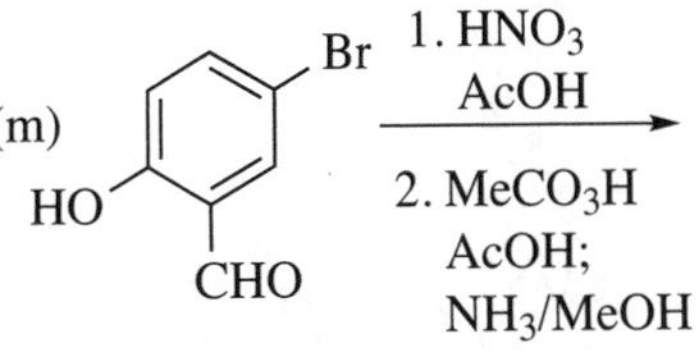
Br
HO
CHO
1. HNO$_3$ / AcOH
2. MeCO$_3$H / AcOH; / NH$_3$/MeOH →

(n)
PMBO
H
O
N 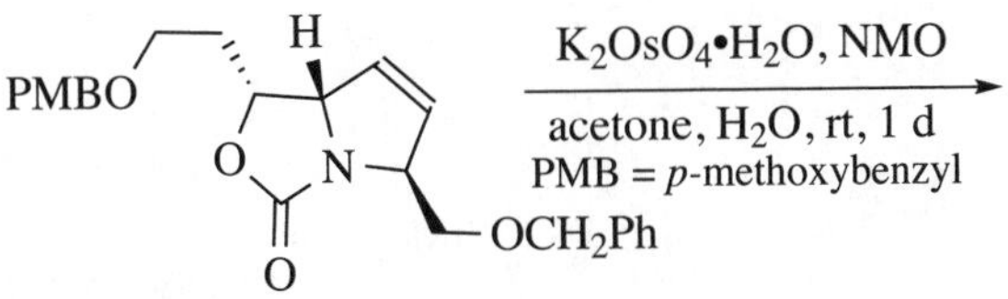
O
OCH$_2$Ph
→ K$_2$OsO$_4$•H$_2$O, NMO / acetone, H$_2$O, rt, 1 d / PMB = p-methoxybenzyl

(o) 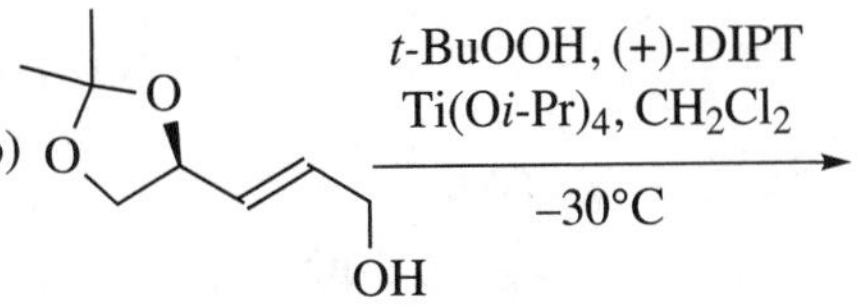
O
O
OH
→ t-BuOOH, (+)-DIPT / Ti(Oi-Pr)$_4$, CH$_2$Cl$_2$ / –30°C →

(p) 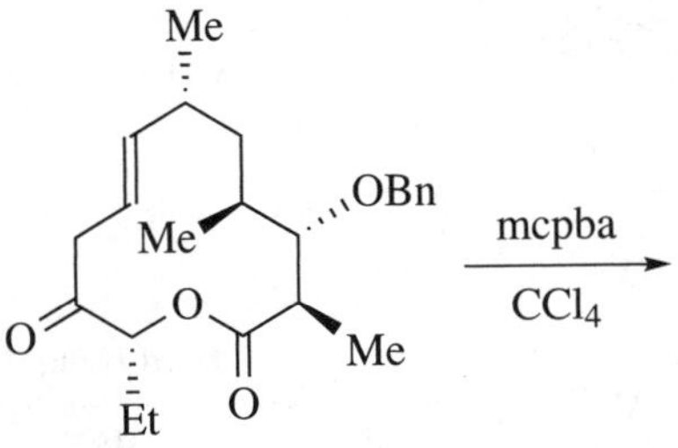
Me
Me
OBn
O
O
Me
Et
O
→ mcpba / CCl$_4$ →

(q) 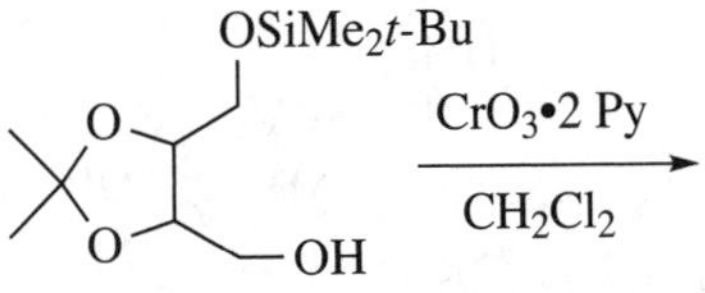
OSiMe$_2$t-Bu
O
O
OH
→ CrO$_3$•2 Py / CH$_2$Cl$_2$ →

(r) 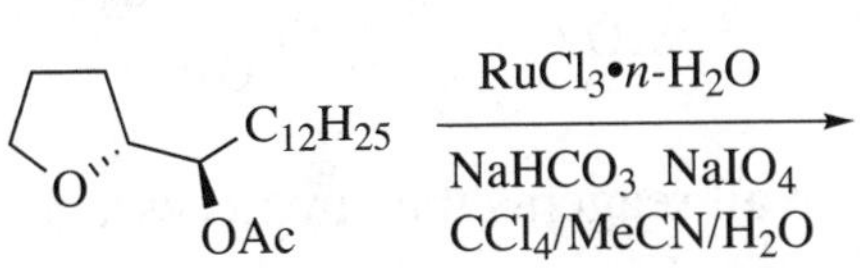
O
C$_{12}$H$_{25}$
OAc
→ RuCl$_3$•n-H$_2$O / NaHCO$_3$ NaIO$_4$ / CCl$_4$/MeCN/H$_2$O →

(s) NMe$_2$
1. 30% aq H$_2$O$_2$ / MeOH
2. Py/KOH / 105°C →

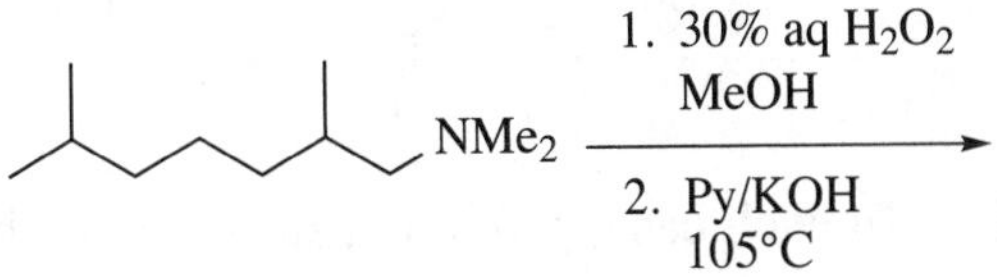

(t)
Me
Me Me
Me
HO
OH
→
AcO OAc
I – OAc
O
O
→

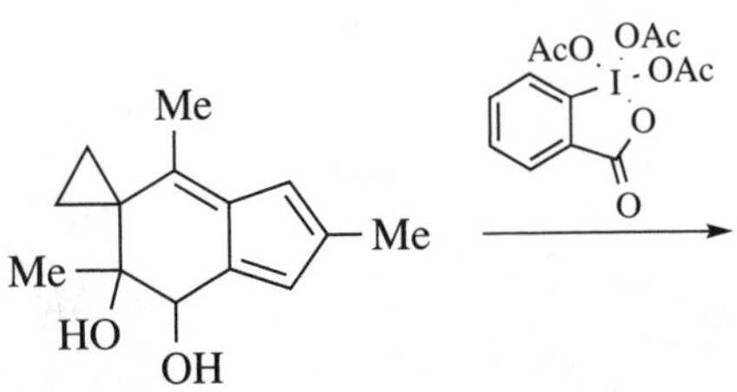

(u)
O
t-BuMe$_2$SiO
HO
AcO'''
→ VO(acac)$_2$ / t-BuOOH, PhH →

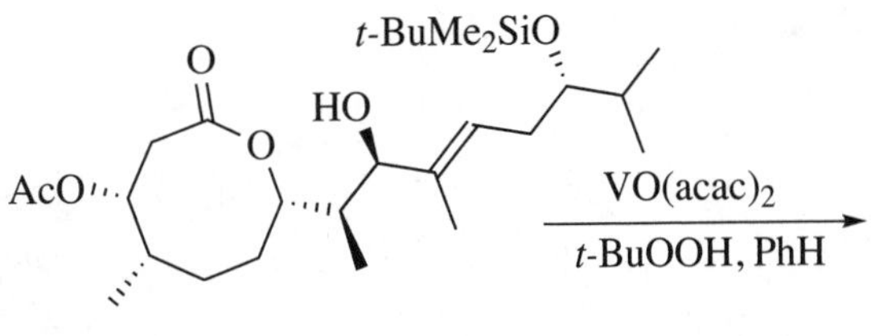

(v) 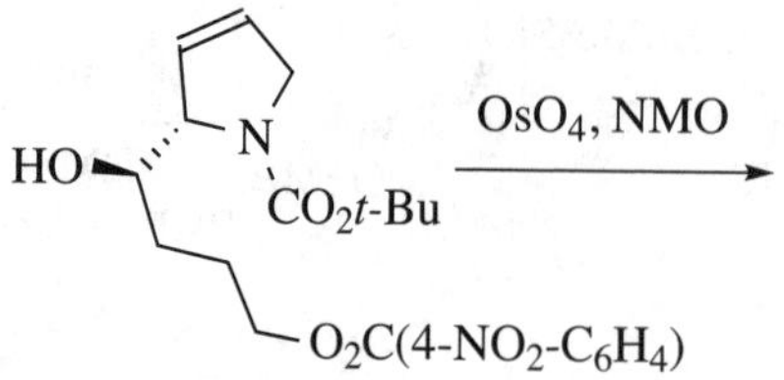 HO— (structure with N–CO$_2$t-Bu, O$_2$C(4-NO$_2$-C$_6$H$_4$))

$\xrightarrow{\text{OsO}_4,\ \text{NMO}}$

(w) 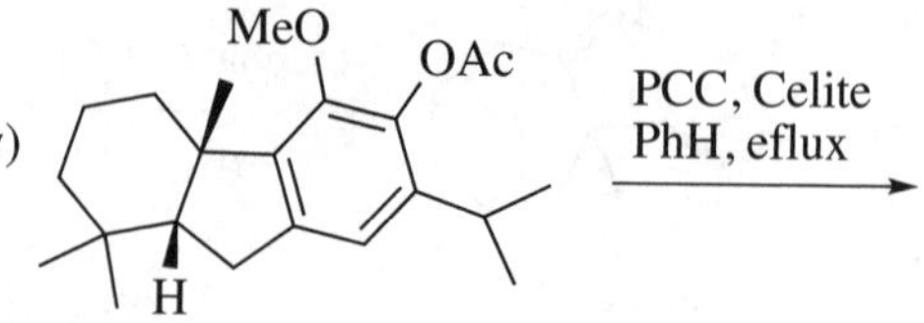 (structure with MeO, OAc, isopropyl, H)

$\xrightarrow[\text{PhH, eflux}]{\text{PCC, Celite}}$

(x) 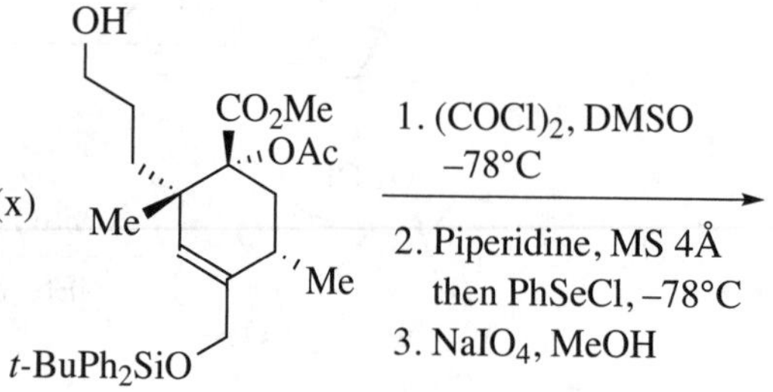 (structure with OH, CO$_2$Me, OAc, Me, Me, t-BuPh$_2$SiO)

1. (COCl)$_2$, DMSO
   −78°C
2. Piperidine, MS 4Å
   then PhSeCl, −78°C
3. NaIO$_4$, MeOH

(y) 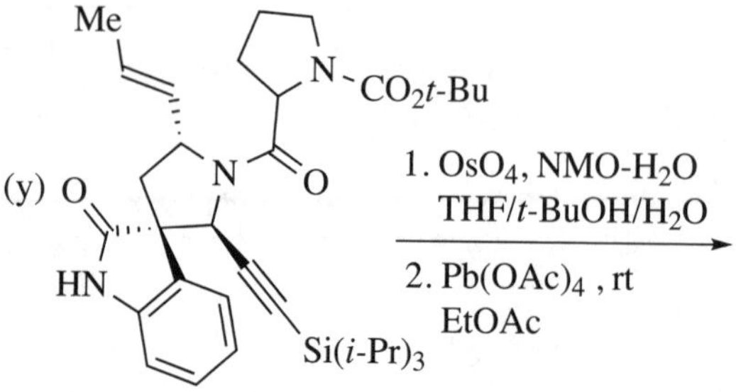 (structure with Me, N–CO$_2$t-Bu, O, HN, Si(i-Pr)$_3$)

1. OsO$_4$, NMO-H$_2$O
   THF/t-BuOH/H$_2$O
2. Pb(OAc)$_4$, rt
   EtOAc

(z) 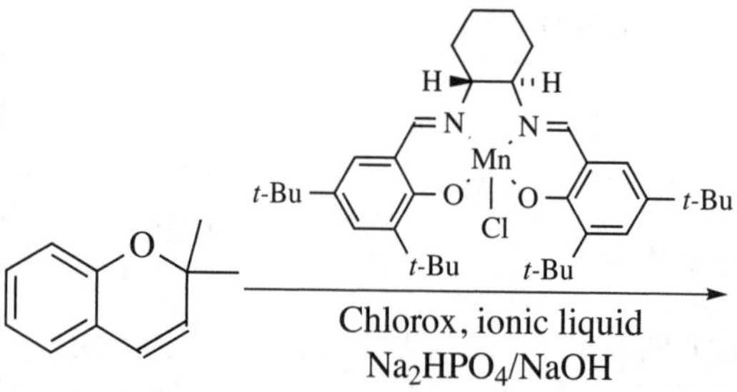 (2,2-dimethyl chromene structure)

(salen Mn catalyst structure: H, H, N, N, Mn, O, Cl, O, t-Bu ×4)

$\xrightarrow[\substack{\text{Chlorox, ionic liquid}\\ \text{Na}_2\text{HPO}_4/\text{NaOH}}]{}$

(aa) 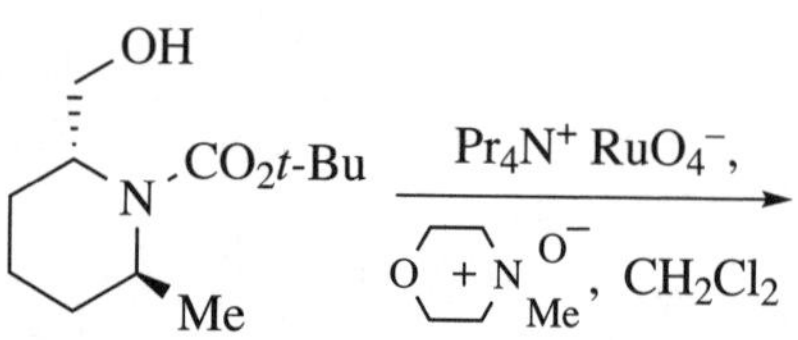 (structure with OH, N–CO$_2$t-Bu, Me)

$\xrightarrow[\substack{\text{O}\ +\ \text{N–O}^-\ \text{Me, CH}_2\text{Cl}_2}]{\text{Pr}_4\text{N}^+\ \text{RuO}_4^-,}$

(ab) 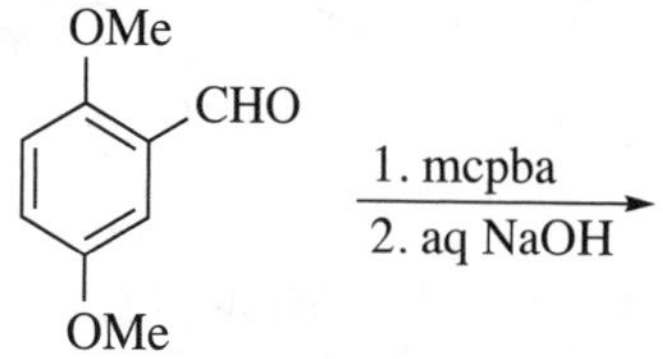 (benzaldehyde with OMe, CHO, OMe)

$\xrightarrow[\text{2. aq NaOH}]{\text{1. mcpba}}$

(ac) 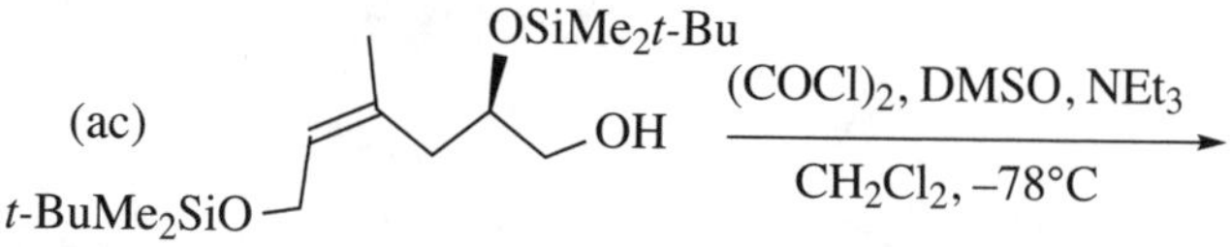 (structure with OSiMe$_2$t-Bu, OH, t-BuMe$_2$SiO)

$\xrightarrow[\text{CH}_2\text{Cl}_2,\ −78°\text{C}]{(\text{COCl})_2,\ \text{DMSO, NEt}_3}$

(ad) 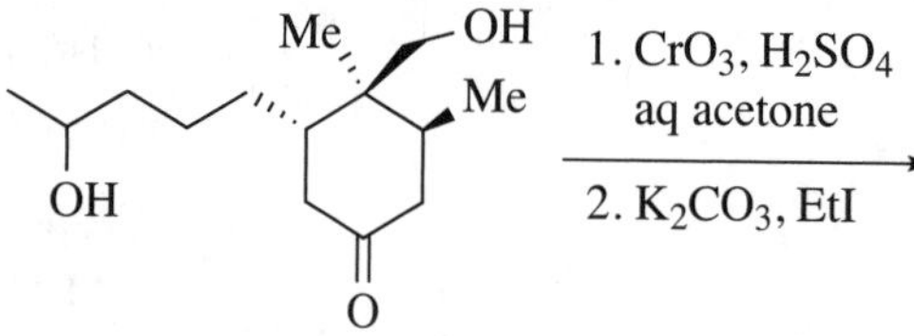 (cyclohexanone structure with Me, OH, Me, OH)

1. CrO$_3$, H$_2$SO$_4$
   aq acetone
2. K$_2$CO$_3$, EtI

(ae) 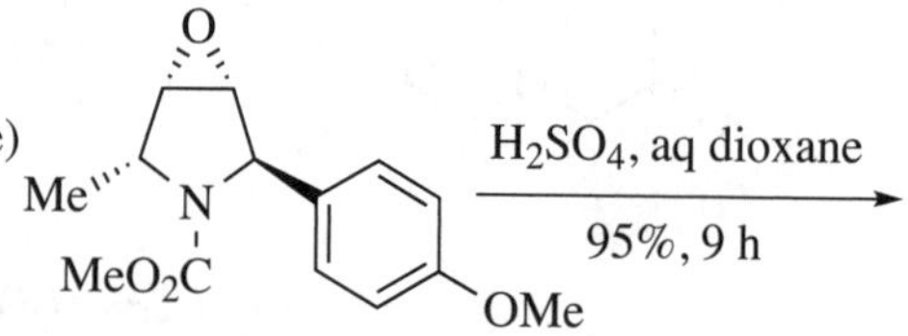 (epoxide pyrrolidine structure with O, Me, N, MeO$_2$C, OMe)

$\xrightarrow[\text{95\%, 9 h}]{\text{H}_2\text{SO}_4,\ \text{aq dioxane}}$

10. In each case, provide a suitable synthesis. Give all reagents that are necessary and show all intermediate products.

(a) 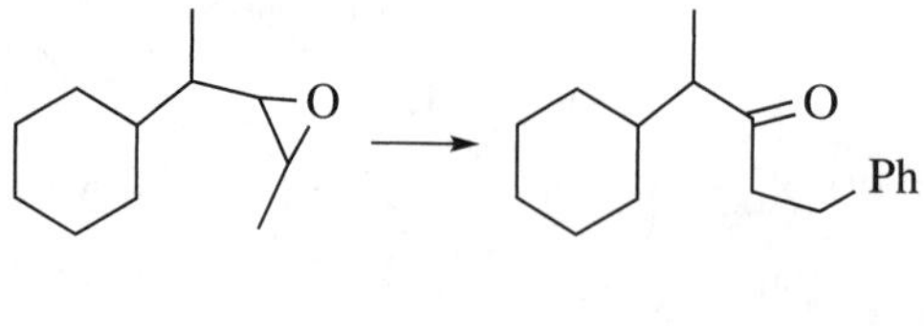 (cyclohexyl epoxide) → (cyclohexyl ketone with Ph)

(b) 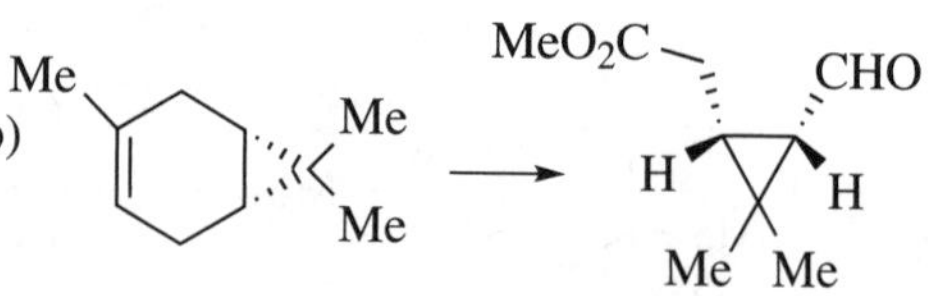 (structure with Me, Me, Me) → (structure with MeO$_2$C, CHO, H, H, Me, Me)

(c)

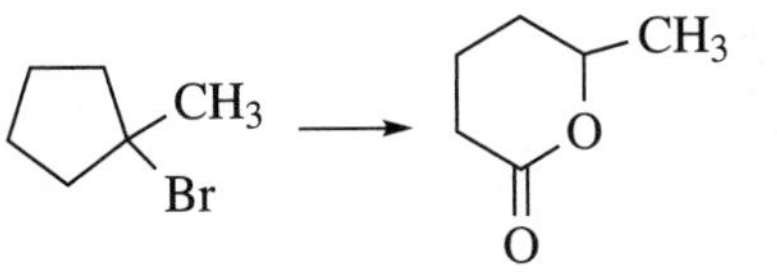

(d)

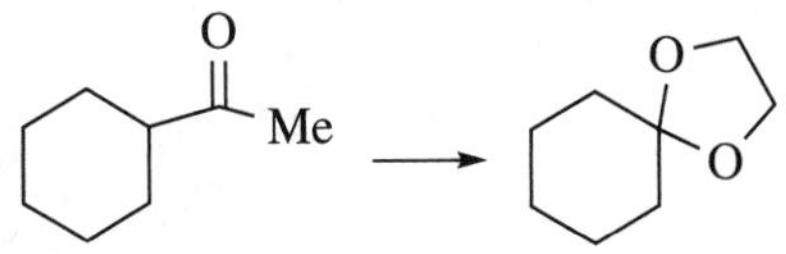

(e)

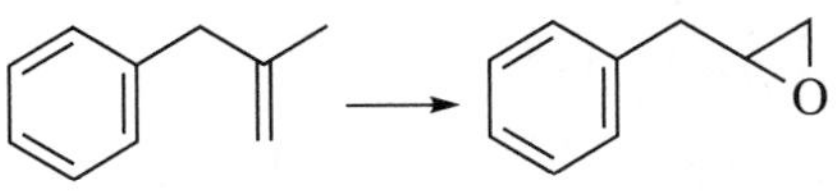

(f)

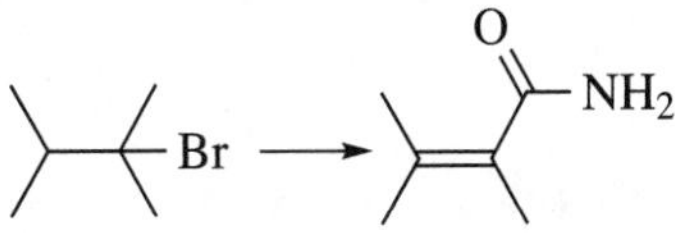

(g)

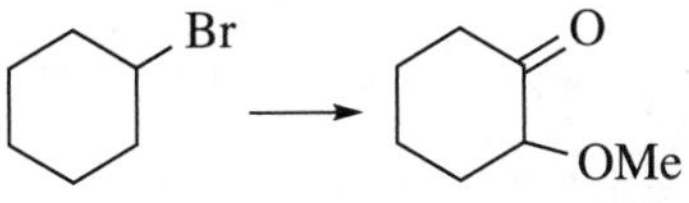

(h)

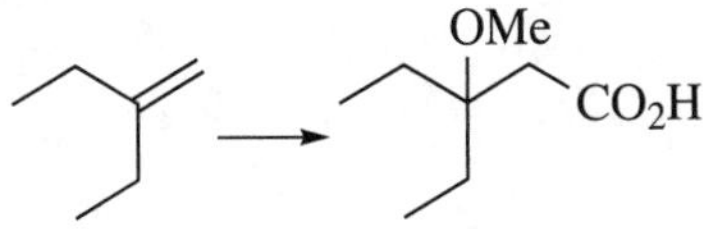

(i)

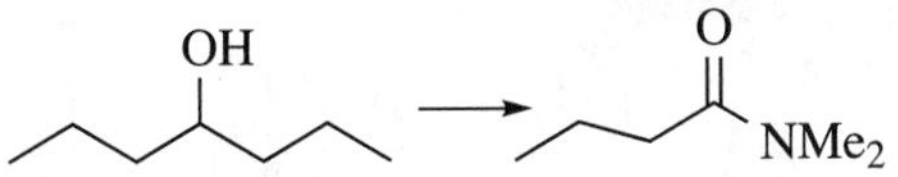

(j)

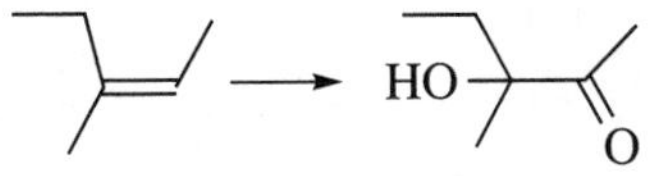

(k)

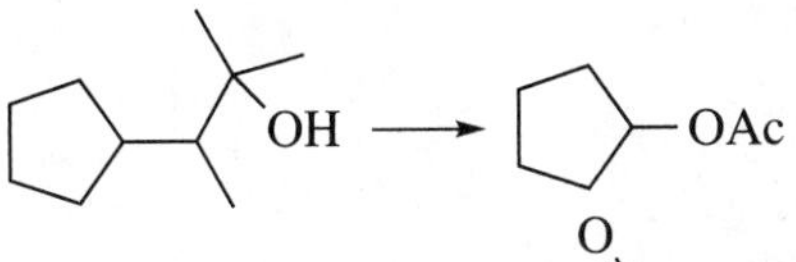

(l)

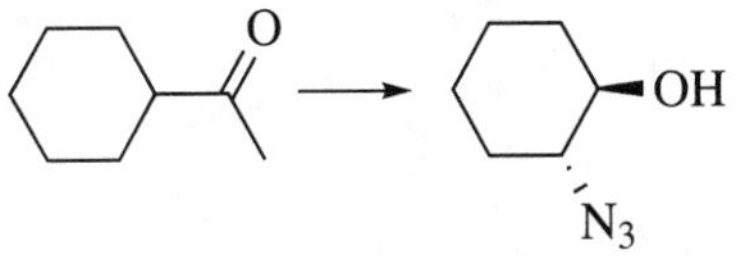

(m)

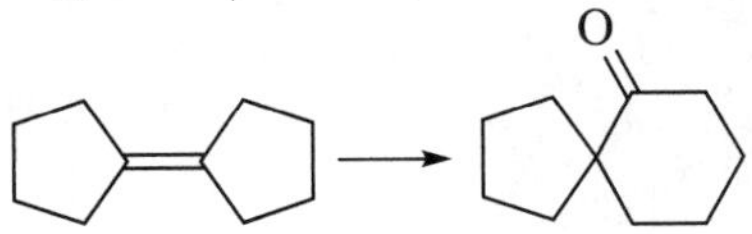

chapter 4

Reduction

4.1. INTRODUCTION

Oxidation reactions were introduced in section 3.1 briefly reviewed. Reduction reactions are equally important, and many different classes of organic molecules can be reduced. If an oxidation reaction proceeds with an increase in oxidation state and a loss of electrons,[1] then a reduction reaction proceeds with a decrease in oxidation state and a gain of electrons. The concept of oxidation stage that was introduced in section 3.1 is also useful for reduction since reduction can be defined as conversion of an atom in a higher oxidation stage to a lower one (III → II, IV → II or II → I), as in the transformation $RCH=NH \rightarrow RCH_2NH_2$.[2] The oxidation numbers of Holleran and Jespersen,[3] and the modified method of Hendrickson, Cram, Hammond[4] and Pine[5] are useful for analyzing reduction reactions. It is also convenient to view organic reductions as the addition of hydrogen to the molecule. This latter definition is not always mechanistically correct, but it is conceptually useful and examples are the reduction of 2-butanone to 2-butanol or propanenitrile to butanamine.

As with oxidation (sec. 3.1), reduction potential can be used to assess the ability of a reagent to reduce molecules. The standard reduction potentials for several metals and metal salts commonly used in synthesis are available.[6] For a metal to behave as a reducing agent, the reverse reaction (Li → Li$^+$, for example) must be facile. The high reduction potential of Li$^+$ → Li (–4.045 V)[6] indicates that the Li → Li$^+$ reaction is favored, and lithium metal is an excellent reducing agent (by contrast, Li$^+$ is a very poor oxidizing agent). Conversely, Sn^{4+} → Sn^{2+} has a positive reduction potential (0.15 V)[6] indicating a facile reaction. On the other hand, Sn^{2+} → Sn° (–0.136 V)[6] indicates that Sn^{2+} a reducing agent. The standard hydrogen electrode is taken as zero V, but this does not really represent the reducing power of hydrogen gas in the presence

1. Yalman, R.G. *J. Chem. Educ.* **1959**, *36*, 215.
2. Soloveichik, S.; Krakauer, H. *J. Chem. Educ.* **1966**, *43*, 532.
3. Holleran, E.M.; Jespersen, N.D. *J. Chem. Educ.* **1980**, *57*, 670.
4. Hendrickson, J.B.; Cram, D.J.; Hammond, G.S. *Organic Chemistry, 3rd ed.,* McGraw-Hill, New York, **1970**.
5. Pine, S.H. *Organic Chemistry*, 5th ed, McGraw-Hill, New York, **1987**.
6. Handbook of Chemistry and Physics, 87th ed., CRC/Taylor and Francis, Boca Raton, Fla., **2006**, pp 8-20-8-32.

of a suitable catalyst. The reduction potentials of hydride reducing agents are not usually listed, although they are among the most potent reducing agents. For these reasons, no attempt will be made to use reduction potentials to correlate the reducing power of each reagent.

Chapter 3 was organized by functional group changes, and the various reagents and techniques for each transformation were discussed. Most of the reduction techniques, despite their differences in mechanism and reagents, accomplish the same or similar functional group changes. There are, of course, differences in functional group reactivity as well as chemoselectivity and stereochemistry for the various reagents. This chapter is organized by type of reduction, with a discussion of applicable functional group transformations. The monograph on reduction by Hudlicky[7] categorized reductions as reduction with complex hydrides, catalytic hydrogenation, electroreduction, reduction with metals, and reductions with non-metals. This excellent work includes an extensive survey of specific organic functional group reductions, to which the reader is referred for detailed information concerning reduction and reduction types. The order of topics in this chapter attempts to correlate a type of reduction with its relative utility in organic synthesis. This chapter will begin with hydride reductions since they are probably the most widely used, followed by catalytic hydrogenation. Reduction by various metals, including reduction by electrolysis and reduction with non-metals will follow. The final sections will deal with miscellaneous nonmetal reductions, including photoreduction and enzymatic or microbial reductions.

Reduction of prochiral centers (sec. 1.4.E) can generate enantiomers and diastereomers. It is therefore important to introduce the models that are used to predict the diastereoselectivity or enantioselectivity of a reduction. These stereochemical explanations are applicable to many reactions in organic chemistry, but it is important that stereoselectivity be viewed as a *part of the reaction* and **not** as a separate topic. For that reason, stereoselective reactions are discussed in each chapter where they are pertinent. It will be seen that reduction includes a wide range of functional group interchanges and important synthetic processes.

4.2. REDUCTION WITH COMPLEX METAL HYDRIDES

4.2.A. Lithium Aluminum Hydride

$$4 \text{ LiH} + \text{AlCl}_3 \xrightarrow{\text{ether}} \text{LiAlH}_4 + 3 \text{ LiCl}$$
$$4 \text{ NaH} + \text{B(OMe)}_3 \longrightarrow \text{NaBH}_4 + 3 \text{ MeO}^- \text{Na}^+$$

Gaylord's monograph with the title shown for this section[8] as well as his contemporary review,[9]

---

7. Hudlicky´, M. *Reductions in Organic Chemistry*, Ellis Horwood Ltd., Chichester *1984*.
8. Gaylord, N.G. *Reduction with Complex Metal Hydrides,* Wiley-Interscience, New York *1956*.
9. Gaylord, N.G. *J. Chem. Educ.* *1957*, *34*, 367.

describes hydride reducing agents that were used until 1957. The two most common hydride reducing agents are lithium aluminum hydride ($LiAlH_4$) and sodium borohydride ($NaBH_4$, to be discussed in sec. 4.4).[10] Schlesinger and co-workers, as well as Brown and co-workers,[11] reported the synthesis and initial applications of $LiAlH_4$ for reduction of organic compounds,[12] in 1947. Lithium aluminum hydride was prepared by reaction of lithium hydride (LiH) with aluminum chloride ($AlCl_3$).[13] Sodium borohydride was first prepared by reaction of sodium hydride (NaH) with trimethylborate, $B(OMe)_3$.[13,14]

Further studies showed that replacement of the hydrogen in $LiAlH_4$ or $NaBH_4$ with alkyl or alkoxy groups led to a variety of reducing agents with modified reducing abilities, chemoselectivity, and stereoselectivity. Lithium aluminum hydride is one of the most powerful reducing agents known for organic substrates bearing a polarizable functional group. This reagent is stable at ambient temperatures, but reacts violently with moisture to give hydrogen gas, lithium hydroxide and aluminum hydroxide (hydrated alumina), as shown:[9]

$$LiAlH_4 \ + \ 4\,H_2O \longrightarrow LiOH \ + \ Al(OH)_3 \ + \ 4\,H_2$$

The moisture does not have to be inadvertently added water. It can arise from "inadequately dried apparatus, fingerprints, and perspiration-soaked gloves", leading to ignition.[15] Although this reagent is commercially available as a powder, lumps are also available. *"Lumps are best crushed by cautious pounding with a hard rubber hammer in a tray with an aluminum foil liner. Grinding with a mortar and pestle should be done in a dry box in an atmosphere of nitrogen or argon.... If ground too vigorously in the presence of air the hydride may take fire."*[15] Theses quotations from Fieser emphasize the reactivity of $LiAlH_4$ and the danger of mishandling this reagent.

Lithium aluminum hydride has some solubility in ether solvents, but tends to form a slurry when relatively large amounts of the reagent are used. The solubility of $LiAlH_4$ in ether is 35-40 g/100 g of ether, but obtaining this solubility requires concentration of a more dilute solution.[14] The solubility is 13g/100 g in THF and 0.1 g in dioxane,[14] which are also commonly used solvents. The reagent is usually represented as $Li^+[AlH_4^-]$ and exists as aggregates of solvated lithium ions and aluminohydride ions in diethyl ether.[16] Such slurries of $LiAlH_4$ are capable of reducing virtually all polarized organic functional groups, as described by Gaylord.[9,8]

10. For a review of 60 years of hydride reductions, see Brown, H.C.; Ramachandran, P.V.. in *Reductions in Organic Synthesis. Recent Advances and Practical Applications*, Abdel-Magid, A.F (Ed.), ACS Symposium Series 641, American Chemical Society, Washington, DC, *1996*. pp 1-30.
11. Nystrom, R.F.; Brown, W.G. *J. Am. Chem. Soc. 1947, 69*, 1197.
12. Finholt, A.E.; Bond, A.C.; Schlesinger, H.I. *J. Am. Chem. Soc. 1947, 69*, 1197.
13. Reference 7, p 14.
14. Schlesinger, H.I.; Brown, H.C.; Finholt, A.E. *J. Am. Chem. Soc. 1953, 75*, 205.
15. Fieser, L.F.; Fieser, M. *Reagents for Organic Synthesis Vol. 1*, Wiley, New York *1967*, pp 581-595.
16. House, H.O. *Modern Synthetic Reactions, 2nd ed,* Benjamin, Menlo Park CA,*1972*, p 49 and citations 5, 12 and 14 therein.

Molecules with an acidic hydrogen such as acids, alcohols, amines, etc. may react with lithium aluminum hydride in an acid-base reaction to generate hydrogen gas with destruction of the active hydride.

4.2.B. Reduction of Carbonyl Compounds

Lithium aluminum hydride rapidly reduces most carbonyl compounds, including aldehydes and ketones,[17] carboxylic acids, acid anhydrides, acid chlorides, esters, lactones, amides, carbamates, imides, and lactams.[18] Most $LiAlH_4$ reductions of carbonyl compounds begin by coordination to the carbonyl oxygen, followed by transfer of the nucleophilic hydride (oxygen is polarized δ- relative to aluminum) to the electrophilic carbonyl (carbon is polarized δ+ relative to oxygen). House illustrated this mechanism for the $LiAlH_4$ reduction of acetone, showing the stepwise formation of alkoxyaluminate intermediates.[16] Initial transfer of hydride to the carbonyl generates **2**, possibly via direct coordination of aluminum to oxygen and transfer of hydride to the carbonyl carbon. Complex **1** has also been proposed as an intermediate. Alkoxide **2** contains three additional active hydrides, so a second equivalent of carbonyl can react to give bis(alkoxide) **3**. Similarly, a third equivalent gives **4** and a fourth gives **5**.

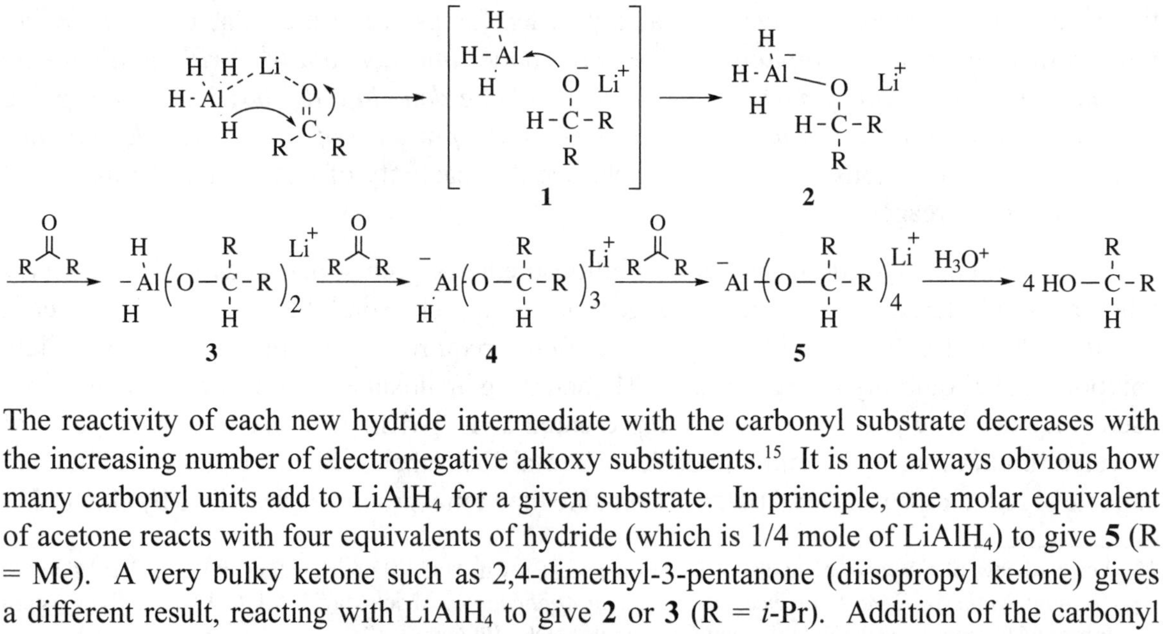

The reactivity of each new hydride intermediate with the carbonyl substrate decreases with the increasing number of electronegative alkoxy substituents.[15] It is not always obvious how many carbonyl units add to $LiAlH_4$ for a given substrate. In principle, one molar equivalent of acetone reacts with four equivalents of hydride (which is 1/4 mole of $LiAlH_4$) to give **5** (R = Me). A very bulky ketone such as 2,4-dimethyl-3-pentanone (diisopropyl ketone) gives a different result, reacting with $LiAlH_4$ to give **2** or **3** (R = *i*-Pr). Addition of the carbonyl compound to $LiAlH_4$ (a 1:1 molar ratio), followed by titration of the complex with water or aqueous acid will release hydrogen gas from the unreacted hydrides. The hydrogen gas can be trapped, allowing an exact determination of the number of carbonyl molecules that reacted with the tetrahydroaluminate. In general, at least one molar equivalent (four equivalents of H^-)

17. Larock, R.C. *Comprehensive Organic Transformations, 2nd ed,* Wiley-VCH, New York, **1999**, p 1077.
18. For reagents used to reduce acid derivatives, see Reference 17, pp 1263-1273.

or more are added to the carbonyl moiety to ensure complete reduction. A complicating factor is that alkoxyaluminate intermediates are subject to disproportionation, as in the reversible interconversion of **6** to **7**,[16,19] which is more of a problem with increasing steric bulk around the carbonyl carbon (ketone > aldehyde; 2° carbon > 1° carbon). Disproportionation regenerates the tetrahydroaluminate anion, which can cause overreduction as well as loss of functional group specificity, and stereoselectivity.[16]

The final alcohol product from the reduction of $R_2C=O$ is generated by treatment of **5** (or any other alkoxyaluminate product) with aqueous acid or with aqueous base via protonation of the alkoxide base. In many cases, titration with water is sufficient to release the alcohol but saturated ammonium chloride or dilute hydrochloric acid is sometimes necessary. Aqueous or acidic workup of a $LiAlH_4$ reaction is exothermic and evolves hydrogen gas, which leads to frothing. Since volatile ether solvents are usually employed, isolation of the alcohol can be difficult and sometimes dangerous with large scale reactions. If excess water is added, the alumina salts formed during the hydrolysis can form a paste that is difficult to filter. Fieser and Fieser described[20a] the basic workup procedure of Mićović and Mhailović's,[20b,c] which has several advantages. It produces a granular precipitate that is easily filtered [for *n* grams of $LiAlH_4$ *n* mL of water, then *n* mL of 15% NaOH followed by 3*n* mL of water]. Although this information is best found in a lab manual, in fact, the choice of a reducing agent is often dictated by the relative ease of isolation of a given product and this discussion is offered with that in mind.

The rate of reduction of aldehydes is faster than that of ketones, because there is less steric hindrance for delivery of hydride to the carbonyl carbon. In extreme cases, reduction of highly hindered ketones may have a half-life that measures in days, but rarely does reduction fail to occur. This observation stands in contrast to results obtained with other reducing agents such as sodium borohydride[21] (sec. 4.4), which reduces only selected functional groups. A typical synthetic use of this reaction is reduction of the aldehyde moiety in **8**, which gave a 92% yield of alcohol **9** in Wiemer and co-worker's synthesis of schweinfurthin B.[22]

19. (a) Kader, M.H.A. *Tetrahedron Lett.* *1969*, 2301; (b) Eliel, E.L.; Senda, Y. *Tetrahedron* *1970*, *26*, 2411.
20. (a) Reference 15, p 584 (b) Mićović, V.M.; Mhailović's, M.L. *J. Org. Chem.* *1953*, *18*, 1190.; (c) Amundsen, L.H.; Nelson, L.S. *J. Am. Chem. Soc.* *1951*, *73*, 242.
21. Brown, H.C.; Ichikawa, J. *J. Am. Chem. Soc.* *1962*, *84*, 373.
22. Treadwell, E.M.; Neighbors, J.D.; Wiemer, D.F. *Org. Lett.* *2002*, *4*, 3639.

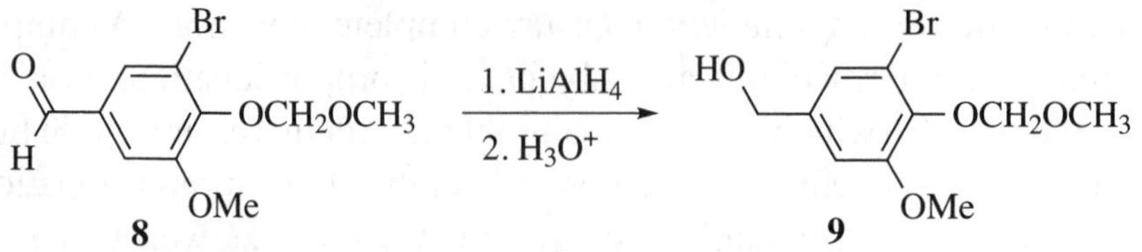

One of the more important characteristics of carbonyl reduction with LiAlH$_4$ is the diastereoselectivity of the reaction when reduction introduces a second stereogenic center. As a general rule, hydride is delivered to the less sterically hindered face of the molecule *unless* special chelation or coordination effects are present. We will present a detailed discussion of this concept in section 4.7.

Reduction of α,β-unsaturated carbonyl derivatives poses a potential problem. Reduction of benzalacetone (**10**) can lead to either the allylic alcohol (**11**) via normal 1,2-addition of hydride to the carbonyl, or to the saturated alcohol (**12**) via 1,4-reduction (delivery of hydride to the alkenyl carbon).[23] Initial 1,2-reduction leads to a bis(alkoxyaluminate) (**13**), which is apparently converted to an intermediate such as **14**.[24] If reduction proceeds via coordination

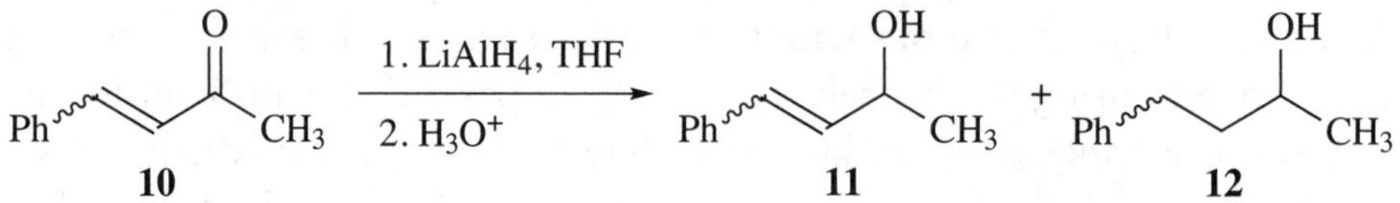

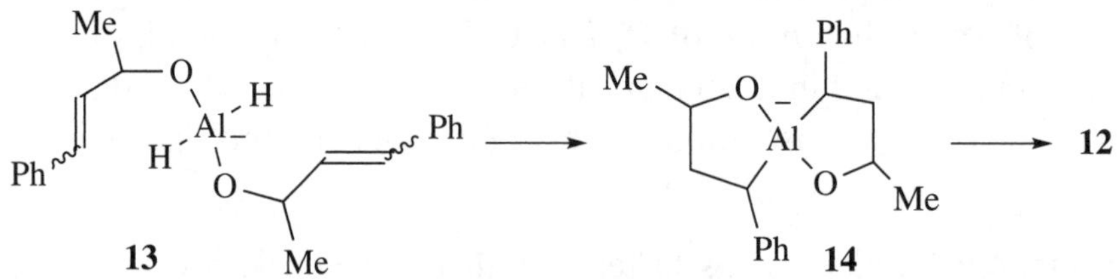

of the aluminum to the carbonyl oxygen (represented by generalized model **16**) 1,2-reduction occurs via path *a* (to give **15**), which is the usual pathway (see **1** → **2**). The β-carbon in **16** is polarized with a positive dipole, however, and hydride can also be delivered to the β-carbon via path *b* (a six-center transition state). This path leads to conjugate (1,4) addition, and enolate **17** is the product. If the R group in **16** is small (e.g., hydrogen in conjugated aldehydes), 1,2-addition predominates. As the R^1 and/or R^2 groups increase in size, delivery to the β-carbon is sterically inhibited (giving more **15**). Conjugate reduction is common for α,β-unsaturated ketones when the β-carbon is relatively unhindered. Conjugate reduction is rare for conjugated aldehydes, although there are a few reagents that give this product in high yield (see the discussion for aluminum hydride in sec. 4.6.B). House pointed out that **14** is

---

23. (a) Reference 7, p 171, 179 and citations 29 and 158 therein (pp. 313, 316); (b) Meek, J.S.; Lorenzi, F.J.; Cristol, S.J. *J. Am. Chem. Soc.* **1949**, *71*, 1830.
24. Reference 16, pp 89-90.

subject to disproportionation, or that another mechanism may be operative in this reduction.[24] Inverse addition of the hydride (a slurry of $LiAlH_4$ is added to the carbonyl compound), short reaction times, and low reaction temperatures generally give moderate to good yields of the conjugate reduction product.[24] Good yields of allylic alcohols can often be obtained from α,β-unsaturated aldehydes, however, when the reaction is done at low temperatures with strict control of the stoichiometry.

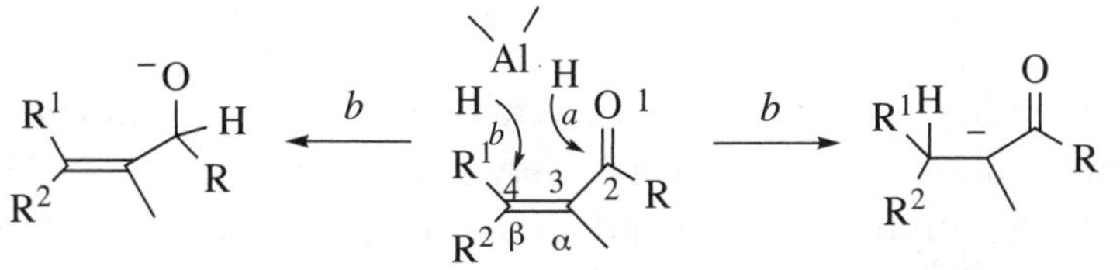

15   **16**   **17**

Reduction of carboxylic acids with one equivalent of $LiAlH_4$ consumes three of the four,[25] and the reduction product is an alcohol. The acidic hydrogen of the COOH unit reacts first. Subsequent reduction of the carbonyl unit leads to the alkoxyaluminate, which is hydrolyzed to the alcohol. In Passaro and Webster's synthesis of the female sex pheromone of the citrus mealybug, *Planococcus citri*, acid **18** was reduced to alcohol **19**.[26] Reduction of acids bearing a heteroatom at the α-position gives the corresponding alcohol.

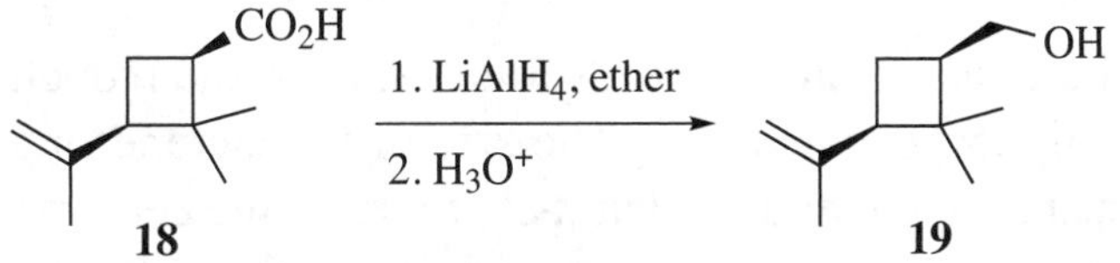

18   **19**

$LiAlH_4$ reduces acid derivatives, including acid chlorides, esters or amides. Reduction of the highly reactive acid chlorides[12] and acid anhydrides[27] is rapid. Partial reduction of aromatic anhydrides[28] to lactones has been reported by controlling the stoichiometry of $LiAlH_4$ and the anhydride. In most cases, however, complete reduction to the diol occurs.[27,12,28]

Reduction of esters (RCO_2R') to hydroxymethyl derivatives (RCH_2OH) along with the alcohol unit R'OH proceeds without loss of stereochemistry at the α-carbon of the ester. Reduction of the *tert*-butyl ester group **20** gave alcohol **21** in 91% yield after hydrolysis (the silyl enol ether moiety was hydrolyzed to the ketone simultaneously), taken from a synthesis of myrmicarin 215A by Movassaghi and Ondrus.[29] A common application that focuses on the R'OH unit is the reduction of acetates, which give the corresponding alcohol ($RCH_2OAc \rightarrow RCH_2OH$) along with ethanol.

25. Reference 8, p 322.
26. Passaro, L.C.; Webster, F.X. *J. Agric. Food Chem.* **2004**, *52*, 2896.
27. Reference 8, p 375 and citation 105 cited therein (p 389).
28. (a) Mahé, J.; Rollet, J.; Willemart, A. *Bull. Chim. Soc. Fr.* **1949**, *16*, 481; (b) Ratouis, R.; Willemart, A. *Compt. Rend.* **1951**, *233*, 1124.
29. Movassaghi, M.; Ondrus, A.E. *Org. Lett.* **2005**, *7*, 4423.

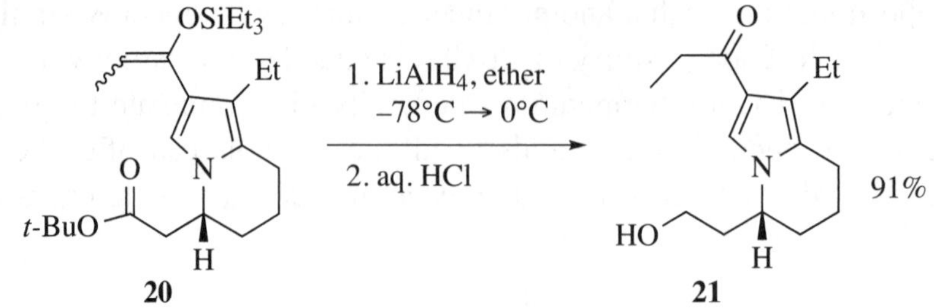

The ester moiety of readily available amino esters (derived from amino acids) can be reduced to give chiral amino alcohols, which have a variety of uses in synthesis.[30] The 1,2-reduction of conjugate esters leads to allylic alcohols under the proper reaction conditions. 1,4-Reduction of conjugated esters is often a problem, as it is with conjugated aldehydes and ketones, but selective 1,2-reduction is easily attained as shown by the conversion of 22 to 23 in >55% yield in Wiemer and co-worker's synthesis of (2E,6Z)-farnesol.[31]

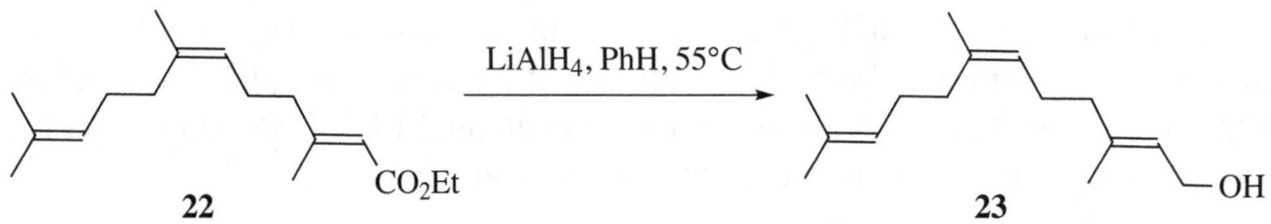

The reduction of cyclic esters (lactones) leads to diols, as in the reduction of δ-valerolactone to give an 85% yield of 1,5-pentanediol.[32] The size of the lactone ring is determined by the number of carbons separating the hydroxyl units.[33] Stereogenic centers adjacent to the oxygen and carbonyl of the lactone[34] are not compromised during the reduction. The conformational bias of the lactone ring can be used to control the relative stereochemistry of substituents as they are incorporated on the ring (Chap. 6). In a synthesis of (−)-enterodiol by Eklund and co-workers, lactone 24 was reduced to diol 25 in 71% yield.[35] When the lactone carbonyl is sterically hindered as in 26, partial reduction is possible. In this case, lactol 27 was formed in > 80% yield.[36] In general, however, partial reduction of lactones with LiAlH$_4$ is *difficult* and one normally expects that LiAlH$_4$ reduction will give a diol.

30. (a) Coppola, G.M.; Schuster, H.F. *Asymmetric Synthesis: Construction of Chiral Molecules Using Amino Acids*, Wiley, New York *1987*; (b) Smith, M.B. *Methods of Non-α-Amino Acid Synthesis*, Marcel Dekker, New York, *1995*.
31. Yu, J.S.; Kleckley, T.S.; Wiemer, D.F. *Org. Lett.* *2005*, 7, 4803.
32. Nystrom, R.F.; Brown, W.G. *J. Am. Chem. Soc.* *1948*, 70, 3738.
33. Lal, K.; Salomon, R.G. *J. Org. Chem.* *1989*, 54, 2628.
34. Corey, E.J.; Ishiguro, M. *Tetrahedron Lett.* *1979*, 2745.
35. Eklund, P.; Lindholm, A.; Mikkola, J.-P.; Smeds, A.; Lehtilä, R.; Sjöholm, R. *Org. Lett.* *2003*, 5, 491.
36. Venkataraman, H.; Cha, J.K. *J. Org. Chem.* *1989*, 54, 2505.

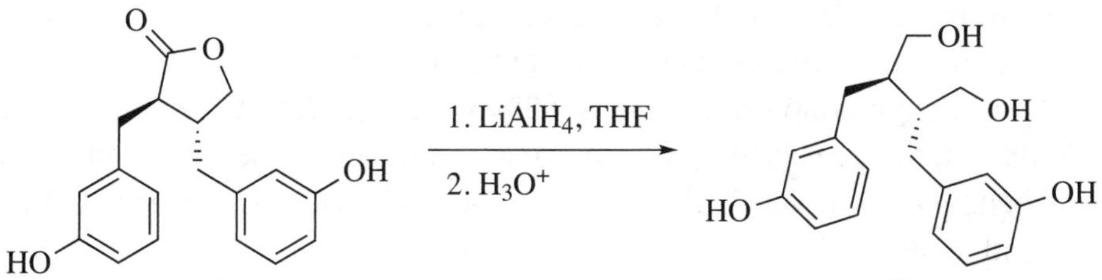

1. LiAlH₄, THF
2. H₃O⁺

24

25

1. LiAlH₄
2. H₃O⁺

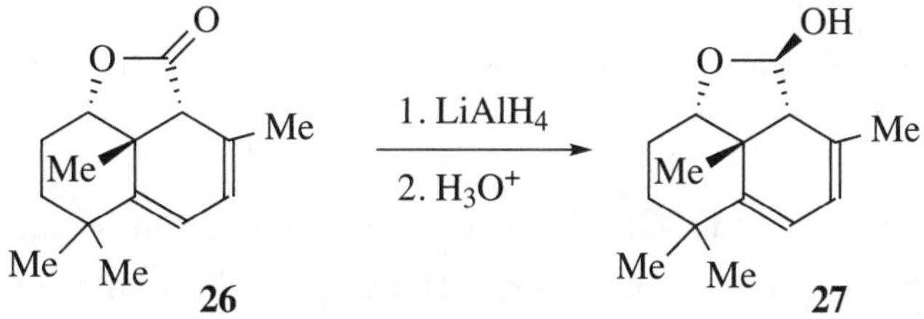

26

27

Amides do not give alcohol products upon hydride reduction, in contrast to other acid derivatives. Reaction of an amide with LiAlH₄ initially gives an intermediate iminium salt, which is further reduced to an amine as the final product. Oxygen is completely removed from the molecule, as illustrated in a synthesis of communesin B by Funk and Crawley,[37] in which amide **28** was reduced to amine **29** in 92% yield. Initial transfer of hydride to the amide forms a complex such as **30**. Further reduction and subsequent elimination of the alkoxyaluminate gives an iminium salt (**31**), and additional hydride (as in **32**) and hydrolysis gives the amine.[38] Increased steric hindrance on nitrogen makes coordination and delivery of hydride more difficult. Although uncommon, C–N bond cleavage can occur to give an aldehyde and the products are an alcohol and an amine. Aldehyde formation is observed when the electron-donating ability of the nitrogen is diminished by substituents, but yields are generally poor.

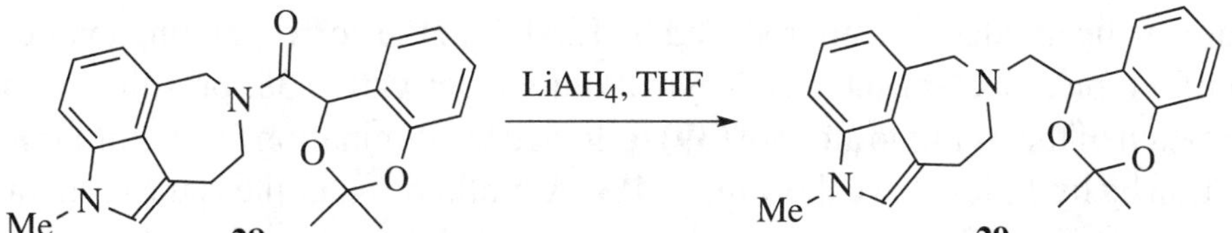

LiAH₄, THF

28

29

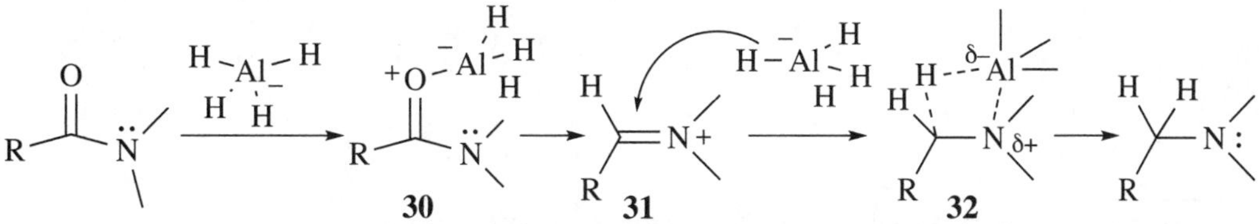

30

31

32

Reduction of derivatives is an effective method for the monomethylation or dimethylation of amines. Reaction of 2-phenylethanamine with formaldehyde and formic acid gave the reactive

37. Crawley, S.L.; Funk, R.L. *Org. Lett.* **2003**, *5*, 3169.
38. Reference 8, pp 544-546.

intermediate iminium salt **33**, which was reduced by formic acid to the *N*-methyl derivative (**34**). In this particular example, **34** was subjected to further reaction with formaldehyde and reduction led to dimethylamino product **35** in 83% yield.[39] A related reaction is the reduction of benzamide derivatives (easily prepared by reaction of an amine with benzoyl chloride or bromide) to *N*-benzyl derivatives.[40] Reduction of an imide usually proceeds with cleavage to give an amine and an alcohol.

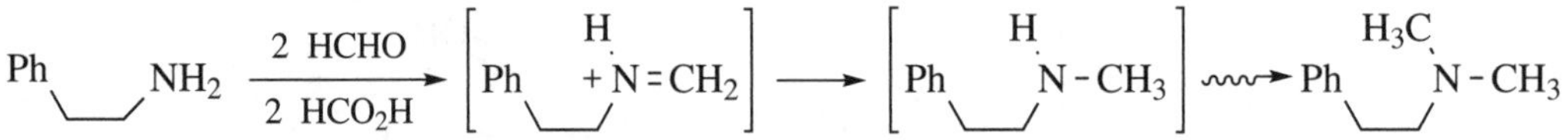

Carbamates are characterized by a O–CO–N group with an electron withdrawing ester unit on the nitrogen. Amines are easily converted to carbamates (sec. 7.3.C.iii) and reduction provides a useful method for the synthesis of methylamines.[41] Indeed, amides and carbamates are common precursors to alkylamines. In Correia and co-worker's synthesis of paroxetine, both the methyl ester and the methyl carbamate units in **36** were reduced with LiAlH$_4$ to give the hydroxymethyl group and the *N*-methylamine **37**, respectively, in 56% yield after hydrolysis.[42] It is also possible to focus on reduction of the alcohol portion of the carbamate. Reduction of a benzyl carbamate gives a benzylic alcohol, for example (this is called a Cbz protecting group; see sec. 7.3.C.iii).[43]

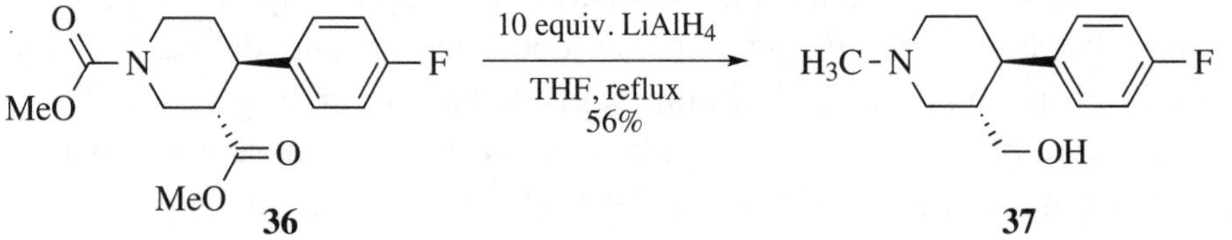

Lactams are cyclic amides, and are reduced by LiAlH$_4$ to the corresponding cyclic amine. The reduction of a typical lactam such as 2-pyrrolidinone derivative **38**, presumably proceeds via initial formation of the α-amino alcohol (**39**) followed by elimination to an iminium salt (**40**) and then reduction by hydride to give the amine (**41**). A synthetic example is the reduction of lactam **42** to crispine A (**43**) in 76% yield, in Allin and co-workers synthesis of (*R*)-(+)-crispine A.[44]

39. Icke, R.N.; Wisegarver, B.B.; Alles, G.A. *Org. Synth. Coll. Vol. 3* **1955**, 723.
40. Hunt, J.H.; McHale, D. *J. Chem. Soc.* **1957**, 2073.
41. Baxter, E.W.; Labaree, D.; Chao, S.; Mariano, P.S. *J. Org. Chem.* **1989**, *54*, 2893.
42. Pastre, J.C.; Duarte Correia, C.R. *Org. Lett.* **2006**, *8*, 1657.
43. Reference 8, p 636.
44. Allin, S.M.; Gaskell, S.N.; Towler, J.M. R.; Page, P.C.B.; Saha, B.; McKenzie, M.J.; Martin, W.P. *J. Org. Chem.* **2007**, *72*, 8972.

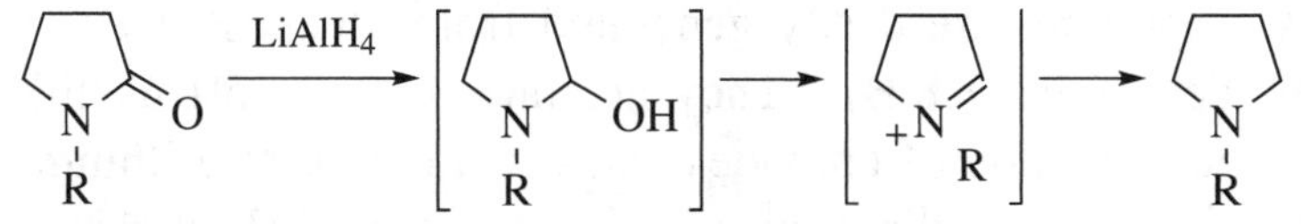

Conjugated amides can undergo both 1,2- and 1,4-addition with hydrides, but there is a competing side reaction. Reductive coupling is possible via generation of an intermediate amide enolate.[45] Reduction of the relatively hindered *N,N*-diethyl-3,3-dimethylacrylamide (**44**) gave a 47% yield of the expected conjugate reduction product **45**, but also 12% of the enolate condensation product **46**.[45] Enolate condensations will be discussed in Chapter 9.

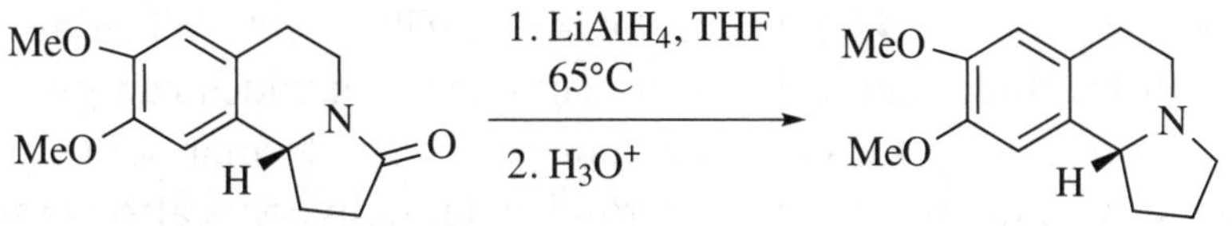

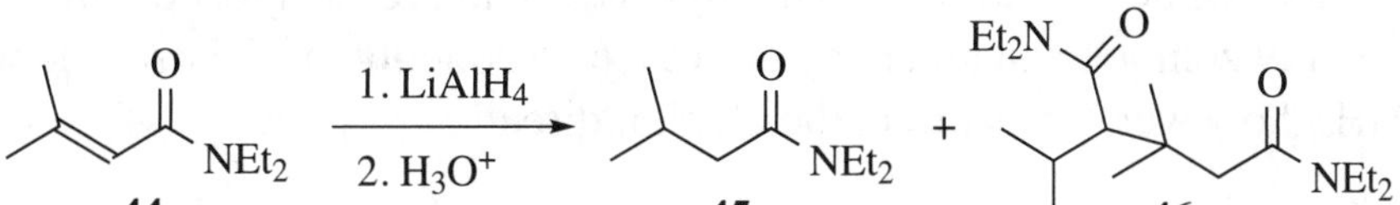

With the exception of amides, it is clear that the principal reduction product of carbonyl derivatives is an alcohol. Succeeding sections will discuss the reduction products observed for functional groups that do not contain a carbonyl group. The functional group transforms for this section include:

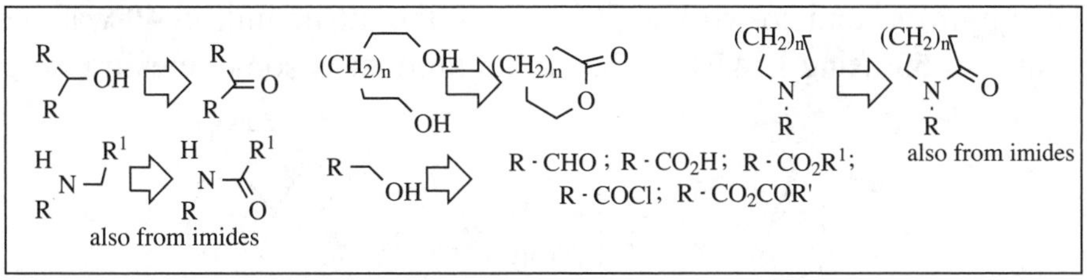

4.2.C. Lithium Aluminum Hydride Reduction of Non-Carbonyl Heteroatom Functional Groups

As shown by Gaylord,[8] lithium aluminum hydride reduces heteroatom-containing functional groups. This section will explore the reduction of several functional groups that are commonly seen in synthesis.

---

45. Snyder, H.R.; Putnam, R.E. *J. Am. Chem. Soc.* **1954**, *76*, 1893.

4.2.C.i. Epoxides. Epoxides are easily generated from alkenes[46] (sec. 3.4), and also from ketones and aldehydes[47] (sec. 8.8.B). They are one of the most useful of all oxygenated functional groups and reduction of epoxides leads to alcohols.[48] Lithium aluminum hydride usually delivers hydride with excellent selectivity to the less substituted carbon of an epoxide.[49] An example is taken from the Yamamoto and co-worker's synthesis of α-bisabolol,[50] where reduction of the epoxide unit in **47** gave alcohol **48** in 93% yield. When the epoxide oxygen can coordinate with the aluminum (via an *ate* complex), epoxides can give products where ydride is delivered to the more substituted carbon.[51] Intramolecular delivery of hydride is to the most electropositive carbon, although intermolecular delivery is also possible.[52]

4.2.C.ii. Ozonides. Ozonolysis is a powerful method for cleaving alkenes to carbonyl products (sec. 3.7.B). Ozonolysis of alkenes initially generates an ozonide that can be reduced by a variety of reagents. Dimethyl sulfide or zinc in acetic acid are the most common reagents for the reduction of an ozonide to an aldehyde or ketone. Reduction of the ozonide with the more powerful LiAlH$_4$, however, gives direct the alcohol directly.

4.2.C.iii. Nitriles. Nitriles are synthetic precursors for both functional group transformation and carbon-carbon bond forming reactions (secs. 8.2, 9.4.B.iv). They are usually formed by a S$_N$2 substitution reaction in which cyanide displaces a leaving group from an alkyl halide or sulfonate ester, or by dehydration of amides (secs. 2.6.A.i, sec. 8.2). Once the nitrile has been formed reduction of the cyano group with LiAlH$_4$ usually gives a primary amine,[53] so nitriles function as protected (latent) aminomethyl groups. This transformation is seen in a synthesis of (–)-tubifoline by Mori and co-workers,[54] in which the nitrile unit in **49** was reduced to an aminomethyl unit in **50** using LiAlH$_4$. The amine product is sometimes trapped (protected,

46. Lewis, S.N. in *Oxidation*, Vol. 1, Augustine, R.L. (Ed.), Marcel Dekker, New York *1969*, p223.
47. Trost, B.M.; Melvin, Jr., L.S. *Sulfur Ylides*, Academic Press, New York, *1975*.
48. For reagents used to reduce epoxides and other cyclic ethers, see Reference 17, p 1019.
49. Reference 8, p 653.
50. Makita, N.; Hoshino, Y.; Yamamoto, H. *Angew. Chem. Int. Ed. 2003, 42*, 941.
51. Azuma, H.; Tamagaki, S.; Ogino, K. *J. Org. Chem. 2000, 65*, 3538.
52. Reference 16, p 104.
53. For reagents used to reduce nitriles, see Reference 17, p 1271.
54. Mori, M.; Nakanishi, M.; Kajishima, D.; Sato, Y. *J. Am. Chem. Soc, 2003, 125*, 9801.

sec. 7.3.C) by reaction with an acyl halide to give an amide, since this allows further synthetic manipulation. Nitrile reduction by hydride proceeds via initial formation of an iminium salt (**51**) and subsequent conversion to a bis(iminoaluminate), **52**.[55] The literature is rather vague concerning the specific structure of **52**, but reduction with additional hydride leads to **53** as reported by Soffer and Katz.[55] The hydrogen atoms in the amine product probably come from hydrolysis, although the nature and substitution pattern of the nitrogen in **53** was not specified. The aluminum moiety in **53** may be coordinated by a solvent molecule (usually ether or THF) or a second aluminum species. In some cases, removal of the α-proton occurs to give the α-lithio nitrile (sec. 9.4.B.iv),[55] especially when a hydride slurry is added to the nitrile solution (inverse addition). A dimeric product is formed that is indicative of an enolate condensation. Addition of butanenitrile to lithium aluminum hydride gave butanamine in 79%, for example, but inverse addition in ether gave only 34% of that amine. Inverse addition also led to 26% of 2-ethyl-1,3-hexanediamine, presumably via a **Thorpe condensation**[56] (sec. 9.4.B.iv) of an intermediate nitrile carbanion.[55]

4.2.C.iv. Azides. Azides are important precursors to amines.[57] The azide anion (N_3^-) is an excellent nucleophile and S_N2 displacement of primary and secondary halides (or alkyl sulfonates) with sodium azide (sec. 2.6.A.i) gives the corresponding alkyl azide. Reduction of the azido group with $LiAlH_4$ gives a primary amino group ($R-N_3 \rightarrow R-NH_2$). An example is taken from the Enders and Müller-Hüwen synthesis of D-*erythro*-sphinganine,[58] in which reduction of the azide group in **54** gave amine **55** in 99% yield (>96% ee).

4.2.C.v. Nitro Compounds. Aliphatic nitro compounds can be reduced to amines,[59] so they also function as amine surrogates. Reduction of nitriles or azides is usually the preferred method to introduce an amino group since reactions that introduce the nitro group often require strong acid. Aromatic nitration (sec. 2.11.B) is relatively easy, however, and reduction to aniline derivatives would be a very attractive synthetic route to this important class of amines.

55. Soffer, L.M.; Katz, M. *J. Am. Chem. Soc.* **1956**, *78*, 1705.
56. (a) Baron H.; Remfry, F.G.P.; Thorpe, Y.F. *J. Chem. Soc.* **1904**, *85*, 1726; (b) Schaefer, J.P.; Bloomfield, J.J. *Org. React.* **1967**, *15*, 1.
57. See Reference 17, p 815.
58. Enders, D.; Müller-Hüwen, A. *Eur. J. Org. Chem.* **2004**, 1732.
59. See Reference 17, p 821.

Hydride reduction is not straightforward because LiAlH$_4$ reduction of nitrobenzene does not give aniline but rather azo compound **56** in 84% yield.[60] Catalytic hydrogenation (see sec. 4.8.D) is the best method for the reduction of aromatic nitro compounds to aniline derivatives. Aliphatic nitro compounds are relatively easy to reduce with LiAlH$_4$. The nitro group in **57** was cleanly reduced to give amine **58** as part of Scanlon and co-worker's synthesis of thyronamines.[61] Note that in this case **57** is a conjugated nitro compound, and both the C=C and the NO$_2$ units were reduced.

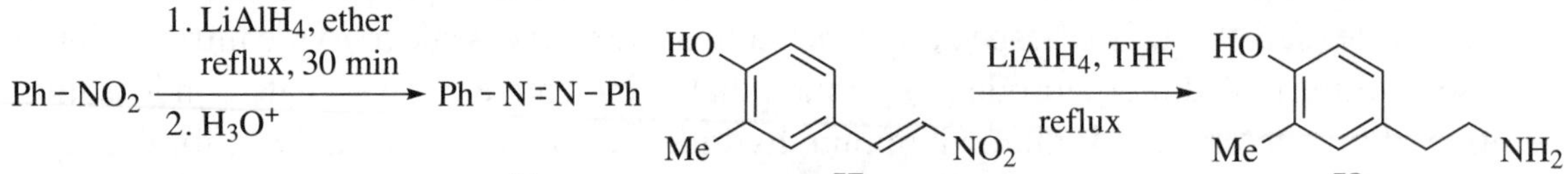

4.2.C.vi. Alkyl Halides. Alkyl halides are reduced by lithium aluminum hydride to give a hydrocarbon in what is known as a **hydrogenolysis** reaction (R$_3$C–X → R$_3$C–H; sec. 4.8.E).[62] Reduction of 1-bromobutane in ether, THF or di-*n*-butyl ether gave yields of *n*-butane in the 30-96% range.[63,32] A synthetic example is taken from the synthesis of (+)-clavosolide A by Lee and co-workers, in which reduction of the primary bromide moiety in **59** gave the corresponding methyl group in 96% yield (see **60**).[64] Benzylic halides[65] are relatively easy to reduce with LiAlH$_4$, when compared to simple aliphatic halides. Electronic effects can influence the yield. The presence of aryl groups leads to more facile reduction, as in the reduction of trityl chloride [Ph$_3$C–Cl] to triphenylmethane [Ph$_3$C–H] in 98% yield. Simple aryl halides are resistant to reduction with LiAlH$_4$,[66] but some substituted halides such as 2-chloro-1-iodobenzene can be reduced (40% yield of chlorobenzene was obtained upon reaction with LiAlH$_4$). Electron-releasing groups such as methyl or methoxy enhance the yield of reduction products.[67] Hydride reduction of allylic halides is sometimes plagued by reductive elimination as a competitive side reaction.

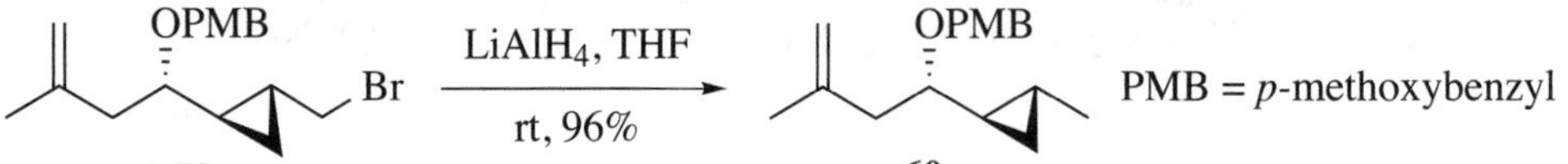

60. Hart, M.E.; Suchland, K.L.; Miyakawa, M.; Bunzow, J.R.; Grandy, D.K.; Scanlan, T.S. *J. Med. Chem.* **2006**, *49*, 1101.
61. Ohta, H.; Kobayashi, N.; Ozaki, K. *J. Org. Chem.* **1989**, *54*, 1802.
62. See Reference 17, p 779.
63. Rassat, A.; Ravet, J.P. *Bull. Soc. Chim. Fr.* **1968**, 3679.
64. Son, J.B.; Kim, S. N.; Kim, N.Y.; Lee, D.H. *Org. Lett.* **2006**, *8*, 661.
65. Trevoy, L.W.; Brown, W.G. *J. Am. Chem. Soc.* **1949**, *71*, 1675.
66. Reference 8, pp 917-922.
67. Feutrill, G.I.; Mirrington, R.N.; Nichols, R.J. *Aust. J. Chem.* **1973**, *26*, 345.

4.2.C.vii. Sulfonate Esters. Sulfonate esters are readily formed by reaction of alcohols with sulfonyl halides and are reduced to the hydrocarbon with $LiAlH_4$, via cleavage of the C–O bond.[62] Reaction of the primary alcohol group in **61** with tosyl chloride generated the tosylate, and subsequent reduction with $LiAlH_4$ led to hydrogenolysis of the C–O bond to give **62** in 66% yield, as part of Harrity and co-worker's synthesis of (–)-nupharolutine.[68] Reduction of sulfonate esters commonly gives yields below 50%, and alkyl sulfonate groups are similarly reduced with $LiAlH_4$.[69] Indeed, cleavage of the C–O bond in primary tosylates ($-CH_2OTs$) or mesylates ($-CH_2OMs$) to a methyl group may be the most common application.[70,71] Formation of an alcohol (via S–O bond cleavage) is often observed when the C–O bond is sterically hindered, accounting for the low yields mentioned above.

1. TsCl, NEt₃ cat DMAP, CH₂Cl₂
2. LiAlH₄, THF, reflux

4.2.C.viii. Sulfonamides. Reduction of sulfonamides does *not* give an alcohol and an amine. Primary sulfonamides resist reductive cleavage with $LiAlH_4$; secondary sulfonamides are cleaved, but the reaction requires a temperature of 120°C in dibutyl ether.[72] Exceptions to this sluggish reactivity occur when the sulfonamide is conjugated to a carbonyl moiety. Oppolzer et al. developed a chiral sulfonamide auxiliary (**Oppolzer's sultam 63**) derived from camphor,[73] and attachment of this auxiliary allows high asymmetric induction in a variety of reactions. In a synthesis of fumagillin by Perlmutter and co-workers, **63** was acylated and converted to **64** in five steps.[74] Reductive cleavage of the sulfonamide with lithium aluminum hydride removed the auxiliary in low yield to give alcohol **65** along with the auxiliary **63**, which could be recycled. Note that complete removal of this group is sometimes a problem, which can severely limit the utility of this auxiliary.

68. Goodenough, K.M.; Moran, W.J.; Raubo, P.; Harrity, J.P.A. *J. Org. Chem.* **2005**, *70*, 207.
69. For an example taken from a synthesis of chrysotricine, see Zhang, J.-X.; Wang, G.-X.; Xie, P.; Chen, S.-F.; Liang, X.-T. *Tetrahedron Lett.* **2000**, *41*, 2211.
70. Rossi, R.; Salvadori, P.A. *Synthesis* **1979**, 209.
71. Fronza, G.; Fuganti, C.; Grasselli, P.; Marinoni, G. *Tetrahedron Lett.* **1979**, 3883.
72. (a) Searles, S.; Nukina, S. *Chem. Rev.* **1959**, *59*, 1077; (b) Klamann, D. *Monatsh Chem.* **1953**, *84*, 651.
73. Oppolzer, W.; Chapuis, C.; Bernardinelli, G. *Helv. Chim. Acta* **1984**, *67*, 1397.
74. Ciampini, M.; Perlmutter, P.; Watson, K. *Tetrahedron: Asymmetry* **2007**, *18*, 243.

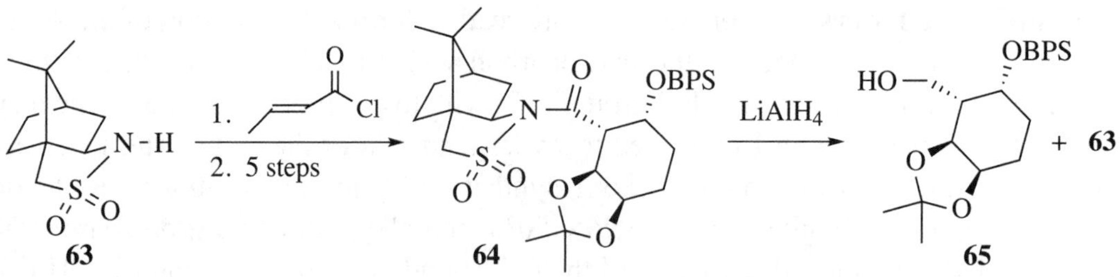

4.2.C.ix. Alkyne-Alcohols. Formation of an alcohol in the presence of another functional group via reduction can sometimes lead to elimination to an alkene. This elimination process can be turned to an advantage when propargylic alcohols are reduced.[75] The so-called **Whiting reaction**[76] reduces an alkynyl diol such as **66** with LiAlH$_4$ to give a diene (**67**, cosmene) via addition of four hydrogen atoms followed by formal elimination of two equivalents of water. In general, LiAlH$_4$ does not reduce alkene or alkynyl moieties, but as noted in the Whiting reaction, propargylic alcohol derivatives (C≡C–CH$_2$–OH) are important exceptions.[77,78] Reduction of alkynyl alcohols, readily formed via condensation of alkyne anions with aldehydes and ketones (sec. 8.3.C), generally gives the trans-allylic alcohol as seen in a synthesis of leucascandrolide A by Carreira and Fettes,[79] in which propargyl alcohol **68** was reduced to **69** and converted to the benzoate ester in 90% overall yield.

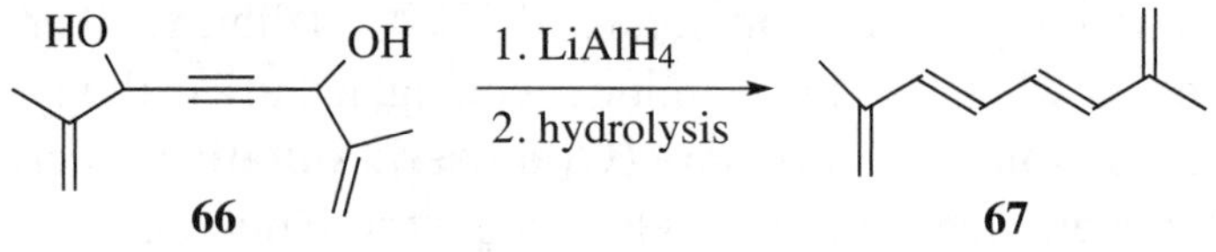

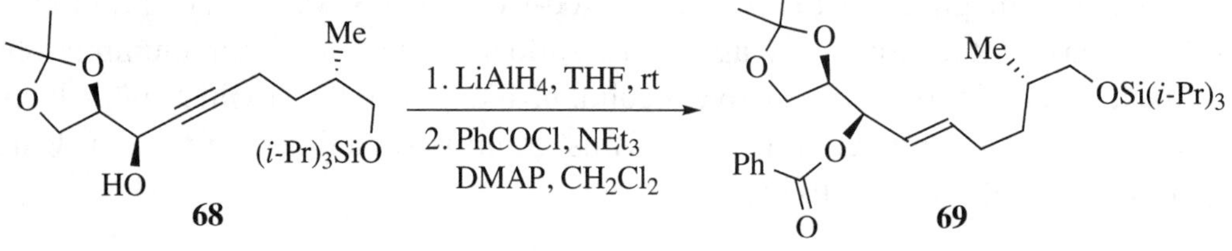

This specialized reduction is not limited to C≡CCHOH systems. Lithium aluminum hydride in refluxing diglyme also reduces homoallylic alkynyl alcohols (C≡CCH$_2$CH$_2$OH). An example is the reduction of 3-decyn-1-ol (**70**) to dec-3*E*-en-1-ol (**71**) in 93% yield, as reported in Vyvyan

75. Karlsen, S.; Frøyen, P.; Skattebøl, L. *Acta Chem. Scand.* **1976**, *30B*, 664; (b) Brown, H.C.; McFarlin, R.F. *J. Am. Chem. Soc.* **1956**, *78*, 252.
76. (a) Nayler, P.; Whiting, M.C. *J. Chem. Soc.* **1954**, 4006; (b) Isler, O.; Montavon, M.; Rüegg, R.; Zeller, P. *Helv. Chim. Acta* **1956**, *39*, 454; (c) The Merck Index, 14th ed., Merck & Co., Inc., Whitehouse Station, New Jersey, **2006**, p ONR-100.
77. Reference 8, pp 968-975.
78. Trost, B.M.; Lee, D.C. *J. Org. Chem.* **1989**, *54*, 2271.
79. Fettes, A.; Carreira, E.M. *J. Org. Chem.* **2003**, *68*, 9274.

and co-worker's synthesis of gibbilimbol A.[80]

The functional group transforms in this section are summarized as

4.3. ALKOXYALUMINATE REAGENTS

Lithium aluminum hydride is a powerful reducing agent that can reduce many functional groups, and it is common for the reduction to span several oxidation stages ($-CO_2Et \rightarrow -CHO \rightarrow -CH_2OH$, for example). There are many instances when a single oxidation stage reduction is desirable (such as $-CO_2Et \rightarrow -CHO$), without overreduction. Selective reduction of one functional group in the presence of others is also important, because multifunctional molecules may contain several reducible functional groups (i.e., lactone + ester + aldehyde). In such a system, lithium aluminum hydride could reduce all of them. An example of this poor selectivity is the reduction of **72** to **73**, in which the ketone, cyano and ester groups were reduced at $-78°C$ (95% yield) in Heathcock and co-worker's synthesis of fawcettimine.[81] A chemoselective (a term introduced by Trost) reduction will reduce a single functional group in high yield, in the presence of the other functional groups. This section will examine reagents that exhibit chemoselectivity for selected functional groups.

Is it possible for a reagent to react faster with one functional group than with another? The answer *may* be yes in some cases, but in **72**, reaction with LiAlH$_4$ rapidly reduced all three functional groups so any rate differences were probably imperceptible. If the hydride reagent were structurally modified so that its reducing power was diminished, however, small

80. Vyvyan, J.R.; Holst, C.L.; Johnson, A.J.; Schwenk, C.M. *J. Org. Chem.* **2002**, *67*, 2263.
81. Heathcock, C.H.; Blumenkopf, J.A.; Smith, K.A. *J. Org. Chem.* **1989**, *54*, 1548.

differences in rates of reduction for various functional groups might be exploited, allowing selective reduction of one functional group in the presence of another. It was shown previously that reaction of lithium aluminum hydride with carbonyls led to alkoxyaluminates, and that each successive reaction with the carbonyl was slower. This is consistent with the observation that the alkoxyaluminates are increasingly less reactive as more OR units replace H in the aluminate. As long as an Al–H unit remains in the molecule, that alkxoyaluminate is a reducing agent. Alkoxyaluminates such as $LiAlH_n(OR)_{4-n}$ can be prepared by the direct reaction of lithium aluminum hydride with an appropriate alcohol. Addition of n equivalents of ROH leads to the alkoxyaluminate with $4-n$ H remaining on aluminum. Brown showed that many alcohols could be used to prepare alkoxyaluminum hydrides. Four equivalents of methanol, for example, reacted with $LiAlH_4$ to form lithium tetramethoxyaluminum [$LiAl(OMe)_4$], but three equivalents (in ether, THF, or glyme) gave $LiAlH(OMe)_3$ [lithium trimethoxyaluminum hydride].[82] Similarly, three equivalents of ethanol gave lithium triethoxyaluminum hydride, $LiAlH(OEt)_3$. Brown first showed that treatment of $LiAlH_4$ with four equivalents of the hindered *tert*-butanol led to the release of only three equivalents of hydrogen gas (i.e., only three equivalents of hydride reacted), generating lithium tri-*tert*-butoxy aluminum hydride, $LiAlH(Ot-Bu)_3$.[83] This hydride failed to react with the excess *tert*-butyl alcohol at ambient temperatures (although it did react upon prolonged heating at elevated temperatures).[83] This reagent is significantly less reactive than $LiAlH_4$ or the other alkoxyaluminum hydrides. Brown went further, and compared the reducing power of lithium aluminum hydride, lithium tri-*tert*-butoxyaluminum hydride and lithium trimethoxyaluminum hydride with a variety of functional groups.[84] Lithium trimethoxyaluminum hydride is about as strong a reducing agent as lithium aluminum hydride.[85] Although the solid reagent is stable and sold commercially, ether solutions are not very stable. It is often preferable to prepare the reagent as needed, by addition of methanol to a THF slurry of lithium aluminum hydride.[85] Reduction of epoxides, nitriles, and tosylates was slow and nitriles gave mixtures of amines and aldehydes in poor yield.[86] Lithium triethoxyaluminum hydride is also unstable in solution and is usually prepared *in situ* by addition of three equivalents of ethanol to $LiAlH_4$ in ether at 0°C.[85] This powerful reducing agent showed good selectivity for reduction of nitriles and amides. Both aliphatic nitriles (such as isobutyronitrile) as well as aromatic nitriles are reduced to aldehydes in good yield.[86] Tertiary amides such as *N,N*-dimethylbutanamide, shown in the reaction, are also reduced to the aldehyde (in this case butanal).[87] As the steric bulk of the alkyl groups attached to the nitrogen of a tertiary amide increased, the yield of aldehyde was diminished. Reduction of the *N,N*-diethyl- analog of this amide, for example, gave only 47% of the aldehyde and reduction

82. Brown, H.C.; Shoaf, C.J. *J. Am. Chem. Soc.* **1964**, *86*, 1079.
83. Brown, H.C.; McFarlin, R.F. *J. Am. Chem. Soc.* **1958**, *80*, 5372
84. Brown, H.C.; Yoon, N.M. *J. Am. Chem. Soc.* **1966**, *88*, 1464.
85. Brown, H.C.; Weissman, P.M. *J. Am. Chem. Soc.* **1965**, *87*, 5614.
86. Brown, H.C.; Garg, C.P. *J. Am. Chem. Soc.* **1964**, *86*, 1085.
87. Brown, H.C.; Tsukamoto, A. *J. Am. Chem. Soc.* **1964**, *86*, 1089.

of the *N,N*-diisopropyl- analog gave no aldehyde.[87] Trimethoxyaluminum hydride and $LiAlH_4$ gave significantly lower yields of aldehydes with both nitriles and tertiary amides.

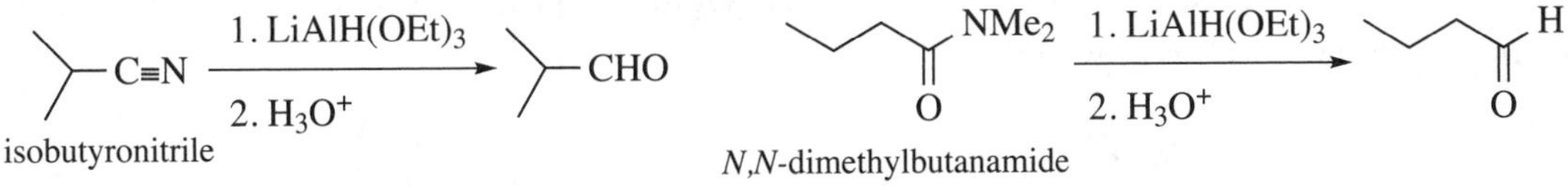

Alkoxyaluminates are relatively mild reducing agents tht can reduce the carbonyl of a lactone to give a lactol. In one example, **75** (an intermediate in Magnus and Hobbs' synthesis of grandisol)[88] was isolated in 97% yield by reduction of **74** with lithium triethoxyaluminum hydride at −20°C. Trialkoxyaluminum hydrides reduce aldehydes, ketones, esters and acid chlorides before reducing nitriles and amides.

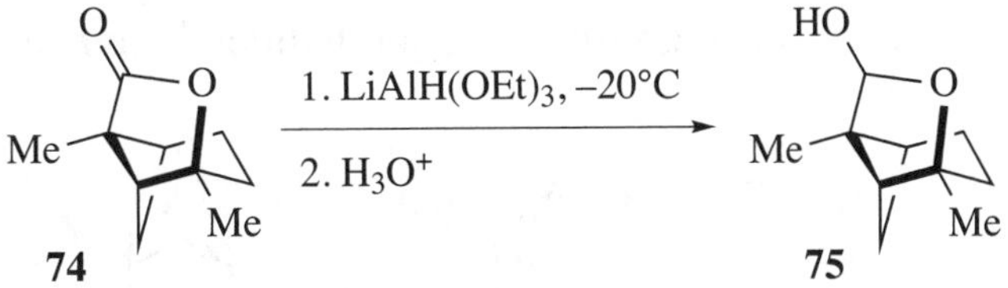

Both lithium diethoxyaluminum hydride $[LiAlH_2(OEt)_2]$[89] and $[LiAlH_3(OEt)]$,[90] lithium ethoxyaluminum hydride have been used in syntheses, but $LiAlH(Ot\text{-Bu})_3$ is used more often. This latter reagent is stable in solution, gives useful concentrations in ether solvents (36 g/100 g of THF, 2 g/100 g of diethyl ether, and 41 g/100 g of diglyme),[83] and reacts with organic substrates faster in solutions where it is in higher concentration. It is thermally stable and heating to 165°C for five hours followed by titration with acid showed that 92% of its active hydrogen was retained.[83] Although lithium tri-*tert*-butoxyaluminum hydride is less reactive than the other alkoxyaluminum hydrides discussed above, it easily reduces aldehydes, ketones, some esters, anhydrides or acid chlorides. Epoxides are reduced very slowly, and nitriles and dimethyl amides are not reduced at all. The use of this reagent is a preferred method for the selective reduction of acid chlorides to aldehydes. The best yields of the aldehyde is realized from aromatic acid chlorides, as in the reduction of *p*-nitrobenzoyl chloride to *p*-nitrobenzaldehyde in 80% yield.[83] The yield of aldehyde from aliphatic acyl chlorides is generally in the 40-60% range. Alkyl esters do not react, allowing some selectivity for the reduction of ketones and aldehydes in the presence of esters and other functional groups. Deprés, Greene, and co-workers reduced the ketone moiety in **76** to alcohol **77** in 83% yield with excellent diastereoselectivity, and without reduction of the lactone unit, in a synthesis

88. (a) Hobbs, P.D.; Magnus, P.D. *J. Chem. Soc. Chem. Commun.* *1974*, 856; (b) Hobbs, P.D.; Magnus, P.D. *J. Am. Chem. Soc.* *1976*, 98, 4594.
89. Schwarz, M.; Oliver, J.E.; Sonnet, P.E. *J. Org. Chem.* *1975*, 40, 2410.
90. (a) Morizur, J.P.; Bidan, G.; Kossanyi, J. *Tetrahedron Lett.* *1975*, 4167; (b) Bidan, G.; Kossanyi, J.; Meyer, V.; Morizur, J.-P. *Tetrahedron* *1977*, 33, 2193.

of homogynolide B.[91] Although alkyl esters are generally unreactive with this reagent, many phenyl esters are reduced to the aldehyde, as in the conversion of methyl *p*-chlorobenzoate to *p*-chlorobenzaldehyde (−22°C, 8 h) in 77% yield.[92] The presence of the *p*-chloro unit undoubtedly enhanced the reactivity.

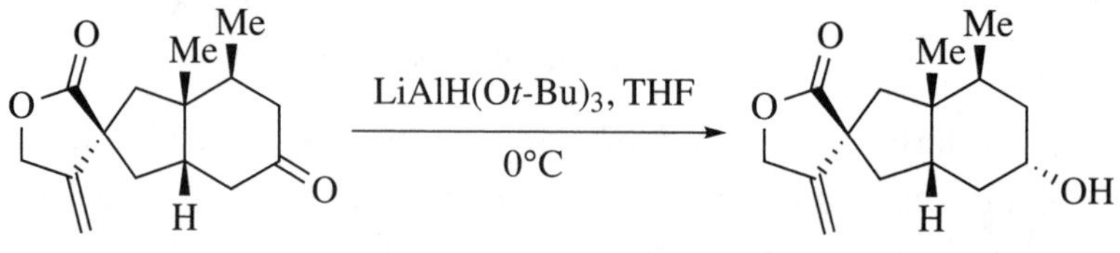

Reaction of conjugated carbonyls with LiAlH(O*t*-Bu)$_3$ gives primarily 1,2-reduction, as in the quantitative reduction of **78** to **79** in Marshall and Ruth's synthesis of globulol.[93] Both sulfonate esters such as the mesylate group in **78**, and halides[94] are resistant to reduction with LiAlH(O*t* Bu)$_3$.

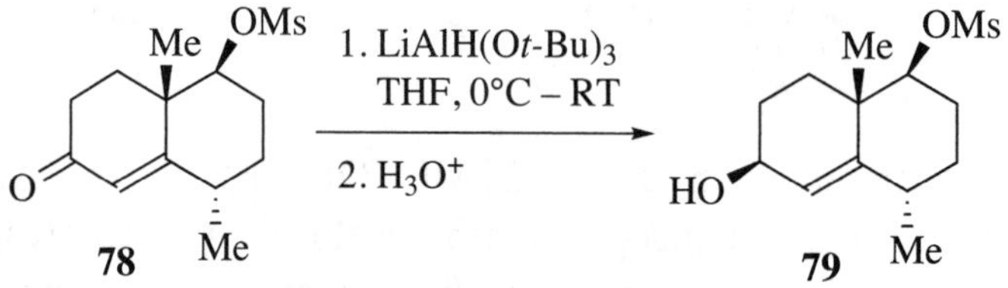

Sodium *bis*(2-methoxyethoxy)aluminum hydride (Red-Al®, Vitride®) was prepared by Vit in 1967.[95c] It was prepared in benzene, as shown, under a hydrogen atmosphere at temperatures > 100°C,[95] and its reducing power is close to that of lithium aluminum hydride. An important advantage of this reagent is its stability to dry air (it does not ignite in even moist air or oxygen), and it is thermally stable up to 200°C unlike LiAlH$_4$, which can detonate at elevated temperatures. Perhaps the greatest utility of Red-Al is its solubility in aromatic hydrocarbon and ether solvents, which allows it to be conveniently used for applications that require inverse addition of hydrides.

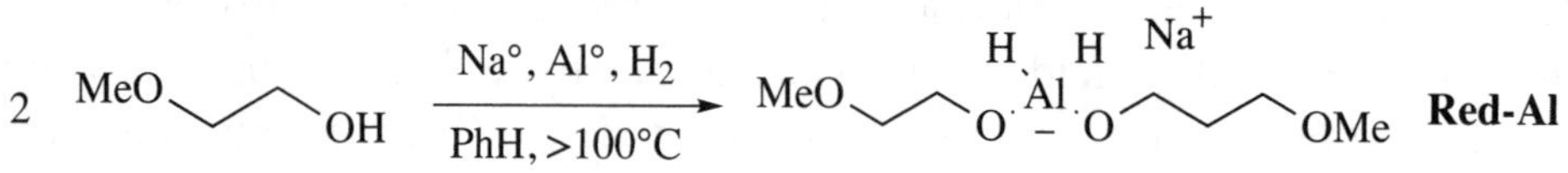

91. Brocksom, T.J.; Coelho, F.; Deprés, J.-P.; Greene, A.E.; Freire de Lima, M.E.; Hamelin, O.; Hartmann, B.; Kanazawa, A.M.; Wang, Y. *J. Am. Chem. Soc.* **2002**, *124*, 15313.
92. Weissman, P.M. Brown, H.C. *J. Org. Chem.* **1966**, *31*, 283.
93. Marshall, J.A.; Ruth, J.A. *J. Org. Chem.* **1974**, *39*, 1971.
94. For a synthetic example, see Daniewski, A.R.; Kiegel, J. *J. Org. Chem.* **1988**, *53*, 5534.
95. (a) Fieser, L.F.; Fieser, M. *Reagents for Organic Synthesis* Vol. 2, Wiley, New York, **1969**, p382; (b) Fieser, L.F.; Fieser, M. *Reagents for Organic Synthesis* Vol. 3, Wiley, New York, **1972**, p260; (c) Vit, J.; Cásensky, B.; Machácek, J. *French Patent*, 1,515,582 **1968** [*Chem. Abstr. 70*: 115009x, *1967*].

The reactivity profile of Red-Al is essentially the same as LiAlH$_4$, reducing aldehydes, ketones[96] and acid derivatives to alcohols. Ketones and aldehydes are reduced with Red-Al at $-78°C$ to the alcohol.[97] The analogy to LiAlH$_4$ reduction also applies to the reduction of carboxylic acids.[98] Reduction of conjugated carbonyls gives primarily 1,2-reduction to an allylic alcohol,[99] as in the reduction of **80** to **81** (80% yield) in the synthesis and structure elucidation of azaspiracid-1 by Nicolaou and co-workers.[100] Red-Al reduces esters to alcohols in a manner identical with LiAlH$_4$.[101] The selectivity for 1,2-reduction can be changed to selectivity for 1,4-reduction by the addition of cuprous bromide (CuBr) to Red-Al, as in the conversion of **82** → **83** in 72% yield.[102] The active reducing agent is probably copper hydride. It is well known that addition of cuprous salts (Cu$^+$) promotes conjugate addition (see organocuprates in sec. 8.7.A.vi).[103] Reaction of cuprous bromide (CuBr) and Red-Al gave a sodium complex whereas reaction of CuBr and LiAlH(OMe)$_3$ gave a lithium complex.[104,105]

Other functional groups can be reduced, such as epoxides to alcohols,[106] and lactams to amines.[107] Epoxides are reduced at the less hindered carbon,[108] but the product distribution can vary with

96. Čapka, M.; Chvalovský, V.; Kochloefl, K.; Kraus, M. *Collect. Czech. Chem. Commun.* **1969**, *34*, 118.
97. Stotter, P.L.; Friedman, M.D.; Minter, D.E. *J. Org. Chem.* **1985**, *50*, 29.
98. Zurflüh, R.; Dunham. L.L.; Spain, V.L.; Siddall, J.B. *J. Am. Chem. Soc.* **1970**, *92*, 425.
99. (a) Markezich, R.L.; Willy, W.E.; McCarry, B.E.; Johnson, W.S. *J. Am. Chem. Soc.* **1973**, *95*, 4414; (b) McCarry, B.E.; Markezich, R.L.; Johnson, W.S. *Ibid.* **1973**, *95*, 4416.
100. Nicolaou, K.C.; Pihko, P.M.; Bernal, F.; Frederick, M.O.; Qian, W.; Uesaka, N.; Diedrichs, N.; Hinrichs, J.; Koftis, T.V.; Loizidou, E.; Petrovic, G.; Rodriquez, M.; Sarlah, D.; Zou, N. *J. Am. Chem. Soc.* **2006**, *128*, 2244.
101. Avery, M.A.; Jennings-White, C.; Chong, W.K.M. *J. Org. Chem.* **1989**, *54*, 1789.
102. Bestmann, H.J.; Schmidt, M. *Tetrahedron Lett.* **1986**, *27*, 1999.
103. Paquette, L.A.; Schaefer, A.G.; Springer, J.P. *Tetrahedron* **1987**, *43*, 5567.
104. Semmelhack, M.F.; Stauffer, R.D.; Yamashita, A. *J. Org. Chem.* **1977**, *42*, 3180.
105. (a) Vedejs, E.; Fedde, C.L.; Schwartz, C.E. *J. Org. Chem.* **1987**, *52*, 4269; (b) Negishi, E.; Akiyoshi, K. *Chem. Lett.* **1987**, 1007.
106. For an example taken from a synthesis of (−)-8-*O*-methyltetrangomycin (MM 47755), see Kesenheimer, C.; Groth, U. *Org. Lett.* **2006**, *8*, 2507.
107. For an example taken from a synthesis of (−)-ibogamine, see White, J.D.; Choi, Y. *Org. Lett.* **2000**, *2*, 2373.
108. (a) Gao, Y.; Sharpless, K.B. *J. Org. Chem.* **1988**, *53*, 4081. For an example taken from a synthesis of δ-*trans*-tocotrienoloic acid, see (b) Maloney, D.J.; Hecht, S.M. *Org. Lett.* **2005**, *7*, 4297.

the solvent, especially when the steric hindrance is similar at each carbon of the epoxide. As with LiAlH₄, Red-Al reduces aromatic nitriles to amines, although aliphatic nitriles typical are difficult to reduce and give poor yields.[109] There are some interesting differences in selectivity, relative to LiAlH₄. Cyanohydrin **84**, for example, was reduced to α-hydroxy-aldehyde **85**,[110] rather than to the amino-methyl compound obtained with lithium aluminum hydride.

Hydrogenolysis of C–X bonds is usually faster than with LiAlH₄, and Red-Al is preferred to LiAlH₄ for reduction of aliphatic halides and many aromatic halides.[111] Primary aliphatic monohalides are reduced to the hydrocarbon (1-bromoheptane was reduced to heptane, for example, in 99% yield).[111] Aromatic halides are also reduced, as in the conversion of bromobenzene to benzene in 99% yield.[111] Under similar conditions, however, chlorobenzene gave only 15% of benzene. Selective mono-reduction is possible when more than one halide is present. Reduction of the *gem*-dibromide moiety in **86**, for example, gave the syn-monobromide (**87**) as the major product.[112]

Sulfonate esters are reduced via C–O cleavage, as with LiAlH₄. Primary tosylates and mesylates are reduced to the methyl derivative.[113] Sterically hindered sulfonate esters often give significant amounts of S–O cleavage during the reduction. Although LiAlH₄ does not readily reduce sulfonamides to the corresponding amine (sec. 4.2.C.viii, *vide supra*),[114] cleavage of the N–S bond with Red-Al is facile, as in the conversion of *N*-tosyl deoxyephedrine (**88**) to deoxyephedrine (**89**) in 64% yield.[114] It was assumed that the sulfonate fragment lost from nitrogen in sulfonamide **88** was reduced to the thiol.

Red-Al can reduce benzylic alcohols to the hydrocarbon at elevated temperatures. The reaction occurs more readily when the aromatic ring contains an electron releasing group in the ortho- or para-position, but the temperatures required can be > 140°C.[115] Even at higher reaction temperatures, reduction to the alcohol can compete with formation of the hydrocarbon, but usually as a minor process. An amino group at a benzylic position is subject to hydrogenolysis. Reduction of 4-(*N,N*-dimethyl-amino)benzaldehyde with Red-Al at 141°C gave a mixture of

109. Černý, M.; Málek, J.; Čapka, M.; Chvalovský, V. *Collect. Czech. Chem. Commun.* **1969**, *34*, 1033.
110. Schlosser, M.; Brich, Z. *Helv. Chim. Acta* **1978**, *61*, 1903.
111. Čapka, M.; Chvalovský, V. *Collect. Czech. Chem. Commun.* **1969**, *34*, 3110.
112. Sydnes, L.; Skattebøl, L. *Tetrahedron Lett.* **1974**, 3703.
113. Zobačová, A.; Hermánková, V.; Jarý J. *Collect. Czech. Chem. Commun.* **1977**, *42*, 2540.
114. Gold, E.H.; Babad, E. *J. Org. Chem.* **1972**, *37*, 2208.
115. Černý, M.; Málek, J. *Tetrahedron Lett.* **1969**, 1739.

4-hydroxymethyl-*N*,*N*-dimethylaniline (63% yield) and 4-methyl-*N*,*N*-dimethylaniline (8%).[116] Similar reduction of 4-aminoacetophenone gave only toluidine, however, in 90% yield.[116] As with LiAlH4, propargylic alcohols are reduced to allylic alcohols.[117,118]

Red-Al behaves as a base in certain applications, *if* a readily reducible functional group is *not* present in the molecule. Kametani used Red-Al to induce a **Steven's rearrangement** (sec. 8.8.C.ii) of berbine methiodide (**90**) to spirobenzylisoquinoline (**91**) (9.4%) plus **92** (22.5%).[119] The yield of rearranged products was poor, which is typical for this application of Red-Al.

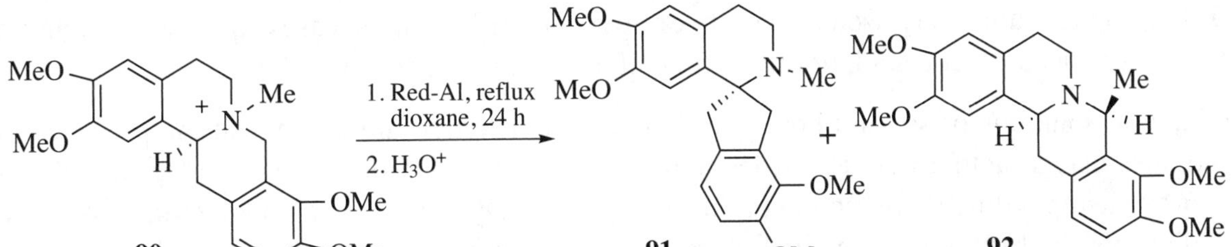

The functional group transforms most useful with Red-Al are:

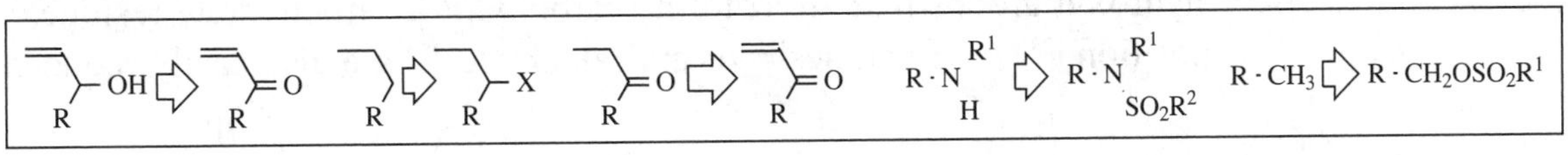

4.4. REDUCTIONS WITH BOROHYDRIDE[120]

4.4.A. Sodium Borohydride

Schlesinger and Brown first prepared sodium borohydride ($NaBH_4$) in 1943,[121] but experimental details were not reported until the 1950s.[122] Its discovery predated that of lithium aluminum hydride, but the security regulations of World War II ($NaBH_4$ was important for the separation of uranium salts as part of the Manhattan project) delayed its disclosure.[122] Sodium borohydride is less reactive than lithium aluminum hydride so it is a selective reagent, often reducing

116. Černý, M.; Málek, J. *Collect. Czech. Chem. Commun.* **1970**, *35*, 1216.
117. (a) Mayer, H.J.; Rigassi, N.; Schwieter, U.; Weedon, B.C.L. *Helv. Chim. Acta* **1976**, *59*, 1424; (b) Kienzle, F.; Mayer, H.; Minder, R.E.; Thommen, W. *Ibid.* **1990**, 524.
118. Jones, T.K.; Denmark, S.E. *Org. Synth. Coll. Vol. 7* **1985**, *64*, 182.
119. Kametani, T.; Huang, S-P.; Koseki, C.; Ihara, M.; Fukumoto, K. *J. Org. Chem.* **1977**, *42*, 3040.
120. For a review of selective reductions using boron reagents, see Burkhardt, E.R.; Matos, K. *Chem. Rev.* **2006**, *106*, 2617.
121. (a) Reference 15, p 1049; (b) Schlesinger, H.I. Brown, H.C. U.S. Patent 2,461,661, 2,461,663 [*Chem. Abstr. 43*: 4684e, 4684i *1949*; U.S. Patent 2,534, 533 [*Chem. Abstr. 45*: 4007i, *1951*].
122. Schlesinger, H.I.; Brown, H.C.; Hoekstra, H.R.; Rapp, L.R. *J. Am. Chem. Soc.* **1953**, 75, 199.

aldehydes, ketones or acid chlorides in the presence of other reducible functional groups.[123] The diminished reactivity is reflected in the solvents used for the reduction. Aqueous and alcohol solvents are preferred due to the excellent solubility of sodium borohydride (25 g/100 g of water at 0°C [88.5 g/100 g at 60°C], 16.4 g /100 g of methanol, 4g /100 g of ethanol).[121a] Sodium borohydride reacts with water to form hydroxyborohydride intermediates, but these are also mild reducing agents. Similar reaction with alcohols is relatively slow so the most common reaction solvents are ethanol and 2-propanol. Sodium borohydride is relatively insoluble in ether solvents (0.1 g/100 g of THF, 5.5 g/100 g of diglyme and 0.8 g/100 g of DME), so these are rarely used for borohydride reductions. In most cases aqueous ammonium chloride, aqueous acetic acid, or dilute mineral acids are used for hydrolysis.

Ethanolic solutions of sodium borohydride reduce aldehydes and ketones in the presence of epoxides, esters, lactones, acids, nitriles or nitro groups.[121a,124] In Cook and co-worker's synthesis of 6-oxoalstophylline,[125] reduction of the aldehyde moiety in **93** gave alcohol **94** in 95% yield. Sodium borohydride often gives the 1,2-reduction product from conjugated carbonyl derivatives, particularly when the terminal carbon of the conjugated system is sterically hindered. In general, sodium borohydride is an excellent reagent for the reduction of ketones or aldehydes in the presence of esters,[126] hydroxyl groups that are α to the carbonyl,[41] a carbohydrate residue,[127] or halogens in the α position.[128] Aryl ketones[129] or aryl aldehydes[130] are also easily reduced.

Unconjugated ketones and aldehydes are usually reduced in preference to a conjugated carbonyl elsewhere in the molecule. Paquette and co-workers treated **95** with NaBH4 and obtained **96** in good yield, without reduction of the conjugated ketone unit, as part of a synthesis of magellanine.[131] Similar reduction with LiAlH4 would reduce both carbonyl groups and give a mixture of 1,2- and 1,4-reduction products from the conjugated ketone.

123. Chaikin, S.W.; Brown, W.G. *J. Am. Chem. Soc.* **1949**, *71*, 122.
124. (a) Brown, H.C.; Krishnamurthy, S. *Aldrichimica Acta* **1979**, *12*, 3; (b) Schenker, R. *Angew. Chem.* **1961**, *73*, 81.
125. Liao, X.; Zhou, H.; Wearing, X. Z.; Ma, J.; Cook, J.M. *Org. Lett.* **2005**, *7*, 3501.
126. Kelley, R.C.; Schletter, I. *J. Am. Chem. Soc.* **1973**, *95*, 7156.
127. Ohrui, H.; Emoto, S. *Agric. Biol. Chem.* **1976**, *40*, 2267.
128. (a) De Amici, M.; De Micheli, C.; Carrea, G.; Spezia, S. *J. Org. Chem.* **1989**, *54*, 2646; (b) Gensler, W.J.; Solomon, P.H. *J. Org. Chem.* **1973**, *38*, 1726.
129. Burka, L.T.; Wilson, B.J.; Harris, T.M. *J. Org. Chem.* **1974**, *39*, 2212.
130. Dai-Ho, G.; Mariano, P.S.; *J. Org. Chem.* **1988**, *53*, 5113.
131. Williams, J.P.; St. Laurent, D.R.; Friedrich, D.; Pinard, E.; Roden, B.A.; Paquette, L.A. *J. Am. Chem. Soc.* **1994**, *116*, 4689.

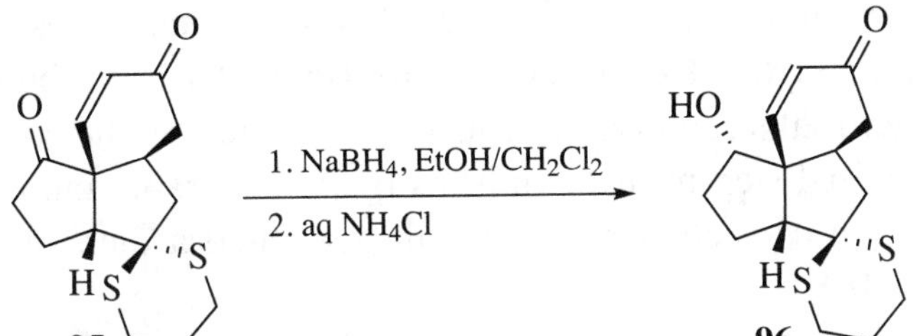

Sodium borohydride reduces ozonides to an alcohol,[132] just as LiAlH$_4$ was used in section 4.2.C.ii. An example is the ozonolysis of **97** where borohydride reduction gave alcohol **98** in 94% yield in a synthesis of (–)-cytoxazone by Jung and co-workers.[133]

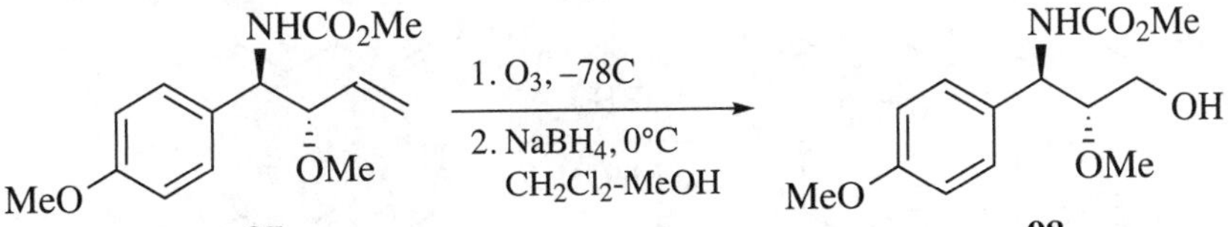

An important use of sodium borohydride is reduction of enamines, imines or iminium salts, which is particularly useful in alkaloid and amino acid syntheses. In a synthesis of (+)-majvinine,[134] Cook and co-workers converted the amine unit in **99** to the imine (**100**) by reaction with benzaldehyde. Subsequent reduction with NaBH$_4$ gave a 92% overall yield of the *N*-benzylamine, **101**. Enamines can also be reduced, as in Werner and co-worker's synthesis of opioid antagonists that used sodium borohydride to selectively reduce enamine **102** to **103**.[135]

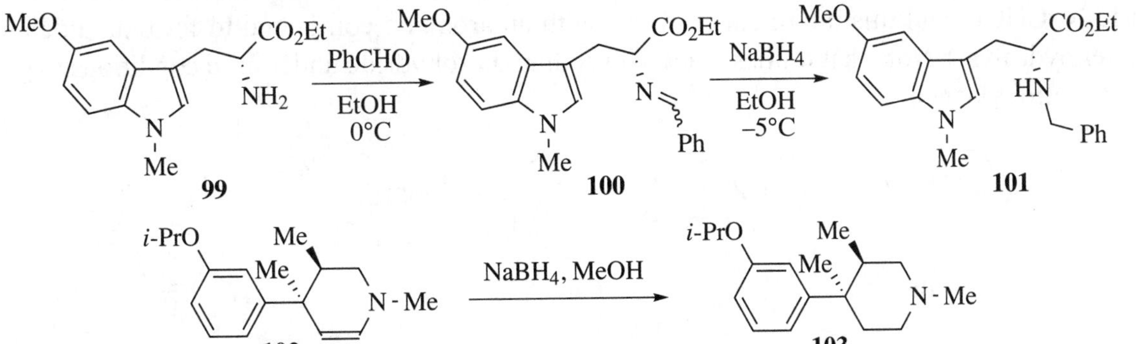

An important variation is called **reductive amination**, in which an aldehyde or ketone is mixed with an amine in the presence of sodium borohydride, generating the corresponding

132. Johnson, C.R.; Senanayake, C.H. *J. Org. Chem.* **1989**, *54*, 735.
133. Kim, J.D.; Kim, I.S.; Jin, C.H.; Zee, O.P.; Jung, Y.H. *Org. Lett.* **2005**, *7*, 4025.
134. Zhao, S.; Liao, X.; Wang,T.; Flippen-Anderson, J.; Cook, J.M. *J. Org. Chem.* **2003**, *68*, 6279.
135. Werner, J.A.; Cerbone, L.R.; Frank, S.A.; Ward, J.A.; Labib, P.; Tharp-Taylor, R.W.; Ryan, C.W. *J. Org. Chem.* **1996**, *61*, 587.

N-alkyl amine. Guillou and co-workers[136] used this technique in a synthesis of maritidine, where the reaction of aldehyde **104**, primary amine **105** and sodium borohydride gave a 78% yield of **106**. Pyridinium salts can also be reduced, illustrated by the conversion of **107** to **108** (haliclamine A) as the final step in that synthesis by Al-Mourabit and co-worker's.[137] In this reaction, initial reduction of each pyridinium salt generated a dienamine, and 1,2-reduction gave the final product, **108**.

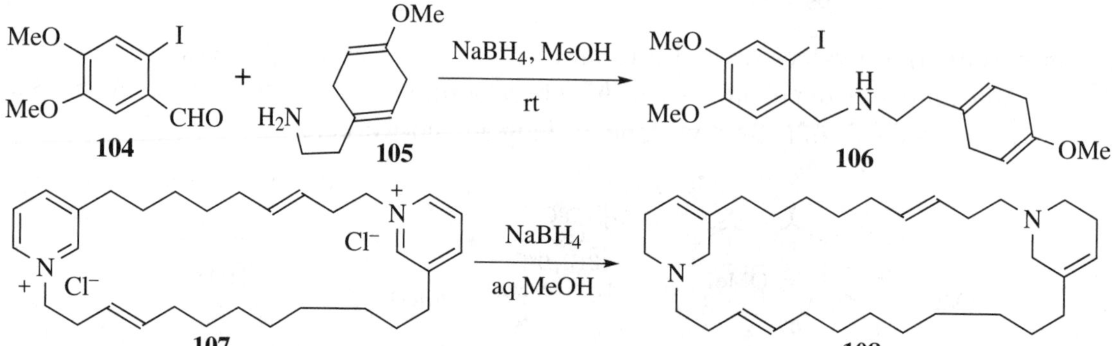

Amides are not reduced directly by $NaBH_4$, but if they are first converted to an iminium derivative reduction to the amine is rapid. Reaction of **109** with phosphorus oxychloride ($POCl_3$) gave the **Vilsmeier-Haack complex**[138] (**110**), and this was reduced to the piperidine derivative **111**.[139] Primary amides are converted to nitriles by this procedure. Vilsmeier complex **110** is related to an intermediate generated in what is known as the **Vilsmeier reaction**,[138] which converts aromatic derivatives to aryl aldehydes. In a Vilsmeier reaction, $POCl_3$, and DMF react to form $Me_2N=CHCl^+$, and this intermediate reacts with an aromatic compound to form an aldehyde after hydrolysis. Note that imides react with sodium borohydride under mild conditions to give a 2-hydroxylactam.[140]

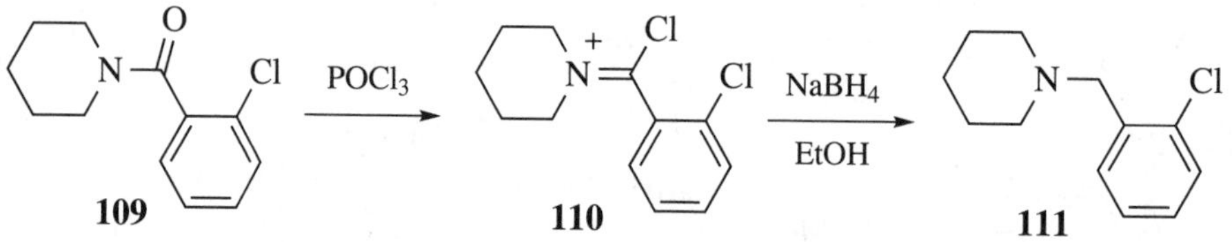

136. Bru, C.; Thal, C.; Guillou, C. *Org. Lett.* **2003**, *5*, 1845.

137. Michelliza, S.; Al-Mourabit, A.; Gateau-Olesker, A.; Marazano, C. *J. Org. Chem.* **2002**, *67*, 6474.

138. (a) Vilsmeier, A.; Haack, A. *Berichte* **1927**, *60*, 119; (b) Mundy, B.P.; Ellerd, M.G.; Favaloro Jr., F.G. *Name Reactions and Reagents in Organic Synthesis 2nd Ed.*, Wiley, New York, **2005**, pp 668-669.;(c) Witiak, D.T.; Williams, D.R.; Kakodkar, S.V.; Hite, G.; Shen, M.-S. *J. Org. Chem.* **1974**, *39*, 1242; (d) de Maheas, M.R. *Bull. Soc. Chim. Fr.* **1962**, 1989; (e) The Merck Index, 14th ed., Merck & Co., Inc., Whitehouse Station, New Jersey, **2006**, p ONR-96.

139. Ur-Rahman, A.; Basha, A.; Waheed, N.; Ahmed, S. *Tetrahedron Lett.* **1976**, 219.

140. For an example taken from a synthesis see Hart, D.J.; Yang, T.K. *J. Chem. Soc. Chem. Commun.* **1983**, 135.

Anhydrides are related to lactones as imides are related to lactams. Partial reduction of an anhydride directly to an hydroxylactone is difficult, but reduction (with ring opening) to a hydroxy acid and subsequent ring closure can generate a lactone when $NaBH_4$ is used. An example is the reduction of anhydride **112** in DMF, to give lactone **114** in 55% yield.[141] Norbornyl systems usually react via the exo face, as discussed in Section 1.5.E, which accounts for the selectivity of this reduction. In this particular example, reduction of the exo carbonyl gave alcohol **113**, which closed to the lactone (**114**) in the presence of the carboxyl group.

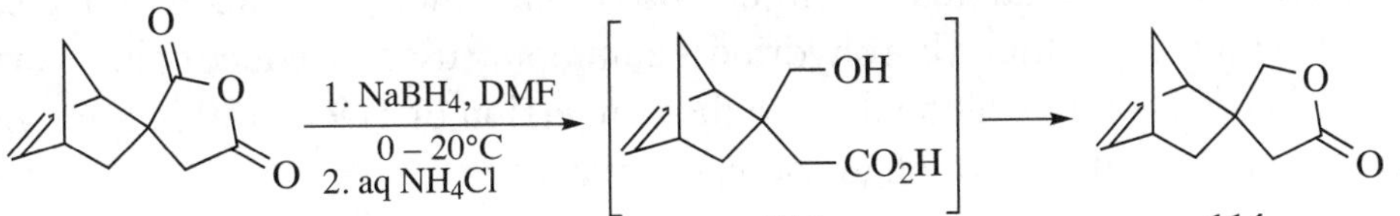

The functional group transforms for this section are

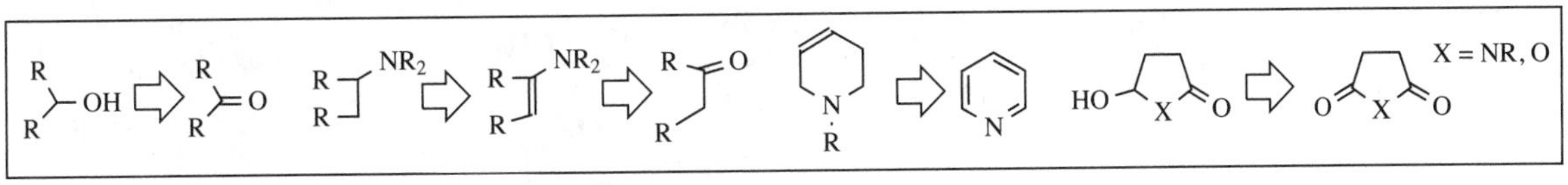

4.4.B. Alternative Metal Borohydrides (Li, Zn, Ce)

In previous sections, the metal counterion for hydride reducing agents was largely restricted to sodium, lithium or potassium. Other metals can modify the course of the reduction. Lithium borohydride ($LiBH_4$), for example,

$$NaBH_4 + LiBr \longrightarrow LiBH_4 + NaBr$$

prepared by reaction of sodium borohydride and lithium bromide,[142] is a more powerful reducing agent than $NaBH_4$. $LiBH_4$ reduces aldehydes, ketones or acid chlorides as well as epoxides, esters or lactones but it does not reduce acids, nitriles or nitro compounds.[143] Lithium borohydride is soluble in ether (4 g/100 g) and THF (21 g/100 g) unlike $NaBH_4$, which is insoluble in the former and sparingly soluble in the latter. The major use of $LiBH_4$ is for the reduction of esters, when the ester is the most reducible group in the molecule. In a synthesis of citreoviral, Woerpel and Peng used $LiBH_4$ to reduce both the ethyl ester and the acetoxy group in **115**, to give an 87% yield of diol **116**.[144] Notice that reductive cleavage of the acetoxy group unmasked the alcohol unit at that carbon. Lithium borohydride usually reduces lactones to the

141. Haslouin, J.; Rouessac, F. *Bull. Soc. Chim. Fr.* ***1977***, 1242.
142. (a) Nystrom, R.F.; Chaikin, S.W.; Brown, W.G. *J. Am. Chem. Soc.* ***1949***, *71*, 3245; (b) Kollonitsch, J.; Fuchs, O.; Gábor, V. *Nature* ***1954***, *173*, 125; (c) Brown, H.C.; Mead, E.J.; Subba Rao, B.C. *J. Am. Chem. Soc.* ***1955***, *77*, 6209.
143. Reference 15, p 603.
144. Peng, Z.-H.; Woerpel, K.A. *Org. Lett.* ***2002***, *4*, 2945.

corresponding diol.[145] Note that LiBH$_4$ reduces an ester in the presence of a carboxylic acid,[146] but there are some cases where LiBH$_4$ fails to reduce esters. For example, attempts to reduce the *N*-allylic pyroglutamate **117** with NaBH$_4$ or LiBH$_4$ failed.[147] The use of **Hojo's reagent** (lithium aluminum hydride pre-adsorbed on silica gel) solved this problem, since the reagent showed great selectivity for reduction of the most reducible group in multifunctional group molecules.[148,149] This reagent was used to reduce **117** to the hydroxymethyl derivative (**118**) in 91% yield, without racemization of the stereogenic center at C5, in Smith and Keusenkothen's synthesis of pyrrolizidine alkaloids.[147] Hojo showed that silica gel is an effective catalyst for several other reactions.[150] Lithium borohydride is commonly used to reduce chiral oxazolidinone auxiliary to the corresponding alcohol, as in the conversion of **119** to **120** in 99% yield, in the synthesis of amphidinolide T3 by Zhao and co-workers.[151]

Metal hydrides with metal ions of higher ionic potential (such as Mg, Ca, Ba, and Sn) are also effective for the reduction of esters.[152a,142b] Magnesium borohydride is prepared *in situ* from NaBH$_4$ and magnesium chloride (MgCl$_2$) in diglyme at 100°C, and readily reduces esters.[150b] Addition of aluminum chloride (AlCl$_3$) to three equivalents of sodium borohydride generates aluminum borohydride [Al(BH$_4$)$_3$], which is a powerful reagent for the reduction lactones, epoxides, carboxylic acids, tertiary amides or nitriles as well as ketones, aldehydes or esters.[153]

145. For an example taken from a synthesis of the C2-C16 segment of laulimalide, see Ghosh, A.K.; Wang, Y. *Tetrahedron Lett.* **2000**, *41*, 2319.
146. For an example taken from a synthesis of (–)-paroxetine, see Yu, M.S.; Lantos, I.; Peng, Z.-Q.; Yu, J.; Cacchio, T. *Tetrahedron Lett.* **2000**, *41*, 5647.
147. Keusenkothen, P.F.; Smith, M.B. *Synth. Commun.* **1989**, *19*, 2859.
148. Kametori, Y.; Hojo, M.; Masuda, K.; Inoue, T.; Izumi, T. *Tetrahedron Lett.* **1982**, *23*, 4585.
149. Kametori, Y.; Hojo, M.; Masuda, R.; Inoue, T.; Izumi, T. *Tetrahedron Lett.* **1983**, *24*, 2575.
150. Hojo, M. in *New Synthetic Methodology and Functionally Interesting Compounds*, Yoshida, Z. (Ed.), Kodensha-Elsevier (Tokyo/Amsterdam), **1986**, pp 43-62.
151. Deng, L.-S.; Huang, X.-P.; Zhao, G. *J. Org. Chem.* **2006**, *71*, 4625.
152. (a) Kollonitsch, J.; Fuchs, O.; Gábor, V. *Nature (London)* **1955**, *175*, 346; (b) Chandrakumar, N.S.; Hajdu, J. *Tetrahedron Lett.* **1981**, *22*, 2949.
153. Brown, H.C.; Subba Rao *J. Am. Chem. Soc.* **1956**, *78*, 2582.

Zinc borohydride [$Zn(BH_4)_2$] is prepared from zinc chloride ($ZnCl_2$) and sodium borohydride in ether. It is most useful[154] for selective reduction of α,β-unsaturated aldehydes and ketones, as in the reduction of 2-cyclohexenone to 2-cyclohexenol in 96% yield (4% of cyclohexanol was formed).[155]

Zinc borohydride is used for the diastereoselective reduction of hydroxy- or alkoxy-carbonyl derivatives (sec. 4.4.B).[156,157] The selectivity is explained by coordination (chelation) of the zinc with the carbonyl oxygen and the proximate hydroxyl oxygen, with delivery of hydride from the less sterically hindered face of the molecule. Typical diastereoselectivity (sec. 4.7) is illustrated by formation of **122** in 77% yield (>98:2 dr) from the zinc borohydride reduction of **121**, taken from the Crimmins and co-worker's[158] synthesis of (–)-laulimalide.

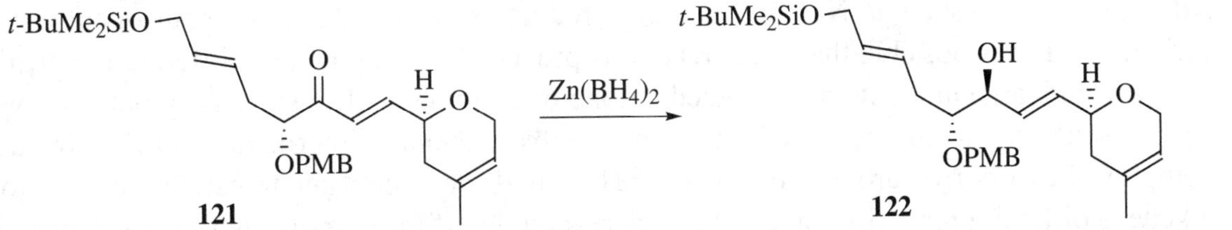

Reduction of epoxides with high selectivity for the more substituted position, is possible using zinc borohydride supported on silica gel.[159a] When 1-nonene oxide (2-heptyloxirane) was treated with this silica-supported reagent, an 88% yield of 1-nonanol was obtained. In general, however, epoxides can be difficult to reduce with $Zn(BH_4)_2$.[159b] When 1-nonene oxide was treated with $Zn(BH_4)_2$ in THF, for example, only a small amount of reduction occurred to give a mixture of regioisomeric products.[159a] Reductive cleavage of a nitrile group has also been observed when a mixture of zinc borohydride and silver tetrafluoroborate ($AgBF_4$) was used, as in the conversion of **123** to **124**, in Husson and co-worker's synthesis of dihydropinidine.[160]

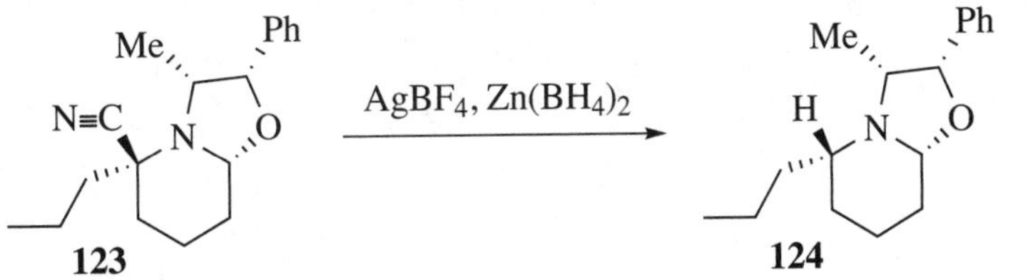

154. Narasimhan, S.; Balakumar, R. *Aldrichimica Acta* **1998**, *31*, 19.
155. (a) Reference 124a, p 5; (b) Yoon, N.M.; Lee, H.J.; Kang, J.; Chung, J.S. Taehan Hwahak Hoechi, *J. Korean Chem. Soc.* **1975**, *19*, 468 [*Chem. Abstr. 84*: 134703z, **1976**].
156. Oishi, T. in *New Synthetic Methodology and Functionally Interesting Compounds*, Yoshida, Z. (Ed.), Kodansha-Elsevier (Tokyo/Amsterdam), **1986**, pp 81-98.
157. (a) Nakata, T.; Fukui, M.; Ohtsuka, H.; Oishi, T. *Tetrahedron Lett.* **1983**, *24*, 2661; (b) *Idem Tetrahedron* **1984**, *40*, 2225.
158. Crimmins, M.T.; Stanton, M.G.; Allwein, S.P. *J. Am. Chem. Soc.* **2002**, *124*, 5958.
159. (a) Ranu, B.C.; Das, A.R. *J. Chem. Soc. Chem. Commun.* **1990**, 1334; (b) Nakata, T.; Tanaka, T.; Oishi, T. *Tetrahedron Lett.* **1981**, *22*, 4723.
160. Guerrier, L.; Royer, J.; Grierson, D.S.; Husson, H.-P. *J. Am. Chem. Soc.* **1983**, *105*, 7754.

Addition of cerium salts to sodium borohydride leads to a reagent, presumably cerium borohydride (known as the **Luche reagent**),[161] that gives very selective 1,2-reduction of conjugated aldehydes and ketones.[162] The carbonyl of a conjugated ketone is reduced faster than that of a normal ketone. Selective 1,2-reduction was observed in the Wipf and co-worker's synthesis of (–)-tuberostemonine,[163] in which reduction of **125** gave alcohol **126** in 71% yield with high diastereoselectivity. In a generalized study, Gemal and Luche examined the extent of 1,2- versus 1,4-reduction of conjugated ketones when $NaBH_4$ was mixed with various metal salts.[164] These results are shown in Table 4.1,[164] and it is clear that the lanthanide salts, including cerium, show a marked preference for 1,2-reduction. Lithium and copper salts give complete reduction of both the carbonyl and the alkene moiety, whereas the other salts give mixtures of 1,2- (**A**) and 1,4-reduction products (**B** and **C**), although they are significantly less reactive. Addition of salts, such as Ni^{2+} or Co^{2+}, give reduction of the alkene without disturbing the carbonyl, but it is possible that these reactions proceed by a completely different mechanistic pathway and large amounts of unreacted ketone often remain. The selectivity obtained when cerium salts are mixed with $NaBH_4$ is explained by a chelated intermediate (**127**). In such a complex, the carbonyl carbon coordinates the borohydride moiety and transfer of hydride to the alkene is difficult unless the complexation is reversible. If the complex is formed irreversibly, hydride transfer to the carbonyl (1,2-addition to give **A** in Table 4.1) is under kinetic control. If the complex forms reversibly (thermodynamic control), 1,4-reduction (to give **B** or **C** in Table 4.1) predominates.

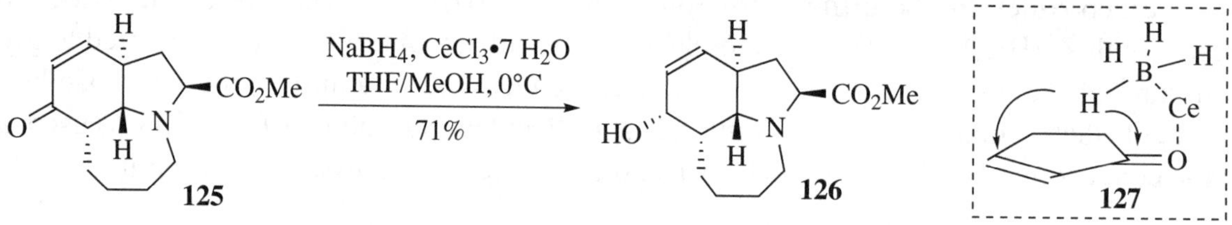

161.   Mundy, B.P.; Ellerd, M.G.; Favaloro Jr., F.G. *Name Reactions and Reagents in Organic Synthesis, 2nd ed.*, Wiley-Interscience, New Jersey, *2005*, p. 805.
162.   Crimmins, M.T.; O'Mahoney, R. *J. Org. Chem.* **1989**, *54*, 1157.
163.   Wipf, P.; Rector, S.R.; Takahashi, H. *J. Am. Chem. Soc.* **2002**, *124*, 14848.
164.   Gemal, A.L.; Luche, J.-L. *J. Am. Chem. Soc.* **1981**, *103*, 5454.

Table 4.1. 1,2- vs. 1,4-Reduction of Cyclopentenone with Sodium Borohydride Admixed with Metal Salts

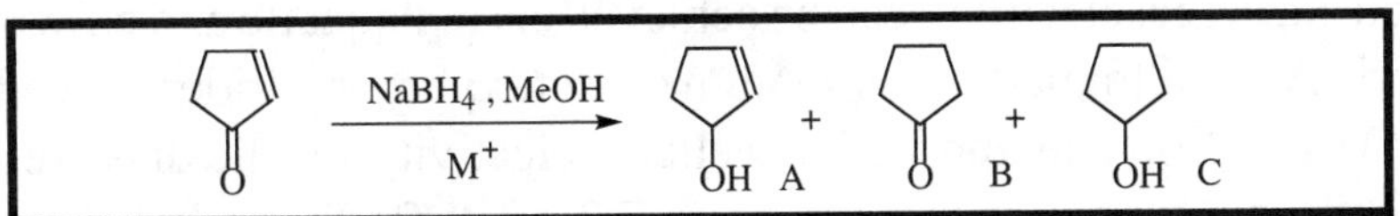

Metal Salt	% Cyclopentanone	% A	% B	% C
La^{3+}	0	90	0	10
Ce^{3+}	0	97	0	3
Sm^{3+}	0	94	0	6
Eu^{3+}	0	93	0	7
Yb^{3+}	0	89	0	11
Y^{3+}	0	86	0	14
Li^+	0	1	0	99
Cu^+	2	6	4	88
Ba^{2+}	2	6	11	76
Zn^{2+}	90		7	3
Fe^{2+}	19	12	9	60
Fe^{3+}	33	5	12	50
Tl^{3+}	86	2	8	4
Ni^{2+}	18	0	76	6
Co^{2+}	27	0	61	12

[Reprinted with permission from Gemal, A.L.; Luche, J.-L. *J. Am. Chem. Soc.* **1981**, *103*, 5454. Copyright © **1981** American Chemical Society.]

4.5. ALKOXY- AND ALKYLBOROHYDRIDES

Alkoxyborohydrides and alkylborohydrides are not used as often as metal alkoxyaluminum hydrides. In general, alkylborohydrides are more reactive than $NaBH_4$. Since $LiAlH_4$ is such a powerful regent, there is little need for alkylalumium hydrides and we will not discuss them. Alkoxyaluminates are less reactive than $LiAlH_4$, and alkoxyborohydrides are less reactive than $NaBH_4$. Acyloxy derivatives are also important. Some derivatives are unstable in ether solvents. For example, sodium trimethoxyborohydride [$NaBH(OMe)_3$] disproportionates in solution,[165] but the derivative triisopropoxyborohydride [$NaBH(OiPr)_3$] is very stable in THF, and is prepared in THF by treatment of $B(Oi\text{-}Pr)_3$ with sodium or potassium hydride.[165,166] This reagent[167] reduces only aldehydes and ketones, reacting with virtually no other functional group.

4.5.A. Acyloxyborohydrides

Acyloxyborohydrides [$NaBH_{4\text{-}n}(O_2CR)_n$] are remarkably selective,[168] as illustrated by the

---

165. Brown, H.C.; Mead, E.J.; Shoaf, C.J. *J. Am. Chem. Soc.* **1956**, *78*, 3616.
166. Brown, C.A. *J. Am. Chem. Soc.* **1973**, *95*, 4100.
167. Brown, C.A.; Krishnamurthy, S.; Kim, S.C. *J. Chem. Soc. Chem. Commun.* **1973**, 391.
168. Gribble, G.W.; Nutaitis, C.F. *Org. Prep. Proc. Int.* **1985**, *17*, 317.

reduction of the aldehyde unit in **128** in the presence of a ketone moiety.[169] Reduction with potassium triacetoxyborohydride (generated by dissolving KBH₄ in acetic acid) gave **129** in 60% yield. Good diastereoselectivity can be achieved, as in Paquette and co-worker's synthesis of (–)-sanglifehrin A.[170] Tetramethylammonium triacetoxyborohydride reduced the ketone unit in **130** to give **131** in 82% yield and with excellent selectivity for the diastereomer shown. In this synthesis, PMB is *p*-methoxybenzyl (see sec. 7.3.A.ii). Sodium triacetoxyborohydride has also been used for selective 1,4-reduction of conjugated esters.[171] Acetoxyborohydride is also used for the reduction of imines and enamines, which is useful for the synthesis of alkaloid precursors. This reagent also reduces imines.[172]

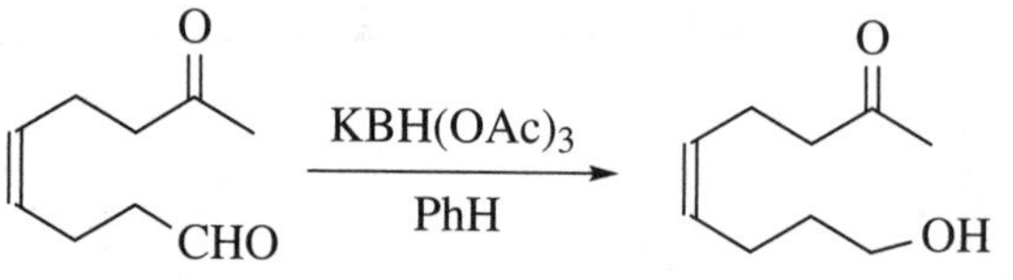

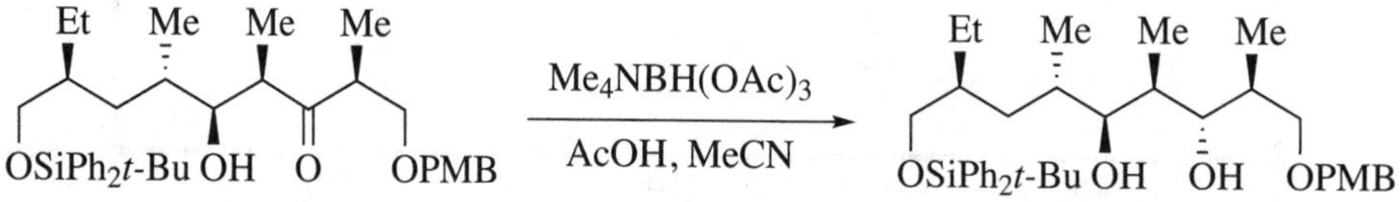

Reductive alkylation is possible when an amine unit is present in the substrate (sec. 4.10.D). The carbon chain of the carboxylic acid fragment is transferred to nitrogen, and an *N*-alkyl amine is the final product. An example is the conversion of tetrahydroquinoline (**132**) to the *N*-propyl derivative (**133**) in 56% yield.[173] When hydroxyl groups are present in the organic substrate, acetoxylation occurs if no other reducible functional group is present. An example is the reduction

of 4-ethoxybenzyl alcohol to the *O*-acetyl derivative, in 95% yield.[174] Although these latter reactions can compete with the reduction of other functional groups, acyloxyborohydrides are important chemoselective reducing agents.

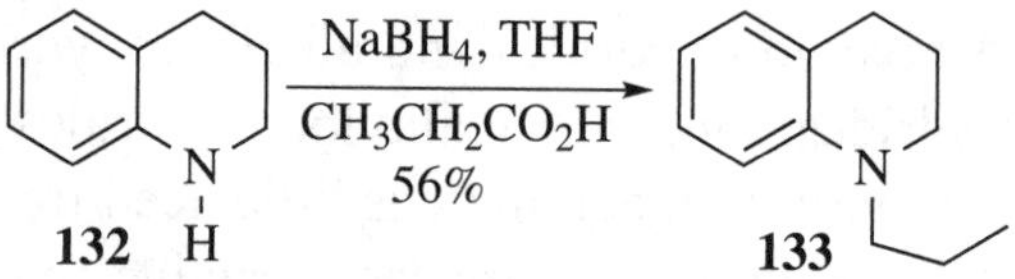

4.5.B. Super-Hydride™

Alkyl groups are known to be electron releasing relative to boron. Therefore, when alkyl groups are attached to the boron of a borohydride reagent the reducing power of that reagent is

169. Tolstikov, G.A.; Odinokov, V.N.; Galeeva, R.I.; Bakeeva, R.S.; Akhunova, V.R. *Tetrahedron Lett.* **1979**, 4851.
170. Paquette, L.A.; Duan, M.; Konetzki, I.; Kempmann, C. *J. Am. Chem. Soc.* **2002**, *124*, 4257.
171. For an example, see Hayahsi, Y.; Rohde, J.J.; Corey, E.J. *J. Am. Chem. Soc.* **1996**, *118*, 5502.
172. Smith, R.G.; Lucas, R.A.; Wasley, J.W.F. *J. Med. Chem.* **1980**, *23*, 952.
173. Gribble, G.W.; Heald, P.W. *Synthesis* **1975**, 650.
174. Prashad, M.; Jigajinni, V.B.; Sharma, P.N. *Indian J. Chem.* **1980**, *19B*, 822.

enhanced.[124a,175] An extremely useful reducing agent has been developed using this idea, lithium triethylborohydride (LiBHEt₃), which is referred to as Super Hydride™. It is prepared by reaction

$$BEt_3 + LiH \xrightarrow{THF,\ 65°C,\ 15\ min} LiBHEt_3$$

of lithium hydride with triethylborane in THF.[176] Super Hydride™ is a more powerful hydride nucleophile than LiBH₄,[177] giving reactions that mimic S_N2 behavior. Super-Hydride™ is commonly used for the reductive dehalogenation of alkyl halides, which illustrates its nucleophilic nataure. As shown in the conversion of 2-*exo*-bromobicyclo[2.2.1]heptane (**134**) to the endo-*d*-derivative (**135**), deuteride displacement proceeded with inversion analogous to an S_N2 process.[177,178] Sulfonate esters such as mesylates, tosylates, and triflates are reduced with LiBHEt₃ via C-O cleavage. In a synthesis of the poison-frog alkaloid (−)-205B, Smith and Kim

converted alcohol **136** to the corresponding mesylate (-CH₂OSO₂Me) and subsequent reduction with LiBHEt₃ gave reduction to the methyl group indicated in **137**, in 83% overall yield.[179] The reagent is highly nucleophilic, so epoxides can be opened regioselectively. Super-Hydride attacks electrophilic centers at the less sterically hindered site, as in the reduction of 1-methyl-1,2-epoxycyclohexane to 1-methyl-cyclohexanol in 99% yield.[180] In reactions with substrates that are sensitive to rearrangement, LiBHEt₃ is superior to LiAlH₄. Reduction of **138** with LiAlH₄ gave 15% of **139** and 85% of the rearranged alcohol **140**.[180] Reduction with Super-Hydride™, however, gave 93% of **139** with < 0.1% of **140**.[180]

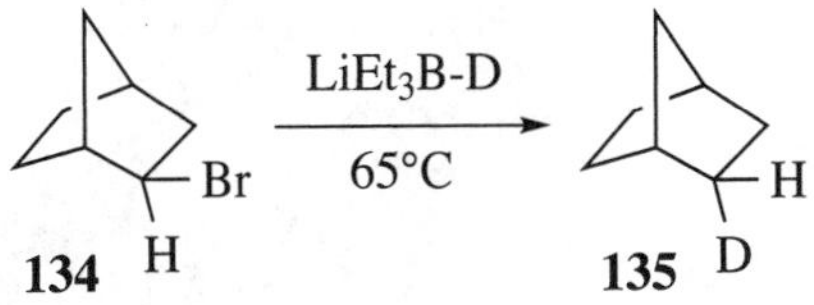

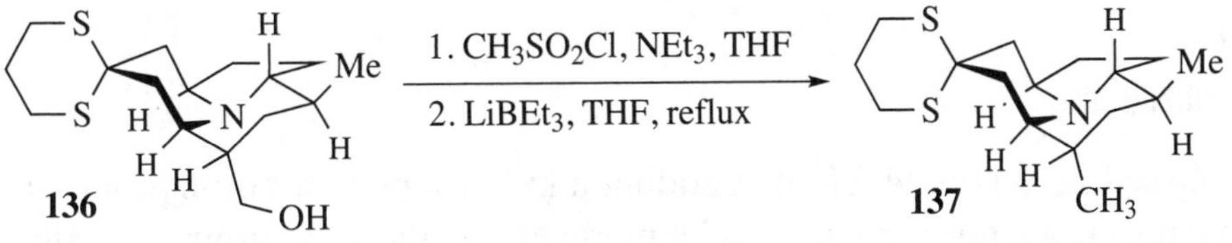

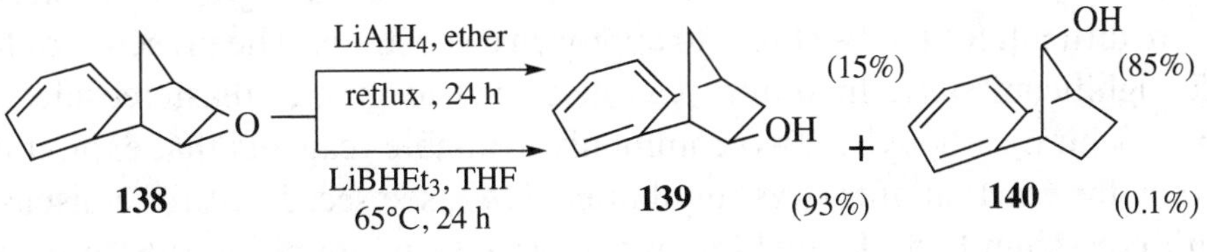

Super-Hydride™ will reduce ketones and aldehydes but it is not used extensively unless the stereochemistry of the alcohol product can be influenced by this reagent, and there are no other groups in the molecule that are reduced by this reagent. In a synthesis of (+)-panepophenathrin

175. Krishnamurthy, S. *Aldrichimica Acta* **1974**, 7, 55.
176. Brown, H.C.; Krishnamurthy, S.; Hubbard, J.L. *J. Am. Chem. Soc.* **1978**, *100*, 3343.
177. Brown, H.C.; Krishnamurthy, S. *J. Am. Chem. Soc.* **1973**, *95*, 1669.
178. Coppi, L.; Ricci, A.; Taddei, M. *J. Org. Chem.* **1988**, *53*, 911.
179. Smith, A.B., III; Kim, D.-S. *Org. Lett.* **2005**, 7, 3247.
180. Krishnamurthy, S.; Schubert, R.M.; Brown, H.C. *J. Am. Chem. Soc.* **1973**, *95*, 8486.

by Porco and co-workers,[181] reduction of ketone **141** with lithium triethylborohydride gave a 98% yield of alcohol **142** with high diastereoselectivity, and without reduction of the epoxide unit. The model of **141** that is provided shows the relative size of the atoms and suggests that path A is more sterically hindered, leading to preferential hydride delivery along path B. Esters are reduced to alcohols, even in the presence of a lactam.[182] Super-Hydride™ usually gives 1,2-reduction of conjugated carbonyl compounds,[183] and it is useful for the reduction of iminium salts. Overman and Heitz.[184] reported the reduction of iminium salt intermediate **144** (derived from thiolactam **143**) to the thioethyl derivative (**145**) in 84% overall yield.

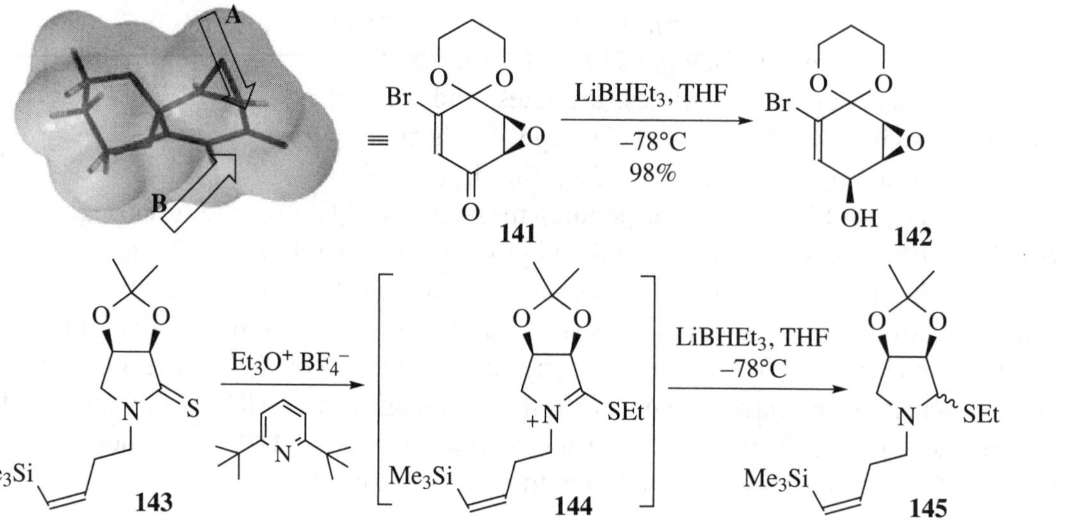

4.5.C. The Selectrides

It was established in Section 4.5.B that adding alkyl groups to a borohydride unit increased the reducing power, but this strength can be modulated if the alkyl groups are sterically bulky since it is more difficult for the B–H unit to approach a substrate. The presence of large groups that provide significant steric hindrance can also lead to greater diastereoselectivity in the reduction of prochiral carbonyls. Two commonly available reagents that exploit this concept are prepared by the reaction of tri-*sec*-butylborane (**147**, see sec. 5.2.A for a discussion of this reagent) with potassium hydride, or lithium trimethoxyaluminum hydride to give potassium tri-*sec*-butyl-borohydride (**146**, known as K-Selectride™) or lithium tri-*sec*-butylborohydride (**148**, known as L-Selectride™).[124a,185,166,176] Both reagents reduce aldehydes and ketones to

181. Lei, X.; Johnson, R.P.; Porco Jr., J.A. *Angew. Chem. Int. Ed.* **2003**, *42*, 3913.
182. Toyooka, N.; Yoshida, Y.; Yotsui, Y.; Momose, T. *J. Org. Chem.* **1999**, *64*, 4914.
183. Perron F.; Albizati, K.F. *J. Org. Chem.* **1989**, *54*, 2044.
184. Heitz, M.-P.; Overman, L.E. *J. Org. Chem.* **1989**, *54*, 2591.
185. (a) Brown, C.A.; Krishnamurthy, S. *J. Organomet. Chem.* **1978**, *156*, 111; (b) Binger, P.; Benedikt, G.; Rotermund, G.W.; Köster, R. *J.L. Ann. Chem.* **1968**, *717*, 21; (c) Corey, E.J.; Albonico, S.M.; Koelliker, U.; Schaaf, T.K.; Varma, R.K. *J. Am. Chem. Soc.* **1971**, *93*, 1491.

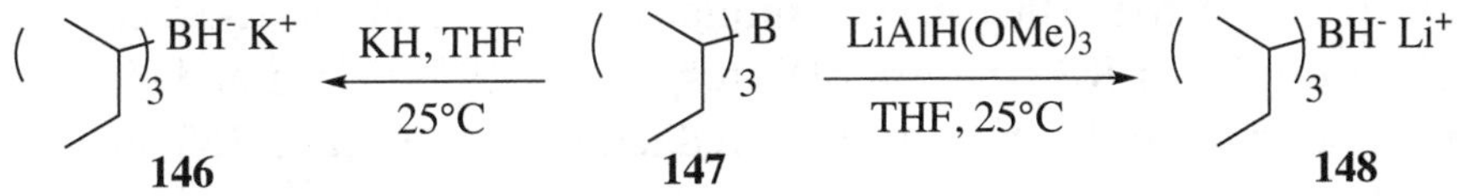

the corresponding alcohol, even at temperatures as low as –78°C. Both the potassium and lithium derivatives are effective for reductions although, by analogy to other borohydrides, the lithium derivative may be somewhat stronger. In one example, L-Selectride™ reduced the ketone moiety of **149** selectively to alcohol **150** in 80% yield as part of Weinreb and co-worker's synthesis of cylindrospermopsin.[186] In a synthesis of (+)-phorboxazole A,[187] Smith and co-workers reduced the ketone moiety in **151** with K-Selectride™ to give alcohol **152** in 95% yield (9:1 dr). Selectride is used rather than another hydride reducing agent for its improved diastereoselectivity.

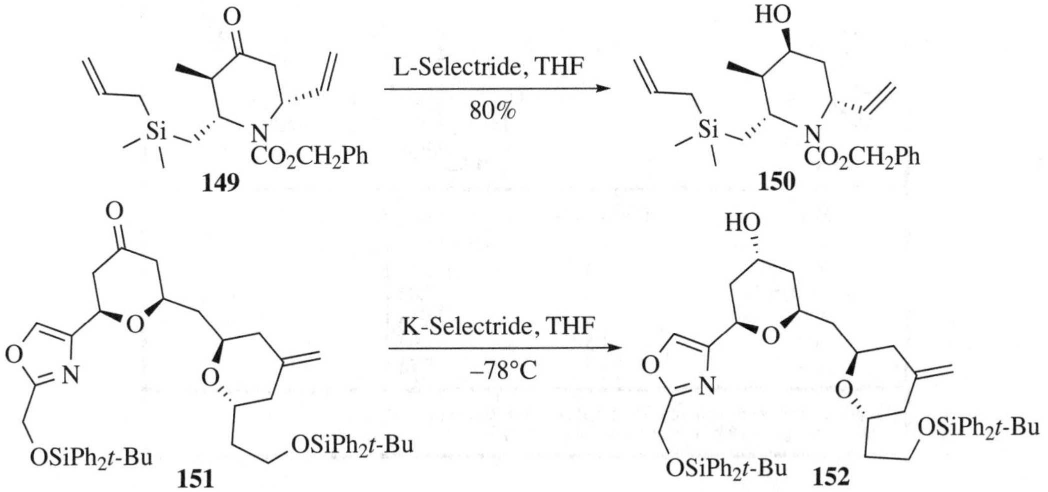

The selectivity of a Selectride™ reagent is influenced by the addition of metal salts, just as the reactivity of $NaBH_4$ was modified in Section 4.4.B. Tanis and co-workers showed that **153** gave a mixture of **154** and **155** upon treatment with L-Selectride™ and added transition metals,[188] via a chelated intermediate complex. Addition of magnesium bromide ($MgBr_2$) favored **154** (8.5:1, 95% yield) and the influence of magnesium bromide, zinc iodide (ZnI_2) and tetravalent titanium is shown in Table 4.2.[188] Magnesium salts gave the best results in Table 4.2, consistent with the chelating properties of that metal. Delivery of hydride from the less hindered face of **153** accounts for the preference for **154** over **155**.

Alkyl borohydrides were prepared from trisiamylborane (siamyl = Me_2CH_2CHMe- ≡ tri-*sec*-isoamyl, sec. 5.2.A) with even greater steric hindrance, but with similar reactivity and selectivity to the Selectrides™. Lithium trisiamylborohydride [$LiBH(CHMeCHMe_2)$] and the

186. Heintzelman, G.R.; Fang, W.-K.; Keen, S.P.; Wallace, G.A.; Weinreb, S.M. *J. Am. Chem. Soc.* **2002**, *124*, 3939.
187. Smith III, A.B.; Razler, T.M.; Ciavarri, J.P.; Hirose, T.; Ishikawa, T. *Org. Lett.* **2005**, *7*, 4399.
188. Tanis, S.P.; Chuang, Y.-H.; Head, D.B. *J. Org. Chem.* **1988**, *53*, 4929.

potassium salt [KBH(CHMeCHMe$_2$)], are similar to the L- and the K-Selectrides™ and are called LS-Selectride™ and KS-Selectride™, respectively.[189] In Ma and Zhu's synthesis of (3S,4aS,6R,8S)-hyperaspine, LS-Selectride™ reduction of **156** gave **157** in 83% yield as a single isomer.[190]

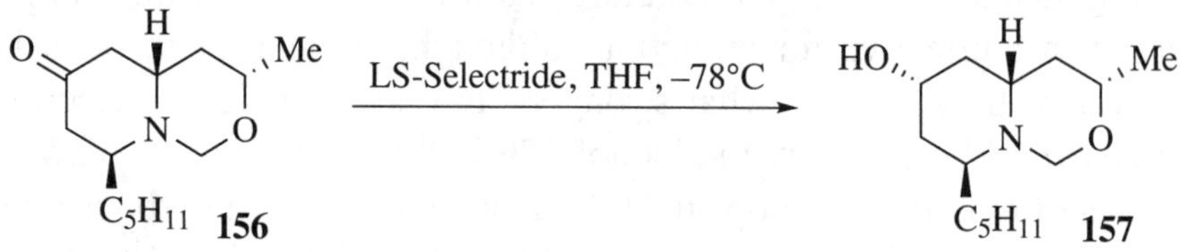

Table 4.2. Influence of Metal Salts on the L-Selectride Reduction of 153

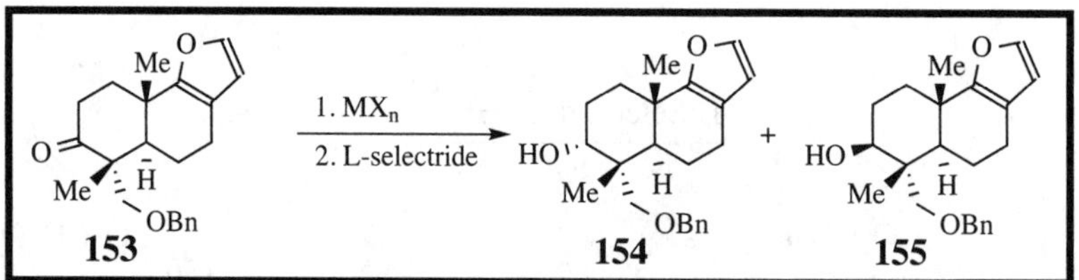

MX$_n$	Equiv. MX$_n$	154 : 155	% Yield
none	0	2.8 : 1	99
ZnI$_2$•OEt$_2$	1.5	1.1 : 1	93
MgBr$_2$•OEt$_2$	1.5	7.0 : 1	95
MgBr$_2$•OEt$_2$	2.0	8.5 : 1	95
Ti(Oi-Pr)$_3$Cl	1.5	4.0 : 1	96
Ti(Oi-Pr)$_4$	1.5	5.7 : 1	93

[Reprinted with permission from Tanis, S.P.; Chuang, Y.-H.; Head, D.B. *J. Org. Chem.* **1988**, *53*, 4929. Copyright © **1988** American Chemical Society.]

4.5.D. Sodium Cyanoborohydride

Alkyl groups are not the only substituents that can be attached to boron in order to generate new borohydride derivatives. Reaction of NaBH$_4$ with HCN gives sodium cyanoborohydride (NaBH$_3$CN),[191] for example, a remarkably stable reagent that it is very selective,[192] and does not decompose in acid solution (the pH should be less acidic than pH 3). It is soluble in THF, MeOH, H$_2$O, HMPA, DMF and sulfolane and they do not react.

$$NaBH_4 + HCN \xrightarrow{THF} NaBH_3CN + H_2$$

---

189. Krishnamurthy, S.; Brown, H.C. *J. Am. Chem. Soc.* **1976**, *98*, 3383.
190. Zhu, W.; Ma, D. *Org. Lett.* **2003**, *5*, 5063.
191. (a) Wittig, G. *J.L. Ann. Chem.* **1951**, *573*, 195 (see pp 209); (b) Wade, R.C.; Sullivan, E.A.; Berchied, Jr., J.R.; Purcell, K.F. *Inorg. Chem.* **1970**, *9*, 2146.
192. Lane, C.F. *Aldrichimica Acta* **1975**, *8*, 3.

Cyanoborohydride readily reduces iodides, bromides or tosylates to the hydrocarbon in excellent yield when HMPA is the solvent,[193] even in the presence of carbonyl groups. The reduction of the iodo group in **158** gave **159** without disturbing the ester or ketone moieties for example.[194] Alcohols are similarly reduced if zinc bromide ($ZnBr_2$) is added to the reagent, as in the reduction of **160** to **161** in Johnson and co-worker's synthesis of longifolene. This reduction presumably proceeds via initial formation of a bromide, although a carbocation intermediate has also been suggested.[195]

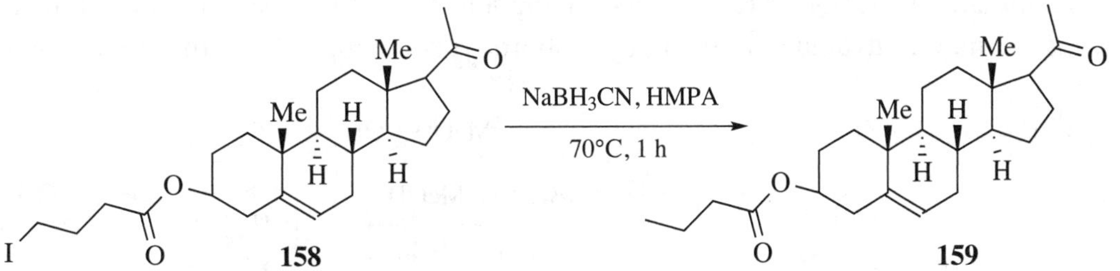

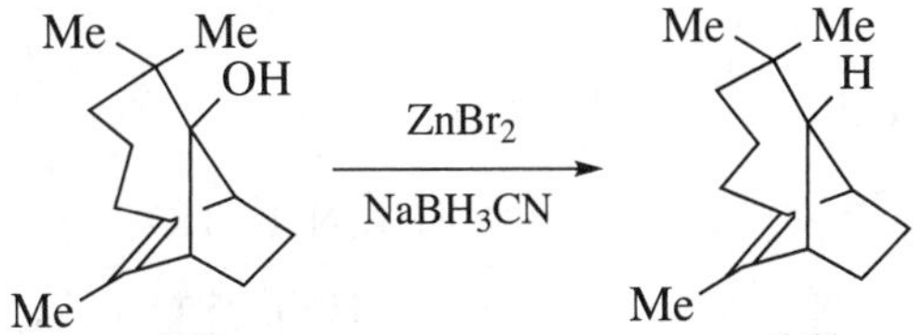

Ketones and aldehydes are reduced in acidic media but *not* at neutral pH. Sodium cyanoborohydride is stable at pH 3-4, where a carbonyl is converted to the protonated form ($C=OH^+$), which is then reduced.[196] An example is the reduction of the aldehyde unit in **162** to give **163** in 89% yield (at pH 4 in this case), in Varela and Di Nardo's asymmetric synthesis of 4-hydroxypipecolic acid.[197] Note that the lactone moiety nor the benzyl carbamate were reduced under these conditions.

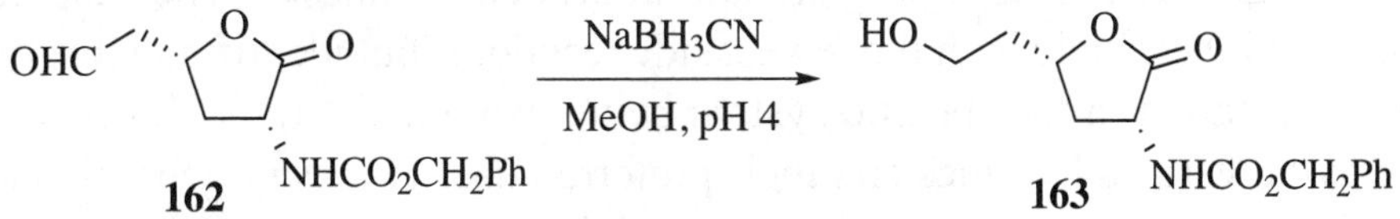

The ability to reduce compounds under acidic conditions is ideal for the reduction of enamines and enamides. When the reduction is done in acidic media protonation of nitrogen gives an

193. Hutchins, R.O.; Milewski, C.A.; Maryanoff, B.E.. *Org. Synth. Coll. Vol. 6* **1988**, 376.
194. Hutchins, R.O.; Kandasamy, D.; Maryanoff, C.A.; Masilamani, D.; Maryanoff, B.E. *J. Org. Chem.* **1977**, *42*, 82.
195. Volkmann, R.A.; Andrews, G.C.; Johnson, W.A. *J. Am. Chem. Soc.* **1975**, *97*, 4777.
196. Borch, R.F.; Bernstein, M.D.; Durst, H.D. *J. Am. Chem. Soc.* **1971**, *93*, 2897.
197. Di Nardo, C.; Varela, O. *J. Org. Chem.* **1999**, *64*, 6119. For reduction of an aryl aldehyde to a mixture of a benzylic alcohol and the hydrogenolysis product in a synthesis of herbertenediol, see Srikrishna, A.; Satyanarayana, G. *Tetrahedron* **2006**, *62*, 2892.

iminium salt, which is reduced *in situ* with cyanoborohydride.[196,198] As mentioned, enamides can be reduced. In a synthesis of nominine, Muratake and co-worker's[199] reduced enamide **164** to the saturated derivative **165** in 91% yield using an acidic solutin of sodium cyanoborohydride. This reagent is excellent for the reduction of iminium salts at neutral pH as well,[200] and it is also useful for the reductive alkylation of amines.[201] Dimethylamino derivatives such as **167** can be prepared from the amine (here, **166**) by treatment with formaldehyde and cyanoborohydride, even in the presence of functional groups, such as the conjugated ketone and the ester in **166**.[202] Apart from the actual $NH_2 \rightarrow NMe_2$ conversion, the selectivity for a specific group shown in this example is the major advantage in using cyanoborohydride rather than other reducing agents.

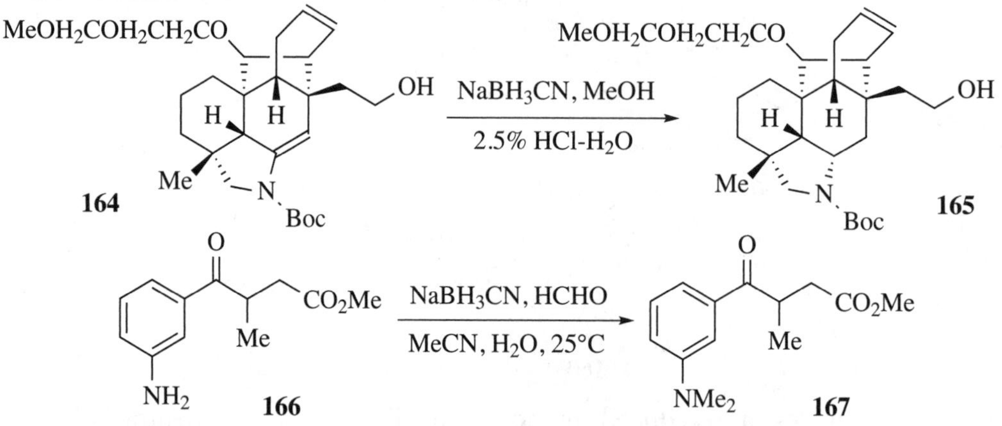

4.5.E.   Typical Applications of Hydride Reducing Agents

Several hydride reducing agents were discussed in preceding sections. Knowing which reducing agent is preferred for a given situation is highly desirable, but making such a choice is not always straightforward, especially in natural product synthesis. The best reagent is the one that gives the highest yield with the fewest side reactions, but identifying this reagent often requires experimentation in the specific system being examined. Table 4.3 is presented as a simple guide to the various hydrides and their preferred use. It is very general, and should be used to only give a *first estimation* of relative reactivity.

198.   Hanaoka, M.; Yoshida, S.; Mukai, C. *J. Chem. Soc. Chem. Commun.* **1984**, 1703.
199.   Muratake, H.; Natsume, M.; Nakai, H. *Tetrahedron* **2006**, *62*, 7093.
200.   Nakagawa, Y.; Stevens, R.V. *J. Org. Chem.* **1988**, *53*, 1871.
201.   Borch, R.F.; Hassid, A.I. *J. Org. Chem.* **1972**, *37*, 1673.
202.   McEvoy, F.J.; Allen, Jr., G.R. *J. Med. Chem.* **1974**, *17*, 281.

Table 4.3. Typical reductions with selected hydride reagents

Hydride	Applications	Special Considerations
LiAlH$_4$	General. reduces most groups: aldehydes, ketones, acids and acid derivatives, nitriles, aliphatic nitro, halides, propargyl alcohols.	Aromatic nitro reduced to diazo. Sulfonate esters give mixtures of C–O and S–O cleavage. Conjugated carbonyls often give mixtures of 1,2- and 1,4-reduction.
LiAlH(OMe)$_3$	Milder than LiAlH$_4$. Generally used for aldehydes, ketones, and conversion of acid chlorides or esters to aldehydes.	Esters are reduced more slowly than ketones or aldehydes. Conjugated carbonyls give mostly 1,2-reduction. Amides and nitriles are reduced to aldehydes.
LiAlH(Ot-Bu)$_3$	Mild and selective. Good for converting acid derivatives to aldehydes.	Esters are reduced more slowly than ketones or aldehydes. Conjugated carbonyls give mostly 1,2-reduction. Amides and nitriles are reduced to aldehydes.
Red-Al	Similar to LiAlH$_4$. Good for carboxylic acids. Nitriles can be converted to aldehydes and aromatic aldehydes to methyl aryls.	Conjugate carbonyls give 1,2-reduction (1,4- when Cu$^+$ is added). Soluble in oranic solvents, so excellent for reverse addition experiments.
NaBH$_4$	General but mild. Good for ketones, aldehydes, acid chlorides, imines, iminium salts.	Esters and most acid derivatives reduce slowly, or not at all. Conjugated carbonyls give mostly 1,2-reduction, but 1,4-reduction is common. Halides are not reduced very well, but epoxides reduce very slowly
LiBH$_4$	Good for esters. Stronger than NaBH$_4^-$, but less selective.	Similar to NaBH$_4$, except for esters.
Zn(BH$_4$)$_2$	Generally similar to NaBH$_4$, but shows higher diastereoselectivity.	Similar to NaBH$_4$.
NaBH$_4$•CeCl$_3$	Gives primarily 1,2-reduction with conjugated carbonyls.	Similar to NaBH$_4$.
MBH(O$_2$CR)$_3$	Good for imines and enamines.	Reducing power is dependent on solvent and the nature of the R group in the RCOOH precursor. Alkylation can occur with amines, and hydroboration can occur with alkenes.
LiBHEt$_3$ (Super Hydride)	Strong reducing agent. Good for halides (S$_N$2-type reductions), halides, and sulfonate esters (C–O cleavage).	Similar to LiAlH$_4$
MBH(sec-Bu)$_3$ (Selectride)	Used primarily with aldehydes and ketones when higher diastereoselectivity is required.	Similar to LiAlH$_4$
NaBH$_3$CN	Excellent for reduction of halides at neutral pH, and alcohols if a Lewis acid is added.	Generally unreactive with most groups at neutral pH. Reduces aldehydes and ketones at pH 4 and lower. Acid derivatives are generally unreactive.

4.6. BORANE, ALUMINUM HYDRIDE, AND DERIVATIVES

4.6.A. Borane

Brown and Schlesinger first reported the preparation and use of diborane as a reducing agent in 1939.[203] This finding began the exploration of hydrides and boranes as reducing agents. Borane reagents are discussed after hydrides only because hydride reagents are now used more often, but borane and its derivatives remain the reagent of choice for many applications. Diborane can be generated by reaction of BF$_3$•OEt$_2$ with LiH[204] or by reaction of NaBH$_4$ and BF$_3$ (sec. 5.2.A).[205]

203.   Brown, H.C.; Schlesinger, H.I.; Burg, A.B. *J. Am. Chem. Soc.* **1939**, *61*, 673.
204.   Schlesinger, H.I.; Brown, H.C.; Hoekstra, H.R.; Rapp, L.R. *J. Am. Chem. Soc.* **1953**, *75*, 199.
205.   Zweifel, G.; Brown, H.C. *Org. React.* **1963**, *13*, 1.

$$6 \text{ LiH} + 8 \text{ BF}_3\text{•OEt}_2 \xrightarrow{\text{ether}} \text{B}_2\text{H}_6 + 6 \text{ LiBF}_4$$

$$3 \text{ NaBH}_4 + 4 \text{ BF}_3\text{•OEt}_2 \xrightarrow{\text{ether}} 2 \text{ B}_2\text{H}_6 + 3 \text{ NaBF}_4$$

Diborane (the stable dimer is B_2H_6, but it is usually represented as BH_3, which is an unstable monomeric form) is a potent reducing agent for aldehydes, ketones, lactones, epoxides, acids, tertiary amides and nitriles. It is ineffective for the reduction of esters, and reacts with alkenes in ether solvents, forming the basis of hydroboration chemistry (sec. 5.2). The primary difference in reactivity between borane and metal hydride reagents is that borane is a good Lewis acid and coordinates to atoms of high electron density, as illustrated by a comparison of the reduction of trichloroacetaldehyde (chloral, 168) to give the alcohol (169), and similar reduction of 2,2-dimethylpropanal (pivalaldehyde, 170) to give 171. Sodium borohydride reduced 168 faster than 170. Diborane, however, reacted faster with 170 due to the more electron-rich carbonyl oxygen.[206]

As mentioned above, borane reduces esters very slowly, and ketones or aldehydes are selectively reduced in the presence of esters. The most widely used application of borane, however, is for the selective reduction of carboxylic acids even in the presence of halides, esters, nitriles or ketones.[207] The dimethyl sulfide•borane complex is commercially available and is often the reagent of choice for this reduction. Although LiAlH_4 reduces both acids and esters and NaBH_4 does not reduce acids (and often reduces esters with difficulty), borane selectively reduces carboxylic acids in the presence of an ester group. Reduction occurs without racemization of adjacent chiral centers, as in the borane reduction of (–)-malic acid to generate (S)-1,2,4-butanetriol in 92% yield.[208] A typical synthetic application is taken from a synthesis of (+)-SCH 351448,[209] in which Lee and co-workers reduced the carboxylic acid unit in 172 to give alcohol 173 using borane-dimethyl sulfide. Borane also reduces epoxides at the less hindered carbon when mixed with catalytic amounts of sodium borohydride.[210] Borane is useful for the conversion of α-amino acids to α-amino alcohols, as in the reduction of phenylalanine to phenylalaninol in 77% yield.[211] Lactams are also reduced to cyclic amines.[212]

206. Reference 124a, p 7.
207. Boger, D.L.; Yohannes, D. *J. Org. Chem.* **1989**, *54*, 2498.
208. (a) Hanessian, S.; Ugolini, A.; Therien, M. *J. Org. Chem.* **1983**, *48*, 4427; (b) Nakata, T.; Nagao, S.; Oishi, T. *Tetrahedron Lett.* **1985**, 26, 75.
209. Kang, E.J.; Cho, E.J.; Lee, Y.E.; Ji, M.K.; Shin, D.M.; Chung, Y.K.; Lee, E. *J. Am. Chem. Soc.* **2004**, *126*, 2680.
210. Klunder, J.M.; Onami, T.; Sharpless, K.B. *J. Org. Chem.* **1989**, *54*, 1295.
211. (a) Lane, C.F. U.S. Patent 3,935 280 1976 [*Chem. Abstr. 84*: 135101p *1976*]; (b) Reference 30a, p 54.
212. For a synthetic example taken from a synthesis of balanol, see Lampe, J.W.; Hughes, P.F.; Biggers, C.K.; Smith, S.H.; Hu, H. *J. Org. Chem.* **1996**, *61*, 4572.

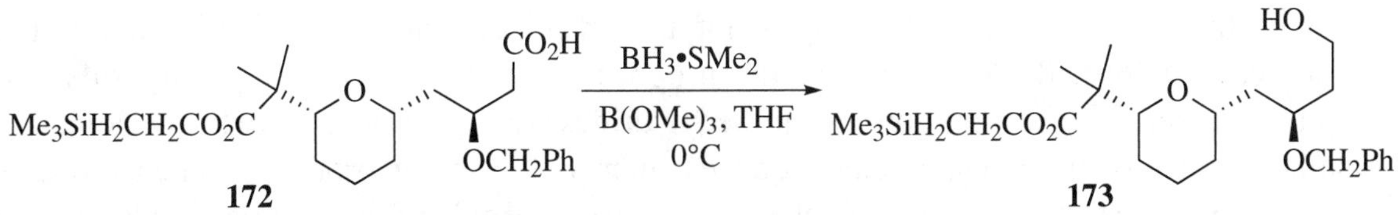

The flammability and difficulties[213] in handling gaseous borane led to development of several stable Lewis base complexes that are easier to use. As noted in the reduction of **172**, one of the more useful reagents is the borane complex of DMS ($H_3B \cdot SMe_2$) where DMS is dimethylsulfide, prepared by bubbling DMS into a THF solution of diborane. This reagent is very soluble in most organic solvents, including benzene and toluene, and gives all the selectivity expected for borane.[213]

Other boranes such as bis(1,2-dimethylpropyl)borane (disiamylborane), and 9-borabicyclo[3.3.1] nonane (usually referred to as simply 9-BBN), both discussed in section 5.2.A, reduce organic functional groups in a manner similar to borane.[214] There are interesting differences in selectivity. Dialkylboranes reduce esters and lactones, whereas similar reduction with borane is very slow. Dicyclohexylborane, for example, reacted with **174** to give 4-deoxyerythropentose (**175**).[215] Tertiary amides are reduced by dialkylboranes to the corresponding aldehyde.[214a] An example is the reduction of *N,N*-dimethylbenzamide with disiamylborane to give benzaldehyde in 89% yield.[214a] The increased steric hindrance of the reagent can have a significant effect on the selectivity in the reduction of prochiral ketones, relative to reduction with $LiAlH_4$ or $NaBH_4$.

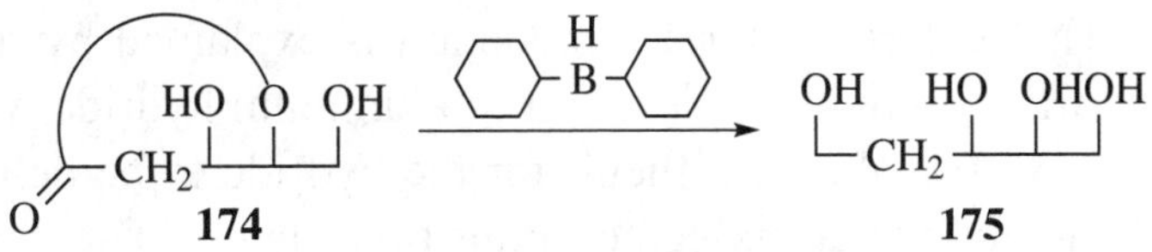

4.6.B. Aluminum Hydride

Borane (BH_3) is considered to be a Lewis acid when compared with borohydride (BH_4^-), which is an *ate* complex of borane. Similarly, AlH_4^- has an electrophilic analog, aluminum hydride (AlH_3), and is capable of selective reduction for many organic functional groups, and is usually prepared by reaction of three equivalents of $LiAlH_4$ with aluminum chloride ($AlCl_3$).[216]

$$3\ LiAlH_4\ +\ AlCl_3\ \longrightarrow\ 4\ AlH_3\ +\ 3\ LiCl$$

213. Lane, C.F. *Aldrichimica Acta* **1975**, *8*, 20.
214. (a) Brown, H.C.; Bigley, D.B.; Arora, S.K.; Yoon, N.M. *J. Am. Chem. Soc.* **1970**, *92*, 7161; (b) Brown, H.C.; Krishnamurthy, S.; Yoon, N.M. *J. Org. Chem.* **1976**, *41*, 1778.
215. Nakaminami, G.; Shoi, S.; Sugiyama, Y.; Isemura, S.; Shibuya, M.; Nakagawa, M. *Bull* Chem. Soc. Jpn. **1972**, *45*, 2624.
216. (a) Eliel, E.L. *Rec. Chem. Progr.,* **1961**, *23*, 129; (b) Doukas, H.M.; Fontaine, T.D. *J. Am. Chem. Soc.* **1951**, *73*, 5917; (c) *Idem, Ibid.* **1953**, *75*, 5355; (d) Wiberg, E.; Schmidt, M. *Z. Naturforsch.* **1951**, *6B*, 333, 460.

Aluminum hydride prepared in this way is not always pristine since mixed chloroaluminum hydrides can be formed, depending on the proportions of lithium aluminum hydride and aluminum chloride.[217] The most common mixed hydrides are Cl_2AlH and $ClAlH_2$, and adjusting the stoichiometry of the reagents can produce either as the major product. When properly prepared, aluminum hydride reduces aldehydes, ketones, acid chlorides, epoxides, acids, tertiary amides or nitriles with ease. Its most common use, however, is the selective 1,2-reduction of conjugated aldehydes, ketones or esters, as in the reduction of the conjugated ester in **176** to the allylic alcohol moiety in **177** (94% yield) in Wilson and Zucker's synthesis of the melon fly pheromone.[218] Similar selectivity is observed with conjugated ketones as shown by the AlH_3 reduction of 2-cyclopentenone to cyclopentenol (**178**, 90% yield), with cyclopentanol as a minor product.[219] It is sometimes possible to control the 1,2- versus 1,4-selectivity of $LiAlH_4$, however, by changing reaction conditions. Reduction of cyclopentenone with $LiAlH_4$ in diethyl ether at -70°C gave an 85% yield of cyclopentenol.[219]

The preference for 1,2-reduction is explained by the Lewis acidity of AlH_3. Initial complexation of aluminum hydride with the carbonyl (as in **179**) makes it difficult for the hydride to be delivered to the alkene moiety via a six-center transition state. The four-center transition state required for delivery of hydride to the carbonyl carbon is, by comparison, very easy to attain. If the complexation is strong, as with AlH_3, then 1,2-reduction predominates. When the coordination is relatively weak, as with $LiAlH_4$, both 1,2- and 1,4-reduction are observed. Aluminum hydride reduces other functional groups as well. In the synthesis of a new phenanthro[9,10,3',4']indolizidine by Wang and co-workers,[220a] lactam **180** was reduced to amine **181** in 80% yield using a mixture of $LiAlH_4$/ $AlCl_3$. In a synthesis of bis(benzofuran) cations with antiprotozoal activity by Tidwell and co-workers, bis(benzofuran-2-yl) ketones were reduced with $LiAlH_4$/$AlCl_3$ to the corresponding bis(benzofuran-2-yl)methane.[220b]

217. Diner, U.E.; Davies, H.A.; Brown, R.K. *Can. J. Chem.* **1967**, *45*, 207.
218. Wilson, S.R.; Zucker, P.A. *J. Org. Chem.* **1988**, *53*, 4682.
219. Brown, H.C.; Hess, H.M. *J. Org. Chem.* **1969**, *34*, 2206.
220. (a) Wang, K.; Wang, Q.; Huang, R. *J. Org. Chem.* **2007**, *72*, 8416; (b) Bakunova, S.M.; Bakunov, S.A.; Wenzler, T.; Barszcz, T.; Werbovetz, K.A.; Brun, R.; Hall, J.E.; Tidwell, R.R. *J. Med. Chem.* **2007**, *50*, 5807.

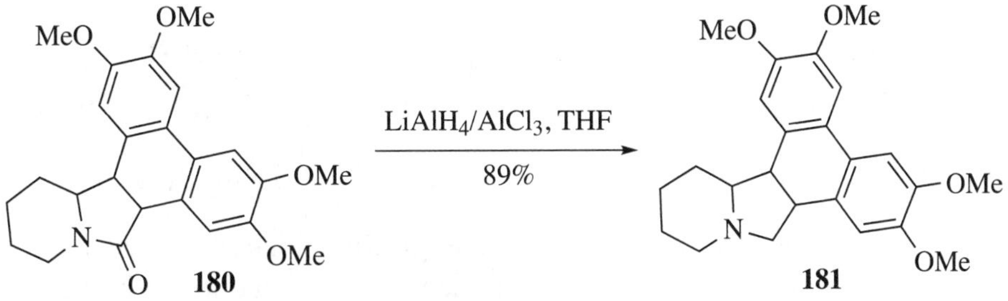

4.6.C. Diisobutylaluminum Hydride

It is often difficult to generate pure AlH_3, which may cause problems in reactions with sensitive substrates. Alkylalanes have been developed that are analogous to the alkylboranes, and these have proven to be quite useful. A commercially available and commonly used alkyl derivative is diisobutylaluminum hydride ($[(CH_3)_2CHCH_2]_2AlH$, **Dibal** or Dibal-H), prepared by refluxing triisobutylaluminum in the solvent heptane, although other solvents can also be used.[221] It is a pyrophoric liquid when used neat, so it is commonly used as a heptane or ether solution. Dibal-H is a strong reducing reagent, reducing most functional groups but the synthetic utility lies in three areas: reduction of esters, reduction of lactones, and reduction of nitriles. Esters are commonly reduced to the alcohol analogous to $LiAlH_4$, as illustrated by reduction of the ester unit in **182** to the hydroxymethyl group in **183** in 86% yield, in Trost and co-worker's synthesis of marcfortine B.[222] Note that the lactam moieties are not reduced under these conditions, and PMB is *para*-methoxybenzyl. Another quite useful reduction involves the reduction of a Weinreb amide (see sec. 2.5.C) to an aldehyde, as in the reduction of **184** to **185**, taken from Leahy and co-worker's synthesis of rhizoxin D.[223]

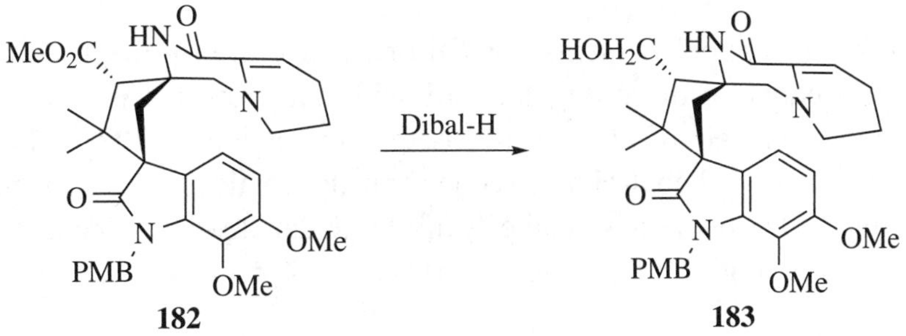

221. (a) Ziegler, K.; Geller, H.G.; Lehmkuhl, H.; Pfohl, W.; Zosel, K. *Annalen,* **1960,** *629,* 1; (b) Reference 15, p 260; (c) Eisch, J.J.; Kaska, W.C. *J. Am. Chem. Soc.* **1966,** *88,* 2213; (d) Gensler, W.J.; Bruno, J.J. *J. Org. Chem.* **1963,** *28,* 1254.
222. Trost, B.M.; Cramer, N.; Bernsmann, H. *J. Am. Chem. Soc.* **2007,** *129,* 3086.
223. Lafontaine, J.A.; Provencal, D.P.; Gardelli, C.; Leahy, J.W. *J. Org. Chem.* **2003,** *68,* 4215.

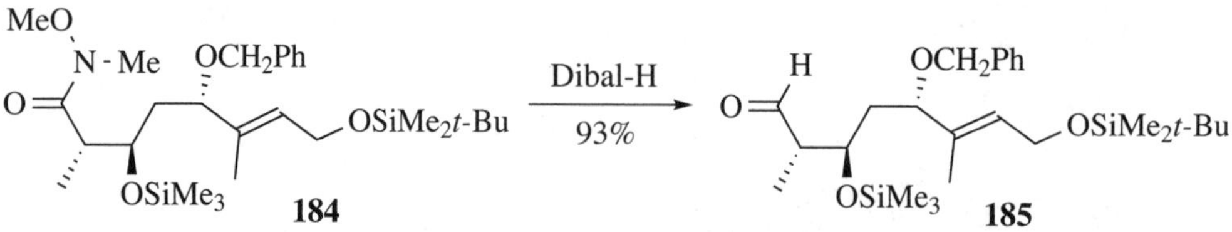

In some cases, it is possible to limit the reduction to give an aldehyde.[224] Generally, reduction of an ester at low temperature leads to the aldehyde whereas reduction at higher temperatures leads to the alcohol. An example is the reduction of the ester unit in **186** in toluene at –78°C to give aldehyde **187** in 93% yield as part of the Dias and Meira synthesis of the C11-C11 fragment of callystatin A.[225] In that same work, but later in the synthesis conjugated ester **188** was reduced to alcohol **189** with Dibal-H in dichloromethane at 0°C (in 95% yield). Clearly, both the solvent and the temperature plays a role in this reduction. In many cases, however, significant amounts of alcohol are formed even at low temperatures.

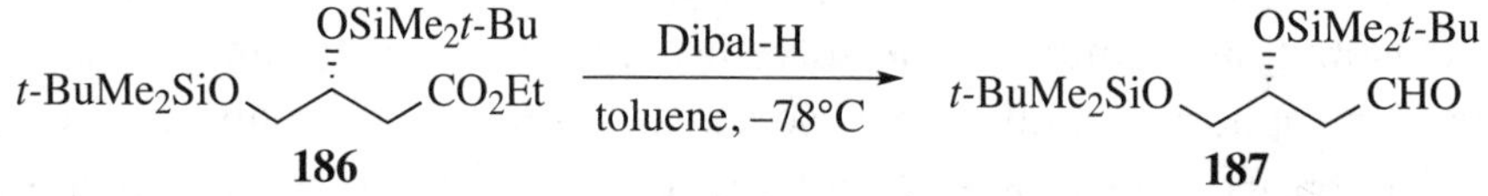

Dibal-H is also used for thereduction of lactones to lactols, which are not always stable. In a synthesis of crispatenine by Parrain and co-workers,[226] lactone **190** was reduced to lactol **191** in 95% yield with Dibal.

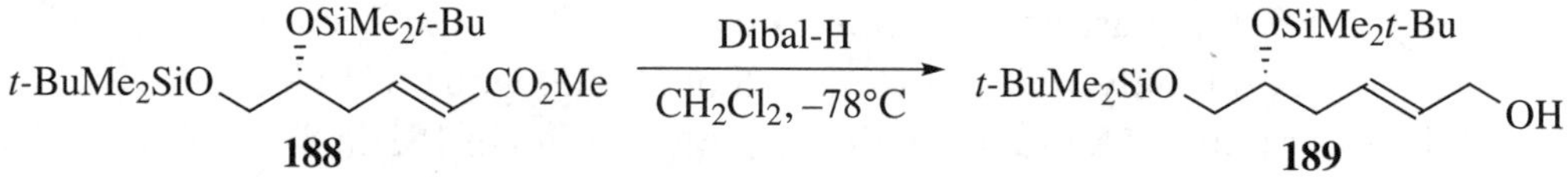

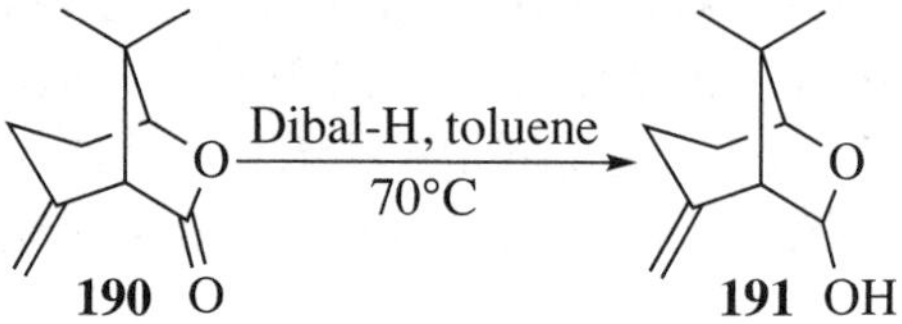

When mixed with alkylcopper reagents (MeCu or *t*-BuCu are the most common), Dibal-H gives selective 1,4-reduction of conjugated carbonyl compounds. An example, taken from Corey and Huang's synthesis of desogestrel, is the conversion of **192** to **193** in 91% yield using HMPA as a solvent.[227] Dibal-H also reduces nitriles to aldehydes. A synthetic example, taken from Andrus' and co-worker's synthesis of (+)-geldanamycin,[228] treated nitrile **194** with Dibal-H and then water to give a 92% yield of **195**.

224. (a) Boger, J.; Payne, L.S.; Perlow, D.S.; Lohr, N.S.; Poe, M.; Blaine, E.H.; Ulm, E.H.; Schorn, T.W.; LaMont, B.I.; Lin, T.-Y.; Kawai, M.; Rich, D.H.; Veber, D.H. *J. Med. Chem.* **1985**, *28*, 1779; (b) Luly, J.R.; Hsiao, C.-N.; BaMaung, N.; Plattner, J.J. *J. Org. Chem.* **1988**, *53*, 6109.
225. Dias, L.C.; Meira, P.R.R. *Tetrahedron Lett.* **2002**, *43*, 8883.
226. Bourdron, J.; Commeiras, L.; Audran, G.; Vanthuyne, N.; Hubaud, J.C.; Parrain, J.-L. *J. Org. Chem.* **2007**, *72*, 3770.
227. Corey, E.J.; Huang, A.X. *J. Am. Chem. Soc.* **1999**, *121*, 710.
228. Andrus, M.B.; Meredith, E.L.; Hicken, E.J.; Simmons, B.L.; Glancey, R.R.; Ma, W. *J. Org. Chem.* **2003**, *68*, 8162.

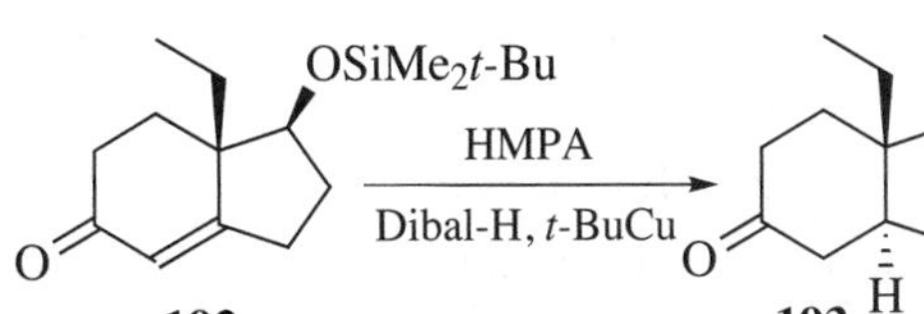

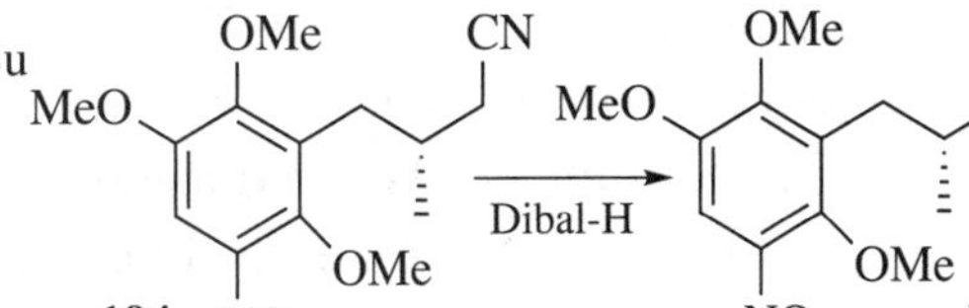

HMPA = hexamethylphosphoramide

It is clear that borane and aluminum hydride, along with their alkyl derivatives, offer significant advantages for the reduction of many functional groups. Borane is often the reagent of choice for reduction of acids in the presence of esters, aluminum hydride for selective 1,2-reduction of conjugated carbonyls, and Dibal-H for partial reduction of lactones and conversion of esters and nitriles to aldehydes. It should be pointed out that Dibal-H also reduces ketones and aldehydes. Reduction of chiral ketone **196** in Table 4.4,[229] which has a methoxy group α- to the carbonyl that is subject to hydrogenolysis, was reduced to a 99:1 mixture of the *anti*-alcohol (**197**) and the *syn*-alcohol (**198**).[229] This example was included to raise a specific question. Why is Dibal-H so selective that **197** is formed as the major product (why is it so diastereoselective)? That question is addressed in Section 4.7.

Table 4.4. Diastereoselectivity with Different Reducing Agents for the Reduction of 196

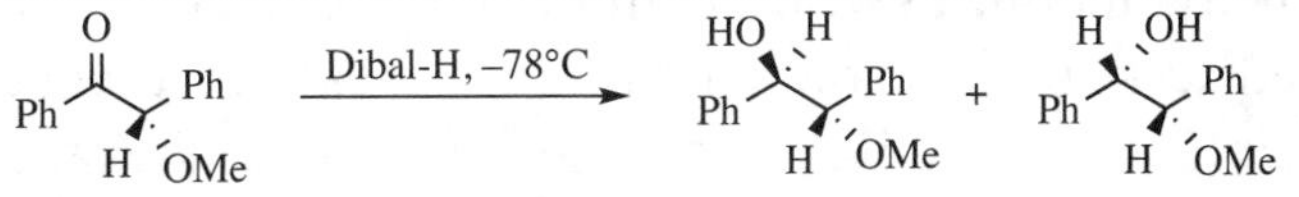

Reducing Agent	Conditions	% Yield	197 : 198
LiAlH₄	THF, 0°C, 5h	89	93.5 : 6.5
	THF, –78°C, 1h	85	95 : 5
	THF-HMPA, 0°C, 5h	82	90.4 : 9.6
	ether, 0°C, 0.5h	84	97 : 3
	ether, –78°C, 1h	83	98 : 2
Dibal-H	toluene, –78°C, 2.5h	78	87 : 13
	THF, –78°C, 1h	85	99 : 1
KBH₄	MeOH (aq), 0°C, 0.5h	86	90 : 10
K-Selectride	THF, –78°C, 1h	79	98 : 2
	ether, –78°C, 3h	68	98 : 2
PhSiMe₂H, Bu₄NF	HMPA, 25°C, 32h	0	
H₂/Pd	EtOH, 25°C, 5h	98	76 : 24

[Reprinted with permission from Davis, F.A.; Serajul Haque, M.; Przeslawski, R.M. *J. Org. Chem.* **1989**, *54*, 2021. Copyright © **1989** American Chemical Society.]

4.7. STEREOSELECTIVITY IN REDUCTIONS

Hydride reduction of prochiral ketones generally gives alcohol products with good diastereoselectivity, but each diastereomer is racemic. An example is the reduction of **196** with

229. Davis, F.A.; Haque, M.S.; Prezeslawski, R.M. *J. Org. Chem.* **1989**, *54*, 2021.

various reducing agents shown in Table 4.4,[230a] where it is clear that the diastereoselectivity (anti/syn selectivity) for reduction of such substrates depends on the reducing agent as well as the reaction conditions. When diastereomers are formed by reduction of carbonyl groups, is it possible to predict the relative stereochemistry of such reductions and also the absolute stereochemistry? The following sections will attempt to answer these questions.

4.7.A. Chiral Additives in Acyclic Systems

The stereochemistry of a product is an important consideration when a functional groups is reduced, as seen in Table 4.4. When 2-pentanone is reduced to 2-pentanol, for example, the carbonyl is a prochiral center (sec. 1.4.E) since addition of hydride from either of the two faces (a or b) of the carbonyl unit in 2-pentanone will generate the two enantiomers of 2-pentanol.[230b] Both faces are equally susceptible to attack, so the alcohol product is racemic and most hydride reagents discussed previously will generate racemic products. Acyclic ketones or aldehydes without a pendant chiral center will also give racemic alcohols. Enantioselectivity is not possible with the simple reagents presented so far, but the use of a chiral reducing agent is expected to give enantioselective reductions.

If a stereogenic center is present in the molecule, reduction will generate diastereomers. Delivery of hydride to the carbonyl unit in **199** can be from both the re and si faces. The presence of the (R)-stereocenter adjacent to the carbonyl, however, means that reduction from the re face gives one diastereomer (**200**) and delivery from the si face gives the other diastereomer (**201**). In some cases, the stereocenter in the ketone can influence the stereochemistry of the newly formed hydroxyl group in the alcohol product. Reductions of this type will be discussed in section 4.7.B.

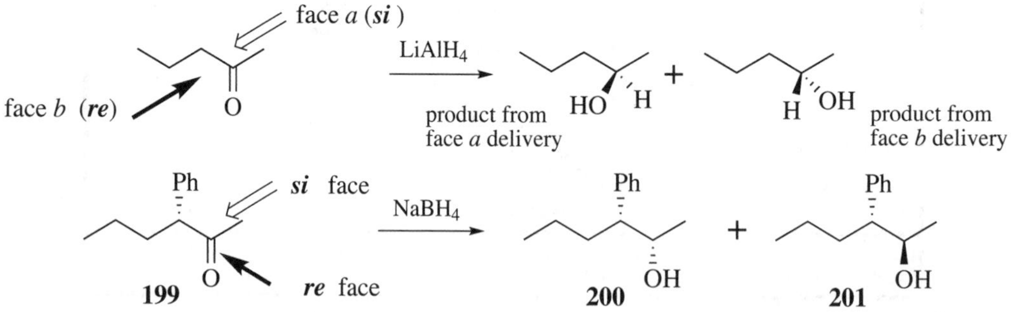

Development of chiral reducing agents led to enantioselective reductions. Morrison and Mosher described this approach in a monograph that covered the literature up to approximately

230. (a) Harding, K.E.; Strickland, J.B.; Pommerville, J. *J. Org. Chem.* **1988**, *53*, 4877; (b) For reagents that lead to asymmetric reduction of aldehydes or ketones, see Reference 17, p 1097.

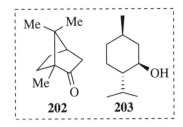

202   203

1970.[231] Early work prepared chiral reagents by reacting hydride reagents with natural products that contained stereogenic centers of known absolute stereochemistry, such as (−)-camphor (**202**)[232,233] or (−)-menthol (**203**).[233b] Reaction of these molecules with LiAlH₄ prior to addition of a substrate with a prochiral carbonyl led to a reagent capable of reducing the carbonyl with modest asymmetric induction.

These reagents were not very effective, however. Reduction of methyl *tert*-butyl ketone (3,3-dimethyl-2-butanone), for example, showed no asymmetric induction when the product (3,3-dimethyl-2-butanol) was analyzed. Other chiral natural products have been attached to LiAlH₄, leading to reductions that proceed with moderate-to-good enantioselectivity. Quinine reacted with LiAlH₄ to give **204**, whereas (+)-(*2S,2R*)-1,2-diphenyl-3-methyl-4-dimethylamino-2-butanol (Darvon alcohol; Darvon is the O₂CEt ester of this alcohol) gave **205**, and ephedrine gave **206**. The lithium aluminum hydride products of both (+)-quinidine (**207**, a diastereomer of quinine) and (+)-cinchonine (the derivative of quinidine without the OMe group on the quinoline unit) are chiral reducing agents.[231] *Cinchona* alkaloids such as these were used by Sharpless as chiral additives in the **Sharpless asymmetric dihydroxylation** reaction, which was discussed in Section 3.5.B.ii. Acetophenone reacted with **204**, for example, to give the (*R*)-phenethyl alcohol in 48% ee.[234] Similar reduction of acetophenone with **205** gave the alcohol with 45% ee, but **206** gave only 13% ee (*R*)-and **207** 23% ee (*S*).[224]

204   205   206   207

Lithium aluminum hydride derivatives of monosaccharides reduce some ketones with moderate % ee. The bis(alkoxy) derivative of LiAlH₄-3-*O*-benzyl-1,2-cyclohexylidene α-D-glucofuranose (**208**), and the analogous 1:1 ethanol adduct (**209**) are examples.[235] Reducing agent **208** reacted with acetophenone to give (*S*)-phenethyl alcohol in 33% ee. Adduct **209** showed greatly improved but opposite selectivity, giving 71% ee in the product (*R*)-phenethyl alcohol when reacted with acetophenone.

231. Morrison, J.D.; Mosher, H.S. *Asymmetric Organic Reactions*, American Chemical Society, Washington, DC *1976*, pp. 202-218.
232. Bothner-By, A.A. *J. Am. Chem. Soc.* *1951*, *73*, 846.
233. (a) Portoghese, P.S. *J. Org. Chem.* *1962*, *27*, 3359; (b) Landor, S.R.; Miller, B.J.; Tatchell, A.R. *Proc. Chem. Soc.* *1964*, 227; (c) *Idem J. Chem. Soc. C 1966*, 1822.
234. Reference 231, p 205 and citations 75 and 74, 76-80 cited therein.
235. (a) Reference 231, p 214; (b) Landor, S.R.; Tatchell, A.R. *J. Chem. Soc. C 1966*, 2280; (c) Landor, S.R.; Miller, B.J.; Tatchell, A.R. *Ibid. 1967*, 197.

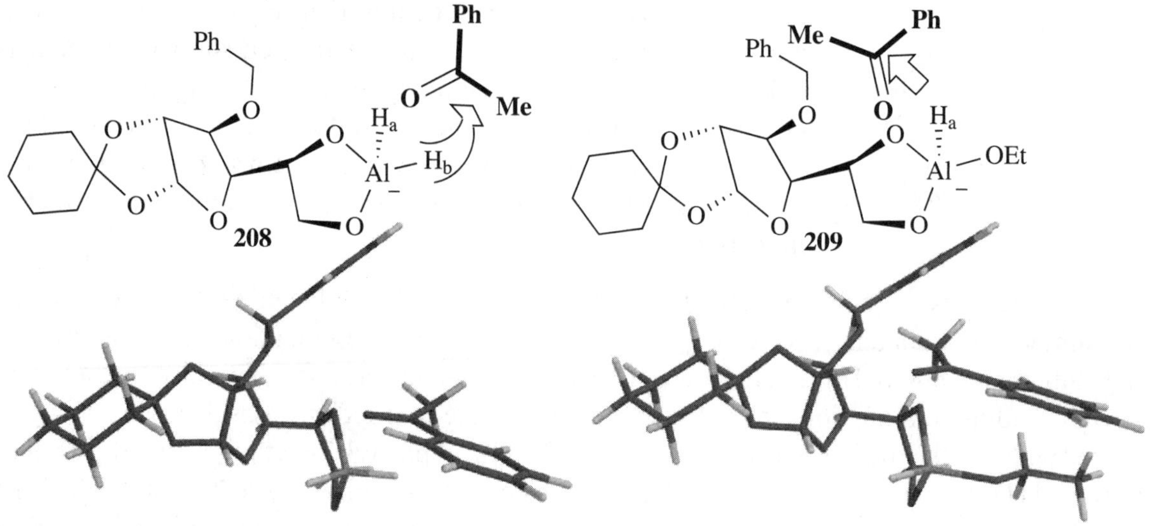

The structural features that lead to asymmetric induction vary with the reagent, making generalization difficult. Landor found that H_b in **208**, for example, was transferred to a carbonyl group during reduction faster than H_a. Acetophenone was reduced to give the (S)-alcohol, and this demands that the ketone approach H_b in **208** away from the protruding O-benzyl group as shown. This model corresponds to approach of the reagent from the *re* face. Presumably, the methyl-oxygen lone-pair interaction is less than the phenyl-oxygen lone pair interaction. This difference is not great and the observed enantioselectivity was only modest. The 3D models of both **208** and **209** are provided, with an acetophenone molecule approaching from the face indicated in the 2D model. When ethanol was incorporated into the reagent (see **209**), H_b was replaced with an ethoxy group so H_a was the only source of hydride. The carbonyl must now approach on the same face of the carbohydrate as before, but to reduce steric interactions the orientation of the ketone relative to H_a places the larger phenyl group up and away from the sugar moiety (the *si* face). This approach leads to reduction that generates the (R)-alcohol.[235]

This steric analysis is attractive, does not apply to reagents such as the LiAlH$_4$-Darvon alcohol reagent (see **205**), prepared by Mosher (Darvon alcohol is **210**).[236] First of all, the hydride reacts to form an alkoxyaluminate (**207**) where the value of *n* may be 1, 2, or 3 (although *n* was shown to be 1 in **211** and is probably 2 after the initial reaction of LiAlH$_4$ with **210**). Second, the coordination state of the aluminum is unknown (both tetrahedral and octahedral coordination complexes are possible). Finally, the age of the reagent is important. Freshly prepared **211** is insoluble in ether solvents and reacts selectively to give the (R)-phenethyl alcohol from acetophenone. If the reagent is allowed to age, it becomes soluble and gives the (S)-alcohol when used for reduction, but this is probably not due to formation of a tris(alkoxy)aluminate (*n* = 3). These ambiguities preclude the simple steric analysis that was possible with the structurally

236. (a) Yamaguchi, S.; Mosher, H.S. *J. Org. Chem.* **1973**, *38*, 1870; (b) Yamaguchi, S.; Mosher, H.S.; Pohland, A. *J. Am. Chem. Soc.* **1972**, *94*, 9254.

well-defined **208** and **209**. When the structure of the active reducing agent is known, steric and facial arguments can predict approach to the carbonyl from the less hindered face with an orientation that minimizes steric interactions (generating the major diastereomer or enantiomer).

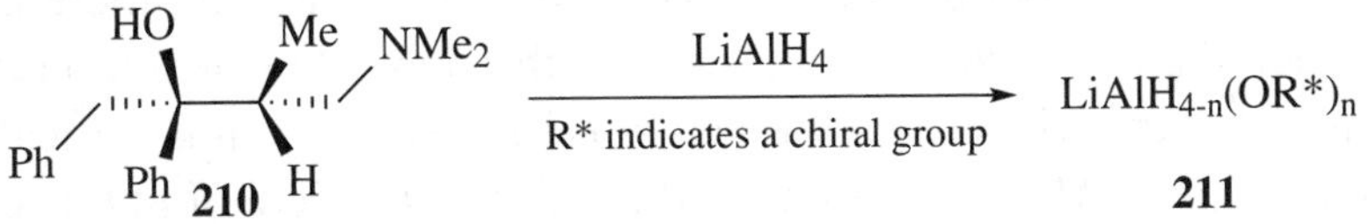

Borane was shown to be a potent reducing agent in section 4.6.A and chiral boranes function as reducing agents for carbonyl compounds.[237] Brown prepared a variety of chiral boranes in connection with extensive studies of hydroboration reactions (sec. 5.4.B). Reaction of diborane and α-pinene gave (–)-(1*R*,2*S*,3*R*,5*R*)-diisopinocampheylborane (**212**), and this chiral borane reduced carbonyl derivatives to alcohols with high asymmetric induction.[238] Trialkylboranes are also capable of reducing carbonyls, but by a different mechanism (i.e. hydride comes from C–H rather than Al–H or B–H). Boranes such as the (–)-(1*R*,2*S*,3*S*,5*S*) complex between pinene and 9-BBN [**213** (R' = H), B-3-pinanyl-9-borabicyclo-[3.3.1]nonane = Alpine borane®] are chiral[239] and were used by Midland and co-workers for the asymmetric reduction of aldehydes and ketones. Alpine borane® reduced *d*-butanal to *d₁*-1-butanol with high enantioselectivity.[240] The source of hydride in Alpine borane® (and generally in trialkylboranes) is the tertiary hydrogen (R' = H) that has been replaced by deuterium in **213** (R' = *d*). Reduction of butanal gave the *(R)*-deuterobutanol (**211**, where R' = *d*) and pinene (**214**).[231] The first important consideration is the observation that tertiary β-hydrogens in these boranes are removed much faster than the secondary or primary β-hydrogens.[241,239a] Approach of the carbonyl to this hydrogen must occur from the face (**216**) that is nearest to the tertiary β-hydrogen (R' = D in **217**). It might

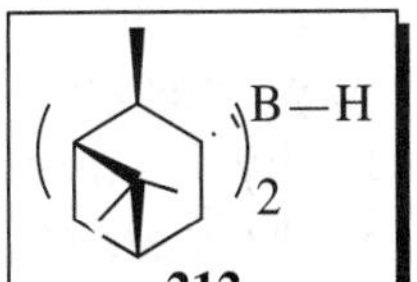

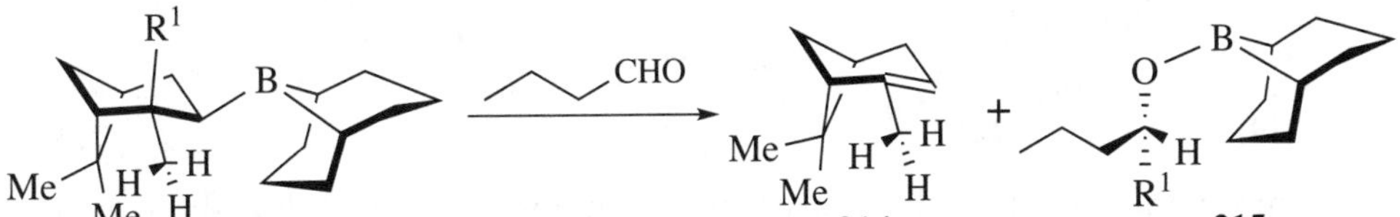

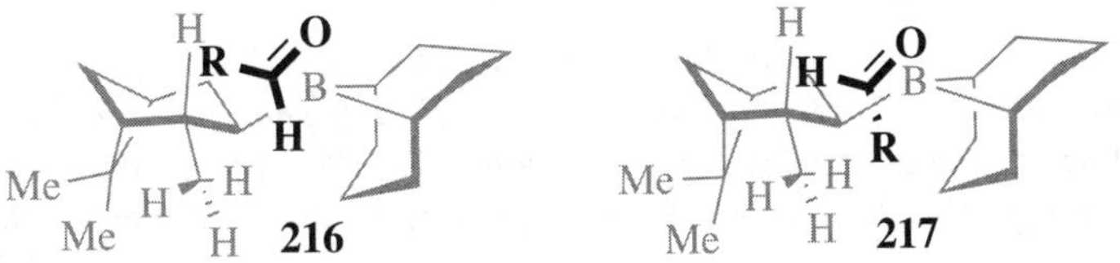

237. For a discussion of factors controlling stereoselectivity in such reductions, see Rogic, M.M. *J. Org. Chem.* **2000**, *65*, 6868.
238. (a) Brown, H.C.; Zweifel, G. *J. Am. Chem. Soc.* **1961**, *83*, 486; (b) Brown, H.C.; Bigley, D.B. *Ibid.* **1961**, *83*, 3166.
239. (a) Midland, M.M.; Tramontano, A.; Zderic, S.A. *J. Organomet. Chem.* **1978**, *156*, 203; (b) Midland, M.M.; Tramontano, A.; Zderic, S.A. *J. Am. Chem. Soc.* **1977**, *99*, 5211.
240. (a) Midland, M.M.; Greer, S.; Tramontano, A.; Zdevic, S.A. *J. Am. Chem. Soc.* **1979**, *101*, 2352; (b) Midland, M.M. McDowell, D.C.; Hatch, R.L.; Tramontano, A. *J. Am. Chem. Soc.* **1980**, *102*, 867.
241. Midland, M.M.; Tramonatano, A.; Zderic, S.A. *J. Organomet. Chem.* **1977**, *134*, C17.

appear at first that the R= group in the aldehyde (R = *n*-propyl) would prefer to be oriented away from the pinanyl ring, but in fact the opposite is true. Examination of **216** shows the R group is over the pinanyl ring but away from the protruding bridgehead hydrogen, which is the required approach for reduction (the *re* face of the aldehyde) and there is minimal steric interaction.[240] When the aldehyde approaches with the opposite orientation (the *si* face as in **217**) the R group interacts with the hydrogens on the methyl group, sterically destabilizing **217**.[240] The orientation in **216** is preferred, leading to the *(R)*-alcohol in this case. The extent of selectivity observed in the reduction of ketones depends on the relative size and nature of the groups attached to the carbonyl. An example is taken from a synthesis of (−)-archazolid B[242] by Trauner and co-workers, in which propargylic ketone **218** was reduced to alcohol **219** in 89% yield (>20:1 dr).

The asymmetric reduction of ketones by borane is catalyzed by chiral oxazaborolidines,[243] and is widely used for the preparation of chiral secondary alcohols.[244] The factors that influence enantioselectivity of the asymmetric reduction have been described,[245] as have the influence of electronic effects and anions on the enantioselectivity of such reductions.[246] In a synthesis of (+)-cryptophycin 52,[247] a potent antimitotic antitumor agent, Ghosh and Swanson reduced keto ester **220** with borane using a catalytic amount of oxazaborolidine **221**, which had been previously prepared by Corey.[248] The chiral alcohol product **222**, formed initially, cyclized via an intramolecular transesterification to lactone **223** in 84% yield and 97% optical purity.

With all reagents generated by addition of chiral materials to LiAlH$_4$, their exact structure

242. Roethle, P.A.; Chen, I.T.; Trauner, D. *J. Am. Chem. Soc.* **2007**, *129*, 8960

243. Itsuno, S.; Sakurai, Y.; Ito, K.; Hirao, A.; Nakashama, S. *Bull. Chem. Soc. Jpn.* **1987**, *60*, 395.

244. (a) Corey, E.J.; Helal, C.J. *Angew. Chem. Int. Ed.*, **1998**, *37*, 1986; (b) Deloux, L.; Srebnik, M. *Chem. Rev.* **1993**, *93*, 763; (c) Singh, V.K. *Synthesis* **1992**, 605; (d) Wallbaum, S.; Martens, J. *Tetrahedron:Asymmetry* **1992**, *3*, 1475.

245. (a) Corey, E.J.; Bakshi, R.K.; Shibata, S. *J. Am. Chem. Soc.* **1987**, *109*, 5551; (b) Corey, E.J.; Link, J.O. *Tetrahedron Lett.* **1989**, *30*, 6275; (c) Puigjaner, C.; Vidal-Ferran, A.; Moyano, A.; Pericas, M.A.; Riera, A. *J. Org. Chem.* **1999**, *64*, 7902; (d) Gilmore, N.J.; Jones, S. *Tetrahedron: Asymmetry* **2003**, *14*, 2115; (e) Xu, J.X.; Wei, T.Z.; Zhang, Q.H. *J. Org. Chem.* **2003, *68*, 10146.

246. Xu, J.; Wei, T.; Zhang, Q. *J. Org. Chem.* **2004**, 69, 6860.

247. Ghosh, A.K.; Swanson, L. *J. Org. Chem.* **2003**, *68*, 9823.

248. Corey, E.J.; Bakshi, R.K.; Shibata, S.; Chen, C.; Singh, V.K. *J. Am. Chem. Soc.* **1987**, *109*, 7925.

must be determined before facial and orientation preferences can be determined for a given substrate. As illustrated by Alpine borane®, the mechanism of hydride transfer must also be known. When these parameters are known, careful examination of models (Dreiding® models or computer generated modeling) should determine the actual steric interactions. Casual inspection of a line drawing is usually insufficient for accurate predictions.

Morrison and Mosher collected data for asymmetric induction in the reduction of several acyclic ketones and their monograph is an excellent source when one must choose an asymmetric reducing agent.[249] The use of (–)-212 for reduction of acetophenone in diglyme, for example, gave 14% ee [(R)-phenethyl alcohol]. Similar reduction of ethyl isopropyl ketone (2-methyl-3-propanone) gave 62% ee of the (S)-alcohol in THF.[249] Many reagents have been developed and used in organic synthesis. Brown compared several chiral boranes with other chiral complexes for the reduction of several cyclic and acyclic ketones.[250] Other chiral reagents used for reduction include enantrane (224),[251] diisopinylcamphenyl chloroborane (Ipc₂BCl, 225),[252] the mixture of diborane and AMDPB (226),[253] (2R,5R)-dimethylborolane (227),[254]

224   **225**   **226**   **227**

NB-enantride (228),[255] (S,S')-N,N'-dibenzoylcystine and (229) and lithium borohydride,[256] 1,2,5,6-di-O-isopropylidene-D-glucofuranose (230) and sodium borohydride,[257] K-glucoride (231),[258] a mixture of LiAlH₄ and N-methylephedrine (232) in the presence of

249. Reference 231, p 216 and citations 84-87 therein.
250. Brown, H.C.; Park, W.S.; Cho, B.T.; Ramachandran, P.V. *J. Org. Chem.* **1987**, *52*, 5406 and references cited therein.
251. Midland, M.M.; Kazubski, A. *J. Org. Chem.* **1982**, *47*, 2814.
252. (a) Chandrasekharan, J.; Ramachandran, P.V.; Brown, H.C. *J. Org. Chem.* **1985**, *50*, 5446; (b) Brown, H.C.; Chandrasekharan, J.; Ramachandran, P.V. *Ibid.* **1986**, *51*, 3394. For a review, see (c) Ramachandran, P.V.; Brown, H.C. in *Reductions in Organic Synthesis. Recent Advances and Practical Applications*, Abdel-Magid, A.F (Ed.), ACS Symposium Series 641, American Chemical Society, Washington, D.C., **1996**. pp 84-97
253. (a) Itsuno, S.; Nakano, M.; Miyazaki, K.; Masuda, H.; Ito, K.; Hirao, A.; Nakahama, S. *J. Chem. Soc. Perkin Trans. 1,* **1985**, 2039; (b) Itsuno, S.; Ito, K.; Hirao, A.; Nakahama, S. *J. Org. Chem.* **1984**, 49, 555; (c) Hirao, A.; Itsuno, S.; Nakahama, S.; Yamazaki, N. *J. Chem. Soc. Chem. Commun.* 1981, 315.
254. Imai, T.; Tamura, T.; Yamamuro, A.; Sato, T.; Wollmann, T.A.; Kennedy, R.M.; Masamune, S. *J. Am. Chem. Soc.* **1986**, *108*, 7402.
255. Midland, M.M.; Kazubski, A. *J. Org. Chem.* **1982**, *47*, 2495.
256. (a) Soai, K.; Oyamada, H.; Yamanoi, T. *J. Chem. Soc. Chem. Commun.* **1984**, 413; (b) Soai, K.; Yamanoi, T.; Hikima, H.; Oyamada, H. *Ibid.* **1985**, 138.
257. (a) Hirao, A.; Nakahama, H.; Mochizuki, D.; Itsuno, S.; Ohowa, M.; Yamazaki, N. *J. Chem. Soc. Chem. Commun.* **1979**, 807; (b) see Morrison, J.D.; Grandbois, E.R.; Howard, S.I. *J. Org. Chem.* **1980**, *45*, 4229.
258. Brown, H.C.; Park, W.S.; Cho, B.T. *J. Org. Chem.* **1986**, *51*, 1934, 3278, 3396.

3,5-dimethylphenol,[259] a mixture of LiAlH$_4$ and (S)-2-(2,6-xylidinomethyl)-pyrrolidine (**236**),[260] the lithium aluminum hydride-(S)-(4-anilino)-3-(methylamino)-1-butanol complex (**233**),[261] Binal-H (**235**),[262] the lithium aluminum hydride-ethanol-(S)-(−)-10,10'-dihydroxy-9,9'-biphenanthryl complex (**234**),[263] the lithium aluminum hydride-**232**-N-ethylaniline complex,[264] the lithium aluminum hydride-**232**-2-(ethylamino)-pyridine complex,[265] and an enzymatic reduction (sec. 4.10.F) with *Thermoanaerobium brockii* alcohol dehydrogenase (**237**).[266] Brown and co-workers collected these data into Table 4.5,[250] to identify the preferred chiral reducing agents for different classes of ketones. Table 4.5 provides general data and it is not exhaustive, but the term effectiveness presumably refers to the order in which the reagents appear. The reagent giving the highest % ee with good yields of the reduction products appears at the top of the list with the following reagents being less effective. This table allows one to choose a chiral additive and reducing agent for reduction of a given type of ketone. Brown also described a series of chiral boronic ester *ate* complexes that showed high diastereoselectivity for the reduction of ketones.[267]

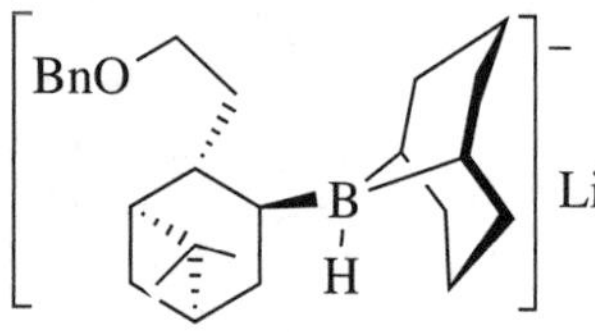

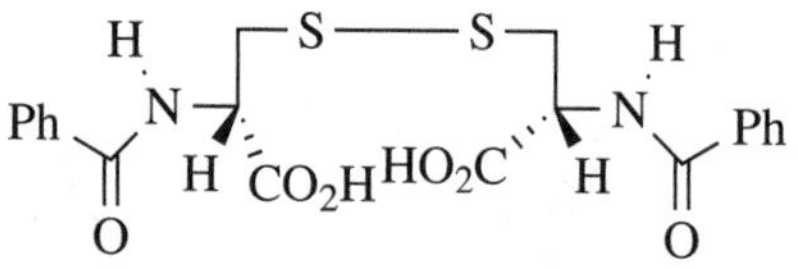

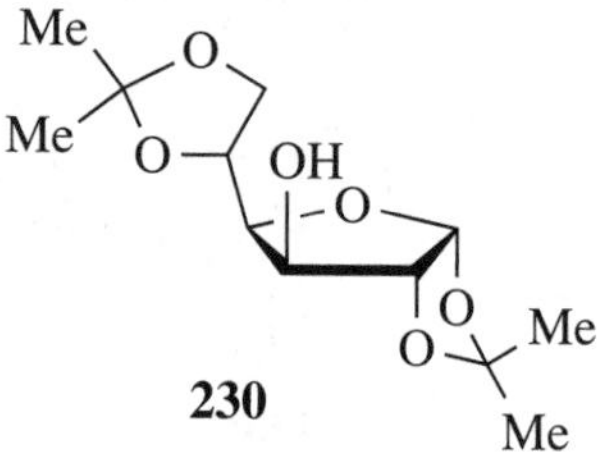

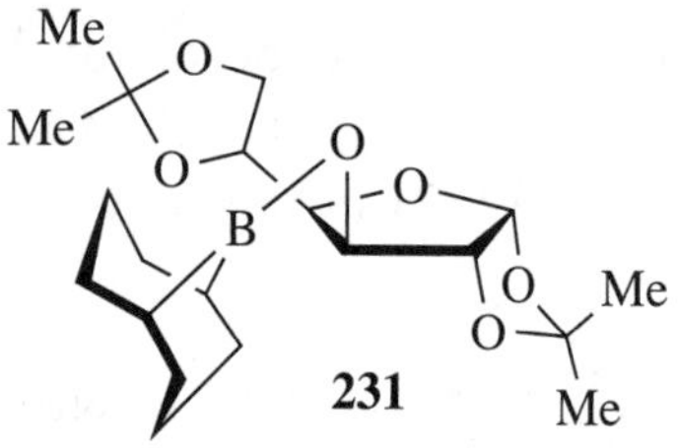

259. (a) Vigneron J.-P.; Bloy, V. *Tetrahedron Lett.* **1979**, 2683; (b) Vigneron J.-P.; Jacquet, I. *Tetrahedron* **1976**, *32*, 939; (c) Jacquet, I.; Vigneron J.-P. *Tetrahedron Lett.* **1974**, 2065.
260. Asami, M.; Mukaiyama, T. *Heterocycles* **1979**, *12*, 499.
261. (a) Sato, T.; Gotoh, Y.; Wakabayashi, Y.; Fujisawa, T. *Tetrahedron Lett.* **1983**, *24*, 4123; (b) Sato, T.; Goto, Y.; Fujisawa, T. *Ibid.* **1982**, *23*, 4111.
262. (a) Noyori, R.; Tomino, I.; Tanimoto, Y.; Nishizawa, M. *J. Am. Chem. Soc.* **1984**, *106*, 6709; (b) Noyori, R.; Tomino, I.; Yamada, M.; Nishizawa, M. *Ibid.* **1984**, *106*, 6717; (c) Noyori, R.; Tomino, I.; Tanimoto, Y. *Ibid.* **1979**, *101*, 3129; (d) Nishizawa, M.; Yamada, M.; Noyori, R. *Tetrahedron Lett.* **1981**, *22*, 247.
263. Yamamoto, K.; Fukushima, H.; Nakazaki, M. *J. Chem. Soc. Chem. Commun.* **1984**, 1490.
264. Terashima, S.; Tanno, N.; Koga, K. *J. Chem. Soc. Chem. Commun.* **1980**, 1026.
265. Kawasaki, M.; Suzuki, Y.; Terashima, S. *Chem. Lett.* **1984**, 239.
266. Kienan, E.; Hafeli, E.K.; Seth, K.K.; Lamed, R. *J. Am. Chem. Soc.* **1986**, *108*, 162.
267. Brown, H.C.; Cho, B.T.; Park, W.S. *J. Org. Chem.* **1987**, *52*, 4020.

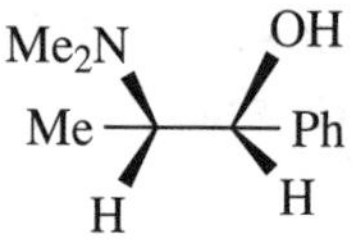

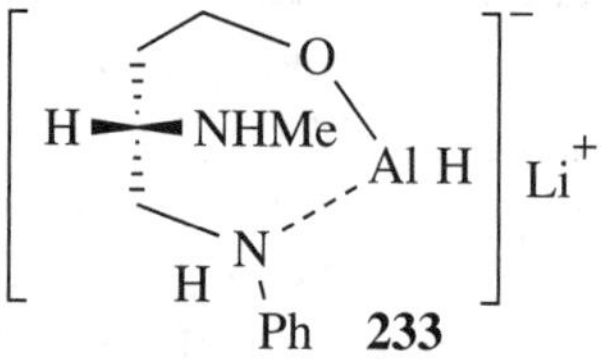

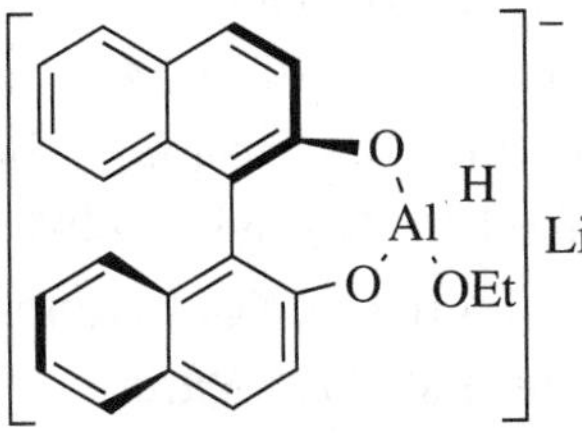

234 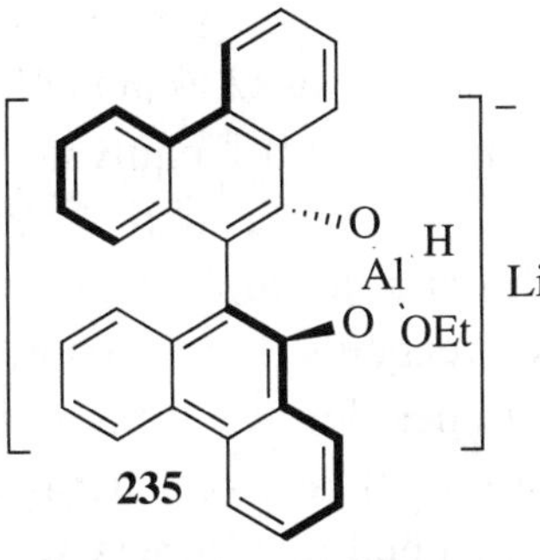   **235** 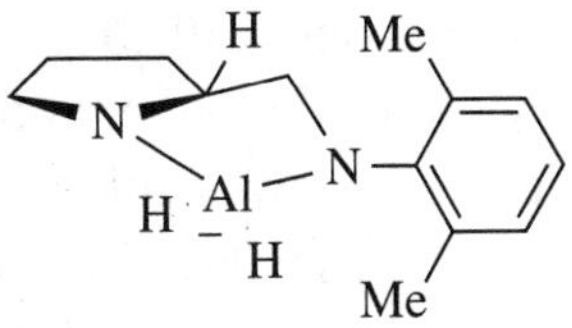   **236**

Table 4.5. Preferred Asymmetric Reducing Agents

Class of Ketones		Preferred Reagents (decreasing order of effectiveness)	
I. Acyclic	a. Unhindered	227	2,5-dimethylborolane
		212	B-Ipc-9-BBN, 6 Kbar
		237	*T. brockii*
	b. Hindered	227	2,5-dimethylborolane
		225	IPc$_2$B–Cl
		226	BH$_3$•AMDPB (2:1)
II. Cyclic		225	IPc$_2$B–Cl
		226	BH$_3$•AMDPB (2:1)
		231	K-Glucoride
III. Aralkyl	a. Unhindered	212	B-Ipc-9-BBN, 6 Kbar
		225	IPc$_2$B–Cl
		233	LiAlH$_4$•DBP•EtOH
		234	Binal-H
		236	LiAlH$_4$•diamine
	b. Hindered	231	K-Glucoride
		236	LiAlH$_4$•diamine
		225	IPc$_2$B–Cl
IV. Heterocyclic		212	B-Ipc-9-BBN, 6 Kbar
		212	B-Ipc-9-BBN, neat
		225	IPc$_2$B–Cl
V. α-Halo		212	B-Ipc-9-BBN, neat
		226	BH$_3$•AMDPB (2:1)
		225	IPc$_2$B–Cl
		234	Binal-H
VI. α-Keto esters		212	B-Ipc-9-BBN, neat
VII. β-Keto esters		229	LiBH$_4$•DBC•t-BuOH
		225	IPc$_2$B–Cl
VIII. Acyclic conjugated enones		234	Binal-H
		A	LiAlH$_4$•MEP•NEA[a]
		236	LiAlH$_4$•aminobutanol
IX. Cyclic conjugated enones		236	LiAlH$_4$•aminobutanol
		B	LiAlH$_4$•MEP•EAP[b]
X. Conjugated ynones		234	Binal-H
		224	NB•Enantrane
		232	LiAlH$_4$•MEP•ArOH
		212	B-Ipc-9-BBN, 6 Kbar

[a]A. **232** + LiAlH$_4$ + 2-ethylaminopyridine[264] [b]B. **232** + LiAlH$_4$ + 2-ethylaminopyridine[265] (*R*)- for 2,2-dimethylcyclopentanone to 16% ee (*S*) for 2-chloroacetophenone to 71% ee (*S*) for 2-cyclohexen-1-one.[267]

[Reprinted with permission from Brown, H.C.; Park, W.S.; Cho, B.T.; Ramachandran, P.V. *J. Org. Chem.*, *1987*, *52*, 5406. Copyright © *1987* American Chemical Society.]

Development of chiral additives for asymmetric reduction is on-going. The, reduction of acetophenone with a **238**-borane complex gave **239** in 88% yield and 92% ee (*R*).[268] Reduction of acetophenone with borane, catalyzed by 0.1 equivalents of oxazaborolidine **240**, gave a >95% yield of the (–)-enantiomer [**241**; 84% ee, (*S*)].[269] When oxazaborolidine-nickel boride complex (**242**) was mixed with borane, acetophenone was reduced to **241** with 94% ee.[270] Note that the use of oxazaborolidine reagents for the enantioselective reduction of ketones was previously described.[271] Oxaborolidines derived from amino alcohols of (1*R*)-camphor have also been used as catalysts.[272] Bolm et al. showed that hyperbranched dendritic amino acids such as **243** (much more highly branched dendrimers were also used) gave reduction when mixed with a borane-dimethyl sulfide complex. In this example, dendritic amino acid **243** reduced acetophenone under these conditions to **239** in 78% yield and 89% ee (*R*).[273]

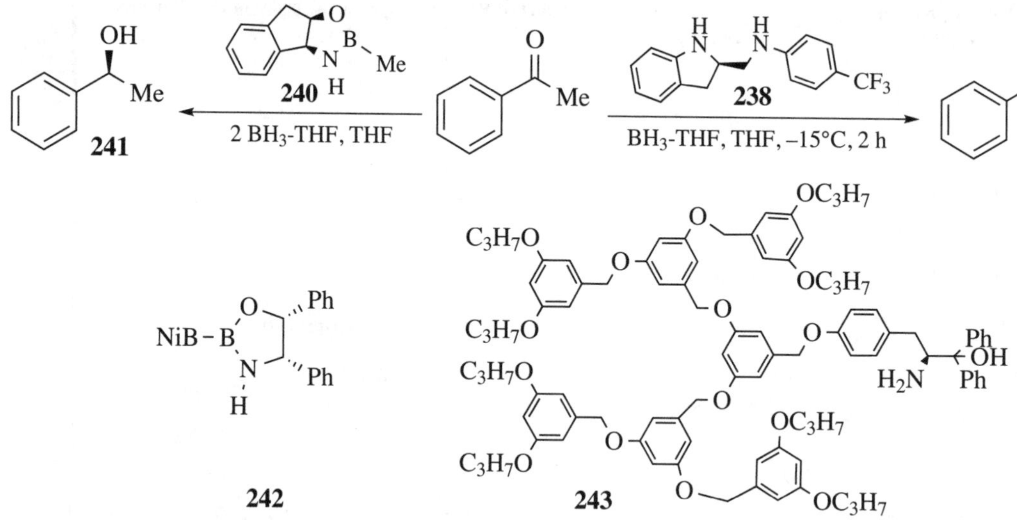

4.7.B. Selectivity in the Reduction of Carbonyl Derivatives Containing a Stereogenic Carbon

As seen in Section 4.7.A, chiral reagents prepared from hydrides gave reasonable diastereoselectivity and the % ee ranged from poor to excellent. When the prochiral center is in a molecule that already contains a stereogenic center, reduction generates diastereomers, as mentioned above with the formation of **200** and **201** from **199**. Two general cases are illustrated by the reduction of **244** or **247**. In the former case, **244** contains a stereogenic

268.   Sato, S.; Watanabe, H.; Asami, M. *Tetrahedron: Asymmetry* **2000**, *11*, 4329.
269.   Jones, S.; Atherton, J.C.C. *Tetrahedron: Asymmetry* **2000**, *11*, 4543.
270.   Molvinger, K.; Lopez, M.; Court, J. *Tetrahedron Lett.* **1999**, *40*, 8375.
271.   (a) King, A.O.; Mathre, D.J.; Tschaen, D.M.; Shinkai, I. in *Reductions in Organic Synthesis. Recent Advances and Practical Applications*, Abdel-Magid, A.F (Ed.), ACS Symposium Series 641, American Chemical Society, Washington, DC, **1996**. pp 98-111; (b) Quallich, G.J.; Blake, J.F.; Woodall, T.M. *Ibid.* pp 112-126.
272.   Santhi, V.; Rao, J.M. *Tetrahedron: Asymmetry* **2000**, *11*, 3553.
273.   Bolm, C.; Derrien, N.; Seger, A. *Chem. Commun.* **1999**, 2087.

center, but it is too far removed from the prochiral center to exert much influence. Note that the absolute configuration of the stereogenic center at C7, in **244**, is unchanged in both **244** and **246**, and C7 does not influence reduction of the carbonyl moiety. Reduction of this compound will be, more or less, the same as reduction of a ketone with no stereogenic center and the resulting mixture should be close to a 1:1 mixture of syn and anti diastereomers (**245** and **246**). Ketone **247**, however, has a stereogenic center adjacent to the carbonyl and its chirality will exert an influence on the approach of the incoming reagent. Reagents will approach from the face presenting the least amount of steric hindrance. In **247A**, approach over the hydrogen atom (the *si* face) will give **249**, but approach over the more hindered methyl group (the *re* face) will give **248** (alcohols **248** and **249** are diastereomers). This analysis assumes that **247A** is frozen into the rotamer shown, which is incorrect of course. This ketone and similar acyclic systems are not locked into a single conformation, and for **247** we can consider two major rotamers with the carbonyl either down (**247A** in the first analysis) or up (**247B**) relative to the methyl group on the stereogenic carbon. In both cases, we have also assumed that the carbonyl bisects the H and Me, although many other rotamers are possible. Reduction from the *si* face in conformation **247B** leads to **248**, and reduction from the *re* face gives **249**. In both rotamers of **247**, the major product of reduction is the one that occurs from the *si* face, over the less sterically demanding hydrogen atom. However, reduction of rotamer **247A** leads to **249** as the major product, whereas reduction of **247B** from the *si* face gives **248** as the major product. If there is no control of the rotamer population at the time of reduction, there is little or no stereoselectivity. Good stereoselectivity in a reaction demands that one face (*re* vs. *si*) is preferred and also that one rotamer is preferred. Prediction of stereochemistry therefore demands a model that makes assumptions about both.

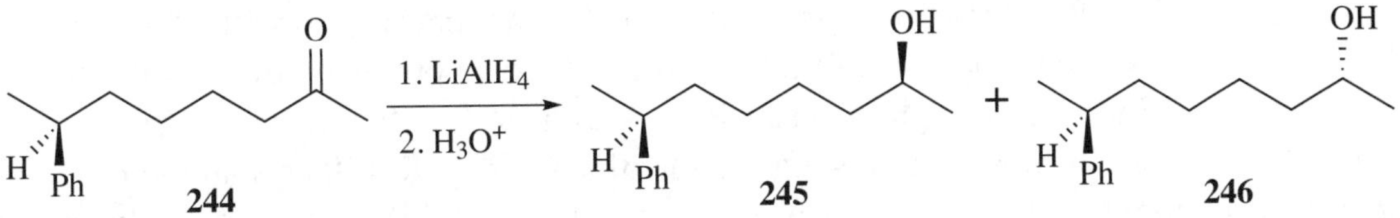

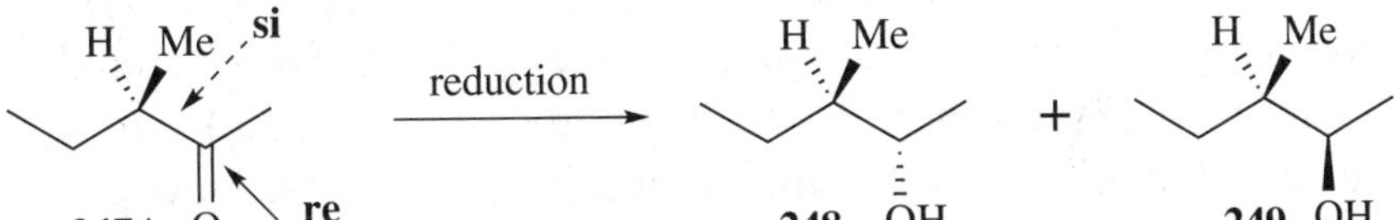

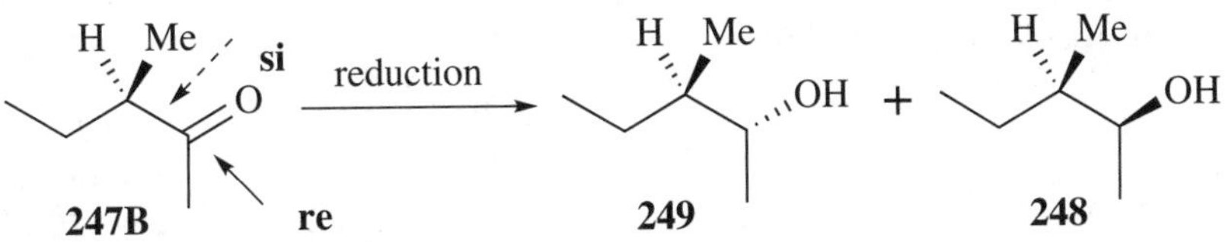

Models for predicting asymmetric induction in systems containing an adjacent stereogenic center have been discussed by Morrison and Mosher.[274] Cram suggested a model for

---

274. Reference 231, pp 84-132.

asymmetric induction in ketones such as **247**, known as **Cram's open chain model (Cram's model)**, or simply **Cram's rule**.[275,276] This model *assumes a kinetically controlled reaction* (nonequilibrating and noncatalytic) for asymmetric 1,2-addition to aldehydes and ketones. The three groups attached to the stereogenic center are R_S (small substituent), R_M (middle-sized substituent), and R_L (large substituent). Determining the relative size of the substituents is not always straightforward, since the *largest* group is defined as the most sterically demanding group relative to the other substituents and the incoming reagent. *Ab initio* calculations by Bienz and co-workers showed that anti-Cram selectivity was favored for compounds that possess very large groups, and that silicon-containing compounds showed the same type of selectivity as the carbon analog.[277] A typical Cram model is shown as its Newman projection (**250**),

and assumes that the predominant rotamer that mimics the transition state for reduction has the large substituent (R_L) syn- to the R^1 group attached to the carbonyl. The hydride is delivered from the less sterically hindered face (over the smallest substituent R_S) to give **251** as the major diastereomer. Stereoselectivity is determined by the extent of the $O \leftrightarrow R_S$ or $O \leftrightarrow R_M$ interaction with the metal hydride. As the steric bulk between R_S and R_M increases, selectivity improves. If R_S and R_M are close in size, this model predicts little or no selectivity. This model also assumes the $R_L \leftrightarrow R^1$ interaction is minimal, which is not entirely correct.[278] As the steric bulk of R^1 increases, the proportion of the anti- product (**253**) produced by $LiAlH_4$ reduction of chiral ketone **252** increases (48% when R^1 = Me; 52% for R^1 = Et; 70% for R^1 = *i*-Pr; 96% for R^1 = *t*-Bu). This observation suggests that the R_S–$R_M \leftrightarrow R^1$ rather than the R_S–$R_M \leftrightarrow O$ interaction is important in this model. In a synthesis of the C1-C12 fragment of the tedanolides by Jung and Yoo, L-Selectride reduction of **254** gave a 75% yield of a single diastereomer (**255**).[279] The Cram model (see **254a**) has the large group (alkyl chain) syn- to the dioxolane moiety, and delivery of hydride over the small group (H) predicts incorrect product. The anti-Cram model must be used since there are oxygen atoms on the α-carbon in the dioxolane ring, and the anti-Cram model predicts **255**.

275. (a) Cram, D.J.; AbdElhafez, F.A. *J. Am. Chem. Soc.* **1952**, *74*, 5828; (b) Cram, D.J.; Kopecky, K.R. *Ibid.* **1959**, *81*, 2748; (c) See also Mengel, A. Reiser, O. *Chem. Rev.* **1999**, *99*, 1191.
276. Eliel, E.L.; Wilen, S.H.; Mander, L.N. *Stereochemistry of Organic Compounds,* Wiley, New York, **1994**, p 879.
277. Smith, R.J.; Trzoss, M.; Bühl, M.; Bienz, S. *Eur. J. Org. Chem.* **2002**, 2770.
278. Reference 231, p 91.
279. Jung, M. E.; Yoo, D. *Org. Lett.* **2007** *9*, 3543.

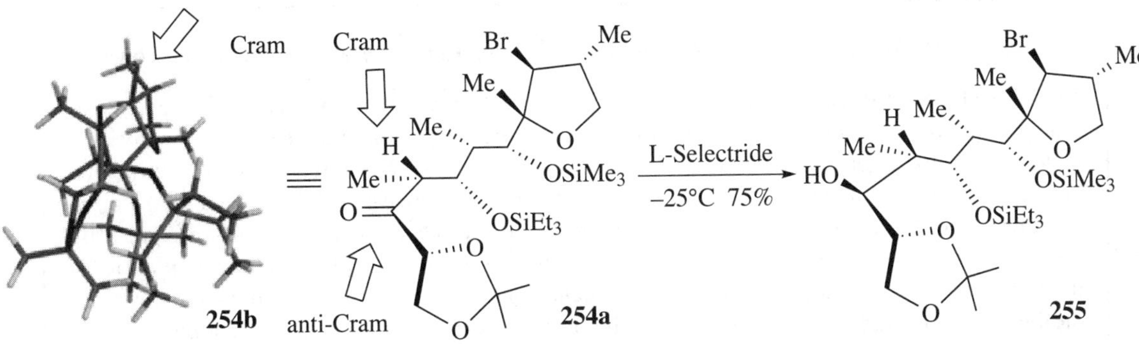

254b ≡ **254a** anti-Cram Cram Cram

L-Selectride
−25°C 75%

255

When the relative difference in size of the groups is negligible, little or no diastereoselectivity is observed. Reduction of **256**, for example, gave a 1:1 mixture of syn- and anti-diastereomers **257** and **258**.[280] The R_M/R_L groups are phenyl and 4-methylphenyl and at the reactive center (the carbonyl), the methyl group of the *p*-tolyl moiety is rather far away. To the approaching hydride reagent phenyl and tolyl are virtually identical in size when analyzed by the Cram model.

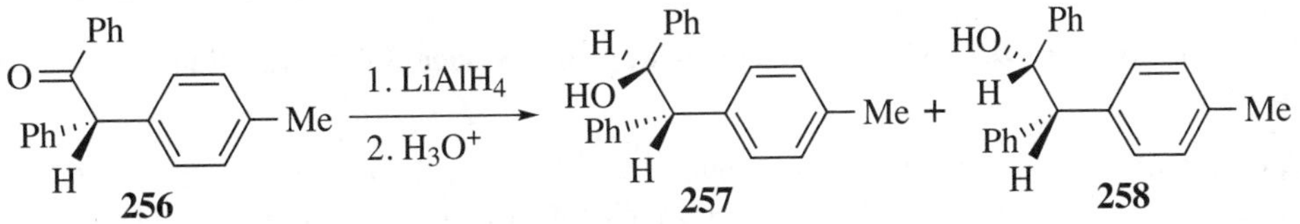

1. LiAlH₄
2. H₃O⁺

256   **257**   **258**

MOLECULAR MODELING: Stereoselectivity of Reductions of Flexible Molecules

Anticipating product distributions in the reduction of flexible (acyclic) ketones is complicated by the need to assign a preferred conformation. A good example is provided by the LAH reduction shown below, which gives a 1:2.5 mixture of the RSR:RSS diastereomers.

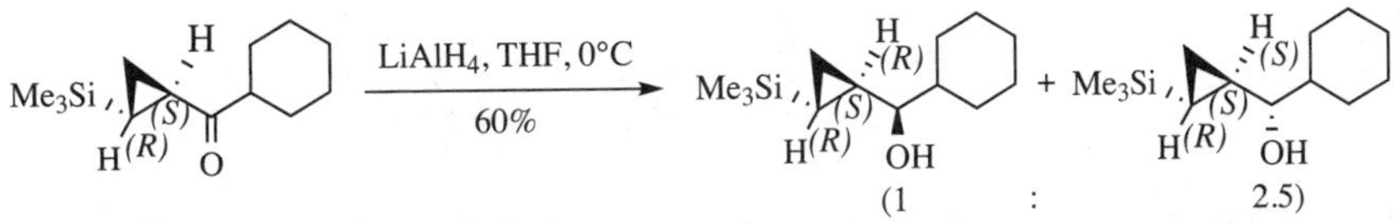

LiAlH₄, THF, 0°C
60%

(1 : 2.5)

The rigid three-member ring attached by a stereogenic center to the carbonyl group may assume one of three possible conformers. In addition, the CC bond connecting the carbonyl to the cyclohexyl ring may be either *equatorial* or *axial* (on the cyclohexyl ring). That is, there are six different (3 x 2) conformers in total. However, quantum chemical calculations suggest that of these six, two nearly-equally-populated conformers account for >90% of the overall distribution. Both of these key conformers must be examined to establish the stereochemical preference for the reduction.

Identifying the lowest-energy conformer(s) of a flexible molecule with more than one degrees of (conformational) freedom is beyond the scope of this problem. This topic is considered in a collection of "mirror problems" included with this text and intended for use with *Spartan Student* or *Spartan*.

---

280. Stocker, J.H.; Sidisunthorn, P.; Benjamin, B.M.; Collins, C.J. *J. Am. Chem. Soc.* **1960**, *82*, 3913.

MOLECULAR MODELING: Stereoselectivity of Reductions of Flexible Molecules, con't.

The LUMO maps for the two important conformers noted above are shown. These are depicted as transparent models, allowing the underlying structures (conformers) to be seen. It is not surprising that both conformers are *equatorially* substituted. Once you are oriented with regard to the structure, you can switch to solid models, allowing the maps to be seen more easily. *Click* on one of the maps and then select *Solid* for the **Style** menu at the bottom of the screen. For each conformer, identify the face of the double bond that is more susceptible to hydride attack by examining the intensity of the blue area on each face, and assign the stereochemistry of the product. By convention, regions on the accessible surface where the LUMO is most concentrated, and most susceptible to attack by an electron donating species such as LAH, are colored "blue", and regions where it is least concentrated and more likely to be electron donating are colored "red". Therefore, the face with the more blue should be the preferred face for delivery of hydride, and lead to the major diastereomeric product. Is the favored product according to the LUMO map the same for both conformers? If it is, is it the product formed in greater amount?

MOLECULAR MODELING: Stereochemistry of Nucleophilic Additions to Flexible Molecules

Assignment of stereochemistry in nucleophilic additions involving rigid molecules (bicyclic, polycyclic, and highly substituted monocyclic molecules) is a straightforward application for a molecular model such as the LUMO map. However, very few molecules are rigid, but instead exist as an equilibrium mixture of two or more different conformers. For reduction of an acyclic ketone, for example, *the appropriate conformer must be identified prior to use of a LUMO map*. It is also known, however, that the presence of a stereogenic center proximal to the reactive center often leads to highly diastereoselective reactions. Consider the LAH reduction of 4-isopropyl-3-heptanone, which gives one diastereomeric alcohol as a major product.

To find the appropriate conformer, it must be recognized that 4-isopropyl-3-heptanone has five rotatable carbon-carbon bonds, which means that there may be as many as 243 (3^5) different conformers. Calculations show that a dozen conformers are required to account for 90% of the room-temperature equilibrium distribution, and three conformers are found to account for 50%. (Identifying the lowest-energy conformer of a flexible molecule with multiple degrees of conformational freedom is beyond the scope of this problem. This topic is considered in a collection of "mirror problems" included with this text, and intended for use with *Spartan Student* or *Spartan*.)

LUMO maps for the three "most important" conformers are provided on screen. Examine each map in turn and identify the face of the double bond that is more susceptible to nucleophilic attack (more blue). Does the preference change with conformer? Can you make a clear "prediction" of the stereochemistry of the product? Is this consistent with the experimental result (see above)?

> **MOLECULAR MODELING: Stereochemistry of Reduction of Diketones**
>
> Sodium borohydride reduction of **95** gives **96**, the product which results from selective reaction at the non-conjugated carbonyl group (see section 4.4.A).
>
>
>
> **95** **96**
>
> The observed preference for **96** would not have been anticipated by examination of a LUMO map, simply because the LUMO in **95** is primarily concentrated on the conjugated carbonyl group. This brings up an important point. The LUMO map is only applicable as an indicator of nucleophilic reactivity where the LUMO corresponds to the "reactive site". Other unoccupied molecular orbitals (LUMO+1, LUMO+2, etc.) may instead be appropriate. The only way to be certain is to examine the first few unoccupied orbitals until the relevant orbital is uncovered. In this particular case, it is the next higher-energy unoccupied orbital (LUMO+1), and the "LUMO+1" map is the appropriate indicator.
>
> First examine the LUMO map for **95** (left). On which carbonyl group is the LUMO primarily localized? Next, examine the LUMO+1 map (right). On which carbonyl group is it primarily localized? Which face of the carbonyl group is should be more susceptible to attack by the nucleophilic hydride, that is, which face is more blue? Is the face that leads to the observed product?

When an oxygen, sulfur, or nitrogen is attached to the carbon α to the carbonyl, the metal of the hydride can coordinate (chelate) to both the heteroatom and the carbonyl oxygen, which effectively locks the molecule into a single rotamer. The Cram designations of large, medium and small no longer have the same meaning. When such a cyclic chelated structure can be formed, the model is changed to the **Cram cyclic model** (more commonly called the **Cram chelation model**).[280,281,275c] When a hydroxyl group, an ether, amino, amide, thiol or thioether group is in the α-position, chelation to that group makes it eclipse the carbonyl, as shown in complex **259**. Delivery of hydride from the less hindered face (over R_S) predicts the major diastereomer, **260**. In one example, the ether-oxygen in the eight-membered ring in **261** along with the carbonyl-oxygen formed the chelated complex **262**, and subsequent quenching with water led to **263** in 48% yield as the single diastereomer shown in Holmes and co-worker's synthesis of (+)-laurencin.[282]

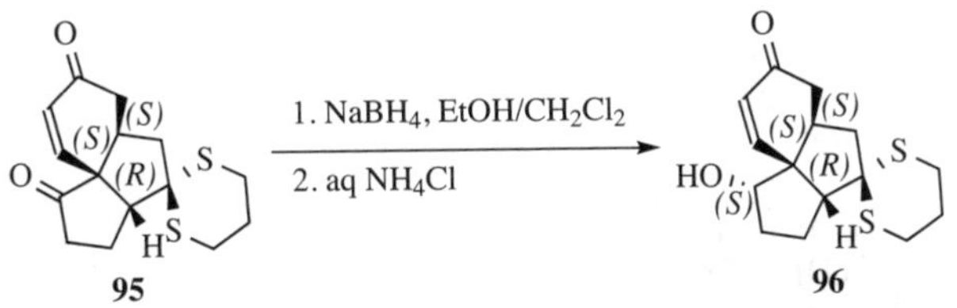

281. Cram, D.J.; Wilson, D.R. *J. Am. Chem. Soc.* ***1963***, *85*, 1245.
282. Burton, J.W.; Clark, J.S.; Derrer, S.; Stork, T.C.; Bendall, J.G.; Holmes, A.B. *J. Am. Chem. Soc.* ***1997***, *119*, 7483.

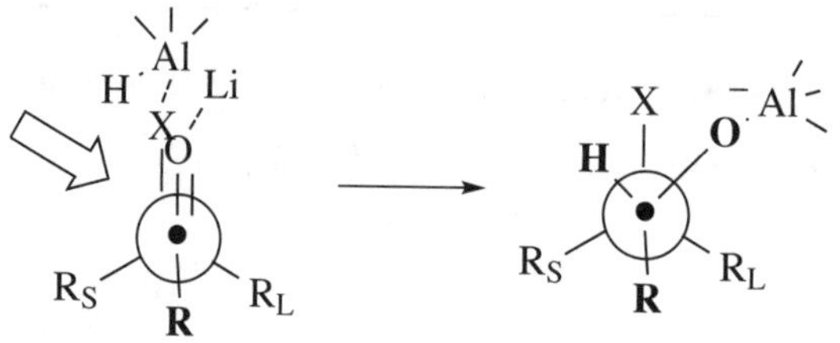

The selectivity in this model, sometimes called **anti-Cram** selectivity, will depend on the reagent, the chelating ability of the substituent, the coordinating ability of the solvent as well as the relative size of the substituents (R_S and R_L). Morrison and Mosher presented data for these systems that included reduction of **263** to an 88:16 mixture of **264** and **265**.[283,280] The presence of the methoxy group in **263** led to a chelated intermediate, and the Cram cyclic model predicts the correct diastereoselectivity. In derivatives where the hydroxyl is blocked (protected, see chap. 7), the reduction proceeds with normal Cram selectivity.[284] In general, a hydroxyl substituent is capable of coordinating via a hydrogen bond and should be more tightly bound than when the hydroxyl is blocked (as in an OMe group). The distance between the chelating group and the carbonyl is an important consideration.

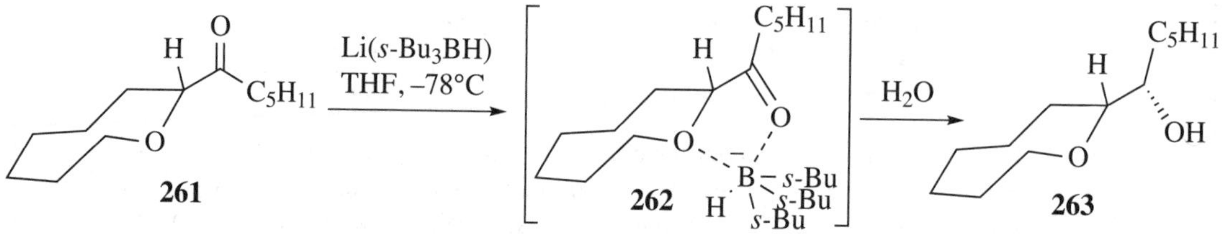

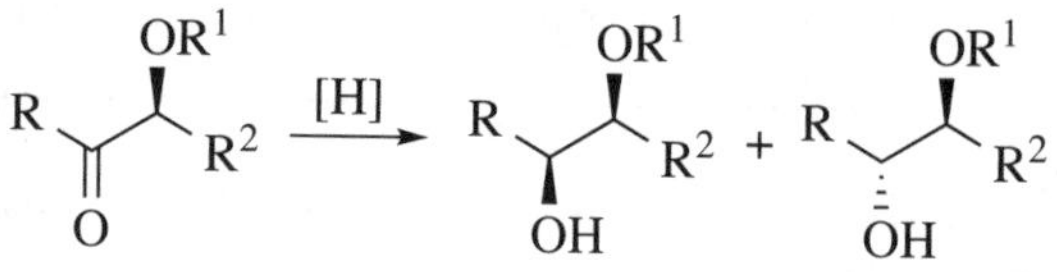

Oishi[285] and co-workers studied the diastereoselectivity of α-alkoxy and α-hydroxy ketones as they reacted with several reducing agents.[285] When $R^1 = H$, there was a distinct preference for the anti (erythro) diastereomer **265** when $Zn(BH_4)_2$ was used for reduction. The anti product was the major one with $LiAlH_4$, but the reaction was less selective. The Cram chelation model for the anti transition state (**266**) to give **265** was favored over the model for syn selectivity (where delivery over the methyl group would give **264**). When hydroxyl was replaced with a bulky silyl ether ($R^1 = SiPh_2t$-Bu), the selectivity changed from anti to syn. Chelation was diminished and Cram's rule (**267**) must be used to predict the correct syn-selectivity. In

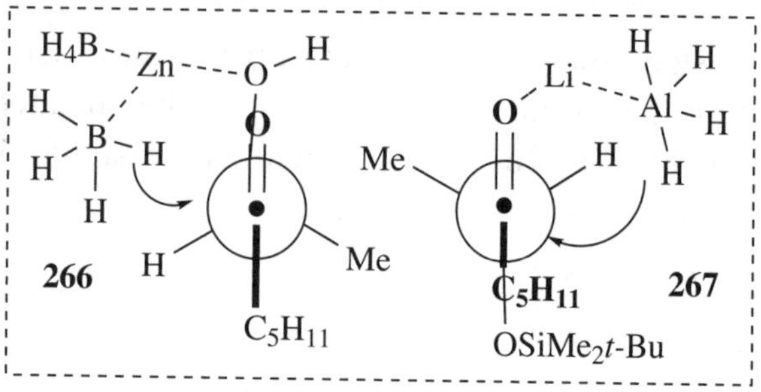

283. Reference 231, pp 100-108.
284. Yamada, S.; Koga, K. *Tetrahedron Lett.* **1967**, 1711.
285. Nakata, T.; Tanaka, T.; Oishi, T. *Tetrahedron Lett.* **1983**, *26*, 2653.

section 4.4.B, excellent diastereoselectivity was obtained in the zinc borohydride reduction of **121**, giving **122** as the major product. Oishi examined the reduction of β-keto esters such as **268** (R^1 = Ph, R^2 = Me) with zinc borohydride and found a preference for the syn-product (**269**, R^1 = Ph, R^2 = Me) over the anti-product (**270**), of more than 99:1 (98% yield).[156] The analogous methyl ketone (**268**, R^1 = Me, R^2 = Bn) gave only a 2:1 mixture of **269/270** under identical reaction conditions.[156] Zinc coordinated both the carbonyl of the ester moiety and the oxygen of the prochiral carbonyl group. Hydride was delivered from the less hindered face (over hydrogen), which is the *si* face when R^1 is phenyl and the *re* face when R^1 is methyl (referring to **268**).

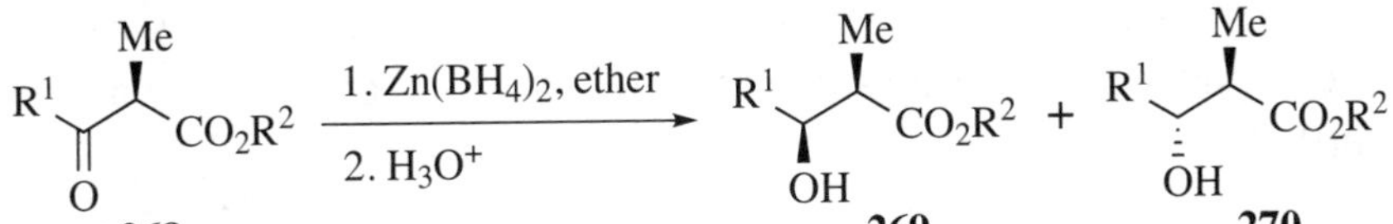

The presence of an α-alkoxy substituent does not always lead to anti-Cram selectivity. Reduction of **271a** gave 82% of the syn diastereomer, **272**.[286] Analysis of the Cram cyclic model for **271** (see **273**) gives an incorrect prediction of the anti-diastereomer (**274**) as the major product. The normal Cram model (see **271b**) predicts the correct syn-diastereomer (**272**) but, in reality, reduction requires that the alkoxy substituent at the C3 position be coordinated with the metal of the reducing agent. In **273**, a five-membered cyclic transition state is possible via coordination with the benzyloxy group. In **275**, coordination of the reagent to the 3-alkoxy substituent leads to a six-centered cyclic transition state, with delivery of hydride from the less hindered methyl face leading to **272**. With hindered reducing agents such as tri-*tert*-butoxyaluminum hydride, **275** is preferred to **273**. As the nature of the alkoxy substituent changes, it is clear that the model used to predict the major diastereomer may change.

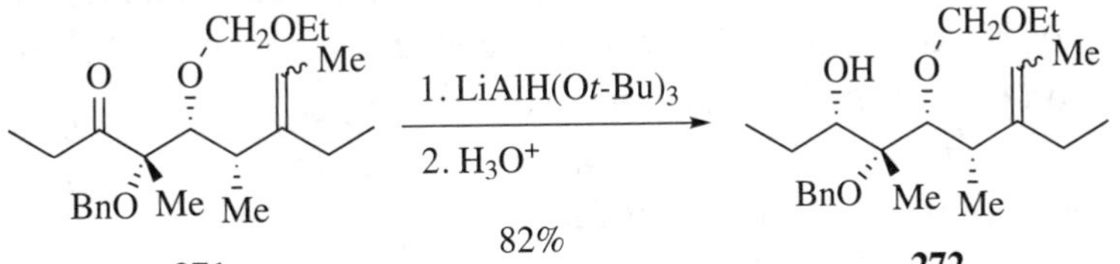

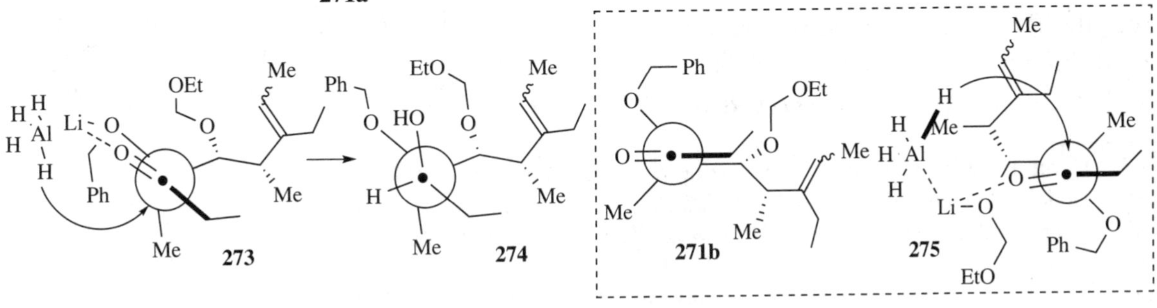

286. Hoagland, D.; Morita, Y.; Bai, D.L.; Märki, H.-P.; Kees, K.; Brown, L.; Heathcock, C.H. *J. Org. Chem.* **1988**, *53*, 4730.

The reduction of **271a** illustrated that chelating substituents at the carbon β to a carbonyl can influence the diastereoselectivity of hydride reductions, which is particularly important for the reduction of 1,3-diketones and β-ketoesters. When the α-substituent is a halogen, however, the open-chain model does not predict the correct results nor does the chelation model. Cornforth et al.[287] suggested that the electron pairs on the carbonyl oxygen and on the halogen repel and assume an anti-conformation, as in **276**. In effect, the halogen becomes R_L with delivery of hydride over the smallest group, which has come to be called the **Cornforth model**. Reduction of **276** gave a 75:25 mixture of **277** and **278**, and the Cornforth model predicted **277** as the major product.[288] Note that Evans and co-workers[289] studied the aldol reaction (sec. 9.4.A) between methyl-substituted Z- and E-enolate anions (sec. 9.2.C) and α-oxygen-substituted aldehydes, and concluded that the diastereoface selectivity (sec 9.5.A) was more consistent with the Cornforth model.

The Cram open-chain model assumed the presence of a reactive species that led to a higher energy eclipsed conformation in the final product. The Cram open-chain model fails to predict the correct diastereoselectivity for many acyclic molecules that contain heteroatom substituents.[290] The Cram model also fails to predict the correct diastereomer when the relative size of R_S and R_M is close. Karabatsos introduced a new model, in an attempt to correct the deficiencies of the Cram model.[291] He assumed a *reactant-like transition state* that showed little bond breaking or bond making. The rotamer chosen for the model showed the groups on the stereogenic carbon to be aligned as in a normal sp^3-sp^2 bond, which led to three rotamers that were considered by the **Karabatsos model (279-281)**. The incoming reagent was assumed to approach over the R_S group.[292] Note that reaction over R_S in all three rotamers gives a product with an anti-conformation. Karabatsos showed that **281** was

287. Cornforth, J.W.; Cornforth, R.H.; Mathew, K.K. *J. Chem. Soc.* **1959**, 112.
288. Bodot, H.; Dieuzeide, E.; Jullien, J. *Bull. Soc. Chim. Fr.* **1960**, 27, 1086.
289. Evans, D.A.; Siska, S.J.; Cee, V.J. *Angew. Chem. Int. Ed.* **2003**, 42, 1761.
290. Reference 231, p 118.
291. Karabatsos, G.J. *J. Am. Chem. Soc.* **1967**, 89, 1367.
292. (a) Karabatsos, G.J.; Hsi, N. *J. Am. Chem. Soc.* **1965**, 87, 2864; (b) Karabatsos, G.J.; Hsi, N. *Tetrahedron* **1967**, 23, 1079; (c) Karabatsos, G.J.; Krumel, K.L. *Ibid.* **1967**, 23, 1097; (d) Karabatsos, G.J.; Fenoglio, D.J.; Lande, S.S. *J. Am. Chem. Soc.* **1969**, 91, 3572. (e) Karabatsos, G.J.; Fenoglio, D.J. *Ibid.* **1969**, 91, 1124, 3577.

favored over **279** by 0.8 kcal mol^{-1} (3.35 kJ mol^{-1}) when $R_S = H$ and $R_M = Me$. The selectivity of a reduction was assumed not to be dependent on the higher energy **279**, but rather on rotamers **280** and **281**. It was then assumed that the destabilizing $O \leftrightarrow R_L$ and $O \leftrightarrow R_M$ energy interactions in these two rotamers would determine the diastereoselection for the reaction. Although this model is inadequate when R_S is large, or when applied to cyclic molecules, it offers advantages over the Cram model in some applications.

Felkin and Anh proposed an alternative model that gave somewhat better results in these systems.[293,294] The model was based on several assumptions. (*1*) the transition states are reactant-like (*2*) the previous eclipsed models introduce significant torsion strain in the partial bonds of these transition states (*3*) the important steric interaction in these transition states involves the group attached to the prochiral center and the incoming group (*4*) polar effects stabilize the transition states in which the separation of the incoming group and any electronegative substituent at the α-carbon is greatest and destabilize the others. Felkin concluded that models **282** and **283** were the most important. The use of these structures to predict diastereoselectivity is called the **Felkin-Anh model**.[293] The correctness of this model has been called into question, based on calculations done by Bientz and co-workers.[295] "According to our calculations, the Felkin-Anh model does not correctly represent the modes of reaction on the course of nucleophilic attack to α-chiral carbonyl compounds. For typical reactions, the two transition states with the M and L groups arranged 'inside' and 'anti' are the key structures, and discrimination is due to the repulsive interactions of the incoming nucleophiles with the 'inside' groups. For compounds with very large groups L, the relevant transition structures are those with this L group placed 'anti' (Felkin-Anh conformations). For the discrimination, the interaction of the incoming nucleophile with the 'inside' rather than the 'outside' group is relevant. This leads to anti-Cram selectivity for such compounds. Acylsilanes should show the same selectivities as the related ketones. The selectivities, however, are expected to be lower and the relevant transition structures might differ from those of the carba analogs."[295]

When R_M and R_S are relatively close in size, the Felkin-Anh model predicts modest to poor selectivity because there is little difference in the destabilizing $R_M \leftrightarrow O$, $R_S \leftrightarrow O$ or $R_S \leftrightarrow R^1$, $R_M \leftrightarrow R^1$ interactions. As R_M or R^1 increases in steric bulk, the increased $R_M \leftrightarrow R^1$ destabilizing interaction in **283** will favor **282**.[296] Reduction of **284** with LiAlH$_4$ was shown to give a 2:1 mixture of **285** and **286**.[297] For

$$282 \qquad 283$$

293.   (a) Chérest, M.; Felkin, H.; Prudent, N. *Tetrahedron Lett.* **1968**, 2199; (b) Chérest, M.; Felkin, H. *Tetrahedron Lett.* **1968**, 2205; (c) Anh, N.T.; Eisenstein, O. *Nov. J. Chem.* **1977**, *1*, 61.
294.   Reference 276, p 881.
295.   Smith, R.J.; Trzoss, M.; Bühl, M.; Bienz, S. *Eur. J. Org. Chem.* **2002**, 2770.
296.   Reference 231, p 116.
297.   Cram, D.J.; McCarty, J.E. *J. Am. Chem. Soc.* **1954**, *76*, 5740.

an analysis of this reduction, the two models of interest are **287** (which predicts **285**) and **288** (which predicts **286**). Note that the analogous 3D model shown above each 2D model). The Me↔Me interaction in **288** is greater than the analogous H↔Me interaction in **287**, making **287** more favorable, and delivery of hydride via **287** predicts **285** as the major product.

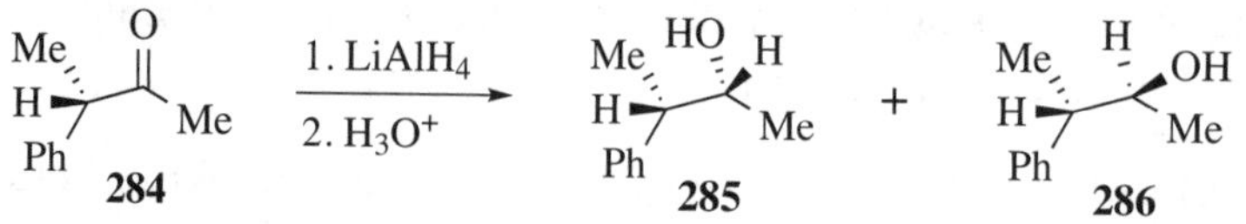

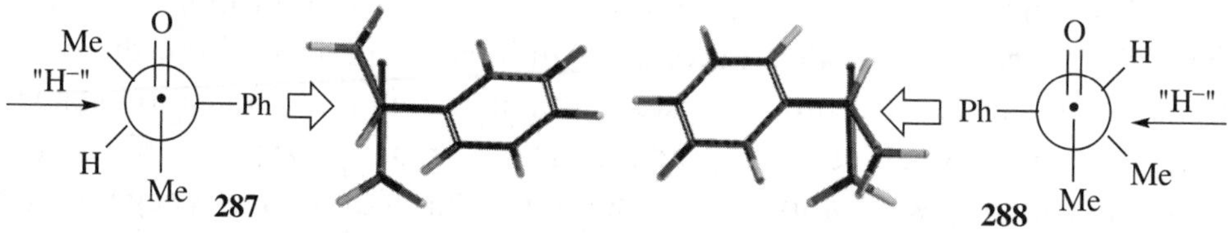

The Felkin-Anh and Cram models are best applied to acyclic systems. Problems arise when any of these models are used to predict the products generated by the reduction of cyclic ketones. These problems will be analyzed and new models for predicting diastereoselectivity in the reduction of cyclic molecules will be discussed in Section 4.7.C. Table 4.6 attempts to compare the two most-used models, the Cram and Felkin-Anh, with the LUMO map approach using calculations. All experimental data are taken from the literature,[275c] and the major diastereomer expected from each substrate (**A** or **B**) is predicted using each model. It is clear that the Cram and Felkin-Anh models give similar predictions and are generally correct, but both fail when there are oxygen substituents at the α-position. In such cases, the anti-Cram or anti-Felkin-Anh model are used to make predictions. Note that when the –NBn₂ group is at the α-position, the anti-models lead to incorrect predictions. The LUMO maps generally give good predictions when compared to the experimental results, with the exception of the α-chloro substituent. For the α-methoxy derivative, the LUMO map predicts the correct diastereomer, but does not predict the outstanding selectivity that was observed in this reaction. LUMO maps are easily obtained, and offer a good model for predicting diastereoselectivity, and can be used in addition to the Cram, anti-Cram or Felkin-Anh models.

Solvent can play a significant role in the selectivity. The reaction of phenylmagnesium bromide with **284**[298] (a Grignard reaction, sec 8.4) led to alcohol products **285** and **286**, and the ratio of *S:R* to *S:S* product was 73:27 in DME, 61:39 in THF, 51:49 in dioxane, and 36:64 in ether. Solvation of the metal in this case is undoubtedly responsible to changes in selectivity, but the result calls into question the efficacy of predictive models. This particular result suggests that model **287** predicts the correct major product in DME and THF, but the anti-Felkin-Anh model predicts the result in ether.[298] Therefore, these models must be used with care, and solvent effects for a given system should be examined by searching the literature making predictions prior to running the experiment.

298. Perez-Ossorio, R.; Perez-Rubalca, A.; Quiroga, M.L. *Tetrahedron Lett.* **1980**, *21*, 1565.

Table 4.6. A Comparison of the Cram and Felkin-Anh Models with LUMO Maps, Against Experimentally Determined Ratios[275c] of Diastereomers Using Various Nucleophiles.

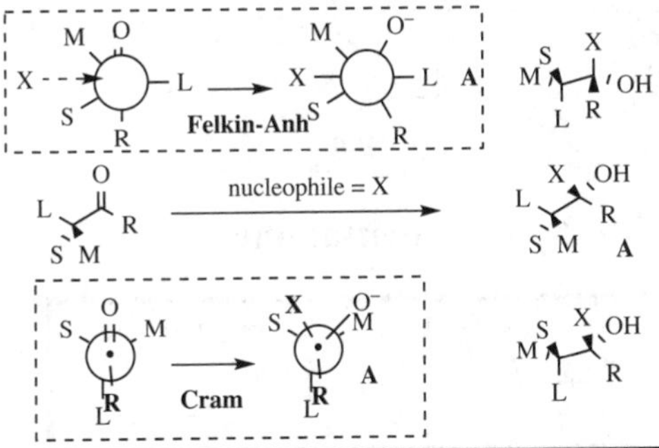

In each case, the table shows a prediction of A or B as the major product, where Cram indicates the use of the Cram model shown, Felkin-Anh indicates the use of the Felkin-Anh model shown, and LUMO indicates the prediction is based on the LUMO map generated for each compound using MacSpartan.

R_S	R_M	R_L	R	Reagent	A:B (exp)	Cram	Felkin-Anh	LUMO map
H	Me	Ph	H	MeLi	80 : 20	A	A	A
H	Me	Ph	Me	LiAlH$_4$	74 : 26	A	A	A
H	Me	Ph	Me	EtMgBr	75 : 25	A	A	A
H	Me	Ph	SiMe$_3$	MeLi	98 : 02	A	A	A
H	Me	Et	H	MeMgBr	60 : 40	A	A	A*
H	Me	c-Hex	H	MeMgI	82 : 18	A	A	A*
H	Me	c-Hex	Me	LiAlH$_4$	62 : 38	A	A	A
H	Me	Cl	Ph	LiAlH$_4$	75 : 25	A	A	B^a
H	C$_7$H$_{15}$	OBn	Me	BuMgI	<01 : 99	A^b	A^b	B
H	Me	NBn$_2$	H	MeMgI	95 : 05	A^c	A^c	A
H	Me	CO$_2$Et	Ph	MeLi	75 : 25	A	A	A
H	Me	OMe	Ph	Me$_2$Mg	<01 : 99	A^b	A^b	B^a
H	Me	OTBS	Et	MeMgCl	40 : 60	A^b	A^b	B
H	Me	OH	Me	LiAlH$_4$	30 : 70	A^b	A^b	B

a Predicts lower selectivity when comparing intensity of blue on each face of the carbonyl LUMO map
b Requires the anti-Cram or anti-Felkin-Anh model for correct predictions.
c Expected anti-Cram mo

4.7.C. Selectivity in Reduction of Monocyclic Molecules

Predictions of stereochemistry for *cyclic* ketones using the Cram or Felkin-Anh models are generally unreliable. Reduction of 2-methylcyclopentanone (see Table 4.7)[299] can be used to illustrate the difficulty. Model **289** is an approximation of the Cram model for 2-methylcyclopentanone, and it predicts the cis isomer to be the major product of hydride reduction. Analysis of cyclic ketones shows that fitting them to the Cram model requires distortion of the ring away from the low energy conformations (chair or envelope for six- and five-membered rings, respectively). The Felkin-Anh model for 2-methylcyclopentanone (**290**) has similar problems but less distortion of the ring is required, although it also predicts the cis product. Reductions of actual molecules (see Table 4.7) show that the trans isomer (**292**) is

---

299. Reference 231, p 119.

favored except when very bulky reducing reagents are used. Both the Cram and Felkin-Anh models gave incorrect predictions.

Table 4.7. Hydride Reduction of 2-Methylcyclopentanone

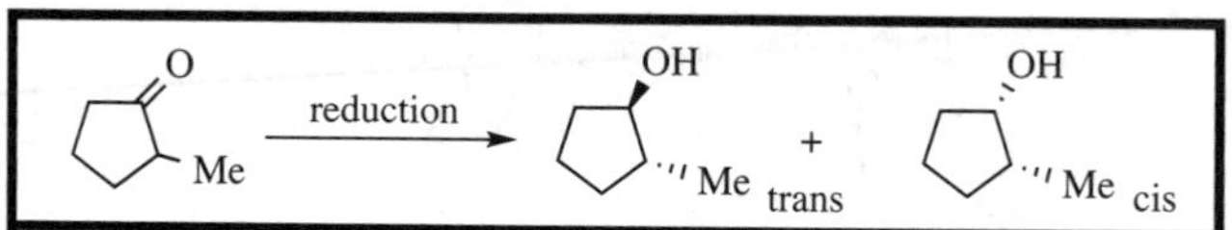

Reagent	% trans	% cis
NaBH$_4$[a]	74	26
LiAlH$_4$[b,c]	76	24
B$_2$H$_6$[e]	69	31
LiAlH(OMe)$_3$[a,b]	56	44
LialH(OEt)$_3$[b]	77	23
LiAlH(Ot-Bu)$_3$[a,b]	72	28
Li(mesityl)$_2$BH$_2$[e]	2	98
L-Selectride[d]	1.5	98.5
Li(siamyl)$_3$BH[d]	0.5	99.5

Reprinted by courtesy of Professor James D. Morrison from Morrison, J.D.; Mosher, H.S. *Asymmetric Organic Reactions*, American Chemical Society, Washington, DC *1976*, p. 119.]

a Ref. [300] b Ref. [301] c Ref. [302] d Ref. [303] e Ref. [304]

The actual experimental results can be explained by using two envelope conformations available to 2-substituted cyclopentanones, **291** and **292** (sec. 1.5.B). In **291**, approach of a reagent to the carbonyl at an angle of ~ 110° (the **Bürgi-Dunitz trajectory**, sec. 6.6.A)[305] leads to an interaction with H$_a$ (path *b*), making path *a* preferred and this leads to the trans product. Remember that this angle assumes a late transition state in which the approach angle approximately mimics the tetrahedral angles of the sp^3-hybridized carbon found in the alcohol products. This result is apparently contradicted by **292**, where approach via path *a* is inhibited by the R group. Since the R group will be pseudo-equatorial in the lowest energy conformation (as in **291**), this model is used to predict the major diastereomer in reactions of cyclopentanone derivatives. Indeed, reduction via path *a* in **291** leads to the correct stereochemical predictions in Table 4.7, except when bulky reducing agents are employed. Structure **293** is a 3D model of **292** (R = Me), showing the steric influence of the group on approach to the acyl carbon.

300. Umland, J.B.; Jefraim, M.I. *J. Am. Chem. Soc.* **1956**, *78*. 2788.
301. Brown, H.C.; Deck, H.R. *J. Am. Chem. Soc.* **1965**, *87*, 5620.
302. Ashby, E.C.; Sevenair, J.P.; Dobbs, F.R. *J. Org. Chem.* **1971**, *36*, 197.
303. Brown, H.C.; Bigley, D.B. *J. Am. Chem. Soc.* **1961**, *83*, 3166.
304. Caro, B.; Boyer, B.; Lamaty, G.; Jaouen, G. *Bull. Soc. Chim. Fr.* **1983**, Pt. 2, 281.
305. Bürgi, H.B.; Dunitz, J.D.; Lehn, J.M.; Wipff, G. *Tetrahedron* **1974**, *30*, 1563.

Inspection of a LUMO map for 2-methylcyclopentanone shows a more intense color for the carbonyl carbon on the same face as the methyl group, suggesting a preference for the *trans*-alcohol, consistent with Table 4.7, although **291** gives an incorrect prediction when bulky reagents such as Selectride are used.

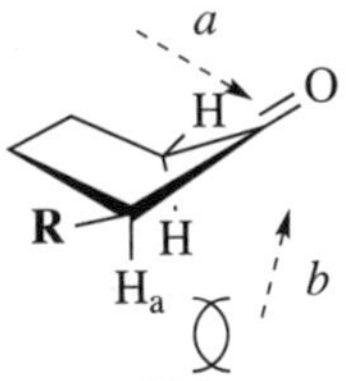

| 291 | 292 | 293 | 294 |

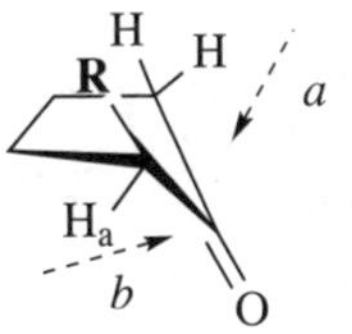

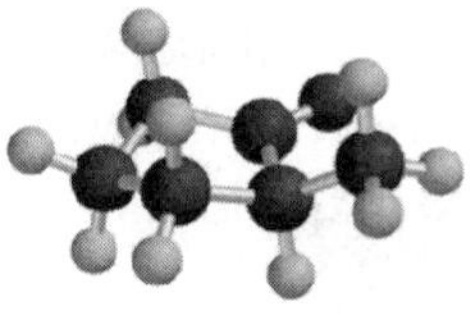

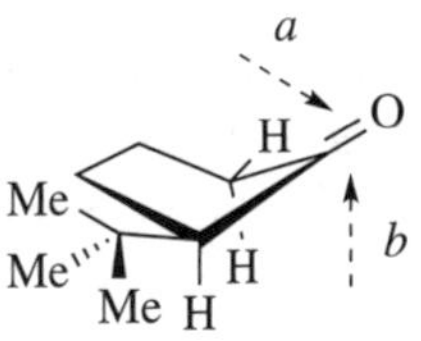

As the α-substituent in 2-substituted cyclopentanones increases in size from methyl to *tert*-butyl (see **295**), path *a* becomes less favored and lower selectivity is observed as seen in Table 4.8.[304] The reason for this is shown for the *tert*-butyl derivative (see **294**), where the methyl groups impose greater hindrance to approach via path *a*. The LUMO map of **294** also indicates less selectivity. Even with this very bulky *tert*-butyl group, essentially a 1:1 mixture of stereoisomers is obtained upon reduction. For clarity, it must be emphasized that **291**, **292**, and **294** are distorted as drawn in a stylized envelope conformation (compare with **293**).

Table 4.8. Lithium Aluminum Hydride Reduction of 2-Substituted Cyclopentanones

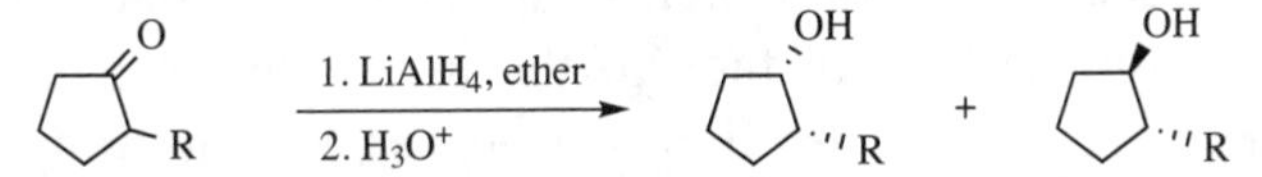

	% cis	% trans
Me	24	76
Et	27	73
i-Pr	47	53
t-Bu	54	46
Ph	40	60
c-C_6H_{11}	42	58

MOLECULAR MODELING: Nucleophilic Attack on Substituted Cyclopentanones

A LUMO map colors a surface of electron density (indicating overall molecular size and shape) with the (absolute) value of the lowest-unoccupied molecular orbital (the LUMO). By convention, blue regions on the surface indicate where the LUMO is most concentrated. Examine maps for 2-methylcyclopentanone (left) and 2-*tert*-butylcyclopentanone (right), and for each assign the favored direction of attack by a nucleophile, that is, the side on which the LUMO is more concentrated. Which compound would be expected to show greater selectivity? Are your results in line with the experimental data shown above?

The Cram model is a poor model for reduction of cyclohexanone derivatives. Antiperiplanar hyperconjugative effects have been invoked to explain the observed results. The Felkin-Anh model suggests an interaction such as that shown in **296**.[306] Cieplak put forth an alternative proposal (see **297**, the **Cieplak model**).[307] The incipient bond (the one being formed) was found to be electron deficient. This model has led to some controversy, and a so-called exterior frontier orbital extension model has been proposed to explain the selectivity.[308]

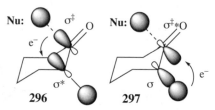

The stereoselectivity associated with substituted cyclohexanones can be shown using the reaction of 4-*tert*-butylcyclohexanone (**298**) with lithium aluminum hydride or sodium borohydride, which gave the trans-alcohol (**299**) as the major product (92% and 80%, respectively) via axial attack (path *a*).[309] The bulky *tert*-butyl group at C4 in **298** effectively locks the molecule into that chair conformation (sec. 1.5.D). The Cram model or the Felkin-Anh model predict path *b* as the major pathway (to give **300**), which is incorrect. A more useful perspective for this analysis is a model represented by **298**, which suggests that approach via path *b* is encumbered by the axial hydrogens on the bottom face of the molecule. An important aspect of these models is the trajectory of approach for the hydride. The angle of approach will be close to 110° (see above)[305] and over the π bond of the carbonyl (as seen in **298**). This angle of approach will exacerbate the steric problems posed by the axial hydrogens at C2 and C6 for path *b* and the axial hydrogens at C3 and C5 for path *a*. In simple cyclohexanone derivatives, path *a* is preferred. When the α-carbon is substituted, as in **301**, path *a* (to give **302**) is preferred to path *b* (which gives **303**) unless R is very large, as shown in Table 4.9.[304,302] Analysis of the intensities of the carbonyl carbon in a LUMO map of 2-methylcyclohexanone indicates that there is a preference for path *a* (see **301**). Structures **304a** and **304b** are 3D-models for 2-methylcyclohexanone, showing the steric impedance imposed by both an axial methyl group (**304b**, blocking path b) and an equatorial methyl group (**304a**). As the steric bulk at C2 is increased, approach from path a is somewhat inhibited, leading to increasing amounts of the cis-product (via path *b*). It is interesting to note that LiAlH(OMe)$_3$ gives more cis-product than LiAlH(O*t*-Bu)$_3$. Generally, smaller reagents give more axial attack (path *a*) and larger reagents give more equatorial attack (path *b*). In their monograph, Morrison and Mosher arranged the

306. (a) Reference 282; (b) Ahn, N.T.; Eisenstein, O.; Lefour, J.-M.; Traàn Huu Daàu, M.E. *J. Am. Chem. Soc.* **1976**, *95*, 6146; (c) Ahn, N.T.; Eisenstein, O. *Nouv. J. Chim.*, **1976**, *1*, 61.
307. (a) Cieplak, A.S. *J. Am. Chem. Soc.* **1981**, *103*, 4540; (b) Reference 276, pp 882-883.
308. Tomoda, S.; Senju, T. *Tetrahedron* **1999**, *55*, 3871.
309. (a) Richer, J.-C. *J. Org. Chem.* **1965**, *30*, 324; (b) Richer, J.-C.; Perrault, G. *Can. J. Chem.* **1965**, *43*, 18.

relative size of hydride reagents as follows.[310]

Table 4.9. Hydride Reduction of 2-Substituted Cyclohexanone Derivatives, 301

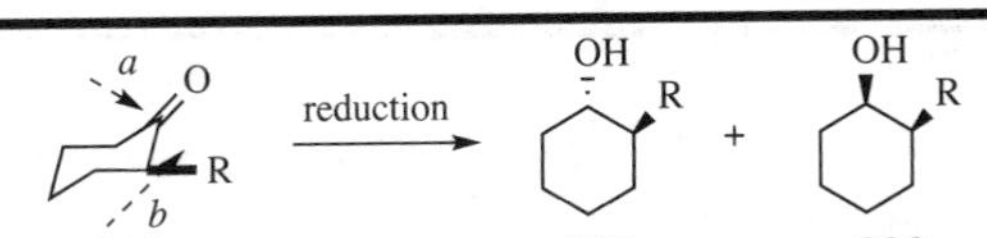

R	Reagent	% 302 (axial attack)	% 303 (equatorial attack)
Me	LiAlH$_4$/THF	76	24
	NaBH$_4$/i-PrOH	70	30
	LiAlH(Ot-Bu)$_3$/THF	70	30
	LiAlH(OEt)$_3$/THF	74	26
	LiAlH(OMe)$_3$/THF	31	69
Et	NaBH$_4$/i-PrOH	65	35
i-Pr	LiAlH$_4$/THF	63	37
	NaBH$_4$/i-PrOH	50	50
t-Bu	LiAlH$_4$/THF	42	58
	NaBH$_4$/i-PrOH	50	50
	LiAlH(OMe)$_3$/THF	36	64
	LiAlH(Ot-Bu)$_3$/THF	46	54

Reprinted in part with permission from Ashby, E.C.; Sevenair, J.P.; Dobbs, F.R. *J. Org. Chem.* **1971**, *36*, 197. Copyright **1971** American Chemical Society.

LiAlH$_4$ < LiAlH(OtBu)$_3$ < NaBH$_4$ < KBH$_4$ < LiAlH(OMe)$_3$ (sec. 4.3) < Ip$_2$BH (sec. 5.4.B)

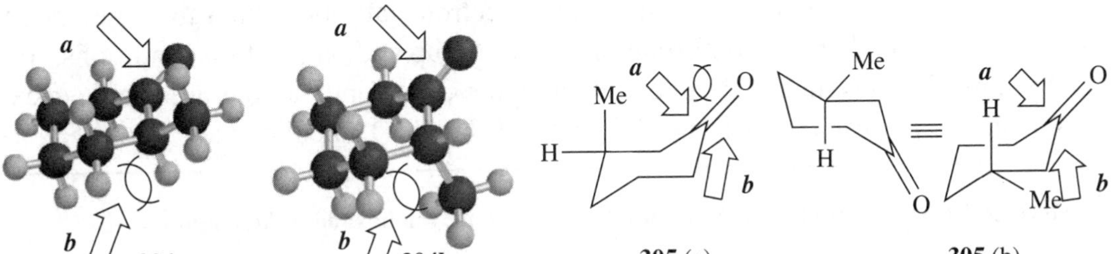

Substituents at C3 or C5 also have a significant effect on the approach of reagents to the carbonyl of cyclohexanone derivatives. Using a model similar to the 2-substituted cyclohexanones, reduction of 3-methylcyclohexanone can proceed via path *a* or path *b* (see **305a**). Since the axial methyl group imposes steric hindrance to path *a*, there is a preference for path *b*. It is cautioned that the lower energy chair conformation of 3-methylcyclohexanone is probably **305b** rather than **305a**. Attack via path *b* in **305a** will give the cis product, but note that attack along the preferred path *a* in **305b** also gives the cis product. If we use **305a** as the standard model, then path *b* is preferred. As shown in Table 4.10[302,311] for 3,3,5-trimethylcyclohexanone (**306**), an *axial* methyl group at C3 of a cyclohexanone blocks path *a* (to give **308**) and the major product is via path *b* to give **307**. The preferred mode of attack is path *b* to give the trans

---

310. Reference 231, pp 123-124 and citations, 38, 55 and 56 therein.
311. Haubenstock, H.; Eliel, E.L. *J. Am. Chem. Soc.* **1962**, *84*, 2363, 2368.

alcohol, **308**. As the size of the hydride reagent increases, path *b* is more highly favored.

Table 4.10. Selectivity in Hydride Reductions of 3,3,5-Trimethylcyclohexanone (306)

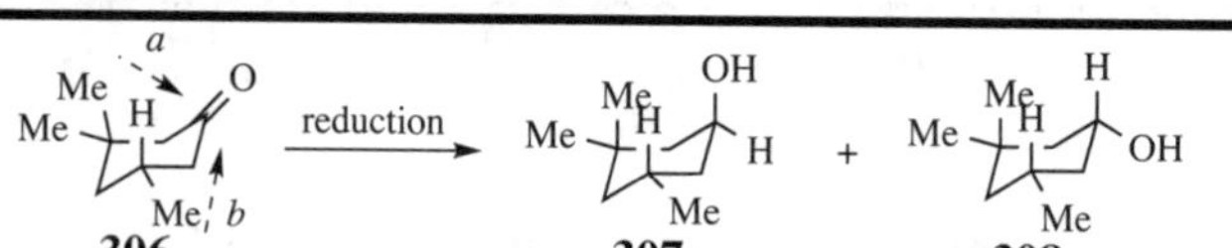

Reagent	% 307	% 308
LiAlH$_4$	52	48
LiAlH(OMe)$_3$	75	25
LiAlH(OEt)$_3$	83	17
LiAlH(O*t*-Bu)$_3$	73	27
NaBH$_4$/EtOH	67	33
NaBH$_4$/aq. MeOH	71	29
KBH$_4$/aq. MeOH	71	29
NaBH(OMe)$_3$/aq/ MeOH	65	35
NaBH(O*i*-Pr)$_3$/diglyme	77	23

[Reprinted in part with permission from Haubenstock, H.; Eliel, E.L. *J. Am. Chem. Soc.* **1962**, *84*, 2363, 2368. Copyright © *1962* American Chemical Society.]

De Maio et al. studied the stereochemistry and relative rates of k_{ax} and k_{eq} for acyl addition reactions of 5-substituted adamantan-2-ones.[312] This kinetic data did not fit current theories of π face diastereoselection. Their studies suggest that k_{ax}/k_{eq} measurements do not adequately indicate what is happening on the two sides of a stereogenic center. In addition, they found that axial reactivity always increases with increasing electronegativity of the substituted at C5 and that k_{eq} changes depend on the conformation of that group and on reaction conditions. In their words: "Simple theories such as those based on dipole-dipole interactions are unable to explain such as behavior, nor can theories based on ground state MO calculations."[312]

MOLECULAR MODELING. Nucleophilic Attack on Cyclohexanone Analogues

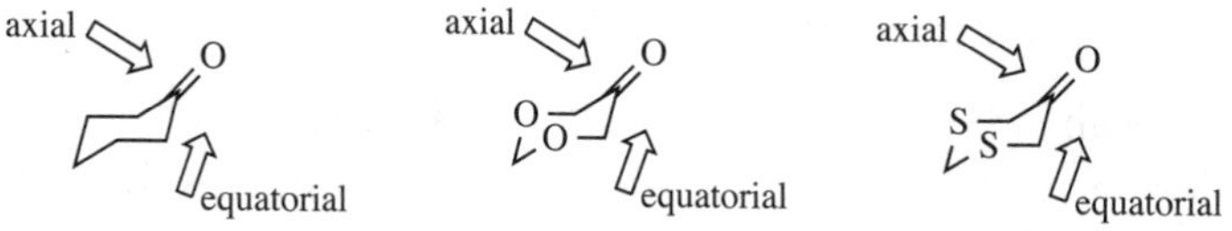

Cyclohexanone derivatives are ideal candidates for stereochemical analysis with molecular models. Nucleophilic addition to a carbonyl requires that the nucleophile approach the carbonyl carbon from an angle of ~110° (the "***Bürgi-Dunitz trajectory***"), that is, either over or under the ring. Examine the LUMO maps for cyclohexanone (top), 1,3-dioxanone (bottom left) and 1,3-dithianone (bottom right). Recall, that the more "blue" the region on the map, the more likely it is to be susceptible to reaction with an electron donating species. For which (if any) of the molecules would nucleophilic attack occur on the *equatorial* face? For which (if any) would it occur on the *axial* face? Which molecule shows the greatest preference and which shows the least preference? Does there appear to be a relationship between the direction and magnitude of the preference and the relative accessibility of the *axial* and *equatorial* faces?

---

312. Di Maio, G.; Solito, G.; Varì, M.R.; Vecchi, E. *Tetrahedron* **2000**, *56*, 7237.

4.7.D. Stereoselectivity in the 1,2-Reduction of Cyclohexenone Derivatives

The conformational problems of cyclohexene derivatives are somewhat different than those observed with cyclohexanone or cyclohexane derivatives. The presence of the alkene moiety effectively flattens four carbons of the ring, making the steric interactions of substituents with an incoming reagent different, relative to cyclohexanone derivatives. The most abundant conformations of cyclohexenone derivatives are **309** and **310**. The first problem that arises in conjugated derivatives is the competition of 1,2- with 1,4-addition for many reducing agents,

discussed in previous sections. In this section only products arising from 1,2-reduction will be discussed. It appears that the C2 group (R^1) offers little interference to either path *a* or *b* unless it is very large. The R^2 group, however, can inhibit approach by path *a* if sufficiently large and a large R^4 may inhibit approach by path *b*.

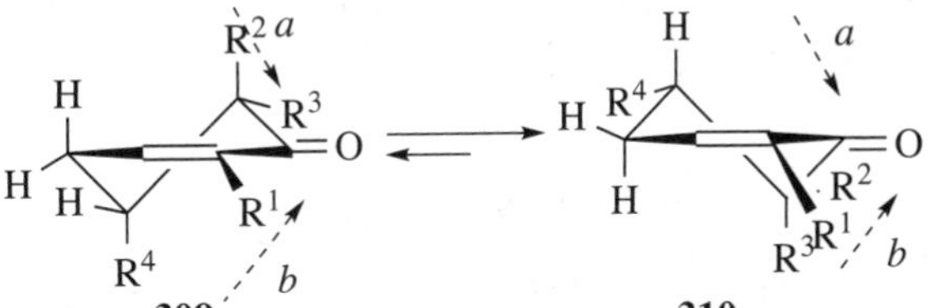

The sodium borohydride-cerium chloride reagent (**Luche reagent**) gave selective 1,2-reduction of conjugated ketones in section 4.4.B. The sensitivity of the cyclohexenyl system to steric blocking when substituents are present at C5 or C6 is shown by reduction of **311**, which showed a preference for **313** over its diastereomer **312**[313]. Analysis of carbonyl region in a LUMO map of **311** shows a more intense blue color on the face marked path *b*, consistent with that face being more accessible and formation of **313** as the major product. A 3D model of **311** is provided to show the stereochemical relationships.

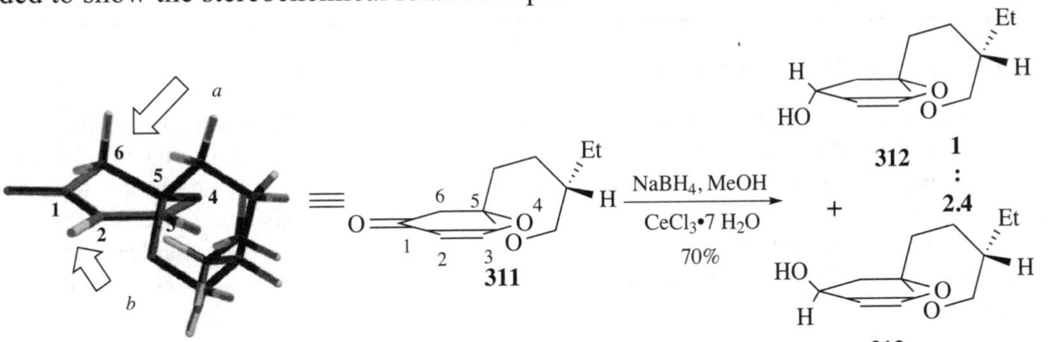

4.7.E. Selectivity in the Reduction of Bicyclic and Polycyclic Derivatives

As with monocyclic ketones, bicyclic ketones show excellent diastereoselectivity upon reduction with hydride reagents. This selectivity is clearly observed in the reduction of bicyclo[2.2.1] heptan-2-one (**314**, norbornanone) and 7,7-dimethylbicyclo[2.2.1]heptan-2-one (**316**). The exo face (path *a* in **314A**) is usually preferred, leading to the endo alcohol **315**. The endo

---

313. Crimmins, M.T.; O'Mahoney, R. *J. Org. Chem.* **1989**, *54*, 1157.

hydrogens (H$_a$ in **314A**) provide steric encumbrance and analysis of a LUMO map of **314B** indicates that path *a* is more accessible, consistent with the observed exo-attack. Approach of a reagent to the carbonyl at an angle of 110° brings the reagent into close proximity to this hydrogen. The exo face appears somewhat hindered but is actually less hindered than the endo face. When a methyl group blocks the exo face (path *a*) (see model **316B**), attack by path *b* is preferred and the exo alcohol (**317**) is the major product. Analysis of a LUMO map of **316B** reveals that path *b* is preferred. In all [2.2.1]-, [2.1.1]-, and [2.2.2]-bicyclic ketones of this type, the major product is predicted by the extent of steric hindrance at the exo face (exo attack with no substituents on the bridge, as in **314**).[314,302,303] When the bridge has substituents (see **316**), endo attack is generally preferred.

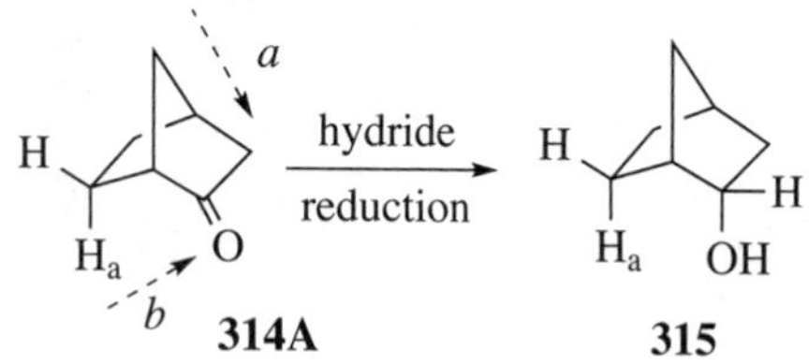

314A   315 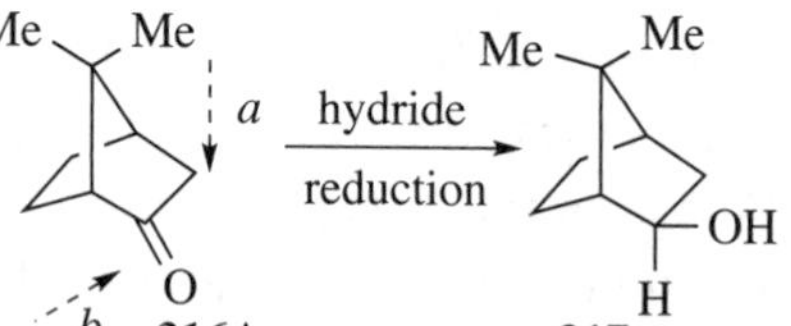   316A   317 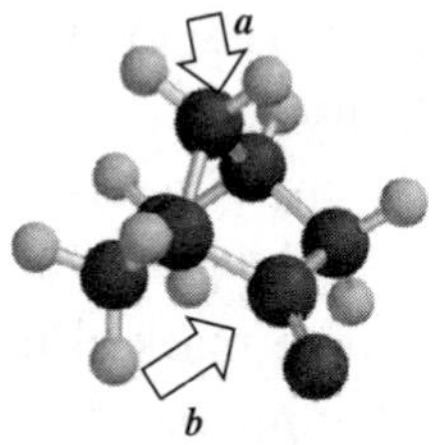   314B 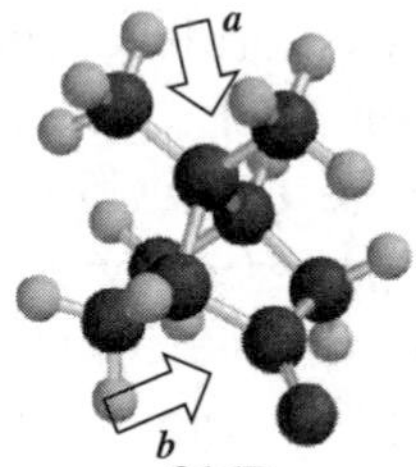   316B

MOLECULAR MODELING: Nucleophilic Attack on Norbornanone

It is known that nucleophilic addition to a carbonyl requires that the nucleophile approach the carbonyl carbon from an angle of ~110° (the "***Bürgi-Dunitz trajectory***"). Where the two faces of the carbonyl group are different, the angle of approach must be considered to determine which is more sterically hindered. Examine a space-filling model for norbornanone (left). Considering the approach angles (*a*) and (*b*), which face of the carbonyl carbon appears to be more encumbered? If steric factors dominate the reaction, what alcohol should result from addition of hydride to norbornanone?

Next, examine the LUMO map for norbornanone (right). This has been displayed in "transparent" mode, allowing you to see the underlying structure. It can be displayed as a solid by *clicking* on the map and selecting **Solid** from the **Style** menu at the bottom of the screen. The "blue regions" of the map indicate maximum concentration of the LUMO on the accessible surface of the molecule, and the "red regions" minimum concentration. The bluest region is most susceptible to attack by the nucleophilic hydride. According to the LUMO map, which face is more likely to be attacked by a nucleophile? Is this the same face as suggested by steric preferences?

When edge-fused bicyclic systems (bicyclo[*m.n.*0]) are examined, as in Table 4.11,[304] they

---

314. Wheeler, O.H.; Mateos, J.L. *Can. J. Chem.* **1958**, *36*, 1431.

behave very similarly to monocyclic ketones. Reduction of decalone (**318**) and indanone (**319**), for example, occurred primarily from path a as shown. In this table, the axial alcohol results from equatorial attack by the hydride and the equatorial alcohol results from axial attack by the hydride. These results show that axial attack (path *a*) is preferred in [m.n.0]-bicyclic ketones **318**, **319** and **320**, as in cyclohexanone since the trajectory of attack is the same. The presence of the bridging carbon inhibits path *a* somewhat in **321** and to a greater extent in **322**. There is also a preference for path *a* in complex compounds such as 3-cholestanone (**323**). Coprostanone (**324**) is an isomer of **323** and it also gives axial attack as shown (via path *a*) although the conformation of the molecule is slightly different.[315] The major diastereomeric product in bicyclic and polycyclic systems can be predicted with little or no modification of the simple cyclohexanone and cyclopentanone models presented above.

Table 4.11. Selectivity in Hydride Reduction of Bicyclic Ketones

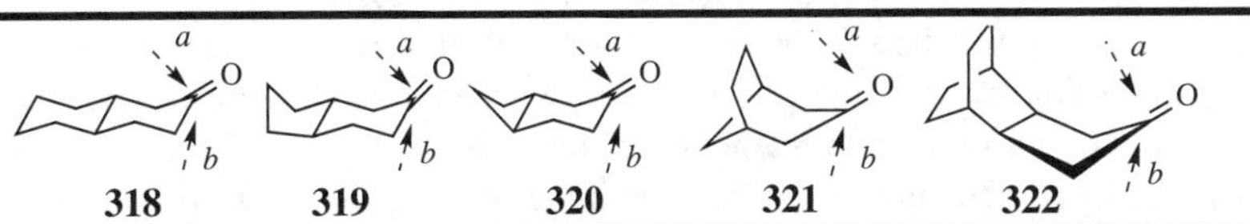

Bicyclic Ketone	Reagent	% ax Alcohol[b]	% eq Alcohol[a]
318	NaBH$_4$	15	85
	LiAlH$_4$	13	88
319	NaBH$_4$	11	89
	LiAlH$_4$	10	90
320	NaBH$_4$	6	94
	LiAlH$_4$	6	94
321	NaBH$_4$	40	60
322	NaBH$_4$	72	28

a Axial attack, path a.   b Equatorial attack, path b.

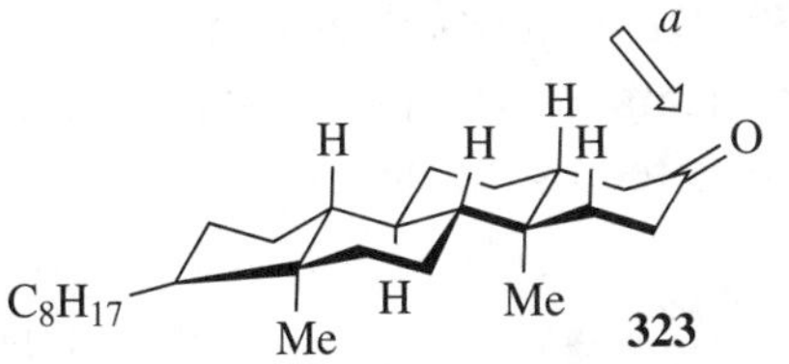

MOLECULAR MODELING: Stereoselectivity of Nucleophilic Additions of Polycyclic Ketones

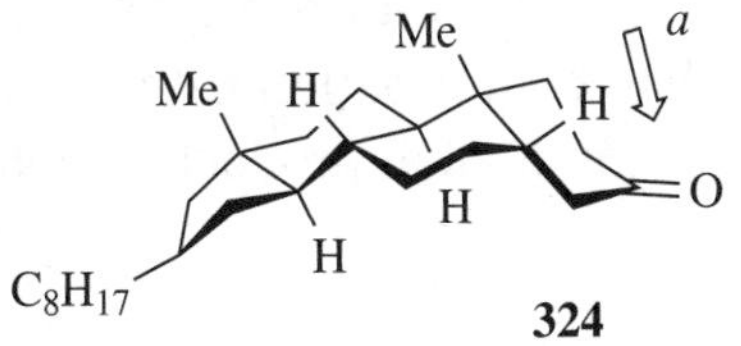

Reduction of bicyclic and polycyclic ketones is similar to that of monocyclic ketones, in that the reducing agent will approach the carbonyl "over" or "under" the ring, and will be subject to steric hindrance imposed by substituents on the ring. Bicyclic, tricyclic and polycyclic ketones such as A-F are generally more rigid that monocyclic ketones, however. Lithium aluminum hydride reduction of *trans*-2-decalone (**A**) occurs primarily from the *axial* face even though this appears to be more crowded than the *equatorial* face. This contrasts with the lack of stereoselectivity generally seen in reactions of unsubstituted steroidal ketones such as **F**. (Remember that the angle

---

315. Reference 231, pp 127-129 and references cited therein.

MOLECULAR MODELING: Stereoselectivity of Nucleophilic Additions of Polycyclic Ketones, con't.

of approach for these and all other compounds considered here is (*a*) or (*b*) as shown for **A** rather than (*c*) or (*d*).) At what point in moving from **A** to **F** does the stereochemical preference diminish? To decide, examine LUMO maps for these and four "intermediate" structures, **B–E**.

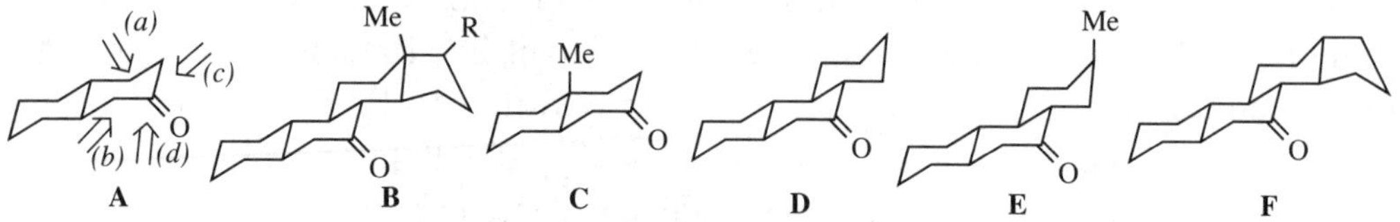

First, examine the map for *trans*-2-decalone (**A**). By convention, regions where the LUMO is concentrated are colored "blue", while regions where it is absent are colored "red". The more blue the region, the more facile will be the reaction with the reducing agent. At which carbonyl face of **A**, *equatorial* or *axial*, is the LUMO more exposed? Next, compare the LUMO map for *trans*-2-decalone with those of the remaining structures, **B–F**. Are any changes observed, either in the preferred face of nucleophilic or in the magnitude of the preference? If there are, identify the structure at which these changes occur.

4.7.F. Selectivity in the Reduction of Natural Products

There are many synthetic examples where hydride reduction proceeds with high diastereoselectivity. Reduction of **325A** to alcohol **326** with lithium tri-*tert*-butoxyaluminum hydride is taken from Donaldson and Greer's synthesis of the C_3-C_{15} bis(oxane) segment of phorboxazole.[316] The conformational drawing (**325B**) suggests that path *a* is preferred in the more stable chair conformation with the two substituents in the equatorial position. This model predicts the major alcohol product, which was converted as the acetate.

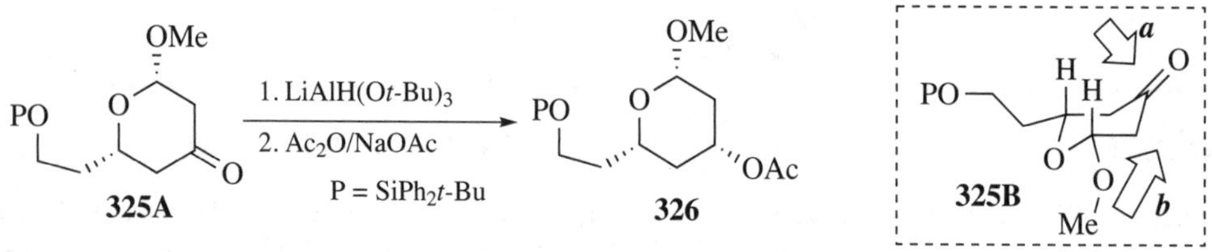

Another example involves a selective reduction of an acyclic ketone, albeit one that has a proximal dioxolane ring. Reduction of **327A** gave **328** in 96% yield with L-Selectride, in Prasad and Gholap's synthesis of (+)7-*epi*-goniofufurone.[317] The 3D model (**327B**) shows the

---

316.   Greer, P.B.; Donaldson, W.A. *Tetrahedron Lett.* **2000**, *41*, 3801.
317.   Prasad, K.R.; Gholap, S.L *J. Org. Chem.* **2007**, *72*, 2.

large substituent is largely out-of-the way for path *a* or path *b*. The conformation of the seven-membered ring, however, leads to some hindrance for path a, making path *b* preferred. This result is consistent with the stereochemistry of the final product, *syn*-diastereomer **328**. Note that the substituent is positioned in a such a way that a "pocket" is available for delivery via path *b*.

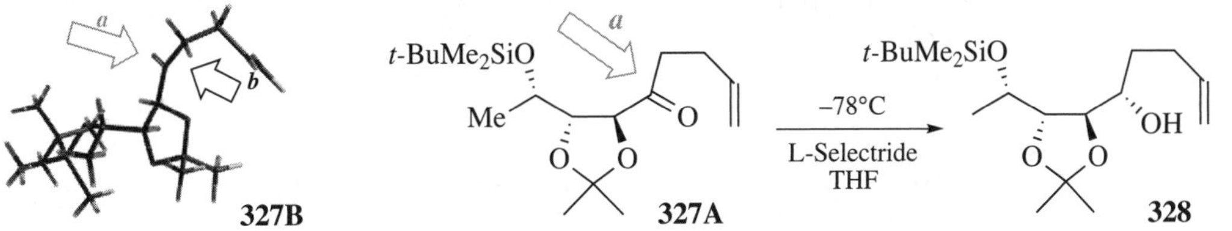

Reduction of norbornane-like precursors to natural products is also predictable. Reduction of the ketone unit in **329**, in Money and co-worker's synthesis of β-santalene for example, gave the exo-alcohol **330** via attack from the less hindered endo face.[318] Conversely, removal of the steric encumbrance at C7 in **331** gave attack via the exo face to give **332** (in about a 4:1 ratio favoring the endo alcohol shown).[319]

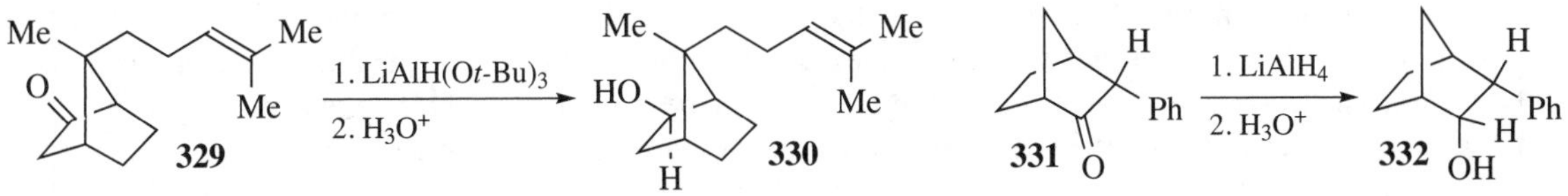

There are many more examples of selectivity in reduction. These few illustrate that the simple principles discussed in this section, and the models used, can be applied to more complex systems with reasonable success.

MOLECULAR MODELING: Stereochemistry of Reduction of Spirocyclic Ketone

Sodium borohydride reduction of the spirocyclic ketone shown below gives a 13:1 mixture of two alcohols [De Shong et. al. *J. Org. Chem.* **1991**, *56*, 3207].

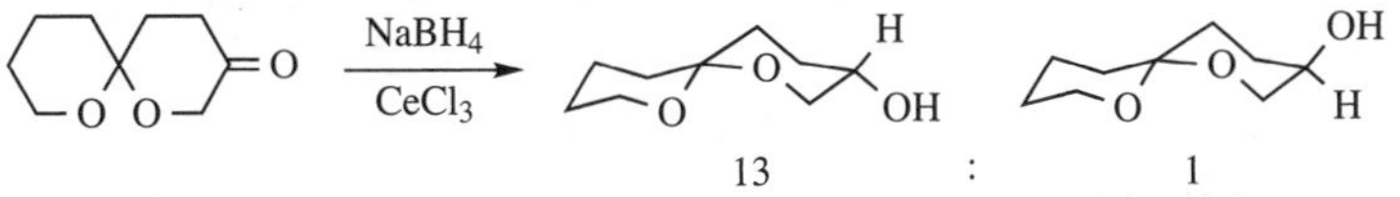

It is reasonable to expect that a LUMO map will be able to anticipate the direction of "hydride" attack (and hence the stereochemistry of the resulting alcohol). Remember that the nucleophile will be delivered to the carbonyl carbon either over or under the ring, based on the Bürgi-Dunitz trajectory. Before it can be used, however, it is first necessary to establish the conformation of the ketone. Assuming that both rings are chairs, there are four possible conformers, as shown, corresponding to whether the C–O bond in the ring that does not contain the carbonyl group is *equatorial* or *axial*. Build and obtain the energy for each of these. For each in turn,

---

318. Hodgson, G.L.; MacSweeney, D.F.; Mills, R.W.; Money, T. *J. Chem. Soc. Chem. Commun.* **1973**, 235.
319. (a) Benjamin, B.M.; Collins, C.J. *J. Am. Chem. Soc.* **1966**, *88*, 1556; (b) Kleinfelter, D.C.; Dye, T.E. *Ibid.* **1966**, *88*, 3174; (c) Kleinfelter, D.C.; Dye, T.E.; Mallory, J.E.; Trent, E.S. *J. Org. Chem.* **1967**, *32*, 1734.

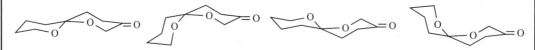
4.8. CATALYTIC HYDROGENATION

Table 4.12.[320] Typical hydrogenation catalysts.

Platinum oxide[a]	Platinum black[b]	Pt/C[c,b]	Pt/Rh oxide[d]	Pd oxide[e]	Pd/BaCO$_3$[f]
Pd/SrCO$_3$[g]	Pd/C[f]	Rh/C[f]	Pd hydroxide/C[h]	Ni boride-P1[i]	Ni(R) W3[q]
Ni boride-P2[j]	Ni (NiC)[k]	Ni-Kieselguhr[m]	Urushibara Ni[n]	Ni(R) W1[p]	Ni-Cu[l]
Raney Ni [Ni(R)][o]	Ni(R) W2[f]	Copper chromite[v]	Ni(R) W4346[a]	Ni(R) W8[t]	Ni(R) W5[t]
Ni(R) W6[s]	Raney cobalt[u]	Ni(R) W7[r]			

---

320. See Reference 323, p 10.

a Ref.[321] b Ref.[322] c Ref.[323] d Ref.[324] e Ref.[325] f Ref.[326] g Ref.[327] h Ref.[328] i Ref.[329] j Ref.[330]
k Ref.[331] l Ref.[332] m Ref.[333] n Ref.[334] o Ref.[335] p Ref.[336] q Ref.[337] r Ref.[338] s Ref.[339] t Ref.[340]
u Ref.[341] v Ref.[342]

Reduction of functional groups with hydrogen gas is a key reaction in organic chemistry, dating to the first hydrogenation of ethene to ethane by von Wilde in 1874.[343] As stated by Hudlicky',[7] "true widespread use of catalytic hydrogenation did not start until 1897 when Sabatier and his co-worker developed the reaction between hydrogen and organic compounds to be a universal reduction method (Nobel Prize, 1912)."[344,345] Other monographs that describe synthetic applications of catalytic hydrogenation are available from Rylander[346] and from

321. (a) Adams, R.; Shriner, R.L. *J. Am. Chem. Soc. 1923, 45*, 2171; (b) Carothers, W.H.; Adams, R. *Ibid. 1923, 45*, 1071; (c) Keenan, C.W.; Giesemann, B.W.; Smith, H.A. *Ibid. 1954, 76*, 229; (d) Frampton, V.L.; Edwards, Jr., J.D.; Henze, H.R. *Ibid. 1951, 73*, 4432.

322. (a) Willstätter, R.; Waldschmidt-Leitz, E. *Berichte 1921, 54*, 113 (see p121); (b) Baltzly, R. *J. Am. Chem. Soc. 1952, 74*, 4586; (c) Theilacker, W.; Drössler, H.G. *Chem. Ber. 1954, 87*, 1676.

323. Baltzly, R. *J. Org. Chem. 1976, 41*, 920.

324. (a) Nishimura, S. *Bull. Chem. Soc. Jpn. 1961, 34*, 32; (b) *Idem, Ibid. 1960, 33*, 566; (c) *Idem Ibid. 1961, 34*, 1544.

325. (a) Shriner, R.L.; Adams, R. *J. Am. Chem. Soc. 1924, 46*, 1683; (b) Starr, D.; Hixon, R.M. *Org. Synth. Coll.. Vol. 2 1943*, 566.

326. Mozingo, R. *Org. Synth. Collect. Vol. 3 1955*, 181.

327. Johnson, W.S.; Rogier, E.R.; Szmuszkovicz, J.; Hadler, H.I.; Ackerman, J.; Bhattacharyya, B.K.; Bloom, B.M.; (b) Stalmann, L.; Clement, R.A.; Bannister, B.; Wynberg, H. *J. Am. Chem. Soc. 1956, 78*, 6289.

328. Pearlman, W.M. *Tetrahedron Lett. 1967*, 1663.

329. Brown, C.A. *J. Org. Chem. 1970, 35*, 1900.

330. (a) Brown, C.A.; Ahuja, V.K. *J. Chem. Soc. Chem. Commun. 1973*, 553; (b) Brown, H.C.; Sivasankaran, K.; Brown, C.A. *J. Org. Chem. 1963, 28*, 214; (c) Brubaker, Jr., C.H. in *Catalysis in Organic Syntheses, 1977*, Smith, G.W. (Ed.), Academic Press, New York, *1977*.

331. Brunet, J.J.; Gallois, P.; Caubere, P. *Tetrahedron Lett. 1977*, 3955.

332. Takai, E. *Sci. Pap. Inst. Phys. Chem. Res. 1968, 62*, 24 [*Chem. Abstr. 69*: 30510h, *1968*].

333. Covert, L.W.; Connor, R.; Adkins, H. *J. Am. Chem. Soc. 1932, 54*, 1651.

334. (a) Taira, S. *Bull. Chem. Soc. Jpn. 1961, 34*, 1294; (b) Urushibara, Y. *Ann. N.Y. Acad. Sci. 1967, 145*, 52 [*Chem. Abstr. 68*: 72708p *1968*]; (c) Motoyama, I. *Bull. Chem. Soc. Jpn. 1960, 33*, 232.

335. (a) Pavlic, A.A.; Adkins, H. *J. Am. Chem. Soc. 1946, 68*, 1471; (b) Mozingo, R. *Org. Synth. Coll. Vol. 3 1955*, 181.

336. Covert, L.W.; Adkins, H. *J. Am. Chem. Soc. 1932, 54*, 4116.

337. Adkins, H.; Pavlic, A.A. *J. Am. Chem. Soc. 1947, 69*, 3039.

338. Adkins, H.; Billica, H.R. *J. Am. Chem. Soc. 1948, 70*, 695.

339. Billica, H.R.; Adkins, H. *Org. Synth. Collect. Vol. 3 1955*, 176.

340. Khan, N.A. *J. Am. Chem. Soc. 1952, 74*, 3018.

341. Reeve, W.; Eareckson II, W.M. *J. Am. Chem. Soc. 1950, 72*, 3299.

342. (a) Connor, R.; Folkers, K.; Adkins, H. *J. Am. Chem. Soc. 1932, 54*, 1138; (b) Adkins, H. *Reactions of Hydrogen*, University of Wisconsin Press, Madison, *1957*.

343. von Wilde, M.P. *Deutsch. Chem. Ber. 1874, 7*, 352.

344. Reference 7, p 3.

345. Sabatier, P.; Senderens, J.B. *Compt. Rend. 1897, 124*, 1358.

346. Rylander, P.N. *Catalytic Hydrogenation in Organic Synthesis* Academic Press, New York *1979*.

Freifelder,[347] Applications of catalytic hydrogenation to organic chemistry are found in the series *Catalysis in Organic Reactions*.[348] Catalytic hydrogenation is used to reduce alkenes,[349] alkynes,[350] ketones and aldehydes,[351] nitriles,[352] nitro compounds[353] or aromatic rings.[354]

4.8.A. Catalytic Activity and Reactivity

Hydrogen gas is added to an organic molecule in the presence of a catalytic amount of a transition metal (see Table 4.12) and the reaction proceeds by adsorption of hydrogen and the substrate on the surface of the metal. The mechanistic rationale for this type of reduction will be discussed in Section 4.8.B. There are two major types of catalysis: **heterogeneous** (catalyst insoluble in the reaction medium) and **homogeneous** (catalyst soluble in the reaction medium). Catalytic heterogeneous hydrogenation can be **supported** (slurry and fixed gel operations)[323] or **unsupported** (primarily solution reactions). The nature and amount of the catalyst and hydrogenation procedure employed influence which functional group that is reduced, the extent of reduction, and the product distribution. Laboratory scale reductions of organic functional groups containing a π bond usually involve a slurry process and an heterogeneous catalysis. Industrial (kilogram or ton scale) applications often use fixed-bed heterogeneous catalysis or homogeneous catalysis. Birch and Williamson reviewed homogeneous hydrogenation in synthesis.[355]

The most commonly used heterogeneous catalysts are platinum, palladium, nickel, rhodium, nickel, and ruthenium. Rylander gave references for the preparation of the most common catalysts, shown in Table 4.12.[332] In some cases, salts of transition metals are used rather than the metal itself although the pure metal adsorbed on a support (see above) is also commonly used. Hudlicky´ presented an order of relative reactivity with propene for groups 8-10 transition metal catalysts,[356] based on the work of Mann and Lien.[357]

---

347. Freifelder, M. *Catalytic Hydrogenation in Organic Synthesis: Procedures and Commentary*, Wiley, New York, *1978*.
348. (a) *Catalysis of Organic Reactions*, Moser, W.R.; Marcel Dekker, New York, *1981*; (b) *Characterization of Heterogeneous Catalysts*, Delannay, F. (Ed.), Marcel Dekker, New York, *1984*; (c) *Catalysis of Organic Reactions*, Kosak, J.R. (Ed.), Marcel Dekker, New York, *1984*; (d) *Catalysis of Organic Reactions*, Augustine, R.L. (Ed.), Marcel Dekker, New York, *1985*; (e) *Catalysis of Organic Reactions*, Rylander, P.N.; Greenfield, H.; Augustine, R.L. (Ed.), Marcel Dekker, New York, *1988*.
349. See Reference 17, p 7.
350. See Reference 17, p 28.
351. See Reference 17, p 1075.
352. See Reference 17, p 875.
353. See Reference 17, p 821.
354. See Reference 17, p 6.
355. Birch, A.J.; Williamson, D.H. *Org. React. 1976, 24*, 1.
356. Reference 7, p 4.
357. Mann, R.S.; Lien, T.R. *J. Catal. 1969, 15*, 1.

	Rh	>	Ir	>	Ru	>	Pt	>	Pd	>	Ni	>	Fe	>	Co	>	Os
	14.0	>	15.0	>	6.5	>	16.0	>	11.0	>	14.0	>	10.0	>	8.1	>	7.4

In this series, rhodium is the most active catalyst. The activation energy for the reaction of hydrogen with propene is also shown, and it is noteworthy that the catalytic activity does not correlate with the activation energy for the hydrogenation reaction (numbers below the metal). These catalysts are used for the hydrogenation of many functional groups, and Table 4.13[358] correlates functional groups with recommended reaction conditions and catalysts. No single catalyst gives excellent results for all functional groups, and there are significant differences in chemo- and stereoselectivity. Note that most hydrogenations are done at ambient temperature and pressure. Only a few of the catalysts listed in Table 4.13 require high pressures, and then only for selected functional groups. In most cases, < 10 mol% of the catalyst is required per mole of the compound. Both Raney nickel and rhodium catalysts appear to require greater amounts of the catalyst. The purpose of this table is to provide a quick guide to the reduction of common functional groups and to show the most common catalysts used. This section will focus on these catalysts and the functional groups they reduce.

Table 4.13. Recommended Reaction Conditions for Catalytic Hydrogenation of Selected Functional Groups

Functional Group	Product	Catalyst	mol %	Temp. (°C)	Pressure (atm)
Alkene	Alkane	A	5-10	25	1-3
		B	0.5-3	25	1-3
		C	30-200	25	1-3
Alkyne	Alkene	D	8	25	1
		E	2 + 2% quinoline	20	1
		F	10 + 4% quinoline	25	1
	Alkane	B	3	25	1
Carboxylic Acid		C	20	25	1-4
Aromtic	Hydroaromatic	B	6-20 + AcOH	25	1-3
		G	40-60	25	1-3
		C	10	75-100	70-100
Heteroaromatic	Hydroaromatic	B	4-7 +AcOH or HCl/MeOH	25	1-4
		H	20 + HCl/MeOH	65-200	130
		C	2	25	1
Aldehyde, Ketone	Alcohol	B	2-4	25	1-4
		A	3-5	25	1
		C	30-100	25	1
Halide	Hydrocarbon	A	1-15 + KOH	25	1
		E	30-100 + KOH	25	1
		C	10-20 + KOH	25	1
RNO_2, Azide	Amine	B	1-5	25	1
		A	4-8	25	1
		C	10-80	25	1-3
Oxime, Nitrile	Amine	B	1-10 + AcOH or HCl/MeOH	25	1-3
		A	5-15 + AcOH	25	1-3
		C	3-30	25	35-70

A = 5% Pd(C); B = PtO$_2$; C = Raney Ni; D =0.3% Pd(CaCO$_3$); E = 5% Pd(BaSO$_4$); F = Lindlar catalyst; G = 5% Rh(Al$_2$O$_3$); H = 5% Rh (C)

[Reproduced with permission from *Reductions in Organic Chemistry* by M. Hudlicky published in *1984* by Ellis Horwood, Chichester.]

358. Reference 7, p 6.

In the context of hydrogenation, the **support** is an inert material on which the catalyst is adsorbed or admixed. Common supports are carbon (platinum black and palladium on carbon, for example) and alumina (rhodium is often adsorbed on alumina) and both calcium sulfate ($CaSO_4$) and barium sulfate ($BaSO_4$) are used. In part, the support dilutes the expensive transition metal catalyst, which allows one to control the rate of reaction as well as making it cost effective. Typically 1-10 mol % of the catalyst (see Table 4.13) is added to the support (sometimes called a carrier). The support can diminish the reactivity of the catalyst. Many other factors influence the reduction. These include the tendency of the catalyst to agglomerate, addition of a promoting substance, synergism between two catalysts, the temperature of the reaction, the ratio of hydrogen to compound being reduced, the pressure of hydrogen gas, the preparation method, the age of the catalyst, and the number of times it has been recycled.[323,7,324]

For many applications, too active a catalyst will lead to overreduction and/or poor selectivity. If the activity of the metal catalyst is too high, addition of compounds that bind to the metal will usually diminish the activity by displacement of hydrogen atoms or the organic substrate being reduced. This process is referred to as **poisoning** of the catalyst, and it can greatly diminish the reducing power of that catalyst. In essence, when the poison is bound to the catalyst, hydrogen and the organic substrate cannot be bound, hence diminished reactivity. Common poisons include metal cations, halides, Hg°, divalent sulfur compounds, carbon dioxide, amines and phosphines.[359] In many cases, the organic molecule being reduced or a minor reaction product can poison the catalyst.[360]

A hydrogenation reaction is rather complex. The surface of the metal catalyst is very porous, allowing adsorption of hydrogen gas, organic substrates and various poisons. As the number of **active sites** (regions of the catalyst that can accommodate and bind hydrogen atoms and/or organic substrates) increases, the rate of hydrogenation increases. A poison will bind to the active site in preference to the substrate or hydrogen, preventing their adsorption.[361] Gases such as carbon dioxide (CO_2) and carbon monoxide (CO) can fill up the active sites as well. Measurement of active sites in a catalyst is difficult. The position of the metal at the surface or just below the surface has a major influence on chemisorption. In many cases, metal salts are used as catalysts and a change in oxidation state during the reaction can deactivate the catalyst.[361]

Reduction of alkenes with homogeneous catalysts will be presented in Section 4.8.B.

---

359. (a) Reference 323, p 4; (b) Maxted, E.B. *Adv. Catal.* **1951**, *2*, 129; (c) Baltzly, R. *J. Org. Chem.* **1976**, *41*, 920, 928, 933.
360. Smutny, E.J.; Caserio, M.C.; Roberts, J.D. *J. Am. Chem. Soc.* **1960**, *82*, 1793.
361. Farranto, R.J.; Hobson, M.C.; Brungard, N.L. in Reference 325e, pp 177-209.

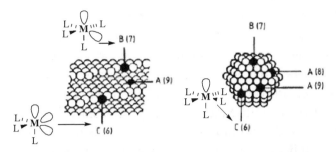

Figure 4.1. The principal surface and particle sites for heterogeneous catalysts. [Reproduced with permission from Maier, W.F. *Angew. Chem. Int. Ed. Engl.* **1989**, *28*, 135. Copyright 1989 VCH Weinheim.]

Heterogeneous catalysis is a surface phenomenon, and it is known that there are different types of metal particles on the surface. Maier suggested the presence of **terrace-, step-** and **kink-**type atoms (in Fig. 4.1)[362] on the surface of a heterogeneous catalyst. These terms refer to different atom types characterized by the number of nearest neighbors,[362] which correspond to different transition metal fragments as well as to different coordination states of that metal.[363] A terrace type atom (A in Fig. 4.1) typically has eight or nine neighbors and corresponds to a coordination model such as ML_5. The step type of atom (B) usually has seven neighbors and can be correlated with coordination model ML_4. Finally, the kink type atom (C) has six neighbors and corresponds to coordination model ML_3. As the number and type of neighboring atoms around each type of atom changes, the binding capacity (active sites) will change. In general, as the particle size increases the relative concentration of terrace atoms will increase, whereas small particle size favors the kink type of surface atoms. Since the **terrace** atoms correspond to ML_5 transition metals, the **step** to ML_4 and the **kink** to ML_3 they should exhibit distinctly different reactivity.[362] Other factors can influence the way each type of atom will interact with an organic substrate, but it is clear the surface of the catalyst is important. Indeed, Maier suggested that the ideal catalyst would be "a material with the right surface structure rather than an empirical mixture of ingredients".[363] It is likely that the use of modifiers, poisons, and special reaction conditions will change the shape, size, and surface structure of the catalyst. This will, of course, influence the selectivity and reactivity. With palladium, both "hydrogenation and hydrogenolysis were found to be sensitive to the surface structure and morphology of the catalyst."[362] It is important to emphasize that Maier states he does not believe this analysis applies to homogeneous catalysis.[363] There are several interesting differences between homogeneous and heterogeneous catalysis. The soluble homogeneous catalysts usually have a dissociated ligand and/or a coordinating solvent that will compete with the organic substrate for active metal sites. Heterogeneous catalysts behave essentially as an insoluble matrix, whereas the homogeneous catalyst is soluble in the reaction solvent and functions as a discreet molecule. Because heterogeneous catalysts are insoluble, only those metal atoms at the surface of the matrix are thought to be important for hydrogenation as noted above. Much of the catalyst is

362. Maier, W.F. *Angew. Chem. Int. Ed.* **1989**, *28*, 135.
363. Maier, W.F. in Reference 325e, pp 211-231, Cf. p 220.

within the matrix and inaccessible. The soluble homogeneous catalyst allows each molecule of catalyst to be accessible (sec. 4.8.B). Thus, homogeneous reactions require less catalyst. The matrix nature of heterogeneous catalysts often means poor chemoselectivity, bond migration, hydrogenolysis and so on (see below). Prediction of stereochemistry in the final product is usually difficult because chemisorption is required for reaction, and the precise nature of the chemisorption is difficult to ascertain. Since heterogeneous catalysts often require polar solvents, problems can arise due to the poor solubility of hydrogen gas in the solvent. The large surface area of heterogeneous catalysts make them susceptible to poisoning, as discussed above, but homogeneous catalysts are soluble and many of the problems arising from chemisorption at a surface disappear. Homogeneous catalysts tend to be significantly more selective for alkenes and alkynes, show less reactivity with heteroatom-containing functional groups, and they are less prone to poisoning.[332] Nonpolar solvents (such as benzene, and methylene chloride) can be used with homogeneous catalysts, which is important because such solvents can dissolve more hydrogen gas, facilitating the reduction. Reaction with an organic substrate occurs with more or less discreet catalyst molecules of known coordination states. The stereochemistry of the reduction product is often predictable by analysis of the geometry of substrate binding around the metal (the binding of hydrogen, solvent and other ligands must also be known). Typical homogeneous catalysts include **Wilkinson's catalyst** $(Ph_3P)_3RhCl$ (**336**)[364] and the iridium complex $[Ir(cod)(Py)PCy_3]^+ PF_6^-$, which is related to **Vaska's catalyst** (see **341**).[365] It is also clear that the main difference between an homogeneous and heterogeneous catalyst is the presence of an organic substrate called a ligand [pyridine (Py), cyclooctadiene (COD), tricyclohexylphosphine (PCy_3), triphenylphosphine, bis(phosphine) ligands, and mono and diamines] in the homogeneous catalyst. These ligands bind to the metal whereas heterogeneous catalysts typically consist of transition metals ($M°$), metal halides or metal oxide salts MX_2, MX_4, and so on where X is Cl or O.

4.8.B. Hydrogenation of Alkenes

Catalytic hydrogenation is commonly used for the reduction of alkenes, alkynes, aromatic hydrocarbons, and aromatic heterocycles, carbonyl derivatives, nitriles or nitro compounds. The reaction with alkenes proceeds on the surface of a heterogeneous metal catalyst, via cleavage of diatomic hydrogen and adsorption of the hydrogen atoms. Horiuti and Polanyi proposed a mechanism in 1934 for the reduction of ethene with hydrogen on nickel.[366] In each step, (*) referred to an undefined nickel atom or complex. It was proposed that hydrogen gas was cleaved to hydrogen atoms that were bound to the nickel (represented by **333**). An

364.   (a) Jardine, F.H., Osborn, J.A.; Wilkinson, G.; Young, G.F. *Chem. Ind. (London)* **1965**, 560; (b) Imperial Chem. Ind. Ltd., *Neth. Appl. 6,602,062* [*Chem. Abstr. 66*: 10556y *1967*]; (c) Bennett, M.A.; Longstaff, P.A. *Chem. Ind., 1965,* 846; (d) Mundy, B.P.; Ellerd, M.G.; Favaloro Jr., F.G. *Name Reactions and Reagents in Organic Synthesis, 2nd ed.*, Wiley-Interscience, New Jersey, *2005*, pp. 868-869.
365.   Suggs, J.W.; Cox, S.D.; Crabtree, R.H.; Quick, J.M. *Tetrahedron Lett. 1981, 22,* 303.
366.   Horiuchi, J.; Polányi, M. *Trans. Faraday Soc. 1934, 30,* 1164.

alkene was also bound to the surface of the metal (represented by **334**), allowing the reaction to proceed, with transfer of hydrogen to the carbon (in **335**). This model is misleading, since the surface is porous and may contain metal atoms in more than one coordination state for heterogeneous catalysts (see above). The exact mechanism of heterogeneous catalysis is not well understood since it is a surface phenomenon. It is clear that the catalyst ruptures the H-H bond to generate a metal-H (H*) species where hydrogen is adsorbed at an active site on the surface of the catalyst. The differing ability of metals to bind hydrogen and to transfer that hydrogen to an organic substrate is a critical factor that leads to differences in reduction for different catalysts.

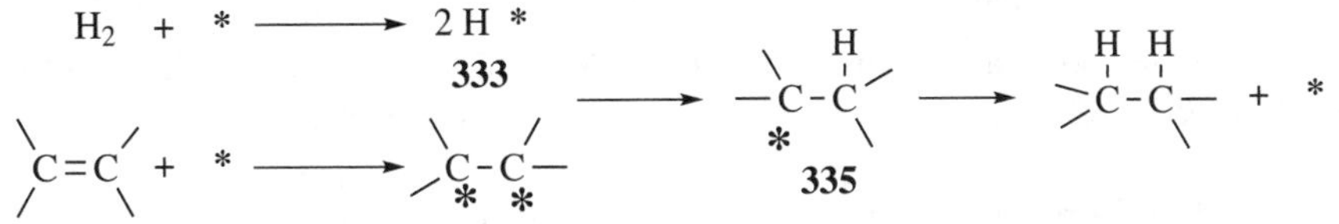

Hydrogen adsorption is best understood for the reduction of alkenes with homogeneous catalysts, because the hydrogenation involves discreet, well-defined particles. Wilkinson's catalyst [**336**, (Ph₃P)₃RhCl] has been studied extensively. Wilkinson's catalyst is prepared from rhodium chloride (RhCl₃) and triphenylphosphine in ethanol (88% is a typical yield of **336**).[367] The mechanism of activation when used for hydrogenation of alkenes is hydrogen activation, substrate activation, and hydrogen transfer.[368] A mechanistic picture of the reaction is shown for the reaction of the square planar catalyst (**336**) with hydrogen gas to form a trigonal bipyramidal species, **337**.[369] Subsequent reaction with the alkene will replace a phosphine ligand to give a species such as **338**. It is also possible that the alkene binds to the catalyst first, followed by hydrogen. Hydrogen is transferred from rhodium to carbon in a stepwise manner, generating a species such as **339**. Release of the alkane generates a trigonal species (**340**) that reacts with additional hydrogen to regenerate **337**. Cotton and Wilkinson used this cycle to illustrate the hydrogenation of alkenes.[370]

367. Osborn, J.A.; Jardine, F.H.; Young, J.F.; Wilkinson, G. *J. Chem. Soc. A* ***1966***, 1711.
368. (a) Collman, J.P. *Acc. Chem. Res.* ***1968***, *1*, 136; (b) Vaska, L. *Ibid.* ***1968***, *1*, 335; (c) Halpern, J. *Quart. Rev.* ***1956***, *10*, 463; (d) *Idem, J. Phys. Chem.* ***1959***, *63*, 398; (e) *Idem, Ann. Rev. Phys. Chem.* ***1965***, *16*, 103.
369. Tolman, C.A.; Meakin, P.Z.; Lindner, D.L.; Jesson, J.P. *J. Am. Chem. Soc.* ***1976***, *96*, 2762.
370. Cotton, F.A.; Wilkinson, G.; Murillo, C.A.; Bochmann, M. *Advanced Inorganic Chemistry, 6th ed*, Wiley, New York, ***1999***, p 1231.

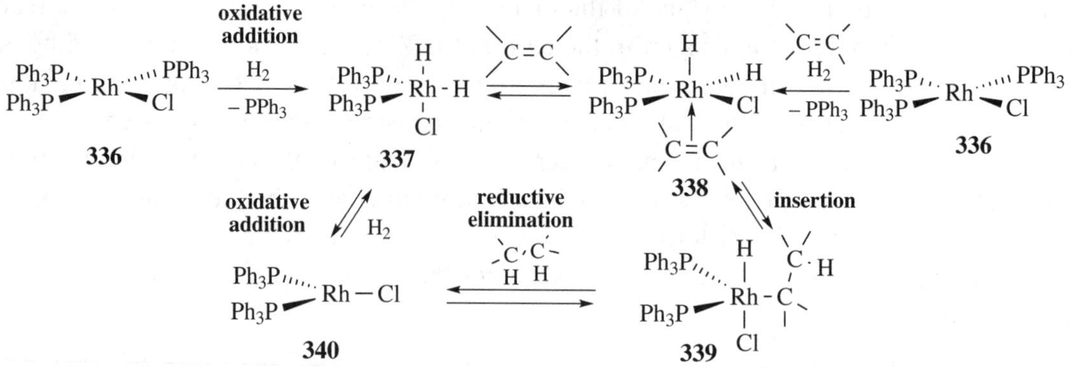

oxidative addition

336   337   338   336

oxidative addition   H₂   reductive elimination   insertion

340   339

[Reprinted in part with permission from Tolman, C.A.; Meakin, P.Z.; Lindner, D.L.; Jesson, J.P. *J. Am. Chem. Soc.* **1976**, *96*, 2762. Copyright © 1976 American Chemical Society.]

There are many other homogeneous catalysts, involving several different transition metals. **Vaska's catalyst (341)** is one of the more important and it also reacts with hydrogen, converting the square planar catalyst to the octahedral intermediate **342**.[371] Reduction of an alkene probably involves replacement of the triphenylphosphine ligand with hydrogen or an alkene in a reversible process.[332]

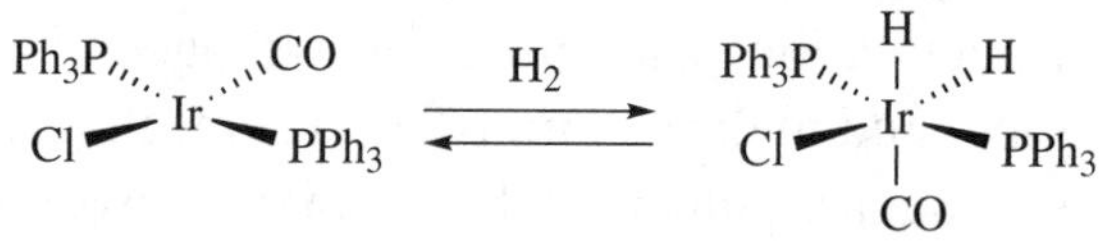

341   342

As shown above, the mechanism of hydrogenation involves coordination of the organic substrate to the metal, usually via π bonds or lone pairs of electrons on heteroatom functional groups. The ability to bind the substrate, how it is bound (π complex or metal-carbon bond) and its reactivity to metal-bound hydrogen are the other factors determining differences in catalysis. Rylander's book[323] examines the reduction of many functional groups, and finds that palladium is preferred for reduction of alkenes, platinum for carbonyl, and rhodium for selective reduction of aromatic rings. The following order of reactivity is useful for general applications:

Pt	C=O	>>	C=C	>	(H)	>	Ar	
Pd	C=C	>	(H)	>	C=O	>	Ar	
Ru	C=O	>	C=C	>	Ar	>	(H)	

where **(H)** = hydrogenolysis

With proper choice of catalyst, high chemoselectivity can be achieved. In the Hagiwara et al. synthesis of (*R*)-(+)-muscopyridine,[372] for example, hydrogenation of **343** with a Pd-C catalyst gave a 98% yield of diketone **344**, with no reduction of the ketone units. Alkenes can be

371. Vaska, L.; Diluzio, J.W. *J. Am. Chem. Soc.* **1962**, *84*, 679.
372. Hagiwara, H.; Katsumi, T.; Kamat, V.P.; Hoshi, T.; Suzuki, T.; Ando, M. *J. Org. Chem.* **2000**, *65*, 7231.

reduced in the presence of benzylic substituents without hydrogenation. In a synthesis of (–)-anaferine[373] by Blechert and Stapper, both alkene units in **345** were hydrogenated to give **346** in 87% yield, without reduction of the benzyl units.

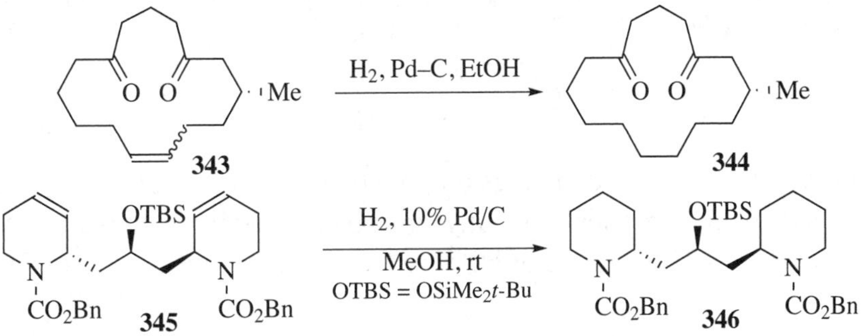

The phenomenon of **hydrogenolysis** (cleavage of a C–X bond, where X is an heteroatom) was introduced with the discussion of LiAlH$_4$ reduction and involves cleavage of carbon-heteroatom bonds, although it did not occur in **345** since reduction of the C=C unit was more facile. This process is particularly facile when the C–X bond is allylic or benzylic to a π-bond. Reaction with hydrogen cleaves the C–X bond (see **347**) to give the hydrocarbon (**348**). The ability of a catalyst to effect hydrogenolysis is indicated by (H) in the comparison of catalyst reactivity for Pt, Pd and Ru. The group X can be halogen, alcohol, ether, amine, and so on. A synthetic example of hydrogenolysis is cleavage of the O-benzyl unit in **349** to give **350**, as part of Lee and co-worker's synthesis of SCH 351448, a novel activator of low-density lipoprotein receptor (LDL–R) promoter.[374] As shown in this example, hydrogenolysis of the PhCH$_2$O–CH$_2$- unit leads to the alcohol and toluene (PhCH$_2$–H).

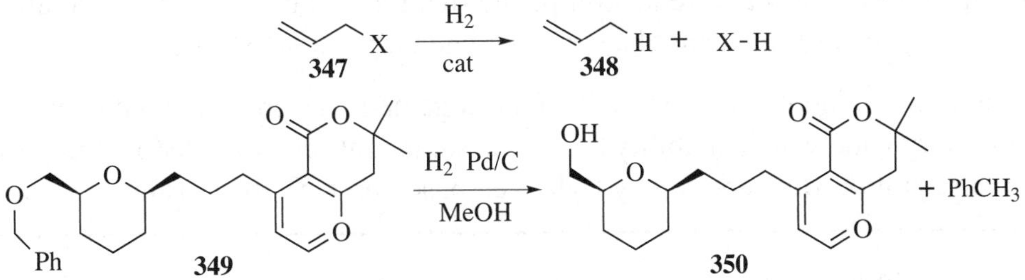

The order of catalytic ability shown above for platinum, palladium, and ruthenium is most useful when a catalyst must be chosen for hydrogenation of a multifunctional molecule. In molecules containing an alkene and a ketone or aldehyde, platinum catalysts selectively reduce the carbonyl. There are also differences in conjugated versus nonconjugated alkenes. An example of this selectivity is the reduction of **351** with palladium and strontium carbonate (SrCO$_3$), which gave selective reduction of one alkene unit to **352** without reducing either the conjugated carbonyl or the cyclopentanone moiety. Changing to a palladium-on-carbon

373. Blechert, S.; Stapper, C. *Eur. J. Org. Chem.* **2002**, 2855.
374. Kang, E.J.; Cho, E.J.; Lee, Y.E.; Ji, M.K.; Shin, D.M.; Chung, Y.K.; Lee, E. *J. Am. Chem. Soc.* **2004**, *126*, 2680.

catalyst led to selective reduction of the conjugated double bond to give **353** in 87% yield.[375]

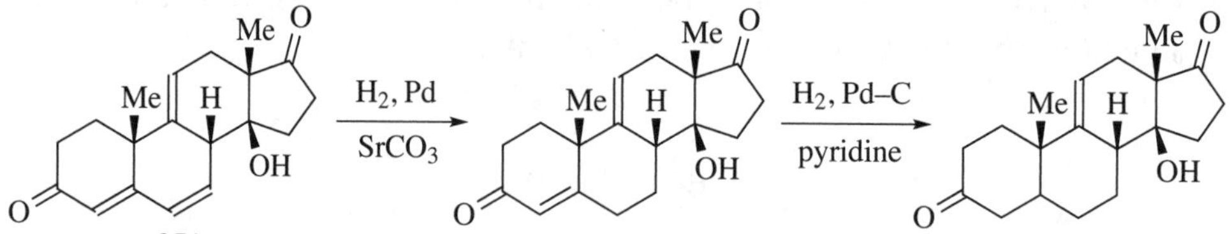

In general, more highly substituted alkenes do not adsorb on the catalyst as effectively and are not readily reduced, which implies that vinyl groups are easily reduced but tetrasubstituted alkenes are very sluggish. The relative order of reduction for mono, di-, tri- and tetrasubstituted alkenes is:

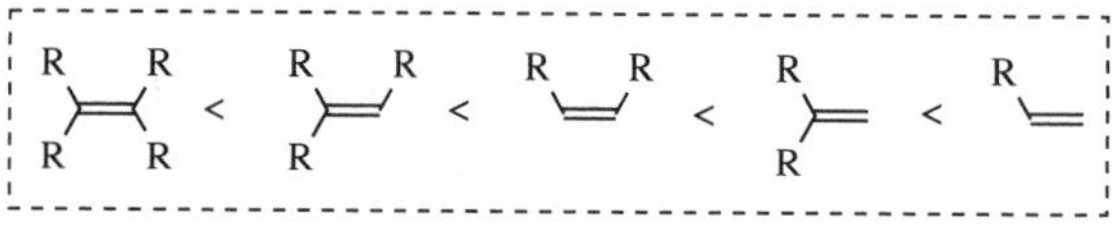

Unfortunately, this selectivity for reduction of the less hindered double bond is not always observed and varies considerably with the catalyst and reaction conditions (see Table 4.13).[358,376] When a double bond is highly strained, it is more easily reduced.[376]

The reduction of the exo-methylene group in **354** to **355** illustrates this selectivity.[377] The exocyclic methylene group is more strained than the internal double bond, which is part of the cyclopentene moiety. Selective reduction of the methylene unit was observed, although the methylene is only disubstituted and the cyclopentene unit is trisubstituted.

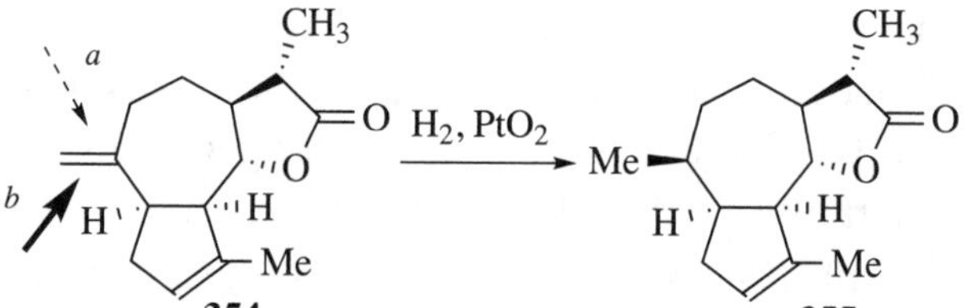

Isomerization of the double bond and even rearrangement are problems with some catalysts, and catalysts vary widely in their ability to induce isomerization. An order of ability to promote double bond migration was reported by Rylander, based on work by Gostunskaya et al.[378]

$$ Pd \ > \ Ni \ \gg \ Rh \ \approx \ Ru \ > \ Os \ \approx \ Ir \ > \ Pt $$

This relative order of bond migration is the same as the relative ability of these metals to absorb hydrogen gas. Rhodium does not show significant isomerization at low temperatures, but

375. Daniewski, A.R.; Kabat, M.M.; Masnyk, M.; Wicha, J.; Wojciechowska, W.; Duddeck, H. *J. Org. Chem.* **1988**, *53*, 4855.
376. Reference 323, p 34.
377. Marx, J.N.; McGaughey, S.M. *Tetrahedron* **1972**, *28*, 3583.
378. (a) Reference 323, p 37; (b) Gostunskaya, I.V.; Petrova, V.S.; Leonova, A.I.; Mironova, V.A.; Abubaker, M.; Kazanskii, B.A. *Neftekhimiya* **1967**, *7*, 3 [*Chem. Abstr. 67*: 21276t, *1967*].

isomerization is facile at temperatures $> 80°C$.[379] Reduction of car-3-ene (**356**) with a platinum catalyst gave carane (**357**), but using a palladium catalyst led to **360**. The initial reaction was a bond isomerization to give **358**, but subsequent opening of the cyclopropane ring gave **359**. The final product (**360**) was generated by reduction of the double bond in **359**.[380]

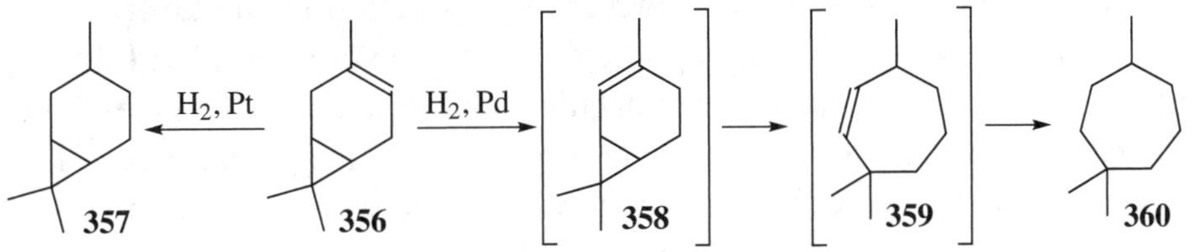

One way to explain the observed results in the reduction of **356** is that palladium binds alkenes reversibly, whereas platinum binds them more tightly and essentially irreversibly. If true, this has consequences for the stereochemistry (cis/trans-) of the final product. Banwell and co-worker's synthesis of γ-lycorane involved reduction of **361** to **362**,[381] where the cis-relationship of the 6-5 ring juncture led to delivery of hydrogen selectively from the bottom of **361** to give the observed stereochemistry in **362**. This experiment suggests that binding hydrogen atoms and the alkene unit to palladium leads to delivery of both hydrogens from the same face.

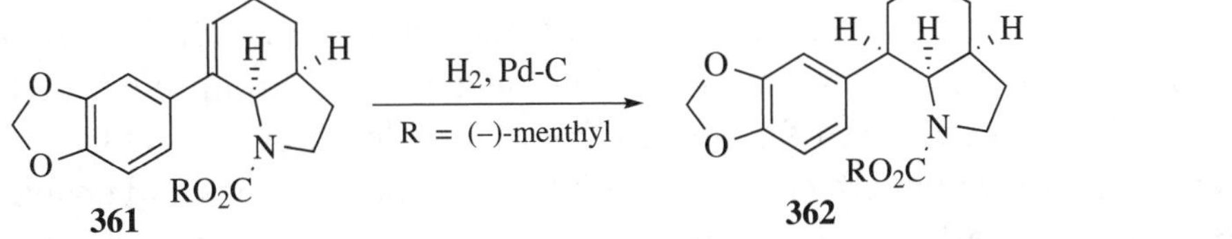

Clean formation of predominantly one stereoisomer also suggests that the alkene is bound tightly to the metal. If the alkene isomerizes prior to delivery of hydrogen, or if the alkyl moiety is not bound tightly to the metal after the first hydrogen is delivered, the alkyl-metal intermediate will equilibrate *prior* to delivery of the second hydrogen. Such an event usually favors the trans-product or, in general, the thermodynamically more stable product. When either a cis- or a trans-product can be produced, catalysts differ significantly in their ability to give the cis-product. The ability of the catalyst to give primarily cis-products on hydrogenation is[382,380]

$$Ir > Os > Pt,Ru,Rh >> Pd$$

Platinum is an excellent choice to give the cis adduct (the reduction is essentially under kinetic control) and palladium will usually give primarily the trans adduct (the reduction is essentially

379. Bond, G.C.; Wells, P.B. *Adv. Catal.* **1964**, *15*, 91.
380. (a) Reference 323, p 41; (b) Cocker, W.; Shannon, P.V.R.; Staniland, P.A. *J. Chem. Soc. C* **1966**, *41*.
381. Banwell, M.G.; Harvey, J.E.; Hockless, D.C.R.; Wu, A.W. *J. Org. Chem.* **2000**, *65*, 4241.
382. Nishimura, S.; Sakamoto, H.; Ozawa, T. *Chem. Lett.* **1973**, 855.

under thermodynamic control). If the selectivity observed in the hydrogenation of **354** is examined more closely, one sees that the platinum probably does not isomerize the alkene prior to delivery of hydrogen nor lead to thermodynamic control. Face *b* is probably not the less hindered face for coordination to the catalyst. The oxygen atoms of the lactone ring may play a role, holding the catalyst and facilitating reduction from that face (a neighboring group effect, see the discussion of haptophilicity below). Note that if **354** gives a reduction product with high stereoselectivity, other molecules with similar structures may show the same stereochemistry. It *may* therefore be possible to predict the major product of hydrogenation reactions by examining the literature for reduction of alkenes with a similar structure. For all reductions, a thorough search of the literature for relevant examples is important.

Predicting the stereochemistry of reduction can be difficult in many cases, and identification of the less hindered face can also be difficult. Rylander pointed out that adsorption may occur to minimize steric interaction but that step may not control the stereochemistry of the final product. The example of **354** seems to reinforce this statement. Control of stereochemistry can vary with the degree of isomerization, the catalyst, reaction conditions, and the purity of the alkene.[383,384] Norbornene systems usually give exclusively exo-(cis)-addition of hydrogen unless there is significant steric encumbrance at the C7 position.[385]

Another phenomenon that influences the stereochemistry of reduction in alkene systems is **haptophilicity**, an anchoring effect in which a functional group binds to the surface of the catalyst and directs delivery of hydrogen from that side of the molecule.[386] This is analogous to the neighboring group effects encountered in previous sections and chapters. As with other neighboring group effects, the extent of its importance greatly depends on the hydrogenation solvent. Hydrogenations that used heterogeneous catalysts are usually explained in terms of "approach, fit and binding of the reducible molecule to the surface of the catalyst".[386,387] This is especially true when the geometry of a molecule or substituent hinders approach or fit.[388] The presence of a chelating group often produces stereochemistry opposite to that predicted by steric effects.[389] Different modes for adsorption of a haptophile on the surface of a catalyst are illustrated in Figure 4.2.[371] Reduction of amine **363** illustrates haptophilicity. The aminomethyl group is bound to the catalyst[390] and directs the hydrogen from the same face as the aminomethyl,

383.   Reference 323, p 43 and references cited therein.
384.   Siegel, S.; Cozort, J.R. *J. Org. Chem.* **1975**, *40*, 3594.
385.   (a) Baird, Jr., W.K.; Surridge, J.H. *J. Org. Chem.* **1972**, *37*, 1182; (b) Baird, Jr., W.C.; Franzus, B.; Surridge, J.H. *Ibid.* **1969**, *34*, 2944.
386.   (a) Reference 323, p 44; (b) Thompson, H.W.; Naipawer, R.E. *J. Am. Chem. Soc.* **1973**, *95*, 6379.
387.   (a) Burwell, Jr., R.L. *Chem. Rev.* **1957**, *57*, 895; (b) Siegel, S. *Adv. Catal.* **1966**, *16*, 123.
388.   (a) Hadler, H.I. *Experientia* **1955**, *11*, 175; (b) Hanaya, K. *Bull. Chem. Soc. Jpn.* **1970**, *43*, 442.
389.   Reference 386b, citations 10 and 6-9 therein.
390.   Thompson, H.W.; Wong, J.K. *J. Org. Chem.* **1985**, *50*, 4270 and references cited therein.

to give **364** with formation of only 10% of the syn diastereomer (**365**).[391]

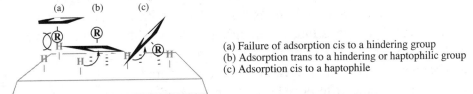

(a) Failure of adsorption cis to a hindering group
(b) Adsorption trans to a hindering or haptophilic group
(c) Adsorption cis to a haptophile

Figure 4.2. Adsorption of a haptophile on the surface of a catalyst. [Reprinted in part with permission from Thompson, H.W.; Naipawer, R.E. *J. Am. Chem. Soc. 1973, 95,* 6379. Copyright © 1973 American Chemical Society.]

The first mode illustrated in Figure 4.2[386] suggests that a neighboring group can interfere with the haptophilic effect, presumably by preventing the required hydrogen bonding or dipole interaction. Hydrogenation of 3β-substituted Δ⁵-steroids [such as 3β-cholesteryl methyl ether (**366**)] yields the saturated derivative with a trans-A/B ring juncture (H is trans to the bridgehead methyl), in most cases.[392] Hydrogenation of 3α-substituted steroids[393] such as epicholesteryl methyl ether (**367**), however, gave a cis-A/B ring juncture. This latter reduction required the addition of catalytic amounts of perchloric or sulfuric acid and failed when done in ethanolic potassium hydroxide.[394] The α-methoxy group prevented adsorption on the α face. Binding occurred on the β face, leading to a cis-A/B ring juncture (H is cis to the bridgehead methyl), illustrates case (a) in Figure 4.2.

The second and third binding modes shown in Figure 4.2 can be illustrated by the reduction of **368**. The ester group in **368** functions as a poor haptophile and does not bind very well to the metal. Steric hindrance leads to the alkene moiety binding to a catalyst particle from the

---

391. (a) For other work involving stereocontrol in hydrogenation reactions, see Thompson, H.W.; McPherson, E. *J. Am. Chem. Soc. 1974, 96,* 6232; (b) Reference 323, p 45.
392. Wagner-Jáuregg, T.; Werner, L. *Z. Physiol. Chem. 1932, 213,* 119.
393. Shoppee, C.W.; Agashe, B.D.; Summers, G.H.R. *J. Chem. Soc. 1957,* 3107.
394. Wilds, A.L.; Johnson, Jr., J.A.; Sutton, R.E. *J. Am. Chem. Soc. 1950, 72,* 5524.

opposite face, and the product (**369**) has the trans-stereochemistry for the ring juncture.[393,386b] Hydrogenation was followed by lithium aluminum hydride reduction of **369** and gave **372**. If the order of reactions is reversed and **368** is first reduced with LiAlH₄ (to give **370**) and then subjected to catalytic hydrogenation, a haptophilic effect of the alcohol (see **373**) delivers hydrogen to that face, giving **371**. The presence of the hydroxymethyl group in **371** led to the opposite stereochemistry upon hydrogenation when compared to that observed with the ester group in **373**, illustrating the difference between the second and third cases in Figure 4.2.

4.8.C. Hydrogenation of Alkynes

Alkynes contain π bonds, of course, and catalytic hydrogenation is usually straightforward. There are two significant problems with hydrogenation of alkynes, however. Hydrogenation of the triple bond first gives an alkene, but overreduction can occur to give an alkane moiety (a two-stage reduction, sec. 4.1). Since the initial product is an alkene product, there is the possibility of stereochemical control problems during the course of the hydrogenation, and hydrogenation of alkynes generally produces a mixture of both cis and trans isomers..

Selective reduction to the alkene without overreduction requires that the alkyne be adsorbed to the catalyst more tightly than the alkene. To selectively generate a cis alkene also requires that double bond migration and cis-trans isomerization of the alkene product be minimized. Palladium is usually a poor choice since both of these processes are faster than transfer of a hydrogen atom to the alkene. An order has been established for selectivity in the reduction of 2-pentyne to 2-pentene[395] without reduction of the C=C moiety in 2-pentene.

$$Pd \quad > \quad Pt \quad > \quad Ru \quad > \quad Rh \quad > \quad Ir$$

Selectivity for formation of the cis alkene (usually 91-98%) was

---

395.   (a) Bond, G.C.; Wells, P.B. *J. Catal.* **1966**, *5*, 419; (b) Reference 323, p 16.

$$ Pd \ > \ Rh \ > \ Pt \ > \ Ru \ \sim \ Ir $$

The best reagent for selective reduction of alkynes to the cis alkene is the **Lindlar catalyst**.[396] Palladium chloride (PdCl$_2$) was precipitated on calcium carbonate (CaCO$_3$) in acidic media and deactivated with lead tetraacetate [Pb(OAc)$_4$] to give the named Pd-CaCO$_3$-PbO catalyst. Reduction of alkynes will stop at the cis alkene with little or no cis-trans isomerization. A variation uses quinoline as a poison, illustrated in Koide and co-worker's synthesis of FR901464, in which alkyne **374** was converted to cis-alkene **375** in excellent yield.[397] Cram and Allinger used a palladium on barium sulfate (BaSO$_4$) catalyst poisoned with quinoline[398a] (sometimes called the **Rosenmund catalyst**),[398b-d] to achieve the same selectivity. The poison is required to deactivate the catalyst (quinoline coordinates to the active sites on the metal), reducing the propensity for overreduction and bond migration. This Cram-Allinger catalyst is widely used in organic synthesis, and is commonly called a Lindlar catalyst. A mixture of nickel boride (Ni$_2$B) on borohydride exchange resin also gives cis alkenes upon hydrogenation of alkynes.[399]

Just as homogeneous catalysts were shown to give superior stereoselectivity with alkenes, relative to homogeneous catalysts,[400] homogeneous catalysts can give improved selectivity with alkynes. This selectivity is illustrated in syntheses of (+)-blastmycinone and (−)-litsenolide C1, by Liu and co-workers[401] in which the hydrogenation of the conjugated alkyne unit **376** with Wilkinson's catalyst gave a 90% yield of **377**.[402] In this case, the selectivity was for the alkyne over the alkene, although the alkyne was completely reduced to the hydrocarbon unit. Homogeneous catalysts are also selective for reduction of an alkene over hydrogenolysis of a C-Br bond, as in the hydrogenation of ω-halo alkenes such as 6-bromo-1-hexene to 1-bromohexane in 78% yield.[403]

The en-yne unit in **376** was selectively reduced, and it is known that dienes are hydrogenated

396. (a) Reference 15, p 566; (b) Lindlar, H. *Helv. Chim. Acta* **1952**, *35*, 446.
397. Albert, B. J.; Sivaramakrishnan, A.; Naka, T.; Koide, K. *J. Am. Chem. Soc.* **2006**, *128*, 2792.
398. (a) Cram, D.J.; Allinger, N.L. *J. Am. Chem. Soc.* **1956**, *78*, 2518; (b) Rosenmund, K.W. *Berichte* **1918**, *51*, 585; (c) Mosettig, E.; Mozingo, R. *Org. React.* **1948**, *4*, 362; (d) Rachlin, A.I.; Gurien, H.; Wagner, D.P. *Org. Synth. Coll. Vol. 6* **1988**, 1007.
399. Choi, J.; Yoon, N.M. *Tetrahedron Lett.* **1996**, *37*, 1057.
400. Milstein, D. *Acc. Chem. Res.* **1984**, *17*, 221.
401. Chen, M.-J.; Lo, C.-Y.; Chin, C.-C.; Liu, R.-S. *J. Org. Chem.* **2000**, *65*, 6362.
402. Evans, D.A.; Morrissey, M.M. *J. Am. Chem. Soc.* **1989**, *106*, 3866.
403. Januskiewicz, K.R.; Alper, H. *Can. J. Chem.* **1984**, *62*, 1031.

to give alkenes, but over-reduction is a problem, as it was with alkynes. Homogenous catalysts can be of value. The diene unit in **378** was hydrogenated to give the non-conjugated alkene unit in **379** (80% yield) in the Serebryakov et al. synthesis of faranal.[404] Both 1,2- and 1,4-reduction are possible, and the major product depends on the catalyst used, as well as the substituents on the C=C units of the diene.

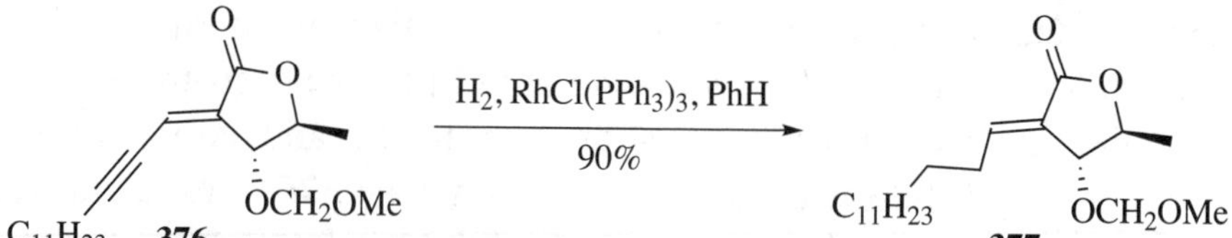

It is possible to reduce cyclopropane rings by catalytic hydrogenation, using even a heterogeneous catalyst such as platinum oxide (PtO$_2$, often called the **Adams' catalyst**). Cyclopropane rings can react with active catalysts, opening during hydrogenation to generate a geminal dimethyl group. In one example, **380** was reduced to **381** to complete Taber and co-worker's synthesis of *trans*-dihydroconfertifolin.[405] The use of a platinum catalyst maximized the hydrogenolysis and minimized reduction of the carbonyl.

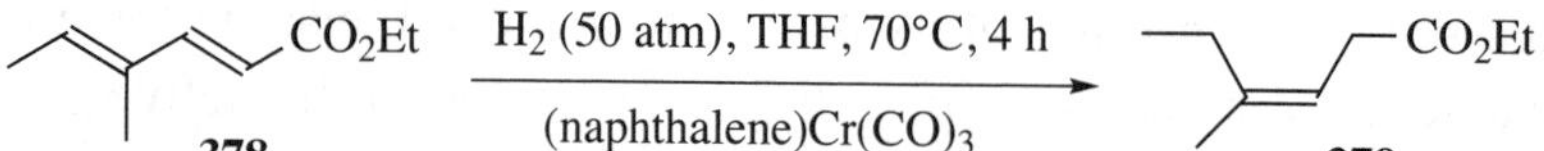

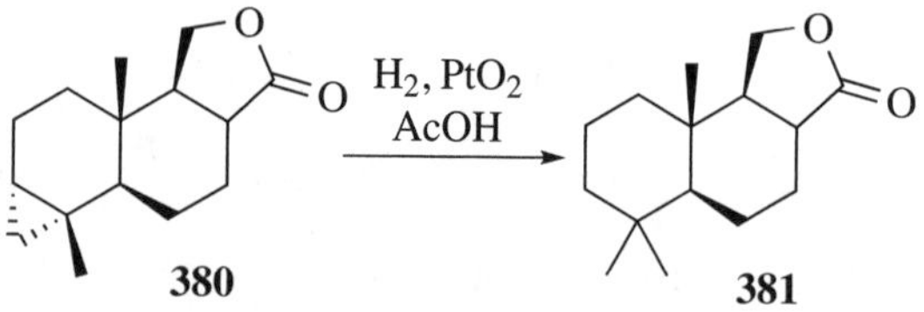

4.8.D. Hydrogenation of Carbonyl Compounds

The π bond of a carbonyl can be reduced by catalytic hydrogenation to give an alcohol. Problems of chemoselectivity and stereoselectivity are common with these compounds as they were with alkenes and alkynes. Hydrogen, in the presence of palladium catalysts, reduces alkenes faster than carbonyls and platinum is usually the preferred catalyst for carbonyls. This preference is apparent in the hydrogenation of **382** to (–)-muscone (**383**), in the final step in an asymmetric synthesis by Hagiwara and co-workers.[406] With alkenyl aldehydes, platinum reduces the aldehyde preferentially when adsorbed on carbon or calcium carbonate, but not when adsorbed on barium sulfate or alumina (Al$_2$O$_3$).[407] In general, catalytic hydrogenation of a ketone unit is straightforward. The proliferation of readily available, easy to use and highly selective hydride and borane reducing agents often makes catalytic hydrogenation a second choice for reducing heteroatom containing functional groups, particular carbonyl derivatives. Many heteroatom functional groups are compatible with hydrogenation of ketones, however.

404. Vasil'ev, A.A.; Engman, L.; Serebryakov, E.P. *J. Chem. Soc. Perkin Trans. 1* **2000**, 2211.
405. Taber, D.F.; Nakajima, K.; Xu, M.; Rheinagold, A.L. *J. Org. Chem.* **2002**, *67*, 4501.
406. Kamat, V.P.; Hagiwara, H.; Katsumi, T.; Hoshi, T.; Suzuki, T.; Ando, M. *Tetrahedron* **2000**, *56*, 4397.
407. Reference 323, p 75.

In a synthesis of plumerinine, Comins and co-workers reduced the ketone unit in **384** to give alcohol **385** in 96% yield using platinum oxide (**Adams' catalyst**).[408]

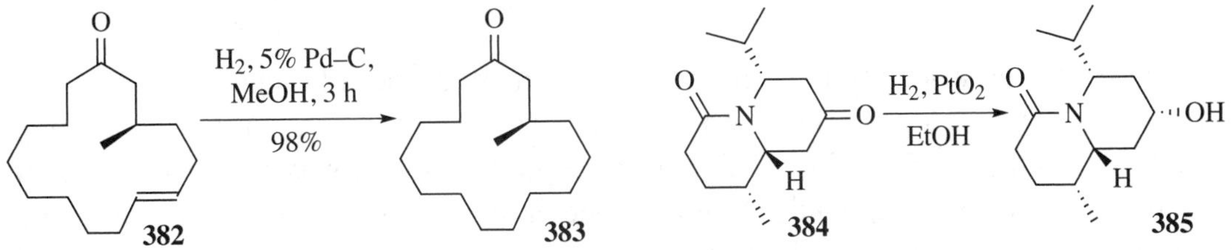

Hydrogenation of conjugated carbonyl compounds can give either 1,2- or 1,4-reduction or a mixture, but changing the catalyst usually controls the product distribution. The palladium-catalyzed hydrogenation of **386**, for example, cleanly reduced the alkene unit to give **387** in 96% yield.[409] Hydrogenation of similar compounds using a platinum catalyst generally leads to preferential reduction of the carbonyl unit.

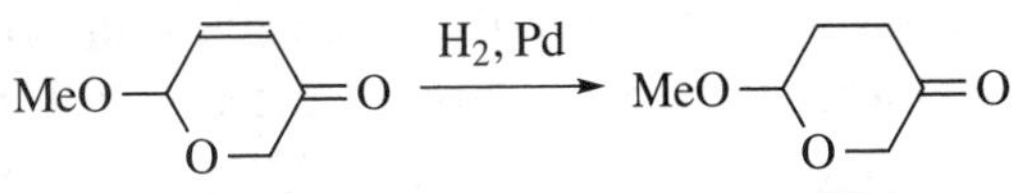

Hydrogenolysis (reductive cleavage) of the C-O bond is rarely a problem in aliphatic carbonyl compounds, and new catalysts[410] are being developed all the time. The presence of an heteroatom substituent may lead to significant hydrogenolysis, however, as in the reduction of the ester ethyl 3-oxobutanoate (**388**).[411] Reduction gave the expected alcohol product (ethyl 3-hydroxybutanoate, **389**), but hydrogenolysis of the C-O bond in **389** gave ethyl butanoate as a side product. It is important to note, however, that it is possible the alcohol moiety in **389** eliminated under the reaction conditions and the resultant alkene moiety was further reduced. Hydrogenolysis *may* not be involved in this transformation at all, but the cited reference suggested that it probably was. In the case of **388**, platinum black gave the greatest amount of reductive cleavage (90% ethyl butanoate and 10% of **389**). The use of additives and of platinum oxide (100% **389**) led to reduction of the ketone without hydrogenolysis.

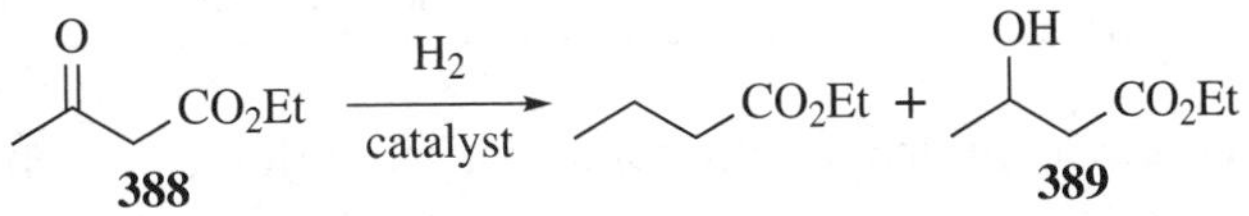

When the carbonyl of ketones or aldehydes is conjugated to aromatic systems, hydrogenolysis products are often the major isolated products. Catalytic hydrogenation of 2-naphthaldehyde (**390**) with palladium-on-barium sulfate gave only 2-methylnaphthalene (**391**).[412] When the **Adams' catalyst** (PtO$_2$) was used, with ferric chloride (FeCl$_3$) as a promoter, an 80% yield of alcohol **392** was obtained, although addition of excess promoter induced further hydrogenolysis to **391**.

408. Comins, D.L.; Zheng, X.; Goehring, R.R. *Org. Lett.* **2002**, *4*, 1611.
409. Tudanca, P.L.L.; Jones, K.; Brownbridge, P. *J. Chem. Soc. Perkin Trans. 1* **1992**, 533.
410. See, for example, Ohkuma, T.; Ooka, H.; Yamakawa, M.; Ikariya, T.; Noyori, R. *J. Org. Chem.* **1996**, *61*, 4872.
411. Reference 323, p 86.
412. Campbell, N.; Anderson, W.; Gilmore, J. *J. Chem. Soc.* **1940**, 819.

Structures: 392 (naphthalene-CH2OH) ← PtO2, EtOH / H2, FeCl3 — 390 (naphthalene-CHO) → Pd°, BaSO4 / H2 — 391 (naphthalene-CH3)

Hydrogenolysis was also observed in the reduction of 2-acetylthiophene (**393**), when the homogeneous catalyst cobalt octacarbonyl [$Co_2(CO)_8$] was used, and 2-ethylthiophene (**394**) was isolated in 52% yield.[413] The homogeneous catalyst was necessary to minimize poisoning by the sulfur compound. As mentioned earlier, homogeneous catalysts are usually less susceptible to poisoning.[414] In general, reduction to the alcohol is rapid with aromatic aldehydes using palladium, but slow with ruthenium. The reverse is observed with aliphatic aldehydes. Hydrogenolysis can occur in acidic media, or with vigorous reaction conditions. Catalysts are available, however, that allow hydrogenation of a ketone unit that is attached to a heterocyclic ring, giving the alcohol without significant hydrogenolysis. An example is a homogeneous ruthenium catalyst used to hydrogenate **395** to **396** in 96% yield, and with high enantioselectivity (see below).[415]

Structures: 393 (2-acetylthiophene) → H2, CO / Co2(CO)8 — 394 (2-ethylthiophene); 395 (2-acetylpyridine) → H2 / (R,R)-Ru catalyst — 396 (pyridyl-CH(OH)CH3)

Some acid derivatives are reduced by catalytic hydrogenation. Relative to alkenes and ketones, several acid derivatives were seen to be less reactive in previous examples. There is a clear difference in reactivity, apart from specificity for the metal catalyst. The ability of hydrogen to reduce acid derivatives follows the expected order, and acid chlorides will be reduced faster than esters, for example. Aldehydes and ketones[416] are reduced significantly faster than all acid derivatives except acid chlorides. The general order of reactivity for hydrogenation is

$$ R\overset{O}{\underset{}{C}}Cl > R\overset{O}{\underset{}{C}}H(R) > R\overset{O}{\underset{}{C}}O\overset{O}{\underset{}{C}}R > R\overset{O}{\underset{}{C}}OR > R\overset{O}{\underset{}{C}}OH > R\overset{O}{\underset{}{C}}NH_2 $$

Esters are difficult to reduce by hydrogenation and the reduction of alkenes, alkynes, aldehydes or ketones in the presence of esters and amides is usually facile. Hydrogenation of carboxylic acids is very difficult. The use of ruthenium oxide (RuO_2) or ruthenium on carbon (Ru/C), at high pressure and temperatures, is a preferred method for reduction of acids to alcohols. Hydrogenolysis can be a problem with esters possessing an allylic or benzylic C–O bond, as in

413. Greenfield, H.; Metlin, S.; Orchin, M.; Wender, I. *J. Org. Chem.* **1958**, *23*, 1054.
414. (a) Marko, L. *Proc. Sympos. Coord. Chem. Tihany Hung.* **1964**, 271 [*Chem. Abstr. 65*: 4707c, **1966**]; (b) Laky, J.; Szabo, P.; Marko, L. *Acta. Chem. Acad. Sci. Hung.* **1965**, *46*, 247; [*Chem. Abstr. 64*: 6511e, **1966**]; (c) Reference 323, p 58.
415. Ohkuma, T.; Koizumi, M.; Yoshida, M.; Noyori, R. *Org. Lett.* **2000**, *2*, 1749.
416. Reference 323, p 69.

the cleavage of one ester linkage in the diolide dicrotaline (**397**) to give **398**.[417]

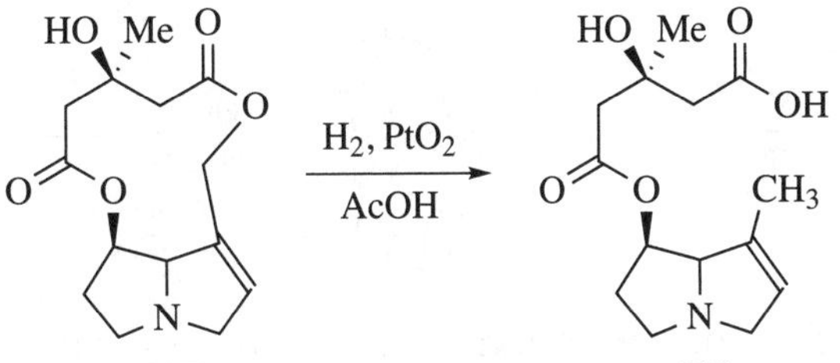

397   **398**

4.8.E. Hydrogenation of Other Heteroatom Functional Groups

Catalytic hydrogenation can be applied to many functional groups, including halides, but there are problems, including poor reactivity, overreduction or competing side reactions, particularly with nitriles and halides. Nitriles are commonly hydrogenated to the amine in good yield via an imine intermediate, and nucleophilic addition of unreacted amine can lead to dimers or polyalkylated species as the final product. Reduction of propanenitrile with a rhodium catalyst gave dipropylamine, for example, after hydrogenolysis of the initially formed condensation product.[418] Reduction of propanenitrile with platinum or palladium catalysts gave the trisubstituted amine (tripropylamine), but rhodium oxide (Rh_2O_3) gave primarily the monoamine. Nickel, nickel boride (NiB) and cobalt are the best catalysts for conversion of low molecular weight nitriles to primary alkyl amines.[419,420] Isolation of the amine product is sometimes difficult, and it is often isolated as its hydrochloride salt. Clive[421] found that hydrogenation of **399**, in the presence of several equivalents of chloroform, which served as a controlled source of HCl,[422] gave amine **400** in high yield as part of a synthesis of brevioxime.[421]

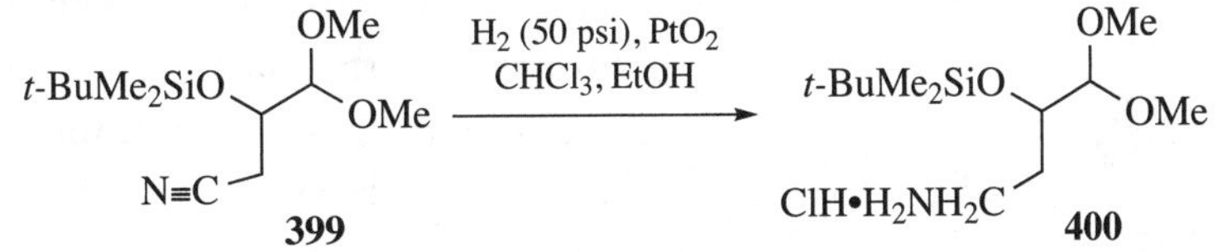

399   **400**

Aromatic nitriles can give mixtures of the primary amine and dibenzylamines (via the condensation reaction), but this depends on the catalyst.[418] Rhodium hydroxide [$Rh(OH)_3$] is very effective for the production of aminomethyl derivatives with repression of the condensation reaction. Hydrogenation of **401** with platinum reduced the nitro group, but not the nitrile to give **402**. Subsequent reduction with palladium gave the aminomethyl group in **403** (and also

417. Devlin, J.A.; Robins, D.J. *J. Chem. Soc. Chem. Commun.* **1981**, 1272.
418. Reference 323, pp 140-141.
419. Zhou, B.; Edmondson, S.; Padron, J.; Danishefsky, S.. *Tetrahedron Lett.* **2000**, *41*, 2039.
420. Dutasta, J.-P.; Gellon, G.; Leuchter, C.; Pierre, J.-L. *J. Org. Chem.* **1988**, *53*, 1817.
421. Clive, D.L.J.; Hisaindee, S. *J. Org. Chem.* **2000**, *65*, 4923.
422. Secrist III, J.A.; Logue, M.W. *J. Org. Chem.* **1972**, *37*, 335.

reduced the chloro moiety) in Folkers and Harris' synthesis of pyridoxine.[423]

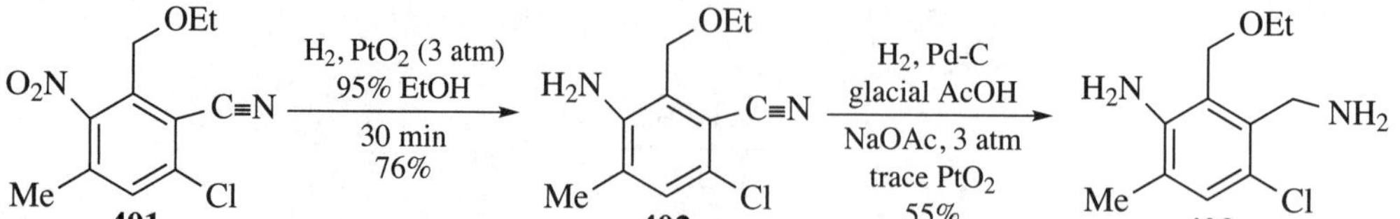

Hydrogenation of nitro compounds is straightforward with palladium, platinum or nickel catalysts. Palladium is the most common catalyst for both aromatic and aliphatic nitro compounds. The reduction of aromatic nitriles with hydride reagents (sec. 4.2.C.iii) gives poor results, and catalytic hydrogenation the preferred method. A synthetic example is taken from Rege and Johnson's synthesis of physostigmine.[424] Hydrogenation of the nitro group in **404** with a palladium-on-carbon catalyst gave a quantitative yield of aniline **405**, under relatively mild conditions. Hydrogenation using platinum oxide converts aromatic nitro compounds to aniline derivatives,[425] even in the presence of other reducible groups.[426]

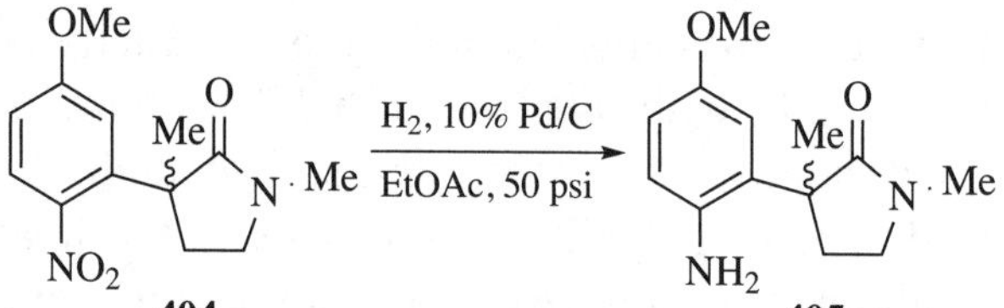

Catalytic hydrogenation is commonly used for reduction of azides, which are important substrates in [3+2]-cycloaddition reactions (sec. 11.11.H) and in rhodium-catalyzed carbene chemistry (sec. 13.9.B.ii). Azides are important precursors to amines because the N_3 moiety can be incorporated into a molecule via an S_N2 reaction in aliphatic substrates, or by aromatic substitution with aromatic substrates. A second step (reduction) is required to give the amine. An example is the palladium-catalyzed reduction of **406** to give a quantitative yield of **407**, in Hecht and co-workers synthesis of deamido-bleomycin A_2, the major catabolite of the antitumor agent bleomycin.[427] Platinum, palladium, and nickel have been used as catalysts for this reduction, but hydrazine is sometimes added to limit the extent of hydrogenolysis. Hydrogenation to the amine can be carried out with retention of the original configuration of the azido group.

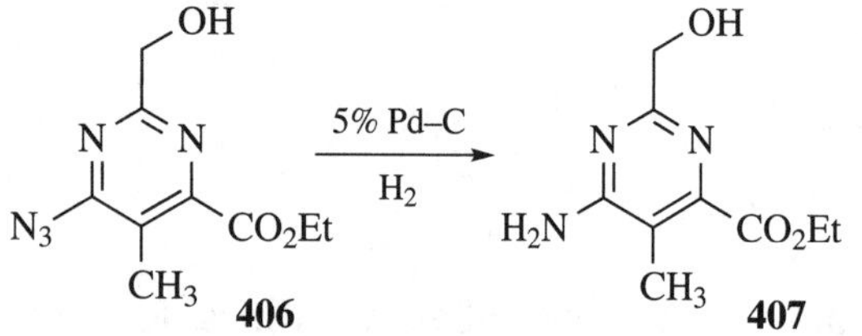

423. (a) Harris, S.A.; Folkers, K. *J. Am. Chem. Soc.* **1939**, *61*, 1245, 3307; (b) Harris, E.E.; Firestone, R.A.; Pfister III, K.;Boettcher, R.R.; Cross, F.J.; Currie, R.B.; Monaco, M.; Peterson, E.R.; Reuter, W. *J. Org. Chem.* **1962**, *27*, 2705.
424. Rege, P.D.; Johnson, F. *J. Org. Chem.* **2003**, *68*, 6133.
425. (a)Bakke,J.;Heikman,H.;Hellgren,E.B.*Acta Chem. Scand.* **1974**,*28B*,393;(b)Bakke,J.*Ibid.* **1974**,*28B*,134.
426. Wender, P.A.; Koehler, K.F.; Wilhelm, R.S.; Williams, P.D.; Keenan, R.M.; Lee, H.Y. *New Synthetic Methodology and Functionally Interesting Compounds*, Yoshida, Z. (Ed.), Kodansha/Elsevier, Tokyo/ Amsterdam, **1986**, pp 163-182 (Cf. p 173).
427. Zou, Y.; Fahmi, N.E.; Vialas, C.; Miller, G.M.; Hecht, S.M. *J. Am. Chem. Soc.* **2002**, *124*, 9476.

Epoxides react with hydrogen and a catalyst by a hydrogenolysis reaction, but it is relatively slow and many functional groups can be reduced preferentially, which means that epoxides can be reduced if there is little or no other functionality. The use of rhodium on carbon (Rh/C)[428] or Wilkinson's catalyst[429] usually avoids reduction of the epoxide. Predicting the direction of ring opening (regioselectivity in the reduction) in epoxides can be difficult. Symmetrical epoxides typically show delivery of hydrogen to both carbons of the oxirane ring, and Rylander discussed the various factors that influence this delivery.[430] Prediction is usually difficult, but it has been observed that addition of hydroxide to the reaction medium influences the regioselectivity of the reduction. Hydrogenation of 1-octyloxirane (**408**) with Raney nickel gave 1-decanol, but hydrogenation using Raney nickel mixed with NaOH gave 2-decanol (**409**).[431]

Hydrogenation of α,β-epoxyketones generally gives the β-hydroxy ketone,[432] as in the conversion of **410** to **411**.[433] If other substituents (electron releasing or withdrawing) are attached to the β-position, the direction of reduction is less predictable. Reduction of aziridines

by hydrogenolysis is similar to that of epoxides and usually occurs at the less hindered C-N bond.[434,435]

4.8.F. Hydrogenolysis Reactions

Halides are commonly incorporated into organic molecules, but in many cases they must be removed. One method is reduction to the corresponding hydrocarbon by hydrogenolysis. Raney nickel can be used, as in reduction of the iodomethyl furanose derivative **412** to the methoxy lactol **413**.[436] Problems arise with many catalysts when the substrate contains alkenyl and

428. Tarbell, D.S.; Carman, R.M.; Chapman, D.D.; Cremer, S.E.; Cross, A.D.; Huggman, K.R.; Kunstmann, M.; McCorkindale, N.J.; McNally, Jr., J.G.; Rosowsky, A.; Varino, F.H.L.; West, R.L. *J. Am. Chem. Soc.* **1961**, *83*, 3096.

429. Sharpless, K.B.; Behrens, C.H.; Katsuki, T.; Lee, A.W.M.; Martin, V.S.; Takatani, M.; Viti, S.M.; Walker, F.J.; Woodard, S.S. *Pure Appl. Chem.* **1983**, *55*, 589.

430. Reference 324, pp 260-265.

431. (a) Newman, M.S.; Underwood, G.; Renoll, M. *J. Am. Chem. Soc.* **1949**, *71*, 3362; (b) Reference 324, p 261; (c) also see Mitsui, S.; Imaizumi, S.; Hisashige, M.; Sugi, Y. *Tetrahedron* **1973**, *29*, 4093 and Sokol'skaya, A.M.; Reshetinikov, S.M.; Bakhanova, E.N.; Kuzembaev, K.K.; Anchevskaya, M.N. *Khim. Khim. Tekhnol. (Alma-ata)* **1966**, *5*, 3 [*Chem. Abstr. 69*: 67027g **1968**].

432. McCurry, Jr., P.M.; Singh, R.K. *J. Org. Chem.* **1974**, *39*, 2316.

433. Strike, D.P.; Smith, H. *Tetrahedron Lett.* **1970**, 4393.

434. Reference 324, p 267.

435. Sugi, Y.; Mitsui, S. *Bull. Chem. Soc. Jpn.* **1969**, *42*, 2984.

436. Schuler, H.R.; Slessor, K.N. *Can. J. Chem.* **1977**, *55*, 3280.

carbonyl functional groups. Hydrogenation with palladium catalysts, for example, commonly removes halogen along with the functional group. If the halogen is to remain in the molecule, platinum is the preferred catalyst.

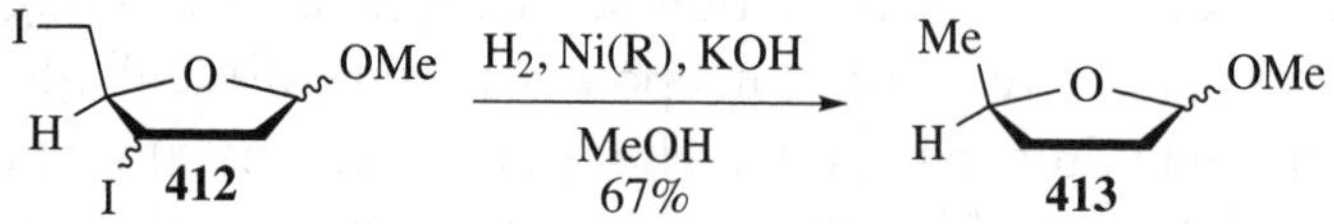

The hydrogenolysis reaction is an important part of many syntheses that involve protection of functional groups (secs. 7.3.A, 7.3.C). Protection of alcohols and amines with a benzyl group is common (secs. 7.3.A, 7.3.C.i).[437] Palladium favors hydrogenolysis, and it is commonly used for PhCH$_2$-//-O or PhCH$_2$-//-N cleavage, leading to toluene and the alcohol or amine. An example of C-N cleavage is the conversion of benzylamine **414** to amine **415** (>90% yield) in a synthesis of 2-epi-perhydrohistrionicotoxin by Hsung and co-workers.[438] Hydrogenolysis is the common method for removal of the benzyloxycarbonyl (Cbz; benzyl carbamate) group, a carbamoyl protecting group often used to protect amines (sec. 7.3.C.iii). This reaction is illustrated by conversion of the PhCH$_2$O$_2$CNH- group in **416** to the amine unit in **417** (80% yield), as part of de Meijere and Larionov's synthesis of belactosin C (**417**)[439] Note that hydrogenolysis of the benzyl ester (-CO$_2$Bn) occurred under these conditions to give the corresponding carboxylic acid moiety. In multi-functional molecules, reduction of other functional groups can compete with hydrogenolysis of the O-Bn bond.[440]

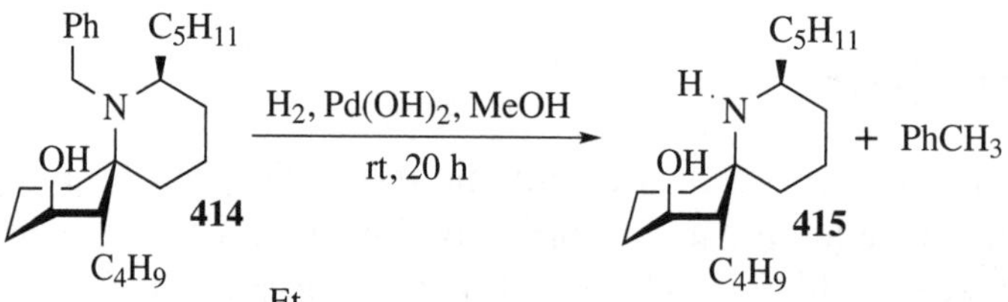

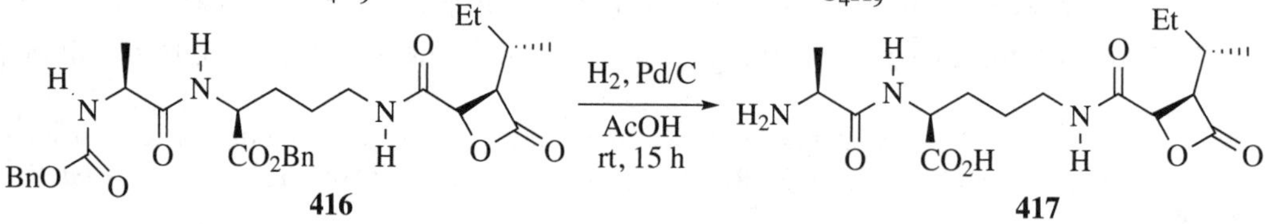

Dithianes and dithiolanes are generated from the reaction of dithiols with aldehydes and ketones, and they are commonly used to generate sulfur stabilized carbanions (sec. 8.6), or as

437. (a) Wuts, P.G.M.; Greene, T.W. *Protective Groups in Organic Synthesis 3rd ed*, Wiley, New York, *1999*; (b) Wuts, P.G.M.; Greene, T.W. *Protective Groups in Organic Synthesis 2nd ed*, Wiley, New York, *1991*; (c) Greene, T.W. *Protective Groups in Organic Synthesis* Wiley, New York *1981*; (d) Wuts, P.G.M.; Greene, T.W. *Protective Groups in Organic Synthesis* 4th ed., Wiley, New Jersey, *2006*.
438. McLaughlin, M.J.; Hsung, R.P.; Cole, K.P.; Hahn, J.M.; Wang, J. *Org. Lett.* *2002*, *4*, 2017.
439. Larionov, O.V.; de Meijere, A. *Org. Lett.* *2004*, *6*, 2153.
440. For an example, see Burgstahler, A.W.; Weigel, L.O.; Sanders, M.E.; Shaefer, C.G.; Bell, W.J.; Vuturo, S.B. *J. Org. Chem. 1977*, *42*, 566.

carbonyl protecting groups (sec. 7.3.B.ii). In addition, sulfur compounds such as sulfoxides and sulfones are important for C-C bond forming reactions via the sulfur carbanions (sec. 8.6.A). In many cases, the sulfur group must be removed from the molecule after its synthetic use has been accomplished. Bivalent sulfur is a common poison of heterogeneous catalysts, and the affinity of sulfur for nickel makes Raney nickel an effective reagent for hydrogenolysis of the C-S bond. Reduction of the dithiolane is usually accomplished by simply heating with Raney nickel in ethanol or methanol, without adding hydrogen gas.[441] An example is the desulfurization of **418** to give **419** in 62% yield, in a synthesis of naphthalene analogs of lignans by Medarde and co-workers.[442]

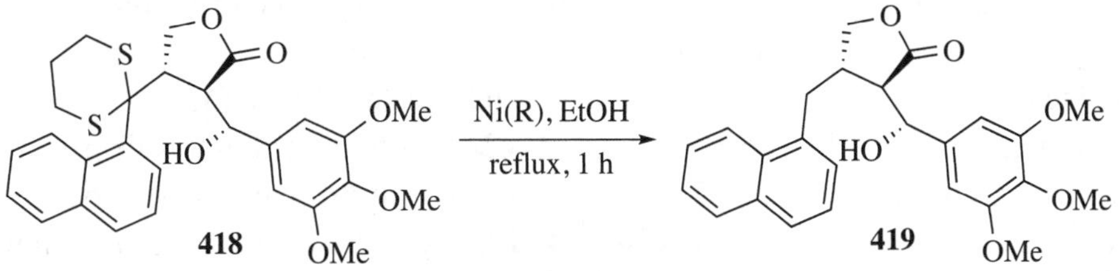

4.8.G.   Hydrogenation of Aromatic and Heteroaromatic Hydrocarbons

Aromatic hydrocarbons and heterocycles are important in many natural products and reduction of the aromatic ring is very important from a synthetic viewpoint. Clearly, the simplest example is reduction of benzene to cyclohexane, and it requires vigorous reaction conditions unless rhodium catalysts are used. Rylander showed the order of various catalysts for the hydrogenation of benzene to be

$$Rh \ > \ Ru \ >> \ Pt \ > \ Pd \ >> \ Ni \ > \ Co$$

Rhodium and ruthenium are used to avoid hydrogenolysis. The catalyst of choice changes somewhat with fused ring aromatic hydrocarbons. The relative order for reduction of naphthalene to tetralin (**420**) is[443]

$$Pd \ > \ Pt \ > \ Rh \ > \ Ir \ > \ Ru$$

This order is due, in part, to the slow reduction of the aromatic ring in **420** with palladium. Rhodium gives initial reduction of naphthalene to **420**, but further reduction to decalin (**421**) is so rapid that **420** is a minor constituent of the final product mixture.

441. For an example, see Davis, F.A.; Chao, B.; Fang, T.; Szewczyk, J.M. *Org. Lett.* **2000**, *2*, 1041.
442. Madrigal, B.; Puebla, P.; Peláez, R.; Caballero, E.; Medarde, M. *J. Org. Chem.* **2003**, *68*, 854.
443. Reference 323, p 180.

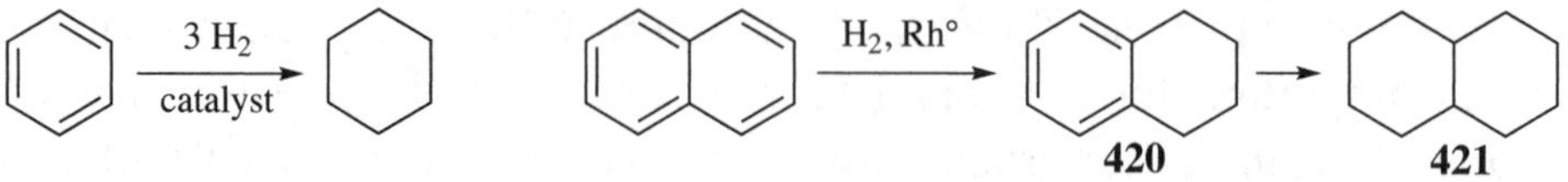

When the aromatic ring is substituted, catalytic hydrogenation can be complicated by side reactions. Reduction of phenols initially gives enol products, but these are usually converted to cyclohexenone or cyclohexanedione derivatives. An example is the reduction of **422** to **423**,[444] which stands in contrast to the reduction of **424**, which in the presence of excess hydrogen gave cyclohexane derivative **425**.[445] Note the selectivity of the reduction for the all cis geometry in **425**. Palladium or platinum catalysts lead to significantly less stereoselectivity in this type of reduction. Catalytic hydrogenation is usually very selective for electron-rich aromatic rings if there is another aromatic ring that bears electron-withdrawing substituents. A simple benzene ring is reduced faster than a ring with an electron-withdrawing group attached to it. Reduction of 1-naphthalenecarboxylic acid (**426**) with Ni(R) led to selective reduction of the more electron-rich ring to give **427**.[446] An example of hydrogenation of a benzene ring in a synthetic sequence is the conversion of **428** to **429** in 74% yield, as part of Denmark and Fu's synthesis of the serotonin antagonist LY426965.[447] Note the selective hydrogenation of only one phenyl group. The first hydrogenation gave a mixture of **429** and the tetrahydro derivative (the cyclohexene compound). Hydrogenation of this mixture in the second step gave **429** in 74% overall yield. Prolonged hydrogenation of **428** gave reduction of both phenyl groups.[443]

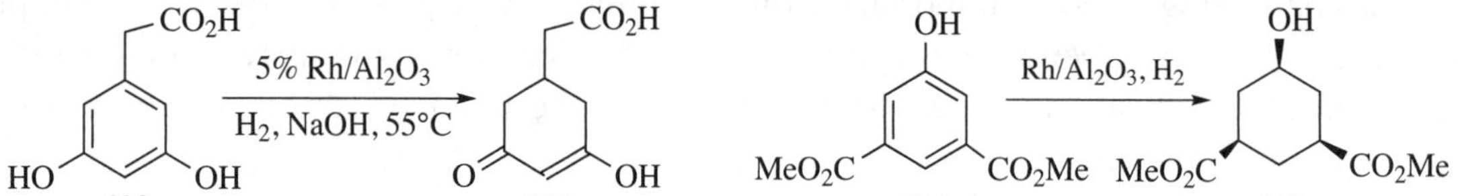

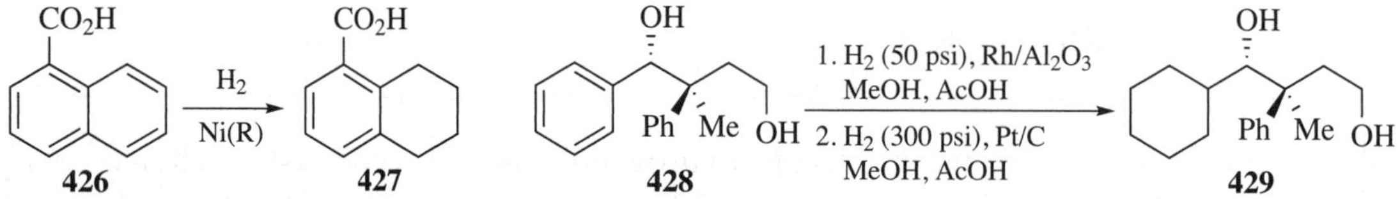

Catalytic hydrogenation of aniline derivatives is often accompanied by disubstituted amine products, as in the formation of dicyclohexylamine (**432**) from aniline.[448] The reactive intermediate in the reduction is an initially formed imine **430**, which is in equilibrium with the enamine under the reaction conditions. Reduction of **430** gives cyclohexylamine, which reacts with **430** to give **431**. Under the reaction conditions, **431** loses cyclohexylamine to give an imine (or an iminium salt). This intermediate then reacts with additional hydrogen to give the

444.   (a) Mokotoff, M.; Cavestri, R.C. *J. Org. Chem.* **1974**, *39*, 409; (b) Reference 323, pp 194-195.
445.   Gensler, W.J.; Solomon, P.H. *J. Org. Chem.* **1973**, *38*, 1726.
446.   Bohlmann, F.; Eickeler, E. *Chem. Ber.* **1979**, *112*, 2811.
447.   Denmark, S.E.; Fu, J. *Org. Lett.* **2002**, *4*, 1951.
448.   Reference 323, p 187.

isolated product, **432**. This process is analogous to the dimerization reactions observed in the

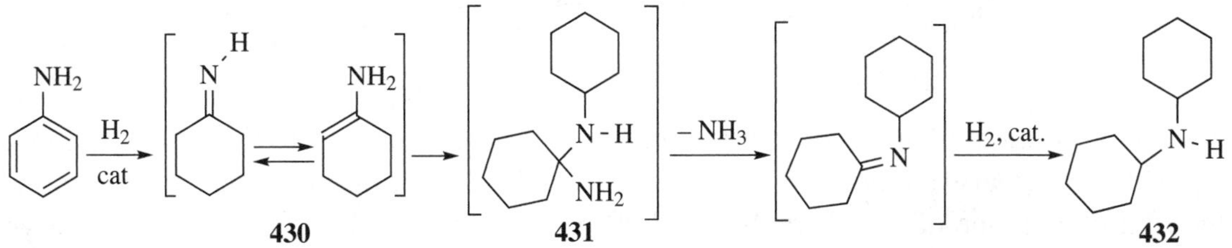

430   **431**   **432**

LiAlH$_4$ reduction of aliphatic nitriles (sec. 4.2.C.iii for the formation of dipropylamine from propanenitrile). This side reaction can usually be limited to a small percentage by the use of ruthenium or rhodium catalysts, allowing isolation of the amine. Note that enamines can be directly reduced to the amine by catalytic hydrogenation.[449] The order of least coupling (with Ru) to most coupling (with Pt) is[448]

$$ Ru \quad < \quad Rh \quad << \quad Pd \quad < \quad Pt $$

The product(s) formed by reduction of heterocyclic aromatic compounds varies with the number of rings, the heteroatom and its position in the molecule, as well as the catalyst. For pyridine and pyrrole derivatives, rhodium is the most common catalyst, and acidic media are used to form the corresponding salt, which is more easily reduced. Hydrogenation of pyrrole is more difficult, which is illustrated by the observation that hydrogenation of 2-vinyl pyrrole (**433**) with 3 atmospheres of hydrogen gave 2-ethyl pyrrole (**434**) in 81% yield without reduction of the pyrrole ring.[450] Hydrogenation of the pyrrole ring is possible under acidic conditions, illustrated by hydrogenation of the previous reaction product (**434**) to give **435**.[450] Rhodium catalysts are usually preferred in low-pressure hydrogenations to reduce heterocyclic aromatic rings.

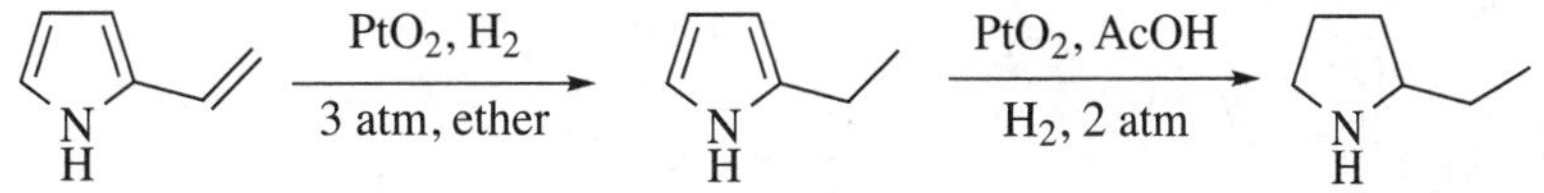

433   **434**   **435**

Both platinum and palladium catalysts can be used for the hydrogenation of heteroaryl derivatives, although platinum is more common. Selective reduction of the heterocyclic ring in 8-methyl-7-hydroxy isoquinoline (**436**) gave **437** in 67% yield, in Woodward and von E. Doering's synthesis of quinine.[451]

449.   Wu, P.-L.; Chu, M.; Fowler, F.W. *J. Org. Chem.* **1981,** *53,* 963.
450.   Cantor, P.A.; Vanderwerf, C.A. *J. Am. Chem. Soc.* **1958,** *80,* 970.
451.   (a) Woodward, R.B.; Doering, W.E. *J. Am. Chem. Soc.* **1944,** *66,* 849; (b) *Idem Ibid.* **1945,** *67,* 860.

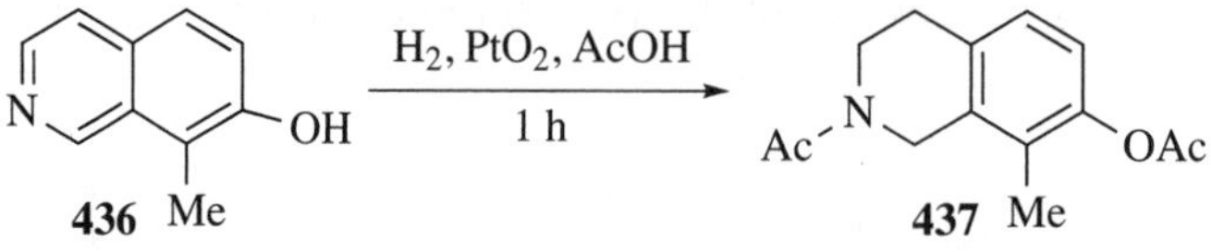

Reductive alkylation of heteroaromatic rings is possible using catalytic hydrogenation.[452] Reduction of heteroaryl ammonium salts is also possible. In an example taken from a synthesis of the trail pheromone of the Pharaoh ant (**440**) by Sonnet and Oliver, treatment of **438** with triphenylphosphine•bromine led to pyridinium salt **439**. Subsequent hydrogenation over PtO$_2$ gave 3-butyl-5-methyloctahydroindolizine **440**.[453]

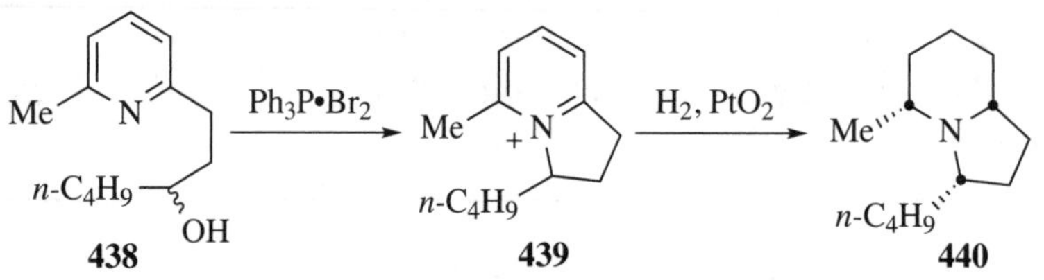

Oxygenated heterocycles are reduced with hydrogen but are prone to hydrogenolysis, especially when platinum is used. Hydrogenation of furan derivatives uses palladium, rhodium, or ruthenium catalysts to minimize this problem. When the reaction is done at higher temperatures and under acidic conditions, however, ring cleavage can occur via hydrogenolysis. Hydrogenation of furan to THF is straightforward when a mixed catalyst of 70% rhodium + 30% platinum is used.[454] The reduction of disubstituted furan **441** to **442**, in the Gerlach et al. synthesis of 8-*epi*-nonactic acid used a rhodium catalyst.[455] Reduction of thiophene derivatives can be difficult because bivalent sulfur is a common poison for heterogeneous catalysts. This problem can be circumvented or diminished by using a large excess of a heterogeneous catalyst (high catalyst loading). As an example, thiophene was reduced to perhydrothiophene in 71% yield, but only when the molar ratio of palladium to thiophene was 2:1 or greater.[456]

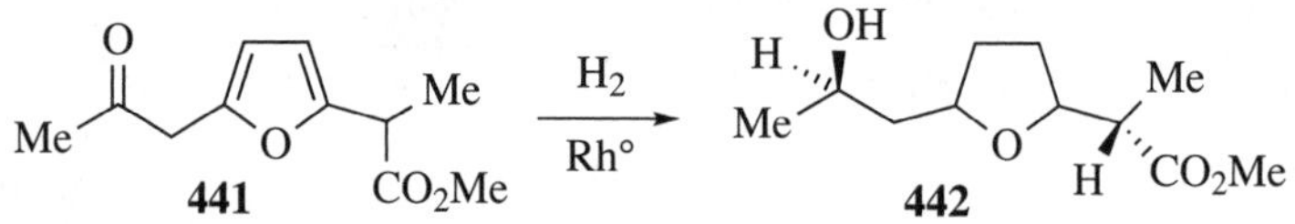

4.8.H. Asymmetric Catalytic Hydrogenation

Unsymmetrical ketones or substituted alkenes have prochiral carbons, and hydrogenation gives

452. Szychowski, J.; MacLean, D.B. *Can. J. Chem.* **1979**, *57*, 1631.
453. Oliver, J.E.; Sonnet, P.E. *J. Org. Chem.* **1974**, *39*, 2662.
454. Reference 323, p 228
455. (a) Gerlach, H.; Wetter, H. *Helv. Chim. Acta* **1974**, *57*, 2306; (b) Gerlach, H.; Oertle, K.; Thalmann, A.; Servi, S. *Ibid.* **1975**, *58*, 2036.
456. Mozingo, R.; Harris, S.A.; Wolf, D.E.; Hoffhine, Jr., C.E.H.; Easton, N.R.; Folkers, K. *J. Am. Chem. Soc.* **1945**, *67*, 2092.

a product that contains one or more stereogenic centers. Therefore, stereoselectivity in the hydrogenation process is an important issue. Asymmetric induction during hydrogenation is possible if a chiral catalyst is used. With heterogeneous catalysts, asymmetric induction usually involves addition of a metal catalyst to a chiral coordinating molecule or a chiral medium[457] such as silk fibroin,[458] (a natural structural protein from silk), which gave products with up to 70% ee (see sec. 1.4.F for a definition of % ee). Other chiral additives include palladium on poly-(L)-leucine,[459] Pd/C with (S)-methionine,[460] palladium on silica gel in the presence of a chiral alkaloid such as narcotine,[461] and platinum with l-menthoxy acetate or tartaric acid.[462] Hydrogenation of a chiral substrate with a heterogeneous catalyst can show enantioselectivity, but problems associated with chemisorption sometimes lead to racemization or bond migration. These secondary reactions are less of a problem with homogeneous catalysts. Indeed, soluble, chiral homogeneous catalysts are usually the best choice in catalytic hydrogenation, especially for alkenes. The occasionally poor activity of homogeneous catalysts with heterocyclic functional groups can limit their utility.

A common way to generate a chiral catalyst involves a modification of Wilkinson's catalyst (**336**) in which an optically active tertiary phosphine, bis- or tris(phosphines) are used as ligands in place of triphenyl-phosphine. If the phosphorous atom of the added phosphine is the stereogenic center, the optical yields are usually 4-22%, as in the conversion of atropic acid (**443**) to hydratropic acid (**444**) with 22% ee.[463] An example of this type of phosphine is (–)-methylpropylphenylphosphine. Chiral bis(phosphines) are commonly used, including **445** (called dipamp)[464] and **446** (called R-camp).[465]

As noted in Section 1.4.A, phosphines are configurationally stable if the temperature is *below* ~ 130°C, which is in sharp contrast to amines which show fluxional inversion even at low temperatures. The configurational stability of chiral phosphines allows their preparation,

457. (a) Reference 333, pp 74-80; (b) Reference 231, pp 292-302 and citations cited therein.
458. (a) Reference 323, pp 292, 297; (b) Reference 333, p 74; (c) Akabou, S.; Sakurai, S.; Izumi, Y.; Fujii, Y. *Nature (London)* **1956**, *178*, 323.
459. Beamer, R.L.; Fickling, C.S.; Ewing, J.H. *J. Pharm. Sci.* **1967**, *56*, 1029.
460. Senoh, S.; Ouchi, S.; Tsunoda, K. *Jpn. Patent* 13, 307, **1963** [*Chem. Abstr. 60*: 3092h **1964**].
461. Padgett, R.E.; Beamer, R.L. *J. Pharm. Sci.* **1964**, *53*, 689.
462. Nakamura, Y. *J. Chem. Soc. Jpn.* **1940**, *61*, 1051 [*Chem. Abstr. 37*: 3772, **1943**].
463. (a) Knowles, W.S.; Sabacky, M.J. *J. Chem. Soc. Chem. Commun.* **1968**, 1445; (b) Cullen, W.R.; Fenster, A.; James, B.R. *Inorg. Nucl. Chem. Lett.* **1974**, *10*, 167.
464. (a) Brown, J.M.; Chaloner, P.A. *J. Chem. Soc. Chem. Commun.* **1980**, 344; (b) *Idem Ibid.* **1978**, 321; (c) *Idem Tetrahedron Lett.* **1978**, 1877; (d) *Idem J. Am. Chem. Soc.* **1980**, *102*, 3040.
465. Knowles, W.S.; Sabacky, M.J.; Vineyard, B.D. *Adv. Chem. Ser,* **1974**, *132*, 274.

isolation and even optical resolution, but they are not always easy to synthesize. A useful alternative is to attach an achiral phosphine to a stereogenic center at carbon, and most are derivatives of 1,2-diphenylphosphinoethane (called diphos or dppe, **447**). When these phosphine ligands are used to form homogeneous catalysts, the enantioselectivity can be very good, if the alkene substrate is properly chosen. Some of the more popular chiral phosphine ligands include **448** (called diop),[466] **449** (called chiraphos),[467] **450** (called prophos),[468] the ferrocenyl derivative **451**,[469] and **452** (called bppm).[470] These bis(phosphine) ligands are often coordinated to rhodium[471] in modified Wilkinson's catalysts.

One of the more important applications of chiral hydrogenation catalysts is the preparation of chiral α-amino acids. Koenig correlated the % ee for the reduction of **453** to **454** with rhodium catalysts prepared from several chiral phosphine ligands (see Table 4.14).[471] In general, reduction of **453** gave the enantiomer of **454** with the absolute configuration opposite that of the chiral phosphine [i.e., (RR) → (S); (SS) → (R); (R) → (S); (S) → (R)]. One complicating factor is that some isomerization (E) → (Z) occurs in **453** during hydrogenation, depending on the solvent and the catalyst. Reduction of **453** in THF with **447** showed 23% isomerization. Reduction with (RR)-diop (**448**) gave 20% isomerization, but dipamp (**445**) showed 0% isomerization. Isomerization was more extensive in ethanol (28% with diphos, 35% with diop, and 6% with dipamp). When benzene was used, no isomerization was observed with any of these catalysts.[472b]

466. (a) Dang, T.P.; Kagan, H.B. *J. Chem. Soc. Chem. Commun.* **1971**, 481; (b) Kagan, H.B.; Dang, T.P. *J. Am. Chem. Soc.* **1972**, *94*, 6429; (c) Levi, A.; Modena, G.; Scorrano, G. *J. Chem. Soc. Chem. Commun.* **1975**, 6.
467. (a) Chan. A.S.C.; Halpern, J. *J. Am. Chem. Soc.* **1980**, *102*, 838; (b) Chua, P.S.; Roberts, N.K.; Bosnich, B.; Okrasinski, S.J.; Halpern, J. *J. Chem. Soc. Chem. Commun.* **1981**, 1278.
468. (a) Fryzuk, M.D.; Bosnich, B. *J. Am. Chem. Soc.* **1978**, *100*, 5491; (b) *Idem Ibid.* **1977**, *99*, 6262.
469. Hayashi, T.; Yamamoto, K.; Kumada, M. *Tetrahedron Lett.* **1974**, 4405.
470. Achiwa, K. *J. Am. Chem. Soc.* **1976**, *98*, 8265.
471. Koenig, K.E. in *Catalysis of Organic Reactions*, Kosak, J.R. (Ed.), Marcel Dekker, New York, **1984**, pp 63-77, cf. pp 72-73.
472. (a) Vineyard, B.D.; Knowles, W.S.; Sabacky, M.J.; Bachmann, G.K.; Weinkauff, D.J. *J. Am. Chem. Soc.* **1977**, *99*, 5946; (b) Koenig, K.E.; Knowles, W.S. *Ibid.* **1978**, *100*, 7561.

Table 4.14. Effect of Chiral Phosphine Ligands on the Asymmetric Hydrogenation of 453

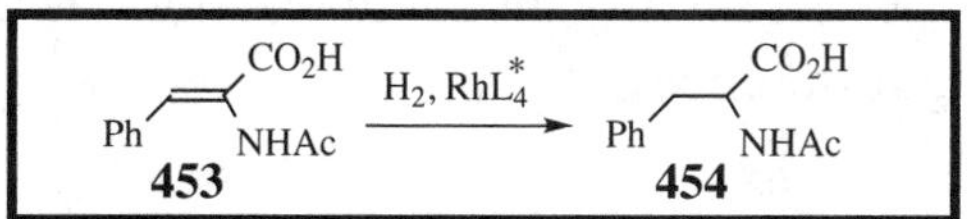

Ligand	% ee	(R) or (S)
(RR)-**448**	85	(S)
(SS)-**449**	95+	(R)
(RR)-**450**	90	(S)
451	93	
(SS)-**452**	91	(R)
(RR)-**445**	94	(S)
(R)-**446**	15	(S)

When these ligands react with rhodium salts, many intermediate metal complexes can be formed, including Rh(diphos)cod$^+$, Rh(diphos)Ac, Rh(dipamp)$_2$, Rh(diop)NBD$^+$, and Rh(camp)$_2$ COD$^+$. In work by Halpern and co-workers, several intermediates in the hydrogenation of **453** with rhodium catalysts were characterized.[467a] An initial square planar Rh(I) coordination complex (**455**) was detected by NMR (^{1}H, ^{13}C, ^{31}P), and its tetrafluoroborate salt was isolated and identified by X-ray crystallography.[473] Lowering the temperature, and using the same NMR techniques allowed observation of a square-planar Rh(III) intermediate (**456**), generated after transfer of hydrogen to the π bond.[467a,473] Halpern presented a mechanistic rationale for the reduction of **457** with this catalyst, based on these intermediates.[474] Halpern and co-workers suggested transformation of a square planar complex to an octahedral complex as the hydrogens are transferred from rhodium to carbon, in a stepwise manner.[467,473] Chiral ligands generate facial selectivity in the complex during the hydrogen transfer.

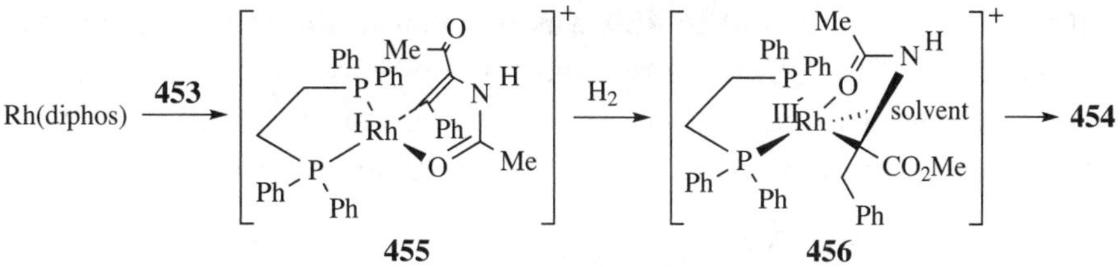

[Reprinted with permission from Chan. A.S.C.; Halpern, J. *J. Am. Chem. Soc.* **1980**, *102*, 838. Copyright © 1980 American Chemical Society.]

4.9. DISSOLVING METAL REDUCTIONS

Reduction involves the gain of electrons by an organic species, as described in this chapter. Many metals and metal salts can transfer electrons to an organic substrate, reducing them with concomitant oxidation of the metal. This section will describe the more common metals used

473. Chan, A.S.C.; Pluth, J.J.; Halpern, J. *Inorg. Chim. Acta* **1979**, *37*, 2477.
474. Halpern, J. in Reference 325d, pp 3-22.

for this purpose as well as the functional groups that are reduced by them. Electron transfer in an electrolytic cell accomplishes essentially the same reactions as treatment with alkali metals, but there are some interesting differences.

4.9.A. Reduction with Alkali Metals

Metals can be used to reduce organic molecules in the presence of a proton transfer agent. Alkali metals can transfer an electron to some organic molecules, but reduction requires delivery of protons from an acidic solvent (with an X–H bond) to reactive intermediates. Transfer of an electron to an organic molecule generates a metal cation along with the radical anion derived from the organic fragment. Water, alcohol or ammonia are common acidic solvents used to quench the radical anion, but primary or secondary amines can also be used. The ability of a metal to reduce organic molecules is directly related to its ability to release electrons (reduction potential), and Hudlick´y tabulated these values for several alkali and transition metals.[475] Based on standard reduction potentials,[6] lithium and potassium followed closely by sodium were the most effective electron transfer metals in this series. It is therefore not surprising that these metals are used most often for reduction of organic molecules. This type of reduction is commonly called a **dissolving metal reduction**.

Metal reduction is most commonly applied to polarized organic substrates containing multiple bonds (C=O, C=N, C≡N). Isolated alkenes are not easily reduced, which is a major difference from the related electrolytic reductions to be discussed in Section 4.9.J. Alkenes that are conjugated to a phenyl ring, another double bond or an aromatic ring are reduced quite easily, however, as in the conversion of **457** to **458** in 97% yield.[476] Alkynes are generally converted to the trans alkene. The alkyne unit in **459** was reduced to the trans-alkene unit in **460**, for example, in a synthesis of (–)-pironetin by Dias and co-workers.[477] Aromatic rings, halides, and sulfur derivatives are also reduced. Complete reduction of the aromatic ring is possible, especially with heterocyclic rings including pyridine and furan derivatives.

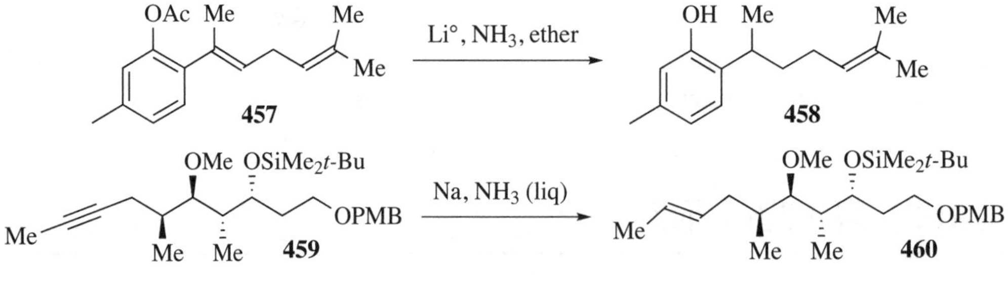

Conjugated dienes are reduced, usually via 1,4-addition to give an alkene. When the conjugated diene unit in **461** was treated with sodium metal in ammonia, a 65% yield of **462** was obtained

475.   Reference 7, p 25.
476.   McEnroe, F.J.; Fenical, W. *Tetrahedron* **1978**, *34*, 1661.
477.   Dias, L.C.; de Oliveira, L.G.; de Sousa, M.A. *Org. Lett.* **2003**, *5*, 265.

as part of Meyers and Lemieux's synthesis of (–)-trichodiene.[478] As expected, a mixture of 1,4-reduction (to give **462**) and 1,2-reduction (to give **463**) was observed (in a ratio of 85:15), which is common with this type of reduction. Note that the nonconjugated *exo*-methylene group of the five-membered ring (in **461**) was not reduced under these conditions.

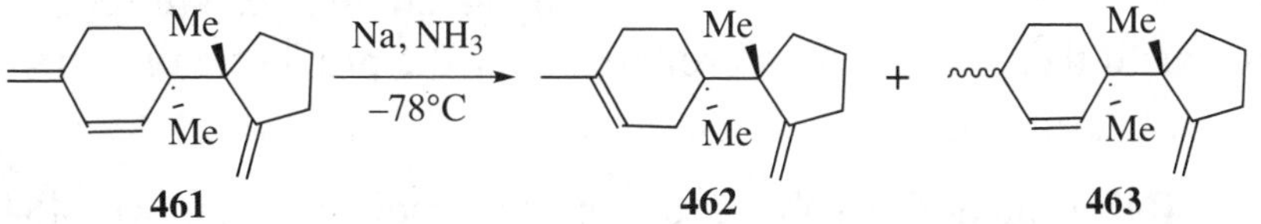

Dissolving metal conditions are capable of reducing many organic functional groups, often with excellent chemo-, regio- and stereoselectivity. The choice of solvent is important, often changing the course of a reaction and having a great influence of the yield and ease of isolation of the products. Liquid ammonia is commonly used, often in conjunction with alcohols. Low molecular weight amines and ethylenediamine can be used in reductions with lithium metal.[479] The most common applications for this reaction include reduction of alkynes, conjugated carbonyls and Birch-type reductions of aromatic rings.

4.9.B. Reduction of Carbonyl Compounds

Metal reduction of ketones and aldehydes begins with electron transfer from the metal to the carbonyl, forming a ketyl (a radical anion). If we use acetone as a typical example, the mechanism of reduction involves transfer of an electron from the metal to the carbonyl π bond giving a radical anion with two canonical forms (**464**).[480] There are two competing pathways for the ketyl, reduction or radical coupling to give **467**, which gives diol **468** after protonation. If we want to facilitate reduction, a protic solvent such as alcohol or water is added to transfer a proton to the oxygen in an acid-base reaction (giving **465**). Protonation, followed by a second electron transfer from the metal, gives the carbanionic species **466**. A second proton is transferred from the solvent to give the final reduction product, 2-propanol.

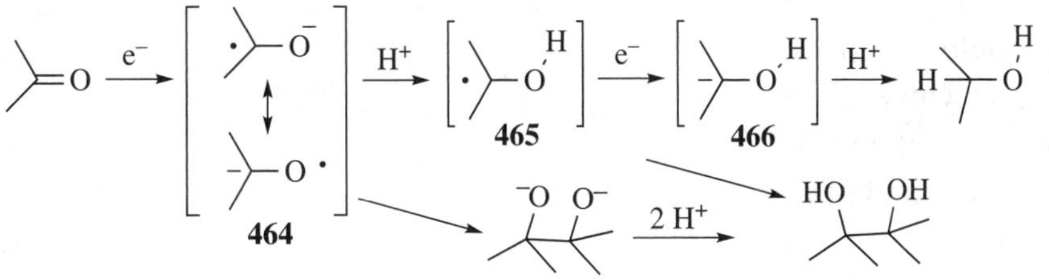

When a ketone or aldehyde is part of a substrate bearing a stereogenic center elsewhere in the

---

478. Lemieux, R.M.; Meyers, A.I. *J. Am. Chem. Soc.* **1998**, *120*, 5453.
479. Garst, M.E.; Dolby, L.J.; Esfandiari, S.; Fedoruk, N.A.; Chamberlain, N.C.; Avey, A.A. *J. Org. Chem.* **2000**, *65*, 7098.
480. Reference 7, p 23.

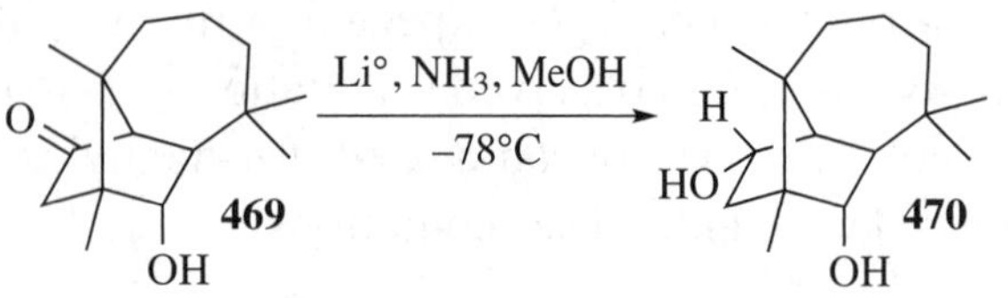

molecule, reduction can proceed with good diastereoselectivity. Reduction of **469** from the exo- face led to a quantitative yield of **470**, in Takasu and Ihara's synthesis of culmorin.[481] Conjugated ketones are reduced selectivity at the alkene moiety and there is usually a preference for the product with a trans-ring juncture (see below).[482]

Formation of diol **468**, as mentioned above, can arise by radical coupling of **464** to give **467** or by radical coupling of **465** to give diol **468** directly. The coupling reaction is called the **pinacol coupling** (see sec. 13.8.D for a more complete description),[567] and is a useful method for the synthesis of 1,2-diols[483] where aromatic ketones give the best yields. Pinacol coupling is best accomplished by using a mixture of magnesium and magnesium iodide ($Mg + MgI_2$), although magnesium amalgam (Mg/Hg) can also be used.[484a] Pinacol coupling of acetone leads to a diol called pinacol (2,3-dimethyl-2,3-butanediol), which gives the reaction its name.

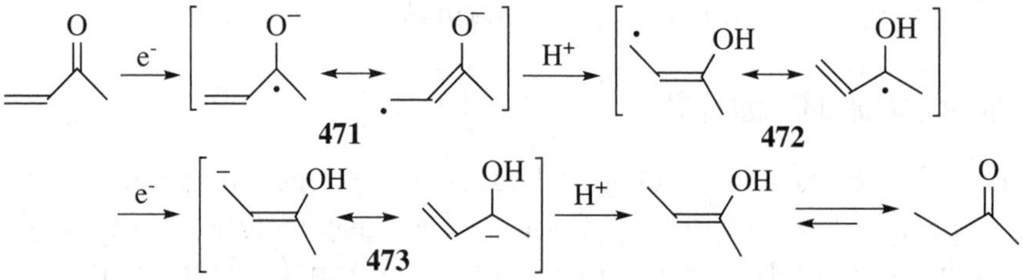

The electron transfer/proton transfer mechanism presented for ketone reduction is under thermodynamic control, and leads to the more stable trans-alcohol.[379b] Conjugated ketones react to give alcohols via 1,4-reduction,[379b] which is also explained by the ketyl mechanism. Transfer of an electron gives a very stable, resonance-delocalized ketyl (**471**). Protonation gives an allylic enol radical (**472**), and transfer of a second electron gives **473**. Proton transfer generates the enol, which tautomerizes to the ketone product. An example is the dissolving metal reduction of **474** to give **475** in quantitative yield, in Covey and co-worker's synthesis of deoxycholic acid.[485] Note that triethylamine was added after quenching with ammonium

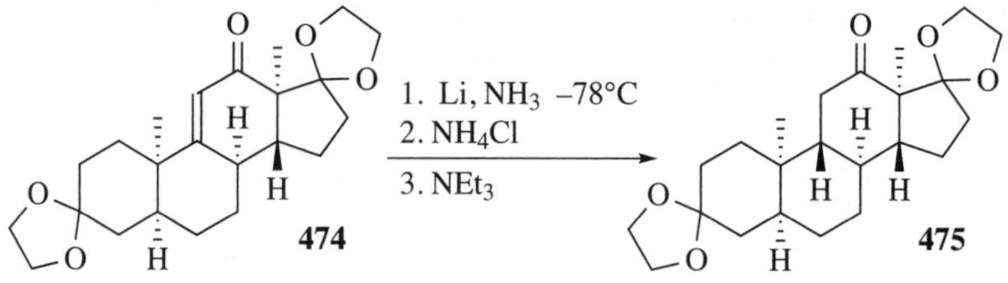

481. Takasu, K.; Mizutani, S.; Noguchi, M.; Makita, K.; Ihara, M. *J. Org. Chem.* **2000,** *65*, 4112.
482. (a) Stork, G.; McMurry, J.E. *J. Am. Chem. Soc.* **1967,** *89*, 5464; (b) Stork, G.; Rosen, P.; Goldman, N.; Coombs, R.V.; Tsuji, J. *Ibid.* **1965,** *87*, 275.
483. Schreibman, A.A.P *Tetrahedron Lett.* **1970,** 4271.
484. (a) Adams, R.; Adams, E.W. *Org. Synth. Coll. Vol. 1* **1932,** 459; (b) Torii, S.; Inokuchi, T.; Kawai, K. *Bull. Chem. Soc. Jpn.* **1979,** *52*, 861.
485. Katona, B.W.; Rath, N.P.; Anant, S.; Stenson, W.F.; Covey, D.F. *J. Org. Chem.* **2007,** *72*, 9298.

chloride to ensure that the ketals were not removed by residual acid (see sec. 7.3.B.). Pinacol coupling is also a competing side reaction with conjugated ketones.[486] Conjugated reduction of α,β-unsaturated esters has been reported using Mg in methanol.[487]

4.9.C. Reduction of Alkynes

An important use of metal reduction is to convert alkynes to trans-alkenes, as illustrated by the conversion of **476** to **477** in 92% yield in Koert and Emde's synthesis of squamocin A.[488]

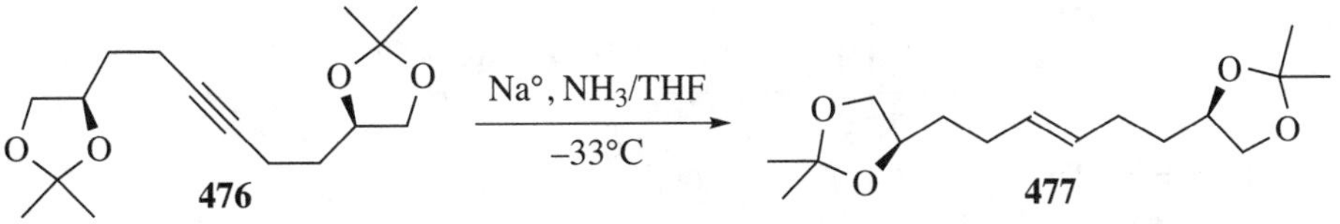

In the general case, the reaction is initiated by electron transfer to the alkyne (see **478**) to give a radical anion (**479**). Electron repulsion of the radical orbital and the carbanion orbital dictates that they be as far apart as possible, which leads to the trans-geometry. Protonation of **479** from either ammonia or an added alcohol co-solvent gives a vinyl radical, **480** and a second electron transfer leads to vinyl carbanion **481**. A second protonation step gives the final trans alkene product. Reduction using either lithium or sodium in liquid ammonia gives primarily the trans alkene,[486b] but increasing amounts of the cis alkene are observed in protonic solvents such as ethanol. Reduction of terminal alkynes is complicated by the acidity of the alkyne hydrogen, which reacts with sodium or lithium to form an acetylide anion (secs. 2.2.A, 8.3.B). Reduction to the alkene is possible, however, by controlling the stoichiometry of the reagent. Reaction of propyne with sodium initially generates the acetylide (**482**), and in the presence of additional alkyne, **482** is reduced to 1-propene as shown.[489] If sodium becomes the limiting reagent, reduction to the alkene is possible. An example is the reduction of 1-octyne to 1-octene in 90% yield (ratio of alkene/sodium = 11.5/1).[486b]

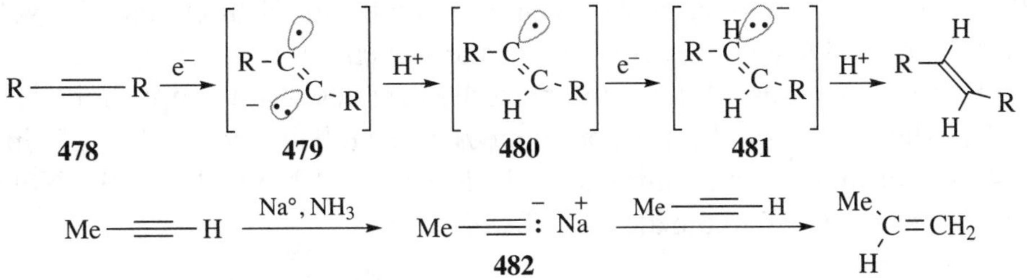

486. (a) Wiemann, J.; Monot, M.R.; Dana, G.; Chuche, J. *Bull. Soc. Chim. Fr.* **1967**, 3293; (b) Kochansky, J.P.; Cardé, R.T.; Taschenberg, E.F.; Roelofs, W.L. *J. Chem. Ecol.* **1977**, 3, 419.
487. For an example taken from a synthesis of (+)-sertraline, see Chandrasekhar, S.; Reddy, M.V. *Tetrahedron* **2000**, *56*, 1111.
488. Emde, U.; Koert, U. *Eur. J. Org. Chem.* **2000**, 1889.
489. Henne, A.L.; Greenlee, K.W. *J. Am. Chem. Soc.* **1943**, *65*, 2020.

Low molecular weight amines such as ethylamine and methylamine are often substituted for ammonia when higher reaction temperatures are required (i.e., $> -33°C$). Changing to an ether solvent also leads to interesting differences in reactivity. As House pointed out,[490] reduction of diphenylacetylene (**483**) with sodium in THF ($-78°C$) gave only 25% of the expected *trans*-stilbene (**487**), 30% of the completely reduced 1,2-diphenylethane and unreacted starting material. Similar reduction with lithium in THF gave a quantitative yield of the deuterated cis alkene **485** upon quenching with *d*-methanol, presumably due to the greater stability of the dimeric dilithio species (**484**) relative to the disodio species **486**.

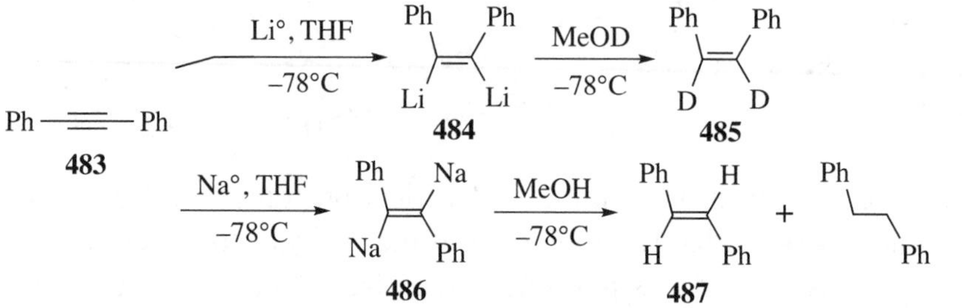

The functional group transforms possible are:

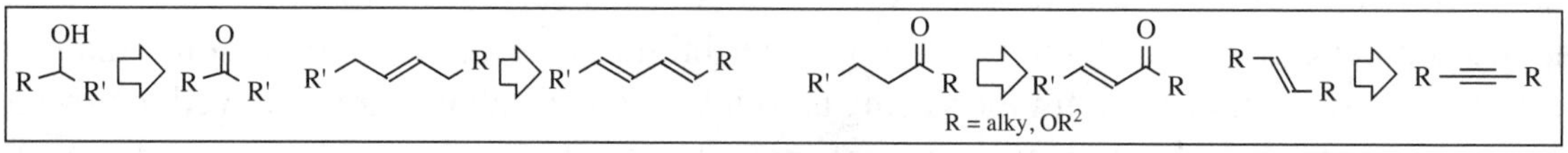

R = alky, OR^2

4.9.D. Hydrogenolysis

Hydrogenolysis may be a competing reaction in dissolving metal conditions, and is particularly useful for the cleavage of *O*-benzyl and *N*-benzyl derivatives. It is widely used as a useful alternative to catalytic hydrogenation. Treatment with sodium in ammonia cleaves an *O*-benzyl group or an *N*-benzyl group. An example of the latter reaction is the conversion of **488** to **489** (in 80% yield), in Passarella and co-worker's synthesis of aloperine.[491] Note that the byproduct is toluene and that ethylenediamine is the solvent. It is also possible to cleave other reactive C–O (or C–N) bonds. Allylic and benzylic acetates, for example, are cleaved to the hydrocarbon, as shown in a synthesis of ottelione A by Katoh and co-worker's,[492] in which the reaction of **490** with lithium and ammonia in THF gave a 98% yield of **491**. Note that both OAc units in **490** are reduced to the alcohol as well.

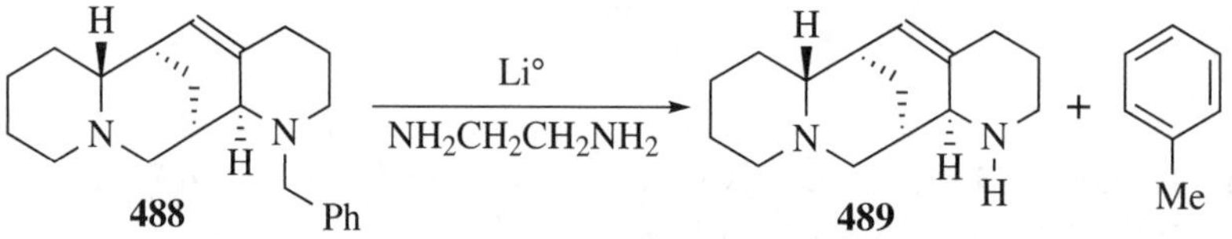

---

490. Reference 16, p 207 and references cited therein.
491. Passarella, D.; Angoli, M.; Giardini, A.; Lesma, G.; Silvani, A.; Danieli, B. *Org. Lett.* **2002**, *4*, 2925.
492. Araki, H.; Inoue, M.; Katoh, T. *Org. Lett.* **2003**, *5*, 3903.

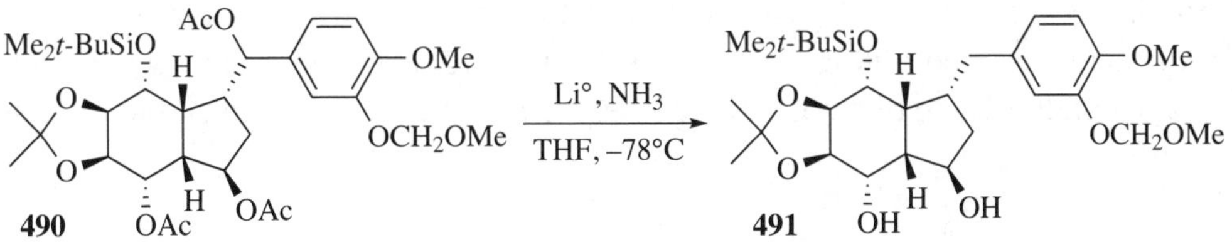

The hydrogenolysis reaction with metals in ammonia involves transfer of an electron to the heteroatom, with cleavage of the C-X bond. Electron transfer to benzyl ether (**492**), for example, generates a radical anion such as **493**. Homolytic cleavage liberates an alkoxide and the benzyl radical (**494**). This benzyl radical removes a hydrogen atom from a hydrogen donor in the medium (solvent, starting ether, etc.) to give toluene as the other reduction product. Cleavage to an alkoxy radical and a benzyl carbanion is also possible, although formation of the resonance stabilized benzyl radical is a more reasonable mechanistic rationale.

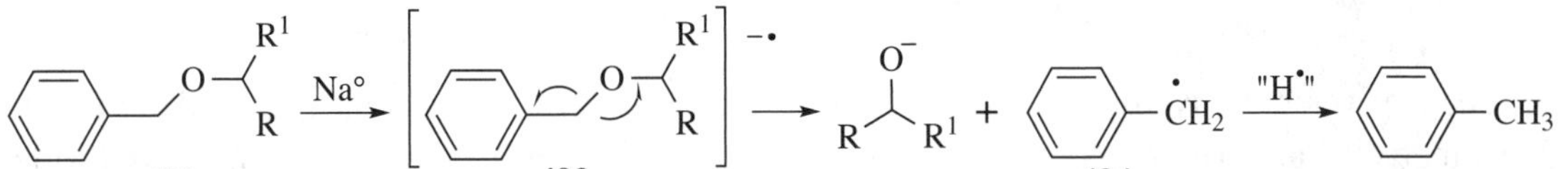

Since the mechanism involves electron transfer to a heteroatom rather than carbon, vinyl or aryl derivatives are easily reduced under these conditions. Many functional groups can be reduced, depending on their ability to accept an electron. Cyclopropane rings are opened under these conditions to give methyl derivatives.[493] The bridging cyclopropane moiety in **495**, for example, was opened with lithium and ammonia to give **496** in 85% yield as part of Srikrishna and Vijaykumar's synthesis of (+)-pinguisenol from (*R*)-carvone.[494] Halides undergo hydrogenolysis with alkali metals, and both alkyl[495] and vinyl halides give good yields of the corresponding alkene.

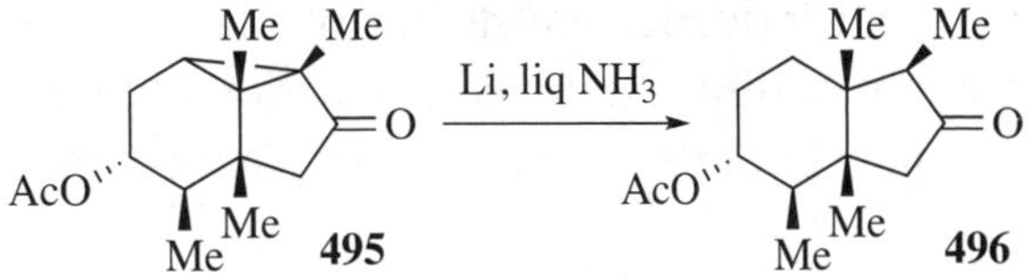

Hydrogenolysis of the C–S bond is rapid, and sulfides, sulfones or sulfoxides are reduced to the hydrocarbon moiety by this method.[496] Since sulfur usually poisons hydrogenation catalysts, reduction with metal is a preferred method for chemoselective cleavage of the C–S bond. In a synthesis of (–)-lasubine II by Back and Hamilton,[497] the sulfone group in **497** (-CHSO₂Tol) was cleaved to the hydrocarbon (-CH₂- in **498**) with lithium in liquid ammonia. Sulfides are similarly reduced.[498] Sodium amalgam is another important reagent for reductive cleavage

493. (a) Packer, R.A.; Whitehurst, J.S. *J. Chem. Soc. Perkin Trans. 1* **1978**, 110; (b) *Idem J. Chem. Soc. Chem. Commun.* **1975**, 757.
494. Srikrishna, A.; Vijaykumar, D. *J. Chem. Soc. Perkin Trans. 1* **2000**, 2583.
495. Eaton, P.E.; Or, Y.S.; Branca, J.S. *J. Am. Chem. Soc.* **1981**, *103*, 2134.
496. For a review that includes reductive desulfonylation, see Nájera, C.; Yus, M. *Tetrahedron* **1999**, *55*, 10547.
497. Back, T.G.; Hamilton, M.D. *Org. Lett,* **2002**, *4*, 1779.
498. Kodama, M.; Matsuki, Y., Itô, S. *Tetrahedron Lett.* **1975**, 3065.

of C–S bond. In a synthesis of gymnocin A,[499] Sasaki and co-workers treated sulfone **499** with Na(Hg), and obtained **500** in 75% yield. Both sulfoxides and sulfones are important in alkylation and condensation reactions of their respective sulfur-stabilized carbanions (sec. 8.6).

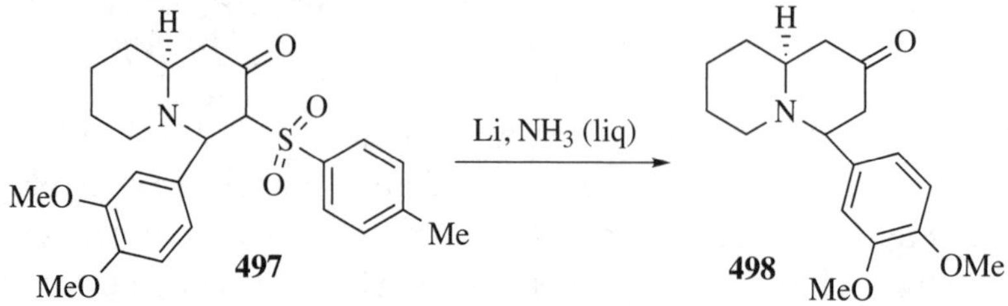

Epoxides are opened by dissolving metal reductions to give an alcohol, usually with good stereoselectivity and regioselectively favoring the product with oxygen at the less hindered carbon. The radical anion intermediate usually leads to the more stable trans alcohol. The presence of substituents at the epoxide-carbon can have a profound influence on the regioselectivity. In a synthesis of formal synthesis of reserpine by Shea and co-workers, reduction of **501** with sodium in ammonia gave a 50% yield of **503**.[500] The regiochemical preference for **503** is driven by formation of the resonance stabilized enolate anion intermediate, **502**. Epoxides are also subject to pinacol type coupling reactions, but the yields are usually low and reduction is the major process.[501] Esters and lactones are usually reduced to the alcohol or diol. Anomalous reactions can occur, however, as in the reduction of a lactone to a lactol,[502] but this is common only with carbohydrates and hindered lactones.

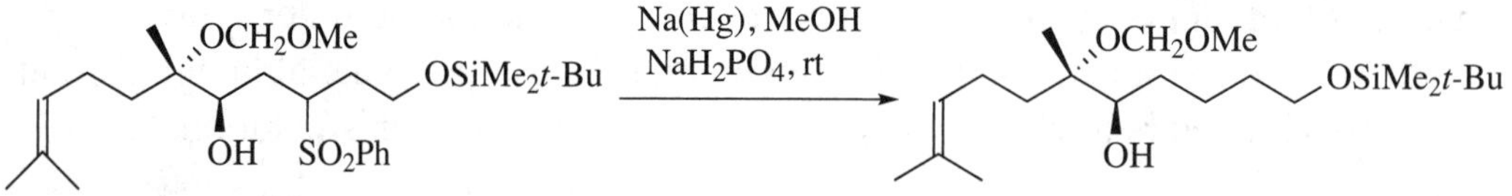

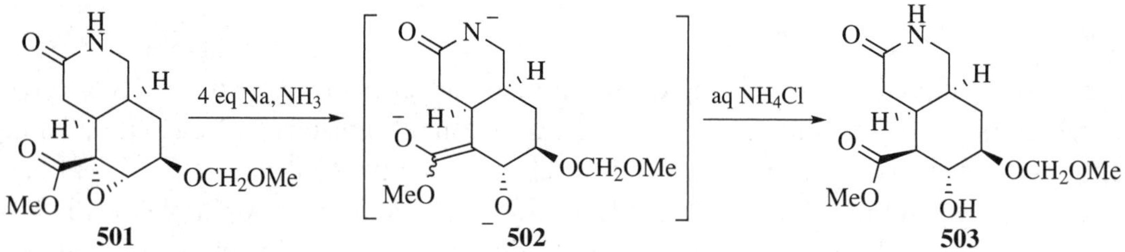

499. Sasaki, M.; Tsukano, C.; Tachibana, K. *Org. Lett.* **2002**, *4*, 1747.
500. Sparks, S.M.; Gutierrez, A.J.; Shea, K.J. *J. Org. Chem.* **2003**, *68*, 5274.
501. Movsumzade, M.M.; Agaev, F. Kh.; Shabanov, A.L. *Azerb. Khim. Zh.* **1966**, *38*, 38 [*Chem. Abstr. 67*: 63653u *1967*].
502. Wong, C.-H.; Whitesides, G.M. *J. Am. Chem. Soc.* **1983**, *105*, 5012.

4.9.E. Birch Reduction

Partial reduction of aromatic rings with alkali metals[503] is an important synthetic method because it allows one to use aromatic substitution reactions to set functionality, and reduction to a nonaromatic system then allows transforms to other compounds. When a dissolving metal reduction is carried out on an aromatic ring containing electron releasing groups (alkyl, alkoxy, etc.) the cyclohexadiene product formed will have those groups on a vinylic carbon (sp^2 carbon),[504] as in the reduction of **504** to **505** in >90% yield as part of Baldwin and co-worker's synthesis of biatractylolide.[505]

Conversely, reduction of an aromatic ring that has an electron-withdrawing group on the ring will generate a product with that group on the sp^3 carbon. Reduction of **506** led to **507**, for example.[506] The electron transfer mechanism invoked for reduction of alkynes and carbonyls also explains these results. Transfer of an electron to anisole is slower due to the OMe and leads to **508** or **509**. The proximity of the negative charge to the electron-rich oxygen in **508** destabilizes it, which leads to a preference for **509**. Protonation and a second electron transfer will lead to vinyl ether **510**. When the substituent on the aromatic ring is changed to an electron-withdrawing group (such as carboxyl), the selectivity changes. Electron transfer to benzoic acid leads to **511** or **512**. The electron-withdrawing carbonyl group stabilizes the developing negative charge in **511**, and it is preferred to **512**. The usual sequence leads to **513** as the final product. This partial reduction of aromatic rings is known as the **Birch reduction**.[507] Complete reduction of the aromatic ring is also possible, especially with those molecules that contain heterocyclic rings such as pyridine or furan derivatives. A recent variation uses lithium and di-*tert*-butylbiphenyl for an ammonia-free Birch reduction.[508]

---

503.   For a brief review of the Birch reduction and its application to the synthesis of natural products, see Schultz, A.G. *Chem. Commun.* **1999**, 1263.

504.   Zimmerman, H.E.; Wang, P.A. *J. Am. Chem. Soc.* **1993**, *115*, 2205.

505.   Bagal, S.K.; Adlington, R.M.; Baldwin, J.E.; Maquez, R.; Cowley, A. *Org. Lett.* **2003**, *5*, 3049.

506.   (a) Caluwe, P.; Pepper, T. *J. Org. Chem.* **1988**, *53*, 1786; (b) Rubin, M.B.; Welner, S. *Ibid.* **1980**, *45*, 1847.

507.   (a) Birch, A.J. *Quart. Rev. (London)*, **1950**, *4*, 69; (b) Birch, A.J.; Smith, H. *Ibid.* **1958**, *12*, 17; (c) Watt, G.W. *Chem. Rev.* **1950**, *46*, 317; (d) The Merck Index, 14th ed., Merck & Co., Inc., Whitehouse Station, New Jersey, **2006**, p ONR-11; (e) Mundy, B.P.; Ellerd, M.G.; Favaloro Jr., F.G. *Name Reactions and Reagents in Organic Synthesis, 2nd ed.*, Wiley-Interscience, New Jersey, **2005**, pp. 94-95.

508.   Donohoe, T.J.; House, D. *J. Org. Chem.* **2002**, *67*, 5015.

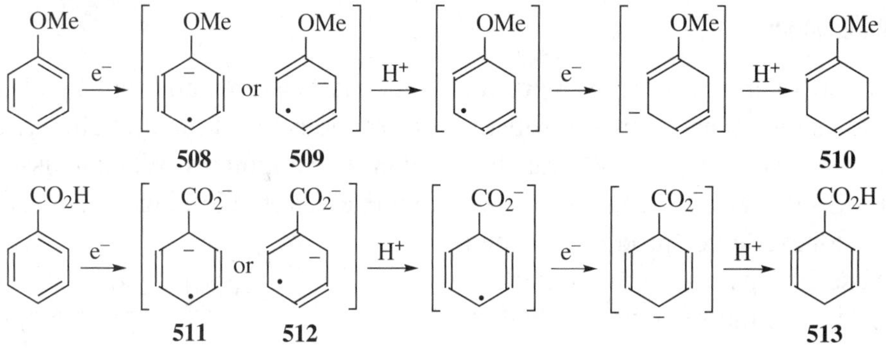

This leads to the functional group transforms:

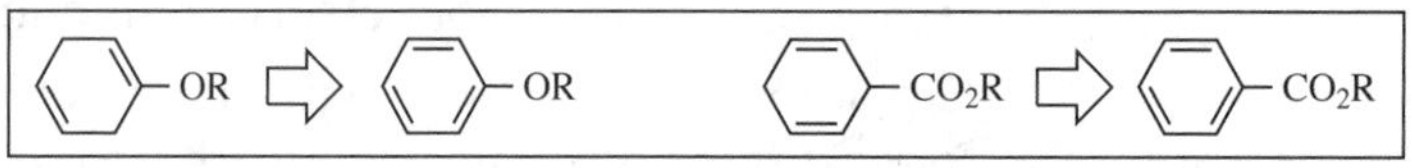

There are a wide variety of synthetic applications of this chemistry, making it one of the more useful reactions in organic chemistry (sec. 10.5.B.vi). Partially reduced bicyclic and tricyclic aromatic derivatives[509] can be formed by this method. It is also possible to reduce anisole derivatives to give cyclohexenone derivatives.

An example is the Birch reduction of methoxybenzene **514**, taken from a synthesis of periplanone C by Saicic and co-workers, which gave the non-conjugated cyclohexenone **515**, in 46% yield

after alkylation with allyl bromide.[510] In some cases, Birch reduction of anisole derivatives followed by treatment with acid, gives the conjugated cyclohexenone derivatives. In a synthesis of aeruginosin 298-A by Bonjoch and co-workers, Birch reduction of **516** gave **517**.[511] Subsequent treatment with aqueous HCl gave **518**, but an intramolecular conjugate addition (sec. 9.) led to **519** as the final product in good yield. Note that it is possible to prevent the double bond of the initially formed vinyl ether from migrating into conjugation,[512]

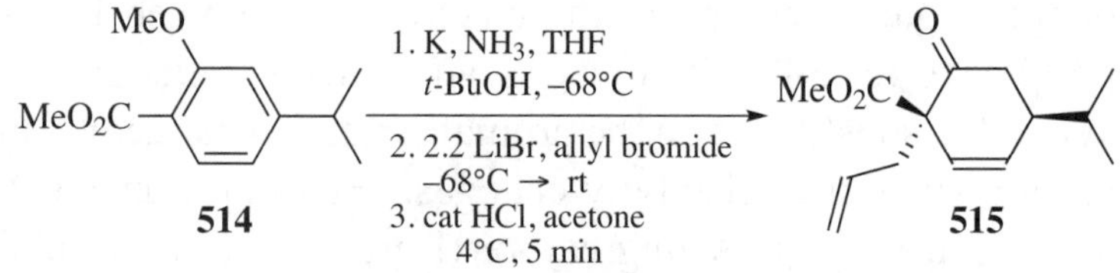

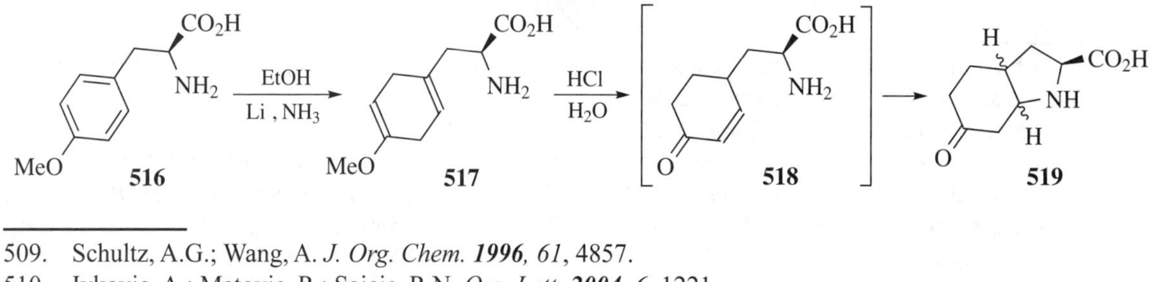

509. Schultz, A.G.; Wang, A. *J. Org. Chem.* **1996**, *61*, 4857.
510. Ivkovic, A.; Matovic, R.; Saicic, R.N. *Org. Lett.* **2004**, *6*, 1221.
511. Valls, N.; López-Canet, M.; Vallribera, M.; Bonjoch, J. *Chem. Eur. J.* **2001**, *7*, 3446.
512. Kasturi, T.R.; Thomas, M. *Indian J. Chem.* **1972**, *10*, 777.

if hydrolysis is done under very mild conditions. Treatment with oxalic acid [(CO$_2$H)$_2$] is commonly used for this purpose. Reduction of anisole derivatives is an important route to cyclohexenone derivatives and the functional group transformation is

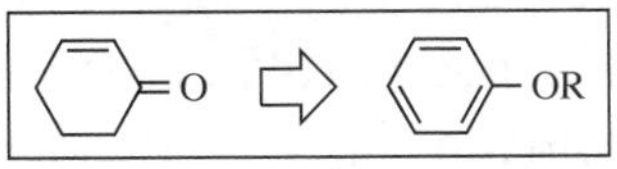

One of the more interesting synthetic features of the Birch reduction is the ability to alkylate the carbanionic intermediate generated by the electron-transfer reaction, as in the synthesis of **515**. In another example, taken from a synthesis of plueromutilin,[513] Zard and co-workers converted **520** to **521** in 72% yield using a Birch reduction and trapping the anion intermediate with the alkyl halide shown.[513]

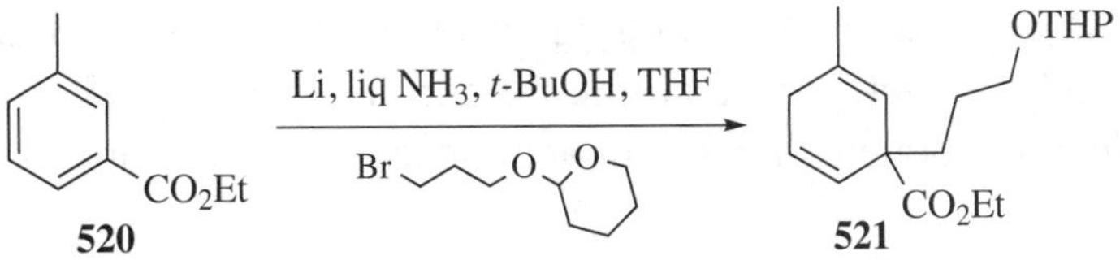

As mentioned briefly above, Birch reduction is an important method for reducing heterocyclic aromatic rings. Although Birch reduction of unsubstituted pyrroles tends to give poor yields, reduction of the *N*-Boc derivatives can give quite good yields,[514] and reduction of pyridine rings to the piperidine derivative has been reported in low yield.[515] Partial reduction of a furan ring is also possible, as observed in the synthesis of secosyrin 1 by Donohoe and co-workers,[516] in which reduction of **522** led to a 68% yield of **523** (>20:1 dr) after reaction of the initially formed anion with the benzylic bromide shown.

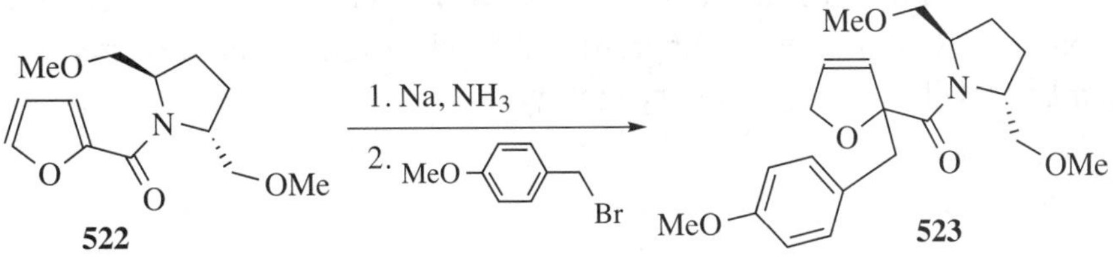

4.9.F. The Benkeser Reduction

Lithium and sodium are the most commonly used metals in metal reductions, but Benkeser and co-workers developed the use of calcium in amines. In the initial work, naphthalene was reduced to an 80:20 mixture of Δ^9-octalin (**524**) and $\Delta^{1(9)}$-octalin (**525**) with lithium in

513. Bacqué, E.; Pautrat, F.; Zard, S.Z. *Org. Lett.* **2003**, *5*, 325.
514. For an example taken from a synthesis of the glycosidase inhibitor (2*R*,5*R*)-bis(hydroxymethyl)-3(*R*),4(*R*) dihydroxy-pyrrolidine (DMDP), see Donohoe, T.J.; Headley, C.E.; Cousins, R.P.C.; Cowley, A. *Org. Lett.* **2003**, *5*, 999.
515. Donohoe, T.J.; Guyo, P.M. *J. Org. Chem.* **1996**, *61*, 7664.
516. Donohoe, T.J.; Fisher, J.W.; Edwards, P.J. *Org. Lett.* **2004**, *6*, 465.

diethylamine-dimethylamine.[517] Replacing lithium with calcium gave a 77:23 mixture of **524** and **525** in 92% yield.[518,519] This method has come to be called the **Benkeser reduction**,[520] and it is a useful modification because Birch reductions with sodium are difficult when run on kilogram scale, whereas the Benkeser reduction can give better yields of products. Addition of the proton source *tert*-butanol allows reduction of benzene rings to stop at the diene stage, which is useful for Birch reduction of aromatic rings.[521] Reduction of anisole with calcium under these conditions gave 86% of **510**, in addition to minor amounts of other reduction products.

Just as lithium and sodium can be used in low molecular weight amines for reduction of alkynes,[522] calcium is an effective reducing agent in amine solvents. Reaction of 2-nonyne with a calcium-methylamine-ethylenediamine mixture gave 87% yield of an 86:8:4 mixture of *trans*-2-nonene/*trans*-3-nonene/*trans*-4-nonene, the latter two products arising by bond isomerization.[523]

This reagent is also effective for reduction of epoxides. Lithium aluminum hydride reduction of **526** was very sluggish and required refluxing in *N*-methylmorpholine to obtain an 84% yield of **527**. Along with **527**, 16% of the rearranged alcohol **528** was isolated.[524] Brown showed that reduction with lithium in diethylamine gave 64% of **527** and 35% of **528** after addition of **526** to the amine solution after one minute.[525] If **526** was added after five minutes, however, only 10% of **527** and 3% of **529** was obtained along with unreacted starting material.[526] When lithium was used with ethylenediamine, 87% of a 99.3:0.2:0.5 mixture of **527/528/529** was obtained.[526] Benkeser showed that reduction of **526** with calcium in ethylenediamine gave an 89% conversion to **527** at ambient temperatures.[526] This method is preferred for reduction of epoxides when LiAlH$_4$ fails or reacts very slowly.

517.   Kaiser, E.M.; Benkeser, R.A. *Org. Synth. Coll. Vol. 6* **1988**, 852.
518.   Benkeser, R.A.; Kang, J. *J. Org. Chem.* **1979**, *44*, 3737.
519.   Benkeser, R.A.; Belmonte, F.G.; Kang, J. *J. Org. Chem.* **1983**, *48*, 2796; (b) Benkeser, R.A.; Belmonte, F.G.; Yang, J. *Synth. Commun.* **1983**, *13*, 1103.
520.   The Merck Index, 14th ed., Merck & Co., Inc., Whitehouse Station, New Jersey, **2006**, p ONR-8.
521.   Benkeser, R.A.; Laugal, J.A.; Rappa, A. *Tetrahedron Lett.* **1984**, *25*, 2089.
522.   (a) Campbell, K.N.; Eby, L.T. *J. Am. Chem. Soc.* **1941**, *63*, 216, 2683; (b) Reference 489; (c) Benkeser, R.A.; Schroll, G.; Sauve, D.M. *Ibid.* **1955**, *77*, 3378.
523.   Benkeser, R.A.; Belmonte, F.G. *J. Org. Chem.* **1984**, *49*, 1662.
524.   Kwart, H.; Takeshita, T. *J. Org. Chem.* **1963**, *28*, 670.
525.   Brown, H.C.; Ikegami, S.; Kawakami, J.H. *J. Org. Chem.* **1970**, *35*, 3243.
526.   Benkeser, R.A.; Rappa, A.; Wolsieffer, L.A. *J. Org. Chem.* **1986**, *51*, 3391.

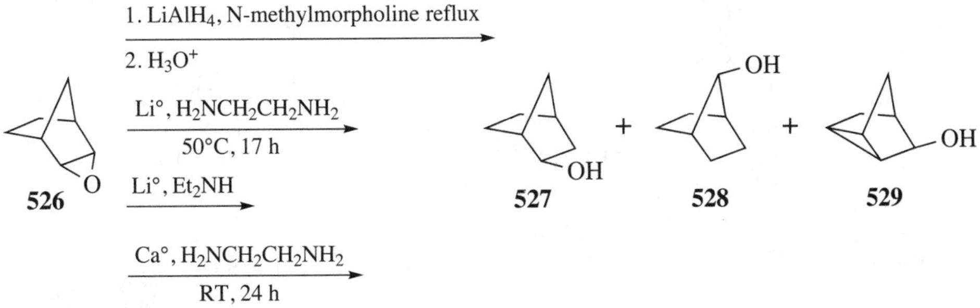

4.9.G. Reduction with Zinc

Zinc is an important metal for reduction of organic molecules,[527] and zinc dust in acetic acid or HCl[528] is probably the most common reagent. Zinc in acetic acid is a very effective reagent for reduction of α-haloketones, as in Deprés, Greene, and co-worker's synthesis of bakkenolide A[529] where dichloride **530** was converted to **531** in 63% yield. Alkyl bromides and chlorides react with zinc and acetic acid very slowly when the halogen is *not* conjugated to a carbonyl. Reduction of primary iodides is relatively fast, and this method is commonly used to insert methyl groups into a synthetic target.[530] Another reaction is observed when a molecule contains two halogen atoms on adjacent carbon atoms (a vicinal dihalide). Reductive dehalogenation to the alkene is a useful process, as seen in the Inoue, Hirama, and Sato synthesis of merrilactone A in which vicinal dichloride **532** was converted to **533**.[531] Similar eliminations are observed when sodium and ammonia[532] or a zinc-titanium tetrachloride (TiCl$_4$) reagent was used.[533]

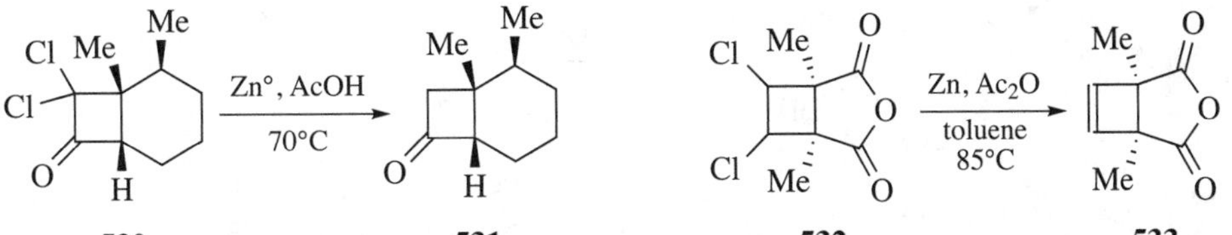

Zinc in acetic acid leads to 1,4-reduction of conjugated ketones such as **534**,[534] in this case giving the saturated ketone **535** in Danishefsky and Mantlo's synthetic studies on calicheamicin. This conjugated system is quite sensitive, but Zn/AcOH gave > 82% of the desired conjugate reduction product **535**.

527. Reference 7, pp 28-29.
528. Reference 15, pp 1276-1284.
529. Brocksom, T.J.; Coelho, F.; Deprés, J.-P.; Greene, A.E.; de Lima, M.E.F.; Hamelin, O.; Hartmann, B.; Kanazawa, A.M.; Wang, Y. *J. Am. Chem. Soc.* **2002**, *124*, 15313.
530. Yamada, K.; Nagase, H.; Hayakawa, Y.; Aoki, K.; Hirata, Y. *Tetrahedron Lett.* **1973**, 4963.
531. Inoue, M.; Sato, T.; Hirama, M. *J. Am. Chem. Soc.* **2003**, *125*, 10772.
532. Schubert, W.M.; Rabinovitch, B.S.; Larson, N.R.; Sims, V.A. *J. Am. Chem. Soc.* **1952**, *74*, 4590.
533. Sato, F.; Akiyama, T.; Iida, K.; Sato, M. *Synthesis* **1982**, 1025.
534. Mantlo, N.B.; Danishefsky, S.J. *J. Org. Chem.* **1989**, *54*, 2781.

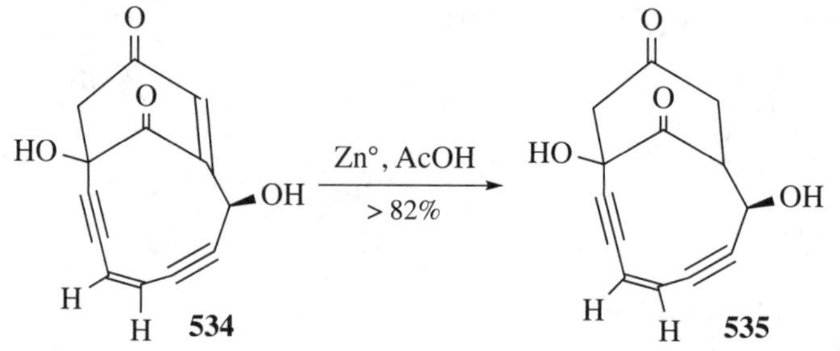

Clemmensen reduction[535] is the classical method for removing a ketone carbonyl via reaction with zinc amalgam (Zn/Hg) in HCl. An example is taken from a synthesis of mutisianthol by Ferraz and co-workers,[536] in which the ketone unit in **536** was removed with zinc-amalgam in HCl to give **537**, in 68% yield. Another use of this reduction is for removal of a hydroxyl group from an acyloin such as **538** (sec. 13.8.E), to give cyclodecanone, **539**.[537] Cram introduced a useful modification of this latter reduction. Reaction of the acyloin **540** with acidic zinc chloride, in the presence of 1,3-propanedithiol,[538] reduced the alcohol via hydrogenolysis, and converted the carbonyl to a dithioketal (**541**) in 82% yield. The dithiane moiety was reduced to **542** in 85% yield with Raney nickel in Cram's example, but hydrolysis to the ketone with aqueous mercuric salts is also possible (sec. 7.3.B.ii).

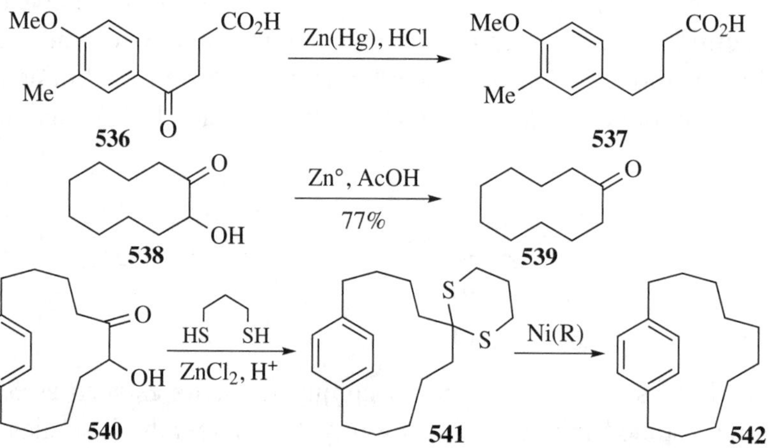

It was previously mentioned that zinc reduces alkynes to alkenes. Näf showed that reduction of the conjugated alkyne (**543**) to the conjugated cis alkene (in **544**) occurred with zinc in the presence of potassium cyanide (KCN). This reduction was not possible by hydrogenation

535. (a) Clemmensen, E. *Berichte* **1913**, *46*, 1837; (b) Vedejs, E. *Org React.* **1975**, *22*, 401; (c) The Merck Index, 14th ed., Merck & Co., Inc., Whitehouse Station, New Jersey, **2006**, p ONR-18; (d) Mundy, B.P.; Ellerd, M.G.; Favaloro Jr., F.G. *Name Reactions and Reagents in Organic Synthesis, 2nd ed.*, Wiley-Interscience, New Jersey, **2005**, pp. 160-161.
536. Ferraz, H.M.C.; Aguilr, A.M.; Silva Jr., L.F. *Tetrahedron* **2003**, *59*, 5817.
537. Prelog, V.; Schenker, K.; Günthard, H.H. *Helv. Chim. Acta* **1952**, *35*, 1598.
538. Cram, D.J.; Cordon, M. *J. Am. Chem. Soc.* **1955**, *77*, 1810.

with a Lindlar catalyst, due to the poor selectivity for reaction with only the alkyne without reduction of any alkenes.[539] The yields obtained by zinc reduction were not always reproducible, however.[539b]

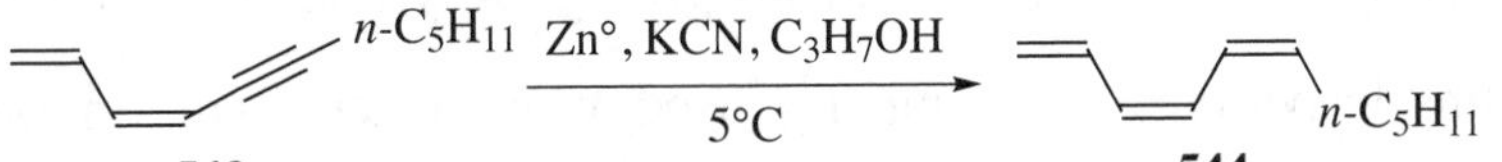

543 → **544**

Zn°, KCN, C₃H₇OH, 5°C

The functional group transforms available with zinc are

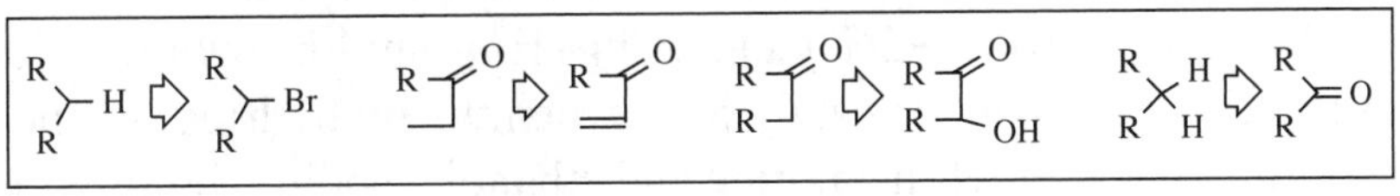

4.9.H.   Reduction with Tin and Tin Compounds

Tin metal is used sparingly as a reducing metal in modern organic synthesis, but tin derivatives such as tributytin hydride are common. Aldehydes, for example, are readily reduced with tributyltin hydride (Bu_3SnH) in aqueous media. An example is the reduction of 2-naphthaldehyde to the alcohol (**392**) in 93% yield, in refluxing methanol.[540] When this reaction was done in water, an 85% yield of **392** was obtained.

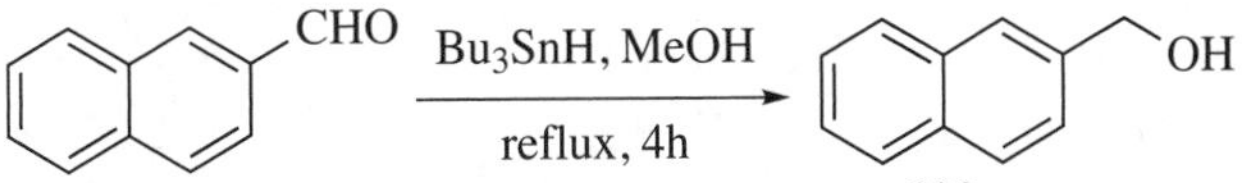

Bu_3SnH, MeOH

reflux, 4h

392

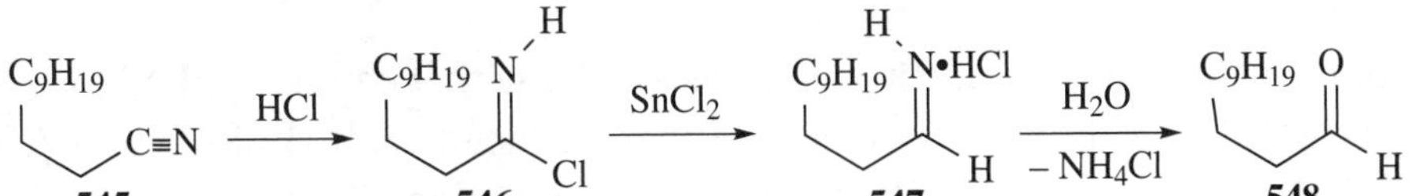

545   **546**   **547**   **548**

For the reduction of other functional groups with tin or a derivative, heating with HCl is often required and in such cases other reducing methods are usually more convenient. Typical applications include reduction of acyloins and aromatic nitro compounds.[541] A classical use of tin that is still used occasionally is the so-called **Stephen reduction**, where a nitrile is converted to the corresponding aldehyde.[542] An example is the reduction of lauronitrile (**545**) to give lauraldehyde (**548**), when treated with a stannous chloride ($SnCl_2$) solution that is saturated with anhydrous HCl.[543] The reaction proceeds by initial addition of HCl to the nitrile to give

539.   (a) Näf, F.; Decorzant, R.; Thommen, W.; Wilham, B.; Ohloff, G. *Helv. Chim. Acta* **1975**, *58*, 1016; (b) Fieser, L.F.; Fieser, M. *Reagents for Organic Synthesis* Vol. 6, Wiley, New York **1977**, p674.

540.   Kamiura, K.; Wada, M. *Tetrahedron Lett.* **1999**, *40*, 9059.

541.   (a) Reference 7, pp 29-30; (b) Reference 15, p1168.

542.   (a) Stephen, H. *J. Chem. Soc.* **1925**, *127*, 1874; (b) Ferguson, L.N. *Chem. Rev.* **1946**, *38*, 227 (see p243); (c) Mosettig, E. *Org. React.* **1954**, *8*, 218 (see pp 246-252).

543.   (a) Lieber, E. *J. Am. Chem. Soc.* **1949**, *71*, 2862; (b) Knight, J.A.; Zook, H.D. *Ibid.* **1952**, *74*, 4560.

546. This intermediate is reduced by stannous chloride to give **547** and SnCl$_2$ is converted to tin tetrachloride (SnCl$_4$). Subsequent treatment with aqueous HCl hydrolyzes imine **547** to aldehyde **548** in 73% overall yield. Stannous bromide (SnBr$_2$) often gives better yields of aldehyde than stannous chloride. Most aromatic nitriles give good yields of the aromatic aldehyde.[544]

As mentioned, tin derivatives are typically used more often than tin metal in synthetic applications. Trialkyltin hydrides, for example, are important for radical hydrogen transfer in reductions of alkyl, alkenyl, and aryl halides (see sec. 13.5.D).[545] In an example, taken from Marino and co-worker's[546] synthesis of (+)-aspidospermidine, the CCl$_2$ moiety of **549** was reduced to -CH$_2$- (see **550**) in 92% yield using tri-*n*-butyltin hydride. Corey and Suggs showed that this reaction can be made catalytic,[547] when the tin hydride is generated *in situ* by photolysis of a mixture of trimethyltin chloride and ethanolic sodium borohydride. Tributyltin chloride can be used as well. The reaction is not restricted to aliphatic derivatives, and vinyl bromides are reduced to the alkene with tin hydride.[548] One of the more common uses of tin hydride is as a hydrogen transfer agent in radical cyclization reactions[549] (sec. 13.7). A number of other functional group reductions have been reported with Bu$_3$SnH, including reduction of nitro groups[550] and azido groups[551] to amines.

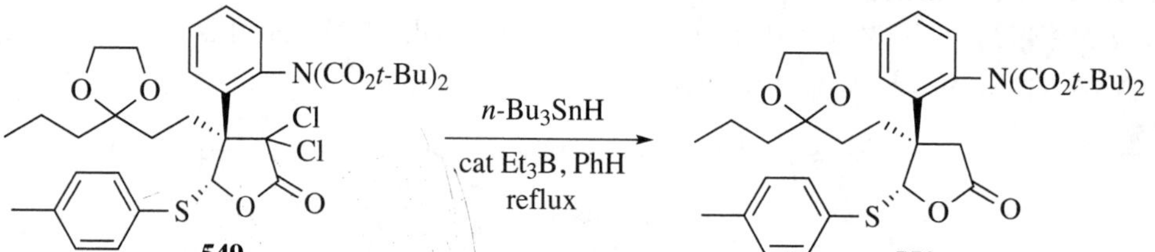

A variation of the tributyltin hydride reduction is available that can be applied to alcohols. The alcohol is converted to a thionocarbonate and then reduced with Bu$_3$SnH and AIBN, which circumvents the need for conversion of the alcohol to a halide. This transformation is known

544. Slotta, K.H.; Kethur, R. *Berichte* **1938**, *71*, 335.
545. Curran, D.P. *Synthesis* **1988**, *417*, 489.
546. Marino, J.P.; Rubio, M.B.; Cao, G.; de Dios, A. *J. Am. Chem.* **2002**, *124*, 13398.
547. Corey, E.J.; Suggs, J.W. *J. Org. Chem.* **1975**, *40*, 2554.
548. For an example taken from a synthesis of cordiachromene, see Bouzbouz, S.; Gougon, J.-Y.; Deplanne, J.; Kirschleger, B. *Eur. J. Org. Chem.* **2000**, 3223.
549. Kraus, G.A.; Hon, Y.-S. *J. Org. Chem.* **1985**, *50*, 4605.
550. Brakta, M.; Lhoste, P.; Sinou, D. *J. Org. Chem.* **1989**, *54*, 1890.
551. Wasserman, H.H.; Brunner, R.K.; Buynak, J.D.; Carter, C.G.; Oku, T.; Robinson, R.P. *J. Am. Chem. Soc.* **1985**, *107*, 519.

as the **Barton deoxygenation** or the **Barton-McCombie reaction**.[552] A synthetic example is the reduction of the OH unit in **551** to give **552** in >90% yield, from Mander and Thomson's synthesis of sordaricin.[553] In the general reaction, treatment with sodium hydride and then formation of a xanthate ester (-O-C(=S)R, a thionocarbonate), is followed by treatment with tributyltin hydride and AIBN to give reductive cleavage of the

C-O bond. Fu and co-workers have shown that the reaction can be done in the presence of polymethylhydrosiloxane using a catalytic amount of tributyltin hydride.[554]

4.9.I. Reduction with Aluminum and Aluminum Compounds

The **Meerwein-Ponndorf-Verley reduction** is a classical reaction that uses metals,[555] and it is the reverse of the **Oppenauer oxidation** discussed in Section 3.2.E.ii.[556] The mechanistic details of this reversible reaction are presented in that section.[557] The major difference between reduction or oxidation is the use of an alcohol solvent rather than a ketone solvent (usually 2-propanol rather than acetone). A ketone is treated with an aluminum alkoxide such as triisopropoxyaluminum, in the presence of a hydrogen-donating solvent such as 2-propanol. Refluxing conditions are usually required to initiate transfer of the hydrogen to the ketone, reducing it and generating acetone from 2-propanol. An uncatalyzed version of this reaction was reported to proceed in supercritical fluids.[558] A synthetic example is the reduction of **553** to give an 84% yield of the *anti*-diastereomer **554**, along with 15% of the *syn*-alcohol in Manaviazar, Hale and co-worker's synthesis of bryostatin 7.[559]

552. (a) Barton, D.H.R.; McCombie, S.W. *J. Chem. Soc. Perkin Trans. 1* **1975**, 1574; Phenyl thionocarbonates were first used by (b) Robins, M.J.; Wilson, J.S.; Hanssla, F. *J. Am. Chem. Soc.* **1983**, *105*, 4059; (c) Robins, M.J.; Wilson, J.S. *J. Am. Chem. Soc.* **1981**, *103*, 932; (d) The Merck Index, 14th ed., Merck & Co., Inc., Whitehouse Station, New Jersey, **2006**, p ONR-6; (e) Mundy, B.P.; Ellerd, M.G.; Favaloro Jr., F.G. *Name Reactions and Reagents in Organic Synthesis, 2nd ed.*, Wiley-Interscience, New Jersey, **2005**, pp. 68-69.
553. Mander, L.N.; Thomson, R.J. *Org. Lett.* **2003**, *5*, 1321.
554. Lopez, R.M.; Hays, D.S.; Fu, G.C. *J. Am. Chem. Soc.* **1997**, *119*, 6949.
555. (a) de Graauw, C.F.; Peters, J.A.; van Bekkum, H.; Huskens, J.; *Synthesis* **1994**, 1007; (b) Wilds, A.L. *Org. React.* **1947**, *2*, 178; (c) The Merck Index, 14th ed., Merck & Co., Inc., Whitehouse Station, New Jersey, **2006**, p ONR-59; (d) Mundy, B.P.; Ellerd, M.G.; Favaloro Jr., F.G. *Name Reactions and Reagents in Organic Synthesis, 2nd ed.*, Wiley-Interscience, New Jersey, **2005**, pp. 418-419.
556. Djerassi, C. *Org. React.* **1951**, *6*, 207.
557. Also see Liu, Y.-C.; Ko, B.-T.; Huang, B.-H.; Lin, C.-C. *Organometallics* **2002**, *21*, 2066.
558. Sominsky, L.; Rozental, E.; Gottlieb, H.; Gedanken, A.; Hoz, S. *J. Org. Chem.* **2004**, *69*, 1492.
559. (a) Manaviazar, S.; Frigerio, M.; Bhatia, G.S.; Hummersone, M.G.; Aliev, A.E.; Hale, K.J. *Org. Lett.* **2006**, *8*, 4477. For a comparison of Meerwein-Pondorff-Verley reduction versus LiAlH4 in a synthesis of dihydrocrinine, see (b) Irie, H.; Uyeo, S.; Yoshitake, A. *J. Chem. Soc. C* **1968**, 1802; (c) *Idem J. Chem. Soc. Chem. Commun.* **1966**, 635.

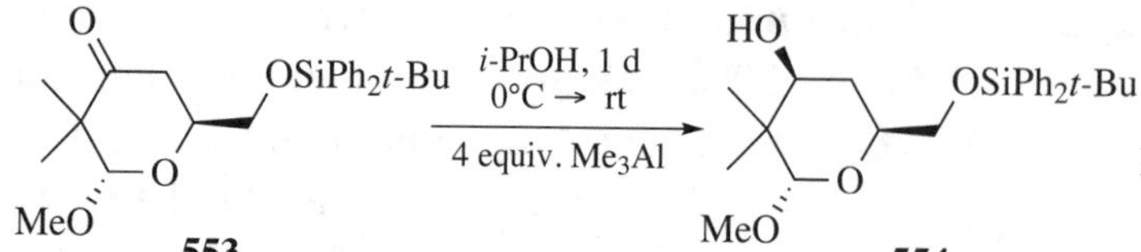

Aluminum amalgam (Al/Hg) is also a useful reagent that is commonly used for hydrogenolysis of sulfur compounds.[560] Reduction of sulfoxides is an important application,[561] as is reduction of sulfones. Aluminum amalgam reduced the α-phenylsulfonyl moiety in **555** to give lactone **556** in 68% yield, for example.[562] Aluminum amalgam has also been used for reduction of nitro compounds to the amine,[563] and reduction of azides to the amine.[564]

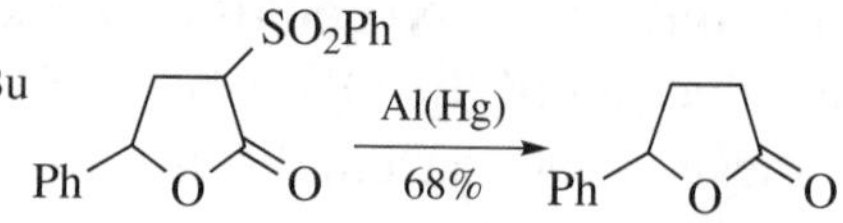

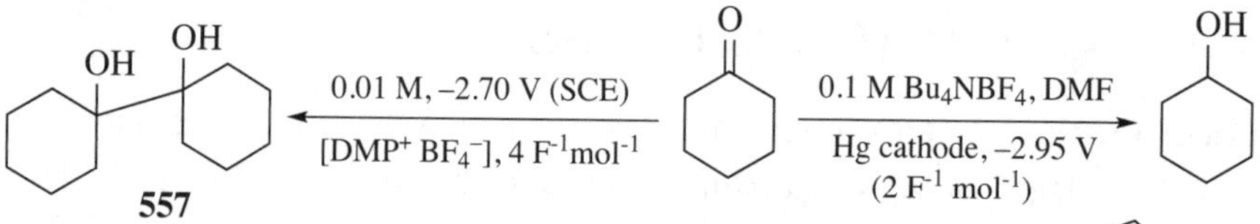

The functional group transforms for reductions with aluminum compounds are

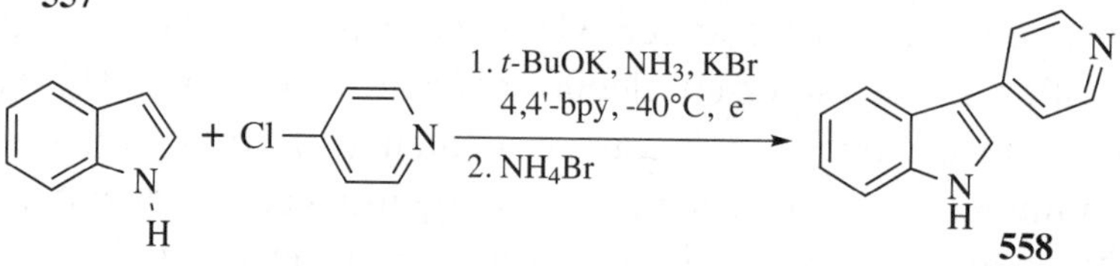

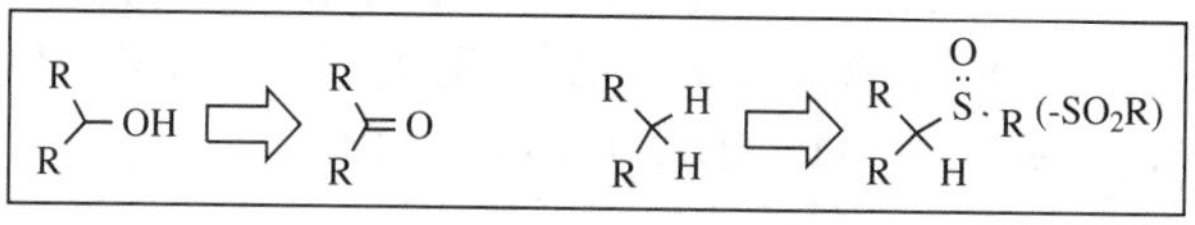

4.9.J. Electrolytic Reductions

The standard reduction potential[6] of alkali metals involves a single electron transfer.[565] Electron transfer can also be done in an electrolytic cell as described in Section 3.1. Electrolysis of acetone is an example where an electron is transferred to the carbonyl, generating a radical anion (a ketyl, see sec. 4.9.B). We have seen that the fate of these reactive species depends on the solvent, and in the case of electrolysis, whether or not the electrode potential allows an additional electron transfer and what reactive species may be present in the reaction medium.

560. Reference 15, pp 20-22.
561. Vig, O.P.; Sharma, S.D.; Matta, K.L.; Sehgal, J.M. *J. Indian Chem. Soc.* **1971**, *48*, 993.
562. Yamamoto, M.; Takemori, T.; Iwasa, S.; Kohmoto, S.; Yamada, K. *J. Org. Chem.* **1989**, *54*, 1757.
563. Fleming, I. in *New Trends in Natural Product Chemistry 1986*, Rahman, A.W.; LeQuesne, P.W. (Ed.), Elsevier, Amsterdam **1986**, p83.
564. (a) Woodward, R.B.; Heusler, K.; Gosteli, J.; Naegeli, P.; Oppolzer, W.; Ramage, R.; Ranganathan, S.; Vorbrüggen, H. *J. Am. Chem. Soc.* **1966**, *88*, 852; (b) Woodward, R.B. *Science* **1966**, *153*, 487.
565. Reference 6, pp 24-25.

There are at least two major types of reaction products available to the ketyl as introduced in Section 4.9.B. It can undergo radical coupling, or it can remove a hydrogen atom from an acidic solvent (such as ethanol or ammonia) and be reduced. The radical coupling of acetone described is formally known as **pinacol coupling** (the **pinacol reduction**)[566],[484a] and is described in Section 13.8.D.

Aldehydes usually undergo reductive dimerization upon electrolysis, as in the formation of 4,5-octanediol in 55% yield from butanal.[567] Small chain aldehydes (such as acetaldehyde) give significant amounts of reduction to the alcohol, but the proportion of dimer increases as the chain length increases. The increased amount of dimer is probably due to increased hydration of the small aldehydes and the alcohol products.

The electrochemical properties of many functional groups have been described in reviews by Steckhan,[568] Degner (industrial uses of electrochemistry),[569] Kariv-Miller,[570] and by Feoktistov.[571] Synthetic applications of anodic electrochemistry have also been reviewed.[572] There are interesting differences between dissolving metal reductions (secs. 4.9.B-G) and electrochemical reactions. By controlling the reduction potential, the nature of the electrode and the reaction medium, cyclohexanone can be reduced to cyclohexanol (sec. 4.9.B) or converted to the 1,2-diol (**557**) via pinacol coupling.[573] Presumably, the more concentrated conditions favor formation of cyclohexanol via reduction of the carbanion. More dilute solutions appear to favor the radical, with reductive dimerization to **557**. More important to this process, however, is the difference in reduction potential (–2.95 versus –2.70 V), and the transfer of two Faradays per mole in the former reaction, and four Faradays per mole in the latter.

Coupling between two different species is possible under some conditions, particularly when an aryl halide is one of the partners. When indole was treated with 4-chloropyridine, for example, under electrolytic conditions, a 60% yield of **558** was obtained[574] when 4,4'-bipyridyl (bpy) was used as a redox mediator.

Reduction of alkynes to the trans-alkene proceeds via a radical anion such as **479** (see sec. 4.9.C), and is virtually identical to similar reduction involving alkali metals (sec. 4.9.C). Electrolytic

566. Schreibman, A.A.P. *Tetrahedron Lett.* *1970*, 4271.
567. Khomyakov, V.G.; Tomilov, A.P.; Soldatov, B.G. *Elektrokhimiya,* *1969*, 5, 853 [*Chem. Abstr.* 71:108403p *1969*].
568. Steckhan, E. in *Electrochemistry I,* Steckhan, E. (Ed.), Springer-Verlag, Berlin, *1987*, pp 31-49.
569. Degner, D. in *Electrochemistry III,* Steckhan, E. (Ed.), Springer-Verlag, Berlin, *1988*, pp 40-91.
570. Kariv-Miller, E.; Palcut, R.I.; Lehman, G.K. in *Electrochemistry III*, Steckhan, E. (Ed.), Springer-Verlag, Berlin, *1988*, pp 97-130.
571. Feoktistov, L.G.; Lund, H.; Eberson, L.; Horner, L. *Organic Electrochemistry*, Baizer, M.M. (Ed.), Marcel Dekker, New York, *1973*, pp 355-366.
572. Moeller, K.D. *Tetrahedron* *2000*, *56*, 9527.
573. Kariv-Miller, E.; Mahachi, T.J. *J. Org. Chem.* *1986*, *51*, 1041.
574. Chahma, M.; Combellas, C.; Thiébault, A. *Synthesis* *1994*, 366.

reduction of but-2-yn-1,4-diol (**559**) gave an 82% yield of *(E)*-but-2-en-1,4-diol (**560**).[575] Electrolytic reduction of alkynes is sensitive to reaction conditions. Although the trans alkene is usually

formed, Hudlicky cited Campbell and Young's work[576] in which electrolysis of alkynes (in 95% EtOH + H_2SO_4) on a spongy nickel cathode gave 65-80% of the cis alkene. Similar reaction in a methylamine solution containing LiCl gave the trans alkene. The sensitivity to reaction conditions is apparent in the reduction of conjugated carbonyl systems, which are reduced electrolytically to give the 1,4-reduction product.[577] Note that electrolysis of cyclohexenone did not give reduction products but rather reductive coupling products. A difference between electrolysis and dissolving metal reductions is the ability to reduce simple double bonds, as in the electrolytic reduction of 1,5-cyclooctadiene to cyclooctene.[578] Another difference is noted in the electrolysis of conjugated acid derivatives, which give reductive coupling products.

Electrolysis of conjugated ketones generates a radical anion by addition of an electron to the conjugated system, but dimerization usually occurs by coupling at the end of the conjugated system.[579] Such dimerizations can occur even with significant steric hindrance at the terminal alkene carbon.[580] Conjugated nitriles and esters show a preference for coupling at the end of the conjugated system. 2-Methylpropenenitrile,[581] for example, was dimerized in high yield to 2,5-dimethyladiponitrile and ethyl acrylate was coupled to an aldehyde to give a γ-hydroxy ester (which cyclized to the lactone).[582]

Electrolytic conditions can be used to achieve Birch reductions (sec. 4.9.E),[583] as in the reduction of 2-methoxynaphthalene (**561**) to give **562** in 95% yield.[584] Electrolysis of heterocycles leads to interesting reduction products. Pyridine, for example,

575. Horner, L.; Röder, H. *Liebigs Ann. Chem.* **1969**, *723*, 11.
576. (a) Campbell, K.N.; Young, E.E. *J. Am. Chem. Soc.* **1943**, *65*, 965; (b) Reference 7, p 45.
577. Hull, D.A.; Heiney, R.E. (Eli Lilly) US 4 082 627 [*Chem. Abstr. 89*: 119769k **1978**].
578. Reference 568, citation 373 therein.
579. (a) Brillas, E.; Ortiz, A. *Electrochem. Acta* **1985**, *30*, 1185; (b) see Zimmer, J.P.; Richards, J.A.; Turner, J.C.; Evans, D.H. *Anal. Chem.* **1971**, *43*, 1000.
580. Bowers, K.W.; Giese, R.W.; Grimshaw, J.; House, H.O.; Kolodny, N.H.; Kronberger, K.; Roe, D.K. *J. Am. Chem. Soc.* **1970**, *92*, 2783.
581. (a) Asahi Chem. J 59 219 485 [*Chem. Abstr. 102*: 228324a **1985**]; (b) *Idem* J 60 050 190 [*Chem. Abstr. 103*: 61495g **1985**]; (c) *Idem* J 60 052 587 [*Chem. Abstr. 103*: 131244a **1985**].
582. (a) Asahi Chem. J 57 108 274 [*Chem. Abstr. 97*: 171368k **1982**]; (b) *Idem* J 58 207 383 [*Chem. Abstr. 100*: 164348e,d **1984**].
583. (a) Asahara, T.; Senö, M.; Kaneko, H. *Bull. Chem. Soc. Jpn.* **1968**, *41*, 2985; (b) Misono, A.; Osa, T.; Yamagishi, T. *Ibid.* **1968**, *41*, 2921.
584. Skatelz, D.H.O. (Hoechst) DE 2 618 276 [*Chem. Abstr. 88*: 62213x **1978**]

gave the reductive dimerization product **563** but air oxidation rapidly converted it to **564**.[585] As well as partial reduction, heteroaromatic rings can be completely reduced, as in the conversion of **565** to **566** in 95% yield in glacial acetic acid.[586]

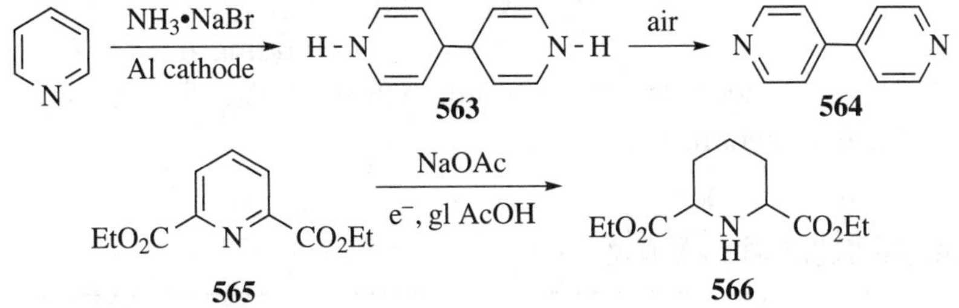

Electrochemical processes can be used for the reduction of many other functional groups. Halogen is removed, as in the reduction of 6-chloro-1-hexene to 1-hexene in 23% yield (in DMF).[587] Aryl halides are similarly reduced electrochemically. 2-Chloroaniline, for example, was reduced to aniline in 85% yield.[588] If the electrodes and solvent are changed, reductive coupling can occur.[589] Elimination is possible in some cases.[590] Epoxides are opened electrochemically, and hydrogen is often delivered to the more substituted carbon, as in the quantitative conversion of **567** to **568**.[591] Azides are reduced to the amine.[592]

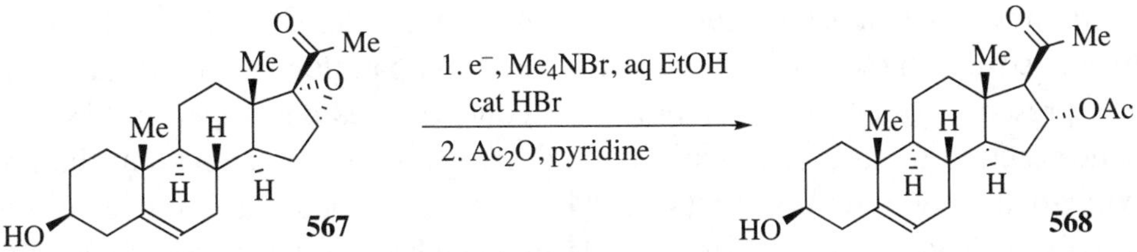

Nitro compounds are reduced to the corresponding amine. Unlike hydride reductions, both alkyl nitro compounds [2-methyl-2-nitro-1,3-propanediol was reduced to 1-amino-2-methyl-1,3-propanediol in 95% yield)][593] and aromatic nitro derivatives [nitrobenzene was reduced to aniline with $Ti(SO_4)_3$ in sulfuric acid and cetyltrimethylammonium bromide] are electrolytically

585. Asahi Chem. *Jpn. JP 81 112,488 1981* [*Chem. Abstr. 96*: 59966e *1982*].
586. (a) Hannebaum, H.; Nohe, H. Müller, H.R. (BASF) DE 2 403 446 [*Chem. Abstr. 83*: 210783x *1975*]; (b) Nohe, H.; Hannebaum, H. (BASF) DE 2 658 951 [*Chem. Abstr. 89*: 109092v *1978*]; (c) Kormasch, V.V.; Kormachev, V.V.; Trentovskaya, L.K.; Abramov, I.A.; Shevnitsyn, L.S.; Danilov, S.D. SU 844 616, [*Chem. Abstr. 95*: 150443x *1981*].
587. Lund, H.; Michel, M.A.; Simonet, J. *Acta Chem. Scand. 1975, 29B*, 217.
588. Seiler, J.N.; Strigham, R.R. (Dow) US 3 677 916 [*Chem. Abstr. 78*: 10920z *1973*].
589. Baldwin, M.M.; Wyant, R.E. (Battelle Dev. Corp) US 4 434 032 [*Chem. Abstr. 100*: 164313; *1984*].
590. Torii, S.; Tanaka, H.; Sasaoka, M.; Kameyama, Y. (Otsuka Chem. Co.) DE 3 507 592 [*Chem. Abstr. 104*: 148628x *1986*].
591. Kamernitskii, A.V.; Reshetova, I.G.; Chernoburova, E.I.; Mairanovskii, S.G.; Lisitsina, N.K. *Izv. Akad. Nauk. SSSR Ser. Khim 1984*, 1901 (Engl. p 1736).
592. Kuwabata, S.; Tanaka, K.; Tanaka, T. *Inorg. Chem. 1986, 25*, 1691.
593. Rignon, M.; Catonne, J.C.; Denisard, F.; Malafosse, J. EP 198722 [FR 2,577,242 *Chem. Abstr. 108*:111810g *1988*].

reduced.[594] If conditions are modified, reductive coupling can give azoxy compounds such as **569** (from **570**)[595] or diazo compounds.[596] A variety of acid derivatives are reduced under electrochemical conditions, including nitriles (to amines),[597] acids (to alcohols),[598,599] esters (to alcohols),[600] and amides (to alcohols).[601] It is possible to selectively reduce a cyclic imide to a lactam.[602]

4.10. NONMETALLIC REDUCING AGENTS

There are several methods for reducing organic functional groups that use nonmetallic reagents. The mechanism of reduction varies with each reagent, and each has a rather specific and useful application. This section is presented as a survey of these methods and is not intended as an in-depth analysis. The intent is to offer alternatives for the reduction of functional groups when more traditional methods fail.

4.10.A. Reduction with Hydrazine (Wolff-Kishner)

When a carbonyl group must be removed from a molecule, the acidic conditions of the **Clemmensen reduction** (sec. 4.9.G) are not always compatible with other functional groups that may be present. An alternative method of reduction was developed in basic media that relied on the reaction of carbonyls with hydrazine to form a hydrazone. In the presence of base, a hydrazone anion intermediate (sec. 9.4.F) can be formed that removes a proton from the acidic solvent leading to reduction. This process has been termed the **Wolff-Kishner**

594. Noel, M.; Anatharaman, P.N.; Udupa, H.U.K. *Electrochem. Acta* **1980**, *25*, 1083.
595. deGroot, H.; van de Henvel, E.; Barendrecht, E., Janssen L.J.J. DE 3 020 846 [*Chem. Abstr. 94*: 54934s **1981**].
596. (a) Kovsman, E.P.; Freidlin, G.N.; Soldatov, B.G.; Motsak, G.V. SU 415 969, 484 745 and 533 590 [*Chem. Abstr. 86*: 23589p, 23588n and 120987z **1977**]; (b) *Idem* SU 655 701 [*Chem. Abstr. 91*: 20129f **1979**].
597. (a) Kirilyus, I.V.; Murzatova, G.K.; Frangulyan, G.A.; Polievktov, M.K. (AS Kaza Chem. Metall.) SU 777 024 [*Chem. Abstr. 94*: 174610e **1981**]; (b) also see Toomey, J.E. (Reilly Tar & Chemical Corp.) US 4 482 437 [*Chem. Abstr. 102*: 53 067h **1985**].
598. Takenaka, S.; Shimakawa, C. (Mitsui Toatsu Chem.) J 60 243 293 and J 60 234 987 [*Chem. Abstr. 104*: 138283s and 98055k **1986**].
599. Beck, F.; Jäger, P.; Guthke, H. (BASF) DE 1 950 282 [*Chem. Abstr. 74*: 140980x **1971**].
600. Horner, L.; Skaletz, D.H.; Hönl, H. (Hoechst) DE 2 428 878 [*Chem. Abstr. 84*: 105228t **1976**].
601. (a) Heitz, E.; Kaiser, U. (Dechema), DE 2 301 032, [*Chem. Abstr. 81*: 130200y **1974**]; (b) Toomey, J.E. (Reily Tar and Chemical Corp.), EP 189 678 [*Chem. Abstr. 105*: 215760v **1986**].
602. (a) Pliva Pharm J 77 116 464 [*Chem. Abstr. 88*: 152418x **1978**]; (b) Crnic, Z.; Djokic, S.; Gaspert, B.; Vajtner, Z.; Maasboel, A. (Pliva Corp.), AU 355 013 [*Chem. Abstr. 93*: 8010u **1980**].

reduction.[603] There are several modifications of this procedure, including isolation of the hydrazone, and using an alkoxide base.[604] Perhaps the most important change in this reaction is the **Huang-Minlon modification**, which uses refluxing diethylene glycol as the solvent and has been shown to be much more efficient for this transformation.[605] Reduction of the aldehyde unit in **571** using this modification gave **572** in 57% yield in Taber and co-worker's synthesis

of (–)-calicoferol B.[606] Alternatives are available. Hydrazine hydrochloride ($N_2H_4 \cdot HCl$) has been shown to be effective for the reduction of sterically hindered ketones.[607] Hydrazine is also used to convert an alkyl phthalimide to an alkyl amine by replacing the H_2C-NR-CH_2 moiety with a $O=C$–$NHNH$–$C=O$ unit, in what is often called the **Ing-Manske procedure**.[608]

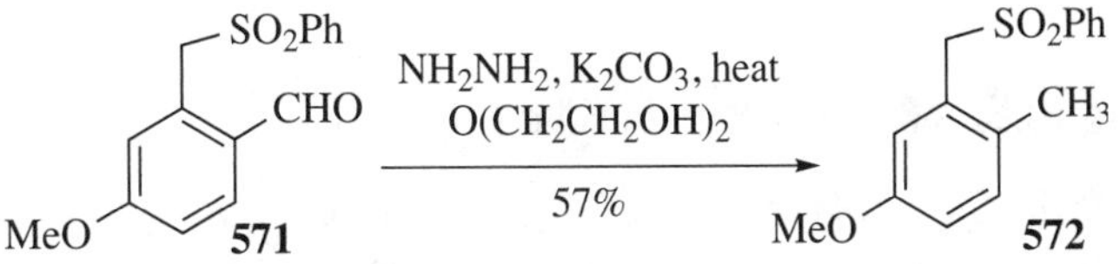

There is an alternative to the Wolff-Kishner or Clemmensen reductions (Sec 4.9.G) of ketones and aldehydes. Reaction of a ketone or aldehyde with 1,3-propanedithiol, in the presence of a Lewis acid, gives the corresponding dithiane (see **573**; Sec 7.3.B.ii). Subsequent treatment with Raney nickel, usually in ethanol or methanol, leads to desulfurization, as shown. Dithiane, dithiolanes, or dithioketals or dithioacetals can be used in this transformation. An example is taken from Pu and Ma's synthesis of (–)-alkaloid 223A,[609] in which **574** was converted to the dithiolane (**575**), and subsequent heating with Raney nickel and hydrogen in isopropanol gave the desulfurized product **576**. See section 8.6 for other reactions of sulfur compounds.

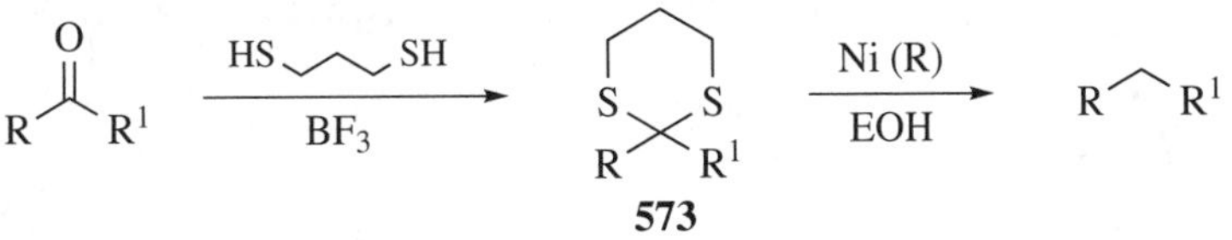

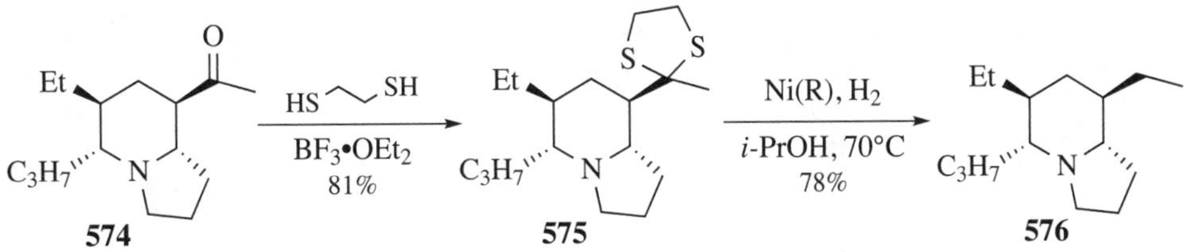

603. (a) Kishner, N. *J. Russ. Phys. Chem. Soc.* **1911**, *43*, 582; (b) Wolff, L. *Annalen*, **1912**, *394*, 86; (c) Todd, D. *Org. React.* **1948**, *4*, 378; (d) The Merck Index, 14th ed., Merck & Co., Inc., Whitehouse Station, New Jersey, **2006**, p ONR-103; (e) Mundy, B.P.; Ellerd, M.G.; Favaloro Jr., F.G. *Name Reactions and Reagents in Organic Synthesis, 2nd ed.*, Wiley-Interscience, New Jersey, **2005**, pp. 704-705.
604. (a) Grudnon, M.F.; Henbest, H.B.; Scott, M.D. *J. Chem. Soc.* **1963**, 1855; (b) Sobti, R.R.; Dev, S. *Tetrahedron* **1970**, *26*, 649.
605. (a) Huang-Minlon *J. Am. Chem. Soc.* **1946**, *68*, 2487; (b) *Idem Ibid.* **1949**, *71*, 3301.
606. Taber, D.F.; Jiang, Q.; Chen, B.; Zhang, W.; Campbell, C.L. *J. Org. Chem.* **2000**. *67*. 4821.
607. Nagata, W.; Itazaki, H. *Chem. Ind.* **1964**, 1194.
608. Ing, H.R.; Manske, R.H.F. *J. Chem. Soc.* **1926**, 2348.
609. Pu, X.; Ma, D. *J. Org. Chem.* **2003**, *68*, 4400.

The functional group transform for reduction with hydrazine or of the dithiane is

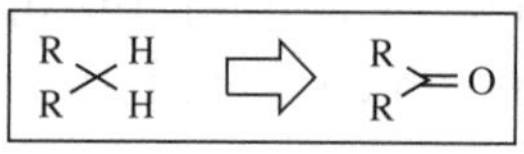

4.10.B. Reduction with Diimide

Diimide (NH=NH) is a transient species generated by reaction of acids with potassium azodicarboxylate,[610] by thermolysis of anthracene-9,10-diimine,[611] from hydrazine derivatives[555] or tosylhydrazone.[612] Diimide reduces isolated C=C units to the corresponding saturated derivatives (i.e., C=C → –CH$_2$CH$_2$–).[613] It is known that diimide gives primarily cis reduction of alkenes[614] and *reduces symmetricall π bonds faster than polarized π bonds*.[615] Generation of diimide with potassium azodicarboxylate gave selective reduction at the least conjugated alkene in **577** to give **578** in 77% yield, an important intermediate in Corey and co-worker's synthesis of ovalicin.[616] The conjugated ketone and exocyclic epoxide were not reduced. In a synthesis of frondosin C, Ovaska and co-worker's reduced a single alkene unit in **579** to give **580** in 70% yield using diimide generated *in situ* from tosylhydrazine.[617] Corey showed that the course of a reduction with diimide could be altered by addition of cupric salts [such as cupric acetate, Cu(OAc)$_2$].[618] This new reagent is highly selective for the reduction of conjugated alkenes, with specificity for the less substituted double bond of the diene.[619]

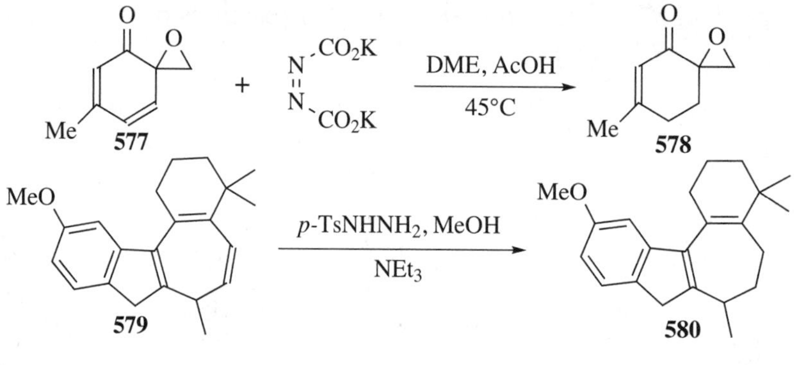

610. Adam, W.; Eggelte, H.J. *J. Org. Chem.* **1977**, *42*, 3987.
611. (a) Reference 7, pp 33-34; (b) Reference 15, pp 257-258.
612. For an example taken from the synthesis of aplyronines, see Marshall, J.A.; Johns, B.A. *J. Org. Chem.* **2000**, *65*, 1501.
613. See Reference 17, p 12.
614. (a) Hünig, S.; Müller, H.-R.; Thier, W. *Tetrahedron Lett.* **1961**, 353; (b) van Tamelen, E.E.; Timmons, R.J. *J. Am. Chem. Soc.* **1962**, *84*, 1067.
615. van Tamelen, E.E.; Dewey, R.S.; Lease, M.F.; Pirkle, W.H. *J. Am. Chem. Soc.* **1961**, *83*, 4302.
616. Corey, E.J.; Dittami, J.P. *J. Am. Chem. Soc.* **1985**, *107*, 256.
617. Li, X.; Kyne, R.E.; Ovaska, T.V. *Org. Lett.* **2006**, *8*, 5153.
618. (a) Corey, E.J.; Mock, W.L.; Pasto, D.J. *Tetrahedron Lett.* **1961**, 347; (b) Corey, E.J.; Hortmann, A.G. *J. Am. Chem. Soc.* **1965**, *87*, 5736.
619. For an example, see Corey, E.J.; Yamamoto, H. *J. Am. Chem. Soc.* **1970**, *92*, 6636.

The functional group transform is

4.10.C. Reduction with Silanes (Hydrosilylation)

Alkyl silanes can be used for the reduction of carbonyls and alkenes. Methylcyclohexene was reduced to methylcyclohexane using a mixture of triethylsilane (Et_3SiH) and trifluoroacetic acid (CF_3CO_2H), in 72% yield.[620] Under the same conditions, however, 1-pentene was not reduced. This reagent is often used for reduction of conjugated carbonyls, probably via formation of a silyl enolate (secs. 9.2, 9.3.B), as in the reduction of cyclohexenone to cyclohexanone in 85% yield with Ph_2SiH_2.[621] Transition metals such as $ZnCl_2$, rhodium derivatives or copper salts can be added to the silane to facilitate the reduction.[622] An example, taken from a synthesis of (+)-manoalide by Kocienski and co-workers, reduced the conjugated double bond in β-ionone (**581**) to give **582** in 90% yield using the two-step procedure shown.[623] Conjugated aldehydes tend to give 1,2-reduction, as in the reduction of 3-phenyl-prop-(2E)-enal to 3-phenylprop-(2E)-en-1-ol in 95% yield.[624] Another example of such selectivity is the reduction of the ketone unit in 2-bromo-1-phenylpropanone to give 2-bromo-1-phenyl-1-propanol in 70% yield.[570] Non-conjugated ketones and aldehydes are reduced by silanes in acid media, as in the reaction of cyclohexanone with Et_3SiH and trifluoroacetic acid to give cyclohexanol in 74% yield.[620] Ketones are usually reduced faster than epoxides.[625,626] The use of a chiral additive gives asymmetric reduction with silanes. Reduction of acetophenone gave (R)-phenethyl alcohol in 99% yield and 84.2% ee, in the presence of **583**.[627]

Reductive cleavage of carbonyls or alcohols, analogous to the Wolff-Kishner or Clemmensen

620. Kursanov, D.N.; Parnes, Z.N.; Bassova, G.I.; Loim, N.M.; Zdanovich, V.I. *Tetrahedron* **1967**, *23*, 2235.
621. Sharf, V.Z.; Freidlin, L.Kh.; Shekoyan, I.S.; Krutii, V.N. *Bull. Akad. USSR Chem.* **1977**, *26*, 995.
622. Ojima, I.; Kogure, T. *Organometallics* **1982**, *1*, 1390.
623. Pommier, A.; Stepanenko, V.; Jarowicki, K.; Kocienski, P.J. *J. Org. Chem.* **2003**, *68*, 4008.
624. Boyer, J.; Corriu, R.J.P.; Perz, R.; Réyé, C. *J. Chem. Soc. Chem. Commun.* **1981**, 121.
625. Mullholland, Jr., R.L.; Chamberlin, A.R. *J. Org. Chem.* **1988**, *53*, 1082.
626. Salomon, R.G.; Sachinvala, N.D.; Raychaudhuri, S.R.; Miller, D.B. *J. Am. Chem. Soc.* **1984**, *106*, 2211.
627. Brunner, H.; Riepl, G.; Weitzer, H. *Angew. Chem. Int. Ed.* **1983**, *22*, 331.

reduction, is illustrated by conversion of cyclohexanone to cyclohexane in 90% yield,[628] using a silane•BF_3 complex. Other modifications, such as addition of cesium fluoride, give reduction of an ester moiety.[629]

The functional group transforms observed in this section are

4.10.D. Reduction with Formic Acid

Formic acid has been used in a variety of rather specific reductions. Heating triphenylcarbinol (584) in formic acid, for example, led to triphenylmethane.[630] The most common application of formic acid reduction involves enamines, presumably via conversion to an iminium salt, which is the actual species reduced by formic acid.[631]

Reduction of iminium salts proceeds by hydride transfer, with formation of carbon dioxide.[632] Reduction of 3-(–)-*erythro*-2-(*N*-methyl)pyrrolidiniminiumbutanoic acid (585) to 587 illustrates that the reaction can be stereoselective. The hydride transfer process is represented by 586, and the stereochemistry of 587 is predicted by analysis with Cram's rule (sec. 4.7.B).[633] The hydride approaches from the more sterically accessible face. A proton catalyzed tautomerization converts enamines to iminium salts, which are then reduced to the corresponding amine.[634] Reduction of iminium salts is apparently responsible for the classical reductive methylation procedure of amines and formaldehyde in the presence of formic acid. The methylation protocol is illustrated by the *N*-methylation of the secondary amine moiety in 588 with formic acid and formaldehyde, to give a 99% yield of 589 in Tran and Kwon's synthesis of alstonerine.[635]

628.   Fry, J.L.; Orfanopoulos, M.; Adlington, M.G.; Dittman, Jr., W.R.; Silverman, S.B. *J. Org. Chem.* **1978**, *43*, 374.
629.   Boyer, J.; Corriu, R.J.P.; Perz, R.; Poirer, M.; Réyé, C. *Synthesis* **1981**, 558.
630.   (a) Kaufmann, H.; Pannwitz, P. *Berichte* **1912**, *45*, 766; (b) Kovache, A. *Ann. Chim.* **1918**, *10*, 184.
631.   Paukstelis, J.V.; Kuehne, M.E. in *Enamines: Synthesis Structure and Reactions*, Cook, A.G. (Ed.), Marcel-Dekker, New York **1969**, pp 169-210 and 313-468.
632.   (a) Lukeš, R.; Jizba, J. *Chem. Listy* **1953**, *47*, 1366 [*Chem. Abstr.* **49**: 323g, **1955**]; (b) *Idem Collect. Czech. Chem. Commun.* **1954**, *19*, 941, 930; (c) Leonard, N.J.; Sauers, R.R. *J. Am. Chem. Soc.* **1957**, *79*, 6210.
633.   Červinka, O. *Collect. Czech. Chem. Commun.* **1959**, *24*, 1880.
634.   (a) Leonard, N.J.; Thomas, P.D.; Gash, V.W. *J. Am. Chem. Soc.* **1955**, *77*, 1552; (b) Bonnett, R.; Clark, V.M.; Giddey, A.; Todd, A. *J. Chem. Soc.* **1959**, 2087.
635.   Tran, Y.S.; Kwon, O. *Org. Lett.* **2005**, *7*, 4289.

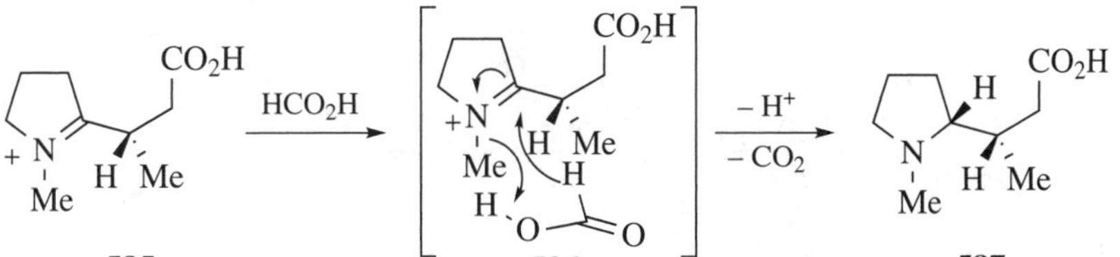

Several modifications of this reduction involve formates and are effective for reduction of various functional groups. A mixture of formic acid and ethyl magnesium bromide was used to reduce decanal to decanol in 70% yield.[636] Decanal was also reduced to decanol in 69% yield by using sodium formate in N-methyl-2-pyrrolidinone as a solvent.[637] Functional groups other than carbonyl derivatives can be reduced under relatively mild conditions with formate derivatives. Ammonium formate, in the presence of palladium on carbon was used to reduce an azide to a primary amine.[638] Aliphatic nitro compounds are also converted to an amine with this reagent.[639]

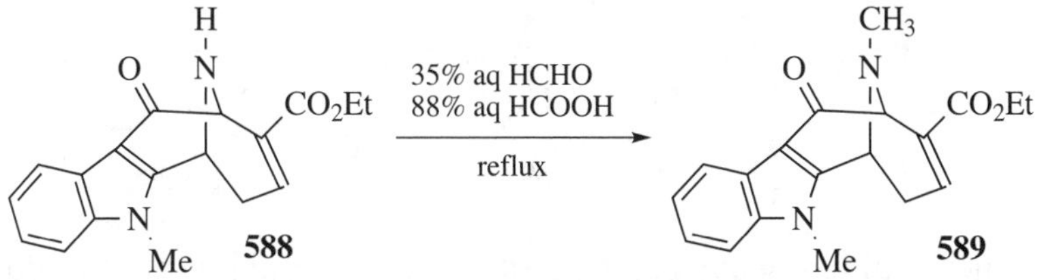

4.10.E. Photoreduction

Electron transfer to certain functional groups can be induced by irradiation with light. When combined with a sensitizer and a hydrogen atom donor, reductions are possible. The basics of photochemical techniques and the use of sensitizers are discussed in Sections 11.10.B and 13.4.

In the presence of a hydrogen atom donor (usually an alcohol or an amine), carbonyl derivatives or halides can be photochemically reduced. Photolysis of ketone **590** gave the benzocyclobutane **591**,[640] but photolysis in the presence of *sec*-butylamine gave alcohol **592** as the product. Photoreduction is often less stereoselective than other techniques. Both metal hydride reduction and catalytic hydrogenation of **593** gave exclusively the endo-alcohol **595**, whereas dissolving metal reduction gave 47% of the exo-isomer **594**.[641] Photoreduction of **593** gave a 54% yield of a 4:1 mixture of **594** and **595**.[641]

636. Babler, J.H.; Invergo, B.J. *Tetrahedron Lett.* **1981**, *22*, 621.
637. Babler, J.H.; Sarussi, S.J. *J. Org. Chem.* **1981**, *46*, 3367.
638. Gartiser, T.; Selve, C.; Delpeuch, J.-J. *Tetrahedron Lett.* **1983**, *24*, 1609.
639. Ram, S.; Ehrenkaufer, R.E. *Tetrahedron Lett.* **1984**, *25*, 3415.
640. Ito, Y.; Kawatsuki, N.; Matsuura, T. *Tetrahedron Lett.* **1984**, *25*, 2801.
641. Momose, T.; Muraoka, O.; Masuda, K. *Chem. Pharm. Bull.* **1984**, *32*, 3730.

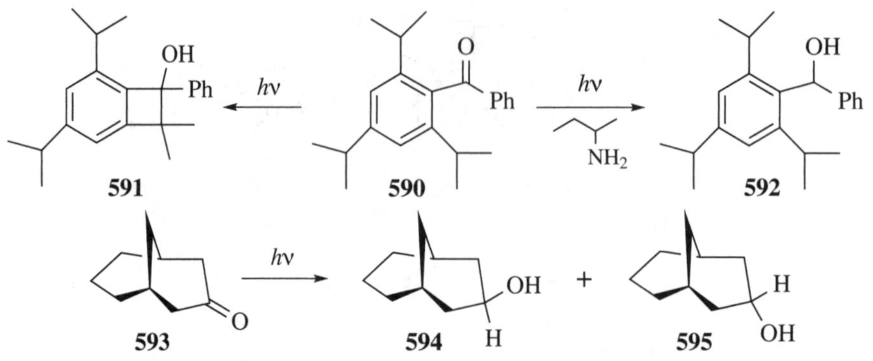

591   590   592

593   594   H   +   595   OH

Competing photochemical pathways can be a problem when using photoreduction (sec. 13.4), including coupling, isomerization or migration of multiple bonds. Functional groups with larger quantum yields will react preferentially but this is very dependent on the wavelength of the light used for irradiation (sec. 11.10.).

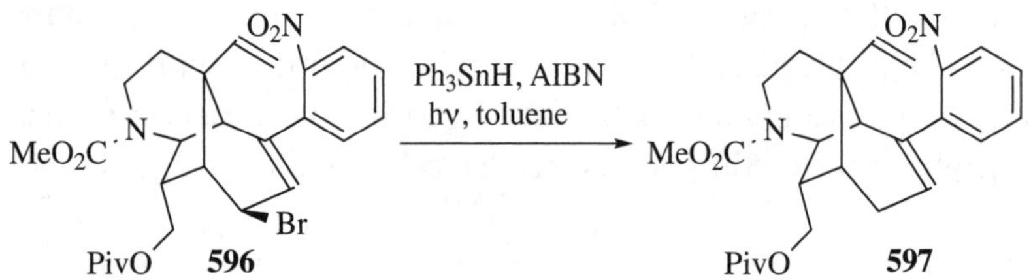

Reduction of halides is a useful application of this process and there are variations that make it more useful. In a synthesis of gelsemine, Danishefsky and co-workers showed that irradiation of bromide **596**, in the presence of the radical initiator AIBN (see sec. 13.4 and 13.3) and triphenyltin hydride led to **597** in 52% yield.[642] The reaction proceeds by formation of a carbon radical from the C-Br unit, and hydrogen transfer from the triphenyltin hydride (see sec. 13.5.D) to give **597**.

Photoreduction of imines has been used in the synthesis of various alkaloids. Irradiation of $\Delta^{1,9}$-octa-hydroquinoline (**598**) led to a 98% yield of *trans*-octahydroquinoline (**600**). Hydrogen is transferred to the nitrogen (2-propanol is the hydrogen transfer agent), leading to the thermodynamically more stable radical (**599**).[643] A second hydrogen transfer delivers the hydrogen to the bottom face of **599** to give **600**.

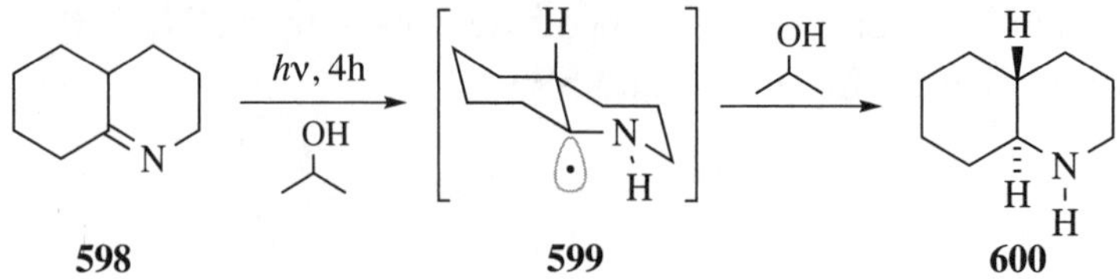

598   599   600

---

642.   Ng, F.W.; Lin, H.; Danishefsky, S.J. *J. Am. Chem. Soc.* **2002**, *124*, 9812.
643.   Hornback, J.M.; Proehl, G.S.; Starner, I.J. *J. Org. Chem.* **1975**, *40*, 1077.

4.10.F. Enzymatic Reductions

Enzymes, and biocatalysts in general, are increasingly available and important tools for the reduction of organic compounds.[644] Enzymatic reductions are often straightforward and highly stereoselective. There are now many enzymatic transformations that are compatible with the use of organic solvents,[645] and the alcohol dehydrogenase obtained from *Geotrichum candidum* is active in supercritical carbon dioxide.[646] Prelog studied the reduction of ketones with several enzymatic systems. Reduction of ketones with *Curvularia fulcata*, for example, gave predictable stereochemical induction based on identifying Large (L) and Small (S) groups around the carbonyl (this method is sometimes called **Prelog's rule**).[647]

If the steric difference between (L) and (S) in **601** is large enough, the enzyme delivers hydrogen from the less hindered face (over S in **601**) to give **602**. Much of the initial work in this area was done with two enzymatic systems, yeast alcohol

dehydrogenase (YAD) and horse liver alcohol dehydrogenase (HLADH). This former reagent is usually obtained from yeast and is a protein with a molecular weight of ~ 80,000, containing four subunits. The exact structure of the enzyme is not always correlated with the that portion of the structure required for reduction (the active site). Such ambiguities of structure make exact mechanistic arguments difficult. An important part of the selectivity observed with these enzymes, however, is "determined by non-bonded interactions of substrate and enzyme in the hydrogen transfer transition state".[647] Prelog argued that compounds capable of assuming a diamond lattice (sec. 1.5.B) gave higher selectivities.

A common and often used enzymatic reagent is baker's yeast (*Saccharomyces cerevisiae*), and it gives selective reduction of β-keto esters and β-diketones. In the Sih and co-workers synthesis of (*L*)-carnitine,[648] a dependence on the size of the ester group was discovered.

Reduction of ethyl acetoacetate (**388**) with baker's yeast gave the *(S)*-alcohol (**603**), but reduction of ethyl β-ketovalerate (**604**) gave the *(R)*-alcohol (**605**).[648] Sih showed the selectivity of reduction changed from *(S)*-selectivity with small chain esters to *(R)*-selectivity with long chain esters.[648] The break point for selectivity occurred at a chain length of four to five carbons [$RC(=O)$-$O(CH_n)H$, $n = 4, 5$]. Baker's yeast has been used in many synthetic applications. Hydroxyester (**603**), for example, was used in Mori's synthesis of (*S*)-(+)-sulcatol.[649] Similar, and highly selective reduction

644. Nakamura, K.; Yamanaka, R.; Matsuda, T.; Harada, T. *Tetrahedron: Asymmetry* **2003**, *14*, 2659.
645. Carrea, G.; Riva, S. *Angew. Chem. Int. Ed., 2000*, *39*, 2226.
646. Matsuda, T. ; Harada, T.; Nakamura, K. *Chem. Commun. 2000*, 1367.
647. Prelog, V. *Pure Appl. Chem. 1964*, *9*, 119.
648. Zhou, B.; Gopalan, A.S.; Van Middlesworth, F.; Shieh, W.-R.; Sih, C.J. *J. Am. Chem. Soc. 1983*, *105*, 5925.
649. Mori, K. *Tetrahedron 1981*, *37*, 1341.

of β-keto esters was observed with *Rhizopus arrhizus* mediated reactions.[650] Carbonyl units in more complex systems can be reduced, as in the reduction of **606** to **607** (84% yield with 95:5 er) in the synthesis of paraherquamide A by Williams and co-workers.[651] Baker's yeast reduction was also used for the preparation of (–)-ethyl (R)-2-hydroxyl-4-phenylbutyrate, an important intermediate for the synthesis of ACE inhibitors.[652] 1,3-Diketones are usually reduced to the β-keto alcohol. Reduction of **608** with baker's yeast, for example, gave **609**, in 50% yield, and 94% ee, in a synthesis of (–)-callystatin A by Enders and co-workers.[653] Highly enantioselective reduction of prochiral ketones was also observed when *Daucus carota* root (carrot root) was used.[654] Cultured cells of *Marchantia polymorpha* have a reductase that leads to the asymmetric reduction of 2-substituted 2-butenolides to the (R)-butenolide.[655]

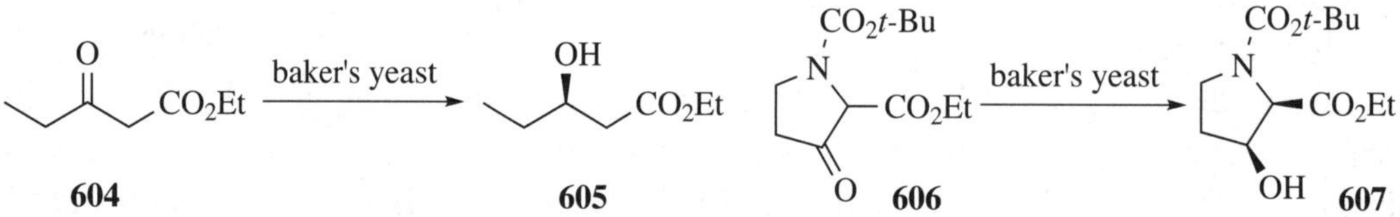

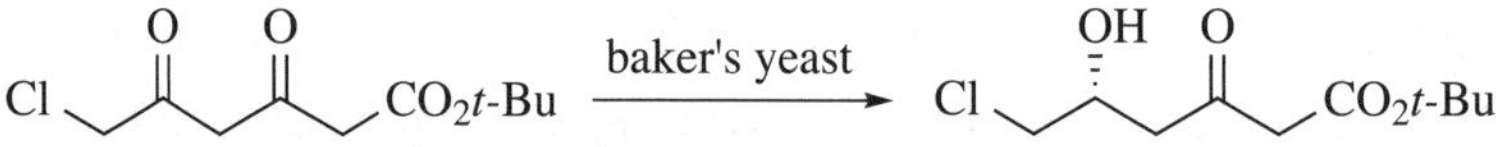

Problems were encountered in the reduction of aryl β-keto esters such as the furan derivative **610**,[656] where baker's yeast was unable to reduce the ketone moiety. Oishi and co-workers examined other selected yeasts including *Kloeckera saturnus*, which converted **610** to a 47:53 mixture of the syn- (**611**) and anti- (**612**) diastereomers, in 47% yield and 87% ee and 53% ee respectively.[656] Another yeast (*Saccharomyces delbrueckii*) gave 43% reduction with a syn/anti ratio of 39:61, and 99% ee of **611** and 94% ee of **612**.[656]

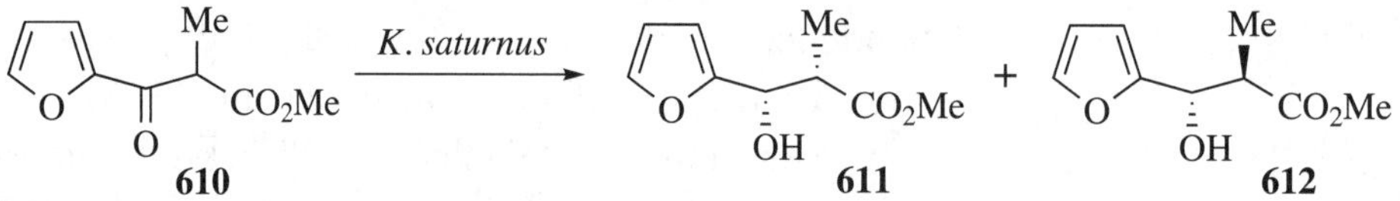

650. Salvi, N.A.; Chattopadhyay, S. *Tetrahedron: Asymmetry* **2004**, *15*, 3397.
651. Williams, R.M.; Cao, J.; Tsujishima, H.; Cox, R. *J. Am. Chem. Soc.* **2003**, *125*, 12172.
652. Fadnavis, N.W.; Radhika, K.R. *Tetrahedron: Asymmetry* **2004**, *15*, 3443.
653. Vicario, J..; Job, A.; Wolberg, M.; Müller, M.; Enders, D. *Org. Lett.* **2002**, *4*, 1023.
654. Yadav, J.S.; Nanda, S.; Reddy, P.T.; Rao, .B. *J. Org. Chem.* **2002**, *67*, 3900.
655. Shimoda, K.; Kubota, N. *Tetrahedron: Asymmetry* **2004**, *15*, 3827.
656. Akita, H.; Furuichi, A.; Koshiji, H.; Horikoshi, K.; Oishi, T. *Tetrahedron Lett.* **1982**, *23*, 4051.

Table 4.15. Diastereoselective Reduction of 613 by Selected Yeasts.

Yeast	% Yield	618 : 619
Endomycopsis fibligera	72	48 : 52
Hansenula anomala	44	43 : 57
Hansenula anomala NI-7572	44	40 : 60
Kloeckera saturnus	46	18 : 82
Lipomyces starkeyi	15	85 : 15
Pichia farinosa	40	15 : 85
Pichia membranaefaciens	51	53 : 47
Rhodotorula glutinis	46	68 : 32
Saccharomyces acidifaciens	14	44 : 56
Saccharomyces debrueckii	63	28 : 72
Saccharomyces fermentati	70	28 : 72
Candida albicans	26	17 : 83

Oishi examined the selectivity for reduction of **613** with a variety of organisms, as shown in Table 4.15.[657] In all cases, the organism produced an excess of either the syn- or the anti-diastereomer (**614** or **615**). The chemical yields were poor in some cases and the reduction proceed with modest to good diastereoselectivity in other cases. The enantioselectivity of the reaction was good to excellent, with 60-99% ee observed for both the *syn-* and the *anti-*diastereomers produced. In other work, Oishi and co-workers found that selected yeasts showed different diastereoselectivity and enantioselectivity when the aryl group was changed from phenyl to thiophene.[657a] Several strains of yeast were used but the yields were poorer than the analogous reduction of **610**. Although the diastereoselectivity was similar, changes in enantioselectivity were very pronounced.

Other organisms have been used to reduce β-keto esters. Indeed, either the *(R)-* or the *(S)-*alcohol can be obtained. Bernardi et al. showed that **388** was reduced to *(S)*-alcohol **603** in 96% yield, 7:93 *(R/S)-* with *G. candidum*.[658] The reaction medium is quite important. Nakamura et al. found, for example, that reduction of ketones with this enzyme in the presence of the hydrophobic polymer Amberlite™ XAD gave the *(S)*-alcohol in high enantiomeric excess, whereas low enantioselectivity was obtained in the absence of the polymer.[659] Reduction of **388** with *Aspergillus niger* gave 98% of a 75:25 mixture favoring the *(R)*-alcohol **616**.[658] α-Keto esters (glyoxalates) are reduced with yeast, as illustrated by the conversion of **617** to *(S)*-alcohol **618** in 47% yield (49% ee).[660] Aryl glyoxalates are usually reduced to the *(R)*-alcohol.[660] It has been shown that using an acetone powder of a microorganism such as *G. candidum* led to improved stereoselectivity.[661] Trifluoromethyl ketones are reduced with

657. (a) Furiuchi, A.; Akita, H.; Koshiji, H.; Horikoshi, K.; Oishi, T. *Chem. Pharm. Bull.* **1984**, *32*, 1619; (b) Akita, H.; Furuichi, A.; Koshiji, H.; Horikoshi, K.; Oishi, T. *Ibid.* **1984**, *32*, 1342; (c) *Idem Ibid.* **1983**, *31*, 4376.
658. Bernardi, R.; Cardillo, R.; Ghiringhelli, D. *J. Chem. Soc. Chem. Commun.* **1984**, 460.
659. Nakamura, K.; Fujii, M.; Ida, Y. *J. Chem. Soc. Perkin Trans. 1* **2000**, 3205.
660. Iriuchijima, S.; Ogawa, M. *Synthesis* **1982**, 41.
661. Matsuda, T.; Harada, T.; Nakajima, N.; Nakamura, K. *Tetrahedron Lett.* **2000**, *41*, 4135.

opposite stereochemistry relative to methyl ketones with this organism.[662] Bioreduction using methanol as a co-solvent and the fungus *Curvularia lunata* CECT 2130 gave chemoselective reduction of aromatic β-keto nitriles to the corresponding (S)-β-hydroxy nitriles with high enantioselectivity.[663]

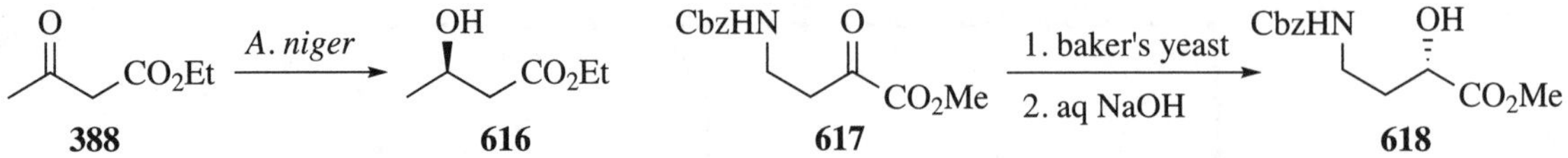

Isolated ketone units can also be reduced with baker's yeast.[664] In a synthetic example, ketone moiety of the piperidone unit in **619** was reduced with baker's yeast to give a 40% yield of **620** (98% ee) in Takeuchi's synthesis of isofebrifugine.[665]

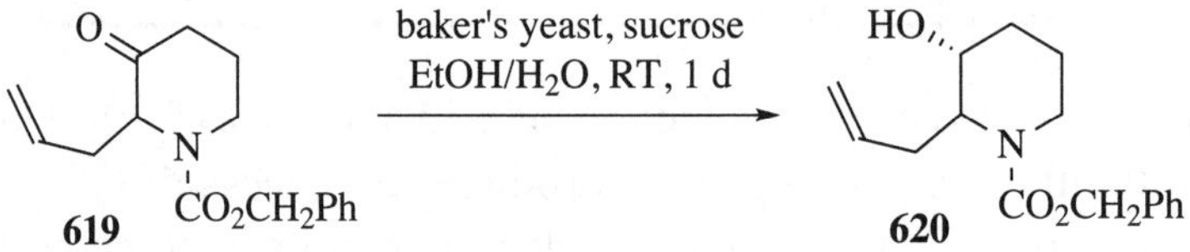

Veschambre used *Beauveria sulfurescens* for the reduction of conjugated carbonyls. *trans*-Crotonaldehyde was reduced to but-(2E)-en-1-ol in 80% yield.[666] 2-Methyl-2-pentenal (**621**), however, gave a mixture of 31% of the conjugated alcohol (**622**) and 69% of the completely reduced alcohol (2-methyl-1-pentanol, **623**).[666] Later work showed this microbial reduction gave the *(S)*-alcohol by a stepwise process in which the conjugated double bond was reduced first, establishing the *(S)*-stereochemistry, followed by reduction of the aldehyde moiety to the alcohol.[666b] Veschambre established a rule for this reduction, **Veschambre's rule**. The hydrogen is delivered to the double bond α to the carbonyl from the front or back depending on the relative bulk of R^1 and R^2, which is simply a modification of Prelog's Rule. Reduction of geranial with baker's yeast gave *(R)*-citronellol, consistent with the prediction. The *(Z)*-isomer neral was predicted to give the *(S)*-isomer, but reduction with baker's yeast gave a 6:4 *(R/S)* mixture, possibly due to isomerization of the double bond in neral prior to delivery of hydrogen.[666b] Shimoda and co-workers found that *Synechococcus sp.* PCC 7942, a cyanobacterium, reduced both the endocyclic double bond and s-*trans*-enones as well as the exocyclic double bond of

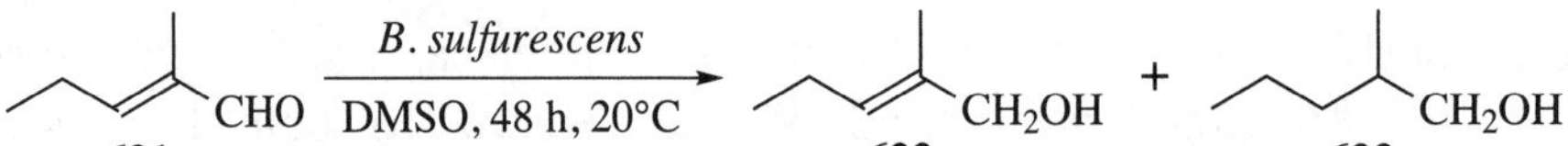

662. Matsuda, T.; Harada, T.; Nakajima, N.; Itoh, T.; Nakamura, K. *J. Org. Chem.* **2000**, *65*, 157.
663. Dehli, J.R.; Gotor, V. *Tetrahedron: Asymmetry* **2000**, *11*, 3693.
664. Lieser, J.K. *Synth. Commun.* **1983**, *13*, 765.
665. Takeuchi, Y.; Azuma, K.; Takakura, K.; Abe, H.; Harayama, T. *Chem. Commun.* **2000**, 1643.
666. (a) Desrut, M.; Kergomard, A.; Renard, M.F.; Veschambre, H. *Tetrahedron* **1981**, *37*, 3825; (b) Bostmembrun-Desrut, M.; Dauphin, G.; Kergomard, A.; Renard, M.F.; Veschambre, H. *Ibid.* **1985**, *41*, 3679.

s-cis-enones with high enantioselectivity.[667]

Another common organism for reduction of the carbonyl group is *Thermoanaerobium brockii* (**238**), first mentioned in Section 4.7.A, in Table 4.5. Keinan et al. showed that small ketones such as 2-butanone were reduced by *T. brockii* to give *(R)*-2-butanol, in 12% yield and 48% ee, but longer chain ketones such as 2-nonanone were reduced to the *(S)*-alcohol [**624**, 85% yield and 96% ee].[266] An alcohol dehydrogenase found in *Thermoanaerobacter ethanolicus* reduces ethynyl ketoses such as **625** to **626** in good yield, and high enantioselectivity [in this case **626** was isolated in 50% yield and >98% ee, *(S)*].[668] These examples illustrate typical enantioselectivity of the reduction and that the selectivity depends on the size and nature of the groups around the carbonyl. Changes in selectivity are explained by knowledge of the active site of the enzyme. The carbonyl is bound close to a nicotinamide moiety in the coenzyme. Two alkyl sites are also present with the smaller alkyl group lying closer to the nicotinamide having the greater affinity.[266] Keinan and Lamed[266] suggested the small alkyl site can accommodate groups smaller than *n*-propyl. For methyl-alkyl ketones, if the alkyl is methyl, ethyl, isopropyl or cyclopropyl, the *(R)*-alcohol predominates but if the alcohol group is larger than *n*-propyl the *(S)*-alcohol predominates. For internal ketones (alkyl-alkyl) the *(S)*-alcohol is usually the major product.[669] It appears that various yeast preparations and *T. brockii* are the most commonly used reducing organisms. New organisms are being used on a regular basis. Lyophilized cells from *Rhodococcus ruber* DSM 44541 for example, reduced ketones to alcohols in good yield, and with excellent enantioselectivity.[670] The low cost and high selectivity of such enzymatic and microbial reagents foretell an increasingly important role in organic synthesis.

4.11. CONCLUSION

It is obvious from this chapter that reductive processes are important in organic synthesis. The diastereoselectivity and enantioselectivity of many reducing agents are important for fixing stereogenic. An attempt has been made to show how reagents were developed as well as their

667. Shimoda, K.; Kubota, N.; Hamada, H.; Kaji, M.; Hirata, T. *Tetrahedron: Asymmetry* **2004**, *15*, 1677.
668. Heiss, C.; Phillips, R.S. *J. Chem. Soc. Perkin Trans. 1* **2000**, 2821.
669. Seebach, D.; Züger, M.F.; Giovannini, F.; Sonnleitner, B.; Friechter, A. *Angew. Chem. Int. Ed.* **1984**, *23*, 151.
670. Stampfer, W.; Kosjek, B.; Faber, K.; Kroutil, W. *J. Org. Chem.* **2003**, *68*, 402.

chemoselectivity.

Chapter 5 will introduce reactions associated with hydroboration. This important synthetic methodology is a logical extension of the borohydride and borane reductions seen in this chapter. Chronologically, the reductive capabilities of boranes preceded the alkene addition reaction, which is the heart of hydroboration. In part, this is why hydroboration is presented after reduction.

It is also noted that the stereochemical models introduced in this chapter will be used continuously in succeeding chapters. Addition of Grignard reagents and other organometallics to carbonyls generally follow Cram's rule or the Felkin-Anh model. Chelation control will be important in many reactions including enolate alkylation reactions. This chapter has therefore been a vehicle for introducing not only reduction but the stereochemical basis for all of organic chemistry.

HOMEWORK

1. Predict the correct diastereomeric product using (a) the appropriate Cram model and (b) the Felkin-Anh model. In each case, draw the model for that specific reaction.

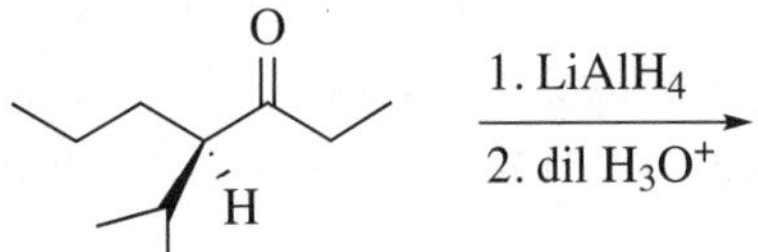

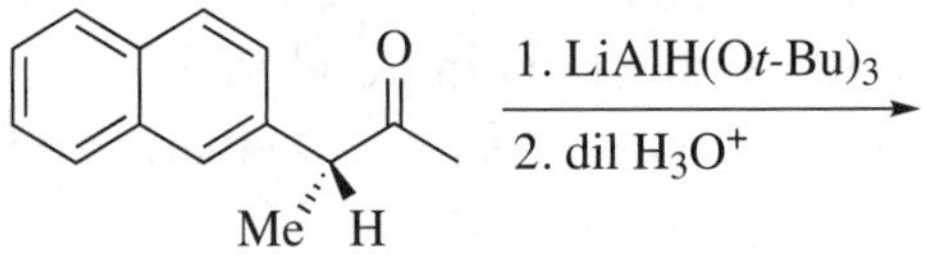

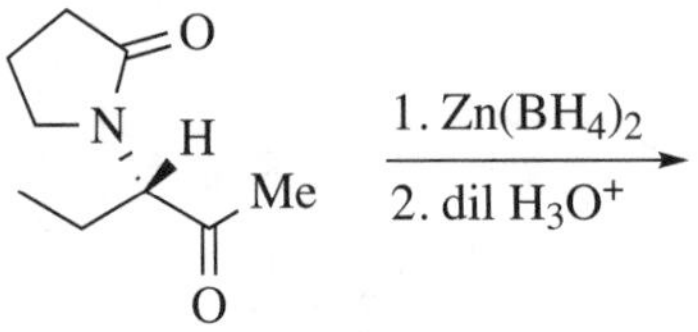

2. Predict the major product and explain the stereochemistry for that product.

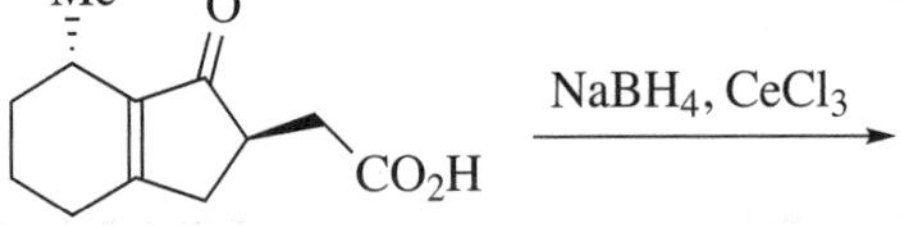

3. In each case, predict the major diastereomer. Explain your answer.

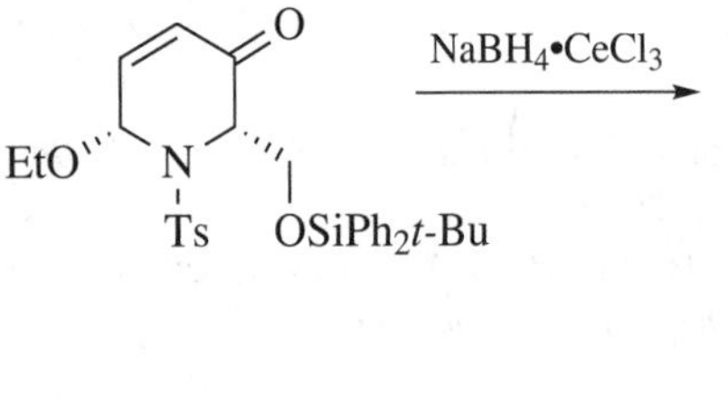

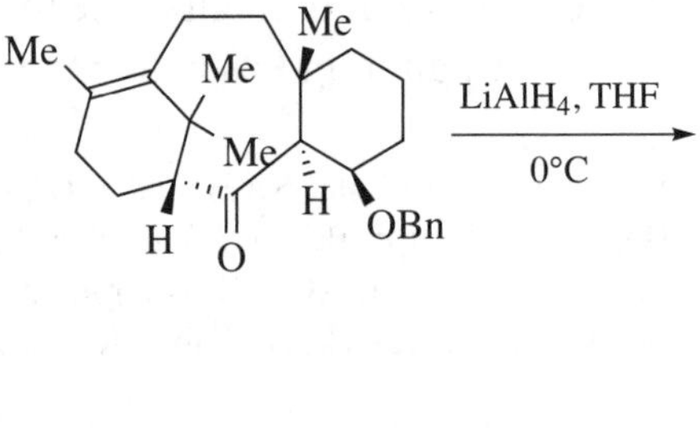

(c) 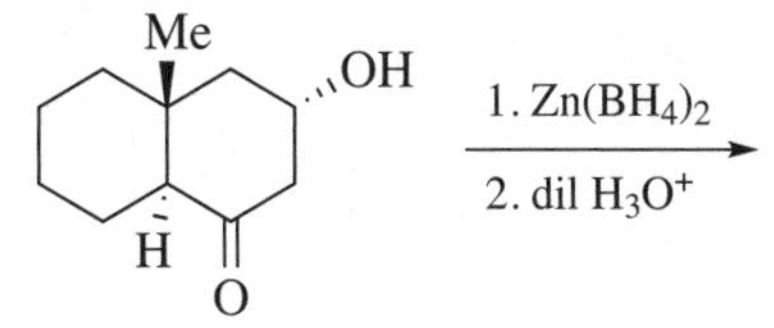 ... (d) 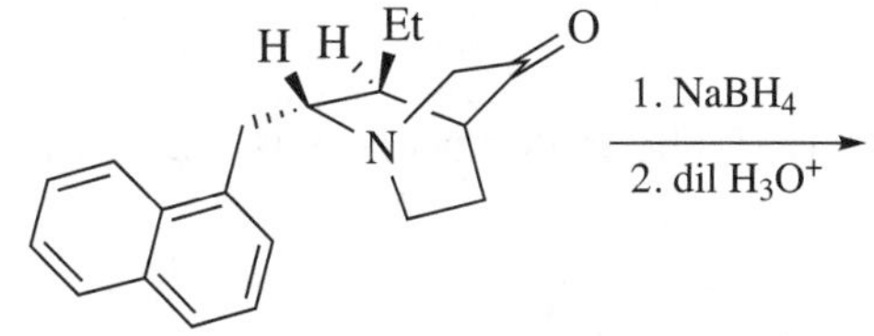 ...

4. Explain why there is such a difference in selectivity for reduction of this conjugated ketone when the reducing agent is changed.

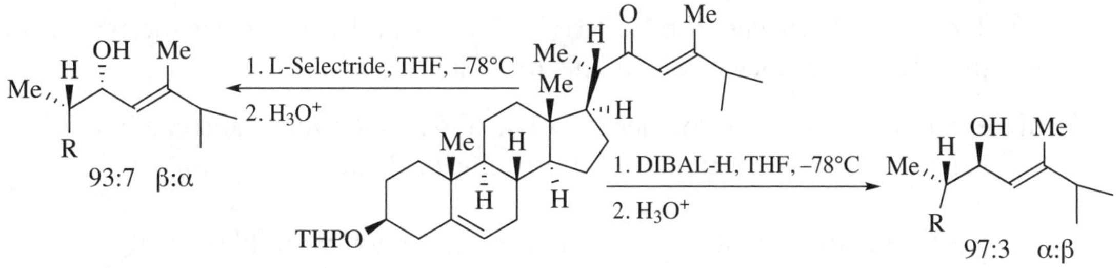

5. Briefly discuss this transformation.

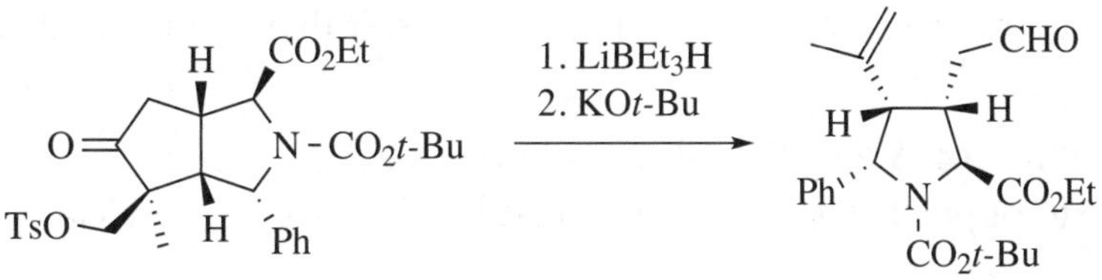

6. Predict the major product of both reactions, (A and B), predict the stereochemistry of each product, and explain your choices.

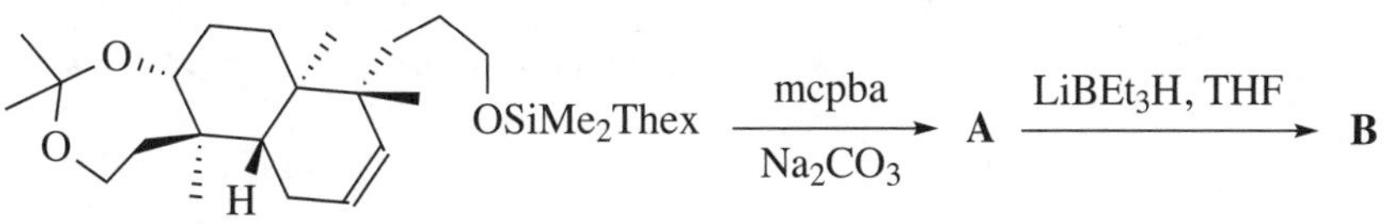

7. Briefly discuss the following transformation, particularly focusing attention on the stereochemistry.

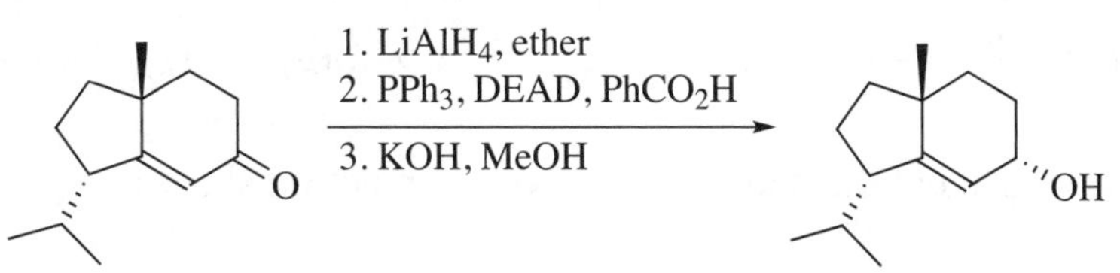

8. Explain each of the following observations or products.

(a) The major process is 1,2-reduction.

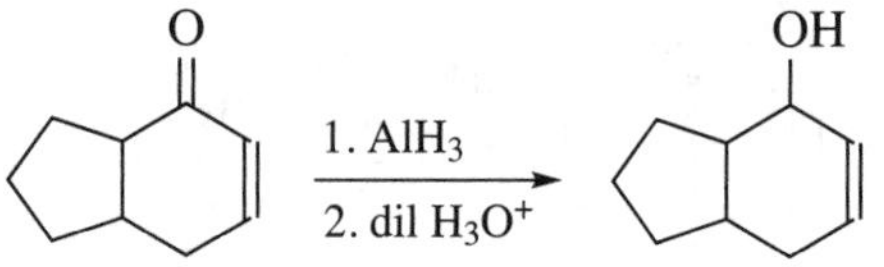

$$\xrightarrow[\text{2. dil } H_3O^+]{\text{1. } AlH_3}$$

(b) L-Selectride gives greater diastereoselectivity in this reduction than does sodium borohydride. Predict the major product.

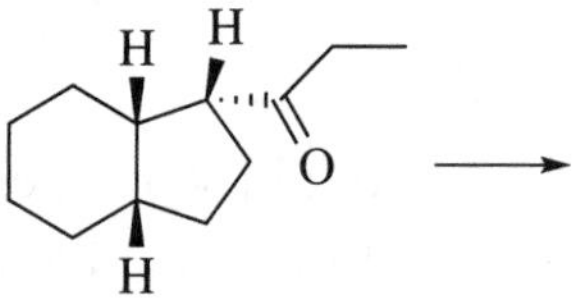

(c) Treatment of benzene with $Na/NH_3/EtOH$ gives 1,4-cyclohexadiene, whereas similar treatment of cyclohexene with this reagent gives no reaction.

(d) Birch reduction of anisole leads to 1-methoxy-1,4-cyclohexadiene but reduction of benzoic acid give 3-carbalkoxy-1,4-cyclohexadiene.

(e) Reduction with $LiAlH_4$ gives a diol but reduction with diborane gives a hydroxy ester.

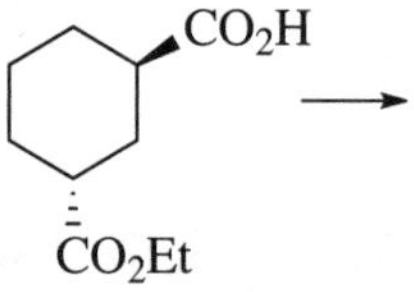

(f) Diimide reduces unconjugated alkenes faster than conjugated alkenes.

(g) Reduction of A and B with zinc borohydride leads to opposite stereochemical results.

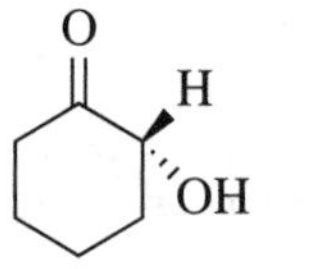

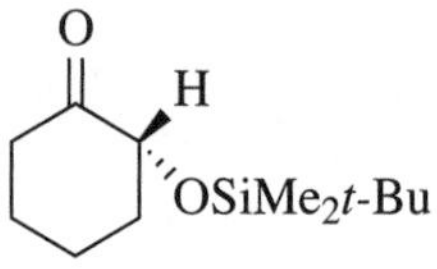

(h) Birch reduction of A, and subsequent acid hydrolysis leads to the conjugated ketone B.

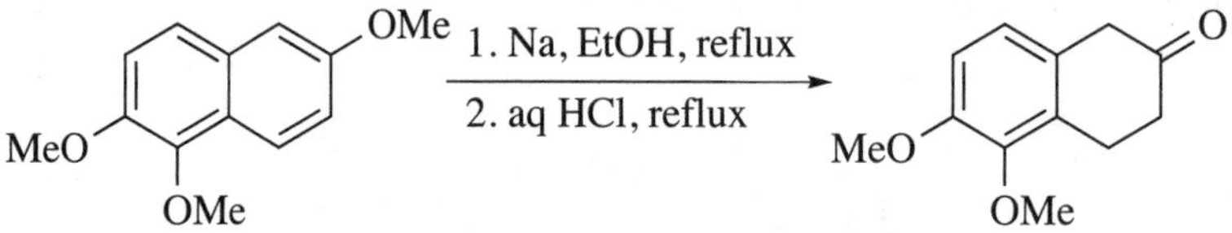

9. For each transformation, give two __different__ reagents that will successfully complete the given conversion.

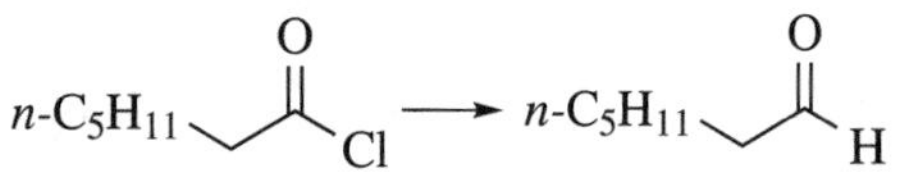

10. In each case, predict the major product of the reaction shown.

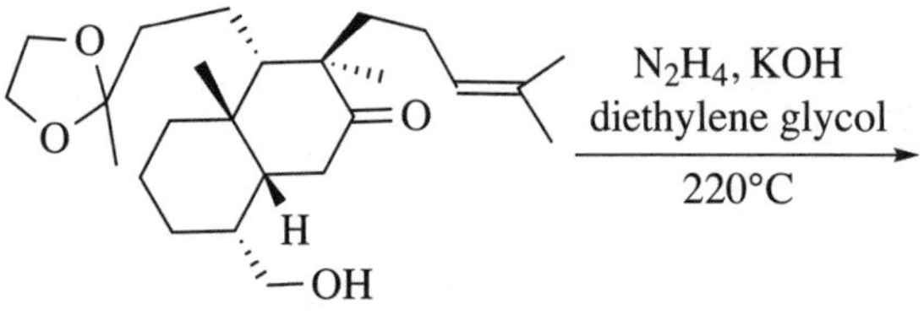

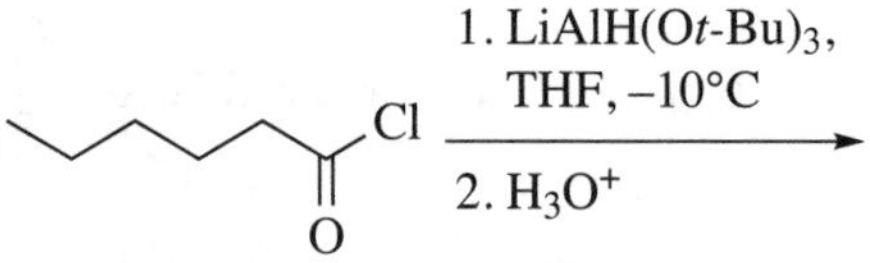

(c) PMB = *p*-methoxybenzyl
Piv = pivaloyl

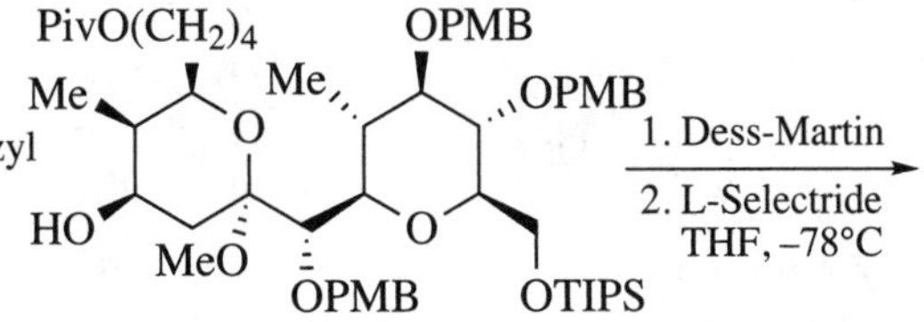

PivO(CH₂)₄ structure with Me, OPMB, OPMB, HO, MeO, OPMB, OTIPS

$$\xrightarrow[\text{2. L-Selectride THF, } -78°C]{\text{1. Dess-Martin}}$$

(d) MeO, MeO, CO₂H, NO₂ benzene structure

$$\xrightarrow[\text{Ni}]{H_2}$$

(e) MeO₂C—N(CO₂*t*-Bu)—CO₂Me (pyrrole)

$$\xrightarrow[\text{2. aq NH}_4\text{Cl}]{\text{1. Li, NH}_3}$$

(f) MeO, MeO, OMe, Br, MeO, CHO benzene structure

MeO—C₆H₄—CH₂NH₂

$$\xrightarrow[\text{ClCH}_2\text{CH}_2\text{Cl}]{\text{NaBH(OAc)}_3}$$

(g) lactone structure with Me, *t*-BuMe₂SiO, CO₂H

$$\xrightarrow{\text{BH}_3\cdot\text{THF, THF}}$$

(h) piperidine with NH₂, Ph, N—CO₂*t*-Bu

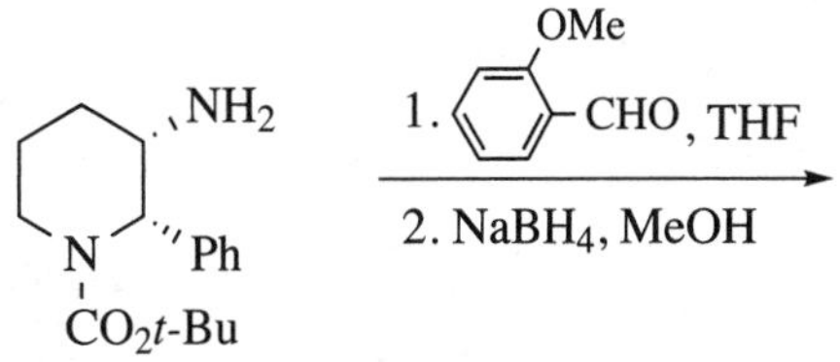

OMe—C₆H₄—CHO, THF

$$\xrightarrow[\text{2. NaBH}_4, \text{MeOH}]{\text{1.}}$$

(i) Br indole-fused structure, N—CO₂Me, HO, H

$$\xrightarrow[\text{rt}]{\text{AlH}_3, \text{THF}}$$

(j) indole with CH₂ CO₂Me, NHCO₂*t*-Bu, NH

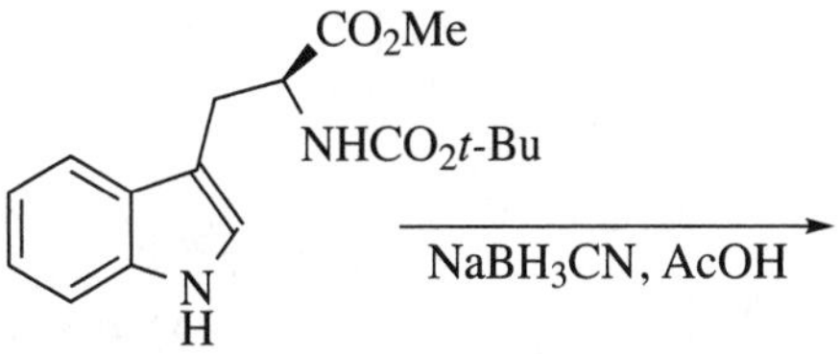

$$\xrightarrow{\text{NaBH}_3\text{CN, AcOH}}$$

(k) Me Me, *t*-BuMe₂SiO, OH, alkyne structure

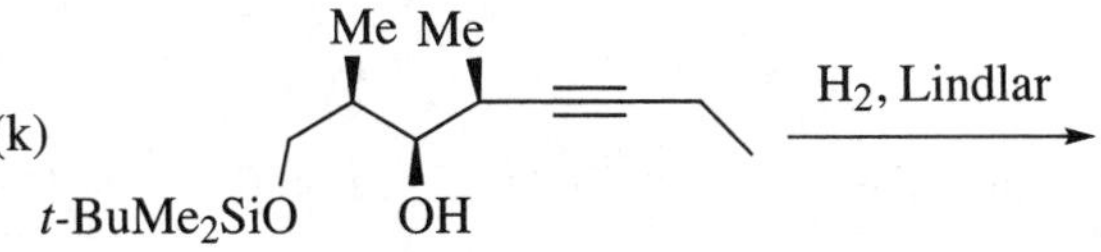

$$\xrightarrow{\text{H}_2, \text{Lindlar}}$$

(l) bicyclic structure H, CO₂Me, N—CO₂*t*-Bu, O=

$$\xrightarrow{\text{NaBH}_4, \text{MeOH}}$$

(m) *t*-Bu, NHPhth, indole structure with N, O, isobutyl, NH

$$\xrightarrow[\text{2. hydrolysis}]{\text{1. LiAlH}_4, \text{THF}}$$

(n) PhMe₂Si, CO₂Me, O, O, O epoxide bicyclic structure

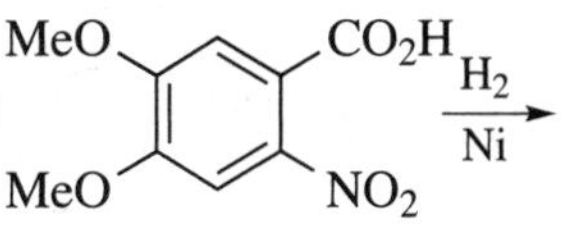

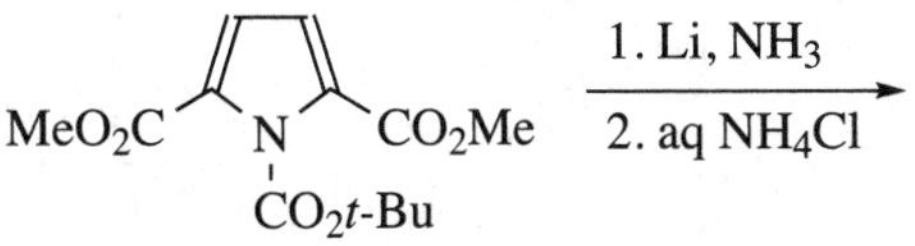

$$\xrightarrow[\text{2. H}_3\text{O}^+]{\text{1. 6 eq LiAlH}_4}$$

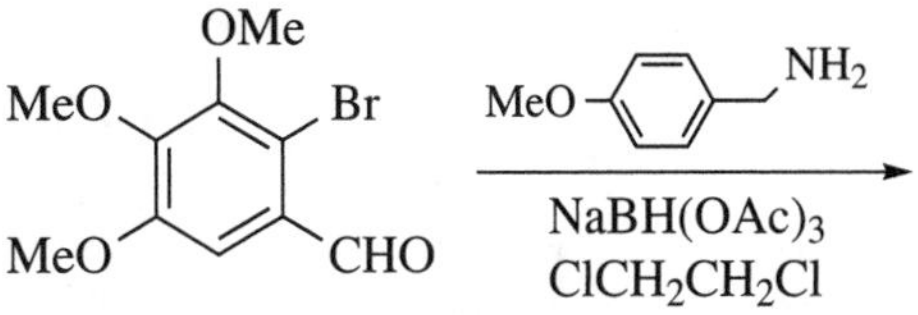

(o) decahydroquinoline structure with OH, H, Me, N—CO₂Bn

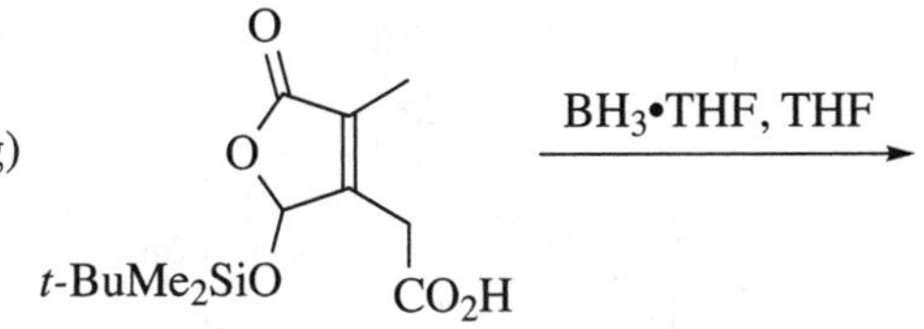

$$\xrightarrow[\text{2. hydrolysis}]{\text{1. LiAlH}_4}$$

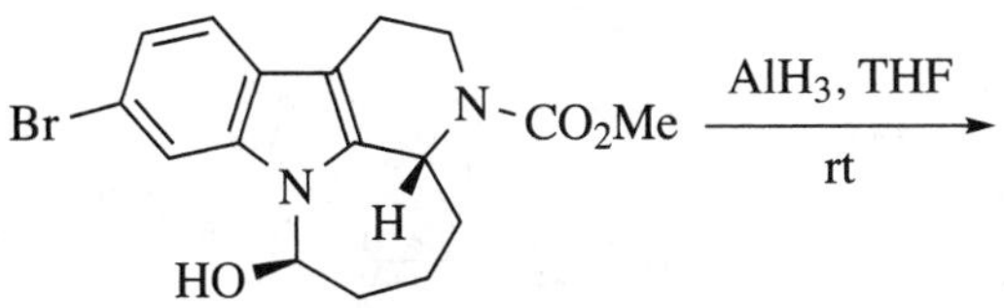

(p) C₁₅H₃₁—C(O)—furan—CH₂CN

$$\xrightarrow[\text{HOCH}_2\text{CH}_2\text{OH}]{\text{NH}_2\text{NH}_2, \text{NaOH}}$$

(q) pyrrolizidine structure with OH, H, N, O, OSiMe₂*t*-Bu

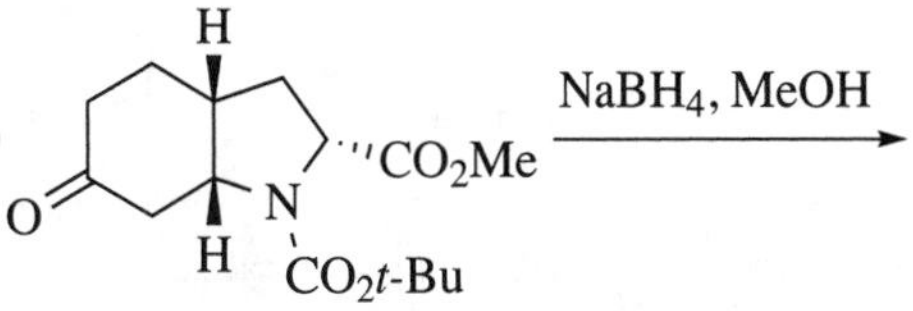

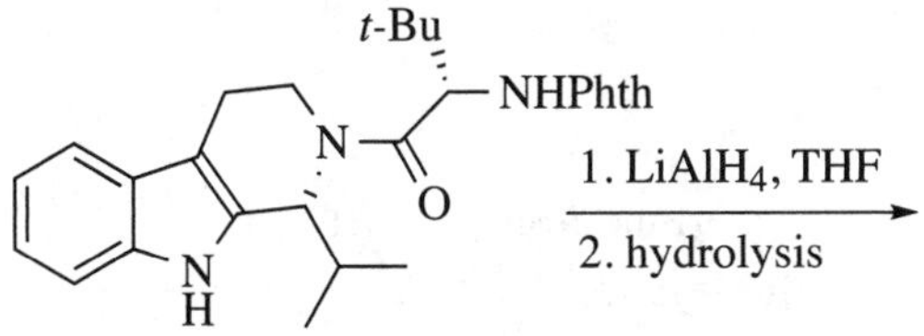

$$\xrightarrow[\text{2. Bu}_3\text{SnH, AIBN}]{\text{1. PhOCSCl, Py}}$$

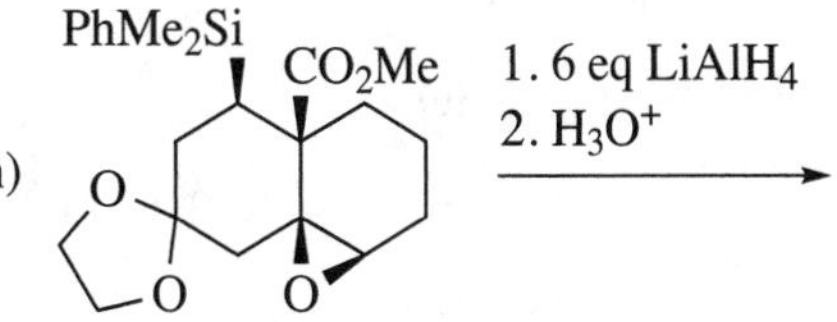

(r) OSiPh₂*t*-Bu, CO₂Me diene-yne structure

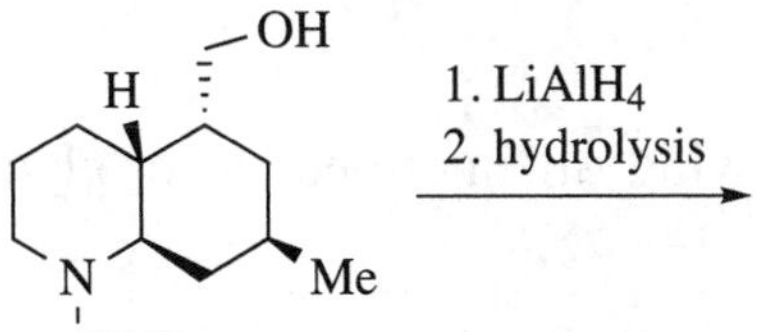

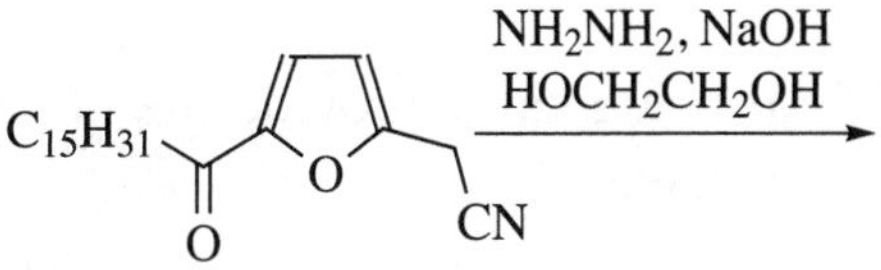

$$\xrightarrow[\text{MeCN, MS 4Å}]{\substack{\text{1. Dibal, CH}_2\text{Cl}_2 \\ -78°C \\ \text{2. TPAP, NMO}}}$$

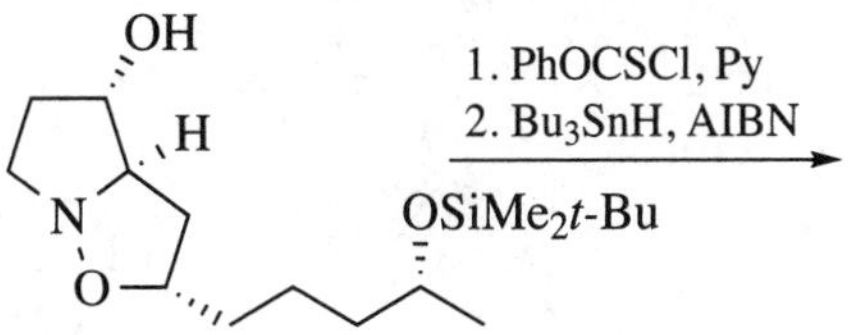

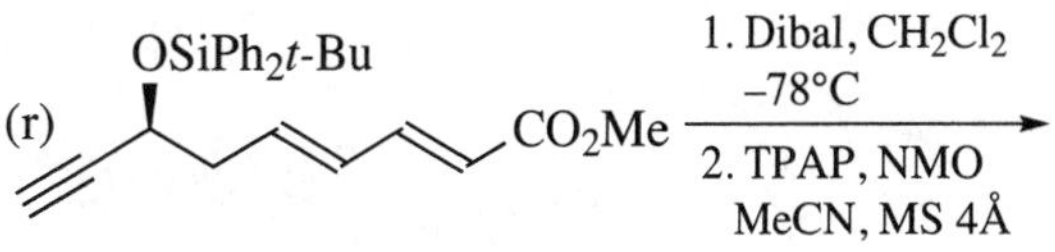

(s) 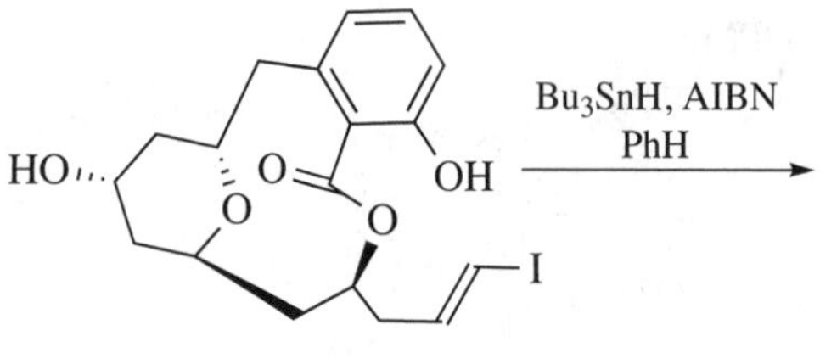 HO,,, O OH O O I — Bu₃SnH, AIBN PhH →

(t) 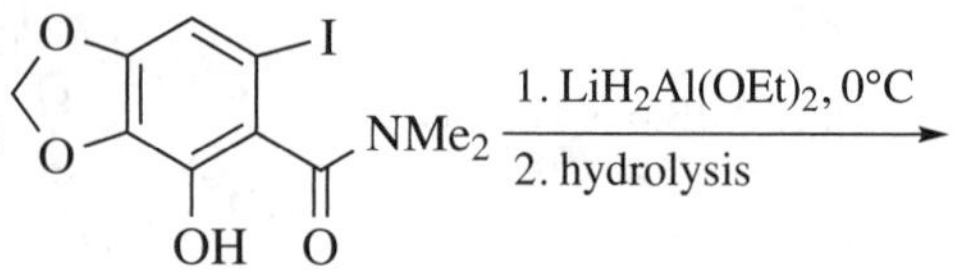 NMe₂ I OH O — 1. LiH₂Al(OEt)₂, 0°C / 2. hydrolysis →

(u) 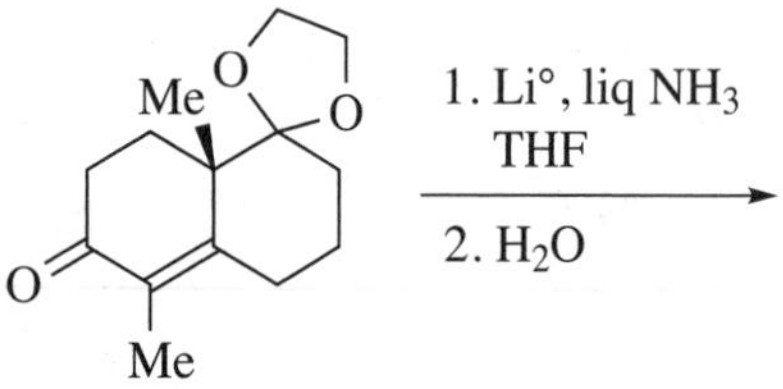 Me, O O, O, Me — 1. Li°, liq NH₃ THF / 2. H₂O →

(v) 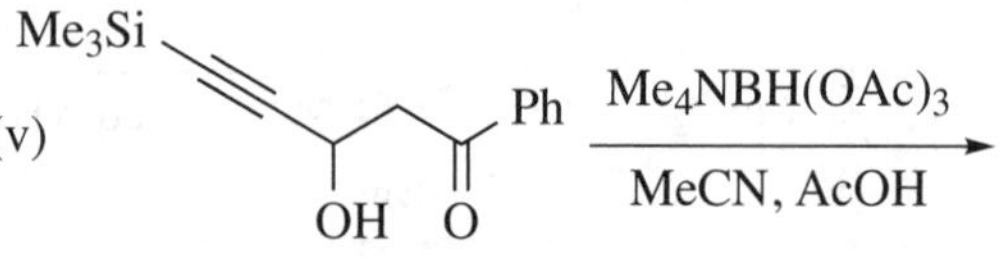 Me₃Si—≡ OH O Ph — Me₄NBH(OAc)₃ / MeCN, AcOH →

(w) 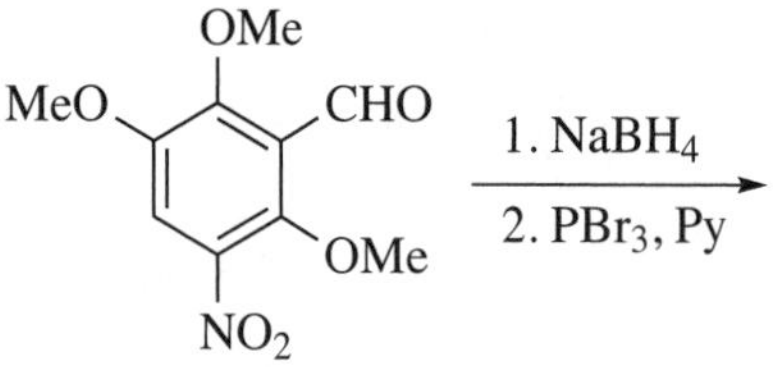 MeO OMe CHO OMe NO₂ — 1. NaBH₄ / 2. PBr₃, Py →

(x) 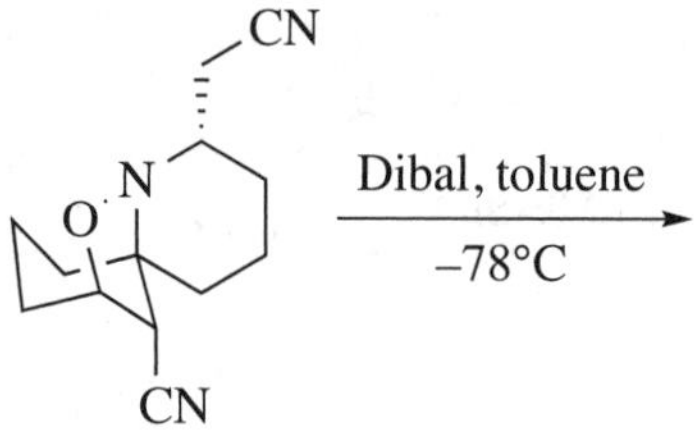 CN N O CN — Dibal, toluene / −78°C →

(y) 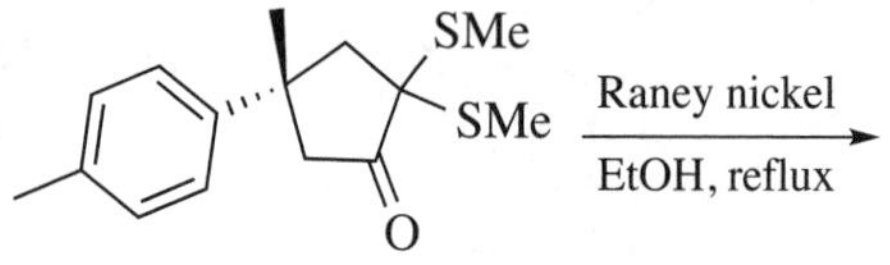 Me SMe SMe O — Raney nickel / EtOH, reflux →

(z) 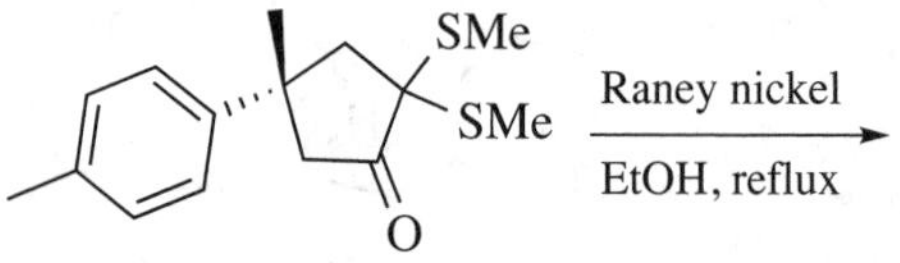 Me SMe SMe O — Raney nickel / EtOH, reflux →

(aa) 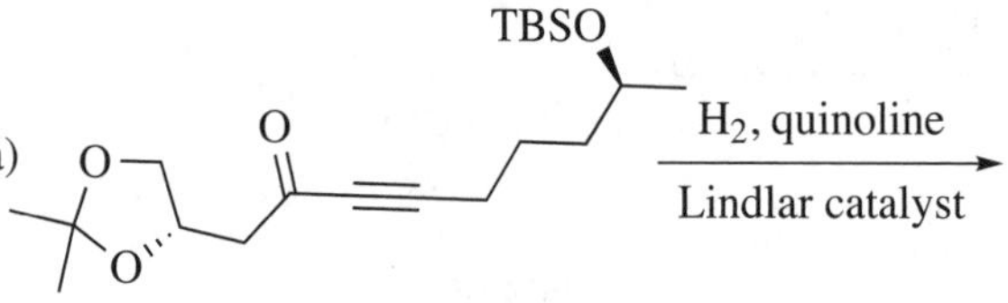 O O O TBSO — H₂, quinoline / Lindlar catalyst →

(ab) 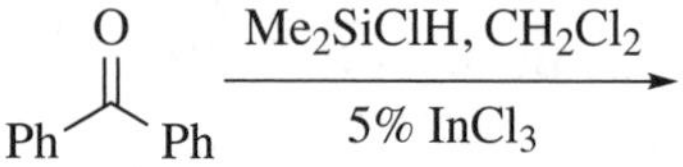 O Ph Ph — Me₂SiClH, CH₂Cl₂ / 5% InCl₃ →

(ac) 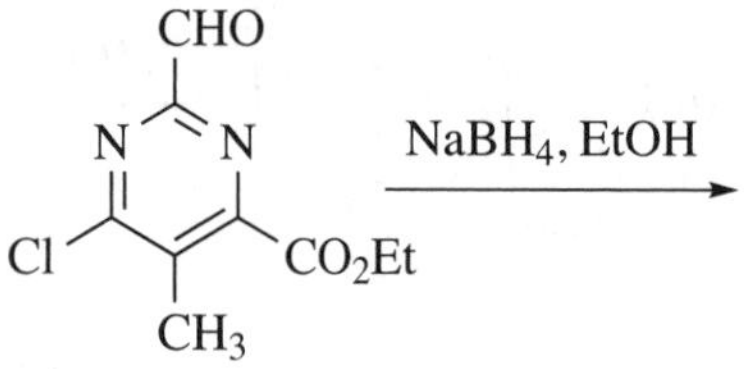 CHO N N Cl CO₂Et CH₃ — NaBH₄, EtOH →

(ad) 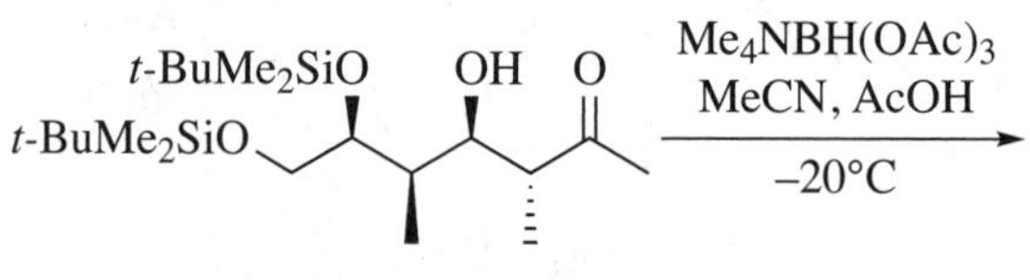 t-BuMe₂SiO OH O t-BuMe₂SiO — Me₄NBH(OAc)₃ / MeCN, AcOH / −20°C →

11. In each case, provide a suitable synthesis showing all intermediate products and all reagents.

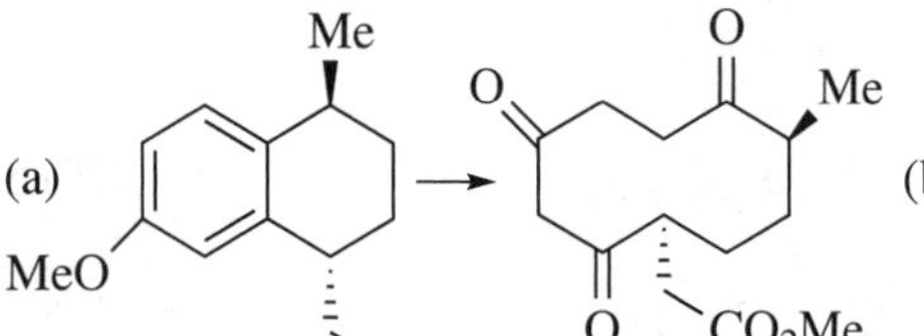

(a) Me MeO CO₂Me → O O Me O CO₂Me

(b) 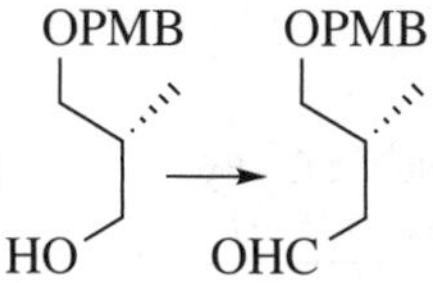 OPMB HO → OPMB OHC

(c) 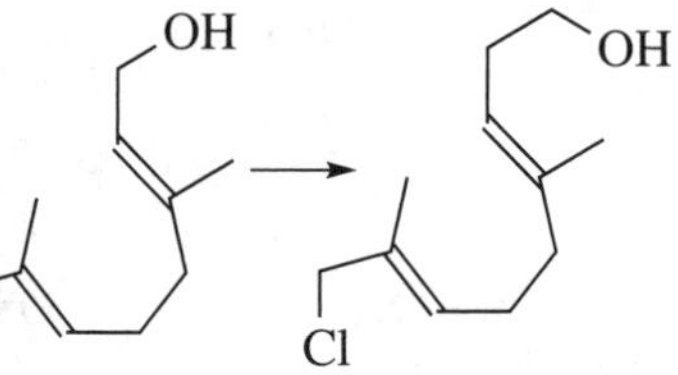 OH → OH Cl

(d)

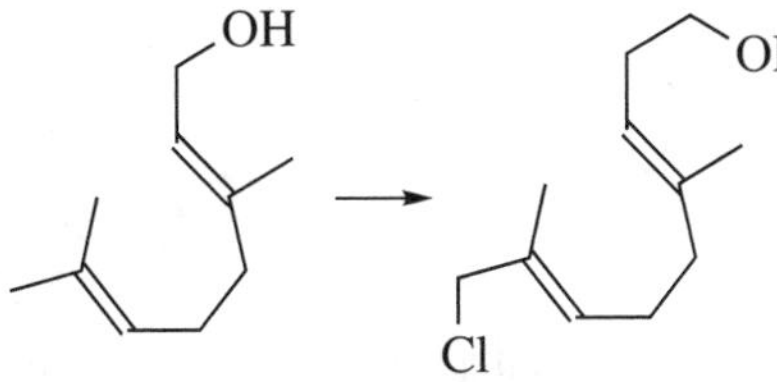

(e)

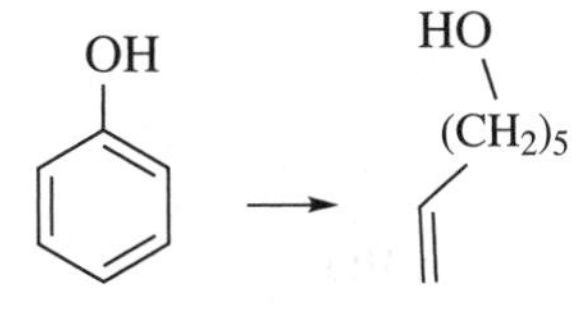

(f)

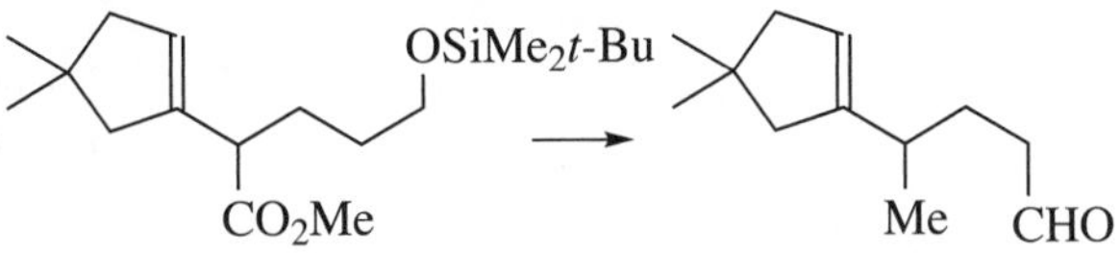

(g)

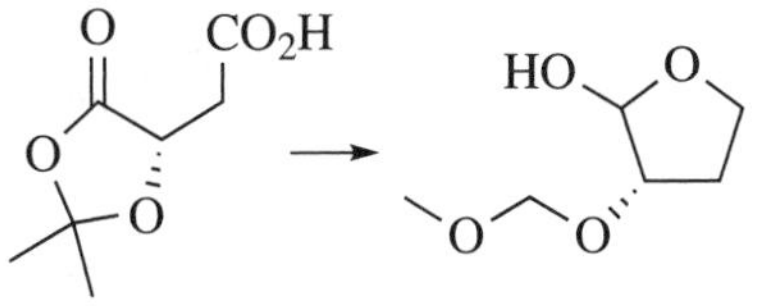

(h)

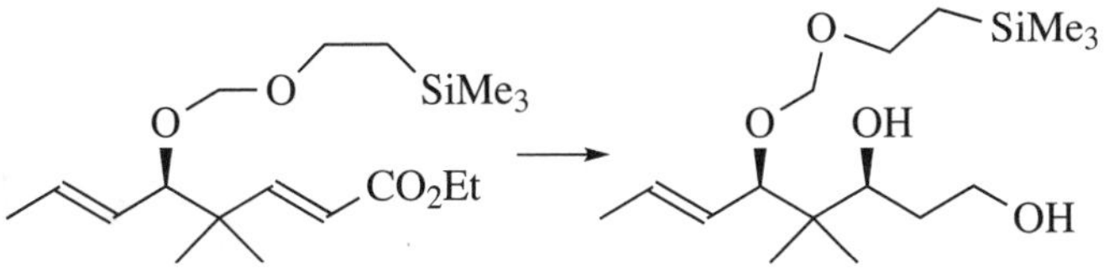

(i)

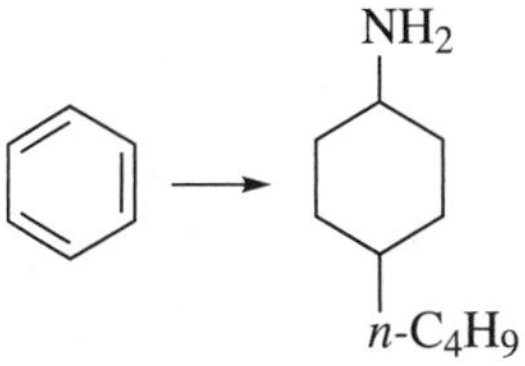

(j)

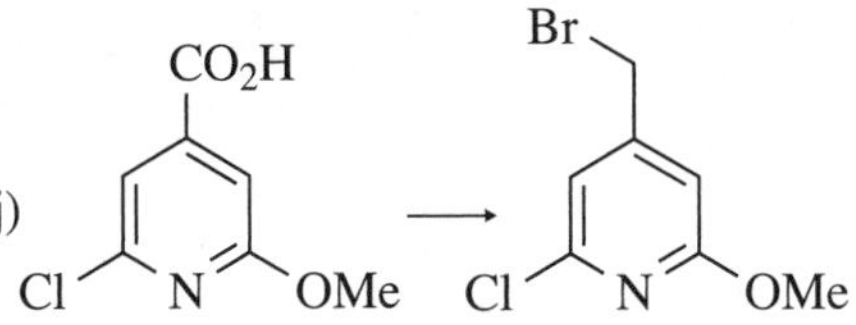

(k)

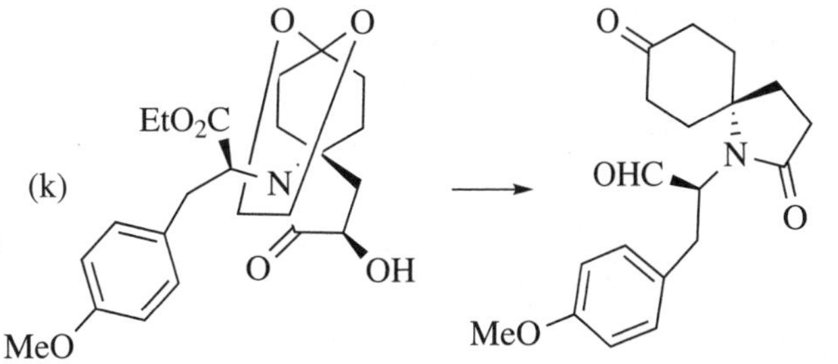

(l)

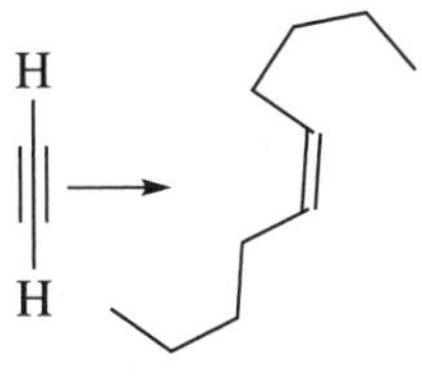

12. Show a complete synthesis for each of the following. Use any starting material of your choosing, that contains six carbons or less, but you must find it in a current catalog from a chemical company (give the source and price of your starting material). Show a retrosynthetic analysis and synthesis for each molecule.

(a) 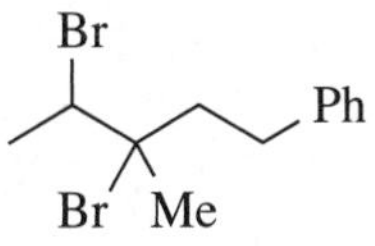 Ph (b) 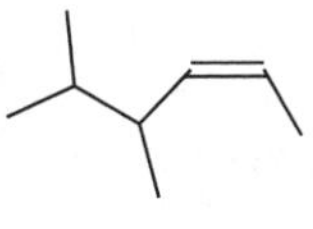 (c) 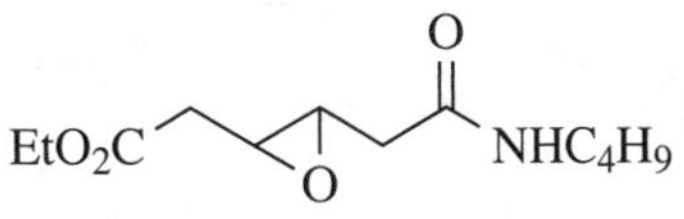 EtO_2C NHC_4H_9 (d) 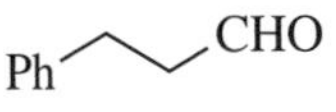 Ph CHO

13. Give reagents that will accomplish each of the following selective transformations.

(a)

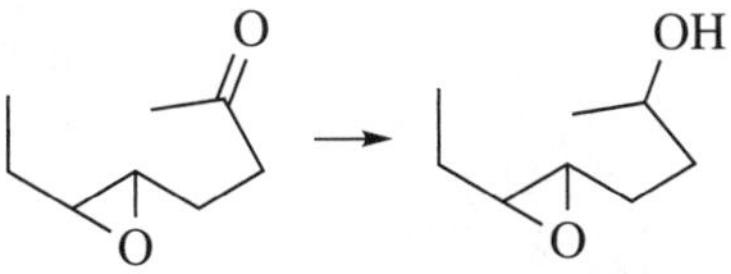

(b)

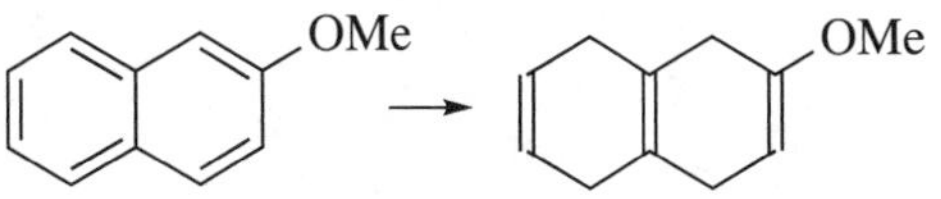

(c)

(d)

(e)

(f)

chapter 5

Hydroboration

5.1. INTRODUCTION

In chapter 4, borane and alkylborane reagents were introduced as powerful and selective reducing agents. The Lewis acidity of borane (BH_3) led to interesting differences in reducing power when compared to $LiAlH_4$ and $NaBH_4$ derivatives. An example of this difference was Brown and co-worker's discovery that diborane (B_2H_6) reacted with alkenes in ether solvents to form trialkylboranes (**1**).[1] Brown's monograph described formation of an alkylborane from an alkene using $NaBH_4/AlCl_3$, $NaBH_4/BF_3$ or diborane itself.[2]

$$6 \overset{R}{\diagdown\!\!=} + B_2H_6 \longrightarrow 2 \left(\underset{\mathbf{1}}{\overset{R}{\diagdown\!\!\diagdown}} \right)_{\mathbf{3}} B$$

Continuing interest in this reaction led to development of a large number of functional group transformations and carbon-carbon bond-forming reactions that utilized mono-, di-, trialkyl, and alkenylboranes. This work was cited in the Nobel Prize for Chemistry awarded to H.C. Brown in 1979. Hydroboration is included in organic synthesis because the technique makes several unique transformations possible, and they are detailed in this chapter. A discussion of hydroboration following the discussion of the reducing properties of boranes (in Chap. 4) is both historically correct and conceptually useful. Where appropriate, the diastereoselectivity and the enantioselectivity of hydroboration reactions will be discussed. In addition to the functional group transformations that are available via hydroboration, the reactions presented in this chapter are a useful review of many reaction types encountered in earlier chapters.

5.2. PREPARATION OF ALKYL AND ALKENYL BORANES

5.2.A. Alkyl Boranes

Diborane reacts with alkenes in ether solvents to form alkylboranes. This important finding apparently resulted from an observation by Brown that reduction of ethyl oleate in ether solvents and in the presence of a Lewis acid, consumed 2.4-2.5 hydrides per equivalents and three equivalents when a larger excess of the reagent was used.[3] Brown discovered that borane had

1. (a) Brown, H.C.; Subba Rao, B.C. *J. Am. Chem. Soc.* **1956,** *78,* 5694; (b) *Idem J. Org. Chem.* **1957,** *22,* 1136.
2. Brown, H.C. *Hydroboration*, W.A. Benjamin, New York, *1962.*
3. Reference 2, p 42.

added to the π bond of the alkene moiety generating a monoalkylborane (**2**), and this is a general reaction. If a sufficient quantity of alkene is present, one of the hydrogen atoms on boron (see **2**) can react with a second equivalent of alkene to give the dialkylborane (**3**) and with a third equivalent to give the trialkylborane (**1**).[2] The reaction of diborane with unsaturated molecules had been reported earlier by Stock and Kuss,[4] in the explosive reaction of acetylene and diborane at 100°C. In 1948, Hurd reported that alkenes reacted with diborane at elevated temperatures to form a trialkylborane,[5] but the reaction was slow at 100°C (24 h) and even slower at ambient temperatures. This earlier work with borane did not use an ether solvent, but Brown's work showed that ethers had a catalytic effect on the addition of borane to alkenes. Although the "reaction of diborane with pure olefins is indeed quite slow, addition of mere traces of ether changed the initially slow reaction to a fast one."[6] This rate acceleration allowed alkylboranes to be readily formed[1a,7] under relatively mild conditions, allowing them to be used in synthesis.

An excellent *in situ* preparation of diborane results from the reaction of boron trichloride (BCl_3) or boron trifluoride (BF_3) with sodium borohydride in the presence of alkenes, as mentioned above. This reagent gave virtually instantaneous conversion of an alkene to a trialkylborane at ambient temperatures, when done in a ether solvent.[8] When diborane was isolated and purified it also reacted with alkenes, rapidly and quantitatively in ether solvents. Homoallylic phosphine boranes undergo intramolecular hydroboration, in the presence of triflic acid.[9]

5.2.A.i. General Properties of Borane Addition to Alkenes. When borane reacts with alkenes, there are several predictable characteristics.[10]

1. Borane reacts with alkenes and alkynes rapidly and quantitatively (in most cases) by addition of H and B to the π- bond of an alkene or an alkyne, to give an alkylborane or an alkenylborane, respectively.[2,11]

4. Stock, A.; Kuss, E. *Berichte* **1923**, *56*, 789.
5. Hurd, D.T. *J. Am. Chem. Soc.* **1948**, *70*, 2053.
6. Reference 2, p 44.
7. Brown, H.C.; Subba Rao, B.C. *J. Am. Chem. Soc.* **1959**, *81*, 6423.
8. Brown, H.C.; Subba Rao, B.C. *J. Am. Chem. Soc.* **1959**, *81*, 6428.
9. Shapland, P.; Vedejs, E. *J. Org. Chem.* **2004**, *69*, 4094.
10. Brown, H.C. *Pure Appl. Chem.* **1976**, *47*, 49.
11. (a) Brown, H.C. *Boranes in Organic Chemistry,* Cornell University Press, Ithica, New York, *1972*; (b) Brown, H.C.; Kramer, G.W.; Levy, A.B.; Midland, M.M. *Organic Synthesis Via Boranes*, Wiley Interscience, New York, *1975*.

2. The reaction proceeds by cis addition of H and B, probably via a four-center transition state such as **4**.

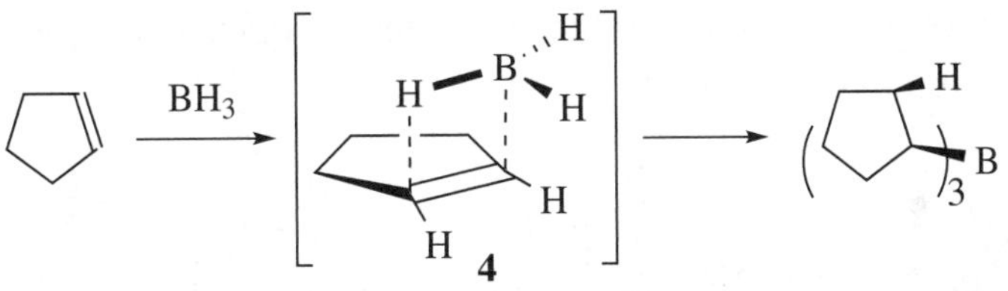

3. The boron adds regioselectively at the less sterically hindered position, as shown in the reaction with 2-methyl-1-butene.

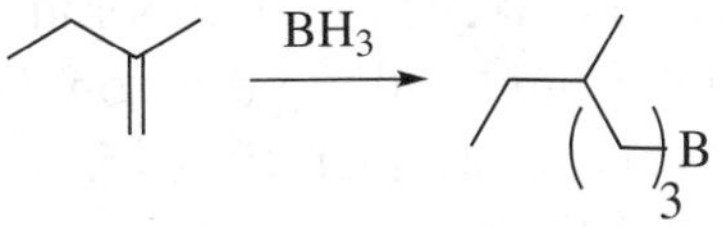

4. The boron adds stereoselectively to the less sterically hindered face of the alkene. Reaction with norbornene, for example, gave 99.6% of the exo-borane (**5**) with only 0.4% of the endo- borane.[12] This is consistent with the normal addition of reagents to the exo face of norbornyl derivatives (sec. 1.5.E).

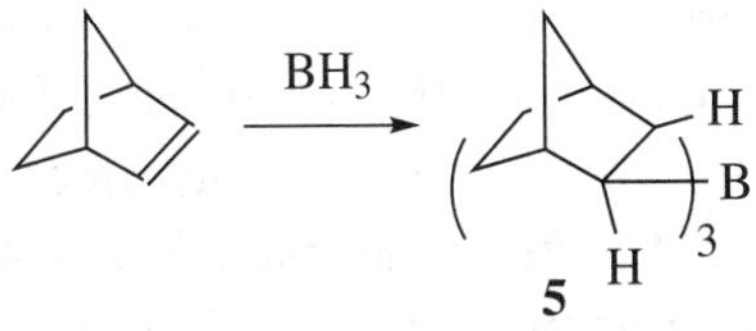

5. Hydroboration proceeds without skeletal rearrangement, at normal temperatures, as in the hydroboration of α-pinene (**6**) to give **7**.[2,11]

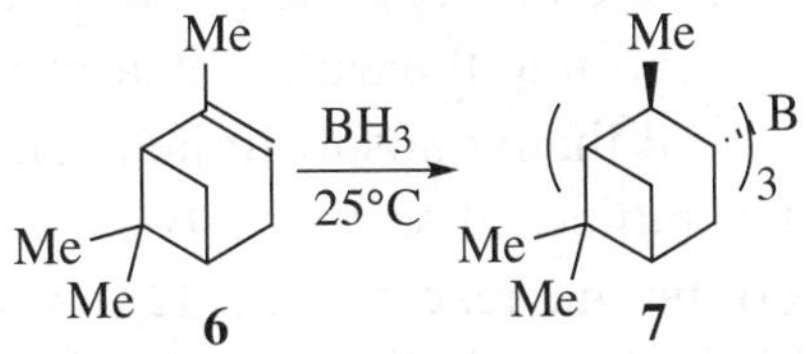

6. In ether solvents, a wide range of other functional groups can be tolerated, even reducible ones such as esters, if the stoichiometry of the reagent is controlled.

7. The organoborane product can be oxidized, eliminated, rearranged, protonated and aminated, leading to a large number of functional group transformations.

As noted in (3), borane adds selectively to the less sterically hindered carbon of the alkene or alkyne, but *mixtures* of Markovnikov product (addition of boron to the most hindered carbon) and anti-Markovnikov (addition of boron to the less hindered carbon) product are usually observed. This selectivity is best under-stood by examining the four-center transition state required for the addition (see **4**). When borane approaches the alkene unit of 2-methylpropene, two orientations are possible (**8** and **9**). In **8**, the BH_2 moiety is positioned over the less hindered

---

12. Brown, H.C. *Chemistry in Britain* **1971**, *7*, 458.

carbon bearing the hydrogen atoms [BH$_2$↔H$_2$C=] because this minimizes the destabilizing steric interactions with the methyl groups on the alkene [BH$_2$↔Me$_2$C=] found in the other orientation (**9**). In other words, this interaction destabilizes **9** and leads to selective delivery of boron to the less hindered carbon via **8**.

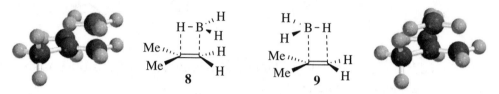

In the reaction of monosubstituted alkenes with diborane, the steric interaction between the BH$_2$ moiety of the borane and the single alkyl group on the π bond is sufficient to give >90% selectivity for the anti-Markovnikov product **10**, over **11** (93:7 for R^1 = Et, R^2, R^3 = H).[13] With di- and trisubstituted alkenes the increased steric interaction between the boron and alkyl groups on the alkene led to extremely high selectivities (R^1 = Ph and R^2 = Et in **10**, for example). Electronic effects are also a factor in the reaction of styrene, which gave only 80% selectivity for **10**. This electronic effect diminishes with distance (as when R^1 is benzyl, which behaves as if it were a simple alkyl group). When there is little steric difference between two groups in disubstituted alkenes (the B–H↔R^1 and B–H↔R^3 interactions are about the same, as when R^1 = Et and R^3 = Me), poor selectivity was observed, giving close to a 1:1 mixture of **10** and **11**. cis Alkenes with large groups attached to the π-bond generally gave good selectivity for the anti-Markovnikov borane, but cis alkenes with small groups showed poor selectivity. In trans alkenes, large differences in the size of substituents were less important, although there was some selectivity for addition of boron to the carbon bearing the smaller substituent. Addition of boron occurred preferentially at the monosubstituted carbon in trisubstituted alkenes. In a synthesis of the bis(spiroacetal) moiety of spirolide B by Brimble and Meilert,[14] for example, the dimethyl sulfide complex of borane reacted with **12** to give a 78% yield of primary alcohol **13** after treatment of the initially formed alkylborane with sodium hydroxide and hydrogen peroxide (see sec. 5.4.A).

There are three B–H bonds in borane and therefore three equivalents of hydride are available

13. Reference 2, pp 113-122.
14. Meilert, K.; Brimble, M.A. *Org. Lett.* **2005,** *7,* 3497.

to react with three equivalents of alkene, which would give a trialkylborane. The boron-alkyl interactions in the four-center transition state (**8** or **9**) largely control the reaction, but destabilizing steric interactions between alkyl substituents attached to the boron in the alkylborane products are also important. Terminal alkenes (such as 1-decene) and relatively unsubstituted alkenes (such as 2-butene or cyclohexene) usually react with borane to give trialkylboranes (R_3B).[15] Relatively unhindered trisubstituted alkenes such as 2-methyl-2-butene usually give a dialkylborane (R_2BH),[15] and hindered alkenes such as *tert*-butylcyclohexene or 2,4,4-trimethyl-2-butene lead to a monoalkylborane (RBH_2).[15] As the steric bulk of the alkyl substituent on the π bond increases, the rate of addition of the initially formed organoborane product to unreacted alkene decreases.

Borane can react with an alkene to give a monoalkylborane. The reactive unit of borane is B–H, and a monoalkylborane has two B–H unit available. It is therefore reasonable that a monoalkylborane can also react with an alkene, forming a dialkylborane. Since the dialkylborane has one B–H unit, it can react with an alkene to form a trialkylborane. Monoalkyl and dialkylboranes are less reactive than diborane itself in reactions with alkenes because alkyl groups are electron releasing, diminishing the Lewis acidity of the boron. The mono-alkylborane derived from 1,2-dimethylcyclopentene, for example, does not react with additional 1,2-dimethylcyclopentene but it reacts rapidly with a less hindered alkene such as 1-butene to give a mixed alkyl trisubstituted borane. Similarly, disubstituted boranes react with 1 equivalent of a new alkene to give a mixed trialkylborane, but they are less reactive than monoalkylboranes. The point of these observations is that mono- and dialkylboranes function as new hydroboration reagents, reacting with alkenes in a manner analogous to borane. no hydrogen remains on boron If a trialkylborane is formed, and it cannot react with additional alkene. Brown used this premise to generate several important organoboranes[16] that have

become commercially available. Reaction of 2-methyl-2-butene gives **14**, diisoamylborane (diisoamylborane = disiamylborane = Sia$_2$BH), 2,3-dimethyl-2-butene gives **15** (*tertiary* hexylborane = thexylborane, ThexBH$_2$)

and α-pinene (**6**) gives **16** (diisopinocampheylborane = Ipc$_2$BH). Dienes form cyclic boranes, as with the reaction of reaction of 2,4-dimethyl-1,4-pentadiene (**17**) to give **18** (3,5-dimethylborinane). A highly useful

borane reagent is prepared by the reaction of 1,5-cyclooctadiene (**19**) and borane, giving **20** (9-borabicyclo-[3.3.1]nonane = 9-BBN).

15. Reference 2, pp 102-112.
16. Lane, C.F. *Aldrichimica Acta* **1973**, *6*, 21.

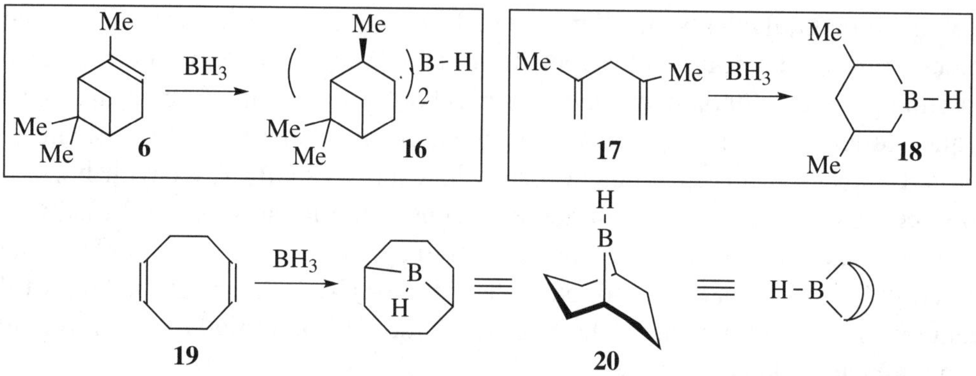

Me   BH$_3$ Me   B–H

Me   Me   **6**   Me   Me   **16**

Me   Me   BH$_3$ Me

Me   **17**   Me   B–H   **18**

BH$_3$ B   H   ≡   H–B

   19   **20**

5.2.A.ii. Regioselectivity and Diastereoselectivity.

There are two stereochemical advantages in using hindered alkylboranes as hydroboration reagents: increased anti-Markovnikov selectivity and enhanced cis selectivity. This selectivity can be seen in Table 5.1,[17] which compares reactions of borane, disiamylborane and 9-BBN with several alkenes. Hindered boranes show greater selectivity for delivery of boron to the less sterically hindered carbon of terminal alkenes (to =CH$_2$) than does diborane. Perhaps the most dramatic increase in selectivity is observed with internal alkenes since they show poor selectivity with borane, whereas the hindered 9-BBN delivers hydrogen selectively to the less hindered carbon in such systems. Borane gave only 80% addition to the terminal carbon of styrene upon reaction, despite the electronic effects of the phenyl. When the bulky 9-BBN added to styrene, however, >98.5% addition to the terminal carbon was observed. It is clear from Table 5.1 that 9-BBN and disiamylborane react with alkenes to give extremely high yields of the anti-Markovnikov addition product. These reagents are more expensive, but 9-BBN is more stable than the pyrophoric diborane, which is almost always used in solution or generated *in situ*.

17. (a) Cragg, G.M.L. *Organoboranes in Organic Synthesi,* Marcel-Dekker, New York, *1973,* pp 63-84; (b) Brown, H.C. *Aldrichimica Acta 1974, 7,* 43.

Table 5.1. Enhanced Anti-Markovnikov Selectivity with Bulky Boranes

Alkene[a]	BH$_3$		Sia$_2$BH		9-BBN	
	a	**b**	**a**	**b**	**a**	**b**
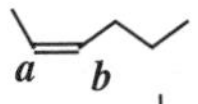 (alkene, a b)	45	45	57	33		
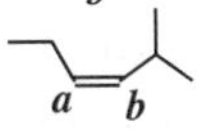 (alkene, a b)	56	43	87	3	90	3
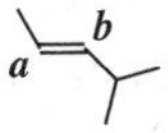 (alkene, b a)	51	39	86	5		
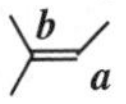 (alkene, b a b)			98	2	94	1
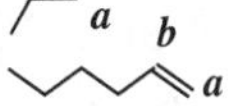 (alkene, a)	94	06	99	1	99.9	0.1
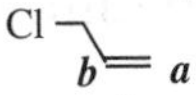 Cl (alkene, b a)	60	40	98	2		
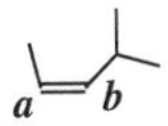 (alkene, a b)	57	43	97	3	99.9	0.1
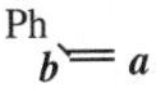 Ph (alkene, b a)	81	19	98.5	1.5		
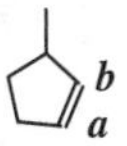 (alkene, b a)	48	40	60	40		

[a] Ca is always the less sterically hindered carbon. For disubstituted alkenes, Ca is the carbon distal (most distant from) a pedant substituent. a and b refer to the points of attachment of the boron atom.

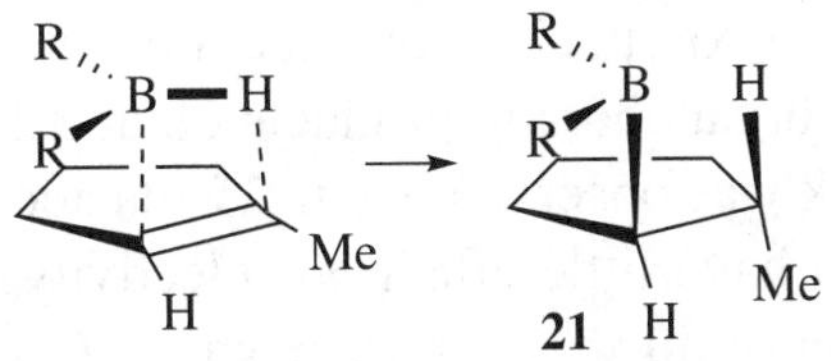

Diborane as well as mono- and dialkylboranes give cis addition in reactions with alkenes, as noted above, which is a result of the four-center transition state in which the boron and hydrogen are delivered from the same face, as illustrated by hydroboration of 1-methylcyclopentene with R$_2$BH to give **21**. The methyl-bearing sp^2 carbon of methylcyclopentene is a prochiral center, and addition of the borane will generate the (R) and (S) enantiomers of the trans-isomer. Hydroboration produces only the *trans*-borane (**21**) via cis-addition of BR$_2$ and H (see **2** in sec. 5.2.A.), making the addition highly diastereoselective. When the compound contains a substituent, diastereoselectivity is determined by approach of the borane from the less sterically hindered face of the alkene (secs. 5.2.A.ii, 5.4.B). In general, boranes containing bulky substituents give greater selectivity. Similar factors were discussed for hydride reductions (sec. 4.7.B). For cyclic molecules[18,17] possessing an exocyclic methylene group, attachment of boron occurs almost exclusively at the terminal alkene carbon. As seen for the reaction of **22** and **23** with diborane (the values in parentheses are percentage yields of each organoborane product obtained by reaction from that face), the preferred face of attack is path **b** (with the major product resulting from attack at the terminal carbon of the alkene). The borane is sufficiently bulky that the diaxial interaction of the axial

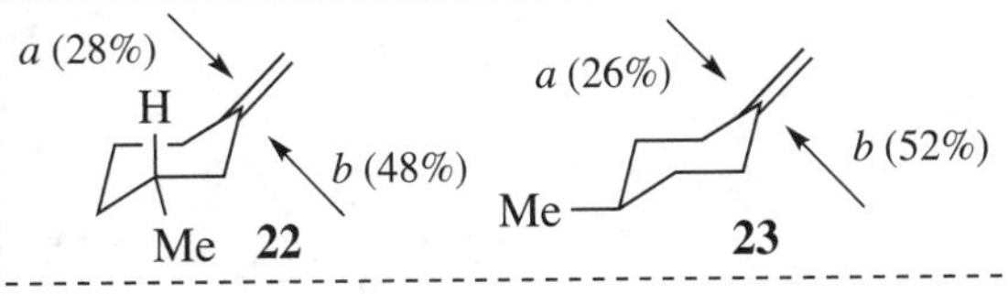

18. Klein, J.; Lichtenberg, D. *J. Org. Chem.* **1970**, *35*, 2654

hydrogens at C3 and C5 inhibits path **a** (opposite the effect seen with hydride reduction of cyclohexanone derivatives, sec. 4.7.C, where path *a* is preferred). As the size of the borane increases, the selectivity for path *b* will also increase.

When the alkene is internal and unsymmetrical, as in 3-methyl-1-cyclohexene (**24**), the reaction is more complicated. In addition to the two faces (top and bottom) of the molecule, there are two different and reactive sp^2 carbons of the alkene. Hydroboration leads to four products **25-28**, although diborane gives little regioselectivity (32:34 for top attack to give **25** and **26**, and selectivity of only 18:15 for bottom attack to give **27** and **28**). There is a preference for delivery of the boron to the face opposite the methyl group (32:18 and 34:15), making **25** and **26** the major products.[17] Similar results were observed when 1,6-dimethyl-cyclohexene reacted with borane.[19]

Reactions of borane with bicyclic alkenes give products with stereochemistry similar to that observed for reduction of bicyclic ketones with hydride reducing agents. Boranes add to norbornene derivatives from the exo face (**29** gave 85% exo attack upon reaction with diborane). In **29-32**, all values are percentage yields of the organoborane products obtained by reaction from that face. Interestingly, only 80% exo attack was observed when **29** reacted with disiamylborane,[17] and the size of the borane appears to have little effect on selectivity. Substituents at the 7-position block the exo face and reaction of **30** with diborane gave 56% of the endo product, with only 10% of the exo product. In this case, reaction of **30** and disiamylborane gave 85% of the endo product, showing enhanced selectivity.[17] Similar results were observed when the C=C unit is within the ring.[20] Reaction of 2-methylbicyclo[2.2.1]hept-2-ene (**31**) with diborane shows the expected exo selectivity and also a preference for attack at the less hindered carbon. The C=C unit in **32** has no steric features to distinguish one carbon from the other, and reaction with diborane gives essentially a 1:1 mixture of regioisomers, although there is a marked preference for exo delivery of boron.

	29	**30**	**31**	**32**
Diborane	(a) 85% exo, 15% endo	(a) 10% exo, 56% endo	(a) 99% exo	(a) 47% exo, 1% endo
Disiamylborane	(b) 80% exo, 8% endo	(b) 10% exo, 85% endo		(b) 50% exo, 2% endo

19. Pasto, D.J.; Klein, F.M. *J. Org. Chem.* **1968**, *33,* 1468.
20. Brown, H.C.; Kawakami, J.H. *J. Am. Chem. Soc.* **1970**, *92* 1990.

When the alkene is part of an edge fused bicyclic system, regio- and stereoselectivity may be influenced by the conformation of the molecule and the relative stereochemistry of the substituents. Hydroboration of **33,** in Heathcock and Kelly's synthesis of eudesmanes,[21,17] led to a 66:34 mixture of boranes with attack from the top face predominating (giving **34**).

MOLECULAR MODELING: Stereochemistry in Hydroboration

Looking at an accurate 3D model of reactants (as opposed to a 2D sketch) is often sufficient to identify the likely product of a reaction. For example, a model for the tricyclodecene derivative (top left) shows distinct "convex" (*exo*) and "concave" (*endo*) faces. This means that there are two different ways for the reagent to approach the carbon-carbon double bond. The model shows that the *exo* face is less crowded, suggesting that hydroboration will occur preferentially here and give the stereochemistry for the alcohol shown in the drawing. This "conclusion" can be amplified by changing to a space-filling model, which reveals the actual size and shape of a molecule.

Of course, "crowded" refers to reactants. While the ***Hammond Postulate*** tells us that the transition state will resemble the reactants if the process is sufficiently *exothermic*, don't forget this is just a postulate. Transition state calculations based on quantum chemical methods are the "gold standard" for modeling selectivity. For the reaction described above, they show that the transition state for hydroboration onto the convex face (top center) is 45 kJ/mol lower in energy than that for reaction onto the concave face (top right). This difference is large enough that only one product will be seen.

Transition state models are not only able to identify the favored direction for approach of a reagent, but are also able to account for changes in selectivity (differences in transition-state energies leading to different products) with change in reactants. For example, calculations show that replacing the methyl ester by hydrogen lowers the difference between the *exo* and *endo* transition states to 25 kJ/mol (*exo* is still favored). Compare the *exo* (bottom center) and *endo* (bottom left) transition states with those of the methyl ester. Has the cavity opened up in response to removal of the ester group?

Although hindered alkylboranes are relatively stable, dialkylboranes can disproportionate, which limits their utility in reactions with alkenes. One solution to this problem was the discovery that boranes containing heteroatoms stabilize the borane and are easy to prepare. Diborane reacts with alcohols such as methanol to form dimethoxyborane [$(MeO)_2BH$].[22]

21. Heathcock, C.H.; Kelly, T.R. *Tetrahedron* **1968**, *24*, 1801.
22. Brown, H.C.; Bigley, D.B. in Reference 11b, p 44

$$2 \text{ MeOH} \xrightarrow{\text{BH}_3, \text{ THF}} (\text{MeO})_2\text{B–H} + 2 \text{ H}_2$$

Similarly, alkyl glycols react to give cyclic dialkoxyborates[23] but these are somewhat unreactive in subsequent reactions with alkenes. Catecholborane (**36**, formed by the reaction of borane with catechol, **35**) reacts with alkenes to give an alkoxyborinate,[24] but the hydroboration reaction requires heating to ~ 100°C. An example is the reaction of **36** with norbornene to form the exo-alkoxyborinate (**37**).[24] An advantage in using catecholboranes is that the boronic acid byproducts are more easily hydrolyzed than the corresponding dialkylboranes (sec. 5.4.A). A synthetic example is taken from a synthesis of apoptolidinone by Crimmins and co-workers,[25] where **38** reacted with **36** and Wilkinson's catalyst [ClRh(PPh$_3$)$_3$][26] with high selectivity for the primary position, and oxidation gave **39** in excellent yield. Note that the monosubstituted alkene reacted preferentially to the trisubstituted alkene in **38**.

Chloroboranes are another class of heteroatom-substituted boranes. Dichloroborane (Cl$_2$BH) and monochloroborane (ClBH$_2$) are formed by the reaction of diborane with varying proportions of boron trichloride etherate (BCl$_3$•OEt$_2$). In ether solvents, both boranes are formed as their etherate complex.[27] An alternative synthesis reacted HCl with borane•THF,[28] but reduction of boron trichloride with lithium borohydride in ether proved to be the most useful route.[27]

In contrast to catecholborane, chloroborane reacts with alkenes at 0°C to give a dialkyl chloroborane such as **40** from cyclopentene.[29] Chloroboranes show great regioselectivity for attachment of boron at the terminal carbon of 1-alkenes (anti-Markovnikov addition).[30] 1-Hexene gave 99.5% of the terminal borane, for example, whereas similar reaction with diborane gave only 94%.[17b] Similarly, styrene reacted to give 96% of the terminal borane

23. Woods, W.G.; Strong, P.L. *J. Am. Chem. Soc.* **1966**, *88*, 4667.
24. Brown, H.C.; Gupta, S.K. *J. Am. Chem. Soc.* **1971**, *93*, 1816.
25. Crimmins, M.T.; Christie, H.S.; Chaudhary, K.; Long, A. *J. Am. Chem. Soc.* **2005**, *127*, 13810.
26. Used to catalyze the hydroboration reaction. See Evans, D.A.; Fu, G.C.; Hoveyda, A.H. *J. Am. Chem. Soc.* **1992**, *114*, 6671.
27. Brown, H.C.; Tierney, P.A. *J. Inorg. Nucl. Chem.* **1959,** *9*, 51.
28. Zweifel, G. *J. Organomet. Chem.* **1967,** *9*, 215.
29. Brown, H.C.; Ravindran, N. *J. Am. Chem. Soc.* **1972,** *94*, 2112.
30. Brown, H.C.; Ravindran, N. *J. Org. Chem.* **1973,** *38*, 182.

compared with 81% from diborane despite the electronic effects of the phenyl group. Internal alkenes show less selectivity as in the reaction of 4-methyl-2-pentene, which gave only 60% delivery of boron to the 2-position. 2,3-Dimethyl-2-butene, however, reacted to give 99.7% of product with the boron moiety at the 2-position. The addition of chloroborane to alkenes is less influenced by steric factors than the similar addition of mono- or dialkylboranes, accounting for the generally poorer selectivity exhibited by chloroborane in its reactions with internal alkenes.

5.2.A.iii. Isomerization of Alkylboranes. One of the early discoveries of organoboranes was their tendency to undergo isomerization if the temperature approached 160°C. This isomerization involves migration of the boron from a more to a less substituted position, usually giving the terminal alkylborane as the predominant isomer.[31] Reaction of 3-hexene with borane initially gave 3-hexylborane **45**, but heating induced rearrangement to 1-hexylborane (**46**). This isomerization can be used to advantage in subsequent reactions of boranes, such as oxidation to alcohols (sec. 5.4.A). If the initial hydroboration occurs *on* a ring (as in **47** from the cyclic alkene), isomerization will put boron at the terminal position of an alkyl side chain, as in the conversion of 1-methylcyclohexene to **49** via heating **47** to form **48**, followed by subsequent heating with 1-decene to give tridecylborane and **49**.[32,33b]

A sacrificial alkene such as 1-decene or 1-hexadecene can be added to drive the overall reaction toward a specific target alkene. Mechanistically, thermal isomerization apparently involves a series of reversible elimination reactions of borane followed by readdition in an anti-Markovnikov manner.[2] If the sacrificial alkene has a significantly higher boiling point, the desired alkene can be distilled from the reaction medium.[31a] The sacrificial alkene must not boil < 160°C, however, which is the temperature required for borane isomerization. This protocol makes it possible to isomerize the double bond of an alkene to a less substituted isomer, a process that Brown termed **contrathermodynamic isomerization**.[33,34] Similar to the preparation of **49**, α-pinene was converted to β-pinene in 54% yield[32] by contrathermodynamic

31. (a) Brown, H.C.; Subba Rao, B.C. *J. Am. Chem. Soc.* **1959**, *81*, 6428; (b) Brown, H.C.; Zweifel, G. *Ibid.* **1966**, *88*, 1433.
32. Brown, H.C.; Zweifel, G. *J. Am. Chem. Soc.* **1967**, *89*, 561.
33. (a) Brown, H.C.; Bhatt, M.V. *J. Am. Chem. Soc.* **1966**, *88*, 1440; (b) Brown, H.C.; Bhatt, M.V.; Munekata, T.; Zweifel, G. *Ibid.* **1967**, *89*, 567 .
34. Brown, H.C., Subba Rao, B.C. *J. Org. Chem.* **1957**, *22*, 1137.

isomerization. Significant amounts of endocyclic isomers often accompany this hydroboration-bond migration process with exocyclic alkenes, and separation of these isomeric by-products can be difficult, limiting the utility of this process.

5.2.A.iv. Hydroboration of Dienes and Heteroatom-Containing Alkenes.

Hydroboration has also been reported with a variety of heteroaryl alkenes.[35] The regioselectivity of hydroboration with 2-vinylpyridine, 3-vinylpyridine, 4-vinylpyridine, as well as 2-vinylfuran, 2-(2-propenyl) furan, 2-vinylthio-phene, 2-(2-propenyl)thiophene were compared with styrene, which showed the expected selectivity for addition of boron to the terminal carbon. In general, the presence of nitrogen in the pyridyl derivatives changed the selectivity for addition from the terminal carbon (the β-carbon) to the α-carbon, even when the reagent was not sterically hindered. Nitrogen releases electrons to the π-bond and the carbon closest to nitrogen receives the most electron density. In reactions with diborane, this electronic effect made attack at C_a preferred when the nitrogen was ortho (2-pyridyl) or para (4-pyridyl). The meta compound (3-pyridyl) does not transmit electrons through the π-system as effectively, and the selectivity was poorer. When excess borane was used, the first equivalent coordinated to nitrogen and the second to the alkene. Addition of BF_3 had the same effect, allowing the subsequent addition of a single equivalent of borane. The bulky reagent 9-BBN showed high selectivity for attack at C_b with 2-pyridyl and 3-pyridyl substituted alkenes, but the electronic effect of the aromatic groups also led to significant amounts of attack at C_a. When the nitrogen was further away (as in 4-pyridyl), the steric and electronic effects are less important and attack at C_a is preferred. Analogous effects were observed with the furan and thiophene derivatives.

Hydroboration of conjugated dienes raises the usual issue, 1,2- versus 1,4-addition. Conjugated dienes such as 1,3-cyclohexadiene (**50**) gave mainly 1,4-addition (85:10 **51:52** with 9-BBN is a typical example). Reaction with dibromoboranes, however, favors the dihydroboration product.[36] In the case of the symmetrical dienes **51** and **52** could result from 1,2-addition to different carbons of one C=C moiety. In this example, however, **51** is considered to be a 1,4-addition product and **52** is considered to be a 1,2-addition product. As the size of the ring increases, dihydroboration increases at the expense of mono-hydroboration. Six-membered rings are less reactive in hydroboration reactions relative to larger and smaller rings.[37] Nonconjugated dienes such as 1,4-cyclohexadiene show the expected independence of the double bonds, giving a mixture of mono-hydroboration and dihydroboration products.[37]

35. (a) Brown, H.C.; Vara Prasad, J.V.N.; Zee, S.-H. *J. Org. Chem.* **1986**, *51*, 439; (b) Brown, H.C.; Vara Prasad, J.V.N.; Gupta, A.-K. *Ibid.* **1986**, *51*, 4296.
36. Brown, H.C.; Bhat, K.S. *J. Org. Chem.* **1986**, *51*, 445.
37. Zweifel, G.; Nagase, K.; Brown, H.C. *J. Am. Chem. Soc.* **1962**, *84*, 190.

5.2.B. Alkenylboranes

53

Boranes add to one π bond of an alkyne just as they add to the π bond of an alkene and the product is the corresponding alkenyl (vinyl) borane, which undergoes many of the same reactions as alkylboranes.[38] Internal alkynes such as 3-hexyne react with diborane to give the trialkenylborane (**53**) in moderate yield, but terminal alkynes such as 1-hexyne give predominantly the 1,1-diboranyl derivative (**54**), plus unreacted alkyne.[39] The 1,1-disubstituted product is formed because hydroboration of the alkene product competes with reaction of the starting alkyne, but hindered boranes such as 9-BBN or Sia_2BH react faster with alkynes than with vinylboranes[40] leading to less of the 1,1-disubstituted product. This rate difference is apparent in the reaction of octa-4-ene-1-yne (**55**) with Sia_2BH to give the (*E,E*)-dienyl-borane, **56**.[39] Unhindered terminal alkenes react with organoboranes faster than do alkynes, or at least competitively, especially with unhindered boranes such as diborane.

Although 9-BBN reacts with terminal alkynes to give the trans-alkenylborane, significant amounts of 1,1-diborane are also formed due to the secondary addition to the alkene mentioned above.[41] Brown suggested this was due to the accessibility of the boron to one face of 9-BBN, which can be corrected by addition of a large excess of the alkyne, which is easily recovered, although this is a poor choice for the hydroboration of expensive alkynes. The alkenylborane derived from 9-BBN is stable enough for isolation and can be conveniently purified, in contrast to most other alkenylboranes. The 9-BBN reagent reacts faster with terminal alkynes[42] than it does with internal alkynes, as shown by the conversion of octa-1,4-diyne (**57**) to **58**.[38] Terminal alkynes react with boranes to give the trans-alkenylborane and 9-BBN reacts with internal alkynes to give the (*Z*)-alkenylborane (with the alkyl groups cis-).[40]

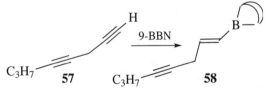

Catecholborane (**36**) reacts with alkynes but requires reaction temperatures that are higher than

38. Brown, H.C.; Campbell, Jr., J.B. *Aldrichimica Acta* **1981,** *14,* 3.
39. Brown, H.C.; Zweifel, G. *J. Am. Chem. Soc.* **1961,** *83,* 3834.
40. Brown, H.C.; Moerikofer, A.W. *J. Am. Chem. Soc.* **1963,** *85,* 2063.
41. Brown, H.C.; Scouten, C.G.; Liotta, R. *J. Am. Chem. Soc.* **1979,** *101,* 96.
42. Brown, C.A.; Coleman, R.A. *J. Org. Chem.* **1979,** *44,* 2328.

those required with alkylboranes, as noted earlier. Phenylethyne, for example, reacts with **36** to give excellent yields of the corresponding vinyl borane in refluxing THF.[43] Alkenylborinates derived from catecholborane are usually stable, allowing them to be isolated by distillation or recrystallization.[38]

With two active hydrogen atoms on boron, thexylborane reacted with two equivalents of 1-nonyne to give primarily (but not exclusively) the *trans-,trans-*dialkenylborane **59**.[44] Terminal alkynes can be used to generate mixed (alkyl-thexyl-alkynyl) boranes that can be used in other transformations (secs. 5.3, 5.4).[44] Reaction of monochloroborane etherate with alkynes gives a dialkenylborane.[45] Significant amounts of the 1,1-diborane are usually formed in reactions with terminal alkynes, but using a 40% excess of the alkyne leads to nearly quantitative yields of the dialkenylborane.[46] A more attractive alternative is to use dibromoborane•methylsulfide (HBBr$_2$•SMe$_2$),[47] which reacts by simple addition with internal alkynes[48] but reacts with terminal alkynes to give the dialkenylborane without formation of the 1,1-diborane byproduct.

Internal alkynes react with boranes to form the cis-alkenylborane. This, of course, is due to the four-center transition state leading to the vinylborane product. The reaction is selective for the less hindered of the two available alkynyl carbons, due to the steric demands of the hydroboration reagent as well as the size and electronic characteristics of the alkynyl substituents. These results are similar to those observed with alkenes when steric factors are the primary influence on the reaction, but the linear nature of alkynes leads to less selectivity than similar reactions with alkenes. When electronic factors dominate, selectivity in the hydroboration reaction is diminished and mixtures of alkenylboranes are formed.[38] The regioselectivity for reaction of several terminal alkynes with hydroboration reagents is shown in Table 5.2.[38] The reagent of choice for selectivity in the hydroboration of alkynes depends largely on steric factors with most hydroboration reagents, as seen in Table 5.2.

43. (a) Brown, H.C.; Gupta, S.K. *J. Am. Chem. Soc.* *1972, 94,* 4370; (b) *Idem Ibid.* *1975, 97,* 5249.
44. Zweifel, G.; Brown, H.C. *J. Am. Chem. Soc.* *1963, 85,* 2066.
45. Brown, H.C.; Ravindran, N. *J. Am. Chem. Soc.* *1976, 98,* 1785.
46. Brown, H.C.; Ravindran, N. *J. Am. Chem. Soc.* *1976, 98,* 1798.
47. (a) Kinberger, K.; Siebert, W. *Z. Naturforsch., B* *1975, 30,* 55; (b) Brown, H.C.; Ravindran, N. *Inorganic Chem.* *1977, 16,* 2938.
48. Brown, H.C.; Campbell, Jr., J.B. *J. Org. Chem.* *1980, 45,* 389.

Table 5.2. Directive Effects for Hydroboration of Substituted Internal Alkynes[a]

Reagent	$n\text{-}C_3H_7\text{-}C\text{≡}C\text{-}H$		$\text{>-}C\text{≡}C\text{-}H$		$\bigcirc\text{-}C\text{≡}C\text{-}H$		$Ph\text{-}C\text{≡}C\text{-}H$	
	a	**b**	**a**	**b**	**a**	**b**	**a**	**b**
Diborane	40	60	25	75				
Thexylborane	39	61	19	81				
Disiamylborane	39	61	7	93	9	91	19	81
Dicyclohexylborane	33	67	8	92	8	92	29	81
HBBr$_2$•SMe$_2$	25	75	4	96	9	91	64	36
9-BBN	22	78	4	96	4	96	65	35

[a] The labels a and b denote the point of attachment of the boron.

An alternative route to vinylborinates adds the dimethyl sulfide complex of HBBr$_2$ to 1-bromo-1-alkynes. Reaction of 1-bromo-1-hexyne (**60**) gave **61** in 87% yield after treatment with an alcohol (2-propanol in this case).[49] Subsequent reduction of the bromide with potassium triisopropoxyborohydride [KBH(Oi-Pr)$_3$] generated cis-vinylborane (**62**) in 89% yield.

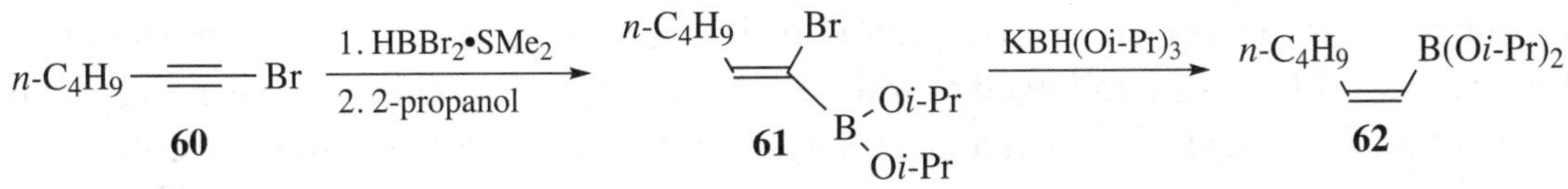

5.3. SYNTHETIC TRANSFORMATIONS

Sections 5.1 and 5.2 described methods for preparing organoboranes and discussed the regio- and stereo-selectivity associated with their formation from alkenes and alkynes. The synthetic value of organoboranes, however, lies in their ability to serve as intermediates for the transformation of alkenes and alkynes into other useful functional groups. The discussion will begin with the reactions of organoboranes to generate functional groups that do not contain heteroatoms. Heteroatom functional groups will be discussed in Sections 5.4-5.5.

5.3.A. Protonolysis

Alkylboranes react with oxygen and are pyrophoric (the lower alkylboranes burn with a green flame). The product of this reaction is an alkylborinate (R$_2$BOR). Although tri-n-butylborane (and higher analogs) do not spontaneously burst into flame, they are sensitive to oxygen and hydroboration reactions should be protected from air.[50] Exceptions are 9-BBN and catecholborane, which are much less reactive with oxygen. This sensitivity to oxygen stands in contrast to the relative insensitivity of alkylboranes to water and alcohol. The reaction of

---

49. Brown, H.C.; Imai, T. *Organometallics* **1984**, *3*, 1392.
50. Reference 2, p 67.

trimethylborane, for example, requires 180°C for 7 h to give only 69% of dimethylborinic acid.[51] Not only is the reaction with water, alcohols or phenols slow, addition of aqueous alkali actually stabilizes trialkylboranes to hydrolysis.[52] Alkylboranes can be hydrolyzed with aqueous mineral acids, but only under vigorous conditions. Conversion of tri-*n*-butylborane to di-*n*-butylboronic acid (**64**) required heating with 48% HBr at reflux,[53] and proceeded through the intermediate bromoborane **63**.

A more useful method for protonolysis is the reaction of alkylboranes and carboxylic acids at temperatures near 100°C.[54] In a typical reaction, two of the alkyl groups of the trialkylborane are removed at ambient temperature on treatment with propionic acid.[55] The yields of hydrocarbon derived from the alkyl portion of the borane are excellent, but protonolysis of the third alkyl group often requires refluxing in diglyme.[55] Primary alkyl groups undergo protonolysis faster than secondary alkyl. Protonolysis proceeds even in the presence of oxygen, nitrogen, chlorine or sulfur substituents. The mechanism of protonolysis begins with coordination of the boron to the carbonyl oxygen of the acid, followed by transfer of hydrogen to the carbon of the alkyl group via six-center transition state **65**.[56] Synfacial transfer of hydrogen leads to retention of configuration for transfer of boron to oxygen and hydrogen to carbon, leading to R–H and **66**. This retention of configuration during the reaction is easily observed in the reaction of tri-*exo*-norbornyl-borane (**67**) with refluxing deuteropropionic acid. The product is *exo*-deuteronorbornane (**68**), where the transfer of deuterium proceeded with complete retention of configuration.[57]

The protonolysis reaction is sensitive to steric effects, presumably in the transition state (see **65**), as seen in organoborane **70**, formed by hydroboration of (+)-Δ^2-carene (**69**).[58] Protonolysis with glacial acetic acid at room temperature *failed* to give carane (**71**), and heating the borane

51. Ulmschneider, D.; Goubeau, J. *Berichte* **1957**, *90*, 2733.
52. Reference 2, p 63.
53. (a) Johnson, J.R.; Snyder, H.R.; Van Campen, Jr., M.G. *J. Am. Chem. Soc.* **1938**, *60*, 115; (b) Reference 2, p 63.
54. (a) Meerwein, H.; Heinz, G.; Majert, H.; Sönke, H. *J. Prakt. Chem.* **1936**, *147*, 226; (b) Goubeau, J.; Epple, R.; Ulmschneider, D.; Lehmann, H. *Angew. Chem.* **1955,** *67*, 710.
55. Brown, H.C.; Murray, K. *J. Am. Chem. Soc.* **1959**, *81*, 4108.
56. Reference 2, p 66.
57. Brown, H.C.; Murray, K.J. *J. Org. Chem.* **1961**, *26*, 631.
58. Cocker, W.; Shannon, P.V.R.; Staniland, P.A. *J. Chem. Soc. C* **1967**, 485.

in refluxing acetic acid gave 16 different products.[59] The failure of the reaction is due to the increased steric bulk around the boron, inhibiting transfer of the hydrogen. This result contrasted with the reaction of the borane of $(+)$-Δ^3-carene (double bond between C3 and C4 leading to the product with the boron at C4), which reacted with propionic acid at 150°C to give carane in 45% yield.[60]

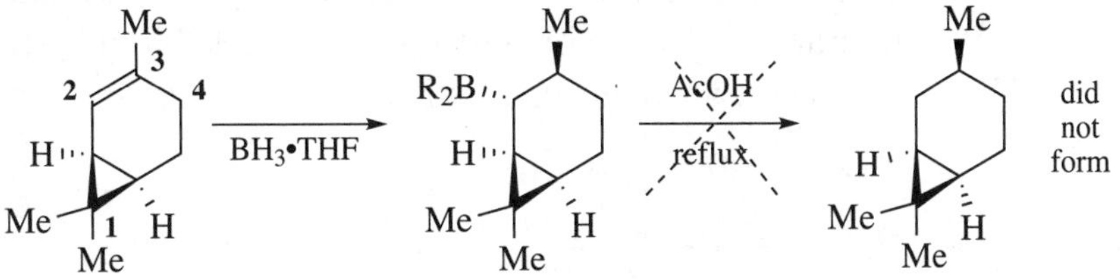

Protonolysis of alkenylboranes to generate an alkene is faster than protonolysis of alkylboranes.[61] Once again, protonolysis proceeds stereospecifically with retention of configuration.   In Unelius and co-worker's synthesis of 9,11-hexadecadienal isomers from female pheromone glands of the sugar cane borer *Diatraea saccharalis*,[62] hydroboration of the alkyne unit in **72** with dicyclohexylborane (THF, -20°C) was followed by protonolysis with glacial acetic acid to give a 68% yield of **73**, after oxidation of the dicyclohexylborinate with NaOH/H$_2$O$_2$. As with alkylboranes, protonolysis of vinylboranes with deuteroacetic acid allows the synthesis of deuterated alkenes.

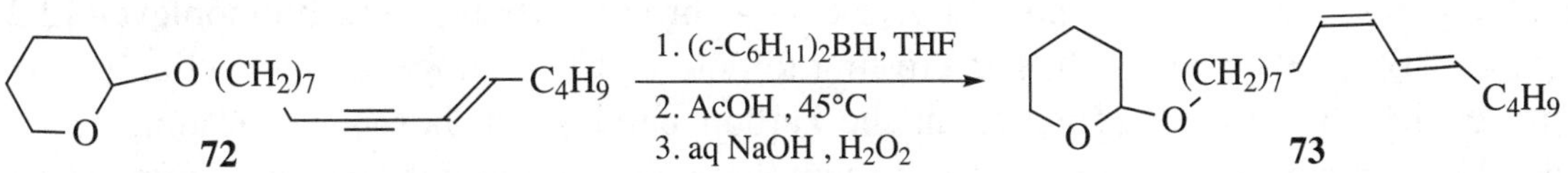

The functional group transforms pertinent to protonolysis are

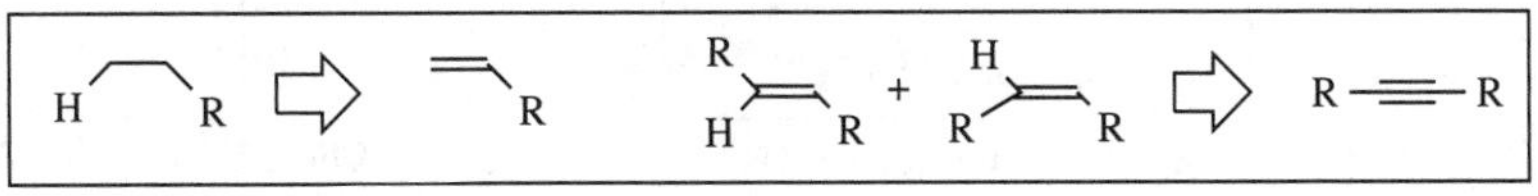

5.3.B. Halogenation

If boron of an alkylborane could be replaced with a halogen, the product would be an alkyl halide. However, reaction of alkylboranes (neat) with chlorine, bromine or iodine is very difficult.[53a] When halogenation is done with bromine or iodine dissolved in dichloromethane, however, the

59. Acharya, S.P.; Brown, H.C. *J. Am. Chem. Soc.* **1967**, *89*, 1925.
60. Brown, H.C.; Suzuki, A. *J. Am. Chem. Soc.* **1967**, *89*, 1933.
61. Brown, H.C.; Zweifel, G. *J. Am. Chem. Soc.* **1961**, *83*, 3834
62. Santangelo, E.M.; Coracini, M.; Witzgall, P.; Correa, A.G.; Unelius, C.R. *J. Nat. Prod.* **2002**, *65*, 909.

reaction is fast and is synthetically useful.[63] A simple example is the reaction of alkenes with a borane followed by addition of bromine, which leads to the alkyl bromide. 2-Bromobutane (**78**) was prepared from 2-butene in 88% yield via hydroboration-bromination.[64] Bromination occurs by a free radical mechanism. Initial reaction of the trialkylborane with bromine generates a bromine radical (Br•), bromoborane (**74**), and 2-butyl radical (**75**). Subsequent reaction with tri-*sec*-butylborane generates the radical at the carbon α to the boron (**76**), and reaction with bromine generates α-bromoborane (**77**). The final reaction between **77** and HBr generates the observed products, **78** and **74**.[63] There are many cases where an organoborane does not react very

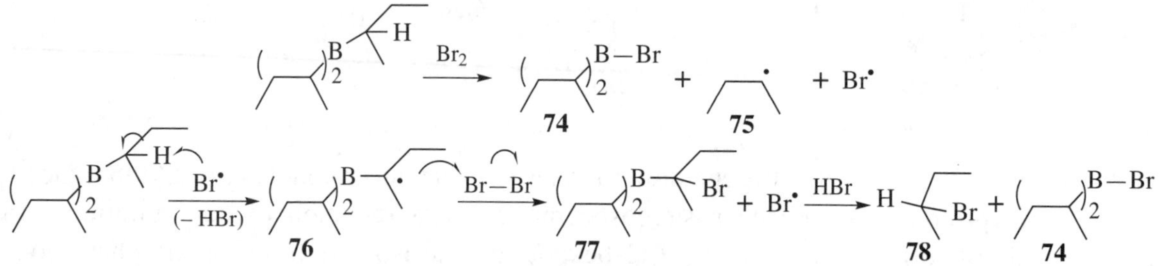

well with bromine, but addition of bases such as sodium hydroxide or sodium methoxide catalyzes the reaction, allowing it to proceed at temperatures as low as 0°C.[65] With this modification, formation of bromides (such as **78**) and iodides via hydroboration is synthetically useful. Hydroboration of norbornene initially gives the exo-borane (**67**), but subsequent reaction with bromine and the borane is slow. Addition of NaOMe converts **67** to an *ate* complex (**79**), which rapidly reacts with bromine to give endo-2-bromonorbornane (2-bromobicyclo[2.2.1] heptane, **80**) in 75% yield.[66] It is interesting to note that the halogenation reaction proceeds with net inversion of configuration at the carbon bearing the boron, suggesting backside attack by bromine. Formation of the *ate* complex and reaction with bromine involves ionic intermediates, not free radical intermediates as noted in reactions of **79**.

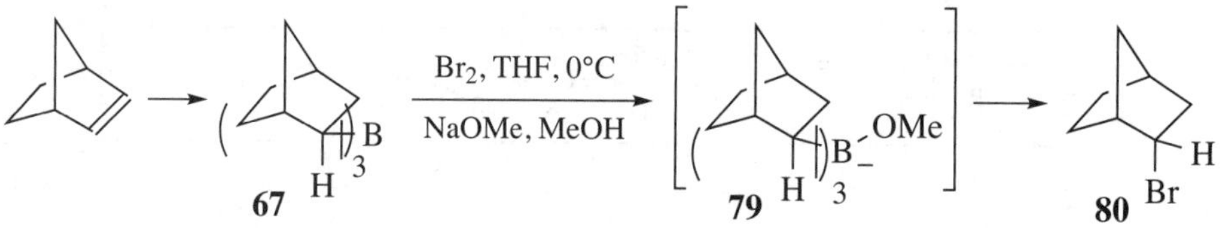

Halogenation of vinylboranes generates vinyl halides, and modification of the reaction conditions can lead to either cis or trans halides. When *E*-alkenyl boronic acid (**81**, derived from 1-octyne) was treated with iodine and sodium hydroxide, *E*-1-iodo-1-octene (**82**) was formed in 90% yield.[67] When the boronic acid was treated with iodine and then with base,

63. Lane, C.F.; Brown, H.C. *J. Am. Chem. Soc.* **1970**, *92*, 7212.
64. Lane, C.F.; Brown, H.C. *J. Organomet. Chem.* **1971**, *26*, C51.
65. (a) Brown, H.C.; Rathke, M.W.; Rogic, M.M. *J. Am. Chem. Soc.* **1968**, *90*, 5038; (b) Brown, H.C.; Lane, C.F. *J. Am. Chem. Soc.* **1970**, *92*, 6660.
66. Brown, H.C.; Lane, C.F. *J. Chem. Soc. Chem. Commun.* **1971**, 521.
67. Brown, H.C.; Hamaoka, T.; Ravindran, N. *J. Am. Chem. Soc.* **1973**, *95*, 5786.

the (Z)-alkenyl iodide (**83**) was produced.[68] Vinylboranes derived from internal alkynes lead to cis-trans mixtures with both of these procedures. Boronic acids derived from alkynes and catecholborane give the (Z)-bromide on addition of bromine followed by sodium hydroxide.[69]

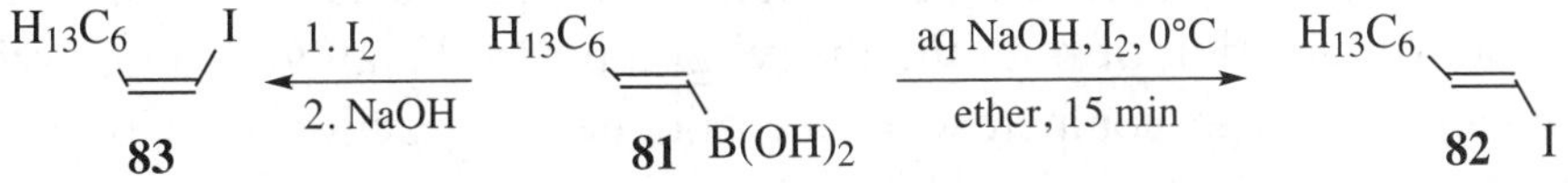

There is an alternative method for converting alkynes to vinyl iodides using B-iodo-9-BBN.[70] This transformation is illustrated in a synthesis of alcyonin by Overman and co-workers, in which the alkyne unit in **84** was treated with this reagent.[71] Subsequent protonolysis gave an 80% yield of **85**. Note that Piv = pivaloyl. This procedure is more direct and avoids some potential problems with previous procedures.

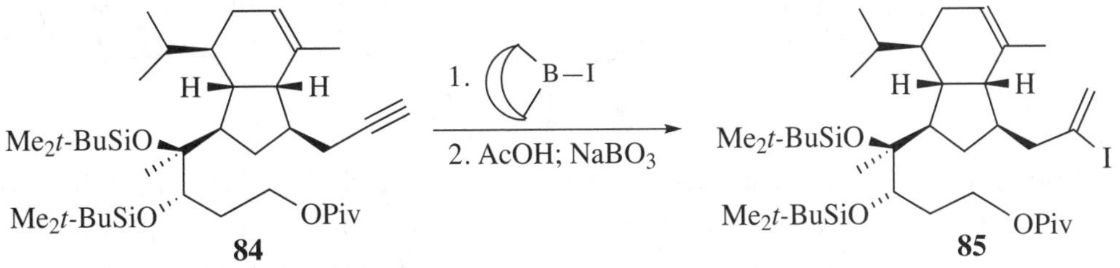

The halogenolysis reaction makes possible the following transformations:

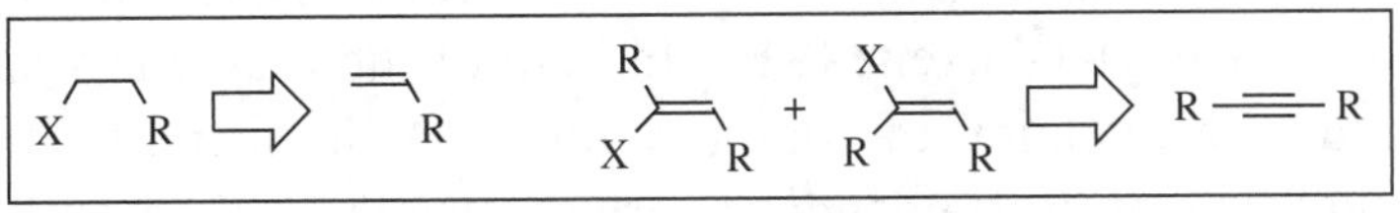

5.3.C. Alkene, Diene, and Alkyne Synthesis

Vinylboranes are versatile intermediates and there are modifications that convert them to alkenes. Hydroboration of a terminal alkyne such as cyclohexylacetylene (**86**) with dicyclohexylborane gave (E)-alkenylborane (**87**). Subsequent treatment with iodine and sodium hydroxide resulted in transfer of an alkyl group from boron to the alkenyl carbon, generating cis-1,2-dicyclohexylethene (**88**) in 77% yield (92% isomeric purity)].[72] Although iodine and base are used, the product is an alkene and not a vinyl iodide. The difference in reaction conditions for halogenation and elimination are subtle, and care must be taken to direct the reaction one way or the other. Conversion to the alkene is believed to proceed via initial coordination of base to

68. Reference 10, p 51.
69. Brown, H.C.; Hamaoka, T.; Ravindran, N. *J. Am. Chem. Soc.* **1973**, *95*, 6456.
70. (a) Hara, S.; Dojo, H.; Takinami, S.; Suzuki, A. *Tetrahedron Lett.* **1983**, *24*, 731; (b) For preparation of B-iodo-9-BBN, see Brown, H.C.; Kulkarni, S.U. *J. Organomet. Chem.* **1979**, *168*, 281.
71. Corminboeuf, O.; Overman, L.E.; Pennington, L.D. *Org. Lett.* **2003**, *5*, 1543.
72. Zweifel, G.; Arzoumanian, H.; Whitney, C.C. *J. Am. Chem. Soc.* **1967**, *89*, 3652.

the boron (to form an *ate* complex), followed by formation of an iodonium (or halonium) ion. This positively charged intermediate facilitates a 1,2-B → C alkyl shift, which is followed by trans elimination of iodine and boron.[10] A major limitation to this procedure is that it is useful only with symmetrical dialkylboranes. Mixed boranes such as those formed from thexylborane and alkynes lead to migration of both the thexyl group and the alkyl group, giving isomeric mixtures. Halogenation can compete with the elimination reaction as expected, which led to development of an alternative cis-alkene synthesis illustrated by the reaction of 4-octyne and 9-BBN. The product of this reaction was *cis*-4-octene in 98% yield, after protonolysis in refluxing methanol.[73] *cis*-4-Octene was also obtained (in 99% yield) in < 15 min at 25°C if 1% acetic acid was added to the reaction.

Manipulation of a vinylborane can lead to trans-alkenes. Hydroboration of a 1-bromoalkyne, followed by treatment with base to induce the B → C alkyl shift and then protonolysis gives the trans alkene.[74] The conversion of 1-bromo-1-hexyne (**60**) to **89** shows that the borane may have alkyl groups different from those found in the alkyne starting material. Treatment with sodium methoxide induced rearrangement to **90**, and protonolysis with acetic acid gave the final product (*E*)-1-cyclohexyl-1-hexene (**91**). In both of these alkene syntheses, the B → C alkyl shift proceeded with retention of configuration.[75] Both alkene-forming reactions have been used for the synthesis of prostaglandin intermediates that relied on the stereoselectivity of the processes.

The dimethyl sulfide complex of chlorothexylborane (see **93**) can be used to generate alkenes from alkynes. Alkenes and alkynes are coupled using this reagent, as illustrated by the reaction of *O*-acetylpent-4-en-1-ol (**92**) and **93**, which gave chloroborane **94** as the initial product. Subsequent reaction with potassium triisopropoxyborohydride (KIPBH) generated an *ate* complex, which reacted with 1-iodo-1-butyne to give **95**. Treatment with sodium methoxide followed by protonolysis gave *trans*-5-octenyl ethanoate (**96**).[76] An alternative synthetic route to

73. (a) Brown, H.C.; Molander, G.A. *J. Org. Chem.* **1986**, *51*, 4512; (b) Brown, H.C.; Wang, K.K. *Ibid.* **1986**, *51*, 4514.
74. Zweifel, G.; Arzoumanian, H. *J. Am. Chem. Soc.* **1967**, *89*, 5086.
75. Zweifel, G.; Fisher, R.P.; Snow, J.T.; Whitney, C.C. *J. Am. Chem. Soc.* **1971**, *93*, 6309.
76. Brown, H.C.; Lee, H.D.; Kulkarni, S.U. *J. Org. Chem.* **1986**, *51*, 5282.

(*E*)-alkenes from alkenyldialkyl-boranes has been developed using cyanogen bromide (BrCN).[77]

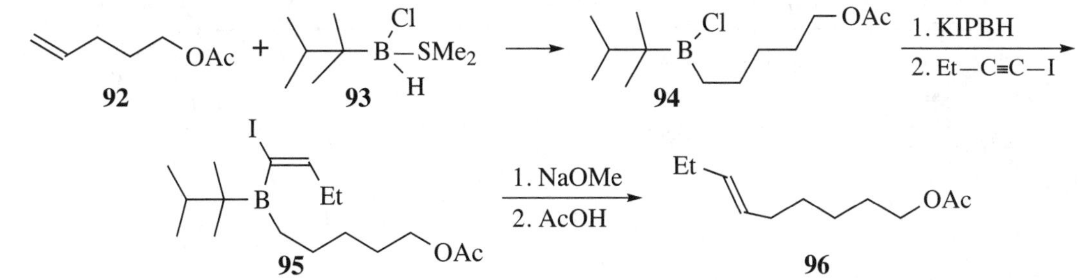

Related techniques have been developed to prepare (*Z,Z*)-, (*Z,E*)-, and (*E,E*)-dienes. Hydroboration of diacetylenes followed by protonolysis gives (*Z-,Z*)-dienes, as in the conversion of **97** to **98**.[78] The requisite symmetrical diacetylenes are prepared by oxidative coupling with oxygen and cuprous chloride, as in the conversion of 1-cyclohexylethyne (**86**) to **97**.[79] Unsymmetrical conjugated dienes are prepared by formation of a diacetylene *ate* complex, prepared from disiamylmethoxyborane by sequential reaction with different acetylides.[80] Unsymmetrical diynes are prepared from dicyclohexyl methylthioborane.[81]

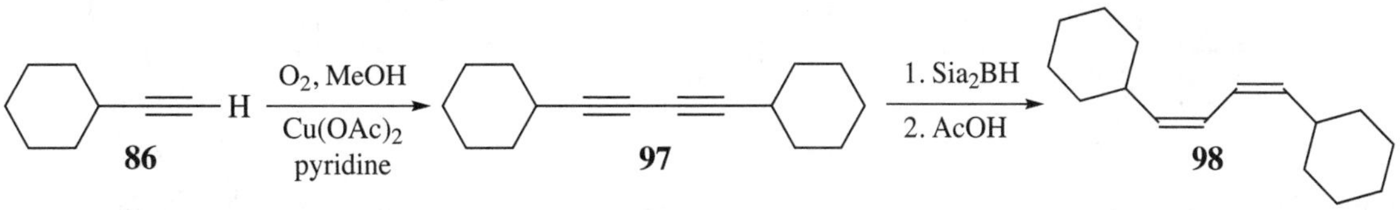

(*E,E*)-Dienes are prepared by a different route. Hydroboration of a 1-chloroalkyne such as 1-chloro-1-hexyne (**99**) with thexylborane gives addition product **100**. Treatment of **100** with 1-hexyne leads to a second addition to give alkyldialkenylborane **101**,[82] and reaction with sodium methoxide leads to a B → C shift and **102**. Protonolysis generated the (*E-,E*)-diene [**103**, (5*E*,7*E*)-dodecadiene].[82] Since two alkyne units are incorporated in different steps, this technique is suitable for the preparation of both symmetrical and unsymmetrical dienes. An alternative route to symmetrical (*E,E*)-dienes involves treatment of dialkenylchloroboranes with methylcopper[83a,48] or by a Suzuki-Miyara-type coupling (sec. 12.7.C) in which a vinylboronic acid is coupled to a vinyl iodide in the presence of a palladium (0) catalyst.[83b] Dienes can also be prepared from dialkyl-alkenylboranes[84] by reaction with sodium methoxide and then cuprous bromide-dimethyl sulfide (CuBr•SMe$_2$).

77. Zweifel, G.; Fisher, R.P.; Snow, J.T.; Whitney, C.C. *J. Am. Chem. Soc.* *1972, 94,* 6560.
78. Zweifel, G.; Polston, N.L. *J. Am. Chem. Soc.* *1970, 92,* 4068.
79. Campbell, I.D.; Eglinton, G. *Org. Synth. Coll. Vol V* *1973,* 517.
80. Sinclair, J.A.; Brown, H.C. *J. Org. Chem.* *1976, 41,* 1078.
81. Pelter, A.; Hughes, R.J.; Smith, K.; Tabata, M. *Tetrahedron Lett.* *1976,* 4385.
82. Negishi, E.; Yoshida, Y. *J. Chem. Soc. Chem. Commun.* *1973,* 606.
83. (a) Campbell, Jr., J.B.; Brown, H.C. *J. Org. Chem.* *1980, 45,* 549; (b) For an example used in a synthesis of (–)-dictyostatin, see Ramachandran, P.V.; Srivastava, A.; Hazra, D. *Org. Lett.* *2007, 9,* 157.
84. Campbell, Jr., J.B.; Brown, H.C. *J. Org. Chem, 1980, 45,* 550.

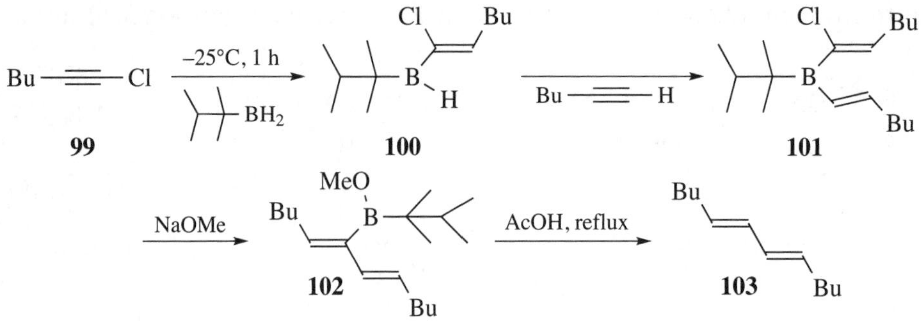

(Z-,E)-Dienes can be prepared by the reaction of alkynes with dibromoborane. Initial reaction of 1-hexyne with the dimethyl sulfide complex of $HBBr_2$ required addition of boron tribromide (BBr_3) to give the borane that was free of dimethyl sulfide.[85] In a subsequent reaction with additional 1-hexyne, divinylborane (**104**) was formed and then treated with iodine/acetic acid to give bromodiene (**105**). Addition of *tert*-butyllithium led to bromine-lithium exchange, leading to the vinyllithium reagent (secs. 8.5.A, 8.5.B), and quenching with MeOH gave **106**.

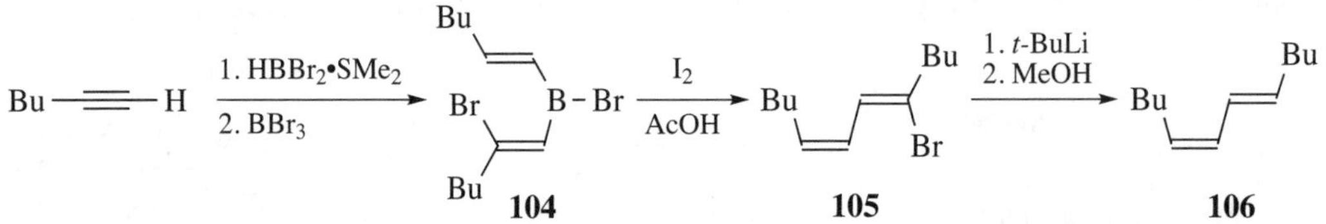

(Z,Z)-Dienes can be formed. Reaction of 1-bromo-1-hexyne (**60**) and the complex $HBBr_2 \cdot SMe_2$ was followed by reaction with potassium triisopropoxyborohydride to form vinylborane **107**. Subsequent coupling with *cis*-1-bromo-1-octene gave the (Z,Z)-diene (**108**) in 87% (>99% ZZ).[86] Internal alkynes can also be coupled to give dienes using this procedure,[87] and terminal alkynes lead to (E,Z)-dienes, although the regioselectivity of this latter reaction is poor.[87,88] Two different alkynes can also be coupled to form unsymmetrical dienes.[89]

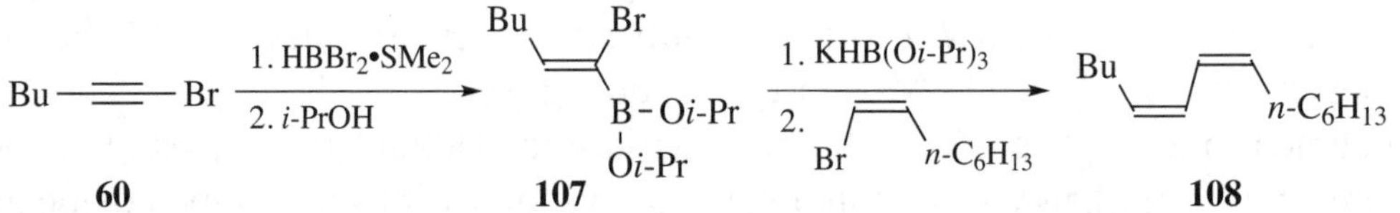

A synthetic route to unsymmetrical alkynes has been reported, which essentially couples an alkene to a terminal alkyne.[90] Reaction of chlorothexylborane (**93**) and 1-octene gave **109** and quenching with methanol gave **110**. An *ate* complex (**111**) was generated when this borane

85. Hyuga, S.; Takinami, S.; Hara, S.; Suzuki, A. T*etrahedron Lett.* **1986**, *27*, 977.
86. Miyaura, N.; Satoh, M.; Suzuki, A. *Tetrahedron Lett.* **1986**, *27*, 3745.
87. Brown, H.C.; Ravindran, N.R. *J. Org. Chem.* **1973**, *38*, 1617.
88. Zweifel, G.; Polston, N.L.; Whitney, C.C. *J. Am. Chem. Soc.* **1968**, *90*, 6243.
89. Zweifel, G.; Backlund, S.J. *J. Organomet. Chem.* **1978**, *156*, 159.
90. Sikorski, J.A.; Bhat, N.G.; Cole, T.E.; Wang, K.K.; Brown, H.C. *J. Org. Chem.* **1986**, *51*, 4521.

reacted with 1-lithio-1-hexyne. Subsequent iodination and oxidation led to a 75% yield of 5-tetradecyne (**112**).[90] Two alkyne byproducts were formed, however, resulting from transfer of different alkyl groups to carbon and causing isolation problems for the alkyne of interest. A limitation[91] is due to the observation that different alkyl groups are transferred from species such as **111** to give mixtures of alkyne products.

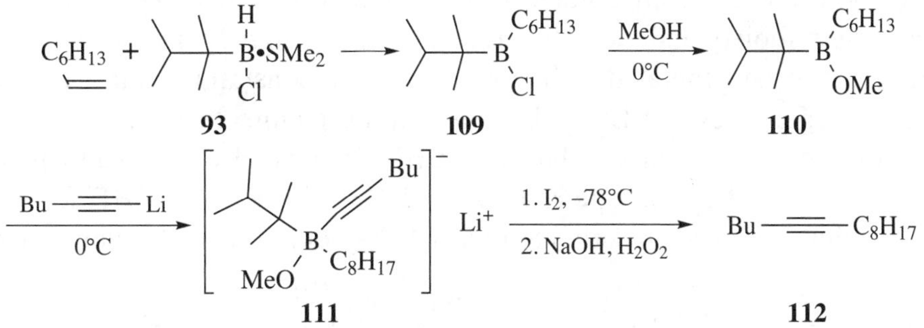

The variety of reactions illustrated in this section shows the range of alkyne-diene transforms.

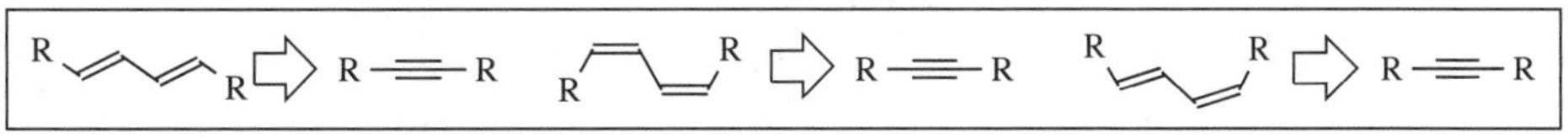

5.3.D. Coupling, Isomerization, and Displacement

Organoboranes can be used as synthetic intermediates to make carbon-carbon bonds. As a practical matter, this reaction couples two alkene moieties. Both symmetrical and unsymmetrical coupling products can be formed, although the method is most useful for symmetrical coupling. Initial hydroboration is followed by treatment with basic silver nitrate ($AgNO_3$) to oxidatively remove boron from the molecule with concomitant coupling of the two alkyl fragments. A typical example is taken from the hydroboration of 2-methyl-1-pentene, which gave the symmetrical coupling product 4,7-dimethyldecane (**113**) in 61% yield after treatment with silver nitrate and NaOH.[92] The reaction probably proceeds via an alkyl silver intermediate. When this reaction is applied to two different alkenes such as $RCH=CH_2$ and $R^1CH=CH_2$, and statistical coupling leads to a mixture of $RCH_2CH_2CH_2CH_2R$, $R^1CH_2CH_2CH_2CH_2R^1$ with only 50% of the cross coupled alkane $RCH_2CH_2CH_2CH_2R^1$.[93] This process is useful for the preparation of large, symmetrical alkanes.

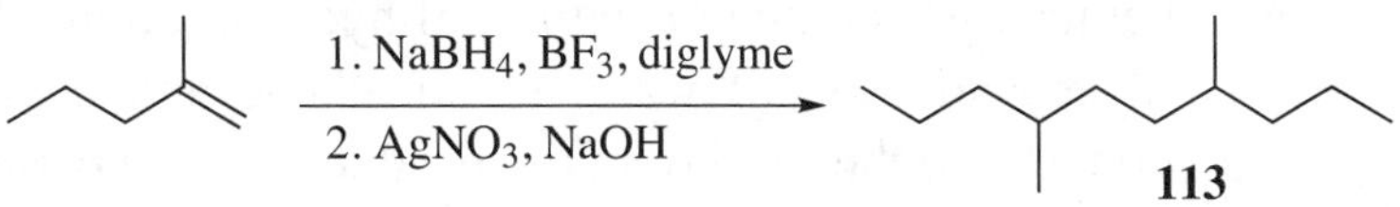

91. Pelter, A.; Drake, R.A. *Tetrahedron Lett.* **1988**, *29*, 4181.
92. Brown, H.C.; Snyder, C.H. *J. Am. Chem. Soc.* **1961**, *83*, 1002.
93. Brown, H.C.; Verbrugge, C.; Snyder, C.H. *J. Am. Chem. Soc.* **1961**, *83*, 1001.

Another variation in the hydroboration sequence uses allylic halides to generate cyclic alkanes. Reaction of 9-BBN with allyl chloride generated the ω-chloroborane **114**. Treatment with hydroxide initially gave **115**, and loss of cyclopropane (formed by internal displacement of chloride) generated the byproduct, **116**.[94] Variations of this process can generate many different substituted and functionalized cyclopropane derivatives. If propargyl bromides are used rather than allylic chlorides, cyclopropylboranes can be formed.[95] Extending the methylene spacer between the alkynyl moiety and the halide (or other leaving group such as tosyl, mesyl, etc.) allows larger rings to be generated in like manner. The constraint against forming rings of eight atoms and larger (secs. 1.5.B, 6.6.B) is a limiting factor. In more recent work, dienes were coupled to form cyclic compound using 9-BBN, followed by treatment with a palladium catalyst (see sec. 12.6 and 12.7). The reaction of **117** with 9-BBN in THF, for example, was followed a palladium catalyzed reaction to give a 42% yield of benzylcyclopentane (**118**).[96]

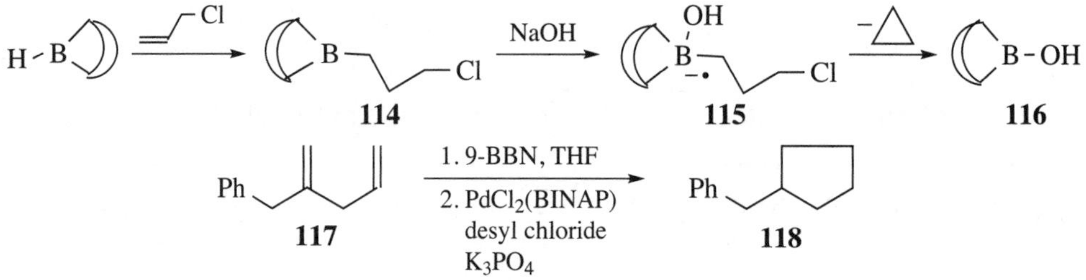

The functional group transformations seen in this section lead to the interesting transforms:

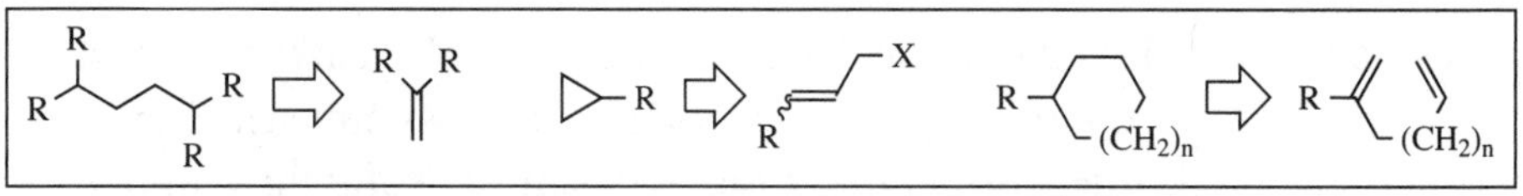

5.4. FORMATION OF OXYGEN-CONTAINING FUNCTIONAL GROUPS

5.4.A. Oxidation of Boranes to Alcohols

Perhaps the most utilized transformation of organoboranes is their oxidation to alcohols, accomplished with basic hydrogen peroxide (H_2O_2, NaOH). The *active* reagent is the hydroperoxide anion (HOO⁻). The overall transformation (alkene → alcohol) occurs with an anti-Markovnikov orientation (the OH group is on the less substituted carbon of the C=C moiety) due to the stereochemical preference of the initial hydroboration. This is important, because both the acid catalyzed addition of water to an alkene and oxymercuration (sec. 2.10.A,B) proceed via a cation to give the Markovnikov product. The two-step hydroboration-oxidation sequence is the preferred method for conversion of alkenes to the less substituted

94. Brown, H.C.; Rhodes, S.P. *J. Am. Chem. Soc.* **1969**, *91*, 2149.
95. Brown, H.C.; Rhodes, S.P. *J. Am. Chem. Soc.* **1969**, *91*, 4306.
96. Lei, A. Zhang, X. *Org. Lett.* **2002**, *4*, 2285.

alcohol.[2] Oxidation of the alkylborane with basic hydrogen peroxide proceeds by initial attack at the boron to give an *ate* complex (**119**), followed by a rapid B → O alkyl shift to form borinate **120**. Additional hydroperoxide anion reacts to transfer the other R groups, generating **121**. Final hydrolysis liberates the alcohol.[97] An example is the conversion of 1-methylcyclopentene to trans-trialkylborane (**122**). Oxidation with basic hydrogen peroxide gave *trans*-2-methylcyclopentanol (**123**) in 86% yield.[98] The conversion of **124** to **125**[99] shows that substituted alkenes react faster than more substituted alkenes.

$$H_2O_2 + OH^- \rightleftharpoons HO_2^- + H_2O$$

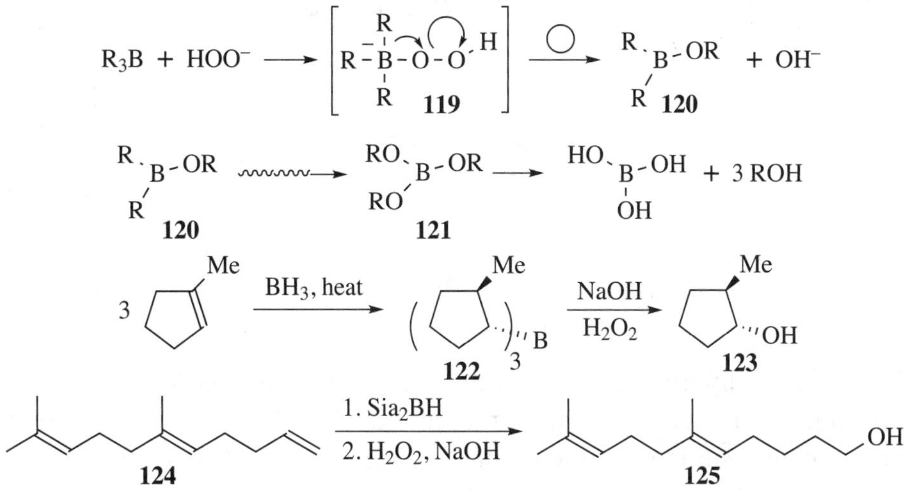

The mechanism shown above implies that the oxidative rearrangement (the B → O shift) proceeds with complete retention of configuration at the boron-bearing carbon, as in the conversion of **122** to **123**. For retention to occur the hydroperoxide anion must attack boron rather than carbon, to be followed by a synfacial migration of the alkyl group to oxygen, without rearrangement, epimerization or racemization. Conversion of an alkene to an alcohol in this manner is most efficient when all three of the groups on the trialkylborane are the same, as with **122**. Oxidation will give three equivalents of the alcohol. When 9-BBN or disiamylborane are used in hydroboration reactions, the trialkylborane being oxidized has at least two sites for formation of a C–OH bond. Reaction of 1-heptene with 9-BBN, for example, generated **126** and oxidation generated not only the desired 1-heptanol, but also 1,5-cyclooctanediol (**127**). Separation of these two alcoholic products can be a problem, which limits the utility of hydroboration reactions. The mono-alcohol may be distilled from the mixture, but chromatographic separation of the mono-alcohol from the diol is usually required.

97. (a) Kuivila, H.G. *J. Am. Chem. Soc.* **1954,** *76,* 870; (b) *Idem Ibid.* **1955,** *77,* 4014; (c) Kuivila, H.G.; Wiles, R.A. *Ibid.* **1955,** *77,* 4830; (d) Kuivila, H.G.; Armour, A.G. *Ibid.* **1957,** *79,* 5659; (e) Wechter, W.J. *Chem. & Ind. (London)* **1959,** 294

98. Zweifel, G.; Brown, H.C. *Org. React.* **1964,** *13,* 1.

99. Chen, L.; Gill, G.B.; Pattenden, G.; Simonian, H. *J. Chem. Soc. Perkin Trans. 1* **1996,** 31.

If disiamylborane were used rather than 9-BBN, the trialkylborane intermediate (**128**) would generate 1-heptanol as well as 2 equivalents of 3-methyl-2-butanol. The separation of the desired alcohol (1-hexanol in this case) from the mono-alcohol byproduct (3-methyl-2-butanol in this case) is potentially more difficult that separating **127** and heptanol. The choice to convert an alkene to an alcohol via hydroboration requires that one weigh the factors of selectivity and reactivity against the ease of isolating the targeted product. When the alcohol product contains many carbons and functional groups, as is common in natural products, the separation problem is probably minimal and selectivity is the major concern. With lower molecular weight alcohol products, ease of isolation may outweigh small increases in selectivity and diborane is probably the reagent of choice.

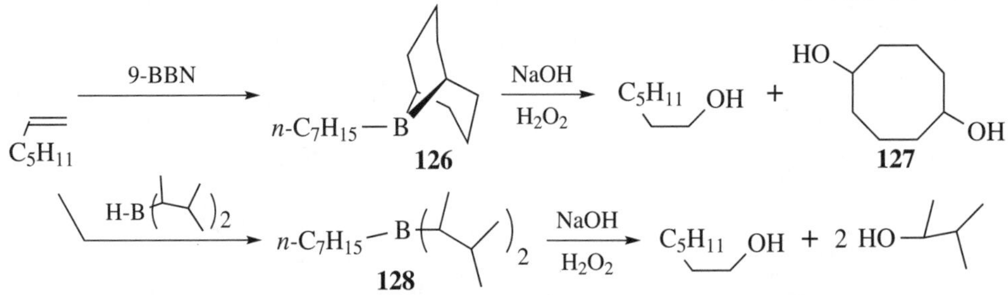

This alcohol-forming hydroboration reaction is highly regioselective, and sometimes diastereoselective.[100] In a synthesis of amphidinolide T3 Zhao and co-workers reacted **129** with a borane•dimethyl sulfide complex, which led to incorporation of the OH group on the less substituted carbon (an 83% yield of alcohol **130**).[101] Note the high regioselectivity, even with borane. Substituted boranes such as disiamylborane, thexylborane or 9-BBN are used in an identical manner, but generally show significantly higher regioselectivity for the less substituted position of the alkene. The alkene unit in **131**, for example, reacted with 9-BBN to give the borane and oxidation gave an 87% yield of **132**, in a synthesis of dolabelide D by Leighton and co-workers.[102] Note that PMB is *p*-methoxybenzyl.

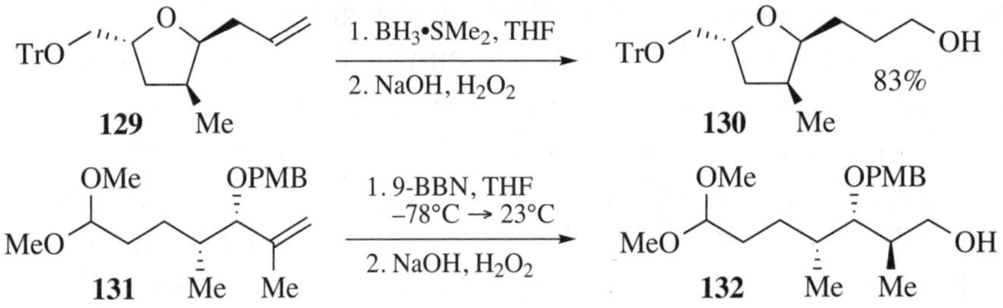

---

100. Evans, D.A.; Sacks, C.E.; Whitney, R.A.; Mandel, N.G. *Tetrahedron Lett.* **1978**, 727.
101. Deng, L.-S.; Huang, X.-P.; Zhao, G. *J. Org. Chem.* **2006**, *71*, 4625.
102. Park, P K.; O'Malley, S.J.; Schmidt, D.R.; Leighton, J.L. *J. Am. Chem. Soc.* **2006**, *128*, 2796.

MOLECULAR MODELING: Hydroboration of Flexible Molecules

As illustrated in the previous problem, there are situations where simply looking at a 3D model of the reactant is sufficient to reliably anticipate the favored (kinetic) product of a reaction. This is something that chemists do all the time using "plastic" models, often with considerable success. There are two underlying assumptions. The first is that the transition state for the reaction will "resemble" the products, or that the Hammond Postulate is valid. The second is that the reaction is sterically controlled.

The situation is complicated and prediction more difficult if the reactant is not rigid. Here it is necessary to first establish its conformation. A good example is provided by the hydroboration reaction shown below, which leads with good regio and stereoselectivity to a single alcohol containing three stereocenters. It is reasonable to assume that the boron adds to the less-substituted carbon, meaning that reactions leading to only two (of four possible products) need to be seriously considered. However, the reactant incorporates several (rotatable) single bonds and may exist in a variety of conformations. Each of the presents district "top" and "bottom" faces of the double bond to the hydroboration reagent.

Steric effects (if indeed these are what is in control) depend on detailed conformation, and without knowledge of conformation. Guessing the conformation of a complex reactant is something that plastic models are ill equipped to do. Conformational energy differences are likely to be very small and may arise because of subtle differences in non-bonded intramolecular interactions. Guessing the preferred conformation of the transition states in a reaction such as hydroboration is likely to be even more difficult, as differences in intermolecular interactions also need to be taken into account. However, quantum chemical calculations are well suited for this purpose, and are able to furnish conformations, detailed geometries and (most important) relative energies of transition states leading to different products.

Transition-state modeling is beyond the capability of ***SpartanModel*** (it is a common application of ***Spartan*** and ***Spartan Student***). However, models generated by these programs can be shown and queried by ***SpartanModel***. Shown on screen are the lowest-energy conformation transition states for addition of borane onto the "top" and "bottom" faces of the alkene shown above. Examine them as space-filling models, and make your best guess as to which is favored. Is this an obvious choice? In fact, the transition state leading to the observed product is 8 kJ/mol lower in energy than the other transition state (corresponding to a room-temperature kinetic product ratio of 95:5). Which transition state shown leads to the major product? Is this the product that is actually observed? However, we can look at results.

Cyclization is an interesting reaction of alkylboranes not discussed in Section 5.3. When the alkyl portion of the borane contains a carbon at the C5 position and when an intermediate dialkylborane possess a B–H bond, cyclization occurs upon vigorous heating (typically 200-

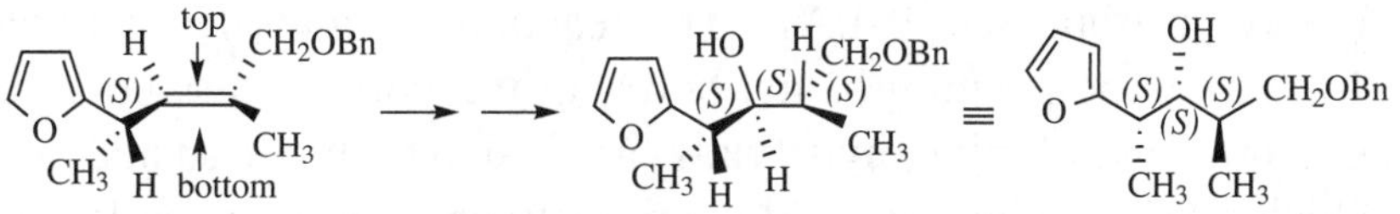

400°C).[103] Hydroboration of 2,4,4-trimethyl-1-pentene gave **133**, and heating to 200°C led to 3,3,5-trimethylborinane (**134**). In general, oxidation of borinanes leads to a diol product. The reaction of **134** with basic H_2O_2 gave an 80% yield of 2,2,4-trimethyl-1,5-pentanediol (**135**).[103]

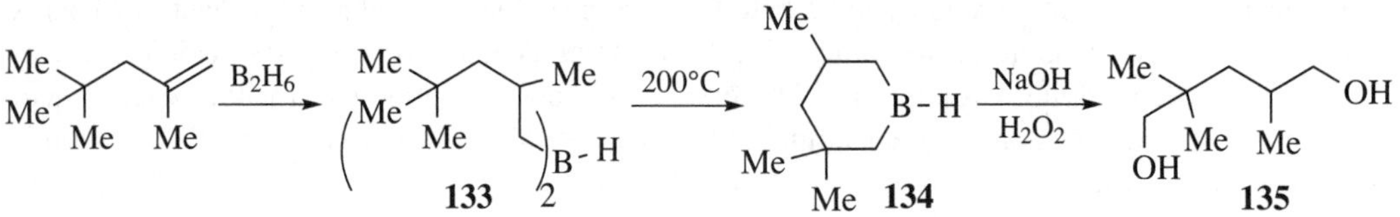

When alkynes are subjected to the hydroboration-oxidation sequence, the alcohol product is actually an enol, and it tautomerizes to a carbonyl derivative (a ketone from internal alkenylboranes and an aldehyde from terminal alkenylboranes). An example is taken from Marshall and Schaaf's synthesis of leptofuranin D,[104] in which the reaction of alkyne **136** with dicyclohexylborane gave vinylborane **137**. Subsequent oxidation gave the enol (**138**), which tautomerized to produce aldehyde **139**, in 95% yield. The synthesis of aldehydes requires that the hydrolysis solution be buffered to pH 7 to minimize condensation reactions of aldehydes (e.g., the **aldol condensation**, sec. 9.4.A). The reaction is regioselective with the position of the carbonyl fixed by the position of the boron in the initially formed alkenylborane. A variation of this reaction uses 1-trimethylsilylalkynes and hydroboration and oxidation leads to a silylketone, which gives a carboxylic acid upon treatment with basic hydrogen peroxide.[105]

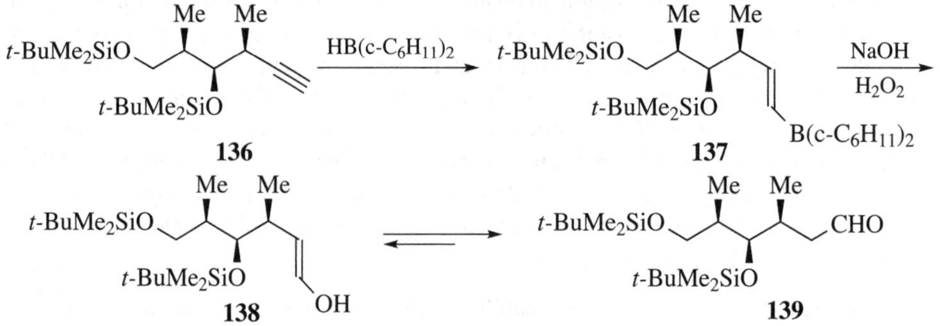

Cyclic boranes can also be used to generate alcohols in a chain extension reaction. The length of the chain is controlled by the ring size of the borinane. B-Methoxyborinane (**140a**) reacted with 1-lithio-1-octyne to give an *ate* complex (**141a**). Iodination and oxidation cleaved the borinane to give 22% of alkynyl alcohol **142**.[106] A similar reaction with B-triphenylmethoxyborinane (**140b**) gave **142** in 85% yield.[106] A variation of this sequence couples cyclic boranes with

103. Reference 2, pp 50-55.
104. Marshall, J.A.; Schaaf, G.M. *J. Org. Chem.* **2003**, *68*, 7428.
105. Zweifel, G.; Backlund, S.J. *J. Am. Chem. Soc.* **1977**, *99*, 3184.
106. Brown, H.C.; Basavaiah, D.; Bhat, N.G. *J. Org. Chem.* **1986**, *51*, 4518.

iodoalkynes, leading to acyclic alkenyl alcohol products.[107] Another chain extension reaction uses diazosilylalknes. The reaction of styrene with catecholborane at 100°C gave **143**. Subsequent reaction with Me_3SiCHN_2 gave **144** and oxidation in the usual manner gave silyl alcohol **145**. Tetrabutylammonium fluoride removes the silyl group, giving 3-phenyl-1-propanol (**146**) in 60% overall yield from styrene.[108]

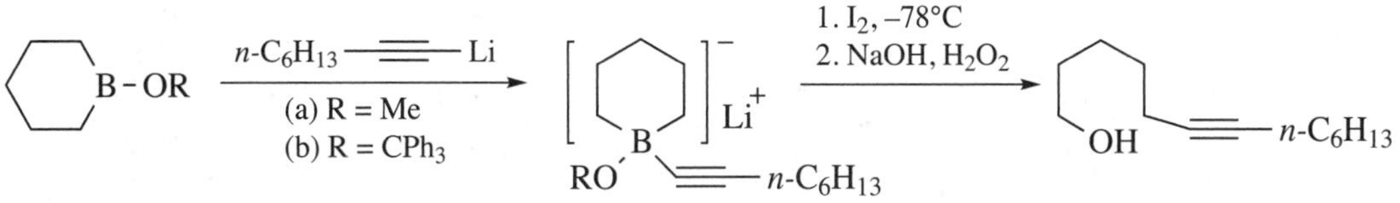

The functional group transformations possible with these reactions are:

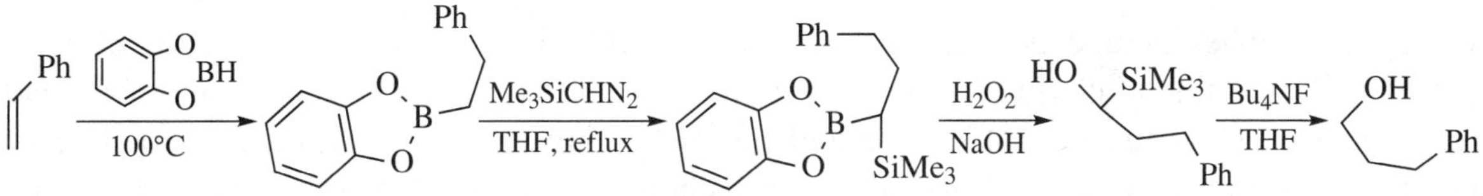

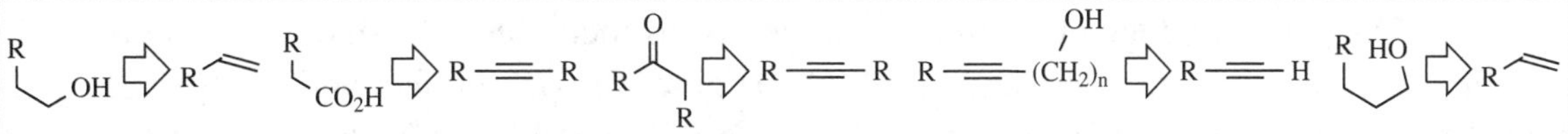

5.4.B. Asymmetric Hydroboration

Optically active products can be produced via hydroboration if an optically active borane is used and if the C=C moiety of the alkene precursor is prochiral (sec. 1.4.E). Brown and others developed several chiral alkylboranes by reaction of borane or achiral organoboranes with naturally occurring chiral alkenes such as pinene or longifolene. These chiral organoboranes react to give hydroboration products with good diastereoselectivity and asymmetric induction. The most common synthetic application is to prepare chiral alcohols from alkenes, using diisopinocampheylborane (**14**, Ipc_2BH) or monoisopinocampheylborane (**147**, $IpcBH_2$) derived from α-pinene (**6**), as well as dilongifolylborane (**148**, Lgf_2BH), which is derived from longifolene **149**.[109] A ligand containing phosphorus and nitrogen groups was used in rhodium-catalyzed asymmetric hydroboration reactions.[110]

107. (a) Basavaiah, D. *Heterocycles* **1982**, *18*, 153; (b) Brown, H.C.; Basavaiah, D.; Singh, S.M.; Bhat, N.G. *J. Org. Chem.* **1988**, *53*, 246; (c) Brown, H.C.; Basavaiah, D.; Kulkarni, S.U.; Bhat, N.G.; Vara Prasad, J.V.N. *Ibid.* **1988**, *53*, 239; (d) Brown, H.C.; Basavaiah, D.; Singh, S.M. *Synthesis* **1984**, 920.

108. Goddard, J.-P.; Le Gall, T.; Mioskowski, C. *Org. Lett.* **2000**, *2*, 1455.

109. Brown, H.C.; Jadhav, P.K.; Mandal, A.K. *Tetrahedron* **1981**, *37*, 3547.

110. Kwong, F.Y.; Yang, Q.; Mak, T.C.W.; Chan, A.S.C.; Chan, K.S. *J. Org. Chem.* **2002**, *67*, 2769

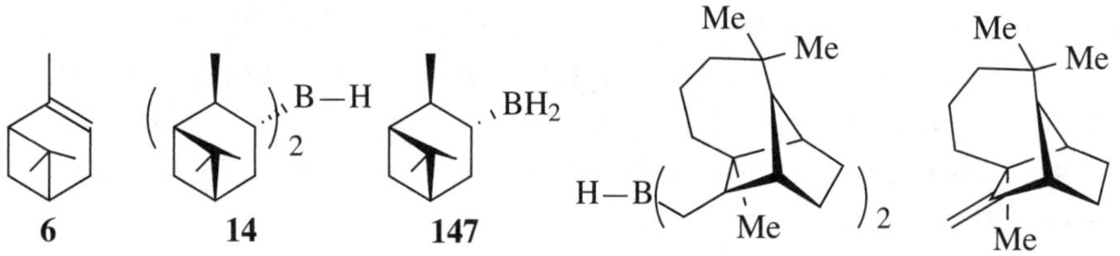

Diisopinocampheylborane is prepared by reaction of either(R)- or (S)-α-pinene (**6**) with diborane,[111] and this dialkylborane product exists largely as a dimer. The product can also be formed by slow reaction of borane and pinene (3 days in THF) with the major isomer precipitating from solution.[107a] This chiral non-racemic reagent reacts with alkenes in the usual manner to give a chiral, non-racemic alcohol after oxidation with basic hydrogen peroxide. Hydroboration of an alkene (**150**) occurs from the less hindered side of the molecule via cis addition (to give **151**), and the oxidation proceeds with complete retention of configuration (to give **152**). The stereogenic center of the newly formed alcohol usually has the *opposite* chirality relative to the organoborane [(−)-IpC$_2$BH gives the (R)-alcohol and (+)-IpC$_2$BH gives the (S)-alcohol].[109] Hydroboration of *cis*-2-butene with (S)-**14**, for example, gave (R)-(−)-2-butanol with 87% ee (in diglyme; 78% ee in THF). Typical percentage asymmetric induction (% ee) for the synthesis of several alcohols using (-)-**14** are: 87% ee (R) when R^1=R^2=Me, R^3=H; 13% (R) when R^1=R^2=R^3=Me; 30% (R) when R^1=H, R^2=i-Pr, R^3=Me; and, 48% (R) when R^1=R^2=H, R^3=propyl.[109]

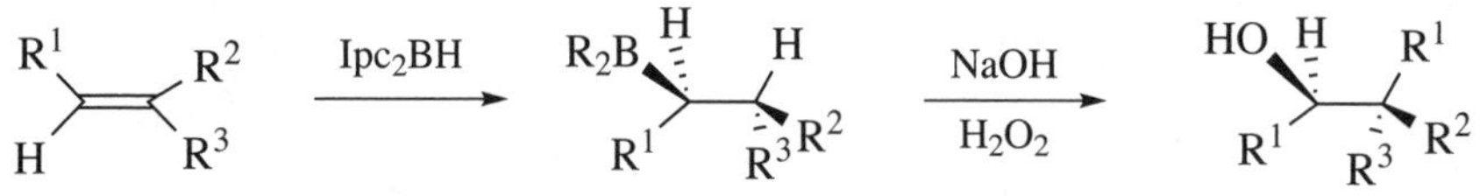

Hydroboration of cis alkenes using **14** leads to an alcohol with higher enantioselectivity than obtained from trans alkenes or hindered terminal alkenes. This can be explained by examination of two proposed structures where the borane approaches the alkene in the hydroboration step (see model **153** for cis alkenes and **154** for trans alkenes.[109] The lowest energy conformation for the reagent positions the two pinanyl rings orthogonal to each other, as shown. When a cis alkene approaches, the favored transition state will minimize all steric interactions. The chirality of the pinanyl rings dictates approach from the right face (as drawn in structures **153**-**156**). If the chirality of this reagent were the opposite of that drawn, approach would be from the left face. The orientation of the alkene is also important in determining enantioselectivity, and this orientation is largely dictated by steric interactions of the alkyl groups on the alkene with one pinanyl ring. The second pinanyl ring is effectively on the other side of the borane, and out of the way of the groups on the alkene. If the alkene approaches from only one face

111.   (a) Brown, H.C.; Yoon, N.M. *Israel J. Chem.* **1977**, *15*, 12; (b) Brown, H.C.; Ayyangar, N.R.; Zweifel, G. *J. Am. Chem. Soc,* **1964**, *86*, 397.

(here the right face), there is a pocket that can accommodate groups on the alkene. In **153**, *cis*-2-butene approaches the right face of the pinanyl reagent and is oriented so that the methyl groups are down. This orientation minimizes steric interactions between the methyl groups and the pinanyl group that is on top of the pinanyl reagent (top and bottom are relative in this discussion since the pinanyl borane reagent possesses C_{2v} symmetry). Approach from the same face but with the methyl groups up as in **155** has a greater steric interaction with the bridging methylene unit, and is disfavored relative to **153**. Therefore, this particular representation suggests the **153** should be the model used to predict stereochemistry for cis alkenes.

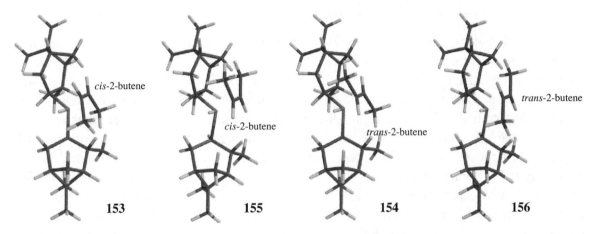

MOLECULAR MODELING: Addition of a Chiral Borane to *cis*-2-Butene

Quantum chemical models are actually at their best when called on to elaborate the effects that subtle changes in structure have on both thermodynamic and kinetic product distributions, most commonly changes in regio and/or stereochemistry. Even though energy differences are likely to be small, comparisons among closely related molecules (or transition states) are most likely to benefit from cancellation of errors. Because practical quantum chemical models involve severe approximations to the Schrödinger equation, the *absolute values* of some calculated quantities may show large errors. Reaction and activation energies, in particular, that result from differencing large numbers (total energies) of dissimilar molecules, are prone to large errors. However, energy differences among products of the reaction that differ only subtly (or energy differences among the transition states leading to these products) are likely to be quite well described, simply because deficiencies in the theoretical model will exist in all of the molecules.

A good example is provided by the hydroboration of *cis*-2-butene using (−)-diisopinocampheylborane as the hydroboration reagent.

> **MOLECULAR MODELING: Addition of a Chiral Borane to *cis*-2-Butene, con't.**
>
> Calculated transition states illustrate the two ways the alkene can approach from a single face, leading eventually to *R* and *S* enantiomers of 2-butanol. The two transition states differ in energy 25 kJ/mol which means that the dominant stereoisomer makes up more than 99% of the total. What is the dominant stereoisomer? Examine both transition state models. Can you point to any differences in the two transition states (steric or otherwise) that would make one better than the other? Elaborate. Suggested reading: Brown, H.C.; Jadhav, P.K.; Mandal, A.K. *Tetrahedron, **1981**, 37,* 3547.

When *trans*-2-butene reacts with the pinanyl borane reagent, approach from one face leads to a situation where one methyl group is up and interacts sterically with the pinanyl ring, but the other methyl group is down and in a pocket (see **154**). The steric interaction of the single methyl group in **154** is greater than the interactions observed in **153**, but somewhat similar to those observed in **155**. The trans alkene should therefore show poorer selectivity than the cis alkene. Rotation of *trans*-2-butene by 180° along the C=C bond axis (to give **156**) exposes the opposite face of the alkene, and also brings one methyl group up and in close proximity to the boranylpinane group that protrudes on top in **156**, which suggests that **156** is not significantly lower in energy in **154**. No orientation that leads to significantly less severe steric interactions for trans alkenes, as was the case with cis alkenes, and enantioselectivity is poorer for the hydroboration-oxidation sequence. If the two substituents on the trans alkene are sufficiently different in size, however, increased selectivity for reaction at the less sterically hindered carbon is possible. With 1,1-disubstituted alkenes, there is little selectivity in the hydroboration-oxidation sequence. Diisopinocampheylborane apparently forms a trimer rather than react with a severely hindered alkene, which further complicates the reaction.[109] When the borane intermediate of the initial alkene reaction is treated with *O*-sulfonylhydroxylamine (sec. 5.5.A), enantioselectivity in the chiral amine product is similar to that observed in the alcohol forming sequence.[109] Partridge and Uskoković used diisopinocampheylborane for the conversion of methylcyclopentadiene to **157**, in a synthesis of loganin.[112]

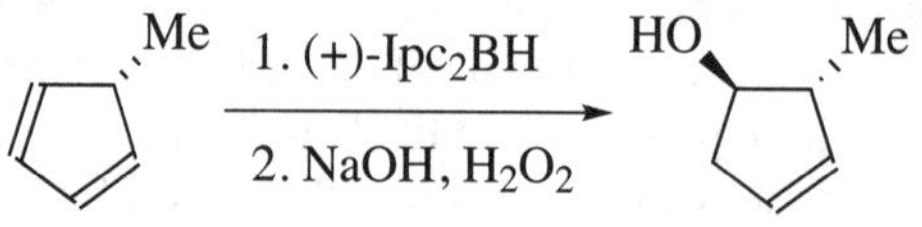

157

112. Partridge, J.J.; Chadha, N.K.; Uskoković, M.R. *J. Am. Chem. Soc, **1973**, 95,* 532.

MOLECULAR MODELING: Stereochemistry of Addition of a Chiral Borane

Previous problems have suggested the difficulty of assigning product distributions using structure models, even models derived from accurate quantum chemical calculations. A further illustration is provided here. Reaction of 5-methylcyclopentadiene with (–)-diisopinocampheylborane, followed by oxidation leads predominately to alcohol **157**. This is one of four possible products and corresponds to approach of boron at C_1 *anti* to the methyl group.

157

The four transition states appear on screen: C_1 *anti* product (top left), C_2 *anti* product (top right), C_2 *syn* (lower right) and C_1 *anti* (lower left). Examine the four both as ball-and-spoke and space-filling models. Based on these models, can you rationalize the fact that the transition state at the top left is 9 kJ/mol lower in energy than any of the other transition states? Hint: Focus your attention on the methyl group of the five-membered ring, and the relationship of the five-membered ring to the pendant pinanyl rings. Elaborate. Guess the identity of the next lowest-energy transition state. From this guess, are you able to conclude that in this case, *syn/anti* selectivity is more or less important than positional selectivity?

Asymmetric induction has been observed in the hydroboration of heterocyclic rings such as dihydrofuran, dihydropyrrole, dihydropyran or dihydrothiophenes.[113] In a typical example, dihydrofuran **158** gave a 92% yield of **159** with 100% ee *R*.[113] Note that these reactions required addition of acetaldehyde to displace α-pinene prior to oxidation of the borane. Dihydropyrrole **160** gave **161** in 92% yield and 89% ee *S*, without the need to use acetaldehyde.[113] These cases show that the hydroboration reaction is highly regioselective for placement of the boron on the alkene carbon distal to the heteroatom. Tetrahydropyridine derivative **162**, however, gave an 85:15 mixture of **163** and **164** in 68% yield, and **163** was formed with 70% ee *R*.[113]

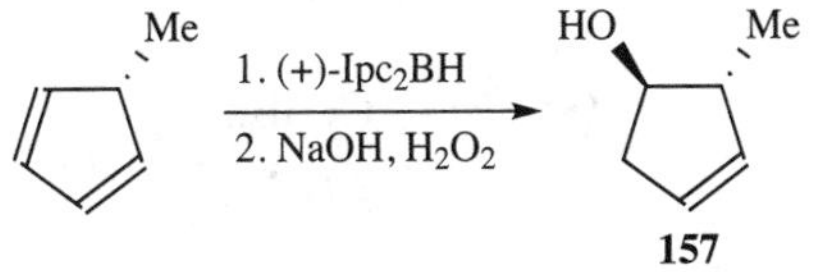

113. Brown, H.C.; Vara Prasad, J.V.N. *J. Am. Chem. Soc.* ***1986***, *108*, 2049.

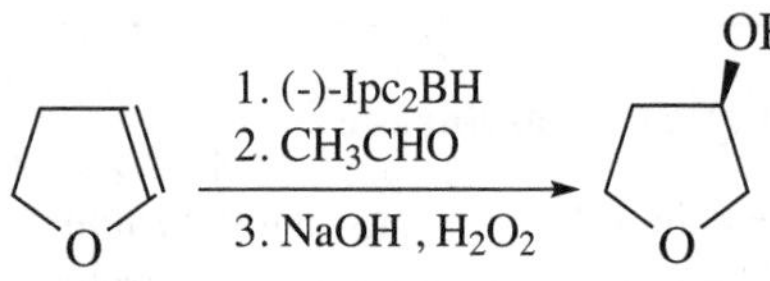

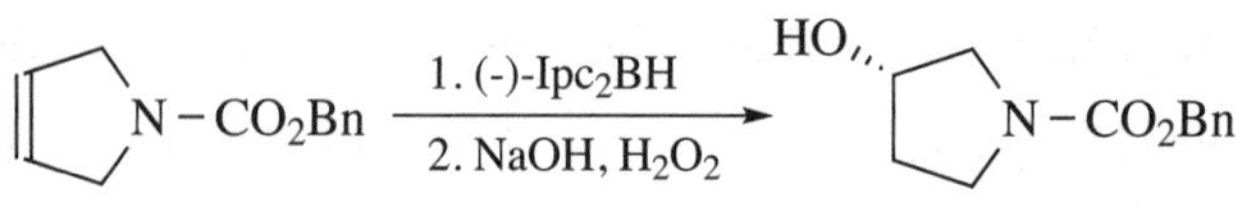

Monoisopinocampheylborane (**147**) is another important chiral borane formed by reaction of α-pinene with the triethylamine complex of thexylborane[114] to give the monoisopinylcamp hylborane•triethyl-amine complex. It is necessary to release monoisopinocampheylborane by treatment with borane•THF or boron trifluoride etherate.[115] This reagent is obviously less hindered than the diisopinocampheyl derivative and reacts faster, even with hindered and polysubstituted alkenes. The highest enantioselectivity is observed with trans alkenes (see **165**). When $R^1=R^3=Me$, $R^2=H$, for example, **167** was formed with 72% ee.[109] The cis alkene ($R^1=R^2=Me$, $R^3=H$,), however, gave **167** in only 24% ee.[109] Initial reaction with **165** generates a chiral organoborane (**166**), and oxidation gives the alcohol (**167**) with retention of configuration.

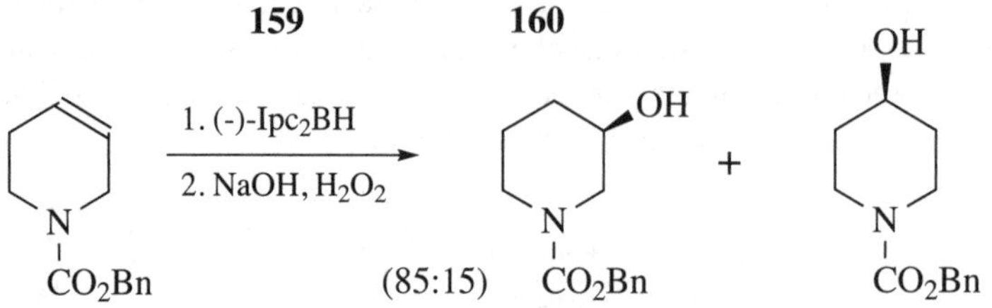

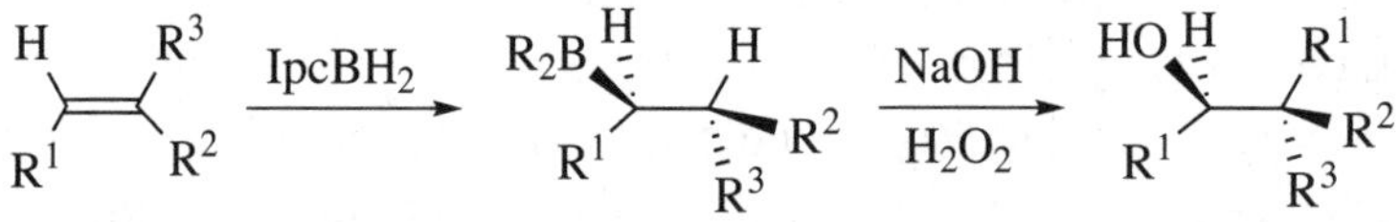

Asymmetric induction increases with increasing steric demands of the alkene, but cis alkenes show the *worst* selectivity. This can be explained by three-dimensional drawing **168**, which shows approach of one face of *cis*-2-butene with the two methyl groups *exo-* to the pinanyl ring (when the methyl groups are *endo-* there is a steric interaction). Rotating *cis*-2-butene by 180° along the axis that *bisects* the C=C bond (C=/=C) exposes the other face of the alkene. This representation clearly suggests no facial bias for the cis alkene and, little or no enantioselectivity in the hydroboration reaction. Examination of **169a** for interaction of one face of *trans*-2-butene with the pinanyl borane reagent shows that the methyl groups occupy "pockets" with minimal steric hindrance. Rotation of 2-butene by 180° along the axis that *bisects* the C=C bond (C=/=C) exposes the opposite face (**169b**), and the steric interaction with the bridging methylene unit is rather severe. These representations suggest a facial bias for the

114. (a) Brown, H.C.; Yoon, N.M. *J. Am. Chem. Soc.* **1977**, *99*, 5514; (b) Brown, H.C.; Schwier, J.R.; Singaram, B. *J. Org. Chem.* **1978**, *43*, 4395; (c) Brown, H.C.; Yoon, N.M.; Mandal, A.K. *J. Organomet. Chem.* **1977**, *135*, C10.
115. Brown, H.C.; Mandal, A.K. *Synthesis* **1978**, 146.

trans-alkene that is not present for the *cis*-alkene, and good enantioselectivity for trans alkenes. 2-Butene shows outstanding selectivity because it is symmetrical. Unsymmetrical alkenes show good selectivity only when there are significant differences in the size of substituents. 1,1-Disubstituted alkenes show good selectivity *if* both groups are large, and particularly if one group is significantly larger than the other.

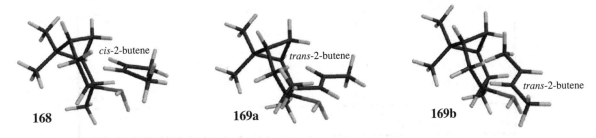

168   **169a**   **169b**

Although **14** shows good selectivity for cis alkenes and **168** shows good selectivity for trans-alkenes, no reagent showed good selectivity with both cis and trans alkenes. A solution to this problem was found when dilongifolylborane (**148**) was prepared by reaction of the dimethyl sulfide complex of borane with (+)-longifolene (**149**), which is readily available [the (–) antipode is rare however].[116] Compound **148** is a dimeric, high-melting white crystalline solid that is stable in ether solvents and can be stored several weeks without decomposition. This reagent gives good enantioselectivity with both cis and trans alkenes, as well as 1,1-disubstituted alkenes.[109] *cis*-2-Butene reacted with **148** to give the alcohol with 78%ee and 2-methyl-2-pentene gave the corresponding alcohol with 75 %ee.[109] The selectivity can be explained with model **170**, which shows the approach of *cis*-2-butene to **148**, and reveals that cis alkenes can be accommodated if small to moderately sized substituents are present. Good selectivity is also observed in reactions with 1,1-disubstituted alkenes. Borane **148** also accommodates trans alkenes, as seen in **171** for approach of *trans*-2-butene, where selectivity increases as the steric size increases.

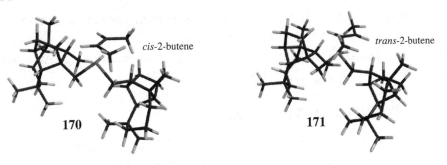

170   **171**

116.   Jahhav, P.K.; Brown, H.C. *J. Org. Chem.* **1981**, *46*, 2988.

Table 5.3. Asymmetric Hydroboration with Carene Based Reagents

Alkene	Reagent (0°C, THF)	Time (h)	Alcohol	% Yield	% ee
cis-2-Butene	2-ICr$_2$BH	12	2-Butanol	63	93 (S)
	4-Icr$_2$BH	72		58	50 (R)
trans--2-Butene	2-ICr$_2$BH	150		69	30 (R)
	4-Icr$_2$BH	72		62	40 (S)
2-Methyl-1-butene	2-ICr$_2$BH	3	2-methyl-1-butanol	61	15 (S)
	4-Icr$_2$BH	4		72	5 (R)
2-Methyl-2-butene	2-ICr$_2$BH	240	3-Methyl-2-butanol	61	37 (S)
	4-Icr$_2$BH	120		75	0
trans-3-Hexene	4-Icr$_2$BH	72	3-Hexanol	65	32 (S)
1-Methylcyclopentene	2-ICr$_2$BH	240	2-methylcyclopentanol	65	3 (RR)
	4-Icr$_2$BH	26		72	3 (RR)

Another class of chiral boranes has been prepared from Δ^2-carene (**69**) and Δ^4-carene (**173**) and the first widely used reagents are (1*S*)-di-2-isocaranylborane (**172**, 2-^{d}Icr$_2$BH) and (1*S*)-di-4-isocaranylborane (**174**, 4-^{d}Icr$_2$BH).[117] These reagents behave in a similar manner to Ipc$_2$BH but are less reactive, and their relative reactivity is: Ipc$_2$BH > 2^{d}Icr$_2$BH > 4-^{d}Icr$_2$BH. Table 5.3[117] illustrates the selectivity and asymmetric induction of **172** and **174** upon reaction with a variety of alkenes. These reagents showed good selectivity with cis alkenes but trans alkenes and terminal alkenes react slowly and show poor selectivity.

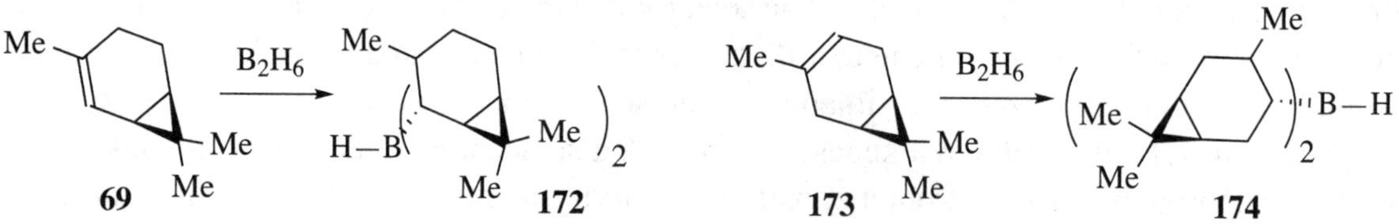

The five reagents discussed in this section show that chiral boranes can induce moderate to good enantioselectivity in the conversion of alkenes to alcohols (or to amines, sec. 5.5.A). These reagents are a powerful addition to the growing list of enantioselective reactions used in organic synthesis.

5.4.C.   Addition of Allylboranes to Carbonyl Compounds

A particularly useful borane is the chiral reagent (−)-B-allyl(diisopinocampheyl)borane.[118] Although the chemistry of acyl addition to aldehydes and ketones to give alcohols will not be formally discussed until Section 8.4.C, the addition of this chiral allylborane to carbonyl compounds will be presented here. In a synthesis of peloruside A by Zhou and Liu ,[119] **175**

117.   Brown, H.C.; Vara Prasad, J.V.N.; Zaidlewicz, M. *J. Org. Chem.* **1988,** *53*, 2911.
118.   Brown, H.C.; Jadhav, P.K. *J. Am. Chem. Soc.* **1983,** *105*, 2092.
119.   Liu, B.; Zhou, W.-S. *Org. Lett.* **2004,** *6*, 71.

was treated with (–)-B-allyl(diisopinocampheyl)borane in ether at –78°C. The usual oxidation gave **176** in good yield.

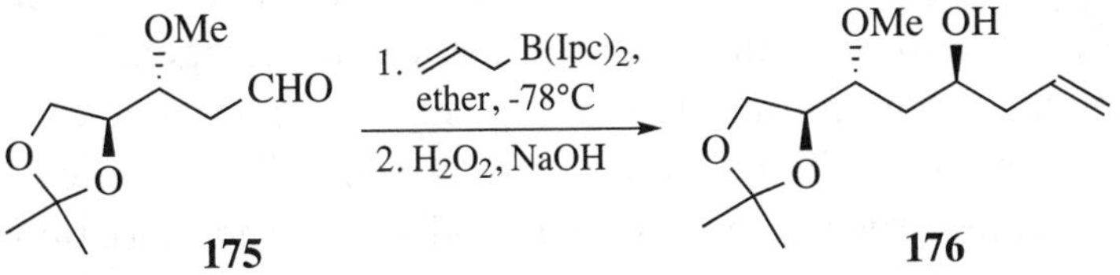

Enantioselectivity in this reaction is usually excellent, but substitution on the alkylborane can diminish diastereoselectivity, as with crotylborane derivatives. This is not always the case, however. In a synthesis of (–)-dictyostatin by Ramachandran, and co-workers,[120] treatment of **177** with two equivalents of (*E*)-(+)-crotyldiisopinocampheylborane followed by an oxidative workup, gave **178** in good yield as a single diastereomer.

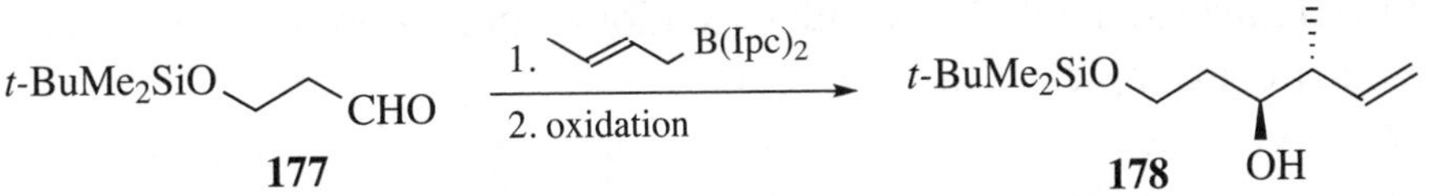

Although this reaction is used most often with aldehydes, other carbonyl compounds can be used. It is known that carboxylic acids,[121] esters,[121a,b,122] anhydrides[122] or amides[122] react with allylic boranes to give the corresponding alcohols. In one example, 2-pyrrolidinone reacted with triallylborane to give a 90% yield of 2,2-diallylpyrrolidine (**179**) after treatment with methanol followed by basic hydrogen peroxide.[123] This transformation is believed to proceed by initial addition to the carbonyl followed by elimination to give an imine. The imine reacts with a second equivalent of triallylborane to give the 2,2-disubstituted derivative. The methanol cleaves the N-B bond and the oxidation step removes the remaining borane derivatives.

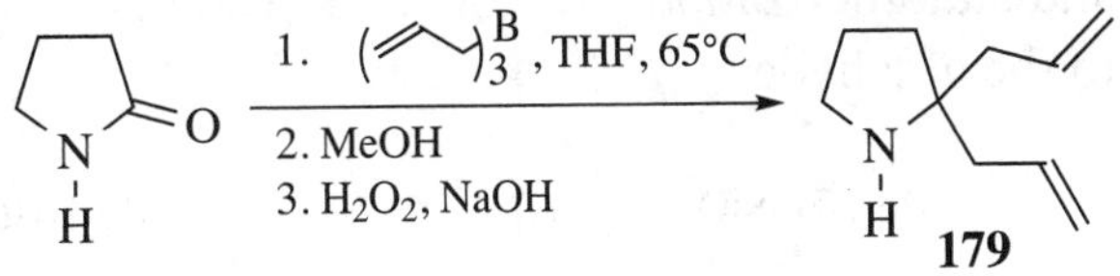

The disconnection for this reaction is

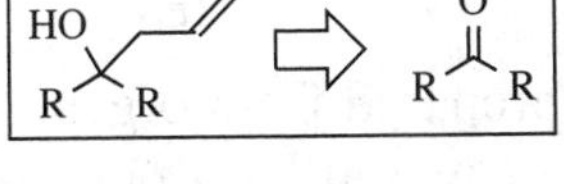

The reader is also directed to Section 12.7.C for a palladium-catalyzed coupling reaction involving boronic acids called **Suzuki-Miyaura coupling**.

---

120. Ramachandran, P.V.; Srivastava, A.; Hazra, D. *Org. Lett.* **2007**, *9*, 157.
121. (a) Mikhailov, B.M.; Bubnov, Yu.N.; Tsyban, A.V.; Grigoryan, M.Sh. *J. Organomet. Chem.* **1978**, *154*, 131; (b) Mikhailov, B.M.; Bubnov, Yu.N.; Tsyban, A.V. *Izv. Akad. Nauk, Ser. khim.* **1978**, 1892; (c) Bubnov, Yu.N.; Demina, E.E.; Bel'sky, V.K.; Zatonsky, G.V.; Ignatenko, A.V. *Izv. Akad. Nauk, Ser. khim.* **1998**, 2320.
122. Kramer, G.W.; Brown, H.C. *J. Org. Chem.* **1977**, *42*, 2292.
123. Bubnov, Y.N.; Pastukhov, F.V.; Yampolsky, I.V.; Ignatenko, A.V. *Eur. J. Org. Chem.* **2000**, 1503.

5.4.D. Oxidation to Aldehydes and Ketones

The alcohols that were formed via hydroboration and oxidation in Section 5.4.A can be oxidized to aldehydes or ketones with an appropriate oxidizing agent (see Chap. 3) after they are isolated from the hydroboration-oxidation sequence. It is possible to combine these steps and oxidize organoboranes directly to ketones by treatment with chromic acid. Hydroboration of cyclohexene, for example, generated tricyclohexylborane and oxidation with chromic acid rather than basic hydroperoxide gave a 65% yield of cyclohexanone directly. Initial oxidation of tricyclohexylborane likely gave the alcohol, which was oxidized to the ketone (cyclohexanone) *in situ.*[124] The oxidation to the alcohol proceeds by a chromate ester of boron and the rate of oxidation is sensitive to steric effects,[125] analogous to the chromic acid oxidation of secondary alcohols (sec. 3.2.A). Using diethyl ether as a solvent facilitates isolation of the final ketone product. When the alkyl substrate is sensitive to acid rearrangement, chromic acid oxidation can facilitate that rearrangement.[126] As with oxidation of organoboranes to form alcohols, there are many synthetic examples. In Harmata and Bohnert's synthesis of sterpurene, hydroboration of alkene **180** followed by oxidation with pyridinium chlorochromate gave a 66% yield of ketone **181**.[127]

A direct synthesis of aldehydes involves hydroboration of terminal alkynes. In a synthesis of bafilomycin,[128] Marshall and Adams first converted the alkyne unit in **182** to the vinylborane unit in **183** using dicyclohexylborane. Subsequent oxidation generated the corresponding enol, and tautomerization gave aldehyde **184** in 81% yield. This sequence converts terminal alkynes to the aldehyde.

Ketones or tertiary alcohols can be prepared from organoboranes using cyano *ate* complexes. Ketone formation requires transfer of two alkyl groups from the boron. Treatment of tri-*n*-butylborane with sodium cyanide (NaCN) gave the cyano *ate* complex (**185**). An electrophilic reagent such as trifluoroacetic anhydride [$(CF_3CO)_2O$] was added to initiate migration of the alkyl groups from boron to carbon, allowing oxidation with basic hydrogen peroxide to give

124. Brown, H.C.; Garg, C.P. *J. Am. Chem. Soc.* **1961,** *83,* 2951, 2952.
125. Ware, J.C.; Traylor, T.G. *J. Am. Chem. Soc. 1963,* **85,** 3026.
126. Lansbury, P.T.; Nienhouse, E.J. *J. Chem. Soc. Chem. Commun. 1966,* 273.
127. Harmata, M.; Bohnert, G.J. *Org. Lett. 2003, 5,* 59.
128. Marshall, J.A.; Adams, N.D. *J. Org. Chem. 2002, 67,* 733.

the symmetrical ketone, 5-nonanone in 94% yield.[129] Treatment of the cyano-*ate* complex with sodium hydroxide and then basic hydrogen peroxide in the usual oxidation sequence leads to a tertiary alcohol. Tri-*n*-butylborane was converted to 5-butyl-5-nonanol in 85% yield along with a 15% yield of 5-nonanone by this procedure.[130]

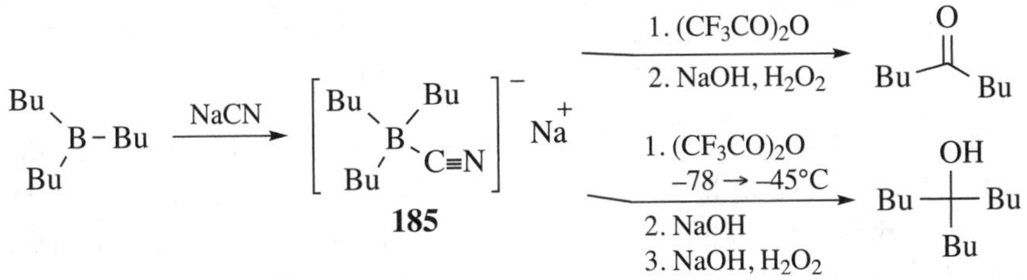

The mechanism of tertiary alcohol formation is postulated to be initial formation of an imine such as **187** from the cyano *ate* complex (**186**). Cyclization via attack at boron in **187** led to **188**, which was opened by reaction with trifluoroacetate anion to give **189**. Oxidation of **189** gave the tertiary alcohol, **190**. The ketone product is formed by oxidation of the intermediate **188**, although oxidative cleavage of **189** is also possible.[131]

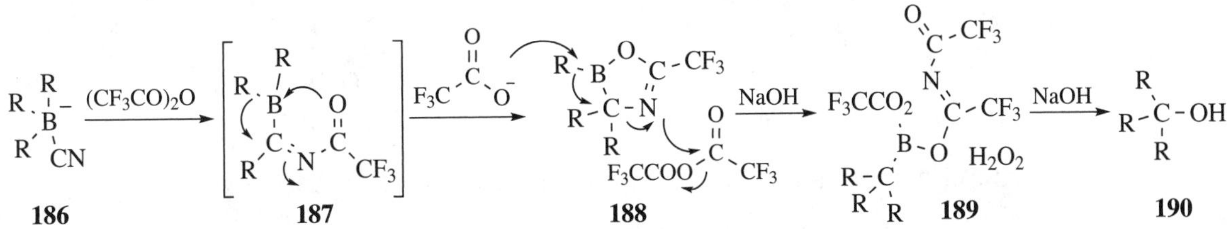

While this method is best suited for the preparation of symmetrical alcohols or ketones, formation of unsymmetrical ketones is difficult unless selective transfer of the alkyl groups can be achieved. Thexylborane provides a solution since it can react sequentially with two different alkenes to generate an unsymmetrical trialkylborane. Reaction of thexylborane with cyclopentene and then arylalkene **191**, for example, gave **192**. Subsequent reaction with sodium cyanide gave the unsymmetrical *ate* complex (**193**), which gave unsymmetrical ketone **194** upon treatment with trifluoroacetic anhydride followed by oxidation.[130b]

129. Pelter, A.; Hutchings, M.E.; Smith, K.G. *J. Chem. Soc. Chem. Commun.* **1970,** 1529.
130. (a) Brown, H.C.; Nambu, H.; Rogić, M.M. *J. Am. Chem. Soc.* **1969,** *91,* 6852; (b) Pelter, A.; Smith, K.; Hutchings, M.G.; Rowe, K. *J. Chem. Soc. Perkin Trans. 1* **1975,** 129; (c) Pelter, A.; Hutchings, M.G.; Rowe, K.; Smith, K. *Ibid.* **1975,** 138.
131. Pelter, A.; Hutchings, M.G.; Smith, K. *J. Chem. Soc. Chem. Commun.* **1971,** 1048.

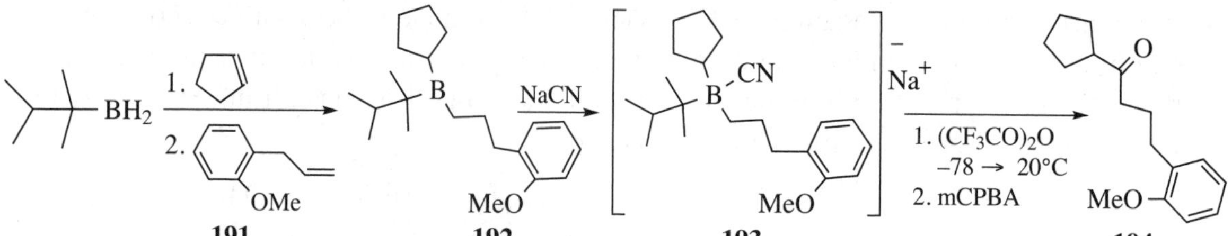

191 **192** **193** **194**

Cyclic ketones can be formed by the hydroboration of dienes, followed by the oxidation sequence just described. In a typical example, taken from Bryson and Welch's synthesis of confertin, **195** was first reacted with thexylborane in the presence of sodium cyanide to give **196**. Subsequent treatment with trifluoroacetic anhydride and then oxidation with basic hydrogen peroxide gave a 70:30 mixture of **197** and **198** in 48% yield.[132] This **boron annulation sequence** has been applied to the synthesis of several bicyclic alkaloids.[133]

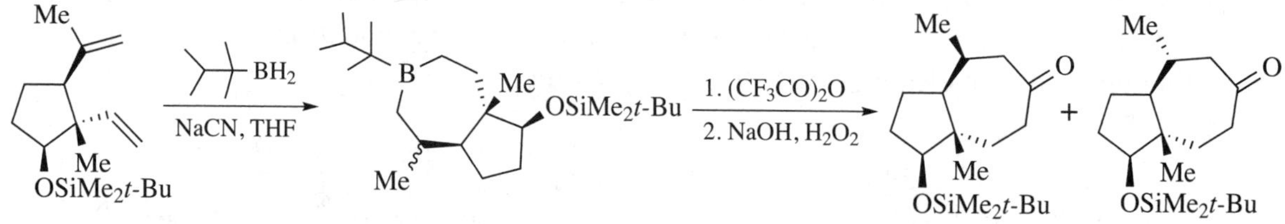

195 **196** **197** **198**

The functional group transforms for the alkene to ketone conversion are:

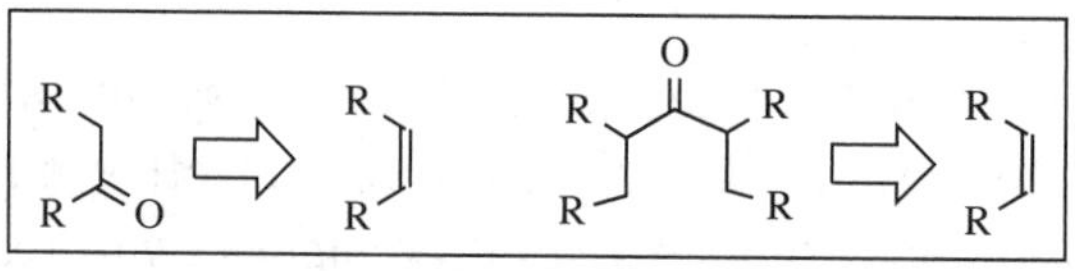

5.4.E. Carbonylation: Formation of Alcohols, Aldehydes and Ketones

There is an alternative method for converting organoboranes into alcohols or carbonyl derivatives that requires more vigorous reaction conditions. Organoboranes react with CO in a reaction that transfers the three alkyl groups of a trialkylborane from boron to carbon,[134] but this transformation requires relatively high pressures of carbon monoxide as well as high temperatures (typically 70 atm and 150°C) to form intermediate **199**. Subsequent oxidation induces the usual B → O shift in **199**, which leads to a symmetrical tertiary alcohol (**200**). A simple example of this reaction involves the reaction of tricyclohexylborane (from cyclohexene) with CO to give **199** (R = cyclohexyl), and oxidation generated tricyclohexylcarbinol (**200**, R

132. Welch, M.C.; Bryson, T.A. *Tetrahedron Lett.* **1988**, 29, 521.
133. Garst, M.E.; Bonfiglio, J.N. *Tetrahedron Lett.* **1981**, 22, 2075.
134. Brown, H.C. *Acc. Chem. Res.* **1969**, 2, 65.

= cyclohexyl) in an overall yield of 86%.[135]

$$R_2B\text{-}R \xrightarrow[150°C]{CO\ (70\ atm)} R\text{-}\underset{\underset{R}{|}}{\overset{\overset{R}{|}}{C}}\text{-}B\text{-}O \xrightarrow[H_2O_2]{NaOH} R\text{-}\underset{\underset{R}{|}}{\overset{\overset{R}{|}}{C}}\text{-}OH$$

$$\textbf{199} \qquad\qquad\qquad \textbf{200}$$

This sequence is useful for the preparation of symmetrical tertiary alcohols, even highly hindered ones that would be difficult to prepare by other routes. It is limited to an organoborane derived from a single alkene, however, since a mixed borane will lead to several alcohol products. Hydroboration of dienes and trienes lead to cyclic, bicyclic or tricyclic tertiary alcohols. The method is illustrated by reaction of diborane with 1,5,9-cyclododecatriene (**201**), which gave tricyclic borane **202** that has a bridgehead boron. Carbonylation and oxidation as before incorporate the carbon group and led to the tricyclic alcohol (perhydrophenalenol, **203**)[136] where the carbinol carbon bridges all three rings. This compound was produced in 70% yield in three chemical steps, and would be difficult to prepare by other synthetic methods.

$$\textbf{201} \qquad\qquad\qquad \textbf{202} \qquad\qquad\qquad\qquad \textbf{203}$$

Another route to tertiary alcohols involves monomethylborane (MeBH$_2$), which can be prepared from lithium methylborohydride (LiMeBH$_3$).[137] Reaction of MeBH$_2$ with two equivalents of 1-hexene gave **204**. Carbonylation and oxidation gave a 1:1 mixture of **205** and **206**.[137] When ten equivalents of 1-hexene were used and the hydroboration was done at –25°C (in THF), carbonylation and oxidation gave **205** in 96% yield.[137] Ketones rather than alcohols can be synthesized if the carbonylation step is done in the presence of water. In this modification, an OH group and two alkyl groups are transferred to carbon. Oxidation leads directly to an intermediate that decomposes to a ketone rather than an alcohol.[138] Hindered boranes, especially thexylborane, react sluggishly at atmospheric pressure, so a pressure of 70 atmospheres of CO is required. Reaction of thexylborane with cyclopentene gave **207**, and subsequent reaction with *O*-acetyl-hept-6-en-1-ol gave the trialkylborane (**208**). Carbonylation in the presence of water followed by oxidation gave the mixed ketone (**209**) in 73% yield.[139] In a similar manner,

135. Brown, H.C.; Ratke, M.W. *J. Am. Chem. Soc.* **1967**, *89*, 2737.
136. Brown, H.C.; Negishi, E. *J. Am. Chem. Soc,* **1967**, *89*, 5478.
137. Brown, H.C.; Cole, T.E.; Srebnik, M.; Kim, K.-W. *J. Org. Chem.* **1986**, *51*, 4925.
138. Brown, H.C.; Rathke, M.W. *J. Am. Chem. Soc,* **1967**, *89*, 2738; (b) Brown, H.C.; Dickason, W.C. *Ibid.* **1969**, *91*, 1226.
139. (a) Brown, H.C.; Negishi, E. *Synthesis* **1972**, 196; (b) *Idem J. Am. Chem. Soc,* **1967**, *89*, 5285.

dienes can be transformed into the corresponding cyclic ketones.[140,141]

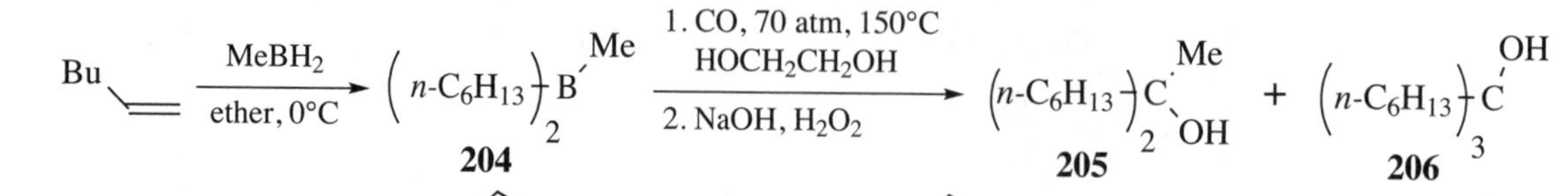

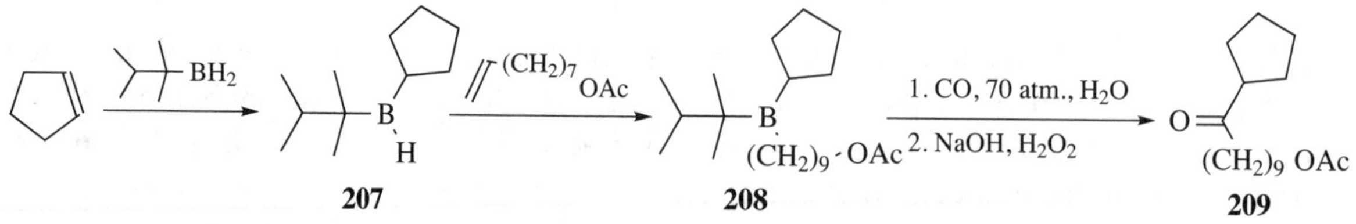

Another modification of the carbonylation sequence adds a single carbon to a substrate and generates an aldehyde. Treatment of 4-vinylcyclohexene (**210**) with 9-BBN, gave the expected borane (**211**). Subsequent carbonylation, in the presence of lithium trimethoxyaluminum hydride [LiAlH(OMe)$_3$, sec. 4.3], gave an α-boranyl alkoxyaluminate (**212**). Treatment with hydroxide liberated the alcohol (**213**), but oxidation of **212** with buffered hydrogen peroxide gave aldehyde **214**.[134,142] The reaction of **211** with a complex hydride increased the rate of CO uptake by the trialkylboranes (this is a general phenomenon).[143] Indeed, a small amount of free trialkylborane must be present in the reaction medium or carbonylation does not occur at a reasonable rate.[144] Trialkylborohydrides are probably the active reducing agents in this reaction.[145] When sodium borohydride or lithium tri-*tert*-butoxyaluminum hydride was used, complex side reactions accompanied carbonylation. For this reaction, lithium trimethoxyaluminum hydride or potassium triisopropoxyborohydride is the preferred reagent.

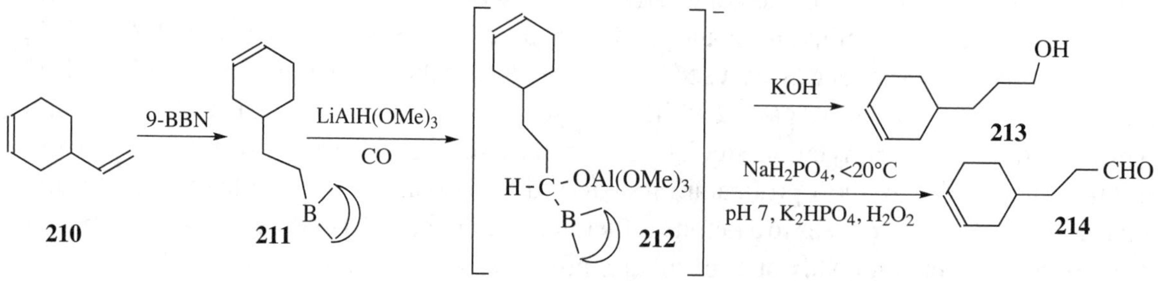

140. Brown, H.C.; Negishi, E. *J. Am. Chem. Soc,* **1967,** *89,* 5477.
141. Brown, H.C.; Negishi, *J. Chem. Soc. Chem. Commun.* **1968,** 594.
142. Brown, H.C.; Knights, E.F.; Coleman, R.A. *J. Am. Chem. Soc,* **1969,** *91,* 2144.
143. (a) Brown, H.C.; Krishnamurthy, S.; Hubbard, J.L. *J. Organomet. Chem.* **1979,** *166,* 271; (b) Brown, H.C.; Krishnamurthy, S.; Hubbard, J.L.; Coleman, R.J. *Ibid.* **1979,** *166,* 281; (c) Brown, C.A.; Hubbard, J.L. *J. Am. Chem. Soc.* **1979,** *101,* 3964
144. Brown, H.C.; Hubbard, J.L. *J. Org. Chem.* **1979,** *44,* 467.
145. Hubbard, J.L.; Smith, K. *J. Organomet. Chem.* **1984,** *276,* C41.

The functional group transforms in this section are quite useful and are outlined below:

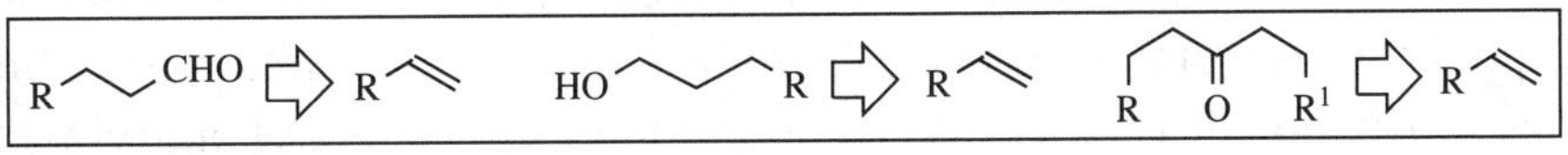

5.4.F.   Conjugate Addition

Borane, monoalkylboranes and dialkylboranes (such as thexylborane and 9-BBN), all with a B-H unit, are known to be powerful reducing agents for carbonyl derivatives (sec. 4.6.A). When no B-H unit is present, as in trialkylboranes, addition to the alkene moiety of α,β-unsaturated ketones or aldehydes occurs. An alkyl group of the borane is transferred to the terminal position of the alkene moiety (1,4-addition), and boron is transferred to the oxygen, to give a boron enolate (sec. 9.4.D). Initial hydroboration of acrolein with tricyclopentylborane gave boron enolate **215**.[146] Hydrolysis of **215** generated an enol and, thereby, aldehyde **216** in 88% yield.[147] Conjugated ketones react in a similar manner to give substituted ketone products.[148] Brown and co-workers noted that the reaction with acrolein was a useful method for the ***three-carbon extension*** of a carbon chain. It was also observed that the reaction was inhibited by addition of free radical inhibitors such as galvinoxyl, which implies that initial addition to the conjugated system probably proceeded by a free radical chain mechanism.[149]

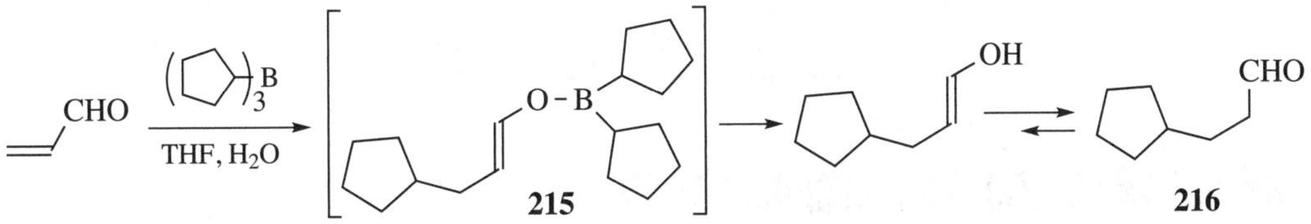

The relatively simple functional group transform for this reaction is:

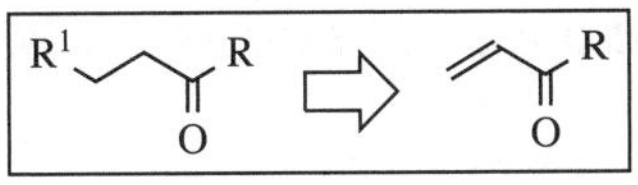

5.4.G.   Carbanions Containing Boron, Boron Enolates, and Enolate-Like Derivatives

Organoboranes react with bromine to form brominated boranes,[150] but they also react with brominated substrates. α-Bromo esters, for example, react with organoboranes to give

146.   Brown, H.C.; Rogić, M.M.; Rathke, M.W.; Kabalka, G.W. *J. Am. Chem. Soc, 1967, 89,* 5709.
147.   Brown, H.C.; Kabalka, G.W.; Rathke, M.W.; Rogić, M.M. *J. Am. Chem. Soc, 1968, 90,* 4165, 4166.
148.   Suzuki, A.; Arase, A.; Matsumoto, H.; Itoh, M.; Brown, H.C.; Rogić, M.M.; Rathke, M.W. *J. Am. Chem. Soc, 1967, 89,* 5708
149.   Kabalka, G.W.; Brown, H.C.; Suzuki, A.; Honma, S.; Arase, A.; Itoh, M. *J. Am. Chem. Soc. 1970, 92,* 710.
150.   (a) Brown, H.C.; DeLue, N.R. *J. Am. Chem. Soc, 1974, 96,* 311; (b) Lane, C.F.; Brown, H.C. *J. Am. Chem. Soc, 1971, 93,* 1025; (c) Pelter, A.; Smith, K.; Brown, H.C. *Borane Reagents,* Academic Press, New York *1988,* p 250.

coupling products if a base is added to the bromination reaction.[151a] Adding a base suppresses radical bromination, allowing an ionic reaction to predominate. This transformation has been successfully applied to several different classes of molecules, including α-bromoketones, α-chloronitriles, α-bromosulfoxides or 4-bromocrotonates.[10] If phenyl or arylboranes are used as the organoborane partner, bromoesters can be arylated.

If a α-bromo carbonyl compound such as ethyl α-bromoacetate is treated with base, an enolate is formed (secs. 9.2, 9.4.B). The enolate can react with a trialkylborane to give an *ate* complex such as **217**, formed from ethyl bromoacetate and trinorbornylborane (**67**). The *ate* complex rearranges via a B → C shift that transfers an alkyl group from boron, replacing the bromine and giving **218**.[151a] The protonated base (the conjugate acid in this case is *tert*-butanol) generated by the initial enolate forming reaction is a sufficiently strong acid to initiate loss of boron. The resulting enol tautomerizes to ester **219** in 85% yield.[151a]

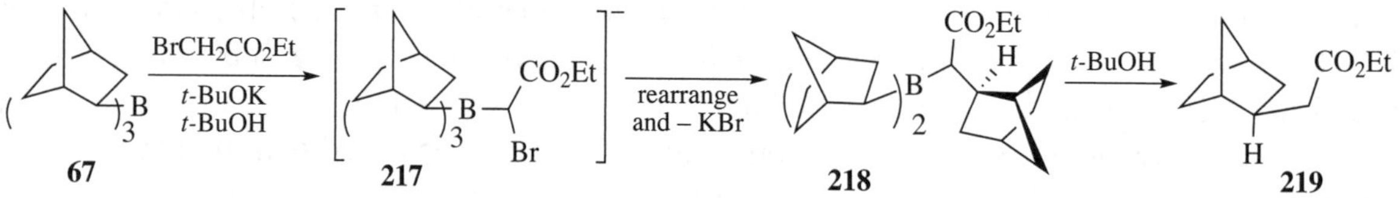

The transforms possible by these reactions are

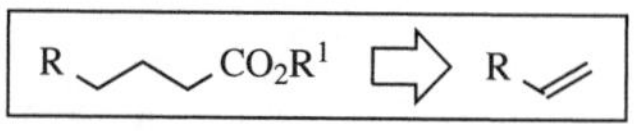

5.5. AMINES AND SULFIDES VIA HYDROBORATION

Both nitrogen and sulfur can be to incorporated into organic molecules by hydroboration. The most common is nitrogen, but there are a few examples where sulfur or even metals have been inserted into organic molecules.

5.5.A. Nitrogen-Containing Compounds

Trialkylboranes are converted to primary and secondary amines in a convenient manner.[152] Chloramine ($ClNH_2$) or hydroxylamine O-sulfonic acid ($H_2N\text{-}OSO_3H$) react with trialkylboranes to give primary amines.[153] Treatment of tributylborane with either reagent, for example, gave 1-aminobutane (**220**). The reaction proceeds via attack by nitrogen at the boron to give *ate* complex **221**. A B → N alkyl shift proceeds with expulsion of the leaving groups (X^- or HSO_4^-), to give **222**, which reacts with a second equivalent of amine to give **223**. Subsequent treatment

151. (a) Brown, H.C.; Rogić, M.M.; Rathke, M.W.; Kabalka, G.W. *J. Am. Chem. Soc,* **1968,** *90,* 818, (b) 1911.
152. For a review of synthetic applications, see Carboni, B.; Vaultier, M. *Bull. Soc. Chim. Fr.* **1995,** *132,* 1003.
153. Reference 150c, p 252.

with NaOH leads to cleavage and formation of the alkylamine (**220**). Similar reaction with

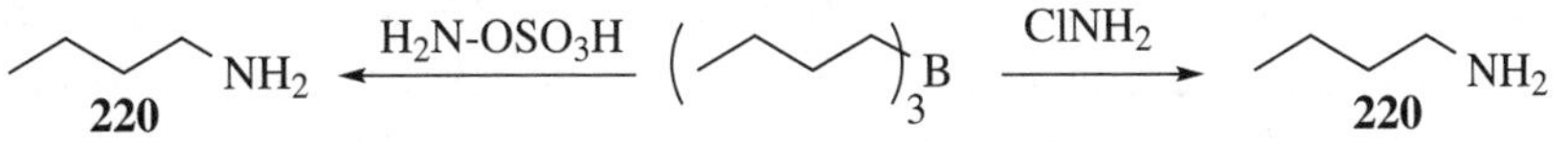

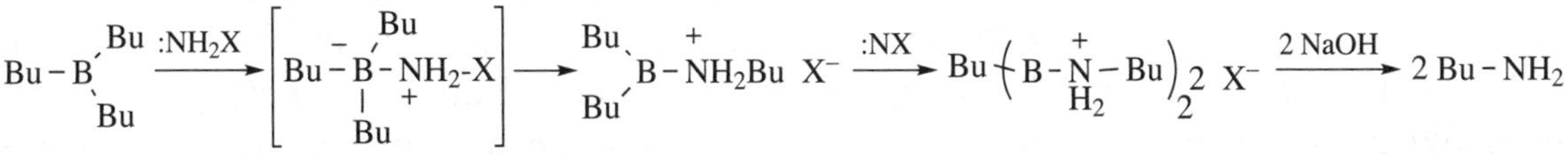

hindered organoboranes requires a temperatures of ~ 100°C.[154] Lewis acid can be added to the reaction to facilitate incorporation of the nitrogen by formation of a more labile *ate* complex, and also to promote the requisite B → N rearrangement. In a synthesis of potentiators of 2-amino-3-(3-hydroxy-5-methyl-isoxazol-4-yl)propanoic acid (AMPA) receptors,[155] Shepherd and co-workers reacted 1-phenylcyclopentene with chloroborane and then trimethylaluminum to form the corresponding *ate* complex. Finally, treatment with hydroxylamine *O*-sulfonic acid gave the *trans*-amine, **224**. Boranes derived from aromatic compounds usually give poor yields of amines, although this can be improved by using Lewis acids as in the example just described. Lewis acids promote the rearrangement (there is a favorable intramolecular reaction) because of the proximity of the aromatic nucleus to the aminating species. Asymmetric hydroboration is possible in these reactions, giving chiral, nonracemic amines. When styrene was treated with catecholborane in the presence of a chiral rhodium catalyst and then treated with MeMgCl (Et$_2$Zn can also be used), a complex was formed that generated a chiral borane. Subsequent treatment with H$_2$NOSO$_3$H gave a 54% yield of **225** in 87% ee.[156]

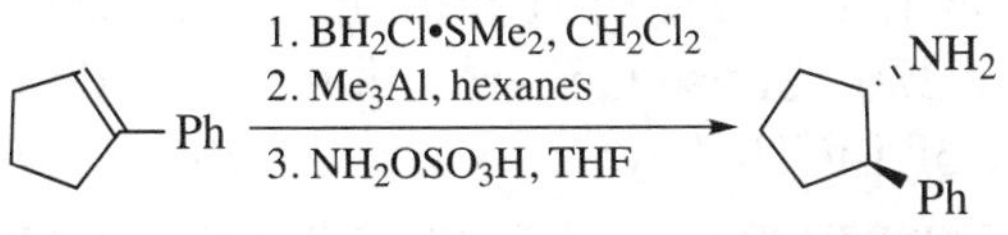

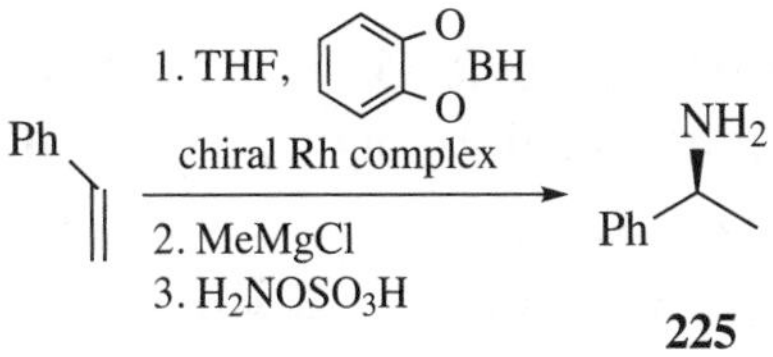

It is also possible for two alkyl groups to migrate from boron to form an intermediate such as **221**.[157] When borane **226** was treated with *N*-chloro-*O*-(dinitrophenyl)hydroxylamine (**227**), a double migration of the alkyl groups led to amino alcohol **228** upon oxidation.[158]

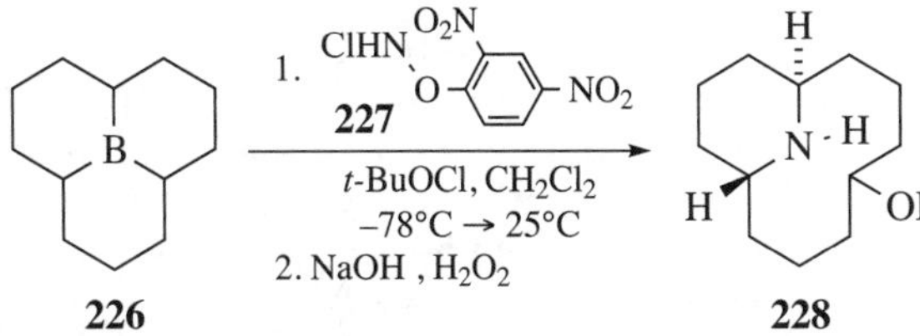

154. Rathke, M.W.; Inoue, N.; Varma, K.R.; Brown, H.C. *J. Am. Chem. Soc. 1966*, *88*, 2870.
155. Shepherd, T.A.; Aikins, J.A.; Bleakman, D.; Cantrell, B.E.; Rearick, J.P.; Simon, R.L.; Smith, E.C.R.; Stephenson, G.A.; Zimmerman, D.M.; Mandelzys, A.; Jarvie, K.R.; Ho, K.; Deverill, M.; Kamboj, R.K. *J. Med. Chem. 2002, 45*, 2101.
156. Fernandez, E.; Hooper, M.W.; Knight, F.I.; Brown, J.M. *Chem. Commun. 1997*, 173.
157. Levy, L.A.; Fishbein, L. *Tetrahedron Lett. 1969*, 3773.
158. Mueller, R.H. *Tetrahedron Lett. 1976*, 2925.

Azides react with trialkylboranes, and the resulting intermediate can be converted to a secondary amine.[159] The initial reaction requires high temperatures, however. In the example shown, the boron in triethylborane is attacked by a nitrogen of phenyl azide (**229**) to form ate complex **230**. A B → N shift of an ethyl group generated **231** via expulsion of nitrogen. Methanolysis and oxidation gave the secondary amine (*N*-ethylaniline), in 78% yield.[159] The oxidation step was required only to destroy excess borane. Both alkyl-dichloroboranes[160]a and dialkylchloroboranes[160]b react with azides to give secondary amines. In one example, dichloroborane reacted with 1-methylcyclopentene to give **232**. Subsequent reaction with cyclohexylazide generated a new chloroborane (**233**), and this was treated with aqueous hydroxide to give an unsymmetrical secondary amine (**234**) in 90% yield. The rearrangement and oxidation proceeded with complete retention of the trans geometry in the cyclopentane ring.[160a] When trialkylboranes react with chlorodimethylamines, the products are tertiary dimethylamines, but a radical scavenger such as galvinoxyl must be added to suppress competing radical reactions.[161]

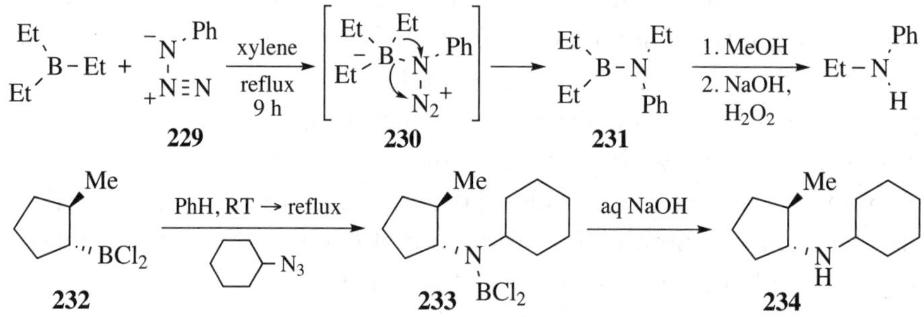

A related azide reaction was used to prepare phenylalanine.[162] Chiral borinane **235** reacted with lithium dichloromethane (LiCHCl$_2$) to give **236**. Subsequent reaction with sodium azide gave **237**. Reaction with a second equivalent of LiCHCl$_2$ was followed by oxidation with NaClO$_2$ to give azido acid **238**. Catalytic hydrogenation (sec. 4.8) gave phenylalanine (**239**) in 63% overall yield.[162]

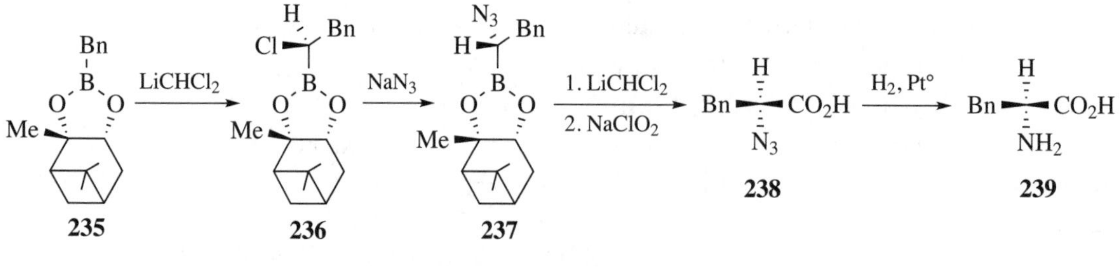

159. Suzuki, A.; Sono, S.; Itoh, M.; Brown, H.C.; Midland, M.M. *J. Am. Chem. Soc.* **1971**, *93*, 4329.
160. (a) Brown, H.C.; Midland, M.M.; Levy, A.B. *J. Am. Chem. Soc.* **1973**, *95*, 2394; (b) *Idem Ibid* **1972**, *94*, 2114.
161. Davies, A.G.; Hook, S.C.W.; Roberts, B.P. *J. Organomet. Chem.* **1970**, *23*, C11.
162. Matteson, D.S.; Beedle, E.C. *Tetrahedron Lett.* **1987**, *28*, 4499.

5.5.B. Sulfur

Trialkylboranes react with organic disulfides to give thioethers (sulfides), but air or light is required because the reaction proceeds by a radical chain process.[163] This reaction is very slow in the dark and ultraviolet (UV) light is required for reasonable reaction rates. Trioctylborane reacts with dimethyl disulfide under these conditions to give methyl 1-octyl sulfide in 95% yield.[163] The mixed borane, *B*-cyclohexyl-3,5-dimethylborinane (**240**) gave 94% of cyclohexyl methyl sulfide (**241**) under the same conditions.[163]

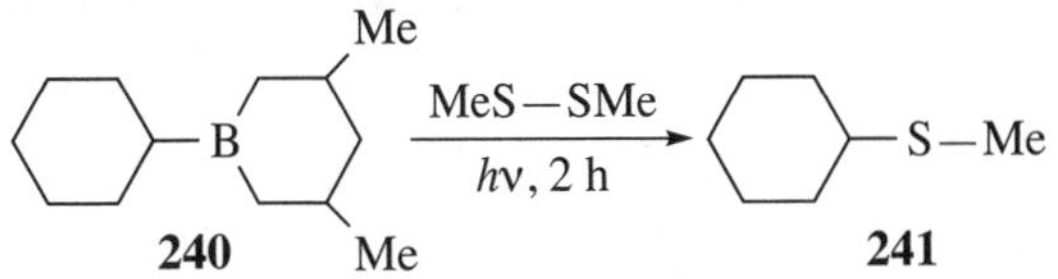

The preparation of unsymmetrical sulfides is increasingly important since chiral sulfoxides can be prepared (sec. 3.9.A) and used in organic synthesis. α-Lithio sulfides give useful carbanion reactions and sulfur ylids can be formed (secs. 8.6, 8.8.B). The difficulties in preparing primary and secondary amines by traditional methods (from halides or tosylates via reaction with imides or azides, sec. 2.6) make the hydroboration-amination sequence a mild and useful alternative for the synthesis of amines.

The functional group transforms available in this section are

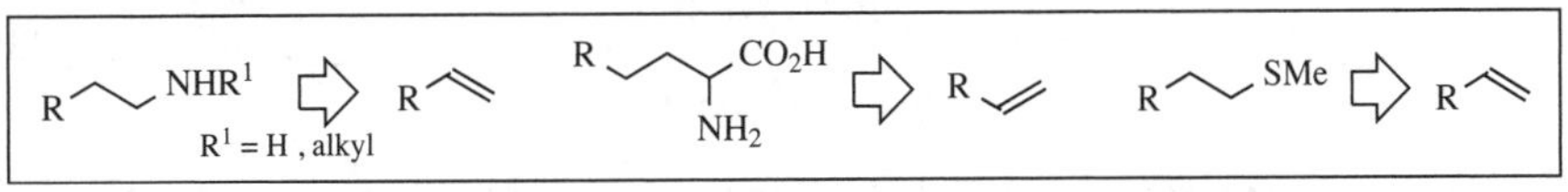

5.6.   CONCLUSION

This chapter concludes the discussion of functional group exchange reactions commonly used in organic synthesis. Hydroboration provides alternatives to many of the transformations described in chap. 1, and adds many unique ones. Hydroboration also provides new examples of controlling asymmetric induction by proper design of reagents. This idea will be reviewed in chap. 6 and it will include general methods for controlling selectivity. Problems unique to forming cyclic compounds will also be discussed.

HOMEWORK

1. Virtually all of the functional group transforms associated with organoboranes involve conversion to an *ate* complex. Explain why this is the case.

2. Give the major product of the reaction shown, and discuss all forms of selectivity that apply to it.

---

163. Brown, H.C.; Midland, M.M. *J. Am. Chem. Soc.* **1971**, *93*, 3291.

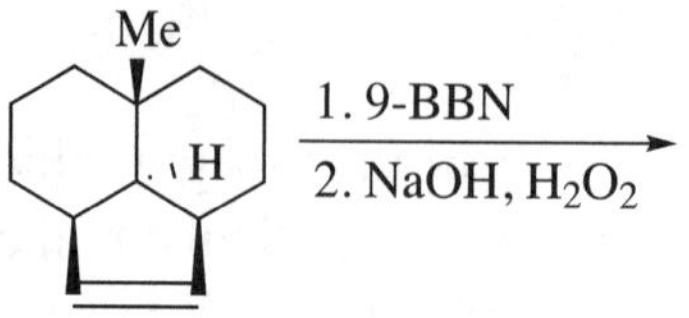

$$\text{1. 9-BBN} \quad \xrightarrow{\text{2. NaOH, H}_2\text{O}_2}$$

3. Predict the structure of each product, A-F.

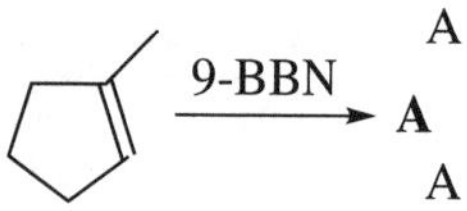

$$\text{A + NH}_2\text{OSO}_3\text{H} \longrightarrow \textbf{B} \quad \text{A + CH}_3\text{CO}_2\text{D, diglyme, 100°C} \longrightarrow \textbf{C}$$

$$\xrightarrow{\text{9-BBN}} \textbf{A}$$

$$\text{A + H}_2\text{O}_2\text{, KOH} \longrightarrow \textbf{D}$$

$$\text{A + I}_2\text{, NaOH} \longrightarrow \textbf{E} \qquad \text{A + (1) CO, H}_2\text{O (2) NaOH, H}_2\text{O}_2 \longrightarrow \textbf{F}$$

4. For each of the following, draw an appropriate transition state and predict the major product(s).

(a) (-)-IpC$_2$BH + (3Z) hexene (b) (-)-IpC$_2$BH + 6-methyl-(3E)-octene (c) (+)-IpCBH$_2$ + (3E)-hexene

5. Give three different examples of carbon-carbon bond-forming reactions involving boranes. Show all starting materials, reagents, and product(s).

6. Provide a mechanistic rationale for the following transformation:

$$\xrightarrow[\text{2. NaOH, H}_2\text{O}_2]{\text{1. B}_2\text{H}_6\text{, 165°C}}$$

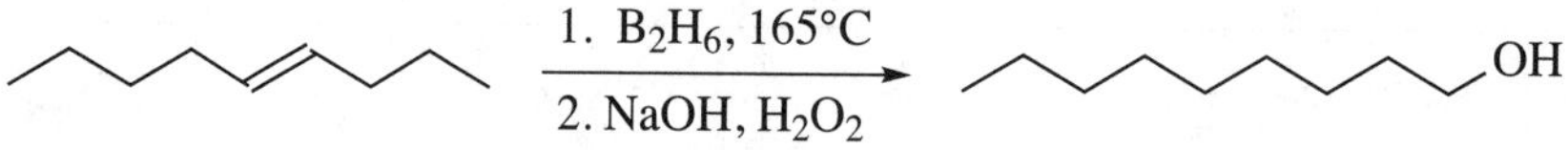

7. Provide a rationale for formation of the indicated product, paying particular attention to the stereochemistry of the secondary alcohol.

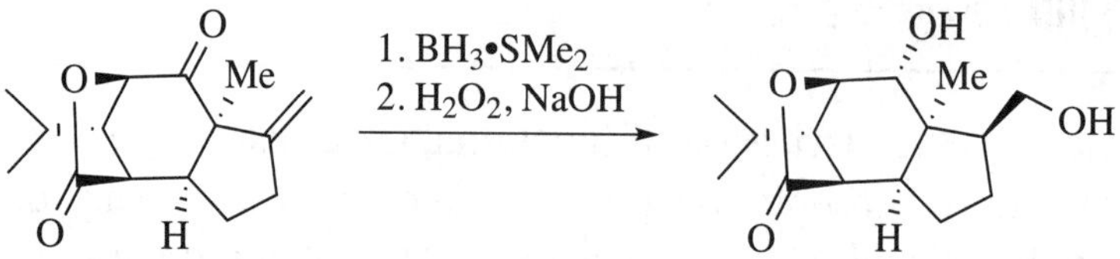

$$\xrightarrow[\text{2. H}_2\text{O}_2\text{, NaOH}]{\text{1. BH}_3\bullet\text{SMe}_2}$$

8. The following transformations are very convenient using a hydroboration strategy. Discuss the possibility of alternative synthetic routes that do <u>not</u> involve hydroboration.

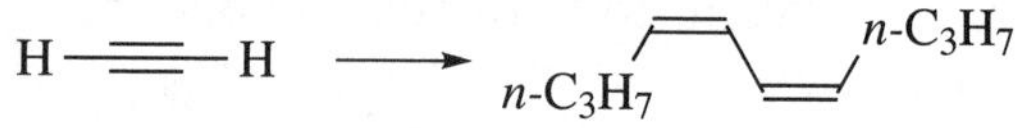

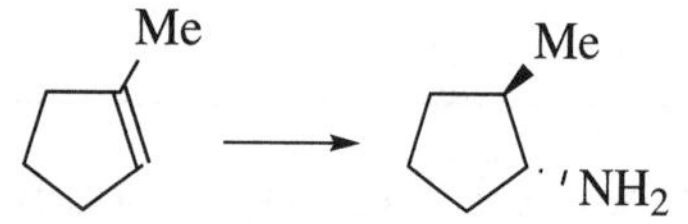

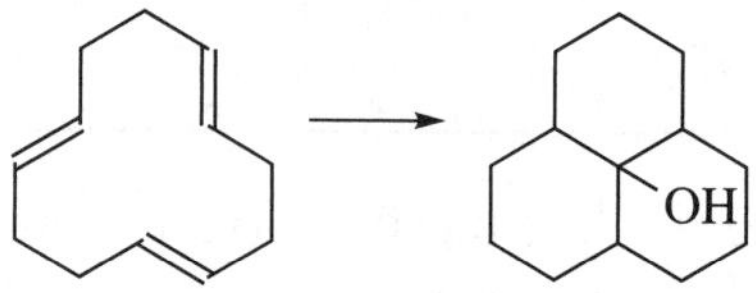

9. Predict the major product of this reaction and justify the stereochemistry of the final product.

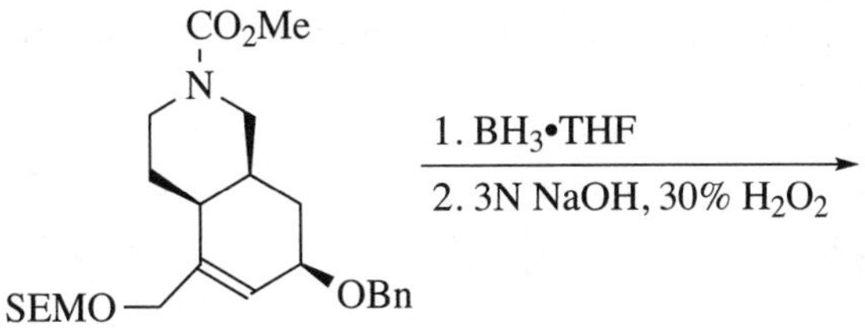

Reactant: N-CO$_2$Me bicyclic with SEMO-CH$_2$ and OBn substituents

1. BH$_3$•THF
2. 3N NaOH, 30% H$_2$O$_2$

10. In each case, give the major product or products. Remember stereochemistry where it is appropriate.

(a) 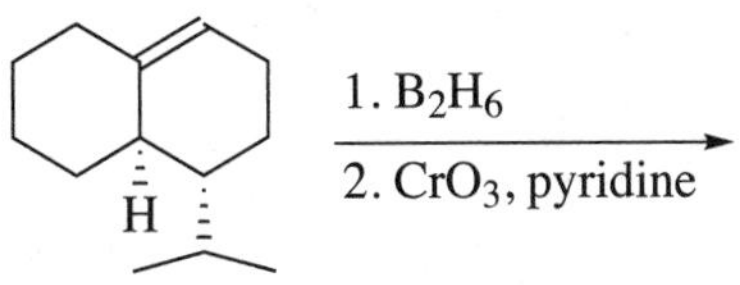
1. B$_2$H$_6$
2. CrO$_3$, pyridine

(b) 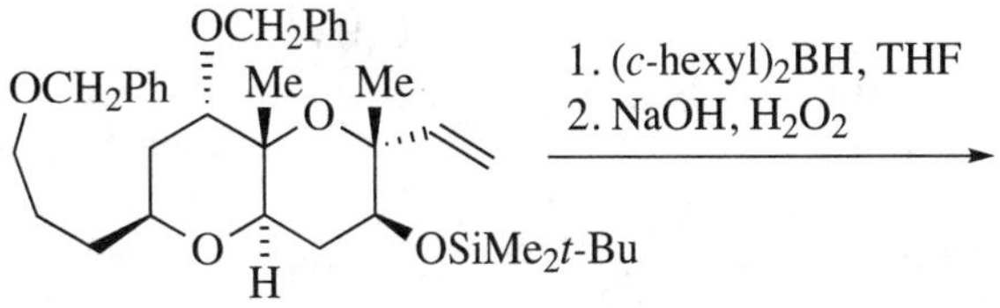
1. (c-hexyl)$_2$BH, THF
2. NaOH, H$_2$O$_2$

(c)
MeO–P(=O)(OMe)–CH=CH–C(CH$_3$)–(p-tolyl)
1. BH$_3$•THF
2. 30% H$_2$O$_2$, NaOH

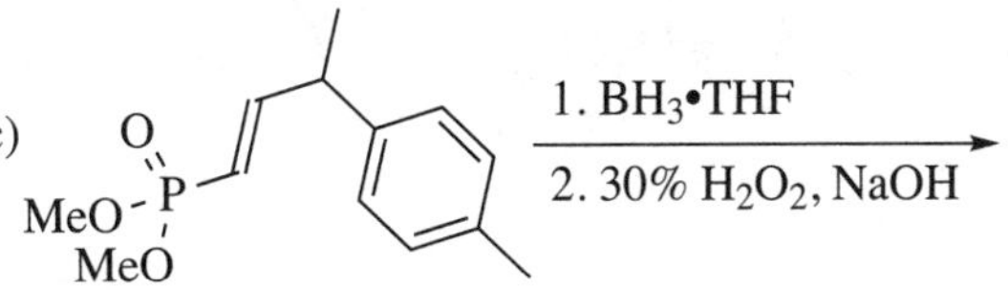

(d) 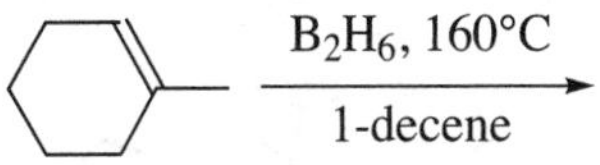
B$_2$H$_6$, 160°C
1-decene

(e) 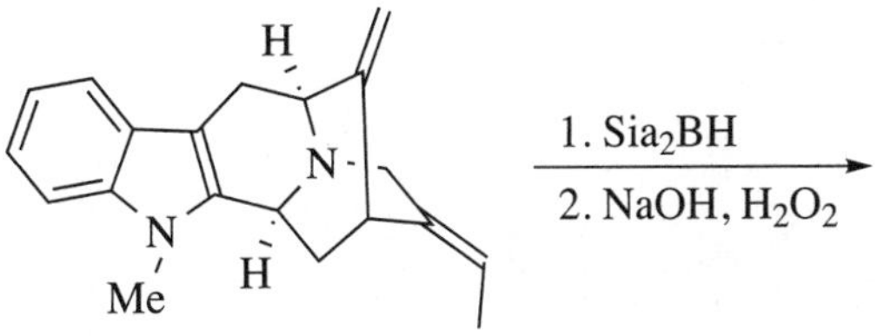
1. Sia$_2$BH
2. NaOH, H$_2$O$_2$

(f) 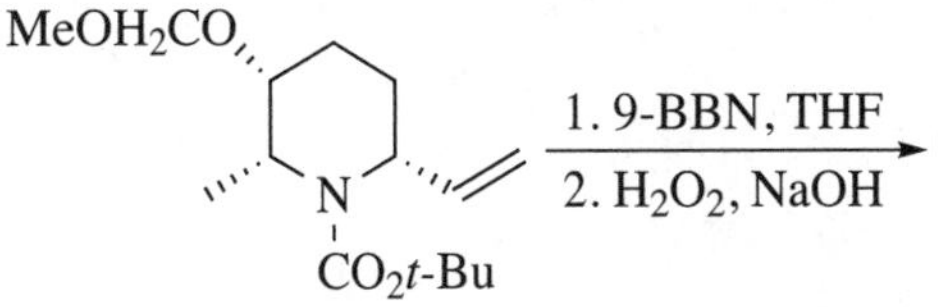
MeOH$_2$CO, piperidine with vinyl, N-CO$_2$t-Bu
1. 9-BBN, THF
2. H$_2$O$_2$, NaOH

(g) 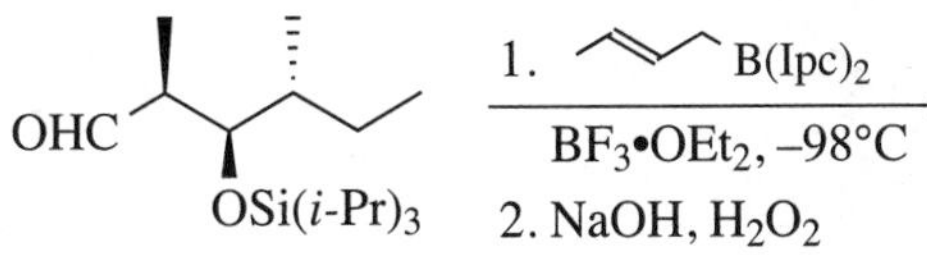 OHC–CH(CH$_3$)–CH(OSi(i-Pr)$_3$)–CH(CH$_3$)–CH$_2$CH$_3$
1. (E)-crotyl–B(Ipc)$_2$, BF$_3$•OEt$_2$, –98°C
2. NaOH, H$_2$O$_2$

(h) 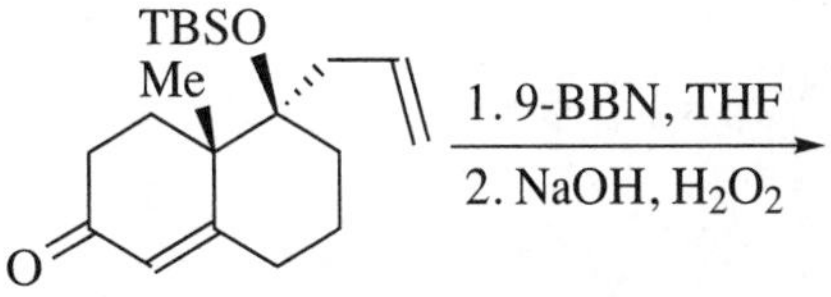
TBSO, Me, allyl decalone with O
1. 9-BBN, THF
2. NaOH, H$_2$O$_2$

(i)
Ph$_2$C=CH$_2$
1. catecholborane, 100°C
2. Me$_3$SiCHN$_2$, THF
3. NaOH, H$_2$O$_2$ 4. Bu$_4$NF, THF

(j) 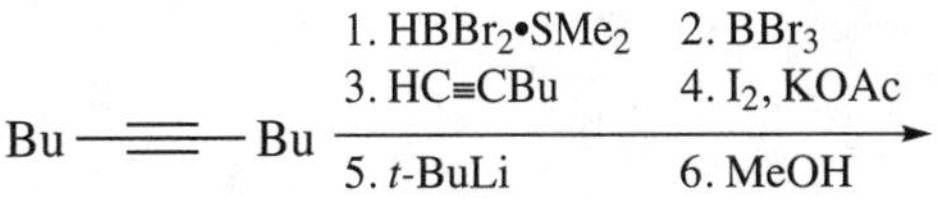 Bu–C≡C–Bu
1. HBBr$_2$•SMe$_2$ 2. BBr$_3$
3. HC≡CBu   4. I$_2$, KOAc
5. t-BuLi   6. MeOH

(k) 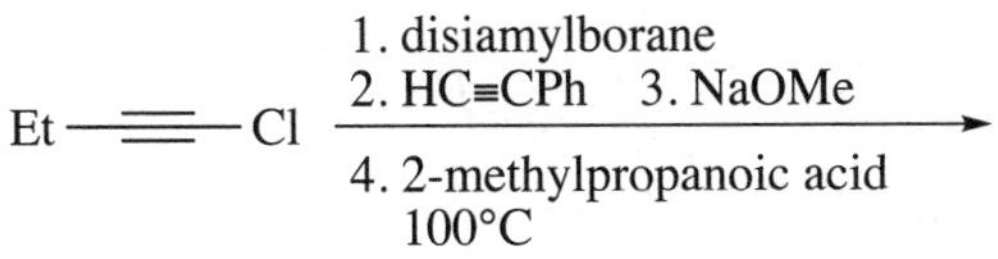 Et–C≡C–Cl
1. disiamylborane
2. HC≡CPh   3. NaOMe
4. 2-methylpropanoic acid
100°C

(l) 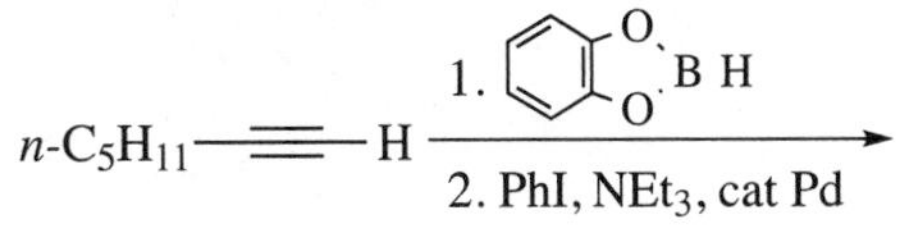 n-C$_5$H$_{11}$–C≡C–H
1. catecholborane (O$_2$C$_6$H$_4$)BH
2. PhI, NEt$_3$, cat Pd

(m) 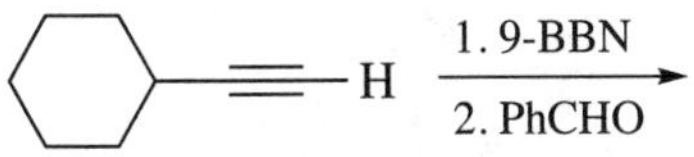 cyclohexyl–C≡C–H
1. 9-BBN
2. PhCHO

(n) CH$_2$=CH–CH$_2$CH$_2$CH$_2$CH$_3$ (hexene)
1. B$_2$H$_6$
2. CO, ethylene glycol heat;
3. NaOH, H$_2$O$_2$

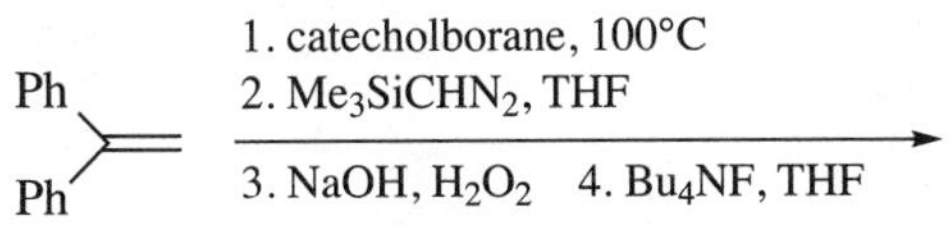

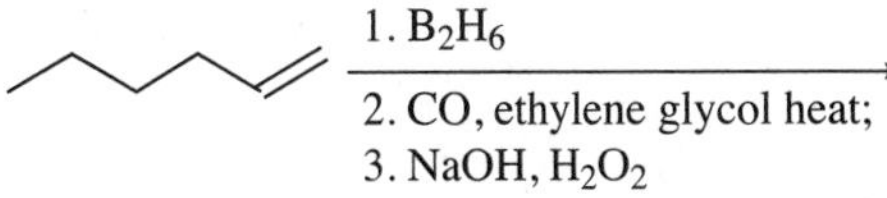

11. Predict the major product. Explain the selectivity of each reaction.

(a)

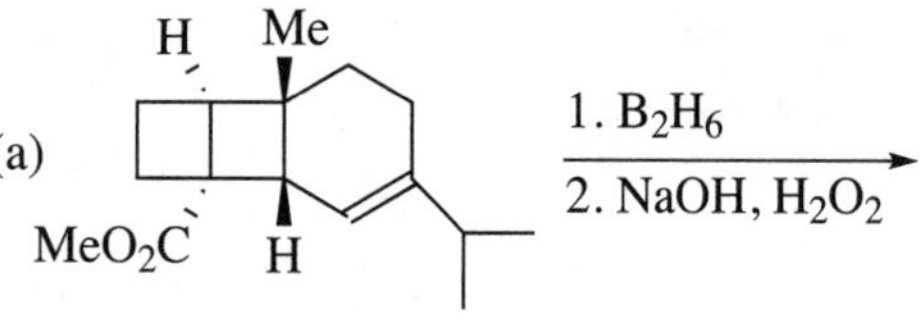

1. B₂H₆

2. NaOH, H₂O₂

$$1.\ B_2H_6$$
$$2.\ NaOH,\ H_2O_2$$

(b)

1. catecholborane
   LiBH₄, THF

2. NaOH, H₂O₂

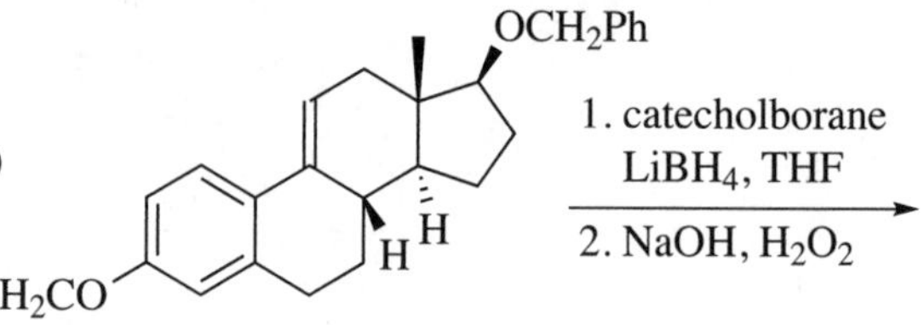

12. In each case, provide a suitable synthesis. Show all reagents and intermediate products.

(a) 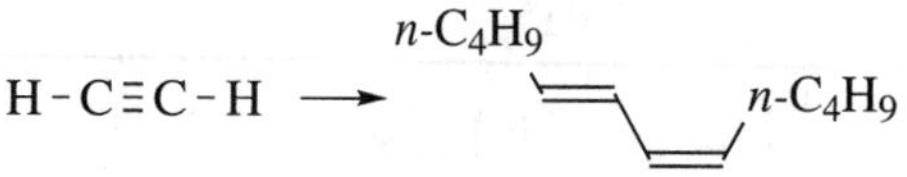 $H-C\equiv C-H$ →

(b)

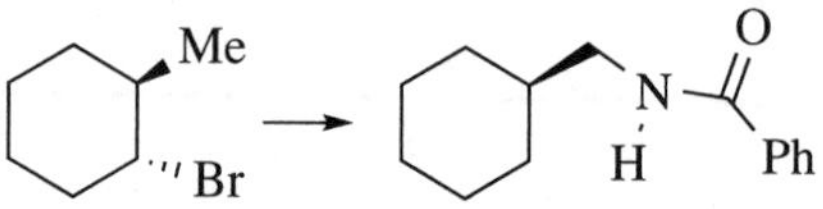

(c)

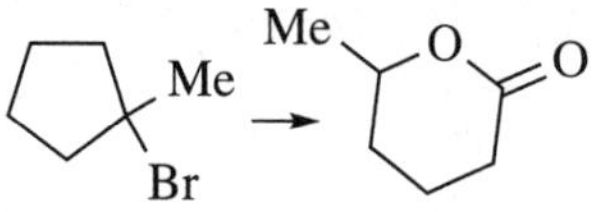

(d)

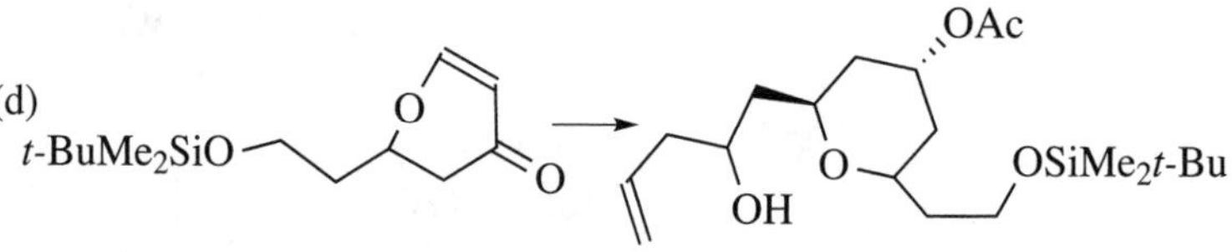

(e)

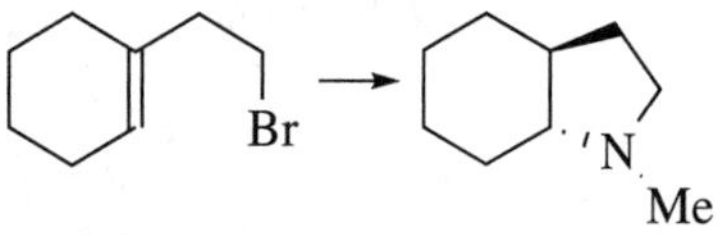

chapter 6

Stereocontrol and Ring Formation

6.1. INTRODUCTION

Chapters 1-5 contain many examples of regioselective, diastereoselective or enantioselective reactions involving both acyclic and cyclic molecules. Chelating effects with certain heteroatom containing substrates was also discussed, with examples of how chelation control could override the effect of non-bonded interactions. There were examples that illustrated the ability to influence, and sometimes control the facial selectivity of a reaction in both acyclic and cyclic systems. Conformational preferences (orientations) in the substrate were also shown to be important. The purpose of this chapter is to focus attention on methods for either controlling the conformational effects in a molecule, or exploiting unalterable conformational effects to achieve a specific synthetic purpose. This chapter provides a review that will summarize key concepts, and focus attention on stereoselectivity prior to discussing carbon-carbon bond-forming reactions.

Section 6.5 in this chapter will summarize methodology for controlling the stereochemistry of acyclic molecules by taking advantage of the conformational bias of cyclic molecules. The cyclic carrier is chemically modified, and then the ring is opened to give a substituted or functionalized acyclic molecule with a certain stereochemical pattern. Another section deals with intramolecular cyclization reactions for generating small, medium, and large ring compounds. The limitations and synthetic value of these ring forming reactions will also be discussed. This chapter will present the fundamental ways an organic chemist recognizes the conformational and topological features of a molecule, and uses that information to give a desired result. This strategy includes recognition of those molecular features that will prevent the desired reaction or lead to incorrect stereochemistry, allowing the chemist to design alternatives.

6.2. STEREOCONTROL IN ACYCLIC SYSTEMS

Reactions of acyclic molecules can generate stereoisomers, and when a chemist refers to "controlling stereochemistry" in such a reaction there is an attempt to generate one stereoisomer in preference to another as the major product of a reaction. Typical situations where control of stereochemistry is important include: (*1*) Regioselectivity (Markovnikov versus anti-Markovnikov), (*2*) Retention versus inversion of configuration, (*3*) cis-trans Selectivity, (*4*), syn-anti Selectivity, and (5) Heteroatom chelation effects.

Reactions where these effects are important were presented in previous chapters. The carbon-carbon bond forming reactions in succeeding chapters contain many additional examples. The purpose of this chapter is not to solve specific synthetic problems (diastereoselective enolate alkylation and condensations, selectivity in the Claisen rearrangement, etc.), but rather to define the problems in a general sense. Approaches for solving these problems will be discussed, but specific solutions are usually presented in other chapters.

6.2.A. Regioselectivity

Regioisomers have the same empirical formula, but groups or atoms are positioned at different carbon atoms. 3-Methylhexanal and 4-methylhexanal are regioisomers, for example. A reaction that generates one regioisomer as the major product can be regioselective or regiospecific. We can take an example from one of the fundamental transformations presented in Chapter 2, addition to an alkene. The reaction of 2-methyl-2-butene with HBr gives 2-bromo-2-methylbutane as the major product, a typical Markovnikov addition, so the reaction is highly regioselective. This reaction proceeds by formation of the more stable carbocation, which reacts with the nucleophilic bromide ion (sec. 2.10.A). If the anti-Markovnikov bromide (1-bromo-2-methylbutane) is desired, a different mechanistic pathway must be followed. Reaction of the alkene with HBr and a peroxide proceeds by a radical mechanism, giving 1-bromo-2-methylbutane, the anti-Markovnikov bromide (secs. 2.10.A, 13.5.B).[1]

Understanding which products are formed and the mechanism of their formation, is essential if one is to control which product is formed. Addition of HBr generates a carbocation, with the charge residing on the more substituted carbon. The three alkyl substituents attached to the positive center provide more electronic stabilization to the p orbital of the cation than is possible in the other possible intermediate, the primary carbocation. Reaction of the tertiary carbocation and the nucleophilic bromide ion generates 2-bromo-2-methylbutane as the major product. When a peroxide is added, Br• is generated *in situ* (sec. 13.3) and this radical adds to the alkene to give the more stable radical. This radical intermediate reacts with more HBr to give the anti-Markovnikov product and more bromine radical. If a reaction generates a cation, Markovnikov orientation is sure to result. If anti-Markovnikov addition is desired, *the reaction must proceed by a different reaction mechanism.* The radical mechanism satisfies this requirement and if the reaction generates a radical, anti-Markovnikov orientation is preferred[2,3] (secs. 13.3-13.6).

---

1. (a) Markovnikoff, V. *Compt. Rend.* *1875, 81*, 668; (b) Isenberg, N.; Grdinic, M. *J. Chem. Educ.* *1969, 46*, 601.
2. Curran, D.P. *Synthesis* *1988*, 489.
3. (a) Keck, G.E.; Yates, J.B. *J. Am. Chem. Soc.* *1982, 104*, 5829; (b) Keck, G.E.; Enholm, E.J.; Yates, J.B.; Wiley, M.R. *Tetrahedron* *1985, 41*, 4079.

Hydroboration of alkenes is another regioselective reaction (see Chap. 5). Addition of borane to the alkene, gives 2-methyl-1-butanol after oxidation of the intermediate alkylborane, a typical anti-Markovnikov addition product (sec. 5.4.A).[4] The hydroboration-oxidation sequence is highly regioselective for the generating the OH unit on the less substituted carbon atom of the alkene. The hydroboration reaction with alkenes to produce alkylboranes[4] (sec. 5.2) proceeds by a four-center transition state rather than a cationic intermediate. The regiochemistry of the final alkylborane product is controlled by the non-bonded steric interactions of the groups attached to boron (in this case *sec*-isoamyl from the disiamylborane) and the groups on the alkene. Oxidation with basic hydrogen peroxide converts the borane to the anti-Markovnikov alcohol. With this alcohol in hand, treatment with PBr_3 gives 1-bromo-2-methylbutane, the anti-Markovnikov bromide mentioned above. Once again, understanding the fundamental mechanism of the hydroboration reaction, allowed us to prepare first the anti-Markovnikov alcohol form the alkene, and then convert it to the bromide. It is clear from these examples, that a product with a certain regiochemistry can be synthesized by modifying the chemical reaction to follow a different mechanistic pathway, or by choosing a different chemical synthesis.

6.2.B. Retention versus Inversion of Configuration

Functional group transformations often proceed with retention, racemization or inversion of a stereochemical center, as in the substitution reactions from Chapter 2 (S_N2 and S_N1, for example). Hydroxylation and dihydroxylation reactions in chapter 3 gave diastereomers or enantiomers, and reduction of ketones and aldehydes in Chapter 4 led to formation of stereoisomeric alcohols. In Chapter 4, Cram's rule or the Felkin-Ahn model allowed one to predict the stereochemistry of the alcohol being formed. Clearly, it is important to predict which stereoisomer is preferred if the reaction is stereoselective, and it is also important to anticipate poor stereocontrol in a reaction. In the S_N2 reaction, we know that the reaction proceeds with complete stereochemical inversion and the S_N1 is predicted to proceed with complete or partial racemization. This section will discuss methodologies that allow one to predict the stereochemical outcome of a reaction, or even to change it after the fact.

A classical example of controlling the configuration of a stereocenter is conversion of a chiral secondary alcohol to the corresponding secondary chloride with thionyl chloride,[5] and it is also another example of changing stereochemistry by modifying the mechanism of the reaction.

---

4. (a) Brown, H.C. *Hydroboration*, W.A. Benjamin, New York, *1962*; (b) *Idem Pure Appl. Chem. 1976, 47*, 49.
5. (a) Lee, C.C.; Finlayson, A.J. *Can. J. Chem. 1961, 39*, 260; (b) Lee, C.C.; Clayton, J.W.; Lee, D.G.; Finlayson, A.J. *Tetrahedron 1962, 18*, 1395.

Both retention of configuration (neat SOCl$_2$) and inversion (SOCl$_2$ + pyridine) are possible (sec. 2.8.A).[5] Thionyl chloride reacts with (S)-2-pentanol to produce a chlorosulfinate ester (**1**). The HCl by-product escapes from the reaction medium, and an intramolecular delivery of chloride in what is called a S$_N$i mechanism (see **1**)[5,6] gives (S)-2-chloropentane (**2**), with concomitant loss of sulfur dioxide (SO$_2$). The synfacial delivery of chlorine generates the chloride with retention of configuration. If pyridine (or another basic tertiary amine) is added, the HCl by-product is trapped as the pyridinium hydrochloride salt (**3**) and nucleophilic chloride ion is in the medium. Displacement of the chlorosulfinate ester (**1**) by chloride ion, in an S$_N$2 reaction, gives chloride **4** with net inversion of configuration.[7] The stereochemistry of the reaction is, therefore, controlled by the presence or absence of a nucleophilic chloride ion in the reaction medium.

Formation of **4** showed that conversion of the OH unit to a leaving group allows an S$_N$2 reaction to occur. Indeed, S$_N$2 displacement is possible with substrates containing many different leaving groups, all with inversion of that stereogenic center. Conversion of (S)-2-pentanol to the corresponding tosylate (**5**), for example, allows reaction by nucleophiles such as azide to give **6** with complete control of the stereogenic center. Changing the poor leaving group OH to the good leaving group tosylate allows greater flexibility in the synthetic sequence. Changing the leaving group may be ineffective, however, if the nucleophile is too weak. As we saw in Chapter 2, there is a great difference in nucleophilic strength for a given substrate and reaction type. If a nucleophile, such as methoxide, reacts with **5**, the S$_N$2 reaction would give a methyl ether. Nucleophiles such as methoxide are strong bases, however, and elimination can occur to give an alkene via the E2 reaction (sec. 2.9.A). The use of highly polar, aprotic solvents such as DMF or DMSO can sometimes increase the yield of substitution product (secs. 2.7.A, 2.6.A.i) and minimize elimination by-products, just as the use of a protic solvent such as methanol usually increases the amount of elimination products. In planning the reaction, the course of a reaction can be modified by changing the structural components of the substrate (good versus poor leaving group), or by changing the reaction partner and/or the solvent. Clearly, competing reactions must be considered, in addition to the factors that control reactivity and stereochemistry.

If one chooses the substrate, reactants or reaction conditions to favor one mechanism over other possibilities, good stereocontrol may be achieved. The ability to control a stereogenic center

6. Lewis, E.S.; Boozer, C.E. *J. Am. Chem. Soc.* **1952**, *74*, 308.
7. Darzens, G. *Compt. Rend.* **1911**, *152*, 1601.

using a S_N2 reaction is illustrated by the reaction of (*S*)-bromide **7** with the potassium salt of benzylthiol to give thioether **8**, in 77% yield and 98% ee as part of the Seki et al. synthesis of an orally active 1-β-methylcarbapenam TA-949.[8] The importance of solvent in this reaction is illustrated by differences in % ee. When THF was used as the solvent, **8** was isolated in only 12% ee. Selectivity was raised to 48% ee in DMF, and to 72% ee when the reaction was done in water.[8]

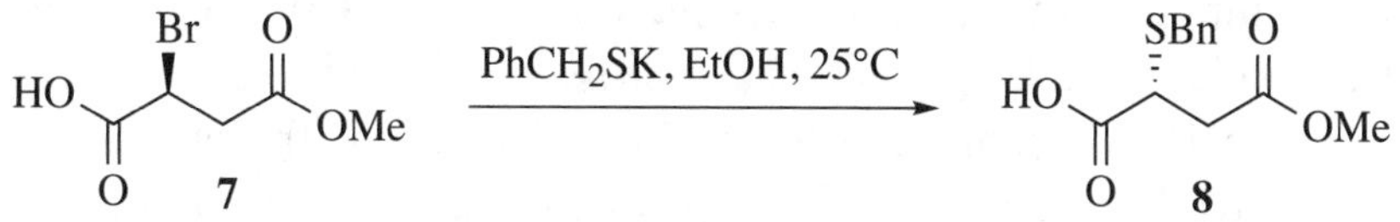

A successful and highly useful S_N2 reaction that is commonly used to change the stereochemistry of a reactive center is the **Mitsunobu reaction** (sec. 2.6.A.ii).[9] The Mitsunobu reaction usually involves mixing diethylazodicarboxylate (DEAD) and triphenylphosphine (Ph$_3$P) with an alcohol in the presence of nucleophiles.[9] Other phosphines can be used as well. The Mitsunobu reaction is typically used to invert a stereochemical center, transforming a (*R*)-hydroxyl unit to a (*S*)-hydroxyl unit for example. This process is illustrated by the conversion of **9** to **11**, taken from Hudlicky, Pettit and co-worker's synthesis of narciclasine.[10] **Luche reduction** (sec. 4.4.B) of **9** gave **10** due to the conformational restraints on delivery of the hydride using this reagent. The relative stereochemistry of the hydroxy unit was incorrect for narciclasine, and was corrected by reaction of **10** with DEAD, tributylphosphine and benzoic acid to give benzoate **11**. The benzoate group was later removed to liberate the alcohol with Amberlyst A21 in methanol.[10] Note that other nucleophiles can be incorporated into a molecule using Mitsunobu conditions.

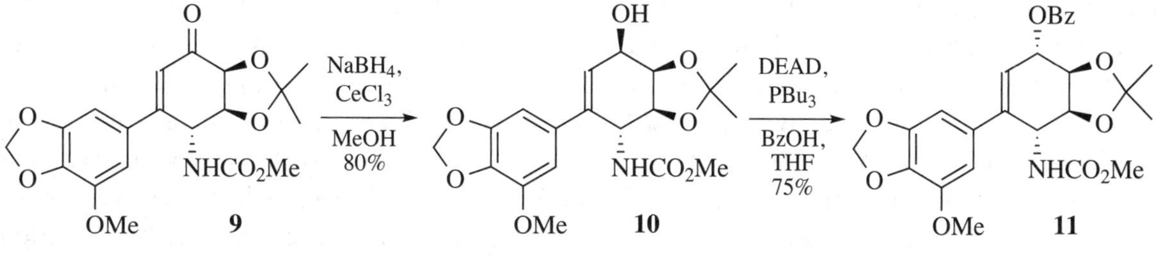

Other methods are available for controlling relative stereochemistry. An example is the NH$_2$ → Br transformation, with complete retention of configuration. Reaction of amino acid **12** with sodium nitrite (NaNO$_2$) and potassium bromide (KBr) converted the amino group to a bromide, and gave α-bromo acid (**13**) with

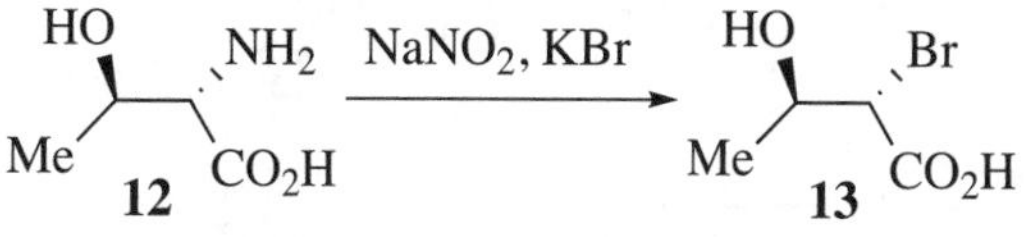

8. Seki, M.; Yamanaka, T.; Kondo, K. *J. Org. Chem.* **2000**, *65*, 517.
9. Mitsunobu, O. *Synthesis* **1981**, 1.
10. Hudlicky, T.; Rinner, U.; Gonzalez, D.; Akgun, H.; Schilling, S.; Siengalewicz, P.; Martinot, T.A.; Pettit, G.R. *J. Org. Chem.* **2002**, *67*, 8726.

complete retention of configuration,[11] a reaction first observed by Walden.[11b-d] This reaction was taken from Salzmann and co-worker's synthesis of the potent carbapenam antibiotic thienamycin.[12] Note that this is a special case where the carboxylate unit participates in the reaction.by conversion of *D*-mannitol to (*R*)- or (*S*)-epichlorohydrin.[13] Tosyl diol **14**, obtained from (*D*)-mannitol, was chlorinated selectively at the more reactive primary alcohol to give **15**. Treatment with sodium gave the epoxide (*S*)-epichlorohydrin (**16**).[13] Reaction of **14** with the basic sodium methoxide, however, generated a hydroxy epoxide via displacement of the tosylate leaving group. Subsequent treatment of the hydroxy epoxide with methanesulfonyl chloride generated the mesylate (**17**). Reaction of the epoxide oxygen with HCl gave chlorohydrin **18**, which gave (*R*)-epichlorohydrin (**19**) upon treatment with sodium methoxide (the OMs in **18** is a better leaving group than Cl). This example shows that understanding reactivity differences in a molecule can lead to control of the stereochemistry in chemical manipulations.

6.2.C. cis-trans Selectivity

Addition of certain reagents to a C≡C unit of an alkyne (hydrogenation, addition of halogens, etc.) can lead to either cis- or trans-alkenes. Control of cis-trans-geometry is well illustrated by catalytic hydrogenation or alkali metal reduction of alkynes, as discussed in Sections 4.8.B and 4.9.C, respectively. The Lindlar catalyst (sec. 4.8.B) allows selective hydrogenation of alkynes to the cis-alkene, as in the conversion of **20** to **21** in 78% yield, in studies directed towards the synthesis of discodermolide by Parker and Katsoulis.[14] This contrasts sharply with treatment of an alkyne with alkali metals (sec. 4.9.C) to give the trans alkene, as in the conversion of **22** to **23**.[15] Once again, a fundamental understanding of the difference in these two reaction mechanisms allowed control of the cis-trans geometry of the final product. Control of cis-trans-geometry (diastereoselectivity) in cyclic systems will be discussed in Section 6.3.D.

11. (a) Shimohigashi, Y.; Waki, M.; Izumiya, N. *Bull. Chem. Soc. Jpn.* **1979**, *52*, 949; (b) Walden, P. *Berichte* **1895**, *28*, 2766; (c) *Idem, Ibid.* **1896**, *29*, 133; (d) *Idem, Ibid.* **1899**, *32*, 1855.
12. (a) Salzmann, T.N.; Ratcliffe, R.W.; Christensen, B.G.; Bouffard, F.A. *J. Am. Chem. Soc.* **1980**, *102*, 6161; (b) Mellilo, D.G.; Liu, T.; Ryan, K.; Sletzinger, M.; Shinkai, I. *Tetrahedron Lett.* **1981**, *22*, 913.
13. Baldwin, J.J.; Raab, A.W.; Mensler, K.; Arison, B.H.; McClure, D.E. *J. Org. Chem.* **1978**, *43*, 4876.
14. Parker, K.A.; Katsoulis, I.A. *Org. Lett,* **2004**, *6*, 1413.
15. Kochansky, J.P.; Cardé, R.T.; Taschenberg, E.F.; Roelofs, W.L. *J. Chem. Ecol.* **1977**, *3*, 419.

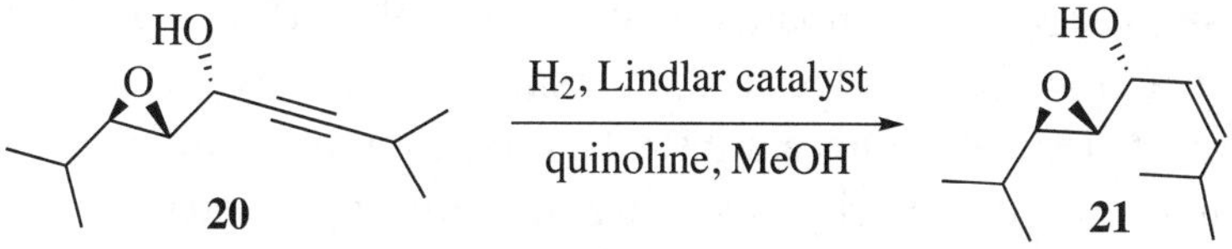

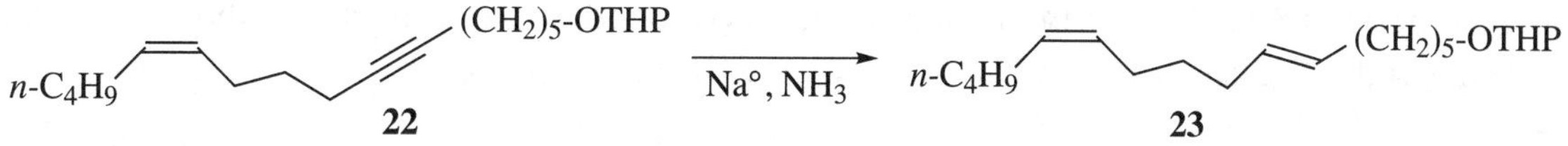

Reactive intermediates in several reactions have a C=C unit, and stereochemistry plays a role in the selectivity of those reactions. Two examples are organometallics such as Grignard reagents derived from vinyl halides, and enolate anions. The configurational stability of these species will be discussed in detail in sections 8.4.F, 8.5.D and in Sections 9.2.C, 9.5, respectively. There are problems associated with cis-trans isomerism. When (1E)-bromo-1-propene (**24**) reacts with magnesium, Grignard reagent 1-propenylmagnesium bromide **25** (R = CH$_3$, R^1

= H, secs. 8.4.A, 8.4.F) is formed. Isomerization led to a mixture of the (E)-isomer (**25**) in 60-70% yield with 30-40% of the (Z)-isomer (**26**),[16] and this is reflected in the subsequent reactions of the Grignard reagent. In other words, when **25** + **26** react with an aldehyde, the corresponding allylic alcohol product will be a mixture of (E)- and (Z)-isomers. In general, reaction of trans alkenyl organometallics tend to give the trans alkenyl product, and cis alkenyl halides give the corresponding cis alkenyl product. Isomerization can be a problem in these reactions, however.[16b]

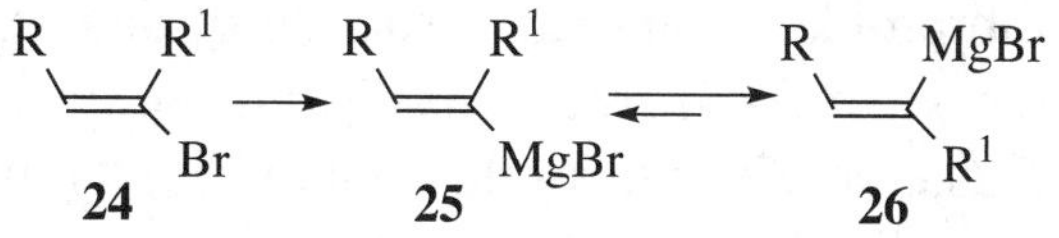

When a ketone reacts with a suitable base (secs. 9.1, 9.2) an enolate anion is formed by removal of the α-proton. In the case of an unsymmetrical ketone such as **27**, a mixture of (Z)-enolate (**28**) and (E)-enolate (**29**) will result (secs. 9.2.E, 9.5.A). The diastereoselectivity and enantioselectivity in products formed by enolate condensation reactions (sec. 9.5) are influenced by the stereochemistry of the enolate anion. When s mixture of geometrical isomers **28** and **29** reacts with aldehydes, a mixture of both syn- and anti-products will be formed, so it is important to

predict or know the geometry of the enolate anion. Several solutions to this problem have been developed, including formation of stable and separable enolate isomers, and controlling reaction conditions to maximize production of one isomer. These techniques will be discussed in Section 9.2, but an example can be presented here to illustrate the approach. Reaction of 3-pentanone (**27**, R = Et, R^1 = Me) with lithium diisopropylamide gave a 77:23 mixture of (E)- and (Z)-enolate anions.[17] Similar reaction with lithium 2,2,6,6-tetramethylpiperidide

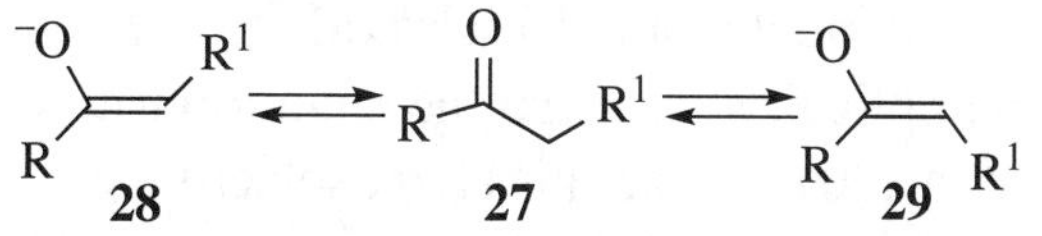

16. (a) Martin, G.J.; Méchin, B.; Martin, M.L. *C.R. Acad. Sci., Ser. C* **1968**, *267*, 986; (b) Martin, G.J.; Martin, M.L. *Bull. Soc. Chim. Fr.* **1966**, 1636.
17. Evans, D.A. in *Asymmetric Syntheses, Vol. 3*, Morrison, J.D. (Ed.), Academic Press, New York, **1984**, p16.

(LTMP), gave only slightly greater amounts of the (*E*)-enolate (86:14 *E/Z*), but addition of HMPA) to this latter reaction medium led to a reversal of selectivity, favoring the (*Z*)-enolate (8:92). Upon subsequent reaction with a substrate such as another aldehyde or ketone, this (*E/Z*)-mixture of enolate anions will give a similar mixture of diastereomers in the products. Some isomerization can occur in the deprotonation or condensation steps, and the product may isomerize under the reaction conditions,[18] as we will see in chapter 9.

6.2.D. syn-anti Selectivity

There are many reactions that generate a mixture of syn and anti diastereomers, and if the reaction is diastereoselective one predominates. Dihydroxylation of alkenes to give a syn-diol is one example from chapter 3, and the addition of bromine to an alkene to give an anti-dibromide is another. Reaction of alkenes with aqueous permanganate or osmium tetroxide (secs. 3.5.A-C) gives a cis diol via the syn hydroxylation mechanism. The reaction is generally diastereoselective for the *cis*-diol, but it is not enantioselective. In synthetic studies aimed at the tricyclic skeleton of natural *Celastraceae* sesquiterpenoids, Barrett and co-workers reacted **30** with a catalytic amount of osmium tetroxide and *N*-methylmorpholine *N*-oxide, giving a 72%% yield of **31a** and **31b** in a 1:1 ratio.[19] Although the reaction was diastereoselective for the *cis*-diol, the lack of enantioselectivity in the presence of the other stereogenic centers led to the mixture.

An alternative method for stereoselectivity generating diols begins with an epoxide (**32**), formed by oxidation of an alkene (sec. 3.4), which is opened to the diol with hydroxide (or another oxygen nucleophile). In this example, syn diol **33** was formed via backside approach of the nucleophilic hydroxide to the less hindered carbon of **32**. Epoxide **32** is assumed to be optically active so the enantiopurity of the diol product can be determined and this is used as a measure of the stereoselectivity of the ring opening. Different mechanistic pathways form diols **31** and **33**. Changing to a different mechanistic process can modify the selectivity. Treatment of this epoxide with a catalytic amount of perchloric acid or sulfuric acid (in 90% aqueous DMSO), for example, gave a mixture of cis and trans diols in roughly equal amounts, although there was a slight excess of the trans.[20] In this latter case, acid catalyzed ring opening proceeds via a carbocation that reacted with water to give a mixture of diastereomers. When ring opening was initiated by KOH in the nonaqueous environment, an S_N2 like mechanism

18. Heathcock, C.H.; Buse, C.T.; Kleschick, W.A.; Pirrung, M.C.; Sohn, J.E.; Lampe, J. *J. Org. Chem.* **1980**, *45*, 1066.
19. Siwicka, A.; Cuperly, D.; Tedeschi, L.; Le Vézouët, R.; White, A.J.P.; Barrett, A.G.M. *Tetrahedron Lett.* **2007**, *48*, 2279.
20. Berti, G.; Macchia, B.; Macchia, F. *Tetrahedron Lett.* **1965**, 3421.

led to formation of a single diastereomer via backside attack. In 85% aqueous DMSO, for example, 1-phenyl cyclohexene oxide reacted with KOH to give the trans diol.

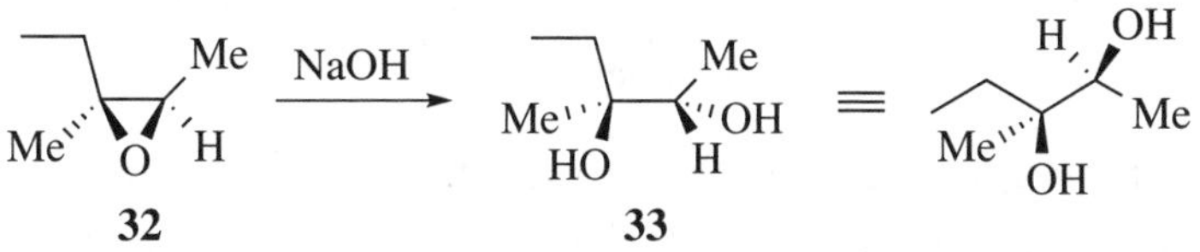

6.2.E. Heteroatom Chelation

One major way to control diastereoselectivity is to take advantage of the chelating effect of neighboring heteroatom groups (neighboring group effects) with certain reagents, which can be illustrated by reaction of chiral allylic alcohol **34** with a peroxyacid. Coordination with the oxygen (see **35**) and delivery of the electrophilic oxygen from that side[21] gave epoxy-alcohol **36** as the major diastereomer, which means that coordination directs the stereochemical course of the reaction. In Chapter 3 and especially in chapter 4, the neighboring group effects of heteroatoms were important for the observed diastereoselectivity in epoxidation, hydride reduction or hydroboration, as well as haptophilic effects in catalytic hydrogenation (sec. 4.8.B).

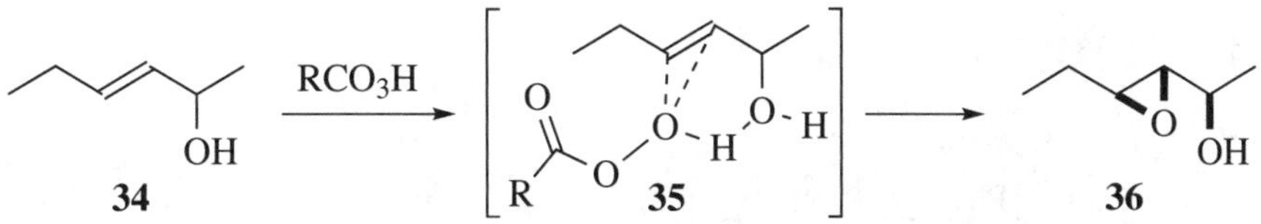

The **Sharpless asymmetric epoxidation** (sec. 3.4.D.i) exploits a chelation effect, because its selectivity arises from coordination of the allylic alcohol to a titanium complex in the presence of a chiral agent. The most effective additive was a tartaric acid ester (tartrate), and its presence led to high enantioselectivity in the epoxidation.[22] An example is the conversion of allylic alcohol **37** to epoxy-alcohol **38** (95% yield, 97% ee), in a synthesis of (–)-apicularen A by Panek and Su.[23] In this reaction, the tartrate, the alkenyl alcohol and the peroxide bind to titanium and provide facial selectivity for the transfer of oxygen from the peroxide to the alkene. Binding of the allylic alcohol to the metal is important for delivery of the electrophilic oxygen and also for controlling orientation of the alkene relative to the epoxidation agent. In this example, coordination with the chiral tartrate led to preferential delivery from the opposite (*si*) face. The reaction is enantioselective and diastereoselective, and syn-anti-selectivity is controlled by the coordination of substrate to the (+) or (–) tartrate.

---

21. Henbest, H.B. *Proc. Chem. Soc.* **1963,** 159.
22. (a) Katsuki, T.; Sharpless, K.B. *J. Am. Chem. Soc.* **1980,** *102,* 5974; (b) Sharpless, K.B.; Verhoeven, T.R. *Aldrichimica Acta* **1979,** *12,* 63.
23. Su, Q.; Panek, J.S. *J. Am. Chem. Soc.* **2004,** *126,* 2425.

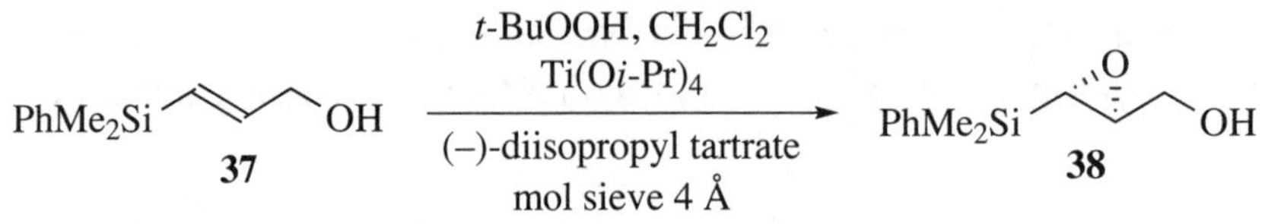

The **Cram chelation model** (sec. 4.7.B) exploits the chelation effects of a heteroatom to influence the rotamer population and, thereby, the selectivity of the reduction. Zinc borohydride [Zn(BH$_4$)$_2$] effectively chelates the carbonyl and alcohol oxygen atoms, in the reduction of **39**, and leads to intermediate **40**. Transfer of hydride to the carbonyl gave primarily the anti diastereomer, **42** (4:96, **41/42**). Lithium aluminum hydride shows less selectivity, due in part to poorer coordination with the heteroatom and reduction of **39** gave a 27:73 mixture of **41** and **42**, respectively.[24] When the chelating hydroxyl group was blocked as a *tert*-butyldiphenylsilyl ether (in **43**, sec. 7.3.A.i), reduction with Red-Al (sec. 4.3) led to a reversal in selectivity (96:4, **44/45**).[24] The chelation ability of a heteroatom varies with the reagent used.

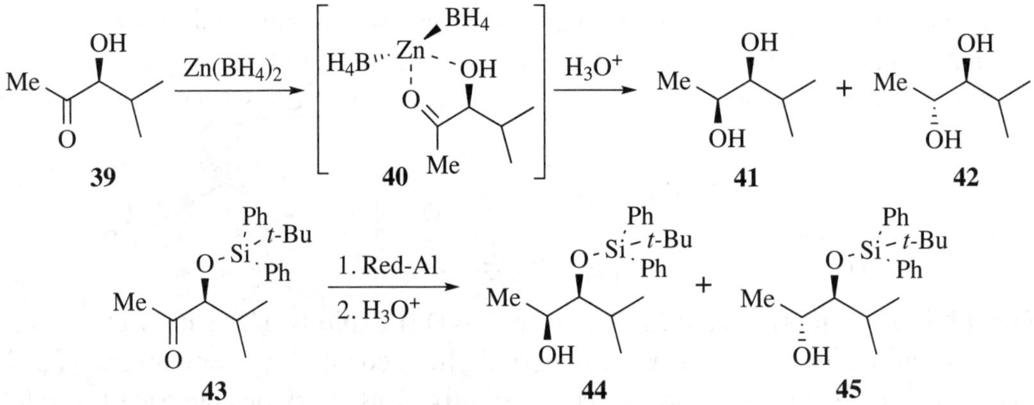

Other heteroatoms can coordinate with reducing agents to afford high selectivity. Reduction of **46** with zinc borohydride, for example, gave a 76:24 mixture of **47** and **48** (61% yield) that was eventually converted by Davis and co-workers to (–)-SS20846A[25a] [(4S)-hydroxy-(2S)-(1-pentadienyl)piperidine], a proposed intermediate in the biosynthesis of the potent antimicrobial agent streptazolin.[25b,c]

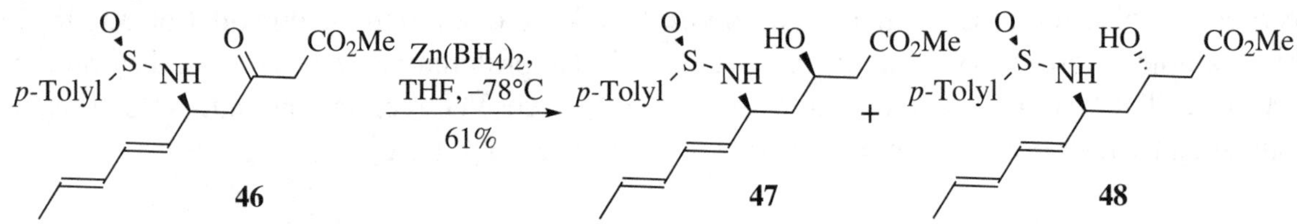

---

24. Nakatu, T.; Yanaka, T.; Oishi, T. *Tetrahedron Lett.* **1983**, *26*, 2653.
25. (a) Davis, F.A.; Chao, B.; Fang, T.; Szewczyk, J.M. *Org. Lett.* **2000**, *2*, 1041; (b) Mayer, M.; Thiericke, R. *J. Org. Chem.* **1993**, *58*, 3486; (c) Grabley, S.; Hammann, P.; Kluge, H.; Wink, J.; Kricke, P.; Zeeck, A. *J. Antibiot.* **1991**, *44*, 797.

The examples in Sections 6.2.D and 6.2.E show modifying the nature of the heteroatom (alcohol versus ether; amine versus amide versus sulfonamide versus azide) can control chelation effects, and the choice of reagent (changing the metal in metal hydrides for example). Such chelating effects have also been used extensively in cyclic systems, and those applications will be discussed in Sections 6.4 and 6.5.

6.3. STEREOCONTROL IN CYCLIC SYSTEMS

Many problems encountered in acyclic systems also arise in cyclic systems. The methods for controlling Markovnikov and anti-Markovnikov regiochemistry, as well as retention or inversion of the configuration are essentially the same as in acyclic systems. It is difficult to separate regiochemical (mode of addition) effects and retention versus inversion effects from cis-trans isomerism in cyclic molecules. Addition to a substituted π bond leads to geometrical isomers, and subsequent reactions give cis-trans isomers and/or diastereomers. The terms syn and anti have no meaning in cyclic systems, and only three discussions fundamental to cyclic molecules will be presented: control of diastereoselectivity, inversion of configuration of a stereogenic center, and the chelating effects of heteroatoms.

An important and differentiating feature of cyclic systems is the relatively rigid conformations assumed by these molecules (sec. 1.5.B). In both Chapters 3 and 4, conformational bias had a profound effect on reaction pathways, and also on the trajectory of attack. These effects provide the tools to control selectivity.

6.3.A. Regioselectivity

The problems associated with regioselective addition to cyclic molecules are essentially the same as those noted in acyclic molecules. Reaction of HBr with methylcyclopentene is highly regioselective for the Markovnikov product, tertiary bromide 1-bromo-1-methylcyclopentane. Methylcyclopentene, however, gave an 86:14 mixture of *trans*-2-methyl-2-cyclopentanol:1-methyl-1-cyclopentanol after hydroboration.[26] As with acyclic systems, regioselectivity improved as the steric bulk of the hydroboration reagent increased. Reaction with 9-BBN followed by oxidation gave an almost quantitative yield of *trans*-2-methyl-2-cyclopentanol.[27] As with acyclic molecules, understanding the mechanism associated with a given transformation is essential for controlling regioselectivity.

For E2 (anti) elimination (sec. 2.9.A) and syn elimination (sec. 2.9.C) reactions were discussed, regiocontrol in the elimination was achieved by allowing the base to attack the molecule in a bimolecular reaction (an external base for anti or E2 elimination) or by binding the base to the molecule (an internal base for syn elimination). Treatment of a halide such as **49** with potassium

26. Zweifel, G.; Brown, H.C. *Org. React.* **1964**, *13*, 1.
27. Brown, H.C. *Organic Synthesis Via Boranes*, Wiley, New York, **1973**, p55.

tert-butoxide in *tert*-butanol led to anti elimination and formation of the more substituted alkene (**50**), which is due to the late transition state of the reaction (the Hammond postulate). Electronic requirements position the leaving group anti to the hydrogen being removed, and effectively lock the conformation of that transition state. The same effect is observed in cyclic halides such as *cis*-2-bromo-1-methylcyclopentane (**51**), which gives methylcyclopentene (**52**) upon treatment with base. The major difference, of course, is the inability of the ring to rotate about the carbon-carbon bonds. Nonetheless, the more highly substituted (more stable) alkene is also produced. For both syn and anti elimination, the β-hydrogen and the leaving group are fixed by the regiochemistry of the halide. *syn*-Elimination involves intramolecular attack of base, dictating removal of the β-hydrogen that is syn to the leaving group. The steric demands of this syn transition state make approach and removal of the less substituted β-hydrogen energetically more favorable. Conversion of **53** to sulfoxide **54** gives the less substituted alkene (3-methyl-1-cyclopentene, **55**) upon thermolysis.

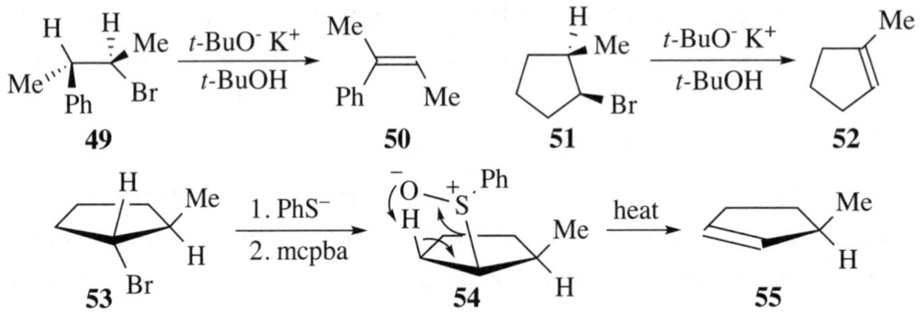

Alkene positional isomerism is occasionally a problem, and it is conceptually related to regiochemistry. Careful choice of synthetic route and reagents is usually required to solve this problem, as illustrated by the Burk and Soffer synthesis of ε-cadinene (**57**) and γ₂-cadinene (**58**).[28] Conversion to the ketone by hydrolysis of the vinyl ether unit in **56** was followed by conversion to the tertiary alcohol by reaction with methyllithium (sec. 8.5.C). When the alcohol was converted to the tertiary chloride, elimination with thionyl chloride/pyridine gave **57**. Conversely, hydrolysis of the vinyl ether unit in **56** to a ketone followed by Wittig olefination (sec. 8.8.A) of the ketone gave the exo methylene derivative, **58**. It is important to note that formation of the C2-C3 C=C unit over the C3-C4 unit in **57**, by elimination of the tertiary chloride precursor is favored for trans decalin derivatives such as this.

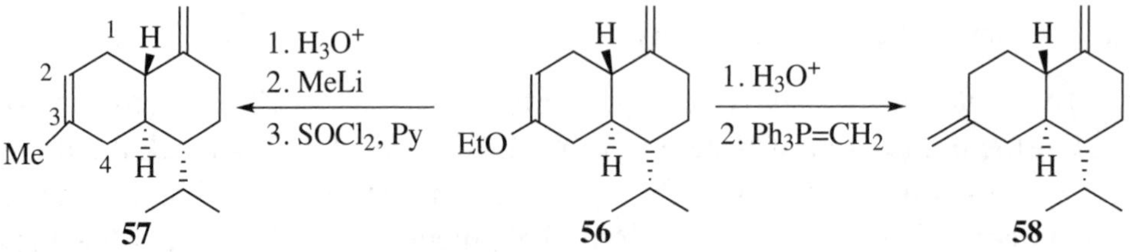

28. Burk, L.A.; Soffer, M.D. *Tetrahedron* **1976**, *32*, 2083.

6.3.B. Bredt's Rule

As we have just seen in the formation of **57** or **58**, regioselectivity is important for the formation of double bonds in elimination reactions. A unique regiochemical problem arises in small bicyclic compounds. Elimination of the bromine and H in 2-bromonorbornane (**59**) gave exclusively the Δ^2-norbornene (**60**) and none of the Δ^1-norbornene (**61**). Formation of an alkene such as **61** requires that bridgehead carbon atoms be compressed toward planarity, but the π overlap required for a planar C=C unit would be diminished due to structure-imposed twisting. The transition state energy required for elimination to a bridgehead carbon in small systems is typically very high, and small bicyclic systems *cannot* accommodate this great increase in strain. The observation that carbon-carbon double bonds cannot be formed to bridgehead atoms in small bicyclic rings is formally called **Bredt's rule**.[29] As originally stated by Bredt, "in systems of the camphane (**62**) and pinane (**63**) series and related compounds, the bridgehead carbons (*a* and *b*) cannot be involved in a carbon double bond."[29a,b] As a corollary, reactions that should lead to such compounds will be hindered or will take a different course.[29] It has been noted that small bridgehead systems that tolerate a trans double bond will violate Bredt's rule.[30,29]

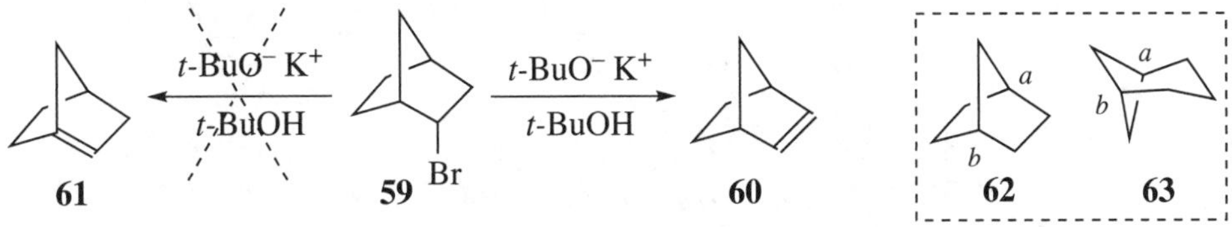

How large must ring systems be before Bredt's rule does not apply? For [2.2.1], [2.1.1], and [1.1.1] bicyclic systems, Bredt's rule is applied. Prelog and co-workers established that cyclization of **64** gave **65** when $n > 6$.[31,29] When $n = 5$, both **65** and **66** were formed, but **65** was not formed when $n < 5$.

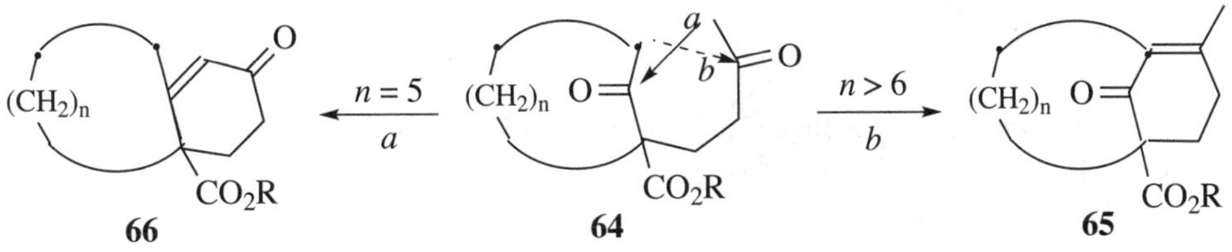

29. (a) Bredt, J.; Houben, J.; Levy, P. *Berichte* **1902**, *35*, 1286; (b) Bredt, J. *Annalen* **1924**, *437*, 1; (c) Fawcett, F.S. *Chem. Rev.* **1950**, *47*, 219; (d) Köbrich, G. *Angew. Chem. Int. Ed.* **1973**, *12*, 464.
30. (a) Weiseman, J.R.; Chong, J.A. *Ibid.* **1969**, *91*, 7775; (b) Wiseman, J.R.; Pletcher, W.A. *Ibid.* **1970**, *92*, 956; (c) Quinn, C.B.; Wiseman, J.R. *Ibid.* **1973**, *95*, 1342, 6120; (d) Quinn, C.B.; Wiseman, J.R.; Calabrese, J.C. *Ibid.* **1973**, *95*, 6121; (e) Krabbenhoft, H.O.; Wiseman, J.R.; Quinn, C.B. *Ibid.* **1974**, *96*, 258; (f) Marshall, J.A.; Fauble, H. *Ibid.* **1970**, *92*, 948.
31. (a) Prelog, V.; Ruzicka, L.; Barman, P.; Frenkiel, L. *Helv. Chim. Acta* **1948**, *31*, 92; (b) Prelog, V.; Barman, P.; Zimmermann, M. *Ibid.* **1949**, *32*, 1284; (c) Prelog, V. *J. Chem. Soc.* **1950**, 420.

Fawcett attempted to correlate the lower limit of Bredt's rule by calculating the sum of the numbers of the bridged atoms [m + n + o = S] in systems such as **63** and **63** with three bridges.[29c] For **62**, $S = 2 + 2 + 1 = 5$, and for **63**, $S = 3 + 1 + 1 = 5$. In Prelog's example (**64**), this [6.3.1] system has $S = 9$ when $n = 5$, and Bredt's rule does not apply. When the values of n are < 5, however, Bredt's rule applies for **64**. Relatively small bridgehead alkenes with a [3.3.1] system such as **67** ($S = 7$) can be prepared by pyrolysis of β-lactone **68** or ammonium salt **69**.[32] Figure 6.1[29d] shows several small bicyclic ring systems for $S = 2$ to $S = 7$, and analysis of these alkenes suggests that the constraints upon Bredt alkenes are important when S is 7 or less.

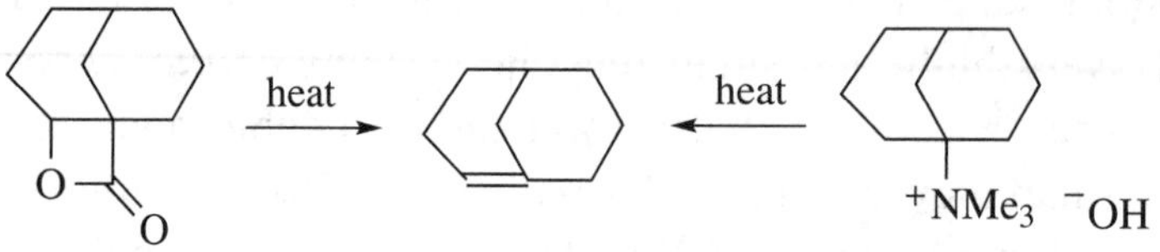

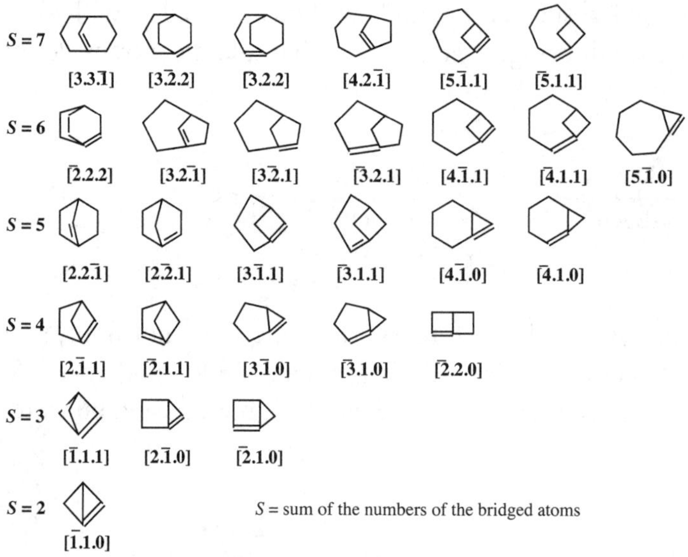

Figure 6.1. Typical Bredt alkenes. [Reproduced with permission from Köbrich, G. *Angew. Chem. Int. Ed. Engl.* **1973**, *12*, 464. Copyright 1973 VCH Weinheim.]

Köbrich established three guidelines[29d] that govern the application of Bredt's rule.

1. For homologs with different S values, the ring strain varies inversely with S.

2. For a given S, the ring strain varies inversely with the size of the larger of the two rings with respect to which bridgehead bond is endocyclic.

---

32. (a) Marshall, J.A.; Faubl, H. *J. Am. Chem. Soc.* **1967**, *89*, 5965; (b) Chong, J.A.; Wiseman, J.R. *Ibid.* **1972**, *94*, 8627 and references cited therein.

3. For a given bicyclic ring skeleton, the ring strain varies inversely with the size of the bridge containing the bridgehead double bond.

For example, these rules predict that **70** is more stable than **71**.[33] These three rules do not allow direct comparison in every case, but are generally applicable. Köbrich presented a list of carbocyclic Bredt's compounds[29d] and stated that many heteroatom and polycyclic analogs can be included in the Figure 6.1 list.[29d]

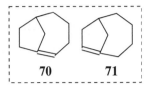

70   **71**

6.3.C. DIASTEREOSELECTIVITY

As mentioned previously, reactions of cyclic molecules that involve formation of stereogenic centers are similar to those of acyclic systems in that they can proceed with clean retention or inversion of configuration, or a mixture of the two. Reaction of sodium azide (NaN_3) and *cis*-4-*tert*-butyl-1-bromocyclohexane is expected to give *trans*-4-*tert*-butylazidocyclohexane with complete inversion of configuration via an S_N2 pathway. In Section 1.5.B, we saw that many cyclic systems exist primarily in one or two low energy conformations, which can be used to control stereochemistry if other groups are present that sterically block approach of a reagent. A typical example is the reduction of a ketone moiety in a cyclic or polycyclic substrate, where the diastereoselectivity of the reduction is controlled by the conformational constraints of the system. In a synthesis of malayamycin A[34] by Hanessian and co-worker's, the ketone unit in **72** was reduced to the alcohol unit in **73** with high diastereoselectivity. The conformation of the ring allowed the diastereoselective reduction to proceed with sodium borohydride, rather than requiring one of the more expensive hydride reducing agents.

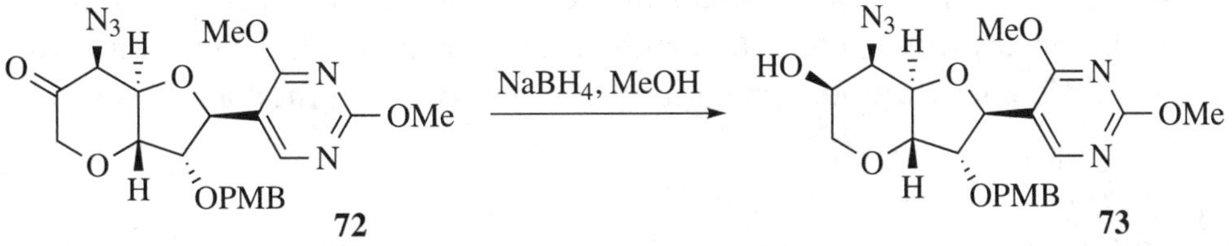

72 → (NaBH$_4$, MeOH) → **73**

In many cases, reaction conditions and strategy combine to give a product with the incorrect stereochemistry at a key center, often an alcohol. In **74**, an intermediate in Philips and Chamberlin's synthesis of dysiherbaine,[35] the stereochemistry of the secondary alcohol is incorrect for the synthetic target. In order to invert the stereochemistry, **74** was oxidized with TPAP and NMO (see Section 3.2.F.i.) to give a 95% yield of ketone **75**. Subsequent reduction with NaBH$_4$ gave a quantitative yield of **76** with the correct stereochemistry. The observed inversion is possible because approach of the hydride reagent to the carbonyl unit in **76** is

33. Wiseman, J.R.; Chan, H.-F.; Ahola, C.J. *J. Am. Chem. Soc.* ***1969***, *91*, 2812.
34. Hanessian, S.; Marcotte, S.; Machaalani, R.; Huang, G. *Org. Lett.* ***2003***, *5*, 4277.
35. Phillips, D.; Chamberlin, A.R. *J. Org. Chem.* ***2002***, *67*, 3194.

restricted by the presence of the tricyclic system and other substituents, as well as the requisite Bürgi-Dunitz trajectory common to hydride reduction of ketone units (see Sections 4.7.C and 6.6.A). Oxidation followed by reduction of a carbonyl is a simple, yet powerful tool for modifying stereochemistry.

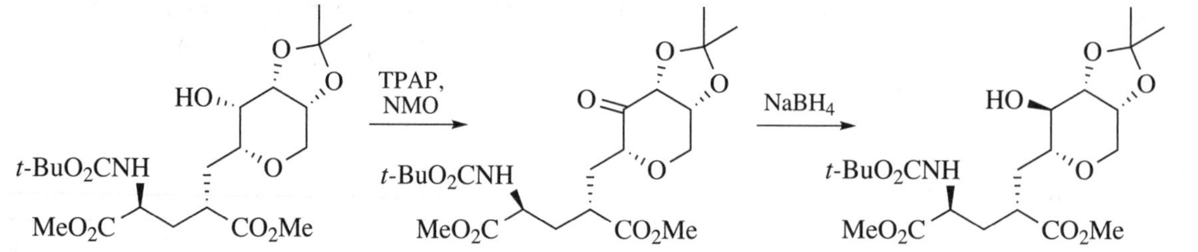

Sometimes, the natural selectivity of a reaction leads to a product with the incorrect stereochemistry required for a given target, but it may be possible to change it. There are at least two options: (*1*) change the stereocenter in the final product., as in **74** → **76**, or (*2*) change the synthetic pathway to generate a compatible intermediate that can produce the desired stereochemistry.

One type of stereocenter that can be corrected is a CHX moiety, where X is an electron-withdrawing group and the H can be removed by treatment with base to form a planar enolate anion (secs. 9.1, 9.2). In a synthesis of quinine and quinidine by Jacobsen and co-workers,[36] lactam **77** (as a 1:1.7 mixture of *cis:trans* isomers) was converted to the enolate anion **78A** with lithium diisopropyl amide (sec. 9.2). Inspection of a model of the enolate anion (**78B**) suggests that protonation will occur from the less hindered top face (A), rather than from face B. The model also suggests that there is only a small difference in steric hindrance of one face relative to the other, and this observation is consistent with the isolated product, a 3:1 mixture favoring **79**. Exploiting such differences in product stability allows one to change the diastereomer population.

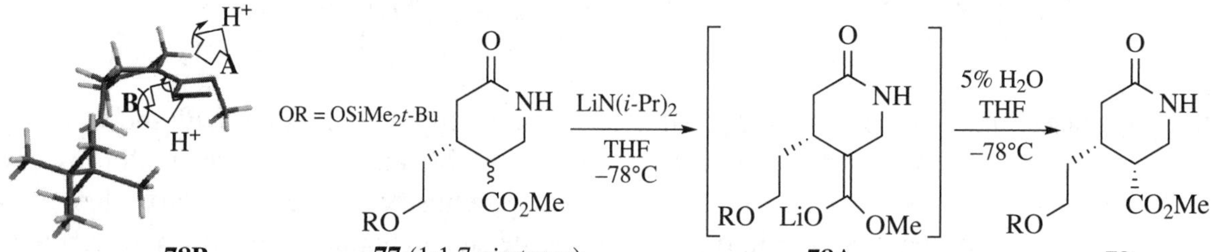

The second approach noted above for correcting stereochemistry is illustrated by formation of epoxides **82** and **84** from conjugated ketone **80**, by Grieco and co-workers.[37] Epoxidation of **80**

36. Raheem, I.T.; Goodman, S.N.; Jacobsen, E.N. *J. Am. Chem. Soc.* **2004**, *126*, 706.
37. (a) Grieco, P.A.; Ohfune, Y.; Majetich, G. *J. Org. Chem.* **1979**, *44*, 3092; (b) Ohfune, Y.; Grieco, P.A.; Wang, C.-L.J.; Majetich, G. *J. Am. Chem. Soc.* **1978**, *100*, 5946.

from the less hindered α face gave **81**, and reduction from that same face gave **82**. The relative stereochemistry was different, however, when initial reduction of **80** (hydride was delivered from the β-face) gave alcohol **83**. Subsequent epoxidation proceeded via coordination of the OH unit in **83** with the epoxidation agent (MCPBA, sec. 3.4.C), and the final product was **84** with the epoxide and hydroxyl unit cis (syn) to each other.[37] Simply changing the order of reactions allowed Grieco to target one diastereomer or the other.

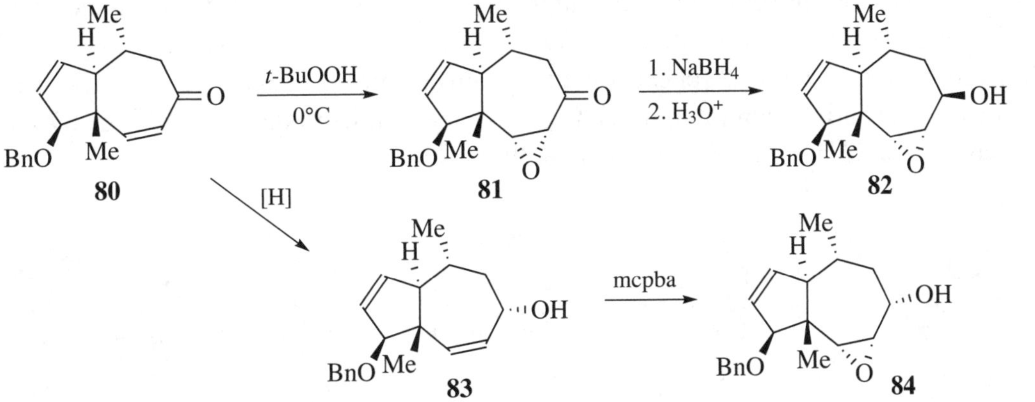

One major difference between cyclic and acyclic systems is the inability of cyclic systems to undergo rotation about carbon-carbon bonds, leading to relatively well-defined conformations. If *trans*-4-*tert*-butyl-1-bromo-cyclohexane is used as an example, there is the usual equilibrating population of the two chair conformations (sec. 1.5.D). The diequatorial conformer is significantly more stable and accounts for the major portion of the equilibrium (sec. 1.5.B). According to Corey's calculations[38] from Section 1.5.D, the equatorial form predominates (>99.99:<1, diequatorial/diaxial). Molecular modeling indicates that the diequatorial conformation is nearly 32 kJ mol^{-1} lower in energy than the diaxial conformation, with a Boltzmann distribution favoring more than 99.99% of the diequatorial conformation. To predict the major product, nucleophilic backside attack on both conformations must be considered, and the less sterically hindered pathway is usually preferred. In this case, the diequatorial conformation is more reactive.

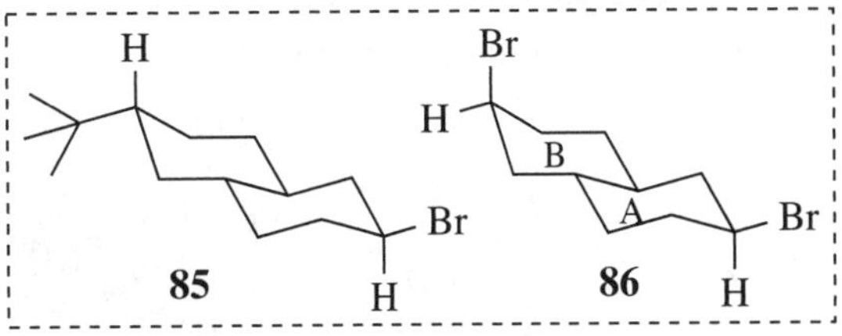

In a conformationally anchored system such as **85**, S$_N$2 displacement of the equatorial bromine may be extremely slow. If the reaction is too slow, the temperature can be raised and a more polar solvent (DMF, DMSO, etc.) can be used (secs. 2.7.A, 2.6.Ai). In some cases, the more vigorous reaction

conditions will lead to an alternative reaction (such as elimination or thermal decomposition) rather than the desired S$_N$2 displacement. Similarly, dibromide **86** is expected to undergo nucleophilic substitution in ring B since displacement of the bromide in ring A is very slow.

38. Corey, E.J.; Feiner, N.F. *J. Org. Chem.* **1980**, *45*, 765.

The relative stereochemistry of the leaving group is very important for intramolecular reactions, as seen in Corey and Watt's synthesis of α-copaene and α-ylangene.[39] Displacement of the tosyl group in **87** by the nucleophilic carbon of the enolate anion (**88A**, formed with dimsyl sodium, sec. 9.2.A and 9.3.A) gave the bridged product (**89**) characteristic of the ylangene natural products. Examination of the conformational drawing for **87** (see **88B**) reveals that the enolate carbon is properly oriented to attack the tosyl carbon from the rear, so displacement of the OTs group is possible only when it is in the equatorial position and is syn to the bridgehead methyl group. Note the letters marking the key atoms. The isomeric axial tosylate (**90**; the tosyl group is anti to the bridgehead methyl group) cannot be displaced via backside attack, since the enolate carbon cannot approach the tosyl carbon from the rear. Ketone **87** will react as shown to give **89**, but the isomeric ketone **90** will not.

Marshall et al. showed that alkenyl aldehydes can be cyclized on acidic silica gel, and the stereochemistry of the bicyclic alcohol product is controlled by the cis or trans geometry of the monocyclic precursor.[40] This reaction is an **ene reaction** (sec. 11.13), which is particularly effective when catalyzed by Lewis acids.[40c,d] The cis precursor **91** was cyclized to **92**, whereas the trans aldehyde **93** was cyclized to **94**. Formation of a seven membered ring in **92** requires a seven-membered chair-like ring for the ene reaction to occur (see **95**). If the oxygen is in the axial position (see **96**) rather than the equatorial position (see **95**), then the alkenyl group is further removed from the acylium cationic center, and transfer of the hydrogen and attack by the π bond for an ene reaction is more difficult. Cyclization of **93** requires transition state **98**. This transition state brings the alkenyl moiety close to the acylium cation, and the oxygen is in the axial position for efficient hydrogen transfer and attack by the alkene, leading to **94**. The opposite conformation with the equatorial oxygen (see **97**) has a greater transannular steric interaction, and is also in a higher energy boat-like transition state that disfavors the ene reaction in this particular system.

39. Corey, E.J.; Watt, D.S. *J. Am. Chem. Soc.* **1973**, *95*, 2303.
40. (a) Marshall, J.A.; Andersen, N.H.; Johnson, P.C. *J. Org. Chem.* **1970**, *35*, 186; (b) Marshall, J.A.; Andersen, N.H.; Schlicher, J.W. *Ibid.* **1970**, *35*, 858; (c) Snider, B.B. *Acc. Chem. Res.* **1980**, *13*, 426.

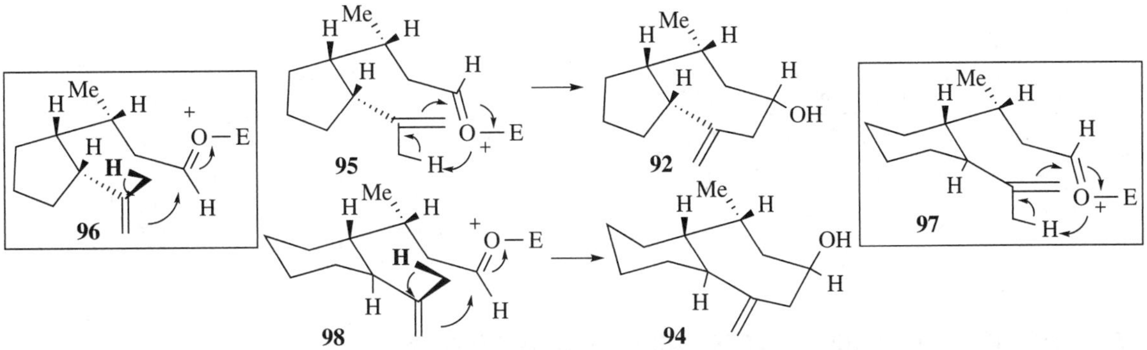

When an alkene unit is the product of a reaction, there is the potential for both (*E*) and (*Z*) isomers. The examples previously cited for E2 and syn elimination are illustrative. In both cases, the stereochemistry (*E*) or (*Z*) of the double bond is important. Stereochemistry may be controlled in acyclic systems by using resolved chiral precursors because the E2 elimination is stereospecific. Controlling cis-trans isomers for alkene derivatives is also possible in medium and large ring molecules, and other reactions are known that introduce the C=C unit. Corey and co-worker's synthesis of caryophyllene (**102**) and isocaryophyllene (**106**) illustrates this type of control.[41] A **Grob fragmentation**[42] (sec. 2.9.E) in **99** led to the (*Z*)-double bond in **100**, required for **102**. The geometry of the double bond was controlled by changing the stereochemistry of the carbon bearing the leaving group in the cyclic precursor (see a discussion of this technique in sec. 6.5). Once the (*Z*)-double bond had been incorporated, the α-hydrogen was equilibrated (via enolization with potassium *tert*-butoxide) from the initially formed cis ring juncture (in **100**) to the more stable trans-ring juncture in **101**. To obtain the other geometric isomer (in **106**) the relative stereochemistry of the OTs group had to be changed to that shown in **103**. A Grob fragmentation of **103** gave the (*E*)-double bond in **104** that is characteristic of **106**.[42] As before, the cis-ring juncture in **104** was equilibrated to the more stable trans ring juncture in **105** by enolization prior to conversion to **106**. These examples show that the (*E/Z*)-stereochemistry of the alkene products was established by the relative stereochemistry of the tosylate precursors, once the reaction to generate the double bond in those alkenes was chosen.

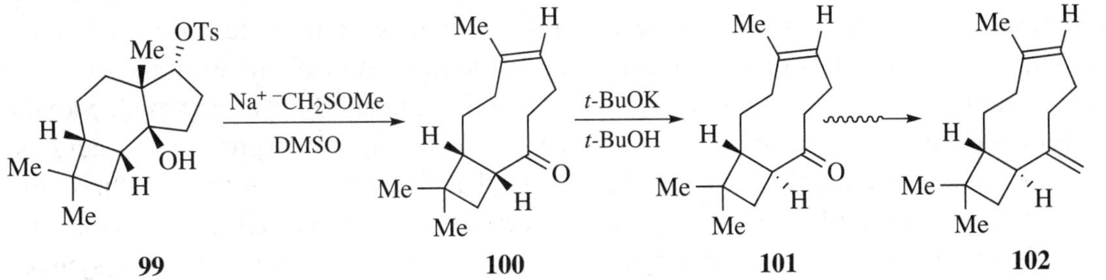

41. (a) Corey, E.J.; Mitra, R.B.; Uda, H. *J. Am. Chem. Soc.* **1963**, *85*, 362; (b) *Idem, Ibid.* **1964**, *86*, 485.
42. Grob, C.A. *Angew. Chemie, Int. Ed.* **1969**, *8*, 535.

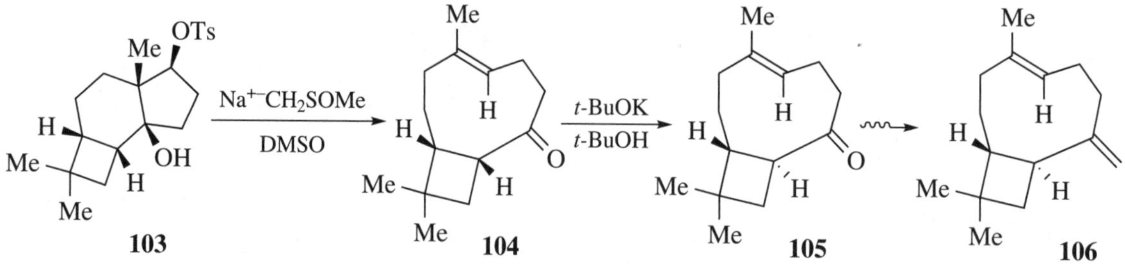

6.4. NEIGHBORING GROUP EFFECTS AND CHELATION EFFECTS

The influence of heteroatom substituents to direct reactivity to one face or another has been noted in this as well as several preceding chapters. In Chapter 3, a neighboring group effect influenced delivery of oxygen to the double bond of an allylic alcohol, described by Henbest and by Sharpless (secs. 3.4.B, 3.4.C). The effect is identical to the effect described for acyclic control in Section 6.2.E (see above). Peroxyacid epoxidation of **107** proceeded via coordination of the peroxyacid to the alcohol (see **108**) to deliver the electrophilic oxygen from that face to give **109**.[22] This result contrasts with epoxidation of allylic acetate **110**, which gave primarily **111**

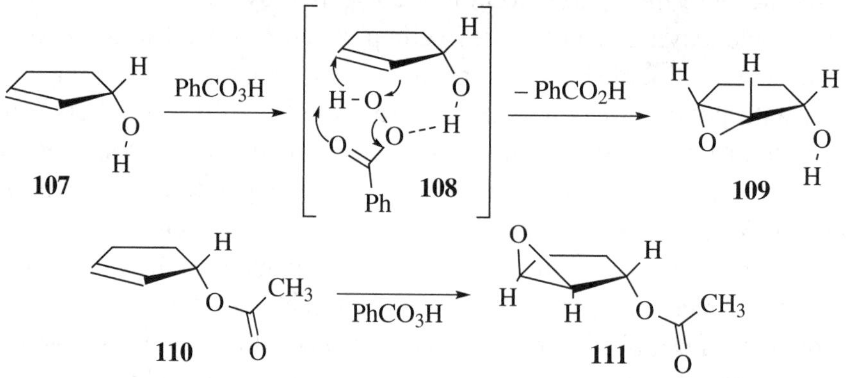

via delivery from the less sterically hindered face. The acetate group inhibits coordination with the peroxyacid, and delivery of the electrophilic oxygen is from the less sterically hindered face, so the epoxide unit is on the opposite side of the ring. A synthetic example of this effect is taken from a synthesis of the A/B ring of ouabain by Jung and Piizzi,[43] in which epoxidation of allylic alcohol **112** occurred from the same face as the hydroxyl group via a neighboring group effect, to give a 96% yield of **113**. Note that this effect was seen in the conjugate epoxidation of ketone **80**, versus the alcohol-directed epoxidation of **83** (see above).[37] In zinc borohydride [Zn(BH$_4$)$_2$] reductions, oxygen atoms can chelate with the zinc.[24] In a synthesis of a C1-C22 fragment of leucascandrolide A, Panek and Dakin treated **114** with zinc borohydride. Coordination of the zinc reagent with the hydroxyl unit led to formation of **115** in 80% yield,

---

43. Jung, M.E.; Piizzi, G. *Org. Lett,* **2003**, *5*, 137.

with a dr of >15:1 favoring the diastereomer shown.[44]

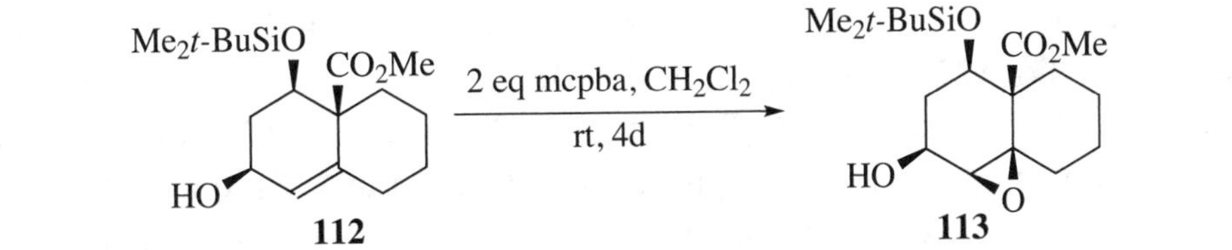

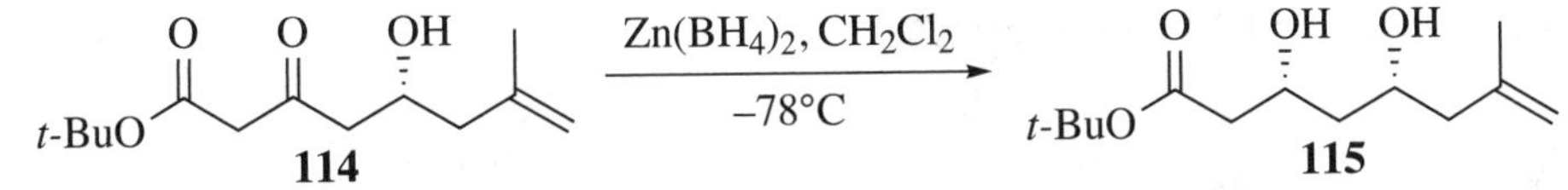

Hydroboration of alkenes to give alcohols typically follows an anti-Markovnikov orientation (sec. 5.2.A). Chelation effects can sometimes control the product distribution, but the mere presence of a heteroatom does not always mean that this will be the case. Hydroboration of **116** (tabersonine) is an example where the hydroboration-oxidation sequence (sec. 5.4.A) gave **117** in

90% yield, as a mixture of diastereomeric secondary alcohols (4:1 β/α mixture favoring **117**). This reaction was part of the Caron-Sigaut et al. synthesis of 14-hydroxyvincadifformine,[45] but they showed that chelation was *not* an important factor contrary to what might

be suspected from a cursory examination of the starting materials. The enamine unit in **116** is part of a bicyclic system where the exo face is more accessible, leading to the stereochemistry in **117** (secs. 1.5.E, sec. 4.7.E). The regiochemistry of the alcohol unit in **117** probably results from steric interactions as borane approaches the C=C unit. Brown showed that electronic and chelation effects can be important in the hydroboration of allylic amines (sec. 5.2.A.iv), but in **116** steric factors control the reaction.

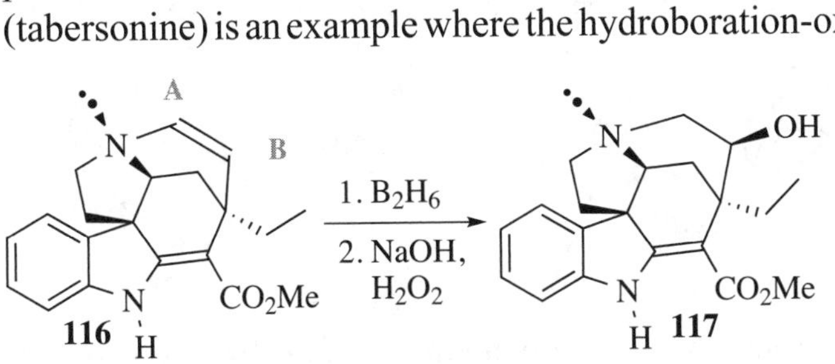

Chelation is important in the dihydroxylation of **118** with osmium tetroxide (sec. 3.5.B), and reduction of the intermediate iminium ion gave **119**, in a synthesis of $\Delta^{20'}$-20'-deoxyvinblastine.[46b] This transformation is compared with the reaction of **118**[46a] with thallium acetate, where reduction of the intermediate iminium ion gave the epimeric alcohol **120**. The nitrogen lone pair

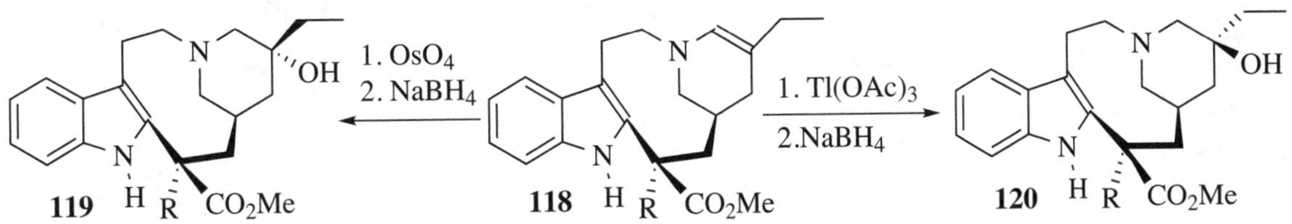

44. Dakin, L.A.; Panek, J.S. *Org. Lett.* **2003**, *5*, 3995.
45. Caron-Sigaut, C.; Le Men-Olivier, L.; Hugel, G.; Lévy, J.; Le Men, J. *Tetrahedron* **1979**, *35*, 957.
46. (a) Mangeney, P.; Andriamialisoa, R.Z.; Langlois, N.; Langlois, Y.; Potier, P. *J. Am. Chem. Soc.* **1979**, *101*, 2243; (b) Langlois, N.; Potier, P. *Tetrahedron Lett.* **1976**, 1099.

is responsible for the reversal of diastereoselectivity. Osmium tetroxide (OsO_4) does *not* coordinate well to the nitrogen and the osmate ester is formed on the less hindered α face of **121**, which leads to diol **122**. Elimination of the α-hydroxylamine gave an iminium salt (**123**) that was subsequently reduced with $NaBH_4$ to give **124** (product **119** above). Thallium (III), however, coordinates quite well to the nitrogen lone pair (as in **125**) to give **126**. Transfer of an oxygen from one acetate unit to the enamine moiety (secs. 9.6.A,B), from the β face, gives **127**. Reduction leads to **128** (product **120** above).

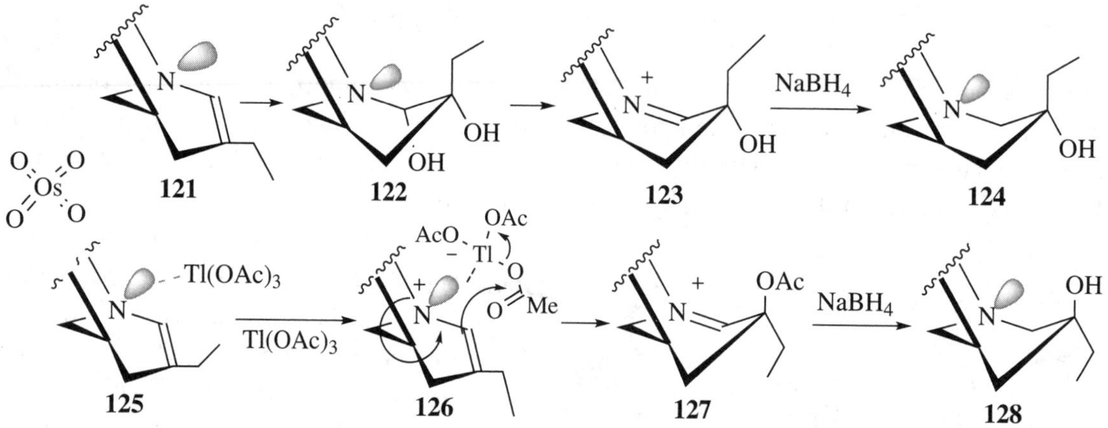

Neighboring group effects usually influence stereo- and regiochemistry of the functional group close to the chelating atom or group. Such effects that can influence a reaction at sites distant from the coordination point.[47] An example of a neighboring group effect with consequences at a remote site is the photochemical isomerization of **129** ($n = 1$) to **130**, in 55% yield (sec. 13.5.E for a brief discussion of this concept).[48] The β-C3 hydroxyl group was converted to an ester, and the benzophenone moiety was positioned on the β face of the molecule by a chain of methylene units. Irradiation generated the benzophenone triplet (see secs. 13.4, 11.10.B for a brief introduction to photochemical techniques), and with an appropriate methylene spacer [$(CH_2)_n$] the hydrogen at C14 was selectively removed to give the $\Delta^{14\text{-}15}$ alkene (**130**, $n = 1$). With this particular spacer, the C8 hydrogen on the α face of the molecule is inaccessible and not removed. Longer or shorter methylene spacers ($n = 0$, $n > 2$) led to a diminished yield of **130**.

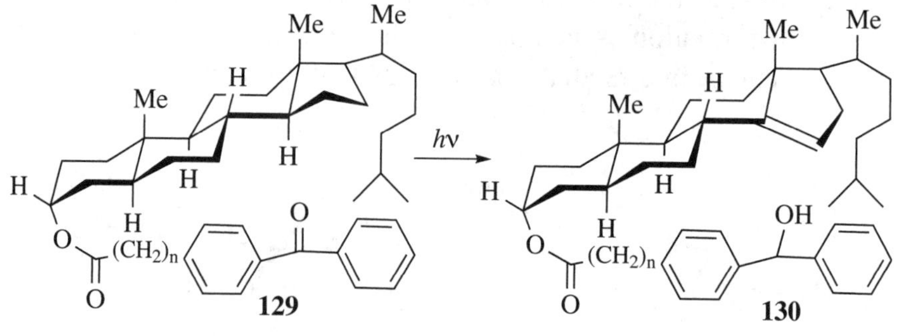

47. For examples, see (a) Sailes, H.; Whiting, A. *J. Chem. Soc. Perkin Trans. 1* **2000**, 1785; (b) Mitchell, H.J.; Nelson, A.; Warren, S. *J. Chem. Soc. Perkin Trans. 1* **1999**, 1899.
48. Breslow, R. *Chem. Soc. Rev.* **1972**, *1*, 553.

Neighboring group effects can compete with steric effects, as illustrated by treatment of oxetane **131** with diethylaluminum-*N*-methylaniline. This elimination reaction gave a 99% yield of (*E*)-alkene **132** in a synthesis of humulene.[49] The aluminum coordinated to the oxetane oxygen on the less hindered β face in **133** and **134**, the two key rotamers. Rotamer **133** is lower in energy due to the severe non-bonded interaction of the propyl group, and the oxetane methyl in **134**. Removal of the indicated hydrogen by the basic nitrogen, in a six-center transition state as seen in **133**, led to the (*E*) isomer **132**. Removal of hydrogen in **134** would give the (*Z*) isomer, but the relatively high energy of that species relative to **133** makes it disfavored.

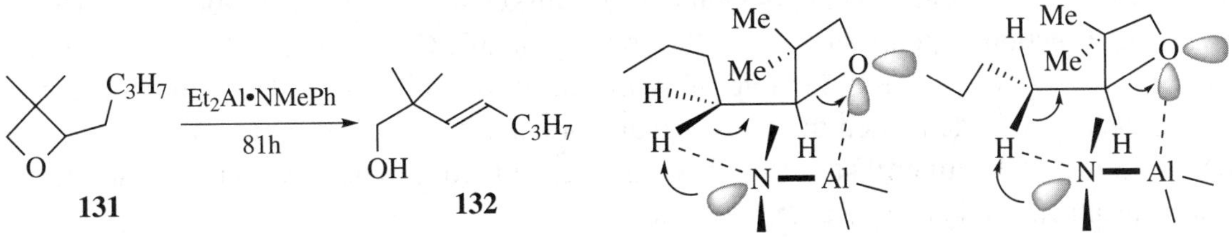

In planning a reaction, not only the obvious effects of α or β hydroxyl, alkoxy or amino groups must be considered, but also the more subtle effects of nearby heteroatom lone pairs. These chelating effects are apparent both in directing stereochemistry or regiochemistry and in slowing or stopping a desired reaction. These effects can be used synthetically to direct attack of a reagent, as shown above. If one wishes to prevent the chelating effect, the offending group must be blocked or removed. Alternatively, the planned synthetic sequence can be accomplished before the chelating atom is present. Careful analysis of all synthetic intermediates is required to use or avoid these effects.

6.5. ACYCLIC STEREOCONTROL VIA CYCLIC PRECURSORS

It is apparent from preceding sections that stereocontrol in cyclic systems is much easier than in acyclic systems, which is due of course, to the conformational bias inherent in cyclic systems. Synthetic chemists have exploited this fact for many years. A cyclic system can be used to position functional groups, often with control of regio- and stereochemistry. The ring is then opened to give an acyclic system. and the regio- and stereochemistry of the substituents has been fixed. There are many examples.

In Chapter 3, ozonolysis (sec. 3.7.B) of cyclic alkenes was shown to generate α,ω-functionalized systems. The utility of the process is demonstrated by the ozonolysis of 1,5-cyclooctadiene to give diol **135** in 85% yield.[50] Conversion of **135** to the racemic sex pheromone of the female

49. Kitagawa, Y.; Itoh, A.; Hashimoto, S.; Yamamoto, H.; Nozaki, H. *J. Am. Chem. Soc.* **1977**, *99*, 3864.
50. (a) Tolstikov, G.A.; Odinokov, V.N.; Galeeva, P.J.; Bekeeva, R.S. *Tetrahedron Lett.* **1978**, 1857; (b) Tolstikov, G.A.; Odinokov, V.N.; Galeeva, R.I.; Bakeeva, R.S.; Rafikov, S.R. *Dokl. Akad. Nauk, SSSR* **1978**, *239*, 1377 (Engl., p. 174).

face fly (**136**) required four steps.[56] Examining the overall sequence shows that the α-and ω-functional groups were incorporated by sequential alkylation reactions at the hydroxyl groups in **135**. Functionalization of each hydroxyl unit and then the alkene unit gave **137**.

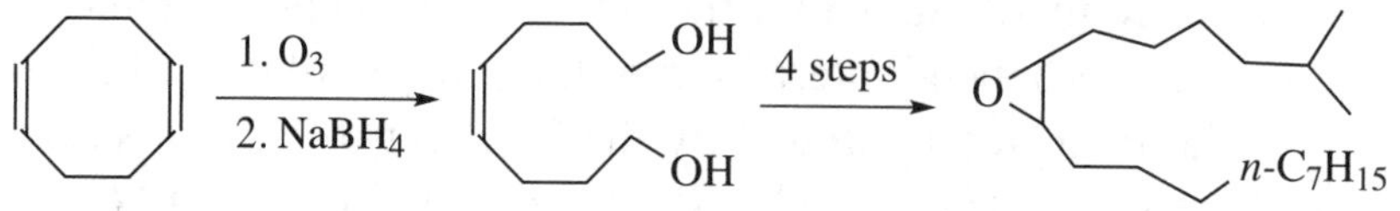

The synthesis of (–)-(R)-muscone by Fehr and co-workers[51] is another example of inserting and positioning functional groups, involving the conversion of **137** to **139**. Treatment of ketone **137** with tosylhydrazine gave an 87% yield of hydrazone **138**. Subsequent reaction with an excess of peroxyacetic acid generated the epoxy-hydrazone *in situ*, and subsequent fragmentation by what is known as **Eschenmoser ring cleavage**[52] gave keto-alkyne **139**. This named reaction transforms an epoxy-ketone to a keto-alkyne.

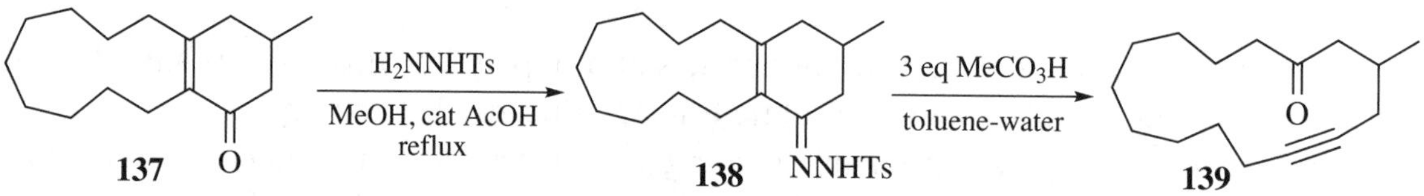

The diastereoselectivity obtained for reactions on a ring makes it possible to use a cyclic molecule to fix a stereocenter in an acyclic target, as seen in the formation of **142** from **140**. In a synthesis of ceramides, Hung and co-workers converted D-Glucosamine **140** to the azide **141** by a four-step sequence, with control of the stereocenters.[53] Further elaboration of the functionality in 7 steps led to ring opening to produce amino-tetraol **142** where the relative stereochemistry of this fragment was fixed by manipulation of the cyclic precursors.

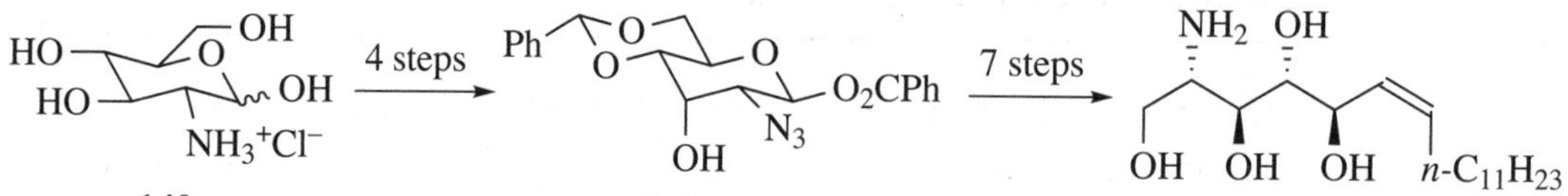

6.6.   RING-FORMING REACTIONS

6.6.A.   Baldwin's Rules

Cyclic compounds play an important role in organic synthesis. The desired compound is not always commercially available, however, and must often be prepared by cyclization reactions

51. Fehr, C.; Galindo, J.; Etter, O. *Eur. J. Org. Chem.* **2004**, 1953.
52. Müller, R.K.; Felix, D.; Schreiber, J.; Eschenmoser, A. *Helv. Chim. Acta* **1970**, *53*, 1479.
53. Luo, S.-Y.; Kulkarni, S.S.; Chou, C.-H.; Liao, W.M.; Hung, S.-C. *J. Org. Chem.* **2006**, *71*, 1226.

from acyclic precursors. This is particularly true for large ring (macrocyclic) compounds and polycyclic molecules. In the latter case, a cyclic molecule acts as a template and the other rings are built onto the template (this is called **annulation**). This section will discuss the salient features of ring-forming reactions commonly encountered in synthesis.

An introduction to cyclization reactions is best begun with a discussion of **Baldwins rules for ring closure**, or simply **Baldwin's rules**. Baldwin studied many nucleophilic, homolytic or cationic ring-closing processes and found a predictable pattern of reactivity. This approach is based on the stereochemical requirements of both reagent and substrate as well as the angles of approach that are allowable when two reactive centers come together.

To form a ring, the two reactive centers are connected by a tether of atoms (usually carbon atoms but not always, and this restriction imposes constraints on the angles from which the reactive centers can approach one another, and on the stereochemistry of the product. If the length and nature of the chain (tether) linking terminal atoms X and Y allows this geometry to be attained, ring formation is possible (**favored**), and we make the predication that the reaction will succeed. If the proper geometry cannot be attained, ring formation is difficult (**disfavored**) and competitive processes often dominate. Baldwin classified ring closing reactions into two categories: **exo** (the electron flow of the reaction is external to the ring being formed [**144** from **143**], and **endo** [the electron flow is within the ring being formed (**146** from **145**)].[54] Baldwin further classified reactions according to the hybridization of the atoms accepting the atom in the ring closing process. If the atom being attacked is sp^3 hybridized, as in **147**, the reaction is termed **tet**, and an exo-tet reaction will generate a ring such as **148**. Attack at an sp^2 atom (**143**) is termed **trig** (forming the ring **144** or **146**). Attack at an sp hybridized atom (**149**) is **dig**, and an exo-dig reaction will generate ring **150**.

It is therefore possible to describe ring forming reactions by the number of atoms in the cyclic product, whether the reaction is exo or endo, and whether it involves **tet**, **trig**, or **dig** intermediates. A **5-exo-tet** reaction represents formation of a five-membered ring by displacement at sp^3 carbon by X, where Y is exo to the ring being formed. The cyclization reactions that form 3-7 membered rings, along with all **exo/endo**, and **tet/trig/dig** possibilities are shown in Figure 6.2.[54]

---

54. (a) Baldwin, J.E. *J. Chem. Soc. Chem. Commun. **1976**,* 734; (b) Baldwin, J.E.; Cutting, J.; Dupont, W. Kruse, L.; Silberman, L.; Thomas, R.C. *Ibid. **1976**,* 736.

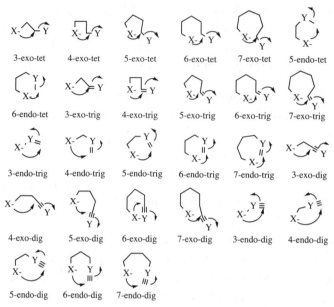

Figure 6.2. Ring closures categorized by Baldwin's rules.

When two reactive ends of a molecule come together, they can approach each other only from certain trajectories called the angle of attack. Before discussing Baldwin's rules, it is first necessary to establish what angles of attack are most likely. Elliot and Graham-Richards described a method for predicting the preferred approach angles based only on the substrate.[55] Note that these models assume a late transition state, so that the

bond angles in the product are close to those in the important transition state (the Hammond postulate). Displacement at a sp³ carbon generally requires backside attack (see **151**) and the incoming group (**X**) must approach the **Y**-bearing carbon at an angle close to 180°. For exo processes, this is usually easy but for endo processes it can be difficult. Reactions at double bonds (trig) are controlled by the planar nature of alkenes, imines or carbonyls. The bond angles are about 120°, but upon reaction the sp² atom is converted to a sp³ atom. Since sp³ atoms are tetrahedral (with bond angles of about 109°, sec. 4.7.C), the best trajectory for attack of a sp² atom (a carbonyl, for example) is about 109° (the Bürgi-Dunitz trajectory;[56] see **152**).

$$\mathbf{X}^- \cdots \overset{\alpha}{\underset{\mathbf{Y}}{\diagup}} \quad \xrightarrow{\alpha=180°} \quad \mathbf{X} \overset{\alpha}{\underset{\mathbf{151}}{\diagup}} \cdots \mathbf{Y}^- \qquad \mathbf{X}^- \cdots \overset{\alpha}{\underset{\mathbf{Y}}{\diagup}} \quad \xrightarrow{\alpha=109°} \quad \overset{\mathbf{X}}{\underset{\mathbf{Y}^-}{\diagup}} \alpha$$

$$\mathbf{151} \qquad\qquad\qquad\qquad \mathbf{152}$$

Treatment of **153** with *azobis*(isobutyronitrile), AIBN, and tributyltin hydride leads to a radical

55. Elliot, R.J.; Graham-Richards, W. *J. Mol. Struct.* **1982**, *87*, 247.
56. Bürgi, H.B.; Dunitz, J.D.; Lehn, J.M.; Wipff, G. *Tetrahedron* **1974**, *30*, 1563.

intermediate (**154**)[57] (see sec. 13.7 for a discussion of radical cyclization). Attack at the alkene unit easily allows approach at 109°, and cyclization gives **155**. Hydrogen transfer to **155** from Bu_3SnH completes the reaction, and generates the product **156**. Formation of large rings is not a problem with respect to Baldwin's rules, because there is sufficient flexibility in the tether to attain the required approach angle.[58]

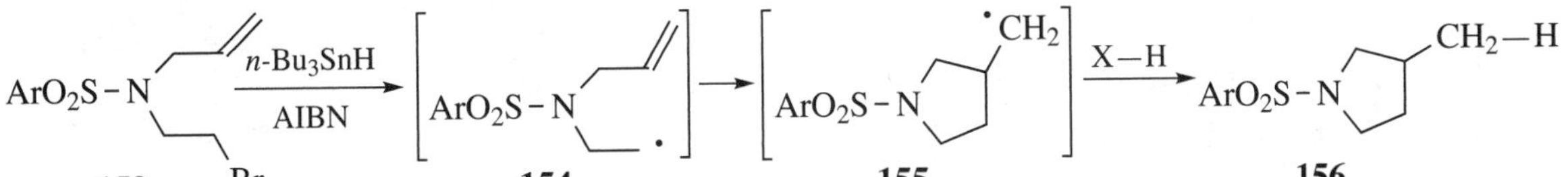

Molecules containing a triple bond have a linear geometry, so the bond angles are 180°. Conversion of the sp atom to an sp^2 atom with the 120° bond angle characteristic of alkenes. Using the same analogy as for the sp^2-sp^3 conversion, an incoming atom should approach the triple bond at an angle of ≈ 120° (see **157**).[59] Ring closure is difficult for reactions where the chain is small (*a* in **158**). With longer chains (*b* and *c* in **158**) the process may be more facile. For intramolecular reactions, attack at an angle of 120° or greater is possible in most cases (see **159**).

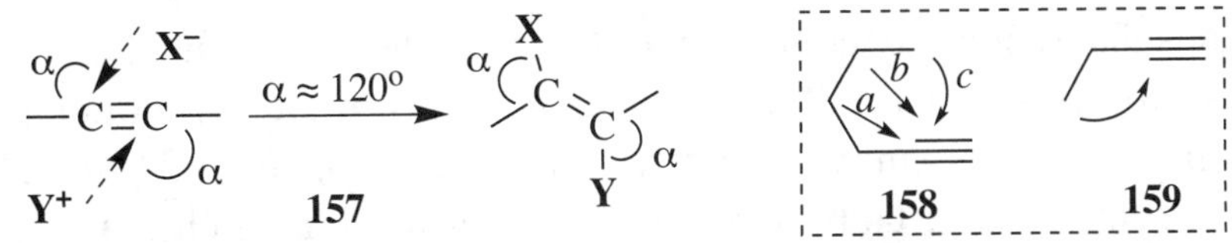

These observations led Baldwin to establish several rules for ring closure, known as **Baldwin's rules**:[54]

1. For **tet** systems: **3-7 exo-tet** are *favored*   **5-6 endo-tet** are *disfavored*
2. For **trig** systems: **3-7 exo-trig** are *favored*   **6-7 endo-trig** are *favored*   **3-5 endo-trig** are *disfavored*
3. For **dig** systems: **5-7 exo-dig** are *favored*   **3-7 endo-dig** are *favored*   **3-4 exo-dig** are *disfavored*

The disfavored reactions are not impossible, **simply more difficult** and usually slower than other competing reactions (inter- or intramolecular). Although some of the reactions show here will not be presented until Chapters 8, 9, 12 and 13, several examples of carbon-carbon bond forming reactions will be given to illustrate these rules.

When **160** reacted with KOH in aqueous acetonitrile, the 5-exo-trig product (**161**) was favored

57. (a) Padwa, A.; Nimmesgern, H.; Wong, G.S.K. *J. Org. Chem.* **1985**, *50*, 5620; (b) *Idem Tetrahedron Lett.* **1985**, *26*, 957; (c) Padwa, A.; Dent, W.; Nimmesgern, H.; Venkatramanan, M.K.; Wong, G.S.K. *Chem. Ber.* **1986**, *119*, 813; (d) Watanabe, Y.; Ueno, Y.; Tanaka, C.; Okawara, M.; Endo, T. *Tetrahedron Lett.* **1987**, *28*, 3953.
58. (a) Porter, N.A.; Magnin, D.R.; Wright, B.T. *J. Am. Chem. Soc.* **1986**, *108*, 2787; (b) Porter, N.A.; Chang, V.H.-T. *Ibid.* **1987**, *109*, 4976.
59. Baughman, R.H. *J. Appl. Phys.* **1972**, *43*, 4362.

over formation of the 6-endo-trig product (**162**), by a ratio of 18:1.[60] The alkynyl derivative (**163**) reacted 10^3 times slower than **160** and gave *only* the 5-exo-dig product (**164**). The 6-endo-dig product (**165**) was ***not*** isolated from the reaction, although the actual product distribution was probably > 100:1 favoring **164** over **165**. It was suggested that sp atoms generally prefer the exo mode of attack to the endo mode.[60]

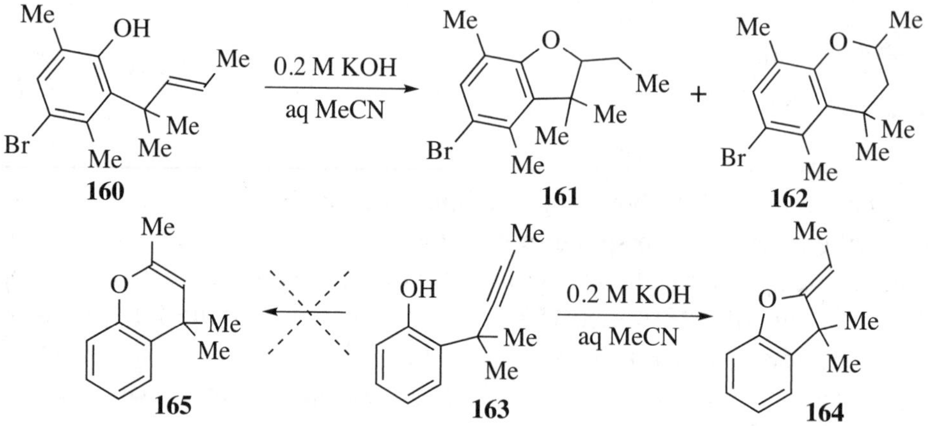

There are reactions that seemingly contradict Baldwin's rules. Presumably, these are reactions with an early transition state that will have different angles of attack. Fountain and Gerhardt showed that *N*-methyl-*O*-cinnamoyl hydroxylamine gave 5-phenyl-3-pyrazolidinone (**166**, which exists as the two valence tautomers shown) under forcing conditions (15 h in refluxing dichlorobenzene), via the disfavored 5-endo-trig pathway.[61] This simply illustrates that the rules say favored or disfavored, not allowed or forbidden. It does point out, however, that factors other than geometry and stereochemistry must be considered. Fountain suggested that electronic factors must be considered.[62] Anselme showed that cyclization of **167** to **168** occurred via a 5-endo-trig process,[61] but under more vigorous conditions than noted by Fountain for the cyclization of **165**. Presumably, the nitrogen lone pair electrons in **167** are more nucleophilic than in CO_2NHMe precursor to **166**.[62]

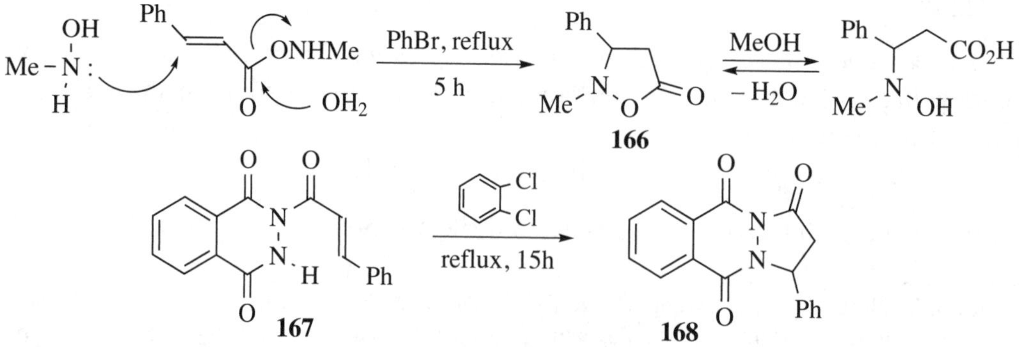

60. Evans, C.M.; Kirby, A.J. *J. Chem. Soc. Perkin 2* ***1984***, 1269.
61. Fountain, K.R.; Gerhardt, G. *Tetrahedron Lett.* ***1978***, 3985.
62. Anselme, J.P. *Tetrahedron Lett.* ***1977***, 3615.

Epoxides and other three-membered rings present special problems since they are in between tet and trig systems. Generally, three membered rings prefer the exo mode of attack (see **169**) to give **170**.[63] The endo mode of attack generates **171**. If we use **172** as an example, protonation of the epoxide oxygen with an acid (camphorsulfonic acid) followed by attack by the OH unit and ring opening gave the observed products.[64] When R was $CH_2CH_2CO_2Me$, the 5-exo product **174** was the only one (94% yield), which is typical. When the R substituent in **172** was changed from alkyl to alkenyl (vinyl), however, the course of ring opening was changed to give only the 6-exo product (**173**) in 90% yield. In general, the 5-exo mode of attack is preferred.

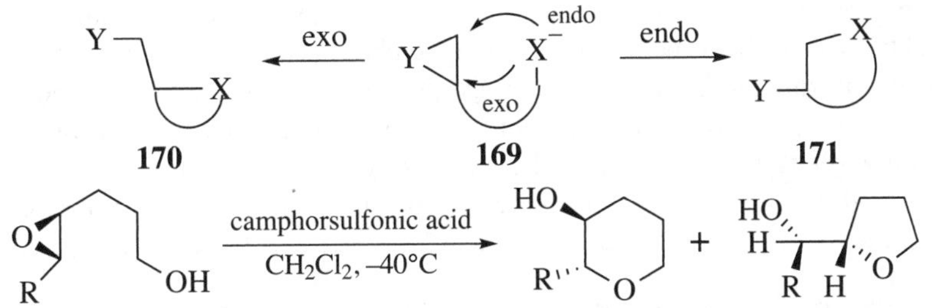

Three-membered rings other than an oxirane ring follow the same ring opening pathways. An example of such an opening is the iodolactonization reaction discussed in Section 2.10.C. A synthetic example is taken from Imanishi and co-worker's synthesis of leustroducsin B, in which alkene-acid **175** was treated with bicarbonate in the presence of iodine and KI, to give carboxylate anion-iodonium intermediate **176**.[65a] Intramolecular displacement of iodide by carboxylate (**path *a***) is a 5-exo process, and is favored over the analogous 6-endo process (**path *b***) to give iodolactone **177** in greater than 80% yield. Note that this reaction may not proceed by the discreet iodination addition product **176**, in which case Baldwin's rules do not apply.

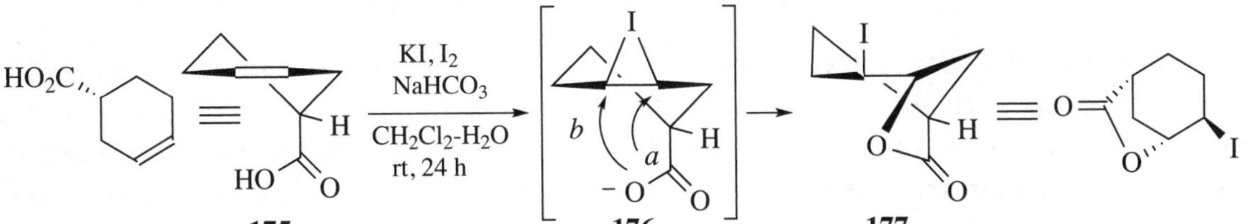

---

63. Stork, G.; Cohen, J.F. *J. Am. Chem. Soc.* **1974**, *96*, 5270.
64. Nicolaou, K.C.; Duggan, M.E.; Hwang, C.-K.; Somers, P.K. *J. Chem. Soc, Chem. Commun.* **1985**, 1359.
65. (a) Miyashita, M.; Tsunemi, T.; Hosokawa, T.; Ikejiri, M. Imanishi, T. *Tetrahedron Lett.* **2007**, *48*, 3829. For iodolactonization, see (b) Klein, J. *J. Am. Chem. Soc.* **1959**, *81*, 3611; (c) van Tamelen, E.E.; Shamma, M. *Ibid.* **1954,** *76*, 2315.

MOLECULAR MODELING: Intramolecular S_N2 Reaction

Ring closure of **A** can proceed via a 7-*endo* tet process to give **B**, or via a 6 *exo*-tet process to give pyran **C**. According to Baldwin's rules, **C** should predominate [see Janda, *et. al.*, *J. Am. Chem. Soc.* ***1995***, *117*, 2659].

Is the six-membered ring also the thermodynamic product (both prior to and following protonation)? To decide, obtain energies for both pairs of compounds. Place each molecule in what you think is a favorable conformation but be careful to use the correct stereochemistry. Following the first calculation, *click* on > at the right of the tab at the top of the screen and select **Continue**. This will bring up a fresh builder palette without removing the text. Does your analysis allow you to conclude that the ring closure is under kinetic or thermodynamic control? Elaborate. Also see homework problem 4.

Third-row elements are difficult to analyze by Baldwin's rules because the atomic radii are larger and the bond lengths are large. Conformations can be attained that are unavailable to molecules that contain only second-row atoms. The constraints on approach angles are therefore less stringent. Thiol **178**, for example, was cyclized to **179** by a 5-endo-trig process (an internal Michael addition).[54b,66] The sulfur atom has 3d orbitals and can receive electrons via back donation from the occupied p orbitals (see **180**). This ability diminishes the requisite angle of attack from 109° to 90° or less, facilitating the endocyclic ring closure.

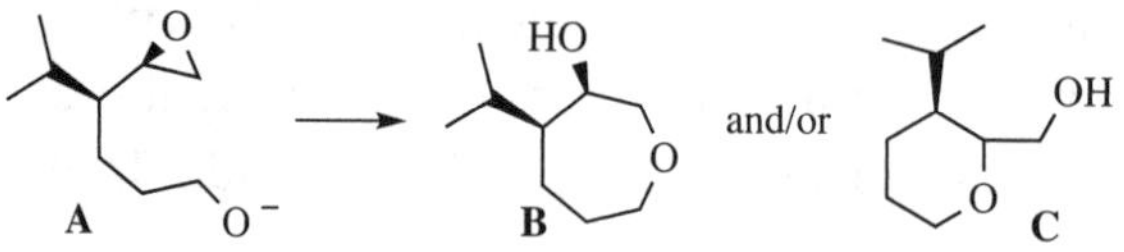

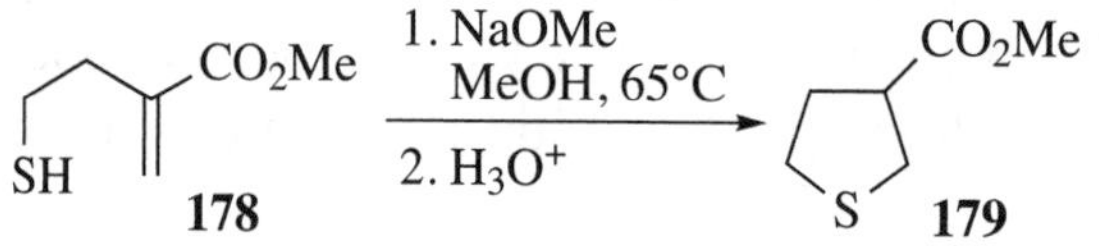

Baldwin's rules also apply to ketone enolate anions,[67] but Baldwin and Lusch[68] modified the rules to make the terminology more specific. The special angular requirements of enolate anions are shown for the exo-tet -process (**181**), and the exo-trig process (**182**). For these two ring-closing reactions, the p orbitals of the enolate anion must approach the reactive center at an angle of about 180° or 109°, respectively. The orientation of the orbital and not just the nucleophilic atom (C or O) must be considered for determining the correct angle of approach (sec. 9.2.B).

66. See (a) Claeson, G.; Jonsson, H.-G. *Arkiv. Kemi* ***1967***, *28*, 167 [*Chem. Abstr.*, *68*: 87083e ***1968***]; (b) *Idem Ibid.* ***1966***, *26*, 247 [*Chem. Abstr. 66*: 85334x ***1967***].
67. Baldwin, J.E.; Kruse, L.I. *J. Chem. Soc. Chem. Commun.* ***1977***, 233.
68. Baldwin, J.E.; Lusch, M.J. *Tetrahedron* ***1982***, *38*, 2939.

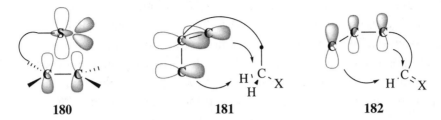

180   181   182

There are two ways to view the ring closure. A kinetic enolate anion (less substituted, derived from removal of the most acidic proton, sec. 9.2.E) such as **183** will close via exo displacement of Y to give **184**. A thermodynamic enolate anion (more substituted, resulting from equilibration to the most stable enolate anion product, sec. 9.2.E) such as **185** also undergoes an exo displacement of Y, but gives **186**. Conversion of **183** to **184** was termed enolendo-exo-tet, and conversion of **185** to **186** was termed enolexo-exo-tet. Similarly, exo-trig reactions can be classified in this manner. The kinetic enolate anion (**187**) gives **188** in an enolendo-exo-trig process. Cyclization of the thermodynamic enolate anion (**189**) gives **190** via an enolexo-exo-trig reaction.

183   enolendo-exo-tet   184   185   enolexo-exo-tet   186

187   enolendo-exo-trig   188   189   enolexo-exo-trig   190

The rules are somewhat different for ring closure with enolates.

6-7	**enolendo-exo-tet**	*favored*		3-5	**enolendo-exo-tet**	*disfavored*
3-7	**enolexo-exo-tet**	*favored*		3-7	**enolexo-exo-trig**	*favored*
6-7	**enolendo-exo-trig**	*favored*		3-5	**enolendo-exo-trig**	*disfavored*

Two competing intramolecular cyclizations (see the aldol condensation, sec. 9.4.A.ii) are illustrated for triketone **191**.[68] In this case, only the favored 6-enolendo-exo-trig product (**192**) was obtained, and not the disfavored 5-enol-endo-exo-trig products (**193** and **194**). Enolate anions

193   194   **5-enolendo exo-trig**   191   **6-enolendo exo-trig**   192

generally show 5-exo-tet reactions at oxygen, and 6-enolendo-exo-tet reactions at carbon. In order to form a five-membered ring from **195**, the oxygen lone pair must be at the correct angle for displacement of bromine (the product is tetrahydrofuran **196**). Another view of **195A** shows the positioning of the orbitals that will allow displacement (see **195B**). In order to generate a six-membered ring, however, the orbitals of the enolate carbon in **197** must be at the correct angle for displacement (see **197B**) and the product will be **198**.

6.6.B. Macrocyclic Compounds

6.6.B.i. Macrocyclic Ring Closures. Baldwin's rules explain most of the cyclization reactions for small- and medium-sized rings encountered in previous chapters, and those that will be seen in succeeding chapters. An exception is the formation of large rings. Formation of carbocyclic rings will be discussed in later chapters in connection with the appropriate carbon-carbon bond-forming reactions. Macrolactonization, however, is a functional group exchange process, and large lactone rings are also an important feature of many natural products. This section will discuss the problems and solutions for macrocyclic cyclizations. The principles discussed here for preparing large ring lactones are applicable to most other macrocyclizations.

Illuminati and Mandolini described the ring-closing reactions of bifunctional chain molecules.[69] In the 1920s and 1930s Ruzicka et al.[70] and Ziegler et al.[71] studied macrocyclic reactions. Formation of macrocyclic rings usually requires an intramolecular cyclization reaction of a bifunctional molecule such as **199**, where cyclization gives the monocyclic product, **200**. In this model, X and Y are reactive functional groups that generate a new bond, represented by Z (which may contain X, Y, or both). An important reaction that competes with cyclization is the intermolecular reaction where initial coupling generates the dimeric species **201**. Repeated intermolecular reactions give oligomers or polymers (**202**). Ruggli discovered that high substrate concentrations favor polymerization, while low concentrations favor cyclization.[72] The rate of cyclization is a function of the structure of the open-chain precursor and that of the product-like transition state. The activation energy for ring closure

69. Illuminati, G.; Mandolini, L. *Acc. Chem. Res.* **1981**, *14*, 95.
70. Ruzicka, L.; Stoll, M.; Schinz, H. *Helv. Chim. Acta* **1926**, *9*, 249.
71. Ziegler, K.; Eberle, H.; Ohlinger, H. *Annalen* **1933**, *504*, 94.
72. (a) Ruggli, P. *Annalen* **1912**, *392*, 92; (b) *Idem, Ibid.* **1913**, *399*, 174; (c) *Idem Ibid.* **1916**, *412*, 1.

is largely determined by the strain energy of the final ring.[73] As shown in Sections 1.5.A, 1.5.D, strain energy is due to "(*1*) bond opposition forces due to imperfect staggering (**Pitzer strain**), (*2*) deformation of ring bond angles (**Baeyer strain**), and (*3*) transannular strain due to repulsive interactions between atoms across the ring when they are forced close to each other."[69,74] As the chain length increases for a cyclization reaction, the probability of the chain termini approaching each other decreases ("negative $\Delta S^{\ddagger}$ due to less freedom of internal rotation around single bonds of the molecular backbone when the disordered, open-chain precursor is converted to the ring shaped transition state").[69] In general, the ring product is a good model of the transition state of cyclization (a ring-shaped transition state) for all but the shortest chains.[75] When large rings are formed that are free of strain, there is minimal strain energy in the transition states. In short chains, an advantage in terms of entropy is offset by an increase in enthalpy due to extremely large strain energies. Ziegler first used this principle of a ring-shaped transition state to generate large membered rings by is known as the **high dilution method**.[76] Cyclization of **203** to **204**, for example, was accomplished by slow addition (6.6 x 10^{-4} mol L^{-1}day^{-1}) of bromoacid to a solvent containing potassium carbonate (K_2CO_3) or hydroxide.[77] For $n = 9$, a 77% yield of **204** was obtained.[77] The main side product of this process was the dimeric ester **205**.

Bromocarboxylate **203** can be cyclized to a lactone and this simple reaction will be used to illustrate cyclization processes in general. An intramolecular cyclization reaction of **203** (governed by k_{intra}) will generate the lactone (**204**), but an intermolecular reaction (k_{inter}) will give the dimeric ester (**205**).[69] The relative preference of a reaction is given by the effective molarity (**EM**) = $k_{\text{intra}}/k_{\text{inter}}$.[78] The EM is supposed to be the first-order rate constant for ring closure times the second order rate constant for reaction between chain ends (if they were not connected), but the EM for five and six-membered rings exceeds the real concentration.[69] Two parameters similar to the EM are useful for predicting conditions under which a ring can be synthesized, free of significant polymerized byproducts.[79] The first is the parameter $C\alpha/M_o$ (see Eqn. 6.1), where k_R is the rate of ring formation of cyclic monomer and k_P is the rate of

73. Liebnan, J.F.; Greenberg, A. *Chem. Rev.* **1976**, *76*, 311.
74. Allinger, N.L.; Tribble, M.T.; Miller, M.A.; Wertz, D.H. *J. Am. Chem. Soc.* **1971**, *93*, 1637.
75. Galli, C.; Mandolini, L. *Eur. J. Org. Chem.* **2000**, 3117.
76. Ziegler, K. in *Methoden der Organischen Chemie (Houben-Weyl), Vol 4/2*, Müller, E. (Ed), Georg Thieme Verlag, Stuttgart, *1955*.
77. (a) Hunsdiecker, H.; Erlbach, H. *Chem. Ber.* **1947**, *80*, 129; (b) Stoll, M. *Helv. Chim. Acta* **1947**, *30*, 1393.
78. Illuminati, G.; Mandolini, L.; Masci, B. *J. Am. Chem. Soc.* **1977**, *99*, 6308.
79. Galli, C.; Mandolini, L. *Gazz. Chim. Ital.* **1975**, *105*, 367.

formation of cyclic dimer. If the initial concentration is less than unity ($\alpha < 1$), the yield is not less than 55%. For medium rings, k_R/k_P should be < 0.1 M.[80]

$$\frac{C_\alpha}{M_o} = \frac{1}{2\alpha} \ln(1 + 2\alpha) \quad \text{where} \quad \alpha = \frac{M_o}{k_R/k_P} \qquad 6.1$$

and k_R and k_P are defined by:

$$\text{monomer} \xrightarrow{k_R} \text{monomeric complex} \qquad 2 \text{ monomer} \xrightarrow{k_P} \text{dimer}$$

Similarly, the parameter η_c/η_m (the yield of cyclic substance obtained by high dilution conditions, Eqn. 6.2) can be used, where η_m is the total number moles of bifunctional substrate, η_c is the number of moles of monomeric ring product produced, and v_f is the constant feed rate in that is measured in mol L^{-1} sec^{-1}.[79] The dimensionless parameter β is a measure of the feed rate, and the relative formation of cyclic monomer is a function of the feed rate.[79]

$$\frac{\eta_c}{\eta_m} = \frac{2}{1 + R(1 + 9n_f k_p/k_R2)} = \frac{2}{1 + R(1 + 8\beta)} \qquad 6.2$$

The dependence on ring size for lactone formation is illustrated by Figure 6.3a (the reactivity for lactone formation), Figure 6.3b (the $\Delta H^\ddagger$ profile for lactone formation) and Figure 6.3c (the $\Delta S^\ddagger$ profile for lactone formation).[69] The quantity $\Delta H^\ddagger_{intra} - \Delta H^\ddagger_{inter}$ is a useful measure of the strain energies involved in ring formation of lactones. As the chain length increases, $\Delta H^\ddagger$ decreases.[81] For very large rings (C_{18}-C_{23}), $\Delta H^\ddagger_{intra} \approx \Delta H^\ddagger_{inter}$ and the intramolecular reaction is nearly strain free. In the C_3-C_{16} region, ring formation is strain dependent with $\Delta H^\ddagger$ maximal at C_3 and C_8 (≈ 8 kcal/mol^{-1} mol^{-1}; 33.5 kJ mol^{-1}). For C_3-C_8, the cis conformation predominates due to the increased strain inherent to a trans conformation.

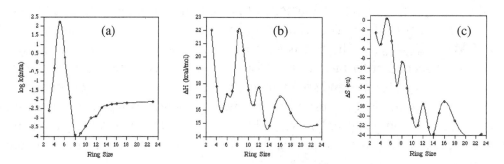

Figure 6.3. For lactone formation: (a) Reactivity profile, (b) $\Delta H^\ddagger$ profile, and (c) $\Delta S^\ddagger$ profile. [Reprinted with permission from Illuminati, G.; Mandolini, L. *Accts. Chem. Res., 1981, 14*, 95. Copyright © 1981 American Chemical Society.]

80. Illuminati, G.; Mandolini, L.; Masci, B. *J. Am. Chem. Soc. 1974, 96*, 1422.
81. Mandolini, L. *J. Am. Chem. Soc. 1978, 100*, 550.

As shown in Figure 6.3(c), $\Delta S^{\ddagger}$ does not show a regular pattern, but more favorable for large rings because of a "looseness in the ring due to low frequency out of plane bending motions".[69,82] The variation in $\Delta S^{\ddagger}$ indicates the importance of individual ring features, with maxima occurring at ring sizes of C_8, C_{12}, and C_{16}. For ring sizes of C_3-C_6, the $\Delta S^{\ddagger}$ values are relatively insensitive to ring size. Solvation around the reactive end plays a major role in C_3 and C_4 rings, but for C_5 rings and larger, such solvation is less important.

MOLECULAR MODELING: Lactone Hydrolysis

In aqueous solution, cyclic lactones are in equilibrium with open chain hydroxycarboxylic acids. Equilibrium favors the five-member ring γ-butyrolactone over 4-hydroxybutanoic acid, but favors 10-hydroxydecanoic acid over the nine-member ring nonalactone.

Obtain structures and energies for both lactone and open chain hydroxycarboxylic acid forms of both compounds as well as for the corresponding tetrahedral intermediates (six calculations in total). For each calculation following the first, *click* on > tab at the top of the screen and select **Continue**. This will give you a fresh builder palette and keep this text on screen. Calculate the energy of loss of hydroxide from the five-member ring intermediate *relative* to that for the ten-member ring intermediate (hydroxide will cancel). Which reaction is favored? Calculate the energy of protonation of the five-member ring intermediate *relative* to that for the ten-member ring intermediate (proton will cancel). Which reaction is favored? Are your results consistent with what is observed? Elaborate.

6.6.B.ii. Synthetic Approaches to Macrocyclic Lactones.[83] There are many examples of biologically important, naturally occurring lactones. High dilution techniques can be used, as mentioned above. In the final step of a synthesis of amphidinolide P, Trost and co-workers simply heated a dilute solution of **206** in hexane to give an 84% yield of amphidinolide P (**207**).[84] In this reaction, the eight-membered ring lactone was opened to give a seco-acid, which closed to the macrocyclic lactone, and the hydroxyl group remaining from the eight-membered ring lactone attacked the ketone moiety to generate the hemi-ketal unit in **207**.

---

82. O'Neal, H.E.; Benson, S.W. *J. Chem. Eng. Data* **1970**, *15*, 266.
83. For a review of macrolactonization in the total synthesis of natural products, See (a) Parenty, A.; Moreau, X.; Campagne, J.-M. *Chem. Rev.* **2006**, *106*, 911. Also see (b) Mundy, B.P.; Ellerd, M.G.; Favaloro Jr., F.G. *Name Reactions and Reagents in Organic Synthesis, 2nd ed.*, Wiley-Interscience, New Jersey, **2005**, pp. 402-403. For a review of syntheses of medium-sized ring lactones, see Shiina, I. *Chem. Rev.* **2007**, *107*, 239.
84. Trost, B. M.; Papillon, J. P. N.; Nussbaumer, T. *J. Am. Chem. Soc.* **2005**, *127*, 17921.

206 → (0.001 M in hexane, reflux, 1 h) → **207**

A variety of cyclization techniques have been developed, but all are based on the idea that the carbonyl end of a ω-substituted acid is activated to facilitate attack by or at the other functionalized end. In one example, trifluoroacetic anhydride reacted with **208** to form mixed anhydride **209**, which activated the carbonyl to attack by the hydroxyl moiety and gave **210** in 31% yield, in the Taub et al. synthesis of zearalenone.[85] Cyclization has also been observed using a mixture of trifluoroacetic acid and trifluoroacetic anhydride.[86] The use of 2-methyl-6-nitrobenzoic anhydride and dimethylaminopyridine with powdered molecular sieves, the so-called **Shiina macrolactonization**, has also been reported.[85cd]

208 → ($(CF_3CO)_2O$, PhH) → **209** → ($-CF_3CO_2H$) → **210**

Another macrocyclization technique has been developed based on formation of a mixed anhydride. In the **Yamaguchi protocol**[87] a seco acid is treated with 2,4,6-trichlorobenzoyl chloride, and the resulting mixed anhydride is heated with DMAP (4-N,N-dimethylamino-pyridine) in toluene. In the Ghosh and Gong synthesis of amphidinolide W, for example, seco-acid **211** was converted to lactone **212** in 50% yield using the Yamaguchi protocol.[88] The methyl group epimerized and a 1:1 mixture of epimers was isolated in this instance. A variation of this procedure prepares the mixed anhydride by reaction of the seco-acid with Boc anhydride (see Chap. 7, Sec. 7.3.C.iii.), followed by heating with DMAP in toluene.[89]

85. (a) Taub, D.; Girotra, N.N.; Hoffsommer, R.D.; Kuo, C.H.; Slates, H.L.; Weber, S.; Wendler, N.L. *Tetrahedron* **1968**, *24*, 2443; (b) *Idem J. Chem. Soc. Chem. Commun.* **1967**, 225; (c) Shiina, I.; Kubota, M.; Oshiumi, H.; Hashizume, M. *J. Org. Chem.* **2004**, *69*, 1822. (d) Wu, Y.; Yang, Y.-Q. *J. Org. Chem.* **2006**, *71*, 4296. For an example taken from a synthesis of lejimalide B, see Schweitzer, D.; Kane, J.J.; Strand, D.; McHenry, P.; Tenniswood, M.; Helquist, P. *Org. Lett.* **2007**, *9*, 4619.
86. Baker, P.M.; Bycroft, B.W.; Roberts, J.C. *J. Chem. Soc. C*, **1967**, 1913.
87. (a) Inanaga, J.; Hirata, K.; Saeki, H.; Katsuki, T.; Yamaguchi, M. *Bull. Chem. Soc. Jpn.*, **1979**, *52*, 1989; (b) Mundy, B.P.; Ellerd, M.G.; Favaloro Jr., F.G. *Name Reactions and Reagents in Organic Synthesis, 2nd ed.*, Wiley-Interscience, New Jersey, **2005**, pp. 710-711.
88. Ghosh, A. K.; Gong, G. *J. Org. Chem.* **2006**, *71*, 1085.
89. Nagarajan, M.; Kumar, V.S.; Rao, B.V. *Tetrahedron* **1999**, *55*, 12349.

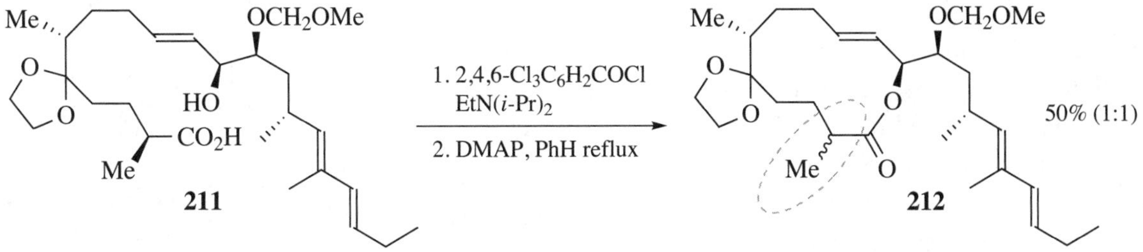

Mercuric trifluoroacetate [Hg(OCOCF$_3$)$_2$] is an effective reagent for the cyclization of hydroxy thio-esters.[90] Masamune et al. used this cyclization procedure for the conversion of **213** to **215** (in 90% yield) in a synthesis of zearalenone dimethyl ether.[91] Masamune developed this procedure (the **Masamune protocol**) to complete the total synthesis of methymycin, where one step involved treatment of a thioester with mercuric salts.[92,91] Mercury(II) has an affinity for bivalent sulfur and it binds the thioester unit as well as the terminal hydroxyl group, as in **214**. This coordination brings the reactive termini in **216** [HO and S(R)C=O] into close proximity, and the hydroxyl oxygen can attack the carbonyl carbon of the thioester to give cyclized product, lactone **213**. Thioester **213** was formed from the corresponding phosphonate ester of the acid. A carboxylic acid is first converted to the phosphonate ester, represented by generic structure **216**. Subsequent reaction of **216** with the thallium salt of 2-methyl-2-propanethiol [TlSC(Me)$_3$] gave **217**. This generic example shows a mild and general procedure for the preparation of thioesters, and was used to prepare **213**.

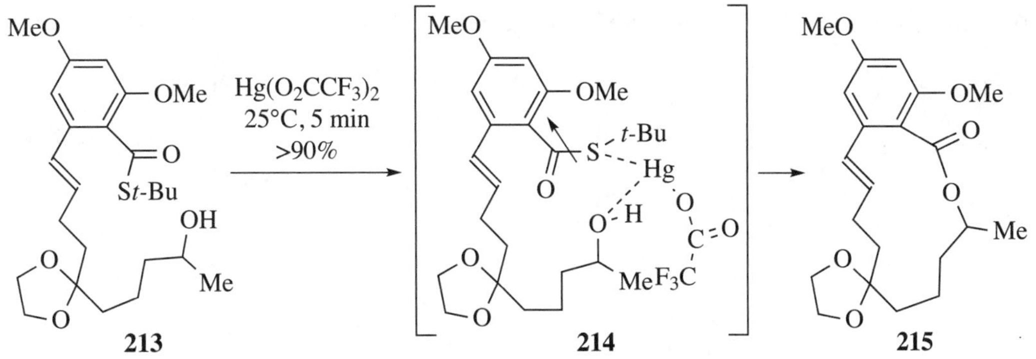

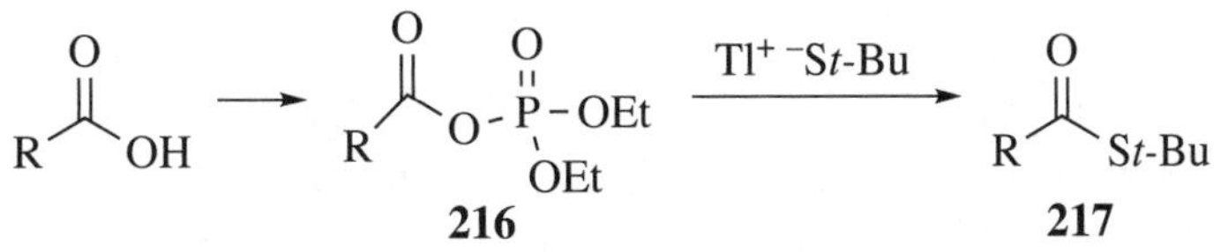

Diethyl azodicarboxylate (EtO$_2$C–N=N–CO$_2$Et, DEAD) is a key reagent in the **Mitsunobu**

90. Masamune, S.; Yamamoto, H.; Kamata, S.; Fukuzawa, A. *J. Am. Chem. Soc.* **1975**, *97*, 3513.
91. Masamune, S.; Kamata, S.; Schilling, W. *J. Am. Chem. Soc.* **1975,** *97*, 3515.
92. Masamune, S.; Kim, C.U.; Wilson, K.E.; Spessard, G.O.; Georghiou, P.E.; Bates, G.S. *J. Am. Chem. Soc.* **1975**, *97*, 3512.

reaction (sec. 2.6.A.ii) and has also been used for macrolactonization.[93] A synthetic example is the reaction of **218** with DEAD, and PPh$_3$, which gave a 73% yield of **219** in Steglich and co-worker's synthesis of ornatipolide.[94] A variety of specialized reagents have

been developed for macrolactonization reactions. Several of these reagents are effective for the cyclization of ω-hydroxy acids, as shown in Table 6.1. For comparison, results for cyclization with DEAD and **222** are included, for conversion of a hydroxy acid to lactone **204**. The **Corey-Nicolaou reagent**[95] (2,2'-dipyridyl disulfide, **220**) and the **Mukaiyama reagent**[96] (2-chloro-1-methyl-pyridinium iodide, **221**) are two of the more important. A number of related reagents have been developed, including imidazole disulfide **222** [2,2'-dithio-(4-*tert*-butyl-1-isopropylimidazole)][97] and imidazole **223** [*N*-(trimethylsilyl)imidazole].[98] As seen in Table 6.1, diolide (**224**) can be formed by reaction of an initial intermolecular coupling product, the ω-substituted carboxylate [X-(CH$_2$)$_n$-CO$_2$-(CH$_2$)$_n$CO$_2^-$]. The nucleophilic oxygen anion of the carboxylate group displaces the X leaving group in a macrolactonization process to give the diolide (see **224**). Diolides are commonly observed during the formation of medium size lactones where transannular interactions lead to poor yields. Cyclization to form a 10-membered lactone is difficult, for example, and under high dilution conditions the formation of the 20-membered diolide is usually favored. Indeed, diolides are common contaminants in reactions that form 6- to 11-membered ring lactones.

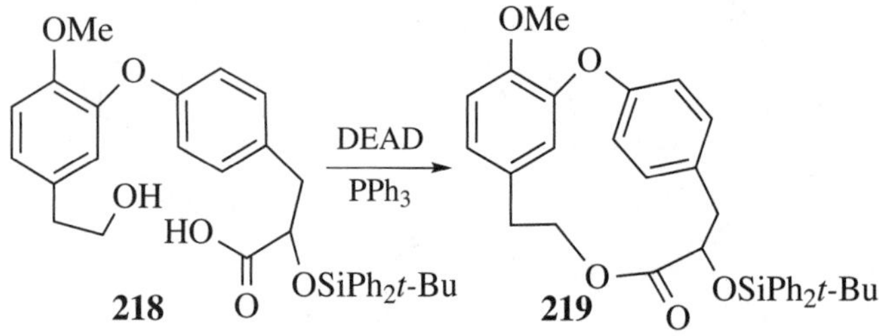

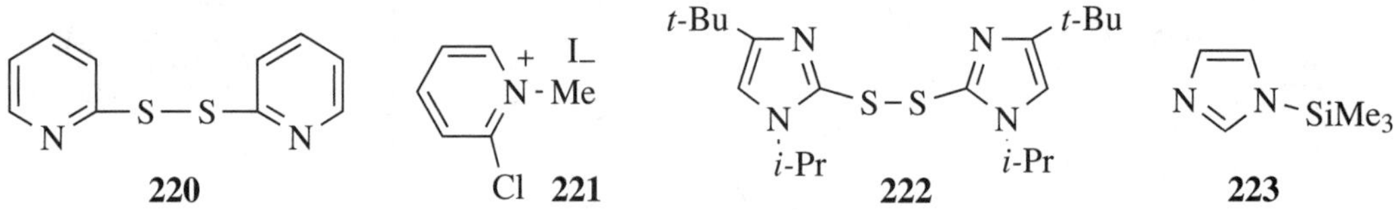

93. Kurihara, T.; Nakajima, Y.; Mitsunobu, O. *Tetrahedron Lett.* **1976,** 2455.
94. Ingerl, A.; Justus, K.; Hellwig , V.; Steglich, W. *Tetrahedron* **2007**, *63*, 6548.
95. (a) Corey, E.J.; Nicolaou, K.C. *J. Am. Chem. Soc.* **1974**, *96*, 5614. The actual cyclization has been called the **Corey-Nicolaou macrocyclization**, (b) Mundy, B.P.; Ellerd, M.G.; Favaloro Jr., F.G. *Name Reactions and Reagents in Organic Synthesis, 2nd ed.*, Wiley-Interscience, New Jersey, **2005**, pp. 184-185.
96. Mukaiyama, T.; Usui, M.; Saigo, K. *Chem. Lett.* **1976,** 49. For a synthetic example taken from a synthesis of the sorangiolides, see Das, S.; Abraham, S.; Sinha, .C. *Org. Lett.* **2007,** *9*, 2273.
97. Corey, E.J.; Brunelle, D.J. *Tetrahedron Lett.* **1976**, 3409.
98. Bates, G.S.; Diakur, J.; Masamune, S. *Tetrahedron Lett.* **1976**, 4423.

Table 6.1. Macrolactonization of ω-Hydroxy Acids

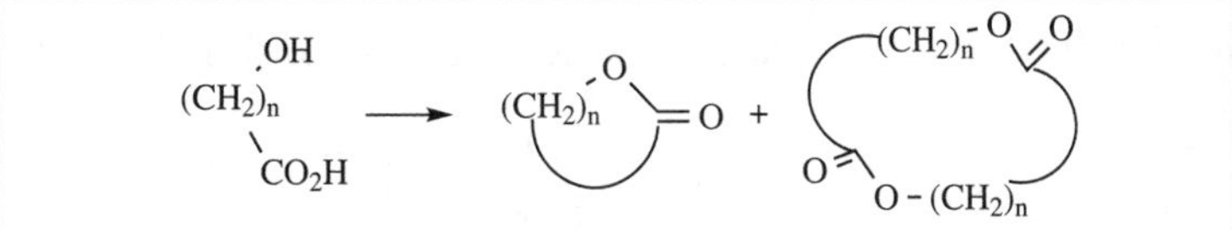

n	Reagent	% 204	% 234	n	Reagent	% 204	% 234
5	220	71	7	11	220	66	7
	221	89	0		221	69	14
	DEAD	40	53		222	87	
6	221	0	93		DEAD	60	17
7	220	8	41	12	220	68	6
	221	13	34	13	222	90	
	DEAD	0	20	14	220	80	5
10	220	47	30		221	84	3
	221	61	24	15	222	96	

The Corey-Nicolaou procedure reacted **220** with a seco acid to form a thioester (**225**), which is in equilibrium with the protonated form. Hydrogen bonding leads to a template effect (see **226**), that allows nucleophilic acyl attack by the hydroxy group to give **227**. Displacement of thiopyridone (**228**) gives the lactone (**204**). A synthetic example that used **220** converted **229** to **230**, in 62% yield, in Kang and co-worker's synthesis of (+)-pamamycin-607.[99] The addition of cupric bromide in a second step facilitated the macrocyclization.[100]

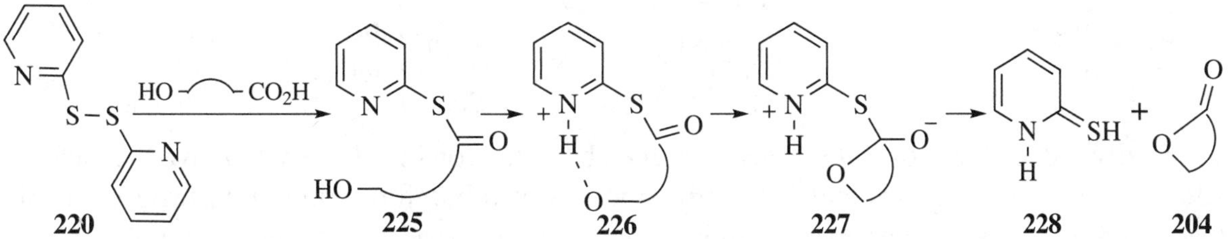

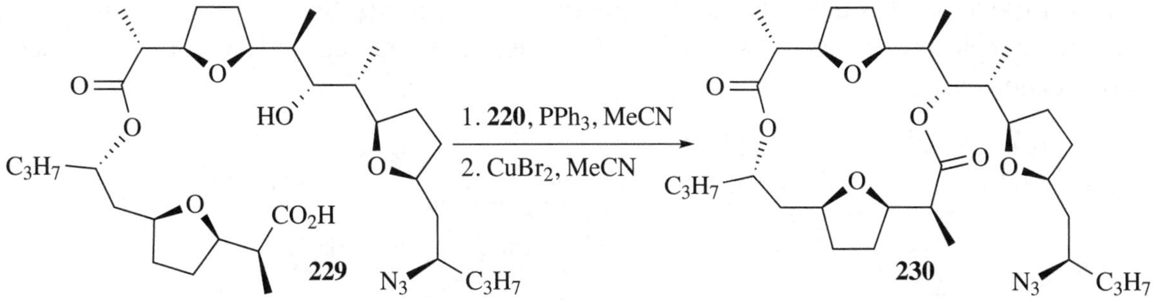

Another useful reagent is 1,1'-carbonyldiimidazole (**232**). Hydroxyacid **231** was converted to **233** (40% yield) in White and co-worker's model studies for a synthesis of erythromycin

99. Kang, S.H.; Jeong, J.W.; Hwang, Y.S.; Lee, S.B. *Angew. Chem. Int. Ed., **2002**, 41*, 1392.
100. Kim, S.; Lee, J.I. *J. Org. Chem. **1984**, 49*, 1712.

B.[101] This reagent promotes formation of an initial imidazolium intermediate to activate the carbonyl to attack. White also cyclized **231** by treatment with tosyl chloride and triethylamine under dilute conditions (0.01-0.02 M), and obtained a 52% yield of **233**.[101]

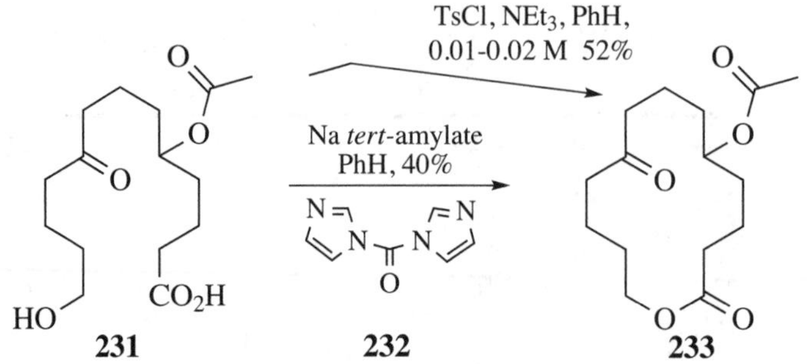

Dibutyltin oxide (*n*-Bu$_2$SnO) was reported to be a tin-template driven extrusion lactonization reagent.[102a,b] This reagent effects cyclization in refluxing mesitylene with an ω-hydroxy acid, giving 22-63% of the corresponding 13-17 membered ring lactone (0% for *n* = 7, 63% for *n* = 14 and 60% for *n* = 15). The reaction works via formation of **234**, where the tin reagent complexes the carbonyl unit and the hydroxyl group (as in **235**).

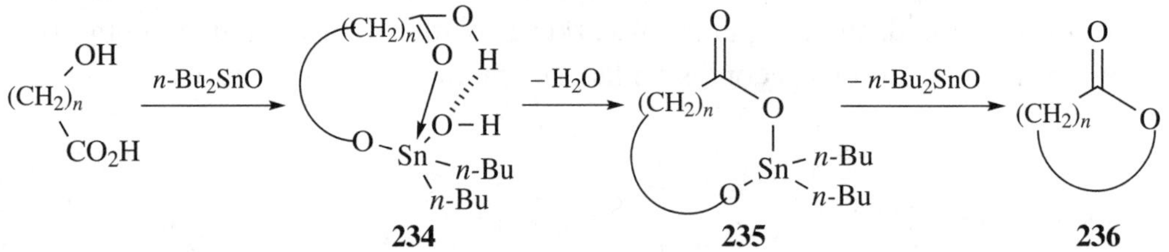

Thermally induced loss of water is accompanied by extrusion of *n*-Bu$_2$SnO to give the lactone, **236**. The reaction requires stoichiometric amounts of dibutyltin oxide to ensure reasonable reaction times. The reagent does not induce racemization at the stereogenic center bearing the hydroxyl group, so it is amenable to asymmetric macrolactonization. Cyclization of **237** to **238**, for example, proceeded in 44% yield and was used in Hanessian and co-workers synthesis of ricinelaidic lactone.[102]

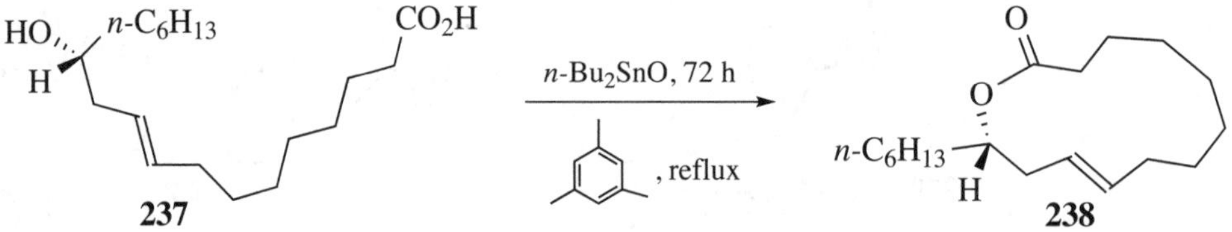

101.   White, J.D.; Lodwig, S.N.; Trammell, G.L.; Fleming, M.P. *Tetrahedron Lett.* **1974,** 3263.
102.   (a) Steliou, K.; Szczygielska-Nowosielska, A.; Favre, A.; Poupart, M.A.; Hanessian, S. *J. Am. Chem. Soc.* **1980,** *102,* 7578; (b) Steliou, K.; Poupart, M.A. *Ibid.* **1983,** *105,* 7130; (c) Shanzer, A.; Mayer-Shochet, N.; Frolow, F.; Rabinovich, D. *J. Org. Chem.* **1981,** *46,* 4662; (d) Shanzer, A.; Libman, J.; Gottlieb, H.E. *Ibid.* **1983,** *48,* 4612.

Trost and co-workers reported another macrolactonization method. Reaction of ω-hydroxy acid **239** with ethoxyethyne and the ruthenium catalyst shown was followed by treatment with camphorsulfonic acid in dilute solution to give a 69% yield of **240**.[103] This two-stage macrolactonization protocol is effective for generating 14-membered rings and larger, and does not involve treatment with base.

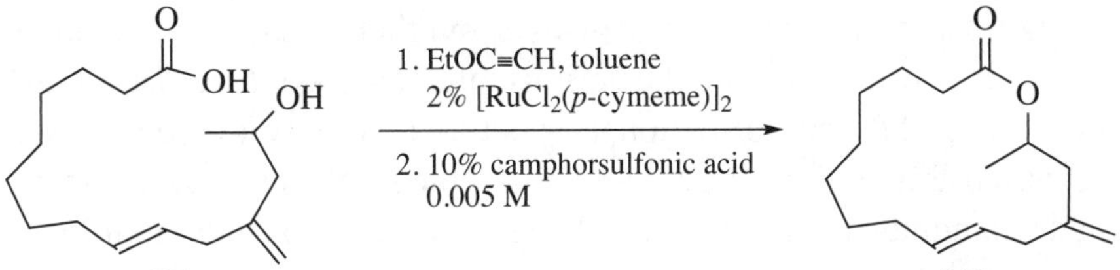

It is useful to note that the yield of macrocyclic product usually depends upon the method used. Clearly, there are several reagents that one can use for this reaction, and the choice is dictated by the ease of reaction and the yield. This is illustrated in a synthesis of (–)-spinosyn A, reported by Frank and Roush.[104] Cyclization of seco-acid **241** with the Mukaiyama reagent (**221**) or under Yamaguchi conditions gave relatively poor yields of macrocycle **242**. Mitsunobu conditions gave better results, using either diethyl azodicarboxylate or diisopropyl azodicarboxylate. Significant amounts of hydrazide were formed using the standard Mitsunobu conditions.[104] The use of DIAD (diisopropylazodicarboxylate) for suppression of this side-product had been reported in the literature,[105] but the yield of **242** was diminished. The use of yet another coupling reagent PyBOP (benzotriazol-1-yloxytripyrrolidinophosphonium hexafluorophosphate)[106] gave the best results, although the related PyBrOP[107] gave significantly poorer results.

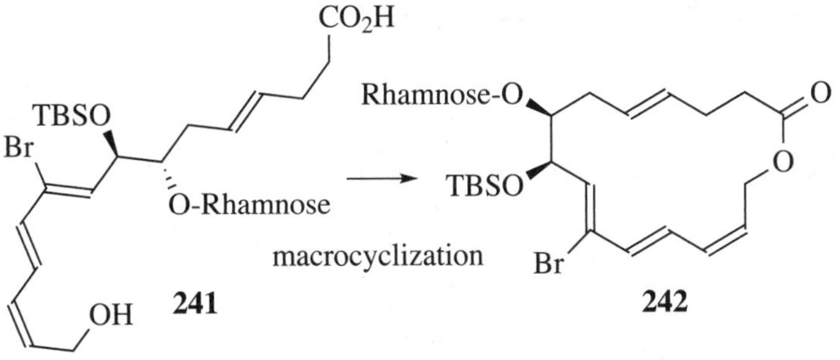

Reagent	% Yield
233, NEt$_3$	33
2,4,6-trichlorobenzoyl chloride, DMAP	38
PPh$_3$, DEAD	54
PPh$_3$, DIAD	48
PyBrOP, DMAP	37
PyBOP, DMAP	70

103.   Trost, B.M.; Chisholm, J.D. *Org. Lett.* **2002**, *4*, 3743.
104.   Frank, S.A.; Roush, W.R. *J. Org. Chem.* **2002**, *67*, 4316.
105.   Evans, D.A.; Ratz, A.M.; Huff, B.E.; Sheppard, G.S. *J. Am. Chem. Soc.* **1995**, *117*, 3448.
106.   Coste, J.; Frérot, E.; Jouin, P.; Castro, B. *Tetrahedron Lett.* **1991**, *32*, 1967.
107.   Frérot, E.; Coste, J.; Pantaloni, A.; Dufour, M.-N.; Jouin, P. *Tetrahedron* **1991**, *47*, 259.

6.7. CONCLUSION

The intent of this chapter was to tie together many of the principles introduced in earlier chapters. In addition, many of the principles used in the important chapters dealing with making carbon-carbon bonds have been discussed. If there is a theme to this chapter, it is that organic chemists can exercise a significant amount of control over synthetic reactions by understanding the mechanism of the trans-formations and the various interactions of heteroatoms and reagents. A thorough understanding of the conformational aspects of reactivity is essential. With knowledge of these principles, a useful plan can be assembled for the synthesis of even complex molecules. Without this understanding, syntheses beyond a few simple steps will be difficult.

One last piece of information is required for synthesis in order to manipulate functional groups. In those cases where a heteroatom functional group interferes with a transformation and cannot be removed, modified, or inserted earlier or later in a synthesis, methods are available to temporarily block it. This method is termed protection and the moiety used to block the reactive fragment is called a protecting group, which will be the focus of Chapter 7.

HOMEWORK

1. In each of the following, give the major product and classify each reaction according to Baldwin's rules:

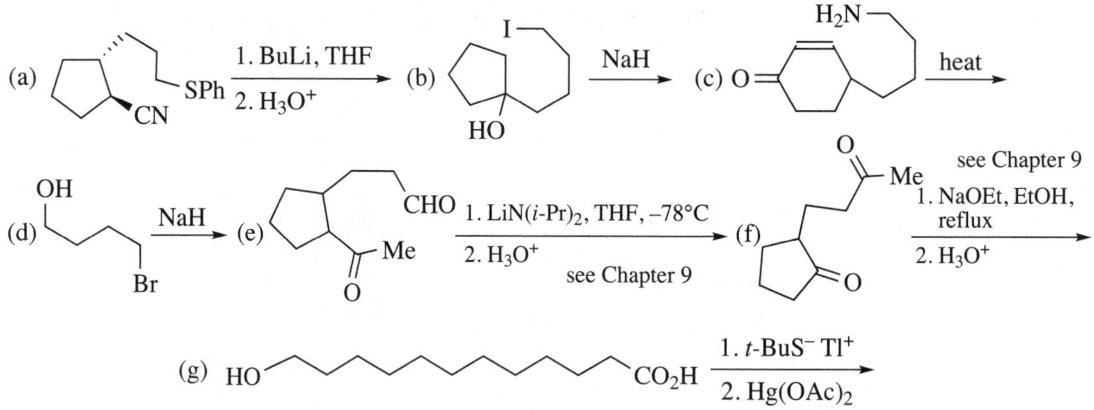

2. Explain why the following sequence results in inversion of configuration of the hydroxyl group:

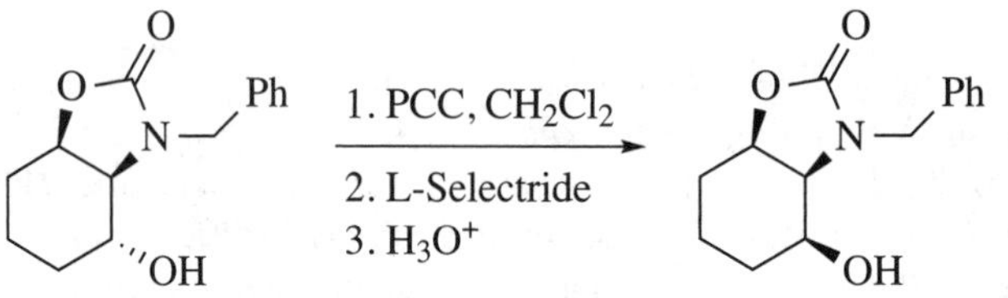

3. Explain this transformation by giving the three intermediate products.

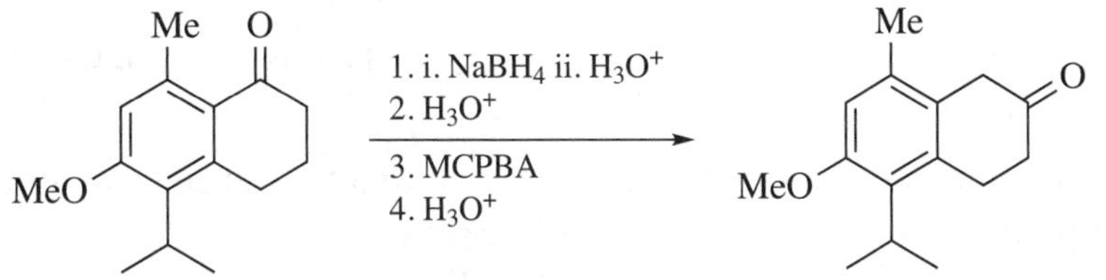
Me
O
MeO
1. i. NaBH4 ii. H3O+
2. H3O+
3. MCPBA
4. H3O+
Me
O
MeO

For related reactions see (a) Fieser, L.F.; Fieser, M. *Advanced Organic Chemistry,* Reinhold, New York, **1961**, pp 897-898; (b) *Technol. Repts. Osaka Univ., **1958**, 8*, 455 [*Chem. Abstr., **1959**, 53*: 18920i].

4. Draw the major product of this reaction and predict the correct stereochemistry. Briefly explain your answer.

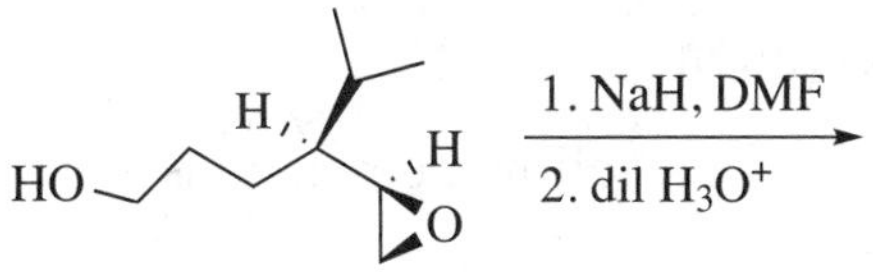
H
HO
H
O
1. NaH, DMF
2. dil H3O+

5. Explain why these two saponification reactions lead to different products (ring closed vs. ring opened).

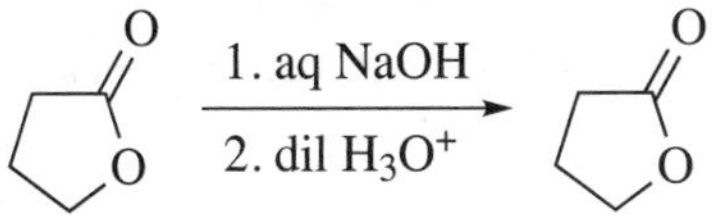
O
O
1. aq NaOH
2. dil H3O+
O
O

but

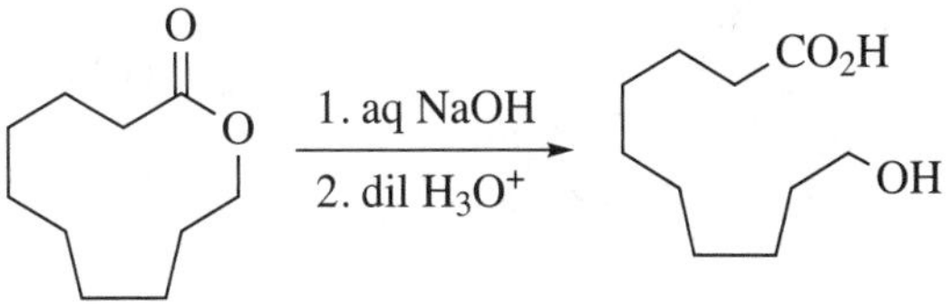
O
O
1. aq NaOH
2. dil H3O+
CO2H
OH

6. Suggest a reason why the carboethoxy group epimerizes upon treatment with potassium carbonate. Why is the second step required?

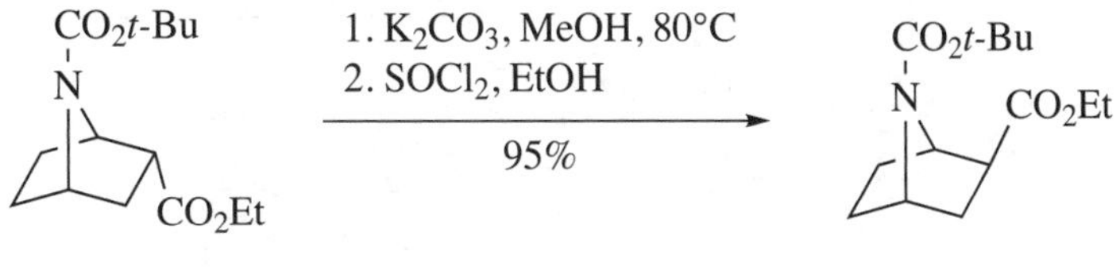
CO2t-Bu
N
CO2Et
1. K2CO3, MeOH, 80°C
2. SOCl2, EtOH
95%
CO2t-Bu
N
CO2Et

7. Offer a mechanistic rationale for the following conversion that explains the stereoselectivity of the reaction.

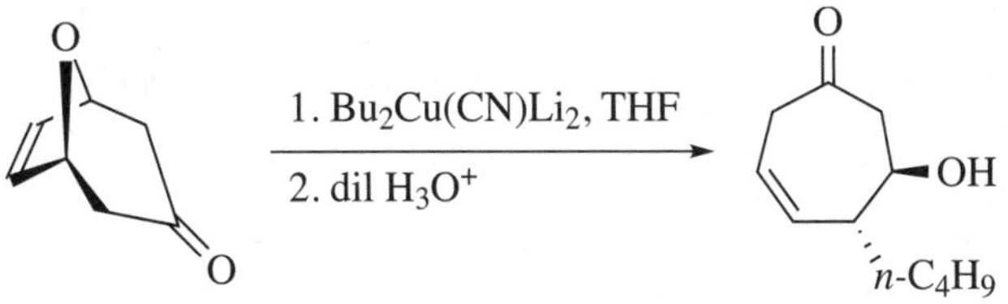
O
O
1. Bu2Cu(CN)Li2, THF
2. dil H3O+
O
OH
n-C4H9

see Chapter 8

8. Show the structure of both A and B. Each of these reactions has a name associated with it. Give the name of each reaction and look up one recent review of each.

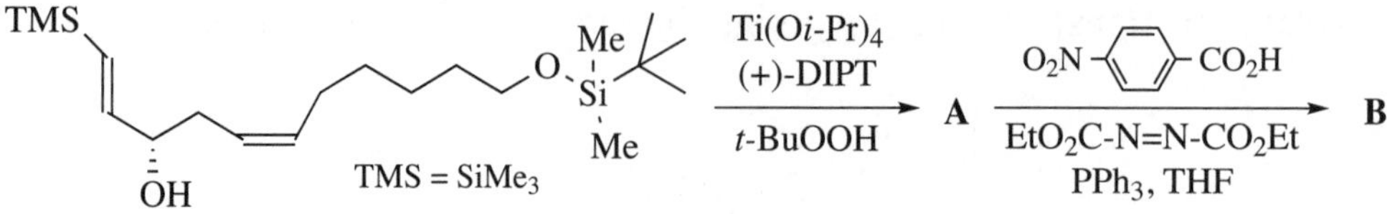

9. Predict the major product of this reaction. Explain your answer in the context of other possible products.

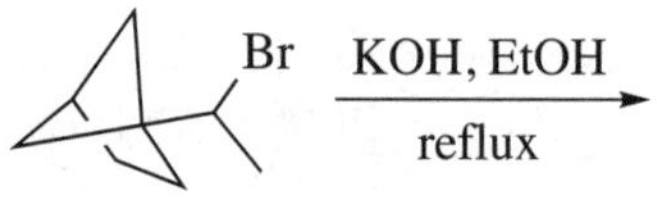

10. Inspection of Table 6.1 suggests that the Mukaiyama reagent gives better yields of lactone with simple ω-hydroxy acids than the Corey-Nicolaou reagent. Offer an explanation that accounts for this observation.

11. Give the major product of the following reactions:

(a)

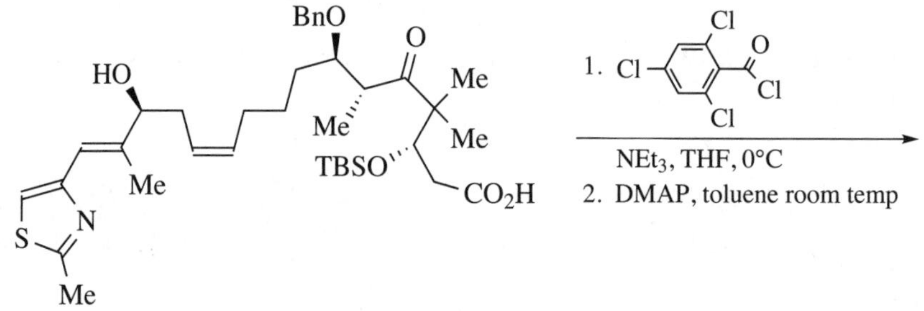

1. Cl—[2,6-dichlorobenzoyl chloride]

NEt₃, THF, 0°C
2. DMAP, toluene room temp

(b)

1. KOH, MeOH
2. 2,4,6-trichlorobenzoyl chloride
   DMAP, NEt₃

(c)

1. TPAP, NMO, CH₂Cl₂
2. NaBH₄, AcOH, THF

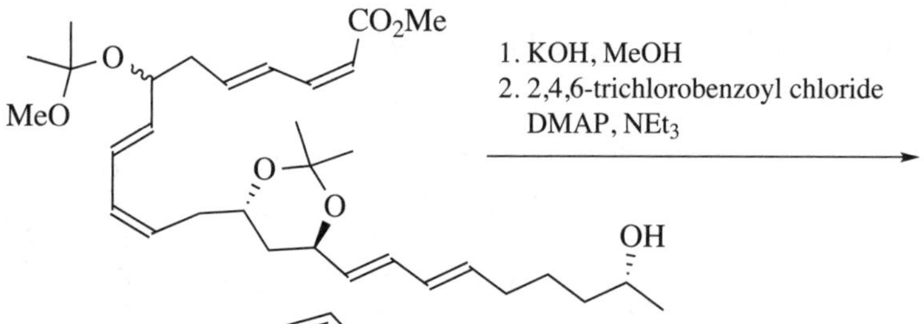

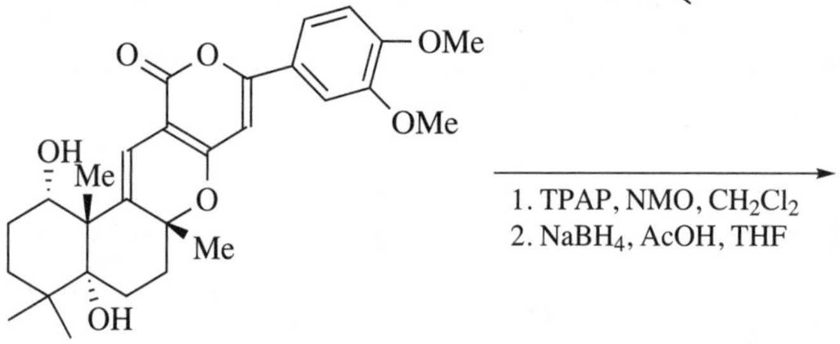

12. In each case provide a suitable synthesis. Show all intermediate products and all reagents.

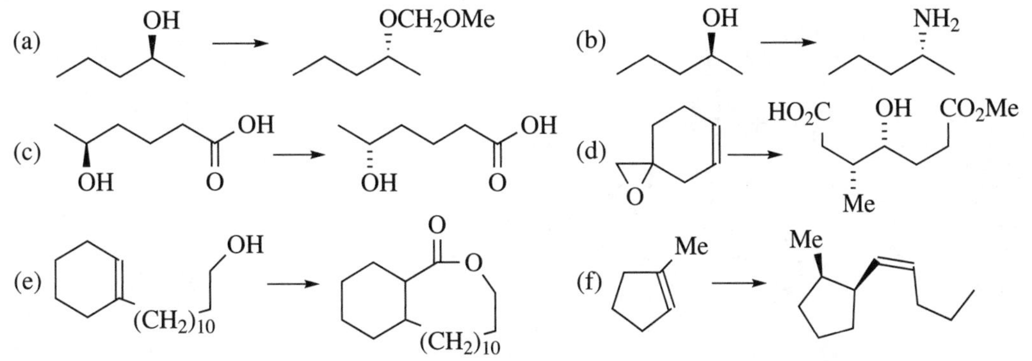

(a)

(b)

(c)

(d)

(e)

(f)

chapter 7

Protecting Groups

7.1. INTRODUCTION

Many problems arise during a synthesis. Some synthetic targets contain more than one functional group, and if they interact with each other or if one group reacts with a reagent competitively with another group, the synthesis can be in serious trouble. The presence of a ketone and an aldehyde in the same molecule at the same time is an obvious point of concern, since they react similarly with nucleophilic species such as a Grignard reagent. Aldehydes are generally more reactive, but the difference in rate may be insufficient to make the reaction chemoselective. The presence of an alcohol and an amine in the same molecule at the same time will also cause problems since reactions with acids, bases, and nucleophiles are possible for both functional groups. One practical solution to such problems is to temporarily block a reactive position by transforming it into a new functional group that will not interfere with the desired transformation. That blocking group is called a **protecting group**. Incorporating a protecting group into a synthesis requires at least two chemical reactions, however. The first reaction transforms the interfering functional group into a different one that will not compete with the desired reaction, and is termed **protection**. The second chemical step transforms the protecting group back into the original group at a later stage of the synthesis. This latter process is termed **deprotection**.

Is the use of a protecting group a desirable feature that should always be incorporated into a synthesis? *No*, it should always be avoided if possible. In many cases, however, it *cannot* be avoided and a discussion of where to anticipate the use of a protecting group in a synthesis is necessary.[1] As mentioned above, the process of protection-deprotection adds two steps to a synthesis. If there are four interfering functional groups and each requires protection, as many as eight steps may be added. Planning the order for insertion of a given functional group can usually minimize the use of protecting groups, but not always. If an aldehyde group can be inserted into a molecule after a ketone-Grignard reaction has been done, for example, there is no need to protect the aldehyde and two synthetic steps are saved. When a protection step is necessary, the first question is what transformation can block the reactivity of a given group but provide a chemical pathway back to the original group in high yield. The protecting group

---

1. For protecting group strategies in organic synthesis see Schelhaas, M.; Waldmann, H. *Angew. Chem. Int. Ed.* **1996**, *35*, 2056.

must be relatively inert and ideally react with a specific reagent to unmask it (i.e., regenerate the original group).

The most important functional groups that cause reactivity problems are alcohols, ketones and aldehydes, and amines. Carboxylic acids contain the COOH group, which can also cause problems, and this is often blocked as an ester. For alcohol groups, the acidic hydrogen of the O-H causes problems, and a protecting group must remove that hydrogen. The acidic hydrogen can be temporarily removed by conversion of the alcohol to an ether, an acetal or an ester. Ketones and aldehydes undergo nucleophilic acyl addition, and the carbonyl moiety is usually protected by conversion to an acetal or a ketal, a thioacetal or a thioketal, or a hydrazone derivative. The interfering part of an amine is the basic, lone pair of electrons. These electrons can be tied up by conversion to a quaternary ammonium salt, but such salts are not always easy to use in synthetic sequences. The lone pair is not actually removed, but electronically delocalized by conversion to an amide or a sulfonamide. In some reactions, the N-H group is acidic, and a protecting group must block the acidic hydrogen.

If a protecting group is to be used, methods must be known to protect the functional group in high yield, and the deprotection must be done with some selectivity, also in high yield. The excellent monographs by Wuts and Greene describe protection for these functional groups as well as many others in great detail.[2] Other books that focus on the nature and use of protecting groups in organic chemistry are also available.[3] The series *Compendium of Organic Synthetic Methods*[4] also provides some examples of recently used protecting groups, and there are reviews,[5] including the use of protecting groups in combinatorial synthesis (see sec. 10.12).[6]

Two essential questions remain. When does one use a protecting group and how is that group

2. (a) Greene, T.W. *Protective Groups in Organic Synthesis* Wiley, New York, *1980*; (b) Wuts, P.G.M.; Greene, T.W. *Protective Groups in Organic Synthesis 2nd ed.*, Wiley, New York, *1991*; (c) Wuts, P.G.M.; Greene, T.W. *Protective Groups in Organic Synthesis* 3rd ed., Wiley, New York, *1999*; (d) Wuts, P.G.M.; Greene, T.W. *Protective Groups in Organic Synthesis* 4th ed., Wiley, New Jersey, *2006*.

3. (a) Kocienski, P.J. *Protecting Groups,* 3rd ed., Georg Thieme Verlag; *2005*; (b) Curti, L.; Czsko, B. *Strategic Applications of Named reactions in Organic Synthesis*, Academic Press, *2005*; (c) Li, J.J. *Name Reactions: A Collection of Detailed Reaction Mechanisms,* 2nd ed., Springer, *2003*; (d) Robertson, J. *Protecting Group Chemistry*, Oxford University Press, *2000*; (e) Hanson, J.R. *Protecting Groups in Organic Synthesis*, Blackwell Pub., *1999*.

4. Smith, M.B. *Compendium of Organic Synthetic Methods, Vol. 11,* Wiley, New York, *2003*; *Vol. 10, 2002*; *Vol. 9, 2000*; *Vol. 8, 1995*; *Vol. 7, 1992*; *Vol. 6, 1988*. Also see Volumes 1-5 of this series, initiated by Harrison, I. and Harrison, S (Wiley).

5. (a) Jarowicki, K.; Kocienski, P. *J. Chem. Soc. Perkin Trans. 1 1999*, 1589; (b) Jarowicki, K.; Kocienski, P. *J. Chem. Soc. Perkin Trans. 1 2000*, 2495; (c) Sartori, G.; Ballini, R.; Bigi, F.; Bosica, G.; Maggi, R.; Righi, P. *Chem. Rev. 2004, 104,* 199.

6. (a) Singh, A.K.; Weaver, R.E.; Powers, G.L.; Rosso, V.W.; Wei, C.; Lust, D.A.; Kotnis, A.S.; Comezoglu, F.T.; Liu, M.; Bembenek, K.S.; Phan, B.D.; Vanyo, D.J.; Davies, M.L.; Mathew, R.; Palaniswamy, V.A.; Li, W.-S.; Gadamsetti, K.; Spagnuolo, C.J.; Winter, W.J. *Org. Process Res. Dev. 2003, 7,* 25; (b) Orain, D.; Ellard, J.; Bradley, M. *J. Comb. Chem. 2002, 4,* 1.

chosen? When to use a protecting group is not an easy question to answer. Wuts and Greene' books are an excellent source for choosing a protecting group, and they provide many examples. Section 7.2 will outline problems where the best solution was to use a protecting group, and Section 7.3 will give a brief review of common protecting groups. It is important to note that there are other uses for the groups labeled here as protecting groups. Such groups can be used to provide directing effects, influence solubility and other physical properties, make it easier to isolate certain products, and they can increase the crystallinity of a compound. In addition, protected functional groups are common in the pharmaceutical industry as prodrugs. They are useful for solid-phase attachment of functional groups (see sec. 10.12), and for tagging molecules (fluorescence tagging for example). Some of these uses will be touched upon in later chapters, but this chapter will focus on classical uses for protecting groups.

7.2. WHEN ARE PROTECTING GROUPS NEEDED?

There is no convenient set of rules that outline when to use protecting groups. Each synthesis is different, and the requirements vary with the reaction sequence and functional groups that are present. In this section, a series of common synthetic problems will be presented where protection-deprotection was a viable solution.

The place to begin this discussion is with a synthesis. An analysis of diol **1** shows that there are several possible disconnections, but we will focus on the CHOH-hexyl bond. Disconnection of the indicated bond leads to disconnect products **2** and **3**. Using the concept of synthetic equivalents introduced in chapter 1, the disconnect products are converted to hexylmagnesium halide (**4**, sec. 8.4.C), and a β-hydroxyaldehyde (**5**). If this synthetic route is chosen, protecting groups are necessary.

Why is it necessary to use a protecting group? Grignard reagent **4** is a base as well as a nucleophile, and attempts to react it with **5** may cause problems because the acidic proton of the hydroxyl group reacts faster than the Grignard reagent adds to the carbonyl. The alkoxide formed by reaction of the OH can react with a second equivalent of Grignard reagent, of course, but there are problems of solubility because the alkoxide is rather polar. In general, it is probably easier to block the OH to prevent the deprotonation. The O-H unit of **5** must be replaced with a functional group that can tolerate the Grignard reaction, but that group must undergo a chemical reaction to convert it back to a hydroxyl. If the hydroxyl group is converted to an ether, as in **6**, the key Grignard reaction can proceed to give **7**. In a subsequent step, the R group can be removed to regenerate the hydroxyl moiety in **1**. In this case, the hydroxyl group was protected, allowing the desired reaction to proceed.

$$6 \xrightarrow[\text{2. H}_3\text{O}^+]{\text{1. } n\text{-C}_6\text{H}_{13}-\text{MgX}} 7 \xrightarrow{\text{Lewis acid}} 1$$

Kitching and co-worker's synthesis of a gland secretion of fruit flies provides a synthetic solution for the preparation of **1**, in which 1,3-propanediol was converted to the tetrahydropyranyl (THP) derivative **8** by reaction with dihydropyran.[7] The identical nature of the two hydroxyls made protection of one very easy, by simply controlling the stoichiometry of the reagent. Oxidation of the unprotected alcohol gave aldehyde **9**, and subsequent addition of hexylmagnesium bromide gave **10**. Deprotection with acidic methanol gave the product, 1,3-nonanediol (**1**). This particular sequence was used to prepare an authentic sample of a synthetic intermediate for purposes of comparison, and no yields were reported. When a disubstituted molecule is symmetrically functionalized (as with 1,3-propanediol), the two groups are chemically equivalent and protection of one of them is rather easy. The reagents used in the remainder of the synthesis will determine the choice of protecting group.

$$\text{diol} \xrightarrow{\text{H}^+} 8 \xrightarrow[\text{NaOAc}]{\text{PCC}} 9 \xrightarrow[\text{2. H}_3\text{O}^+]{\text{1. } n\text{-C}_6\text{H}_{13}\text{MgBr}} 10 \xrightarrow[\text{H}^+]{\text{MeOH}} 1$$

In many syntheses, protection must be used or the sequence will not work, but what are the consequences of using protecting groups in synthesis? One price to be paid is at least two synthetic steps (insertion and removal of the protecting group) that must be added to the sequence, and the overall yield of the synthesis will suffer accordingly. The ability to generate a target directly must be compared to the loss of material that accompanies the use of protecting groups, and each route evaluated in the context of the specific synthesis. Usually, protecting groups are used because there is no direct way to solve the problem without them.

In complex synthetic targets, it is common for two or more identical functional groups to be present. It maybe difficult to incorporate all of those groups simultaneously, and there is a chance that once two groups are incorporated, one will interfere with reactions aimed at another. A synthesis of sigialomycin D by Danishefsky and Geng provides an example where several alcohol moieties are incorporated in a molecule.[8] A description of the specific protecting groups presented in this example will be presented later in this chapter, but an analysis of the strategy will illustrate how their use led to the desired target. The synthesis begins with D-2-deoxyribose **11**. Protection of the diol unit allowed ring-opening to a hydroxy-aldehyde, protection of the primary alcohol as the pivaloyl ester, Wittig reaction to extend the carbon chain and hydroboration gave alcohol **12**. Protection of the diol unit in **11** was essential,

7. Kitching, W.; Lewis, J.A.; Perkins, M.V.; Drew, R.; Moore, C.J.; Schuring, V. König, W.A.; Francke, W. *J. Org. Chem.* **1989**, *54*, 3893.
8. Geng, X.; Dainishefsky, S.J. *Org. Lett.* **2004**, *6*, 413.

because **12** contains four OH units and three are protected so that only one will participate in the next sequence. A series of five reactions converted **12** to **13**, and the OH unit was protected as a silyl ether, allowing selective deprotection of the pivaloyl ester to give **14**. In three steps, **14** was converted to alkyne-ester **14**. Again, careful choice of protecting groups and the order in which groups are incorporated allowed first one side of the molecule to be elaborated, and then the other. Ester **15** was converted to lactone **16** in three steps, and another three steps were required to incorporate the aromatic ring system (see **17**) that contained two protected phenolic units. The choice of the silyl ether protecting group allowed that alcohol to be unmasked in **17**, and Martin's sulfurane conditions[9] converted the benzylic alcohol in **17** to the C=C unit in **18**. Finally, acid hydrolysis converted the dioxolane group in **18** to the diol, a structural component of sigialomycin D (**19**). This sequence illustrates that a sugar derivative can be manipulated in such a way that two of the hydroxyl groups are carried throughout the synthesis, in protected form, to the final product **19**. Other alcohol units are incorporated and used along the way, and each is protected with a group that allows that OH group to be deprotected and used a the proper

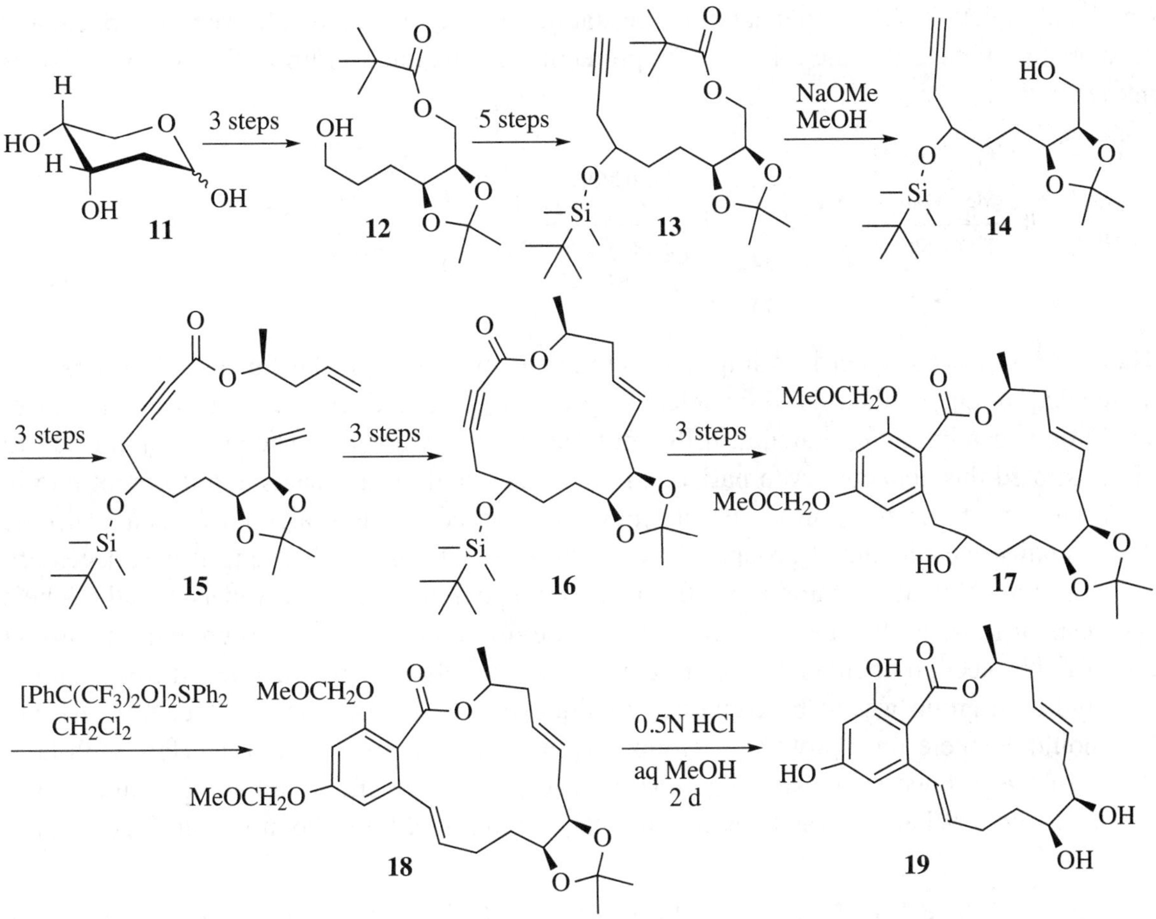

---

9. Martin, J.C.; Arhart, R.J. *J. Am. Chem. Soc.* **1971**, *93*, 4327.

time. Despite the need for protection and deprotection, this sequence allowed Danishefsky to incorporate specific functional groups in a controlled manner and in the desired sequence, one of the positive aspects of using protecting groups.

If protecting groups must be used in a synthesis, the goal is to be efficient and keep the number of steps to a minimum. It is often necessary to protect a group, deprotect it, and then later protect it a second time. If these steps could somehow be combined, it could save on the total number of steps in the synthesis and make the overall process more efficient. An example that illustrates this type of efficiency is the conversion of **20** to **23** in Silverstein's synthesis of α-cubebane.[10] In this case, the starting material is **20**, which contains two primary hydroxyl groups that are first protected as a ketal in **21**. This choice allows manipulation of **21** to give **22** where a third hydroxyl group has been incorporated. The next step in the sequence required manipulation of the –CH$_2$OH unit that is adjacent to the methyl group. Rather than deprotect **22** and then re-protect the resulting triol, **22** was treated with acid and that induced deprotection of the seven-membered ring ketal and formation of the more stable five-membered ring ketal moiety in **23**. Understanding the stability of the two ketals allowed Silverstein and co-workers to maximize the utility of the protecting group, and minimize the number of steps in the synthesis.

There are occasions when it is impossible to avoid placing two like functional groups in a molecule, yet one of them is to be selectively transformed and the other must be protected. The groups can be incorporated at different times in the synthesis. The conversion of **11** to **19** illustrated this concept. Alternatively, *if* one functional group reacts at a different rate in the protecting group reaction selectivity may be achieved. Amino alcohol **24** is a molecule that contains two functional groups that can interfere with each other, and it was necessary to incorporate both at the same time (from an amino acid precursor) in Andersen and Globe's synthesis of kessanol.[11] The functional groups are different, an amino group and a hydroxyl group and protection of either the OH or the NH$_2$ selectively requires chemical differentiation. The hydroxyl group was to be converted to aldehyde **26**. Apart from the fact that OH and NH$_2$ could compete for a protecting group reagent, the amine unit would interfere with later stages of the planned synthesis. The goal was to protect the amine, allowing manipulation of the alcohol. When **24** was treated with acetic anhydride, both the amine and the alcohol

10. Pearce, G.T.; Gore, W.E.; Silverstein, R.M.; Peacock, J.W.; Cuthbert, R.A.; Lanier, G.N.; Simeone, J.B. *J. Chem. Ecol.* **1975**, *1*, 115.
11. Andersen, N.H.; Golec, Jr, F.A. *Tetrahedron Lett.* **1977**, 3783.

formed their respective acetates in **25**. Knowledge that an ester is more easily hydrolyzed than the amide, however, allowed both groups to be protected and the OH unit to be selectively deprotected. Treatment of **25** with aqueous KOH gave the alcohol, which was oxidized to the aldehyde in **26** by **Collins oxidation** (sec. 3.2.B).

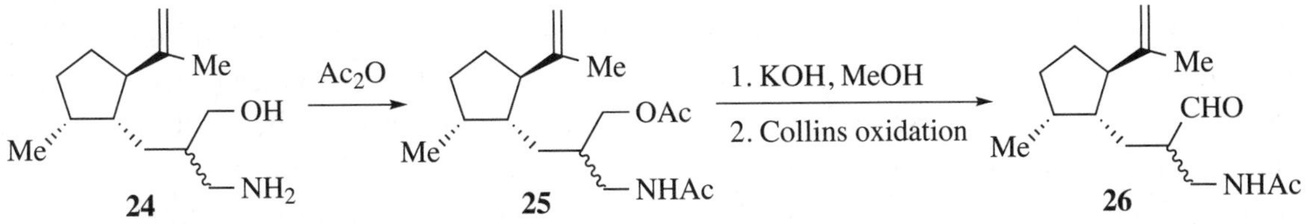

A final example illustrates planning a synthesis in such a way that groups that may interfere with each other are protected. The synthesis of **30** from **27**, taken from the Chiang et. al. synthesis of inhibitors of cholesterol biosynthesis,[12] involves conversion of the $Me_2C=C$ group

to an aldehyde that can undergo an olefination reaction. Clearly, the presence of the ketone moiety in **27** would interfere with this sequence. For this reason, a precursor to the aldehyde group in **29** was incorporated early in the sequence, in the form of an alkene because the aldehyde could be unmasked later by oxidative cleavage. Before this step, however, the ketone unit was

protected as the ketal (**28**) so that ozonolysis of the alkene would give the aldehyde unit in **29**. **Horner-Wadsworth-Emmons olefination** (sec. 8.8.A.iii) of this aldehyde gave **30**. Knowing that aldehyde and ketone units must be present in the same molecule allowed them to be synthetically incorporated at different times, and in a manner where one would not interfere with the chemistry of the other.

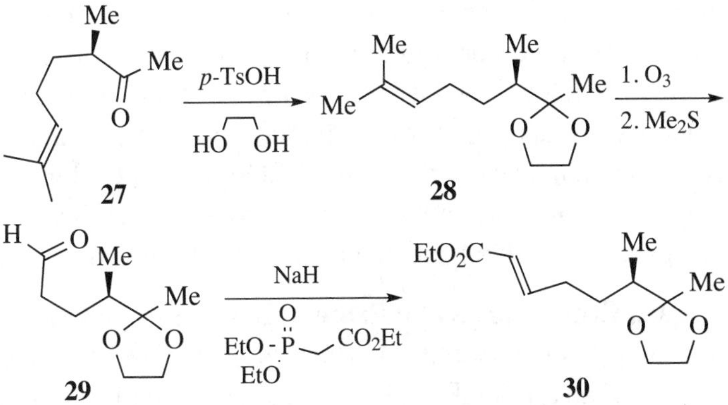

7.3. PROTECTING GROUPS FOR ALCOHOLS, CARBONYLS, AND AMINES

This section will introduce the actual functional group modifications used for protecting groups. The monographs by Wuts and Greene[2] describe protecting groups for many different functional groups. This section is not exhaustive, but rather representative. Most of the protecting groups used in synthesis involve alcohols, ketones, aldehydes or amines, and only common protecting groups will be examined. The intent is to introduce protecting groups, and to provide a few simple tools that may solve a synthetic problem.

---

12. Chiang, Y.-C.P.; Yang, S.S.; Heck, J.V.; Chabala, J.C.; Chang, M.N. *J. Org. Chem.* **1989**, *54*, 5708.

7.3.A. Protection of Alcohols

Before a protecting group can be developed for a given functional group, the interfering part of that group must be identified. In the case of alcohols, this is almost always the acidic hydrogen of the O–H unit. If an alcohol moiety interferes with a planned reaction, the OH must be converted to O–X (another functional group), but this reaction must proceed in high yield and under mild conditions. The new functional group (now called the **protecting group**) must be impervious to all reagents used in succeeding synthetic steps but must be easily removed in high yield, under conditions that do not disturb other functionality. All of this chemistry must be done within the limitations of reactivity imposed by the given functional group.

Alcohols give two major reactions that largely fit these criteria. Conversion of alcohols to ethers usually proceeds in high yield to give a relatively inert product. Due to this lack of reactivity, the reaction conditions required for cleavage of the ether may be very harsh. Specialized ether protecting groups have therefore been developed that can be removed under mild and selective conditions. Another solution is to convert alcohols to acyclic acetals or ketals. These groups possess much of the inert character of ethers, but are easily converted back to the alcohol. It is also possible to convert alcohols to esters under mild conditions, and hydrolysis easily converts them back to the alcohol. There is a problem if an ester group is used, however, since many reagents react with its acyl carbon. A few examples of ester protecting groups will be discussed, but ethers and acetals constitute the vast majority of alcohol protecting groups.

7.3.A.i. Ether and Acetal Protecting Groups. A simple way to protect the OH of an alcohol is by conversion to its **methyl ether (–OMe)**.[13] Reaction of an alcohol with a base [often sodium hydride (NaH) in THF or DMF] gives an alkoxide. Subsequent reaction with iodomethane gives the methyl ether via an S_N2 reaction (sec. 2.6.A.i) called the **Williamson ether synthesis**.[14] Methyl ethers can also be prepared by treatment of an alcohol with trimethyloxonium tetrafluoroborate (**Meerwein's reagent**),[15] or with dimethyl sulfate (Me_2SO_4),[16] a base and a phase-transfer agent (such as tetrabutylammonium iodide).[17] Methyl ethers are stable to strong bases, nucleophiles, organometallics, ylids, hydrogenation [except for benzylic ethers (see below)], oxidizing agents, and hydride reducing agents. They are stable in the pH range 1→14. Methyl ethers can be cleaved with concentrated HI, but trimethylsilyl iodide unmasks the alcohol in chloroform at ambient temperatures.[18] Lewis acids such as boron tribromide

13. Reference 2a, pp 14-16, 296-298; Reference 2b, pp 15-17; Reference 2c, pp 23-27; Reference 2d, pp 25-30.
14. (a) Williamson, A.W. *J. Chem. Soc.* **1851**, *4*, 229; (b) Dermer, O.C. *Chem. Rev.* **1934**, *14*, 409.
15. (a) Meerwein, H.; Hinz, G.; Hofmann, P.; Kroning, E.; Pfeil, E. *J. Prakt. Chem.* **1937**, *147*, 257; (b) Earle, M.J.; Fairhurst, R.A.; Giles, R.G.; Heaney, H. *Synlett* **1991**, 728.
16. (a) Merz A. *Angew. Chem. Int. Ed.* **1973**, *12*, 846; (b) Reuvers, J.T.A.; de Groot, A. *J. Org. Chem.* **1986**, *51*, 4594.
17. Merz, A. *Angew. Chem. Int. Ed.* **1973**, *12*, 846.
18. Jung, M.E.; Lyster, M.A. *J. Org. Chem.* **1977**, *42*, 3761.

(BBr_3)[19] also cleave methyl ethers to alcohols. The use of this group is illustrated in Williams and co-worker's synthesis of paraherquamide A,[20] in which vanillin (**31**) was used as a starting material and has the OMe unit present. After six synthetic steps, lactam **32** was treated with boron tribromide to remove the methyl protecting group, giving a 99% yield of **33**.

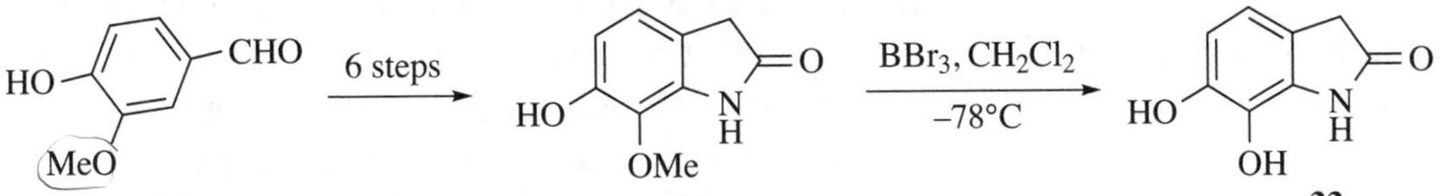

Another common protecting group is the benzyl ether ($-OCH_2Ph$, **OBn**),[21] prepared by reaction of the nucleophilic alkoxide of an alcohol with benzyl bromide or chloride.[22] Benzyl ethers are stable to a wide range of reagents including pH 14, carbanions, and organometallics, nucleophiles, hydrides and some oxidizing agents. The two best methods for removing the benzyl group involve cleavage of the $O–CH_2Ph$ bond by hydrogenolysis (secs. 4.8.E, 4.9.D). The most common method is probably catalytic hydrogenation with a palladium catalyst,[23] but hydrogenolysis with sodium or potassium in ammonia (sec. 4.9.D) is also used.[24] Wood and co-workers used the *O*-benzyl group in a synthesis of kalihinol C,[25] in which the free alcohol in **34** was converted to benzyl ether (**35**) in 93% yield by treatment with sodium hydride (KH) and benzyl bromide. Seven synthetic steps were required to give **36**, which contained the original *O*-benzyl group. Treatment with sodium and ammonia led to deprotection, and formation of **37** in 91% yield.

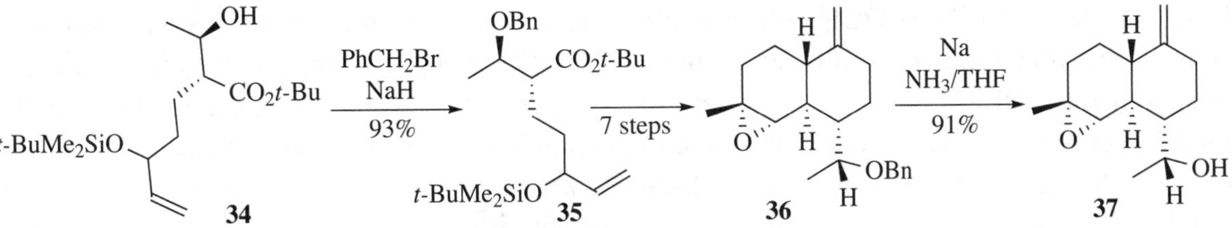

19. (a) Niwa, H.; Hida, T.; Yamada, K. *Tetrahedron Lett.* **1981**, *22*, 4239; (b) Kuehne, M.E.; Pitner, J.B. *J. Org. Chem.* **1989**, *54*, 4553; (c) Grieco,P.A.; Nishizawa, M.; Oguri, T.; Burke, S.D.; Marinovic, N. *J. Am. Chem. Soc.* **1977**, *99*, 5773.
20. Williams, R.M.; Cao, J.; Tsujishima, H.; Cox, R.J. *J. Am. Chem. Soc.* **2003**, *125*, 12172.
21. Reference 2a, pp 29-31, 296-298; Reference 2b, pp 47-53; Reference 2c, pp 76-84; Reference 2d, pp 102-106.
22. (a) Dueno, E.E.; Chu, F.; Kim, S.-I.; Jung, K.W. *Tetrahedron Lett.* **1999**, *40*, 1843; (b) Iwashige, T.; Saeki, H. *Chem. Pharm. Bull.,* **1967**, *15*, 1803.
23. (a) Heathcock, C.H.; Ratcliffe, R. *J. Am. Chem. Soc.* **1971**, *93*, 1746; (b) Hartung, W.H.; Simonoff, C. *Org. React.* **1953**, *7*, 263.
24. (a) McClosky, C.M. *Adv. Carbohydr. Chem.* **1957**, *12*, 137; (b) Reist, E.J.; Bartuska, V.J.; Goodman, L. *J. Org. Chem.* **1964**, *29*, 3725; (c) Schön, I. *Chem. Rev.* **1984**, *84*, 287; (d) Zakarian, A.; Batch, A.; Holton, R.A. *J. Am. Chem Soc.* **2003**, *125*, 7822.
25. White, R.D.; Keaney, G.F.; Slown, C.D.; Wood, J.L. *Org. Lett.* **2004**, *6*, 1123.

Alcohols can be converted to a *tert*-butyl ether[26] [O–C(CH$_3$)$_3$, **-O-*t*-Bu**] by reaction of the alcohol with 2-methyl-2-propene (isobutylene) and an acid catalyst (sulfuric acid or BF$_3$).[27] This procedure takes advantage of the facile formation and relative stability of a tertiary cation, which allows subsequent reaction with the alcohol in a S$_N$1 type reaction (sec. 2.6.B) to give the ether. A limitation on this reaction is that other functionality in the molecule must be compatible with acidic conditions. As a protecting group, *tert*-butyl is stable to pH 1→14, as well as to nucleophiles, organometallics, hydrides, catalytic hydrogenation, oxidations and dissolving metal reductions. The group can be removed with strong aqueous acid (< pH 1), as well as with Lewis acids. Two common methods for removal of the *tert*-butyl group are treatment with anhydrous trifluoroacetic acid (CF$_3$COOH, TFA)[27a] or with trimethylsilyl iodide.[18]

Commonly, incorporation or removal of methyl, benzyl, *tert*-butyl and other simple ether protecting groups is incompatible with the presence of sensitive functional groups elsewhere in the molecule. In the case of alcohols, a molecule may contain several hydroxyl moieties that require protection and subsequent deprotection in a specific order. Several ether-like protecting groups have been developed that show greater selectivity in reactions with acids or Lewis acids, allowing more latitude in the choice of a method to remove them. Wuts and Greene classified these alternatives as substituted methyl ethers, but their RO–CHR^1–O–R^2 structure shows them to be acetals. There are many possible derivatives, but only a few are used with great frequency.

One is methoxymethyl ether (O-CH$_2$OMe **OMOM**),[28] formed by reaction of the alcohol with a base, and then with ClCH$_2$OCH$_3$ (chloromethyl methyl ether).[29] The MOM group is somewhat sensitive to acid, but is generally stable in the pH 4-12 range. It is important to note that HCl in an aqueous THF solution (pH ≈ 1) does not cleave the BnO group, whereas it does cleave the OMOM group. The MOM group is stable to nucleophiles, organometallic reagents, hydrides, catalytic hydrogenation (except with platinum catalysts in acidic media) and to oxidation, but reacts with acidic media to give the alcohol. Sensitivity to acid is the key to removing this group, and both 50% aqueous acetic acid with a sulfuric acid catalyst[30] and HCl in MeOH[31] are commonly used for deprotection. In a synthesis of spiculoic acid A,[32] Mehta and Kundu prepared alcohol **38**, and subsequent reaction with MOM chloride gave **39** in 85% yield. Eleven synthetic steps converted **39** to **40**, which incorporateed the other alcohol protecting groups *O*-benzyl (see above) and *O*-diphenyl-*tert*-butylsilyl (see sect. 7.3.A.ii). Treatment

26. Reference 2a, pp 26-27, 296-298; Reference 2b, pp 41-42; Reference 2c, pp 65-67; Reference 2d, pp 82-84.
27. (a) Beyerman, H.C.; Bontekoe, J.S. *Proc. Chem. Soc.* **1961**, 249; (b) Beyerman, H.C.; Heiszwolf, G.J. *J. Chem. Soc.* **1963**, 755.
28. Reference 2a, pp 16-17, 296-298; Reference 2b, pp 17-21; Reference 2c, pp 27-33; Reference 2d, pp 30-38.
29. Kluge, A.F.; Untch, K.G.; Fried, J.H. *J. Am. Chem. Soc.* **1972**, *94*, 7827.
30. Laforge, F.B. *J. Am. Chem. Soc.* **1933**, *55*, 3040.
31. Auerbach, J.; Weinreb, S.M. *J. Chem. Soc. Chem. Commun.* **1974**, 298.
32. Mehta, G.; Kundu, U.K. *Org. Lett.* **2005**, *7*, 5569.

with BF$_3$•etherate in the presence of 1,3-dithiane removed the MOM group to give alcohol **41** in 84% yield.

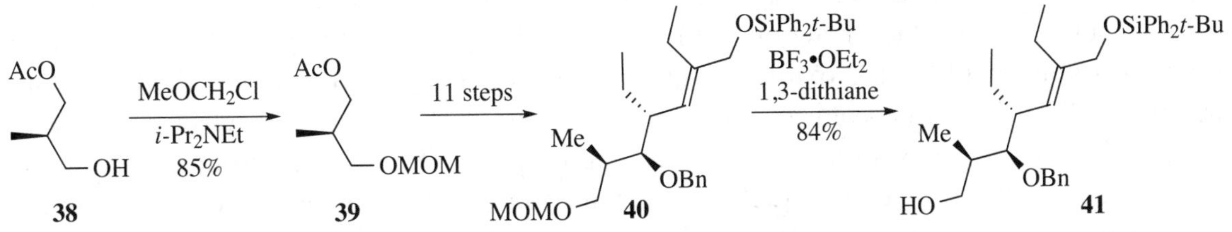

A variation on the MOM protecting group replaces the oxygen with sulfur to generate the methylthio-methyl ether (O–CH$_2$SMe, **OMTM**).[33] The target alcohol is converted to the alkoxide with sodium hydride or another suitable base, and then treated with chloromethylmethyl sulfide (ClCH$_2$SMe).[34a] An alternative method uses DMSO and acetic anhydride in acetic acid.[34b] This thioether protecting group is stable to pH 1-12 but it is sensitive to very strong aqueous acid and to Lewis acids (especially mercuric salts). It reacts with many hydrogenation catalysts (bivalent sulfur is a poison, sec. 4.8.A) and can be oxidized to a sulfoxide with many oxidizing agents (sec. 3.9.A). The group is stable to reaction with nucleophiles, organometallics or hydrides. It can be removed by treatment with Lewis acids, and mercury salts are especially effective.[34] This group can also be removed by treatment with iodomethane in mildly basic solution in some cases.[33]

The MTM group was used to protect several of the alcohol groups in Corey and co-worker's synthesis of erythronolide A.[35] Alcohol **42** contains a sensitive OH group and was converted to the MTM derivative (**43**). This allowed the construction of the *tris*(MTM) lactone **44**. Treatment with potassium carbonate (K$_2$CO$_3$) and CH$_3$I (aq. acetone, 40°C, 15 h) gave the triol (**45**) in 80% yield, which was converted to erythronolide A.

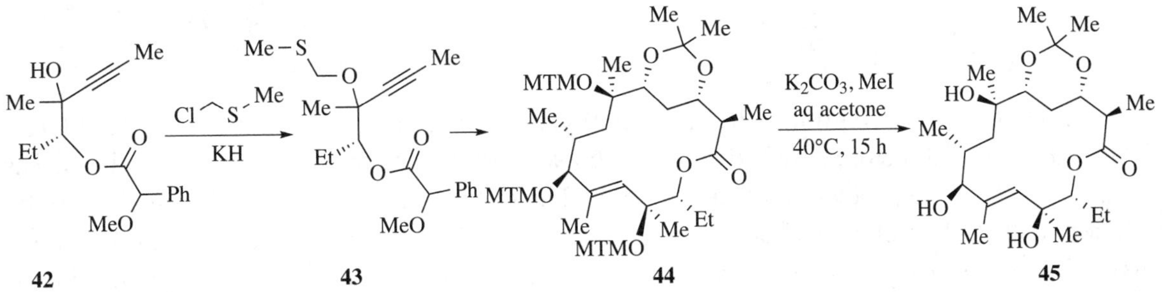

33. Reference 2a, pp 17-18, 296-298; Reference 2b, pp 21; Reference 2c, pp 33-35; Reference 2d, pp 38-41.
34. (a) Corey, E.J.; Bock, M.G. *Tetrahedron Lett.* **1975**, 3269; (b) Pojer, P.M.; Angyal, S.J. *Aust. J. Chem.* **1978**, *31*, 1031.
35. (a) Corey, E.J.; Trybulski, E.J.; Melvin, Jr., L.S.; Nicolaou, K.C.; Secrist, J.A.; Lett, R.; Sheldrake, P.W.; Falck, J.R.; Brunelle, D.J.; Haslanger, M.F.; Kim, S.F.; Yoo, S. *J. Am. Chem. Soc.* **1978**, *100*, 4618, 4620; (b) Corey, E.J.; Hopkins, P.B.; Kim, S.; Yoo, Y.; Nambiar, K.P.; Falck, J.R. *Ibid.,* **1979**, *101*, 7131.

Another common acetal protecting group is the 2-methoxyethoxymethyl ether (–O–CH$_2$OCH$_2$CH$_2$OMe, **OMEM**).[36] The MEM group is attached by reaction of the alcohol with NaH in THF or DME (although diisopropylethylamine [**Hünig's base**] can be used in dichloromethane), and then reaction with methoxyethoxy chloromethyl ether.[37] The MEM group is stable to pH 1-12, oxidizing agents, nucleophiles, hydrogenation, hydrides, and organometallics, but is cleaved in strong aqueous acid (pH < 1) and is sensitive to Lewis acids. This latter observation is the key to deprotection. Either zinc bromide (ZnBr$_2$) or titanium tetrachloride (TiCl$_4$) in dichloromethane can be used for deprotection,[37] although other deprotection methods are known. In a synthesis of ristocetin aglycone, Boger and co-workers converted alcohol **46** to the OMEM derivative **47** in 99% yield.[38] After 14 steps, **48** was obtained and treatment with *B*-bromocatecholborane gave **49** in > 90% yield.

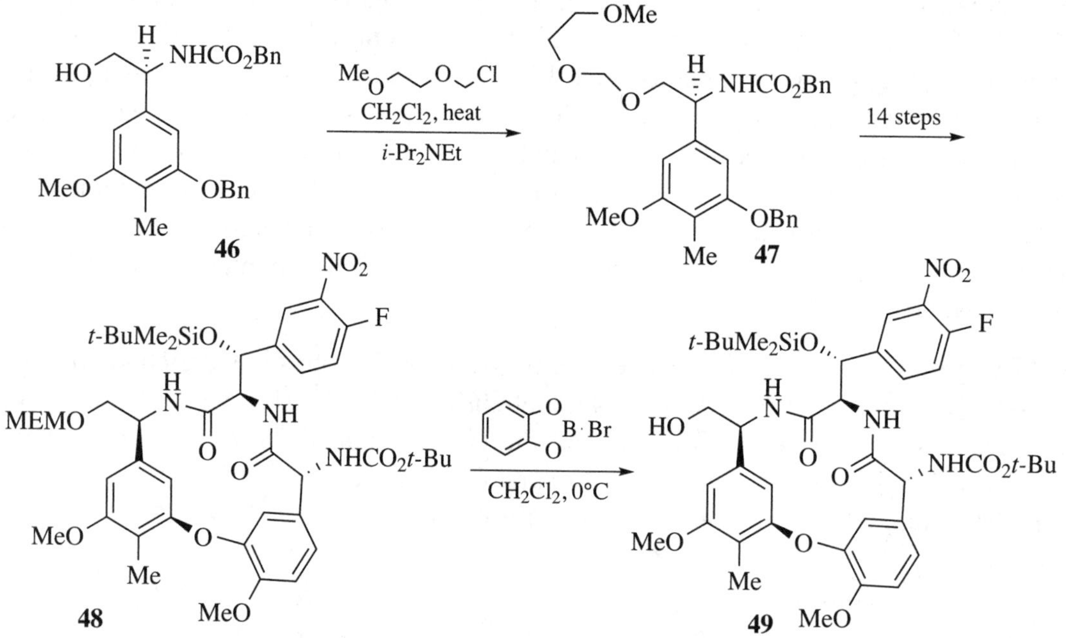

A variation of the OMEM group is the 2-(trimethylsilyl)ethoxymethyl ether (O–CH$_2$OCH$_2$CH$_2$–SiMe$_3$, **OSEM**),[39] developed by Lipshutz and co-workers.[40] The trimethylsilyl group is used as a trigger for facile deprotection of the alcohol. The OSEM group is attached to an alcohol by reaction with 2-(trimethylsilyl)ethoxymethyl chloride (Me$_3$SiCH$_2$CH$_2$OCH$_2$Cl) and Hünig's base. The SEM group is stable in a pH range of 4-12. Some nucleophiles and many Lewis acids react with this group, but it is generally stable to hydrides and to many organometallics

36. Reference 2a, pp 19, 296-298; Reference 2b, pp 27-29; Reference 2c, pp 41-44; Reference 2d, pp 49-53.
37. Corey, E.J.; Gras, J.-L.; Ulrich, P. *Tetrahedron Lett.* **1976**, 809.
38. Crowley, B.M.; Mori, Y.; McComas, C.C.; Tang, D.; Boger, D.L. *J. Am. Chem. Soc.* **2004**, *126*, 4310.
39. Reference 2a, pp 20-21, 296-298; Reference 2b, pp 30-31; Reference 2c, pp 45-48; Reference 2d, pp 54-58.
40. (a) Lipshutz, B.H.; Pegram, J.J. *Tetrahedron Lett.* **1980**, *21*, 3343; (b) Lipshutz, B.H.; Moretti, R.; Crow, R. *Ibid.*, **1989**, *30*, 15; (c) Lipshutz, B.H.; Miller, T.A. *Ibid.*, **1989**, *30*, 7149.

(such as butyllithium), as well as most oxidizing agents. The presence of silicon makes the group sensitive to fluoride [as with silyl protecting groups, see below], and treatment with tetrabutylammonium fluoride (n-Bu$_4$N$^+$F$^-$, TBAF) usually regenerates the alcohol, but not always. Indeed, the OSEM group is difficult to remove in some cases. In the Keaton and Phillip's synthesis of (–)-7-demethylpiericidin A$_1$,[41] the pyridyl OH unit in **50** was protected as the SEM derivative (**51**) in 97% yield using silver carbonate to effect the coupling. This fragment was modified and then coupled to another fragment in a convergent synthesis to give **52**. Final treatment with tetrabutylammonium fluoride removed the SEM group as well as the cyclic silyloxy unit to give **53** in 70% yield.

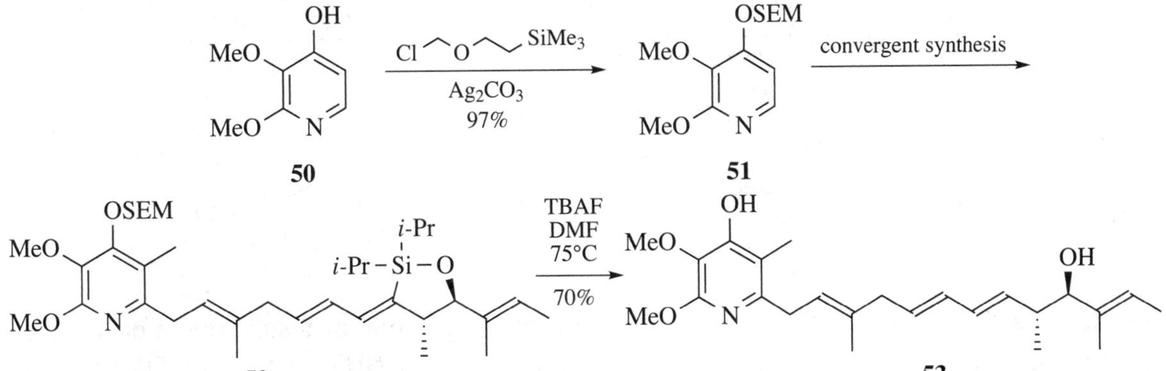

Tetrahydropyranyl ethers (OTHP, see **55**)[42] are acetal protecting groups that can be found in many older organic syntheses. An alcohol reacts with dihydropyran (**54**) in the presence of an acid, such as p-toluenesulfonic acid, to give the THP derivative (**55**) via an oxygen stabilized cation.[43] One problem with this protecting group is the fact that a new stereogenic center is created when **55** is formed. If ROH is chiral, **55** will exist as diastereomers, which may make identification, isolation, and purification difficult if this protecting group is carried through many synthetic steps. This group is stable to base (pH 6-12) but unstable to aqueous acid and to Lewis acids.

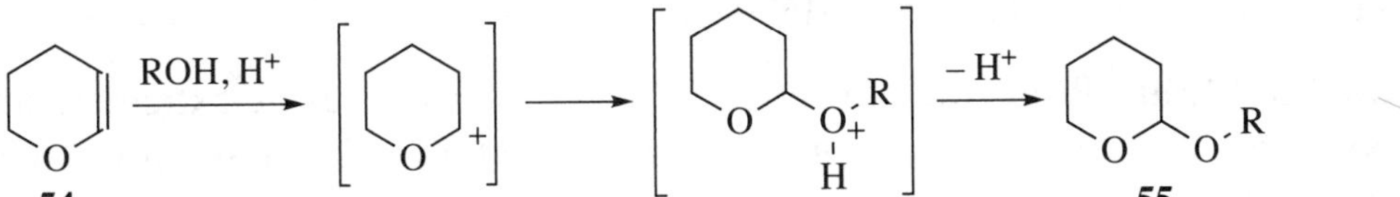

It is relatively stable to nucleophiles and organometallics. It is also stable to hydrogenation, hydrides, and oxidizing agents.[43] It is removed with aqueous acid or methanolic tosic acid.[44] Greene reported[43] that explosions have occurred when tetrahydropyranyl ethers are treated

41. Keaton, K.A.; Phillips, A.J. *J. Am. Chem. Soc.* ***2006**, 128,* 408.
42. Reference 2a, pp 21-22, 296-298; Reference 2b, pp 31-34; Reference 2c, pp 49-52; Reference 2d, pp 59-68.
43. This report is taken from (a) Bernady, K.F.; Floyd, M.B.; Poletto, J.F.; Weiss, M.J. *J. Org. Chem.* ***1979**, 44,* 1438;(b) Miyashita, M.; Yoshikoshi, A.; Grieco, P.A. *Ibid.* ***1977**, 42,* 3772.
44. Corey, E.J.; Niwa, H.; Knolle, J. *J. Am. Chem. Soc.* ***1978**, 100,* 1942.

with diborane:basic hydrogen peroxide[45] or with 40% peroxyacetic acid.[44] An example involving this protecting group is taken from von Zezschwitz and co-worker's synthesis of okaspirodiol.[46] The initial synthetic steps that incorporated the OTHP unit were taken from earlier work of White and co-workers during their degradation and absolute configurational assignment to C34-botryococcene.[47] Alkynyl alcohol **56** was protected as the THP derivative **57** in 90% yield. The synthesis of **58** required only two steps, but converged with a fragment that contained a second THP group. Final treatment with methanolic HCl liberated the two hydroxyl groups in **59** (>80% yield), which was used to complete the synthesis.

Greene classified another generic type of acetal protecting group as a substituted ethyl ether. A common member of this group is 1-ethoxyethyl ether [O–CH(OCH$_2$Me)Me, **OEE**].[48] As with the OTHP group, diastereomers can be generated when the OEE group is incorporated in a molecule with the same problems. The protected derivative will exist as diastereomers, since there are two stereogenic centers. The group is usually attached to the alcohol via reaction with ethyl vinyl ether under acid conditions (HCl or tosic acid).[49] The group is stable to a relatively narrow pH range of 6-12, but it is compatible with reactions of nucleophiles, organometallics, hydrogenation, hydrides, and most oxidants. It is sensitive to acid conditions and to Lewis acids. The most common methods for cleavage to the alcohol are treatment of the OEE ether with aqueous acetic acid,[49a] or with aqueous HCl in THF.[49b] Fukuda and Okamoto used this protecting group in a synthesis of AM6898A.[50]

7.3.A.ii. Silyl Ether Protecting Groups. Silyl ethers have become extremely important for the protection of alcohols with the generic structure OSiR$_3$. Variation in the R group leads to significant differences in the stability of the protecting group. In general, silyl ethers can be cleaved with aqueous base or acid, but the rate of hydrolysis for a secondary silyl ether

45. See Meyers, A.I.; Schwartzman, S.; Olson, G.L.; Cheung, H.-C. *Tetrahedron Lett.* **1976**, 2417.
46. Bender, T.; Schuhmann, T.; Magull, J.; Grond, S.; von Zezschwitz, P. *J. Org. Chem.* **2006**, *71*, 7125.
47. White, J.D.; Somers, T.C. G. Reddy, N. *J. Org. Chem.* **1992**, *57*, 4991.
48. Reference 2a, pp 25, 296-298; Reference 2b, pp 414-415; Reference 2c, pp 709-711; Reference 2d, pp 74-75.
49. (a) Chládek, S.; Smrt, J. *Chem. Ind (London)* **1964**, 1719; (b) Meyers, A.I.; Comins, D.L.; Roland, D.M.; Henning, R.; Shimizu, K. *J. Am. Chem. Soc.* **1979**, *101*, 7104.
50. Fukuda, Y.; Okamoto, Y. *Tetrahedron* **2002**, *58*, 2513.

is significantly slower than that of a primary silyl ether.[51] In addition, alkyl alcohols are more reactive than phenolic derivatives, and the order of reactivity for other silane-protected functional groups is $COOSiR_3 > NHSiR_3 > CONHSiR_3 > SSiR_3$.[52] Steric factors play a major role in the relative stability of the $OSiR_3$ group in cleavage reactions. The prototype of this class of protecting groups is the trimethylsilyl ether [$O–Si(Me)_3$, **OTMS**].[53] The group is usually attached to the alcohol by reaction with chlorotrimethylsilane (Me_3SiCl) in the presence of an amine base such as triethylamine or pyridine,[54] although Wuts and Greene list several other methods. This group is sensitive to many aqueous conditions, and the TMS group is sometimes cleaved in an aqueous workup procedure.[55] The group is not very stable to organometallics such as Grignard reagents, or to nucleophiles, hydrogenation or hydrides, but it can be used with some oxidants. The sensitivity of OTMS to these reagents is often the result of the aqueous or acid workup procedures that are required for a given reaction. If strictly anhydrous conditions are used, the TMS group can be used for short-term protection. Under aprotic conditions, fluoride ion (such as tetrabutylammonium fluoride) readily attacks silicon and cleaves the O-Si bond.[51a] A synthetic example is the protection of the alcohol moiety in **60**, using trimethylsilyl triflate and 2,6-lutidine to give **61** in 86% yield, in Hiersemann and co-worker's synthesis of (–)-15-acetyl-3-propionyl-17-norcharaciol.[56] This group was sufficiently stable to allow the next three steps to give fragment **62** that also contained *O-tert-*butyldimethylsilyl (OTBDMS, see below) and *O*-triethylsilyl (OTES, see below) protecting groups. Treatment with tetrabutylammonium fluoride removed both the OTMS group and the OTES group to give **63**.

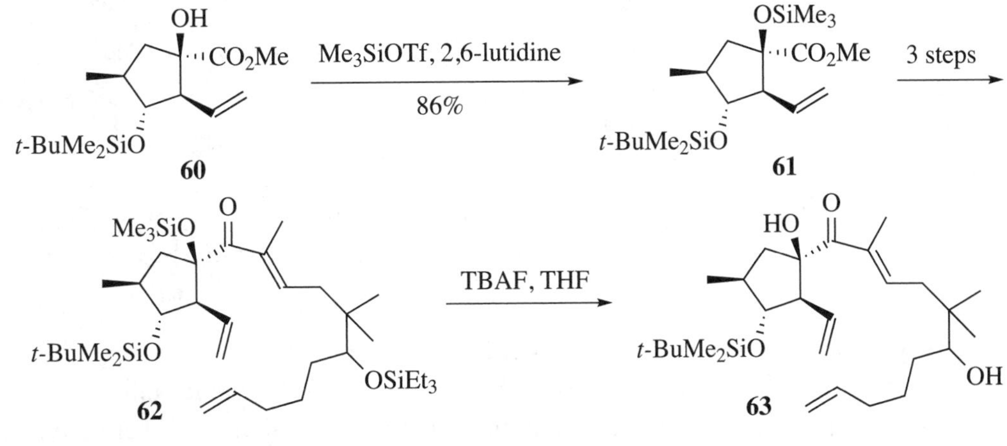

51. (a) Reference 2a, pp 39; (b) McInnes, A.G. *Can. J. Chem.* **1965**, *43,* 1998.
52. (a) Reference 51a; (b) Cooper, B.E. *Chem. Ind. (London)* **1978**, 794.
53. Reference 2a, pp 40-42; Reference 2b, pp 68-73; Reference 2c, pp 116-121; Reference 2d, pp 171-178.
54. (a) Corey, E.J.; Snider, B.B. *J. Am. Chem. Soc.* **1972**, *94*, 2549; (b) Sweely, C.C.; Bentley, R.; Makita, M.; Wells, W.W. *Ibid.,* **1963***, 85,* 2497.
55. Reference 2a, pp 296-298; Reference 2b, pp 414-415; Reference 2c, pp 709-711; Reference 2d, pp 171-178.
56. Helmboldt, H.; Kohler, D.; Hiersemann, M. *Org. Lett.* **2006**, *8*, 1573.

The fact that hydrolysis of the OTMS group is so facile makes it unsuitable for most alcohol moieties that require long-term protection. A more stable silyl derivative is the triethylsilyl group (O-SiEt$_3$, **OTES**).[57] This group is attached by reaction of the alcohol with chlorotriethylsilane (Et$_3$SiCl) and pyridine.[58] The OTES group is generally more stable than OTMS, and in particular it is less sensitive to water. Aqueous acetic acid cleaves OTES and also alcoholic tosic acid,[53] but fluoride can also be used for removing it. In a synthesis of (+)-dolastatin 19 by Paterson and co-workers,[59] fragment **64** was converted to the tetra(triethylsilyloxy) or TES derivative **65**, in 72% yield using triethylsilyl triflate and 2,6-lutidine. After four synthetic steps that included a convergent step, compound **66** was obtained. Treatment with tosic acid in methanol deprotected the OTES alcohol but not the dimethyl-*tert*-butylsilyl alcohol (see below) to give **67** but this diol was not isolated but treated *in situ* with *tert*-butyldimethylsilyl triflate and 2,6-lutidine (see below) to give **68** in 74% yield. Note that treatment with tosic acid liberated the OTES alcohol, but that hydroxyl group reacted with the ketone moiety to give a ketal unit in **68**, via reaction of the intermediate hemi-ketal with the methanol.

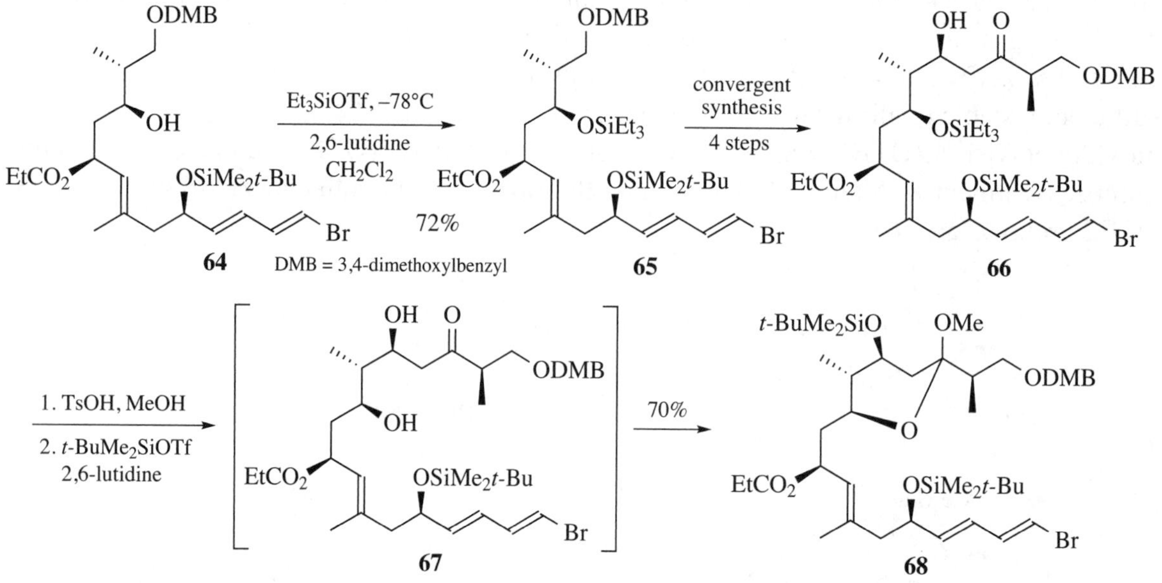

Triisopropylsilyl ether [O–Si(CH(Me)$_2$)$_3$, **OTIPS**],[60] is prepared via triisopropylchlorosilane (*i*-Pr$_3$SiCl)[61] and imidazole. The group is stable in the pH 2-14 range and it is generally impervious to nucleophiles, organometallics, hydrogenation (except in acetic acid), hydride and oxidizing agents. It is usually removed with fluoride (TBAF or aqueous HF). In an example

57. Reference 2a, p 44; Reference 2b, pp 72-73; Reference 2c, pp 121-123; 709-711; Reference 2d, pp 178-183.
58. Hart, T.W.; Metcalfe, D.A.; Scheinmann, F. *J. Chem. Soc. Chem. Commun.* **1979**, 156
59. Paterson, I.; Findlay, A.D.; Florence, G. J. *Org. Lett.* **2006**, *8*, 2131.
60. Reference 2a, pp 50, 296-298; Reference 2b, pp 74-75; Reference 2c, pp 125-127; 709-711; Reference 2d, pp 183-187.
61. Ogilvie, K.K.; Thompson, E.A.; Quilliam, M.A.; Westmore, J.B. *Tetrahedron Lett.* **1974**, 2865.

taken from the Cook and co-worker's synthesis of (–)-macroine,[62] the primary alcohol unit in diol **69** was selectively protected with triisopropylsilyl chloride in the presence of 2,6-luitidine to give the OTIPS derivative **70** in 90% yield. Four synthetic steps gave the OTIPS protected intermediate **71** and deprotection of the OTIPS groups with TBAF gave **72** in 86% yield.

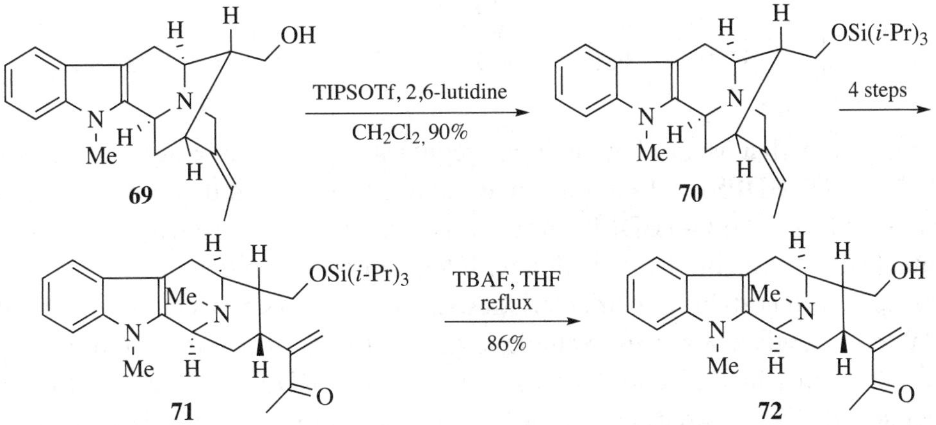

One of the most used silyl ether protecting groups is *tert*-butyldimethylsilyl [O-Si(Me)₂*t*-Bu, abbreviated as **OTBDMS** although it is commonly abbreviated as **OTBS**].[63] The most common method for attaching this group reacts the alcohol with *tert*-butyldimethylsilyl chloride using imidazole[64a] or dimethylaminopyridine[64b] as the base. The analogous silyl triflate (*t*-BuMe₂Si-OSO₂CF₃) is also useful for generating this ether from alcohols, with an amine base.[64c] The group is stable to base and reasonably stable to acid (pH 4 → 12) than the other silyl protecting groups, and also to nucleophiles, organometallics, hydrogenation (except in acidic media), hydrides or oxidizing agents. It can be removed with aqueous acid, but fluoride ion is the most common cleavage method (fluoride (tetrabutylammonium fluoride,[64] tetrabutylammonium chloride, and KF,[65] and also aqueous HF[66]). Groth and Kesenheimer used this protecting group in a synthesis of (–)-8-*O*-methyltetrangomycin (MM 47755),[67] for a key alcohol in **73**. Conversion to the silyl ether by reaction the silyl triflate gave **74** in 96% yield. Nine synthetic steps were required to form **75**, and removal of the TBS group with aqueous HF gave **76** in 49% yield.

62. Liao, X.; Zhou, H.; Yu, J.; Cook, J.M. *J. Org. Chem.* **2006**, *71*, 8884.
63. Reference 2a, pp 44-46, 296-298; Reference 2b, pp 77-83; Reference 2c, pp 127-141; 709-711; Reference 2d, pp 189-211.
64. (a) Corey, E.J.; Venkateswarlu, A. *J. Am. Chem. Soc.* **1972**, *94*, 6190; (b) Chaudhary, S.K.; Hernandez, O. *Tetrahedron Lett.* **1979**, 99; (c) Corey, E.J.; Cho, H.; Rücker, C.; Hua, D.H. *Ibid., **1981**, 22*, 3455.
65. Carpino, L.A.; Sau, A.C. *J. Chem. Soc. Chem. Commun.* **1979**, 514.
66. Newton, R.F.; Reynolds, D.P.; Finch, M.A.W.; Kelly, D.R.; Roberts, S.M. *Tetrahedron Lett.* **1979**, 3981.
67. Kesenheimer, C.; Groth, U. *Org. Lett.* **2006**, *8*, 2507.

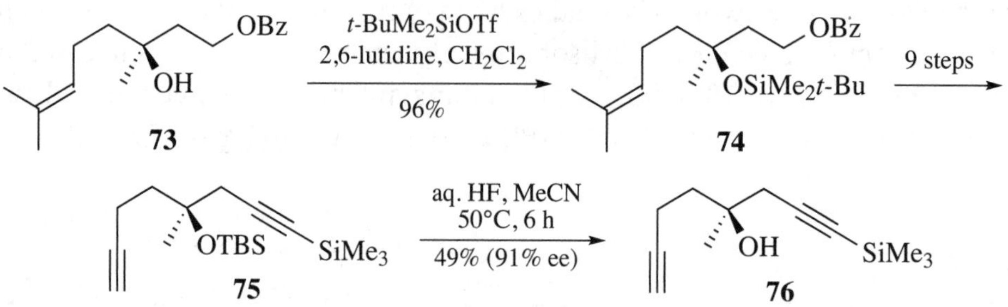

t-BuMe$_2$SiOTf
2,6-lutidine, CH$_2$Cl$_2$
96%

OBz
OH
73

OBz
OSiMe$_2$t-Bu
74

9 steps

OTBS SiMe$_3$
75

aq. HF, MeCN
50°C, 6 h
49% (91% ee)

OH SiMe$_3$
76

A protecting group that is used with great regularity is the *tert*-butyldiphenylsilyl group [*O*-SiPh$_2$C(Me)$_3$, **OTBDPS**].[68] Hanessian and Lavallee developed this protecting group when it was discovered that the O-TBDMS group was sensitive to hydrogenation.[69] The TBDPS group is attached to the alcohol by reaction with *tert*-butyldiphenylsilyl chloride (*t*-BuPh$_2$SiCl) and imidazole.[69] It has stability characteristics similar to dimethyl-*tert*-butyl derivatives, but is more stable to Lewis acids. An example is the synthesis of macroviracin A by Takahashi and co-workers,[70] in which alcohol **77** was converted to TBDPS derivative **78** in 96% yield. The conversion to **79** required several steps, in a convergent synthesis (sec. 10.3.C), and deprotection with TBAF gave **80** in 82% yield.

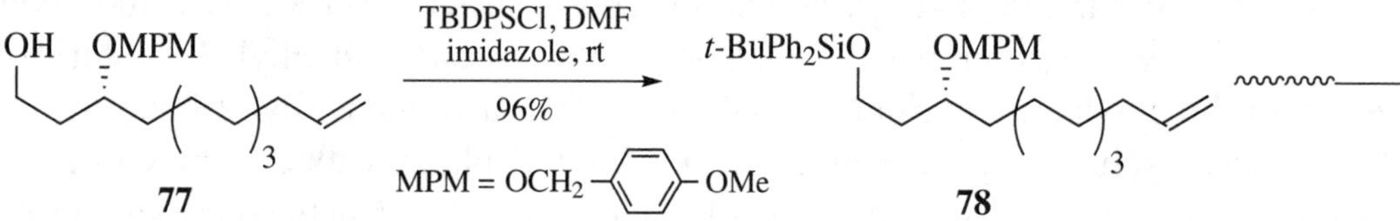

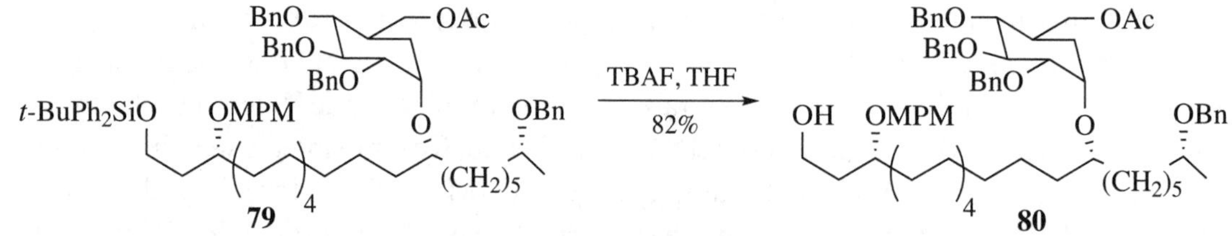

OH OMPM
3
77

TBDPSCl, DMF
imidazole, rt
96%

MPM = OCH$_2$—⟨⟩—OMe

t-BuPh$_2$SiO OMPM
3
78

BnO—⟨⟩—OAc
BnO
BnO
O
OBn

t-BuPh$_2$SiO OMPM
(CH$_2$)$_5$
79 4

TBAF, THF
82%

BnO—⟨⟩—OAc
BnO
BnO
O
OBn

OH OMPM
(CH$_2$)$_5$
4 **80**

7.3.A.iii. Ester Protecting Groups. Another class of alcohol protecting groups are the *O*-esters (O-COR), where COR is an acyl group. Ester protecting groups are limited in scope due to their susceptibility to nucleophilic acyl substitution, hydrolysis or reduction. Several esters are commonly used in synthesis, however, including acetates, benzoates or mesitoates. Although many esters may be used in synthesis, this discussion will focus only on the more common derivatives.

68. Reference 2a, pp 47-48, 296-298; Reference 2b, pp 83-84; Reference 2c, pp 141-145; 709-711; Reference 2d, pp 211-215.
69. Hanessian, S.; Lavallee, P. *Can. J. Chem.* **1975**, *53*, 2975; (b) *Idem Ibid.*, **1977**, *55*, 562.
70. Takahashi, S.; Souma, K.; Hashimoto, R.; Koshino, H.; Nakata, T. *J. Org. Chem.* **2004**, *69*, 4509.

Acetates (**OAc**)[71] are formed by reaction of an alcohol with acetic anhydride or acetyl chloride and pyridine or triethylamine. Acetates are stable to pH from 1 up to 8, ylids and organocuprates, catalytic hydrogenation, borohydrides, Lewis acids or oxidizing agents. Hydrolysis with acid or base (saponification) cleaves the ester to the alcohol and an acid. Reaction with LiAlH$_4$ reductively cleaves the acetate to the alcohol and ethanol. Acetates have found widespread use in synthesis. In the Sha and co-workers synthesis of the hexahydrobenzofuran subunit of avermectin,[72] diol **81** was acetylated to give **82** in 81% yield. Subsequent elimination of the tertiary alcohol introduced the alkene unit, and deprotection liberated the alcohol in **83** in 60% and 87% respectively.

Alcohols react with benzoyl chloride (PhCOCl) or benzoic anhydride [(PhCO)$_2$O], with pyridine or NEt$_3$, to give the benzoyl derivative (benzoate ester, **OBz**).[73] Benzoates are more stable to hydrolysis than acetates, with a pH stability from pH 1 up to 10-11. Treatment with 10% K$_2$CO$_3$ (pH 11), for example, will hydrolyze the O–Bz group. Benzoyl esters are more stable to reactions with nucleophiles such as cyanide and acetate, ylids or organocuprates. Benzoate esters also resist catalytic hydrogenation, and reaction with borohydrides or with oxidizing agents. The group is usually cleaved by basic hydrolysis or by reduction with LiAlH$_4$.

Three additional protecting groups in this category are the mesitoate ester (OCO-C$_6$H$_2$-2,4,6-trimethyl, **OMes**),[74] the pivaloyl ester (OCO-*t*-Bu, **OPiv**),[75,76] and the *para*-methoxybenzoyl ester [OCO-C$_6$H$_4$-(4-Me)]. The mesitoate ester is formed by reaction of an alcohol with mesitoyl chloride in the presence of pyridine or triethylamine.[77] This ester is stable to hydrolysis (pH 1-12), nucleophilic attack, organometallics, catalytic hydrogenation; reaction with boranes, borohydrides, Lewis acids or oxidation. This stability is due to the two ortho methyl groups, which block the acyl carbon to attack. Both LiAlH$_4$[74] and concentrated alcoholic potassium

71. Reference 2a, pp 53-55, 300-302; Reference 2b, pp 88-92; Reference 2c, pp 150-160; 713-716; Reference 2d, pp 223-239.
72. Sha, C.-K.; Huang, S.-J.; Zhan, Z.-P. *J. Org. Chem.* **2002**, *67*, 831.
73. Reference 2a, pp 61-62, 300-302; Reference 2b, pp 100-103; Reference 2c, pp 173-178; 713-716; Reference 2d, pp 255-262.
74. Reference 2a, pp 63-64, 300-302; Reference 2b, pp 103-104; Reference 2c, pp 178-179; 713-716; Reference 2d, p 263.
75. (a) Reference 2c, pp 170-173; 713-716; (b) Reference 2d, pp 250-254.
76. For a synthetic example, taken from a synthesis of elutherobin, see Chen, X.-T.; Bhattacharya, S.K.; Zhou, B.; Gutteridge, C.E.; Pettus, T.R.R.; Danishefsky, S.J. *J. Am. Chem. Soc.* **1999**, *121*, 6563.
77. Corey, E.J.; Achiwa, K.; Katzenellenbogen, J.A. *J. Am. Chem. Soc.* **1969**, *91*, 4318.

tert-butoxide[78,74] will cleave the group. Cleavage of the C–O bond is occasionally a problem with LiAlH$_4$.[79] Similar reactivity is observed with OPiv, but the precursor is pivaloyl chloride, *t*-BuCOCl.

The *p*-methoxybenzoyl group (**OPMB** or sometimes **OMPM** as in **83**)[80] is a useful protecting group. The precursor to this ester is *p*-methoxybenzoyl chloride. Cleavage of the *p*-methoxybenzyl ether unit is usually accomplished by oxidation with DDQ (2,3-dichloro-5,6-dicyano-1,4-benzoquinone). An example is taken from the Perkins and Lister synthesis of dolabriferol,[81] in which alcohol **84** was treated with trichloromethyl imidate (**85**) to give the O-PMB derivative **86**, in 84% yield. Three synthetic steps generated **87**, and deprotection by treatment with 2,3-dichloro-5,6-dicyano-1,4-benzoquinone (DDQ) gave **88**.

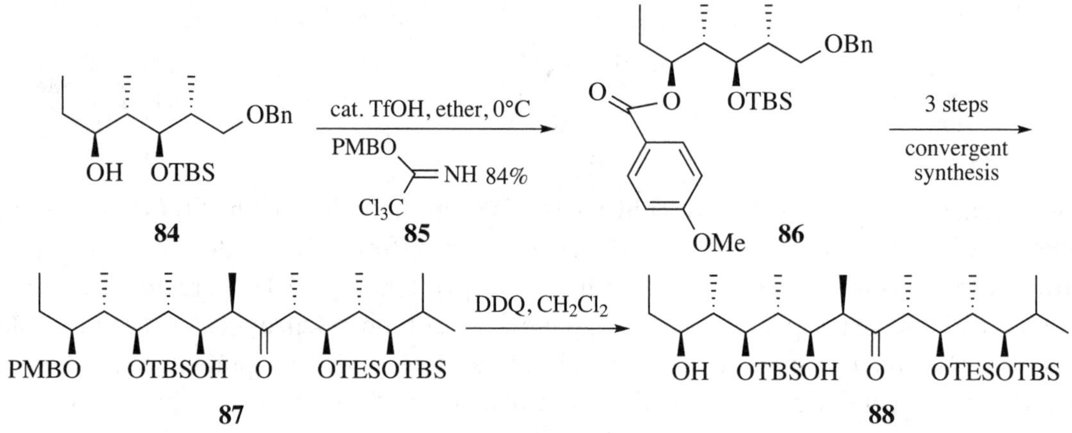

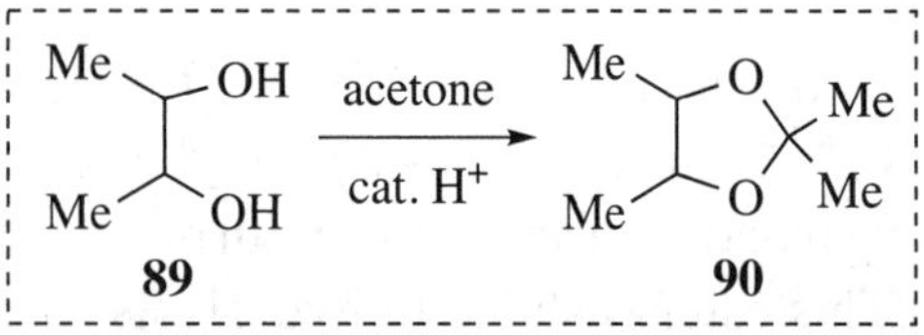

7.3.A.iv. Protection of Diols. 1,2-Diols are obviously alcohols, but the vicinal nature of the hydroxyls allows them to be protected as cyclic ketals. When a 1,2-diol such as 2,3-butanediol (**89**) reacts with a ketone such as acetone in the presence of an acid catalyst, a 1,3-dioxolane (**90**) is formed. Ketals such as **90** that are derived from acetone are called **acetonides** (isopropylidene ketals).[82] A 1,3-diol will generate a six-membered ring acetonide, which is a 1,3-dioxane derivative. Common methods for acetonide formation are reaction of the diol with 2-methoxy-1-propene in the presence of an acid such as anhydrous HBr[83] or reaction

78. (a) Reference 74; (b) Gassman, P.G.; Schenk, W.N. *J. Org. Chem.* **1977**, *42*, 918.
79. For cleavage of esters, see Robins, D.J.; Sakdarat, S. *J. Chem. Soc. Perkin Trans. 1* **1981**, 909.
80. (a) Reference 2c, p 36; (b) Stork, G.; Isobe, M. *J. Am. Chem. Soc.* **1975**, *97*, 6260.
81. Lister, T.; Perkins, M.V. *Org. Lett.* **2006**, *8*, 1827.
82. Reference 2a, pp 76-78, 304-306; Reference 2b, pp 123-127; Reference 2c, pp 207-215; 717-719; Reference 2d, pp 306-318.
83. Corey, E.J.; Kim, S.; Yoo, S.; Nicolaou, K.C.; Melvin, Jr., L.S.; Brunelle, D.J.; Falck, J.R.; Trybulski, E.J.; Lett, R.; Sheldrake, P.W. *J. Am. Chem. Soc.* **1978**, *100*, 4620.

of acetone with an acid catalyst.[84] The group is stable to base, but not to acid (pH 4 → 12). It is stable to nucleophiles, organometallics, catalytic hydrogenation, hydrides, and oxidizing agents. It can be cleaved with aqueous HCl, with acetic acid[85] or with *p*-toluenesulfonic acid in methanol.[86] This group has been used extensively in the manipulation of carbohydrates. Acetonides are useful in other synthetic applications, as in the protection of the 1,2-diol unit found in **91**, in Danishefsky and Siegel's synthesis of garsubellin A,[87] which was converted to **92** by reaction with dimethoxypropane and tosic acid in acetone, as part of This protection step was followed by elaboration of the molecule to **93** in three steps, and deprotection with aqueous perchloric acid gave **94**.

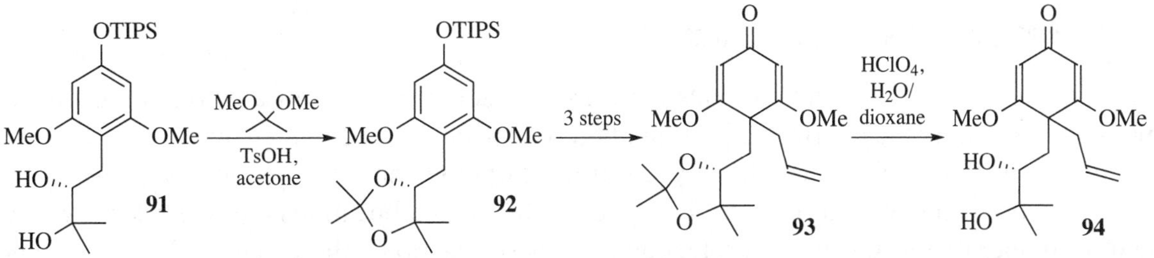

Diols react with other aldehydes or ketones to give a wide array of ketals or acetals. Two of the more common derivatives are benzaldehyde (benzylidene acetals) and cyclohexanone (cyclohexylidene ketals). In an initial study by Lee and co-workers aimed at the synthesis of rolliniastatin 1,[88] diol **95** was converted to the benzylidene acetal (**96**) with benzaldehyde dimethyl acetal, and then converted to **97** in eight synthetic steps. The diol was deprotected to give **98** using hydrogen with a Pd/C catalyst, via hydrogenolysis of the benzylic position. This route proved to be problematic, however, and another route was used to complete the synthesis.

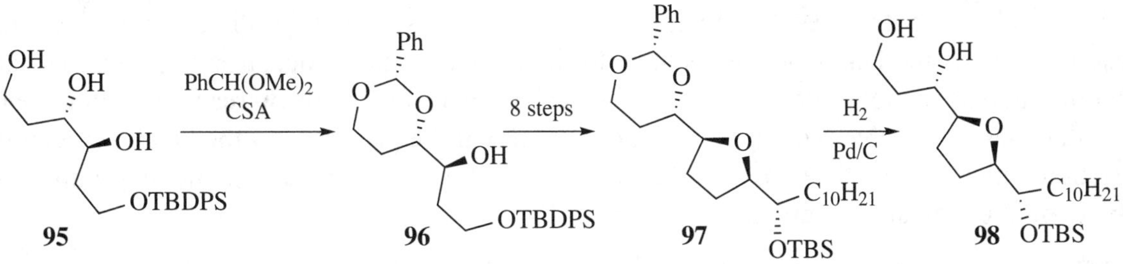

Wuts and Greene[2] discuss many other ketal protecting groups that have been used in synthesis,

84. (a) Schmidt, O.Th. *Methods Carbohydr. Chem.* **1963**, *2*, 318; (b) de Belder, A.N. *Adv. Carbohydr. Chem.* **1965**, *20*, 219.
85. Lewbart, M.L.; Schneider, J.J. *J. Org. Chem.* **1969**, *34*, 3505.
86. (a) Ichihara, A.; Ubukata, M.; Sakamura, S. *Tetrahedron Lett.* **1977**, 3473; (b) Kimura, J.;Mitsonobu, O. *Bull. Chem. Soc. Jpn.* **1978**, *51*, 1903.
87. Siegel, D.R.; Danishefsky, S.J. *J. Am. Chem. Soc.* **2006**, *128*, 1048.
88. Keum, G.; Hwang, C.H.; Kang, S.B.; Kim, Y.; Lee, E. *J. Am. Chem. Soc.* **2005**, *127*, 10396.

but acetonide is probably the most common. Cyclopentylidene or cyclohexylidene[89] and methylene[90] have been used, as well as a benzophenone ketal.[91] Discussions of methods for protecting alcohols in this section have been extensive, due to the wide range of alcohol protecting groups. Since the hydroxyl is capable of being transformed into a wide range of other functional groups, its protection is of particular importance. More than one hydroxyl moiety is commonly incorporated into a molecule, so the requirement for several selective protecting groups is obvious. The following sections will deal with protection of the other two key functional groups, the carbonyl group found in ketones and aldehydes and also the nitrogen found in amines.

7.3.B.   Protection of Aldehydes and Ketones

As with alcohols, developing methods for the protection of ketones or aldehydes must begin with understanding the interfering portion of these molecules. The main reaction of a carbonyl is nucleophilic acyl addition. The oxygen of the carbonyl can also function as a base in the presence of a suitable acid, leading to oxygen-stabilized carbocations (secs. 2.6.B.i, 12.2.A) and a number of reactions. To protect against these reactions, the electrophilic carbon and the π bond of the carbonyl must be removed by conversion to a new functional group that is easily converted back to the carbonyl. One method is to reduce the C=O unit to an alcohol (see Chap. 4), and then protect the OH unit as described in sec. 7.3.A. When it is appropriate in the synthesis, the alcohol is deprotected, and then oxidized back to the aldehyde or ketone. Such a protocol requires four synthetic steps. When feasible, protection procedures should require only two steps: protection and deprotection. With this limitation in mind, the major method for protecting ketones and aldehydes is conversion to a ketal or acetal, using alcohol or diol reactants. A variation of this reaction uses thiol or dithiol reactants to generate dithioketals or dithioacetals. In both cases, treatment with acid (or Lewis acids with the dithio derivatives) readily converts the ketals or acetals back to the ketone or aldehyde. It is important to note that in previous sections, alcohols were protected as ketals or acetals by reaction with ketones or aldehydes, but in this section the carbonyl group will be protected as the ketal or acetal by reaction with alcohols. The group is the same. The difference is which functional group is the object of attention and isolation.

7.3.B.i.   Ketals and Acetals. Ketals and acetals are formed by reaction of the carbonyl with alcohols such as methanol or ethanol under anhydrous conditions, in the presence of an acid catalyst. It is obvious that many alcohols can be used to generate acetals and ketals, but methanol and ethanol are probably the most common ones used. The yields of product with these reagents are high, and the lower molecular weight alcohol by-products are easily removed after deprotection. Another attractive feature of the methoxy unit is that it gives simple singlets

89. Reference 2d, p 318.
90. Reference 2d, pp 300-302.
91. Reference 2d, p 344.

in the ^{1}H NMR spectrum that usually do not obscure important signals from the remainder of protected substrate. In general, dry HCl (gas), sulfuric acid,[92] $BF_3 \cdot OEt_2$ or *p*-toluenesulfonic acid (PPTS) are the catalysts of choice. The addition of molecular sieves to adsorb water usually improves the yield of ketal or acetal.[93] **Diethyl and dimethyl ketals and acetals** are generally stable in a pH range of ~ 4-12, but they are sensitive to strong aqueous acid and to Lewis acids. They are stable to nucleophiles, organometallics, catalytic hydrogenation, hydrides and most oxidizing agents (they do react with ozone).[94] Conversion of the ketal or acetal back to the ketone or aldehyde is accomplished by treatment with aqueous acids such as trifluoroacetic acid[95] or oxalic acid (HOOC–COOH).[96] A popular alternative for regenerating the carbonyl from a ketal or acetal is transketalization, whereby acetone is added along with PPTS (or another acid catalyst). Acetone generates a new ketal (2,2-dimethoxypropane) in this reaction, liberating the originally protected carbonyl derivative.[97] Other acetals or ketals can be used as well.

In Kuwahara and co-worker's synthesis of a sesquiterpene isolated from the pheromone gland of a stink bug, *Tynacantha marginata* Dallas,[98] conjugated aldehyde **99** was treated with methanol in the presence of an acidic resin to give the dimethyl acetal **100**. Under the reaction conditions, the O-EE unit was deprotected to the alcohol. Diethoxy acetals are also quite useful. A synthetic example using an diethyl acetal is taken from Tanino and co-workers synthesis of (–)-coriolin.[99] Formation of the anion from 2-methylpropanenitrile and reaction with the diethyl acetal of 2-bromoacetaldehyde gave **101**. The nitrile unit was elaborated to the methylthio-ketone unit in **102**, and treatment with sulfuric acid regenerated aldehyde **103**.

The 1,2- and 1,3-diols, which form cyclic ketals or cyclic acetals [1,3-dioxolanes or 1,3-dioxanes

92. Cameron A.F.B.; Hunt, J.S.; Oughton, J.F.; Wilkinson, P.A.; Wilson, B.M. *J. Chem. Soc.* **1953**, 3864.
93. Roelofsen, D.P.; Wils, E.R.J.; Van Bekkum H. *Recl. Trav. Chim. Pays-Bas* **1971**, *90*, 1141.
94. Reference 2a, pp 116-120, 312-314; Reference 2b, pp 178-183; Reference 2c, pp 297-304; 725-727; Reference 2d, pp 435-444, 1009-1016.
95. Ellison, R.A.; Lukenbach, E.R.; Chiu, C.-W. *Tetrahedron Lett.* **1975**, 499.
96. Huet, F.; Lechevallier, A.; Pellet, M.; Conia, J.M. *Synthesis* **1978**, 63.
97. Colvin, E.W.; Raphael, R.A.; Roberts, J.S. *J. Chem. Soc. Chem. Commun.* **1971**, 858.
98. Kuwahara, S.; Hamade, S.; Leal, W.S.; Ishikawa, J.; Kodama, O. *Tetrahedron* **2000**, *56*, 8111.
99. Mizuno, H.; Domon, K.; Masuya, K.; Tanino, K.; Kuwajima, I.. *J. Org. Chem.* **1999**, *64*, 2648.

as discussed for protection of diol, see above], are particularly important for the protection of aldehydes and ketones. 1,3-Dioxolane derivatives are formed by reaction of the carbonyl with 1,2-ethanediol, and reaction with 1,3-propanediol gives 1,3-dioxanes. Dioxane protected ketones are usually hydrolyzed back to the carbonyl faster than the analogous dioxolane protected derivatives, but dioxolane protected aldehydes are hydrolyzed faster than dioxane protected.[100] Despite this difference in reactivity, dioxolanes appear to be the most commonly used protecting group. A carbonyl group reacts with 1,2-ethanediol (ethylene glycol) and PPTS, $BF_3 \cdot OEt_2$, or oxalic acid to give the **ethylenedioxy** ketal or acetal (a 1,3-doxolane).[101] Reaction of the carbonyl with 1,3-propanediol (propylene glycol) under similar conditions generates the 1,3-dioxane, a **propylenedioxy** ketal or acetal.[102] Cyclic ketals are sensitive to acid and are generally stable to pH 4-12 and to reactions with nucleophiles, organometallics, hydrogenation (except in acetic acid), hydrides and oxidizing agents, but they do react with Lewis acids. The most common method for cleavage of the dioxane or dioxolane derivatives is treatment with aqueous acid, including HCl in THF[103] and aqueous acetic acid.[104] In a synthesis of securinine by Ailbés, de March and co-worker's,[105] a protected aldehyde was incorporated into Grignard reagent **105**, which reacted with the aldehyde unit in **104** to give **106**. After four synthetic steps, the dioxolane group in **107** was converted to the aldehyde in **108** with DDQ. Note the use of DDQ as another method for deprotection.[106]

100. (a) Newman, M.S.; Harper, R.J. *J. Am. Chem. Soc.* **1958**, *80,* 6350; (b) Smith, S.W.; Newman, M.S. *Ibid* **1968**, *90,* 1249, 1253.
101. Reference 2a, pp 124-125; Reference 2b, pp 188-195, 185-186; Reference 2c, pp 307-308; 725-727; Reference 2d, pp 454-466.
102. Reference 2a, pp 121-126, 312-314; Reference 2b, pp 188-195, 185-186; Reference 2c, pp 307-308; 725-727; Reference 2d, pp 449-452.
103. (a) Grieco, P.A.; Nishizawa, M.; Oguri, T.; Burke, S.D.; Marinovic, N. *J. Am. Chem. Soc.* **1977**, *99,* 5773; (b) Grieco, P.A.; Yokoyama, Y.; Withers, G.P.; Okuniewicz, F.J.; Wang, C.-L.J. *J. Org. Chem.* **1978**, *43,* 4178.
104. Babler, J.H.; Malek, N.C.; Coghlan, M.J. *J. Org. Chem.* **1978**, *43,* 1821.
105. Ailbés, R.; Ballbé, M.; Busqué, F.; de March, P.; Elias, L.; Figueredo, M.; Font, J. *Org. Lett.* **2004**, *6,* 1813.
106. Tanemura, K.; Suzuki, .; Horaguchi, T. *J. Chem. Soc. Chem. Connun.* **1992**, 979.

A variation in ketal protection introduces alkyl substituents on the diol moiety. Barrett and co-workers, for example, used 2,2-dimethyl-1,3-propanediol to protect one of the ketone units of the quinone in **109** (forming **110**), in studies aimed at a synthesis of lactonamycin.[107] Seven synthetic steps led to **111**, and deprotection with aqueous acetic acid regenerated the carbonyl in quinone **112**.

7.3.B.ii. Dithioketals and Dithioacetals. The use of cyclic ketals and acetals in conjunction with acyclic carbonyl derivatives offers some selectivity for protection of more than one carbonyl. Expanding the methodology to include the sulfur analogs [dithioketals and dithioacetals, $R_2C(SR')_2$] gives even more flexibility. Both cyclic and acyclic dithio derivatives are known and used, but cyclic dithioketals and acetals are more common. A ketone or aldehyde reacts with two equivalents of a thiol (mercaptan) and an acid (typically HCl and BF_3).[108,111a] The main advantage of these groups is their stability to acid. They are stable from pH 1-12 and compatible with nucleophiles, organometallics, hydrides and oxidizing agents, although CrO_3 can oxidize the sulfur.[109] Dithioacetals and ketals are particularly reactive with mercuric and silver salts, and are relatively unreactive with many other Lewis acids.[102] Mercuric chloride under aqueous conditions is the most commonly used method for conversion back to the carbonyl.[110] As with the oxygenated analogs, the use of cyclic dithioketals and acetals is very prevalent. Reaction with 1,2-ethanedithiol generates 1,3-dithiolanes, and 1,3-propanedithiol gives 1,3-dithianes.

Boron trifluoride etherate ($BF_3 \cdot OEt_2$) is commonly used as an acid catalyst to form dithianes and dithiolanes from ketones and aldehydes.[111] These protecting groups are similar in their stability to the acyclic derivatives. Cyclic dithioketal and dithioacetal derivatives are relatively insensitive to acid, showing stability from pH 1-12. They are stable to nucleophiles, organometallics, hydrides and many oxidizing agents.[109] They are subject to hydrogenolysis

107. Henderson, D.A.; Collier, P.N.; Pave, G.; Rzepa, P.; White, A.J.P.; Burrows, J.N.; Barrett, A.G.M. *J. Org. Chem.* **2006**, *71*, 2434.

108. Zinner, H. *Chem. Ber.,* **1950**, *83*, 275.

109. Reference 2a, pp 129-132, 312-314; Reference 2b, pp 198; Reference 2c, pp 333-340; 725-727; Reference 2d, pp 477-500.

110. English, Jr., J.; Griswold, Jr., P.H. *J. Am. Chem. Soc.* **1945**, *67*, 2039.

111. (a) Fujita, E.; Nagao, Y.; Kanelo, K. *Chem. Pharm. Bull.* **1978**, *26*, 3743; (b) Corey, E.J.; Bock, M.G. *Tetrahedron Lett.* **1975**, 2643.

and the bivalent sulfur atoms in dithianes can poison heterogeneous hydrogenation catalysts (sec. 4.8.A). They are also sensitive to strong Lewis acids such as aluminum chloride (AlCl$_3$) and mercuric salts. This latter reaction is the basis of the most common cleavage reaction, the **Corey-Seebach procedure**.[112] Other cleavage reagents include Vedejs' use of boron trifluoride etherate in aqueous THF containing mercuric oxide (HgO),[103c] N-Bromosuccinimide,[113] iodine in DMSO,[114] ceric ammonium nitrate [Ce(NH$_4$)$_2$(NO$_3$)$_6$, **CAN**],[115] and iodomethane in aqueous media.[116] One of the best cleavage procedures is Corey's use of N-chlorosuccinimide and silver nitrate in aqueous acetonitrile.[117] There are many synthetic examples that use dithiane or dithiolane protecting groups, in combination with many deprotection protocols. In Kadota, Yamamoto and co-worker's synthesis of gambierol,[118] aldehyde **113** was converted to dithiane **114** by reaction with 1,3-propanedithiol and BF$_3$•etherate. This sequence required relatively short-term protection, and after two synthetic steps that produced **115**, the dithiane group was removed with iodomethane and bicarbonate in aqueous acetonitrile to give **116**.

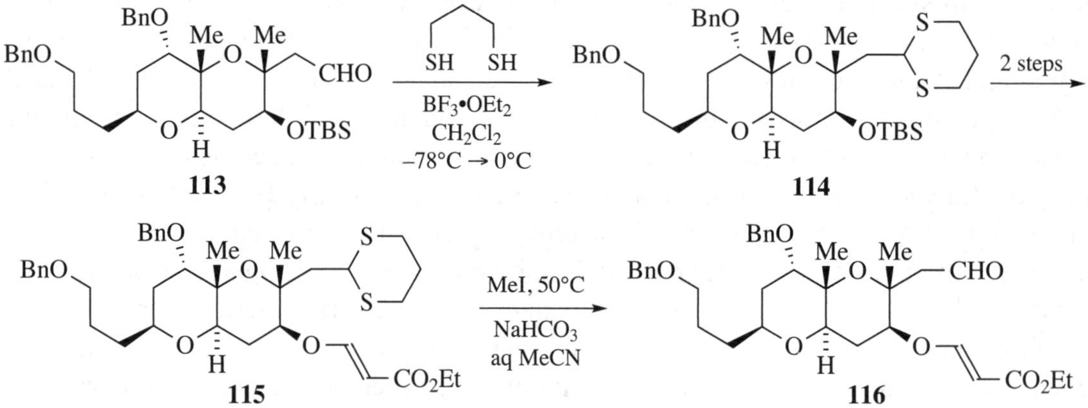

Other 1,2- and 1,3-dithiols have been used as protecting groups, but those discussed here are the most common. Wuts and Greene[2] discuss other methods for protection of ketones and aldehydes, including cyanohydrins, hydrazones, oximes, oxazolidines or imidazolidines.[119] Most of these are rather specialized and will not be discussed in this general presentation.

112. (a) Seebach, D.; Corey, E.J. *J. Org. Chem.* **1975**, *40*, 231; (b) Seebach, D. *Synthesis* **1969,** 17; (c) Vedejs, E.; Fuchs, P.L. *J. Org. Chem.* **1971**, *36*, 366.

113. (a) Cain, E.N.; Welling, L.L. *Tetrahedron Lett.* **1975**, 1353; (b) Corey, E.J.; Erickson, B.W. *J. Org. Chem.* **1971**, *36*, 3553.

114. Chattopadhyaya, J.B.; Rama Rao, A.V. *Tetrahedron Lett.* **1973**, 3735.

115. Ho, T.-L.; Ho, H.C.; Wong, C.M. *J. Chem. Soc. Chem. Commun.* **1972**, 791.

116. (a) Fétizon, M.; Jurion, M. *J. Chem. Soc. Chem. Commun.* **1972**, 382; (b) Takano, S.; Hatakeyama, S.; Ogasawara, K. *Ibid.,* **1977**, 68.

117. Corey, E.J.; Erickson, B.W. *J. Org. Chem.* **1971**, *36*, 3553.

118. Kadota, I.; Takamura, H.; Sato, K.; Ohno, A.; Matsuda, K.; Satake, M.; Yamamoto, Y. *J. Am. Chem. Soc.* **2003**, *125*, 11893.

119. Reference 2a, pp 141-147, 312-314; Reference 2b, pp 210-223; Reference 2c, pp 348-364; Reference 2d, pp 506-528.

7.3.C. Protection of Amines

The last major type of functional group to be discussed will be amines. The interfering feature of this group is the lone pair of electrons on nitrogen, which makes the molecule basic and nucleophilic. If the amine reacts with alkyl groups, the electron pair is used to forms a bond to carbon to generate an ammonium salt. Alkylation, basicity or oxidation are not possible for ammonium salts, but they are not often used as protecting groups since these charged species are incompatible with many reactions, and solubility can be a problem. In addition, ammonium salts are good leaving groups for substitution and elimination reactions. A more useful protection method generates a derivative in which the electron pair is completely or partially delocalized, reducing the basicity and nucleophilicity of the nitrogen. The three most common methods of amino group protection involve conversion of primary and secondary amines to a tertiary amine (usually benzyl or trialkylsilyl), conversion to an amide or a carbamate, and to a lesser extent, conversion to specialized sulfonamides.[120] Only the first two of these methods will be discussed.

7.3.C.i. *N*-Alkyl and *N*-Silyl Protecting Groups. Benzyl (*N*-CH$_2$Ph, **N-Bn**)[121] is a common alkyl protecting group for nitrogen. An amine is treated with benzyl chloride or benzyl bromide, usually in the presence of a base such as potassium carbonate (K$_2$CO$_3$) or hydroxide.[122] This group is stable to acid and base (pH 1-12) and to nucleophiles, organometallics, hydrides, but reacts with Lewis acids (as a Lewis base). The N-C bond is subject to hydrogenolysis by catalytic hydrogenation or dissolving metals. Hydrogen and palladium on carbon[123] (sec. 4.8.E) or sodium in liquid ammonia[124] (sec. 4.9.D) are the two most common methods of cleavage. Most of the other *N*-alkyl groups used are less general.[125]

A synthetic example using the *N*-benzyl protecting group is taken from Comins and Kuethe's synthesis of (+)-cannabisativine.[126] Formation of amide **118** in 82% yield from acid chloride **117**, under basic conditions, used *N*-benzyl 4-hydroxylbutylamine prepared from 4-hydroxylbutylamine and benzyl chloride.[127] After four synthetic steps, to give **119**, treatment with sodium in ammonia deprotected the *N*-benzyl, the *O*-benzyl, and the *N*-tosyl groups to give **120** [(+)-cannabisativine] in 76% yield. It has also been reported that *N*-benzyl can be removed in the presence of *O*-benzyl by catalytic transfer hydrogenation in the presence of a

120. For a review of nitrogen protecting groups, see Theodoridis, G. *Tetrahedron* **2000**, *56*, 2339.
121. Reference 2a, pp 272-273, 332-334; Reference 2b, pp 335-338; Reference 2c, pp 620-621; 745-747; Reference 2d, pp 814-818.
122. Velluz, L.; Amiard, G.; Heymès, R. *Bull. Soc. Chim. Fr.* **1954**, 1012.
123. Hartung, W.H.; Simonoff, R. *Org. React.* **1953,** *7,* 263.
124. du Vigneaud, V.; Behrens, O.K. *J. Biol. Chem.* **1937,** *117,* 27.
125. Reference 2a, pp 268-275, 332-334; Reference 2b, pp 321-341; Reference 2c, pp 573-586; 619-620; Reference 2d, pp 803-809.
126. Kuethe, J.T.; Comins, D.L. *J. Org. Chem.* **2004,** *69*, 5219.
127. Lesher, G.Y.; Surrey, A.R. *J. Am. Chem. Soc.* **1955,** *77*, 636.

hydrogen donor, 1,4-cyclohexadiene.[128]

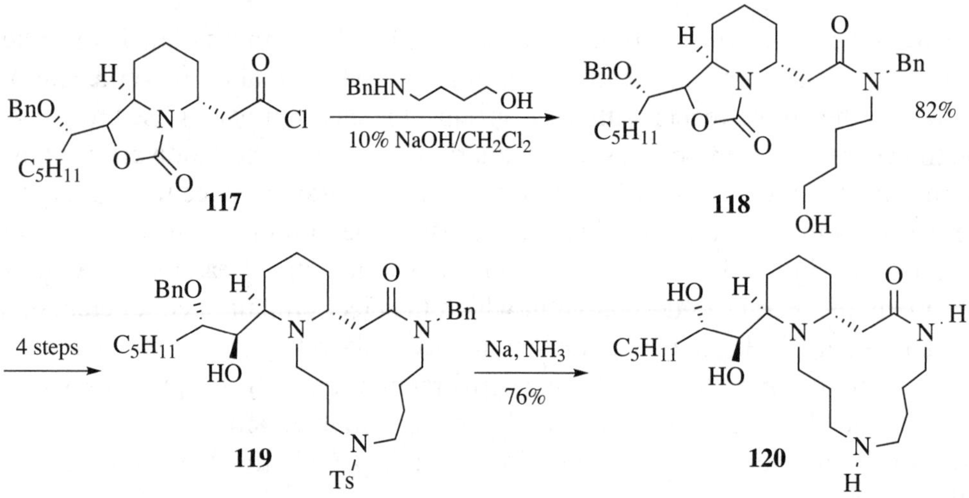

The benzyl protection group can be applied to primary or secondary amines. Scammells and Gathergood incorporated a primary amine precursor in the synthesis of psilocin,[129] by reaction of tosylate **121** with dibenzylamine to give **122** in 56% yield. The dibenzylated species functions as a primary amine surrogate, where both hydrogens of the nitrogen has been removed. A palladium-catalyzed coupling reaction with aryl iodide **123** (sec. 12.7) gave **124**, and catalytic hydrogenation with a palladium on carbon catalyst removed both benzyl groups to give **125** in 83% yield.

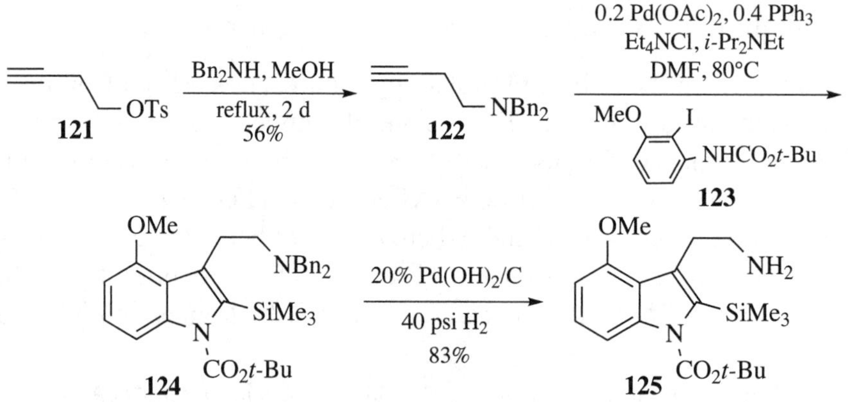

A simple silyl protecting group is trimethylsilyl (*N*-SiMe₃, ***N*-TMS**), which is formed by reaction of the amine with chlorotrimethylsilane and triethylamine or pyridine. The trimethylsilyl group is sensitive to water or alcohol solvents, and the reaction of Me₃SiNEt₂ and alcohols

---

128.   Bajwa, J.S.; Slade, J.; Repic, O. *Tetrahedron Lett.* **2000,** *41,* 6025.
129.   Gathergood, N.; Scammells, P.J. *Org. Lett.* **2003,** *5,* 921.

gives ROSiMe$_3$. The *N*-TMS group is therefore used under anhydrous conditions.[130] In a synthesis of (–)-21-isopentenylpaxilline by Smith and Cui[131] reaction of aniline derivative **126** with MeLi (sec. 8.5) and excess chlorotrimethylsilane gave an 84% yield of **127**, blocking both hydrogens on the primary amine. Alkylation (sec. 8.5) gave **128**, and treatment with ethanolic HCl gave **129** in 82% yield for the last two steps.

7.3.C.ii. ***N*-Acyl Protecting Groups.** Amides are common protecting groups for amines. *N*-Acetyl is the best known of the amide protecting groups, and *N*-acylamines are known as acetamide derivatives (*N*-COMe, ***N*-Ac**).[132] Reaction of acetic anhydride or acetyl chloride with an amine, in the presence of a base such as pyridine or triethylamine will generate the acetamide.[133] Acetamides are sensitive to strong acid and base ,but are stable in the pH range 1→12. Nucleophiles and many organometallics react with the *N*-acetyl group. Organolithium reagents are unreactive, although reactive Grignard reagents are. This group can be reduced by catalytic hydrogenation, borane or borohydride reducing agents (not LiAlH$_4$), and can be oxidized by many oxidizing agents. Reactions that use Lewis acids, such as glycosylation, may cause problems if this protecting group is present. The two major methods for conversion of *N*-acetamides to the amine are treatment with aqueous acid[134] and treatment with triethyloxonium tetrafluoroborate[135] (Et$_3$O$^+$BF$_4^-$, **Meerwein's reagent**[136]). Reduction with LiAlH$_4$ gives the ethylamine derivative, but cleavage is sometimes observed.[137] In a synthesis of (–)-securinine,[138] by Honda and co-worker's, the *N*-acetyl group was used to protect **130** as acetamide **131**. Conversion to **132** in five steps was followed by cleavage with trifluoroacetic

130.   (a) Reference 2a, pp 283; Reference 2b, pp 69-71; (c) Pratt, J.R.; Massey, W.D.; Pinkerton, F.H.; Thames, S.F. *J. Org. Chem.* **1975**, *40*, 1090.

131.   Smith III, A.B.; Cui, H. *Org. Lett.* **2003**, *5*, 587.

132.   Reference 2a, pp 251-252, 328-330; Reference 2b, pp 351-352; Reference 2c, pp 552-555; 741-743; Reference 2d, pp 775-779.

133.   Barrett, A.G.M.; Lana, J.C.A. *J. Chem. Soc. Chem. Commun.* **1978**, 471.

134.   Dilbeck, G.A.; Field, L.; Gallo, A.A.; Gargiulo, R.J. *J. Org. Chem.* **1978**, *43*, 4593.

135.   Hanessian, S. *Tetrahedron Lett.* **1967**, 1549.

136.   (a) Meerwein, H. *Org. Synth. Coll. Vol. 5* **1973**, 1080; (b) For reactions with amides and lactams see Borch, R.F. *Tetrahedron Lett.* **1968**, 61.

137.   Gaylord, N.G. *Reductions with Complex Metal Hydrides*, Ellis Horwood Ltd., Chichester, *1964*, pp 544-546.

138.   Honda, T.; Namiki, H.; Kaneda, K.; Mizutani, H. *Org. Lett.* **2004**, *6*, 87.

acid to give the amine group in **133**.

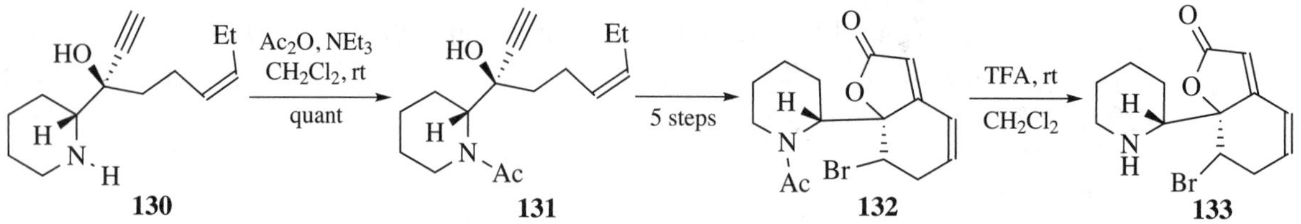

A simple modification of the structure has dramatic effects on the stability of the protecting group. If trifluoroacetyl is used rather than acetyl, the trifluoroacetamide (*N*-COCF$_3$, *N*-**TFA**)[139] is generated. The trifluoroacetyl group is usually attached by reaction of the amine with trifluoroacetic anhydride [(CF$_3$CO)$_2$O] in the presence of triethylamine or pyridine.[133,140] This group retains its stability to acid, but is more sensitive to base (pH 1-10) than the acetamide analog. It is stable to anhydrous bases or tetraalkylammonium hydroxides. The group reacts with most nucleophiles, organolithium reagents, LiAlH$_4$ and borohydrides via nucleophilic acyl substitution, but is stable to oxidation, Lewis acids, borane, and most catalytic hydrogenation conditions. It is usually removed with potassium carbonate in aqueous methanol,[141] by reduction with sodium borohydride[142] or by ammonia in methanol.[143] In a synthesis of microsclerodermin E by Ma and Zhu,[144] the primary amine unit in **134** was converted to the trifluoroacetamide (**135**), in 62% yield. After 7 steps, **136** was deprotected with ethanoic sodium borohydride to give a quantitative yield of **137**.

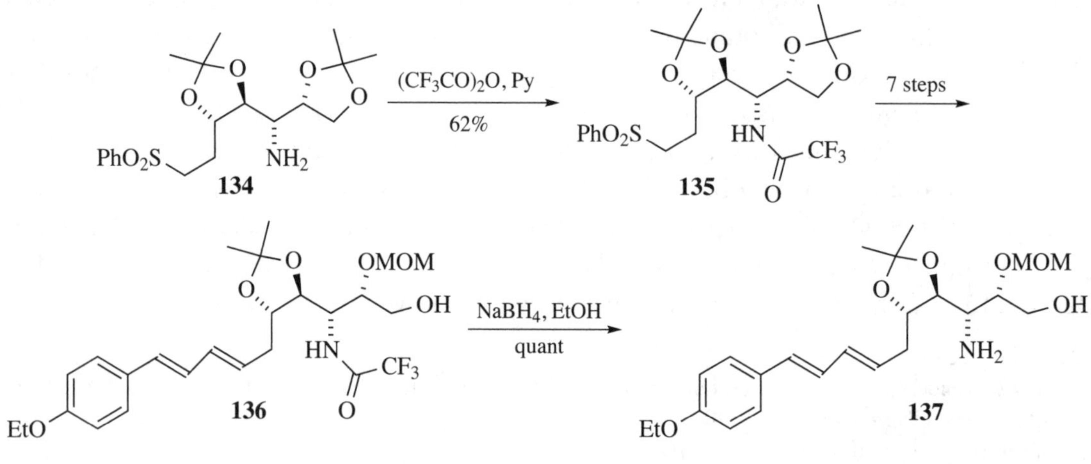

139. Reference 2a, pp 254-255, 328-330; Reference 2b, pp 353-354; Reference 2c, pp 556-558; 741-743; Reference 2d, pp 781-783.
140. Green, J.F.; Jham, G.N.; Neumeyer, J.L.; Vouros, P. *J. Pharm. Sci.* **1980**, *69*, 936.
141. (a) Quick, J.; Meltz, C. *Ibid.,* **1979**, *44*, 573; (b) Schwartz, M.A.; Rose, B.F.; Vishnuvajjala, B. *J. Am. Chem. Soc.* **1973**, *95*, 612.
142. Weygand, F.; Frauendorfer, E. *Chem. Ber.* **1970**, *103*, 2437.
143. Imazawa, M.; Eckstein, F. *J. Org. Chem.* **1979**, *44*, 2039.
144. Zhu, J.; Ma, D. *Angew. Chem. Int. Ed.* **2003**, *42*, 5348.

Benzamides (*N*-COPh, **N-Bz**)[145] are formed by reaction of an amine with benzoyl chloride in pyridine or triethylamine.[146] This group is stable to pH 1-14, nucleophiles, organometallics (not RLi), hydrogenation, hydrides (except LiAlH$_4$, and borane) and oxidizing agents. Cleavage is accomplished with 6 N HCl or HBr in acetic acid,[147] hot and concentrated aqueous NaOH,[148] and reduction with DIBAL-H (sec. 4.6.C).[149]

7.3.C.iii. N-Carbamate Protecting Groups. The carbamates (*N*-COOR) are a related class of protecting groups for nitrogen.[150] Several different carbamates have been used for the protection of amino acids in peptide synthesis. One of the most popular is the *tert*-butyl carbamate (*tert*-butoxycarbonyl, [NCOC(Me)$_3$], **N-Boc**).[151] Commercially available BOC-ON® [(Me)$_3$C-O-CO-CO$_2$N=C(CN)Ph] reacts with amines, in the presence of another base such as triethylamine to give the N-Boc derivative.[152] Similarly, reaction with di-*tert*-butyldicarbonate ([(CH$_3$)$_3$CCO]$_2$CO) under basic conditions gives the BOC-protected amine.[153] This group is sensitive to strong anhydrous acid (stable to pH 1 → 12) and trimethylsilyl triflate (Me$_3$Si-OTf), but it is stable to nucleophiles, organometallics (including organolithium reagents although Grignard reagents can react by nucleophilic acyl addition), hydrogenation (except under acidic conditions), hydrides, oxidizing agents (not Jones' conditions), aqueous acid, and mild Lewis acids. The Boc group is usually removed by treatment with aqueous HCl[154] or with anhydrous trifluoroacetic acid.[155] In a synthesis of the C(21)-C(26) fragment of superstolide A by Nishida and co-workers[156] the amide in **138** was converted to the N-Boc derivative **139**, in 98% yield. After six synthetic steps to give **140**, treatment with trifluoroacetic acid cleaved the Boc group to give the free amine in **141**.

145. Reference 2a, pp 261-263, 328-330; Reference 2b, pp 355-356; Reference 2c, pp 560-561; 741-743; Reference 2d, pp 785-787.
146. White, E. *Org. Synth. Collect. Vol. 5* **1973**, 336.
147. Ben-Ishai, D.; Altman, J.; Peled, N. *Tetrahedron* **1977**, *33*, 2715.
148. Khorana, H.G.; Turner, A.F.; Vizsolyi, J.P. *J. Am. Chem. Soc.* **1961**, *83*, 686.
149. Gutzwiller, J.; Uskoković, M. *J. Am. Chem. Soc.* **1970**, *92*, 204.
150. Reference 2a, pp 222-248, 324-326; Reference 2b, pp 327-330; Reference 2c, pp 518-525; 737-739; Reference 2d, pp 707-718.
151. Reference 2a, pp 324-326; Reference 2d, pp 725-735.
152. Itoh, M.; Hagiwara, D.; Kamiya, T. *Bull. Chem. Soc. Jpn.* **1977**, *50*, 718.
153. Tarbell, D.S.; Yamamoto, Y.; Pope, B.M. *Proc. Natl. Acad. Sci.* **1972**, *69*, 730.
154. Stahl, G.L.; Walter, R.; Smith, C.W. *J. Org. Chem.* **1978**, *43*, 2285.
155. Lundt, B.F.; Johansen, N.L.; Vølund, A.; Markussen, J. *Int. J. Pept. Protein Res.* **1973**, *12*, 258.
156. Nagata, T.; Nakagawa, M.; Nishida, A. *J. Am. Chem. Soc.* **2003**, *125*, 7484.

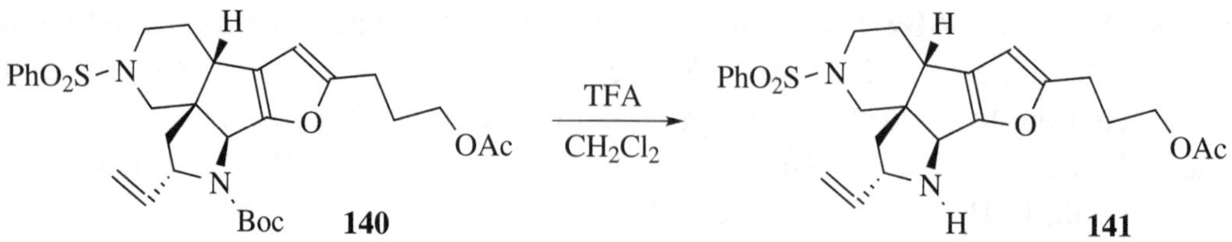

Another widely used carbamate is the benzyl carbamate (*N*-CO$_2$CH$_2$Ph, *N*-**Cbz**, benzyloxy carbonyl),[157] which Greene points out has been used since 1932.[158] The group is attached by reaction of an amine with benzyl chloroformate (PhCH$_2$O$_2$CCl) in the presence of base (aqueous carbonate or triethylamine). The group is very stable to acid and base (pH 1-12), to nucleophiles, to milder organometallics (it reacts with organolithium reagents and Grignard reagents), to milder Lewis acids, and to most hydrides (it reacts with LiAlH$_4$). The group is sensitive to hydrogenolysis by catalytic hydrogenation, which is the primary means of cleaving this group (Pd on carbon is the usual catalyst).[152,159]

In a synthesis of diazonamide A,[160] Nicolaou and co-workers protected amine **142** as the Cbz derivative **143** in 94% yield. This group remained intact through 29 steps, including a convergent synthetic step, until **144** was obtained. Catalytic hydrogenation removed the Cbz group to give **145**, in >80% yield, which was one step away from diazonamide A.

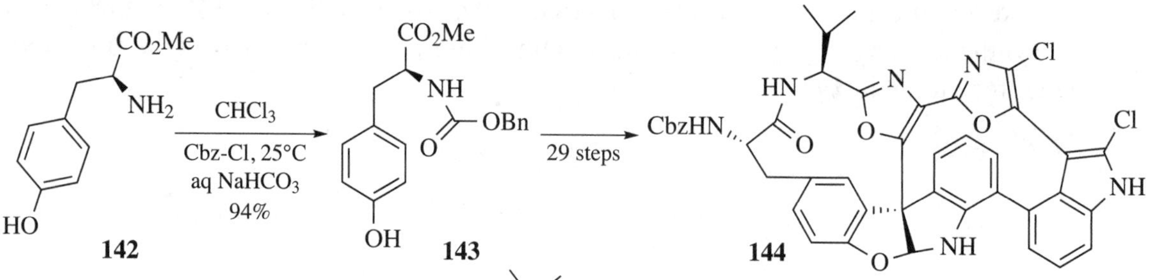

157. Reference 2a, pp 239-241, 324-326; Reference 2b, pp 335-338; Reference 2c, pp 531-537; 737-739; Reference 2d, pp 748-756.
158. (a) Reference 2a, p239; (b) Bergmann, M.; Zervas, L. *Berichte* **1932**, *65*, 1192.
159. Meienhofer, J.; Kuromizu, K. *Tetrahedron Lett.* **1974**, 3259.
160. Nicolaou, K.C.; Rao, P.B.; Hao, J.; Reddy, M.V.; Rassias, G.; Huang, X.; Chen, D.Y.-K.; Snyder, S.A. *Angew. Chem. Int. Ed.* **2003**, *42*, 1753.

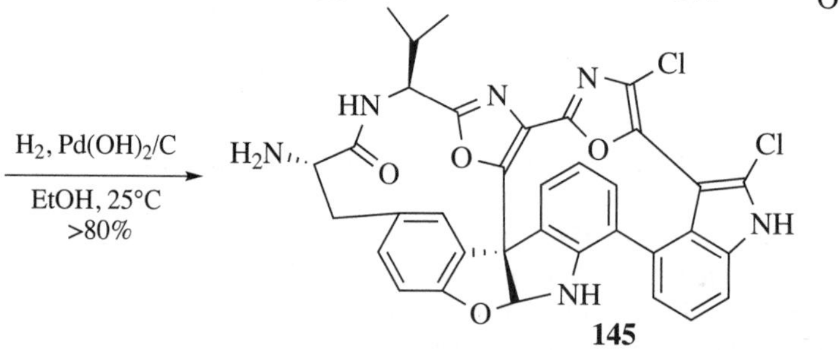

7.4. CONCLUSION

It is obvious that the preceding discussions are not exhaustive, but they are representative of the methods used by synthetic chemists to protect key functional groups. The monographs by Wuts and Greene[2] also discuss methods for protection of alkynes,[161] phosphates,[162] carboxylic acids,[163] amides[164] and thiols.[165] These monographs also contain detailed discussions of other protecting groups that can be used for alcohols, amines, ketones, and aldehydes.

There are obviously other situations that demand a protecting group, but the cases presented above give a useful arsenal of protecting groups as well as a sample of the situations that require their use. One should always be cautious with protecting groups, remembering that each protection-deprotection sequence adds two steps to a synthesis that diminishes the overall yield. Protecting and deprotecting steps often involve the use of an excess of reagents. These reagents are also often cheap, and the yields are usually high. Also, one-pot and simultaneous deprotection saves steps. New protecting groups for specific and general situations are always being developed, and a search of the current literature is always useful.

With this chapter, the first part of this book is complete. Functional groups have been introduced, and the manipulation and interchange of functional groups and their protection have been discussed. All of the fundamental concepts for organic synthesis have also been introduced: stereochemistry, chirality, conformational control, chelation effects and various problems in selectivity. The remaining chapters will deal with formation of carbon-carbon bonds, but the basic principles introduced in Chapters 1-7 will be invoked throughout.

161. Reference 2d, pp 927-933.
162. Reference 2d, pp 934-985.
163. Reference 2b, pp 224-276; Reference 2c, pp 369-453; 729-731; Reference 2d, pp 533-646.
164. Reference 2a, pp 152-192, 315-317; Reference 2b, pp 349-362; Reference 2c, pp 632-647.
165. Reference 2a, pp 195-222, 319-321; Reference 2b, pp 277-308; Reference 2c, pp 454-493; 733-735; Reference 2d, pp 647-695.

HOMEWORK

1. In each of the following, discuss the relative merits of protecting functional groups or finding a chemoselective reagent for: (i) reduction with hydrides, (ii) oxidation with Jones or PCC, (iii) catalytic hydrogenation, and (iv) reaction with MeMgBr (Chap. 8).

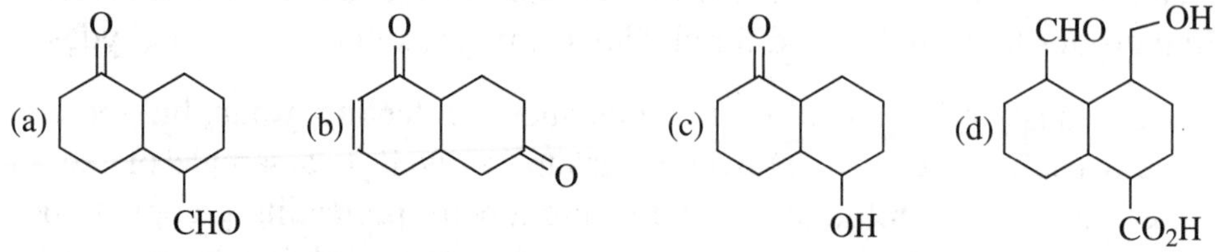

2. In the transformation of **A** to **B**, and then to **C**, taken from the synthesis of 7-deoxypancratistatin, deprotection was followed by conversion of a lactone to a lactam. Provide a mechanism for the deprotection of both the O-MOM group and the dioxolane, giving a triol, and also for the lactone-lactam conversion in the second step.

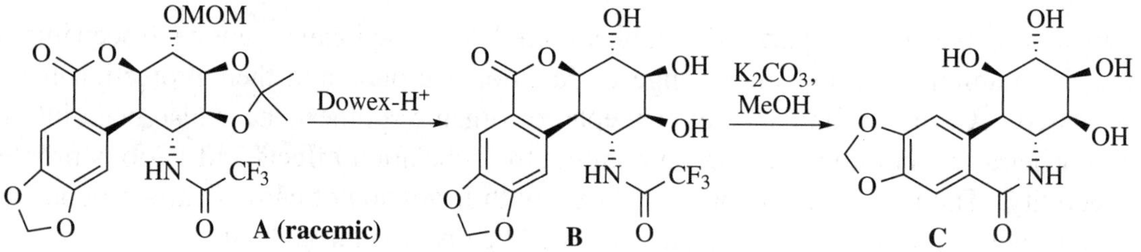

3. Explain the following transformation.

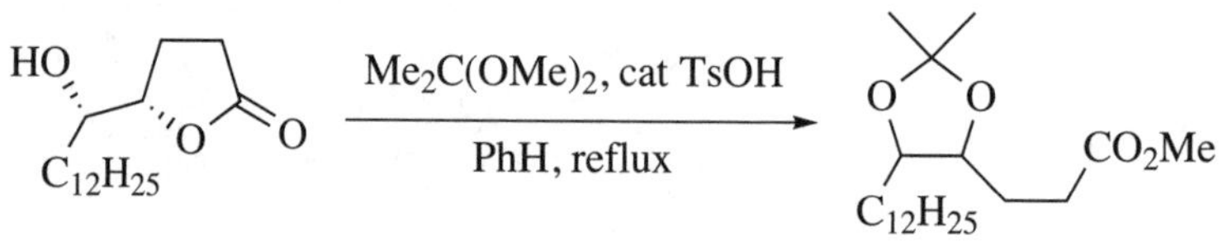

4. In each of the following, give the major product. Show stereochemistry where it is appropriate.

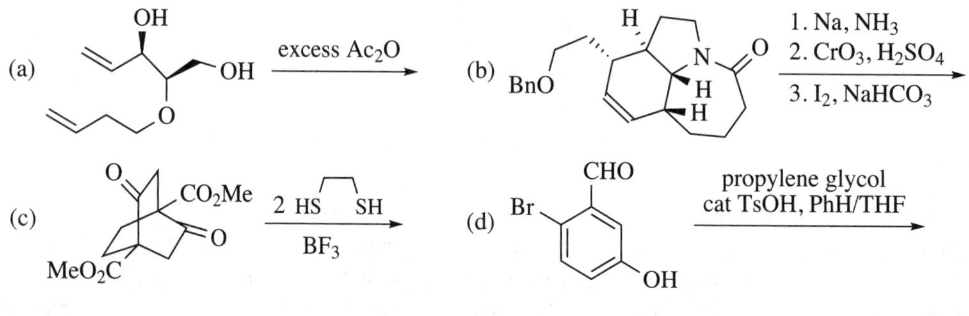

(e) 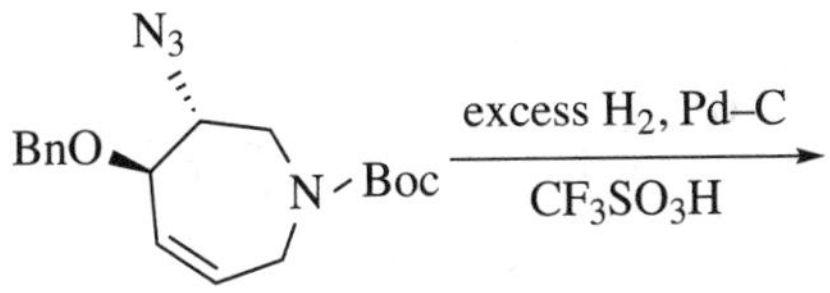 BnO, N₃, N–Boc [seven-membered N-heterocycle bearing azide, OBn, and alkene]

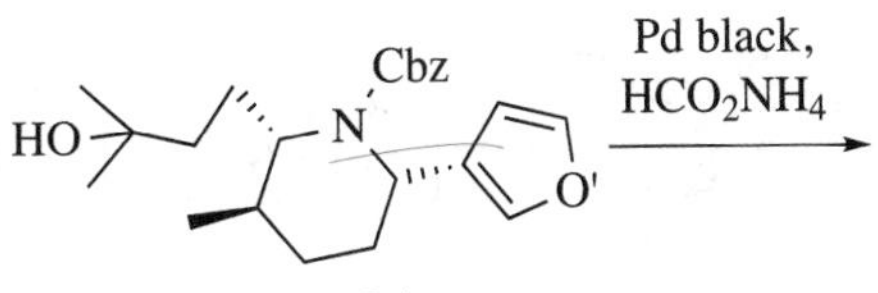

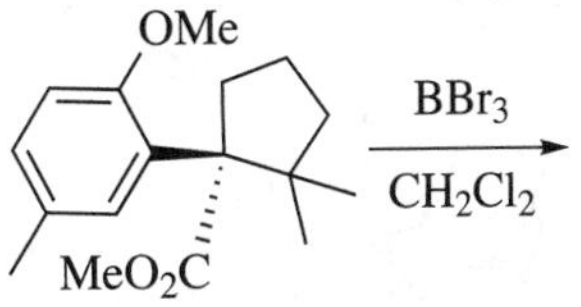

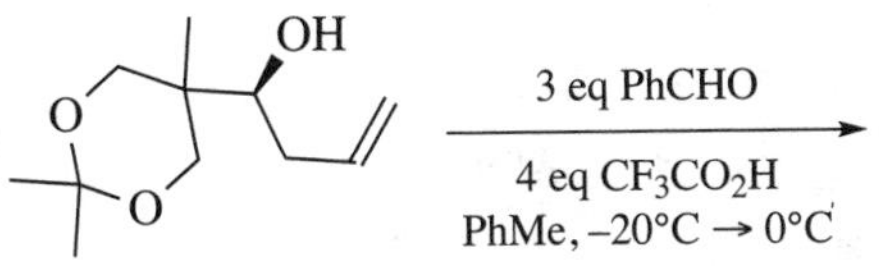

$\xrightarrow[\text{CF}_3\text{SO}_3\text{H}]{\text{excess H}_2, \text{Pd–C}}$

(f) HO [piperidine bearing N–Cbz, methyl, and a furan ring]

$\xrightarrow[]{\text{Pd black,}\ \text{HCO}_2\text{NH}_4}$

(g) 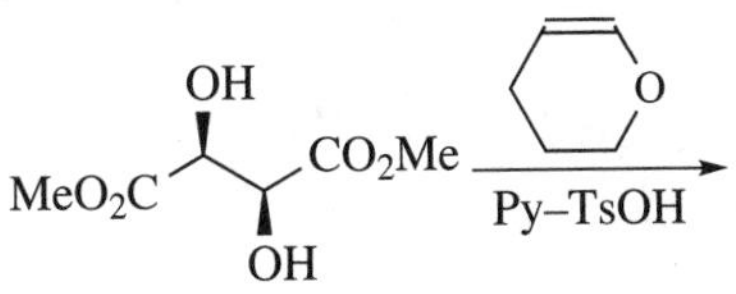 MeO₂C, OH, OH, CO₂Me [dimethyl tartrate-type diol]

$\xrightarrow[\text{Py–TsOH}]{\text{[dihydropyran]}}$

(h) 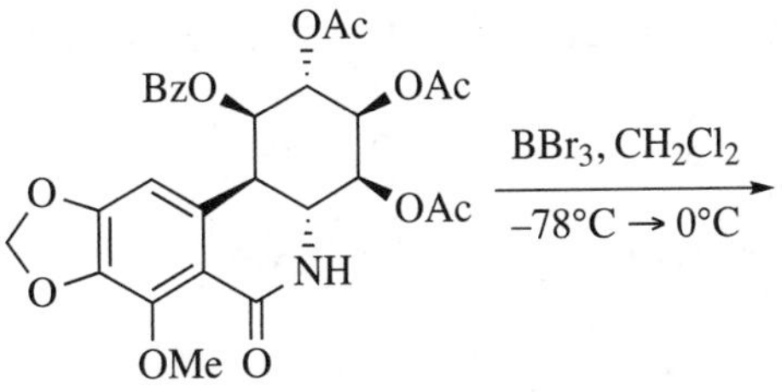 [methylenedioxy aromatic lactam with OMe, NH, C=O, OMe; cyclohexane ring bearing BzO, OAc, OAc, OAc]

$\xrightarrow[-78°C \to 0°C]{\text{BBr}_3, \text{CH}_2\text{Cl}_2}$

(i) 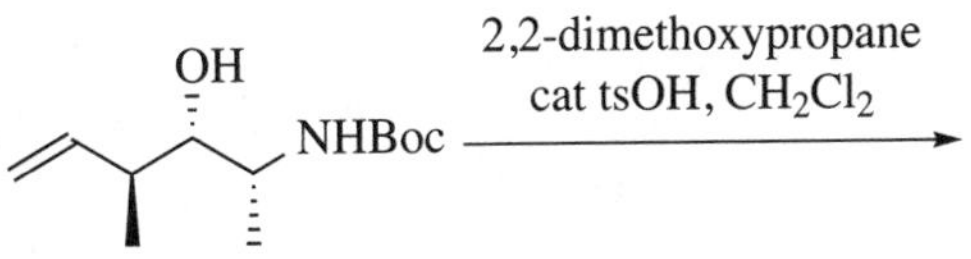 OH, NHBoc [chain with vinyl and two methyl stereocenters]

$\xrightarrow[\text{cat tsOH, CH}_2\text{Cl}_2]{\text{2,2-dimethoxypropane}}$

(j) OMe [aromatic bearing a 1,1-dimethylcyclopentane with MeO₂C]

$\xrightarrow[\text{CH}_2\text{Cl}_2]{\text{BBr}_3}$

(k) [1,3-dioxane bearing OH and allyl group]

$\xrightarrow[\text{PhMe, } -20°C \to 0°C]{\begin{array}{c}\text{3 eq PhCHO}\\ \text{4 eq CF}_3\text{CO}_2\text{H}\end{array}}$

(l) 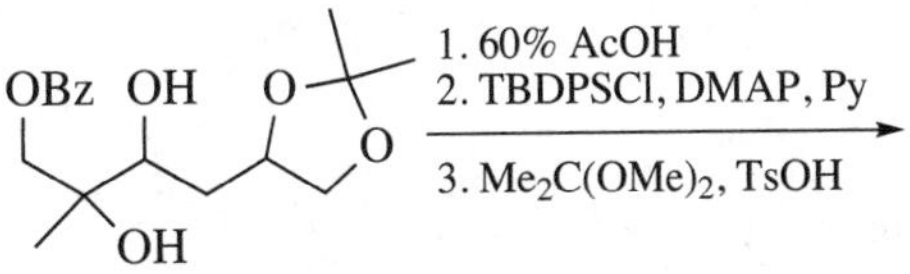 OBz, OH [chain with acetonide]

$\xrightarrow{\begin{array}{l}1.\ 60\%\ \text{AcOH}\\ 2.\ \text{TBDPSCl, DMAP, Py}\\ 3.\ \text{Me}_2\text{C(OMe)}_2, \text{TsOH}\end{array}}$

(m) 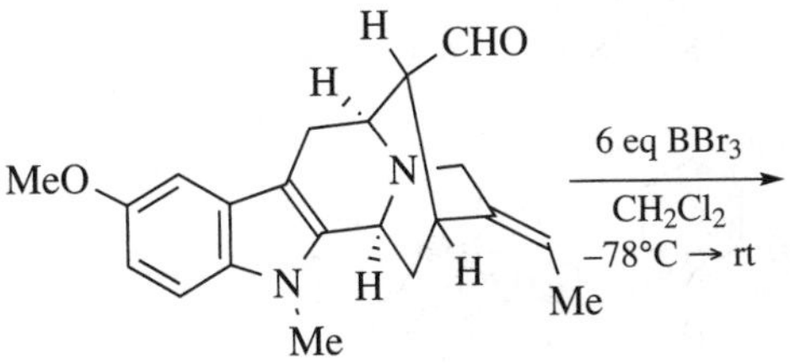 MeO [indole alkaloid skeleton bearing H, CHO, N, N–Me, Me, vinyl]

$\xrightarrow[\begin{array}{c}\text{CH}_2\text{Cl}_2\\ -78°C \to \text{rt}\end{array}]{\text{6 eq BBr}_3}$

(n) 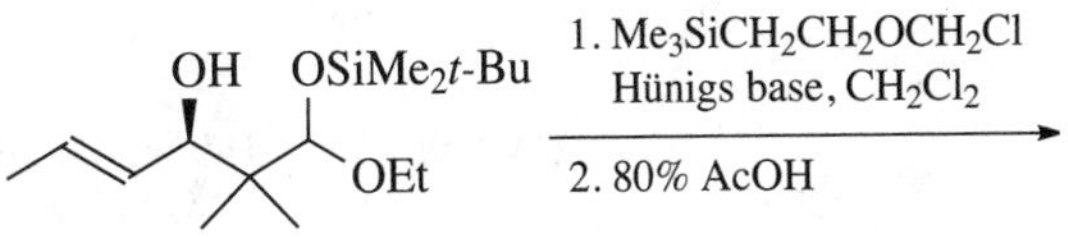 OH, OSiMe₂t-Bu, OEt [chain]

$\xrightarrow{\begin{array}{l}1.\ \text{Me}_3\text{SiCH}_2\text{CH}_2\text{OCH}_2\text{Cl}\\ \quad\ \text{Hünigs base, CH}_2\text{Cl}_2\\ 2.\ 80\%\ \text{AcOH}\end{array}}$

(o) [bicyclic amine bearing vinyl, NH, and CH₂OH]

$\xrightarrow[\text{THF}]{\text{NaH, BnBr}}$

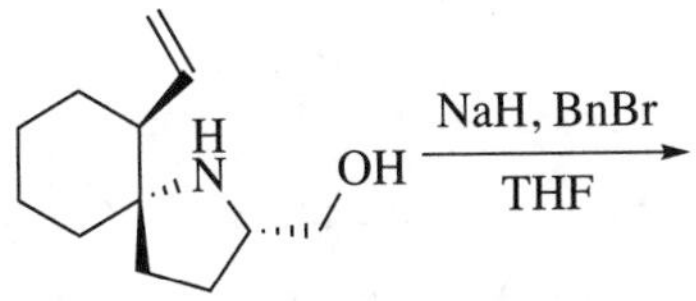

5. For each of the following, determine if protection is required in the given sequence. If so, choose the protecting group(s), explain your choice(s), and provide all necessary reagents for the entire sequence.

(a) 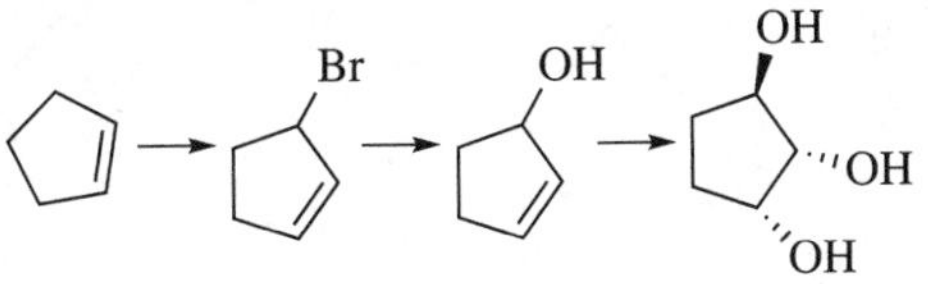 [cyclopentene] → [bromo-cyclopentadiene] → [hydroxy-cyclopentadiene] → [cyclopentane triol, OH, OH, OH]

(b) 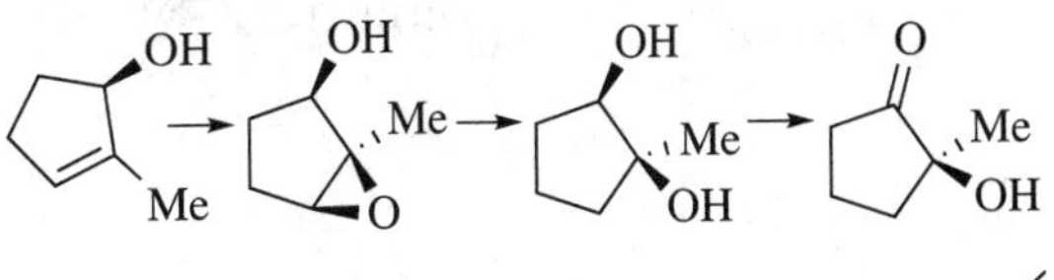 [cyclopentene with OH, Me] → [epoxide with Me, O] → [cyclopentane OH, Me, OH] → [cyclopentanone, O, Me, OH]

(c) 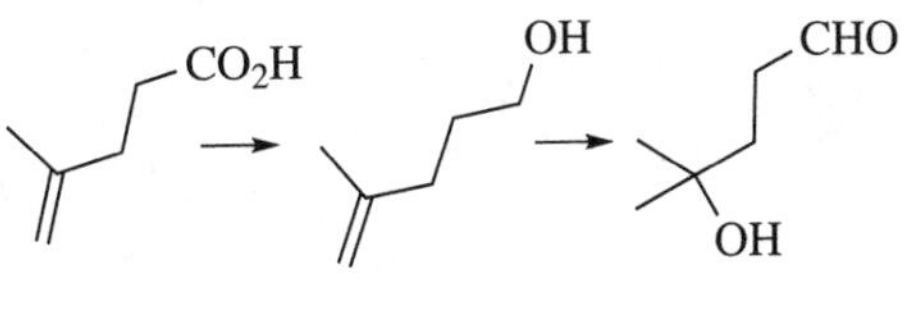 CO₂H [chain with terminal alkene] → OH [chain with terminal alkene] → CHO [chain with terminal alkene and OH]

(d) 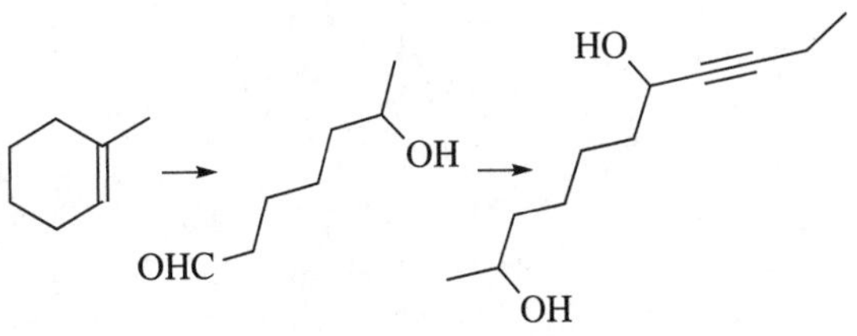 [methylcyclohexene] → [chain with OHC, OH] → [chain with HO, alkyne, OH]

(e)

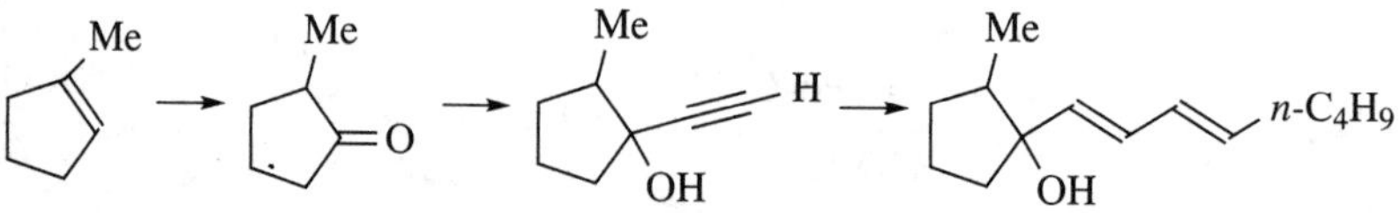

6. In each case, complete the synthesis showing all reagents and all intermediate products.

(a)

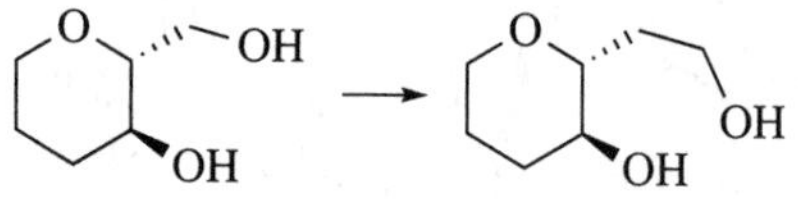

(b)

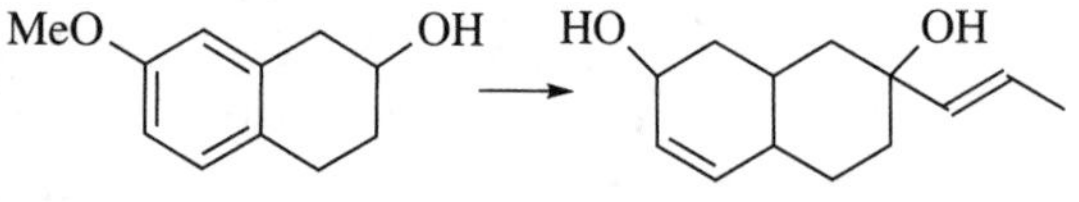

(c)

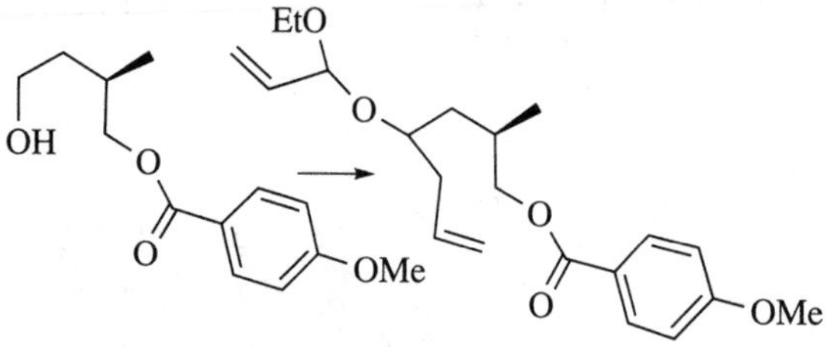

(d)

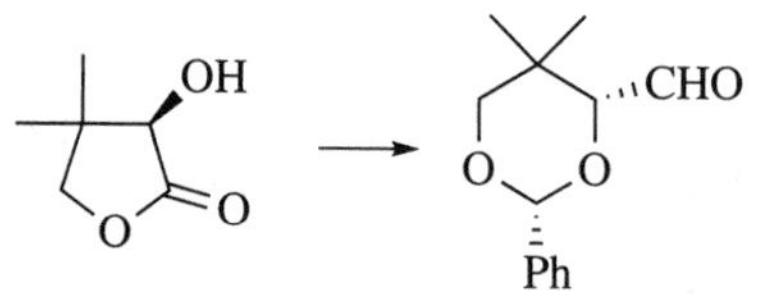

(e)

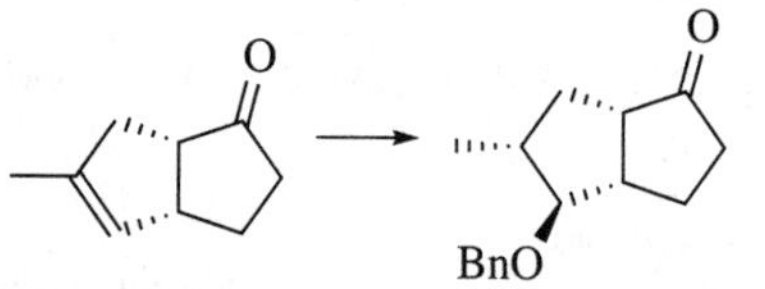

(f)

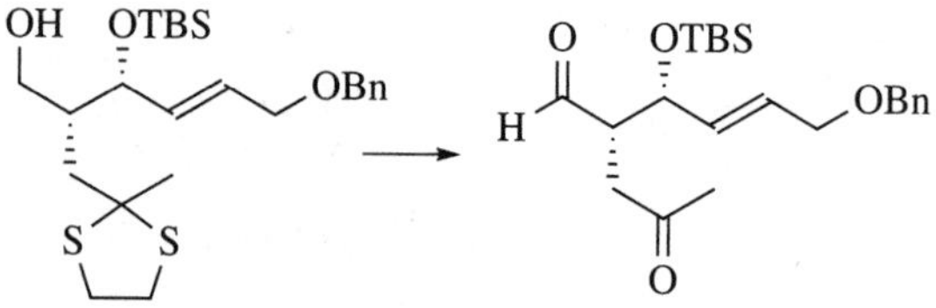

(g)

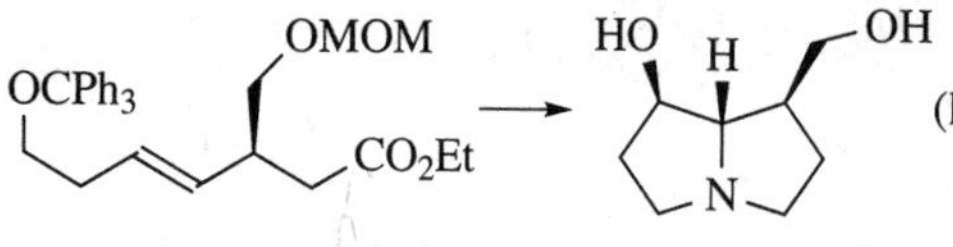

(h)

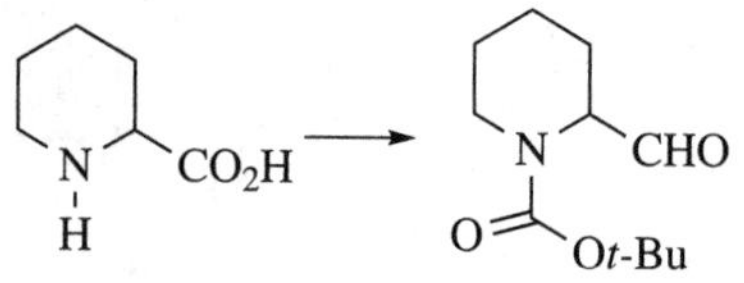

(i)

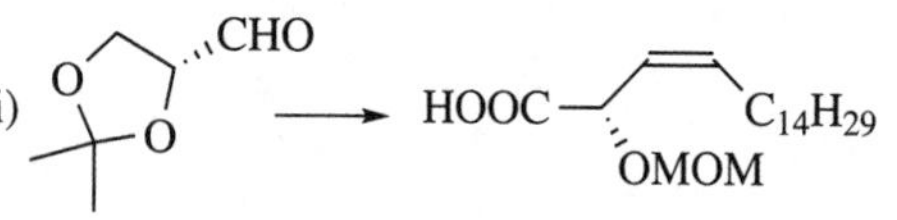

(j)

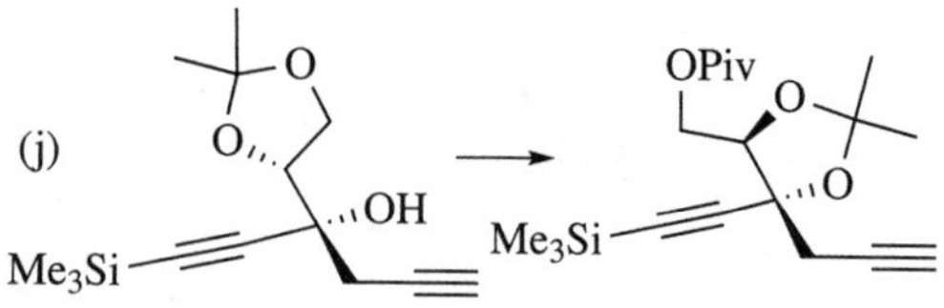

(k)

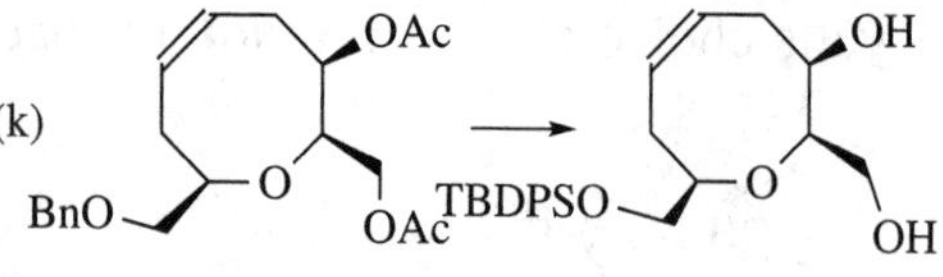

(l)

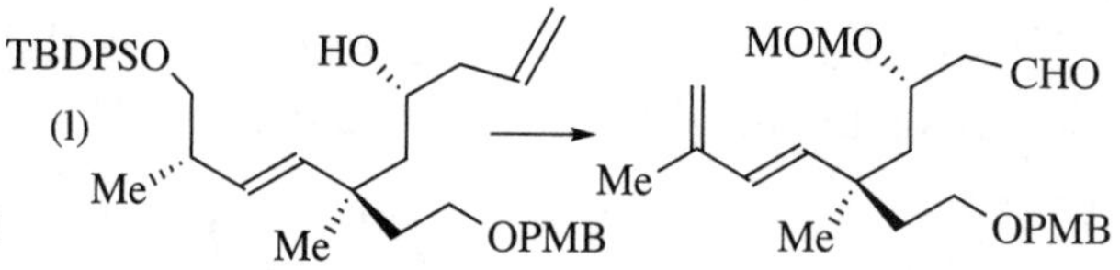

chapter 8

C^d Disconnect Products: Nucleophilic Species that Form Carbon-Carbon Bonds

8.1. INTRODUCTION

The formation of new carbon-carbon bonds poses a major challenge in a synthesis of a complex molecule. Introduction of new functional groups that are compatible with existing functional groups, and also with the reagents that are necessary to complete the transformation further complicates the problem. Finally, the problems of regioselectivity, stereoselectivity, chemoselectivity, and absolute configuration of the stereogenic centers must be addressed in a reasonable manner. As an introduction, the major reactions that build new carbon-carbon bonds have been classified into four major types: nucleophilic, electrophilic, pericyclic and radical reactions. This chapter will focus on those reactions that employ a nucleophilic carbon species. Referring back to Chapter 1, disconnection of a bond generated two fragments (R-C^a and R-C^d), where R-C^d was a donor or nucleophilic fragment and R-C^a was an acceptor or electrophilic fragment (sec. 1.2). The main purpose of this chapter is to identify those types of molecules that function as C^d, and to expand on the number of reactions that were presented in Chapter 1. In most cases the nucleophilic carbon is an organometallic derivative, where the metal is magnesium, lithium, sodium, potassium or a transition metal.

8.2. CYANIDE

8.2.A. Formation of Nitriles

Cyanide ion is a powerful nucleophile that offers a simple but convenient method for generating carbon-carbon bonds. When cyanide displaces a leaving group in a S_N2 reaction, the carbon chain is extended by one carbon, and the nitrile can be converted to a variety of other functional groups.[1]

$$R-X + NaCN \xrightarrow{DMSO} R-C\equiv N + NaX \quad X = Br, Cl, I, OSO_2R$$

Williamson reported the displacement reaction of ethyl halides or sulfonate esters by cyanide ion to give propanenitrile ($CH_3CH_2CH_2CN$)[1] in 1854,[2] but Wöhler and Liebig synthesized the

---

1. Mowry, D.T. *Chem. Rev.* **1948**, *42*, 189.
2. Williamson, A.E. *J. Prakt. Chem.* **1854**, *61*, 60.

first nitrile in 1832.[3] Since the reaction with an alkyl halide is an S_N2 process (sec. 2.6.A.i), the best yields of nitrile are obtained with primary and secondary substrates, whereas tertiary halides such as 2-chloro-2-methylbutane sometimes react via elimination to give the alkene (sec. 2.9.A).[4] Cyanide displacement of secondary halides often gives poor yields, however. The use of polar, aprotic solvents such as DMSO, DMF or THF give the best yields of nitriles such as hexanenitrile (capronitrile, 97% yield after 20 minutes (sec. 2.6.A.i).

Sulfonate ester leaving groups can also be used, as in the conversion of alcohol **1** to the mesylate group and then to nitrile **2** with NaCN in DMSO,[5] in a synthesis of dipeptide nitriles as reversible and potent cathepsin S inhibitors by Ward, Spero and co-workers. The S_N2 reactions proceed with the expected inversion of configuration. Another common reagent that generates a nitrile is tosylmethylisocyanide (TsCH$_2$N≡C, also called TosMIC),[6] which reacts with ketones or aldehydes to generate a nitrile ($R_2C=O \rightarrow R_2CH–C\equiv N$).[6]

Reactions that incorporate a cyano group add one carbon to that molecule. Nitriles are versatile synthetic intermediates since they can be converted to several other functional groups. Three important reactions of nitriles are hydrolysis to acid derivatives,[7] reduction to amines (secs. 4.2.C.iii, 4.8.D), and reduction to aldehydes[8] (secs. 4.3, 4.6.C, 4.9.H). In this chapter, we will see that nitriles react with both Grignard reagents (sec. 8.4.C.iii) and organolithium reagents (sec. 8.5.E) to form ketones. Treatment of nitriles containing an α-proton with a strong base, such as lithium diisopropylamide, generates the α-carbanion, which can react with alkyl halides, aldehydes, or ketones or another nitrile (see the Thorpe reaction in sec. 9.4.B.iv). Reactions of cyanide are not limited to alkylation of halides and sulfonate esters. Epoxides are also opened at the less sterically hindered carbon by cyanide via an S_N2-like process.[9]

Disconnection of a C–CN bond is particularly attractive when the CN unit is attached to a primary or a secondary carbon, which reflects the relative ease of formation of this bond by the

3. Wöhler, F.; Liebig, J. *Annalen* **1832**, *3*, 249, 267.
4. Hass, H.B.; Marshall, J.R. *Ind. Eng. Chem.* **1931**, *23*, 352.
5. Ward, Y.D.; Thomson, D.S.; Frye, L.L.; Cywin, C.L.; Morwick, T.; Emmanuel, M.J.; Zindell, R.; McNeil, D.; Bekkali, Y.; Givadot, M.; Hrapchak, M.; DeTuri, M.; Crane, K.; White, D.; Pav, S.; Wang, Y.; Hao, M.-H.; Grygon, C.A.; Labadia, M.E.; Freeman, D.M.; Davidson, W.; Hopkins, J.L.; Brown, M.L.; Spero, D.M. *J. Med. Chem.* **2002**, *45*, 5471.
6. Oldenziel, O.H.; van Leusen, D.; van Leusen, A.M. *J. Org. Chem.* **1977**, *42*, 3114.
7. For an example taken from a synthesis of *R*-(+)-myrmicarin 217, see Sayah, B.; Pelloux-Léon, N.; Vallée, Y. *J. Org. Chem.* **2000**, *65*, 2824.
8. For an example taken from a synthesis of laulimalide, see Mulzer, J.; Hanbauer, M. *Tetrahedron Lett.* **2000**, *41*, 33.
9. Kergomard, A.; Veschambre, H. *Tetrahedron Lett.* **1976**, 4069.

S_N2 reaction and the variety of functional group transformations available. The disconnection for nitriles can be simplified as:

$$R - C \equiv N \implies R - X$$

8.2.B. Isonitriles: Formation and Reactions

Both potassium and sodium cyanide react with alkyl halides to give excellent yields of the nitrile in the solvent DMSO. An isomeric product is often observed (an isonitrile, also called an isocyanide, **4**)[10] when the reaction is done in a refluxing alcohol solvent, or when certain metal cyanides are used. Gautier and Hofmann were the first to report isonitriles.[11] Cyanide is a bidentate nucleophile, and in general C^- is a better nucleophile than N^-. Nucleophilic strength depends on the metal counterion, however. If M^+ in MCN is sodium or potassium, which favor formation of ionic compounds, cyanide reacts primarily at carbon. If M^+ is a metal such as silver that forms relatively covalent bonds, nucleophilic strength at carbon is diminished and cyanide reacts primarily at nitrogen. These competing pathways are illustrated by the reaction of MCN with 1-bromopentane to give **3** or **4**. Reaction with potassium cyanide (KCN) generates hexanenitrile (**3**), whereas reaction with silver cyanide (AgCN) gives the isomeric isonitrile **4**. If the reaction is done in alcoholic solvents, which minimizes the nucleophilicity of ionic species, the yield of isonitrile is improved. Table 8.1.[1] shows the relationship between formation of isonitriles when ionic and covalent metal cyanides react with alkyl halides. The best yields of isonitrile are observed with silver cyanide and copper cyanide, which form highly covalent metal cyanides. The observed reactivity is due, in part, to the formation and stability of intermediates such as **5**. Reaction of an alkyl halide with silver cyanide is a preparative route to isonitriles, **6**, but they can also be formed by dehydration of formamide derivatives.[12]

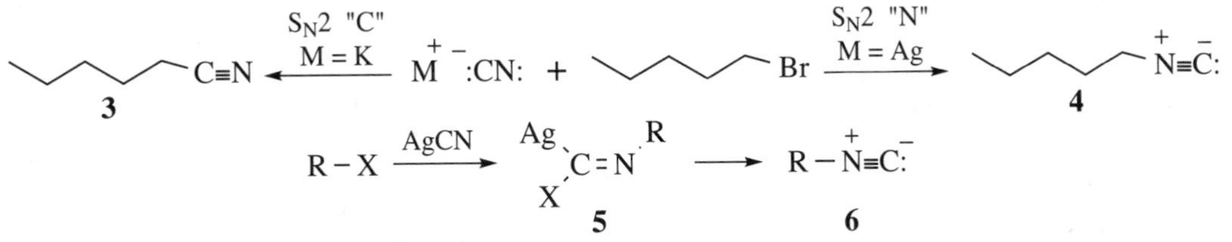

10. (a) Guillemard, H. *Compt. Rend.* **1906**, *143*, 1158; (b) Guillemard, H. *Ibid.* **1907**, *144*, 141; (c) Guillemard, H. *Ibid.* **1907**, *144*, 326; (d) Guillemard, H. *Bull. Soc. Chim. Fr.* **1907**, *1*, 269; (e) Guillemard, H. *Ibid.* **1907**, *1*, 530.
11. (a) Gautier, A. *Ann. Chim. Paris* **1869**, *17*, 103; (b) Gautier, A. *Annalen* **1867**, *142*, 289; (c) Hofmann, A.W. *Ibid.* **1867**, *144*, 114.
12. Ugi, I. *Isonitrile Chemistry*, Academic Press, New York, ***1971***.

Table 8.1. Preparation of isonitriles from alkyl halides and metal cyanides.

$R-X + MCN \longrightarrow R-\overset{+}{N}\equiv\overset{-}{C}:$	
Metal Cyanide	**% $R-N\equiv C$:**
AgCN	quantitative
CuCN	56
$Cd(CN)_2$	11
$Ni(CN)_2$	8
$Zn(CN)_2$	2.6
KCN	trace

Hydrolysis of isonitriles with cold, dilute aqueous acid leads to an amine and formic acid. If the reaction of cyanide with an alkyl halide gives a mixture of nitrile and isocyanide, mild hydrolysis with aqueous acid affects only the isocyanide, allowing isolation of the nitrile. Isocyanides can also be converted to the nitrile by thermal rearrangement, but the reaction requires temperatures of between 140° and 240°C.[1] Heating 3-α-isocyanocholestane **7** to 270°C, for example, gave 3-α-cyano-cholestane (**8**). Interestingly, the reaction proceeded with >99% retention of configuration.[13] If the isonitrile does not contain functionality that would be damaged by heating, good yields of the nitrile can be obtained. Isonitriles are also formed by formylation of a primary amine followed by dehydration, as in the conversion of **9** to **10** in a synthesis of kalihinol C (**10**) by Wood and co-workers.[14] Note that isocyanates[15] are easily formed from isonitriles by reaction with halogens.[15b]

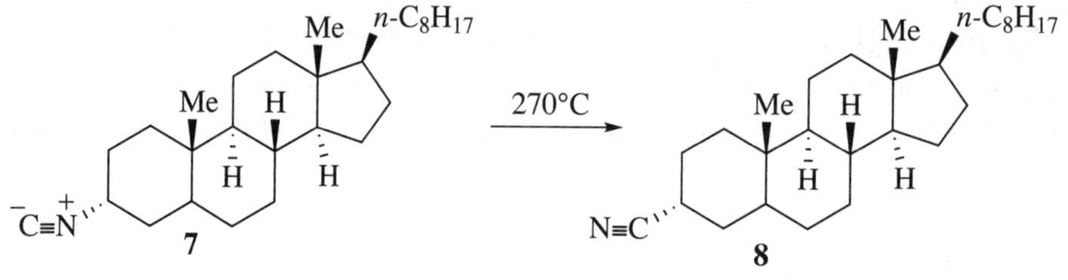

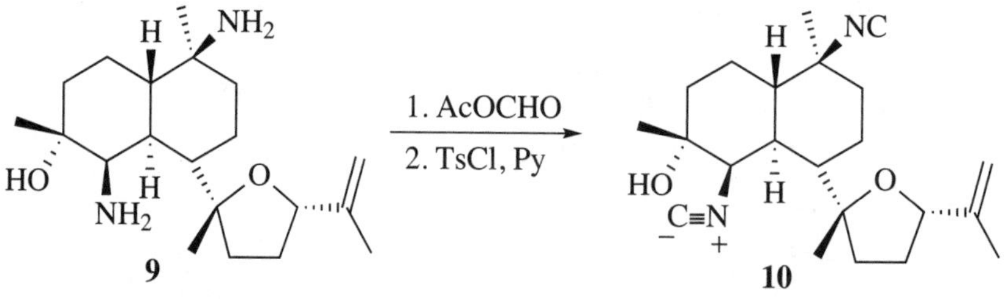

13. Horobin, R.W.; Khan, N.R.; McKenna, J.; Hutley, B.G. *Tetrahedron Lett.* **1966**, 5087.
14. White, R.D.; Keaney, G.F.; Slown, C.D.; Wood, J.L. *Org. Lett.* **2004**, *6*, 1123. For another example using this approach, in a synthesis of aspidophytine, see Sumi, S.; Matsumoto, K.; Tokuyama, H.; Fukuyama, T. *Org. Lett.* **2003**, *5*, 1891.
15. (a) Nef, J.U. *Annalen* **1892**, *270*, 267; (b) Johnson, Jr., H.W.; Daughhetee, Jr., P.H. *J. Org. Chem.* **1964**, *29*, 246.

Isonitriles can be used for a few useful synthetic transformations.[10,16] The **Passerini reaction**[17] is a convenient route to amide esters and is considered to be a **three-component coupling reaction**.[18a] When propionic acid was heated with acetone and *tert*-butylisonitrile, for example, the product was α-propanoyloxy amide **11**.[17a] A useful modification of the Passerini reaction used trifluoroacetic acid rather than a normal aliphatic carboxylic acid. When α-phenoxyacetaldehyde (**12**) reacted with *tert*-butylisonitrile and trifluoroacetic acid (TFA), **13** was formed. Hydrolysis of this trifluoroacetyl ester with aqueous sodium carbonate led to isolation of the α-hydroxyamide (**14**) in 69% yield.[19] This modification has the advantage that hydrolysis conditions for trifluoroacetate esters are milder than for normal esters. Note that the four-component coupling of an isonitrile, an amine, an acid and an aldehyde or ketone to generate an amido amide is called the **Ugi reaction**.[18b-d]

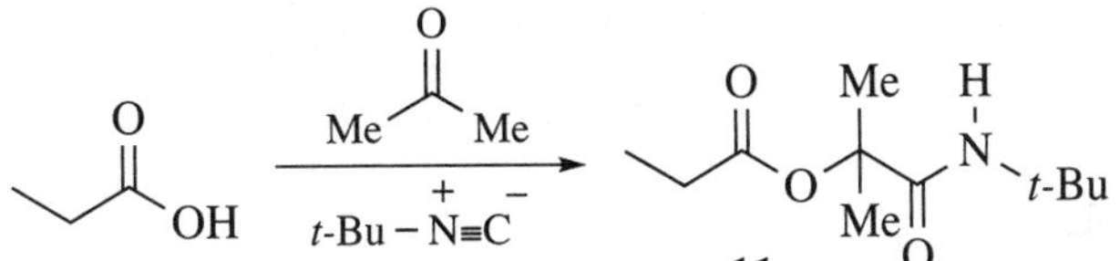

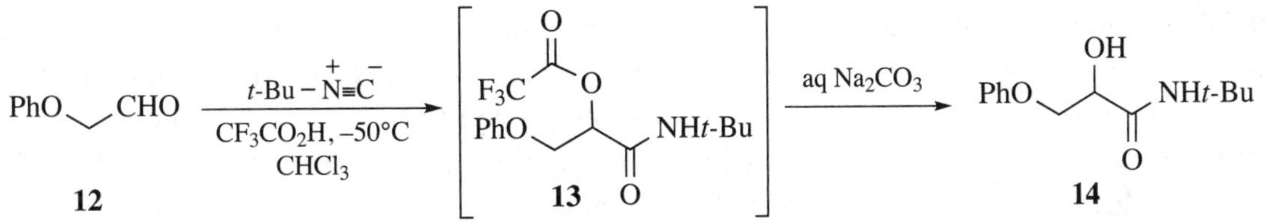

Isonitriles are somewhat limited in their synthetic utility, but they have some interesting disconnections:

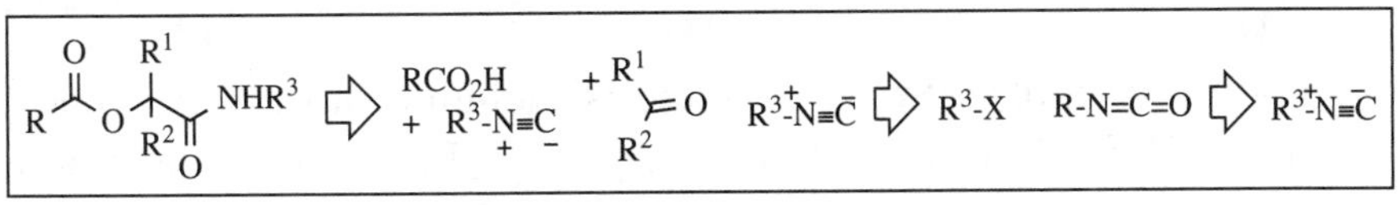

16. (a) Ferosie, I. *Aldrichimica Acta* **1971**, *4*, 21; (b) Dömling, A.; Ugi, I. *Angew. Chem. Int. Ed.* **2000**, *39*, 3168.

17. (a) Passerini, M. *Gazz. Chim. Ital.* **1921**, *51 II*, 126; (b) Passerini, M. *Ibid.* **1924**, *54*, 529; (c) The Merck Index, 14th Ed., Merck & Co., Inc., Whitehouse Station, New Jersey, **2006**, p ONR-69; (d) Mundy, B.P.; Ellerd, M.G.; Favaloro Jr., F.G. *Name Reactions and Reagents in Organic Synthesis, 2nd ed.*, Wiley-Interscience, New Jersey, **2005**, pp. 482-483.

18. (a) For a review, of isocyanide based multi-component coupling reactions, see Dömling, A. *Chem. Rev.* **2006**, *106*, 17. For information concerning the Ugi reaction, see (b) Lindhorst, T.; Bock, H.; Ugi, I. *Tetrahedron* **1999**, *55*, 7411; (c) Dömling, A.; Ugi, I. *Angew. Chem. Int. Ed.* **2000**, *39*, 3168. See pp 3183-3184 and references therein. Also see(d) Tye, H.; Whittaker, M. *Org. Biomol. Chem.* **2004**, *2*, 813.

19. Lumma, Jr., W.C. *J. Org. Chem.* **1981**, *46*, 3668.

8.2.C. Other Nitrile-Forming Reactions

Aryl halides are inert to reaction with cyanide ion under normal S_N2 conditions (chap. 2, sec. 2.11), but they do react when heated with cuprous salts like cuprous cyanide (CuCN).[20] This transformation is called the **Rosenmund-von Braun reaction**,[21] and it can be applied to a wide range of aryl halides, and the products can subsequently be used in a variety of reactions.[22] This transformation does *not* work when potassium or sodium cyanide is used. An amine base such as *N*-methylpyrrolidine is often added to facilitate the reaction. A synthetic example is taken from a synthesis of stachybotrylactam by Kende and co-workers,[23] in which aryl bromide **15** was treated with cuprous cyanide in hot DMF to give a 92% yield of aryl nitrile **16**.

Aliphatic or aromatic carboxylate salts are also nitrile precursors. For example, fusion (250-300°C) of sodium propionate with cyanogen bromide[24] gave propanenitrile.[25,24b] Cyanide is a nucleophile that reacts with carbonyl derivatives via acyl addition, although this is usually a reversible reaction. The classical **Strecker synthesis**[1,26] used to prepare racemic α-amino acids is based on controlled addition to give a cyanohydrin [RCHCN(OH)]. When phenylacetaldehyde (**17**) was treated with NH_3 and HCN, for example, the product was the amino nitrile, **18**. Hydrolysis of the nitrile generated the amino acid, phenylalanine (**19**).[26]

There are many modifications of the Strecker synthesis, including those by Erlenmeyer, Tiemann, Zelinsky and Stadnikoff or Knoevenagel and Bucherer.[1] In all cases, the reaction produces

20. Newman, M.S.; Boden, H. *J. Org. Chem.* **1961**, *26*, 2525.
21. (a) Rosenmund, K.W.; Struck, E. *Berichte* **1916**, *52*, 1749; (b) The Merck Index, 14th ed., Merck & Co., Inc., Whitehouse Station, New Jersey, **2006**, p ONR-81.
22. (a) Ito, T.; Watanabe, K. *Bull. Chem. Soc. Jpn.* **1968**, *41*, 419; (b) Akanuma, K.; Amamiya, H.; Hayashi, T.; Watanabe, K.; Hata, K. *Nippon Kagakai Zashi* **1960**, *81*, 333 [*Chem. Abstr. 56*: 406i **1962**)]; (c) Cohen, T.; Lewin, A.H. *J. Amer. Chem. Soc.* **1966**, *88*, 4521.
23. Kende, A.S.; Deng, W.-P.; Zhong, M.; Guo, X.-C. *Org. Lett.* **2003**, *5*, 1785.
24. (a) Douglas, D.E.; Eccles, J.; Almond, A.E. *Can. J. Chem.* **1953**, *31*, 1127; (b) Douglas, D.E.; Burditt, A.M. *Ibid.* **1958**, *36*, 1256; (c) Barltrop, J.A.; Day, A.C.; Bigley, D.B. *J. Chem. Soc.* **1961**, 3185.
25. Payot, P.H.; Dauben, W.G.; Replogle, L. *J. Am. Chem. Soc.* **1957**, *79*, 4136.
26. (a) Strecker, A. *Annalen* **1850**, *75*, 27; (b) The Merck Index, 14th ed., Merck & Co., Inc., Whitehouse Station, New Jersey, **2006**, p ONR-90; (c) Mundy, B.P.; Ellerd, M.G.; Favaloro Jr., F.G. *Name Reactions and Reagents in Organic Synthesis, 2nd ed.*, Wiley-Interscience, New Jersey, **2005**, pp. 632-633.

racemic amino acids but newer methodology introduces enantioselectivity into the process. One modification of the original Strecker procedure uses a chiral amine to produce amino acids with good asymmetric induction.[27] Another example of a stereoselective Strecker reaction uses (R)-phenylglycinol as chiral auxiliary for the diastereoselective synthesis of (R)-amino acids.[28] In one example by Irie and co-workers, taken from a synthesis of 8-octylbenzolactam-V9, a selective activator for protein kinase Cε and η, aldehyde **20** (containing the chiral auxiliary on nitrogen) reacted with phenylglycinol and trimethylsilyl cyanide to give **21**, in >70% yield after refluxing in methanol.[29] Catalytic enantioselective Strecker reactions are also known.[30]

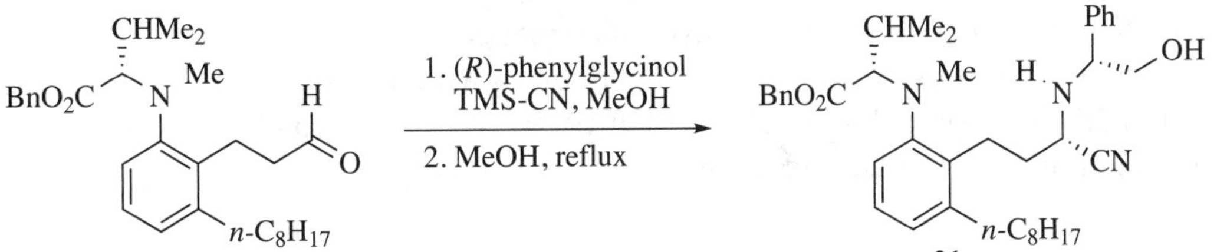

Other acid derivatives can also be prepared with HCN and carbonyl derivatives. Addition of HCN to a ketone or aldehyde gives a cyanohydrin in a reversible reaction, and hydrolysis gives the α-hydroxy acid.[1,31] This process is an equilibrium process favoring the cyanohydrin for aliphatic aldehydes and ketones. The yields are usually poor with aryl and diaryl ketones, however. These reactions add the following disconnections to those previously shown for nitriles:

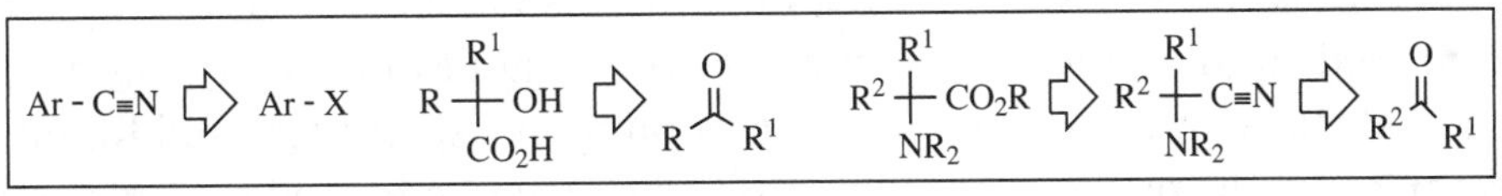

8.3. ALKYNE ANIONS (R–C≡C:⁻)

The conjugate base of an alkyne is an alkyne anion (older literature refers to them as acetylides), and it is generated by reaction with a strong base and is a carbanion. It functions as a nucleophile (a source of nucleophilic carbon) in S_N2 reactions with halides and sulfonate esters. Acetylides react with ketones, with aldehydes via nucleophilic acyl addition and with acid derivatives via nucleophilic acyl substitution. Acetylides are important carbanion synthons for the creation of

27. Weinges, K.; Stemmle, B. *Chem. Ber.* **1973**, *106*, 2291.
28. (a) Chakraborty, T.K.; Hussain, K.A.; Reddy, G.V. *Tetrahedron* **1995**, *51*, 9179. Also see (b) Ma, D.; Tian, H.; Zou, G. *J. Org. Chem.* **1999**, *64*, 120; (c) Ferraris, D.; Young, B.; Cox, C.; Dudding, T.; Drury III, W.J.; Ryzhkov, L.; Taggi, A.E.; Lectka, T. *J. Am. Chem. Soc.* **2002**, *124*, 67; (d) Kobayashi, S.; Ishitani, H. *Chem. Rev.* **1999**, *99*, 1069; (e) France, S.; Guerin, D.J.; Miller, S.J.; Lectka, T. *Chem. Rev.* **2003**, *103*, 2985.
29. Nakagawa, Y.; Irie, K.; Yanagita, R.C.; Ohigashi, H.; Tsuda, K-i.; Kashiwagi, K.; Saito, N. *J. Med. Chem.* **2006**, *49*, 2681.
30. For a review, see Groger, H. *Chem. Rev.* **2003**, *103*, 2795.
31. Reference 1, pp 231-246.

new carbon-carbon bonds. Some of the chemistry presented here will deal with the synthesis of alkynes (see Chapter 2) in order to provide continuity with the discussion of acetylides.

8.3.A. Preparation of Alkynes

Alkynes have been thoroughly studied over many years[32] and numerous methods are available for their preparation.[33] One of the most useful is dehydrohalogenation of dihalides. This method works best for aryl derivatives when trans elimination is possible.[34] Treatment of tetrabromide **22** with potassium *tert*-butoxide gave the usual E2 elimination (sec. 2.9.A) to give **23**. Subsequent base induced dehydrohalogenation gave mono-alkyne **24**, since only one trans elimination was possible.[35] Treatment of **24** with a second equivalent of base gave bis(alkyne) **25**, but required a cis elimination pathway. This second elimination was more difficult, requiring excess base and more vigorous conditions.

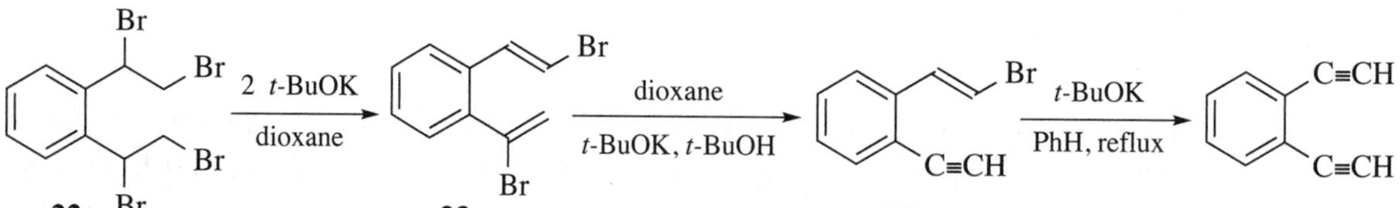

As the chain length of the vinyl halide increases, the yield of terminal alkyne product generally decreases.[23,36] Vinyl bromide **26** ($n = 0$) gave 61% of alkyne **27** ($n = 0$), for example, but 1-bromo-1-hexene (**26**, $n = 4$) gave only 18% of the corresponding alkyne (**27**, $n = 4$).[36] Alkynes can also be formed from non-terminal (internal) vinyl halides (such as 3-bromo-3-heptene), but when alkoxide bases are used, the vigorous conditions usually required for elimination of the halogen can cause the triple bond of the product to migrate to the terminal position.[37]

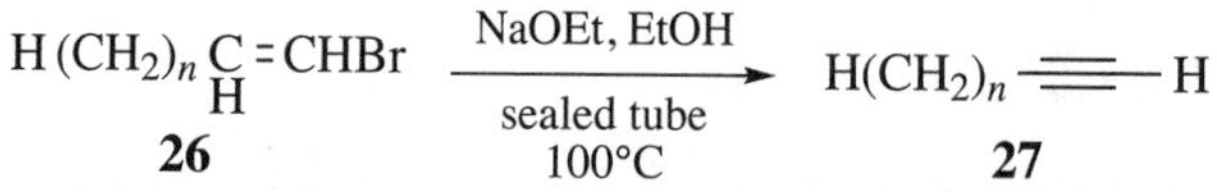

1,1-Dibromoalkenes are a useful source of terminal alkynes, which give an alkyne anion when treated with *n*-butyllithium. In a synthesis of (+)-phomactin A by Halcomb and Mohr,[38] the **Corey-Fuchs procedure**,[39] which is related to the Wittig olefination reaction (see sec. 8.8.A), was used to convert aldehyde **28** to **29**. Subsequent treatment with butyllithium gave **30**, in 92% overall yield.

32. Viehe, H.R. (Ed.) *Chemistry of Acetylenes*, Marcel Dekker, New York, *1969*.
33. Larock, R.C. *Comprehensive Organic Transformations, 2nd Ed.* Wiley-VCH, New York, *1999*, pp 569-581.
34. (a) Reference 32 , pp 100-168; (b) Fiesselmann, H.; Sasse, K. *Chem. Ber. 1956, 89*, 1775.
35. Gaudemar-Bardone, F. *Ann. Chim. (Paris) 1958 , 3*, 52.
36. Loevenich, J.; Losen, J.; Dierichs, A. *Berichte 1927, 60*, 950.
37. Vaughn, T.H. *J. Am. Chem. Soc. 1933, 55*, 3453.
38. Mohr, P.J.; Halcomb, R.L. *J. Am. Chem. Soc. 2003, 125*, 1712.
39. Corey, E.J.; Fuchs, P.L. *Tetrahedron Lett. 1972, 13*, 3769.

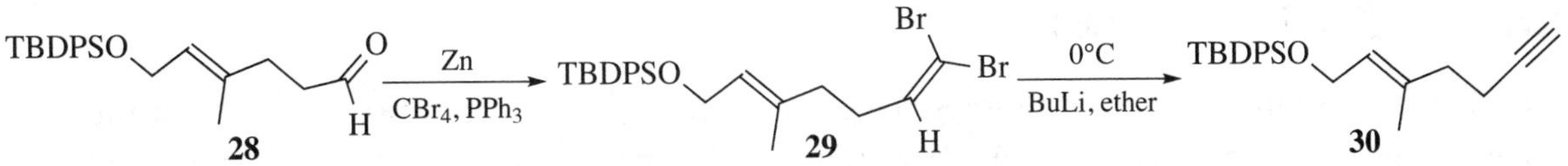

Coupling is an important reaction of alkynes that leads to di-ynes. Two classical alkyne-coupling reactions involve copper derivatives, but a variety of conditions can be used.[40] In the **Glaser reaction**,[41] an alkyne such as phenylacetylene reacts with basic cupric chloride ($CuCl_2$), and subsequent air oxidation gives a diyne (in this case **31**, in 90% yield). This reaction has also been reported to give a quantitative yield of **31** in the presence of $CuCl_2$ and NaOAc, using supercritical CO_2 as the solvent.[42] Glaser-type coupling of alkynes can be accomplished with copper derivatives and bases other than hydroxide. Cupric acetate and pyridine is also an effective reagent.[43] The second important reaction is the **Cadiot-Chodkiewicz coupling**,[44] in which a bromoalkyne such as reacts with a monoalkyne in the presence of cupric chloride and an amine to give the diyne.[44a] In a synthesis of panaxytriol by Danishefsky and Yun,[45] Cadiot-Chodkiewicz cross-coupling of bromo-alkyne **32** and alkyne **33** gave a 63% yield of diyne **34**. Internal coupling reactions are also possible. Indium metal[46] was used to couple alkynes with allyl bromide with an anti-Markovnikov regiochemistry.

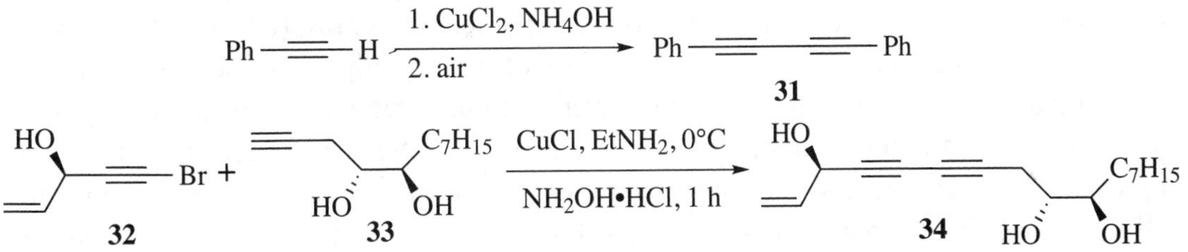

40. Cadiot, P.; Chodkiewicz, W. in *Chemistry of Acetylenes*, Viehe, H.G. (Ed.), Marcel-Dekker, New York, *1969*, pp 597-648.
41. (a) Glaser, C. *Berichte* *1869*, *2*, 422; (b) *Idem Annalen 1870, 154*, 137; (c) The Merck Index, 14th ed., Merck & Co., Inc., Whitehouse Station, New Jersey, *2006*, p ONR-37; (d) Mundy, B.P.; Ellerd, M.G.; Favaloro Jr., F.G. *Name Reactions and Reagents in Organic Synthesis, 2nd ed.*, Wiley-Interscience, New Jersey, *2005*, pp. 276-277.
42. Li, J.; Jiang, H. *Chem. Commun.* *1999*, 2369.
43. Brown, A.B.; Whitlock, Jr., H.W. *J. Am. Chem. Soc.* *1989*, *111*, 3640.
44. (a) Chodkiewicz, W.; Cadiot, P. *Compt. Rend.* *1955*, *241*, 1055; (b) Mundy, B.P.; Ellerd, M.G.; Favaloro Jr., F.G. *Name Reactions and Reagents in Organic Synthesis, 2nd ed.*, Wiley-Interscience, New Jersey, *2005*, pp. 128-129.
45. Yun, H.; Danishefsky, S.J. *J. Org. Chem.* *2003*, *68*, 4519.
46. (a) Araki, S.; Imai, A.; Shimizu, M.; Yamada, A.; Mori, Y.; Butsugan, Y. *J. Org. Chem.* *1995*, *60*, 1841; (b) Ranu, B.C. *Eur. J. Org. Chem.* *2000*, 2347.

8.3.B. Acidity of Terminal Alkynes

Acetylene and other terminal alkynes have an acidic hydrogen atom (C≡C-H), and they are weak acids. A strong base is required to remove that proton to give an alkyne anion such as **35**, and this carbanion is a carbon nucleophile that reacts with alkyl halides or sulfonate esters (R^1-X where R^1 = Me, 1°, 2° alkyl, and X = halogen or OSO_2R) via an S_N2 sequence to give disubstituted alkynes such as **36**. In a synthesis of (–)-neodysiherbaine A[47] by Hatakeyama and co-workers, triflate **37** was treated with the alkyne anion from butyllithium/HC≡CCH$_2$OPMB to give a 96% yield of **38**.

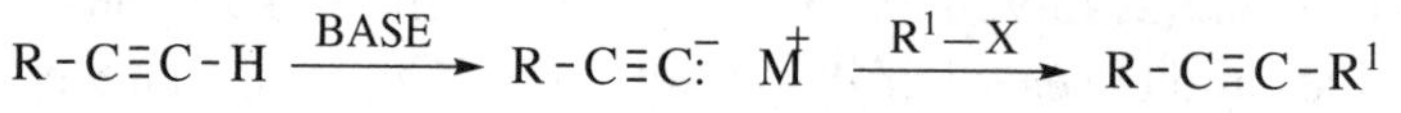

As shown in Table 8.2,[48] the acidity of a proton is enhanced (lower pK_a) as the **s character** of the carbon to which it is attached increases.[49] A variety of bases (sodium hydride, lithium, sodium, or potassium amide, lithium dialkyl amides, organolithium reagents or Grignard reagents) can be used to generate an alkyne anion.[50] 1-Octyne reacted with methylmagnesium bromide, for example, to give methane and the magnesium salt of the acetylide.[50] This reaction must be considered if a Grignard reaction (secs. 8.4.A, 8.4.F.i) is one step of a planned synthesis that involves a molecule containing a terminal alkyne.[51]

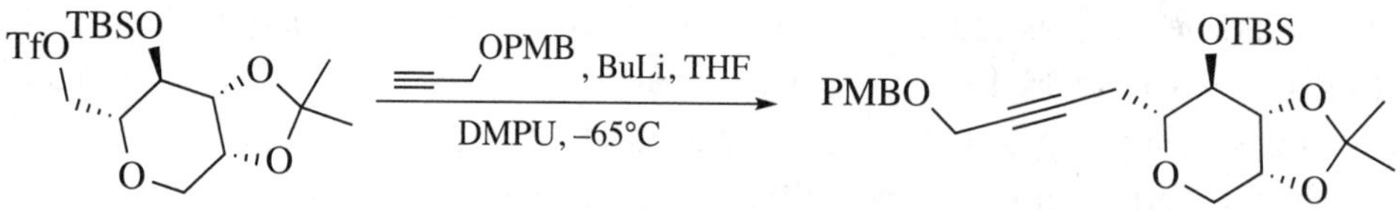

47. Takahashi, K.; Matsumura, T.; Corbin, G.R.; Corbin, R.M.; Ishihara, J.; Hatakeyama, S. *J. Org. Chem.* **2006,** *71,* 4227.
48. (a) Hopkinson, A.C. in *The Chemistry of the Carbon-Carbon Triple Bond,* Patai, S. (Ed. John Wiley, Chichester-New York, **1978,** pp 75-136; (b) Matthews, W.S.; Bares, J.E.; Bartmess, J.E.; Bordwell, F.G.; Cornforth, F.J.; Drucker, G.E.; Margolin, Z.; McCallum, R.J.; McCollum, G.J.; Vanier, N.R. *J. Am. Chem. Soc.* **1975,** *97,* 7006; (c) McEwen, W.K. *Ibid.* **1936,** *58,* 1124; (d) Streitwieser, Jr., A.; Reuben, D.M.E. *Ibid.* **1971,** *93,* 1794.
49. (a) Cram. D. *Fundamentals of Carbanion Chemistry,* Academic Press, New York, **1965**; (b) Maksić, Z.B.; Eckert-Maksić, M. *Tetrahedron* **1969,** *25,* 5113; (c) Maksić, Z.B.; Randic, M. *J. Am. Chem. Soc.* **1973,** *95,* 6522; (d) Emsley, J.W.; Feenez, J.; Sutcliffe, C.H. *High Resolution NMR,* Pergamon Press, New York, **1966.**
50. Kriz, J.; Beneš, M.J.; Peška, J. *Tetrahedron Lett.* **1965,** 2881.
51. (a) Prevost, C.; Gaudemar, M.; Honigberg, J. *Compt. Rend.* **1950,** *230,* 1186; (b) Wotiz, J.H.; Matthews, J.S.; Lieb, J.A. *J. Am. Chem. Soc.* **1951,** *73,* 5503.

Table 8.2. Relative Acidity of Common Weak Acids

Acid	Conjugate Base	pK$_a$
RC≡CH	RC≡C:$^-$	25
R$_2$C=CH$_2$	R$_2$CH3=CH:$^-$	36
H$_3$C–C(=O)–CH$_3$	H$_3$C–C(=O)–CH$_2^-$	19
ROH	RO$^-$	17

The acidity of an alkynyl hydrogen is strongly influenced by the nature of the group on the other side of the triple bond, as shown in Table 8.3.[52] When the R group is electron releasing (e.g., an alkyl group such as butyl) the acidity of the proton decreases. Likewise, an electron withdrawing R group increases the acidity.[52] In Table 8.3, the alkyne anion was trapped with D$_2$O to give the *d*-alkyne (**39**). Strong bases such as MNH$_2$ (M = Li, K, Na), or *n*-butyllithium in ether (sec. 8.5.G) abstract the proton rapidly at –80°C. Ethylmagnesium bromide is less basic so alkyne deprotonation is slower (30-90 minutes at reflux).[52] Conjugation enhances the acidity[53] (see Table 8.3) and *n*-PrC≡C–C≡CH reacts instantaneously with ethylmagnesium bromide in refluxing ether. This contrasts with *n*BuC≡CH, which shows no reaction after 15 min under the same reaction conditions, although it reacts slowly when longer reaction times are used.[52] At higher reaction temperatures, dilithiation [HC≡CH → HC≡C–Li → Li-C≡C–Li] can be a problem with acetylene. Attempts to form HC≡C–M with either *n*-butyllithium (M = Li) or Grignard reagents (M = MgBr) in ether led to M–C≡C–M due to the instability of M–C≡C–H. In THF, however, both HC≡CMgBr and HC≡CLi are stable enough below 35°C to be used as nucleophiles in an S$_N$2 reaction. Therefore, *n*-butyllithium can be used without problem under the proper conditions (ether, < -10°C).

Table 8.3. Effect of Substituents on Alkyne Acidity

$$\text{R–C≡C–H} \xrightarrow[\text{D}_2\text{O}]{\text{NEt}_3,\ \text{DMF}} \text{R–C≡C–D} \quad \mathbf{39}$$

R	Relative Acidity
Ph-	1.0
H-	0.73
n-Bu-	0.058
Cl(CH$_2$)$_4$-	0.033
HC≡C(CH$_2$)$_4$-	0.076
MeO-	2.0

The acidity of a terminal alkyne hydrogen leads to another problem when one attempts to form the Grignard reagent of propargyl bromide (**40**). The initial magnesium insertion reaction generates a Grignard reagent, but migration of the MgBr moiety is possible because of the

---

52. (a) Reference 32 , p 171; (b) Wotiz, J.H.; Hollinsworth, C.A.; Dessy, R.E. *J. Org. Chem.* **1955**, *20*, 1545.
53. Reference 32 , p 15.

terminal alkyne moiety to form an allene structure (**41**). Allene Grignard **41** is converted to the propynylmagnesium salt (**42**) at temperatures below 0°C.[54] Attempts to use the bromomethyl carbon atom in alkyne **40** as a nucleophile should therefore be approached with caution.

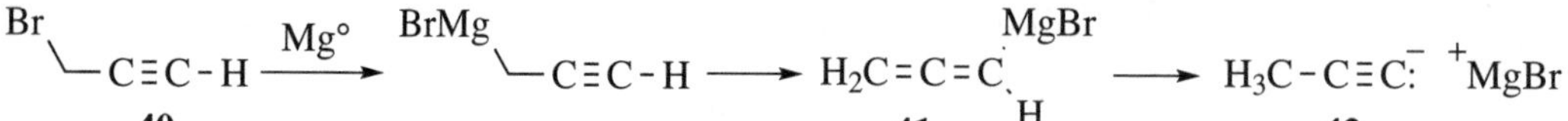

When an alkyne possesses a α-proton (like 1-butyne), heating can induce a rearrangement to an internal alkyne (2-butyne) under basic conditions.[55] In a synthesis of (–)-terpestacin by Jamison and Chan,[56] for example, heating alkyne **43** with *tert*-butoxide in DMSO gave a 95% yield of **44**. This rearrangement is usually not a problem in S_N2 alkylation reactions involving alkyl halides, at reaction temperatures from ambient to ≈ 120°C (rearrangement requires much higher temperatures). If the alkylation is sluggish and the reaction requires refluxing in DMF or DMSO however, such rearrangements may be competitive.

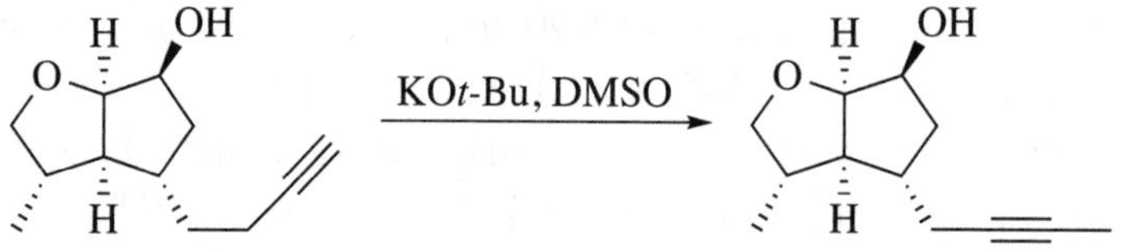

8.3.C. Reactions of Alkyne Anions

As noted above, alkyne anions are very useful in S_N2 reactions with alkyl halides and in acyl addition reactions to a carbonyl.[57] Alkyl halides[58] and sulfonate esters (tosylates and mesylates primarily) are electrophilic substrates for alkyne anions, as in the conversion of **37** to **38** above. Another example that uses a sulfonate ester leaving group is taken from Sasaki and co-worker's synthesis of the EFGH ring system of ciguatoxin, in which alcohol **45** was first treated triflic anhydride to generate the triflate.[59] Subsequent reaction with the alkyne anion resulting from the reaction of trimethylsilylacetylene with butyllithium, gave a 61% yield of **46**.

54. (a) Zakharova, A.I. *Zh. Obshch. Khim.* **1947**, *17*, 1277; (b) Zakharova, A.I. *Ibid.* **1949**, *19*, 1297; (c) Pasternak, Y. *Compt. Rend.* **1962**, *255*, 3429; (d) Jacobs, T.L.; Prempree, P. *Ibid.* **1967**, *89*, 6177.
55. (a) Wotiz, J.H.; Billups, W.E.; Christian, D.T. *J. Org. Chem.* **1966**, *31*, 2069; (b) Jacobs, T.L.; Dankner, D. *Ibid.* **1957**, *22*, 1424; (c) Bainvel, J.; Wojtkowiak, B.; Romanet, R. *Bull. Soc. Chim. Fr.* **1963**, 978.
56. Chan, J.; Jamison, T.F. *J. Am. Chem. Soc.* **2003**, *125*, 11514.
57. For reactions of this type, see Larock, R.C. *Comprehensive Organic Transformations, 2nd Ed.* Wiley-VCH, New York, **1999**, pp 585-605.
58. Porter, N.A.; Chang, V.H.-T.; Magnin, D.R.; Wright, B.T. *J. Am. Chem. Soc.* **1988**, *110*, 3554.
59. Sasaki, M.; Noguchi, T.; Tachibana, K. *J. Org. Chem.* **2002**, *67*, 3301.

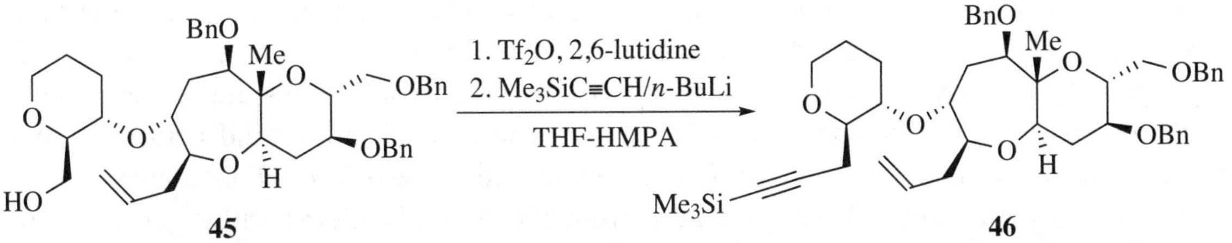

As with cyanide, S_N2 reactions of alkyne anions can be done with substrates other than halides or sulfonate esters. Epoxides are opened by acetylides at the less sterically hindered carbon to give an alkynyl alcohol. A synthetic example is taken from Kishi and co-worker's synthesis of mycolactones A and B, in which reaction of epoxide **47** with the indicated lithium alkyne anion was facilitated by addition of boron trifluoride etherate, to give a 96% yield of **48**.[60]

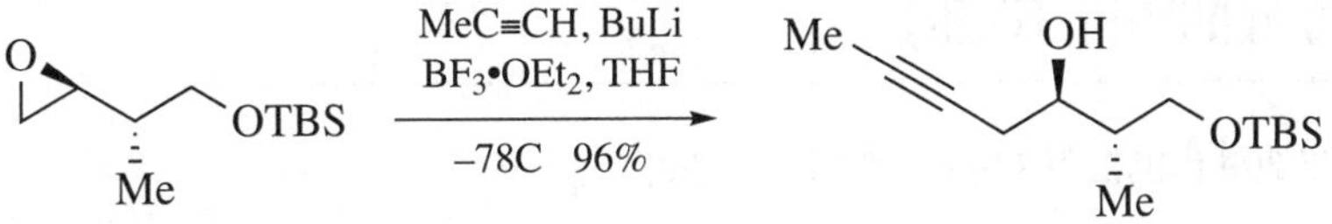

Nucleophilic acyl addition to ketones or aldehydes to give an alkynyl alcohol is the other major synthetic use of alkyne anions. A synthetic example is taken from a synthesis of the C10-C21 segment of lycoperdinoside by Chakraborty and co-workers in which aldehyde **49** reacted with a lithio-alkyne generated form the alkyne and *n*-butyllithium to give a 1:1 mixture of diasteromeric alcohols **50** in >90% yield.[61] The reaction is selective for ketones and aldehydes in the presence of acid derivatives, if the acetylide is not present in large excess.[62]

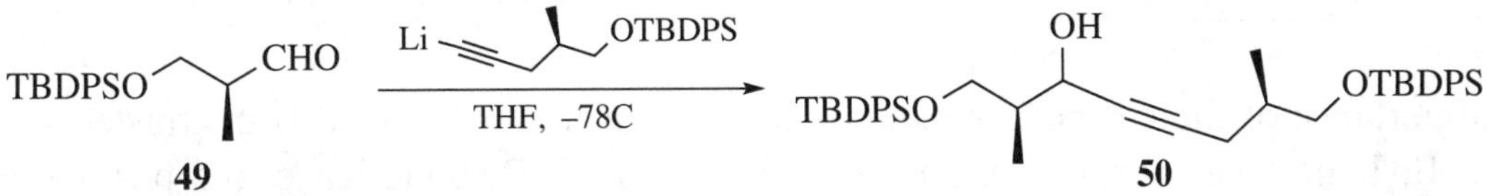

A variety of polar substituents can be attached to the C≡C unit.[63] Alkyne anions show some propensity for 1,2-addition in conjugated systems, especially if the β-carbon of the double bond is hindered.

The reactions of acetylides with carbonyls and halides lead to the following disconnections:

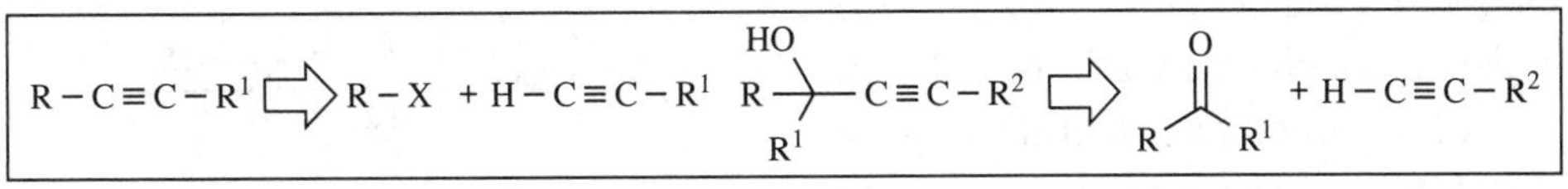

---

60. Song, F.; Fidanze, S.; Benowitz, A.B.; Kishi, Y. *Tetrahedron* ***2007**, *63*, 5739.
61. Chakraborty, T.K.; Goswami, R.K.; Sreekanth, M. *Tetrahedron Lett.* ***2007**, *48*, 4075.
62. Butlin, R.J.; Holmes, R.B.; Mcdonald, E. *Tetrahedron Lett.* ***1988**, *29*, 2989.
63. Ohkuma, K. *Agr. Biol. Chem. (Tokyo)* ***1965**, *29*, 962.

Disconnection of the C-//-C≡C bond is a high priority when the C≡C unit is attached to a primary or secondary carbon, reflecting the facile formation of that bond via the S_N2 reaction. The alkyne moiety is easily converted to a variety of other functional groups, including reduction to alkenes, oxidation to ketones and aldehydes or addition of halides and acid derivatives. Two-carbon or multiple-carbon chain extensions are characteristic of this reaction, and the functional group transforms that are available from alkynes make them highly useful synthons. The ability to reduce alkynes to either the (E) or (Z) alkene is also of great synthetic value, as discussed in Section 4.9.B for cis reduction, and in Section 4.10.D for trans reduction. Addition reactions (with HX or X_2) were discussed in Sections 2.10.A, 2.10.C. There are several copper-catalyzed coupling reactions of aliphatic and aromatic alkynes, but these are primarily radical mediated reactions and will be discussed in Chapter 13.

8.4 GRIGNARD REAGENTS (C-MG)

8.4.A. Preparation and Properties of Grignard Reagents

Victor Grignard reported what we now call the **Grignard reaction** in 1900,[64a] and he was awarded the Nobel Prize in 1912. The Grignard reaction is now one of the most important, and powerful in all of organic chemistry, and it set the stage for much of the organometallic chemistry that is so important in modern organic chemistry.[64b] The so-called **Grignard reagent** (RMgX) is formed by the reaction of magnesium (Mg(0)) with an alkyl or aryl halide, usually in an ether solvent. A simple example is the reaction of bromoethane with magnesium in ether to give **51**.

$$\boxed{^{\delta+}C-Br^{\delta-}} \quad CH_3CH_2Br + Mg^{\circ} \xrightarrow{\text{ether}} CH_3CH_2MgBr \quad \boxed{^{\delta-}C-Mg^{\delta+}-Br}$$
$$\textbf{51}$$

The importance of the Grignard reagent is seen when the C–Br bond is contrasted with the C–Mg–Br bond. The normal bond polarization in the C–Br bond leads to a positive dipole on carbon, so it is an electrophilic carbon. The C–Mg bond in the Grignard reagent, however, generates a negative dipole on carbon, so it is a nucleophilic carbon. This ability to *invert* the polarity of a carbon atom means the Grignard reagent can function as a carbon nucleophile, usually in reactions with a carbonyl derivative. The reaction was first discovered in 1899 by Barbier, who was Grignard's mentor, although its nature was not well understood at the time. Barbier discovered that a ketone (such as **52**) reacted with magnesium metal, Mg(0), in the presence of iodomethane in ether to give an alcohol (**53**, after hydrolysis) that resulted

64. (a) Grignard, V. *Compt. Rend.* **1900**, *130*, 1322; (b) Richey, Jr., H.G. *Grignard Reagents. New Developments*, John Wiley, Chichester, **2000**; (c) The Merck Index, 14th ed., Merck & Co., Inc., Whitehouse Station, New Jersey, **2006**, p ONR-38; (d) Mundy, B.P.; Ellerd, M.G.; Favaloro Jr., F.G. *Name Reactions and Reagents in Organic Synthesis, 2nd ed.*, Wiley-Interscience, New Jersey, **2005**, pp. 284-287.

from coupling of the two organic fragments.[65] This conversion is now known as the **Barbier reaction** or **Barbier coupling**.[65,66] Its utility is somewhat limited by the requirement that all reagents be mixed together. In subsequent work, Grignard took the reaction much further. He premixed magnesium metal and the halide, characterized the resulting product as RMgX, and showed how RMgX reacted with many functional groups (particularly with ketones and aldehydes).

Note that there are several modern versions of the Barbier reaction that use transition metals or their derivatives.[67] Examples include indium metal, samarium iodide, and lead, although lithium and magnesium can be used. An example, taken from a synthetic approach to the guanacastepenes by Srikrishna and Dethe,[68] reacted ketone **54** with lithium metal and 4-bromo-1-butene with ultrasound irradiation to give **55**. A retro-Barbier fragmentation has also been reported

in which an alcohol was treated with ten equivalents of bromine and fifteen equivalents of potassium carbonate in chloroform to give a bromo ketone.[69]

When organomagnesium compounds are formed using the Grignard protocol, the reagent is not isolated. In solution, a Grignard reagent (see **56**) is not the simple monomeric species RMgX. Ashby reviewed the composition and discussed the mechanism for formation of Grignard reagents.[70] Ashby's conclusion was that the reaction proceeds by a single electron transfer (SET) process. Other work involving the formation of aromatic Grignard reagents, however, indicates that no freely diffusing radicals were detectable.[71] The RMgX structure (**56**) usually drawn for a Grignard reagent is in equilibrium with dimethylmagnesium (**57**) and $MgBr_2$

$$\text{trimer} \rightleftharpoons \text{dimer} \rightleftharpoons 2\ \underset{\mathbf{56}}{\text{RMgX}} \rightleftharpoons \underset{\mathbf{57}}{R_2Mg} + MgX_2 \rightleftharpoons \text{dimer} \rightleftharpoons \text{trimer}$$

65. Barbier, P. *Compt. Rend.* **1899**, *128*, 110.
66. (a) The Merck Index, 14th ed., Merck & Co., Inc., Whitehouse Station, New Jersey, **2006**, p ONR-5; (b) Mundy, B.P.; Ellerd, M.G.; Favaloro Jr., F.G. *Name Reactions and Reagents in Organic Synthesis, 2nd ed.*, Wiley-Interscience, New Jersey, **2005**, pp. 58-59.
67. For the use of indium catalyzed reactions in organic synthesis see (a) Chauhan, K.K.; Frost, C.G. *J. Chem. Soc. Perkin Trans. 1* **2000**, 3015. For a list of reagents used in this type of reaction, see (b) Smith, M.B.; March, J. *March's Advanced Organic Chemistry, 6th ed.*; Wiley, New York, **2007**, pp 1302,1313; (c) For reactions of this type, see Larock, R.C. *Comprehensive Organic Transformations, 2nd Ed.* Wiley-VCH, New York, **1999**, pp 1126-1133. Also see (d) Blomberg, C. *Reactions and Structure. Concepts in Organic Chemistry, Vol. 31. The Barbier Reaction and Related One-Step Processes,* Springer Verlag, Berlin, **1993**.
68. Srikrishna, A.; Dethe, D.H. *Org. Lett.* **2004**, *6*, 165.
69. Zhang, W.-C.; Li, C.-J. *J. Org. Chem.* **2000**, *65*, 5831.
70. Ashby, E.C. *Quart. Rev. Chem Soc.* **1967**, *21*, 259.
71. Walter, R.I. *J. Org. Chem.* **2000**, *65*, 5014.

as well as **58**, in what is called the **Schlenk equilibrium**.[72] In most common solvents the term Grignard reagent really refers to this equilibrium. It was later reported that dimers such as **59** and **60** were present in ether solutions. The Grignard reagent equilibrium is more complex in ether than it is in THF.[73] Smith and Becker verified that the equilibrium favored RMgBr in ether, although there is a mixture.[74] In THF,[75] all species (**56-60**) were present and the RMgBr concentration was diminished relative to similar solutions in ether. A Grignard reagent (RMgX) can associate into dimers and trimers in many solvents via a solvent dependent interaction with the halogen.[76]

In other work, Ashby and co-workers concluded[77] that the Grignard reagent exists primarily as the RMgX species in ether, THF or triethylamine but the composition varies in each solvent. In ether, the monomeric species is largely of RMgX with lesser amounts of R_2Mg and MgX_2. Association to non-monomeric species is extensive at concentrations > 0.3 M, and the dimers are assumed to be **59** and **60**. The trimeric structure can be represented as **61**. Ether solutions of Grignard reagents are stable if protected from moisture and air. A 2 N solution of CH_3MgI in ether was stored in a sealed tube for 20 years, and shown to have virtually the same concentration of Grignard reagent as when originally sealed.[78]

In THF, relatively little association occurs and the most abundant monomeric species include both RMgX and $R_2Mg + MgX_2$, as mentioned above. In triethylamine (rarely used as a solvent for Grignard reactions), the only species observed is RMgX for simple alkyl magnesium bromides and chlorides. The amine functions as a Lewis base and coordinates with RMgX, leading to a monomeric species. The solution is more complex for alkyl magnesium iodides or aryl magnesium halides. In hydrocarbon solvents, alkyl halides and Mg(0) react at temperatures >100°C[79] to give a mixture of insoluble organomagnesium and inorganic magnesium compounds

72. Schlenk, W.; Schlenk, W. *Berichte* **1929**, *62*, 920.
73. (a) Whitesides, G.M.; Kaplan, F.; Nagarajan, K.; Roberts, J.D. *Proc. Natl. Acad. Sci.* **1962**, *48*, 1112; (b) Ashby, E.C.; Smith, M.B. *J. Am. Chem. Soc.* **1964**, *86*, 4363.
74. Smith, M.B.; Becker, W.E. *Tetrahedron Lett.* **1965**, 3843.
75. (a) Smith, M.B.; Becker, W.E. *Tetrahedron* **1966**, *22*, 3027; (b) Smith, M.B.; Becker, W.E. *Ibid.* **1967**, *23*, 4215.
76. (a) Walker, F.W.; Ashby, E.C. *J. Am. Chem. Soc.* **1969**, *91*, 3845; (b) Ashby, E.C.; Walker, F.W. *J. Organomet. Chem.* **1967**, *7*, P17.
77. (a) Reference 69, p 278; (b) Parris, G.E.; Ashby, E.C. *J. Am. Chem. Soc.* **1971**, *93*, 1206.
78. Gilman, H.; Esmay, D.L. *J. Org. Chem.* **1957**, *22*, 1011.
79. Ashby, E.C.; Reed, R. *J. Org. Chem.* **1966**, *31*, 971.

$(R_2Mg + MgX_2 +$ highly associated combinations).[80,79]

Grignard reagents can be formed from primary, secondary and tertiary halides. Alkyl chlorides, bromides or iodides can be used. In all cases, an ether solvent stabilizes the Grignard reagent by forming a Lewis acid-Lewis base charge-transfer complex such as **62**. Coordination with ether also assists in the initial magnesium insertion reaction, and coordination minimizes decomposition of the Grignard reagent via disproportionation. Aryl and vinyl halides react with magnesium to form aryl magnesium halides such as phenylmagnesium bromide (PhMgBr), or vinyl Grignard reagents such as 1-propenylmagnesium bromide ($CH_3CH=CHMgBr$). In both cases, the C–Br bond of the aryl or vinyl bromide is stronger than the analogous C–Br bond of an alkyl halide. A stronger Lewis base is required, both to assist the insertion and to stabilize the organometallic (see **63**). Therefore, a more basic solvent THF is used when aryl or vinyl Grignard reagents must be prepared. Grignard formation occurs in ether, but may be sluggish, and the yields can be poor.

Grignard reagents react as nucleophilic carbanions, and they are also very basic. The conjugate acid of the Grignard is an alkane (MeMgX → Me–H), which is a very weak acid (pK_a = 25-40).[81] Formation of such a weak conjugate acid suggests that the Grignard reagent is a very strong base. Chevrot et al. showed that methylmagnesium halides are more basic than other straight-chain alkyl or aryl Grignard reagents.[81] Vinyl Grignard reagents and allylic Grignard reagents are slightly more basic than alkylmagnesium halides, and organomagnesium chlorides (RMgCl) are more basic than organomagnesium bromides (RMgBr). This finding may be due to displacement of the Schlenk equilibrium towards RMgX.

The basicity of Grignard reagents is an important factor in their reactions with electrophilic compounds. Grignard reagents react with the slightly acidic hydrogen of water[82] to give the hydrocarbon (R–H) and a magnesium hydroxide. Alcohols react in a similar manner. Another electrophilic reagent that reacts with Grignard reagents is molecular oxygen, which gives a hydroperoxide anion (**64**) as an initial product. This anion reacts with additional Grignard reagent to give an alkoxide (**65**),[82] and hydrolysis liberates the alcohol as the final product. The use of anhydrous solvents and exclusion of air is essential for optimum yields in the formation and reactions of Grignard reagents on small scale. On large scale with relatively simple reactants, less care need be taken, because the Grignard reagent will scavenge small amounts of water or other reactive compounds.

80. (a) Bryce-Smith, D.; Cox, G.F. *J. Chem. Soc.* **1961**, 1175; (b) Zakharkin, L.I.; Okhlobystin, O. Yu.; Strunin, B.N. *Tetrahedron Lett.* **1962**, 631.
81. (a) Chevrot, C.; Folest, J.C.; Troupel, M.; Cachelou, C.; Perichon, J. *J. Electroanal. Chem. Interfac. Electrochem.* **1974**, *55*, 263; (b) Chevrot, C.; Folest, J.C.; Troupel, M.; Périchon, J. *Ibid.* **1974**, *54*, 135; (c) Chevrot, C.; Folest, J.C.; Troupel, M.; Périchon, J. *C.R. Acad. Sci. Ser. C* **1971**, *273*, 613.
82. Reference 67a, pp 794-795 and references cited therein.

$$RMgX + HOH \longrightarrow R\!-\!H + HOMgX$$

$$RMgX + O_2 \longrightarrow ROOMgX \xrightarrow{RMgX} 2\ ROMgX \longrightarrow 2\ ROH$$

$$\underset{\textbf{64}}{} \qquad\qquad \underset{\textbf{65}}{}$$

8.4.B. The Reaction with Alkyl and Aryl Halides

By analogy to cyanide or acetylides, a Grignard reagent might be expected to displace an alkyl halide in an S_N2 reaction ($RMgX + R^1Y \rightarrow R\!-\!R^1$). For Grignard reagents derived from simple aliphatic, aryl, or vinyl halides that react with aliphatic alkyl halides, the yield of coupling product ($R\!-\!R^1$) is usually poor. Indeed, if Grignard reagents readily coupled with alkyl halides, Grignard reagents would be difficult to prepare. Only reactive Grignard reagents such as allylmagnesium halides react with alkyl halides that are also highly reactive (iodomethane, allyl bromide, benzyl bromide) to give good yields of a coupling product.[83] It is important to note that highly reactive Grignard reagents intended for uses other than coupling are usually prepared in dilute solution with excess magnesium, and via slow addition of the halide to avoid coupling. An example of a successful coupling reaction involves reaction of allyl chloride (**66**) with allylmagnesium chloride to give **67**.[84] Intramolecular coupling reactions of mesylates at the C3 of indole systems via the *N*-magnesium indole have been reported.[85]

8.4.B.i. The Kharasch Reaction. Alkyl halides are not very reactive in coupling reactions with Grignard reagents. With a few exceptions, good yields of coupling are obtained only by addition of transition metal catalysts.[86] Kharasch showed the effectiveness of several transition metals that promoted the coupling reaction of phenylmagnesium bromide and chlorodiphenylmethane (**68**, Table 8.4)[87]. It is clear that cobalt chloride ($CoCl_2$) is very effective and it has been widely used. As shown in Table 8.4, the products of this reaction are the cross coupling product (triphenylmethane, **69**) as well as the symmetrical coupling product (the Wurtz product, **70**, sec. 8.5.B.ii). This transition metal catalyzed coupling reaction is known as the **Kharasch reaction**. Ferric chloride is a very effective catalyst for the cross coupling of alkyl halides with aryl Grignard reagents, when tetramethylethylenediamine (TMEDA) is

83. Kharasch, M.S.; Reinmuth, O. *Grignard Reactions of Nonmetallic Substances*, Prentice-Hall, New York, *1954*, pp 1054-1056.
84. Chen, L.; Gill, G.B.; Pattenden, G.; Simonian, H. *J. Chem. Soc. Perkin Trans. 1 1996*, 31.
85. (a) Smith III, A.B.; Cui, H. *Org. Lett. 2003, 5,* 587; (b) Smith III, A.B.; Davulcu, A.H.; Kurti, L. *Org. Lett. 2006, 8,* 1665.
86. Reference 83, pp 1054-1132.
87. Sayles, D.C.; Kharasch, M.S. *J. Org. Chem. 1961, 26,* 4210.

used as a stoichiometric additive.[88] The use of Fe(acac)$_3$ allows coupling with alkyl halides that posses a β-hydrogen.[89]

Table 8.4. Transition Metal Catalyzed Coupling of Phenylmagnesium Bromide and Benzal Bromide

<table>
<tr><td colspan="3" align="center">PhMgBr + Ph$_2$CH–Cl $\xrightarrow{\text{M}}$ Ph$_2$CH–Ph + Ph$_2$CH–CHPh$_2$
         68   **69**   **70**</td></tr>
<tr><th>M</th><th>% 69</th><th>% 70</th></tr>
<tr><td>No catalyst</td><td>0</td><td>90</td></tr>
<tr><td>CoCl$_2$</td><td>82</td><td>6</td></tr>
<tr><td>FeCl$_3$</td><td>63</td><td>17</td></tr>
<tr><td>Cu$_2$Cl$_2$</td><td>30</td><td>47</td></tr>
<tr><td>MnCl$_2$</td><td>0</td><td>82</td></tr>
</table>

[Reprinted with permission from Sayles, D.C.; Kharasch, M.S. *J. Org. Chem.* **1961**, *26*, 4210. Copyright © **1961** American Chemical Society

Disproportionation is a serious problem in transition metal catalyzed coupling reactions. Metal-hydrogen exchange followed by Wurtz coupling (sec. 8.5.B.ii) is also a problem. Iron, cobalt, manganese or nickel salts catalyze disproportionation, as shown below for the reaction of cyclopentylmagnesium bromide (**71**) and ethylmagnesium bromide.[90] The latter reagent disproportionated to give ethane and ethene, and similar disproportionation of cyclopentylmagnesium bromide gave cyclopentane and cyclopentene. Only a 10% yield of a coupling product (**72**) was observed, and there no cross-coupling was observed. The reaction of phenylmagnesium and ethylmagnesium bromide behaved similarly, giving an 81% yield of biphenyl.[91]

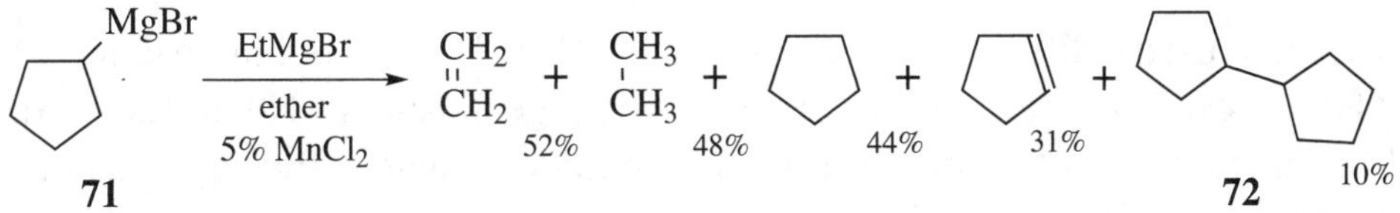

Silver salts are effective catalysts for coupling alkyl Grignard reagents and alkyl halides. The reaction of ethylmagnesium bromide and 1-bromopropane, for example, gave 47% of butane, 12% of hexane, and 37% of the cross-coupled product, pentane.[92] Disproportionation is less of a problem but the statistical coupling seen in this example is typical, which clearly limits the synthetic utility of the reaction if the cross-coupling product is the desired target. When the target is the symmetrical coupling product (the alkyl groups are the same in the halide and the

88. Nakamura, M.; Matsuo, K.; Ito, S.; Nakamura, E. *J. Am. Chem. Soc.* **2004**, *126*, 3686.
89. Nagano, T.; Hayashi, T. *Org. Lett.* **2004**, *6*, 1297.
90. (a) Kharasch, M.S. Urry, W.H. *J. Org. Chem.* **1948**, *13*, 101; (b) Tamura, M.; Kochi, J.K. *J. Am. Chem. Soc.* **1971**, *93*, 1485, 1487.
91. Kharasch, M.S.; Fields, J.K. *J. Am. Chem. Soc.* **1941**, *63*, 2316.
92. Tamura, M.; Kochi, J.K. *J. Am. Chem. Soc.* **1971**, *93*, 1483.

Grignard reagent), however, product yields can be good. Indeed, reaction of ethylmagnesium bromide with bromoethane gave a 92% yield of butane.[92]

$$RMgBr + Cu^{I}Br \longrightarrow RCu + MgBr_2 \qquad RCu + R^{1}Br \xrightarrow{\text{slow}} R\text{-}R^{1} + Cu^{I}Br$$

Cuprous [Cu(I)] salts are readily available, and when mixed with Grignard reagents give excellent yields of cross-coupled products with very little disproportionation.[92,93] This reaction will be discussed in greater detail in Section 8.7.A. However, a simple example is the reaction of isopropylmagnesium bromide with bromoethane in the presence of cuprous bromide (CuBr), which gaave 2-methylbutane.[92] The origin of this reaction dates to 1936, and work by Gilman and Straley,[94] with later work by Cotton and by Kochi,[93] and by Noller and co-workers.[95] Kochi and Tamura studied the mechanism of the reaction,[92] and proposed an alkylcopper(I) species (RCu) as an intermediate that reacted with the alkyl halide to give the coupled product (R-R') in the rate-determining (slowest) step of the sequence.[92] The reaction is straightforward with primary alkyl halides,[96] but secondary and tertiary halides do not react as well. If the concentration of Cu(I) is too high with secondary and tertiary halides, disproportionation occurs. The Grignard reagent, however, can be primary, secondary, tertiary, aryl, or vinyl. A useful example is the conversion of the Grignard reagent of 2-bromo-2-butene (73) to (3Z)-methyl-2-hexene (74) by the Cu(I) catalyzed reaction with iodopropane.[97] Grignard reagent 73 was enriched in the (Z) isomer, and the reaction proceeded with high stereoselectivity to give predominantly the (Z)-isomer of 74. Organocuprates of this type will be discussed in Section 8.7.A.

Kochi and Tamura used a different copper catalyst (Li₂CuCl₄, prepared by reaction of LiCl and CuCl₂ in THF) to catalyze the coupling of Grignard reagents and alkyl halides.[98] An example is taken from a synthesis of bioactive dihydroceramides isolated from the dental pathogen *Porphyromonas gingivalis*, by Smith, Nichols and co-workers,[99] in which bromide 75 was coupled to Grignard reagent 76 using Li₂CuCl₄ to give a 95% yield of 77. Note that the coupling occurred with the unprotected hydroxyl group in 75.

93. (a) Cotton, F.A. *Chem. Rev.* **1955**, *55*, 551; (b) Tamura, M.; Kochi, J. *J. Organomet. Chem.* **1971**, *31*, 289.
94. Gilman, H.; Straley, J.M. *Recl. Trav. Chim. Pays-Bas* **1936**, *55*, 821.
95. (a) Parker, V.D.; Piette, L.H.; Salinger, R.M.; Noller, C.R. *J. Am., Chem. Soc.* **1964**, *86*, 1110; (b) Parker, V.D.; Noller, C.R. *Ibid.* **1964**, *86*, 1112.
96. Schlosser, M. *Angew. Chem.* **1974**, *86*, 751.
97. Derguini-Boumechal, F.; Linstrumelle, G. *Tetrahedron Lett.* **1976**, 3225.
98. Tamura, M.; Kochi, J. *Synthesis* **1971**, 303.
99. (a) Mun, J.; Onorato, A.; Nichols, F.C.; Morton, M.D.; Saleh, A.I.; Welzel, M.; Smith, M.B. *Org. Biomol. Chem.* **2007**, *5*, 3826; and see (b) Takikawa, H.; Nozawa, D.; Kayo, A.; Muto, S.; Mori, K. *J. Chem. Soc., Perkin Trans. 1* **1999**, 2467.

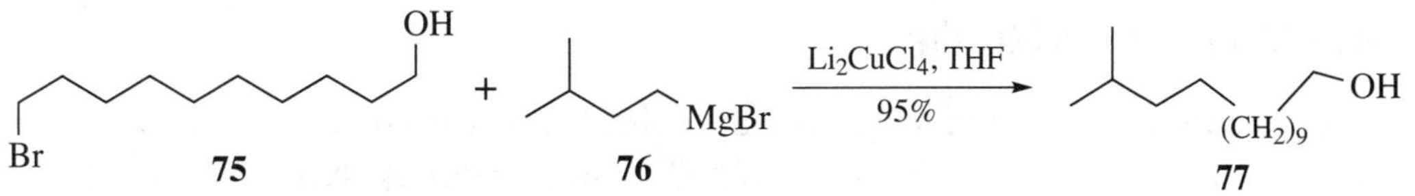

Coupling of aryl halides and alkylmagnesium halides under these conditions is also possible. Fürstner and Leitner reported that aryl halide couple with Grignard reagents, in the presence of a catalytic amount of an iron catalyst such as Fe(acac)$_2$ to give the arene.[100] An example is the reaction of 2-chloropyridine with hexylmagnesium bromide to give **78** in 90% yield.

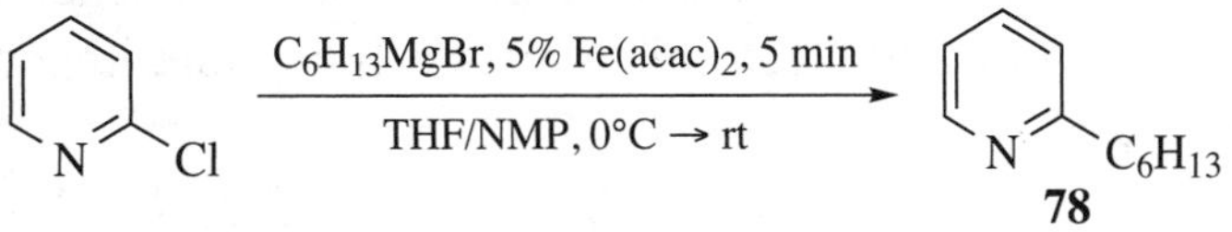

Metal-catalyzed coupling reactions lead to the disconnection

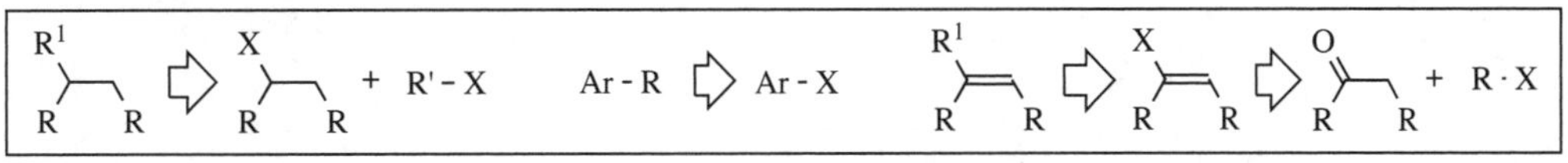

The coupling reaction of Grignard reagents and cuprous salts does not work well with secondary and tertiary halides, as mentioned above. Similar reagents can be prepared with cuprous salts and organolithium reagents that are much more reactive and give better yields of coupling products. These organocuprate reagents will be discussed in Section 8.7.

8.4.B.ii. Reaction with α-Halocarbonyls. Treatment of a ketone with NBS, NCS, Cl$_2$, SOCl$_2$ and so on[101] gives an α-chloro ketone, **79**. α-Haloketones give good yields of coupling products with Grignard reagents, even without a transition metal catalyst. There are two electrophilic sites, the carbonyl carbon and the halogen-bearing carbon, but the two are easily differentiated. The halogen-bearing carbon is more susceptible to attack by Grignard reagents. Subsequent reaction with phenylmagnesium bromide gave **80**.[102] This method is superior for the preparation of α-arylketones, but the yields are often poor if the carbonyl is not conjugated to an aromatic nucleus (i.e., with simple alkyl ketone derivatives).

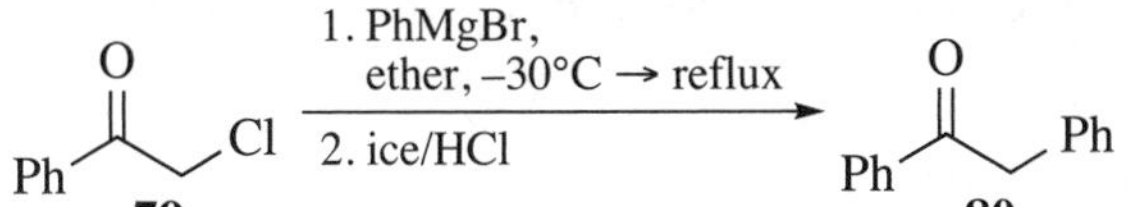

The disconnection for substitution of a α-halo-ketone is

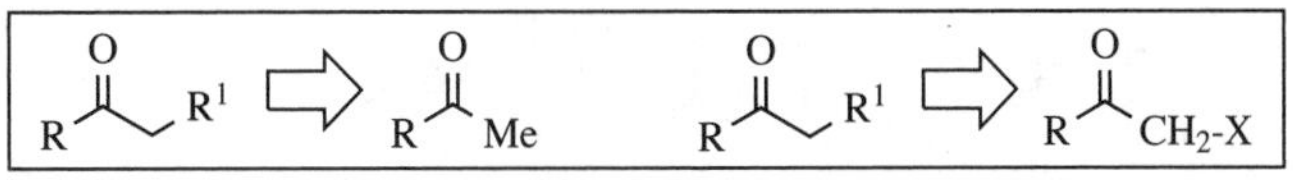

---

100. Fürstner, A.; Leitner, A. *Angew. Chem. Int. Ed.* **2002**, *41*, 609.
101. (a) Reed, Jr., S.F. *J. Org. Chem.* **1965**, *30*, 2195; (b) Rappe, C.; Kumar, R. *Arkiv. Kemi.* **1965**, *23*, 475 [*Chem. Abstr. 63*: 5521c **1965**].; (c) Zbiral, E.; Rasberger, M. *Tetrahedron* **1969**, *25*, 1871; (d) Wyman, D.P.; Kaufman, P.R. *J. Org. Chem.* **1964**, *29*, 1956.
102. Ando, T. *Yuki Gosei Kagaku Kyokaishi* **1959**, *17*, 777 [*Chem. Abstr. 54*: 4492b **1960**].

8.4.C. Reactions with Carbonyl Derivatives

8.4.C.i. Acyl Addition. Grignard reagents react with aldehydes and ketones to give alcohols, in what is commonly called the Grignard reaction.[103] This reaction is shown for cyclopentanone, which reacts with the nucleophilic isopropylmagnesium bromide at the electrophilic carbonyl carbon to give alkoxide **82**. Hydrolysis is required to liberate the alcohol product, **83**. Ashby suggested the mechanism shown below.[104] It invokes a cyclic transition state (**81**), in which the alkyl group is transferred from magnesium to oxygen, forming the alkoxide and may involve a single electron transfer (SET). This cyclic transition state has stereochemical implications for the alcohol product, since steric and electronic effects will be more important in a cyclic than an acyclic transition state.

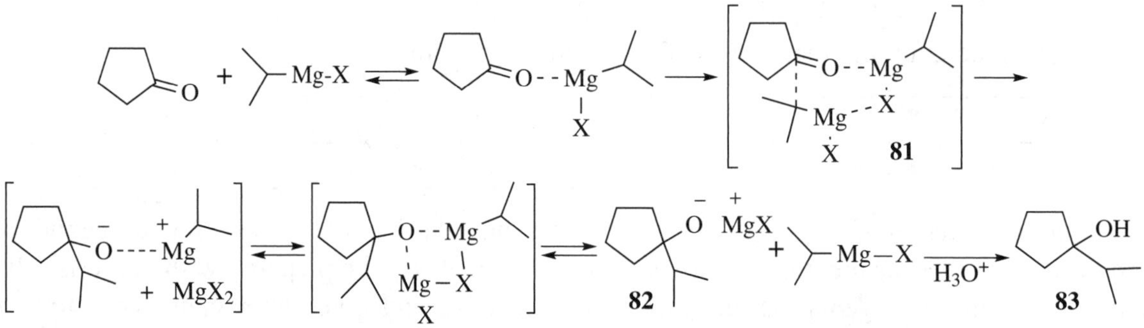

The Grignard reaction is one of the best methods to prepare secondary alcohols from aldehydes, since it generates a carbon-carbon bond during the reaction (RMgX + R^{1}CHO). A simple example is the reaction of 4-methoxyphenylmagnesium bromide and aldehyde **84** to give alcohol **85** in good yield and good diastereoselectivity, in Rozwadowska and Grajewska's synthesis of cytoxazone.[105] With ketones, the reaction generates tertiary alcohols (RMgX + R$^1_2$C=O → RR$^1_2$CHOH). In McMorris and co-worker's synthesis of (–)-irofulven,[106] **86** reacted with methylmagnesium chloride, to give an 87% yield of **87**. Note the selectivity for addition to the non-conjugated ketone, and delivery of methyl from the top face of **86**, as the molecule is drawn.

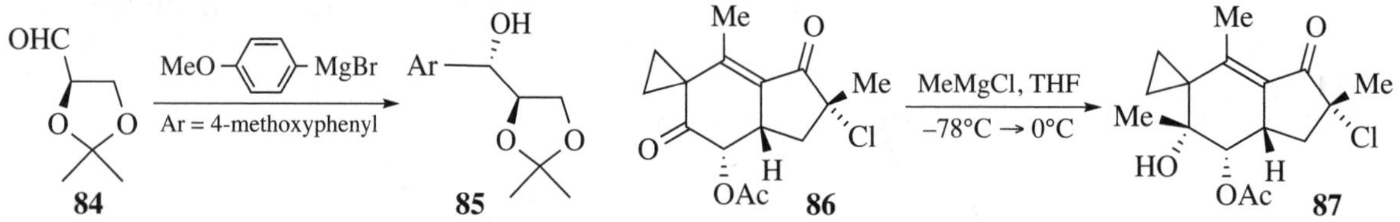

103. For reactions of this type, see Larock, R.C. *Comprehensive Organic Transformations, 2nd Ed.* Wiley-VCH, New York, **1999**, pp 1125-1176.
104. Reference 70, p 283
105. Grajewska, A.; Rozwadowska, M.D. *Tetahedron: Asymmetry* **2007**, *18*, 803.
106. McMorris, T.C.; Staake, M.D.; Kelner, M.J. *J. Org. Chem.* **2004**, *69*, 619.

MOLECULAR MODELING: Unoccupied Molecular Orbital Maps and Stereoselectivity of Grignard Reactions Involving Diketones

A Grignard reagent is characterized by a δ carbon atom, which reacts as a nucleophile with carbonyl groups. Many Grignard reactions with ketones or aldehydes proceed with good-to-excellent diastereoselectivity. One way to model this is to examine the (unoccupied) molecular orbital on the substrate into which the pair of electrons on the nucleophile ("CH$_3$⁻", in the case of methylmagnesium chloride) will go, and identify the most electrophilic (electron poor) site for reaction with the nucleophilic (electron rich) carbon of the Grignard reagent. While the LUMO is commonly the "relevant" orbital, in the case of **86** which contains two carbonyl groups, it is not. Here the relevant unoccupied orbital is the next higher-energy orbital ("LUMO+1"). This orbital can be mapped on an electron density surface, and the resulting map employed to see which carbonyl face presents the better target for the nucleophile.

Examine both LUMO (left) and LUMO+1 (right) maps for **86**. On which carbonyl group is the LUMO primarily localized? (By convention, regions where the orbital is most heavily concentrated are colored "blue".) Next, examine the LUMO+1 map. On which carbonyl group is this orbital primarily localized? Which carbonyl face, *syn* or *anti* to the OAc group, is likely to be more susceptible to attack by the nucleophilic Grignard reagent, that is, for which face is the map more blue? Is the product anticipated by the model that which is actually observed?

MOLECULAR MODELING: Transition States and Stereoselectivity of Grignard Reactions of Conjugated Carbonyl Compounds

A more accurate (and more difficult approach) to assigning selectivity Grignard reactions is to compare the energies of the transition states leading from reactants to different isomeric products. For example, addition of methylmagnesium chloride to **86** could conceivably occur onto either the conjugated or non-conjugated carbonyl groups and, for each, *syn* or *anti* to the OAc substituent. The major product (**87**) follows from addition to the non-conjugated carbonyl *anti* to OAc. Application of transition state models, which requires that the transition state for each of these four possibilities be found, and the lowest-energy transition state be identified, is beyond the scope of *SpartanModel*. We will limit ourselves to examining results obtained using *Spartan*.

Transition states for addition of methyl magnesium chloride *syn* and *anti* to the non-conjugated carbonyl (top left and right), and *syn* and *anti* to the conjugated carbonyl (bottom left and right) appear on screen. Relative energies are as follows: *anti* to non-conjugated carbonyl, 0 kJ/mol; *syn* to conjugated carbonyl, 8; *syn* to non-conjugated carbonyl, 33; *anti* to conjugated carbonyl, 39. The lowest-energy transition state is for addition *anti* to non-conjugated carbonyl group and leads to the observed major product. The model shows a switch in the preferred stereochemistry of reaction between addition to the non-conjugated carbonyl and conjugated carbonyl groups. *Anti* addition is favored for the non-conjugated carbonyl group *syn* addition is favored for the conjugated carbonyl group. The preference for (*anti*) addition to the conjugated carbonyl group over (*syn*) addition to the conjugated carbonyl group is only 8 kJ/mol. Which transition state for addition to the non-conjugated carbonyl group (*syn* or *anti* to the reagent) is more similar to the geometry of the reactants? (Look at the lengths of the CC bond that is being formed and the CO double bond that is being converted to a single bond.) Is this the result you expect? Elaborate.

The disconnection for this process is

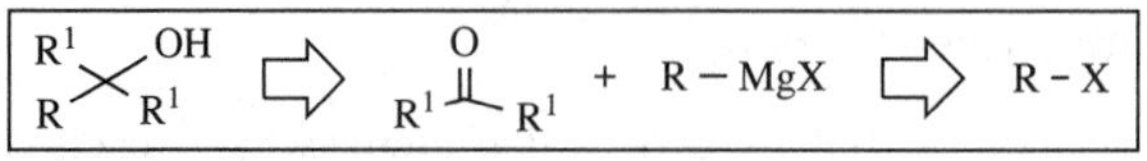

8.4.C.ii. Acyl Substitution. Although the reaction with aldehydes and ketones is the most common, Grignard reagents react with most other carbonyl derivatives.[107] As we saw in Section 2.5.C, acid derivatives contain a leaving group such as Cl, O_2R, OR, or NR_2, and react with nucleophiles via acyl substitution. Initial addition of a Grignard reagent to the acyl carbon of an acid derivative generates an alkoxide intermediate (**88**), called a **tetrahedral intermediate**, and displacement of the leaving group (X) leads to a ketone. If the initially formed ketone product is more reactive than the acid derivative starting material, further reaction with the Grignard reagent can give tertiary alcohol **90** via alkoxide **89**.[108] The competition between the ketone product and the acyl starting material can lead to mixtures of products, and there may be significant amounts of unreacted starting material.

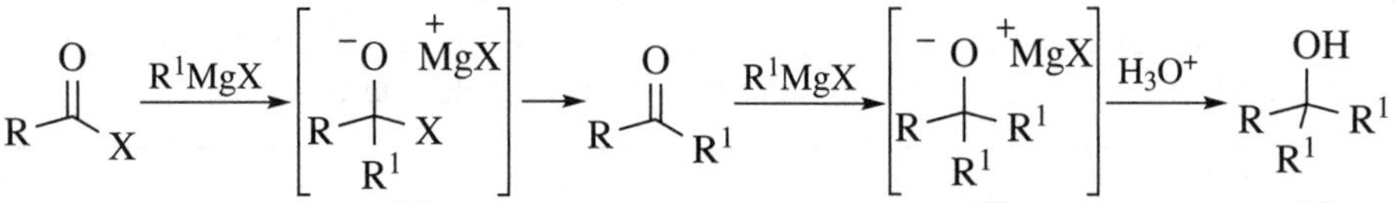

Acid chlorides are the most reactive acid derivatives, where chlorine is the leaving group.[109] Reaction of Grignard reagents with acid chlorides gives excellent yields of the tertiary alcohol when an excess of the Grignard reagent is used, especially under vigorous reaction conditions. When the reaction is done at low temperatures and/or when metal catalysts are added, isolation of the ketone product is often possible.[110] In some cases, the ketone product can be isolated. Reaction of hexanoyl chloride and **91**, taken from Dussalt and co-worker's synthesis of peroxyacarnoate A,[111] gave a 78% yield of ketone **92**.

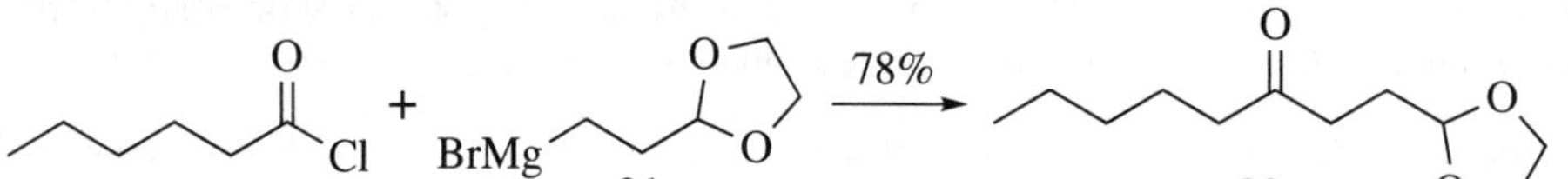

An excellent method for converting an acid chloride to a ketone employs transition metal catalysts such as ferric chloride, in conjunction with low reaction temperatures.[112] Reaction

---

107. For reactions of this type, see Larock, R.C. *Comprehensive Organic Transformations, 2nd Ed.* Wiley-VCH, New York, *1999*, pp 1389-1411.
108. (a) Reference 82, pp 567-568; (b) Reference 83
109. For a review of reactions of acyl chlorides with organometallic reagents, see Dieter, R.K. *Tetrahedron* *1999*, *55*, 4177.
110. Stowell, J.C. *J. Org. Chem.* *1976*, *41*, 560.
111. Xu, C.; Raible, J.M.; Dussault, P.H. *Org. Lett.* *2005*, *7*, 2509.
112. Cason, J.; Kraus, K.W. *J. Org. Chem.* *1961*, *26*, 1768, 1772.

of butylmagnesium bromide with octanoyl chloride at –60°C gave a mixture of 13% of **93**, 4% of **94** (from the secondary reaction of butyl-magnesium bromide and **93**, and 60% of **95**.[112] When a catalytic amount (2 mol %) of ferric chloride (FeCl₃) was added, however, a 76% yield of ketone **93** was obtained along with 3% of the over alkylation product, **94**.[112] If the intermediate ketone is unreactive to nucleophilic substitution due steric hindrance or peculiar electronic factors (diisopropyl ketone and phenyl-*tert*-butyl ketone are both sterically hindered), the ketone can usually be isolated (sec. 8.4.G for problems with hindered ketones). In many instances, however, even dry ice temperatures and reverse addition techniques give poor yields of the ketone.[113]

	93	**94**	**95**
catalyst = NONE, –60°C	13%	4%	60%
catalyst = 2% FeCl₃, –60°C	76%	3%	15%

Changing the metal from magnesium to one that gives a less reactive organometallic provides an alternative method for isolating the ketone. Gilman and co-workers[114] discovered that a Grignard reagent reacts with cadmium chloride (CdCl₂) to give a dialkylcadmium reagent (R₂Cd), and this less reactive organometallic is more selective in its reactions with carbonyl derivatives.[115] This reaction is rather slow, however, and in many cases it is not clear that dialkylcadmium is the active reagent.[116] Dialkylcadmium reagents are readily formed from primary Grignard reagents, but secondary and tertiary organocadmium reagents are relatively unstable.[117] The likely decomposition pathway is dissociation to radicals, which disproportionate to give alkane and alkene products. Reagents such as di-*n*-pentylcadmium (**96**) are formed by reaction of a Grignard reagent (pentylmagnesium bromide) and anhydrous cadmium chloride (CdCl₂). Such reagents react rapidly with acyl halides but very slowly or not at all with ketones, aldehydes, esters or amides. The yields of ketone products obtained by reaction with acid chlorides are moderate to good.[118] In general, dialkylcadmium reagents do not react with the ketone products, but there are exceptions when highly reactive ketone products are formed.[115] When **96** reacted with α-chloroacetyl chloride, a 46% yield of **97** was obtained.[118]

---

113. Newman, M.S.; Smith, A.S. *J. Org. Chem.* **1948**, *13*, 592.
114. (a) Gilman, H.; Nelson, J.F. *Rec. Trav. Chim.* **1936**, *55*, 518; (b) Cason, J. *J. Org. Chem.* **1948**, *13*, 227; (c) Le Guilly, L.; Tatibouët, F. *Compt. Rend. Ser. C* **1966**, *262*, 217; (d) LeGuilly, L.; Chenault, J.; Tatibouët, F. *Ibid.* **1965**, *260*, 6634.
115. (a) Cason, J. *Chem. Rev.* **1947**, *40*, 15; (b) Shirley, D.A. *Org. React.* **1954**, *8*, 28 (see pp 35-38).
116. Cason, J. *J. Am. Chem. Soc.* **1946**, *68*, 2078.
117. Cason, J.; Fessenden, R.J. *J. Org. Chem.* **1960**, *25*, 477.
118. (a) Miyano, M.; Dorn, C.R.; Mueller, R.A. *J. Org. Chem.* **1972**, *37*, 1810; (b) Miyano, M.; Dorn, C.R. *Tetrahedron Lett.* **1969**, 1615; (c) Archer, S.; Unser, M.J.; Froelich, E. *J. Am. Chem. Soc.* **1956**, *78*, 6182; (d) Bindra, J.S.; Bindra, R. *Prostaglandin Synthesis,* Academic Press, New York, **1973**, pp 52-53.

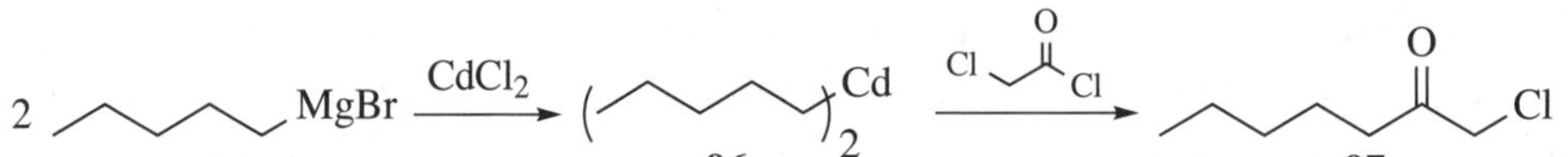

Although esters are much less reactive than acid chlorides (ethoxy is a poorer leaving group than Cl), they are only slightly less reactive than the ketone products resulting from reaction with a Grignard reagent. A Grignard reaction with esters rarely gives the ketone in good yield. A mixture of ester, ketone and tertiary alcohol is often observed when only one equivalent each of Grignard reagent and ester are used. Two or more equivalents of Grignard reagent give the alcohol as the only product.[119] The tendency for over-reaction of some acid derivatives can have useful synthetic applications. The reaction of **98** with an excess of methylmagnesium bromide[120] illustrates the reaction at hand, but introduces an interesting feature. If we focus on the benzyl ester, addition of two equivalents of Grignard reagent leads to the tertiary alcohol in **99**. The acetate group in **98** is also an ester, and reaction with MeMgBr leads to the alcohol unit in **99** and *tert*-butanol (via loss of acetone

from initial addition of MeMgBr and subsequent reaction of acetone to give the alcohol). We have previously seen another application of this reaction in the **Barbier-Wieland degradation** (sec. 3.7.A.iv.),[121] used to decrease the chain length of carboxylic acid esters by one carbon.

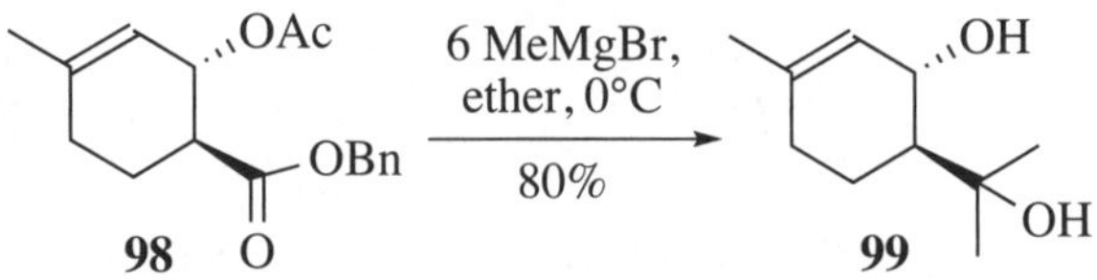

The usual disconnections for reaction with esters and acid chlorides are

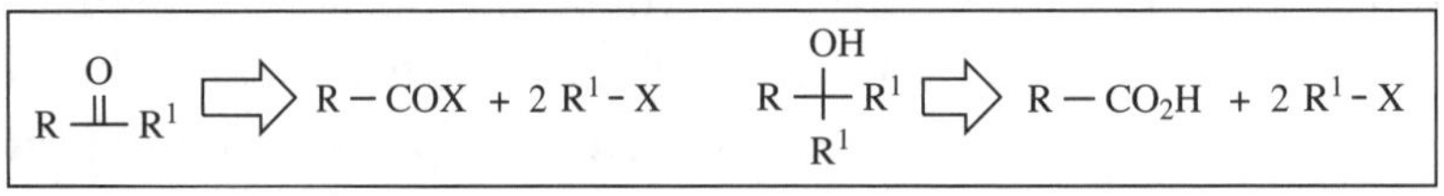

Alkyl esters such as *n*-butyl mesitoate (**100**) resist the acyl addition reaction because the ortho methyl groups sterically hinder the acyl carbon. When **100** reacted with phenylmagnesium iodide, for example, the products were mesitoic acid (**101**) and iodobutane.[122] Acyl addition is inhibited by the presence of the methyl groups, but iodide (assisted by Mg^{2+} complexation with the carbonyl) attacks the O–C group (O–Bu in **100**) via an S_N2 reaction that displaces the

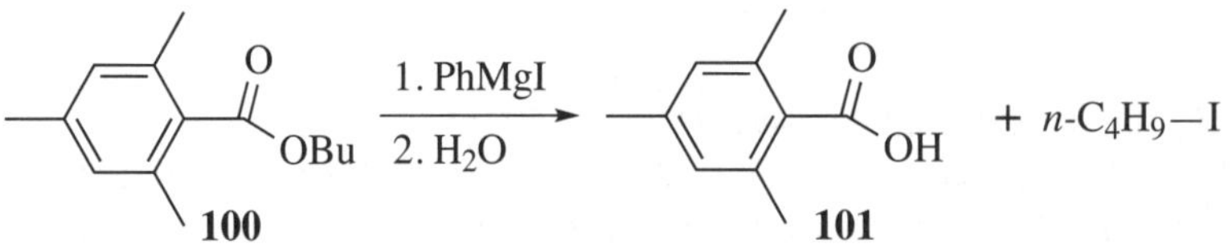

---

119. Fieser, L.F.; Heymann, H. *J. Am. Chem. Soc.* **1942**, *64*, 376.
120. Evans, D.A.; Barnes, D.M.; Johnson, J.S.; Lectka, T.; von Matt, P.; Miller, S.J.; Murry, J.A.; Norcross, R.D.; Shaughnessy, E.A.; Campos, K.R. *J. Am. Chem. Soc.* **1999**, *121*, 7582.
121. (a) Wieland, H. *Berichte* **1912**, *45*, 484; (b) Barbier, P.; Locquin, R. *Compt. Rend.* **1913**, *156*, 1443.
122. (a) Reference 83, p 567; (b) Fuson, R.C.; Bottorff, E.M.; Speck, S.B. *J. Am. Chem. Soc.* **1942**, *64*, 1450.

carboxyl moiety (which functions as a leaving group). Grignard reactions with lactones are interesting in that ring opening accompanies acyl addition, giving a diol in which the two ends are differentiated. Valerolactone, for example, reacted with butylmagnesium iodide to give 5-methyl-1,5-hexanediol. Amides are less reactive than esters (NR_2 is a poorer leaving group than OR) and the ketone product can often be isolated, although the yields are usually poor. Reaction of *N*-methylacetamide with two equivalents of phenylmagnesium bromide gave alkoxide **102**.[123] Hydrolysis led to loss of the amine (via **103**) and formation of the ketone product. Since **102** cannot react with additional Grignard, the product is the ketone. *N,N*-Dialkyl amides also react with Grignard reagents to give a ketone, but as mentioned above, the yields are usually poor.[124] An exception is a Weinreb amide, which we have seen previously (see sec. 2.5.C). This activated amide –CONMe(OMe) is an excellent leaving group, and the reaction stops at the ketone. In a synthesis of naamidine A by Watson and co-workers, amide **104** was converted to ketone **105** in 62% yield, by reaction with the Grignard reagent shown.[125]

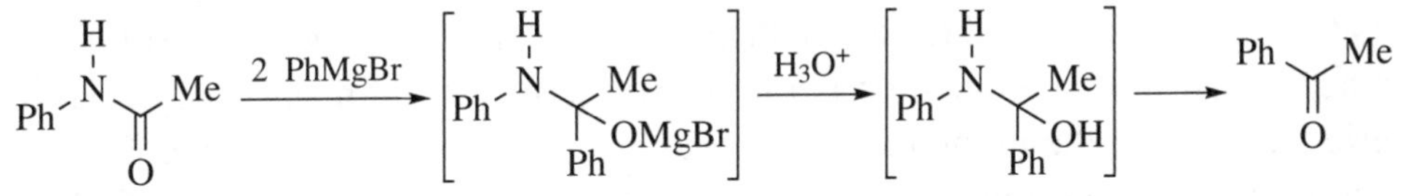

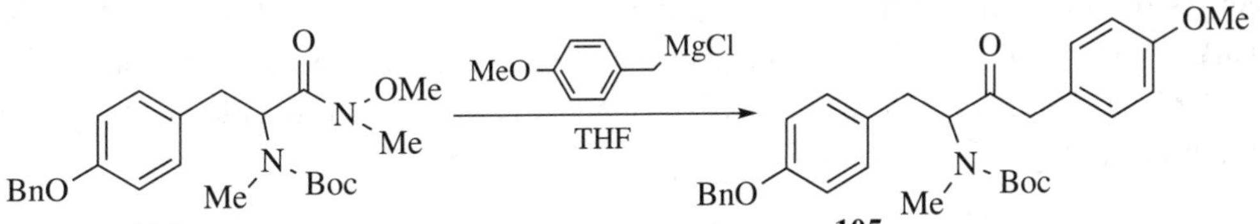

The reaction with lactams follows a somewhat different pathway than that just described for amides. The final product is an amine or an amino ketone. 2-Pyrrolidinone and 2-piperidone derivatives give primarily the amine, with minor amounts of ring cleavage products (amino ketones). At low temperature, however, the ring-opened ketone can be obtained, as in Martin and Simila's synthesis of the azatricyclic skeleton of FR901483 in which *N*-Boc-2-pyrrolidinone (**106**) reacted with the butenylmagnesium bromide to give iminium salt **107** which opened to give **108** in 98% yield after hydrolysis.[126a] Larger ring lactams give mainly the amino ketone, as in the reaction of *N*-methyl caprolactam with ethylmagnesium bromide to give a mixture of dialkylamine **109** in 42% yield and the ring-opened amino ketone in 23% yield .[126b]

123. Heyns, K.; Pyrus, W. *Chem. Ber.* **1955**, *88*, 678.
124. Busch, M.; Fleischmann, M. *Berichte* **1910**, *43*, 2553.
125. Aberle, N.S.; Lessene, G.; Watson, K.G. *Org. Lett.* **2006**, *8*, 419.
126. (a) Simila, S.T.M.; Martin, S.F. *J. Org. Chem.* **2007**, *72*, 5342; (b) Lukes, R.; Dudek, V.; Sedláková, O.; Korán, J. *Coll. Czech. Chem. Commun.* **1961**, *26*, 1105.

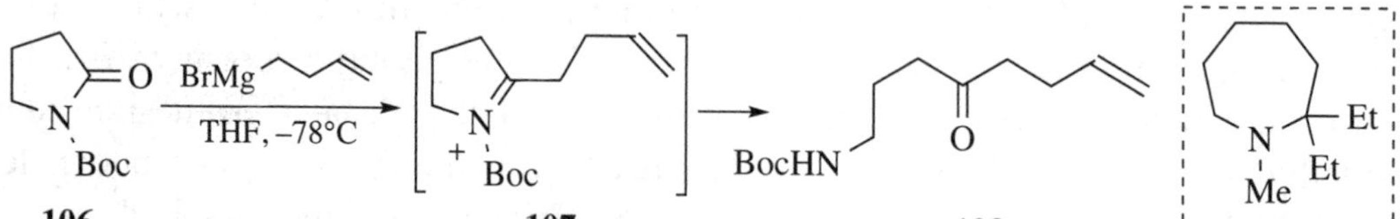

106　　　　**107**　　　　**108**　　　　**109**

8.4.C.iii.　Reaction with Nitriles.

Nitriles are carboxylic acid derivatives and react with Grignard reagents (also sec. 8.2.C).[127] Initial attack generates an intermediate iminium salt that can be hydrolyzed (with loss of ammonia) to the corresponding ketone.[128] The best yields are obtained with aryl nitriles, as illustrated by the reaction of 3-methoxybenzonitrile with the Grignard reagent derived from 4-bromo-1-butene. The initial reaction generates iminium salt **110**,[129] and subsequent hydrolysis produced ketone **111**. When an alkyl nitrile reacts with a Grignard derived from an hindered alkyl halide, the yield of ketone can be poor, but this is not always the case. Reaction of cyclopropanecarbonitrile (**112**) and isopropylmagnesium bromide gave > 80% of ketone **113** (R = *i*-Pr).[130] The intermediate iminium salt is not usually isolated since hydrolysis to the ketone is rapid. Hindered iminium salts can resist hydrolysis due to steric blocking of the imino carbon, however, and treatment with aqueous acid[131] under vigorous conditions can be required to give even modest yields. Reaction of **112** with *tert*-butylmagnesium bromide, followed by heating at reflux with 1 N HCl, gave only 35% of **113** (R = *t*-Bu).[130a] Hindered nitriles often give no reaction at all when they are reacted with very hindered Grignard reagents. Reaction of *tert*-butylmagnesium chloride with cyanomethyl-2,4,6-trimethylbenzene, for example, gave 0% of the ketone.[132]

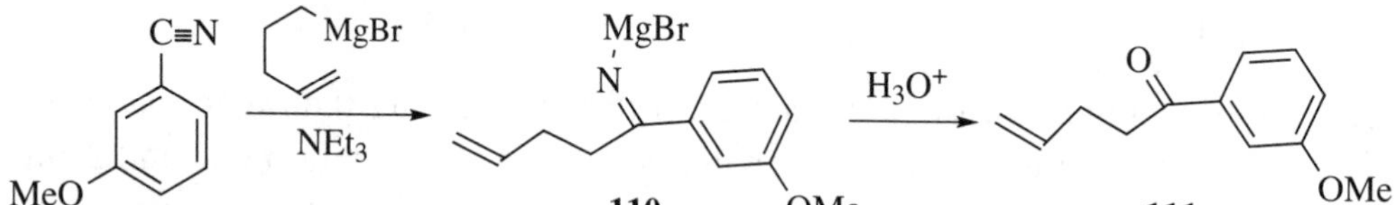

110　　　　**111**

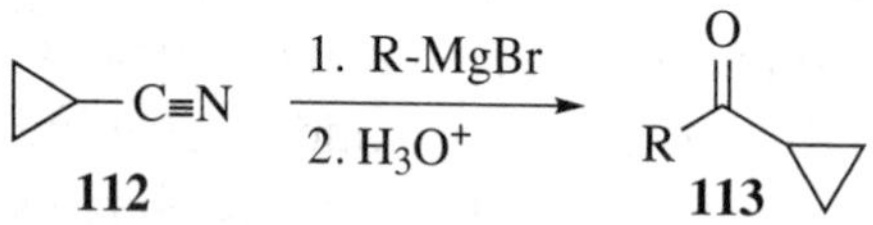

112　　　　**113**

127.　For reactions of this type, see Larock, R.C. *Comprehensive Organic Transformations, 2nd Ed.* Wiley-VCH, New York, *1999*, pp 1420-1422.
128.　Reference 83, pp 767-845.
129.　Taber, D.F.; Wang, Y.; Pahutski, Jr., T.F. *J. Org. Chem.* **2000**, *65*, 3861.
130.　(a) Hrubiec, R.T.; Smith, M.B. *J. Org. Chem.* **1984**, *49*, 431; (b) Hanack, M.; Ensslin, H.M. *Annalen* **1966**, *697*, 100; (c) Hanack, M.; Bocher, S.; Herterich, I.; Hummel, K.; Vött, V. *J.L. Ann. Chem.* **1970**, *733*, 5.
131.　Citron J.D.; Becker, E.I. *Can. J. Chem.* **1963**, *41*, 1260.
132.　Bruylants, A. *Bull. Soc. Chem. Fr.* **1958**, 1291.

The disconnection for the nitrile-Grignard reagent reaction is:

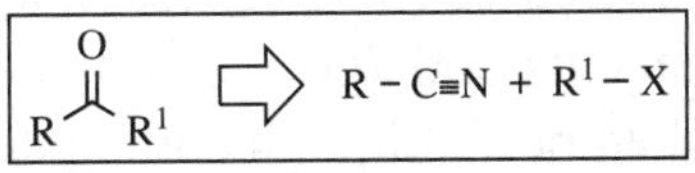

8.4.C.iv. Reaction with Carbon Dioxide. Carbon dioxide (CO_2, O=C=O) is actually a carbonyl derivative, and reacts with a Grignard reagent[133] to give a carboxylate salt, that gives the corresponding carboxylic acid upon hydrolysis.[134] Conversion of benzylmagnesium bromide to phenylacetic acid[135] is an example of this transformation. This functional group interchange of a halide to an acid proceeds with a one-carbon extension of the chain. An interesting example is taken from a synthesis of (+)-biotin by Seki and co-workers,[136] in which nitrile **114** was generated *in situ*, and reacted with the bis(Grignard) reagent shown. Initial reaction with the nitrile led to formation of the ketone unit in **115** after hydrolysis (see sec 8.4.C.iii above), but the second C–MgBr unit reacted with carbon dioxide and hydrolysis with aqueous citric acid generated the carboxylic unit. The yield of **115** was 79% in the solvent shown, but the yield was significantly less in THF and other solvents.

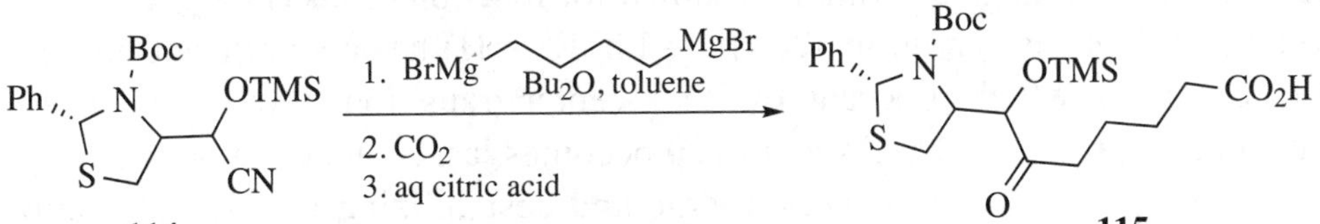

The disconnection for the carboxylation reaction is

8.4.D. Conjugate Addition

α,β-Unsaturated carbonyl derivatives have two reactive sites, and nucleophilic addition of a Grignard reagent can occur at either the acyl carbon (1,2-addition) or the alkenyl carbon (1,4-addition). When a Grignard reagent reacts with a conjugated carbonyl such as **116**, the 1,4-addition product is an enolate anion (**117**, see sec. 9.2). The 1,2-addition product is the usual alkoxide **118**. The course of this reaction varies with the steric bulk of R^1 in the Grignard reagent, and R in the carbonyl compound. As the size of R increases, the amount of 1,4-addition

133. (a) Oppolzer, W.; Kündig, E.P.; Bishop, P.M.; Perret, C. *Tetrahedron Lett.* **1982**, *23*, 3901; (b) Rowsell, D.G. Brit. Patent 1,392,907 [*Chem. Abstr. 83*: P114682t *1975*].
134. (a) Sneeden, R.P.A. in *The Chemistry of Carboxylic Acids and Esters*, Patai, S. (Ed.), Interscience, New York, *1969*, pp. 137-175; (b) Reference 83, pp 913-948.
135. Eberson, L. *Acta Chem. Scand.* **1962**, *16*, 781.
136. Seki, M.; Mori, Y.; Hatsuda, M. Yamada, S. *J. Org. Chem.* **2002**, *67*, 5527.

increases, although α,β-unsaturated aldehydes usually give only 1,2-addition.[137] Conversely, as the R^1 group of the Grignard increases in size, less conjugate addition is observed for a given R. In general, Grignard reagents undergo 1,2-addition with conjugated aldehydes and relatively unhindered conjugated ketones, although conjugate alkylation often competes in the latter case. In a synthesis of (+)-methynolide,[138] Cossy and co-worker's reacted methacrolein with ethylmagnesium bromide, and obtained a 68% yield of the 1,2-addition product, **119**.

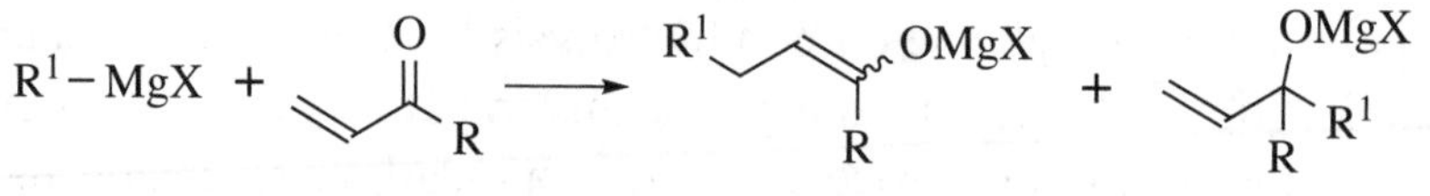

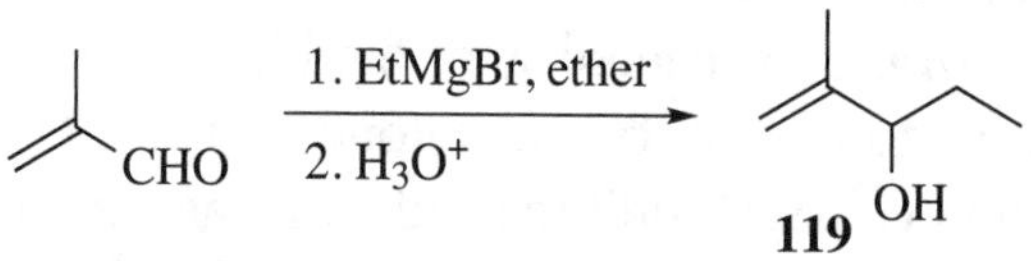

In Table 8.5[139] the influence of R^2 (from the Grignard reagent) and R^1 (attached to the carbonyl) on the percentage of conjugate addition is shown for reaction of alkyl Grignard reagents with **120**. Note that the phenyl group in **120** (R^3 = Ph, R^4 = H) has a significant influence on the reaction (see below), which proceeds by a six-center transition state (see **121**) to give the conjugate addition product, **122**. As R^1 in **120** becomes larger, attack at the carbonyl (path *a*) in a four-center transition state induces a large and destabilizing $R^2 \leftrightarrow R^1$ interaction leading to preferred attack at the end of the planar conjugated system (path *b*). When R^1 is small, R^2 is easily transferred to the carbonyl by the usual four-center transition state (path *a*).[141] Increasing the size of R^3 or R^4 increases the magnitude of the $R^3(R^4) \leftrightarrow R^2$ interaction in the six-center transition state (path *b*), diminishing the amount of 1,4-addition. Removing the Lewis acid byproduct (MgX_2) facilitates formation of **121** with an increase in 1,4 addition product **122**.[141] With alkylidene malonates and related compounds, virtually no 1,2-addition is observed,[140] although two electron-withdrawing groups attached to the alkene moiety make conjugate addition rather facile.

137.   Hauser, F.M.; Hewawasam, P.; Rho, Y.S. *J. Org. Chem.* **1989**, *54*, 5110.
138.   Cossy, J.; Bauer, D.; Bellosta, V. *Tetrahedron* **2002**, *58*, 5909.
139.   Maroni-Barnaud, Y.; Maroni, P.; Fualdès, A.M. *Compt. Rend.* **1962**, *254*, 2360.
140.   Mane, R.B.; Krishna Rao, G.S. *J. Chem. Soc. Perkin Trans. 1* **1973**, 1806.

Table 8.5. **Effect of Steric Hindrance on 1,4- vs. 1,2-Addition to α,β-Unsaturated Ketones**

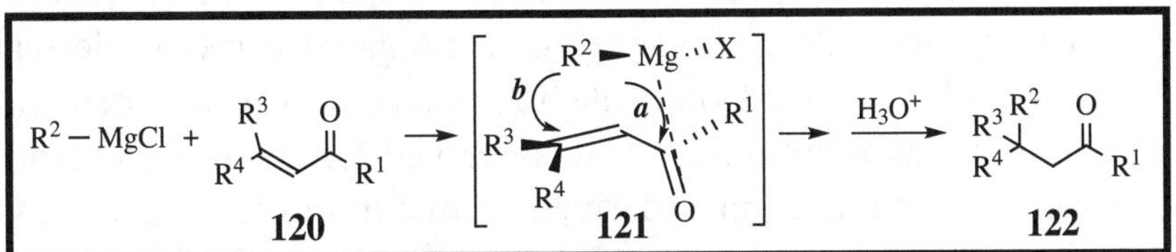

R^1	R	% 122
Me	*i*-Pr	50
Me	*t*-Bu	33
t-Bu	*i*-Pr	60
t-Bu	*t*-Bu	48
Ph	*i*-Pr	66
Ph	*t*-Bu	67
PhCH=CH-	*i*-Pr	70
PhCH=CH-	*t*-Bu	93

A similar transition state can be drawn for the reaction of Grignard reagents and α,β-unsaturated esters, where steric hindrance in the ester moiety partially blocks the acyl carbon and leads to more 1,4-addition.[141] An example is the reaction of butylmagnesium bromide with *sec*-butylcrotonate to give *sec*-butyloctanoate.[142] Asymmetric induction is possible in these reactions, as illustrated by the reaction of the (–)-menthyl ester[143] of crotonic acid **123** to give the (*S*)-(+) enantiomer of 3-phenyl derivative **124** in 46% yield.[144] Interestingly, addition of Cu_2Cl_2 to the reaction gave the (*R*)-(–) enantiomer.

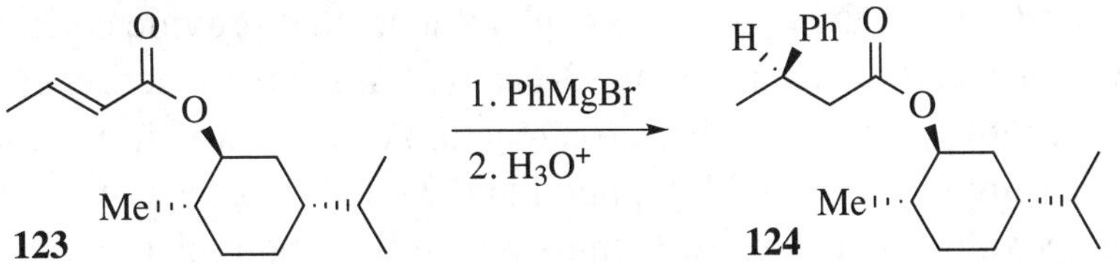

In general, addition of cuprous [Cu(I)] salts to a Grignard facilitates conjugate addition, partly because a more highly reactive species (R_2MgCu or RCu) is formed, and partly because copper coordinates better to the carbonyl, which facilitates the six-center transition state. The organocuprates derived from organolithium reagents (R_2CuLi, sec. 8.7.A) are extremely useful in conjugate addition reactions.[145] Organocopper reagents derived from the reaction of a Grignard reagent with a cuprous salt, however, are also very useful. In a synthesis of *N*-deacetyllappaconitine by Taber and co-workers, butenylmagnesium bromide reacted with

141. Munch-Petersen, J. *J. Org. Chem.* **1957**, *22*, 170.
142. Munch-Petersen, J.; Jacobsen, S. *Compt. Rend.* **1962**, *255*, 1355.
143. Seiji, S.; Yumiko, S.; Sawada, S.; Sejima, Y.; Ohi, S.; Shunsuke, O.; Inouye, Y. *Bull Kyoto Univ. Educ. Ser. B,* **1979**, *55*, 33 [*Chem. Abstr. 92*: 129098s *1980*].
144. Inouye, Y.; Walborsky, H.M. *J. Org. Chem.* **1962**, *27*, 2706.
145. (a) Bernady, K.F.; Weiss, M.J. *Prostaglandins* **1973**, *3*, 505; (b) Reference 118d, p 111.

conjugated ketone **125** in the presence of CuBr–SMe$_2$ and triethylsilyl chloride to give a 78% yield of the conjugate addition product **126**.[146] This example is typical in that cuprous salts generally lead to 1,4-addition, despite the usual propensity of conjugated aldehyde to give 1,2-addition with nucleophiles. A variety of conjugated carbonyl derivatives undergo 1,4-addition with Grignard reagents. Conjugate addition to α,β-unsaturated amides is known and proceeds with little variation from the results observed for conjugated ketones or esters.[147] As mentioned above, the organocuprates derived from lithium will be discussed in greater detail in Section 8.7.A.

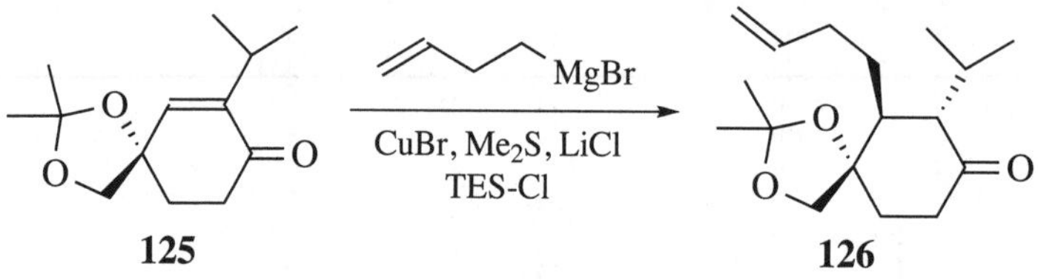

125   126

8.4.E.   Reaction with Epoxides[148]

Grignard reagents react with epoxides to form a new carbon-carbon bond with opening of the three-membered ring to give an alcohol after hydrolysis. There are two electrophilic carbon atoms, and the Grignard reagent attacks the less hindered carbon in an S$_N$2 like reaction.[149] If the carbons of the epoxide moiety are primary or secondary, attack can occur at either carbon leading to a mixture of alcohol products. When 2-ethyloxirane reacts with methylmagnesium bromide, attack at the primary carbon gave 3-pentanol with only trace amounts of attack at the secondary carbon. When the epoxide has substituents that provide equal or close to equal steric hindrance at each carbon, as in 2-ethyl-3-propyloxirane the reaction produces a mixture of regioisomeric alcohol products. In this case, reaction of this epoxide and methylmagnesium bromide gave a mixture of 4-methyl-3-heptanol and 3-methyl-4-heptanol. When a Grignard reagent opens an epoxide, the new substituent is on the β-carbon relative to the OH.

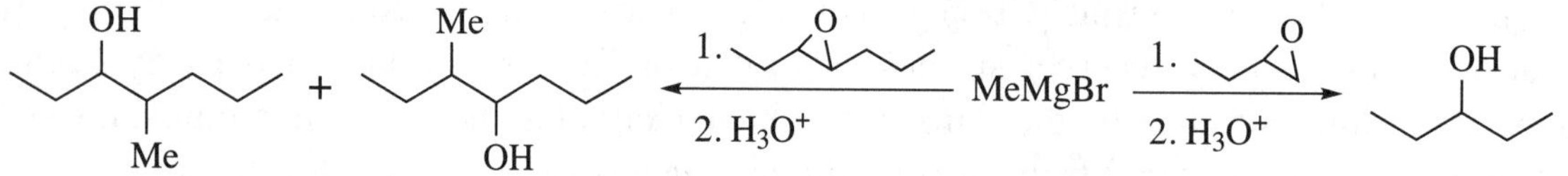

Metal salts such as cuprous ion promotes the ring opening reaction of epoxides, which is particularly useful for directing the incoming Grignard reagent to a particular carbon (to make

---

146.   Taber, D.F.; Liang, J.-L.; Chen, B.; Cai, L. *J. Org. Chem.* **2005**, *70*, 8739.
147.   Gilbert, G.; Aycock, B.F. *J. Org. Chem.* **1957**, *22*, 1013.
148.   (a) Taylor, S.K.; Haberkamp, W.C.; Brooks, D.W.; Whittern, D.N. *J. Heterocyclic Chem.* **1983**, *20*, 1745; (b) Linstrumelle, G.; Lorne, R.; Dang, H.P. *Tetrahedron Lett.* **1978**, 4069; (c) Boireau, G.; Namy, J.L.; Abenhaïm, D. *Bull. Soc. Chim. Fr.* **1972**, 1042.
149.   Sano, M.; Kodama, H.; Matsuda, H.; Matsuda, S. *Nippon Kagaku Kaishi* **1974**, 1716 [*Chem. Abstr. 82*: 42560f *1975*].

the reaction more regioselective). An example is taken from Martín and co-worker's synthesis of (+)-muconin in which the reaction of epoxide **127** with 1-undecylmagensium bromide (formed *in situ*) and cuprous iodide,[150] gave **128** in 70% yield. Note that the Grignard reagent added to the less substituted carbon. Allylic epoxides can be opened in a conjugated addition manner using Grignard reagents, analogous to an S_N2' reaction (sec. 2.6.A.iii). Isopropylmagnesium bromide reacted with **129**, for example, attacking the alkene unit with concomitant opening of the epoxide to give a 76% yield of **130** with a 22.1:1.0 Z:E ratio.[151]

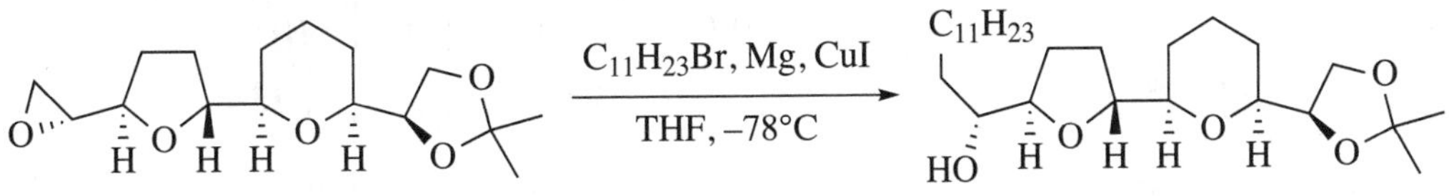

This fundamental reaction leads to useful synthetic applications. Similar alkylation of aryl epoxides, followed by oxidation of the resulting alcohol, makes the Grignard-epoxide reaction an effective route for the synthesis of α-aryl ketones. Epoxidation of cyclohexene with *m*-chloroperoxybenzoic acid (sec. 3.4.B.iii) gave cyclohexene oxide. Reaction with phenylmagnesium bromide and hydrolysis led to 2-phenylcyclohexanol. Oxidation of the secondary alcohol moiety with chromium trioxide (or with another oxidizing agent as in sec. 3.2) gave the targeted 2-phenylcyclohexanone.

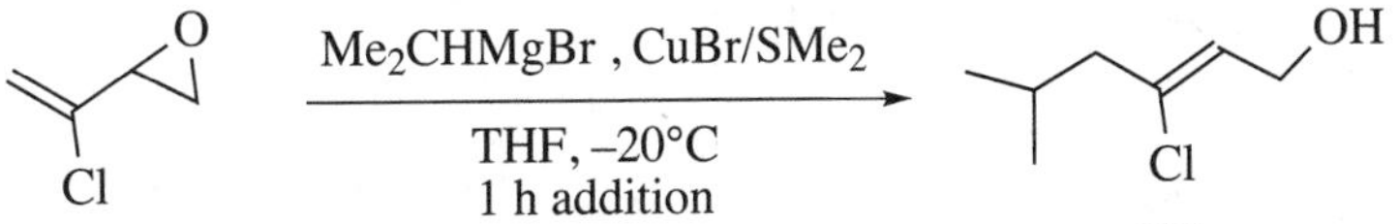

This sequence generates the useful disconnections:

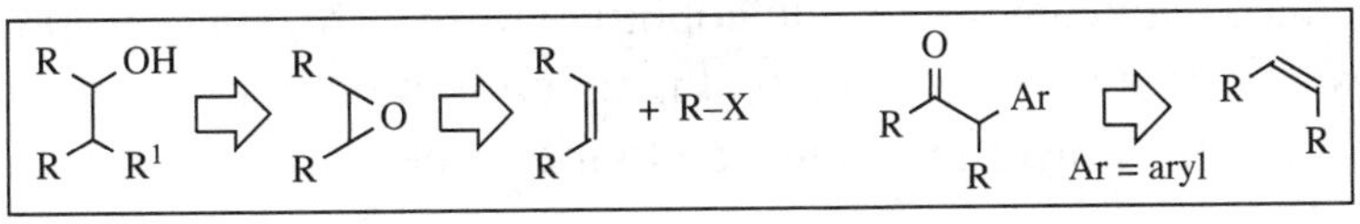

Looking back to the Schlenk equilibrium (sec 8.4.A), Grignard reagents contain the mild Lewis acid MgX_2 even if no other reagent is added. This Lewis acid is part of the reagent, and it can induce acid catalyzed rearrangements of acid-sensitive epoxides, complicating the normal product distribution. When 5,6β-epoxycholestane (**131**) was treated with methylmagnesium iodide, the product was 4aα-methyl-A-homo-β-norcholestan-4aβ-ol, **134**,[152] which is explained by reaction of the epoxide with MgI_2 to give **132**. A 1,2-hydride shift (secs. 2.6.B.iii, sec. 12.2.B) generated ketone derivative **133**, which reacted with additional methylmagnesium

150. Pinacho Crisostomo, F.R.; Carrillo, R.; Leon, L.G.; Martin, T.; Padron J.M.; Martin, V.S. *J. Org. Chem.* **2006**, *71*, 2339
151. Taber, D.F.; Mitten, J.V. *J. Org. Chem.* **2002**, *67*, 3847.
152. Frankel, J.J.; Julia, S.; Richard-Neuville, C. *Bull. Soc. Chim. Fr.* **1968**, 4870.

iodide to give the alcohol (**134**). The involvement of MgI_2 in the conversion of **131** to **134** was later substantiated by similar reaction of a cholestan-3-one derivative with methylmagnesium iodide.[153] The yield was low, since cationic rearrangement was slower than nucleophilic attack in the aprotic solvent used for this Grignard reaction.

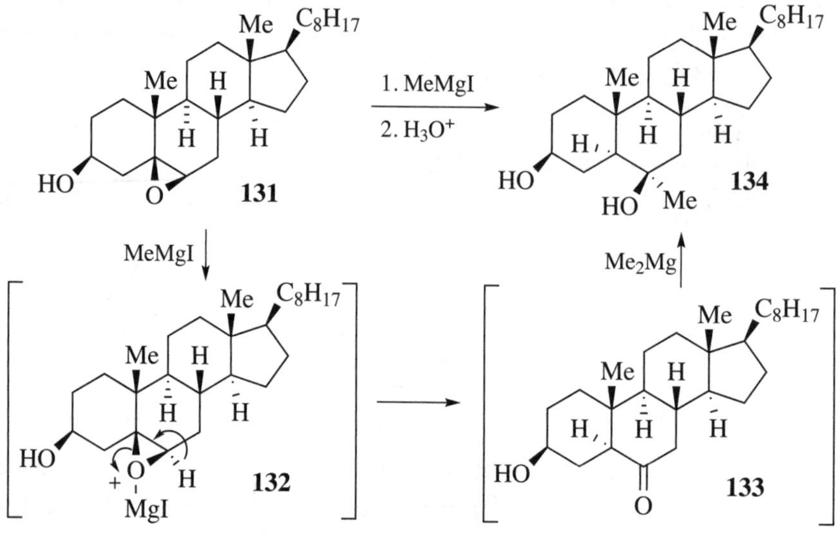

If the epoxide moiety is attached to a tertiary carbon, cationic rearrangements can lead to significant amounts of rearranged product. The cationic intermediate generated from the epoxide and MgX_2 does not always rearrange to a ketone derivative, but can give S_N1 type reactions if a tertiary center is present. Reaction of phenylmagnesium bromide with epoxide **135**, for example, led to a coordination complex of the magnesium and oxygen in **136**. The oxonium salt opened to generate an oxygen-stabilized cation (**137**), which is particularly stable. When this intermediate reacted with the nucleophilic PhMgBr reagent, the product was alcohol **138** after hydrolysis.[154] It should be noted that in epoxy-ketones or epoxy-aldehydes, opening the epoxide ring is usually slower than addition to the carbonyl.[155]

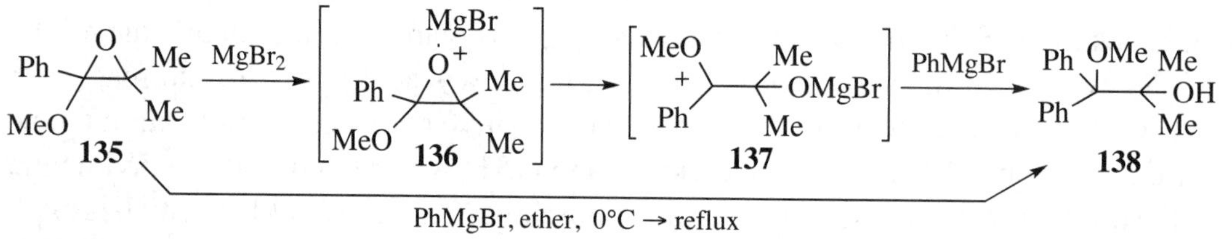

PhMgBr, ether, 0°C → reflux

153. (a) Rao, P.N.; Uroda, J.C. *Tetrahedron Lett.* **1964**, 1117; (b) Hall, N.D.; Just, G. *Steroids* **1965**, *6*, 111; (c) Julia, S.; Lavaux, J.P.; Lorne, R.; Riz, J.-C. *Bull. Soc. Chim. Fr.* **1967**, 3218.
154. Stevens, C.L.; Holland, W. *J. Org. Chem.* **1958**, *23*, 781.
155. Sepúlveda, J.; Soto, S.; Mestres, R. *Bull. Soc. Chim. Fr., Part 2* **1983**, 233.

8.4.F. Selectivity in Grignard Reactions

8.4.F.i. Configurational Stability of Grignard Reagents.

In principle, the carbon atom bearing a MgX unit can be a stereogenic center, which has important stereochemical implications for any reaction. *When the C–Mg bond of a Grignard reagent is formed, however, the stereochemical integrity of that bond is poor.* For this reason, the generation and use of chiral Grignard reagents is *not* a synthetically useful option in most systems. The NMR studies by Roberts and Whitesides [156] and also Fraenkel and co-workers[157] on Grignard regents in ether, established that the C–Mg has significant ionic character and undergoes rapid inversion at the magnesium-bearing carbon atom. 3-Cyclohexenylmagnesium bromide (**139**), for example, undergoes inversion of configuration at the C–Mg position. The rate of inversion decreased when the solvent was changed from ether to THF.[158]

Fraenkel and co-workers also showed that 2-butylmagnesium bromide (**140**) undergoes rapid inversion at ordinary temperatures,[157,159] thereby making a Grignard reagent derived from chiral 2-butyl bromide effectively racemic. However, Hoffmann and Hölzer used chiral Grignard reagent **141**, which was configurationally stable at –78°C.[160] In a nickel-catalyzed reaction of **141** with vinyl bromide, the so-called **Kumada-Corriu coupling**,[161] an 80% yield of **142** was obtained in 89% ee, along with 11% of 1-phenyl-1-butene. Grignard reagent **141** was generated by the reaction of a chiral chlorosulfone with an excess of ethylmagnesium chloride at –78°C.

Since the carbon-magnesium bond is stereochemically unstable, a Grignard reagent derived from a chiral halide equilibrate (**143a** → **143b**),[162] so subsequent Grignard reaction with an aldehyde or ketone will give racemic products. There are a few exceptions to this configurational instability. Walborsky et al. reported that the Grignard reagent derived from **144** reacted with

---

156. Whitesides, G.M.; Roberts, J.D. *J. Am. Chem. Soc.* **1965**, *87*, 4878.
157. (a) Fraenkel, G.; Dix, D.T. *J. Am. Chem. Soc.* **1966**, *88*, 979; (b) Fraenkel, G.; Dix, D.T.; Adams, D.G. *Tetrahedron Lett.* **1964**, 3155.
158. Maercker, A.; Geuss, R. *Angew. Chem. Int. Ed.* **1971**, *10*, 270.
159. Fraenkel, G.; Dix, D.T.; Adams, D.G. *Tetrahedron Lett.* **1964**, 3155.
160. (a) Hölzer, B.; Hoffmann, R.W. *Chem. Commun.* **2003**, 732. See also (b) Hoffmann, R.W.; Hölzer, B.; Knopff, O.; Harms, K. *Angew. Chem. Int. Ed.* **2000**, *39*, 3072.
161. (a) Tamao, K.; Hiyama, T.; Negishi, E.-i. *J. Organomet. Chem.* **2002**, *653*, 1; (b) Corriu, R.J.P.; Masse, J.P. *J. Chem. Soc. Chem. Commun.* **1972**, 144a; (c) Tamao, K.; Sumitani, K.; Kumada, M. *J. Am. Chem. Soc.* **1972**, *94*, 4374; (d) Mundy, B.P.; Ellerd, M.G.; Favaloro Jr., F.G. *Name Reactions and Reagents in Organic Synthesis, 2nd ed.*, Wiley-Interscience, New Jersey, **2005**, pp. 386-387.
162. Morrison, J.D.; Mosher, H.S. *Asymmetric Organic Reactions*, ACS, Washington, **1976**, pp 414-415.

carbon dioxide to give the corresponding carboxylic acid with some retention of configuration

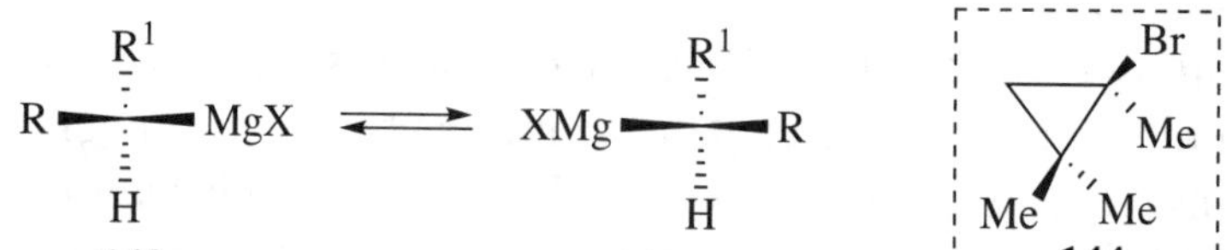

143a   **143b**   **144**

(14% ee) in 38% yield.[163] The % ee can be as high as 80%, however, when the reactions are carried out in ether.[163] Better selectivity was observed with menthol-based Grignard reagents.[164] Cyclopropyl derivatives tend to give better selectivity than acyclic or larger ring Grignard reagents. α-Halo-Grignard reagents have been shown to be excellent precursor to chiral Grignard reagents. Hoffmann showed that ethylmagnesium chloride reacted with α-halosulfoxide **145** to give **146**, along with 4-chloropenyl ethyl sulfoxide.[165] When **146** was treated with more ethylmagnesium chloride between −50 to −30°C, Grignard reagent **147** was formed in solution. When this reacted with phenylisothiocyanate, thioamide **148** was isolated in 56% yield and 83% ee. Hoffmann observed that racemization of the Grignard reagent **147**[166] was slowest in the presence of chloride, an anion of low nucleophilicity. The use of ethylmagnesium chloride was important because the reaction was least complicated by formation of a rearranged Grignard product.[167]

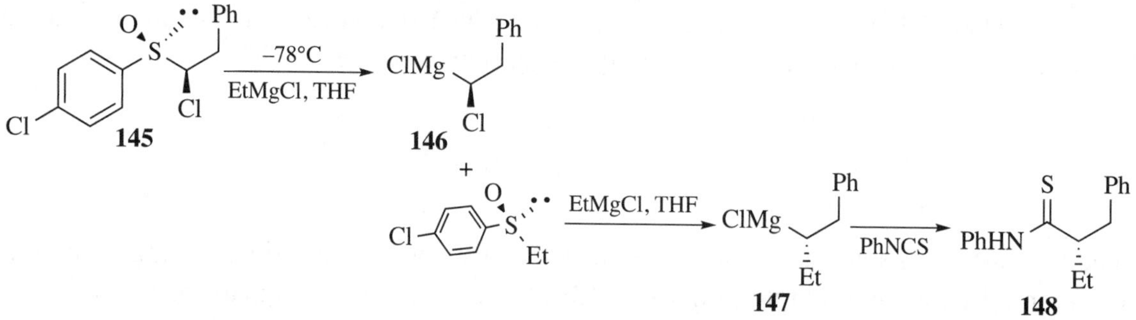

When a α-halo Grignard reagent is treated with another organometallic reagent (such as another Grignard reagent), a new Grignard reagent is generated by a carbon-carbon bond-forming reaction.[167] *This appears to be a rearrangement product, but actually arises by formation of a carbenoid species* (sec. 13.8.D) that undergoes a C–H insertion reaction. Reaction of diiodide **149** with isopropylmagnesium chloride, for example, generated the iodo-Grignard reagent **150** at −78°C. As **150** warmed up, it reacted with additional isopropylmagnesium chloride to give

163. Walborsky, H.M.; Impastato, F.J.; Young, A.E. *J. Am. Chem. Soc.* **1964**, *86*, 3283.
164. (a) Tanaka, M.; Ogata, I. *Bull. Chem. Soc. Jpn.* **1975**, *48*, 1094; (b) Schumann, H.; Wassermann, B.C.; Hahn, F.E. *Organometallics* **1992**, *11*, 2803; (c) Dakternieks, D.; Dunn, K.; Henry, D.J.; Schiesser, C.H.; Tiekink, E.R. *Organometallics* **1999**, *18*, 3342.
165. Hoffmann, R.W.; Hölzer, B.; Knopff, O.; Harms, K. *Angew. Chem. Int. Ed.* **2000**, *39*, 3072.
166. Hoffmann, R.W.; Nell, P.; Leo, R.; Harms, K. *Chem. Eur. J.* **2000**, *6*, 3359.
167. Hoffmann, R.W.; Knopff, O.; Kusche, A. *Angew. Chem. Int. Ed.* **2000**, *39*, 1462.

151, and quenching with methanol gave the observed product **152**.[167] For most of the reactions listed in this section, such carbenoid reactions should not be a major problem.

8.4.F.ii. Diastereoselectivity and Enantioselectivity. Reaction of a Grignard reagent reacts with unsymmetrical ketones or aldehydes that contain a prochiral carbonyl carbon, leads to a new stereogenic center. Diastereomers result if another stereogenic center is present in either the Grignard reagent (but not at the carbon bearing the Mg) or the carbonyl substrate. If 2-pentanone reacts with the Grignard reagent derived from 3-phenyl-1-bromobutane (**153**), the resulting alcohol will be a mixture of syn and anti diastereomers (**154** and **155**, respectively). Each diastereomer will be racemic. In light of the previous discussion in section 8.4.F.i., it may be possible for one diastereomer to predominate (the reaction would be diastereoselective) but it should not be enantioselective.

The products from such reactions can be organized into four categories that reflect simple combinations of chiral and achiral Grignard reagents reacting with chiral or achiral carbonyl derivatives. (*1*) Mixing an achiral Grignard reagent and a carbonyl derivative that does not possess a substituent attached to a stereogenic center generates one new stereogenic center. (*2*) Mixing a chiral Grignard reagent and a carbonyl derivative that does not possess a substituent attached to a stereogenic center generates diastereomers. (*3*) Mixing an achiral Grignard reagent and a carbonyl derivative that has a substituent attached to a stereogenic center leads to diastereomers, with possible diastereoselectivity. (4) Mixing a chiral Grignard reagent with a carbonyl derivative that has a substituent attached to a stereogenic center will lead to good diastereoselectivity.

If there is steric bias in a molecule, and if the addition of a Grignard reagent generates diastereomers, the products may be formed with good diastereoselectivity. The rationale for diastereoselectivity discussed in section 4.8 for reduction generally applies to reaction of Grignard reagents and carbonyl derivatives. In a synthesis of vinigrol, by Paquette and co-workers,[168] ketone **156** reacted with vinylmagnesium chloride to give 54% of **157** and 24% of **158**. The observed diastereoselectivity is typical of the exo-selectivity observed similar bicyclic systems (sec. 1.4.E).[169]

168.   Paquette, L.A.; Guevel, R.; Sakamoto, S.; Kim,I.H.; Crawford, J. *J. Org. Chem.* **2003**, *68*, 6096.
169.   (a) Jung, M.E.; Hudspeth, J.P. *J. Am. Chem. Soc.* **1980**, *102*, 2463; (b) Idem *Ibid.* **1978**, *100*, 4309.

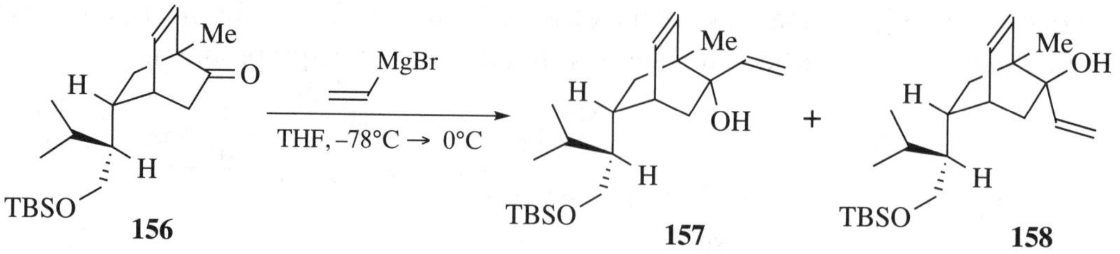

MOLECULAR MODELING: Stereochemistry of Vinyl Magnesium Bromide Additions

One synthetic route to vinegrol involves addition of vinyl magnesium bromide to **156.** This occurs from the *exo* carbonyl face (leading to **157**) rather than from the *endo* carbonyl face (leading to **158**). Remember that the addition of nucleophiles to the carbonyl carbon must approach from an angle of 110° (the "Bürgi-Dunitz trajectory"). This fact must be considered when determining the preferred face for reaction, and any possible steric hindrance to that addition.

Examine a space-filling model of **156** (left). Assume that the Grignard reagent approaches from either (*a*) or (*b*), and determine if both *exo* and *endo* faces accessible to the reagent? Is there a clear steric preference for addition to one over the other? Next, examine the LUMO map for **156** (right). This maps the (absolute) value of the lowest-unoccupied molecular orbital onto an electron density surface, which like a space-filling model, indicates overall molecular size and shape. By convention, regions on the surface where the LUMO is most concentrated are colored "blue", while those where it is absent are colored "red". The reagent (a nucleophile) should prefer to attack **156** on the face with the greatest blue area. Does the LUMO map show a clear preference for attack? If it does, which product (**157** or **158**) should result? Note that the LUMO map is provided as a transparent model to allow you to see the underlying structure. You can change it to a solid model (making comparison of the *exo* and *endo* faces easier), by clicking on the map and selecting *Solid* from the **Style** menu at the bottom of the screen.

Good diastereoselectivity can be observed in less sterically demanding systems for selected Grignard reagents. Hoffmann reported good diastereoselectivity when α-haloalkyl Grignard reagents reacted with benzaldehyde.[170] Grignard reagent **159**, for example, gave an 82% yield of **160 + 161** in a 92:8 ratio, showing the reaction clearly favored the syn diastereomer.

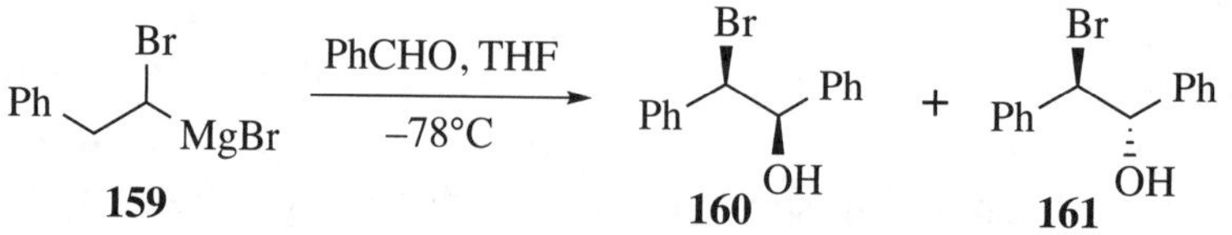

Assuming that Grignard reagent addition to a carbonyl is a diastereoselective reaction that produces racemic products, it is important to ask if it is possible for the reaction to be

---

170. Schulze, V.; Nell, P.G.; Burton, A.; Hoffmann, R.W. *J. Org. Chem.* **2003**, *68*, 4546.

enantioselective. If we rely on the carbon bearing the magnesium atom, the answer is usually no. If we incorporate a stereogenic center elsewhere in the molecule, however, the relative merits of the four cases listed above must be discussed.

Case 1 involves an achiral Grignard reagent reacting with a prochiral ketone to give a racemic alcohol. Some asymmetric induction[171] has been achieved by forming asymmetric complexes of the Grignard reagent with solvents such as (2R, 3R)-(+)-dimethoxybutane.[172] The two oxygens of this chiral ether coordinate with the Grignard reagent to form a monomeric species such as 162.[162] In 162, coordination with the oxygens renders the magnesium asymmetric, and reaction with a prochiral ketone will generate a diastereomeric transition state for the acyl addition[162] so an excess of one enantiomer will result. Since R² in 162 exists as an equilibrating mixture of stereoisomers (see 144 and 145), complex 162 represents four stereochemically distinct solvated species, not necessarily present in equal amounts, and each can react with a prochiral substrate at different rates.[162] The extent of induced stereoselectivity is generally low, often 2-3%,[162] even in an asymmetric solvent. When phenylmagnesium bromide reacted with 2-octanone in the optically active solvent 2-methyltetrahydrofuran, for example, 2-phenyl-2-octanol was formed in only 11-18% ee.[173]

In some cases, it is possible to form a chiral, nonracemic alcohol from a prochiral ketone and a Grignard reagent, by adding an asymmetric ligand (a chiral additive) to the reaction mixture. This case is differentiated from the reaction of Grignard reagent with a carbonyl species that has a stereogenic center present in the same molecule (case *3* above, a Cram's rule type system: see sec. 4.8.B for a discussion of Cram's rule). Presumably, the Grignard reagent forms a chiral complex with these additives, promoting facial selectivity in the addition reaction with the carbonyl. In work by Nozaki and co-workers, addition of (−)-sparteine (163) to the reaction of ethylmagnesium bromide and benzaldehyde gave (R)-(+)-1-phenyl-1-propanol in 15% yield and 22% ee.[174] Under the same conditions, reaction with acetophenone gave racemic 2-phenyl-2-butanol in 11% yield. Seebach and co-workers showed that addition of 164 to benzaldehyde gave 1-phenyl-1-propanol in 6-8% ee.[175] Inch et al. added furanose 165[176] to a mixture of methylmagnesium bromide and 1-phenyl-1-heptanone, and obtained a 95% yield of (R)-(+)-2-phenyl-2-octanol with 70% ee. Under similar conditions, however, ethylmagnesium bromide reacted with acetophenone to give a 48% yield of (S)-(−)-2-phenyl-

171. Solladie, G. in *Asymmetric Synthesis Vol. 2*, Morrison, J.D. (Ed.), Academic Press, New York, *1983*, pp 157-183.
172. (a) Cohen, H.L.; Wright, G.F. *J. Org. Chem. 1953, 18*, 432; (b) Allentoff, N.; Wright, G.F. *Ibid. 1957, 22*, 1.
173. Iffland, D.C.; Davis, J.E. *J. Org. Chem. 1977, 42*, 4150.
174. (a) Nozaki, H.; Aratani, T.; Toraya, T.; Noyori, R. *Tetrahedron 1971, 27*, 905; (b) Nozaki, H.; Aratani, T.; Toraya, T. *Tetrahedron Lett. 1968*, 4097.
175. (a) Seebach, D.; Langer, W. *Helv. Chim. Acta 1979, 62*, 1701; (b) Seebach, D.; Crass, G.; Wilka, E.-M.; Hilvert, D.; Brunner, E. *Ibid. 1979, 62*, 2695.
176. Inch, T.D.; Lewis, G.J.; Swainsbury, G.L.; Sellers, D.J. *Tetrahedron Lett. 1969*, 3657.

2-butanol with 27% ee.[176] Chiral amino alcohols were used as additives by Battioni and Chodkiewicz, but gave optical yields of only 0-20%.[177] Mukaiyama and co-workers obtained good results with the chiral ligand **166** in reactions of organolithium, dialkylmagnesium and Grignard reagents with benzaldehyde.[178] When **166** was added to the reaction, the optical yield of alcohol **167** depended on both the solvent and the temperature of the reaction. In case (*2*), the stereogenic center of the Grignard might be expected to give some selectivity for the syn or anti diastereomeric product. The configurational instability of the C-Mg bond[158,159,162] at the putative stereogenic carbon, however, leads to an equilibrating mixture that is effectively a chiral, racemic Grignard reagent, with poor syn-anti selectivity in the diastereomeric alcohol mixture.

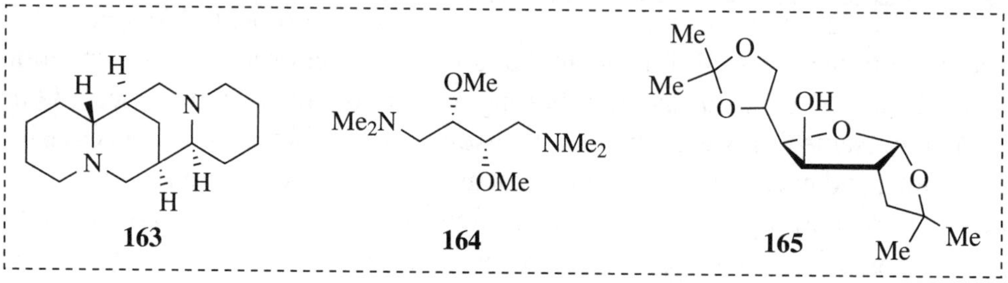

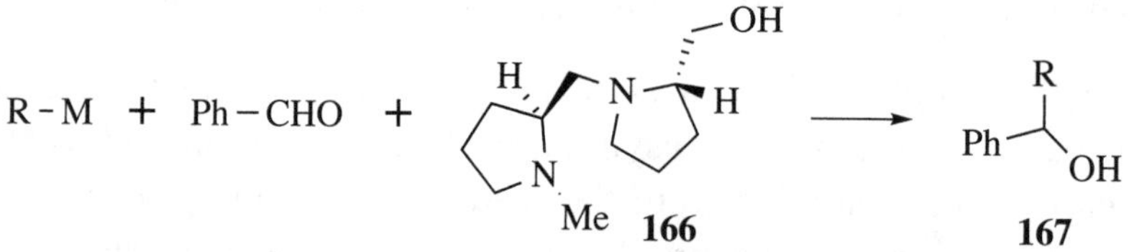

If the carbonyl partner contains a stereogenic carbon, as in case (*3*), the asymmetry of that center should exert an influence when the Grignard reagent reacts with the prochiral carbonyl (the faces of the carbonyl are diastereotopic), leading to some diastereoselectivity in the alcohol product. This is the system that formed the basis of the Cram model,[179] the Cornforth model,[180] the Felkin-Anh model[181] and the Karabatsos model[182] discussed in Section 4.8.B. The Cram model, the Cram chelation model or the Felkin-Anh model are generally used to predict selectivity in addition of Grignard reagents to a prochiral carbonyl with stereogenic α- or β- carbons. The presence of a stereogenic center in one of the reactive partners induces facial bias on approach of the two reagents, leading to moderate to good diastereoselection for either the syn or the anti isomer. Examples taken from Mosher and Morrison's compilation are

177. Battioni, J.P.; Chodkiewicz, W. *Bull. Soc. Chim. Fr.* **1972**, 2068.
178. (a) Mukaiyama, T.; Soai, K.; Sato, T.; Shimizu, H.; Suzuki, K. *J. Am. Chem. Soc.* **1979**, *101*, 1455; (b) Soai, K.; Mukaiyama, T. *Chem. Lett.* **1978**, 491; (c) Sato, T.; Soai, K.; Suzuki, K.; Mukaiyama, T. *Ibid.* **1978**, 601; (d) Mukaiyama, T.; Suzuki, K.; Soai, K.; Sato, T. *Chem. Lett.* **1979**, 447.
179. (a) Cram, D.J.; Abd Elhafez, F.A *J. Am. Chem. Soc.* **1952**, *74*, 5828; (b) Cram, D.J.; Kopecky, K.R. *Ibid.* **1959**, *81*, 2748.
180. Cornforth, J.W.; Cornforth, R.H.; Mathew, K.K. *J. Chem. Soc.* **1959**, 112.
181. Cherest, M.; Felkin, H.; Prudent, N. *Tetrahedron Lett.* **1968**, 2199.
182. Karabatsos, G.J. *J. Am. Chem. Soc.* **1967**, *89*, 1367.

shown in Table 8.6,[183] and Cram's open chain model predicts the diastereoselectivity of case (3) reactions. In Table 8.6, ketone **168** reacted with various Grignard reagents (R^2MgX) to give a mixture of diastereomeric alcohols, **169** and **170**. When the LUMO map (see sec. 4.7.B) for **168** (R_S =H, R_M = Me, R_L = Ph, R^1 = H) is prepared, the blue intensity about the carbonyl carbon indicates a preference for **169**.

Table 8.6.   Cram Selectivity for Carbonyls with an Adjacent Stereogenic Center (168)

R_S	R_M	R_L	R^1	R^2	X	169 : 170
H	Me	Ph	H	Me	Br	2.4 : 1
				Me	I	2.0 : 1
				Et	Br	3.0 : 1
				Ph	Br	>4.0 : 1
H	i-Pr	Ph	H	i-Pr	Br	1.9 : 1
Me	Et	Ph	H	Ph	Br	2.9 : 1
H	Et	Ph	H	Me	I	2.5 : 1
				Et	Br	3.0 : 1
H	Me	Et	H	Me	Br	1.5 : 1
H	Me	Et	Me	Me_2CHCH_2	Br	1.8 : 1

[Reprinted by courtesy of Professor James D. Morrison from Morrison, J.D.; Mosher, H.S. *Asymmetric Organic Reactions*, American Chemical Society, Washington, D.C. *1976*, pp. 92-93.]

As discussed in Chapter 4, the Cornforth dipolar model (sec. 4.7.B) is better able to predict the stereoselectivity for reactions of α-chloroketones, and the Cram cyclic (chelation) model (**171**)[184] is preferred for reactions of ketones that have heteroatoms in the α-position. The chelation model is very common in Grignard reactions, and it assumes that the magnesium is coordinated to both the heteroatom and the carbonyl oxygen as shown in **171**. For this reason, coordination of an organometallic with carbonyls that have a heteroatom in the α- and β- positions has been termed **chelation control** when it results in formation of products with high diastereoselectivity.[185] Part of this model assumes that the incoming group in **171** (R attached to magnesium) is delivered from the less hindered face (over the smallest group, R_S). In this model, the R^1 group could be H or alkyl. An example of chelation control is the reaction of α-benzyloxy ketone (**172**) with vinylmagnesium bromide, a key step in the Williams and White synthesis of (±)-citreoviridin.[186] In this case, diastereomer **174** was formed exclusively, and the chelated model (**173**) predicted the observed stereochemistry.

183.   Reference 162, pp 92-93.
184.   (a) Reference 162, pp 94-98; (b) Cram, D.J.; Wilson, D.R. *J. Am. Chem. Soc.* **1963**, *85*, 1245.
185.   Still, W.C.; Schneider, J.A. *Tetrahedron Lett.* **1980**, *21*, 1035.
186.   Williams, D.R.; White, F.H. *J. Org. Chem.* **1987**, *52*, 5067.

172 → **173** → **174**

Case (*4*) from above should give the best diastereoselectivity. The instability of the C-Mg bond, however, leads to poor diastereoselectivity for the carbon bearing the magnesium as observed in case (*2*). The stereogenic center on the carbonyl substrate remains intact, and it influences the reaction, leading to good diastereoselectivity in the alcohol product upon reaction with a prochiral carbonyl derivative. For all practical purposes this is a case (*3*) situation, except that there is loss of diastereoselectivity due to the inability to control the configuration of the stereogenic center in the Grignard reagent.

8.4.F.iii. Stereochemical Integrity of Alkenyl Grignard Reagents. A different stereochemical problem arises when vinylmagnesium halides are formed from alkenyl halides. Both (*E*) and (*Z*) isomers can be formed. Reaction of stereochemically pure (*E*) or (*Z*) alkenyl halides with magnesium gave a mixture of (*E*) and (*Z*) vinyl Grignard reagents.[187] Subsequent reaction with carbonyl derivatives gives an allylic alcohol product as an *E/Z* mixture. Martin et al.[187] reported that reaction of pure (*E*)- or (*Z*)-1-bromo-1-alkenes with magnesium metal and then with CO_2 gave stereospecific conversion to the corresponding acid. Although (*E*) halide gave mainly the (*E*) product, this is *misleading*. In fact, severe loss of stereochemical integrity was observed and a mixture of both (*E*) and (*Z*) isomers was always obtained, as shown in Table 8.7.[188] It is important to note that there are significant differences between the stereochemical ratio of the Grignard reagent and that of the acid product. Table 8.7 shows a comparison of bromopropenes and bromohexenes, where loss of stereochemical integrity is more pronounced in the longer chain alkenes. In addition, greater loss of stereochemical integrity is apparent upon reaction with the electrophile (CO_2). A partial solution to this problem is reaction of the vinyl Grignard reagent with alkyl halides in the presence of cuprous iodide [Cu(I)I], which does little for the initial loss of stereochemistry upon formation of the Grignard, but in subsequent reactions, greater retention of stereochemistry is observed.[189] An example is the reaction of **175** with iodopropane in the presence of cuprous iodide, giving a 97% yield of **176**. The original 90:10 (*Z/E*) mixture in **175** was retained (88:12 *Z/E*) in the final product **176**.

187. Martin, G.J.; Mèchin, B.; Martin, M.L. *C.R. Acad. Sci. Ser. C* **1968**, *267*, 986.
188. Martin, G.J.; Martin, M.L. *Bull. Soc. Chim. Fr.* **1966**, 1636.
189. Derguini-Boumechal, F.; Linstrumelle, G. *Tetrahedron Lett.* **1976**, 3225.

Table 8.7. **Stereoselectivity in the Formation of Grignard Reagents from Stereo-chemically Pure Vinyl Halides and Subsequent Reaction with CO_2**

Halide	(Z : E)	RMgBr (Z : E)	RCOOH (Z : E)
(Z)-1-Bromo-1-propene	98.5 : 1	80-90 : 20-10	85-95 : undetected
(E)-1-Bromo-1-propene	5 : 95	30-40 : 70-60	15-5 : 85-95
(Z)-1-Bromo-1-hexene	97 : 1	75-85 : 25-15	45-55 : 55-45
(E)-1-Bromo-1-hexene	1 : 99	40-50 : 60-80	35-45 : 65-55

Another problem arises with allylic Grignard reagents. When the magnesium halides derived from (E)-1-bromo-2-butene were quenched with water (see Table 8.8),[190] the double bond had isomerized and a mixture of 1-butene, (Z)-2-butene and (E)-2-butene was obtained. This allylic rearrangement is well known in reactions of Grignard reagents.[191] Increasing the solvating power of the solvent, as in changing from diethyl ether to THF, increased the lability of the C–Mg bond and gave more isomerization.

Table 8.8. **Isomerization of Crotylmagnesium Halides**

Crotyl-MgX (X)	Solvent	1-Butene	(Z)-2-Butene	(E)-2-Butene
Br	ether	75	11	14
Cl	ether	80	9	11
I	ether	80	10	10
Cl	THF	60	15	25
I	THF	55	20	25

Reaction of an allylic magnesium halide with an electrophile can give a new bond at either carbon of the allylic carbanion, so two isomeric products are possible.[192] Felkin and co-workers described the reaction with ketones as a non-cyclic S_E2' rearrangement, and it is sensitive to steric encumbrance at the carbonyl carbon.[193] Allylic Grignard reagent **177**, when generated from either the (E)- or the (Z)- bromide, formed an equilibrating mixture (**177** and **180**). When this mixture of Grignard reagents reacted with a ketone (via **178** and **181**, respectively), isomeric alcohols **179** and **182** were formed. Increasing the steric bulk of R^1 and R destabilizes **178** relative to **181**, favoring **182** over **179**. The 2-butenyl derivative (R^1 = Me) reacted primarily as the crotyl derivative (**177**),[194] but **177** was in equilibrium with a small amount of **180**.[195]

190. Agami, C.; Andrac-Taussig, M.; Prévost, C. *Bull. Soc. Chim. Fr.* **1966**, 1915.
191. Agami, C. *Bull. Soc. Chim. Fr.* **1967**, 4031.
192. Hoffmann, R.W. *Angew. Chem. Int. Ed.* **1982**, *21*, 555.
193. Chérest, M.; Felkin, H.; Frajerman, C. *Tetrahedron Lett.* **1971**, 379.
194. Gaudemar, M. *Bull. Soc. Chim. Fr.* **1958**, 1475.
195. Nordlander, J.E.; Young, W.G.; Roberts, J.D. *J. Am. Chem. Soc.* **1961**, *83*, 494.

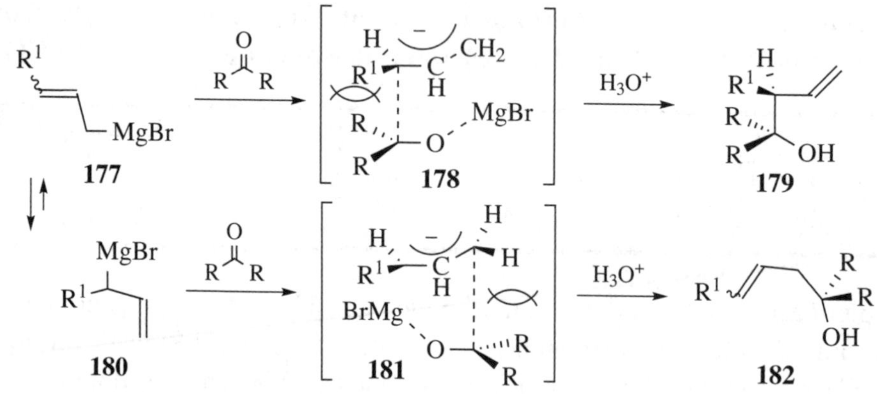

When **177/180** reacted with aldehydes[196] or unhindered ketones,[197] α-methallyl alcohol **179** was formed (**182/179** ratio is < 0.01). With diisopropyl ketone (R = *i*-Pr) the **182/179** ratio was 0.5[198] and with di-*tert*-butyl ketone **182** was the major product.[198a,199] *tert*-Butylalkylmagnesium bromide (R¹ = *t*-Bu)[194] increased the strain in **178** to the extent that even the unhindered substrate acetone gave predominantly **182** (R = Me), although similar reaction with acetaldehyde gave only **179** (**182179** ratio is <0.01). In all cases **182** was isolated predominantly as the (*E*) isomer [(*E/Z*) ratio is > 20:1]. Cyclic ketones such as cyclopentanone showed the greatest selectivity for **182** [cyclopentanone: **182:179** = 65:1, R = R¹ = -(CH₂)₄-] and cyclohexanone gave a 3:1 ratio (R=R¹ = -(CH₂)₅-). Miginiac, Prévost and co-workers noted that 1-bromo-2-pentene reacted with magnesium and various electrophiles to give primarily the 3-alkyl derivatives.[200] A minor side reaction is sometimes observed when attempting to prepare the Grignard reagent of a ω-haloalkene. The Grignard reagent derived from reaction of 6-bromo-1-hexene and magnesium gave ~ 5% of 1-methylcyclopentane via addition of the Grignard reagent to the alkene double bond.[201] The analogous organolithium reagents can give excellent yields of cyclization under similar conditions (RX + Li°), as discussed in Section 8.5.F.ii.

8.4.G. Reduction, Organocerium Reagents, and Enolization

8.4.G.i. Reduction. Although Grignard reagents react with ketones and aldehydes to give alcohols in good yield, there are a few reactions that compete with the normal reactions. Two common side-reactions are reduction and enolization. Reduction of carbonyl compounds occur when the carbonyl unit and the magnesium-bearing carbon of the Grignard reagent are sterically

196. Oh, Kiun-Houo *Ann. Chim (Fr.)* **1940**, *13*, 175.
197. Roberts, J.D.; Young, W.G. *J. Am. Chem. Soc.* **1945**, *67*, 148.
198. (a) Benkeser, R.A.; Young, W.G.; Broxterman, W.E.; Jones, Jr., D.A.; Piaseczynski, S.J. *J. Am. Chem. Soc.* **1969**, *91*, 132; (b) Young, W.G. Roberts, J.D. *Ibid.* **1945**, *67*, 319; (c) Gross, B.; Prevost, C. *Bull Soc. Chim. Fr.* **1967**, 3610.
199. Footnote 6 in Benkeser, R.A.; Broxterman, W.E. *J. Am. Chem. Soc.* **1969**, *91*, 5162.
200. Miginiac-Groizeleau, L.; Miginiac, P.; Prévost, C. *Compt. Rend.* **1965**, *260*, 1442.
201. Drozd, V.N.; Ustynyuk, Yu.A.; Tsel'eva, M.A.; Dmitriev, L.B. *Zh. Obshch. Khim.* **1969**, *39*, 1991 (Engl., p 1951).

hindered,[202] and the Grignard reagent must possess a β-hydrogen. Reaction of **183** with *tert*-butylmagnesium chloride, for example, gave >90% of the reduction product, alcohol **185** with none of the carbonyl addition product.[203] The reduction is thought to proceed by a coordination complex that leads to a transition state close to **184**, which is similar to the transition state observed in the **Meerwein-Ponndorf-Verley reduction**[204] (sec. 4.10.J) and the **Oppenauer oxidation** (sec. 3.2.E). Coordination between the carbonyl oxygen and magnesium leads to complex **184**, where hydrogen transfer gives the alcohol **185**, accompanied by elimination of H and MgX from the Grignard to give an alkene (in this case isobutylene). Hamelin[203b] showed that even an unhindered aldehyde such as propanal can be reduced with a small Grignard reagent such as ethylmagnesium bromide. A Grignard reaction between hindered ketones and aldehydes often leads to reduction as a synthetically useful process. Camphor (**186**) is reduced to borneol (**187**) and isoborneol (**188**) upon reaction with Grignard reagents, with selectivity for isoborneol as seen in Table 8.9.[205]

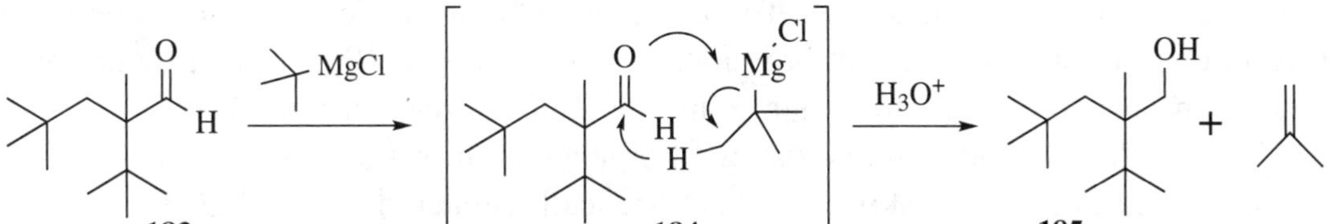

Table 8.9. Reduction of Camphor with Grignard Reagents

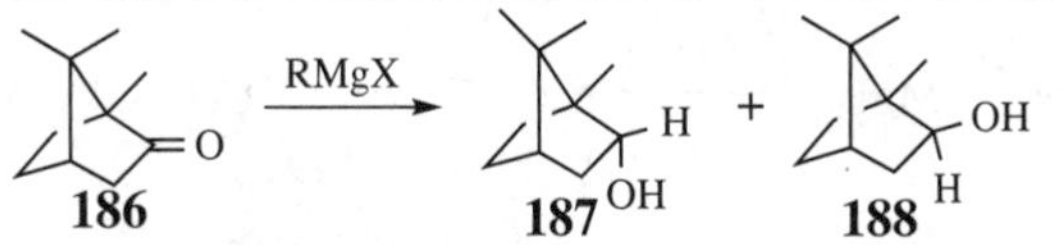

RMgX	% 187	% 188
EtMgBr	35	29
i-PrMgBr	45	45
n-PrMgBr	34	48
Et₂CHMgBr	29	55

Varying degrees of asymmetric induction can be achieved in the reduction process when chiral Grignard reagents and asymmetric solvents are employed.[206] In most cases, the %

202. Reference 83, p 138.
203. (a) Whitmore, F.C.; Whitaker, J.S.; Mosher, W.A.; Breivik, O.N.; Wheeler, W.R.; Miner, Jr., C.S.; Sutherland, L.H.; Wagner, R.B.; Clapper, T.W.; Lewis, C.E.; Lux, A.R.; Popkin, A.H. *J. Am. Chem. Soc.* **1941**, *63*, 643; (b) Hamelin, A. *Bull. Soc. Chim. Fr.* **1961**, 926.
204. (a) Meerwein, H.; Schmidt, R. *Ann.* **1925**, *444*, 221; (b) Ponndorf, W. *Angew. Chem.* **1926**, *39*, 138; (c) Verley, A. *Bull. Soc. Chim. Fr.* **1925**, *37*, 537, 871; (d) Wilds, A.L. *Org. React.* **1944**, *2*, 178.
205. Malkonen, P.J. *Suomen Kemistilehti* **1965**, *38B*, 89 [*Chem. Abstr. 63*: 8411b **1965**].
206. (a) Birtwistle, J.S.; Lee, K.; Morrison, J.D.; Sanderson, W.A.; Mosher, H.S. *J. Org. Chem.* **1964**, *29*, 37 and references cited therein; (b) Reference 162, pp 177-202.

ee for reduction is small. Nasipuri and coworkers[207] described the asymmetric reduction of phenylalkyl ketones.[208] A cyclic model[209] and an acyclic model[210] were proposed. Asymmetric induction was higher with phenylalkyl ketones than with cyclohexylalkyl ketones or *tert*-butylalkyl ketones.[211]

Since the Grignard reduction can be used synthetically, the appropriate functional group transform is

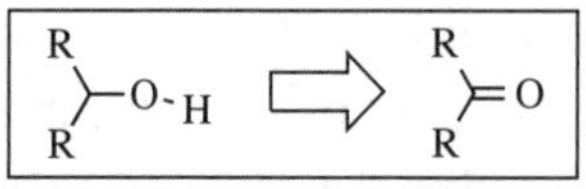

8.4.G.ii. Organocerium Reagents. Reduction in Grignard reactions of hindered ketones can severely limit a carbonyl alkylation reaction. This limitation can be circumvented, however, by the converting the Grignard reagent to an organocerium reagent. Imamoto and coworkers showed that methyl mesityl ketone (**190**) gave only 10% of the acyl addition product (**191**) upon reaction with butylmagnesium bromide.[212] Organocerium **189**, formed by the reaction of butylmagnesium bromide and cerium chloride ($CeCl_3$), reacted with **190** to give a 57% yield of **191**[212] A synthetic example using an organocerium reagent is taken from Majetich and Zhang's synthesis of perovskone, in which sterically hindered ketone **192** reacted with the organocerium reagent derived from vinylmagnesium bromide to give the 1,2-addition product **193**.[213] In this example, mild acid hydrolysis led to elimination of water and formation of the extended conjugated ketone **194** in an overall yield of 90%.

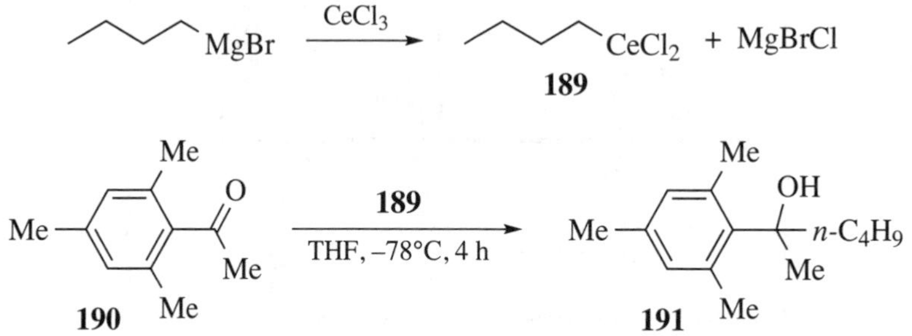

207. Nasipuri, D.; Ghosh, C.K.; Mukherjee, P.R.; Venkataraman S. *Tetrahedron Lett.* **1971**, 1587 and references cited therein.
208. Reference 162, pp 182-187.
209. Mathieu, J.; Weill-Raynal, J. *Bull. Soc. Chim. Fr.* **1968**, 1211.
210. Cabaret, D.; Welvart, Z. *Chem. Commun.* **1970**, 1064.
211. Morrison, J.D. *Survey of Progress in Chemistry, Vol. 3*, Academic Press, New York, **1966**, p 147.
212. (a) Imamoto, T.; Sugiura, Y.; Takiyama, N. *Tetrahedron Lett.* **1984**, *25*, 4233; (b) Imamoto, T.; Sugiura, Y. *J. Organomet. Chem.* **1985**, *285*, C21; (c) Imamoto, T.; Takiyama, N.; Nakamura, K *Tetrahedron Lett.* **1985**, *26*, 4763.
213. Majetich, G.; Zhang, Y. *J. Am. Chem. Soc.* **1994**, *116*, 4979.

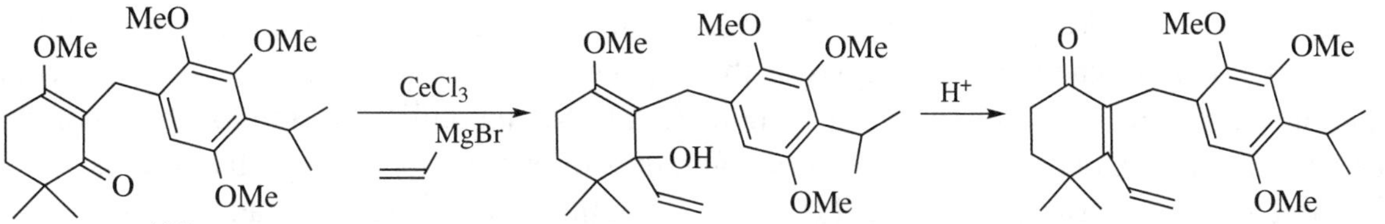

Organolithium reagents (sec. 8.5) are also useful precursors. Vinylcerium reagent **196** was prepared from vinyl bromide **195**, in Ovaska and Roses' synthesis of fused polycyclic ring systems.[214] When **196** reacted with the cyclopentanone derivative shown, alcohol **197** was isolated in 61% yield after hydrolysis.

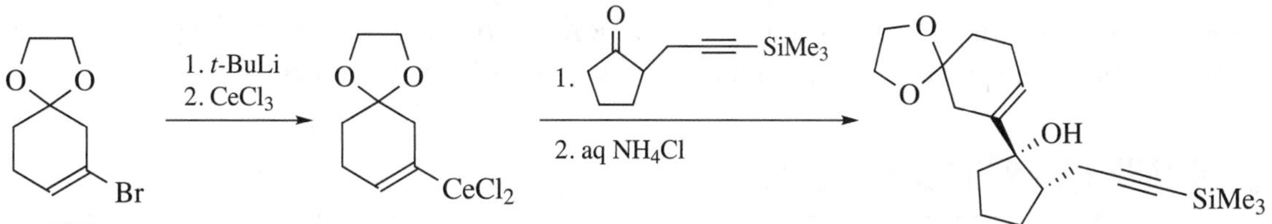

Table 8.10. Competition for Nucleophilic Addition, Reduction, and Enolate Formation in Reactions of Diisopropyl Ketone with Grignard Reagents

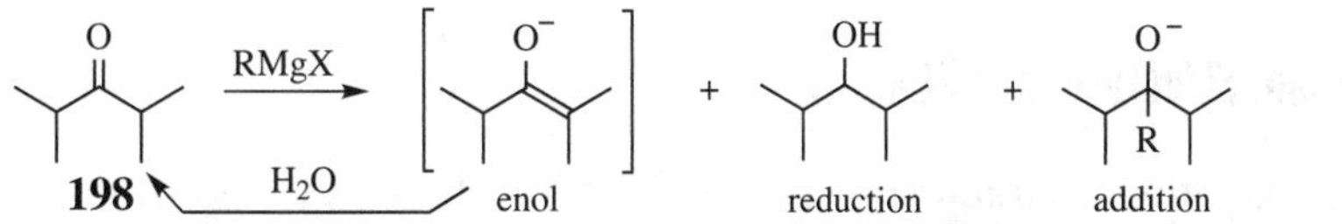

RMgX	RMgX/Ketone	% Enolate	% Reduction	% Addition
EtMgBr	1.3	1	21	78
	1.5	1	19	80
	2.5	1	15	80
	1.2	2	21	77
n-PrMgCl	1.2	2	51	46
	2.5	1	37	62
n-PrMgBr	1.2	1	64	35
	1.4	2	60	36
n-PrMgI	1.2	2	69	30
i-PrMgCl	1.2	28	72	0
i-PrMgBr	1.2	29	65	0
	1.4	30	65	0
i-PrMgI	1.4	30	70	0

[Reprinted with permission from Cowan, D.O.; Mosher, H.S. *J. Org. Chem.* **1962**, *27*, 1. Copyright © **1962** American Chemical Society.]

8.4.G.iii. Enolization. Aldehydes and ketones have acidic protons on the α-carbon, which can be removed by the basic Grignard reagent to generate an enolate anion (sec. 9.2.A). Formation

---

214. Ovaska, T.V.; Roses, J.B. *Org. Lett.* **2000**, *2*, 2361.

of the enolate anion can sometimes compete with nucleophilic addition to the carbonyl. If there are no secondary reactions, hydrolysis of the enolate product regenerates the starting ketone and the net result is isolation of the starting material. Nucleophilic addition to the carbonyl is usually faster than deprotonation, but small amounts of enolization can be observed as shown in Table 8.10 for diisopropyl ketone (198).[215] With unhindered Grignard reagents, the amount of enolate anion formed is insignificant, and reduction is the major reaction that competes with nucleophilic addition. With the hindered isopropylmagnesium halide, steric interactions completely suppressed addition, and reduction accounted for 69% of products with only 2% enolization. Hamelin observed similar reactivity in reactions with ethylmagnesium bromide,[216] in which the effect of solvent and temperature on reduction and enolization was probed. Increasing the size of the groups attached to the carbonyl increased the amount of reduction but only slightly increased the amount of enolization. Increasing the size of the group attached to magnesium, in the presence of a hindered ketone, significantly increased enolate formation.

8.5. ORGANOLITHIUM REAGENTS (C–Li)

Grignard reagents are clearly important in synthesis. If MgX is replaced with Li, another class of organometallic reagents is available, the organolithium reagents. Organolithium reagents are also potent nucleophiles and they are more basic than Grignard reagents. Both features can be exploited in synthesis.[217]

8.5.A. Preparation, Structure, and Stability

Organolithium reagents, characterized by a C-Li bond, are as important in organic synthesis as the Grignard reagents. Lithium is less electronegative than carbon, and the carbon of the C–Li bond is polarized δ–, as in Grignard reagents.[218] Organolithium reagents are expected to behave both as a nucleophile and as a base. It is important to understand the chemical properties of organolithium reagents, and to note differences with Grignard reagents before discussing their reactions.[219]

Ethyllithium was discovered by Schlenk and Holtz in 1917,[220] but it was not until 1930[221] that Ziegler and Colonius discovered that organolithium reagents could be prepared by direct

215. Cowan, D.O.; Mosher, H.S. *J. Org. Chem.* **1962**, *27*, 1.
216. Hamelin, R. *Bull Soc. Chim. Fr.* **1961**, 915 and references cited therein.
217. *Organolithiums in Enantioselective Synthesis* (Series: Topics in Organometallic Chemistry, Vol. 5), Hodgson, D.M. (Ed.), Springer Verlag, Heidelberg, **2003**.
218. (a) Wakefield, B.J. *The Chemistry of Organolithium Compounds*, Pergamon Press, Oxford, **1974**; (b) Clayden, J. *Organolithiums: Selectivity for Synthesis,* Pergamon, Amsterdam, The Netherlands, **2002**.
219. Mallan, J.M.; Bebb, R.L. *Chem. Rev.* **1969**, *69*, 693.
220. Schlenk, W.; Holtz, J. *Berichte* **1917**, *50*, 262.
221. Ziegler, K.; Colonius, H. *Annalen* **1930**, *479*, 135.

reaction of lithium metal and an alkyl halide,[222] as in the reaction of lithium and chlorobutane to give *n*-butyllithium ($CH_3CH_2CH_2CH_2Li$, or BuLi). Other halides can be used, as seen by the conversion of chloride **199** to **200**, taken from a synthesis of (−)-frontalin by Yus and co-workers.[223] The naphthalene added to this reaction may assist formation of **200** via formation of lithium naphthalenide. Two years prior to the work of Ziegler and Colonius, Schlenk and Bergmann had reported that ethyllithium (CH_3CH_2Li, or EtLi) reacted with fluorene (**201**), with exchange of the acidic hydrogen with Li to give fluorenyllithium (**202**).[224] These two reactions (lithium-halogen exchange, and lithium-hydrogen exchange) describe the main methods for preparing organolithium reagents and the latter also describes the most useful synthetic application, removal of an acidic hydrogen to generate a new organolithium species.

Organolithium reagents are usually written as R–Li, but this simple representation does not begin to describe their actual structure.[219,222,225] Some physical data[222] suggest that the C–Li bond in an organolithium is highly covalent. It is known that organolithium reagents exist as associated aggregates, due in part to this covalent character. Both methyllithium[226] and ethyllithium[227] have been obtained in crystalline form and their crystal structure determined by X-ray crystallography. The structure in Figure 8.1[228] is that of crystalline isopropyllithium reported by Siemeling and co-workers, which exists as a hexamer, (*i*-PrLi)$_6$. The X-ray crystal structure of 9,9-dilithiofluorene has also been reported,[229] the first for a dilithiated hydrocarbon.

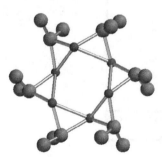

Figure 8.1. X-ray crystal structure of isopropyllithium (hydrogen atoms are hidden for clarity). [Reprinted with permission from Siemeling, U.; Redecker, T.; Neumann, B.; Stammler, H.-G. *J. Am. Chem. Soc.* **1994**, *116*, 5507. Copyright © 1994 American Chemical Society.] Drawn with Spartan software, Wavefunction, Inc. Hydrogen atoms omitted for clarity.

222. Deberitz, J. *Janssen Chimica Acta* **1984**, *2*, 3.
223. Yus, M.; Ramón, D.J.; Prieto, O. *Eur. J. Org. Chem.* **2003**, 2745.
224. Schlenk, W.; Bergmann, E. *Annalen* **1928**, *463*, 1 (see p 98).
225. (a) Brown, T.L. *Advan. Organomet. Chem.* **1965**, *3*, 365; (b) Brown, J.M. *Chem. Ind. (London)* **1972**, 454.
226. Weiss, E.; Lucken, E.A.C. *J. Organomet. Chem.* **1964**, *2*, 197.
227. Dietrich, H. *Acta Cryst.* **1963**, *16*, 681.
228. Siemeling, U.; Redecker, T.; Neumann, B.; Stammler, H.-G. *J. Am. Chem. Soc.* **1994**, *116*, 5507.
229. Linti, G.; Rodig,, A.; Pritzkow, H. *Angew. Chem. Int. Ed.* **2002**, *41*, 4503

Methyllithium and ethyllithium are tetrameric in the solid state,[227,230] and most organolithium reagents are highly associated in solution.[218a,219] The degree of association is related to the solvent and structure of the organolithium, but tends to be higher for straight-chain than for branched (secondary and tertiary) organolithium reagents. The extent of association of the organolithium is important since it can affect the rate of metal-halogen or metal-hydrogen exchange, as well as the product distribution. Table 8.11[222,231,225a] shows the degree of association of several common organolithium reagents in various solvents, which decreases as the coordinating ability of the solvent increases. This observation can loosely be compared with increasing solvent polarity. The most commonly used solvents in reactions of organolithium reagents are ether, THF, pentane or hexane.[232] Increased solvation reduces the ability of the organolithium reagent to associate, and thereby form solvated oligomers.[232] Williard and co-workers determined the X-ray structures of butyllithium•TMEDA, •THF, and •DME complexes, and showed that they are tetramers.[233] Complexing also increases the ionic character of the relatively covalent C–Li bond, lowering the energy requirements for the transition state leading to reactions.[234] (−)-Sparteine is often added to organolithium reagents, and the resulting reagent can lead to asymmetric induction in certain reactions. Strohmann and co-workers determined the X-ray crystal structure of *tert*-butyllithium•(−)-sparteine,[235] which has the monomeric structure **203**.

Table 8.11. Association of Organolithium Reagents in Common Solvents

R–Li $\longrightarrow$ (R–Li)$_n$		
R	**Solvent**	**n**
Et	cyclohexane	6
	hexane	6
	ether	2
	benzene	6,2
n-Bu	cyclohexane	6
	benzene	6,2
	ether	6
n-Bu (+ TMEDA)	hexane	1
t-Bu	hexane	4
	benzene	4
Ph	ether	2

[Reprinted with permission from Mallan, J.M.; Bebb, R.L. *Chem. Rev., 1969, 69,* 693. Copyright © *1969* American Chemical Society.]

230. Dietrich, H. *Z. Naturforsch.* **1959**, *14B*, 739.
231. (a) Reference 219, p 696; (b) Brown, T.L. *Acc. Chem. Res.* **1968**, *1*, 23; (c) West, P.; Waack, R. *J. Am. Chem. Soc.* **1967**, *89*, 4395.
232. Miginiac-Groizeleau, L. *Bull. Soc. Chim. Fr.* **1963**, 1449.
233. Nichols, M.A.; Williard, P.G. *J. Am. Chem. Soc.* **1993**, *115*, 1568.
234. (a) Waack, R.; Doran, M.A. *Chem. Ind. (London)* **1962**, 1290; (b) Zakharkin, L.I.; Okhlobystin, O. Yu.; Bilevitch, K.A. *Tetrahedron* **1965**, *21*, 881.
235. Strohmann, C.; Seibel, T.; Strohfeldt, K. *Angew. Chem. Int. Ed.* **2003**, *42*, 4531.

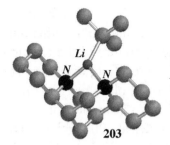

203

[Reprinted with permission from Strohmann, C.; Seibel, T.; Strohfeldt, K. *Angew. Chem. Int. Ed.* **2003**, *42*, 4531. Copyright © **2003** Wiley-VCH Verlag GmbH & Co, KGaA, Weinheim.] Structure drawn with Spartan by Wavefunction, Inc. Hydrogen atoms omitted for clarity.

Although ethers are used as solvents for reactions of organolithium reagents, they also react with them. Ethers are Lewis bases and coordinate to lithium, stabilizing the organometallic reagent and participating in subsequent reactions. In most cases, the ether fragment does not appear in the final product.[236,218] The α-hydrogen of an ether is readily removed by the basic organolithium reagents, via Li–H exchange (this is an acid-base reaction, and the conjugate acid is an alkane, RLi → R–H). Disproportionation of the lithio-ether derivative (**204**) generates an

204

alkoxide and an alkene when tetrahydrofuran (THF) is the solvent.[237] The acidity of the α-proton varies with the ether and the organolithium reagent. It is not surprising, therefore, that organolithium reagents have different stabilities in different solvents. The half-life of *n*-butyllithium in diethyl ether at 25°C is 153 hours, whereas that of ethyllithium is 54 hours, and that of cyclohexyllithium is 30 minutes.[238] This stability to reaction of *n*-butyllithium can be used as an approximate reaction order for Li–H exchange. Note that the relative stability of organolithium reagents can be determined, based on the tin-lithium exchange equilibrium reaction.[239] Primary organolithium reagents exhibit better stability (they are less reactive) when compared to secondary or tertiary reagents. It has also been determined that *n*-butyllithium decomposes in THF in 2 h at 25°C.[240] Stability can be increased by the addition of a less acidic co-solvent such as diethyl ether or tetrahydropyran (THP). Solutions of *n*-butyllithium are

236. (a) Reference 219, p 696; (b) Applequist, D.E. O'Brien, D.F. *J. Am. Chem. Soc.* **1963**, *85*, 743; (c) Curtin, D.Y.; Koehl, Jr., W.J. *Ibid.* **1962**, *84*, 1967; (d) Eastham, J.F.; Gibson, G.W. *J. Org. Chem.* **1963**, *28*, 280; (e) Eastham, J.F.; Gibson, G.W. *J. Am. Chem. Soc.* **1963**, *85*, 2171; (f) Mulvaney, J.E.; Gardlund, Z.G.; Gardlund, S.L. *Ibid.* **1963**, *85*, 3897; (g) Waack, R.; Doran, M.A.; Stevenson, P.E. *J. Organomet. Chem.* **1965**, *3*, 481.
237. (a) Reference 219, p 697; (b) Bartlett, P.D.; Friedman, S.; Stiles, M. *J. Am. Chem. Soc.* **1953**, *75*, 1771; (c) Rembaum, A.; Siao, S.P.; Indictor, N. *J. Polymer. Sci.*, **1962**, *56*, S17; (d) Gilman, H.; Gaj, B.J. *J. Org. Chem.* **1957**, *22*, 1165.
238. (a) Reference 219, pp 697, 716; (b) Seyferth, D.; Cohen, H.M. *J. Organomet. Chem.* **1963**, *1*, 15.
239. Graña, P.; Paleo, M.R.; Sardina, F.J. *J. Am. Chem. Soc.* **2002**, *124*, 12511.
240. (a) Reference 219, p 716; (b) Gilman, H.; Schwebke, G.L. *J. Organomet. Chem.* **1965**, *4*, 483.

stable for 24 hours in an ether-THF mixture, and up to one week in an ether-THP solution.[240] Addition of a hydrocarbon cosolvent such as pentane, hexane or cyclohexane to the ether also extends the lifetime of the organolithium reagent. Lowering the temperature of the reaction from 25 to 0 or to –78°C significantly reduces the rate of the cleavage reaction, extending the lifetime of the organolithium in solution, and as noted above the solvent plays a critical role. In DME, the half-life of *tert*-butyllithium is 11 minutes at –70°C but <2 minutes at -40°C, whereas in THF the half life is 338 minutes at –40°C and only 42 minutes at –20°C.[241] In THF, *n*-butyllithium has a half life of 1039 minutes at 0°C but only 38 minutes at 20°C, and in DME at –20°C the half life is 11 minutes. In ether, however, *n*-butyllithium has a half-life of 9180 minutes (153 hours) at 20°C but 1860 minutes (31 hours) at the reflux temperature of 35°C.[241] The general reactivity of organolithium reagents in cleavage reactions of diethyl ether via Li-H exchange is[242]

The reactivity of organolithium reagents in ethers can be increased by addition of a Lewis base that is stronger than diethyl ether or THF. When dimethoxyethane is used as a solvent, for example, the reactivity is greatly increased.[243] The two oxygens in dimethoxyethane coordinate to the lithium of the organolithium reagent, and generate what is essentially a monomeric organolithium species (**RLi-DME**). Such coordination may also increase the polarity of the C–Li bond, and thereby the carbanionic character. Similar effects are observed when stoichiometric amounts of nitrogen bases such as DABCO (1,4-diazabicyclo[2.2.2]octane[244] and TMEDA[244,218] are added. Alkoxides

such as potassium *tert*-butoxide have also been used,[245] but are less effective. The effects of the additives are apparent in the reaction of benzene with *n*-butyllithium.[245] A refluxing solution of *n*-butyllithium in benzene is rather stable, giving < 5% of phenyllithium. Addition of TMEDA, however, led to a 76% yield of phenyllithium (three h at ambient temperatures).[245] The structure of a 1:1 complex between allyllithium and TMEDA was determined by NMR at low temperature.[246] With unsymmetrical alkyl substitution at the termini, the allyl C–C bond to the more substituted terminus is of higher bond order than to the less substituted terminus.

241. Stanetty, P.; Mihovilovic, M.D. *J. Org. Chem.* **1997**, *62*, 1514.
242. Gilman, H.; Haubein, A.H.; Hartzfeld, H. *J. Org. Chem.* **1954**, *19*, 1034.
243. Schlosser, M. *J. Organomet. Chem.* **1967**, *8*, 9.
244. Reference 222, p 4 and reference 10 therein.
245. Eberhardt, G.G.; Butte, W.A. *J. Org. Chem.* **1964**, *29*, 2928.
246. Fraenkel, G.; Qiu, F. *J. Am. Chem. Soc.* **2000**, *122*, 12806.

Many of the organolithium reagents used for reactions are commercially available, usually packaged in bulk ether or hydrocarbon solvents and the latter often contains many hydrocarbon impurities. Impurities such as these are not a problem for most applications unless the reaction products are alkanes or alkenes, since many products are isolated by chromatography. If the reagent is allowed to age in an ether solvent, there may be contamination caused by the deprotonation reaction, which reduces the amount of organolithium and also generates alkoxides. Once a bottle of commercial reagent has been opened, it reacts with moisture and oxygen to further reduce the molarity of the solution and generate additional alkoxide. For these reasons organolithium reagents should always be standardized prior to each use. Several methods have been developed for this purpose, each taking advantage of the basicity or complexing ability of the organolithium reagent.

Gilman titration[247] gives the total alkali (organolithium + alkoxide) when one aliquot is titrated first with water, and then with standardized HCl. Quenching a second aliquot with benzyl chloride, followed by titration with H_2O and then standardized HCl gives the alkoxide concentration. The difference in these values is the molarity of the organolithium. 1,3-Diphenylacetone tosylhydrazone (**A**),[248] 2,5-dimethoxybenzyl alcohol (**B**)[249] and diphenylacetic acid (**C**)[250] have all been utilized for standardization, each relying on removal of an acidic proton to generate a colored anion. The use of 1,10-phenanthroline (phen, **D**) as a titration reagent relies on complexation between the organolithium and the two nitrogens, leading to a reversible charge transfer complex that has a distinct color.[251] Titration with 2-butanol destroys the complex, and the resultant color change allows the molarity to be determined. Salicylaldehyde phenylhydrazone (**E**) has been used as an indicator, where deprotonation of the phenolic OH leads to a yellow alkoxide anion.[252] More than one equivalent of the organolithium reagent removes the hydrazone NH to give a red dianion, which is the endpoint for the titration.

247. (a) Gilman, H.; Haubein, A.H. *J. Am. Chem. Soc.* **1944**, *66*, 1515; (b) Gilman, H.; Cartledge, F.K. *J. Organomet. Chem.* **1964**, *2*, 447.
248. Lipton, M.F.; Sorensen, C.M.; Sadler, A.C.; Shapiro, R.H. *J. Organomet. Chem.* **1980**, *186*, 155.
249. Winkle, M.R.; Lansinger, J.M.; Ronald, R.C. *J. Chem. Soc. Chem. Commun.* **1980**, 87.
250. Kofron W.G.; Baclawski, L.M. *J. Org. Chem.* **1976**, *41*, 1879.
251. Watson, S.C.; Eastham, J.F. *J. Organomet. Chem.* **1967**, *9*, 165.
252. Love, B.E.; Jones, E.G. *J. Org. Chem.* **1999**, *64*, 3755.

8.5.B. Metal-Halogen Exchange

For applications where the organolithium reagent is not commercially available, the reagent can be prepared by direct synthesis (by reaction of lithium metal with the appropriate alkyl halide). The reaction between one organolithium reagent and an alkyl halide generates a new organolithium reagent, along with a new halide. This reaction can be used to prepare many reagents that can be used in subsequent reactions. Methods have been reported for the preparation and synthetic use of functionalized aromatic and heteroaromatic organolithium reagents that do **not** rely on deprotonation methods.[253]

8.5.B.i. Preparation of Organolithium Reagents. When an alkyl halide reacts with lithium metal, a new organolithium reagent is produced along with a new alkyl halide. This process is an exchange reaction, and it probably proceeds by a one-electron transfer from lithium to the halide as shown.[218,219] Using highly purified lithium leads to a dramatic decrease in formation of the organolithium reagent,[254] and it was discovered that the lithium must contain at least 0.02% of sodium for the reaction to proceed normally.

$$R-X + |Li-Li|_n \longrightarrow \left[\begin{array}{cc} R\text{---}\bullet\text{---}X^- \\ |Li^\bullet|_{n-1} \text{-----} Li^+ \end{array} \right] \longrightarrow \left[\begin{array}{cc} R\bullet & X^- \\ |Li\bullet|_{n-1} & Li^+ \end{array} \right]$$

A related and useful process is the **Wittig-Gilman reaction**, in which an organolithium (R'–Li) reacts with an alkyl halide to produce a new organolithium (R–Li) via metal-halogen exchange.[255] Several mechanisms have been proposed for this exchange.[256] The reaction is reversible, and the equilibrium favors the more stable organolithium reagent. The reactivity of the alkyl halide partner is also important. If R–X is more susceptible to nucleophilic displacement than R'–X, the reaction proceeds to the right. An alternative view is to recognize that when the pK_a of RH is less than that of R'H (RH is more acidic than R'H), R'H is favored.

$$R-X + R^1-Li \rightleftharpoons R-Li + R^1-X$$

Exchange of the newly formed organolithium with the newly formed alkyl halide can lead to a mixture of organolithium reagents.[257] Proper choice of the initial halide and the organolithium reagent can usually produce a good yield of the desired organolithium reagent, and minimize

253. Nájera, C.; Sansano, J.M.; Yus, M. *Tetrahedron* **2003**, *59*, 9255.
254. (a) Beel, J.A.; Koch, W.G.; Tomasi, G.E.; Hermansen, D.E.; Fleetwood, P. *J. Org. Chem.* **1959**, *24*, 2036; (b) Kamienski, C.W.; Esmay, D.L. *Ibid.* **1960**, *25*, 1807; (c) Seyferth, D.; Suzuki, R.; Murphy, C.J.; Sabet, C.R. *J. Organomet. Chem.* **1964**, *2*, 431.
255. (a) Jones, R.G.; Gilman, H. *Org. React.* **1951**, *6*, 339; (b) Gilman, H.; Langham. W.; Jacoby, A.L. *J. Am. Chem. Soc.* **1939**, *61*, 106; (c) Wittig, G.; Pockels, U.; Dröge, H. *Chem. Ber.* **1938**, *71*, 1903.
256. (a) Schlosser, M. *Struktur und Reaktivität Polarer Organometallic*, Springer-Verlag, New York, *1973*, p 187; (b) Russel, G.A.; Lamsun, D.W. *J. Am. Chem. Soc.* **1969**, *91*, 3967.
257. Applequist, D.E; O'Brien, D.F. *J. Am. Chem. Soc.* **1963**, *85*, 743.

the reverse reaction. This exchange reaction is very rapid when *tert*-butyllithium is used with primary alkyl halides. Bailey showed that *tert*-butyllithium reacted with neopentyl iodide (**205**) in a mixture of pentane and ether, to give **206** in < 5 min at −78°C.[258] In this case, **206** was treated with butanal, to give an 89% yield of alcohol **207** (sec. 8.5.C). It should be noted, however, that this particular reaction gives good results only with primary iodides. Halides that react with the organolithium reagent should generate a new organolithium that is much less reactive. This observation is important when an aryl halide or a vinyl halide is used to generate an aryllithium (such as 4-lithiotoluene)[259] or a vinyllithium reagent.[260,258] Heteroaryllithium reagents such as 3-lithiothiophene (**208**)[261] generated from 3-bromothiophene, are formed rapidly in ether solvents.[255a] Coupling with alkyl halides is an important reaction of the relatively stable aryllithium reagents, as in the reaction of **208** with geranyl bromide (**209**) to give **210**.[262]

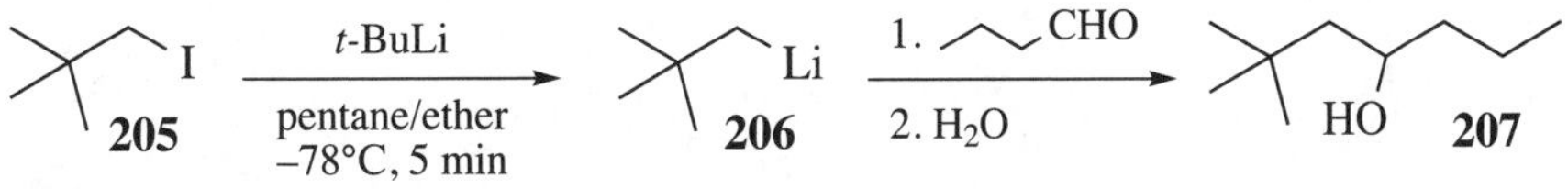

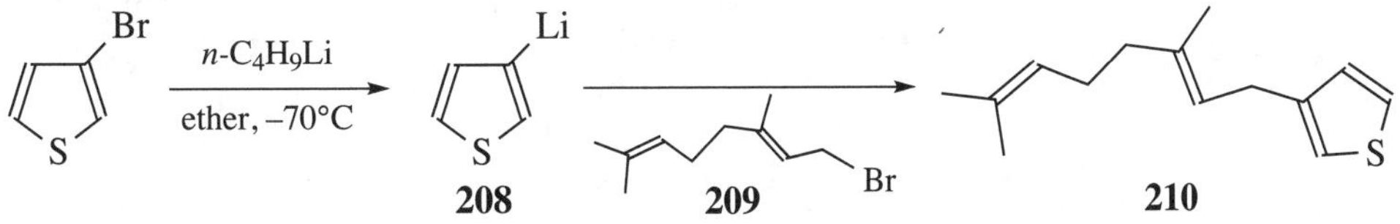

The disconnection for aryl alkylation is

The lithium-halogen exchange is particularly useful for the preparation of stereodefined alkenyllithium reagents.[263] Two equivalents of an organolithium reagent are necessary. When

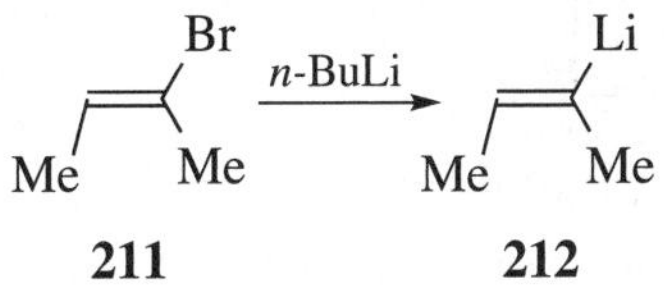

n-butyllithium reacts with **211**, one equivalent of *n*-butyllithium exchanges to form **212** plus 1-bromobutane and the second equivalent of *n*-butyllithium reacts with 1-bromobutane to form butene and butane. In this particular example,[264] the geometric

258. Bailey, W.F.; Punzalan, E.R. *J. Org. Chem.* **1990**, *55*, 5404.
259. Gilman, H.; Langham, W.; Moore, F.W. *J. Am. Chem. Soc.* **1940**, *62*, 2327.
260. Merrill, R.E.; Negishi, E. *J. Org. Chem.* **1974**, *39*, 3452.
261. Moses, P.; Gronowitz, S. *Arkiv. Kemi* **1962**, *18*, 119 [*Chem. Abstr. 56*: 10173c **1962**].
262. Semenovskii, V.V.; Emel'yanov, M. *Izvest. Akad. Nauk. SSSR, Ser. Khim.* **1980**, 2578 (Engl., p 1833).
263. (a) Negishi, E. *Organometallics In Organic Synthesis* Vol. 1, Wiley-Interscience, New York, **1980**, pp 39-40; (b) Kluge, H.F.; Untch, K.G.; Fried, J.H. *J. Am. Chem. Soc.* **1972**, *94*, 7827; (c) Cahiez, G.; Bernard, D.; Normant, J.F. *Synthesis* **1976**, 245.
264. (a) Curtin, D.Y.; Crump, J.W. *J. Am. Chem. Soc.* **1958**, *80*, 1922; (b) Bordwell, F.G.; Landis, P.S. *Ibid.* **1957**, *79*, 1593; (c) Seyferth, D.; Vaughn, L.R. *J. Organomet. Chem.* **1963**, *1*, 201.

configuration of the 2-bromo-2-butenes **211** and **213** is retained upon metal-halogen exchange, giving the corresponding lithiobutenes (**212** and **214**, respectively). There is less isomerization of the alkenyl group during Li–H exchange than is observed with formation of the analogous Grignard reagent with magnesium. Stereodefined organolithium reagents are important for coupling reactions with alkyl halides. **Wurtz coupling** (*vide infra*, sec. 8.5.B.ii) is minimized when aryllithium or vinyllithium reagents are formed by metal-halogen exchange. An example is the reaction of vinyl bromide **213** with lithium, and subsequent reaction with iodooctane to give a 77% yield of **215** (94% *Z*).[265] Alkenyllithium reagents react with aldehydes and ketones in the usual manner. Vinyllithium reagents, as well as other Grignard and organolithium reagents react with DMF to produce conjugated aldehydes. In Kende and co-worker's synthesis of *Stachybotrys* spirolactams,[266] vinyl iodide **216** was treated with *tert*-butyllithium to initiate I↔Li exchange, and the resulting vinyllithium reagent reacted with an excess of DMF to give an 83% yield of conjugated aldehyde **217**.

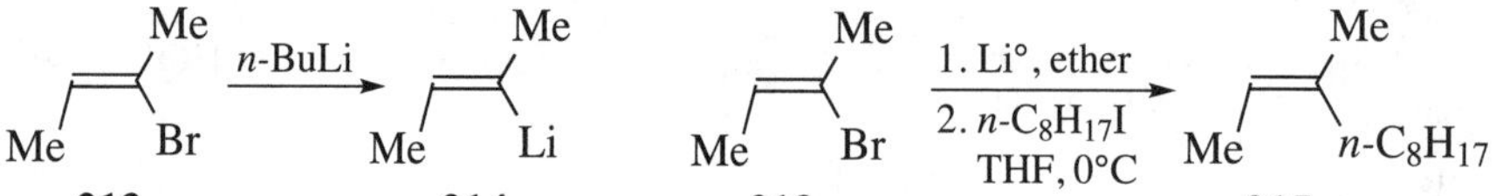

The alkenyllithium-alkyl halide disconnection is:

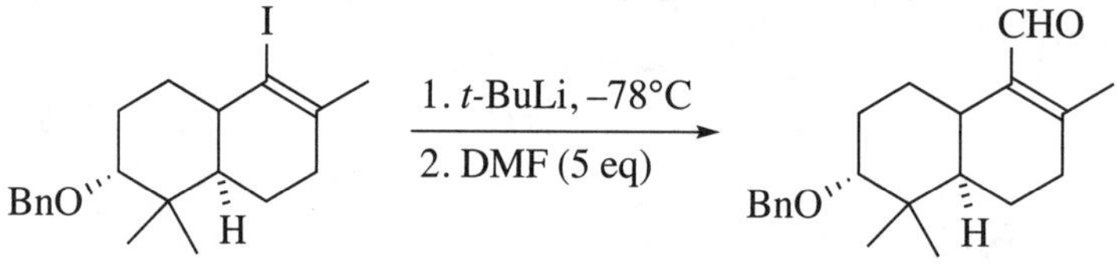

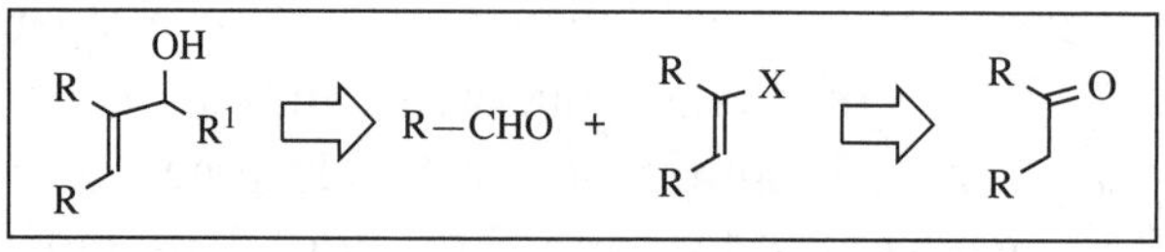

8.5.B.ii. Wurtz Coupling. There are two major side reactions in a metal-halogen exchange reaction between an organolithium reagent and an alkyl halide, coupling (the **Wurtz reaction,**[267,268] which generates R–R, R^1–R^1, and R–R^1) and α-metalation. The halogen is

265. (a) Stowell, J.C. *Carbanions in Organic Synthesis* Wiley Interscience, New York, *1979*, p 50; (b) Millon, J.; Lorne, R.; Linstrumelle, G. *Synthesis 1975*, 434; (c) Cahiez, G.; Masuda, A.; Bernard, D.; Normant, J.F. *Tetrahedron Lett. 1976*, 3155; (d) Cahiez, G.; Bernard, D.; Normant, J.F. *Synthesis 1977*, 130; (e) Cahiez, F. Normant, J.F. *Tetrahedron Lett. 1977*, 3383.
266. Deng, W.-P.; Zhong, M.; Guo, X.-C.; Kende, A.S. *J. Org. Chem. 2003*, *68*, 7422.
267. (a) Wurtz, A. *Ann. Chim. Phys. 1855*, *44*, 275; (b) Idem *Annalen 1855*, *96*, 364.
268. (a) The Merck Index, 14th ed., Merck & Co., Inc., Whitehouse Station, New Jersey, *2006*, p ONR-104; (b) Mundy, B.P.; Ellerd, M.G.; Favaloro Jr., F.G. *Name Reactions and Reagents in Organic Synthesis, 2nd ed.*, Wiley-Interscience, New Jersey, *2005*, pp. 708-709.

electron withdrawing, making the proton on the α-carbon acidic, which facilitates its removal by the basic organometallic. As noted for the **Kharasch reaction** with Grignard reagents (sec. 8.4.B.i), this sort of exchange leads to two organolithium reagents and two halides, with direct coupling products R–R and R^1–R^1 as well as a cross coupling product R–R^1. Wurtz coupling also occurs during metal-halogen exchange reactions, forming R–R.

$$R-Li + R^1-X \longrightarrow R-R + R^1-R^1 + R-R^1$$
$$M + R-X \longrightarrow R-R + M-X + R-M$$

Wurtz coupling can be avoided by using an organotin compound as a precursor to the organolithium reagent.[269] Allyllithium (**219**) is often difficult to prepare by the usual metal-halogen exchange methods. An alternative synthesis first reacted allyl bromide with triphenyltin hydride (Ph$_3$SnH) to give allyltriphenyltin (**218**). Subsequent metal-metal (Sn–Li) exchange with phenyllithium gave an excellent yield of **219** along with tetraphenyltin (Ph$_4$Sn).[267] This route is useful when the organolithium reagent is very reactive, or when the halide does not undergo exchange at a reasonable rate (as with vinyl bromide to give vinyllithium).[270]

Reactive halides, such as allyl or benzyl halides, usually give Wurtz coupling during lithium-halogen exchange, leading to low concentrations of the organolithium reagent. Less reactive alkyl halides often fail to react at all. Addition of cuprous salts greatly enhances the reaction with alkyl halides, but this generates a new reagent, a lithium dialkylcuprate. This important class of reagents will be discussed in Section 8.7.A.

8.5.C. Reactions with Aldehydes and Ketones

The reactions of organolithium reagents (RLi) parallel those of Grignard reagents (RMgX) in many ways.[103] Williard and co-worker's *ab initio* study of the reaction of methyllithium aggregates with formaldehyde in aqueous media showed one formaldehyde and one water around lithium ions, coordinating the reactive methyl group via polar solvation.[271] Addition occurred via a concerted mechanism involving a four-membered ring transition state, but hydrophobic solvents and low substrate concentrations promoted a non-concerted mechanism involving ring opening of the dimers and the trimer. Organolithium reagents that are not commercially available can be prepared, such as the organolithium reagent derived from **216**. Commercially available organolithium reagents can be added to carbonyl derivatives, and often provide the most efficient route to substituted alcohols. Organolithium reagents react

269. Gröbel, B.-T.; Seebach, D. *Chem. Ber.* **1977,** *110*, 867.
270. (a) West, R.; Glazer, W.H. *J. Org. Chem.* **1961,** *26*, 2096; (b) Seyferth, D.; Weiner, M.A. *Chem. Ind. (London)* **1959,** 402; (c) Juenge, E.C.; Seyferth, D. *J. Org. Chem.* **1961,** *26*, 563.
271. Hæffner, F.; Sun, C.; Williard, P.G. *J. Am. Chem. Soc.* **2000,** *122*, 12542.

with ketones and aldehyde via acyl addition, as expected of a carbon nucleophile. In a synthesis of briarellin E,[272] Overman and co-worker's converted vinyl iodide **220** to the corresponding vinyl iodide, and subsequent reaction with aldehyde **221** gave **222** as a 3:1 mixture of *anti:syn* diastereomers, in 62% yield.

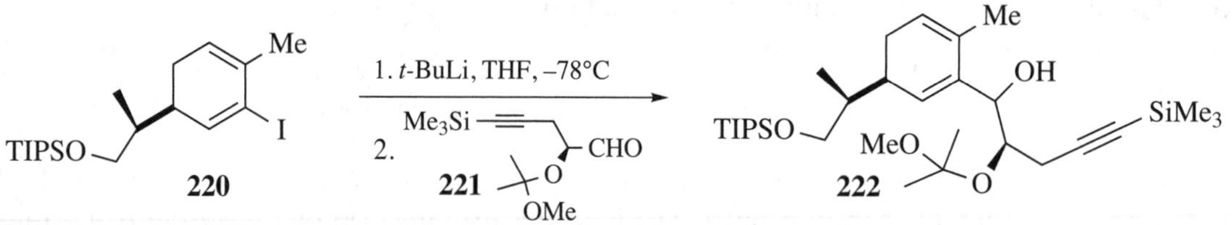

Unlike similar reactions with Grignard reagents, reduction is not a serious problem.[273] *n*-Butyllithium, for example, added readily to the hindered ketone di-*tert*-butyl ketone (**223**) to give alcohol **224** by direct addition to the carbonyl.[273b] Organolithium reagents usually prefer 1,2-addition in conjugated systems due primarily to their enhanced nucleophilicity and poor coordinating ability relative to Grignard reagents. The reaction of methyllithium with **225**, for example, gave an 88% yield of **226** in Sha and co-worker's synthesis of (–)-pinguisenol.[274] Although the C=C unit in **225** is sterically hindered, 1,2-addition is the normal reaction. Indeed, reaction of methyllithium with 3-penten-2-one gave < 1% of conjugate addition, with 2-methyl-3-penten-2-ol being the major product.[275] Note that addition of cuprous iodide to the reaction led to > 99% of conjugate addition[275] (sec. 8.7.A.vi).

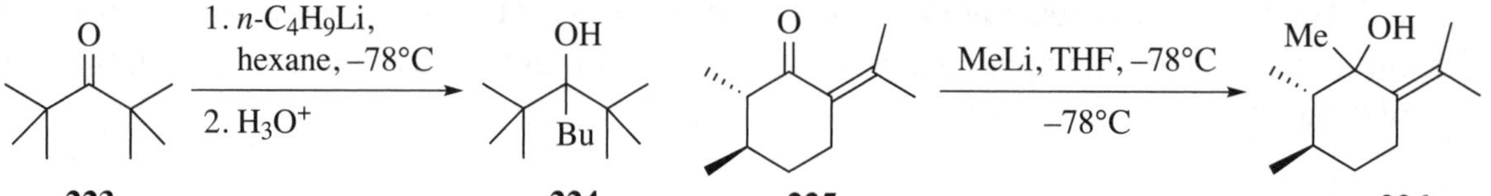

The disconnection for organolithium addition is

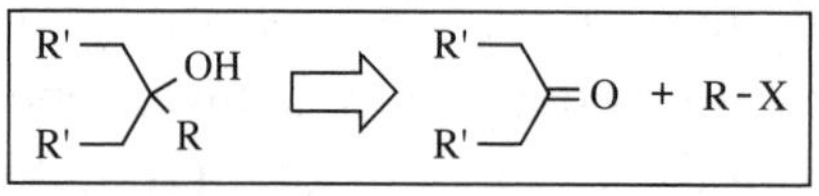

8.5.D. STEREOSELECTIVITY IN ORGANOLITHIUM REACTIONS

8.5.D.i. Configurational Stability. As with Grignard reagents, it is important to know if an organolithium reagent with a stereogenic carbon at the C–Li unit is configurationally

272. Corminboeuf, O.; Overman, L.E.; Pennington, L.D. *J. Am. Chem. Soc.* **2003**, *125*, 6650.
273. (a) Huet, F.; Emptoz, G. *J. Organomet. Chem.* **1975**, *101*, 139; (b) Buhler, J.D. *J. Org. Chem.* **1973**, *38*, 904.
274. Sha, C.-K.; Liao, H.-W.; Cheng, P.-C.; Yen, S.-C. *J. Org. Chem.* **2003**, *68*, 8704.
275. (a) House, H.O.; Respess, W.L.; Whitesides, G.M. *J. Org. Chem.* **1966**, *31*, 3128; (b) Braude, E.A.; Timmons, C.J. *J. Chem. Soc.* **1950**, 2007; (c) Gilman, H.; Kirby, R.H. *J. Am. Chem. Soc.* **1941**, *63*, 2046.

stable in subsequent reactions. In general, such stereogenic centers are not very stable and undergo racemization[276] unless cyclopropyl halides are used as precursors, similar to the diastereoselectivity noted with cyclopropyl Grignard reagents (sec. 8.4.F.i). Reaction of halide **227** (X = Cl, Br, I) with lithium metal, followed by quenching with CO_2, gave carboxylic acid **228**.[163] The amount of racemization varied as the halide was changed. Table 8.12[163] shows the % of racemization observed when the organolithium reagent is generated from R-X. It follows the order: I > Br > Cl, where R is a cyclopropyl group. Reaction of **227** [X = I, 1% sodium (25 microns)] gave 64% racemization, but decreasing the sodium concentration to 0.002% gave 87% racemization. An increase in the particle size of the sodium increased the amount of racemization. Both results are consistent with a complex heterogeneous surface reaction. The nature of the halide, the percentage of Na, and the particle size all contribute to the relative rate of formation of the organolithium as well as the stereoselectivity of the reaction.[277] Note that the choice of solvent can also influence the stereochemistry of organolithium reactions. There is some evidence that organolithium reagents have greater stereochemical stability in hydrocarbon solvents than they do in ether.[278]

Table 8.12. Racemization of Cyclopropyllithium with Changes in Halide

X	Time (min)	% 228	Optical Purity (%)	Racemization (%)
Cl	40	73	71	29
Cl	60	76	66	34
Br	32	60	46	54
Br	42	70	42	58
I	33	42	36	64
I	41	60	36	64

[Reprinted with permission from Walborsky, H.M.; Impastato, F.J.; Young, A.E. *J. Am. Chem. Soc.* *1964*, *86*, 3283. Copyright © *1964* American Chemical Society.]

8.5.D.ii. Diastereoselective Addition Reactions. If the ketone or aldehyde contains a stereogenic center α- to the carbonyl, addition of the organolithium reagent can give a product with good diastereoselectivity. The stereoselectivity of addition for Grignard reagents and organolithium reagents with aldehydes and ketones is predicted in a manner similar to that observed with reducing agents in section 4.8. Acyl addition of organolithium reagents to ketones and aldehydes usually follows Cram's rule (sec. 4.8.B), for example, as seen in Table

276. (a) Witanowski, M.; Roberts, J.D. *J. Am. Chem. Soc.* *1966*, *88*, 737; (b) Lardicci, L.; Lucarini, L.; Palagi, P.; Pino, P. *J. Organomet. Chem.* *1965*, *4*, 341; (c) Buncel, E. *Carbanions: Mechanistic and Isotopic Aspects*, Elsevier, *1975*.
277. Walborsky, H.M.; Aronoff, M.S. J. *Organomet. Chem.* *1965*, *4*, 418.
278. (a) Curtin, D.Y.; Koehl, Jr., W.J. *J. Am. Chem. Soc.* *1962*, *84*, 1967; (b) Reference 216, pp 697-698.

8.13.[279] Since lithium does not coordinate as readily as magnesium in reactions with carbonyls, chelation effects are somewhat diminished relative to Grignard reagents, and the Cram cyclic model is not as useful for predicting the selectivity. When complexing agents are added to the reaction, however, the cyclic model does predict the major diastereomer. Table 8.13, taken from Mosher and Morrison's work, shows the usual Cram selectivity for ketone **229** with a variety of organolithium reagents. The products are diastereomeric alcohols **230** and **231**.

Table 8.13. Cram Selectivity in Reactions of Organolithium Reagents with Ketones and Aldehydes

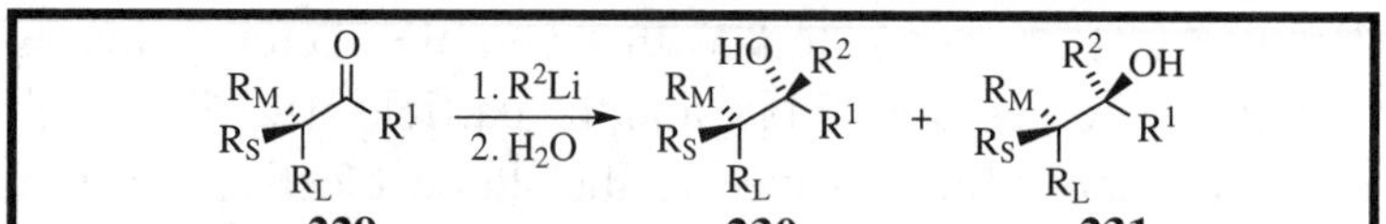

R_S	R_M	R_L	R^1	R^2	230 : 231
H	i-Pr	Ph	H	i-Pr	1 : 1
Me	Et	Ph	Me	Et	2.3 : 1
Me	Et	Ph	Et	Me	10 : 1
H	Me	cyclohexyl	H	Me (in pentane)	1.5 : 1
H	Me	cyclohexyl	H	Me (in ether)	1.2 : 1

[Reprinted by courtesy of Professor James D. Morrison from Morrison, J.D.; Mosher, H.S. *Asymmetric Organic Reactions*, American Chemical Society, Washington, D,. 1976, pp. 92-93.]

Reaction with cyclic ketones usually favors formation of the syn diastereomer as shown in Table 8.14.[280] As with Grignard reagents, attack at a carbonyl occurs at the less sterically hindered face of the molecule. Reaction of various organolithium and Grignard reagents with 2-methylcyclopentanone, for example, gave a mixture of cis- (**232**) and trans- (**233**) alcohols, generally favoring **232**.

Table 8.14. Selectivity in Reactions of 2-Methylcyclopentanone with Organolithium Reagents and Grignard Reagents

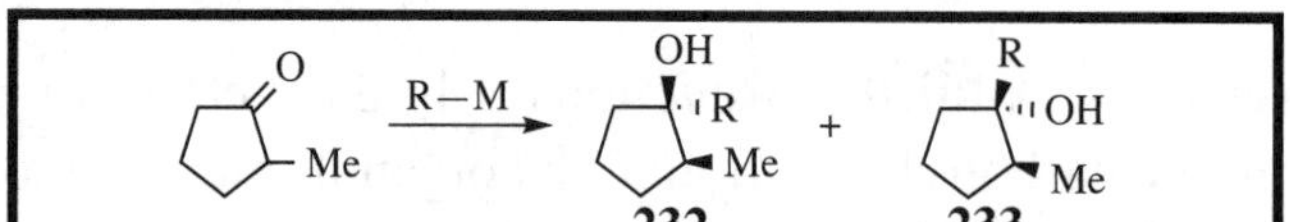

RM	Solvent	232 : 233	% Yield
MeLi	ether	70 : 30	53
MeMgI	ether	45 : 55	95
EtMgBr	THF	75 : 25	68
n-PrMgBr	ether	80 : 20	50
$C_6H_{13}MgBr$	ether	80 : 20	53
PhLi	ether	95 : 5	53
PhMgBr	THF	99 : 1	60

279. Reference 162, pp 92-93.
280. Battioni, J.-P.; Capmau, M.-L.; Chodkiewicz, W. *Bull. Soc. Chim. Fr.* **1969**, 976.

8.5.E. Reactions of Organolithium Reagents with Other Functional Groups

Organolithium reagents react with carbonyl compounds other than aldehydes and ketones analogous to Grignard reagents.[107] Esters usually react to give a tertiary alcohol,[281] and acid chlorides[282] give mixtures of unreacted starting material and tertiary alcohol. Adding an excess of the organolithium reagent gives the tertiary alcohol exclusively, which is formally analogous to the acyl substitution reactions noted for Grignard reagents (sec. 8.4.C.ii). If an ester is treated with excess methyllithium, a tertiary dimethyl carbinol results as in the conversion of **234** to **235** (95%) in Venkateswaran and co-worker's synthesis of heliannuol B.[283] Careful selection of RLi and the reaction conditions often allows the synthesis of ketones rather than tertiary alcohols, however.[284]

Reaction with amides can give a ketone,[123,285] but removal of the proton on the α-carbon of amides or lactams by the basic organolithium to form an enolate anion is a common side reaction. For this reason, non-nucleophilic bases such as lithium diisopropylamide are commonly used (sec. 9.2.B).[286] Organolithium reagents react with Weinreb's amide (see sec. 2.5.C) to give ketones, just as it was noted for Grignard reagents in Section 8.4.C.ii.[287] Primary amides react with excess organolithium to give a nitrile, as in the conversion of phenylacetamide (**236**) to benzonitrile in 72% yield.[288] Reaction of **236** with three equivalents of butyllithium gave trilithiated species **237**, fragmentation gave **238** and hydrolysis gave benzonitrile.[288]

281. (a) Nelson, P.H.; Strosberg, A.M.; Untch, K.G. *J. Med. Chem.* **1980**, *23*, 180; (b) Kleemann, A.; Hesse, J.; Engel, J. *Arzneim-Forsch.* **1981**, *31*, 1178 [*Chem. Abstr. 95*: 186978q *1981*].

282. Locksley, H.D.; Murray, I.G. *J. Chem. Soc. C* **1970**, 392.

283. Roy, A.; Biswas, B.; Sen, P.K.; Venkateswaran, R.V. *Tetrahedron Lett.* **2007**, *48*, 6933.

284. (a) Corey, E.J.; Kim, S.; Yoo, S.-e.; Nicolaou, K.C.; Melvin, Jr., L.S.; Brunelle, D.J.; Falck, J.R.; Trybulski, E.J.; Lett, R.; Sheldrake, P.W. *J. Am. Chem. Soc.* **1978**, *100*, 4620; (b) Corey, E.J.; Trybulski, E.J.; Melvin, Jr., L.S.; Nicolaou, K.C.; Secrist, J.A.; Lett. R.; Sheldrake, P.W.; Falck, J.R.; Brunelle, D.J.; Haslanger, M.F.; Kim, S.; Yoo, S.-e. *Ibid.* **1978**, *100*, 4618; (c) Corey, E.J.; Melvin, Jr., L.S.; Haslanger, M.F. *Tetrahedron Lett.* **1975**, 3117.

285. (a) Evans, E.A. *J. Chem. Soc.* **1956**, 4691; (b) Jones, E.; Moodie, I.M. *J. Chem. Soc. C,* **1968**, 1195.

286. (a) Hullot, P.; Cuvigny, T.; Larchevêque, M.; Normant, H. *Can. J. Chem.* **1976**, *54*, 1098; (b) Durst, T.; Van Den Elzen, R.; Legault, R. *Ibid.* **1974**, *52*, 3206; (c) Crouse, D.N.; Seebach, D. *Berichte* **1968**, *101*, 3113.

287. For a synthetic example taken from a synthesis of (+)-frontalin, see Kanada, R.M.; Taniguchi, T.; Ogasawara, K. *Tetrahedron Lett.* **2000**, *41*, 3631.

288. Kaiser, E.M.; Vaulx, R.L.; Hauser, C.R. *J. Org. Chem.* **1967**, *32*, 3640.

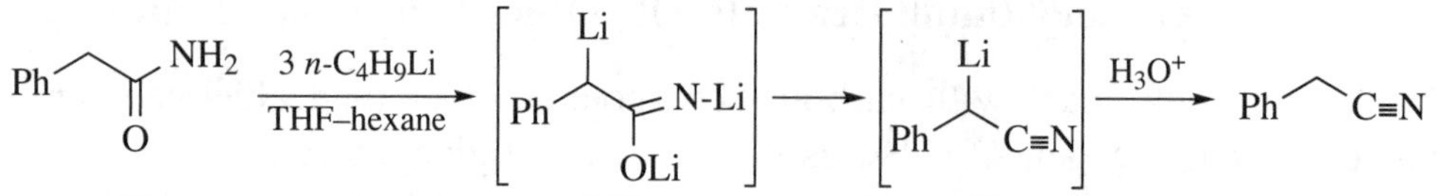

In section 8.5.B.i the reaction of a Grignard reagent with DMF (an amide) gave an aldehyde, and organolithium reagents react similarly. An example is taken the synthesis of geldanamycin by Andrus and co-workers,[289] in which the reaction of **239** with *n*-butyllithium gave an aryllithium reagent (see sec. 8.5.F for directed ortho metalation reactions), which reacted with DMF *in situ* to give aryl aldehyde **240** in 79% yield after hydrolysis.

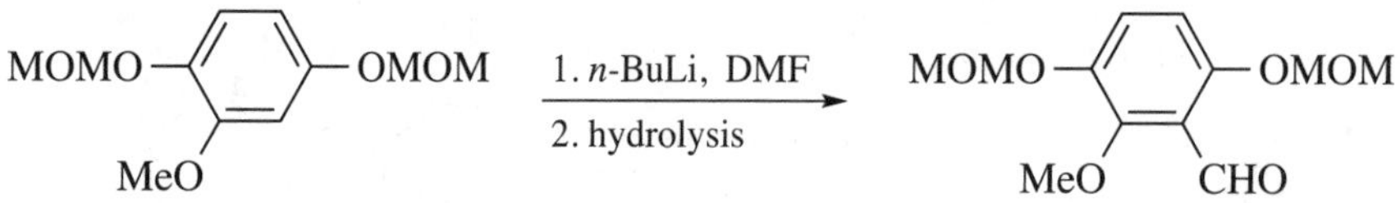

Organolithium addition to acid derivatives gives the disconnections:

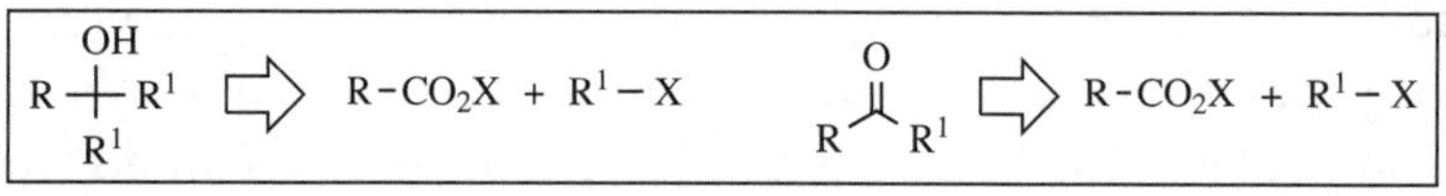

Organolithium reagents are more reactive than the corresponding Grignard reagent. A comparison with Grignard reagents readily demonstrates this statement. The reaction of a Grignard reagent and a carboxylic acid gives the carboxylate salt. Reaction of an organolithium reagent with a carboxylic acid gives the expected lithium carboxylate salt, but a second equivalent adds to the carboxylate to give a ketone. 2-Furyllithium (**241**) reacted with acetic acid, for example, to give the dianion (**242**).[290] Hydrolysis of **242** gave the hydrate, which was converted to the ketone (**243**) under the reaction conditions. The reaction of **242** with acetic acid *in situ*, however generated some ketone product prior to the hydrolysis step. In the presence of an excess of the furyllithium reagent, further reaction gave alcohol **244** as a byproduct.[290]

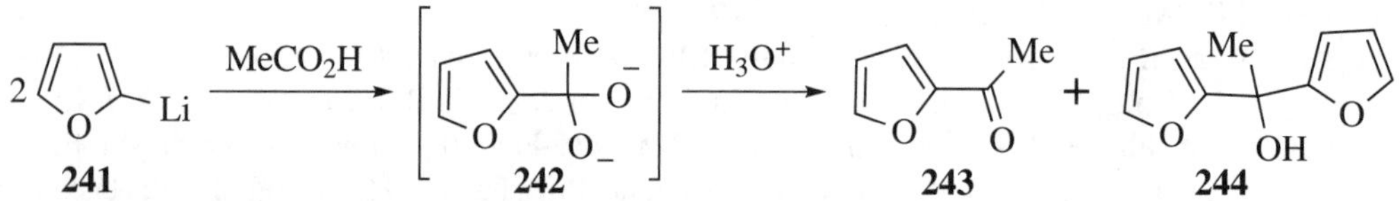

The disconnection for ketone formation from the acid is:

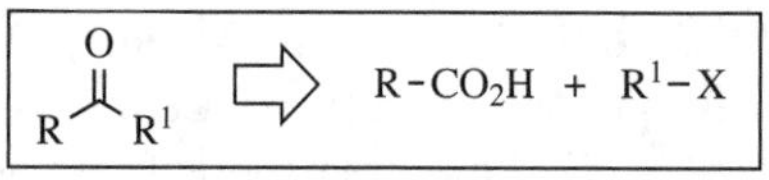

---

289. Andrus, M.B.; Hicken, E.J.; Meredith, E.L.; Simmons, B.L.; Cannon, J.F. *Org. Lett.* **2003**, *5*, 3859.
290. Heathcock, C.H.; Gulik, L.G.; Dehlinger, T. *J. Heterocyclic Chem.* **1969**, *6*, 141.

Organolithium reagents react vigorously with carbon dioxide to give the corresponding carboxylic acid after hydrolysis. Typically, the organolithium solution is carefully poured onto dry ice. An example is taken from Maier and Kühnert's synthesis of apicularen A,[291] in which carboxylic acid **246** was prepared from **245** in 77% yield. We saw this reaction earlier when cyclopropyllithium reagent **227** was quenched with CO_2.

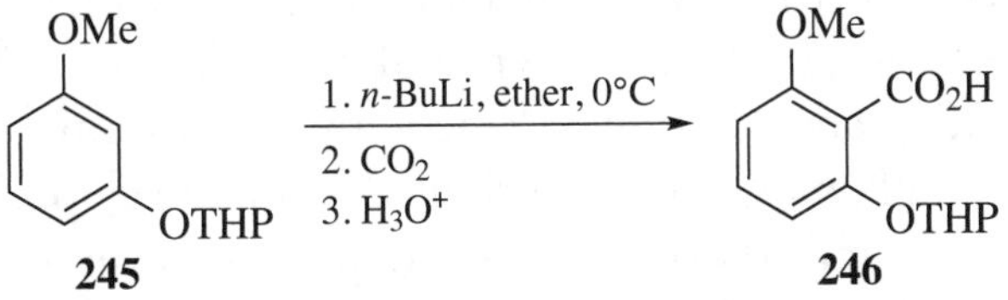

Nitriles can be hydrolyzed to or prepared from carboxylic acids, and nucleophiles[292] can attack the electrophilic carbon, so they are considered to be acid derivatives. 3-Thienyllithium (**208**) reacted with 4-methylpyridine carbonitrile (**247**), for example, to give a *N*-lithio imine (**248**) analogous to the reaction of Grignard reagents. Subsequent acid hydrolysis converted the initially formed imine to the ketone **249**.[293] Addition of an organolithium reagent initially generates an imine, so reduction to an amine (secs. 4.2.C.iii, 4.4.A, 4.5.A) is also possible. Reaction of butyllithium with nitrile **250**, for example, gave an 84% yield of lithiated imine **251**,[294] and reduction with methanolic sodium borohydride gave a 90% yield of amine **252**.

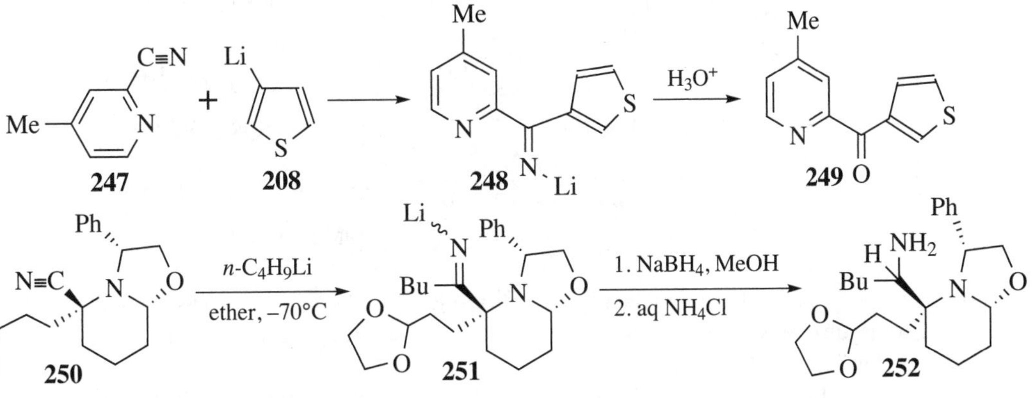

The disconnections for these two reactions are:

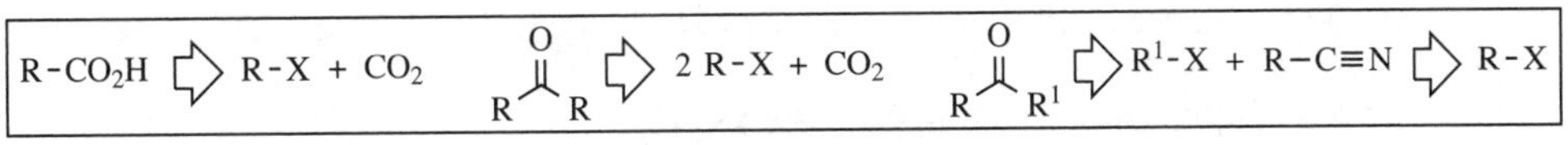

291. Kühnert, S.M.; Maier, M.E. *Org. Lett.* **2002**, *4*, 643.
292. (a) Reference 219, p 698; (b) O'Sullivan, W.I.; Swamer, F.W.; Humphlett, W.J.; Hauser, C.R. *J. Org. Chem.* **1961**, *26*, 2306.
293. (a) Granados Jarque, R.; Bosch Cartes, J.; Lopez Calahorra, F. *Spanish Patent* 453,484 [*Chem. Abstr. 90*: P152156b *1979*]; (c) Koshinaka, E.; Ogawa, N.; Yamagishi, K.; Kato, H.; Hanaoka, M. *Yakugaku Zasshi* **1980**, *100*, 88 [*Chem. Abstr. 93*: 71507b *1980*].
294. (a) Zhu, J.; Quirion, J.-C.; Husson, H.-P. *Tetrahedron Lett.* **1989**, *30*, 6323; (b) Arseniyadis, S.; Huang, P.Q.; Husson, H.-P. *Ibid.* **1988**, *29*, 1391.

8.5.F. Directed Ortho Metalation

When an aromatic ring has a heteroatom or a heteroatom-containing substituent, reaction with a strong base such as an organolithium reagent usually leads to an ortho lithiated species.[295] Subsequent reaction with an electrophilic species gives the ortho substituted product. Gilman and by Wittig discovered this ortho selectivity independently in 1939-1940, when anisole was found to give ortho deprotonation in the presence of butyllithium.[296] This reaction generated ortho-lithiated anisole (**253**), and subsequent reaction with CO_2 gave 2-methoxybenzoic acid (**254**).[296a] Hauser and co-workers greatly contributed to this reaction.[297] Snieckus[298] provides a table of relative directing abilities of various groups, and several categories of functional groups are capable of this ortho-directing effect. When two directing metalation groups have a 1,3-relationship on the aromatic ring, they work cooperatively to enhance the ortho selectivity.[299] An example is taken from Moriarty and coworker's synthesis of treprostinil.[300] Treatment of anisole derivative **255** with butyllithium and then allyl bromide, gave a 60% yield of **256**. Collum and co-worker's rate studies of the butyllithium-TMEDA mediated lithiation reaction of five arenes suggest that the lithiation reactions do not necessarily rely on a complex-induced proximity effect.[301] They[301] proposed a triple-ion based model that depends largely on inductive effects proposed by Schlosser.[302]

295. For a review of directed ortho metalation, see (a) Snieckus, V. *Chem. Rev.* **1990**, *90*, 879; (b) Gschwend, H.W.; Rodriguez, H.R. *Org. React.* **1979**, *26*, 1. Also see (c) Green, L.; Chauder, B.; Snieckus, V. *J. Heterocyclic Chem.* **1999**, *36*, 1453.

296. (a) Gilman, H.; Bebb, R.L. *J. Am. Chem. Soc.* **1939**, *61*, 109; (b) Wittig, G.; Fuhrman, G. *Chem. Ber.* **1940**, *73*, 1197.

297. (a) Puterbaugh, W.H.; Hauser, C.R. *J. Org. Chem.* **1964**, *29*, 853; (b) Slocum, D.W.; Sugarman, D.I. *Adv. Chem. Ser.*, **1974**, *No. 130*, 227.

298. See reference 295a, and Scheme 7 therein (pp. 884-885).

299. See reference 295a, and Table 3 therein (p 885) for cooperative effects of several 1,3-related directed metalation groups.

300. Moriarty, R.M.; Rani, N.; Enache, L.A.; Rao, M.S.; Batra, H.; Guo, L.; Penmasta, R.A.; Staszewski, J.P.; Tuladhar, S.M.; Prakash, O.; Crich, D.; Hirtopeanu, A.; Gilardi, R. *J. Org. Chem.* **2004**, *69*, 1890.

301. (a) Chadwick, S.T.; Rennels, R.A.; Rutherford, J.L.; Collum, D.B. *J. Am. Chem. Soc.* **2000**, *122*, 8640. Also see (b) Collum, D.B. *Acc. Chem. Res.* **1992**, *25*, 448

302. (a) Maggi, R.; Schlosser, M. *Tetrahedron Lett.* **1999**, *40*, 8797; (b) Schlosser, M. *Angew. Chem. Int. Ed.* **1998**, *37*, 1497; (c) Büker, H.H.; Nibbering, N.M.M.; Espinosa, D.; Mongin, F.; Schlosser, M. *Tetrahedron Lett.* **1997**, *38*, 8519.

As a rule, the powerful organolithium bases should be used with organic solvents in which they are highly soluble, due to their association into aggregates (see sec. 8.5.A).[295a] Additives such as TMEDA are usually important because they effectively break down the aggregates (see sec. 8.5.A) to generate monomers and dimers in solution.[301] Roberts and Curtin proposed that ortho lithiated species are stabilized by coordination many years ago,[303] and Snieckus summarized the early evidence pointing to the idea that the directed ortho-lithiation process is a three-step sequence.[295a] The three steps are coordination of the organolithium aggregate to the heteroatom-containing aromatic compound, deprotonation to give the coordinated ortho-lithiated species, and finally, reaction with an electrophile to give the final product as with any organolithium reagent.

As mentioned, many groups are capable of directed ortho metalation, which is illustrated by the reaction of *N*-cumylbenzamide (**257**) with 3.2 equivalents of *sec*-butyllithium in the presence of TMEDA to give **258**[304] as a reactive intermediate. Subsequent treatment with benzaldehyde gave an 86% yield of **259**. The reaction is not limited to simple aromatic rings. In a synthesis of (+)-aigialospirol by Hsung and co-worker's, aryl amide **260** was treated with an excess of *sec*-butyllithium and then with an excess of TMEDA to give the *ortho*-lithiated amide **261**.[305] Subsequent treatment with aldehyde **262** gave **263** in 50% overall yield. In previous examples, the conversion of **239** to **240** and **245** to **246** exploited this chemistry. This regioselective approach to preparing functionalized aromatic rings is an important addition to synthetic methodology.

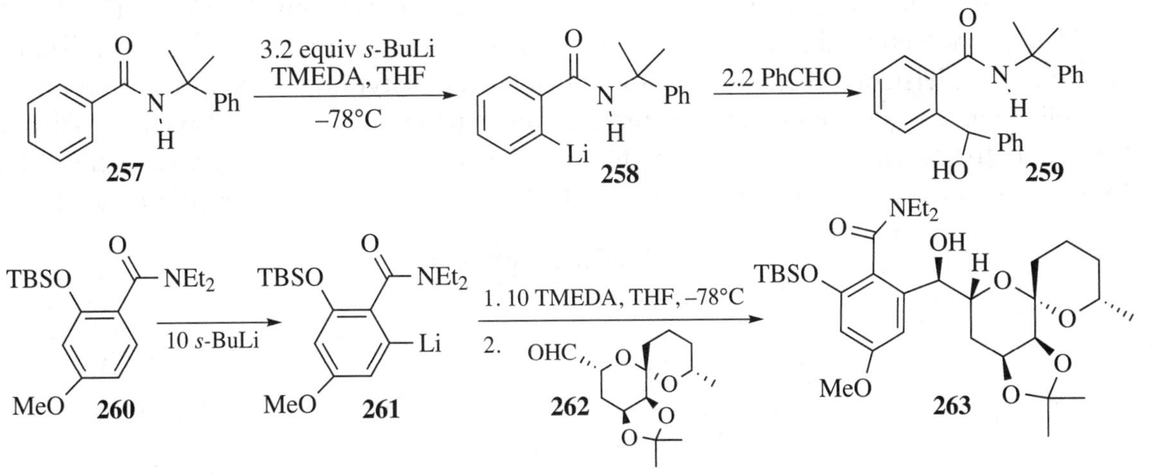

A disconnection for these reactions is generalized as:

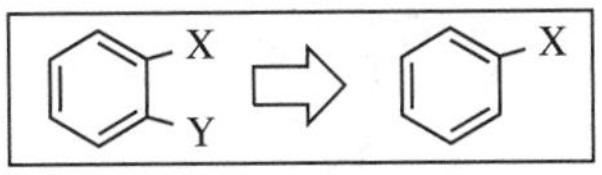

303. (a) Roberts, J.D.; Curtin, D.Y. *J. Am. Chem. Soc. **1946**, 68*, 1658; (b) Morton, A.A. *J. Am. Chem. Soc. **1947**, 69*, 969; (c) Chatgilialoglu, C.; Snieckus, V. in *Chemical Synthesis Gnosis to Prognosis; NATO ASI Series E. Applied Sciences*, Vol. 320, p 191. Kluwer Academic Publishers, Netherlands, **1994**.
304. Chauder, B.; Green, L.; Snieckus, V. *Pure Appl. Chem. **1999**, 71*, 1521.
305. Figueroa, R.; Hsung, R.P.; Guevarra, C.C. *Org. Lett., **2007**, 9*, 4857.

8.5.G. Addition to Epoxides, Alkenes, and Alkynes

8.5.G.i. Reactions with Epoxides. Epoxides are opened to the alcohol as expected, untroubled by the MgX_2 catalyzed rearrangements often observed with Grignard reactions. Methyllithium reacted with epoxide **264** to give the alcohol (**265**), in 68% yield.[306] The methyllithium added selectively to the allylic position to give the (*E*)-carbinol. In general, the organolithium attacks the epoxide at the less sterically hindered (less substituted) carbon, as was noted with Grignard reagents (8.4.E.).

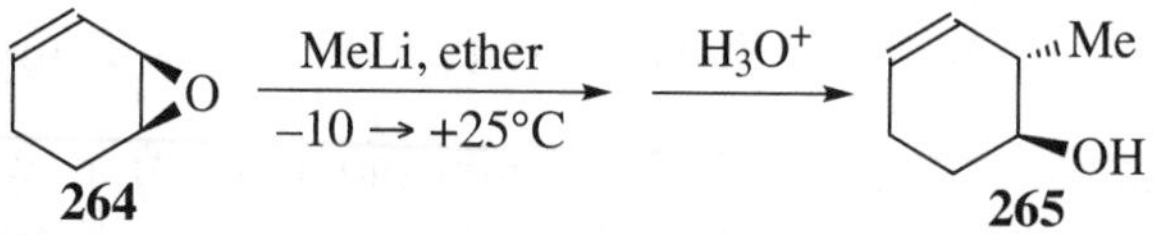

264 **265**

The epoxide ring opening gives the disconnection:

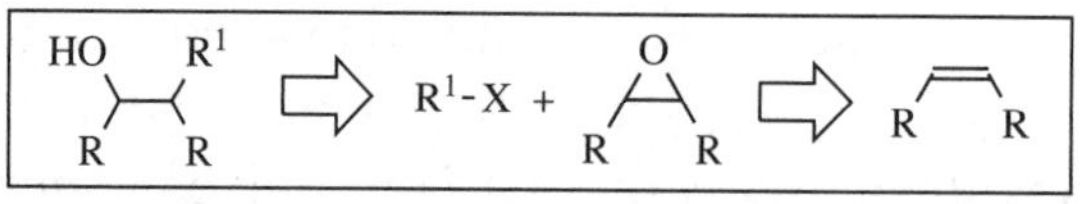

8.5.G.ii. Intramolecular Addition to Alkenes and Alkynes. While organolithium reagents react with alkenes or alkynes in an intramolecular reaction, both are relatively inert to the analogous intermolecular reaction. Ward and Lawler treated alkynyl bromide **266** with *n*-butyllithium, which gave a 60% yield of the alkylidene cyclopentane **267**.[307] There were several other products produced in this reaction, and the mechanistic pathway for the cyclization was not well understood. Bailey et al. showed that the *tert*-butyllithium exchange reaction and the cyclization reaction are rapid at low temperatures. Reaction of 6-iodo-3-methyl-1-hexene (**268**) gave **269**, which cyclized to a mixture of **270** and **271**. Quenching with methanol at −78°C gave a mixture of 76% of **273**, 5.6% of **274** and 14.2% of uncyclized product **272**.[308]

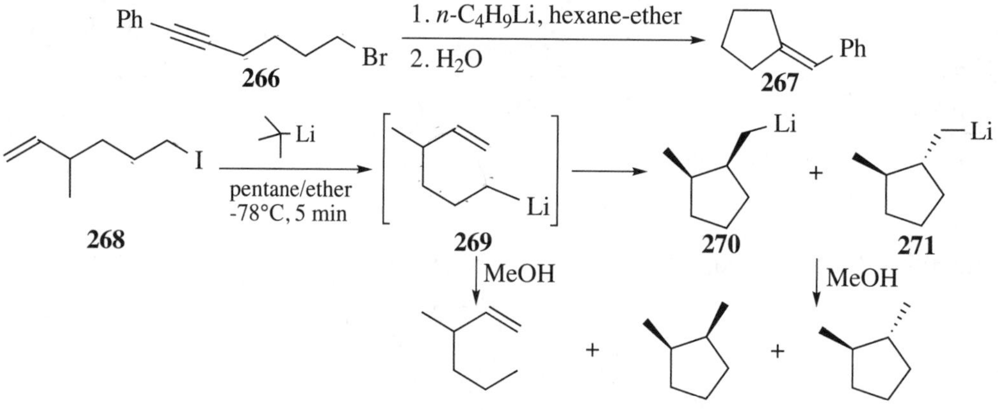

266 **267** **268** **269** **270** **271** **272** **273** **274**

306. Edward, D.E.; Ho, P.T. *Can. J. Chem.* ***1977**, 55*, 371.
307. Ward, H.R.; Lawler, R.G. *J. Am. Chem. Soc.* ***1967**, 89*, 5517.
308. Bailey, W.F.; Nurmi, T.T.; Patricia, J.J.; Wang, W. *J. Am. Chem. Soc.* ***1987**, 109*, 2442.

Bailey et al. also showed the initially formed organolithium product could be trapped with a variety of electrophiles, including aldehydes.[309] Cooke has shown that alkyllithium and alkenyllithium derivatives containing an ester moiety can be cyclized to form ketone products by direct addition or by Michael addition.[310] Bicyclic and spirocyclic molecules can be prepared by addition of TMEDA to the initially formed organolithium, because it induces a second cyclization reaction. The organolithium products were trapped with various electrophiles.[311] There is good enantioselectivity in the presence of (–)-sparteine. The reaction of **275** with *tert*-butyllithium and (–)-sparteine (**163**) gave **276**, which cyclized to give **277**. Quenching with water gave **278** in 60% yield (65% ee), in ether, and 85% yield (87% ee) in toluene.[312] When the reaction was done in THF, however, an 80% yield of **278** was obtained but with 0% ee. The extent of asymmetric induction is dependent on the structure of the added ligand.[313]

8.5.H. Basicity: Metal-Hydrogen Exchange

Organolithium reagents are stronger bases than Grignard reagents so removal of the acidic α-hydrogen to form an enolate anion can be a problem in reactions with carbonyl derivatives. Addition to the carbonyl of ketones and aldehydes is the faster process with common organolithium reagents. Specialized organolithium reagents can give different results, however. Lithium triphenylmethide, for example, often gives selective deprotonation with aldehydes and ketones. The extent of enolate formation with organolithium is dependent on many factors.[314] Because organolithium reagents are powerful bases,[315] the metal-hydrogen exchange reaction with functionalized substrates plays an important role in many syntheses.[316] This exchange reaction is believed to proceed via a four-center transition state such as **279**, generating a new organolithium (R–Li) from the acid (R–H + R¹–Li).[220,317] For this reaction to occur, R–H

309. Bailey, W.F.; Ovaska, T.V.; Leipert, T.K. *Tetrahedron Lett.* **1989**, *30*, 3901.
310. (a) Cooke, Jr., M.P. *J. Org. Chem.* **1992**, *57*, 1495; (b) Cooke, Jr., M.P.; Widener, R.K. *Ibid.* **1987**, *52*, 1381; (c) Cooke, Jr., M.P. *Ibid.* **1984**, *49*, 1144; (d) Cooke, Jr., M.P.; Houpis, I.N. *Tetrahedron Lett.* **1985**, *26*, 4987.
311. Bailey, W.F.; Khanolkar, A.D. *J. Org. Chem.* **1990**, *55*, 6058.
312. (a) Gil, G.S.; Groth, U.M. *J. Am. Chem. Soc.* **2000**, *122*, 6789. For the identical reaction with the *N,N*-diallyl derivative, see (b) Bailey, W.F.; Mealy, M.J. *J. Am. Chem. Soc.* **2000**, *122*, 6787.
313. Mealy, M.J.; Luderer, M.R.; Bailey, W.F.; Sommer, M.B. *J. Org. Chem.* **2004**, *69*, 6042.
314. House, H.O.; Kramar, V. *J. Org. Chem.* **1963**, *28*, 3362.
315. Roberts, J.D.; Curtin, D.Y. *J. Am. Chem. Soc.* **1946**, *68*, 1658.
316. (a) Reference 219, pp 698-705; (b) Reference 218.
317. (a) Ingold, C.K. *Helv. Chim. Acta* **1964**, *47*, 1191; (b) Batalov, A.P.; Rostokin, G.A.; Korshunov, I.A. *J. Gen. Chem. USSR* **1965**, *35*, 2146 (Engl., p 2135).

must be a stronger acid than R^1–H, and R^1–Li must be a stronger base than R–Li. The rate-determining step[318] in cleavage of the C–H bond is initial coordination of the lithium atom with the atom bearing the most acidic hydrogen,[287,319] In general, the more acidic the proton being removed in R–H, the more facile the reaction with R^1–Li. Six factors influence the acidity of a C–H bond.[320]

$$R-H + R^1\text{-Li} \longrightarrow \left[\begin{array}{cc} R-H \\ | \quad | \\ Li \cdot R^1 \end{array} \longleftrightarrow \begin{array}{cc} R\text{--------}H \\ | \qquad | \\ Li\text{-------}R^1 \end{array} \right] \longrightarrow R-Li + R^1\text{-H}$$

279

Table 8.15. The pK_a as a Function of % s Character

Compound	Bond Hybridization	% s-Character	pK$_a$
ethyne	sp	50	25
ethene	sp^2	33	36.5
cyclopropane	sp$^{2.28}$	30	39
ethane	sp^3	25	42

1. The first factor is the percentage of **s character**.[320a] As the s character of a bond increases, the attached proton becomes more acidic, because the 2s orbital lies closer to the nucleus than the 2p orbitals. When the proton is removed, increased electron density in the 2s orbital gives greater stability to the carbanion. As shown in Table 8.15,[320a] this leads to enhanced acidity of the alkynyl C–H relative to the analogous alkenyl C–H. Diminishing the C–C–C bond angle (i.e., cyclopropane) increases the % s character, and a cyclopropane C–H is more acidic than a normal alkane hydrogen. The percentage of s character[321] can be determined from ^{13}C NMR nuclear spin-spin coupling constants.

2. Conjugative effects enhance the acidity of certain covalently bound hydrogen atoms.[320a] The presence of a substituent that can delocalize electron density toward a more electronegative atom makes that proton more acidic. In addition, the conjugate base (the carbanion) formed after its removal is more stable because the electrons can be delocalized over several atoms. Carbonyl groups provide delocalization in the resonance stabilized enolate anion **280**, for example. The alkene bond linkage in enolate anion **280** is usually planar, but steric interactions, reduction of bond angles in small rings, and solvent interactions can inhibit formation of the planar enolate. The

318. (a) Shatenshtein, A.I.; Kamrad, A.G.; Shapiro, I.O.; Ranneva, Yu.I.; Zvyagintseva, E.N. *Dokl. Akad. Nauk, SSSR* **1966**, *168*, 364; (b) Barnes, R.A.; Nehmsmann, L.J. *J. Org. Chem.* **1962**, *27*, 1939; (c) Benkeser, R.A.; Trevillyan, A.E.; Hooz, J. *J. Am. Chem. Soc.* **1962**, *84*, 4971; (d) Bryce-Smith, D. *J. Chem. Soc.* **1963**, 5983.

319. (a) Morton, A.A. *J. Am. Chem. Soc.* **1947**, *69*, 969; (b) Gilman, H.; Morton, Jr., J.W. *Org. React.* **1954**, *8*, 258.

320. (a) Reference 49a, pp 48-85; (b) Bent, H.A. *Chem. Rev.* **1961**, *61*, 275; (c) Walsh, A.D., *Trans. Faraday Soc.* **1949**, *45*, 179.

321. Closs, G.L.; Closs, L.E. *J. Am. Chem. Soc.* **1963**, *85*, 2022.

stabilizing effects of conjugation are diminished in an enolate anion as a result of these factors[322] (sec. 9.2). The proton H_a in diketone **281**, for example, is much less acidic when compared to H_a in dione **282**.[322] This effect has been attributed to poor overlap of the carbanion orbital with the p orbitals of the carbonyl in **281**, due to the steric constraints of the bicyclo[2.2.2]octane ring system. The diminished acidity is therefore not entirely due to the difficulty that arises when forming a planar enolate. Aryl and alkenyl groups also provide conjugative stabilization, as in the benzyl anion (**283**). Conjugating groups such as nitro and cyano are excellent stabilizing substituents, as shown in **284** and **285**, respectively.

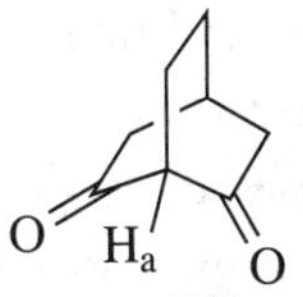

3. Inductive effects[320a] involve both through bond transmission of electrons, and field (through space) effects (sec. 2.2.A). As the electronegativity of a substituent increases, the central atom diverts increasing amounts of s character to the orbital occupied by the unshared electron pair.[320a] Methoxyacetic acid (pK_a 3.57) is more acidic, for example, than acetic acid (pK_a 4.75).[323] Most carbanion stabilizing groups act through a mixture of inductive and conjugative effects that are difficult to separate. The fluoride anion and quaternary ammonium cations are the only common substituents that stabilize ions by pure inductive effects.

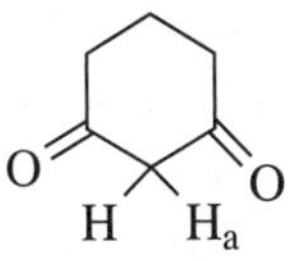

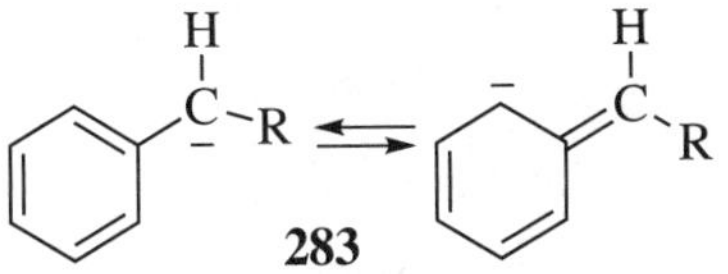

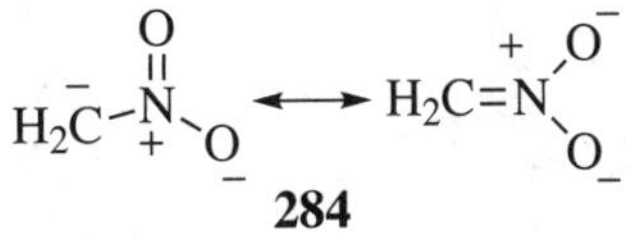

4. Homoconjugative effects[320a] occur when the α-carbon (the position adjacent to a group that induces acidity) is blocked, but the β-carbon bears a hydrogen. Removal of the β-hydrogen of certain ketones (called β-enolization) can give a stabilized carbanion, when its p orbital is held rigidly at the proper angle and distance to the p orbital of the carbonyl. The effect[324] is usually limited to a compound where an acidic proton is disposed β to the carbonyl, and the α-position does not possess hydrogen atoms. A **homoenolate** (first reported by Nickon and co-workers) is believed to be resonance stabilized by species such as **287** and **288**,[325] derived from treatment of **286** with base.[326]

322. Bartlett, P.D.; Woods, G.F. *J. Am. Chem. Soc.* **1940**, *62*, 2933.
323. King, E.J. *J. Am. Chem. Soc.* **1960**, *82*, 3575.
324. Johnson, A.L.; Petersen, W.W.; Rampersad, M.B.; Stothers, J.B. *Can. J. Chem.* **1974**, *52*, 4143.
325. Nickon, A.; Kwasnik, H.; Swartz, T.; Williams, R.O.; DiGiorgio, J.B. *J. Am. Chem. Soc.* **1965**, *87*, 1615.
326. Werstiuk, N.H. *Tetrahedron* **1983**, *39*, 205.

The acidity of H_a in **286** is thereby enhanced.[327]

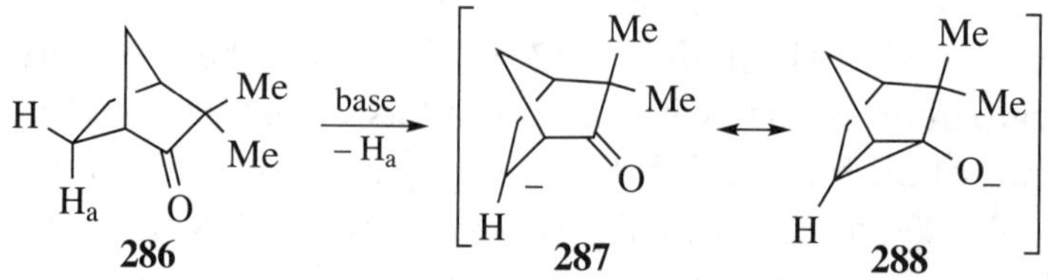

5. Aromatic character[320a] is an important consideration when comparing the conjugate bases of several weak acids. If one of the conjugate bases is a resonance stabilized anion, the acid precursor will be more acidic when compared to an acid that generates a non-aromatic conjugate base. The pK_a of cyclopentadiene is ~ 15, due in part to removal of a proton that leads to the stable aromatic cyclopentadienyl anion **289**. This acidity sharply contrasts with cycloheptatriene (pK_a of ~ 36), where loss of the hydrogen would give the antiaromatic 8π electron cycloheptatrienyl anion **290**.

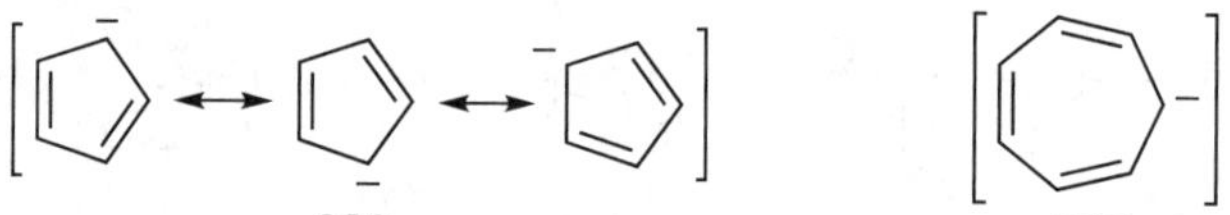

6. d-Orbital effects[320a] are important only when atoms are present that are not in the second row of the periodic chart, such as sulfur, phosphorus or transition metals. Carbanions can be stabilized by overlap of the electron pair with 3d orbitals of third row elements such as phosphorus and sulfur.[328] Dithioacetal **291** is more acidic than acetal **292** by 5-6 pK_a units. The carbanionic p orbital aligns itself to minimize electrostatic repulsion, as in **293**.[329] Forcing these orbitals into close proximity by confining them to a polycyclic system significantly decreases the acidity. Tris(sulfone) **294** is less acidic than **295**, but in this case solvation effects contribute to the difference in acidity.[328,329]

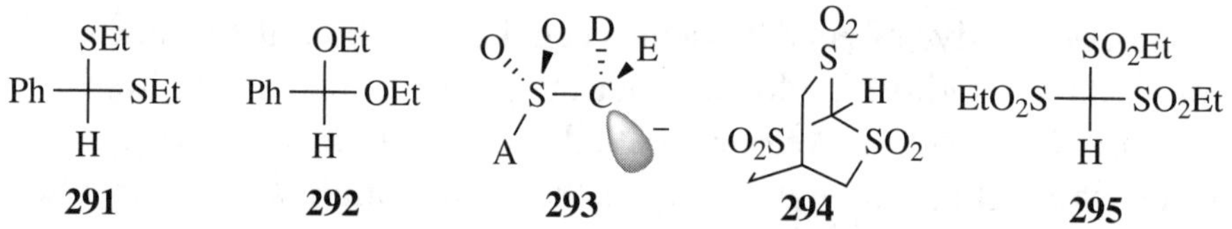

In summary, all six factors described above contribute to the following order of acidity, based on the ability of that functional group to enhance the acidity of

$$NO_2 > CO > SO_2 > CO_2H > CO_2R \sim CN \sim C(=O)NH_2 > X > H > R$$

The pK_a values of several common functional groups are presented in Table 8.16 to show the

327. (a) Nickon, A.; Lambert, J.L. *J. Am. Chem. Soc.* **1962**, *84*, 4604; (b) *Idem Ibid.* **1966**, *88*, 1905.
328. von E. Doering, W.; Hoffmann, A.K. *J. Am. Chem. Soc.* **1955**, *77*, 521.
329. von E. Doering, W.; Levy, L.K. *J. Am. Chem. Soc.* **1955**, *77*, 509.

effect on acidity of carbonyl substituents in the context of other organic acids. In reactions with an appropriate acidic molecule, the organolithium reagent removes the acidic proton to give the new organolithium reagent. In general, an organolithium reagent will exchange with an acidic hydrogen of another molecule if the conjugate acid of the newly-formed organolithium reagent is a *weaker* acid than the original organolithium reagent. According to this rationale, methyllithium (pK_a of methane is 40) will remove the acidic hydrogen in phenylacetylene (pK_a is 21) (see Table 8.16). Metal-hydrogen exchange therefore generates a carbanion that can subsequently be used as a nucleophile in addition reactions, or for substitution reactions. Enolization of carbonyl derivatives is usually accomplished with lithium dialkylamide bases rather than alkyllithium reagents, as discussed in Section 9.2.B.

Table 8.16. pK^a Values of Organic Acids

ACID	pK$_a$	ACID	pK$_a$	ACID	pK$_a$	ACID	pK$_a$
H–C(NO$_2$)$_2$–H	3.6[a]	H–C(CN)$_2$–H	12[a]	Me$_2$C(CHO)–H (O=C with H,H)	20.8[d,e]	cyclic S,S with Ph,H	29.6[c]
Ph–CO$_2$H	4.2[a]	MeC(O)–C(CO$_2$Et)(Et)–H	12.7[a]	fluorene (H,H)	21[b]	cyclic S,S with H,H	31.3[c], 36.5[i]
MeC(O)CH(SO$_2$Me)C(O)Me / H–C(SO$_2$Me)	4.7[b]	H–C(CO$_2$Et)$_2$–H	13.3[a]	Ph–C≡C–H	21[b]	Ph$_3$C–H	32.5[b]
Me–CO$_2$H	4.8[b]	H–C(SO$_2$Me)$_2$–H	14[a]	Me$_2$C(C(O)Me)–H	21.3[d]	cyclic S,S with Ph,H	34[c]
H–C(CHO)$_2$–H	5.0[a]	cyclopentadiene (H,H)	14-15[b]	Me$_2$C(C(O)Et)–H	21.6[d]	H–C(Ph)$_2$–H (with Ph,Ph)	35[b]
CH$_3$C(O)CH(CHO)–H	5.9[a]	EtO$_2$C–CH(Et)–CO$_2$Et	15[a]	Ph$_2$N–H	23[b]	H–C(Ph)(H)–H	35[b][41][f]
acetylacetone type (H,H)	5.85[a]	H–OH	15.7[b]	Et$_2$N–H	23[b]	cycloheptatriene (H,H)	36[b]
H–C(NO$_2$)(H)–H	8.6[a]	H$_3$C–OH	16[b]	H$_3$C–SO$_2$Me	23[a]	cis H$_2$C=CH$_2$ (H,H)	36.5[b][44][f]
PhC(O)CH$_2$C(O)–H,H	9.4[a]	CH$_3$C(O)–Cl / H	16.5[a]	t-Bu$_2$CH–H	23.5[d,e]	benzene H	37[b][44][f]
Me$_3$N⊕–H	9.8[a]	Et–OH / t-Bu–OH	18[c]	H$_3$C–CO$_2$H	24[g]	cyclic S,S with Me,H	38.3[c]
Ph–OH	9.9[a,k]	PhC(O)–CH$_2$–H (H,H)	18.6[b,d]	H–C≡C–H	25[b]	cyclic S,S with Et,H	38.6[c]
H–CH$_2$NO$_2$	10.2[a]	PhC(O)–CH(Me)–H (H,H)	19.1[d]	H–CH$_2$CO$_2$Et	25.6[h]	cyclopropane H	30[b]
=/–O-H	10.5[j]	PhC(O)–C(Me)(Me)–H	19.5[d]	H$_3$C–C(O)–NH$_2$	25[a]	H–CH$_3$	40[b]
CH$_3$C(O)CH$_2$C(O)Me / H,Me	11[a]	CH$_3$C(O)–H (H,H)	20[a]	Ph–NH$_2$	27[b]	cyclohexane H,H	45[b][51][f]

a Ref. [330] b Ref. [331] c Ref. [332] d Ref. [333] e Ref. [334] f Ref. [335] g Ref. 279 h Ref. [336] i Ref. [337] j Ref. [338] k Ref. [339]

The acidities of several ketones have also been measured in DMSO.[340]

Metal-hydrogen exchange is useful for the preparation of allylic and alkyl[330b], aryl[341a] or heteroaryl lithio derivatives. Several examples were used in previous syntheses, including the conversion of **239**→**240** and **245**→**246**. Asymmetric deprotonation is possible using *sec*-butyllithium and (–)-sparteine (**163**, and see the reaction of **275** above). The reaction of Boc pyrrolidine derivative **296**, for example, gave **297** and subsequent reaction with the methylating agent dimethyl sulfate gave a 45% yield of **298** in a ratio of 93:7 syn/anti diastereomers.[342] It is also possible to generate allyllithium derivatives from allylic hydrocarbons.[343]

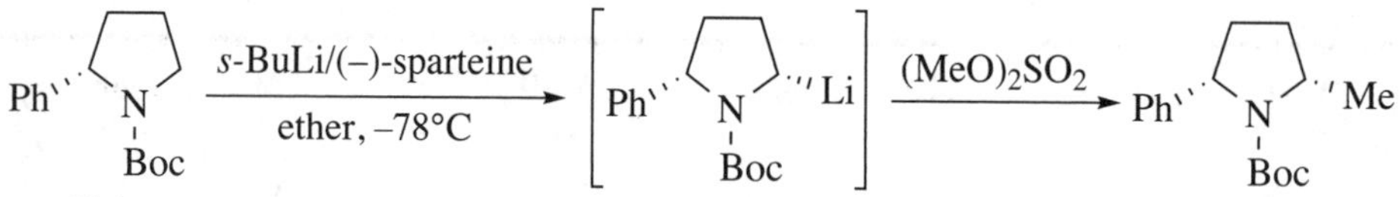

The disconnections for the aryl substitution reactions are

$$Ar{-}CO_2H \Rightarrow Ar{-}H \quad ArCR_2OH \Rightarrow Ar{-}H + R_2C{=}O \quad Ar{-}CH_2R \Rightarrow Ar{-}H + R{-}CH_2X \quad RCH{=}C(R)CH_2R \Rightarrow RCH{=}C(R)CH_3$$

330. Pearson, R.G.; Dillon, R.L. *J. Am. Chem. Soc.* **1953**, *75*, 2439

331. Reference 49a, pp 4, 10, 13, 14, 43, 48.

332. Streitwieser, Jr., A.; Ewing, S.P. *J. Am. Chem. Soc.* **1975**, *97*, 190..

333. Zook, H.D.; Kelly, W.L.; Posey, I.Y. *J. Org. Chem.* **1968**, *33*, 3477.

334. House, H.O. *Modern Synthetic Reactions, 2nd Ed.* W.A. Benjamin, Menlo Park, CA., **1972**, p 494.

335. (a) Streitwieser, Jr., A.; Caldwell, R.A.; Granger, M.R. *J. Am. Chem. Soc.* **1964**, *86*, 3578; (b) Streitwieser, Jr., A.; Maskornick, M.J.; Ziegler, G.R. *Tetrahedron Lett.* **1971**, 3927.

336. Amyes, T.L.; Richard, J.P. *J. Am. Chem. Soc.* **1996**, *118*, 3129.

337. For acidity of 1,3-dithiane in DMSO, see Xie, L.; Bors, D.A.; Streitwieser, A. *J. Org. Chem.* **1992**, *57*, 4986.

338. Werstiuk, N.H.; Andrew, D. *Can. J. Chem.* **1990**, *68*, 1467.

339. Liptak, M.D.; Gross, K.C.; Seybold, P.G.; Feldgus, S.; Shields, G.C. *J. Am. Chem. Soc.* **2002**, *124*, 6421.

340. Bordwell, F.G.; Harrelson, Jr., J.A. *Can. J. Chem.* **1990**, *68*, 1714.

341. (a) Letsinger, R.L.; Schnizer, A.W. *J. Org. Chem.* **1951**, *16*, 869; (b) Breslow, R.; Grant, J.L. *J. Am. Chem. Soc.* **1977**, *99*, 7745.

342. Wu, S.; Lee, S.; Beak, P. *J. Am. Chem. Soc.* **1996**, *118*, 716.

343. For a synthetic example taken from a synthesis of β-bisabolene , see Crawford, R.J.; Erman, W.F.; Broaddus, C.D. *J. Am. Chem. Soc.* **1972**, *94*, 4298.

MOLECULAR MODELING: Acidities of Carboxylic Acids

The electrostatic potential is the energy felt by a point positive charge at a location near a molecule. It balances the repulsion between the charge and the nuclei and the attraction between the charge and the electrons. Of more practical value is an electrostatic potential map. This presents the electrostatic potential at all locations on an electron density surface which, like a space-filling model, depicts the accessible area of a molecule. As such, an electrostatic potential map provides a measure of the distribution of charge on the surface. By convention, negatively charged regions on the surface are colored "red", positively charged regions are colored "blue" and neutral regions are colored "green".

There are many times in organic chemistry when knowing the relative acidity of two or more acids may be important issue, and pK_a values are not available in the literature. Here, electrostatic potential maps may provide a viable alternative. Do these correlate with acidity for series of related compounds? For example, does the (maximum) potential on the acidic hydrogen in carboxylic acids correlate with measured pK_a? One after another, build four (or more) of the carboxylic acids from the following list. (After the first model, *click* on > at the top of the screen and select **Continue** to bring up a fresh builder without removing this text.) Calculate an electrostatic potential map for each of the four carboxylic acids and use a graphing program to make a plot of the value of the electrostatic potential at the acidic hydrogen vs. pK_a (given after the name). To "measure" the value of the potential at a location on the map, *click* on the location. It will be reported at the bottom of the screen.

trichloroacetic acid	0.7
oxalic acid	1.23
dichloroacetic acid	1.48
cyanoacetic acid	2.45
chloroacetic acid	2.85
fumaric acid	3.10
terephthalic acid	3.51
formic acid	3.75
(E)-3-chloroacrylic acid	3.79
benzoic acid	4.19
3-(*p*-chlorophenyl)-*trans*-acrylic acid	4.41
crotonic acid	4.70
acetic acid	4.75
pivalic acid	5.03

Based on the few compounds you have examined, are you able to conclude that electrostatic potentials at the acidic hydrogen correlate with the pKa's of carboxylic acids? Try to identify a carboxylic acid that is more acidic (smaller pKa) than trichloroacetic acid and use the best fit line you obtained to predict its pKa. Try to identify a carboxylic acid that is less acidic (larger pKa) than pivalic acid and predict its pKa.

8.6. SULFUR STABILIZED CARBANIONS AND UMPOLUNG

Introducing a heteroatom (X) into an organic molecule enhances the acidity of the proton adjacent to the heteroatom (H–C–X). A sulfur group makes the proton more acidic due to electron-withdrawing effects, as well as by stabilization of the resulting carbanion by the d orbitals (see above).[328,329] Several different sulfur-stabilized carbanions are known, and have been used extensively in organic synthesis.[344,345] Sulfur stabilized carbanions react with alkyl halides or with aldehydes and ketones in the same manner as other carbanions.

8.6.A. Sulfur Carbanions

Thioethers (sulfides)[348] can be converted to their α-carbanion derivative (**299**, M = Li, Na, K) by treatment with strong bases such as *n*-butyllithium, sodium amide (NaNH$_2$) or potassium amide (KNH$_2$). The hydrogen atom α-to the sulfur atom of sulfides is a weaker acid than that the hydrogen atom α-to a carbonyl, nitrile or nitro group, and stronger bases are required for deprotonation. Sulfides are stronger acids, than the allylic, vinyl or aromatic hydrocarbons discussed above, however. The various oxides of sulfur, including sulfoxides (**300**), sulfones (**301**) or sulfonate esters (**302**) can be converted to α-carbanions. Increasing the number of oxygens on the sulfur atom generally increases the acidity of the α-proton. Bordwell and co-workers studied the acidity of several sulfur derivatives using DMSO as a solvent.[346,48b] Anders and co-workers studied the structure of sulfur-stabilized allyllithium compounds in solution, and found that they were monomers in THF.[347]

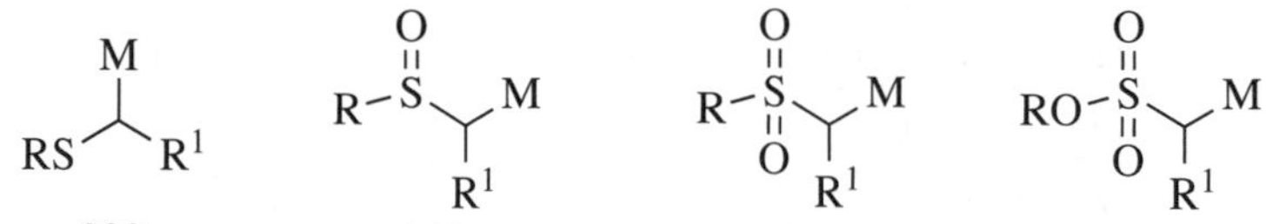

344. Reference 265a, pp 94-103.
345. Block, E. *Aldrichimica Acta* **1978**, *11*, 51.
346. (a) Reference 344, p 9; (b) Bordwell, F.G.; Algrim, D. *J. Org. Chem.* **1976**, *41*, 2507; (c) Bordwell, F.G.; Bares, J.J.; Bartmess, J.E.; McCollum, G.J.; van der Puy, M.; Vanier, N.R.; Matthews, W.S. *Ibid.* **1977**, *42*, 321; (d) Bordwell, F.G.; Bares, J.E.; Bartmess, J.E.; Drucker, G.E.; Gerhold, J.; McCollum, G.J.; van der Puy, M.; Vanier, N.R.; Matthews, W.S. *Ibid.* **1977**, *42*, 326; (e) Bordwell, F.G.; Matthews, W.S.; Vanier, N.R. *J. Am. Chem. Soc.* **1975**, *97*, 442; (f) Bordwell, F.G.;Bartmess, J.E.; Drucker, G.E.; Margolin, Z.; Matthews, W. *Ibid.* **1975**, *97*, 3226.
347. Piffl, M.; Weston, J.; Günther, W.; Anders, E. *J. Org. Chem.* **2000**, *65*, 5942.

Oxidation of a sulfoxide to a sulfone places a second oxygen on sulfur (RSO_2R, sec. 3.9.A.iii), and this increases the relative acidity of the α-hydrogen. Dimethyl sulfone is more acidic than the corresponding sulfoxide (RSOR, DMSO) by 6.1 pK_a units. The presence of an electron-releasing methyl group ($PhSO_2CH_3$ versus $PhSO_2CH_2CH_3$)[317c] causes a reduction of acidity by two pK_a units. As the group attached to the sulfur becomes more electron withdrawing, the acidity increases ($Me_3CSO_2CH_3$ = 30.3, $MeSO_2CH_3$ = 31.1, $PhSO_2CH_3$ = 29.0, $CF_3SO_2CH_3$ = 18.8).[346c,g] A similar trend is observed for the thio-analogs of acetonitrile (Me_3CSCH_2CN = 22.9, Me_2CHSCH_2CN = 23.6, $CH_3CH_2SCH_2CN$ = 24.0, CH_3SCH_2CN = 24.3, $PhSCH_2CN$ = 20.9).[346g] Two sulfur groups significantly increase the acidity of the proton ($CH_3SO_2CH_3$ = 31.1, $CH_3CH_2SO_2CH_2SO_2CH_2CH_3$ = 14.4, $CH_3SCH_2SOCH_3$ = 29.0),[346g] and another electron withdrawing group leads to an increase in acidity ($CH_3SO_2CH_2CN$ = 13.6, CH_3SCH_2CN = 24.3).[346g]

As mentioned, the proton on the α-carbon of a sulfide is a very weak acid and its removal requires a strong base.[348] Organolithium reagents are commonly used for this purpose. Phenylisopropyl sulfide (**303**) reacted with *tert*-butyllithium (THF, HMPA) to give the α-lithio derivative **304**. Alkyl sulfides often require addition of HMPA or TMEDA to facilitate Li-H exchange. Organolithium reagent **304** reacted with allyl bromide to give **305** in 87% yield, or with benzaldehyde to give **306** in 92% yield.[349]

There are many synthetic applications for sulfide carbanions. Sulfide **307** required the addition of DABCO to facilitate deprotonation by butyllithium, but the resulting anion coupled readily with allylic chloride **308** under these conditions to give **309**. The sulfide moiety was removed by a dissolving metal reduction (for example, see secs. 4.9.D, 4.9.F) to give dendrolasin **310**.[349b] The ability to remove the sulfur after activating a carbon for nucleophilic attack is one of the most attractive features of organosulfur chemistry. Coupling of allylic halides and allylic thiocarbanions is sometimes called **Biellmann alkylation**.[350] Intramolecular reactions occur readily and reaction with epoxides or alkyl halides is possible.

348. (a) Gilman, H.; Beaber, N.J. *J. Am. Chem. Soc.* **1925**, *47*, 1449; (b) Ipatieff, N.N.; Pines, H.; Friedman, B.S. *Ibid.* **1938**, *60*, 2731; (c) Hurd, C.D.; Gershbein, L.L. *Ibid.* **1947**, *69*, 2328; (d) Campbell, J.R. *J. Org. Chem.* **1964**, *29*, 1830; (e) Kharasch, M.S.; Fuchs, C.F. *Ibid.* **1948**, *13*, 97.

349. (a) Dolak, T.M.; Bryson, T.A. *Tetrahedron Lett.* **1977**, 1961; (b) Kondo, K.; Matsumoto, M. *Ibid.* **1976**, 391.

350. (a) Biellmann, J.F.; Ducep, J.B. *Tetrahedron* **1971**, *27*, 5861; (b) Altman, L.J.; Ash, L.; Marson, S. *Synthesis* **1974**, 129; (c) Grieco, P.A.; Masaki, Y. *J. Org. Chem.* **1974**, *39*, 2135.

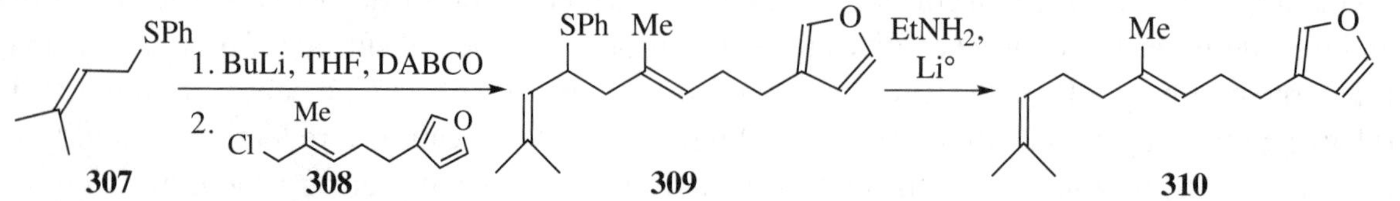

When a sulfide group is attached to the α-carbon of a molecule an electron-withdrawing group such as a carbonyl (the sulfide is β to the carbonyl), the acidity of the α-proton is greatly enhanced (see Table 8.16). When α-phenylthio-γ-butyrolactone (**311**) was treated with lithium diisopropylamide (LDA, sec. 9.2.B), the 2-lithio derivative was formed. Subsequent reaction with iodomethane gave **312**.[351] The sulfide was especially useful in this synthesis, since facile oxidation to the corresponding sulfoxide (**313**) and subsequent heating induced syn-elimination to alkene **314** (sec. 2.9.C.v). The sulfide-alkylation sequence can be applied to many systems, and in some cases highly functionalized molecules can be prepared.

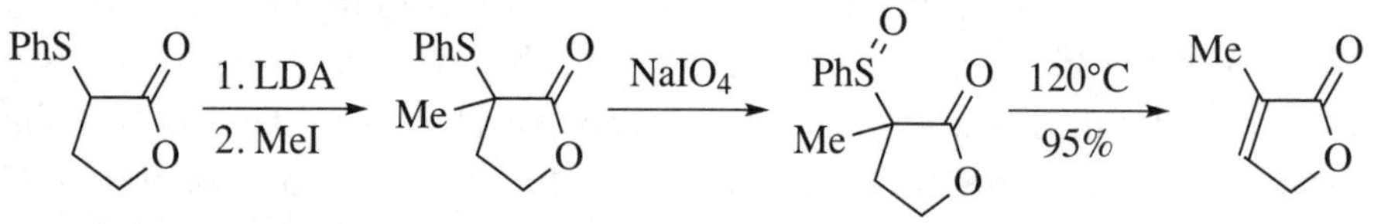

Sulfoxide carbanions formed by deprotonation of sulfoxides are common in alkylation reactions.[352] Chiral sulfoxides (sec. 3.9.A.ii) are possible, and reacton of the corresponding organolithium derivatives with aldehydes or ketones generates diastereomeric products. Once the organolithium derivative is generated, the stereochemical integrity of the C–Li bond is remarkably stable, possibly due to interactions with the adjacent functional groups. Conversion of the asymmetric keto-sulfoxide (**315**) to the carbanion, followed by condensation with propanal gave an alcohol-sulfoxide. The sulfoxide group was removed by treatment with aluminum amalgam to complete the asymmetric synthesis of 3-hydroxyketone **316** in 67% yield and 64% ee.[353] Note that *tert*-butylmagnesium bromide was used as the base. The reductive cleavage of the sulfoxide with aluminum amalgam (sec. 4.9.I) is one of the most common methods for removing a sulfur moiety.

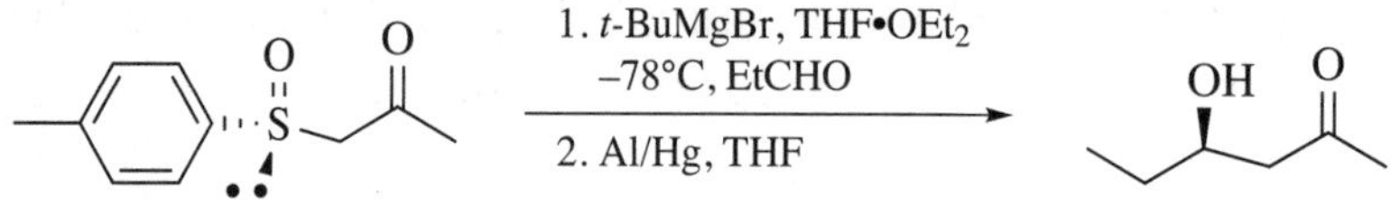

351. (a) Grieco, P.A.; Reap, J.J. *Tetrahedron Lett.* **1974**, 1097; (b) for a different example, see Trost, B.M.; Salzmann, T.N. *J. Org. Chem.* **1975**, *40*, 148.
352. Reference 344, pp 94-97.
353. Schnedier, F.; Simon, R. *Synthesis* **1986**, 582.

Another variation in sulfoxide reactivity is shown after an initial conversion of **317** to **318** via reaction with LDA and alkylation with methyl iodide.[354b] In this case, removal of the sulfur moiety allowed formation of an allylic alcohol. Heating allylic sulfoxide **318** induced a facile [2,3]-sigmatropic rearrangement (sec. 11.12.A) to give **319**, and hydrolysis of **319** gave cyclohexenol derivative **320**. Trimethyl phosphite [P(OMe)$_3$] was added to trap a minor isomer of the equilibrium. This reaction is called the **Mislow-Evans rearrangement**.[354,355] The [2,3]-sigmatropic rearrangement that occurs with allylic sulfoxides is useful and predictable, since it proceeds with high stereoselectivity.[354c] This rearrangement is useful for introducing functional groups other than allylic alcohols into a molecule, as in Otera and co-worker's use of this reaction to generate conjugated aldehydes.[356] Selenoxides also give this reaction.[357]

Sulfones are useful organosulfur compounds.[358] In Yamada and co-worker's synthesis of cladocoran A,[359] sulfone **321** was converted to the α-lithio compound by treatment with *n*-butyllithium. Subsequent reaction with 4-iodo-2-methyl-1-butene gave an 89% yield of sulfone **322**. Reduction of the sulfone with sodium amalgam gave a 91% yield of **323**. Note that sodium amalgam or Raney nickel [Ni(R)] are commonly used to remove sulfur from organic molecules (sec. 4.9.D).[360] Reaction of a sulfone carbanion with an aldehyde, acetylation of the resulting alcohol, followed by elimination to an alkene is called **Julia olefination**.[361] In a synthesis of (–)-siccanin,[362] Trost and co-workers used this technique to convert sulfone **324** to **325** in 93% overall yield.

354. (a) Bickart, P.; Carson, F.W.; Jacobus, J.; Miller, E.G.; Mislow, K. *J. Am. Chem. Soc.* **1968**, *90*, 4869; (b) Evans, D.A.; Andrews, G.C. *Acc. Chem. Res.* **1974**, *7*, 147; (c) Hoffmann, R.W. *Angew. Chemie. Int. Ed.* **1979**, *18*, 563.

355. (a) The Merck Index, 14th ed., Merck & Co., Inc., Whitehouse Station, New Jersey, **2006**, p ONR-62; (b) Mundy, B.P.; Ellerd, M.G.; Favaloro Jr., F.G. *Name Reactions and Reagents in Organic Synthesis, 2nd ed.*, Wiley-Interscience, New Jersey, **2005**, pp. 430-431.

356. Sato, T.; Otera, J.; Nozaki, H. *J. Org. Chem.* **1989**, *54*, 2779.

357. For an example taken from a synthesis of pseudocodeine, see Kshirsagar, T.A.; Moe, S.T.; Portoghese, P.S. *J. Org. Chem.* **1998**, *63*, 1704.

358. (a) Magnus, P.D. *Tetrahedron* **1977**, *33*, 2019; (b) Reference 344, pp 97-103.

359. Miyaoka, H.; Yamanishi, M.; Kajiwara, Y.; Yamada, Y. *J. Org. Chem.* **2003**, *68*, 3476.

360. Welch, S.C.; Gruber, J.M. *J. Org. Chem.* **1982**, *47*, 385.

361. (a) Julia, M.; Paris J.M. *Tetrahedron Lett.* **1973**, *14*, 4833; (b) The Merck Index, 14th ed., Merck & Co., Inc., Whitehouse Station, New Jersey, **2006**, p ONR-50; (c) Mundy, B.P.; Ellerd, M.G.; Favaloro Jr., F.G. *Name Reactions and Reagents in Organic Synthesis, 2nd ed.*, Wiley-Interscience, New Jersey, **2005**, pp. 356-357.

362. Trost, B.M.; Shen, H.C.; Surivet, J.-P. *Angew. Chem. Int. Ed.* **2003**, *42*, 3943.

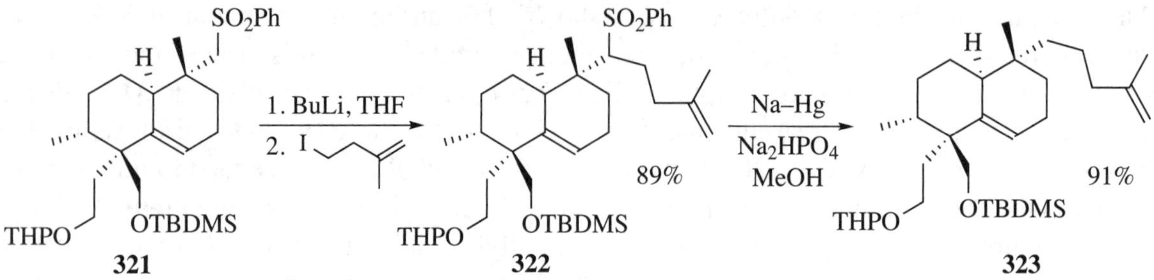

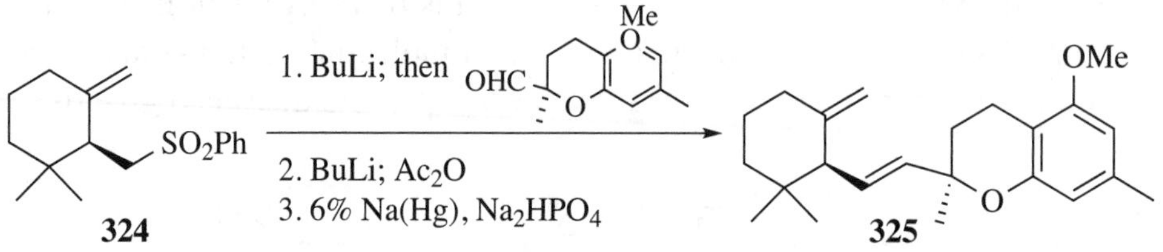

In 1-lithio vinyl sulfides (sec. 8.6) the sulfur atom stabilizes the carbanionic center, allowing either alkylation or condensation reactions. Lithio vinyl sulfones are similarly stabilized. Vinyl sulfone (**326**), for example, was converted to the 1-lithio derivative with methyllithium.[363] This organometallic reacted in the usual manner with alkyl halides (methyl iodide) to give the coupling product, **327**.

Vinyl sulfoxides undergo an unusual elimination reaction when a leaving group such as chloride is also present on the double bond. When **328** was treated with *tert*-butyllithium, exchange occurred to give **329**.[364] In the presence of additional *tert*-butyllithium, chloride is lost and a lithio alkyne (**330**) is formed in what is known as the **Fritsch-Buttenberg-Wiechell rearrangement**.[365] Hydrolysis led to the final product, alkyne **331**, in 74% yield. A similar reaction was observed when **328** was treated with Grignard reagents.

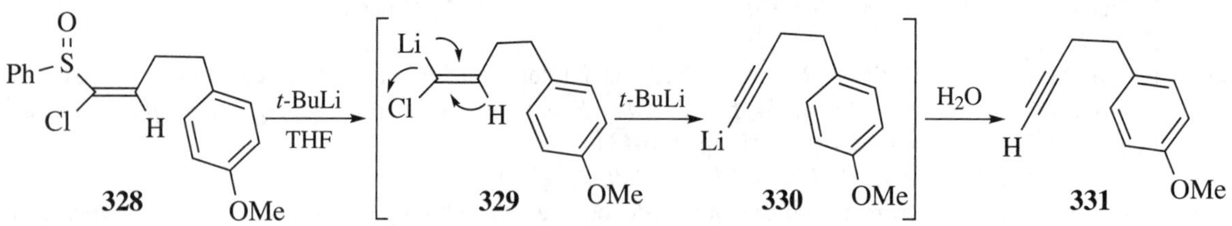

363. Eisch, J.J.; Galle, J.E. *J. Org. Chem.* **1979**, *44*, 3277, 3279.
364. Satoh, T.; Hayashi, Y.; Yamakawa, K. *Bull. Chem. Soc. Jpn.* **1993**, *66*, 1866.
365. (a) Fritsch, P. *Annalen* **1894**, *279*, 319; (b) Buttenberg, W.P. *Annalen* **1894**, *279*, 327; (c) Wiechell, H. *Annalen* **1894**, *279*, 332; (d) Stang, P.J. *Chem. Rev.* **1978**, *78*, 383; (e) Stang, P.J.; Fox, D.P. *J. Org. Chem.* **1978**, *43*, 364; (f) The Merck Index, 14th ed., Merck & Co., Inc., Whitehouse Station, New Jersey, **2006**, p ONR-35; (g) Mundy, B.P.; Ellerd, M.G.; Favaloro Jr., F.G. *Name Reactions and Reagents in Organic Synthesis, 2nd ed.*, Wiley-Interscience, New Jersey, **2005**, pp. 262-263.

Formation of a carbanion is often accompanied by loss of the SO_2 moiety in α-halo sulfones, via the so-called **Ramberg-Bäcklund reaction**,[366] as in the conversion of 1-bromoethyl ethyl sulfone (**332**) to 2-butene.[367] Initial reaction with KOH generated the sulfone carbanion (**333**), which displaced bromide intramolecularly to give an episulfone (**334**). Under the reaction conditions, **334** lost sulfur dioxide (SO_2) to give the alkene. Meyers introduced a modification of this reaction in which the sulfone was converted directly to the alkene without isolating the α-halosulfone, using KOH in CCl_4.[368,369] Sulfuryl chloride and bromine can also be used.[369]

Sulfonate carbanions have been used less extensively than sulfone, sulfoxide or sulfide carbanions. Sulfonates are good leaving groups and subject to nucleophilic displacement or elimination in the presence of the bases required to generate the carbanion. Cyclic sulfonates (sultones, **335**) have been alkylated[370] via initial reaction with *n*-butyllithium and then an alkyl halide. Sultone **335** was used in Wolinsky and co-worker's synthesis of the terpene β-santalene (**337**), via alkylation with 1-bromo-3-methyl-2-butene to give **336**.[371] The sulfur was removed by reduction with aluminum hydride (AlH_3, sec. 4.6.B), and subsequent treatment with $POCl_3$ and pyridine generated the exo-methylene group (secs. 2.8.A, 2.9.A).[371] Note that reductive cleavage of the sultone can give the thioalcohol via S–O cleavage and/or the alcohol via C–O cleavage.[370a] Sultones and sulfonate esters are rarely used in synthesis, but sulfides and sulfones are common.

366. (a) Ramberg, L.; Bäcklund, B. *Arkiv. Kemi Mineral. Geol.* **1940**, *13A*, 50 [*Chem. Abstr. 34:* 47255, **1940**]; (b) Paquette, L.A. *Acc. Chem. Res.* **1968**, *1*, 209; (c) The Merck Index, 14th ed., Merck & Co., Inc., Whitehouse Station, New Jersey, **2006**, p ONR-77; (d) Mundy, B.P.; Ellerd, M.G.; Favaloro Jr., F.G. *Name Reactions and Reagents in Organic Synthesis, 2nd ed.*, Wiley-Interscience, New Jersey, **2005**, pp. 540-541.
367. (a) Block, E. *Reactions of Organosulfur Compounds*, Academic Press, New York, **1978**; (b) Block, E. *J. Chem. Educ.* **1971**, *48*, 814; (c) Paquette, L.A., *Org. React.* **1977**, *25*, 1; (d) Bordwell, F.G.; Jarvis, B.B.; Corfield, P.W.R. *J. Am. Chem. Soc.* **1968**, *90*, 5298; (e) Paquette, L.A.; Philips, J.C. *Ibid.* **1969**, *91*, 3973; (f) Paquette, L.A.; Houser, K.W. *Ibid.* **1969**, *91*, 3870; (g) Corey, E.J.; Block, E. *J. Org. Chem.* **1969**, *34*, 1233.
368. Meyers, C.Y.; Malte, A.M.; Matthews, W.S. *J. Am. Chem. Soc.* **1969**, *91*, 7510.
369. (a) Pommelet, J.-C.; Nyns, C.; Lahousse, F.; Morény, R.; Viehe, H.G. *Angew. Chemie. Int. Ed.* **1981**, *20*, 585; (b) Martel, H.J.J.-B.; Rasmussen, M. *Tetrahedron Lett.* **1975**, 947.
370. (a) Smith, M.B.; Wolinsky, J. *J. Org. Chem.* **1981**, *46*, 101; (b) Durst, T.; Tin, K.C. *Can. J. Chem.* **1970**, *48*, 845.
371. (a) Wolinsky, J.; Marhenke, R.L.; Eustace, E.J. *J. Org. Chem.* **1973**, *38*, 1428; (b) Coffen, D.L.; Grant, B.D.; Williams, D.L. *Int. J. Sulfur Chem., Part A* **1971**, *1*, 13.

MOLECULAR MODELING: Relative CH Acidities of Sulfur Compounds

The absolute acidity of a molecule is the energy required to remove a proton.

$$R\text{–}H \rightarrow R^- + H^+$$

As it turns out, it is very difficult to accurately calculate absolute acidity (and impossible to actually measure such a quantity). However, even simple quantum chemical methods provide a reasonable account of the relative acidities of closely related compounds.

Obtain an electrostatic potential map for phenyl methyl sulfoxide. This colors an electron density surface (revealing the overall size and shape) with the energy that a positive charge "feels" at various locations on the surface. By convention, positive regions (where the charge is repelled) are colored "blue" and negative regions (where the charge is attracted) are colored "red". Acidic sites will be blue. Identify these regions on the map and record the value of the potential. (*Click* on the region and the value of the potential will be displayed at the bottom of the screen.) Repeat your calculations for phenyl methyl sulfone and phenyl methyl sulfide. For each new structure, *click* on > to the right of the tab at the top of the screen and select ***Continue***. According to electrostatic potential maps, which compound is most acidic and which is least acidic?

Next calculate energies for the two reactions shown below (comparing acidities of phenyl methyl sulfoxide and phenyl methyl sulfone with that of phenyl methyl sulfide as a standard).

$$Ph\text{–}SO\text{–}CH_3 + Ph\text{–}S\text{–}CH_2^- \rightarrow Ph\text{–}SO\text{–}CH_2^- + Ph\text{–}S\text{–}CH_3$$
$$Ph\text{–}SO_2\text{–}CH_3 + Ph\text{–}S\text{–}CH_2^- \rightarrow Ph\text{–}SO_2\text{–}CH_2^- + Ph\text{–}S\text{–}CH_3$$

You have already calculated energies for the three neutral molecules, so you only need to obtain energies for the three anions. Work out energies of the two reactions. Are both the sulfoxide and the sulfone stronger acids than the sulfide? Which is the stronger acid, the sulfoxide or the sulfone?

Repeat your calculations for 1,3-dithiolane and for 1,3-dithiane. Which protons in 1,3-dithiolane and which in 1,3-dithiane are more acidic?

1,3-dithiolane

1,3-dithiane

The disconnections for these reactions found in this section are

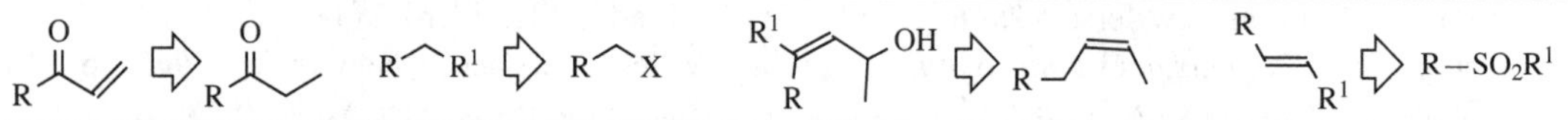

8.6.B. Umpolung

We are used to seeing a certain kind of reactivity associated with a given functional group. In some cases, it is possible to reverse the reactivity, in effect making an electrophilic center in to a nucleophilic one and vice versa. This section will explore reactions that allow this reversal.

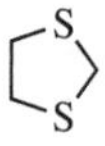

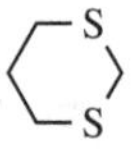

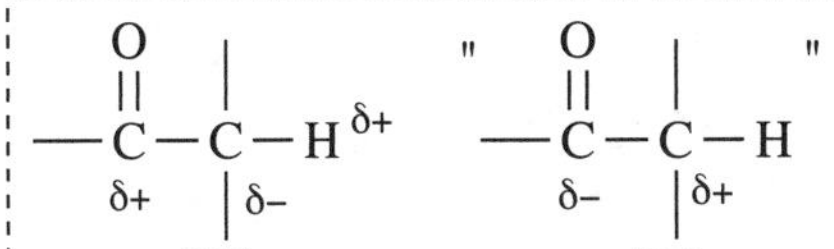

338 **339**

8.6.B.i. Definition of Umpolung. Under normal conditions, a carbonyl group is polarized as shown in **338**, with the carbonyl carbon electrophilic and the a-carbon nucleophilic. Facile formation of an enolate anion (see **280**) demonstrates this bond polarization. In a hypothetical **339** the acyl carbon would be nucleophilic, but to attain this reversal the carbonyl must be transformed into another functional group that allows the carbon to become nucleophilic. The ability to regenerate the carbonyl after completion of the desired reactions is another requirement. Seebach termed this process **umpolung**.[372] There is umpolung (reversal of reactivity) whenever one observes a 1,2-*n* relationship of the functional groups in a carbonyl compound, where *n* is the number of carbon atoms separating the functional groups. For Umpolung to occur, an even number of carbons should separate the carbons bearing the functional groups (0, 2, 4, ... as in **345**, where no carbons separate the groups, and in **346** where two carbons separate the groups).[372a] This relationship contrasts with normal reactivity, where an odd number of carbons separate the carbonyl bearing functional groups (1, 3, 5, ... as **347** where one carbon separates the groups, and in **348** where three carbons separate the groups).[372a] Table 8.17[372c] shows several umpolung equivalents, along with the reactivity of the normal fragment and typical reagents that give the umpolung reactivity.

Table 8.17. Typical Umpolung Equivalents and Reagents

Normal		Desired Umpolung		Umpolung Reagents
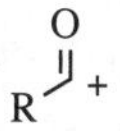	a_1	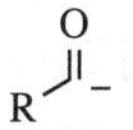 **340**	d_1	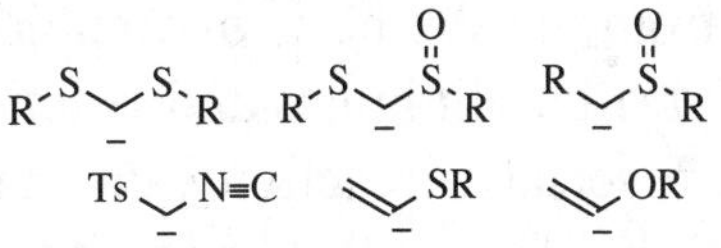
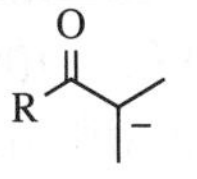	d_2	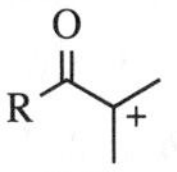 **341**	a_2	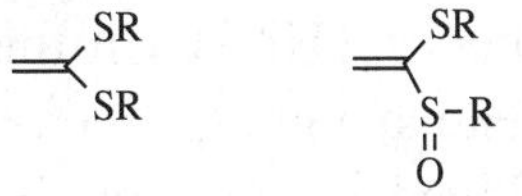
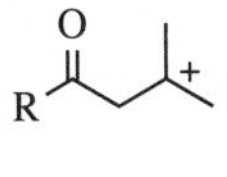	a_3	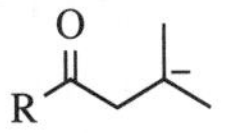 **342**	d_3	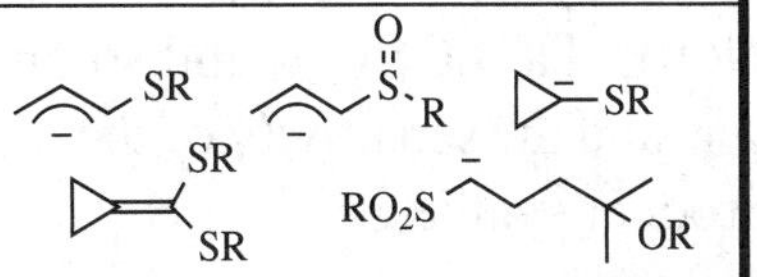
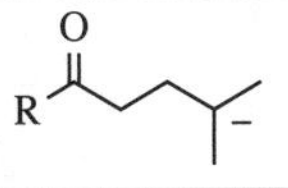	d_4	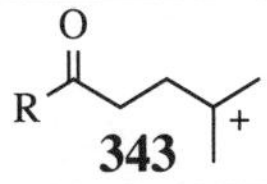 **343**	a_4	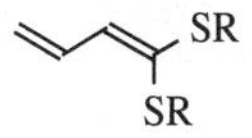
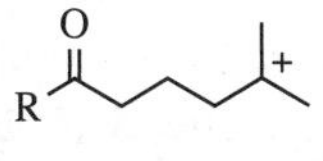	a_5	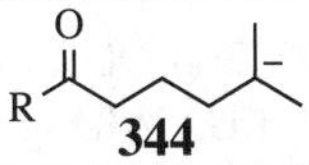 **344**	d_5	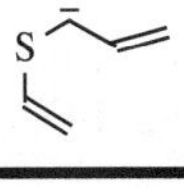

[Reprinted from Gröbel, B.-T.; Seebach, D. *Synthesis* *1977*, pp. 357. Copyright *1977* by Georg Thieme Verlag.]

372. (a) Seebach, D. *Angew. Chem. Int. Ed.* ***1979**, 18,* 239; (b) Seebach, D. *Ibid.* ***1969**, 8,* 639; (c) Gröbel, B.-T.; Seebach, D. *Synthesis* ***1977***, 357; (d) Seebach, D.; Kolb, M. *Chem. Ind. (London)* ***1974**,* 687.

345   **346**   **347**   **348**

8.6.B.ii. Acyl Anion Equivalents. The most common umpolung equivalent is that for acyl anion **340**, and dithianes derived from aldehydes (**349**) are the most common reagent used for this purpose. Dithianes are available from ketones, and this is a common ketone protecting group (sec. 7.3.B.ii), but no hydrogen is available for removable. 1,3-Propanedithiol is the usual precursor for the preparation of **349**. Note that a nonthiolic, odorless equivalent has been reported.[373] Removal of the hydrogen adjacent to the sulfur in **349** requires a base such as *n*-butyllithium, and the product is α-lithiodithiane (**350**). The

acidity of this hydrogen is largely due to greater polarizability of the sulfur, and the greater C–S bond length, rather than to the presence of the d orbitals.[374] The pK_a of 1,3-dithiane (**349**, R = H) is 31.1.[335] 2-Methyl-1,3-dithiane (**349**, R = Me) has a pK_a of 38.3, however, whereas 2-phenyl-1,3-dithiane (**349**, R = Ph) has a pK_a of 29.6.[307] The acidity is clearly dependent on the electronic effect of the substituent. There are many dithioacetal derivatives (remember that dithioketals do not posses an acidic hydrogen), including both acyclic and cyclic species. For reasons of stability, availability and ease of preparation, dithianes are used most often.

A typical application is the reaction of 1,3-dithiane with an aldehyde to give **351** in 92% yield, taken from the synthesis of amphidinolide T3 by Zhao and co-workers.[375] Subsequent treatment of *tert*-butyllithium, and reaction with iodide shown gave **352** in 83% yield. To complete the reversal of reactivity, the carbonyl must be unmasked, which is usually done by reaction with a Lewis acid that has a high affinity for sulfur, such as mercuric oxide with boron trifluoride in aqueous THF.[376] In this particular example, treatment of **352** with iodine and sodium bicarbonate in aqueous acetone gave an 82% yield of **353**.[375] Many different oxidants have been employed to unmask dithianes, including Cl$_2$, Br$_2$, *t*-BuOCl, NBS, Pb(OAc)$_4$, Ce(NH$_3$)$_6$(NO$_3$)$_4$, Tl(O$_2$CCF$_3$)$_3$, and so on.[377] The choice of reagent depends on the facility of the hydrolysis and the sensitivity of the molecule to further reaction. Dithianes are widely used in natural product synthesis.[378]

373. Liu, Q.; Che, G.; Yu, H.; Liu, Y.; Zhang, J.; Zhang, Q.; Dong, D. *J. Org. Chem.* **2003**, *68*, 9148.
374. Bernardi, F.; Csizmadia, I.G.; Mangini, A.; Schlegel, H.B.; Whangbo, M.H.; Wolfe, S. *J. Am. Chem. Soc.* **1975**, *97*, 2209.
375. Deng, L.-S.; Huang, X.-P.; Zhao, G. *J. Org. Chem.* **2006**, *71*, 4625.
376. (a) Seebach, D.; Corey, E.J. *J. Org. Chem.* **1975**, *40*, 231; (b) Seebach, D. *Synthesis* **1969**, 17.
377. Romanet, R.F.; Schlessinger, R.H. *J. Am. Chem. Soc.* **1974**, *96*, 3701.
378. Yus, M.; Nájera, C.; Foubelo, F. *Tetrahedron* **2003**, *59*, 6147.

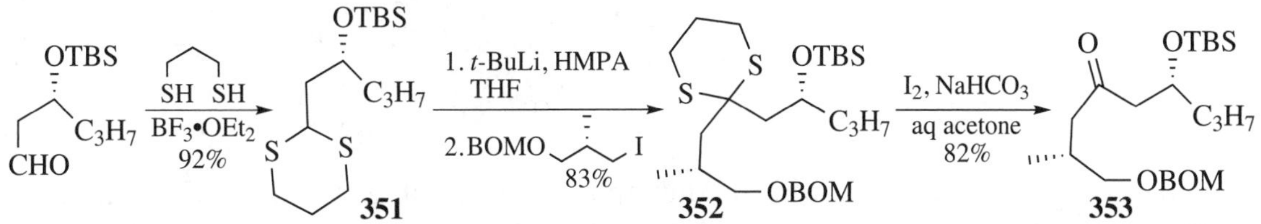

Lithiated dithianes react best with primary and secondary alkyl halides,[379] acyl halides, ketones and aldehydes,[380] as well as epoxides.[381] Nitriles and amides react only if there is no acidic proton. An important synthetic use of this reaction is the preparation of highly functionalized aldehydes using **349** (R = H) as a starting material.[382] This is an important route to β-hydroxy and alkoxy aldehydes, and many substituents can be incorporated.

The disconnection for the acyl anion umpolung is

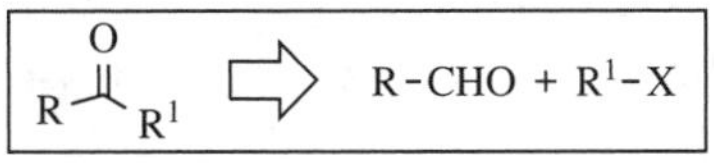

Dithiane anions can be used in more creative ways, and with electrophilic species such as epoxides. In the construction of an advanced intermediate of spongistatin,[383] Smith and co-worker's reacted 2-silyl dithiane **354** with *tert*-butyllithium to form the 2-lithio species. Subsequent reaction with epoxide **355** and then epoxide **356** led to initial coupling, followed by *in situ* loss of the silyl unit to allow coupling of the second epoxide giving a 69% yield of **357**. Smith termed this sequence a "linchpin assembly",[384] and the dithiane unit was converted to the ketone later in the synthesis. This two-directional approach was used with halides in Stockman and co-worker's synthesis of perhydrohistrionicotoxin.[385] 1,3-Dithiane was treated with butyllithium and then with halide **358**. This sequence was followed by a second round of butyllithium and then **358**, giving **359** in an overall yield of 70%. Later in this synthesis, the dithiane moiety was converted to the ketone by treatment with NCS and silver nitrate in aqueous acetonitrile.[385]

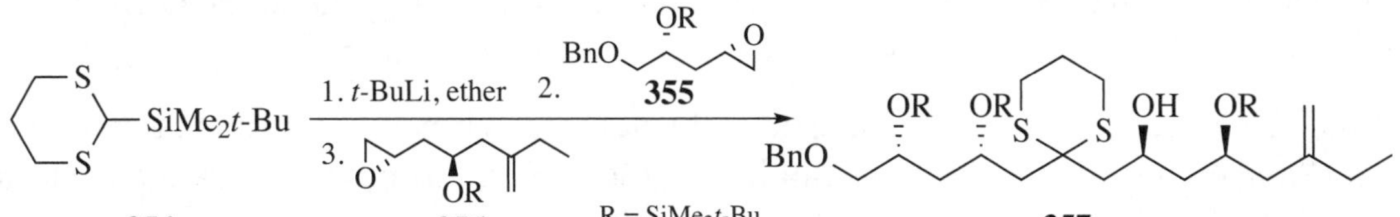

379. Corey, E.J.; Andersen, N.H.; Carlson, R.M.; Paust, J.; Vedejs, E.; Vlattas, I.; Winter, R.E.K. *J. Am. Chem. Soc.* **1968**, *90*, 3245.
380. (a) Corey, E.J.; Weigel, L.O.; Chamberlin, A.R.; Lipshutz, B. *J. Am. Chem. Soc.* **1980**, *102*, 1439; (b) Corey, E.J.; Weigel, L.O.; Floyd, D.; Bock, M.G. *Ibid.* **1978**, *100*, 2916.
381. Corey, E.J.; Bock, M.G. *Tetrahedron Lett.* **1975**, 2643.
382. Vedejs, E.; Fuchs, P.L. *J. Org. Chem.* **1971**, *36*, 366.
383. Smith III, A.B.; Doughty, V.A.; Sfouggatakis, C.; Bennett, C.S.; Koyanagi, J.; Takeuchi, M. *Org. Lett.* **2002**, *4*, 783.
384. Smith III, A.B.; Pitram, S.M.; Boldi, A.M.; Gaunt, M.J.; Sfouggatakis, C.; Moser, W.H. *J. Am. Chem. Soc.* **2003**, *125*, 14435.
385. Stockman, R.A.; Sinclair, A.; Arini, L.G.; Szeto, P.; Hughes, D.L. *J. Org. Chem.* **2004**, *69*, 1598.

Although conjugate addition is known, dithiane anions most often give 1,2-addition with conjugated systems. Cyclohexenone reacted with **350** (R = Me) to give the 1,2-addition product **360** in 99% yield.[386] In this case, treatment with aqueous acid led to a rearrangement that gave **361** in 83% yield. Unmasking the carbonyl with mercuric salts led to an 80% yield of **362**.[386] With stabilized carbanions of this type, 1,4-addition products are observed in some cases, but this may be due to a kinetic-thermodynamic phenomenon in which the initially formed 1,2-addition product equilibrates to a 1,4-adduct.[387] This equilibration can be induced by treatment with base. For example, Wilson and co-workers showed that treatment of **360** with 1-1.5 equivalents of KH in HMPA/THF led to a 23% yield of the 1,4-adduct (the 3-dithianylcyclohexanone).[388] The extent of 1,2- versus 1,4-addition observed is dependent on the reaction conditions employed.

As mentioned above, many reagents have been used for hydrolysis, which in part is due to variations in reactivity of the dithiane as the nature of the 2-alkyl group changes. Several problems can arise during hydrolysis. The reaction becomes acidic so $BaCO_3$, $CaCO_3$ or HgO are often added to maintain neutrality. Hydrolysis of dithianes derived from aldehydes often give acetals in neutral media. When an alkenyl group is present, addition of mercuric ion can give a competitive oxymercuration reaction (sec. 2.10.B). In general, $HgO/BF_3 \cdot OEt_2$ in aqueous organic solvents is used for hydrolysis of sensitive substrates.[386]

Acyl anion equivalents are not limited to dithianes, and virtually any dithioacetal can be used. Another common reagent is dithiolane (**363**),[389] prepared from an aldehyde and 1,2-ethanedithiol. A dithiolane such as **363** reacts with organolithium reagents to give 2-lithio-1,3-dithiolane derivative **364**, and reaction with an alkyl halide in the usual manner generates

386. (a) Corey, E.J.; Seebach, D. *Angew. Chem. Int. Ed.* **1965**, *4*, 1075, 1077; (b) Corey, E.J.; Crouse, D. *J. Org. Chem.* **1968**, *33*, 298.
387. (a) Schultz, A.G.; Yee, Y.K. *J. Org. Chem.* **1976**, *41*, 4044; (b) Deschamps, B.; Anh, N.T.; Seyden-Penne, J. *Tetrahedron Lett.* **1973**, 527; (c) Luchetti, J.; Krief, A. *Ibid.* **1978**, 2697; (d) Still, W.C.; Mitra, A. *Ibid.* **1978**, 2659.
388. (a) Wilson, S.R.; Misra, R.N.; Georgiadis, G.M. *J. Org. Chem.* **1980**, *45*, 2460; (b) Wilson, S.R.; Misra, R.N. *Ibid.* **1978**, *43*, 4903.
389. Herrmann, J.L.; Richman, J.E.; Schlessinger, R.H. *Tetrahedron Lett.* **1973**, 2599.

365. The reaction conditions are important since lithio dithiolanes such as **364** can fragment (see below). For this reason, dithiane derivatives are more commonly used as umpolung reagents as mentioned above. The same reagents that unmasked a dithiane to give back a carbonyl will convert a dithiolane to a carbonyl. Other reagents that function as acyl anion equivalents include **366**,[390] **367**,[391] and **368**.[392]

Schlessinger and co-workers used a mixed sulfide-sulfoxide derivative (**369**) to achieve the umpolung. The advantages of this reagent include ease of generation (the proton is more acidic and NaH, KH, LDA can be used), greater stability of the anion and lower cost. Since the anion is more stable it is less reactive, and the alkylation reactions generally require higher temperatures.[393] The diethyl derivative is superior to the dimethyl.[394] Acid hydrolysis is more facile, which is important since dithiane hydrolysis can be sluggish with highly substituted derivatives. The lithio-derivative **369** gave primarily Michael addition (1,4- or conjugate addition) in conjugated systems rather than the 1,2-addition usually observed with dithiane anions.[395] These examples are meant to show there are many possible acyl anion equivalents. The exotic reagents are used to fine-tune an umpolung reaction when other reagents are unsatisfactory. In general, relatively simple dithiane and dithiolane derivatives are used.

A major drawback to using dithiolane anions is their relative instability when compared with dithiane anions (see discussion of **360**, see above). Lithio-dithiolanes are stable only when R in **364** is an anion stabilizing group such as COR or CO_2R, but when used in conjunction with such electron-withdrawing groups, the dithiolane becomes very useful.

Enol ethers such as methyl 1-propenyl ether (1-methoxypropene, **370**) constitute a second major class of acyl anion equivalents. When **370** was treated with *tert*-butyllithium (note the need for a stronger base with the less acidic vinyl hydrogen) and then condensed with benzaldehyde, the product was **371**.[396] Lithiation of vinyl derivatives was described in Section 8.5. Facile

390. (a) Hackett, S.; Livinghouse, T. *J. Org. Chem.* **1986**, *51*, 879, 1629; (b) Trost, B.M.; Miller, C.H. *J. Am. Chem. Soc.* **1975**, *97*, 7182.
391. Trost, B.M.; Salzmann, T.N. *J. Am. Chem. Soc.* **1973**, *95*, 6840.
392. (a) Iriuchijima, S.; Maniwa, K.; Tsuchihashi, G. *J. Am. Chem. Soc.* **1975**, *97*, 596; (b) Carey, F.A.; Hernandez, O. *J. Org. Chem.* **1973**, *38*, 2670.
393. (a) Richmann, J.E.; Herrmann, J.L.; Schlessinger, R.H. *Tetrahedron Lett.* **1973**, 3267; (b) Ogura, K.; Tsuchihashi, G. *Ibid.* **1971**, 3151.
394. Carlson, R.M.; Helquist, P.M. *J. Org. Chem.* **1968**, *33*, 2596.
395. Herrmann, J.L.; Richman, J.E.; Schlessinger, R.H. *Tetrahedron Lett.* **1973**, 3271.
396. Baldwin, J.E.; Höfle, G.; Lever, Jr., O.W. *J. Am. Chem. Soc.* **1974**, *96*, 7125.

hydrolysis with aqueous acid liberated the corresponding ketone (**372**), completing the acyl anion equivalency. Schlosser and co-workers found that a mixture of *sec*-butyllithium and potassium *tert*-butoxide could be used to generate the lithium anion of *O*-tetrahydropyranyl enol ethers.[397] This modification generates a product that is more easily hydrolyzed to the ketone.

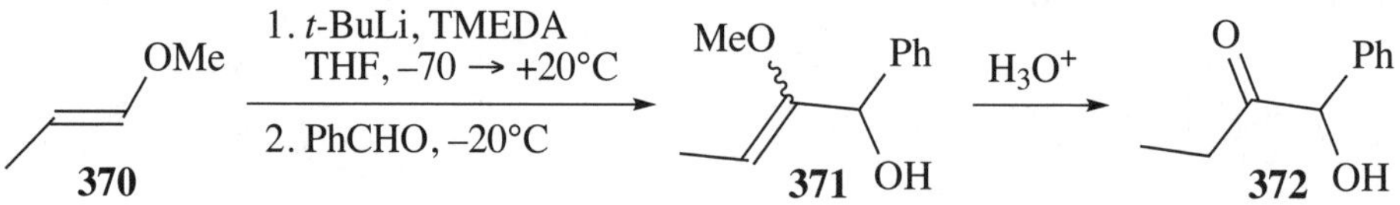

Vinyl sulfides can also function as acyl anion equivalents, but their preparation has not as yet been discussed. Wittig reagent such as **373** (sec. 8.8.A) reacted with benzenesulfenyl chloride to produce a new ylid (**374**), which condensed with benzaldehyde to give a 79% yield of the vinyl sulfide, **375**.[398] When **375** was treated with mercuric salts, a 73% yield of phenylacetone was obtained. In the second approach, a vinyltin compound (**376**)[399] was treated with methyllithium, and lithium-metal exchange provided **377**.[269] Alkylation with iodomethane gave a 91% yield of **378**, and treatment with titanium tetrachloride led to a 75% yield of 1,1-diphenylacetone.[269]

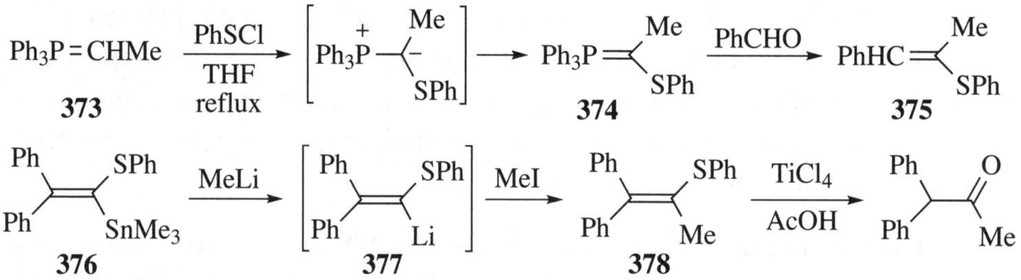

In both cases, sulfur stabilized the lithium derivative and reaction with alkyl halides proceeded in a manner analogous to other vinyllithium reagents. Hydrolysis required mercuric salts or another Lewis acid to give the ketone, as in the preparation of $Ph_2CHCOMe$ from **378**. Another preparation involves direct treatment of an alkylvinyl sulfide, such as ethylvinyl sulfide, with *sec*-butyllithium to give alkenyllithium (**379**). Subsequent treatment with an alkyl halide such as 1-bromooctane led to **380**, and hydrolysis with mercuric[400] (see sec. 7.3.B.ii for a related reaction with dithioketals and dithioacetals) gave ketone **381** in 90% yield.

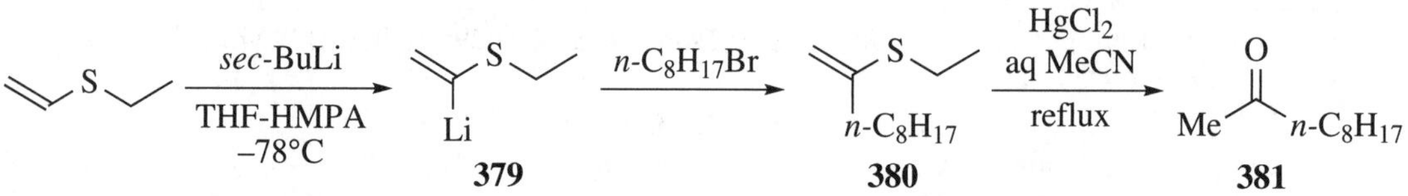

397. Hartmann, J.; Stähle, M.; Schlosser, M. *Synthesis* **1974**, 888.
398. (a) Corey, E.J.; Seebach, D. *J. Org. Chem.* **1966**, *31*, 4097; (b) Carey, F.A.; Court, A.S. *Ibid.* **1972**, *37*, 939.
399. For another synthetic example using stannane derivatives, see Hanessian, S.; Martin, M.; Desai, R.C. *J. Chem. Soc. Chem. Commun.* **1986**, 926.
400. Oshima, K.; Shimoji, K.; Takahashi, H.; Yamamoto, H.; Nozaki, H. *J. Am. Chem. Soc.* **1973**, *95*, 2694.

The disconnection for vinyl sulfides and enol ethers is

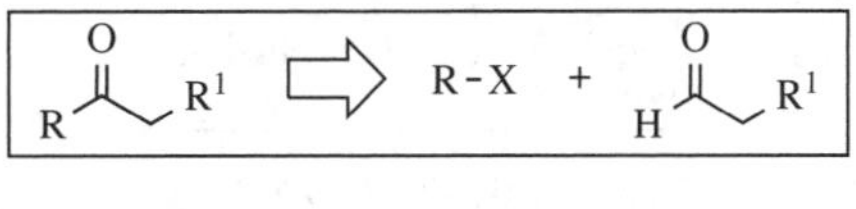

8.6.B.iii. A R(CO)CH₂⁺ Umpolung Equivalent. Reagents that reverse the usual nucleophilic character of the α-carbon to give **341** (see Table 8.17) are ketene dithioacetals (see **386**, **388**, and **392** below). These compounds can be prepared by several synthetic routes One of the more useful preparations is the **Peterson olefination**,[401] which is an alternative to the Wittig reaction (sec. 8.8.A) in which either trimethylsilylmethyllithium (**382**, X = Li) or its Grignard analog (X = MgBr, MgCl, etc.) reacts with aldehydes or ketones to give a β-silyl alcohol (**383**). Subsequent treatment with base induces elimination of the hydroxyl and the silyl group to give an alkene (in this case 1,1-diphenylethene, **384**).[401] Ketene dithioacetals can be prepared by Peterson olefination, as in the reaction of 2-trimethylsilyl-1,3-dithiane (**385**) with *n*-butyllithium, followed by reaction with an aldehyde (such as isobutyraldehyde to give **386** in 44% yield).[402]

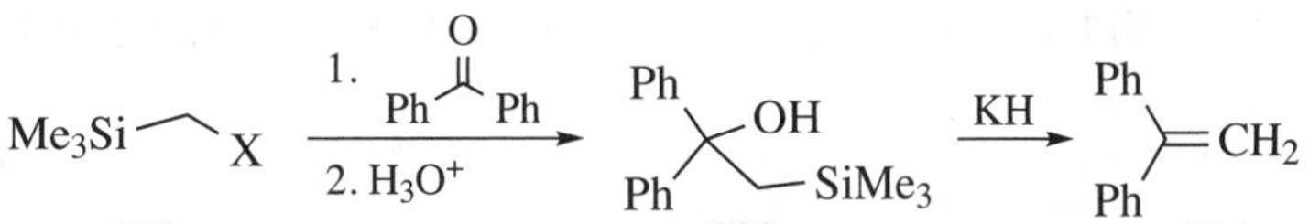

Dithiocarboxylic acid derivatives also provide a useful route to dithioketene acetals (another name for ketene dithioacetals).[403] Treatment of dithiopropionic acid (**387**) with excess lithium diisopropylamide (LDA, for *O*-alkylation reactions see sec. 9.3.B), followed by reaction with excess iodoethane led to a 97% yield of ketene dithioacetal **388**.[404] Ketene dithioacetals can be prepared by reaction of an aldehyde or ketone with dithioalkyl phosphonate esters such as **389**, which gave excellent yields of the dithioketene acetal upon reaction with aldehydes or ketones.[405] Reaction of **389** with acetophenone, for example, gave **390** in 81% yield. Corey and Märkl showed that phosphonate ylids such as **391** also react with compounds such as

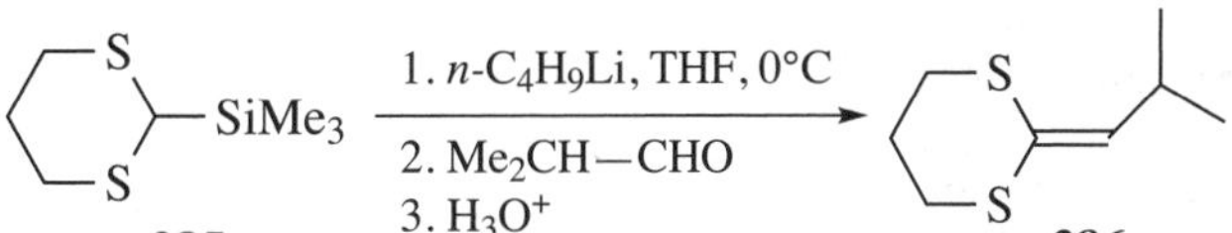

401. (a) Peterson, D.J. *J. Org. Chem.* **1968**, *33*, 780; (b) The Merck Index, 14th ed., Merck & Co., Inc., Whitehouse Station, New Jersey, **2006**, p ONR-71; (c) Mundy, B.P.; Ellerd, M.G.; Favaloro Jr., F.G. *Name Reactions and Reagents in Organic Synthesis, 2nd ed.*, Wiley-Interscience, New Jersey, **2005**, pp. 496-497.
402. (a) Carey, F.A.; Court, A.C. *J. Org. Chem.* **1972**, *37*, 1926; (b) Trost, B.M.; Kunz, R.A. *J. Am. Chem. Soc.* **1975**, *97*, 7152;
403. Beiner, J.M.; Thuillier, A. *C.R. Acad. Sci., Ser. C* **1972**, *274*,642.
404. Ziegler, F.E.; Chan, C.M. *J. Org. Chem.* **1978**, *43*, 3065.
405. Mikolajczyk, M.; Grzejszczark, S.; Zatorski, A.; Mlotkowska, B.; Gross, H.; Costisella, B. *Tetrahedron* **1978**, *34*, 3081.

benzaldehyde to give the desired product, in that case **392**.[406]

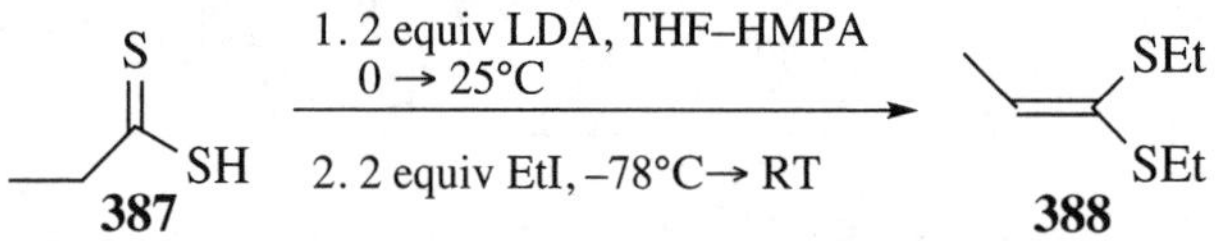

The actual umpolung reaction that allows a ketene dithioacetal to function as if it had an electrophilic carbon α to a carbonyl is achieved by conjugate addition of nucleophiles to the ketene dithioacetal, followed by hydrolysis. Both sulfide and sulfoxide ketene derivatives can be used. Conjugate addition of an ester enolate derived from *tert*-butyl acetate (secs. 9.2, 9.4.B, 9.7.A) to the ketene dithioacetal [CH_2=$C(SOMe)_2$] gave the stable dithioacetal anion (**393**). Conversion of the dithioacetal to the bis(sulfoxide) enhanced the ability of that species to function as a Michael acceptor. Subsequent transformation of **393** gave the aldehyde-ester (**394**).[407]

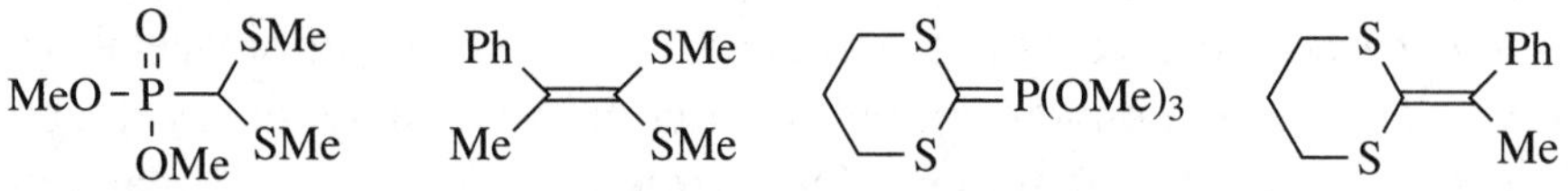

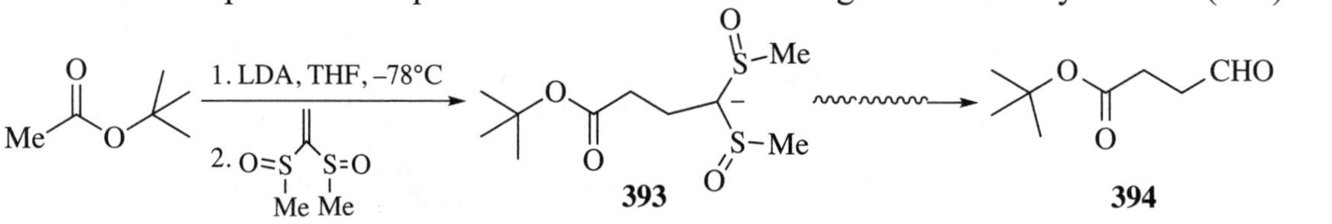

The disconnection for this process is:

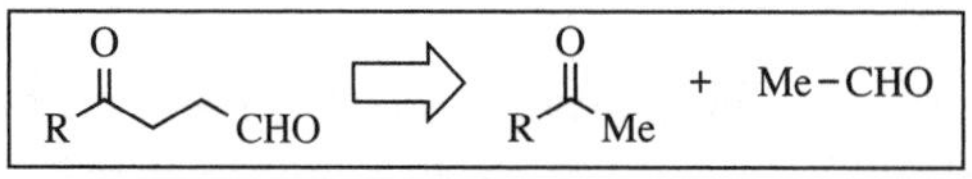

8.6.B.iv. R(CO)CH$_2$CH$_2$⁻ Umpolung Equivalents. In this umpolung equivalent (**342**, see Table 8.17), the reactive center is more distant from the carbonyl, making it more difficult to generate the umpolung reagent. Ketene dithioacetals can be used for the umpolung equivalent of **342**, although the yields tend to be low. When the ketene dithioacetal possesses an allylic hydrogen (as in **395**), treatment with a strong base generates a resonance-stabilized anion (**396**). Alkylation with an appropriate reagent (E–X) generates two products (**398** and **399**) from each of the two carbanionic centers. Two sulfur atoms stabilize the carbanion in **396**, so **399** usually predominates unless R² in **395** is an electron withdrawing group. In that case, **396A** may be stabilized to a greater extent giving **398** as the major product. Hydrolysis of **399** generates **400**. Only **398** will yield the umpolung product (**397**) upon hydrolysis, suggesting that the umpolung is observed as the major process only when R² is an electron-withdrawing

406. (a) Corey, E.J.; Märkl, G. *Tetrahedron Lett.* **1967**, 3201; (b) Lemal, D.M.; Banitt, E.H. *Ibid.* **1964**, 245.
407. Herrmann, J.L.; Kieczykowski, G.R.; Romanet, R.F.; Wepplo, P.J.; Schlessinger, R.H. *Tetrahedron Lett.* **1973**, 4711.

group. These principles are illustrated by the reaction of the ketene dithioacetal (**395**) with *n*-butyllithium to give a resonance-stabilized anion, which could be alkylated at either the α- (**399**) or the γ-position (**398**).[372c,408] When R^2 = alkyl, the anion adjacent to the sulfur atoms is the most prevalent and dominates alkylation reactions. As mentioned above, the desired umpolung equivalent is obtained only when the anion (which gives **398**) is stabilized by electron-withdrawing groups at R^2, or when a hindered electrophile is used.[408] When R^2 is cyano, the major product was **398**, but when R^2 was *n*-propyl, the major product was **399**.[408]

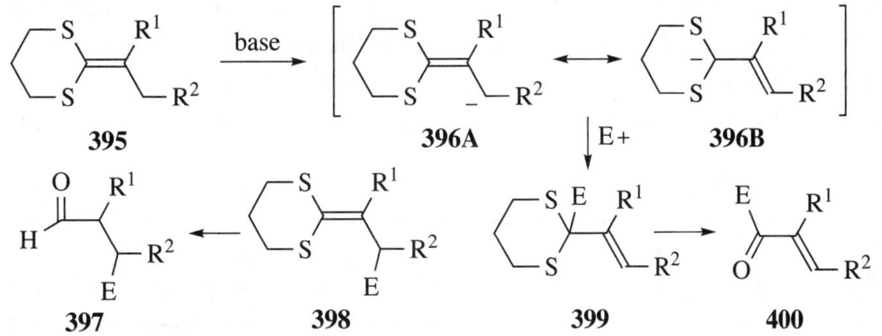

A more efficient method to generate this umpolung equivalent is via the dianion of allylic thiols. Treatment of allylthiol (2-propene-1-thiol, **401**) with base removes the acidic S–H hydrogen, and a second equivalent of strong base removes the hydrogen from the α-carbon to give **402**. The course of the subsequent alkylation reaction depends on the metal counterion of the base.[409] When M = Li (from *n*-butyllithium) alkylation occurred at both the α and γ carbanionic centers (to give **404** and **405**, respectively), but γ-alkylation predominated (a 68:31 mixture favoring **404** in 93% yield).[409,410] When M = Mg (magnesium *tert*-butoxide was added to effect the Li–Mg exchange reaction), substitution occurred predominantly at the carbon α to sulfur, giving **403**.[411] Hydrolysis of **404** with mercuric salts liberated the aldehyde. There is a great preference for reaction at the α-position unless that position is sterically hindered. The solvent, the nature of the R group on sulfur and the nature of the electrophile play a role. Allylic sulfides are used more often than allylic thiols. The monoanion derived from an allylic sulfide also gives α and γ substitution.[412,354b]

408. Meyers, A.I.; Nolen, R.L.; Collington, E.W.; Narwid, T.A.; Strickland, R.C. *J. Org. Chem.* **1973**, *38*, 1974.
409. Geiss, K.-H.; Seuring, B.; Pieter, R.; Seebach, D. *Angew. Chem. Int. Ed.* **1974**, *13*, 479.
410. Hartmann, J.; Muthukrishnan, K.; Schlosser, M. *Helv. Chim. Acta* **1974**, *57*, 2261.
411. Seebach, D.; Geiss, K.-H.; Pohmakotr, M. *Angew. Chem. Int. Ed.* **1976**, *15*, 437.
412. Atlani, P.M.; Biellmann, J.F.; Dube, S.; Vicens, J.J. *Tetrahedron Lett.* **1974**, 2665.

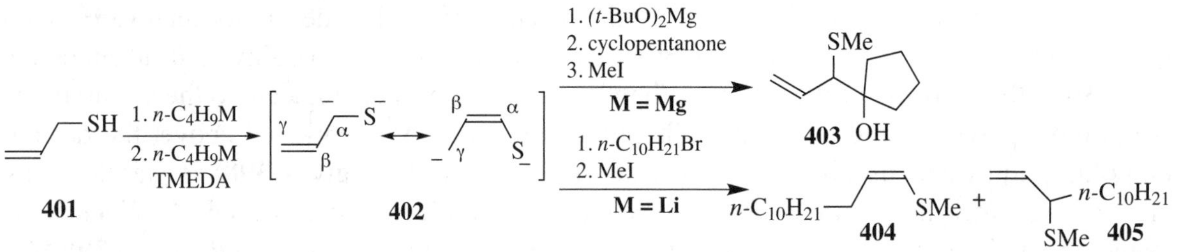

In the solvent THF, γ-substitution predominated in reactions of **402**, and addition of DABCO increased that yield, presumably by coordination near the sulfur to sterically block that position. Addition of HMPA favors more α-substitution, and addition of a strongly coordinating species such as a cryptate blocks the γ-position, which leads to α-substitution as the sole product.

Reasonable disconnections for this process are

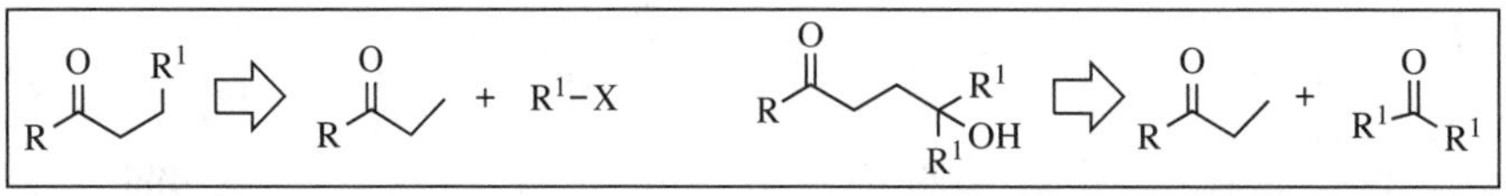

8.6.B.v. A R(C=O)CH$_2$CH$_2$CH$_2^+$ Equivalent. In this umpolung equivalent, the potentially positive center is even further removed from the carbonyl (see **343** in Table 8.17). This umpolung equivalent is best achieved by reaction of a nucleophile with a vinylogous ketene dithioacetal such as **406**.[413] Initial reaction of **406** with *n*-butyllithium proceeds via Michael type addition rather than deprotonation (there are no acidic hydrogens) to give the ketene dithioacetal anion (**407**). In order to achieve this umpolung equivalent, the ketene dithioacetal *cannot* have an acidic hydrogen present in the dithioacetal unit, since it would be selectively deprotonated. Addition of an alkyl halide such as iodomethane leads to alkylation at the dithiane carbon (to give **408**), and hydrolysis of the dithiane with mercuric chloride leads to the conjugated ketone **409**.[413] As mentioned earlier, this umpolung reagent is somewhat limited in utility because it cannot have acidic hydrogen atoms elsewhere in the molecule. It is important to mention what is becoming a pattern. As the umpolung center is removed further from the carbonyl, preparation of the requisite reagents becomes more difficult.

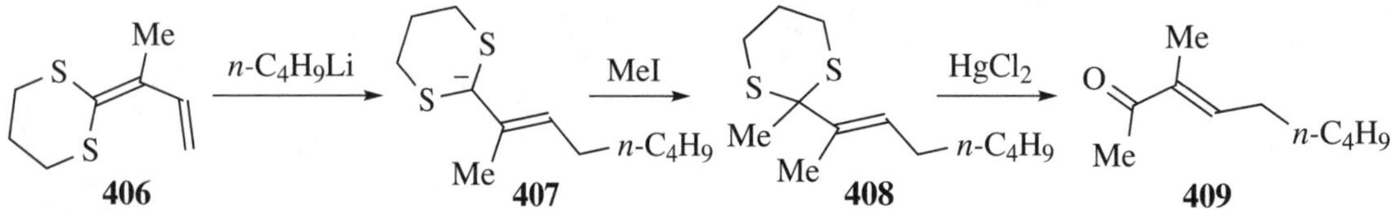

The disconnection is

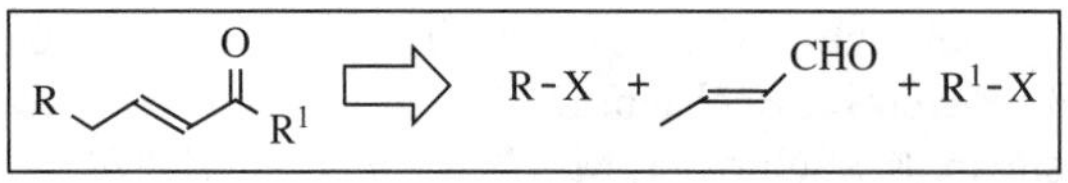

---

413. Seebach, D.; Kolb, M.; Gröbel, B.-T. *Angew. Chem. Int. Ed.* **1973**, *12*, 69.

8.6.B.vi. An R(C=O)CH₂CH₂CH₂CH₂⁻ Equivalent. The final umpolung equivalent in Table 8.17 is **344**, where the reactive center is separated from the carbonyl by three carbons. This umpolung equivalent is relatively easy to obtain, however, by the use of an allyl vinyl sulfide such as **410**. Upon heating, a thio-Claisen rearrangement (sec. 11.12.D.iii) accomplishes the desired umpolung transformation.[414] A generalized example is treatment of **410** with base, followed by quenching with an electrophile (E⁺) to give **411**. Heating leads to a thio-Claisen rearrangement and formation of thiocarbonyl derivative **412**. Subsequent hydrolysis converts the thiocarbonyl to the carbonyl unit in **413**. A simple example is the reaction of allyl vinyl sulfide **414** with *sec*-butyllithium to give the reactive α-lithio sulfide, and this reacted with benzyl bromide to give **415**. Heating in aqueous dimethoxyethane led to the thio-Claisen product **416**, and hydrolysis provided aldehyde **417** in 62% overall yield.[415] The specificity for the trans alkene in **416** arises from the chair-like transition state (see **415**, secs. 11.12.C, 11.12.D), in which the benzyl group is equatorial in the lowest energy conformation leading to product.

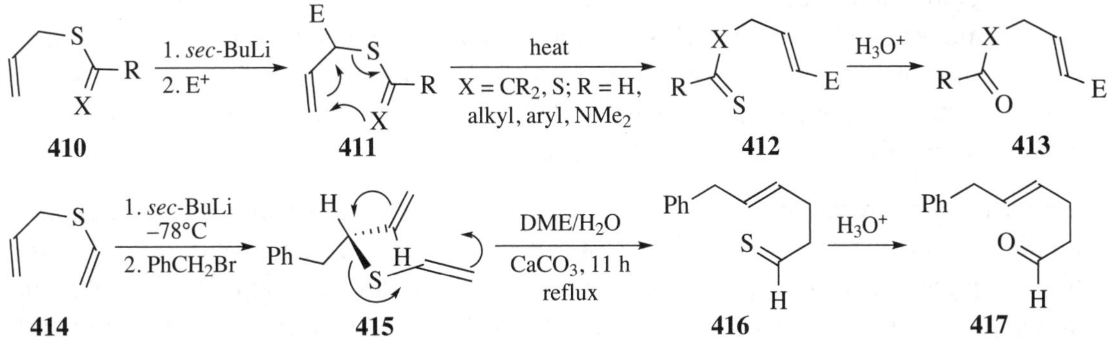

The disconnection for this Umpolung is:

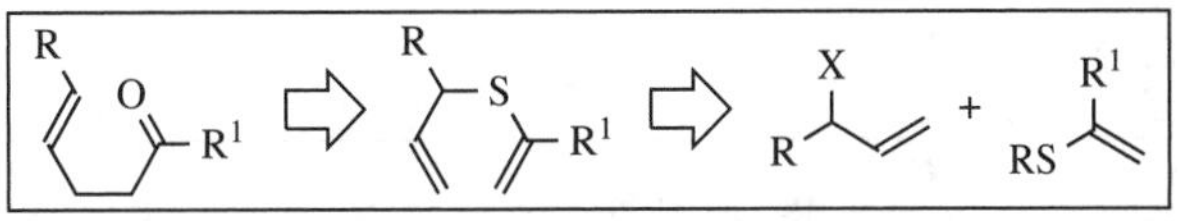

The ability to alter normal reactivity patterns of carbonyl derivatives allows a variety of carbon-carbon bond forming processes to proceed. The products derived from these reactions can often be obtained by no other route and provide great flexibility in synthetic planning.

8.7   ORGANOCOPPER REAGENTS (C–Cu)

In all previous sections, the Cᵈ reagent used as a nucleophilic source of carbon was an organometallic derived from sodium, potassium, magnesium or lithium reagents. In this section the focus will be on a class of reagents that contains a carbon-copper bond, the lithium

---

414.   Corey, E.J.; Shulman, J.I. *J. Am. Chem. Soc.* **1970**, *92*, 5522.
415.   Oshima, K.; Takahashi, H.; Yamamoto, H.; Nozaki, H. *J. Am. Chem. Soc.* **1973**, *95*, 2693.

dialkylcuprates (R'$_2$CuLi).

$$2 \; n\text{-BuLi} + \text{Cu}^{\text{I}}\text{I} \xrightarrow[-20°C]{\text{ether}} \text{Bu}_2\text{CuLi}$$

8.7.A. Lithium Dialkylcuprates

8.7.A.i. Preparation of Gilman Reagents. Gilman et al. first observed that reaction of cuprous salts with two equivalents of an organolithium reagent generated an organocuprate such as lithium di-*n*-butylcuprate (Bu$_2$CuLi).[416,94] The reagent formed is usually called a **Gilman reagent**, and is drawn as R$_2$CuLi.[416,94] The mechanism[417] of the reaction probably involves an intermediate and conversion of a Cu(I) species to a transient Cu(III) species, and it may proceed via one-electron transfer.[418] The stability of organocopper species varies considerably with their structure. Whitesides and House[419] showed that dimethylcuprate (Me$_2$CuLi) was stable in ether for hours, at 0°C under nitrogen. Secondary and tertiary cuprates, however, rapidly decompose in ether above –20°C.[419] Disproportionation is the primary decomposition route for organocuprates, and the halide also plays a role in the stability of the cuprate. Cuprous iodide gives better results than cuprous bromide,[419] but the dimethyl sulfide-cuprous bromide complex (Me$_2$S•CuBr) has been used with success.[420]

There are two major reactions of organocuprates as a carbon nucleophile:[421] (*1*) reaction with alkyl halides and (*2*) conjugate addition with α,β-unsaturated ketones. With α,β-unsaturated ketones, conjugate addition is promoted when ether is used as a solvent.[418] The substitution reaction is promoted by the use of THF or ether-HMPA as a solvent.[418] As mentioned earlier, the mechanism of these reactions probably involves a one-electron transfer, although other mechanistic proposals are in the literature.[422] The general reactivity of organocuprates with electrophiles follows the order:[418]

416. Gilman, H.; Jones, R.G.; Woods, L.A. *J. Org. Chem.* **1952**, *17*, 1630.
417. For an extensive discussion of the mechanism of reaction between organocuprates and alkyl haldies or epoxides, see Mori, S.; Nakamura, E.; Morokuma, K. *J. Am. Chem. Soc.* **2000**, *122*, 7294.
418. Posner, G.H. *An Introduction to Synthesis Using Organocopper Reagents*, Wiley, New York, **1980**.
419. Whitesides, G.M.; Fischer, Jr., W.F.; San Filippo, Jr., J.; Bashe, R.W.; House, H.O. *J. Am. Chem. Soc.* **1969**, *91*, 4871.
420. House, H.O.; Wilkins, J.M. *J. Org. Chem.* **1978**, *43*, 2443.
421. (a) Ashby, E.C.; DePriest, R.N.; Tuncay, A.; Srivastava, S. *Tetrahedron Lett.* **1982**, *23*, 5251; (b) Posner, G.H.; Ting, J.-S.; Lentz, C.M. *Tetrahedron* **1976**, *32*, 2281; (c) Johnson, C.R.; Dutra, G.A. *J. Amer. Chem. Soc.* **1973**, *95*, 7777, 7783; (d) Smith, R.A.J.; Hannah, D.J. *Tetrahedron* **1979**, 35, 1183; (e) Castro, C.E.; Havlin, R.; Honwad, V.K.; Malte, A.; Mojé, S. *J. Am. Chem. Soc.* **1969**, 91, 6464; (f) House, H.O.; Koespell, D.G.; Campbell, W.J. *J. Org. Chem.* **1972**, *37*, 1003.
422. Nakamura, E.; Mori, S. *Angew. Chem. Int. Ed.* **2000**, *39*, 3750.

The R groups can be primary, secondary, tertiary alkyl, aryl, or heteroaryl and can bear remote functionality such as ethers, acetals or ketals, sulfides, or other labile groups.[418]

8.7.A.ii. Coupling Reactions. In previous sections, Grignard reagents and organolithium reagents usually gave poor yields of coupling products in a reaction with an alkyl halide unless catalyzed by transition metals. Cuprous salts were used with Grignard reagents to catalyze coupling reactions (sec. 8.4.B.i). Addition of cuprous salts to organolithium reagents gave organocuprates (R'_2CuLi), the active species in the coupling reaction, and this has become the

$$R-X \ + \ R^1_2CuLi \ \xrightarrow[-20°C]{ether} \ R-R^1$$

preferred method for the formation of coupled products (R–R') from alkyl halides.[423] The reactivity of the cuprate is due in large part to the low ionic character of the Cu–C bond; the low oxidation potential (0.15 eV) separating Cu(I) and Cu(II) ions, the tendency for Cu to form polynuclear Cu clusters, and the formation of mixed-valence Cu compounds.[418] The coupling reacation of organocuprates with alkyl halides, tosylates or acetates is straightforward, but the mechanistic details have been a source of some controversy. Primary alkyl halides generally give good yields of the coupling product (R–R^1) in a Wurtz-type process[424] (sec. 8.5.B.ii). Table 8.18[419] shows the relative reactivity of di-*n*-butylcuprate with various organohalides to give the alkane R–R'. The reaction proceeded faster and in higher yield in THF than in ether. From this study the order of reactivity of the halide substrate was determined to be RI > RBr > RCl. The secondary halide gave reduced yields, and tertiary halides gave virtually no reaction under conditions that produced large amounts of product from primary halides. Sulfonate esters such as mesylates or tosylates and also acetates are susceptible to displacement by organocuprates.[425]

Table 8.18. Relative Reactivity of Lithium Di-*n*-butyl Cuprate with Alkyl Halides

$$R-X \ + \ \textit{n}-Bu_2CuLi \ \xrightarrow{ether, -20°C} \ R-\textit{n}-Bu$$

R–X	Solvent	Time (h)	% R–*n*-Bu
1-Iodopentane	ether	1	68
	ether	26	70
	THF	1	98
1-Bromopentane	THF	1	98
1-Chloropentane	THF	1	80
	ether	1	10
	ether	26	10
2-Bromopentane	THF	1	12
	THF	26	12
2-Bromo-2-methylpentane	THF	1	<10

[Reprinted with permission from Whitesides, G.M.; Fischer Jr., W.F.; San Filippo Jr., J.; Bashe, R.W.; House, H.O. *J. Am. Chem. Soc.* **1969**, *91*, 4871. Copyright © **1969** American Chemical Society.]

---

423. For reactions of this type, see Larock, R.C. *Comprehensive Organic Transformations, 2nd Ed.* Wiley-VCH, New York, **1999**, pp 118-122.

424. House, H.O. *Acc. Chem. Res.* **1976**, *9*, 59.

425. (a) Posner, G.H.; Ting, J.S.; Lentz, C.M. *Tetrahedron* **1976**, *32*, 2281; (b) Posner, G.H.; Ting, J.S. *Tetrahedron Lett.* **1974**, 683.

The coupling reaction is extremely useful for joining alkyl fragments or alkyl-aryl, aryl-aryl, or heteroaryl fragments by reaction with alkyl halides or sulfonate esters.[426] In a typical synthetic application, the cuprate derived from **418** (via Li–Br exchange with *tert*-butyllithium and reaction with Li_2CuCl_4 to form the aryl cuprate) reacted with the allylic bromide shown to give a 74% yield of **419** in the Overman and co-workers synthesis of the kinesin motor protein inhibitor adociasulfate 1.[427] Note that organocuprates react with secondary bromides with inversion whereas the identical reaction with a secondary iodide proceeds with racemization. In this synthesis, TBDMS is *tert*-butyldimethylsilyl (see sec. 7.3.A.i).

If there is excessive steric encumbrance when an organocuprates reacts with an allylic system, the coupling may proceed by a S_N2' like pathway (sec. 2.6.A.iii).[428] A synthetic example is taken from a synthesis of paroxetine (PAXIL) by Krische and Koech in which allylic phosphonate **420** reacted with the cuprate derived from $ClSiMe_2(Oi\text{-}Pr)$ to give the S_N2' product **421** in 96% yield.[429] Cyclopropylcarbinyl halides react with organocuprates[430] such as di-*n*-butylcuprate to give the (*E*)-homoallylic compound.

In addition to the expected displacement of alkyl halides, organocuprates react with vinyl halides and aryl halides to give good yields of the coupled product.[431] This alkenyl halide coupling reaction proceeds with high stereoselectivity, in contrast to the Grignard and organolithium coupling reactions. (*E*)-2-Bromostyrene (**422**) reacted with diphenylcuprate to give a 90:<2 mixture of (*E/Z*) stilbenes (**423/424**).[424] Similarly, the (*Z*)-bromide (**425**) gave a <1:73 mixture of **423/424**.[424] It is apparent that the coupling proceeded with little or no loss of stereochemical

426. Kojima, Y.; Wakita, S.; Kato, N. *Tetrahedron Lett.* **1979**, 4577.
427. Bogenstätter, M.; Limberg, A.; Overman, L.E.; Tomasi, A.L. *J. Am. Chem. Soc.* **1999**, *121*, 12206.
428. (a) Magid, R.M. *Tetrahedron* **1980**, *36*, 1901; (b) DeWolfe, R.H.; Young, W. *Chem. Rev.* **1956**, *56*, 753; (c) Goering, H.L.; Singleton, Jr, V.D. *J. Org. Chem.* **1983**, *48*, 1531.
429. Koech, P.K.; Krische, M.J. *Tetrahedron* **2006**, *62*, 10594.
430. (a) Hrubiec, R.T.; Smith, M.B. *Tetrahedron* **1984**, *40*, 1457; (b) Hrubiec, R.T.; Smith, M.B. *J. Org. Chem.* **1984**, *49*, 385.
431. (a) Bowlus, S.B.; Katzenellenbogen, J.A. *J. Org. Chem.* **1973**, *38*, 2733; (b) Cooke, Jr., M.P. *Tetrahedron Lett.* **1973**, 1983.

integrity of the vinyl bromide. Vinyl sulfonate esters such a triflate derivatives [triflate = OTf = -SO$_2$CF$_3$] also react with organocuprates to give the coupling product.[432] A mechanism of "oxidative" addition between a cuprate and an alkenyl halide has been proposed.[433] The initially formed π-complex is thought to behave as a cuprio(III)cyclopropane[434] rather than a simple CuI/alkenyl halide complex, where charge transfer takes place from the 3d orbital of the bent R$_2$Cu moiety to the π*/σ*-mixed orbital of a deformed alkenyl halide. Subsequent C–Br bond cleavage in this complex may go through the "three-centered" or the "eliminative" pathway, of which the latter is preferred.

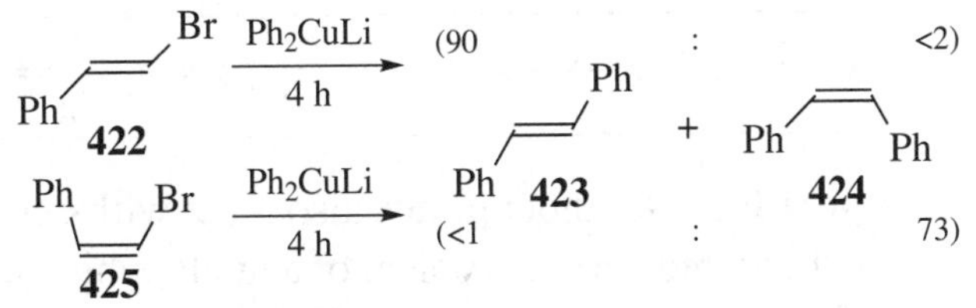

Coupling reactions with alkyl halides often use a large excess of cuprate as mentioned above, which can be a severe limitation if the halide precursor to the organocuprate is difficult to prepare or unavailable from commercial sources. A solution is to prepare a cuprate with a ligand that is less reactive, allowing selective transfer of the substituent of choice. An example is the use of **428** in a reaction with **429** that gave **430**, an intermediate in Corey and co-worker's synthesis of *N*-methylmaytansine.[435,381] The alkynyl-copper bond in species such as **428** is very stable relative to alkyl-copper and alkenyl-copper bonds. When an alkyl-alkynyl cuprate or an alkenyl-alkynyl cuprate is prepared, the alkyl or alkenyl group will be transferred selectively. Cuprate **428** was prepared by reaction of the Cu(I) species **426** with vinyl bromide **427** in the presence of *n*-butyllithium.[435] It is noteworthy that only the alkenyl group was transferred to the benzylic iodide, and not the alkynyl moiety. In general, this method allows a synthetically prepared halide such as **427**, which may be difficult to obtain in pure form, to be coupled to another halide with greater efficiency.

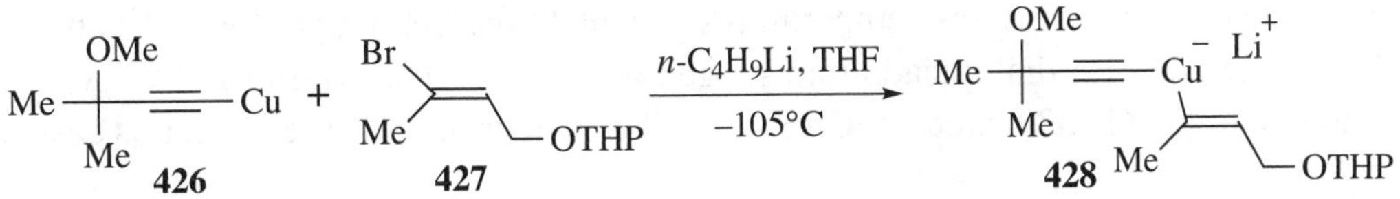

432. Corey, E.J.; Das, J. *J. Am. Chem. Soc.* **1982**, *104*, 5551.
433. Yoshikai, N.; Nakamura, E. *J. Am. Chem. Soc.* **2004**, *126*, 12264.
434. Mori, S.; Nakamura, E.; Morokuma, K. *Organometallics* **2004**, *23*, 1081.
435. (a) Corey, E.J.; Wetter, H.F.; Kozikowski, A.P.; Rao, A.V.R. *Tetrahedron Lett.* **1977**, 777; (b) Corey, E.J.; Bock, M.G.; Kozikowski, A.P.; Rao, A.V.R.; Floyd, D.; Lipshutz, B. *Ibid.* **1978**, 1051.

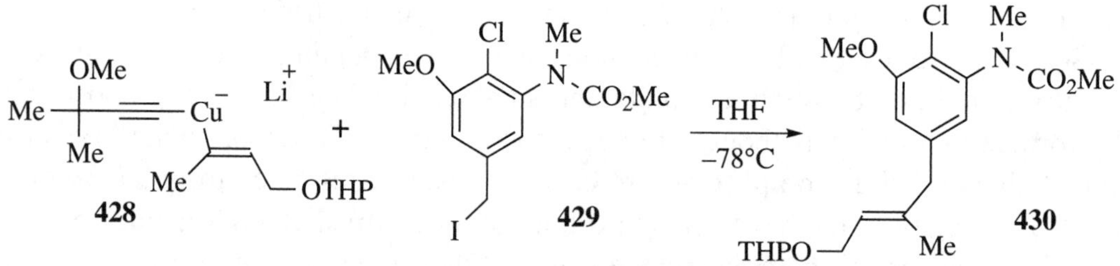

The alkyl or aryl halide -organocuprate coupling disconnection is

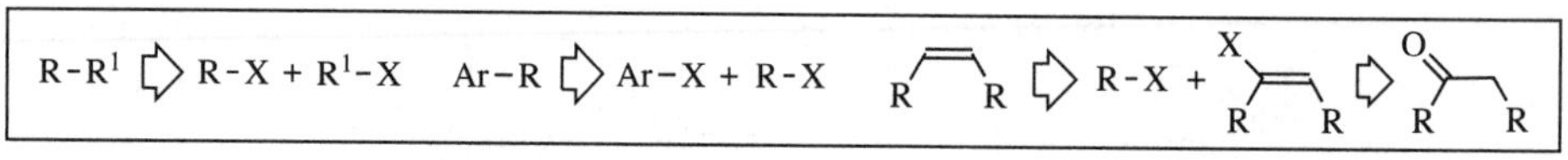

8.7.A.iii. Reaction with Epoxides. Organocuprates also react with epoxides,[436] with addition occurring at the less sterically hindered carbon, typical of a nucleophilic addition. In the Toyota and co-worker's synthesis of (+)-mycalamide A, dimethylcuprate reacted with epoxide **431** to give **432** in >90% yield.[437] The ring opening was highly selective for the alcohol shown.

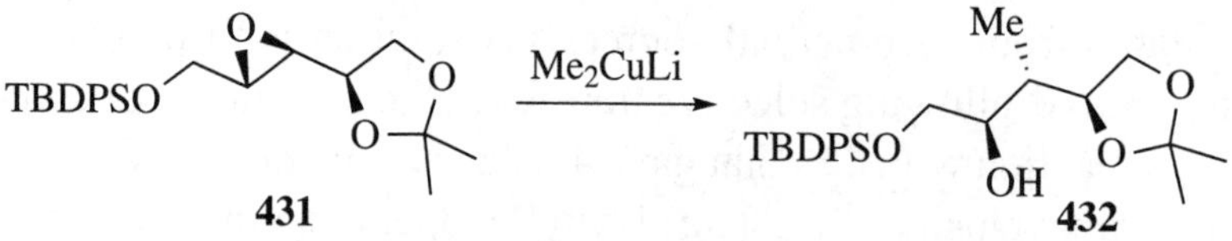

The epoxide-organocuprate disconnection is

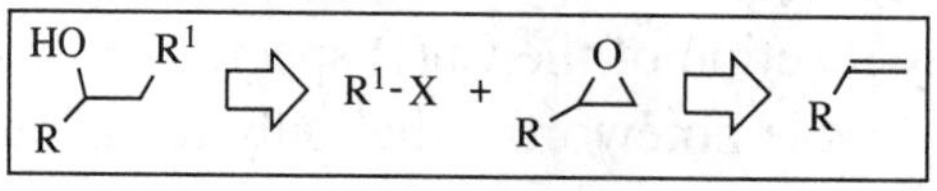

8.7.A.iv. Reaction with Acid Chlorides. As briefly mentioned in Section 8.4.C.ii, the reaction of a dialkyl cuprate and an acid chloride is an excellent method for the synthesis of ketones.[438,439] The reaction usually requires low temperatures to isolate the ketone product[425,424] and is more general than the use of a dialkylcadmium reagent (sec. 8.4.C.ii). Masamune et al. used this reaction to convert acid **433** to ketone **434** in a synthesis of erythronolide.[440] In this case, oxalyl

436. (a) Herr, R.W.; Wieland, D.M.; Johnson, C.R. *J. Am. Chem. Soc.* **1970**, *92*, 3813; (b) Staroscik, J.; Rickborn, B. *Ibid.* **1971**, *93*, 3046; (c) Wieland, D.M.; Johnson, C.R. *Ibid.* **1971**, *93*, 3047; (d) Anderson, R.J. *Ibid.* **1970**, *92*, 4978; (e) Herr, R.W.; Johnson, C.R. *Ibid.* **1970**, *92*, 4979; (e) Anderson, R.J.; Adams, K.G.; Chinn, H.R.; Henrick, C.P. *J. Org. Chem.* **1980**, *45*, 2229.
437. Kagawa, N.; Ihara, M.; Toyota, M. *J. Org. Chem.* **2006**, *71*, 6796.
438. Fex, T.; Froborg, J.; Magnusson, G.; Thorén, S. *J. Org. Chem.* **1976**, *41*, 3518.
439. For reactions of this type, see Larock, R.C. *Comprehensive Organic Transformations, 2nd Ed.* Wiley-VCH, New York, **1999**, pp 1389-1400.
440. (a) Masamune, S.; Choy, W.; Kerdesky, F.A.J.; Imperiali, B. *J. Am. Chem. Soc.* **1981**, *103*, 1566; (b) Masamune, S.; Hirama, M.; Mori, S.; Ali, Sk.A.; Garvey, D.S. *Ibid.* **1981**, *103*, 1568.

chloride was used to prepare the requisite acid chloride.

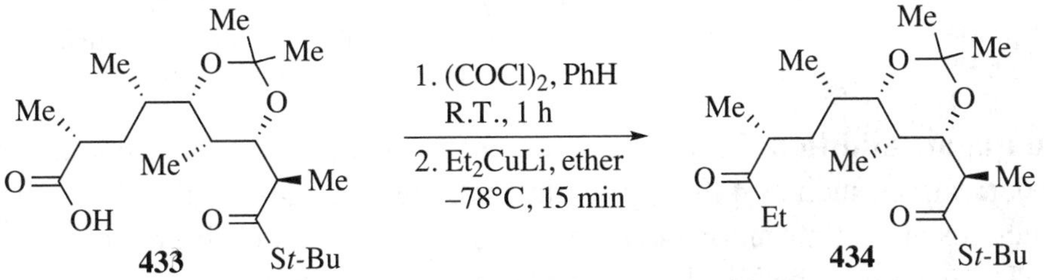

The acid chloride-organocuprate disconnection is

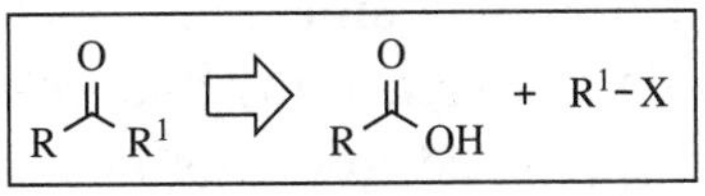

8.7.A.v. Reaction with Aldehydes and Ketones.[441] The reaction of organocuprates with aldehydes is fast. Dimethylcuprate reacted with benzaldehyde to give phenylethyl alcohol at < –90°C. Ketones react more slowly, as illustrated by the reaction of dimethylcuprate and 5-nonanone which occurred at about –10°C to give 5-methyl-5-nonanol.[442] No reaction occurred at lower temperatures. Higher temperatures are often required for reaction of hindered aldehydes, leading to significant decomposition of the organocuprate via disproportionation before the addition reaction occurs. Still showed a typical cuprate addition to an aldehyde when dimethylcuprate reacted with **435** (at –78°C) to give **436**.[443] In this case the excellent diastereoselectivity (30:1 anti:syn **436**) was the result of chelation control (secs. 4.7.B, sec. 6.2.E). The cuprate addition usually follows Cram's rule (sec. 4.7.B), in the absence of chelation effects. Di-*n*-butylcuprate reacted with **437** in the presence of chlorotrimethylsilane to give a 6.4:1 mixture of **438/439** (**439** is predicted by Cram's rule, the Cram product).[444] As mentioned above, ketones react very slowly with cuprates. Addition of chlorotrimethylsilane to the reaction can alleviate this problem (this was used in the reaction of **437**). With conjugated ketones, addition of chlorotrimethylsilane enhances the rate of 1,2-addition. It also enhances Cram diastereoselectivity when an organocuprate reacts with a chiral aldehyde.[444] There is a clear enhancement of rate, and this is probably not just a consequence of trapping the enolate anion that results from conjugate addition. The precise mechanism is unclear but may be related to an equilibration reaction for organocuprate conjugate additions.[443]

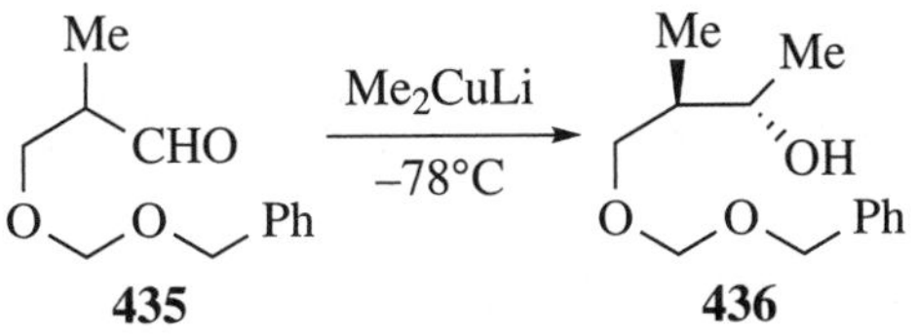

441. For reactions of this type, see Larock, R.C. *Comprehensive Organic Transformations, 2nd Ed.* Wiley-VCH, New York, *1999*, pp 1150-1151.
442. Posner, G.H.; Whitter, C.E.; McFarland, P.E. *J. Am. Chem. Soc. 1972, 94,* 5106.
443. Still, W.C.; McDonald III, J.H. *Tetrahedron Lett. 1980, 21,* 1031.
444. Matsuzawa, S.; Isaka, M.; Nakamura, E.; Kuwajima, I. *Tetrahedron Lett. 1989, 30,* 1975 and references cited therein.

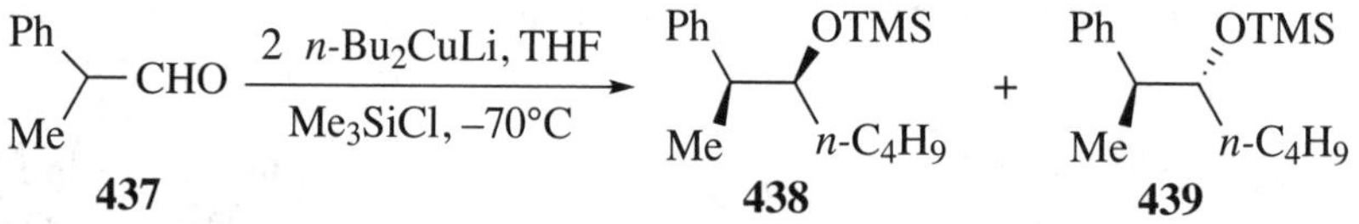

8.7.A.vi. Conjugate Addition. Organocuprates react with α,β-unsaturated conjugated aldehyde and ketone derivatives such as **440** to give almost exclusive 1,4-addition.[445] Initial conjugate addition generates an enolate anion (**441**, secs. 9.2.A, 9.7.A), and subsequent hydrolysis gives ketone **442**. The reaction is believed to proceed via reductive elimination of a σ-allylcopper(III) compound that leads to a C–C bond.[446] A density functional study has shown that the C–C bond formation occurs directly from the π-allyl complex via an enyl[σ + π]-type transition state.[447] A synthetic example is taken from Ollivier, Piras and co-worker's synthesis of grandisol, inwhich conjugated aldehyde **444** (prepared from conjugated ester **443**) reacted with lithium dimethylcuprate to give a 97% yield of **445**.[448] In addition to the dialkyllithium cuprates, it is also possible to form magnesium cuprates that react in an identical manner. A synthetic example is taken from Harris and Padwa's synthesis of 6-epi-indolizidine 223A,[449] in which conjugated piperidone **446** reacted with diethylmagnesium cuprate to give **447** in 90% yield.

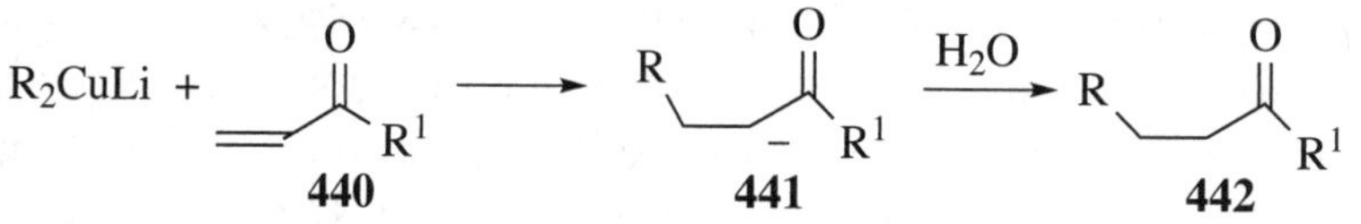

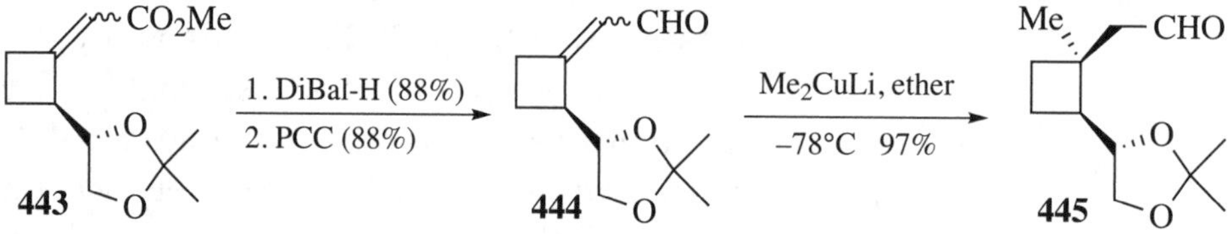

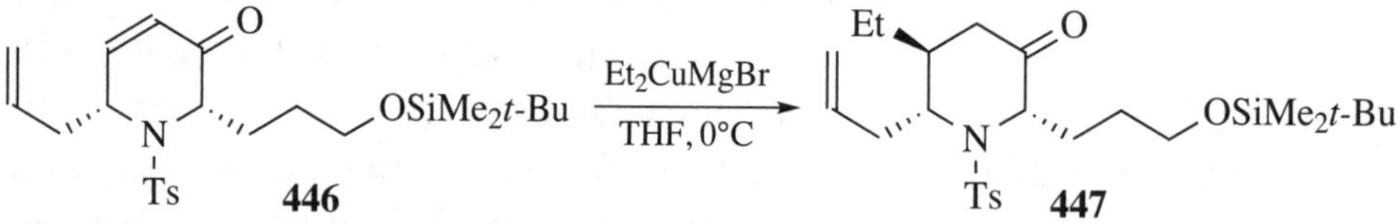

After the initial addition of a cuprate to a conjugated system, an enolate anion is formed, which can react in a subsequent reaction with an alkyl halide, an aldehyde or another electrophilic species. In a synthesis of *dendrobatid* alkaloid 251F by Aubé and co-worker's,[450]

445. For reactions of this type, see Larock, R.C. *Comprehensive Organic Transformations, 2nd Ed.* Wiley-VCH, New York, *1999*, pp 1599-1613

446. See (a) Goering, H. L.; Singleton Jr., V. D. *J. Org. Chem. 1983, 48*, 1531; (b) Goering, H. L.; Tseng, C. C. *J. Org. Chem. 1983, 48*, 3986; (c) Goering, H. L.; Kantner, S. S. *J. Org. Chem,. 1984, 49*, 422.

447. Yamanaka,, M.; Kato, S.; Eiichi Nakamura, E. *J. Am. Chem. Soc. 2004, 126*, 6287.

448. Bernard, A.M.; Frongia, A.; Ollivier, J.; Piras, P.P.; Secci, F.; Spiga, M. *Tetrahedron 2007, 63*, 4968.

449. Harris, J.M.; Padwa, A. *J. Org. Chem. 2003, 68*, 4371.

450. Wrobleski, A.; Sahasrabudhe, K.; Aubé, J. *J. Am. Chem. Soc. 2002, 124*, 9974.

conjugate addition of dimethyl cuprate to α,β-unsaturated ketone **448** gave enolate anion **449**. Subsequent addition of the aldehydes shown gave **450**, after elimination of water to form the conjugated ketone unit. The ability to add two alkyl substituents via conjugate addition and then alkylation with an alkyl halide or another electrophile (a double alkylation) is an important variation of conjugate addition with organocuprates. In general, the two groups incorporated by such a double-alkylation have a trans relationship,[451] but the cis-trans ratio varies with the organocuprate and the substrate.

The reaction is not limited to the use of simple organocuprates. Mixed cuprate **428** was one example. Another example that used an unsymmetrical organocuprate is Sih and co-worker's reaction of (–)-dialkenylcuprate (**451**) with **452** to give **453**, an intermediate in a prostanoid synthesis.[452,453] As with all Michael-type additions, the reaction of organocuprates is somewhat sensitive to steric effects at the reactive π bond. Addition usually occurs to the less sterically hindered position. Steric hindrance also plays a role in the diastereoselectivity of the reaction. Heathcock et al. showed that the position of the substituent relative to conjugated system is a significant factor in the formation of (*E*) or (*Z*) isomers from cycloheptenones.[454] The stereoselectivity increases as the group on the ring increases in size and when it is closer to the β-carbon.

Because of the reactivity of aldehydes with cuprates (sec. 8.7.A.v), 1,2-addition to α,β-unsaturated aldehydes can compete with 1,4-addition, although the former usually predominates. α,β-Unsaturated ketones normally give 1,4-addition. In Table 8.19,[455] Marshall showed how the copper reagent and the proportion of RLi to Cu influenced the ratio of 1,2- to 1,4-addition with **454** and di(4-pentenyl)cuprate. Reaction with this cuprate gave the 1,2-addition product **455**, the 1,4-addition products **456** and **457**, and the dialkylation product **458** when the initial

451. Paczkowski, R.; Maichle-Mössmer, C.; Maier, M.E. *Org. Lett.* **2000**, *2*, 3967.
452. (a) Sih, C.J.; Salomon, R.G.; Price, P.; Peruzzoti, G.; Sood, R. *J. Chem. Soc. Chem. Commun.* **1972**, 240; (b) Sih, C.J.; Salomon, R.G.; Price, P.; Sood, R.; Peruzzotti, G. *J. Am. Chem. Soc.* **1975**, *97*, 857.
453. Kluge, A.F.; Untch, K.G.; Fried, J.H. *J. Am. Chem. Soc.* **1972**, *94*, 9256.
454. Heathcock, C.H.; Germroth, T.C.; Graham, S.L. *J. Org. Chem.* **1979**, *44*, 4481.
455. Marshall, J.A.; Audia, J.E.; Shearer, B.G. *J. Org. Chem.* **1986**, *51*, 1730.

addition product was quenched with iodomethane. Diminishing the RLi/Cu ratio to <2:1 suppressed 1,2-addition. The best yields were obtained when the intermediate enolate was trapped with chlorotrimethylsilane, which was mentioned above with the reaction of **437** with di-*n*-butylcuprate. The improvement in yield is presumably due to rate acceleration.[456,457] The enhanced diastereoselectivity can be seen in the addition of divinylcuprate to **459**, which gave two racemic silyl enol ethers, **460** and **461**. In the absence of chlorotrimethylsilane the ratio of enolates was 56:44 (this is also the diastereomeric ratio of **460/461**, although the silyl enol ether cannot be formed without addition of chlorotrimethylsilane).which also increased the product ratio to >99:1 (**460/461**).[457] Isolation of the silyl enol ethers was followed by hydrolysis to give **462** as the major product.

Table 8.19. Organolithium/Cuprous Halide Ratio in 1,2 versus 1,4-Addition of Organocuprates to 454

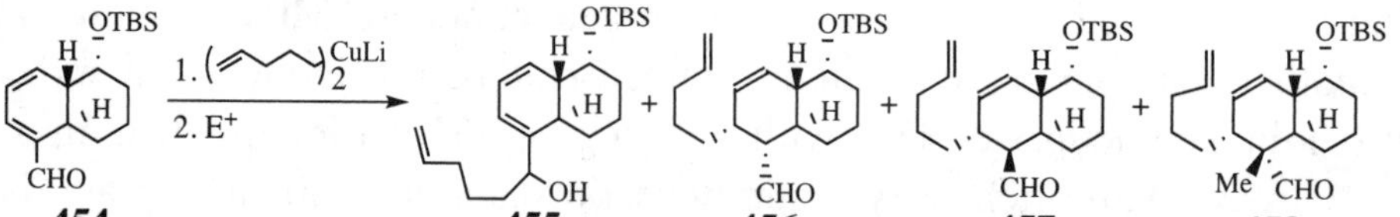

Conditions	E⁺	% 455	% 456	% 457	% 458
2 RLi, CuI, Me₂S, 2:1 ether:THF, 14h	MeI	17	37	19	7
2 RLi, CuBr•SMe₂, 2:1 ether:THF, 14h	MeI	17ᵃ	20	16	8
1.8 RLi, CuI, Me₂S, 1.6:1 ether:THF, 14h	MeI	0	30	22	33
1.8 RLi, CuI, Me₂S, 1.8:1 ether:THF, 10h	H₃O⁺	0	65		
1.6 RLi, CuI, Me₂S, 1.6:1 ether:THF, 14h	H₃O⁺	0ᵇ	0		
1.8 RLi, CuI, Me₂S, 1.6:1 ether:THF, 14h	1. TSCl 2. H₃O⁺	0	91		

ᵃ + 10% of **454**.   ᵇ + 92% of **454**.

[Reprinted with permission from Marshall, J.A.; Audia, J.E.; Shearer, B.G. *J. Org. Chem.* **1986**, *51*, 1730. Copyright © **1986** American Chemical Society.

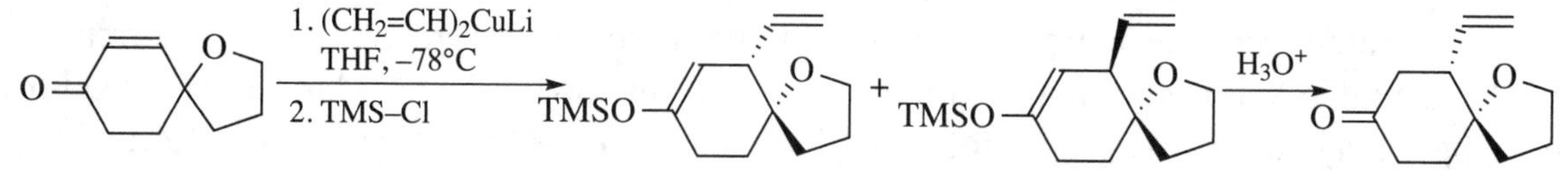

Conjugated esters are less reactive than the corresponding ketones.[458] Conjugate addition to ester **463** gave only 8% methylated product **464** after reaction with dimethylcuprate for 44 hours.

456. Taylor, R.J.K. *Synthesis* **1985**, 364.
457. (a) Corey, E.J.; Boaz, N.W. *Tetrahedron Lett.* **1985**, *26*, 6015, 6019; (b) Alexakis, A.; Berlan, J.; Besace, Y. *Ibid.* **1986**, *27*, 1047; (c) Johnson, C.R.; Marren, T.J. *Ibid.* **1987**, *28*, 27.
458. (a) Posner, G.H. *Organic. React.* **1972**, *19*, 1; (b) Normant, J.F.; *Synthesis* **1972**, 63.

Despite the steric hindrance of the hexyl group, the analogous ketone (**465**) gave a quantitative yield of **466** under the same conditions. Although conjugated esters usually react slower than conjugated ketones, they do give conjugate addition products unless a more reactive group is present in the molecule. Addition of the diallylcuprate shown to ester **467**, for example, gave the alkylated ester (**468**) an intermediate in Nicolaou and co-worker's synthesis of tylosin.[459] Note that conjugated esters show greatly enhanced reactivity when RCu•BF$_3$ is used as the organocopper reagent.[460] With highly hindered systems, especially β,β-disubstituted conjugate esters, RCu•BF$_3$ is the reagent of choice.[461]

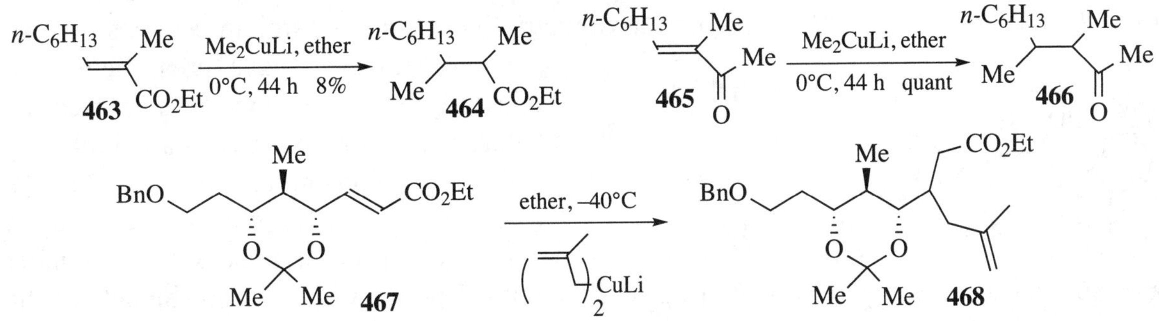

The disconnection for cuprate conjugate addition is

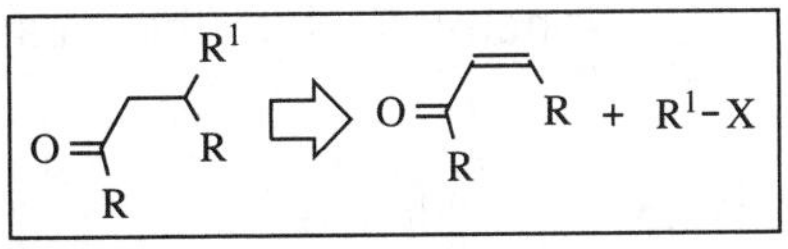

8.7.A.vii. Asymmetric Induction. Addition of an organocuprate to a conjugated system often creates a new stereogenic center, so a discussion of diastereoselectivity and possible enantioselectivity is appropriate. Controlling asymmetric induction in organocuprate addition reactions involves the use of organocuprates whose chirality has been modified.[462] Typical chiral groups include 1,2,5,6-di-*O*-isopropylidene-α-*D*-glucofuranose (**469**), (*S*)-proline (**470**), (*S*)-*N*-methylproline (**471**), and neomenthyl thiol (**472**).[462] Addition of the organocuprate to cyclohexenone in the presence of these chiral ligands gave the chiral ketone, **473**, but with generally poor %ee (4.7-26% ee).[462]

459. Nicolaou, K.C.; Pavia, M.R.; Seitz, S.P. *J. Am. Chem. Soc.* **1982**, *104*, 2027, 2030.
460. (a) Yamamoto, Y. *Angew. Chem. Int. Ed.* **1986**, *25*, 947; (b) Lipshutz, B.H.; Ellsworth, E.L.; Siahaan, T.J. *J. Am. Chem. Soc.* **1988**, *110*, 4834; (c) Idem, *Ibid.* **1989**, *111*, 1351.
461. Yamamoto, Y.; Maruyama, K. *J. Am. Chem. Soc.* **1978**, *100*, 3240.
462. Tomioka, K.; Koga, K. in *Asymmetric Synthesis, Vol. 2*, Morrison, J.D. (Ed.), Academic Press, New York, **1983**, pp 219-221.

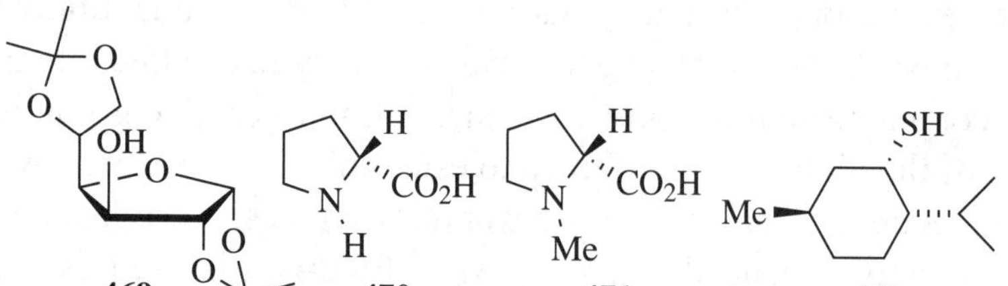

In acyclic systems, Mukaiyama and co-workers found that addition of chiral magnesium cuprate R(Z*)CuLi derived from **471** gave good asymmetric induction.[463] Lithium cuprates derived from **474**, however, gave poor yields and modest % ee.[464] Kretchmer[465] also used the alkaloid (–)-sparteine (**163**) as a complexing agent with magnesium cuprates (RCuMgX$_2$), but observed < 10% optical purity upon addition to α,β-unsaturated ketones. Langer and Seebach observed products with up to 27% optical purity when the co-solvent for the reaction was (*R,R*) or (*S,S*)-1,4-(dimethylamino)-2,3-dimethoxybutane.[466] Gustafsson used an asymmetric ligand (RR*CuLi) where R* = (–)-[1-(dimethylamino)-ethyl] phenyl[467] or *o*-[cyclohexyl(dimethylamino)methyl]-phenyl,[468] but they gave < 5% ee in organocuprate reactions.

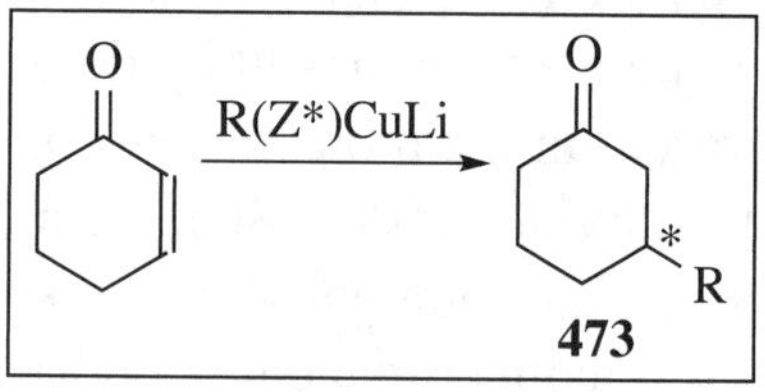

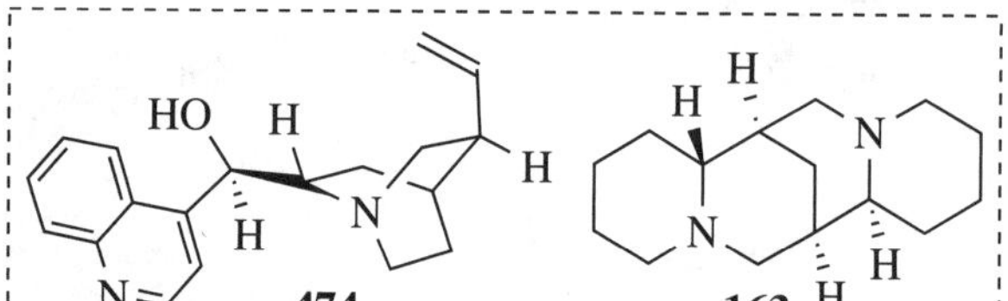

Bertz and co-workers showed that cuprates added to cyclohexenone to give the 3-phenyl derivative (**475**) in optical yields of 0-50%.[469] Typical examples are R^{1}R^{2}N = (4*S*,5*S*)-(+)-5-amino 2,2-dimethyl-4-phenyl-1,3-dioxane (the reaction proceeded in 62% yield, 50% ee); (*R*) or (*S*) α-methylbenzylamine (70% yield, 30% ee); (*R*) or (*S*) α-(1-naphthyl) ethylamine (70% yield, 30% ee); and (-) α-2-naphthylethyl-amine (50% yield, 40% ee).[469]

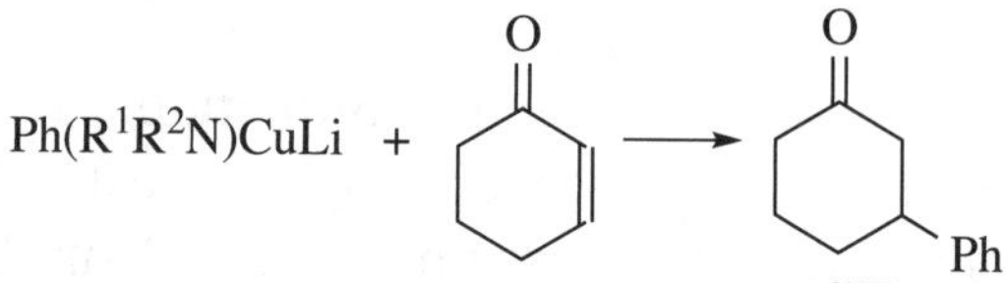

Good-to-excellent asymmetric induction has been achieved with a new generation of chiral additives or chiral ligands.[470] Corey et al. achieved enantioselectivities in the 75-95% range with the chiral controller ligand **476**, obtained from (+) or (–)-ephedrine.[471] This chiral ligand

463. Imamoto, T.; Mukaiyama, T. *Chem. Lett.* ***1980***, 45.
464. Huché, M.; Berlan, J.; Pourcelot, G.; Cresson, P. *Tetrahedron Lett.* ***1981***, 1329.
465. Kretchmer, R.A. *J. Org. Chem.* ***1972***, *37*, 2744.
466. Langer, W.; Seebach, D. *Helv. Chim. Acta* ***1979***, *62,* 1710.
467. Gustafsson, B.; Nilson, M.; Ullenius, C. *Acta Chem. Scand. Ser. B* ***1977***, *B31*, 667.
468. Gustafsson, B. *Tetrahedron* ***1978***, *34*, 3023.
469. Bertz, S.; Dabbagh, G.; Sundararajan, G. *J. Org. Chem.* ***1986***, *51*, 4953.
470. For a review of enantioselective conjugate additions see Sibi, M.P.; Manyem, S. *Tetrahedron* ***2000***, *56*, 8033.
471. Corey, E.J.; Naef, R.; Hannon, F.J. *J. Am. Chem. Soc.* ***1986***, *108*, 7114.

was converted to the asymmetric copper complex (477), and reaction with cyclohexenone with selective transfer of the butyl group gave 3-butylcyclohexanone (478) in 80% yield (80% ee).[471] Similar reaction with ethyllithium gave the ethyl derivative in 66% yield with an optical purity > 95% ee. If the organolithium reagent was heavily contaminated with alkoxides, however, the optical purity was as low as 9% ee.

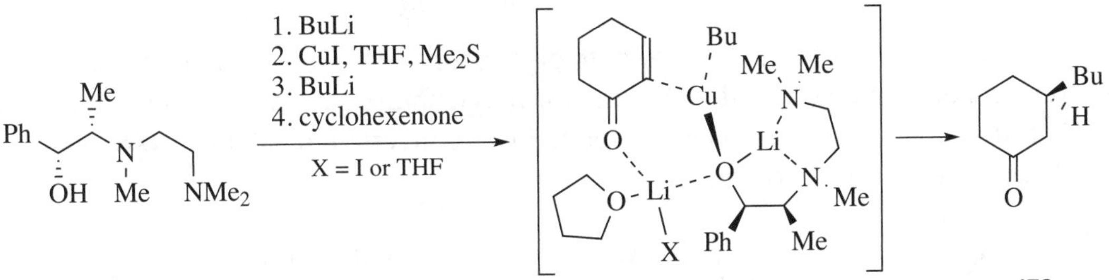

476   477   478

A different approach to solving this problem is to attach a chiral auxiliary to the enone substrate. An auxiliary is a chiral material that is attached to an achiral substrate. The asymmetry of the auxiliary controls the selectivity of the reaction and is then removed in a separate chemical step. Posner and co-workers studied the use of optically pure α-carbonyl-α,β-ethylenic sulfoxides (479, see sec. 3.9.A.ii for the preparation of optically active sulfoxides).[472] Conjugate addition of di-(2,2,-dimethylpropyl)magnesium cuprate and then reductive cleavage of the C–S bond of the sulfoxide with aluminum amalgam gave 3-(2,2-dimethylpropyl)cyclohexanone (480) in 99% yield as a 7:1 (R/S) mixture.[472] Oppolzer and Löher prepared (–)-(8E)-phenylmenthyl acrylates such as 481.[473] The conformational stability of the cyclohexyl ring in 481 ensured that the phenyl group blocked one face of the α,β-unsaturated ester, leading to excellent facial selectivity during the addition. Conjugate addition (here with phenylcopper-boron trifluoride- see the previous discussion in sec. 8.7.A.vi) followed by saponification of the ester (482 was formed in 76% yield) gave the acid 483 in > 99% ee. Oppolzer et al. also reported excellent diastereoselectivity for addition of organocopper derivatives to α,β-unsaturated sulfonamides derived from camphor (see sec. 11.9.A for further discussion of chiral auxiliaries).[474]

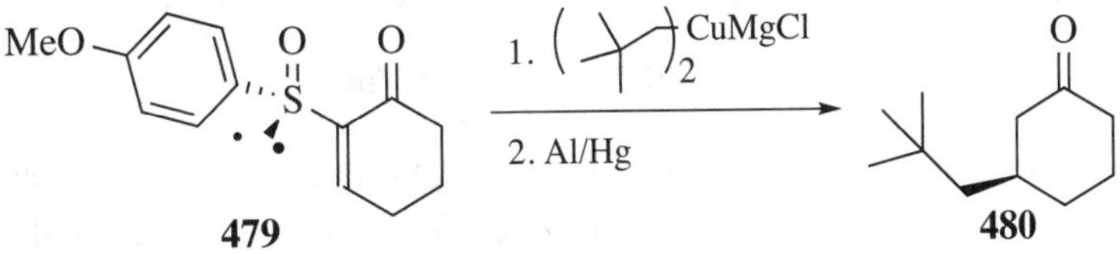

479   480

---

472.   Posner, G.H.; Frye, L.L.; Hulce, M. *Tetrahedron* **1984**, *40*, 1401.

473.   Oppolzer, W.; Löher, H.J. *Helv. Chim. Acta* **1981**, *64*, 2808.

474.   (a) Oppolzer, W.; Mills, R.J.; Pachinger, W.; Stevenson, T. *Helv. Chim. Acta* **1986**, *69*, 1542; (b) Oppolzer, W.; Schneider, P. *Ibid.* **1986**, *69*, 1817.

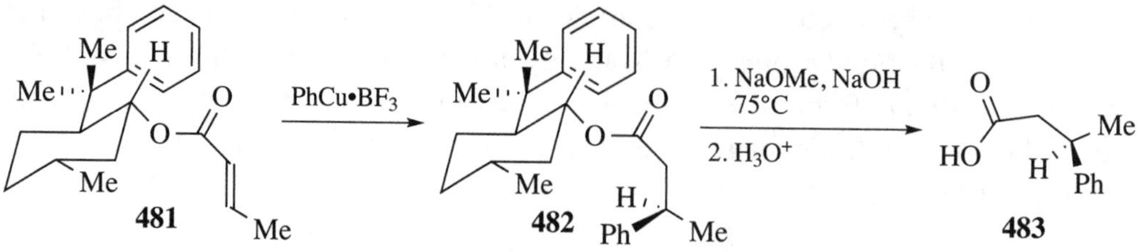

Other asymmetric substrates can be used with organocuprates. One example is Scolastico and co-worker's report that reaction of **484** with di-*n*-butylcuprate gave a 70% yield of **485** (78-98 % ee).[475] Using this auxiliary, alkylated ester **485** was converted to aldehyde **486** in 80% yield with 91% asymmetric induction.[475]

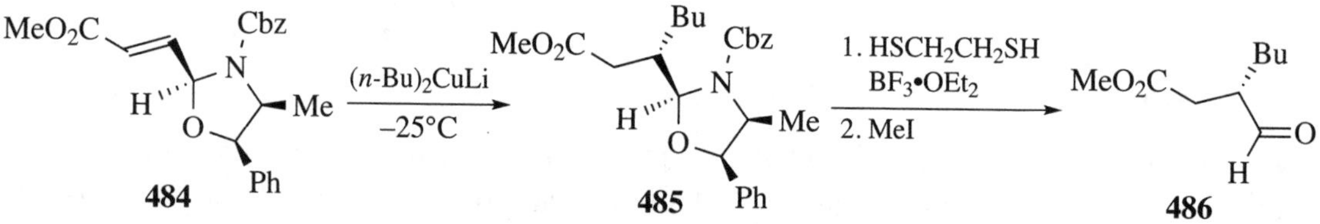

A less general method involves transfer of a chiral group from the cuprate to the enone. The chiral Gilman reagent [R*$_2$CuLi, R = (*R*)-2-(1-dimethylamino)phenyl] reacted with (4*E*)-methyl-3-penten-2-one (**487**) to give **488** in 30% yield and 80% ee.[476] Reaction of the mixed cuprate [ThR*CuLi where Th = 2-thienyl and R* = (*R*)-2-(1-dimethylamino)phenyl] gave **488** in 70% yield (82% ee).[476]

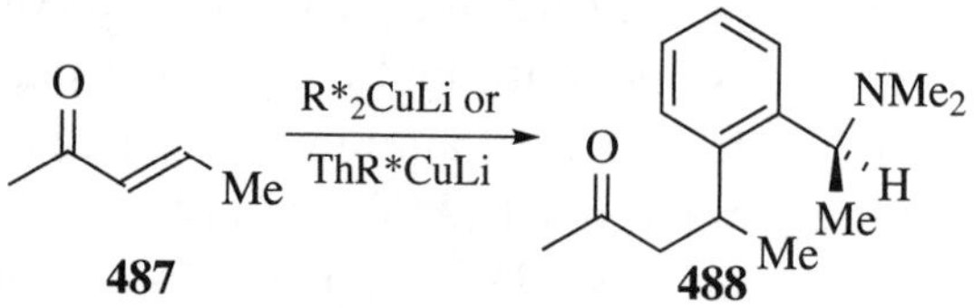

8.7.B. Higher Order Cuprates

Cuprates are very useful but there are a few problems associated with them, including reactivity and thermal stability. In an attempt to circumvent these problems, Lipshutz and co-workers developed the so-called *higher order mixed cuprates* (R$_2$Cu(CN)Li$_2$),[477] prepared by reaction of two equivalents of organolithium reagent with cuprous cyanide (CuCN).[478] Mixed cuprates

$$2 \text{ R−Li} + \text{CuCN} \longrightarrow \text{R}_2\text{Cu(CN)Li}_2$$

react faster than Gilman reagents with alkyl halides, even secondary halides. Mixed cuprates are generally the reagent of choice for this purpose. Table 8.20[479a] shows the reaction of alkyl

475. Bernardi, A.; Cardani, S.; Poli, G.; Scolastico, C. *J. Org. Chem.* **1986**, *51*, 5041.
476. Malmberg, H.; Nilsson, M.; Ullenius, C. *Tetrahedron Lett.* **1982**, *23*, 3823.
477. Lipshutz, B.H.; Wilhelm, R.S.; Kozlowski, J.A. *Tetrahedron* **1984**, *40*, 5005.
478. Lipshutz, B.H.; Wilhelm, R.S. *J. Am. Chem. Soc.* **1982**, *104*, 4696.
479. (a) Lipshutz, B.H.; Wilhelm, R.S.; Kozlowski, J.A.; Parker, D. *J. Org. Chem.* **1984**, *49*, 3928; (b) Yamamoto, K.; Iijima, M.; Ogimura, Y.; Tsuji, J. *Tetrahedron Lett.* **1984**, *25*, 2813; (c) Takahashi, T.; Okumoto, H.; Tsuji, J. *Ibid.* **1984**, *25*, 1925.

halides with the higher order cuprate derived from butyllithium and cuprous cyanide to give the alkane, R-n-Bu. Note that Bertz questioned the validity of structure of cuprates such as R$_2$Cu(CN)Li$_2$, citing ^{13}C NMR evidence that the reagent actually existed as R$_2$CuLi•LiCN in THF.[480] Lipshutz contradicted this conclusion,[481] when the ^{1}H and ^{13}C NMR spectra of the higher order cuprate Me$_2$Cu(CN)Li$_2$ were compared with those of Me$_2$CuLi•LiBr and Me$_2$CuLi•LiI, and he concluded that the spectra were not at all similar. Lipshutz concluded that the spectra proved the presence of a Cu–CN bond, and a higher order cuprate structure. *Ab initio* evidence,[482a] additional NMR evidence[483] as well as kinetic evidence[482b,c] by Bertz again called these structures into question, and he suggested their structure to be cyano-Gilman reagents. Several studies concluded there is no higher order cyanocuprate, but that the active species is best described as R$_2$CuLi•LiCN.[482d] A theoretical study suggests that a minor species R(CN)CuLi•LiR may play an important role due to its greatly enhanced reactivity.[484]

Table 8.20. Reaction of Higher Order Cyanocuprates with Alkyl Halides

R–X + 2 n-Bu$_2$Cu(CN)Li$_2$ ⟶ R–n-Bu

Halide	Temp. (°C)	Time (h)	% R–n-Bu
Iodocyclopentane	−78	2	82
Bromocyclopentane	0	6	86
Iodocyclohexane	−78	1	100
Bromocyclohexane	25	6	41
2-Iodopentane	−50	2	99
2-Bromopentane	0→25	2	94
2-Chloropentane	25	11	28
2-Tosyloctane	25	8	>80
(10 equivalents cuprate)			

[Reprinted with permission from Lipshutz, B.H.; Wilhelm, R.S.; Kozlowski, J.A.; Parker, D. *J. Org. Chem.* **1984**, *49*, 3928. Copyright © **1984** American Chemical Society.]

Higher order organocuprates react with chiral halides to give chiral coupling products. When (*R*)-2-bromooctane (**489**) reacted with the mixed-cuprate [EtMeCu(CN)Li$_2$], a 72% yield of (*R*)-3-methylnonane (**490**) was obtained,[479b,c] which is the result expected of a nucleophilic S$_N$2

like displacement of the bromide. As mentioned in Section 8.7.A, the reaction probably proceeds via single electron-transfer process, but the stereochemistry of this reaction mimics nucleophilic substitution. The extent of inversion is very dependent on the nature of the reacted organocuprate, however. 2-Iododecane derivatives

480. Bertz, S.H. *J. Am. Chem. Soc.* **1990**, *112*, 4031.
481. Lipshutz, B.H.; Sharma, S.; Ellsworth, E.L. *J. Am. Chem. Soc.* **1990**, *112*, 4032.
482. (a) Snyder, J.P.; Spangler, D.P.; Behling, J.R.; Rossiter, B.E. *J. Org. Chem.* **1994**, *59*, 2665; (b) Snyder, J.P.; Bertz, S.H. *J. Org. Chem.* **1995**, *60*, 4312; (c) Bertz, S.H.; Miao, G.; Eriksson, M. *Chem. Commun.* **1996**, 815. Also see (d) Krasuse, *Angew. Chem. Int. Ed.* **1999**, *38*, 79.
483. Bertz, S.H.; Nilsson, K.; Davidsson, Ö.; Snyder, J.P. *Angew. Chem. Int. Ed.* **1998**, *37*, 314.
484. Nakamura, E.; Yoshikai, N. *Bull. Chem. Soc. Jpn.* **2004**, *77*, 1.

showed virtually no inversion of configuration when with Gilman-type reagent or with higher order cuprates.[479a] Bromides, on the other hand, gave virtually complete inversion with both reagents.[479a] The smallest amount of inversion was obtained with symmetrical cuprates, and the largest amount with mixed cuprates.

This displacement reaction is not limited to allyl derivatives or to substrates a bearing halide leaving group. Both vinyl and aryl derivatives react with higher order cuprates, similar to the Gilman reagents. Lipshutz and Elworthy found that vinyl triflates are particularly useful in cuprate coupling reactions. The reaction of vinyl triflate **491** reacted with the mixed cuprate **492**, gave **493** in 87% yield.[485] An interesting feature of cuprate **492** is the presence of 2-thienyl as an unreactive ligand (see **428**). Lipshutz et al. found this to be most effective for the selective transfer of the group other than thienyl.[486] This is analogous to the use of alkynyl groups as unreactive substituents in Gilman reagents. The mixed-alkyl cuprates are easily prepared by sequential addition of two different organolithium reagents to cuprous cyanide. In general, for reagents such as $R(Me)Cu(CN)Li_2$ (R > Me), the R group is transferred selectively for both the halide displacement and the conjugate addition.

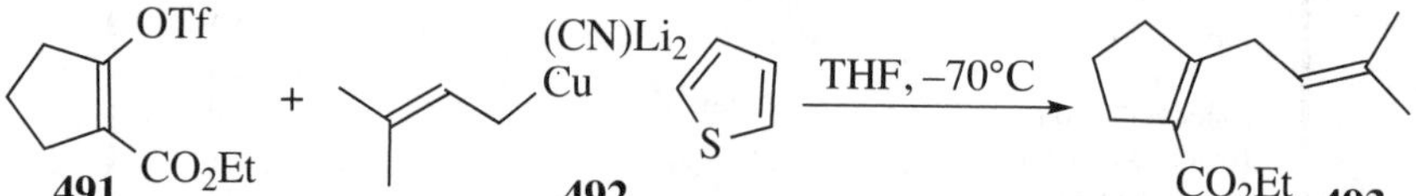

Epoxide ring opening is more efficient with the higher order cuprate than with Gilman reagents, and attack is at the less sterically hindered carbon.[479a] Reaction of **494** with di-*n*-propylcuprate, for example, gave only 15-30% of **495**. Similar reaction with $Pr_2Cu(CN)Li_2$ gave an 86% yield of **495**.[487,488] Higher order cuprates give excellent yields of conjugate addition with α,β-unsaturated compounds, as do Gilman reagents.[489] In the Reddy and co-worker's synthesis of (+)-cyanthiwigin AC,[490] conjugated ketone **496** reacted with the higher order methyl cuprate derived from 2-propenyllithium and CuCN to give an 85% yield of **497**.

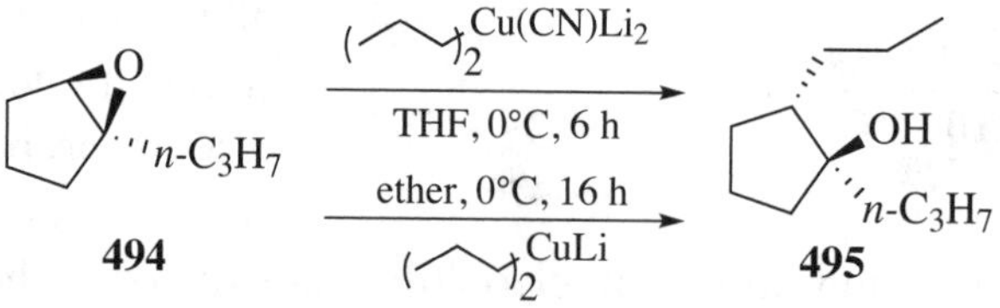

485. Lipshutz, B.H.; Elworthy, T.R. *J. Org. Chem.* **1990**, *55*, 1695.
486. Lipshutz, B.H.; Kozlowski, J.A.; Parker, D.A.; Nguyen, S.L.; McCarthy, K.E. *J. Organomet. Chem.* **1985**, *285*, 437.
487. Lipshutz, B.H.; Kozlowski, J.; Wilhelm, R.S. *J. Am. Chem. Soc.* **1982**, *104*, 2305.
488. Lipshutz, B.H.; Kotsuki, H.; Lew, W. *Tetrahedron Lett.* **1986**, *27*, 4825.
489. Lipshutz, B.H.; Wilhelm, R.S.; Kozlowski, J.A. *J. Org. Chem.* **1984**, *49*, 3938.
490. Reddy, T.J.; Bordeau, G.; Timble, L. *Org. Lett.* **2006**, *8*, 5585.

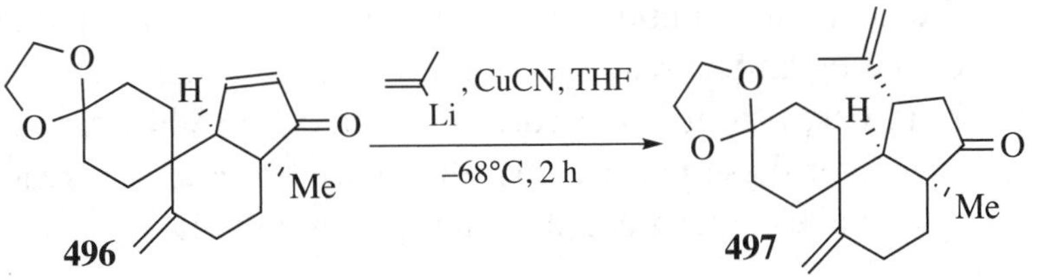

Dieter et al. developed an α-aminoalkylcuprate that allows incorporation of nitrogen. Cyanocuprate **498** reacted with cyclohexenone in the presence of chlorotrimethylsilane to give a 98% yield of the conjugate addition product **499**.[491a]

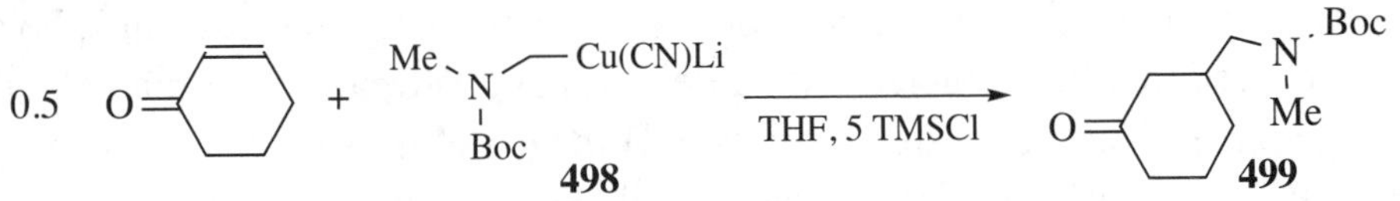

The disconnections for the higher order cuprate are identical to those observed with Gilman reagents.

8.8. YLIDS

Ylids are defined as compounds in which a carbanionic carbon is immediately adjacent to an atom bearing a positive center. Many atoms can support a positive charge but phosphorus, sulfur or nitrogen lead to particularly useful and interesting compounds. This section will deal with ylids and their reactions, focusing on phosphorus ylids such as **500**, sulfur ylids such as **501** or nitrogen ylids such as **502**. Of these three classes of ylids,[492] the phosphonium ylids (**500**) and other phosphorus ylids are the most studied due to their relative stability and ability to generate alkenes from carbonyl compounds. The sulfur ylids (**501**) are also well known,[493] and react as carbanions with carbonyl derivatives to give oxiranes or cyclopropane derivatives. The nitrogen ylids (**502**) are not as well known as isolable entities, but there are a few synthetic applications when stabilizing groups are present.[494]

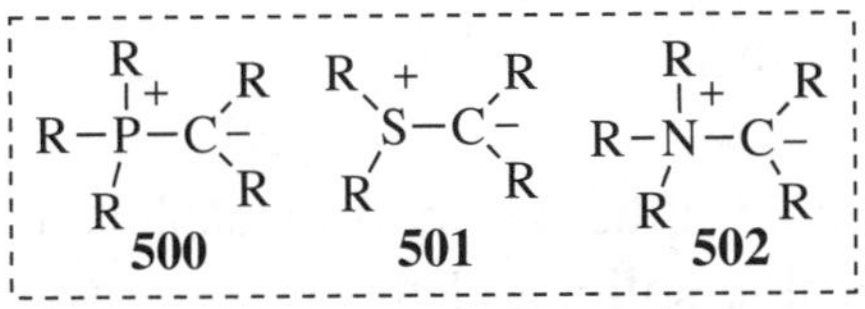

8.8.A. Phosphorus Ylids

8.8.A.i. Wittig Reagents. Phosphorus ylids have a history dating to the 1890s and to the

---

491. (a) Dieter, R.K.; Alexander, C.W.; Nice, L.E. *Tetrahedron* **2000**, *56*, 2767. Also see (b) Dieter, R.K.; Lu, K.; Velu, S.E. *J. Org. Chem.* **2000**, *65*, 8715.
492. Johnson, A.W. *Ylid Chemistry,* Academic Press, New York, *1966*
493. Trost, B.M.; Melvin, Jr., L.S. *Sulfur Ylides*, Academic Press, New York, *1975*.
494. Zugravescu, I.; Petrovanu, M. *Nitrogen-Ylid Chemistry*, McGraw Hill, New York, *1976*.

pioneering work of Michaelis and Gimborn.[495] Wittig and co-workers found that phosphonium salts (**503**) were generated by an S_N2 reaction of a trialkylphosphine with an alkyl halide, and they reacted with strong bases to give the corresponding ylid (**504**), a resonance stabilized species.[496a] The discovery of these phosphorus ylids may have been serendipitous, because Wittig was attempting to make pentavalent phosphorus compounds.

In earlier work, Staudinger and Hauser found that phosphazines ($R_3P=N-N=CR_2$) generated $R_3P=CR_2$ upon heating and they examined some of the chemistry of these phosphorus ylids,[497] but Wittig discovered that they reacted with aldehydes and ketones to form an alkene product. This latter reaction has come to be called the **Wittig olefination reaction**, or simply the **Wittig reaction**.[496] In early papers, it was postulated that initial reaction of ylid **504** with a ketone or aldehyde generated a labile intermediate called a betaine **505**, which collapsed to an oxaphosphetane (**506**). Either of these intermediates can decompose to give an alkene (reaction of **504** with benzophenone, for example, led to 1,1-diphenylethene, **384**), and a trisubstituted phosphine oxide (in the case of **504**, triphenylphosphine).[496] The P–O bond is particularly strong, with a bond dissociation energy of 130-140 kcal mol^{-1} (544.2-586.0 kJ mol^{-1}).[498] Formation of this strong bond is largely responsible for decomposition of the intermediate betaine or oxaphosphetane to give the alkene.

Betaine or oxaphosphetane intermediates are not usually isolable, although an oxaphosphetane derived from a highly specialized ylid was isolated as a stable solid, and its structure was determined by X-ray crystallography.[499] The existence of betaine derivatives as discrete intermediates has been called into question, and this issue will be addressed in connection with mechanistic discussions of the Wittig reaction in section 8.8.A.ii.

Examination of the generalized structure **508** (generated from phosphonium salt **507**) shows

495. Michaelis, A.; Gimborn, H.V. *Berichte* **1894**, *27*, 272.
496. (a) Wittig, G.; Rieber, M. *Annalen* **1949**, *562*, 187; (b) Wittig, G.; Geissler, G. *Ibid.* **1953**, *580*, 44; (c) Wittig, G.; Schöllkopf, U.; *Chem. Ber.* **1954**, *87*, 1318; (d) Gensler, W.J. *Chem. Rev.* **1957**, *57*, 191 (see p 218); (e) The Merck Index, 14th ed., Merck & Co., Inc., Whitehouse Station, New Jersey, **2006**, p ONR-101; (f) Mundy, B.P.; Ellerd, M.G.; Favaloro Jr., F.G. *Name Reactions and Reagents in Organic Synthesis*, *2nd ed.*, Wiley-Interscience, New Jersey, **2005**, pp. 696-697.
497. (a) Stauadinger, H.; Hauser, E. *Helv. Chim. Acta* **1921**, *4*, 861; (b) Staudinger, H.; Meyer, J. *Ibid.* **1919**, *2*, 612.
498. Hartley, S.B.; Holmes, W.S.; Jacques, J.K.; Mole, M.F.; McCoubrey, J.C. *Quart. Rev. Chem. Soc.* **1963**, *17*, 204.
499. Bestmann, H.J.; Roth, K.; Wilhelm, E.; Böhme, R.; Burzlaff, H. *Angew. Chem. Int. Ed.* **1979**, *18*, 876.

the formal charge on phosphorus is +1. When R^1 and/or R^2 are alkyl, the positive charge on phosphorus is diminished due to the inductive effect of the alkyl groups. Such ylids are more reactive with a ketone or aldehyde, but decomposition of the resulting oxaphosphetane is more difficult. The R groups on the phosphorus should not be a group that contains α-protons, since deprotonation would lead to a mixture of isomeric ylids. Reaction of **509** with *n*-butyllithium, for example, gives two ylids, **510** and **511**. To prevent this problem, all but one of the R groups should *not* possess α-hydrogens, which accounts for the use of Ph_3P and other triaryl phosphines in the Wittig reaction. If R^1 in **507** is electron-withdrawing (COR, CO_2R, CN, etc.), the negative charge of the ylid is delocalized by that group, decreasing both the nucleophilicity and reactivity of the carbanionic carbon. Such ylids usually react well with aldehydes, but ketones are less reactive.

The strength of the base required for deprotonation of the phosphonium salt to form the ylid depends on the relative acidity of the α proton, which depends on the alkyl or aryl group attached to that carbon ($-CHR^1R^2$ in **508**) as well as the nature of the groups attached to the phosphorus. The α-hydrogen of a phosphonium salt such as **507** is much more acidic than that of the analogous structure without the phosphorus atom (2,2-dimethylpropane). The pK_a for H_a of fluorene (**201**), for example, is 25[500] but H_a in the triphenylphosphonium derivative **512** has a pK_a of <10.[501] The proton (H_a) in the

corresponding ammonium salt (**513**) has a pK_a slightly less than that of fluorene.[502] Electron-releasing groups such as alkyl ($R^1 = R^2$ = alkyl in **507**) make the α-hydrogen less acidic, whereas the presence of electron-withdrawing groups (R^1 and/or $R^2 = COR$, CO_2R, CN, etc. in **507**) make that hydrogen more acidic. The wide range of acidity for hydrogens in phosphonium salts leads to a range of different bases that can be used. Both sodium hydride and potassium hydride have been used as bases, as has dimsyl sodium (sodium salt of DMSO) in DMSO. It is more common to use *n*-butyllithium, *tert*-butyllithium or *sec*-butyllithium in ether or THF. Guanidine bases also promote the Wittig reaction, as well as the Horner-Wadsworth-Emmons reaction discussed in section 8.8.A.ii, under mild conditions, with high efficiency and isolation of products is facile.[503] Guanidine bases include 1,5,7-triazabicyclo[4.4.0]dec-5-ene (TBD,

500. McEwen, W.K. *J. Am. Chem. Soc.* **1936**, *58*, 1124.
501. Johnson, A. Wm.; LeCount, R.B. *Tetrahedron* **1960**, *9*, 130.
502. Wittig, G.; Felletschin, G. *Annalen* **1944**, *555*, 133.
503. Simoni, D.; Rossi, M.; Rondanin, R.; Mazzali, A.; Baruchello, R.; Malagutti, C.; Roberti, M.; Invidiata, F.P. *Org. Lett.* **2000**, *2*, 3765.

514) and its methyl derivative MTBD (**515**), which were ≈ 100 times more effective than tetramethylguanidine (TMG, **516**). It is also noted that Wittig reactions can be done using stable phosphorus ylids in the presence of silica gel in hexane.[504]

Triarylphosphonium ylids[505] react with aldehydes and ketones by what is known as the **Wittig reaction**,[496c,506] to form alkenes. The reaction is particularly useful for incorporating exocyclic methylene groups.[507,508] For this and other work, Wittig was awarded the Nobel prize in 1979. Poor stereocontrol is sometimes a problem in Wittig olefination reactions (see sec. 8.8.A.ii).[509] The synthetic applications are generally straightforward. In Crimmins and co-worker's synthesis of astrogorgin,[510] ketone **517** reacted with triphenylphosphonium ylid $Ph_3P=C(Me)CO_2Et$ to give a near quantitative yield of **518**. A variety of substituents (R^1 and R^2 in **508**) can be attached to the carbanionic carbon in

phosphorus ylids.[511] Arylalkyl ylids ($Ph_3P=CRAr$) are commonly used in synthesis. Even reactive groups such as esters and nitriles can be used. Other functionality is compatible with this reaction. The functionalized phosphonium salt **519** reacted with the NaH and THF, and was then treated with thiophene-aldehyde **520** to give a 55% yield of **521** in Al-Mourabit and co-worker's synthesis of haliclamine A.[512] Otherwise reactive functional groups such as aldehydes, ketones, esters, etc. can be incorporated into the ylid. In a synthesis of perophoramidine by Weinreb and Artman, the lactone unit of ylid **522** was successfully incorporated in **523** in 98% yield.[513] In general, an ylid that contains an electron-withdrawing group at the carbanionic carbon is less reactive, but they still react with carbonyl derivatives. The phosphorane can even possess another acidic proton, although two equivalents of base are required to generate the anion ylid. An example is $Ph_3P(CH_2)_3CO_2H$, which was converted first to the carboxylate

504. Patil, V.J.; Mävers, U. *Tetrahedron Lett.* **1996**, *37*, 1281.
505. For reactions of this type, see Larock, R.C. *Comprehensive Organic Transformations, 2nd Ed.* Wiley-VCH, New York, **1999**, pp 327-332.
506. (a) Wittig, G.; Haag, W. *Berichte* **1955**, *88*, 1654; (b) Wittig, G. *Acc. Chem. Res.* **1974**, *7*, 6; (c) Schöllkopf, U. *Angew. Chem.* **1959**, *71*, 260; (d) Trippett, S. *Quart. Rev. Chem. Soc.* **1963**, *17*, 406;
507. (a) Corey, E.J.; Mitra, R.B.; Uda, H. *J. Am. Chem. Soc.* **1963**, *85*, 362; (b) Idem *Ibid.* **1964**, *86*, 485.
508. Ley, S.V.; Murray, P.J. *J. Chem. Soc. Chem. Commun.* **1982**, 1252.
509. For a computational study of reactivity and selectivity in the Wittig reaction, see Robiette, R.; Richardson, J.; Aggarwal, V.K.; Harvey, J.N. *J. Am. Chem. Soc.* **2006**, *128*, 2394.
510. Crimmins, M.T.; Brown, B.H.; Plake, H.R. *J. Am. Chem. Soc.* **2006**, *128*, 137.
511. Reference 492, pp 119-129.
512. Michelliza, S.; Al-Mourabit, A.; Gateau-Olesker, A.; Marazano, C. *J. Org. Chem.* **2002**, *67*, 6474.
513. Artman III, G.D.; Weinreb, S.M. *Org. Lett.* **2003**, *5*, 1523.

anion, and then to the ylid by treatment with base.[514] The resulting ylid reacted with aldehydes in the usual manner to give an alkene, but several carbons should separate the anionic centers. Protected alcohols can also be incorporated as part of the ylid.[514] It is even possible to do the Wittig reaction in aqueous media with water-soluble phosphonium salts developed by Warren and co-workers.[515] There are other nonclassical applications of the Wittig reaction,[516] and intramolecular versions are known.[517]

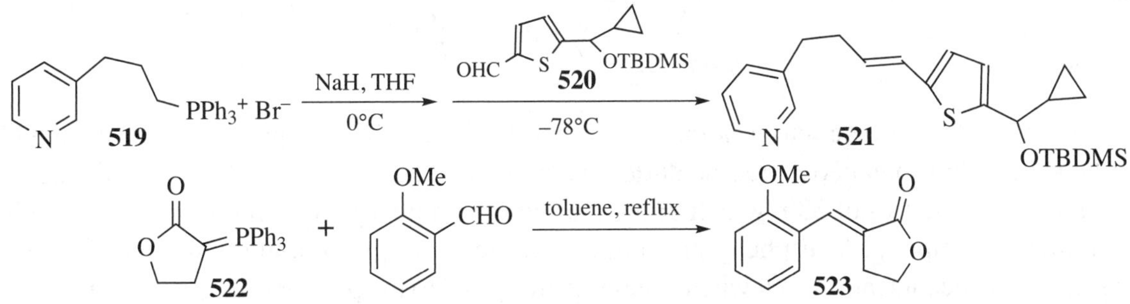

Incorporation of electron rich groups other than alkyl on the ylid is also possible. A useful application of this idea is the chain extension of ketones or aldehydes. In the synthesis of (+)-N_a-methylsarpagine,[518] Cook and co-worker's treated ketone **524** with a methoxymethyl ylid to generate a vinyl ether product (**525**). Subsequent acid hydrolysis converted the vinyl ether to aldehyde **526** in 90% overall yield. This sequence converted a ketone to an aldehyde with a one-carbon chain extension. Vinyl sulfides can also be formed by Wittig methodology (see **375** and **378** in sec. 8.6.B.ii).[519]

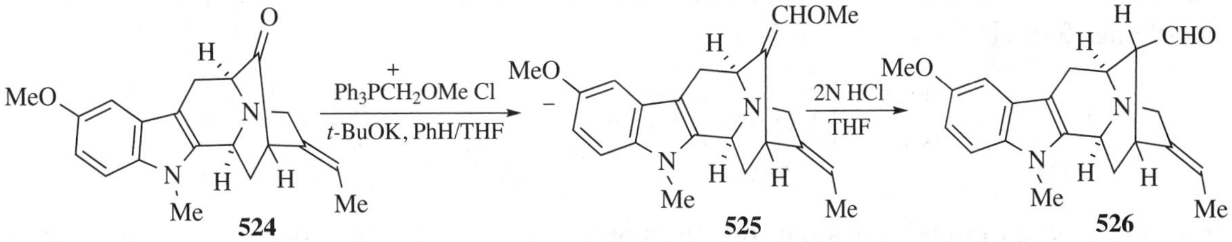

The presence of halogen atoms in a molecule is compatible with formation of an ylid and subsequent Wittig reaction. Carbon tetrabromide and triphenylphosphine have been used to give vinyl dibromide products, which are rapidly converted to the corresponding alkyne. Indeed, this is an important synthetic route to alkynes called the **Corey-Fuchs procedure**

514. Marinier, A.; Deslongchamps, P. *Tetrahedron Lett.* **1988**, *29*, 6215.
515. Russell, M.G.; Warren, S. *J. Chem. Soc. Perkin Trans. 1* **2000**, 505.
516. Murphy, P.J.; Lee, S.E. *J. Chem. Soc. Perkin Trans. 1* **1999**, 3049.
517. For reactions of this type, see Larock, R.C. *Comprehensive Organic Transformations, 2nd Ed.* Wiley-VCH, New York, **1999**, p 328.
518. Zhao, S.; Liao, X.; Cook, J.M. *Org. Lett.* **2002**, *4*, 687.
519. Corey, E.J.; Carpino, P. *J. Am. Chem. Soc.* **1989**, *111*, 5472.

previously mentioned in chapter 2.[520] Double elimination is possible to generate an alkyne. The reaction of **527** with this reagent gave vinyl dibromide **528** in 88% yield, as part of Ellman and co-worker's synthesis of (+)-lithospermic acid.[521] When **528** was treated with an excess of butyllithium and then methyl chloroformate, alkyne **529** was isolated in 93%% yield.

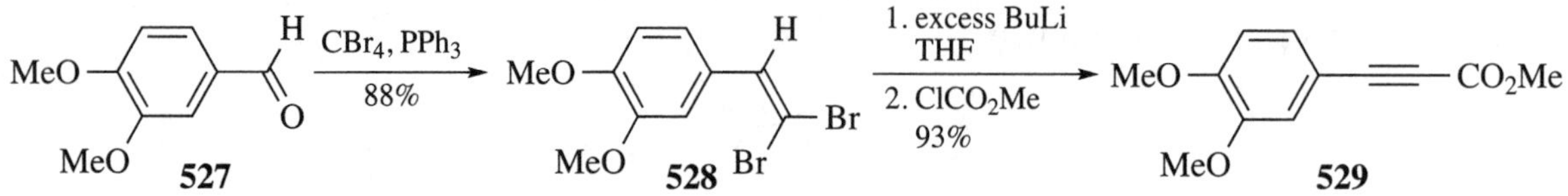

For use in the Wittig reaction, it is important that the R[1] and R[2] groups on the ylid (see **508**) not be subject to spontaneous fragmentation when the carbanion is formed. This is illustrated by bis(phosphonium) salt **530**, which upon treatment with phenyllithium did not give the ylid but rather decomposed to triphenylphosphine and the vinyl phosphonium salt **531**.[522] Such fragmentation occurs most often when a leaving group is β to the phosphorus atom. Conversion to the bis(ylid) is straightforward when the two phosphorus-containing groups are separated by three or more methylene groups.[522]

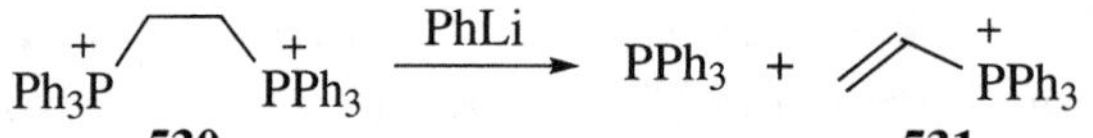

Ylids with a heteroatom substituent can decompose to a carbene intermediate (sec. 13.9.B). Butoxymethyl ylid **532** spontaneously fragmented to triphenylphosphine and carbene **533**, for example. This carbene reacted with an additional molecule of the ylid (**527**) to give the 1,2-dienolether **535** via the zwitterion **534**.[523]

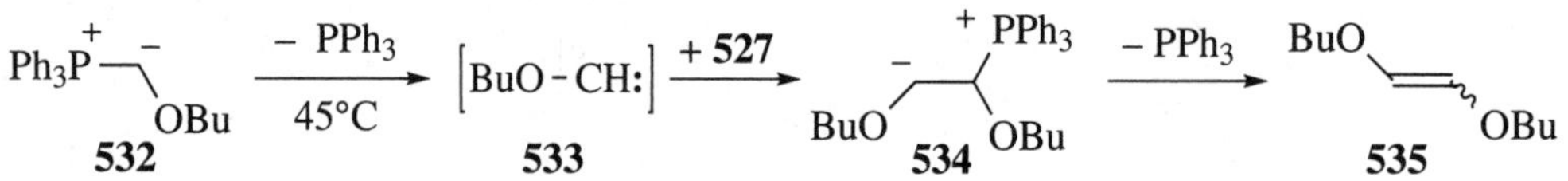

Phosphorus ylids rarely rearrange, but the presence of a reactive group such as a halide[524] or ester[525] can lead to anomalous products. In one example, the carbanionic carbon of **536** displaced the bromide to give triphenylcyclobutylphosphonium bromide (**537**), via an internal S_N2 reaction.[522] In ester **538**, nucleophilic attack of the carbanionic carbon generated α-keto ylid **539** via acyl substitution.[523]

520. Corey, E.J.; Fuchs, P.L. *Tetrahedron Lett.* **1972**, 3769.
521. O'Malley, S.J.; Tan, K.L.; Watzke, A.; Bergman, R.G.; Ellman, J.A. *J. Am. Chem. Soc.* **2005**, *127*, 13496.
522. (a) Wittig, G.; Eggers, H.; Duffner, P. *Annalen* **1958**, *619*, 10; (b) Trippett, S. *Chem. & Ind. (London)* **1956**, 80; (c) Dicker, D.W.; Whiting, M.C. *J. Chem. Soc.* **1958**, 1994.
523. (a) Wittig, G.; Böll, W. *Chem. Ber.* **1962**, *95*, 2526; (b) Trippett, S. *Proc. Chem. Soc.* **1963**, 19.
524. (a) Mondon, A. *Annalen* **1957**, *603*, 115; (b) Scherer, K.V.; Lunt III, R.S. *J. Org. Chem.* **1965**, *30*, 3215.
525. Bergel'son, L.D.; Vaver, V.A.; Barsukov, L.I.; Shemyakin, M. *Izvest. Akad. Nauk. SSSR* **1963**, 1134.

Ph₃P⁺ ~~~ Br ⟶ Ph₃P⁺ ◻   Ph₃P⁺ ~~~~ (OEt)=O ⟶ ⁺PPh₃ =O

536   **537**   **538**   **539**

8.8.A.ii. (E/Z) Isomers in Wittig Reactions. When a phosphorus ylid reacts with an aldehyde or ketone, the alkene that is formed is a mixture of (E) and (Z) isomers.[526] Initial postulations suggested that the stereochemistry of the alkene products was controlled by the stereochemistry of the betaine, which rapidly collapsed to an oxaphosphetane. Subsequent fragmentation to the alkene occurred via syn-elimination.[492] An example of this postulate is shown by the reaction of ylid **540** and 2-butanone to generate betaine **541**, and the anti-oxaphosphetane, **543**.[492] Syn elimination led to the (E)-alkene, **545**. The reaction mixture also contained betaine **542**, the precursor to the syn oxaphosphetane (**544**), which led to the (Z) alkene, **546**.

540 + (ketone) ⟶ **541** + **542**

545 ⟵ **543** + **544** ⟶ **546**

Resonance stabilized ylids such as Ph₃P=CHCO₂Et and Ph₃P=CHPh give predominantly the (E) alkene, whereas non-stabilized ylids such as Ph₃P=CHEt give predominantly the (Z) alkene.[527] These results are also rationalized in terms of a betaine intermediate. For resonance-stabilized ylids, formation of the betaine is reversible and equilibration favors formation of the thermodynamically more stable anti-betaine. The anti-betaine may also collapse faster than the syn-betaine.[528] If elimination of triphenylphosphine oxide is slow, the possibility of equilibration to the anti-betaine is maximized, and there will be increased amounts of the (E) alkene. Tricyclohexylphosphine ylids are less electrophilic than triphenyl ylids, and they show greater selectivity for formation of the (E) alkene.[529] Addition of a protic solvent or a Lewis acid to coordinate the oxygen atom of the betaine leads to more (Z) isomer.[526,529] If the metal coordinates with the betaine (or the oxaphosphetane) after reaction of an aldehyde and Ph₃P=CHCO₂Me, this coordination slows down the interconversion of **549** and **550** and diminishes the stereoselectivity. The equilibrium concentration of **547** and **548** is not the same as the concentration of **549** and **550**. The complexed form shows some preference for the anti-conformation **550**, which gives a slight preference for the (Z) isomer. These arguments refer only to stabilized phosphoranes and ignore the stereochemistry of the groups around the phosphorus

526.   Reucroft, J.; Sammes, P.G. *Quart. Rev. Chem. Soc.* **1971**, *25*, 135.
527.   Schlosser, M.; Christman, K.F. *Annalen* **1967**, *708*, 1.
528.   Speziale, A.J.; Bissing, D.E. *J. Am. Chem. Soc.* **1963**, *85*, 1888, 3878.
529.   Bestmann, H.J.; Kratzer, O. *Berichte* **1962**, *95*, 1894.

atom, which are stable throughout the reaction.[530] When the ylid is not stabilized (as when it possesses alkyl substituents), betaine and oxaphosphetane formation is essentially irreversible. Reaction of unsaturated ylids with aromatic aldehydes or α,β-unsaturated aldehydes shows some reversibility, however.[531] Lithium salts are usually present in the reaction medium, and this can lead to poor yields of alkene products. Under salt free conditions, the reaction appears to be under kinetic control, leading to the syn-betaine and the (Z) alkene.[532]

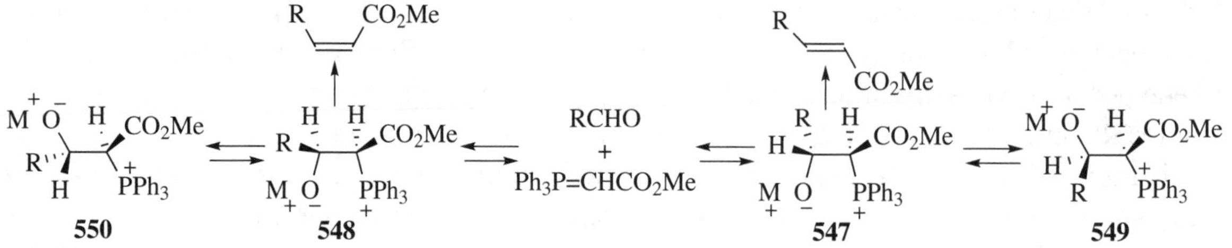

Schlosser et al. found that manipulating the solvent and the salt concentration allows some control of the (E/Z) selectivity, and rationalized his observations in terms of a betaine intermediate.[533] Betaine **551** can react with unreacted *n*-butyllithium to form a dilithio species (**552**), which rapidly establishes an equilibrium with **553**.[533] If lithium salts are added to this equilibrium, oxaphosphetane formation and elimination from the kinetic syn betaine **551** is inhibited. Further addition of excess butyllithium leads to **552** and **553**. This equilibrium favors the anti species **553**. Protonation leads to betaine **555** (or the potassium salt, **554**) and thereby to an oxaphosphetane which collapses to the anti alkene. This study suggests that excess base and excess salt promotes equilibration and formation of the anti (E) alkene.

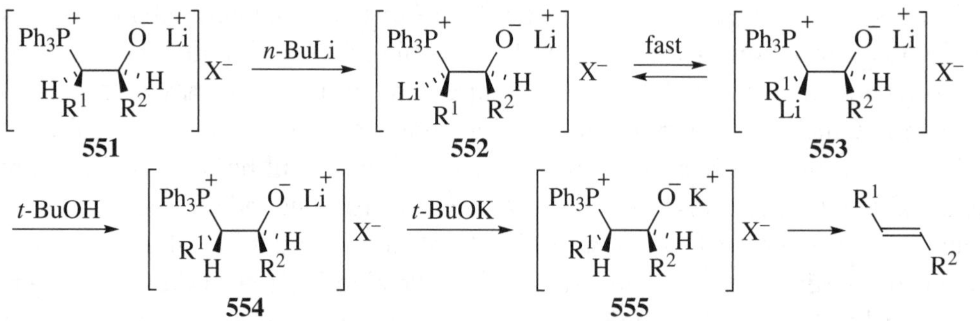

To obtain the (Z) alkene, salt free conditions are required. The syn betaine is relatively hindered, and as the carbonyl reactant becomes more hindered there is a greater preference for the (Z)

530. Bladé-Font, A.; McEwen, W.; VanderWerf, C.A. *J. Am. Chem. Soc.* **1960**, *82*, 2646, 2396.
531. Schlosser, M.; Christmann, K.F. *Angew. Chem. Int. Ed.* **1965**, *4*, 689; *Ibid.* **1964,** *3*, 636.
532. (a) Bestmann, H.J. *Angew. Chem. Int. Ed.* **1965**, *4*, 583; (b) Wittig, G.; Eggers, H.; Duffner, P. *Annalen* **1958***, 619*, 10.
533. (a) Schlosser, M.; Müller, G.; Christmann, K.F. *Angew. Chem. Int. Ed.* **1966**, *5*, 667; (b) Schlosser, M.; Christmann, K.F. *Ibid.* **1966**, *5*, 126.

alkene.[534] Schneider explained this effect by considering the importance of the substituent configuration around the phosphorus atom during reaction.[535] The intermediate **556** (apical oxygen and equatorial ylid carbon) must rotate at the O-C bond to form the betaine (**557** or **558**). The Ph↔R^2 steric interaction is maintained throughout, but the R^1↔Ph interaction in **556** inhibits formation of **558** and favors **557** (the syn betaine), leading to the (Z) alkene. With protic solvents or polar additives, the configuration of the phosphorane may change, and this model does not apply. For stabilized ylids, the pre-betaine intermediate is represented by **559** and is oriented so that the positive end is attracted to the nucleophilic acyl carbon, and forms the anti betaine and the (E) alkene.[529] The R^1 group rotates to form the betaine, and this rotation minimizes the R^1↔CO_2Et interaction.

The intermediacy of the betaines in the Wittig reaction has been questioned, and Vedejs has proposed an alternative explanation. The Wittig reaction is subject to solvent effects that indicate a nonpolar transition state for stabilized ylids.[536] There appears to be no direct evidence for the presence of betaines, and none have been isolated. Alternatively, Vedejs and Snoble detected oxaphosphetanes as the only observable intermediates in several Wittig reactions of non-stabilized ylids,[537] using ^{31}P NMR.[538] Vedejs devised a test for the betaine mechanism based on changes in phosphorus stereochemistry of betaines versus oxaphosphetanes.[539] The results of this test suggested that "the conventional betaine mechanism[540,505e] can play at most a minor role in the Wittig reaction".[539] Vedejs points out that the "stereochemical test does not necessarily disprove mechanisms via intermediates with lifetimes that are short compared to the time scale of bond rotation."[539]

Vedejs proposes two oxaphosphetane models to predict the stereochemistry in the Wittig reaction. For systems that give primarily cis alkenes, a cis-oxaphosphetane is required based on the assumption that the reaction proceeds by an early transition

534. Axen, U.; Lincoln, F.H.; Thompson, J.L. *J. Chem. Soc. Chem. Commun.* **1969**, 303.
535. Schneider, W.P. *J. Chem. Soc. Chem. Commun.* **1969**, 785.
536. (a) Frøyen, P. *Acta Chem. Scand.* **1972**, *26*, 2163; (b) Aksnes, G.; Khalil, F.Y. *Phosphorus* **1972**, *2*, 105 [*Chem. Abstr.* 78:70865v **1973**]; (c) Idem, *Ibid.* **1973**, *3*, 79 [*Chem. Abstr.* 79:145749s **1973**].
537. Vedejs, E.; Snoble, K.A.J. *J. Am. Chem. Soc.* **1973**, *95*, 5778.
538. Vedejs, E.; Meier, G.P.; Snoble, K.A.J. *J. Am. Chem. Soc.* **1981**, *103*, 2823.
539. Vedejs, E.; Marth, C.F. *J. Am. Chem. Soc.* **1990**, *112*, 3905.
540. (a) Maercker, A. *Org. React.* **1965**, *14*, 270; (b) Schlosser, M. *Topics Stereochem.* **1973**, *5*, 1.

state (like starting materials), and the oxaphosphetane has a puckered conformation (**560**).[541] Similarly, for systems that give trans-alkenes as the major product, a *trans*-oxaphosphetane is required, the reaction is assumed to proceed by a late transition state (product-like) and the oxaphosphetane has a planar structure (**561**).

A trans-oxaphosphetane is more stable than the cis-oxaphosphetane, and equilibrium conditions favors the trans-product. Kinetic control conditions appear to dominate Wittig reactions and lead to the cis-product using this analogy.[537] Vedejs and co-workers suggest that there is no single, dominant Wittig transition state geometry but rather a continuum of related mechanistic variants.[538] Examination of **560** and **561** suggests that the interplay of 1,2-steric interactions (interactions of substituents on P and C2 in **560** and also on C2-C3) and 1,3-steric interactions (interactions of substituents on P and C3 in **561**) will determine cis or trans selectivity.[539] Puckered transition states are more important when large groups at the α-carbon provide increasing 1,3-steric interactions. For stabilized ylids, 1,2-interactions dominate the transition state, leading to trans selectivity.

Maryanoff et al. also studied this problem, and found a prevalence of (*Z*)-alkene and cis-oxaphosphetane in the salt-free Wittig reaction, which forms the alkene by syn elimination of $Ph_3P=O$.[542] His work also suggests a very slow equilibration of cis- and trans-oxaphosphetanes relative to the rate of alkene formation. There is a significant concentration dependence on the stereochemistry of the reaction. Increasing concentration favored the trans-oxaphosphetane.[542] At high dilution, alkenes were generally formed with (*Z*)-selectivity in THF with LiBr present. The concentration effect may be associated with sequestration of lithium by THF. The ability of lithium soluble salts to promote formation of (*E*)-alkenes from non-stabilized ylids is well known (see above), but it is concentration dependent in THF. Based on studies with many types of phosphorus ylids, Maryanoff et al. concluded that betaines may have a "meaningful, albeit transient, existence,[542]" despite the lack of direct evidence for their existence. Since kinetic experiments demonstrated that cis-oxaphosphetanes undergo reversal to the ylid and an aldehyde faster than does the trans-oxaphosphetane (see above), the degree of stereochemical drift in Wittig reactions is probably due to the relative rates of oxaphosphetane reversal.

Stereochemical control in a Wittig reaction is seen in Wu and co-worker's synthesis of annonacin,[543] in which aldehyde **562** was treated with ylid **563** to give an 81% yield of **564**. Pan and co-worker's[544] synthesis of (–)-6,7-dehydroferruginyl methyl ether is a trans specific reaction, in contrast to formation of **564**. Aldehyde **565** reacted with the ylid generated from phosphonium chloride **566** to give the trans alkene (**567**) in 60% yield.

---

541. Vedejs, E.; Marth, C.F. *J. Am. Chem. Soc.* **1988**, *110*, 3948.
542. (a) Maryanoff, B.E.; Reitz, A.B.; Mutter, M.S.; Inners, R.R.; Almond, Jr., H.R.; Whittle, R.R.; Olofson, R.A. *J. Am. Chem. Soc.* **1986**, *108*, 7664; (b) Maryanoff, B.E.; Reitz, A.B.; Duhl-Emswiler, B.A. *Ibid.* **1985**, *107*, 217.
543. Hu, T.-S.; Wu, Y.-L.; Wu, Y. *Org. Lett.* **2000**, *2*, 887.
544. Gan, Y.; Li, A.; Pan, X.; Chan, A.S.C.; Yang, T.-K. *Tetrahedron: Asymmetry* **2000**, *11*, 781.

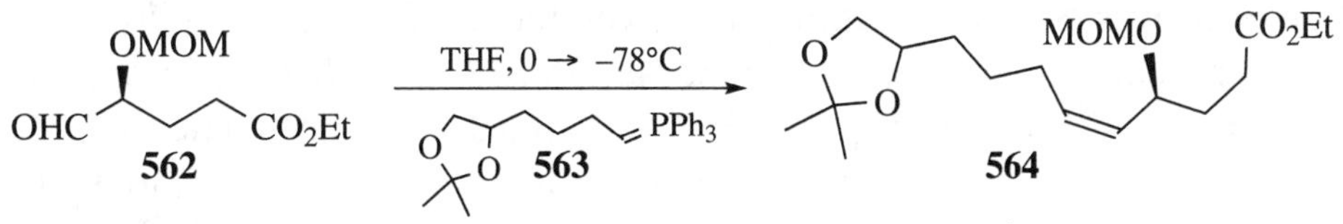

8.8.A.iii. Phosphine Oxides and Phosphonate Esters.[545] Many extensions of the Wittig reaction have been introduced that improve or modify the reactivity and/or stereoselectivity of the ylid. Horner et al. showed that α-lithiophosphine oxides such as that derived from **568** react with aldehydes or ketones to give a β-hydroxy phosphine oxide (**569**) as an isolable species.[546] Subsequent treatment with base liberates the alkene (**570**). Wadsworth and Emmons modified the **Horner reaction** to use phosphonate ester derivatives such as **571**. Reaction with aldehydes or ketones (such as benzaldehyde) in the presence of base gave the olefination product **572** in 84% yield.[547] These variations have come to be called the **Horner-Wadsworth-Emmons modification** of the Wittig reaction, or simply the **Horner-Wadsworth-Emmons olefination**.[548] It is sometimes called **Horner-Emmons olefination**.

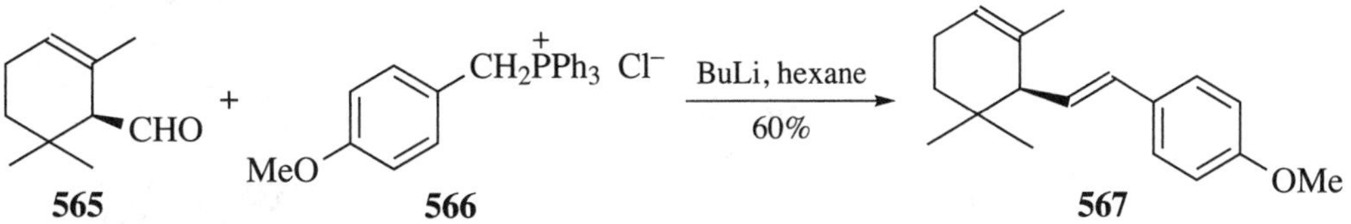

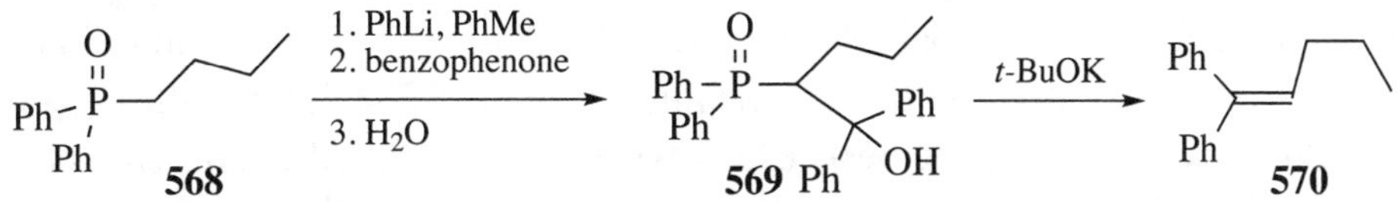

The major product of olefination with phosphonate carbanions is usually the (E)-isomer.[549] However, a (Z)-selective reaction has been developed that uses sodium iodide and DBU as the base.[550] Speziale and Ratts suggested that increased amounts of (Z) alkene were obtained by increasing the steric bulk of the L and L^1 groups in **573** and **576**, formed by reaction of the phosphonate ester ylid with a carbonyl compound. This equilibrium favors the anti

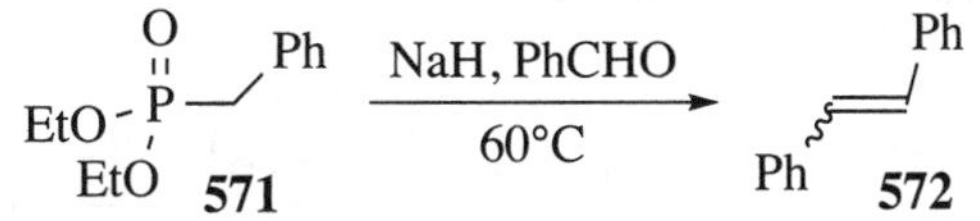

545. For reactions of this type, see Larock, R.C. *Comprehensive Organic Transformations, 2nd Ed.* Wiley-VCH, New York, *1999*, pp 332-335.
546. Horner, L.; Hoffmann, H.; Wippel, J.H.G.; Klahre, G. *Berichte* *1959*, *92*, 2499.
547. Wadsworth, Jr, W.S.; Emmons, W.D. *J. Am. Chem. Soc.* *1961*, *83*, 1733.
548. (a) Boutagy, J.; Thomas, R. *Chem. Rev.* *1974*, *74*, 87; (b) Mundy, B.P.; Ellerd, M.G.; Favaloro Jr., F.G. *Name Reactions and Reagents in Organic Synthesis, 2nd ed.*, Wiley-Interscience, New Jersey, *2005*, pp. 334-335.
549. Wadsworth, D.H.; Schupp III, O.E.; Seus, E.J.; Ford, Jr., J.A. *J. Org. Chem.* *1965*, *30*, 680.
550. Ando, K.; Oishi, T.; Hirama, M.; Ohno, H.; Ibuka, T. *J. Org. Chem.* *2000*, *65*, 4745.

conformation **573** over **576**, and elimination gives more (Z) product, **575**.[551] This model assumes that steric encumbrance is more important in **573** and **576** than in oxaphosphetanes **574** and **577**, respectively, that are required for syn elimination to the alkene. This equilibrium generally favors **575** over **578** due to the stereochemical preferences in the initially formed ylid products.

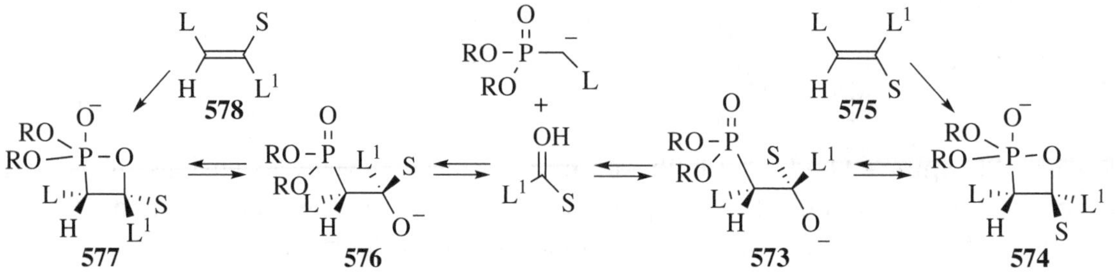

Bases such as *n*-butyllithium, potassium *tert*-butoxide or sodium hydride are usually required to generate the phosphonate carbanion.[552] The nucleophilicity of that anion is influenced by the nature of the metal ion. Addition of a metal ligand to the reaction mixture also influences the relative ease of removal of the α-proton. With some metals, a weaker base can be used to generate the carbanion.[553] Addition of metal salts to **579** (R = Et) led to formation of carbanion **580**, which was stabilized by chelation to the metal.[554] Addition of complexing Lewis acids such as LiBr or MgBr$_2$ allowed the ylid to be formed with a base as weak as triethylamine. Magnesium bromide was the most efficient metal additive for producing this product.[554] Subsequent reactions were done with cyclohexanone to give **581** (also with benzaldehyde to give the =CHPh derivative), and the best results were obtained in ether solvents or acetonitrile with lithium halides, although magnesium bromide was also effective.

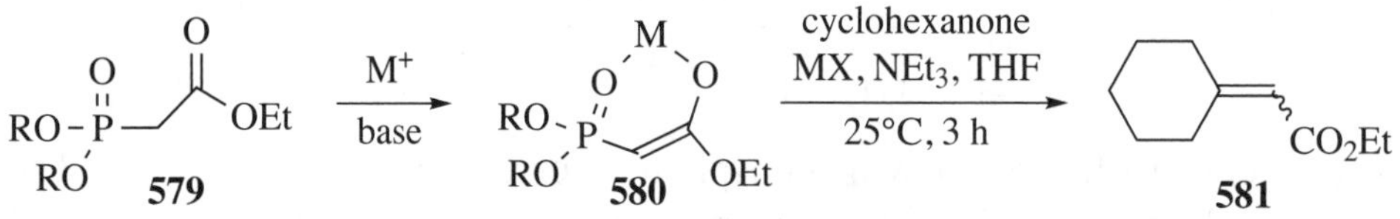

Still and Gennari[555] showed that octanal reacted with the carbanion derived from **579** to give a mixture of **582** and **583**. Using KN(TMS)$_2$ (potassium hexamethyldisilazide) as a base (sec. 9.2.B) in THF in the presence of 18-crown-6, gave an 8:1 mixture of **582/583** in 81% yield when OR = OMe in **579**. Changing OR to OCH$_2$CF$_3$ led to isolation of a 12:1 mixture of **582/583** in 90% yield.[555] Reaction of **579** (R = CH$_2$CF$_3$) with benzaldehyde, under the

551. Speziale, A.J.; Ratts, K.W.O. *J. Am. Chem. Soc.* **1963**, *85*, 2790.
552. Wadsworth, Jr., W.S. *Org. React.* **1977**, *25*, 73.
553. (a) Bottin-Strzalko, T.; Corset, J.; Froment, F.; Ponet, M.J.; Seyden- Penne, J; Simmonin, M.J. *J. Org. Chem.* **1980**, *45*, 1270; (b) Blanchette, M.A.; Choy, W.; Davies, J.T.; Essenfeld, A.P.; Masamune, S.; Roush, W.R.; Sakai, T. *Tetrahedron Lett.* **1984**, *25*, 2183.
554. Rathke, M.W.; Nowak, M. *J. Org. Chem.* **1985**, *50*, 2624.
555. Still, W.C.; Gennari, C. *Tetrahedron Lett.* **1983**, *24*, 4405.

same conditions, gave a >50:1 mixture of (Z/E) alkenes in >95% yield.[555] When Triton B (PhCH$_2$NMe$_3$$^+OH^-$) was used in THF with **579** (R = CH$_2$CF$_3$) a 7:1 mixture of **582/583** was obtained (84%), but the use of KO*t*-Bu in THF gave a 2:5 mixture of **582/583** in 70% yield.[555] These studies show that the preference for the (Z)-alkene product is strongly influenced by the reaction conditions and phosphonate carbanion chosen for the reaction. An asymmetric olefination using a benzopyranoisoxazolidine auxiliary was reported, generating alkylidene cyclohexane derivatives with good asymmetric induction.[556]

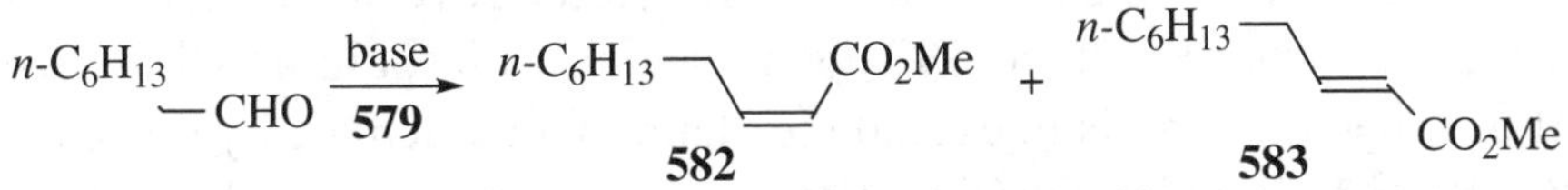

There are many applications of phosphonate ester methodology in organic synthesis. In a synthesis of (+)-himgaline,[557] Evans and Adams reacted phosphonate ester **585** with aldehyde **584** in the presence of Hünig's base and lithium perchlorate to incorporate the chiral auxiliary for an 82%% yield of conjugated carbonyl **586**.

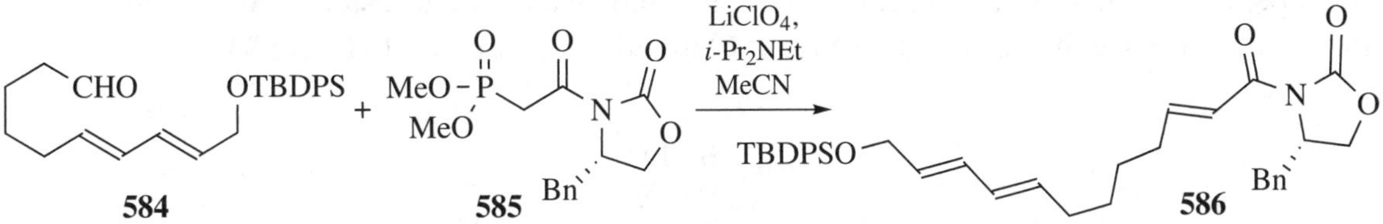

Intramolecular olefination reactions are possible using phosphonate carbanions. This type of cyclization is also possible with Wittig reagents, but it is more facile with the more stable phosphonate carbanions. An example is taken from a synthesis of palmerolide A by De Brabender and co-worker's, in which **587** was cyclized to macrolide **588** containing the conjugate ketone moiety in 70% yield from the primary alcohol precursor to **587**.[558]

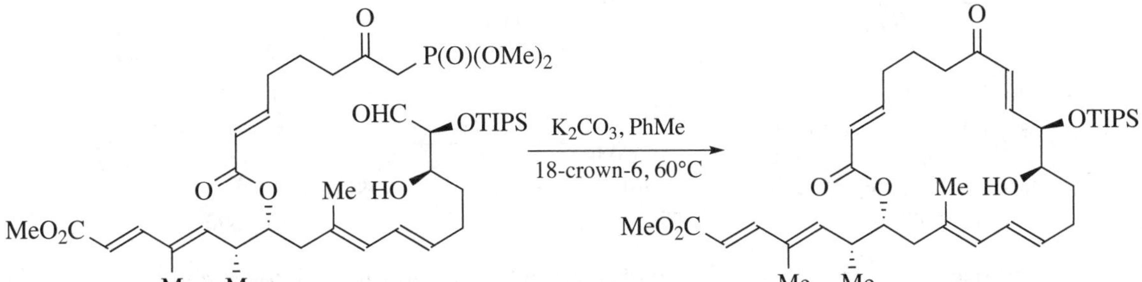

Warren showed that olefination of ketones and aldehydes, or even esters with phosphine oxide carbanions, allowed some control of stereochemistry in the alkene products. Reaction with

556. Abiko, A.; Masamune, S. *Tetrahedron Lett.* **1996**, 37, 1077.
557. Evans, D.A.; Adams, D.J. *J. Am. Chem. Soc.* **2007**, *129,* 1048.
558. Jiang, X.; Liu, B.; Lebreton, S.; De Brabander, J.K. *J. Am. Chem. Soc.* **2007**, *129*, 6386

an aldehyde or ketone generates diastereomeric β-hydroxyphosphine oxide products, which can be isolated and then chromatographically separated. Reaction of **589** with *n*-butyllithium was followed by addition of 3,4-methylenedioxybenzaldehyde (**590**, known as piperonal) to give a 9:1 mixture of **591/592**, the syn and anti alcohol, respectively.[559] The pure syn alcohol **591** was isolated in 75% yield by chromatography on silica gel. Subsequent treatment with sodium hydride in DMF gave the (*Z*)-alkene, α-isosafrole (**593**). If **589** reacted first with *n*-butyllithium and then with the analogous ester (**594**), the keto-phosphine oxide **595** was produced. Reduction of the ketone with sodium borohydride gave predominantly (≈ 6:1) the anti isomer **596**, which gave the (*E*)-alkene **597** upon reaction with sodium hydride. An interesting application of this methodology is to extend the chain of an aldehyde by one carbon (R–CHO → RCH$_2$CHO).[560] Indeed, many functional groups can be incorporated into the phosphonate ester, including nitriles, esters, carboxylic acids, etc. An interesting variation was used by Sorensen and co-worker's in a synthesis of FR182877,[561] in which Weinreb amide **598** was treated with LiCH$_2$P(O)(OMe)$_2$ (**599**) to give phosphonate ester **600**. Weinreb amides have been seen to react with Grignard reagents (Sec. 8.4), and reduced to aldehydes (Sec. 4.6.C), and phosphonate anions give highly functionalized phosphonate esters. Such phosphonate esters can, of course, be used in subsequent olefination reactions (also see **584**).

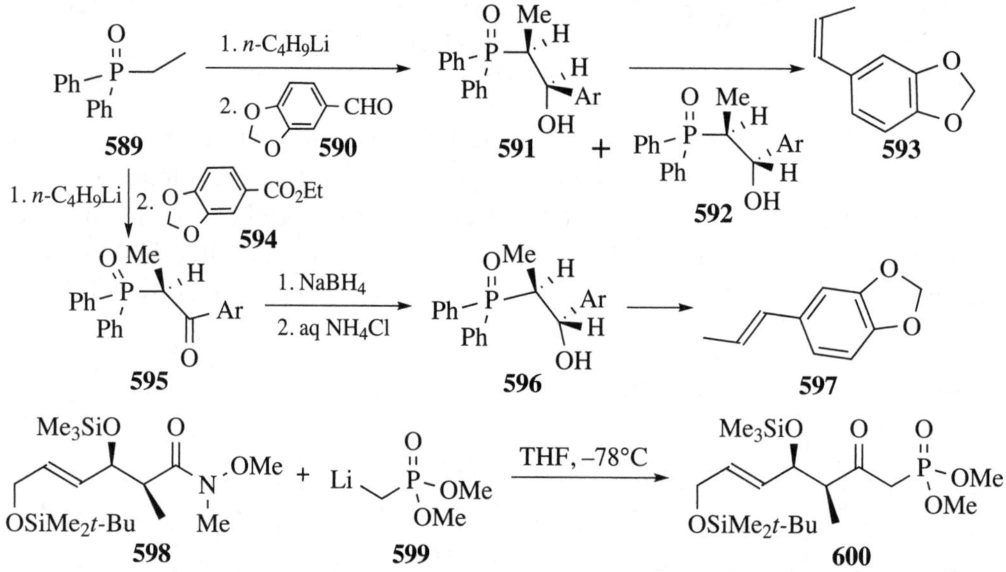

The use of phosphine oxide ylids and phosphonate ester carbanions just described is very similar to Corey's use of α-lithiophosphonic acid bis(amides) for olefination reactions. Corey showed

559. Buss, A.D.; Warren, S. *J. Chem. Soc. Chem. Commun.* **1981,** 100.
560. (a) Earnshaw, C.; Wallis, C.J.; Warren S. *J. Chem. Soc. Chem. Commun.* **1977,** 314; (b) Earnshaw, C.; Wallis, C.J.; Warren, S.T., *J. Chem. Soc. Perkin Trans. 1* **1979,** 3099; (c) Davidson, A.H.; Warren, S. *Ibid.* **1976,** 639.
561. Vanderwal, C.D.; Vosburg, D.A.; Welle, S.; Sorensen, E.J. *J. Am. Chem. Soc.* **2003,** *125,* 5393.

that reaction of **601** with *n*-butyllithium gave the lithio derivative **602**.[562] Subsequent reaction with methyl 2,2-dimethylpropanoate (methyl pivaloate) gave the corresponding ketone, **603**. Reduction with LiAlH$_4$ and quenching with water led to isolation of alcohol **604**. Heating in refluxing benzene with silica gel, liberated the alkene **605** in 71% yield.[560]

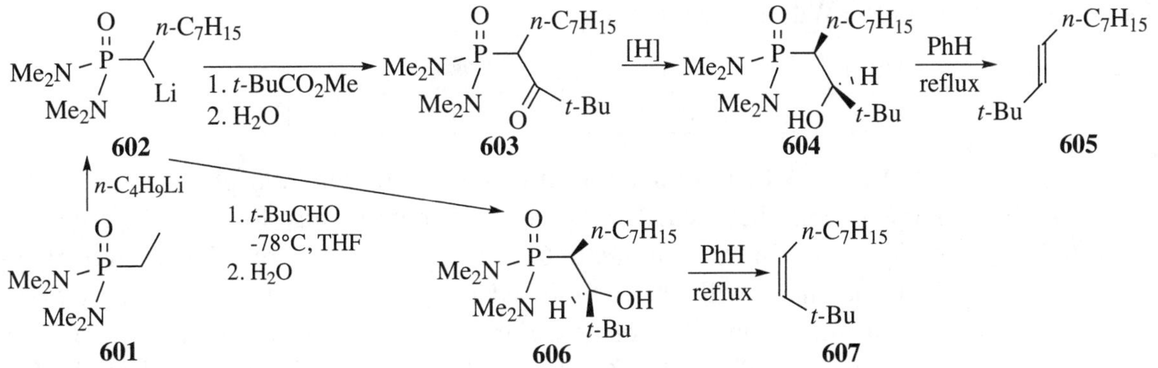

Corey also showed that the stereochemistry of the alkene could be controlled. Reaction of **601** with butyllithium and then pivalaldehyde, for example, gave **606**. Thermal elimination gave a 3:1 (*Z/E*) mixture of **607** and **605** favoring **607**.[563] By comparison, the triphenylphosphonium carbanion from 1-bromooctane reacted with pivalaldehyde to give a 98.5:1.5 mixture of **607/605**, which clearly favored the (*Z*)-isomer.[563] Good yields of (*E*) or (*Z*) alkenes were obtained by oxidation of the alcohol intermediate derived from reaction with aldehydes or ketones, with manganese dioxide in chloroform. Sodium borohydride was used for reduction of this ketone, as shown above, allowing control the relative stereochemistry of the intermediate product.

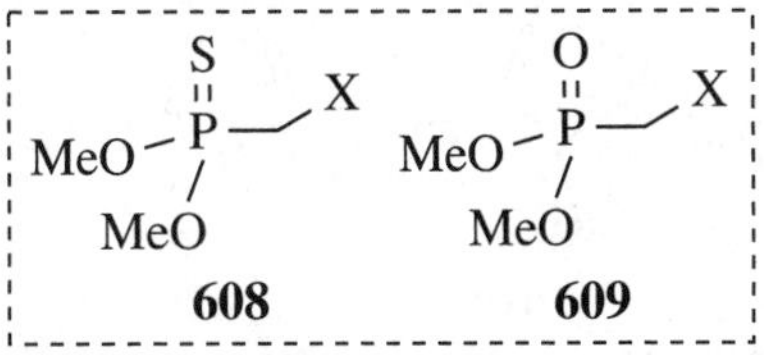

Corey and Kwiatkowski developed phosphonothioate derivative **608**[564] and phosphonate derivative **609**, and showed the corresponding ylids gave good yields of alkene products upon reaction with aldehydes and ketones when X = Cl, C(=O) R, or Ar. When X = H or alkyl in **609** (the ylid is not stabilized), formation of the lithio derivative was difficult and elimination to the alkene was also sluggish. The thio derivative (**608**), however, reacted with carbonyl derivatives and decomposed to alkene products at temperatures ranging from ambient to 65°C. An example is the reaction of **610** with *n*-butyllithium and then 1-bromobutane to give the alkylated product, **611**. This result is unusual since Wittig-type reagents do not undergo alkylation reactions in good yield. Addition of more base to **611** gave a new ylid and condensation with benzophenone at 25°C, gave **612**. Elimination to the alkene required higher temperatures (65°C) when the carbonyl component was a nonconjugated aldehyde or ketone.

562. Corey, E.J.; Kwiatkowski, G.T. *J. Am. Chem. Soc.* **1966,** *88,* 5652.
563. Corey, E.J.; Kwiatkowski, G.T. *J. Am. Chem. Soc.* **1966,** *88,* 5653.
564. Corey, E.J.; Kwiatkowski, G.T. *J. Am. Chem. Soc.* **1966,** *88,* 5654.

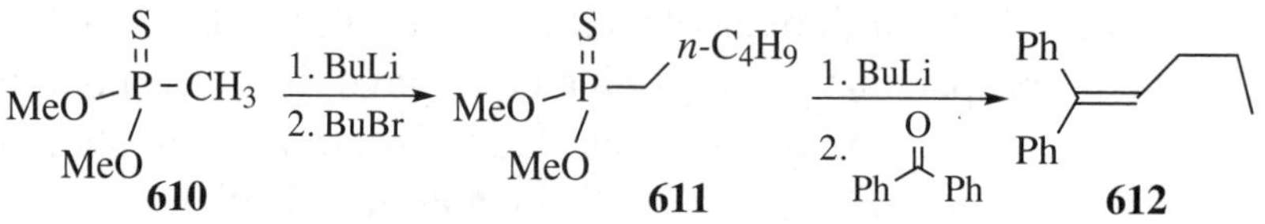

The disconnection for the Wittig type reactions is

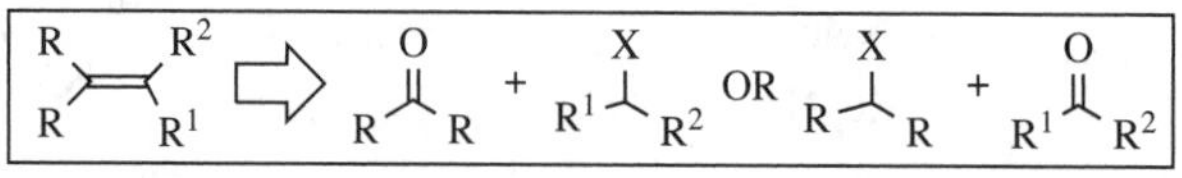

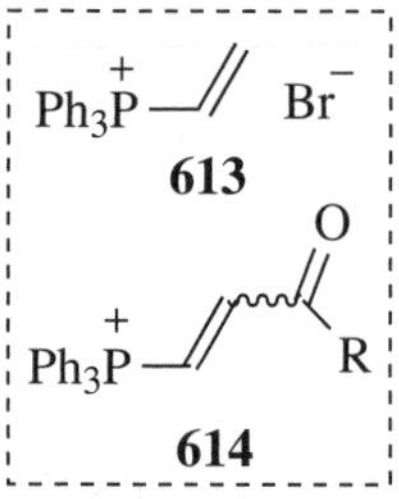

8.8.A.iv. Alkenylphosphonium Salts. Alkenylphosphonium salts are known, and they react with carbonyl compounds to form heterocyclic compounds. Phosphonium salt (**613**) was generated by reaction of 1-phenoxy-2-bromoethane with PPh₃, followed by heating.[565] Phosphonium salt **614** was formed from the corresponding β-chloro-α,β-unsaturated ketone by reaction with PPh₃ and **613**, followed by reaction with aqueous KBr.[566] The most common use of these phosphonium salts is in a reaction with carbonyl compounds that possess a nucleophilic atom at the β- or γ-position. Treatment of **615** with sodium hydride generated an alkoxide, and addition to **613** gave ylid **616** and intramolecular olefination reaction gave dihydrofuran derivative **617** in 50% yield.[567] α-Thioketones react with vinyl phosphonium salts to give dihydrothiophenes. When 3-mercapto-1-phenyl-2-propanone reacted with Ph₃P=CMe₂, for example, the product was **618**.[568] In a similar manner, pyrrole-2-carboxaldehyde (**619**) reacted with vinylphosphonium salts to give a 1,2*H*-pyrrolizine derivative (**620**).[569]

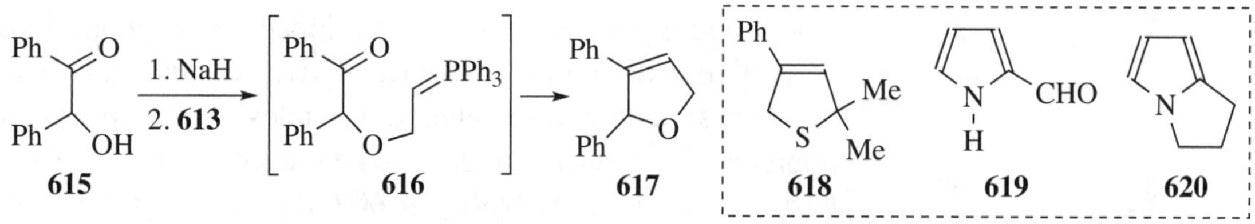

An interesting modification of this reaction generates a normal Wittig reagent by reaction of **613** with an organocuprate (sec. 8.7.A). Addition of dibutyl cuprate to **613** generated ylid **621**. Subsequent reaction with hexanal gave the expected alkene **622** as a 1:4 mixture of (*Z/E*) isomers.[570]

565. Schweizer, E.E.; Bach, R.D. *Org. Synth.* **1968**, *48*, 129.
566. Zbiral, E.; Rasberger, M. Hengstberger, H. *Annalen* **1965**, *725*, 22.
567. (a) Schweizer, E.E.; Liehr, J.C. *J. Org. Chem.* **1968**, *33*, 583; (b) Schweizer, E.E.; Creasy, W.; Liehr, J.C.; Jenkins, M.E.; Dalrymple, D.L. *Ibid.* **1970**, *35*, 601; (c) Schweizer, E.E. *J. Am. Chem. Soc.* **1964**, *86*, 2744.
568. McIntosh, J.M.; Goodbrand, H.B. *Tetrahedron Lett.* **1973**, 3157.
569. Schweizer, E.E.; Light, K.K. *J. Am. Chem. Soc.* **1964**, *86*, 2963.
570. Just, G.; O'Connor, B. *Tetrahedron Lett.* **1985**, *26*, 1799.

8.8.B. Sulfur Ylids

8.8.B.i. Sulfonium and Sulfoxonium Ylids. Just as phosphines react with alkyl halides to give the corresponding phosphonium salt, sulfides (thioethers) react with alkyl halides to give the corresponding sulfonium salt.[493,571] Ingold and Jessop reported sulfonium ylid **624** in 1930 (from the fluorene dimethylsulfonium salt **623**), but this initially formed ylid rearranged to give **625**[572] in a sulfur analog of the **Sommelet-Hauser rearrangement**[573] (sec. 8.8.C.ii). It is now known that sulfur ylids are formed by reaction with a strong base[574] (such as RO⁻/ROH, NaH/DMSO, RLi, or LiNR₂ in THF or ether) with the appropriate sulfonium salt. The sulfonium salt is formed by reaction of a dialkyl sulfide or a diaryl sulfide (R_2S) with primary, allyl or benzyl halides[575] The reaction of diphenyl sulfide and iodoethane gives sulfonium salt **626**.[576] Subsequent reaction with *tert*-butyllithium generated the ylid (**627**), which contains the Ph_2S^+ moiety, a good leaving group for S_N2 displacement by a nucleophile. The gegenion of the sulfonium salt is iodide, which can displace Ph_2S. In other words, the reaction that generates the sulfonium salt is reversible.[577] The use of nonnucleophilic gegenions such as tetrafluoroborate or perchlorate is usually a necessity to minimize this side reaction. When highly electrophilic species such as silver salts are used, other complicating reactions are possible. Rearrangements can occur by reaction of $AgBF_4$ with the alkyl halide to form the secondary cation (sec. 12.2.B). The straight-chain sulfonium salt is usually the major product.[578]

571. Reference 492, pp 304-366.
572. (a) Ingold, C.K.; Jessop, J.A. *J. Chem. Soc.* **1930**, 713; (b) Agami, C. *Bull. Soc. Chim. Fr.* **1965**, 1021.
573. (a) Hilbert, G.E.; Pinck, L.A. *J. Am. Chem. Soc.* **1938**, *60*, 494; **1946**, *68*, 751; (b) Pine, S.H. *Org. React.* **1970**, *18*, 403.
574. Reference 493, pp 13-23, 29-33.
575. (a) Corey, E.J.; Chaykovsky, M. *J. Am. Chem. Soc.* **1965**, *87*, 1353; (b) Hatch, M.J. *J. Org. Chem.* **1969**, *34*, 2133; (c) LaRochelle, R.W.; Trost, B.M.; Kiepski, L. *Ibid.* **1971**, *36*, 1126; (d) Hauser, C.R.; Kantor, S.W.; Brasen, W.R. *J. Am. Chem. Soc.* **1953**, *75*, 2660; (e) Ratts, K.W.; Yao, A.N. *J. Org. Chem.* **1966**, *31*, 1185; (f) Speziale, A.J.; Tung, C.C.; Ratts, K.W.; Yao, A.N. *J. Am. Chem. Soc.* **1965**, *87*, 3460.
576. Corey, E.J.; Oppolzer, W. *J. Am. Chem. Soc.* **1964**, *86*, 1899.
577. (a) Ray, F.E.; Farmer, J. *J. Org. Chem.* **1943**, *8*, 391; (b) Ray, F.E.; Levine, I. *Ibid.* **1938**, *2*, 267.
578. (a) Trost, B.M.; Bogdanowicz, M.J. *J. Am. Chem. Soc.* **1973**, *95*, 5298; (b) Tang, C.S.F.; Rapoport, H. *J. Org. Chem.* **1973**, *38*, 2806.

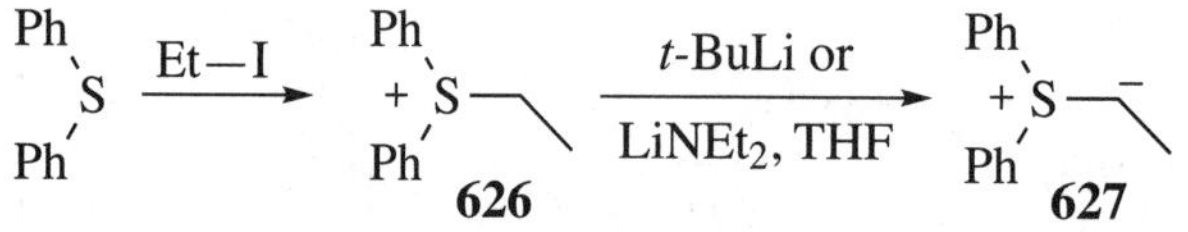

Dimethyl sulfide is a useful precursor to sulfur ylids, reacting with iodomethane to give trimethylsulfonium iodide (**628**). When this salt was treated with *n*-butyllithium, deprotonation of the hydrogen on the α-carbon led to dimethylsulfonium methylid (**629**). Dimethyl sulfoxide also reacts with alkyl halides to give a sulfoxonium salt. When DMSO reacted with iodomethane, trimethylsulfoxonium iodide (**630**) was formed. As with the dimethyl sulfide derivative, treatment with a strong base such as *n*-butyllithium generated the corresponding ylid, dimethylsulfoxonium methylid (**631**).

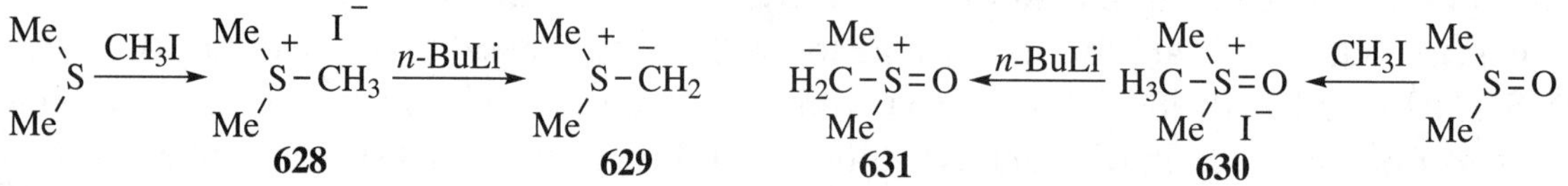

When **629** reacts with carbonyl derivatives, it essentially behaves as a carbanion rather than an ylid. In a synthesis of aphidicolin by Toyota, Sasaki, and Ihara,[579] the reaction of dimethylsulfonium methylid with ketone **632** gave the acyl addition product **633** as the initial product. Dimethyl sulfide (Me_2S) is a good leaving group, and the nearby alkoxide unit displaced Me_2S to give oxirane **634**, in 89% yield. Sulfur ylids normally react with aldehydes or ketones to generate an epoxide, and *not* an alkene.[580] Dimethylsulfoxonium ylid **631** also reacts with carbonyl compounds to give an intermediate (**635**) that contains a leaving group, DMSO, and displacement by the adjacent alkoxide also gives an oxirane (**636**).[575] An intermediate such as **633** or **635** was generated independently and shown to form an epoxide.[581]

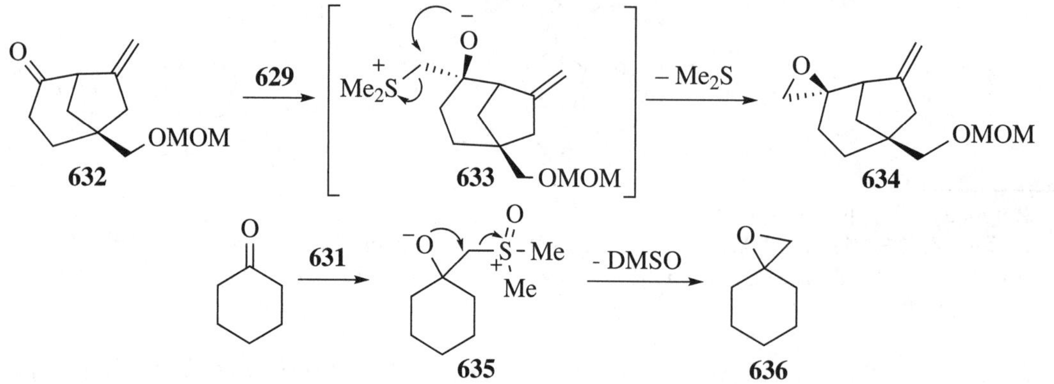

It is known that dimethylsulfonium methylid (**629**) adds irreversibly to ketones and aldehydes.

579.   Toyota, M.; Sasaki, M.; Ihara, M. *Org. Lett.* **2003**, *5*, 1193.
580.   For reactions of this type, see Larock, R.C. *Comprehensive Organic Transformations, 2nd Ed.* Wiley-VCH, New York, **1999**, pp 944-946.
581.   Johnson, C.R.; Schroeck, C.W.; Shanklin, J.R. *J. Am. Chem. Soc.* **1973**, *95*, 7424.

On the other hand, dimethylsulfoxonium methylid (**631**) adds reversibly. This difference is reflected in their reactivity with cyclic ketones such as 4-*tert*-butyl cyclohexanone (**637**). Dimethylsulfonium methylid (**629**) reacted with **637** to give the oxirane with an axial exocyclic C–C bond (**638**), which is less stable than the product with an axial C–O bond (**639**).[582] When **631** reacted with **637**,[583] however, the oxirane with the more stable equatorial exocyclic C–C bond (**639**) was formed. This is attributed to the reversibility of the addition of **631** to the ketone leading to the more stable oxirane whereas **629** formed the intermediate salt irreversibly in what is essentially a kinetic process.

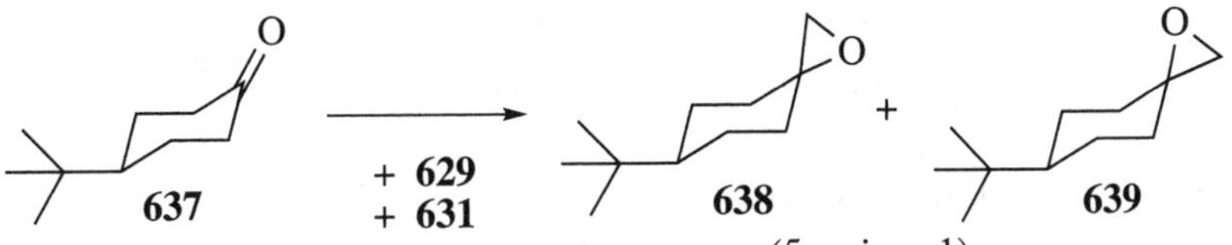

Another difference in the reactivity of these two ylids is reflected in their reactions with α,β-unsaturated ketones. The carboethoxy dimethylsulfonium methylid **640** reacted with cyclohexenone to give the usual alkoxy sulfonium salt **641**. Intramolecular displacement of dimethyl sulfide led to the expected oxirane (**642**), analogous to the formation of **634**. This product represents a generalized reaction, and dialkylsulfonium ylids react via 1,2-addition with displacement of the sulfide to give an allylic oxirane.[581] When carboethoxy dimethylsulfoxonium ylid **643** reacted with cyclohexenone, however, the 1,2-addition adduct was formed reversibly and 1,4-addition becomes competitive. The conjugate addition product **644** is an enolate anion (secs. 9.2, 9.3, 9.7) and intramolecular displacement of DMSO generated a cyclopropane ring, as in **645**.[583] The latter reaction is more facile when the conjugate system becomes more

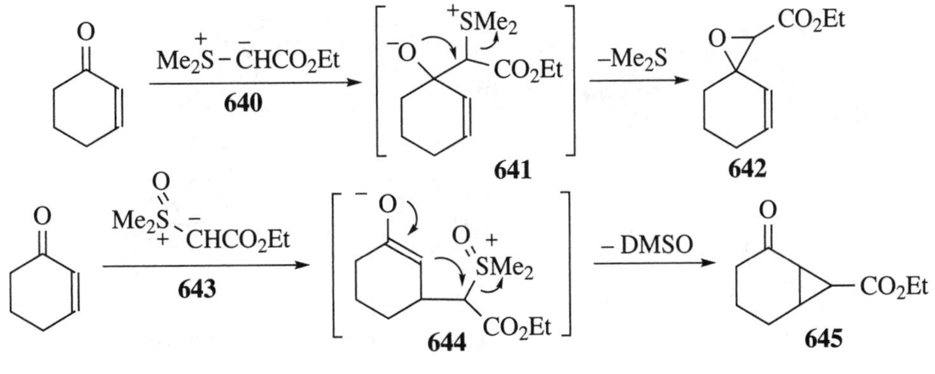

amenable to Michael addition. Acrylate **646**, for example, gave only a 9% yield of **647** after 1 h, whereas **648** gave a 91% yield of **649** under the same reaction conditions.[584] The reaction is

582. (a) Carlson, R.G.; Behn, N.S. *J. Org. Chem.* **1967**, *32*, 1363; (b) Johnson, C.R.; Katekar, R.A. *J. Am. Chem. Soc.* **1970**, *92*, 5753; (c) Johnson, C.R. *Acc. Chem. Res.* **1973**, *6*, 341.
583. Reference 493, p 40.
584. Landon, S.R.; Punja, N. *J. Chem. Soc. C* **1967**, 2495.

synthetically useful despite the limitations, as shown by the conversion of *R*-(–)-carvone (**650**) to **651** in 96% yield as part of Brocksom and co-worker's synthesis of guaiane sesquiterpenes.[585]

646 **647** **648** **649** **650** **651**

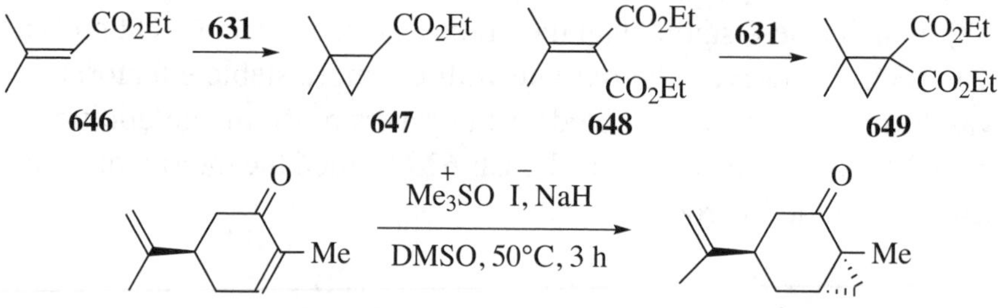

Disconnections for reactions of these sulfur ylids are

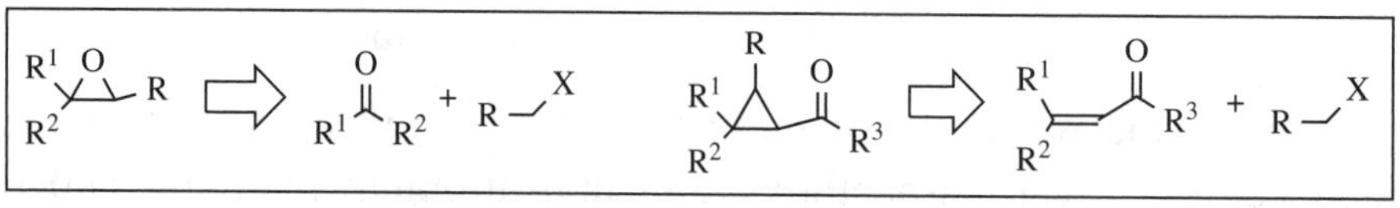

This methodology has been applied to a variety of synthetic applications.[586] Epoxide **652** reacted with **629** to give the allylic alcohol **653**, in 77% yield, a step in Kadota and co-worker's synthesis of hemibrevetoxin B.[587] Generating an oxirane in the presence of a nucleophile allows further reaction. Initial reaction of the ketone moiety of **654** with **631** gave the oxirane **655**. The nucleophilic nitrogen of imidazole opened the oxirane, however, to give **656**.[588] The cyclopropanation reaction is also quite useful for derivatization of key compounds, as in the cyclopropanation of the *N*-methyluridine moiety in **657** with **631** generated the **658** in 80% yield (7:3 β/α).[589]

652 **653**

654 **655** **656**

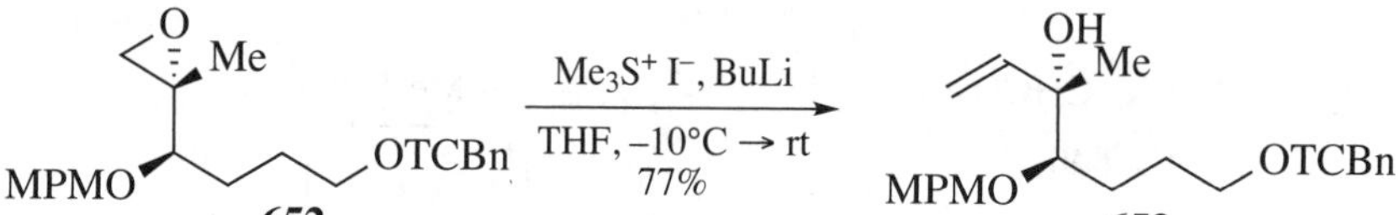

585. de Faria, M.L.; de A. Magalhaes, R.; Silva, F.C.; de O. Matias, L.G.; Ceschi, M.A.; Brocksom, U.; Brocksom, T.J. *Tetrahedron: Asymmetry* ***2000**, 11*, 4093.
586. Reference 493, pp 51-107.
587. Kadota, I.; Abe, T.; Ishitsuka, Y.; Touchy, A.S.; Nagata, R.; Yamamoto, Y. *Tetrahedron Lett.* ***2007**, 48*, 219.
588. Chaykovsky, M.; Benjamin, L.; Fryer, R.I.; Meltesics, W. *J. Org. Chem.* ***1970**, 35*, 1178.
589. Kunieda, T.; Witkop, B. *J. Am. Chem. Soc.* ***1971**, 93*, 3478.

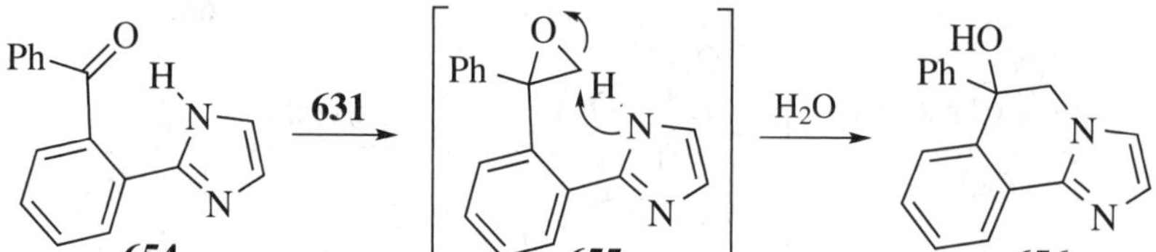

8.8.B.ii. Diphenylcyclopropylsulfonium Derivatives. Trost and Bogdanowicz showed that diphenyl-cyclopropylsulfonium ylids such as **659** reacted with ketones and aldehydes in a manner similar to **629**. Attack at the carbonyl generated the expected intermediate **660**, and intramolecular displacement of diphenyl sulfide by alkoxide gave the oxirane **661**. This oxirane, however, is an oxaspiropentane derivative and is susceptible to several additional synthetic manipulations.[590,593a] A useful transformation is the conversion of the oxaspiropentane moiety to cyclobutanone derivative (**662**) by treatment with Lewis acids such as HBF_4, $LiClO_4$, or $Eu(fod)_3$, where fod = tris(6,6,7,7,8,8,8)-heptafluoro-2,2-dimethyl-3,5-octanedionate.[591] Such products arise via a cationic rearrangement of an oxonium ion (sec. 12.2.B), formed by protonation of (or coordination to) the oxiranyl oxygen. Ring opening generated a cyclopropylcarbinyl cation, which rearranged to the oxo-stabilized cyclobutyl cation (sec. 12.2.B), and this gave the cyclobutanone product. When better coordinating Lewis acids are used [$LiClO_4$ or $Eu(fod)_3$], there is a chelation effect that further stabilizes the intermediate cation.

An alternative reaction pathway is available when a carbon adjacent to the oxiranyl moiety possesses a hydrogen. Reaction of **663** with lithium diisopropylamide (sec. 9.2.B) removed the hydrogen on the carbon α to the oxiranyl carbon and formed **664** via opening of the oxaspiro moiety. When this was done in the presence of chlorotrimethylsilane, the alkoxide was trapped as the trimethylsilyl ether **665** (sec. 7.3.A.i). Subsequent heating (flash vacuum pyrolysis at ~ 300-500°C) led to a rearrangement that generated **666**.[592] This so-called **vinyl cyclopropane rearrangement** is a sigmatropic rearrangement that will be discussed in sections 11.12.B and 13.9.C.iii. Silyl enol ether **666** is simply a protected enolate (secs. 9.2.B, 9.2.D) and aqueous hydrolysis liberated the unstable enol, **667** (sec. 7.3.A.i). Tautomerization favored the final product, ketone **668**.[592] Trost et al. applied this methodology to a synthesis of allamandin.[593]

590. (a) Trost, B.M.; Bogdanowicz, M.J. *J. Am. Chem. Soc.* **1971**, *93*, 3773; (b) Trost, B.M.; LaRochelle, R.; Bogdanowicz, M.J. *Tetrahedron Lett.* **1970**, 3449.
591. Trost, B.M.; Keeley, D.; Bogdanowicz, M.J. *J. Am. Chem. Soc.* **1973**, *95*, 3068.
592. Trost, B.M.; Bogdanowicz, M.J. *J. Am. Chem. Soc.* **1973**, *95*, 289.
593. (a) Trost, B.M.; Mao, M.K.-T.; Balkovec, J.M.; Buhlmayer, P. *J. Am. Chem. Soc.* **1986**, *108*, 4965; (b) Trost, B.M.; Balkovec, J.M.; Mao, M.K-T. *Ibid.* **1986**, *108*, 4974.

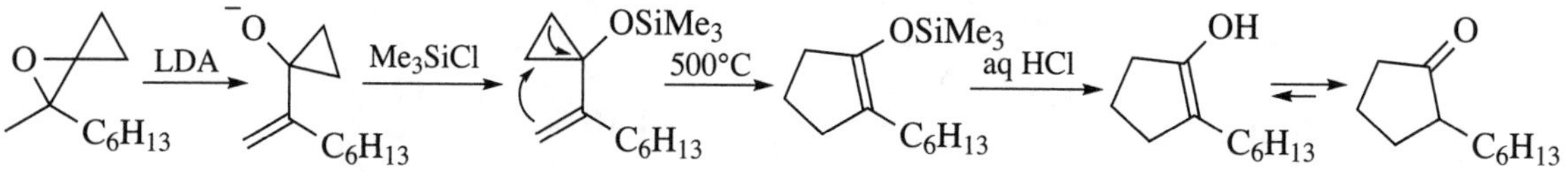

The disconnections for the cyclopropyl sulfide reactions are

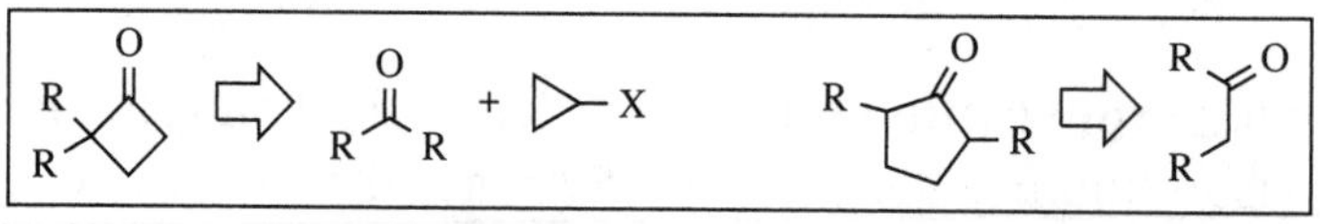

8.8.C. Nitrogen Ylids[494,594]

8.8.C.i. Formation of Nitrogen Ylids. The first report of a nitrogen ylid was probably by Krohnke in 1935, who prepared the resonance stabilized pyridinium ylid **669** by reaction of pyridinium salt **670** with potassium carbonate (K$_2$CO$_3$).[595] Wittig and Wetterling[596] found that phenyllithium reacted with tetramethylammonium bromide (**671**) to give the corresponding ylid (**672**). The ylid was shown to be the lithium bromide complex. Complexation with added or generated metal salts or other reagents is very important for the stability of nitrogen ylids. Ylid **672** decomposed to carbene (CH$_2$:) (see sec. 13.9) in solvents that strongly coordinate LiBr, and this carbene polymerized to polymethylene under the reaction conditions. In ether (ambient temperatures for 90 h), reaction of tetramethylammonium bromide (**671**) and phenyllithium gave 20% trimethylamine and 12% polymethylene. In dimethoxyethane (DME), **672** formed a strong complex with LiBr and polymethylene was formed in 74% yield. This result suggested that the free (uncomplexed) ylid was rather unstable.

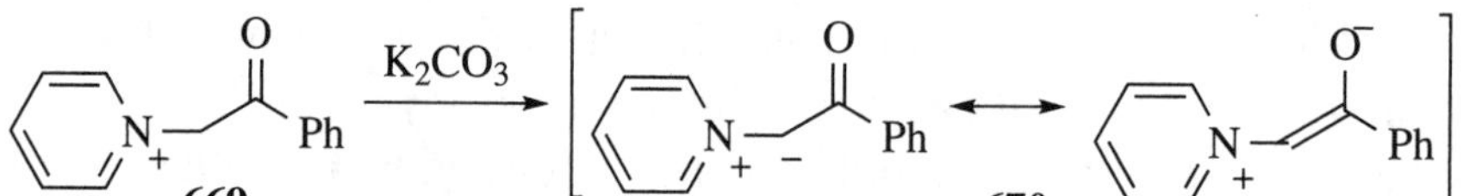

Wittig showed that tetramethylammonium bromide (**671**) reacted with two equivalents of phenyllithium to form a species (**672** and/or **673**) that gave 18% of monosubstituted product (**674**) and 20% of the bis-adduct (**675**) on reaction with benzophenone followed by hydrolysis.[597]

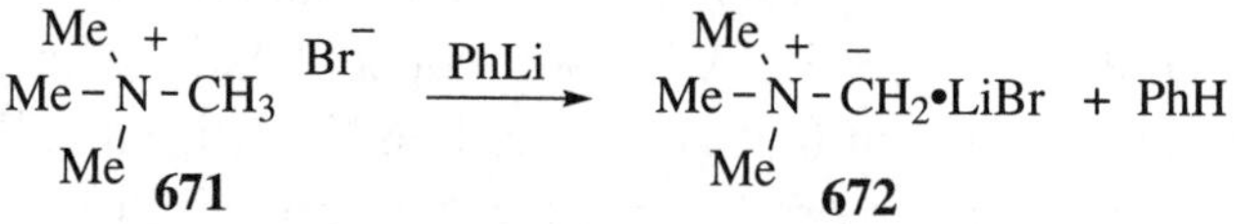

---

594. Reference 494, pp 251-283.
595. Kröhnke, F. *Berichte* **1935**, *68*, 1177.
596. Wittig, G.; Wetterling, M. *Annalen* **1947**, *557*, 193.
597. (a) Wittig, G.; Rieber, M. *Annalen* **1949**, *562*, 177; (b) Wittig, G.; Polster, R. *Ibid.* **1956**, *599*, 1.

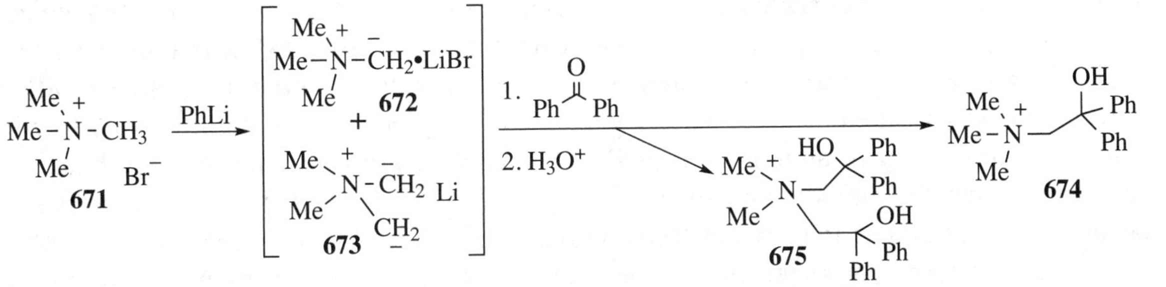

Although nitrogen ylids behave as carbanions, alkylation of **672** in THF is difficult because the carbanion is also basic and induces an elimination reaction with an alkyl halide (loss of trimethylamine). Reaction of **672** with bromocyclohexane, for example, gave a 92% yield of cyclohexene.[598] Ylid **672** does behave as a carbanion if the substrate does not possess a leaving group. Reaction with either benzonitrile or ethyl benzoate gave the ammonium ketone **676**.[598] Similar reaction with benzoyl chloride gave a mixture of the usual ketone product (**676**) and a secondary product resulting from *O*-alkylation of the enol form of the acid chloride. Nitrogen ylids behave as both a stronger base and a stronger carbanion than any of the phosphorus or sulfur ylids previously encountered.

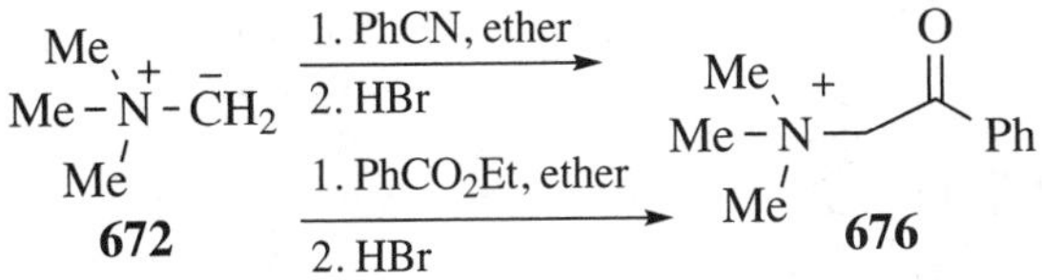

Pyridinium ylids such as those discovered by Krohnke (see above) and simple *N*-alkyl derivatives generally undergo nucleophilic reactions as carbanions. Pyridinium ylids react with aldehydes in a Knoevenagel reaction (sec. 9.4.B.iii) rather than as an ylid. Carbanion **677**, for example, reacted with 4-nitrobenzaldehyde to give **678** after elimination of water from the initial alcohol product.[596,597a,599] Pyridinium ylids[600] undergo Michael addition on reaction with conjugated carbonyl systems. Another class of nitrogen ylids is azomethine ylids, which are useful synthetic intermediates and often formed by thermal ring opening of aziridine derivatives.[601] The main synthetic use of azomethine ylids is in [3+2]-cycloadditions and this will be discussed in Section 11.11.F.[602]

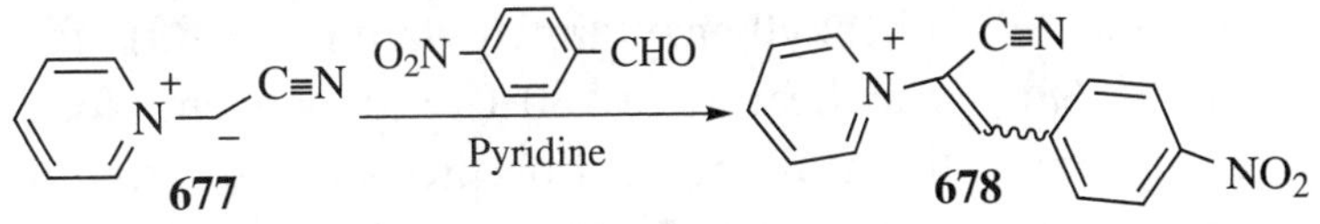

598. (a) Weygand, F.; Daniel, H.; Schroll, A. *Berichte* **1964**, *97*, 1217; (b) Weygand, F.; Daniel, H. *Ibid.* **1961**, *94*, 3147.

599. Also see (a) Kröhnke, F. *Chem. Ber.* **1951**, *84*, 388; (b) Idem, *Ibid.* **1950**, *83*, 253.

600. (a) Zecher, W.; Kröhnke, F. *Berichte* **1961**, *94*, 690; (b) Kröhnke, F.; Zecher, W.; Curtze, J.; Drechsler, D.; Pfleghar, K.; Schnalke, K.E.; Weis, W. *Angew. Chem.* **1962**, *74*, 811.

601. Huisgen, R.; Scheer, W.; Mäder, H.; Brunn, E. *Angew. Chem. Int. Ed.* **1969**, *8*, 604.

602. (a) Lown, J.W. *Recent Chem. Progr.* **1971**, *32*, 51 [*Chem. Abstr.* 76:3599g **1972**]; (b) Kellogg, R.M.; *Tetrahedron* **1976**, *32*, 2165; (c) Huisgen, R. *J. Org. Chem.* **1976**, *41*, 403; (d) Hermann, H.; Huisgen, R.; Mäder, H. *J. Am. Chem. Soc.* **1971**, *93*, 1779; (e) Fleming, I. *Frontier Orbitals and Organic Chemical Reactions*, Wiley-Interscience, London, **1976**, pp 109-110, 148-160.

8.8.C.ii. The Stevens' Rearrangement and the Sommelet Rearrangement.

The intermediacy of nitrogen ylids has been suggested for two classical reactions, the **Stevens' rearrangement**[603,573c] and the **Sommelet-Hauser rearrangement** (sometimes called the **Sommelet rearrangement**).[604] Stevens found that treatment of phenacylbenzyldimethylammonium bromide (**679**) with aqueous hydroxide gave amino ketone **682** The reaction probably proceeds via hydrogen abstraction from the ammonium salt (**679**) to give the ylid **680**. A N→C 1,2-benzyl shift occurs via a carbanionic intermediate **681** to give amine **682**.[603e] The mechanism[605] involves a [2,3]-sigmatropic rearrangement of the ylid, followed by proton migration to restore aromaticity. The preparation of complex alkaloids is a common synthetic application of the Stevens' rearrangement. The reaction of bis(α,α-o-xylylene) ammonium bromide (**683**) with phenyllithium led to the α-lithio compound **684**. This nitrogen ylid was unstable and rearranged with loss of LiBr, to **685**.[606]

Formation of o-benzylbenzyldimethylamine (**689**) from benzhydryltrimethylammonium bromide (**686**),[604] on heating to 180°C with concentrated hydroxide, illustrates the Sommelet rearrangement. Initial deprotonation probably occurred at the benzylic site, but equilibrium conditions generated ylid **687**. Nucleophilic attack at the proximal benzene ring gave **688** via cleavage of the C–N bond, and subsequent aromatization gave the final product **689**. Evidence suggests the mechanism is a [1,2]-shift of the ylid via a caged radical pair intermediate.[605]

603. (a) Stevens, T.S.; Creighton, E.M.; Gordon, A.B.; MacNicol, M. *J. Chem. Soc.* **1928**, 3193; (b) Stevens, T.S. *Ibid.* **1930**, 2107; (c) Thomson, T.; Stevens, T.S. *J. Chem. Soc.* **1932**, 55; (d) The Merck Index, 14th ed., Merck & Co., Inc., Whitehouse Station, New Jersey, **2006**, p ONR-89; (e) Mundy, B.P.; Ellerd, M.G.; Favaloro Jr., F.G. *Name Reactions and Reagents in Organic Synthesis, 2nd ed.*, Wiley-Interscience, New Jersey, **2005**, pp. 618-619.

604. (a) Sommelet, M. *Compt. Rend.* **1937**, *205*, 56; (b) Hauser, C.R.; van Eenam, D.N. *J. Am. Chem. Soc.* **1956**, *78*, 5698; (c) The Merck Index, 14th ed., Merck & Co., Inc., Whitehouse Station, New Jersey, **2006**, p ONR-88; (d) Mundy, B.P.; Ellerd, M.G.; Favaloro Jr., F.G. *Name Reactions and Reagents in Organic Synthesis, 2nd ed.*, Wiley-Interscience, New Jersey, **2005**, pp. 608-609.

605. (a) Tanaka, T.; Shirai, N.; Sugimori, J.; Sato, Y. *J. Org. Chem.* **1992**, *57*, 5034; (b) Lepley, A.R.; Giumanini, A.G. in *Mechanism of Molecular Migrations, Vol. 3;* Thyagarajan, B.S. (Ed.), Wiley-Interscience, New York, **1971**, p 297.

606. Wittig, G.; Tenhaeff, H.; Schoch, W.; Koenig, G. *Annalen* **1951**, *572*, 1.

Steric effects appear to be less significant than electronic effects for ylid stability.[607] The **ortho substitution rearrangement**[608] involves reaction of benzyltrimethylammonium salt **690** with $NaNH_2$ to give 2-(dimethylaminomethyl)toluene (**691**). This rearrangement is either the same reaction as the Sommelet rearrangement, or closely related.[608] Hauser and Jones showed that this rearrangement could be used for a ring expansion,[609] where reaction of dimethylammonium salt **692** with sodium amide gave an 83% yield of **693**.

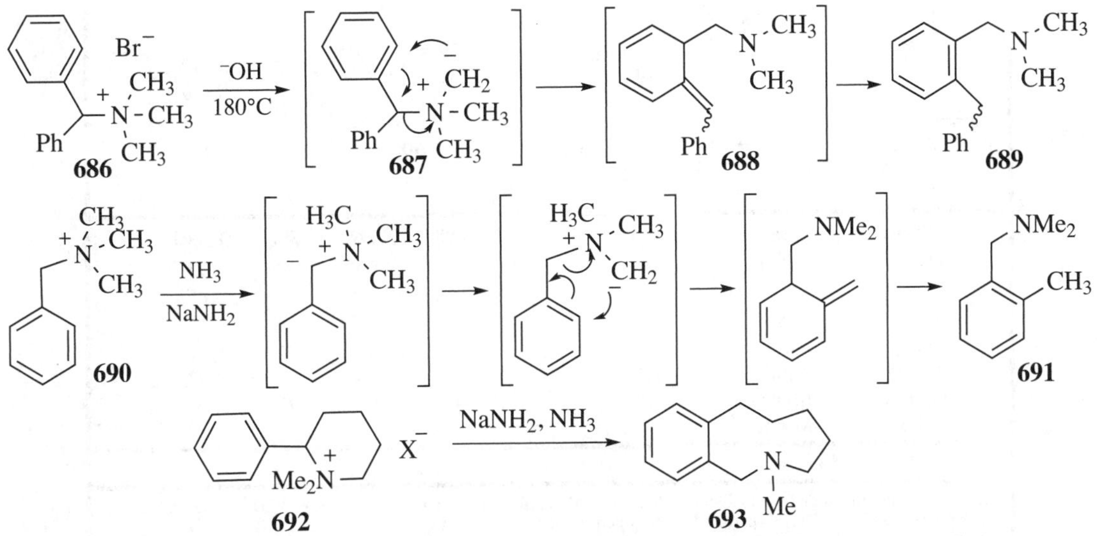

The Sommelet and Stevens rearrangements compete with each other in some cases, as shown in Table 8.21.[610] The major products from reaction of **694** with base are the Stevens products **696** and **697**, and the Sommelet-Hauser product **698**. Small amounts of the para rearrangement product (**699**) and the direct displacement product (**700**) are also present. The ammonium salt precursor to the ylid (**694**) can generate either ylid (**695** or **701**), but formation of **701** is sterically inhibited relative to **695**. The Stevens product of **701** (amine **702**) is more sterically crowded than **696** or **697**. In polar aprotic solvents the major product is the Sommelet-Hauser product **698**, but in nonpolar solvents (hexane) the Stevens product **696** predominates. In DMSO, the base is more nucleophilic. leading to the displacement product **700** in high yield. Alkoxide bases are not strong enough to generate the ylid from **694**, and the displacement reaction (to give **700**) dominates in those cases.

607. Heard, G.L.; Yates, B.F. *Aust. J. Chem.* **1994**, *47*, 1685.
608. (a) Kantor, S.W.; Hauser, C.R. *J. Am. Chem. Soc.* **1951**, *73*, 4122; (b) Puterbaugh, W.H.; Hauser, C.R. *Ibid.* **1964**, *86*, 1105.
609. Jones, G.C.; Hauser, C.R. *J. Org. Chem.* **1962**, *27*, 3572.
610. Pine, S.H.; Munemo, E.M.; Phillips, T.R.; Bartolini, G.; Cotton, W.D.; Andrews, G.C. *J. Org. Chem.* **1971**, *36*, 984.

Table 8.21. Competition Between the Sommelet and Stevens Rearrangements

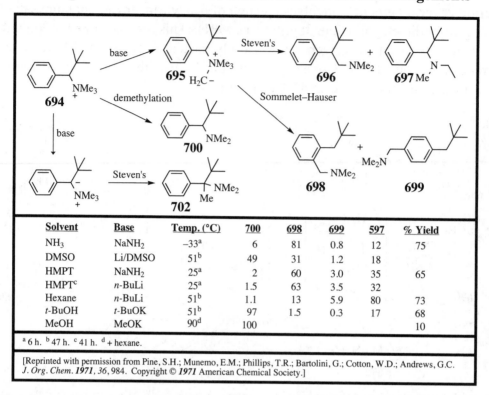

Solvent	Base	Temp. (°C)	700	698	699	597	% Yield
NH$_3$	NaNH$_2$	–33[a]	6	81	0.8	12	75
DMSO	Li/DMSO	51[b]	49	31	1.2	18	
HMPT	NaNH$_2$	25[a]	2	60	3.0	35	65
HMPT[c]	n-BuLi	25[a]	1.5	63	3.5	32	
Hexane	n-BuLi	51[b]	1.1	13	5.9	80	73
t-BuOH	t-BuOK	51[b]	97	1.5	0.3	17	68
MeOH	MeOK	90[d]	100				10

[a] 6 h. [b] 47 h. [c] 41 h. [d] + hexane.

[Reprinted with permission from Pine, S.H.; Munemo, E.M.; Phillips, T.R.; Bartolini, G.; Cotton, W.D.; Andrews, G.C. *J. Org. Chem.* **1971**, *36*, 984. Copyright © **1971** American Chemical Society.]

Note that a nitrogen ylid has been invoked in the **Hofmann elimination reaction** (sec. 2.9.C.i).[611] This ylid mechanism is highly questionable for many substrates[612] that proceed via β-elimination with a coordinated hydroxide. In those cases, removal of the hydrogen directly by hydroxide appears more likely. The ylid mechanism probably operates when PhLi is used as a base[613] and there is steric inhibition of the usual β-elimination process.[614]

The useful disconnections of nitrogen ylids are

8.8.D. Transition Metal Olefination Reagents

Other organometallic derivatives give Wittig-type olefination reactions, and titanium derivatives

611. Cope, A.C.; Ciganek, E.; LeBel, N.A. *J. Am. Chem. Soc.* **1959**, *81*, 2799.
612. Reference 494, p 277-281.
613. Wittig, G.; Polster, R. *Annalen* **1956**, *599*, 13.
614. Cope, A.C.; Mehta, A.S. *J. Am. Chem. Soc.* **1963**, *85*, 1949.

are the most commonly used.[615] The initial addition reaction to the carbonyl gives a transient metalated alcohol that leads to an alkene. An example is the **Tebbe reagent**, which exists as a bridged methylene species (see **703**) where Cp is cyclopentadienyl,[616] and the reaction with carbonyl compounds that leads to alkene derivatives is called **Tebbe olefination** The aluminum species can be varied

to include chloride ($AlCl_3$), triperdeuteromethyl aluminum $[Al(CD_3)_3]$ or trimethylsilyl $[AlCl(CH_2SiMe_3)_2]$. The most common reagent is the aluminum dimethylaluminum chloride compound shown. The Tebbe reagent reacts with ketones or aldehydes to give an alkene, analogous to the Wittig reagent. In a synthesis of (+)-hygrine, Jew, Park and co-worker's converted aldehyde **704** to alkene **705** in 82% yield using the Tebbe reagent in THF.[617]

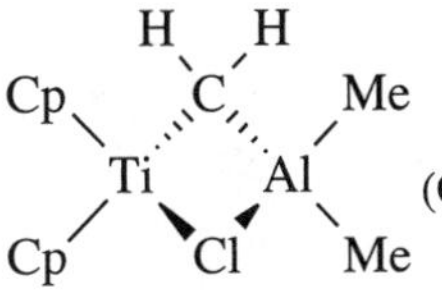

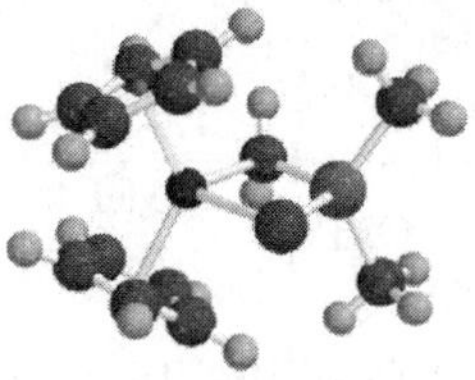

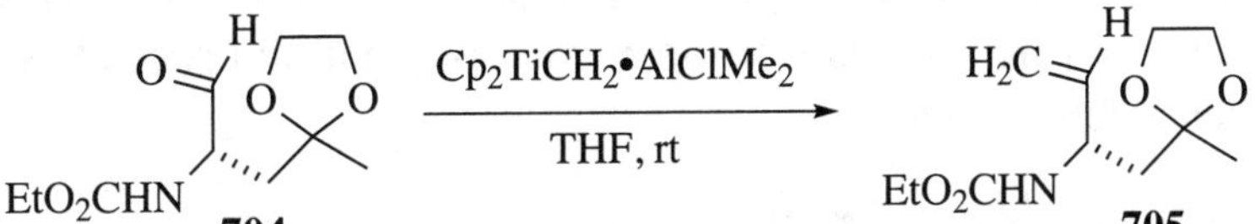

The Tebbe reagent is quite useful in that it reacts with the carbonyl of esters or lactones to give vinyl ethers, in contrast to common Wittig reagents. For example, in a synthesis of the C17-C28 fragment of spongistatin 1, Roush and Holson converted ester **706** to vinyl ether **707** in 74% yield.[618] The Tebbe reagent also reacts with conjugated esters via 1,2-addition to give the vinyl ether.[616a]

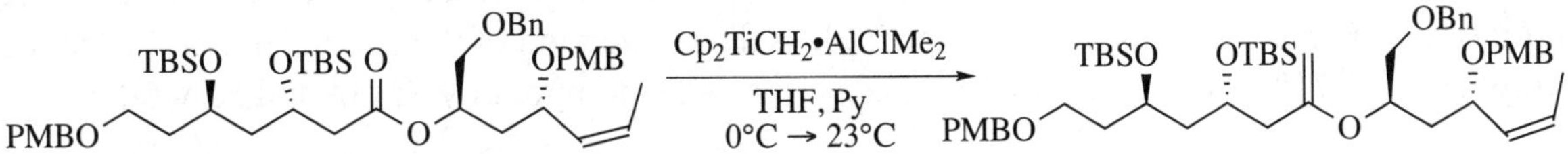

Other titanium-based olefination reagents have been developed. Eisch used a zinc analog of the Tebbe reagent (**708**) in a reaction with benzophenone to give 1,1-diphenylethene in 78% yield.[619] Similarly, Clawson, Buchwald and Grubbs used **709**

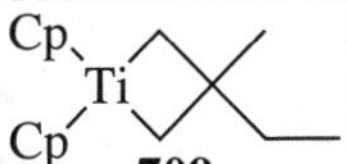

615. For reactions of this type, see Larock, R.C. *Comprehensive Organic Transformations, 2nd Ed.* Wiley-VCH, New York, *1999*, pp 341-350.

616. (a) Tebbe, F.N.; Parshall, G.W.; Reddy, G.S. *J. Am. Chem. Soc. 1978, 100,* 3611; (b) The Merck Index, 14th ed., Merck & Co., Inc., Whitehouse Station, New Jersey, *2006*, p ONR-93; (c) Mundy, B.P.; Ellerd, M.G.; Favaloro Jr., F.G. *Name Reactions and Reagents in Organic Synthesis, 2nd ed.*, Wiley-Interscience, New Jersey, *2005*, p. 856.

617. Lee, J.-H.; Jeong, B.S.; Ku, J.-M.; Jew, S.-s.; Park, H-g. *J. Org. Chem. 2006, 71,* 6690.

618. Holson, E.B.; Roush, W.R. *Org. Lett. 2002, 4,* 3719.

619. Eisch, J.J.; Piotrowski, A. *Tetrahedron Lett. 1983, 24,* 2043.

in olefination reactions with ketones and aldehydes.[620] Alkoxy-titanium reagents such as **710** have been employed, as in the conversion of cyclohexane carboxaldehyde to 1-cyclohexyl-1,3-butadiene (**711**) in 86% yield (96:4 Z/E).[621]

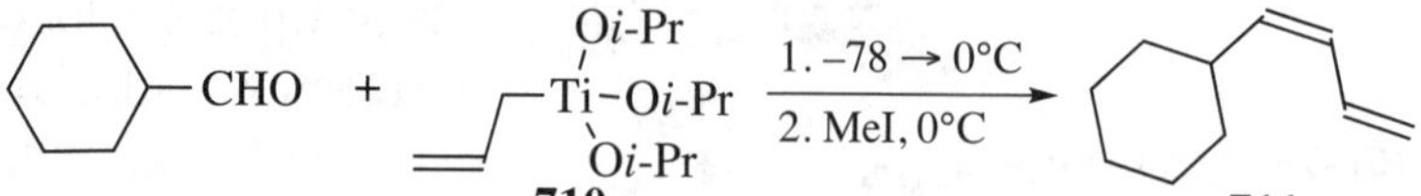

Other titanium reagents have been used for olefination reactions. Petasis developed Cp_2TiMe_2 (**713**), now called the **Petasis reagent**.[622] The Petasis reagent is prepared by reaction of methyllithium with titanocene dichloride (Cp_2TiCl_2)], which avoids some of the difficulties of **703** (high cost, long preparation times, short shelf-lives, extreme sensitivity to air and water, residual aluminum reagents). Aldehydes and ketones react with the Petasis reagent to give an alkene, illustrated by the conversion of **712** to **714** in 76% yield, taken from Smith and co-worker's synthesis of (+)-phorboxazole A.[623] Note that the ketone moiety reacted in preference ot the carbony of the thioester. Petasis proposed that olefination proceeds via methyl transfer from an intermediate such as **715**.[624]

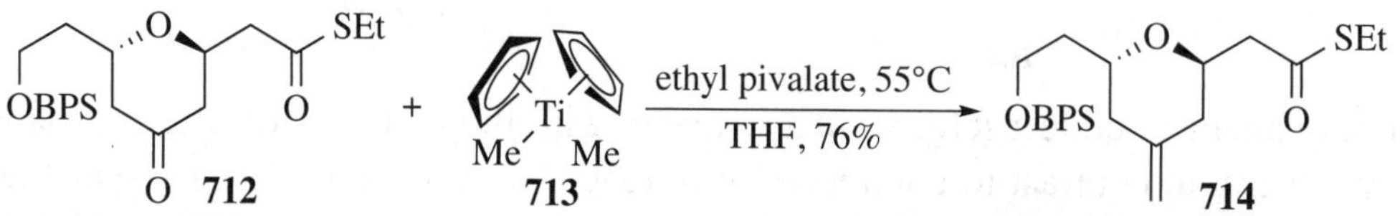

Hughes proposed that olefination of esters proceeds via a titanium carbene such as **716**, and an oxatitanacycle **717**.[625] This latter proposal has independent support, based on mass spectrometry evidence provided by Pilli, Eberlin and co-workers.[626] In the absence of an aldehyde or ketone moiety, an ester moiety can react. An example is the conversion of the lactone unit in **718** to the vinyl ether in **719** (85% yield) in a synthesis of (−)-kendomycin by Smith and co-workers.[627]

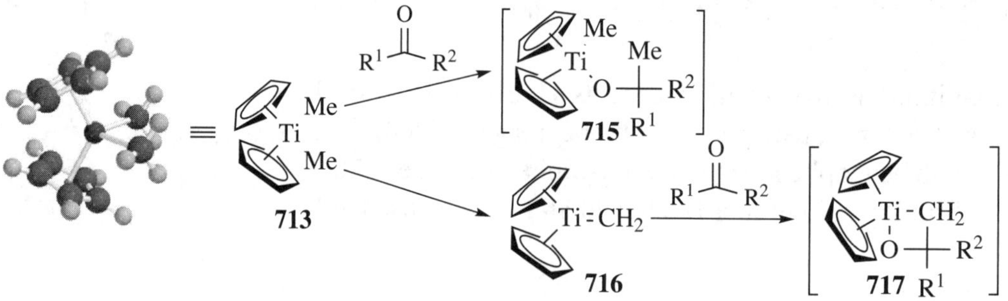

620. Clawson, L.; Buchwald, S.L.; Grubbs, R.H. *Tetrahedron Lett.* **1984**, *25*, 5733.
621. Ukai, J.; Ikeda, Y.; Ikeda, N.; Yamamoto, V. *Tetrahedron Lett.* **1983**, *24*, 4029.
622. Petasis N.A.; Bzowej, E.I. *J. Am. Chem. Soc.* **1990**, *112*, 6392.
623. Smith III, A.B.; Razler, T.M.; Ciavarri, J.P.; Hirose, T.; Ishikawa, T. *Org. Lett.* **2005**, *7*, 4399.
624. Petasis N.A.; Bzowej, E.J. *J. Org. Chem.* **1992**, *57*, 1327.
625. Hughes, D.L.; Payack, J.F.; Cai, D.; Verhoeven, T.R.; Reider, P.J. *Organometallics,* **1996**, *15*, 663.
626. Meurer, E.C.; Santos, L.S.; Pilli, R.A.; Eberlin, M.N. *Org. Lett.* **2003**, *5*, 1391.
627. Smith III, A.B.; Mesaros, E.F.; Meyer, E.A. *J. Am. Chem. Soc.* **2006**, *128*, 5292.

718 → (Cp₂TiMe₂, THF, 60-63°C 85%) → **719**

A Petasis-like reagent has also been reported that gives the same reaction: $CH_2(ZnI)_2$, $TiCl_2$, and TMEDA in THF.[628] Oshima and co-workers developed a $TiCl_4$ reagent (Zn-CH_2Br_2-$TiCl_4$)[629] that gave olefination. It was electrophilic and generally suppressed the tendency of a ketone substrate to enolize. Lombardo developed a more active form of this reagent by dropwise addition of $TiCl_4$ to a stirred suspension of zinc dust in CH_2Br_2 and THF at −40°C, followed by warming the mixture to 5°C and *stirring for 3 days*.[630] Lombardo used this reagent in synthetic studies of C_{20} gibberellins (GA₃₈).[630] Mehta and Islam used Lombardo's reagent to convert **720** to **721**, in 71% yield, in a synthesis of ottelione.[631] Lombardo's modified reagent is particularly important when the ketone or aldehyde to be used in the olefination reaction is sensitive to basic reagents. Another variation of this reagent uses CH_2Br_2, Zn, $TiCl_4$ and a catalytic amount of $PbCl_2$.[632]

720 → (Zn-$TiCl_4$-CH_2Br_2, CH_2Cl_2, 0°C) → **721**

The **Nysted reagent (722)**[633] is commercially available, and it reacts with aldehydes or ketones in the presence of BF₃•etherate to give an alkene. Reaction with 2-phenylpropanal, for example, gave an 82% yield of **723**.[634] The **Takai reaction**[635] uses an alkyl di-or triiodide in the presence of $CrCl_2$ to generate the alkene. An example is taken from Huang and co-worker's synthesis of psymberin, in which aldehyde **724** was converted to the vinyl iodide **725** in 90% yield as a 5:1 *E/Z* mixlure.[636]

628. Matsubara, S.; Ukai, K.; Mizuno, T.; Utimoto, K. *Chem. Lett.* *1999*, 825.
629. Takai, K.; Hotta, Y.; Oshima, K.; Nozaki, H. *Tetrahedron Lett.* *1978*, 2417.
630. (a) Lombardo, L. *Tetrahedron Lett.* *1982*, *23*, 4293; (b) Lombardo, L. *Org. Synth. Coll. Vol. 8* *1993*, 386.
631. Mehta, G.; Islam, K. *Angew. Chem. Int. Ed.* *2002*, *41*, 2396.
632. (a) Takai, K.; Kakiuchi, T.; Kataoka, Y.; Utimoto, K. *J. Org. Chem.* *1994*, *59*, 2668; (b) Takai, K.; Kataoka, Y.; Miyai, J.; Okazoe, T.; Oshima, K.; Utimoto, K. *Org. Synth.*, *1995*, *73*, 73. For a synthetic example taken from a synthesis of (−)-periplanone B, see Hodgson, D.M.; Foley, A.M.; Boulton, L.T.; Lovell, P.J.; Maw, G.N. *J. Chem. Soc. Perkin Trans. 1* *1999*, 2911.
633. (a) Nysted, L.N. *US Patent* 3 865 848, *1975* [*Chem. Abstr.* *1975*, 83:10406q]; (b) Mundy, B.P.; Ellerd, M.G.; Favaloro Jr., F.G. *Name Reactions and Reagents in Organic Synthesis, 2nd ed.*, Wiley-Interscience, New Jersey, *2005*, p. 826.
634. Matsubara, S.; Sugihara, M.; Utimoto, K. *Synlett* *1998*, 313.
635. (a) Takai, K.; Nitta, K.; Utimoto, K. *J. Am. Chem. Soc.* *1986*, *108*, 7408; (b) Okazoe, T.; Takai, K.; Utimoto, K. *J. Am. Chem. Soc.* *1987*, *109*, 951; (c) Mundy, B.P.; Ellerd, M.G.; Favaloro Jr., F.G. *Name Reactions and Reagents in Organic Synthesis, 2nd ed.*, Wiley-Interscience, New Jersey, *2005*, pp. 640-641.
636. Huang, X.; Shao, N.; Palani, A.; Aslanian, R.; Buevich, A. *Org. Lett.* *2007*, *9*, 2597.

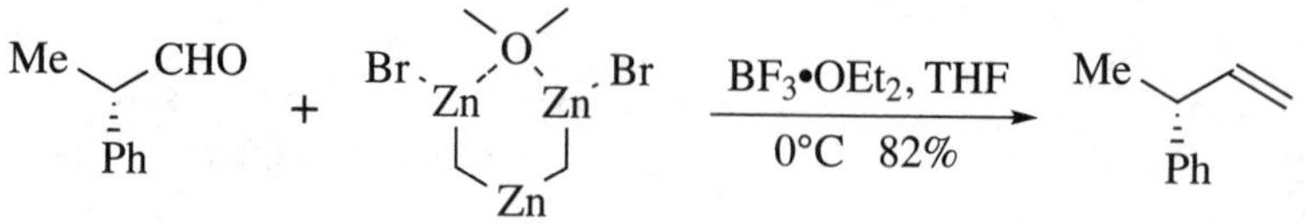

Me—CHO + Br—Zn—O—Zn—Br / Zn **722** →[BF₃•OEt₂, THF, 0°C 82%] Me—CH=CH₂ / Ph **723**

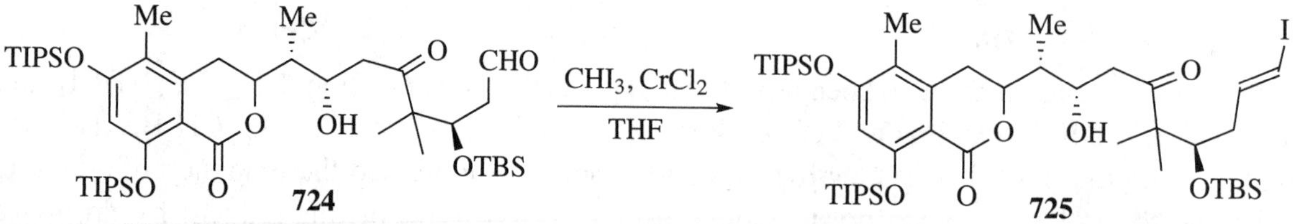

Structure **724** →[CHI₃, CrCl₂, THF]→ Structure **725**

Other organometallic reagents have been reported to give Wittig-type olefination reactions with ketones and aldehydes. Typical examples are **726**,[637] **727**,[638] **728**,[639] **729**,[640] and **730**.[641] This list is certainly not exhaustive, and is probably not completely representative. It does, however, illustrate some of the reagent types that have appeared, and continue to appear.

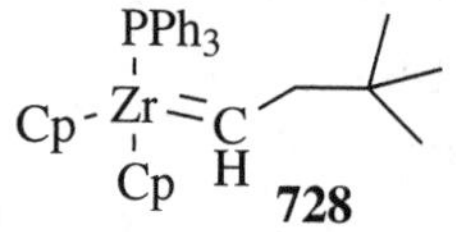

$(MeS)_2B=CH_2$ $Cl_3Mo=CH_2$ Cp₂Zr=C(H)(t-Bu) [**728**]   W-cyclopentylidene [**729**] $Me_2Te^+-CHCO_2Et^-$

726   **727**   **728** 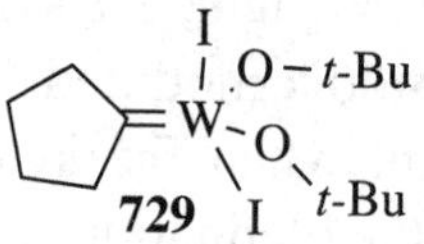   **729**   **730**

The disconnections for these reagents are identical to those in sec. 8.8.A for the Wittig reagents.

8.9. OTHER ORGANOMETALLIC CARBANIONIC COMPOUNDS

8.9.A. Organoiron Compounds

8.9.A.i. Formation and Stability. A few organoiron compounds have found their way into organic synthesis.[642] In general, Fe–C bonds are sensitive to homolytic cleavage, producing organic radicals and the metal in a lower oxidation state. This instability is due to small energy differences between the filled *d* orbitals, and the valence *s* and *p* anti-bonding orbitals of the Fe–C bond. This bond can be stabilized in at least two ways. The first involves addition of a ligand possessing acceptor properties (CO, cyclopentadienyl, phosphines, amines, etc.). The second involves alteration of the effective electronegativity of the carbon (induce a different hybridization state), or attachment of strongly electronegative groups such as fluorine to the carbon.

637. Pelter, A.; Singaram, B.; Wilson, J.W. *Tetrahedron Lett.* **1983**, *24*, 635.
638. Kauffmann, T.; Ennen, B.; Sander, J.; Wieschollek, R. *Angew. Chem. Int. Ed.* **1983**, *22*, 244.
639. Clift, S.M.; Schwartz, J. *J. Am. Chem. Soc.* **1984**, *106*, 8300.
640. (a) Aguero, A.; Kress, J.; Osborn, J.A. *J. Chem. Soc. Chem. Commun.* **1986**, 531; (b) Freudenberger, J.H.; Shrock, R.R. *Organometallics* **1986**, *5*, 398.
641. Osuka, A.; Mori, Y.; Shimizu, H.; Suzuki, H. *Tetrahedron Lett.* **1983**, *24*, 2599.
642. Davies, S.G. *Organotransition Metal Derivatives: Applications to Organic Synthesis,* Pergamon Press, Oxford, *1982*.

The iron must eventually be removed from the molecule to be useful in organic synthesis. Cleavage[643] by a proton source is one way to break the Fe–C bond. Reaction with either water or an alcohol is usually sufficient, although a catalytic amount of acid is sometimes required.[643] This process simply protonates the alkyl fragment in an organoiron compound such as **731** to give methane and **732**.[643] Some organoiron complexes are thermally labile, generating an alkene via β-elimination.[644] Thermal disproportionation is also observed and the thermal stability of simple alkyl groups in $RFe(CO)_2Cp$ was reported to be R = Me ≈ Ph >> Et > i-Pr. An example of the disproportionation reaction is the fragmentation of **733** to the alkene. Reaction of the organoiron complex with an halogen[643] also cleaves the C–Fe bond. When **735** was treated with iodine, an alkyl halide (R–I) and the iron iodide (**735**) were formed. The two most common methods for converting an organoiron to the alkyl fragment are protonolysis and halogenation. Both of these techniques will be used in the following sections.

$$MeFe(CO)_2Cp \xrightarrow{H^+} CH_4 + [CpFe(CO)_2(H_2\overset{+}{O})] \qquad L_nM-CH_2CH_2R \xrightarrow{heat} L_nM-H + CH_2=CHR$$

$$\textbf{731} \qquad\qquad\qquad\qquad \textbf{732} \qquad\qquad\qquad\qquad \textbf{733}$$

$$RFe(CO)_2Cp \xrightarrow{I_2} R-I + I-Fe(CO)_2Cp$$

$$\textbf{734} \qquad\qquad \textbf{735}$$

8.9.A.ii. Cyclopentadienylirondicarbonyl (Fp) Compounds.
Cyclopentadienylbis(carbonyliron) (Fp, **737**)A is a common organoiron reagent.[645] This derivative can be converted to an alkyl derivative (**739**, where R = alkyl) by reaction with an alkyl halide, or to the protio derivative (**739**, R = H) by reaction with an acid.[646] The anionic reagent **738** is prepared by reaction of iron pentacarbonyl with the dimer of cyclopentadiene (**736**, sec. 115.B) at 200°C to give the dimeric species **737**. Treatment of **737** with base leads to loss of cyclopentadiene and formation of **738**. A byproduct of this process is ferrocene (**740**) formed by extrusion of carbon monoxide from **737**.[646]

The protonated form of Fp can add to conjugated carbonyl derivatives such as acrylonitrile[647] and also with conjugated dienes.[648] Both SO_3[649] and carbon monoxide can be inserted into the Fe-C bond of FpR.[598] Protonolysis provides a synthetic route to substituted nitriles or alkenes.

643. Piper, T.S.; Wilkinson, G. *J. Inorg. Nucl. Chem.* **1956**, *3*, 104.
644. Braterman, P.S.; Cross, R.J. *J. Chem. Soc. Dalton Trans.* **1972**, 657.
645. Eisch, J.J.; King, R.B. *Organomet. Synth.* **1965**, *1*, 114, 152.
646. Cotton, F.A.; Wilkinson, G.; Murillo, C.A.; Bochmann, M. *Advanced Inorganic Chemistry, 6th. Ed.,* Wiley, New York, **1999**. pp.808-812.
647. Green, M.L.H.; Nagy, P.L.I. *J. Chem. Soc.* **1963**, 189
648. Ariyaratne, J.K.P.; Green, M.L.H. *J. Chem. Soc.* **1963**, 2976
649. Bibler, J.P.; Wojcicki, A. *J. Am. Chem. Soc.* **1964**, *86*, 5051

In each case the Fp unit can also be removed photochemically, thermally, or by treatment with triphenylphosphine. Hydride extraction to give an alkene or a diene is also known.[650]

8.9.A.iii. Sodium Tetracarbonyl Ferrate. An extremely useful organoiron reagent is sodium tetracarbonyl ferrate [$Na_2Fe(CO)_4$],[651] usually prepared by reduction of iron pentacarbonyl. The synthetic utility of this reagent lies in its ability to react with alkyl halides in a stepwise manner, including the reaction with two different alkyl halides as reported independently by Cooke and by Collman and co-workers.[651,652] When sodium tetracarbonyl ferrate reacted with 1-bromooctane, a trigonal bipyramidal complex (**741**) was formed by insertion of the alkyl fragment.

$$Fe(CO)_5 \xrightarrow{1\% \ Na(Hg)} Na_2Fe(CO)_4$$

Subsequent treatment with iodoethane led to a second insertion and expansion of the coordination shell to an octahedral species (**742**). When the reaction was pressurized with carbon monoxide, CO was inserted in the C–Fe bond to give an acyl derivative **743**. Protonolysis with acetic acid cleaved the Fe–C bond, and concomitant migration led to coupling of the alkyl and acyl fragments to produce 3-undecanone, along with iron tricarbonyl [$Fe(CO)_3$]. When **741** was treated with triphenylphosphine, CO insertion occurred via an intramolecular migration and protonolysis led to 3-undecanone.

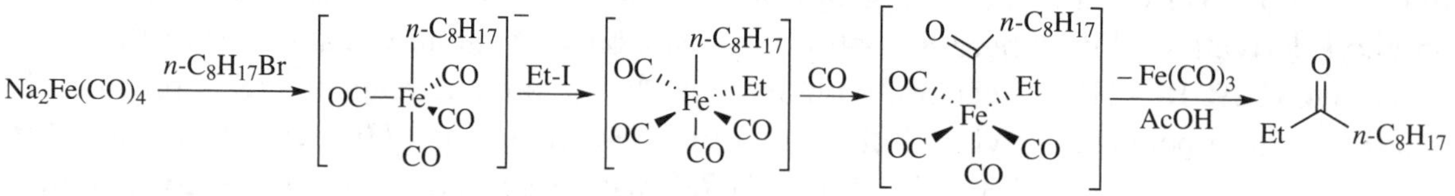

The sequence shown above generates an unsymmetrical ketone. A modification of this reaction involves addition of only one alkyl halide to sodium tetracarbonyl ferrate, followed by pressurization with carbon monoxide.

Reaction with 1-bromohexane gave the expected complex, **744**. When treated with carbon monoxide, carbonyl insertion gave **745** and subsequent oxidation with oxygen in the presence of a base led to heptanoic acid.[653] Iodination in the presence of water also gave the acid but when the reaction was done in ethanol, ethyl heptanoate was formed. Similarly, *N,N*-diethylheptanamide was generated by reaction with iodine and diethylamine.[653] In general, the

650. Green, M.L.H.; Smith, M.J. *J. Chem. Soc. A,* **1971**, 3220
651. Cooke, Jr., M.P. *J. Am. Chem. Soc.* **1970**, *92*, 6080
652. (a) Collman, J.P.; Winter, S.R.; Clark, D.R. *J. Am. Chem. Soc.* **1972**, *94*, 1788; (b) Johnson, B.F.G.; Lewis, J.; Thompson, D.J. *Tetrahedron Lett.* **1974**, 3789; (c) Cooke, Jr., M.P.; Parlman, R.M. *J. Am. Chem. Soc.* **1975**, *97*, 6863
653. (a) Collman, J.P.; Winter, S.R.; Komoto, R.G. *J. Am. Chem. Soc.* **1973**, *95*, 249; (b) Watanabe, Y.; Yamashita, M.; Mitsudo, T.; Tanaka, M.; Yakegami, Y. *Tetrahedron Lett.* **1973**, 3535; (c) Yamashita, M.; Watanabe, Y.; Mitsudo, T.; Takegami, Y. *Ibid.* **1976**, 1585

yields of ketone were higher with primary aliphatic halides and tosylates. Secondary substrates gave significant amounts of elimination, although this could be minimized if THF was used as a solvent.

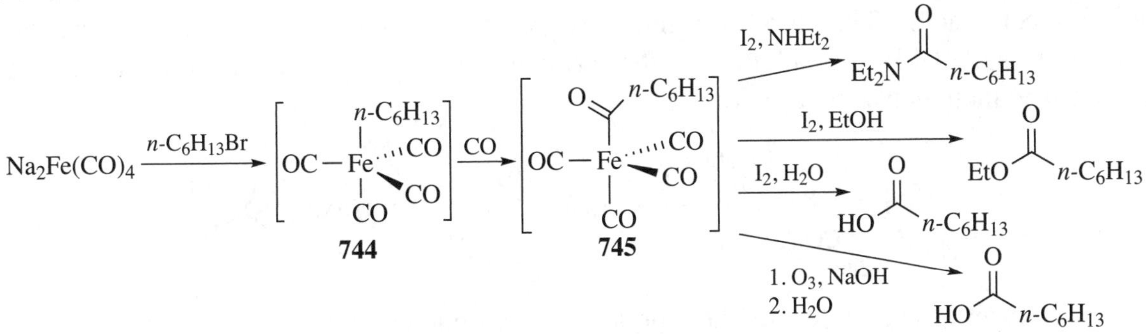

The disconnection for this process is

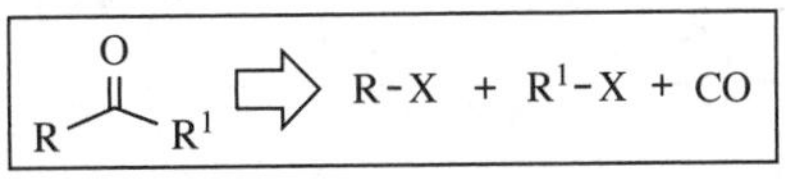

8.9.A.iv. Chiral Organoiron Species. In work reported independently by Davies and by Liebeskind, iron reagents such as **746** were prepared, shown to be chiral and resolved into the (R) and (S) antipodes. One of the phenyl groups in **746** effectively blocks one face of the carbonyl moiety, inducing high diastereoface selection in reactions at that carbonyl or at the adjacent carbon. The carbonyl partner of the iron species can undergo enolate reactions similar to those of an acid derivative (secs. 9.2, 9.3). Oxidation with iodine or bromine, as shown above,[653] cleaves the C–Fe bond and generates the acid derivative. Davies et al.[654] showed that the asymmetric acetyl derivative **746** can be deprotonated to form an enolate and condensed with an aldehyde (such as propanal) or alkylated (as with iodomethane). Reaction of **746** with butyllithium and then propanal led to a second reaction with excess butyllithium, followed by alkylation with iodomethane and oxidation of the C–Fe bond with bromine gave **747** with high

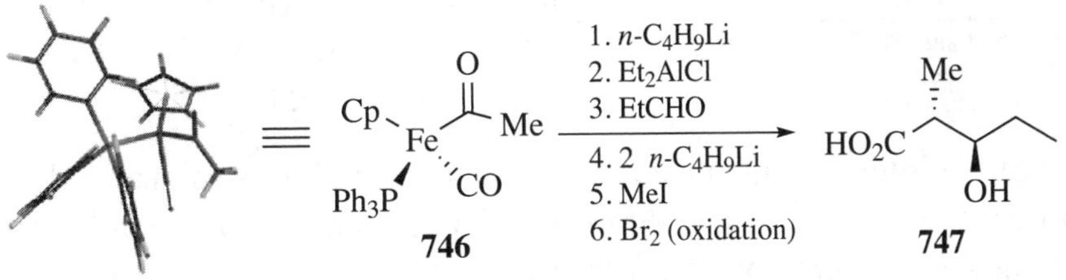

asymmetric induction.[654,655] Liebeskind and Welker reacted **746** with lithium diisopropylamide (sec. 9.2.A) and propanal, and showed that the enantioselectivity in the final β-hydroxy acids (**748**

---

654. (a) Davies, S.G.; Dordor-Hedgecock, I.M.; Warner, P.; *Tetrahedron Lett.* **1985**, *26*, 2125; (b) Ambler, P.W.; Davies, S.G. *Ibid.* **1985**, *26*, 2129; (c) Davies, S.G.; Dordor, I.M.; Walker, J.C.; Warner, P. *Ibid.* **1984**, *25*, 2709; (d) Davies, S.G.; Dordor, I.M.; Warner, P. *J. Chem. Soc. Chem. Commun.* **1984**, 956.
655. Davies, S.G.; Walker, J.C. *J. Chem. Soc. Chem. Commun.* **1985**, 209.

and **749**) was dependent on the conditions used to form the enolate.[656] If diisobutylaluminum chloride (*i*-Bu$_2$AlCl) was used, a 5.2:1 mixture of **748** and **749** was reversed (to 1:11.6), in 66% yield. The iron group is a chiral auxiliary,[657] where the acyl iron derivative behaves essentially as a protected acid. The chiral iron moieties are useful variation of enolate condensation chemistry (sec. 9.4.B). In addition to the formation of the condensation product, the high asymmetric induction will prove valuable.

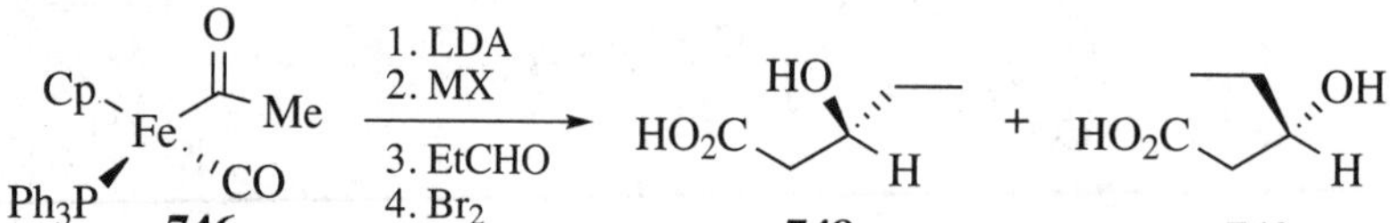

The disconnections possible with these organoiron compounds are

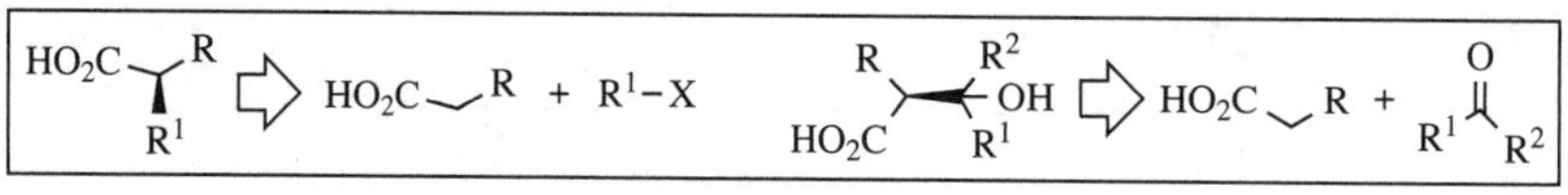

8.9.B. Organoaluminum Compounds

The chemistry of organoaluminum compounds is electrophilic in nature. Zweifel and Whitney, however, showed that conversion of an organoaluminum to its *ate* complex[658] changed it into a nucleophilic species. Examples of this reaction have been largely confined to the vinyl alanates, which react with electrophiles to give the (*E*) or the (*Z*) isomer, depending on the reaction conditions. Treatment of an alkyne such as 3-hexyne with diisobutylaluminum hydride (DIBAL–H) gave vinyl alanate **750**. When **750** reacted with methyllithium, aluminate **751** formed in 73% yield. Subsequent reaction with an electrophile such as carbon dioxide led to a transfer of the vinyl group to the carbonyl, forming the (*E*)-acrylic acid derivative **752**.

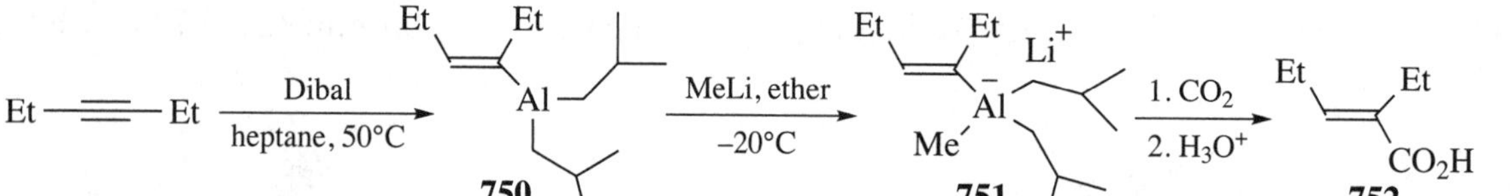

The geometry of the intermediate alanate, and the final product was changed by first reacting methyllithium with Dibal to give the aluminate **753**. When this *ate* complex reacted with the alkyne (3-hexyne), a new *ate* complex (**754**) was formed. Subsequent reaction with carbon dioxide gave (*Z*)-acrylate **755**. The aluminum *ate* complex also reacted with alkynes, and

656. Liebeskind, L.S.; Welker, M.E. *Tetrahedron Lett.* **1984**, *25*, 4341.
657. (a) Liebeskind, L.S.; Welker, M.E. *Tetrahedron Lett.* **1985**, *26*, 3079; (b) Davies, S.G.; Easton, R.J.C.; Gonzalez, A.; Preston, S.C.; Sutton K.H.; Walker, J.C. *Tetrahedron* **1986**, *42*, 3987.
658. (a) Zweifel, G.; Whitney, C.C. *J. Am. Chem. Soc.* **1967**, *89*, 2753; (b) Zweifel, G.; Steele, R.R. *Ibid.* **1967**, *89*, 2754, 5085.

hydrolysis converted the intermediate into substituted 1,3-butadienes.[659] Reaction of the *ate* complex with halogens gave the corresponding vinyl halide, with retention of the geometry observed in the complex.[658] Similar reaction with alkyl halides gave the substitution product.[660] Vinyl alanates generated in this way can be quenched with other electrophiles. In a synthesis of (+)-testudinariol A, Amarasinghe and Montgomery converted a silyl-alkyne to a vinyl bromide by reacting the alkyne unit with Dibal (sec. 4.6.C) followed by bromination with Br_2.[661]

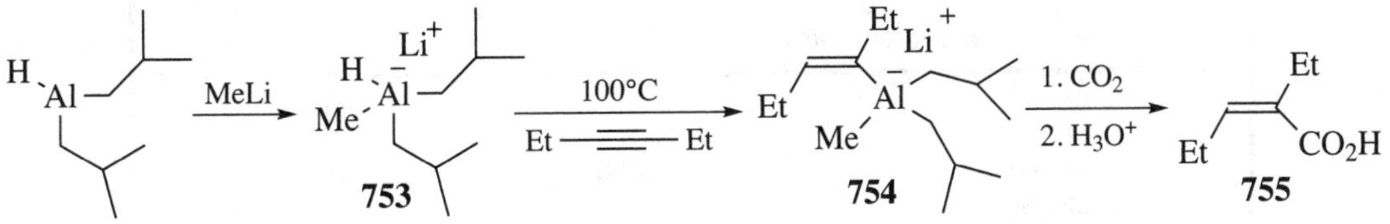

Vinyl alanes are also generated by the transition metal catalyzed addition of trialkylaluminum reagents to alkynes. In Hoye and co-worker's synthesis of elenic acid,[662] 1-dodecyne reacted with Me_3Al and Cp_2ZrCl_2 and then MeLi. The resulting *ate* complex **756** then reacted with triflate **757** to give a 96% yield of the alkene, **758**.

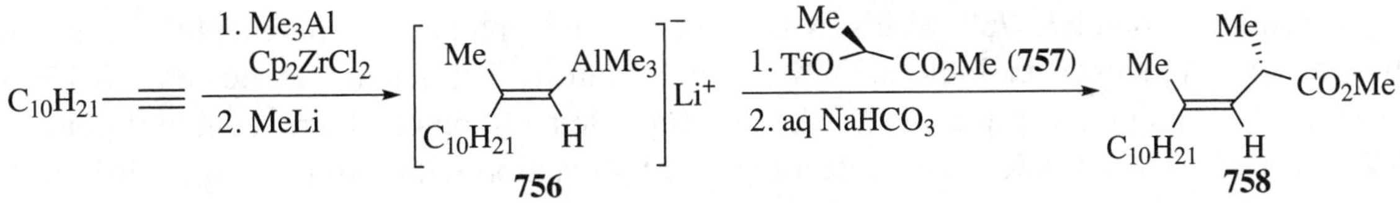

The disconnection for the alkyne-alanate reaction is

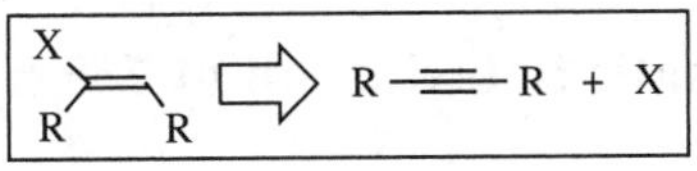

8.9.C.   Organochromium Reagents

Benzene and its derivatives react with chromium hexacarbonyl $[Cr(CO)_6]$[663] to give an arylchromium complex, **759**. Chromium resides in a position perpendicular to the plane of the ring, as shown. Semmelhack and co-workers found that the chromium activated the benzene ring to metal-hydrogen exchange with organolithium reagents (sec. 8.5.G), and these new lithium reagents reacted with alkyl halides to give the alkyl substitution product. As shown in Table 8.22,[664] reaction of **759** with butyllithium generated **760** via lithium-hydrogen exchange. Subsequent reaction with an electrophile gave **761** in good yield.

659.   (a) Zweifel, G.; Polston, N.L.; Whitney, C.C. *J. Am. Chem. Soc.* **1968**, *90*, 6243; (b) Zweifel, G.; Miller, R.L. *Ibid.* **1970**, *92*, 6678.
660.   Yamamoto, Y.; Yatagai, H.; Maruyama, K. *J. Org. Chem.* **1980**, *45*, 195.
661.   Amarasinghe, K.K.D.; Montgomery, J. *J. Am. Chem. Soc.* **2002**, *124*, 9366.
662.   Hoye, R.C.; Baigorria, A.S.; Danielson, M.E.; Pragman, A.A.; Rajapakse, H.A. *J. Org. Chem.* **1999**, *64*, 2450.
663.   (a) Strohmeier, W. *Chem. Ber.* **1961**, *94*, 2490; (b) Rausch, M.D. *J. Org. Chem.* **1974**, *39*, 1787.
664.   Semmelhack, M.F.; Bisaha, J.; Czarny, M. *J. Am. Chem. Soc.* **1979**, *101*, 768.

Table 8.22. Li–H Exchange in Arylchromium Complex 759 and Substitution Reactions with Electrophiles

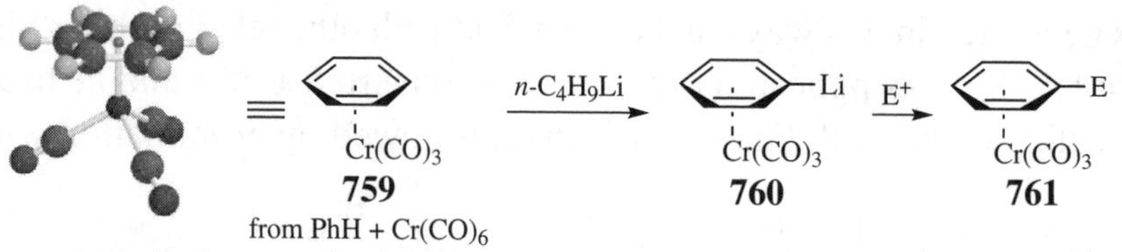

E⁺	E	% 761
CO_2	$-CO_2Me$	72
acetone	$-C(OH)Me_2$	29
$MeOSO_2F$	$-CH_3$	91
PhCHO	$-CH(OH)Ph$	0
Me_3SiCl	$-SiMe_3$	94
I_2	$-I$	76

[Reprinted with permission from Semmelhack, M.F.; Bisaha, J.; Czarny, M. *J. Am. Chem. Soc.* *1979*, *101*, 768. Copyright © *1979* American Chemical Society.]

Arylchromium complex **759** reacts with carbon nucleophiles via nucleophilic aromatic substitution. As shown in Table 8.23,[665] organolithium reagents and enolates add to **759** to generate a carbanionic complex **762**. The nucleophilic addition can be reversible if complex **762** is heated. The complex can be decomposed by reaction with iodine (to give **763**) or by protonolysis (with trifluoroacetic acid, TFA) to give cyclohexadiene derivative **764**. The reagents used to remove the metal are similar to those used with organoiron complexes (sec. 8.9.A).

Table 8.23. Nucleophilic Aromatic Substitution of 759 by Organolithium Reagents and Enolate Anions

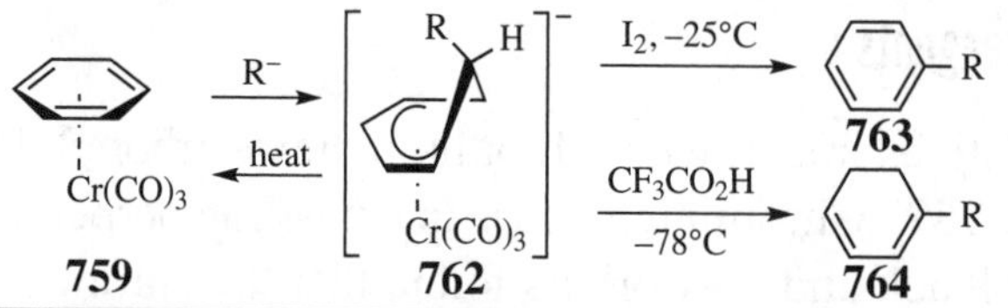

R⁻	% 763
$LiCH_2CN$	68
$LiCMe_2CN$	94
2-Lithiodithiane	93
$LiCMe_3$	97
$LiCH_2CO_2CMe_3$	87

[Reprinted with permission from Semmelhack, M.F.; Hall, H.T.; Yoshifuji, M.; Clark, G. *J. Am. Chem. Soc.* *1975*, *97*, 1247 and Semmelhack, M.F.; Hall Jr., H.T.; Yoshifuji, M. *Ibid* *1976*, *98*, 6387. Copyrights © *1976* and © *1975* American Chemical Society.]

When the benzene ring is functionalized, the chromium complex can react with nucleophiles to

---

665. (a) Semmelhack, M.F.; Hall, H.T.; Yoshifuji, M.; Clark, G. *J. Am. Chem. Soc.* *1975*, *97*, 1247; (b) Semmelhack, M.F.; Hall, Jr., H.T.; Yoshifuji, M. *Ibid.* *1976*, *98*, 6387.

give ortho, meta or para isomers. In general, the meta substitution product predominates[666,667] but all three isomeric products are formed. These transformations have been used in synthesis. Semmelhack and co-workers,[668,666] used an arylchromium complex in a synthesis of acorenoneand also in a synthesis of (±)-frenolicin that exploited the lithium-hydrogen exchange reaction.[669] Chromium complexes of polynuclear aromatic molecules undergo nucleophilic substitution, but not on the ring complexed to chromium.[670] Asymmetric induction is possible when chiral ligands are used in conjunction with the organolithium addition reactions. The reaction of **765** with phenyllithium in toluene, in the presence of (–)-sparteine gave complex **766**, which was treated with propargyl bromide to give a 72% yield of **767** in 54% ee.[671]

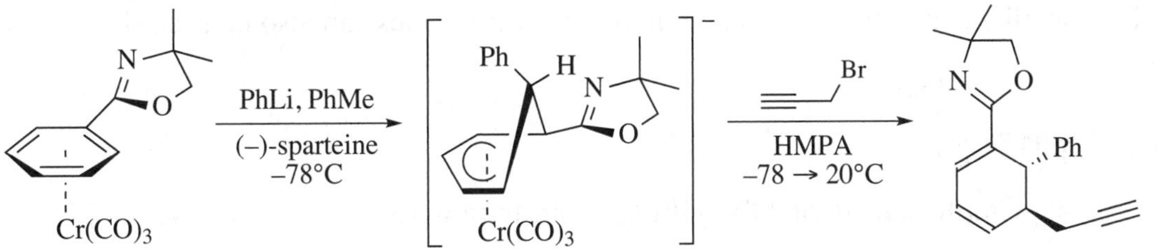

A different asymmetric application allows formation of CH_2Li units from *ortho*-methyl groups in chromium-arene complexes, which then react as carbon nucleophiles. Amide **768** reacted with butyllithium[672] in the presence of a chiral amine, and subsequent reaction with benzyl bromide gave **769** in 63% yield and 67% ee. The amide unit clearly activates the substituent at *ortho*-position to substitution, and the presence of the chiral amine led to asymmetric induction.

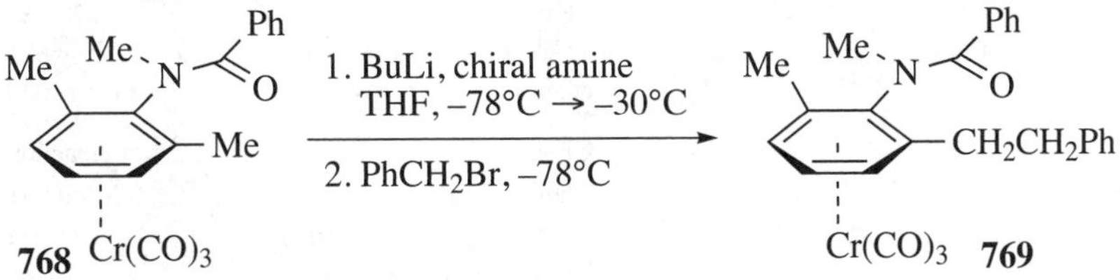

The disconnections available from these organochromium reactions are:

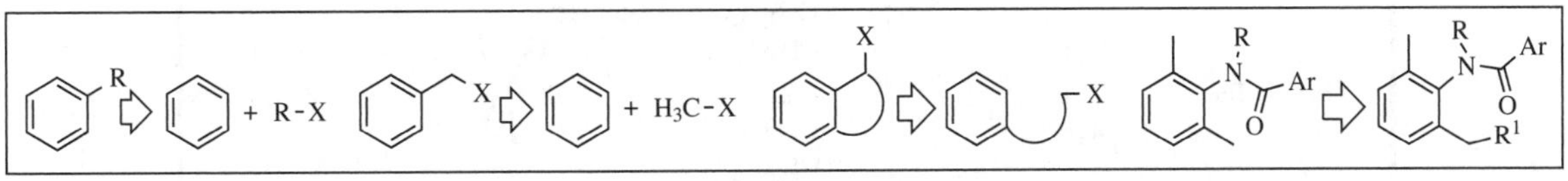

666. Semmelhack, M.F.; Harrison, J.J.; Thebtaranonth, Y. *J. Org. Chem.* **1979**, *44*, 3275.
667. (a) Semmelhack, M.F.; Clark, G.R.; Farina, R.; Saeman, M. *J. Am. Chem. Soc.* **1979**, *101*, 217; (b) Semmelhack, M.F.; Hall, Jr., H.T.; Farina, R.; Yoshifuji, M.; Clark, G.; Bargar, T.; Hirotsu, K.; Clardy, J. *Ibid.* **1979**, *101*, 3535.
668. Semmelhack, M.F.; Yamashita, A. *J. Am. Chem. Soc.* **1980**, *102*, 5924
669. Semmelhack, M.F.; Zask, A. *J. Am. Chem. Soc.* **1983**, *105*, 2034.
670. Semmelhack, M.F.; Seufert, W.; Keller, L. *J. Am. Chem. Soc.* **1980**, *102*, 6584.
671. Amurrio, D.; Khan, K.; Kündig, E.P. *J. Org. Chem.* **1996**, *61*, 2258.
672. Koide, H.; Hata, T.; Uemura, M. *J. Org. Chem.* **2002**, *67*, 1929.

In general, nucleophilic aromatic substitution reactions are rather difficult with unsubstituted aryl derivatives, or when the aromatic ring contains a strongly electron-releasing group. Formation of the chromium complex activates such aromatic compounds to nucleophilic substitution. Since the nucleophiles are carbon nucleophiles, this technique offers a route to carbon bonds that would be very difficult to form by other methods.

8.10. ALLYLIC TIN, ALKYLTITANIUM, AND ALLYLIC SILANE COMPLEXES

Other metal derivatives can be used for coupling. Among the more popular reagents are allyl- or alkyl-metallic compounds of tin and titanium. Allylsilanes can also be used if a Lewis acid is added.

8.10.A. Allyltin Reagents

Table 8.24. Condensation of Allyltributyl Stannane with Chiral Aldehydes (770)

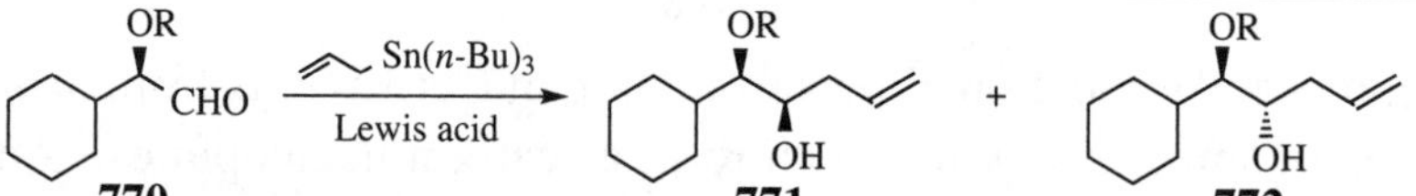

R	Lewis Acid	Solvent	Temp. (°C)	771 : 772
CH$_2$Ph	BF$_3$•OEt$_2$	CH$_2$Cl$_2$	−78	39 : 61
	MgBr$_2$	ether		94 : 6
		THF		20 : 80
		CH$_2$Cl$_2$	−23	>250 : 1
	MgCl$_2$	ether		no reaction
		THF		22 : 78
		CH$_2$Cl$_2$		no reaction
	Mg(ClO$_4$)$_2$	ether		69 : 31
	ZnBr$_2$	THF	67	27 : 73
		CH$_2$Cl$_2$		77 : 23
		PhMe		41 : 59
	ZnI$_2$	ether		35 : 64
		CH$_2$Cl$_2$		97 : 3
	TiCl$_4$	CH$_2$Cl$_2$	−78	>250 : 1
SiMe$_2$t-Bu	TiCl$_4$	CH$_2$Cl$_2$	−78	36 : 64
	MgBr$_2$	CH$_2$Cl$_2$		21 : 79
	ZnI$_2$	CH$_2$Cl$_2$		53 : 47
	BF$_3$•OEt$_2$	CH$_2$Cl$_2$	−78	9 : 91
	BF$_3$•OEt$_2$ (2 equiv)	CH$_2$Cl$_2$	−78	5 : 95

[Reprinted with permission from Keck, G.E.; Boden, E.P. *Tetrahedron Lett.* **1984**, *25*, 265, Copyright **1984**, with permission from Elsevier Science.]

Tetravalent tin complexes add to aldehydes and ketones in the presence of a Lewis acid. Allyltin complexes are, by far, the most widely used of these compounds.[673] A typical example

---

673. For reactions of this type, see Larock, R.C. *Conmprehensive Organic Transformations, 2nd ed.*, Wiley-VCH, New York, **1999**, pp 373-378.

is taken from the work of Keck and Boden, in which a chiral aldehyde (**770**) was treated with allyltributyltin and various Lewis acids.[674] As shown in Table 8.24,[674] a mixture of syn (**771**) and anti (**772**) products was obtained. The ratio of **771**/**772** was dependent on the structure of the R group in **770**, the solvent and the Lewis acid.[674] The anti product (**772**) was obtained by using the *tert*-butyldimethylsilyloxy derivative (sec. 7.3.A.i) of **770** with two equivalents of boron trifluoride in dichloromethane. The syn product is obtained preferentially when the benzyloxy derivative of **770** is used with titanium tetrachloride in dichloromethane.[674]

Allyltin and crotyltin coupling is highly diastereoselective. In a synthesis of (+)-roxaticin, Evans and Connell reacted aldehyde **773** with allyltributytin in the presence of $SnCl_4$, and obtained a 90% yield of **774** with a 35:1 diastereoselectivity.[675] As the size of the group attached to the α-carbon of the aldehyde increased, the diastereoface selectivity increased for the syn diastereomer relative to allyltin compounds. In the example cited, the chelation product exceeded the product predicted by the Felkin-Anh model (sec. 4.7.B), being formed in a ratio of >200:1. In general, the Felkin-Anh model (predicting the syn diastereomer) gave the greatest success when BF_3 was used as a catalyst in the presence of diphenyl-*tert*-butylsilyloxy aldehydes.[676] Chelation control with silyloxy derivatives did not contribute significantly to any condensation product,[676a] although $MgBr_2$ gave 91:9 diastereofacial selectivity and 89:11 syn selectivity.[676a] Enantioselective addition of allyltributytin to aldehydes is also possible, using $Ti(Oi\text{-}Pr)_4$ and a chiral binaphthol derivative such as *(S)*-BINOL [*(S)*-(−)-1,1'-bi-2-naphthol]. In a synthesis of mucocin, Takahashi and co-workers reacted aldehyde **775** with allyltributytin[677] in the presence of *(S)*-BINOL, and obtained a 76% yield of **776** (>98% ee).

Homoallylic alcohols can be prepared. When benzaldehyde reacted with 2-methylbut-3-en-1-ol,[678] a 78% yield of **777** was obtained as a 49/1 (*E/Z*) mixture.

674. Keck, G.E.; Boden, E.P. *Tetrahedron Lett.* **1984**, *25*, 265. Allyltributyl tin can undergo conjugate addition, as in a synthesis of *ent*-Sch 47554 by Morton, G.E.; Barrett, A.G.M. *Org. Lett.* **2006**, *8*, 2859.
675. Evans, D.A.; Connell, C.T. *J. Am. Chem. Soc.* **2003**, *125*, 10899.
676. (a) Keck, G.E.; Abbott, D.E. *Tetrahedron Lett.* **1984**, 1883; (b) Keck, G.E.; Abbott, D.E.; Boden, E.P.; Enholm, E.J. *Ibid.* **1984**, *25*, 3927.
677. Takahashi, S.; Kubota, A.; Nakata, T. *Angew. Chem. Int. Ed.* **2002**, *41*, 4751.
678. Sumida, S.-i.; Ohga, M.; Mitani, J.; Nokami, J. *J. Am. Chem. Soc.* **2000**, *122*, 1310.

Tin-catalyzed addition of the alcohol substrate to the aldehyde, followed by a [3,3]-sigmatropic rearrangement, led to the products. Clearly, this is not the same reaction, but it is useful to compare it with the other reactions.

The allyltin disconnections are

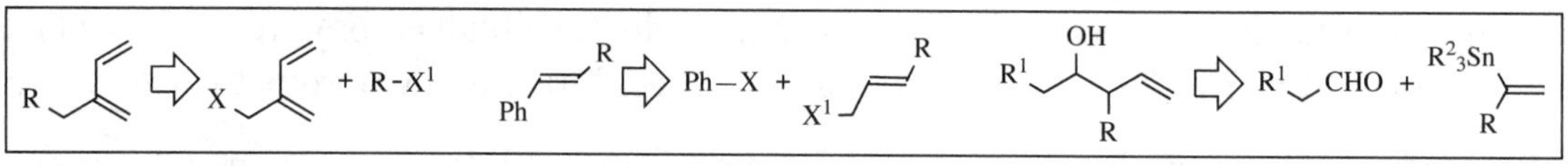

Other allyl metal complexes can be used to generate carbon bonds, including allyl zinc complexes. Stille showed that 3-methyl-1-bromo-2-butene reacted with zinc chloride ($ZnCl_2$) to generate the π-allyl zinc species. Coupling required the use of an organotin species such as α-trimethyltin isoprene (**778**). Reaction in refluxing THF led to a 94% yield of myrcene (**779**).[678]

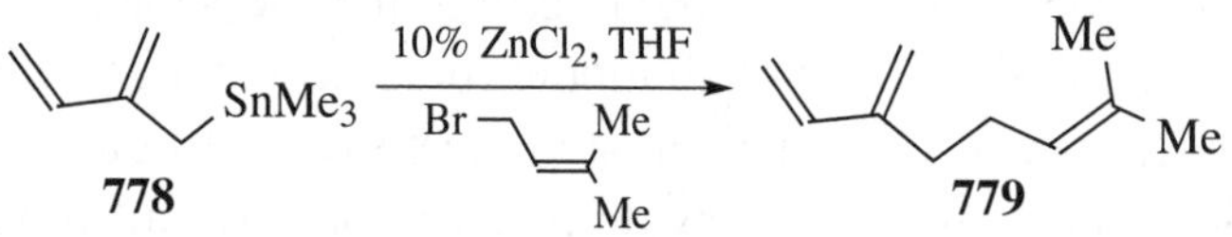

8.10.B. Alkyltitanium Reagents

A related coupling reaction is the condensation of aldehydes with alkyl trichlorotitanium compounds ($RTiCl_3$). When methyltrichlorotitanium ($MeTiCl_3$) was coupled with aldehyde **780** in dichloromethane at –78°C, a 91:9 mixture of **782/783** was formed via the chelated complex **781**.[679] Allyltitanium compounds also couple to aldehydes. In a synthesis of FR66979,[680] Cuifolini and Ducray reacted aldehyde **784** with allyltitanium anion **785** and obtained alcohol **786** in good yield.

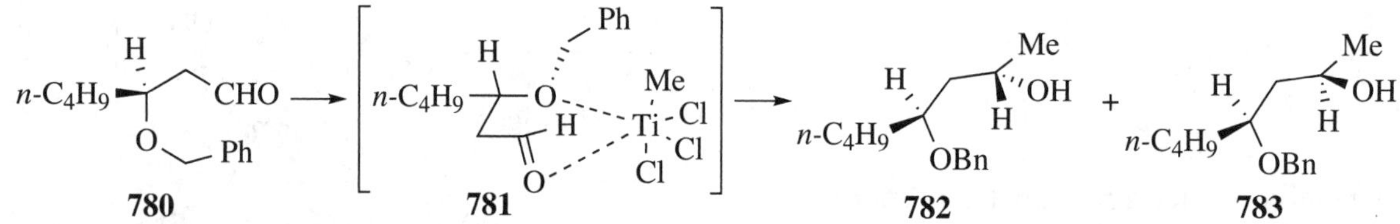

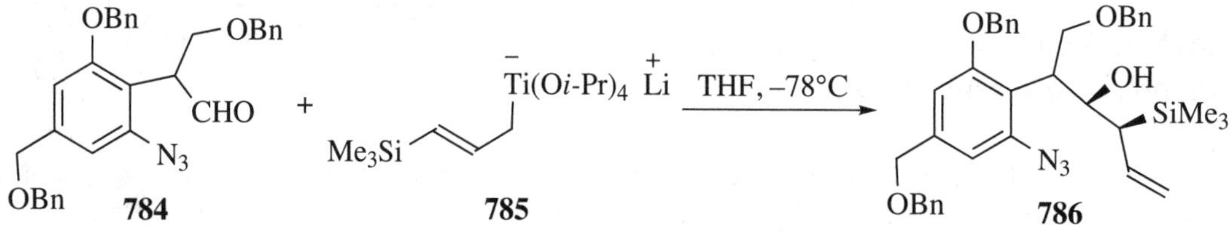

679. Reetz, M.T.; Jung, A. *J. Am. Chem. Soc.* **1983**, *105*, 4833.
680. Ducray, R.; Ciufolini, M.A. *Angew. Chem. Int. Ed.* **2002**, *41*, 4688.

8.10.C. Allylsilane Reagents

Another variation of this reaction couples an aldehyde with an allylsilane (rather than an allyltitanium species), in the presence of a Lewis acid. Sakurai and Hosomi had previously noted the coupling of allylsilanes to aldehydes in the presence of $TiCl_4$.[681] Reetz et al. used $TiCl_4$, $SnCl_4$, BF_3 or $AlCl_3$ in reactions of allyltrimethylsilane, and observed good diastereoselectivity.[682] In this study, the solubility of the Lewis acid, and the temperature of the reaction the Lewis acid employed were important factors for both reactivity and selectivity. A high degree of stereoselectivity is possible, if a chiral catalyst is used. In a synthesis of fostriecin,[683] Shibasaki and co-workers reacted conjugated aldehyde **787** with allyltrimethoxysilane in the presence of AgF and (R)-p-tolyl-BINAP, and obtained an 80% yield of **788** (dr = 28:1).

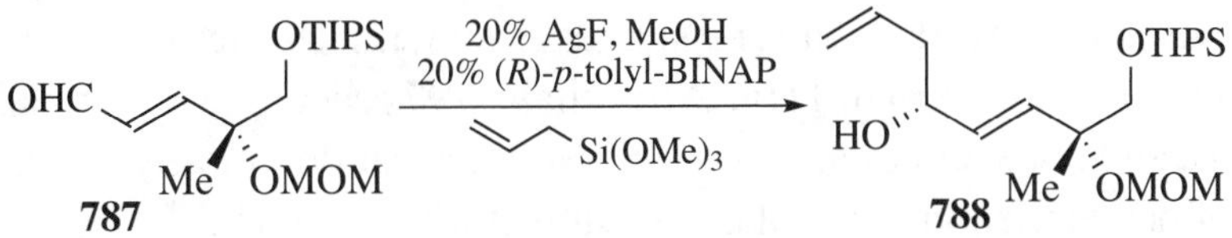

Heathcock and Kiyooka examined a similar coupling with aldehydes having a stereogenic α-carbon (as in **789**) rather than the β-carbon (as in Reetz's **787**). Upon reaction with allyltrimethylsilane, aldehyde **789** gave a mixture of **790** and **791**.[684] Using tin tetrachloride ($SnCl_4$) as a catalyst provided better syn selectivity in the products (**790** and **791**) than did BF_3

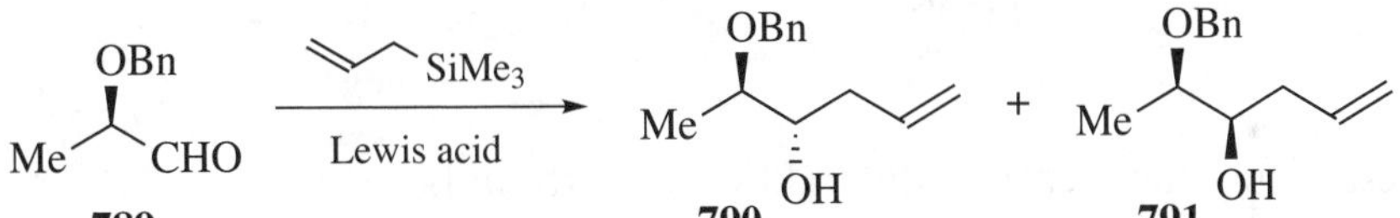

or $TiCl_4$ ($TiCl_4$ did not catalyze this coupling reaction).[684] As with the allyltin complexes, these results are explained by chelation control. A synthetic example, taken from Fürstner and co-worker's synthesis of herbarumin I, also illustrates an interesting variation.[685] Furanose derivative **792** reacted with allyltrimethylsilane in the presence of TMSOTf to give **793** (58% of the β-anomer + 5% of

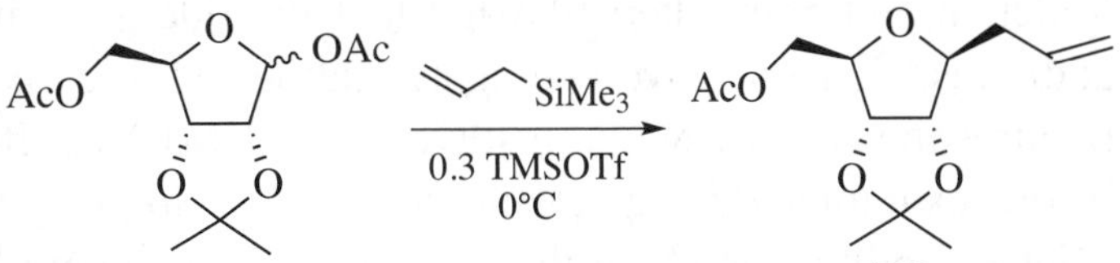

681. (a) Hosomi, A.; Sakurai, M. *Tetrahedron Lett.* **1976**, 1295; also see (b) Trost, B.M.; Coppola, B.P. *J. Am. Chem. Soc.* **1982**, *104*, 6879.
682. Reetz, M.T.; Kesseler, K.; Jung, A. *Tetrahedron Lett.* **1984**, *25*, 729.
683. Fujii, K.; Maki, K.; Kanai, M.; Shibasaki, M. *Org. Lett.* **2003**, *5*, 733.
684. (a) Kiyooka, S.; Heathcock, C.H. *Tetrahedron Lett.* **1983**, *24*, 4765; (b) Heathcock, C.H.; Kiyooka, S.; Blumenkopf, T.A. *J. Org. Chem.* **1984**, *49*, 4214.
685. Fürstner, A.; Radkowski, K.; Wirtz, C.; Goddard, R.; Lehmann, C.W.; Mynott, R. *J. Am. Chem. Soc.* **2002**, *124*, 7061.

the α-anomer). The allylsilane reacted via the aldehyde generated *in situ* from the protected hemi-acetal.

8.10.D. Silane Carbanions

Allylic silanes are synthetically useful compounds[686] that are usually prepared by reaction of chlorotrialkylsilanes and allylic Grignard reagents, as with the preparation of the *tert*-butyldimethylsilane **796**.[687] When a silane such as **794** is treated with fluoride ion, fluoride attacks the silicon and displaces the carbanion **795**. If the anion is generated in the presence of an electrophile, typical carbanion reactions occur. It is unlikely that the carbanion (**795**) has a significant lifetime as a discrete entity. Indeed, it need not exist at all, since the carbanionic character of that carbon increases as the fluoride departs, forming a transient hypervalent silicon species. Sakurai showed that allylic silanes react with ketones or aldehydes and tetra-*n*-butylammonium fluoride (TBAF) in THF. Allyl silane **797** reacted with butanal in the presence of TBAF[688] to give an 83% yield of alcohol **798**. Similar reaction of allylic silanes with TBAF, in the presence of benzophenone gave the acyl substitution product.[688]

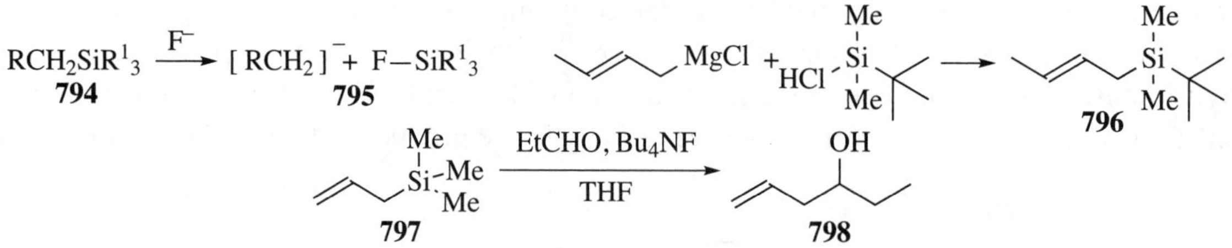

The disconnection for silane carbanion addition to carbonyls is

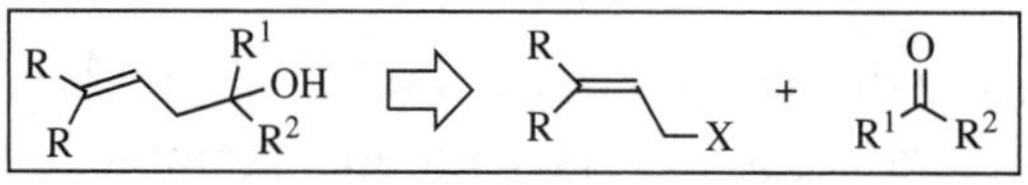

Majetich and co-workers showed that fluoride-generated carbanions undergo Michael-type additions.[689] To achieve conjugate addition with carbanions generated from silanes usually requires the use of DMF as a solvent, and HMPA is often added to obtain satisfactory yields. In this example, **799** was treated with TBAF to generate the carbanion **800**. Intramolecular Michael addition and hydrolysis led to the bicyclic product **801**.[689]

686. Chabaud, L.; James,P.; Landais, Y. *Eur. J. Org. Chem.* **2004**, 3173
687. Hosomi, A.; Sakurai, H. *Tetrahedron Lett.* **1978**, 2589.
688. (a) Hosomi, A.; Shirahata, A.; Sakurai, H. *Tetrahedron Lett.* **1978**, 3043; (b) Sakurai, H. *Pure Appl. Chem.* **1982**, *54*, 1.
689. (a) Majetich, G.; Desmond, Jr., R.W.; Soria, J.J. *J. Org. Chem.* **1986**, *51*, 1753; (b) Majetich, G.; Casares, A.; Chapman, D.; Behnke, M. *Ibid.* **1986**, *51*, 1745.

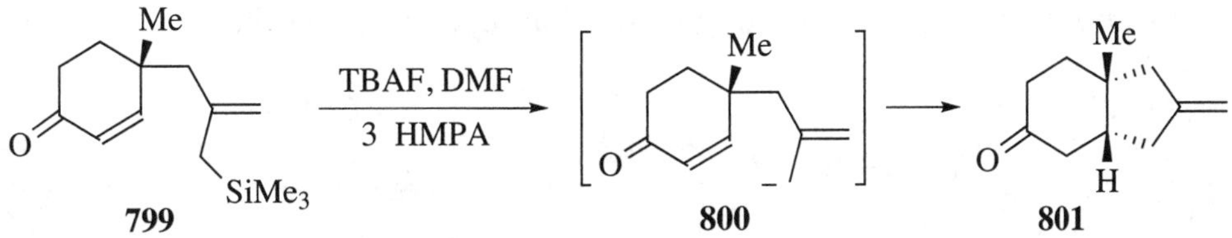

799 → TBAF, DMF / 3 HMPA → [**800**] → **801**

Note that Vedejs used the fluoride/silane carbanion reaction to generate a novel imidate ylid that was used in a [3+2]-cycloaddition reaction with ethyl acrylate (sec. 11.11.F).[690]

The disconnections possible with these latter reactions are

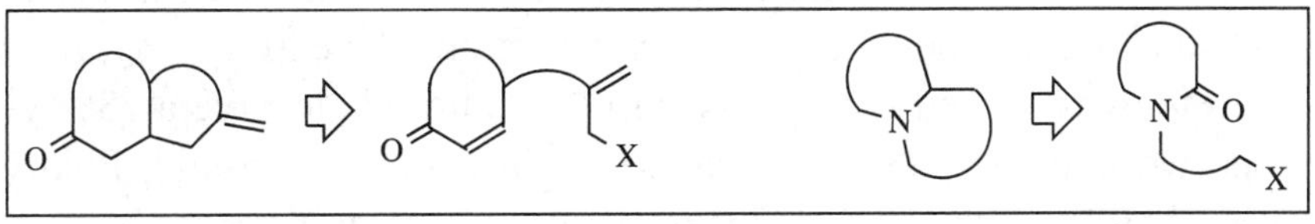

8.11. PHENOLIC CARBANIONS

There are additional methods for generating carbanionic centers that are not classified in previous sections. Two of the more useful are treatment of organosilanes with fluoride and formation of resonance stabilized aromatic carbanions. These two methods will be presented in this section, with an emphasis on the more synthetically useful versions of each reaction.

When phenols are treated with base, the resulting phenoxide ion (**802**) has four resonance contributors as shown. If an electrophilic species were added to **802**, the major product is usually

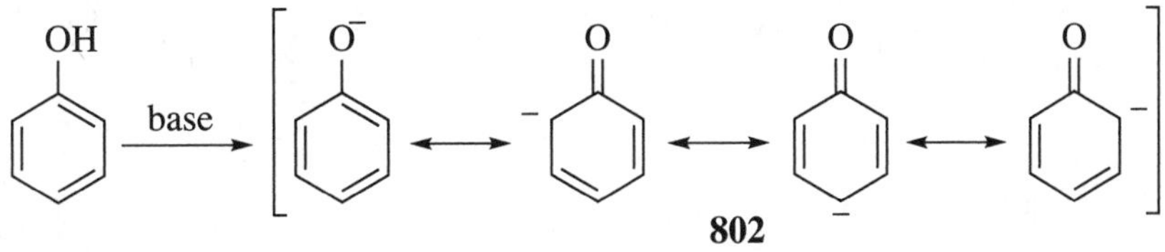

802

the ether (*O*-alkylation) via Williamson ether synthesis (sec. 2.6.A.i). If an electrophilic center is tethered to the phenolic moiety, however, it is possible to trap the carbanion center intramolecularly. Winstein showed that *p*-bromophenylsulfonyl (OBs) derivatives such as **803** reacted with potassium *tert*-butoxide to give the reactive carbanion intemediate **804**.[691] In the absence of an exogenous electrophile, carbanionic resonance contributors can displace an attached leaving group such as bromine, iodine, or a sulfonate ester to form a carbocyclic ring. When **803** was treated with *tert*-butoxide, **804** is one resonance contributor of the intermediate. Displacement of the bromobenzenesulfonyl leaving group in **804** by the aromatic carbanion

690. (a) Vedejs, E.; Larsen, S.; West, F.G. *J. Org. Chem.* **1985**, *50*, 2170; (b) Vedejs, E.; Martinez, G.R. *J. Am. Chem. Soc.* **1979**, *101*, 6452; **1980**, *102*, 7993; (c) Vedejs, E.; West, F.G. *J. Org. Chem.* **1983**, *48*, 4773.
691. Baird, R.; Winstein, S. *J. Am. Chem. Soc.* **1962**, *84*, 788.

proceeded via an Ar_1^--5 mechanism[692] to give the spirocyclic ketone (**805**).

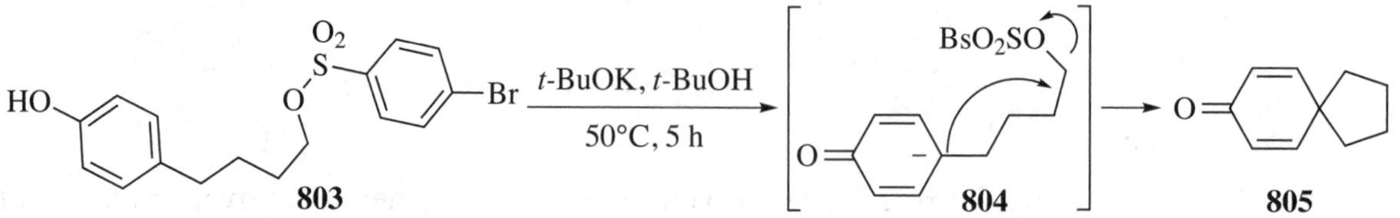

A phenol need not be used directly, as long as a phenoxide anion can be formed easily. Spirocyclization of a bis(methanesulfonate) has been reported, where the requisite phenolic anion is generated *in situ*.[693] Corey and co-workers[694] as well as Crandall and Lawton[695] used spirocyclization to construct the spirocyclic portion of cedrene. In another example, phenol derivative **806** was treated with potassium *tert*-butoxide to give a 73% yield of **807** in Mukherjee and co-workers synthesis of zizaene.[696] In another example, Fukuyama and co-workers[697] converted **808** to **809** (77% yield) in a synthesis of duocarmycin.

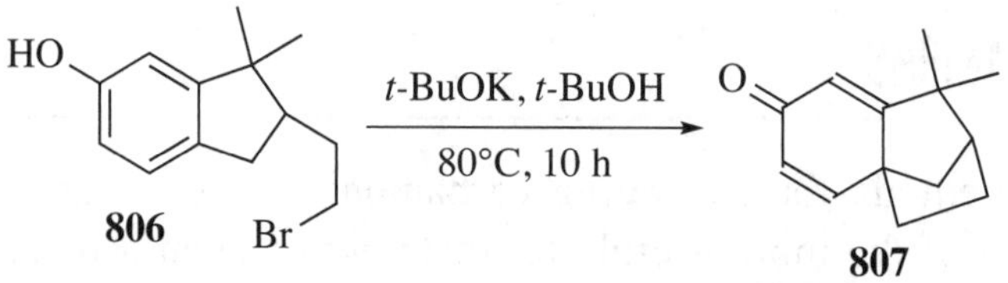

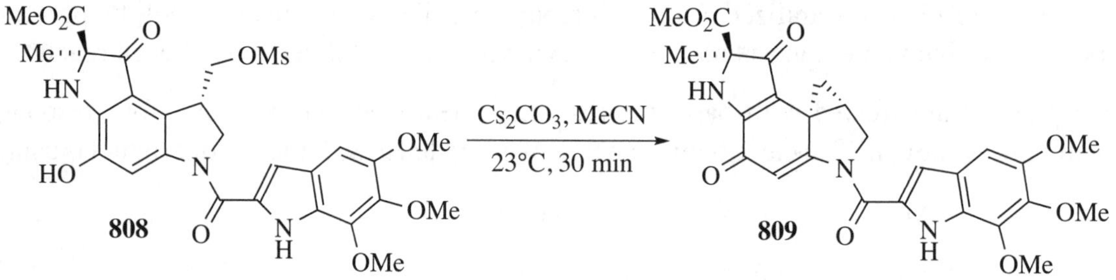

This spirocyclic disconnection is

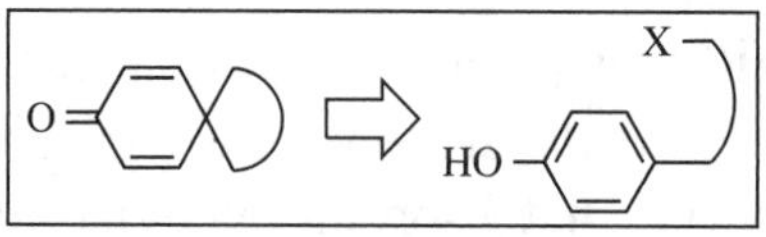

692. (a) Winstein, S.; Heck, R.; Lapporte, S.; Baird, R. *Experientia* **1956**, *12*, 138; (b) Heck, R.; Winstein, S. *J. Am. Chem. Soc.* **1957**, *79*, 3105; (c) Winstein, S.; Baird, R. *Ibid.* **1957**, *79*, 756; (d) Scott, F.L.; Glick, R.E.; Winstein, S. *Experientia* **1957**, *13*, 183.
693. Marx, J.N.; Bih, Q.-R. *J. Org. Chem.* **1987**, *52*, 336.
694. Corey, E.J.; Girotra, N.N.; Mathew, C.T. *J. Am. Chem. Soc.* **1969**, *91*, 1557.
695. Crandall, T.G.; Lawton, R.G. *J. Am. Chem. Soc.* **1969**, *91*, 2127.
696. Pati, L.C.; Roy, A.; Mukherjee, D. *Tetrahedron* **2002**, *58*, 1773.
697. Yamada, K.; Kurokawa, T.; Tokuyama, H.; Fukuyama, T. *J. Am. Chem. Soc.* **2003**, *125*, 6630.

8.12. CONCLUSION

This chapter has shown how incorporation of a metal or an appropriate electron withdrawing group or atom gives carbanions that can be used to form new carbon-carbon bonds. The great variety of nucleophilic species, the range of electrophilic species with which they react, and the high degree of stereoselectivity which accompanies many of the reactions show why these are among the most powerful disconnections in organic synthesis. The useful carbonyl stabilized carbanions, the enolates will be discussed in Chapter 9.

HOMEWORK

1. Predict the major product and explain the stereochemistry of the following reaction.

$$t\text{-BuMe}_2\text{SiO} \quad \xrightarrow[\text{CuBr}\bullet\text{Me}_2\text{S}]{\text{BF}_3\bullet\text{OEt}_2,\ \text{MeMgBr}}$$

with CO$_2$Et, N–Boc, OMe substituents

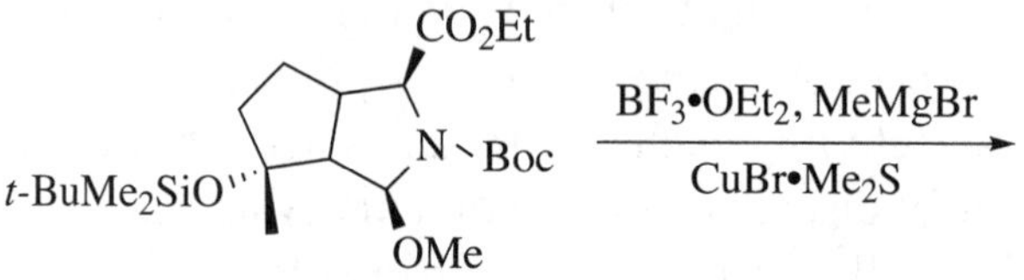

2. The Grignard reaction of compound **86** in the text gave alcohol **87**. Offer an explanation for this selectivity.

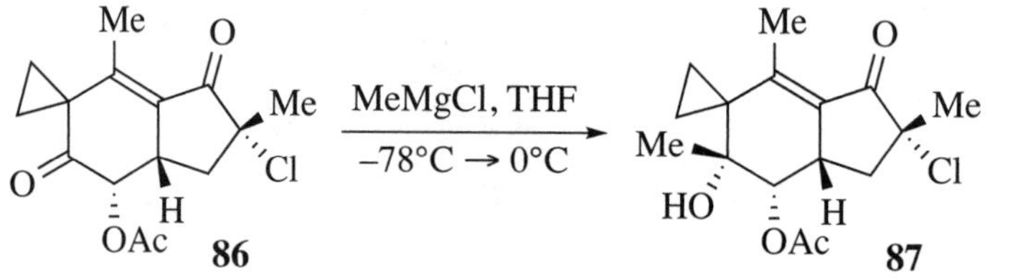

$$\xrightarrow[-78°\text{C} \rightarrow 0°\text{C}]{\text{MeMgCl, THF}}$$

86 → **87**

3. Given the reaction sequence, give a mechanistic rationale for formation of the cyclohexanone product.

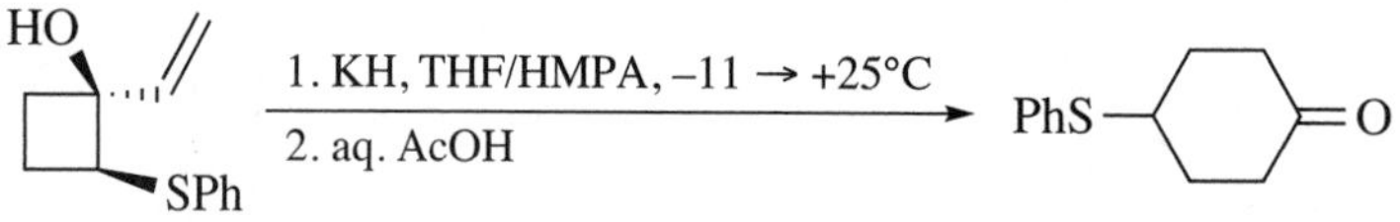

$$\xrightarrow[\text{2. aq. AcOH}]{\text{1. KH, THF/HMPA, } -11 \rightarrow +25°\text{C}}$$

4. In each case, draw the transition state for the Cram, Karabatsos and Felkin-Anh models of the reaction of each molecule with (1) MeMgBr (2) PhMgCl (3) CH$_3$C≡C:$^-$Na$^+$ (4) EtLi (5) 2-lithio-1,3-dithiane.

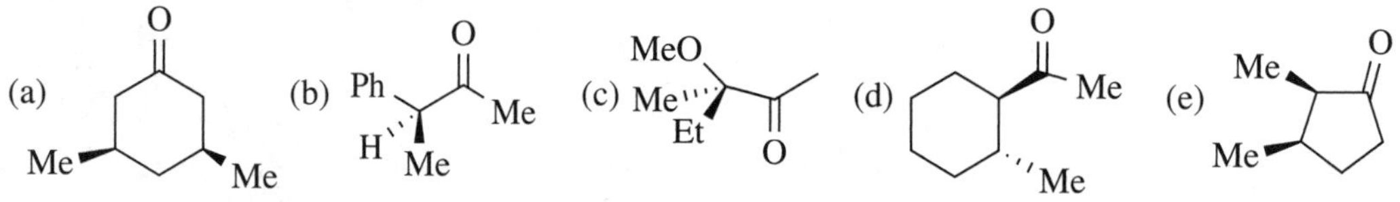

(a) (b) Ph (c) MeO (d) (e) Me

Draw the major product(s) that result from each reaction, after hydrolysis.

5. The following reaction gives a diastereomeric ratio of about 3:1. Draw both of these products, and predict the major product.

$$\text{OHC} \quad \xrightarrow{\text{BuLi}} \quad$$

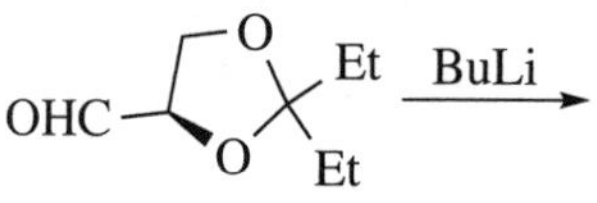

6. Give a mechanistic explanation for the formation of the product shown from the designated starting material.

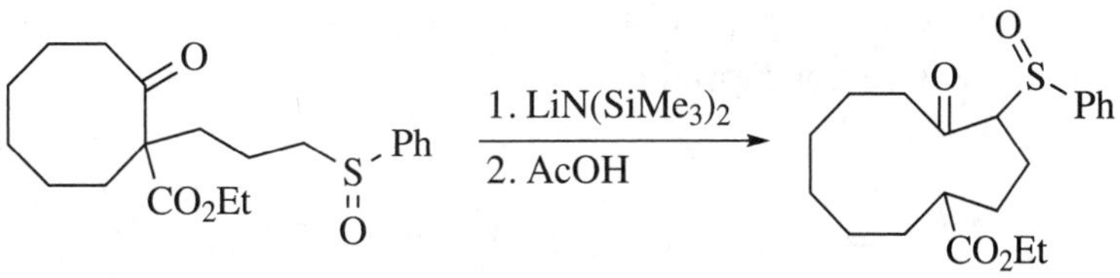

7. Give the major product for (a-e) when treated with (1) *n*-BuLi/THF/-78°C; (2) MeI/-78 → 0°C; (3) aq NH₄Cl.

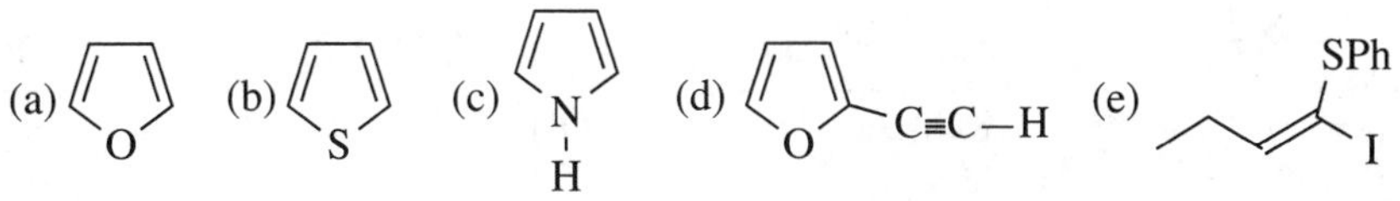

8. For each of the following give a complete reaction that illustrates its use for the formation of a carbon-carbon bond:

(a) $CdCl_2$; (b) Cp_2TiMe_2; (c) $CeCl_3$ (d) lithium 2,2,6,6-tetremethylpiperidide; (e) CO; (f) CuCN; (g) *N,N,N',N'*-tetramethylethylenediamine; (h) $FeCl_3$; (i) NaCN; (j) $NaNH_2$; (k) BF_3; (l) $Na_2Fe(CO)_4$; (m) *i*-Bu₂AlH; (n) PPh_3; (o) NaH; (p) $(MeO)_2POMe$; (q) DMSO.

9. Rationalize formation of the indicated product in this reaction.

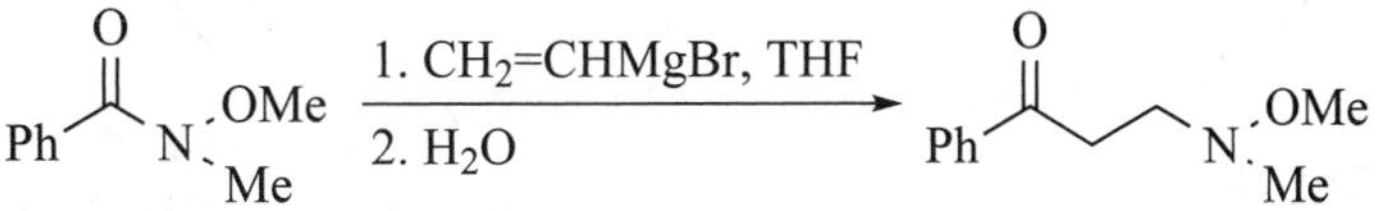

10. Each of the following molecules can be used to standardize a solution of an organolithium reagent. Describe, with reactions, the acid-base chemistry involved in each reaction.

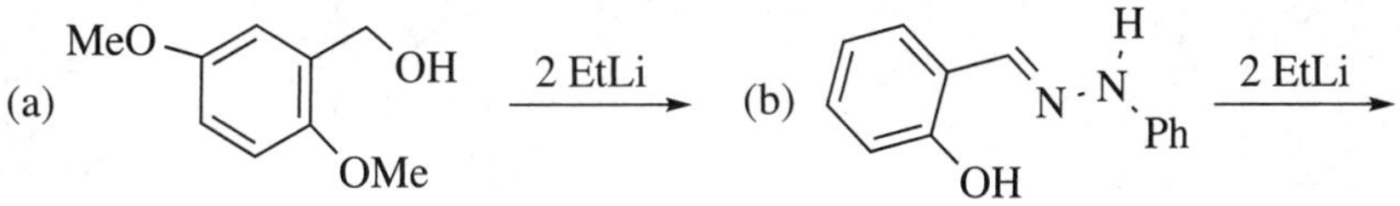

11. Provide a suitable synthetic sequence for each transformation (a-d).

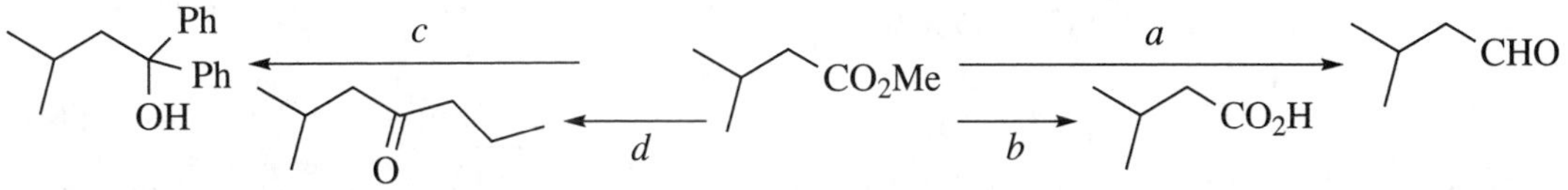

12. For (a-c) give the major product of each reaction, with correct stereochemistry, and justify your choice.

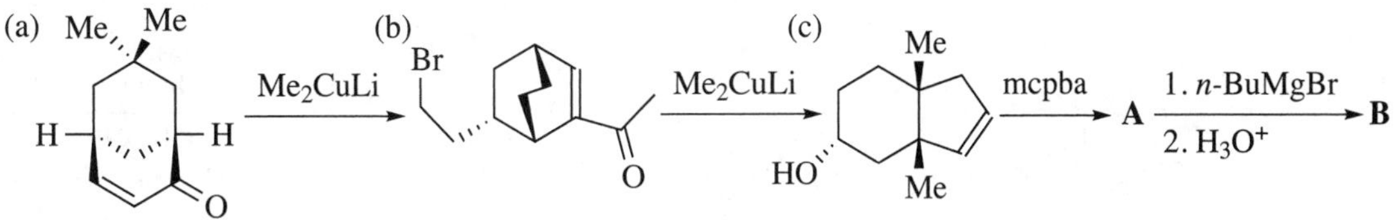

13. In each of the following reactions, predict the major product, with the correct stereochemistry where appropriate:

(a) 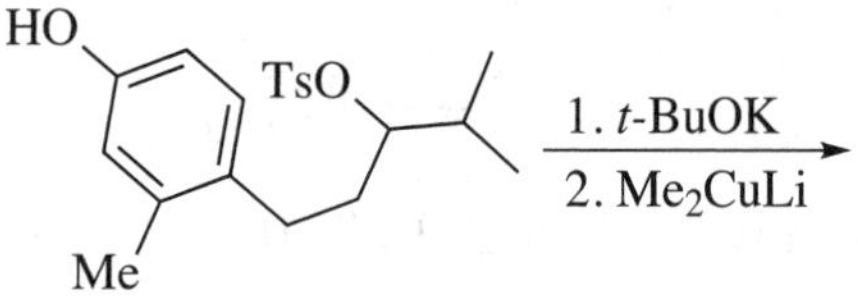
1. t-BuOK
2. Me₂CuLi

(b) 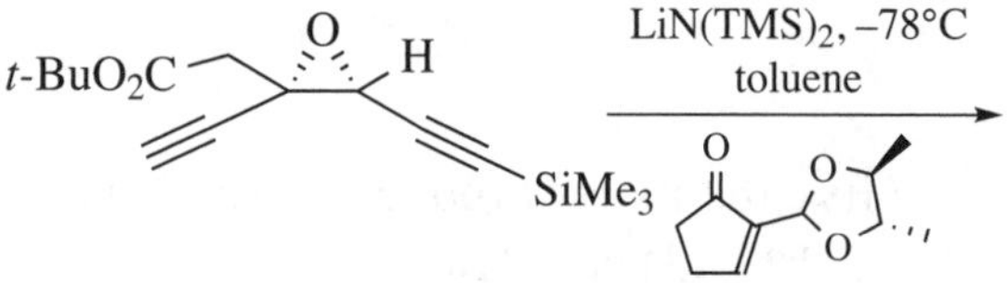
LiN(TMS)₂, −78°C
toluene

(c) 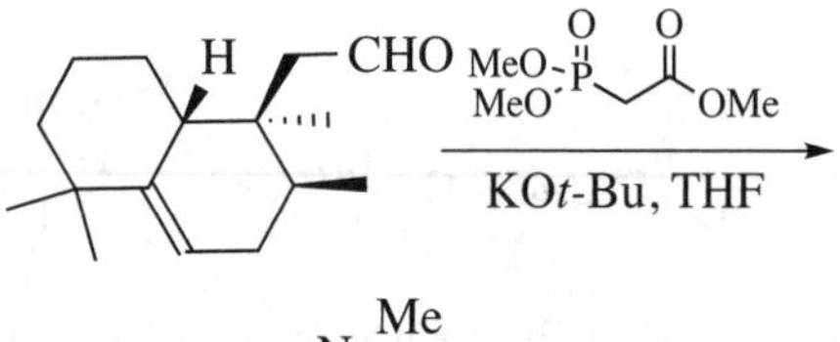
MeO–P(O)(OMe)–CH₂–C(O)–OMe
KOt-Bu, THF

(d) 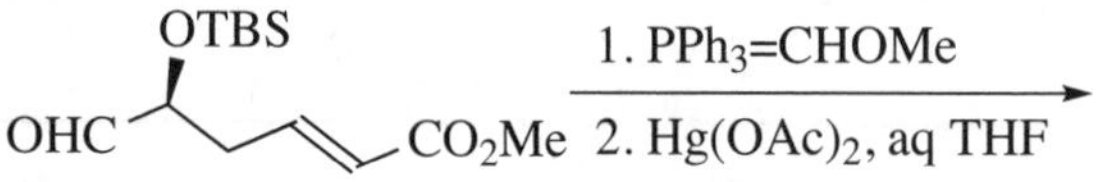
1. PPh₃=CHOMe
2. Hg(OAc)₂, aq THF

(e) 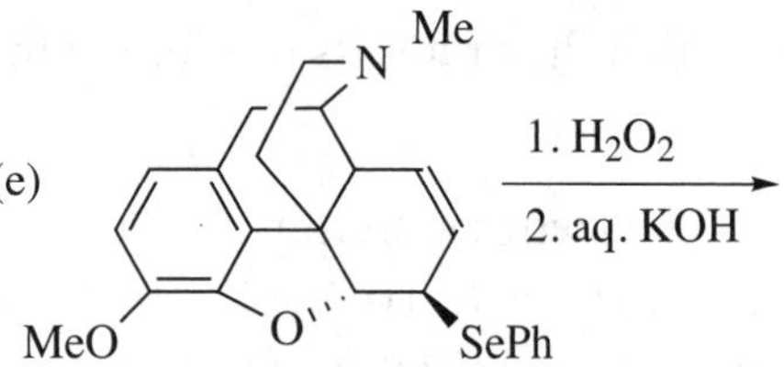
1. H₂O₂
2. aq. KOH

(f) 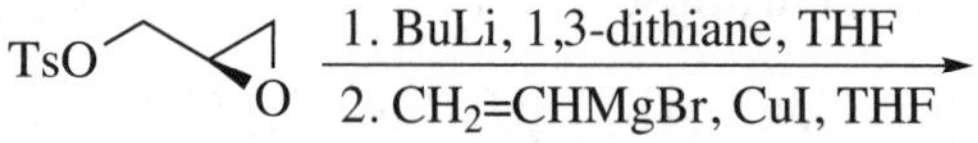 TsO–CH₂–epoxide
1. BuLi, 1,3-dithiane, THF
2. CH₂=CHMgBr, CuI, THF

(g) 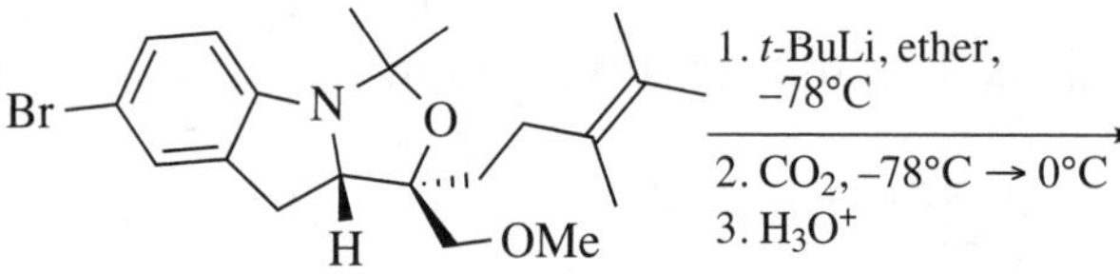
1. t-BuLi, ether, −78°C
2. CO₂, −78°C → 0°C
3. H₃O⁺

(h) 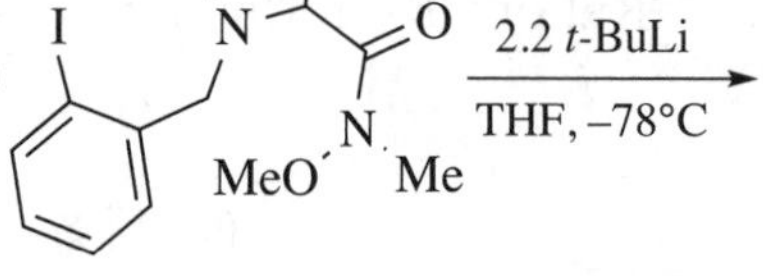
2.2 t-BuLi
THF, −78°C

(i) 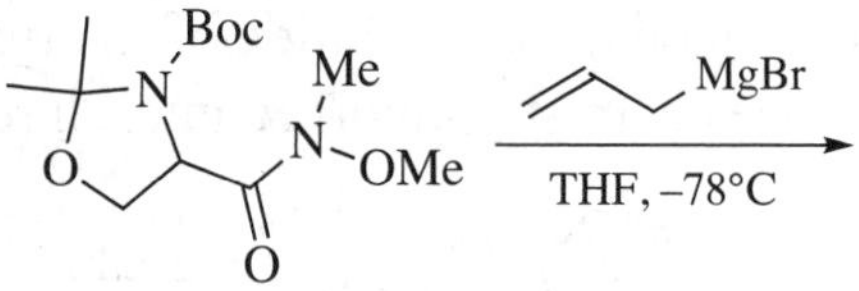
allyl–MgBr
THF, −78°C

(j) 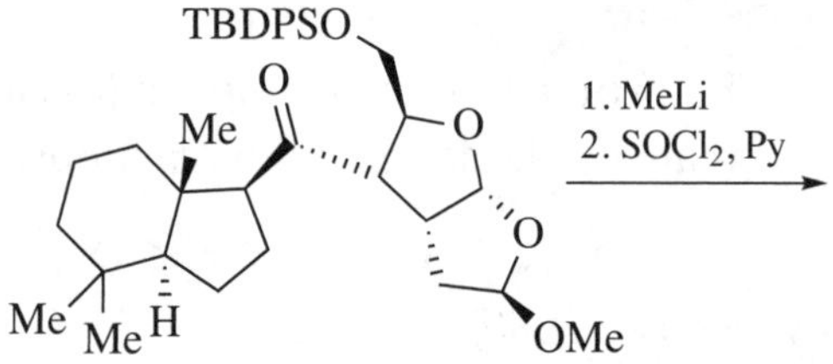
1. MeLi
2. SOCl₂, Py

(k) 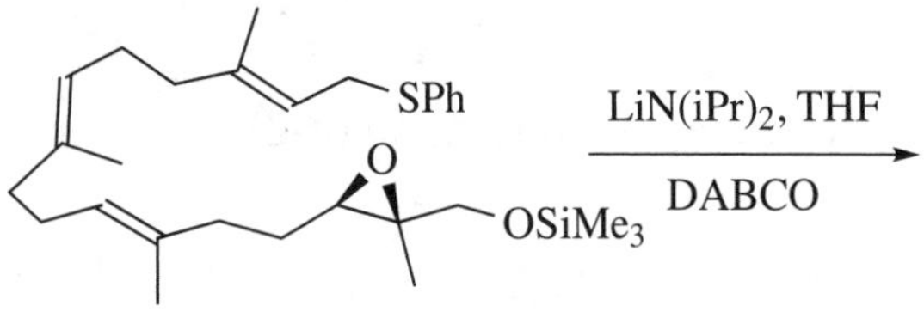
LiN(iPr)₂, THF
DABCO

(l) 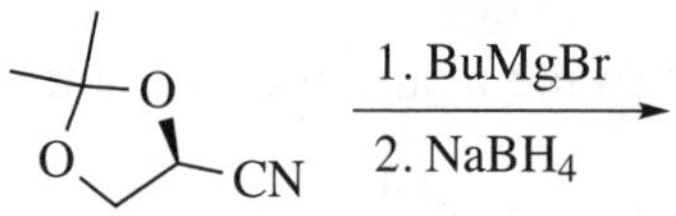
1. BuMgBr
2. NaBH₄

(m) 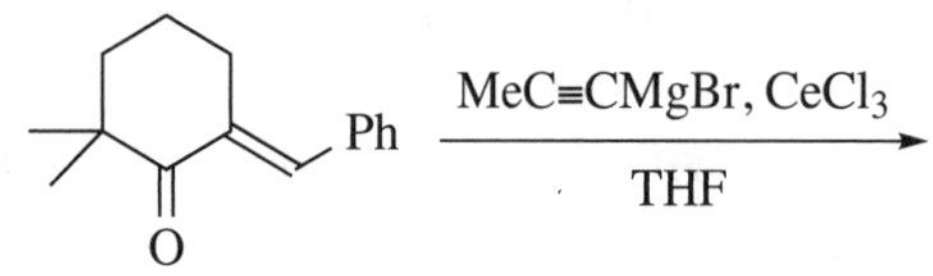
MeC≡CMgBr, CeCl₃
THF

(n) 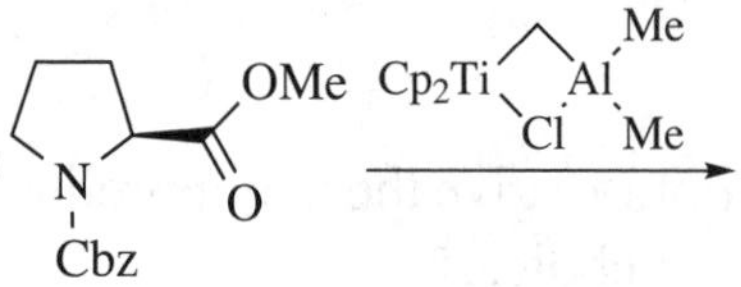
Cp₂Ti(Cl)(Me)Al Me

(o)
Me₂CuLi

(p)
1. 2 sec-BuLi, TMEDA
allyl bromide, THF
2. BCl₃, Bu₄NI
CH₂Cl₂, −78°C → rt

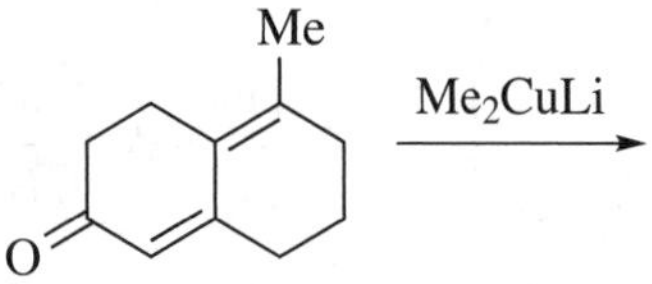

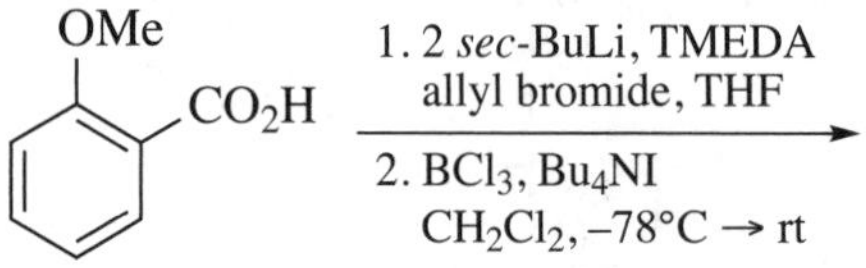

(q)

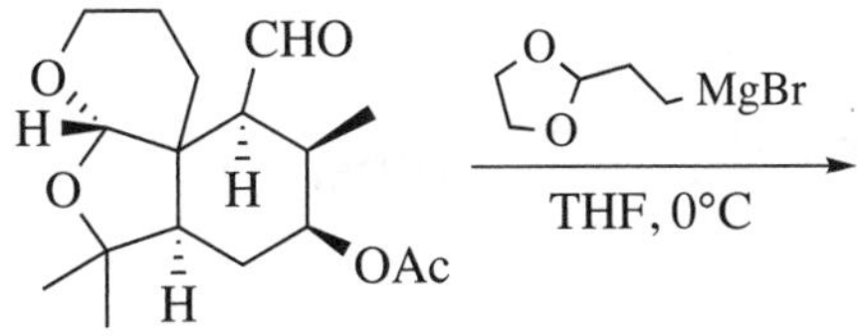

$\dfrac{\text{MgBr}}{\text{THF, 0°C}}$ (with 1,3-dioxolane ethyl MgBr reagent)

(r) 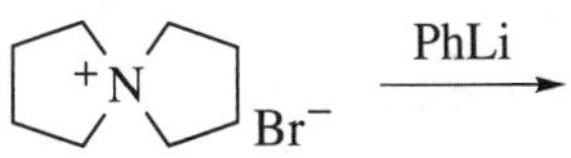 Br⁻ $\xrightarrow{\text{PhLi}}$

(s) t-BuMe$_2$Si (1,3-dithiane) $\xrightarrow[\substack{\text{2. } -78°C \rightarrow -45°C \\ 2.5\ \text{eq} \quad \text{BnO} \diagdown \diagup \text{O}}]{\text{1. } t\text{-BuLi, THF-HMPA, } -78°C}$

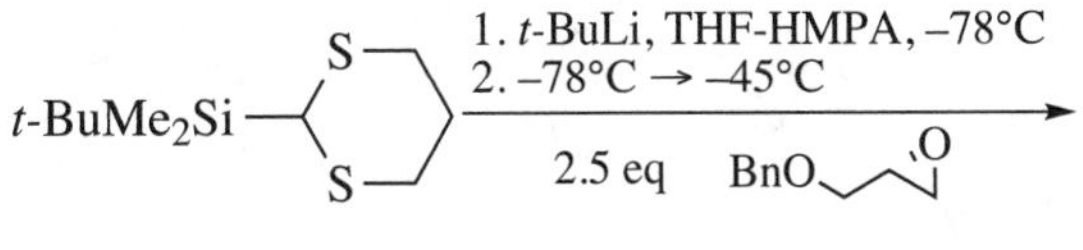

(t) HO $\diagup$ (epoxide, Me) $\diagdown$ OTIPS, OTIPS $\xrightarrow[\text{2.} \diagup \text{MgBr}]{\text{1. TPAP, NMO}}$

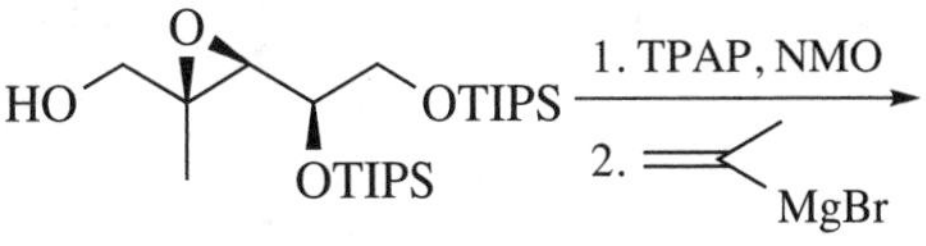

(u) 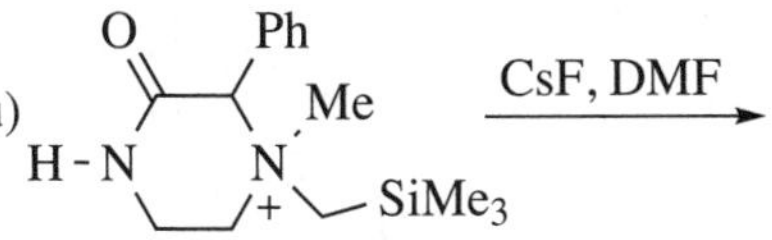 (piperazinone, Ph, Me, SiMe$_3$) $\xrightarrow{\text{CsF, DMF}}$

(v) 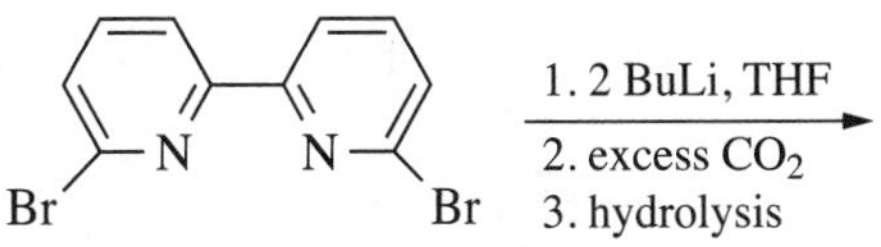 (6,6′-dibromo-2,2′-bipyridine) $\xrightarrow[\substack{\text{2. excess CO}_2 \\ \text{3. hydrolysis}}]{\text{1. 2 BuLi, THF}}$

(w) 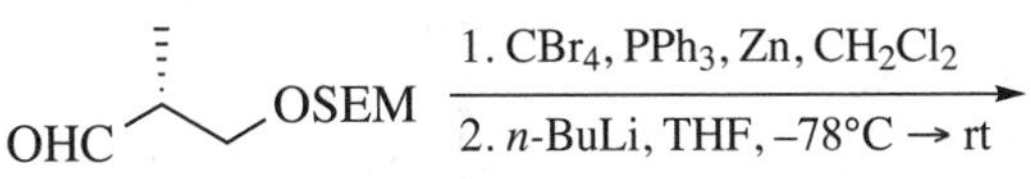 OHC $\diagdown$ (Me) $\diagdown$ OSEM $\xrightarrow[\text{2. } n\text{-BuLi, THF, } -78°C \rightarrow \text{rt}]{\text{1. CBr}_4\text{, PPh}_3\text{, Zn, CH}_2\text{Cl}_2}$

(x) 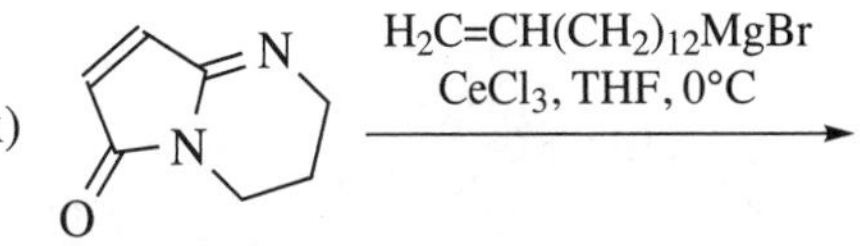 (bicyclic amidine, O) $\xrightarrow[\text{CeCl}_3\text{, THF, 0°C}]{\text{H}_2\text{C=CH(CH}_2)_{12}\text{MgBr}}$

(y) 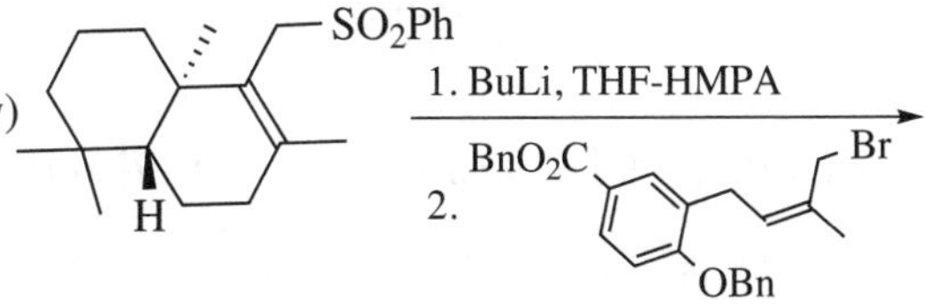 (decalin with SO$_2$Ph) $\xrightarrow[\substack{\text{BnO}_2\text{C} \quad \text{Br} \\ \text{2.} \quad \text{OBn}}]{\text{1. BuLi, THF-HMPA}}$

(z) 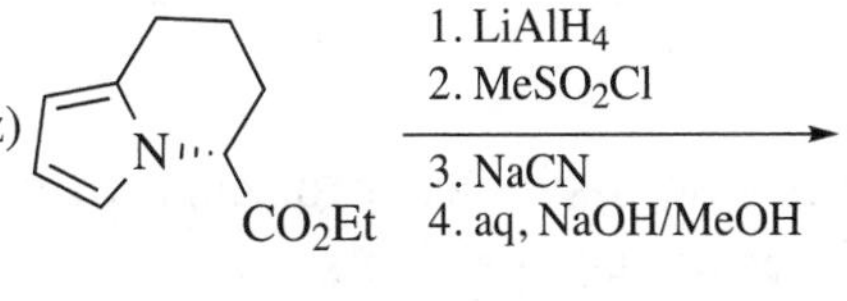 (indolizidine, CO$_2$Et) $\xrightarrow[\substack{\text{2. MeSO}_2\text{Cl} \\ \text{3. NaCN} \\ \text{4. aq, NaOH/MeOH}}]{\text{1. LiAlH}_4}$

(aa) 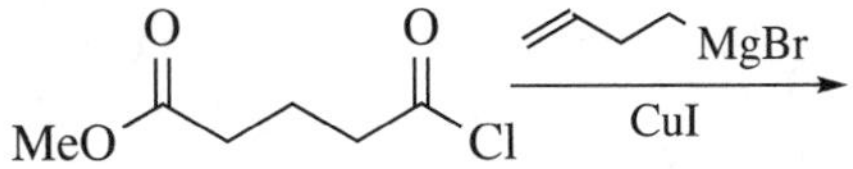 MeO$_2$C $\diagdown\diagup\diagdown$ C(=O)Cl $\xrightarrow[\text{CuI}]{\diagup\diagdown \text{MgBr}}$

(ab) 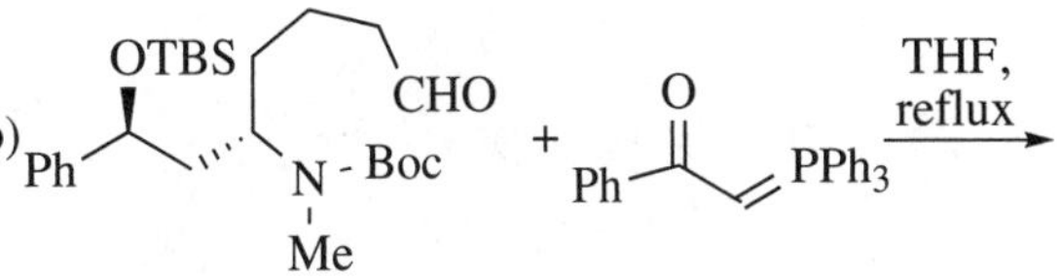 Ph $\diagdown$ (OTBS) $\diagdown$ N(Boc)(Me) $\diagdown$ CHO $+$ Ph-C(=O)-CH=CH-PPh$_3$ $\xrightarrow[\text{reflux}]{\text{THF,}}$

(ac) 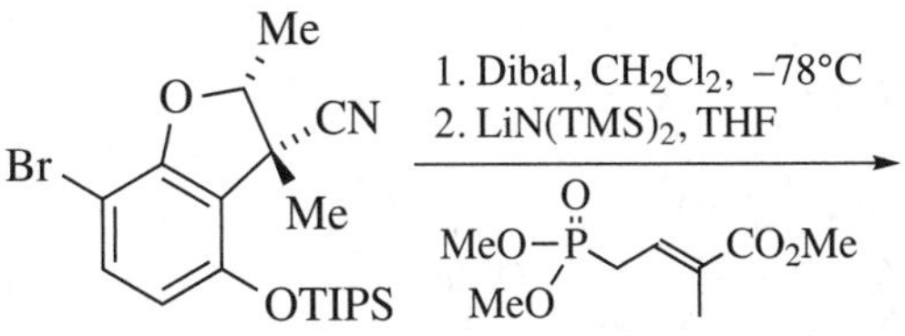 Br $\diagup$ (benzofuran, Me, CN, Me, OTIPS) $\xrightarrow[\substack{\text{2. LiN(TMS)}_2\text{, THF} \\ \text{MeO-P(=O)(OMe)-CH=C(Me)-CO}_2\text{Me}}]{\text{1. Dibal, CH}_2\text{Cl}_2\text{, } -78°C}$

(ad) 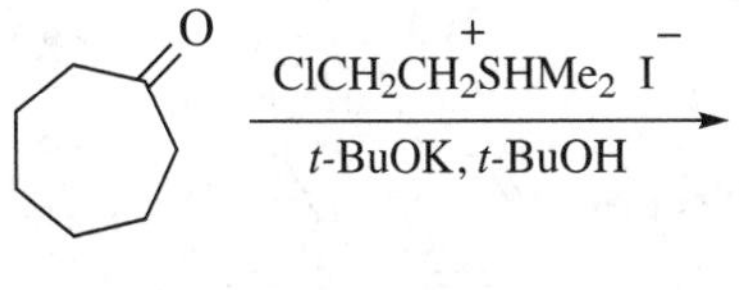 (cycloheptanone) $\xrightarrow[\text{$t$-BuOK, t-BuOH}]{\text{ClCH}_2\text{CH}_2\text{SHMe}_2^+ \ \text{I}^-}$

(ae) 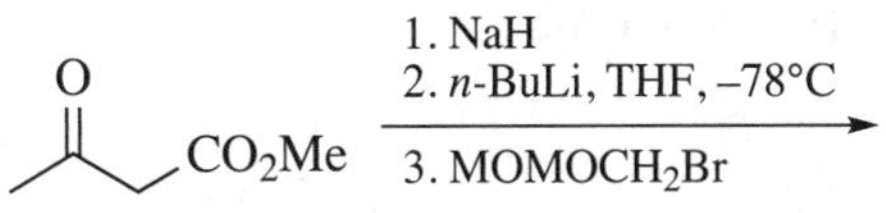 O=C(CH$_3$)-CH$_2$-CO$_2$Me $\xrightarrow[\substack{\text{2. } n\text{-BuLi, THF, } -78°C \\ \text{3. MOMOCH}_2\text{Br}}]{\text{1. NaH}}$

(af) 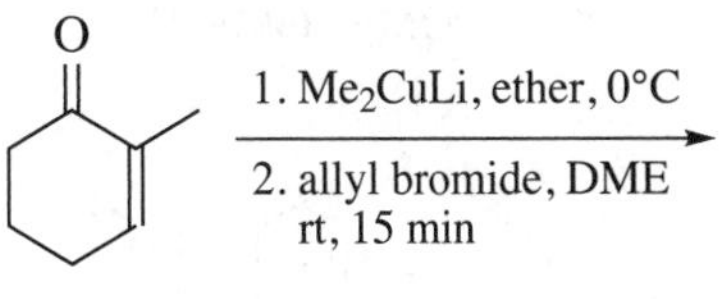 (2-methylcyclohexenone) $\xrightarrow[\substack{\text{2. allyl bromide, DME} \\ \text{rt, 15 min}}]{\text{1. Me}_2\text{CuLi, ether, 0°C}}$

(ag) 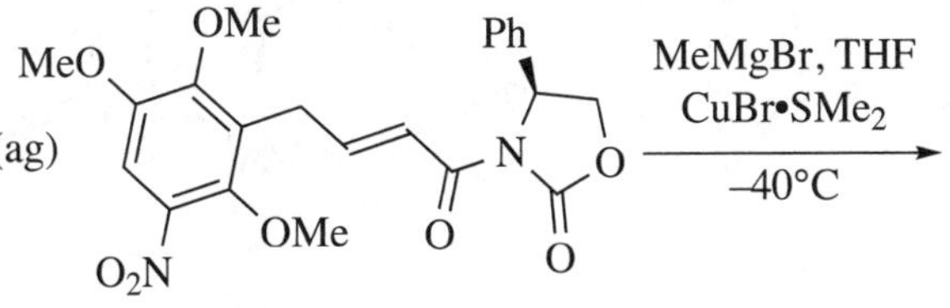 (trimethoxy-nitro-aryl enoyl oxazolidinone, Ph) $\xrightarrow[-40°C]{\substack{\text{MeMgBr, THF} \\ \text{CuBr·SMe}_2}}$

(ah) 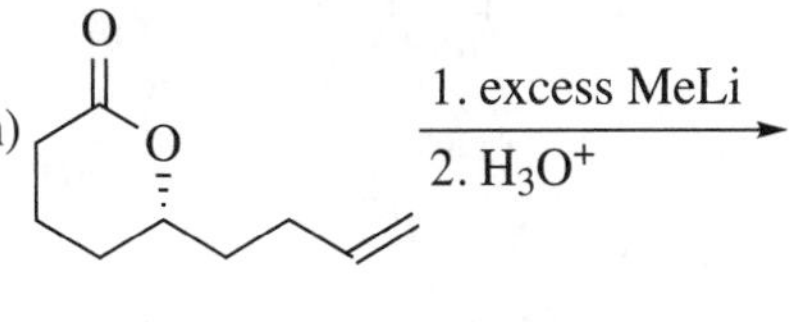 (lactone with allyl side chain) $\xrightarrow[\text{2. H}_3\text{O}^+]{\text{1. excess MeLi}}$

(ai) 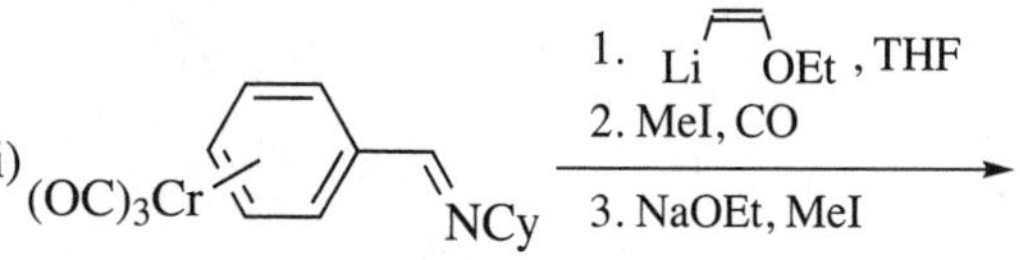 $(OC)_3Cr$ [arene]—CH=NCy

1. Li [—OEt], THF
2. MeI, CO
3. NaOEt, MeI

(aj) 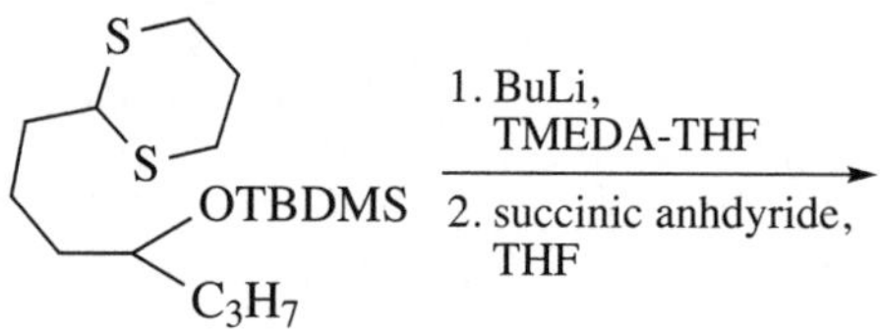 [1,3-dithiane structure]—OTBDMS, C_3H_7

1. BuLi, TMEDA-THF
2. succinic anhydride, THF

(ak) 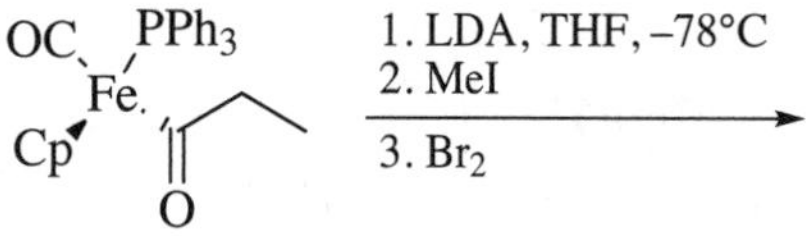 Cp, OC, PPh_3, Fe, [acyl O]

1. LDA, THF, −78°C
2. MeI
3. Br_2

(al) 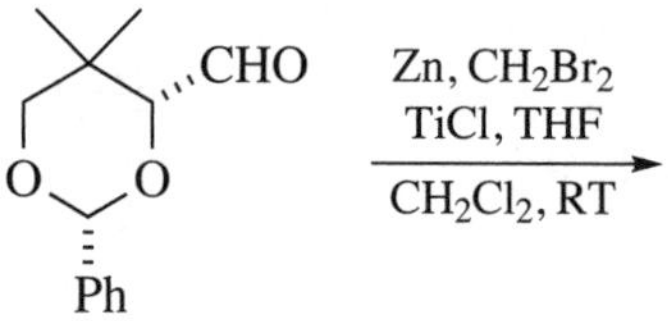 [1,3-dioxane with Ph]—CHO

Zn, CH_2Br_2
TiCl, THF
CH_2Cl_2, RT

(am) 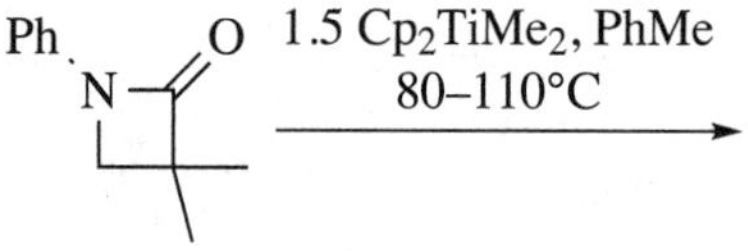 Ph—N—[β-lactam with O and methyls]

1.5 Cp_2TiMe_2, PhMe
80–110°C

(an) 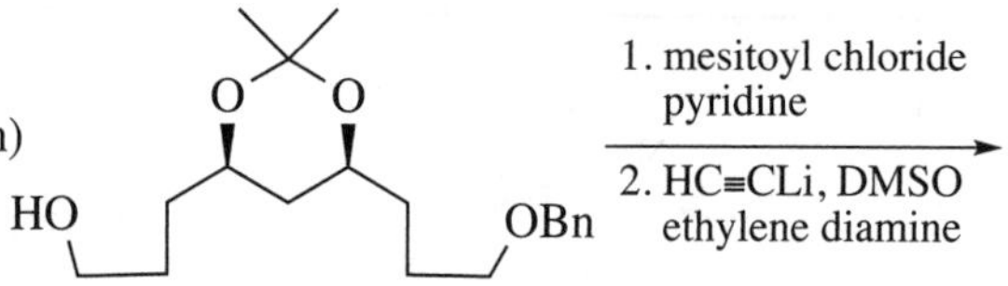 [1,3-dioxane, HO— and —OBn chains]

1. mesitoyl chloride pyridine
2. HC≡CLi, DMSO ethylene diamine

(ao) 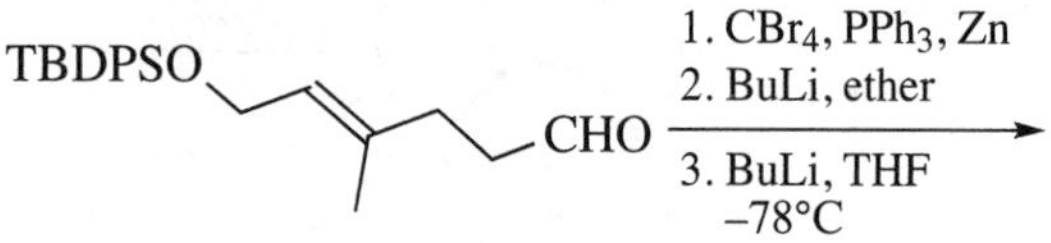 TBDPSO—[trisubstituted alkene chain]—CHO

1. CBr_4, PPh_3, Zn
2. BuLi, ether
3. BuLi, THF −78°C

(ap) 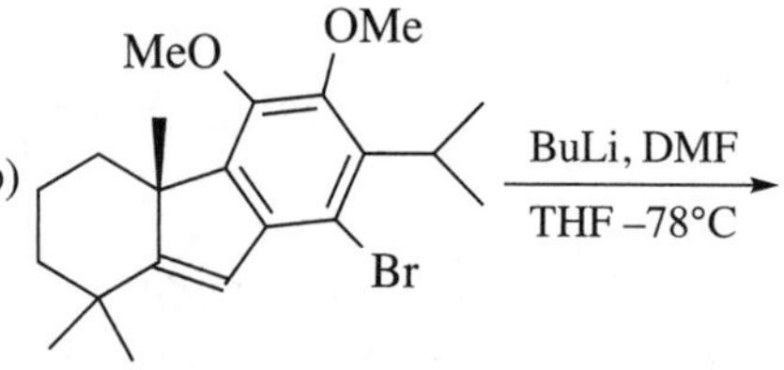 [fused tricyclic aromatic, MeO, OMe, isopropyl, Br, gem-dimethyl]

BuLi, DMF
THF −78°C

14. Provide a synthesis for each of the following transformations. Show all reagents and intermediate products.

(a) 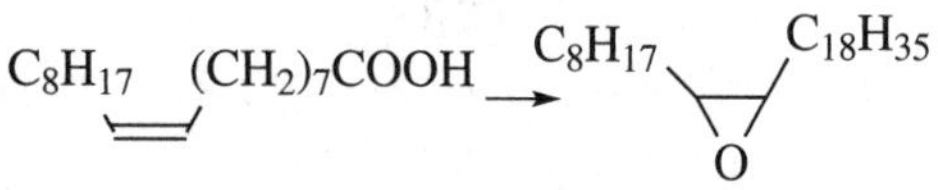 C_8H_{17}—[cis alkene]—$(CH_2)_7COOH$ → C_8H_{17}—[epoxide]—$C_{18}H_{35}$

(b) [1,3-dioxane]—$OSiMe_2t$-Bu, —OAc → [1,3-dioxane]—OH, allyl, [alkene]—OTIPS

(c) HO, OHC, MeO, OMe, MeO, OMe OMe OMe [biaryl] → O=, MeO, OMe, [allyl], MeO, OMe OMe OMe [biaryl]

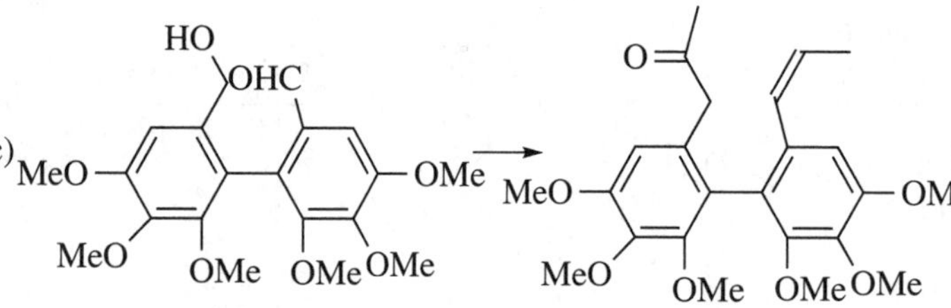

(d) 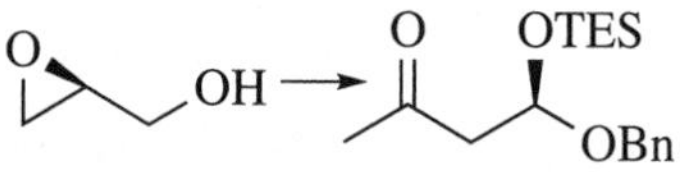 [epoxide]—OH → O=, OTES, OBn

(e) 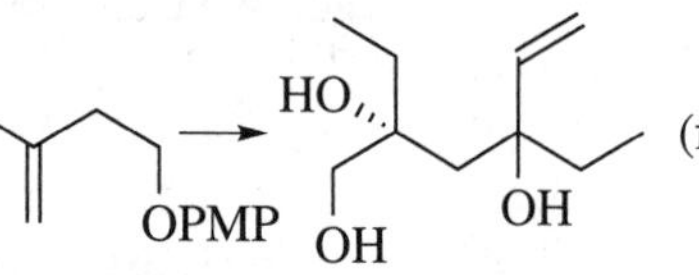 [diene, OPMP] → HO,,, [alcohol], OH, vinyl, OH

(f) 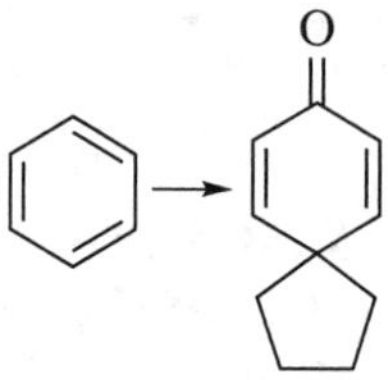 [benzene] → [spiro cyclohexenone-cyclopentane]

(g) 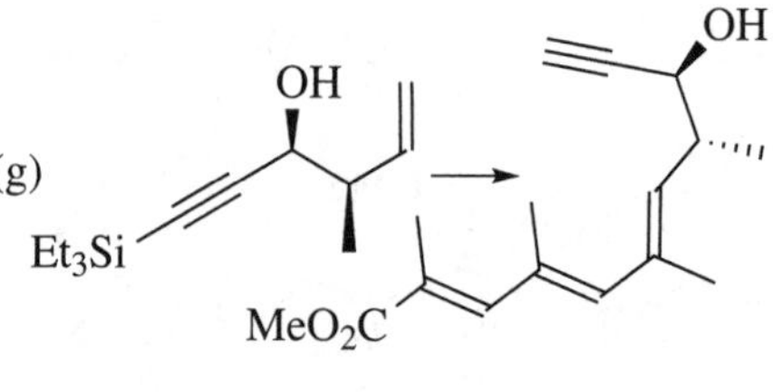 Et_3Si—[alkyne]—OH, [diene, MeO_2C] → [alkyne]—OH, [diene chain]

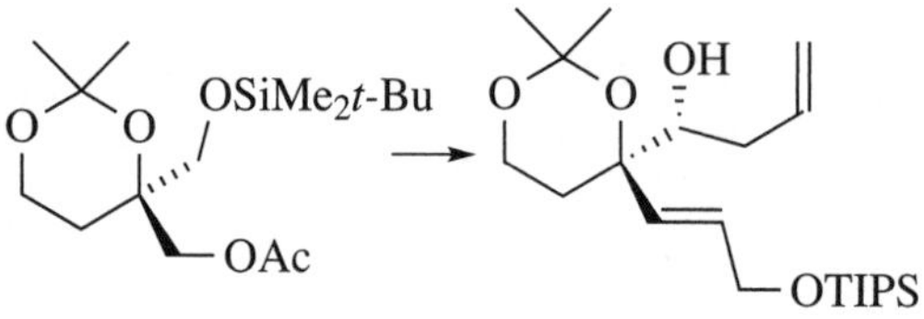

(h)

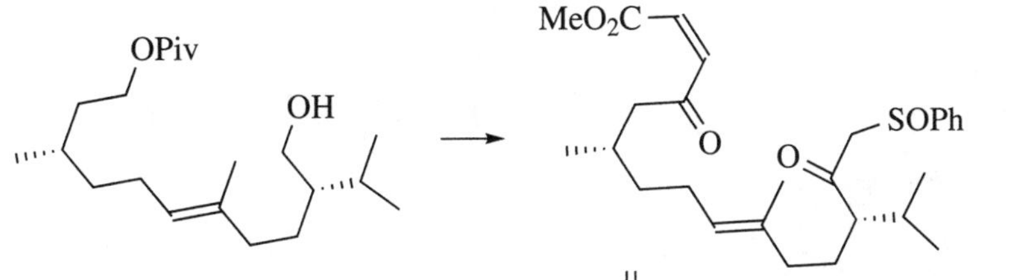

(i)

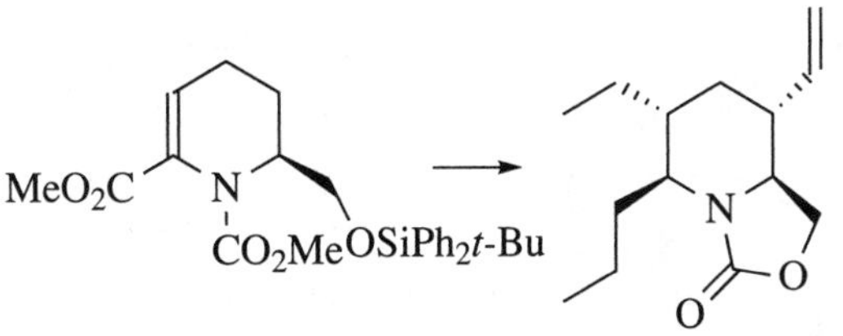

(j)

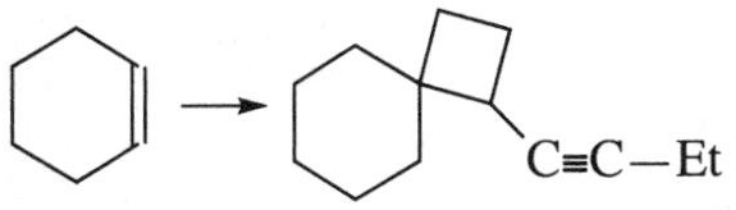

(k)

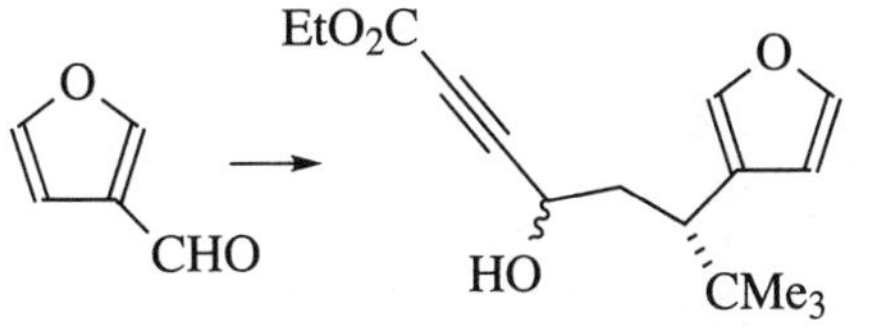

(l)

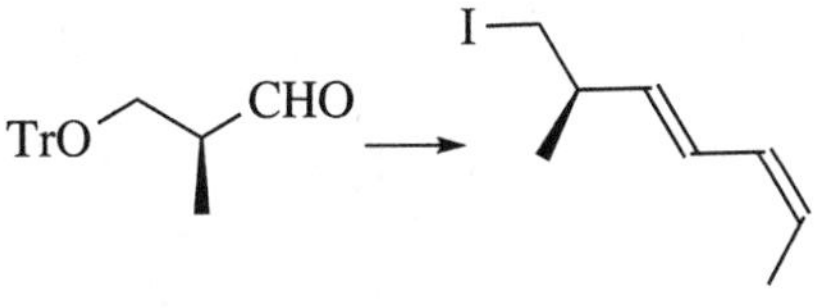

(m)

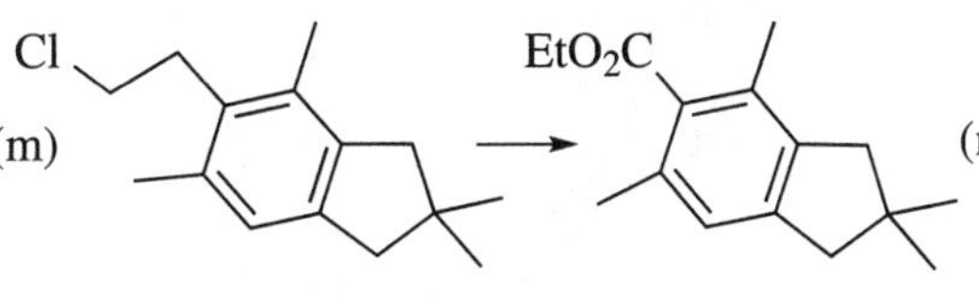

(n)

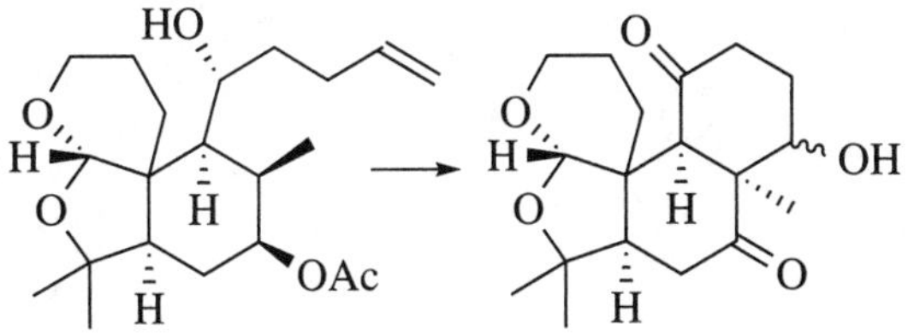

(o)

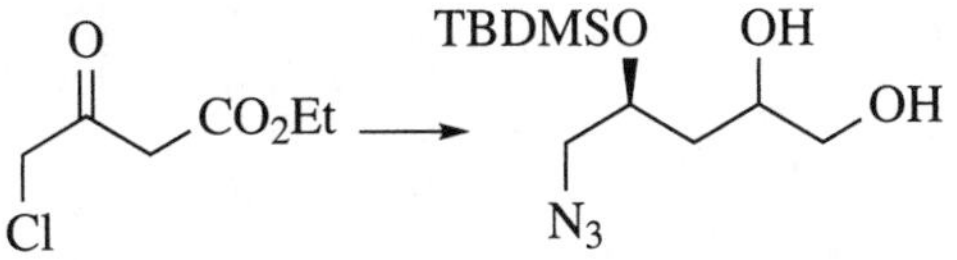

(p)

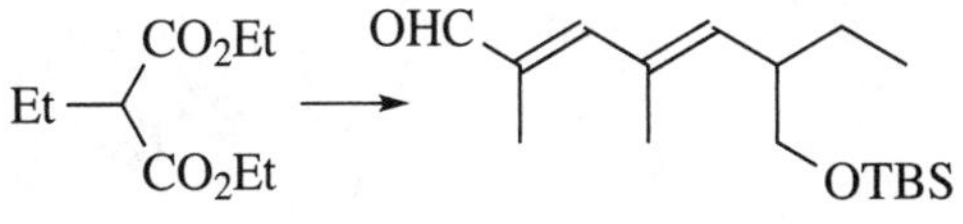

(q)

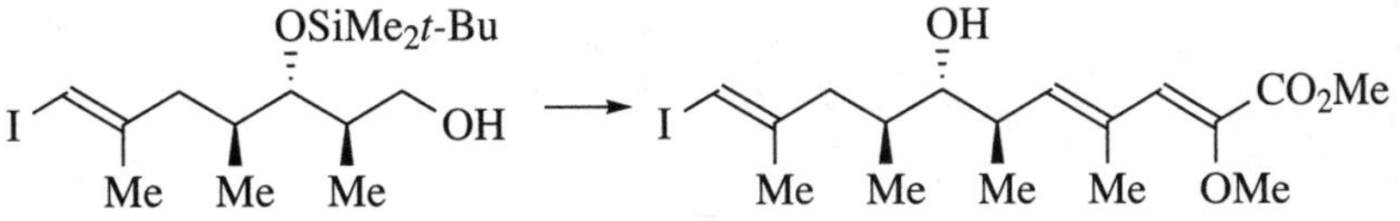

15. Synthesize each of the following molecules from a starting material of no more than six carbons. That starting material must be commercially available from a chemical company. Show your retrosynthetic analysis and all reagents and intermediate products of the synthesis.

(a)

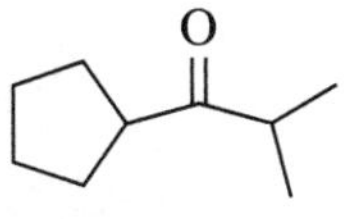

(b)

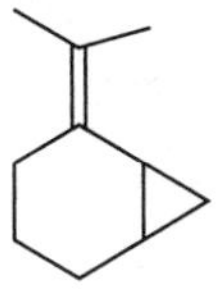

(c)

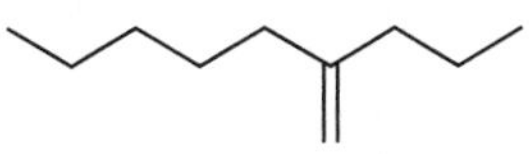

(d)

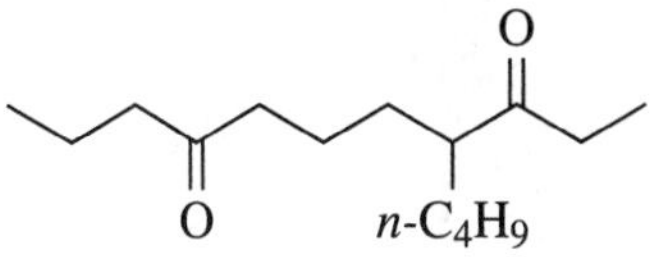

(e)

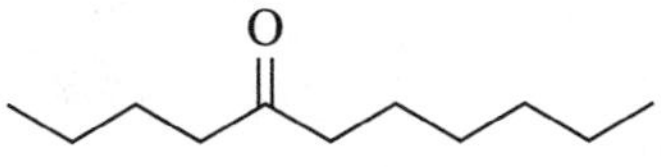

(f)

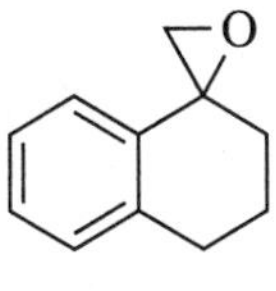

(g)

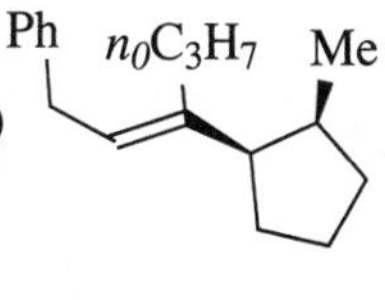

(h)

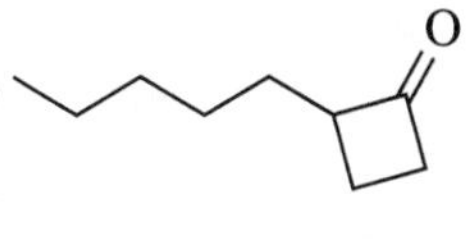

chapter 9

Cᵈ Disconnect Products: Nucleophilic Species that Form Carbon-Carbon Bonds: Enolate Anions

9.1. INTRODUCTION

In Chapter 8, a hydrogen atom on a carbon α to an electron-withdrawing group was shown to be a weak acid. Removal of that proton by a base generated several synthetically useful reagents that contained a carbanion center, and functioned as a C^d disconnect product (sec. 1.2). When the electron-withdrawing group is a single carbonyl group (ketones, aldehydes, esters, etc.), the pK_a of the adjacent protons are generally in the range pK_a 19-24.[1] As discussed in Section 8.5.G, the effects of the carbonyl group are due to (*1*) the inductive electron-withdrawing ability of the unsaturated substituent, but mainly due to (*2*) the ability of these substituents to delocalize the negative charge remaining after a proton has been removed.[2] With the enhancement in acidity induced by the carbonyl, weaker bases can be used for deprotonation when compared to sulfides, sulfoxides and so on (introduced in Chapter 8).[2] Deprotonation generates a resonance stabilized **enolate anion**. The bases used in previous chapters are all rather nucleophilic, and when dealing with aldehydes, ketones or carboxylic acid derivatives, acyl addition to the carbonyl may compete with deprotonation. Acyl addition must be suppressed or diminished if an enolate anion is to be generated, and used in subsequent reactions. This result can be achieved by decreasing the nucleophilicity of the base while maintaining or enhancing its basicity. This chapter will introduce enolate anions, and discuss solutions to the problems posed, as well as synthetically useful reactions.

9.2. FORMATION OF ENOLATE ANIONS

9.2.A. Preparation and Properties

The attachment of an electron-withdrawing carbonyl on a C–H moiety enhances the acidity of that hydrogen, the so-called α-hydrogen. Reaction of a carbonyl derivative such as 3-methyl-2-butanone (**1**) and a suitable base leads to an acid-base reaction in which H_a is removed (as H^+), making the α-carbon a carbanion (see **2**). This carbanion (called an **enolate anion**)

1. House, H.O. *Modern Synthetic Reactions, 2nd Ed.*, W.A. Benjamin, Menlo Park, CA., *1972*, p 494 and references 1, 2b cited therein.
2. (a) d'Angelo, J. *Tetrahedron* *1976*, *32*, 2979; (b) Stowell, J.C. *Carbanions in Organic Synthesis* Wiley-Interscience, New York, *1979*, pp 127-216.

is resonance stabilized, with the electrons delocalized on the carbon and on the oxygen. If the carbanion center is the preferred site for reaction, an enolate anion will behave like any other nucleophilic center upon reaction with an electrophile, so an enolate anion is a carbon nucleophile. Table 9.1 reproduces a portion of Table 8.16 in Section 8.5.H, and several things are apparent. In **1**, there are two chemically different types of α-hydrogen atoms (H_a and H_b). There are three identical hydrogen atoms on the methyl group (labeled H_b), and one on C3 (labeled H_a). The presence of the electron-releasing methyl groups on the α-carbon diminishes the acidity of H_a because the C–H_a bond is less polarized.

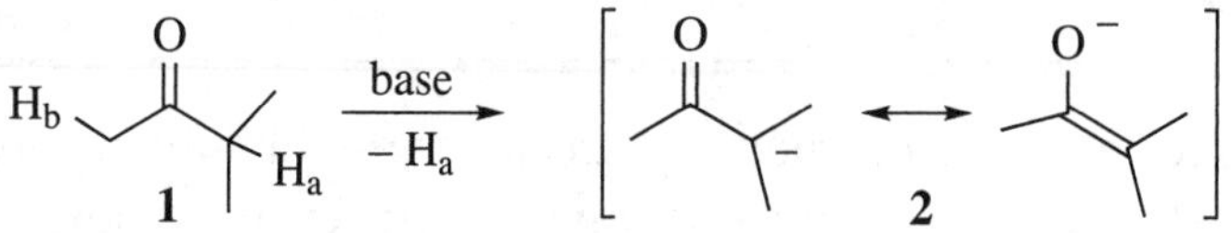

Table 9.1.   pK_a values of Typical Carbon Acids

Carbon Acid	pKa	Carbon Acid	pKa	Carbon Acid	pKa	Carbon Acid	pKa
N≡C–H	25[a]	CO₂Et / CO₂Et (H)	15[a]	**4**	20.8[b,c]	Cl (H)	16.5[a]
HO–H (O)	24[a]	Me Me (H, O) **7**	11[a]	**6**	23.5[b,c]	CO₂Et / CO₂Et (H)	13.3[a]
5	21.3[b]	CHO / CHO (H) **9**	5.9[a]	**3**	20[a]	Me / Me (H, O) **8**	9.0[a]
Ph (H)	19.5[b]	NO₂ / NO₂ (H)	3.6[a]	Ph–H (O)	19.1[d,b]	H Me / MeO₂S Me **10**	4.7[a]
Ph (H)	18.6[b]						

a Ref. [3] b ref. [4], c Ref. [5] d Ref. [6]

This trend can be seen in Table 9.1 by comparing **3** (acetone) with **4** that has a pK_a of 20.8. They are obviously close. However, when **4** is compared to **5**, with a pK_a of 21.3, and **6** with a pK_a of 23.5, the presence of the alkyl substituents leads to a higher pK_a (less acidic proton). Since deprotonation is an acid-base reaction, and we know that alkyl groups are electron-releasing, the effect on the equilibrium is that the carbanionic center in the enolate anion is

3.   Pearson, R.G.; Dillon, R.C. *J. Am. Chem. Soc.* **1953**, *75*, 2439.
4.   Zook, H.D.; Kelly, W.L.; Posey, I.Y. *J. Org. Chem.* **1968**, *33*, 3477.
5.   Reference 1, p 494.
6.   Cram, D.J. *Fundamentals of Carbanion Chemistry*, Academic Press, New York, **1965**, pp 4, 10, 13, 14, 43, 48.

destabilized and the C–H$_a$ bond is stronger (see **1**). These factors shift the acid-base equilibrium to the left (toward **1**) and contribute to decreasing acidity of the hydrogen. The α-hydrogen of an aldehyde is more acidic than that of a ketone, as seen by comparing malonaldehyde (**9**, pK_a of 5.9) with acetylacetone (**8**, pK_a of 9.0). Comparing the α-hydrogen of acetone to the α-hydrogen of acetic acid (pK_a of 20 versus 24) shows that the α-hydrogen of the ketone is more acidic. If acetone is compared with acetylacetone (**8**, pK_a of 9.07 in Table 9.1) it is clear the presence of the second carbonyl group greatly enhances the acidity of H$_a$. The presence of an electron-releasing methyl group in **7** diminishes the acidity (pK_a of 11). An electron withdrawing sulfonyl group (see **10**) enhances the acidity (pK_a of 3.6).

The enol content of carbonyl derivatives correlates with the acidity of the α proton. In ketone **11** the acidic proton of the enol (**12**) is attached to oxygen rather than carbon (from the keto form, **11**). Removal of H$_a$ from **12** generates enolate anion **13**. Removal of H$_a$ from **11** simply generates the other resonance contributor. In general, monoalkylated ketone derivatives such as **11** exist primarily in this keto form, but 1,3-dicarbonyl derivatives (such as **7-10** for example) have relatively high concentrations of enol (see **12**), and this contributes to the greater acidity of the proton on the α-carbon. The relative enol content of a molecule is strongly dependent on the solvent, as observed with ethyl acetoacetate.[7] For ethyl acetoacetate, the enol content is 0.4% in water, 2.2% in 50% aqueous ethanol, 6.9% neat, 11% in ethanol, 27% in ether, and 46% in hexane.[7] There is a clear dependence on hydrogen-bonding ability and polarity. The extensive hydrogen bonding that is possible in water minimizes intramolecular stabilization of the enol, whereas aprotic solvents such as ether or hexane maximize the intramolecular hydrogen bonding, which stabilizes the enol. Removal of a proton connected to oxygen is expected to be more facile than similar removal from a carbon, and one expects that as the enol content increases, the acidity should increase.

MOLECULAR MODELING: Hydrogen Bonding and Keto-Enol Tautomerism

It is known that propanal exists almost exclusively as the keto form, whereas there is evidence to suggest that *cis*-3-hydroxyacrolein (the enol form) is actually more stable than its keto form, malonaldehyde (CH$_2$-(CHO)$_2$, A).

---

7.   Stewart, R. *The Investigation of Organic Compounds,* Prentice-Hall, Englewood Cliffs, NJ, *1966*, p.12.

There is a clear correlation between enhanced acidity of a ketone or aldehyde and increased enol content, but inductive effects also play a role in the increased acidity of H_a for 1,3-dicarbonyl derivatives such as **11** (where $R^1 = CO_2Et$; sec. 2.2.A). The carbonyl group is a typical electron-withdrawing substituent but there are many others. In chapter 8, the ability to increase the acidity of a proton was shown to follow the order:

$$NO_2 > \overset{O}{\underset{C}{\|}} > SO_2R > \overset{O}{\underset{C\sim OR}{\|}} \approx C\equiv N > \overset{O}{\underset{C\sim R}{\|}} > Ph$$

and this is reflected in Table 9.1. Note that recent studies of X-ray structures of enolate anions and amide bases suggest that the enolate anion may be formed by direct removal of the proton from the α-carbon via an intermediate tetrameric lithium complex involving the base and the carbonyl starting material (sec. 9.2.D). Whether the α-hydrogen is removed directly from carbon or from the O–H of an enol, the presence of an electron-withdrawing group enhances the acidity of that hydrogen. The presence of two or more groups enhances the acidity of an α-proton even more, but the effect is not strictly additive, as shown in Table 9.2.[8,3]

Table 9.2. Enhancement of Acidity with Increasing Number of Electron-Withdrawing Groups

Compound	pK_a	Compound	pK_a
CH_3NO_2	11	CH_3COCH_3	20
$CH_2(NO_2)_2$	4	$CH_2(COCH_3)_2$	9
$CH(NO_2)_3$	0	$CH(COCH_3)_3$	6
$CH_3SO_2CH_3$	23	CH_3CN	25
$CH_2(SO_2CH_3)_2$	14	$CH_2(CN)_2$	12
$CH(SO_2CH_3)_3$	0	$CH(CN)_3$	0

[Reprinted with permission from Pearson, R.G.; Dillon, R.C. *J. Am. Chem. Soc.* *1953*, *75*, 2439. Copyright © *1953* American Chemical Society.]

---

8.   (a) Buncel, E. *Carbanions: Mechanistic and Isotopic Aspects*, Elsevier, Amsterdam, *1975*, p 5; (b) Reference 6, p. 12.

MOLECULAR MODELING: Electrostatic Potential Maps and Relative CH Acidities

The electrostatic potential is the energy of interaction of a point positive charge with the nuclei and electrons that make up a molecule. As such, it provides a broad indicator of excess positive or negative charge at any location. A more useful indicator, termed an electrostatic potential map, is provided by coloring different locations on an electron density surface (designating overall molecular size and shape) with the value of the electrostatic potential. By convention, locations with excess positive charge are colored "blue", while those with excess negative charge are colored "red". Neutral regions are colored "green".

Electrostatic potential maps may be used to rank the acidities of closely-related compounds, for example, the CH acidities of alkanes (substituted methanes). It is known that replacing H with an electron-withdrawing group such and CN or NO_2 leads to significant increase in the acidity of the hydrogen atoms remaining on that carbon. In other words, nitromethane is more acidic than methane. One after another, obtain geometries and electrostatic potential maps for acetonitrile, malonitrile and tricyanoethylene. Start the second and third calculations by *clicking* on > at the right of the tab at the top of the screen and selecting **Continue**. Compare the three maps, focusing on the region around the acidic hydrogen(s). The more positive ("blue") the electrostatic potential in this region, the more acidic the hydrogen (and the compound). You can switch between transparent displays, allowing you to see the structures underneath, and solid displays, allowing easier comparisons, *clicking* on the map and selecting either ***Transparent*** or ***Solid*** from the **Style** menu at the bottom of the screen. Which compound is the most acidic and which is the least acidic?

Repeat your calculations and analysis for nitromethane, dinitromethane and trinitromethane. Is the ordering of acidities with increasing number of electron-withdrawing groups the same as noted for the cyano compounds? Is the nitro group more or less effective in increasing CH acidity? Elaborate.

MOLECULAR MODELING: Site of Protonation in Pyrrole and Indole

Quantum chemical models may be used to compare base strengths (measured as their proton affinities) of related molecules. Such models may also be used to identify the most basic site within a given molecule and to rank the strengths of alternative sites. Such an analysis is important to determine if there may be more than one site for protonation in an acid-base reaction and, if there is, which site is likely to be most basic. Even a molecule as simple as pyrrole offers three different protonation sites: the nitrogen, C_2 and C_3. Use two different modeling tools to assign which site is favored.

The simpler approach is to examine the electrostatic potential map for pyrrole itself. Build pyrrole and *click* on the electrostatic potential map icon. In turn, *click* on regions of the map in the vicinity of N, C_2 and C_3 to obtain the electrostatic potential (a numerical value is displayed at the bottom of the screen). You might find it helpful to switch from a solid to a transparent model in order to see the underlying structure. *Click* on the map and select ***Transparent*** from the **Style** menu at the bottom right of the screen. Remember that blue regions indicate electron acceptor sites, such as acidic protons, whereas red regions indicate electron-donating (basic) sites. Based on this analysis, which atom is the most basic, and where does pyrrole protonate?

MOLECULAR MODELING: Site of Protonation in Pyrrole and Indole, con't.

The more complicated but usually more reliable approach to assess relative basicity is to compare the energies of the positive ions resulting from protonation. *Click* on > and select ***Continue*** from the screen to bring up a fresh build palette. Build N-protonated pyrrole starting from the NH_4^+ icon and obtain its energy. Repeat the process for C_2 and C_3 protonated pyrrole (build each starting from the CH_3^+ icon). Which protonated form is most stable (lowest in energy)? Is it reasonable to expect that two or more different protonated forms of pyrrole will be seen in an equilibrium mixture at room temperature? Elaborate. Is your conclusion the same as that you arrived at by comparing electrostatic potential maps?

Repeat both electrostatic potential map and energy calculations for indole and its protonated forms. Consider only protonation sites that are analogous to those in pyrrole (in the five-membered ring, and not those associated with the benzene ring). Is the preferred protonation site in indole the "same" as it is in pyrrole? Is the ordering of the nitrogen and two carbon proton affinities the same in the two molecules? If not provide an explanation as to why not.

9.2.B. Non-Nucleophilic Bases

Although the carbonyl group enhances the acidity of the α-proton, it also reacts with the base used for the deprotonation via acyl addition. Both organolithium or Grignard reagents are powerful bases, but they react with carbonyl derivatives primarily by nucleophilic acyl addition to the carbonyl, as in the reaction of 3-hexanone and *n*-butyllithium to give **14**. This acyl addition reaction can be suppressed somewhat by lowering the reaction temperature to –78°C, –100°C or even lower, but nucleophilic addition is usually quite facile. Other bases must be used to remove the α-proton in the acid-base reaction.
Ideally, the base would not be very nucleophilic, to suppress acyl addition. However, the base must be sufficiently basic to remove a relatively weak acidic proton.

14

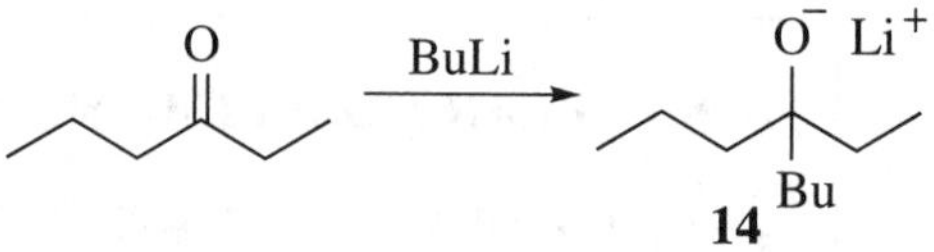

Hydroxide and alkoxides (formed from reaction of alcohols with NaH, NaOH, etc.) are strong enough bases to deprotonate most ketones. Sodium hydroxide (NaOH) in water or alcohol can be used, as well as $MeOH/MeO^-$, $EtOH/EtO^-$ or *t*-BuO^-/*t*-BuOH. These are nucleophilic reagents, however, and their use depends on the reversibility of the nucleophilic addition reaction to allow the deprotonation to compete. The bases NaH and KH are poor nucleophiles and they are very useful for generating enolates, although not kinetically active enough for a wide range of applications. Sodium and potassium hydride are used most often with malonate derivatives, acetoacetic esters, or 1,3-diketones.

Amines are weak acids, and are deprotonated only in the presence of a strong base. When a secondary amine is treated with a strong base, a strong conjugate base (R_2NH + base → $R_2N^-\ M^+$) is generated, the amide anion (R_2N^-). This conjugate base is more basic than the alkoxide obtained from similar reaction with an alcohol, and is easily capable of removing the α-proton of a ketone or aldehyde. Generation of $NaNH_2$ from ammonia is well known, and reaction of a secondary amine with a basic molecule such as *n*-butyllithium gives the analogous lithium dialkylamide. Lithium diisopropylamide (**15**, LDA)[9] is formed by reaction of diisopropylamine and *n*-butyllithium.[10] At first glance, it appears that LDA has a nucleophilic nitrogen atom. In reactions with carbonyl derivatives, however, nucleophilic attack at the acyl carbon is sterically blocked. This is illustrated in Figure 9.1, which represents the approach of LDA from the top face to 2-butanone, with the carbonyl oxygen projected to the front.

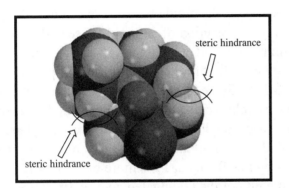

Figure 9.1. Steric hindrance that inhibits nucleophilic acyl addition of LDA (*top*) as it approaches 2-butanone (*bottom*). The N–Li is aligned with the C=O unit in the front of the diagram.

As the nitrogen approaches the carbonyl group at the required Bürgi-Dunitz angle of about 110°[11] (sec. 6.6.A), the methyl groups of the isopropyl units are repelled by the methyl and ethyl groups, which inhibits nucleophilic attack by nitrogen at the carbonyl carbon. In other words, steric hindrance prevents close contact of the nitrogen and the acyl carbon. Therefore, nitrogen is categorized as a poor nucleophile in this reaction. Since approach to an α hydrogen is not sterically encumbered, LDA reacts as a base without any problem. Dialkylamide bases such as LDA do not react well via acyl addition due to steric hindrance but deprotonate ketones or aldehydes, so dialkylamides are categorized as **non-nucleophilic bases**. All this really means is that nucleophilic acyl addition is slow relative to the acid-base reaction that removes the α-proton. Virtually any secondary amine can be converted to the corresponding amide base (R_2NLi), including lithium diethylamide (**16**), lithium tetramethylpiperidide

9. Hamell, M.; Levine, R. *J. Org. Chem.* **1950**, *15*, 162.
10. (a) Albarella, J.P. *J. Org. Chem.* **1977**, *42*, 2009; (b) Sasson, I.; Labovitz, J. *Ibid.* **1975**, *40*, 3670.
11. (a) Bürgi, H.-B.; Shefter, E.; Dunitz, J.D. *Tetrahedron* **1975**, *31*, 3089. Also see (b) Polt, R.; Seebach, D. *J. Am. Chem. Soc.* **1989**, *110*, 2622.

(17)[12] or lithium hexamethyldisilazide (18),[13,14] although 15 is probably the most commonly used amide base. In 17, the four methyls at C2 and C6 inhibit approach of nitrogen to an acyl carbon. In 18 [a bis(silyl)amide], the bulky trimethylsilyl groups sterically block the nitrogen and in this case steric hindrance and also diminished base strength leads to slow deprotonation for many carbonyl species. Relatively unhindered ketones, aldehydes or esters usually react to form the enolate anion without major problems. The hindered chelating base 1,8-bis(dimethylamino)naphthalene (19) is a very poor nucleophile, but extremely efficient at removing protons.[15] It is sold under the name Proton Sponge™.

It is noted that there are differences in base strength for the amide bases. We can use pK_a for a comparison of the amines, the conjugate acid of the reaction of an amide base and an acid. A stronger base has a conjugate acid with a higher pK_a. If we compare the acetylide anion with hydroxide, the conjugate acid acetylene with a pK_a of about 25 can be compared with water, with a pK_a of 15.7 (see chap. 8, Table 8.16). Since acetylene has the higher pK_a, we can conclude that acetylide is a much stronger base relative to hydroxide. Fraser and Mansour reported that the pK_a of $(Me_3Si)_2NH$ is 29.5, diisopropylamine is 35.7, dicyclohexylamine is 35.7, and 2,2,6,6-tetramethylpiperidine is 37.3, all reported in THF.[16] Therefore, we can compare the base strength of 15, 17 and 18. Using the pK_a data, 17 is more basic than 15, and 18 is the least basic in this series. In water,[17] the pK_a of the conjugated acid of 19 is reported to be 12, but it is 7.47 in DMSO.[18] As a comparison, ammonia has a pK_a of 9.24 (ammonium ion, $NH_3 \rightarrow NH_4^+$),[19] whereas amide (H_2N^-) has a pK_a of 38 (ammonia, $H_2N^- \rightarrow NH_3$),[20] both measured in water. Note that this data is consistent with amide being a much stronger base than ammonia.

The structure of LDA is not the monomeric species suggested by structure 15, but rather an aggregate. Jackman and co-workers,[21a] and later Seebach and co-workers studied the colligative

12. Olofson, R.A.; Dougherty, C.M. *J. Am. Chem. Soc.* **1973**, *95*, 581.
13. Amonoo-Neizer, E.H.; Shaw, R.A.; Skovlin, D.O.; Smith, B.C. *J. Chem. Soc.* **1965**, 2997.
14. For a discussion of the structure of lithium hexamethyldisilazide, see Lucht, B.L.; Collum, D.B. *Acc. Chem. Res.* **1999**, *32*, 1035.
15. (a) Ishida, A.; Mukaiyama, T. *Chem. Lett.* **1976**, 1127; (b) Diem, M.J.; Burow, D.F.; Fry, J.L. *J. Org. Chem.* **1977**, *42*, 1801.
16. Fraser, R.R.; Mansour, T.S. *J. Org. Chem.* **1984**, *49*, 3443.
17. Kresge, A.J. *Pure. Appl. Chem.* **1981**, *53*, 189.
18. Hess, A.S.; Yoder, R.A.; Johnston, J.N. *Synlett* **2006**, 147.
19. Bruckenstein, S.; Kolthoff, I.M., in Kolthoff, I.M.; Elving, P.J. *Treatise on Analytical Chemistry, Vol. 1, part. 1, Wiley, NY,* **1959**, pp. 432-433.
20. Bunbcel, E.; Menon, B. *J. Am. Chem. Soc.* **1977**, *99*, 4457.
21. Jackman, L.M.; De Brosse, C.W. *J. Am. Chem. Soc.* **1983**, *105*, 4177; (b) Bauer, W.; Seebach, D. *Helv. Chim. Acta* **1984**, *67*, 1972; (c) Williard, P.G.; Salvino, J.M. *J. Org. Chem.* **1993**, *58*, 1. Also see (d) Shobatake, K.; Nakamato, K. *Inorg. Chem. Acta* **1970**, *4*, 485.

properties of LDA as a solution in THF, and concluded that dilute solutions exist as both dimers (**20**) and monomers (**21**).[21b] The structure of LDA in solution is assumed to be the THF solvated dimer shown by

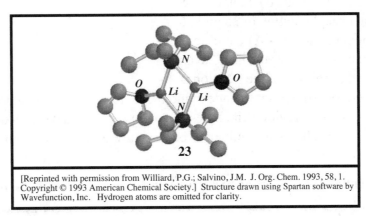

Williard and Salvino's X-ray structure **23**,[21c] that shows the nitrogen may be more sterically encumbered than is suggested in the representation in Figure 9.1. Note that **23** is the same representation as **20** where n = 1. Solution structures of chiral lithium amides having internal sulfide coordination have also been determined, and mixed dimeric complexes are generally formed.[22]

[Reprinted with permission from Williard, P.G.; Salvino, J.M. J. Org. Chem. 1993, 58, 1. Copyright © 1993 American Chemical Society.] Structure drawn using Spartan software by Wavefunction, Inc. Hydrogen atoms are omitted for clarity.

Collum and Rutherford studied the solution structure of lithium diethylamide (**16**) in THF and in ether, where dimers and trimers, and also four-, five- and six-rung ladder structures such as **22** were detected.[23] Collum and Galiano-Roth investigated the aggregation state of LDA in solution during the metalation of *N,N*-dimethylhydrazones (sec. 9.4.F.ii) using [6]Li and ^{15}N NMR.[24] The NMR data also suggested a solvated, dimeric structure such as **20**. Collum and DePue found evidence the structure of lithium isopropylcyclohexylamine (LICA) was an equilibrating set of stereoisomeric dimers in solution, cis-**24** and trans-**24**.[25] Collum's studies found no evidence for monomeric species in solution, and concluded that **21** was not a significant species in ether solutions of LDA.[24,25] Collum and co-worker's showed that enolization of ketones with LiN(TMS)$_2$, in the presence of diisopropyl

22. Sott, R.; Faranander, J.; Dinér, P.; Hilmersson, G. *Tetrahedron Asymmetry* **2004**, *15*, 267.
23. Rutherford, J.L.; Collum, D.B. *J. Am. Chem. Soc.* **1999**, *121*, 10198.
24. Galiano-Roth, A.S.; Collum, D.B. *J. Am. Chem. Soc,* **1989**, *111*, 6772.
25. DePue, J.S.; Collum, D.B. *J. Am. Chem. Soc.* **1988**, *110*, 5518, 5524.

ether, substituted tetrahydrofurans, and cineole proceed via dimer-based transition structures.[26] The hindered ethers accelerate the enolization by sterically destabilizing the reactants and stabilizing the transition structures. Pratt and co-workers[27] found that lithium tetramethylpiperidide (LiTMP) in THF gave the best selectivity for formation of the (E)-enolate anion of ketones. Selectivity was improved further by using a LiTMP-butyllithium mixed aggregate. Less polar solvents led to diminished selectivity.

9.2.C. (E/Z) Geometry in Enolate Formation

Formation of an enolate anion from a ketones is often accompanied by formation of (E) and (Z) isomers. Treatment of 2-methyl-3-pentanone with LDA (THF, –78°C), for example, gave a 60:40 mixture of the (Z) and (E) enolates (25 and 26)[28] Subsequent acyl addition of 25 and 26 to a carbonyl compound (sec. 9.4.A) could lead to different stereochemical consequences, even if the orientation and facial bias of the reaction is controlled. It is therefore important to understand the factors that control selectivity in enolate forming reactions. Evans' review[29] includes a compilation of several studies of selective enolization. Table 9.3.[30] shows the influence of structure and base on enolate geometry for simple ketones.

Table 9.3. The Influence of Ketone Structure and Base on Enolate Geometry

R^1	$LiNR_2$[a]	(Z) : (E)	R^1	$LiNR_2$[a]	(Z) : (E)
Et	LTMP	14 : 86	i-Pr	LHDS	59 : 41
Et	LTMP/HMPA	92 : 8	i-Pr	LTMP	>98 : 2
Et	LDA	23 : 77	t-Bu	LDA	32 : 68
Et	LICA	35 : 65	Ph	LDA	>98 : 2
Et	LHDS	66 : 34	mesityl	LDA	>98 : 2
Et	(Me₂PhSi)₂NLi	100 : 0	mesityl	LICA	5 : 95
i-Pr	LDA	60 : 40	mesityl	LHDS	4 : 96
i-Pr	LICA	59 : 41			87 : 13

[a] **LDA** = lithium diisopropylamide (15). **LICA** = lithium isopropylcyclohexylamide (24). **LTMP** = lithium 2,2,6,6-tetramethylpiperidide (17). **LHDS** = lithium hexamethyldisilazide (18)

26. Zhao, P.; Lucht, B.L.; Kenkre, S.; Collum, D.B. *J. Org. Chem.* **2004**, *69*, 242.
27. Pratt, L.M.; Newman, A.; St. Cyr, J.; Johnson, H.; Miles, B.; Lattier, A.; Austin, E.; Henderson, S.; Hershey, B.; Lin, M.; Balamraju, Y.; Sammonds, L.; Cheramie, J.; Karnes, J.; Hymel, E.; Woodford, B.; Carter, C. *J. Org. Chem.* **2003**, *68*, 6387.
28. Heathcock, C.H.; Buse, C.T.; Kleschick, W.A.; Pirrung, M.A.; Sohn, J.E.; Lampe, J. *J. Org. Chem.* **1980**, *45*, 1066.
29. Evans, D.A. in *Asymmetric Syntheses, Vol. 3*, Morrison, J.D. Ed., Academic Press, New York, *1984*, pp 1-110.
30. Reference 29, p 16.

Only moderate variations in kinetic (E/Z) ratios are observed, except when LHDS was used as the base, which favored the thermodynamically more stable (Z) enolate. The sterically demanding base LTMP favored formation of the kinetic E enolate.[31] This equilibration may involve an aldol addition–retro-aldol process as suggested by Rathke and co-workers (via **27**).[31] The **aldol condensation** is discussed in Section 9.4.A. Addition of HMPT (hexamethylphosphorus triamide) destabilizes and deaggregates aldolate **27**, thus promoting equilibration.

The (E/Z) trends observed for ketones is also seen in enolate anions of esters and amides, as shown in Table 9.4.[32] The kinetic (E) ester enolate usually predominates with esters unless HMPT is added,[33] but amides[34] give the thermodynamic (Z) isomer. It is unlikely that the equilibration observed with ketones is responsible for reversal in enolate geometry in esters on addition of HMPT. Such equilibration would show a preference for formation the (Z)-enolate with both LDA and *sec*-butyllithium, and would lead to significant amounts of **Claisen condensation** (sec. 9.4.B.i). The influence of the dialkylamide base and the carbonyl structure on the (E/Z) enolate ratio can be rationalized by the so-called **Ireland model**,[33,35] where the two important steric factors are taken to be R^1 versus Me in **29** and Me versus L in **31** (see Table 9.5)[36] and the Zimmerman-Traxler model in sec. 9.5.A.iii. This model was developed to predict stereochemistry in the anionic accelerated Claisen rearrangement (sec. 11.12.D.ii), but is used for predicting stereochemistry in ester enolates. Enolate anions actually exist as aggregate dimers, tetramers or hexamers (sec. 9.2.D), so this model has some limitations. It generally gives good predictions, and is also a useful tool for predicting (E) and (Z) isomers for enolate anions. When R^1 is not sterically demanding (as in OR) the H $\leftrightarrow$ L interaction in **28** is much smaller than the Me $\leftrightarrow$ L interaction in **30** and more of the (E) isomer (**29**) is produced. When R^1 is large (such as *tert*-butyl) the large Me $\leftrightarrow$ R^1 interaction in **28** is larger than the H $\leftrightarrow$ R^1 interaction and usually larger than the Me $\leftrightarrow$ L interaction in **30**. In this case, **30** is favored, leading to an increased amount of the (Z) isomer (**31**). It is possible that addition of HMPT disrupts the pericyclic transition states represented by **28** and **30**.

31. (a) Fataftah, A.Z.; Kopka, I.E.; Rathke, M.W. *J. Am. Chem. Soc.* **1980**, *102*, 3959; (b) also see Nakamura, E.; Hashimoto, K.; Kuwajima, I. *Tetrahedron Lett.* **1978**, 2079.
32. Reference 29, p 17 and references 39, 40, 45-47 therein.
33. Ireland, R.E.; Mueller, R.H.; Willard, A.K. *J. Am. Chem. Soc.* **1976**, *98*, 2868.
34. Reference 29, p 17.
35. Reference 29, pp 18-20.
36. (a) Reference 29, p 19; also see (b) Meyers, A.I.; Snyder, E.S.; Akerman, J.J.H. *J. Am. Chem. Soc.* **1978**, *100*, 8136; (c) Hoobler, M.A.; Bergbreiter, D.E.; Newcomb, M. *Ibid.* **1978**, *100*, 8182; (d) Davenport, K.G.; Eichenauer, H.; Enders, D.; Newcomb, M.; Bergbreiter, D.E. *Ibid.* **1979**, *101*, 5654.

Table 9.4. Enolate Geometry of Esters and Amides

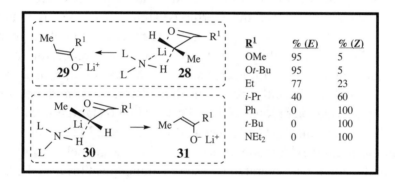

R^1	R^2	LiNR$_2$	(Z):(E)	R^1	R^2	LiNR$_2$	(Z):(E)
OMe	Me	LDA	5:95	Ot-Bu	Et	LDA, HMPT	77:23
OCH$_2$OMe	Me	LDA	<2:98	St-Bu	Me	LDA	10:90
Ot-Bu	Me	LDA	5:95	NEt$_2$	Me	LDA	>97:3
OMe	Et	LDA	9:91	NEt$_2$	Me	sec-BuLi	75:25
OMe	Et	LDA, HMPT	84:16	Pyrrolidino	Me	LDA	>97:3
Ot-Bu	Et	LDA	5:95	Pyrrolidino	Me	sec-BuLi	80:20

Table 9.5. Formation of (E) and (Z) Enolates and the Ireland Model

R^1	% (E)	% (Z)
OMe	95	5
Ot-Bu	95	5
Et	77	23
i-Pr	40	60
Ph	0	100
t-Bu	0	100
NEt$_2$	0	100

9.2.D. Structure and Aggregation State of Enolate Anions

Enolate geometry must is a function of the structure of the enolate anion. Metal enolates exist as dimers[37a] or other aggregates in ether solvents[32] (see Section 9.2.C. for LDA aggregates).[37d] Jackman and Szeverenyi suggested that the lithium enolate of isobutyrophenone exists as a tetramer (**32**) in THF solution,[38] but as a dimer (**33**) in DME.[39] House et al proposed these aggregates,[40] and found that ketone enolates of groups 1, 2 and 3 metals exist primarily as the *O*-metal enolate.[41,39] For lithium enolates, increasing the donor properties of the solvents

37. (a) Bernstein, M.P.; Collum, D.B. *J. Am. Chem. Soc.* **1993**, *115*, 789; (b) Bernstein, M.P.; Romesberg, F.E.; Fuller, D.J.; Harrison, A.T.; Collum, D.B.; Liu, Q.Y.; Williard, P.G. *Ibid.* **1992**, *114*, 5100; (c) Collum, D.B.*Acc. Chem. Res.* **1992**, *25*, 448; (d) Reference 29, p 21 and citations 59-64 therein.
38. (a) Jackman, L.M.; Szeverenyi, N.M. *J. Am. Chem. Soc.* **1977**, *99*, 4954; (b) Jackman, L.M.; Lange, B.C. *Ibid.* **1981**, *103*, 4494.
39. Jackman, L.M.; Lange, B.C. *Tetrahedron* **1977**, *33*, 2737.
40. House, H.O.; Gall, M.; Olmstead, H.D. *J. Org. Chem.* **1971**, *36*, 2361.
41. (a) House, H.O.; Auerbach, R.A.; Gall, M.; Peet, N.P. *J. Org. Chem.* **1973**, *38*, 514; (b) Orsini, F.; Pelizzoni, F.; Ricca, G. *Tetrahedron Lett.* **1982**, *23*, 3945; (c) House, H.O.; Prabhu, A.V.; Phillips, W.V. *J. Org. Chem.* **1976**, *41*, 1209.

from ether → THF → DME increases the Li–O bond polarization, enhancing the α-carbon π-electron density.[42] An increase in π-electron density is also observed for the series: Li → Na → K. Increasing π-electron density is associated with greater reactivity of carbon in the enolate alkylation reactions.[43]

R = (Me₂C=C(Ph)—)   S = solvent

[Reprinted with permission from Jackman, L.M.; Szeverenyi, N.M. J. Am. Chem. Soc. 1977, 99, 4954. Copyright © 1977 American Chemical Society.]

Lithium enolate anions of ketones exist as aggregates in solution.[38-40,43d,44] Mixed aggregates between the enolate anion and the amide base are also possible.[45] In 1981, Seebach and co-workers confirmed by X-ray crystallography that the lithium enolates of pinacolone and cyclopentanone form a tetrameric aggregate in the solid state, and it was assumed that a similar species exited in solution.[46] A THF solvated tetramer of lithium pinacolonate is shown (see 34), as reported by Seebach.[41] Williard et al. reported the X-ray structure of the unsolvated lithium enolate of pinacolone to be a hexamer (35).[47] The structure of the potassium enolate of pinacolone was a hexamer, although the sodium enolate was a tetramer. The tetramer was composed of a cubic cluster of four enolized ketone molecules, four sodium atoms, and four *unenolized* solvating ketones units, which contrasts with 35 in which each oxygen atom is coordinated to three lithium atoms, and each enolate double bond is almost antiperiplanar to one lithium atom. A variety of X-ray crystal structures for organolithium derivatives confirm the presence of aggregates in the solid state.[48] There is also evidence that the aggregate structure is preserved in solution and is probably the actual reactive species.[2b]

42. Reference 29, p 22.

43. (a) Zook, H.D.; Gumby, W.L. *J. Am. Chem. Soc.* **1960**, *82*, 1386; (b) Zook, H.D.; Russo, T.J. *Ibid.* **1960**, *82*, 1258; (c) Zook, H.D.; Russo, T.J.; Ferrand, E.F.; Stotz, D.S. *J. Org. Chem.* **1968**, *33*, 2222; (d) Zook, H.D.; Kelly, W.L.; Posey, I.Y. *Ibid.* **1968**, *33*, 3477; (e) Zook, H.D.; Miller, J.A. *Ibid.* **1971**, *36*, 1112.

44. Stork, G.; Hudrlik, P.F. *J. Am. Chem. Soc.* **1968**, *90*, 4464.

45. Sun, C.; Williard, P.G. *J. Am. Chem. Soc.* **2000**, *122*, 7829.

46. (a) Amstutz, R.; Schweizer, W.B.; Seebach, D.; Dunitz, J.D. *Helv. Chim. Acta* **1981**, *64*, 2617; (b) Seebach, D.; Amstutz, D.; Dunitz, J.D. *Ibid.* **1981**, *64*, 2622.

47. (a) Williard, P.G.; Carpenter, G.B. *J. Am. Chem. Soc.* **1986**, *108*, 462; (b) Williard, P.G.; Carpenter, G.B. *Ibid.* **1985**, *107*, 3345.

48. For example, see (a) Bauer, W.; Laube, T.; Seebach, D. *Chem. Ber.* **1985**, *118*, 764; (b) van Koten, G.; Jastrzebski, J.T.B.H. *J. Am. Chem. Soc.* **1988**, *107*, 697; (c) Polt, R.L.; Stork, G.; Carpenter, G.B.; Williard, P.G. *Ibid.* **1984**, *106*, 4276; (d) Amstutz, R.; Laube, T.; Schweizer, W.B.; Seebach, D.; Dunitz, J.D. *Helv. Chim. Acta* **1984**, *67*, 224.

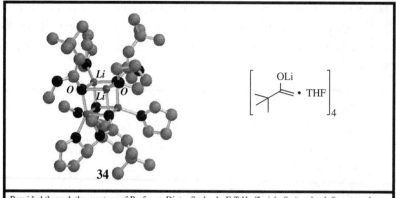

Provided through the courtesy of Professor Dieter Seebach, E.T.H., Zurich, Switzerland. Structure drawn using Spartan software by Wavefunction, Inc. Hydrogen atoms are omitted for clarity.

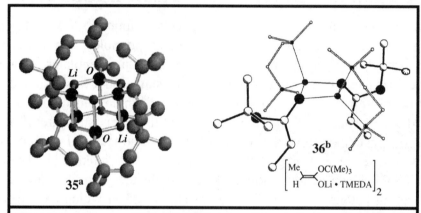

[a]Provided through the courtesy of Professor Paul Williard, Brown University, Providence, RI. Structure 35 is drawn using Spartan software by Wavefunction, Inc. [b] Provided through the courtesy of Professor Dieter Seebach, E.T.H., Zurich, Switzerland.

Seebach et al. showed that lithium enolates derived from esters are aggregates. The lithium enolates of (*Z*)-*tert*-butyl propionate (**36**) and of *tert*-butyl-2-methylpropionate are dimers in the solid state.[49] In both cases, the crystal contains two coordinated TMEDA molecules (see **36**). The lithium enolate of (*Z*)-methyl 3,3-dimethylbutanoate was found to be a dimer containing four THF molecules. Focusing on colligative properties rather than X-ray analysis, Collum and co-workers determined that lithium 2-carbomethoxycyclohexanone dimethylhydrazone (sec. 9.4.F.ii) is a dimeric THF solvate in the solid state.[50] *Perhaps the most important implication of this work is that the reactivity of enolate anions in alkylation and condensation reactions (see below) will be influenced by the aggregate state of the enolate.* If the enolate anion is are attached to (or a part of) a tetrameric or hexameric complex, then simple steric

49. Seebach, D.; Amstutz, R.; Laube, T.; Schweizer, W.B.; Dunitz, J.D. *J. Am. Chem. Soc.* **1985**, *107*, 5403.
50. (a) Wanat, R.A.; Collum, D.B.; Van Duyne, G.; Clardy, J.; De Pue, R.T. *J. Am. Chem. Soc.* **1986**, *108*, 3415;
    (b) Wanat, R.A.; Collum, D.B. *Ibid.* **1985**, *107*, 2078.

and electronic arguments for reactivity and selectivity are flawed if they are based only on examining monomeric species. The relative proportions of (E) and (Z) enolate anions will be influenced by the extent of solvation and the aggregation state.

The solvent plays a significant role in the aggregation state of enolate anions, and this directly influences their reactivity. Streitwieser et al. found that selected lithium enolate anions have lower aggregation states in dimethoxyethane than in THF, and that aggregation is much higher in methyl *tert*-butyl ether.[51] He found that alkylation and acylation of the enolate anions was extremely slow in methyl *tert*-butyl ether. These reactions apparently require additional solvation of the lithium cation, which is ineffective in this solvent.

Once formed, an enolate anion can react with a variety of electrophiles. The approach of the electrophile is usually assumed to be perpendicular (as in 37), in reactions with alkyl halides and epoxides, to maximize overlap of participating orbitals.[52] This model also applies to the deprotonation reaction. The LUMO of the electrophile is important, and Dunitz and co-workers[53] showed that a non-vertical approach (represented by 38 or 39)[54] is preferred when nucleophiles attack carbonyl centers. The angle of approach (as discussed in secs. 4.7.C, sec. 6.6.A. and 9.2.B) is ~ 110° and is called the **Bürgi-Dunitz trajectory**.[11,53] Models 38 and 39 imply that as the enolate nucleophile approaches the carbonyl carbon, the approach angle is somewhat restricted, which enhances the steric effects between R and the approaching electrophile. Since the geometry of the enolate

will influence the approach of the electrophile, geometric constrictions must be considered when examining stereoselectivity. It appears that enolate-alkyl halide transition states are essentially reactant-like, and that the product stereochemistry is determined largely by steric factors.

51. Streitwieser, A.; Juaristi, E.; Kim, Y.-J.; Pugh, J.K. *Org. Lett.* **2000**, *2*, 3739.
52. (a) Reference 29, p 25; (b) Velluz, L.; Valls, J.; Nominé, G. *Angew. Chem. Int. Ed.* **1965**, *4*, 181 (see p 189); (c) Valls, J.; Toromanoff, E. *Bull Soc. Chim. Fr.* **1961**, 758; (d) Toromanoff, E. *Ibid.* **1962**, 708, 1190; (e) Bucourt, R. *Ibid.* **1964**, 2080; (f) Toromanoff, E.; Bucourt, R. *Tetrahedron Lett.* **1976**, 3523.
53. (a) Bürgi, H.B.; Dunitz, J.D.; Lehn, J.M.; Wipff, G. *Tetrahedron* **1974**, *30*, 1563; (b) Bürgi, H.B.; Dunitz, J.D.; Shefter, E. *J. Am. Chem. Soc.* **1973**, *95*, 5065; (c) Bürgi, H.B.; Lehn, J.M.; Wipff, G. *Ibid.* **1974**, *96*, 1956; (d) Bürgi, H.B. *Angew. Chem. Int. Ed.* **1975**, *14*, 460.
54. (a) Agami, C. *Tetrahedron Lett.* **1977**, 2801; (b) Agami, C.; Chauvin, M.; Levisalles, J. *Ibid.* **1979**, 1855; (c) Agami, C.; Levisalles, J.; Lo Cicero, B. *Tetrahedron* **1979**, *35*, 961.

Lithium enolate anions exist as large aggregates, and their approach to the electrophile is restricted by steric and electronic considerations, as well as by the relative geometry of the molecule. Nonetheless, good predictions for reactivity and diastereoselectivity can be made based on the steric requirements that would be present in a monomeric system. For example, in most reactions of enolate anions the electrophile will be delivered to the less hindered face of the enolate to give the major product. In all models used to describe reactivity (secs. 9.5.A.iii-9.5.A.v), a monomeric enolate will be shown but the facial and orientational bias of the enolate is clearly influenced by the state of aggregation in solution.

9.2.E. Kinetic Versus Thermodynamic Control

In an unsymmetrical ketone, the protons on both sides of the carbonyl are acidic (H_a and H_b in **1**), and the pK_a values of the protons are different. When such a ketone is treated with an appropriate base, removal of the different protons will lead to different regioisomeric enolates. 3-Hexanone (**40**), for example, reacted with LDA to give a mixture of **41** and **42**, and each enolate anion is a mixture of *E* and *Z* isomers. In **40**, each α-carbon has one alkyl substituent (methyl and ethyl, respectively). In many cases, however, the

α- and α'-carbon can be monosubstiuted or disubstituted. The α-protons at each carbon will have slightly different pK_a values, with the less substituted position having the more acidic proton. As noted in Table 9.1, 3,3-dimethyl-2-butanone (**4** , O=C–C$\underline{H}_3$) has a pK_a of 20.8 and 4,4-dimethyl-3-pentanone (**5**, O=CC$\underline{H}_2$CH$_3$) has a pK_a of 21.7. In general, the difference in pK_a for monosubstituted versus disubstituted positions is typically only about 1 pK_a unit. 2-Methyl-3-pentanone (**43**) has two acidic protons, H_a and H_b. As discussed previously, the methyl group releases electrons to the carbon bearing H_b, reducing the acidity of H_b by about 1 pK_a unit relative to H_a. Removal of H_a leads to enolate **44**, and loss of H_b leads to enolate **45**. Enolate

44 results from loss of the most acidic proton, which is removed faster in an acid-base reaction to give the **kinetic** enolate anion (less stable). Enolate **45** is the more thermodynamically stable, but it is formed only after equilibration of the initially formed enolate anion and is known as the **thermodynamic** enolate anion. This observation raises the question if it is possible to form one enolate or the other regioselectively. It is noted that **44** can be formed as a mixture of (*E*) and (*Z*) isomers. The (*E/Z*) ratio is one of the parameters that can be controlled to influence selectivity in diastereoselective aldol condensation reactions, discussed in section 9.5.

MOLECULAR MODELING: Acidity of Hydrogens Attached to α Carbons in Ketones

It is well known that the pK_a of the proton on the α-carbon of a ketone is influenced by substituents on that carbon. Alkyl substituents are electron releasing and halogens, alkoxy groups, amino groups and carbonyl groups are electron withdrawing.

Two different types of proton are identified for each of compounds **A-D** shown below.

A **B** **C** **D**

Construct **A**, and obtain an electrostatic potential map. This displays the energy that a positive charge feels at a location on the surface of a molecule. Areas for which the charge is repelled are colored blue and are associated with more electrophilic atoms. Compare the surface around H_a and H_b. Which hydrogen appears to be the more acidic (more blue)? *Click* on > to the right of the tab at the top of the screen and select **Continue** to bring up a fresh builder palette without removing this text. Build **B**, obtain its electrostatic potential map and determine the relative acidities of H_a and H_b. Which is more acidic? What is the influence of the methyl group in **B** when compared with **A**? Repeat the sequence (*click* on > and select **Continue**) to build **C** and obtain its electrostatic potential map. Which of the two hydrogens, H_a or H_b, is more acidic? What does this tell you about the influence of the chlorine? Repeat one last time to make **D**, obtain its electrostatic potential map and determine the relative acidities of H_a and H_b. What does this tell you about the influence of the second carbonyl group on acidity?

Display all four compounds simultaneously on screen. *Check* the box to the left of each of the four tabs at the top of the screen. Which of the four compounds shown is likely to have the most acidic proton? Elaborate.

Enolate formation is an acid-base reaction (sec. 2.2.A), and control of this equilibrium is the key to regioselectivity. As with any acid-base reaction, the equilibrium is determined by the relative strengths of the acid-base and conjugate acid-conjugate base pairs. Reaction of a strong acid and a strong base generates a weak conjugate acid-weak conjugate base, and K_a is large. If the conjugate acid and conjugate base are stronger than the initial acid-base pair, K_a is small. A typical acid-base equilibrium reaction is shown for the reaction of **46** to give **47**.

46 **47**

$$K_a = \frac{k_1}{k_{-1}} \text{ and } pK_a = -\log K_a$$

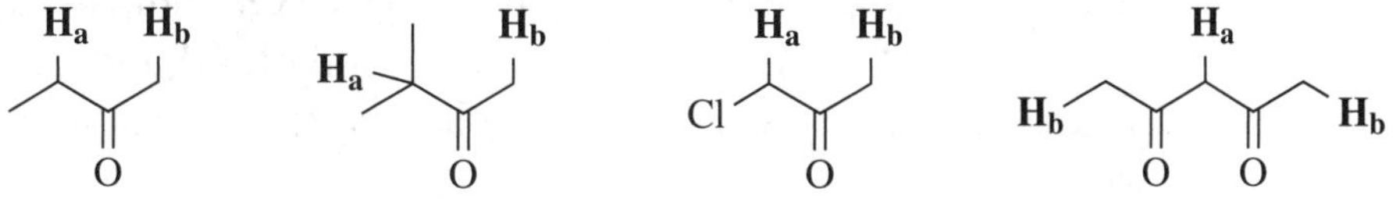

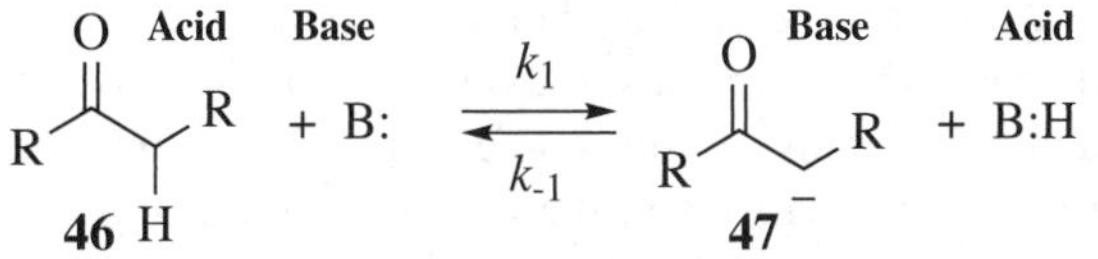

Kinetic control means that the kinetic enolate is the major product (K_a is large and $k_1 >> k_{-1}$). Thermodynamic or equilibrium control means the thermodynamic enolate is the major product, and K_a is near unity. The choice of solvent, base, cation and temperature equilibrium control K_a. ***Kinetic control*** is favored by several factors,[2] including (*1*) aprotic solvents, (*2*) strong bases that are weakly nucleophilic and **generate a conjugate acid that is weaker than the ketone or aldehyde**, (*3*) cations such as Li which complex oxygen: (not Na^+ or K^+), (*4*) low temperatures, and (*5*) short reaction times. The conditions that favor thermodynamic control must be the opposite of those described for kinetic control.[2] ***Thermodynamic control*** conditions include (*1*) protic solvents, (*2*) nucleophilic bases that **generate a conjugate acid that is stronger than the ketone or aldehyde**, (*3*) cations that form ionic M–O bonds, (*4*) higher temperatures, and (*5*) long reaction times. These parameters can be explained by examining the three species involved in the equilibrium, **43-45**. For kinetic control (formation of **44**), the reaction should be essentially irreversible and the reaction conditions should suppress the equilibrium.

Solvents can be grouped according to whether or not they have an acidic proton. Protic solvents (solvents containing an acidic hydrogen), usually involve O–H, S–H or N–H and promote equilibrium conditions (small pK_a). Remember than a $K_a = 1$ represents a 50:50 equilibrium. Aprotic solvents favor a large K_a (an equilibrium that is shifted to the right). In the reaction of **46**, the ketone is the acid ($pK_a \sim 21$) and a protic solvent such as ethanol ($pK_a \sim 17$) or water ($pK_a \sim 15.8$) will be a stronger acid. Once the conjugate base (the enolate anion, **47**) is formed, it reacts with the protic solvent to regenerate the ketone, driving the equilibrium to the left (small K_a). Aprotic solvents such as THF do not have an acidic proton and the enolate anion cannot react with them, meaning that once formed the enolate anion will be relatively stable (equilibrium shifted to the right, large K_a). In other words, aprotic solvents such as ether or THF favor kinetic control, and protic solvents such as ethanol, water, methanol or *tert*-butanol favor thermodynamic control.

MOLECULAR MODELING: Solvent Effects on Enolate Formation

In solution, the reaction of 2-methylcyclopentanone with a base leads to the enolate anion that results from deprotonation at the less substituted C_5 position, and not the more substituted C_2 position. In other words, the protons on C_5 are more acidic than the proton on C_2. This observation suggests that the proton on the carbon bearing the methyl group, the more substituted C_5 carbon, is less acidic. The usual explanation for this observation is that the methyl group is an electron donor, and it acts to destabilize the anionic center to which it is attached. However, the fact that a methyl group is larger than hydrogen suggests that it might also act to disperse the negative charge and actually lead to stabilization of the carbanion center.

Such interpretations can be tested by quantum chemical calculations. Given that the calculations strictly correspond to measurements carried out in the gas phase, do they necessarily lead to the same results as measurements carried out in solution? In other words, they may not conclude that the methyl group actually has a destabilizing effect on the carbon to which it attached. To explore this, compare energies for the two possible enolate anions resulting from deprotonation of 2-methylcyclopentanone. First build the enolate anion corresponding to deprotonation at the less substituted carbon (C_2). You can easily do this by using icons for sp^2, sp^3 and anionic carbons and sp^2 oxygen. Obtain its energy using the properties icon. Now make the enolate anion for deprotonation at C_5 (the more substituted carbon), and obtain the energy. *Click on > at the right of the tab at the top of the screen and select **Continue** to bring up a fresh builder palette while maintaining this text. Compare energies of the two. Which enolate anion is more stable? Is this result consistent with destabilization of the anionic center by the methyl group? If it is not, provide a plausible interpretation of the experimental result.

Whether or not the methyl group is actually an electron donor pushing more negative charge onto the carbanion center or just a "bulky" group dispersing the negative charge, it's overall effect on a carbanion center would be expected to be modest. Change it for a methoxy group (presumably a much stronger electron donor). Start each by *clicking* on > to the right of the tab at the top of the screen and selecting **Continue**. Which enolate anion is favored here and by how much? Is there a significant change from what you observed for methyl?

Finally, repeat your calculations and analysis for the two enolate anions formed by deprotonation of 2-phenylcyclopentanone. Start each by *clicking* on > to the right of the tab at the top of the screen and selecting **Continue**. Which enolate anion is favored? What does your result say about the effect of a phenyl group on a carbanion center? Is the magnitude of the preference smaller, larger or about the same as that for 2-methylcyclopentanone? Rationalize your result.

The base plays two roles in this reaction. First, it must react with the acid (see **43**) quickly and efficiently (it must be a strong base). The base also generates a conjugate acid after deprotonation that plays a significant role in the position of the overall equilibrium. If the conjugate acid is more acidic than **43**, it will reprotonate the conjugate base (the enolate anion) and drive the equilibrium back toward **43**. When LDA deprotonates **43** to form **44** *in an aprotic solvent,* the only acids in the equilibrium are **43** ($pK_a \sim 21$) and diisopropylamine ($pK_a \sim 25$), the conjugate acid formed by reaction with LDA. Since diisopropylamine is a weaker acid than the ketone, reaction with the enolate anion (the conjugate base) is relatively slow, which shifts the overall equilibrium to the right (large K_a) and favors kinetic enolate formation. Note that diisopropylamine is a weak acid, and given sufficient time it will react with the enolate anion to establish an equilibrium (see below). If the reaction is done in THF with sodium ethoxide as the base, enolate **44** is also formed since H_a is the most acidic proton. The pK_a of this ketone acid is ~ 21 (as mentioned, see Table 9.1), but as ethoxide reacts with **43** it generates ethanol as the conjugate acid with a $pK_a \sim 17$. Ethanol easily reacts with the enolate anion, shifting the equilibrium to the left so that **43**, ethoxide, ethanol and some **44** are all in solution. Although

H_b in **43** is less acidic than H_a, it is acidic enough to be removed by ethoxide. If the conversion of **43** to **44** is slow and/or reversible, the less acidic H_b will be removed to generate enolate anion **45**. The conjugate acid (ethanol) will reprotonate **45**, establishing a second equilibrium between **44** and **45**. If the solvent is ethanol, which is a stronger acid than **43**, this equilibrium is favored and as mentioned, **43-45**, ethanol and NaOEt will all be in solution. The C=C unit of enolate anion **45** is more highly substituted and assumed to be thermodynamically more stable. Therefore, once is established, the equilibrium will shift to favor the more stable enolate anion **45** (hence the term thermodynamic control). This discussion can be summarized by saying that strong bases such as LiNH$_2$ or LiNR$_2$ favor kinetic control, since they generate weak conjugate acids, but strong bases such as NaOH, NaOMe, NaOEt or NaO*t*-Bu will favor thermodynamic control because they generate conjugate acids that are stronger than the starting ketone.

The counterion (M$^+$) of the base (M$^+$NR$_2^-$) will be the counterion of the enolate anion unless another metal ion is added. The covalent character of the M–O bond in the enolate plays a role in the kinetic versus thermodynamic equilibrium. If the M–O bond is relatively covalent (as with the Li–O bond), the initially formed enolate (**44**) will form more easily and be less likely to react with an acid to give **43**, which favors the kinetic process. If the M–O bond is more ionic (M = Na, K), the ionic character of the enolate increases and it becomes increasingly easier to reprotonate the anionic oxygen to give **44**. This promotes the equilibrium. This relationship is summarized and reinforced by noting that sodium and potassium enolates equilibrate readily, whereas magnesium, lithium, zinc, copper or aluminum enolates favor kinetic control.[55]

Reaction temperature is another important variable. If the reaction is kept cold, all reactions (including the forward and reverse reactions that constitute the equilibrium) are slowed. If there is enough energy at low temperatures to convert **43** to **44**, the subsequent acid-base reactions (that will promote the equilibrium) will be slower, which favors kinetic control. Typical reaction temperatures are –78°C (CO$_2$ in acetone or 2-propanol) and –100°C (ether in CO$_2$).[56] Conversely, high reaction temperatures promote the equilibrium process. In most cases, the temperatures observed with thermodynamically controlled reactions are the reflux temperatures of the solvent (refluxing ethanol, water, methanol, *tert*-butanol).

The final parameter in this process is the length of time the molecules are allowed to react. If the reaction time is short (typically min to h), it is more difficult for the equilibration process to develop. If the reaction is allowed to go for a long time (typically hours to days, although

55. (a) Stork, G.; Rosen, P.; Goldman, N.; Coombs, R.V.; Tsuji, J. *J. Am. Chem. Soc.* **1965**, *87*, 275; (b) Patterson, Jr., J.W.; Fried, J.H. *J. Org. Chem.* **1974**, *39*, 2506.

56. Gordon, A.J.; Ford, R.A. *The Chemist's Companion*, John Wiley & Sons, London, **1972**. p 451.

minutes can be a long time in some reactions), equilibration and thermodynamic control is more likely. The time factor is linked to the relative strength of all acids in the system. When **44** is formed by reaction of LDA and **43**, its conjugate acid is diisopropylamine. Diisopropylamine is a weak acid, and when given a sufficiently long contact time with a strong base (the enolate anion is a strong base) some acid-base equilibration can occur. Few reactions are quantitative, and virtually none are instantaneous. Remember that about six half-lives are required for a reaction to be >98% complete. Therefore, unreacted starting material will be available during the course of the reaction. Further, if the conversion of **43** to **44** occurs in only 90%, then 10% of unreacted ketone remains. With a pK_a of ~ 21, **43** is acidic enough to reprotonate **44** and establish the equilibrium required for thermodynamic control, if given sufficient time. If the reaction itself is slow, enolate **44** is being formed in the presence of unreacted **43** and there is the possibility of a reaction. The terms short and long are obviously relative and vague, and vary with the electrophile for a particular enolate anion. In general, short reaction times favor kinetic control and long reaction times favor thermodynamic control.

The intramolecular cyclization reaction of bromoketone **48**[57] illustrates how changing reaction conditions can influence formation of the kinetic or thermodynamic enolate. The reaction of **48** with LDA (ether, 0°C) generates the kinetic enolate (**49**). Subsequent intramolecular displacement of the bromide by the enolate anion gives cyclohexanone **50**. Some elimination of **49** under the reaction conditions occurred, giving **52**. A complicating factor in this example is that the oxygen in **49** is also a nucleophile, and O-alkylation (discussed in sec. 9.3.B) gave the pyran product **51**, although **50** was the major product. When **48** was treated with *tert*-butoxide in *tert*-butanol, the thermodynamic enolate (**53**) was formed and this cyclized to the dihydropyran **51** via O-alkylation. A four-membered ring product (**54**) would be the normal enolate alkylation product that is analogous to **50**, but its formation has a relatively high activation barrier and it was a minor product (secs. 1.5.B, 6.6.B).

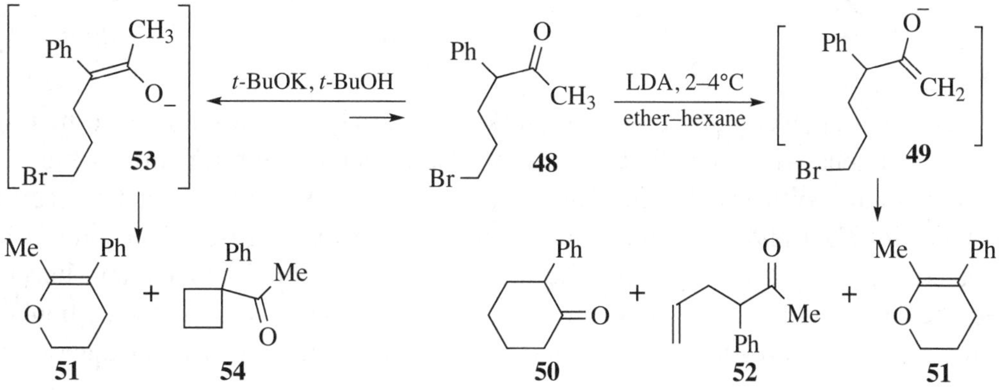

57. (a) House, H.O.; Philips, W.V.; Sayer, T.S.B.; Yau, C.-C. *J. Org. Chem.* **1978**, *43*, 700; (b) Etheredge, S.J. *Ibid.* **1966**, *31*, 1990.

Carbonyl compounds other than ketones readily form enolate anions. It was noted in Table 9.1 that the α-hydrogen of an aldehyde was more acidic than that of a ketone. When aldehydes are used, however, there is only one acidic proton, and in principle, either kinetic or thermodynamic control conditions can be used. Formation of (E) or (Z) isomers of **55** remeîns a problem, however. It may be desirable to use kinetic

control conditions to control geometry, or minimize subsequent reactions such as nucleophilic addition of the conjugate base to the carbonyl (an aldol-type condensation), which is more facile with aldehydes than with ketones (sec. 9.4.A). A nonnucleophilic base such as LTMP often maximizes formation of the enolate. Acid derivatives such as esters and *N,N*-disubstituted amides have only one acidic α-proton, and deprotonation gives a single regioisomeric enolate. Methyl butanoate reacted with LDA, for example, to give enolate **56** as a mixture of (E) and (Z) isomers. Self-condensation is a problem with esters (see section 9.4.B.i).

When primary or secondary amides are treated with a base, there is a complicating reaction that was not possible with esters, ketones or aldehydes. The N–H moiety is acidic enough to react with the bases used for deprotonation. Treatment of **57** with base gave the *N*-lithio derivative, but the α-lithio derivative (**58**) can be generated by addition of two equivalents of base.[58] Enolate anion formation is straightforward with tertiary amides such as dimethylisobutyramide (**57**, R = Me), and the resultant enolate anion (**59**) reacted with butanal to give amido-alcohol **60** in 68% yield[53] (see sec. 9.4.B).

When an ester has an acetyl at the β-position (keto-ester **61**), it is possible to form a dianion. Subsequent reactions will occur first at the site of the second deprotonation, which is more nucleophilic. In the synthesis of stemonamide by Kende and co-workers,[59] methyl acetoacetate was treated with NaH and then with butyllithium to give dianion **61**. The sodium hydride deprotonated the most acidic hydrogen between the two carbonyl groups, and butyllithium removed the less acidic proton adjacent to the methyl group. This latter carbon is more nucleophilic, and reaction with allyl bromide followed by quenching with an aqueous workup

58. (a) Cuvigny, T.; Hullot, P.; Larchevêque, M.; Normant, H. *C.R. Hebd. Seances Acad. Sci., Ser. C* **1974**, *279*, 569; (b) Hullot, P.; Cuvigny, T.; Larchevêque, M.; Normant, H. *Can. J. Chem.* **1976**, 54, 1098; (c) Crouse, D.N.; Seebach, D. *Chem. Ber.* **1968**, *101*, 3113.
59. Kende, A.S.; Hernando, J.I.M.; Milbank, J.B.J. *Tetrahedron* **2002**, *58*, 61.

led to **62** in >90% yield.

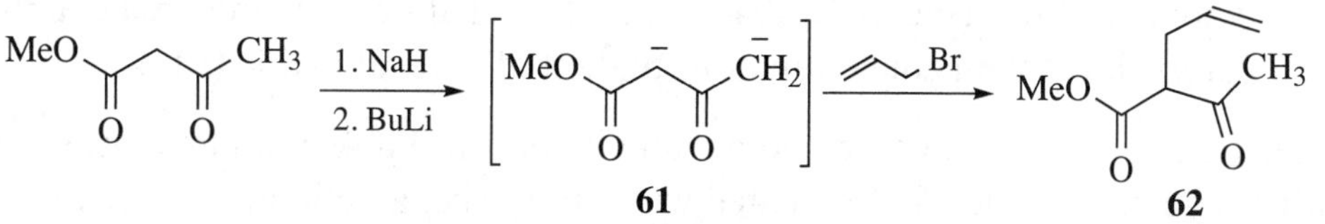

All ketones, aldehydes, and esters mentioned so far have had an α-carbon with a hydrogen that could be removed by base. When a ketone does not have such a proton (it is a non-enolizable ketone), treatment with strong base can lead to a C–C bond cleavage reaction. Discovered by Semmeler,[60] Haller and Bauer[61] developed the reaction illustrated by reaction of **63** with NaNH$_2$ to give amide **65**. Acyl addition of the amide anion to the carbonyl of the ketone led to **64**, and loss of the anion Ph⁻ gave the amide product. Treatment with aqueous acid (the workup) will protonate the anion to give benzene. This reaction is known as the **Haller-Bauer reaction**.[62]

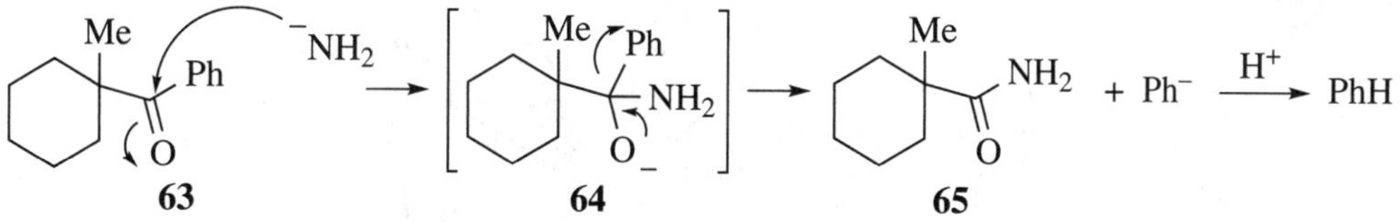

9.3.   REACTIONS OF ENOLATE ANIONS WITH ELECTROPHILES

There are two major reactions of enolates: (*1*) displacement reactions with alkyl halides or other suitable electrophiles and (*2*) nucleophilic acyl addition to carbonyl compounds. Reaction of enolate anion **61** with allyl bromide to give **62**, and reaction of **59** with butanal to give **60** are simple examples. Enolate anions are carbon nucleophiles, and react similarly to the acetylides discussed in section 8.3.C. The reactions discussed in this section are solution-phase enolate anion reactions. Solid-phase enolate anion reactions[63] have been investigated[64] by high resolution-magic angle spinning NMR spectroscopy.

9.3.A.   Enolate Alkylation

An enolate anion is a nucleophile, and both the oxygen and the carbon are nucleophilic (sec. 9.3.B), but the reaction at carbon usually predominates. Unlike Grignard reagents and

60. Semmler, F.W. *Berichte* **1906**, *39*, 2577.
61. Haller, A.; Bauer, E. *Compt. Rend.* **1908**, *147*, 824.
62. (a) Hamlin, K.E.; Weston, A.W. *Org. React* **1957**, *9*, 1; (b) Mehta, G.; Venkateswaran, R.V. *Tetrahedron* **2000**, *56*, 1399; (c) Paquette, L.A.; Gilday, J.P. *Org. Prep. Proceed. Int.* **1990**, *22*, 167.
63. See Dörwald, F.Z. *Organic Synthesis on Solid Phase, Supports, Linkers, Reactions*, Wiley-VCH, New York, **2000**, Chapters 12-13.
64. Fruchart, J.-S.; Lippens, G.; Kuhn, C.; Gras-Masse, H.; Melnyk, O. *J. Org. Chem.* **2002**, *67*, 526.

organolithium reagents enolate anions react easily with the electrophilic carbon of an alkyl halide.[65] As with acetylides (sec. 8.3), reaction an alkyl halide with an enolate anion forms a new carbon-carbon bond with concomitant formation of a metal halide (M–X).

Many synthetic applications involve enolate alkylation. In Funk and Greshocks synthesis of welwistatin,[66] cyclohexanone **66** was treated with lithium hexamethyldisilazide to generate the enolate anion, and addition of allylic bromide **67** gave the alkylated ketone **68** in 75% yield, with >10:1 selectivity for the diastereomer shown. Note that HMPA [hexamethylphosphoramide, $(Me_3N)_3P=O$] was added to facilitate formation of the anion and also to assist the alkylation step. Another example is the reaction of **69** with LDA to give enolate **70**, where the lithium is chelated with the nitrogen lone pair electron.[67] The facial bias imposed by the stereogenic centers, and the relatively rigid conformation of **70** led to reaction from the bottom face (the α face in steroid nomenclature). Addition of 1-bromo-2-cyclohexene from this less sterically hindered face led to **71** with high diastereoselectivity (>95:<5 favoring the stereochemistry shown for **71**).[67]

In some alkylation reactions, HMPA (hexamethylphosphoramide) is used as an additive to facilitate the alkylation. Hexamethylphosphorus triamide [HMPT, $(Me_2N)_3P$] also coordinates with the enolate anion, thereby diminishing the aggregate state and increasing reactivity. This additive also enhances solvent polarity and enhances the facility of an S_N2 alkylation step (sec. 2.6.A.i). The use of additives such as HMPA or HMPT is very common when the enolate anion reacts slowly, and/or the halide is relatively unreactive. Other acid derivatives such as lactones can react with LDA to give an enolate anion, which then reacts with alkyl halides in the usual manner.

---

65. For reactions of this type, see Larock, R.C. *Comprehensive Organic Transformations, 2nd ed.,* Wiley-VCH, New York, *1999*, pp 1476-1483.
66. Greshock, T.J.; Funk, R.L. *Org. Lett.* ***2006***, *8*, 2643.
67. Thottathil, J.K.; Moniot, J.L.; Mueller, R.H.; Wong, M.K.Y.; Kissick, T.P. *J. Org. Chem.* ***1986***, *51*, 3140.

The general disconnections for enolate alkylation are

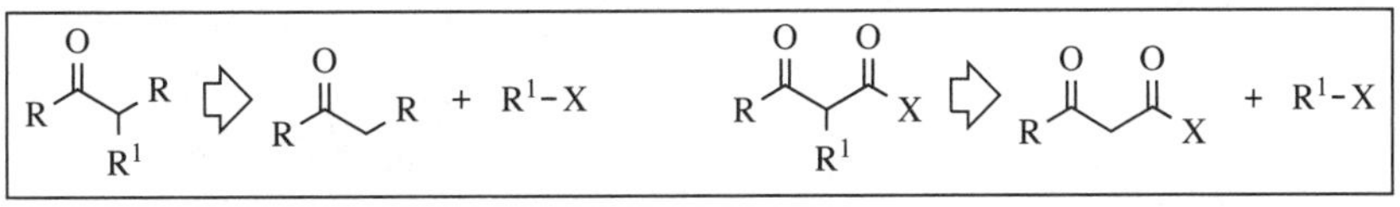

Alkylation reactions with unsymmetrical ketones that have essentially the same pK_a for both α-hydrogens often show poor regioselectivity, and over-alkylation can be a problem when reactive alkyl halides are used. Partial equilibration can occur under some reaction conditions, as shown for reactions of **72**, leading to poor regioselectivity.[68] When $LiNH_2$ was used as a base in THF, and then quenched with iodobutane, a single alkylation product was obtained in 95% yield (**73**, the kinetic product since equilibration is minimal under these conditions). With excess base and a complexing agent such as HMPA, however, equilibration occurred and alkylation gave a mixture of **73** and **74** (52% and 14% yield, respectively, along with

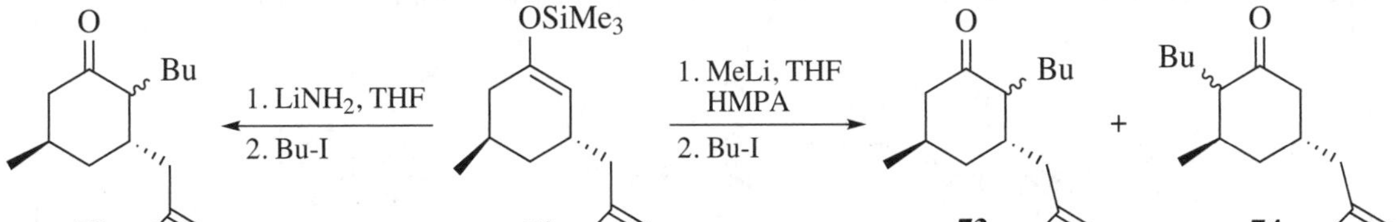

dialkylation products).[68] Polyalkylation may be a problem when a very reactive halide such as iodomethane is used under equilibrating conditions. Treatment of cyclohexanone with sodium amide and then iodomethane gave the expected 2-methylcyclohexanone (**75**), but this α-alkyl derivative was deprotonated under the reaction conditions to give **76**, and a second methylation gave **77**. If an excess of iodomethane was present, the tetramethylated cyclohexanone derivative (**78**) was formed. The 2,6-dimethyl and the 2,2,6-trimethylcyclohexanone derivatives are also possible products from this reaction, although it is usually difficult to prepare and isolate, **77** by this method.

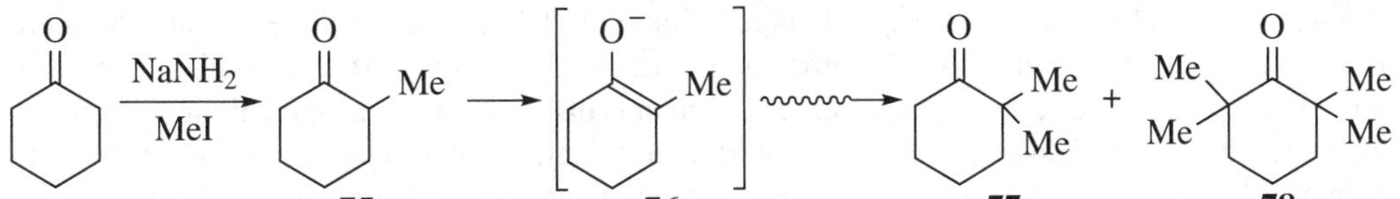

Enolates derived from acid derivatives can also be alkylated. Esters, lactones, amides, lactams or carboxylic acids can all be directly alkylated via an enolate anion intermediate. An example is the treatment of lactone **79** with NaH in DMF (NaH can be used because of the enhanced acidity of the 1,3-dicarbonyl moiety in **79**) followed by reaction with an excess of iodomethane

68. (a) Binkley, E.S.; Heathcock, C.H. *J. Org. Chem.* **1975**, *40*, 2156; (b) for a related reaction, see Stork, G.; Rosen, P.; Goldman, N.L.*J. Am. Chem. Soc.* **1961**, *83*, 2965.

to give an 89% yield of **80** as part of Hale and Li's synthesis of (+)-eremantholide.[69] Note that alkylation proceeded with inversion of configuration.

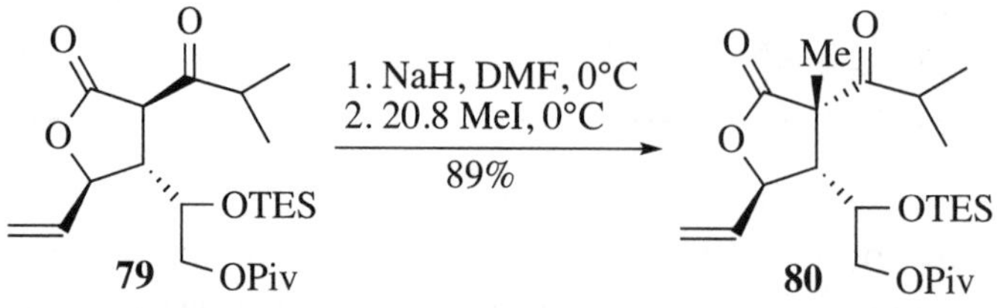

MOLECULAR MODELING: Stereochemistry of Reactions of Enolates

Examination of an accurate 3D model will often provide sufficient insight into the stereochemistry of the product resulting from attack of an electrophile on an enolate anion. This, of course, assumes that steric factors are the controlling influence. Consider for example, the enolate formed by deprotonation of the polycyclic lactone shown below (see also *homework problem 1*).

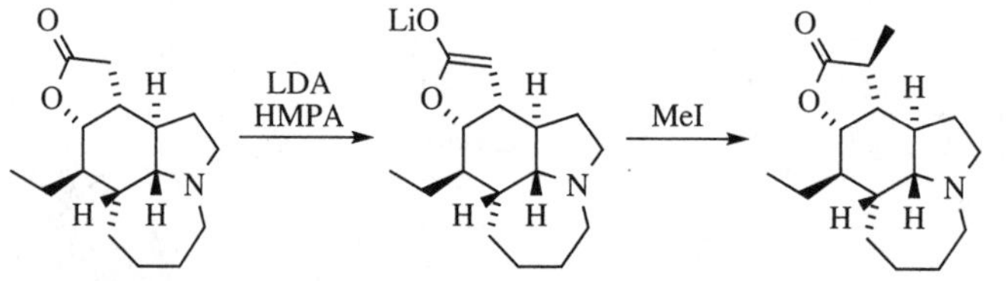

First, examine a space-filling model for the enolate anion (left). Is the "convex" or "concave" face more accessible, or are they equally accessible? In other words, when the enolate anion donates electrons to the methyl carbon of MeI, is the methyl group likely to attach to the "top" face of the product, as shown? Does this rule in favor of steric control of the stereochemistry of reaction? Elaborate.

Next, examine a local ionization potential map for the enolate anion (right). This colors an electron density surface which gives the overall molecular size and shape (where a reagent can approach) with the value of the local ionization potential (the energy required to expel an electron from a given location on the surface). By convention, regions on the surface where the ionization potential is small (electrons are loosely bound) are colored "red", while regions where the ionization is large (electrons are tightly bound) are colored "blue". The more loosely bound the electrons, the more likely they are to be donated to an approaching electrophile. In summary, red regions are more likely the target for attack by an electrophile. The map is displayed as a transparent model to allow you to see the structure underneath. Once you become oriented, you can switch to a solid model to allow the colors to be seen more clearly (*click* on the map and select **Solid** from the **Style** menu at the bottom of the screen). Based on the local ionization potential map, what product is favored? Is this the observed product? Is this the same product that is suggested by inspection of the space-filling model? If not, suggest why not.

---

69. Li, Y.; Hale, K.J. *Org. Lett.* **2007,** *9,* 1267.

MOLECULAR MODELING: Selectivity in Methylation of Enolates

Enolate anions react with alkyl halides to give the corresponding alkylated carbonyl derivative. Enolate anions from ketones, aldehydes, esters, lactones, lactams, etc., are all known to undergo reaction with alkyl halides such as iodomethane. Such alkylation reactions are often highly regio and stereoselective. While steric preferences may be obvious from inspection of a 3D model (or even a 2D sketch), other factors may contribute to or even dominate product selection. Here, graphical models such as the local ionization potential map may prove of value. Consider for example, methylation of the enolate anion formed from each of the three bicyclo[4.4.0]decanones, **A-C**. Each leads to a single product.

Examine local ionization potential maps for the three enolate anions for **A-C** (left to right on screen) and assign the products. These correspond to deprotonation at the carbon substituted with the electron-withdrawing group. (Note that the model for **A** has been simplified by replacing the OTHP group by hydrogen; this is far removed from the reaction site and should be of little consequence.) These maps show whether electrons are tightly bound ("blue") or loosely bound ("red") as a function of location on an electron density surface (depicting overall molecular size and shape). The smaller (more red) the ionization potential for a region on the map, the more likely it is to donate electrons to the electrophile, so you want to identify the red regions. The maps are displayed as transparent models in order for you to see the structure underneath. Once you are properly oriented, you can switch to solid models, by *clicking* on one of the maps and selecting *Solid* from the **Style** menu at the bottom of the screen. This will allow you to more easily distinguish subtle differences in the local ionization potential for different regions. You can also "measure" the value of the ionization potential at a particular location on the surface by *clicking* on this location. The value will be displayed at the bottom of the screen. Do some or all the maps reveal a preference for methylation onto one face of the enolate over the other? Which enolate shows the greatest preference and which shows the least? Draw each of the resulting products.

There are two classical reaction sequences in organic chemistry that rely on enolate alkylation. One is the **malonic ester synthesis**.[70,71] In a synthetic example taken from Sulikowski and co-worker's studies on the biosynthesis of phomoidride B,[72] di-*tert*-butyl malonate was treated with sodium hydride in the presence of 6-iodo-2(*E*)-hexene to give the alkylated product **81** in >95% yield. In this study, the deuterated acid was the target, and hydrolysis of **81** to the diacid with trifluoroacetic acid was followed by α-deuteration by treatment of the diacid with NaOD in D$_2$O, giving **82**. Heating **82** in the NaOD/D$_2$O solution led to decarboxylation as well as

70. (a) Meyer, K.H. *Berichte* **1912**, *45*, 2864; (b) Hauser, C.R.; Hudson, Jr., B.E. *Org. React.* **1942**, *1*, 266.
71. (a) The Merck Index, 14th ed., Merck & Co., Inc., Whitehouse Station, New Jersey, **2006**, p ONR-57; (b) Mundy, B.P.; Ellerd, M.G.; Favaloro Jr., F.G. *Name Reactions and Reagents in Organic Synthesis, 2nd ed.*, Wiley-Interscience, New Jersey, **2005**, pp. 406-407.
72. Sulikowski, G.A.; Agnelli, F.; Spencer, P.; Koomen, J.M.; Russell, D.H. *Org. Lett.* **2002**, *4*, 1447.

incorporation of a second α-deuterium, and isolation of **83** in 83% yield from **81**.

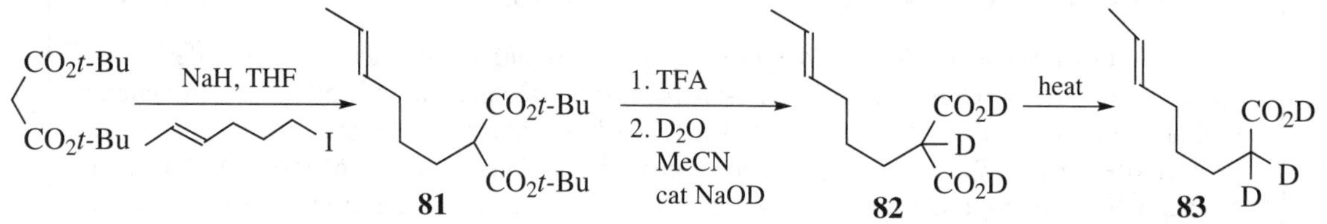

The second classical reaction mentioned above is the **acetoacetic ester synthesis**.[68b,73] In this sequence, an ester of acetoacetic acid (3-oxobutanoic acid) such as ethyl acetoacetate is treated with base under thermodynamic control conditions and alkylated, as with the malonic ester synthesis. Reaction with sodium ethoxide in ethanol (an ethyl ester is being used) gave the enolate anion, and quenching with benzyl bromide led to **84**. Saponification and decarboxylation gave a substituted ketone (**85**). Although the malonic ester synthesis and the acetoacetic ester synthesis are fundamentally similar, the different substrates lead to formation of either a highly substituted acid or a ketone. The reaction is not restricted to acetoacetate derivatives, and any β-keto-ester can be used.[74]

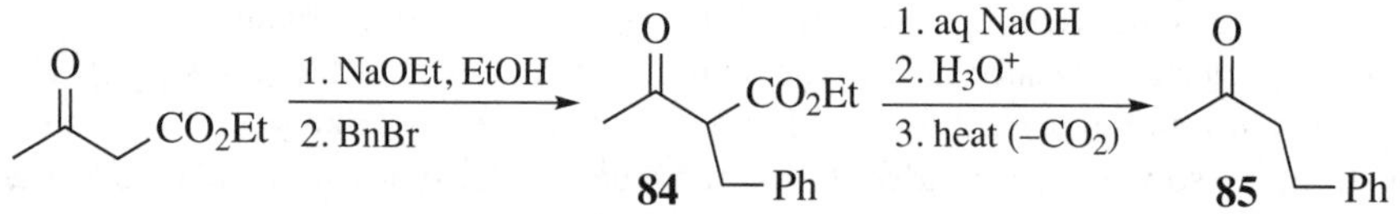

The key to all of these reactions is the enhanced acidity of the hydrogen on the carbon between the two carbonyls. As noted in Table 9.1, 1,3-dicarbonyl derivatives have a pK_a in the range of 8-12 for that proton. The enolate anion formed from a 1,3-dicarbonyl compound is somewhat less reactive because the enolate is more stable due to resonance delocalization. For this reason, higher reaction temperatures are usually (but not always) required. A typical experiment will use the reflux temperature of the alcohol solvent being used. Another alternative is to use a complexing agent such as HMPT or HMPA (see above), or DABCO (diazabicyclo[2.2.2] octane).

Intramolecular cyclization is possible using a doubly activated enolate anion. In the Deslongchamps and co-workers synthesis of (+)-martimol,[75] keto-ester **86** was treated with the basic cesium carbonate and intramolecular displacement of the allylic chloride gave the macrocycle **87** in 75% yield. There is a related ring-forming process called the **Perkin alicyclic**

73. (a) Krauch, H.; Kunz, W. *Organic Name Reactions*, John Wiley, New York, *1964*, pp 1-2; (b) The Merck Index, 14th ed., Merck & Co., Inc., Whitehouse Station, New Jersey, *2006*, p ONR-1; (c) Mundy, B.P.; Ellerd, M.G.; Favaloro Jr., F.G. *Name Reactions and Reagents in Organic Synthesis, 2nd ed.*, Wiley-Interscience, New Jersey, *2005*, pp. 2-3.
74. Cochrane, J.S.; Hanson, J.R. *J. Chem. Soc. Perkin Trans. 1 1972*, 361.
75. Toró, A.; Nowak, P.; Deslongchamps, P. *J. Am. Chem. Soc. 2000, 122*, 4526.

synthesis[76] (see sec. 9.4.B.i). In general, this useful modification of the malonic ester synthesis condenses an α,ω-dihalide with a malonic ester (although other 1,3-dicarbonyl derivatives can be used). This particular reaction converts malonic acid derivatives to cyclic compounds having a carboxylic acid substituent (cyclopentanecarboxylic acid or cyclohexanecarboxylic acid, for example). Notice that this is an intramolecular alkylation reaction, which is quite common. An example of an intramolecular alkylation reaction that employs a ketone rather than a keto-ester treated **88** with *tert*-butoxide in THF to give a 92% yield of **89**, in Coate's and co-worker's synthesis of cameroonan-7α-ol.[77]

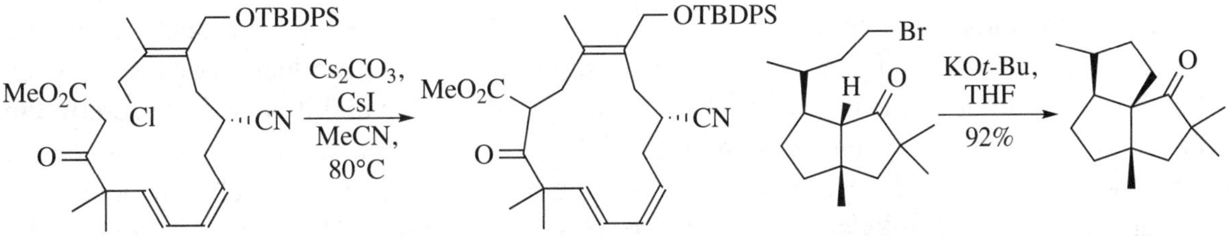

Variations of the malonic ester and acetoacetic ester sequences lead to many useful synthetic opportunities. In the examples quoted, the base-solvent pair used was ethanol-sodium ethoxide where the alkoxide is the conjugate base of the solvent. If NaOEt–EtOH were used with a methyl ester, transesterification[78] would give a mixture of methyl and ethyl esters as products. For both malonic ester and acetoacetic ester removal of the most acidic proton (α to both carbonyls) also gives the more thermodynamically stable enolate. Either NaOEt–EtOH or LDA-THF will generate the desired enolate. The malonic ester synthesis is most useful for the synthesis of highly substituted monoacids, and the acetoacetic ester synthesis is used to prepare substituted methyl ketones.

The disconnections for the malonic and acetoacetic ester syntheses are

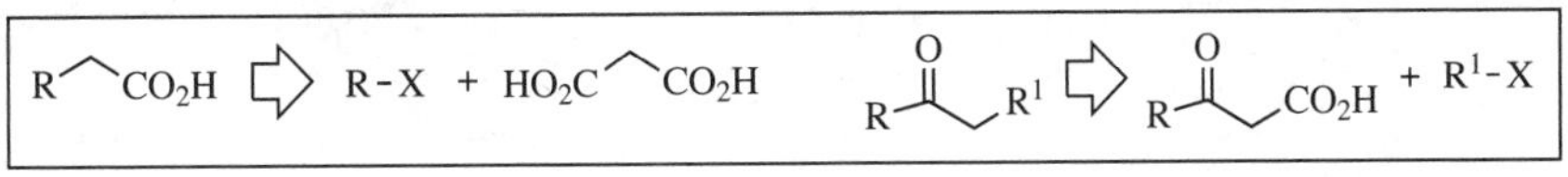

76. (a) Perkin, Jr., W.H. *Berichte* **1883**, *16*, 1787 (see p 1793); (b) Perkin, Jr., W.H. *J. Chem. Soc.* **1885**, *47*, 801; (c) van Heyningen, E. *J. Am. Chem. Soc.* **1954**, *76*, 2241; (d) The Merck Index, 14th ed., Merck & Co., Inc., Whitehouse Station, New Jersey, **2006**, p ONR-71; (e) and see Mundy, B.P.; Ellerd, M.G.; Favaloro Jr., F.G. *Name Reactions and Reagents in Organic Synthesis, 2nd ed.*, Wiley-Interscience, New Jersey, **2005**, pp. 492-493.
77. Davis, C.E.; Duffy, B.C.; Coates, R.M. *J. Org. Chem.* **2003**, *68*, 6935.
78. (a) Smith, M.B.; March, J. *March's Advanced Organic Chemistry, 6th ed.*, Wiley, New York, **2007**, pp 1419-1421; (b) Koskikallio, J. in Patai, S. Ed. *The Chemistry of Carboxylic Acids and Esters*, Interscience, New York, **1969**.

9.3.B.   O Versus C Alkylation of Enolate Anions

As discussed in Section 9.2.E, cyclization of **48** gave **50**, via an intramolecular alkylation reaction of enolates. However, vinyl ether **51** was also formed during this reaction, which illustrates an important competing reaction in the alkylation of enolates. An enolate anion is a bidentate nucleophile because it contains *two* nucleophilic centers, the carbanionic center and the oxygen anion as illustrated by resonance contributors **90**. Reaction of the *O*-nucleophile in **90** with an alkyl halide leads to the vinyl ether **91**. Reaction at the C nucleophile of **90** generates the usual alkylation product **92**. Since a carbanion is usually more nucleophilic than an alkoxy anion for common electrophiles, **92** tends to be the major product. There are several factors that can influence the relative proportion of these products, however. When **92** is the major product, the process is referred to as C-alkylation and it is greatly preferred in ether solvents when lithium enolates are used. Ether solvents promote the nucleophilic nature of the enolate, favoring the carbanion. The lithium enolate forms a relatively covalent O–Li bond, again favoring *C*-alkylation. When very reactive halides (iodomethane, allyl bromide) react with sterically unhindered enolates (usually with a potassium or sodium gegenion), *O*-alkylation (formation of product **91**) is preferred. The highly ionic O–Na or O–K bond effectively increases the nucleophilic character of the oxygen. When the alkylation occurs intramolecularly, *O*-alkylation can be a much more serious problem. Approach of the carbon nucleophile to the electrophilic center is sometimes restricted due to the nature of the tether (see sec. 6.6.A for **Baldwin's rules**). When the highly polar solvent DMSO is used, the enolate is largely monomeric, which increases reactivity and favors *O*-alkylation. The product distribution depends on the substitution pattern of the enolate and the nature of the counterion, as in Table 9.6.[79]

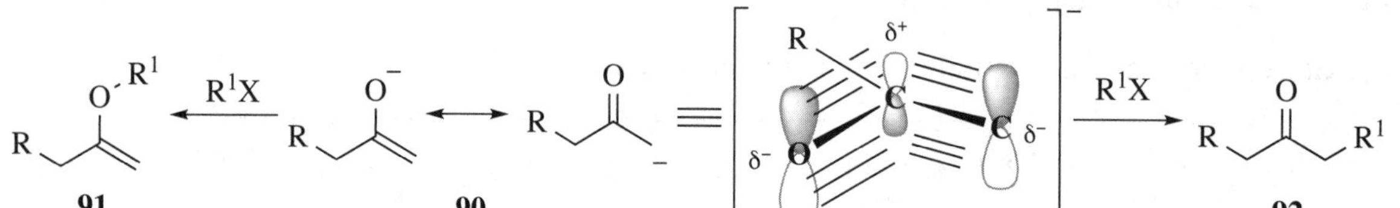

---

79. Zook, H.D.; Miller, J.A. *J. Org. Chem.* **1971**, *36*, 1112.

Table 9.6. *O*- Versus *C*-Alkylation for Phenyl Ketones in DMSO

$$\text{Ph-CO-CR R}^1 \xrightarrow[\text{DMSO, 30°C}]{\text{Me-S(=O)-M}} \mathbf{94} \xrightarrow[\text{30°C}]{\text{R}^2\text{X}} \mathbf{95} + \mathbf{96}$$

93 → 94 → 95 + 96

R	R^1	M	R^2–X	95 : 96	R	R^1	M	R^2–X	95 : 96
H	Ph	Na	n-PrCl	1 : 3.7	Me	Me	Li	i-BuCl	1 : 1.4
			i-BuCl	1 : 4.8				n-Amyl Cl	1 : 0.45
			n-Amyl Cl	1 : 3.6			Na	n-PrCl	1 : 0.77
			n-Amyl Br	1 : 7.1				i-BuCl	1 : 1.5
H	Et	Li	n-PrCl	1 : 0.9				n-Amyl Cl	1 : 0.48
			n-PrBr	1 : 2.3	Et	Et	Na	n-PrCl	1 : 0.14
			i-BuCl	1 : 1.9				i-BuCl	1 : 0.28
			n-Amyl Cl	1 : 0.77				n-Amyl Cl	1 : 0.11
			n-Amyl Br	1 : 1.2	Ph	Ph	Li	n-PrCl	1 : <0.01
			n-Amyl I	1 : 5				n-Amyl Cl	1 : <0.01
		Na	n-PrCl	1 : 0.77					
			i-BuCl	1 : 1.9					
			n-Amyl Cl	1 : 0.8					
			n-Amyl Br	1 : 1.6					
			n-Amyl I	1 : 4.4					
		Cs	n-Amyl Cl	1 : 0.77					

[Reprinted with permission from Zook, H.D.; Miller, J.A. *J. Org. Chem.* **1971**, *36*, 1112. Copyright © *1971* American Chemical Society.]

Treatment of ketone **93** with dimsyl anion generates enolate **94**, but this enolate anion can potentially form a mixture of (*E*)- and (*Z*)-isomers. Reaction of **94** with various alkyl halides led to the *O*-alkylation product (**95**) and the *C*-alkylation product (**96**). In DMSO, the ratio of *O*- to *C*-alkylation is relatively insensitive to changes in the gegenion (the metal), suggesting that the enolate is an unencumbered anion in this cation solvating medium. Note that *C*-alkylation is preferred except when the steric bulk at the nucleophilic carbon becomes too great (R, R^1 = Ph, Ph or Et, Et). For PhCOCHPh$_2$ (**93**, R^1 = R^2 = Ph), *O*-alkylation is effectively the only process.

Enolate anions react with acyl halides or anhydrides to give the acylated product. Both *C*-acylation and *O*-acylation are possible. In general *O*-acylation predominates[80] (section 9.3.C). If *C*-acylation is desired, the acyl unit must be modified. One solution is to use acyl cyanides. In an example taken from Nicolaou and co-worker's synthesis of coleophomone B,[81] acyl cyanide **98** was prepared from aldehyde **97** by treatment with Et$_2$AlCN (**Nagata's reagent**)[82] followed by oxidation of the resulting alcohol with PCC. When **98** was treated with

80. See Krapcho, A.P.; Diamanti, J.; Cayen, C.; Bingham, R. *Org. Synth. Coll. Vol. V 1973*, 198.
81. Nicolaou, K.C.; Vassilikogiannakis, G.; Montagnon, T. *Angew. Chem. Int. Ed. 2002, 41*, 3276.
82. Nagata, W.; Yoshioka, M.; Murakami, M. *Org. Synth. Coll. Vol. VI 1988*, 307.

5-methyl-1,3-cyclohexanedione and triethylamine, C-acylation proceeded smoothly to give **99** in 98% yield.

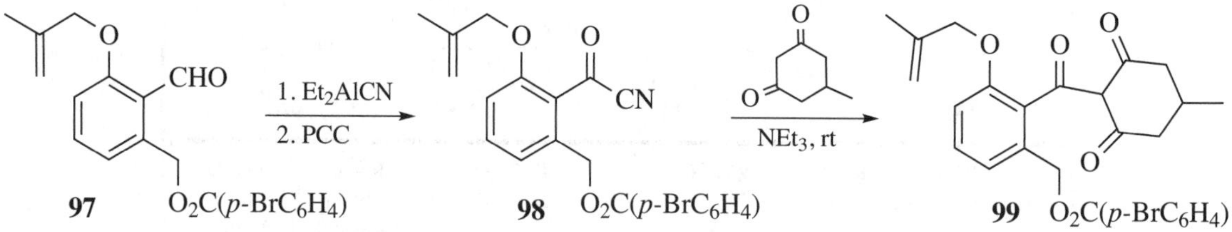

This type of reaction can sometimes be useful in synthesis, but the structural units on the substrate influence *C*- versus *O*- reactivity. In the Honda et al. formal synthesis of securinine,[83] *N*-benzylamino ketone (**100**) was treated with lithium hexamethyldisilazide and then sorbic anhydride. This sequence gave dienyl ester **101** by *O*-alkylation of the initially formed enolate anion. When the nitrogen-protecting group was changed to *N*-Boc (see **102**), however, the identical reaction gave *C*-acylation and the product was **103**. Neither of these products was suitable for the synthesis, but the reaction with 2-acetyl pyridine (**104**) gave **105**, which was transformed into securinine. This example shows that *O*- versus *C*-acylation is very dependent on the local environment and electronic effects within the enolate anion.

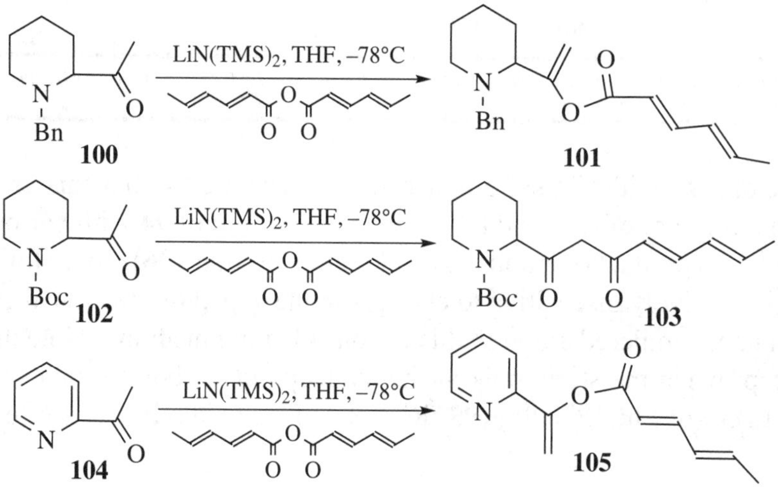

The synthesis of (+)- and (−)-saudin by Boeckman and co-worker's[84] employed an interesting variation that involved *O*-alkylation. Treatment of **106** with potassium hexamethyldisilazide removed a proton from the distal methyl group (acidic due to vinylogy caused by the conjugated enone system), and subsequent reaction with the triflate shown gave enol ether **107** in good yield. Apart from the use of a vinylogous enolate anion, this example illustrates that triflates also give *O*-alkylated product.

83. Honda, T.; Namiki, H.; Kudoh, M.; Watanabe, N.; Nagase, H.; Mizutani, H. *Tetrahedron Lett.* **2000**, *41*, 5927.
84. Boeckman Jr., R.K.; Ferreira, M.d R.R.; Mitchell, L.H.; Shao, P. *J. Am. Chem. Soc.* **2002**, *124*, 190.

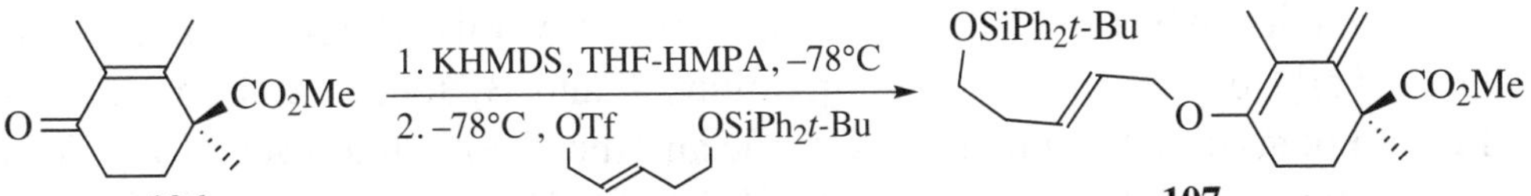

To put this problem into perspective, when ether or THF are used as the solvent, *O*-alkylation is a much less serious problem in reactions with alkyl halides. With the possible exception of iodomethane, the use of lithium enolates in ether solvents leads to *C*-alkylation as the major product in virtually all cases. When the enolate carbanion center is sterically hindered, *O*-alkylation can be a problem even in THF or ether. Some reagents such as silyl halides and anhydrides show a preference for *O*-alkylation. Both of these *O*-alkylation reactions will be discussed in Section 9.3.C.

9.3.C. Enolate Reactions with Non-Carbonyl Electrophiles

Enolate anions react with many electrophilic species other than alkyl halides, analogous to the reactions of other carbanions (Chap. 8). Nucleophilic ring opening of an epoxide to give alcohols is very common,[85] occurring at the less sterically hindered carbon. Grieco et al. reacted the dilithio salt of acetic acid ($LiCH_2CO_2Li$, sec. 9.4.B.vii) with epoxide **108** to give **109**, in a synthesis of bigelovin.[86] The regioselectivity in the ring opening is probably due to a combination of neighboring group effects with the adjacent hydroxyl group, and steric blocking of one carbon by the bridgehead methyl.

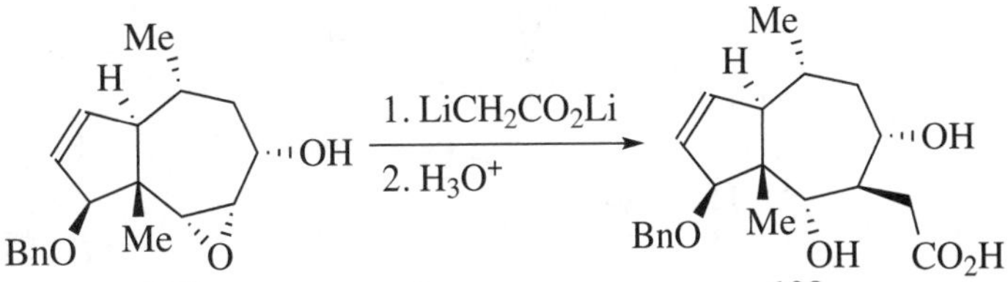

The epoxide ring opening gives the interesting disconnection,

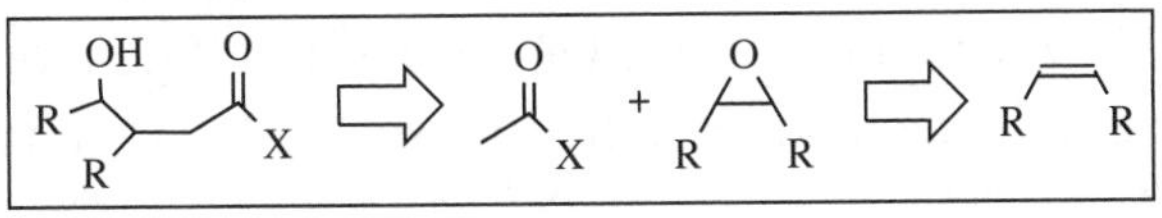

It is also possible to oxygenate an enolate to give the corresponding acyloin (an α-hydroxy ketone) or alkoxy derivative. In Danishefsky and co-worker's synthesis of myrocin C,[87] reaction of an enol ether such as **110** with 3,3-dimethyldioxirane (DMDO) gave the protected α-hydroxy ketone (**111**). Reaction of an enolate anion with oxygen and trimethylphosphite [$(MeO)_3P$] gives an α-hydroxy derivative.[88] MoOPh is an effective oxygenation reagent for

85. For a review of such reactions, see Taylor, S.K. *Tetrahedron* **2000**, *56*, 1149.
86. Grieco, P.A.; Ohfune, Y.; Majetich, G. *J. Org. Chem.* **1979**, *44*, 3092.
87. Chu-Moyer, M.Y.; Danishefsky, S.J.; Schulte, G.K. *J. Am. Chem. Soc.* **1994**, *116*, 11213.
88. For an example taken from a synthesis of brefeldin A, see Corey, E.J.; Wollenberg, R.H.; Williams, D.R. *Tetrahedron Lett.* **1977**, 2243.

enolate anions, and it can be used with enolate anions of several different carbonyl compounds. In a synthetic work towards the trichodermamides A and B, Joullié and co-workers treated lactone **112** with potassium hexamethyldisilazide and then MoOPh to give **113** in 70% yield.[89] The actual structure of MoOPh is $MoO_5•Py•HMPA$.[90] Nozoe and co-workers showed that 2-phenyl-2-*p*-toluene-sulfonyloxazidine (sec. 3.9.A.ii) could convert the enolate derived from a 2-pyrrolidinone derivative to the 3-hydroxy derivative (α-hydroxylation) in good yield.[91] Another reagent that reacts with enolate anions to generate α-hydroxy ketones is the **Davis reagent** (2-phenylsulfonyl-3-phenyl-oxaziridine).[92] A variety of enolate anions, including those derived from carboxylic acid derivatives, can be hydroxylated in this manner. In a synthesis of jiadifenin by Danishefsky and co-workers, the lactone enolate anion generated from sodium hexamethyldisilazide and lactone **114** was treated with the Davis reagent to give a 42% yield of α-hydroxy lactone **115**.[93]

Enolates react with sulfenyl halides to give an α-alkylthio ketone. As shown for cyclohexanone, reaction with LDA followed benzenesulfenyl chloride (PhSCl) produced phenylthio derivative **116**. The α-proton adjacent to both the carbonyl and the sulfur was more acidic, and deprotonation with potassium hydride gave the enolate. Subsequent alkylation with iodobutane gave a 91% yield of **117**.[94] As mentioned in Chapter 2, the sulfide moiety can be oxidized to a sulfoxide and thermally eliminated (syn elimination, sec. 2.9.C.v) to give an alkene.[95] Similar

89. Wan, X.; Doridot, G.; Joullie, M.M. *Org. Lett.* **2007**, *9*, 977. Also see Vedejs, E. *J. Am. Chem. Soc.* **1974**, *96*, 5944.

90. (a) Mimoun, H.; Seree de Roch, L.; Sajus, L. *Bull. Soc. Chim. Fr.* **1969**, 1481; (b) Regen, S.L.; Whitesides, G.M. *J. Organomet. Chem.* **1973**, *59*, 293.

91. Ohta, T.; Hosoi, A.; Nozoe, S. *Tetrahedron Lett.* **1988**, *29*, 329.

92. (a) Davis, F.A.; Chen, B.-C. *Chem. Rev.* **1992**, *92*, 919. (b) Also see Nagasaka, T.; Imai, T. *Chem. Pharm. Bull.* **1995**, *43*, 1081; (c) Mundy, B.P.; Ellerd, M.G.; Favaloro Jr., F.G. *Name Reactions and Reagents in Organic Synthesis, 2nd ed.*, Wiley-Interscience, New Jersey, **2005**, p. 755.

93. Cho, Y.S.; Carcache, D.A.; Tian, Y.; Li, Y.-M.; Danishefsky, S.J. *J. Am. Chem. Soc.* **2004**, *126*, 14358.

94. See Coates, R.M.; Pigott, H.D.; Ollinger, J. *Tetrahedron Lett.* **1974**, 3955.

95. (a) Trost, B.M.; Salzmann, T.N. *J. Org. Chem.* **1975**, *40*, 148; (b) Grieco, P.A.; Reap, J.J. *Tetrahedron Lett.* **1974**, 1097.

methodology can be applied to formation and syn-elimination of the corresponding selenides, -CO-CH(R)-Se-R[1].[96]

In addition to the methods described for controlling the ratio of thermodynamic and kinetic products, other techniques have been developed to trap the enolate based on reactions with reagents that prefer *O*-alkylation. Reaction of a ketone with acetic anhydride, usually in the presence of a catalytic amount of perchloric acid, generates the thermodynamic enol acetate **118**.[97] When **118** was treated with methyllithium, acyl addition to the acetyl group led to the thermodynamic enolate (**76**) in high yield. Once available, **76** was treated with an alkyl halide or a carbonyl derivative in the usual manner. Some isomerization occurs on treatment with methyllithium, but the major product of this process is that resulting from alkylation of the thermodynamic enolate.

Trialkylsilyl halides show a great propensity to react with the oxygen rather than the carbon of enolate anions. Stork and Hudrlik showed that enolates can be trapped as the trialkylsilyl enol ether via *O*-alkylation, which is most useful for kinetic enolates in which a lithium enolate (such as the kinetic enolate derived from 2-methyl-cyclohexanone and LDA) is reacted with trimethylsilyl chloride to give an isolable intermediate, **119**.[98] The enolate is trapped with high efficiency, and conversion to the enolate is readily accomplished by treatment with methyllithium to generate kinetic enolate **120** and the volatile trimethylsilane (Me_3SiH).[99] This latter reaction is subject to some equilibration, but gives the kinetic alkylation product in good yield. *O*-Silyl enolates[100] can be generated by other reagents, reacting with a variety of starting materials. With kinetic enolates, the trapping procedure is especially useful since the isomeric silyl enol ethers can be isolated and separated from amine by-products and unreacted ketone. Unmasking and alkylation can be accomplished under controlled conditions that minimize equilibration, maximizing the yield of kinetic product.

96. (a) Grieco, P.A.; Miyashita, M. *J. Org. Chem.* **1974**, *39*, 120; (b) Reich, H.J.; Renga, J.M.; Reich, I.L *J. Org. Chem.* **1974**, *39*, 2133.
97. (a) House, H.O.; Kramar, V. *J. Org. Chem.* **1963**, *28*, 3362; (b) House, H.O.; Phillips, W.V.; VanDerveer, D. *Ibid.* **1979**, *44*, 2400.
98. Stork, G.; Hudrlik, P.F. *J. Am. Chem. Soc.* **1968**, *90*, 4464, 4462.
99. House, H.O.; Czuba, L.J.; Gall, M.; Olmstead, H.D. *J. Org. Chem.* **1969**, *34*, 2324.
100. Rasmussen, J.K. *Synthesis* **1977**, 91.

119 (74% yield, 99% pure)   120

9.4. ENOLATE CONDENSATION REACTIONS

The condensation reaction of enolate anions and carbonyl derivatives is one of the most useful in organic chemistry. The condensation is nothing more than an acyl addition reaction of the nucleophilic enolate to an electrophilic carbonyl carbon.[101] The generalized reaction is represented by enolate 121 reacting with carbonyl derivative 122 (an aldehyde or a ketone in this case) to give the acyl addition product 123. The reaction is straightforward unless R^1 and/or R^2 are very large. Acyl addition may also be slow if R^3 and/or R^4 in the carbonyl partner are large. If the carbonyl is an aldehyde or a ketone, the reaction gives 123. If 122 is an acid derivative, however, R^3 or R^4 is a leaving group (Cl, O_2CR, OR, NR_2) and 123 reacts further to produce a 1,3-dicarbonyl compound via acyl substitution (sec. 9.4.B). There are many synthetic variations, and to further complicate matters the enolate may be generated under kinetic or thermodynamic conditions. The more common variations of this reaction are well known, and usually have a name associated with them (i.e. named reactions).[102,103,104,105]

9.4.A. The Aldol Condensation[106]

9.4.A.i. Intermolecular Reactions. The **aldol condensation**[107] is one of the classical reactions of Organic chemistry, apparently first reported by Chiozza in 1856[107a] and later expanded by

101. For reactions of this type, see Larock, R.C. *Comprehensive Organic Transformations, 2nd ed.,* Wiley-VCH, New York, *1999*, pp. 1317-1324.
102. *The Merck Index, 14th Edition,* Merck and Co, Inc., Whitehouse, New Jersey, *2006*, pp ONR-1-ONR-105.
103. Mundy, B.P.; Ellerd, M.G.; Favaloro Jr., F.G. *Name Reactions and Reagents in Organic Synthesis, 2nd ed.,* Wiley-Interscience, New Jersey, *2005*.
104. Krauch, H.; Kunz, W. *Organic Name Reactions,* John Wiley, New York, *1964*.
105. Surrey, A.R. *Name Reactions in Organic Chemistry, 2nd Ed.,* Academic Press, New York, *1961*.
106. (a) *Modern Aldol Reactions,* Mahrwald, R. (Ed.), Wiley, NY, *2004*; (b) The Merck Index, 14th ed., Merck & Co., Inc., Whitehouse Station, New Jersey, *2006*, p ONR-1; (c) Mundy, B.P.; Ellerd, M.G.; Favaloro Jr., F.G. *Name Reactions and Reagents in Organic Synthesis, 2nd ed.,* Wiley-Interscience, New Jersey, *2005*, pp. 26-27.
107. (a) Chiozza, L. *Annalen 1856,* 97, 350; (b) Wurtz, A. *Compt. Rend. 1872, 74,* 1361; (c) Perkin, W. *Berichte, 1882, 15,* 2802; (d) Nielsen, A.T.; Houlihan, W.J. *Org. React. 1968, 16,* 1; (e) Mundy, B.P.; Ellerd, M.G.; Favaloro Jr., F.G. *Name Reactions and Reagents in Organic Synthesis, 2nd ed.,* Wiley-Interscience, New Jersey, *2005*, pp. 8-9; (f) Reference 104, pp 6-10.

Wurtz[107b] and by Perkin.[107c] In its earliest version, an aldehyde was mixed with a base and an aldehyde that did not contain an enolizable hydrogen (such as benzaldehyde), under what we now categorize as thermodynamic conditions. Ketones were also used. If acetophenone reacted with benzaldehyde in the presence of sodium ethoxide (in ethanol), the initial product was alkoxide **124**. Hydrolysis provided the β-hydroxy ketone product (an aldol or aldolate) **125**. In general, the hydrolysis step provides the aldolate, but sometimes elimination of water (dehydration) accompanies the hydrolysis to give a conjugated ketone (**126**). Note that heating is usually required to induce dehydration unless there is a conjugating substituent. When an aromatic aldehyde (such as benzaldehyde) is the reaction partner, the aqueous acid workup almost always leads to elimination of water from the aldol product. If an aliphatic aldehyde is used, the aldol product is usually easy to isolate. When an aromatic aldehyde is condensed with an enolate of an aliphatic aldehyde or ketone to give the α, β-unsaturated compound, the reaction is called the **Claisen-Schmidt reaction**.[108,107d] Note that the aldol condensation can be done using high-intensity ultrasound,[109] where good yields of aldol products were isolated that normally give significant elimination under non-ultrasound conditions.

A major drawback of the reaction as originally formulated is self-condensation. The enolate derived from 3-pentanone (see **127**) exists in equilibrium with the ketone under these conditions, and benzaldehyde is present. Enolate **127** can react with benzaldehyde to give the aldolate product **128**, and hydrolysis gives the alcohol as a mixture of diastereomers **129** and **130**. At this point, we cannot predict one or the other as a major product. For the time being, we will assume that the reaction produces close to a 1:1 mixture of the two products. If this mixture is heated, conjugated ketone **131** is formed. The problem with this approach is clear when we recognize that enolate **127** can also react with unenolized 3-pentanone to produce the self-condensation aldol product (**132**), and hydrolysis gives **133**. Therefore, this reaction could produce three different aldol products.

108.   (a) Claisen, L.; Claparède, A. *Berichte,* **1881**, *14*, 2460; (b) Schmidt, J.G. *Ibid.* **1881**, *14*, 1459; (c) The Merck Index, 14th ed., Merck & Co., Inc., Whitehouse Station, New Jersey, **2006**, p ONR-18.
109.   Cravotto, G.; Demetri, A.; Nano, G.M.; Palmisano, G.; Penoni, A.; Tagliapietra, S. *Eur. J. Org. Chem.* **2003**, 4438.

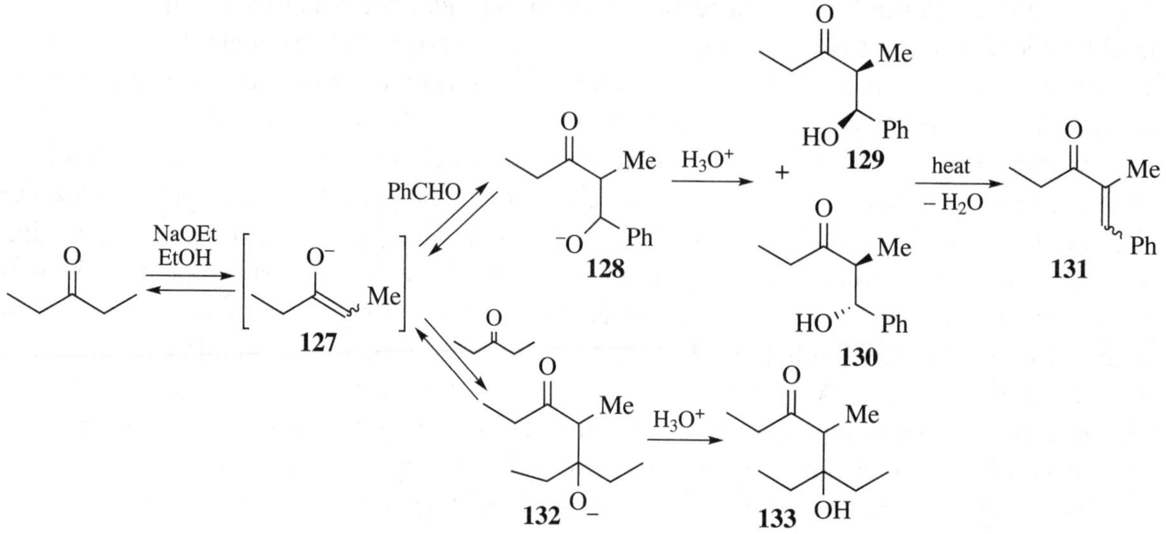

This latter point is one disadvantage of the traditional aldol condensation, but there are also advantages. The main advantage is the construction of a relatively complex carbon skeleton from simple precursors. In addition, this carbon-carbon bond forming reaction generates new stereogenic (but racemic) centers. As will be seen in Section 9.5.A, good-to-excellent diastereoselectivity is possible in many cases. The reaction conditions used for the aldol condensation of 3-pentanone described above are rather harsh, and this is a disadvantage if elimination of water from the aldol product occurs, or if self-condensation occurs under these equilibrating conditions. Aldehydes that do not contain an enolizable proton can be added to the enolate of a ketone to give a cross-coupling product, but a large excess of the aldehyde is required to minimize self-condensation.

A crossed-aldol condensation is often desirable between two reactive partners that have an enolizable position, as in 3-pentanone with cyclopentanone. If 3-pentanone and benzaldehyde, which has no enolizable protons, can lead to three products, what will happen in this new case? If an aldol condensation occurs under thermodynamic conditions, 3-pentanone reacts with sodium ethoxide to give enolate **127**. This enolate can condense with either unenolized 3-pentanone (to produce **133**) or with unenolized cyclopentanone (to produce **135**). Both of these ketones are symmetrical, and there is no opportunity for additional enolates, which would further complicate the reaction (see below). The pK_a of 3-pentanone and cyclopentanone are close. Under equilibrium conditions, both enolate anions (**127** and **134**) will be formed, and will be available for reaction. In addition to the formation of **133** and **135**, enolate anion **134** can condense with either the unenolized 3-pentanone (to produce **136**) or with unenolized cyclopentanone (to produce **137**), giving a total of four aldol products in this reaction. Diastereomers are possible for **133** and **135**, increasing the actual number of products. The low yield of the desired aldolate, and problems associated with separation and isolation make this reaction unattractive. Even if one aldolate is produced with some selectivity, the problems of poor yields and isolation of the desired product remain. It is clear that this mixed aldol condensation is not a synthetically useful reaction as described above.

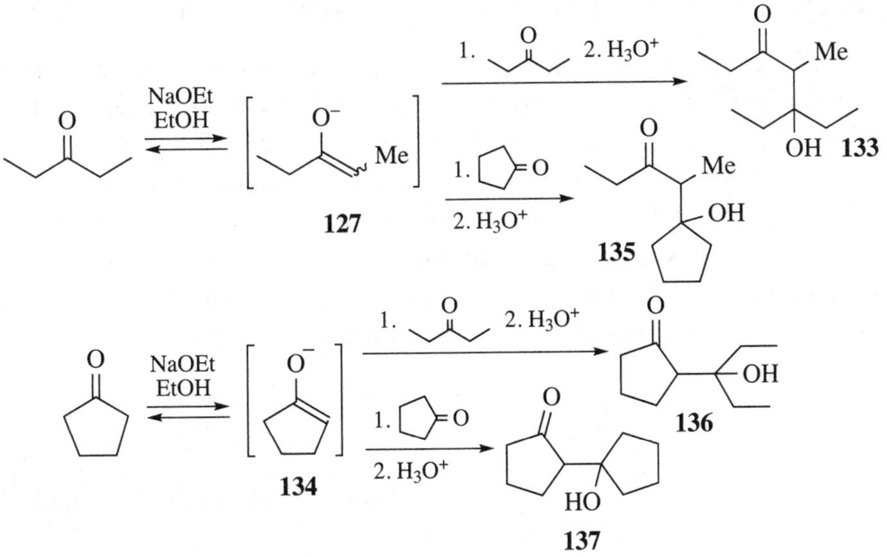

If an unsymmetrical ketone is used in this reaction, the problem is exacerbated. Reaction of **43** with sodium ethoxide, under thermodynamic control conditions, generates two different enolate anions. When reacted with an aldehyde with no α-hydrogens (benzaldehyde), two aldol products are formed (**138** and **139**) if there is no self-condensation of **43**. When **43** reacted with sodium ethoxide under thermodynamic conditions in the presence of an unsymmetrical ketone such as 2-butanone, the kinetic and thermodynamic enolates of both ketones were formed, so four different enolate anions are formed and each one reacts with two different ketones. Therefore, the attempted mixed aldol condensation of 2-butanone and **43** can produce eight different aldolate products.

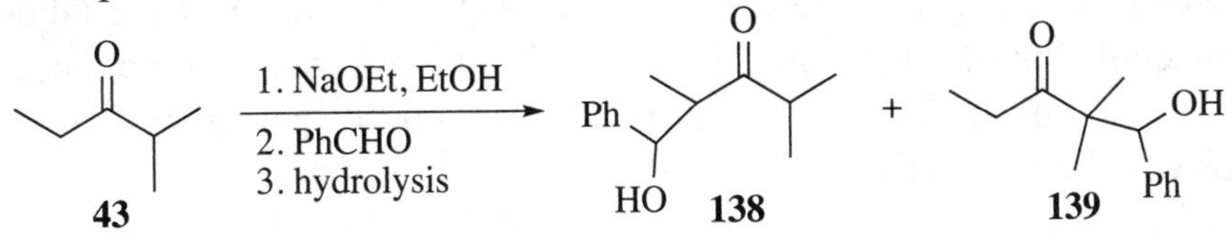

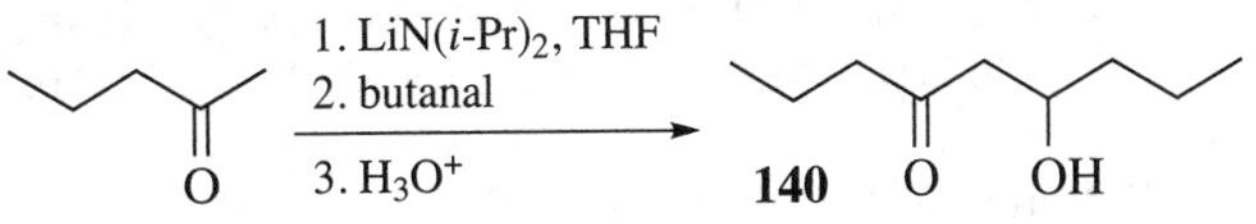

One solution to this problem lies in changing from thermodynamic cto kinetic control conditions. If 3-pentanone, for example, were treated with LDA in THF at –78°C, enolate **127** would be produced as the major enolate anion and in high yield. In a subsequent step, cyclopentanone (or **43**) could be added to give a single aldol condensation product. The advantage of kinetic control is the formation of primarily one enolate anion under non-equilibrating conditions, which minimizes self-condensation products. A kinetically controlled mixed aldol condensation is the reaction of 2-pentanone with LDA and subsequent condensation with butanal, which gave **140** as the major product.[110] If the thermodynamic aldolate is desired, the thermodynamic enolate must be trapped as the enol acetate (see above) and then treated with a carbonyl compound after unmasking with methyllithium.

There are countless synthetic examples of the aldol condensation. In the Perkins and Lister synthesis of dolabriferol[111] treated the ethyl ketone unit in **141** with an excess of LDA to generate the enolate anion. Subsequent addition of aldehyde **142** gave a 78% yield of aldolate product **143** with 85% ds. Enantioselective[112] aldol condensation reactions have been reported. An enantioselective crossed-aldol reaction of aldehydes has been reported. In work by MacMillan and Northrup,[113] *L*-proline was used as a chiral base. The cross-coupling of propanal with cyclohexanecarboxaldehyde, for example, gave an 87% yield of **144** as a 14:1 *anti:syn* mixture, and the *anti*-diastereomer was obtained in 99% enantiomeric purity.

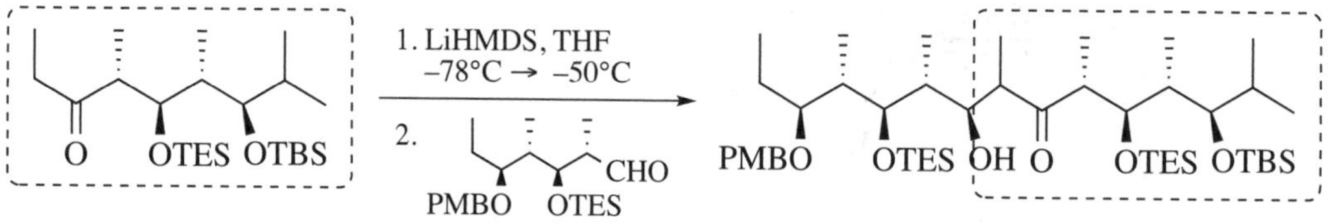

Aldol condensation reactions can be done under acidic conditions. In a synthesis of the immunosuppressant FR901483 by Fukuyama and co-workers,[114] keto-aldehyde **145** was treated with camphorsulfonic acid in refluxing benzene, and subsequent protonation of the aldehyde led to the protonated intermediate in the presence of the enol form of the ketone (see **146**). Reaction of the enol with the protonated aldehyde via an intramolecular aldol cyclization (also see Sec. 9.4.A.ii), produced aldolate product **147**, and loss of the proton gave an excellent yield of alcohol **148** with high selectivity.

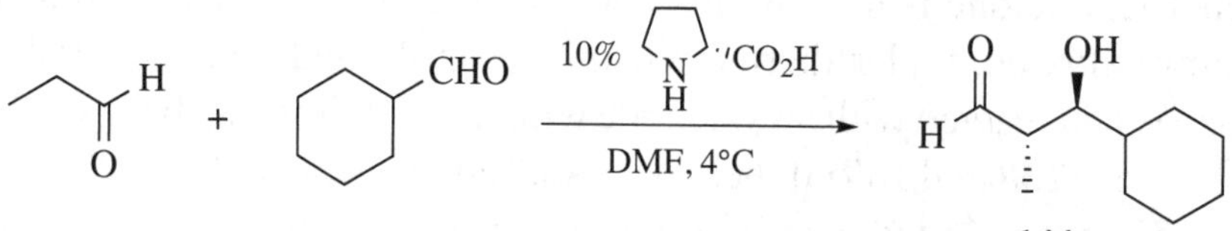

110. (a) Stork, G.; Kraus, G.A.; Garcia, G.A. *J. Org. Chem. 1974, 39*, 3459; (b) Gaudema, M. *C.R. Acad. Sci. Ser. C 1974, 279*, 961.
111. Lister, T.; Perkins, M.V. *Org. Lett. 2006, 8*, 1827.
112. (a) Cordova, A.; Notz, W.; Barbas III, C.F. *J. Org. Chem. 2002 67*, 301; (b) List, B.; Pojarliev, P.; Castello, C. *Org. Lett, 2001, 3*, 573; (c) Yoskikawa, N.; Kumagai, N.; Matsunaga, S.; Moll, G.; Oshima, T.; Suzuki, T.; Shibasaki, M. *J. Am. Chem. Soc. 2001, 123*, 2466; (d) Trost, B.M.; Silcoff, E.R.; Ito, H. *Org. Lett. 2001, 3*, 2497.
113. Northrup, A.B.; MacMillan, D.W.C. *J. Am. Chem. Soc. 2002, 124*, 6798.
114. Kan, T.; Fujimoto, T.; Ieda, S.; Asoh, Y.; Kitaoka, H.; Fukuyama, T. *Org. Lett. 2004, 6*, 2729.

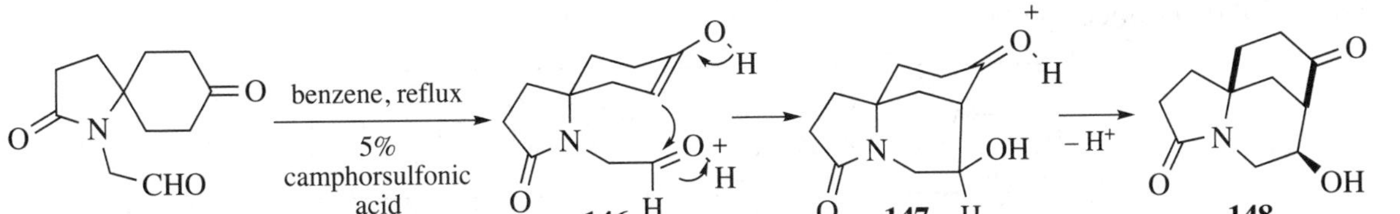

Lewis acids also catalyze the aldol condensation. In presence of a catalytic amount of TiCl$_4$,[115] 2-butanone was condensed with benzaldehyde in toluene to give an 83% yield of the aldolate product. Under these conditions, the reaction was highly regioselective for the thermodynamic product. In another example, taken from the synthesis of siphonarin B by Patterson and co-workers,[116] aldehyde **149** was condensed with ketone **150** in the presence of Sn(OTf)$_2$ to give a 92% yield of **151**. The aldolate product formed diastereoselectively, generating a 2.7:1 mixture of syn,syn:(syn,anti + anti,anti) diastereomers.

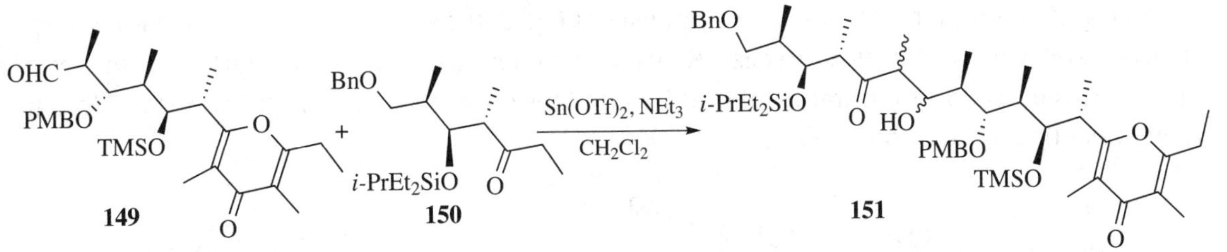

Nevalainen and Simpura reported another variation of this classic reaction called the aldol-transfer reaction. In the presence of a suitable catalyst, usually an aluminum compound, an aldolate product reacts with an aldehyde to give a ketone and a new aldolate.[117] An example is the reaction of benzaldehyde with aldolate **152** in the presence of 5% of aluminum catalyst (**153**). In dichloromethane at ambient temperature, a 62% yield of aldolate **154** was obtained after a reaction time of 43 h. The other product of this reaction was acetone, which was readily removed. This transformation involves a retro-aldol reaction of **152** (see sec. 9.5.A.vi.) and the resultant enolate anion reacts with benzaldehyde. Several aldehydes were used and **152** is particularly attractive (the aldol condensation product of acetone) because acetone is the second product.

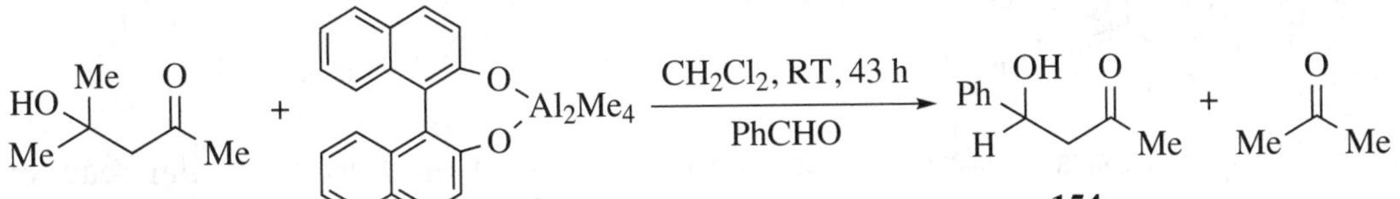

115. Mahrwald, R.; Gündogan, B. *J. Am. Chem. Soc.* ***1998***, *120*, 413.
116. Paterson, I.; Chen, D.Y.-K.; Franklin, A.S. *Org. Lett.* ***2002***, *4*, 391.
117. Simpura, I.; Nevalainen, V. *Angew. Chem. Int. Ed.* ***2000***, *39*, 3422.

The aldol disconnection is

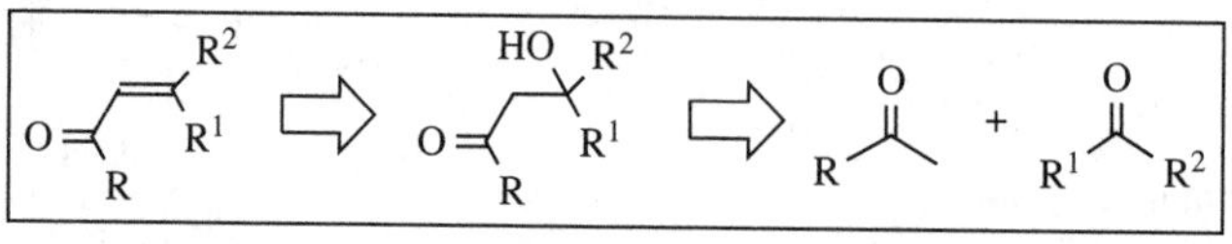

9.4.A.ii. Intramolecular Reactions. The intramolecular aldol condensation is particularly useful, and is illustrated by the simple reaction of 2,5-hexanedione with LDA to first give kinetic enolate **155**. This enolate could condense with a second molecule of dione to give an aldolate product, but intramolecular attack of the second carbonyl group is much faster and leads to the cyclic aldolate product **156**, which gives **157** upon hydrolysis. In Skrirsihna and Ramasastry's synthesis of (+)-2β-hydroxysolanascone,[118] intramolecular aldol condensation of **158** with piperidine as the base gave a 90% yield of **159**. The course of the intramolecular aldol condensation can be influenced by steric and conformational factors, as well as the rate of competing cyclization reactions. In a synthesis of (+)-dichroanone by Stoltz and McFadden,[119] diketone **160** was heated with excess KOH in xylenes, using a Dean-Stark trap to remove water, and the resulting intramolecular aldol condensation led to **161** in 80% yield. Note that water was eliminated upon workup.

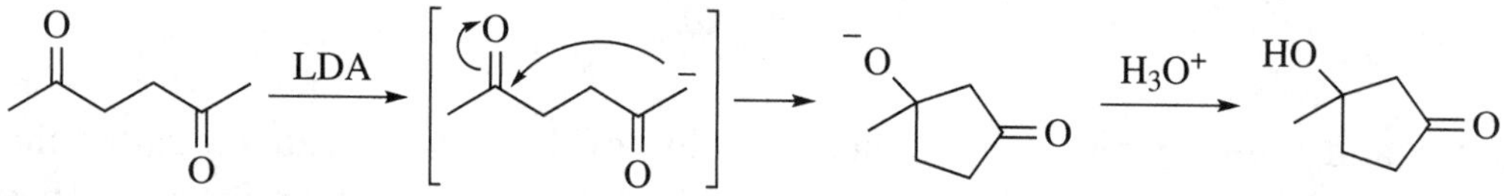

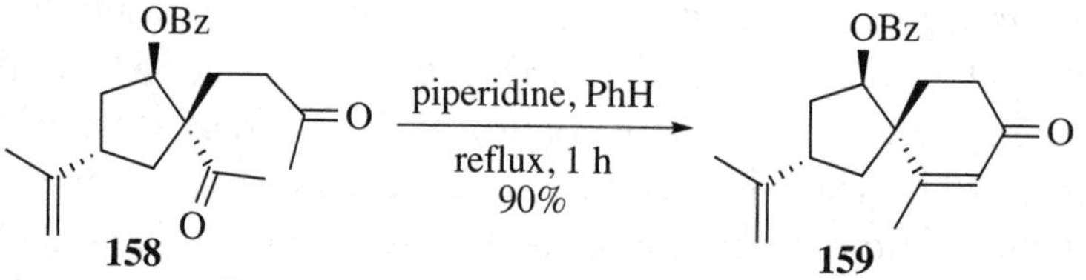

Another example illustrates some the problems inherent in the intramolecular reaction that do not arise with the intermolecular version. Deprotonation of **162** with sodium ethoxide under thermodynamic control conditions gave enolate anion **163** and isolated product **167** (water was lost from the initial aldolate product), which was used in Rouessac and Alexandre's

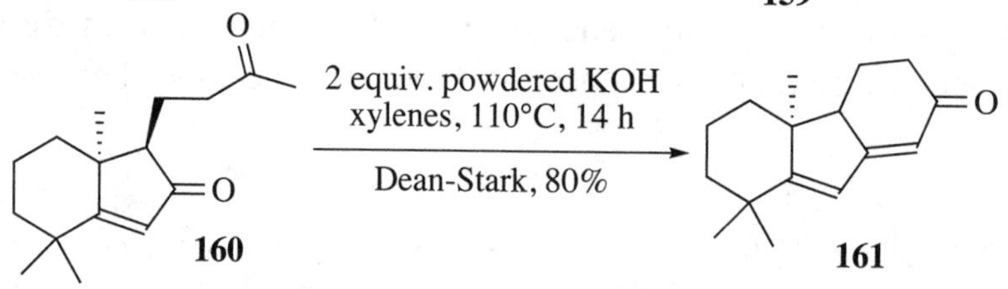

118. Srikrishna, A.; Ramasastry, S.S.V. *Tetrahedron Lett.* **2006,** *47,* 335.
119. McFadden, R.M.; Stoltz, B.M. *J. Am. Chem. Soc.* **2006,** *128* 7738.

synthesis of isoshyobunone.[120]   Under these reaction conditions, four different enolate anions (**163** and **168-170**) are possible via **162**. The intramolecular aldol condensation reaction with **168** and **170** would generate four-membered rings, which is energetically less favorable (secs. 1.5.B, 6.6.B) than formation of the six-membered rings that can be formed from the other enolates, **169** and **163**. Another product was formed in the enolate condensation reaction of **162**, dienone **167**, the kinetic product, relative to cyclization via **169**. When enolate **163** closed to form **164**, hydrolysis generated the aldolate **165**. Dehydration gave the ketone with the C=C unit conjugated to the carbonyl group (**166**), but the isopropenyl double bond remained out of conjugation. Under the acidic conditions of the workup, that double bond moved into conjugation to give the final product **167**. Despite several possible reaction pathways, this intramolecular aldol cyclization proceeded with high selectivity for a single product. The cyclization is governed by the factors discussed in section 6.6 for ring closures, which poses problems not encountered with the intermolecular aldol condensation.

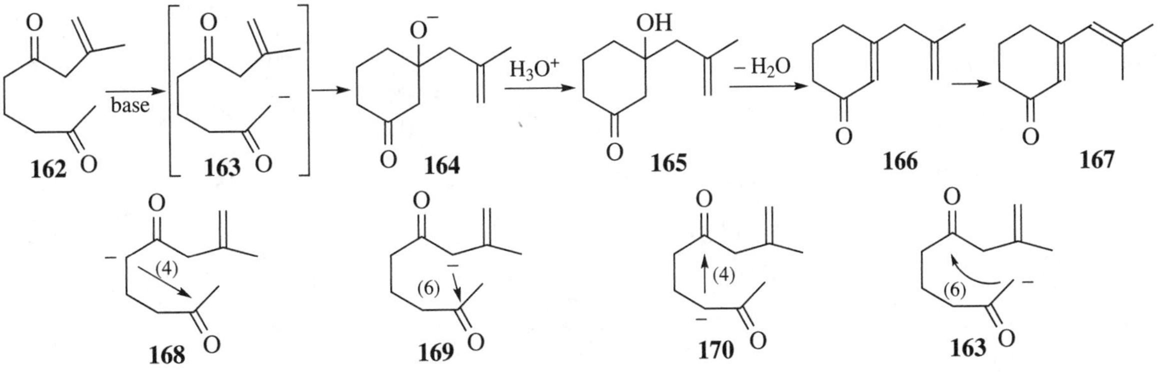

The disconnection for an intramolecular aldol condensation is:

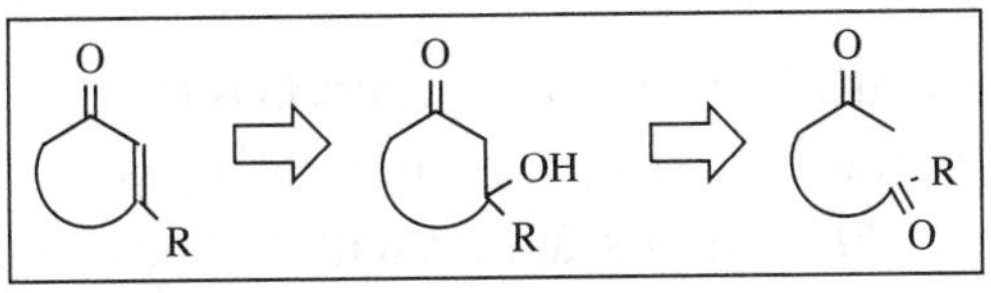

9.4.B.   Condensation Reactions of Acid Derivatives

Just as ketone and aldehyde enolates react with ketones and aldehydes, they can also react with esters and other acid derivatives.[121] Similarly, esters and other acid derivatives can be converted to their enolate anions, and then react with another acid derivative, or with an aldehyde or ketone. There are many variations of this reaction. When an acid derivative is attacked by a nucleophile, the alkoxide product contains a leaving group and this leads to acyl

120.   (a) Alexandre, C.; Rouessac, F. *J. Chem. Soc. Chem. Commun.* **1975**, 275; (b) *Idem Bull. Chim. Soc. Fr.* **1977**, 117.
121.   For reactions of this type, and related reactions, see Larock, R.C. *Comprehensive Organic Transformations, 2nd ed.,* Wiley-VCH, New York, **1999**, pp 1528-1546.

substitution as described for reactions of acid derivatives and Grignard reagents (sec. 8.4.C.ii) and organolithium reagents (sec. 8.5.E). This chemistry is summarized by the conversion of carbonyl derivative **171** to the enolate (**172**), and subsequent reaction with an acid derivative (**173**). Attack at the acyl carbon produces the usual alkoxide (**174**), but the X group is a leaving group (Cl, O$_2$CR, OR, NR$_2$) and expulsion of X leads to a 1,3-dicarbonyl compound **175**. This is, of course, the normal acyl substitution sequence. This section will examine those reactions where the carbonyl partner is an acid derivative and, in many cases, the enolate partner will also be an acid derivative. Esters, lactones, anhydrides, amides or lactams can be used as the enolate anion precursor, reacting with aldehydes, ketones or esters. Both intermolecular and intramolecular reactions are known. The remainder of this section will focus on a variety of carbonyl enolate precursors. To illustrate that virtually all acid derivatives can be used in this reaction, amide **176** was treated with LDA, and intramolecular condensation with the ketone unit gave a 92% yield of lactam-alcohol **177**, in McWhorter and Liu's synthesis of 8-desbromohinckdentine A.[122]

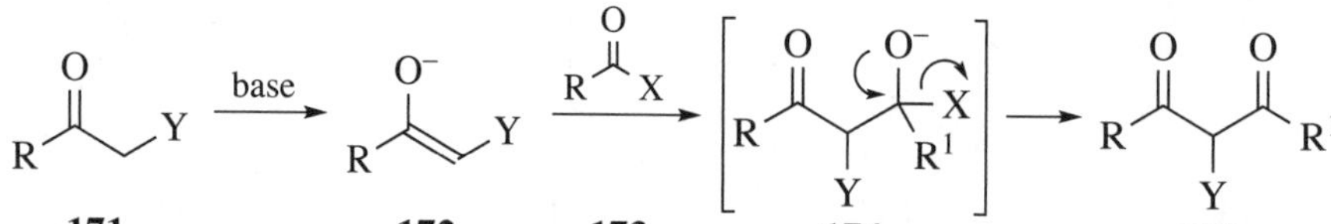

9.4.B.i. The Claisen Condensation. A classical reaction is the condensation of an ester enolate with an ester, illustrated by the self-condensation of ethyl butanoate in the presence of sodium ethoxide to give β-keto-ester **180**. Initial reaction with the base under thermodynamic control in this case, generated the enolate anion (**178**). This anion attacked the carbonyl of a second molecule of ethyl butanoate to give the tetrahedral intermediate **179**. Displacement of ethoxide generated ketone **180**. As shown here, this reaction is known as the **Claisen condensation**.[123]

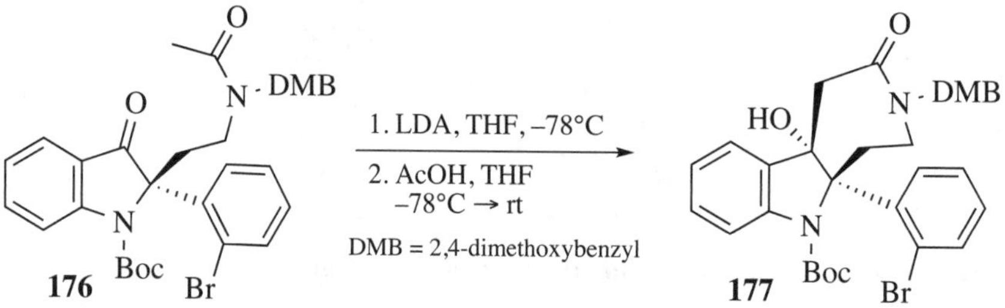

122. Liu, Y.; McWhorter Jr., W.W. *J. Am. Chem. Soc.* **2003**, *125*, 4240.
123. (a) Hellon, R.; Oppenheim, A. *Berichte,* **1877**, *10*, 699; (b) Israel, A. *Annalen* **1855**, *231*, 197; (c) Claisen, L. *Berichte,* **1887**, *20*, 655; (d) Beyer, C.; Claisen, L. *Ibid.* **1887**, *20*, 2178; (e) Claisen, L.; Stylos, N. *Ibid.* **1887**, *20*, 2188; (f) Claisen, L. *Annalen* **1894**, *281*, 306; (g) Reference 104, pp 94-96; (h) Reference 105, pp 49-51; (i) The Merck Index, 14th ed., Merck & Co., Inc., Whitehouse Station, New Jersey, **2006**, p ONR-18; (j) Mundy, B.P.; Ellerd, M.G.; Favaloro Jr., F.G. *Name Reactions and Reagents in Organic Synthesis, 2nd ed.*, Wiley-Interscience, New Jersey, **2005**, pp. 150-151.

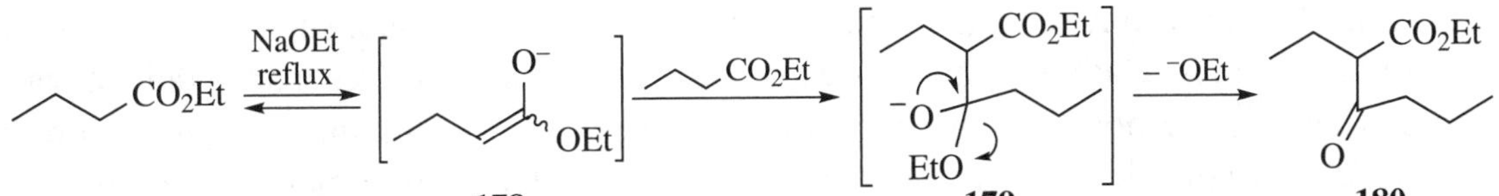

The reaction described for ethyl butanoate is a self-condensation, but as with the aldol condensation, Claisen condensation of two different esters can result in a mixture of products under thermodynamic control conditions. The reaction of two different esters is called the **crossed-Claisen** (or a **mixed Claisen**) condensation. A generalized reaction involving RCO_2Et and R^1CO_2Et can lead to at least four different condensation products. One ester (RCO_2Et) can condense with itself to give **181** or with R^1CO_2Et to give **182**. Similarly, ester R^1CO_2Et can condense with itself to give **183**, or with RCO_2Et to give **184**. As with the aldol condensation, kinetic control conditions are preferred if a particular crossed Claisen product is desired. A synthetic example is taken from Panek and co-worker's synthesis of leucascandrolide A,[124] in which the lithium enolate anion of *tert*-butyl acetate reacted with β-epoxy ester **185** in THF, to give a 70% yield of **186**. Self-condensation of the lithium enolate with the parent ester is sometimes a problem when LDA is used as a base,[125] but using the *tert*-butyl ester or LICA minimizes this competing process.[126] In general, forming the ester enolate under kinetic control conditions gives greater flexibility in the condensation, allowing the lithium enolate to be condensed with a variety of other esters (as in formation of **186**). It is noted that a full equivalent of base is required to form the enolate anion. The Claisen condensation can be used in incorporate an aldehyde unit into a molecule. In Padwa and Danca's synthesis of jamtine,[127] ester **187** was treated with LDA to form the enolate anion, and subsequent reaction with ethyl formate gave aldehyde **188** via the acyl substitution reaction.

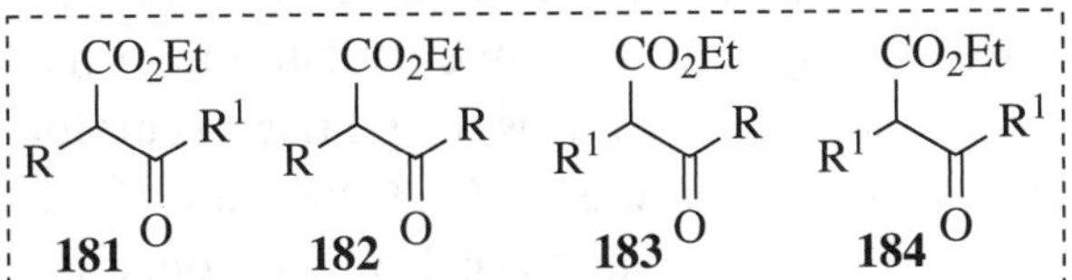

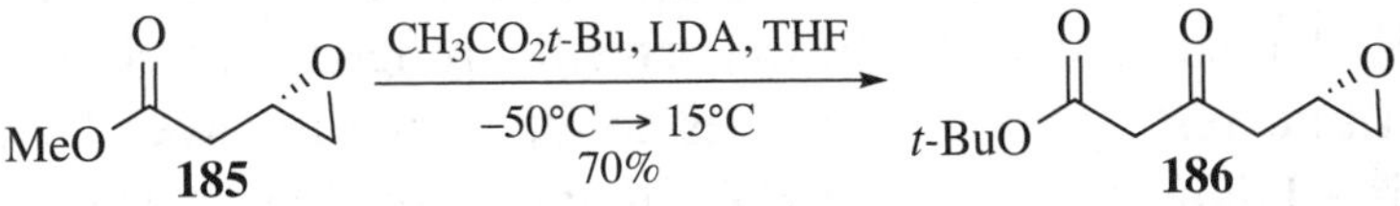

The α-proton in a normal Claisen product is more acidic than in the starting ester since it is adjacent to two carbonyls (see Table 9.1). Reaction with base under thermodynamic conditions

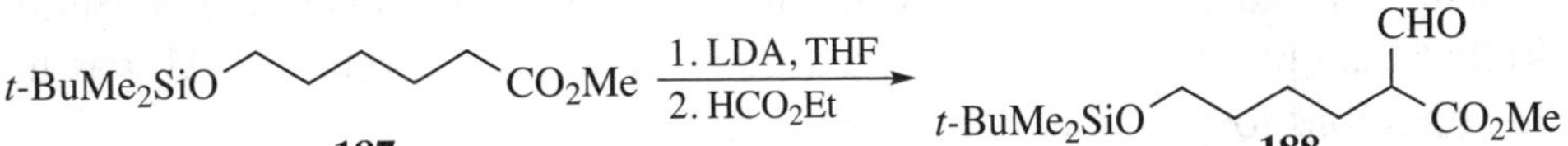

124. Su, Q.; Dakin, L.A.; Panek, J.S. *J. Org. Chem.* ***2007****, 72*, 2.
125. (a) Rathke, M.W.; Sullivan, D.F. *J. Am. Chem. Soc.* ***1973****, 95*, 3050; (b) Lochmann, L.; Lím, D. *J. Organomet. Chem.* ***1973****, 50*, 9; (c) Sullivan, D.F.; Woodbury, R.P.; Rathke, M.W. *J. Org. Chem.* ***1977****, 42*, 2038.
126. Rathke, M.W.; Lindert, A. *J. Am. Chem. Soc.* ***1971****, 93*, 2318.
127. Padwa, A.; Danca, M.D. *Org. Lett.* ***2002****, 4*, 715.

will give the 1,3-dicarbonyl enolate, where deprotonation of a product such as **186** is certain. The resulting anion is relatively stable, however, due to the increased resonance delocalization possible when two carbonyl groups are present (sec. 9.2.A). This means that the product enolate is *less reactive* than the enolate derived from the ester starting material, and generally does not give significant amounts of condensation product. The thermodynamic conditions that employ an alkoxide base require that the alcohol solvent be the conjugate acid of that base (methanol with sodium methoxide, for example), or transesterification will give two different esters for all products.

There are two important variations of this condensation. In the first, an ester enolate is condensed with a ketone or aldehyde. This has been called the **Claisen reaction**.[128] Pons and co-worker's synthesis of goniothalamin[129] condensed ethyl acetate with cinnamaldehyde to give an 85% yield of **189**. When the reaction is modified to use an aldehyde or ketone enolate rather than an ester enolate but the reactive partner is an ester, it is still referred to as the **Claisen reaction**. In this variant, a ketone or aldehyde enolate reacts with an ester to give a 1,3-diketone. In the Hua et al. synthesis of tetrahydro-1-oxpyranobenzopyrans, the dianion of **190** was condensed with ethyl pyridine-3-carboxylate gave an 88% yield of **191**,[130] which exists largely in the enol form shown.

Reaction of an ester enolate with an acid chloride will also generate a β-keto-ester and is a useful alternative to the Claisen condensation. Ketone enolates can also be condensed with acid chlorides.[131] An ester enolate can be trapped with trimethylsilyl chloride, as with aldehydes and ketones. An interesting variation of this condensation involved lactone **192**, which was condensed with a lithio-acetate (R = *t*-Bu, Et) in a modified Claisen condensation to give **193**.[132] Rather than the normal loss of alkoxide to give the β-diketone, **193** was hydrolyzed under mild conditions to give alkoxy hydroxy-ester **194**.

128. Claisen, L. *Berichte,* **1890**, *23,* 976.
129. Fournier, L.; Kocienski, P.; Pons, J.-M. *Tetrahedron* **2004**, *60,* 1659.
130. Hua, D.H.; Chen, Y.; Sin, H.-S.; Maroto, M.J.; Robinson, P.D.; Newell, S.W.; Perchellet, E.M.; Ladesich, J.B.; Freeman, J.A.; Perchellet, J.-P.; Chiang, P.K. *J. Org. Chem.* **1997**, *62,* 6888.
131. Beck, A.K.; Hoekstra, M.S.; Seebach, D. *Tetrahedron Lett.* **1977**, 1187.
132. Duggan, A.J.; Adams, M.A.; Brynes, P.J.; Meinwald, J. *Tetrahedron Lett.* **1978**, 4323.

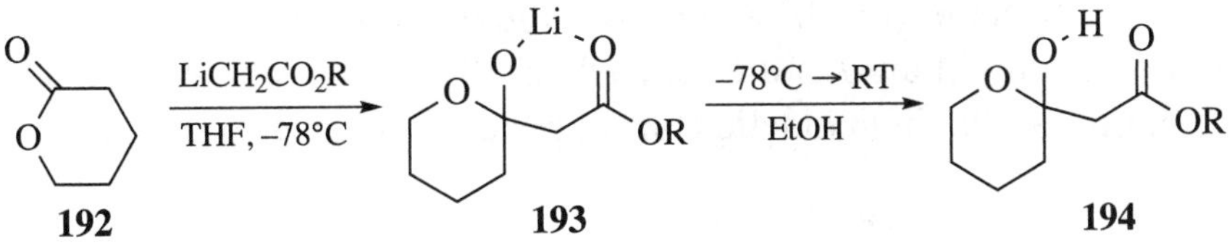

The Claisen disconnection is

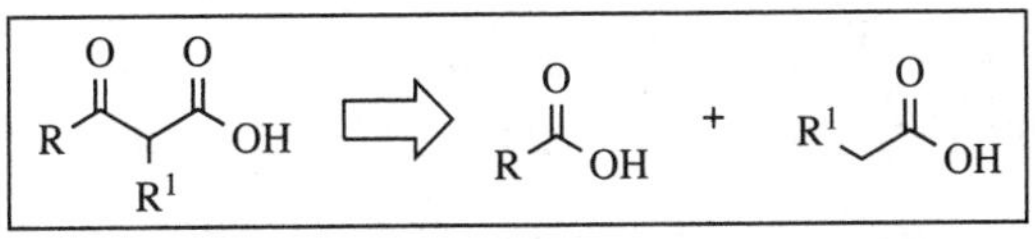

A variation of this condensation involves reaction with aldehydes, and it is called the **Perkin reaction**[133] (also see sec. 9.3.A). Condensation of an aldehyde (having no enolizable protons) with the enolate of an acid anhydride leads to an acetoxy ester such as **195**.[133a] Internal acyl substitution by the alkoxide forms the *O*-acetyl ester and liberates the carboxylate anion (**196**). Subsequent reaction with more acetic anhydride generates a new mixed anhydride **197**. Saponification leads to the β-hydroxy acid, which eliminates water in the acid hydrolysis step to give the final product, the aryl acrylic acid derivative (**198**, cinnamic acid).

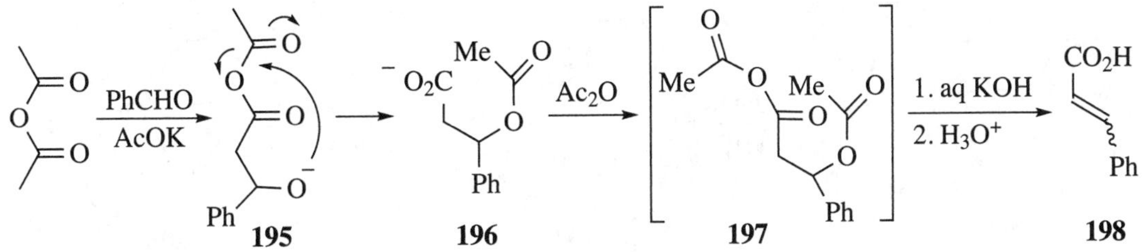

The Perkin disconnection is

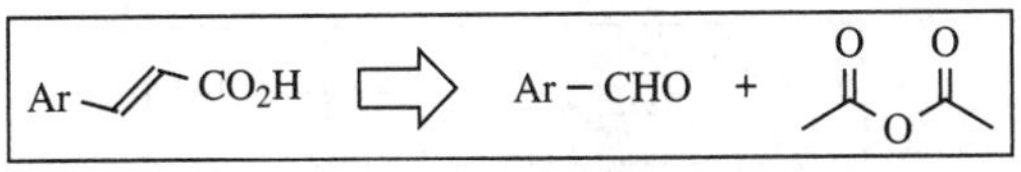

9.4.B.ii. The Dieckmann Condensation. Just as there is an intramolecular version of the aldol condensation, there is an intramolecular version of the Claisen condensation but it has

133. (a) Perkin, W.H. *J. Chem. Soc.* *1869*, *21*, 53, 181; (b) Johnson, J.R. *Org. React.* *1942*, *1*, 210; (c) Reference 104, pp. 344-346; (d) Reference 105, pp 184-186; (e) The Merck Index, 14th ed., Merck & Co., Inc., Whitehouse Station, New Jersey, *2006*, p ONR-71; (f) Mundy, B.P.; Ellerd, M.G.; Favaloro Jr., F.G. *Name Reactions and Reagents in Organic Synthesis, 2nd ed.*, Wiley-Interscience, New Jersey, *2005*, pp. 492-493.

been given a different name (the **Dieckmann condensation**).[134,135] This reaction involves intramolecular cyclization of an α,ω-diester such as **199**. The reaction is usually done under equilibrating conditions, although kinetic control conditions can also be used.

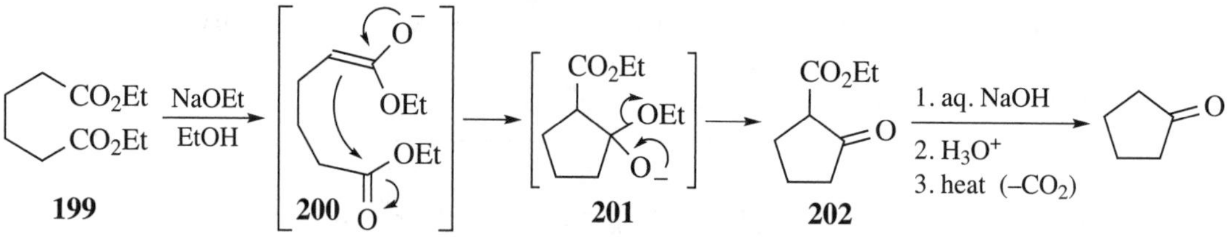

The initially formed enolate (**200**) attacked the ester group on the opposite side of the molecule to form a ring (tetrahedral intermediate **201**). Loss of ethoxide in the usual manner gave β-keto-ester **202**. Hydrolysis and thermally induced decarboxylation gave cyclopentanone.[134] A synthetic example is taken from Zhai and co-worker's synthesis of 19,20-dihydrosubincanadine B, in which diester **203** was treated with potassium *tert*-butoxide in THF and cyclization gave **204**. Subsequent heating with acetic acid-HCl led to formation of the carboxylic acid *in situ*, and decarboxylation to give the ketone **205** in 81% overall yield.[136] There are many synthetic applications of the Dieckmann cyclization.

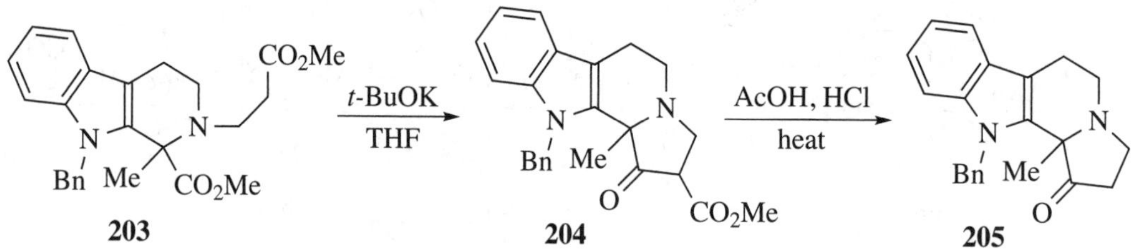

The Dieckmann disconnection is

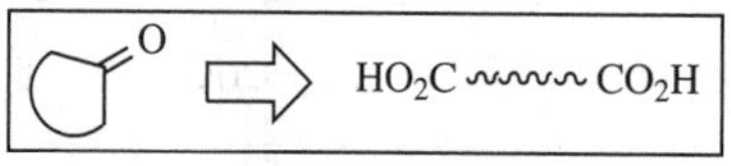

The size of the ring being formed has a great influence on the course of the reaction. When **206** was treated with sodium ethoxide, three isomeric Dieckmann products were possible, **207-209**. Reaction of the enolate via path *a* may generate a three-membered ring, which is energetically

134. (a) Dieckmann, W. *Berichte*, **1894**, *27*, 102, 965; (b) *Idem, Ibid.* **1900**, *33*, 2670; (c) Dieckmann, W.; Groenveld, A., *Ibid.* **1900**, *33*, 595; (d) Hauser, C.R.; Hudson, B.E. *Org. React.* **1942**, *1*, 266 (see p 274); (e) Schaefer, J.P.; Bloomfield, J.J. *Ibid.* **1967**, *15*, 1; (f) Reference 104, pp 124-125; (g) Reference 105, pp 75-77; (h) The Merck Index, 14th ed., Merck & Co., Inc., Whitehouse Station, New Jersey, **2006**, p ONR-23; (i) Mundy, B.P.; Ellerd, M.G.; Favaloro Jr., F.G. *Name Reactions and Reagents in Organic Synthesis, 2nd ed.*, Wiley-Interscience, New Jersey, **2005**, pp. 204-205.
135. For reactions of this type, see Larock, R.C. *Comprehensive Organic Transformations, 2nd ed.,* Wiley-VCH, New York, **1999**, pp 1316-1317.
136. Liu, Y.; Luo, S.; Fu, X.; Fang, F.; Zhuang, Z.; Xiong, W.; Jia, X.; Zhai, H. *Org. Lett.* **2006**, *8*, 115.

less favorable. Attack of the enolate via path *b* would generate **209**, but this is disfavored due to the sterically hindered center adjacent to the enolate carbon. This leaves reaction paths *c* (to give **207**) and *d* (to generate **208**) as viable possibilities. What is the product? Reaction of **206** under these conditions gave **208** as the only product.[137] The yield of this product varied with the base used to initiate enolate formation.[137] The presence of the methyl group in **206** influences the course of the reaction and similar cyclization with the triester lacking that methyl gave a different major product.[137]

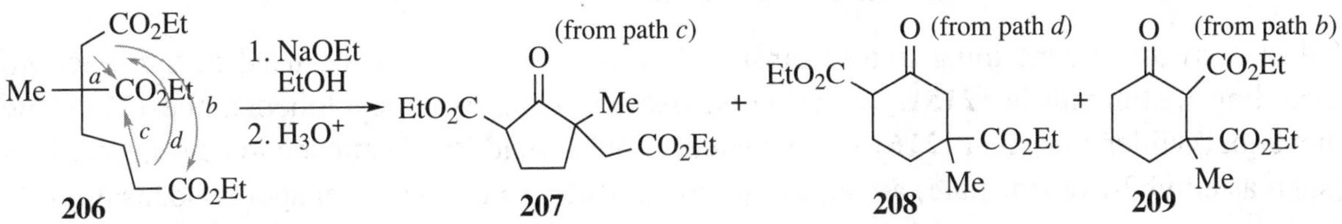

9.4.B.iii.   The Knoevenagel Condensation.

Another classical reaction, called the **Knoevenagel condensation**[138,133b] involves malonate enolates in a condensation reaction with aldehydes, usually a non-enolizable aldehyde. An aldehyde or ketone can be condensed with an active methylene compound (H_2CX_2 or $HRCX_2$) such as a malonic ester using a primary or secondary amine as the base. To avoid a competing aldol condensation, the aldehyde should have no α-protons. Diethyl malonate and benzaldehyde were condensed to form **210**, for example. Spontaneous elimination of water gave the alkylidene product (**211**), saponification followed by decarboxylation gave the final Knoevenagel product, conjugated acid **198**.

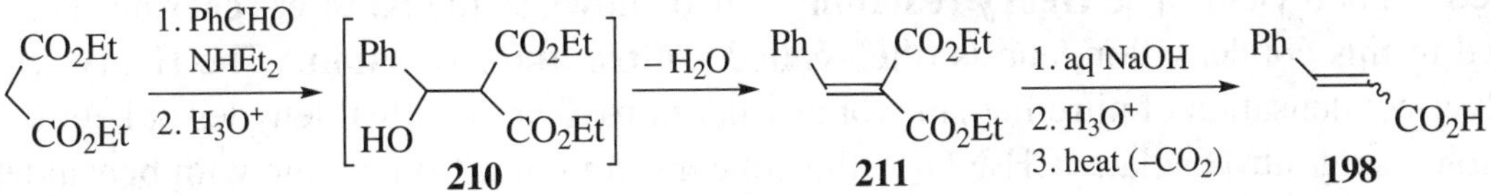

Contrary to aldol condensation without conjugation in the products, elimination commonly occurs upon hydrolysis of the Knoevenagel products. This is a facile route to substituted methylene malonic acids [$RCH=C(CO_2H)_2$] and acrylic acid derivatives. When the base is pyridine with a trace of piperidine (rather than diethylamine) and malonic acid is the enolate

137. (a) Goldberg, M.W.; Hunziker, F.; Billeter, J.R.; Rosenberg, H.R. *Helv. Chim. Acta* **1947**, *30*, 200; (b) Banerjee, D.K. *J. Indian Chem. Soc.* **1940**, *17*, 453; (c) Chakravarty, N.K.; Banerjee, D.K. *Ibid.* **1946**, *23*, 377; (d) Dutta, J.; Biswas, R.N. *Ibid.* **1961**, *38*, 335.
138. (a) Japp, F.R.; Streatfeild, F.W. *J. Chem. Soc.* **1883**, *43*, 27; (b) Knoevenagel, F. *Berichte* **1896**, *29*, 172; (c) *Idem, Ibid.* **1898**, *31*, 730; (d) Jones, G. *Org. React.* **1967**, *15*, 204; (e) Reference 104, pp 261-262; (f) Reference 105, pp 148-149; (g) The Merck Index, 14th ed., Merck & Co., Inc., Whitehouse Station, New Jersey, **2006**, p ONR-51; (h) Mundy, B.P.; Ellerd, M.G.; Favaloro Jr., F.G. *Name Reactions and Reagents in Organic Synthesis, 2nd ed.*, Wiley-Interscience, New Jersey, **2005**, pp. 364-365.

212 Ts   **213** Ts

precursor, the reaction is called the **Doebner condensation** or the **Doebner modification**.[139] In a synthesis of (+)-hapalindole Q by Kerr and Kinsman,[140] malonic acid reacted with aldehyde **212** in refluxing pyridine to give an 86% yield of **213**. Note that decarboxylation occurred under the reaction conditions. Kwon and co-workers showed that microwave irradiation facilitates the Knoevenagel condensation with aldehydes.[141]

Other derivatives containing two electron-withdrawing groups can be used in the reaction, including malononitrile (**215**). When condensed with 2,3-benzocycloheptanone (**214**), the dicyanoalkylidene product (**216**) was isolated in 46% yield.[142] Cyanoacetic acid derivatives (such as ethyl 2-cyanoacetate) are also common partners in the Knoevenagel condensation.[143]

214   **215**   **216**

As noted in Table 9.1, dinitromethane (**217**) has a pK_a of 3.6. Amines or even sodium carbonate are excellent bases for the deprotonation of **217** to give the resonance stabilized nitro-enolate **218**. When this enolate anion reacted with piperonal, the dinitroalkylidene product **219** was formed in good yield. The **Henry reaction**[144] (also known as the **Kamlet reaction**) is closely related to this condensation, and is often called a **nitro aldol reaction**. The Henry reaction involves condensation of nitromethane (or another nitroalkane) with aldehydes or ketones. An example using nitro enolate ⁻CH_2NO_2 is the condensation of nitromethane with benzaldehyde to give a 75% yield of 1-nitro-2-phenylethene.[145] Mono-nitro compounds such as nitrobutane

139. (a) Doebner, O. *Berichte, 1900, 33,* 2140; (b) The Merck Index, 14th ed., Merck & Co., Inc., Whitehouse Station, New Jersey, *2006,* p ONR-25; (c) Mundy, B.P.; Ellerd, M.G.; Favaloro Jr., F.G. *Name Reactions and Reagents in Organic Synthesis, 2nd ed.,* Wiley-Interscience, New Jersey, *2005,* pp. 212-213.
140. Kinsman, A.C.; Kerr, M.A. *J. Am. Chem. Soc. 2003, 125,* 14120.
141. (a) Kim, S.-Y.; Kwon, P.-S.; Kwon, T.-W. *Synth. Commun. 1997, 27,* 533; (b) Kwon, P.-S.; Kim, Y.-M.; Kang, C.-J.; Kwon, T.-W. *Synth. Commun. 1997, 27,* 4091.
142. Campaigne, E; Subramanya, R; Maulding, D.R. *J. Org. Chem. 1963, 28,* 623.
143. McElvain, S.M.; Lyle, Jr., R.E. *J. Am. Chem. Soc. 1950, 72,* 384.
144. (a) Henry, L. *Compt. Rend. 1895, 120,* 1265; (b) Kamlet, J. *U.S. Patent* 2,151,171, *1939* [*Chem. Abstr. 33:* 50039 *1939*]; (c) Hass, H.B.; Riley, E.F. *Chem. Rev. 1943, 32,* 373 (see p 406); (d) Lichtenthaler, F.W. *Angew. Chem. Int. Ed. 1964, 3,* 211. (e) For a review, see Luzzio, F.A. *Tetrahedron 2001, 57,* 915; (f) The Merck Index, 14th ed., Merck & Co., Inc., Whitehouse Station, New Jersey, *2006,* p ONR-43; (g) Mundy, B.P.; Ellerd, M.G.; Favaloro Jr., F.G. *Name Reactions and Reagents in Organic Synthesis, 2nd ed.,* Wiley-Interscience, New Jersey, *2005,* pp. 300-301.
145. Knoevenagel, E.; Walter, L. *Berichte, 1904, 37,* 4502.

can also react with a suitable base to generate an enolate anion, and then react with aldehydes or with ketones.[146] Several variations of this reaction include using microwave irradiation to assist the reaction[147] or Mg-Al hydrotalcites to induce condensation,[148] and using powdered KOH without solvent[149] or proazaphosphatranes [P(RNCH$_2$CH$_2$)$_3$N] to promote the reaction.[150] Asymmetric induction was observed in a synthesis of (–)-denopamine by Trost and co-workers[151] when aldehyde **220** reacted with nitromethane, in the presence of diethylzinc and a chiral 2,6-diaminophenol derivative, to give nitro-aldol product **221** in 88% yield (90% ee).

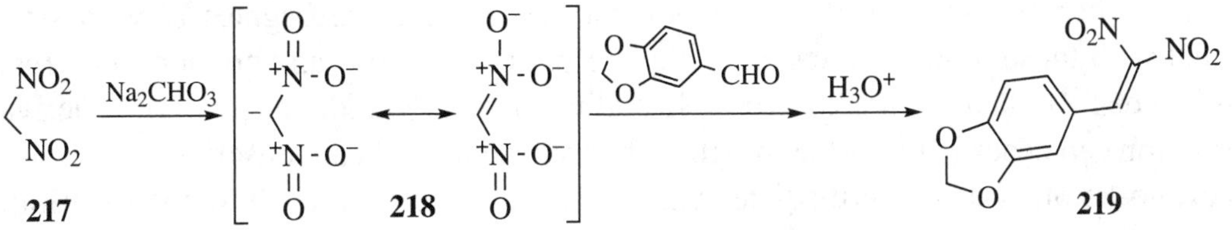

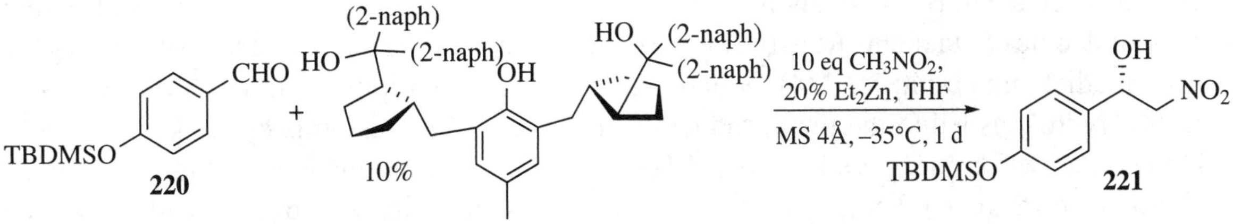

Note that nitro enolates have other synthetic uses. When nitrobutane was treated with sodium hydroxide, nitro-enolate **222** was formed. Rather than addition of an aldehyde or a ketone, **222** was treated with concentrated sulfuric acid to form butanal with loss of N$_2$O, in what is known as the **Nef reaction**.[152] Modern versions of this reaction use bases such as LDA and less vigorous oxidizing agents such as MoOPh.[153]

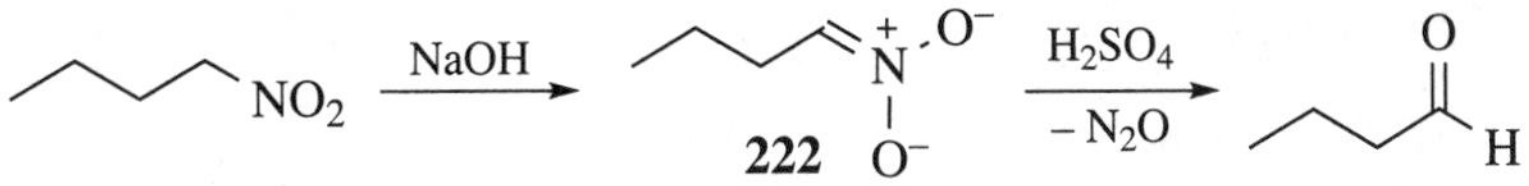

The Knoevenagel disconnection is:

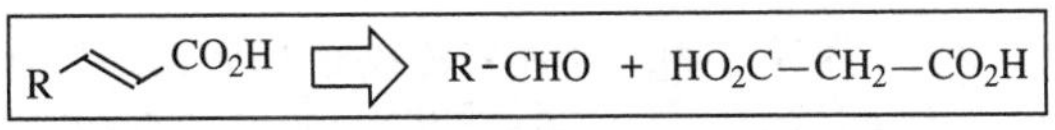

146. Hass, H.B.; Susie, A.G.; Heider, R.L. *J. Org. Chem.* **1950**, *15*, 8.
147. Varma, R.S.; Dahiya, R.; Kumar, S. *Tetrahedron Lett.* **1997**, *38*, 5131.
148. Bulbule, V.J.; Deshpande, V.H.; Velu, S.; Sudalai, A.; Sivasankar, S.; Sathe, V.T. *Tetrahedron* **1999**, *55*, 9325.
149. Ballini, R.; Bosica, G.; Parrini, M. *Chem. Lett.* **1999**, 1105.
150. Kisanga, P.B.; Verkade, J.G. *J. Org. Chem.* **1999**, *64*, 4298.
151. Trost, B.M.; Yeh, V.S.C.; Ito, H.; Bremeyer, N. *Org. Lett.* **2002**, *4*, 2621.
152. (a) Nef, J.U. *Annalen* **1894**, *280*, 263; (b) Noland, W.E. *Chem. Rev.* **1955**, *55*, 137; (c) The Merck Index, 14th ed., Merck & Co., Inc., Whitehouse Station, New Jersey, **2006**, p ONR-64; (d) Mundy, B.P.; Ellerd, M.G.; Favaloro Jr., F.G. *Name Reactions and Reagents in Organic Synthesis, 2nd ed.*, Wiley-Interscience, New Jersey, **2005**, pp. 452-453.
153. Gabobardes, M.; Pinnick, H. *Tetrahedron Lett.* **1981**, *22*, 5235.

An interesting variation of the Knoevenagel condensation generates aldol condensation products (see Sec. 9.4.A). Condensation of β-keto esters and aldehydes at pH 7.8 in aqueous media gives the aldolate. The condensation of ethyl 3-oxobutanoate and pentanal gave 4-hydroxy-2-octanone in 80% yield.[154] This new methodology constitutes a regioselective alternative to the aldol condensation. Note that heating the β-hydroxy-ketone at pH 1 liberates a conjugated ketone, prepared from the keto-ester and aldehyde in a one-pot procedure.

9.4.B.iv. Nitrile Enolates. Nitrile enolates are formed by reaction of a nitrile with LDA or another suitable base. Nitrile enolates exist and react as aggregates,[155] with structures consistent with those found in the solid state.[156] Both alkylation[157] and condensation reactions with aldehydes[158] or ketones are known.[159] In addition to alkyl halides and carbonyl derivatives, condensation can occur with another nitrile. The base-catalyzed condensation of two nitriles to give a cyano-ketone via an intermediate cyano enolate, is known as the **Thorpe reaction**.[160,134e] Reaction of butanenitrile with sodium ethoxide gave a nitrile enolate, which reacted with a second molecule of butanenitrile at the electrophilic cyano carbon to give **223**. Hydrolysis gave an intermediate imine-nitrile (**224**), which is in equilibrium with the enamine form (**225**, sec. 9.6.A). Hydrolysis led to the final product of the Thorpe reaction, an α-cyano ketone **226**.[160] Mixed condensations are possible when LDA and kinetic conditions are used to generate the α-lithionitrile (a mixed Thorpe reaction). When pentanenitrile was treated with LDA and condensed with benzonitrile, 2-cyano-1-phenyl-1-pentanone was the isolated product after acid hydrolysis. Nitrile enolates can also be alkylated with a variety of alkyl halides.[161]

154. Kourouli, T.; Kefalas, P.; Ragoussis N.; Ragoussis V. *J. Org. Chem.* **2002**, *67*, 4615.
155. Fleming, F.F.; Shook, B.C. *J. Org. Chem.* **2002**, *67*, 2885, and references 2b,c therein.
156. Carlier, P.R.; Lo, K.M. *J. Org. Chem.* **1994**, *59*, 4053.
157. For a synthetic example taken from a synthesis of (+)-neosymbioimine, see Varseev, G.N.; Maier, M.E. *Org. Lett.* **2007**, *9*, 1461.
158. For a synthetic example taken from a synthesis of AM6898D, see Fukuda, Y.; Sakurai, M.; Okamoto, Y. *Tetrahedron Lett.* **2000**, *41*, 4173.
159. For reactions of this type, see Larock, R.C. *Comprehensive Organic Transformations, 2nd ed.,* Wiley-VCH, New York, **1999**, p 1317.
160. (a) Reference 104, p 449; (b) Baron H.; Remfry, F.G.P.; Thorpe, J.F. *J. Chem. Soc.* **1904**, *85*, 1726; (c) The Merck Index, 14th ed., Merck & Co., Inc., Whitehouse Station, New Jersey, **2006**, p ONR-93; (d) Mundy, B.P.; Ellerd, M.G.; Favaloro Jr., F.G. *Name Reactions and Reagents in Organic Synthesis, 2nd ed.,* Wiley-Interscience, New Jersey, **2005**, pp. 646-647.
161. Larchevêque, M.; Cuvigny, T. *Tetrahedron Lett.* **1975**, 3851.

The Thorpe disconnection is:

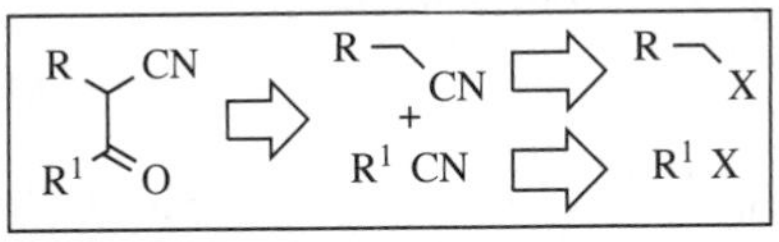

There is an intramolecular version of the Thorpe reaction called the **Thorpe-Ziegler reaction**.[162] When an α,ω-dinitrile is treated with base, formation of the enolate is followed by cyclization. When adiponitrile (**227**) was treated with sodium ethoxide, cyclization gave the usual imine-enamine mixture (**228**). Hydrolysis gave the cyano-ketone (2-cyanocyclopentanone, **229**).[162b] In Deslongchamp's synthesis of (+)-martimol,[75] dinitrile **230** was treated with potassium *tert*-butoxide in *tert*-butanol at 85°C, and after heating to 115°C with acetic acid-phosphoric acid ketone **231** was obtained in 68% yield.

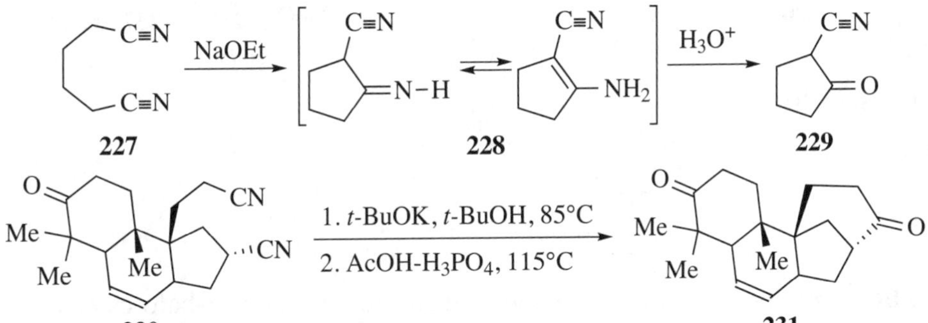

The Thorpe-Ziegler disconnection is

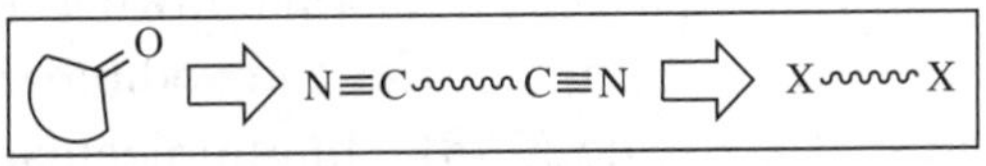

9.4.B.v. The Stobbe Condensation. Condensation of succinic ester derivatives (such as diethyl succinate, **232**) with non-enolizable ketones or aldehydes and a base gives the condensation product **233**. The alkoxide reacts with the distal ester via acyl substitution to give a lactone intermediate (**234**). In the original version of this reaction, saponification of **234** gave the α-alkylidene monoester **235**. The reaction is limited to those α,ω-diesters for which the Dieckmann condensation is not a competitive reaction so succinic acid derivatives are commonly used. This transformation is known as the **Stobbe condensation**.[163] The aldehyde or ketone substrate is not limited to non-enolizable derivatives. An example is the double-

162. (a) Reference 104, pp 514-516; (b) Ziegler, K.; Eberle, H.; Ohlinger, H. *Annalen* **1933**, *504*, 94.
163. (a) Stobbe, H. *Berichte* **1893**, *26*, 2312; (b) Stobbe, H. *Annalen* **1894**, *282*, 280; (c) Johnson, W.S.; Daub, G.H. *Org. React.* **1951**, *6*, 1; (d) Reference 104, pp 439-440; (e) Reference 105, pp 228-230; (f) The Merck Index, 14th ed., Merck & Co., Inc., Whitehouse Station, New Jersey, **2006**, p ONR-90; (g) Mundy, B.P.; Ellerd, M.G.; Favaloro Jr., F.G. *Name Reactions and Reagents in Organic Synthesis, 2nd ed.*, Wiley-Interscience, New Jersey, **2005**, pp. 626-627.

Stobbe condensation of aldehyde **236** with diethyl succinate (sodium ethoxide was used as a base) to give **237**, taken from the Robinson and co-worker's synthesis[164] of dicaffeoyltartaric acid analogs.

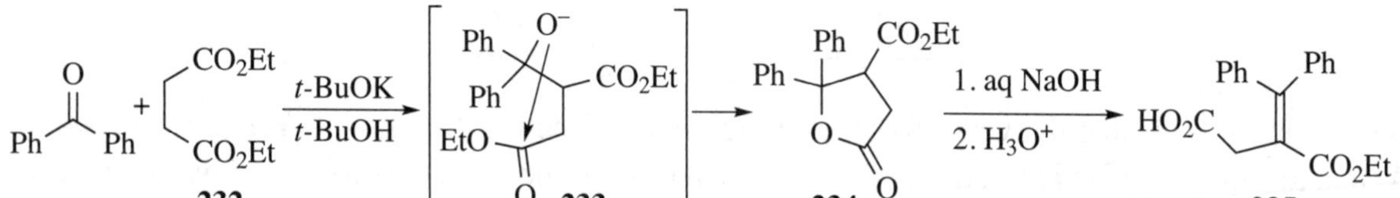

The Stobbe disconnection is

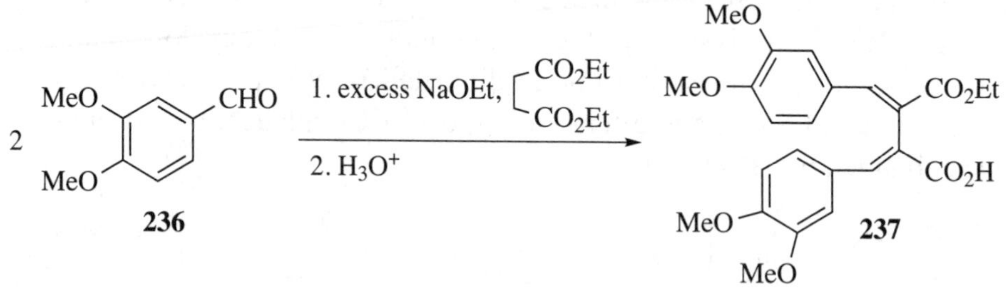

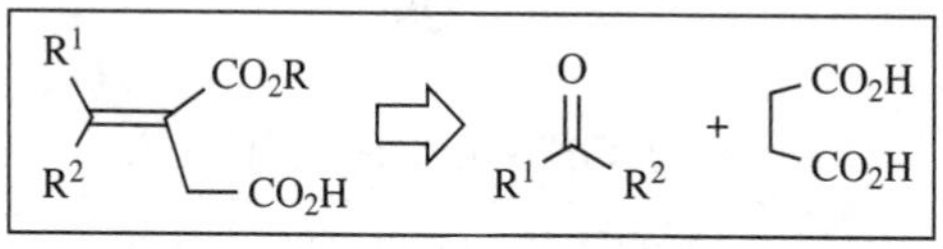

9.4.B.vi. The Darzens' Glycidic Ester Condensation. When a α-halo ester is treated with base and the resulting enolate anion condensed with a carbonyl derivative, the product is a halo-alkoxide. This nucleophilic species can displace the halogen intramolecularly to produce an epoxide, which forms the basis of a classical reaction known as the **Darzens' glycidic ester condensation.**[165] Reaction of ethyl α-chloroacetate and sodium ethoxide in the presence of benzaldehyde generated the usual alkoxide (**238**). Intramolecular displacement of chloride, however, gave the glycidic ester (**239**). Saponification gave the epoxy acid (**240**), which was

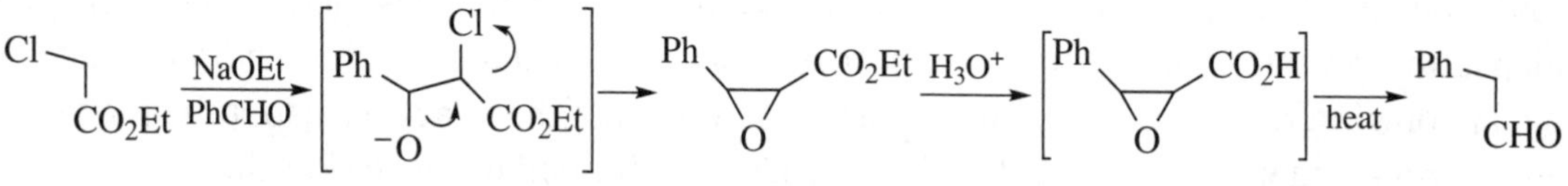

unstable and lost carbon dioxide to generate a substituted aldehyde or ketone such as phenyl

---

164. Reinke, R.A.; King, P.J.; Victoria, J.G.; McDougall, B.R.; Ma, G.; Mao, Y.; Reinecke, M.G.; Robinson, Jr., W.E. *J. Med. Chem.* **2002**, *45*, 3669.

165. (a) Darzens, G. *Compt. Rend.* **1904**, *139*, 1214; (b) *Idem Ibid.* **1906**, *142*, 214; (c) Ballester, M. *Chem. Rev.* **1955**, *55*, 283; (d) Reference 104, pp 116-118; (e) Reference 105, pp 68-70; (f) The Merck Index, 14th ed., Merck & Co., Inc., Whitehouse Station, New Jersey, **2006**, p ONR-22; (g) Mundy, B.P.; Ellerd, M.G.; Favaloro Jr., F.G. *Name Reactions and Reagents in Organic Synthesis, 2nd ed.*, Wiley-Interscience, New Jersey, **2005**, pp. 198-199.

acetaldehyde in this case. This sequence is a useful **chain extension reaction** in which an aldehyde (PhCHO) is converted to a longer chain aldehyde homolog (PhCH$_2$CHO). A ketone is converted to a longer chain aldehyde. An enantioselective synthesis of glycidic amides using camphor-derived sulfonium salts has been reported.[166] Cyclohexanone, for example, reacted with ethyl α-chloroacetate in the presence of *tert*-butoxide to give cyclohexane carboxaldehyde via hydrolysis of the initially formed glycidic ester.[167] Using a α-alkyl-α-chloroester in this sequence leads to a ketone as the final product. An asymmetric version of this reaction isknown, in which (–)-8-phenylmenthyl ester of chloroacetic acid (**241**) reacted with acetone to give a 64% yield of **242**, in 87% de.[168] A catalytic asymmetric Darzens condensation[169] has been reported using a chiral phase transfer agent. Note that a Wittig reaction that produces a vinyl ether (Ph$_3$P=CHOR + ketone → R$_2$C=CHOR) that can be hydrolyzed to a chain extended aldehyde (sec. 8.8.A.i). It is an attractive alternative to the Darzens' glycidic ester condensation.

The Darzens' disconnection is

9.4.B.vii. Acid Dianions. All of the named reactions discussed in Section 9.4 constitute relatively minor variations of the fundamental condensation reaction of aldehydes, ketones or acid derivatives with another aldehyde, ketone, or acid derivative. The ability to produce kinetic enolates from acid derivatives has made possible another useful modification of the enolate reaction. Carboxylic acids have an acidic proton that is removed by one equivalent of base to first give a carboxylate (see **243**). Addition of a second equivalent of a powerful base such as a dialkylamide leads to the dianion (**244**). Subsequent reaction with an electrophilic

$$H_3C-CO_2H \xrightarrow[\text{THF}]{\text{LDA}} \left[H_3C-CO_2^-\right] \xrightarrow[\text{HMPA, 50°C}]{\text{LDA, THF}} \left[H_2\bar{C}-CO_2^-\right] \xrightarrow[\text{2. aq HCl}]{\text{1. } n\text{-C}_4\text{H}_9\text{Br}} n\text{-C}_4\text{H}_9\text{-HC}-CO_2H$$
$$\mathbf{243}\mathbf{244}$$

species, in this case 1-bromobutane, occurred first at the more nucleophilic α-carbon to give hexanoic acid.[170] The carboxylate is usually generated with *n*-butyllithium and the enolate with LDA, although two equivalents of LDA can be used. As discussed in Chapter 8, treatment

166. Aggarwal, V.K.; Charmant, J.P.H.; Fuentes, D.; Harvey, J.N.; Hynd, G.; Ohara, D.; Picoul, W.; Robiette, R.; Smith, C.; Vasse, J.-L.; Winn, C.L. *J. Am. Chem. Soc.* **2006**, *128*, 2105.
167. Hunt, R.H.; Chinn, L.J.; Johnson, W.S. *Org. Synth., Coll. Vol. 4* **1963**, 459.
168. Ohkata, K.; Kimura, J.; Shinohara, Y.; Takagi, R.; Hiraga, Y. *Chem. Commun.* **1996**, 2411.
169. Arai, S.; Shioiri, T. *Tetrahedron Lett.* **1998**, *39*, 2145.
170. (a) Reference 2b, p 158; (b) Pfeffer, P.E.; Silbert, L.S.; Chirinko, Jr., J.M. *J. Org. Chem.* **1972**, *37*, 451.

of a carboxylic acid with an organolithium reagent usually gives the ketone, via reaction of the carboxylate (**243**) with a second equivalent of an organolithium reagent (sec. 8.5.E).[171] For this reason, dialkylamides are commonly employed for the second deprotonation.[172] An acid dianion gives the typical reactions of enolates derived from acid derivatives.[173] When 3-methylbutanoic acid was treated first with sodium hydride and then with LDA, dianion **245** was formed.[174] The more acidic carboxylic acid proton was removed by the first equivalent of base, and least acidic hydrogen (the α-hydrogen) was removed with the send equivalent of base, and that position is the more nucleophilic center in **245**. Therefore, subsequent reaction with 4-bromo-1-butene occurred exclusively at the α-position, and hydrolysis gave an 95% yield of **246**, as reported in Koskinen and co-worker's synthesis of nor-1,6-germacradien-5-ols.[174]

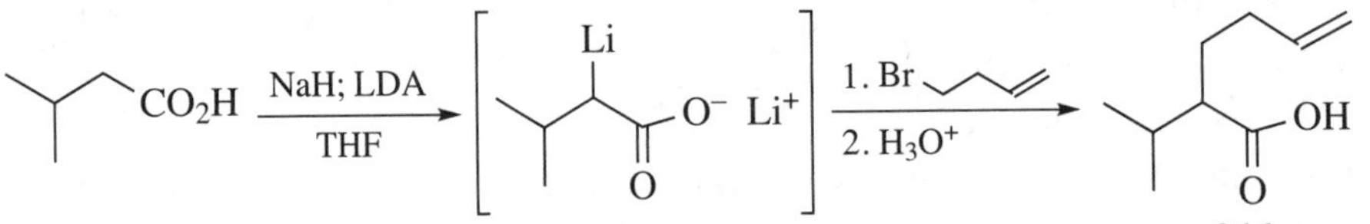

Dianions of this type react with ketones,[175] epoxides[176] or esters[177] as well as a wide variety of other electrophiles. The dilithio derivative of 2-methylpropionic acid was condensed with the epoxide moiety in **247** to form an hydroxy acid, which cyclized to form the lactone ring in **248**.[176] Since most of the enolates of acid derivatives contain a leaving group, the alkoxide resulting from reaction with an epoxide often displaces that leaving group to give the lactone.

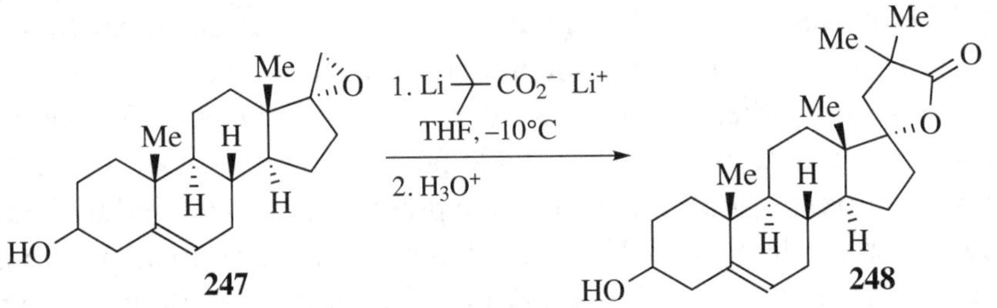

The condensation of enolates with alkyl halides or other carbonyl derivatives allows a wide variety of synthetic and functional group transformations in the carbon-carbon bond-forming process. Enolates are, therefore, among the most powerful synthetic intermediates known. In addition to generating a new carbon-carbon bond, the reaction proceeds with high diastereoselectivity in most cases, making it even more useful.

171. Heathcock, C.H.; Gulik, L.G.; Dehlinger, T. *J. Heterocyclic Chem.* **1969**, *6*, 141.
172. For a convenient preparation, see Parra, M.; Sotora, E.; Gil, S. *Eur. J. Org. Chem.* **2003**, 1386.
173. Petragnani, N.; Yonashiro, M. *Synthesis* **1982**, 521.
174. Nevalainen, M.; Koskinen, A.M.P. *J. Org. Chem,* **2002**, *67*, 1554.
175. Moersch, G.W.; Burkett, A.R. *J. Org. Chem.* **1971**, *36*, 1149.
176. (a) Creger, P.L. *J. Org. Chem.* **1972**, *37*, 1907; (b) Creger, P.L. *J. Am. Chem. Soc.* **1967**, *89*, 2500.
177. (a) Kuo, Y.N.; Yahner, J.A.; Ainsworth, C. *J. Am. Chem. Soc.* **1971**, *93*, 6321; (b) Pfeffer, P.F.; Silbert, L.S. *Tetrahedron Lett.* **1970**, 699.

to the reaction medium, good diastereoselectivity can be obtained (sec. 9.5.A.v, see below), and some asymmetric induction is possible when chiral catalysts are used. Stereochemically pure aldehydes such as **258** lead to good diastereoselectivity. The reaction of benzaldehyde with **260** gave a 74% yield of **262** (3.2:1 syn/anti),[185] in the presence of cupric triflate and bis(oxazolidine) catalyst **261**. Interestingly, this reaction was done in aqueous media. A water-accelerated reaction of ketene silyl acetals has been reported.[186]

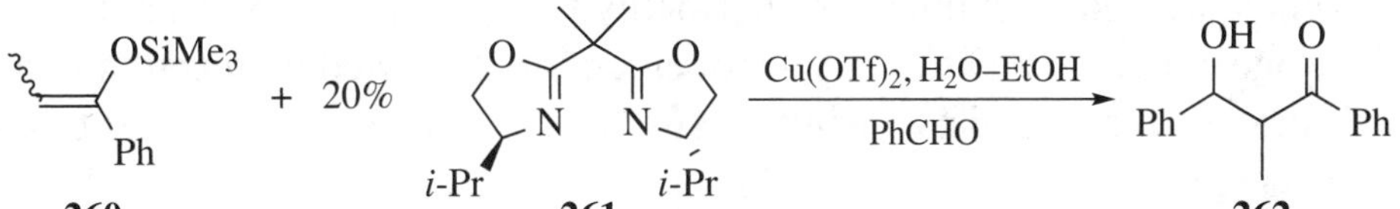

The bis(oxazolidine) catalyst can be applied to silyl enol ethers as well, and good asymmetric induction can be achieved. (*R*)-BINOL-Titanium catalysts have been used in solvents such as toluene or CH_2Cl_2,[187] and also in supercritical fluoroform and in supercritical carbon dioxide. Other reactions have been reported using catalysts closely related to **261** and dichloromethane as the solvent.[188] Chiral oxazaborolidinone derivatives are effective catalysts for enantioselective Mukaiyama aldol reactions.[189] In a synthesis of halipeptins A and D by Nicolaou and co-workers,[190] aldehyde **263** was condensed with $Me_2C=C(OMe)OTMS$ in the presence of **264** to give **265** in 87% yield and about 95:5 diastereoselectivity.

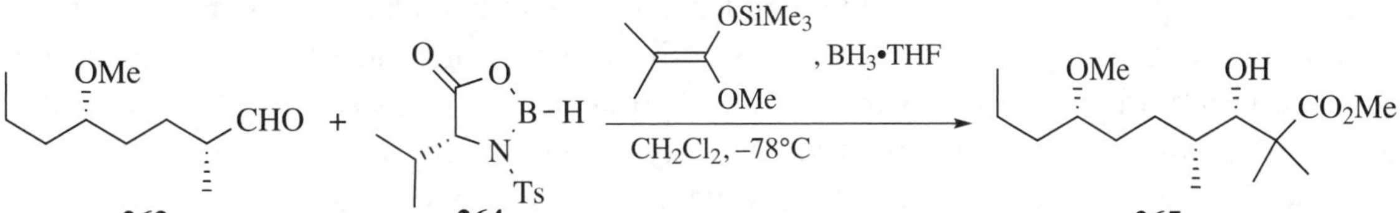

A variation of the Mukaiyama aldol reaction involves the reaction of silyl enol ethers with acetals,[191] in the presence of $TiCl_4$. An example is the reaction of silyl-enol ether **266** with 1,1-dimethoxycyclohexane (**267**, a protected cyclohexanone, sec. 7.3.B.i) to give **268** in 91% yield.[192,193,194]

185. (a) Kobayashi, S.; Nagayama, S.; Busujima, T. *Chem. Lett.* **1999**, 71; (b) *Idem Tetrahedron* **1999**, *55*, 8739.
186. Loh, T.-P.; Feng, L.-C.; Wei, L.-L. *Tetrahedron* **2000**, *56*, 7309.
187. Mikami, K.; Matsukawa, S.; Kayaki, Y.; Ikariya, T. *Tetrahedron Lett.* **2000**, *41*, 1931.
188. Evans, D.A.; Burgey, C.S.; Kozlowski, M.C.; Tregay, S.W. *J. Am. Chem. Soc.* **1999**, *121*, 686.
189. Ishihara, K.; Kondo, S.; Yamamoto, H. *J. Org. Chem.* **2000**, *65*, 9125.
190. Nicolaou, K.C.; Lizos, D.E.; Kim, D.W.; Schlawe, D.; de Noronha, R.G.; Longbottom, D.A.; Rodriquez, M.; Bucci, M.; Cirino, G.*J. Am. Chem. Soc.* **2006**, *128*, 4460.
191. Mukaiyama, T.; Murakami, M. *Synthesis* **1987**, 1043.
192. Mukaiyama, T.; Hayashi, M. *Chem. Lett.* **1974**, 15.
193. Ishihara, H.; Inomata, K.; Mukaiyama, T. *Chem. Lett.* **1975**, 531.
194. Mukaiyama, T.; Ishihara, H.; Inomata, K. *Chem. Lett.* **1975**, 527.

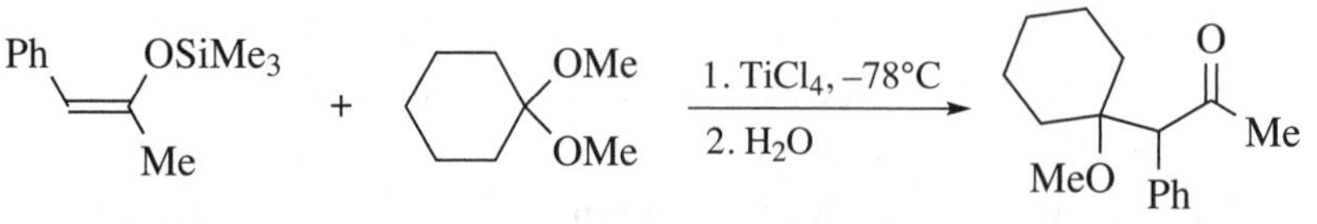

266 + **267** → **268**

Mukaiyama conditions were used intramolecularly for the successful closure of an 11-membered ring in Smith and co-worker's synthesis of normethyljatrophone.[195] Conversion of the alkynyl conjugated yn-one moiety in **269** to a silyl-enol ether, was followed by reaction with TiCl₄ and reaction with the dioxolane moiety. Ring closure (sec. 7.3.B.i for formation of dioxolanes) gave **270** in 47% yield.[195]

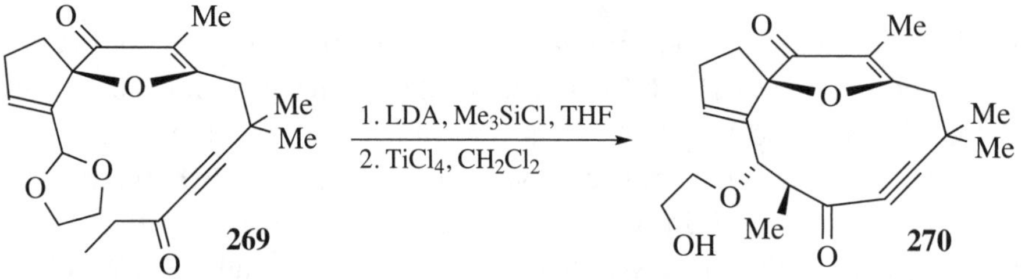

Conjugated ketones react with silyl-enol ethers under Mukaiyama conditions to give the 1,4-addition product in high yield. Reaction of **254** with **271**, for example, gave a 95% yield of the 1,4-product (**272**).[196] Aldol condensations with conjugated systems often lead to 1,4-addition of the enolate anion (sec. 9.7.A). Since the aldol condensation is usually done under basic conditions, further reaction of the products including self-condensation of the starting materials is commonly observed. Mukaiyama conditions are relatively mild, and the 1,4-addition proceeds in good yield with few deleterious side reactions.

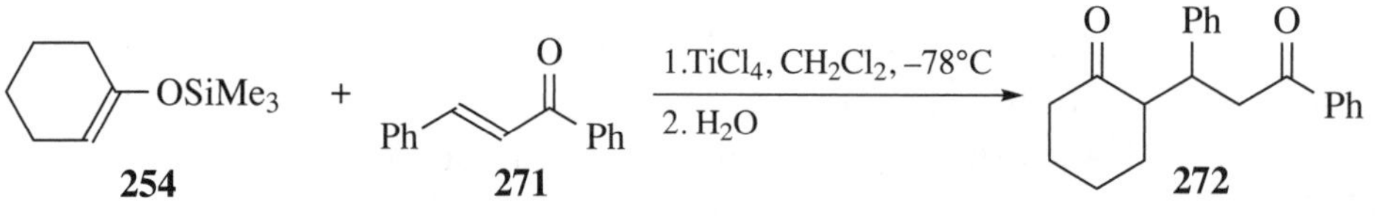

The Mukaiyama reaction can also be applied to ketene acetals such as **274**, derived from the enolate anion of an ester such as that in **273**. In a synthesis of a novel angiogenesis inhibitor (–)-azaspirene by Hayashi and co-workers,[197] **273** reacted with LDA to give the enolate anion and reaction with with *tert*-butyldimethylsilyl triflate gave **274**. Subsequent condensation with phenylpropargyl aldehyde, mediated by magnesium bromide, gave **275** in 69% yield (about 92% de). Other Lewis acids were used in this reaction, but they gave **275** with much poorer

195. Smith III, A.B.; Guaciaro, M.A.; Schow, S.R.; Wovkulich, P.M.; Toder, B.H.; Hall, T.W. *J. Am. Chem. Soc.* **1981**, *103*, 219.
196. (a) Narasaka, K.; Soai, K.; Mukaiyama, T. *Chem. Lett.* **1974**, 1223; (b) Narasaka, K.; Soai, K.; Aikawa, Y.; Mukaiyama, T. *Bull. Chem. Soc. Jpn.* **1976**, *49*, 779.
197. Hayashi, Y.; Shoji, M.; Yamaguchi, J.; Sato, K.; Yamaguchi, S.; Mukaiyama, T.; Sakai, K.; Asami, Y.; Kakeya, H.; Osada, H. *J. Am. Chem. Soc.* **2002**, *124*, 12078.

diasteroselectivity. This variation is very useful since esters can be used as ketene acetal precursors, greatly expanding the utility of the reaction.

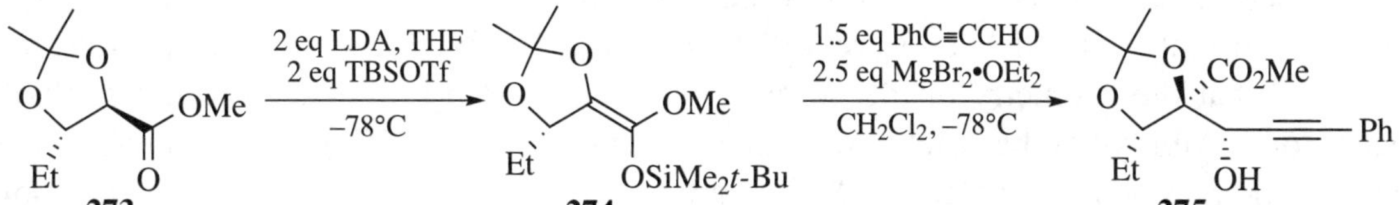

The disconnections are identical to those shown for the aldol condensation.

9.4.D. Boron Enolates

Vinyl boronates (boronic esters such as **276**) are important intermediates in directed aldol condensations, particularly when derived from aldehydes. Trapping the enolate as the boronic ester, and separating the desired isomer, can solve regiochemical problems in enolate formation. Subsequent condensation proceeds with excellent diastereoselectivity (sec. 9.5). Mukaiyama and co-workers showed that conjugate addition of tri-*n*-propylborane (sec. 5.2.A) with methyl vinyl ketone gave **276**, and reaction with benzaldehyde gave a 91% yield of **277:278** as a 33:1 mixture of diastereomers.[198,199,178] Mukaiyama's conjugate addition method for the preparation of **276** was first reported by Brown.[200] Masamune also prepared **276** via reaction of diazoketone **279** with tributylborane (sec. 13.9.C for the preparation and other reactions of diazoketones).[198] Other methods for the preparation of boron enolates include the reaction of borane derivatives with ketones, thioesters or ketone enolates (see below).

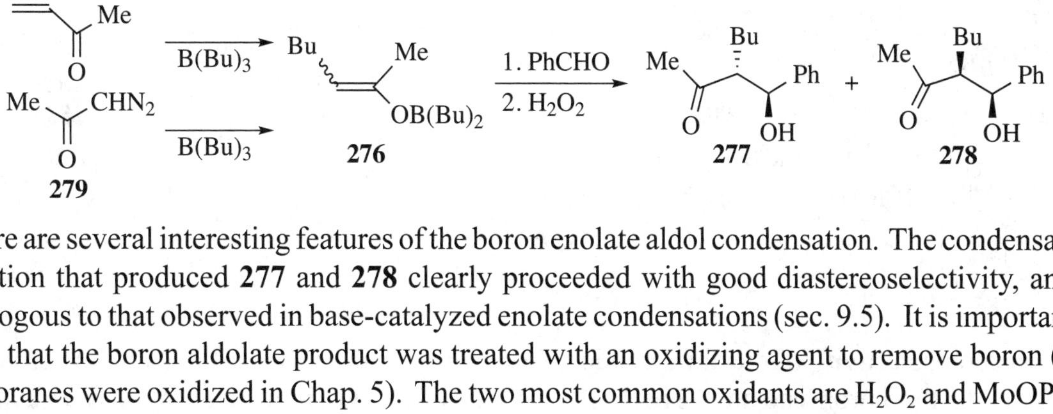

There are several interesting features of the boron enolate aldol condensation. The condensation reaction that produced **277** and **278** clearly proceeded with good diastereoselectivity, and is analogous to that observed in base-catalyzed enolate condensations (sec. 9.5). It is important to note that the boron aldolate product was treated with an oxidizing agent to remove boron (just as boranes were oxidized in Chap. 5). The two most common oxidants are H_2O_2 and MoOPh,[203] which did not degrade the diastereoselectivity of the initial condensation. Several factors help to control or at least influence the diastereoselectivity of the reaction. The first is the geometry

---

198.   Masamune, S.; Mori, S.; Van Horn, D.; Brooks, D.W. *Tetrahedron Lett.* **1979**, *1665.*

199.   Mukaiyama, T.; Inomata, K.; Muraki, M. *J. Am. Chem. Soc.* **1973**, *95*, 967.

200.   Suzuki, A.; Arase, A.; Matsumoto, H.; Itoh, M.; Brown, H.C.; Rogić, M.M.; Rathke, M.W. *J. Am. Chem. Soc.* **1967**, *89*, 5708.

of the boron enolate. In most preparations, the (E)-isomer of boron enolate **276** predominates. This isomer led to **277** with high diastereoselectivity. Masamune found that treatment of **276** with lithium phenoxide led to isomerization of the (E)-isomer to the (Z)-isomer,[198] and subsequent condensation with benzaldehyde gave a 90% yield of **277** and **278** (>2:1 ratio). Masamune also found that the geometry of the boron enolate was greatly influenced by its method of preparation. Tributylborane reacted with methyl vinyl ketone (MVK) to give predominantly the (E)-isomer, as shown. Ketene **280** reacted with sulfide Bu₂BSt-Bu to give the (E)-enolate (**281**) as the major isomer. Condensation with isobutyraldehyde gave a 75% yield of **282** and **283**, in a ratio of 5:>95 favoring **283**.[201] Masamune also showed that boron enolates are generated by reaction of a thioester (**284**) with dibutylboryl triflate and diisopropylethylamine (Hünigs base). The product of this reaction was boron enolate **285**, and subsequent condensation with isobutyraldehyde gave **282** and **283**, although the diastereomeric ratio was reversed to give a >95:5 ratio (79% yield) favoring **282**.[196] A similar reaction occurs with enolizable ketones such as **286**, where the syn:anti ratio of the aldolate products resulting from reaction of the boron enolate (**287**) varied as Hünigs base or 2,6-lutidine was used as the base.[202] The best selectivity for the syn-diastereomer (**287**) was observed with 9-BBN derivatives, and the best anti-diastereoselectivity (**288** > **289**) was with dicyclopentylboranes and Hünigs base.

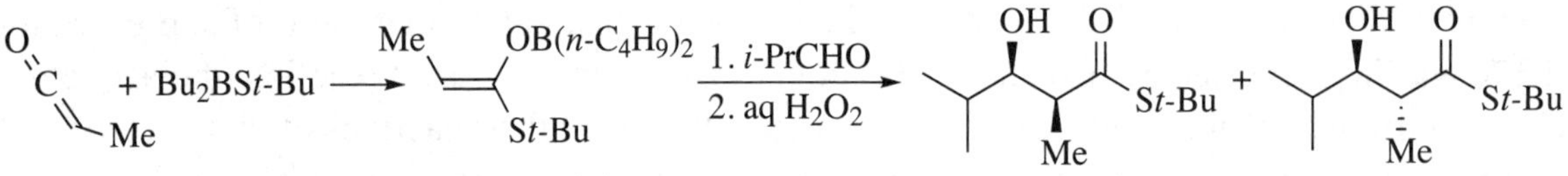

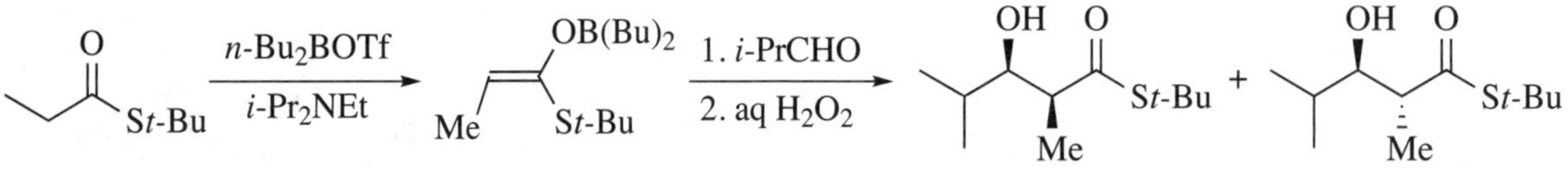

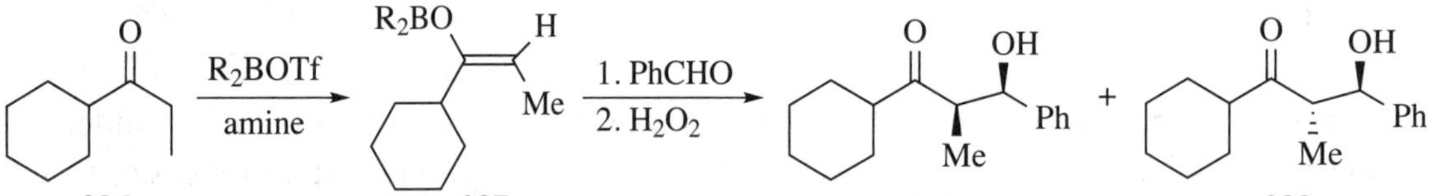

The diastereoselectivity of the aldol condensation varies with the substituents on the carbonyl precursor to the boron enolate. The boron enolates derived from ketones such as **290** are shown in Table 9.8.[203] The lithium enolate of **290** reacted with a boron triflate to give the (Z) enolate **291** and the (E) enolate **292**. Reaction with an aldehyde such as benzaldehyde gave the syn (**293**) and the anti boron aldolate (**294**). The boron aldolates were converted to the

201. Hirama, M.; Masamune, S. *Tetrahedron Lett.* **1979**, 2225.
202. Van Horn, D.E.; Masamune, S. *Tetrahedron Lett.* **1979**, 2229.
203. Evans, D.A.; Nelson, J.V.; Vogel, E.; Taber, T.R. *J. Am. Chem. Soc.* **1981**, *103*, 3099.

aldolate by treatment with H_2O_2 or MoOPH.[203] The results were similar to those observed by Masamune. The best selectivity was obtained in an ether solvent with the enolate and product ratios a function of the base. Hünigs base gave a (Z/E) ratio of >99:1 whereas lutidine gave a 69:31 ratio. The ligand (L) was important and L = n-Bu led to a (Z/E) ratio of 99:1, whereas L = cyclopentyl gave 84:16.[203,204]

Table 9.8.   **Selectivity in Boron Enolate Aldol Condensations**

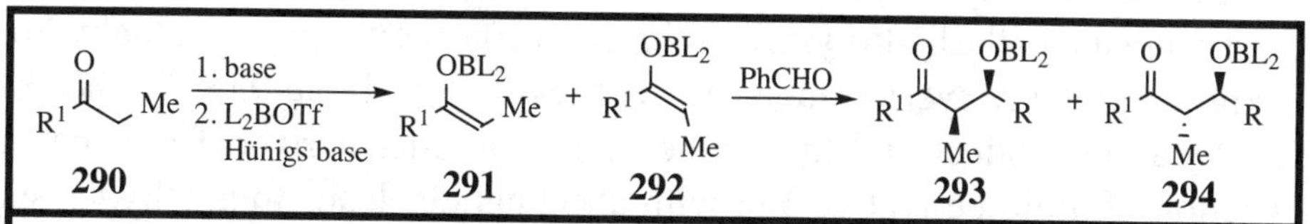

R^1	291 : 292	293 : 294	% Yield
Et	>99 : 1	>97 : 3	77
Ph	>99 : 1	>97 : 3	82
i-Bu	>99 : 1	>97 : 3	82
i-Pr	45 : 55	44 : 56	92
i-Bu	>99 : 1	>97 : 3	65

[Reprinted with permission from Evans, D.A.; Nelson, J.V.; Vogel, E.; Taber, T.R. *J. Am. Chem. Soc.* *1981, 103,* 3099 and Evans, D.A.; Bartroli, J.; Shih, T.L. *J. Am. Chem. Soc. 1981, 103,* 2127. Copyright © *1981* American Chemical Society.]

Modern versions of this condensation react dialkylchloroboranes with ketones or aldehydes, in the presence of an amine base. A synthetic example of this reaction is taken from Paterson and co-worker's studies toward the synthesis of spirastrellolide A.[205] Ketone **295** was treated with the chiral B-chloro diisopinocampheylborane (see Chap. 5) in the presence of triethylamine to give the boron enolate *in situ*, which reacted with aldehyde **296** to give an 87% % yield of **297** with a diastereoselectivity of >20:1.

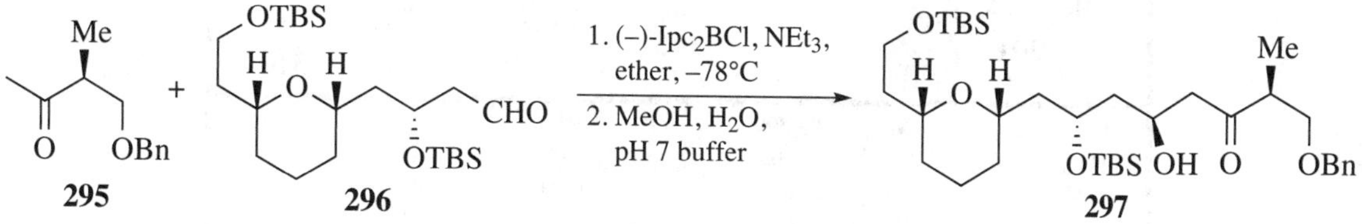

9.4.E.   The Meyers Aldehyde Synthesis

Groups other than a carbonyl can stabilize a carbanion. Meyers and co-workers[206] described the preparation of α-lithio derivatives of dihydro-1,3-oxazines (**298**) and 2-oxazolines (**300**), and subsequent alkylation and condensation

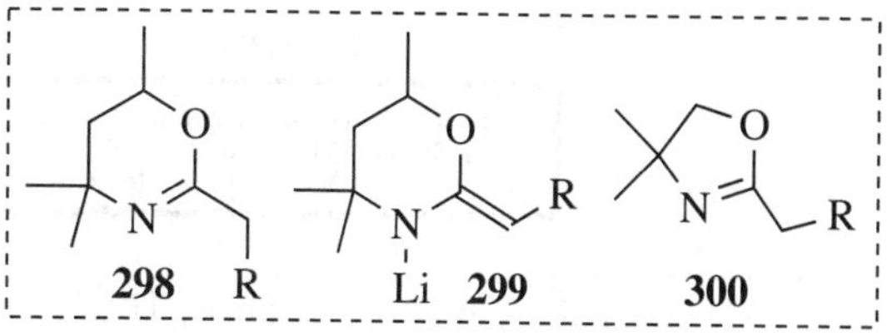

204.   Evans, D.A.; Bartroli, J.; Shih, T.L. *J. Am. Chem. Soc. 1981, 103,* 2127.
205.   Paterson, I.; Anderson, E.A.; Dalby, S.M.; Loiseleur, O. *Org. Lett. 2005, 7,* 4125.
206.   Meyers, A.I.; Nabeya, A.; Adickes, H.W.; Politzer, I.R. *J. Am. Chem. Soc. 1969, 91,* 763.

reactions via α-lithio derivatives such as (**299**). Meyers and co-workers treated the commercially available **301** with *n*-butyllithium to form the α-lithio derivative (**302**), stabilized by chelation with the nitrogen. When iodomethane was added, a carbanion displacement reaction occurred to give **303**.[206] Reduction of the imine unit in **303** with borohydride (sec. 4.4.A) gave amine **304**, and hydrolysis produced the aldehyde (propanal). When **302** was treated with benzophenone, the initial condensation product after hydrolysis was the alcohol (**305**). Reduction of the imine moiety provided **306**[207] and hydrolysis led to conversion of the oxazolidine to aldehyde **307**, but this eliminated water under the hydrolysis conditions to give the conjugated aldehyde (**308**). This sequence is known as the **Meyers' aldehyde synthesis**.[206,208] Table 9.9[206,207] shows formation of both the alkylation product and the condensation product derived from the nucleophile **302**. The reaction of lithio derivative **302** with alkyl halides leads to aldehydes such as **309**, and reaction with aldehydes or ketones leads to conjugated carbonyl derivatives such as **310**.

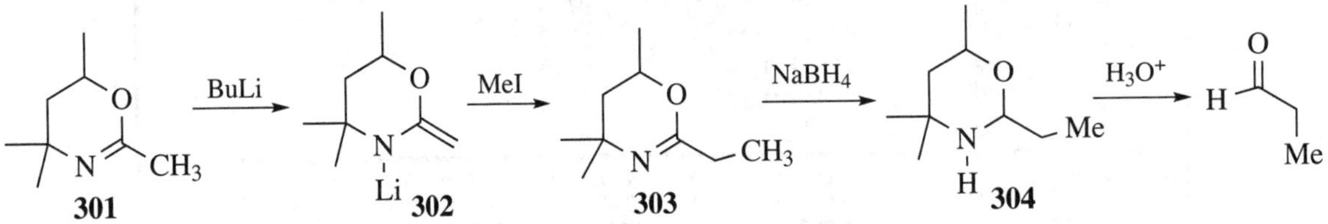

301 → (BuLi) → **302** (Li) → (MeI) → **303** (CH₃) → (NaBH₄) → **304** → (H₃O⁺) → aldehyde (H, Me)

Table 9.9. Aldehyde Formation via Alkylation and Condensation with Dihydro-1,3-Oxazine (**301**)

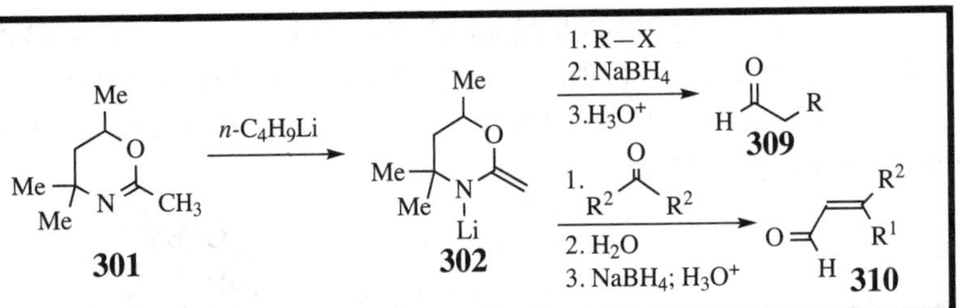

301 → (*n*-C₄H₉Li) → **302**

1. R—X
2. NaBH₄
3. H₃O⁺ → **309**

1. R²C(=O)R²
2. H₂O
3. NaBH₄; H₃O⁺ → **310**

R	X	% 309[a]	R¹	R²	% 310[a]
Me	I	60	C_3H_7	H	61
n-Pr	I	65	C_6H_{13}	H	48
n-Bu	Br	67	Ph	H	64
CH₂=CHCH₂	Br	53	Me	Me	50
i-Pr	I	47	$-(CH_2)_5-$		53
PhCH₂	Br	54	Et	Et	62
			Ph	Me	50
			Ph	Ph	62

[a] overall yield from **301**

207. Meyers, A.I.; Nabeya, A.; Adickes, H.W.; Fitzpatrick, J.M.; Malone, G.R.; Politzer, I.R. *J. Am. Chem. Soc.* **1969**, *91*, 764.
208. (a) Meyers, A.I.; Nabeya, A.; Adickes, H.W.; Politzer, I.R.; Malone, G.R.; Kovelesky, A.C.; Nolen, R.L.; Portnoy, R.C. *J. Org. Chem.* **1973**, *38*, 36; (b) The Merck Index, 14th ed., Merck & Co., Inc., Whitehouse Station, New Jersey, **2006**, p ONR-61.

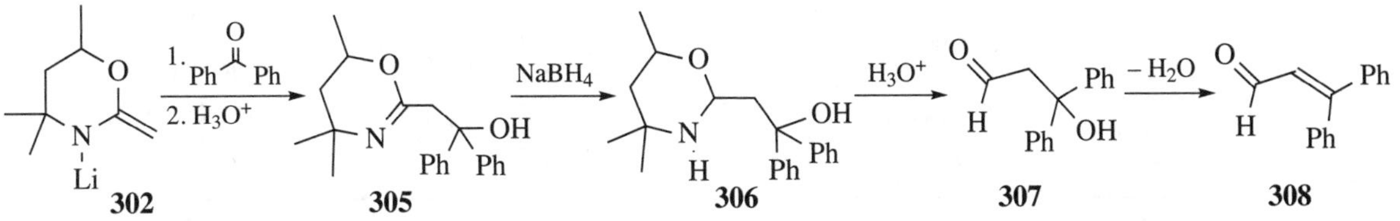

Meyers and others[209] showed that 2-oxazolines such as **311** are carbanionic synthons when they are converted to their α-lithio derivative. Reaction of **311** with *n*-butyllithium, for example, gave **312**. Addition of benzyl bromide gave the alkylated product **313**, and heating that product in ethanolic sulfuric acid gave an 84% yield of the ethyl ester, ethyl 3-phenylpropanoate (**314**).[210] Ketones react similarly to give a conjugated acid derivative.

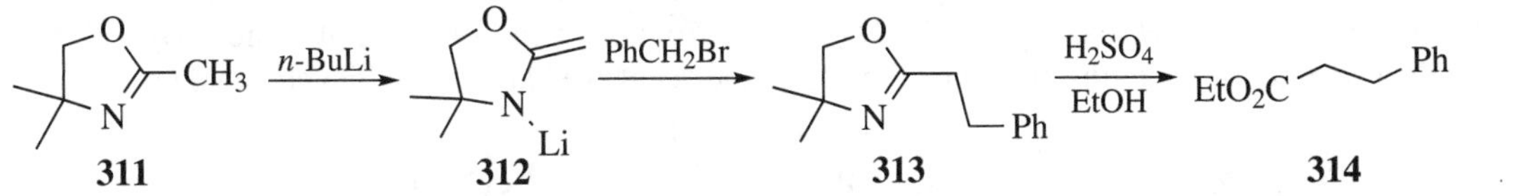

These reactions are not limited to the simple oxazine and oxazolines described above. They are readily prepared from aldehydes and amino alcohols. Meyers used this methodology to prepare chiral precursors for use in asymmetric synthesis. When keto-acid **315** was treated with valinol (**316**), chiral bicyclic lactam **317** was formed.[211] Sequential alkylations with iodomethane and then 4-methoxybenzyl bromide gave **318** in 85% yield (30:1 selectivity). Upon hydrolysis, keto acid **319** was produced in 88% yield with > 98% ee.[179] Alkylation of **317**, or related analogs, proceeds with excellent diastereoselectivity, as seen in Meyers and co-worker's asymmetric synthesis of (+)-mesembrine.[212]

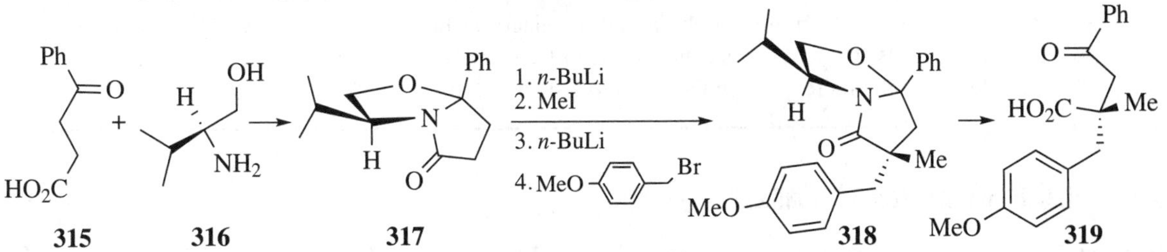

It is clear that the Meyers aldehyde synthesis provides a facile route not only to a variety of aldehydes but also to esters, acids, and chiral derivatives. The disconnections available by this methodology include

209. For related work, see (a) Wehrmeister, H.L. *J. Org. Chem.* **1962**, *27*, 4418; (b) Allen, P.; Ginos, J. *Ibid.* **1963**, *28*, 2759.
210. Meyers, A.I.; Temple, Jr., D.L. *J. Am. Chem. Soc.* **1970**, *92*, 6644, 6646.
211. Meyers, A.I.; Harre, M.; Garland, R. *J. Am. Chem. Soc.* **1984**, *106*, 1146.
212. Meyers, A.I.; Hanreich, R.; Wanner, K. Th. *J. Am. Chem. Soc.* **1985**, *107*, 7776.

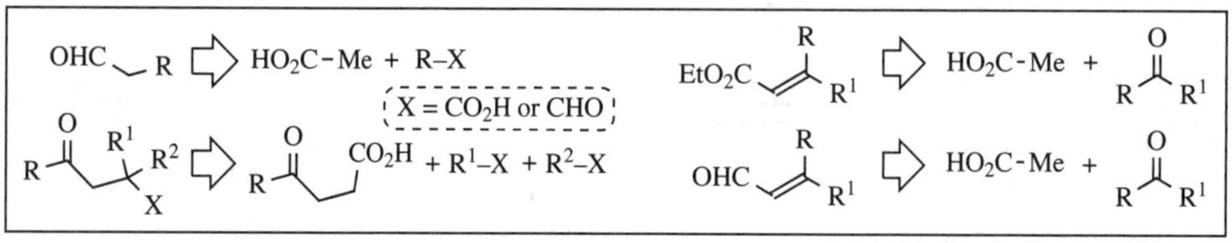

MOLECULAR MODELING: Where Steric Considerations May Not Be Sufficient

In some cases steric effects revealed by 2D drawings or even accurate 3D models may not lead to a clear assignment of product, or even to a prediction of the wrong product. Either steric effects are not the controlling factor, or, if they are, they may not be easy to anticipate from a structural model. An example is provided by reaction of 4-methoxybenzyl bromide with the enolate anion formed from **A**. The major product, **318**, follows from reaction of the enolate anion (**B**) with the electrophilic carbon of the benzylic bromide on the "concave" face of the bicyclic enolate.

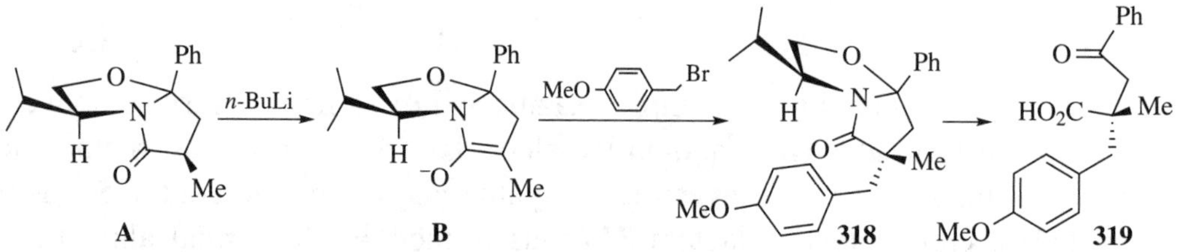

Examine a space-filling model of **B**. Is the convex face (lower left) visually more crowded than the concave face (upper left)? Are you able to predict the observed product?

Next examine the local ionization potential map for **B** (right). This shows whether electrons are tightly bound ("blue") or loosely bound ("red") as a function of location on the accessible surface of the enolate. The more loosely bound the electrons the more likely they will be donated to an electrophile. Is the convex or concave face of the enolate more susceptible to attack by an electrophile? Does the more reactive face lead to the observed product?

9.4.F. Imine and Hydrazone Carbanions

In Section 9.4.A, it was noted that there were problems with aldol-type reactions, especially with the directed aldol condensation. In particular, aldehydes with an α-hydrogen have great difficulty adding to ketones due to their propensity for self-condensation. The ability to use kinetic control conditions in enolate reactions of ketones and aldehydes often solves this problem. There are also several alternative approaches that involve the use of carbanions derived from imines and hydrazones and these can be very useful.[213]

9.4.F.i. Imine Carbanions. Wittig prepared the ethylamine imine of acetaldehyde (see **320**), and subsequent reaction with lithium diethylamide gave α-lithio derivative **321**. When benzophenone was added, the expected carbanion condensation reaction occurred to give a

---

213. Wittig, G.; Reiff, H. *Angew. Chem. Int. Ed.* **1968**, *7*, 7.

27% yield of **322**.[214,215] With these experiments, Wittig and co-workers demonstrated several important features of imine carbanions. The imine moiety functioned as a protected carbonyl (see sec. 7.3.B for protecting groups used with ketones and aldehydes) in this reaction, and the carbonyl could be regenerated after the alkylation or condensation reaction. The imine clearly provides activation to the α-position to produce the carbanion on treatment with a suitably strong base. For Schiff bases (imines derived from aldehydes) deprotonation can occur only at the α-carbon, with no possibility of self-condensation or dimerization. The sequence shown above was one of the first examples of an aldehyde directed aldol condensation.

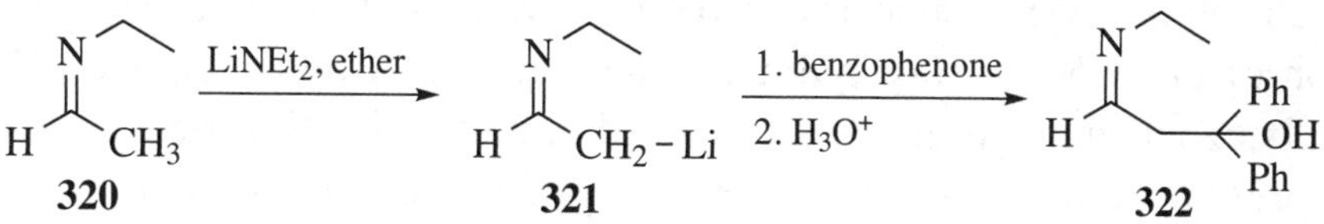

In a synthesis of the phomoidrides by Nicolaou and co-workers,[216] imine **323** was treated with LDA, and then the aldehyde to generate a substituted imine. Subsequent acid hydrolysis converted the imine to an aldehyde, with accompanying elimination of water to give a 50% overall yield of conjugated aldehyde **324**. When imine anions of this type are condensed with an ester, the product is an enamino ketone.[217] There is a potential problem when this approach is used. When imine carbanions react with ethyl chloroformate, two products are usually formed; *C*-acylation as the major product and *N*-acylation as a minor product.[218]

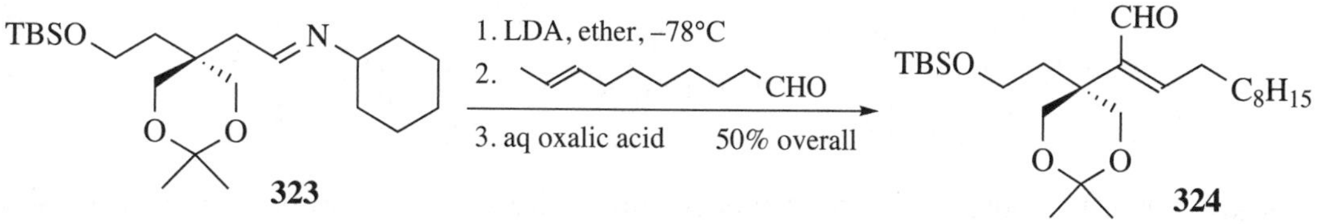

The imine-carbanion method was modified to accomplish the interesting ketone → conjugated aldehyde transformation. The α-lithioimine (**325**) reacted with cyclohexanone to give **326**. Hydrolysis at a carefully controlled pH of 4.5 led to a conjugated aldehyde **327** in 90% yield.[219]

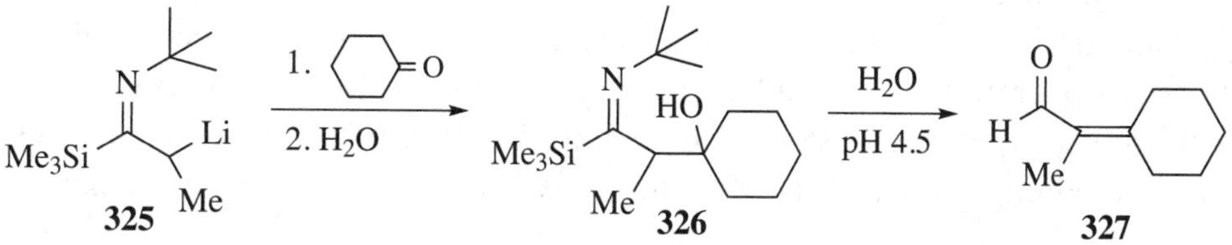

214. Wittig, G.; Frommeld, H.-D. *Chem. Ber.* **1964**, *97*, 3541.
215. Wittig, G.; Frommeld, H.-D. *Chem. Ber.* **1964**, *97*, 3548.
216. Nicolaou, K.C.; Jung, J.; Yoon, W.H.; Fong, K.C.; Choi, H.-S.; He, Y.; Zhong, Y.-L.; Baran, P.S. *J. Am. Chem. Soc.* **2002**, *124*, 2183.
217. Wittig, G.; Suchanek, P. *Tetrahedron Suppl. 8, Part I* **1966**, *22*, 347.
218. Reference 213, p 12 and citation 6 therein.
219. Corey, E.J.; Enders, D.; Bock, M.G. *Tetrahedron Lett.* **1976**, 7.

9.4.F.ii. Hydrazone Carbanions. Corey and Enders[220] introduced an alternative to Wittig's imine-carbanion methodology discussed in Section 9.4.F.i. Modification of the reaction to use a hydrazone rather than an imine made it more controllable and useful. In a synthesis of (+)-leucascandroide A by Paterson and Tudge,[221] hydrazone **328** (PMB = p-methoxybenzl) was converted to the anion with LDA and subsequent reaction with the propargylic bromide gave **329**. Subsequent treatment with tetrabutylammonium fluoride (TBAF) initiated conversion of the hydrazone to the ketone and the O-silyl group to the alcohol (sec. 7.3.A.i), giving **330** in 65% yield. The hydrazone carbanion reacts like other carbanions, not only with alkyl halides but with other electrophilic species such as carbonyl derivatives or epoxides.[222] This is illustrated by the reaction of the N,N-dimethylhydrazone derivative of cyclohexanone (**331**) with LDA and then with benzophenone, to give a 90% yield of **332**.[223] Hydrazone carbanions generally show a great preference for 1,2-addition to conjugated aldehydes such as crotonaldehyde,[223] which suggests that a α-lithiohydrazone is a highly carbanionic species and that the reaction with aldehydes may be under chelation control. N-Tosyl hydrazone derivatives are readily converted to a dianion and undergo the same type of reactions.[224]

Oximes are a useful source of carbanions in aldol-like condensations. Jung et al. showed that oxime **333** reacted with two equivalents of n-butyllithium to give the dianion **334**, which reacted with 1-bromopropane.[225,226] Subsequent treatment with water resulted in a 53% yield of **335**. O-Tetrahydropyranyl oximes such as **336** can be used in this sequence, and the fact that the oxygen is blocked removes the necessity for generating a dianion, which may have solubility problems that diminish the yield of subsequent reactions. Conversion of 2-heptanone to the O-tetrahydropyranyl oxime produced a 61:39 mixture of (E/Z) isomers (**336**). Treatment with LDA followed by condensation with acetone led to a mixture of oximes (**337/338**) in 94% yield, maintaining the 61:39 (E/Z) ratio.[227] Warming **336** to –50°C (from –78°C) led

220. (a) Corey, E.J.; Enders, D. *Tetrahedron Lett.* **1976**, 3; (b) *Idem Chem. Ber.* **1978**, *111*, 1337.
221. Paterson, I.; Tudge, M. *Angew. Chem. Int. Ed.,* **2003**, *42*, 343.
222. Corey, E.J.; Enders, D. *Chem. Ber.* **1978**, *111*, 1362.
223. Corey, E.J.; Enders, D. *Tetrahedron Lett.* **1976**, 11.
224. Lipton, M.F.; Shapiro, R.H. *J. Org. Chem.* **1978**, *43*, 1409.
225. Jung, M.E.; Blair, P.A.; Lowe, J.A. *Tetrahedron Lett.* **1976**, 1439.
226. (a) Adlington, R.M.; Barrett, A.G.M. *J. Chem. Soc. Chem. Commun.* **1979**, 1122; (b) *Idem Ibid.* **1978**, 1071.
227. Ensley, H.E.; Lohr, R. *Tetrahedron Lett.* **1978**, 1415.

to isomerization, giving exclusively the (*E*)-isomer. When the condensation of (*E*)-**336** with acetone was done at –50°C, a single product (**337**) was obtained.[227]

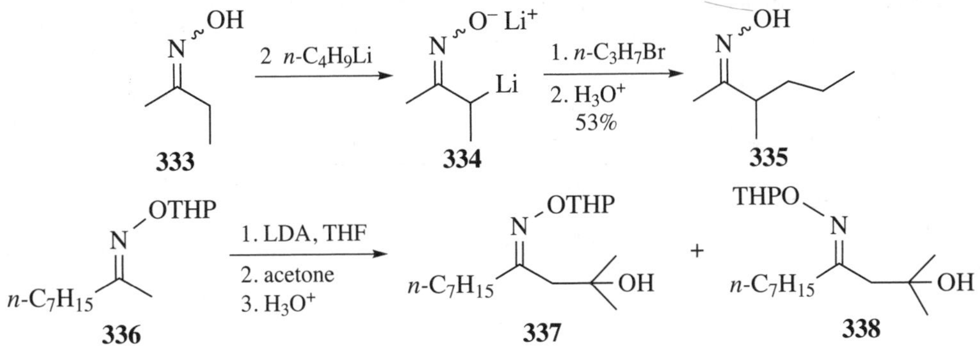

9.4.G. Nozaki-Hiyama-Kishi Coupling

Another acyl addition reaction involves an organochromium intermediate. Although it is not formally an enolate anion condensation, it is related to the acyl addition reactions presented in this chapter. As originally formulated by Nozaki and co-workers, vinyl triflates such as **339** were treated with an aldehyde and $CrCl_2$ in the presence of a nickel catalyst.[228] Initial formation of a vinyl nickel species was followed by *in situ* transmetalation to give a vinyl Cr(III) species, which added to the aldehyde. Hydrolysis gave the alcohol product. When **339** reacted with benzaldehyde, an 83% yield of **340** was obtained.[228] Kishi and co-workers employed a similar reagent for coupling activated alkenes with aldehydes, during synthetic studies toward palytoxin,[229] and later work involving the synthesis of taxanes[230] expanded the reaction. Hiyama et al. studied reactions of bromobutenes with aldehydes in the presence of Cr(II) reagents.[231] This transformation is the **Nozaki-Hiyama-Kishi reaction**,[232] although it has also been referred to as the **Takai-Utimoto reaction**. Fürstener and Shi employed a Cr(II)-Mn(0) redox couple to give essentially the same coupling reaction.[233] Boeckman and Hudack used this reagent to prepare anti-diols with high stereoselectivity.[234] A slight modification of

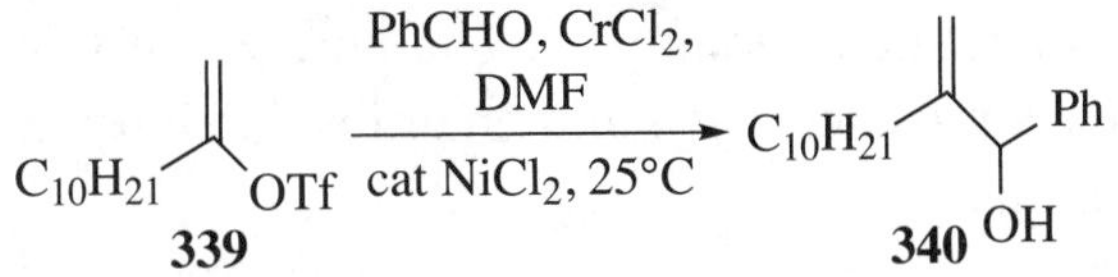

228. Takai, K.; Tagashira, M.; Kuroda, T.; Oshima, K.; Utimoto, K.; Nozaki, H. *J. Am. Chem. Soc.* **1986**, *108*, 6048.
229. Jin, H.; Uenishi, J.; Christ, W.J.; Kishi, Y. *J. Am. Chem. Soc.* **1986**, *108*, 5644.
230. Kress, M.H.; Ruel, R.; Miller, L.W.H.; Kishi, Y. *Tetrahedron Lett.* **1993**, *34*, 5999.
231. (a) Hiyama, T.; Kimura, K.; Nozaki, H. *Tetrahedron Lett.* **1981**, *22*, 1037; (b) Hiyama, T.; Okude, Y.; Kimura, K.; Nozaki, H. *Bull. Chem. Soc. Jpn.* **1982**, *55*, 561.
232. (a) Cintas, P. *Synthesis* **1992**, 248; (b) Wessohann, L.A.; Scheid, G. *Synthesis* **1999**, 1 (see pp 16-18); (c) The Merck Index, 14th ed., Merck & Co., Inc., Whitehouse Station, New Jersey, **2006**, p ONR-67; (d) Mundy, B.P.; Ellerd, M.G.; Favaloro Jr., F.G. *Name Reactions and Reagents in Organic Synthesis, 2nd ed.*, Wiley-Interscience, New Jersey, **2005**, pp. 466-467.
233. Fürstner, A.; Shi, N. *J. Am. Chem. Soc.* **1996**, *118*, 12349.
234. Boeckman, Jr., R.K.; Hudack, Jr., R.A. *J. Org. Chem.* **1998**, *63*, 3524.

this fundamental reaction leads to vinyl iodides. [235] An intramolecular example, taken from a synthesis of bipinnatin J by Trauner and Reothle, showed that treatment of **341** with CrCl₂/NiCl₂ gave **342** (bipinnatin) in 59% yield, with a diastereoselectivity of >9:1.[236]

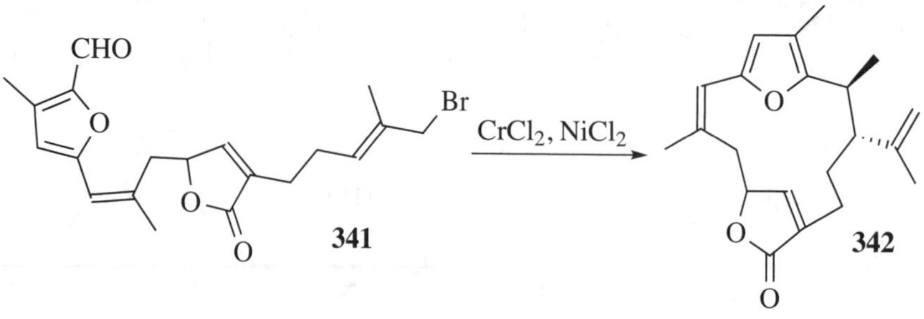

9.5. STEREOSELECTIVE ENOLATE REACTIONS

When a prochiral enolate reacts with an aldehyde or an unsymmetrical ketone to give a aldolate product (**343**), two new stereogenic centers are created (when R ≠ H). When an enolate reacts

with an alkyl halide, one new stereogenic center is created as in **344** (when R ≠ H). There have been many advances in the area of asymmetric enolate methodology.[237] The purpose of this section is to discuss those factors that influence diastereoselectivity, and enantioselectivity in alkylation and condensation reactions of enolate anions.

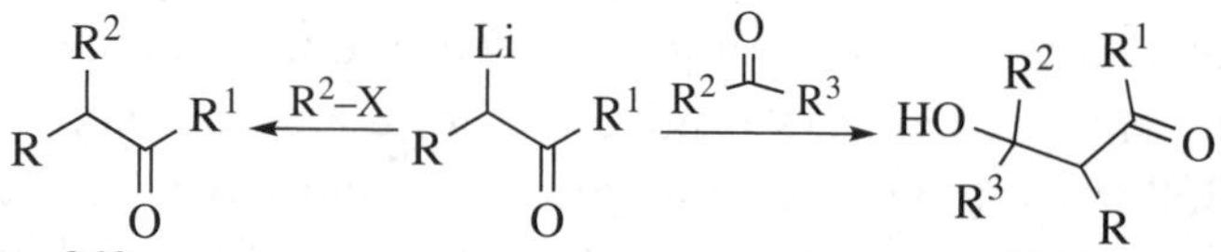

Note that several factors influence both the stereoselectivity of hydrogen exchange and enolate formation in base-promoted reactions. Houk, Ando and co-workers found that differing conjugative stabilization by CH *p*-orbital overlap does not directly influence stereoselectivity.[238] Steric effects dominate only in exceptionally crowded transition structures, but torsional strain involving vicinal bonds contributes significantly to the stereoselectivity of all cases studied.

9.5.A. Simple Diastereoselection

9.5.A.i. Alkylation. If the aldol condensation reaction that produces **344** could be controlled, one diastereomer might be formed in preference to the other, making the reaction diastereoselective. There are two essential factors that control diastereoselectivity in this reaction: the *face* from which the two reagents approach and the relative *orientation* of the two molecules.

235. Lee, K.-Y.; Oh, C.-Y.; Ham, W.-H. *Org. Lett.* **2002**, *4*, 4403.
236. Roethle, P.A.; Trauner, D. *Org. Lett.* **2006**, *8*, 345.
237. Arya, P.; Qin, H. *Tetrahedron* **2000**, *56*, 917.
238. Behnam, S.M.; Benham, S.E.; Ando, K.; Green, N.S.; Houk, K.N. *J. Org. Chem.* **2000**, *65*, 8970.

Enolate alkylation with simple aldehydes and ketones does not generally lend itself to enantioselective control due to the planar nature of the enolate π system.[239] Inspection of **345** shows that the *si-re* face (face *a*) has no more steric hindrance than the *re-si* face (face *b*). When this enolate reacts with iodomethane, therefore, no facial selectivity is anticipated and the methylated product will be racemic at the newly-formed stereogenic center. In general, enolate alkylation reactions produce chiral, racemic products. The reaction can be diastereoselective, however, when substituents attached to the molecule provide facial bias. In general, enolate alkylation proceeds by approach of the enolate to the halide from the less sterically hindered face of the enolate anion. In the Molander and Haas synthesis of davanone,[240] ketone **346** was treated with LDA and then with iodomethane. Three products were formed, 58% of **348**, 8% of **349** and <2% of dialkylation product **350**. About 20% of unreacted **346** was recovered. Inspection of the enolate anion formed from **346** (see **347**) shows that the exo face of the bicyclic system is the least hindered for attack of iodomethane, and accounts for the stereochemistry of the methyl group(s) in all three products The regioselectivity for **348** is attributed to formation of the enolate anion and alkylation away from the bridgehead methyl group. Another example is the conversion of **351** to **353** in Paquette and Heidelbaugh's synthesis of xyloside derivatives.[241] Reaction with LDA led to enolate anion **352**, and reaction with formaldehyde gave hydroxymethyl derivative **353**, with reaction occurring from the less hindered exo face of the enolate anion.

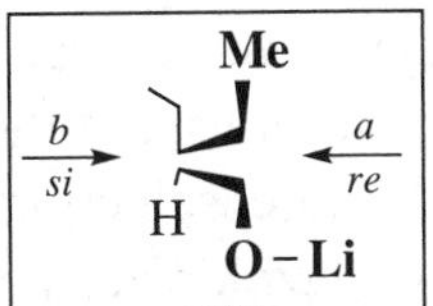

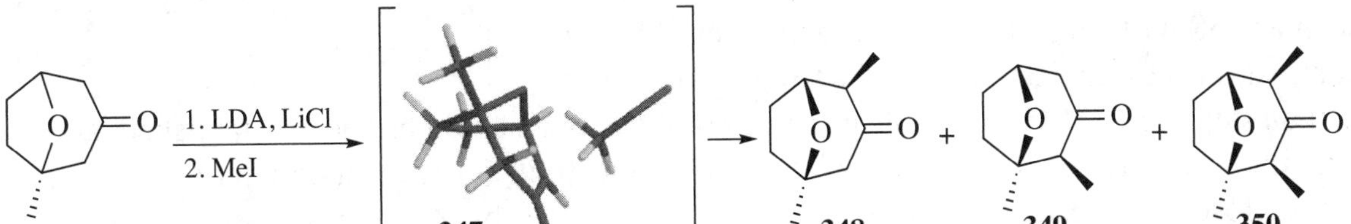

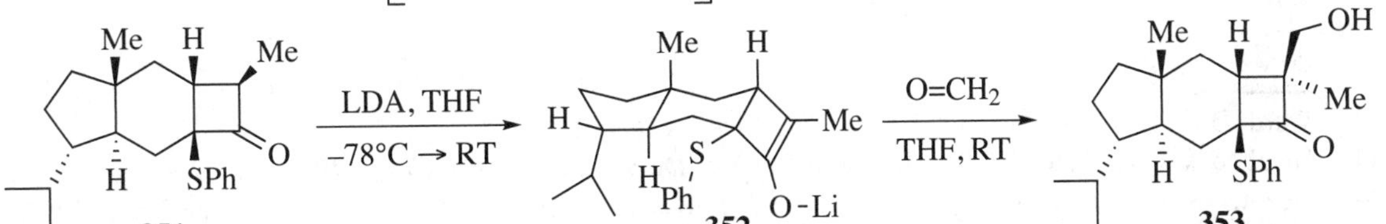

9.5.A.ii. Diastereoselectivity in the Aldol Condensation.
Enolate anions react with other carbonyl compounds to give aldol or Claisen condensation products, as discussed in previous sections. An aldol condensation with the enolate of 1-phenyl-1-propanone and benzaldehyde generates two new stereocenters, and gives two racemic diastereomers (four stereoisomers). These two diastereomers are the racemic anti diastereomer (**354** and **357**), and the racemic

239. Meyers, A.I.; Williams, D.R.; Erickson, G.W.; White, S.; Druelinger, M. *J. Am. Chem. Soc.* **1981**, *103*, 3081.
240. Molander, G.A.; Haas, J. *Tetrahedron* **1999**, *55*, 617.
241. Paquette, L.A.; Heidelbaugh, T.M. *Synthesis* **1998**, 495.

syn diastereomer (**355** and **356**). Diastereoselectivity in this reaction depends on the reaction conditions and the enolate and aldehyde partners, and this section will explore the origins of diastereoselection. Many aldol products undergo syn-anti isomerization in the presence of imidazole via an enolization mechanism,[242] so in some cases the syn-anti ratio can be modified.

Heathcock et al. described processes that formed diastereomers in the aldol condensation, from precursors that do not contain a chiral center, as **simple diastereoselectivity**.[243,244] Ther naming protocols for the diastereomers produced in the aldol condensation are erythro/threo[245,246,247,248] and the syn/anti[249] (see sec. 1.4.B). The latter convention designates diastereomers **354** and **357** as anti, and diastereomers **355** and **356** as syn.

9.5.A.iii. The Zimmerman-Traxler Model.

It would be useful to have a simple model that would predict the major diastereomer produced in an aldol condensation with confidence. In Section 9.2.C, the geometry of the enolate was shown to be important and the geometry could be controlled to some extent. In general, a (Z)-enolate such as **358** will generate a syn diastereomer (**359**). In contrast, an (E) enolate such as **360** will generate the anti diastereomer (**361**). Once the geometry of the enolate anion has been established, a cyclic transition state model is used to predict diastereoselectivity. The most popular transition state is the closed or chelated transition state,[250] first proposed by Zimmerman and Traxler[250b] for

242. Ward, D.E.; Sales, M.; Sasmal, P.K. *J. Org. Chem.* **2004**, *69*, 4808.
243. Heathcock, C.H.; White, C.T.; Morrison, J.J.; Van Derveer, D. *J. Org. Chem.* **1981**, *46*, 1296.
244. Heathcock, C.H. in *Asymmetric Synthesis Vol. 3*, Morrison, J.D. Ed., Academic Press, New York, **1983**.
245. Reference 244, citations 5-14 therein. For example, see (a) Cahn, R.S.; Ingold, C.; Prelog, V. *Angew. Chem. Int. Ed.* **1966**, *5*, 385; (b) Noyori, R.; Nishida, I.; Sakata, J.; Nishizawa, M. *J. Am. Chem. Soc.* **1980**, *102*, 1223; (c) Evans, D.A.; Vogel, E.; Nelson, J.V. *Ibid.* **1979**, *101*, 6120; (d) Seebach, D.; Golinski, J. *Helv. Chim. Acta* **1981**, *64*, 1413; (e) Seebach, D.; Prelog, V. *Angew. Chem. Int. Ed.* **1982**, *21*, 654.
246. Winstein, S.; Lucas, H.J. *J. Am. Chem. Soc.* **1939**, *61*, 1576, 2845.
247. (a) Lucas, H.J.; Schlatter, M.J.; Jones, R.C. *J. Am. Chem. Soc.* **1941**, *63*, 22; (b) Cram, D.J. *Ibid.* **1952**, *74*, 2149; (c) Curtin, D.Y.; Kellom, D.B. *Ibid.* **1953**, *75*, 6011; (d) House, H.O. *Ibid.* **1955**, *77*, 5083.
248. (a) Eliel, E.L. *Stereochemistry of Carbon Compounds*, McGraw Hill, New York, **1962**; (b) Mislow, K. *Introduction to Stereochemistry*, W.A. Benjamin, New York, **1965**.
249. Masamune, S.; Ali, Sk.A.; Snitman, D.L.; Garvey, D.S. *Angew. Chem. Int. Ed.* **1980**, *19*, 557.
250. (a) Reference 244, p 154; (b) Zimmerman, H.E.; Traxler, M.D. *J. Am. Chem. Soc.* **1957**, *79*, 1920.

the Ivanov condensation of phenylacetic acid and benzaldehyde. The **Ivanov condensation**[251] treats a carboxylic acid with an excess of a Grignard reagent to give a bis(magnesium) ketene acetal (**362**).[251a] Subsequent reaction with a carbonyl compound (benzaldehyde) led to an aldol-like product (**363**). The cyclic model proposed to describe the Ivanov reaction (model **364**) can be used for aldol condensations, if magnesium is replaced with lithium to give cyclic model **365**. When **365** is used to predict stereochemistry in the aldol condensation, it is referred to as the **Zimmerman-Traxler model**.[250]

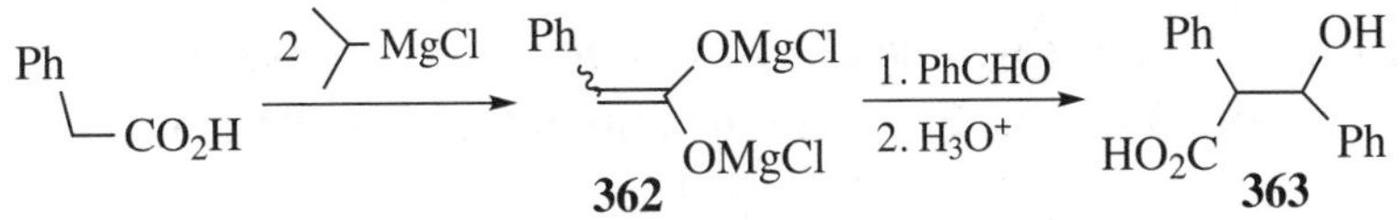

There is *no* evidence that the Zimmerman-Traxler model represents the actual transition state for aldol-like reactions. Nonetheless, this model is an extensively used mnemonic that makes reasonable predictions in many cases. It is used to predict structure-selectivity relationships for lithium, boron, zinc or magnesium enolates when they react with aldehydes and ketones. There are four possible transition states, indicated by for an aldol condensation (**366, 367, 368**, and **369**), reflecting the different orientations of enolates with respect to the aldehyde (or ketone), as well as the (*E/Z*) geometry of the enolate. For simplicity, approach of the aldehyde to the enolate from a single face[252] is shown in all cases (the back face is shown, as well as the front face), and the reaction generates enantiomers.

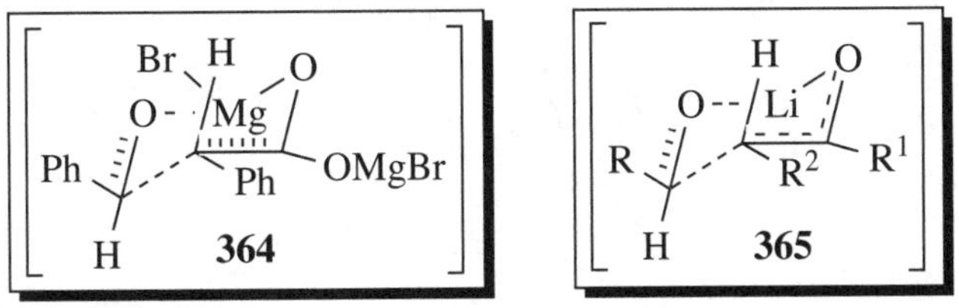

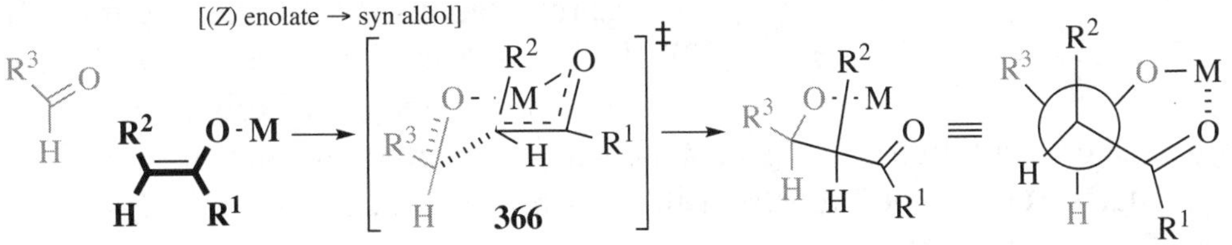

---

251. (a) Ivanoff, D.; Spassoff, A. *Bull. Chim. Soc. Fr.* **1931**, *49,* 19; (b) Ivanoff, D.; Mihova, M.; Christova, T. *Ibid.* **1932**, *51,* 1321; (c) Ivanoff, D.; Nicoloff, N.I. *Ibid.* **1932**, *51*, 1325, 1331; (d) The Merck Index, 14th ed., Merck & Co., Inc., Whitehouse Station, New Jersey, **2006**, p ONR-47; (e) Mundy, B.P.; Ellerd, M.G.; Favaloro Jr., F.G. *Name Reactions and Reagents in Organic Synthesis, 2nd ed.*, Wiley-Interscience, New Jersey, **2005**, pp. 342-343.
252. Reference 244, p 155.

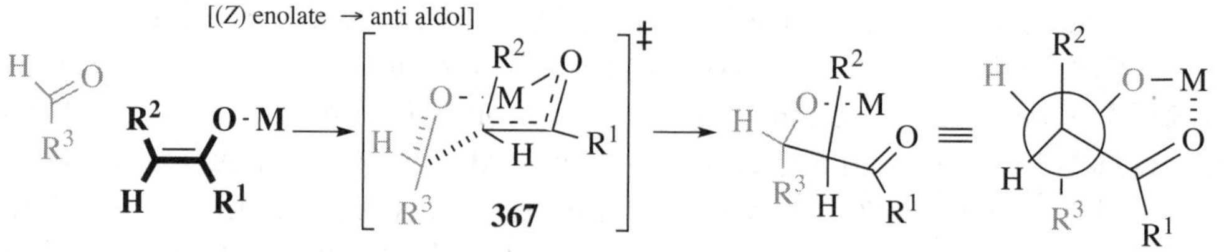

[(Z) enolate → anti aldol]

367

Referring to these generalized models, when R^2 is small, the chair transition state explains the stereochemical trend. For (Z)-enolates, the R^1↔R^3 interaction dominates, and **366** is favored over **367** as the steric bulk increases, leading to more of the syn isomer. For (E) enolates, the R^1↔R^3 interaction favors **368** over **369**, and the anti aldolate is the major product.

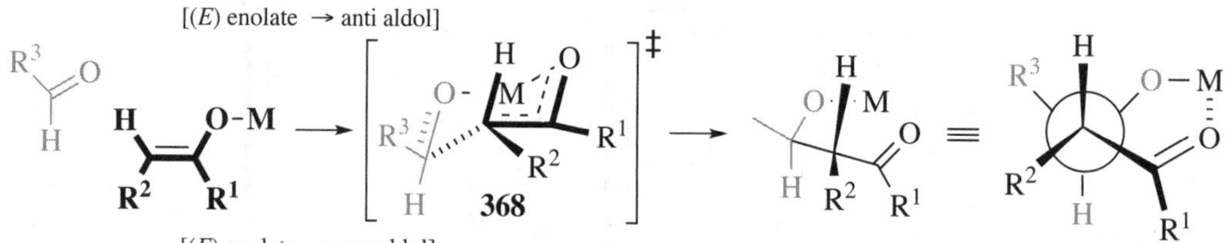

[(E) enolate → anti aldol]

368

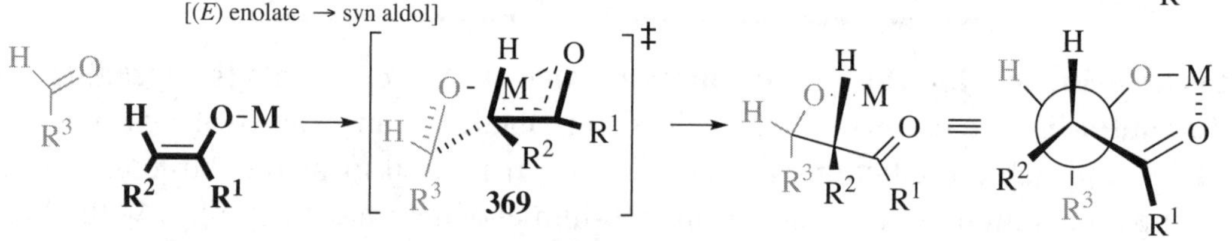

[(E) enolate → syn aldol]

369

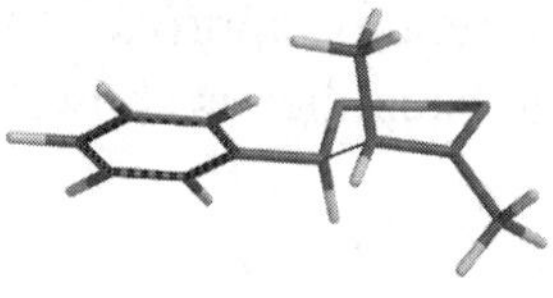

370 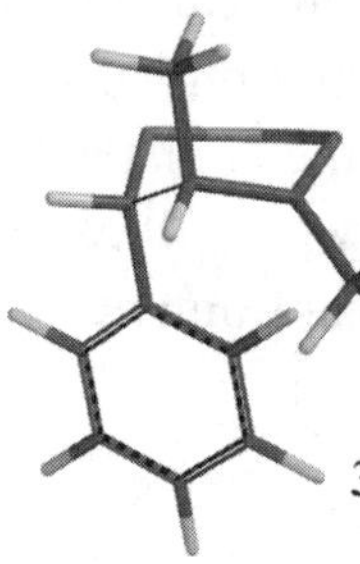   **371**

Model **370** is shown for transition state **366**, for the specific condensation of benzaldehyde (R^3 = Ph) with the (E)-enolate of 2-butanone (R = R^2 = Me). A similar model (**371**) is shown for the opposite orientation (transition state **367**). These representations are not the true transition states, but they show the pseudo-axial and pseudo-equatorial interactions of the phenyl and methyl groups for the condensation of benzaldehyde with the thermodynamic enolate of 2-butanone. Transition states **366** and **367** represent the generalized reaction with a (Z)-enolate with an aldehyde for two different orientations. For the (E)-enolate, the steric interactions in **367** (see **371**) appear to be greater than those in **366** (see **370**) due to the pseudo-equatorial phenyl and methyl groups in **367** (**371**). In **366**, one group is pseudo-axial and the other is pseudo-equatorial, minimizing the steric interaction.

Transition state **366** would be used to predict the major product of the aldol reaction, where the (E)-enolate predicts the syn-diastereomer. Model **372** represents **368** and **373** represents **369** as transition states for the reaction of benzaldehyde with the (Z) enolate of 2-butanone. Pseudo-diaxial groups in **372** (**368**) have minimal steric interactions. Conversely, the pseudo-diequatorial placement of the phenyl and methyl groups in **373** (**369**) brings those groups into relatively close proximity. Transition state **368** is therefore lower in energy, and predicts the anti diastereoselectivity of the final product. Using the Zimmerman-Traxler model for the benzaldehyde-2-butanone system, transition state **374** for this system predicts the initial aldolate product (**375**) and the final syn product, **376**. Similarly, the (*Z*) enolate uses transition state **377** to give the aldolate product (**378**) and the final anti product **379**.

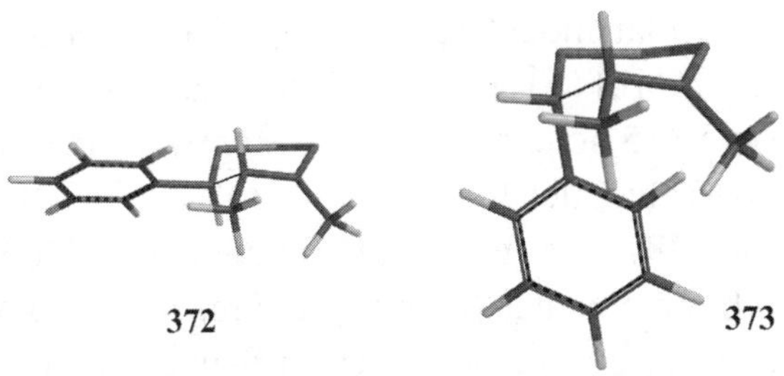

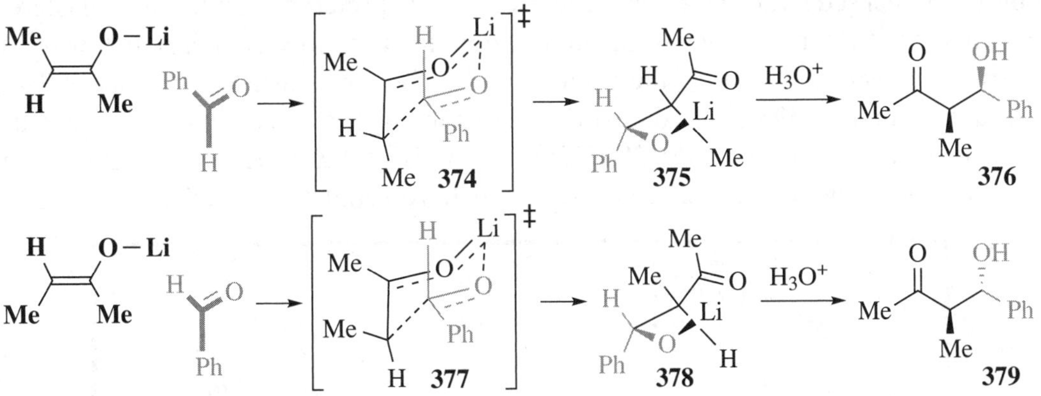

Identical conclusions can be drawn when the Zimmerman-Traxler model is applied to ester enolates. The (*Z*) enolate **380** (derived from the corresponding ester in Table 9.10) reacted with various aldehydes to give a mixture of the anti **381** and the syn (**382**) product, with the anti product predominating.[253]

Table 9.10. The Syn/Anti Selectivity of Ester Enolates with Benzaldehyde[253]

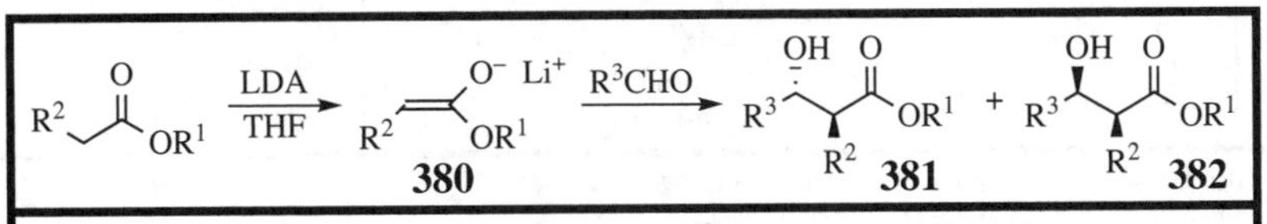

R^1	R^2	R^3	381 : 382
Me	Me	Me	57 : 43
		Ph	55 : 45
		i-Pr	55 : 45
CH$_2$OMe	Me	Me	67 : 33
		i-Pr	90 : 10
DMP[a]	Me	Ph	88 : 12
		n-C$_5$H$_{11}$	86 : 14
		i-Pr	>98 : 2
		t-Bu	>98 : 2
	Et	*i*-Pr	>98 : 2

[a] DMP = 2,6-dimethylphenyl

253. Reference 244, pp 123-124.

A similar model has been applied to aldol condensation reactions of boron enolates.[254] The reaction of (*E*) enolate **383**, for example, can lead to either transition state **385** or **386**. Transition state **385** generates the threo (anti) diastereomer **389**. Transition state **386** leads to the erythro (syn) aldolate **390**. Likewise, (*Z*)-enolate **384** has two transition states, **387** and **388**. Transition state **387** leads to the anti- diastereomer (**389**), and **388** leads to the syn diastereomer **390**. With lithium enolates (M = Li), the reaction is diastereoselective if R^1 is sterically demanding (e.g., R = *tert*-butyl). When R^1 is relatively small, the diastereoselectivity is diminished. Examining boron enolates shows the B–O bond to be shorter than the O–Li bond, which maximizes stereocontrol in a kinetic aldol process. The (*Z*)-enolate of 3-pentanone (**391**, R^1 = ethyl in **384**, L = *n*-butyl) condensed with benzaldehyde to give a >97:3 preference for the syn- diastereomer, **392**.[223] This result clearly suggests transition state **388** is the lowest energy pathway. When R^1 is *tert*-butyl (**384**, L = butyl), the preference for the syn diastereomer was >97:1. For cyclohexanone enolate **393** (locked into the (*E*) geometry), the reaction with benzaldehyde gave primarily the anti diastereomer (**394**, 33:67 syn/anti).[254] A Zimmerman-Traxler-like model also is useful for predicting the diastereoselection in boron enolates as well as lithium enolates.

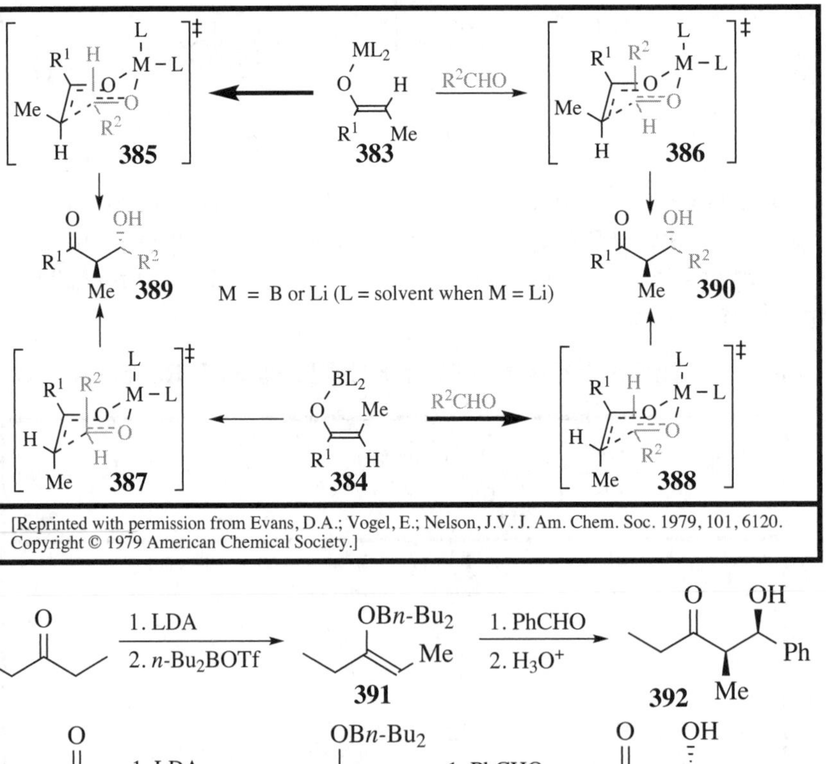

[Reprinted with permission from Evans, D.A.; Vogel, E.; Nelson, J.V. J. Am. Chem. Soc. 1979, 101, 6120. Copyright © 1979 American Chemical Society.]

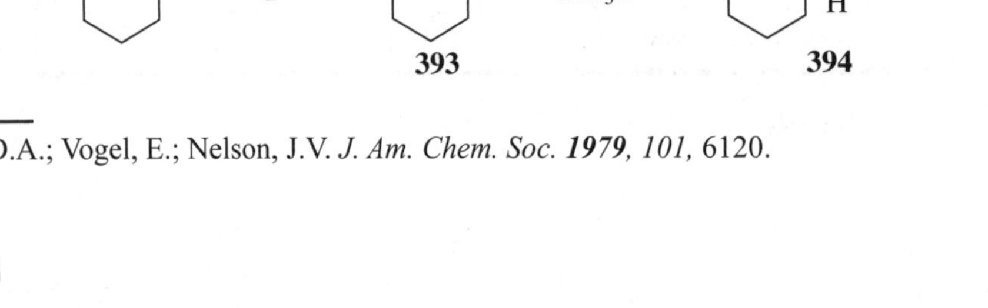

---

254. Evans, D.A.; Vogel, E.; Nelson, J.V. *J. Am. Chem. Soc.* **1979**, *101*, 6120.

9.5.A.iv. The Evans' Model. (Z) Enolates are more stereoselective than (E) enolates, even when R^1 is not large. The Zimmerman-Traxler model transition states **366-369** do not account for this observation. It has been suggested that the transition states are not chair-like, but skewed as in **395-398**.[255] In this representation, (Z) enolate **395** leads to the syn aldolate. Similarly, (Z) enolate **396** gives the anti aldolate, (E) enolate **397** gives the anti aldolate and (E) enolate **398** is the precursor to the syn aldolate. The major steric interactions in this model are those for $R^1 \leftrightarrow R^3$ and $R^2 \leftrightarrow R^3$. For both (Z) and (E) enolates, the $R^1 \leftrightarrow R^3$ interaction favors **395** and

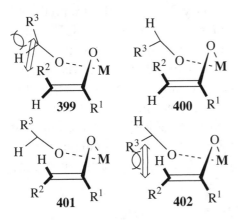

395 [(Z) - syn]
396 [(Z) - anti]
397 [(E) - anti]
398 [(E) - syn]

397, respectively. The $R^2 \leftrightarrow R^3$ interaction is more important for the (E) enolate and leads to **396** and **398** when R^1 is large. The angle at which the reagents approach is obviously important for these transition states. It is assumed that the angle of attack of a nucleophile is 110° as it attacks a carbonyl (sec. 6.6.A), the **Bürgi-Dunitz trajectory** (sec. 9.2.D).[11] This angle of attack brings R^2 and R^3 in **397** into close proximity and relieves the $R^1 \leftrightarrow R^3$ interaction. Similar results are observed for **396** and **398**, but to a lesser degree, which explains why (E) enolates show comparable $R^2 \leftrightarrow R^3$ and $R^1 \leftrightarrow R^3$ interactions unless R^1 is very large, and also accounts for the large effect of the R^2 group on the diastereoselectivity. As R^2 increases in size, transition states **395** and **397** become less important.

399 **400** **401** **402**

Evans et al. considered boat conformations **399-402**[256] as possible transition states, in addition to the chair conformations used in the Zimmerman-Traxler model. In these boat conformations, reaction of an aldehyde with (Z) enolate via orientation **399** gives the syn aldolate, but reaction via the opposite orientation of the aldehyde (see **400**) gives the anti aldolate. Likewise, reaction of an aldehyde with (E) enolate via the orientation shown in **401** gives the anti aldolate, whereas the opposite orientation (see **402**) gives the syn aldolate. A given (Z) or (E) enolate, therefore, has one or two chair transition states, or one or two boat transition states that are energetically favorable for an aldol condensation. Only transition states **400** and **401** are considered important, because of the significant $R^2 \leftrightarrow R^3$ interactions in **399** and **402**. As R^2 becomes larger, however, the (Z) enolate might favor **400**

255. (a) Heathcock, C.H. *Comprehensive Carbanion Chemistry Vol. 2*, Durst, T.; Buncel, E. Eds., Elsevier, Amsterdam, *1985*; (b) Heathcock, C.H.; Buse, C.T.; Kleschick, W.A.; Pirrung, M.A.; Sohn, J.E.; Lampe, J. *J. Org. Chem. 1980, 45*, 1066; (c) Fellmann, P.; Dubois, J.E. *Tetrahedron 1978, 34*, 1349; (d) Dubois, J.E.; Fellmann, P. *Tetrahedron Lett. 1975*, 1225.

256. Evans, D.A.; Nelson, J.V.; Taber, T.R. *Topics in Stereochemistry 1982, 13*, 1.

or the chair transition state. Boat transition states predict that increasing the steric bulk of R^2 in an (E) enolate will either have no effect on stereochemistry or increase selectivity for the anti aldolate. There is some evidence for a modest increase in anti selectivity with increasing the size of R^2.[257]

9.5.A.v. The Noyori Open-Chain Model.

In the Mukaiyama reaction, the Zimmerman-Traxler and Evans' models are not satisfactory for predicting diastereoselectivity. Several open (nonchelated) transition states have been considered as useful models. The condensation reaction of carboxylic acid dianions with aldehydes indicated that anti selectivity increased with increasing dissociation of the gegenion (the cation, M^+).[258] When analyzing an aldol condensation that does not possess the bridging cation required for the Zimmerman-Traxler model, an aldehyde and enolate adapt an eclipsed orientation as they approach. Noyori reported syn selectivity for the reaction of a mixture of (Z)-silyl enol ether **403** and (E)-silyl enol ether **404** with benzaldehyde, in the presence of the cationic reagent tris(diethylamino) sulfonium (TAS).[259] This reaction is clearly a variation of the Mukaiyama reaction, which usually proceeds with poor diastereoselectivity (sec. 9.4.C). The reaction of the (Z) enolate **403** and the (E) enolate (**404**), led to formation of **405** as the major product. The 86:14 (**405/406**) preference[260]

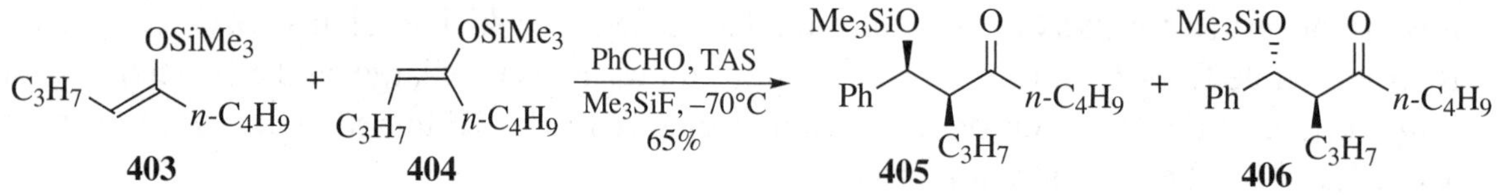

for the syn diastereomer was independent of the enolate geometry. These results are predicted with the open-chain model. The (Z) enolate will form a transition state such as **407**, which

leads to the anti diastereomer. The (E) enolate will react via **408** to give the syn enolate. In both cases, the opposite orientation will lead to a significant $R^2 \leftrightarrow R^3$ interaction that destabilizes that transition state. This does not explain why both (E- and Z) enolates lead to the syn-diastereomer (**405**), and some equilibration or isomerization must occur during the course of the reaction, which will be discussed in Section 9.5.A.vi.

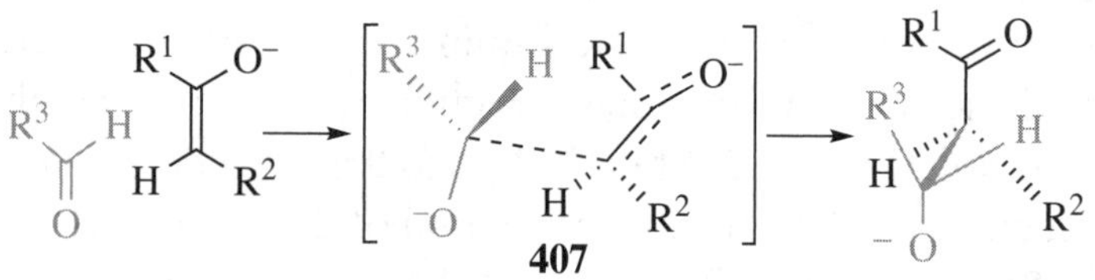

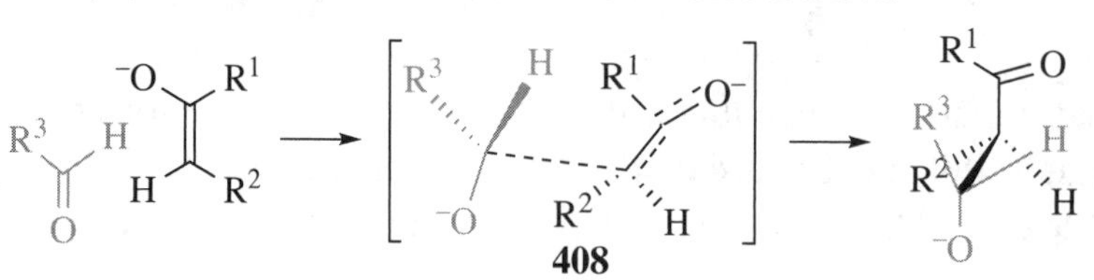

---

257. Reference 244, p 151.
258. Mulzer, J.; Zippel, M.; Brüntrup, G.; Segner, J.; Finke, J. *Liebigs Ann. Chem.* **1980**, 1108.
259. Noyori, R.; Yokoyama, K.; Sakata, J.; Kuwajima, I.; Nakamura, E.; Shimizu, M. *J. Am. Chem. Soc.* **1977**, *99*, 1265.
260. Noyori, R.; Nishida, I.; Sakata, J. *J. Am. Chem. Soc.* **1981**, *103*, 2106.

9.5.A.vi. Isomerization in the Aldol Condensation. It is clear from the previous discussion that steric effects control the orientation of the molecule as it approaches the enolate anion. In simple diastereoselection processes, one observes moderate to good syn/anti selection depending on the steric demands of the enolate and carbonyl partner. However, other processes are at work in this condensation that lead to variations in the syn/anti ratio. Heathcock reported the selectivity for a variety of lithium enolates derived from **409** with benzaldehyde (see Table 9.11).[28,261] The (E) and (Z) enolates (**411** and **410**, respectively) can be trapped with trimethylsilyl chloride to give silyl-enol ethers **413** and **412**, respectively. The observed **412/413** ratio is taken to be the ratio of (Z) and (E)-enolates originally generated from **419**. When the enolate mixture reacted with benzaldehyde, the syn aldolate **414** and the anti aldolate **415** were formed.[261] It is clear from Table 9.11 that the (E/Z) enolate ratio (taken to be **414/415**) is not the same as observed in the aldolate products. One explanation is that the enolate anions equilibrate under the reaction conditions. Isomerization may occur during or after the aldolate-forming step, and the enolate ratio may be the same but the aldehyde adds from different faces of the enolate anion (see above).

Table 9.11. Stereoselectivity in Enolate Formation and Diastereoselectivity in the Aldol Condensation

R	Base	412 : 413	414 : 415	R	Base	412 : 413	414 : 415
LiO	LDA		45 : 55	i-Pr	LDA	60 : 40	82 : 18
MeO	LDA	5 : 95	62 : 38		LICA	59 : 41	75 : 25
t-BuO	LDA	5 : 95	49 : 51		LHMDS	>98 : 2	90 : 10
i-Pr₂N	LDA	81 : 19	63 : 57		LTMP	32 : 68	58 : 42
	LTMP	52 : 48	68 : 32	i-Bu	LDA	>98 : 2	>98 : 2
H	MeLi	100 : 0	50 : 50		LHMDS	>98 : 2	>98 : 2
		0 : 100	65 : 35	Ph	LDA	>98 : 2	88 : 12
Et	LDA	30 : 70	64 : 36		LICA	>98 : 2	87 : 13
	LICA	35 : 65	62 : 38		LHMDS	>98 : 2	88 : 12
	LHMDS	66 : 34	77 : 23		LTMP	>98 : 2	83 : 17
	LTMP	20 : 80	66 : 34				

LDA = lithium diisopropylamide   LICA = lithium isopropylcyclohexylamide
LHMDS = lithium hexamethyldisilazide   LTMP = lithium tetramethylpiperidide

[Reprinted with permission from Heathcock, C.H.; Buse, C.T.; Kleschick, W.A.; Pirrung, M.A.; Sohn, J.E.; Lampe, J. J. Org. Chem. **1980**, *45*, 1066. Copyright © **1980** American Chemical Society.]

The nature of the amide base had a substantial effect on the (E/Z) ratio of the enolate. In general, LTMP gave mainly the (E) enolate and LHMDS the (Z) enolate, which is probably related to the aggregation state of the base. As the size of R increases, the use of LTMP and LDA increases

261. Reference 244, pp 122-132.

the amout of (Z) enolate, which is related to the conformation of the ketone immediately prior to deprotonation, using the Felkin-Anh model (sec. 4.7.B). Conformation **416** is more stable than **418** for propanal by 0.8 kcal mol^{-1} (3.35 kJ mol^{-1}).[262] This difference is probably greater in ketones (R = alkyl or aryl) and esters (R = O-alkyl). This observation suggests that as R becomes larger the (Z) enolate is preferred, regardless of the base. The observation that LTMP gives more (E) enolate can be explained by a steric interaction between the base and the methyl group (see **416**). The methyl↔R interaction is greater, but the interaction of the methyl group and the base is minimized in **419**, This model assumes approach of the base and ketone is not along the axis of the C–H bond, but over the face of the incipient enolate plane.[263] With benzaldehyde, (E) and (Z) enolates may exhibit different degrees of stereoselectivity.

The results in Table 9.11 can be explained by the Zimmerman-Traxler-like transition state (**420**). The R^2↔R^5 and R^1↔R^4 interactions are very important. When R^5 is large (such as adamantyl) the (Z) enolate gives the syn aldolate, and the (E) enolate the anti aldolate. When R^5 is small (Ph, Et), the (Z) enolate gives primarily the syn aldolate. When R^5 = H, reactions of the (Z) enolate become stereorandom and lead to diastereomeric mixtures of products. When R^5 is small (Ph, Et), the (E) enolate usually gives a 50:50 mixture of syn and anti aldolates, probably because the R^1↔R^4 interaction balances the R^2↔R^5 interaction. When R^5 = H or Me, the (E) enolate is selective for the syn diastereomer.

A complicating factor is the observation that aldolates can undergo syn/anti equilibration by enolization or by reverse aldolization. Aldolates such as **421** can be deprotonated to the dianion (**422**), which undergoes alkylation with iodomethane to give the anti product (**423**), as shown.[264] This equilibration is clearly the basis of the aldol-transfer reaction discussed in **152** to **154** in Section 9.4.A.i. If **423** forms a new enolate, equilibration can lead to a mixture of syn and anti products. The primary mechanism for syn/anti equilibration appears to be reverse aldolization.[265] A retro-aldol condensation will convert the syn diastereomer (**424**) into the aldehyde and enolate components, which can regenerate **424** or form the anti diastereomer **425**.

262. Karabatsos, G.J.; Hsi, N. *J. Am. Chem. Soc.* **1965**, *87*, 2864.
263. (a) Ireland, R.E.; Willard, A.K. *Tetrahedron Lett.* **1975**, 3975; (b) Nakamura, E.; Hashimoto, K.; Kuwajima, I. *Ibid.* **1978**, 2079.
264. (a) Fráter, G. *Helv. Chim. Acta* **1979**, *62*, 2825, 2829; (b) Seebach, D. Wasmuth, D. *Ibid.* **1980**, 63, 197.
265. Reference 244, pp 161-162.

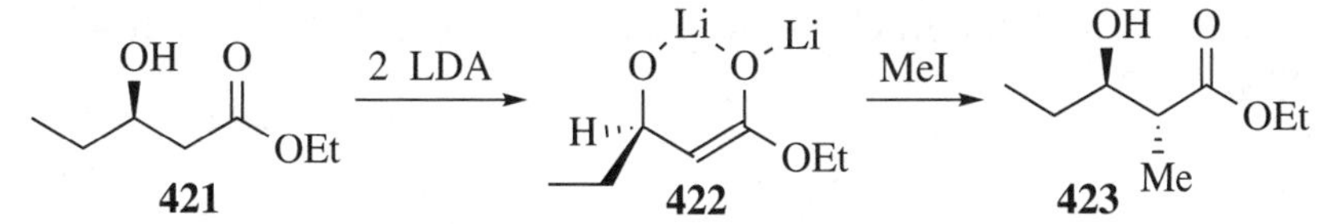

Syn/anti equilibration can be much slower than reverse aldolization, as with the (Z) enolate of 2,2-dimethyl-3-pentanone).[261] Aldolates derived from the more basic ketone enolates are more likely to suffer reverse aldolization than aldolates derived from the less basic enolates of esters, amides or carboxylate salts. Steric crowding in an aldolate promotes reverse aldolization. The metal is very important, and some metals form stable chelates that generate aldolates resistant to reverse aldolization. Boron enolates, for example, do not undergo equilibration even at elevated temperatures. Lithium enolates equilibrate more slowly than other alkali metal enolates and potassium enolates equilibrate rapidly.

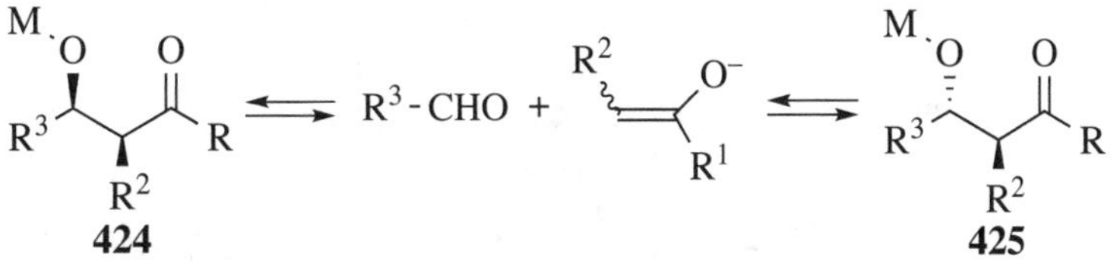

9.5.B. Selectivity with Chiral, Nonracemic Reactants

9.5.B.i. Diastereoface Selectivity[243,266] Heathcock defined three kinds of stereoselection for the aldol condensation: simple diastereoselection, diastereoface selection, and double stereodifferentiation.[243,258] Simple diastereoselection occurred when an achiral aldehyde and an achiral ketone reacted to give a mixture of syn and anti aldol products **426** and **427**. This example illustrates the type of reaction discussed previously.

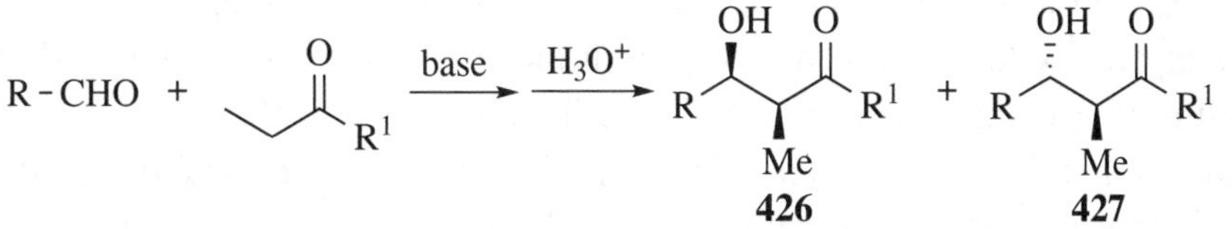

If either the enolate or carbonyl partner has a stereogenic center near the reactive center, it will influence both orientation and facial selectivity.[243,255] Reaction of an achiral enolate such as **429** with an aldehyde or a ketone that possesses a (_S_) stereogenic center (**428**) will give the (_SRS_) diastereomer (**430**) and the (_SRR_) diastereomer (**431**) aldolate products. The (S) stereogenic center in **428** can induce high diastereoselectivity, with **430** as the major product. An aldehyde with a (R) stereocenter will similarly influence the reaction, giving **431** as the major aldolate product. Several factors influence the extent of diastereoselectivity. A second variation reacts an enolate possessing a stereogenic center (such as **432**) with an achiral aldehyde or ketone.

266. Also see (a) White, C.T.; Heathcock, C.H. *J. Org. Chem.* **1981**, *46*, 191; (b) Heathcock, C.H.; Young, J.P.; Hagen,M.C.; Pirrung, M.C.; White, C.T.; VanDerveer, D. *J. Org. Chem.* **1980**, *45*, 3846.

In ($\underline{R}$)-**432**, the reaction with an aldehyde leads to the ($RS\underline{R}$) diastereomer **433** and the ($SR\underline{R}$) diastereomer, **434**.

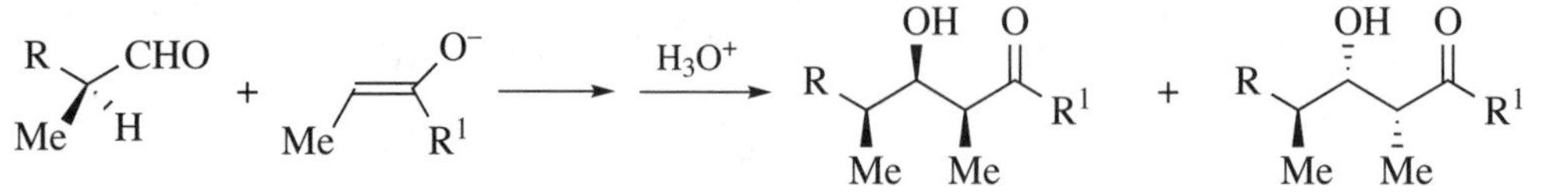

428 (S)   **429**   **430** [(S,R,S) - syn, syn]   **431** [(S,R,R) - syn, anti]

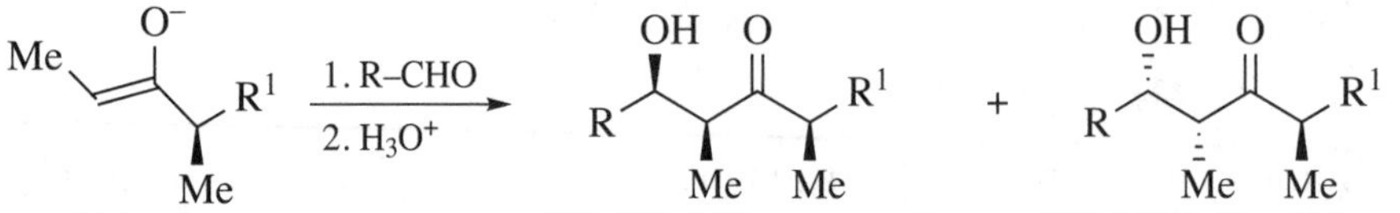

432 (R)   **433** [(R,S,R) - syn, syn]   **434** [(S,R,R) - anti, syn]

Double stereodifferentiation[267] results from the reaction of an enolate possessing a stereogenic center and an aldehyde or ketone that also has a stereogenic center. An example, taken from Calter and Liao's synthesis of siphonarienedione,[268] condensed chiral aldehyde **435** with enolate anion **436**. Here, stereoisomer **437** was produced in 65% yield as a result of double stereodifferentiation.

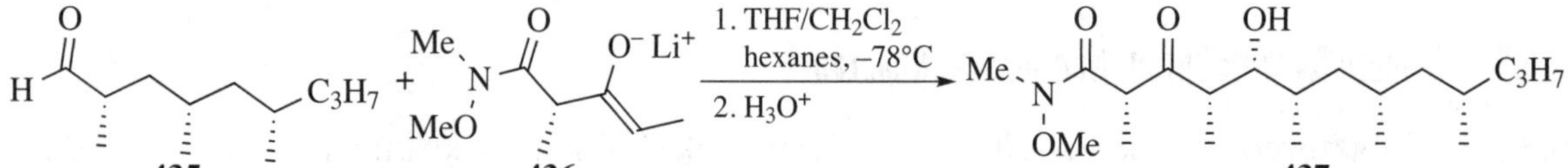

435   **436**   **437**

Improvements in diastereoselection arise from factors that influence the orientation of the reaction, and the face from which the two reactants approach. For simple diastereoselectivity, steric factors influence the orientation of the aldehyde and enolate, leading to good syn/anti selectivity using the models previously described. When a stereogenic center is incorporated in the enolate or aldehyde, increased steric hindrance affects the approach and transition state energies, leading to greater diastereoselectivity. The presence of the stereogenic center makes one face of the molecule more hindered than another when comparing the possible transition states, and a preference for one over another is called facial selectivity. As observed in Table 9.12,[269] when an aldehyde possessing a stereogenic center (such as **438**) was condensed with various enolate anions (**439**), the syn-diastereomer (**440**) predominated over the anti-diastereomer (**441**) in every case.

267. (a) Horeau, A.; Kagan, H.-B.; Vigneron J.-P. *Bull. Soc. Chim. Fr.* **1968**, 3795; (b) Izumi, Y.; Tai, A. *Stereodifferentiating Reactions*, Kodanshar Ltd., Tokyo, Academic Press, New York, *1977*.
268. Calter, M.A.; Liao, W. *J. Am. Chem. Soc.* **2002**, *124*, 13127.
269. Reference 244, p 166.

Table 9.12. Influence of an Asymmetric Aldehyde on Facial Selectivity in the Aldol Condensation with a Metal Enolate

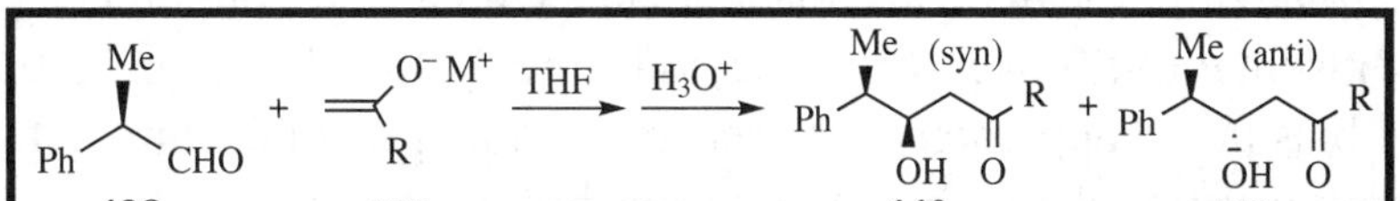

R	M	Temp. (°C)	440 : 441	% Yield
MeO	ZnBr	80 (in PhH)	74 : 26	78
MeO	Li	-78	75 : 25	91
t-BuO	Li	-78	80 : 20	70
Me	Li	-78	76 : 24	98
Me	c-(C$_5$H$_9$)$_2$B	0	65 : 35	
t-Bu	Li	-78	80 : 20	80
Me$_2$N	Li	-78	78 : 22	67

In most cases, Crams rule (sec. 4.7.B) predicts the major isomer if the reaction partner (or partners) contain a stereogenic center.[270] To understand how this rule applies to orientational and facial selectivity, we must understand the transition state of the reaction. The Zimmerman-Traxler model is used most often, and if it is applied to **438** and **439**, the syn selectivity can be predicted. The facial selectivity shown in **442** and **443** arises from the methyl group. In **443**, the enolate approaches from the face opposite the methyl, leading to diminished steric interactions and syn product (**444**). If the enolate approaches via **442**, the steric impedance of the methyl group destabilizes that transition state relative to **443**. In both **442** and **443**, a Cram orientation is assumed (see above), although other rotamers are possible. The appropriate rotamer for reaction is that where R$_L$ is anti to the carbonyl oxygen. Since the phenyl group is R$_L$, **442** and **443** are assumed to be the appropriate orientation for the aldehyde. If an aldehyde or ketone follows anti-Cram selectivity, this aldehyde orientation must be adjusted.

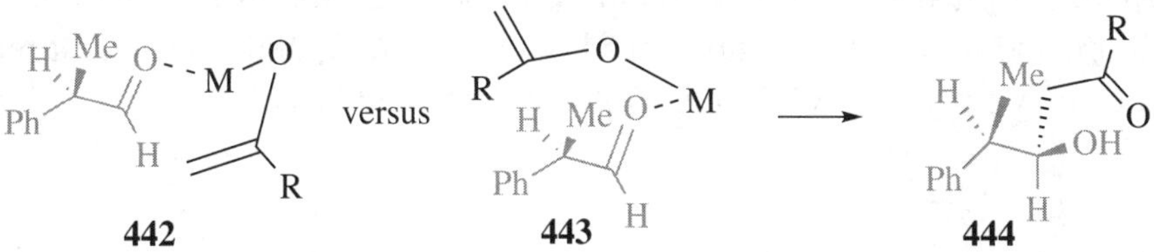

Cram selectivity (sec. 4.7.B) results from approach of the enolate over the less hindered face of the stereogenic center of the aldehyde (over the H, R$_S$) rather than over the larger methyl group (R$_M$). The phenyl group is taken to be R$_L$, and it is anti to the carbonyl oxygen. This model assumes a reasonable steric difference between R$_M$ and R$_S$ (as in the Cram model), and if the steric difference between the two groups is small, there is little facial selectivity in the transition state and poor enantioselectivity in the final product.

270. (a) Cram, D.J.; AbdElhafez, F.A. *J. Am. Chem. Soc.* **1952**, *74*, 5828; (b) Cram, D.J.; Kopecky, K.R. *Ibid.* **1959**, *81*, 2748.

The presence of a stereogenic center in the enolate imparts facial selectivity, and reaction of (S)-3-methyl-2-pentanone (**445**) was treated with LDA, chiral enolate **446** was formed.[271] Subsequent reaction with propanal gave the aldolates with a slight preference (57:43 to 63:36) for the anti (S,R) isomer (**447**) over the syn (S,S) isomer (**448**). Once again, the methyl group induced a facial bias in the approach of the reagent, leading to attack on the face opposite the methyl. The results were similar for both lithium and boron enolates. If the transition states for the boron enolates are examined (**449** and **450**),[272] the observed facial bias favors **449** as the least hindered transition state, *assuming an orientation of R_L in the transition state according to Cram's rule*, which leads to a preference for the (S,R) product.

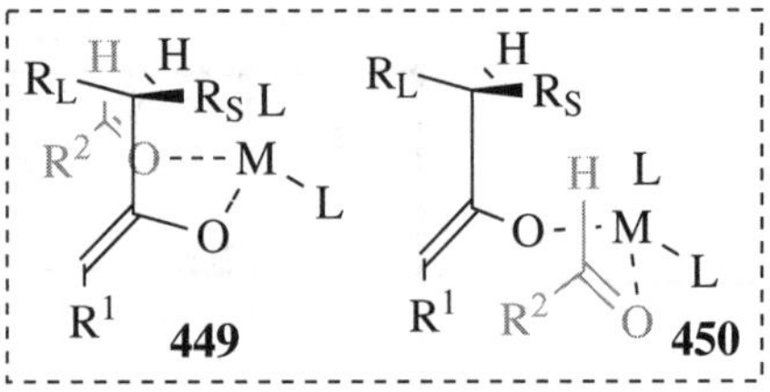

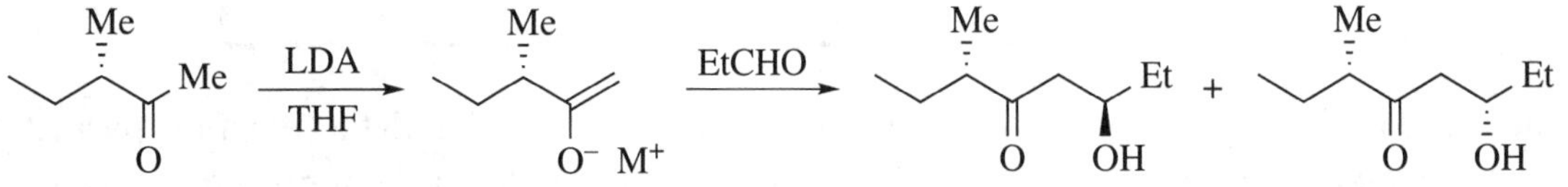

Diastereoface selection is observed when a chiral auxiliary is attached to either the enolates or the carbonyl substrate. A chiral auxiliary is an asymmetric group attached to an achiral substrate. The chirality of the auxiliary influences the course of the reaction (an aldol condensation) by inducing a chiral transition state, which influences the chirality of the final product. The auxiliary is cleaved to give the normal aldolate product. Oppolzer et al. developed a sulfonamide alcohol auxiliary based on camphor (**Oppolzer's sultam**) in which an acid derivative is attached as a chiral ester.[273] The auxiliary was removed by reduction of the ester group with LiAlH₄. A related sultam auxiliary also gave good results. In Mori's synthesis of (1S,3S,7R)-3-methyl-α-himachalene,[274] the sex pheromone of the sand fly *Lutzomyia longipalpis*, amide **451** was prepared from the appropriate acid chloride and Oppolzer's sultam.[275] Treatment with butyllithium and iodomethane gave an 81% yield of **452**, and subsequent removal of the auxiliary by basic hydrolysis gave **453** in 67% yield and >99% ee.

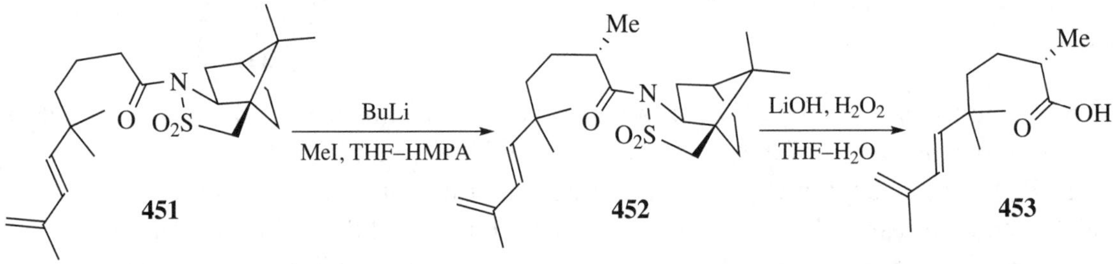

271. Seebach, D.; Ehrig, V.; Teschner, M. *Liebigs Ann. Chem.* **1976**, 1357.
272. Evans, D.A.; Taber, T.R. *Tetrahedron Lett.* **1980**, *21*, 4675.
273. Oppolzer, W.; Dudfield, P.; Stevenson, T.; Godel, T. *Helv. Chim. Acta* **1985**, *68*, 212.
274. Tashiro, T.; Bando, M.; Mori, K. *Synthesis* **2000**, 1852.
275. Oppolzer, W.; Chapuis, C.; Dupuis, D.; Guo, M. *Helv. Chim. Acta* **1985**, *68*, 2100.

In Leahy and co-worker's synthesis of rhizoxin D, a so-called **Evans' auxiliary** (the chiral, substituted oxazolidinone unit) was used in **454**.[276] Generation of the enolate anion with sodium hexamethyldisilazide was followed by reaction with allyl bromide to give a 73% yield of **455**. Helmchen and co-workers[277] achieved good asymmetric induction in the alkylation of ester enolates using a camphor based sulfonamide auxiliary.

MOLECULAR MODELING: Enolates that Hold Onto Metals

Up to this point, molecular modeling problems relating to enolate anions have not considered the possible role of the metal counterion. Were the metal to remain attached to an enolate, it could coordinate with other electron-rich centers and restrict conformational freedom. The net effect would be to limit the range of shapes available to the enolate, making it much easier to anticipate the product of electrophilic addition. For example, the conformational flexibility of compound **A** would be greatly reduced upon enolization to **B**, were the lithium to remain and coordinate with the carbonyl oxygen on the oxazolidinone ring.

Metal complexation might also be expected to affect the steric demands of the free enolate, perhaps changing the favored direction of approach of an electrophile. Examine space-filling models for the free and lithium-complexed enolates (top left and right). Is one face of the free enolate more crowded than the other face? Is there a clear "steric" preference for one product over the other? If there is, does this preference lead to the observed product? Does your conclusion change for the lithium-complexed enolate?

What effect does "holding onto to metal" have on the reactivity of an enolate anion. Is the lithium-complexed enolate **B** likely to be more or less susceptible to attack by an electrophile than the corresponding free enolate? Compare the local ionization potential map for the free enolate (bottom left) with that of the corresponding lithium complex **B** (bottom right). The maps are displayed as transparent models in order for to see the structure underneath. After you have oriented yourself, you can switch to solid models. *Click* on the map and select *Solid* from the **Style** menu at the bottom of the screen. Recall that the lower the local ionization potential (colored "red" in the map), the more likely the molecule will undergo reaction with the electrophile. According to the local ionization potential maps, is the free enolate or its lithium complex more reactive?

---

276. Lafontaine, J.A.; Provencal, D.P.; Gardelli, C.; Leahy, J.W. *J. Org. Chem.* **2003**, *68*, 4215.
277. Schmierer, R.; Grotemeier, G.; Helmchen, G.; Selim, A. *Angew. Chem. Int. Ed.* **1981**, *20*, 207.

Using a chiral additive often generates a reactive species that functions as a transient auxiliary, and constitutes another method for inducing enantioselectivity in aldol-like reactions. Koga and co-workers[278] reacted the achiral ketone acetophenone with a chiral, nonracemic base (**456**) in the presence of *N,N,N',N'*-tetramethylenediamine (TMEDA), to form an enolate coordinated chiral complex. Reaction of this complex with benzaldehyde at –100°C led to asymmetric induction in the final aldolate product (**457**), which was produced in 73% yield with high enantioselectivity for the (*R*)-enantiomer.[278] List and Notz showed that the addition of 30 mol% of L-proline induced an asymmetric aldol condensation between hydroxyacetone and cyclohexanecarboxaldehyde to give **458** in 60% yield, with >20:1 regioselectivity, >20:1 diastereoselectivity and >100:1 enantioselectivity.[279] The use of chiral Lewis bases in the aldol reaction has been reviewed,[280] as have catalytic aldol reactions.[281]

The presence of a chiral auxiliary can influence the stereochemistry of products formed in a Mukaiyama aldol reaction. A stereogenic center derived from a chiral amino alcohol was incorporated in the enol ether moiety (see **459**). When reacted with benzaldehyde and TiCl₄, the alcoholate products were **460** and **461** (R* = the chiral auxiliary). The diastereoselectivity was quite good (95:5 favoring **460**), and each product was formed with high enantioselectivity.[282] It is also possible to incorporate the stereogenic center into the starting material (a **chiral template**, as discussed in sec. 10.9).[283]

278. Muraoka, M.; Kawasaki, H.; Koga, K. *Tetrahedron Lett.* **1988**, *29*, 337.
279. Notz, W.; List, B. *J. Am. Chem. Soc.* **2000**, *122*, 7386.
280. Denmark, S.E.; Stavenger, R.A. *Acc. Chem. Res.* **2000**, *33*, 432.
281. Machajewski, T.D.; Wong, C.-H. *Angew. Chem. Int. Ed.* **2000**, *39*, 1352.
282. Gennari, C.; Molinari, F.; Cozza, P.; Oliva, A. *Tetrahedron Lett.* **1989**, *30, 5163.*
283. Woodward, R.B.; Logusch, E.; Nambiar, K.P.; Sakan, K.; Ward, D.E.; Au-Yeung, B.-W.; Balaram, P.; Browne, L.J.; Card, P.J.; Chen, C.H.; Chênevert, R.B.; Fliri, A.; Frobel, K.; Gais, H.-J.; Garratt, D.G.; Hayakawa, K.; Heggie, W.; Hesson, D.P.; Hoppe, D.; Hoppe, I.; Hyatt, J.A.; Ikeda, D.; Jacobi, P.A.; Kim, K.S.; Kobuke, Y.; Kojima, K.; Krowicki, K.; Lee, V.J.; Leutert, T.; Malchenko, S.; Martens, J.; Matthews, R.S.; Ong, B.S.; Press, J.B.; Rajan Babu, T.V.; Rousseau, G.; Sauter, H.M.; Suzuki, M.; Tatsuta, K.; Tolbert, L.M.; Truesdale, E.A.; Uchida, I.; Ueda, Y.; Uyehara, T.; Vasella, A.T.; Vladuchick, W.C.; Wade, P.A.; Williams, R.M.; Wong, H.N.-C. *J. Am. Chem. Soc.* **1981**, *103,* 3210.

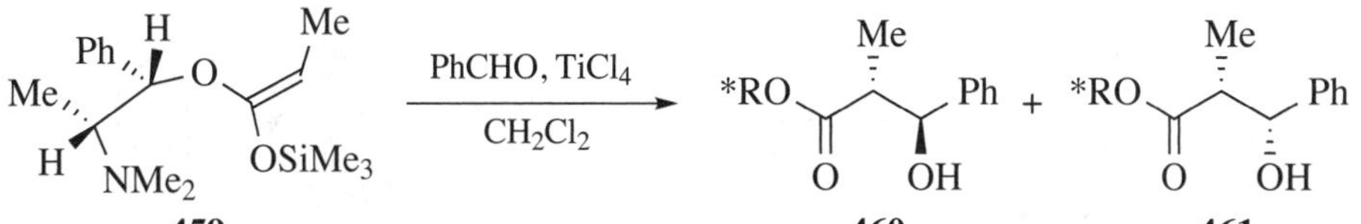

459

9.5.B.ii. Double Stereodifferentiation. Double stereodifferentiation employs a stereogenic center on both the enolate and the aldehyde or ketone (see **435** above). The two reactants may have incipient chirality, but the nature of the stereogenic centers lead to a lower energy transition state during the reaction (**consonant** double stereodifferentiation, matched). Alternatively, that chirality can lead to a higher energy transition state (**dissonant** double stereodifferentiation, mismatched).[257,284] This concept of matched and mismatched was introduced in the discussion of the **Sharpless asymmetric epoxidation** in section 3.4.D.i. The reaction of chiral ketone **462** with LDA generates the corresponding enolate anion, and subsequent reaction with benzaldehyde occurs with high selectivity for the syn isomer (**463** rather than **464**).

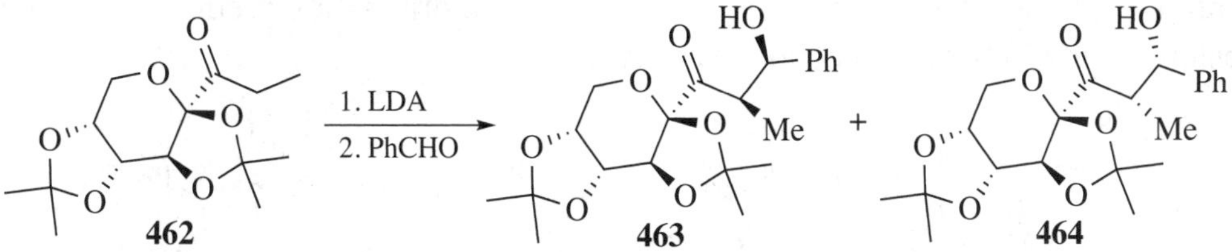

462   463   464

This reaction demonstrates the incipient selectivity of an aldol condensation with chiral ketone **465**. If we replace benzaldehyde with chiral aldehydes, we have an example of matched versus mismatched reactions in an aldol condensation. The condensation of (*R*)-aldehyde **465** or the S aldehyde **470** with enolates derived from ketone **463** is illustrative.[284,243] When **463** reacted with the mismatched aldehyde **466**, the selectivity was poor due to dissonance in the transition state, with formation of a 62:38 mixture of **467** and **468**. Some anti diastereomers were also produced, which is quite different from the reaction of **462** with **470**, the matched aldehyde. Consonant stereodifferentiation led to excellent selectivity for a single isomer (**471**), produced in > 97% yield. The steric effects of the consonance versus the dissonance approach can be described in terms of the generalized Zimmerman-Traxler models **472** and **473**. In **472**, the R and R^2 groups are oriented away from each other (matched), minimizing the steric interaction. In **473**, however, the R↔R^2 interaction is maximized (mismatched), destabilizing that transition state.

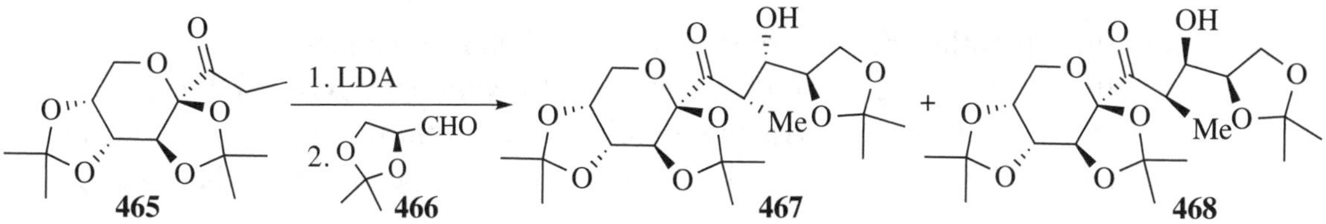

465   466   467   468

284. Heathcock, C.H.; White, C.T. *J. Am. Chem. Soc.* **1979**, *101*, 7076.

469 470 471

472 473

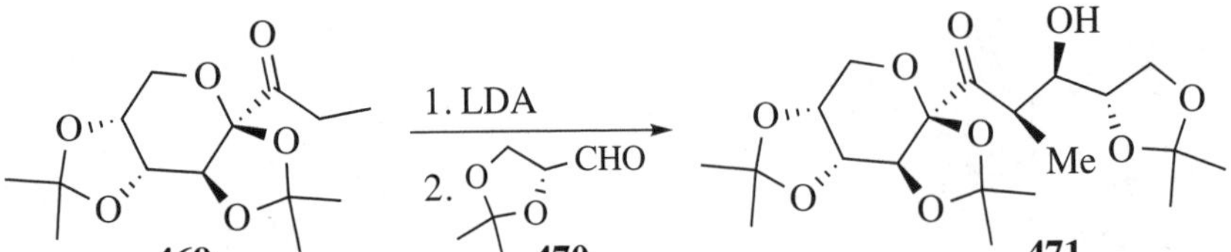

This selectivity is seen in the condensation of chiral aldehyde **474** with the chiral enolate anion **475**. The matched transition state has the methyl groups oriented away from each other, which is lower in energy than the opposite case in which the methyl groups interfere with each other (both options assume a Cram orientation). The matched orientation predicts formation of the anti diastereomer **476** for this reaction, as shown.

474 475 476

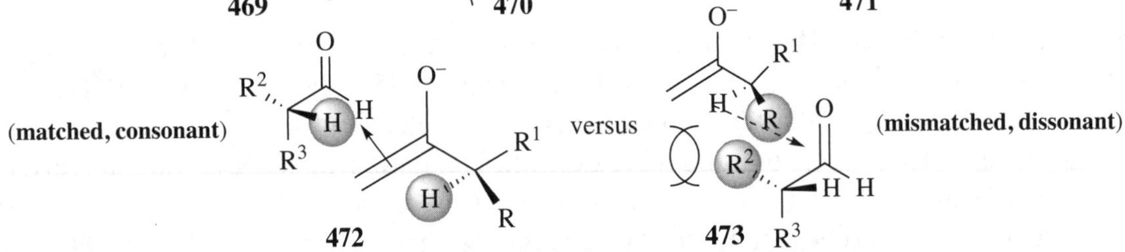

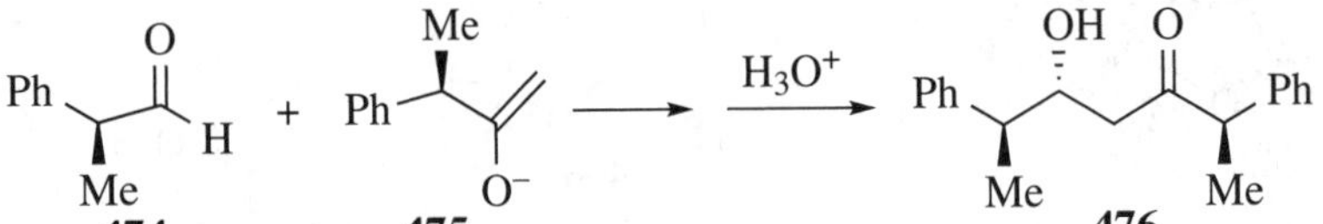

When applied to the Mukaiyama reaction, stereodifferentiation leads to excellent diastereoselection and enantioselection. When silyl ketene acetal **477** was reacted with chiral aldehyde **478** in the presence of $TiCl_4$, a 75% yield of **479** and **480** was obtained as a 98:2 mixture favoring **479**.[285]

477 478 479 480

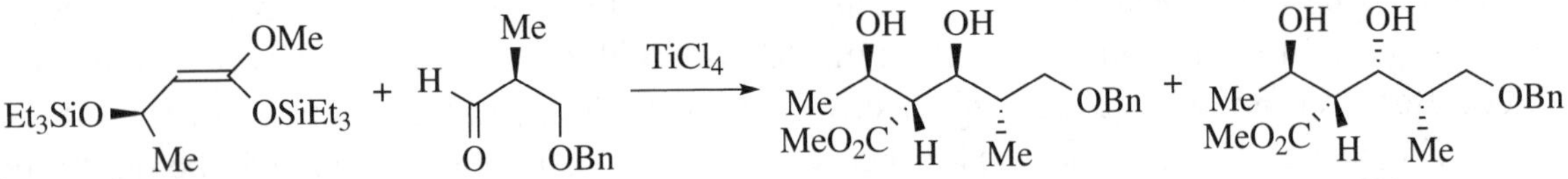

9.5.C. Chelation Control

Chelating substituents influence the stereoselectivity of enolate condensation and alkylation reactions, usually by generating a complex that is conformationally rigid. When β-hydroxy-ester **481** was treated with two equivalents of LDA, the dilithio derivative was generated. Lithium was chelated to both oxygens of the dianion, and this type of intermediate was seen in

285. Shirai, F.; Nakai, T. *Tetrahedron Lett.* ***1988***, *29,* 6461.

previous sections (see **422** and **456**).[286,29] Reaction of the (*Z*) enolate (**482**) with an electrophile leads to its delivery from the less hindered face, giving the anti product **484**. In the (*E*)-enolate (**483**), chelation again leads to delivery of the electrophile from the less hindered face (away from R[1]), and **484** is again the major product.[286,29] If the hydroxyl group is blocked (protected), the chelating effect is greatly diminished or removed. Reaction of $OSiR_3$ derivatives of **481**, where the chelating effect is removed because there is no O–H group, with LDA/THF and an electrophile gave the opposite aldolate ratio of stereoisomers with an (*E/Z*) ratio of > 90:10.[286,29] Kraus and Taschner used this analysis (structures **482** and **483**)[286] to predict the stereoselectivity of an epoxidation reaction that gave glycidic esters (sec. 9.4.B.vi). The model was later applied to enolate alkylations and condensations, as shown.[29]

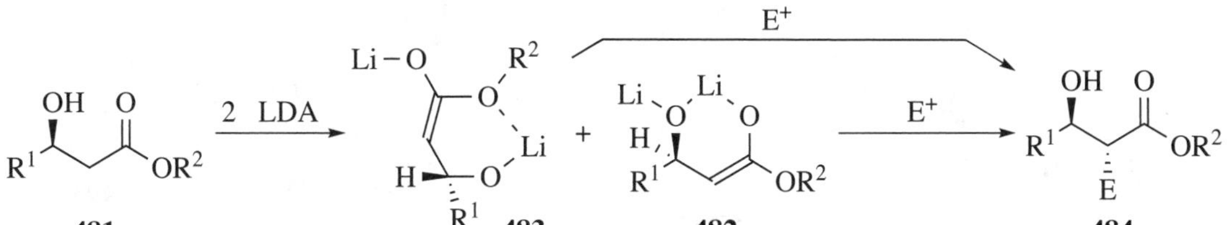

Evans and Takacs prepared several chiral auxiliaries derived from amino alcohols such as valinol or prolinol, know known as **Evans' auxiliaries**. Prolinol amides such as **485** preferentially form the (*Z*)-enolate (**486**) over the (*E*)-enolate (**487**). Alkylation proceeds with chelation control and good diastereoselectivity (from the *si* face) to give the alkylated products **488** and **489**, favoring **488**.[287,29] The (*Z*) enolate (**486**) is preferred over **487**, since **486** is chelated and **487** is not. Addition of the electrophile from the less hindered face of each enolate anion gives the observed products. Evans and co-workers also prepared asymmetric carbamates that showed opposite diastereoface selectivity.[288] In both cases; small alkyl halides such as methyl iodide showed less stereoselectivity than the more sterically demanding benzyl bromide.

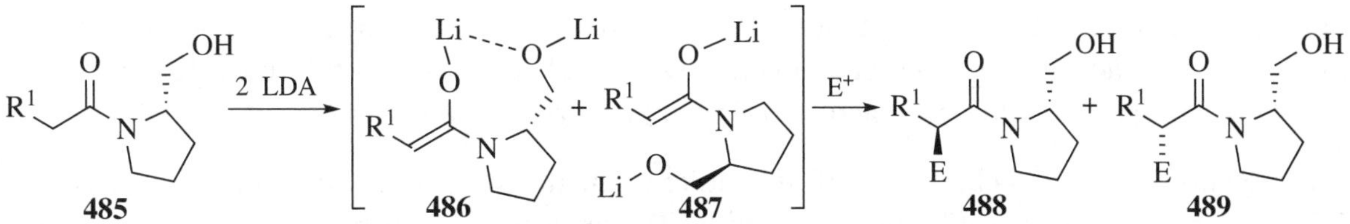

Diastereoselectivity is observed in reactions of carbanions derived from imines and hydrazones when those species contain a stereogenic center or a chiral auxiliary (sec. 9.4.F). Asymmetric imines can be used, and chiral oxazoline derivatives have also been prepared and used in the alkylation sequence (sec. 9.3.A). Meyers and co-workers showed that chiral oxazoline

286. Kraus, G.A.; Taschner, M.J. *Tetrahedron Lett.* **1977**, 4575.
287. (a) Evans, D.A.; Takacs, J.M. *Tetrahedron Lett.* **1980**, *21*, 4233; (b) Sonnett, P.E.; Heath, R.R. *J. Org. Chem.* **1980**, *45,* 3138.
288. Evans, D.A.; Ennis, M.D.; Mathre, D.J. *J. Am. Chem. Soc.* **1982**, *104*, 1737.

490 could be alkylated to give the ethyl derivative **491**.[289] A second alkylation generated the diastereomeric product **492**, and hydrolysis provided the chiral lactone (**493**) in 58% yield and with a selectivity of 70% ee for the (*R*) enantiomer.[289] As pointed out in Section 9.4.F.ii, hydrazone carbanions can be used for alkylation or condensation reactions. In a synthesis of (–)-callystatin A,[290] Enders and co-worker's prepared the chiral hydrazone **495** (by reaction of the aldehyde with the hydrazone known as **SAMP** [**494**, *N*-amino-(2*S*)-(methoxymethyl) pyrrolidine)]),[291] and showed that treatment with LDA and then iodomethane gave the alkylated product. Ozonolysis cleaved the SAMP group to give methylated aldehyde **496** in 73% overall yield from **495**. The enantiomer of **494** is known as **RAMP**, *N*-amino-(2*R*)-(methoxymethyl) pyrrolidine) and both enantiomers have been applied in organic synthesis.[292]

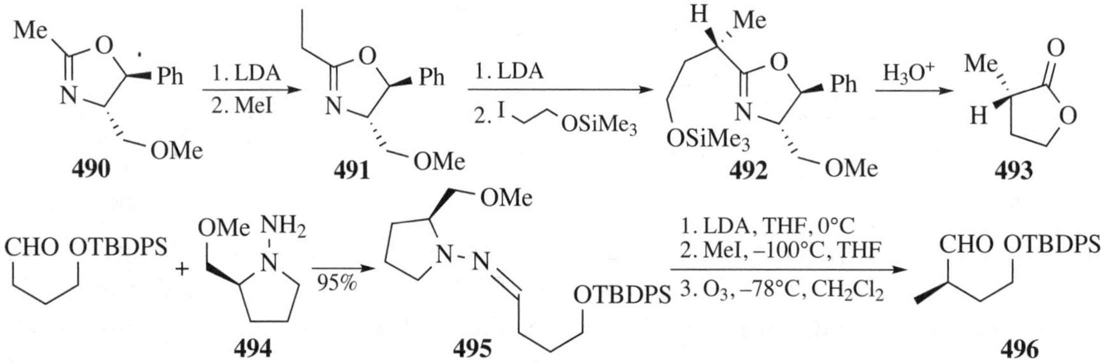

9.5.D. Diastereoselectivity in Alkylidene Enolates

Enolates derived from cyclic compounds cyclohexane carboxylic acid or cyclohexanecarboxaldehyde generate enolate anions that are unique. These enolates have an exocyclic double bond that can exist as (*E*) and (*Z*) isomers. The facial and orientational bias in alkylation and condensation reactions of such enolates is influenced by the conformation of the ring it is attached to. Alkylidene cyclohexane enolates show a preference for equatorial attack, as observed in addition reactions of cyclohexanone derivatives (sec. 4.7.C,D).

The position and stereochemistry of substituents on the ring influences the selectivity. In 1,2-chiral systems such as **497**, treatment with LDA leads to an equilibrating mixture of enolate anions (**498A** and **498B**). The destabilizing Me↔OMe interaction in **498A** is absent in **498B**,

289. (a) Meyers, A.I.; Yamamoto, Y.; Mihelich, E.D.; Bell, R.A. *J. Org. Chem.* **1980**, *45*, 2792; (b) Hoobler, M.A.; Bergbreiter, D.Z.; Newcomb, M. *J. Am. Chem. Soc.* **1978**, *100*, 8182.
290. Vicario, J.L.; Job, A.; Wolberg, M.; Müller, M.; Enders, D. *Org. Lett.* **2002**, *4*, 1023.
291. (a) Enders, D.; Eichenauer, H. *Chem. Ber.* **1979**, *112*, 2933; (b) Enders, D.; Frey, P.; Kipphardt, H. *Org. Synth.* **1987**, *65*, 173.
292. See (a) Job, A.; Janeck, C.F.; Bettray, W.; Peters, R.; Enders, D. *Tetrahedron* **2002**, *58*, 2253; (b) Enders, D.; Lenzen, A.; Backes, M.; Janeck, C.; Catlin, K.; Lannou, M.-I.; Runsink, J.; Raabe, G. *J. Org. Chem.* **2005**, *70*, 10538.

and reaction with CH₃I proceeds from the less sterically hindered path *b* to give **499** rather than **500**.[293,28] When R in **497** was hydrogen a 70:30 mixture of **499/500** was obtained, but when R was methyl an 80:20 mixture was obtained, favoring **499**. This result is opposite to results observed in the reduction of cyclohexanone derivatives (sec. 4.7.C). One explanation for this difference is that enolate anions such as **498A** exist as aggregates (dimers or tetramers) that increases the blockage to attack via path *a*.

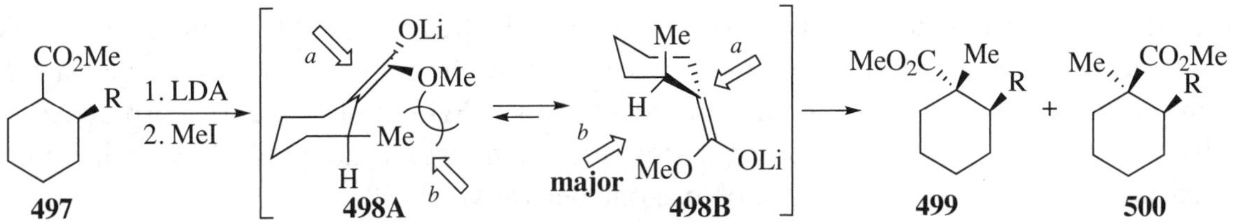

A 1,3-disubstituted system is shown with ester **501**, where treatment with LDA generates the equilibrating enolates **502A** and **502B**, favoring **502B**. When the enolate precursor was the acid (R = H in **501**), alkylation with iodomethane showed a slight preference for formation of **504** over **503**. Note that reaction of the acid proceeded via formation of a dianion. When R¹ was methyl (R = H in **501**), however, a 52:48 mixture of **503/504** was obtained. The methyl ester of **501** (R = methyl) showed a distinct preference for **503** when R¹ was OMe, where a 78:22 mixture of **503/504** was obtained. When R¹ = methyl and R = OMe, a 90:10 mixture of **503/504** was obtained. This mixture is explained by delivery of iodomethane from path *a* in **502B** (equatorial attack).[292b] Presumably, steric interactions with the OR group of the enolate moiety are minimal (or equal) in both conformations, making the conformation with the equatorial R¹ group (**502B**) preferred. In the 1,4-disubstituted system (**505**), reaction with LDA generated the alkylidene enolate **506**. Alkylation with iodomethane proceeded via path *b* (interaction with the 1,3-diaxial hydrogens in **506** destabilized attack along path *a*) to give the major product (**508**).[292b,294] An 84:16 mixture of **507/508** was obtained when R was either OMe or *t*-Bu. Addition of the electrophile favored equatorial alkylation (path *b* to give **507**), but a mixture of **507** and **508** was obtained.

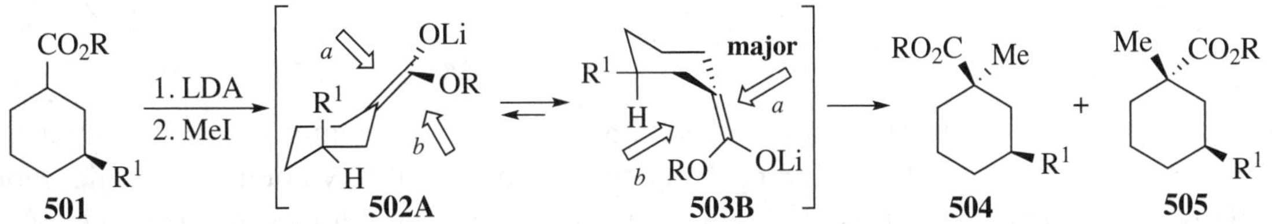

---

293. (a) Krapcho, A.P.; Dundulis, E.A. *J. Org. Chem.* **1980**, *45*, 3236; (b) Johnson, F.; Malhotra, S.K. *J. Am. Chem. Soc.* **1965**, *87*, 5492, 5493.
294. House, H.O.; Bare, T.M. *J. Org. Chem.* **1968**, *33*, 943.

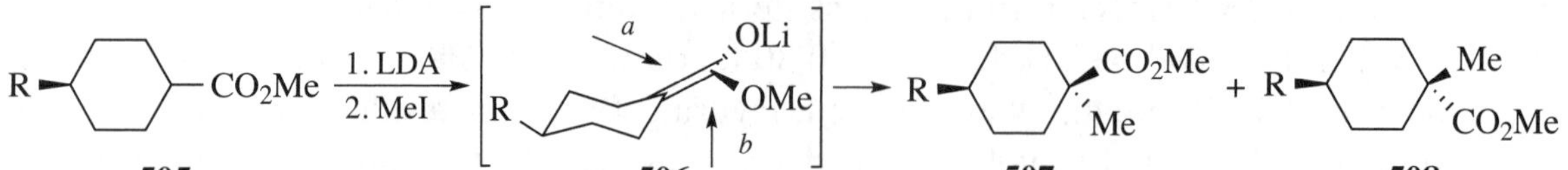

Five- and six-membered ring enolates derived from cyclopentanone and cyclohexanone derivatives have an endocyclic double bond, and reactions with an electrophile occur from the sterically less hindered face. Enolate **510** was obtained by treatment of **509** with lithium amide. Subsequent reaction with an alkyl halide led to delivery of the halide from the face opposite the alkenyl group (path *a*), to give the trans product shown (**511**) in 60% yield.[295] Approach via path *b* would have serious steric consequences, and that transition state is destabilized. Similar effects are observed with 3-alkyl-cyclohexanone derivatives.[296,294,295]

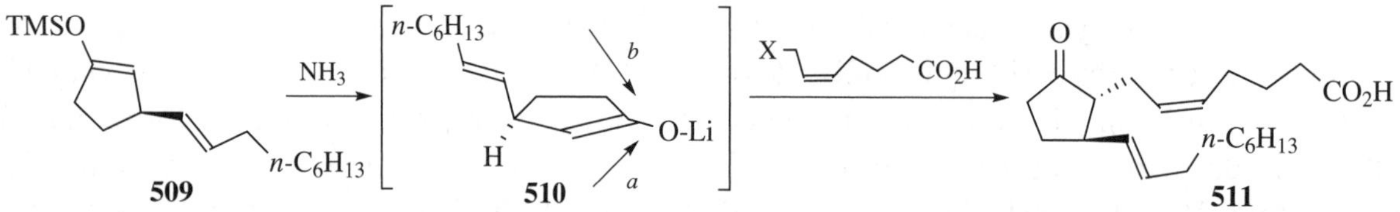

Delivery of an electrophile to the less hindered face of an enolate also occurs in intramolecular alkylation reactions. When **512** was treated with potassium *tert*-butoxide, a mixture of (*E*) and (*Z*) enolates (**513** and **514**, respectively) was obtained. Intramolecular displacement of bromide generated a single isomer (**515**).[297] In this case, the electrophile can approach the enolate from only one face (the bottom or α face). Because of this conformational constraint, both (*E*) and (*Z*) enolates lead to the same product. In cyclopentanone and cyclohexanone enolates, an increase in the size of a facial blocking group increases selectivity. When that group was small, the selectivity decreased.

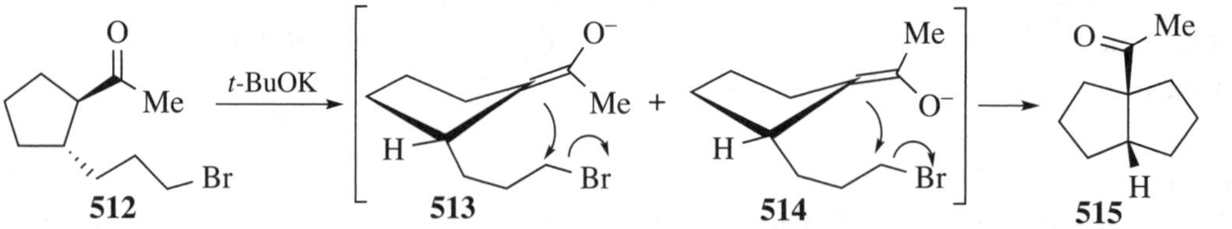

The preference for the lowest energy conformation of the enolate anion is seen in larger ring systems as well (sec. 1.5.B,C), leading to good selectivity in alkylation and condensation reactions. The methyl group provides only small steric encumbrance to approach of the electrophile in enolate **517** (derived from lactone **516** and LDA). The preferred mode of attack for this relatively stable conformation was from the top face (path *a*, pseudo-equatorial attack)

---

295. Patterson, Jr., J.W.; Fried, J.H. *J. Org. Chem.* **1974**, *39*, 2506.
296. Posner, G.H.; Sterling, J.J.; Whitten, C.E.; Lentz, C.M.; Brunelle, D.J. *J. Am. Chem. Soc.* **1975**, *97*, 107.
297. House, H.O.; Sayer, T.S.B.; Yau, C.-C. *J. Org. Chem.* **1978**, *43*, 2153.

and gave the syn diastereomer (**518**) with >99:1 selectivity.[298]

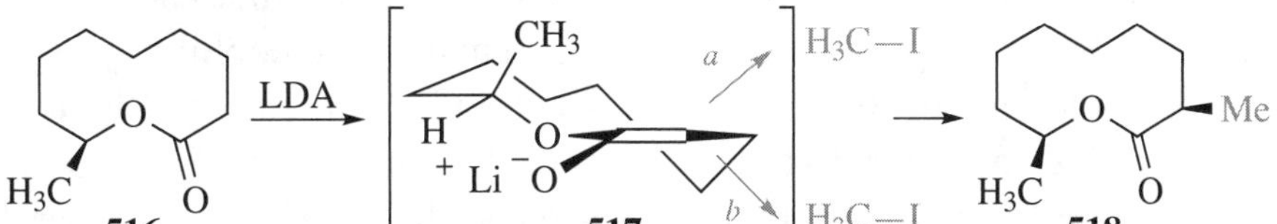

9.6 ENAMINES

9.6.A. Preparation of Enamines

An older method is very effective in controlling the kinetic-thermodynamic product. In 1954, Stork and co-workers[299] reported that cyclohexanone reacted with pyrrolidine to give the corresponding enamine (**519**),[300] a term introduced by Wittig and Blumenthal.[301] An enamine is essentially a nitrogen enolate, and it can react with alkyl halides such as iodomethane to give an equilibrium mixture of the iminium salt (**510**) and the alkylated enamine (**511**).[299] Hydrolysis with aqueous acid led to loss of pyrrolidine and formation of 2-methylcyclohexanone. A more general sequence shows the behavior of a generic enamine (**522**) reacting with a halide via the β-carbon to give α-alkyl iminium salt, via an S_N2 process. The intermediate iminium salt (**523**) can isomerize under the reaction conditions to the less substituted enamine **524** in an equilibrium process, and hydrolysis yields the corresponding α-alkyl carbonyl compound. Alkylation gives the best yields with reactive primary halides, since it is essentially an S_N2 reaction. This sequence is referred to as the **Stork enamine synthesis**.[299,302] Enamines are usually formed by reaction of a secondary amine with a ketone, in the presence of an acid catalyst.[303]

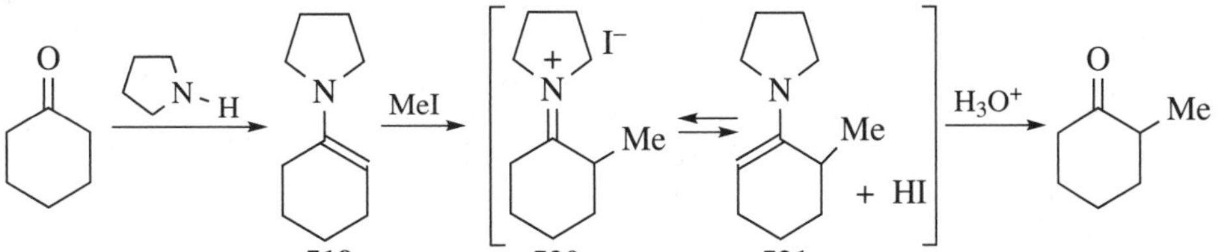

298. Still, W.C.; Galynker, I. *Tetrahedron* **1981**, *37*, 3981.
299. Stork, G.; Terrell, R.; Szmuszkovicz, J. *J. Am. Chem. Soc.* **1954**, *76*, 2029.
300. For an older review of the chemistry of enamines see Cook, A.G. Ed., *Enamines: Synthesis Structure and Reactions*, Marcel Dekker, New York, **1969**.
301. Wittig, G.; Blumenthal, H. *Berichte,* **1927**, *60*, 1085.
302. (a) The Merck Index, 14th ed., Merck & Co., Inc., Whitehouse Station, New Jersey, **2006**, p ONR-90; (b) Mundy, B.P.; Ellerd, M.G.; Favaloro Jr., F.G. *Name Reactions and Reagents in Organic Synthesis, 2nd ed.*, Wiley-Interscience, New Jersey, **2005**, pp. 628-629.
303. (a) Hayne, L.W. in Reference 300, pp 55-100; (b) Herr, M.E.; Heyl, F.W. *J. Am. Chem. Soc.* **1952**, *74*, 3627; (c) Heyl, F.W.; Herr, M.E. *Ibid.* **1953**, *75*, 1918; (d) Herr, M.E.; Heyl, F.W. *Ibid.* **1953**, *75*, 5927.

Enamines such as **522** are bidentate nucleophiles, with both nitrogen and carbon functioning as nucleophiles.[304] Alkylation at nitrogen is a problem with reactive halides such as methyl iodide or allyl bromide, and when the groups on nitrogen are small. Nucleophilicity usually parallels that of the unsubstituted amine. In general, the kinetic nitrogen enolate (the less substituted enamine) is formed due to reduced steric hindrance.[305] The enamine exists primarily as a planar species (**523** or **524**) in which the lone pair electrons on nitrogen are parallel with the π bond. If the steric interaction (R↔R¹) in **525** is greater than the R↔Me interaction in **526**, there is a preference for that kinetic nitrogen enolate. When diethylamino or dimethyl-amino moieties are used, these interactions are small, and a mixture of **525** and **526** is obtained. As R and R¹ increase in size, the steric encumbrance increases, favoring **526**. Pyrrolidine, piperidine, morpholine or diethylamine are the most common amine precursors to enamines. It is noted that when aldehydes such as **527** react with secondary amines, initial reaction gives an iminium salt. Reaction with excess amine leads to aminal **528** rather than to the enamine. Treatment with a base such as sodium carbonate is required to induce elimination of the amine (pyrrolidine in this case), to give enamine **529**.[306]

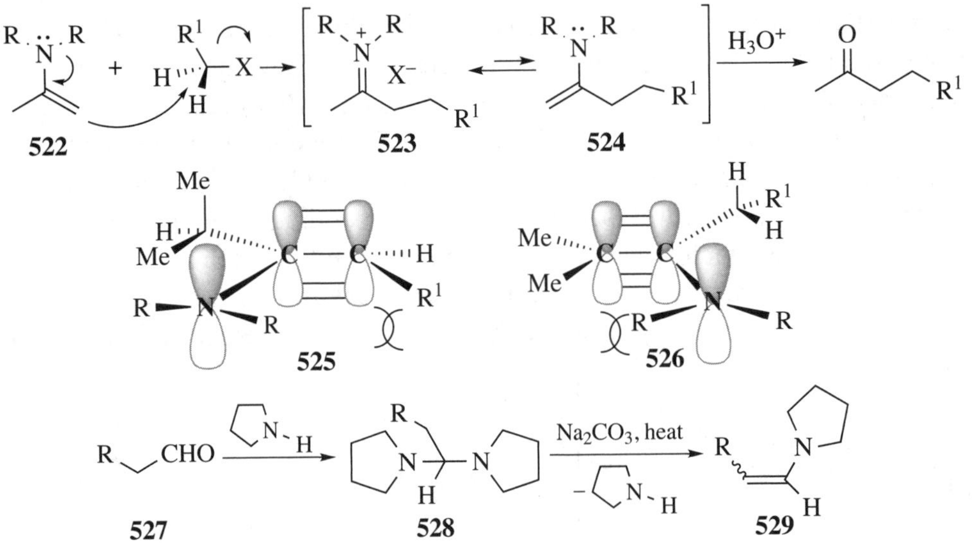

9.6.B. Reactions of Enamines

As noted above, enamines behave as nitrogen enolates in their reactions with alkyl halides, generating an iminium salt (see **523**) that can isomerize to the less substituted enamine (see **524**). Although this new enamine can react with additional halide, it is present in very small

304. For a review of enamine chemistry, as well as that of imines and oximes, see Adams, J.P. *J. Chem. Soc. Perkin Trans. 1*, **2000**, 125.
305. (a) Stork, G.; Brizzolara, A.; Landesman, H.; Szmuszkovicz, J.; Terrell, R. *J. Am. Chem. Soc.* **1963**, *85*, 207; (b) Kuehne, M. *Ibid.* **1959**, *81*, 5400; (c) Kuehne, M. *Ibid.* **1962**, *84*, 837.
306. Mannich, C.; Davidsen, H. *Berichte* **1936**, *69*, 2106.

quantities in most cases, and overalkylation is not a serious side reaction. If an excess of the halide is used, however, the second alkylation can be used to advantage. A typical enamine synthesis is taken from Carpenter and Davis' work, in which ketone **530** reacted with pyrrolidine to give **531**.[307] Reaction with ethyl bromoacetate, followed by hydrolysis gave **532**. Enamines can react with aryl halides in addition to alkyl halides, but the aryl halide must be activated by the presence of electron-withdrawing groups since this coupling reaction is formally a nucleophilic aromatic substitution (sec. 2.11.C).[308,305c] As noted above, pyrrolidine enamines are usually too reactive at carbon for *N*-acylation to compete.

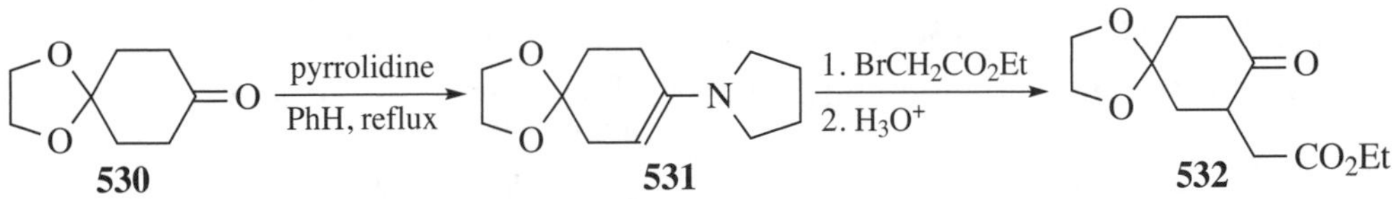

Enamines react with α,β-unsaturated compounds almost exclusively by conjugate addition to give the corresponding substituted ester, ketone or nitrile. Reaction of **519** with acrylonitrile, for example, gave a new enamine (**533**). Subsequent hydrolysis liberated the substitute ketone (**534**) in 80% overall yield from enamine **519**.[305a,309]

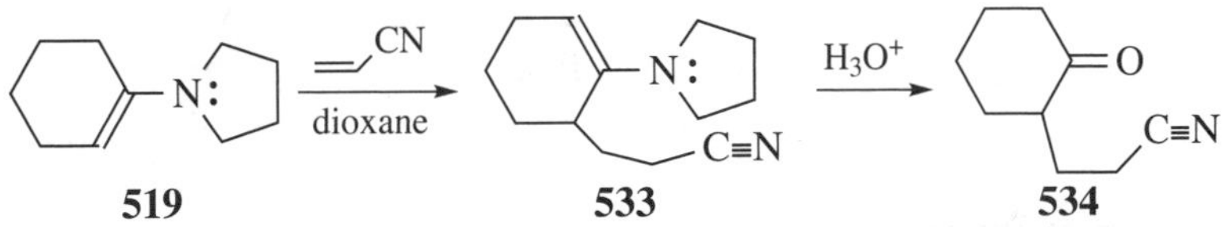

With the exception of pyrrolidine enamines noted above, enamines react very efficiently with acyl halides and the product after hydrolysis is a dicarbonyl compound. The morpholine enamine of isobutryaldehyde (**535**) reacted with acetyl chloride to give keto-iminium salt **536**. Hydrolysis gave keto aldehyde **537** in 66% yield.[310] Enamines react with acid chlorides in the

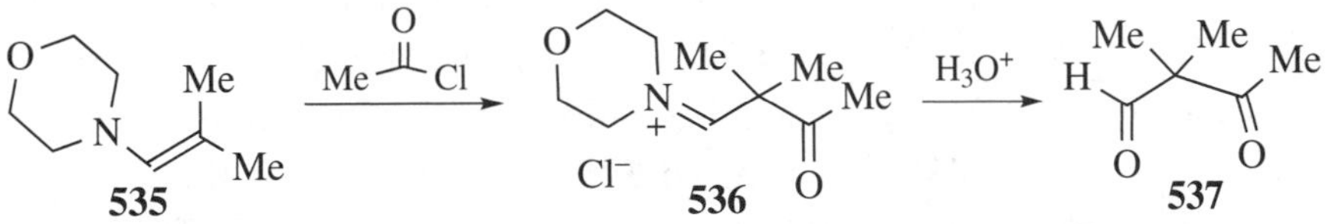

presence of a base to give a ketene, and this reacts via a thermal [2+2]-cycloaddition to give a cyclobutanone derivative (see sec. 11.10.A). Ketene formation is usually not a significant problem in acylation reactions, since the enamine is the only base in the system.[311] Morpholino enamines of cyclic ketones usually give the best yield of ketone product on reaction with acyl halides. A side reaction that is sometimes important is an aldol-like reaction between enamines

307. Davis, K.M.; Carpenter, B.K. *J. Org. Chem.* **1996**, *61*, 4617.
308. Alt, G.H. in Reference 300, p 131.
309. Williamson, W.R.N. *Tetrahedron* **1958**, *3*, 314.
310. (a) Alt, G.H. in Reference 300, pp 135-136; (b) Inukai, T.; Yoshizawa, R. *J. Org. Chem.* **1967**, *32*, 404.
311. Hünig, S.; Benzing, E.; Lücke, E. *Berichte* **1957**, *90*, 2833.

and aldehydes, which are both in solution,[312] but it is not a serious problem when the enamine is derived from a ketone.

Prior to the development of kinetic control in enolate anion condensation reactions, enamines were used extensively in synthesis.[313,314] An example is the synthesis of lupinine (**541**) using the optically active chlorocarbonate **539** and octahydroquinolizine **538**.[315] The enamine product of the initial reaction (**540**) was isolated in this case and selectively reduced to lupinine (**541**). Other reactions are possible with enamines. Cossy and Belotti reported that cyclohexene-enamines can be converted to aromatic amines.[316] In their work, cyclohexanone and pyrrolidine gave 1-pyrrolidinecyclohexene (**542**) and this was converted to 1-phenylpyrrolidine (**543**) in 83% yield by heating with palladium on carbon in the presence of nitrobenzene.

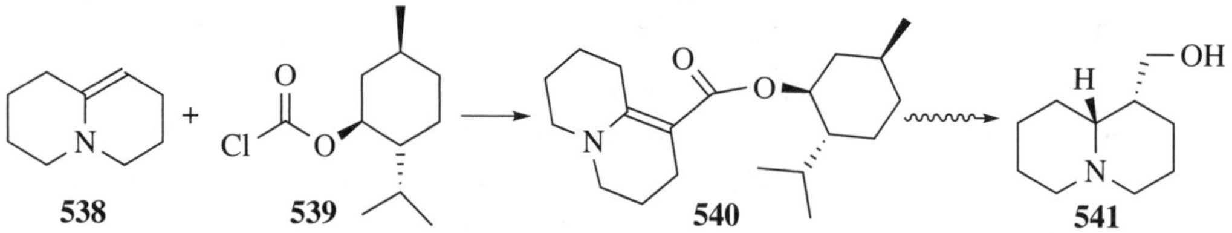

9.6.C. Asymmetric Enamine Syntheses

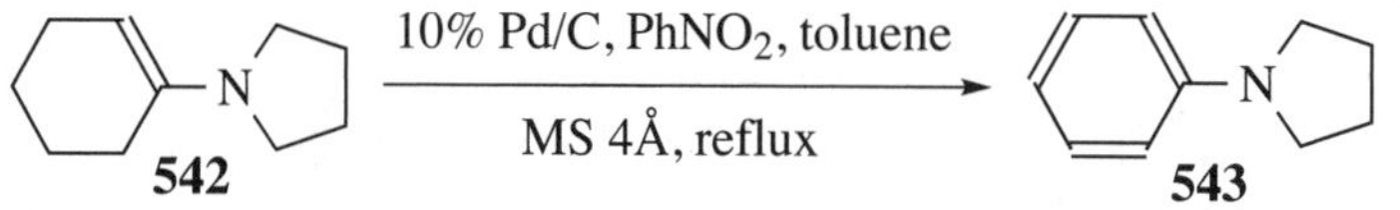

Asymmetric enamine derivatives generally give moderate-to-good diastereoselectivity.[317] The synthesis of lupinine (**541**) provided one example where an acid chloride substrate possessed a chiral auxiliary. An alternative approach is to prepare the enamine from a chiral amine.

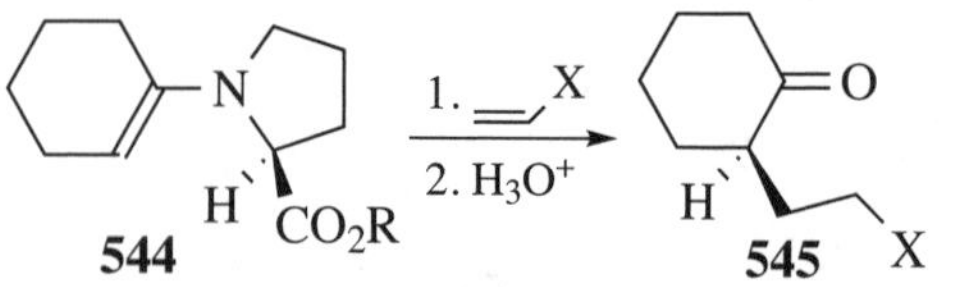

Yamada et al. prepared proline ester enamines such as **544**. In this case, the chiral enamine was used in a Michael addition with acrylonitrile or methyl acrylate to give **545**.[318] The yield of the ketone product was rather poor, and the asymmetric induction (% ee) was in the range 15-59% with acrylonitrile and methyl acrylate.[318] The chiral amine precursor to the enamine was synthesized from proline derivatives. When this chiral amine was distilled extensive racemization occurred, which is a major drawback to a practical use of this procedure. In addition, some racemization

312. (a) Alt, G.H. in Reference 300, pp 156-165; (b) Birkofer, L.; Kim, S.M.; Engels, H.D. *Berichte* **1962**, *95*, 1495.
313. Kuehne, M.E. in Reference 300, pp 313-468
314. Lansbury, P.T.; Wang, N.Y.; Rhodes, J.E. *Tetrahedron Lett.* **1972**, 2053.
315. Goldberg, S.I.; Ragade, I. *J. Org. Chem.* **1967**, *32*, 1046.
316. Cossy, J.; Belotti, D. *Org. Lett.* **2002**, *4*, 2557.
317. Seebach, D.; Imwinkelried, R.; Weber, T. in *Modern Synthetic Methods 1986*, Scheffold, R. Ed., Springer-Verlag, Berlin, **1986**, pp 217-246.
318. (a) Yamada, S.; Hiroi, K.; Achiwa, K. *Tetrahedron Lett.* **1969**, 4233; (b) Yamada, S.; Otani, G. *Ibid.* **1969**, 4237.

occurred during isolation of the enamine product (**544**). Failure to resolve these problems led to the use of crude amines and enamine products in all subsequent reactions.[319] It is possible to obtain good enantioselectivity with chiral enamines, if the chiral amine precursor is carefully chosen. Whitesell and Felman found that enamine **546** reacts with alkyl halides to give chiral ketones such as **547** with good optical yields (+ MeI = 83% ee; + *n*-PrI = 93% ee; + CH_2=$CHCH_2Br$ = 87% ee).[320]

The disconnections for enamine reactions are the same as those shown for enolates.

9.7. MICHAEL ADDITION AND RELATED REACTIONS

9.7.A. Michael Addition

The 1,4- (conjugate) addition[321e] of a carbon nucleophile to an α,β-unsaturated carbonyl system is usually reversible, and referred to as **Michael addition**.[321] The term Michael addition is also applied to the conjugate addition of amines to unsaturated carbonyl compounds. The addition of dimethylamine to **548** in water,[322] for example, generated **549** and illustrates the Michael addition of an amine. The rate acceleration effect observed when the reaction is done in water makes this

particular reaction less common. Enolates and carbanions are common partners in Michael additions.[323] The anion of 1-phenyl-1,3-butanedione (**550**) generated *in situ* by reaction of the diketone with DBU, reacted with ethyl acrylate to give the Michael addition product, enolate anion **551**. Quenching the enolate anion with aqueous acid generated diketo ester **552** in 90% yield, in Taber and co-worker's synthesis of the ethyl ester of the major urinary metabolite of prostaglandin E_2.[324] Conjugate addition of most enolates is often reversible, and the relative stability of the enolate starting material (**550**) versus the enolate product (**551**) will determine

319. (a) Bergbreiter, D.E.; Newcomb, M. in *Asymmetric Synthesis Vol. 2*, Morrison, J.D. Ed., Academic Press, New York, *1980*, pp 220; (b) Hiroi, K.; Yamada, S. *Chem. Pharm. Bull., 1973, 21*, 47; (c) Otani, G.; Yamada, S. *Ibid. 1973, 21*, 2112.

320. Whitesell, J.K.; Felman, S.W. *J. Org. Chem. 1977, 42*, 1663.

321. (a) Michael, A. *J. Prakt. Chem. 1887, 35*, 379; (b) Bergmann, E.D.; Gingberg, D.; Pappo, R. *Org. React. 1959, 10*, 179; (c) Reference 104, pp 315-316; (d) Reference 105, pp 173-174; (e) Perlmutter, P. *Conjugative Addition Reactions in Organic Synthesis,* Pergamon Press, Oxford, *1992*; (f) The Merck Index, 14th ed., Merck & Co., Inc., Whitehouse Station, New Jersey, *2006*, p ONR-61; (g) Mundy, B.P.; Ellerd, M.G.; Favaloro Jr., F.G. *Name Reactions and Reagents in Organic Synthesis, 2nd ed.*, Wiley-Interscience, New Jersey, *2005*, pp. 428-429.

322. Naidu, B.N.; Sorenson, M.E.; Connolly, T.P.; Ueda, Y. *J. Org. Chem. 2003, 68*, 10098.

323. For reactions of this type, see Larock, R.C. *Comprehensive Organic Transformations, 2nd ed.,* Wiley-VCH, New York, *1999*, p 1317.

324. Taber, D.F.; Teng, D. *J. Org. Chem. 2002, 67*, 1607.

the position of the equilibrium. Solvent is important, and alcoholic solvents usually promote the equilibrium. Many examples of conjugate additions with Grignard reagents, organolithium reagents, or organocuprates were given in Chapter 8, and enolates also add to conjugated systems in a Michael fashion. Enantioselective Michael additions are well known,[325a] and a lanthanide-linked BINOL complex as been used effectively for the addition of malonate anions to cyclohexenone.[325b] One synthetic example is taken from the Takasu et al. synthesis of culmorin,[326] in which treatment of **553** with LHDMS led to a double Michael addition to give **554** in 87% yield. Initial formation of enolate anion **554** allows a Michael addition to the conjugated ester to form a seven-membered ring in **555**. This newly formed ester enolate reacts with the conjugated ketone by a second Michael addition to give enolate anion **556**, which is drawn a second time to show the bicycloheptanone structure. Hydrolysis gave the final product **557**.

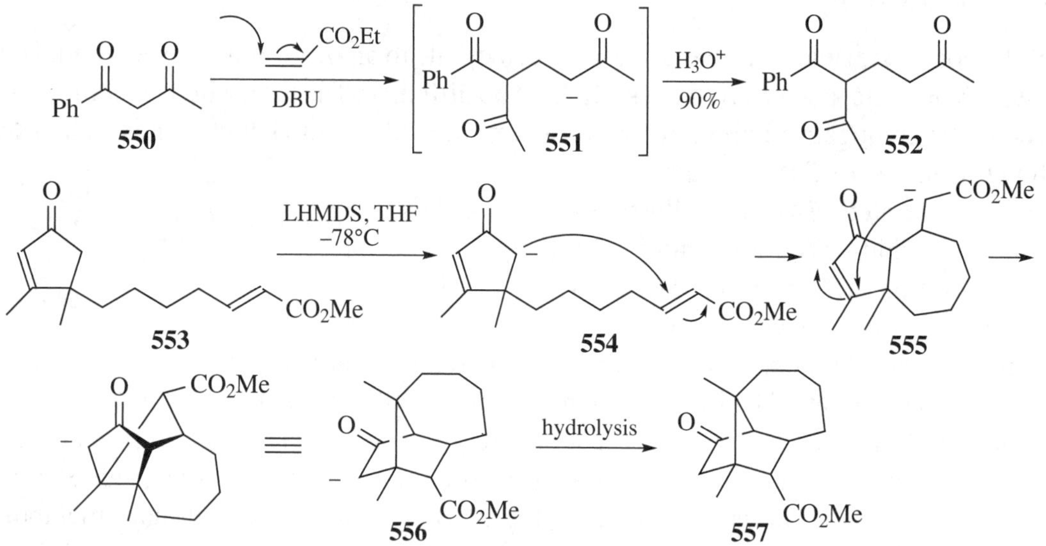

9.7.B. Baylis-Hillman Reaction

An interesting variation of the Michael addition has been used to prepare highly functionalized acylate derivatives. In its fundamental form, an acrylate ester reacts with an aldehyde in the presence of an amine or phosphine (such as tributylphosphine) catalyst. Presumably, Michael addition generates an enolate anion such as **558**, which condenses with the aldehyde and then loses the phosphine or amine to give the final product **559**. This

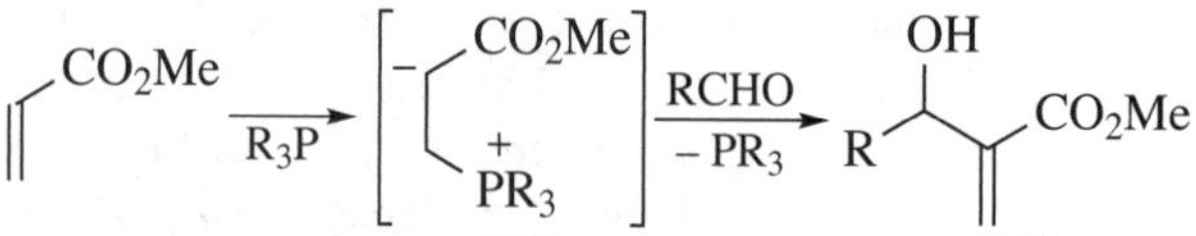

325. (a) Sibi, M.P.; Manyem, S. *Tetrahedron* **2000**, *56*, 8033; (b) Kim, Y.S.; Matsunaga, S.; Das, J.; Sekine, A.; Ohshima, T.; Shibasaki, M. *J. Am. Chem. Soc.* **2000**, *122*, 6506.
326. Takasu, K.; Mizutani, S.; Noguchi, M.; Makita, K.; Ihara, M. *J. Org. Chem.* **2000**, *65*, 4112.

transformation is known as the **Baylis-Hillman reaction**[327] (also known as the **Morita-Baylis-Hillman reaction**).[328] Aldol condensation, induced by the amine catalyst, can be a competing side reaction. Coupling is often slow, but Leahy and Rafel found that using DABCO as the catalyst at 0°C led to faster formation of Baylis-Hillman products, and in good yield.[329] An example is the reaction of **560** with ethyl acrylate in the presence of DABCO, giving an 86% yield of **561** as a 1:1 mixture of diastereomers in the Tandano and co-worker's synthesis of (+)-tubelactomicin A.[330]

Chloromethyl derivatives formed from $TiCl_4$ and triethylamine in CH_2Cl_2 at $-78°C$ are treated with triethylamine or DBU to give a Baylis-Hillman product.[331] Samarium iodide (SmI_2) mediates the Baylis-Hillman reaction.[332] Asymmetric Baylis-Hillman reactions are known.[333] Using Oppolzer's sultam[334] as an auxiliary, Leahy and co-workers[335] reacted amide **562** with propanal to give an 85% yield of **563** in >99% ee.[334] Treatment of **563** with camphorsulfonic acid in methanol in a second step gave an 85% yield of (+)-**564**. Another camphor-based auxiliary has also been used with success.[336] A chiral acetylenic ester titanium alkoxide complex has been used to generate chiral Baylis-Hillman products,[337] and the use of lanthanide triflates complexed with chiral diamine ligands derived from camphor gave Baylis-Hillman products with high enantioselectivities.[338] Intramolecular Baylis-Hillman reactions have also been reported.[339]

327. (a) Ciganek, E. *Org. React.* **1997**, *51*, 201; (b) Basavaiah, D.; Rao, P.D.; Hyma, R.S. *Tetrahedron* **1996**, *52*, 8001; (c) Drewes, S.E.; Roos, G.H.P. *Tetrahedron* **1988**, *44*, 4653; (d) Shi, M.; Li, C.-Q.; Jiang, J.-K. *Tetrahedron* **2003**, *59*, 1181; (e) The Merck Index, 14th ed., Merck & Co., Inc., Whitehouse Station, New Jersey, **2006**, p ONR-7; (f) Mundy, B.P.; Ellerd, M.G.; Favaloro Jr., F.G. *Name Reactions and Reagents in Organic Synthesis, 2nd ed.*, Wiley-Interscience, New Jersey, **2005**, pp. 74-75.
328. (a) Morita, K.; Suzuki, Z.; Hirose, H. *Bull. Chem. Soc. Jpn.* **1968**, *41*, 2815. (b) Baylis, A.B.; Hillman, M E.D. *German Patent* 2155113, **1972** (*Chem. Abstr.* **1972**, *77*: 34174q). For a recent review, see (c) Basavaiah, D.; Rao, A.J.: Satyanarayana, T. *Chem. Rev.* **2003**, *103*, 811.
329. Rafel, S.; Leahy, J.W. *J. Org. Chem.* **1997**, *62*, 1521.
330. Motozaki, T.; Sawamura, K.; Suzuki, A.; Yoshida, K.; Ueki, T.; Ohara, A.; Munakata, R.; Takao, K.-i.; Tadano, K.-i. *Org. Lett.* **2005**, *7*, 2265.
331. Shi, M.; Jiang, J.-K.; Feng, Y.-S. *Org. Lett.* **2000**, *2*, 2397.
332. Youn, S.W.; Park, H.S.; Kim, Y.H. *Chem. Commun.* **2000**, 2005.
333. Langer, P. *Angew. Chem. Int. Ed.* **2000**, *39*, 3049.
334. (a) Oppolzer, W. *Pure Appl. Chem.* **1990**, *62*, 1241; b) Kim, B.H.; Curran, D.P. *Tetrahedron* **1993**, *49*, 293; (c) Oppolzer, W. *Tetrahedron* **1987**, *43*, 1969; (d) Oppolzer, W. *Pure Appl. Chem.* **1988**, *60*, 39.
335. Brezinski, L.J.; Rafel, S.; Leahy, J.W. *J. Am. Chem. Soc.* **1997**, *119*, 4317.
336. Yang, K.-S.; Chen, K. *Org. Lett.* **2000**, *2*, 729.
337. Suzuki, D.; Urabe, H.; Sato, F. *Angew. Chem. Int. Ed.* **2000**, *39*, 3290.
338. Yang, K.-S.; Lee, W.-D.; Pan, J.-F.; Chen, K. *J. Org. Chem.* **2003**, *68*, 915.
339. Keck, G.E.; Welch, D.S. *Org. Lett.* **2002**, *4*, 3687. Also see Krishna, P.R.; Kannan, V.; Sharma, G.V.M. *J. Org. Chem.* **2004**, *69*, 6467.

$$562 \quad \xrightarrow[\text{CH}_2\text{Cl}_2, 0°\text{C}]{\text{EtCHO, DABCO}} \quad 563 \quad \xrightarrow[\text{CSA = camphorsulfonic aid}]{\text{CSA, MeOH}} \quad 564$$

9.7.C. Robinson Annulation

A major synthetic challenge in the 1930s through the 1950s was construction of various steroid molecules. One important route that emerged from this effort was the use of a Michael addition protocol to produce a bicyclic ketone. This sequence soon came to be known as the **Robinson annulation**.[340,321d] The reaction can be illustrated by treatment of 3-pentanone with sodium ethoxide (equilibrating conditions) in the presence of MVK. Other bases have been used, including lanthanoid triisopropoxides that behave as catalysts in this reaction.[341] The initial product is a cyclic alkoxy ketone (**565**), which gives alcohol **566** upon hydrolysis. This alcohol usually loses water with heating to give enone **567** as the final Robinson annulation product.

The reaction conditions favor formation of a thermodynamic enolate, although this is irrelevant with 3-pentanone. Conversion of 3-pentanone to its enolate (**127**) was followed by Michael addition to MVK to give a new enolate product (**568**). This enolate could react with the carbonyl at C6 to give a four-membered ring (path *a*), but as discussed in Sections 6.6.A,B, the energy requirements for this reaction are relatively high and the process is energetically unfavorable. Under the equilibrium conditions of the reaction, a small portion of kinetic enolate **569** can be formed. An intramolecular aldol condensation at C6 is possible (path *b*) that generates a six-membered ring (**565**), and this is energetically more favorable than forming a four-membered ring from **568**. When a cyclic ketone such as cyclohexanone reacts with MVK, Robinson annulation generates a bicyclic ketone.[340b-e]

---

340. (a) Annulation is the act or process of forming a ring; the formation of rings or ring-shaped parts. Annelation is an alternative but less desirable term, which means to anneal one ring onto another. The term annelation refers to the phenomenon whereby some rings in fused systems give up part of their aromaticity to adjacent rings, as on p. 61 in Smith, M.B.; March, J. *March's Advanced Organic Chemistry, 6th ed.*; Wiley, New York, *2007*; (b) DuFeu, E.C.; McQuillin, F.J.; Robinson, R. *J. Chem. Soc.* **1937**, 53; (c) Balasubramanian, K.; John, J.P.; Swamnathan, S. *Synthesis* **1974**, 51; (d) The Merck Index, 14th ed., Merck & Co., Inc., Whitehouse Station, New Jersey, *2006*, p ONR-80; (e) Mundy, B.P.; Ellerd, M.G.; Favaloro Jr., F.G. *Name Reactions and Reagents in Organic Synthesis, 2nd ed.*, Wiley-Interscience, New Jersey, *2005*, pp. 556-557.
341. Okano, T.; Satou, Y.; Tamura, M.; Kiji, J. *Bull. Chem. Soc. Jpn.* **1997**, 70, 1879.

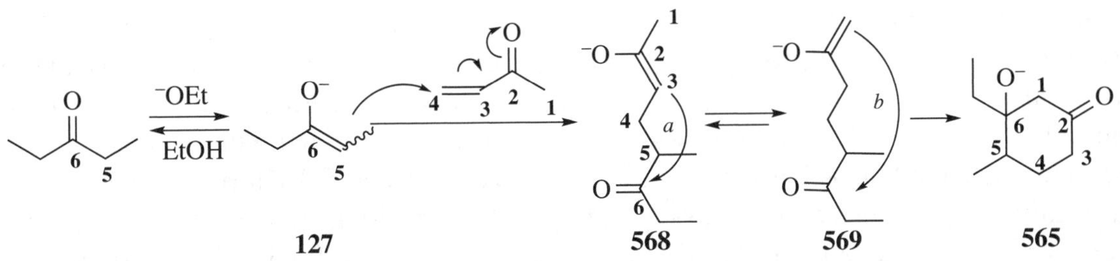

127   **568**   **569**   **565**

A synthetic example is taken from a synthesis of ouabain by Jung and Piizzi,[342] in which β-keto ester **570** reacted with methyl vinyl ketone and ethanolic potassium *tert*-butoxide to give **571** in good yield. There are many variations of the Robinson annulation that can be used to prepare synthetically useful products,[343,344] including some using a conjugated ketone surrogate known as a **Mannich base** such as **572**. Under the basic conditions of the Robinson annulation, the β-amino ketone eliminates diethylamine to form the conjugated ketone *in situ*, which may avoid the deleterious side reactions caused by the equilibrium conditions. A Mannich base is produced by a **Mannich reaction**,[345] where dimethylamine, for example, reacts with formaldehyde and acetone to give **572**.[345a,346] Acid-catalyzed Robinson annulation reactions are also known. An example is the conversion of **573** to **574** in good yield.[347]

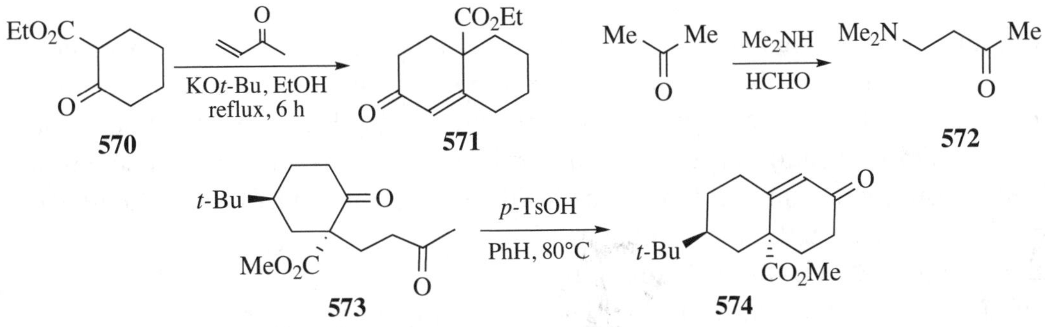

570   **571**   **572**

573   **574**

The Robinson Annulation disconnection is

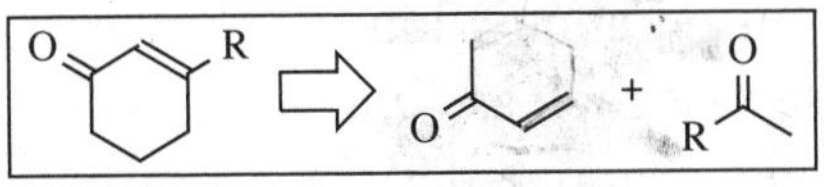

342. Jung, M.E.; Piizzi, G. *Org. Lett.* **2003**, *5*, 137.
343. Scott, W.L.; Evans, D.A. *J. Am. Chem. Soc.* **1972**, *94*, 4779.
344. Hackett, S.; Livinghouse, T. *J. Org. Chem.* **1986**, *51*, 1629.
345. (a) Mannich, C.; Krösche, W. *Arch. Pharm.* **1912**, *250*, 647; (b) Blicke, F.F. *Org. React.* **1942**, *1*, 303; (c) The Merck Index, 14th ed., Merck & Co., Inc., Whitehouse Station, New Jersey, **2006**, p ONR-57; (d) Mundy, B.P.; Ellerd, M.G.; Favaloro Jr., F.G. *Name Reactions and Reagents in Organic Synthesis, 2nd ed.*, Wiley-Interscience, New Jersey, **2005**, pp. 408-409.
346. For a synthetic example using a Manich reaction for saframycin alkaloids, see Zhou, B.; Guo, J.; Danishefsky, S.J. *Tetrahedron Lett.* **2000**, *41*, 2043.
347. Busch-Petersen, J.; Corey, E.J. *Tetrahedron Lett.* **2000**, *41*, 6941.

The reversible nature of the initial Michael addition is a problem with the Robinson annulation. A solution is to use a conjugated system that is particularly prone to Michael addition and forms the product, essentially irreversibly. α-Silyl vinyl ketones have been shown to be powerful Michael acceptors.[348] The lithium enolate of cyclohexanone reacted with conjugated ketone **575** to produce the Michael product, **576**.[348a] In this case, the initially formed Michael adduct was stabilized by the presence of the silyl group at the α-position, driving the reaction toward the product. Hydrolysis produced **577**, which was converted to the Robinson product (**578**) in 80% overall yield by treatment with NaOMe/MeOH under the requisite thermo-dynamic conditions.[348a] When compared to this sequential process, normal treatment of cyclohexanone enolate with MVK under Robinson conditions, which gave adduct **578** in < 5% yield.

9.7.D. Selectivity in the Robinson Annulation

Many modifications of the fundamental Robinson annulation sequence have appeared over the years, one of which uses silyl derivatives such as **575**. An important modification uses chiral precursors or chiral additives to give asymmetric induction either during the Michael addition, during the aldol cyclization or in both reactions. Excellent asymmetric induction has been obtained by using an asymmetric base such as (S)-(−)-proline (shown in the reaction) to generate the enolate. Hajos and Parrish obtained bicyclic ketone **580** (**Hajos diketone**) by treating **579** with MVK and proline. The Robinson product (**580**) was isolated with an optical purity of 93.4%.[349] Proline acts as a

catalyst in an asymmetric Robinson annulation.[350] Similarly, Wynberg and Helder showed that addition of asymmetric catalysts such as *Cinchona* alkaloids gave optical yields of 5-25% ee for Michael additions of cyclohexanone derivatives,[351] but up to 71% ee in some cases.[352] Langstrom and Bergson first employed asymmetric catalysts in the Michael addition.[353] An antibody-catalyzed enantioselective Robinson annulation has also been reported.[354]

348. (a) Stork, G.; Ganem, B. *J. Am. Chem. Soc.* **1973**, *95*, 6152; (b) Stork, G.; Singh, J. *Ibid.* **1974**, *96*, 6181.
349. Hajos, Z.G.; Parrish, D.R. *J. Org. Chem.* **1974**, *39*, 1612, 1615
350. Bui, T.; Barbas III, C.F. *Tetrahedron Lett.* **2000**, *41*, 6951.
351. Hermann, K.; Wynberg, H. *J. Org. Chem.* **1979**, *44*, 2238.
352. Wynberg, H.; Helder, R. *Tetrahedron Lett.* **1975**, 4057.
353. Langstrom, B.; Bergson, G. *Acta Chem. Scand.* **1973**, *27*, 3118.
354. Zhong, G.; Hoffmann, T.; Lerner, R.A.; Danishefsky, S.; Barbas III, C.F. *J. Am. Chem. Soc.* **1997**, *119*, 8131.

Solvent influences the stereochemistry of the Robinson annulation.[355] The condensation of methylcyclohexanone and **581** with sodium hydride in dioxane led to >95% of **582** with syn relationship for the methyl groups. Similar reaction in DMSO, however, led to >95% of **583** with had an anti relationship for those methyl groups. The change in stereochemistry was explained by a hydrogen transfer, that is facile in the highly polar DMSO but not in dioxane.[350] In general, polar protic solvents tend to give the trans product whereas less polar aprotic solvents favor the cis product.

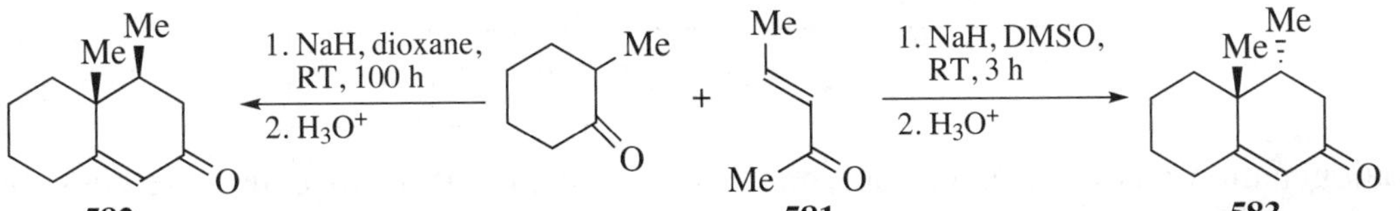

The metal also plays a role. The presence of a highly covalent O–Li bond favors the trans product, but a more ionic O–Na or O–K bond in alcoholic solvents favors the cis.[355] The lithium species is less bulky, due to poorer solvation in the aprotic solvent. The protic solvent hydrogen bonds extensively with the ionic sodium species, leading to a large increase in the relative size of the alkoxide moiety. This bulky group influences the geometry of the intermediate and leads to an increase in the cis product.[355]

9.8. ENOLATE REACTIONS OF α-HALO CARBONYL DERIVATIVES

9.8.A. α-Halogenation

Several reactions formally involve enolates, but the course of the reaction is altered by the presence of reactive functionality. The **Darzens' glycidic ester synthesis** discussed in Section 9.4.B.vi is one example. Another involves the formation of α-halogenated ketones. Treatment of a ketone with X_2 or NXS (X = Cl, Br, I) usually gives the α-halo ketone, as in the conversion of cyclopentanone to 2-bromocyclopentanone.[356] When there is the possibility of forming kinetic and thermodynamic products, the thermodynamic product usually predominates. In the synthesis of this α-bromoketone, $KClO_3$ oxidized the HBr byproduct back to Br_2. A serious problem in the halogenation is further reaction of the product with the halogen to give an α,α-dihalo compound. The presence of the first halogen promotes further enolization and polyhalogenation is often a serious problem. Halogenation of silyl enol ethers

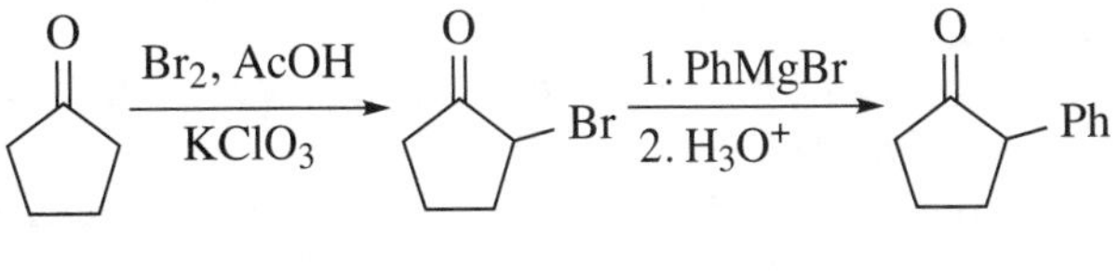

355. (a) Marshall, J.A.; Warne, Jr., T.M. *J. Org. Chem.* **1971**, *36*, 178; (b) Scanio, C.J.V.; Starrett, R.M. *J. Am. Chem. Soc.* **1971**, *93*, 1539.
356. (a) Reed, Jr., S.F. *J. Org. Chem.* **1965**, *30*, 2195; (b) Rappe, C.; Kumar, R. *Arkiv. Kemi.* **1965**, *23*, 475 [*Chem. Abstr. 63*: 5521c **1965**]; (c) Zbiral, E.; Rasberger, M. *Tetrahedron* **1969**, *25*, 1871; (d) Wyman, D.P.; Kaufman, P.R. *J. Org. Chem.* **1964**, *29*, 1956.

gives a good yield of monohaloketone.[357] Cupric (Cu^{2+}) salts such as $CuCl_2$ and $CuBr_2$ are often added to convert a ketone to α-chloro or α-bromo ketones.[358] The presence of the halogen modifies the bond polarity of the molecule so that displacement of halogen by nucleophiles is faster than acyl addition to the carbonyl. The reaction of 2-bromocyclopentanone with phenylmagnesium bromide gave 2-phenylcyclopentanone,[359] for example. This variation of the Grignard reaction is an excellent route to α-aryl ketones[353] (sec. 8.4.B.ii).

This leads to the interesting disconnection:

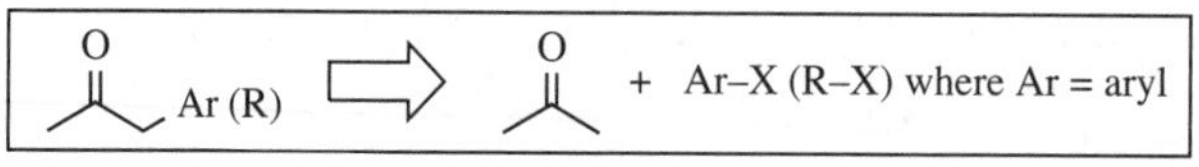

Polyhalogenation can be a severe side reaction in the formation of α-halo ketones, but can be used to synthetic advantage in the **haloform reaction** (this has been called the **Lieben iodoform reaction**).[360] Reaction of 2-butanone with sodium hydroxide gave the enolate, and in the presence of bromine the enolate displaced bromide from Br–Br to form **584**. Incorporation of the bromine made the α-hydrogen more acidic, and a second bromination occurred to give **585**, followed by a third bromination to give **586**. Nucleophilic acyl addition of hydroxide to all carbonyl derivatives (e.g., 2-butanone, **584** and **585**) was reversible up to this point.

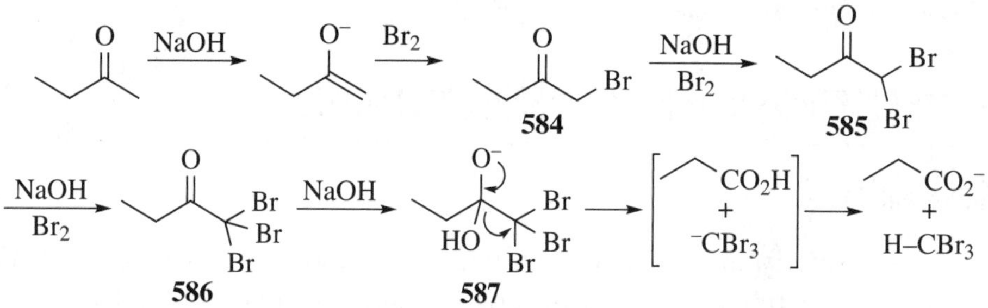

The Br_3C^- group is a good leaving group, however, so nucleophilic attack at the carbonyl leads to cleavage of the $O=C-CBr_3$ bond (via **587**) to produce butanoic acid and Br_3C^-. This carbanion rapidly deprotonated butanoic acid to produce the two final products, butanoic acid and a haloform (bromoform, $CHBr_3$). This sequence is a synthetic method for oxidative cleavage of a methyl group attached to the carbonyl, complementing the known cleavage of carbonyls by Cr(VI) (sec. 3.8.C). This reaction also cleaves the H_3C-C bond of methyl

357. (a) Reuss, R.H.; Hassner, A. *J. Org. Chem.* **1974**, *39*, 1785; (b) Blanco, L.; Amice, P.; Conia, J.M. *Synthesis* **1976**, 194, 196.
358. Smith, M.B.; March, J. *March's Advanced Organic Chemistry, 6th ed.,* Wiley-Interscience, Hoboken, New Jersey, **2007**, pp 776-780.
359. Ando, T. *Yuki Gosei Kagaku Kyokaishi* **1959**, *17*, 777 [*Chem. Abstr. 54*: 4492b **1960**].
360. (a) Fuson, R.C.; Bull, B.A. *Chem. Rev.* **1934**, *15*, 275; (b) Seelye, R.N.; Turney, T.A.; *J. Chem. Educ.* **1959**, *36*, 572; (c) The Merck Index, 14th ed., Merck & Co., Inc., Whitehouse Station, New Jersey, **2006**, p ONR-56; (d) Mundy, B.P.; Ellerd, M.G.; Favaloro Jr., F.G. *Name Reactions and Reagents in Organic Synthesis, 2nd ed.*, Wiley-Interscience, New Jersey, **2005**, pp. 392-393.

carbinols [RCH(OH)Me] to give the corresponding acid and iodoform. When iodine is used rather than bromine, the cleavage of methyl ketones or methyl carbinols is referred to as the **iodoform reaction**. The iodoform reaction constitutes a classical test for the presence of a methyl ketone moiety or a methyl carbinol moiety in an unknown molecule.

The haloform disconnection is

9.8.B. The Reformatsky Reaction

Another enolate-like reaction is the **Reformatsky reaction**,[361] which employs a nucleophilic organozinc intermediate, generated from an α-halo carbonyl and zinc metal. This condensation reaction is widely used in synthesis.[362] The organozinc reagent, in this case derived from ethyl 2-bromopropanoate (see **588**), attacks the carbonyl of an aldehyde or ketone (such as acetophenone) to give **589**. Hydrolysis gives the condensation product, β-hydroxyester **590**.

Preparation of the organozinc complex in the Reformatsky reaction can be a problem, and it often requires special preparation of the zinc (activated zinc). Activated zinc has been prepared by various procedures.[363] The use of ultrasound techniques produces a finely dispersed zinc that also facilitates the Reformatsky reaction.[364] As noted above, hydrolysis gives the hydroxy ester (**590**) to complete this two-carbon chain extension process. The organozinc reagent is generally less reactive than a Grignard or organolithium reagent, and condensation reactions proceed well with aldehydes and ketones but are sluggish with esters. A synthetic example is taken from a synthesis of rhinocerotinoic acid by Rivett and co-workers,[365] in which ketone **591** was treated with Zn and ethyl bromoacetate to give a 93% yield of **592** as a 1:1 mixture of epimers. A SmI_3 reaction has been reported that gives the Reformatsky product (see sec. 9.8.B)

361. (a) Reformatsky, S. *Berichte* **1887**, *20*, 1210; (b) Diaper, D.G.M.; Kuksis A. *Chem. Rev.* **1959**, *59*, 89; (c) Rathke, M.W. *Org. React.* **1975**, *22*, 423; (d) Reference 104, pp 370-371; (e) Reference 105, pp 200-202; (f) The Merck Index, 14th ed., Merck & Co., Inc., Whitehouse Station, New Jersey, **2006**, p ONR-78; (g) Mundy, B.P.; Ellerd, M.G.; Favaloro Jr., F.G. *Name Reactions and Reagents in Organic Synthesis, 2nd ed.*, Wiley-Interscience, New Jersey, **2005**, pp. 544-545.
362. Ocampo, R.; Dolbier Jr., W.R. *Tetrahedron* **2004**, *60*, 9325.
363. (a) Rieke, R.D.; Li, P.T.-J.; Burns, T.P.; Uhm, S.T. *J. Org. Chem.* **1981**, *46*, 4323; (b) Rieke, R.D.; Uhm, S.J. *Synthesis* **1975**, 452.
364. Han, B.H.; Boudjouk, P. *J. Org. Chem.* **1982**, *47*, 5030.
365. Gray, C.A.; Davies-Coleman, M.T.; Rivett, D.E.Q. *Tetrahedron* **2003**, *59*, 165.

with high diastereoselectivity in aqueous media.[366] Vedejs and Ahmad used an intramolecular version of this reaction aimed at the cytochalasin ring system.[367] In this reaction, **593**, was cyclized to give **594** as a 1:1 mixture of epimeric alcohols. The use of highly reactive **Rieke zinc**[368] led to a 75% yield of **594**. When samarium iodide (SmI$_2$) in trifluoroacetic acid was used for the cyclization,[369] a 46% yield of **592** was obtained as one diastereomeric alcohol.

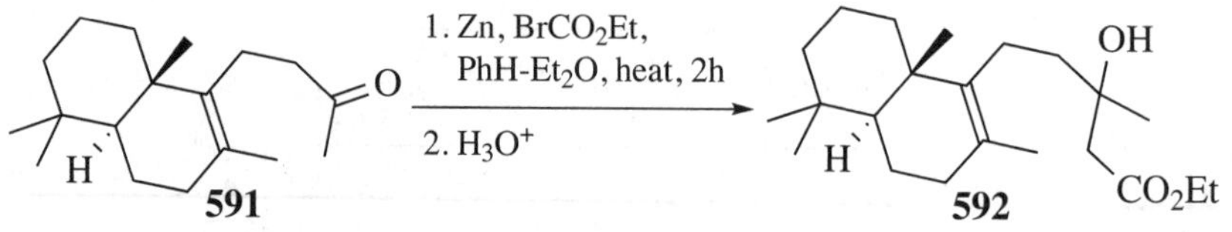

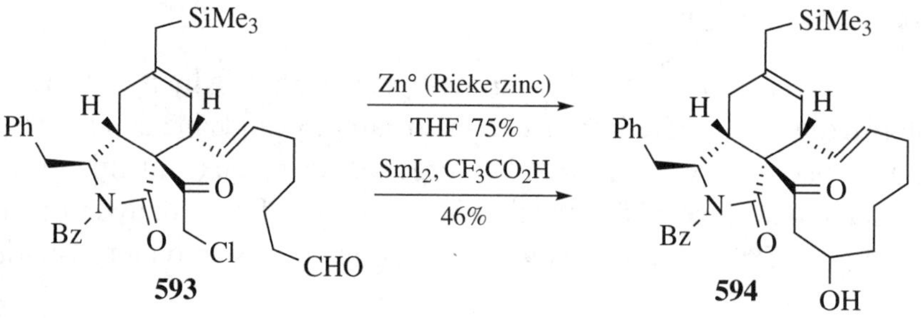

The Reformatsky disconnection is

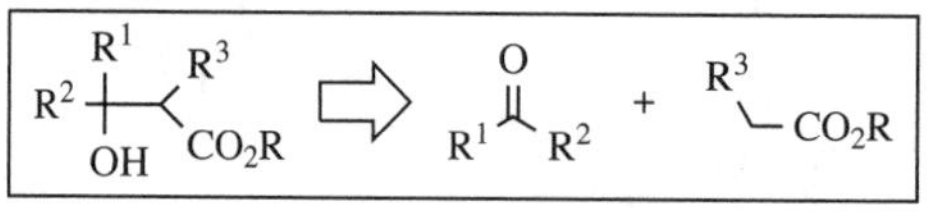

The selectivity of the Reformatsky reaction is apparent in Table 9.13,[370,371] taken from Heathcock's work, which shows the syn/anti selectivity for the hydroxy ester products (anti product **595** and syn product **596**) resulting from reaction of α-haloesters with ketones and aldehydes.[371] Aldehydes generally show poorer selectivity than do ketones for formation of the anti product, **595**. The bromozinc aldolate products from ketones were shown to equilibrate under reaction conditions, but those from aldehydes did not. In general, increasing the size of R^2 in the α-halo-carbonyl derivative led to greater anti selectivity.

366. Hayakawa, R.; Shimizu, M. *Chem. Lett.* **1999**, 591.
367. Vedejs, E.; Ahmad, S. *Tetrahedron Lett.* **1988**, *29*, 2291.
368. Arnold, R.T.; Kulenovic, S.T. *Synth. Commun.* **1977**, *7*, 223.
369. Tabuchi, T.; Kawamura, K.; Inanaga, J.; Yamaguchi, M. *Tetrahedron Lett.* **1986**, *27*, 3889.
370. Reference 244, pp 144-152.
371. (a) Reference 244, p 146; (b) Canceill, J.; Basselier, J.J.; Jacques, J. *Bull. Soc. Chim. Fr.* **1967**, 1024; (c) Canceill, J.; Jacques, J. *Ibid.* **1970**, 2180.

Table 9.13. *syn/anti* Selectivity in the Reformatsky Reaction

R^1	R^2	593 : 594	R^1	R^2	593 : 594
H	Me	37 : 63	H	Ph	76 : 24
H	Et	46 : 54	Me	Me	67 : 33
H	*i*-Pr	53 : 47	Me	Et	70 : 30
H	c-C_6H_{11}	50 : 50	Me	*i*-Pr	83 : 17
H	*t*-Bu	69 : 31			

In more recent work, germanium catalysts such as $GeCl_4$ gave a highly diastereoselective Reformatsky reaction favoring the syn diastereomer.[372] Another syn selective reaction was reported using $TiCl_2$ and Cu.[373] Enantioselective Reformatsky reactions have also used chiral amino alcohols as additives.[374]

9.8.C. The Favorskii Rearrangement

There is a classical reaction of enolate anions derived from α-halo ketones that involves a rearrangement. In the **Favorskii rearrangement**,[375] also called the **Wallach degradation**, a α-chloroketone such as 2-chlorocyclohexanone is treated with NaOH under thermodynamic conditions to form the enolate anion **597**. Intramolecular displacement of chlorine leads to **598**, which is attacked by hydroxide to give the acyl addition product, **599**. Cleavage of the adjacent C–C bond leads to enolate **600**, which is protonated by the adjacent carboxylic acid to give carboxylate **601**. Hydrolysis gives the final product, cyclopentanecarboxylic acid. This reaction can be useful synthetically, as in Ley and co-worker's synthesis of trilobolide.[376] Treatment of **602** with sodium methoxide gave a 95%% yield of the Favorskii rearrangement product (**603**), obtained as the methyl ester since the reaction was done in methanol.

372. Kagoshima, H.; Hashimoto, Y.; Oguro, D.; Saigo, K. *J. Org. Chem.* **1998**, *63*, 691.
373. Mukaiyama, T.; Kagayama, A.; Igarashi, K.; Shiina, I. *Chem. Lett.* **1999**, 1157.
374. (a) Andrés, J.M.; Martín, Y.; Pedrosa, R.; Pérez-Encabo, A. *Tetrahedron* **1997**, *53*, 3787; (b) Mi, A.; Wang, Z.; Zhang, J.; Jiang, Y. *Synth. Commun.* **1997**, *27*, 1469.
375. (a) Favorskii, A.E. *J. Prakt. Chem.* **1913**, *88*, 658; (b) Wallach, O. *Annalen*, **1918**, *414*, 296; (c) Kende, A.S. *Org. React.* **1960**, *11*, 261; (d) Turro, N.J. *Acc. Chem. Res.* **1969**, *2*, 25; (e) The Merck Index, 14th ed., Merck & Co., Inc., Whitehouse Station, New Jersey, **2006**, p ONR-30; (f) Mundy, B.P.; Ellerd, M.G.; Favaloro Jr., F.G. *Name Reactions and Reagents in Organic Synthesis, 2nd ed.*, Wiley-Interscience, New Jersey, **2005**, pp. 238-239.
376. (a) Oliver, S.F.; Högenauer, K.; Simic, O.; Antonello, A.; Smith, M.D.; Ley, S.V. *Angew. Chem.* **2003**, *42*, 5996; (b) Also see Lee, E.; Yoon, C.H. *J. Chem. Soc. Chem. Commun.* **1994**, 479.

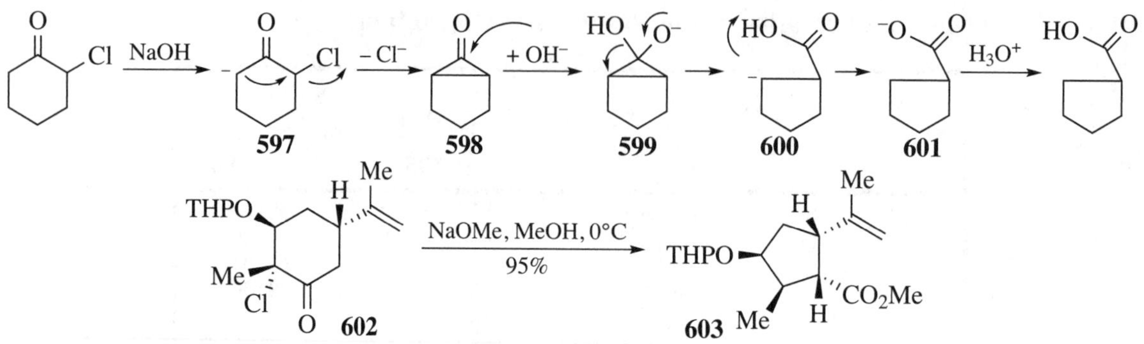

597 **598** **599** **600** **601**

NaOMe, MeOH, 0°C
95%

602 **603**

9.9. CONCLUSION

In conclusion, the reaction of enolates with alkyl halides and carbonyl compounds provides one of the most powerful and useful of all carbon-carbon bond-forming processes. Not only are a variety of synthetic products available, but the reaction can be made to proceed with excellent diastereoselectivity. In conjunction with the reactions of carbanions, nucleophilic methods constitute the largest single type of methodology for making new carbon-carbon bonds.

HOMEWORK

1. Predict the major product of this reaction, and explain the stereochemistry.

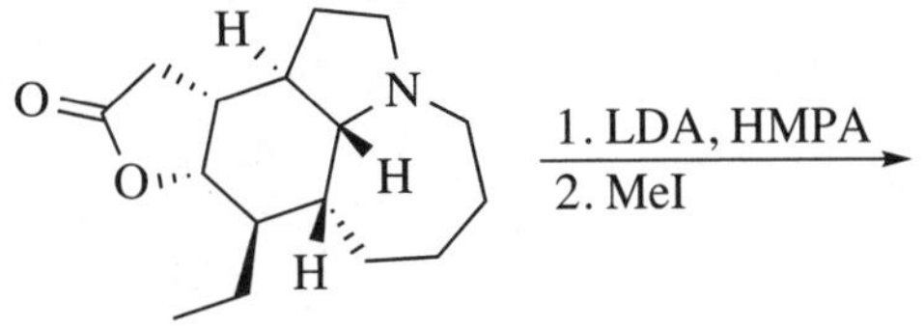

H
O
N
O
H
H
1. LDA, HMPA
2. MeI

2. Explain formation of the product shown given the reaction conditions.

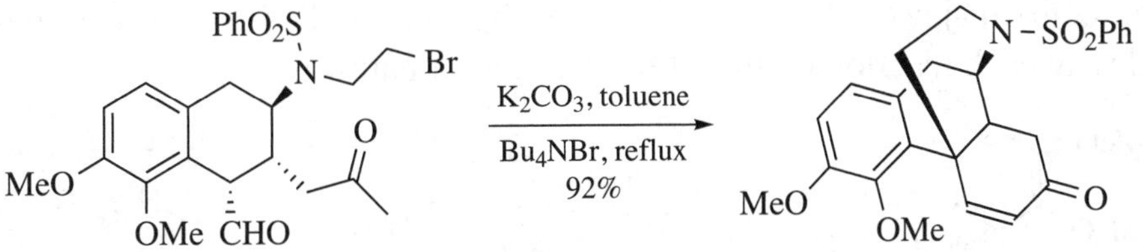

PhO2S
N
Br
O
MeO
OMe CHO
K2CO3, toluene
Bu4NBr, reflux
92%
N - SO2Ph
MeO
OMe
O

3. Give a mechanistic rationale for the following reaction.

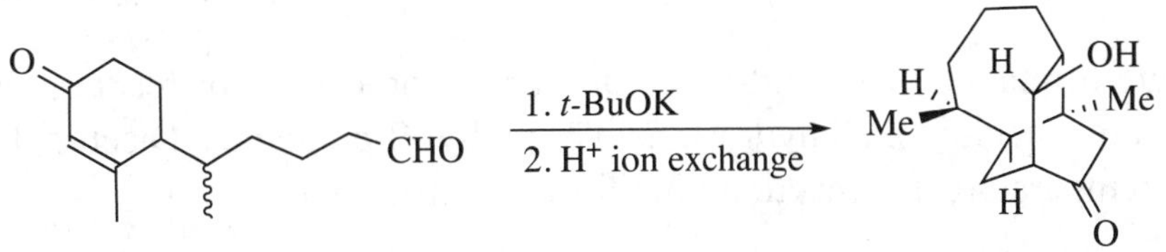

O
CHO
1. t-BuOK
2. H+ ion exchange
H
H
Me
OH
Me
H
O

4. In the following sequence, pulegone is transformed into 2,3-dimethylcyclohexanone. Explain this transformation.

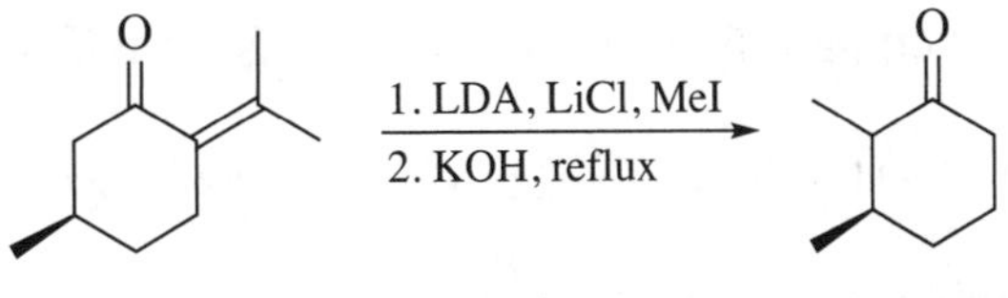

O
1. LDA, LiCl, MeI
2. KOH, reflux
O

5. For the reactions shown, draw the syn and anti products. Explain why there is a difference in selectivity and why the anti product predominates with PhCHO but there is essentially a 1:1 mixture with cyclohexanone.

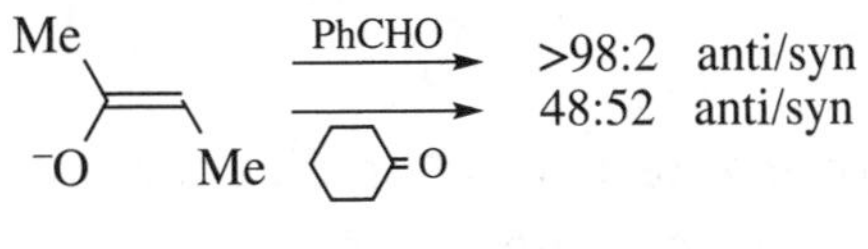

Me
-O
Me
PhCHO
O
>98:2 anti/syn
48:52 anti/syn

6. For the reaction of 3-pentanone and lithium diisopropylamide (-78°C, THF):

   (a) Draw both (*E*) and (*Z*) enolates.

   (b) Give the major product of the (*Z*) enolate when PhCHO approaches from the *re* face;

from the *si* face.

Show the Zimmerman-Traxler model for both approaches.

(c) Give the major product when the (*E*) enolate reacts with *A*, assuming *A* is chiral and nonracemic.

(d) Give the major product when the (*Z*) enolate reacts with chiral, non-racemic A.

7. Predict the major product (A or B) for each reaction. Discuss formation of each product and why each reaction might follow a different pathway.

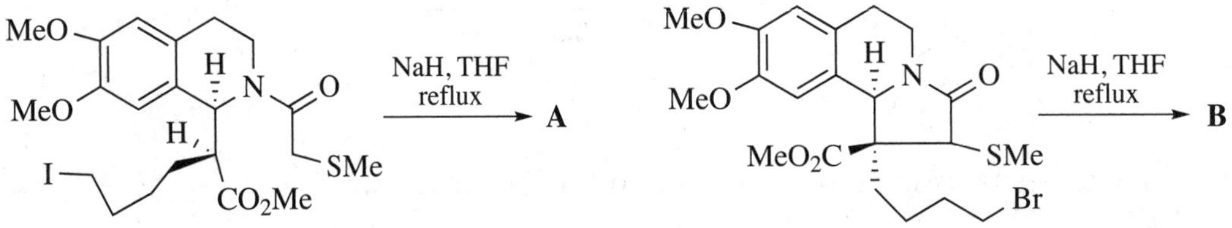

8. Enolate formation is an acid-base equilibrium process. For reactions A and B, discuss the relative acidity of all hydrogens in 3-methyl-2-butanone. Discuss the role of solvent, base, temperature and reaction time for reactions A and B.

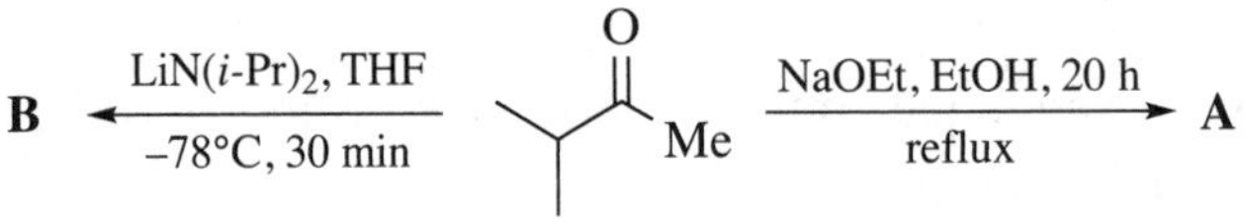

Write the acid-base equations for both A and B using the standard reaction shown.

$$\text{acid} + \text{base} \xrightleftharpoons{K_a} \text{conjugate acid} + \text{conjugate base}$$

9. Explain the following transformation.

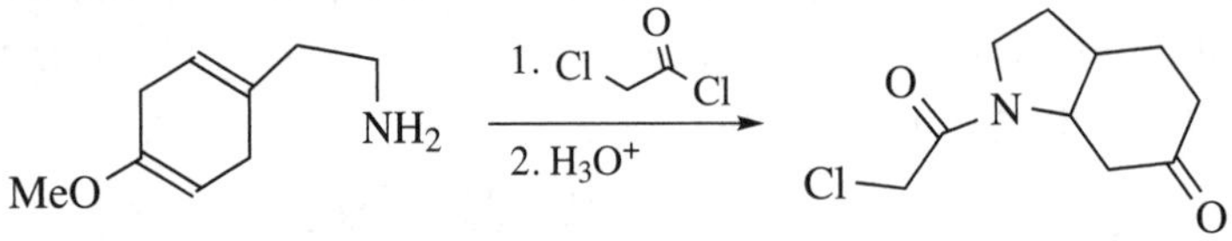

10. In each of the following reactions, predict the major diastereomer and show the Zimmerman-Traxler transition state:

(a) 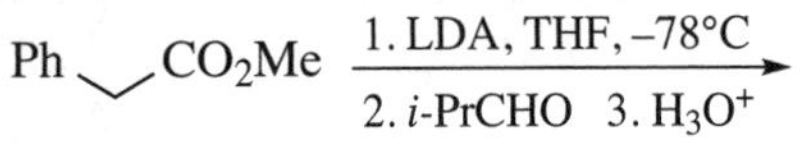 Ph⁀CO₂Me $\xrightarrow[\text{2. } i\text{-PrCHO \quad 3. H}_3\text{O}^+]{\text{1. LDA, THF, }-78°\text{C}}$

(b) 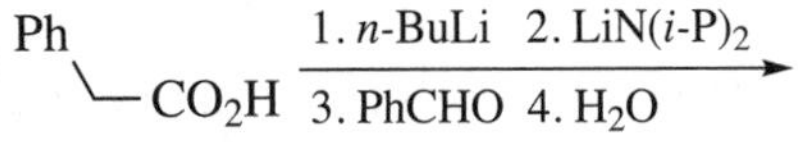 Ph⁀CO₂H $\xrightarrow[\text{3. PhCHO \quad 4. H}_2\text{O}]{\text{1. } n\text{-BuLi \quad 2. LiN}(i\text{-P})_2}$

(c) 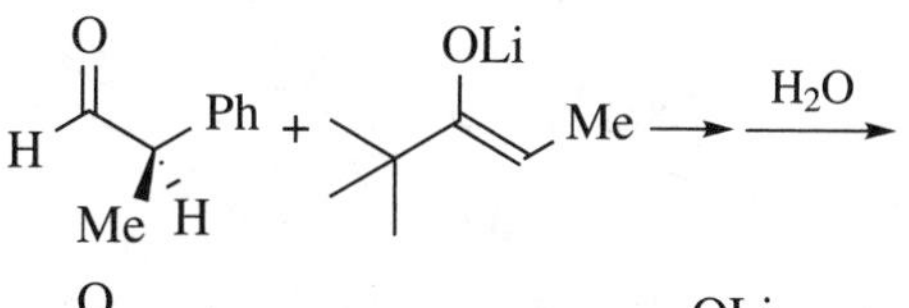
O
Ph
Me H
OLi
Me
H_2O

(d) 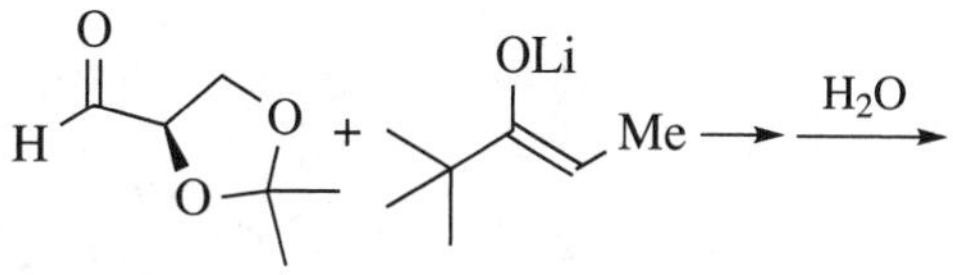
O
OLi
Me
H_2O

(e) 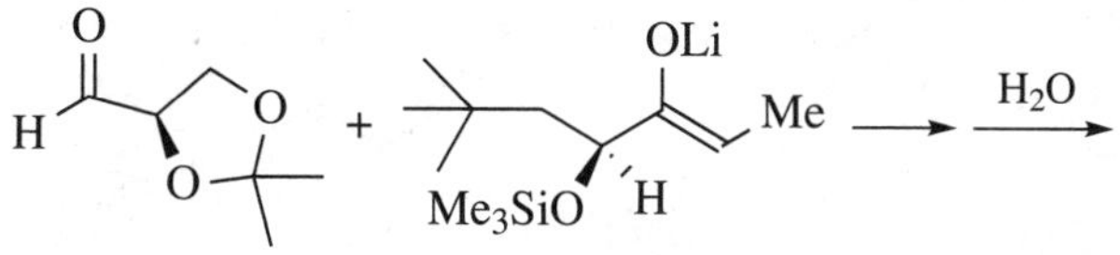
OLi
Me
Me_3SiO H
H_2O

11. Use models to predict the major diastereomer formed during this methylation reaction.

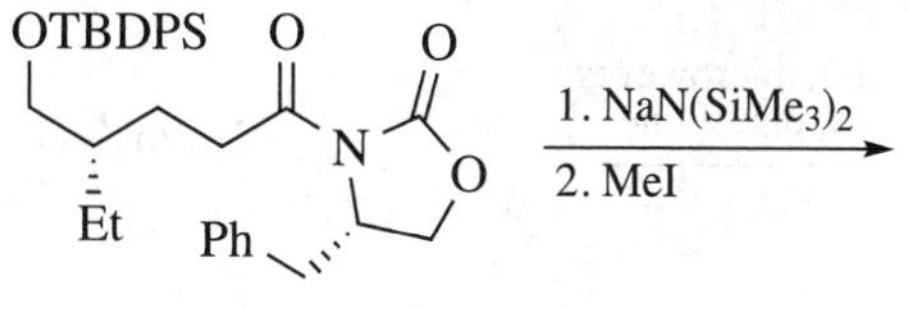
OTBDPS
Et
Ph
1. $NaN(SiMe_3)_2$
2. MeI

12. Provide a mechanism for the following transformation.

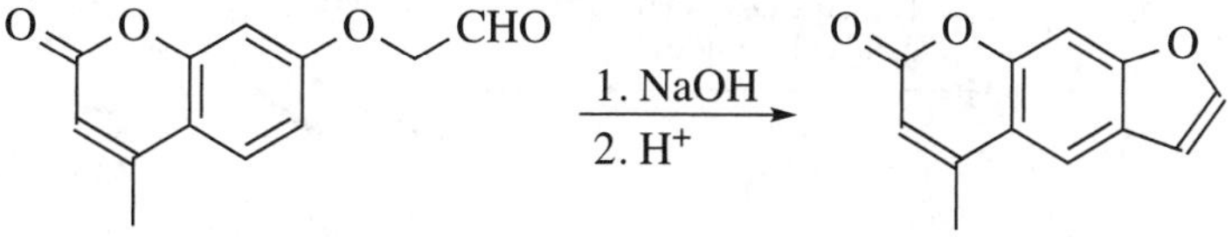
CHO
1. NaOH
2. H^+

13. Provide a mechanistic rationale for the following transformation.

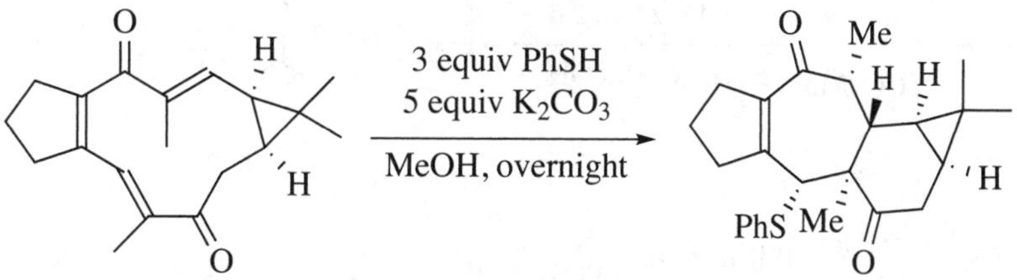
3 equiv PhSH
5 equiv K_2CO_3
MeOH, overnight
PhS Me

14. For each of the following, give all products. What is the strategic purpose of the first condensation with benzaldehyde? Compare and contrast routes A and B:

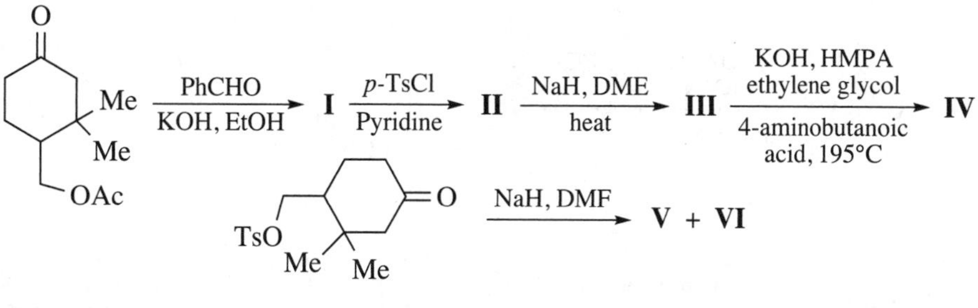
Me
Me
OAc
PhCHO
KOH, EtOH
I
p-TsCl
Pyridine
II
NaH, DME
heat
III
KOH, HMPA
ethylene glycol
4-aminobutanoic acid, 195°C
IV
TsO
NaH, DMF
V + VI

15. Show a full synthetic sequence, with all intermediate products and reagents, for conversion of 1-methyl-1-cyclohexene into each of the molecules shown.

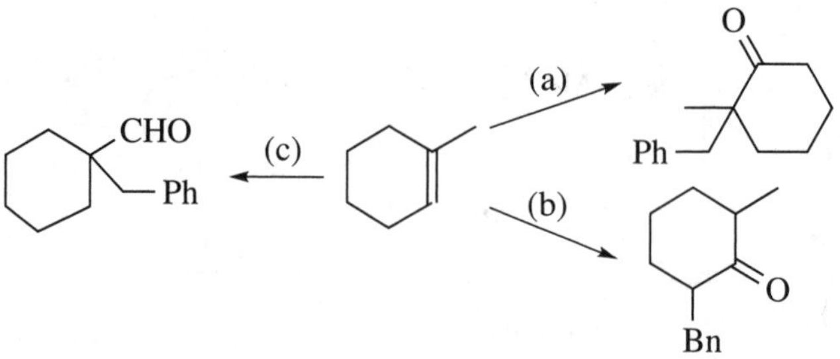

16. In each case, give the major product, with correct stereochemistry where appropriate.

(a) 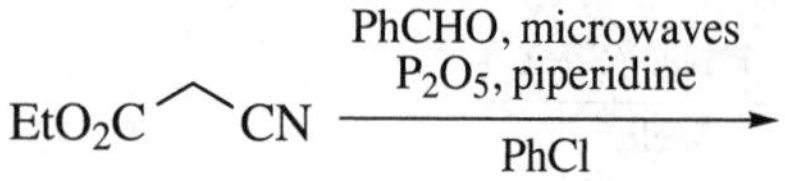 EtO₂C—CN → $EtO_2C\text{—}CN$

PhCHO, microwaves
P₂O₅, piperidine
————————→
PhCl

(b) 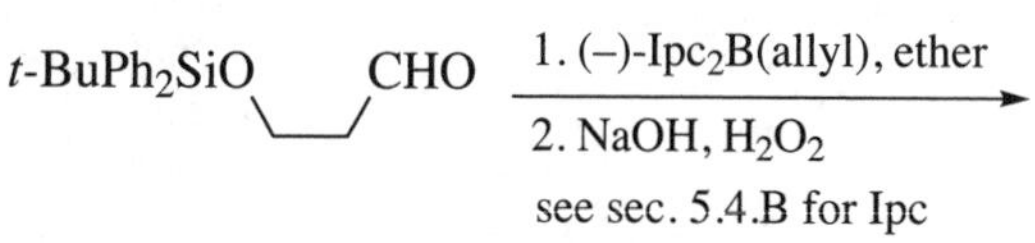 $t\text{-BuPh}_2SiO$ ... CHO
1. (–)-Ipc₂B(allyl), ether
2. NaOH, H₂O₂
see sec. 5.4.B for Ipc

(c) 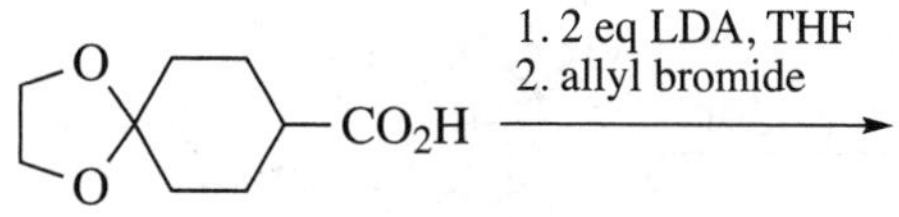 [structure] —CO₂H
1. 2 eq LDA, THF
2. allyl bromide
————————→

(d) 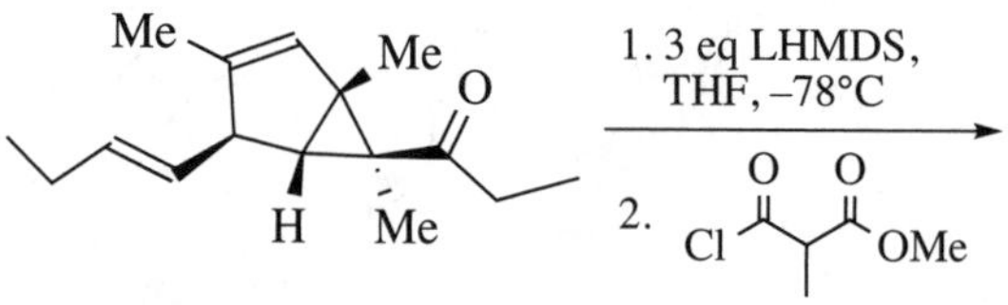 [structure with Me, Me, O, H, Me]
1. 3 eq LHMDS,
THF, –78°C
2. [structure] Cl ... OMe

(e) 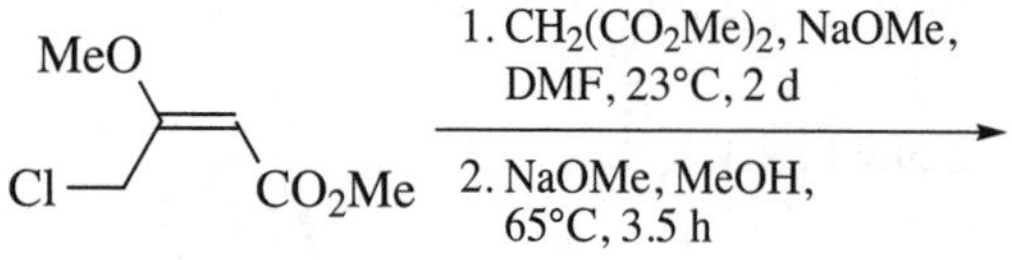 MeO / Cl—CO₂Me
1. CH₂(CO₂Me)₂, NaOMe,
DMF, 23°C, 2 d
2. NaOMe, MeOH,
65°C, 3.5 h
————————→

(f) 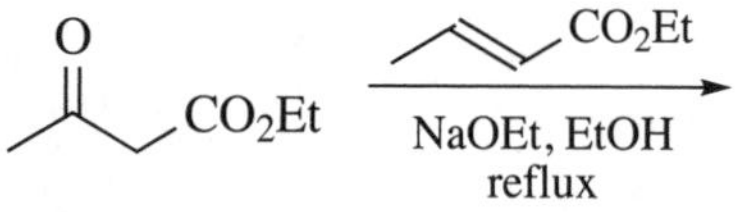 [structure] CO₂Et
/=\—CO₂Et
NaOEt, EtOH
reflux
————————→

(g) 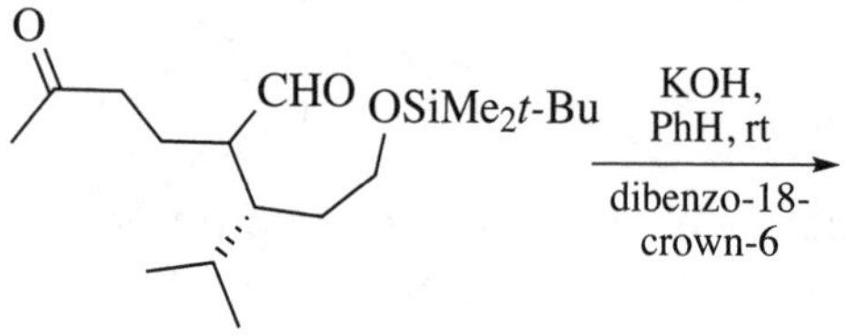 [structure with CHO OSiMe₂t-Bu]
KOH,
PhH, rt
————————→
dibenzo-18-
crown-6

(h) 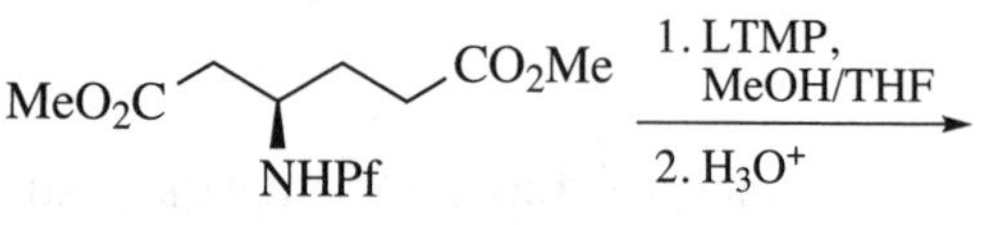 MeO₂C ... CO₂Me
NHPf
Pf = 9-phenyl-9-fluorenyl
1. LTMP,
MeOH/THF
2. H₃O⁺
————————→

(i) 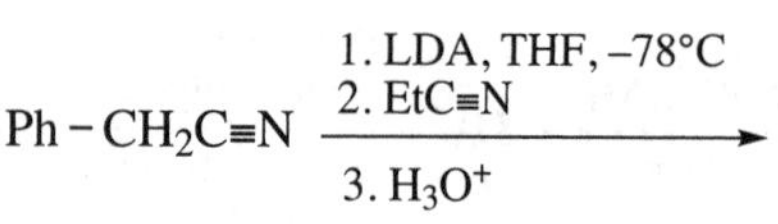 Ph—CH₂C≡N
1. LDA, THF, –78°C
2. EtC≡N
3. H₃O⁺
————————→

(j) 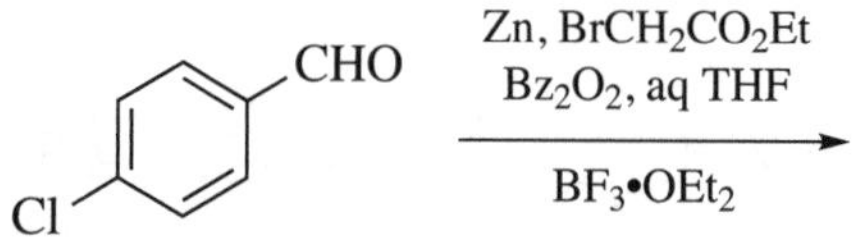 [structure] CHO, Cl
Zn, BrCH₂CO₂Et
Bz₂O₂, aq THF
————————→
BF₃•OEt₂

(k) 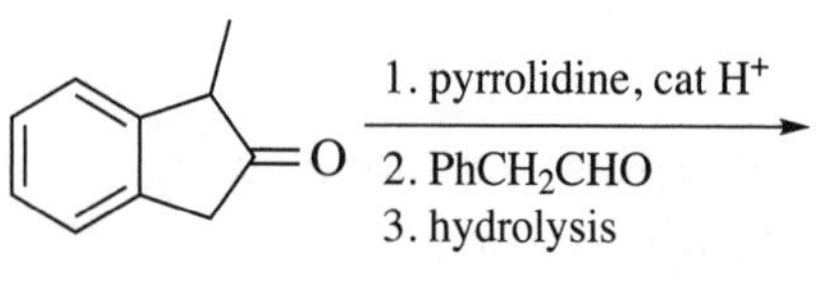 [structure] O
1. pyrrolidine, cat H⁺
2. PhCH₂CHO
3. hydrolysis
————————→

(l) 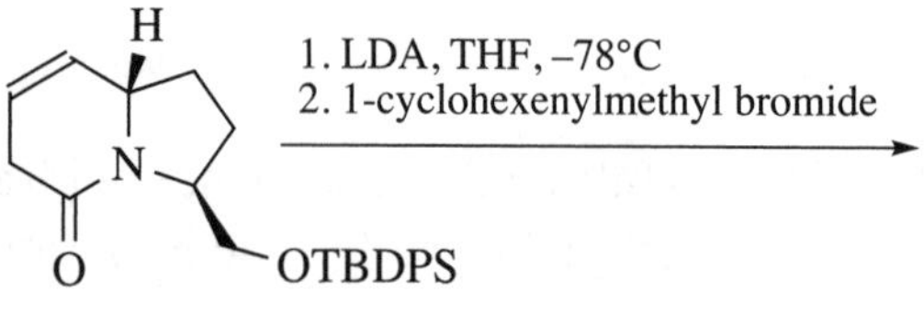 [structure with H, N, O, OTBDPS]
1. LDA, THF, –78°C
2. 1-cyclohexenylmethyl bromide
————————→

(m) 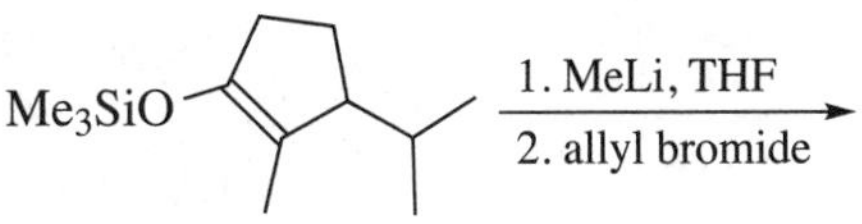 Me₃SiO [structure: cyclopentene with isopropyl and methyl] → 1. MeLi, THF 2. allyl bromide

(n) 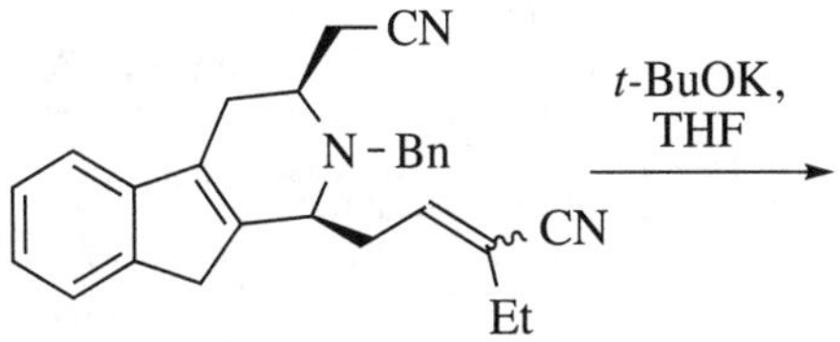 [structure: tricyclic indene with CH₂CN, N–Bn, and side chain with CN, Et] → t-BuOK, THF

(o) 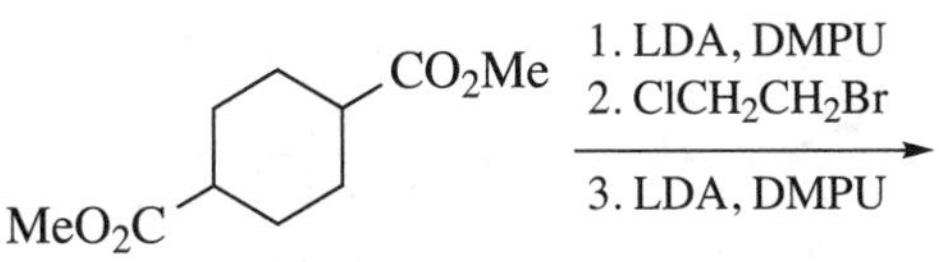 [cyclohexane with CO₂Me and MeO₂C] → 1. LDA, DMPU 2. ClCH₂CH₂Br 3. LDA, DMPU

(p) [bicyclic indolizidinone with H, Me, OH, N, =O] → 1. 2 eq LDA, THF 2. Me₂CHCHO

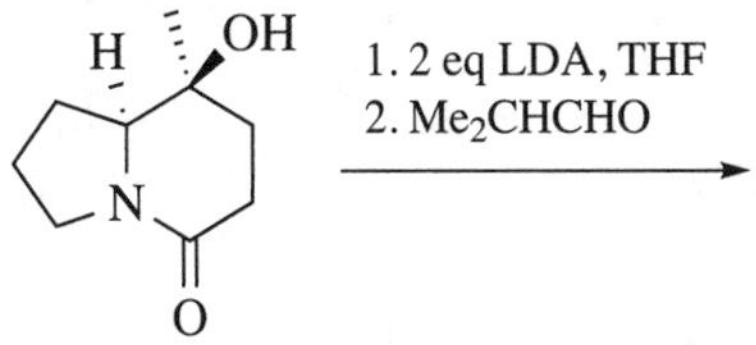

(q) 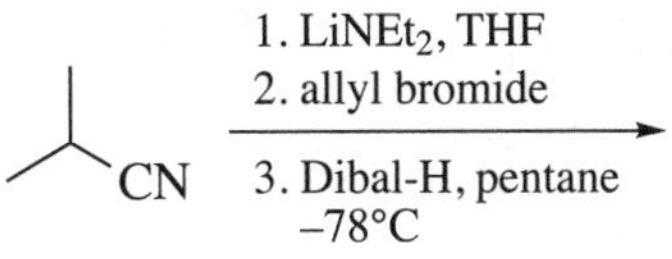 [isopropyl CN] → 1. LiNEt₂, THF 2. allyl bromide 3. Dibal-H, pentane −78°C

(r) 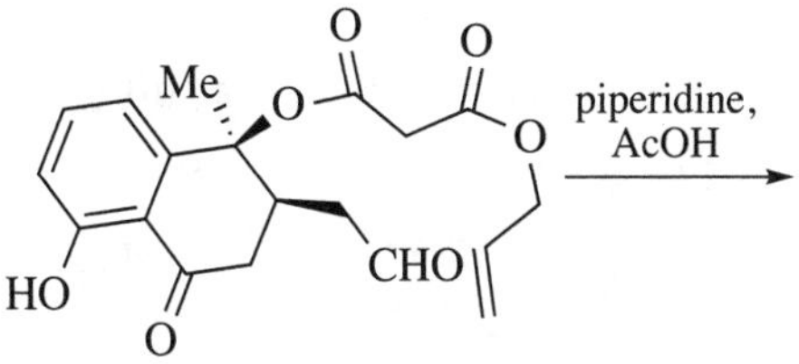 [structure: naphthalenone with Me, O, O, malonate allyl ester, CHO, HO, O] → piperidine, AcOH

(s) 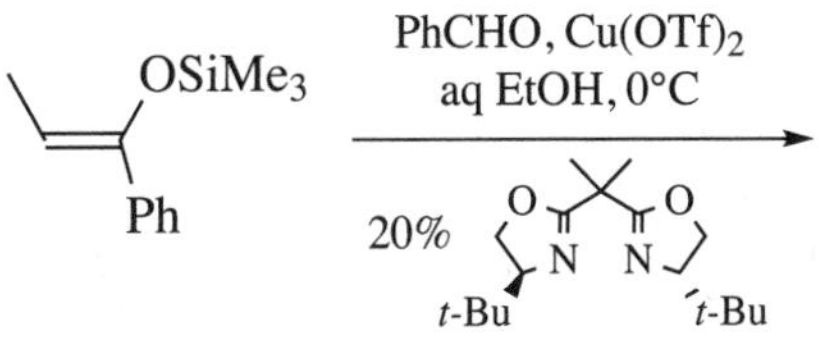 [structure: OSiMe₃, Ph, vinyl] → PhCHO, Cu(OTf)₂ aq EtOH, 0°C 20% [bis-oxazoline with t-Bu groups]

(t) [polycyclic structure with dioxolane, Me, Me, O, O, OH, H] → 20% aq HCl THF, 50°C

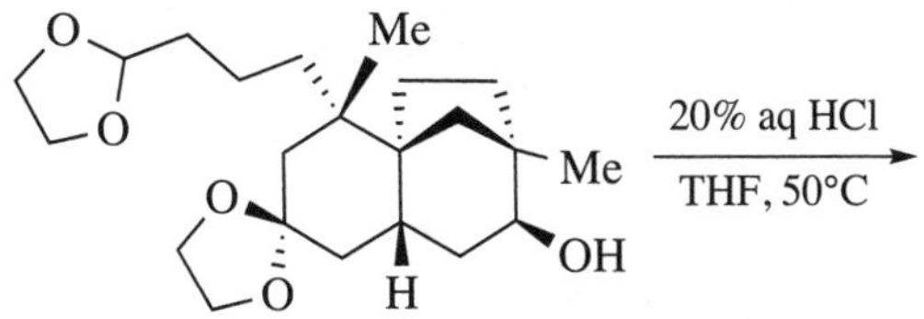

(u) 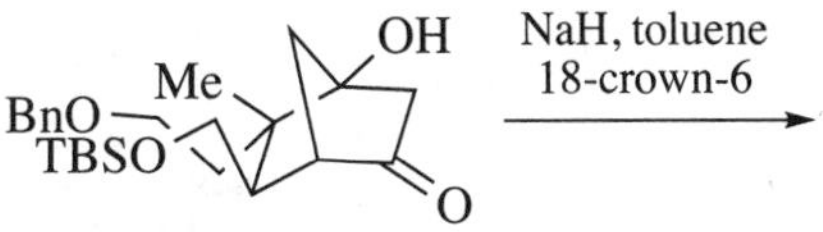 BnO, Me, TBSO [bicyclic structure with OH, O] → NaH, toluene 18-crown-6

(v) 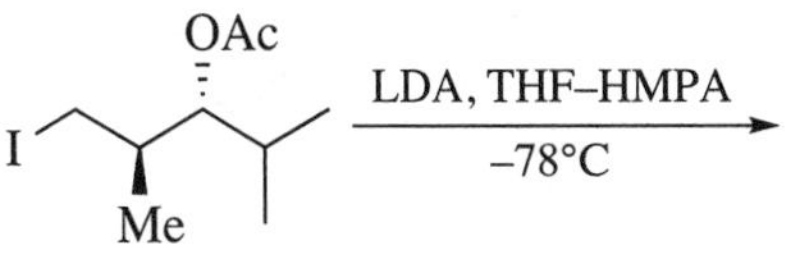 I [structure with OAc, Me, isopropyl] → LDA, THF–HMPA −78°C

(w) 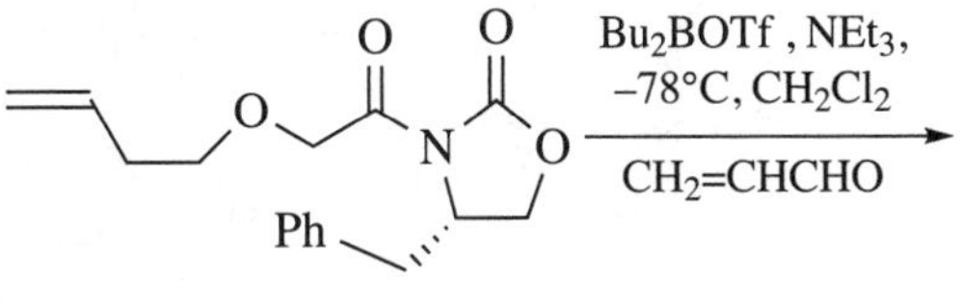 [oxazolidinone with O, O, O, N, Ph, ethyl, butenyl] → Bu₂BOTf, NEt₃, −78°C, CH₂Cl₂ CH₂=CHCHO

(x) 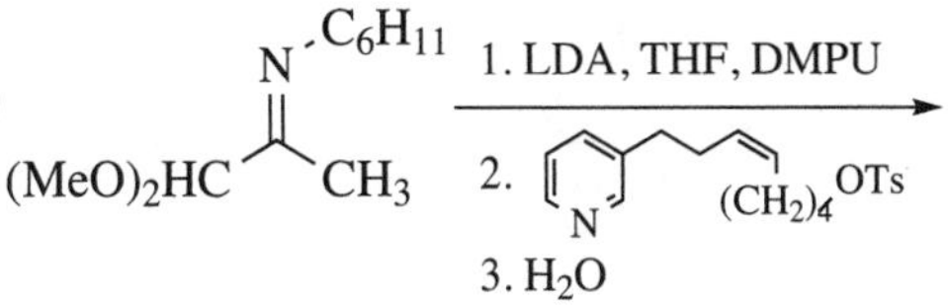 (MeO)₂HC [structure: N–C₆H₁₁, CH₃] → 1. LDA, THF, DMPU 2. [pyridine (CH₂)₄OTs] 3. H₂O

(y) 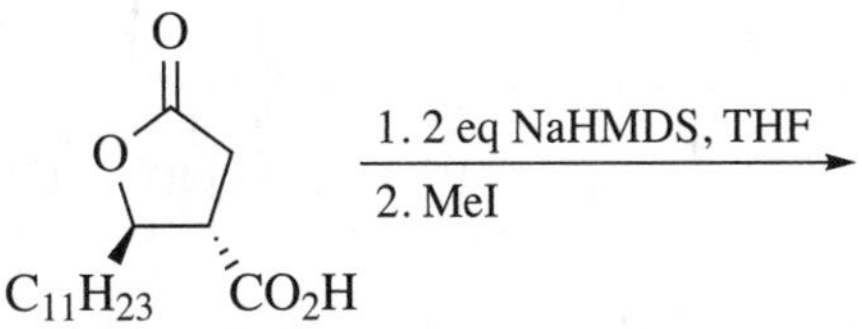 [lactone: O, O, C₁₁H₂₃, CO₂H] → 1. 2 eq NaHMDS, THF 2. MeI

(z) 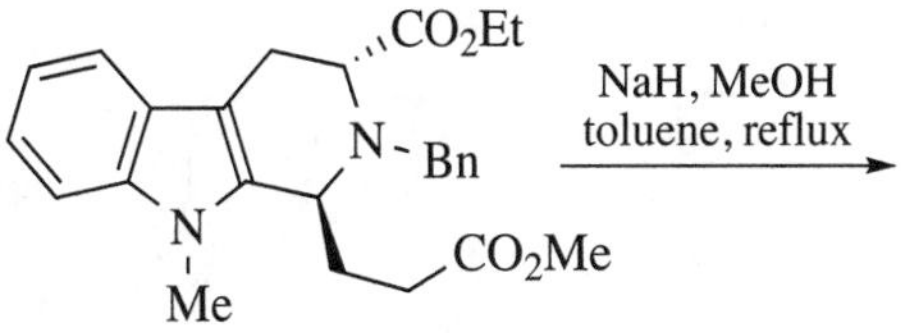 [structure: indole tricyclic with CO₂Et, N–Bn, N–Me, CO₂Me] → NaH, MeOH toluene, reflux

(aa) 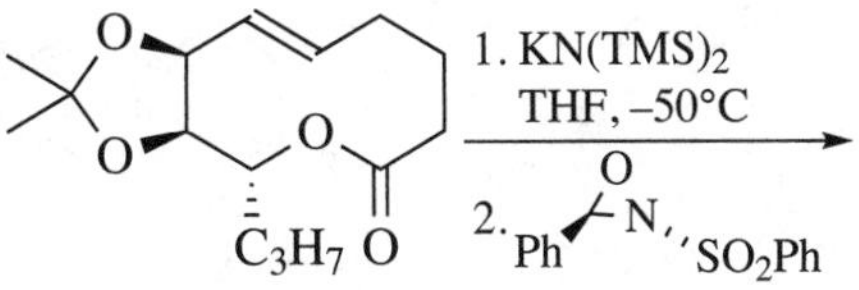 [structure: dioxolane, O, O, O, C₃H₇, O] → 1. KN(TMS)₂ THF, −50°C 2. Ph–N–SO₂Ph

(ab) 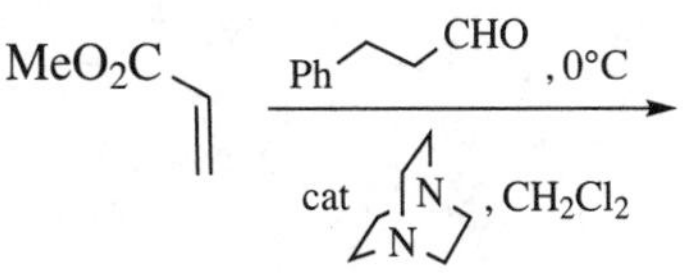 MeO₂C [vinyl] + Ph [CHO] → , 0°C cat [bicyclic diamine], CH₂Cl₂

(ac) 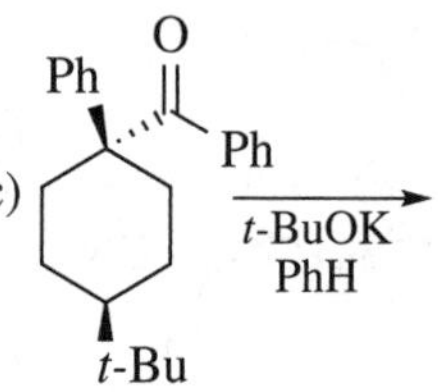 → (ad) 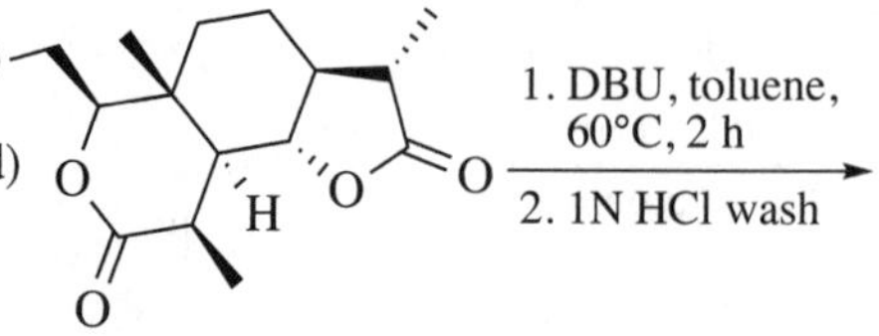 t-BuOK, PhH →

1. DBU, toluene, 60°C, 2 h
2. 1N HCl wash

(ae) 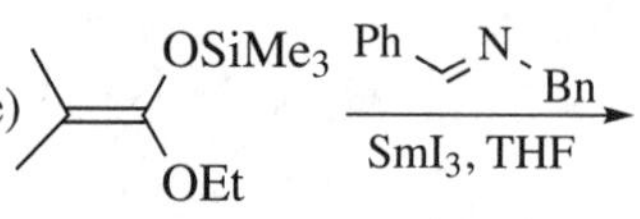 → $\begin{array}{c}\text{Ph} \diagdown \text{N} \diagup \text{Bn}\end{array}$ SmI$_3$, THF

(af) 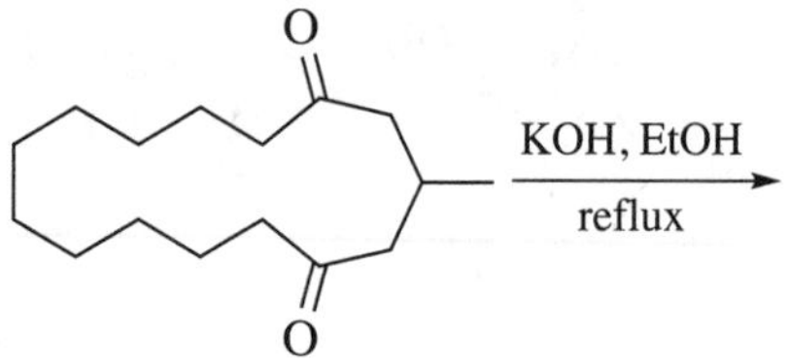 → KOH, EtOH reflux → (ag) 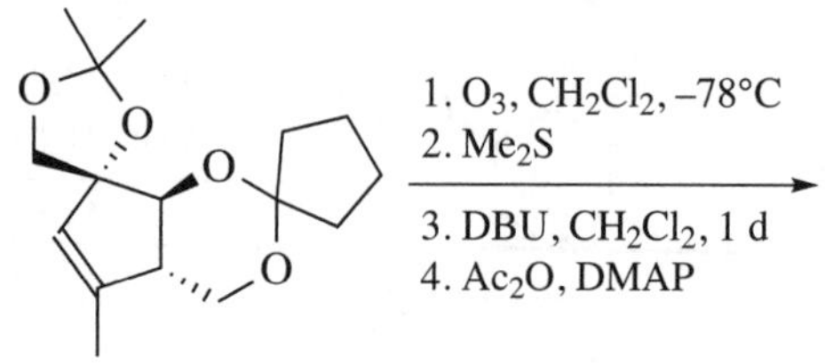 →

1. O$_3$, CH$_2$Cl$_2$, −78°C
2. Me$_2$S
3. DBU, CH$_2$Cl$_2$, 1 d
4. Ac$_2$O, DMAP

17. In each case, provide a suitable synthesis. Provide all reagents and show all intermediate products.

(a)

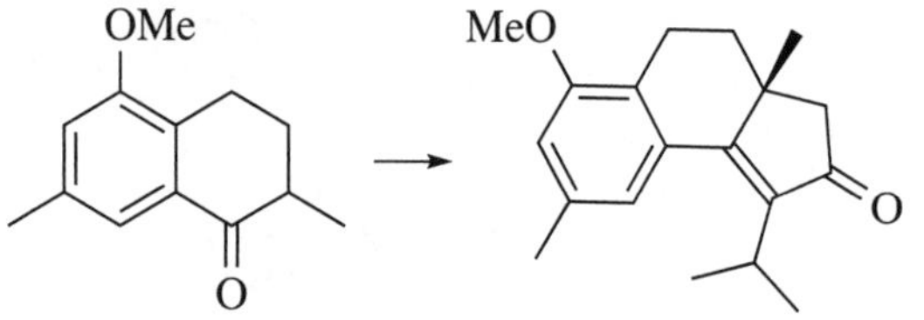

(b)

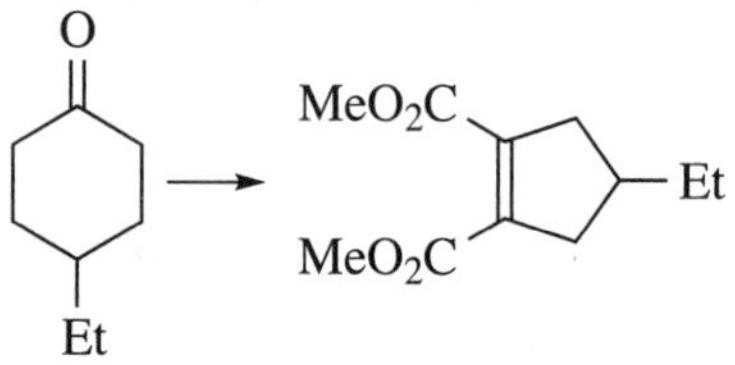

(c)

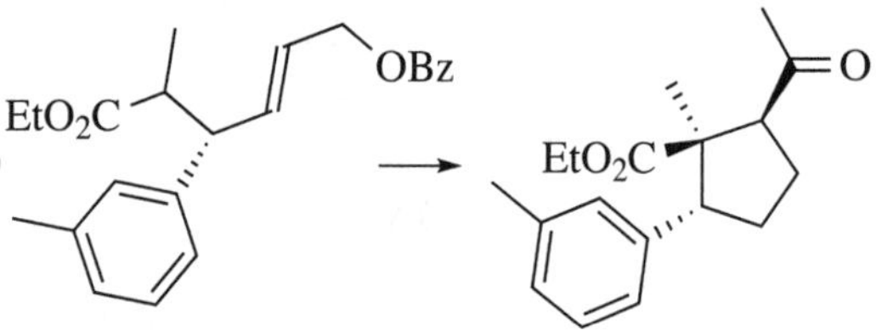

(d)

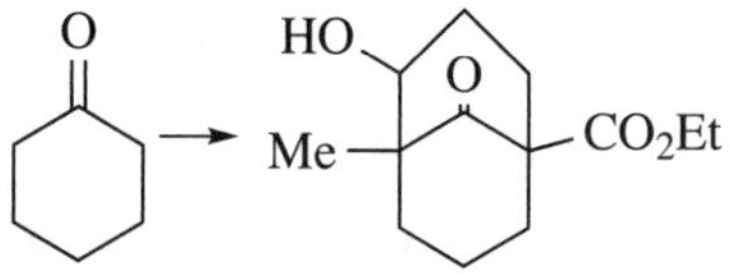

(e)

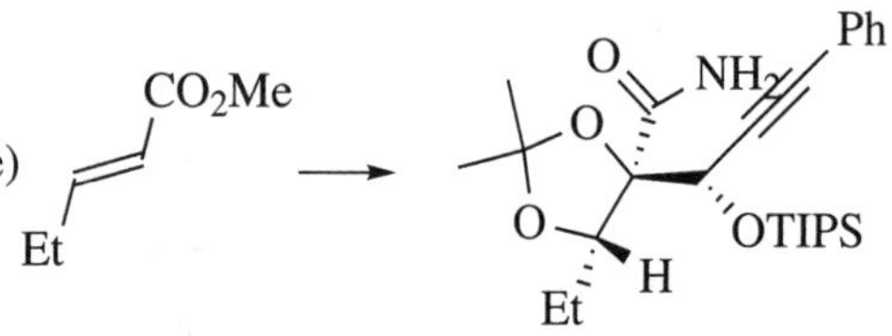

(f)

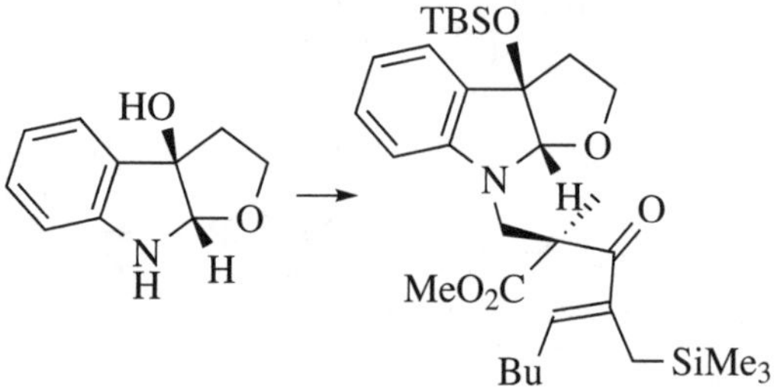

(g)

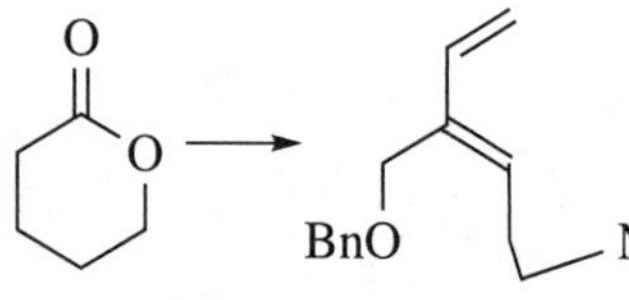

(h)

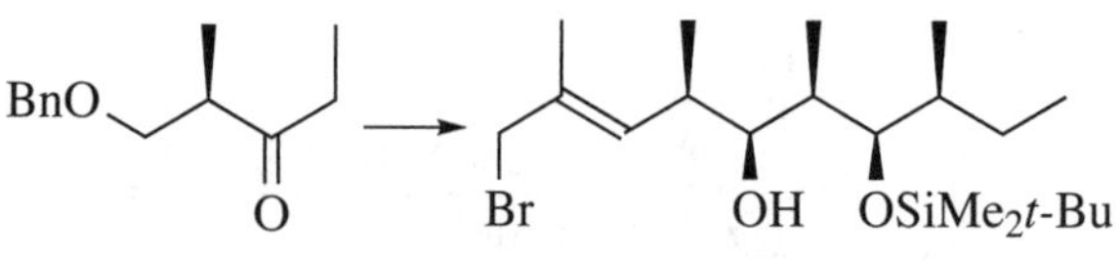

(i)

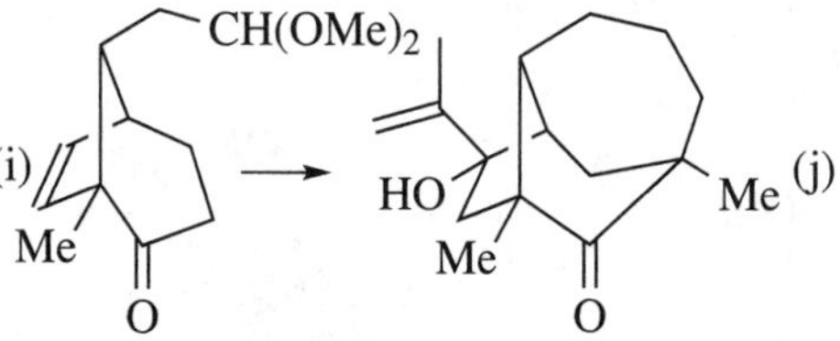

(j)

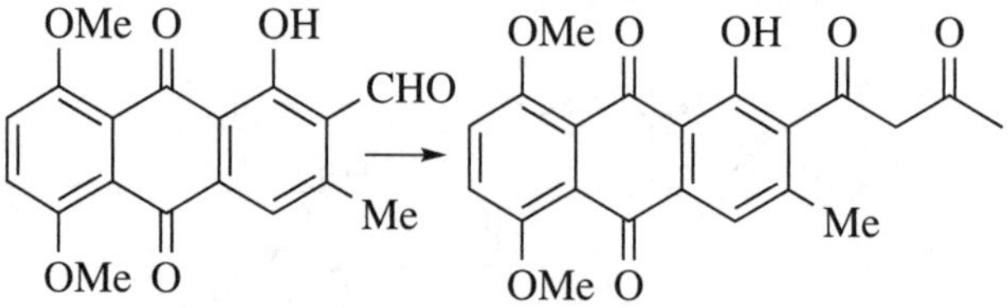

(k) 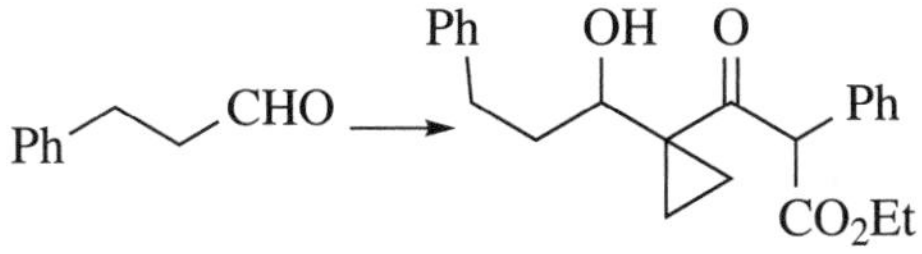 Ph—CHO → (structure with Ph, OH, O, Ph, CO₂Et)

(l) 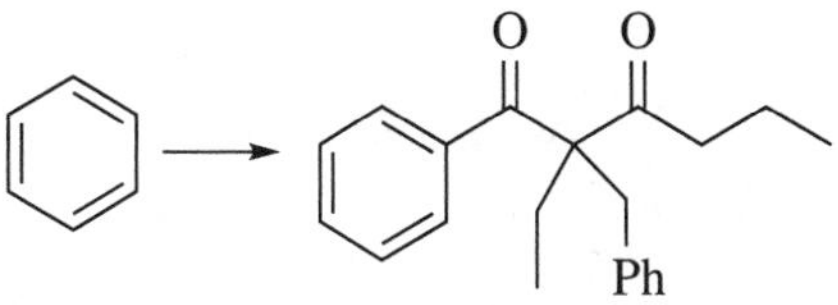 (benzene) → (structure with O, O, Ph)

(m) 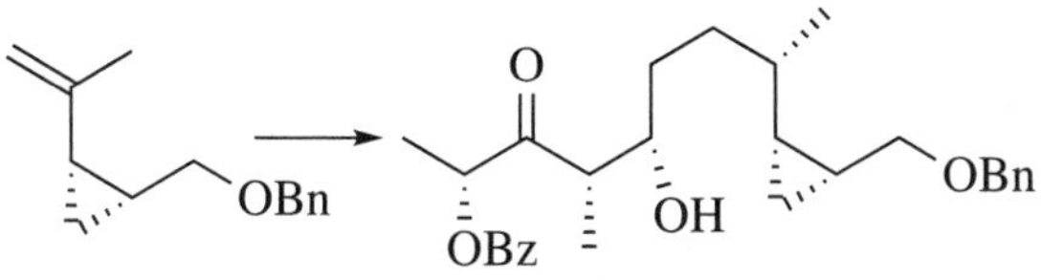 (structure) → (structure with O, OBn, OBz, OH, OBn)

(n) 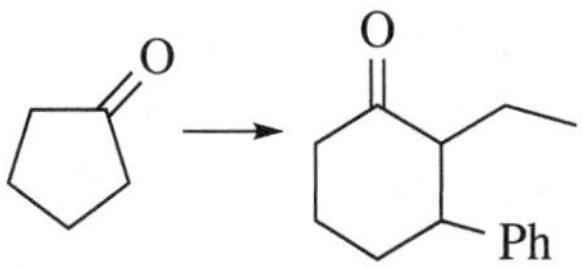 (cyclopentanone O) → (cyclohexanone with Ph)

(o) (structure with O, CO₂Et) → (structure with O)

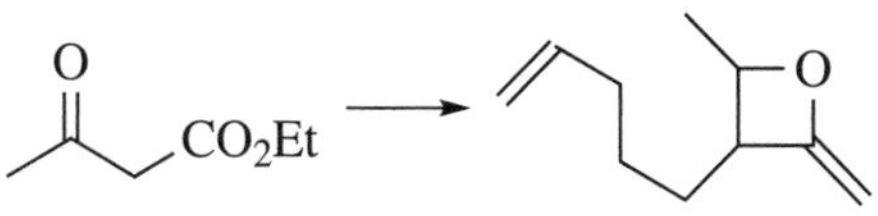

(p) 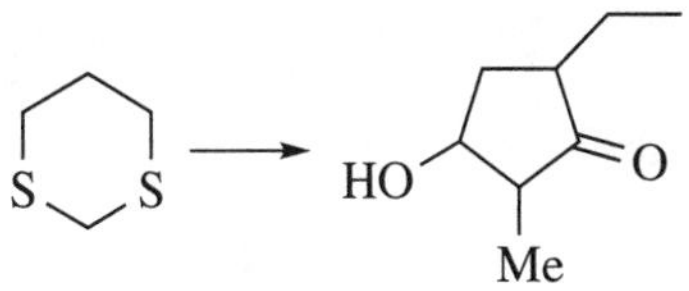 (dithiane, S, S) → (HO, Me, O)

18. For each of the following, choose a commercially available starting material (give the commercial source and the price) of six carbons or less. Convert that starting material into the target shown, showing all intermediate products and all reagents. Give the commercial price of all reagents used. Show your retrosynthetic analysis:

(a) 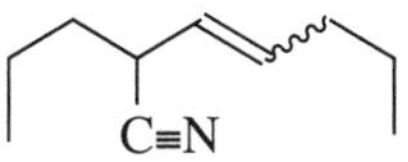 (structure with C≡N)

(b) 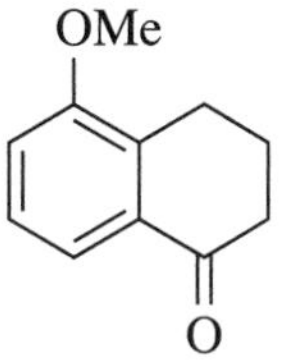 OMe (structure with O)

(c) O Ph (structure)

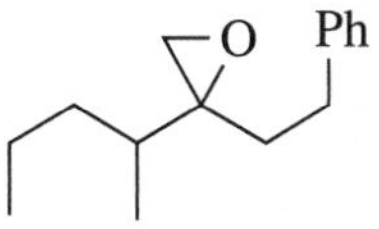

(d) 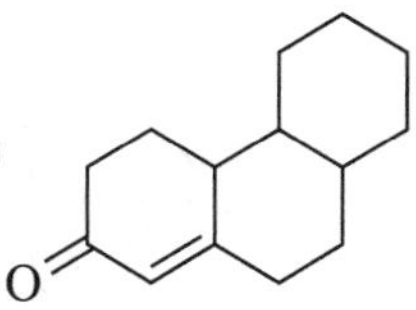 (structure with O)

(e) 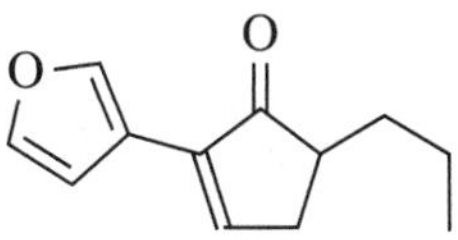 O (structure with O)

(f) 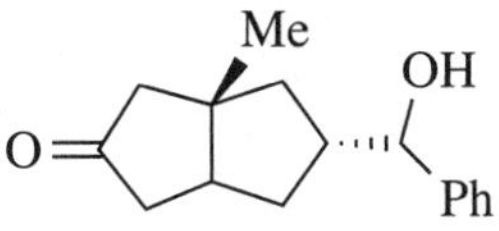 Me (structure with O, OH, Ph)

(g) 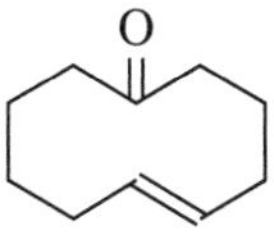 O (structure)

(h) 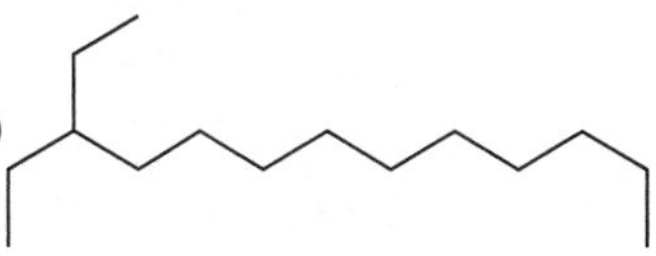 (structure)

chapter 10

Synthetic Strategies

10.1. INTRODUCTION

Throughout this book, functional group transforms for each new reaction were shown at the end of the discussion. In addition, disconnections for carbon-carbon bond-forming reactions have also been given. These transforms and disconnections form the basis of the modern approach to total synthesis. Complex molecules may require alternative strategies for their total synthesis rather than the simple ones presented earlier in this book.

The logic and planning behind Corey's many syntheses spanning over 30 years were summarized in a monograph.[1a] That monograph detailed several strategic approaches to total syntheses where those principles were applied. Nicolaou and co-authors have also published collections of total syntheses.[1b,c] The science of total synthesis dates to the nineteenth century, and Corey cited many examples that illustrate its history. In 1904 Perkin synthesized terpineol (**1**),[2] in part to ascertain the actual structure of this natural product. In the synthesis of tropinone (**2**) by Robinson in 1917,[3] he envisioned that an "imaginary hydrolysis of the substance (tropinone) may be resolved into succinaldehyde, methylamine and acetone."[4b] In this analysis we can apply modern strategy terminology to saw that he is disconnecting **2** into simple disconnect fragments. The synthesis of equilenin (**3**) by Bachmann et al. in 1939[4] was the first multi-step synthesis of a steroid precursor to estrone. Fischer and Kirstahler synthesized hemin (**4**)[5]

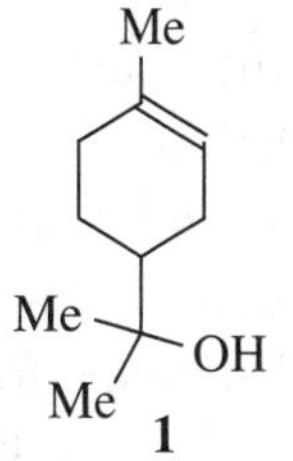

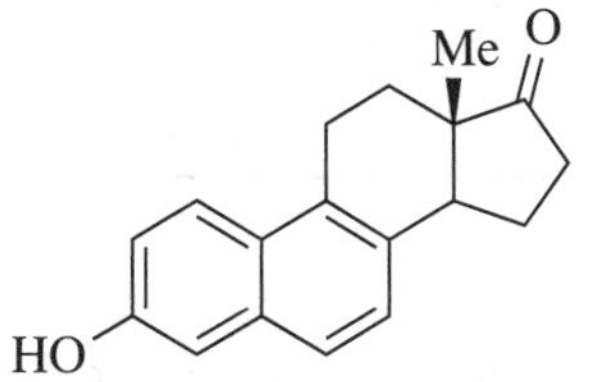

1. (a) Corey, E.J.; Cheng, X.-M. *The Logic of Chemical Synthesis*, Wiley, New York, *1989*; (b) Nicolaou, K.C.; Snyder, S.A. *Classics in Total Synthesis II : More Targets, Strategies, Methods*, Wiley, New York, *2003*; (c) Nicolaou, K.C.; Sorenson, E.J. *Classics in Total Synthesis, 1996*, VCH, Weinheim.
2. Perkin, Jr., W.H. *J. Chem. Soc. 1904, 85*, 654.
3. (a) Robinson, R. *J. Chem. Soc. 1917, 111*, 762; (b) Fleming, I. *Selected Organic Syntheses*, Wiley, London, *1973*, p 18.
4. Bachmann, W.E.; Cole, W.; Wilds, A.L. *J. Am. Chem. Soc. 1939, 61*, 974.
5. (a) Fischer, H.; Kirstahler, A. *Annalen 1928, 466*, 178; (b) Fischer, H.; Zeile, K. *Ibid. 1929, 468*, 98.

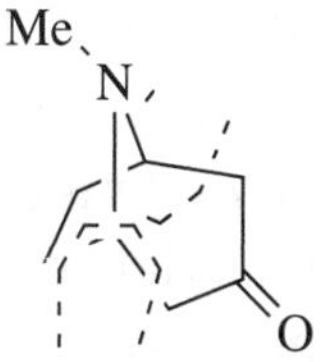

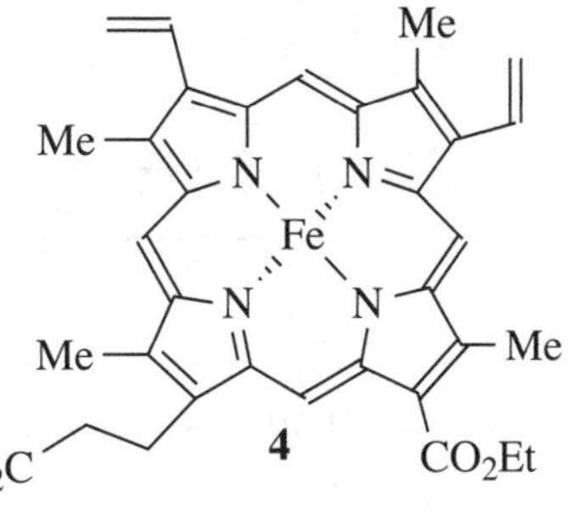

and recognized that four similar pyrrole units composed the structure, requiring the synthesis of each fragment prior to assembling **4**.

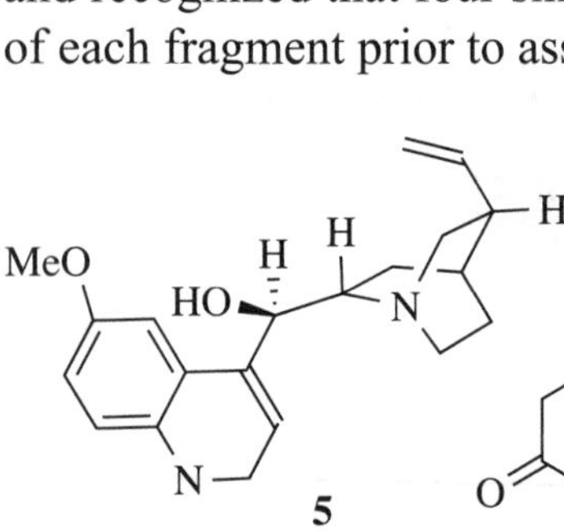

With Woodward, synthesis assumed a sophistication previously unknown,[6] and targets such as quinine (**5**, with von E. Doering),[7] cortisone (**6**),[8] strychnine (**7**)[9] or reserpine (**8**)[10] were synthesized (along with many other molecules). Over these years, the structural and stereochemical features of the targets increased in complexity. As more complex structures were targeted, the newly acquired ability to control stereochemical features and stereogenic centers led to greater synthetic achievements. Corey formalized the terminology for this approach to synthesis, making it an understandable process. This approach is called **retrosynthesis**, and it allows one to dissect a molecule and arrive at relatively simple starting materials in what is now called the **disconnection (or synthon) approach**.[11] Most chemists now use the disconnection approach when they undertake a synthetic problem, but there are other strategies that can be applied to planning a synthesis. Corey included several strategies for synthesis in a computer program called LHASA (sec. 10.4.B), with the goals of assisting chemists in synthetic planning and actually suggesting syntheses for complex molecules. This approach to synthesis will be presented later in this chapter, as will other techniques for assembling a retrosynthetic plan.

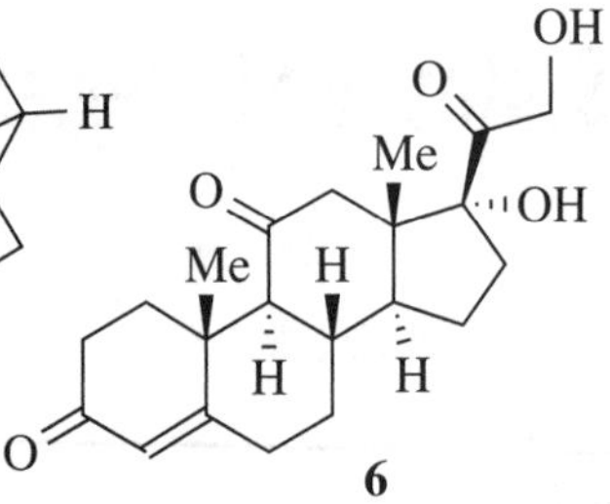

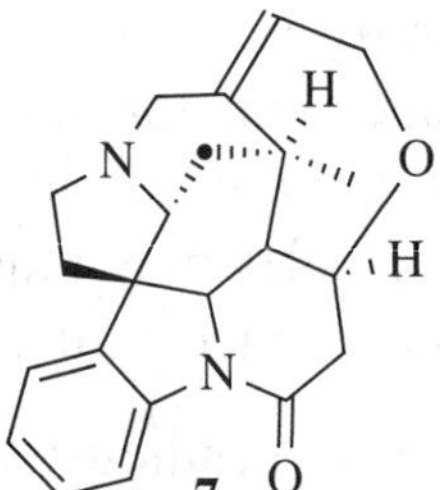

10.2. TARGET SELECTION

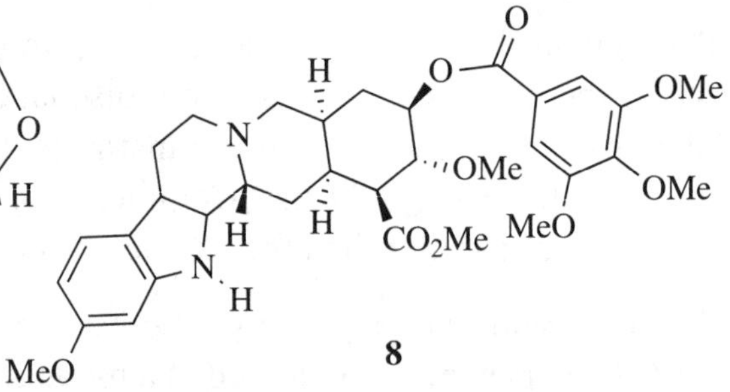

10.2.A. What Is the Rationale for Total Synthesis?

The choice of a target is obviously the point of departure for a synthesis. The synthesis of an

6. Woodward, R.B. in *Perspectives in Organic Chemistry*, Todd, A.R. (Ed., Interscience, New York, *1956*, pp 159-184.
7. Woodward, R.B.; Doering, W.E. *J. Am. Chem. Soc.* *1945*, 67, 860.
8. (a) Woodward, R.B.; Sondheimer, F.; Taub, D.; Heusler, K.; McLamore, W.M. *J. Am. Chem. Soc.* *1952*, 74, 4223; (b) Woodward, R.B.; Sondheimer, F.; Taub, D. *Ibid.* *1951*, 73, 4057, 3547.
9. Woodward, R.B.; Cava, M.P.; Ollis, W.D.; Hunger, A.; Daeniker, H.U.; Schenker, K. *Tetrahedron* *1963*, 19, 247.
10. Woodward, R.B.; Bader, F.E.; Bickel, H.; Frey, A.J.; Kierstead, R.W. *Tetrahedron* *1958*, 2, 1.
11. Reference 1, pp 81-91.

organic molecule usually begins with two questions: (*1*) why was this molecule chosen as a target? and (*2*) where do I begin? The answer to (*1*) often lies in the needs and interests of the synthetic chemist. It may be a challenging stereochemical problem or functional group combination. A molecule may possess unique chemical or biological properties. The answer to (*2*) is the basis for this chapter.

Many reasons are cited for a synthesis, and the challenging problems encountered in a total synthesis provide many opportunities to find new reactions, processes or strategies that may be of value to other organic chemists. Corey and Wipke[12] stated a few of the useful applications that may be found in a retrosynthetic analysis-synthesis problem. Many of these may be useful in solving problems for another target, or in applications to a completely different area of organic chemistry.

"As a synthetic tree (secs. 10.3, 10.4) is developed, findings may emerge which suggest new relationships or simplifications that can lead to even more effective analysis of the original target structure. In this sense, it can be said that imbedded in each problem is a learning opportunity, which if seized can lead to extraordinarily simple or elegant situations. Learning opportunities are equally great in the subsequent stages of synthetic work which involve (i) the selection of specific chemical reactions for transforming one synthetic intermediate to the next in the sequence, (ii) the selection of specific reagents, (iii) the design of experiments, and (iv) experimental execution and analysis. During these stages observations, discoveries, inventions, or theorems of great importance may result."[12]

A similar fundamental basis for total synthesis was described by Deslongchamps.[13]

"Nowadays, for synthetic chemists interested in making fundamental contributions to their field of research, the choice of a given target is not which should be considered as the most important decision; the choice of a specific target is simply an excuse to put in practice a new strategy or to demonstrate the value of, either a new reaction, or a new set of reaction conditions. Also, when one chooses a very difficult target with these principles in mind, it creates a must to innovate in order to succeed. In other words, since the researcher has put himself or herself in a situation that organic chemists have not faced before, the chances of discovering something new and original are then quite high."[13]

"The preceding suggests that what is most important is not the choice of a given target, but how the goal is going to be achieved. In other words, it is the chemistry that one discovers along the way that is the important parameter, not the fact that one succeeds in the synthesis of a given compound, natural or non-natural. Indeed, when peers eventually have to evaluate one's work, either on a short-term or even more so on a long-term basis, it is only the value of the chemistry that will be deemed important. The fact that a total synthesis is reported for the

12. Corey, E.J.; Wipke, W.T. *Science* **1969**, *166*, 178.
13. Deslongchamps, P. *Aldrichimica Acta* **1984**, *17*, 59.

first time or that the compound has important biological properties or is of theoretical interest is of secondary importance from a chemical point of view".[13]

It is often true that synthesis of a particular target is of paramount importance to an individual or organization. The criterion for that importance is always obvious to the researcher(s). To one beginning a study of organic reactions and synthesis, however, it may be of value to examine several commonly quoted criteria for synthesis. These criteria, along with Corey's and also Deslongchamps' evaluations, provide a reasonable answer to question (1). The disconnection approach in its several forms will be used to address question (2).

In addition to the eight molecules shown in the introduction, six molecules are shown as examples of the wide variety of organic molecules that have been synthetic targets. Dysidiolide (9), isolated from the marine sponge *Dysidea etheria* de Laubenfels is a natural inhibitor of a protein phosphatase called csc25A, which is involved in a signaling system for cell division and a potential target for anticancer therapy.[14] Not all targets come from nature. Viagra (sildenafil, 10)[15] was developed by Pfizer Inc. to treat impotence, and Merck Inc. developed indinavir (11)[16] as a HIV-1 protease inhibitor. Epothilone (12) is a compound with the same mechanism of action as the important anticancer drug taxol, but has a less complex structure that makes it easier to prepare.[17] Pancratistatin (13) has been an important target for organic synthesis due to

14. (a) Gunaskera, G.P.; McCarthy, P.J.; Kelly-Borges, M.; Lobkovsky, E.; Clardy, J. *J. Am. Chem. Soc.* *1996*, *118*, 8759; (b) Blanchard, J.L.; Epstein, D.M.; Boisclair, M.D.; Rudolph, J.; Pal, K. *Bioorg. Med. Chem. Lett.* *1999*, *9*, 2537.

15. (a) Bell, A.S.; Brown, D.T.; Nicholas, K. *Eur. Pat. Appl. EP 463 756* [*Chem. Abstr. 1992, 116*:P255626q]. For biological activity, see (b) Boolell, M.; Gepi-Attee, S.; Gingell, J.C.; Allen, M.J. *Br. J. Urol 1996, 78,* 257; (c) Terrett, N.K.; Bell, A.S.; Brown, D.; Ellis, P. *Bioorg. Med. Chem. Lett. 1996, 6,* 1819.

16. (a) Vacca, J.P.; Holloway, M.K.; Dorsey, B.D.; Hungate, R.W.; Guare, J.P. Eur. Pat. Appl. 541,168 [*Chem. Abstr. 120*: 54552w *1994*]; (b) Vacca, J.P.; Dorsey, B.D.; Guare, J.P.; Holloway, M.K.; Hungate, R.W.; Levin, R.B. US Patent 5,413,999 [*Chem. Abstr. 123*: 47889v *1995*]; (c) Dorsey, B.D.; Levin, R.B.; McDaniel, S.L.; Vacca, J.P.; Guare, J.P.; Darke, P.L.; Zugay, J.; Emini, E.A.; Schleif, W.A.; Quintero, J.C.; Lin, J.H.; Chen, I.-W.; Holloway, M.K.; Fitzgerald, P.M.D.; Axel, M.G.; Ostovic, D.; Anderson, P.S.; Huff, J.R. *J. Med. Chem. 1994, 37,* 3443; (d) Askin, D.; Eng, K.K.; Rossen, K.; Purick, R.M.; Wells, K.M.; Volante, R.P.; Reider, P.J. *Tetrahedron Lett. 1994, 35,* 673; (e) Maligres, P.E.; Upadhyay, V.; Rossen, K.; Cianciosi, S.J.; Purick, R.M.; Eng, K.K.; Reamer, R.A.; Askin, D.; Volante, R.P.; Reider, P.J. *Tetrahedron Lett. 1995, 36,* 2195; (f) Vacca, J.P.; Dorsey, B.D.; Schleif, W.A.; Levin, R.B.; McDaniel, S.L.; Darke, P.L.; Zugay, J.; Quintero, J.C.; Blahy, O.M.; Roth, E.; Sardana, V.V.; Schlabach, A.J.; Graham, P.I.; Condra, J.H.; Gotlib, L.; Holloway, M.K.; Lin, J.; Chen, I.-W.; Vastag, K.; Ostovic, D.; Anderson, P.S.; Emini, E.A.; Huff, J.R. *Proc. Natl. Acad. Sci. 1994, 91,* 4096.

17. For a synthesis, see (a) Meng, D.; Bertinato, P.; Balog, A.; Su, D.-S.; Kamenecka, T.; Sorensen, E.J.; Danishefsky, S.J. *J. Am. Chem. Soc, 1997, 119,* 10073. For isolation and characterization, see (b) Gerth, K.; Bedorf, N.; Höfle, G. Irschik, H.; Reichenbach, H. *J. Antibiotics 1996, 49,* 560; (c) Höfle, G.; Bedorf, N.; Steinmetz, H.; Schomburg, D.; Gerth, K.; Reichenbach, H. *Angew. Chem. Int. Ed. 1996, 35,* 1567. For biological activity, see Bollag, D.M.; McQueney, P.A.; Zhu, J.; Hensens, O. Koupal, L.; Liesch, J. Goetz, M.; Lazarides, E.; Woods, C.M. *Cancer Res. 1995, 55,* 2325.

its promising anticancer activity.[18] The preparation of **14** (the common name of this compound is [4]-phenylene) helped show that the delocalized bonds of a benzene ring were effectively localized in these compounds, generating the elusive cyclohexatriene unit.[19] This discovery caused scientists to reexamine the underlying principles of aromaticity.

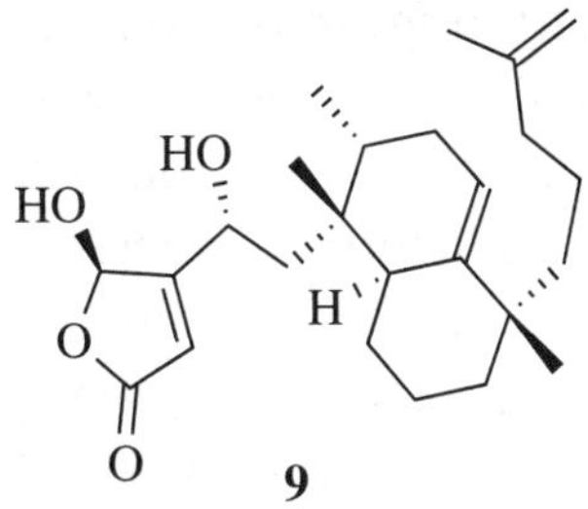

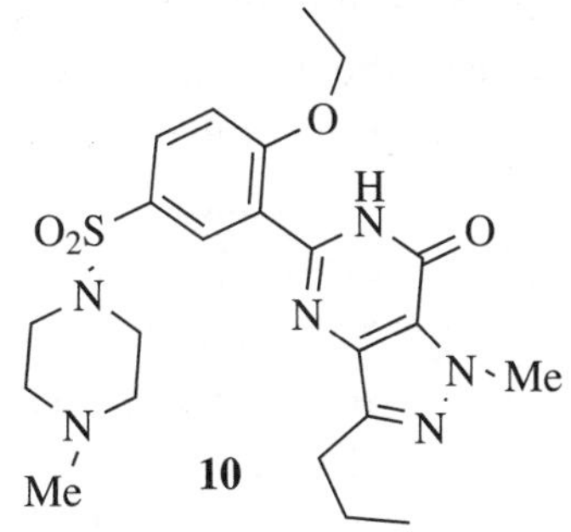

The differing structural features offer unique synthetic problems. When such syntheses are published, the step-by-step reaction sequence used to construct each molecule is reported and, occasionally, a summary of the rationale behind the synthesis. This summary sometimes provides insight into why a target was chosen, but not always. To a chemist searching for a target, inspection of previous work may not be of great value, although a close study of these syntheses will be of enormous value for understanding and devising strategies. The five criteria listed below are often cited as characteristic of important synthetic targets. These may be of value to a chemist who is being introduced to synthesis for the first time. Much of the information gained from examining these criteria may be useful for disconnecting the target. **(1) Structural verification; (2) important biological activity; (3) analog generation and studies; (4) structural or topological challenges-fundamental understanding of the nature of bonding and molecules; (5) development of new reactions or reagents.**

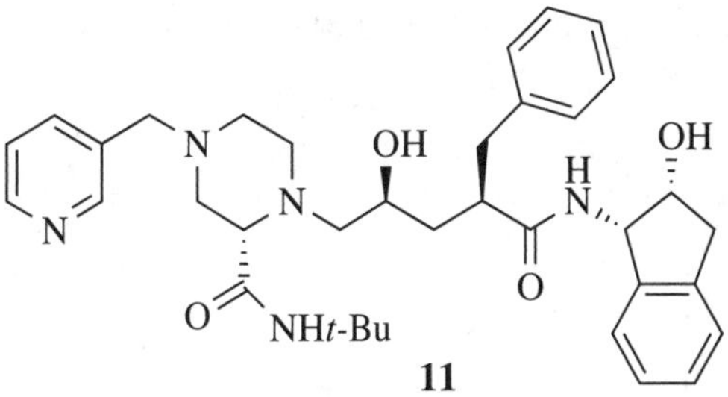

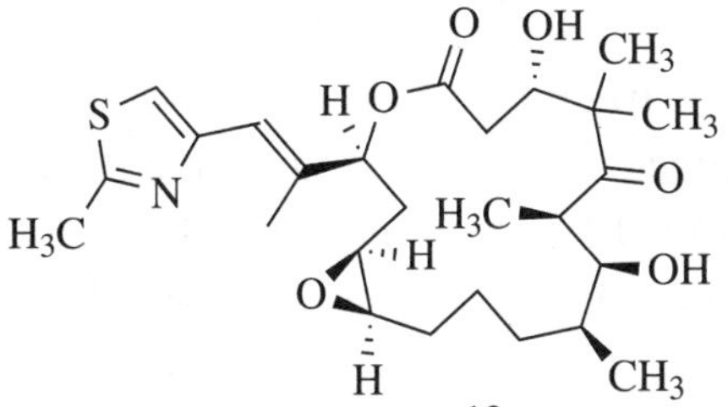

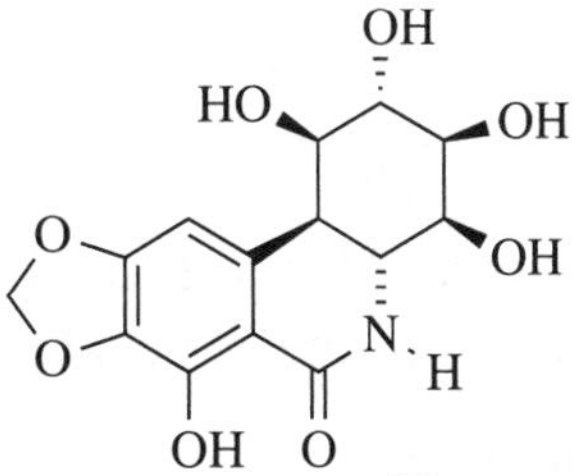

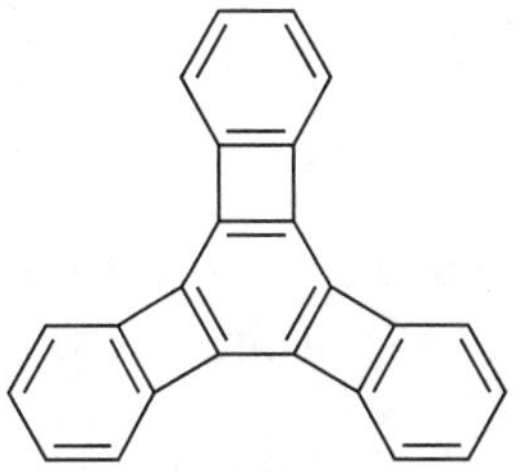

18. (a) Pettit, G.R.; Gaddamidi, V.; Cragg, G.M. *J. Nat. Prod.* **1984**, *47*, 1018; (b) Pettit, G.R.; Gaddamidi, V.; Cragg, G.M.; Herald, D.L.; Sagawa, Y. *J. Chem. Soc. Chem. Commun.* **1984**, 1693; (c) Pettit, G.R.; Gaddamidi, V.; Herald, D.L.; Singh, S.B.; Cragg, G.M.; Schmidt, J.M.; Boettner, F.E.; Williams, M.; Sagawa, Y. *J. Nat. Prod.* **1986**, *49*, 995 (d) *Antitumor Alkaloids*, Suffness, M.; Cordell, G.A. *The Alkaloids, Vol. XXV*, Academic Press, Orlando, FL, **1985**, pp 198-212, 338-340 (see p 207). For a review of total syntheses, see (e) Manpadi, M.; Korienko, A. *Org. Prep. Proceed. Int.* **2008**, *40*, 107.

19. The chemical name of **14** is bis-benzo[3,4]cyclobuta[1,2-a:1',2'-c]biphenylene. See (a) Diercks, R.; Vollhardt, K.P.C. *J. Am. Chem. Soc.* **1986**, *108*, 3150; (b) Vollhardt, K.P.C.; Mohler, D.L. *Adv. Strain Org. Chem.* **1996**, *5*, 121; (c) Schulman, J.M.; Disch, R.L. *J. Phys. Chem. A* **1997**, *101*, 5596.

10.2.B. Structural Verification

Despite the power of X-ray crystallography and high field NMR, the structures of many complex natural products were incorrectly reported. Examples, as reported by Weinreb,[20] include cylindrospermopsin,[21] the sclerophytins[22] and batzelladine F,[23] where the correct structures were determined by the total synthesis. Weinreb uses the example of lepadiformine to demonstrate the value of total synthesis for determination of structure of a natural product where the original report gave an incorrect structure.[20] Synthesis is used for proof of structure, or sometimes revision of an originally proposed structure as in Yamada and co-worker's synthesis of cladocoran A (15),[24] which was originally published by Fontana, et al as 16.[25] Structure 15 corresponded exactly to the spectral data of Fontana, and the original proposed structure 16 differs from 15 in the stereochemical assignments at C7, C12, and C16. Synthesis of the compound allowed the correct structure to be defined, and revised.

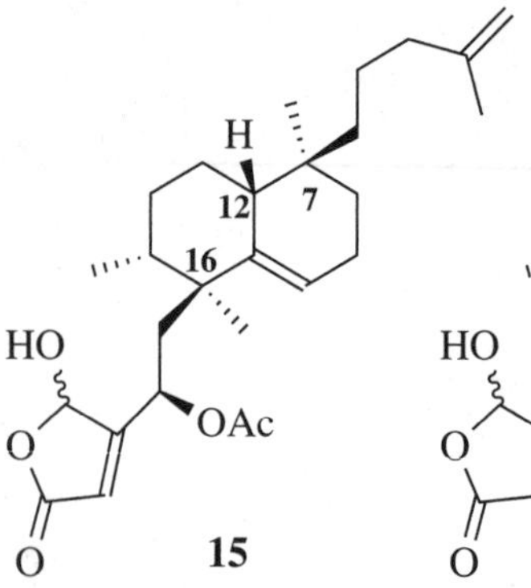

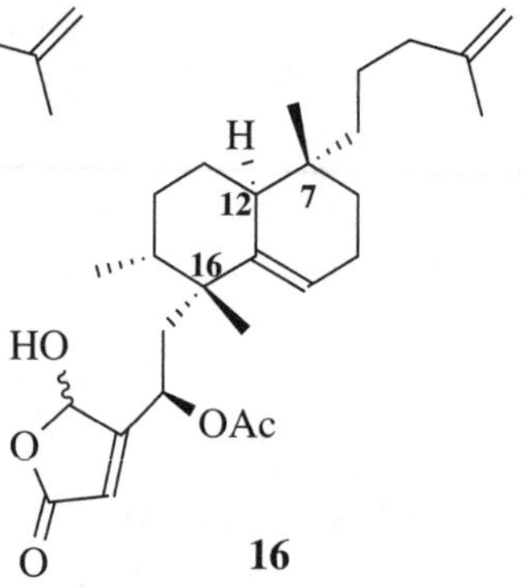

Single crystal X-ray crystallography has become a powerful method for determining the structures of complex organic molecules.[26] An example where X-ray analysis was invaluable for the structure proof is taken from Qian-Cutrone and co-workers isolation of new antitumor alkaloids, stephacidin A (17) and B (18), produced by *Aspergillus*

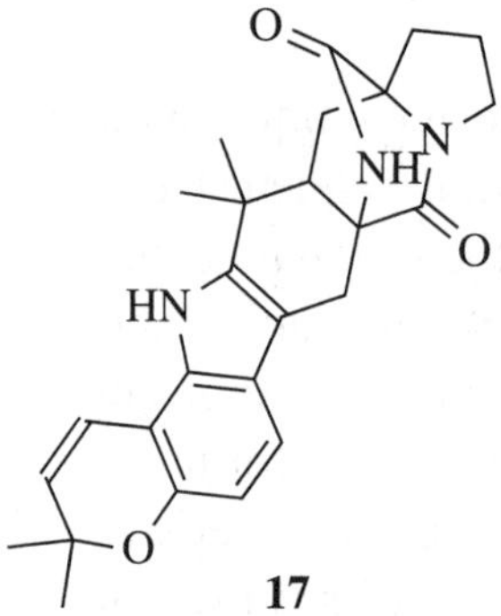

20. Weinreb, S.M. *Acc. Chem. Res.* **2003**, *36*, 59.
21. Heintzelman, G. R.; Fang, W.-K.; Keen, S. P.; Wallace, G. A.; Weinreb, S. M. *J. Am. Chem. Soc.* **2002**, *124*, 3939.
22. (a) Bernardelli, P.; Moradei, O. M.; Friedrich, D.; Yang, J.; Gallou, F.; Dyck, B. P.; Doskotch, R. W.; Lang, T.; Paquette, L. A. *J. Am. Chem. Soc.* **2001**, *123*, 9021; (b) Overman, L. E.; Pennington, L. D. *Org. Lett.* **2000**, *2*, 2683.
23. Cohen, F.; Overman, L. E. *J. Am. Chem. Soc.* **2001**, *123*, 10782.
24. Miyaoka, H.; Yamanishi, M.; Kajiwara, Y.; Yamada, Y. *J. Org. Chem.* **2003**, *68*, 3476.
25. Fontana, A.; Ciavatta, M..; Cimino, G. *J. Org. Chem.,* **1998**, *63*, 2845.
26. For example see (a) Powell, H.M.; Prout, C.K.; Wallwork, S.C. *Ann. Rept. Progr. Chem.* **1964**, *60*, 593; (b) Seemann, V. AD 626597, *U.S. Govt. Res. Develop. Rept.* **1966**, *41*, 99 [*Chem. Abstr. 67*: R15845m *1967*]; (c) Schredt, E.; Weitz, G. (Ed.) *Landoet-Boernster, Group III: Crystal and Solid State Physics, Vol. 5, Pts a & b: Structure Data of Organic Crystals*, Springer, New York, **1971**; (d) Koyama, H.; Okada, K.; Itoh, C. *Acta Crystallogr. Sect. B*, **1970**, *26*, 444.

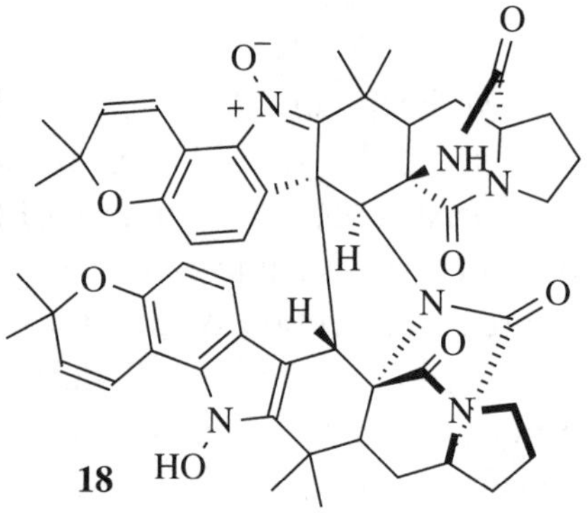

ochraceus WC76466.[27] Extensive analysis using NMR suggested that **18** was a dimeric structure related to **17**. Although all structural fragments were determined, the point at which the monomeric units of **17** were linked to form the dimer could not be established. Single-crystal X-ray analysis established the final structure as **18**, with the relative stereochemistry shown as well as the presence of the nitrone and hydroxylamine units.[28]

It is reasonable to ask why a lengthy synthetic verification might be necessary for a given target when such techniques are available. In many cases, X-ray analysis is not possible due to an unsuitable crystalline form, the unavailability of single crystals, poor morphology of the molecule, or a derivative. In many cases, of course, the molecule does not form a crystalline derivative at all, although crystalline derivatives can sometimes be prepared. As noted for volubilide, even high-field NMR techniques often fail to give the absolute stereochemistry of all chiral centers, although the relative stereochemistry may be discerned. Chemical degradation used in conjunction with spectroscopy is also an important identification protocol but is not successful in every case. Total synthesis and comparison with an authentic sample therefore remains an important method for final structure determination.

10.2.C. Biological Activity

An obvious reason for total synthesis is the importance of the target in medicine, agriculture or other commercial and humanitarian ventures. An important or interesting molecule is usually isolated from a natural source in very small quantity, as with pancratistatin (**13**). If the molecule is subsequently shown to possess significant biological activity, additional material may be required for further testing or determination of the structure. The small quantities isolated or the difficulties and expense inherent in gathering natural specimens often prevent collection of sufficient amounts of the natural product. If the molecule is to undergo further biological screening, or be used clinically, total synthesis may be the only means to obtain sufficient material.

An example of a synthesis used to verify a structure or to generate sufficient material for useful purposes is seen in pancratistatin (**13**), an *Amaryllidaceae* isoquinoline alkaloid that was isolated by Pettit et al. from *Pancratium littorale* in 0.039% of the dry weight of the bulbs of *P. littorale*.[18a,b] It was also isolated from the bulbs of *Zephyranthese grandiflora*, but in only 0.0019% yield.[18a,b] The natural abundance of **13** reported from these plant sources clearly show that harvesting these plants is a poor source to obtain quantities suitable for testing. In biological testing, pancratistatin is believed to behave similarly to narciclasine, which inhibits protein synthesis in eukaryotic cells[29] and exhibits antineoplastic activity in ovarian sarcoma

27. Qian-Cutrone, J.; Krampitz, K. D.; Shu, Y. Z.; Chang, L. P. U.S. Patent 6,291,461, 2001.
28. Qian-Cutrone, J.; Huang, S.; Shu, Y.-Z.; Vyas,D.; Fairchild, C.; Menendez, A.; Krampitz, K.; Dalterio, R.; Klohr, S.E.; Gao, Q. *J. Am. Chem. Soc.* **2002**, *124*, 14556.
29. (a) Rivera, G.; Gosalbez, M.; Ballestra, J.P.G. *Biochem. Biophys. Res. Commun.* **1980**, *94*, 800; (b) Baez, A.; Vazquez, D. *Biochim. Biophys. Acta* **1978**, *518*, 95; (c) Barbacid, M.; Vazquez, D. *J. Mol. Biol.* **1974**, *84*, 603.

and lymphatic leukemia[18c] as well as being an antimitotic.[18c] Pancratistatin is in preclinical development at the National Cancer Institute, "based on its high activity against the M5076 sarcoma system."[18d] It also exhibited good activity against the P388 leukemia.[18c] The significant antitumor activity coupled with the small amounts of material available from natural sources have prompted many total syntheses, including those by Danishefsky and Lee,[30] Hudlicky and co-workers,[31] Trost and Pulley,[32] Rigby et al.[33] and Kim and co-worker's.[34]

Synthesis of a complex natural product by a multi-step synthetic route can be very expensive. To produce even one gram of active product can require large quantities of both starting material and time. If the synthetic goal is to produce usable quantities of material at a reasonable cost, this must be factored into the retrosynthetic scheme. If one requires only milligram quantities, greater latitude in the choice of starting materials and reagents is possible.

10.2.D. Analog Studies

A target that has interesting or commercially attractive properties may exhibit deleterious side effects, or be unstable to storage and handling. Subsequent studies often show that structural modification to the basic skeleton gives a molecule with different characteristics and the new molecule becomes the synthetic target.

Chlorothiazide is 2*H*,1,2,4-benzothiadiazine-7-sulfonamide, 6-chloro-1,1-dioxide (**19**).[35] It is a diuretic also used to treat hypertension. Studies "determined that significant alterations in the pharmacological profile would follow changes in the heterocyclic ring rather than the ring bearing the sulfonamide group."[36] Structural modification led to the synthesis of chlorexolone (**20**).[37] Conversion of the heterocyclic ring in **19** to the *N*-cyclohexyl lactam

30. Danishefsky, S.; Lee, J.Y. *J. Am. Chem. Soc.* **1989**, *111*, 4829.
31. Tian, X.; Hudlicky, T.; Königsberger, K. *J. Am. Chem. Soc.* **1995**, *117*, 3643.
32. Trost, B.M.; Pulley, S.R. *J. Am. Chem. Soc.* **1995**, *117*, 10143.
33. Rigby, J.H.; Maharoof, U.S.M.; Mateo, M.E. *J. Am. Chem. Soc.* **2000**, *122*, 6624.
34. (a) Ko, H.; Kim, E.; Park, J.E.; Kim, D.; Kim, S. *J. Org. Chem.* **2004**, *69*, 112; (b) Kim S.; Ko, H.; Kim, E.; Kim, D. *Org. Lett.* **2002**, *4*, 1343.
35. (a) Novello, F.C.; Sprague, J.M. *J. Am. Chem. Soc.* **1957**, *79*, 2028; (b) Novello, F.C. *U.S. Patent* 2,809,194 [*Chem. Abstr.* 52: 2939h **1958**]; (c) Hinkley, D.F. *U.S. Patent* 2,937,169 [*Chem. Abstr.* 54: 18565i **1960**]; (d) Dupont, D.; Dideberg, O. *Acta Crystallogr., Sect. B* **1970**, *26*, 1884; (e) Wolff, F.W.; Basabe, J.; Grant, A.; Krees, S.; Lopez, N.; Vicktora, J. *Med. Ann. DC*, **1971**, *40*, 98 [*Chem. Abstr. 75* : 47093w, **1971**]; (f) Tubaro, E., *Boll. Chim. Farm.*, **1963**, *102*, 505 [*Chem. Abstr.* 60: 2198b, **1964**]; (g) Ford, R.V. *Ann. New York Acad. Sci.* **1960**, *88*, 809 [*Chem. Abstr.* 61:12493d, **1964**]; (h) Sorice, F., *Policlinico (Rome), Sez. Prat.* **1960**, *67*, 625 [*Chem. Abstr.* 55: 14692cd **1961**].
36. Lednicer, D.L.; Mitscher, L.A., *Organic Chemistry of Drug Design*, Vol. 1, Wiley, **1977**, p 321.
37. May & Baker Ltd., *Belg. Patent* 620,654 [*Chem. Abstr.* 59: P11436c **1963**].

moiety in **20** resulted in improved hypotensive action[38] and diuretic action, but did not increase urinary pH or urinary bicarbonate excretion relative to **19**.[39] Synthetic modification of a known structure can identify a synthetic target.

Another rationale for preparing analogs of a given target is to study the structure-biological activity profile of a molecule. Flavone 8-acetic acid (**21**) was believed to have potential as an antitumor agent.[40] Denny and co-workers showed that 9-oxo-9H-xanthene-4-acetic acid (**22a**) is as active

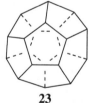

(a) X = H (b) X = Me (c) X = Cl

21   **22**

as **21** against colon-38 tumors in mice, and is more dose potent.[41] It was also shown that small lipophilic substituents at the 5-position (see **22b** and **22c**) enhanced dose potency.[42] A systematic study[43] of synthetic derivatives of **22** showed that many related derivatives possessed enhanced antitumor activity, which is a typical case in which a known compound was structurally modified based on structure-activity properties of a related compound. The impetus for this synthesis was therefore a search for a more potent and efficacious drug.

10.2.E. Structural Challenges

Occasionally, a molecule presents such a structural and chemical challenge that its potential as a target is irresistible. Such a synthesis often pushes back the limits of known chemistry for making carbon-carbon bonds, and gives insight into structure, bonding, or fundamental reaction properties of organic molecules.

Paquette and co-workers synthesized dodecahedrane (**23**),[44a] and the topology (shape, structural features) of **23** is essentially that of a ball. Having the form of a dodecahedron, first described as one of five regular polyhedra in 400-350 B.C.E. in Plato's *Timaeus*, **23** has a spherical superpolycyclopentanoid topology[44b] that possesses the highest known point group symmetry (I_n, icosahedral).[44a] It also

23

38. (a) Maxwell, D.R.; McLusky, J.M. *Nature* **1964**, *202*, 300; (b) Bayeli, P.F.; Montagnani, M.; Zampetti, L.P.; Antonelli, A. *Boll. Soc. Ital. Biol. Sper.* **1969**, *45*, 406 [*Chem. Abstr. 73:* 2550k **1970**]; (c) Patterson, R.R.; Macaraeg, Jr., P.V.J.; Schrogie, J.J. *Clin. Pharmacol. Ther.* **1969**, *10*, 265 [*Chem. Abstr. 70:* 95348b **1969**].
39. Baba, W.L.; Lant, A.F.; Wilson, G.M. *Clin. Pharmacol. Therap.* **1966**, *7*, 212 [*Chem. Abstr. 64:* 18220h **1966**].
40. (a) Smith, G.P.; Calveley, S.B.; Smith, M.J.; Baguley, B.C. *Eur. J. Cancer Clin. Oncol.* **1987**, *23*, 1209 [*Chem. Abstr. 108*:15894s **1988**]; (b) Ching, L.-M.; Baguley, B.C. *Ibid.* **1989**, *25*, 821 [*Chem. Abstr. 111*:33243v **1989**]; (c) *Idem. Ibid.* **1989**, *25*, 1513 [*Chem. Abstr. 112*:15992e **1990**]; (d) Zwi, L.J.; Baguley, B.C.; Gavin, J.B.; Wilson, W.R. *J. Natl. Cancer Inst.* **1989**, *81*, 1005.
41. Rewcastle, G.W.; Atwell, G.J.; Baguley, B.C.; Calveley, S.B.; Denny, W.A. *J. Med. Chem.* **1989**, *32*, 793.
42. Atwell, G.L.; Rewcastle, G.W.; Baguley, B.C.; Denny, W.A. *J. Med. Chem.* **1990**, *33*, 1375.
43. Rewcastle, G.W.; Atwell, G.J.; Zhuang, L.; Baguley, B.C.; Denny, W.A. *J. Med. Chem.* **1991**, *34*, 217.
44. (a) Ternasky, R.J.; Balogh, D.W.; Paquette, L.A. *J. Am. Chem. Soc.* **1982**, *104*, 4503; (b) Mehta, G. *J. Sci. Ind. Res.* **1978**, *37*, 256.

shows a unique encapsulation of a cavity incapable of solvation, negligible angle strain but great torsion strain,[45] twenty symmetry equivalent methine units and an absence of structurally allied substances. Paquette was the first to synthesize this molecule, and the creative and sometimes novel chemistry arising from the synthetic solution to targets such as **23** can usually be applied to other synthetic endeavors.

10.2.F. New Reactions and New Reagents

During a total synthesis, completion of a key transformation via known chemical reactions may be impossible. Development of new methodology or modification of existing reactions must be accomplished to give the desired transformation and complete the synthesis. Once developed, later work often shows the reagents developed for that synthesis can be utilized for a variety of other purposes (recall Corey's and Deslongchamps' analyses). An example is the total synthesis of vitamin B-12 (**24**) reported by Woodward and Eschenmoser.[46] Late in this synthesis, a key step required hydrolysis of an amide in an intermediate possessing six ester moieties. The reagent used in this transformation was an α-chloronitrone (**25**), which gave a vinyl nitronium ion (**26**) when treated with silver ion.

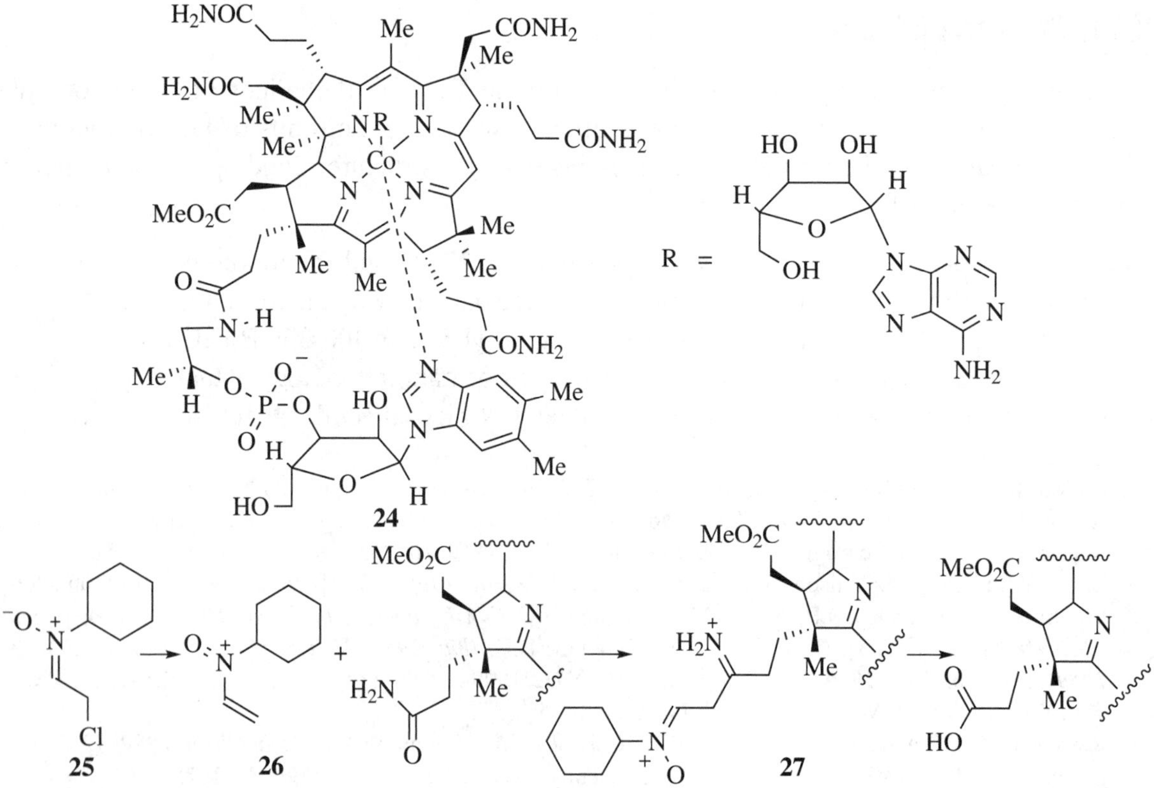

45. Ermer, O. *Angew. Chem. Int. Ed.* **1977**, *16*, 411.
46. Woodward, R.B. *Pure Appl. Chem.* **1973**, *33*, 145 (pp 171-173).

Reaction with the amide gave **27**. Hydrolysis of **27** gave the acid, allowing the synthesis of **24** to be completed. Some years later, Eschenmoser developed a series of transformations using derivatives of the α-chloronitrone described above. Nitrone **28** was used as a diene partner in a Diels-Alder reaction (secs. 11.4.A, 11.7), and the resultant cycloadduct (**31**) was transformed into several synthetically useful molecules. The secondary product that competed with the cycloaddition was the addition product (**29**), which was also converted to synthetically useful products. Note that the end-products include furan and alkyne derivatives[47] when the addition partner is changed from alkenes to alkynes to other derivatives. In the specific example shown, cycloaddition of methylcyclohexene and **28** gave the addition product (**29**) in 59% yield, and subsequent hydrolysis led to a 90% yield of **30**. The cycloadduct (**31**) was formed in 30% yield. This sequence illustrates how a reagent developed for one purpose can be a powerful reagent in a completely different application. Indeed, the discovery of such new reagents and transformations is commonly the result of a total synthesis.

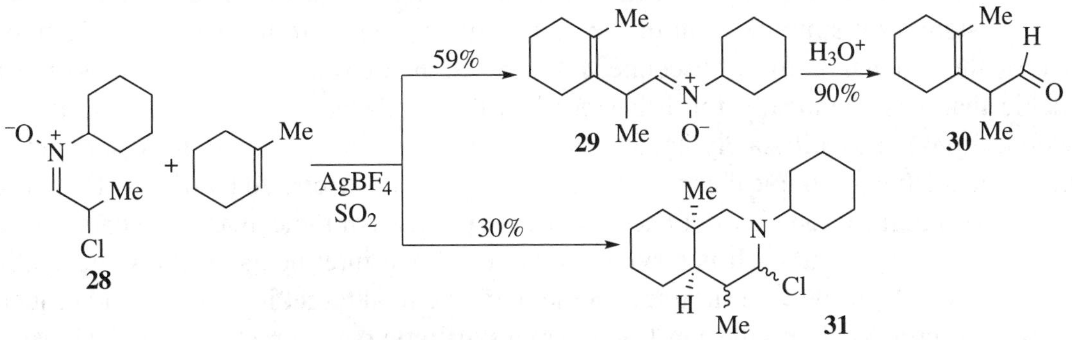

To summarize, the criteria discussed for target selection can be useful during a search for possible molecules to synthesize. They may also offer insights into possible applications to chemical problems that may arise. An understanding of the history of a target and how it is used often provides clues to its chemistry and physical properties. These are essential to execution of a synthesis, if not the actual retrosynthetic analysis.

10.3. RETROSYNTHESIS

When a target is chosen, there are guidelines that allow its systematic disconnection to a starting material. The process that develops this roadmap of synthetic intermediates is called retrosynthesis. This section will describe several ways to analyze a molecule, with the goal of identifying key bonds for disconnection.

47. (a) Shatzmiller, S.; Gygax, P.; Hall, D.; Eschenmoser, A. *Helv. Chim. Acta* **1973**, *56*, 2961; (b) Kempe, U.M.; Das Gupta, T.K.; Blatt, `K.; Gygax, P.; Felix, D.; Eschenmoser, A. *Ibid.* **1972**, *55*, 2187 (c) Gygax, P.; Das Gupta, T.K.; Eschenmoser, A. *Ibid.* **1972**, *55*, 2205; (d) Das Gupta, T.K.; Felix, D.; Kempe, U.M.; Eschenmoser, A. *Ibid.* **1972**, *55*, 2198; (e) Petrzilka, M.; Felix, D. Eschenmoser, A. Ibid. **1973**, *56*, 2950; (f) Shatzmiller, S. Eschenmoser, A. *Ibid.* **1973**, *56*, 2975.

10.3.A. The Disconnection Approach

Once a synthetic target has been chosen, reactions must be developed that will allow the total synthesis to be accomplished. The choice of a **starting material** (the molecule, purchased or readily prepared, that is used in the first reaction of the synthetic sequence) is critical. How does one analyze a target in order to determine the best starting material? The answer to that question is the essence of total synthesis, and its answer requires: (*1*) a detailed analysis of the structure of the target, (*2*) an excellent knowledge of chemical reactions, (*3*) a good understanding of stereochemistry, bonding, and reactivity, and (*4*) a well-developed chemical intuition. This information is then used to **disconnect** bonds in the target, simplifying the structure along reasonable chemical reaction pathways.[1] Wipke and Howe[48] defined the important bonds that are disconnected, as approached by the computer program LHASA (sec. 10.4.B) developed by Corey and co-workers for analysis of retrosynthetic pathways:[49,69] Their definition was "There are usually certain bonds in a molecule whose disconnection in the retrosynthetic direction leads to a significant simplification of the ... structure. These are termed **strategic bonds**." If one of these bonds is to be disconnected, a reaction (or sequence of reactions) must be available that will chemically form that bond in the synthesis. Subsequent simplifications (disconnections) lead ultimately to a molecule that can be recognized as commercially available, available by simple chemical techniques or already prepared by others. This process of structural simplification via disconnection leads to a series of molecular fragments that serve as key intermediates, and each is a synthetic target. Such intermediates allow the synthetic chemist to mentally bridge the starting material with the final target in a logical and sequential manner. This process allows the construction of a **synthetic tree**, for which chemical reactions must be provided to accomplish the planned transformations.

An online synthetic tree for a very simple molecule is presented by the Helsinki University of Technology, as shown in Figure 10.1,[50] similar to more general synthetic trees presented elsewhere.[51,159] The target (T_0) is disconnected to "a logically restricted set of structures which may be converted in a single synthetic operation (a chemical step) to the synthetic target."[12] This gives the disconnection product in the first retro-reaction. The subtree represents further disconnection of one branch of the synthetic tree. After several retro-reactions, the synthetic tree will yield several starting materials from one or more of the first branches, which can then be used to construct the molecule. Working *backwards* toward the starting material in the manner illustrated in Figure 10.1 is termed **retrosynthesis**.[1] Each succeeding structure has been modified and always simplified. Bonds have been disconnected, and functional

48. Wipke, W.T.; Howe, W.J. *Computer Assisted Organic Synthesis*, American Chemical Society, Washington, *1971*.
49. (a) Corey, E.J.; Howe, W.J.; Pensak, D.A. *J. Am. Chem. Soc.* *1974*, *96*, 7724; (b) Corey, E.J.; Wipke, W.T.; Cramer, III, R.D.; Howe, W.J. *J. Am. Chem. Soc.* *1972*, *94*, 421; (c) Corey, E.J.; Jorgensen, W.L. *Ibid.* *1976*, *98*, 189.
50. Helsinki University of Technology, Laboratory of Organic Chemistry. See http://www.hut.fi/Yksikot/Orgaaninen/Opetus/
51. Ottenheijm, H.C.J. *Janssen Chimica Acta* *1984*, *2*, 3.

groups transformed until the relatively simple starting materials shown were obtained. The term **disconnection**[52,12] therefore refers to "mentally breaking bonds to give successively more simple precursor molecules, but always in a manner in which those bonds can be reformed by known or reasonable chemical reactions." Warren has described this approach in great detail.[52]

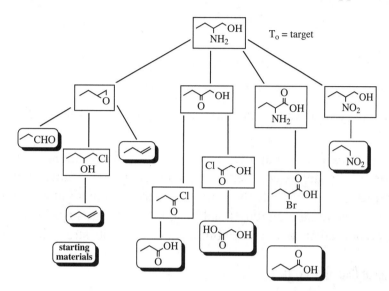

Figure 10.1. A Synthesis Tree.

Massanet and co-workers provided an example of the disconnection analysis as applied to a real synthetic problem, in a synthesis of thapsigargin (**32**).[53] In Figure 10.2, the retrosynthesis analysis begins with clipping the labile lactone ring to give **33**. The next disconnection leads to fragment **34**, in which an isopropenyl unit is the precursor to the lactone-diol moiety. This fragment is, in turn, disconnected to give fragment **35** indicates that no new carbon atoms are incorporated, but rather indicates a reorganization (in this case a rearrangement reaction). The final disconnection of **35** leads to two fragments (**36** and **37**), which are either commercially available or readily prepared. This analysis is outlined in the box and shows how a molecule is simplified into smaller, and more manageable fragments by a series of disconnections. Each fragment is prepared and the molecule assembled, following what is essentially the reverse of the retrosynthetic sequence shown. Analysis and simplification of a complex target using this approach is the essence of retrosynthesis.

52. Warren, S., *Designing Organic Syntheses: A Programmed Introduction to the Synthon Approach*, John Wiley, New York, *1978*.
53. Manzano, F.L.; Guerra, F.M.; Moreno-Dorado, F.J.; Jorge, Z.D.; Massanet, G.M. *Org. Lett.* *2006*, *8*, 2879.

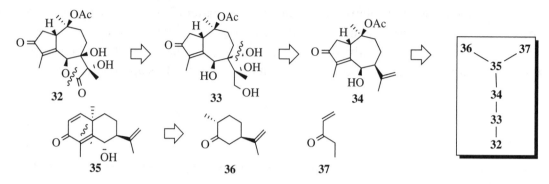

Figure 10.2. Massanet and Co-worker's Retrosynthetic Analysis of Thapsigargin, **32**.

A decision must be made concerning the relative priority of simplification versus controlling the stereochemistry of the stereogenic center, which was not made clear in the above analysis. The analysis of target **32** must include the reality of choosing a method to chemically form that bond in the context of the remaining functionality and stereochemistry. There is not necessarily one correct answer, but rather several possibilities, each with its strengths and weaknesses.

10.3.B. The Problem of Complex Targets

The disconnection of truly complex targets raises the level of difficulty to a higher level of analyses. Isobe and Hamajima's synthesis of the right hand segment of ciguatoxin (**38**) makes the point, where the retrosynthesis shown in Scheme 10.1 was given in the paper.[54] For the salient reactions of the synthesis, the reader referred to the paper, but several features are clear. First, the initial disconnection of ring F generates two fragments (**39** and **40**). Kira and Isobe had previously synthesized fragment **39**,[55] and the reader is referred to that work for the retrosynthetic analysis and the actual synthesis. Refunctionalization of fragment **40** gave **41**, and disconnection of ring H led to **42**. Disconnection of ring I gave **43**, which set up the disconnection to **44** and **45**. In this paper, **44** was prepared and coupled with **45**, which was prepared from a key intermediate that was synthesized in a previous paper by Isobe and co-workers[56] The retrosynthetic analysis involves many parts, detailed analysis of protecting groups and specific reactions, and requires independent syntheses of the fragments. The complexity of the target is reflected in the retrosynthetic analysis, which includes not only reactions and functionality, but also stereochemistry.

54. Hamajima, A.; Isobe, M. *Org. Lett.* **2006**, *8*, 1205.
55. Kira, K.; Isobe, M. *Tetrahedron Lett.* **2001**, *42*, 2821.
56. Baba, T.; Huang, G.; Isobe, M. *Tetrahedron* **2003**, *59*, 6851.

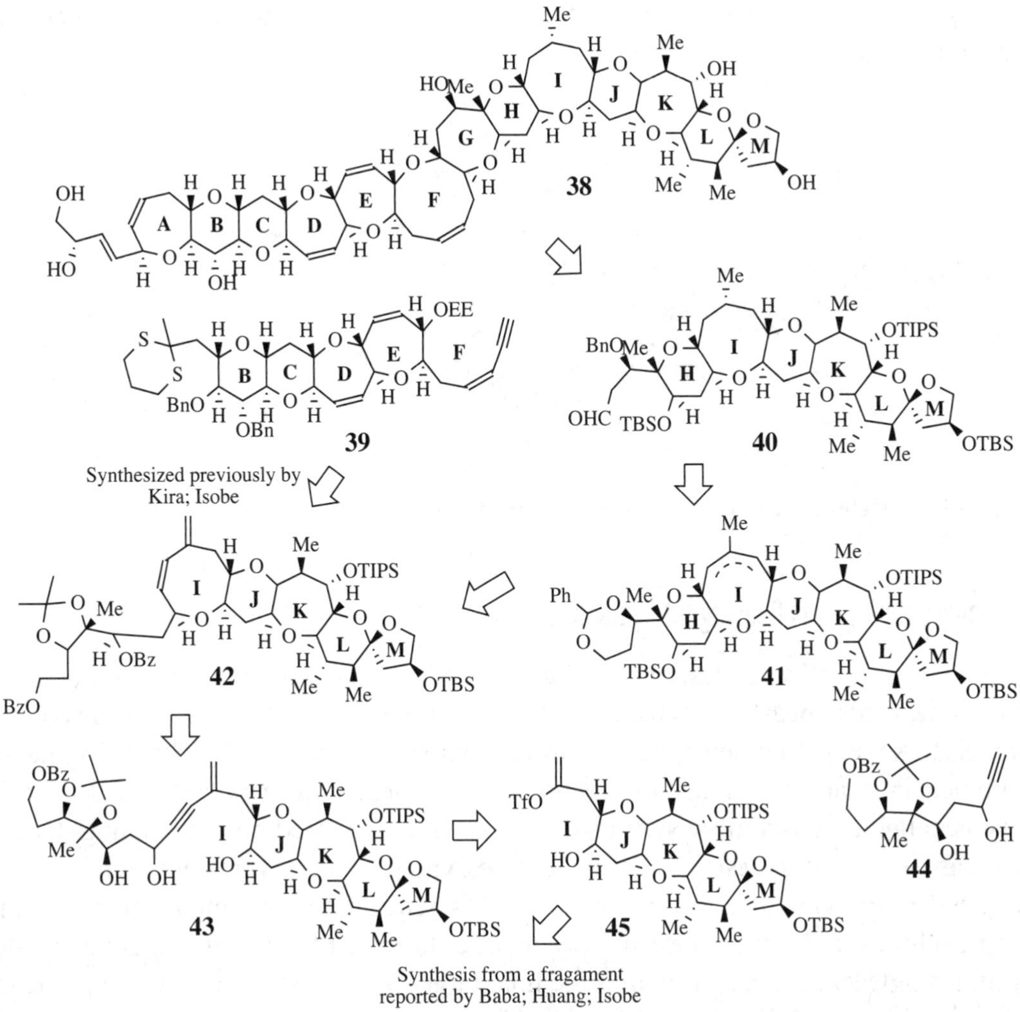

Scheme 10.1. Isobe's Retrosynthetic Analysis of a Portion of Ciguatoxin, **38**.

Stereocenters are usually quite important in the total synthesis of complex targets, but not always. Indeed, some targets have no stereocenters, but construction of the molecule by making key bonds and incorporation of substituents and/or functional groups can be quite challenging. A relatively simply example in Scheme 10.2 is taken from Kelly and co-worker's synthesis of pterocellin A (**46**),[57] and the first disconnection generates **47** by clipping the five-membered ring (an intramolecular coupling reaction for the synthesis). The next disconnection generates three fragments (**48-50**), two pyridine fragments and a functionalized acyl fragment. Fragments **48** and **49** are combined to give **51** as the next disconnection fragment, which can be prepared from the commercially available kojic acid (**52**). The fragment **50** is prepared from the commercially available 2-bromo-3-pyridinol, **53**. Note that **46** has no stereocenters, but

57. O'Malley, M.M.; Damkaci, F.; Kelly, T.R. *Org. Lett.* **2006**, *8*, 2651.

the synthesis must accommodate making the key bonds as well as positioning the heteroatoms and substituents at the proper positions.

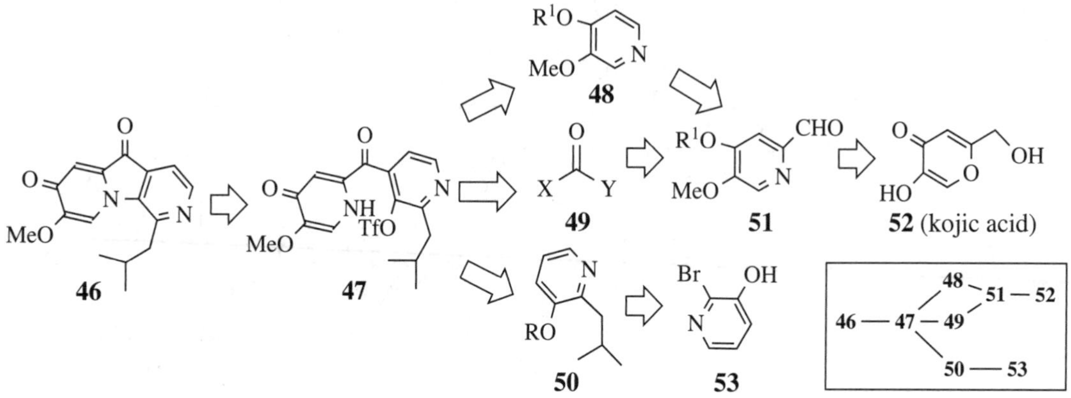

Scheme 10.2. Kelly and Co-worker's Retrosynthetic Analysis of Pterocellin A, **46**.

10.3.C. Consecutive versus Convergent Syntheses

Any synthetic strategy must describe how to proceed from starting material to product. One can work backwards linearly from the target to a starting material. Alternatively, disconnection can branch to several fragments, each with its own starting material. These fragments can be combined in a non-linear fashion to prepare the target. A synthesis based on a linear pathway is called a **consecutive synthesis**, and a synthesis based on a branching pathway is called a **convergent synthesis**. In a convergent synthesis the several pieces of a molecule are synthesized individually, and the final target is sequentially assembled from the pieces. This approach differs from a consecutive synthesis, in which the target is assembled step by step from a single starting material until the final target is reached. These two approaches are illustrated in Figure 10.3,[58] which was used by Velluz et al. to describe their synthetic approach. It appears that the convergent synthesis should have fewer steps in the overall synthesis. In this example, if each step has a mean yield of 90%, the yield of final product after 15 steps is $(0.9)^{15}$ ≈ 0.21 (21%). If the largest sequence is split and divided into convergent solutions, there would be only four consecutive steps $(A_1 \rightarrow B_1 \rightarrow C_1 \rightarrow D_1 \rightarrow Z)$ in a *perfectly convergent pathway*. The yield of the final product would be $(0.9)^4 \approx 0.66$ (66%). It is difficult to achieve such a perfect convergence since functional group manipulation, steric considerations and asymmetric structural features lead to imperfections.[58] Hendrickson's SYNGEN approach to synthesis (sec. 10.10.A) relies on a convergent synthetic strategy, and will be described later.

---

58. Velluz, L.; Valls, J.; Mathieu, J. *Angew. Chem. Int. Ed.* **1967**, *6*, 778.

CONSECUTIVE
CONVERGENT
= end product (target)
= starting material(s)

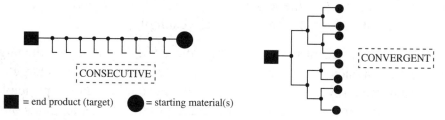

Figure 10.3. Consecutive versus convergent syntheses. [Reproduced with permission from Velluz, L.; Valls, J.; Mathieu, J. *Angew. Chem. Int. Ed.* ***1967***, *6*, 778. Copyright 1967 VCH Weinheim.]

Massanet and Co-worker's retrosynthetic analysis of thapsigargin, **32**, in Figure 10.2 is one example of a consecutive synthesis. Another consecutive synthesis is shown in Scheme 10.3, in which Selvakumar and Rajulu[59] converted 2-fluoronitrobenzene **54** to paniculidine B (**59**), in 8 linear steps with an overall yield of 22.7%. Nucleophilic aromatic substitution (sec 2.11.C) followed by a Knoevenagel-type condensation (sec. 9.4.B.iii) converts **54** to the diester **55**. A previously reported rearrangement to **56**[60] was followed by decarboxylation (sec. 2.9.D) and reduction of the cyano group to the aldehyde in **57** with diisobutylaluminum hydride (sec. 4.6.C). A Barbier-type addition (sec. 8.4.C) gave alcohol **58**, and hydroboration-oxidation (sec. 5.4.A) gave paniculidine B, **59**. Although **59** is not structurally complex, this consecutive synthesis illustrates the fundamental idea of beginning with a single compound and transforming it directly to the target. Another consecutive synthesis is Corey and co-worker's synthesis of longifolene in Scheme 10.4.

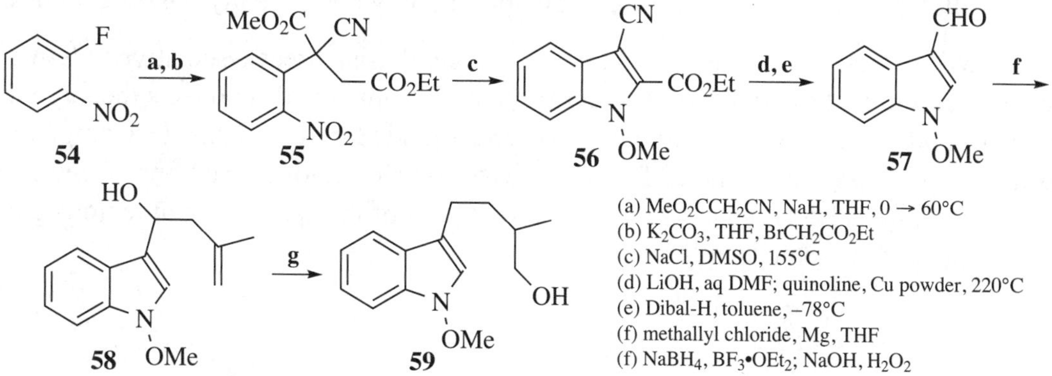

54
55
56
57
58
59
a, b
c
d, e
f
g
(a) MeO_2CCH_2CN, NaH, THF, 0 → 60°C
(b) K_2CO_3, THF, $BrCH_2CO_2Et$
(c) NaCl, DMSO, 155°C
(d) LiOH, aq DMF; quinoline, Cu powder, 220°C
(e) Dibal-H, toluene, –78°C
(f) methallyl chloride, Mg, THF
(f) $NaBH_4$, BF_3•OEt_2; NaOH, H_2O_2

Scheme 10.3. Selvakumar's Consecutive Synthesis of Paniculidine B (**59**).

Both Isobe's and co-worker's synthesis of ciguatoxin (**38**, Scheme 10.1) and Kelly and co-worker's synthesis of pterocellin A (**46**, Scheme 10.2) illustrate convergent syntheses. Disconnection to separate fragments requires that each fragment be synthesized independently, and then those fragments are joined in one or more key steps to complete the synthesis. This

59. Selvakumar, N.; Rajulu, G.G. *J. Org. Chem.* ***2004***, *69*, 4429.
60. Selvakumar, N.; Reddy, B.Y.; Azhagan, A.M.; Khera, M.K.; Babu, J.M.; Iqbal, J. *Tetrahedron Lett.* ***2003***, *44*, 7065.

fact is quite clear in the retrosynthetic analyses presented for **38** and for **46**. The reader is referred to the cited papers for more details of these examples.

Both consecutive and convergent strategies can be attractive, depending on the target and only careful consideration of each will yield the best approach for an individual target. Obviously, once it is completed a given synthesis can be continually reexamined and, in principle, improved upon. In general, a convergent strategy is expected to produce larger amounts of the target in the fewest steps if a scheme can be devised that converges at a useful point. A symmetrical convergent strategy is often the most attractive[61] (also see sec. 10.10.A).

10.4. SYNTHETIC STRATEGIES[62]

10.4.A. Defining Various Strategic Approaches

Over many years, E.J. Corey and co-workers have described an elegant approach to the synthesis of complex targets that was incorporated in the computer program LHASA (sec. 10.4.B). The purpose of that program is to assist the analysis of complex synthetic targets.[48,12,63] The monograph by Corey and Cheng[1] describes approaches for constructing a synthetic tree, and also the logic used for that construction. These modifications include:[12] (*1*) interconversion, removal, or introduction of functional groups, (*2*) extension of the atomic chains or appendages, (*3*) generation of atomic rings, (*4*) rearrangement of chain or ring members, and (*5*) cleavage of chains or rings. Each chemical fragment in the synthetic tree must possess chemical properties that are predictable and allow selective combination in only one of many possible modes.[1,12]

Corey and Wipke described three methodologies for synthesis: direct associative, intermediate associative, and logic centered.[12] The direct associative approach disconnects the target (such as **60**)[12] at bonds that generate a structure easily recognized by the chemist. Compound **60**, for example, gives easily recognized fragments based on well-known reactions. A retro Diels-Alder reaction (sec. 11.5.B) gave **61** and **62**. Hydrolysis of the ester and amide units gave **63**-**65**. This approach is somewhat limited.

61. Bertz, S.A. *J. Chem. Soc. Chem. Commun.* **1984**, 218.
62. See Tatsuta, K.; Hosokawa, S. *Chem. Rev.* **2005**, *105*, 4707.
63. Corey, E.J.; Long, A.K.; Rubenstein, S.D. *Science* **1985**, *228*, 408.

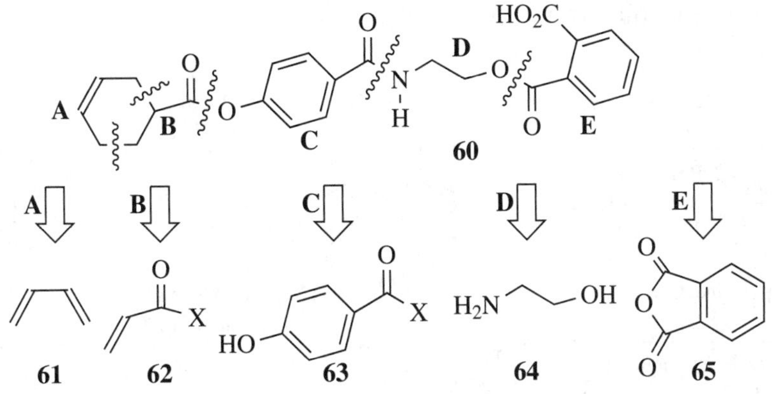

The intermediate associative approach recognizes a major subunit in the target that corresponds to a known, or potentially available starting material. The choice of this starting material directs the retrosynthetic analysis along a specific pathway.[12] The scope and rigor of solutions to the problem are sometimes very limited, however.[12] Sarett's synthesis of cortisone (6)[64] used deoxycholic acid as a starting material, available as an intermediate, taken from Kendall and co-worker's previous synthesis of 11-keto-etiolithocholic acid.[65] This earlier synthesis required a total of 30 steps, but Sarett's second synthesis was shorter, and introduced a general method for introduction of the 17α hydroxyl group.[66]

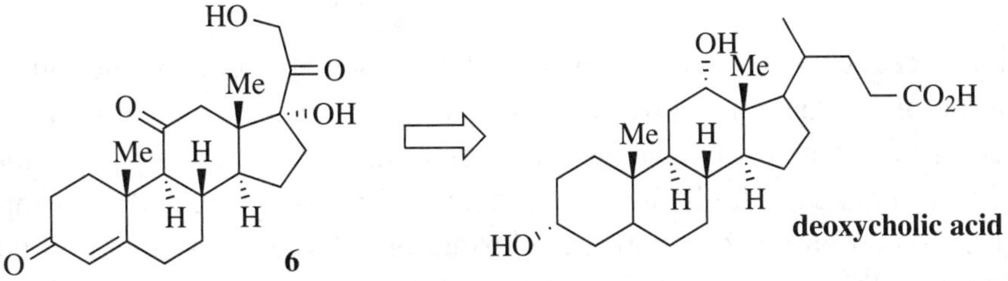

The logic centered approach leads to a restricted set of structures that may be converted in a single step to the target. Each new structure can, in turn, be converted to a new structure. The purpose of the retrosynthetic analysis is to generate synthetic intermediates (the synthesis tree) that terminate with a number of starting materials.[63] Corey et al. outlined this approach in 1964, in the synthesis of longifolene (67), which is shown in Scheme 10.4.[67] This synthesis used the **Wieland-Miescher ketone (66)** as the starting material.[68] Analyses of targets such as longifolene begin with a search for key structural features that may be structurally significant. These key features are[12] (1) individual molecular chains, rings, and appendages, (2) individual

64. (a) Sarett, L.H. *J. Biol. Chem.* **1946**, *162*, 601; (b) Fieser, L.F. Fieser, M. *Steroids*, Van Nostrand Reinhold, New York, **1959**, pp 640-650.
65. McKenzie, B.F.; Mattox, V.R.; Engel, L.L.; Kendall, E.C. *J. Biol. Chem.* **1948**, *173*, 271.
66. (a) Sarett, L.H. *J. Am. Chem. Soc.* **1948**, *70*, 1454; (b) Sarett, L.H. *Ibid.* **1949**, *71*, 2443.
67. Corey, E.J.; Ohno, M.; Mitra, R.B.; Vatakencherry, P.A. *J. Am. Chem. Soc.* **1964**, *86*, 478; (b) Corey, E.J.; Mitra, R.B.; Uda, H. *Ibid.* **1964**, *86*, 485.
68. Wieland, P.; Miescher, K. *Helv. Chim. Acta* **1950**, *33*, 2215.

functional groups, (3) asymmetric centers and attached groups, and (4) chemical reactivity.

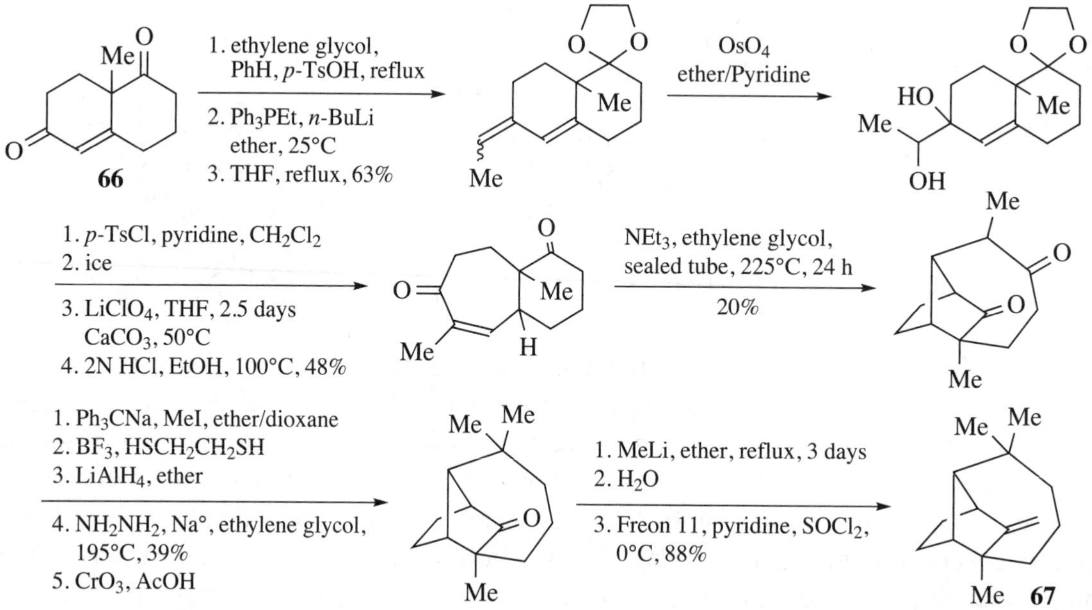

Scheme 10.4. Corey's synthesis of longifolene (**67**).

The next goal is to reduce molecular complexity. This is done by meeting the following goals:[12] (1) simplification of internal connectivity by scission of rings, (2) reduction of molecular size by disconnection of chains or appendages, (3) removal of functionality, (4) modification or removal of sites of unusually high chemical reactivity of instability, and (5) simplification of stereochemistry. There are also certain transformations that are important adjuncts to these goals but do not simplify the molecule. Examples of these transformations are:[12] (1) functional group interchanges, (2) introduction of functional groups, (3) modification of functional groups to control the level of chemical reactivity, (4) introduction of groups that permit stereochemical or positional control, and (5) internal rearrangement. Above all, the disconnections must lead to chemical reactions that will allow construction of the target. The basic mechanistic aspects of these reactions must be well understood, since the reaction pathway is the same in the forward and reverse directions (principle of microscopic reversibility).[12] If a set of disconnections is found to correlate with a mechanistically sound set of chemical reactions in the forward directions, those steps are taken as part of the synthetic tree. In all steps, stereochemical and topological aspects must be considered.

In an analysis of a target, the vast body of Organic chemistry may lead to the generation of retrosynthetic trees that are too large to be manageable, and include numerous dead ends.[63] It is essential that simplification techniques, search strategies, screening procedures and logical analysis be employed as the tree is constructed. If subunits of the molecule that are potential

synthons (the disconnection products often taken as the chemical building blocks) can be found, a strong measure of control over tree branching is possible.[63,69] The retrosynthetic transform and synthetic reaction to accomplish that transform are intricately linked. Just as the power of a transform to simplify a structure is critically important, the most powerful synthetic reactions are those that reliably increase molecular complexity.[63] Transforms that decrease or increase the molecular complexity of a target are important in the retrosynthetic strategy. Bertz described the importance of molecular complexity in the context of synthetic analysis, particularly as applied to convergent syntheses,[70] a topic also addressed by Merrifield and Simmons.[70d] Bertz suggested that "topological complexity, as well as other sources of molecular complexity, be examined *in addition* to the classical considerations of synthetic efficiency when evaluating synthetic routes."[70a]

From the preceding discussions it is clear that there are many approaches for a disconnection approach that constriucts a synthesis tree. Corey et al. described five different strategies for retrosynthetic analysis:[63]

(*1*) **Transform based strategies**. Identify a powerful transform for a specific target that produces a reasonable line in the synthetic tree. Two or three powerful simplifying transforms can be applied successively to the target structure. Alternatively, the same simplifying transforms can be repeatedly applied.[71]

(*2*) **Structure goal strategies**. Identify a potential starting material, building block, retrosynthetic subunit, or initiating chiral element (see the chiral template approach in sec. 10.9).

(*3*) **Topological strategies**. Identify one or more bonds whose disconnection can lead to major molecular simplification, which is the essence of the strategic bond approach[72] to be discussed in Section 10.5. These bonds may be in bridged or fused ring cyclic systems or appear as connectors to appendages at rings, functional groups, or stereocenters.

(*4*) **Stereochemical strategies**. Heuristically derived procedures for reducing stereochemical complexity in the retrosynthetic direction, these procedures remove stereocenters and take advantage of steric screening or proximity to a functional group. The latter may allow simplification of the target by application of functional group removal transforms.[73] Corey defines heuristic as "a noun to mean heuristic principle, a 'rule-of-thumb', which may lead by a shortcut to the solution of a problem or may lead to a blind alley."[49b]

69. Corey, E.J. *Quart. Rev. Chem. Soc.* **1971**, *25*, 455.
70. (a) Bertz, S.H. *J. Am. Chem. Soc.* **1982**, *104*, 5801; (b) *Idem, bid.,*, **1981**, *103*, 3599; (c) *Idem J. Chem. Soc. Chem. Commun.* **1981**, 818; (d) Merrifield, R.E.; Simmons, H.E. *Proc. Natl. Acad. Sci, USA* **1981**, *78*, 1329, 692.
71. Corey, E.J.; Long, A.K. *J. Org. Chem.* **1978**, *43*, 2208.
72. Corey, E.J.; Howe, W.J.; Orf, H.W.; Pensak, D.A.; Petersson, G. *J. Am. Chem. Soc.* **1975**, *97*, 6116.
73. For a synthetic example, see Corey, E.J.; Arnett, J.F.; Widiger, G.N. *J. Am. Chem. Soc.* **1975**, *97*, 430.

(5) Functional group oriented strategies. When one or more functional groups are related to an interconnecting atom path, a simplifying transform can often be found. Interconversion, addition, or removal of functionality can pave the way for further retrosynthetic simplification. Both internal protection of functional groups and replacement of a highly reactive group by another less reactive equivalent can be useful.[74]

10.4.B. LHASA

Corey and co-workers developed a computer program that uses the strategies described above in a way that emulates the most effective problem solving approaches of a synthetic chemist, with an emphasis on complex rather than routine synthetic problems. This computer program is called **LHASA** (**L**ogic and **H**euristics **A**pplied to **S**ynthetic **A**nalysis; see sec. 1.5.D). It analyzes a molecule, develops retrosynthetic schemes and supplies appropriate chemical reactions. Wipke and Howe provide a useful description of this program:[48]

"One important aspect of the project has been the writing of a general purpose computer program which will aid the laboratory chemist and will employ both the basic and more complex techniques for synthetic design as elucidated by this study. The program (hereafter also called **LHASA**) is intended to propose a variety of synthetic routes to whatever molecule it is given. The responsibility for final evaluation of the merit of the routes lies with the chemist. The program is to be an adjunct to the laboratory chemist as much as any analytical tool."[48]

LHASA is an interactive program that displays the target, allowing the chemist to select a synthetic strategy (or strategies). The program then suggests synthetic routes that satisfy the goals of the selected strategy.[63] The analysis generates retrosynthetic precursors and generates a synthetic tree (see Figure 10.4).[63] The chemist takes responsibility for choosing strategies and tactics and for deciding which precursors should be submitted to LHASA for further simplifications. LHASA selects the actual transforms, based on a database of known reactions.

Portions of a typical LHASA analysis are shown in Figure 10.4.[75] A LHASA transform is shown in part (*a*), in this case based on the Fries rearrangement (see sec. 12.5.E). In part (*b*), a target molecule is shown (**Target 3**), and a retrosynthesis is provided that is based toward acetophenone. Note that a description of the projected chemistry is shown over each disconnect arrow. In part (*c*), a retrosynthetic tree is shown as part of the analysis of a target, where the numbers represent chemical structures generated by LHASA. In part (*d*), a specific retrosynthetic analysis is shown in which one unit, the hydroxymethyl group, is removed from Target 3, with the projected reaction being an organometallic transform. The data provided by Figure 10.4 is meant to illustrate the type of information that is available from the LHASA program. A sample retrosynthesis for aphidicolin is shown in Figure 10.5, taken from the work

74. Corey, E.J.; Orf, H.W.; Pensak, D.A. *J. Am. Chem. Soc.* **1976**, *98*, 210.
75. (a) Johnson, A.P.; Marshall, C.; Judson, P.N. *J. Chem. Inf. Comput. Sci.* **1992**, *32*, 411. Also see (b) Johnson, A.P.; Marshall, C. *J. Chem. Inf. Comput. Sci.* **1992**, *32*, 418.

of Corey and co-workers.[76]

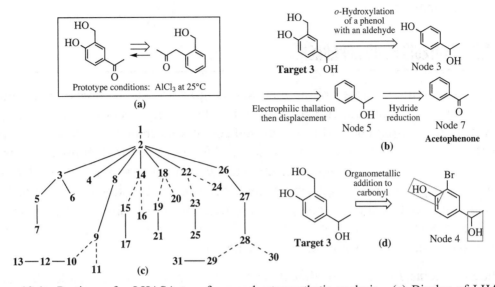

Figure 10.4. Portions of a LHASA transform and retrosynthetic analysis. (a) Display of LHASA transform. (b) Retrosynthesis of **Target 3** to acetophenone. (c) Part of the tree from the retrosynthetic analysis of target. (d) Retrosynthetic removal of the hydroxymethyl group from **target 3** by a route involving an organometallic addition transform. [Reprinted from Johnson, A.P.; Marshall, C.; Judson, P.N. *J. Chem. Inf. Comput. Sci. 1992, 32*, 411. Copyright© 1992 by the Amer. Chem. Soc.]

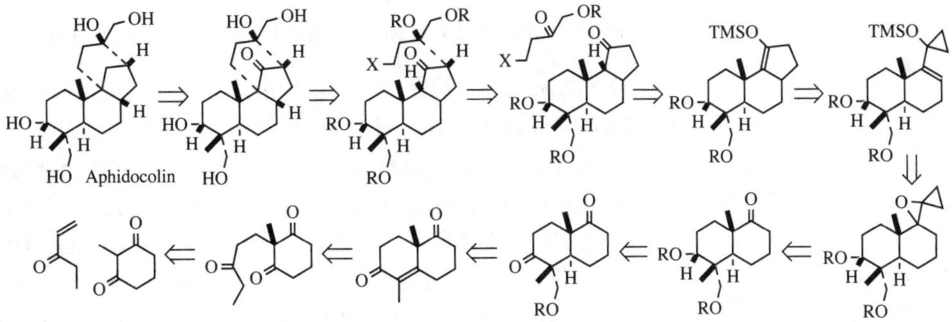

Figure 10.5. A LHASA-generated retrosynthetic pathway for the synthesis of aphidicolin. [Reprinted from Corey, E.J.; Long, A.K.; Lotto, G.I.; Rubenstein, S.D. *Recl. Trav. Chim. Pays-Bas 1992, 111*, 304.]

Note that a variation of this computer technique has been developed, with the goal of synthesis simplification. "Mathematics Applied to Synthetic Analysis (MASA) is a useful addition to Logic and Heuristics Applied to Synthetic Analysis (LHASA), as it can be used to calculate the simplification afforded by alternative disconnections of a target molecule. One-bond and two-bond disconnections that are more efficient as far as simplification is concerned than those

---

76. Corey, E.J.; Long, A.K.; Lotto, G.I.; Rubenstein, S.D. *Recl. Trav. Chim. Pays-Bas 1992, 111*, 304.

identified by the LHASA criteria have been found by using MASA."[77] This approach is based on identifying topological strategic bonds.

10.5 THE STRATEGIC BOND APPROACH

Several available strategies can be applied to the specific task of breaking the first (and subsequent) bond(s) in the target. For the student beginning a study of synthesis, a more formalized approach would be useful. Corey and co-workers enumerated several rules that tie together the strategies and goals of retrosynthetic analysis into a **strategic bond analysis**. These rules provide insight into the first disconnection and the complete synthetic tree. The analysis begins with an inspection of the target molecule to determine if the structure is compatible with simple solutions. Before the strategic bonds are determined, simplification may be possible if the molecule possesses symmetry, is structurally similar to another molecule that has been previously synthesized, or has repeating units within the structure.

The following items are only guidelines. These guidelines are a reasonable place to begin planning a synthesis, but by no means the only way to approach a problem. If a disconnection is reasonable and there is either known chemistry or planned chemistry to reconstitute that bond in a synthesis, proceed.

10.5.A. The Preliminary Scan

The analysis inspects the molecule to determine if it can be simplified. There are three criteria.[78]

68

10.5.A.i. Is There Symmetry or Near Symmetry in Two Parts of the Molecule? If so, the synthesis may be simplified by synthesizing and then joining together the two or more identical pieces. A good example is the molecule C-toxiferin I (**68**),[79] which is essentially a dimer. In **68**, one-half of the molecule is identical to the other, and it has been shown that treatment with sulfuric acid cleaves the molecule into a single product that regenerates the dimer on treatment with hot sodium acetate.[79] Synthesis of one piece and joining the two identical pieces together greatly simplifies an overall synthesis with a built-in convergence.

The symmetry may not be obvious and the target should be carefully analyzed. Usnic acid (**73**) is an example of a molecule that possesses potential symmetry.[80] The molecule may not

77. Bertz, S.H.; Rücker, C.; Rücker, G.; Sommer, T.J. *Eur. J. Org. Chem.* **2003**, 4737.
78. Bersohn, M.; Esack, A. *Chem. Rev.* **1976**, *76*, 269.
79. (a) Bernauer, K.; Berlage, F.; von Philipsborn, W.; Schmid, H. Karrer, P. *Helv. Chim. Acta* **1958**, *41*, 2293; (b) Berlage, F. Bernauer, K.; von Philipsborn, W.; Waser, P.; Schmid, H. Karrer, P. *Ibid.* **1959**, *42*, 394.
80. Corey, E.J. *Pure Appl. Chem.* **1967**, *14*, 19.

be obviously symmetrical, but the symmetry appears upon disconnection and suggests a retro phenolic oxidative coupling (see sec. 13.8.B). The synthesis of usnic acid is shown in Scheme 10.5,[81] with a phenolic oxidative coupling of a single phenol derivative (**69**) to give **71** via coupling of the initial oxidation product **70**. The **phenolic oxidative coupling** reaction (sec. 13. 8.B) is a radical process by which two phenol moieties are joined.[82] The two phenolic moieties reacted via an acid-catalyzed Michael reaction to give a dimer (**71**), which cyclized via the phenol hydroxyl group to give **72**. Treatment with concentrated sulfuric acid completed the two-step synthesis of **73**. Recognizing the potential symmetry greatly shortened the synthesis.

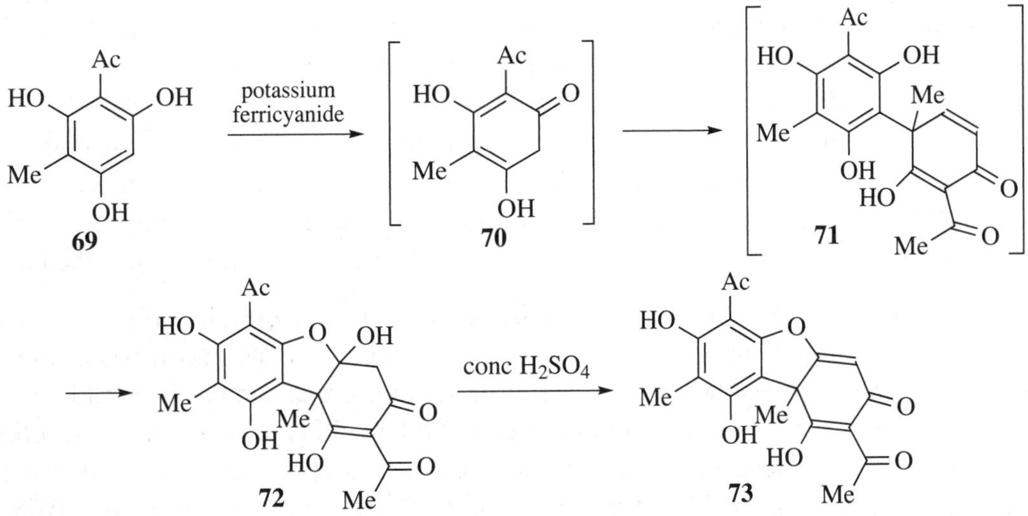

Scheme 10.5. Phenolic oxidative coupling strategy for the synthesis of usnic acid (**73**).

10.5.A.ii. Is the Problem Like One Already Solved? Woodward' and co-workers synthesis of reserpine (**8**) is shown in Scheme 10.6.[10] Compound **74** was important intermediate in that synthesis. Deserpidine (**79**)[83] is a natural product that is structurally similar to **8**, but lacking the methoxy group on the indole moiety. It closely mimics the activity of **8**,[84] and a total

81. Barton, D.H.R.; Deflorin, A.M.; Edwards, O.E. *J. Chem. Soc. 1956*, 530.
82. (a) Kotani, E.; Miyazaki, F.; Tobinaga, S. *J. Chem. Soc. Chem. Commun. 1974*, 300; (b) Barton, D.H.R.; James, R.; Kirby, G.W. Widdowson, D.A. *Ibid. 1967*, 266; (c) Kametani, T.; Fukumoto, K. *Synthesis 1972*, 657; (d) Kametani, T.; Yamaki, K. Terui, T.; Shibuya, S.; Kukumoto, K. *J. Chem. Soc. Perkin I 1972*, 1513.
83. (a) Wenkert, E.; Liu, L.H. *Experientia 1955, 11*, 302; (b) Schlittler, E.; Ulshafer, P.R.; Pandow, M.L.; Hunt, R.M.; Dorfman, L. *Ibid. 1955, 11*, 64; (c) Huebner, C.F.; MacPhillamy, H.B.; Schlittler, E.; St. André, A.F. *Ibid. 1955, 11*, 303; (d) MacPhillamy, H.B.; Dorfman, L.; Huebner, C.F.; Schlittler, E.; St. André, A.F. *J. Am. Chem. Soc. 1955, 77*, 1071; (e) Yamamoto, T.; Yamanaka, M. *Shinyaku to Rinsho 1955, 4*, 19 and pp 175 [*Chem. Abstr. 51*: 14984h *1957*] and see the other articles cited in this abstract.
84. Lednicer, D.L.; Mitscher, L.A.; *Organic Chemistry of Drug Design*, Vol. 1, John Wiley, *1977*, pp 320-321.

synthesis of **79** was developed by Velluz[85] and later by Protiva.[86] Intermediate **75**, prepared from **74** from the Woodward synthesis, was used as the starting material to give the alternative synthesis also shown in Scheme 10.6.[86] Reaction of **75** with 6-methoxytryptamine (obtained with great difficulty) gave **76**.[8] Intermediate **76** was converted to **77**, and then to **8**. In the synthesis of **79**, key intermediate was **78** prepared by reaction of tryptamine with **75**. This simple modification of Woodward's synthesis obviously saved Velluz and Protiva the time and effort required to develop a new synthetic approach. This example suggests that thorough familiarization with the target, including knowledge of all previous syntheses of the target and also related compounds (i.e., thorough search of the literature) is always beneficial. One does not always want to mimic a literature approach, but when the target is structurally similar to one that is known, the use of a key intermediate or transformation sequence in designing a retrosynthesis is prudent and usually desirable. Obviously, if one has an improved method or route to a target, the retrosynthesis is biased to include this new method. It is common for a published total synthesis to be followed by several later syntheses that improve or expand upon it. Syntheses developed later are often useful for producing large quantities of the natural product, as the initial sequence may produce only small amounts (milligrams) of the target.

10.5.A.iii. Is the Molecule a String of Available or at Least Simple Pieces? The synthesis[87] of the macrolide antibiotic nonactin (**83**)[88] is a useful example. Nonactin consists of four units of nonactic acid (**83**),[88] joined together in a tetrameric ring. A total synthesis would be greatly simplified by preparing **82**, since the final synthesis of **83** would require only macrocyclization (formation of a large lactone ring, sec. 6.6.B).[89] Gerlach and co-worker's synthesis of **83**, which is shown in Scheme 10.7,[87] utilized this approach. The synthetic precursor of **82** was **81**, produced in several steps from furanyl derivative **80**. Note that when two remote, reactive ends of a molecule must be brought together so reaction can occur, intermolecular processes often dominate (sec. 6.6.B) and the low yield of **83** obtained by Gerlach is a result of such problems, which does not diminish the simplicity of the synthetic approach that recognized

85. Velluz, L.; Muller, G.; Joly, R.; Nominé, G.; Mathieu, J.; Allais, A. Warnant, J.; Valls, J.; Bucourt, R.; Jolly, J. *Bull. Soc. Chim. Fr.* **1958**, 673.
86. (a) Weichet, H.; Pelz, K.; Bláha, L. *Collect. Czech. Chem. Commun.* **1961**, *26*, 1537; (b) Protiva, M.; apek, A.; Jílek, J.O.; Kakác, B.; Tadra, M. *Ibid.* **1961**, *26*, 1537; (c) Ernest, I.; Protiva, M. *Ibid.* **1961**, *26*, 1137; (d) Protiva, M.; Rajser, M.; Jilek, J.O. *Monatsh. Chem.* **1960**, *91*, 703.
87. (a) Gerlach, H.; Wetter, H. *Helv. Chim. Acta* **1974**, *57*, 2306; (b) Gerlach, H.; Oertle, K.; Thalmann, A.; Servi, S. *Ibid.* **1975**, *58*, 2036.
88. (a) Keller-Schierlein, W.; Gerlach, H. *Fortschr. Chem. Org. Naturstoffe* **1968**, *26*, 161; (b) Ando, K.; Oishi, H.; Hirano, S.; Okutomi, T. Suzuki, K.; Okazaki, H.; Sawada, M.; Sagawa, T. *J. Antibiot. (Tokyo)* **1971**, *24*, 347; (c) Dobler, M.; Dunitz, J.D. Kilbourn, B.T. *Helv. Chim. Acta* **1969**, *52*, 2573; (d) Dobler, M. *Ibid.* **1972**, *55*, 1371; (e) Iitaka, Y.; Sakamaki, T. Nawata, Y. *Chem. Lett.* **1972**, 1225.
89. (a) Taub, D.; Girotra, N.N.; Hoffsommer, R.D.; Kuo, C.H.; Slates, H.L.; Weber, S.; Wendler, N.L. *Tetrahedron* **1968**, *24*, 2443; (b) Corey, E.J.; Nicolaou, K.C. *J. Am. Chem. Soc.* **1974**, *96*, 5614; (c) Masamune, S.; Kim, C.U.; Wilson, K.E.; Spessard, G.O. Georghiou, P.E.; Bates, G.S. *Ibid.* **1975**, *97*, 3512; (d) Masamune, S.; Yamamoto, H.; Kamata, S.; Fukuzawa, A. *Ibid.* **1975**, *97*, 3513.

nonactin as a string of repeating fragments.

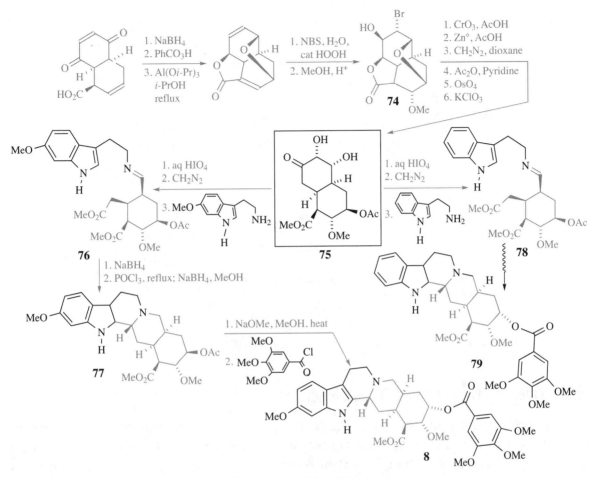

Scheme 10.6. Syntheses of reserpine (**8**) and deserpidine (**79**).

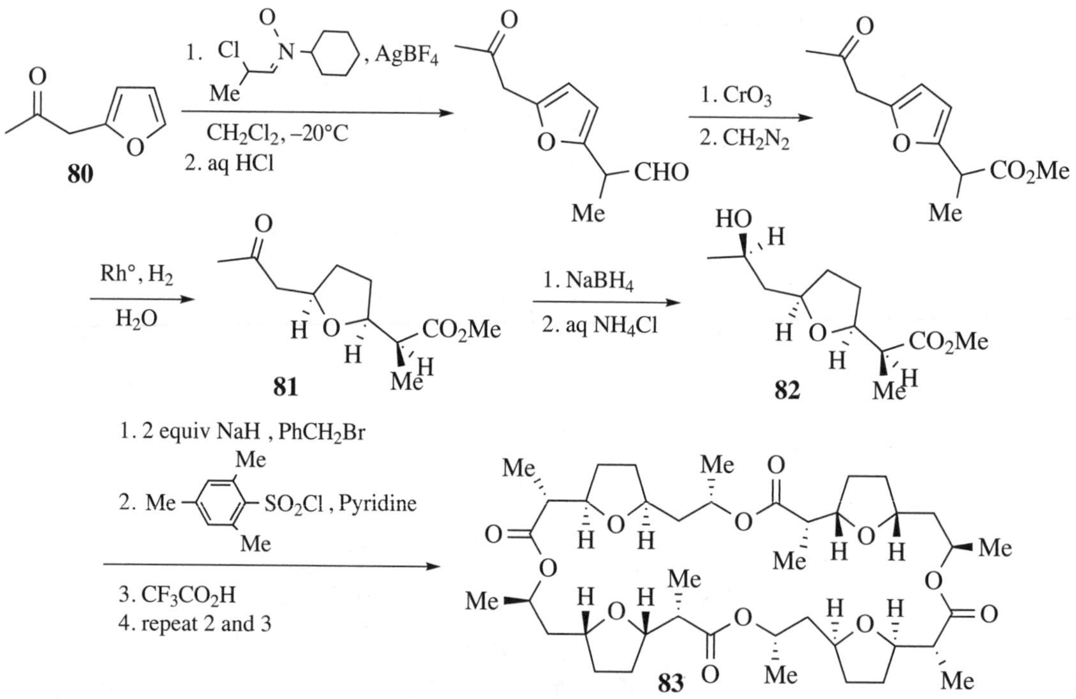

Scheme 10.7. Gerlach and co-Worker's Synthesis of Nonactin (**83**)

10.5.B. Criteria for Disconnection of Strategic Bonds

Once the preliminary scan is complete, the next step is to disconnect the molecule by breaking bonds that can be reconnected by known or reasonable chemical steps, to generate the synthetic tree. As discussed above, these important bonds are called strategic bonds. Bersohn has restated Corey's rules to tie together the strategies outlined in Section 10.4.[78,49] One should not be a slave to these statements but rather treat them as excellent suggestions for how to begin planning a synthesis:

1. Labile groups should be removed first.

2. Transforms should remove as much functionality as possible and should remove stereochemical centers where possible.

3. Favor transforms that generate closely related intermediates.

4. If a certain substructure interferes with a key process, then use a transform to remove that substructure.

5. Where possible, use transforms to build bridges between the goal atoms.

6. Convert the molecule under consideration into the product of one of the powerful

reactions.

7. Disconnect appendages attached to atoms bearing certain functional groups such as OH or C=O.

8. Consider all pairs of functional groups in the molecule to ascertain whether known transforms can disconnect any of the intervening bonds.

9. In a ring system preferentially disconnect the strategic bonds.

The examples that follow in Section 10.5.B.i-10.5.B.viii will illustrate an application of each of these rules.

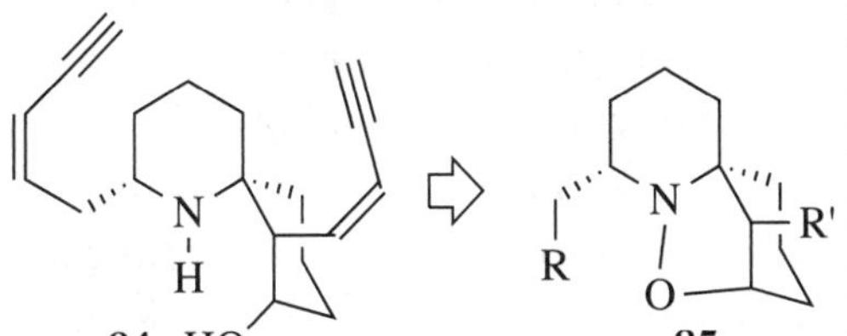

10.5.B.i. Remove Labile Groups First. Removing a labile group is illustrated by a retrosynthetic step in Holmes and co-worker's synthesis of (–)-histrionicotoxin, **84**.[90] The alcohol and conjugated ene-yne units in **84** are quite reactive. Both units are removed in the initial disconnection and replaced with R and R' groups, which represent functionality that can be elaborated to the conjugated system. In addition, the labile amine and alcohol units are converted to the isoxazolidine unit in **85**. This disconnection allows one to focus on constructing the fundamental spirocyclic ring system, knowing that highly reactive groups will be added last.

10.5.B.ii. Remove Stereochemical Centers. The vast majority of natural products and important synthetic targets are characterized by several stereogenic centers that are often contiguous. A major task of a total synthesis is the generation of these stereogenic centers, in a controlled and predictable manner. It is therefore important to plan all reactions, and often the entire retrosynthesis around stereoselective reactions. Chemical reactions that involve formation of bonds *at* a chiral carbon allow a measure of control over stereochemical induction. The stereogenic center and the reaction center usually have a 1,2-, 1,3-, or occasionally a 1,4-relationship. It is also possible that a remote center is held in close proximity due to a particular conformation, and exerts an influence on the reaction. Molecular or computer modeling is usually necessary to observe the correct influence of remote centers on a disconnection. Disconnections are, in general, made at a bond connected to a stereogenic center (1,2-relationship) rather than to a bond containing a pendant group that contains a stereogenic center. In

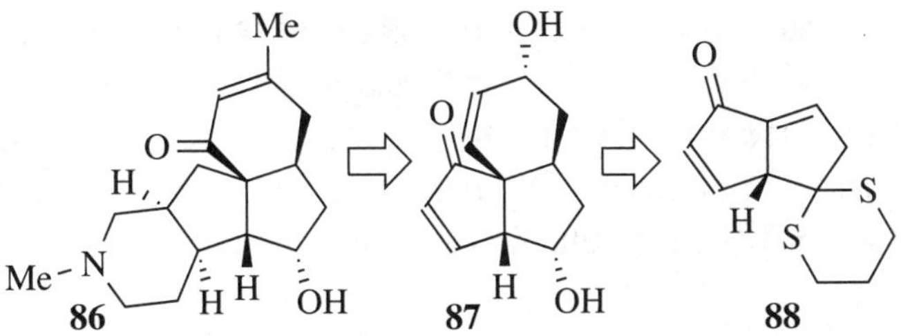

---

90. Williams, G.M.; Roughley, S.D.; Davies, J.E.; Holmes, A.B. *J. Am. Chem. Soc.* **1999**, *121*, 4900.

Paquette and co-worker's synthesis of magellanine (**86**),[91] disconnection of the piperidine ring removed two stereogenic centers, although the allylic alcohol unit introduced another (see **87**). The next disconnection removed the cyclohexenol ring along with two stereogenic centers, and the use of a protected carbonyl (the dithiane) effectively removed the stereogenic center at the secondary alcohol, giving **88**.

10.5.B.iii. Favor Closely Related Intermediates. It is usually preferable to simplify the target molecule in small bites (i.e., prune small branches of the synthesis tree rather than large branches). Moving too far down the synthetic tree with large disconnections will mean that several carbon-carbon bond-forming reactions or functional group exchanges are required to convert one disconnect product to another. It may be difficult to find transforms where several bonds can be formed simultaneously in the synthesis, although such transformations do exist in some cases. For a first analysis, simplification should be done so that a minimum number of carbon-carbon bond-forming reactions and functional group exchanges are required for each disconnect product. In this way, it is easy to understand the simplification and the implications for the requisite chemical reactions, which is illustrated by the relatively simple disconnection sequence used by Kibayashi and co-workers in a synthesis of (–)-(3*R*,6*S*,9*R*)-decarestrictine C_2, **89**.[92] The target was first converted to the seco-acid (**90**), removing the labile ester unit, and disconnection to aldehyde **91** suggests a condensation reaction to make that bond. The last disconnection leads to tetraol **92**, again suggesting a condensation reaction. In each case, one key bond is disconnected and the differences between target and disconnect product are structurally close and predictable. Generally, one disconnects one or two bonds per disconnection, but more may be disconnected if the pieces are closely related.

10.5.B.iv. Remove Interfering Substructures. An example provided by Corey and co-workers[74] is the disconnection of **93** and **95** to **94**. Path *a* is a simple epoxidation (sec. 3.4), and path *b* is a Baeyer-Villiger oxidation (sec. 3.6). A cyclic cis alkene moiety such as that in **94** shows moderate reactivity with peroxyacids, but the ketone moiety in **94** is very reactive, since it is strained. Disconnection of **95** to **94** is viable, since the synthesis will proceed without interference from the alkene moiety. Disconnection of **93** to **94**, however, requires that the interfering carbonyl be removed in any synthetic attempt. The best retrosynthesis of **93** will probably involve protection of the ketone (sec. 7.3.B), as a ketal for example. In Figure 10.4,[75] the

91. Williams, J.P.; St. Laurent, D.R.; Friedrich, D.; Pinard, E.; Roden, B.A.; Paquette, L.A. *J. Am. Chem. Soc.* **1994**, *116*, 4689.
92. Arai, M.; Morita, N.; Aoyagi, S.; Kibayashi, C. *Tetrahedron Lett.* **2000**, *41*, 1199.

LHASA program generated a retrosynthetic sequence with boxes around reactive or interfering functional groups that can be protected and dashed boxes around unstable (interfering) groups that are not easily protected. Quite often, this rule refers to protection of interfering structure.

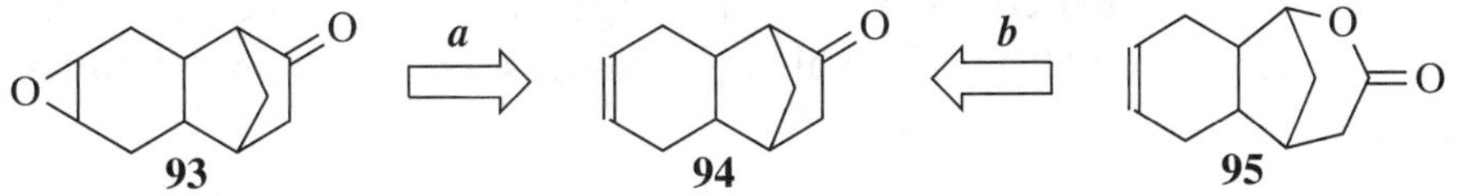

10.5.B.v. Use Transforms to Bridge Goal Atoms. If a disconnection results in cleavage of the target into separate pieces, chemical reconnection usually requires formation of one to four bonds. Two-bond disconnections are well known, although there are not many of them, and if three or four bonds must be formed it is usually very difficult to rejoin all the bonds in one synthetic step. The Diels-Alder cycloaddition (sec. 11.4.A) is an example in which two bonds are connected. If possible, it is useful to bridge the two pieces together by formation of one of the requisite bonds in the retrosynthesis, bringing the two pieces close enough to allow subsequent formation of the other bond(s). An illustration of this principle is Trost and co-worker's retrosynthetic analysis of ibogamine (**100**)[93] in Scheme 10.8 ($\Leftarrow$ represents a retrosynthetic disconnection and $\rightarrow$ represents the synthetic step). The first disconnection of the nine-membered ring gave **99** in which the two disconnected fragments are attached. The proximity of these fragments allows the usually difficult nine-membered ring to be formed in the synthesis. The bridged ring was disconnected to give **98**, in which the two fragments are still bridged. Disconnection to **97** gave a fragment (**96**) that was recognized as a precursor to the starting diene. In the actual synthesis (also in Scheme 10.8), reaction of intermediate **97** with tryptamine gave **98**, which bridged the indole and the azabicyclo[2.2.2]octane moieties, allowing formation of **99**. The organopalladium reagent used for this coupling will be discussed in Section 12.7. Final joining of the bridged pieces gave **100**, completing the total synthesis.

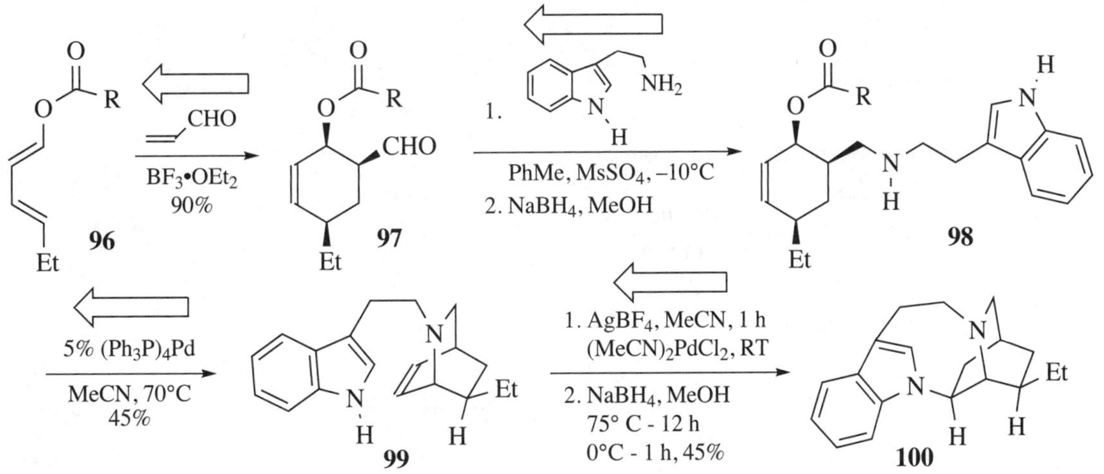

Scheme 10.8. Trost and co-Worker's Synthesis of Ibogamine (**100**).

93. Trost, B.M.; Godleski, S.A.; Genet, J.P. *J. Am. Chem. Soc.* **1978**, *100*, 3930.

10.5.B.vi. Convert the Molecule into the Product of One of the Powerful Reactions.

Powerful reactions have been defined as those synthetic reactions that reliably increase molecular complexity. Disconnections that are based on powerful reactions will, therefore, greatly *decrease* the molecular complexity of the target, which is, of course, a major goal of a disconnection strategy. Corey listed several examples of such reactions.[63] Synthetic examples of these reactions are shown in Table 10.1.[12]

Table 10.1. Powerful Reactions[a]

1. Carbocyclic Diels–Alder

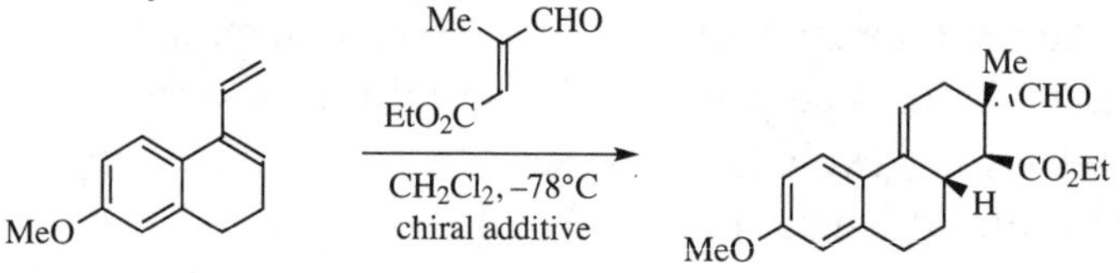

EtO$_2$C, Me, CHO; CH$_2$Cl$_2$, −78°C, chiral additive → product (MeO, Me, CHO, CO$_2$Et, H)   92%[94]

2. Quinone and Related Diels–Alder

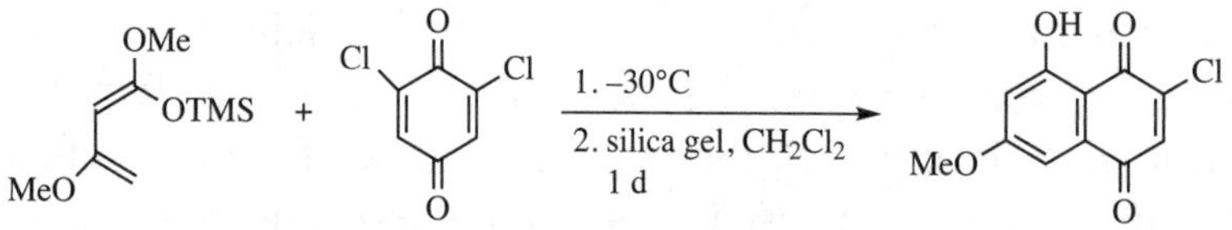

OMe, OTMS, MeO + Cl, Cl quinone → 1. −30°C; 2. silica gel, CH$_2$Cl$_2$, 1 d → product (OH, O, Cl, MeO, O)   82%[96]

3. Heteroatom Diels–Alder

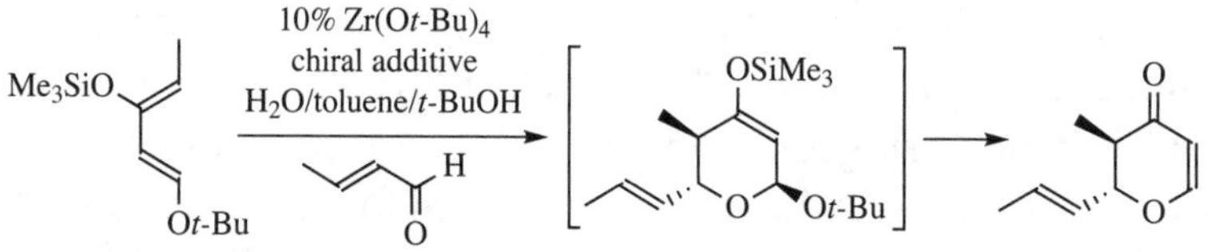

Me$_3$SiO, Ot-Bu + CHO → 10% Zr(Ot-Bu)$_4$, chiral additive, H$_2$O/toluene/t-BuOH → [OSiMe$_3$, O, Ot-Bu] → product (O, O)   89%[97]

4. Robinson Annulation

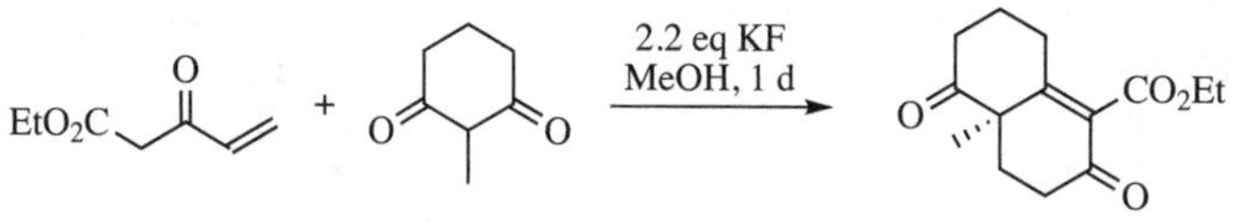

EtO$_2$C, O + O, O → 2.2 eq KF, MeOH, 1 d → product (O, CO$_2$Et, O)   74%[98]

5. Position Selective Partial Aromatic Reduction

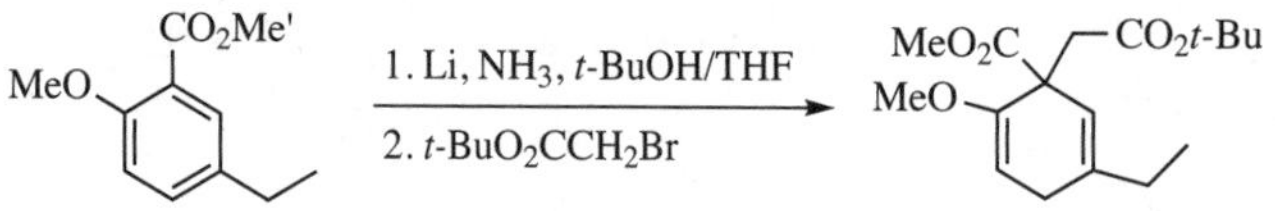

MeO, CO$_2$Me' → 1. Li, NH$_3$, t-BuOH/THF; 2. t-BuO$_2$CCH$_2$Br → product (MeO$_2$C, CO$_2$t-Bu, MeO)   85%[99]

6. Cation ≠ Cyclization

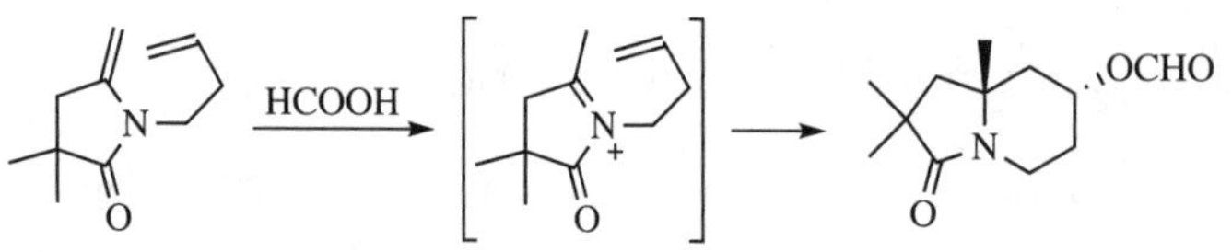

→ HCOOH → [N$^+$, O] → product (OCHO, N, O)   _[101]

7. Radical π Cyclization

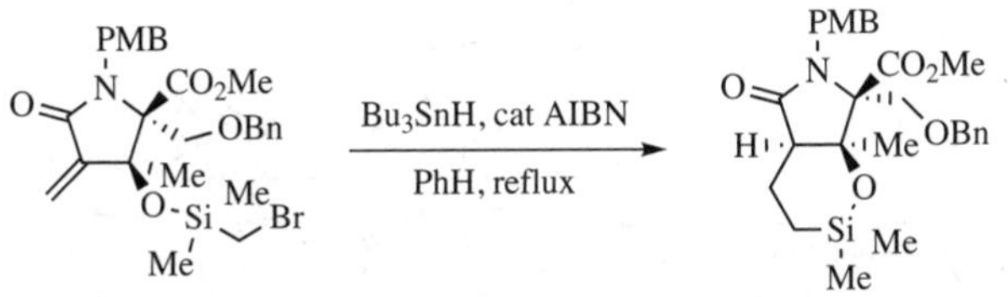

Bu₃SnH, cat AIBN

PhH, reflux

89%[102]

8. Aldol Condensation

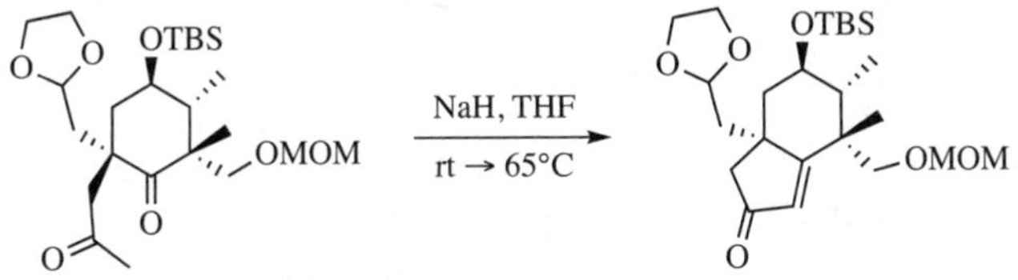

NaH, THF

rt → 65°C

80%[103]

9. Silyl-acyloin Cyclization

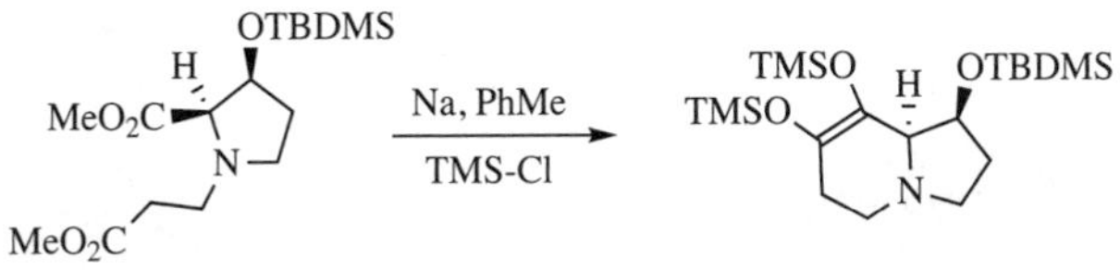

Na, PhMe

TMS-Cl

_[104]

10. Internal S_N2 Cyclization

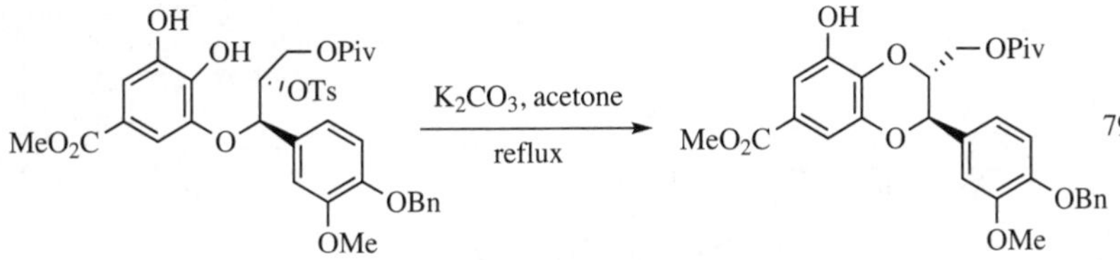

K₂CO₃, acetone

reflux

79%[105]

11. Friedel–Crafts Type Cyclization

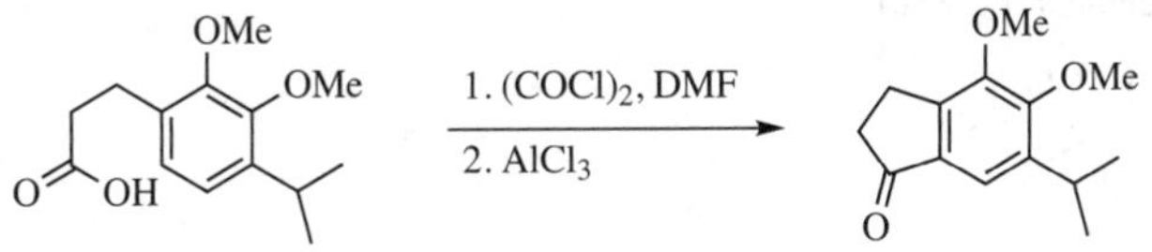

1. (COCl)₂, DMF

2. AlCl₃

84[106]

12. Internal Ene Reaction

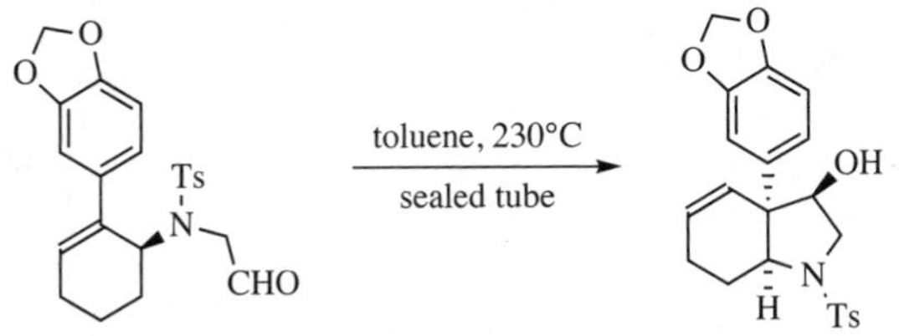

toluene, 230°C

sealed tube

59%[107]

13. Cationic Rearrangement

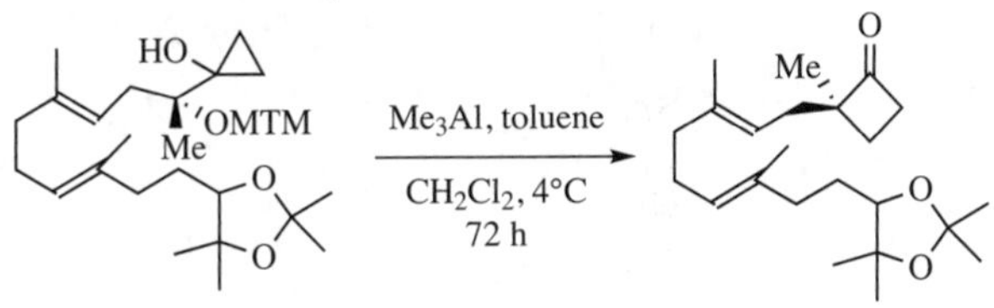

Me₃Al, toluene

CH₂Cl₂, 4°C

72 h

90[108]

14. Photocyclization

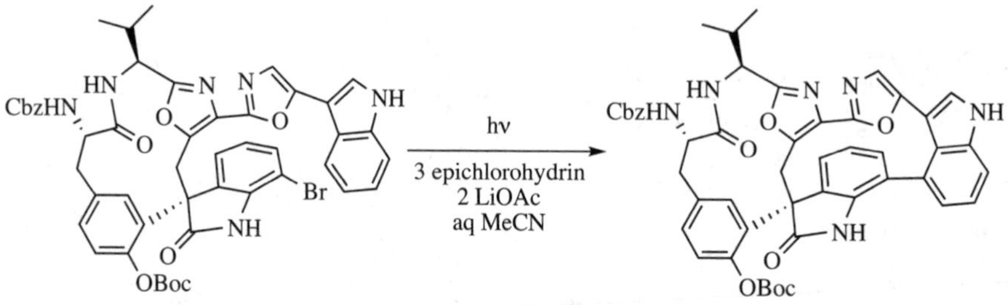
CbzHN
HN
NH
Br
OBoc
hν
3 epichlorohydrin
2 LiOAc
aq MeCN

33[109]

15. Fischer Indole Synthesis

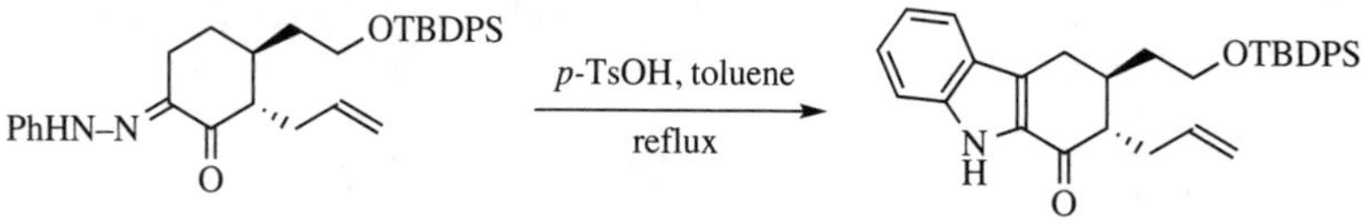
OTBDPS
PhHN–N
p-TsOH, toluene
reflux

72%[111]

16. Knorr Pyrrole Synthesis

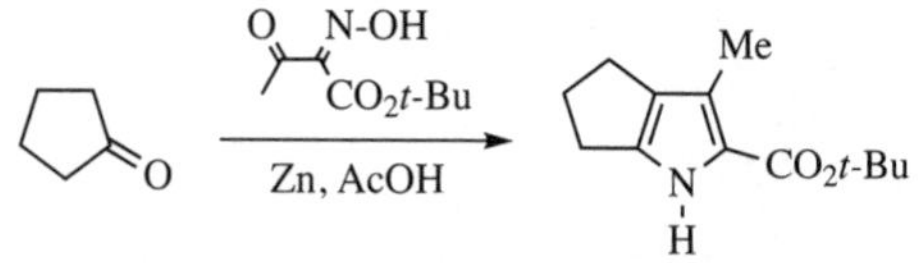
N-OH
CO2t-Bu
Zn, AcOH
Me

—[112]

17. [m,n]-Sigmatropic Rearrangement (anionic Oxy-Cope)

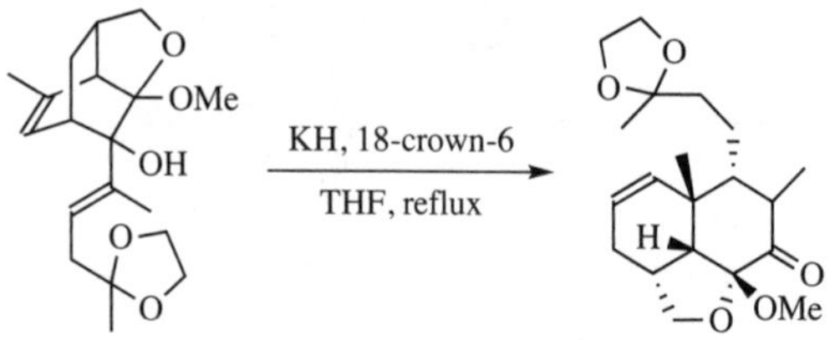
OMe
OH
KH, 18-crown-6
THF, reflux

97%[113]

18. [*m,n*]-Sigmatropic Rearrangement (Claisen)

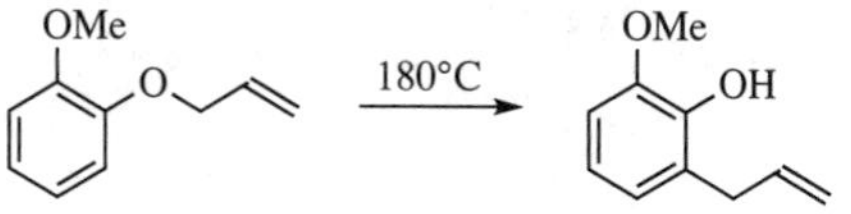
OMe
180°C
OH

80%[114]

19. Ring-closing metathesis (RCM)

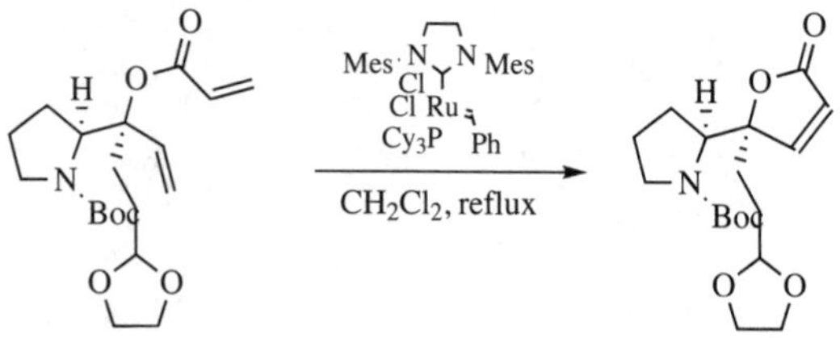
Mes
Cy3P
Ph
CH2Cl2, reflux
Boc

65%[115]

20. Heck Reaction and other Palladium-catalyzed coupling reactions

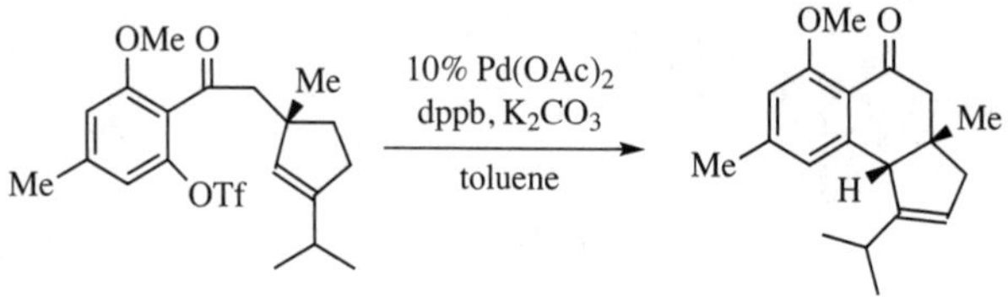
OMe O
Me
OTf
10% Pd(OAc)2
dppb, K2CO3
toluene

58%[116]

The intermolecular **Diels-Alder reaction** (sec. 11.4) is shown first. Example 1 from Table 10.1 is a carbo-Diels-Alder reaction taken from the catalytic enantioselective synthesis of estrone by Corey and co-workers in 92% yield (94% ee).[94] Intermolecular and intramolecular versions of this important reaction are known. Example 2 is a quinone Diels-Alder reaction (see Mikami et al. for another example),[95] taken from the Dallavalle and co-worker's synthesis of topopyrone in which Diels-alder cycloaddition was followed by hydrolysis and elimination to give the product shown in 82% yield.[96] Example 3 is a hetero Diels-Alder reaction (sec. 11.7), taken from a synthesis of (+)-prelactone C by Kobayashi and co-workers.[97] A Lewis acid catalyzed cyclization of an aldehyde with the functionalized diene shown, gave the dihydropyran product, which was converted to the enol under the reaction conditions and gave the pyrone in 89% yield and 93% ee. Example 4 is a **Robinson annulation** (sec. 9.7.C) induced by the basic potassium fluoride in a synthesis of norzoanthamine by Theodorakis and co-workers.[98]

Example 5 is taken from Zard and Sharp's synthesis of aspidospermidine,[99] in which **Birch reduction** (see sec. 4.9.E) of the aromatic ring was followed by alkylation of the intermediate anion and then hydrolysis to give the cyclohexadiene in 85% yield for two steps. Example 6 is a cationic π cyclization reported by Grubbs and Cannizzo,[100] that uses Speckamp's procedure[101] (sec. 12.2.C). In Example 7 a radical π cyclization (sec. 13.7), reported by Corey and co-workers in a synthesis of salinosporamide A,[102] generated the silyloxy ring in 89% yield. Example 8 is an **aldol condensation** (sec. 9.4.A) reported by Mehta and Shinde in a synthesis of (+)-minwanenone.[103] Example 9 used the silyl-acyloin cyclization (sec. 13.8.E) to close the ring in the Sih and co-workers synthesis of castanospermine.[104] Chan and co-workers synthesis of 5'-methoxyhydnocarpin-D used the intramolecular S_N2 displacement (sec. 2.6.A) of the secondary tosylate by the phenolic oxygen nucleophile in Example 10.[105] This transformation was included because it represents those reactions that give one the ability to control relative and absolute stereochemistry because the S_N2 reaction proceeds with inversion

94. Hu, Q.-H.; Rege, P.D.; Corey, E.J. *J. Am. Chem. Soc. 2004, 126*, 5984.
95. Mikami, K.; Motoyama, Y.; Terada, M. *J. Am. Chem. Soc. 1994, 116*, 2812.
96. Gattinoni, S.; Merlinni, L.; Dallavalle, S.D. *Tetrahedron Lett. 2007, 48*, 1049.
97. Yamashita, Y.; Saito, S.; Ishitani, H.; Kobayashi, S. *J. Am. Chem. Soc. 2003, 125*, 3793.
98. Ghosh, S.; Rivas, F.; Fischer, D.; González, M.A.; Theodorakis, E.A. *Org. Lett. 2004, 6*, 941.
99. Sharp, L.A.; Zard, S.Z. *Org. Lett. 2006, 8*, 831.
100. Cannizzo, L.F.; Grubbs, R.H. *J. Org. Chem. 1985, 50*, 2316.
101. (a) Dijkink, J.; Speckamp, W.N. *Tetrahedron Lett. 1975*, 4047; (b) Shoemaker, H.E.; Speckamp, W.N. *Ibid. 1978*, 1515; (c) Speckamp, W.N. *Recl. Trav. Chim. Pays-Bas 1981, 100*, 345.
102. Reddy, L.R.; Saravanan, P.; Corey, E.J. *J. Am. Chem. Soc. 2004, 126*, 6230.
103. Mehta, G.; Shinde, H.M. *Tetrahedron Lett. 2007, 48*, 8297.
104. Bhide, R.; Martezaei, R.; Scilimati, A.; Sih, C.J. *Tetrahedron Lett. 1990, 31*, 4827.
105. Chan, K.-F.; Zhao, Y.; Chow, L.M.C.; Chan, T.H. *Tetrahedron 2005, 61*, 4149.

of configuration (secs. 2.6.A, 6.2, 6.3). Example 11 shows a **Friedel-Crafts acylation reaction** (sec. 12.4.D) in which the carboxylic acid was first converted to the acid chloride and then treatment with $AlCl_3$ gave the ketone in 84% yield as part of Sarpong and Simmons synthesis of salviasperanol.[106] Example 12 is an internal carbonyl **ene reaction** (sec. 11.13) taken from the Mori and co-worker's synthesis of (+)-crinamine,[107] which gave the indicated product in 59% yield. In Example 13, Corey and Kingsbury treated the cyclopropylcarbinyl alcohol starting material with trimethylaluminum and observed the rearrangement product shown as part of a synthesis of β-araneosene.[108] In this case, treatment with *p*-toluenesulfonic acid also led to rearrangement but with much lower selectivity. Nicolaou and co-workers A used a **Witcop-type photocyclization**[109a,b] (secs. 13.4, 13.7) used a synthesis of diazonamide A, as shown in Example 14.[109c] In this case, the hexatriene-cyclohexene isomerization is reversible, allowing the cyclization to proceed in the presence of an oxidizing agent such as air.[110] Blechert and Jiricek's synthesis of (–)-gilbertine provides an illustration of the **Fischer indole synthesis** (sec. 12.5.C),[111] shown in Example 15. In Example 16, Lash and Zhang used a **Knorr pyrrole synthesis** procedure in the synthesis of porphyrin mineral abelsonite and related petroporphyrins.[112] An anionic **oxy-Cope rearrangement** (sec. 11.12.C.ii) is shown in Example 17 to illustrate **sigmatropic rearrangement** processes (sec. 11.12.A), taken from the Liao and Hsu synthesis of bilosespene A.[113] Example 18 is another sigmatropic rearrangement, the **Claisen rearrangement** (sec. 11.12.D), in this case used by Brimble and co-workers in a synthesis of γ-rubromycin.[114] Increased interest in the ring-closing metathesis reaction is illustrated by Example 19 in the Alibé, de March and co-worker's synthesis of securinine.[115] The ability to generate two alkene units in different parts of the molecule and then use them to form a ring certainly makes this reaction worthy of addition to the list of powerful reactions. Finally, palladium-catalyzed coupling reactions have become extremely important in organic synthesis. The Heck reaction, Stille coupling and Suzuki coupling are important variations that will be discussed in chapter 12 (sec. 12.7). Example 20 shows a Heck coupling, taken from a synthesis of hamigeran B by Trost and co-workers.[116]

106. Simmons, E.M.; Sarpong, R. *Org. Lett.* **2006**, *8*, 2883.
107. Nishimata, T.; Sato, Y.; Mori, M. *J. Org. Chem.* **2004**, *69*, 1837.
108. Kingsbury, J.S.; Corey, E.J. *J. Am. Chem. Soc.* **2005**, *127*, 13813.
109. (a) Yonemitsu, O.; Cerutti, P.; Witkop, B. *J. Am. Chem. Soc.* **1966**, *88*, 3941; (b) Theuns, H.G.; Lenting, H.B.M.; Salemink, C.A.; Tanaka, H.; Shibata, M.S.; Ito, K.; Lousberg, R.J.J. *Heterocycles* **1984**, *22*, 2007; (c) Nicolaou, K.C.; Chen, D.Y.-K.; Huang, X.; Ling, T.; Bella, M.; Snyder, S.A. *J. Am. Chem. Soc.* **2004**, *126*, 12888.
110. Stermitz, F.R. *Organic Photochemistry*, Vol. 1, Chapman, O.L. (Ed., Marcel-Dekker, New York, *1967*, p 247.
111. Jiricek, J.; Blechert, S. *J. Am. Chem. Soc.* **2004**, *126*, 3534.
112. (a) Zhang, B.; Lash, T.D. *Tetrahedron Lett.* **2003**, *44*, 7253; and see (b) Harbuck, J. ; Rapoport, H. *J. Org. Chem.* **1971**, *36*, 853; (c) Knorr, L. *Berichte*, **1884**, *17*, 1635; (d) Knorr, L. *Annalen* **1886**, *236*, 290; (e) Knorr, L.; Lange, H. *Berichte*, **1902**, *35*, 2998.
113. Hsu, D.-S.; Liao, C.-C. *Org. Lett.* **2003**, *5*, 4741.
114. Tsang, K.Y.; Brimble, M.A.; Bremner, J.B. *Org. Lett.* **2003**, *5*, 4425.
115. Alibés, R.; Ballbé, M.; Busqué, F.; de March, P.; Elias, L.; Figueredo, M.; Font, J. *Org. Lett.* **2004**, *6*, 1813.
116. Trost, B.M.; Pissot-Soldermann, C.; Chen, I.; Schroeder, G.M. *J. Am. Chem. Soc.* **2004**, *126*, 4480.

The reactions in Table 10.1 are classified as powerful due to the variety of transformations (formation of rings, molecular reorganization, generation of reactive functional groups from relatively unreactive functionality, or for functional group insertion) that were achieved all in essentially one synthetic step. Other reactions could easily be termed powerful, but these are sufficient to illustrate that if a reaction induces extensive and useful structural modifications, the synthetic tree should be biased to take advantage of that powerful chemistry.

10.5.B.vii. Disconnect from Atoms Bearing Certain Functional Groups. Certain functional groups are tied together along a reaction pathway. In many cases, the functional group transforms (CH–OH → C=O and C=O → CH–OH for example) involve ionic or polarized intermediates or reactants. Ionic functional group transformations are a useful illustration of this rule. Both anionic and cationic reactions involve a (positive) or (δ+) end, and a (negative) or (δ–) end. The functional groups corresponding to these reactive fragments have a special reactivity relationship which can be exploited. When a bond is disconnected, identification of each end of the disconnect product as a donor or acceptor can lead to identification of possible reaction pathways to reconnect the bond (discussed in sec. 1.2). The disconnection of **101** with the appropriate

donor and acceptor sites clearly suggests an **aldol condensation** (sec. 9.4.A), where **101** can be synthesized from a synthetic equivalent of **102**.

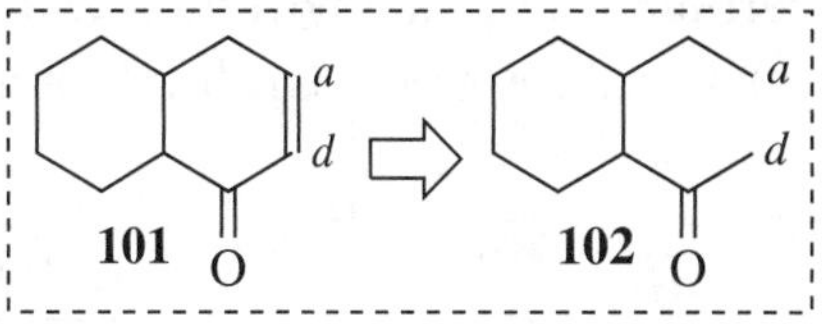

The disconnection **103** ⇒ **104** suggests a relationship between the tertiary amine in **103** and the dienyl azide moieties in **104**.[117] When **104** was heated in toluene, a [3+2]-cycloaddition occurred to give a triazoline (**105**) (sec. 11.11.H), but this unstable molecule decomposed (with loss of nitrogen) to form vinyl aziridine **106** as a mixture of stereoisomers.[117] When **106** was heated to 480°C (using flash vacuum pyrolysis, FVP), the nitrogen analog of a **vinylcyclopropane rearrangement** (sec. 11.12.B) occurred, and the resulting product was subjected to catalytic hydrogenation to give **103**. This reaction sequence (azide addition, vinylcyclopropane rearrangement) led to a bicyclic amine, linking the amine in **103** with the azido diene moieties in **104**. Clearly, the special relationship of these groups is exploited by a combination of reactions.

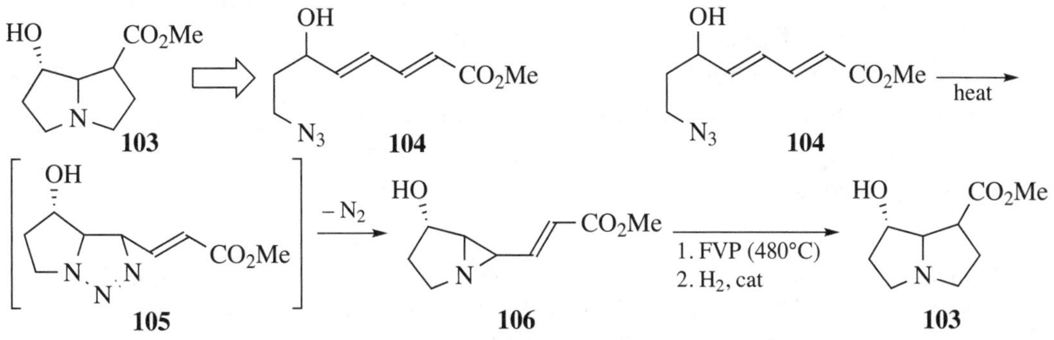

117.   Hudlicky, T.; Seoane, G.; Lovelace, T.C. *J. Org. Chem.* **1988**, *53*, 2094.

10.5.B.viii. Consider Bonds Related by Known Transforms. Some functional groups are related by the reactions that transform one to the other. They may be as obvious as alcohol → ketone. There are other transformations that connect or disconnect several bonds and/or functional groups at one time. The relationship is not always obvious, but its recognition can lead to great simplification of a target in one disconnection.

An important and useful example is the **oxy-Cope rearrangement** (see Example 17 in Table 10.1) in which the diene structure is related to the alkene-ketone functionality in the final product. Another example is the well known 1,3 elimination reaction known as the **Grob fragmentation** (sec. 2.9.E).[118] In a synthesis of periplanone C by Saicic and co-workers,[119] treatment of **107** with methanesulfonyl chloride generated the primary mesylate. Subsequent treatment with KOH led to the alkoxide and Grob fragmentation as shown gave **108** in 75% yield from **107**. In this case, the alkoxide and the leaving group (mesylate) are related to the carbonyl and alkenyl unit via the 1,3-elimination. Recognition of the spatial and structural relationships of the Grob fragmentation allows one to plan the disconnection, **108⇒106**.

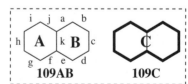

10.6. STRATEGIC BONDS IN RINGS

10.6.A. Carbocyclic Rings

In addition to the eight criteria in Section 10.5.B, there is a ninth criterion[78] that identifies strategic bonds in rings. Bonds are strategic because their disconnection maximizes the simplification of a polycyclic molecule. Corey devised a set of guidelines that accommodate the special problems encountered in monocyclic and polycyclic structures.[72]

1. **Strategic bonds must be in a primary ring**. A **primary ring** cannot be expressed as the envelope of two or more smaller rings bridged or fused to one another.[72,120] Five-, six- and seven-membered rings are usually easy to form, whereas other rings (especially C-8 and larger) are difficult to form by cyclization techniques (secs. 1.5.C, sec. 6.6.B). The two

118. (a) Grob, C.A. *Experientia* **1957**, *13*, 126; (b) Cherbuliez, E. Baehler, Br.; Rabinowitz, J. *Helv. Chim. Acta* **1961**, *44*, 1820 (c) Zurflüh, R.; Wall, E.N.; Siddall, J.B.; Edwards, J.A. *J. Am. Chem. Soc.* **1968**, *90*, 6224.
119. Ivkovic, A.; Matovic, R.; Saicic, R.N. *Org. Lett.* **2004**, *6*, 1221.
120. See Reference 72, footnotes 8-10 cited therein.

six-membered rings A and B in **109AB** are primary rings, but the 10-membered ring C (see **109C**) that encloses A and B is not (i.e., bond *k* is *not* strategic). It is important to note that bonds *a-i* are all strategic by this rule since every one is a part of a primary ring. The presence of ring C as an envelope of rings A and B does not change this.

2. **Strategic bonds must be directly attached to another ring.** The strategic bond should be exo to another ring. A ring disconnection that produces two functionalized appendages leads to a more complex overall synthesis. Three disconnections are shown for **110**. Paths I and II disconnect a bond exo to ring A to give **112** and **113**, respectively. Synthetically, this ring-closure is straightforward, and may potentially proceed with stereoselectivity. Disconnection III leads to **111**, which is a more difficult intermediate to prepare. Although the ring closure for III is probably facile, the disconnections in I or II lead to greater simplification and have a greater potential for stereocontrol. There are very few ring closure methods in which bonds are fused to preexisting three-membered rings and strategic bonds may not be exo to rings of that size.[72,120]

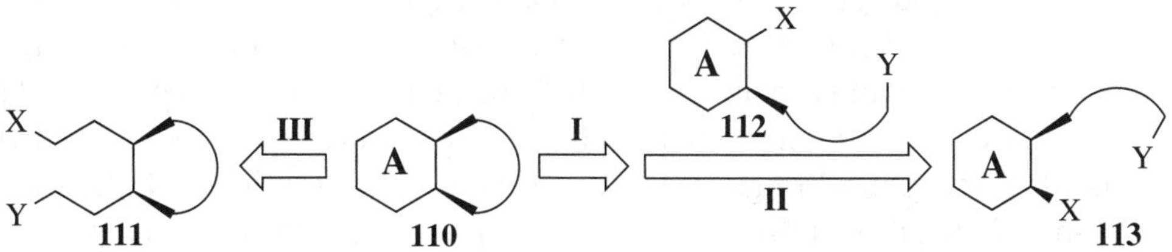

3. **Strategic bonds must be in rings with the greatest degree of bridging**. Disconnection of a bond that is highly bridged (connected to several rings) leads to greater simplification of that structure. Disconnection **A** in the highly bridged four-membered ring of **114** leads to **115**, with a bicyclo[4.4.0]decane unit, whereas disconnection **B** leads to **116**. The bicyclo[3.1.1]heptane ring system in **116** is less accessible, and provides less simplification when compared to **115**.

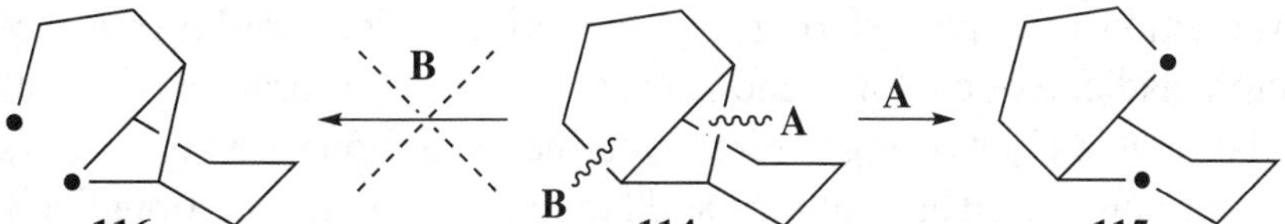

A **maximum bridging ring** is defined as one that is bridged at the greatest number of sites. The maximum bridging ring(s) is(are) selected from the set of "synthetically significant rings that is defined as the set of all primary rings and all secondary rings less than eight membered."[72,120] Structure **117** has six significant rings: **A, B, C, D, E,** and **F**.[120] The highlighted rings in **A → D** are primary rings, but those in **E** and **F** are secondary. Ring structure **C** is bridged at four sites, more than any other, and is the maximal bridging ring.[72] Ring **B** contains as many bridgehead sites as **C**, but it is not a maximal bridging ring, because it is bridged to other rings at only two of these sites (***b*** and ***d***). Therefore, the number of times a ring is bridged is not a valid criterion for determining maximal bridging character. Ring **E** is bridged to as many rings as ring **A** but has one less bridgehead site than **C**.

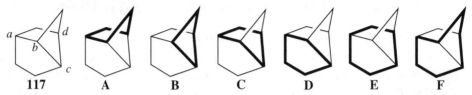

117   A   B   C   D   E   F

4. **Bonds common to a pair of bridged primary rings are not strategic.** Any bond common to a *pair* of bridged or fused primary rings whose envelope is eight membered or larger cannot be strategic.[72] Such a disconnection would generate a ring of more than seven members. Disconnection of the central bond in **109** would result in a 10-membered ring compound (**118**), and rings of this size are often difficult to synthesize. For that reason, this type of

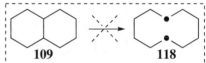

disconnection should be avoided.[120] If the two fused or bridged rings are *directly* joined elsewhere by another bond (as in **114**), the bond *may* be strategic. As shown in **114**, disconnection of the indicated bond does not produce a 10-membered ring but two fused six-membered rings (see **115**). Similar disconnection in **109**, however, gives a 10-membered ring (**118**), as mentioned above, and is not strategic.

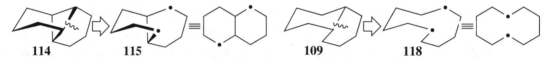

114   115   109   118

5. **Bonds within aromatic rings are not strategic.** Aromatic nuclei are common fragments in natural products but they are usually incorporated as intact units of another fragment, although aromatic rings can be reduced and converted to other functionality. Although introduction of a phenyl ring and reduction (**Birch reduction**, see sec. 4.9.E) to a cyclohexadiene or cyclohexenone is common, incorporation of a cyclohexene and later oxidation to a phenyl ring is rare. Notable exceptions are quinones (sec. 3.3.A), which are commonly used as synthons and later converted to functionalized aromatic rings.[121] Incorporation of a carbocyclic ring and subsequent aromatization often requires harsh conditions and several synthetic steps. For these reasons, it is rarely useful to create an aromatic ring in a disconnection step unless aromatization is facile and non-injurious to other functionality.

6. **In a cyclic arc, cleavage of the bond should not leave stereocenters in the side chain**. Reactions that join two fragments to generate a chiral center allow greater

---

121.   (a) Fu, P.P.; Harvey, R.G. *Chem. Rev. **1978**, 78,* 317; (b) Walker, D.; Hiebert, J.D. *Ibid. **1967**,* 67, 153; (c) Jackman, L.M. *Adv. Org. Chem. **1960**, 2,* 329.

control than those reactions that join fragments containing a chiral center remote to the site of bond formation. Disconnection of bond **B** in **119** would lead to **120**, requiring the synthesis of the alcohol with a stereogenic center on the chain prior to ring closure. It is not clear there would be any control over the stereochemistry of that center as it is formed. The reaction resulting from disconnection of bond **A** (to **121**), however, forms the ring and generates the stereogenic center simultaneously. Disconnection of **119** to **121** will lead to a shorter synthetic sequence, and there is the possibility that it can proceed with reasonable stereoselectivity.

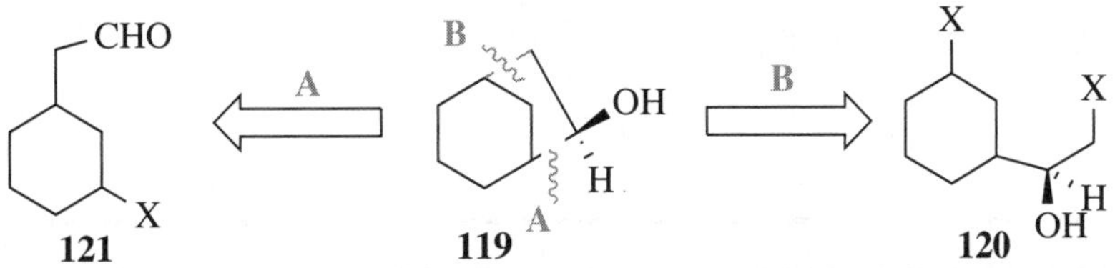

A special case arises when the bond is attached to a stereogenic center, but that center is not the only one on the arc linking the two common atoms.[72] Such a bond may be strategic. In **122**, bond *a* is strategic, since the arc in the disconnect product (**123**) does not contain an asymmetric center. In **124**, the arc contains a stereogenic center, and bond *a* is not strategic, since that disconnection gives **125**.

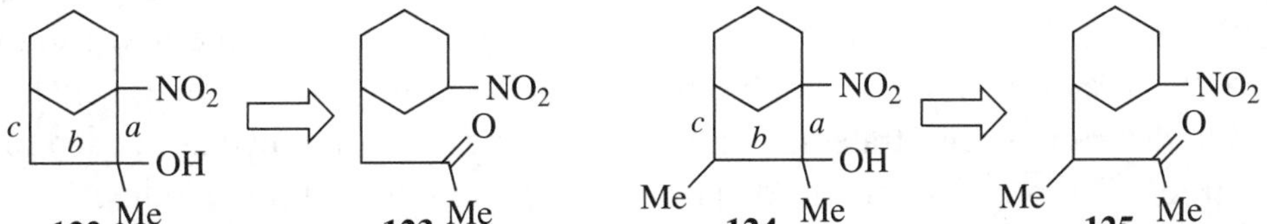

These rules can be applied to a carbocyclic molecule such as **126**. An analysis based on these six rules is shown in Table 10.2.[72] Rule 1 shows that all bonds are part of a primary ring and rule 5 does not apply since there are no aromatic rings. Rule 2 shows that several bonds are exo to primary rings, indicated in **A**. The primary rings are highlighted in **B-D**. There are larger rings forming an envelope of these smaller rings, but bonds exo to those highlighted rings will take into account all possibilities. Rule 3 suggests that bonds *i, j, l, f,* and *n* are part of bridging rings. Disconnection of bond *m* is ruled out since the disconnection is away from the stereogenic center and does not involve an exo bond. Disconnection of bond *k* leads to a large ring and is ruled out. Bonds *b-d, f,* and *g-i* are not disconnected since it would violate Rule 6.

Table 10.2. Strategic Bond Analysis for Carbocyclic Compound 126

BOND	1	2	3	4	5	6	STRATEGIC	BOND	1	2	3	4	5	6	STRATEGIC
		RULE[a]								**RULE**[a]					
a		Y		Y	X			h					X		
b					X			i	Y	Y			X	Y	YES
c					X			j	Y	Y			X	Y	YES
d					X			k	Y			X	X	Y	
e		Y			X	Y		l	Y	Y			X	Y	YES
f		Y	Y		X	Y	YES	m					X		
g					X			n	Y	Y			X	Y	YES

Y = rule applies, and X = rule is not applicable.

10.6.B. Heterocyclic Rings

The six rules discussed in Section 10.6.A apply to many carbocyclic ring systems. If a bond in the ring is connected to an heteroatom, that bond can be considered strategic if it satisfies rules 2, 4, 5 and 6. Corey generated the strategic bonds for lycopodine (**127**) in Figure 10.6, using these rules.[72] Bonds 3, 12, 14, and 19 were found to be strategic for the *first* disconnection. Thereafter, these bonds may or may not be strategic in the disconnect product, which should be reexamined to determine new strategic bonds. This process is continued as needed to generate a synthetic tree. The process used to determine strategic bonds in lycopodine followed five steps. (*1*) Structure of lycopodine (arbitrary bond numbering in **127**). (*2*) Primary rings. There are four primary rings, highlighted in **A-D**. (*3*) No synthetically significant (< eight-membered) secondary rings. (*4*) Bridgehead atoms. There are two, marked with (•) in **127**. (*5*) Strategic bond determination by the carbocyclic procedure.

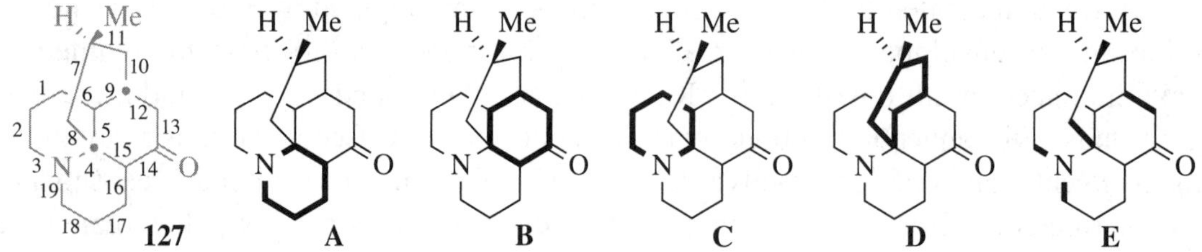

Bond Number as in 127																			
Rule 1:	1	2	3	4	5	6	7	8	9	10	11	12	13	14	15	16	17	18	19
Rule 2:			3	4	5	6		8	9	10		12		14	15	16			19
Rule 3:					5			8	9	10		12	13	14	15				
Rule 4:	1	2	3			6	7	8		10	11	12	13	14		16	17	18	19
Rule 5:	1	2	3	4	5	6	7	8	9	10	11	12	13	14	15	16	17	18	19
Rule 6:	1	2	3	4	5	6	7		9		11	12	13	14	15	16	17	18	19

Strategic bond determination: carbon-heteroatom procedure
Bond Number

Rule 2b:			3	4															19
Rule 4:	1	2	3			6	7	8		10	11	12	13	14		16	17	18	19
Rule 5:	1	2	3	4	5	6	7	8	9	10	11	12	13	14	15	16	17	18	19
Rule 6:	1	2	3	4	5	6	7		9		11	12	13	14	15	16	17	18	19

The Strategic bonds are: 3, 12, 14, 19 (see **E**)

Figure 10.6. Strategic bond analysis of lycopodine, **127**. [Reprinted with permission from Corey, E.J.; Howe, W.J.; Orf, H.W.; Pensak, D.A.; Petersson, G. *J. Am. Chem. Soc. 1975, 97,* 6116. Copyright © 1975 American Chemical Society.]

Disconnection of bonds *e* and *a* leads to complex ring systems relative to the other disconnections, bonds *l, n, f, j,* and *i*. These fragmentations show reasonable simplification of the ring system and can be reformed by reasonable synthetic routes. Rule 4 (bridgehead atoms) discounts only bond *k* since it would open to a 10-membered ring, which leads to bonds *l, f, n, j,* and *i* as strategic (see the highlighted bonds in **E**).

A retrosynthetic analysis of Stork's synthesis of lycopodine is outlined in Scheme 10.9.[122] The numbering system used in Corey's analysis (Figure 10.6) is retained, with bonds 3, 12, 14 and 19 being strategic. Bond 19 of **127** is disconnected to give **128**. It is noted that Corey's rules do not formally list manipulation of functional groups prior to disconnection of the target, but such flexibility is expected in any plan. Disconnection of **128** to **129** involves cleavage of bond 19, strategic by Corey's analysis, and the novel use of an aromatic ring as the equivalent of bonds 16-19. Further disconnection shows that bond 15 in **129** is now strategic, although it is not in **127**. This observation emphasizes that each disconnect product should be analyzed separately, in order to make full use of the strategic bond approach and maximize one's synthetic options. Bond 3, which is strategic in **127**, is disconnected to give **130**, and disconnection of bond 20 gives **131**. The final disconnection (to **132** and ethyl 3-oxobutanoate) disconnects bonds 8 and 10. These bonds are not strategic in **127**, but they are strategic in a molecule such as **131**.

Corey provided several additional analyses for a variety of polycyclic molecules (see Figure 10.7).[72] The strategic bonds are shown in bold lines offering several disconnections for each target. Disconnection of each strategicbond will lead to a different synthetic approach (a different branch on the synthetic tree). Indeed, disconnection of one strategic bond can

---

122. (a) Stork, G.; Kretchmer, R.A.; Schlessinger, R.H. *J. Am. Chem. Soc. 1968, 90,* 1647; (b) Stork, G., *Pure Appl. Chem. 1968, 17,* 383.

generate several completely different syntheses (see Section 10.7).

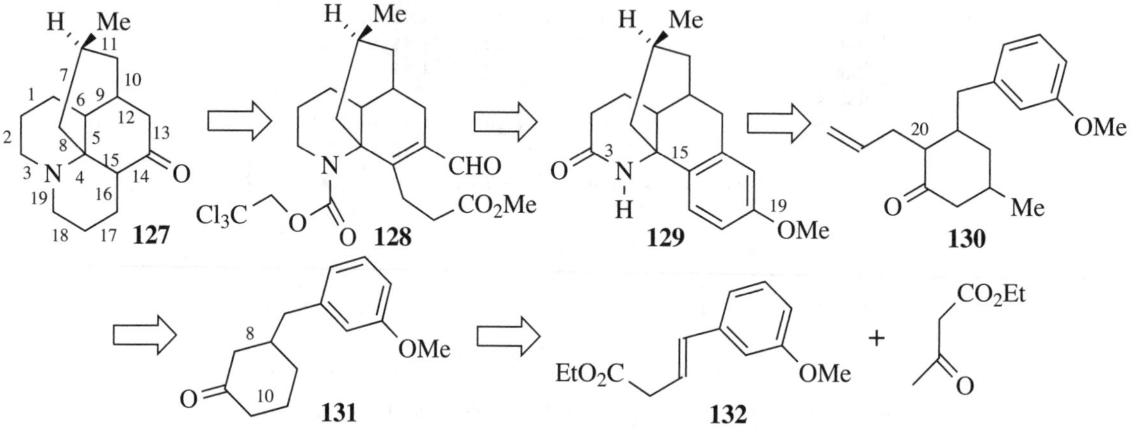

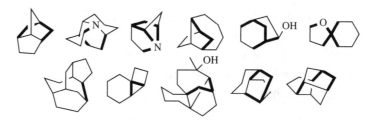

Scheme 10.9. Stork and co-worker's synthesis of lycopodine (**127**).

Figure 10.7. Strategic bonds in selected ring systems. [Reprinted with permission from Corey, E.J.; Howe, W.J.; Orf, H.W.; Pensak, D.A.; Petersson, G. *J. Am. Chem. Soc.* **1975**, *97*, 6116. Copyright © 1975 American Chemical Society.]

10.6.C. An Intramolecular Strategy

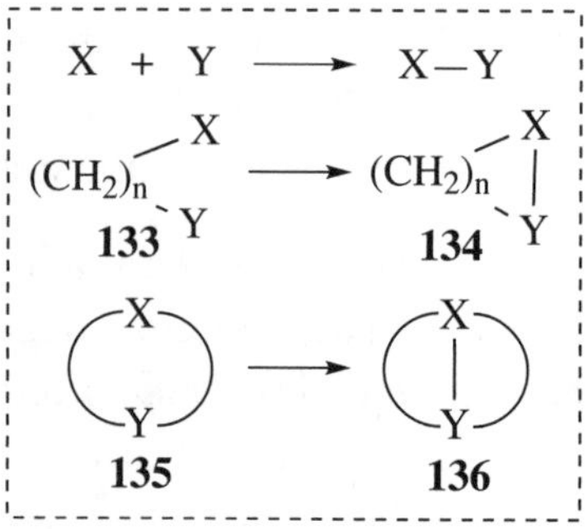

Deslongchamps developed a strategy[13] that relies on an intramolecular approach using unsymmetrical molecules with a rigid conformation and a needed absolute configuration. In a cyclization reaction, two ends of a molecule react and the transition state will assume many if not all of the conformational restrictions imposed by the ring being formed, as discussed in sections 1.5.B, 6.5, 6.6.B. This is a cyclization process, illustrated by the reactions **133** → **134**, where X and Y are tied together by a chain of intervening methylene units. When compared with X+Y → X–Y, there are more constraints on conformation and reactivity in **133**. If two units of a cyclic molecule react, a polycyclic system is generated where reactivity and stereochemistry are controlled by the rigid conformation of the cyclic precursor. This intramolecular reactions illustrated by **135** → **136**. Indeed, there are more conformational and reactivity constraints on **135** than on **133**. It is possible to tie together a reaction center and a reagent, which will limit the angles from which a reagent can approach the reaction center with a great influence on the stereochemistry of the product.[13] The chain length (n in **133**) must allow generation of a ring with a low barrier to its formation (3-7, $n = 1$-5 in common5.B). Increasing chain rigidity by incorporating a cis double bond or a ring will increase the conformational restrictions on the reaction and will favor the cyclization process. Deslongchamps defined the conversion of **133** to **134** as a **Level 1 Process** (an intramolecular process having one chain).[13] Compare this reaction with formation of **136** in which the reactive center and reagent in the precursor (**135**) are held together by tethering chains, and a higher degree of stereochemical control is observed. This transannular reaction is defined as a **Level 2 Process** (an intramolecular process with two chains).[13] A synthetic plan based on a Level 2 process requires the use of medium and large rings. The ultimate degree of stereochemical restriction occurs when the reagent and substrate are held rigidly in space in an appropriate orientation for the desired reaction to take place.[13]

Scheme 10.10 outlines Deslongchamps' synthesis of triquinacene derivatives.[123] Manipulation of **Thiele's acid** (**137**) gave the key product, **138**. A photochemical fragmentation gave **139**, and an intramolecular aldol (sec. 9.4.A.ii) gave **140**. With the basic skeleton formed, **140** was converted to triquinacene (**141**), and also to triquinacenecarboxylic acid (**142**). Scheme 10.10 provides an analysis of the synthesis, which involved a Level 1 intramolecular aldol (**139** → **140**) and a reverse Level 1 process in the conversion of **138** → **139**.

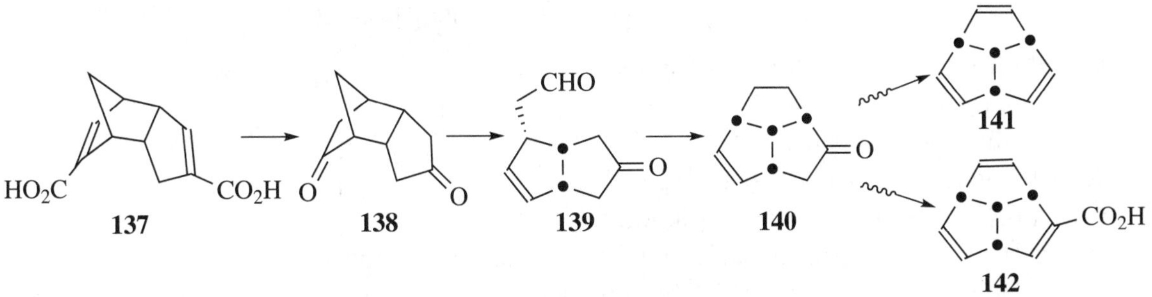

Scheme 10.10. Deslongchamps' syntheses of triquinacene (**141**) and triquinacenecarboxylic acid (**142**).

A synthesis involving a Level 2 process is Deslongchamps' synthesis of twistane (**146**), outlined in Scheme 10.11.[124] Treatment of **143** with sodium hydride in dioxane gave **144**, and intramolecular displacement of the mesylate gave 4-twistanone, **145**. The two reacting groups are brought into proper position by the transannular nature of the cyclization.

123. Deslongchamps, P.; Cheriyan, U.O.; Lambert, Y.; Mercier, J.-C.; Ruest, L.; Russo, R.; Soucy, P. *Can. J. Chem.* **1978**, *56*, 1687.
124. Gauthier, J.; Deslongchamps, P. *Can. J. Chem.* **1967**, *45*, 297.

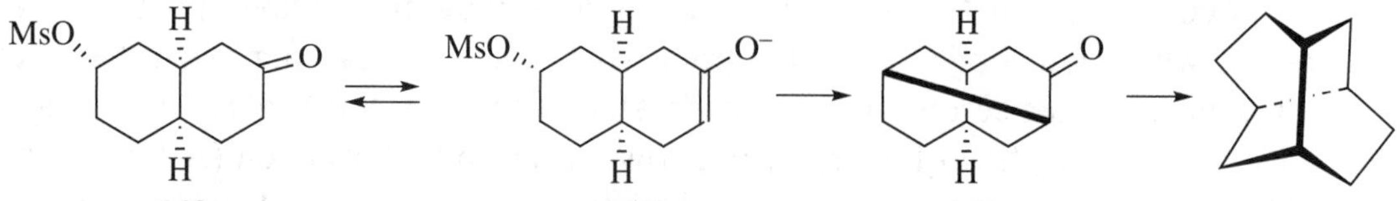
MsO
H
O
H
MsO
H
O⁻
H
H
O
H

143 144 145 146

Scheme 10.11. An outline of Deslongchamps' synthesis of twistane, **146**.

10.7. SELECTED SYNTHETIC STRATEGIES: PANCRATISTATIN

As mentioned in Section 10.6.C, disconnection of a single strategic bond can lead to several different synthetic schemes. The nature of this approach maximizes individuality in a given synthesis and, as shown with lycopodine (**127**), many different strategies are possible. This diversity is shown explicitly by several syntheses of pancratistatin (**13**), first introduced in Section 10.2.A.

A few retrosynthetic analyses are illustrated in Schemes 10.12, 10.13, 10.14 and 10.15 of the several syntheses of **13** that appear in the literature.[18e] In addition, Keck and co-workers synthesis of (+)-7-deoxypancratistatin (**172**)[125] is shown for a comparison of hat general strategy. The four retrosyntheses for **13** shown in Schemes 10.12,[125] 10.13,[126] 10.14[127] and 10.15,[128] respectively, are based on the actual published synthesis rather than a pre-synthesis analysis. A retrosynthesis for the closely related 7-deoxypancratistatin (**172**) is shown in Scheme 10.16.[129] An analysis of **13** via Corey's rules (sec. 10.6.A) suggests that bonds 1, 2, 6, and 8 are strategic, with the amide bond 9 considered to be labile (see sec. 10.5.B.i). Bonds 13 and 14 could also be considered labile since they are part of an acetal unit.

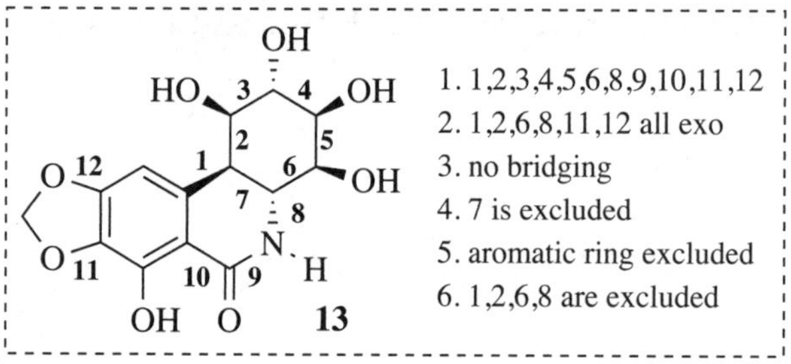
OH
HO
OH
OH
O
O
N
H
OH
O
1 2 3 4 5 6 7 8 9 10 11 12
13
1. 1,2,3,4,5,6,8,9,10,11,12
2. 1,2,6,8,11,12 all exo
3. no bridging
4. 7 is excluded
5. aromatic ring excluded
6. 1,2,6,8 are excluded

125. Keck, G.E.; McHardy, S.F.; Murry, J.A. *J. Am. Chem. Soc,* ***1995***, *117*, 7289.
126. (a) Iwao, M.; Reed, J.N.; Snieckus, V. *J. Am. Chem. Soc.* ***1982***, *104*, 5531; (b) Watanabe, M.; Sahara, M.; Kubo, M.; Furukawa, S.; Billedeau, R.J.; Snieckus, V. *J. Org. Chem.* ***1984***, *49*, 742.
127. For example, see (a) Manitto, P. *Biosynthesis of Natural Products*, Halsted Press, John Wiley, New York, ***1981***; (b) Steyn, P.S. (Ed.) *The Biosynthesis of Mycotoxins. A Study in Secondary Metabolism*, Academic Press, New York, ***1980***; (c) Arigoni, D. *Pure Appl. Chem.* ***1968***, *17*, 331; (d) Seiler, M.; Acklin, W.; Arigoni, D. *J. Chem. Soc. D,* ***1970***, 1394; (e) Hamberg, M.; Swensson, J.; Wakabayaski, T.; Samuelsson, B. *Proc. Natl. Acad. Sci. USA,* ***1974***, *71*, 345; (f) Hamberg, M. Samuelsson, B. *J. Biol. Chem.* ***1967***, *242*, 5344.
128. Stocking, E.M.; Williams, R.M. *Angew. Chem. Int. Ed.,* ***2003***, *42*, 3078.
129. (a) de la Torre, M.C.; Sierra, M.A. *Angew. Chem. Int. Ed.,* ***2004***, *43*, 160; (b) Johnson, W.S. *Acc. Chem. Res.* ***1968***, *1*, 1; (c) Hendrickson, J.B., *The Molecules of Nature* W.A. Benjamin, New York, ***1965***, pp 12-57; (d) Bernfeld, P. *Biogenesis of Natural Compounds*, Pergamon Press, ***1963***.

In Scheme 10.12,[125] the first disconnection of Danishefsky's synthesis involved cleavage of bond 9 (the amide), converting the carboxyl unit to a lactone and the amine unit to a trichloroimidate. This disconnection also changed the trans stereochemistry found in 13 to the cis stereochemistry shown for 147, allowing a Diels-Alder-based strategy that produced the cyclohexene ring in 148. Disconnection to 149 mainly involved functional group manipulation that allowed the hydroxyl groups to be incorporated with the correct relative stereochemistry. This functional group manipulation was continued in the disconnection to 150, which contained the requisite cyclohexene unit. constructed via a Diels-Alder reaction from diene disconnect product 151, which disconnects bonds 4 and 6. This diene unit was prepared from aldehyde 152 via disconnection of bond 1. Disconnection of bond 10 in 152 as well as bond 1 led to pyrogallol (153) as the starting material. The actual synthesis involved a total of 26 steps that produced racemic 13. It appears that the key disconnection was bond 9, the amide unit, bonds 4 and 6 for the Diels-Alder reaction, and bonds 1 and 10 for attachment of groups to the aromatic ring.

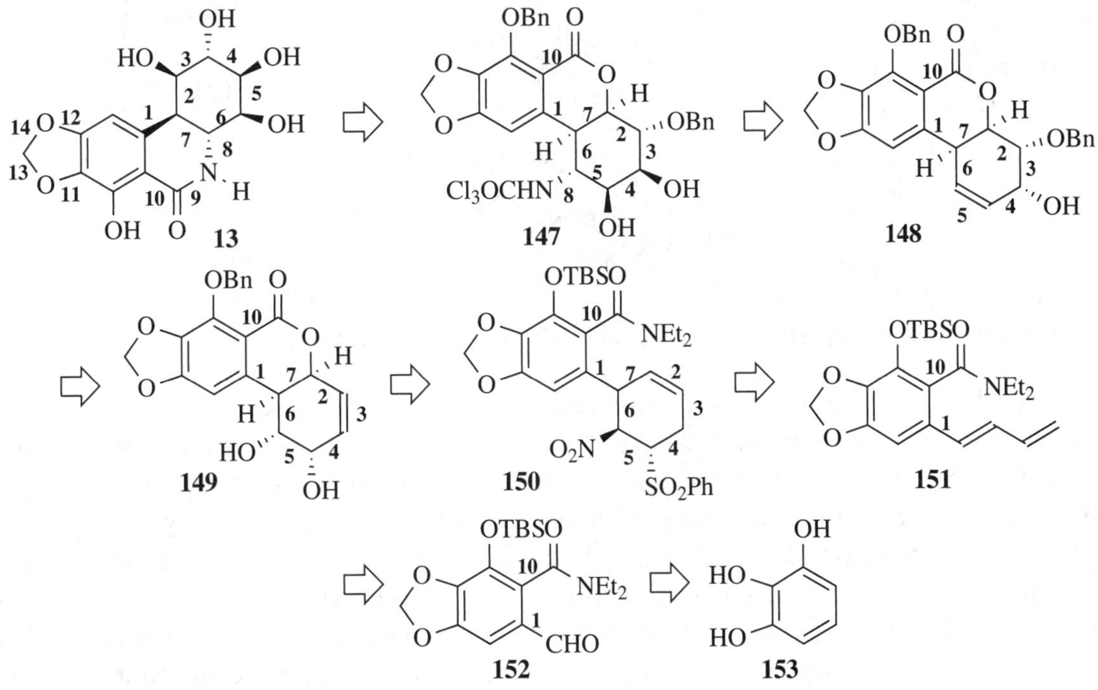

Scheme 10.12. Retrosynthetic analysis of Danishefsky's pancratistatin synthesis.

In Hudlicky's retrosynthetic analysis (Scheme 10.13),[126] bonds 9 and 1 were utilized. The amide bond (9) was again disconnected but to a N-Boc ester (154) rather than the lactone seen in Danishefsky's synthesis. Manipulation of the functional groups was also different, proceeding via the epoxide shown in 154. Disconnection to 155 illustrates the manner in which the functional groups were manipulated, setting up disconnection of bond 1 to give two products, 156 and 157. Amide 156 was prepared from commercially available materials, and contains bonds 10-12. Fragment **157** contains one entire ring of **13**, and this differs significantly from Danishefsky's synthesis, where this ring was put together via a Diels-Alder reaction that required extensive manipulation (and steps) to obtain the desired functionality. In Hudlicky's approach, this ring is incorporated intact, and contains functionality that allows the tetraol unit to be incorporated more efficiently. The key step is the disconnection of **155** to **156** and **157**, where the two fragments were attached by an ortho-metalation procedure described by Snieckus and co-workers (see sec. 8.5.F).[126] Fragment **157** is derived from **158**, which is commercially available, but is generated by reaction of bromobenzene with *Pseudomonas putida*, as described in sec. 3.5.D. This synthesis produced (+)-pancratistatin, and required 16 steps from the commercially available diol **158**, and 17 steps if **158** was prepared from bromobenzene.

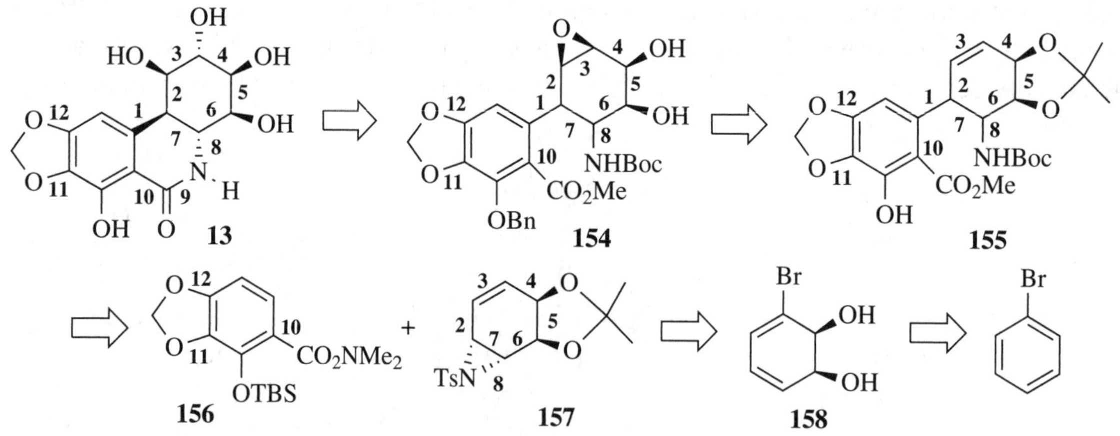

Scheme 10.13. Retrosynthetic analysis of Hudlicky's pancratistatin synthesis.

Trost used bonds 9 and 10 as well as 8 and 1 as outlined in Scheme 10.14.[127] The first disconnections involve functional group manipulation to set up the ring for the hydroxyl units as in **159**. The next disconnection is a key step in which bond 10 is disconnected to give **160**, rather than bond 9 in the other syntheses. This disconnection implies generation of an aryllithium from the bromide in **160** adding to the isocyanate group, generating the lactam and the requisite ring. The next disconnection to **161** gives the precursors to the isocyanate as well as the hydroxyl precursors, which sets up the next key disconnection to give fragments **162** and **163**. The precursor to Grignard reagent **162** was prepared using a literature procedure, and azide **163** was derived from the next disconnection to **164**, a diol readily available from benzoquinone in five steps. With **64** as the starting material, this synthesis produced (+)-pancratistatin in ~ 16 steps. The bromide precursor to **162** was prepared separately. Since **164** was prepared from benzoquinone, the true starting material, five more steps are added to bring the total to 21.

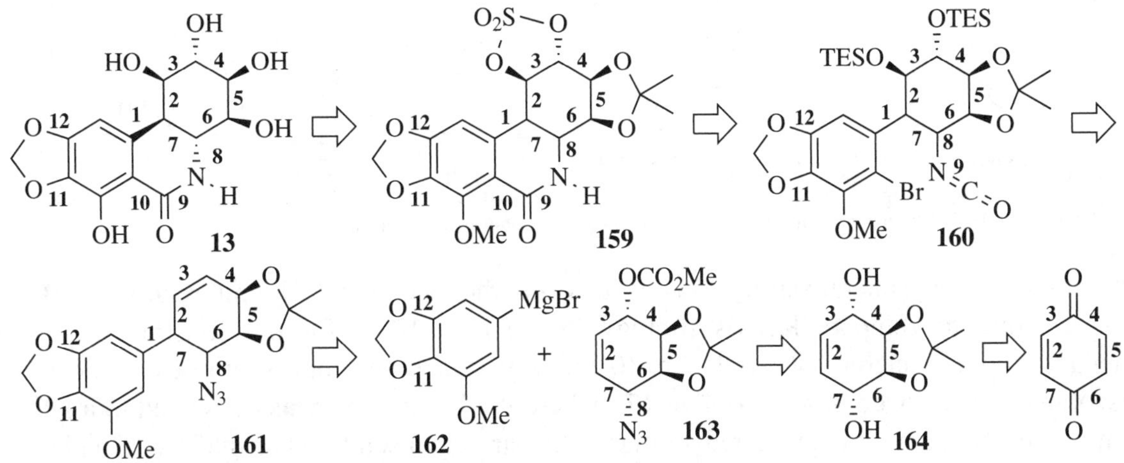

Scheme 10.14. Retrosynthetic analysis of Trost's synthesis of pancratistatin.

Rigby's synthetic plan is outlined in Scheme 10.15[128] and used bonds 1 and 10. As in the previous synthesis, the first two disconnections involve manipulation of the hydroxyl groups, first a disconnection to give **165** and then to **166**. It is noted that these disconnections represent several steps and a change in the stereochemistry of the ring juncture stereochemistry, which is not shown. Disconnection of bond 1 in **166** to **167** is a key step that uses a photoaddition reaction to set that bond in the synthesis. Disconnection of bond 10 in **167** leads to two fragments, **168** and **170**. As with Trost's synthesis, bond 10 will be set via an isocyanate in **168**, this time via a Curtius rearrangement. Fragment **168** is disconnected to give **169** as one of the starting materials. The ring is used as an intact entity, with functional group manipulation to give the hydroxyl groups. Fragment **170** is derived from the commercially available 2,3-dihydroxybenzaldehyde (**171**). The conversion of **169** to **168** required six steps. Counting the coupling of **168** to **170** as one step, and using **169** as the starting material, the overall synthesis produced (+)-pancratistatin in 24 steps. The synthesis involved an enzymatic resolution using cholesterol esterase, in the sequence that converted **169** to **168**, to prepare optically active material.

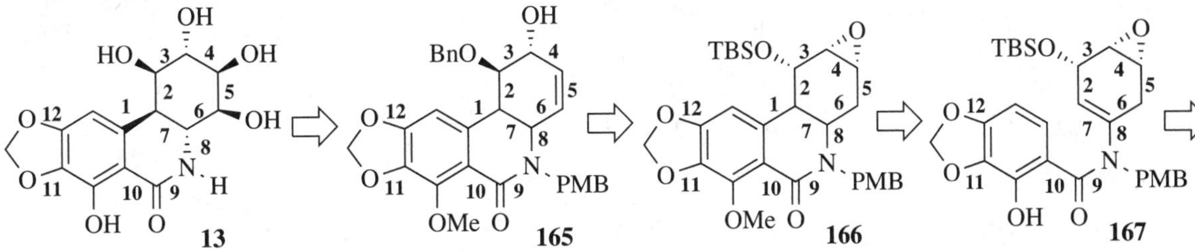

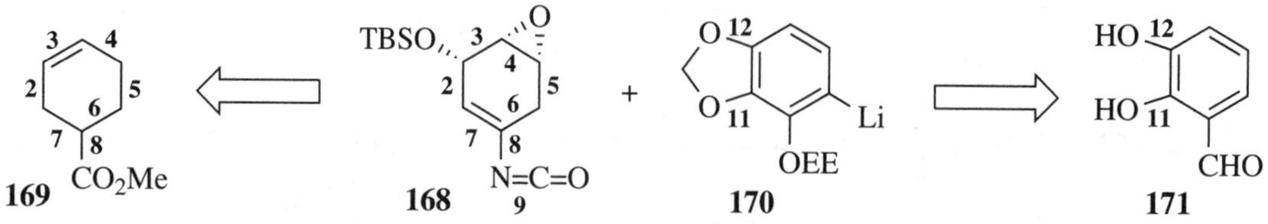

Scheme 10.15. Retrosynthetic analysis of Rigby's synthesis of pancratistatin.

Although 7-deoxypancratistatin (**172**) is missing the phenolic OH unit relative to **13**, it is interesting to compare Keck's strategy (Scheme 10.16)[129] for the construction of the phenanthridone skeleton (see Scheme 10.15) with the other four syntheses. The initial disconnection removed bond 9 to give **173** where the lactam unit was converted to a lactone unit and the amine unit is protected similar to the strategy used by Danishefsky. Bond 9 was actually converted to a trifluoroacetyl amide unit and the last step in the synthesis is the key lactone to lactam reorganization. The next disconnection removes bond 7, which was *not* strategic using analysis by Corey's rules, to give **174**. This is a key disconnection because the six-membered ring was made in the synthesis by a radical cyclization process (see sec. 13.7). It also illustrates that bond 7 is not strategic in **13** or **172**, but it *is* strategic in **173** because now it is part of a primary ring. *Remember, each new disconnect product must be reexamined for strategic bonds.* Disconnection of **174** to **175** essentially sets up the ring for manipulation of functional groups, and the functionality necessary for the radical cyclization. Disconnection to **176** is also required to set the functionality and this allows a key disconnection to **177**, which utilizes formation of an aryllithium to attack the ester carbonyl of **177** leading to the reorganized aldehyde-ketone **176** in the synthesis. Disconnection of **177** breaks the labile ester bond to give fragment **178**, which is prepared by known methods, and **179** that contains the carbons of the tetraol ring in **172** along with appropriate functionality. This precursor is derived from the available D-gulonolactone (**180**). This synthesis produced (+)-7-deoxypancratistatin in 20 steps.

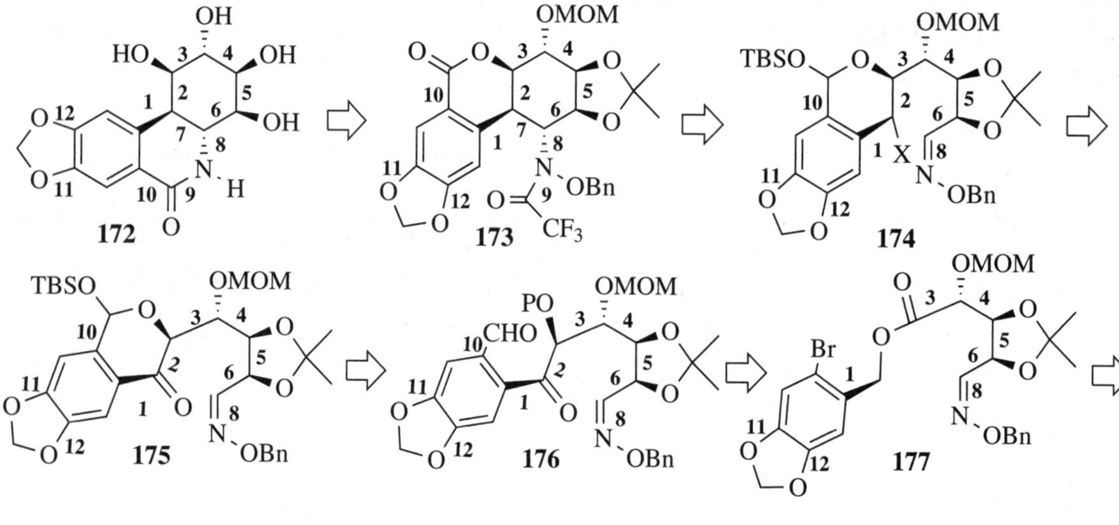

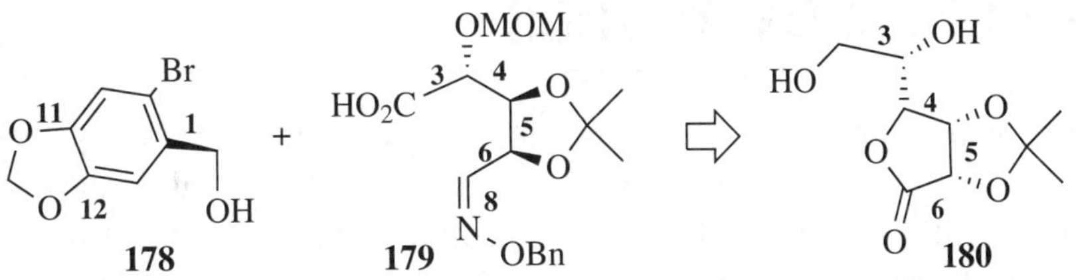

Scheme 10.16. Retrosynthetic analysis of Keck's synthesis of 7-deoxypancratistatin.

The syntheses in schemes 10.12-10.16 rely on disconnection of the amide bond (9), essentially a functional group interchange, although it is clearly not that simple since either bonds 1 or 10 are also disconnected. The three key rings, the benzene ring, the lactam ring and the tetraol ring are also targeted. Although the disconnected bonds are similar, the strategies differ based on which of the three rings is key. In the syntheses of Danishefsky, Hudlicky and Keck, the lactam ring was formed close to the end or at the end of the synthesis. In these cases, the other two functionalized rings formed atropisomers that made formation of the lactam difficult. Steps were added and the sequences modified to get around this problem. In Rigby's synthesis, the lactam ring was formed in the middle of the synthesis by linking the aromatic ring to the tetraol ring by an amid linkage, with formation of the lactam ring by a photoaddition reaction. In Hudlicky and Trost's syntheses the aromatic ring was joined to the tetraol ring, in one case by an aryl anion opening of an aziridine and in the other by a S_N2' type addition of an aryl anion to a functionalized cyclohexene derivative. The carbonyl and amine units were intact, allowing the lactam ring to be formed. In the syntheses of Danishefsky and Keck, the lactam ring was formed by a lactone-to-lactam reorganization, where the lactone unit linked the aryl ring to the tetraol ring. In Danishefsky's synthesis, the tetraol ring was generated by a Diels-Alder reaction and functional group manipulation. In the syntheses of Hudlicky, Trost, and Rigby the tetraol ring was incorporated as an intact ring, from different precursors, requiring functional group manipulation. In Keck's synthesis, the tetraol ring was constructed by manipulating a carbohydrate precursor that contained all the requisite carbons and functionality (a chiral template driven synthesis-see sec. 10.9 for other examples of this strategy). The actual key bonds used to generate the carbon skeleton are actually very similar, but the strategies for making those bonds generated vastly different total syntheses. Which is best? All methods produce the target and all have their strengths. The most efficient route that produces the best yield of the target is obviously one of the more important considerations, but *best is always relative to the needs of those asking the question.*

10.8. BIOMIMETIC APPROACH TO RETROSYNTHESIS

In previous sections, a target was analyzed and the disconnection analysis led to reactive fragments. The disconnection process was continued until a suitable starting material was found. The actual synthesis followed the disconnection sequence to construct the molecule. If there was a bias in the retrosynthesis or the actual synthesis, it was usually due to the special interests of the chemist, recognition of an attractive substructure, or special knowledge of one area of organic chemistry.

There is an important alternative to the disconnection approach in which the retrosynthetic plan is biased to follow the known biosynthetic pathway for a particular target. In effect, the biosynthetic route becomes the synthetic tree, and appropriate reagents must be found to mimic the processes that Nature accomplishes with enzymes. Knowledge about formation of the molecule in Nature, how its functional groups interact and which bonds are cleaved with what reagents is usually gathered during the isolation and identification process. In many cases, intermediates of the biosynthetic pathway (also called the biogenetic pathway) for the formation of a natural product are known. Targeting those intermediates to generate the synthetic tree will effectively mimic the biosynthetic pathway and is referred to as a **biomimetic** or a **biosynthetic approach**. Although many biopathways have been determined,[127] the pathways remain unknown for many important classes of natural products. Generalized pathways are known, such as biosynthetic Diels-Alder reactions,[128] and can be used to develop a biomimetic strategy for total synthesis. This method offers an important alternative in synthetic planning, however. A number of reviews are available that describe this approach when applied to total synthesis.[129]

10.8.A. Polyene Cyclization

A good example of this approach is the **Johnson polyene cyclization reaction**, or just **polyene cyclization**.[82c,130,133] This polyene cyclization reaction was termed the **Stork-Eschenmoser hypothesis** by Johnson, named after the scientists who made the proposal. The biogenetic pathway for production of the steroid lanosterol from acetyl coenzyme A (acetyl Co-A) is shown in Figure 10.8.[131] From this Figure it is clear that squalene is the direct biogenetic precursor of squalene-2,3-epoxide (**181**), which is the direct precursor to lanosterol. Lanosterol is ultimately converted to cholesterol,[132,135] where enzymes close the ring in the biological process. The Stork-Eschenmoser hypothesis reasoned that treatment of a squalene-like polyene (all trans double bonds) with an appropriate acid would lead to a carbocation and the proximity

130. Mundy, B.P.; Ellerd, M.G.; Favaloro Jr., F.G. *Name Reactions and Reagents in Organic Synthesis, 2nd ed.*, Wiley-Interscience, New Jersey, **2005**, pp. 348-349.
131. (a) Clayton, R.B. *Quart. Rev. Chem. Soc.* **1965**, *19*, 168; (b) van Tamelen, E.E.; Willett, J.D.; Clayton, R.B.; Lord, K.E. *J. Am. Chem. Soc.* **1966**, *88*, 4752; (c) Corey, E.J.; Russey, W.E. Ortiz de Montellano, P.R. *Ibid.* **1966**, *88*, 4750; (d) Corey, E.J.; Russey, W.E. *Ibid.* **1966**, *88*, 4751.
132. Wendt, K.U.; Schulz, G.E.; Corey, E.J.; Liu, D.R. *Angew. Chem. Int. Ed.,* **2000**, *39*, 2812.

of the C=C units would lead to ring formation.[133] If this occurred in a sequential manner, several rings would be formed to build the polycyclic system via carbocation intermediates.[137]

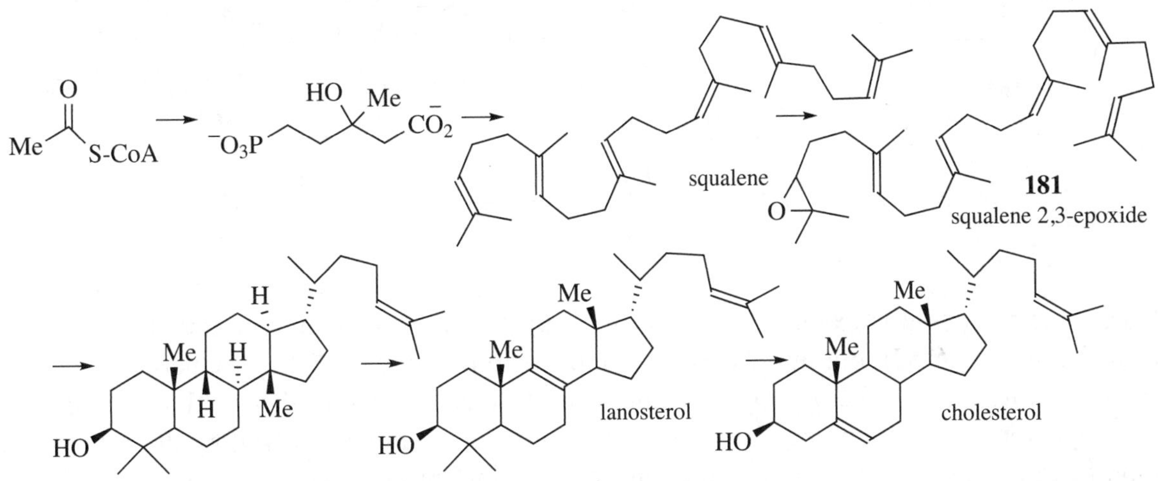

Figure 10.8. The biogenesis of cholesterol from acetyl coenzyme-A.

An all-chair conformation was assumed to be preferred for this polyene system, and an all-trans geometry was required for the double bonds to be in the proper position for cyclization to the proper stereochemistry. In work by Johnson and co-workers, addition of an acid catalyst to the terminal π bond generated a cation from **181**. Each π bond in turn acts as a nucleophile, and attacks the neighboring carbocation generated by the previous addition. Each ring closing reaction proceeds via a *six-center transition state*. In the final step, formation of a five-membered ring should be favored and the process is terminated by an E1 reaction (sec. 2.9.B) to give **182**. If the alkene precursor has an (*E*) geometry, the cyclic product should be the more stable trans-fused ring due to the thermodynamic (equilibrium) nature of the cyclization, which was confirmed when treatment of trienes **183** and **185** with acid gave the same trans-fused decalin, **184**.[134] The cis-fused decalin was not observed in either case.

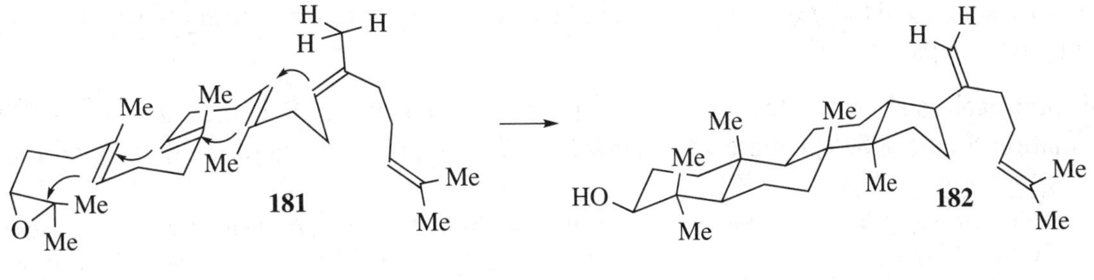

133. (a) Stork, G.; Burgstahler, A.W. *J. Am. Chem. Soc.* **1955**, *77*, 5068; (b) Eschenmoser, A.; Ruzicka, L.; Jeger, O.; Arigoni, D. *Helv. Chim. Acta* **1955**, *38*, 1890.
134. Stadler, P.A.; Nechvatal, A.; Frey, A.J.; Eschenmoser, A. *Helv. Chim. Acta* **1957**, *40*, 1373.

When this process was extended to tetraene (**186**),[135] treatment with acid gave the expected tricyclic alcohol (**187**), but in only 5-10% yield. The remainder of the product mixture was an intractable oil. As the number of rings being formed increased, the yield and efficiency of the polyene cyclization decreased. This problem was eventually solved by three synthetic modifications: (*1*) incorporation of one pre-formed ring into the polyene system to stabilize the conformation, (*2*) incorporation of a more reactive center at the terminus, and (*3*) use of a milder and more selective acid. The planned cyclization is a Level 1 intramolecular cyclization, as described by Deslongchamps (sec. 10.6.C).[13] In acyclic systems, the conformation required for cyclization may not be the low energy conformation assumed by the polyene. When the polyene is attached to a preformed ring, the conformational stability of that ring must stabilize the entire polyene system in the lowest energy chair conformations required for efficient cyclization. Cyclopentanoid derivatives were found to be very effective. Once the cyclization process has begun, it is important to terminate the cationic process in an efficient manner that minimizes competitive side reactions. Incorporation of an alkyne or a vinyl fluoride at the reaction terminus will give a vinyl cation or an α-fluoro cation, respectively.[136] In both cases, the highly reactive cation product will trap a nucleophile in an S_N1 process, stopping the cyclization process. Finally, acids such as hydrofluoric acid and trifluoroacetic acid were found to be effective catalysts for the cyclization.[133] They are strong acids, but are not oxidizing acids, and the gegenions do not induce secondary reactions during the cyclization or termination steps.

This approach is shown in the synthesis of progesterone (**193**) in Scheme 10.17.[137] A Wittig olefination[138] (sec. 8.8.A) coupled **188** with **189**. The improved technology developed for the

135. (a) Reference 133b, p 3 and reference 6 cited therein; (b) Eschenmoser, A.; Felix, D.; Gut, M.; Meier, J.; Stadler, P. in *CIBA Foundation Symposium on the Biosynthesis of Terpenes and Steroids*, Wolestenholme, G.E.W.; O'Connor, M. (Eds.), J and A Churchill Ltd., London, *1959*.
136. (a) Johnson, W.S.; McCarry, B.E.; Markezich, R.L.; Boots, S.G. *J. Am. Chem. Soc.* *1980*, *102*, 352; (b) van Tamelen, E.E. Loughhead, D.G. *Ibid.* *1980*, *102*, 869; (c) Hoye, T.R.; Kurth, M.J. *Ibid.* *1979*, *101*, 5065.
137. Johnson, W.S.; Gravestock, M.B.; McCarry, B.E. *J. Am. Chem. Soc.* *1971*, *93*, 4332.
138. (a) Wittig, G.; Schöllkopf, U. *Berichte* *1954*, *87*, 1318; (b) Wittig, G.; Haag, W. *Ibid.* *1955*, *88*, 1654; (c) Trippett, S. *Quart. Rev. Chem. Soc.* *1963*, *17*, 406; (d) Wittig, G. *Acc. Chem. Res.* *1974*, *7*, 6.

cyclization led to a good yield of **192** from **191**. Polyene **191** was prepared by an intramolecular aldol condensation (sec. 9.4.A.ii) after hydrolysis of the dioxolane protecting groups in **190** (sec. 7.3.B.i).

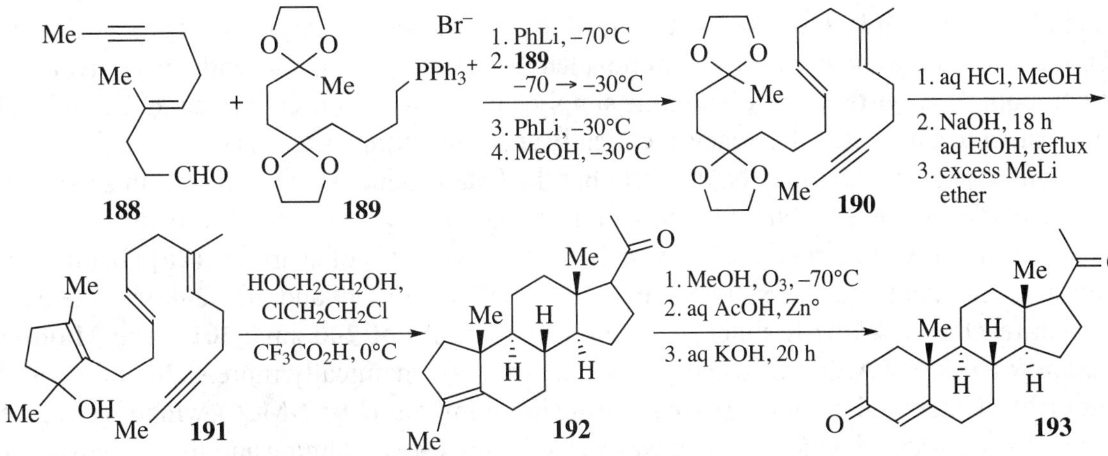

Scheme 10.17. Johnson and Co-Worker's Synthesis of Progesterone.

Choice of reagents and reaction conditions are critical for success in polyene cyclizations. The exact cyclization conditions are often specific to the particular target. It is also clear that to accomplish chemically what enzymes do routinely presents many problems, often in the areas of regio- and stereocontrol. Only if these problems can be overcome by suitable choice of reagents and conditions, can a biogenetic route be used successfully. An enantioselective polyene cyclization usedg $SnCl_4$ mixed with BINOL derivatives[139] in which (–)-ambrox (**195**), the most important commercial substituent for ambergris used in the perfume industry, was prepared from triene **194**. Tricyclic compound **195** was obtained in 30% yield, but with only modest enantiopurity (42% ee). Three other tricyclic compounds were obtained as well (total yield of cyclized products was 54%). When the polyene was anchored to an aromatic ring, however, this approach gave good yields of polycyclic compounds with good enantioselectivity (62-87% ee).

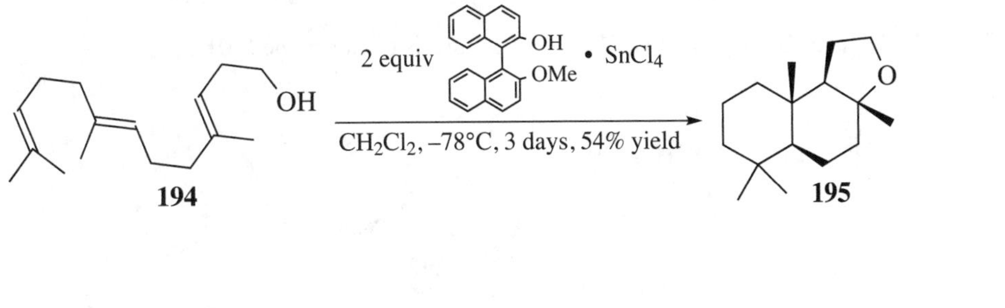

---

139.   Ishihara, K.; Nakamura, S.; Yamamoto, H. *J. Am. Chem. Soc.* **1999**, *121*, 4906.

10.8.B. Sparteine

van Tamelen and co-worker's synthesis of sparteine (**201**), outlined in Scheme 10.18,[140] is another example that follows a known biosynthetic pathway. In this case, conversion of diaminoketoaldehyde **198** to 8-ketosparteine (**200**) was the key biosynthetic step. van Tamelen did not use the labile and unavailable aminoaldehyde (**198**), but rather a synthetic equivalent of this compound. A **synthetic equivalent** is simply a molecule that is converted to the molecule of interest, or to a molecule that reacts identically and is later unmasked. The chemical structure of the synthetic equivalent may be different but the final product is the same. In this example, the synthetic equivalent of **198** is the bis(iminium) ketone **197**, generated from amino-ketone **196** by treatment with mercuric salts. van Tamelen's synthesis of sparteine (**201**) begins with piperidine. Generation of the synthetic equivalent **197** led to **199** and then **200** via sequential Mannich reactions. Wolff-Kishner reduction (sec. 4.10.A) of **200** gave **201**. The **Mannich reaction**[141] (secs. 9.7.A, 9.7.C) usually gives the thermodynamically more stable product via addition of an intermediate enolate (or enol) to an iminium salt ($H_2C=NMe_2^+$), which is produced *in situ*. When 2-methylcyclohexanone was treated with dimethylamine and formaldehyde, for example, both the kinetic product (**203**) and the thermodynamic product (**202**) were formed, with **202** predominating.

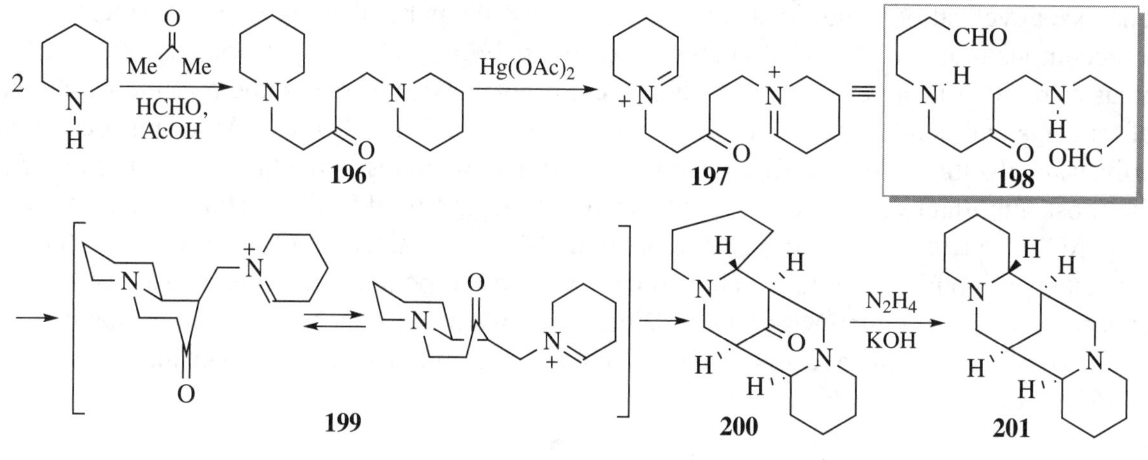

Scheme 10.18. van Tamelen and co-Worker's Synthesis of Sparteine (**201**).

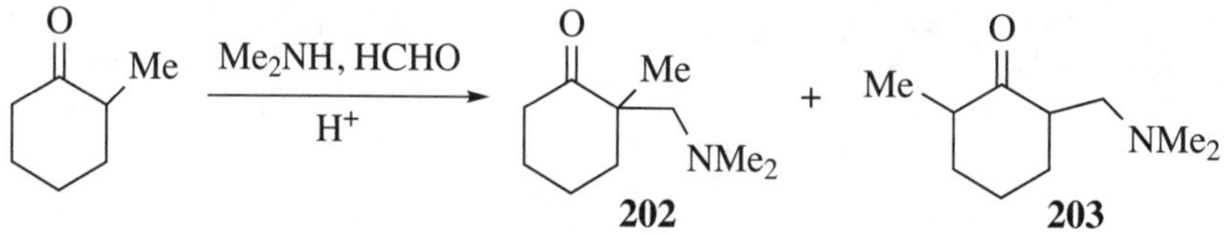

140. van Tamelen, E.E.; Foltz, R.L. *J. Am. Chem. Soc.* **1960**, *82*, 2400.
141. (a) Blicke, F.F. *Org. React.* **1942**, *1*, 303; (b) Maxwell, C.E. *Org. Syn. Coll. Vol. 3* **1958**, 305.

10.8.C. Deoxyerythronolide B

The final example illustrates yet another use of the biomimetic approach. Figure 10.9[142] shows a retrosynthetic representation for Masamune's synthesis of deoxyerythronolide B (**204**). The biosynthetic building blocks of this and other macrolide antibiotics are known to be acetate and/or propionate units combined head-to-tail, as seen in **204**.[143] The stereocenters in **204** are clearly shown in the acyclic (seco acid) form of the macrolide (**206a**). The specific biopathway is not utilized but rather modified to include the basic building blocks, seven propionate units (**bold lines** in **204**).[146] Seco acid **207** (X = OH) was constructed by sequential aldol condensation reactions (sec. 9.4.A) of propionaldehyde units, as shown by the disconnections in Figure 10.9. The chiral boron enolates derived from propanal control the symmetric induction. The diastereoface selectivity of the aldol condensation was discussed in Section 9.5.B.[144] Both (R) and (S) boron enolates of propionaldehyde (**205a** and **205b**) were prepared and used as a fundamental biomimetic building block. Sequential aldol condensations generate the erythronolide skeleton, with the correct stereochemistry predominating in each aldol product. The targeted macrolide[146] was generated from the acyclic seco acid thioester (**206b**) via treatment with copper triflate and Hünig's base (diisopropylethyl amine). This and other techniques for macrolactonization[145] were discussed in Section 6.6.B.

204   **205** (a) (R) (b) (S)   **206** (a) X = OH (b) X = t-Bu

142. (a) Masamune, S.; Choy, W.; Kerdesky, F.A.J.; Imperiali, B. *J. Am. Chem. Soc.* **1981**, *103*, 1566; (b) Masamune, S.; Hirama, M.; Mori, S.; Ali, Sk.A.; Garvey, D.S. *Ibid.* **1981**, *103*, 1568 (c) Kaneda, T.; Butte, J.C.; Taubman, S.B.; Corcoran, J.W. *J. Biol. Chem.* **1962**, *237*, 322.

143. (a) Birch, A.J.; Djerassi, C.; Dutcher, J.D.; Majer, J.; Perlman, D. Pride, E.; Rickards, R.W.; Thomson, P.J. *J. Chem. Soc.* **1964**, 5274; (b) Manwaring, D.J.; Rickards, R.W.; Guadiano, G.; Nicolella, V. *J. Antibiot.,* **1969**, *22*, 545; (c) Corcoran, J.W.; Chick, M. *Biochemistry of the Macrolide Antibiotics*, Ed. Snell, J.F. Academic Press, **1966**, p 159;(d) Birch, A.J.; Pride, E.; Rickards, R.W.; Thomson, P.J.; Dutcher, J.D.; Perlman, D.; Djerassi. C. *Chem. Ind., (London) 1960*, 1245.

144. (a) Evans, D.A.; Takacs, J.M.; McGee, L.R.; Ennis, M.D.; Mathre, D.J. Bartroli, J. *Pure Appl. Chem.* **1981**, *53*, 1109; (b) Evans, D.A.; McGee, L.R. *J. Am. Chem. Soc.* **1981**, *103*, 2876; (c) Evans, D.A.; Nelson, J.V.; Vogel, E.; Taber, T.R. *Ibid.* **1981**, *103*, 3099 and references cited therein; (d) Heathcock, C.H. *Science* **1981**, *214*, 395 and references cited therein.

145. (a) Back, T.G. *Tetrahedron* **1977**, *33*, 3041; (b) Masamune, S.; Bates, G.S.; Corcoran, J.W. *Angew. Chem. Int. Ed.* **1977**, *16*, 585.

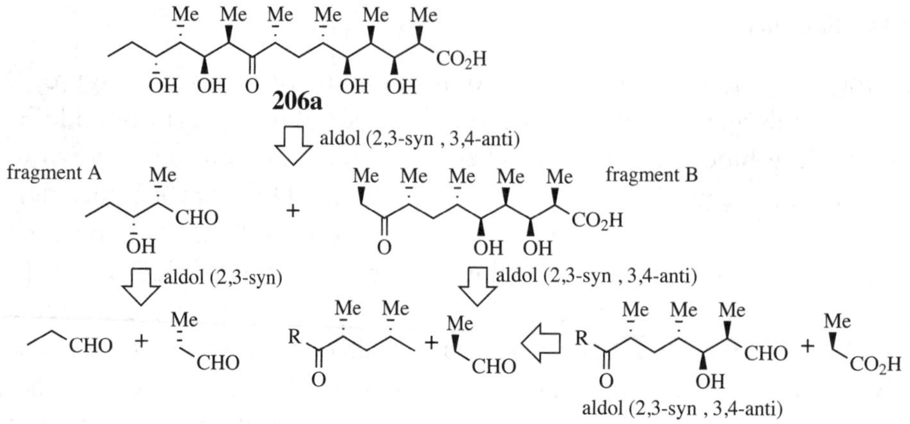

Figure 10.9. Retrosynthetic Analysis of the Seco Acid of Deoxyerythronolide B.

The synthetic examples in this section show that a biosynthetic pathway can be used to generate a synthetic tree, which can be of great utility and simplifies the task of growing the synthesis tree. If the biosynthesis is unknown one can turn to other processes for guidance.

10.9.   THE CHIRAL TEMPLATE APPROACH

In the *synthon approach*,[1] a target was disconnected to give a suitable starting material, usually without prior knowledge of the identity of that compound. An alternative strategy generates strategic bond disconnections by "locating segments containing a number of chiral centers".[146] The fragments containing these stereogenic centers are part of a structure (a **chiron**, which is a **chir**al synth**on**), which will be a fragment of or related to a naturally occurring and/or readily available chiral material.[147] Such synthons are now commonly referred to as chiral templates. A **chiral template** is chosen by examination of the structure of a target for symmetry, chirality and functionality. This information is then *decoded* and transposed onto a carbon framework of a suitable synthetic precursor (the chiron or template). This has been called the **chiron approach**, but is more commonly referred to as the **chiral template approach**. The chiron is obtained from chiral starting materials (chiral templates) by systematic functionalization.[151a] This process is illustrated in Figure 10.10,[151a] where a target (**207**) is disconnected to a suitable chiron (**208**). The chiron is prepared from one of several chiral templates. Hanessian et al. analyzed a chiron-based synthesis in Figure 10.11,[151a] in a general sense, and many of the elements found in Corey's various strategies are also found here. Hannessian also discusses the fundamental difference between Corey's synthon approach and the chiron approach. In Figure 10.12[151a] the synthon approach was applied to a target, which was disconnected to a diene and a carbene.[151a] The chiron disconnection led to a chiron (**210**) that is available from

146.   Hanessian, S., *Total Synthesis of Natural Products: The 'Chiron' Approach*, Pergamon Press, *1983*, p 22.
147.   (a) Hanessian, S.; Franco, J.; Larouche, B. *Pure Appl. Chem. 1990, 62*, 1887; (b) Reference 150, p 21.

a suitable chiral template (**209**).[151a] The synthon approach can be correlated with "the type of functionality present in the target molecule and the chemical feasibility or precedent that dictates the strategy. The chiron approach to synthesis involves disconnection of strategic bonds in a target molecule, with minimal perturbation of stereogenic centers. A maximum overlap of functionality, as well as stereochemical and carbon framework relationships between target (or substructure) and the chiron is ideally sought. It is the type of substructure, and its possible chiral progenitor that dictate the strategy and chemistry to be carried out.[148]

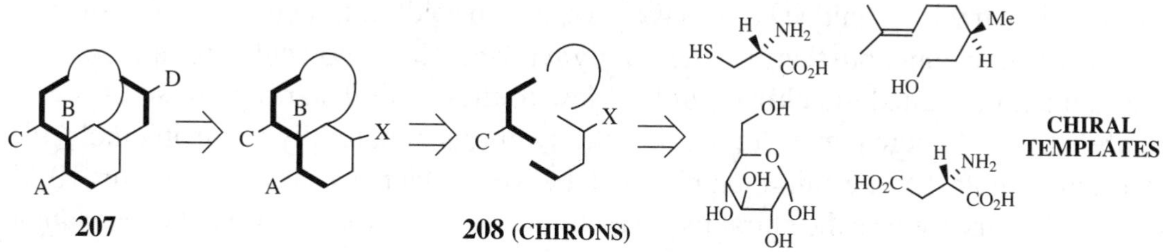

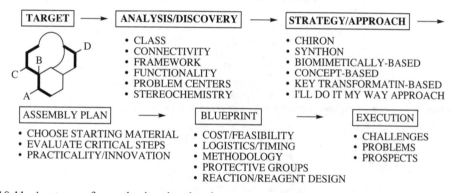

Figure 10.10. Model for total synthesis using the *chiron approach*. [Reprinted from Hanessian, S.; Franco, J.; Larouche, B. *Pure. Appl. Chem.* ***1990****, 62*, 1887. Copyright ***1990*** IUPAC.]

TARGET →	ANALYSIS/DISCOVERY →	STRATEGY/APPROACH →
	• CLASS • CONNECTIVITY • FRAMEWORK • FUNCTIONALITY • PROBLEM CENTERS • STEREOCHEMISTRY	• CHIRON • SYNTHON • BIOMIMETICALLY-BASED • CONCEPT-BASED • KEY TRANSFORMATIN-BASED • I'LL DO IT MY WAY APPROACH

ASSEMBLY PLAN →	BLUEPRINT →	EXECUTION
• CHOOSE STARTING MATERIAL • EVALUATE CRITICAL STEPS • PRACTICALITY/INNOVATION	• COST/FEASIBILITY • LOGISTICS/TIMING • METHODOLOGY • PROTECTIVE GROUPS • REACTION/REAGENT DESIGN	• CHALLENGES • PROBLEMS • PROSPECTS

Figure 10.11. Anatomy of a synthesis using the *chiron approach*. [Reprinted from Hanessian, S.; Franco, J.; Larouche, B. *Pure. Appl. Chem.* ***1990****, 62*, 1887. Copyright ***1990*** IUPAC.]

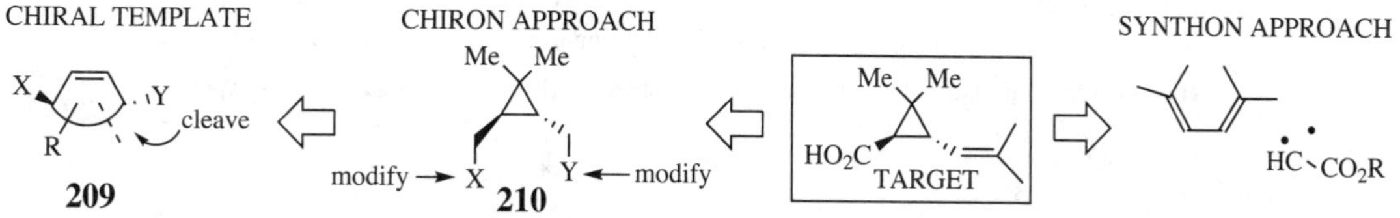

Figure 10.12. A comparison of the *synthon approach* and the *chiron approach*. [Reprinted from Hanessian, S.; Franco, J.; Larouche, B. *Pure. Appl. Chem.,* ***1990****, 62*, 1887. Copyright ***1990*** IUPAC.]

The distinguishing feature of the chiral template approach is recognition of the structural elements of a particular compound within the target structure. The retrosynthetic scheme is

---

148.   Reference 150, p 4.

then biased so that the selected molecule will be the starting material. This leads to a single branch of the synthesis tree rather than analysis of several branches.

There may be several chiral templates for a given chiron, so there is flexibility in the choice of which branch is chosen. This approach is analogous to Corey's starting material based strategy previously discussed in Section 10.4.[63] If the recognized starting material is readily available as a single enantiomer (see Figure 10.13),[151a] the chirality of this compound can be transferred to the final target without a resolution step. This is clearly of great value and the main driving force for exploitation of this technique. Carbohydrates, amino acids or terpenes are a few of the many classes of organic compounds that provide readily available chiral molecules, usually from natural sources sometimes called the **chiral pool**. A few of these chiral starting materials are shown in Figure 10.13,[151a] where they were used for the synthesis of naturally occurring molecules. It is important to note that the chiral template can be well hidden, as shown in Figures 10.13 and 10.14.[151a] (S)-Carvone was the chiral template for a disconnection analysis of eucannabinolide (**211**), and (+)-camphor was the chiral template for pheromone **212**.[151a] Some searching is required to discover the 'hiding place' of these chiral templates in the target (see Figure 10.14).

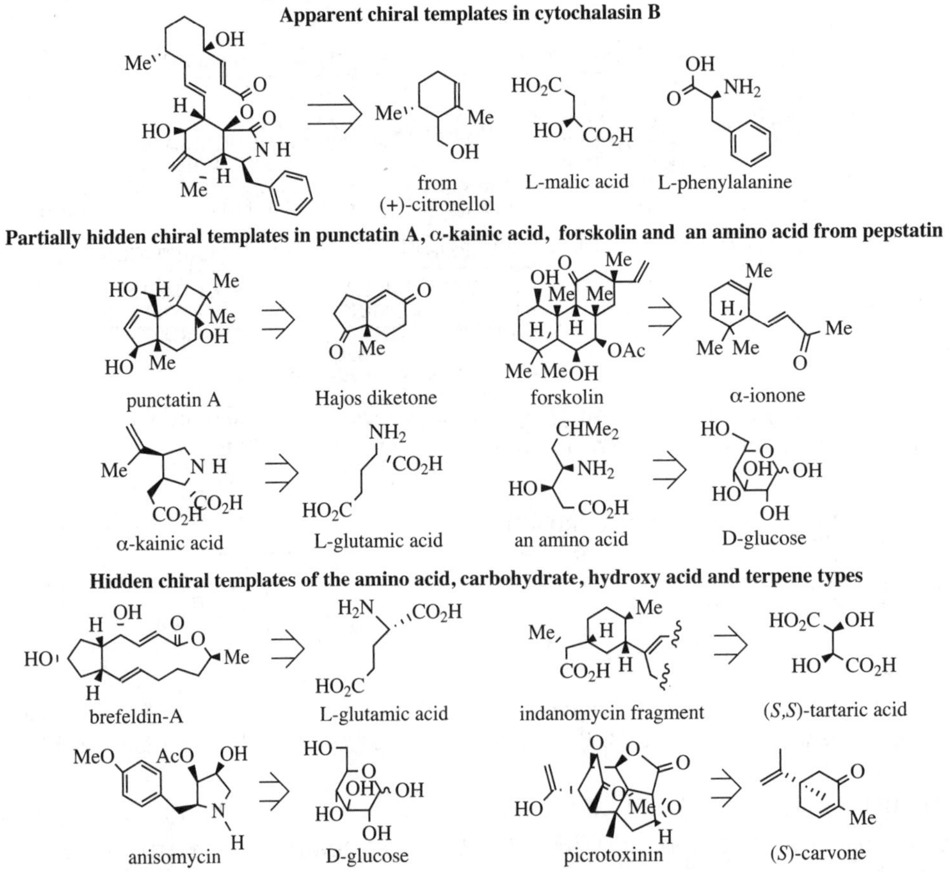

Figure 10.13. Hidden chiral templates. [Reprinted from Hanessian, S.; Franco, J.; Larouche, B. *Pure. Appl. Chem.*, *1990*, *62*, 1887. Copyright *1990* IUPAC.]

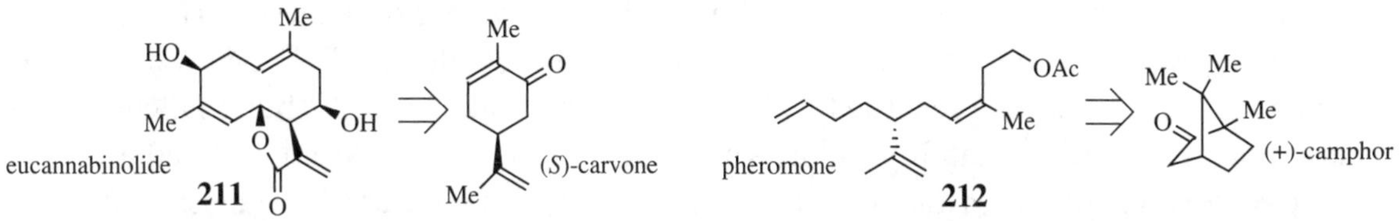

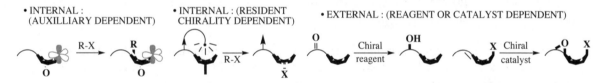

Figure 10.14. Very hidden chiral templates. [Reprinted from Hanessian, S.; Franco, J.; Larouche, B. *Pure. Appl. Chem.*, *1990*, *62*, 1887. Copyright *1990* IUPAC.]

In general, Hanessian discusses three types of asymmetric induction that are pertinent to all syntheses, not only those to which the chiron approach will be applied (see Figure 10.15).[151a] The first type of asymmetric induction is auxiliary dependent (a chiral auxiliary is attached to an achiral substrate, used to transfer asymmetry during the reaction and later removed). The internal (resident chirality) dependent form of asymmetric induction relies on a chiral center being part of the reactant. The external type of asymmetric induction makes use of a chiral reagent or a chiral catalyst. Auxiliaries were used in Chapter 9 for asymmetric aldol condensations (sec. 9.5). Internal asymmetric induction was a key element of Cram's rule as applied to reductions (sec. 4.7.B) and Grignard addition to carbonyl derivatives (sec. 8.4.F.ii), as well as stereodifferentiation reactions related to the **aldol condensation** (see sec. 9.5.B). Chiral reagents and chiral catalysts were used in asymmetric reductions (sec. 4.7.A), in asymmetric Grignard reagents (sec. 8.4.F), and in organocuprate reactions (sec. 8.7.A.vii).

• INTERNAL :
(AUXILLIARY DEPENDENT)   • INTERNAL : (RESIDENT
CHIRALITY DEPENDENT)   • EXTERNAL : (REAGENT OR CATALYST DEPENDENT)

Figure 10.15. Types of asymmetric induction. [Reprinted from Hanessian, S.; Franco, J.; Larouche, B. *Pure. Appl. Chem.*, *1990*, *62*, 1887. Copyright *1990* IUPAC.]

Carbohydrates are particularly useful chiral templates.[149] The pyranose or furanose form of a sugar can be opened to give a chiral acyclic fragment. Functional groups can be systematically replaced with others, with a high degree of stereocontrol. In Figure 10.16[150] an analysis of erythronolide A (**213**) reveals two carbohydrate chirons, **214** and **215**. Various synthetic manipulations of the sugar moiety are also shown, eventually leading to a recognizable sugar (D-glucose) protected as the *O*-methyl derivative.

Fraser-Reid suggested that the two most valuable advantages offered by carbohydrate derivatives for organic synthesis are (*1*) highly stereoselective reactions resulting from conformational bias and (*2*) ready proof of molecular stereochemistry.[151] Synthetic targets with two to four

149.   Hanessian, S. *Acc. Chem. Res.* *1979*, *12*, 159.
150.   Reference 150, p 34.
151.   Anderson, R.C.; Fraser-Reid, B. *J. Am. Chem. Soc.* *1975*, *97*, 3870.

contiguous chiral centers can be constructed from chiral pyranose derivatives. Higher carbon sugars (with greater than four chiral centers) offer formidable challenges.[152,153] Many macrolides, or segments thereof, may be regarded as higher carbohydrate sugars,[154] but challenges arise in stereoselectivity and availability in known configurations. Fraser-Reid outlined a concept termed **pyranoside homologation** to fill this need.[155,156] The monograph by Hanessian provides an excellent description of this approach, using carbohydrates and many other chiral starting materials.[150] This monograph is strongly recommended to provide a detailed explanation of this approach. In this section, the approach can only be outlined in the context of beginning the construction of a synthetic tree based on the chiral template approach.

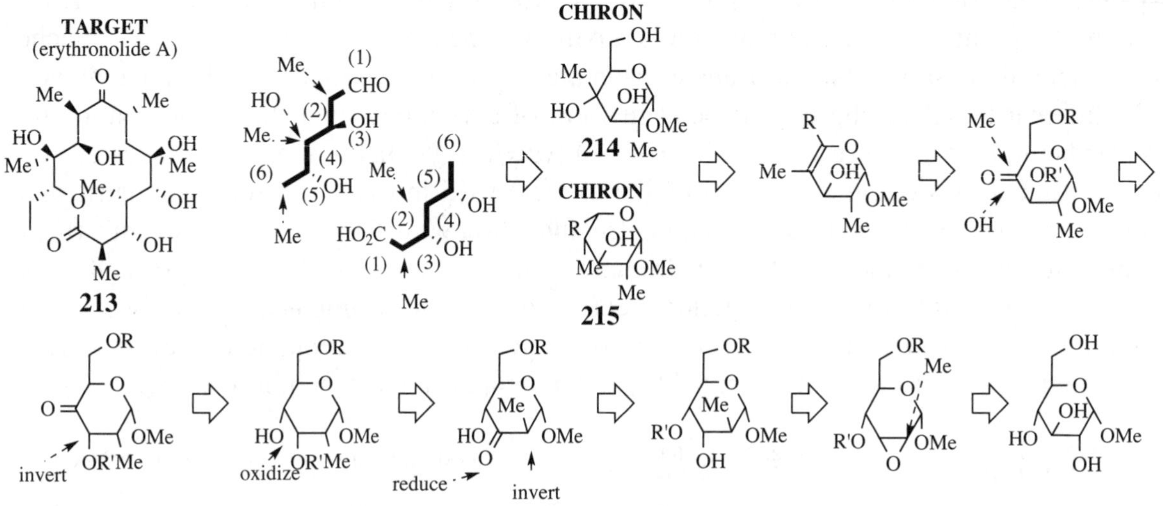

Figure 10.16. Chiron analysis of erythronolide A (**213**). [Reprinted from Hanessian, S., *Total Synthesis of Natural Products: The Chiron Approach*, Pergamon Press, *1983*, p. 34 by courtesy of Professor Stephen Hannesian. Copyright 1983, with permission from Elsevier Science]

Takahashi and Nakata's retrosynthesis of mucocin (**216**) in Scheme 10.19[157] clearly shows the disconnection roadmap has been biased toward three carbohydrates, D-galactose (**221**), 2,5-anhydro-D-mannitol (**222**) or L-rhamnose (**224**). Initial disconnection to two fragments (**217** and **218**) shows the convergent nature of the synthesis. Fragment **217** is disconnected to give two fragments, **219** and **220**, and these fragments are derived from two of the sugar

152. Fraser-Reid, B.; Magdzinski, L.; Molino, B.F.; Mootoo, D.R. *J. Org. Chem.* *1987*, *52*, 4495.
153. (a) Secrist, J.A.; Wu, S.-R. *J. Org. Chem.* *1979*, *44*, 1434; (b) Secrist III, J.A.; Barnes, K.D. *Ibid.* *1980*, *45*, 4526; (c) Brimacombe, J.S.; Kabir, A.K.M.S. *Carbohydr. Res.* *1986*, *152*, 329, 335; (d) Danishefsky, S.J.; Larson, E.; Springer, J.P. *J. Am. Chem. Soc.* *1985*, *107*, 1274.
154. Celmer, W.D. *Pure Appl. Chem.* *1971*, *28*, 413.
155. (a) Molino, B.F.; Magdzinski, L.; Fraser-Reid, B. *Tetrahedron Lett.* *1983*, *24*, 5819; (b) Magdzinski, L.; Cweiber, B.; Fraser-Reid, B. *Ibid.* *1983*, *24*, 5823; (c) Fraser-Reid, B.; Magdzinski, L.; Molino, B. in *Current Trends in Organic Synthesis*, Nozaki, H. (Ed.), Pergamon, New York, *1983*.
156. Fraser-Reid, B.; Magdzinski, L.; Molino, B. *J. Am. Chem. Soc.* *1984*, *106*, 731.
157. Takahashi, S.; Nakata, T. *J. Org. Chem.* *2002*, *67*, 5739.

precursors. Fragment **218** is derived from **223** by a series of steps, and this key fragment (**223**) is prepared from rhamnose. In the actual synthesis, significant functional group modification was required to convert each sugar into the appropriate synthetic fragment, which is both the major advantage and the major disadvantage of the method. It is a disadvantage in that the chiron approach does not necessarily lead to shorter syntheses, since the synthetic tree is pruned to give a single option. Persistence in a single approach may add several extra steps in functional group transformations. It may also lead to more difficult carbon-carbon bond forming reactions. There is usually sufficient pliability in the synthetic tree, however, that a reasonable pathway can be found that will accommodate an available chiral template without undue cost in time or synthetic difficulties. The advantage is that the asymmetry in the starting materials leads to increased enantioselectivity in all reactions, which usually means that no resolution steps are necessary.

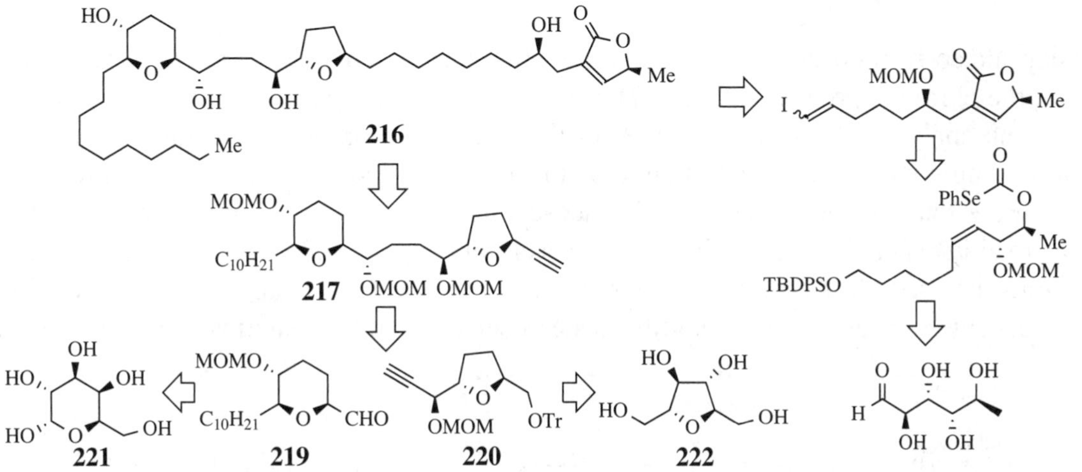

Scheme 10.19. Chiral template-based retrosynthesis of mucocin (**216**). [Reprinted with permission from Takahashi, S.; Nakata, T. *J. Org. Chem.* **2002**, *67*, 5739, Copyright 2002, with permission from the American Chemical Society.]

Any chiral molecule available in Nature or other sources can, in principle, be used in a synthesis. Scheme 10.20[158] shows Srikrishna's synthesis of (–)-neopupukean-10-one (**229**) from the chiral template (*R*)-carvone (**225**). Formation of the enolate anion at C6 and 1,4-addition to the conjugated ester, followed by a second conjugate addition of the resulting anion to the conjugated ketone at C3 leads to **226** after saponification of the ester. Formation of the acid chloride and reaction with diazomethane gave diazoketone **227**, and rhodium catalyzed insertion (see sec. 13.9.C.iii) closed the five-membered ring, and hydrogenation of the alkene unit gave **228**. Conversion of the less hindered ketone to the dithiolane and reductive cleavage with Raney nickel gave the target **229**. The carbon atoms in the template (**225**) are numbered in **226** and in **229** for convenience.

---

158. Srikrishna, A.; Gharpure, S.J. *Chem. Commun.* **1998**, 1589.

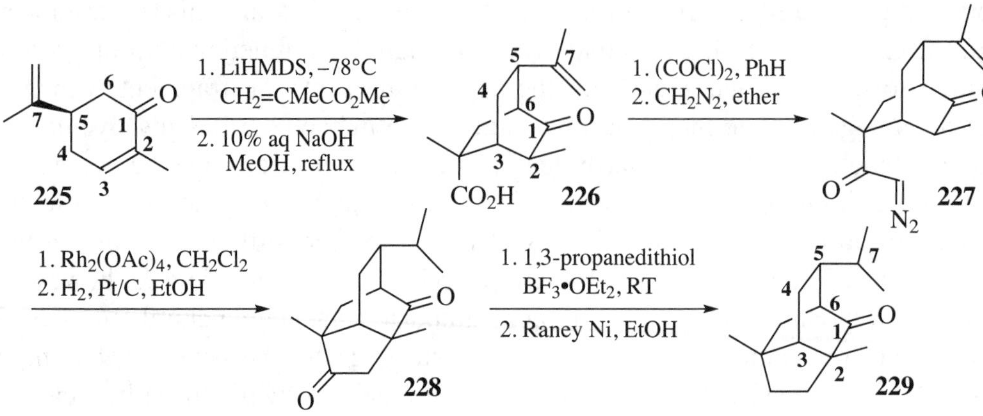

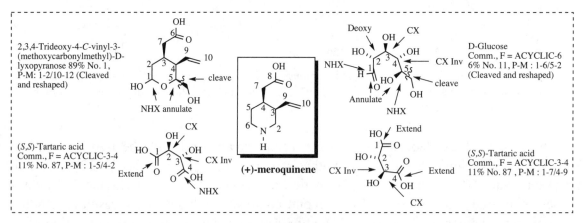

Scheme 10.20. Srikrishna and Gharpure's synthesis of (−)-neopupukean-10-one (**229**).

Hanessian and co-workers developed a computer program that will execute retrosynthetic plans based on the chiron approach, called **CHIRON**.[159] Hanessian synthesized (+)-meroquinene based on the analysis shown in Figure 10.17,[163c] where the computer generated four chiral templates along with notations as to how they must be modified for conversion to a requisite chiron. In the total synthesis, tetraacetyl glucose was chosen as the chiral template starting material, readily available from D-glucose.[160] In a chiral template approach, not all of the chiral centers must be used. It is possible to use only one or two without changing the target, while other chiral centers are generated by highly diastereoselective and/or enantioselective reactions.

Figure 10.17. CHIRON analysis of meroquinene. [Reprinted with permission from Hanessian, S.; Faucher, A.-M.; Léger, S. *Tetrahedron* **1990**, *46*, 231, Copyright 1990, with permission from Elsevier Science.]

159. (a) Reference 150, pp 9-22; (b) Hanessian, S.; Franco, J.; Gagnon, G.; Laramée, D.; Larouche, B. *J. Chem. Int. Compt. Sci*, **1990**; *30*, 413; (c) Hanessian, S.; Faucher, A.-M.; Léger, S. *Tetrahedron* **1990**, *46*, 231; (d) Hanessian, S.; Sakito, Y.; Dhanoa, D.; Baptistella, L. *Ibid.*. **1989**, *45*, 6623.
160. Rao, D.R.; Lerner, L. *Carbohydr. Res.* **1972**, *22*, 345.

10.10. COMPUTER GENERATED STRATEGIES

In Section 10.4.B a computer based method for generating retrosyntheses was presented called LHASA, and Section 10.9 mentioned the computer program CHIRON. Computer methods are generically known as computer-assisted organic synthesis. This section will present two such approaches, the SYNGEN approach of Hendrickson, and the MARSEIL-SOS approach of Barone. Later generation programs are also available[161] but this section will use the initially reported programs to introduce the fundamental approach.

10.10.A. Hendrickson's SYNGEN Approach

The strategic bond approach often leads to consecutive type syntheses, although disconnection of two or more strategic bonds can easily generate convergent pieces. The strategic bond approach generates intermediates on the synthetic tree in an interactive manner that leaves selection to the chemist.[162] The interactive-decision making aspects of the strategic bond approach maximizes opportunities to create new methodology and strategy. If the aim, as stated by Deslongchamps,[13] is to develop new methodology and chemistry as a lasting achievement then this is an attractive approach. If direct synthesis of a specific target by the most efficient route is more important than one that develops new methodology, then alternative strategies may be desirable. The interactive approach tends to focus on the first levels of the synthetic tree and does not allow the chemist to see which retrosynthetic start leads to early discovery of starting materials and so the shortest pathway.[166]

Hendrickson pointed out that the stepwise backward protocol described in the previous sections suffers from several problems:[163] (*1*) The procedure creates too many intermediates to manage and most must be deleted without knowing if they will lead to more efficient syntheses. (*2*) There is no method for predicting yield. (*3*) The procedure requires an extensive database of transforms and reactions. (*4*) The focus of a transform is on the functional groups and those reactions that will completely remove a functional group are not discernible by this method. Estrone (**232**), for example, is easily obtained from **230** via stepwise reduction.[164,165] Initial reduction of **230** gives **231**, which was converted to **232**. This sequence is not likely to be

161. Ihlenfeldt, W.-D.; Gasteiger, J. *Angew. Chem. Int. Ed.* **1995**, *34*, 2613.
162. Hendrickson, J.B. *Acc. Chem. Res.* **1986**, *19*, 274.
163. (a) Hendrickson, J.B.; Berstein, Z.; Miller, T.M.; Parks, C.; Toczko, A.G. Chapter 6 in ACS Symposium Series, No. 408, Hohne, B.; Pierce, T. (Eds.), American Chemical Society, Washington, DC, **1989**, pp 62-81; (b) Hendrickson, J.B. *J. Chem. Inf. Compt. Sci.* **1979**, *19*, 129.
164. Ananchenko, S.N.; Torgov, I.V. *Tetrahedron Lett.* **1963**, 1553.
165. (a) Smith, H.; Hughes, G.A.; McLoughlin, B.J. *Experientia* **1963**, *19*, 177; (b) Smith, H.; Hughes, G.A.; Douglas, G.H.; Hartley, D.; McLoughlin, B.J.; Siddall, J.B.; Wendt, G.R.; Buzby, Jr., G.C.; Herbst, D.R.; Ledig, K.W.; McMenamin, J.R.; Pattison, T.W.; Suida, J.; Tokolics, J.; Edgren, R.A.; Jansen, A.B.A.; Gadsby, B.; Watson, D.H.R.; Phillips, P.C. *Ibid.* **1963**, *11*, 394; (c) Douglas, G.H.; Graves, J.M.H.; Hartley, D.; Hughes, G.A. McLoughin, B.J.; Siddall, J.; Smith, H. *J. Chem. Soc.* **1963**, 5072.

derived from LHASA or the strategic bond approach since there are no functional groups that could be correlated with the pertinent transforms.

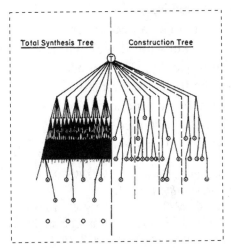

In a complex synthetic tree (see Figure 10.18),[166] which pathway is the best? As stated by Hendrickson, "selection, not generation, is the central problem."[167] The tree must be simplified and then subdivided. Assume that the retrosynthesis to a given target has five levels (five disconnect products). One disconnection may generate 30 reactions that can synthesize the target. If each of the disconnect targets is just as complex, the second disconnection will also generate ~ 30 reactions. If the synthesis is taken to five levels in the synthesis tree, each with the same 30-reaction complexity, there are 30^5 (24.3 million) routes.[167] If the reaction sequence has 25 steps, an already unmanageable problem becomes worse. If the number of synthetic routes at each disconnection stage were only three, a 25-step synthesis would have 25^3 (15,625) possibilities, still rather unmanageable. Hendrickson reported of four methods to simplify the synthesis tree:[170] (1) Systematize structures and reactions into digital form for a computer. (2) Simplify the tree by "condensing trivial disconnections into familial generalizations". (3) Subdivide the tree into independent subtrees that have no reactions in common, and analyze each subtree separately. (4) Select the optimal pathways.

Figure 10.18. The synthesis tree. [Reprinted with permission from Hendrickson, J.B. *Acc. Chem. Res.* *1986*, *19*, 274. Copyright © 1986 American Chemical Society.]

166.   Hendrickson, J.R. *J. Am. Chem. Soc.* *1977*, *99*, 5439.

The process begins with a search for key reactions, the construction reactions (carbon-carbon bond-forming reactions) that will assemble the target skeleton from the skeletons of the starting materials. Initial consideration affords only a major simplification of the total synthesis tree to generate a construction tree. The construction tree contains only the key constructions that assemble the target skeleton from starting skeletons.[166] Each full synthetic sequence is a separate construction plan,[166] separated by the dashed lines in Figure 10.18, and each plan is an independent smaller tree that may be examined separately for its detailed chemistry.[166] This is systematized by breaking the target into smaller fragments. The simplest gross definition of a particular synthesis is the **bondset**-the set of all bonds constructed in a synthesis (γ in number).[167] The concept of bondset is illustrated by three different disconnection schemes[171] for the same steroid: **233A** ($\gamma = 3$),[168] **233B** ($\gamma = 6$),[169] and **233C** ($\gamma = 8$).[170]

241A   **241B**

241C

[Reprinted with permission from Hendrickson, J.B. *Accounts Chem. Res.* *1986*, *19*, 274. Copyright © 1986 American Chemical Society.]

Such a bondset, once defined, implies a multistep sequence passing through the tree from starting material to target, seeing both at once.[171] The bondset is a skeletal conception and immediately defines the skeleton of the starting material as well as the sites on each synthon at which construction is to occur (heavy dots), that is, the construction sites.[171] Each bondset defines an independent subtree of synthetic sequences, all of which construct only that particular set of γ bonds in the target. In general, there are 8! ordered bondsets for a given molecule, although relatively few are efficient. Hendrickson's analysis indicated that the typical total synthesis constructs 25-33% of all skeletal bonds in the target using starting skeletal fragments averaging about three-to-four carbons. That is, for every three or four bonds that are made in

167.   (a) Hendrickson, J.B. *Topics in Current Chemistry, 1976, 62,* 51 (b) Hendrickson, J.B. *J. Am. Chem. Soc. 1975, 97,* 5763.
168.   Anand, N.; Bindra, J.S.; Ranganathan, S. *Art in Organic Synthesis,* Holden Day, *1970,* p 181.
169.   Reference 172, pp 3,4.
170.   Reference 172, p 130.
171.   (a) Velluz, L.; Nominé, G.; Amiard, G.; Torelli, V.; Cérède, J. *C.R. Hebd. Seances Acad. Sci. 1963, 257,* 3086; (b) Velluz, L. Nominé, G.; Mathieu, J.; Toromanoff, E.; Bertin, D.; Vignau, M. Tessier, J. *Ibid. 1960, 250,* 1510; (c) Velluz, L.; Nominé, G. Mathieu, J.; Toromanoff, E.; Bertin, D.; Tessier, J.; Pierdet, A. *Ibid. 1960, 250,* 1084.

a synthesis only one skeletal bond is made. Synthons actually used are predominantly acyclic unbranched skeletons of four carbons or less, or aromatic derivatives. The average size of starting materials can be defined as n_o/k for an overall average of 4.0 carbons. The bondset size as a proportion of the total bonds is therefore $\gamma/b_o \approx 0.24$ (the construction ratio).

Hendrickson used two different syntheses of estrone (232) to illustrate this approach and Figure 10.19[171] shows the bondsets, construction plans, and weights (see below) for both. Velluz and co-worker's convergent retrosynthetic strategy is shown in Figure 10.20[58,171] and compred with the closely related syntheses of Torgov and Ananchenko[168] and Smith et al.[169] With Hendrickson's bondset analysis, the 18 carbons of 232 should give an average of six pieces, each of three or more carbons. If a molecule is composed of 15-25 carbons and the construction ratio of ≈ 0.21, a synthesis should require 4-7 constructions.[171] For this example, there are nine bond constructions ($\gamma = 9$). A skeleton of n atoms and r rings has $n_o + r_o - 1$ bonds and $b_o = n_o + r_o - 1$. If there are k pieces then $\gamma = k + r_o - r - 1$. The following definitions apply: b_o = number of C-C single (σ) bonds in the target; n_o = number of carbons in the target; r_o = number of carbocyclic rings in the target; r = number of carbocyclic rings in the starting material; k = number of synthons; γ = number of bonds constructed (bondsets). For Velluz's synthesis, $b_o = 21$, $n_o = 18$, $r_o = 4$, $r = 0$ and $k = 6$. Therefore, $b_o = 18 + 4 - 1 = 21$ and $\gamma = 6 + 4 - 0 - 1 = 9$.

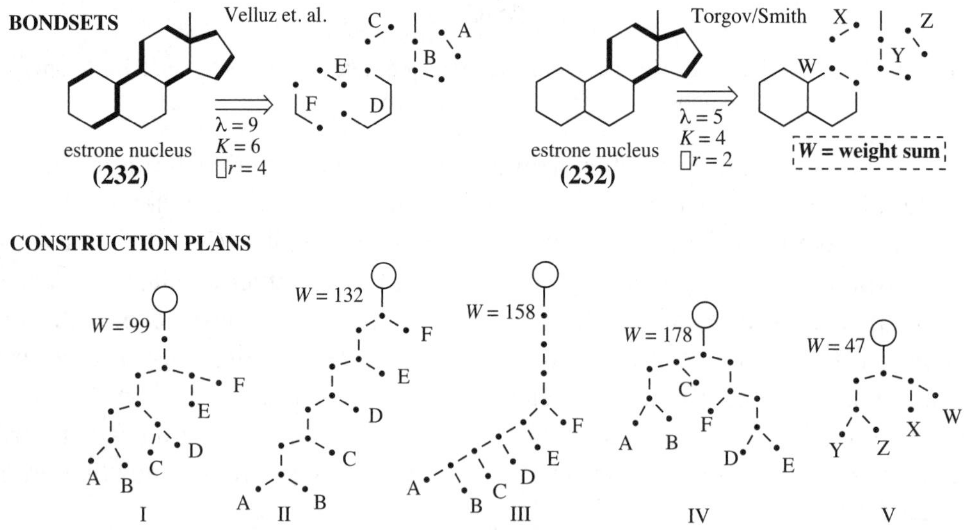

Figure 10.19. Bondsets and Construction Plans for Estrone (232). [Reprinted with permission from Hendrickson, J.B. *J. Am. Chem. Soc.* **1975**, *97*, 5763. Copyright © 1975 American Chemical Society.]

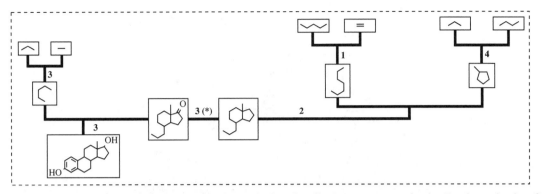

Figure 10.20. A convergent retrosynthetic analysis of Velluz's estrone synthesis. Total synthesis of estradiol, schematic. The figures between the individual stages indicate the numbers of reaction steps. (*) denotes optical resolution at that step. [Reproduced with permission from Velluz, L.; Valls, J.; Mathieu, J. *Angew. Chem. Int. Ed.* **1967**, *6*, 778. Copyright 1967 VCH Weinheim.]

Convergent syntheses disconnect the target into at least two pieces, each of which is further disconnected, which usually leads to a more efficient strategy as with the Torgov-Smith approach. Rather than proceeding backwards one step at a time, the convergent disconnection goes down many levels into the tree, closer to the starting materials to be used. After restricting the disconnection to the convergent type to maximize simplification, the next step is to examine the various functional groups and reactions necessary for the construction sequences. The number of steps should be kept to a minimum. In the estrone synthesis, there are 24 bonds. If these are to be made from one-carbon pieces (the bondset = 1), there are 24! (= 6 x 10²³ different pathways for assembly). The convergent strategy will increase the number of bondsets and decrease the number of convergent routes that can be applied.[167] If the bondset is increased to five members ($\gamma = 5$), there are 20,349 ways to cut estrone, but there are γ! routes (2.5 million ordered bondsets), which is an enormous number, but better than 6 x 10²³.

Hendrickson devised a weighting scheme for each construction tree based on the weights of the several starting materials for each level of the synthesis tree.[166,170] The fundamental idea is to calculate the overall weight of starting materials that will be required to generate a gram of the target. Since these molecular weights are unknown at the time the synthetic fragments are assembled, the number of skeletal carbons in each piece (η_i) and the weight sum (W) gives a measure of the efficiency of the reaction. The reciprocal of the average yield for each step is X, and the number of steps each fragment (i) passes through is l (the level of the starting material in the synthetic tree is i). The quantity W is then given by the expression[167]

$$W = (\Sigma\, \eta_i)(X^{li})$$

Hendrickson provided a table of X^{li} values that are correlated to the number of steps (l, see Figure 10.21).[167] Table 10.3[167] reproduces Hendrickson's table, but the yield has been inserted, remembering that X is the *reciprocal* of the yield. The value of W assumes *no* refunctionalization in the synthesis.

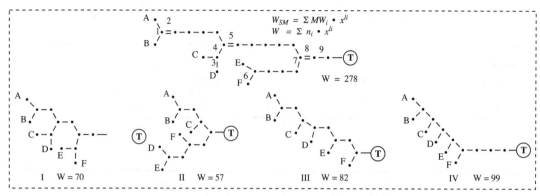

Figure 10.21. Weighting scheme for synthetic tree analyses.

Table 10.3. Values of X^a and the Yield for Syntheses of Steps l

l (No. Steps)	X^1 Approximate[a]	X^1 Actual[a]	% Yield	l (No. Steps)	X^1 Approximate[a]	X^1 Actual[a]	% Yield	
1	1.25	1.25	80	9	7.5	7.45	13.4	
2	1.5	1.56	64.1	10	9.5	9.31	10.7	
3	2	1.95	51.3	11	11.5	11.64	8.6	$^a X = \dfrac{1}{\% \text{ yield}}$
4	2.5	2.44	41.0	12	14.5	14.55	6.9	
5	3	3.05	32.8	13	18	18.19	5.5	
6	4	3.81	26.2	14	23	22.7	4.4	
7	5	4.77	21.0	15	28	28.4	3.5	
8	6	5.96	16.8					

A simple analysis of **234** shows a three-level synthetic tree. The first disconnection ($l = 1$) to **235** (η = number of carbons = 11) and **236** ($\eta = 5$) requires a Diels-Alder disconnection (sec. 11.4.A). The level 2 disconnection to **237** ($\eta = 7$) and **238** ($\eta = 4$) is also a Diels-Alder disconnection. The final disconnection (level 3) generates cyclopentanone (**239**, $\eta = 5$) suggesting a condensation reaction with an organometallic. For this example, the weighted average (W) is given by the following calculation:

$$W = \eta_{235} \cdot X^1 + \eta_{236} \cdot X^1 + \eta_{237} \cdot X^2 + \eta_{238} \cdot X^2 + \eta_{239} \cdot X^3$$
$$W = (11)(1.25) + (5)(1.25) + (7)(1.5) + (4)(1.5) + (5)(2)$$
$$W = 46.5$$

The weighted averages shown in Figures 10.21 and 10.19 were calculated in this manner. In Figure 10.21, synthesis tree II is taken to be the most efficient (lowest W) and represents the *fully convergent* synthesis.

An ideal synthesis in the context of this method requires only sequential construction reactions, with no refunctionalization needed to repair functional groups between constructions.[166] This ideal synthesis was assumed in the weighting scheme discussed above, although it is very difficult to attain as a practical matter. Hendrickson's approach was to develop a numerical description of functionality.[172] The structures become a list of simple numbers, ordered by the numbering of the skeletal carbons. A reaction is expressed by its net structural change, the arithmetic change in the functionality from substrate to product or vice versa.[166] The reaction is represented by a number list for the pertinent carbons that generate the substituent functionality when added to the product list (or *vice versa*).[173] Hendrickson chose to use half-reactions in the actual analysis,[166] and a construction reaction may be seen as two linked half-reactions on each side of the bond-formed.[174,175] Each partial synthon undergoes a half-reaction and each construction half-reaction is characterized by the net structural change relating substrate and product for that particular synthon.[166] The skeletal format for construction of a half-reaction is shown by the relationship of a substrate (**240**) and a product (**241**). Functionality appears on

these carbons and changes from substrate to product in a manner characteristic of a given half-reaction.[167b] A numerical system was developed to describe structures,[156,158] based on four kinds of attachments to any carbon: to hydrogen (H); a σ bond (R); a π bond (Π); and a carbon-heteroatom bond (Z). The functionality is then $\pi + z$ for a given skeleton of known σ bonds and $\underline{h}$ derived by difference. Functionality is, therefore, expressed by z and π.

These concepts led to the computer program **SYNGEN**[167] (**SYN**thesis **GEN**eration).[176,177] As with Corey's LHASA program, inspection of the main principles of the program can offer useful information for a synthesis. It is emphasized that this section has only touched the surface of Hendrickson's detailed and comprehensive analysis. Once understood, any target can be analyzed in detail to provide a synthetic tree. The purpose of this discussion is to give general concepts that may be applied without a computer analysis or dependence on any one approach. With this in mind, SYNGEN has two parts. The first is a skeletal dissection of the target and the second is generation of functionality necessary to assemble the synthesis. There are several general rules for design of an optimal synthesis, as described by Hendrickson:[167]

1. Minimize refunctionalization.

2. Refunctionalize early.

172. (a) Hendrickson, J.B. *J. Am. Chem. Soc.* **1971**, *93*, 6847, 6854; (b) Hendrickson, J.B. *J. Chem. Educ.* **1978**, *55*, 216.
173. For retrieval of reactions from databases, see Hendrickson, J.B.; Miller, T.M. *J. Org. Chem.* **1992**, *57*, 988.
174. Hendrickson, J.B.; Grier, D.L.; Toczko, A.G. *J. Am. Chem. Soc.* **1985**, *107*, 5228.
175. For a description of reactions, see Hendrickson, J.B. *Recl. Trav. Chim. Bay-Bas* **1992**, *111*, 323.
176. Hendrickson, J.B.; Braun-Keller, E.; Toczko, A.G. *Tetrahedron (Suppl.)* **1981**, *37*, 359.
177. Hendrickson, J.B. *CHEMTECH* **1998**, *28(9)*, 35.

3. Minimize constructions (γ), which implies the use of large structural pieces of several constructions in a single step (sec. 10.5.B.vi for powerful reactions).

4. Join large pieces last to minimize their large weight loss.

5. Fully convergent routes are preferred.

6. Cyclize early. The structure weight will be minimized if cyclizations are carried out early.

7. Chiral resolution should come early. Since resolution discards > 50% of the material, the rationale is obvious.

There are nine characteristics of the SYNGEN approach.[180]

1. It is an executive program not interactive with the chemist.

2. It assesses all possible routes within clearly defined constraints.

3. It uses digital expressions for molecules and reactions.

4. It dissects the initial skeleton for efficient assembly.

5. It draws on a pool of several thousand available starting materials .

6. It attempts to weight each route to economize the number of steps.

7. It limits the analysis primarily to construction reactions.

8. It generates reactions from mechanistic logic.

9. It does not predict yields.

Hendrickson summarized seven reductive criteria for the basic protocol of the approach.[178]

1. Select from the target structure prime bonds for construction.

    (a) Functionality criteria: pairwise consideration of functionality and their relative positions on the skeleton.

    (b) Skeletal (and stereochemical) criteria.

2. Combine these bonds into bondsets, adequate to dissect reasonable synthon skeletons but minimal in γ.

3. From each bondset isolate the synthon skeletons dissected, noting the pattern of construction sites on each.

4. For each synthon skeleton, locate in the general sequence list corresponding to its pattern of construction sites the sequences applicable to the skeletal (σ) levels of those sites. This defines for each synthon a set of all self-consistent sequences of half-reactions, their particular starting materials and their consequent product functionality.

5. Eliminate, for each synthon, undesirable sequences:

---

178. Hendrickson, J.B. *J. Am. Chem. Soc.* **1975**, *97*, 5784.

(a) Sequences unsuitable for the synthon skeleton (at sites other than construction sites), which must bear functionality, $f = 4 - \sigma$.

(b) Unavailable starting materials.

(c) Unacceptable product functionality for the target structure.

(d) Exclusion of certain functionality in rings (e.g., triple bonds) and certain reactions (cf. Grignard) for cyclizations.

(e) other criteria, such as implicit refunctionalization or regiospecificity in sequences.

6. Select a prime synthon, with the most construction sites, and match successive constructions serially with the other synthons as dictated by the bondset, accepting only matching half-reactions of opposite polarity, to create full synthetic routes.

7. Eliminate matchings in which subsequent constructions on one synthon are not compatible with existing functionality on any other synthon already linked to it.

Hendrickson provided details for several of these criteria, which are instructive to a chemist beginning the disconnection process.

1a: A particular skeletal bond is identified, defined by the functional group (or pair of groups) examined.[171] These may be summarized (Table 10.4)[171] as a function of the position of the functionalized sites on the target skeleton. Some bonds will have a higher priority if the target functionality is exactly correct for certain constructions, or if more constructions are available to yield the particular sites of functionality.[178] A more detailed table of construction bond data is available.[182]

Table 10.4. Construction of Bonds Dictated by Product Functionality

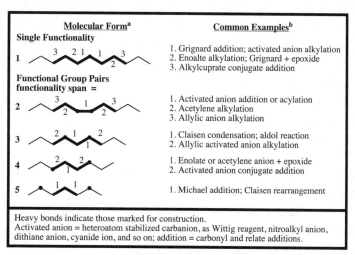

Heavy bonds indicate those marked for construction.
Activated anion = heteroatom stabilized carbanion, as Wittig reagent, nitroalkyl anion, dithiane anion, cyanide ion, and so on; addition = carbonyl and relate additions.

1b: Certain considerations of the skeleton define particular bonds to be considered for construction. (*1*) The first is the presence of the skeleton in available starting materials of recognizable large skeletal units such as aromatic rings or monoterpenes for sesquiterpene synthesis. (*2*) The second is division into two or three similar sized skeletons for convergent synthesis efficiency (*divide* each *target into two disconnect fragments at each retrosynthetic step*). Two or three synthons of similar size may be separately constructed in parallel, then linked. Bonds at or near the center of the skeleton can be used to disconnect the skeleton into separate component synthons. (*3*) Choose bond pairs for annelation. (*4*) Inspect for the presence of several like groups that could be introduced simultaneously. (*5*) Inspect for the presence of skeletal features with limited synthetic choices: quaternary or tertiary centers and small (three- and four-membered) rings. (*6*) Consider strategic ring disconnections in polycyclic skeletons. (*7*) Consider stereochemical features.

Hendrickson restated this principle[171] in a manner that can be used for planning the first disconnection.

"Fully convergent bondsets of any target skeleton are readily created by dissection of the target into two pieces, then cutting each of these in two again. When all the pieces so obtained also corresponds to skeletons of available starting materials, this creates an ordered bondset for a potential synthesis using those starting materials."

2. In the bondsets γ, which is the number of construction reactions in the synthetic sequence, should be minimized to make the synthesis more economical (shorter).

3. The pattern of construction sites on any synthon can be defined as to their relative placement or location and the order in which they undergo construction half-reactions.[171] The synthon skeletons are marked as construction sites on which is placed the appropriate functionality for actually constructing half-reactions in a self-consistent sequence. The numerical codification for individual half-reactions[182] defines a reactive strand of up to three functionalized carbons out from the construction site. Except for a terminal (primary) construction site ($\sigma = 1$), there is a question as to which strand of carbons from the site bears the activating functionality. Secondary construction sites ($\sigma = 2$) have two choices and tertiary ($\sigma = 3$) have three. A synthesis with three steps having secondary sites to undergo construction ($\alpha\beta\gamma$) has $2^3 = 8$ ways to attach the reactive strand of functionality on the skeleton required for construction.[171] Hendrickson[179] provided a table of sequence lists for use in this procedure.

4. The process is illustrated in Figure 10.22[166] for testrone (**195**). The skeletal fragmentations lead to bicyclo[4.4.0]decane (F), the two-carbon fragment (E), and the cyclopentane fragment (D). The appropriate functionalization is also shown in which vinyl chloride (J) is the synthetic equivalent of (E), ketone (K) is the equivalent of F, and 2-methyl-1,3-cyclopentanedione (H) substitutes for (D).

---

179.   Reference 171, pp 101-169.

5-7:These are self-explanatory.

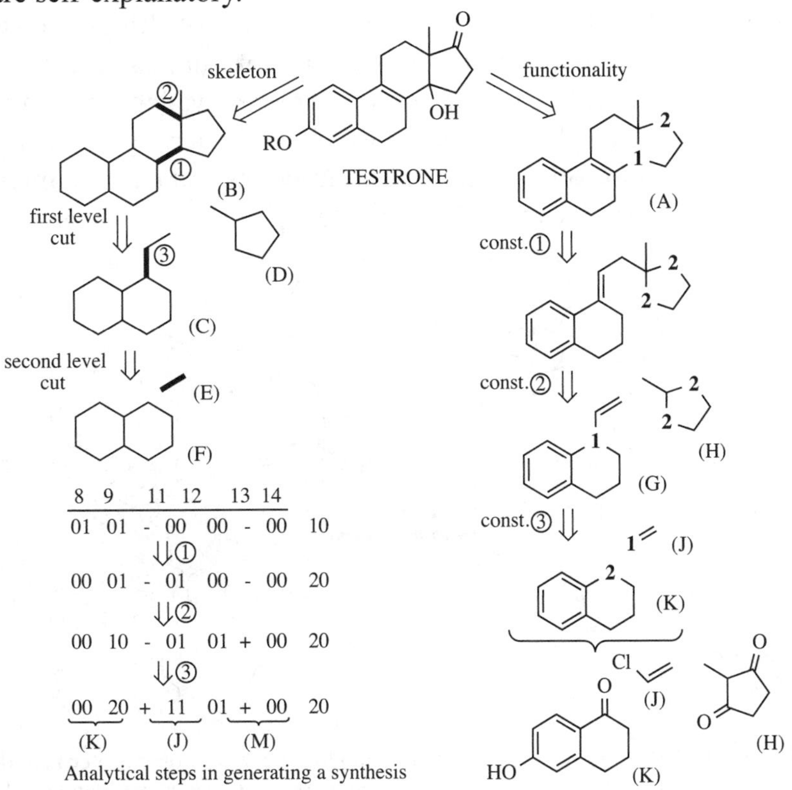

Figure 10.22. A SYNGEN-Generated Retrosynthesis for Testrone. [Reprinted with permission from Hendrickson, J.B. *Accounts Chem. Res.* **1986**, *19*, 274. Copyright © 1986 American Chemical Society.]

Hendrickson's approach can be applied without a computer. However, a beginner does not have the extensive background of reactions or a mental pool of reasonable starting materials and so, without a computer, may not be able to take full advantage of this methodology. The main lesson from this section will point a beginner toward a search for convergent disconnections and the use of an approach that breaks the molecule into two fragments at each disconnection level. Some rationale must be provided as to which two fragments are the best.

10.10.B. Barone's MARSEIL/SOS System

Barone developed an approach that relies on a microcomputer and the program **MARSEIL/ SOS** (**S**imulated **O**rganic **S**ynthesis).[180] It aims to be a usable electronic lab notebook for organic synthesis. The fundamental basis of this program is[184] "about 350 reactions which

180. Azario, P.; Barone, R.; Chanon, M. *J. Org. Chem.* **1988**, *53*, 720.

constitute about 33% of March and Smith,"[181] and it follows a retrosynthetic approach.[182] The structural features of the target are determined in two parts. (*1*) Rings and standard features such as nucleophilic centers, electron-withdrawing groups, and so on are determined using classical organic chemistry; (*2*) A file of substructures is available that are matched with the target. In a general sense, both parts are excellent advice to the chemist beginning a disconnection. Attempting to correlate known substructures and reactive fragments can point to which fragmentation makes the most chemical sense.

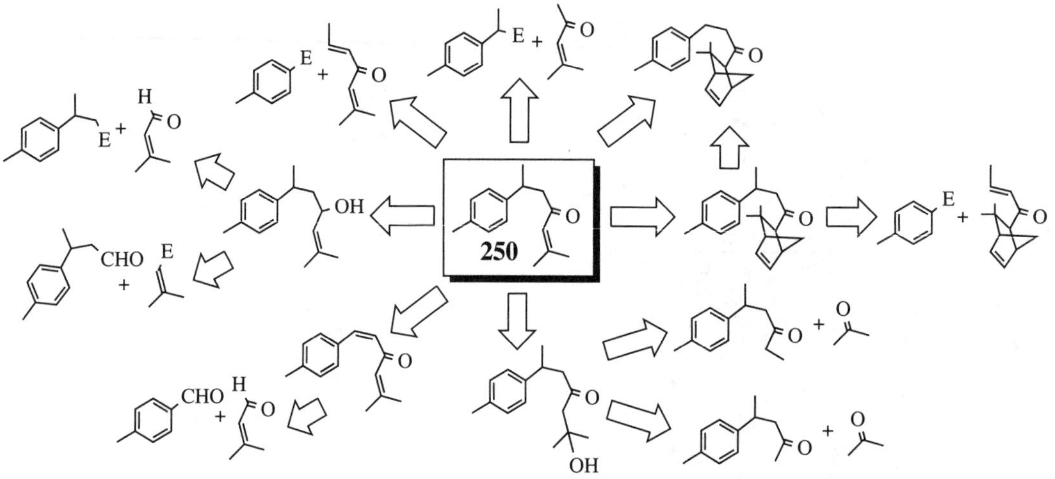

Figure 10.23. MARSEIL-SOS Generated Retrosynthetic Scheme for Turmerone. [Reprinted with permission from Bertrand, M.P.; Monti, H.; Barone, R. and the Journal of Chemical Eductation, Vol. 63, **1986**, 624. Copyright 1986, Divison of Chemical Education, Inc.

The actual search for a particular transform requires that one look for a characteristic substructure, evaluating that substructure and providing reactions to build it. The lack of attention to stereochemistry in the target or disconnect products is a major drawback to the program. In one example (Figure 10.23),[183] this program generated several synthesis pathways for (±)-ar-turmerone (**242**).[184] It was pointed out that the students who used this program must choose the more reasonable path, "after discussion with the instructor."[187] The instructor chose path 6 (Figure 10.24), and it is clear that this program relies on the chemical intuition and knowledge of reactions of an experienced chemist. In this case, the novice will not be able to reasonably choose the best approach. Barone used a retrosynthetic approach to define the logic of Marseil/SOS, and this was contained in another program, **REKEST** (**RE**search for the **KE**y **ST**ep).[185]

181. Smith, M.B.; March, J. *March's Advanced Organic Chemistry, 6th ed.*; Wiley-Interscience, Hoboken, New Jersey, *2007*.
182. Barone, R.; Chanon, M. *Nouv. J. Chim. 1978, 2,* 659.
183. Bertrand, M.P.; Monti, H.; Barone, R. *J. Chem. Educ. 1986, 63,* 624.
184. (a) Grieco, P.A.; Finkelhor, R.S. *J. Org. Chem. 1973, 38,* 2909; (b) Ho, T.L. *Synthetic Commun. 1974, 4,* 189; (c) Park, O.S. Grillasca, Y.; Garcia, G.A.; Maldonado, L.A. *Ibid. 1977, 7,* 345 and references cited therein.
185. Barone, R.; Chanon, M. *Chimia 1986, 40,* 436.

The fundamental logic for disconnection is based on a library of stylized reactions, shown in Figure 10.24.[189] Each reaction is a pattern that allows deletion or addition of one or several bonds in the target. The labels A, B, C, and D are any atoms and the disconnections generally involve deletion of one or two bonds. An analysis for **243** is shown in Figure 10.24.[189] Once the choice of reaction has been made, appropriate atoms for A, B, C, and D are determined within the boundaries of atoms that are compatible with real chemical reactions. Clearly, this general approach is of value because the chemist can examine a target for disconnections that follow the reactions in Figure 10.24, which leads to better decisions as to which disconnections should be made. The bonds chosen for disconnection maximize simplification of the target. Bonds are chosen that can easily be chemically formed. A pool of known starting materials and chemical reactions is essential. Once the disconnection process begins, the chemical experience and interest of the chemist will usually determine which pathway is best suited for their purpose.

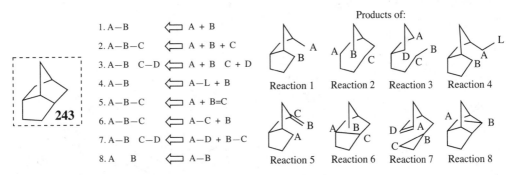

Figure 10.24. Reaction Types for REKEST and MARSEIL/SOS.

10.11. DEGRADATION TECHNIQUES AS A TOOL FOR RETROSYNTHESIS

10.11.A. Chemical Degradation

Structural information about complex molecules is often obtained by degrading a target into simple, and identifiable pieces. Common methods for degrading a molecule include hydrolysis, thermolysis, photolysis or treatment with chemical reagents. The bonds that break in a breaks are correlated with known reactions by which these bonds can be formed by chemical means. Analysis of degradation products can then be used to construct a complete or partial synthetic tree, or it can point to a key reaction for a potential synthesis.

Bleomycin A_2 (**244**) was isolated by Umezawa and co-workers,[186] and shown to be a potent anticancer agent. The total synthesis of bleomycin reported by Hecht and co-workers is a

---

186.   (a) Ikekawa, T.; Iwami, F.; Hiranaka, H.; Umezawa, H. *J. Antibiot. (Tokyo)* **1964**, *17A*, 194; (b) Umezawa, H.; Maeda, K.; Takeuchi, T.; Okami, Y. *Ibid.* **1966**, *19A*, 200; (c) Umezawa, H. Suhara, Y.; Takita, T.; Maeda, K. *Ibid.* **1966**, *19A*, 210.

superb example of the chemical degradation approach. While determining the structure, it was shown that hydrolysis and other chemical manipulations of **242** gave nine key fragments: aminopropyldimethyl sulfonium (**245**); 2-aminoethyl-2',4-bithiazole-4'-carboxylic acid (**246**); L-threonine (**247**); (2*R*,3*S*,4*R*)-4-amino-3-hydroxy-2-methylpentanoic acid (**248**); L-*erythro*-β-hydroxy-histidine (**249**); β-amino-β-(4-amino-6-carboxy-5-methylpyrimidin-2-yl)propionic acid (**250**); a rearrangement product, L-β-aminoalanine (**251**); gulose (**252**); and, carbamoyl mannose (**253**).

Hecht's synthetic approach was to first synthesize each of the individual pieces. The total synthesis required connection of the nine molecules to give bleomycin.[187] The disconnected pieces are shown in Scheme 10.21[187] and the plan to synthesize bleomycin from these pieces constitutes an outline of the synthetic tree. Hecht and c-worker's synthesis of bleomycin A_2 is also an excellent example of a convergent synthesis (sec. 10.3.C). Note that Boger synthesized bleomycin using a different strategy.[188]

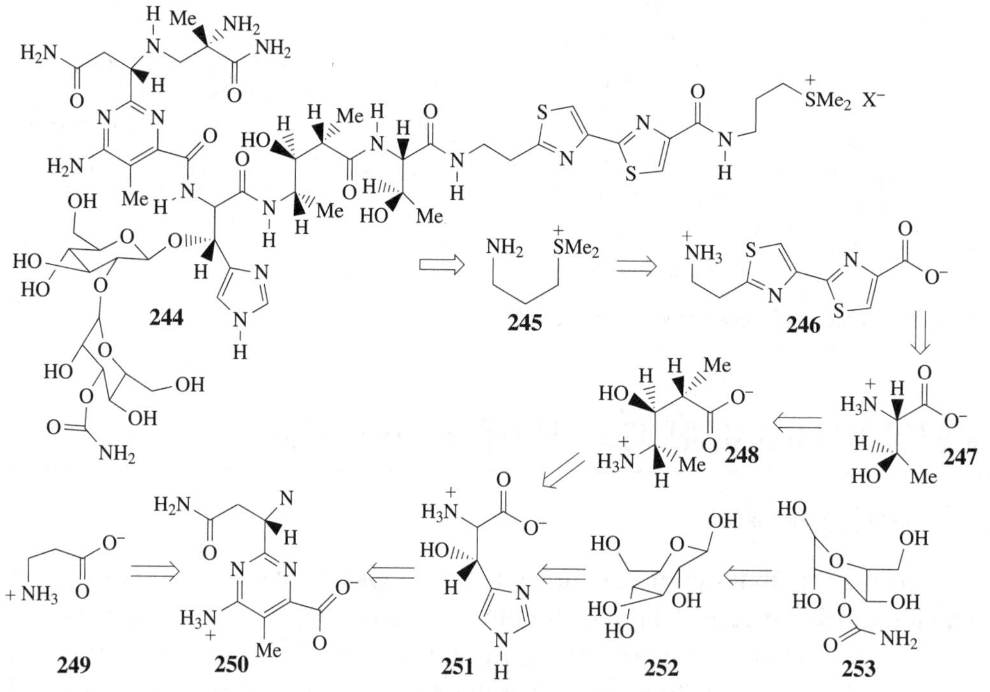

Scheme 10.21. Retrosynthetic Analysis of Hecht's Bleomycin A_2 Synthesis.

187. (a) McGowan, D.A.; Jordis, U.; Minster, D.K.; Hecht, S.M. *J. Am. Chem. Soc.* ***1977***, *99*, 8078; (b) Arai, H.; Hagman, W.K.; Suguna, H.; Hecht, S.M. *Ibid.* ***1980***, *102*, 6631; (c) Levin, M.D.; Subrahamanian, K.; Katz, H.; Smith, M.B.; Burlett, D.J.; Hecht, S.M. *Ibid.* ***1980***, *102*, 1452; (d) Hecht, S.M.; Rupprecht, K.M.; Jacobs, P.M. *Ibid.* ***1979***, *101*, 3982; (e) Ohgi, T.; Hecht, S.M. *J. Org. Chem.* ***1981***, *46*, 1232; (f) Pozsgay, V.; Ohgi, T.; Hecht, S.M. *Ibid.* ***1981***, *46*, 3761; (g) Aoyagi, Y.; Suguna, H.; Murugesan, N.; Ehrenfeld, G.M.; Chang, L-H.; Ohgi, T.; Shekhani, M.S.; Kirkup, M.P.; Hecht, S.M. *J. Am. Chem. Soc.* ***1982***, *104*, 5237; (h) Aoyagi, Y.; Katano, K.; Suguna, H.; Primeau, J.; Chang, L.-H.; Hecht, S.M. *Ibid.* ***1982***, *104*, 5537.
188. See Boger, D.L.; Cai, H. *Angew. Chem. Int. Ed.* ***1999***, *38*, 448 and references cited therein.

Reactions other than hydrolysis can fragment a molecule, and the degradation of securinine (**254**)[189] in Scheme 10.22 uses other chemical reactions to generate a synthetic tree . Reductive cleavage of the C-N bond yielded **255**. The alkene unit was reduced by catalytic hydrogenation to **258**, but the lactone unit could also be reduced with LiAlH$_4$ to give **256**. Allylic alcohol **256** was oxidatively degraded to **257** with permanganate. Reduction of **258** liberated **259**. Hydrogenation in the presence of base, however, gave **260**. Reduction of the lactam carbonyl in **260** with LiAlH$_4$ gave **261**. With this degradation scheme in mind, a retrosynthesis is shown in Scheme 10.23,[190] based on Horii's total synthesis. Note the similarity of **262** (in Scheme 10.23) to **255** and **258**, and also the relationship of **263** to **256**. Disconnection product **264** yielded the two starting materials, the monoethyleneketal of 1,2-cyclohexanedione (**265**) and 2-lithiopyridine.

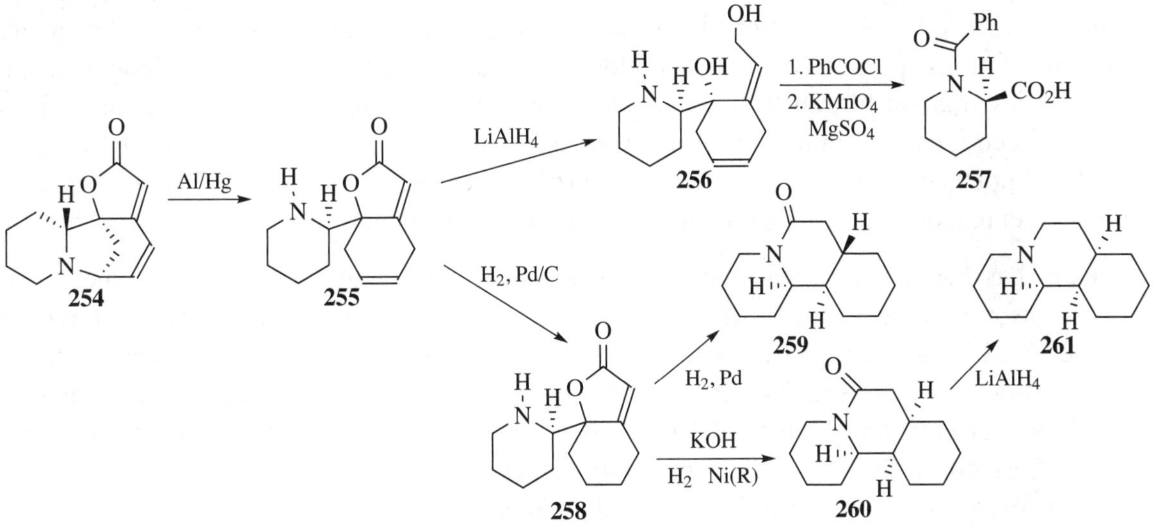

Scheme 10.22. Chemical Degradation of Securinine (**254**).

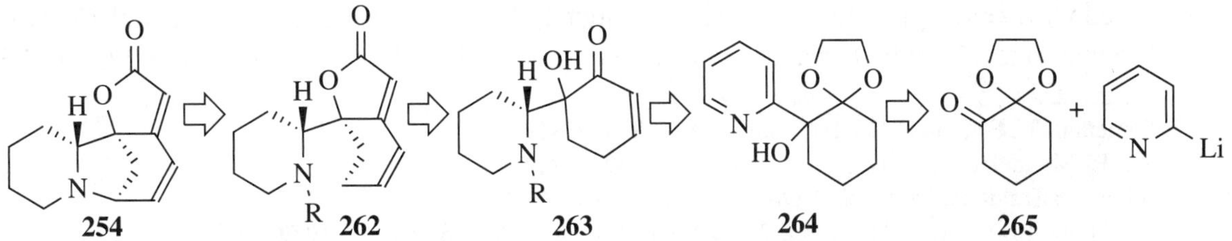

Scheme 10.23. Retrosynthetic Analysis of Horii's Securinine Synthesis.

189.   (a) Nakano, T.; Yang, T.H.; Terao, S. *J. Org. Chem.* **1963**, *28*, 2619; (b) Satoda, I.; Murayama, M.; Tsuji, J.; Yoshii, E. *Tetrahedron Lett.* **1962**, 1199; (c) Saito, S.; Kotera, K.; Shigematsu, N.; Ide, A.; Sugimoto, N.; Horii, Z.; Hanaoka, M. Yamawaki, Y.; Tamura, Y. *Tetrahedron* **1963**, *19*, 2085; (d) Nakano, T.; Yang, T.H.; Terao, S. *Tetrahedron Lett.* **1963**, 665 (e) Saito, S.; Kodera, K.; Sugimoto, N.; Horii, Z.; Tamura, Y. *Chem. Ind., (London), 1962*, 1652.

190.   Horii, Z.; Hanaoka, M.; Yamawaki, Y.; Tamura, Y.; Saito, S. Shigematsu, N.; Kotera, K.; Yoshikawa, H.; Sata, Y.; Nakai, H. Sugimoto, N. *Tetrahedron* **1967**, *23*, 1165.

10.11.B. Retro-Mass Spectral Degradation

A target may resist hydrolysis or chemical degradation, or the degradation products may not yield useful information. An alternative degradation technique is available that uses the ionizing electron beam of a mass spectrometer. The ionization pathways available from electron impact in the mass spectrum are bond fission processes that occur by known and predictable pathways. Indeed, each fragmentation pathway usually follows an analogous chemical reaction pathway *in a retrosynthetic manner*. It therefore follows that an examination of mass spectral ionization patterns can give clues for suitable disconnections and a synthetic tree.

This basic premise was outlined by Kametani and Fukumoto,[191] and used for the synthesis of several alkaloids. As mentioned, the key feature of the analysis is the fact that many chemical reactions appear as their retro analog in the mass spectrum. These retro reactions are disconnections that can be translated directly to a synthetic tree. Some examples of retro reactions that have been identified in the mass spectrum are (*1*) retro-Diels-Alder (sec. 11.5.B),[192] (*2*) a retro-1,3-dipolar addition (sec. 11.11),[193] (*3*) a retro Wagner-Meerwein (sec. 12.2.B),[194,195] and (*4*) a retro-Ritter reaction.[196] The **Ritter reaction**[197] couples an alkene and a nitrile to give an amide.

Kametani has applied retro-mass spectral analysis to total synthesis. The mass spectra of tetrahydroisoquinoline alkaloids show fragmentations by the three major pathways shown in Scheme 10.24.[195] Kametani observed that path *c* (a retro Diels-Alder reaction) was the major fragmentation pathway (sec. 11.5.B).[195] Xylopinene (**266**) gave this specific fragmentation (to **267** and **268**),[198] and Scheme 10.25[195] shows the resulting total syntheses reported by Kametani, based on the Diels-Alder reaction suggested by path *c* in Scheme 10.24 and the observed fragmentation pattern. Two related synthetic routes involve thermal conversion of benzocyclobutane derivatives to a diene (sec. 11.10.F) that reacts with the dihydroisoquinoline starting material (**269**). In one case, heating cyano derivative **270** gave a transient diene (**271**) that reacted with **269** to give the Diels-Alder adduct **272**.[199] In the second case, alcohol **273** was heated to generate **274**, which reacted with **269** to give **275**.[200] Reduction of either **272** or **275**

191.   Kametani, T.; Fukumoto, K. *Acc. Chem. Res. 1976, 9*, 319.
192.   Bel, P.; Mandelbaum, A. *Org. Mass. Spectrom. 1981, 16*, 513.
193.   Selva, A.; Traldi, P.; Fantucci, P. *Org. Mass Spectrom. 1978, 13*, 695.
194.   Wolfshütz, R.; Schwarz, H.; Blum, W.; Richter, W.J. *Org. Mass. Spectrom. 1979, 14*, 462.
195.   Smith, M.B.; March, J. *March's Advanced Organic Chemistry, 6th ed.*, Wiley-Interscience, Hoboken, New Jersey, *2007*, pp 1580-1585 and references cited therein.
196.   Nagubandi, S. *Org. Mass. Spectrom. 1980, 15*, 535.
197.   (a) Krimen, L.I.; Cota, D.J. *Org. React. 1969, 17*, 213; (b) Ritter, J.J.; Minieri, P.P. *J. Am. Chem. Soc. 1948, 70*, 4045; (c) Ritter, J.J.; Kalish, J. *Ibid. 1948, 70*, 4048.
198.   Budzikiewicz, H.; Djerassi, C.; Williams, D.H. *Structure Elucidation of Natural Products by Mass Spectrometry*, Vol. 1, Holden-Day, San Francisco, *1964*.
199.   (a) Kametani, T.; Takahashi, T.; Honda, T.; Ogasawara, K.; Fukumoto, K. *J. Org. Chem. 1974, 39*, 447; (b) Kametani, T.; Kajiwara, M. Takahashi, T.; Fukumoto, K. *J. Chem. Soc. Perkin Trans. 1 1975*, 737.
200.   Kametani, T.; Katoh, Y.; Fukumoto, K. *J. Chem. Soc. Perkin Trans. 1 1974*, 1712.

gave **266** in good yield. The carbocyclic analog of this reaction[201,202] had been used previously in synthesis, and Oppolzer had previously used the amido analog of this condensation reaction for alkaloid synthesis (sec. 11.10.F).[203] Its use by Kametani in this case, however, was the direct result of construction of a synthesis tree from the mass spectrum.

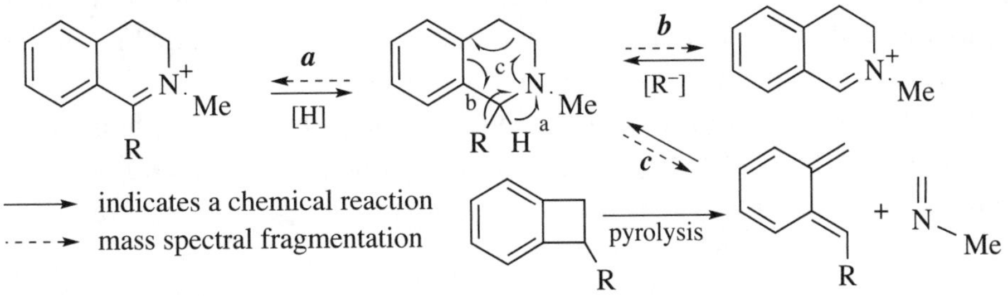

indicates a chemical reaction
mass spectral fragmentation

Scheme 10.24. Kametani's Mass Spectral Analysis of Tetrahydroisoquinoline Alkaloids. [Reprinted with permission from Kametani, T.; Fukumoto, K. *Accts. Chem. Res.* **1976**, *9*, 319. Copyright © 1976 American Chemical Society.]

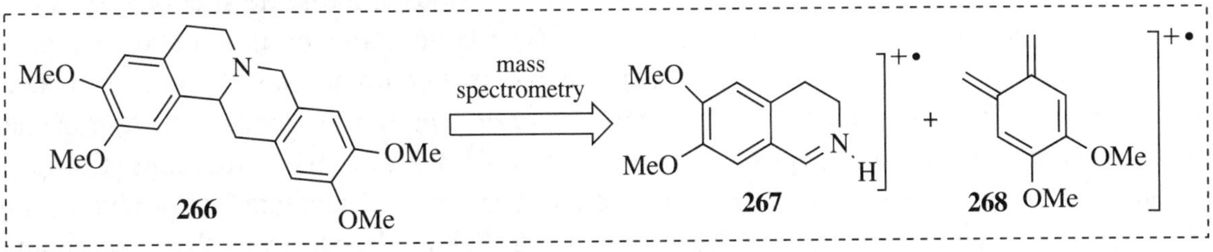

Scheme 10.25. Synthesis of Xylopinene (**266**) Based on Retro-Mass Spectral Analysis.

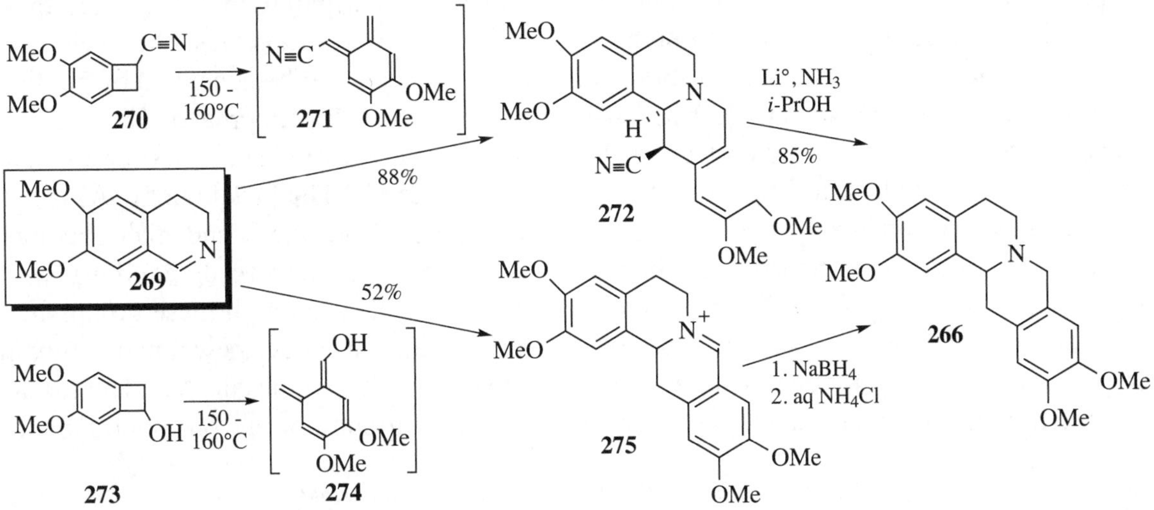

201. Klundt, I.L. *Chem. Rev.* **1970**, *70*, 471.
202. Kametani, T.; Takahashi, T.; Ogasawara, K.; Fukumoto, K. *Tetrahedron* **1974**, *30*, 1047.
203. (a) Oppolzer, W. *J. Am. Chem. Soc.* **1971**, *93*, 3834, 3833; (b) Oppolzer, W.; Keller, K. *Ibid.* **1971**, *93*, 3836; (c) Oppolzer, W. *Tetrahedron Lett.* **1974**, 1001.

The availability of mass spectrometers, and variety of mass spectrometric techniques and detectors clearly makes this an attractive tool for retrosynthetic analysis. The major drawback is that structural identification of the mass spectral fragment is a necessity, which is often unknown and its determination can require fragmentation studies on isotopically labeled compounds. Modern mass spectrometry allows the use of *mass spectrometry of mass spectrometry* to identify fragments, and with the high-resolution mass spectrometers capable of identifying the mass of individual fragments, identification of key fragments is often possible. This analytical technique brings retro-mass spectral analysis into the fold of modern methods for determining a retrosynthesis tree. If the structure of a fragment is not easily determined, however, this approach will be of little value for constructing a synthetic tree. If the fragment is known, retro-mass spectral analysis can be a useful tool.

10.12. COMBINATORIAL CHEMISTRY

Combinatorial chemistry is a term used to describe various microscale methods of solid-state synthesis and testing. It involves the synthesis of large numbers of compounds (called libraries), by doing reactions in a manner that produces large combinations of products, usually as mixtures.[204] This approach has been called *irrational drug design*, since early approaches involved making a large vat of all possible chemical combinations of several reactants. Compare this with parallel synthesis, where the same reactions are repeated separately to produce many individual but related products. In other words, a parallel synthesis means that a compound library is constructed by synthesizing many compounds in parallel, keeping each compound in a separate reaction vessel. When the final compounds are kept separate in this manner, their identity is easily discerned. The pin-method of combinatorial synthesis, described below, is one such parallel synthesis method. Libraries of compounds can also be generated by pooling strategies, and the teabag method described below illustrates this type of approach. The methods of combinatorial synthesis will be described below, but there are several strategies for efficient synthesis of a target illustrated by the cartoons in Figure 10.25.[208c] The first strategy (A) is one where doubly functionalized monomer units are linked in peptide synthesis and other oligomer synthesis. In strategy B, the synthesis begins with a template that is already equipped with several functional groups. Different building blocks are attached through these groups in a series of separate steps or in one pot. In strategy C, the reactions introduce new functionality in the product. As each new functional group is added, that product becomes the template for the next reaction. Combinatorial syntheses can be performed both in solution and on a solid support.

204. (a) Czarnik, A.W.; DeWitt, S.H. (Eds.), *A Practical Guide to Combinatorial Chemistry*, Amer. Chem. Soc. Washington, D.C., *1997*; (b) Chaiken, I.N.; Janda, K.D. (Eds.), *Molecular Diversity and Combinatorial Chemistry: Libraries and Drug Discovery*, American Chemical Society, Washington, D.C., *1996*; (c) Balkenhol, F.; von dem Bussche-Hünnefeld, C.; Lansky, A.; Zechel, C. *Angew. Chem. Int. Ed.* *1996*, *35*, 2289; (d) Thompson, L.A.; Ellman, J.A. *Chem. Rev.* *1996*, 96, 555; (e) Pavia, M.R.; Sawyer, T.K.; Moos, W.H. (Eds.), *Bioorg. Med. Chem. Lett. Symposia-in-print no. 4*, *1993*, *3*, 381.

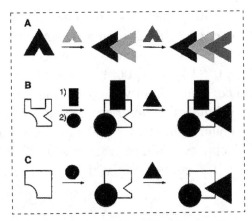

Figure 10.25. Strategies in Combinatorial Synthesis. [Reprinted with permission from Balkenhol, F.; von dem Bussche-Hünnefeld, C.; Lansky, A.; Zechel, C. *Angew. Chem. Int. Ed.* **1996**, *35*, 2289.]

In 1963 Merrifield introduced solid-state synthesis for the synthesis of peptides.[205] This technique involves chemical functionalization of a polystyrene bead (or another polymeric bead) that reacts with the carboxylic acid portion of a *N*-protected amino acid to give a polymer-bound amino ester such as **276**. When **276** is treated with a reagent to deprotect the amine, it can react with another N-protected amino acid, activated at the carbonyl, to give a dipeptide. This procedure can be repeated to generate the desired polypeptide, and when the target has been attained a reagent is added to cleave the polypeptide from the bead (hydrolysis). This solid-state synthesis can be applied to other types of chemical transformations.[206]

The fundamental idea of the Merrifield solid-state synthesis can be extended, using several beads where each is attached to a different amino acid (V, I, S, etc.)[207] Subsequent reaction can generate a "pool" of peptides. If this mixture of beads was treated with three activated amino acids (V, I, S), a total of nine products would be generated, each one attached to a bead, as shown in Figure 10.26. This approach can be expanded to generate a large number of peptides. If the bead containing V were made to react with 20 different amino acids, a mixture of 20 dipeptides would be generated. If each of these 20 dipeptides were to react with 20

205. Merrifield, R.B. *J. Am. Chem. Soc.* **1963**, *85*, 2149.
206. (a) Crowley, J.I.; Rapoport, H. *Acc. Chem. Res.* **1976**, *9*, 135; (b) Leznoff, C.C. *Acc. Chem. Res.* **1978**, *11*, 327.
207. The 3-letter codes and the appropriate 1-letter code for the essential amino acids are given here. glycine (gly, G), alanine (ala, A), valine (val, V), leucine (leu, L), isoleucine (ile, I), phenylalanine (phe, F), serine (ser, S), threonine (thr, T), tyrosine (tyr, Y), cysteine (cys, C), methionine (met, M), asparagine (asn, N), glutamine (gln, Q) aspartic acid (asp, D), glutamic acid (glu, E), lysine (lys, K), tryptophan (trp, W) histidine (his, H), arginine (arg, R), proline (pro, P).

different amino acids, 20 x 20 or 400 tripeptides would be generated as a mixture. If this 400 member tripeptide library were to react with 20 amino acids, a mixture of 20 x 20 x 20, or 8000 tetrapeptides, would be generated. If we continued this process, 20^4 (160,000) pentapeptides, 20^5 (3,200,00) hexapeptides, 20^6 (1,280,000,000) heptapeptides, and 20^7 (25,600,000,000) octapeptides would be generated; a huge library of compounds all mixed together.[208] It is easy to imagine variations that would build other libraries. A mixture of the three amino acids shown in Figure 10.26, for example, could be mixed with 20 amino acids and this would generate 60 dipeptides. This library of dipeptides could be reacted with 20 amino acids to generate 60^{20} different tetrapeptides (3.656×10^{35}) compounds, which would be a library from a library.

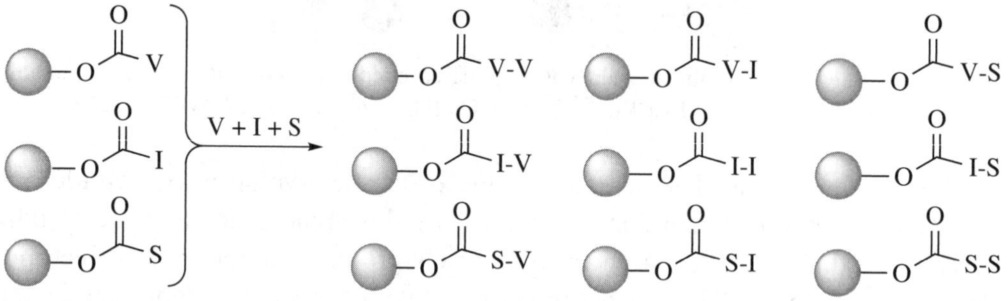

Figure 10.26. Generation of a Nine-Component Dipeptide Library.

Early in the development of combinatorial chemistry it was discovered that when chemical reactions occurred close to the polymer bead there were problems, both in chemical reactivity and in releasing the final product from the bead. An important variation added a *spacer* between the bead and the functional group used for chemistry.

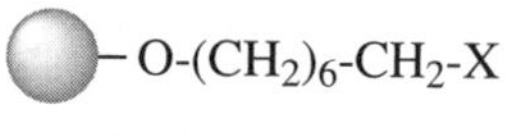

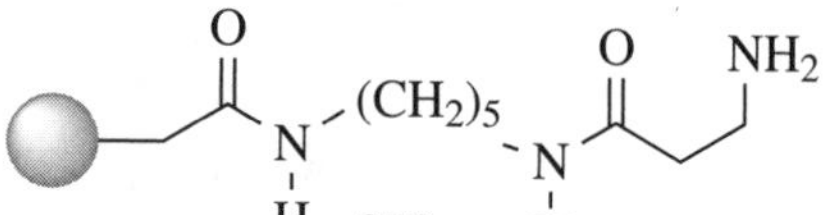

Hydrocarbon chains connected to the ether linkage (as in **277**) diamide groups, and polyamide chains such as **278** are commonly used. The group is usually be attached to the bead by an acid-labile unit, or one that is susceptible to enzymatic cleavage. Two common, and highly acid-labile anchors are the aryl derivatives 4-(2',4'-dimethoxyphenylhydroxylmethyl)phenoxy (**279**) or 2-methoxy-4-alkoxybenzyl (**280**).[209]

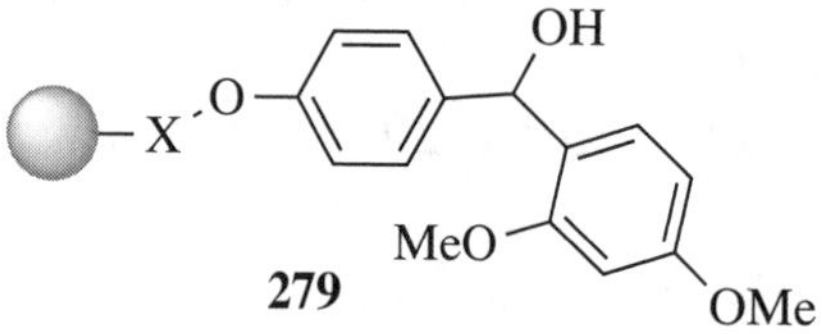

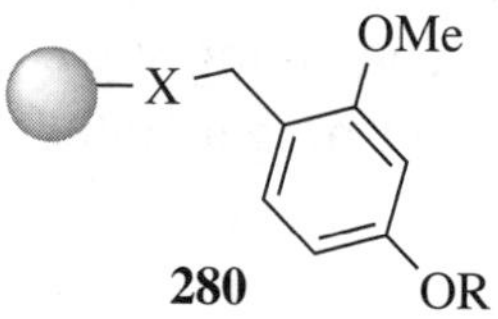

208. (a) Nielsen, J. *Chem. Ind.* **1994**, 902; (b) Sepetov, N.F.; Krchák, V.; Stanková, M.; Wade, S.; Lam, K.S.; Lebl, M. *Proc. Natl. Acad. Sci. USA* **1995**, *92*, 5426.
209. (a) Rink, H. *Tetrahedron Lett.* **1990**, *28*, 3787; (b) Barlos, K.; Gratos, D.; Kallitsis, J.; Papaphotiu, G.; Sotiriu, P.; Wenqing, Y.; Schäfer, W. *Tetrahedron Lett.* **1989**, *30*, 3943.

A reasonable question is how is the combinatorial technique done experimentally? Two methods will be presented here to illustrate the fundamental techniques. Houghten described a teabag method[210] that could prepare as many as 150 different peptides with a yield of up to 50 mg.[211] A polypropylene bag (see Figure 10.27)[215] is charged with ≈ 100 mg of a polymeric support (usually polystyrene-1% divinylbenzene). Each bag contains a label and is immersed in a polyethylene bottle containing piperidine/DMF. A computer-automated device adds amino acids, and the peptide is generated within each teabag. Treatment of the bags with a deprotection reagent and subsequent washing liberates the peptides.

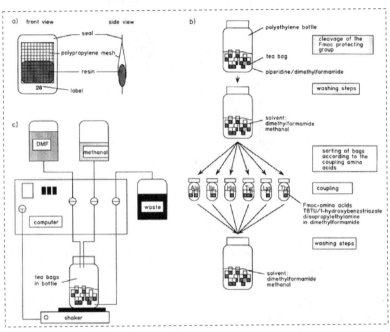

Figure 10.27. The Teabag Approach to Combinatorial Peptide Synthesis. [Reprinted with permission from Jung, G.; Beck-Sickinger, A.G. *Angew. Chem., Int. Ed.* **1992**, *31*, 367.]

A second technique that accomplishes peptide synthesis on polyethylene rods was developed by Geyson et al.[212] A typical array of rods (called pins) consists of 96 pins (typically 4 mm in diameter and 40 mm in length),[215] arranged in 8 rows of 12 rods (see Figure 10.28b). The pins fit into the wells of plates (see Figure 10.28),[215] which were developed for the enzyme-linked immunosorbent assay (ELISA). The polyethylene rods are functionalized with acrylic acid, and N^{β}-Fmoc-β-alanyl-1,6-diaminohexane is attached as a spacer (see Figure 10.28a). The idea is to attached an amino acid to each of the pins, and place the solution containing the reactants into the wells. The pins are dipped into the wells and allowed to react. To deprotect the terminal

210. Houghten, R.A. *Proc. Natl. Acad. Sci. USA* **1985**, *82*, 5131.
211. Jung, G.; Beck-Sickinger, A.G. *Angew. Chem. Int. Ed.* **1992**, *31*, 367.
212. Geysen, H.M.; Meloen, R.H.; Barteling, S.J. *Proc. Natl. Acad. Sci. USA* **1984**, *81*, 3998. Also see Kerr, J.M.; Banville, S.C.; Zukermann, R.N. *Bioorg. Med. Chem. Lett.* **1993**, *3*, 463

position, deprotection reagent is added to a new plate of wells, and the pins are dipped into that. Since the "polyethylene rods cannot swell or shrink, and adsorbed molecules can only be removed with difficulty, thorough washing after each reaction step of the synthesis cycle is of utmost importance".[215] Recent work has shown that soft, but mechanically strong polymer rods derived from styrene-divinylbenzene copolymers prepared by adjusting the reaction conditions,[213] could be cut into disks with good swelling characteristic in various solvents, and that resist osmotic shock. Such disks are an alternative support for solid phases synthesis.

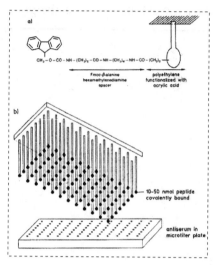

Figure 10.28. The Pin Approach to Combinatorial Peptide Synthesis. [Reprinted with permission from Jung, G.; Beck-Sickinger, A.G. *Angew. Chem., Int. Ed.* **1992**, *31*, 367.]

Chemical reactivity is an important consideration. If a bead reacts with many different reactants (*n* coupling partners), there will be *n* products, which is also a problem in solution combinatorial synthesis. Individual substances can be obtained by fast parallel syntheses, that is, *n* substances are synthesized in *n* reaction vessels.[208c] Since it is reasonable that some reactants will react faster, and in higher yields than others, there may be different *amounts* of the *n* products. In other words, some coupling reactions occur at faster rates than others. This problem of different reaction rates can be avoided by the so-called *split method*.[214,228] As shown in Figure 10.29,[208d] the solid support is divided into three equal parts in this case, and each of these is treated with a component A. The resulting products are mixed and divided again, giving three mixtures in which the resin-linked components (A^1, A^1, A^3) are present in equimolar amounts. When these are treated with new reagents (B^1, B^2, B^3), nine defined products are produced. Since there is only one reactant in solution in each reaction vessel at any one time, all reactions proceed to completion, despite possible differences in kinetics.[208c] Repetition of the cycle of division, reaction, and mixing gives large compound libraries in

213. Hird, N.; Hughes, I.; Hunter, D.; Morrison, M.G.J.T.; Sherrington, D.C.; Stevenson, L. *Tetrahedron* **1999**, *55*, 9575.
214. Sebestyén, F.; Dibó, G.; Kovács, A.; Furkua, A. *Bioorg. Med. Chem. Lett.* **1993**, *3*, 413.

which all compounds are present in equimolar amounts. Ley and co-workers developed an approach for the rapid automated optimization of polymer-supported reagents in synthesis using an optimized set of reaction conditions with an array of 80 compounds.[215]

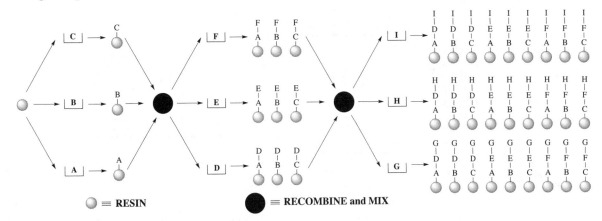

Figure 10.29. Split Synthesis Method. [Reprinted with permission from Thompson, L.A.; Ellman, J.A. *Chem. Rev.* **1996**, 96, 555. Copyright © 1996 American Chemical Society.]

An important component of the split method is ***deconvolution***. The library formed by these methods will be a mixture, and when the mixture must be tested in solution structure determination can be very difficult. An example of deconvolution is shown in Figure 10.30[208c] for a molecule with four variable substituents (A-D), where five different possibilities for each substituent lead to a total library of $5^4 = 625$ compounds. In the first deconvolution, 25 sublibraries are synthesized, each containing 25 compounds, where the substituents A and B are defined. Using a predetermined measure of activity (which compounds are being targeted), the most active sublibrary contains the optimal combination of substituents A and B, A^2B^1 in Figure 10.30. Based on this, five further sublibraries in the form $A^2B^1C^nC^{1-5}$ are prepared,[208c] each containing five components in which C is now defined. The optimal residue D is determined in the last step by synthesis of the five individual compounds in the form $A^2B^1C^3D^n$, and in Figure 10.30, the most active compound was $A^2B^1C^3D^5$.[208c] To get around the need for a new synthesis of another sublibrary for each assay of the total library, Han, et al. used a procedure called recursive deconvolution.[216] An aliquot is held back after each cycle of the split synthesis. From active mixtures, similarly in one or more steps, the single active compound is identified.[208c] No new split syntheses are involved since only the intermediates retained during the first split synthesis need to be elaborated further.[208c]

215. Jamieson, C.; Congreve, M.S.; Emiabata-Smith, D.F.; Ley, S.V. *Synlett* **2000**, 1603.
216. (a) Han, H.; Wolfe, M.; Brenner, S.; Janda, K.D. *Proc. Natl. Acad. Sci. USA* **1995**, *92*, 6419; (b) Erb, E.; Janda, K.D.; Brenner, S. *Proc. Natl. Acad. Sci. USA* **1994**, *91*, 11422.

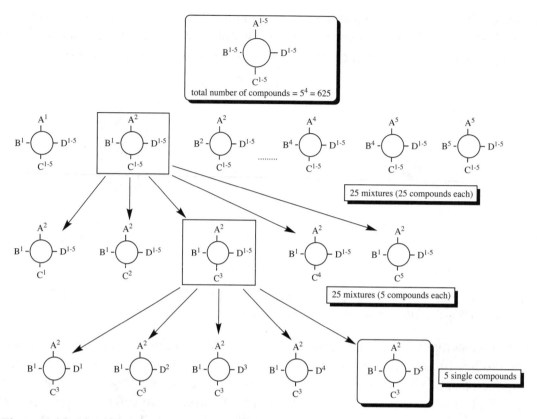

Figure 10.30. Identification of the Most Active Library Component by Deconvolution. [Reprinted with permission from Balkenhol, F.; von dem Bussche-Hünnefeld, C.; Lansky, A.; Zechel, C. *Angew. Chem. Int. Ed.* **1996**, *35*, 2289.]

It is reasonable to ask is how each product of a massive library can be identified as a unique entity, and isolated? There are several techniques by which this can be accomplished. All products can be detached from the beads and isolated as a mixture. This mixture can be analyzed using high performance liquid chromatography (HPLC) and gas chromatography (GC) techniques, particularly when combined with mass spectrometry (HPLC/mass spectrometry and GC/mass spectrometry). In principle, each component can be separated and identified.[214,217] Direct monitoring of solid-phase reactions, on the bead itself without prior cleavage from the resin, is possible using soft laser desorption time-of-flight mass spectrometry, if photocleavable phenacyl ester or *o*-nitroveratryl linker groups are used.[218] Lam and co-workers reported a "one-bead-

217. For examples of peptide analysis using these techniques, see (a) Hean, M.T.W.; Regnies, F.E.; Wehr, C.T (Eds.), *High Performance Liquid Chromatography of Proteins and Peptides*, Academic Press, New York, **1983**; (b) Hean, M.T.W. *Adv. Chromatogr.* **1982**, *20*, 1; (c) Bruins, A.P.; Covey, T.R.; Henion, J.D. *Anal. Chem.* **1987**, *59*, 2642; (d) Bruins, A.P.; Weidolf, O.G.; Henion, J.D.; Buddle, W.L. *Anal. Chem.* **1989**, *59*, 2647; (e) Grolo, M.; Takeuchi, T.; Ishii, D. *Adv. Chromatogr.* **1989**, *30*, 167.
218. Gerdes, J.M.; Waldmann, H. *J. Comb. Chem.* **2003**, *5*, 814.

one-compound" encoding approach, based on mass spectrometry.[219] As stated by Lam, prior to library synthesis the inner core of each bead is derivatized with 3-4 different coding blocks on a cleavable linker.[223] An individual coding block that contains a functional group with the same chemical reactivity encodes each functional group on the scaffold.[223] During the library synthesis, the same chemical reactions take place on the scaffold (outer layer of the bead) and coding blocks (inner core of the bead) concurrently.[223] After screening, the coding tags in the positive beads are released, followed by mass determination of the moelcuels using matrix-assisted laser desorption ionization (MALDI) Fourier transform mass spectrometry.[223]

It would be useful to know the bead to which each product is attached. A solution to this problem is the use of a molecular tag, work pioneered by Brenner and Lerner in 1992,[220] and by Nielsen.[221,212] The idea is to give a binary chemical tag for each reagent. In the work of Still and co-workers, a constant segment of a library is attached to a polymeric bead using standard solid-phase methods.[222] A tag for the first reagent is attached, and then the amino acid is attached. After processing, a second tag is attached, followed by the second amino acid in the sequence. Still and co-worker's coding strategy used haloaromatic tags such as **281**.[226,208d] When attached as an amide, the chain length (n) can be varied along with the haloaryl ether unit, to create a binary synthesis code. The reagents can be designated in binary as 001 (reagent 1), 010 (reagent 2), 011 (reagent 3), to 111 (reagent 7). If reagent 3 were used in the first step and reagent 1 in the second and reagent 6 in the third, the description would be 001 011 011.[226] In **281**, nine different haloaryl units could be attached to create nine tags, T^1-T^9 and each could be linked to one reagent. Therefore, aryl halide A might be 001 (T^1), B 002 (T^2), and C 011 (T^6) and so on. Two possibilities are pentachlorophenyl and 2,4,6-trichlorophenyl. Using different haloaromatic units results in a large number of unique tags. As the peptide chain grows, each tag gives the exact sequence of that chain. In this case, any attached tag can be detached by UV and decoded by electron capture GC. Still and co-workers related the tagging molecules to the binary bits of the synthesis code by arranging them (T^1-T^9) in order of their GC elution order where

219.   Song, A.; Zhang, J.; Lebrilla, C.B.; Lam, K.S. *J. Am. Chem. Soc.* **2003**, *125*, 6180.
220.   Brenner, S.; Lerner, R.A. *Proc. Natl. Acad. Sci. USA* **1992**, *89*, 5381.
221.   Nielsen, J.; Brenner, S.; Janda, K.D. *J. Am. Chem. Soc.* **1993**, *115*, 9812.
222.   Ohlmeyer, M.H.J.; Swanson, R.N.; Dillard, L.W.; Reader, J.C.; Asouline, G.; Kobayahsi, R.; Wigler, M.; Still, W.C. *Proc. Natl. Acad. Sci. USA* **1993**, *90*, 10922.

T^1 is retained the longest on the GC column used and designates the rightmost bit of the binary synthesis code.[226] In this way, the tags are easily identified and the product can be identified.

An interesting solution is to color code the beads and the vessel in which the reaction is done.[223] If there were eight subunits, for example, each one could be partitioned into different containers with different color caps. If each subunit were attached to a different color bead, one bead of each color could be added to each color-coded vessel. When the next subunit is attached, the compounds formed can be sorted individually by cap and by color. This process can be contained as each new subunit is attached. Another color-coded approach can be used to identify products that are susceptible to a particular chemical reaction. The approach couples the acceptor molecule to an enzyme such as alkaline phosphatase or to a fluorescent, and then adds these in soluble form to generate the peptide-bead library.[224] Those beads that contain compounds active for this analysis will be stained and are usually visible to the naked eye or under a low-power microscope. The stained beads can then be removed for analysis. In one reported study, a monoclonal antibody active against β-endorphin was of interest because it had a high affinity for the epitope sequence UCCF,[227] and six reactive beads were obtained from 2 million screened beads from a pentapeptide library.

A powerful method to identify individual components of a combinatorial library is known as encoded combinatorial chemistry. Brenner and Lerner[224] designed a method where each chemical sequence is labeled with a so-called genetic tag (nucleotide sequences), and each tag is constructed by chemical synthesis. A monomeric chemical unit is attached to a polymeric structure and this is followed by addition of an oligonucleotide sequence. The sequence of that oligonucleotide constitutes a tag. Active products from the library are selected by binding to a receptor. After the chemical entity is bound to a target, the genetic tag can be amplified by replication, and utilized for enrichment of the bound molecules by serial hybridization to a subset of the library.[224] The nature of the chemical structure bound to the receptor is decoded by sequencing the nucleotide tag.[224] Another encoding technique is illustrated by synthesizing a peptide and a nucleotide chain in an alternating, bidirectional manner.[212] Sequencing the nucleotide chain determines the sequence of the peptide chain.

This chapter concerns synthesis. How is combinatorial chemistry pertinent to this discussion? Many small molecules can be synthesized using combinatorial techniques and evaluated for biological activity. One example is the synthesis of substituted benzodiazepines such as **282**, shown in Scheme 10.26.[225] This approach has become important ro the pharmaceutical industry

223. Guiles, J.W.; Lanter, C.L.; Rivero, R.A. *Angew. Chem. Int. Ed.* **1998**, *37*, 926.
224. Lam, K.S.; Salmon, S.E.; Hersh, E.M.; Hruby, V.J.; Kazmierski, W.M.; Knapp, R.J. *Nature* **1991**, *354*, 82.
225. (a) Bunin, B.A.; Plunkett, M.J.; Ellman, J.A. *Proc. Natl. Acad. Sci. USA* **1994**, *91*, 4708. Also see (b) DeWitt, S.H.; Kiely, J.S.; Stankovic, C.J.; Schroeder, M.C.; Cody, D.M.R.; Pavia, M.R. *Proc. Natl. Acad. Sci. USA* **1993**, *90*, 6909.

for drug discovery.[226]

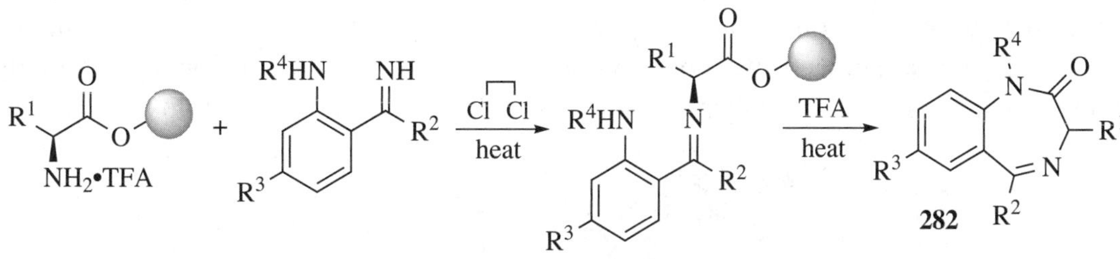

Scheme 10.26. Combinatorial Synthesis of Benzodiazepines.

Many combinatorial-based syntheses have been reported.[227] Schreiber and co-workers described a synthesis strategy that yielded diverse small molecules by combinatorial techniques.[228] Andrus and co-workers, with the goal of multidrug resistance reversal, have prepared a solution-phase indexed combinatorial library of nonnatural polyenes such as **283**.[229] Varying R and R^1 in **283** generated this library. Ellman and co-workers reported a combinatorial library of synthetic receptors targeting vancomycin-resistant bacteria,[230] and Paterson and co-workers prepared polyketide-type libraries by iterative asymmetric aldol reactions on a solid support.[231] Rieser and co-workers used combinatorial liquid-phase synthesis to prepare [1,4]-oxazepine-7-ones by the Baylis-Hillman reaction (see sec. 9.7.B).[232] Schreiber and co-workers reported the synthesis and evaluation of a library of polycyclic small molecules for use in chemical genetic assays.[233] Bauer and co-workers reported a library of *N*-substituted 2-pyrazoline compounds such as **284**, by parallel solution-phase synthesis, for screening as therapeutic agents as antibacterials, antivirals, and anti-inflammatory compounds.[234] Libraries of trisaccharides have been prepared by combinatorial methods.[235] Han and co-workers reported highly substituted thiophene derivatives such as **285** as novel phosphodiesterase-4 (PDE-4) inhibitors,[236] and Ley

226. (a) Gallop, M.A; Barrett, R.W.; Dower, W.J.; Fodor, S.P.A.; Fordon, E.M. *J. Med. Chem.* **1994**, *37*, 1233; (b) Gordon, E.M.; Barrett, R.W.; Dower, W.J.; Fodor, S.P.; Gallop, M.A. *J. Med. Chem.* **1994**, *37*, 1385; (c) Alper, J. *Science, 1994, 264*, 1399; (d) Houghten, R.A.; Pinilla, C.; Blondelle, S.E.; Appel, J.R.; Dooley, C.T.; Cuervo, J.H. *Nature (London)* **1991**, *354*, 84.

227. For reviews, see Dolle, R.E.; Le Bourdonnec, B.; Morales, G.A.; Moriarty K.J.; Salvino, J.M. *J. Comb. Chem.* **2006**, *8*, 597; Dolle, R.E. *J. Comb. Chem.; Idem* **2005**, *7*, 623; **2004,** Idem *6, 623; 2003*, 5, 693; Idem *2002, 4*, 369; Idem *2001, 3, 477*; Idem *2000*, 2, 383.

228. Burke, M.D.; Berger, E.M.; Schreiber, S.L. *J. Am. Chem. Soc.* **2004**, *126*, 14095.

229. Andrus, M.B.; Turner, T.M.; Asgari, D.; Li, W. *J. Org. Chem.* **1999**, *64*, 2978.

230. Xu, R.; Greiveldinger, G.; Marenus, L.E.; Cooper, A.; Ellman, J.A. *J. Am. Chem. Soc.* **1999**, *121*, 4898.

231. Paterson, I.; Donghi, M.; Gerlach, K. *Angew. Chem. Int. Ed.* **2000**, *39*, 3315.

232. Räcker, R.; Döring, K.; Reiser, O. *J. Org. Chem.* **2000**, *65*, 6932.

233. Tan, D.S.; Foley, M.A.; Stockwell, B.R.; Shair, M.D.; Schreiber, S.L. *J. Am. Chem. Soc.* **1999**, *121*, 9073.

234. Bauer, U.; Egner, B.J.; Nilsson, I.; Berghult, M. *Tetrahedron Lett.* **2000**, *41*, 2713.

235. Takahshi, T.; Adachi, M.; Matsuda, A.; Doi, T. *Tetrahedron Lett.* **2000**, *41*, 2599.

236. Han, Y.; Giroux, A.; Lépine, C.; Laliberté, F.; Huang, Z.; Perrier, H.; Bayly, C.I.; Young, R.N. *Tetrahedron* **1999**, *55*, 11669.

and co-workers reported the solution phase synthesis of functionalized bicyclo[2.2.2]octanes.[237] Libraries of heterocyclic compounds have been prepared from peptides and polyamides.[238] Wolf and Hawes also showed that combinatorial techniques are useful for high-throughput screening of enantioselective catalysts, allowing rapid evaluation of their utility.[239]

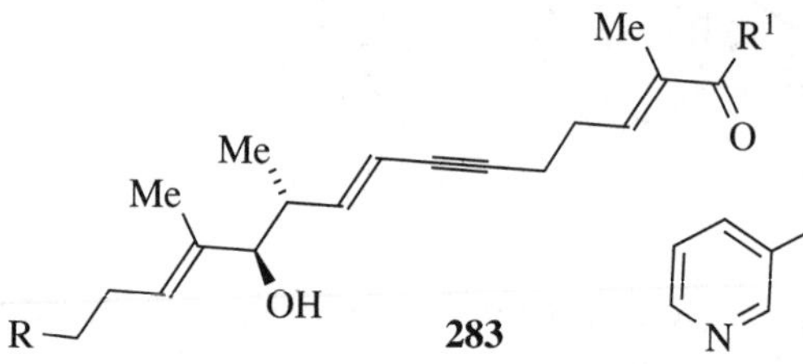

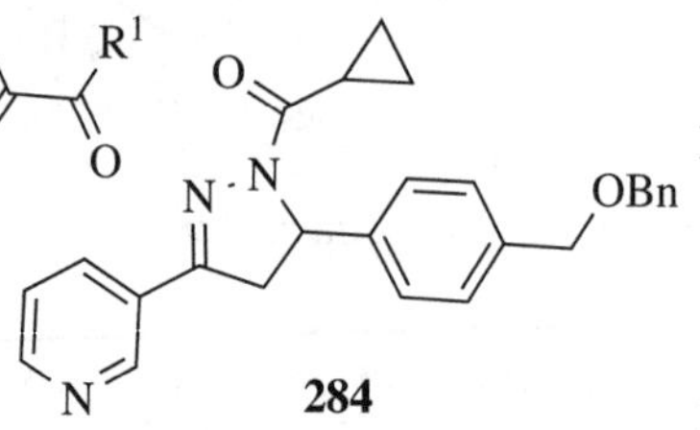

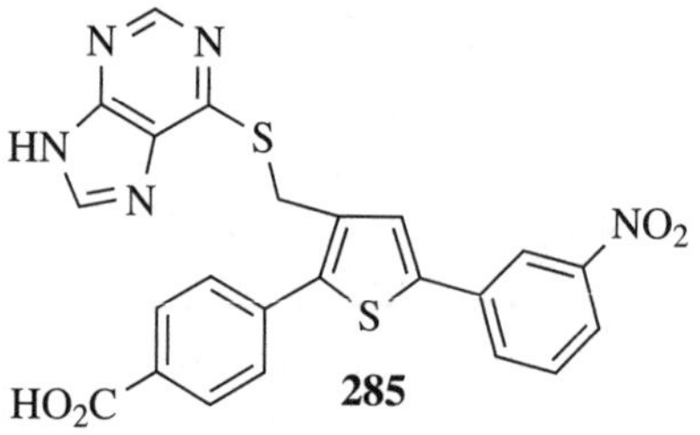

237. Ley, S.V.; Massi, A. *J. Chem. Soc. Perkin Trans. 1* **2000**, 3645.
238. Nefzi, A.; Ostresh, J.M.; Yu, J.; Houghten, R.A. *J. Org. Chem.* **2004**, *69*, 3603.
239. Wolf, C.; Hawes, P.A. *J. Org. Chem.* **2002**, *67*, 2727.

10.13. CONCLUSION

This chapter summarized general approaches to the total synthesis of complex molecules. These techniques can just as easily be applied to simple molecules, and to short synthetic sequences. One must remember that the concepts of conformational control, regio- and stereochemical control, and steric considerations (discussed in all previous chapters) must be an integral part of the planning at each step of a synthesis. These concepts should be applied to every disconnection. Planning is essential.

The key steps of a synthesis usually involve formation of carbon-carbon bonds. In Chapters 8 and 9, nucleophilic methods involving the disconnect product C^d were discussed in detail. In the chapters that follow, methods of forming carbon-carbon bonds by pericyclic reactions (disconnect products that arise from breaking more than one bond), and by reactions that have cationic intermediates (C^a disconnect products) or radical intermediates ($C^·$ disconnect products-carbenes are included in this chapter) will be presented. In each chapter, additional examples of synthetic strategy using those particular reactions will be presented.

NOTE: **For some problems you may need to consult chapters 11-13 for reactions not yet covered. The focus is on strategy and planning as much as on chemical reactions. Consult your instructor if necessary.**

1. For each of the following literature syntheses. (*1*) Determine the total number of steps and the percentage yield for each of the carbon-carbon bond-forming steps. (*2*) Identify all powerful reactions-justify each choice in the context of that synthesis. (*3*) Determine the number and percentage yield for all protection-deprotection steps (if any). (*4*) Discuss why each protecting group was chosen for that particular application.

(a)

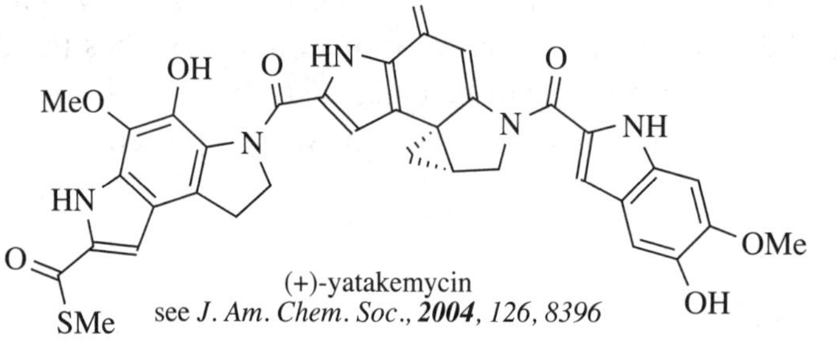

(+)-yatakemycin
see *J. Am. Chem. Soc.*, **2004**, *126*, 8396

(b)

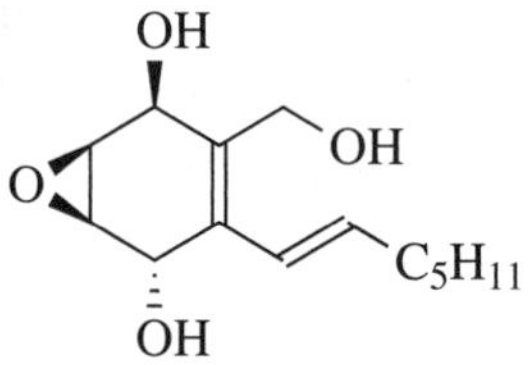

(+)-eupenoxide
see *Org.Lett.* **2004**, *6*, 2389

(c)

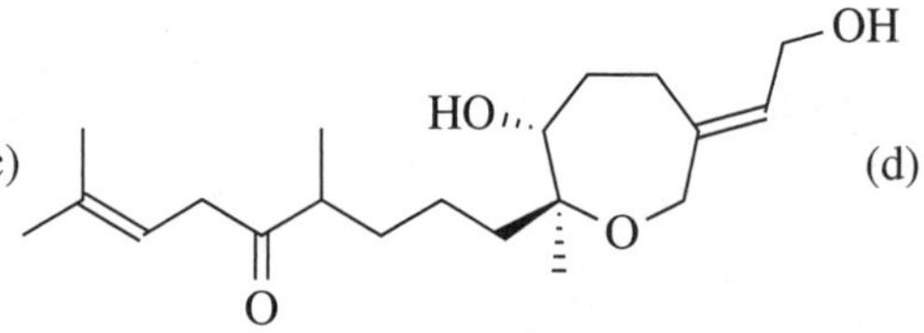

(+)-zoapatanol
see *Org. Lett.* **2004**, *6*, 2149

(d)

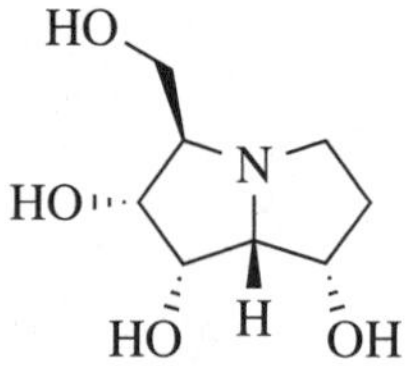

1-epiaustraline
see *Org. Lett.* **2004**, *6*, 2003

(e)

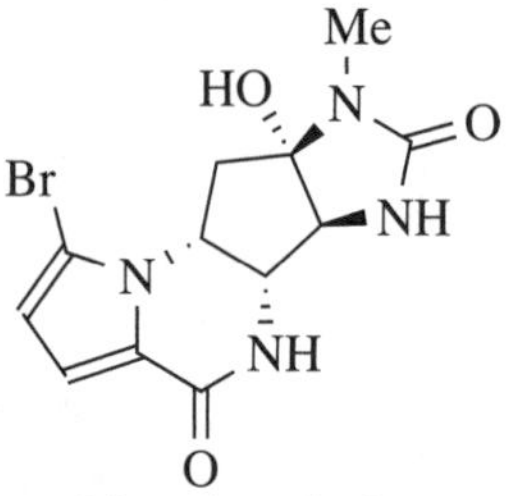

(–)-agelastatin A
see *Org. Lett.* **2004**, *6*, 2615

(f)

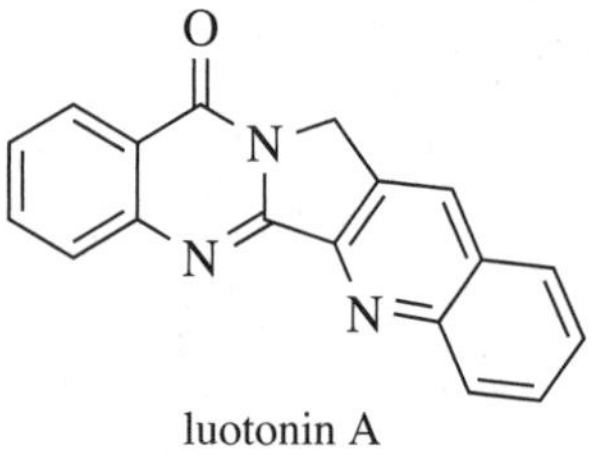

luotonin A
see *J. Org. Chem.*, **2004**, *69*, 4563

(g)

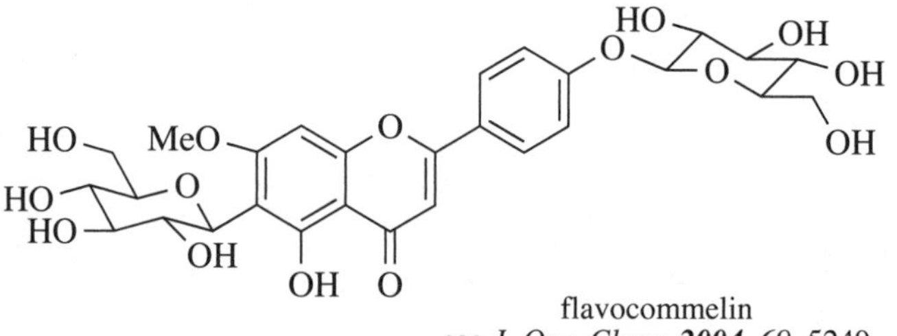

flavocommelin
see *J. Org. Chem.* **2004**, *69*, 5240

2. For each of the following: (*1*) Determine the strategic bonds. (*2*) Give two different retrosyntheses based on two different first disconnections. (*3*) Show all donor and acceptor sites in each disconnect fragment. (*4*) Give a synthesis for each molecule based on <u>one</u> of your strategic bonds. Choose a starting material that is commercially available with at

most five carbons. (Give the cost per gram of that starting material.).

(a)

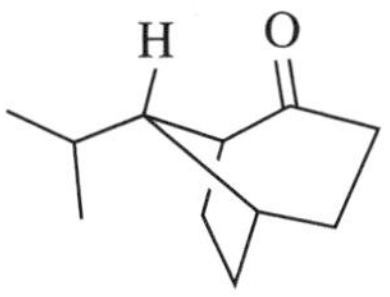

(b)

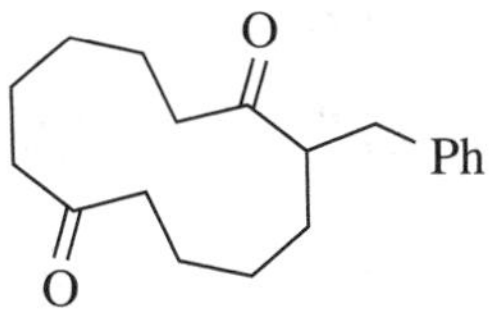

(c)

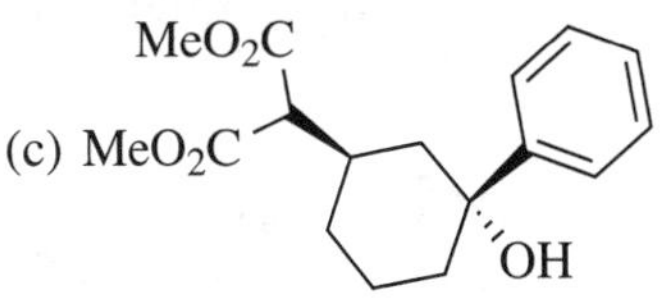

3. For each of the following: (*1*) Determine the strategic bonds. (*2*) Devise a retrosynthetic scheme based on a consecutive strategy. (*3*) Devise a retrosynthetic scheme based on a convergent strategy. (*4*) Show a total synthesis based on <u>both</u> (*2*) and (*3*).

(a)

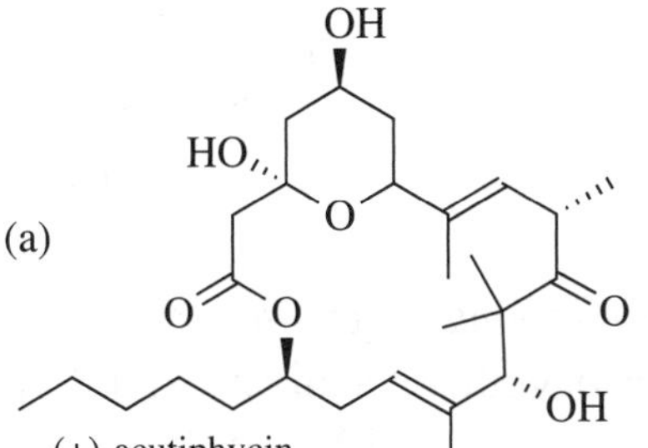

(+)-acutiphycin

(b)

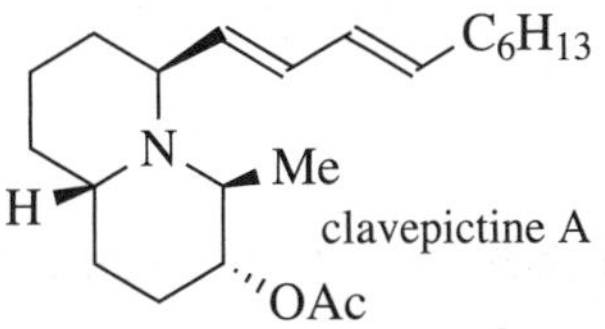

clavepictine A

(c)

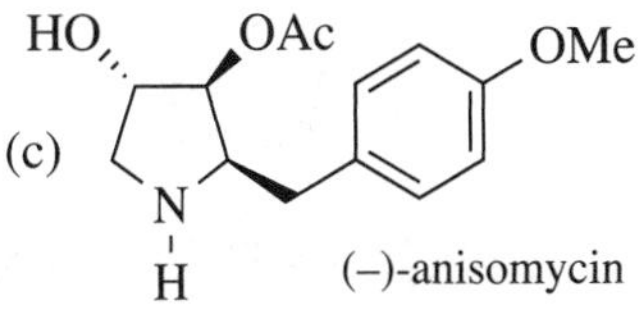

(–)-anisomycin

4. For each of the following, show a reasonable retrosynthetic analysis that gives a starting material with a reasonable cost (current chemical catalog). Show the complete synthesis. Assume an average yield of 75% for each step. What is the overall percentage yield for your synthesis? How many grams of your starting material do you need in order to synthesize 1 gram of the target?

(a)

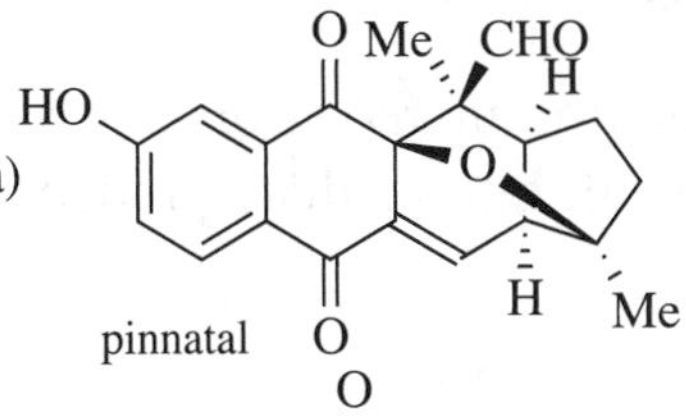

pinnatal

(b)

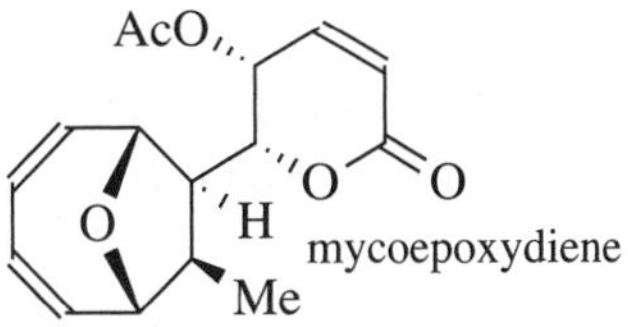

mycoepoxydiene

(c)

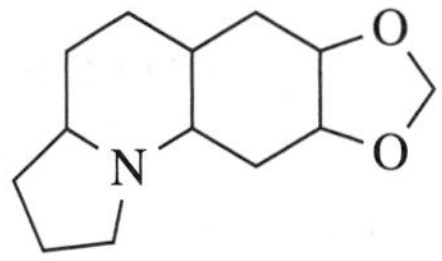

(d)

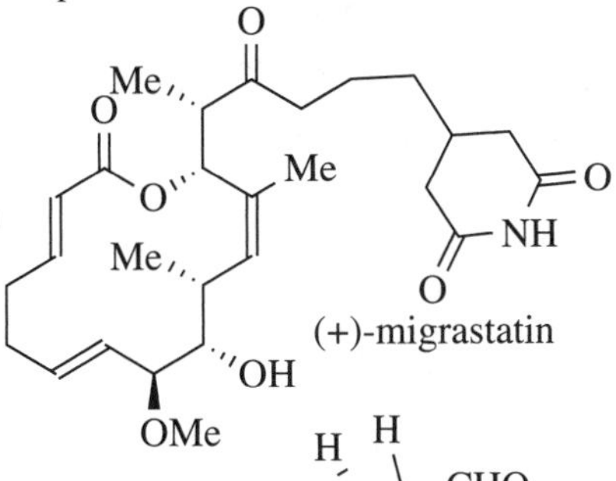

(+)-migrastatin

(e)

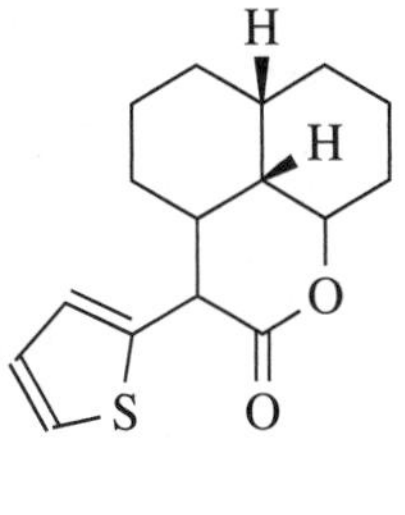

(f)

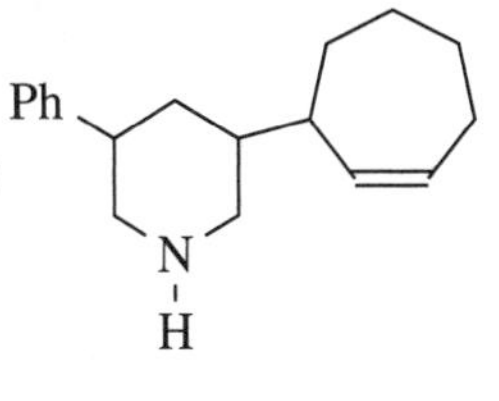

(g)

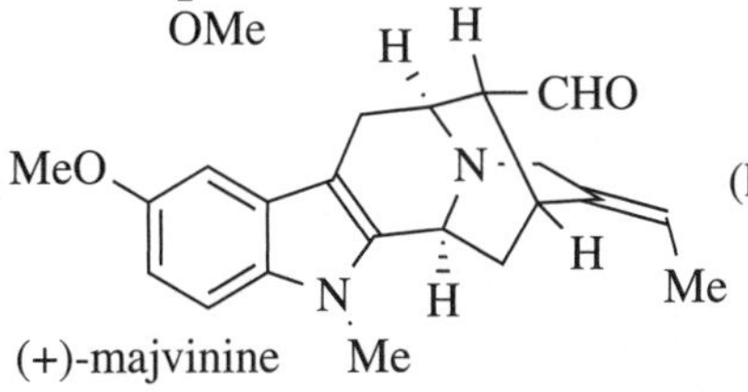

(+)-majvinine

(h)

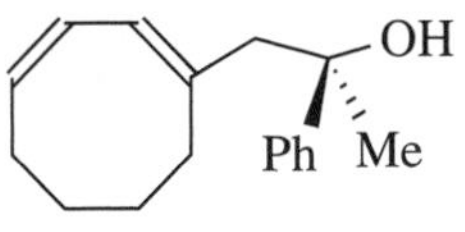

(i)

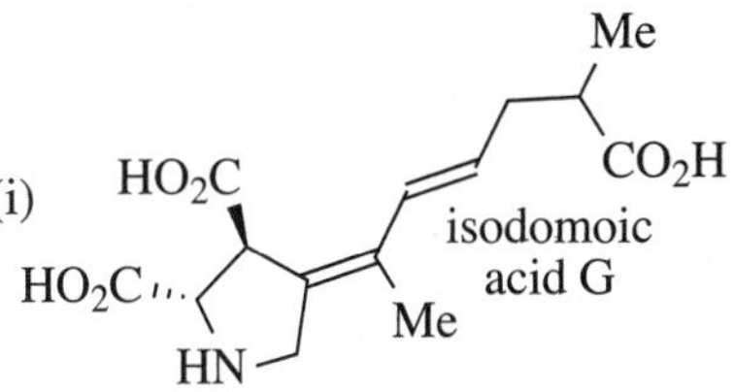

isodomoic acid G

5. Synthesize each of the following from starting materials of at most six carbons (cite chemical catalog and price per gram):

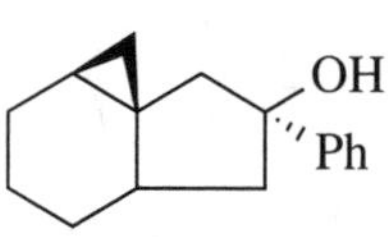

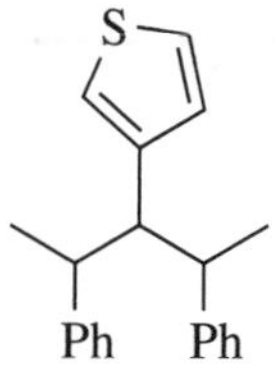

6. Suggest a complete synthesis of the molecule shown (spongidipepsin) from any starting material of no more than six carbons. Show your retrosynthetic analysis, all intermediate products and all reagents.

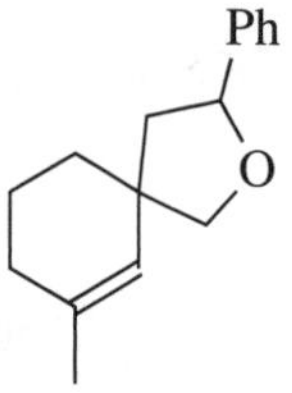

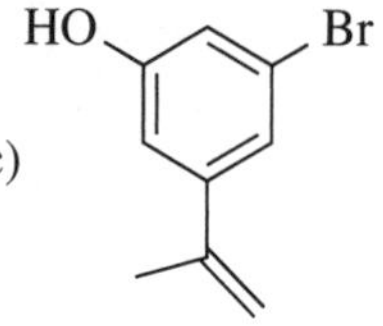

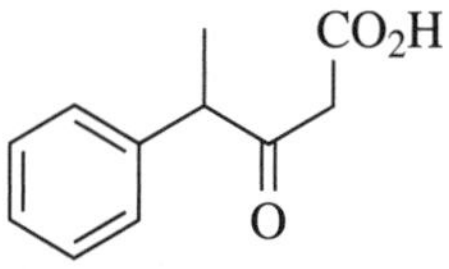

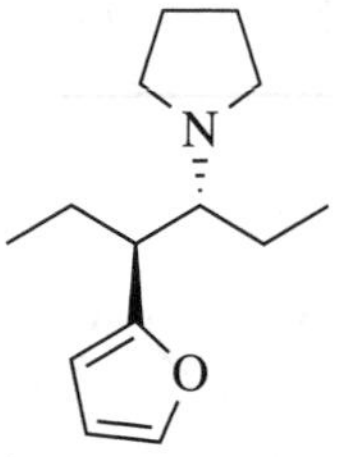

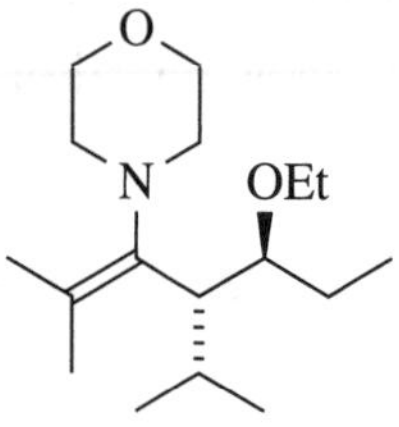

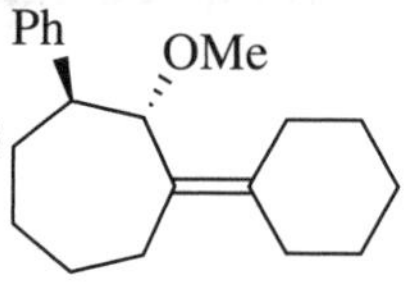

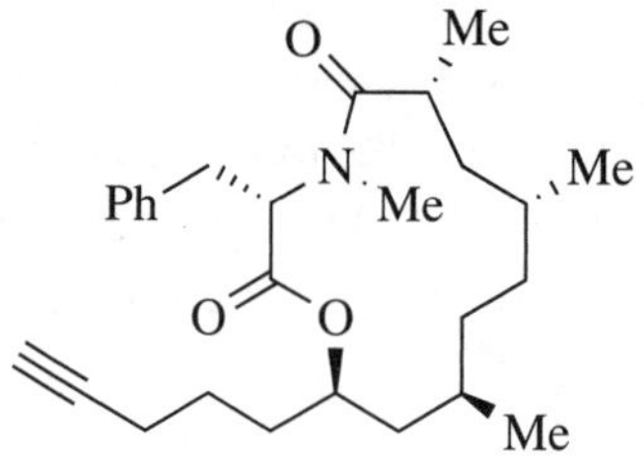

7. Suggest a complete synthesis of the molecule shown (spirofungin A) from any reasonable starting material. Show your retrosynthetic analysis, all intermediate products and all reagents.

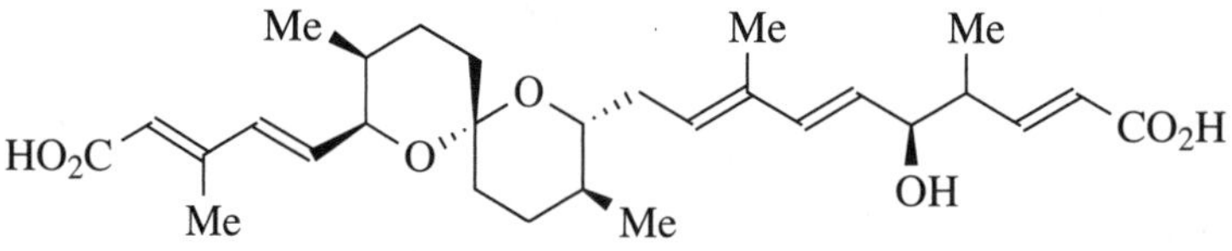

8. For each of the following show a synthesis of the designated starting material to the indicated target. Show all reagents and intermediate products.

(a)

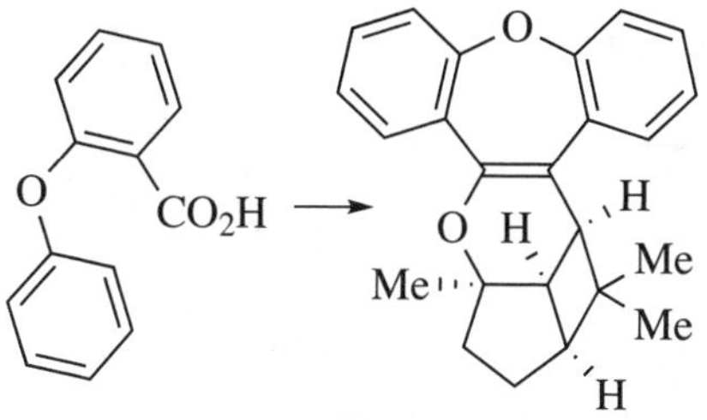

(b)

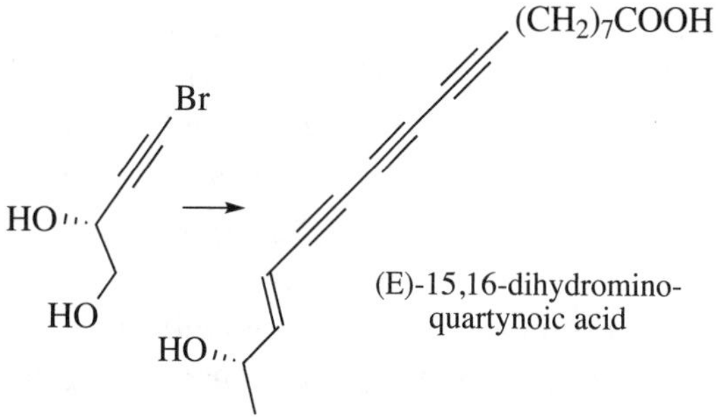

(E)-15,16-dihydromino-quartynoic acid

(c)

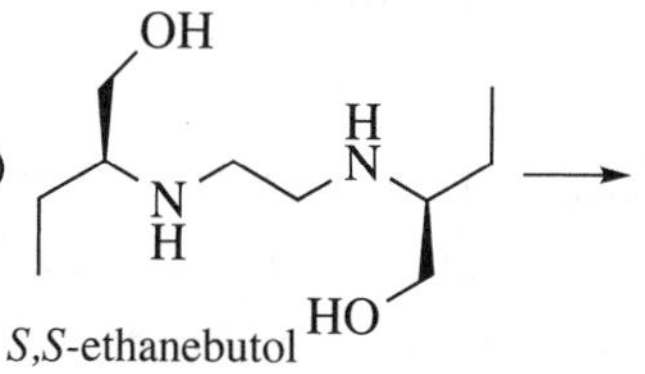

S,S-ethanebutol → L-methionine

(d)

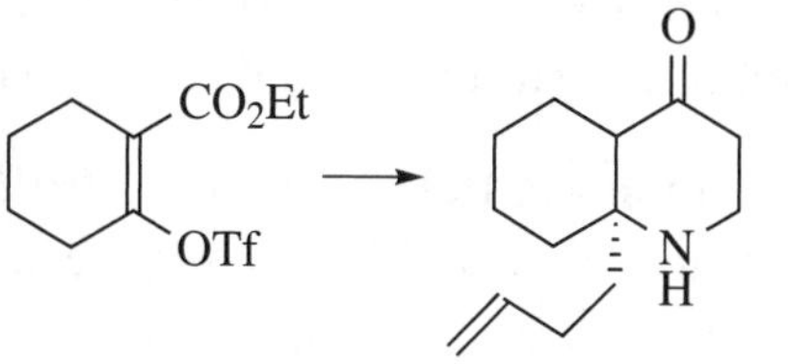

(e)

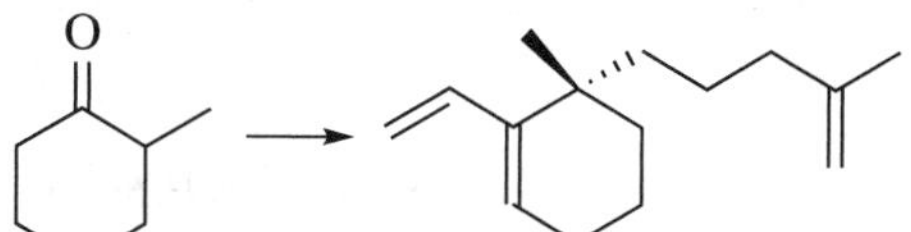

(f)

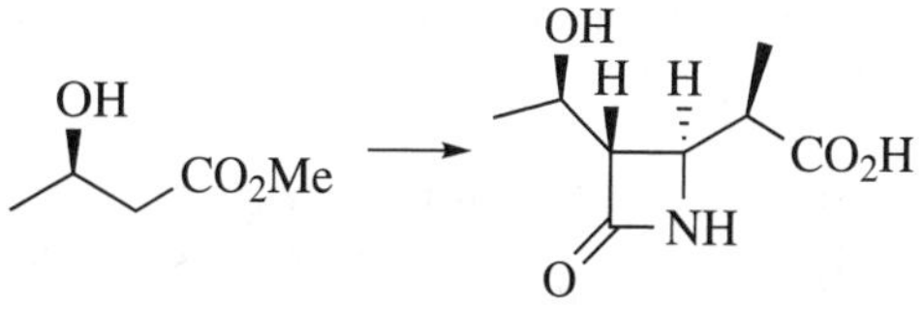

(g)

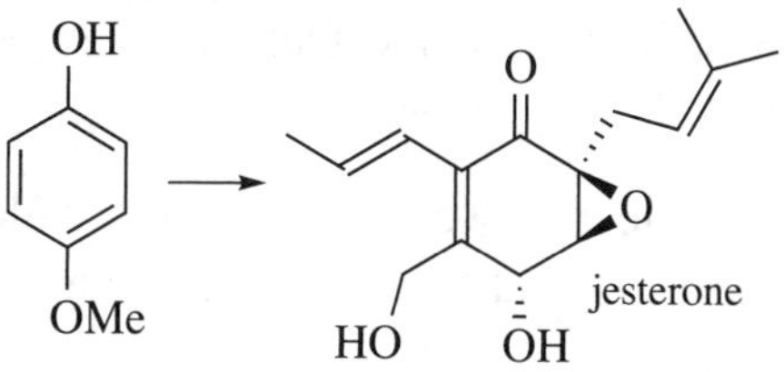

jesterone

(h) diethyl L-tartrate →

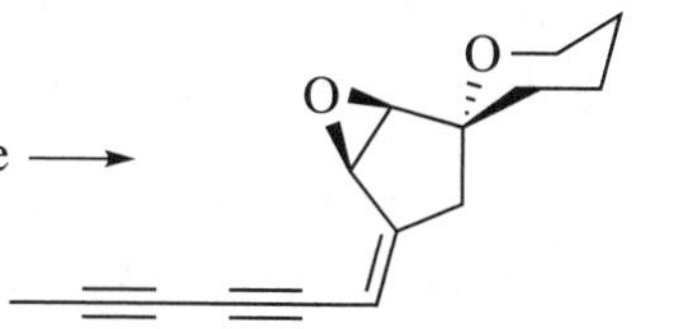

9. Construct a retrosynthetic tree based on the given template as a starting material.

(a)

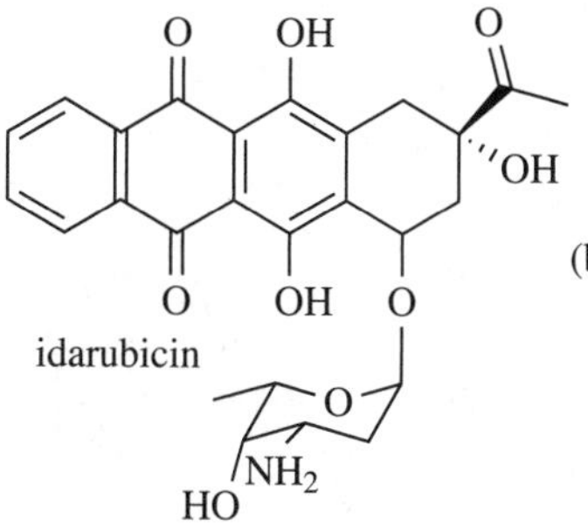

idarubicin

(b)

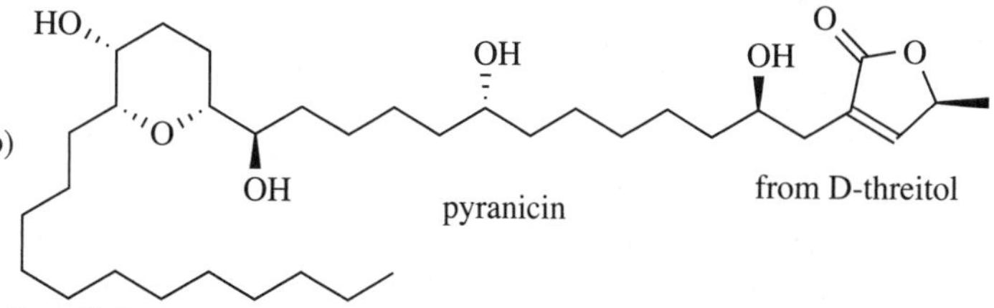

pyranicin

from L-rhamnose

from D-threitol

(c) 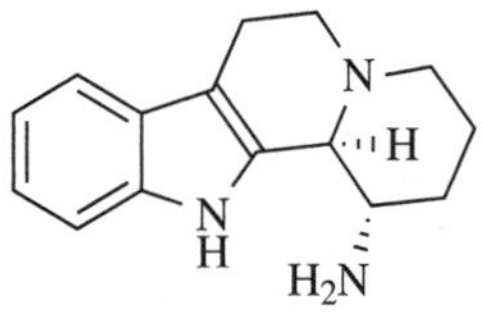 from L-glutamic acid

(d) 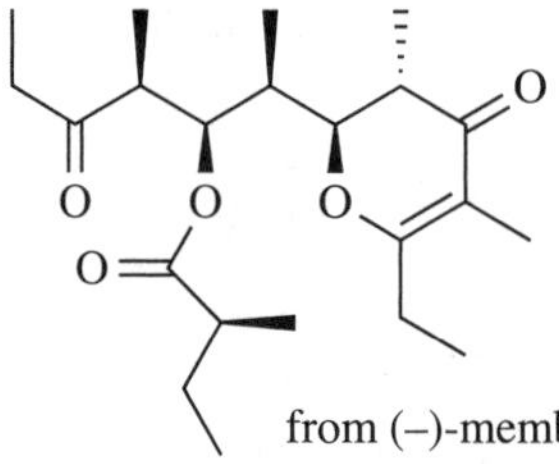 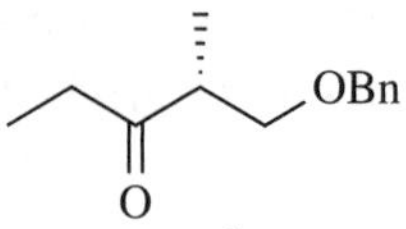 from (−)-membrenone-A

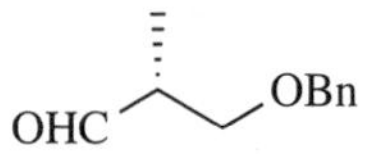

and

10. Plumet published a synthesis of (+)-7-deoxypancratistatin, starting with furan as a starting material (see *Org. Lett.* **2000**, *2*, 3683). Draw the retrosynthesis from this paper and compare and contrast it with those syntheses given in Section 10.7. Pay particular attention to strategic bonds and to strategy.

11. In a review article (*Synthesis* **1998**, 1559), Chanon presents 8 different syntheses of silphinene and 18 different syntheses of hirsutene. Based on your definitions of efficiency, cost of starting materials, control of stereochemistry, and so on, compare and contrast these syntheses. Draw a conclusion as to which you believe to be the best for each target.

12. In 1992, Sir John Cornforth presented a lecture entitled *The Trouble with Synthesis* (see *Aust. J. Chem.*, **1993**, *46*, 157). Read the article and consider the points made in the context of this chapter, particularly those quoted by Corey and Deslongchamps early in the chapter. Also see *Aldrichimica Acta*, **1994**, *27*, 71.

13. Each reference discusses a proposed biosynthetic pathway for the molecule shown. Construct a retrosynthetic tree based on that pathway and then show the total synthesis.

(a) 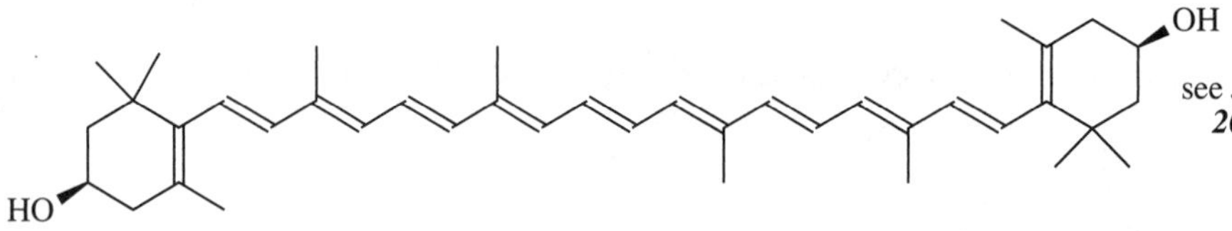 see *J. Org. Chem.*, **2002**, *67*, 871

(b) 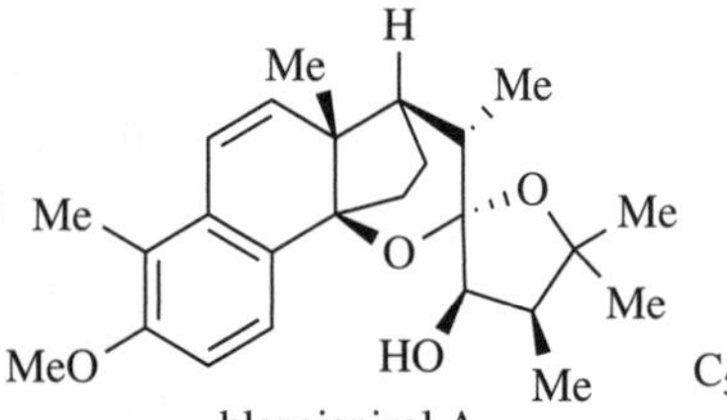 blazeispirol A
see *Tetrahedron*, **2002**, *58*, 10251

(c) 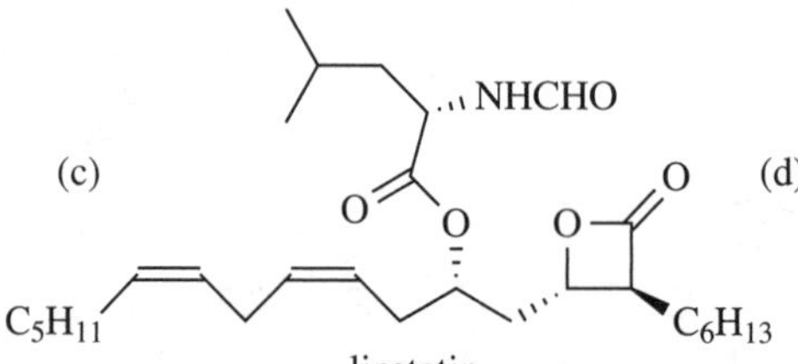 lipstatin
see *J. Org. Chem.*, **2002**, *67*, 2257

(d) 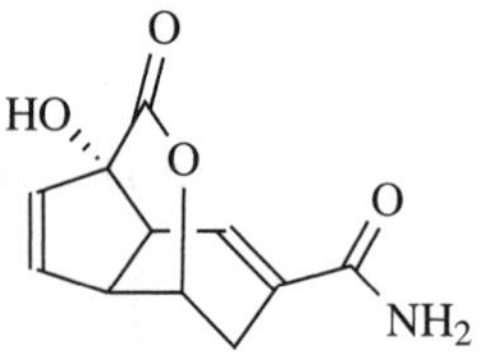 echnisosporin
see *Eur. J. Org. Chem.*, **2002**, 983

(e) 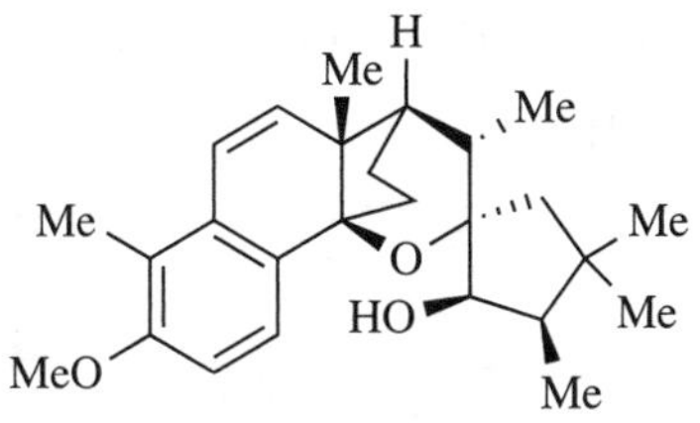 blazeispirol A
see *Tetrahedron Lett.*, **2000**, *41*, 6101

(f) 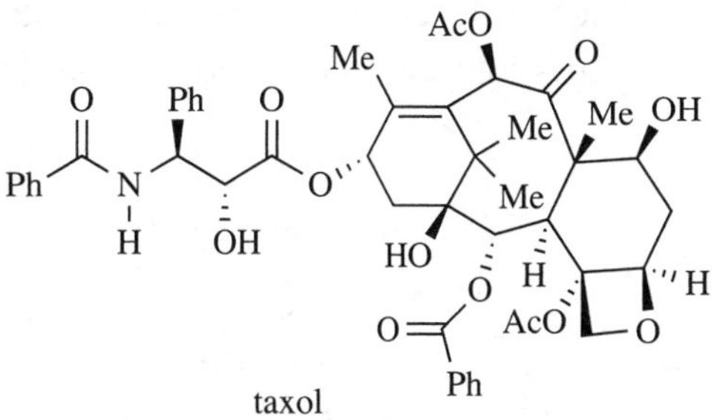 taxol
see *J. Org. Chem.*, **2000**, *65*, 7865

14. Each of the following references give mass spectral fragmentation information for the molecule shown. Construct a retrosynthetic tree based on this information and then give the total synthesis.

(a) 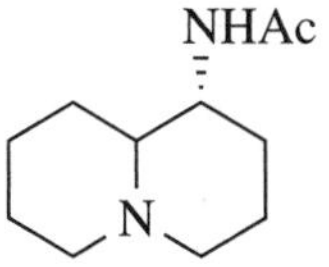 epiquinamide
see *J. Nat. Prod.*,
2003, *66*, 1345

(b) 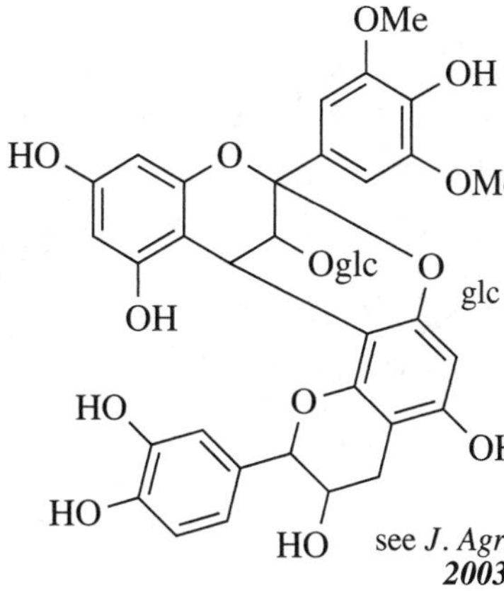 glc = glucose
see *J. Agric. Food Chem.*
2003, *51*, 3592

(c) 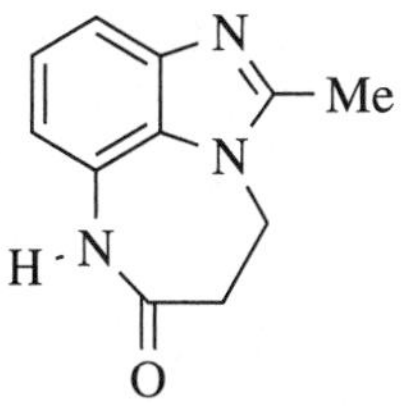 *Org. Mass Spectrom.*,
1978, *13*, 141

(d) ciguatoxin congener
CTX3C ($C_{60}H_{86}O_{19}$)
see *J. Am. Chem. Soc.*,
2000, *122*, 4988

(e) 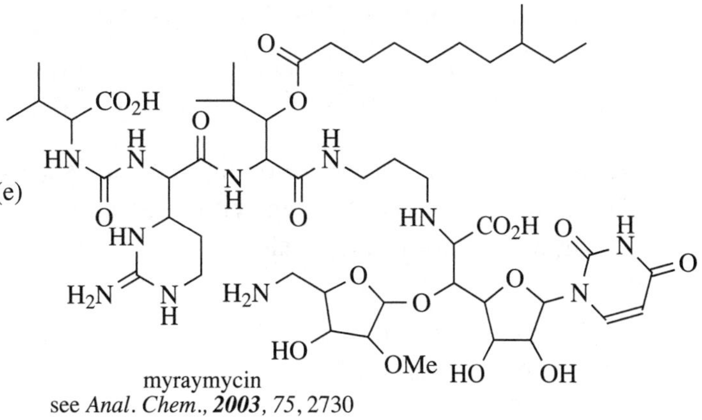 myraymicin
see *Anal. Chem.*, **2003**, *75*, 2730

15. Each reference describes a chemical degradation of that structure. Construct a retrosynthetic tree based on that degradation scheme and then show the total synthesis.

(a)

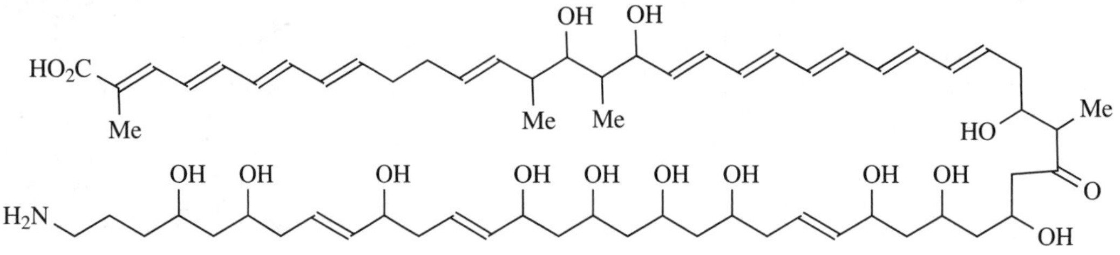

linearmycin A
Org. Lett., **2003**, 5, 93

(b)

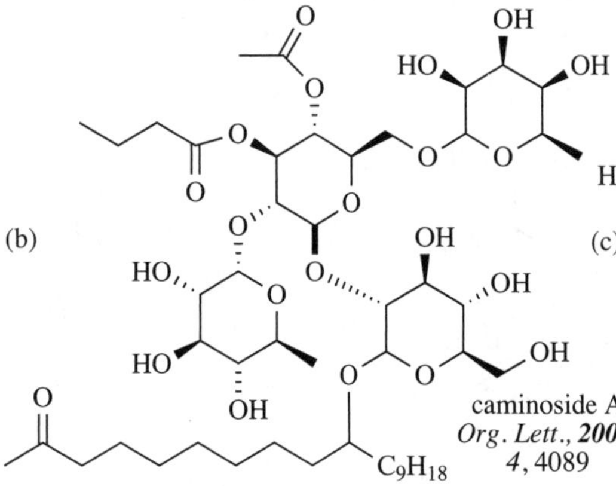

caminoside A
Org. Lett., **2002**,
4, 4089

(c)

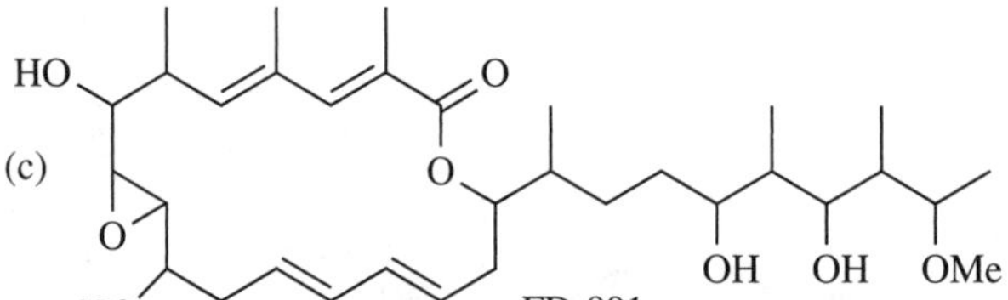

FD-891
Org. Lett., **2002**, 4, 3383

(d)

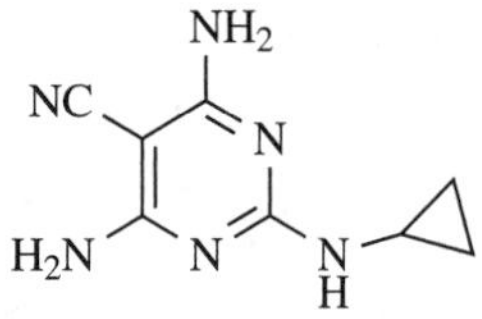

dicyclanil
J. Agric. Food. Chem., **2002**, 50, 5115

(e)

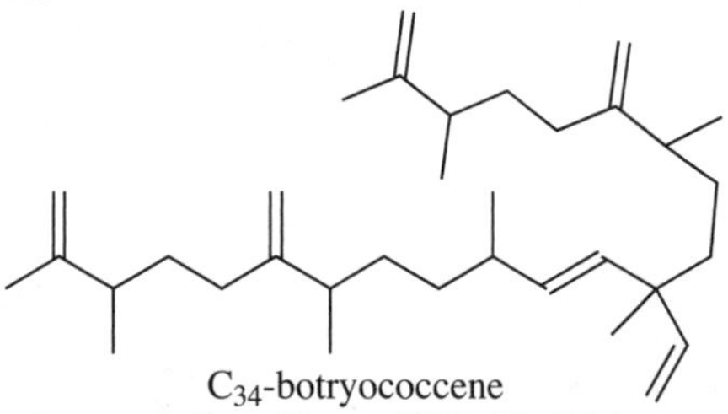

C$_{34}$-botryococcene
see *J. Org. Chem.*, **1992**, 57, 4991

16. Devise a synthesis for the following molecules showing all chemical steps and reagents. The literature reference given with each molecule refers to its isolation and structure determination, not its synthesis.

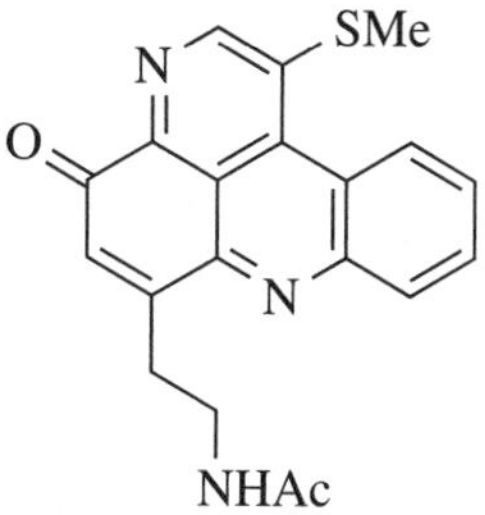

isodiplamine
Tetrahedron, **2002**, *58*, 9779

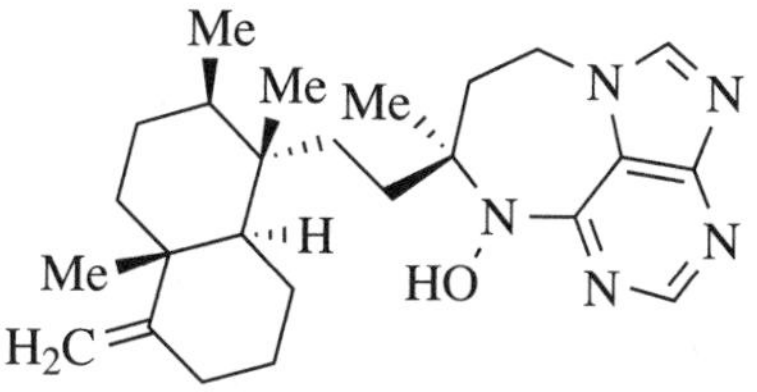

asmarine A
J. Nat. Prod., **2000**, *63*, 299

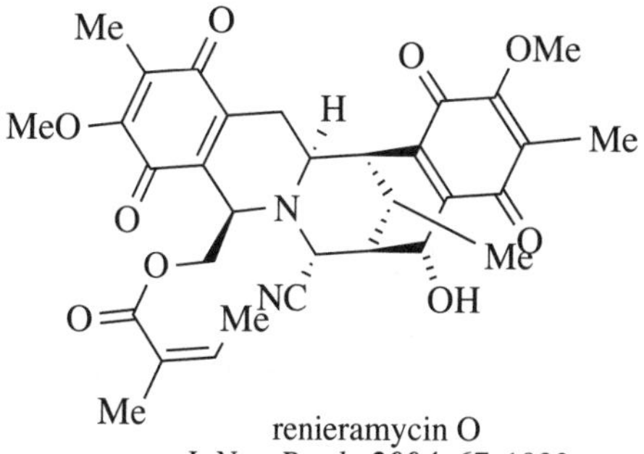

renieramycin O
J. Nat. Prod., **2004**, *67*, 1023

17. For each of the following, use a convergent strategy to devise a synthetic scheme. Show the total synthesis based on that scheme.

(a) (−)-lemonomycin

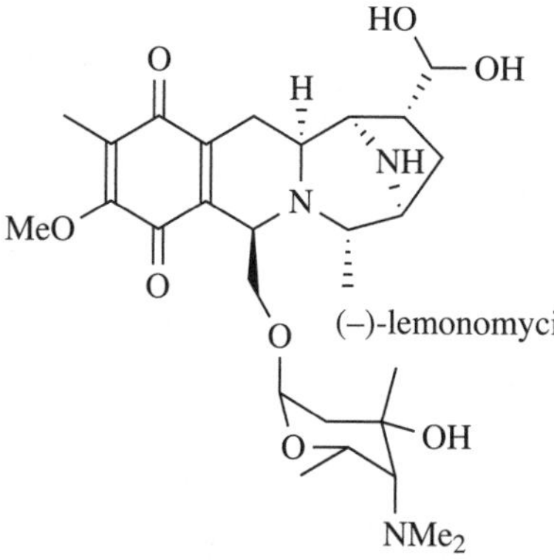

(b) *cis*-3,4-dihydrohamacanthin B

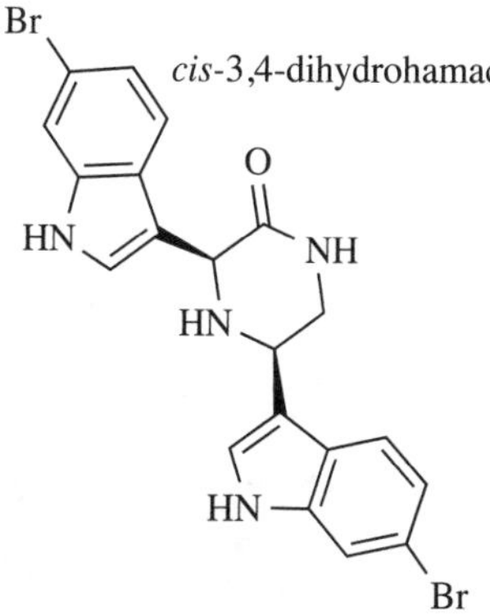

(c) (−)-subersic acid

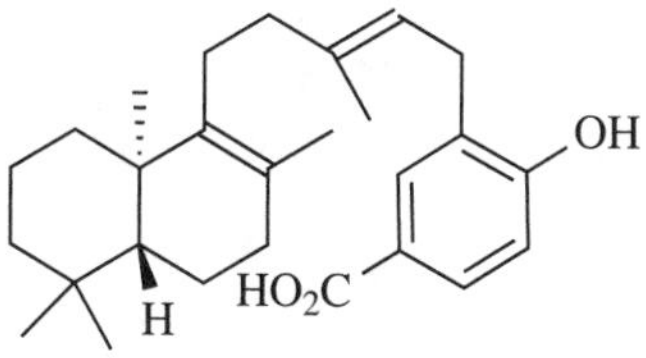

(d) (+)-macquarimicin A

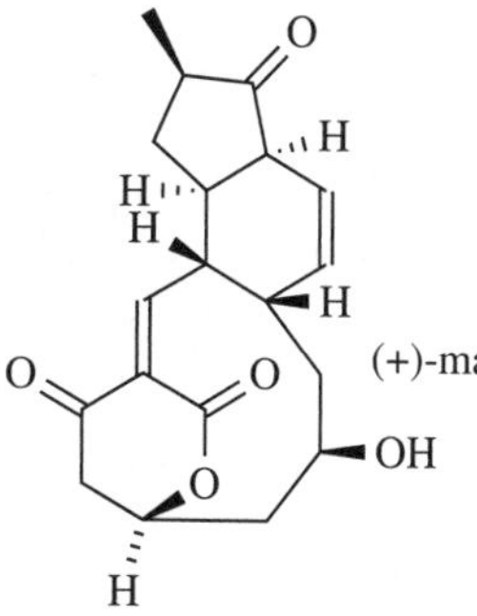

18. Search the literature to find four <u>different</u> syntheses of the molecule epothilone. Compare and contrast them. Can you devise a new approach?

chapter 11

Pericyclic Carbon-Carbon Bond Forming Reactions: Multiple Bond Disconnections

11.1. INTRODUCTION

Reactions involving ionic or highly polarized compounds comprise the majority of organic reactions, as seen in other chapters. Chapters 8 and 9 discussed methods for making carbon bonds via nucleophilic carbon species. Chapter 12 will discuss carbon-carbon bond formation with reagents involving electrophilic carbon, and chapter 13 will discuss making carbon-carbon bonds via radical, carbene, and related processes. The reactions in these chapters involve one-carbon disconnections for the most part. There are reactions that involve two-bond disconnections, however, which lead to disconnect products that are readily assembled by known reactions. Many of these reactions generate carbon-carbon bonds by electronic reorganization of molecules, rather than via ionic or radical species in what are generically cal led **pericyclic reactions**. The **Diels-Alder reaction**, **Claisen rearrangement**, and the **Cope rearrangement** fall into this category. Without question, the synthetic utility of these reactions ranks them among the highest in importance. This chapter will focus on methods that use pericyclic reactions, and it will present a brief overview of the mechanistic basis for each reaction. As in previous chapters, however, the emphasis will be on the synthetic utility of each reaction.

11.2. FRONTIER MOLECULAR ORBITAL THEORY

11.2.A. Frontier Orbitals

This chapter deals with molecules containing π bonds: alkenes, dienes, carbonyls, imines, nitriles, and so on. In a typical alkene, bonding is described in terms of the σ- and π-orbitals. Electronic interactions are represented by the molecular orbitals, using σ-orbitals (sp^2 hybridized

1   **2**

for alkenes, **1**) and π-orbitals (**2**). These orbitals take on a directional character [represented by (+ or dark shading) and (– or light shading)] when they are in close proximity to another atom, orbital, or polarized species, as is the case in a covalent bond. There are, therefore, two possibilities when two sp^2 hybridized orbitals come together. Orbitals of the same sign can be directed toward each other, as in **3**, or orbitals of opposite sign can be directed toward each other, as in **4**. In **3**, the electron density between the two nuclei is maximized (electronic attraction for the positive nucleus), but in **4**, it is minimized. Since a strong covalent bond (mutual sharing of electron density between nuclei) is associated with significant electron density between the nuclei, **3** represents a **bonding** interaction, and **4** is usually viewed as an **antibonding** interaction. These terms have meaning only when considering the outermost orbital containing electrons, the valence electrons (valence orbitals).

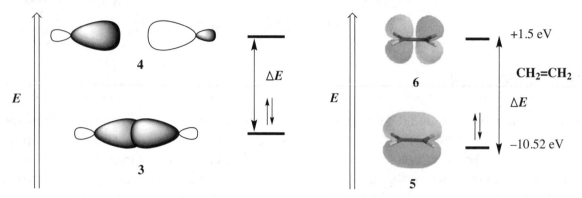

Figure 11.1. Bonding and Antibonding σ- and π-Orbitals.

Similar arguments can be made for the overlap of the p-orbitals that constitute a π bond. In **5**, the overlap of orbitals with the same sign represents the maximum bonding interaction for the π-bond of ethylene (a bonding molecular orbital), and **6** represents an antibonding molecular orbital. As with a σ-bond, the net energy of the bonding interaction in a π bond is lower than that of the antibonding interaction, as shown in Figure 11.1, and it is more stable. The energy of each molecular orbital can be measured, and this energy difference (ΔE) will vary with substituents on the alkene. This observation has important implications for the reactivity of these compounds, and this concept will be discussed below.

The π-orbitals are the key to understanding pericyclic reactions, and we will focus on the π bond of alkenes and dienes in this section. The difference in energy between the two π molecular orbitals (**5** and **6**) can be represented in terms of their relative energies, as shown in Figure 11.1, taken from Houk's work.[1] The lowest energy orbital containing the bonding electrons is called the **highest occupied molecular orbital**, the **HOMO**. Experimentally, the energy of the HOMO is the **negative** of the **ionization potential (IP)** of the molecule.[2] In terms of chemical reactivity, the HOMO is the orbital that donates electrons, and the IP is a reasonable measure of its energy. The higher energy orbital does not contain electrons, but is the next available energy level if electrons are available. The energy that describes accepting electrons is the **electron affinity (EA)**, and this orbital is called the **lowest unoccupied molecular orbital**, the **LUMO**. Experimentally, the energy of the LUMO is taken to be the **negative** of the EA of the alkene.[1,2] *In a simplistic reaction model, the LUMO would be expected to accept electrons from the HOMO*

1. Houk, K.N.; Sims, J.; Watts, C.R.; Luskus, L.J. *J. Am. Chem. Soc. 1973, 95,* 7301.
2. Houk, K.N. in *Pericyclic Reactions, Vol. II*, Marchand, A.P.; Lehr, R.E. (Eds.), Academic Press, New York, *1977*, p. 203.

of another π bond. The difference in energy between the HOMO and LUMO (ΔE) is determined by the difference of the IP (-10.52 eV for ethene[2] where 1 eV = **electron volt** = 23.06 kcal mol^{-1} = 96.53 kJ mol^{-1}), and the EA is +1.5 eV (for a ΔE of 12.02 eV = 277.2 kcal mol^{-1} = 1160.4 kJ mol^{-1}).

The HOMO and LUMO of ethene (or of any other molecule) are referred to as **frontier molecular orbitals (FMO)**, after the pioneering work of Fukui.[3] Analyzing several different types of reactions with FMO theory, the HOMO was believed to deliver electrons to the LUMO of a reactive center that could accept the charge (as mentioned above). The FMO theory has now been shown to be "a first approximation to a perturbation treatment of chemical reactivity".[4] Perturbation theory,[5] as first described by Coulson and Longuet-Higgins,[6] treats the molecular orbital of two interacting components as a perturbation of the product of their individual orbitals.[7] *The theory fails to explain large perturbations, and does not predict the transition state, but gives "an estimation of the slope of an early part of the path along the reaction coordinate leading to the transition state".*[16] This theory usually allows one to identify the higher and the lower energy transition state, and this information is useful for predicting reactivity. The transition state energies can be described as reactant-like (early transition state) or product-like (late transition state). The well-known **Hammond postulate** states "that transition states for exothermic reactions are reactant-like, and for endothermic reactions are product-like".[8] Frontier orbital effects are therefore particularly important in exothermic reactions.[9]

Analysis of a reaction by frontier orbital theory has additional benefits, particularly for predicting reactivity and stereochemistry. Woodward and Hoffman[10] pointed out "that electrocyclic reactions followed the stereochemistry dictated by the symmetry, or nodal properties of the HOMO of the polyene".[11] This concept of orbital symmetry will be important for discussions of all pericyclic reactions. Of particular importance is the difference in energy between the HOMO of one π system, and the LUMO of a second π-system, because this energy difference (ΔE) will be used to predict reactivity in pericyclic reactions (see below). A cautionary note was raised by Spino and co-worker's,[12] who showed that the HOMO$_{\text{diene}}$–LUMO$_{\text{alkene}}$ values should be used with caution when attempting to predict inverse electron demand Diels-Alder

3. (a) Fukui, K.; Yonezawa, T.; Nagata, C.; Shingu, H. *J. Chem. Phys.* **1954**, *22*, 1433; (b) Fukui, K. in *Molecular Orbitals in Chemistry, Physics and Biology*, Löwdin, P.-O.; Pullman, B. (Eds.), Academic Press, New York, **1964**, p. 513.
4. Reference 2, p 183.
5. (a) Dewar, M.J.S.; Dougherty, R.C. *The P.M.O. Theory of Organic Chemistry*, Plenum Press, New York, **1975**; (b) Dewar, M.J.S. *Molecular Orbital Theory for Organic Chemists*, McGraw Hill, New York, **1969**.
6. Coulson, C.A.; Longuet-Higgins, H.C. *Proc. Roy. Soc, A* **1947**, *192*, 16.
7. Fleming, I. *Frontier Molecular Orbitals and Organic Chemical Reactions*, Wiley, London, **1976**, p 23.
8. Hammond, G.S. *J. Am. Chem. Soc.* **1955**, *77*, 334.
9. Reference 7, p 24.
10. (a) Woodward, R.B.; Hoffman, R. *Angew. Chem. Int. Ed.* **1969**, *8*, 781; (b) *Idem The Conservation of Orbital Symmetry*, Verlag-Chemie GmbH/Academic Press, Weinheim, **1971**.
11. Reference 2, pp 183-184.
12. Spino, C.; Rezaei, H.; Dory, Y.L. *J. Org. Chem.* **2004**, *69*, 757.

reactions (see sec. 11.5), where the asynchronicity of the transition state should be taken into account. Orbital coefficients (see sec. 11.2.B) can be used to understand both reactivity and regiochemistry, however.[13]

11.2.B. HOMO, LUMO Energies and Orbital Coefficients

Figure 11.1 showed the molecular orbitals for ethylene, and these are shown again in Figure 11.2. The magnitude of each orbital, called the **orbital coefficient**, is now shown. The + and − signs correlate with the symmetry of the orbital,[2] which is somewhat arbitrary, but useful for examining the directionality of two reactive orbitals. When two π systems differ in their substitution pattern, there will be differences in the magnitude of orbital densities (larger or smaller orbitals). Predictions concerning reactivity must, therefore, indicate the magnitude of electron density in each orbital.

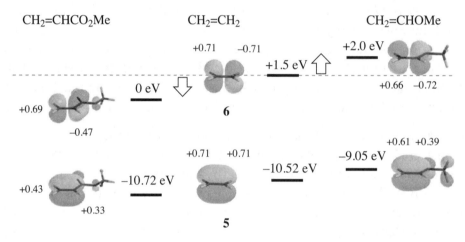

Figure 11.2. Changes in Alkene Homo and Lumo Energies Upon Introduction of Substituents.

The orbital coefficients in the HOMO of a symmetrical molecule such as ethene are identical in magnitude and sign. The LUMO coefficients are equal in magnitude but opposite in sign, also a result of this symmetry, and this leads to no node in **5**, and one node in **6** (a **node** is a *point of zero electron density* relative to the orbital lobes). The importance of orbital coefficients will be discussed in Section 11.4.C. Low energy orbitals are associated with a minimal number of nodes, and higher energy orbitals with increasing numbers of nodes. If we change the structure of ethene by replacing a hydrogen atom with an electron-withdrawing group, as is the case for methyl acrylate, electron density is distorted away from the π bond of the alkene, toward the carbonyl. This distortion will influence the energies of the HOMO and LUMO as well as the magnitude of the orbital coefficients. The energy of the HOMO of methyl acrylate is -10.72 eV and the LUMO energy is 0 eV.[13] These energy values are lower than the frontier orbitals of ethene (see Figure 11.2), which is shown in Figure 11.2,[2] and the dashed line marks

---

13. Reference 2, pp 203-205.

the energy of the LUMO of ethene as the reference point. The LUMO energy is affected to a greater extent than the HOMO energy. As seen in Figure 11.2, the electron-withdrawing group diminishes the size (orbital coefficient) of the orbitals closest (*proximal*) to the δ+ carbonyl carbon, leaving the most distant (*distal*) orbital with the largest coefficient.

The opposite effect is expected when a hydrogen of ethene is replaced with an electron releasing group such as OMe, as is the case with methyl vinyl ether (see Figure 11.2). The oxygen atom releases electrons toward the π bond, raising the energy of the orbitals relative to ethene (to -9.05 eV for the HOMO and +2.0 eV for the LUMO).[2] In Figure 11.2,[2] the HOMO coefficients of the carbon proximal to the methoxy are diminished (0.39 vs. 0.61), whereas the orbital density of the distal carbon is increased. In the LUMO, however, the orbital on the proximal carbon is slightly larger than that on the distal carbon. In general, the HOMO coefficients for methyl vinyl ether are typical of other electron-releasing substituents, but the LUMO coefficients are often closer to each other in magnitude. Table 11.1 shows the HOMO and LUMO energies for several common alkenes, along with the HOMO and LUMO coefficients, where available.

Table 11.1   HOMO and LUMO Energies and Orbital Coefficients of Common Alkenes

Alkene	HOMO (eV)	c1	c2	LUMO (eV)	c3	c4
$H_2C=CH_2$[a]	−10.52	0.71 [.63]	0.71 [.63]	1.5	0.71 [.86]	−0.71 [.86]
$H_2C=CHCl$[a]	−10.15	0.44	0.30	0.5	0.67	−0.54
H_2CHMe[a]	−9.88	0.67	0.6	1.8[a,b]	0.67	−0.65
MeHC=CHMe	−9.13[c]	[−.61]	[−.61]	2.22[d]	[−.78]	[.78]
EtHC=CH_2	−9.63[e]	[−.64]	[−.59]	2.01[e]	[−.80]	[.80]
cyclohexene	−8.94[f,g]	[−.60]	[−.60]	2.1[g]	[−.79]	[.79]
$H_2C=CHPh$	−8.48	0.49[−.43]	0.32[−.29]	0.8	0.48[−.51]	−0.33[.35]
$H_2C=CHCH_2OH$[a,d]	−10.1	−0.31	−0.36	1.1	−0.64	−0.56
$H_2C=CHOMe$[a]	−9.05(−8.93)[c]	0.61[−.60]	0.39[−.42]	2.0	0.66[−.79]	−0.72[.83]
$H_2C=CHOEt$	−9.03[a,c]			5.53[a,c]		
$H_2C=C(OEt)_2$	−8.68[a,c]			6.28[a,c]		
$H_2C=CHSMe$[a]	−8.45	0.34	0.17	1.0	0.63	−0.48
$H_2C=CHNMe_2$[a]	−9.0	0.50	0.20	2.5	0.62	−0.69
$H_2C=CHCO_2H$[b]	−10.93	[−.49]	[−.47]	2.91	−0.63[−.65]	0.46[.47]
$H_2C=CHCO_2Me$	−10.72(−10.64)[a,c]	0.43	0.33	0(3.12)[a,c]	0.69	−0.47
$MeO_2CHC=CHCO_2Me$[a,d]	−10.95[a,c]			1.94[a,c]		
$H_2C=C(CO_2Et)_2$	−10.96[a,c]			2.43[a,c]		
$H_2C=CHCHO$	−10.89[b]	0.58[a][−.59]	0.48[a][−55]	0.60[b]	−0.58[b][−.62]	−0.40[b][.43]
$H_2C=CHCHO•BF_3$[b]	−12.49			0.43	−0.53	0.25
methyl vinyl ketone	−10.16[s]	[−.10]	[−.18]		[−.61]	[.44]
cyclopentenone	−9.34[c,q]	0.56[a,b]	0.70[a,b]	−0.64[q]	−0.74[a,b]	0.52[a,b]

Table 11.1 HOMO and LUMO Energies and Orbital Coefficients of Common Alkenes (con't).

Alkene	HOMO (eV)	c1	c2	LUMO (eV)	c3	c4
p-benzoquinone	−10.29[h]	[−.34]	[−.34]	−1.91[i]	[−.33]	[.33]
maleic anhdyride	−11.95[h](−12.13)[a,c]	[−.34]	[−.34]	−0.57[j](1.03)[a,c]	[−.51]	[.51]
N-phenylmaleimide	−10.64[o]	[−.10]	[.10]	−0.49[i]	[.46]	[−.46]
$H_2C=CHCN$	−10.92	0.60(.64)[a,b]	0.49(0.6)	0[a,b]	0.68(.55)[a,b]	−0.54(−.67)[a,b]
$H_2C=C(CN)_2$[r]	−11.38	0.61	0.45	−1.54	0.66	−0.49
$(NC)_2C=C(CN)_2$	−11.8[k]			−1.80[l](−2.03)[m]		
$H_2C=CHNO_2$[a]	−11.4	0.62[.005]	0.60[−.002]	0.7	0.54[.50]	−0.32[.30]
$HC\equiv CH$	−11.4[g]	[−.62]	[−.62]	2.6[g]	[−.86]	[.86]
$MeC\equiv CMe$	−9.9[c]			3.43[n]		
$HC\equiv CCN$[r]	−11.81	0.56	0.43	0	0.57	−0.41
$MeO_2CC\equiv CCO_2Me$	−11.5[p]	[.37]	[.37]	−0.60[f]	[.42]	[−.42]
acetaldehyde[z]	−11.49	0.44[aa]		4.21	[.44]	[−.44]
thioacetaldehyde[z]	−9.25	0.62[aa]		1.81	0.64	−0.62[.09][u]
phenanthrene (C9-C10)	−8.1[c,q]	[−.37]	[.37]	−0.31[q]	0.65	−0.66[.07][u]
$H_2C=C=CH_2$[t]	−10.14	−0.47	−0.56(.20)[u]	2.4	0.67	−0.63(.10)[u]
$H_2C=C=CMe_2$[t]	−9.67	−0.66	−0.63(.09)[u]	0.78	0.54	−0.65(.05)[u]
$H_2C=C=CHOMe$[t]	−9.33	−0.53	−0.63(.12)[u]	1.01	0.48	−0.63(.04)[u]
$H_2C=C=CHCN$[t]	−10.45	−0.63	−0.55(.22)[u]	−0.01	0.22	0.57(.75)[w]
$H_2C=C=CHCO_2Me$[t]	−10.62	−0.67	−0.53(.23)[u]	−0.07	−0.51	0.71(−.32)[w]
$H_2C=C=O$[v]	−12.55(−12.7)[x]	−0.73	−0.27(.61)[v]	2.55	−0.16	0.45(−.23)[w]
$MeHC=C=O$[v]	−11.52(−8.95)[y]	−0.67	−0.33(.55)[w]			
$PhHC=C=O$	−10.61(−10.56)[x]	−0.53	−0.30(.43)			
$Me_2C=C=O$	−8.45[y]					
$ClHC=C=O$	−9.24[y]					
$Cl_2C=C=O$	−9.15[y]					
$NCHC=C=O$	−10.07[y]					

Values in [ ] calculated using MacSpartan, v2.0.2, Hartree-Fock calculations at STO-3G level. [a] All data From Ref. 13 unless otherwise noted. [b] From Ref. 14. [c] From Ref. 15. [d] From Ref. 16. [e] From Ref. 17. [f] From Ref. 18. [g] From Ref. 19. [h] From Ref. 20. [i] From Ref. 21. [j] From Ref. 22. [k] From Ref. 23. [l] From Ref. 24. [m] From Ref. 25. [n] From

14. Kahn, S.D.; Pau, C.F.; Overman, L.E.; Hehre, W.J. *J. Am. Chem. Soc.* **1986**, *108*, 7381.
15. *Handbook of Chemistry and Physics, 87th ed.*, CRC/Taylor and Francis, Boca Raton, Fla., **2006**, pp 10-203-10-223.
16. Jordan, K.D.; Michejda, J.A.; Burrow, P.D. *Chem. Phys. Lett.* **1976**, *42*, 227.
17. Kadifachi, S. *Chem. Phys. Lett.* **1984**, *108*, 233.
18. Sauer, J.; Wiest, H.; Mielert, A. *Chem. Ber.* **1964**, *97*, 3183.
19. Pearson, R.G. *J. Org. Chem.* **1989**, *54*, 1423.
20. Dougherty, D.; Brint, P.; McGlynn, S.P. *J. Am. Chem. Soc.* **1978**, *100*, 5597.
21. Heinis, T.; Chowdhury, S.; Scott, S.L. Kebarle, P. *J. Am. Chem. Soc.* **1988**, *110*, 400.
22. Samuilov, Ya.D.; Uryadov, V.G.; Uryadova, L.F.; Konovalov, A.I. *Zh. Org. Khim.* **1985**, *21*, 1249 (Engl. 1137).
23. Dinur, U.; Honig, B. *J. Am. Chem. Soc.* **1979**, *101*, 4453.
24. Konovalov, A.I.; Kiselev, V.D.; Vigdorovich, O.A. *Zh. Org. Khim.* **1967**, *3*, 2085 (Engl. 2034).
25. Dewar, M.J.S.; Rzepa, H.S. *J. Am. Chem. Soc.* **1978**, *100*, 784.

Ref. [26]. [o] From Ref. [27]. [p] From Ref. [28]. [q] From Ref. [29]. [r] all data From Ref. [30]. [s] From Ref. [31]. [t] all data From Ref. [32]. [u] third coefficient (C1 for CR_2; C2 for $=C=$; C3 for CH_2); [v] all data From Ref. [33] unless otherwise noted. [w] third coefficient (C1 for RCH; C2 for $=C=$; C3 for $=O$). [x] From Ref. [34]. [y] From Ref. [35] [z] From Ref. [36] [aa] For O=C, where C_1 is for O and C_2 is for C [ab] From Ref. [37]; [ac] From Ref. [38] ; [ad] *trans*; [ae] From Ref. [39]

The data in Table 11.1 came from a variety of sources, including calculations[40] and experimentally determined ionization potential and EA data. These numbers vary with the experimental method, and the sophistication level of the calculation. They should be viewed as useful first estimates when comparing ΔE's using values from two different sources (and/or methods). Caution should be exercised. The purpose of this table is to use the HOMO-LUMO energies to predict chemical reactivity (*vide infra*, sec. 11.4.B). The orbital coefficients will be used to predict regioselectivity (*vide infra*, sec. 11.4.C.ii). When two or more electron-withdrawing groups are present (tetracyanoethylene, benzoquinone, maleic anhydride, etc.), greater lowering of the HOMO is apparent and the LUMO energy may fall below 0 eV. Ethyne is generally higher in energy than ethene, but electron-releasing and -withdrawing substituents have a similar effect (note −11.4 eV for acetylene vs. −11.5 eV for dimethyl acetylene dicarboxylate, pointing to the problems of using different methods to determine the ionization potential).

Data are also shown for allene and ketene derivatives, which react similarly to alkenes in pericyclic reactions. The HOMO and LUMO energies are generally lower than those of ethene and the orbital coefficients of interest are those at C2 and C3 in the LUMO of allenes and C1 and C2 in ketenes. As will be seen in Sections 11.4.B and 11.11, the LUMO energies and coefficients are important for predicting reactivity and selectivity in several types of pericyclic reactions.

As seen above, a conjugating substituent (C=O or C≡N, Ph) lowers the energy of both the HOMO and the LUMO, and its presence also perturbs the orbital coefficients. Table 11.1 shows an entry for styrene in which the HOMO is higher in energy, but the LUMO is lower (relative to ethene), and this result is also seen in Figure 11.3. The orbital coefficients are smaller closest to the phenyl, and are significantly smaller in magnitude. These effects appear in dienes where

26. Ng, L.; Jordan, K.D.; Krebs, A.; Rüger, W. *J. Am. Chem. Soc.* **1982**, *104*, 7414.
27. El-Basil, S.; Said, M. Ind. *J. Chem.* **1980**, *19B*, 1071.
28. Bihlmaier, W.; Huisgen, R.; Reissig, H.-U.; Voss, S. *Tetrahedron Lett.* **1979**, 2621.
29. Michl, J.; Becker, R.J. *J. Chem. Phys.* **1976**, *46*, 3889.
30. Houk, K.N. *Acc. Chem. Res.* **1975**, *8*, 361.
31. Hentrich, G.; Gunkel, E.; Klessinger, M. *J. Mol. Struct.*, **1974**, *21*, 231.
32. Padwa, A.; Bullock, W.H.; Kline, D.N.; Perumattam, J. *J. Org. Chem.* **1989**, *54*, 2862.
33. Meslin, J.C.; N'Guessan, Y.T.; Quiniou, H.; Tonnard, F. *Tetrahedron* **1975**, *31*, 2679.
34. Kuzuya, M.; Miyake, F.; Okuda, T. *J. Chem. Soc. Perkin Trans. 2* **1984**, 1471.
35. Bock, H.; Hirabayashi, T.; Mohmand, S. *Chem. Ber.* **1981**, *114*, 2595.
36. Yu, Z.-X.; Wu, Y.-D. *J. Org. Chem.* **2003**, *68*, 412.
37. Suishu, T.; Shimo, T.; Somekawa, K. *Tetrahedron* **1997**, *53*, 3545.
38. Spino, C.; Rezaei, H.; Dory, Y.L. *J. Org. Chem.* **2004**, *69*, 757.
39. Yoshitake, Y.; Yamaguchi, K.; Kai, C.; Akiyama, T.; Handa, C.; Jikyo, T.; Harano, K. *J. Org. Chem.* **2001**, *66*, 8902.
40. Using MacSpartan, v.2.0.2, Wavefunction, Inc., Irvine, CA.

the conjugating substituent is another ethene moiety (1,3-butadiene, for example). In Figure 11.3, the HOMO energy for butadiene is –9.03 eV (with coefficients of 0.57 and 0.41). The LUMO energy is lowered to +1.0 eV with coefficients of 0.56 and -0.42, relative to ethene,[41] which is a net electron-withdrawing effect, but the influence of the HOMO and LUMO is similar to the effects noted for styrene.

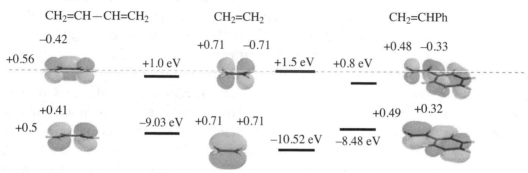

Figure 11.3. Influence of a Conjugating Substituent on Alkene Homo and Lumo Energies.

In Figure 11.3, 1,3-butadiene is treated as a substituted alkene, which is a contrivance used to illustrate the significant interaction of the conjugated π system, as reflected in the HOMO and LUMO energies. Although the HOMO-LUMO coefficients for all four atoms of the conjugated system must be considered, only those at the termini will interact with the alkene in a Diels-Alder reaction. 1,3-Butadiene is symmetrical, and the coefficients reflect not only the equal distribution of electron density in the orbitals (see Figure 11.3), but also the electron withdrawing effect invoked in Figure 11.3. The largest concentration of electron density occurs at the terminal carbons, and these are expected to be the major sites of reactivity. In butadiene, there are four π orbitals and four π electrons, so there are four molecular orbitals as shown. The HOMO and LUMO energy levels in Figure 11.4[41] are only two of four frontier orbitals. The lowest energy orbital has no node, the HOMO has one node, the LUMO has two nodes, and the highest energy orbitals has three nodes. The orbitals are shown using a convenient drawing convention with p-orbitals at each carbon. Orbital pictures calculated with MacSpartan are also shown for the HOMO and LUMO.[24] In this diagram, the two lowest energy orbitals contain electrons, with the HOMO being the higher of the two. For analysis of cycloaddition reactions, the HOMO will be more important than the other orbitals, since it contains electrons that can be donated. Similarly, the LUMO is the lowest energy of the unfilled orbitals, and is more important for reactions since it will preferentially accept electrons. For dienes, attention will be focused entirely on the HOMO and LUMO.

---

41. Reference 2, p 204.

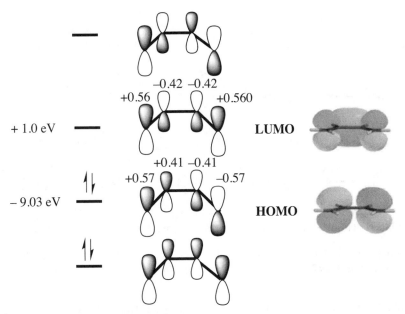

Figure 11.4. Orbital Diagram for 1,3-Butadiene.

When substituents are attached to a diene, effects on the frontier orbitals are similar to those observed with alkenes. Electron withdrawing groups decrease the energy of the HOMO and LUMO relative to butadiene, and electron releasing groups increase the HOMO and LUMO energies. There is a new problem of regiochemistry, however, since the substituent can be attached to C1 or C2 of the diene. The Z group is a generic electron-withdrawing group, and the X group is a generic electron-releasing group, as used by Fleming[42] and Houk.[43] The energies for the HOMO and LUMO are mostly taken from the work of Houk.[43] As expected, the electron withdrawing groups lower the HOMO and the LUMO (see Figure 11.5).[43] The C1 substituent lowers the energy more than a C2 substituent and the orbital coefficients are larger on the orbitals distal to the Z group for C1 substituents, although they are larger on the proximal carbon for C2 substituents. The orbital on the carbon distal to the X group is smallest for C1 substituents, and largest for C2 substituents in the LUMO. In the HOMO, the orbital on the carbon distal to the Z group attached to C1 is largest for the HOMO, and smallest for the LUMO, which is reversed when the Z group is attached to C2 (see Figure 11.5). Orbital coefficients are largest for C2 substituents in the LUMO. Typical substituted dienes are presented in Table 11.2, along with the HOMO and LUMO values and the available coefficients.

42. Reference 7, p 128.
43. Houk, K.N. *J. Am. Chem. Soc.* **1973**, *95*, 4092.

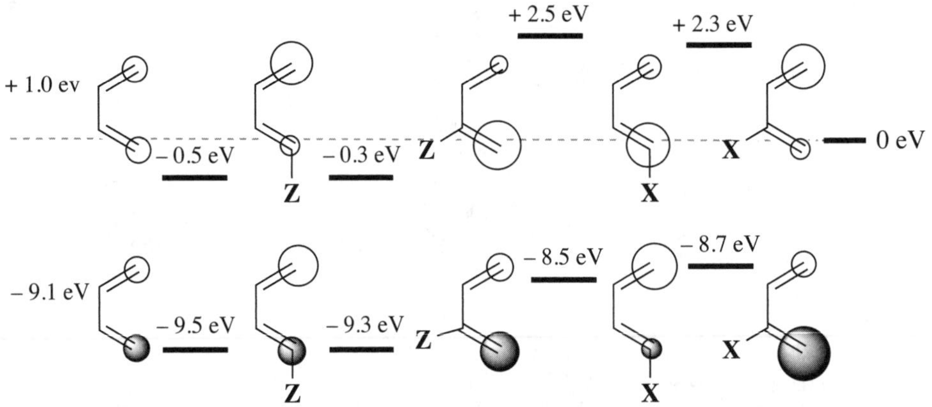

Figure 11.5. Relative Homo and Lumo Energies of Electron-Rich and Electron-Poor Dienes.

[Reprinted with permission from Fleming, I. *Frontier Molecular Orbitals and Organic Chemical Reactions,* Wiley, London, *1976*, p. 128. Copyright © *1976* by John Wiley and Sons, Inc.]

As with the energy levels and coefficients presented in Table 11.1, the values in Table 11.2 vary with the experimental method used to obtain them, and with the sophistication level of the calculation. Two HOMO and LUMO values are given for butadiene to illustrate this. It is clear, however, that the electron-releasing and -withdrawing effects are as predicted. Cyclopentadiene is interesting in that it shows significant lowering of the LUMO, as does cyclohexadiene relative to butadiene, and especially relative to the acyclic 2,4-pentadiene. Introduction of an electron-releasing heteroatom in the ring (furan, thiophene, and pyrrole) leads to a LUMO of lower energy relative to methoxy or methylthiobutadiene, but the LUMOs are higher in energy than cyclo-pentadiene. The HOMO energies (as well as the c_1 and c_2 orbital coefficients) will prove useful for analyzing Diels-Alder and [3+2]-cycloaddition reactions.

Table 11.2   HOMO, LUMO Energies and Orbital Coefficients for Substituted Dienes

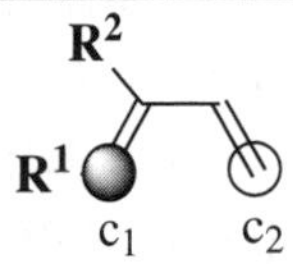

Diene	HOMO (eV)	c1	c2	LUMO (eV)	c3	c4
 [a]	−9.07(−8.85)[b]	0.57	−0.57	1.0(3.38)[b]	0.56	0.56
 Me [b]	−9.78[a](−8.54)	0.314	0.315	3.51	0.629[c]	0.617[c]
 Me [b]	−8.81°					
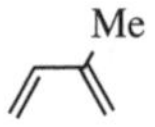 Me [b]	−9.04a(−8.72)	0.340	0.296	3.38	0.56[d]	0.55[d]

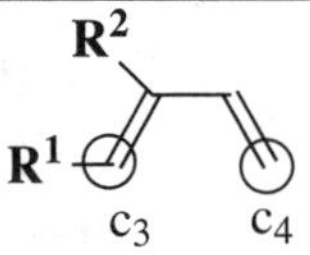

Diene	HOMO (eV)	c_1	c_2	LUMO (eV)	c_3	c_4
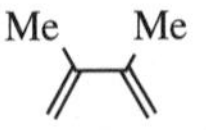 Me, Me (2,3-dimethylbutadiene)	$-8.76^{a}(-8.62)^{o}$			2.18^{e}		
 Me —/=\— Me	$-8.39^{a}(-8.09)^{o}$					
 Ph —/=\	$-8.16^{a,o}$	0.408^{f}	0.416^{f}			
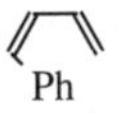 Ph (cyclic)	-8.38^{o}					
 Ph —/=\— Ph	-8.09^{o}					
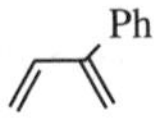 Ph (2-phenyl)	$-8.77^{a,o}$	0.572^{g}	0.335^{g}			
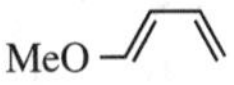 MeO —/=\	$-8.21^{a,o}(-8.24)^{b}$	0.235^{b}	0.313^{b}	3.77^{b}	0.644^{c}	0.609^{c}
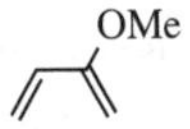 OMe (2-methoxy)	$-8.62^{a,o}$	0.352^{b}	0.103^{b}	3.60^{b}		
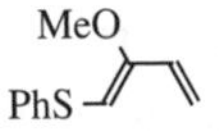 MeO, PhS	-9.94^{p}	-0.296^{p}	0.231^{p}	2.41^{p}	-0.399^{p}	-0.502^{p}
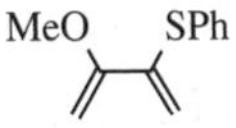 MeO, SPh	-10.50^{p}	-0.079^{p}	0.329^{p}	2.18^{p}	-0.431^{p}	-0.615^{p}
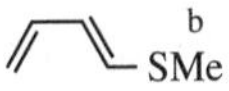 SMe [b]	-7.94	$0.24-$	0.256	3.25		
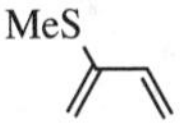 MeS [b]	-8.37	0.399	0.201	3.25		
 CO_2H	-9.41^{b}	0.483^{f}	0.460^{f}	1.99^{b}		
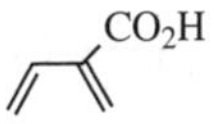 CO_2H [b]	-9.24	0.279	0.326	2.39		
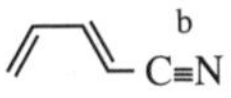 $C{\equiv}N$ [b]	-9.45			1.86		
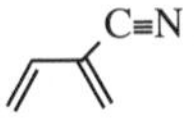 $C{\equiv}N$ [b]	-9.58	0.595^{g}	0.490^{g}	2.12		
 (cyclopentadiene)	$-8.97^{h}(-8.58)^{o}$			0.75^{i}		
 (cyclohexadiene)	$-8.2^{j,k}$			$0.73^{k}(0.80)^{j,l}$		
 (pyrrole, N-H)	$-8.2^{h,m}$			$2.38^{l,n}$		
 (furan, O)	$-8.8^{h,m}$			$1.76^{l,n}$		
 (thiophene, S)	$-8.9^{h,m}$			$1.17^{l,n}$		

[a] taken from Ref.[44] unless otherwise noted. [b] taken from Ref. 13 unless otherwise noted. [c] from Ref.[45]. [d] from Ref. 29. [e] from Ref.[46]. [f] from Ref.[47]. [g] from Ref.[48] [h] from Ref. 14. [i] from Ref. 17. [j] from Refs. 15,[49]. [k] from Ref.[50]. [l] from Ref.[51]. [m] from Ref.[52]. [n] from Ref.[53] [o] from Ref.[54] [p] from Ref.[55]

11.3. ALLOWED AND FORBIDDEN REACTIONS

Most of the reactions involving multiple π bonds can be classified by the number of π-electrons in each reaction partner. A single π bond (2 π electrons) can interact with another single π bond of an alkene or a diene (2 more π electrons). The reaction of two alkenes (see Figure 11.6) is therefore classified as a **2 π + 2 π** reaction. When a diene with four π electrons reacts with an alkene having two π electrons (Figure 11.6), there are a total of 6 π electrons and this is a **4 π + 2 π** reaction. Reaction of an allyl anion with an alkene (Figure 11.6) is classified as a **3 π + 2 π** reaction, and is a simple example of the so-called dipolar addition discussed in Section 11.11. Woodward and Hoffman[10b,56] correlated each reaction type with the symmetry of their frontier orbitals, allowing predictions to be made concerning their reactivity. They classified reactions as **allowed** or **forbidden**. In an allowed reaction, a four-center transition state is stabilized more than a two-center transition state; in a forbidden reaction, vice versa. For reaction of a diene and an alkene (see Figure 11.6), the four-center transition state is represented by **7** and the two-center transition state by **8**.[57] This problem can be further defined in terms of the number of π electrons in the polyene. In **4 n + 2 n π** systems (*n* = the number of π-electrons), the 4 π-electrons in 1,3-butadiene, for example, react with the 2 π electrons of an alkene. The HOMO of the diene contains terminal atoms with orbitals of opposing symmetry (HOMO$_{diene}$ in **7**) and the orbital coefficients have opposite signs,[58] which contrasts with a 4n π electron system (n = 0, 1, 2, 3, ..., with 0, 4, 8, 12, ..., π electrons, such as the 2 π-electrons of one alkene reacting with the 2 π electrons of a carbonyl π bond). The HOMO contains orbitals of the same symmetry and sign in symmetrical alkenes, but the LUMO contains orbitals of

44. Sustmann, R.; Schubert, R. *Tetrahedron Lett.* *1972*, 2739.
45. Kakushima, M. *Can. J. Chem.* *1979, 57*, 2564.
46. Kroner, J.; Bock, H. *Theoret. Chim. Acta* *1968, 12*, 214.
47. Alston, P.V.; Ottenbrite, R.M.; Shillady, D.D. *J. Org. Chem.* *1973, 38*, 4075.
48. Alston, P.V.; Ottenbrite, R.M. *J. Org. Chem.* *1975, 40*, 1111.
49. Jordan, K.D.;Burrow, P.D. *Chem. Phys. Lett.* *1975, 36*, 594.
50. Giordan, J.C.; McMillan, M.R.; Moore, J.H.; Staley, S.W. *J. Am. Chem. Soc.* *1980, 102*, 4870.
51. Jordan, K.D.; Burrow, P.D. *Acc. Chem. Res.* *1978, 11*, 341.
52. Lazzaroni,R.;Boutique,J.P.;Riga,J.;Verbist,J.J.;Fripiat,J.G.;Delhalle,J.*J. Chem. Soc. Perkin Trans. 2* *1985*,97.
53. Van Veen, E.H. *Chem. Phys. Lett.* *1976, 41*, 535.
54. Rücker, C.; Lang, D.; Sauer, J.; Friege, H.; Sustmann, R. *Chem. Ber.* *1980, 113*, 1663.
55. Alston, P.V.; Gordon, M.D.; Ottenbrite, R.M.; Cohen, T. *J. Org. Chem.* *1983, 48*, 5051.
56. Hoffmann, R.; Woodward, R.B. *J. Am. Chem. Soc.* *1965, 87*, 2046.
57. Reference 2, p 216.
58. Reference 2, p 209.

opposing symmetry and sign (see LUMO$_{alkene}$ in **7**). The symmetry of the alkene LUMO is the same as that found in the HOMO of the diene. In a reaction, maximum stability in the transition state is observed when the terminal carbons of the HOMO$_{diene}$ interact with the termini of the LUMO$_{alkene}$. These orbitals have the same symmetry and such interactions are allowed. If the orbitals have the opposite symmetry (and sign) the interaction is forbidden.[59]

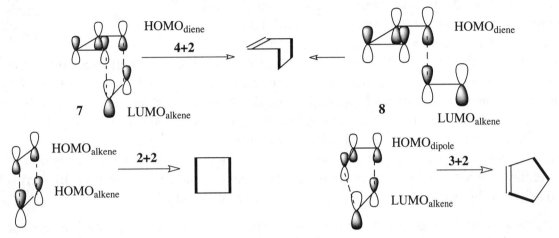

Figure 11.6. Common Pericyclic [*m+n*] Reactions.

Woodward and Hoffman described addition modes, based on the facial approach of the reactive termini, as **suprafacial** and **antarafacial**.[60] "A suprafacial process is one in which bonds made or broken lie on the same face of the system undergoing reaction", as in **9**.[60] An antarafacial process has the "newly formed or broken bonds on opposite sides of the reaction system," as in **10**.[60] Woodward and Hoffman then described several specific π interactions, and classified

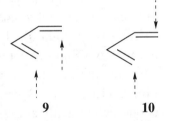

them as **symmetry allowed** or **symmetry forbidden** (the so-called **Woodward-Hoffman rules**). Each π process could then be assigned an *s* (**suprafacial**) or an *a* (**antarafacial**) to describe the facial approach of the reactive termini. Lehr and Marchand use addition of bromine to *cis*-2-butene to illustrate these modes of addition.[61] Antarafacial addition leads to the *d,l*-dibromide (**11**), whereas a suprafacial addition would give the meso-dibromide (**12**). These modes of addition for electrocyclic reactions will be further illustrated in later discussions (see 11.11).

59. Reference 2, p 210.
60. Reference 10b, p 65.
61. Lehr, R.E.; Marchand, A.P. *Orbital Symmetry*, Academic Press, New York, *1972*, p 16.

Finally, it is imperative that we categorize the energy source used to effect the reaction as either heat (a thermal reaction) or light (a photochemical reaction - sec. 11.10.B is a brief introduction to photochemistry). The lowest energy arrangement of electrons (shown in **13** for an alkene) is referred to as the **ground state**. If this orbital is used for reaction, the HOMO will react with an orbital of like symmetry, but the reaction does not require promotion of an electron to the LUMO. This type of reaction only requires heating, since absorption of a photon of light will promote an electron to the antibonding orbital (convert **13** → **14**). If the alkene reacts photochemically, a singly occupied molecular orbital (a SOMO) is formed, represented by **14**, and it is the reactive orbital. Under thermal conditions, the ground state HOMO will dominate. If, however, the HOMO cannot react, heating will not affect the desired reaction, and it will occur only under photochemical conditions where an electron is promoted to the higher energy SOMO. It is, therefore, possible to characterize reactions as thermally allowed or forbidden, and as photochemically allowed or forbidden.

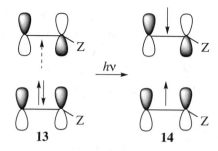

With these tools in hand, the **Woodward-Hoffman rules** of cycloaddition describe two π systems of $m \pi$ and $n \pi$ electrons. With these rules, $4q$ and $4q+2$ systems ($q = 0, 1, 2, 3,....$) are classified as thermally and photochemically allowed or forbidden, as presented in Table 11.3.[62] "For suprafacial interactions, the thermally allowed reactions involve one $4q + 2$ molecule and one $4q$ molecule"[10] [an alkene ($q = 1$ for $4q$) and butadiene ($q = 1$ for $4q + 2$)]. "This gives a total of ($4q+2$) π-electrons and the reaction is allowed. Forbidden reactions involve two $4q$ or two $4q+2$ systems for a total of $4n\pi$ electrons".[10] In the following section, the three major types of cycloaddition reactions shown in Figure 11.6 will be examined. In Figure 11.6, the [4+2]-cycloaddition ($4n + 2n$) is thermally allowed, the [2+2]-cycloaddition is photochemically allowed and the [3+2]-cycloaddition is thermally allowed (although it is a $3n + 2n$ reaction).

Table 11.3. Rules for [*m+n*] Cycloadditions

[*m + n*]	Allowed (ground state, thermal) Forbidden (excited state, hν)	Forbidden (ground state, thermal) Allowed (excited state, hν)
$4q$	$m_s + n_a$	$m_s + n_s$
	$m_a + n_s$	$m_s + n_a$
$4q + 2$	$m_s + n_s$	$m_s + n_a$
	$m_a + n_a$	$m_a + n_s$

---

62. Reference 10b, p 70.

11.4. [4 + 2]-CYCLOADDITIONS

11.4.A. The Diels-Alder Reaction

In 1906, Albrecht reported a thermal reaction between cyclopentadiene and *p*-benzoquinone that gave a dimeric complex, later identified to be the [4+2]-cycloadduct, **15**.[63] In 1928, Diels and Alder reported this reaction, along with those of several dienes and alkenes, including maleic anhydride and acrolein, and characterized the products.[64,68] From these initial reactions, countless variations have been reported, and this has become one of the most important methods for constructing molecules in all of organic chemistry. This thermal reaction of a diene and an alkene (or other two π-electron system, generically known as a **dienophile**) is now called the **Diels-Alder reaction**.[65,66] Diels and Alder were awarded the Nobel Prize for this reaction in 1950. The reaction is explained in terms of molecular orbital theory, and is recognized as a [4 *n* + 2 *n*]-cycloaddition reaction. In Section 11.3, a [4+2]-reaction was shown to be thermally allowed, and Figure 11.7 shows that the C1 and C4 orbitals of the HOMO of butadiene have the same symmetry as the orbitals of the LUMO of ethene. The reaction proceeds thermally, presumably by donation of electrons from the HOMO$_{diene}$ to the LUMO$_{alkene}$. Orbitals of the same symmetry (see **7**) react with transfer of electrons in a six-center transition state (**16**) to form two σ bonds and a new π bond in the cyclohexene product (**17**), as shown in Figure 11.8.

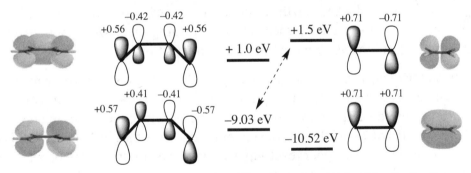

Figure 11.7. HOMO$_{butadiene}$–LUMO$_{ethene}$ Interactions That Control the Diels-Alder Cyclization.

63. Albrecht, W. *Annalen* **1906**, *348*, 31.
64. (a) Diels, O.; Alder, K.; Lübbert, W., Naujoks, E.; Querberitz, F.; Röhl, K.; Segeberg, H. *Annalen* **1929**, *470*, 62; (b) Diels, O.; Alder, K. *Ber.* **1929**, *62*, 2081; (c) *Idem Ibid.* **1929**, *62*, 2087.
65. Larock, R.C. *Comprehensive Organic Transformations, 2nd Ed.*, Wiley-VCH, New York, **1999**, pp 523-544.
66. (a) The Merck Index, 14th ed., Merck & Co., Inc., Whitehouse Station, New Jersey, **2006**, p ONR-2624 (b) Mundy, B.P.; Ellerd, M.G.; Favaloro Jr., F.G. *Name Reactions and Reagents in Organic Synthesis, 2nd ed.*, Wiley-Interscience, New Jersey, **2005**, pp. 206-209.

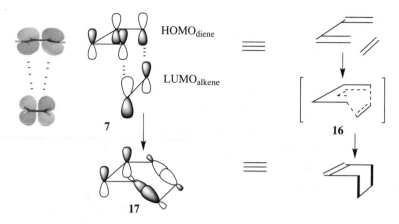

Figure 11.8. The Diels-Alder Reaction of Butadiene and Ethene.

The terminal orbitals in the *cisoid* conformation (sec. 1.5.A) bear the proper symmetry for reaction with the alkene, and only in this conformation can the HOMO orbitals come within bonding distance of the LUMO orbital of the alkene. Acyclic dienes are in a dynamic equilibrium consisting of both cisoid (**18c**) and transoid (**18t**) conformations, but the lower energy transoid conformation is the major one. The equilibrium mixture can be used because as the cisoid conformation reacts with the alkene, the equilibrium is shifted to give more cisoid as the reaction proceeds. If a diene is locked into a transoid geometry, as in **19**, the Diels-Alder reaction does not occur. If the diene is locked into the cisoid form, as in cyclopentadiene (**20**), the reaction is expected to proceed faster. It is important to point out that this simple analysis ignores any steric interaction between diene and dienophile.

Substituents on the diene can increase the energy barrier to the cisoid form, which diminishes the rate of the Diels-Alder reaction. Attempts to react 2-methyl-2,4-pentadiene (**21**) with maleic anhydride at ambient temperature gave no reaction. The more reactive alkene tetracyanoethylene (TCNE, **22**) gave 11% of **23** in 20 h, but the major product was the [2+2]-cycloaddition product (**24**, sec. 11.10) in 69%.[67] This poor reactivity was attributed to the energy barrier required for **21** to assume a cisoid conformation.

67. Stewart Jr., C.A. *J. Am. Chem. Soc.* *1962*, *84*, 117.

11.4.B. Reactivity in the Diels-Alder Reaction

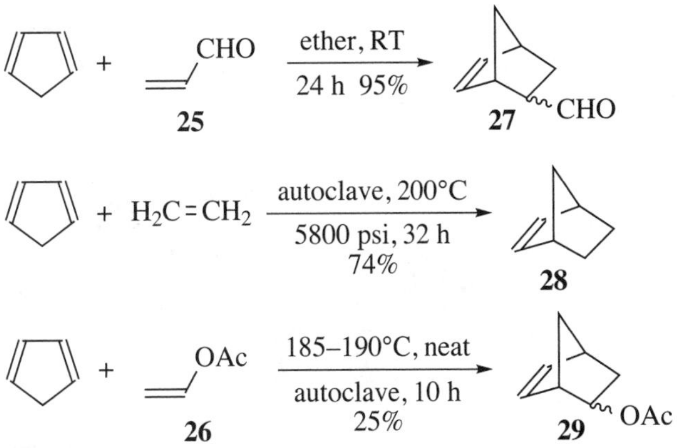

Diels-Alder reactions of cyclopentadiene with different alkenes proceed at different rates. Cyclopentadiene reacts with acrolein (**25**) in ether at room temperature (24 h) to give a 95% yield of **27**.[68] This contrasts with the reaction of cyclopentadiene and ethene in ether to give a 74% yield of norbornene (**28**), which required heating to 200°C in an autoclave at 5800 psi pressure or 32 h. The presence of an electron releasing group on the alkene causes the reaction to proceed with greater difficulty. Vinyl acetate (**26**) reacted with cyclopentadiene at 190°C in an autoclave (10 h, neat) to give only 25% of **29**.

Generally, simple alkenes and electron-rich alkenes (bearing an electron-releasing group) will react with butadiene or cyclopentadiene only under very vigorous conditions. Molecular orbital theory explains these results by comparing the relative energy levels of the HOMO and LUMOs. If we use butadiene as an example (shown in Figure 11.9), the $HOMO_{diene}$-$LUMO_{alkene}$ energy difference (ΔE) is 9.07 eV (209.15 kcal mol^{-1}, 875.5 kJ mol^{-1}) for a reaction between butadiene and ethyl acrylate. The ΔE for ethene-butadiene is 10.57 eV (243.74 kcal mol^{-1}, 1020.3 kJ mol^{-1}) and 11.07 eV (255.27 kcal mol^{-1}, 1068.6 kJ mol^{-1}) for methyl vinyl ether-butadiene. The smaller ΔE for ethyl acrylate is associated with a faster reaction.[69] These energy effects are correlated with the energy expression [Equation (11.1)][70] derived by Klopman[71] and Salem[72] from perturbation theory. Fleming noted that the energy separation term of the HOMO and LUMO (E_r–E_s) is in the denominator of the equation, and a larger energy gap will slow the overall reaction.[70] The equation is

$$\Delta E = -\sum (q_a + q_b)\, \beta_{ab}S_{ab} + \sum_{k<l} \frac{Q_k Q_l}{e_R k_l} + \sum_{r}^{occ}\sum_{s}^{unocc} - \sum_{s}^{unocc}\sum_{r}^{occ} \frac{2\,(\Sigma_{ab}c_{ra}c_{sb}\beta_{ab})^2}{E_r - E_s} \qquad 11.1$$

where q_a and q_b are electron populations of atomic orbitals a and b; β and S are the resonance and overlap integrals, respectively; the energy of the bonding orbital is

$$E_r = \frac{\alpha_1 + \beta}{1 + S} \qquad 11.2$$

68. Diels, O.; Alder, K. *Annalen* **1928**, *460*, 98.
69. Reference 7, pp 114-121.
70. Reference 7, p 27.
71. Klopman, G. *J. Am. Chem. Soc.* **1968**, *90*, 223.
72. Salem, L. *J. Am. Chem. Soc.* **1968**, *90*, 543, 553.

and the energy of the antibonding orbital is

$$E_s = \frac{\alpha_2 + \beta}{1 + S} \qquad 11.3$$

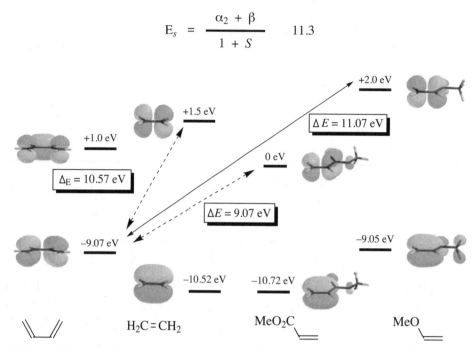

Figure 11.9. HOMO$_{diene}$-LUMO$_{alkene}$ Interactions of Butadiene and Representative Alkenes

The first term in Equation 11.1 is the Coulomb integral,[72b] and Q_k and Q_l are the total charges on atoms k and l; e is the local dielectric constant; R_{kl} is the distance between atoms k and l; C_{ra} is the coefficient of atomic orbitals (a) in molecular orbital (r), where r refers to the molecular orbitals on one molecule and s refers to those on the other; and E_r and E_s are the energies of the molecular orbitals.

A comparison of ΔE for the HOMO$_{diene}$ and LUMO$_{alkene}$ of electron rich dienes and electron poor dienes upon reaction with ethyl acrylate (see Figure 11.9), shows that ΔE is smaller for electron rich dienes. For example, 1-ethoxy-butadiene reacts faster than ethyl 1,3-pentadienoate. A similar trend is observed with ethene and methyl vinyl ether, but the overall trend that the diene reacts faster with the electron poorer alkene is retained, as observed in the reaction of 1,3-butadiene with *trans*-cinnamic acid to give **30**, which required heating to 140°C for 20 h.[73] Similar reaction with 2-ethoxy-1,3-butadiene gave a mixture of two cycloadducts after heating to 170°C for 48 h in an autoclave, but **31** was the major product.[74] The reaction of maleic anhydride and tetracyanoethylene showed similar reactivity.[75]

73. Alder, K.; Vagt, H. Vogt, W. *Annalen* **1949**, *565*, 135.
74. House, H.O.; Gannon, W.F.; Ro, R.S.; Wluka, D.J. *J. Am. Chem. Soc.* **1960**, *82*, 1463.
75. (a) Sustmann, R.; Schubert, R. *Angew. Chem. Int. Ed.* **1972**, *11*, 840; (b) see Reference 7, p 113.

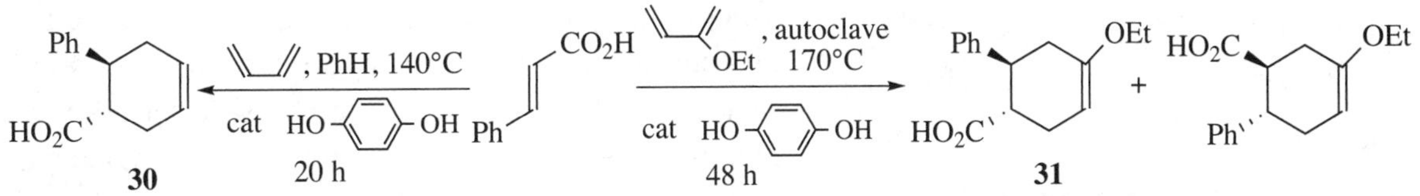

Electron-releasing or -withdrawing substituents have a significant effect on the reactivity of the diene. Comparing the reaction of cyclopentadiene and acrolein to give **25** with cyclopentadiene and vinyl acetate to give **27** illustrates this point. Substituent effects on the HOMO of the diene will increase or decrease the HOMO$_{diene}$-LUMO$_{alkene}$ interaction (ΔE). Electron-withdrawing groups will lower the HOMO of the diene and electron-releasing groups will raise the LUMO of the diene.

The conclusions drawn from these analyses show that the HOMO$_{diene}$-LUMO$_{alkene}$ ΔE correctly predicts the *relative* reactivity of a given diene with a variety of alkenes, or of an alkene with a variety of dienes. As seen above, cyclopentadiene reacted faster with ethyl acrylate and slowest with methyl vinyl ether. The HOMO of cyclopentadiene is -9.87 eV (Table 11.2), the LUMO of methyl acrylate is 0 eV, and the LUMO of methyl vinyl ether is +2.0 eV (Table 11.1). The ΔE for methyl acrylate and cyclopentadiene is 9.87 eV, but the ΔE for methyl vinyl ether and cyclopentadiene is 11.87 eV. Methyl acrylate is expected to react faster. In general, alkenes with electron-withdrawing substituents react faster than alkenes with electron releasing substituents. This point is illustrated in Woodward's synthesis of reserpine, which began with the Diels-Alder reaction of benzoquinone **32** and 1,3-butadiene, to give **33** in 86%.[76]

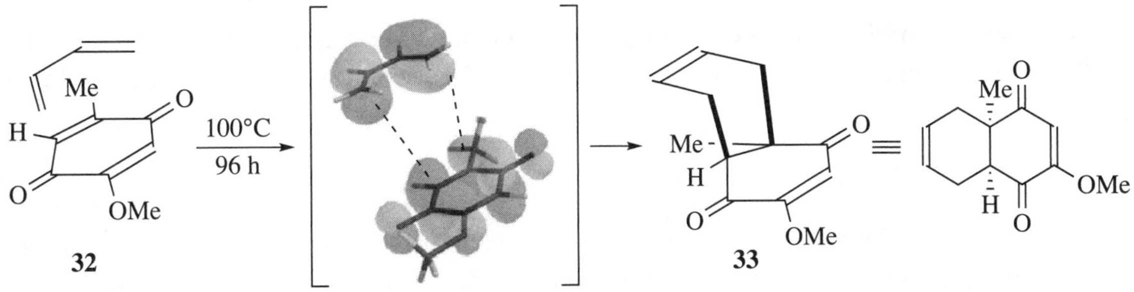

Reaction occurred preferentially at the double bond bearing the methyl group rather than at the more electron-rich vinyl ether. The HOMO$_{diene}$-LUMO$_{alkene}$ ΔE for the methoxy side is larger than that for the methyl side, and butadiene reacted preferentially with the methyl side rather than the methoxy side. The Diels-Alder reaction forms two bonds in a single reaction and from a retrosynthetic standpoint, the two-bond disconnection for the Diels-Alder reaction is

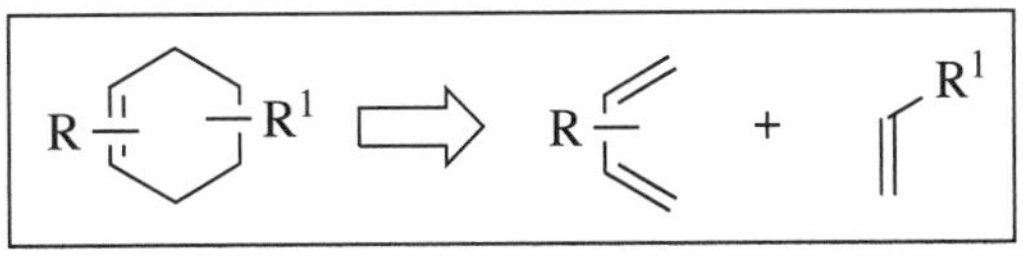

76. Woodward, R.B.; Bader, F.E.; Bickel, H.; Frey, A.J.; Kierstead, R.W. *Tetrahedron* **1958**, 2, 1.

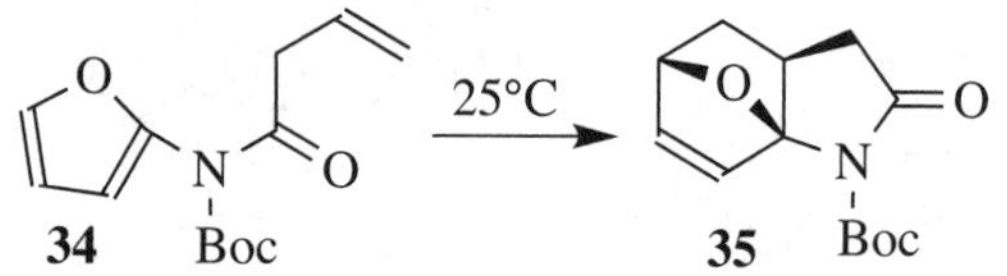

To conclude this section, it is important to note that alkenes are not the only hydrocarbon dienophiles that are useful partners in the Diels-Alder reaction. Heterocyclic compounds, alkynes and even benzyne are quite useful. In a synthesis of *epi*-zephyranthine by Padwa and Wang[77] an intramolecular Diels-Alder reaction of the alkene unit with the furan ring **34** gave a 55% yield of **35**. A benzyne derivative of pyridine was shown to undergo Diels-Alder reactions.[78] When 3-chloro-2-methoxypyridine (**36**) was treated with *tert*-butyllithium in THF at −78°C and then with furan, for example, cycloadduct **39** was obtained in 74% yield. The reaction is believed to proceed by aryl lithiation to give **37**, which loses LiCl to give the benzyne, **38**, which traps furan in a Diels-Alder reaction to give **39**.

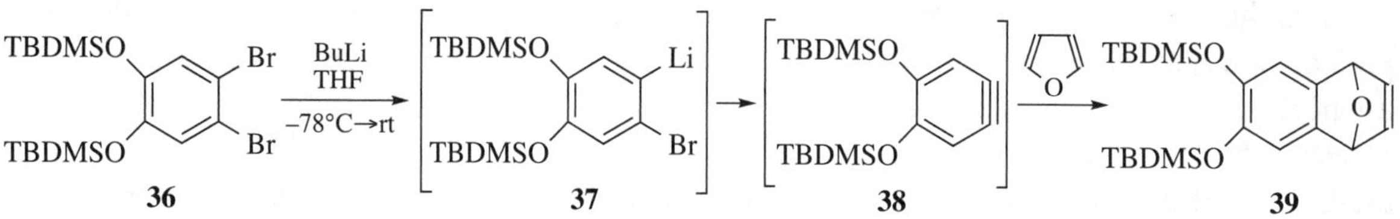

Note that an alternative explanation has been presented, based on the relative position of the diene and the dienophile in the electrophilicity scale.[79] This analysis led to characterization of the $HOMO_{Nu-} - LUMO_{E+}$ interactions along a polar DA reaction and characterization of what the authors believe is a more favorable, two-center interaction along an asynchronous bond-formation process.

The **Arrhenius equation** indicates that the rate of a reaction depends on its activation energy, which is the difference in energy between the reactants and the transition state. The Arrhenius equation is

$$\text{rate} = k\,[A]^a\,[B]^b \quad \text{and} \quad k = Ae^{-\frac{E_A}{RT}} \quad \text{or} \quad \ln k = \ln A - \frac{E_A}{RT}$$

where rate is in mol dm^{-3}s^{-1}, k is the rate constant, [A] and [B] are concentrations in mol dm^{-3}, the superscript a is the order of the reaction with respect to A and the superscript b is the order of the reaction with respect to B. The rate constant k is calculated as shown, where A is the frequency factor (pre-exponential factor), E_a is the activation energy in J mol^{-1} (not kJ mol^{-1}), R is the gas constant and T is the temperature in Kelvin. All other factors being equal, the lower the activation energy the faster the reaction, but the Arrhenius equation is most successful for comparisons of closely related reactions. A Diels-Alder cycloaddition of different dienophiles with a common diene, for example, should lend itself to such a rate analysis.

77. Wang, Q.; Padwa, A.. *Org. Lett.* **2004**, *6*, 2189.
78. Connon, S.J.; Hegarty, A.F. *J. Chem. Soc. Perkin Trans. 1* **2000**, 1245.
79. Domingo, L.R. *Eur. J. Org. Chem.* **2004**, 4788.

MOLECULAR MODELING: HOMO-LUMO Gaps and the Rates of Diels-Alder Reactions

Diels-Alder cycloaddition reactions most commonly involve electron-rich dienes and electron-deficient dienophiles (usually alkenes). According to the Frontier Molecular Orbital (FMO) model, the reaction is driven by the interaction of the most available occupied and unoccupied molecular orbitals. Specifically, electron transfer from the highest-occupied molecular orbital (HOMO) on the diene to the lowest-unoccupied molecular orbital (LUMO) on the dienophile stabilizes the transition state. This implies that the smaller the HOMO-LUMO gap, the faster will be the reaction. Therefore, it is reasonable to expect that the energy difference between HOMO of the diene and LUMO of the dienophile will relate to the activation energy for the reaction. While the HOMO and LUMO energies (and the HOMO-LUMO gap) cannot be measured (orbitals are not "real"), both are closely related to quantities that can be. Specifically, the HOMO energy is related to the ionization potential (the higher the energy of the orbital, the easier it is to expel an electron), and the LUMO energy is related to the electron affinity (the lower the energy of the orbital, the easier it is to capture an electron).

Build cyclopentadiene (the diene) and calculate the HOMO energy using the molecular orbitals menu item. Build each of the alkenes listed in the table (dienophiles), and calculate its LUMO energy. After the first calculation, *click* on > at the right of the tab at the top of the screen and select **Continue**. This will bring up a fresh builder palette while keeping this text on screen.

<u>Dienophile</u>	<u>log Relative Rate</u>
acrylonitrile	0
trans-1,2-dicyanoethylene	1.89
cis-1,2-dicyanoethylene	1.94
1,1-dicyanoethylene	4.64
tricyanoethylene	5.66
tetracyanoethylene	7.62

You can't make plots with *SpartanModel*, although you can with *Spartan Student* or *Spartan*. Use an external graphing program (or graph paper) to make a plot of the HOMO-LUMO gap ($E_{LUMO} - E_{HOMO}$) vs. the log of the rate of Diels-Alder cycloaddition reactions between cyclopentadiene and each of the dienophiles (tabulated above as relative to the log of the rate for reaction of cyclopentadiene and acrylonitrile). Does a decrease in HOMO-LUMO gap lead to a corresponding increase in reaction rate?

Optional: Try to identify a common diene with a HOMO energy that is higher (more positive) than the HOMO energy of cyclopentadiene. Try to identify a common dienophile with a LUMO energy that is lower (less positive) than the LUMO energy of tetracyanoethylene. Calcualate the HOMO-LUMO gap and, based on the correlation obtained above, predict a log of the rate of cycloaddition (relative to that of cyclopentadiene and acrylonitrile).

There is an alternative way to look at differences in rate of a Diels-Alder reaction. The most common Diels-Alder reactions are between electron rich dienes and electron-deficient dienophiles, for example, between cyclopentadiene and methyl vinyl ether or acrolein. This fact suggests that (negative) charge may be transferred from the diene to the dienophile, and may contribute to what is driving the reaction.

MOLECULAR MODELING: Geometries of Transition States and the Hammond Postulate

The energy required to reach the transition state in a reaction is known as the activation energy, E_{act}. Organic chemists, therefore, think of a transition state as being "at the top" of the pathway commonly referred to as a "reaction coordinate" connecting reactants and products. In part, this is true. A transition state is an energy maximum along the reaction coordinate. However, a chemical reaction where the reactant contains N atoms actually involves movement over an energy surface with 3N-6 dimensions, and a transition state is actually an energy minimum in each of the remaining (3N-7) dimensions. Liken a chemical reaction to a drive across a mountain range. The "goal" is to go from one valley to another, just as the goal of a chemical reaction is to go from one minimum on the energy surface (the reactants) to another minimum (the product). The road moves about a two-dimensional surface whereas the energy surface for a chemical reaction may involve dozens or even hundreds of dimensions. *There is no need to climb to the top of a mountain (a maximum in both dimensions) to go from one valley to another. Rather, is sufficient to go through a pass between mountains (a maximum in one dimension but a minimum in the other). Similarly, it is only necessary to pass through an energy maximum in one coordinate (the reaction coordinate) for a chemical reaction.*

Because a transition state is an energy maximum (albeit only in one of many dimensions) it cannot function to "trap" a molecule, let alone collect a population of "trapped" molecules. *This means that a transition state cannot even be observed experimentally let alone isolated or characterized. Quite simply, a transition state does not exist in the sense that a molecule exists.* However, it is possible to calculate the geometry and other properties of a transition state and to verify that in fact it is on a pathway connecting reactants and product. We shall refer to this as *transition state modeling.*

It is not possible to actually draw a multi-dimensional energy surface, and locating a transition state relies on its mathematical description. Reactants, products and transition states are so-called stationary points on an energy surface, that is, locations for which the first derivative of the energy with respect to each and every coordinate is zero. Think of them as points on the surface where the energy is neither heading up or down but rather is "flat". While the vast majority of stationary points have no "chemical meaning", there are two classes of points that are of interest. The *first* are points for which the second derivative of energy is positive for each and every coordinate. The second derivative gives the curvature of the coordinate at the point on the surface. A positive value for the second derivative indicates that the surface is curved up and corresponds to a minimum in the coordinate, whereas a negative value indicates that the surface is curved down and corresponds to a maximum. Points for which the second derivative is positive for all dimensions correspond to "stable" molecules, although not necessarily molecules that can actually be isolated. The *second* are points for which the second derivative of the energy is positive for all coordinates except for one (the reaction coordinate) for which the second derivative of the energy is negative. These *may* correspond to transition states.

MOLECULAR MODELING: Geometries of Transition States and the Hammond Postulate, con't.

Even though they cannot be determined experimentally, the geometries of transition states are often brought up in the discussion of chemical reactions. For example, the choice of which regio or stereochemical product is preferred often comes down to which results from the "least crowded" transition state. Several examples of this appear in the succeeding problems in this chapter.

Another relationship that is commonly made is that between the geometry of the transition state and that of the reactants. According to the **Hammond Postulate**, the more *exothermic* a reaction the more its transition state will resemble the reactants. Said in reverse, the more the transition state resembles the reactants, the more *exothermic* a reaction is likely to be. The Hammond Postulate provides justification for use of the properties of reactants, for example, its steric demands, to anticipate the properties of the transition state.

Transition states obtained from quantum chemical calculations for Diels-Alder cycloadditions of cyclopentadiene and ethylene and for cyclopentadiene and tetracyanoethylene appear at the left and right of the screen, respectively. The first corresponds to a reaction that is very slow, that is, has a high energy barrier, while the second corresponds to a reaction that is very fast, that is, has a low energy barrier. According to the Hammond Postulate, which transition state should more closely resemble its reactants? Elaborate. Examine the two transition states to see if can find evidence supporting or refuting the suitability of the Hammond Postulate in this instance. You can calculate equilibrium geometries for the reactants (ethylene, tetracyanoethylene and cyclopentadiene) if you think that this will help.

MOLECULAR MODELING: Arrhenius Equation

While the geometries of transition states obtained from quantum calculations are certainly "interesting", in that they show the "making" and "breaking" of bonds on the way from reactants to products. Will they tell the chemist what really matters: "will a reaction go" and if it does "what products will be formed"? Put another way, do changes in structure fortell changes in activation energies and if they do, will they be large enough to discern?

Compare transition states for the Diels-Alder reaction of 2-methoxybutadiene and styrene leading to the *para* regioisomer (top left) and for the corresponding reaction of 2-methoxybutadiene and 2-butyne (top right). Can you see any difference in their geometries that would tell you that the first reaction is fast while the second reaction is slow? Specifically, is one of the transition states "more delocalized" and presumably more stable than the other transition state? A measure of this might be that the difference between formal, single, and double bonds (in the reactants) is more "blurred" in one of the transition states than in the other.

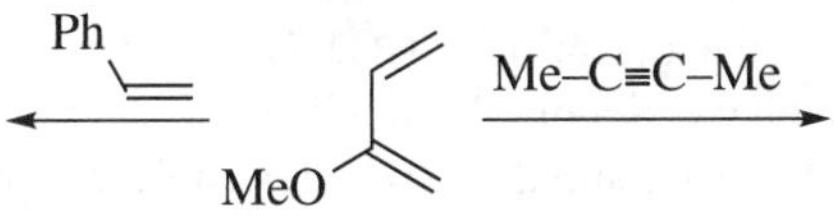

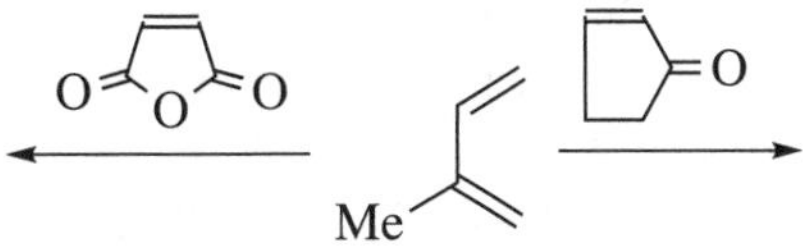

Next, compare transition states for Diels-Alder reactions involving 2-methylbutadiene as the diene, first with maleic anhydride (bottom left) and then with 2-cyclopenten-1-one leading to the *para*

regioisomer (bottom right) as dienophiles. Here, reaction with maleic anhydride is known to be much faster than that with 2-cyclopenten-1-one. Is one of the transition states more delocalized than the other (use the same criterion as before)? For these examples, do you find it easy or difficult to tell from examining the transition states, that one reaction is likely to be much faster than the other?

Questions about whether a particular reaction will actually occur and if it does, what its product will be may be answered by connecting the rates of chemical reactions to the energies of their transition states. This requires a model, the most common of which is known as the *Arrhenius equation*. According to the Arrhenius equation, the rate of a chemical reaction depends on the concentration of reactants, [R], the so-called pre-exponential or "A factor", which accounts for the fact that not all encounters between reactants will lead to reaction, and the energy barrier or *activation energy*, $\Delta E^{\ddagger}$, that the reactants must surmount to get over the transition state. The activation energy is simply the difference the energy of the transition state and that of the reactants.

$$\text{rate} \propto [R] \, A \, \exp[-\Delta E^{\ddagger}/RT]$$

R is the gas constant and T is the temperature (in K). Where the concern is which product is favored for a particular reaction, that is, where the reactants are the same, the experimental conditions (correlations) will be identical and the A factor may safely be assumed to be the same for all competing reactions. In this case, the dominant product will be that formed from the transition state with the lowest energy. So the goal is to identify the lowest-energy transition state among all possible transition states and link it to a product.

Back to the two examples referred to earlier in this problem. Based on an analysis of the activation energy, the results are clear. Diels-Alder reaction of 2-methoxybutadiene and styrene occurs with an activation energy that is 32 kJ/mol lower than that for reaction of 2-methoxybutadiene and 2-butyne, and the Diels-Alder reaction of 2-methylbutadiene and maleic anhydride occurs with an activation energy that is 44 kJ/mol lower than that for reaction of 2-methylbutadiene and 2-cyclopenten-1-one. These activation energy differences are large enough that competition between dienophiles styrene and 2-butyne in the case of 2-methoxybutadiene as a diene and maleic anhydride and 2-cyclopenten-1-one in the case of 2-methyl butadiene as a diene would not be close.

In summary, while comparison of the geometries of transition states for related reactions may not always lead to a clear assignment of relative reaction rates, comparison of activation energies will likely provide an reliable indicator.

Most synthetically-important Diels-Alder reactions involve electron-rich dienes and electron-deficient dienophiles. According to the picture first introduced by Woodward and Hoffmann, favorable interaction between the highest-occupied molecular orbital (the HOMO) on the diene and the lowest-unoccupied molecular orbital (the LUMO) on the dienophile stabilizes the transition state and lowers the activation energy. As a consequence, electrons should be transferred from filled to empty molecular orbitals, and be reflected in the charge distribution of transition state. In the case of the Diels-Alder reactions between electron-rich dienes and electron-deficient dienophiles, electrons should move from the HOMO (on the diene) into the LUMO (on the dienophile). Does significant charge transfer actually occur? If it does, is the extent of transfer related to the degree to which the diene is electron rich and the dienophile is electron poor? Does greater charge transfer necessarily imply greater stabilization of the transition state and lower activation energy?

One measure of molecular charge distribution is provided by an electrostatic potential map. This colors each location on an electron density surface, indicating overall size and shape, with the value of the electrostatic potential, indicating the energy that a positive charge "feels" at this location. By convention, electron-rich regions where the charge is attracted are colored "red", whereas electron-poor regions where the charge is repelled are colored "blue". Compare electrostatic potential maps for the transition state for the Diels-Alder reaction of cyclopentadiene and acrolein (top right) with those of the reactants, cyclopentadiene (top left) and acrolein (top center). The maps are displayed as transparent models to allow you to see the underlying structures. Once you are properly oriented, switch to solid models to allow the extent of any differences to be ascertained. *Click* on one of the maps and select *Solid* from the **Style** menu at the bottom of the screen. Is there evidence of charge transfer between the reactants and their fragments in the transition state? Elaborate. If there is, is it in the direction supposed by the Woodward-Hoffmann model, that is, has the diene become "less red" and the dienophile "less blue"?

Next compare electrostatic potential maps for the transition state for the Diels-Alder reaction of cyclopentadiene and tetracyanoethylene (bottom right) with those of the reactants, cyclopentadiene (top left) and tetracyanoethylene (bottom center). This reaction is known to be faster (lower activation energy) than that involving acrylonitrile as a dienophile. Is the direction of charge transfer the same as that for the reaction involving acrylonitrile? If it is, is the magnitude of the transfer, smaller, larger or about the same? Does there appear to be a link between the extent of charge transfer and rate for Diels-Alder reactions? Elaborate.

11.4.C. Selectivity in the Diels-Alder Reaction

The Diels-Alder reaction is one of the powerful reactions, as categorized by Corey in the LHASA program (sec. 10.4.B). One of the powerful features of the Diels-Alder reaction is the stereochemical selectivity inherent to the cycloaddition, and many examples exist in the total synthesis of natural products.[80]

11.4.C.i. Cis-Trans Selectivity. In the reaction of **32**, the cycloadduct (**33**) showed a cis relationship at the ring juncture. The methyl and the hydrogen at the bridgeheads were derived from the respective alkene and it is clear that the relative stereochemistry of these groups in the alkene reactant was retained in the Diels-Alder cycloadduct (**33**). As seen in the transition state for that reaction, as butadiene approached the alkene, the hydrogen and the methyl were

---

80. (a) Nicolaou, K.C.; Snyder, S.A.; Montagnon, T.; Vassilikogiannakis, G. *Angew. Chem. Int. Ed.,* ***2002***, *41*, 1669; (b) Desimoni, G.; Tacconi, G.; Barco, A.; Pollini, G.P. *Natural Product Synthesis Through Pericyclic Reactions*, American Chemical Society, Washington, ***1983***, pp 119-254.

pushed away, and no isomerization occurred during the reaction. The product (**33**) reflected this stereochemistry, with the methyl and hydrogen cis to each other. In general, reactions requiring relatively low reaction temperatures and fast reaction times show little epimerization (*cis* → *trans*). Maleic anhydride reacted with butadiene in refluxing xylene to give the cis cycloadduct (**40**), which was opened under the reaction conditions to the cis-diacid (**41**).[81] The cis configuration of the alkene reaction partner (maleic anhydride) is locked by the cyclic nature of the anhydride, and retained in the product. When the stereochemistry is not locked, it can be controlled in some cases. When 1-phenyl-1,3-butadiene reacted with acryloyl chloride, for example, the cis adduct (**42**) was formed, but at refluxing temperatures, a mixture of **42** and **43** was formed, with **43** predominating.[82] This example illustrates that epimerization to a more stable product is possible when the product contains an α-hydrogen that is subject to enolization.

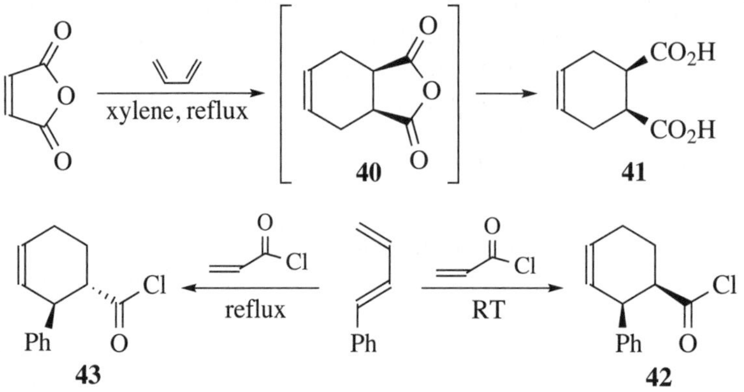

In the monograph by Desimoni et al.,[83] four rules are listed for Diels-Alder reactions, taken from Sauer's work[84] and based on molecular orbital arguments.

1. Electron-withdrawing substituents (Z) on dienophiles and electron releasing substituents (X) on dienes increase the rate of reaction. The reverse substituent effects (inverse electron demand) likewise increase the rate.

2. The diene and dienophile configurations are retained in the adduct (the **cis principle**).

3. The endo transition state is favored over the exo transition state (the **endo rule**).

4. The Z-substituted dienophiles react with 1-substituted butadienes (in normal electron demand Diels-Alder reactions) to give 3,4-disubstituted cyclohexenes, independent of the nature of diene substituents (the **ortho effect**).

The effects of electron-withdrawing and releasing substituents on a diene (Rule 1) were previously discussed, and the stereochemistry of the alkene (dienophile, Rule 2) was retained

81. Korolev, A.; Mur, V. *Dokl. Akad. Nauk. SSSR* **1948**, *59*, 251.
82. Meek, J.S.; Poon, B.T.; Merrow, R.T.; Christol, S.J. *J. Am. Chem. Soc.* **1952**, *74*, 2669.
83. Reference 80b, pp 128-130.
84. Sauer, J. *Angew. Chem. Int. Ed.* **1966**, *5*, 211; (b) *Idem Ibid.* **1967**, *6*, 16.

in the cycloadduct. An illustration is the reaction of *E,E*-diene **44** with maleic anhydride to give **45**.[85,86] Similarly, *E,Z*-butadienes **46** reacted with maleic anhydride to give cycloadduct **47**.[87,88] With a wide range of substrates, the stereochemistry of the substituents in the diene was retained in the cycloadduct.

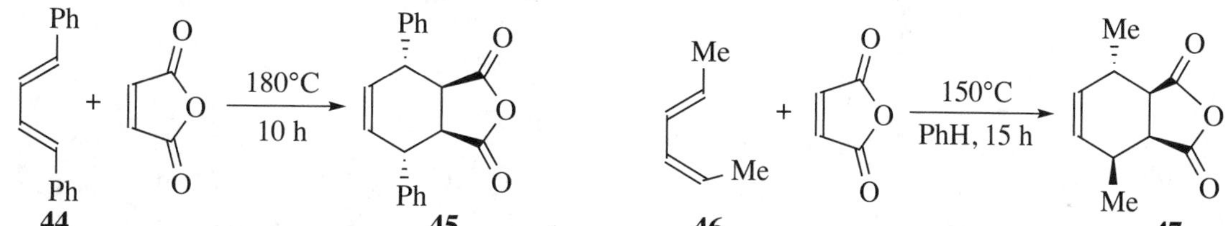

In order for the stereochemistry of the diene substituents to be maintained in **45** and **47**, the relative motions of these substituents during ring formation must be as shown in **50**. The outer substituents on the diene move toward each other, and the inner substituents move away from each other producing **51**. Such motion is described as **conrotatory** (orbital motion in the same direction; **50 → 51**). Alternatively, the outer substituents can move in opposite directions, as indicated in **48**, and is termed **disrotatory** (orbital motion in opposite directions, **48 → 49**).[89] It is clear from previous examples that the Diels-Alder reaction proceeded with a disrotatory motion in those cases, leading to predictable stereochemistry in the cycloaddition. The Merck cortisone synthesis, performed by Sarett et al.[90] began with reaction of **52** and benzoquinone to give **53**. The cis stereochemistry at the ring juncture is predictable by examination of transition state shown. Disrotatory motion of the methyl and hydrogen led to the stereochemistry shown in **53**.

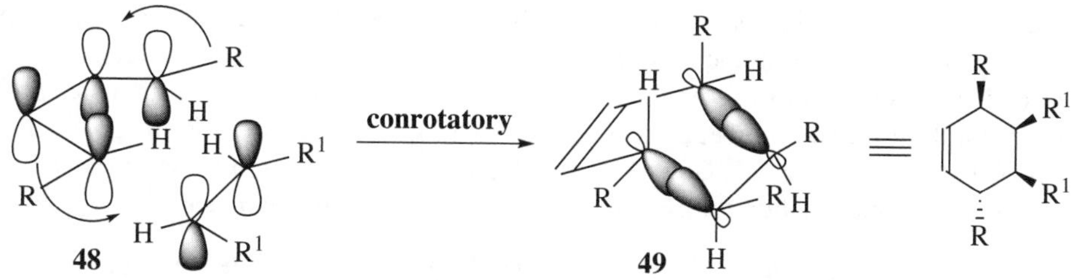

85. Martin, J.G.; Hill, R.K. *Chem. Rev.* **1961**, *61*, 537 (see p 544).
86. Alder, K.; Schumacher, M. *Annalen* **1951**, *571*, 87.
87. Reference 86, pp 544-545.
88. Alder, K.; Vogt, W. *Annalen* **1951**, *571*, 137.
89. Also see David, C.W. *J. Chem. Ed.* **1999**, *76*, 999.
90. (a) Sarett, L.H.; Arth, G.E.; Lukes, R.M.; Beyler, R.E.; Poos, G.I.; Johns, W.F.; Constantin, J.M. *J. Am. Chem. Soc.* **1952**, *74*, 4974; (b) Sarett, L.H.; Lukes, R.M.; Poos, G.I.; Robinson, J.M.; Beyler, R.E.; Vandergrift, J.M.; Arth, G.E. *Ibid.* **1952**, *74*, 1393.

disrotatory

50   51

52   53

MOLECULAR MODELING: Conformation and the Rates of Diels-Alder Reactions

Calculations indicate that the activation energy for the Diels-Alder reaction of cyclopentadiene and acrylonitrile is 29 kJ/mol less than that for the corresponding reaction of (*trans*) 1,3-butadiene and acrylonitrile. One possible explanation is that the lowest-energy conformer of 1,3-butadiene is s-*trans* and not s-*cis*. Thus, energy needs to be expended to get into the "reactive" s-*cis* conformer. An alternative explanation is that the π system of cyclopentadiene may be inherently more electron rich than that of 1,3-butadiene, that is, cyclopentadiene is a better diene.

Obtain energies for both s-*cis* and s-*trans* conformers of 1,3-butadiene. Note that the *cis* diene may not be planar, so start with a non-planar structure. You can rotate around the carbon-carbon single bond by *clicking* on it, holding down both the left mouse button and the **Alt** key (**Option** key on Macintosh) and moving the mouse up and down. What is the energy difference between the "reactive" *cis* conformer of 1,3-butadiene and the *trans* conformer? Is conformation a possible explanation for the difference in reactivity of the two dienes? Compare electrostatic potential maps for *cis*-1,3-butadiene and cyclopentadiene. Which diene appears to be more electron rich? Is this difference a possible explanation for the difference in reactivity of the two dienes?

Next, try to explain why the Diels-Alder reaction of 2,3-dimethylcyclopentadiene and acrylonitrile is faster that that between 2,3-dimethyl-1,3-butadiene and acrylonitrile. Identify the lower-energy conformer for 2,3-dimethyl-1,3-butadiene. If it is not the "reactive" (s-*cis*) conformer, what is the energy that must be expended before the Diels-Alder reaction can take place? Is the "penalty" smaller, larger or about the same as for 1,3-butadiene? Given that the activation energy for the reaction of 2,3-dimethyl-1,3-butadiene and acrylonitrile is 29 kJ/mol lower than that for corresponding reaction of 2,3-dimethylcyclopentadiene, what if anything does your result suggest about the importance of conformation on relative rate. Compare electrostatic potential maps for the two methyl-substituted dienes. Which appears to be the more electron rich? What if anything does your result suggest about the importance of the "electron richness" of the diene on relative rate?

11.4.C.ii. Regioselectivity of Cycloaddition Reactions. It is apparent from several of the preceding examples that more than one regioisomeric cycloadduct is possible with certain substituent patterns on the diene and/or alkene. When the 2-substituted diene **54** reacts with alkene CH$_2$=CHX, two orientations for the X group are possible relative to R: the meta orientation in **55** and the para orientation in **56**. Dienes with substituents at C1, such as **57**, react with CH$_2$=CHX to give a mixture of the ortho product (**58**) and the meta product (**59**). Although a mixture of regioisomers is formed, the Diels-Alder reaction is highly regioselective in most cases. The reaction of 1,3-pentadiene and methyl acrylate, for example, gave a 68% yield of the ortho adduct, **60**.[91] Similarly, 1-phenyl-1,3-butadiene reacted with acrolein to give **61**, in 64% yield.[74] The endo products are cis **60** and **61**. In the Diels-Alder reaction of 1-alkyl butadienes (**57**, R = Me, i-Pr, n-Bu, t-Bu) with CH$_2$=CHX (X = CHO, CN, CO$_2$Me), the ortho regioisomer **58** was favored over the meta cycloadduct (**59**),[92] with a preference between 4:1 and 10:1.

The propensity for ortho, meta, or para selectivity will vary with the diene (electron-releasing or withdrawing substituents) and the alkene (electron-releasing or withdrawing substituents). Fleming assembled a table to predict regioselectivity for the generalized alkenes CH$_2$=CH–R, CH$_2$=CH-Z, and CH$_2$=CH–X (where R = an alkyl group, Z = an electron-withdrawing group, and X = an electron-releasing group) using frontier orbital considerations.[93]

Examining the orbital coefficients of the HOMO and LUMO for both diene and alkene predicts selectivity. The transition state for the reaction of methyl acrylate and 2-phenyl-1,3-butadiene (see **62**) shows the orbitals with largest coefficients (HOMO$_{diene}$-LUMO$_{alkene}$) combine (the absolute value of the orbital coefficient is used, |-0.625| $\leftrightarrow$ |0.69| and |-0.475| $\leftrightarrow$ |-0.471|) to predict the cycloadduct produced in the greatest amount (the para product, **63**). The HOMO$_{62}$-LUMO$_{alkene}$ ΔE is [0 - (-8.77) = 8.77], and it predicts the relative reactivity of these reactants. The magnitude of the orbital coefficients in each partner is important, and overlap of the larger orbital coefficients of the diene and alkene predict the selectivity. In this case, the larger orbital coefficients are 0.625 for the diene and 0.69 for the alkene, and their overlap predicts the observed selectivity for the para product. Similarly, the orbital coefficients for 1-methoxy-

91. Meek, J.S.; Ragsdale, J.W. J. Am. Chem. Soc. **1948**, 70, 2502.
92. (a) Reference 80b, p 137; (b) Titov, Y.A. Usp. Khim. **1962**, 31, 529.
93. Reference 7, pp 133-138.

1,3-butadiene and acrolein correctly predict the major product is the ortho adduct. For simple cases, the orbital coefficients can be calculated, using a program such as Spartan™.[40]

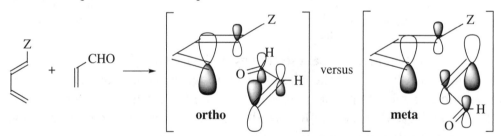

This discussion allows an explanation of the **ortho effect** presented earlier, in connection with the Diels-Alder reaction. When a C=X or a C≡X unit (where X is a heteroatom) reacts with a 1-substituted diene, the ortho product is generally preferred. Ortho selectivity is explained by the size of the orbital coefficients. When coefficient size is combined with secondary orbital interactions (sec. 11.4.C),[47,43,94] as indicated in the transition state shown in Figure 11.10, the carbonyl orbital is responsible for a secondary interaction that stabilizes the ortho transition state. If the meta transition state is involved, then orbital symmetry dictates an exo mode, and there is no possibility for secondary orbital interactions. Secondary orbital interactions such as this are possible only in an endo transition state. The application of secondary orbital interactions to this problem has been questioned.[95]

Figure 11.10. Frontier Orbital Control of Regiochemistry For Electron-Rich Dienes and Acrolein.

The ability to predict the regioselectivity of Diels-Alder reactions is a cornerstone of their use in synthesis. Methyl vinyl ketone (MVK), for example, reacted with 2-methyl-1,3-butadiene to give a 3:1 mixture of **64** and **65**. Gutsche et al. used the para product (**64**) to synthesize α-bisabolol.[96] They also reported that using stannic chloride as a catalyst (to be discussed in sec. 11.6.A) increased the ratio of **64/65** to 93:7.[96]

---

94. Also see Eisenstein, O.; Lefour, J.-M.; Ahn, N.T. *Chem. Commun.* **1971**, 969.
95. Fleming, I.; Michael, J.P.; Overman, L.E.; Taylor, G.F. *Tetrahedron Lett.* **1978**, 1313.
96. (a) Gutsche, G.D.; Maycock, J.R.; Chang, C.T. *Tetrahedron* **1968**, *24*, 859; (b) Rittersdorf, W.; Cramer, F. *Ibid.* **1968**, *24*, 43.

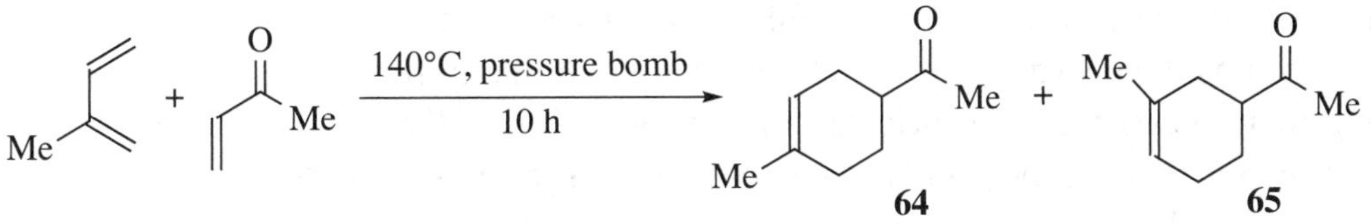

The diene and/or the alkene components can be part of a amore complex structure. An example is the reaction of the C=C unit in conjugated lactam **66** with the conjugated diene-yne **67**, taken from a synthesis of (–)-gymnodimine by Romo and co-workers.[97] A chiral *bis*-oxazolone **68**-hexafluoroantimonate complex catalyzes the reaction[98] to give **69** in 85% yield, and 95% ee.

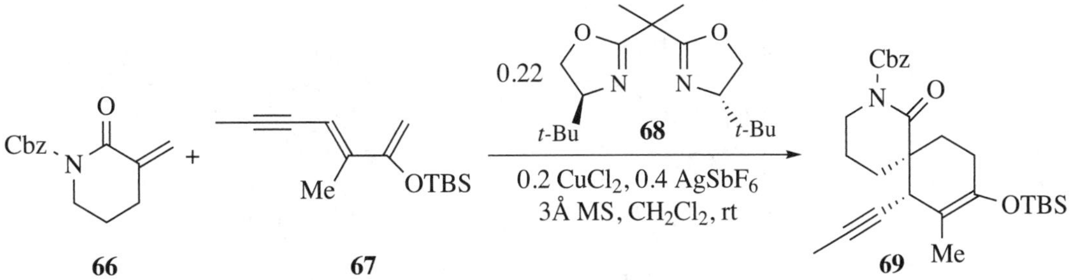

Hehre et al. proposed an alternative method to predict the regioselectivity of Diels-Alder reactions. The FMO approach does not predict "that a phenyl substituent is a significantly better regiodirector than methyl".[14,99] Hehre suggested using the average hydride potentials for the carbons of substituted ethylenes,[14] and the electrostatic (proton) potentials for the carbons of substituted dienes.[14] Correlating the largest coefficients of each reactive pair of molecules can accurately predict the regioselectivity of the resulting cycloadduct.

11.4.C.iii. Endo Selectivity. In the reaction that generated **53**, the observed stereochemistry demanded that the benzoquinone be under the diene at the time of reaction (see that transition state in sec. 11.4.C.i), which is the normal reaction of dienophiles bearing a substituent with a π-bond and is called an **endo mode** of addition (the group of interest is under or down, relative to the diene, rather than out or up). This endo addition is seen again in the reaction of diene **70** with methacrolein to give **71**, via the endo transition state shown, taken from Rawal and Kozmin's synthesis of (–)-α-elemene.[100] In this case, the disrotatory motion of the pyrrolidine unit places it down in the transition state. The carbonyl moiety is endo in the transition state, leading to the observed cis geometry (CHO and pyrrolidino in **71**). This endo orientation is seen again in the transition state for a reaction of acrylic acid with a generic diene (**72**) and

97. Kong, K.; Moussa, Z.; Romo, D. *Org. Lett.* **2005,** *7,* 5127.
98. (a) Evans, D.A.; Miller, S.J.; Lectka, T.; Matt, P.V. *J. Am. Chem. Soc.* **1999,** *121,* 7559. (b) Evans, D.A.; Barnes, D.M.; Johnson, J.S.; Lectka, T.; von Matt, P.; Miller, S.J.; Murry, J.A.; Norcross, R.D.; Shaughnessy, E.A.; Campos, K R. *J. Am. Chem. Soc.* **1999,** *121,* 7582.
99. Also see Hehre, W.J.; Pau, C.F.; Hout Jr., R.F.; Francl, M.M. *Molecular Modeling - Computer Aided Descriptions of Molecular Structure and Reactivity*, Wiley, New York, **1987**.
100. Kozmin, S.A.; Rawal, V.H. *J. Am. Chem. Soc.* **1997,** *119,* 7165.

it contrasts with the exo orientation in **74**, which gives the opposite relative stereochemistry for the ring substituents (trans- in **75** vs. cis- in **73**). In many reports of Diels-Alder reactions spanning many years the endo orientation is preferred, usually in a ratio of 2→4:1, which is the basis of rule 3 listed previously, the so-called **endo rule** or **Alder endo rule**.[101] Martin and Hill define the rule as follows: "Endo addition involves the tendency for dienophile substituents to be so oriented in the favored transition state that they lie directly above the residual unsaturation of the diene.... The transition state that is best stabilized by spatial orbital overlap and simultaneously least destabilized by unfavorable steric repulsion has the lowest free energy of all possible transition states, and consequently predominates in the kinetically determined product".[85]

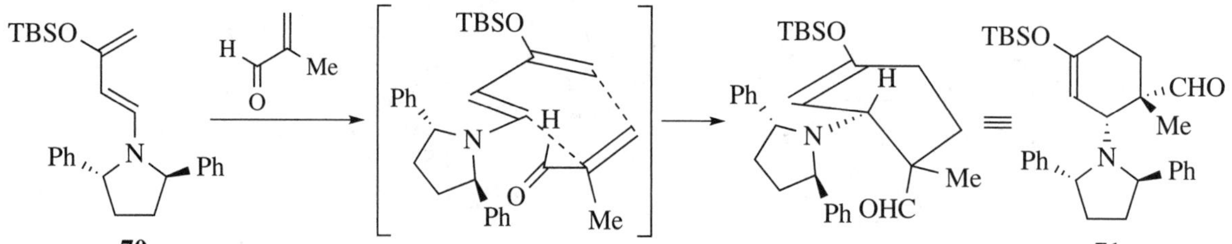

endo transition state

70   **71**

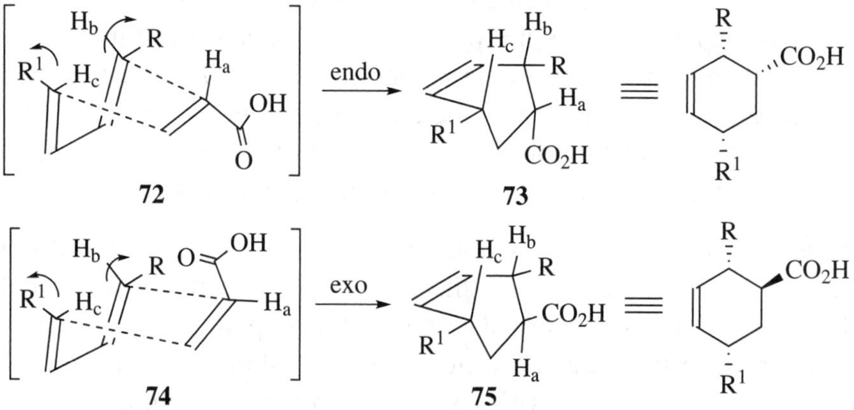

72   **73**

74   **75**

An endo transition state is best observed in reactions with cyclic dienes such as cyclopentadiene, where the exo/endo orientation is easily detected in the cycloadduct. Cyclopentadiene, for

76   **77**

example, reacts to give an endo product (X group is under the ring in **76**) and an exo product (X group is up as in **77**).[102] Both electron-rich and -poor substituents (X) on the alkene exhibit endo selectivity upon reaction with cyclopentadiene

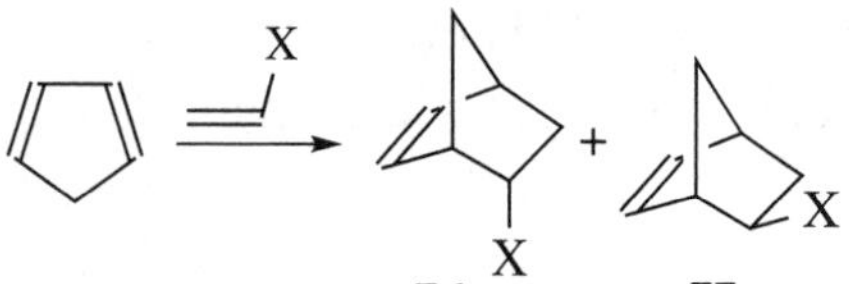

101.   Alder, K.; Stein, G. *Angew. Chem.* **1937**, *50*, 510.
102.   Reference 85, p 546.

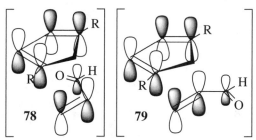

The **Alder endo rule** is explained by FMO theory, invoking **secondary orbital interactions**.[103] A comparison of the endo transition state (**78**) with the exo transition state (**79**) for the reaction of cyclopentadiene and acrolein shows the orbitals of the carbonyl are properly aligned in **78** and of the correct symmetry for stabilization by the orbitals of the $HOMO_{diene}$. This interaction is absent in the exo transition state (**79**). This secondary orbital interaction stabilizes **78** and lowers the energy of that transition state relative to **79**. Maleic anhydride[104] reacts with cyclopentadiene, for example, to give only the endo product, and in like manner *cis*-methacrylic acid reacts to give a 90:10 mixture of endo/exo cycloadducts.[105] As mentioned in section 11.4.C.ii for the ortho effect, the existence of secondary orbital interactions for explaining endo selectivity has been called into question.[106] Indeed, Ogawa and Fujimoto[107] demonstrated by the application of perturbation theory that the effect of secondary orbital interactions should be much less significant than has been assumed within the frontier orbital scheme. Somekawa and co-workers reported calculations that concluded selectivity arises from interactions of the diene π and n orbitals, as well as the diene-π orbitals and dienophile-π orbitals.[108]

Examination of such interactions is important for the understanding of regioselectivity in Diels-Alder reactions. Houk formulated a general approach for secondary orbital interactions.[109]

1. For all allowed cycloadditions involving the reaction of a 4n electron polyene with an alkene with extended conjugation, the endo transition state can be stabilized by secondary orbital interaction (see **80** and **81**).

2. For the allowed 4n + 4m+2 cycloaddition of two polyenes, where m ≠ 0, the endo transition state will be destabilized (see **82**).[110]

103. Hoffmann, R.; Woodward, R.B. *J. Am. Chem. Soc.* **1965,** *87*, 4388, 4389.
104. Alder, K.; Stein, G.; von Buddenbrock, F.; Eckardt, W.; Frercks, W.; Schneider, S. *Annalen* **1934,** *514*, 1.
105. Alder, K.; Günzl, W.; Wolff, K. *Chem. Ber.* **1960,** *93*, 809.
106. García, J.I.; Mayoral, J.A.; Salvatella, L. *Acc. Chem. Res.* **2000,** *33*, 658.
107. Ogawa, A.; Fujimoto, H. *Tetrahedron Lett.* **2002,** *43*, 2055.
108. Kiri, S.; Odo, Y.; Omar, H.I.; Shimo, T.; Somekawa, K. *Bull. Chem. Soc. Jpn.* **2004,** *77*, 1499.
109. Houk, K.N. *Tetrahedron Lett.* **1970,** 2621.
110. Reference 2, pp 228-229.

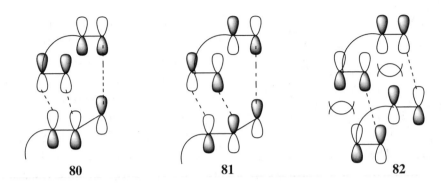

80   **81**   **82**

11.5. INVERSE ELECTRON DEMAND AND THE RETRO DIELS-ALDER

Two variations for the Diels-Alder reaction are synthetically useful. In one, the HOMO-LUMO interactions that drive the reaction change due to heteroatoms and/or substituents on the diene and/or alkene units. The reaction is driven by the $LUMO_{diene}$-$HOMO_{alkene}$ interaction rather than the $HOMO_{diene}$-$LUMO_{alkene}$ interaction. The second uses the reversible nature of the Diels-Alder reaction. Manipulation of the cycloadducts followed by a retro-Diels-Alder reaction gives a new diene or alkene. Both variations will be examined in this section.

11.5.A. Inverse Electron Demand Diels-Alder Reactions

The $LUMO_{diene}$-$HOMO_{alkene}$ interaction for butadiene and methyl vinyl ether is lower than the comparable $HOMO_{diene}$-$LUMO_{alkene}$ interaction in dienes bearing an electron-withdrawing group. In some cases this leads to the **inverse electron demand** reaction mentioned above. Fleming uses the example of azonia salt **83**,[111] which reacted with diethyl ketene acetal to give **84**. Similar reaction with allyl alcohol gave **85**, and acrylonitrile gave **86**.[111a] In all examples, the reaction proceeded to 75% completion and the relative rates of reactions provided with each transformation indicates that electron-rich alkenes reacted faster. Inverse electron demand Diels-Alder reactions are controlled by the $LUMO_{diene}$ $HOMO_{alkene}$ interaction, which usually means that an electron-rich alkene reacts faster than an electron-poor alkene, the opposite of what is normally observed (see above).

111. (a) Reference 7, pp 111-112; (b) Fields, D.L.; Regan, T.H.; Dignan, J.C. *J. Org. Chem.* **1968**, *33*, 390; (c) Porter, N.A.; Westerman, I.J.; Wallis, T.G.; Bradsher, C.K. *J. Am. Chem. Soc.* **1974**, *96*, 5104.

[Reprinted with permission from Fleming, I. *Frontier Molecular Orbitals and Organic Chemical Reactions,* Wiley, London, ***1976***, pp. 112 and 129. Copyright © ***1976*** by John Wiley and Sons, Inc.]

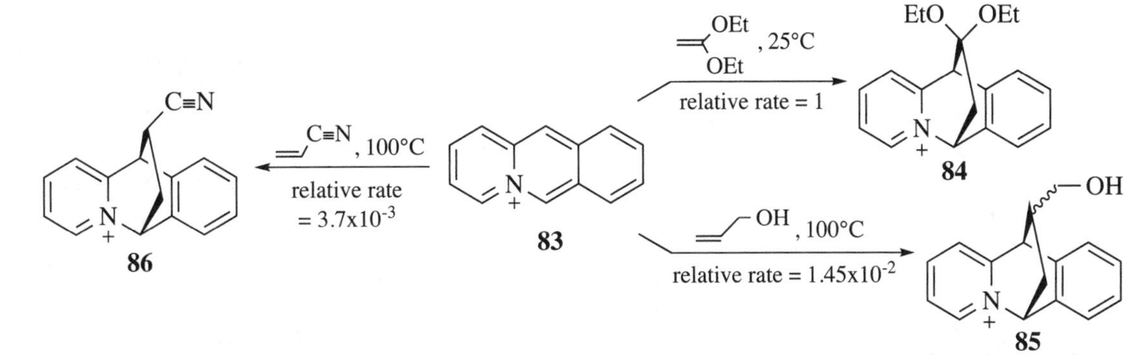

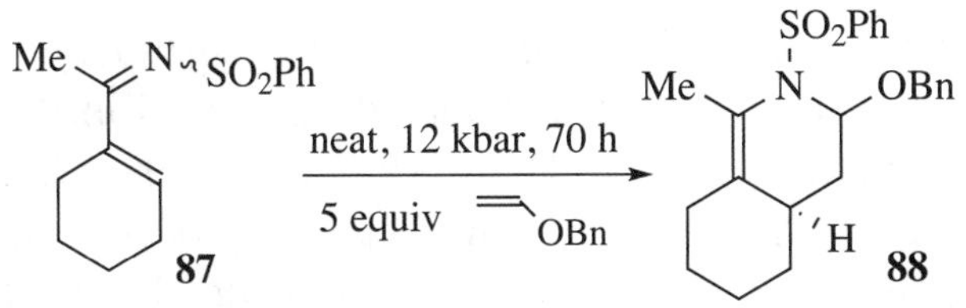

In another example, Boger and Kasper prepared 1-azadiene **87** (azadienes are discussed in sec. 11.7.C.v) and showed that the LUMO of the diene controlled the subsequent Diels-Alder reaction. Reaction of **87** with butyl vinyl ether (n-BuOCH=CH_2) gave a 74% yield of **88** (>20:1 endo/exo).[112] This reaction was sluggish and high pressure (sec. 11.6.C) was required to force the reaction to product with a reasonable reaction time and in good yield. Boger and co-workers also showed that tetrazines react with alkynes, with inverse electron demand, to give 1,2-diazines, after loss of nitrogen from the original Diels-Alder cycloadduct.[113] It is apparent from these few examples that the structures of most reverse electron demand Diels-Alder reactions involve heteroatoms, or another unique feature. They are less prevalent than the normal electron demand Diels-Alder reactions discussed elsewhere in this chapter. A synthetic example is taken from Dang and co-worker's synthesis of purine nucleosides[114] for which 2,4,6-tris(ethoxycarbonyl)-1,3,5-triazine (**91**), a useful diene used by Boger for the synthesis of pyrimidines,[115] reacted with 5-amino-1-benzyl-4-imidazolecarboxylic acid (**89**) to give **94** in 83% yield. Decarboxylation of **89** occurred during the reaction to give the reactive species **90** prior to the Diels-Alder reaction, which gave cycloadduct **92**.[114] This compound was not isolated, but rather decarboxylated to give **93**, and this compound lost ammonia to give the observed product. Imidazole **90** is so reactive, that this reaction occurs at the relatively mild conditions shown, which is somewhat unusual for inverse-electron demand Diels-Alder reactions.

112. Boger, D.L.; Kasper, A.M. *J. Am. Chem. Soc.* ***1989,*** *111*, 1517.
113. Soenen, D.R.; Zimpleman, J.M.; Boger, D.L. *J. Org. Chem.* ***2003,*** *68*, 3593.
114. Dang, Q.; Liu, Y.; Erion, M.D. *J. Am. Chem. Soc.* ***1999,*** *121*, 5833.
115. See (a) Boger, D.L.; Dang, Q. *J. Org. Chem.* ***1992,*** *57*, 1631; (b) Boger, D.L.; Menezes, R.F.; Dang, Q. *J. Org. Chem.* ***1992,*** *57*, 4333; (c) Boger, D.L.; Kochanny, M.J. *J. Org. Chem.* ***1994,*** *59*, 4950.

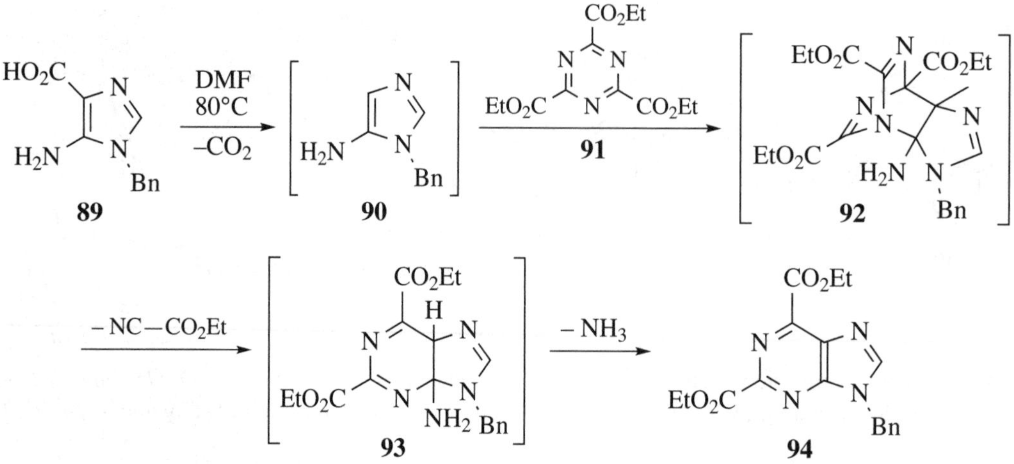

Enantioselective inverse-electron demand hetero-Diels-Alder reactions have been reported. Jorgensen and Juhl reported that 3-methybutanal reacted with conjugated ketone **95** in the presence of a catalytic amount of a chiral pyrrolidine to give the Diels-Alder adduct thast led to **96** in 93% overall yield, and 89% ee after oxidation with PCC.[116] This catalytic procedure is believed to proceed by initial formation of the pyrrolidine enamine of the aldehyde, followed by an inverse electron demand Diels-Alder reaction. In the presence of water and silica, the *N,O*-acetal is converted to the hemi-acetal, which is then oxidized with PCC. Jacobsen and co-workers reported that chromium complex **97** showed catalyzed the inverse electron Diels-Alder reaction of ten equivalents of ethyl vinyl ether and 2-butenal, to give pyran **98**, in 75% yield and 94% ee.[117]

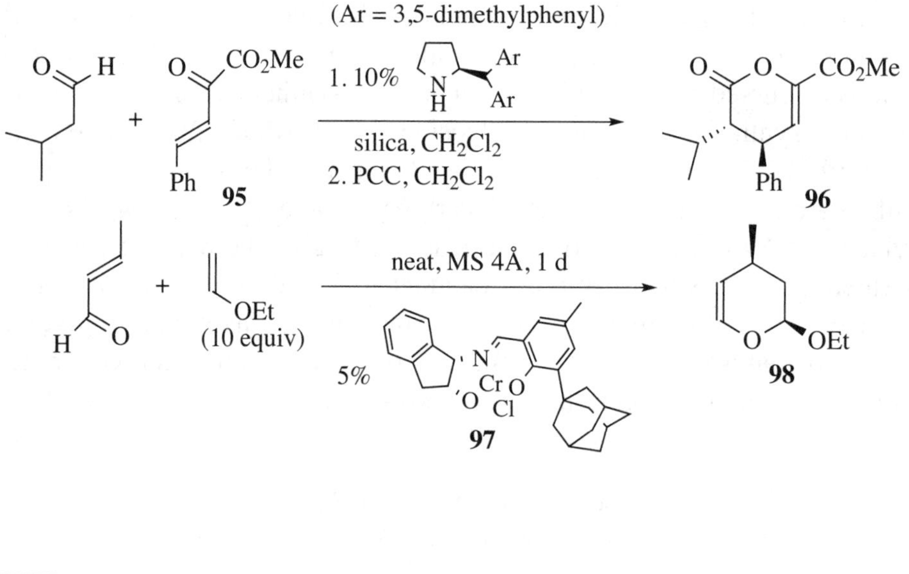

116.   Juhl, K.; Jørgensen, K.A. *Angew. Chem. Int. Ed.,* **2003**, *42*, 1498.
117.   Gademann, K.; Chavez, D.E.; Jacobsen, E.N. *Angew. Chem. Int. Ed.* **2002**, *41*, 3059.

A typical inverse electron demand disconnection is:

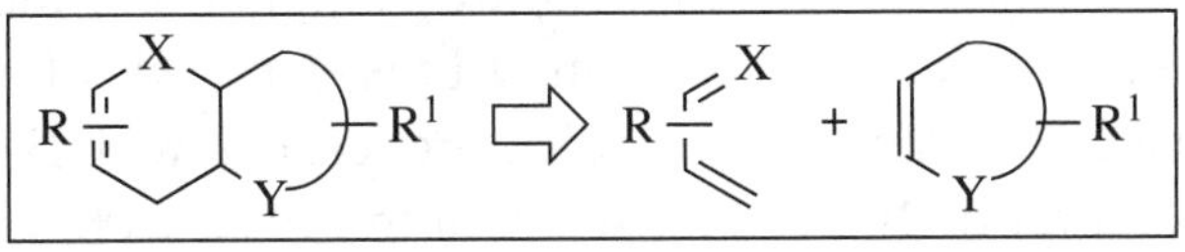

11.5.B. Retro Diels-Alder Reaction

The Diels-Alder reaction is reversible under certain circumstances.[118] This reversibility can be exploited synthetically with the correct choice of diene, dienophile and reaction conditions because the equilibrium usually favors the thermodynamically more stable cycloadduct as the major product. Woodward and Baer showed that maleic anhydride reacted with fulvalene **99** at ambient temperatures to give **100** the endo product (see below).[119] When heated to 50°C for 10 min., however, **101** was the major product with only 20.5% of **100**, and this dropped to 3.1% upon heating at 50°C for 60 min.[119] This result is consistent with a reversible reaction pathway favoring the more stable exo product.[119] This reverse reaction (cycloadduct → diene + dienophile) is called a **retro Diels-Alder reaction**.[120]

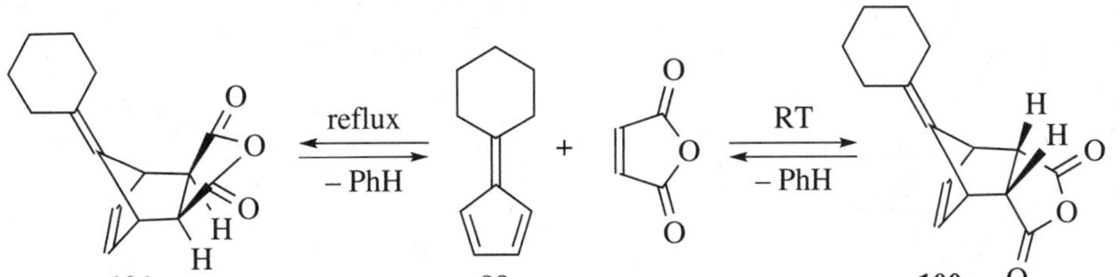

A classical example of a retro Diels-Alder reaction is the self condensation of cyclopentadiene (**20**) to give the cyclopentadienyl dimer **102**. This reaction occurs quickly at temperatures > 25°C, but is slow at low temperatures such as –78°C. Many Diels-Alder reactions involving cyclopentadiene are done at low temperatures with Lewis acid catalysts (sec. 11.6.A) to suppress

cyclopentadiene dimerization. When heated to 160-240°C, a retro-Diels-Alder reaction of **102** gives 2 equivalents of monomeric cyclopentadiene.[121]

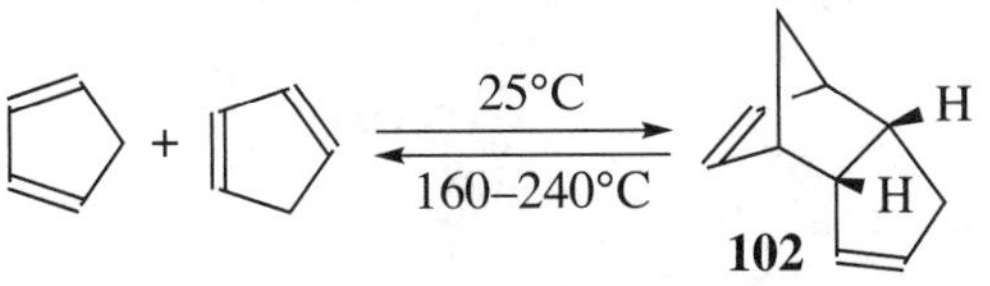

118. (a) Kwart, H.; King, K. *Chem. Rev.* **1968**, *68*, 415; (b) Smith, G.G.; Kelly, F.W. *Progr. Phys. Org. Chem.* **1971**, *8*, 75 (see p 201).

119. Woodward, R.B.; Baer, H. *J. Am. Chem. Soc.* **1944**, *66*, 645.

120. (a) Ripoll, J.L.; Rouessac, A.; Rouessac, F. *Tetrahedron* **1978**, *34*, 19; (b) Lasne, M.-C.; Ripoll, J.L. *Synthesis* **1985**, 121; (c) Ichihara, A. *Ibid.* **1987**, 207.

121. (a) Moffett, R.B. *Org. Synth. Coll. Vol. 4* **1963**, 238; (b) Korach, M.; Nielsen, D.R.; Rideout, W.H. *Org. Synth. Coll. Vol. 5* **1973**, 414.

Most of the synthetic applications of retro-Diels-Alder reactions involve cyclopentadiene or furan diene partners, but alkyne dienophiles are common and there are many potential substrate. One example is illustrated in Baran and Burns' synthesis of haouamine A in which a dilute solution of pyrone **103** was heated in *o*-dichlorobenzene with BHT to give cycloadduct **104**, which lost CO_2 via a retro-Diels-Alder reaction to give a 21% yield of **104** (+ 30% of **103**) after hydrolysis of the acetoxy groups.[122] The retro-Diels-Alder usually requires higher temperatures than the Diels-Alder reaction, and the normal Diels-Alder product can be obtained without competition from the retro reaction. When the retro Diels-Alder reaction is desired, flash vacuum pyrolysis is a common technique used in synthesis.[123] Lewis acids also catalyze retro-Diels-Alder reactions (sec. 11.6.A).[124] In a synthesis of pentenocin B, Sugahara, Fukuda and Iwabuchi utilized a retro Diels-Alder reaction beginning with ketodicyclopentadiene **106**.[125] Compound **106** was converted in ten steps to the highly functionalized compound **107**. When **107** was heated to 280°C, the retro-Diels-Alder reaction liberated cyclopentadiene and the targeted conjugated ketone **108** in 93% yield.

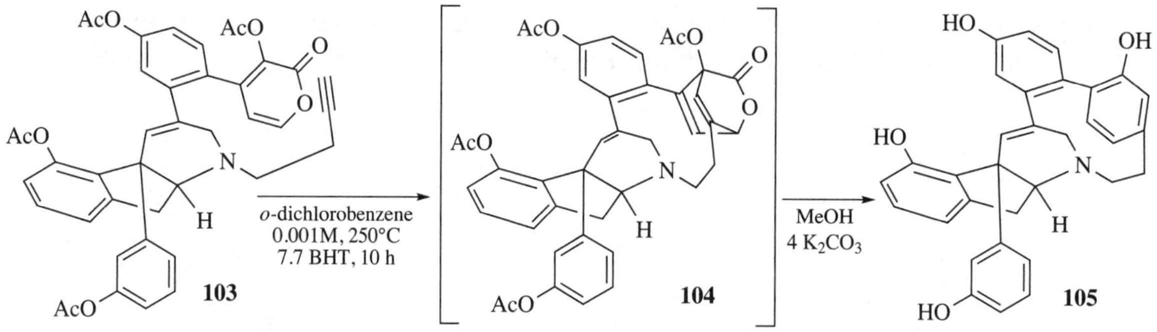

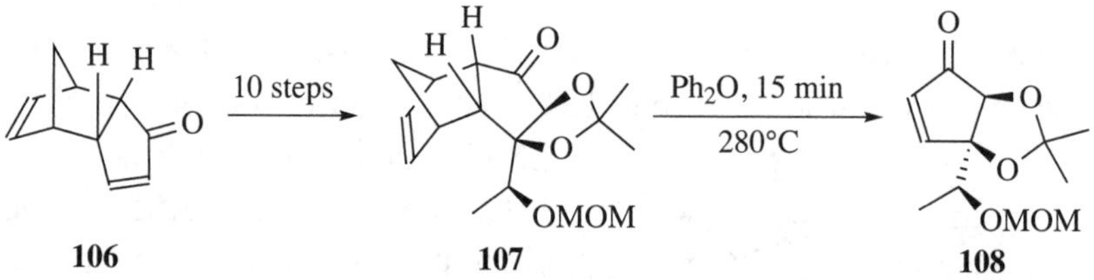

A variety of interesting and useful products are available by using the Diels-Alder reaction to insert functionality and the retro-Diels-Alder to transform that product into another. The retro-Diels-Alder reaction therefore leads to several interesting disconnections, including

122.   Baran, P.S.; Burns, N.Z. *J. Am. Chem. Soc.* ***2006***, *128*, 3908.
123.   Stork, G.; Nelson, G.L.; Rouessac, F.; Gringore, O. *J. Am. Chem. Soc.* ***1971***, *93*, 3091.
124.   Grieco, P.A.; Abood, N. *J. Org. Chem.* ***1989***, *54*, 6008.
125.   Sugahara, T.; Fukuda, H.; Iwabuchi, Y. *J. Org. Chem.* ***2004***, *69*, 1744.

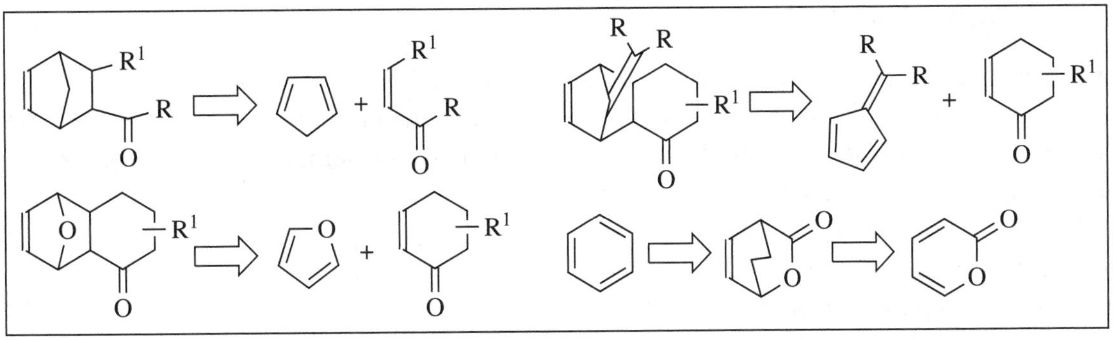

11.6. RATE ENHANCEMENT IN DIELS-ALDER REACTIONS

In many instances, Diels-Alder reactions give poor yields and/or require excessively harsh reaction conditions and long reaction times. There are several fairly straightforward techniques that increase the rate and, in some cases, the selectivity of Diels-Alder reactions. The three most important methods are catalysis by Lewis acids, rate enhancement in aqueous media, and the use of high pressure (usually greater than 5 kbar). The exploration of the Diels-Alder reaction in environmentally friendly solvents such as ionic liquids is also of great interest. Two common ionic liquids are [bmim] [BF$_4$] or [bmim][ClO$_4$], and [emim][PF$_6$], where [bmim]$^+$ and [emim]$^+$ are 1-butyl-3-methylimidazolium (**109**) and 1-ethyl-3-methylimidazolium (**110**), respectively.[126] Both hydrogenbutylimidazolium tetrafluoroborate (HBuIm, **111**) and 1,3-dibutylimidazolium, tetrafluoroborate (DiBuIm, **112**)[127] have been used in Diels-Alder reactions,[128] where these compounds were the ionic liquid medium. An asymmetric Diels-Alder reaction was reported using ionic liquids.[129] Pyridinium based ionic liquids such as ethylpyridinium tetrafluoroborate (**113**) have also been used.[130] The use of green solvents such as these is an important, and rapidly expanding area of research.

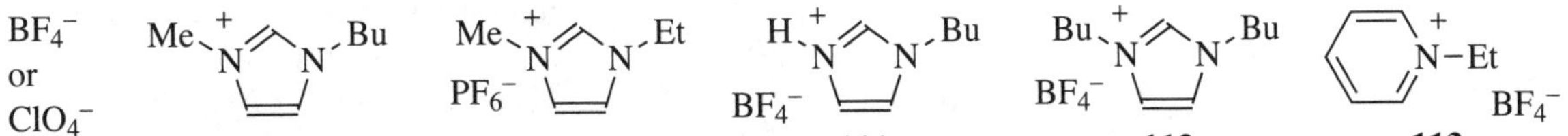

126. Kumar, A. *Chem. Rev.* **2001**, *101*, 1 (see p. 8).
127. For discussion of HBuIM and DiBuIm, see (a) Harlow, K. J.; Hill, A. F.; Welton, T. *Synthesis* **1996**, 697; (b) Holbrey, J. D.; Seddon, K. R. *J. Chem. Soc. Dalton Trans.* **1999**, 2133; (c) Larsen, A. S.; Holbrey, J. D.; Tham, F. S.; Reed, C. A. *J. Am. Chem. Soc,.* **2000**, *122*, 7264.
128. Jaegar, D. A.; Tucker, C. E. *Tetrahedron Lett.* **1989**, *30*, 1785.
129. Meracz, I.; Oh, T. *Tetrahedron Lett.* **2003**, *44*, 6465.
130. See Xiao, Y.; Malhotra, S.V. *Tetrahedron Lett.* **2004**, *45*, 8339.

11.6.A. Catalysis by Lewis Acids

Lewis acids are, in general, excellent catalysts for the Diels-Alder reaction.[131] Their catalytic activity lies in the ability of the Lewis acid to complex a heteroatom component of the diene or dienophile (usually the dienophile). Complexation with acrolein, for example, generates an ionic intermediate such as **114**,[132] and this zwitterion will influence the course of the Diels-Alder reaction.[133] Formation of **114** lowers the net energy of both the HOMO and the LUMO relative to the uncomplexed dienophile.[134] Protonated acrolein was used to mimic **114,** and the energy lowering of these frontier molecular orbitals was significant.[134] Ottenbrite and co-workers calculated the HOMO/LUMO energies, as well as the orbital coefficients, for acrolein and the boron trifluoride complex, as shown in Figure 11.11.[135] In the reaction of acrolein with a suitable diene, ΔE for the $\text{HOMO}_{\text{diene}}$-$\text{LUMO}_{\text{alkene}}$ will be lowered when comparing the acrolein reaction with the protonated acrolein reaction. The HOMO/LUMO energies and orbital coefficients were calculated[134] and are different from those presented in Table 1.1. Using the LUMO values of 3.68 eV and 1.05 eV reported by Ottenbrite,[135] the ΔE ($\text{HOMO}_{\text{diene}}$-$\text{LUMO}_{\text{alkene}}$) for reaction with cyclopentadiene is lowered from 12.65 for acrolein to 10.02 eV for the BF_3 complex in a reaction with cyclopentadiene (HOMO = –8.97 eV from Table 1). The lower ΔE predicts acceleration of the Diels-Alder reaction upon addition of the Lewis acid. An alternative view of this effect sees that the denominator in Equation (11.1) is lowered, which lowers the activation of the reaction when the Lewis acid lowers the HOMO and LUMO energies.

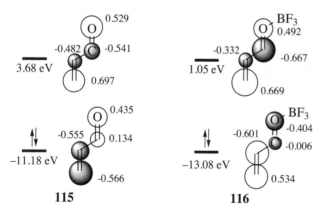

Figure 11.11. The HOMO-LUMO Diagram for Acrolein and Acrolein•BF_3. [Reprinted with permission from Guner, O.F.; Ottenbrite, R.M.; Shillady, D.D.; Alston, P.V. *J. Org. Chem.* **1987**, *52*, 391. Copyright © 1987 American Chemical Society.]

131. (a) Yates, P.; Eaton, P. *J. Am. Chem. Soc.* **1960**, *82*, 4436; (b) Fray, E.I.; Robinson, R. *Ibid.* **1961**, *83*, 249.
132. Reference 7, p 162.
133. Thompson, H.W.; Melillo, D.G. *J. Am. Chem. Soc.* **1970**, *92*, 3218.
134. (a) Houk, K.N.; Strozier, R.W. *J. Am. Chem. Soc.* **1973**, *95*, 4094; (b) For a computational study of the BF_3-catalyzed cycloaddition of furan and methyl vinyl ketone, see Avalos, M.; Babiano, R.; Bravo, J.L.; Cintas, P.; Jiménez, J.L.; Palacios, J.C.; Silva, M.A. *J. Org. Chem.* **2000**, *65*, 6613.
135. Guner, O.F.; Ottenbrite, R.M.; Shillady, D.D.; Alston, P.V. *J. Org. Chem.* **1987**, *52*, 391.

The orbital coefficients are influenced by the extent of coordination with the Lewis acid.[132] All orbital coefficients of acrolein (**115**) are affected (see the protonated form **116**), but those of the alkene moieties are changed less than those of the carbonyl, bringing about rather large effects in the HOMO and in the LUMO. The carbonyl carbon is most affected, as expected.[132,136] In the reaction of acrolein with 2-phenylbutadiene, the para product (**117**) is preferred in the uncatalyzed reaction. The difference in orbital coefficients is much greater in the catalyzed reaction, and this leads to a greater preference for the para product in the catalyzed reaction, due to more efficient overlap of the larger coefficients. Therefore, a common benefit of Lewis acid catalysis is greater regioselectivity in the cycloadduct.[137] In general, the Lewis acid decreases the reaction time, and increases product yield and selectivity. Reaction with cyclopentadiene derivatives generally show increased endo selectivity when done in the presence of aluminum chloride.[138] Aluminum chloride increases the endo/exo ratio to favor the endo-product and the anti/syn ratio to favor the anti product. The syn/exo product was not formed in reportable amounts, presumably due to the severe steric interaction between the anti-chlorine and an exo-carbomethoxy. Stronger Lewis acids ($AlCl_3 > TiCl_4 > BF_3$, see sec. 2.3) have a greater effect on this reaction. Increased selectivity and rate of reaction are also observed with acyclic dienes. Additional examples of Lewis acid catalysis will be presented throughout this chapter. There is also a report that amidinium ions catalyze the Diels-Alder reaction.[139]

MOLECULAR MODELING: Catalysis of Diels-Alder Reactions by Lewis Acids

The "important" orbital interaction for Diels-Alder cycloadditions involving electron-rich dienes and electron-poor dienophiles is that between the diene HOMO and the dienophile LUMO. Changes that diminish the separation of those orbitals (the ***HOMO-LUMO gap***), either by raising the energy of the HOMO or lowering the energy of the LUMO (or both), will increase this interaction and should accelerate the reaction. One way to accomplish this is to add a Lewis acid such as BF_3 which complexes to the dienophile and lowers the energy of the LUMO. For example, quantum chemical calculations show a reduction in the activation energy for cycloaddition of cyclopentadiene and acrylonitrile of 43 kJ/mol as a result of complexing BF_3 end-on to the nitrogen in the dienophile. This suggests an increase in reaction rate, consistent with experimental observation.

A Lewis acid should also change the shape of the dienophile LUMO, leading to changes in regioselectivity of Diels-Alder reactions. For example, quantum chemical transition-state

136. Reference 7, p 163.
137. Williamson, K.L.; Hsu, Y.-F.L. *J. Am. Chem. Soc.* **1970**, *92*, 7385.
138. Sauer, J.; Kredel, J. *Tetrahedron Lett.* **1966**, 731.
139. Schuster, T.; Kurz, M.; Göbel, M.W. *J. Org. Chem.* **2000**, *65*, 1697.

calculations show that the "uncatalyzed" reaction of 1-methylcyclopentadiene and acrylonitrile leads to a room-temperature ratio of *ortho* to *meta* products of 81:19, whereas the BF_3 catalyzed reaction increases the ratio to 99:1. Similarly, while the uncatalyzed reaction of 2-methylcyclopentadiene and acrylonitrile is non-regioselective (yielding a 55:45 ratio of *para* to *meta* products), but the BF_3 catalyzed reaction leads to a ratio of 70:30 (*para:meta*).

Examine and compare LUMO maps for acrylonitrile (left) and the acrylonitrile-BF_3 complex (right). These color a surface of electron density according to the value of the LUMO at a particular point on the surface. By convention, regions where the LUMO is most concentrated are colored "blue", while those where it is least concentrated are colored "red". The maps are displayed as transparent models allowing you to see the structure underneath. Once you are oriented, you can switch to solid models, by *clicking* anywhere on the map and selecting **Solid** from the **Style** menu at the bottom of the screen. For which carbon of acetonitrile is the LUMO more concentrated? Is the direction of the preference the same for the BF_3 complex? If it is, is the magnitude of the preference diminished or amplified, or does it not change significantly.

The HOMO in 1-methylcyclopentadiene is more concentrated on the unsubstituted double bond and the HOMO in 2-methylcyclopentadiene is more concentrated on the substituted double bond. Which regioisomer should be favored for each cycloaddition with the alkenes noted above? Assume that the preferred product will be that where the carbon on the diene with the higher HOMO concentration will position itself over the carbon on the dienophile with the higher LUMO concentration. Are your "predictions" consistent with the results of the transition-state calculations? Are any changes in regioselectivity resulting from BF_3 complexation consistent with the calculations?

There are many synthetic applications of Lewis acid catalyzed Diels-Alder reactions, and several have been presented previously in this chapter. Formation of **69** in Sect. 11.4.C.ii and **98** in Sect. 11.5.A are two examples. Another example is taken from the synthesis of the ABC ring of micrandilactone A by Chen, Yang and co-workers,[140] in which *E*-alkene **118** reacted with 1,3-butadiene **119** in the presence of the titanium tetrachloride bis(tetrahydrofuran) complex to give a 79% yield of **120**.

11.6.B. Rate Enhancement in Aqueous Media

For a long time, solvent polarity was believed to have no effect on the course of a Diels-Alder reaction. Berson, however, showed a clear relationship between the endo/exo product ratio

---

140. Zhang, Y.-D.; Tang, Y.-F.; Luo, T.-P.; Shen, J.; Chen, J.-H.; Yang, Z. *Org. Lett.* **2006**, *8*, 107.

and solvent polarity in the Diels-Alder reaction of cyclopentadiene and acrylates.[141] A patent by Hopff and Rautenstrauch in 1939[142] reported the first water accelerated reaction when they showed that yields in the Diels-Alder reaction were enhanced in aqueous detergent solutions. In 1980, Breslow showed there was a hydrophobic acceleration for an intermolecular Diels-Alder reaction in which cyclopentadiene reacted with methyl vinyl ketone to give **121**.[143] When non-polar compounds are suspended in water their relative insolubility causes them to associate, diminishing the water-hydrocarbon interfacial area (a **hydrophobic effect**). This association is greater in water than in methanol, which brings the reactive partners into close proximity, increasing the rate of reaction. Any additive that increases the hydrophobic effect will increase the rate, as shown in Table 11.4[143] for the preparation of **121**. Lithium chloride (LiCl) increases

Table 11.4. Hydrophobic Effects in the Diels-Alder Reaction of Cyclopentadiene and Methyl Vinyl Ketone in Aqueous Media

Solvent	Additive	$k_2 \times 10^5\ M^{-1}\,s^{-1}$	Solvent	Additive	$k_2 \times 10^5\ M^{-1}\,s^{-1}$
isooctane		5.94		LiCl	10,800
methanol		75.5		$(NH_2)_3C^+\,Cl^-$	4,300
water		4,410		β-cyclodextrin	10,900
				α-cyclodextrin	2,610

[Reprinted with permission from Rideout, D.; Breslow, R. *J. Am. Chem. Soc.* **1980**, *102*, 7816. Copyright © 1980 American Chemical Society.]

the hydrophobic effect by salting-out nonpolar material, but guanidinium chloride decreases hydrophobic interactions. β-Cyclodextrin (cycloheptaamylose) possesses a hydrophobic cavity and if the reactive intermediates can fit within this cavity there is a significant rate enhancement. The smaller α-cyclodextrin cannot accommodate the reactive species, and the rate is significantly diminished. Microwave-assisted Diels-Alder reactions have been reported in both water and in ionic liquid media using Lewis acid catalysts.[144]

There are many examples of Diels-Alder reactions that exploit this rate enhancement effect. Grieco and co-worker's synthesis of quassinoids led to the reaction of **122** and **123**, in benzene at 25°C, which gave a 1:1.18 mixture of **124** and **125** in 52% yield but required a reaction time of 288 hours.[145] When the same reaction was done in water, an 82% yield was obtained (1.3:1

141. Berson, J.A.; Hamlet, Z.; Mueller, W.A. *J. Am. Chem. Soc.* **1962**, *84*, 297.
142. Hopff, H.; Rautenstrauch, C.W. *U.S. Patent* 2,262,002, **1939** [*Chem. Abstr., 36*: 10469, **1942**.
143. Rideout, D.; Breslow, R. *J. Am. Chem. Soc.* **1980**, *102*, 7816.
144. Chen, I.-H.; Young, J.-N.; Yu, S.J. *Tetrahedron* **2004**, *60*, 11903.
145. Grieco, P.A.; Garner, P.; He, Z. *Tetrahedron Lett.* **1983**, *24*, 1897

124/125) after 168 hours. Diels-Alder cyclization with the free acid rather than the ester, in water, gave an 85% yield (1.5:1 **124/125**) in 17 hours. Similar reaction in 1:1 aqueous methanol gave a quantitative yield after 97 hours (1:1.25 **124/125**). When the sodium salt of the acid was used, a quantitative yield was obtained in 5 hours (3:1 **124/125**).[145] In all cases, the reactions were carried out at room temperature with a fivefold excess of diene over dienophile. Breslow et al. showed that addition of guanidinium perchlorate to a Diels-Alder reaction slowed the reaction, which was taken as evidence that the special ability of water to enhance the rate of the Diels-Alder reaction was due to a hydrophobic effect.[146]

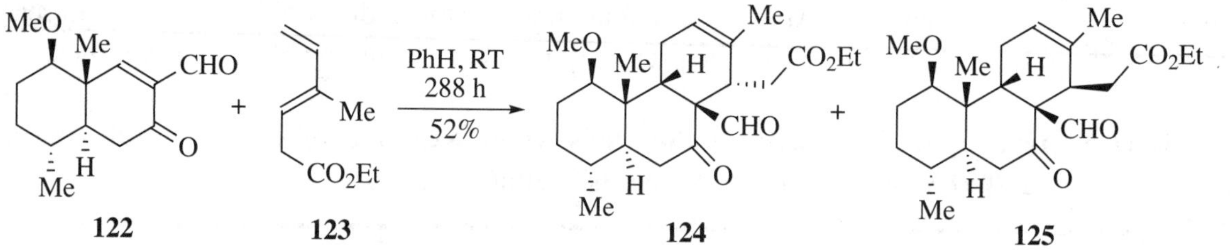

An aqueous environment creates a medium where the "reactants orient themselves within a micelle, thereby increasing their effective molarity and, consequently, the rate."[147] Liotta and co-workers, however, point out that reaction of 2,6-dimethylbenzoquinone (**127**) and diene carbonate **126** showed no rate enhancement in water relative to the same reaction in benzene. The carbonate was essentially insoluble in water and rate enhancement requires at least partial solubility of the diene and/or dienophile in the aqueous medium. Other solvents with the ability to enhance molecular aggregation while maintaining the solubility of the reactants showed an increase in rate. Ethylene glycol, for example, showed a large rate enhancement (relative to methanol) for the reaction of **126** and **127** to give a quantitative yield of **128**. The influence of several solvents (relative dielectric constants and relative rates with reaction in benzene as the standard) was examined for several substrates.[147] Liotta explains this effect by comparing **129** and **130** in Figure 11.12,[147] where the π-stacked arrangement in the micelle (**129**) assumes a smaller volume and is probably preferred. The larger micelle volumes in the end-on arrangement (**130**) will disrupt hydrogen bonding and be higher in energy. The smaller volume of **129** will retain the maximum hydrogen bonding and minimize disruption.[147]

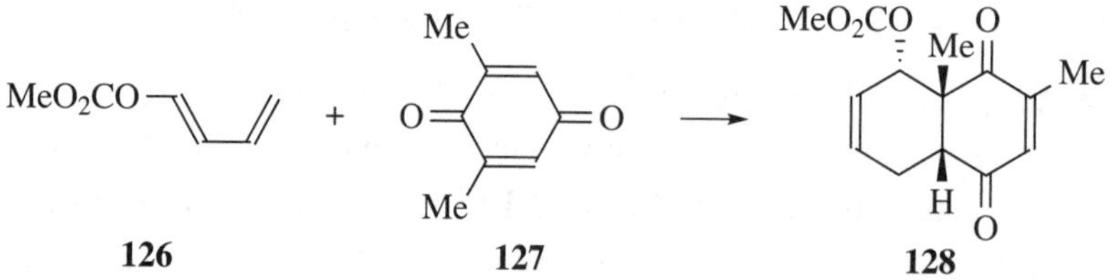

146. Breslow, R.; Maitra, U.; Rideout, D. *Tetrahedron Lett.* **1983**, *24*, 1901.
147. Dunams, T.; Hoekstra, W.; Pentaleri, M.; Liotta, D. *Tetrahedron Lett.* **1988**, *29*, 3745.

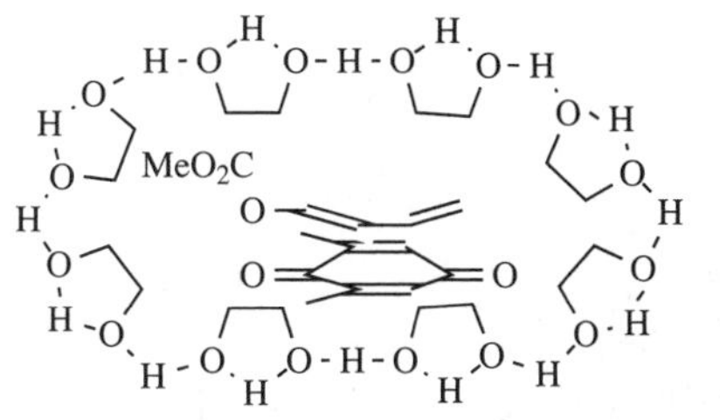

129 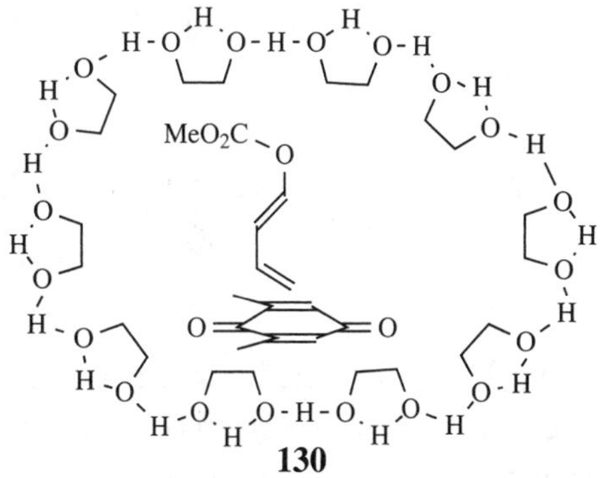　　　　　　　　**130**

Figure 11.12. π Stacked versus end-on arrangement in the micelle of aqueous Diels-Alder reactions. [Reprinted with permission from Dunams, T.; Hoekstra, W.; Pentaleri, M.; Liotta, D. *Tetrahedron Lett.* *1988, 29,* 3745, Copyright 1988, with permission from Elsevier Science.]

MOLECULAR MODELING: Selectivity in Cycloaddition of Quinone Carbonate

Cycloaddition of dienyl carbonate **126** with quinone **127** can lead to four possible products (*syn* and *anti* orientation of OCO_2Me and the quinone methyl groups, each with two *endo*-regioisomers and two *exo*-regioisomers). Previously, anticipating the regiochemical preference for cycloaddition reactions has relied on aligning the largest orbital coefficients associated with the HOMO on the diene and LUMO on the dienophile. While such procedure may be satisfactory where the positions on the diene and dienophile are markedly different, it is not reliable where the differences are subtle. Consider cycloaddition of **126** and **127**, leading to four possible products. The molecular orbitals of the reactants (especially 127) are delocalized, and seems unlikely that the observed product **128** could have been anticipated from examination of their orbital coefficients.

Calculations offer a more reliable method for determining the regiochemical preferences. Transition states corresponding to each possible product are generated and their energies compared. The lowest-energy transition state leads to the dominant product. Examine the transition states leading to the four products. From left to right and top to bottom on screen, these correspond to the *syn-endo*, *anti-endo*, *syn-exo* and *anti-exo* products. Can you pick out which among them is "least crowded"? If you can, is this the transition state leading to **128**?

All four transition states have very similar energies. In accord with observation, the transition state leading to the *syn-endo* product, **128**, is favored and leads to 82% of the total at room temperature. *Anti-endo* and *syn-exo* products each account for 9% and the *anti-exo* product for 2%. All four products arise from the same reactants, meaning that differences in the energies of the transition states are the same as differences in activation energies. Therefore, the kinetic produce is that which follows from the lowest energy transition state.

In summary, while comparison of the geometries of transition states for related reactions looking for which is "least crowded" may not always lead to a clear assignment of the favored product, comparison of activation energies will likely provide an reliable indicator.

Grieco and Larsen also showed that iminium salts such as **131**, formed by reaction of an amine and formaldehyde (sec. 4.4) reacted with cyclopentadiene in aqueous media to give a near-quantitative yield of the Diels-Alder adduct.[148] An internal Diels-Alder reaction was also possible[148] (sec. 11.8).

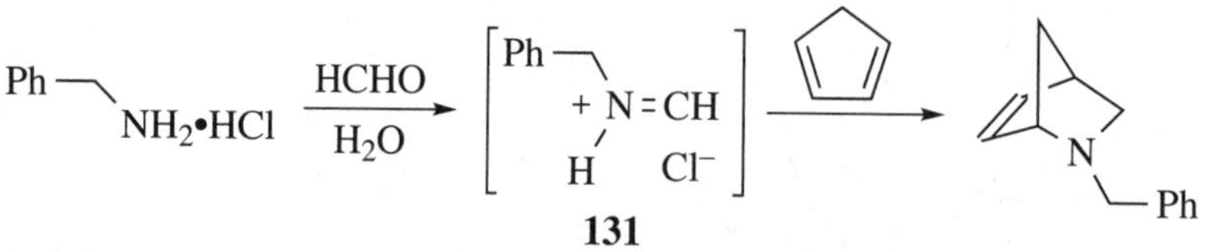

131

The use of highly polar media such as water clearly assists the Diels-Alder reaction, and similar rate enhancement has been observed in other reactions. Grieco et al. showed that 5.0 M lithium perchlorate in diethyl ether is an effective medium for these reactions.[149] In a synthesis of (–)-elisapterosin B,[150] Rychnovsky and Kim used the reagent to facilitate the Diels-Alder reaction of diene **132** and quinone **133**. The two products were formed in 75% yield, as a 1.7:1 mixture of **134** and **135**.

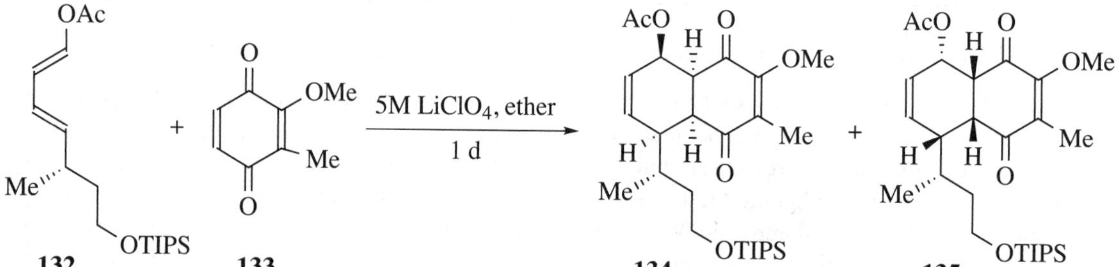

When highly polar media such as this are used, weak acids become strong acids, and protonation of dienes has been observed in intramolecular reactions, that can lead to diene isomerization. Such isomerization may be competitive with cycloaddition.[151] Grieco and Kaufman found that for such reactions, water is the polar medium of choice.[151] Recently, Augé found that lithium trifluoromethane-sulfonate in ether or acetonitrile is a useful substitute for lithium perchlorate in cycloaddition reactions.[152]

11.6.C. Rate Enhancement under High-Pressure Conditions

The rate of Diels-Alder reactions can be increased by the application of high pressure. High pressure is but one of a number of methods for activating reactions, including microwave and

148. Larsen, S.D.; Grieco, P.A. *J. Am. Chem. Soc.* **1985**, *107*, 1768.
149. For examples, see (a) Grieco, P.A.; Nunes, J.J.; Gaul, M.D. *J. Am. Chem. Soc.* **1990**, *112*, 4595; (b) Grieco, P.A.; Moher, E.D. *Tetrahdron Lett.* **1993**, *34*, 5567; (c) Grieco, P.A.; Beck, J.P. *Ibid.* **1993**, *34*, 7367; (d) Grieco, P.A.; Beck, J.P.; Handy, S.T.; Saito, N.; Daeuble, J.F.; *Ibid.* **1994**, *35*, 6783; (e) Grieco, P.A.; Piñeiro-Nuñez, M.M. *J. Am. Chem. Soc.* **1994**, *116*, 7606; (f) Grieco, P.A.; Dai, Y. *J. Am. Chem. Soc.* **1998**, *120*, 5128; (g) Grieco, P.A.; Kaufman, M.D. *Tetrahedron Lett.* **1999**, *40*, 1265.
150. Kim, A.I.; Rychnovsky, S.D. *Angew. Chem. Int. Ed.* **2003**, *42*, 1267.
151. Grieco, P.A.; Kaufman, M.D. *J. Org. Chem.* **1999**, *64*, 6041.
152. Augé, J.; Gil, R.; Kalsey, S.; Lubin-Germain, N. *Synlett* **2000**, 877.

ultrasound enhancement.[153] Microwave-assisted Diels-Alder reactions have also been shown to be more effective when pressurized reaction vessels are used.[154] An early example by Dauben and Kozikowski[155] demonstrated that methyl 1-pentadienoate reacted with **136** to give **137** in 88% yield. Matsumoto and co-workers reviewed applications of high-pressure techniques to organic synthesis,[156] and also published a monograph.[157] The second part of the review[156B] deals with high-pressure pericyclic reactions, including Diels-Alder reactions.

The thermodynamic properties of solutions are well known, and the rate of a reaction can be expressed in terms of the activation volume, $\Delta V^{\ddagger}$

$$\frac{\delta \ln k}{\delta p} = \frac{V^{\ddagger}}{RT} \qquad 11.4$$

"The activation volume is the difference in partial molal volume between the transition state and the initial state. From a synthetic point of view this could be approximated by the molar volume."[156] Equation (11.4) suggests that the rate of the reaction will be accelerated with increasing pressure if the volume of activation is negative. As the pressure increases, the value of $\Delta V^{\ddagger}$ decreases and the system does not strictly obey equation (11.4) > 10 kbar (1 bar = 0.986924 atm. = 1.1019716 kg cm^{-2}). Considerable information has been collected concerning $\Delta V^{\ddagger}$, principally from work by le Noble.[158] If the transition state of a reaction involves bond formation, concentration of charge, or ionization, a negative volume of activation often results. Cleavage of a bond, dispersal of charge, or neutralization of the transition state and diffusion control lead to a positive volume of activation. Matsumoto summarizes the reactions for which rate enhancement is expected at high pressure.[156] (*1*) Reactions in which the molecularity number (number of molecules) decreases when starting materials are converted to products: cycloadditions, condensations. (*2*) Reactions that proceed via cyclic transition states: Claisen and Cope rearrangements (sec. 11.12.D and 11.11.C, respectively). (*3*) Reactions with dipolar transition states: **Menschutkin reaction**[159] (tertiary amines with alkyl halides to produce

153. Jenner, G. *Tetrahedron* **2002**, *58*, 5185.
154. Kaval, N.; Dehaen, W.; Kappe, C.O.; Van der Eycken, E. *Org. Biomol. Chem.* **2004**, *2*, 154.
155. Dauben, W.G.; Kozikowski, A.P. *J. Am. Chem. Soc.* **1974**, *96*, 3664.
156. (a) Matsumoto, K; Sera, A.; Uchida, T. *Synthesis* **1985**, 1; (b) Matsumoto, K.; Sera, A. *Ibid.* **1985**, 999
157. Matsumoto, K.; Morris, A.R. *Organic Synthesis at High Pressure*, Wiley, New York, **1991**.
158. (a) le Noble, W.J. *Progr. Phys. Org. Chem.* **1967**, *5*, 207; (b) Isaacs, N.S. *Liquid Phase High Pressure Chemistry*, Wiley, Chichester, **1981**; (c) Asano, T.; le Noble, W.J. *Chem. Rev.* **1978**, *78*, 407.
159. (a) Menschutkin, N. *Z. Physik. Chem.* **1890**, *5*, 589; (b) *Idem Ibid.* **1890**, *6*, 41; (c) Ingold, C.K. *Structure and Mechanism in Organic Chemistry, 2nd Ed.*, Cornell Univ. Press, New York, **1969**, p 435; (d) T*he Merck Index, 14th ed.*, Merck &Co., Inc., Whitehouse Station, New Jersey, **2006**, p ONR-60.

quaternary ammonium halides), electrophilic aromatic substitution. (*4*) Reactions with steric hindrance.

Table 11.5[156] shows the variation of rate constants with pressure. Many high-pressure reactions are done neat, but if a solvent is used, the influence of pressure on that solvent is important. The melting point generally increases at elevated pressures, and this influences the viscosity of the medium (the viscosity of liquids increases approximately two times per kilobar increase in pressure). Controlling the rate of diffusion of reactants in the medium is also important, leading to another influence of high pressure on reactivity.[156,160] In most reactions, pressure is applied (5-20 kbar) at room temperature and then the temperature is increased until reaction takes place. The temperature is lowered and the pressure is reduced to isolate the products.

Table 11.5. Variations of Rate Constants with Pressure

$\Delta V^{\ddagger}$ (cm^3 mol^{-1})	Temperature (°C)	$k_p/k_1{}^a$ [$\Delta GG^{\ddagger}$ (kJ mol^{-1})]					
		5 kbar	**10 kbar**	**15 kbar**	**20 kbar**	**30 kbar**	**50 kbar**
-10	25	7.5	57	430	3200	1.8×10^5	7.5×10^8
	50	6.4	41	270	1700	7.1×10^4	1.5×10^8
	100	5.0	25	130	630	1.6×10^4	1.2×10^7
		(-5)	(-10)	(-15)	(-20)	(-30)	(-50)
-20	25	57	3200	1.8×10^5	1.0×10^7	3.3×10^{10}	5.6×10^{17}
	50	41	1700	7.1×10^4	2.9×10^6	5.0×10^9	2.4×10^{16}
	100	25	630	1.6×10^4	4.0×10^5	2.5×10^8	1.5×10^{14}
		(-10)	(-20)	(-30)	(-40)	(-60)	(-100)
-30	25	430	1.8×10^5	7.7×10^7	3.3×10^{10}	6.0×10^{15}	4.2×10^{26}
	50	270	7.1×10^4	1.9×10^7	5.0×10^9	3.6×10^{14}	3.7×10^{24}
	100	130	1.6×10^4	2.0×10^6	2.5×10^8	4.0×10^{12}	1.9×10^{21}
		(-15)	(-30)	(-45)	(-60)	(-90)	(-150)

^{a} k_p = rate constant under high pressure; k_1 = rate constant at atmospheric pressure

[Reprinted from Matsumoto, K; Sera, A.; Uchida, T. *Synthesis* **1985**, pp. 1. Copyright **1985** by Georg Thieme Verlag.]

Diels-Alder reactions with furans can be very problematic due to the thermal sensitivity of the cycloadduct, and their tendency for cycloreversion. These problems are due to the strain inherent to the 7-oxabicycloheptane cycloadduct (**138**), as well as the aromatic character of the furan. Raising the pressure of the reaction to 13.5 to 15 kbar in dichloromethane as a solvent, leads to fast reactions and yields approaching those obtained with Lewis acid catalysts and long reaction times.[156] 2-Methylfuran is more reactive than furan but substituents on the alkene (dienophile) lower the yield of the cycloadduct, even under high pressure. One synthetic example is taken from Funk and Maeng's synthesis of fasicularin, in which diene **139** reacted

160.   Firestone, R.A.; Vitale, M.A. *J. Org. Chem.* **1981**, *46*, 2160.

with **140** in dichloromethane at 12 kbar to give a 93% yield of **141**.[161]

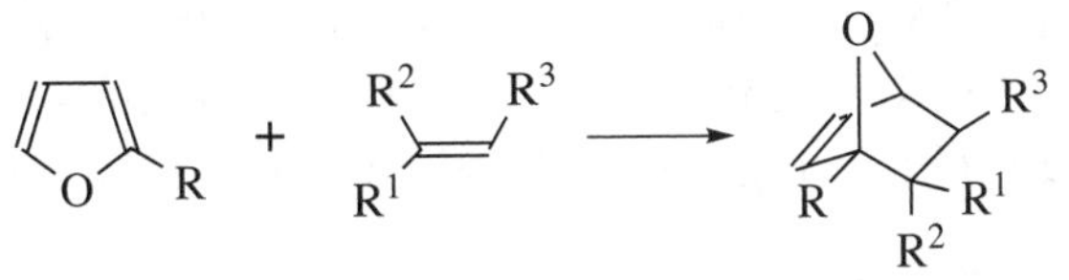

138

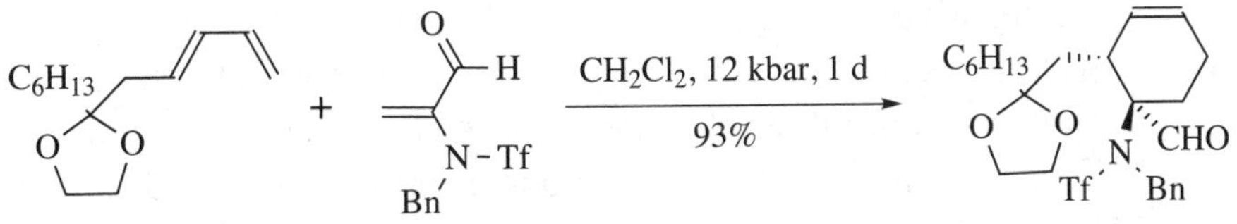

139   **140**   **141**

Furan also reacted with benzoquinone (at 20 kbar) to give a 14% yield of the endo adduct and a 15% yield of the exo adduct, along with 71% of unreacted benzoquinone.[162] Better results were obtained with functionalized furans such as **142**, which reacted with 4-methyl-2-cyclopentenone at 15 kbar to give a 1:2 mixture of **143/144** in 95% yield (after 2 days), in Ghosez' and co-worker's synthesis of polyfunctionalized cis-hydrindanones.[163] Thiophenes react with maleic anhydride under high pressure conditions, where it was observed that solvent-free conditions led to a significant lowering of the requisite reaction pressure, improvement in yield of products, and allowed less reactive dienophiles such as methyl acrylate to be used.[164] 2-Pyrones are another candidate for high pressure reactions since the normal Diels-Alder reaction usually requires high temperatures, and the products often lose carbon dioxide in a retro-Diels-Alder reaction.[165]

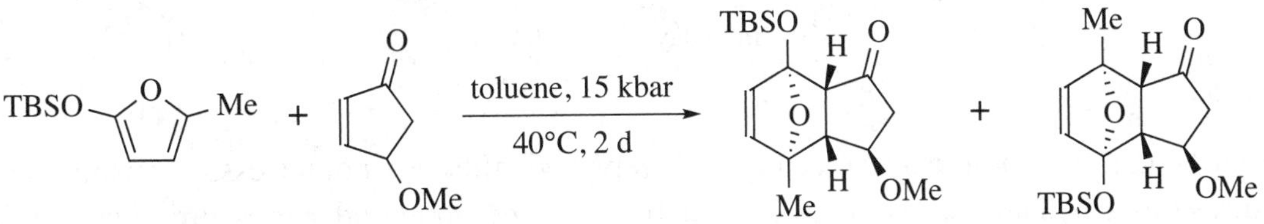

142   **143**   **144**

Pyridones react at high pressure to give the cycloadduct,[166] whereas heating at atmospheric pressure usually gives no reaction at all.[166] A variety of relatively unreactive dienes and/or dienophiles give enhanced yield of cycloadducts under pressure. Naphthalene reacted with maleic anhydride at 9.5 kbar and 100°C.[167] Acetaldehyde reacted with 1-methoxybutadiene

161. Maeng, J.-H.; Funk, R.L. *Org. Lett.* **2002**, *4*, 331.
162. Jurczak, J.; Koz`luk, T.; Filipek, S.; Eugster, C.H. *Helv. Chim. Acta* **1983**, *66*, 222.
163. Trembleau, L.; Patiny, L.; Ghosez, L. *Tetrahedron Lett.* **2000**, *41*, 6377.
164. Kumamoto, K.; Fukada, I.; Kotsuki, H. *Angew. Chem. Int. Ed.* **2004**, *43*, 2015.
165. Pfaff, E.; Plieninger, H. *Chem. Ber.* **1982**, *115*, 1967.
166. Gisby, G.P.; Royall, S.E.; Sammes, P.G. *J. Chem. Soc. Chem. Commun.* **1979**, 501.
167. (a) Jones, W.H.; Mangold, D.; Plieninger, H. *Tetrahedron* **1962**, *18*, 267; (b) Plieninger, H.; Wild, D.; Westphal, J. *Ibid.* **1969**, *25*, 5561.

(sealed tube, 160-180°C, 6 hours) to give only 2% of the dihydropyran cycloadduct.[168] When the reaction was done neat at 14 kbar and 80°C for 6 hours, 50% of pyran was obtained[168] and reaction in ether at 20 kbar (65°C, 5 hours)[169] gave a 62% yield of pyran. Dienamines reacted with conjugated esters at 13.9 kbar to give a 70% yield of an enamine cycloadduct.[170,155]

Intermolecular Diels-Alder reactions have a large negative volume of activation[171] ($\sim$ -25 to -45 cm^3 mol^{-1}) and a large negative volume of reaction. The activation volume for intramolecular Diels-Alder reactions was measured as -25 cm^3 mol^{-1}.[172] The effects on the intermolecular Diels-Alder can be dramatic, but intramolecular Diels-Alder reactions do not always respond to high pressure.[155,170] There are exceptions.[173] In one example, **145** could be cyclized to give a 1:1 mixture of **146/147** in refluxing toluene, at ambient pressure.[174] In order to suppress side reactions, a small amount of the radical scavenger BHT (2,6-di-*tert*-butyl-4-phenylphenol) was used. When done at 13 kbar in dichloromethane at room temperature, an 88% yield of a 1:2.3 mixture of **146/147** was obtained without the need for added BHT. It is noted that attempts to cyclize **145** using several different Lewis acids were not successful. Intramolecular Diels-Alder reactions will be discussed in Section 11.8.

A variation in this technique has recently been reported, although not necessarily using the high pressures described above. Diels-Alder reactions can be done under pressure in supercritical

168. Makin, S.M.; El'yanov, B.S.; Raifel'd, Yu.E. *Izv. Akad Nauk. SSSR Ser. Khim.* **1976**, 831 (Engl. p 810).
169. Jurczak, J.; Chmielewski, M.; Filipek, S. *Synthesis* **1979**, 41.
170. Darling, S.D.; Subramanian, N. *J. Org. Chem.* **1975**, *40*, 2851.
171. (a) vor der Brück, D.; Bühler, R.; Hück, C.-Ch.; Plieninger, H.; Weale, K.E.; Westphal, J.; Wild, D. *Chem.-Ztg.* **1970**, *94*, 183 [*Chem. Abstr., 73*:3128r *1970*]; (b) McCabe, Jr.; Eckert, C.A. *Acc. Chem. Res.* **1974**, *7*, 251.
172. Isaacs, N.S.; der Beeke, P.V. *Tetrahedron Lett.* **1982**, *23*, 2147.
173. (a) Klärner, F.-G.; Diedrich, M.K.; Wigger, A.E. *Chemistry under Extreme or Non-Classical Conditions*, van Eldik, R.; Hubbard,. C.D. (Eds.), Wiley, New York, **1997**, pp 103-161; (b) Jurczak, J.; Gryko, D.T. *Chemistry under Extreme or Non-Classical Conditions*, van Eldik, R.; Hubbard,. C.D. (Eds.),, Wiley, New York, **1997**, pp 163-188; (c) Ibata, T. *Organic Synthesis at High Pressures*, Matsumoto, K.; Archeson, R.M. (Eds), Wiley, New York, **1991**, pp 213-285; (d) Uyehara, T.; Yamamoto, Y. *Organic Synthesis at High Pressures*, Matsumoto, K.; Archeson, R.M. (Eds), Wiley, New York, **1991**, pp 409-422; (e) Diedrich, M.K.; Klärner, F.-G. *J. Am. Chem. Soc.* **1998**, *120*, 6212; (f) Ciobanu, M.; Matsumoto, K. *Liebigs Ann.-Recl.* **1997**, 623.
174. Plietker, B.; Seng, D.; Frölich, R.; Metz, P. *Tetrahedron* **2000**, *56*, 873.

carbon dioxide, although the synthetic application may be limited by slow reaction rates.[175] Addition of a Lewis acid can compensate, as illustrated by Kobayashi's use of scandium triflate[176a] in the reaction of methyl vinyl ketone with 2-methyl-1,3-butadiene to give a 93:7 mixture of **148/149**. Rayner and co-workers had previously used scandium triflate in supercritical CO_2,[176b] and reported that the maximum selectivity was obtained at a density of 1.12 g mL^{-1}.

11.7. HETEROATOM DIELS-ALDER REACTIONS

The Diels-Alder reaction is quite remarkable in its scope. One of the more attractive features is the ability to incorporate heteroatoms into either the diene, or the alkene unit, generating heterocyclic compounds. This section will examine several of the more useful variations.

11.7.A. Heteroatom Substituents

In previous sections, several examples were shown using alkoxydienes or alkoxyalkenes. Acyloxy dienes are very useful in organic synthesis. One example is Bohlmann and co-worker's synthesis of liglularone, which reacted acetoxydiene **150** with quinone **151** to give **152**.[177] An attractive feature of dienes that contain a vinyl ether or vinyl ester moiety is the facile hydrolysis of the cycloadduct (the vinyl ester unit in **152**) to give a ketone (see **153**). The hydrolysis of vinyl ethers was previously shown with the Birch reduction (sec. 4.9.E).

150   **151**   **152**   **153**

Danishefsky used a variety of silyloxybutadienes in Diels-Alder reactions[178] in place of alkoxy and acetoxy derivatives. Treatment of the resulting siloxy cycloadducts with fluoride ion (sec. 7.3.A.i) or simply with aqueous acid will generate an enol, which tautomerizes to the carbonyl under very mild conditions. In Nishida and co-worker's synthesis of (-)-nakadomarin A,[179]

175.   (a) Renslo, A.R.; Weinstein, R.D.; Tester, J.W.; Danheiser, R.L. *J. Org. Chem.* **1997**, *62*, 4530; (b) Weinstein, R.D.; Renso, A.R.; Danheiser, R.L.; Harris, J.G.; Tester, J.W. *J. Phys. Chem.* **1996**, *100*, 12337; (c) Hyatt, J.A. *J. Org. Chem.* **1984**, *49*, 5097; (d) Paulaitis, M.E.; Alexander, G.C. *Pure Appl. Chem.* **1987**, *59*, 61; (e) Ikushima, Y.; Saito, N.; Arai, M. *Bull. Chem. Soc. Jpn.*, **1991**, *64*, 282; (f) Isaacs, N.S.; Keating, N. *J. Chem. Soc. Chem. Commun.* **1992**, 876.
176.   (a) Matsuo, J.-i.; Tsuchiya, T.; Odashima, K.; Kobayashi, S. *Chem. Lett.* **2000**, 178; (b) Oakes, R.S.; Heppenstall, T.J.; Shezad, N.; Clifford, A.A.; Rayner, C.M. *Chem. Commun.* **1999**, 1459.
177.   Bohlmann, F.; Förster, H.-J.; Fischer, C.H. *J.L. Ann. Chem.* **1976**, 1487.
178.   (a) Danishefsky, S. *Acc. Chem. Res.* **1981**, *14*, 400; (b) *Idem Aldrichimica Acta* **1986**, *19*, 59.
179.   Ono, K.; Nakagawa, M.; Nishida, A. *Angew. Chem. Int. Ed.* **2004**, *43*, 2020.

154 reacted with diene **155** to give **156**. Subsequent reaction with trifluoroacetic acid gave a 52% overall yield of conjugated ketone **157**. Preparation of silyl enol ethers is usually easier than the preparation of the corresponding enol ether.[180] Diene **155** is a variation of the commercially available OSiMe$_3$ derivative (**159**, known as **Danishefsky's diene**).[181] This diene, or analogs, have been very useful for a variety of natural product syntheses.[151a] Many structural variations are possible for the alkoxybutadiene partner in the Diels-Alder reaction. Chiral heterosubstituted 1,3-butadienes[182] are common in [4+2]-cycloaddition reactions and chiral dienophiles are used. Avenoza, Peregrina and co-workers used an asymmetric hetero-Diels-Alder reaction in a synthesis of carbacephams.[183] Chiral non-racemic imine **158**[184] reacted with Danishefsky's diene to give a 17:83 mixture of **160:161**, in 65% yield. Chiral Diels-Alder reactions will be discussed in section 11.9.

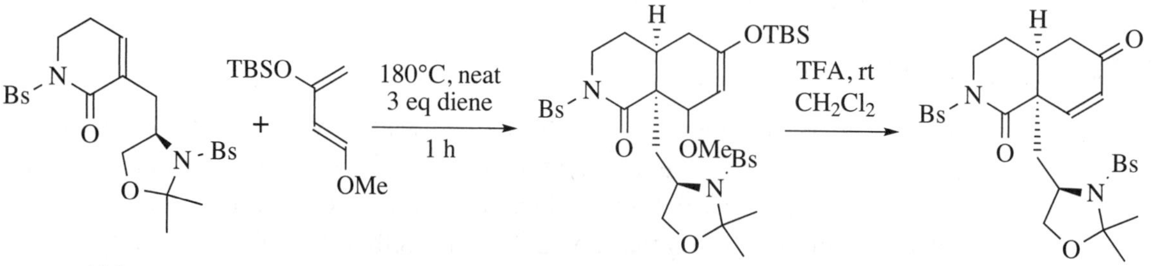

It is clear from these examples that oxygenated dienes are convenient and useful precursors to cyclohexenones and 7-oxabicycloheptane derivatives. Typical disconnections for oxygenated dienes are

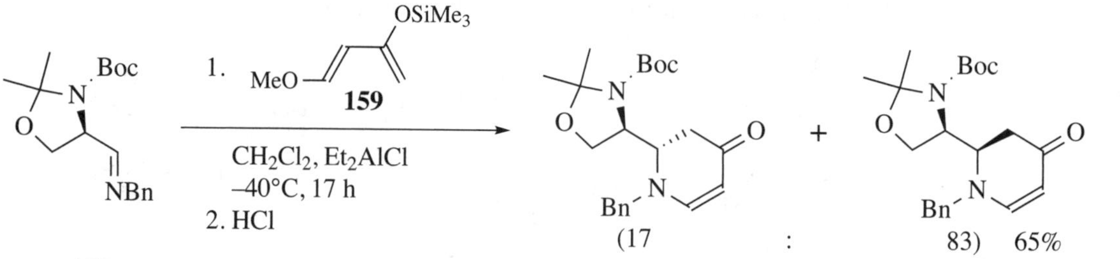

180. (a) Stork, G.; Hudrlik, P. *J. Am. Chem. Soc.* **1968**, *90*, 4462; (b) House, H.O.; Czuba, L.; Gall, M.; Olmstead, H.D. *J. Org. Chem.* **1969**, *34*, 2324; (c) House, H.O.; Gall, M.; Olmstead, H.D. *Ibid.* **1971**, *36*, 2361.
181. (a) Danishefsky, S.; Kitahara, T.; Schuda, P.F.; Etheredge, S.J. *J. Am. Chem. Soc.* **1976**, *98*, 3028; (b) Danishefsky, S.; Kitahara, T.; McKee, R.; Schuda, P.F. *Ibid.* **1976**, *98*, 6715; (c) Danishefsky, S.; Schuda, P.F.; Kitahara, T. Etheredge, S.J. *Ibid.* **1977**, *99*, 6066; (c) Mundy, B.P.; Ellerd, M.G.; Favaloro Jr., F.G. *Name Reactions and Reagents in Organic Synthesis*, 2nd ed., Wiley-Interscience, New Jersey, **2005**, p. 753.
182. Barluenga, J.; Suárez-Sobrino, A.; López, L.A. *Aldrichimica Acta* **1999**, *32*, 4.
183. Avenoza,A.;Busto,J.H.;Cativiela,C.;Corzana,F.;Peregrina,J.M.;Zurbano,M.M.*J. Org. Chem.***2002**,*67*,598.
184. (a) Langenbeck. W.; Gödde, O.; Weschky, L.; Schaller, R. *Chem. Ber.* **1942**, *75*, 232; also see (b) Hünig, S.; Kahanek, H. C. *Ibid.* **1957**, *90*, 238; (c) Farmer, M.L.; Billups, W.E.; Greenlee, R.E.; Kurtz, A.N. *J. Org. Chem.* **1966**, *31*, 2885; (d) Mannich, C.; Handke, K.; Roth, K. *Chem. Ber.* **1936**, *69*, 2112.

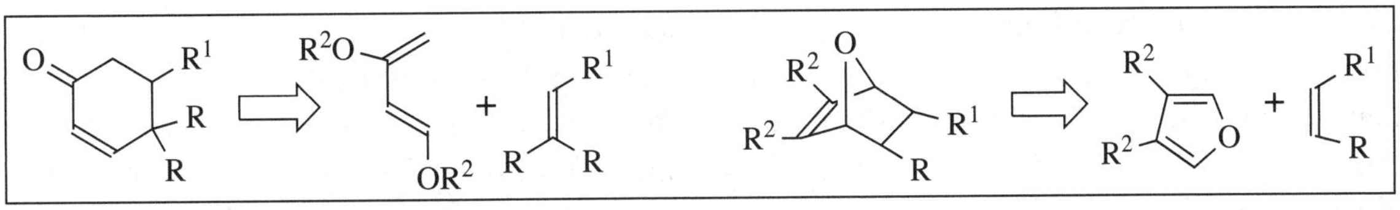

11.7.B. Amino and Amido Dienes

Azadienes (amino or amido dienes) constitute another important class of hetero-substituted dienes. There are four major types, amino dienes such as **162**, amido dienes such as **163**, carbamoyl dienes such as **164**, and pyrrole derivatives such as **165**. Both 1- and 2-substituted aminodienes are possible, and they function as diene partners in Diels-Alder reactions.

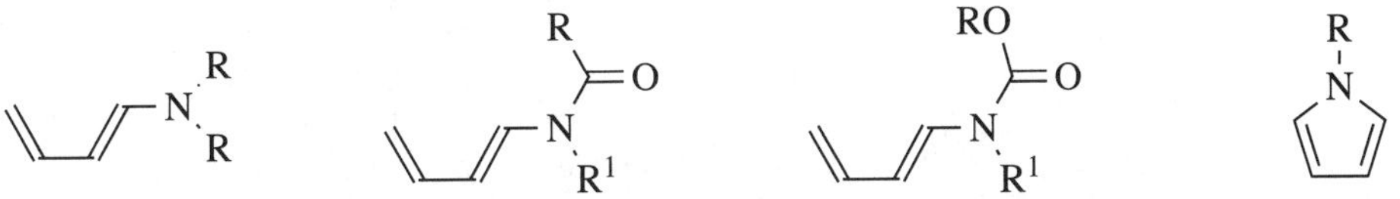

Enamines have been used as dienophiles in previous examples. There are a few examples of Diels-Alder reactions with 1-(dialkylamino)-1,3-butadienes.[184] It is noted that a density functional study of the Diels-Alder reaction of 1,2-diaza-1,3-butadiene with alkenes showed the transition state to be concerted, but asynchronous.[185] The allylic amine cycloadducts of this reaction show a propensity for elimination, and this has led to the use of dienyl amides (**163**) and carbamates (**164**), whose cycloadducts are significantly more stable and easier to isolate.[186] The conversion of **70** to **71** in Section 11.4.C.iii used a diene of this type (see above). In another example, Oppolzer et al. showed that **166** reacted with methyl acrylate to give a 73% yield of **167** and 18% of **168**.[187] The endo- (cis-) selectivity is apparent in this example and is general for this class of dienes. Amido dienes such as **166** can be prepared by reaction of conjugated aldehydes such as crotonaldehyde with an amine, a base, and an acid chloride.[188] In a synthesis of asparagamine A,[189] Overman and co-worker's showed that pyrrole **169** reacted with the nitro-acrylate shown to give **170**, which was hydrogenated to give **171** in 73% yield. Similarly, Overman prepared dienyl carbamate **172** via a **Curtius rearrangement**[190] (a thermal rearrangement of a diazoketone to an isocyanate; dienyl azido ketone **172** → **173**; see sec.

185. Avalos, M.; Babiano, R.; Clemente, F.R.; Cintas, P.; Gordillo, R.; Jiménez, J.L.; Palacios, J.C. *J. Org. Chem. 2000, 65*, 8251.
186. Smith, M.B. *Org. Prep. Proceed. Int. 1990, 22*, 315.
187. Oppolzer, W.; Bieber, L.; Francotte, E. *Tetrahedron Lett. 1979*, 4537.
188. (a) Oppolzer, W.; Fröstl, W. *Helv. Chim. Acta 1975, 58*, 587; (b) Oppolzer, W.; Bieber, L.; Francotte, E. *Tetrahedron Lett. 1979*, 981.
189. Brüggemann, M.; McDonald, A.I.; Overman, L.E.; Rosen, M.D.; Schwink, L. Scott, J.P. *J. Am. Chem. Soc. 2003, 125*, 15284.
190. (a) Curtius, T. *J. Prakt. Chem. 1894, 50*, 275; (b) Smith, P.A.S. *Org. React. 1946, 3*, 337; (c) Saunders, J.H.; Slocombe, R.J. *Chem. Rev. 1948, 43*, 203.

2.11.C). Heating with benzyl alcohol gave carbamate **174**.[191] Overman and co-workers also found thermolysis of trichloroacetimidic esters of propargyl alcohols generated trichloromethyl dienyl carbamates.[192],[193] Dienyl carbamate reacted with styrene to give a quantitative yield of a mixture of **175** and **176** in a 93:7 ratio.[194] 1-Amino-3-siloxy-1,3-butadienes also react with unactivated aldehydes under mild conditions, at room temperature and in the absence of Lewis acids.[195] Dienyl lactams such as **177** have also been prepared and shown to undergo Diels-Alder reactions.[196] The same ortho and endo selectivity that was observed with dienyl amides was observed when *N*-(1,3-butadienyl)-2-pyrrolidinone reacted with ethyl acrylate and other alkenes.[196c] When chiral lactam **177** reacted with ethyl acrylate, Smith and co-workers showed that the cycloadduct was formed with excellent enantioselectivity.[196d] Rawal and co-workers prepared **178**, and showed Diels-Alder reactions proceed in good yield and with reasonable stereoselectivity.[197]

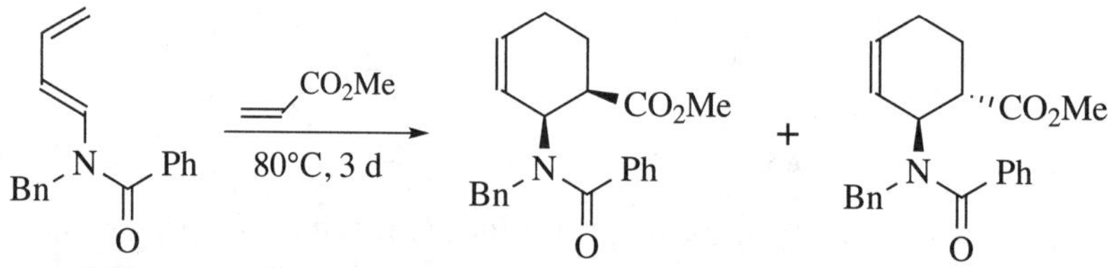

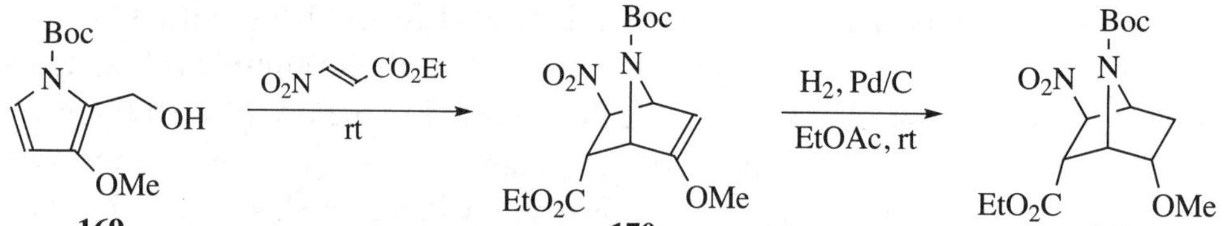

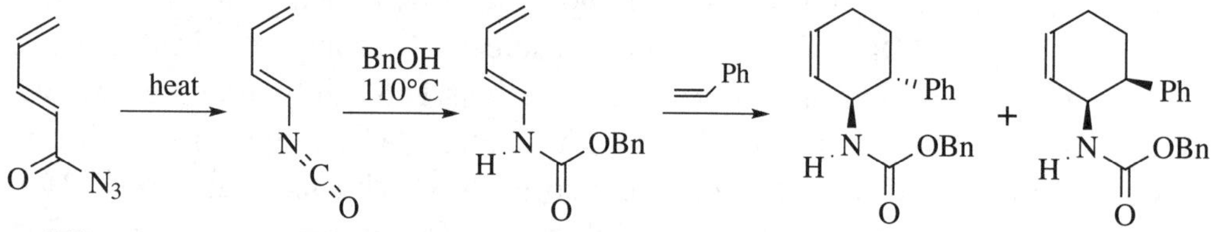

191. (a) Overman, L.E.; Taylor, G.F.; Jessup, P.J. *Tetrahedron Lett.* **1976,** 3089; (b) Overman, L.E.; Taylor, G.F.; Petty, C.B.; Jessup, P.J. *J. Org. Chem.* **1978,** *43*, 2164.
192. (a) Overman, L.E.; Clizbe, L.A. *J. Am. Chem. Soc.* **1976,** *98*, 2352; (b) Overman, L.E. *Ibid.* **1976,** *98*, 2901.
193. Overman, L.E.; Clizbe, L.A.; Freerks, R.L.; Marlowe, C.K. *J. Am. Chem. Soc.* **1981,** *103*, 2807.
194. (a) Overman, L.E.; Freerks, R.L.; Petty, C.B.; Clizbe, L.A.; Ono, R.K.; Taylor, G.F.; Jessup, P.J. *J. Am. Chem. Soc.* **1981,** *103*, 2816; (b) Overman, L.E.; Taylor, G.F.; Houk, K.N.; Domelsmith, L.N. *Ibid.* **1978,** *100*, 3182.
195. (a) Huang, Y.; Rawal, V.H. *Org. Lett.* **2000,** *2*, 3321; (b) Huang, Y.; Iwama, T.; Rawal, V.H. *J. Am. Chem. Soc.* **2000,** *122*, 7843.
196. (a) Murata, K.; Terada, A. *Bull Chem. Soc. Jpn.* **1967,** *40*, 414; (b) Terada, A.; Murata, K. *Ibid.* **1967,** *40*, 1644; (c) Zezza, C.A.; Smith, M.B. *J. Org. Chem.* **1988,** *53*, 1161; (d) Menezes, R.F.; Zezza, C.A.; Sheu, J.L.; Smith, M.B. *Tetrahedron Lett.* 1989, *30*, 3295.
197. Janey, J.M.; Iwama, T.; Kozmin, S.A.; Rawal, V.H. *J. Org. Chem.* **2000,** *65*, 9059.

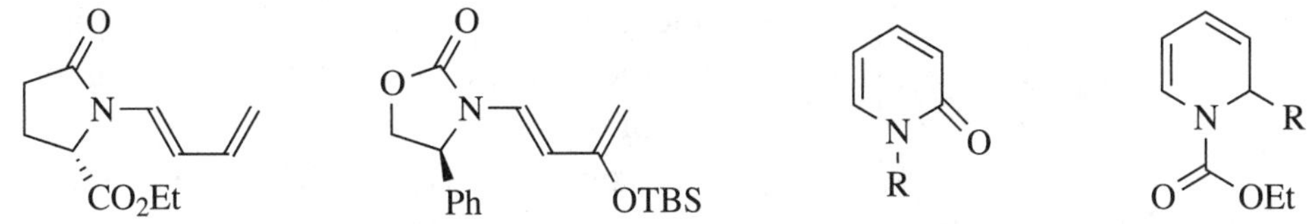

177 **178** **179** **180**

There are two other classes of dienyl lactams that have been used synthetically, *N*-alkyl-2-pyridones (**179**), and *N*-acyl-1,2-dihydropyridines (**180**). Pyridones tend to be relatively unreactive, requiring high temperatures, often giving low yields of the cycloadduct.[198] *N*-Acyl pyrrole derivatives[199] also function as diene partners in the Diels-Alder reaction.

Nitrogen containing dienes lead to useful synthetic disconnections such as

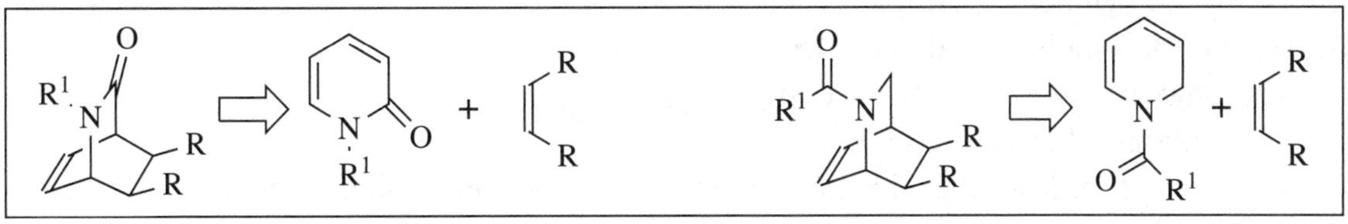

11.7.C. Heteroatom Dienes and Heteroatom Alkene Dienophiles

Heteroatoms can be incorporated into the π framework of either the alkene partner or the diene partner of Diels-Alder reactions. This section will focus on two structural types that function as dienophiles: ketones or aldehydes, and imines (**181**). We will also focus on three types that function as dienes: conjugated carbonyls (**182**), 1-azadienes (**183**), and 2-azadienes (**184**). The use of heteroatom dienes and alkenes in organic synthesis was thoroughly reviewed by Boger and Weinreb,[200] and by Fringuelli and Taticchi.[201]

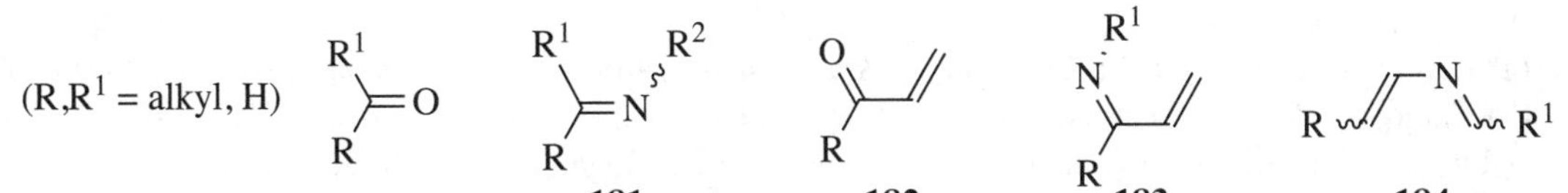

(R,R^1 = alkyl, H)

181 **182** **183** **184**

11.7.C.i. Aldehydes and Ketones. Most reactions that use carbonyl compounds as dienophiles involve aldehydes, which are more reactive than ketones. Hetero-Diels-Alder reactions of ketones are known, however.[202] Simple aldehydes and ketones do not react very well in

198. Raucher, S.; Lawrence, R.F. *Tetrahedron Lett.* **1983**, *24*, 2927.
199. Groves, J.K.; Cundasawmy, N.E.; Anderson, H.J. *Can. J. Chem.* **1973**, *51*, 1089.
200. Boger, D.L.; Weinreb, S.M. *Hetero Diels-Alder Methodology in Organic Synthesis*, Academic Press, San Diego, **1987**.
201. Fringuelli, F.; Taticchi, A. *Dienes in the Diels Alder Reaction*, Wiley, New York, **1990**.
202. Jørgensen, K.A. *Eur. J. Org. Chem.* **2004**, 2093.

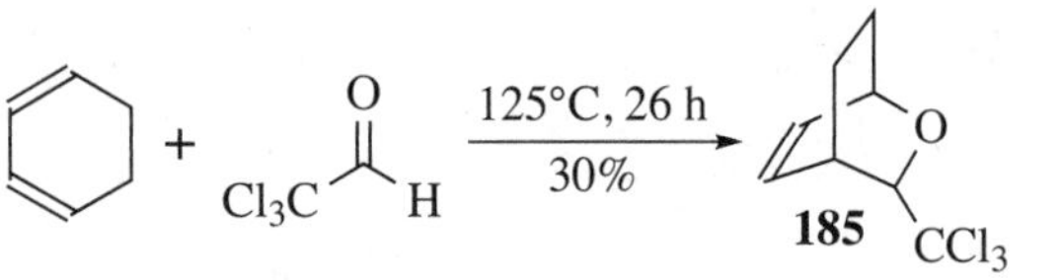

uncatalyzed reactions,[203] although electron-deficient aldehydes and ketones are good partners. Chloral reacts with cyclohexadiene to give **185**.[204] Glyoxalates[205] (RO_2CCHO) also add to dienes. The carbonyl moiety of electron deficient ketones functions as a dienophile,[206a] but ketone dienophiles may require elevated temperatures and or high pressure to obtain good reactivity.[206] Unactivated ketones react in hydrogen bond-promoted reactions[207] with activated conjugated dienes.[208]

As mentioned, relatively simple aliphatic and aromatic aldehydes are unreactive except with very reactive dienes, and the reaction usually requires Lewis acid catalysis.[209] Danishefsky's diene reacts with aldehydes to give pyrones.[210] A synthetic example is the reaction of Danishefsky's diene (**159**) with **158** shown above,[181] catalyzed by diethyl aluminum chloride. In Rainier

and co-worker's synthesis of the hemibrevetoxin B ring system,[211] diene **155**, which is obviously related to Danishefsky's diene, reacted with *O*-benzyl 4-hydroxybenzaldehyde in the presence of zinc chloride to give **186** in 92% yield. Pyrones and pyrans are particularly useful for the preparation of acyclic precursors to macrocyclic antibiotics. Exploiting the selectivity inherent in the cyclic pyran (sec. 6.5) controls the stereochemistry of the highly oxygenated acyclic fragment.[212,213] The disconnection is

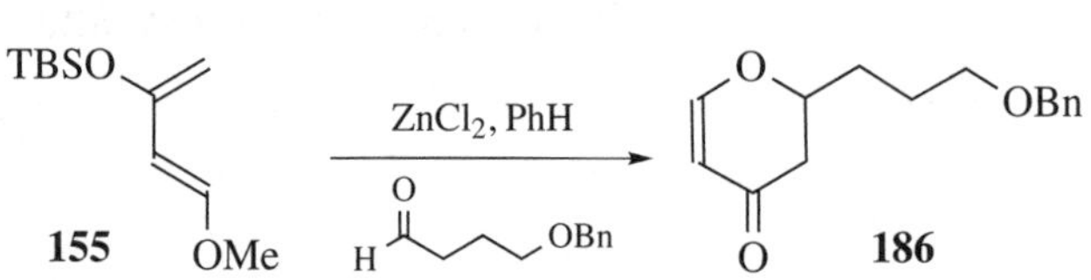

203. (a) Tietze, L.F.; Kettschau, G. *Top. Curr. Chem.* **1997**, *189*, 1; (b) Kumar, A. *Chem. Rev.* **2001**, *101*, 1; (c) Jørgensen, K.A. *Angew. Chem. Int. Ed.* **2001**, *39*, 3558.

204. (a) Begley, M.J.; Benner, J.P.; Gill, G.B. *J. Chem. Soc. Perkin Trans. 1* **1981**, 1112; (b) Smushkevich, Y.I.; Belov, V.N.; Kleev, B.V.; Akimova, A.Y. *Zh. Org. Khim.* **1967**, *3*, 1036 (Engl. p 997); (c) Dale, W.J.; Sisti, A.J. *J. Org. Chem.* **1957**, *22*, 449.

205. (a) Kanowal, A.; Jurczak, J.; Zamojski, A. *Rocz. Chem.* **1968**, *42*, 2045 [*Chem. Abstr. 71*: 3209b, *1969*]; (b) Zamojski, A.; Konowal, A.; Jurczak, J. *Ibid.* **1970**, *44*, 1981 [*Chem. Abstr., 75*: 35008j, *1971*]; (c) Yablonovskaya, S.D.; Shekhtman, N.M.; Antonova, N.D.; Bogatkov, S.V.; Makin, S.M.; Zefirov, N.S. *Zh. Org. Khim.* **1970**, *6*, 871 (Engl. p 871).

206. (a) Daniewski, W.M.; Kubak, E.I.; Jurczak, J. *J. Org. Chem.* **1985**, *50*, 3963; (b) Bonjouklian, R.; Ruden, R.A. *Ibid.* **1977**, *42*, 4095.

207. Huang, Y.; Rawal, V.H. *J. Am. Chem. Soc.* **2002**, *124*, 9662.

208. Jørgensen, K.A. *Eur. J. Org. Chem.* **2004**, 2093.

209. Jurczak, J.; Golebiowski, A.; Rahm, A. *Tetrahedron Lett.* **1986**, *27*, 853.

210. (a) Reference 200, p 108; (b) Danishefsky, S.; Kerwin Jr., J.F.; Kobayashi, S. *J. Am. Chem. Soc.* **1982**, *104*, 358; (c) Danishefsky, S.; Kerwin Jr., J.F. *J. Org. Chem.* **1982**, *47*, 3183; (d) Danishefsky, S.; Webb II, R.R. *Ibid.* **1984**, *49*, 1955.

211. Rainier, J.D.; Allwein, S.P.; Cox, J.M. *Org. Lett.* **2000**, *2*, 231.

212. (a) Danishefsky, S.; Pearson, W.H.; Harvey, D.F. *J. Am. Chem. Soc.* **1984**, *106*, 2456; (b) Kishi, Y. *Pure Appl. Chem.* **1981**, *53*, 1163.

213. Danishefsky, S.J. *Aldrichimica Acta* **1986**, *19*, 59 (see p 62).

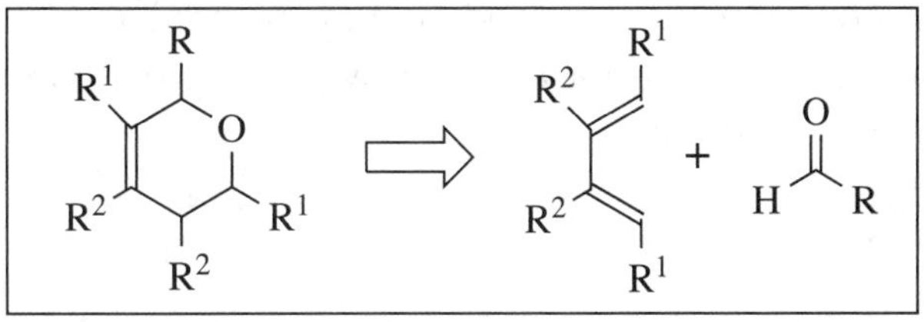

11.7.C.ii. Imines. Imines are used less often as dienophiles, but there are examples. Iminium salts are also known dienophiles. Simple imines are usually generated *in situ*, although more useful imines can be stabilized by the presence of electron-withdrawing groups on the nitrogen. *N*-Tosyl imines such as **188** can be isolated. When **188** reacted with diene **187**, a 58% yield of tetrahydropyridine (**189**)[214] was obtained as part of Weinreb and Heintzelman's synthesis

of cylindrospermopsin. *N*-Acyl imines react similarly via the corresponding iminium salt. Bis(carbamates) are precursors to iminium salts that undergo Diels-Alder reactions.[215] Böhme and co-workers generated iminium salt **191** from chloromethylamine **190**, and subsequent reaction with 2,3-dimethylbutadiene gave **192** in 69% yield.[216] Grieco and co-worker's previously cited preparation and Diels-Alder reaction of iminium salt **126**, and the one derived from amine **128** in aqueous media, are related to this type of cycloaddition (sec. 11.6.B). Formation of nitrogen-containing cycloadducts via Diels-Alder reactions is an attractive route for the synthesis of alkaloids. Even simple imines can function as dienophiles. Danishefsky and Vogel reacted pyrroline (**193**) with siloxydiene **194** to give **195**, in a synthesis of ipalbidine.[217]

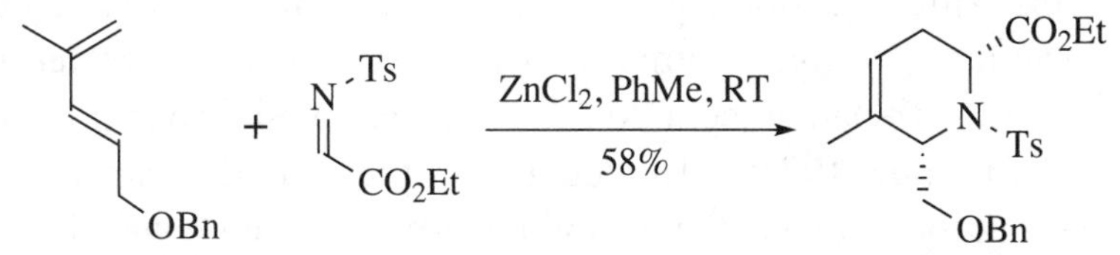

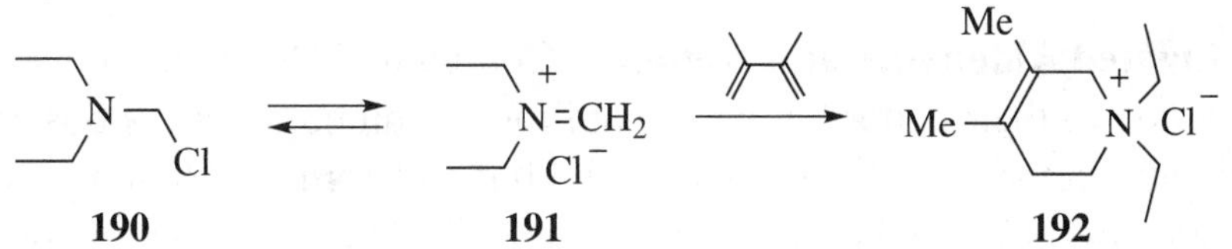

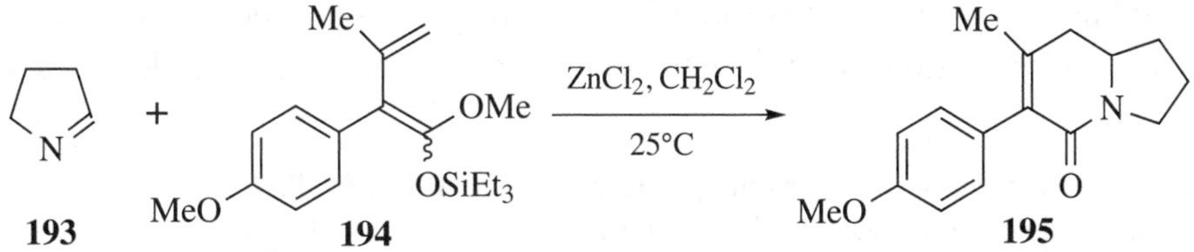

A large variety of stable imines are easily prepared, making this technique valuable for the

214. Heintzelman, G.R.; Weinreb, S.M. *J. Org. Chem.* **1996**, *61*, 4594.
215. Merten, R.; Muller, G. *Angew. Chem.* **1962**, *74*, 866.
216. Böhme, H.; Hartke, K.; Müller, A. *Chem. Ber.* **1963**, *96*, 607.
217. Danishefsky, S.; Vogel, C. *J. Org. Chem.* **1986**, *51*, 3915.

preparation of nitrogen-containing cyclic and polycyclic molecules. Catalytic asymmetric hetero-Diels-Alder reactions using imines are known.[218] Two important disconnections for this type of Diels-Alder reactions are

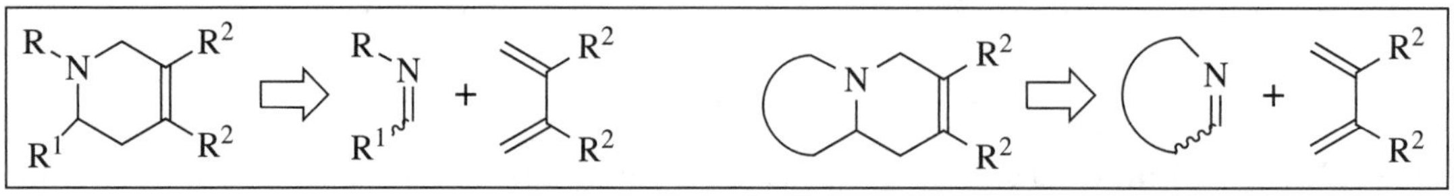

11.7.C.iii. Nitroso-Type Compounds. Nitroso compounds (R–N=O) are dienophiles in Diels-Alder reactions, giving heterocyclic rings. In Kibayashi and co-worker's synthesis of fasicularin,[219a] hydroxamic acid **196** was treated with tetrapropylammonium periodate and 9,10-dimethylanthracene to give transient acylnitroso compound **197**, and the resultant Diels Alder product **198** was formed in 84% yield. In this particular example, the Diels-Alder adduct essentially "protected" the acyl nitroso unit, which was used in a subsequent reaction.

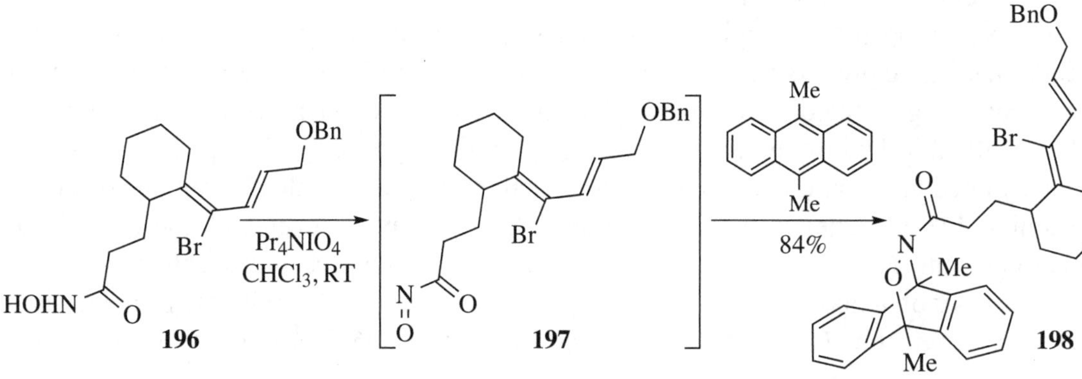

11.7.C.iv. Conjugated Aldehydes and Ketones. Conjugated aldehydes and ketones function as dienes, in addition to their better known capabilities as dienophiles, as observed by Sherlin et al.[220] Methyl vinyl ketone reacted with acrolein, but gave only 15% of a pyran cycloadduct in Mundy and co-worker's synthesis of brevicomin.[221] However, pyran formation by this reaction is well known,[222] and 2-substituted 3,4-dihydropyrans are the major products in almost all cases. This cycloaddition is obviously related to the reaction of aldehydes with dienes,

218. For a review see Jørgensen, K.A. *Angew. Chem. Int. Ed.* **2000**, *39*, 3558.
219. (a) Abe, H.; Aoyagi, S.; Kibayashi, C. *J. Am. Chem. Soc.* **2000**, *122*, 4583. For the use of nitroso dienophiles for the synthesis of mannosidase and fucosidase inhibitors, see (b) Joubert, M.; Defoin, A.; Tarnus, C.; Streith, J. *Synlett* **2000**, 1366. Also see (c) Bach, P.; Bols, M. *Tetrahedron Lett.* **1999**, *40*, 3461.
220. Sherlin, S.M.; Berlin, A.Y.; Serebrennikova, T.A.; Rabinovich, R.F. *J. Gen. Chem. USSR* **1938**, *8*, 7,22.
221. (a) Mundy, B.P.; Otzenberger, R.D.; DeBernardis, A.R. *J. Org. Chem.* **1971**, *36*, 2390, 3830; (b) Lipkowitz, K.B.; Mundy, B.P.; Geeseman, D. *Synth. Commun.* **1973**, *3*, 453; also see (c) Bhupathy, M.; Cohen, T. *Tetrahedron Lett.* **1985**, *26*, 2619.
222. Desimoni, G.; Tacconi, G. *Chem. Rev.* **1975**, *75*, 651.

but provides pyran products with regiochemical differences relative to those prepared in sec. 11.7.C.i. Both Lewis acid catalysis and high-pressure techniques facilitate the cycloaddition. Electron deficient alkenes react as dienophiles with conjugated carbonyl derivatives,[223] but electron rich alkenes (such as methyl vinyl ether) also react. An example is the reaction of the conjugated aldehyde **199** with the C=C unit of vinyl ether **200**, to give a 42% yield of **201** in Korte and co-worker's synthesis of *dl,iso*-iridomyrmecin.[224] Boger and Weinreb.[225] show several reactions of this type

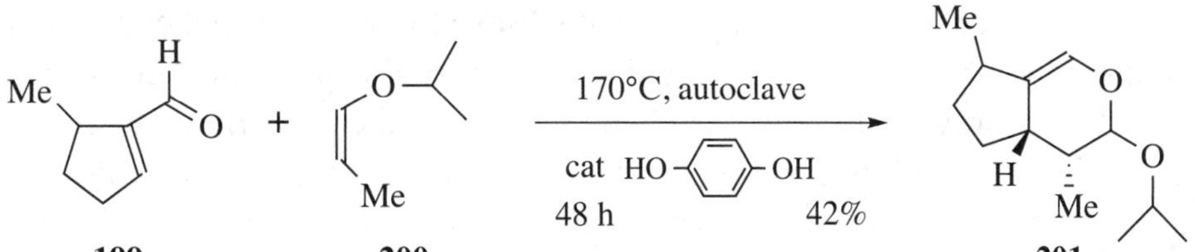

The disconnection for this heteroatom Diels-Alder reaction is

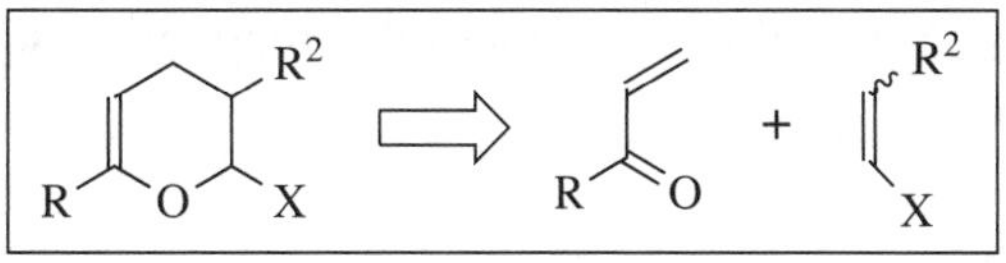

11.7.C.v. Azadienes. Azadienes such as **182** and **183**[226] are useful diene partners that can be used for the preparation of many heterocycles and alkaloids. The utility of 1-azadienes in the Diels-Alder reaction is limited by the fact that isomerization occurs to give a conjugated imine, probably via a proton transfer. An example is the equilibration of **202** and **203**. Addition of maleic anhydride to this mixture gave the Diels-Alder adduct **204**, derived from aminobutadiene **202**.[227] Incorporation of tertiary alkyl substituents are used to stabilize the 1-azadiene. Even when isomerization is not possible, the aminobutadiene is relatively unreactive and the yields can be poor. An important side reaction is a competitive [2+2]-cycloaddition (sec. 11.10.A), further contributing to poor yields of the [4+2]-cycloadduct.[228,229]

223. Childers Jr., W.E.; Pinnick, H.W. *J. Org. Chem.* **1984**, *49*, 5276.
224. Korte, F.; Büchel, K.H.; Zschocke, A. *Chem. Ber.* **1961**, *94*, 1952.
225. Reference 200, pp 167-213.
226. For a review, see Behforouz, M.; Ahmadian, M. *Tetrahedron* **2000**, *56*, 5259.
227. (a) Snyder, H.R.; Robinson Jr., J.C. *J. Am. Chem. Soc.* **1941**, *63*, 3279; (b) Snyder, H.R.; Cohen, H.; Tapp, W.J. *Ibid.* **1939**, *61*, 3560.
228. (a) Reference 200, p 242 and references 18a-c therein; (b) Taylor, E.C.; Eckroth, D.R.; Bartulin, J. *J. Org. Chem.* **1967**, *32*, 1899.
229. Garashchenko, Z.M.; Skvortsova, G.G.; Shestova, L.A. *USSR Patent* 370,208, **1973** [*Chem. Abstr.* 79: 31900d, *1973*]

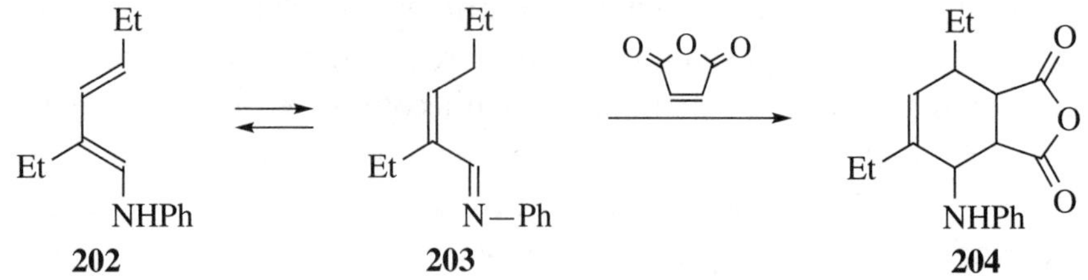

2-Azadienes such as **205** were prepared by Ghosez and co-workers and then reacted with conjugated alkynes [such as the methyl propiolate (**206**)] to give an initial cycloadduct **207**, but facile loss of dimethylamine led to aromatization and formation of pyridine derivative **208** as the final product.[230] The previously noted azadiene-aminobutadiene rearrangement is impossible in **205**, and the dimethylamino moieties further stabilize the azadiene structure. Ghosez and co-workers also used 2-azadienes that were stabilized by the presence of silyloxy groups.[230] Boger and Weinreb discuss several examples of 1- and 2-azadienes in Diels-Alder reactions.[231] Kende and co-workers used a less obvious azadiene in a synthesis of alantrypinone.[232] When 6H-pyrazino[2.1-b]quinazoline-6-one **209** was heated with 3-methyleneoxindole, a mixture of **210** and **211** was formed in 55% yield and 18% yield, respectively.

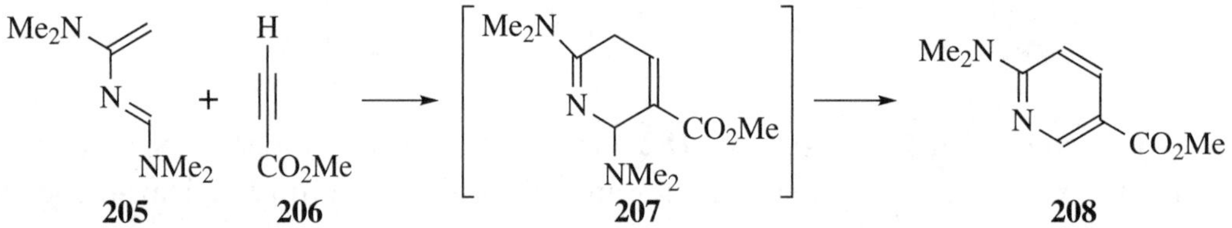

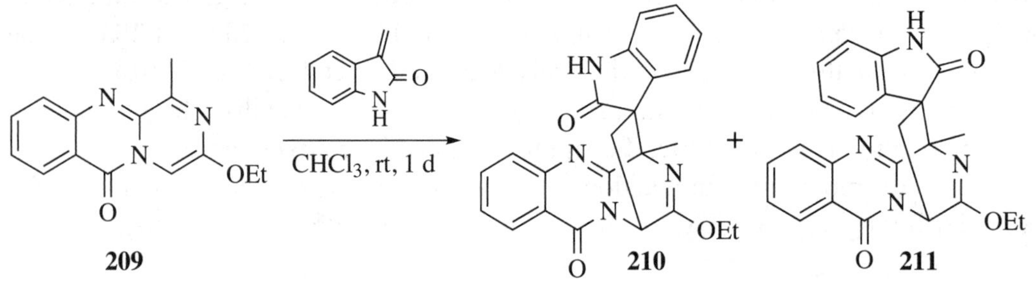

The synthetic utility of azadienes Diels-Alder reactions lies in the synthesis of heterocycles and alkaloids. Three disconnections related to this section are

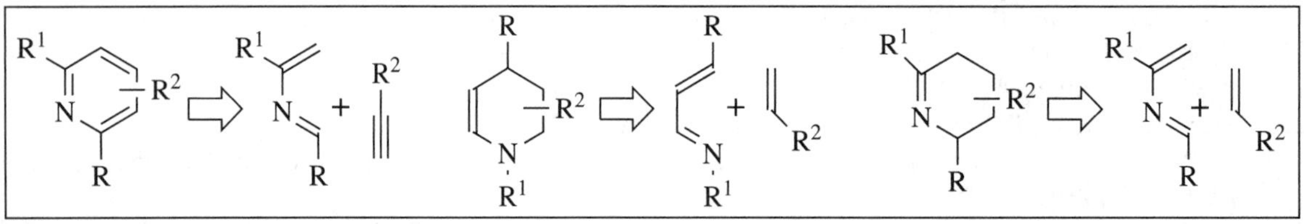

---

230. Sainte, F.; Serckx-Poncin, B.; Hesbain-Frisque, A.-M.; Ghosez, L. *J. Am. Chem. Soc.* **1982**, *104*, 1428.
231. Reference 200, pp 239-358.
232. Kende, A.S.; Fan, J.; Chen, Z. *Org. Lett.* **2003**, *5*, 3205.

11.8. INTRAMOLECULAR DIELS-ALDER REACTIONS[233]

In previous discussions of the Diels-Alder reaction, the reactions of **34** and of **145** involved an intramolecular cycloaddition. The Diels-Alder reaction can occur intramolecularly if the diene and alkene are connected by an intervening chain of atoms, illustrated by the conversion of **212** to **213**. This version of the reaction has become one of the most powerful synthetic methods (sec. 10.5.B.vi) for constructing cyclic compounds with high regio- and stereoselectivity.[234] There is much interest

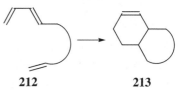

212   **213**

in intramolecular Diels-Alder reactions and many reviews, including those by Carlson in 1974,[235] Oppolzer in 1977,[236] Brieger and Bennett in 1980,[237] Fallis in 1984[238] and Smith in 1990.[186] Virtually all of the intermolecular reactions shown in previous sections have intramolecular analogs. The atoms that link the diene and alkene (the **tether**) can sometimes inhibit close approach of those moieties. Indeed, the ground state conformation of a molecule can greatly influence the reaction. Padwa and co-worker's studied the intramolecular Diels-Alder reaction

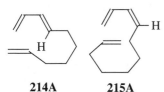

214A   **215A**

of *N*-alkenyl amidofurans,[239] and found that placement of a carbonyl in the tether can cause the molecule to adopt a conformation that is closer to the reactive conformation. The geometry of the diene and alkene (*E*) or (*Z*) also plays a role, which can be illustrated by comparing the intramolecular Diels-Alder reactions of **214A** and **215A**. *trans*-Diene **214A** readily attains the proper geometry for reaction (see **214B**) with the methylene groups in the tether assuming a relatively normal conformation for a small ring. *cis*-Diene **215A**, however, requires significant distortion of the methylene groups in the tether to even approach the proper orientation for the cycloaddition. In model **215B**, severe distortion is

214B   **215B**

apparent in the tether, where one or more bonds must be elongated to achieve the conformation shown. Such bond elongation and distortion of the tether is energetically unfavorable for this molecule.

233. See Takao, K.-i.; Munakata, R.; Tadano, K.-i. *Chem. Rev.* **2005**, *105*, 4779.
234. Fallis, A.G. *Acc. Chem. Res.* **1999**, *32*, 464.
235. Carlson, R.G. *Ann. Rep. Med. Chem.* **1974**, *9*, 270.
236. Oppolzer, W. *Angew. Chem. Int. Ed.* **1977**, *16*, 10 (see p 10-18).
237. Brieger, G.; Bennett, J.N. *Chem. Rev.* **1980**, *80*, 63 (see p 67).
238. Fallis, A.G. *Can. J. Chem.* **1984**, *62*, 183.
239. Bur, S.K.; Lynch, S.M.; Padwa, A. *Org. Lett.* **2002**, *4*, 473.

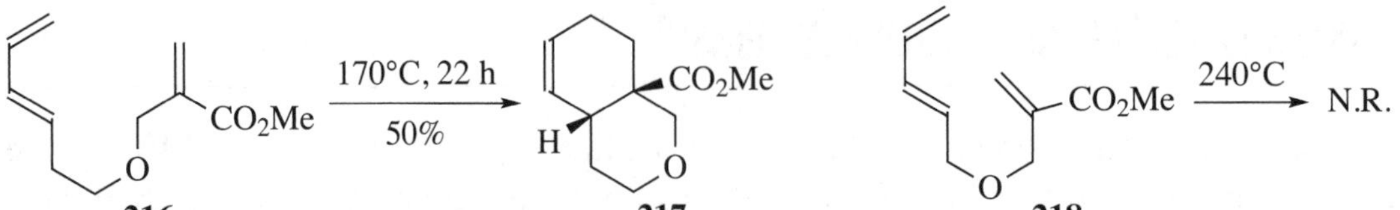

The *length* of the tether is very important.[240,241] Boeckman and Demko showed that Diels-Alder cyclization of **216** gave 50% of a 70:30 mixture (βH/αH = cis/trans) of **217**.[242] Attempted cyclization of **218**, which has fewer carbons in the tether, gave no cyclized product. This latter example shows a preference for the transoid conformation and has a higher energy of activation for interconversion of the two rotamers than does **216**.[242,243a]

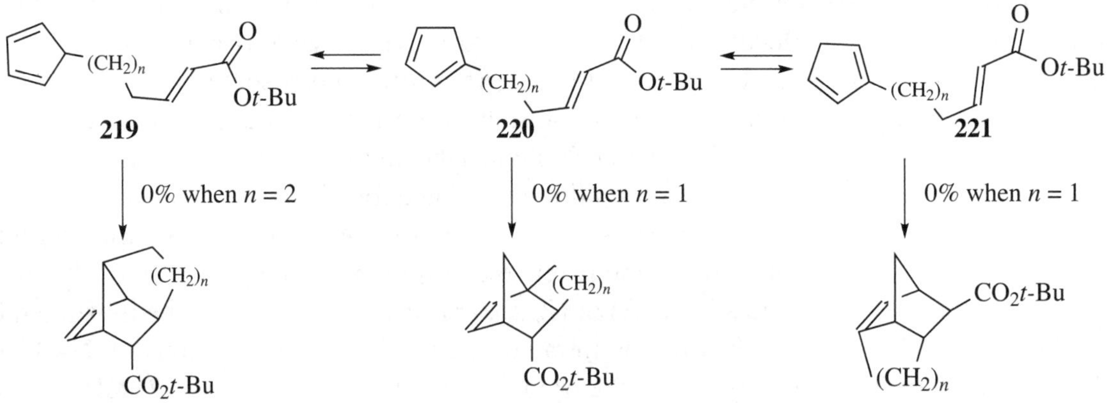

[Reprinted with permission from Stille, J.R.; Grubbs, R.H. *J. Org. Chem.* **1989**, *54*, 434. Copyright © 1989 American Chemical Society.]

The influence of the tether length is particularly striking in the substituted cyclopentadienes examined by Stille and Grubbs.[244] Each cyclopentadiene derivative is in equilibrium with three isomers, the 1-substituted, 2-substituted and 3-substituted cyclopentadienes. A two-carbon tether ($n = 1$), as in **219** → **221**, gave 53% of **222** when heated in benzene.[244] The 2-substituted derivative (**220**) and 3-substituted derivative (**223**) gave no cycloadduct (**223** and **224**, respectively, where $n = 1$), presumably due to the difficulty in forming the four-membered ring in **223**, and attaining the proper geometry for the bridgehead π bond in **224** (see **Bredt's rule** in sec. 6.6.A). The endo-transition state required to generate **222** is relatively easy to achieve. Adding one carbon to the tether (**219**, $n = 2$), however, and heating led to a 6.4:1

240. Kraus, G.A.; Bougie, D.; Jacobson, R.A.; Su, Y. *J. Org. Chem.* **1989**, *54*, 2425.
241. Tagmazyan, K.Ts.; Mkrtchyan, R.S.; Babayan, A.T. *Zh. Org., Khim.* **1974**, *10*, 1642. (Engl. p 1657).
242. Boeckman Jr., R.K.; Demko D.M. *J. Org. Chem.* **1982**, *47*, 1789.
243. (a) For a related reaction with furan derivatives, see Parker, K.A.; Adamchuck, M.R. *Tetrahedron Lett.* **1978,** 1689; (b) Reference 80b, p 167.
244. Stille, J.R.; Grubbs, R.H. *J. Org. Chem.* **1989**, *54*, 434.

mixture of **223** and **224** ($n = 2$).[244] The cyclization of **219** → **222** ($n = 2$) was *not* observed. The 2-substituted derivative gave the highest yield of cyclized product when $n = 2$.[244] Adding a total of four carbons to the tether gave results similar to when $n = 2$ in **219-221**.[244] These examples clearly show that the length and nature of the tether is important, and inspection of all possible transition states is essential for predicting the cycloadducts arising from intramolecular Diels-Alder reactions.

In addition to the problems concerning reactivity and the nature of the tether, at least two different regioisomers are possible for internal Diels-Alder reactions (**226** and **227** from **225**), depending on the orientation of the alkene as it approaches the diene. If the tether is short, only **227** is possible, but with longer tethers **226** can be a side product or even the major product. In Thomas and co-worker's studies[245] on the synthesis of cytochalasin, an intramolecular Diels-Alder reaction of **228** gave a 27:5% mixture of **229** and **230**. In general, (*E*)-dienes do not give products such as **220**, although (*Z*)-dienes do. As the reactive termini approach, the alkene arm can assume an exo or an endo mode. The exo transition state leads to the trans-product, and the endo transition state leads to the cis product. The final outcome depends on the conformation assumed in the transition state, based on the structures of the reactive species (sec. 1.5.D).[237] In general, the exo transition state is preferred.

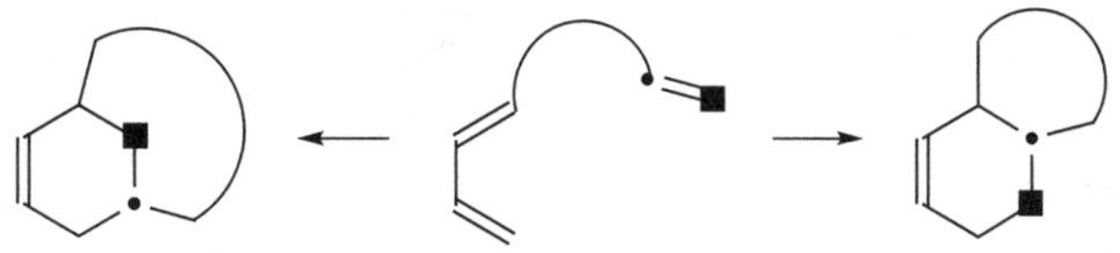

226   225   227

[Reprinted with permission from Desimoni, G.; Tacconi, G.; Barco, A.; Pollini, G.P. *Natural Product Synthesis Through Pericyclic Reactions*, American Chemical Society, Washington, *1983*. Copyright © 1983 American Chemical Society.]

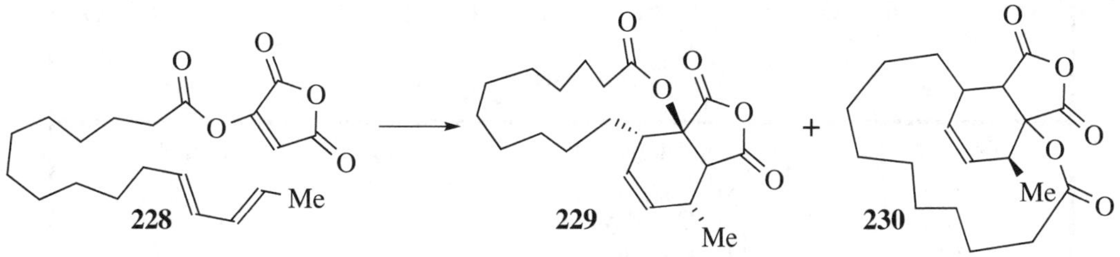

228   229   230

As mentioned previously, the length of the tether, reaction conditions and catalyst, if any, will influence the cis/trans ratio. Houk and co-workers compared the cyclization of trienes (**231**) with several tether lengths to give the bicyclic cycloadduct **232**.[246] The cis adduct was preferred except when $n = 5$ (to form a nine-membered ring). Gras initially reported that when

---

245.   Bailey, S.J.; Thomas, E.J.; Turner, W.B.; Jarvis, J.A. *J. Chem. Soc. Chem. Commun.* **1978**, 474.
246.   Smith, D.A.; Sakan, K.; Houk, K.N. *Tetrahedron Lett.* **1986**, *27*, 4877.

$n = 2$ the trans product was preferred[247] (later corrected), but Houk found that the cis product was preferred. When $n = 4\text{-}7$, the product yields were decreased, reflecting the difficulty in generating the transition state for an 8-11-membered ring in **232** (secs. 1.5.B,C, 6.6.B). There are many applications of the intramolecular Diels-Alder reaction to natural product synthesis. In a synthesis of sordaricin,[248] Mander and Thomson stirred **233** at 40°C for 3 days, and obtained quantitative yield of **234**. Using an intramolecular Diels-Alder strategy that involved a transannular cyclization, Tadano and co-worker's converted **235** to **236**, in a synthesis of (+)-macquarimicin A.[249]

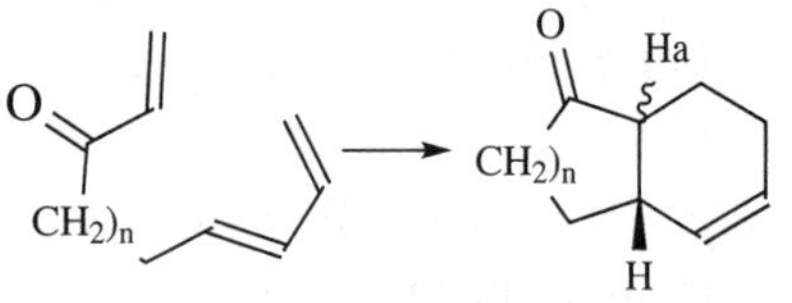

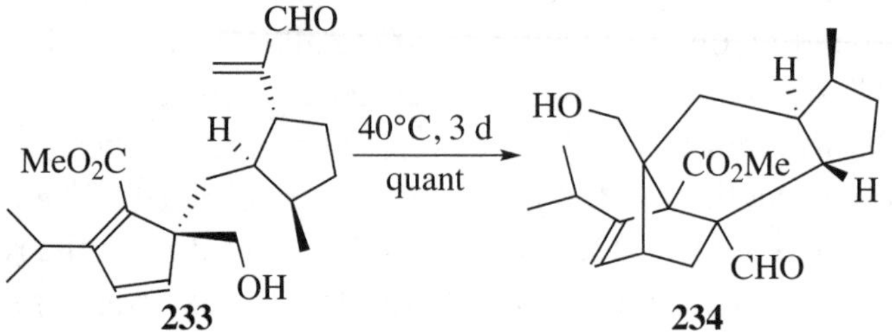

231   **232**   **233**   **234**

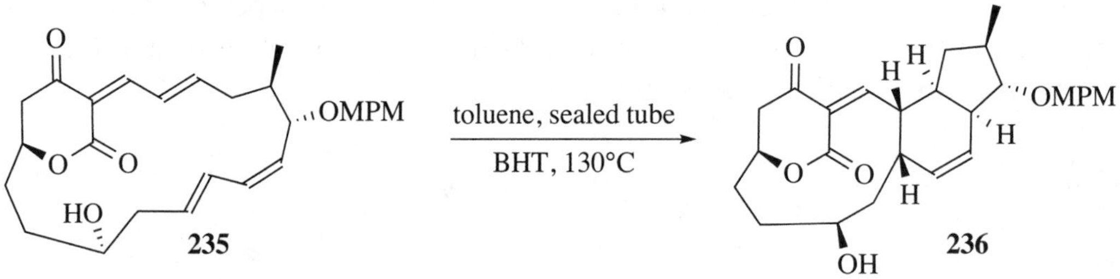

235   **236**

Table 11.6. Intramolecular Cycloaddition of Selected Trienes[250]

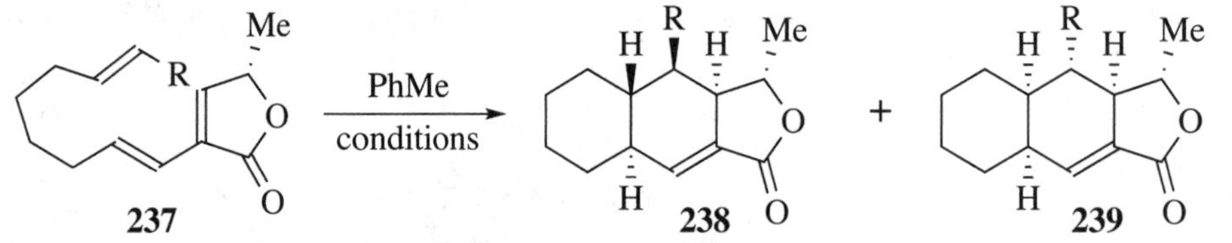

237   **238**   **239**

R	Conditions	% Yield	238:239
CO$_2$Et	200°C (3 h)	77	1 : 4
CO$_2$Et	110°C (1 day)	58	1 : 1
CO$_2$Et	SiO$_2$•Et$_2$AlCl	31	3 : 1
CH$_2$OTBS	110°C (16 h)	0	1 : 1
CH$_2$OTBS	SiO$_2$•Et$_2$AlCl	75	20 : 1

[Reprinted with permission from Hart, D.J.; Li, J.; Wu, W.-L.; Kozikowski, A.P. *J. Org. Chem.* **1997**, *62*, 5023. Copyright © **1997** Amer. Chem. Soc.]

247.   Gras, J.-L.; Bertrand, M. *Tetrahedron Lett.* **1979**, 4549.
248.   Mander, L.N.; Thomson, R.J. *Org. Lett.* **2003**, *5*, 1321.
249.   Munakata, R.; Katakai, H.; Ueki, T.; Kurosaka, J.; Takao, K.; Tadano, K. *J. Am. Chem. Soc.* **2003**, *125*, 14722.

Even when a single substrate is used, the ratio of diastereomeric products can vary with reaction conditions. In Hart, Kozikowski and co-workers synthesis of (+)-himbeline and (+)-himbacine,[250] **237** was cyclized to a mixture of **238** and **239** using the conditions in Table 11.6. At higher temperatures, the reaction favored **239** but with an acid promoter at lower temperatures, **238** was favored. When the ester group in **237** was changed to the reduced form (a TBS protected alcohol), heating to 110°C gave a 1:1 mixture whereas the acid promoted lower temperature reaction gave a 20:1 mixture favoring **238**.

MOLECULAR MODELING: Thermodynamic vs. Kinetic Control in Diels-Alder Reactions

The distribution of products for a reaction may reflect either their relative energies (***thermodynamic control***), or the relative energies of the transition states (***kinetic control***). High temperatures and long reaction times generally favor thermodynamic control, although in practice it is often difficult to say with certainty whether kinetics or thermodynamics is in control. On the other hand, quantum chemical calculations provide the means to access both relative product and transition state energies and, therefore, can provide an unambiguous account. A good example is provided by comparison of intramolecular Diels-Alder reactions for E-1,3,9-decatriene, which yields an equal amount of *cis* and *trans* products, and of its 3-methyl derivative, which shows a preference for a *cis* product.

R = H at 230°C 48 : 52 (*trans : cis*) 95%
R = Me at 160°C 94 : 6 (*trans : cis*) 95%

Click on > to the right of the tab at the top of the screen and select ***Continue*** to bring up the model-building palette. Build the *trans* product A for R = H and obtain the energy. Repeat the sequence (*click* on >, select ***Continue***, build and obtain the energy) for the corresponding *cis* product B, and for the *trans* and *cis* products for R=Me. Are the calculated relative energies consistent with the experimental data? In particular, do you find that reaction of the unsubstituted compound occurs with little if any stereochemical preference, while that of the methyl substituted compound shows a modest preference for *exo* addition? Overall, do the calculations suggest that the reaction is or is not under thermodynamic control?

Click on the rightmost tab at the top of the screen to bring back the screen containing the transition state models that opened this problem. Shown are transition states for *exo* (left) and *endo* (right) cycloadditions of E-1,3,9-decatriene (top) and 3-methyl-E-1,3,9-decatriene (bottom). Calculated energies (not provided) for the *exo* and *endo* transition states for reaction of E-1,3,9-decatriene are nearly the same, suggesting that the reaction will produce an equal mixture of *trans* and *cis* products. On the other hand, calculations show that the *exo* transition state for reaction of 3-methyl-E-1,3,9-decatriene is 12 kJ/mol lower in energy than the corresponding *endo* transition state. This translates into a 96:4 *trans:cis* mixture (favoring for *exo* addition) at 358 K (160°C). These results are close to what is observed experimentally, and they support the notion that the reaction is kinetically controlled. Examine the individual transition state geometries. Can you find any evidence that reaction of the methyl-substituted triene should be more selective than the reaction of the unsubstituted compound?

---

250. Hart, D.J.; Li, J.; Wu, W.-L.; Kozikowski, A.P. *J. Org. Chem.* **1997**, *62*, 5023

In the cyclization of dienyl amides, Oppolzer noted a difference in regiochemistry between dienes having the carbonyl *external to the ring being formed* (**240** → **241**),[251,252] and dienes having the carbonyl *in the ring being formed* (**242** → **243** + **244**).[251a] This difference was rationalized by a preference for the transition state that maximized overlap of the amide π-orbitals, and the diene π-orbitals. This secondary orbital interaction is analogous to the endo effect noted in sec. 11.4.C).

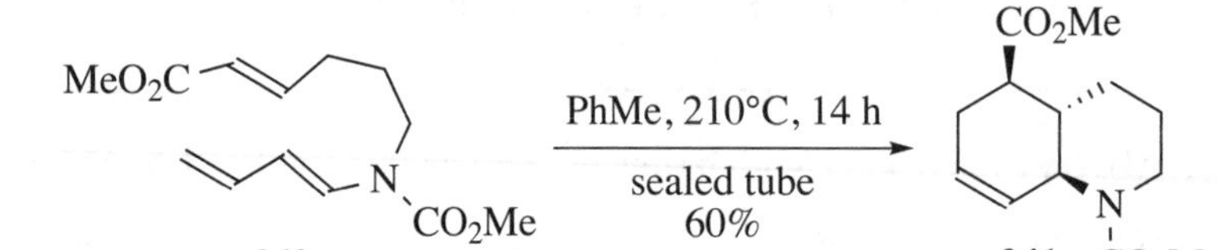

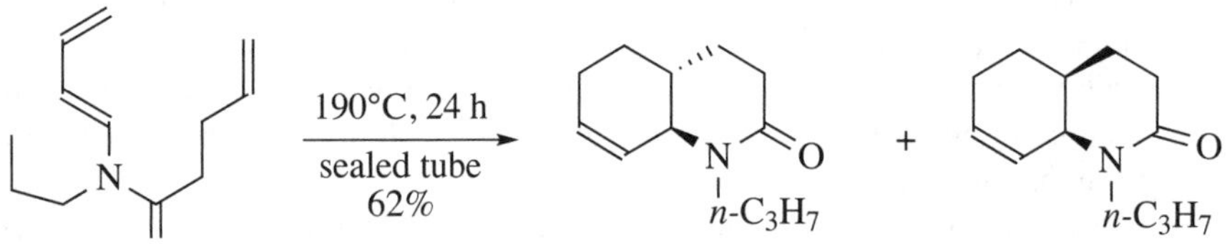

There are many synthetic examples carbocyclic, intramolecular Diels-Alder reactions that involve heteroatom substituents. In Martin and co-worker's synthesis of manzamine A,[253] imine-diene **245** was heated to 150°C to give a 74% yield of **246**.

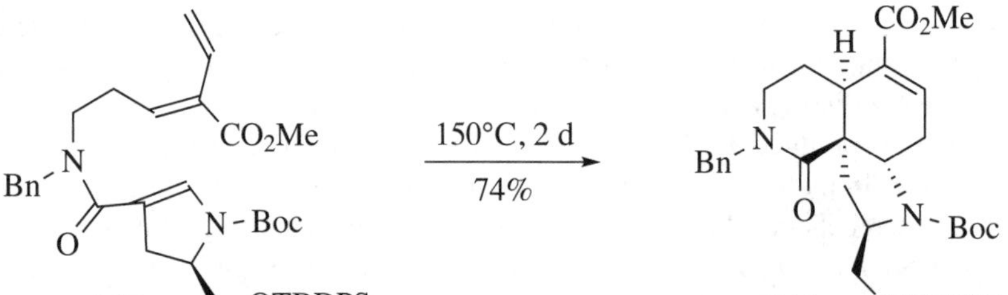

The concept of tethered intramolecular reactions has become popular. The basis of this concept is that the diene and alkene units are linked together by a species that allows an intramolecular Diels-Alder but can be chemically removed later. In the work of Batey and co-workers,[254a] reaction of boronic acid (**247**) with sorbic alcohol (**248**) gave **249**, and heating in a sealed tube at 190°C in the presence of the radical inhibitor BHT (2,6-di-*tert*-butyl-4-phneylpheonol) gave the

251. (a) Oppolzer, W.; Fröstl, W. *Helv. Chim. Acta* **1975**, *58*, 590; (b) Oppolzer, W.; Fröstl, W.; Weber, H.P. *Ibid.* **1975**, *58*, 593.
252. Witiak, D.T.; Tomita, K.; Patch, R.J. *J. Med. Chem.* **1981**, *24*, 788.
253. Humphrey, J.M.; Liao, Y.; Ali, A.; Rein, T.; Wong, Y.-L.; Chen, H.-J.; Courtney, A.K.; Martin, S.F. *J. Am. Chem. Soc.* **2002**, *124*, 8584.
254. (a) Batey, R.A.; Thadani, A.N.; Lough, A.J. *J. Am. Chem. Soc.* **1999**, *121*, 450; (b) Boeckman, R.K., Jr.; Shao, P.; Wrobleski, S.T.; Boehmler, D.J.; Heintzelman, G.R.; Barbosa, A.J. *J. Am. Chem. Soc.* **2006**, *128*, 10572.

cycloadduct **250**. Oxidation with trimethylamine-*N*-oxide and hydrolysis gave a 90:10 mixture of **251/252** in 84% yield. The boronic ester linkage in **249** was a tether leading to the regio- and stereoselectivity of an intramolecular process. Treatment with the amine oxide and hydrolysis removed the tether to give the cycloadducts. In a synthesis of (+)-tetronolide by Boeckman and co-workers, an acetal tether was used to anchor an intramolecular Diels-Alder reaction.[254b]

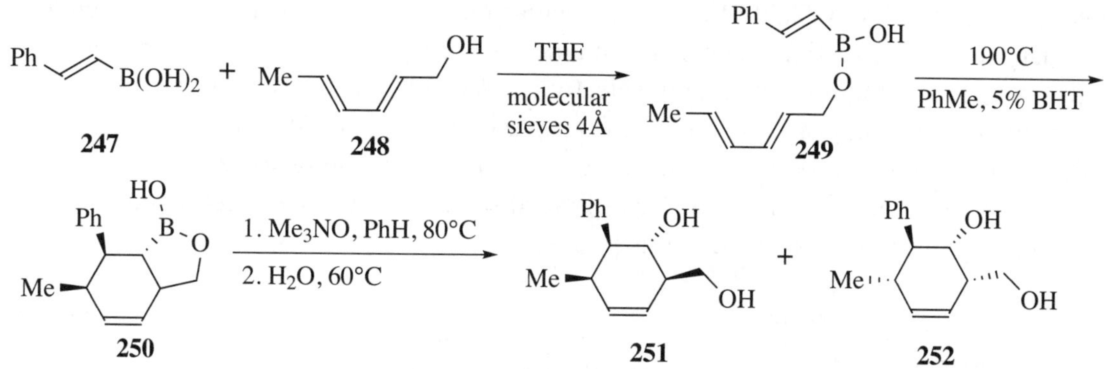

Silylalkyl tethers are also quite useful. In a synthesis of C-aryl glycosides by Martin and co-workers,[255] treatment of aryl dichloride **253** with *tert*-butyllithium generated the benzyne intermediate. An intramolecular Diels-Alder reaction with the furan moiety led to **254** in 61% yield. Subsequent treatment with tetrabutylammonium fluoride and then trifluoroacetic acid cleaved the silane tether, generating an OH moiety, and opening the oxabicycloheptane unit to give **255** in 72% yield.

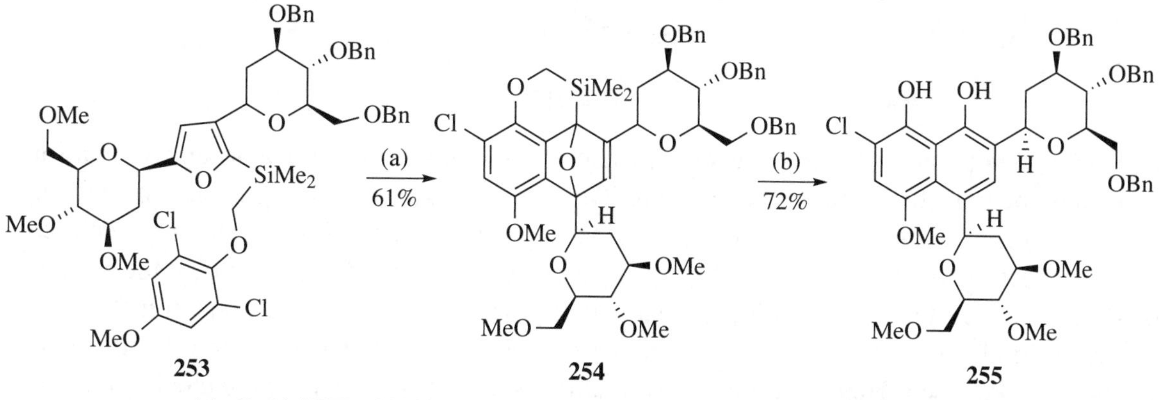

(a) *t*-BuLi, THF, −90°C → −10°C   (b) 1. Bu₄NF, DMF, rt   2. TFA, CH₂Cl₂

255.   Kaelin Jr., D.E.; Sparks, S.M.; Plake, H.R.; Martin, S.F. *J. Am. Chem. Soc.* **2003**, *125*, 12994.

11.9. ENANTIOSELECTIVE DIELS-ALDER REACTIONS

The diastereoselectivity inherent to the Diels-Alder reaction is apparent in several examples in preceding reactions. The reaction is not inherently enantioselective since there is no facial control (biasing approach of alkene to diene from a single face) for intermolecular reactions (some facial control is available for intramolecular reactions). The ortho rule, the endo rule (secondary orbital interactions) and steric interactions provide some orientational control, but facial control is also required for enantioselectivity. When ethyl acrylate reacts with 2-methyl-1,3-pentadiene, endo approach can occur from the bottom as in **256A**, or from the top as in **256B**. Clearly, the two products resulting from attack at opposite faces (**257A** and **257B**) are mirror images and enantiomers. This lack of facial selectivity leads to racemic mixtures in all Diels-Alder cyclizations discussed to this point.

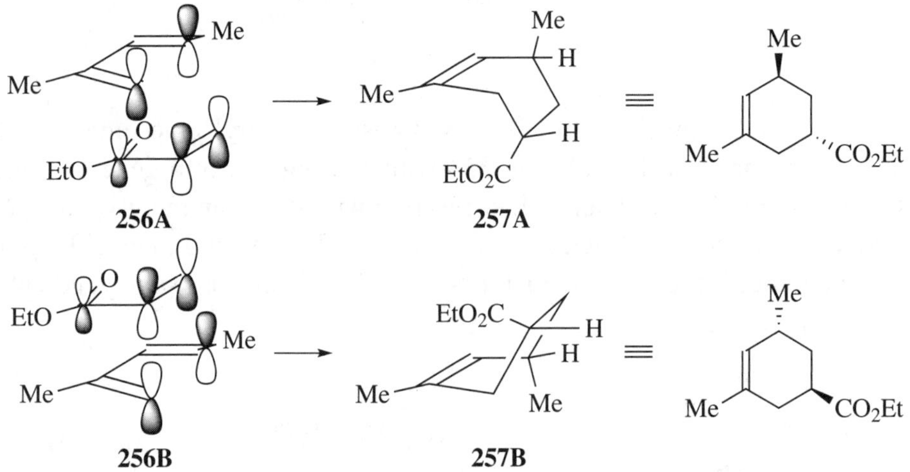

As in Chapters 1-5 and 8-9, the key to enantioselectivity is providing both facial and orientational control in the transition state of the reaction. The orientational control (regioselection and endo selectivity) for the diene and alkene is provided by orbital interactions, including secondary orbital effects. As in previous chapters, facial selectivity will be provided by the presence of stereogenic centers in the molecule that usually make one face in the transition state more sterically hindered than the other. The diene can be constructed from a chiral precursor (a **chiral template**, sec. 10.9). Alternatively, a chiral auxiliary (X_c) can be attached to the diene or alkene. A third approach is related to using a chiral auxiliary in that a chiral material (usually a chiral catalyst) is added to the reaction, usually providing a transient chiral auxiliary (sec. 4.7.A). Catalytic, enantioselective Diels-Alder reactions are now well known, and many catalysts and auxiliaries are available.[256] The rare earth (III) salt catalyzed asymmetric Diels-Alder reaction with a chiral dienophile in supercritical carbon dioxide has been reported, and

---

256. Corey, E.J. *Angew. Chem. Int. Ed.* **2002**, *41*, 1651.

the use of alternative solvents is an on-going area of study.[257]

11.9.A   Chiral Auxiliaries

A chiral auxiliary is a chiral molecule, usually derived from Nature, which is attached to the reaction substrate (in high yield and under mild conditions), and provides facial selectivity (and hopefully orientation control as well) in a reaction.[258] After the reaction of interest, the auxiliary must be removed (in high yield and under mild conditions) to regenerate the original functional group. For a Diels-Alder reaction, the chiral auxiliary must be attached to either an achiral diene (as in **258**) or an achiral dienophile (as in **261**) to generate a chiral molecule. A limitation, of course, is the availability of a functional group in the alkene or diene that can react with a chiral auxiliary. A Diels-Alder reaction with **258**, for example, would generate **259** as a mixture of diastereomers, each in high enantiomeric purity (% ee, sec. 1.4.F). The auxiliary is then removed to give the chiral target **260**. A similar sequence with **261** generated **262**, and removal of X_c gave **263**. The auxiliary provides facial selectivity in the Diels-Alder reaction, transferring that chirality to the cycloadduct. The Diels-Alder reactions of both **258** and **261** *may* proceed with excellent enantioselectivity, but may proceed with moderate diastereoselectivity. This problem of diastereoselectivity and enantioselectivity must be addressed in all asymmetric reactions, and was seen in reductions (Chap. 4) as well as nucleophilic carbon bond forming reactions (Chap. 8), and especially in enolate reactions (Chap. 9).

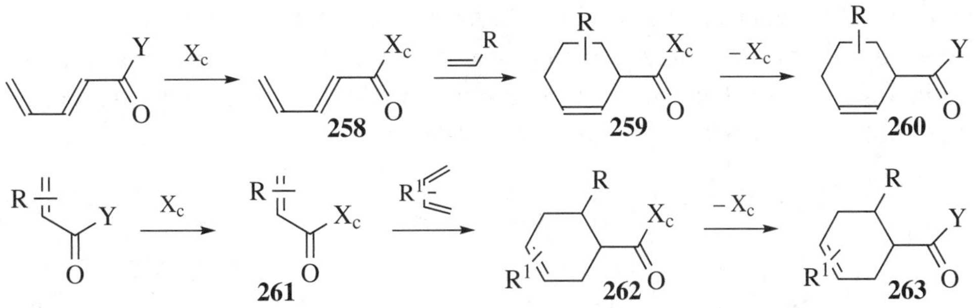

Morrison and Mosher[259] showed that menthyl was an early auxiliary, and when an acrylic acid derivative reacts with menthol to form a menthyl ester, **264**, asymmetric induction is possible upon reaction with cyclopentadiene to give diastereomers **265** and **266**,[260] as shown in Table 11.7.[259] In the absence of a Lewis acid, however, the % ee is rather poor.

257.   Fukuzawa, S.; Metoki, K.; Esumi, S. *Tetrahedron* **2003**, *59*, 10445.
258.   For a review of bicyclic lactam chiral auxiliaries, see Groaning, M.D.; Meyers, A.I. *Tetrahedron* **2000**, *56*, 9843.
259.   Morrison, J.D.; Mosher, H.S. *Asymmetric Organic Reactions*, Amer. Chem. Soc., Washington, D.C., **1976**, p256.
260.   (a) Sauer, J.; Kredel, J. *Tetrahedron Lett.* **1966**, 6359; (b) Farmer, R.F.; Hamer, J. *J. Org. Chem.* **1966**, *31*, 2418; (c) Cervinka, O; Kríz, O. *Collect. Czech. Chem. Commun.* **1968**, *33*, 2342.

Table 11.7. Selectivity in the Diels-Alder Reaction of Cyclopentadiene and Menthyl Acrylate

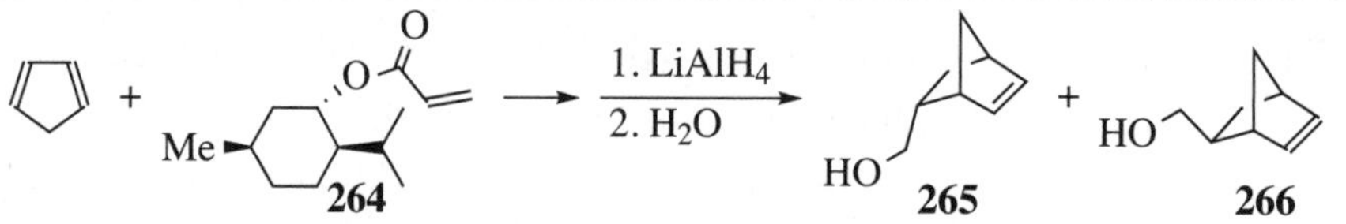

Catalyst	Solvent	Temperature (°C)	265 [%ee]	266 [%ee]
None	CH$_2$Cl$_2$	0	(+) [9]	
None		35	(+) [7]	(+) [3.1]
AlCl$_3$•OEt$_2$		0	(+) [49]	(-) [36]
		-70	(+) [67]	
BF$_3$•OEt$_2$		0	(+) [74]	(-) [43]
		-20	(+) [82-85]	
SNCl$_4$	PhMe	48	(+) [41]	
None	Neat	160	(+) [6]	

[Reprinted by courtesy of Professor James D. Morrison from Morrison, J.D.; Mosher, H.S. *Asymmetric Organic Reactions*, American Chemical Society, Washington, DC *1976*, pp. 256.]

One of the first examples of an asymmetric Diels-Alder reaction is the work of Walborsky et al.[261] in the reaction of (–)-dimenthyl fumarate (**267**) and 1,3-butadiene to give **268** and **269**, after reduction of the ester products with lithium aluminum hydride.[262] Reaction in toluene using a Lewis acid catalyst generally provided the best asymmetric induction, in good yield. Hydride reduction of esters (sec. 4.2.B) is a common method for removing ester auxiliaries. In general, menthol type auxiliaries provide only moderate selectivity.[263]

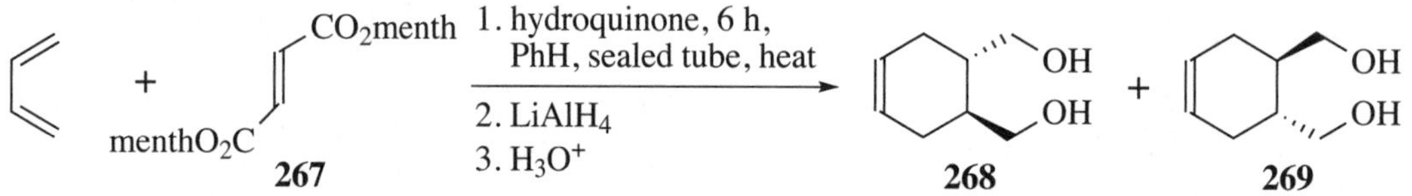

In Oppolzer's review,[264] several chiral auxiliaries related to menthol and other chiral alcohols were presented, giving the diastereomeric endo adducts **270** and **271**, as shown in Table 11.8.[265] In all cases, the chiral ester in Table 11.8 provided good-to-excellent selectivity. Oppolzer discussed the source of this selectivity by invoking Figure 11.13[265] (from 5-benzyloxymethyl cyclopentadiene) in which the derivatized menthyloxy chiral auxiliary **272** provided facial selectivity to the acrylate, for attack from either the *re* face or the *si* face. The *si* face is less

261. Walborsky, H.M.; Barash, L.; Davis, T.C. *Tetrahedron* **1963**, *19*, 2333.
262. Reference 200, p 253.
263. Roush, W.R.; Gillis, H.R.; Ko, A.I. *J. Am. Chem. Soc.* **1982**, *104*, 2269.
264. Oppolzer, W. *Angew. Chem. Int. Ed.* **1984**, *23*, 876.
265. Oppolzer,W.;Kurth,M.;Reichlin,D.;Chapuis,C.;Mohnhaupt,A.;Moffatt,F.*Helv. Chim.Acta***1981**,*64*,2802.

hindered and **273** [the (*R*) cycloadduct] is preferred to **274**, consistent with the results in Table 11.8. Figure 11.13 can be used to illustrate the source of facial selectivity for other chiral auxiliaries. The poor-to-moderate asymmetric induction with menthol and related auxiliaries contrasts with Trost's use of **276** with napthoquinone **275**, which gave 98% of **277** with >97% ee.[266]

Table 11.8. Enantioselectivity in the Diels-Alder reaction of Cyclopentadiene with Conjugated Esters Derived from Chiral Alcohols

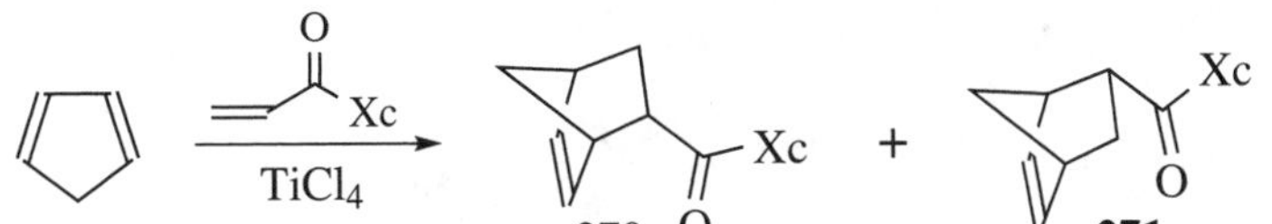

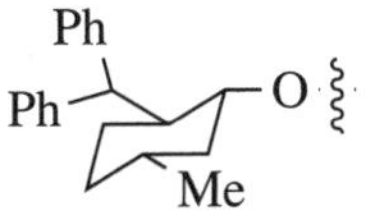

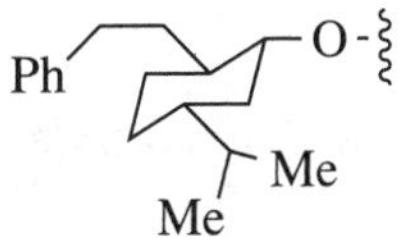

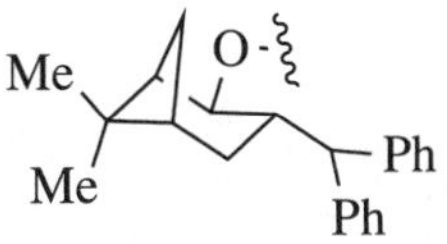

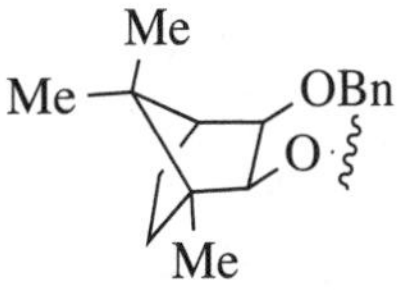

Other auxiliaries can be attached to either the dienophile or the diene. In Rawal and Kozmin's synthesis of (–)-α-elemene, a chiral pyrrolidine unit attached to the diene unit of **70** and served as a chiral auxiliary (see sec. 11.4.C.iii). In that synthesis, the Diels-Alder adduct was **71**, but the auxiliary was removed by elimination to an alkene unit later in the synthesis.[100]

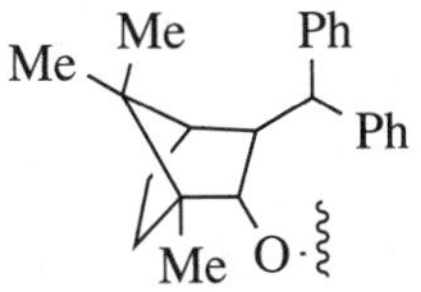

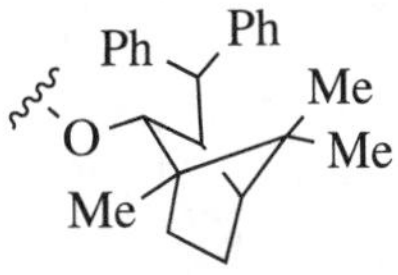

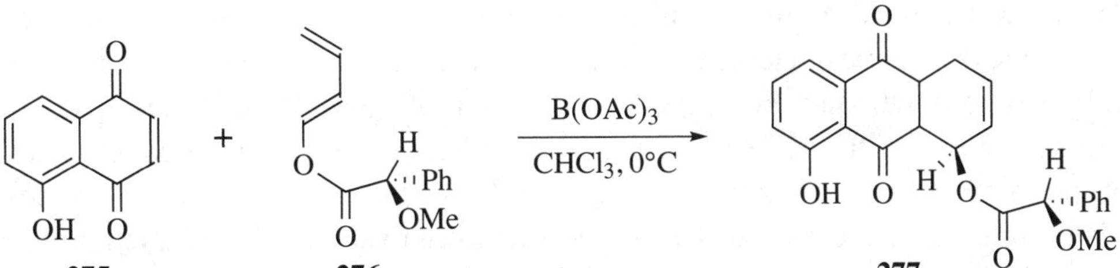

266. Trost, B.M.; O'Krongly, D.; Belletire, J.L. *J. Am. Chem. Soc.* **1980**, *102*, 7595.

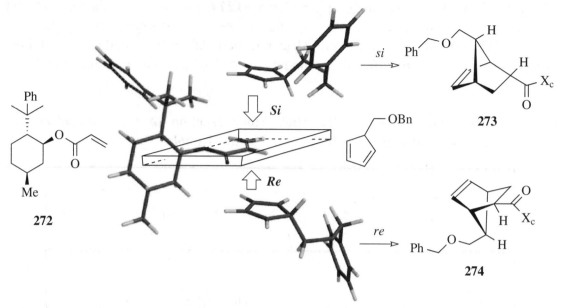

Figure 11.13. Facial selectivity of chiral acrylates in the Diels-Alder reaction. [Reproduced with permission from Oppolzer, W. *Angew. Chem. Int. Ed. 1984, 23*, 876. Copyright © 1984 VCH Weinheim.]

Evans developed a new generation of chiral auxiliaries, based on conversion of amino alcohols to oxazolidinone derivatives. The use of these auxiliaries to give the corresponding conjugated amide **278** led to excellent selectivity in the intramolecular Diels-Alder reaction, which generated mixtures of **279** and **280**.[267] Coordination with the Lewis acid catalyst generated a complex in which facial selectivity is enhanced relative to the uncomplexed species. When dimethylaluminum chloride was used as a Lewis acid catalyst with **281**, for example, chelated intermediate **282** was formed. This provided orientational control for the dienophile, and the chiral nature of the auxiliary provided the requisite facial control, with approach of the diene from the less sterically hindered face. Good selectivity was observed with the intermolecular Diels-Alder reaction as well.[268] In a synthesis of (+)-eutipoxide B by Okamura and co-

267. Evans, D.A.; Chapman, K.T.; Bisaha, J. *Tetrahedron Lett. 1984, 25*, 4071.
268. Evans, D.A. Chapman, K.T.; Bisaha, J. *J. Am. Chem. Soc. 1984, 106*, 4261.

workers,[269] the Cinchonine-catalyzed reaction of pyrone **283** and acrylamide **284** (containing the **Evan's auxiliary** shown) proceeding good yield, with greater than 95% selectivity for **285**.

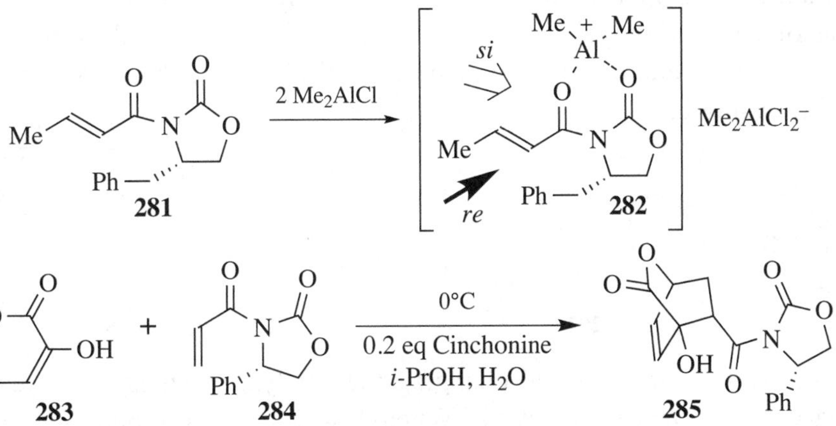

Oppolzer et al. developed a different set of auxiliaries based on a camphor precursor,[270] typified by sultam **286** (**Oppolzer's sultam**) and used in a chiral synthesis of the aglycone of loganin.[271] Sultam **286** was also the precursor to acrylamide **287**, which reacted with diene **288** to give a 66% yield of **289** (7:1 *exo:endo*), in Roush and co-worker's synthesis of the spirotetronate subunits of quartromicin.[272] Selectivity in this Diels-Alder reaction is due to chelation of the sultam unit with the catalyst ($MeAlCl_2$).

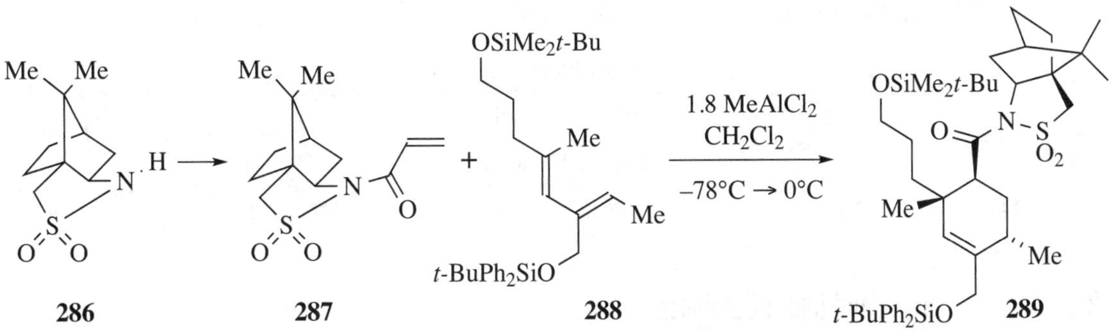

Many auxiliaries give excellent asymmetric induction in the Diels-Alder reaction. Helmchen and Schmierer developed an auxiliary based on camphor that is quite useful.[273] Helmchen[274] also prepared the lactone auxiliary **290** and the imide auxiliary **292**, and showed they give

269. Shimizu, H.; Okamura, H.; Iwagawa, T.; Nakatani, M. *Tetrahedron* **2001**, *57*, 1903.
270. Oppolzer, W.; Chapuis, C.; Dupuis, D.; Guo, M. *Helv. Chim. Acta* **1985**, *68*, 2100.
271. Vandewalle, M.; van der Eycken, J.; Oppolzer, W.; Vullioud, C. *Tetrahedron* **1986**, *42*, 4035.
272. Roush, W.R.; Limberakis, C.; Kunz, R.K.; Barda, D.A. *Org. Lett.* **2002**, *4*, 1543.
273. (a) Helmchen, G.; Schmierer, R. *Angew. Chem. Int. Ed.* **1981**, *20*, 205; (b) Schmierer, R.; Grotemeier, G. Helmchen, G.; Selim, A. *Ibid.* **1981**, *20*, 207.
274. Poll, T.; Abdel Hady, A.F.; Karge, R.; Linz, G.; Weetman, J.; Helmchen, G. *Tetrahedron Lett.* **1989**, *30*, 5595.

opposite enantioface selectivity in the Diels-Alder reaction. In the presence of a Lewis acid such as TiCl$_4$, differences in coordination site led to a different chiral face in the transition state of the cycloaddition and opposite stereochemistry. Helmchen used an auxiliary based on **290** in a synthesis of cyclosarkomycin.[275] Other auxiliaries include **291**, developed by Marchand-Brynaert[276] and co-workers for 1-amino dienes, and the *exo*-selective auxiliary **293** developed by Kudo and Kawamura,[277]

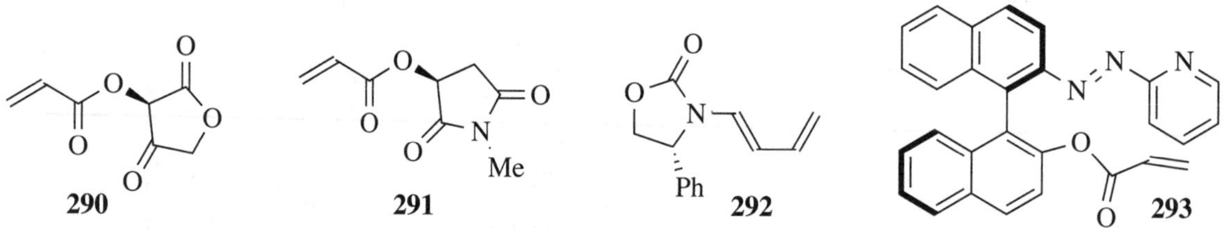

Ghosez and co-workers showed that changing the catalyst in Diels-Alder reactions using a chiral auxiliary can influence selectivity. When **294** reacted with diene **295**, in the presence of a silyl triflate catalyst, an 88:12 mixture of **296/297** was obtained in 96%.[278] Changing the catalyst to europium [Eu(fod)$_3$], however, gave 98% of **297** (0:100 **296/297**). A variety of auxiliaries have been developed that can be very effective in catalyzing Diels-Alder reactions with high asymmetric induction. It is clear that the best results are obtained when the substrate-auxiliary complex can chelate to the catalyst. A major drawback with these auxiliaries is that they can sometimes be difficult to remove (vigorous aqueous acid or base or treatment with LiAlH$_4$ may be required). Dienes and dienophiles must have heteroatom substituents for these auxiliaries or other methodology must be developed.

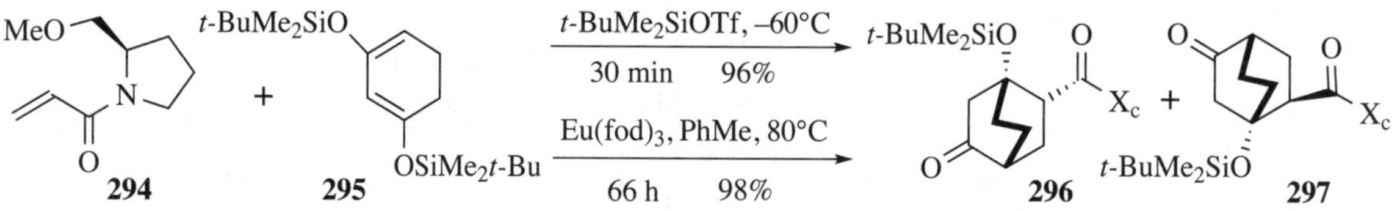

11.9.B. Chiral Additives and Chiral Catalysts

Clearly related to chiral auxiliaries are chiral additives, which usually form transient auxiliaries to provide the requisite facial selectivity. In Section 4.7.A, we saw that naturally occurring chiral amines were added to reducing agents to achieve asymmetric induction. Protected proline and abrine esters (*L*-abrine is *N*-Methyl-*L*-tryptophan) and related compounds have

275. Linz, G.; Weetman, J.; Abdel Hady, A.F.; Helmchen, G. *Tetrahedron Lett.* **1989**, *30*, 5599.
276. Robiette, R.; Cheboub-Benchaba, K.; Peeters, D.; Marchand-Brynaert, J. *J. Org. Chem.* **2003**, *68*, 9809.
277. Kawamura, M.; Kudo, K. *Chirality* **2002**, *14*, 727.
278. (a) Lamy-Schelkens, H.; Ghosez, L. *Tetrahedron Lett.* **1989**, *30*, 5891; (b) Lamy-Schelkens, H.; Giomi, D.; Ghosez, L. *Ibid.* **1989**, *30*, 5887.

been used as chiral catalysts.[279] These amines can also be used in Diels-Alder reactions, with variable results. One example is the reaction of **298** with *N*-methylmaleimide to give **299**,[280] in the presence of a chiral additive. The use of various chiral amine additives [quinine, quinidine, cinchonine, cinchonidine, (*S*)-prolinol, and (*1R,2S*)-*N*-methylephedrine], gave enantiomeric excesses between 16-61%) in the cycloadduct. In this particular example, quinidine proved the most effective additive with a 61% ee. Presumably, coordination between the chiral amine and **298** via hydrogen-bonding (or a dipole-dipole interaction) provides diastereoselection in the subsequent reaction with *N*-methylmaleimide.

Preparation of a chiral catalyst will provide facial selectivity when it coordinates to the substrate (usually the dienophile), forming a chiral transition state containing the transient auxiliary. Roush and co-workers used menthoxyl and bornyloxyaluminum catalysts,[263] providing an example in which traditional chiral materials are attached to aluminum chloride. Significantly more effective catalysts have been developed based on coordination of chiral molecules [usually bis(amines) or diols] with aluminum or titanium reagents. Corey et al. prepared chiral catalyst **301**, which reacted with **300** to form chiral complex **302**.[281] This complex was formed *in situ*, and reacted with cyclopentadiene to give an 88% yield of **303**, with 94% ee (in a ratio of 96:4 endo/exo).

Corey and co-workers developed other catalysts for enantioselective Diels-Alder reactions. One is a C_2-symmetric chiral bis(oxazoline)-Fe(III) complex (**304**).[282] Similar copper-coordinated catalysts have been developed and give good-to-excellent enantioselectivity in Diels-Alder reactions.[283] Analogs of **300** reacted with cyclopentadiene in the presence of a **304**•$FeCl_2I$

279. Ahrendt, K.A.; Borths, C.J.; MacMillan, D.W.C. *J. Am. Chem. Soc.* **2000**, *122*, 4243.
280. Riant, O.; Kagan, H.B. *Tetrahedron Lett.* **1989**, *30*, 7403.
281. Corey, E.J.; Imwinkelried, R.; Pikul, S.; Xiang, Y.B. *J. Am. Chem. Soc.* **1989**, *111*, 5493.
282. Corey, E.J.; Imai, N.; Zhang, H.-Y. *J. Am. Chem. Soc.* **1991**, *113*, 728.
283. For copper coordinated catalysts, see (a) Evans, D.A.; Barnes, D.M.; Johnson, J.S.; Lectka, T.; von Matt, P.; Miller, S.J.; Murry, J.A.; Norcross, R.D.; Shaughnessy, E.A.; Campos, K.R. *J. Am. Chem. Soc.* **1999**, *121*, 7582. For copper and zinc coordinated catalysts see (b) Yao, S.; Roberson, M.; Reichel, F.; Hazell, R.G.; Jørgensen, K.A. *J. Org. Chem.* **1999**, *64*, 6677.

catalyst to give cycloadducts with good enantioselectivity (93:7, with 99:1 endo/exo-selectivity) at –50°C in dichloromethane (15, 10 mol% of the catalyst).[282] The FeI_3 complex that is shown with **304**, as well as the $FeCl_3$ complex gave slightly poorer selectivity than the $FeCl_2I$ complex. The (S)-tryptophan derived oxazoborolidine (**305**) catalyst was designed to facilitate intramolecular interactions in highly enantioselective Diels-Alder reactions.[284] Oxazolidinone catalyst **68**, as the hexafluoroantimonate complex, in Sect. 11.4.C.ii, is another example of an extremely useful catalyst for the Diels-Alder reaction related to **304**. In a synthesis of estrone,[285] Corey and co-workers reacted **307** with **308**, in the presence of chiral catalyst **306** (an oxazaborolidinium salt), to give a 92% yield of **309**, in 94% ee. Catalysts related to **306** have been shown to be broad spectrum catalysts for enantioselective Diels-Alder reactions.[286]

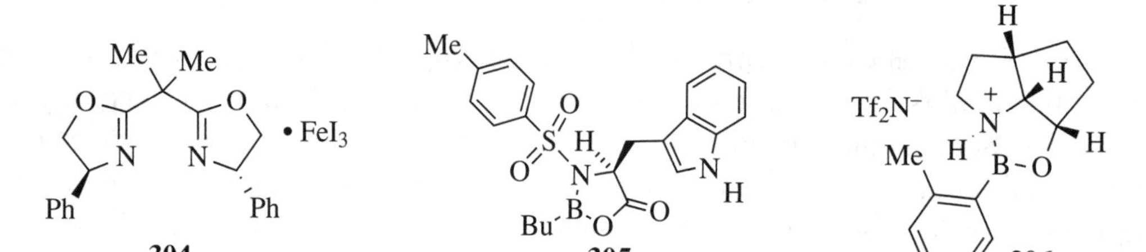

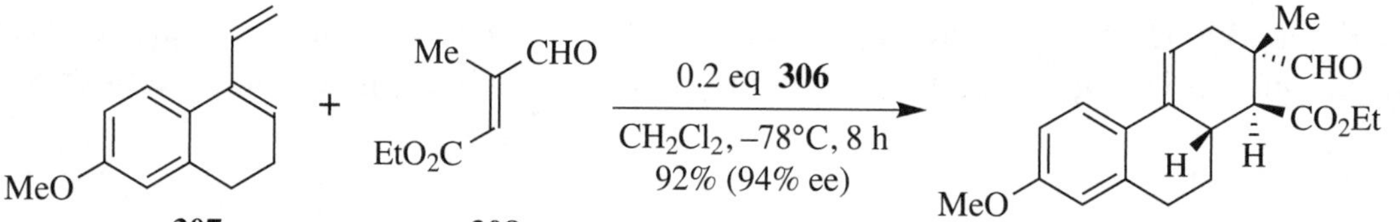

Yamamoto and co-workers reported an asymmetric hetero-Diels-Alder reaction that was catalyzed by a chiral organoaluminum reagent (**311**).[287] When diene **310** reacted with benzaldehyde in the presence of 10 mol% of **311**, a mixture of *cis*-dihydropyrone (**312**) and *trans*-dihydropyrone (**313**) was obtained in 77 and 7% yield, respectively. The optical activity of **312** was 95% ee, and that of **313** was 53% ee.[287] In the complex that mediates this reaction, the diene approaches benzaldehyde with an endo alignment that minimizes steric repulsion between the incoming diene and the triphenylsilyl moiety on that face.

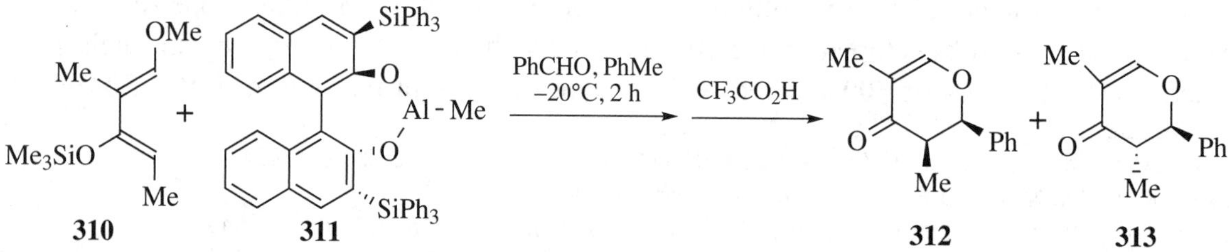

284. Corey, E.J.; Loh, T.-P. *J. Am. Chem. Soc.* **1991**, *113*, 8966.
285. Hu, Q.-Y.; Rege, P.D.; Corey, E.J. *J. Am. Chem. Soc.* **2004**, *126*, 5984.
286. Ryu, D.H.; Lee, T.W.; Corey, E.J. *J. Am. Chem. Soc.* **2002**, *124*, 9992.
287. Maruoka, K.; Itoh, T.; Shirasaka, T.; Yamamoto, H. *J. Am. Chem. Soc.* **1988**, *110*, 310.

Narasaka et al. developed a titanium catalyst generated by complexation with chiral diols.[288] The dienophile must contain functionality that will coordinate with the metal catalyst to form a chiral complex, and these catalysts are less effective with dienes and dienophiles that do not contain heteroatoms. A related titanium-BINOL complex has been used to catalyze Diels-Alder reactions.[289] Kelly prepared a transient boron catalyst, prepared *in situ* with borane and a chiral diol.[290] It is clear that chiral additives, particularly chiral catalysts can be effective. The asymmetric induction is usually good to excellent with properly designed dienophiles. Effective chiral catalysts continue to be reported and will undoubtedly be a major force in chiral synthesis involving Diels-Alder reactions.

11.9.C. Chiral Templates

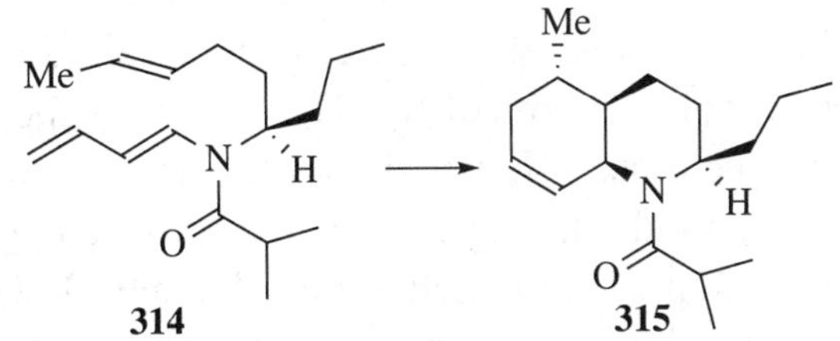

The last approach to chiral synthesis begins with a chiral precursor whose components will be integrated into the final product, which is the **chiral template approach** (sec. 10.9). The examples of asymmetric syntheses involving Diels-Alder **templates** are as varied as the

314   **315**

targets. One is by Oppolzer, who reported an asymmetric synthesis of pumiliotoxin C via Diels-Alder cyclization of **314**, prepared from (R)-norvaline, to give **315**.[291]

Carbohydrates are efficient chiral templates and have been used in the Diels-Alder reaction. Sherburn and Lilly used L-ascorbic acid as a template to synthesize **316**.[292] Heating this compound in refluxing toluene led to a 96:4 mixture of **317** and **318** (68% yield). The chirality inherent in the carbohydrate precursor provides the needed facial selectivity that is transferred to the cycloadduct product. The use of chiral templates for preparing Diels-Alder precursors will undoubtedly increase in importance. As the need for enantiomerically pure material increases, the chiral template driven Diels-Alder reactions will be increasingly effective.

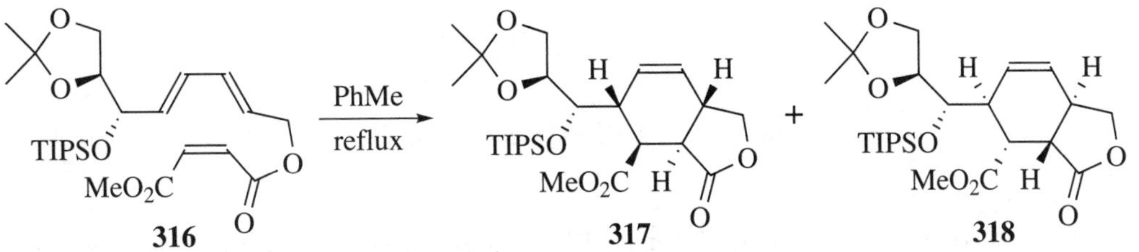

316   **317**   **318**

288. (a) Narasaka, K.; Iwasawa, N.; Inoue, M.; Yamada, T.; Nakashima, M.; Sugimori, J. *J. Am. Chem. Soc.* **1989**, *111*, 5340; (b) Narasaka, K.; Inoue, M.; Yamada, T.; Sugimori, J.; Iwasawa, N. *Chem. Lett.* **1987**, 2409.
289. Mikami, K.; Motoyama, Y.; Terada, M. *J. Am. Chem. Soc.* **1994**, *116*, 2812.
290. Kelly, T.R.; Whiting, A.; Chandrakumar, N.S. *J. Am. Chem. Soc.* **1986**, *108*, 3510.
291. Oppolzer, W.; Flaskamp, E. *Helv. Chim. Acta* **1977**, *60*, 204.
292. Lilly, M.J.; Sherburn, M.S. *Chem. Commun.* **1997**, 967.

11.10. [2+2]-CYCLOADDITION REACTIONS

The Diels-Alder reaction occupies a prominent place in organic synthesis, but other pericyclic reactions are quite useful. One is the [2+2]-cycloaddition that occurs between two alkenes, or between an alkene and a carbonyl or imine.[293] Cyclobutanes and oxetanes are prepared by this reaction, and β-lactams and β-lactones have been prepared. This section will focus on both thermal and photochemical [2+2]-cycloaddition reactions.

11.10.A. Thermal [2+2]-Cycloadditions

In Section 11.3, thermal [2+2]-cycloaddition reactions that form cyclobutane derivatives such as **321**[294,10a] were shown to be symmetry forbidden, but photochemically allowed. There are, however, several examples of thermal [2+2]-cycloadditions that probably occur by a radical or dipolar intermediate (**319** or **320**, respectively), or by HOMO-HOMO or LUMO-LUMO interactions.[295] Lewis-acid catalyzed reactions of alkynes with alkenes or conjugated carbonyl compounds are also known,[296] proceeding via a metal-stabilized cationic intermediate.[296b] Reaction of dimethylaminoisobutene (**322**) with methyl acrylate gives cyclobutane **324** by what is believed to be a zwitterionic intermediate (**323**).[297] As seen in the reaction of **322**, enamines are partners and alkynes react with ketenes to form cyclobutene derivatives.[298] Ionic intermediates need not be invoked in all cases, since some reactions proceed via diradical intermediates. The HOMO of **322** and LUMO of methyl acrylate can be used to predict the regiochemistry in **324**.[299] In general, the presence of an electron-withdrawing group on a ketene, such as a chlorine, an alkoxy group, an aryl group or a vinyl group greatly enhances the rate of the [2+2]-cycloaddition reaction with alkenes, particularly intramolecular cycloadditions.[300] It is also known that cis alkenes are more reactive than trans alkenes in reactions with ketenes.[301] Cycloaddition reactions with cis alkenes are stereospecific, but some loss of stereochemistry is observed in reactions with trans alkenes.[299,300,43]

293.   (a) Reference 65, pp 161-165; (b) Lee-Ruff, E.; Mladenova, G. *Chem. Rev.* **2003**, *103*, 1449.
294.   Reference 80b, pp 33-63.
295.   Reference 80b, p 35.
296.   (a) Dopper, J. H.; Greijdanus, B.; Oudman, D.; Wynberg, H. *J. Chem. Soc. Chem. Commun.* **1975**, 972; (b) Meyers, A. I.; Tschantz, M. A.; Brengel, G. P. *J. Org. Chem.* **1995**, *60*, 4359; (c) Groaning, M. D.; Brengel, G. P.; Meyers, A. I. *Tetrahedron* **2001**, *57*, 2635.
297.   Brannock, K.C.; Bell, A.; Burpitt, E.D.; Kelly, C.A. *J. Org. Chem.* **1961**, *26*, 625.
298.   For a synthetic example, see Fex, T.; Froborg, J.; Magnusson, G.; Thorén, S. *J. Org. Chem.* **1976**, *41*, 3518.
299.   Reference 80b, p 33.
300.   (a) Snider, B.B. *Chem. Rev.* **1988**, *88*, 793; (b) Snider, B.B.; Walner, M. *Tetrahedron* **1989**, *45*, 3171.
301.   (a) Ghosez, L.; O'Donnell, M.J. in *Pericyclic Reactions, Vol. II*, Marchand, A.P.; Lehr, R.E. (Eds), Academic Press, New York, *1977*, pp 79-140; (b) Frey, H.M.; Isaacs, N.S. *J. Chem. Soc. B* **1970**, 830.

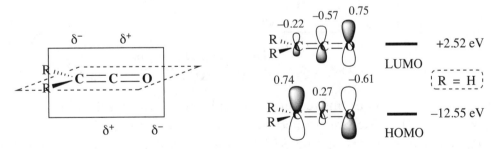

319 **320** **321**

322 **323** **324**

Ketenes ($R_2C=C=O$) react at the C=C unit via thermal [2+2]-cycloaddition reactions to give cyclobutanone derivatives. Ketenes are highly electrophilic due to a low-lying LUMO (π^* C=O), and the HOMO is relatively high in energy (see Table 11.1). Ketene itself has a HOMO of -12.55 eV and a LUMO of +2.55 eV,[43,302] and those values along with those of the orbital coefficients are shown in Figure 11.14 (also see Table 11.1). Polarization of the C=C unit in ketene (the combination of electrophilic and nucleophilic character) is responsible for the facile [2+2]-cycloaddition reactions. As mentioned above, the presence of an electron-withdrawing group enhances the rate of the [2+2]-cycloaddition reaction. Inspection of Table 11.1 shows that the HOMO for ketene is –12.55 eV, but the HOMO for PhCH=C=O is –10.61 eV, that for ClCH=C=O is –9.24 eV and that for $Cl_2C=C=O$ is –9.15 eV. Clearly, the presence of these groups raises the energy of the HOMO, and would be expected to increase the rate of a cycloaddition driven by the LUMO of an alkene. This effect has been noted in many intermolecular reactions of ketenes and alkenes.[303] The attempted intramolecular cyclization (see below) of **325** (X = H) failed. When the α-chloro analog (**325**, X = Cl) was prepared, however, treatment with triethylamine led to **326**, which cyclized to give a 55% yield of the [2+2]-cycloadduct (**327**) along with 19% of the ene adduct (**328**, see sec. 11.13) where X = Cl.

Figure 11.14. Orbital Diagram for Ketene.

302.   (a) Reference 7, p 143; (b) Reference 80b, p 33-42.
303.   (a) Krepski, L.R.; Hassner, A. *J. Org. Chem.* **1978**, *43*, 2879; (b) Bak, D.A.; Brady, W.T. *Ibid.* **1979**, *44*, 107; (c) Martin, P.; Greuter, H.; Bellusˇ, D. *Helv. Chim. Acta.* **1984**, *64*, 64; (d) Brady, W.T. *Synthesis* **1971**, 415.

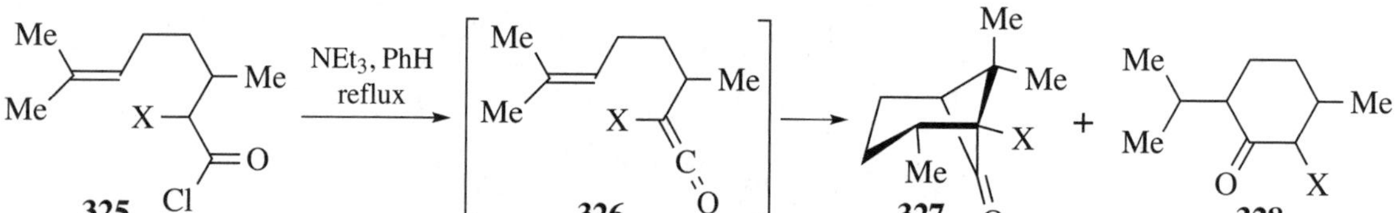

In a typical reaction, dimethylketene (**329**) and ethene were heated to give cyclobutanone **330** in a reaction controlled by the HOMO$_{alkene}$-LUMO$_{ketene}$ interaction.[304] Inspection of the orbital picture for this reaction (see **331**) shows that [2+2]-cycloadditions are allowed when one component is suprafacial and the other is antarafacial. In **331**, the HOMO of the alkene is of the correct symmetry to interact with the low-lying LUMO of the ketene.[305a] Reaction of an alkene and a ketene (see **331**) yields a cyclobutanone (**332**).[306] Ketenes such as **329** are readily generated by reaction of an acid chloride with an amine,[306] or by treatment with Zn-Cu in the case of α-chloro acid chlorides. Since carboxylic acids and acid chlorides are readily available, ketenes are convenient partners for reaction with both alkenes and dienes.[307] In Delair and co-worker's synthesis of (+)-amphorogynine A,[308] the acid chloride of trichloroacetic acid was treated with Zn-Cu in the presence of POCl$_3$ to give dichloroketene (**333**), which reacted with **334** to give dichlorocyclobutanone **335** in >90%% yield.

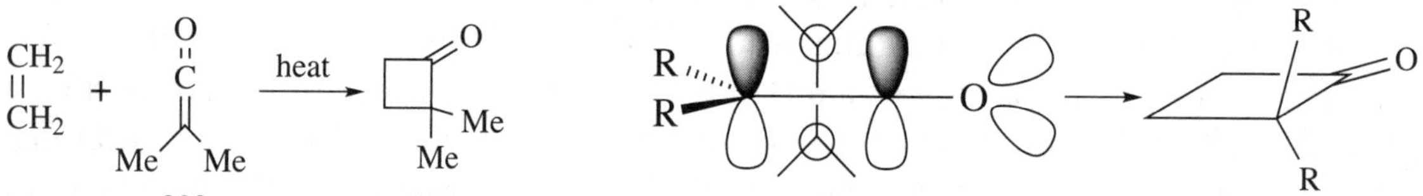

[Reprinted with permission from Tidwell, T.T. *Ketenes*, Wiley, New York, *1995*, p 487. Copyright © *1995* by John Wiley and Sons, Inc.]

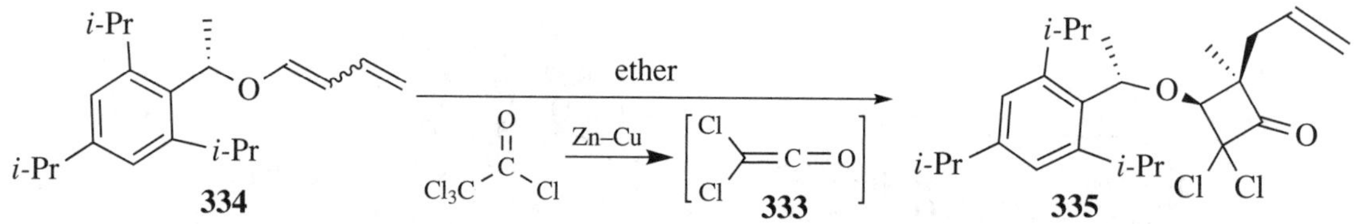

The intramolecular version of this reaction is usually more efficient than the intermolecular version. When **336** (R = Me) was treated with triethylamine (0.2% solution of acid chloride in refluxing dichloromethane), ketene **337** (R = Me) was formed and an intramolecular

304. (a) Reference 80b, pp 38-40; (b) Sustmann, R.; Ansmann, A.; Vahrenholt, F. *J. Am. Chem. Soc. 1972, 94,* 8099.
305. (a) Tidwell, T.T. *Ketenes*, Wiley, New York, *1995*, p 487; (b) Also see Reference 80b, p 39.
306. Brady, W.T. *Tetrahedron 1981, 37,* 2949.
307. For a reaction with cyclopentadiene, see Grieco, P.A. *J. Org. Chem. 1972, 37,* 2363.
308. Roche, C.; Kadlecˇiková, K.; Veyron, A.; Delair, P.; Philouze, C.; Greene, A. E.; Flot, D.; Burghammer, M. *J. Org. Chem. 2005, 70,* 8352.

[2+2]-cycloaddition gave an 80% yield of **338**.[309] Ghosez reported, however, that reaction of **336** (R = H) under the same conditions gave only 3% of **338** (R = H). Many unactivated ketones (see above) show similar poor reactivity, and to enhance the reactivity of such molecules Ghosez and co-workers developed a ketene iminium salt[310] cyclization procedure. Pyrrolidine amide **339** was treated with triflic anhydride [(CF$_3$SO$_2$)$_2$O] and collidine to generate the ketene iminium salt (**340**). Subsequent internal [2+2]-cycloaddition gave **341**, and after hydrolysis of the iminium salt product **338** was obtained in 75% yield. Ghosez and Marchand-Brynaert showed that ketene iminium salts were more electrophilic than ketenes and did not dimerize.[311]

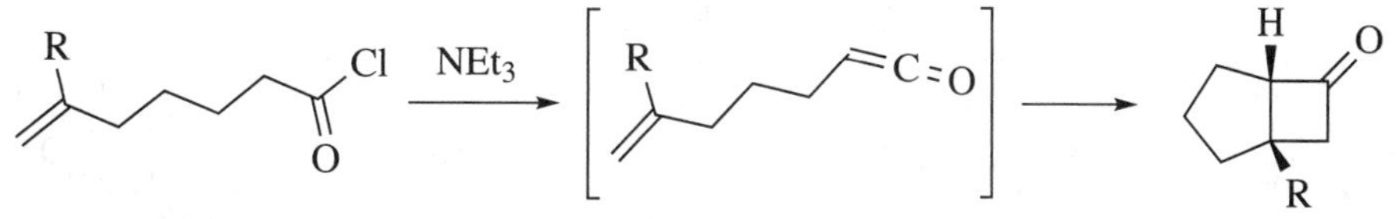

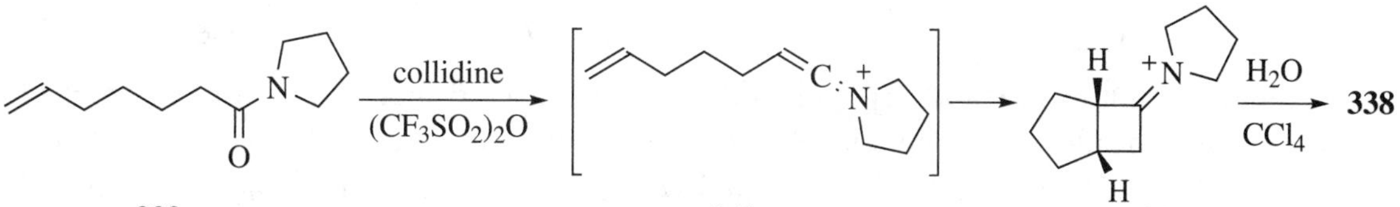

Another example of this approach is the conversion of **342** to **343** in ~ 65% yield,.[312] There are many examples of internal [2+2]-cycloaddition of **344** to **345**),[313] and Brady and Giang.[314] In both Snider's and Brady's examples, the presence of an α-alkoxy substituent led to a more facile cycloaddition. As mentioned above, vinyl ketenes also react well in the cycloaddition. For intramolecular [2+2]-cycloadditions involving vinyl ketenes, Snider defined three different reaction types: Type I (the ketene carbon has a side chain),[315] Type II (a side chain is attached α- to the ketene carbon),[316] and Type III (a side chain is attached β- to the ketene

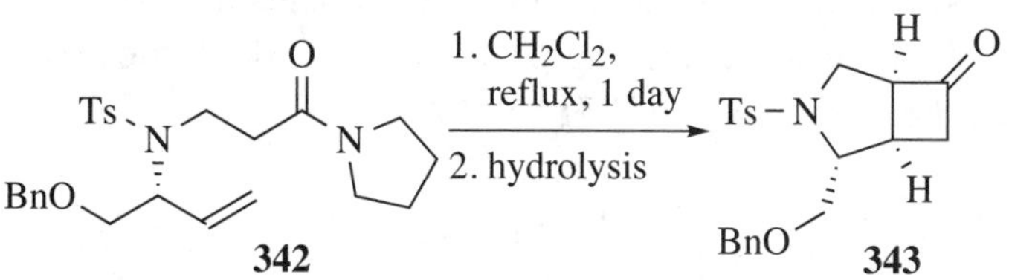

309. Markó, I.; Ronsmans, B.; Hesbain-Frisque, A.-M.; Dumas, S.; Ghosez, L. *J. Am. Chem. Soc.* **1985**, *107*, 2192.
310. Falmagne, J.B.; Escudero, J.; Taleb-Sahraoui, S.; Ghosez, L. *Angew. Chem. Int. Ed.* **1981**, *20*, 879.
311. Ghosez, L.; Marchand-Brynaert, J. *Iminium Salts in Organic Chemistry, Part I*, Böhme, J.; Viehe, H.G. (Eds), Wiley, New York, **1976**, pp 421-532.
312. Monache, G.D.; Misiti, D.; Salvatore, P.; Zappia, G.; Pierini, M. *Tetrahedron: Asymmetry* **2000**, *11*, 2653.
313. (a) Snider, B.B.; Hui, R.A.H.F.; Kulkarni, Y.S. *J. Am. Chem. Soc.* **1985**, *107*, 2194; (b) Snider, B.B.; Hui, R.A.H.F. *J. Org. Chem.* **1985**, *50*, 5167.
314. Brady, W.T.; Giang, Y.F. *J. Org. Chem.* **1985**, *50*, 5177.
315. Kulkarni, Y.S.; Burbaum, B.W.; Snider, B.B. *Tetrahedron Lett.* **1985**, *26*, 5619.
316. Kulkarni, Y.S.; Snider, B.B. *J. Org. Chem.* **1985**, *50*, 2809

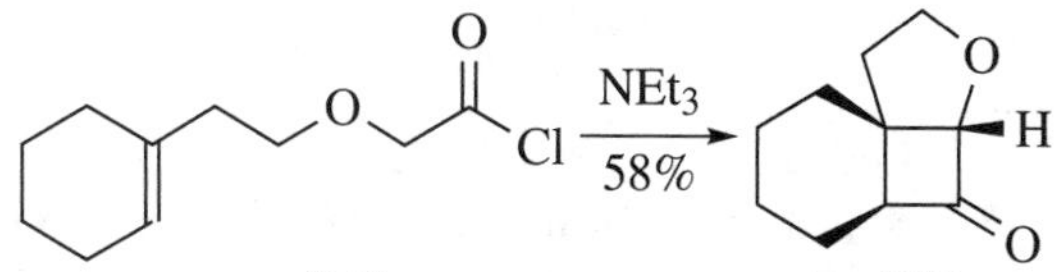

carbon).[313a] In reactions of ketenes, primarily from the work of Ghosez and co-workers,[309] Snider et al. (as in the conversion example of a Type I cyclization is the conversion of acid **346** to the ketene (**347**) and cyclization to **348** in 97% yield.[317] With vinyl ketenes, [1,5]-sigmatropic hydrogen shifts are sometimes a problem[317] (sec. 11.12.A). Snider used a Type III cyclization of this type in a synthesis of isocomene.[318] Funk et al. synthesized clovene using this methodology,[319] and Ernst and co-workers showed this method was effective for the synthesis of triquinane derivatives.[320] In another extension of this methodology, Halcomb and co-worker reported an intramolecular ketene-allene cycloaddition.[321]

Ketenes react with imines via [2+2]-cycloaddition to produce β-lactams.[322] An example is the reaction of the acid chloride of phenylacetic acid with Proton Sponge™ [1,8-bis(dimethylamino) naphthalene] to give ketene **349**, which reacted with the tosyl imine shown and a quinuclidine catalyst to give a 65% yield of β-lactam **350** in 96% ee.[323] Chiral ammonium salts derived from chiral quinidine compounds and Cinchona alkaloids have been used to catalyze the [2+2] cycloaddition of ketenes and imines, producing β-lactams with excellent enantioselectivity.[324] Ketenes are useful in many asymmetric syntheses.[325] *N*-Substituted isocyanates also undergo thermal [2+2]-cycloaddition reactions with alkenes, generating β-lactams.[326] Aldehydes react with silylketenes to give β-lactones, as in the conversion of **351** to **352**.[327]

317. Lee, S.Y.; Kulkarni, Y.S.; Burbaum, B.W.; Johnston, M.I.; Snider, B.B. *J. Org. Chem.* **1988**, *53*, 1848.
318. Snider, B.B.; Beal, R.B. *J. Org. Chem.* **1988**, *53*, 4508.
319. Funk, R.L.; Novak, P.M.; Abelman, M.M. *Tetrahedron Lett.* **1988**, *29*, 1493.
320. Veenstra, S.J.; De Mesmaeker, A.; Ernst, B. *Tetrahedron Lett.* **1988**, *29*, 2303.
321. McCaleb, K.L.; Halcomb, R.L. *Org. Lett.* **2000**, *2*, 2631.
322. For reviews of the formation of β-lactams, see (a) Brown, M.J. *Heterocycles* **1989**, *29*, 2225; (b) Isaacs, N.S. *Chem. Soc. Rev.* **1976**, *5*, 181; (c) Mukerjee, A.K.; Srivastava, R.C. *Synthesis* **1973**, 327. For a review of cycloaddition reactions of imines, see (d) Sandhu, J.S.; Sain, B. *Heterocycles* **1987**, *26*, 777.
323. Taggi, A.E.; Hafez, A.M.; Wack, H.; Young, B.; Drury III, W.J.; Lectka, T. *J. Am. Chem. Soc.* **2000**, *122*, 7831.
324. Taggi, A.E.; Hafez, A.M.; Wack, H.; Young, B.; Ferraris, D.; Lecktka, T. *J. Am. Chem. Soc.* **2002**, *124*, 6626.
325. Orr, R.K.; Calter, M.A. *Tetrahedron* **2003**, *59*, 3545.
326. (a) Johnston, D.B.R.; Schmitt, S.M.; Bouffard, A.F.; Christensen, B.G. *J. Am. Chem. Soc.* **1978**, *100*, 313; (b) Bouffard, A.F.; Johnston, D.B.R.; Christensen, B.G. *J. Org. Chem.* **1980**, *45*, 1130.
327. Fournier, L.; Kocienski, P.; Pons, J.-M. *Tetrahedron* **2004**, *60*, 1659.

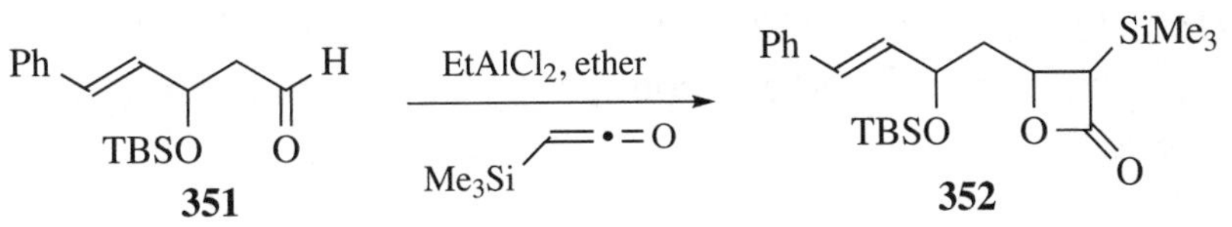

Ph ⟍⟍⟍ H   EtAlCl₂, ether   Ph ⟍⟍⟍   SiMe₃
TBSO   O   Me₃Si ⟍=•=O   TBSO   O
351   **352**

Thermal [2+2]-cycloadditions lead to the following disconnections:

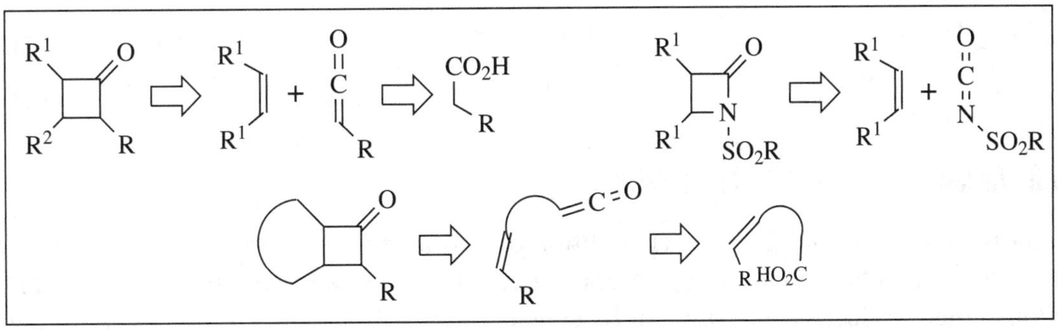

The intramolecular ketene-alkene cycloaddition has also been used to prepare highly substituted phenols. As originally formulated, a cyclobutenone such as **353** was converted to a phenol derivative (**357**). When **353** was heated in benzene, ketene **354** was formed. When done in the presence of an alkyne,[328] an *in situ* [2+2]-cycloaddition led to a new cyclobutenone (**355**), but a four-electron electrocyclic cleavage reaction occurred under the reaction conditions to give **356**. This compound also reacted, via a six-electron electrocyclic ring closure to form a six-membered ring, and tautomerization gave the aromatic final product, phenol **357**.

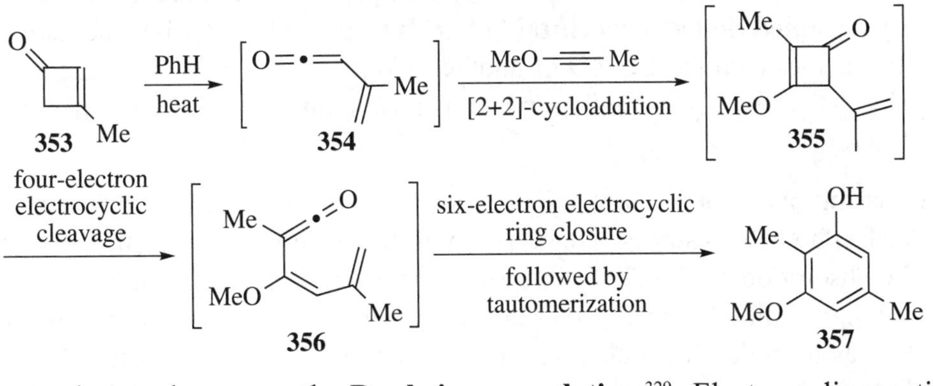

353 PhH/heat → **354** MeO≡Me [2+2]-cycloaddition → **355**

four-electron electrocyclic cleavage → **356** six-electron electrocyclic ring closure followed by tautomerization → **357**

This sequence is now known as the **Danheiser annulation**.[329] Electrocyclic reactions such as this are discussed in section 11.11. In a synthetic application taken from Danheiser and co-worker's synthesis of (−)-ascochlorin, alkyne **358** was photolyzed in the presence of **359**,

328. (a) Danheiser, R.L.; Gee, S.K. *J. Org. Chem.* **1984**, *49*, 1672; (b) Danheiser, R.L.; Brisbois, R.G.; Kowalczyk, J.J.; Miller, R.F. *J. Am. Chem. Soc.* **1990**, *112*, 3093.

329. Mundy, B.P.; Ellerd, M.G.; Favaloro Jr., F.G. *Name Reactions and Reagents in Organic Synthesis, 2nd ed.*, Wiley-Interscience, New Jersey, **2005**, pp. 196-197

and then refluxed in toluene to give **360** in 65% yield.[330] Resorcinols[331] have been prepared by this method. Liebeskind and co-workers reported a related method for the synthesis of highly substituted quinones.[332]

11.10.B. General Principles of Photochemistry

It is known that a normal [2+2]-cycloaddition of two alkene moieties does not occur readily under thermal reaction conditions. In one sense, therefore, the thermal [2+2]-cycloaddition reactions noted in Section 11.10.A can be considered highly useful special cases. In contrast to the thermal reaction, photochemically induced cycloaddition occurs with great ease. In the photochemical [2+2]-cycloaddition, promotion of an electron from the populated HOMO of the alkene to the unpopulated LUMO to generate an excited state. Fleming's generic orbital pictures[333] can be used in a simple illustration of this process. The HOMO and LUMO for substituted alkenes is shown in Figure 11.2. For a reaction to occur between two alkenes, the HOMO of one alkene must react with the LUMO of the second, but the $HOMO_{alkene}$-$LUMO_{alkene}$ interaction is symmetry forbidden. When a photon of light, is absorbed by an alkene, an electron is promoted from the filled HOMO to the empty LUMO. The higher energy orbital is now a **singly occupied molecular orbital** (a **SOMO**, see section 2.4.B), and is of the correct symmetry to interact with the LUMO of another alkene. In other words, reaction occurs by the $LUMO^{*}_{alkene}$-$LUMO_{alkene}$ interaction, which is now symmetry allowed (where * represents the excited state).

Before discussing photochemical reactions, some general principles of photochemistry will be introduced. Photochemistry is "concerned with the chemical change that may be brought about by the absorption of light".[334] Chemical change is not a requirement, however, since **fluorescence** (light emitted from a species that has absorbed radiation) and **chemiluminescence** (light emitted as a product of a chemical reaction) can occur.[334] A molecule can absorb a photon of light, and the photoactivated species can undergo [2+2]-cycloaddition or a variety of other reactions (sec. 13.4).

---

330. Dudley, G.B.; Takaki, K.S.; Cha, D.D.; Danheiser, R.L. *Org. Lett.* ***2000***, *2*, 3407.
331. Danheiser, R.L.; Nishida, A.; Savariar, S.; Trova, M.P. *Tetrahedron Lett.* ***1988***, *29*, 4917.
332. Liebeskind, L.S.; Iyer, S.; Jewell, Jr., C.F. *J. Org. Chem.* ***1986***, *51*, 3065.
333. Reference 7, pp 128, 219.
334. Wayne, R.P. *Principles and Applications of Photochemistry,* Oxford Univ. Press, Oxford, ***1988***, p 1.

Application of light increases the energy of a molecule according to

$$E = h\nu \qquad \text{where} \qquad \nu = \frac{c}{\lambda} \qquad\qquad 11.5$$

where h is Planck's constant (6.63×10^{-34} J sec), c is the speed of light (3×10^8 m sec^{-1}), ν is its frequency in sec^{-1}, and λ is the wavelength of the light in nm, which is inversely proportional to frequency. As E is usually expressed in kcal mol^{-1}, the molecular value of E calculated from Eqn 11.5 must be multiplied by Avogadro's number, $N = 6.02 \times 10^{23}$ mol^{-1}. This expression gives $E = 2.86 \times 10^4/\lambda_{nm}$ kcal mol^{-1}, including conversion factors. An energy of 1 kcal mol^{-1} corresponds to a wavelength (λ) of 2.86×10^5 Å, and in general:[335]

$$1 \text{ kcal mol}^{-1} = \frac{2.86 \times 10^5 \text{ kcal mol}^{-1}}{\lambda \text{ [Å]}} \quad \text{and} \quad \lambda = 2.86 \times 10^5 \text{ Å} \qquad\qquad 11.6$$

This wavelength corresponds to a wavenumber (ν) given by

$$\nu = \frac{c}{\lambda} = \frac{10_8 \text{ Å cm}^{-1}}{2.86 \times 10^5 \text{ Å}} = 353 \text{ cm}^{-1}, \text{ also } = 1 \text{ kcal mol}^{-1} \qquad\qquad 11.7$$

Therefore, 1 kcal mol^{-1} = 2.86×10^5 Å = 353 cm^{-1}, which is in the infrared (IR) region of the electromagnetic spectrum. An energy of 10 kcal mol^{-1} corresponds to 3530 cm^{-1}, also in the IR region. Light of wavelength 2000 Å is at the lower end of the UV region (see Figure 11.15) and corresponds to 143 kcal mol^{-1}. Light in the UV and visible region can excite molecules to higher electronic states. Wavelengths in the UV are usually expressed in nanometers (1 nm = 1 mμ = 1×10^{-7} cm), and the UV region is 200-400 nm.

		Violet	Red			
UV	Near UV	Vis.		IR		

10	200	400	800	2860	28600	ν (mμ = nm)
2860	143	71.5	35.75	10	1	E (kcal mol^{-1})
11972	598.6	299.3	149.7	41.86	4.186	E (KJ mol^{-1})
100	2000	4000	8000	2.86×10^4	2.86×10^5	λ (Å)
1000×10^3	50×10^3	25×10^3	12.5×10^3	3530	353	ν (cm^{-1})

Figure 11.15. A Portion of the Electromagnetic Spectrum.

How does one know what wavelength of light to use for a particular molecule? Each functional group absorbs light according to its individual structure. If that absorption is in the UV or the IR, a UV spectrum or an IR spectrum can be obtained to determine the maximum wavelength of

335. DePuy, C.H.; Chapman, O.L. *Molecular Reactions and Photochemistry*, Prentice-Hall, Englewood Cliffs, New Jersey, *1972*, p 5.

light absorbed. Ultraviolet light is probably used most often, and the wavelength of maximum absorption (λ_{max}) is a good guide as to which wavelength of light should be chosen for irradiation. Table 11.9[336] shows the energy of absorption in the UV (λ_{max}) for several common types of organic molecules, along with the energy in kcal mol^{1} and kJ mol^{-1}.

Table 11.9. Absorption Maxima for Representative Molecules and Functional Groups

Molecule	Transition	λ_{max} (nm)	E (kcal mol^{-1})	E (kJ mol^{-1})
C_4H_9I	n → σ*	224	127.7	534.6
$CH_2=CH_2$	π → π*	165	173.3	725.4
HC≡CH	π → π*	173	165.3	691.9
acetone	π → π*	150	190.7	798.3
	n → σ*	188	152.1	636.7
	n → π*	279	102.5	429.1
$CH_2=CHCH=CH_2$	π → π*	217	131.8	551.7
$CH_2=CHCHO$	π → π*	210	136.2	570.1
	n → π*	315	90.8	380.1
benzene	π → π*	180	158.9	665.2
		200	143.0	598.6
		255	112.2	469.7

Functional Group	λ_{max} (nm)	E (kcal mol^{-1})	E (kJ mol^{-1})
RCH=CHR	165	173.3	725.4
	193	148.2	620.4
RC≡CR	173	165.3	691.9
R(C=O)R'	188	152.1	636.7
	279	102.5	429.1
RCHO	290	98.6	412.7
RCOOH	<208	<137.5	<575.6
-C≡N	<160	<178.8	<748.5
-N=N-	347	82.4	344.9
$R_2C=N-$	190	150.5	630.0

[Reprinted with permission from Silverstein, R.M.; Bassler, G.C.; Morill, T.C. *Spectrophotometric Identification of Organic Compounds*, 4th Ed., Wiley, New York, *1981*, pp. 308 and 312. Copyright © *1981* by John Wiley and Sons, Inc.]

From this table it is apparent that organic molecules absorb energy in the region of ~ 80-200 kcal mol^{-1}, easily accessible by exposure of the molecule to a UV light source. Some molecules can react with far less amounts of energy, undergoing photochemical transformations on exposure to visible light or even IR light.

The wavelength of light absorption and corresponding energy for several common functional

336. Silverstein, R.M.; Bassler, G.C.; Morill, T.C. *Spectrophotometric Identification of Organic Compounds, 4th Ed.*, Wiley, New York, *1981*, pp 308 and 312.

groups is also shown in Table 11.10,[337] along with the bond dissociation energies of that functional group. This table illustrates that the wavelength of light used for irradiation closely matches the amount of energy necessary to rupture a given bond (bond dissociation energy). In this section, the main focus will be on using alkenes and carbonyls in [2+2]-photocycloaddition reactions, where the functional groups generally absorb in the 200-300 nm range (140-90 kcal mol^{-1}, 586-377 kJ mol^{-1}). Many light sources will generate light in this region, usually as a broad band of several wavelengths, although it is easy to obtain a light source where one or two intense wavelengths predominate. A low-pressure sodium lamp, for example, generates light of primarily 589 nm (48.6 kcal mol^{-1}, 203.3 kJ mol^{-1}). Mercury lamps usually generate light at $\approx$ 253.7 nm (112.7 kcal mol^{-1}, 471.9 kJ mol^{-1}). A low pressure mercury lamp tends to give only this wavelength whereas a high-pressure mercury lamp gives several other intense lines as well.[338] If a molecule is irradiated with a light source generating several intense wavelengths of light, other functional groups in a molecule may absorb the light and react, including the photoproduct. A convenient method for limiting the wavelength of light (and thereby the energy of irradiation) is to filter out higher energy wavelengths (and lower energy if necessary). A filter is inserted between the light source and the reactants and sometimes the reaction vessel, made out of Pyrex™ glass or quartz glass, is the filter. The filter may be attached to the light source. Table 11.11[339,340] shows several common filters and their approximate range of light transmission.

Table 11.10. Bond Dissociation Energies and Spectral Regions for Irradiation

Bond	$\underline{E^a}$ (kcal mol^{-1})	$\underline{E^a}$ (kJ mol^{-1})	Wavelength (nm)	Spectral Region	$\underline{E}$ (kcal mol^{-1})	$\underline{E}$ (kJ mol^{-1})
C=O	145	607.0	200		140	586.0
O–H	118	494.0	250		110	460.5
C=C	102	427.0	250		110	460.5
C–H	94	393.5	300		90	376.7
S–H	87	364.2	350		80	334.9
C–O	69	288.8	400	Visible, violet	70	293.0
S–S	63	263.7	400		70	293.0
C–C	58	242.8	50	Visible, blue	62	259.5
			500	Visible, green	54	226.0
			550	Visible, yellow	50	209.0
^{a} Bond Dissociation Energy			600	Visible, red	48	200.9
			650	Visible, red	44	184.2
			1000		29	121.4

337. Johnson, E. in *The Physiology and Pathophysiology of the Skin*, Vol. 8 of *The Photobiology of the Skin: Lasers and the Skin*, Jarrett, A. (Ed.), Academic Press, *1984*, p 2401.
338. Calvert, J.G.; Pitts Jr., J.N. *Photochemistry*, Wiley, New York, *1966*, p 474.
339. Murov, S.L. *Handbook of Photochemistry*, Marcel Dekker, New York, *1973*, pp 107-108.
340. Reference 338, p 748.

Table 11.11. Photochemical Filters and Their Effective Transmittance Range

Filter	Range (nm)[a]	Range (kcal mol^{-1})	[kJ mol^{-1}]	E_{max}, nm	(kcal mol^{-1})	[kJ mol^{-1}]
Quartz crystal[b]	>200	143	[598.6]	>230	124.3	[520.2]
Vycor[b]	>200	143	[598.6]	~280	102.1	[427.4]
Pyrex$^{(TM)}$[b]	>280	102.1	[427.4]	~360	79.4	[332.4]
Lead glass[b]	>290	98.6	[412.4]	~380	75.2	[314.8]
Window glass[c]	>310	92.3	[386.4]	~350	81.7	[342.0]
Schott UV R-310[c]	275-310	104 - 92.3	[435.3 - 386.4]	310	92.3	[386.4]
Schott UV R-280[c]	240-320	119.2 - 89.4	[499.0 - 374.2]	280	102.1	[427.4]
Schott UV R-250[c]	225-300	127.1 - 95.3	[532.0 - 398.9]	250	114.4	[478.9]
Suprasil[c]	>160	178.8	[748.2]	180	158.9	[665.2]
Corning 5850[c]	280→450	102.1 - 63.6	[427.4 - 266.2]	~360	79.4	[332.4]
Corning 5970[c]	310→420	92.3 - 68.1	[386.4 - 285.1]	~360	79.4	[332.4]

[a] Estimated at 50% transmittance from an absorption curve   [b] reference 340   [c] Reference 339

[Reprinted in part with permission from Calvert, J.G.; Pitts Jr., J.N. *Photochemistry*, Wiley, New York, *1966*, pp. 748. Copyright © *1966* by John Wiley and Sons, Inc.]

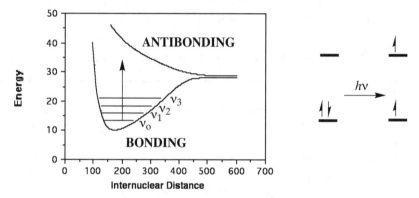

Figure 11.16. Energy Transition in an Electronically Excited Molecule. [C.H. Depuy/O.L. Chapman, MOLECULAR REACTIONS AND PHOTOCHEMISTRY, ©1972, pp. 6. Reprinted by permission of Prentice Hall, Englewood Cliffs, New Jersey.]

When a molecule absorbs a photon of light, an electron is promoted from the bonding molecular orbital to an anti-bonding molecular orbital ($n \rightarrow \sigma^*$, $n \rightarrow \pi^*$, or $\pi \rightarrow \pi^*$) generating a SOMO. Figure 11.16 illustrates this energy transition.[341] There are several ways a molecule can dissipate excess energy from an excited state,[342] as shown in a Jablonski diagram (Figure 11.17). When the excess energy is lost by emission of light (**radiative deactivation**), the process is termed **fluorescence** (occurs with spin conservation) or **phosphorescence** (occurs with spin inversion). If the deactivation is nonradiative, the transition can occur between two states of the same spin

---

341. Reference 335, p 6.
342. Margaretha, P. *Preparative Organic Photochemistry*, Lehn, J.-M. (Ed.), Springer-Verlag, Berlin, *1982*, pp 6, 7.

(**internal conversion**) or with inversion of spin (**intersystem crossing**). The excess energy is liberated as heat in both cases. Deactivation must occur in such a way that there is no motion of the atoms during the electronic transition (the **Franck-Condon principle**). Transition occurs when the geometries of the excited and ground state are similar as an absorption (solid curves in Figure 11.18),[342] or an emission (dashed curves in Figure 11.18).[342]

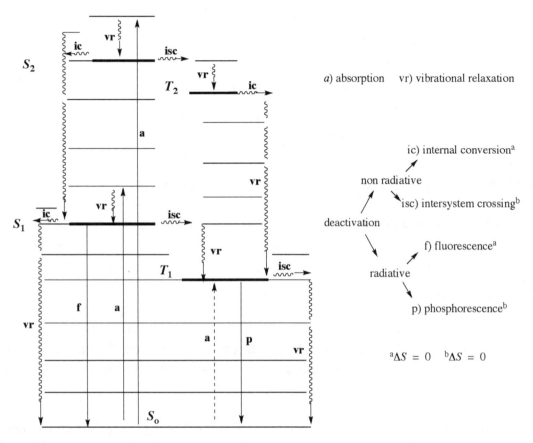

Figure 11.17. Jablonski Diagram.

In Figure 11.18, (I) represents the 0→0 transition, which is the most intense for emission and absorption (most probable transition). In Figure 11.18, (II) represents a case where the excited and ground states have different geometries, and the most intense transitions are 0→2' and 0'→2 (the 0→0 transition is indicated as a point of reference for case II). In both cases, the absorption and emission spectra are approximately mirror images. Therefore, it is possible to estimate the energy difference (ΔE) between the excited and ground states.

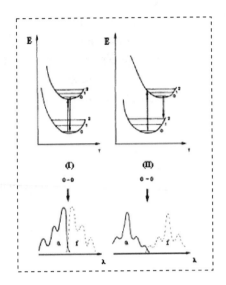

Figure 11.18. Correlation Between Absorption (a) and Emission (Fluorescence, f) Spectrum.

The excited state can also interact with another molecule in a bimolecular process[343] via energy transfer

$$A^* + B \longrightarrow B^* + A \qquad \qquad 11.8$$

or via exciplex formation

$$A^* + B \longrightarrow A^* : B \qquad \qquad 11.9$$

Energy transfer occurs when ΔE_{A^*A} is $> \Delta E_{BB^*}$. In this case, an excited molecule in a triplet state can transfer energy to the other molecule which is in the singlet state:

$$^3B_1 + {}^1A_0 \longrightarrow {}^1B_2 + {}^3A_1 \qquad \qquad 11.10$$

This process is important if the rate constant of intersystem crossing $k_{si} \rightarrow_{tr}$ for a given molecule is small, or if direct irradiation of these molecules is inconvenient. This process is the foundation of a sensitized reaction.

In many cases, photon absorption is not facile and a sensitizer (S*) is added to facilitate electron transfer between two molecules,[344] as shown for the conversion of D-Y to D• and A-X to A•.[345] In these reactions,

$$S^* + D\text{-}Y \longrightarrow \overset{\bullet}{S^-} + D\bullet + Y^+$$

$$S^* + A\text{-}X \longrightarrow \overset{\bullet}{S^+} + A\bullet + X^-$$

343. Reference 342, p 8.
344. Eaton, D.F. *Pure App. Chem.* **1984**, *56*, 1191.
345. Lazár, M.; Richly´, J.; Klimo, V.; Pelikán, P.; Valko, L. *Free Radicals in Chemistry and Biology*, C.R.C. Press, **1989**, p 22.

D–Y is an electron-rich species that generates the **sensitizer** (a molecule that donates an electron to S). Reductive fragmentation (S donates an electron to A–X) allows fragmentation of the electron poor (acceptor) species. As seen in Figure 11.19,[346] the use of a sensitizer involves transfer of excitation energy from an electronically excited molecule (the sensitizer) to the ground state of another molecule, usually providing a triplet excited state. Singlet excitation energy can be transferred, but the lifetime of single excited states is remarkably short (10^{-8} vs. $>10^{-6}$ s for triplets).[346] Triplet energy transfer requires that the triplet energy of the donor be higher than that of the acceptor molecule by 3 kcal mol^{-1} (12.6 kJ mol^{-1}) or more. Benzophenone is a common triplet sensitizer. The electron transfer occurs by intersystem crossing ($S_1 \rightarrow T_1$, where S is the singlet and T is the triplet), which is spin forbidden. Nonetheless, it occurs with great efficiency in molecules such as benzophenone. These processes are illustrated in Figure 11.19.[346]

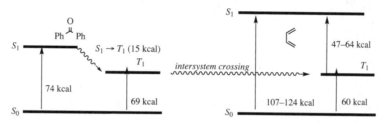

Figure 11.19.   Singlet to Triplet Intersystem Crossing For Sensitized Reactions.   [C.H. Depuy/O.L. Chapman, MOLECULAR REACTIONS AND PHOTOCHEMISTRY, ©1972, pp. 36. Reprinted by permission of Prentice Hall, Englewood Cliffs, New Jersey.]

In the photoinduced electrocyclization reaction of 1,3-butadiene, cyclobutene and bicyclobutane (**361**) are produced via the singlet diene. Photolysis in the presence of triplet excited benzophenone gave the dimers (**362** and **363**) via the triplet diene.[347] Irradiation at 366 nm

361   **362**   **363**

[C.H. Depuy/O.L. Chapman, MOLECULAR REACTIONS AND PHOTOCHEMISTRY, ©1972, pp. 37. Reprinted by permission of Prentice Hall, Englewood Cliffs, New Jersey.]

(78.1 kcal mol^{-1}, 327.1 kJ mol^{-1}) of a mixture of butadiene and benzophenone led to exclusive absorption by benzophenone. The triplet energy of benzophenone (69 kcal mol^{-1}, 288.8 kJ mol^{-1}) is transferred with 100% efficiency via intersystem crossing ($S_1 \rightarrow T_1$), due to this small energy gap.[348] Triplet butadiene is ≈ 60 kcal mol^{-1} (251.2 kJ mol^{-1}) above the ground state, and the 69 kcal mol^{-1} (288.8 kJ mol^{-1}) of benzophenone is more than adequate for diffusion

---

346.   Reference 335, p 36.
347.   Reference 335, p 37.
348.   Reference 335, p 38

controlled energy transfer to butadiene, as shown on the Jablonski diagram in Figure 11.19.[346] Since the $S_0 \rightarrow S_1$ energy gap of benzophenone is lower than the $S_0 \rightarrow S_1$ transition for butadiene, transfer between singlet benzophenone and singlet butadiene does not occur.[346] Benzophenone is one of many commonly used sensitizers. Others include acetophenone and naphthalene. More exotic sensitizers are known and used, but rose bengal (**364**), chlorophyll a (**365**) and methylene blue (**366**) are readily available and commonly used.

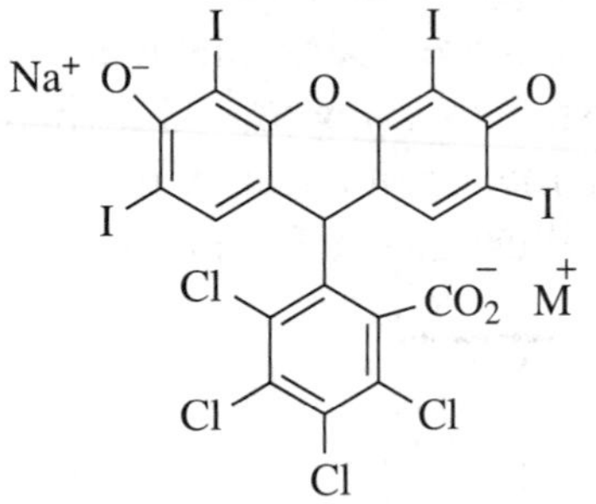

364 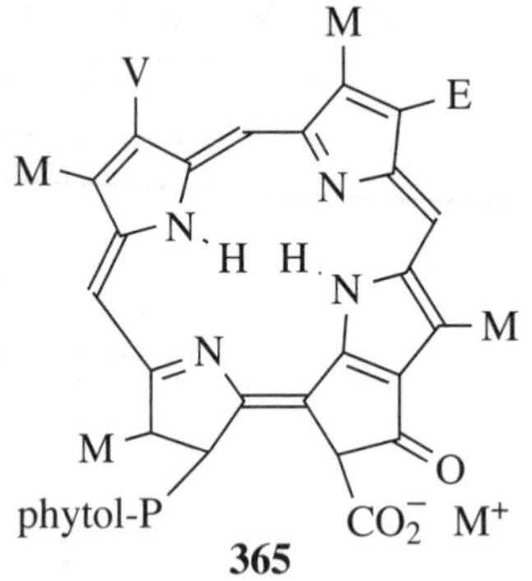   **365** 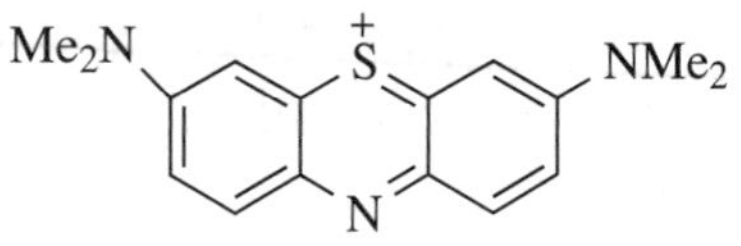   **366**

In photochemical reactions, it is important to measure the efficiency of the photolysis, which is done by the parameter known as **quantum yield** (or **quantum efficiency**) defined as the number of molecules reacted or formed per light quanta absorbed.[349] The quantum yield for formation of a product (Φ_{form}) is

$$\Phi_{form} = \frac{\text{No. of molecules of product formed}}{\text{No. of quanta absorbed}} \qquad 11.11$$

Quantum yield can also be expressed in terms of the number of molecules of starting material (SM) which disappear per quantum of light (Φ_{dis}):[350]

$$\Phi_{dis} = \frac{\text{No. of molecules of SM destroyed}}{\text{No. of quanta absorbed}} \qquad 11.12$$

For photochemical reactions that give a single product, $\Phi_{form} = \Phi_{dis}$. If several products are formed, however, $\Phi_{dis} = \Sigma\,\Phi_{form}$. If a reaction does *not* go through a chain mechanism, Φ will have a value between 0 and 1. Free radical chain mechanisms have quantum yields as high as several thousand,[292] which reflects the photochemical initiation step leading to a radical chain carrier molecule or molecules.

---

349.   Reference 335, p 20.
350.   Reference 335, p 76.

11.10.C. Photochemical [2+2]-Cycloadditions

[2+2]-Photocycloaddition reactions are quite useful in synthesis, and variations include the cycloaddition of alkenes, alkynes or conjugated carbonyl compounds with other alkenes, alkynes or conjugated systems.[351] In general, photolysis of enones requires wavelengths > 3000Å (300 nm, 95.3 kcal mol^{-1}, 399.1 kJ mol^{-1}), corresponding to the n→π* transition.[352] This reaction is usually accomplished by photolysis in a Pyrex container, which is opaque to light < 300 nm. The n→π* transition for acrolein occurs at 315 nm (90.8 kcal mol^{-1} = 380.1 kJ mol^{-1}; the λ_{max} for acrolein).[353] Ethene, however, has a strong π→π* transition at ≈ 165 nm. In general, irradiation of alkene must use higher energy (165 nm, 173.3 kcal mol^{-1} = 725.4 kJ mol^{-1}) to populate the LUMO. There are countless synthetic examples. In one, taken from Sorensen and Shipe's synthesis of guanacastepene,[354] irradiation of **367** in a mixture of Hünig's base and ether led to a 76% yield of cyclobutane derivative **368**.

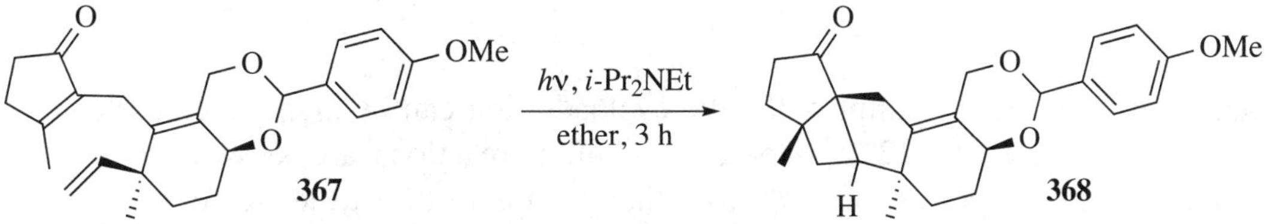

Hammond and Liu studied this type of photocyclization and showed that myrcene (**369**) cyclized to **370** in the presence of a benzophenone sensitizer.[355] Direct (unsensitized) photocyclization of **369** gave primarily **371**, and a small amount of β-pinene (**372**).[356] Photolysis in the presence of the benzophenone sensitizer leads to a triplet diradical intermediate, whereas the direct photolysis generated an excited diene that does not cross to the triplet manifold (see above) prior to cyclization.

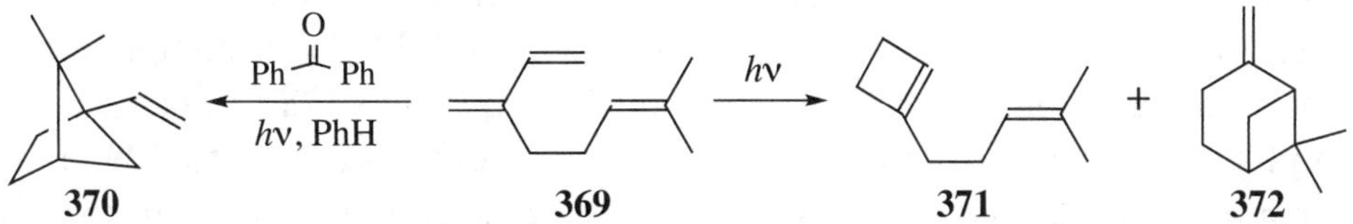

There are many examples of photochemical [2+2]-cycloadditions involving enones and alkenes, or alkenes with alkenes. Solvent-controlled intramolecular [2+2] photocycloadditions of

351. For a review of intermolecular reactions in synthesis, see Bach, T. *Synthesis* **1998**, 683.
352. Eaton, P.E. *Acc. Chem. Res.* **1968**, *1*, 50.
353. Reference 335, p 308.
354. (a) Shipe, W.D.; Sorensen, E.J. *Org. Lett.* **2002**, *4*, 2063; (b) Shipe, W.D.; Sorensen, E.J. *J. Am. Chem. Soc.* **2006**, *128*, 7025.
355. Liu, R.S.H.; Hammond, G.S. *J. Am. Chem. Soc.* **1964**, *86*, 1892.
356. Crowly, K.J. *Proc. Chem. Soc.* **1962**, 245, 334.

α-substituted enones have been studied.[357] The conversion of **367** to **368** is an example of the reaction with conjugated carbonyl compounds, and an example of the latter is taken from Hsung and co-worker's synthesis of rhododaurichromanic acid A,[358] in which **373** was photolyzed to give a 79% yield of **374** as a 1:1 mixture of isomers (at methyl). Williams and Callahan used this reaction in a synthesis of shyobunone,[359] and Wender and Lechleiter synthesized 10-epijuneol by a similar route.[360] It should be noted that selectivity may be an issue for intermolecular reactions, and a cyclohexenone derivative reacting with 1- or 2-butene will give a mixture of diastereomers.

It is apparent from these examples that the cycloaddition can be highly stereoselective, and asymmetric intramolecular [2+2]-photocycloaddition reactions are known. One example used chiral α- and β-hydroxy acids[361] as tether groups to direct the stereochemistry of the cycloaddition. The relative stereochemistry observed in many [2+2]-photocycloadditon reactions generally corresponds to a suprafacial addition for both alkene fragments.[362]

For intermolecular reactions, free rotation around the original carbon-carbon double bond can take place in the excited state of the molecule, although not in the intramolecular reaction that formed **374**. Because of such rotation, photocyclization of 2-butene leads to four tetramethylcyclobutane isomers (**375-378**).[362] Similar results are observed when unsymmetrical alkenes are photocyclized. Selectivity is improved with stereochemically pure alkenes. Reaction of pure *cis*-2-butene gave only **375** and **376** with complete retention of configuration. Similarly, pure *trans*-2-butene gave only **377** and **378**. In general, reaction of an alkene and a conjugated carbonyl derivative gives a derivative such as **379**, and coupling of two nonconjugated alkenes gives **380**. Despite the selectivity shown in the reaction of 2-butene, the relative proportions of cis and trans cycloadduct may depend on the reaction conditions. Eaton[352,363] showed that irradiation of cyclopentene gave a mixture of **381** (the head-to-head dimer) and **382** (the head-to-tail dimer). Both the molarity of cyclopentene and the solvent were varied (see Table 11.12).[352,263] The head-to-tail cycloadduct (**382**) was the

357. Ng, S. M.; Bader, S. J.; Snapper, M. L. *J. Am. Chem. Soc.* **2006**, *128*, 7315.
358. Kurdyumov, A.V.; Hsung, R.P.; Ihlen, K.; Wang, J. *Org. Lett.* **2003**, *5*, 3935.
359. Williams, J.R.; Callahan, J.F. *J. Chem. Soc. Chem. Commun.* **1979**, 404.
360. Wender, P.A.; Lechleiter, J.C. *J. Am. Chem. Soc.* **1978**, *100*, 4321.
361. Faure, S.; Piva-Le-Blanc, S.; Bertrand, C.; Pete, J.-P.; Faure, R.; Piva, O. *J. Org. Chem.* **2002**, *67*, 1061.
362. Yamazaki, H.; Cvetanović, R.J. *J. Am. Chem. Soc.* **1969**, *91*, 520.
363. Eaton, P.E. *J. Am. Chem. Soc.* **1962**, *84*, 2344, 2454.

major product in all cases, but the selectivity varied with reaction conditions. The ratio of dimers was independent of the method of triplet generation, independent of the concentration of the sensitizer and quencher and was unaffected by the presence of excess reactant. The solvent had an effect, and the concentration effect was more pronounced in nonpolar solvents. The regiochemistry of the cycloaddition was controlled by the orbital coefficients.[364] The interaction of the largest coefficients predicted the regiochemistry, and chemoselectivity can also be predicted.

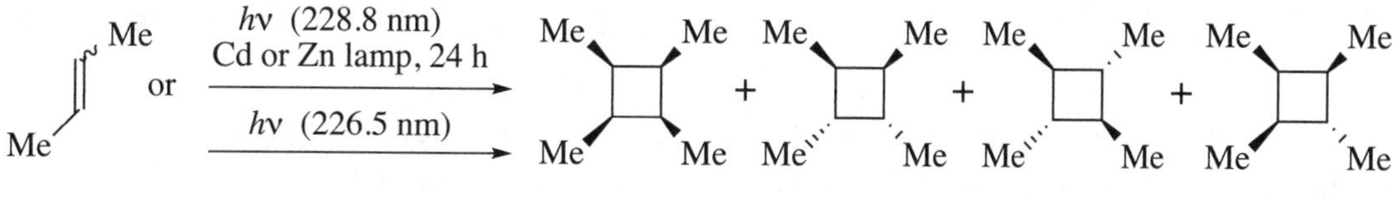

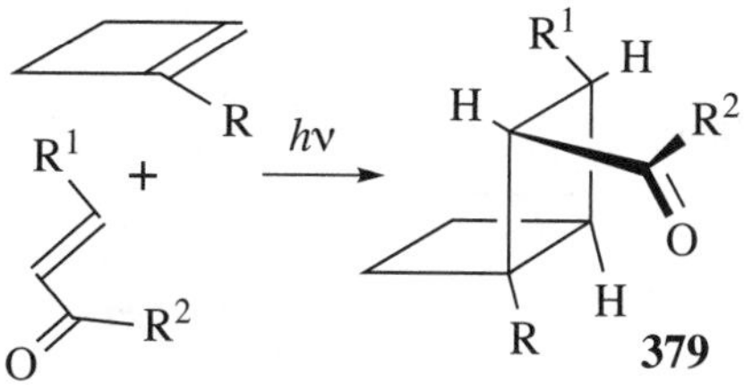

Table 11.12. Influence of Solvent and Concentration on the Photocyclization of Cyclopentenone

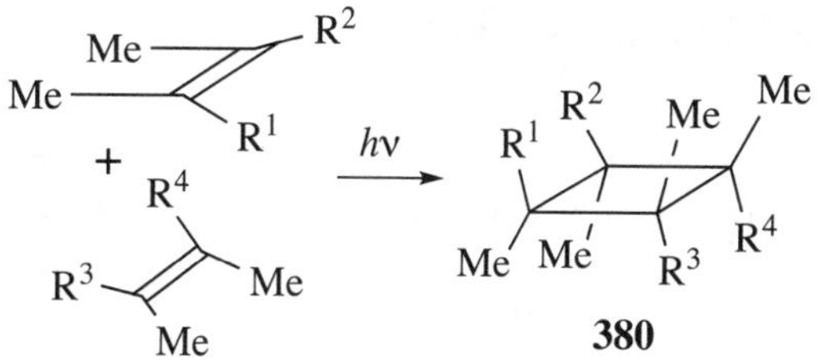

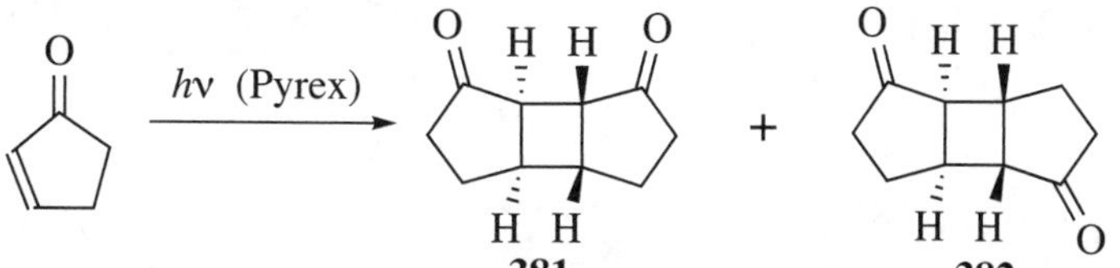

Solvent	Molarity	370:371	Solvent	Molarity	370:371
neat	12.4	43 : 57	t-BuOH	1.0	24 : 76
PhH	10.0	41 : 59	EtOAc	1.0	30 : 70
	5.0	35 : 65	dimethyl 1,4-dcyclohexane		
	2.0	27 : 73	dicarboxylate	1.0	33 : 67
	1.1	23 : 77	MeOH	10.5	41 : 59
	0.093	15 : 85		1.0	41 : 59
PhH-cyclopentanone	1.0	27 : 73		0.1	39 : 61

[Reprinted with permission from Eaton, P.E. *Accts. Chem. Res.* **1968**, *1*, 50. Copyright © *1968* American Chemical Society.]

---

364.   Servé, P.; Rosenberg, H.M.; Rondeau, R. *Can. J. Chem.* **1969**, *47*, 4295.

Intramolecular alkene cycloadditions can be used to generate quaternary centers and for construction of highly bridged molecules. We saw examples in the photocyclization of **367** and of **373**. Another example is taken from Sulikowski and Dohroh's synthesis of the tetracyclic core of bielschowskysin A, in which conjugated lactone **383** was photolyzed to give a 50% yield of **384**.[365] Note that the stereochemistry of the spirocyclic lactone is determined by bond rotation in triplet biracial intermediates.

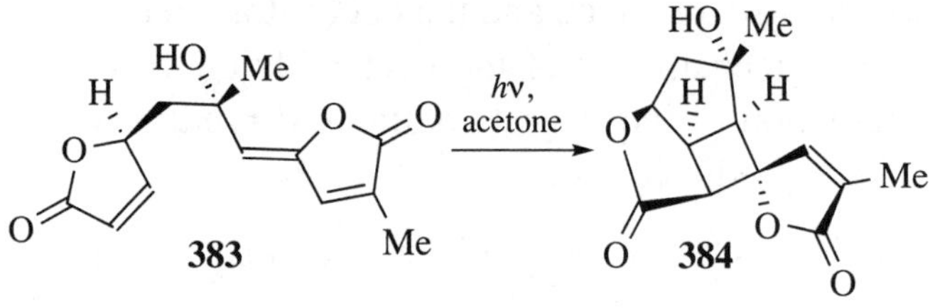

Electrocyclic ring closure (see section 11.11) can occur in diene systems, even with small rings. Photolysis of 2-pyridone **385**, for example, gave **386**. Unmasking the enol with fluoride gave the carbonyl,[366] and reduction with cyanoborohydride (sec. 4.5.D) gave the interesting and useful β-lactam **387**.

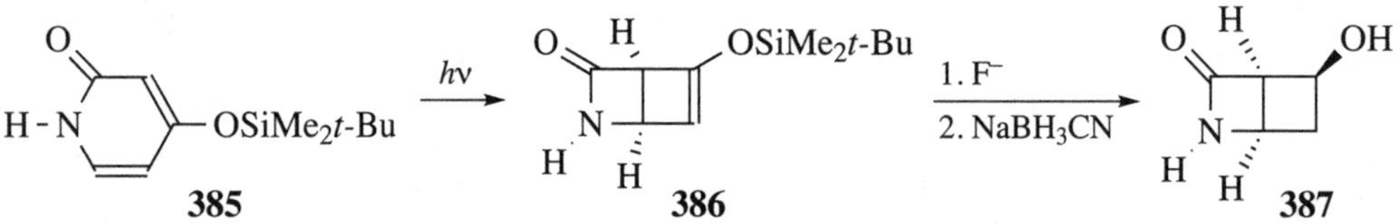

[2+2]-photocycloaddition and electrocyclic reactions of alkenes allow the following disconnections:

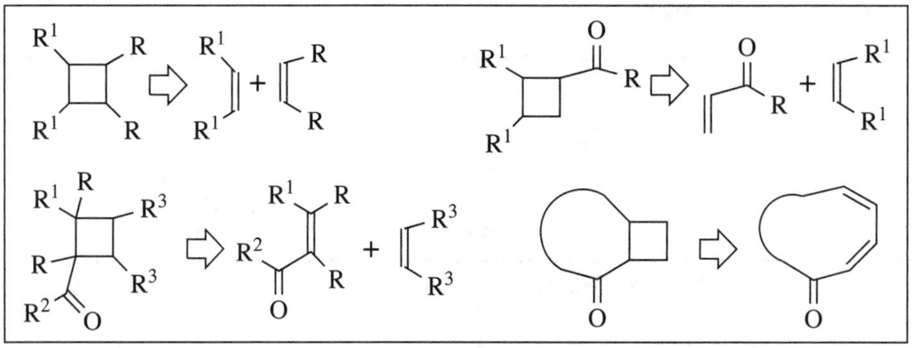

11.10.D. The Paternò-Büchi Reaction

[2+2]-Photocyclization is not limited to two alkene partners, and one important reaction involves cycloaddition of a carbonyl and an alkene. This reaction was discovered by Paternò

365. Doroh, B.; Sulikowski, G.A. *Org. Lett.* **2006**, *8*, 903.
366. (a) Sato, M.; Katagiri, N.; Muto, M.; Haneda, T.; Kaneko, C. *Tetrahedron Lett.* **1986**, *27*, 6091; (b) Kaneko, C.; Katagiri, N.; Sato, M.; Muto, M.; Sakamoto, T.; Saikawa, S.; Naito, T.; Saito, A. *J. Chem. Soc. Perkin Trans. 1* **1986**, 1283.

and Chieffi in 1909[367] and expanded by Büchi et al.;[368] thus it is called the **Paternò-Büchi reaction**.[369] The product is an oxetane (**389**), and the reaction is believed to proceed via a diradical intermediate (**388**). The regioselectivity of the reaction can be explained by molecular orbital considerations, however. Electron- rich carbonyls of aldehydes or ketones

388 389

can be used. The alkene moiety is usually electron rich but this is not a requirement. A synthetic example that also illustrates the intramolecular version of the reaction is Greaney and co-worker's synthesis of merrilactone A, in which **390** was photolyzed to give a 93% yield of oxetane **391** where the ketone carbonyl cyclized to the C=C unit of the conjugated ketone moiety.[370] Irradiation of aldehydes such as this often uses light at 290 nm (98.6 kcal mol^{-1}; 412.8 kJ mol^{-1}), generally corresponding to the n→π* transition of the carbonyl. Lower energy irradiation can also be used, however, such as laser irradiation in the visible region (445-515 nm, 64.3-55.5 kcal mol^{-1}, 269.2-232.3 kJ mol^{-1}).[371] Intramolecular Paternò-Büchi reactions are used in other natural product synthesis.[372]

This reaction can be done in the presence of a rather wide range of functionality. It has been used with aromatic compounds such as 2-furylmethanol derivatives, as in the reaction of **392**[373] with benzophenone to give a 65% yield of **393** along with a 20% yield of **394**. The photocycloaddition product of the Paternò-Büchi reaction with these substrates is an acetal or a ketal, and it can be hydrolyzed to produce acyl derivatives or other functionality. This synthetic route is attractive since achiral fragments generate cycloadducts with several contiguous chiral centers whose relative stereochemistry can usually be predicted with great accuracy.

367.   Paternò, E.; Chieffi, C. *Gazz. Chim. Ital.* *1909*, *39*, 341.
368.   Büchi, G.; Inman, C.G.; Lipinsky, E.S. *J. Am. Chem. Soc.* *1954*, *76*, 4327.
369.   (a) Reference 65, pp 912-913; (b) The Merck Index, 14th ed., Merck & Co., Inc., Whitehouse Station, New Jersey, *2006*, p ONR-69; (c) Mundy, B.P.; Ellerd, M.G.; Favaloro Jr., F.G. *Name Reactions and Reagents in Organic Synthesis, 2nd ed.*, Wiley-Interscience, New Jersey, *2005*, pp. 484-485.
370.   Iriondo-Alberdi, J.; Perea-Buceta, J.E.; Greaney, M.F. *Org. Lett.* *2005*, *7*, 3969.
371.   Adam, W.; Kliem, U.; Mosandl, T.; Peters, E.M.; Peters, K.; von Schnering, H.G. *J. Org. Chem.* *1988*, *53*, 4986.
372.   de la Torre, M.C.; García, I.; Sierra, M.A. *J. Org. Chem.* *2003*, *68*, 6611.
373.   D'Auria, M.; Racioppi, R.; Romaniello, G. *Eur. J. Org. Chem.* *2000*, 3265.

The Paternò-Büchi disconnection is

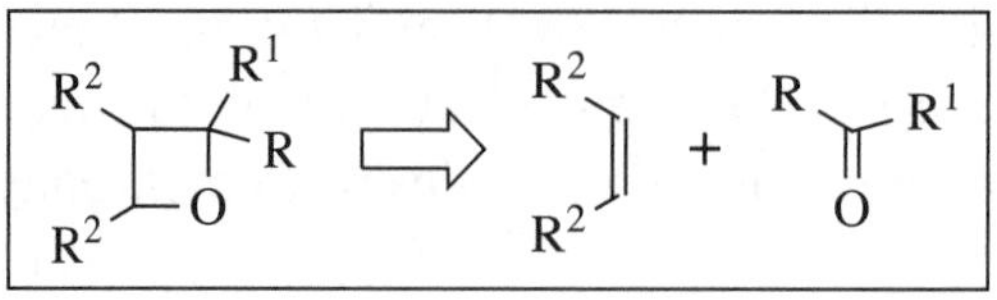

11.10.E. [2+2]-Cycloreversion Reactions

The reverse of the [2+2]-cycloaddition is a thermally allowed σ2s + σ2a cycloreversion.[374] Conversion of **395** to alkenes **397** and **398** illustrates this process. Note that the geometry of one alkene precursor is retained (suprafacial ring opening), but the other is reversed (antarafacial ring opening represented by **396**).[375] The thermal process requires interaction of a HOMO and a LUMO, which must proceed via the antarafacial ring opening represented by **396**.

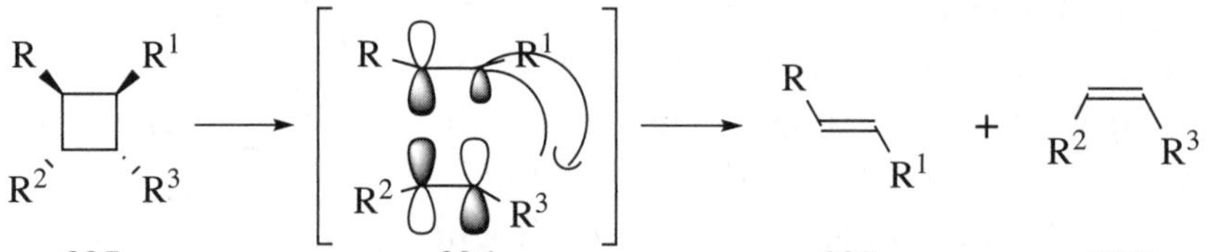

[Reprinted with permission from Desimoni, G.; Tacconi, G.; Barco, A.; Pollini, G.P. *Natural Product Synthesis Through Pericyclic Reactions,* American Chemical. Society, Washington, DC, *1983*. Copyright © 1983 American Chemical Society.]

An example is the photochemical [2+2]-cycloaddition to form **399**, which is followed by thermal ring opening with cleavage of either cyclobutane moiety (**path *a*** or **path *b***). **Path *a*** leads to a 10-membered ring diene (**400**), and path *b* to the monocyclic diene (**401**).[376]

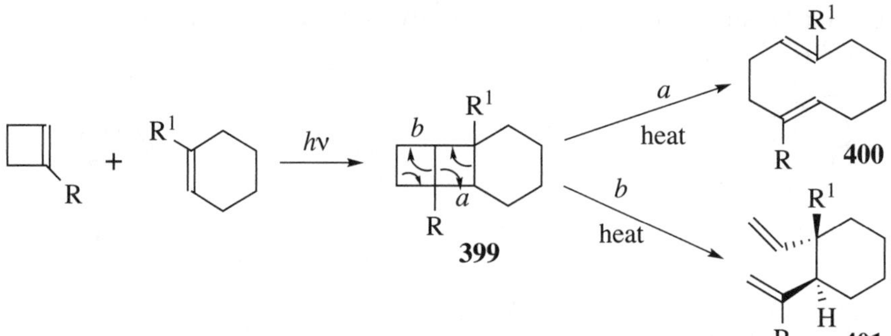

[Reprinted with permission from Desimoni, G.; Tacconi, G.; Barco, A.; Pollini, G.P. *Natural Product Synthesis Through Pericyclic Reactions,* American Chemical. Society, Washington, DC, *1983*. Copyright © 1983 American Chemical Society.]

---

374.   Reference 80b, pp 77-83.
375.   Reference 80b, p 77.
376.   (a) Reference 80b, p 77; (b) see Wender, P.A.; Lechleiter, J.C. *J. Am. Chem. Soc.* *1977, 99,* 267.

The pathway taken by the reaction appears to depend on the substituents. Cycloaddition of ketone **402** with methylcyclobutene generates derivatives such as **403**, with alkyl substituents at the bridgehead positions. Upon heating, cycloadducts of this type open to give **404** (shyobunone), an elemane sesquiterpene, although the yield here was only 5% yield of **404**,[377,259] however, which contrasts with **406** that was formed from methyl cyclobutenecarboxylate and cyclohexenone **405**. Cycloadduct **406** contains an ester moiety, and upon heating it opened to give the germacradiene sesquiterpene isabelin (**407**) in the synthesis reported by Wender and Lechleiter.[378]

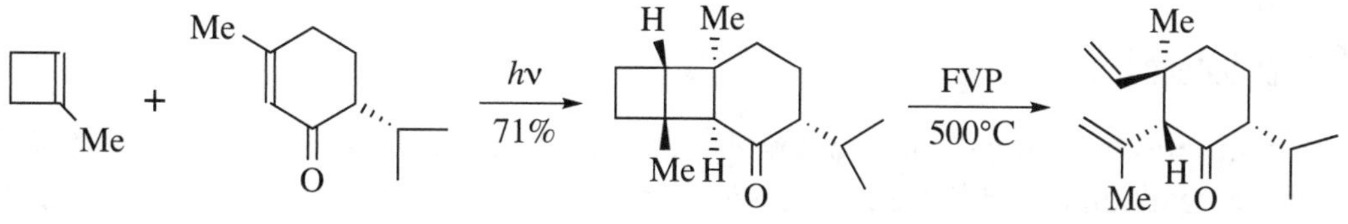

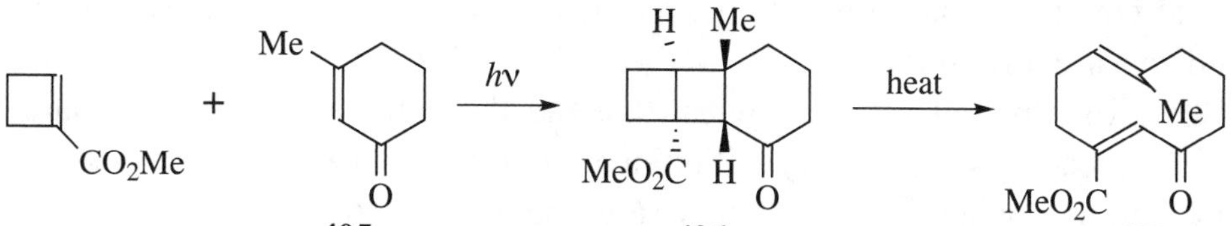

Typical cycloreversion disconnections are

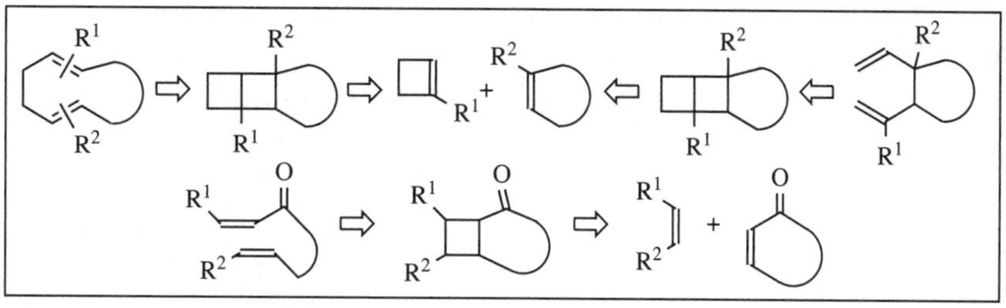

11.11 ELECTROCYCLIC REACTIONS

Woodward and Hoffmann defined electrocyclic transformations[379] as "the formation of a single bond between the termini of a linear system containing k π-electrons (**408** → **409**), and the converse process".[380] Electrocyclic reactions[381] involve an equilibrium between a ring-closed

377. Williams, J.R.; Callahan, J.F. *J. Org. Chem.* **1980**, *45*, 4475.
378. Wender, P.A.; Lechleiter, J.C. *J. Am. Chem. Soc.* **1980**, *102*, 6340.
379. For a review of biosynthetic and biomimetic electrocyclizations, see Beaudry, C.M.; Malerich, J.P.; Trauner, D. *Chem. Rev.* **2005**, *105*, 4757.
380. Woodward, R.B.; Hoffmann, R. *J. Am. Chem. Soc.* **1965**, *87*, 395.
381. See *Electrocyclic Reactions*, Ansari, F.L.; Quereshi, R.; Quereshi, M.L. Wiley-VCH, Weinheim, New York, *1999*.

and a ring-opened molecule that are structurally related. The reaction, which occurs within one molecule, involves the interconversion of a σ-electron system to a π-electron system. This section will explore synthetically useful electrocyclic reactions.

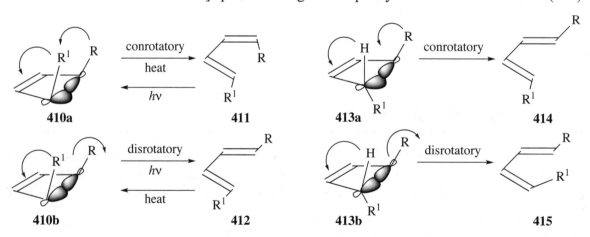

408   409

[Reprinted with permission from Woodward, R.B.; Hoffmann, R. *J. Am. Chem. Soc.* **1965**, *87*, 395. Copyright © 1965 American Chemical Society.]

11.11.A. Cyclobutene Ring Opening

Thermal and photochemical ring opening reactions of cyclobutenes (**410** and **413**) are electrocyclic reactions. Although both (*E,E*) and (*E,Z*) dienes are possible by opening a cyclobutene ring, cis disubstituted cyclobutenes (**410a**) give only the (*E,Z*) diene (**411**) under thermal conditions, and only the (*E,E*) diene (**412**) photochemically. Opening of **410** can proceed via a conrotatory process (**410a**) to give **411**, or a disrotatory process (**410b**) to give **412**. Experimental results suggest that the thermal process is conrotatory and the photochemical process is disrotatory. Similarly, thermal conrotatory ring opening of **413a** leads to **414**, and photochemical disrotatory ring opening (**413b**) leads to **415**.[382] An example is the pyrolysis of cyclobutene **417** to give substituted butadiene derivative **418**. Cyclobutene derivatives can be prepared by several routes. In this example, cyclobutene **417** was prepared from cyclopropylcarbinyl alcohol **416** by treatment with $SOBr_2$ at 100°C,[383] or aqueous HBr/ $ZnBr_2$[384] followed by heating with copper triflate and Hünig's base. Heating **417** [with thionyl bromide at 100°C[383] or via FVP[384]] opened the ring to the *S*-phenylthiobutadiene derivative (**418**).

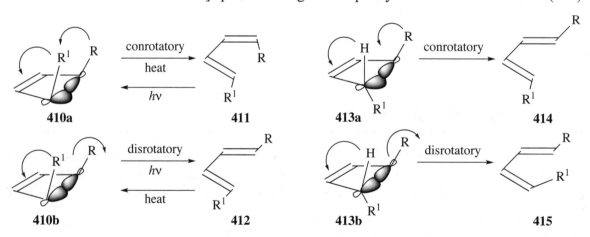

382. Reference 10b, pp 38-46.
383. Trost, B.M.; Jungheim, L.N. *J. Am. Chem. Soc.* **1980**, *102*, 7910.
384. Kwon, T.W.; Smith, M.B. *Chem. Lett.* **1989**, 2027.

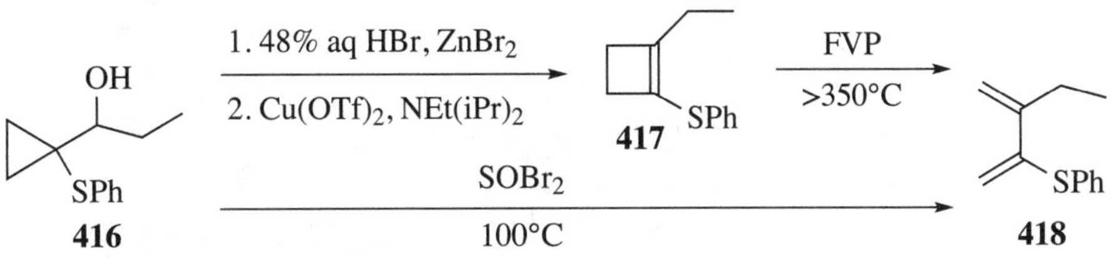

A related reaction has been applied to several natural product syntheses, primarily in work reported by Oppolzer[385] and Kametani Nemoto.[386] In these cases benzocyclobutene derivatives (**419**) open upon heating to form a transient diene (**420**) that is highly reactive in Diels-Alder reactions because formation of the cycloadduct regenerates the aromatic ring. This method is powerful for constructing polycyclic ring systems, and the

use of amide cyclobutane derivatives leads to heterocyclic alkaloids. The structure of the ring bearing the heteroatom is dependent on the type of amide (as seen previously in sec. 11.7). Heating **421**, for example, gave the trans-adduct **422** in 90% yield but **423** gave the cis-adduct **424** in 78% yield.[387] Amide **421** has the carbonyl within the ring being formed, whereas **423** has the carbonyl external to the ring being formed. This methodology was used in Oppolzer and Keller's synthesis of chelidonine.[387a] The presence of a nitrogen atom attached to the cyclobutane unit is not required for the reaction, and other alkaloids can be produced when the nitrogen atom is moved away from the cyclobutane unit.[388] Nitrogen does not have to be present in the substrate. In a synthesis of viridin,[389] Sorensen and co-workers heated benzocyclobutene **425** in degassed xylenes, to give an 83% yield of **426**. Sorensen suggested that the silyl ether substituent conferred a high degree of torquoselectivity[390] to the ring-opening process.

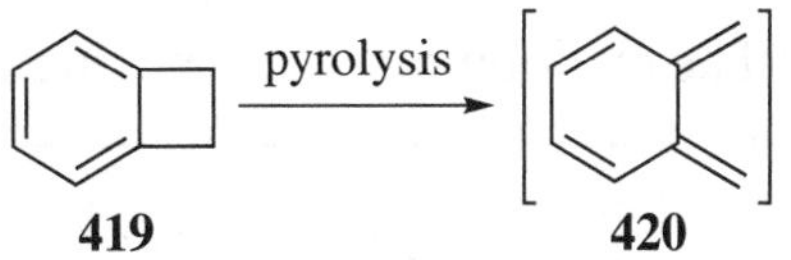

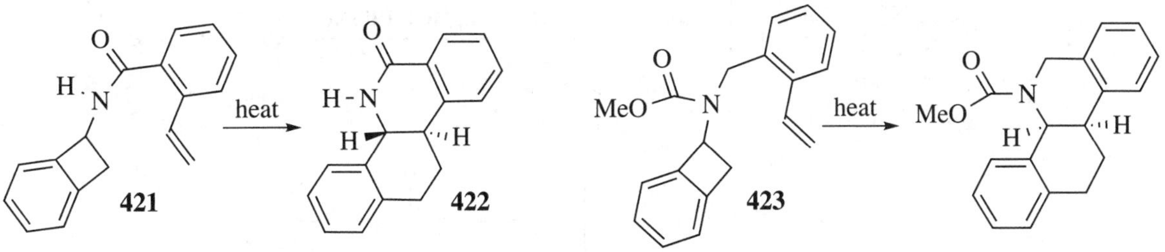

385. Oppolzer, W. *Synthesis* ***1978***, 793.
386. Kametani, T.; Nemoto, H. *Tetrahedron* ***1981***, *37*, 3.
387. (a) Oppolzer, W.; Keller, K. *J. Am. Chem. Soc.* ***1971***, *93*, 3836; (b) Oppolzer, W. *Tetrahedron Lett.* ***1974***, 1001.
388. Verrat, C.; Hoffmann, N.; Pete, J.P. *Synlett* ***2000***, 1166.
389. Anderson, E.A.; Alexanian, E.J.; Sorensen, E.J. *Angew. Chem. Int. Ed.* ***2004***, *43*, 1998.
390. (a) Rondan, N.G.; Houk, K.N. *J. Am. Chem. Soc.* ***1985***, *107*, 2099; (b) Houk, K.N.; Spellmeyer, D.C.; Jefford, C.W.; Rimbault, C.G.; Wang, Y.; Miller, R.D. *J. Org. Chem.* ***1988***, *53*, 2125.

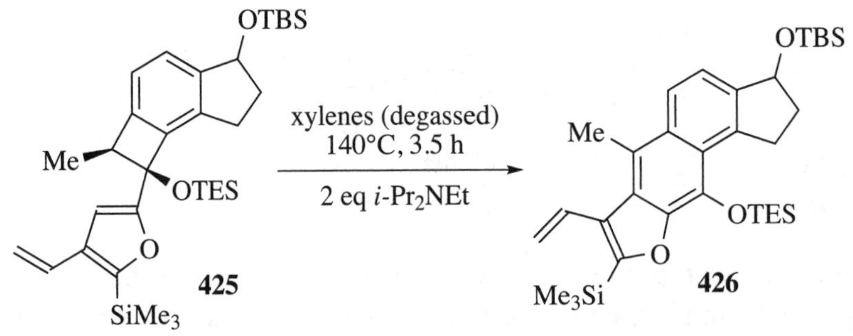

The cyclobutene ring opening disconnections are

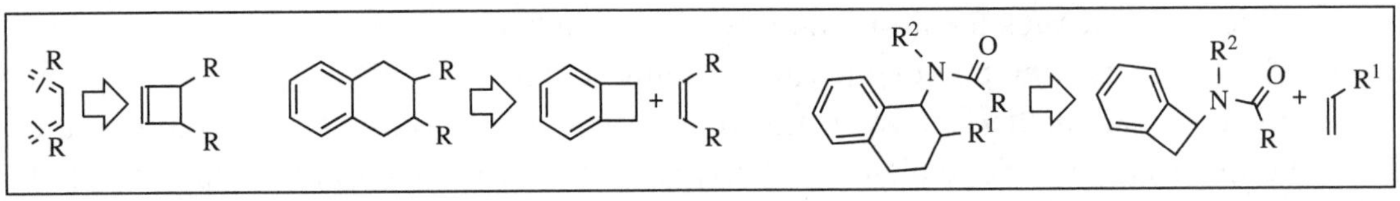

11.11.B. Hexatriene Derivatives and Related Compounds

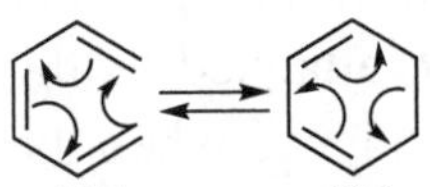

Another useful electrocyclic reaction is the interconversion of 1,3,5-hexatriene (**427**) to cyclohexadiene (**428**). Woodward and Hoffmann categorized the thermal cyclization of hexatrienes as uniquely disrotatory, but a stereospecific conrotatory process is observed under photochemical conditions.[380] This is supported by orbital and state correlation diagrams of Longuet-Higgins and Abrahamson.[391]

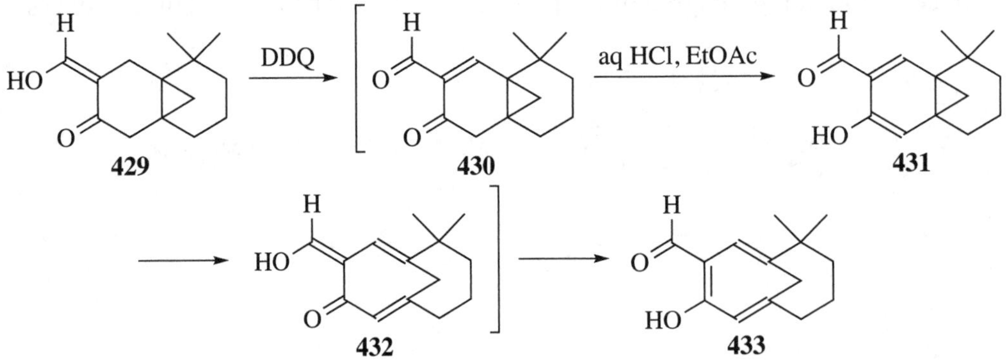

6π-Electrocyclic reactions have been used many times in synthesis,[392] often combined with other structural features to generate interesting structures. In a synthesis of dihydrospiniferin-1 by Marshall and Conrow,[393] treatment of **429** with DDQ gave the dehydrogenated product **430**,

391. Longuet-Higgins, H.C.; Abrahamson, E.W. *J. Am. Chem. Soc.* **1965**, *87*, 2045.
392. Nicolaou, K.C.; Sasmal, P.K.; Xu, H. *J. Am. Chem.Soc.* **2004**, *126*, 5493. For an example taken from a synthesis of 9,10-deoxy-tridachione, see Rodriguez, R.; Adlington, R.M.; Eade, S.J.; Walter, M.W.; Baldwin, J.E.; Moses, J.E. *Tetrahedron* **2007**, *63*, 4500.
393. Marshall, J.A.; Conrow, R.E. *J. Am. Chem. Soc.* **1980**, *102*, 4274.

which upon treatment with aqueous acid generated enol **431**. Electrocyclic ring opening gave **432**, which led to **433**, isolated in 74% yield.

This type of electrocyclic ring close is not limited the carbocyclic systems. Parker and Mindt showed that enolizable vinyl quinone imides were converted to dihydroisoquinolines by thermal 6π-electrocyclization.[394] Specifically, quinone imide **434** enolized to **435** *in situ*, and 6π-electrocyclization gave **436** in 55% yield.

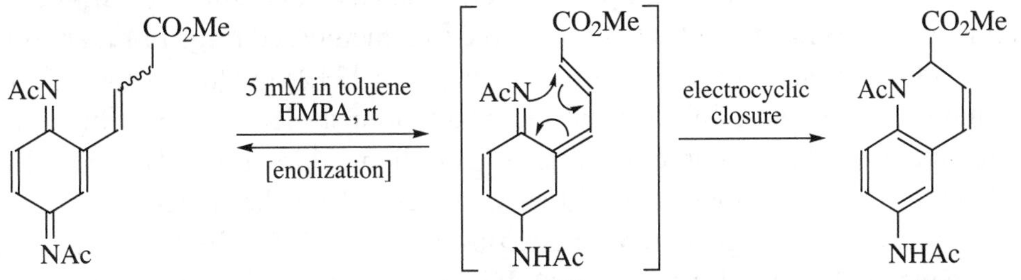

The electrocyclization has been extended to other π-systems. Okamura and co-worker's reported an 8π-electron electrocyclization reaction in the synthesis of an analog of 1α,25-dihydroxyvitamin D₃.[395] Lindlar hydrogenation of the triple bond in **437** generated **438**, which was a transient product immediately undergoing the 8π electrocyclization to give **439** in 50% yield.

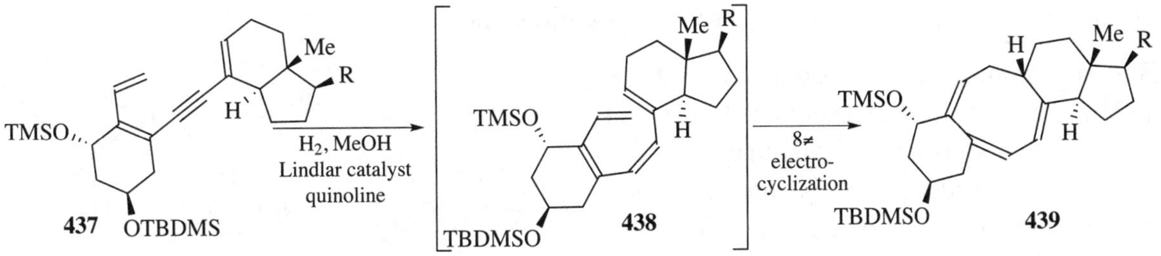

11.12   [3+2]-CYCLOADDITION REACTIONS

We have seen both [4+2]- and [2+2]-cycloaddition reactions in previous sections. There is another class of cycloaddition reactions that involve a molecule containing a π bond (an alkene, alkyne, etc., see **441**) and a highly polarized, and often ionic compound, that is called a 1,3-dipole. 1,3-Dipoles usually take the form of **440**. When these two react, a five-membered ring is generated. Depending on the nature of *a-e*, carbocyclic or heterocyclic rings can be

394. Parker, K.A.; Mindt, T.L. *Org. Lett.* **2002,** *4*, 4265.
395. Hayashi, R.; Fernández, S.; Okamura, W.H. *Org. Lett.* **2002,** *4*, 851.

formed, and a variety of substituents can be incorporated on the ring.[396] This section will explore this class of reactions, which are particularly useful for the synthesis of heterocyclic compounds and natural products.[397]

11.12.A. Dipoles and Dipolarophiles

Cycloaddition reactions that involve dipolar species such as **440** (called **1,3-dipoles**), reacting with π-bonds [called **dipolarophiles (441)**] to give five-membered rings (**442**) are often called 1,3-dipolar cycloadditions. Formally, the reaction is a [3+2]-cycloaddition. A 1,3-dipole usually contains heteroatoms, and it usually contains four π-electrons distributed over three atoms. The dipolarophile is commonly an alkene or alkyne derivative. Huisgen showed that the rate of bimolecular reactions is not greatly influenced by the dielectric constant of the solvent, and proceeds with high stereo- and regioselectivity.[398,399] The reaction is believed to proceed by a concerted, thermal cyclization.[400] It is noted that Firestone has proposed a two step diradical mechanism.[401] Cossío, Schleyer and co-workers studied these cycloaddition reactions computationally, and found that the transition structures associated with concerted supra-supra processes are in-plane aromatic, but aromaticity does not determine regiochemistry.[402] Electrostatic interactions and solvent effects can modify the regiochemical and stereochemical outcome.[402] In a computational study of the 1,3-dipolar cycloaddition of nitrones to alkenes, Di Valentin et al.[403] found that the concerted path (Huisgen's mechanism)[398] should overwhelm the stepwise diradical mechanism proposed by Firestone.[401]

Both intermolecular cycloadditions (**440** → **442**) and intramolecular reactions (**443** → **444**) are possible. An isomeric bridged cycloadduct (**445**) can be produced in this latter reaction if the tether is long enough. Padwa reviewed the internal [3+2]-cycloaddition reaction,[404] as did Oppolzer.[405]

396.   For a review of pyrrolizidine and indolizidine alkaloid syntheses involving 1,3-dipolar cycloadditions, see Broggini, G.; Zecchi, G. *Synthesis, **1999**,* 905.

397.   *Synthetic Applications of 1,3-Dipolar Cycloaddition Chemistry Toward Heterocycles and Natural Products,* Padwa, A.; Pearson, W.H. (Eds), Wiley, New York, Chichester, **2003**.

398.   (a) Huisgen, R. *J. Org. Chem.* **1968**, *33*, 2291; (b) *Idem Angew. Chem. Int. Ed.* **1963**, *2*, 565, 633.

399.   (a)Reference 7, pp 148-161; (b) Reference 80b, pp 89-117.

400.   (a) Hoffman, R.; Woodward, R.B. *Acc. Chem. Res.* **1968**, *1*, 17; (b) Eckell, A.; Huisgen, R.; Sustmann, R.; Wallbillich, G.; Grashey, D.; Spindler, E. *Chem. Ber.* **1967**, *100*, 2192.

401.   (a)Firestone,R.A.*J.Org.Chem.***1968**,*33*,2285;(b)*IdemIbid.***1972**,*37*,2181;(c)*IdemJ.Chem.Soc.A***1970**,1570.

402.   Cossío, F.P.; Morao, I.; Jiao, H.; Schleyer, P.v.R. *J. Am. Chem. Soc.* **1999**, *121*, 6737.

403.   Di Valentin, C.; Freccero, M.; Gandolfi, R.; Rastelli, A. *J. Org. Chem.* **2000**, *65*, 6112.

404.   Padwa, A. *Angew. Chem. Int. Ed.* **1976**, *15*, 123

405.   Reference 236 (see p 18-22).

This type of dipolar addition reaction was introduced in chapter 3 in connection with ozonolysis (sec. 3.7.B), as well as permanganate (sec. 3.5.A)[406] and osmium (sec. 3.5.B) dihydroxylation of alkenes. Ozone is a classical example of a dipolar molecule (see canonical forms of **446**, $^+$O-O-O$^-$ and $^-$O-O-O$^+$). The [3+2]-cycloaddition of ozone with ethene is controlled by interaction of the frontier molecular orbitals.[407] Interactions of the HOMO of ozone with the LUMO of the alkene, and of the LUMO of ozone with the HOMO of the alkene are symmetry allowed.[408] There are three possible orbital interactions, proposed by Sustmann.[409,400b] Substituents that raise the dipole HOMO energy or lower the LUMO energy of the dipolarophile will accelerate the reaction controlled by HOMO$_{dipole}$-LUMO$_{dipolarophile}$.[400b] Substituents that lower the dipole LUMO energy or raise the HOMO energy of the dipolarophile will accelerate the reaction controlled by LUMO$_{dipole}$-HOMO$_{dipolarophile}$.[400b]

As discussed in Section 3.7.B, ozone adds to an alkene to generate a 1,2,3-trioxolane. Addition of ozone to 1-pentene, for example, generated **447**. Rearrangement of this initially formed cycloadduct is facile, even at temperatures as low as –78°C, and results in cleavage of the carbon-carbon bond in **447** with formation of a 1,2,4-trioxolane (**448**). This product is usually called an ozonide and can be either oxidized or reduced to give the carbonyl compounds characteristic of the oxidative or reductive cleavage reactions.

There are many types of 1,3-dipoles other than ozone (see Table 11.13).[410] A few synthetically useful examples are diazoalkanes (**449**, secs. 13.9.B.ii, 13.9.C), nitrile ylids (**450**), nitrile oxides (**451**), azides (**452**), azomethine ylids (**453**), nitrones (**454**) and carbonyl ylids (**455**). The HOMO-LUMO values for these dipolar compounds, and their orbital coefficients at the terminal atoms as shown in Table 11.13,[410] taken largely from Houk's work but also from work by Bastide and Henri-Rouseau.[411] When the energies and orbital coefficients of the alkene LUMO (from Table 11.1) are used with Table 11.13, reactivity and regioselectivity can be predicted.

---

406. Houk, K.N.; Strassner, T. *J. Org. Chem.* **1999**, *64*, 800.
407. (a) Caramella, P.; Gandour, R.W.; Hall, J.A.; Deville, C.G.; Houk, K.N. *I. Am. Chem. Soc.* **1977**, *99*, 385; (b) Houk, K.N.; Sims, J.; Watts, C.R.; Luskus, L.J. *Ibid.* **1973**, *95*, 7301.
408. Reference 7, p 93.
409. Sustmann, R. *Tetrahedron Lett.* **1971**, *12*, 2717.
410. Reference 7, pp 148-149.
411. (a) Houk, K.N.; Sims, J.; Duke Jr., R.E.; Strozier, R.W.; George, J.K. *J. Am. Chem. Soc.* **1973**, *95*, 7287; (b) Bastide, J.; Ghandour, N.E.; Henri-Rousseau, O. *Bull. Chim. Soc. Fr.* **1973**, 2290, 2294; (c) Bastide, J.; Henri-Rousseau, O. *Ibid.* **1973**, 2294; (d) *Idem Ibid.* **1974**, 1037.

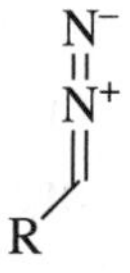

N=N⁺ / R (449)   ⁻CH₂–N⁺≡N / R (450)   O⁻–N⁺≡N / R (451)   N⁻–N⁺≡N / R (452)   R₁ / N⁺ / R R (453)   R–N⁺–O⁻ / R R (454)   O / R R / R X / O⁺ (455)

449    450    451    452 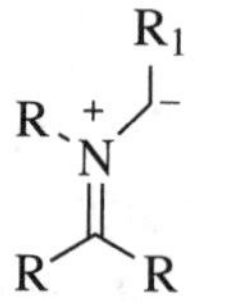   453 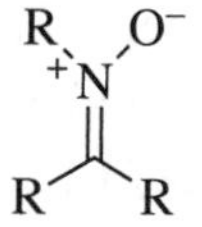   454 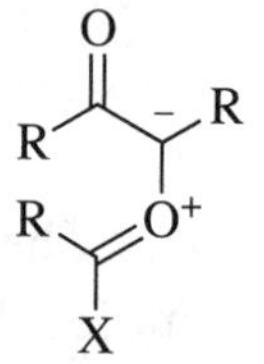   455

Table 11.13 HOMO and LUMO Energies and Orbital Coefficients of Common 1,3-Dipoles

	R^1–X—Y—Z		
HOMO		c_1	c_2
LUMO		c_3	c_4

Dipole	X–Y–Z	HOMO	c1	c2	HOMO	c3	c4
ozone	^+O–O–O^-	-13.5			-2.2		
nitrile ylid	HC≡N⁺–CH₂⁻	-7.7	1.07	1.50	0.9	0.69	0.64
nitrile oxide	HC≡N⁺–O⁻	-11.0	0.81	1.24	-0.5	1.18	0.17
diazoalkanes	H₂C=N⁺=N⁻	-9.0	1.57	0.85	1.8	0.66	0.56
azides	HN=N⁺=N⁻	-11.5	1.56	0.72	0.1	0.37	0.76
azomethine ylids	H₂C=NH⁺–CH₂⁻	-6.9	1.28	1.28	1.4	0.73	0.73
nitrones	H₂C=NH⁺–O⁻	-9.7	1.11	1.06	-0.5	0.98	0.32

[Reprinted with permission from Fleming, I. *Frontier Molecular Orbitals and Organic Chemical Reactions*, Wiley, London, *1976*, pp. 148-149. Copyright © *1976* by John Wiley and Sons, Inc.]

11.12.B. Nitrile Ylids

Frontier molecular orbital analysis is illustrated by reaction of nitrile ylid **456** [412] with ethene, methyl acrylate or methyl vinyl ether as diagrammed in Figure 11.20[412] (all HOMO and LUMO values and coefficients are from Table 11.1). The ΔE values from Figure 11.20 suggest that nitrile ylids should react fastest with electron-deficient alkenes and slowest with electron-rich alkenes, similar to the normal electron demand found in Diels-Alder reactions. If so, there are two possible regioisomers for reaction of **450** with methyl acrylate, **457** and **458**. The orbital coefficients (see **456** and Table 11.1) suggest that **458** is the preferred product. Estimation of the values from Table 11.13 (for H–C≡N⁺–CH₂⁻), and the observation that Ph–C≡N⁺–CH₂⁻ has a HOMO of –6.4 eV and a LUMO of +0.6 eV,[413] shows that these values are sensitive to structural modification. Therefore, general predictions of either reactivity or regioselectivity are risky and each case should be considered individually. Electron deficient alkenes have the lowest ΔE for all of the dipoles shown in Table 11.13, and should react fastest.

---

412.   Reference 7, p 93.
413.   Reference 7, p 148.

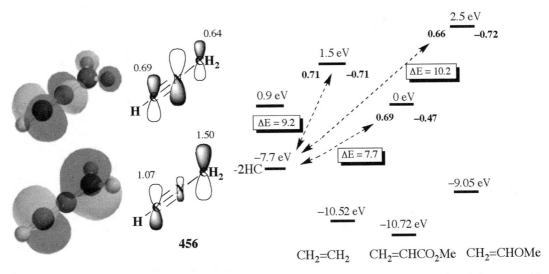

Figure 11.20. The HOMO-LUMO Interactions of Nitrile Ylid With Alkenes. [Reprinted, in part, with permission from Fleming, I. *Frontier Molecular Orbitals and Organic Chemical Reactions*, Wiley, London, *1976*, p. 151. Copyright © *1976* by John Wiley and Sons, Inc.]

The inverse electron demand ΔE is lower for reaction of ozone, azides or nitrile oxides and suggests that the fastest reactions are with electron-rich alkenes. A synthetic example, taken from a synthesis of *ent*-alantrypinone by Hart and Magomedov,[414] involved thermal fragmentation of **459** to **460**, and rapid proton transfer to give nitrile ylid **461** *in situ*. Subsequent intramolecular [3+2]-cycloaddition with the distal carbonyl gave an 80% yield of **462**. This example not only illustrates an intramolecular variation, but also that carbonyls can serve as the dipolarophile.

---

414. Hart, D.J.; Magomedov, N.A. *J. Am. Chem. Soc. 2001, 123,* 5892.

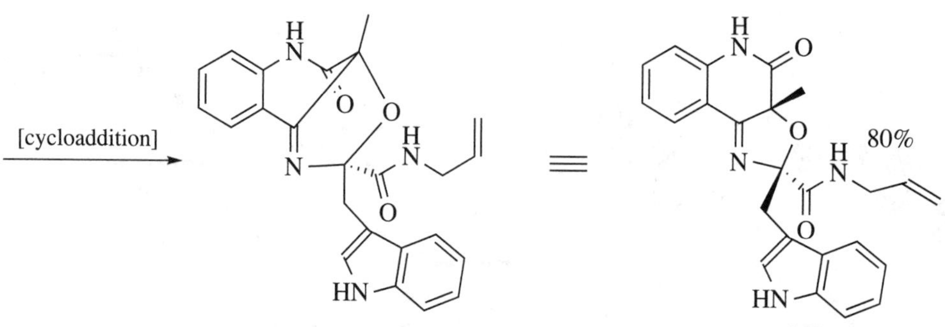

[cycloaddition] →

462 80%

The typical nitrile ylid disconnection is

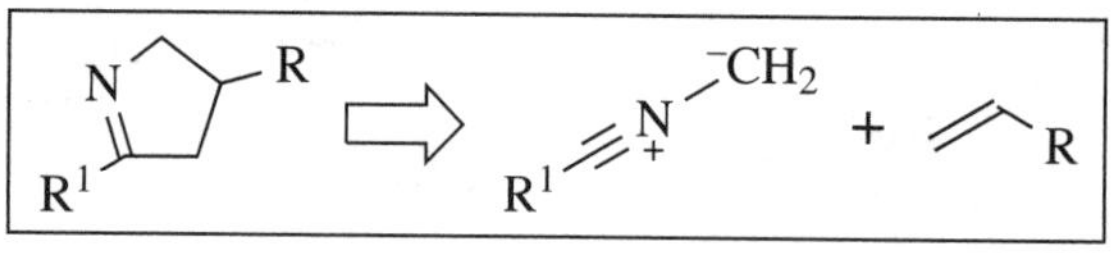

11.12.C. Nitrile Oxides

Frontier molecular orbital theory predicts that nitrile oxides should react faster with electron-rich alkenes (vinyl ethers, enamines or even simple alkenes). The orbital coefficients suggest (see Tables 11.1 and 11.22) that propene reacts with **463** ($HOMO_{dipole}$-$LUMO_{dipolarophile}$) to give **464**. The regiochemistry of the product depends on the substrate, but substituents on the alkene are usually adjacent to the oxygen in the product. Products in which an alkyl substituent is adjacent to a carbon that bears an electron-releasing group (such as methyl vinyl ether to give **465**), or an electron-withdrawing group are usually preferred. In a synthesis of (+)-calafianin,[415] Porco and co-workers generated nitrile oxide **467** *in situ* from the acyl oxime shown[415] (generated from ethyl chlorooximinoacetate and diisopropylethylamine via $EtO_2CC(Cl)=NOH$).[416] The [3+2]-cycloaddition with **466** gave a 1.8:1 mixture of isoxazolines **468** and **469** in 85-93% yield.[415] The use of bases such as DBU or $KHCO_3$ led to very poor yields of isoxazolines, but good yields were obtained using Kobayashi and co-worker's procedure with zirconium (IV) alkoxides.[417]

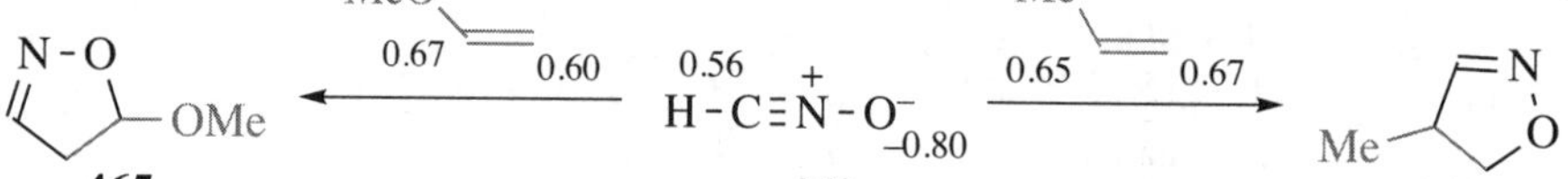

---

415. Bardhan, S.; Schmitt, D.C.; Porco Jr., J.A. *Org. Lett.* **2006,** *8,* 927.

416. Blanchette, M.A.; Choy, W.; Davis, J.T.; Essenfeld, A.P.; Masamune, S.; Roush, W.R.; Sakai, T. *Tetrahedron Lett.* **1984,** *25,* 2183.

417. (a) Kobayashi, S.; Shimizu, H.; Yamashita, Y.; Ishitani, H.; Kobayashi, H. *J. Am. Chem. Soc.* **2002,** *124,* 13678; (b) Yamashita, Y.; Kobayashi, S. *J. Am. Chem. Soc.* **2004,** *126,* 11279.

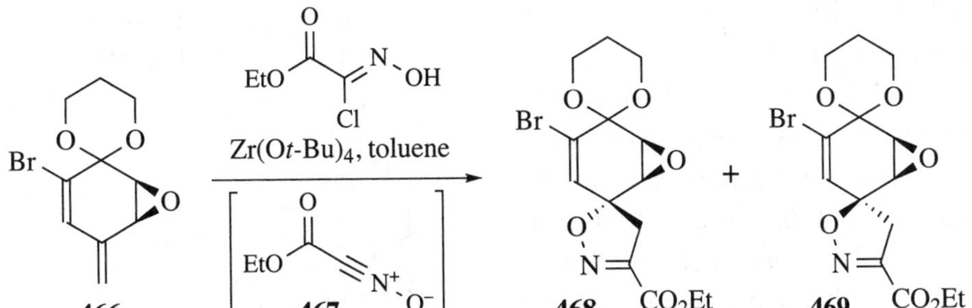

Apart from the nitro compound-phenylisocyanate route, two other procedures are used to prepare nitrile oxides. These two methods are (*1*) treatment of a hydroxamic acid with base and (*2*) oxidation of substituted oximes.[418] Nitrolic acids [R(NO$_2$)C=NOH] also give nitrile oxides upon heating.[419] Relatively complex applications of this reaction are possible.[420,421] A [3+2]-cycloaddition reaction of nitrile oxide was catalyzed by antibodies.[422] Nitrile oxide cycloadditions have also been done in supercritical carbon dioxide as the reaction medium.[423]

The nitrile oxide disconnection is

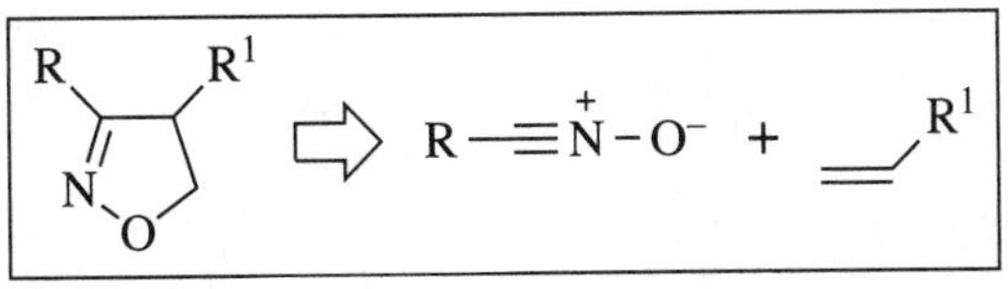

11.12.D. Nitrones

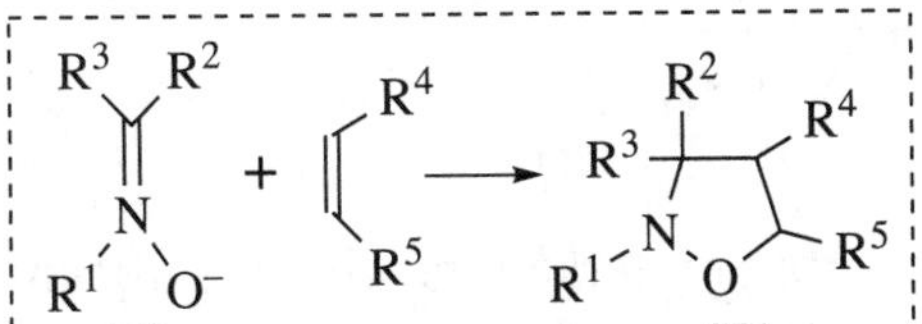

Nitrones (**470**) are potent dipoles in cycloaddition reactions.[424] Cycloaddition of an alkene to a nitrone produces an isoxazolidine (**471**), and internal cycloaddition gives a fused-ring isoxazolidine (**472 →**

473). Nitrones are commonly prepared

in situ via oxidation of an *N*-alkenyl hydroxylamine or condensation of an aldehyde with *N*-methyl hydroxylamine

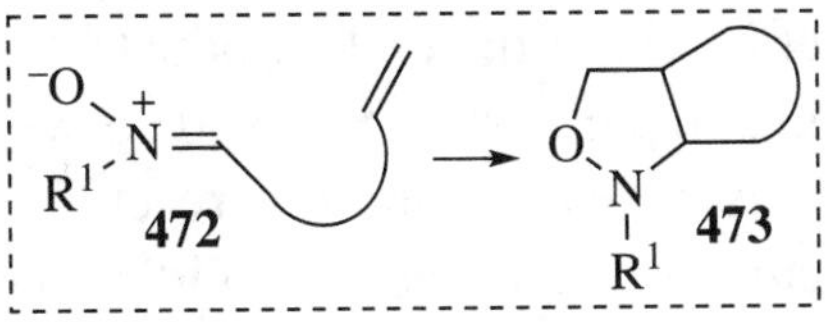

418. Grundmann, C.; Grünanger, P. *The Nitrile Oxides*, Springer-Verlag, Berlin, *1971*.
419. Matt, C.; Gissot, A.; Wagner, A.; Mioskowski, C. *Tetrahedron Lett. 2000, 41*, 1191.
420. See Tufariello, J.J. *Acc. Chem. Res. 1979, 12*, 396.
421. (a) Cornforth, J.W. in Jackson, A.H.; Smith, K.M. *The Total Synthesis of Natural Products, Vol 1*, ApSimon, J. (Ed.), Wiley, New York, *1973*, p.143; (b) Traverso, G.; Barco, A.; Pollini, G.P.; Anastasia M.; Sticchi, V.; Pirillo, D. *Il Farmaco Ed. Sci. 1969, 24*, 946 [*Chem. Abstr. 72*: 78342u, *1970*].
422. Toker, J.D.; Wentworth Jr., P.; Hu, Y.; Houk, K.N.; Janda, K.D. *J. Am. Chem. Soc. 2000, 122*, 3244.
423. Lee,C.K.Y;Holmes,A.B.;Al-Duri,B.;Leeke,G.A.;Santos,R.C.D.;Seville,J.P.K.*Chem.Commun.2004*,2622.
424. See reference 65, pp 777-778.

or any secondary hydroxylamine.[404] The HOMO$_{\text{nitrone}}$-LUMO$_{\text{alkene}}$ ΔE also suggests that nitrones react best with electron-deficient alkenes (such as methyl acrylate) The HOMO coefficients for **474** (Table 11.13) and the LUMO coefficients of methyl acrylate (Table 11.1) suggests that the electron-withdrawing substituent is aligned with the C terminus of the dipole, giving regioisomer **475**. When a nitrone reacts with a simple alkene such as propene (see Table 11.1), the same regiochemical

outcome is predicted. The small difference in the magnitudes of the orbitals suggests poor regioselectivity, but this is not always the case and the presence of an electron-withdrawing group leads to good regioselectivity. There is a report in which nitrones react with phenyl alkynes to give β-lactams.[425] In a synthesis of stemonidine,[426] Figueredo and coworkers reacted nitrone **476** and a conjugated ester to give a 78% yield of **477** along with 5% of the diastereomer. Nagasawa and co-workers used an alkene without an electron-withdrawing substituent in a synthesis of crambescidin 359.[427] Nitrone **478** reacted with alkene **479** in refluxing toluene to give a 67% yield of the desired stereoisomer **480**, again with good regioselectivity.

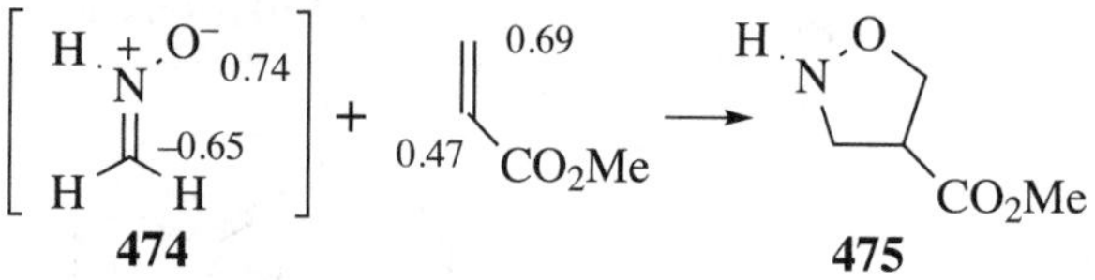

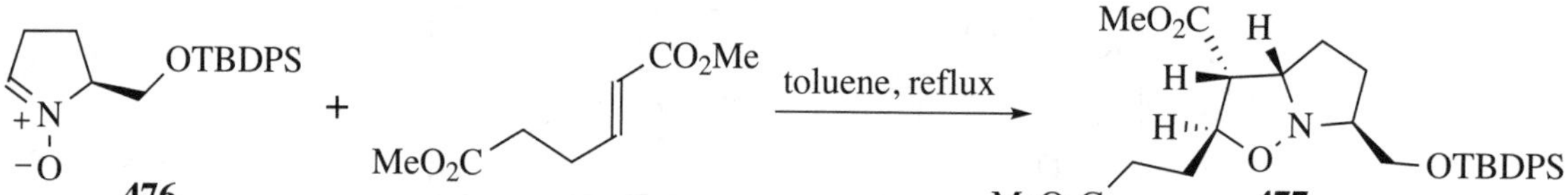

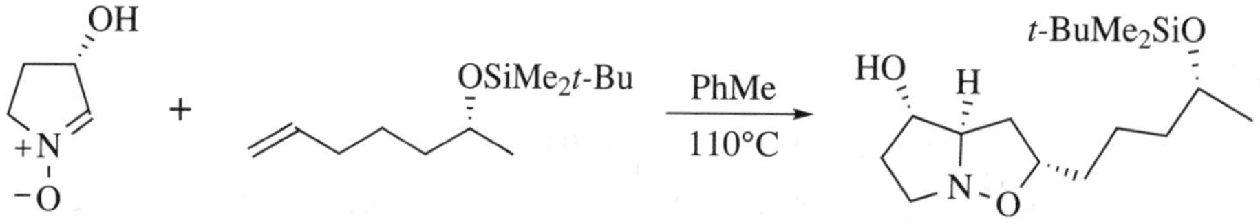

Alkyl dipolarophiles with a tendency for strong secondary orbital interactions give significant amounts of the endo product.[428] Racemization can occur in the cycloadduct after it is formed in some cases.[429] Sibi and co-workers developed a chiral catalyst that leads to *exo*-selectivity in nitrone cycloaddition reactions.[430] Reaction of **481** with the nitrone shown, catalyzed by the copper triflate-bis(oxazolidine) complex **482**, gave a 94% yield of **483** (96:4 *exo:endo*) and in 98% ee for the *exo* product shown.

425. Ye, M.-C.; Zhou, J.; Tang, Y. *J. Org. Chem* **2006**, *71*, 3576.
426. Sanchez-Izquierdo, F.; Blanco, P.; Busque, F.; Alibes, R.; de March, P.; Figueredo, M.; Font, J.; Parella, T. *Org. Lett.* **2007**, *9*, 1769.
427. Nagasawa, K.; Georgieva, A.; Koshino, H.; Nakata, T.; Kita, T.; Hashimoto, Y. *Org. Lett.* **2002**, *4*, 177.
428. Tufariello, J.J.; Ali, Sk.A. *Tetrahedron Lett.* **1978**, 4647.
429. Boyle, L.W.; Peagram, M.J.; Whitham, G.H. *J. Chem. Soc. B* **1971**, 1728.
430. Sibi, M.P.; Ma, Z.; Jasperse, C.P. *J. Am. Chem. Soc.* **2004**, *126*, 718.

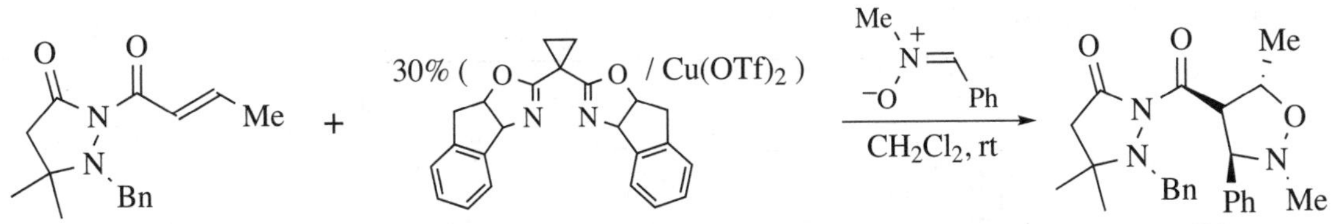

Nitrones can be prepared by reaction of an aldehyde or ketone and a *N*-alkylhydroxylamine, as in the reaction of ketone **484** with hydroxylamine **485** to give nitrone **486**. Treatment with ethyl acrylate resulted in a 9:1 mixture of isoxazolidine (**487**) as well as a diastereomer in 74% yield, as part of the Snider and Lin synthesis of (–)-FR901483.[431] An alternative preparation of nitrones[432] is via reduction of a nitro-ketone with zinc, which gave the nitrone *in situ*. Intramolecular cycloadditions are possible with nitrones, giving the expected cycloadduct.[433] If C2 of the alkene and the α-position of the nitrone were hindered, however, a bridged-bicyclo[3.2.1] product was produced in significant amounts.[433] In a synthesis of neplanocin A by Gallos and co-workers,[434] reaction of ketone **488** and MeNHOH in the presence of pyridine led to nitrone **489**. Subsequent heating initiated the intramolecular cycloaddition that gave **490** in 75% overall yield from **488**. Enantioselective nitrone cycloaddition reactions have been reported, using chiral imidazolidinones[435a] or chiral aluminum BINOL complexes.[435b] Examples of enantioselective cycloadditions were given in a review.[436]

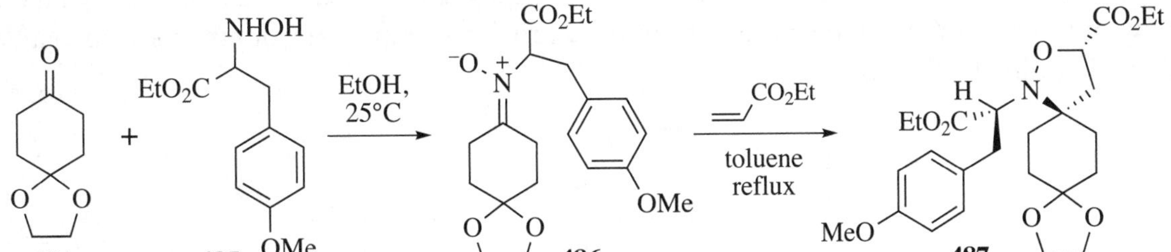

431. Snider, B.B.; Lin, H. *J. Am. Chem. Soc.* **1999**, *121*, 7778.
432. (a) Tufariello, J.J.; Trybulski, E.J. *J. Org. Chem.* **1974**, *39*, 3378; (b) Gössinger, E.; Imhof, R.; Wherli, H. *Helv. Chim. Acta* **1975**, *58*, 96.
433. (a) LeBel, N.A. *Trans. N.Y. Acad. Sci.* **1965**, *27*, 858 [*Chem. Abstr. 64*:548h **1966**]; (b) Reference 404, p 124; (c) LeBel, N.A.; Whang, J.J. *J. Am. Chem. Soc.* **1959**, *81*, 6334.
434. Gallos, J.K.; Stathakis, C.I.; Kotoulas, S.S.; Koumbis, A.E. *J. Org. Chem.* **2005**, *70*, 6884.
435. (a) Jen, W.S.; Wiesner, J.J.M.; MacMillan, D.W.C. *J. Am. Chem. Soc.* **2000**, *122*, 9874; (b) Jensen, K.B.; Roberson, M.; Jørgensen, K.A. *J. Org. Chem.* **2000**, *65*, 9080.
436. Gothelf, K.V.; Jørgensen, K.A. *Chem. Commun.* **2000**, 1449.

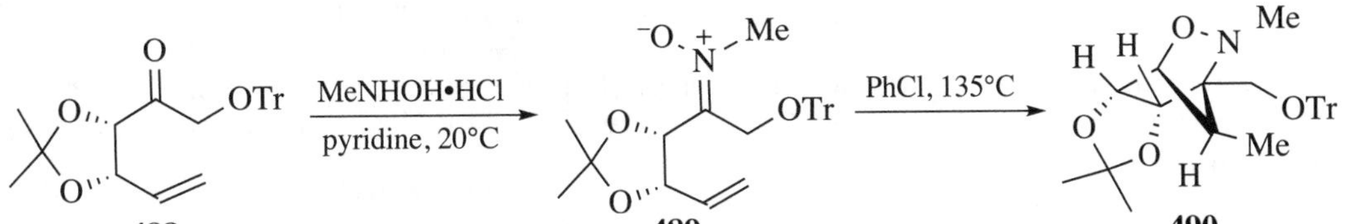

The nitrone disconnection is

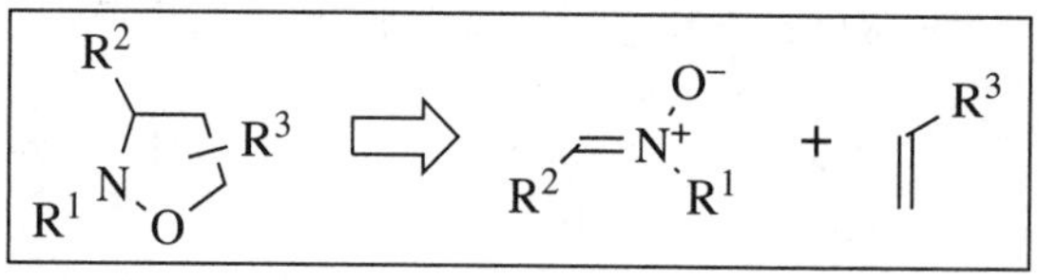

11.12.E. Diazoalkanes

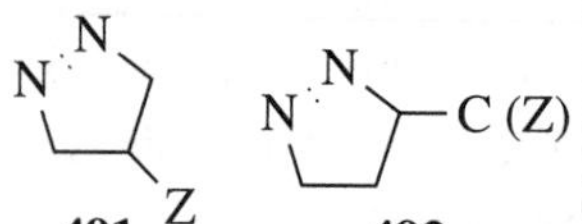

Diazoalkanes are common 1,3-dipoles used in organic synthesis (see sec. 13.9.C for other reactions of diazoalkanes). As diagrammed in Figure 11.21, diazomethane should react fastest with electron deficient alkene (C=C–Z) and slowest with electron-rich alkenes (C=C–X). Orbital coefficients indicate the regiochemical preferences as shown in **491** and **492** for various alkenes. A synthetic example is the [3+2]-cycloaddition of vinyl diazoalkane **493** with the terminal alkene moiety of (1,3,5Z)-octatriene, to give pyrazoline **494** in 55% yield.[437] Photolysis (sec. 11.10.B) of the pyrazoline led to extrusion of molecular nitrogen (N_2) via a nonsynchronous cleavage of the C–N bond.[438] This reaction is thought to proceed via discreet diradical intermediates[439] that close to form a cyclopropane ring.[440] In this example, photolysis of **494** gave a mixture of **495** and **496**. Cyclopropane **496** was isolated and then converted to ectocarpene via a Cope rearrangement (sec. 11.12.C). The synthetic utility of the pyrazoline product arises from the extrusion of nitrogen from the molecule under relatively mild conditions.

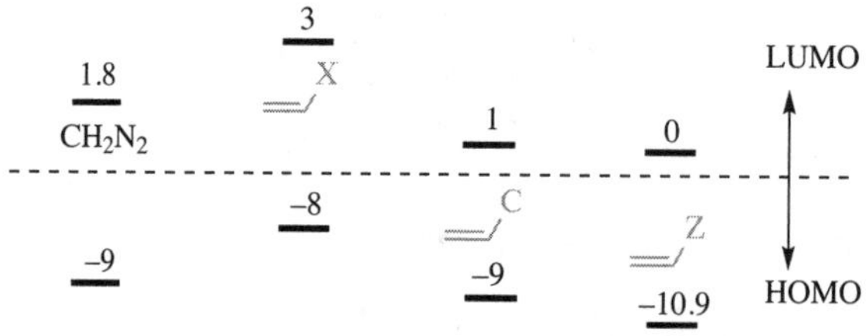

Figure 11.21. Reactivity of diazomethane with alkenes. [Reprinted with permission from Houk, K.N.; Sims, J.; Watts, C.R.; Luskus, L.J. *J. Am. Chem. Soc.* **1973**, *95*, 7301. Copyright © 1973 American Chemical Society.]

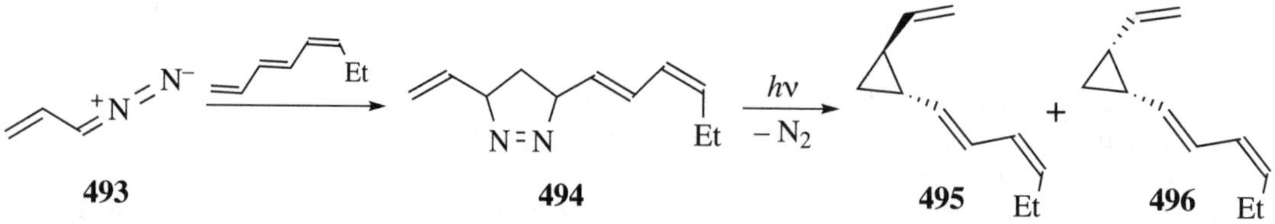

437. Schneider, M.P.; Goldbach, M. *J. Am. Chem. Soc.* **1980**, *102*, 6114.
438. Hiberty, P.C.; Jean, Y. *J. Am. Chem. Soc.* **1979**, *101*, 2538.
439. Givens, R.S.; Oettle, W.F. *J. Am. Chem. Soc.* **1971**, *93*, 3301.
440. See reference 65, pp 137-139.

The reaction is amenable to intramolecular [3+2]-cycloadditions.[441] There are application in which extrusion of nitrogen is not a desirable pathway. In a synthesis of manzacidin A by Maruoka and co-workers, [3+2]-cycloaddition of ethyl diazoacetate to methacrolein (497) gave pyrazoline 498 in 54% yield and 90% ee.[442] Subsequent reduction of the aldehyde with sodium borohydride and treatment with trimethyl orthoformate and tosic acid led to 499, a key intermediate in the synthesis. Later in the synthesis, the pyrazoline unit was reduced to a 1,3-diamine and the [3+2]-cycloaddition was used to set the 1,3-diamine subunit.

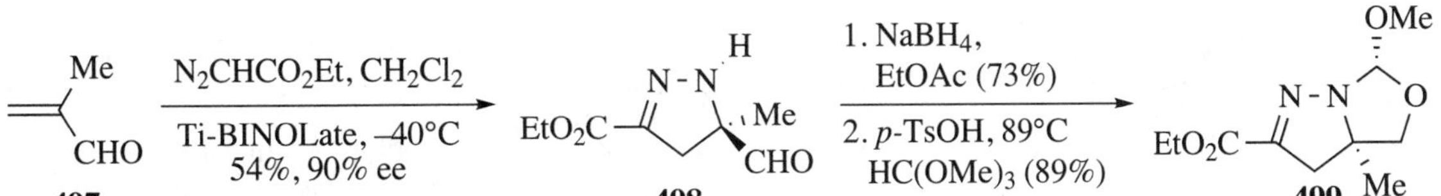

A general diazoalkane disconnection is

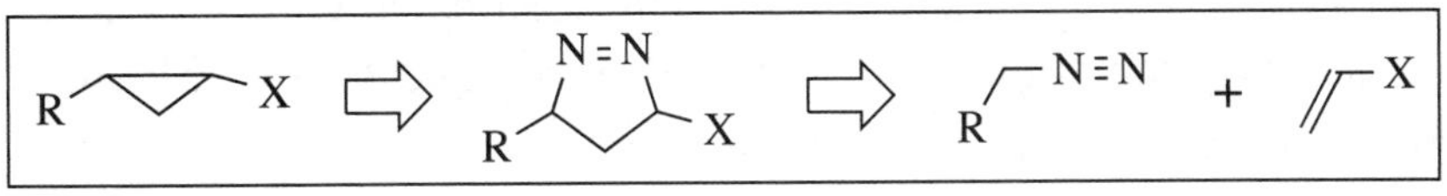

11.12.F. Azomethine Ylids

Azomethine ylids are useful for the synthesis of pyrrolidine-containing compounds. Table 11.13 indicated electron-deficient alkenes and simple alkenes react fastest (in that order), but the orbital coefficients for the simple case predicts near symmetrical orbital density with simple azomethine ylids (see 500) and little regioselectivity in the cyclic imine products. Reaction of 500 with methyl acrylate is therefore expected to give a mixture of pyrrolidines 501 and 502 and reaction with propene would also lead to a mixture of regioisomers. Clearly, substituents at either carbon or on the nitrogen will influence the regioselectivity. An alkyl group attached to the imino carbon leads to formation of different reactive dipoles, including 503a and its resonance contributor 503b. The presence of the electron-releasing alkyl group will change the orbital coefficients, leading to some regioselectivity in the cycloaddition. The relative stability of the iminium salt moiety ($C=NR_2^+$) will also play a role. Resonance contributor 503a is favored over 503b, and is expected to dominate the cycloaddition.

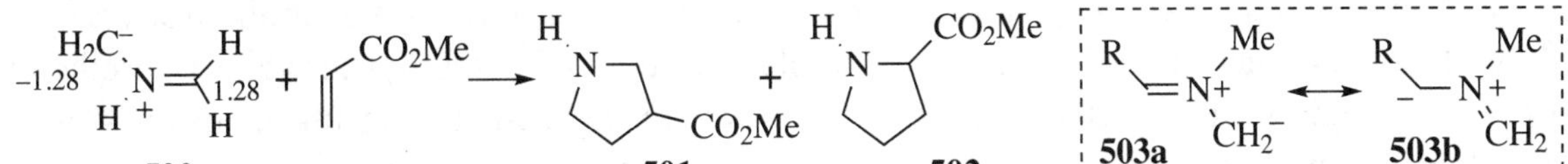

---

441. Piers, E.; Geraghty, M.B.; Soucy, M. *Synth. Commun.* **1973**, *3*, 401.
442. Kano, T.; Hashimoto, T.; Maruoka, K. *J. Am. Chem. Soc.* **2006**, *128*, 2174.

The two resonance forms in azomethine ylids lead to formation of a mixture of regioisomeric cycloadducts in reactions with alkenes. Resonance stabilizing substituents usually enhance the regioselectivity, and intramolecular reactions generally favor only one mode of addition. An example of an intermolecular reaction is taken from a synthesis of (–)-kainic acid by Fukuyama and co-workers,[443] in which conjugated lactone (**504**) reacted with azomethine ylid **505** (generated *in situ* from **506** and trifluoroacetic acid) to give **507** in 83% yield and a diastereoselectivity of 20:1.

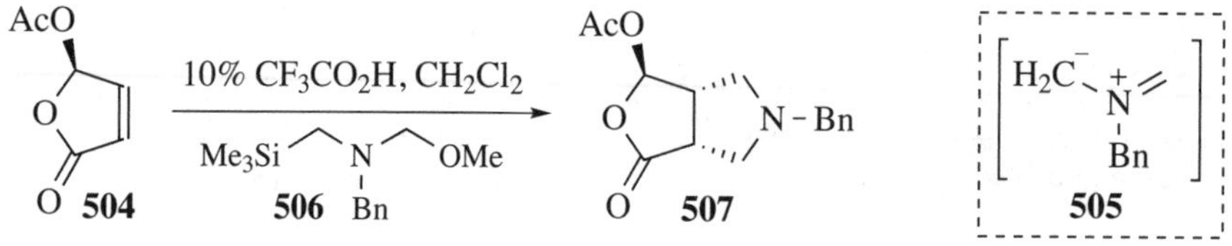

Intramolecular versions of the reaction form polycyclic systems. In a synthesis of pancracine by Pandey and co-workers,[444] silylmethylamine **508** was treated with 2.5 equivalents of silver fluoride to generate the azomethine ylid **509**, which cyclized to give **510** in 56% yield as a single diastereomer.

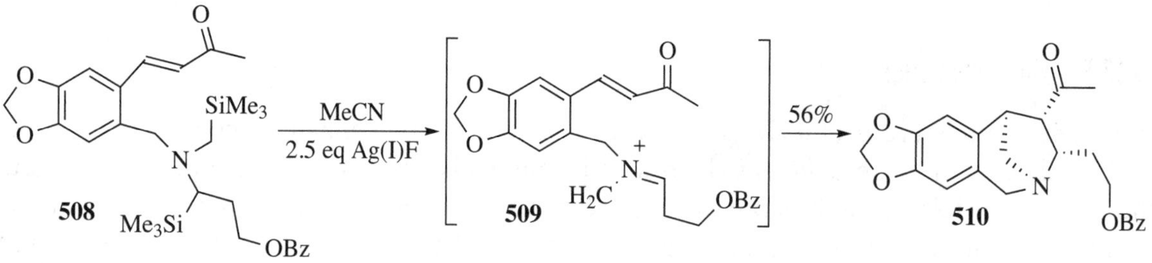

The azomethine ylid disconnection is

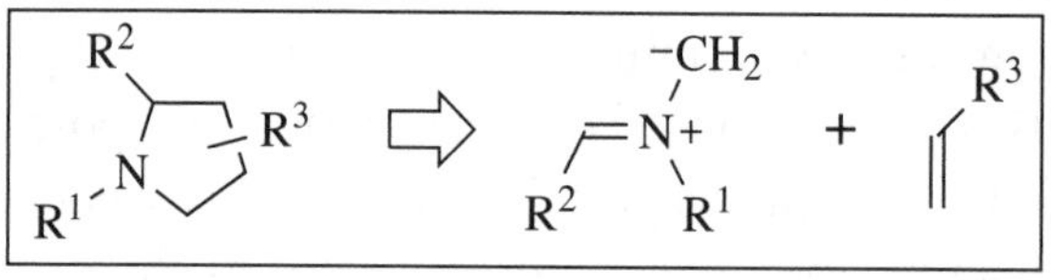

11.12.G. Azomethine Imines

[3+2]-Dipolar additions of azomethine imines (see **511**) and alkenes give tetrahydropyrazines (pyrazolizidines) such as **512**. In this example, reaction with methyl acrylate is predicted to give the regioisomer with the carbomethoxy group adjacent to the nitrogen. A similar regiochemical preference is predicted upon reaction with simple alkenes such as propene. Azomethine imines are usually generated from substituted hydrazone derivatives, which are prepared by reaction of the corresponding aldehydes or ketones with substituted hydrazine

443. Morita, Y.; Tokuyama, H.; Fukuyama, T. *Org. Lett.* **2005**, *7*, 4337.
444. Pandey, G.; Banerjee, P.; Kumar, R.; Puranik, V.G. *Org. Lett.* **2005**, *7*, 3713.

derivatives.[445] Overman and co-workers used an intramolecular 1,3-dipolar cycloaddition of azomethine imines to prepare triazacyclopenta[cd]pentalenes.[446] When **513** was treated with thiosemicarbazide, initial formation of **514** was followed by formation of the azomethine imine *in situ*, and cycloaddition gave an 89% yield of **515**. Note that the carboethoxy unit in **514** was converted to an amide unit during the course of the reaction.

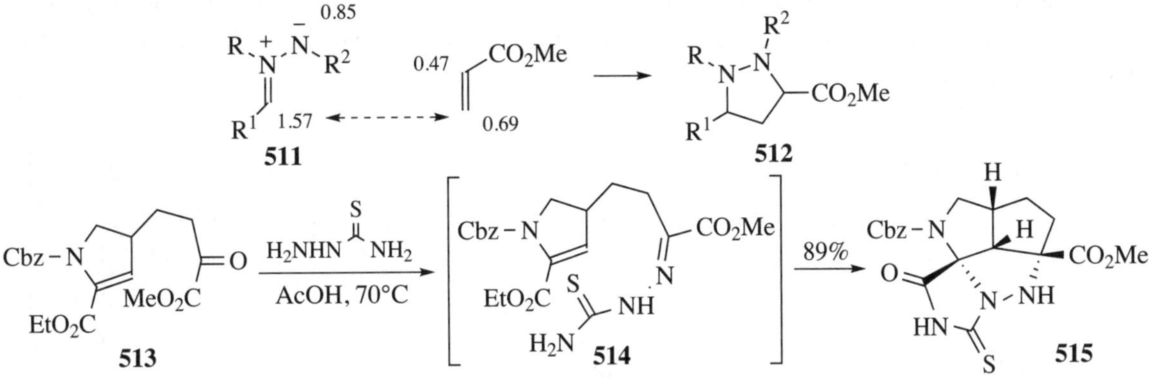

The azomethine imine disconnection is

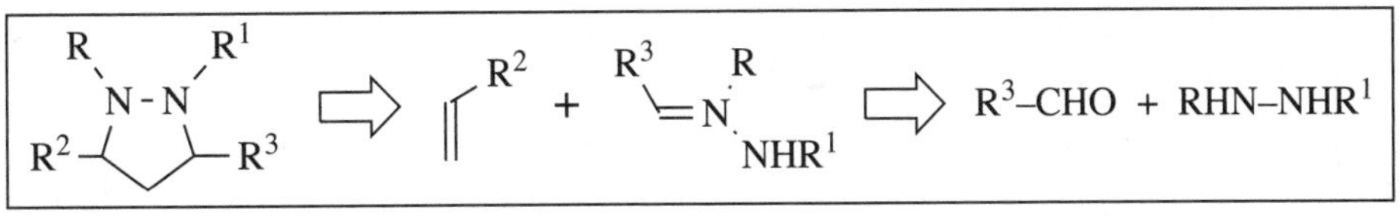

11.12.H Alkyl Azides

Alkyl azides have the structure $R-N_3$ ($R-N^--^+N\equiv N$) and undergo [3+2]-cycloaddition reactions. The first report of the reaction of an azide with an alkene was by Wolff in 1912,[447] and its scope and stereochemistry were later examined by Alder and Stein.[448] This [3+2]-cycloaddition reaction is believed to be concerted[449] and it gives a triazoline product such as **516**. In Table 11.13, hydrazoic acid (HN_3) was shown to have a HOMO of –11.5 eV, with orbital coefficients of 1.56 and 0.72. In a reaction with methyl acrylate, the [3+2]-cycloadduct is triazoline **516**, based on correlation of the largest orbital coefficients. This heterocycle usually decomposes thermally to give pyrroline derivatives,[450] and photochemically to give

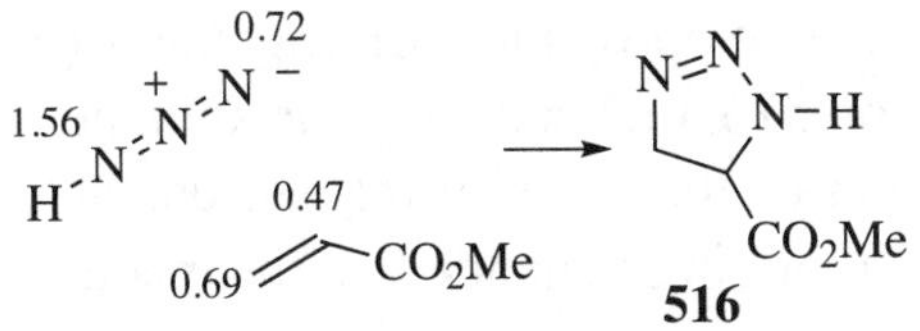

445. (a) Oppolzer, W. *Tetrahedron Lett.* **1970,** 2199; (b) *Idem Ibid.* **1972,** 1707.
446. Bélanger, G.; Hong, F.-T. Overman, L.E.; Rogers, B.N. Tellew, J.E.; Trenkle1, W.C. *J. Org. Chem.* **2002,** *67,* 7880.
447. Wolff, L. *Annalen* **1912,** *394,* 23, 59, 68.
448. (a) Alder, K.; Stein, G. *Annalen* **1931,** *485,* 211; (b) *Idem Ibid.* **1933,** *501,* 1.
449. Scheiner, P.; Schomaker, J.H.; Deming, S.; Libbey, W.J.; Nowack, G.P. *J. Am. Chem. Soc.* **1965,** *87,* 306.
450. Logothetis, A.L. *J. Am. Chem. Soc.* **1965,** *87,* 749.

aziridines.[451,452] An example, taken from Marazano and co-workers' synthesis of the ABC ring of manzamine A,[453] reacted mesylate **517** with sodium azide in hot DMF to give **518**. The [3+2]-cycloaddition produced triazole **519**, which extruded nitrogen to give **520**. In this example, proton transfer occurred to give **521** along with **522** (resulting from epimerization of the proton adjacent to the ester unit) in a 3:2 ratio, in 67% combined yield.[453] Photochemical decomposition of the triazoline gives the aziridine[452] but pyrrolines can also be formed.[412a] An aziridine is not necessarily an intermediate in the pathway that forms five-membered ring derivatives, however.[454]

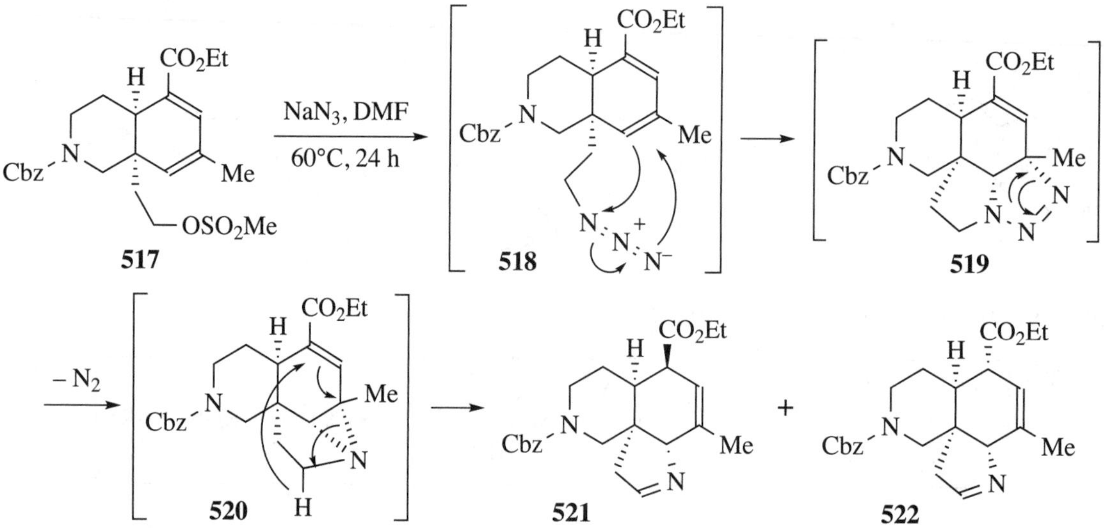

An interesting synthetic application of this process is Hudlicky and co-worker's preparation of several pyrrolizidine alkaloids, via treatment of conjugated ketone **523** with sodium azide (NaN_3) to give keto-azide **524** in 88% yield.[455] Addition of the **524** to the diene apparently generated a triazoline, which was not isolated but decomposed to a transient vinyl aziridine.[456] Reduction of the carbonyl unit in **524** to give alcohol **525** (in 64% yield) used cerium borohydride ($NaBH_4/CeCl_3$, sec. 4.4.B). Subsequent heating in toluene gave **526** and **527** in about 50% yield (as a 64:36 mixture) via cycloaddition followed by extrusion of nitrogen ($N \equiv N$).[455] Hudlicky combined this reaction with a nitrogen analog of the vinyl cyclopropane rearrangement (to be discussed in sec. 11.12.B). Flash vacuum pyrolysis of **526** induced the rearrangement to give pyrrolizidine **528** in 43% yield (after hydrogenation of the initially formed unstable enamine). The extent of nitrogen extrusion to form an aziridine ring depended on the reaction

451. (a) Scheiner, P. *J. Am. Chem. Soc.* **1968**, *90*, 988; (b) Huisgen, L.; Möbius, G.; Müller, H.; Stangl, G.; Szeimies, G.; Vernon, J.M. *Chem. Ber.* **1965**, *98*, 3992.
452. See reference 65, p 819.
453. Herdemann, M.; Al-Nourabit, A.; Martin, M.-T.; Marazano, C. *J. Org. Chem.* **2002**, *67*, 1890.
454. Leeson, P.D.; Carling, R.W.; James, K.; Baker, R. *J. Org. Chem.* **1990**, *55*, 2103.
455. Hudlicky, T.; Seoane, G.; Lovelace, T.C. *J. Org. Chem.* **1988**, *53*, 2094.
456. Hudlicky, T.; Frazier, J.O.; Kwart, L.D. *Tetrahedron Lett.* **1985**, 26, 3523.

conditions.[457] Alkyl azides of the type shown in these examples can be generated by reaction of a suitable derivative with tosyl azide[458] or mesyl azide.[459]

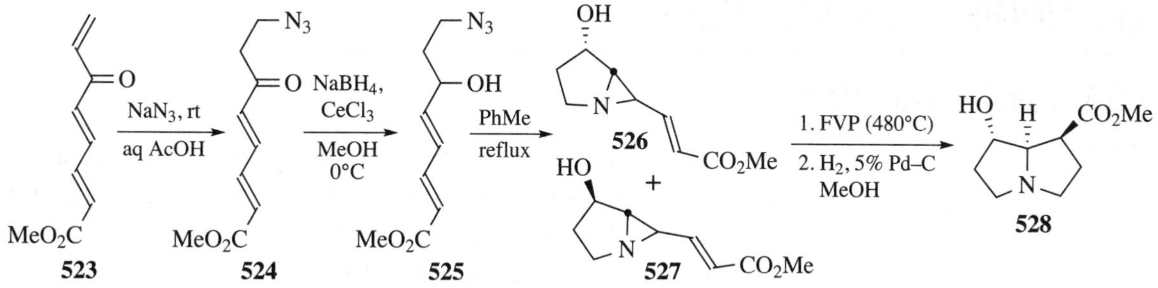

The alkyl azide disconnection is:

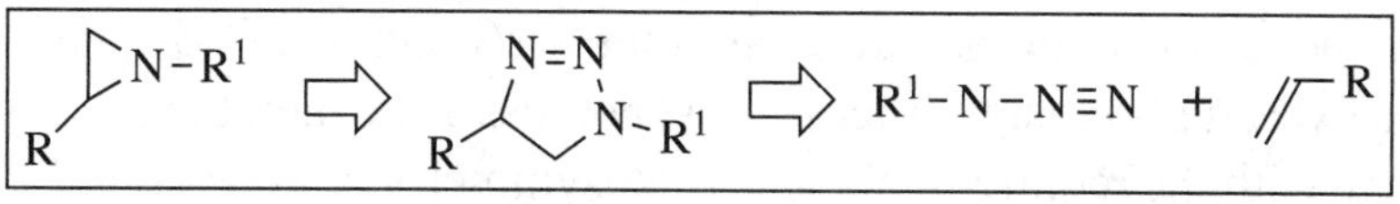

11.12.1. Carbonyl Ylids

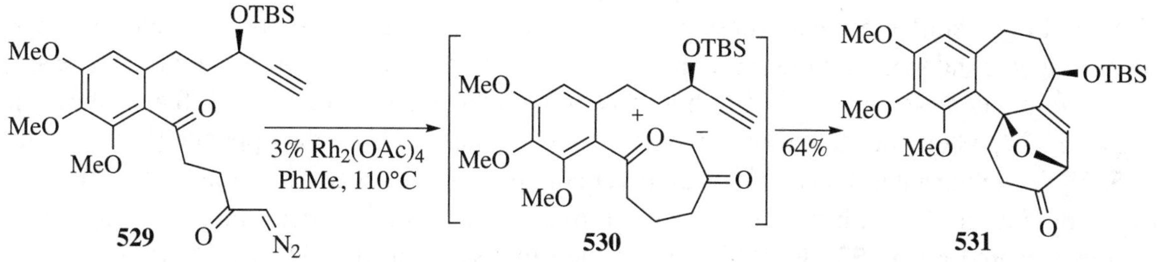

The final class of dipoles discussed here are the carbonyl ylids. Padwa et al. generated these reactive intermediates from a rhodium (II)-catalyzed reaction with 1-diazopentanediones.[460] A synthetic example of this reaction is the formation of carbonyl ylid **530** from diazoketone **529** upon treatment with dirhodium tetraacetate, taken from Schmalz and co-worker's synthesis of (−)-colchicine.[461] Once **530** was generated, an intramolecular [3+2]-cycloaddition gave **531** in 64% yield and >98% de. Note that in cases reported by Padwa,[460a] rhodium catalysts other than

457. Hudlicky, T.; Frazier, J.O.; Seoane, G.; Tiedje, M.; Seoane, A.; Kwart, L.D.; Beal, C. *J. Am. Chem. Soc.* **1986**, *108*, 3755.
458. (a) Ledon, H.T. *Org. Synth. Coll. Vol. 6* **1988**, 414; (b) Hudlicky, T.; Govidan, S.V.; Frazier, J.O. *J. Org. Chem.* **1985**, *50*, 4166.
459. (a) Taber, D.F.; Ruckle Jr., R.E.; Hennessy, M.J. *J. Org. Chem.* **1986**, *51*, 4077; (b) Boyer, J.H.; Mack, C.H.; Goebel, W.; Morgan Jr., L.R. *Ibid.* **1959**, *23*, 1051.
460. (a) Padwa, A.; Carter, S.P.; Nimmesgern, H. *J. Org. Chem.* **1986**, *51*, 1157; (b) Padwa, A.; Carter, S.P.; Nimmesgern, H.; Stull, P.D. *J. Am. Chem. Soc.* **1988**, *110*, 2894; (c) Padwa, A.; Fryxell, G.E.; Zhi, L. *J. Org. Chem.* **1988**, *53*, 2875; (d) *Idem J. Am. Chem. Soc.* **1990**, *112*, 3100.
461. Graening, T.; Bette, V.; Neudorfl, J.; Lex, J.; Schmalz, H.-G. *Org. Lett.* **2005**, *7*, 4317.

Rh$_2$(OAc)$_4$ were required to generate the carbonyl ylid.

11.13. SIGMATROPIC REARRANGEMENTS

11.13.A. [m, n]-Sigmatropic Shifts

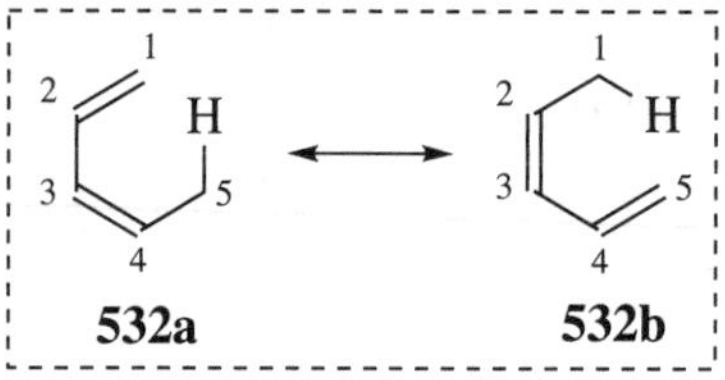

Fleming defines a sigmatropic rearrangement as a "reaction in which a σ bond (a substituent) moves across a conjugated system to a new site", as in the transfer of hydrogen across the face of **532a** to generate **532b** (which in this simple case is identical to **532a**)[462] The nomenclature used for this process is a [m,n]-sigmatropic shift, where an atom or group migrates from position *m* to position *n*. The shift for **532** moves the hydrogen atom at bond 1 to bond 5, a [1,5]-sigmatropic shift. "A [1,5] sigmatropic hydrogen-transfer process must possess a six-membered pericyclic transition state in accordance with the requirements of orbital symmetry conservation".[463] The migrating group is hydrogen in **532**, but carbon and other atoms or groups can migrate. Migration can occur from the same face of the π-system or the opposite face, as illustrated in Figure 11.22.[464] Migration from C1 to C5 across the top face of the molecule as drawn is called a suprafacial [1,5] hydrogen shift. In other words, an atom moves from C1, across the π-face of **532**, and forms a new bond to C5 on the same face. "The σ bond is made and broken on the same side of the conjugated system".[465,462] Another possibility is the antarafacial [1,5] shift also shown in Figure 11.22, where the atom moves from C1, but attaches to C5 on the opposite (bottom) face of **532**. While the antarafacial shift is an inversion, it does not cause inversion of the geometry of a single atom, but rather moves the migrating group to the opposite face.[461] A synthetic example heated allene **533** to 100°C, which led to formation of **534** in 59% yield.[466] This transformation was essential to Okamura and co-worker's synthesis of 9,11-dehydrovitamin D$_3$ analogs.

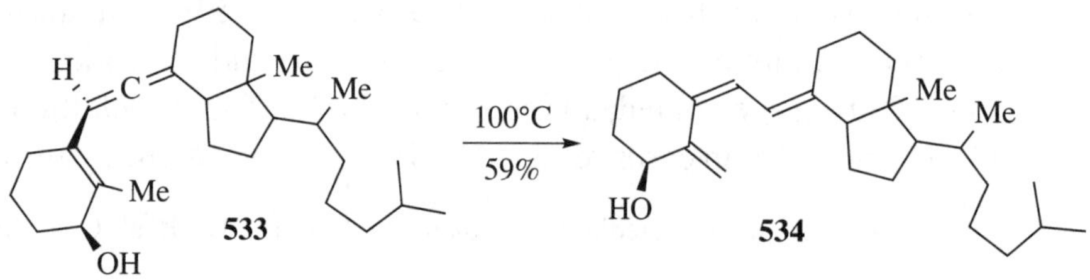

462. Reference 7, p 98.
463. Kwart, H.; Brechbiel, M.W.; Acheson, R.M.; Ward, D.C. *J . Am. Chem. Soc.* **1982**, *104*, 4671.
464. Miller, B. *Advanced Organic Chemistry: Reactions and Mechanisms, 2nd Ed.*, Pearson/Prentice Hall, Upper Saddle River, NJ, **2004**, p. 91.
465. Roth, W.R.; König, J.; Stein, K. *Chem. Ber.* **1970**, *103*, 426.
466. Okamura, W.H.; Aurrecoechea, J.M.; Gibbs, R.A.; Norman, A.W. *J. Org. Chem.* **1989**, *54*, 4072.

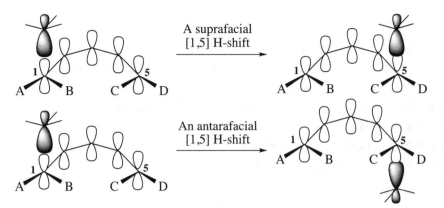

Figure 11.22. Suprafacial[1,5] and antarafacial [1,5] shifts. [Reprinted with permission from Miller, B. *Advanced Organic Chemistry: Reactions and Mechanisms, 2nd Ed.*, Pearson/Prentice Hall, Upper Saddle River, NJ, *2004*, p. 91. Copyright © *2004*.]

Sigmatropic shifts do not always require high temperatures. The [1,5]-sigmatropic rearrangement of substituted cyclopentadienes (**535**) occurs readily at room temperature,[467] to generate three different species (**535-537**). Note that the rearrangement produces two forms of **537** (**a** and **b**) that are identical when R = H, but with disubstituted cyclopentadienes, may be different isomers. The presence of several isomeric compounds leads to regiochemical control problems in Diels-Alder reactions of substituted cyclopentadienes, as discussed in Section 11.8[468] (see **219**).[244] Those rearrangements can now be defined as a series of [1,5]-sigmatropic hydrogen shifts.

It is possible to trap the 5-substituted cyclopentadiene, however, as in the Diels-Alder reaction of **539** to give **540** in 55% yield.[469] Diene **539** was prepared via alkylation of the thallous salt of cyclopentadienyl anion (**538**). Note that the isolation technique for **539** avoided heating above 0°C, in order to suppress the [1,5]-sigmatropic rearrangement. The Diels-Alder reaction that produced **540** used a Lewis acid to catalyze the reaction, allowing the temperature to be

467. Alder, K.; Ache, H.-J. *Chem. Ber. 1962, 95*, 503.
468. Unpublished results, cited as reference 22 in Oppolzer, W. *Angew. Chem. Int. Ed. 1977, 16*, 10, see p 13.
469. (a) Corey, E.J.; Ravindranathan, T.; Terashima, S. *J. Am. Chem. Soc. 1971, 93*, 4326; (b) Corey, E.J.; Keolliker, U.; Neuffer, J. *Ibid. 1971, 93*, 1489.

maintained below 0°C.[469]

Migration of carbon atoms around 1,3,5-cycloheptatrienes (tropilidenes) is a sigmatropic rearrangement involving an electrocyclization to a norcaradiene, followed by 1,5-shifts of a cyclopropane bond. When **541**, **542**, or **543** are heated to >300°C, the methyl group appears to "walk" around the ring, and this rearrangement is called the "walk rearrangement".[470] Initial electrocyclization to **544** was followed by sequential 1,5-shifts-electrocyclizations. It is noted that **545** and **546** are isolated, and since they result from a hydrogen shift but there is no cyclization an alternative mechanism is possible.[470a,471]

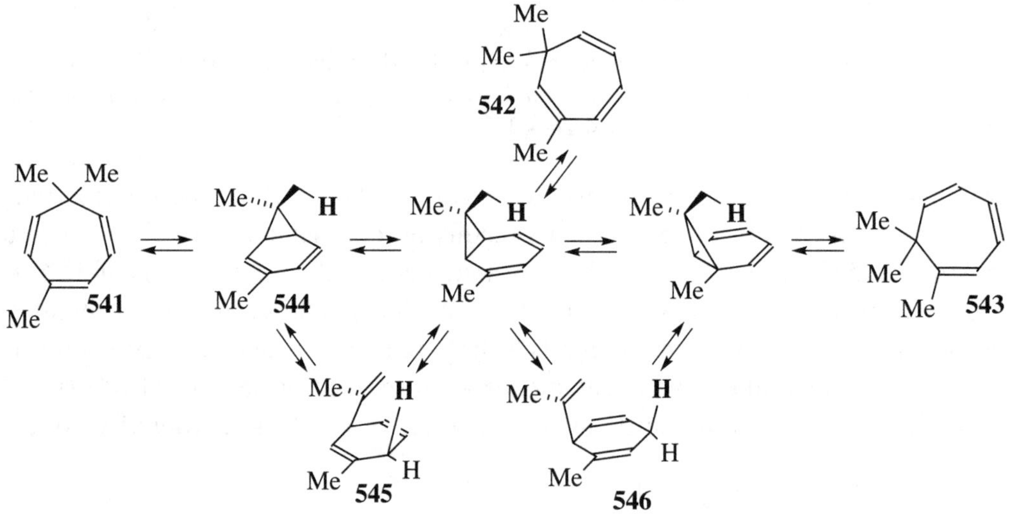

[Reprinted with permission from Berson, J.A. *Accs. Chem. Res.* **1968**, *1*, 152. Copyright © 1968 American Chemical Society.]

Sigmatropic rearrangements can occur in other π systems. In an allylic system, hydrogen migration requires an antarafacial shift for a [1,3]-sigmatropic rearrangement (see **547**),[472] and the reaction is rare.[473] With longer conjugated systems such as 1,3,5-heptatriene, [1,7]-sigmatropic rearrangement is thermally allowed, although it is an antarafacial shift[474] (see **548**). The longer carbon backbone can twist, which allows antarafacial delivery of the C1 hydrogen to C7, as shown. This rearrangement can be difficult, and rarely occurs in cyclic trienes. There are exceptions, however, as in quinol **549** where heating in benzene led to **550**,

470. (a) Berson, J.A. *Accs. Chem. Res.* **1968**, *1*, 152; (b) Berson, J.A.; Willcott III, M.R. *J. Am. Chem. Soc.* **1966**, *88*, 2494.

471. (a) Ter Borg, A.P.; Razenberg, E.; Kloosterziel, H. *Rec. Trav. Chim.*, **1965**, *84*, 1230; (b) Egger, K.W. *J. Am. Chem. Soc.* **1967**, *89*, 3688.

472. Miller, B. *Advanced Organic Chemistry: Reactions and Mechanisms, 2nd Ed.*, Pearson/Prentice Hall, Upper Saddle River, NJ, **2004**, p. 94.

473. Baldwin, J.E.; Fleming, R.H. *J. Am. Chem. Soc.* **1972**, *94*, 2140.

474. Näf, F.; Decorzant, R.; Thommen, W.; Willhalm, B.; Ohloff, G. *Helv. Chim. Acta* **1975**, *58*, 1016.

and a subsequent [1,7]-H shift gave trihydroxybenzene **551** as part of Nicolaou and coworker's synthesis of bisorbicillinoids.[475]

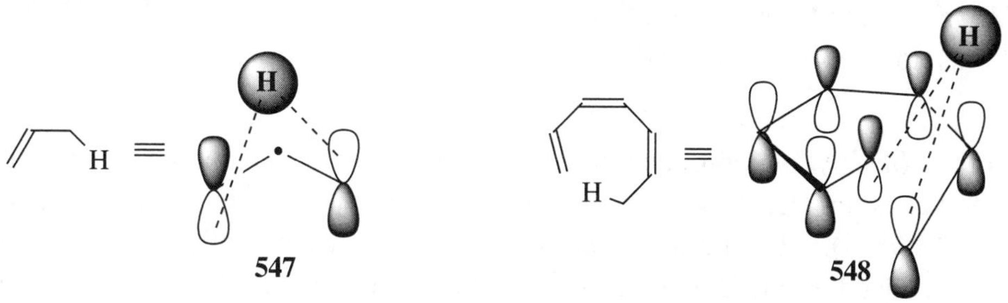

[Reprinted with permission from Miller, B. *Advanced Organic Chemistry: Reactions and Mechanisms, 2nd Ed.*, Pearson/Prentice Hall, Upper saddle River, NJ, **2004**, p. 94. Copyright © **2004**.]

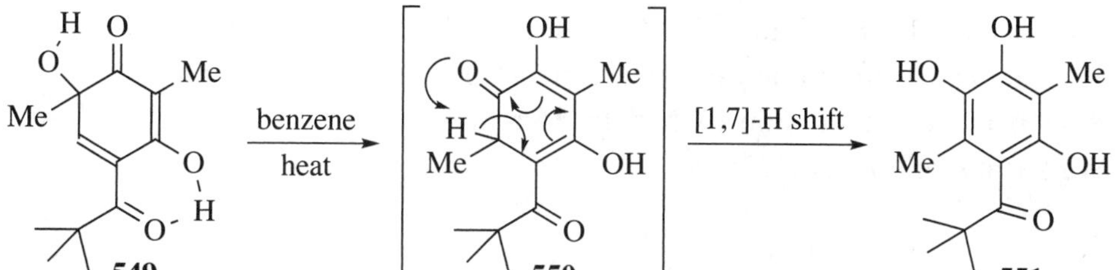

11.13.B. Vinylcyclopropane Rearrangements

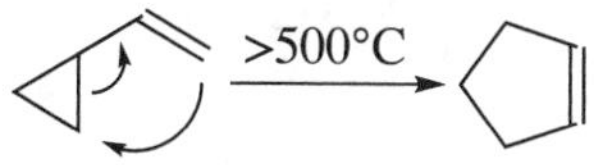

Cyclopropylcarbinyl systems generates cyclopentenes via a [1,3]-alkyl sigmatropic rearrangement.[476] Vigorous heating of vinylcyclopropane **552** (usually >500°C) gives rearrangement to a cyclopentene, as shown. The first reports of this rearrangement were by Vogel[477] and by Overberger and Borchert.[478] Both concerted[479] and diradical[480] pathways have been proposed for this reaction. An alternative proposal suggests a continuous diradical transition state.[481] A

475. Nicolaou, K.C.; Vassilikogiannakis, G.; Simonsen, K.B.; Baran, P.S.; Zhong, Y.-L.; Vidali, V.P.; Pitsinos, E.N.; Couladouros, E.A. *J. Am. Chem. Soc.* **2000**, *122*, 3071.

476. (a) Baldwin, J.E. *Chem. Rev.* **2003**, *103*, 1197; (b) Mil'vitskaya, E.M.; Tarakanova, A.V.; Plate, A.F. *Russ. Chem. Rev.* **1976**, *45*, 469; (c) Frey, H.M. *Adv. Phys. Org. Chem.* **1966**, *4*, 147 (see pp 155-163 and 175-176).

477. Vogel, E. *Angew. Chem.* **1960**, *72*, 4.

478. Overberger, C. G.; Borchert, A. E. *J. Am. Chem. Soc.* **1960**, *82*, 1007.

479. (a) Shields, T.C.; Billups, W.E.; Lepley, A.R. *J. Am. Chem. Soc.* **1968**, *90*, 4749; (b) Berson, J.A.; Dervan, P.B.; Malherbe, R.; Jenkins, J.A. *Ibid.* **1976**, *98*, 5937; (c) Gajewski, J.J.; Olson, L.P. *Ibid.* **1991**, *113*, 7432.

480. (a) Willcott, M.R.; Cargle, R.L. *J. Am. Chem. Soc.* **1967**, *89*, 723; (b) Doering, W. von E.; Schmidt, D. *Tetrahedron* **1971**, *27*, 2005; (c) Roth, W.R.; Schmidt, E.K.G. *Tetrahedron Lett.* **1971**, 3639; (d) Zimmerman, H.E.; Fleming, S.A. *J. Am. Chem. Soc.* **1983**, *105*, 622; (e) McGaffin, G.; de Meijere, A.; Walsh, R. *Chem. Ber.* **1991**, *124*, 939.

481. (a) Doering, W. von E.; Sachdev, K. *J. Am. Chem. Soc.* **1974**, *96*, 1168; (b) *Idem Ibid.* **1975**, *97*, 5512; (c) Roth, W.R.; Lennartz, H.; Doering, W. von E.; Birladeanu, L.; Guyton, C.A.; Kitagawa, T. *Ibid.* **1990**, *112*, 1722.

synthetic example is Trost and Parquette's synthesis of 11-hydroxyjasionone,[482] where **553** was heated to 600°C at 0.06 mmHg (FVP) to give a 97% yield of **554**.

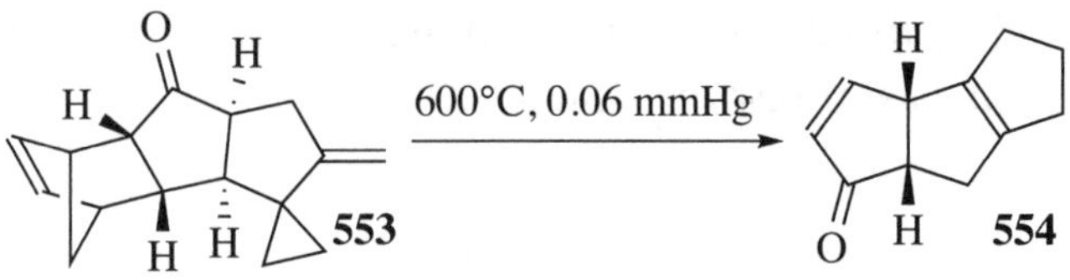

A useful source of vinylcyclopropane derivatives is the cyclopropanation reaction that occurs when diazoalkanes are decomposed to form a carbene-type intermediate in the presence of a diene (sec. 13.9.C.i). Heating diazoketone **555** in the presence of a cupric salt generates a carbene, which added to one C=C unit of the diene moiety to give the bicyclo[3.1.0]hexanone derivative **556**. Heating to 580°C induced the vinylcyclopropane rearrangement, which gave a cyclopentene ring in **557**. Hudlicky et al. converted **557** to hirsutene (**558**) in two steps.[483] Additional examples of cyclopropanation of alkenes with diazoalkane derivatives will be discussed in sections 13.9.B.ii and 13.9.C. This vinylcyclopropane rearrangement is general in its scope, and very useful in organic synthesis.[484] There is a distinct preference for the trans isomer, as seen in the conversion of **555** to **556**, where one cyclopropyl hydrogen and the methyl have a trans-relationship.[485]

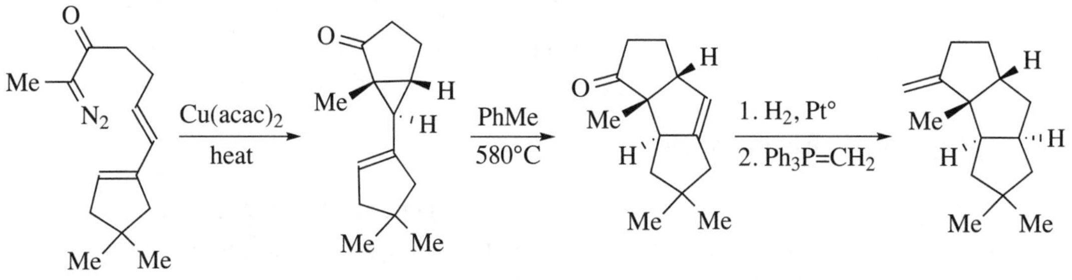

The vinylcyclopropane rearrangement is key step in Trost's **seco-alkylation** sequence (see sec. 8.8.B.ii). In this protocol, trimethylsilyloxyvinylcyclopropanes such as **559** are generated by reaction with diphenylcyclopropyl sulfide, and then with lithium diisopropyl amide (LDA). These steps open the oxaspiropentane, and the resulting alkoxide anion is trapped with chlorotrimethylsilane. Thermal rearrangement of the vinylcyclopropane[486bc] gave the five-membered ring silyl enol ether (**560**), which gave ketone **561** upon hydrolysis.[486a]

482. Trost, B.M.; Parquette, J.R. *J. Org. Chem.* **1994**, *59*, 7568.
483. Hudlicky, T.; Kutchan, T.M.; Wilson, S.R.; Mao, D.T. *J. Am. Chem. Soc.* **1980**, *102*, 6351.
484. For another synthetic example, see Wender, P.A.; Eissenstat, M.A.; Filosa, M.P. *J. Am. Chem. Soc.* **1979**, *101*, 2196.
485. For another synthetic example, see Marino, J.P.; Kaneko, T. *J. Org. Chem.* **1974**, *39*, 3175.
486. (a) Trost, B.M.; Bogdanowicz, M.J. *J. Am. Chem. Soc.* **1973**, *95*, 289; (b) Wilcott, M.R.; Cargle, V.H. *Ibid.* **1967**, *89*, 723; (c) *Idem Ibid.* **1969**, *91*, 4310.

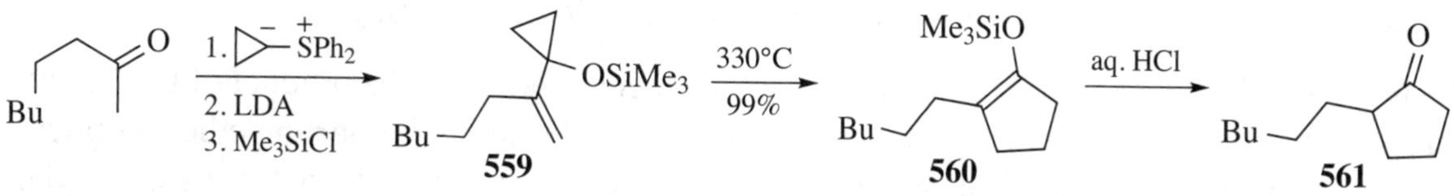

The sigmatropic hydrogen shift and vinylcyclopropane disconnections are

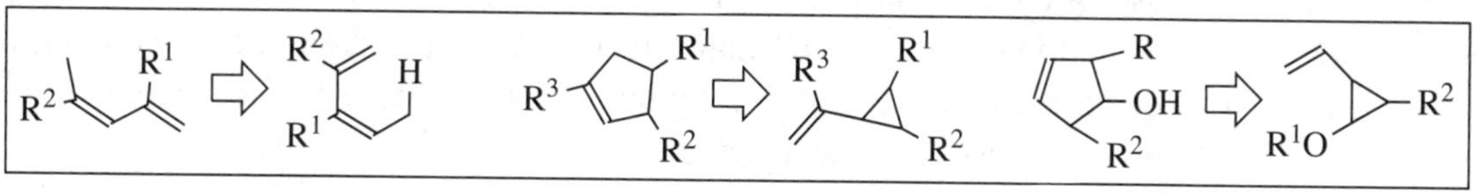

11.13.C. [2.3]-Sigmatropic Rearrangement (Wittig Rearrangement)

Other [m,n]-sigmatropic rearrangements are known. An important rearrangement that is useful in synthesis is the [2.3]-sigmatropic rearrangement known as the Wittig rearrangement.

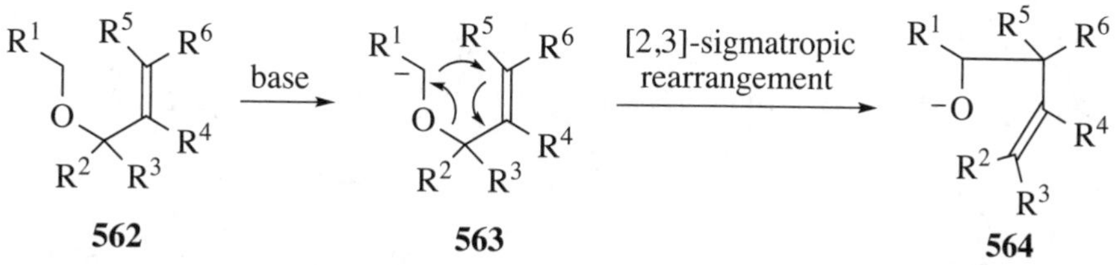

11.13.C.i. Wittig Rearrangement. Heating sulfur ylids (sec. 8.8.B) bearing an allylic group converts them to unsaturated sulfides[487] via a concerted [2,3] sigmatropic rearrangement.[488] Nitrogen ylids[489] give similar rearrangements, as do the conjugate bases of allylic ethers (**563**). This latter rearrangement involves treatment of an allylic ether (**562**) with a suitable base to give the anion (**563**), which undergoes a [2,3]-sigmatropic rearrangement to give the homoallylic alcohol **564**. This transformation is known as the **[2,3]-Wittig rearrangement.**[490] It is noted that the [2,3]-Wittig rearrangement may involve severe deformation of the molecule in order to

487. For example, see (a) Blackburn, G.M.; Ollis, W.D.; Plackett, J.D.; Smith, C.; Sutherland, I.O. *Chem. Commun.* **1968**, 186; (b) Trost, B.M.; LaRochelle, R. *Tetrahedron Lett.* **1968**, *9*, 3327; (c) Kirmse, W.; Kapps, M. *Chem. Ber.* **1968**, *101*, 994, 1004; (d) Ceré, V.; Paolucci, C.; Pollicino, S.; Sandri, E.; Fava, A. *J. Org. Chem.* **1981**, *46*, 3315; (e) For a review as applied to ring expansions, see Vedejs, E. *Acc. Chem. Res.* **1984**, *17*, 358.
488. Hoffmann, R.W. *Angew. Chem. Int. Ed.* **1979**, *18*, 563.
489. (a) Jemison, R.W.; Ollis, W.D. *J. Chem. Soc., D: Chem. Commun.* **1969**, 294; (b) Honda, K.; Inoue, S.; Sato, K. *J. Am. Chem. Soc.* **1990**, *112*, 1999.
490. (a) Cast, J.; Stevens,. T. S.; Holmes, J. *J. Chem. Soc.* **1960**, 3521; (b) Makisumi, Y.; Notzumoto, S. *Tetrahedron Lett.* **1966**, *7*, 6393; (c) Nakai, T.; Mikami, K. *Org. React.* **1994**, *46*, 105; (d) Marshall, J.A. *Comp. Org. Synth.* **1991**, *3*, 975; (e) Nakai, T.; Mikami, K. *Chem. Rev.* **1986**, *86*, 885; (f) The Merck Index, 14th ed., Merck & Co., Inc., Whitehouse Station, New Jersey, **2006**, p ONR-102; (g) Mundy, B.P.; Ellerd, M.G.; Favaloro Jr., F.G. *Name Reactions and Reagents in Organic Synthesis, 2nd ed.*, Wiley-Interscience, New Jersey, **2005**, pp. 700-701.

proceed.[491] It has been shown that SmI$_2$ induces a [2,3] Wittig rearrangement.[492]

Enantioselective [2,3]-Wittig rearrangements have been reported. The product of this reaction has potential stereogenic centers at C-3 and C-4 (if R$^5 \neq$ R^6), and if the starting ether is optically active because of a stereogenic center at C-1 the product may be optically active as well. Many examples are known in which an optically active ether was converted to a product that was optically active, because of chirality at C-3, C-4 or both.[493] If a suitable stereogenic center is present in R^1 (or if a functional group in R^1 made to incorporate a stereogenic center), then stereocontrol over three contiguous stereogenic centers can be achieved. Stereocontrol of the new double bond (E or Z) has also been accomplished.

A synthetic example of the [2,3]-Wittig rearrangement is taken from Martin and co-worker's synthesis of (+)-ambruticin S.[494,495] In this synthesis, the OH unit in alcohol **565** was O-alkylated with trimethyltin iodomethane to give **566**. Subsequent reaction with butyllithium led to formation of the anion and rapid Wittig rearrangement to give alcohol **567** in 70% overall yield after workup, with high diastereoselectivity (>10:1). A second example, taken from Tsubuki, Takahashi and Honda's synthesis of one unit of kallolide A,[496] illustrates a useful intramolecular variation of the Wittig rearrangement, where stereocontrol in the product was a function of the stereochemistry of the alkene unit in the starting allylic ether. Treatment of E-**568** with sec-butyllithium generated a 95:5 mixture of **569** and **570**, in 73% yield, via the [2,3]-Wittig rearrangement. Similar treatment of Z-**568** with sec-butyllithium, however, gave

491. You, Z.; Koreeda, M. *Tetrahedron Lett.* **1993**, *34*, 2597.
492. (a) Kunishima, M.; Hioki, K.; Kono, K.; Kato, A.; Tani, S. *J. Org. Chem.* **1997**, *62*, 7542. Also see (b) Hioki, K.; Kono, K.; Tani, S.; Kunishima, M. *Tetrahedron Lett.* **1998**, *39*, 5229.
493. (a) Mikami, K.; Nakai, T. *Synthesis* **1991**, 594; (b) Fujimoto, K.; Nakai, T. *Tetrahedron Lett.* **1994**, *35*, 5019; (c) Kusche, A.; Brückner, R.; Harms, K. *Chem. Ber.* **1990**, *123*, 917; (d) Wu, Y.; Houk, K.N.; Marshall, J.A. *J. Org. Chem.* **1990**, *55*, 1421; (e) Marshall, J.A.; Wang, X.J. *J. Org. Chem.* **1990**, *55*, 2995.
494. Kirkland, T.A.; Colucci, J.; Geraci, L.S.; Marx, M.A.; Schneider, M.; Kaelin Jr., D.E.; Martin, S.F. *J. Am. Chem. Soc.* **2001**, *123*, 12432.
495. Also see Davidson, A.H.; Wallace, I.H. *J. Chem. Soc. Chem. Commun.* **1986**, 1759.
496. Tsubuki, M.; Takahashi, K.; Honda, T. *J. Org. Chem.* **2003**, *68*, 10183.

a 5:95 mixture of **569:570** in 84% yield.

Other [2,3] sigmatropic rearrangements are known. In chapter 8 (sec. 8.6.A) the **Mislow-Evans rearrangement** involved a rearrangement of allylic sulfoxides to allylic alcohols by treatment with a thiophilic reagent such as trimethyl phosphite. This migration is from sulfur to oxygen, and [2,3] oxygen-to-sulfur migrations are also known.[497] The **Sommelet-Hauser rearrangement** (Sec. 8.8.C.ii) is also a [2,3] sigmatropic rearrangement.

11.13.C.ii. aza-Wittig Rearrangement. An aza-Wittig rearrangement is known that involves nitrogen ylids (sec. 8.8.C).[498] The rearrangement is fundamentally similar to the other Wittig rearrangements just discussed. Amides are the most common precursor because the ylid derived from amide are somewhat more stable. A synthetic example is taken from Anderson and Whiting and Anderson's synthesis of kainic acid.[499] The requisite amide is produced by simple alkylation of allylic bromide **571** with the anion of BocNHCH$_2$CONMe$_2$ (91% yield of **572**). Subsequent reaction with LDA generated the anion, and [2,3]-sigmatropic rearrangement occurred spontaneously to give **573** in 78% yield.

11.13.D. The Cope Rearrangement

11.13.D.i. The Normal Cope Rearrangement. Sigmatropic rearrangements are not confined to hydrogen shifts occurring within a π framework. Thermolysis of *1,5-dienes* results in a six π-electron reorganization, which is a [3,3]-sigmatropic rearrangement illustrated by the interconversion of **574** and **575**. This reaction is known as the **Cope rearrangement**[500] and both the reactivity and selectivity of this reaction can be explained

497. (a) Braverman, S.; Mechoulam, H. *Isr. J. Chem.* **1967**, *5*, 71, (b) Braverman, S.; Stabinsky, Y. *J. Chem. Soc., D: Chem. Commun.* **1967**, 270; (c) Rautenstrauch, V. *Chem. Commun.* **1970**, 526; (d) Tamaru, Y.; Nagao, K.; Bando, T.; Yoshida, Z. *J. Org. Chem.* **1990**, *55*, 1823.

498. (a) Coldham, I. *J. Chem. Soc. Perkin Trans. 1* **1993**, 1275; (b) Anderson, J.C.; Siddons, D.C.; Smith, S.C.; Swarbrick, M.E. *J. Chem. Soc. Chem. Commun.* **1995**, 1835; (c) Ahman, J.; Somfai, P. *J. Am. Chem. Soc.* **1994**, *116*, 9781.

499. Anderson, J.C.; Whiting, M. *J. Org. Chem.* **2003**, *68*, 6160.

500. (a) Cope, A.C.; Hardy, E.M. *J. Am. Chem. Soc.* **1940**, *62*, 441; (b) Rhoads, S.J.; Raulins, N.R. *Org. React.* **1975**, *22*, 1; (c) Schröder, G.; Oth, J.F.M.; Merény, R. *Angew. Chem. Int.* **1965**, *4*, 752; (d) The Merck Index, 14th ed., Merck & Co., Inc., Whitehouse Station, New Jersey, **2006**, p ONR-19; (e) Mundy, B.P.; Ellerd, M.G.;Favaloro Jr., F.G. *Name Reactions and Reagents in Organic Synthesis, 2nd ed.*, Wiley-Interscience, New Jersey, **2005**, pp. 170-171.

by frontier molecular orbital theory.[501] In practical terms, a 1,5-diene is converted to another 1,5-diene by a reversible sigmatropic rearrangement.[502] The reverse-Cope rearrangement is useful in many applications.[503] Calculations suggest two transition states for the Cope rearrangements, a chair (**576**) and a boat (**577**).[504] In **576**, the orbitals of interest are at the termini (C1 and C6) that reacts via a chair-like conformation (**576**), Mechanistic work also suggests that the transition state[505,504] for the Cope rearrangement is predominantly aromatic in nature. A boat transition state is also possible (**577**), which is close in energy to **576** according to calculations (37.1 kcal mol^{-1} for the chair versus 47.8 kcal mol^{-1} for the boat; 155.3 versus 200.1 kJ mol^{-1}),[504] which is in reasonable agreement with experimental data.[506] In other work, the chair transition state was estimated to be 5.7 kcal mol^{-1} (23.8 kJ mol^{-1}) lower in energy relative to the boat transition state,[507] and there is a preference for the chair conformation.[508] When substituents are present on the molecule, a destabilizing orbital interaction in the boat transition state can lead to the chair transition state being favored. In general terms, if the 1,5-diene cannot attain a chair conformation, the Cope rearrangement is often slow and higher reaction temperatures are required.

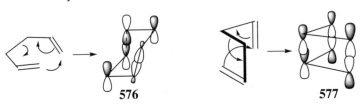

[Reprinted with permission from Dewar, M.J.S.; Jie, C. *J. Am. Chem. Soc.* **1987**, *109*, 5893. Copyright © 1987 American Chemical Society.]

It is known that (*E,E*)-diene **578** rearranges faster than (*Z,Z*)-diene **580** in the Cope rearrangement.[507] The (*E,E*)-diene is more stable than the (*E,Z*)-diene by about 2 kcal mol^{-1} (8.37 kJ mol^{-1}), due to the pseudo-1,3-diaxial methyl-hydrogen interactions in transition state **581**. These interactions are absent in transition state **579**, which has two pseudo-equatorial methyl groups.[507] As

501. Fukui, K. *Accs. Chem. Res.* **1971**, *4*, 57.
502. See reference 65, pp 464-465.
503. Cooper, N.J.; Knight, D.W. *Tetahedron* **2004**, *60*, 243.
504. Dewar, M.J.S.; Jie, C. *J. Am. Chem. Soc.* **1987**, *109*, 5893.
505. Staroverov, V.N.; Davidson, E.R. *J. Am. Chem. Soc.* **2000**, *122*, 186.
506. (a) von E. Doering, W; Toscano, V.G.; Beasley, G.H. *Tetrahedron* **1971**, *27*, 299; (b) von E. Doering, W.; Troise, C.A. *J. Am. Chem. Soc.* **1985**, *107*, 5739.
507. von E. Doering, W.; Roth, W.R. *Tetrahedron* **1962**, *18*, 67.
508. Shea, K.J.; Stoddard, G.J.; England, W.P.; Haffner, C.D. *J. Am. Chem. Soc.* **1992**, *114*, 2635, and references cited therein.

the steric bulk of the substituents on the diene increases, more vigorous reaction conditions (higher temperatures) are required, illustrated by syn-3,4-disubstituted hexadienes such as **582**.[507] Two chair-like transition states (**583a** and **583b**) are in equilibrium, and each has a pseudo-equatorial and a pseudo-axial group, leading to similar energies for each conformation. Cope rearrangement of **583a** gave **584**, the (*E,Z*)-diene, but **583b** also gave the (*E,Z*)-isomer (**585**). trans-3,4-Disubstituted hexadiene **586**, however, generated **587a** (with two pseudo-equatorial groups) and **587b** (with two pseudo-axial groups).[507] Since the 1,3-diaxial interaction (R-R) in **587b** will be greater than the analogous interaction (H-H) in **587a**, **587a** will be lower in energy. Conformation **587a** leads to the (*E,E*)-isomer **588** and **587b** leads to the (*Z,Z*)-isomer **589**. Since **587a** is the lowest energy transition state, the (*E,E*)-diene should predominate. These predictions are borne out with the reaction of meso-2,3-dimethyl-1,5-hexadiene (**582**, R = Me), which gave 99.7% of **584** (R = Me) and 0.3% of **585** (R = Me) at 225°C after 6 h. The (*dl*)-isomer (**586**, R = Me) gave 90% of **588**, 10% of the cis-,cis-diene product and <1% of **589** (R = Me).[507]

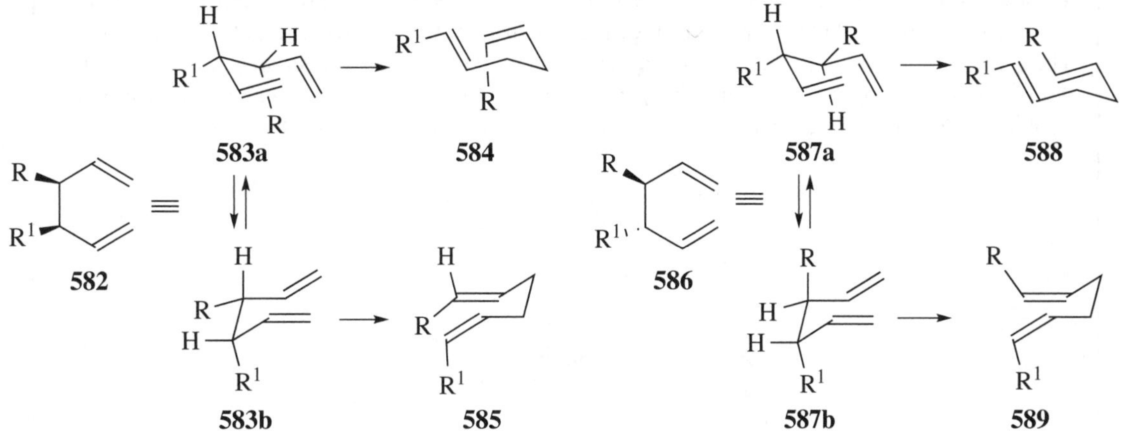

von E. Doering and Roth[507] determined that the reaction gave a >300:1 preference for products derived from the chair transition state over the boat transition state, leading to the energy preference of 5.7 kcal mol^{-1} (23.9 kJ mol^{-1}) for products derived from the chair transition state mentioned above. The influence of the pseudo equatorial versus the pseudo axial groups can be seen in the Cope rearrangement of **590**, also drawn as the pseudo-diequatorial rotamer. Hexadiene **590** gave only the (*E,E*)-diene (**591**) rather than **592** when heated to 80°C for 8 h. The diaxial rotamer[509] would lead to the unobserved (*Z,Z*)-isomer (**592**). The obvious implication is a marked conformational preference for the pseudo-diequatorial rotamer (see the conformational drawing of **590**). This compares with the 3-phenyl-4-methyl derivative, which gave a 90% yield of the (*E,E*) diene and 10% of the (*Z,Z*)-diene.[509]

509. Lutz, R.P.; Bernal, S.; Boggio, R.J.; Harris, R.O.; McNicholas, M.W. *J. Am. Chem. Soc.* **1971**, *93*, 3985.

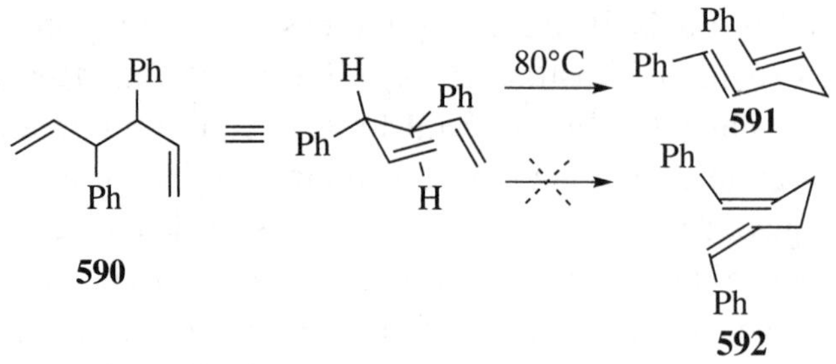

590 **591** **592**

MOLECULAR MODELING: Woodward-Hoffmann Rules

The ***Woodward-Hoffmann rules*** suggest that if given the choice, an intramolecular rearrangement will prefer a path that allows high-energy molecular orbitals (in particular, the HOMO) to change smoothly from reactants to products rather than an alternative route that leads to a change in sign. Woodward and Hoffmann referred to the former as an "allowed reactions" and the latter as a "forbidden reactions". Sigmatropic rearrangements such as the Cope rearrangement are one example. The Cope rearrangement of 2E,4Z,6E-octatriene can either occur via a ***disrotatory*** motion (terminal double bonds turning toward each other) to give a *cis* diene, or a ***conrotatory*** motion (terminal double bonds turning away from each other) to give the less crowded *trans* diene. isomer.

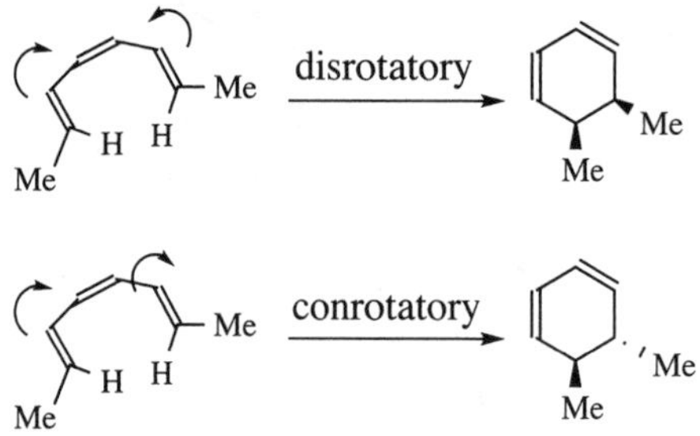

The highest-occupied molecular orbitals for conformers of 2E,4Z,6E-octatriene that are "poised" to *cis* and *trans* 1,5-dienes are displayed on screen (left and right, respectively. Describe the difference between the two orbitals. Which if either corresponds to an allowed reaction?

3,4-Disubstituted dienes (such as **590**) undergo Cope rearrangement more easily (at lower temperatures) than the corresponding 1,6-disubstituted dienes (see **578**). Substituents at the C2 and C5 positions, however, have little effect. These observations can best be explained by identifying the Cope rearrangement as an equilibration process (as seen in **583a → 583b**). Under the thermal conditions of the reaction, the thermodynamically most

593   **594**

stable diene will accumulate. Heating 3-methyl-1,5-hexadiene (**593**) gave ~ 60:40 mixture of **593** and **594**.[510] In this case, the energy difference between **593** and **594** is rather small, and **593** is only slightly favored. The Cope rearrangement is most efficient when the product diene

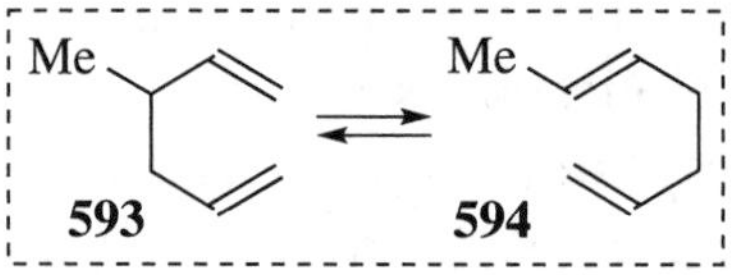

---

510. Levy, H.; Cope, A.C. *J. Am. Chem. Soc.* **1944**, *66*, 1684.

of the equilibrium is significantly more stable than the starting diene.

Introducing strain into the molecule lowers the activation barrier for the [3,3]-sigmatropic shift, as does introduction of an oxygen atom. The relative rate of each reaction is inversely proportional to the highest activation barrier:[511] 33.5 kcal mol^{-1} (140.2 kJ mol^{-1}) for **595**,[512] 23.1 kcal mol^{-1} (96.7 kJ mol^{-1}) for **596**,[513] 19.4 kcal mol^{-1} (81.2 kJ mol^{-1}) for **597**,[514] and 18.2 kcal mol^{-1} (76.2 kJ mol^{-1}) for **598**.[515]

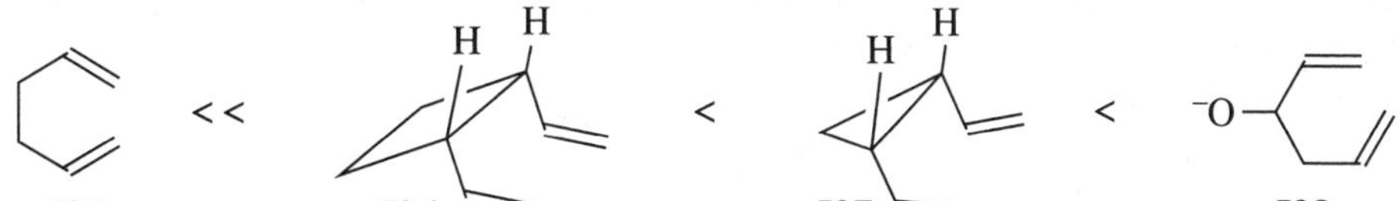

[Reprinted with permission from Desimoni, G.; Tacconi, G.; Barco, A.; Pollini, G.P. *Natural Product Synthesis Through Pericyclic Reactions*, American Chemical Society, Washington, DC, *1983*. Copyright © 1983 American Chemical Society.]

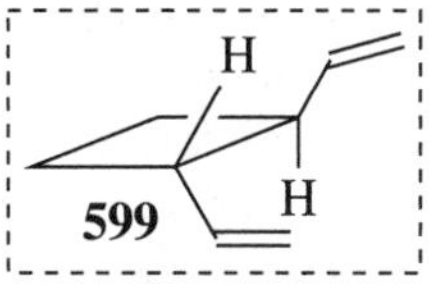

When both strain and stereochemical bias are present in the diene, remarkable rate accelerations can be observed. The activation energy for **596** ($\Delta E \sim 23.1$ kcal mol^{-1}, 96.7 kJ mol^{-1}) is lower than that for the trans isomer, **599** ($\Delta E = 34.0$ kcal mol^{-1}, 142.3 kJ mol^{-1}).[513] Diene **599** cannot assume either a chair or a boat conformation without distortion of the molecule, making the Cope rearrangement slow. The cis compound can assume either a chair or a boat conformation and will easily undergo the Cope rearrangement. The strain imparted by a cyclobutane ring is seen in the 3,4 bond of the diene (see **596**), but this strain is diminished in the Cope rearrangement product (**600**), which is an eight-membered ring with some inherent strain (sec. 1.5.B), but not enough to preclude opening the four-membered ring. Dictyopterene C (**602**) was prepared via Cope rearrangement of the cis-diene, **601**.[516] The strain of the cyclopropane ring accelerated the reaction, which required heating to only 80°C. The process is effectively irreversible due to significant differences in energy between product and starting material.

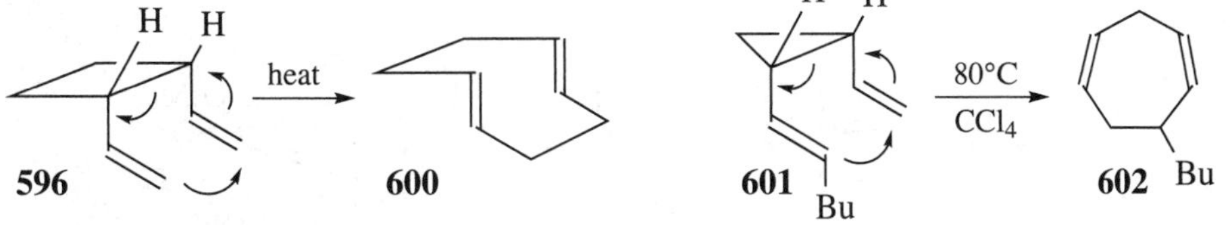

511.   Reference 80b, pp 277-278.
512.   von E. Doering, W.; Toscano, V.G.; Beasley, G.H. *Tetrahedron 1971, 27,* 5299.
513.   Hammond, G.S.; DeBoer, C.D. *J. Am. Chem. Soc. 1964, 86,* 899.
514.   Brown, J.M.; Golding, B.T.; Stofko Jr., J.J. *J. Chem. Soc. Chem. Commun. 1973,* 319.
515.   Evans, D.A.; Golob, A.M. *J. Am. Chem. Soc. 1975, 97,* 4765.
516.   Billups, W.E.; Chow, W.Y.; Cross, J.H. *J. Chem. Soc. Chem. Commun. 1974,* 252.

MOLECULAR MODELING: Cope Rearrangements of *cis*-Divinylcyclopropane and *cis*-1,3-Divinylcyclobutane

The Cope rearrangement is a reversible [3,3]-sigmatropic rearrangement of a 1,5-diene. It is known that introduction of steric strain at the C_3-C_4 position of the diene, such as observed with divinylcyclopropane, leads to acceleration in the rate of rearrangement. Therefore, Cope rearrangements of *cis*-divinylcyclopropane (to *cis,cis*-1,3-cycloheptadiene) and *cis*-1,3-divinylcyclobutane (to *cis,cis*-1,3-cyclooctadiene) are faster than the corresponding Cope rearrangement of 1,5-hexadiene (to itself).

Quantum chemical calculations show that the energy barriers for rearrangement of the three and four-member ring compounds are, respectively, 77 and 40 kJ/mol lower than for rearrangement of 1,5-hexadiene. (An energy barrier is obtained by subtracting the energy of the transition state from that of the reactant.) The same calculations show that both ring opening reactions are *exothermic* (by 96 kJ/mol for divinylcyclopropane and 70 kJ/mol for divinylcyclobutane), consistent with the notion that opening a small ring leads to relief of strain. Cope rearrangement of 1,5-hexadiene leads back to 1,5-hexadiene, meaning that there is no thermodynamic driving force. Such a reaction is said to be ***thermoneutral***.

Transition state structures for all three reactions are shown on screen, in addition to those for the corresponding reactants. Compare geometries of the Cope transition states for *cis*-divinylcyclopropane (top left) and *cis*-1,3-divinylcyclobutane (top center) with that of 1,5-hexadiene (bottom). In terms of carbon-carbon bond distances, which transition state is most similar to its reactant and which least similar (geometries for reactants are provided below those for transition states)? Is there a relationship between the *exothermicity* of a reaction and the degree to which the transition state resembles its reactant? Elaborate. What chemical principle are you "testing"?

Synthetic applications of the Cope rearrangement often involve interesting structural changes. The conversion of cyclopropyl ketone **603** to **604** in 58% yield is part of Ito, Iguchi and co-worker's[517] synthesis of clavubicyclone.

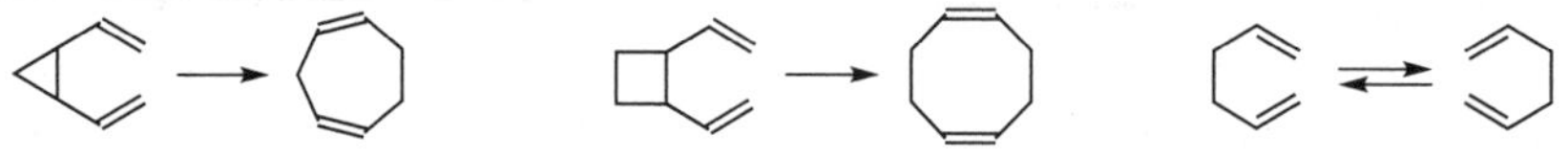

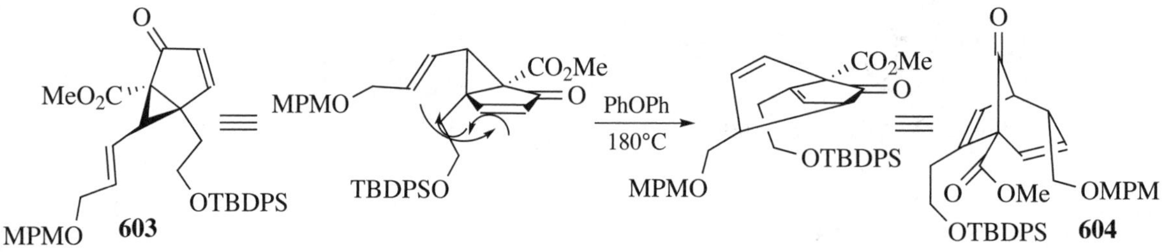

517. Ito, H.; Takeguchi, S.; Kawagishi, T.; Iguchi, K. *Org. Lett.* ***2006***, *8,* 4883.

The Cope rearrangement is also effective when used in conjunction with other reactions. Davies and co-workers[518] used a cyclopropyl-accelerated Cope rearrangement in conjunction with a rhodium-catalyzed diazoalkane cyclopropanation (sec. 13.9.C.iii). In a synthesis of the tricyclic core of vibsanin E, treatment of **605** with a dirhodium catalyst with chiral ligands led to cyclopropanation and formation of **606**, which gave the Cope rearrangement under the reaction conditions to give **607** in 69% yield and 64% ee.[519] The rearrangement of **606** could also be viewed as a vinylogous vinyl cyclopropane rearrangement, but the Cope mechanistic view is probably best and constitutes an effective synthesis of seven-membered ring compounds.

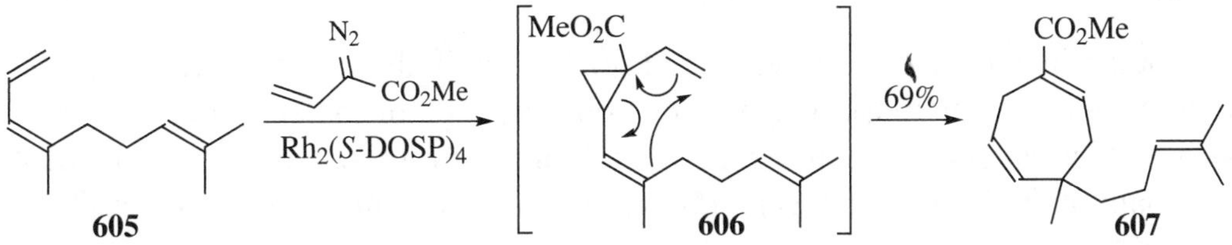

Majetich and Hull also showed that the presence of an enolate moiety in conjunction with the diene can facilitate the Cope rearrangement. This variation was called an **enolate accelerated Cope rearrangement**.[520] When **608** was heated to 180°C, **610** was formed in 45% yield via bond migration from the initial Cope product, **609**. When **608** was treated with fluoride, however, enolate anion **611** was formed, and Cope rearrangement to **612** occurred at 25°C (92% yield). Hydrolysis gave **610**.

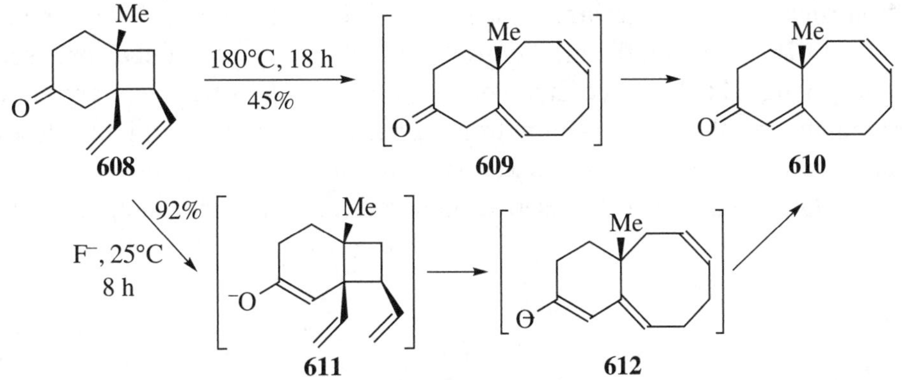

11.13.D.ii. oxy-Cope Rearrangement. Problems caused by the equilibrium inherent to the Cope rearrangement can be circumvented by the presence of an oxygen group on C3 or C4 of the diene (**598**), as mentioned above.[521] Rearrangement of **613**, for example, required heating to only 65°C (refluxing THF) to give the Cope product, which in this case is an enol (**614**) that

518.   Davies, H.M.L.; McAfee, M.J.; Oldenburg, C.E.M. *J. Org. Chem.* **1989,** *54,* 930.
519.   Davies, H M.L.; Loe, Ø.; Stafford, D.G. *Org. Lett.* **2005,** *7,* 5561.
520.   Majetich, G.; Hull, K. *Tetrahedron Lett.* **1988,** *29,* 2773.
521.   See reference 65, pp 1306-1307.

tautomerized to the ketone (**615**)[522] in Jung and co-worker's synthesis of coronafacic acid. When the OH unit is at C3 of the diene, the enol product is subject to keto-enol tautomerism and this shifts the equilibrium toward the final ketone product, facilitating the overall transformation.

A major factor in the rate enhancement observed for hydroxy and alkoxy dienes is the irreversibility of the reaction, assisted by formation of an enol (or enolate under basic conditions, *vide infra*), which favors the ketone form and inhibits the reverse Cope rearrangement. This alcohol variant of the Cope rearrangement is called the **oxy-Cope rearrangement**.[523]

Conversion of the allylic alcohol to the potassium alkoxide enhances the rate of rearrangement by a factor of up to 10^{10}-10^{17}, in what is called the **anionic oxy-Cope rearrangement**. This rearrangement generates an enolate anion directly, and hydrolysis gives the ketone or aldehyde. An example is taken from Castle and co-worker's synthesis of hasubanonine,[524] where **616** was treated with *tert*-butoxide to give the alkoxide which rearranged to **617**. Subsequent hydrolysis led to **618** in 74% yield. The presence of a sulfur substituent, in addition to the alcohol unit, leads to an even greater rate acceleration.[525] An interesting application of this reaction is **Wender's macroexpansion reaction**, in which an initial anionic oxy-Cope rearrangement of **619** generated **620**, in which the allylic groups are properly positioned for a second Cope rearrangement, giving **621**.[526] Since this work was published, it has been argued that **620** is not an intermediate and that a [5,5]-sigmatropic rearrangement leads directly to **621**.[526c] An antibody-catalyzed oxy-Cope reaction is known, and mechanistic studies suggest a diradical intermediate.[527] Other bases can be used to generate the alkoxide, including phosphazene bases.[528]

522. Jung, M.E.; Hudspeth, J.P. *J. Am. Chem. Soc.* **1980**, *102*, 2463.
523. Mundy, B.P.; Ellerd, M.G.; Favaloro Jr., F.G. *Name Reactions and Reagents in Organic Synthesis, 2nd ed.*, Wiley-Interscience, New Jersey, **2005**, pp. 472-473.
524. Jones, S.B.; He, L.; Castle, S L. *Org. Lett.* **2006**; *8*, 3757.
525. Paquette, L.A.; Reddy, Y.R.; Vayner, G.; Houk, K.N. *J. Am. Chem. Soc.* **2000**, *122*, 10788.
526. (a) Wender, P.A.; Sieburth, S.McN. *Tetrahedron Lett.* **1981**, 2471; (b) Wender, P.A.; Sieburth, S.McN.; Petraitis, J.J.; Singh, S.K. *Tetrahedron* **1981**, *37*, 3967; (c) Wender, P.A.; Ternansky, R.J.; Sieburth, S.McN. Tetrahedron Lett. 1985, 26, 4319.
527. Black, K.A.; Leach, A.G.; Kalani, Y.S.; Houk, K.N. *J. Am. Chem. Soc.* **2004**, *126*, 9695.
528. Mamdani, H.T.; Hartley, R.C. *Tetrahedron Lett.* **2000**, *41*, 747.

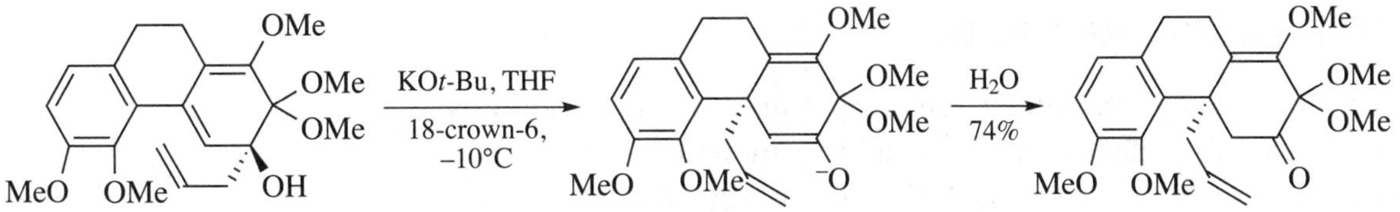

11.13.D.iii. The aza-Cope Rearrangement. It is well established that the presence of an oxygen atom adjacent to the π bond accelerates the Cope rearrangement. Overman and co-workers[529] reported that nitrogen compounds (iminium salt fragments in the diene) undergo an **aza-Cope rearrangement**.[530] In a synthesis of gelsemine by Overman and co-workers,[531] cyanomethylamine (**622**) was treated with KOH to form an imine (**623**) that gave the aza-Cope rearrangement to **624** (the initially formed imine isomerized to the enamine under the reaction conditions). Also under the reaction conditions, the vinyl carbonate was converted to the ketone and **625** was obtained in 81% yield. This reaction can also be viewed as an aza-Claisen rearrangement (sec. 11.12.D.iii). A recent report showed that a 3-azo-Cope rearrangement occurred using Zn^{2+}-Montmorillonite clay catalysis, with microwave irradiation.[532]

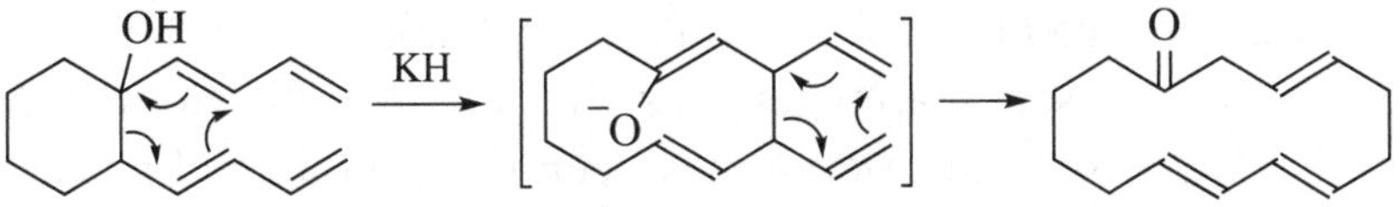

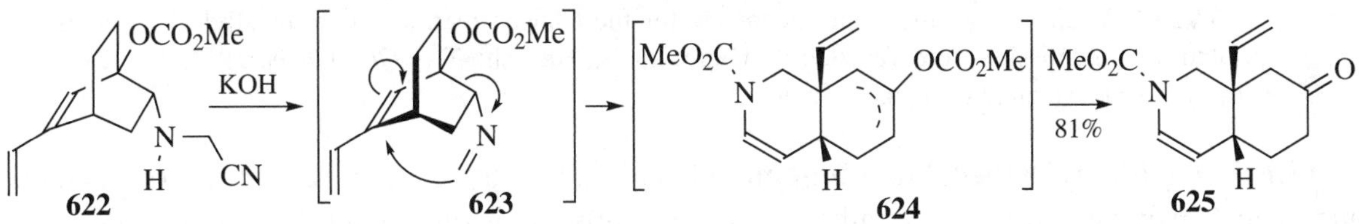

The Cope disconnections are

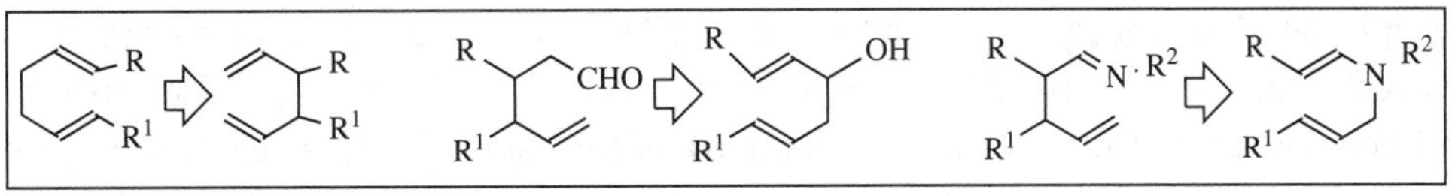

---

529. Jacobsen, E.J.; Levin, J.; Overman, L.E. *J. Am. Chem. Soc.* **1989**, *110*, 4329.
530. See reference 65, p 1308.
531. Earley, W.G.; Jacobsen, J.E.; Madin, A.; Meier, G.P.; O'Donnell, C.J.; Oh, T.; Old, D.W.; Overman, L.E.; Sharp, M.J. *J. Am. Chem. Soc.* **2005**, *127*, 18046.
532. Yadav, J.S.; Subba Reddy, B.V.; Rasheed, M.A.; Kumar, H.M.S. *Synlett* **2000**, 487.

11.13.E. The Claisen Rearrangement

[3,3]-Sigmatropic rearrangements are possible when heteroatoms are part of the chain linking the alkene moieties. Overman's aza-Cope rearrangement of **623** was an illustration. When that heteroatom is an oxygen as in allyl vinyl ether **626**, the [3,3]-sigmatropic rearrangement to **627** is called the **Claisen rearrangement**.[533,500b]   Claisen rearrangements are thermal reactions[534] usually done in solution, but a microwave-assisted reaction has been done on a solid phase.[535] The Claisen rearrangement has been a mainstay of organic synthesis.[536]

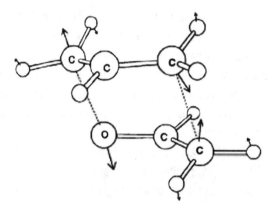

Figure 11.23. Calculated transition state geometry for the Claisen rearrangement of allyl vinyl ether [Reprinted with permission from Yamabe; S.; Okumoto, S.; Hayashi, T. *J. Org. Chem.* **1996**, *61*, 6218. Copyright © 1996 American Chemical Society.]

11.13.E.i. Allyl-Vinyl Ether Rearrangements. This reaction predated the Cope rearrangement and is obviously very similar. An alkenyl carbonyl product is obtained just as in the oxy-Cope rearrangement, but the carbonyl is generated directly rather than via initial enol formation. As with the Cope rearrangement, the Claisen rearrangement proceeds via a chair-like transition state (see calculated geometry shown in Figure 11.23).[537] The chair transition state was reported to be favored over the boat by 2.5-3.0 kcal mol^{-1} (10.5-12.6 kJ mol^{-1}).[538] In different work, Houk and co-workers estimated that the chair transition state for this reaction is more stable

533. (a) Claisen, L. *Berichte* **1912**, *45*, 3157; (b) Claisen, L.; Tietze, E. *Ibid.* **1925**, *58*, 275; (c) *Idem Ibid.* **1926**, *59*, 2344; (d) Tarbell, D.S. *Org. React.* **1944**, *2*, 1; (e) The Merck Index, 14th ed., Merck & Co., Inc., Whitehouse Station, New Jersey, **2006**, p ONR-18; (f) Mundy, B.P.; Ellerd, M.G.; Favaloro Jr., F.G. *Name Reactions and Reagents in Organic Synthesis, 2nd ed.*, Wiley-Interscience, New Jersey, **2005**, pp. 156-159.
534. See reference 65, pp 1302-1303.
535. Sampath Kumar, H.M.; Anjaneyulu, S.; Subba Reddy, B.V.; Yadav, J.S. *Synlett* **2000**, 1129.
536. Martín Castro, A.M. *Chem. Rev.* **2004**, *104*, 2939.
537. Yamabe; S.; Okumoto, S.; Hayashi, T. *J. Org. Chem.* **1996**, *61*, 6218.
538. (a) Vittorelli, P.; Winkler, T.; Hansen, J.J.; Schmid, H. *Helv. Chim. Acta* **1968**, *51*, 1457; (b) Hansen, H.J.; Schmid, H. *Tetrahedron* **1974**, *30*, 1959.

than the boat transition state by about 6.6 kcal mol⁻¹ (27.6 kJ mol⁻¹).[539] Ziegler reviewed the synthetic applications of the Claisen rearrangement, and the stereoselectivity that characterizes this reaction is apparent in those applications.[540] A generic Claisen rearrangement is illustrated by allyl vinyl ether **628**, which rearranges to **630** via chair transition state **629**. The two methyl groups are equatorial in the transition state, a result of the (*E,E*)-geometry in **628**.[540,538]

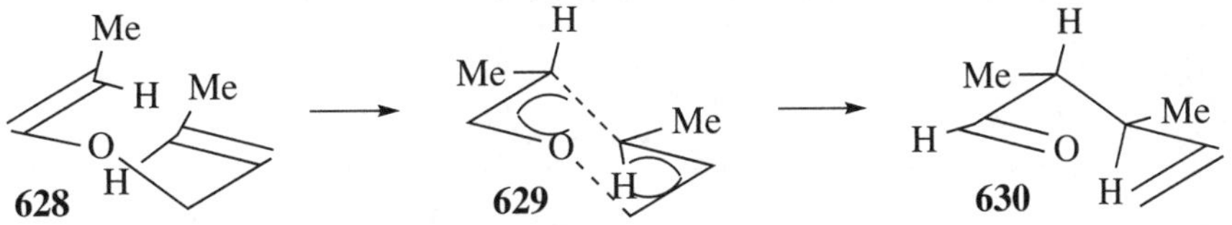

The (*E,Z*)- and (*Z,Z*)-derivatives will rearrange more slowly since one or both of the methyl groups will be axial in the transition state (**629**). When a substituent is present at C4 (see **626**), the substituent prefers the equatorial position in the transition state. The selectivity is about 99% when an alkyl group is present rather than a hydrogen (greater selectivity is observed when the substituents are -OR or -NR₂).[540,541] The energy requirements and stereochemical arguments are virtually identical with those discussed for the Cope rearrangement. A synthetic example is taken from a synthesis of garsubellin A,[542] in which Shibasaki and co-workers heated allyl vinyl ether **631** to 200°C in the presence of sodium acetate and obtained **632** in 96% yield. The allyl vinyl ether is a precursor to molecules containing both an aldehyde (or ketone) and an alkene.

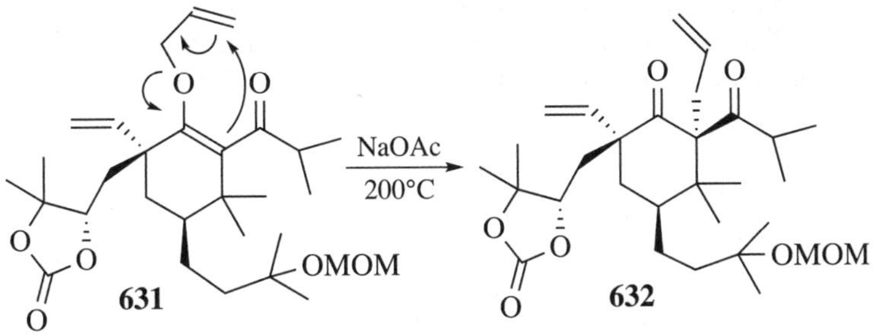

The Claisen rearrangement is not restricted to aliphatic derivatives. Rearrangement occurs with aryl allyl ethers such as **633**, which gave **634** in 80% yield, in Theodorakis and co-worker's synthesis of the cyclic core of *Garcinia* natural products.[543] These rearrangements can be sluggish, and a Lewis acid catalyst such as TiCl₄ is sometimes required.[544]

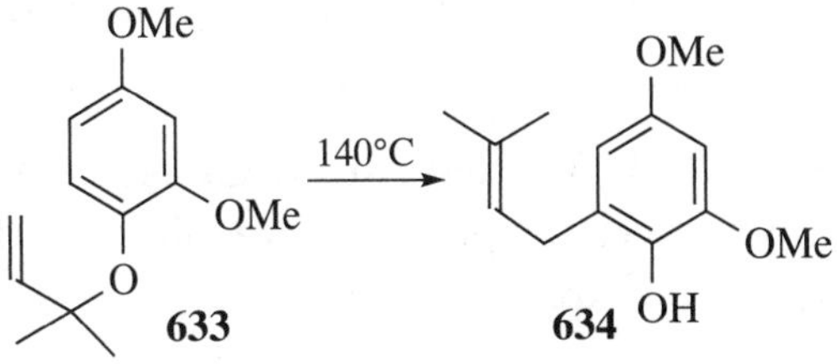

539. Vance, R.L.; Rondan, N.G.; Houk, K.N.; Jensen, F.; Borden, W.T.; Komornicki, A.; Wimmer, E. *J. Am. Chem. Soc.* **1988**, *110*, 2314.
540. Ziegler, F.E. *Acc. Chem. Res.* **1977**, *10*, 227.
541. Faulkner, D.J.; Petersen, M.R. *Tetrahedron Lett.* **1969**, 3243.
542. Kuramochi, A.; Usuda, H.; Yamatsugu, K.; Kanai, M.; Shibasaki, M. *J. Am. Chem. Soc.* **2005**, *127*, 14200.
543. Tisdale, E.J.; Chowdhury, C.; Vong, B.G.; Li, H.; Theodorakis, E.A. *Org. Lett.* **2002**, *4*, 909.
544. For an example, see Saidi, M.R. *Heterocycles* **1982**, *19*, 1473.

A catalyst is not always necessary for the Claisen rearrangement to occur.[545]

Lewis acids catalyze the Claisen rearrangement, often with improved stereoselectivity. Acyclic allyl vinyl ethers give a Claisen rearrangement upon treatment with a catalytic amount of Pd(II).[546] A Ho(fod) catalyst has been used as well.[547] An example of this type of reaction is the conversion of **635** to a 22:71 mixture of **636** and **637** in 99% yield.[548]

i-PrO$_2$C   O   C$_3$H$_7$ $\xrightarrow[\substack{\text{CH}_2\text{Cl}_2,\,\text{MS 3 Å}\\ \text{1 day}}]{0.05\ \text{Cu(OTf)}_2,\ 25°\text{C}}$ C$_3$H$_7$ O   CO$_2$*i*-Pr   +   C$_3$H$_7$ O   CO$_2$*i*-Pr

Bn   **635**   **636**   Bn   **637**   Bn

MOLECULAR MODELING: Stereochemistry of Claisen Rearrangements

Two enantiomeric products are typically formed in a Claisen rearrangement. In other words, the reaction gives a racemic product. When there is stereogenic center close to the reactive atoms, however, the substituent on that center acts as a "controller group" to generate a product enantioselectivity. For example, while Claisen rearrangement of (*S*)-4-(6-methyl-*E*-2-heptenyl) vinyl ether may lead to two different enantiomeric products, in fact, a single product is formed in >96% abundance.

H   O   Me   Me   O

    Me $\xrightarrow{\text{heat}}$ or   H

                   Me   O

                       H

Shown on screen are the two chair transition states that lead to the two different enantiomeric products. Choose the one that you think should be the lower in energy, and justify your selection. Hint. Consider the conformational analysis of six-membered rings in **Chapter 1**. Which of the two products shown corresponds to the product that results from the transition state you have chosen?

11.13.E.ii. Variants of the Claisen Rearrangement. Many methods are used to prepare the allyl vinyl ether precursors required for the Claisen rearrangement. Many of these preparations are rather specialized variations, usually named after the scientist(s) who developed them. One method used for the preparation of allyl vinyl ethers (**640**) is the acid catalyzed reaction of vinyl ethers such as **638** and allyl alcohols such as **639**. Vinyl ether **640** is formed *in situ* but usually not isolated since a Claisen rearrangement occurs under the reaction conditions to

545. Subramanian, R.S.; Balasubramanian, K.K. *Tetrahedron Lett.* **1988**, *29*, 6797.
546. (a) van der Baan, J.L.; Bickelhaupt, F. *Tetrahedron Lett.* **1986**, *27*, 6267; (b) Sugiura, M.; Nakai, T. *Chem. Lett.* **1995**, 697; (c) Hiersemann, M. *Synlett* **1999**, 1823.
547. Trost, B.M.; Schroeder, G.M. *J. Am. Chem. Soc.* **2000**, *122*, 3785.
548. Hiersemann, M.; Abraham, L. *Org. Lett.* **2001**, *3*, 49.

give **641**. This transformation has been termed the **Marbet-Saucy variant**[549] of the Claisen rearrangement (**Marbet-Saucy or Saucy-Marbet reaction**). A synthetic example is the treatment of **642** with 3-methoxyisoprene to generate **643** as a reactive intermediate. Subsequent Claisen rearrangement gave **644** in Faulkner and co-worker's synthesis of squalene.[550] Hsung and co-workers applied this reaction to the synthesis of highly substituted chiral homoallenyl alcohols using chiral ynamides.[551]

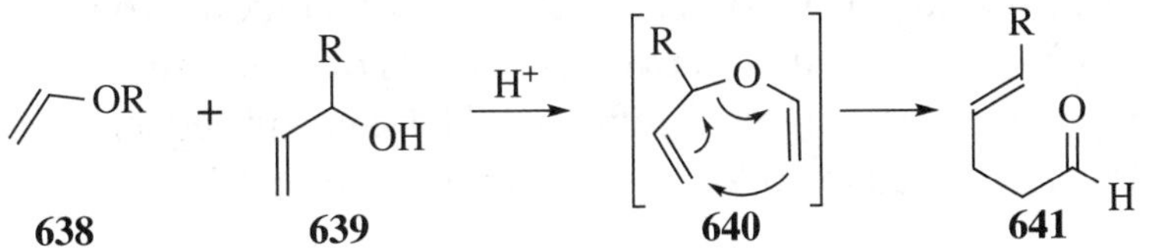

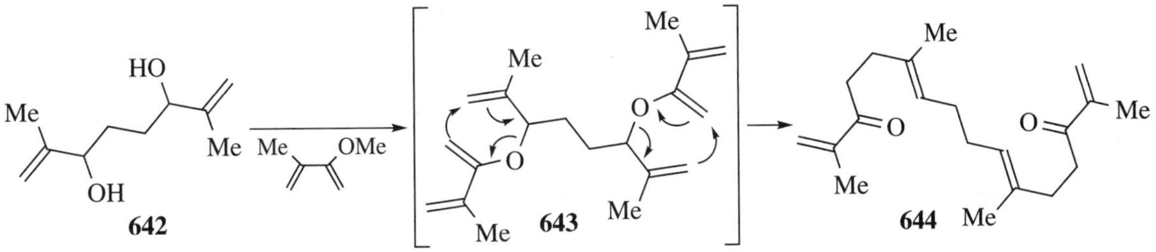

A similar variant of the Claisen rearrangement reacts allyl alcohols such as **645** with β-bromoselenides (**646**) to give **647**. Oxidation to the transient selenoxide (**648**) and *in situ* syn-elimination (sec. 2.9.C.vi) gave **649**. This ether possessed the proper functionality for Claisen rearrangement to the final product, ester **650**.[552] Petrzilka used a variation of this method for the synthesis of phoracantholide J.[553]

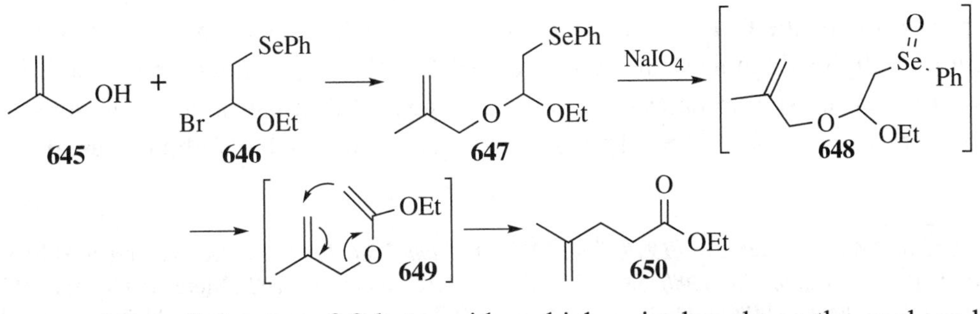

Carrol prepared the allyl ester of β-keto acids, which exist largely as the enol, and give a Claisen rearrangement to a β-keto acid.[554] This reaction is usually called the **Carroll variant**,

549. Marbet Saucy, G. *Helv. Chim. Acta* ***1967**, 50*, 2091, 2095.
550. Faulkner, D.J.; Petersen, M.R. *J. Am. Chem. Soc.* ***1973**, 95*, 553.
551. Frederick, M.P. Hsung, R.P.; Lambeth, R.H.; Mulder, J.A.; Tracey, M.R. *Org. Lett.* ***2003**, 5*, 2663.
552. Petrzilka, M. *Helv. Chim. Acta* ***1978**, 61*, 2286.
553. Petrzilka, M. *Helv. Chim. Acta* ***1978**, 61*, 3075.
554. Tanabe, M.; Hayashi, K. *J. Am. Chem. Soc.* ***1980**, 102*, 862.

the **Carroll-Claisen rearrangement** or the **Carroll reaction**,[555] although it has also been called the **Kimel-Cope rearrangement**.[556] Decarboxylation is facile at this reaction temperature. Heating **651** on neutral alumina led to Carroll reaction that was accompanied by loss of CO_2 to give a 72% yield of **652**, in Cossy, Meyer and co-worker's synthesis of zincophorin methyl ester,[557] Wilson introduced a variation of the Carrol reaction with mild conditions for the rearrangement.[558] Dianions of allylic acetoacetates rearrange at room temperature or in refluxing THF. When **653** was heated to 170°C for 2 h (neat), 50% of a mixture (67:33) of **654** and **655** was obtained. When **653** was treated with two equivalents of LDA (sec. 9.2.B and 9.4.B.vii), however, the enolate dianion rearranged in refluxing THF and subsequent decarboxylation in refluxing carbon tetrachloride led to an 84% yield of **654** with no **655**.[558]

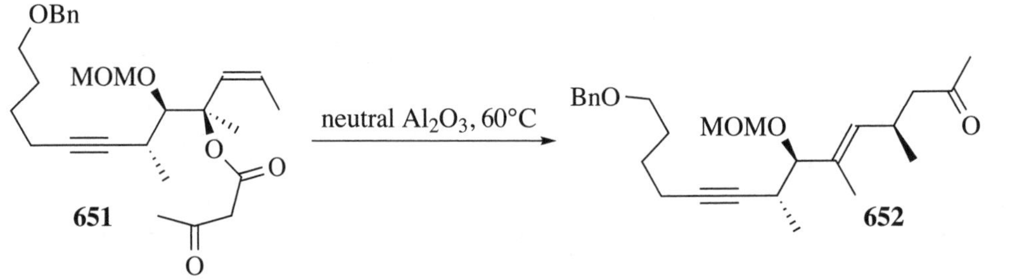

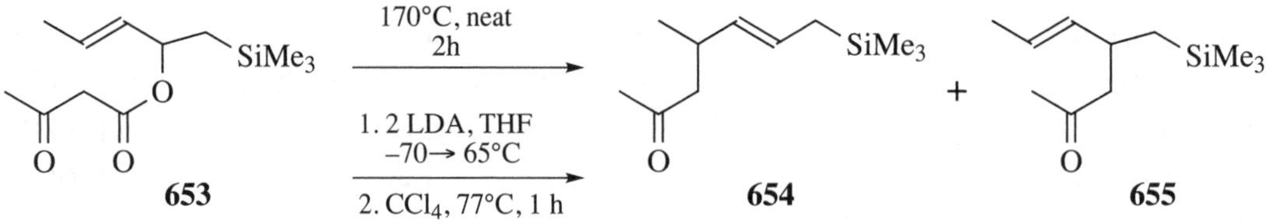

Another variation of the Claisen rearrangement involves ketene acetals, which are generated by reaction of allylic alcohols with ortho esters [$R^1CH_2C(OEt)_3$] under acidic conditions. The resulting product (**656**) eliminates ethanol to give **657**, and this product undergoes a Claisen rearrangement to give ester **658**. This sequence is known as the **Johnson-ester variant**[559]

555. (a) Carroll, M.F. *J. Chem. Soc.* **1940**, 704, 1266; (b) *Idem Ibid.* **1941**, 507; (c) Reference 80b, p 300-302; (d) Ziegler, F.E. *Chem. Rev.* **1988**, *88*, 1423; (e) The Merck Index, 14th ed., Merck & Co., Inc., Whitehouse Station, New Jersey, **2006**, p ONR-16; (f) Mundy, B.P.; Ellerd, M.G.; Favaloro Jr., F.G. *Name Reactions and Reagents in Organic Synthesis, 2nd ed.*, Wiley-Interscience, New Jersey, **2005**, pp. 134-135.
556. Kimel, W.; Cope, A.C. *J. Am. Chem. Soc.* **1943**, *65*, 1992.
557. (a) Defosseux, M.; Blanchard, N.; Meyer, C.; Cossy, J. *Org. Lett.* **2003**, *5*, 4037; (b) Idem *J. Org. Chem.* **2004**, *69*, 4626.
558. Wilson, S.R.; Price, M.F. *J. Org. Chem.* **1984**, *49*, 722.
559. (a) Johnson, W.S.; Werthemann, L.; Bartlett, W.R.; Brocksom, T.J.; Li, T.T.; Faulkner, D.J.; Petersen, M.R. *J. Am. Chem. Soc.* **1970**, *92*, 741; (b) Johnson, W.S.; Gravestock, M.B.; Parry, R.J.; Myers, R.F.; Bryson, T.A.; Miles, H. *Ibid.* **1971**, *93*, 4330; (c) Reference 80b, p 302-308; (d) Mundy, B.P.; Ellerd, M.G.; Favaloro Jr., F.G. *Name Reactions and Reagents in Organic Synthesis, 2nd ed.*, Wiley-Interscience, New Jersey, **2005**, pp. 350-351.

of the Claisen rearrangement, or the **ortho ester Claisen rearrangement.**[560] An example is taken from a synthesis of (–)-cucumin H by Srikrishna and Dethe, in which allylic alcohol **659** was heated with triethyl orthoacetate in propionic acid to give an 80% yield of the Claisen rearrangement product **661** via **660**.[561]

656 **657** **658**

659 **660** **661**

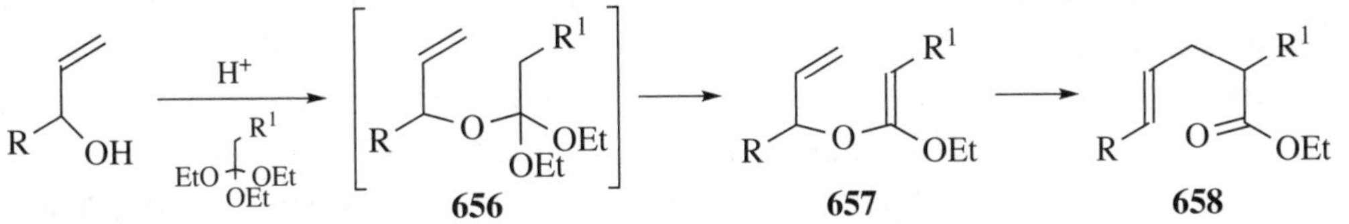

A related variation was reported by Eschenmoser, who showed that heating an allylic alcohol with *N,N*-dimethylacetamide dimethyl acetal (**662**) gave **663**. This transient intermediate eliminated methanol to give **664**, and subsequent Claisen rearrangement generated amide **665** in a sequence that is known as the **Eschenmoser variant** or the **Eschenmoser-Claisen rearrangement.**[562] Corey and co-worker's synthesis of thromboxane B_2 utilized this methodology.[563] Another synthetic example is the conversion of **666** to **667** in good yield, as part of Kobayashi and co-worker's synthesis of tuberonic acid.[564]

662 **663** **664** **665**

666 **667**

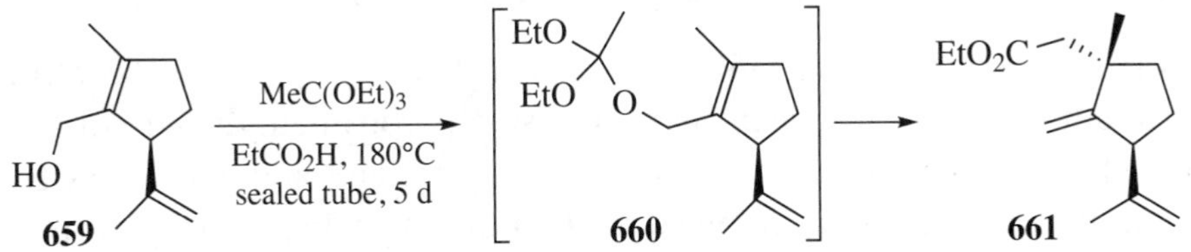

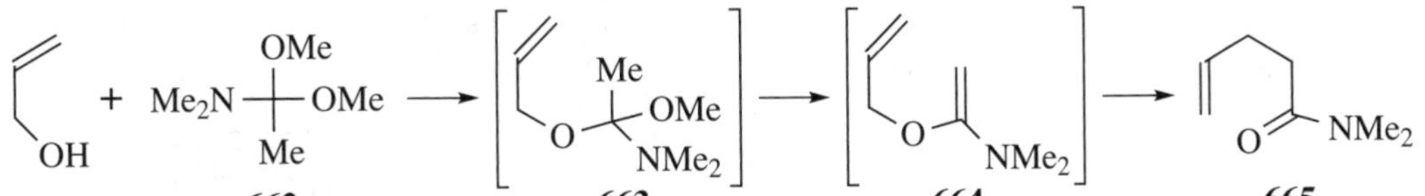

560. For a discussion of acyclic stereoselection in this reaction, see Daub, G.W.; Edwards, J.P.; Okada, C.R.; Allen, J.W.; Maxey, C.T.; Wells, M.S.; Goldstein, A.S.; Dibley, M.J.; Wang, C.J.; Ostercamp, D.P.; Chung, S.; Cunningham, P.S.; Berliner, M.A. *J. Org. Chem.* ***1997***, *62*, 1976.
561. Srikrishna, A.; Dethe, D.H. *Org. Lett.* ***2003***, *5*, 2295.
562. (a) Felix, D.; Gschwend-Steen, K.; Wick, A.E.; Eschenmoser, A. *Helv. Chim. Acta* ***1969***, *52*, 1030; (b) Reference 80b, pp. 309-311.
563. Corey, E.J.; Shibasaki, M.; Knolle, J. *Tetrahedron Lett.* ***1977***, 1625.
564. Nonaka, H.; Wang, Y.-G.; Kobayashi, Y. *Tetrahedron Lett.* ***2007***, *48*, 1745..

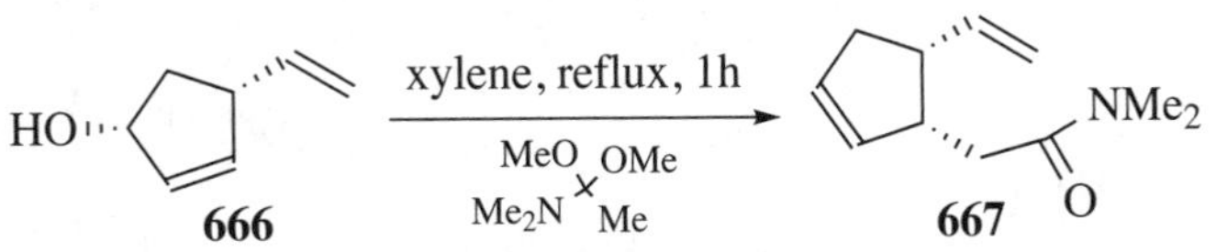

Esters can be converted to the corresponding enolate via treatment with base and can rearrange, as observed by Arnold and Hoffman.[565] When the enolate was trapped as the O-trialkylsilyl ether, a 1,5-diene was generated that gave a Claisen rearrangement under milder conditions. The solvent has a great influence on the selectivity of the reaction, and THF is commonly used.[566] The enolate Claisen rearrangement occurs at ambient temperatures, whereas most Claisen rearrangements require temperatures of 120-200°C.[567] The enolate accelerated Claisen rearrangement is obviously similar to the enolate accelerated Cope rearrangement described by Majetich in sec. 11.12.C.[520] If hexamethylphosphoramide (HMPA) is added, the lithium ion (Li⁺) is solvated. Lithium coordination with oxygen is poor and the syn-isomer predominates. If THF is used, O↔Li coordination is strong and the anti-isomer is the major product.

In section 9.4.B,C, an ester such as **668** reacted with LDA under kinetic control conditions to form enolate anion **669**, which then reacted with chlorotrimethylsilane to give silyl enol ether **670**. Compounds such as **670** were prepared by Ireland and Mueller, who found that subsequent Claisen rearrangement followed by hydrolysis of the initially formed silyl ester gave acid **671**, in what has come to be known as the **Ireland variant** or the **Ireland-Claisen rearrangement**.[568] There are several stereoselective variations of this reaction.[569] An example of the reaction is taken from a synthesis of galbonolide B by Parsons and co-workers,[570] in which formation of the silyl enol ether of ester **672** under the conditions shown led to a 75% yield of the corresponding acid **673**.

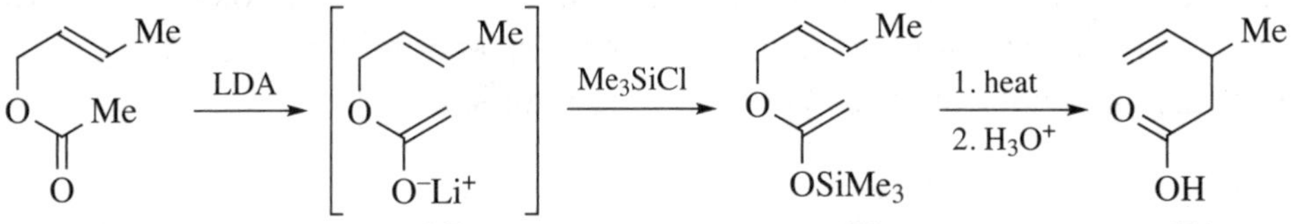

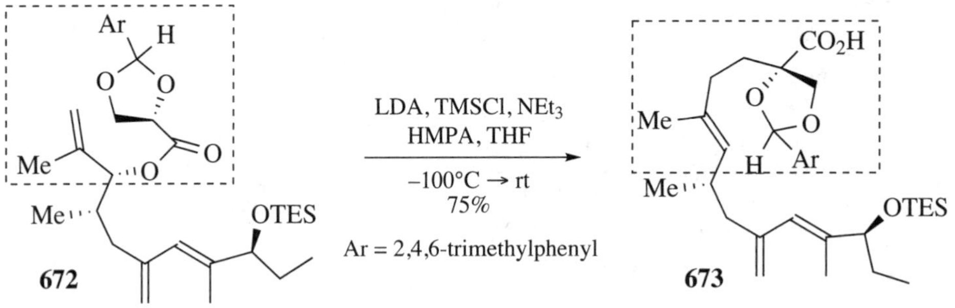

565. Arnold, R.T.; Hoffmann, C. *Synth. Commun.* **1972**, *2*, 27.
566. Ireland, R.E.; Mueller, R.H.; Willard, A.K. *J. Am. Chem. Soc.* **1976**, *98*, 2868.
567. Reference 80b, p 312.
568. (a) Ireland, R.E.; Mueller, R.H. *J. Am. Chem. Soc.* **1972**, *94*, 5897; (b) Pereira, S.; Srebnik, M. *Aldrichimica Acta* **1993**, *26*, 17; (c) Reference 80b, p 311-319; (d) Mundy, B.P.; Ellerd, M.G.; Favaloro Jr., F.G. *Name Reactions and Reagents in Organic Synthesis, 2nd ed.*, Wiley-Interscience, New Jersey, **2005**, pp. 340-341.
569. Chai, Y.; Hong, S-p.; Kubdsaym H.A.; McFarland, C.; McIntosh, M.C. *Tetrahedron* **2002**, *58*, 2905.
570. Parsons, P.J.; Pennicott, L.; Eshelby, J.; Goessman, M.; Highton, A.; Hitchcock, P. *J. Org. Chem.* **2007**, *72*, 9387.

MOLECULAR MODELING: Ireland-Claisen Rearrangement

The Ireland-Claisen rearrangement involves formation of an enolate anion of an allylic ester. Sigmatropic rearrangement leads to a carboxylic acid product. The enolate anion of **672** is trapped by reaction with trimethylsilyl chloride to give a mixture of E and Z stereoisomers of the intermediate **A**. Each may undergo a Claisen rearrangement (presumably via a chair transition state), leading first to the intermediate **B** and finally to **673**. The stereochemistry of the product is determined in the reaction between intermediates **A** and **B**.

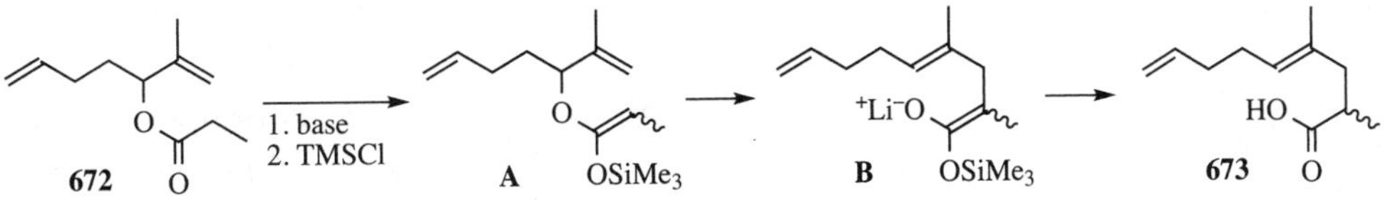

Quantum chemical calculations show that the transition state arising from the Z stereoisomer of **A** is 10 kJ/mol lower in energy than the transition state arising from the E stereoisomer. Examine the two transition states and, based on conclusions reached in previous molecular modeling problems, provide a reason for the preference.

Rate enhancement is possible, especially the enolate Claisen rearrangement. Denmark et al.[571] showed that other carbanionic centers accelerate the Claisen rearrangement. Generation of the anion of sulfone **674** (sec. 8.6.A) with various bases led to acceleration of the reaction relative to the thermal reaction of **674** and also influenced the syn/anti ratio (**675/676**), generally favoring **675**. In general, a donor group at the allyl position accelerates the rate and the presence of an amino stabilizing group increases the rate even more.

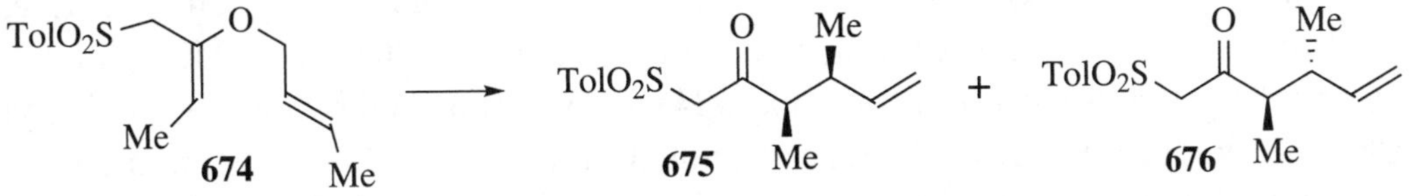

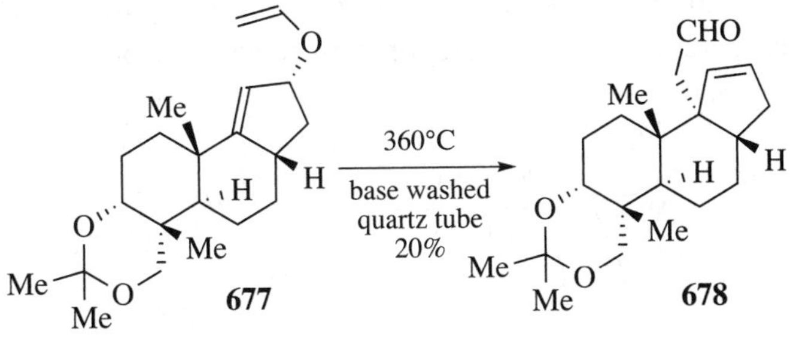

Aqueous solvents can accelerate the Claisen rearrangement in many cases, as observed in the Diels-Alder reaction (sec. 11.6.B). McMurry et al. showed that heating **677** to 360°C gave only a 20% yield of **678** in a synthesis of aphidicolin.[572] The difficulty lay in the sensitivity of the product to

571. (a) Denmark, S.E.; Harmata, M.A.; White, K.S. *J. Am. Chem. Soc.* **1989,** *111,* 8878; (b) Denmark, S.E.; Harmata, M.A. *Ibid.* **1982,** *104,* 4972.
572. McMurry, J.E.; Andrus, A.; Ksander, G.M.; Musser, J.H.; Johnson, M.A. *Tetrahedron* **1981,** *37 (Suppl 1),* 319.

the reaction conditions and the steric hindrance to formation of the correct chair transition state. When Grieco and co-workers heated the water soluble diol **679** to 80°C an 85% yield of **680** was obtained,[573] which dramatically demonstrated the rate enhancement possible in aqueous solvents.

As with Diels-Alder reactions, solubility of the Claisen precursor in the aqueous medium is required.

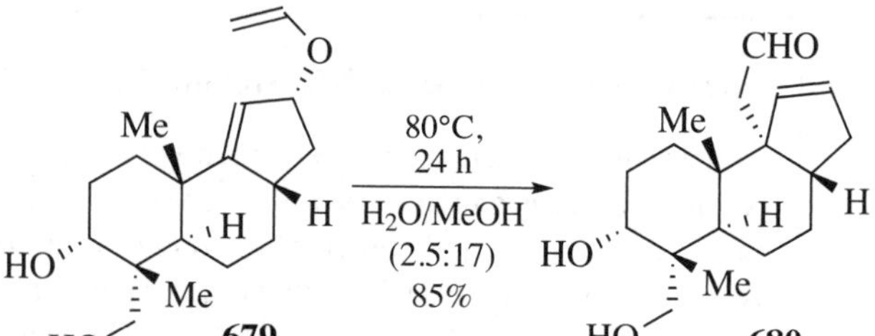

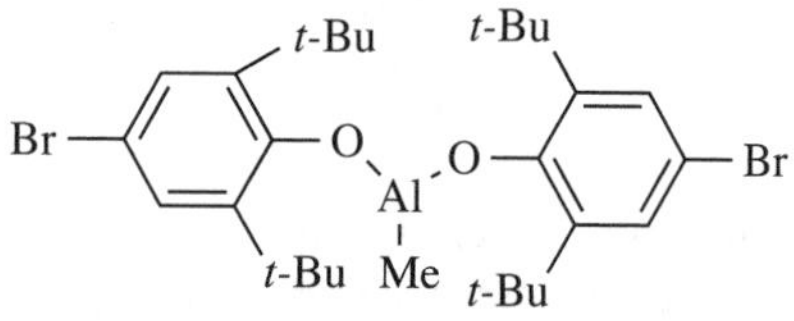

Many transition metal compounds catalyze the Claisen rearrangement[544] (see the reaction of **637** with copper triflate cited above). Aluminum reagents such as **681** or **682** can also catalyze the Claisen rearrangement with opposite stereochemistry, as reported by Yamamoto and co-workers.[574]

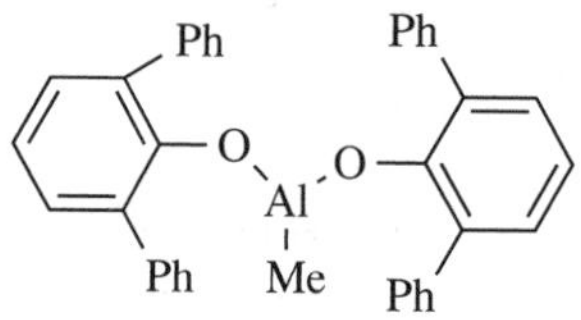

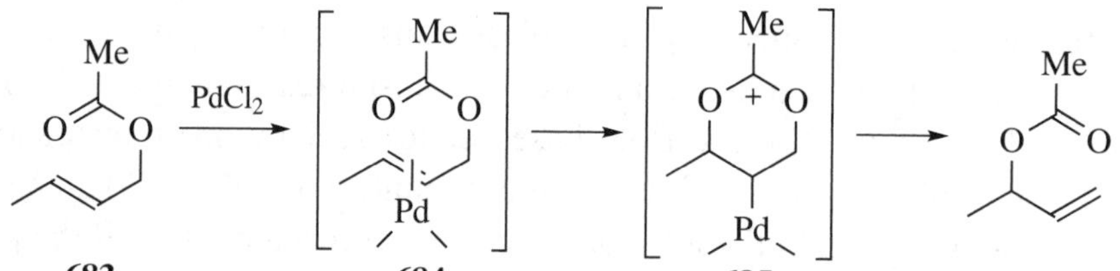

Palladium chloride (PdCl₂) is known to catalyze the rearrangement of allyl ethers in a reaction that looks very much like a Claisen transition state. Treatment of **683** with PdCl₂, for example, initially generated the π-palladium complex (**684**) that was attacked by the carbonyl oxygen to give **685**, and ring opening gave the rearrangement product **686**.[575,576] It is important to note, however, that this reaction proceeds in a step-wise manner via a π-allyl palladium intermediate and not via a concerted [3,3]-sigmatropic rearrangement (sec. 12.6 for a discussion of palladium-catalyzed reactions). Therefore, one must exercise care in categorizing reaction types. It is included here only because of the similarity in products obtained by this process and is not meant to imply that it proceeds by an identical mechanism to the Claisen rearrangement. Grieco

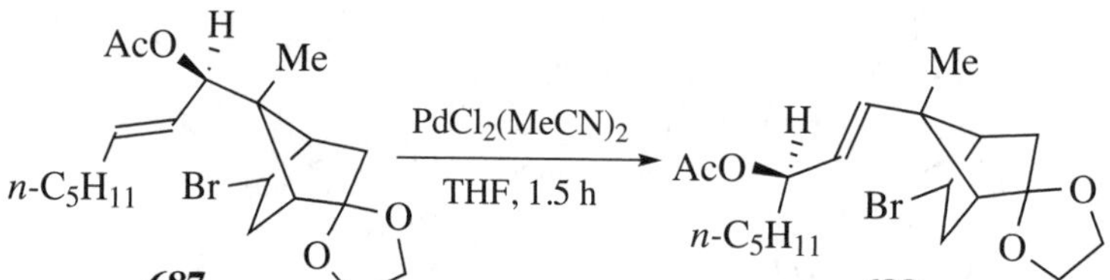

573. Grieco, P.A.; Brandes, E.B.; McCann, S.; Clark, J.D. *J. Org. Chem.* **1989**, *54*, 5849.
574. Nonoshita, K.; Banno, H.; Maruoka, K.; Yamamoto, H. *J. Am. Chem. Soc.* **1990**, *112*, 316.
575. Henry, P.M. *J. Am. Chem. Soc.* **1972**, *94*, 5200.
576. Kitching, W.; Rappoport, Z.; Winstein, S.; Young, W.G. *J. Am. Chem. Soc.* **1966**, *88*, 2054.

and co-workers used a similar palladium catalyzed Claisen rearrangement to convert **687** to **688** (93% yield) in a synthesis of 15-epi-12-methyl prostaglandin $F_{2\alpha}$.[577]

11.12.E.iii. thio-Claisen and aza-Claisen Rearrangements.

Heteroatoms other than oxygen can be used. The two most common are sulfur and nitrogen (giving thio-Claisen and aza-Claisen rearrangements, respectively). Heating allyl vinyl sulfides (**689**) leads to a

[3,3]-rearrangement that gives thioaldehyde **690**, and subsequent hydrolysis generates an aldehyde **691**.[578] This variant is known as the **thio-Claisen rearrangement**.[579]

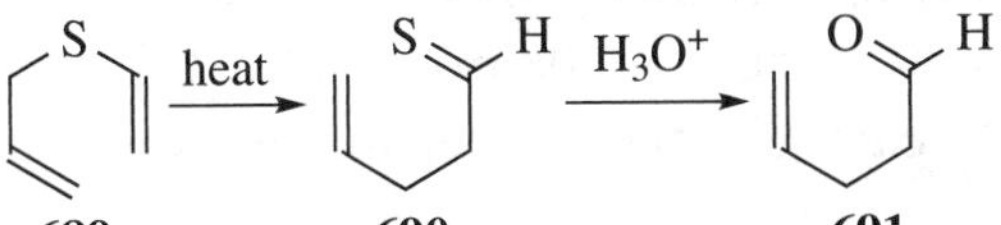

The thio-Claisen rearrangement was the basis of an **umpolung** reagent discussed in Section 8.6.B.vi. Treatment of **689** with *sec*-butyllithium gave the α-lithio sulfide (**692**), and condensation with benzyl bromide gave the alkylated sulfide, **693**. Heating led to a thio-Claisen rearrangement and hydrolysis of the resulting product gave aldehyde **694** in 62% overall yield.[580] This sequence was correlated with umpolung equivalent **695**.[580]

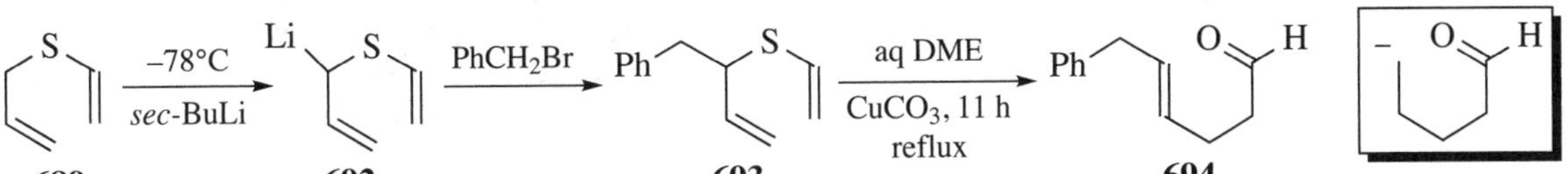

Iminium salts have been used in **aza-Claisen rearrangement** reactions (also see the aza-Cope rearrangement in sec. 11.12.C.iii).[530,581] In the generic reaction, heating *N*-allylic enamines led to an imine that could be hydrolyzed to an aldehyde,[582] or converted to another nitrogen-containing moiety. Lewis acids such as titanium tetrachloride (TiCl₄) catalyze the aza-Claisen rearrangement. When aldehyde **696** reacted with 3-(*N*-phenyl)-1-butene (**697**) in the presence of TiCl₄, the initially formed enamine (**698**) rearranged to **699** and was hydrolyzed to aldehyde **700** in 61% overall yield.[583]

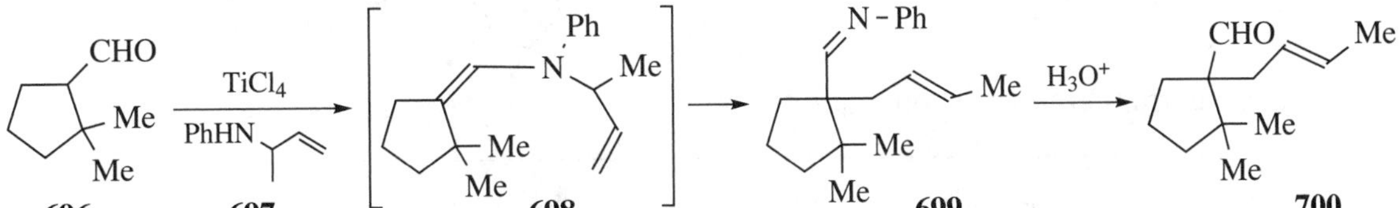

577. Grieco, P.A.; Takigawa, T.; Bongers, S.L.; Tanaka, H. *J. Am. Chem. Soc.* **1980**, *102*, 7587.
578. Oshima, K.; Takahashi, H.; Yamamoto, H.; Nozaki, H. *J. Am. Chem. Soc.* **1973**, *95*, 2693.
579. Majumdar, K.C.; Ghosh, S.; Ghosh, M. *Tetrahedron* **2003**, *59*, 7251.
580. Takano, S.; Hirama, T.; Araki, T.; Ogasawara, K. *J. Am. Chem. Soc.* **1976**, *98*, 7084.
581. Makisumi, Y. *Tetrahedron Lett.* **1966,** 6413.
582. Hill, R.K.; Gilman, N.W. *Tetrahedron Lett.* **1967**, 1421.
583. Hill, R.K.; Khatri, H.N. *Tetrahedron Lett.* **1978**, 4337.

Stille and Cook developed an attractive route to aza-Claisen precursors. Allylamine reacted with 2-methylpropanal (isobutyraldehyde), for example, to give allyl imine 701.[584] When 701 reacted with 2-methylpropanoyl chloride, N-acyl derivative 702 was produced (94% overall yield from allylamine) and reduction with LiAlH$_4$ give a 98% yield of tertiary amine 703. Stille found that conversion of the amine to the ammonium salt by treatment with HCl led to clean conversion to 704 in 82% yield via a [3,3]-sigmatropic rearrangement (an aza-Claisen rearrangement). Reduction of the iminium salt led to an 81% yield of 705 from 704. Stille and co-workers also found that Lewis acids as well as HCl react with the allyl enamine precursor to facilitate the aza-Claisen rearrangement.[585]

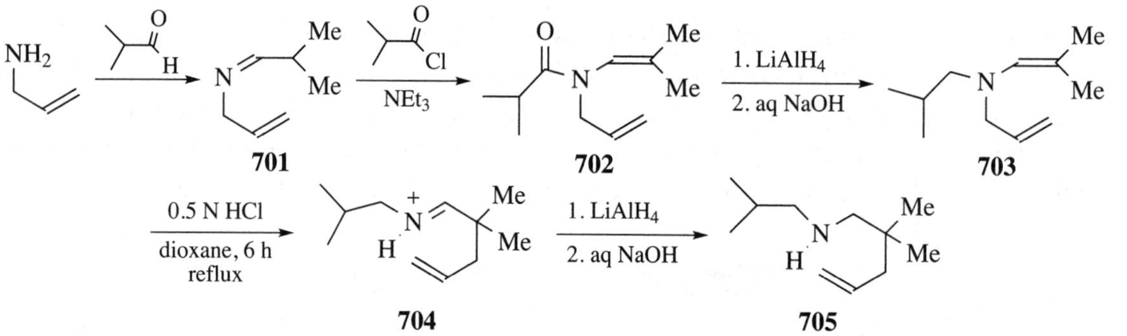

Other types of iminium salts can be used this reaction. β-Carboline 706 was reduced and converted to iminium salt 707 by treatment with formaldehyde (secs. 4.4.B, 4.10.D). The aza-Claisen rearrangement (sometimes called the **2-azonia rearrangement** in this system) was used by Winterfeldt and co-workers[586] to generate 15β-methoxyyohimban-17-ol).

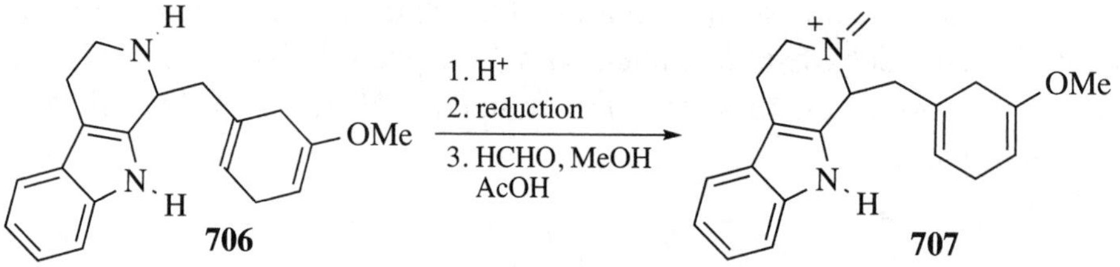

The Claisen rearrangement disconnections are

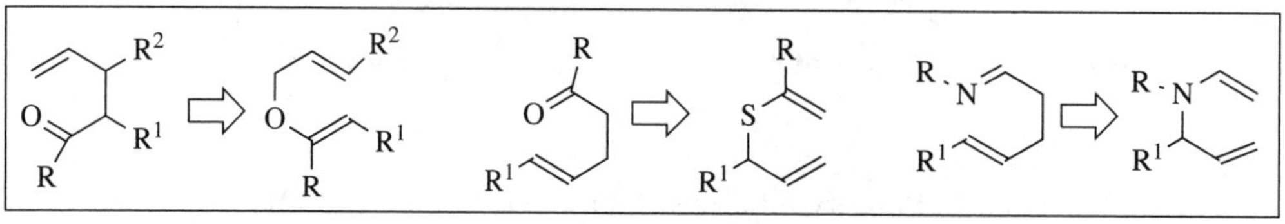

584. (a) Cook, G.R.; Stille, J.R. *J. Org. Chem.* **1991**, *56*, 5578; (b) Cook, G.R.; Barta, N.S.; Stille, J.R. *Ibid.* **1992**, *57*, 461.
585. Barta, N.S.; Cook, G.R.; Landis, M.S.; Stille, J.R. *J. Org. Chem.* **1992**, *57*, 7188.
586. Ahmad, V.U.; Feuerherd, K.H.; Winterfeldt, E. *Chem. Ber.* **1977**, *110*, 3624.

11.14. THE ENE REACTION

The Diels-Alder reaction is a thermal cycloaddition involving 1,3-dienes and alkenes, and [3+2]-cyclo-addition reactions involve a π bond and a 1,3-dipole. The Cope, oxy-Cope and Claisen rearrangements are thermal, intramolecular reactions of 1,5-dienes. A [2+2]-cycloaddition usually involves reaction between two alkenes, or certainly two π bonds. A reaction that is different from any seen so far occurs with certain alkenes and allylic systems. In its fundamental form, it is "the indirect substituting of a compound with a double bond (**enophile**) to an alkene that has an allylic hydrogen (the **ene**)"[587] to give a new, substituted alkene. This transformation is called the **ene reaction**.[588] Although several related examples of this reaction were reported early in the 20th century,[587] it was not until the work of Alder and co-workers in 1943[589] that the ene reaction was recognized as a synthetically useful tool. Because of this work, this transformation is often referred to as the **Alder ene reaction**. Oppolzer often used magnesium-ene cyclizations in syntheses.[590] Oppolzer and co-workers also developed a Pd-catalyzed Zn-ene cyclization using allyl acetate precursors.[591]

11.14.A. Enes and Enophiles

The ene reaction involves an alkene fragment (the ene) that removes a hydrogen from an allylic fragment with formation of a new carbon bond, as in **708**.[592,264] As with the other reactions in this chapter, reactivity and stereochemistry in the ene reaction can be explained by frontier orbital theory. The reaction proceeds via interaction of the HOMO of the alkene (ene) and the LUMO of the allylic partner (enophile),[593] illustrated by the reaction of propene and ethene in Figure 11.24, with the transition structure calculated by Houk and Loncharich.[594] The intermolecular reaction usually requires very high temperatures (typically 250-600°C). 2-Methylpropene reacts with acetylene to give a 40% yield of 2-methyl-1,4-pentadiene, for example, but the reaction requires a large excess of isobutylene at 170 atmospheres.[595] The fact

587. Hoffmann, H.M.R. *Angew. Chem. Int. Ed.* **1969**, *8*, 556.
588. (a) See reference 65, p 236; (b) The Merck Index, 14th ed., Merck & Co., Inc., Whitehouse Station, New Jersey, **2006**, p ONR-28; (c) Mundy, B.P.; Ellerd, M.G.; Favaloro Jr., F.G. *Name Reactions and Reagents in Organic Synthesis, 2nd ed.*, Wiley-Interscience, New Jersey, **2005**, pp. 226-227.
589. Alder, K.; Pascher, F.; Schmitz, A. *Berichte* **1943**, *76*, 27.
590. (a) Oppolzer, W. *Angew. Chem. Int. Ed.* **1989**, *28*, 38; (b) Oppolzer, W. *Pure Appl. Chem.* **1990**, *62*, 1941; (c) Oppolzer, W. in *Comprehensive Organic Synthesis*, Trost, B.M., Ed., Pergamon Press, New York, **1991**, Vol. 5, pp 29-61.
591. (a) Oppolzer, W.; Schröder, F. *Tetrahedron Lett.* **1994**, *35*, 7939; (b) See Yasui, K.; Goto, Y.; Yajima, T.; Taniseki, Y.; Fugami, K.; Tanaka, A.; Tamaru, T. *Tetrahedron Lett.* **1993**, *34*, 7619.
592. Taber, D.F. *Reactivity and Structure Concepts in Organic Chemistry, Vol. 18*, Springer-Verlag, Berlin, **1984**.
593. Oppolzer, W.; Snieckus, V. *Angew. Chem. Int. Ed.* **1978**, *17*, 476.
594. Loncharich, R.J.; Houk, K.N. *J. Am. Chem. Soc.* **1987**, *109*, 6947.
595. Cywinski, N.F. *J. Org. Chem.* **1965**, *30*, 361.

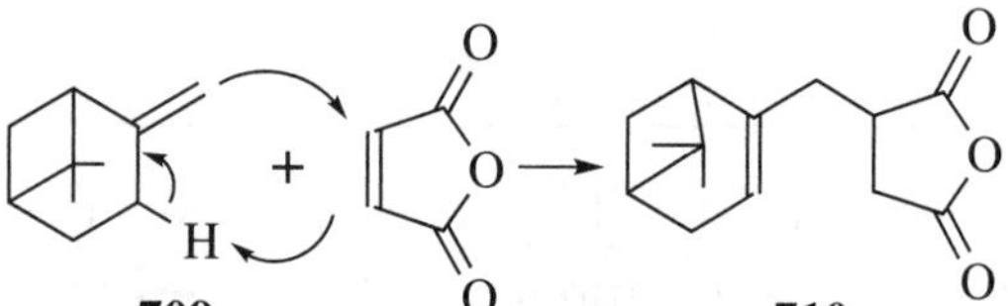

that the reaction is concerted (rather than involving a cationic intermediate) can be illustrated by the ene reaction of β-pinene (**709**) and maleic anhydride to give **710**, without skeletal rearrangement of the pinene ring.[596] In general, allylic alkenes react faster than allylic alkynes in the ene reaction.[587] This point is illustrated by the vigorous conditions of the previously cited reaction of acetylene. If the alkene contains a strained double bond, the ene reaction is accelerated (lower reaction temperatures). Caryophyllene (**711**) contains a strained trans-double bond and gave an ene reaction with maleic anhydride under very mild conditions (refluxing benzene) to give **712**, but the yield was only 19%.[597]

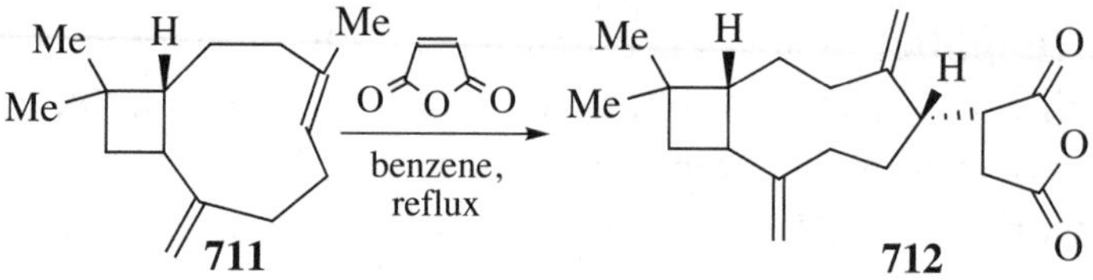

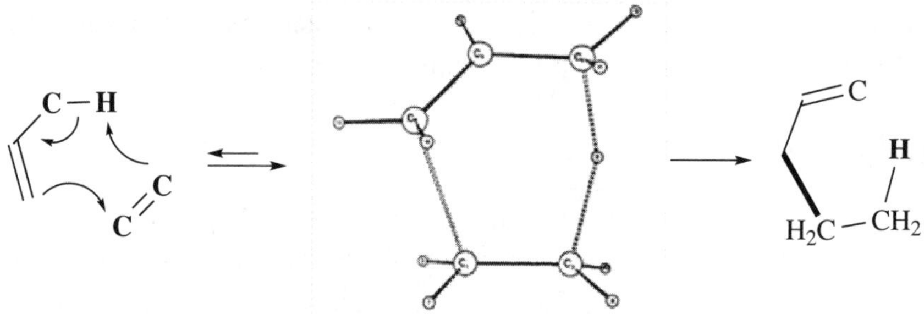

Figure 11.24. Calculated Transition State For the Alder Ene Reaction of Propene And Ethylene. [Reprinted with permission from Loncharich, R.J.; Houk, K.N. *J. Am. Chem. Soc.* **1987**, *109*, 6947. Copyright © 1987 American Chemical Society.]

MOLECULAR MODELING: Transition State for an Ene Reaction

Transition states for chemical reactions cannot be observed experimentally, simply because they are not energy minima and do not form collections on which measurements may be made. However, the structures, energies and "properties" of transition states may be calculated. Propene and maleic anhydride undergo an ene reaction to yield allylsuccinic anhydride.

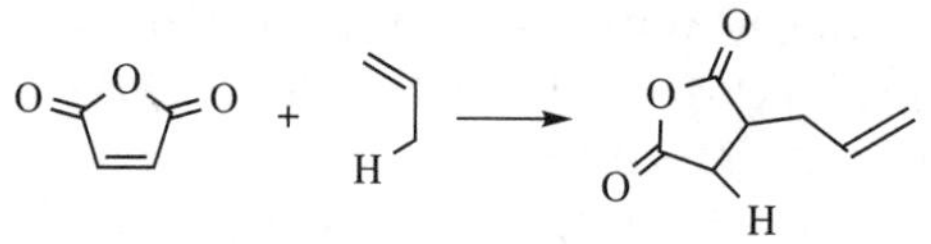

The transition state for this reaction is displayed on screen. Measure bond distances for the C–C bonds that are broken and formed and compare these to the corresponding distances in the reactants. Build propene. *Click* on > and then select **Continue** to get a model-building palette. Repeat the process for

---

596. Arnold, R.T.; Showell, J.S. *J. Am. Chem. Soc.* **1957**, *79*, 419.
597. (a) Nickon, A. *J. Am. Chem. Soc.* **1955**, *77*, 1190; (b) Ruzicka, L.; Plattner, P.A.; Balla, G. *Helv. Chim. Acta* **1941**, *24*, 1219; (c) Ruzicka, L.; Zimmerman, W. *Ibid.* **1935**, *18*, 219; (d) Rydon, H.N. *J. Chem. Soc.* **1939**, 537.

maleic anhydride. Describe changes to C–C bond distances in going from reactants to transition state. Are the changes to the C–O single and double bonds as large as those to the C–C bonds? Is the migrating hydrogen bound more tightly to propene or to the product (or is it roughly midway in between)?

The "infrared spectrum" of a transition state cannot be observed, but it can be calculated. Each line in the spectrum corresponds to a vibration that a molecule undergoes. The frequency of the line is proportional to the **square root** of the so-called force constant, that is, the difficulty of moving ("up") from an energy well. Motions that are "easy" (for example, changes in torsional angles) give rise to low frequencies, whereas motions that are "difficult" (for example, CH bond stretches) give rise to high frequencies. One vibration in the infrared spectrum of a transition state corresponds to a "downhill" motion, that is, the force constant is negative. Taking the square root gives rise to an imaginary frequency. *Click* on this frequency in the spectrum (designated as "negative" 1500 cm^{-1}) to see the motion. Is it almost entirely associated with the migrating hydrogen or do other atoms move as well?

Carbonyl compounds are enophiles in the ene reaction, particularly when the enophilic carbonyl is conjugated to another carbonyl moiety. Diethyl mesoxalate (**713**), for example, reacted with 1-pentene to give **714**.[598]

Intramolecular ene reactions have been studied in great detail.[599,405] The most common ene precursors are 1,6- and 1,7-dienes, with 1,6-dienes the more reactive and more useful. The

product of the ene reaction of a 1,6-diene is a five-membered ring derivative. Cyclization of 1,6-octadiene (**715**, a cis-trans mixture) at 457°C[600] gave exclusively cis-isomer **717** in 35% yield, via a transition state that is believed to resemble **716**. Note that internal ene reactions typically give a mixture of diastereomers.[601]

Oppolzer et al. described the stereochemistry for cis- and trans-enophiles in the ene reaction. When amide **718** containing a cis- alkene unit and amide **720** containing a trans-alkene unit reacted via an ene reaction, the same product (**719**) was formed.[602] The temperature used for the reaction had a large influence on product distribution and stereochemistry. Cyclization of

598. Achmatowicz, O.; Achmatowicz, Jr., O. *Roczniki Chim.* **1962,** *36,* 1791 [*Chem. Abstr. 59:* 8610b **1963**].
599. Reference 592, p 61-94.
600. Huntsman, W.D.; Solomon, V.C.; Eros, D. *J. Am. Chem. Soc.* **1958,** *80,* 5455.
601. Pines, H.; Hoffman, N.E.; Ipatieff, V.N. *J. Am. Chem. Soc.* **1954,** *76,* 4412.
602. Oppolzer, W.; Pfenninger, E.; Keller, K. *Helv. Chim. Acta* **1973,** *56,* 1807.

721 at 450°C, for example, gave a mixture of **722**, **723** and **724** (9:25:41). Ene reaction of **721** at 290°C gave the same products, but in a ratio of 2:7:40.[603] Good selectivity is possible in the ene reaction, as in the conversion of **725** to **726** in 60% yield at 250°C, in Stork's explorations toward the synthesis of prostaglandins.[604] In this case, the silyl group on the secondary hydroxyl was selectively removed by treatment with dilute HCl in THF (sec. 7.3.A.i). This selectivity may be due to increased steric hindrance at that position or to a rate-determining oxygen protonation step, or both.[604] Oppolzer and Bättig used an ene reaction to convert **727** to **729** via an endo-transition state (see **728**) in a synthesis of modhephene, with good selectivity.[605] The conformational constraints imposed by the molecular structure of the ene led to high selectivity in the rearrangement products. Conjugated alkynes are also enophiles.[606]

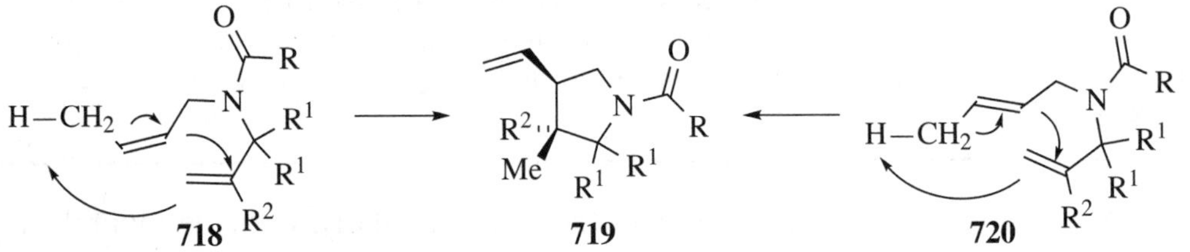

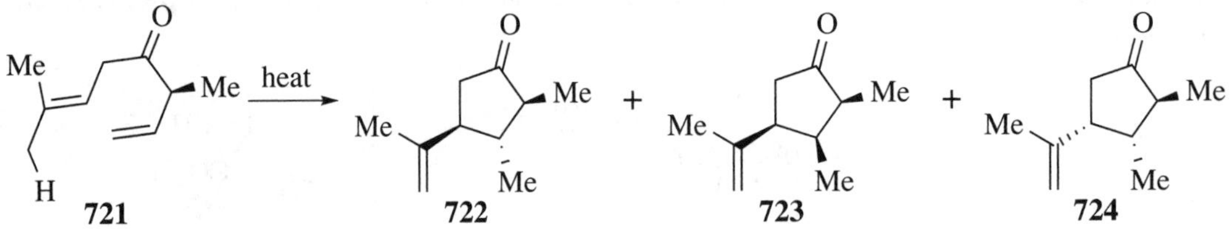

	722		723		724
450°C	(9	:	25	:	41)
290°C	(2	:	7	:	40)

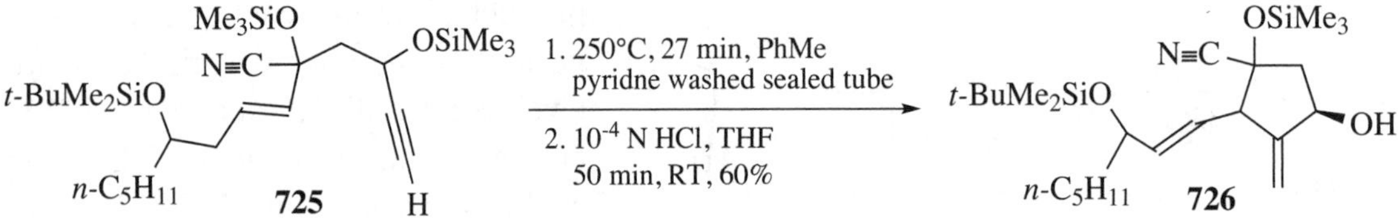

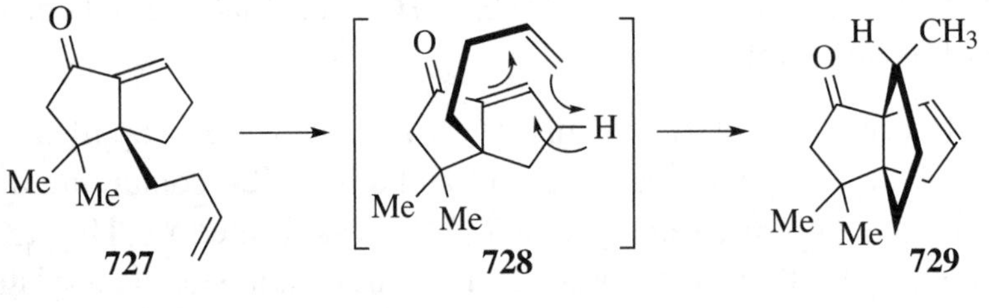

603. Schulte-Elte, K.H.; Gadola, M.; Ohloff, G. *Helv. Chim. Acta* **1971**, *54*, 1813.
604. Stork, G.; Kraus, G. *J. Am. Chem. Soc.* **1976**, *98*, 6747.
605. Oppolzer, W.; Bättig, K. *Helv. Chim. Acta* **1981**, *64*, 1575.
606. Takahashi, K.; Mikami, K.; Nakai, T. *Tetrahedron Lett.* **1988**, *29*, 5277.

MOLECULAR MODELING: Transition-State Structures Anticipate Kinetic Product Distributions

Some ene reactions may give rise to two or more isomeric products. If the reaction is under kinetic control, the preferred product (and more generally, the distribution of products) may be anticipated by comparing the energies of transition states leading to the different products. A good example is provided by the intramolecular ene reaction of diene **A**, which is known to give primarily **B** rather than **C**.

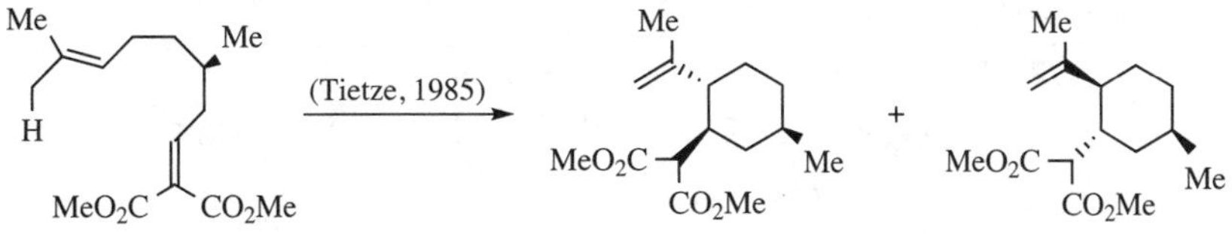

A **B** (>99%) **C**

The two pertinent transition state models obtained from quantum chemical calculations are provided on screen. The structure on the left is 8 kJ/mol lower in energy that that on the right. Draw the product that results from this transition state and confirm that it leads to the dominant product. Why do you think that the transition state on the left more stable?

Lewis acids are effective catalysts for the ene reaction in many ene-ene or ene-yne reactions, although they are most effective with carbonyl ene reactions (see below). Buchwald and co-workers,[607] for example, reported a titanocene-catalyzed intramolecular ene reaction in which ene-yne **730** was treated with $Cp_2Ti(CO)_2$ in hot toluene to give a 97% yield of ene product **731**.

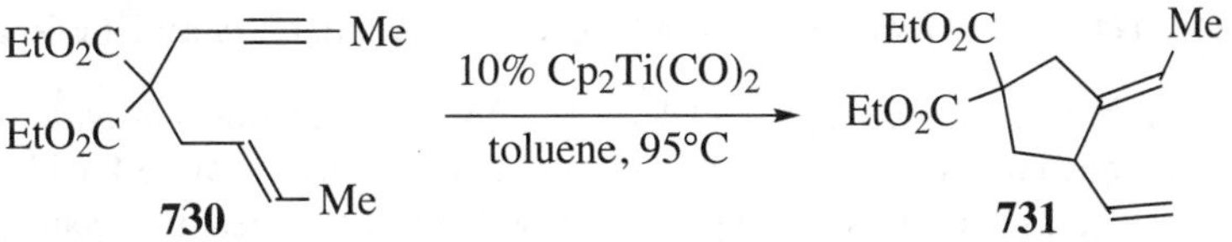

730 **731**

11.14.B. Ene Reactions of 1,7-Dienes

Ene reactions of 1,7-dienes are less common. Both six-membered ring products (such as **732** via path *a*) and eight-membered ring products (such as **733** via path *b*) are possible.[587] The first ene reaction with a 1,7-diene was probably reported by Hunstman et al.[608] Six-membered rings are usually preferred to eight- membered rings, but ene reactions of 1,7-dienes require higher reaction temperatures. Heating 8-methyl-1,7-nonadiene (**734**, R = H) to 440°C, for example, gave no reaction although a good yield of **735** was obtained at 490°C.[604] The

607. Sturla, S.J.; Kablaoui, N.M.; Buchwald, S.L. *J. Am. Chem. Soc.* ***1999**, 121*, 1976.
608. Hunstman, W.D.; Lang, P.C.; Madison, N.L.; Uhrick, D.A. *J. Org. Chem.* ***1962**, 27*, 1983.

presence of a carbomethoxy group led to a more facile ene reaction in which **734** (R = CO$_2$Me) was converted to **735** in about 25% yield, but still required a temperature of 400°C.[608]

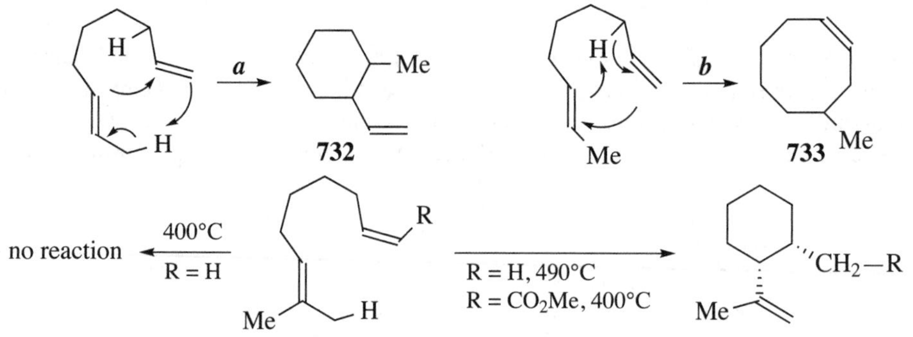

The presence of an electron-withdrawing group increases the rate of the ene reaction by lowering the LUMO energy. Lewis acids greatly enhance the rate of ene reactions involving 1,7-dienes.[609] When **736** was treated with zinc bromide (ZnBr$_2$) the ene reaction occurred at room temperature to give an 88% yield of desmethyl **737/738** (>99:1 favoring **737**).[609a,610]

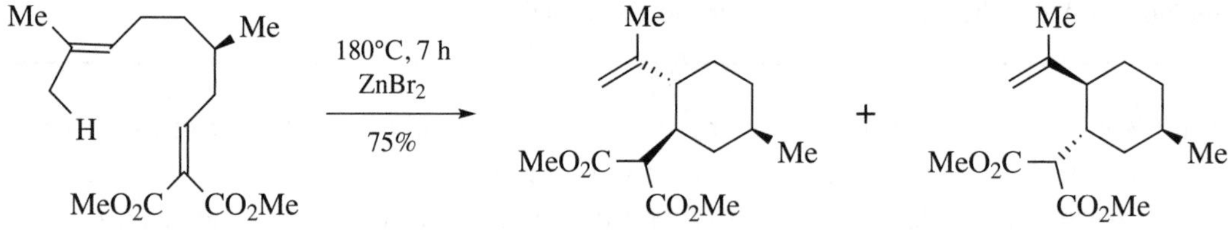

MOLECULAR MODELING: Nature of the Migrating Hydrogen in an Ene Reaction

Observed variation in rate among closely-related reactions can sometimes provide insight into the nature of the reaction. For example, the intramolecular ene reaction of **734** with R=H requires heating to 490°C, whereas the corresponding reaction with R=CO$_2$Me readily occurs at 400°C. This observation suggests that the migrating hydrogen bears "negative charge", that is, it is stabilized by an electron-accepting group at the "receiving end".

Examine the electrostatic potential map for transition state for reaction with R=H (left). Recall that this map colors an electron density surface which gives to overall molecular size and shape with the value of the electrostatic potential (the energy of a point positive charge with the nuclei and electrons of a molecule). By convention, electron-rich regions (negative) on are "red", electron-deficient regions (positive) are "blue" and neutral regions are "green". The maps are displayed as transparent models to allow you to see the underlying structure. Once you are oriented with regard to the position of the migrating hydrogen, switch to solid models by *clicking* on one of the maps and

---

609. (a) Tietze, L.F.; Beifuss, U. *Angew. Chem. Int. Ed.* **1985**, *24*, 1042; (b) *Idem Liebigs Ann. Chem.* **1988**, 321; (c) Nakatani, Y.; Kawashima, K. *Synthesis* **1978**, 147; (d) Sakane, S.; Maruoka, K.; Yamamoto, H. *Tetrahedron* **1986**, *42*, 2203.
610. Tietze, L.F.; Beifuss, U.; Ruther, M. *J. Org. Chem.* **1989**, *54*, 3120.

MOLECULAR MODELING: Nature of the Migrating Hydrogen in an Ene Reaction, con't.

selecting ***Solid*** from the **Style** menu at the bottom of the screen. Is the migrating hydrogen negatively charged, positively charged or neutral? What does this suggest about the ability of an electron-withdrawing group such as acetate to stabilize the transition state? Next, examine the electrostatic potential map for transition state for reaction with R= CO_2Me (right). Is either the direction and/or magnitude of the charge on the migrating hydrogen significantly different than that for the reaction with R=H? Are your results consistent with the observed rate change? Elaborate.

11.14.C. Carbonyl and Imino Ene Reactions

When a carbonyl moiety is the ene component of a 1,6- or 1,7-diene system, the ene reaction is more favorable. Heating **739** to 110°C, for example, led to a 94% yield of **740** in White and Somers' synthesis of 2-desoxystemodinone.[611] In **739**, the 1,6-diene is the π-bond of the aldehyde carbonyl and the π bond of the cyclohexene unit, which is more easily seen in the conformational drawing. The reaction occurred in refluxing toluene in good yield, and with excellent stereoselectivity due to the rigid nature of the tricyclic system.

MOLECULAR MODELING: Models vs. 2D Sketches

The design of new total syntheses requires the ability to reliably "guess" the products of the individual steps in the overall sequence of reactions. This is why we study organic reactions, and place so much emphasis on three-dimensional structure (stereochemistry). It seems odd then that chemists use 2D drawings and not 3D models to display structures. An example of the limitations of 2D drawings is provided by the ene reaction of **739**, where the stereochemistry of the product depends on the "reactive" conformation of the aldehyde. Even though there are only 9 conformers (threefold rotation around the hydroxyl group and threefold rotation of the aldehyde), merely looking at the drawing does not easily reveal the stereochemistry of the product.

Obtain energies for different conformers of **739** to identify the one with the lowest energy. To rotate around a single bond, first *click* on the bond and then move the mouse up and down while simultaneously holding down the left button and the **Alt** key (**Option** key on Macintosh). Speculate why this conformer is preferred. Assuming that this is also the reactive conformer, predict the product of the ene reaction.

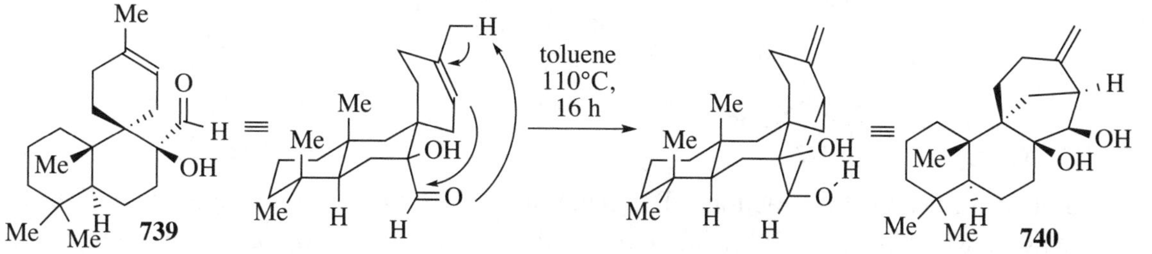

611. White, J.D.; Somers, T.C. *J. Am. Chem. Soc.* ***1994**, 116*, 9912.

The carbonyl ene reaction is catalyzed by Lewis acids.[612] Isobutylene reacted with methyl propiolate at 220°C to give 47% of **741** and 3% of **742**. The presence of aluminum chloride (AlCl₃), however, allowed reaction to occur at 25°C and gave a 61% yield of **741**.[613] Coordination of the Lewis acid and enophile lowers the LUMO, accelerating the reaction due to the diminished ΔE, just as in the Diels-Alder reaction. The C=O of an aldehyde reacts as an ene. The EtAlCl₂-catalyzed ene reaction of **743** to give **744** in up to 68% yield was part of Paquette and co-workers synthesis of dumsin.[614] Alkenyl ketones also undergo ene reactions. As with the conversion of **743** to **744**, dimethylaluminum chloride (Me₂AlCl) and diethylaluminum chloride (Et₂AlCl) are excellent catalysts for the carbonyl ene reaction and are often the catalyst of choice. When catalyzed by chiral chromium (III) complexes, high asymmetric induction is possible in hetero-ene reactions of this type.[615]

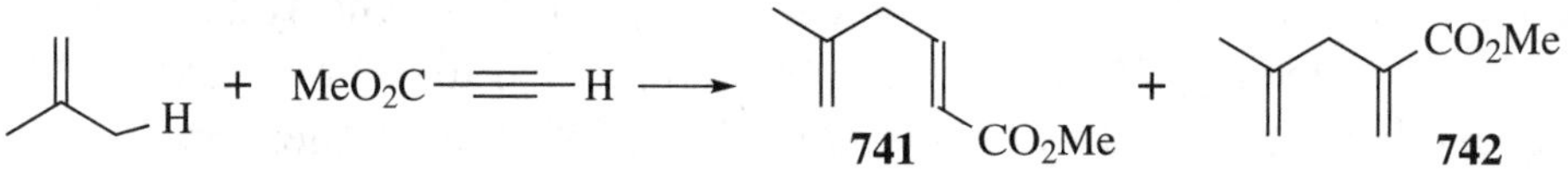

230°C		47%	3%
AlCl₃, 25°C		61%	–

The catalyst influences the reaction. The reaction of **745** with formaldehyde gave **746**, but when ZnCl₂, Me₃Al, AlCl₃, EtAlCl₂, and Et₂AlCl were examined, only diethylaluminum chloride (Et₂AlCl) gave good yields (61%).[616] The selectivity in the reaction is due to the more favorable interaction of the aluminum and aryl moieties in the transition state. Good selectivity was also observed in the reaction of *trans*-2-butene with **747**. Anti-selectivity to give **748** was observed when tin chloride (SnCl₄) catalyzed the reaction with either *cis*- or *trans*-2-butene, but syn selectivity (to **749**) was observed when alkyl aluminum catalysts were used (Me₂AlCl, Me₂AlOTf, MeAl(OTf)₂).[617]

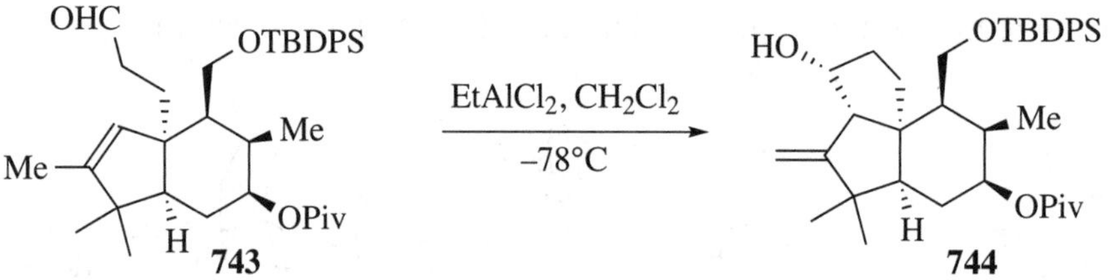

612. Snider, B.B. *Acc. Chem. Res.* **1980**, *13*, 426.
613. (a) Snider, B.B. *J. Org. Chem.* **1974**, *39*, 255; (b) Greuter, H.; Bellus, D. *Synth. Commun.* **1976**, *6*, 409; (c) Åkermark, B.; Ljungqvist, A. *J. Org. Chem.* **1978**, *43*, 4387; (d) Snider, B.B. *J. Org. Chem.* **1976**, *41*, 3061; (e) Snider, B.B.; Rodini, D.J.; Conn, R.S.E.; Sealfon, S. *J. Am. Chem. Soc.* **1979**, *101*, 5283.
614. Paquette, L.A.; Hu, Y.; Luxenburger, A.; Bishop, R.L. *J. Org. Chem.* **2007**, *72*, 209.
615. Ruck, R.T.; Jacobsen, E.N. *J. Am. Chem. Soc.* **2002**, *124*, 2882.
616. Prashard, M.; Tomesch, J.C.; Shapiro, M.J. *Tetrahedron Lett.* **1989**, *30*, 4757.
617. Mikami, K.; Loh, T.-P.; Nakai, T. *Tetrahedron Lett.* **1988**, *29*, 6305.

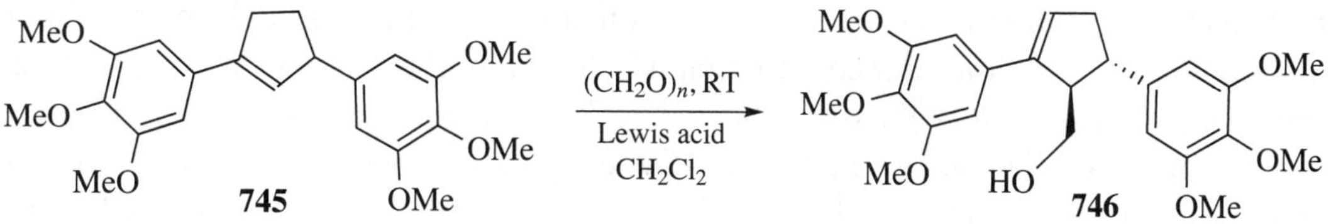

Imino-ene reactions are known.[618] In one report, lanthanum derivatives catalyzed the reaction. When imine **750** reacted with α-methyl styrene in the presence of Yb(OTf)$_3$, for example, a 90% yield of the ene product (**751**) was obtained when chlorotrimethylsilane was used as an additive.[619] Catalytic, enantioselective imino-ene reactions of α-imino esters have also been reported.[620]

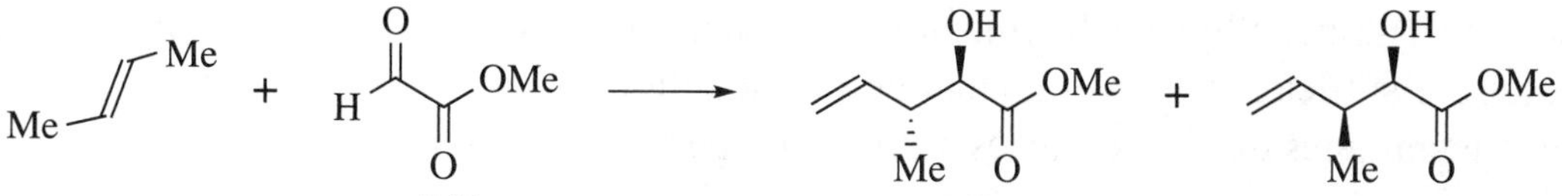

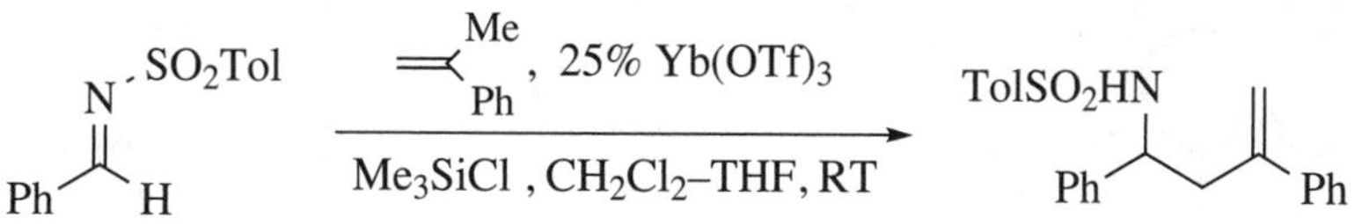

11.14.D. Chiral Ene Reactions

The fundamental principles of asymmetric synthesis employed in previous chapters and reactions in this chapter also apply to ene reactions. Chiral additives, chiral auxiliaries or the preparation of chiral templates all lead to good enantioselectivity when applied to the ene reactions. Most chiral ene reactions seem to involve addition of a chiral catalyst or the use of a chiral auxiliary. Oppolzer et al. utilized a chiral auxiliary built into **752** to prepare **753** (90% de)

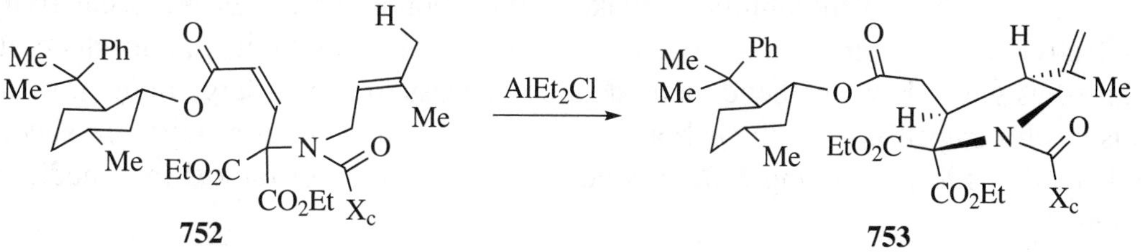

618. For examples, see (a) Achmatowicz Jr., O.; Pietraszkiewwicz, M. *J. Chem. Soc. Chem. Commun.* **1976**, 484; (b) Borzilleri, R.M.; Weinreb, S.M. *Synthesis* **1995**, 347; (c) Laschat, S.; Grehl, M. *Angew. Chem. Int. Ed.* **1994**, *33*, 458; (d) Mikami, K.; Kaneko, M.; Yajima, T. *Tetrahedron Lett.* **1993**, *34*, 4841.
619. Yamanaka, M.; Nishida, A.; Nakagawa, M. *Org. Lett.* **2000**, *2*, 159.
620. (a) Drury, III, W.J.; Ferraris, D.; Cox, C.; Young, B.; Lectka, T. *J. Am. Chem. Soc.* **1998**, *120*, 11006; (b) Yao, S.; Fang, X.; Jørgensen, K.A. *Chem. Commun.* **1998**, 2547.

in a synthesis of (+)-α-allokainic acid.[621] Yamamoto and co-workers used the chiral aluminum catalyst **754** for the intermolecular ene reaction of 1,6-dichlorobenzaldehyde and 2-phenylthio-1-propene to give **755** in 96% yield (65 % ee).[622] These catalysts are similar to those used in Section 11.9.B. A similar titanium catalyst (**756**) coupled methylenecyclohexane and methyl glyoxalate to give an 89% yield of **757** (98 % ee). Catalyst **756**[623] was prepared *in situ* by reaction of (*R*)-binaphthyl and bis(diisopropoxytitanium) dibromide [(iPrO)$_2$TiBr$_2$]. These few examples are sufficient to illustrate that chiral induction is possible in the ene reaction, using the same techniques that were successful in other pericyclic reactions.

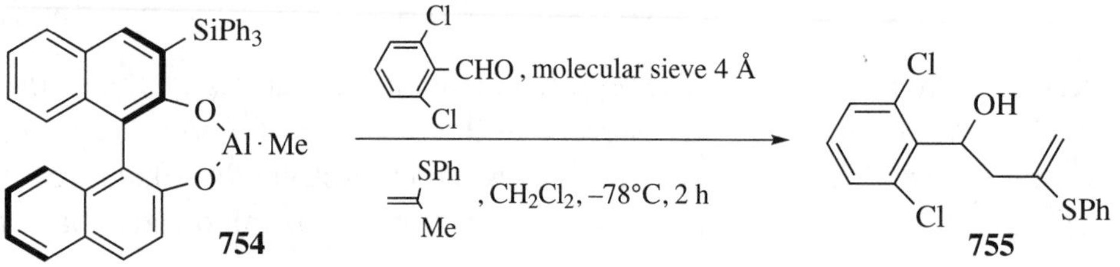

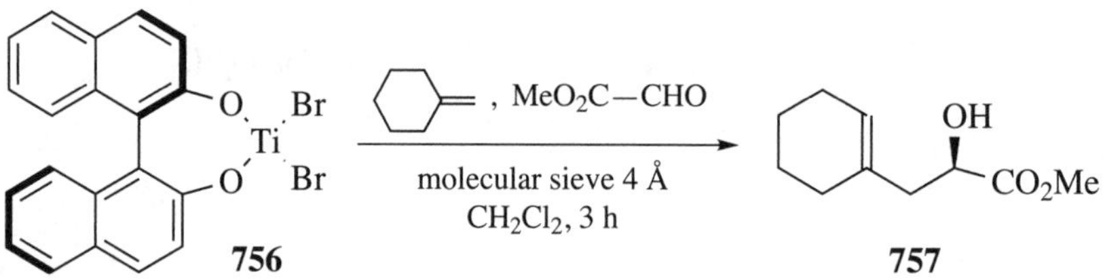

11.15. CONCLUSION

It is clear from the preceding sections that a wide range of cyclic and polycyclic molecules can be prepared by pericyclic reactions. In addition, stereocontrolled syntheses of acyclic molecules are possible via initial cycloaddition followed by cleavage of the ring. The power of these reactions lies in the ability to make carbon bonds with high regioselectivity and stereoselectivity. If the proper chiron or chiral auxiliary is used, high enantioselectivity can be obtained as well. Few synthetic methods allow access to such a large number of natural products, with compatibility with such a large number of functional groups. In succeeding chapters, methods for generating carbon bonds via electrophilic or radical intermediates will be presented.

621. (a) Oppolzer, W.; Robbiani, C.; Bättig, K. *Helv. Chim. Acta* **1980**, *63*, 2015; (b) *Idem Tetrahedron* **1984**, *40*, 1391; (c) Oppolzer, W.; Mirza, S. *Helv. Chim. Acta* **1984**, *67*, 730.
622. Maruoka, K.; Hoshino, Y.; Shirasaka, T.; Yamamoto, H. *Tetrahedron Lett.* **1988**, *29*, 3967.
623. Mikami, K.; Terada, M.; Nakai, T. *J. Am. Chem. Soc.* **1989**, *111*, 1940.

HOMEWORK

1. Using the tables in this chapter (where possible), particularly Tables 11.1 and 11.2, evaluate the relative rate of reaction for each of the following. For the *faster* reaction, predict the structure, and where data is available predict the regiochemistry and stereochemistry of the major product(s).

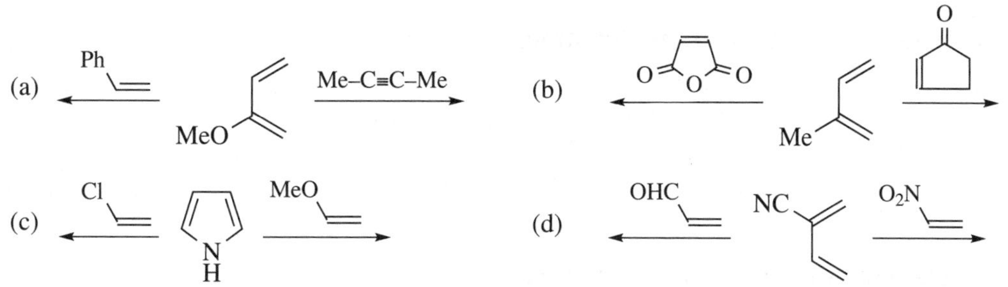

(a)

(b)

(c)

(d)

2. Give a mechanistic explanation for the following transformation:

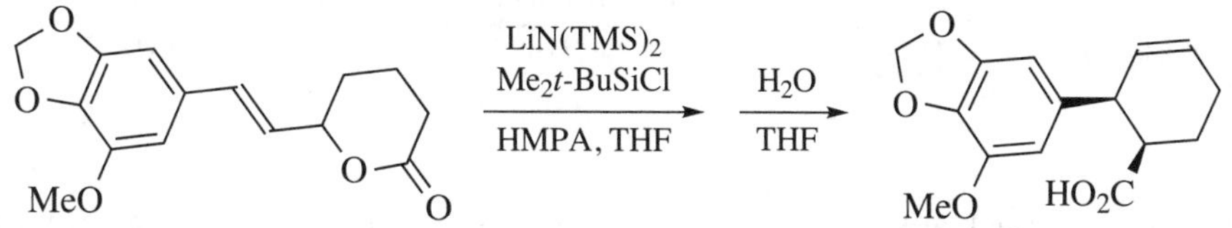

3. In section 11.5.A, Boger and co-worker's conversion of tetrazines to diazines via inverse electron demand Diels-Alder reactions was mentioned. Give a mechanistic rationale for the transformation shown, based on this paper, that accounts for formation of the indicated product.

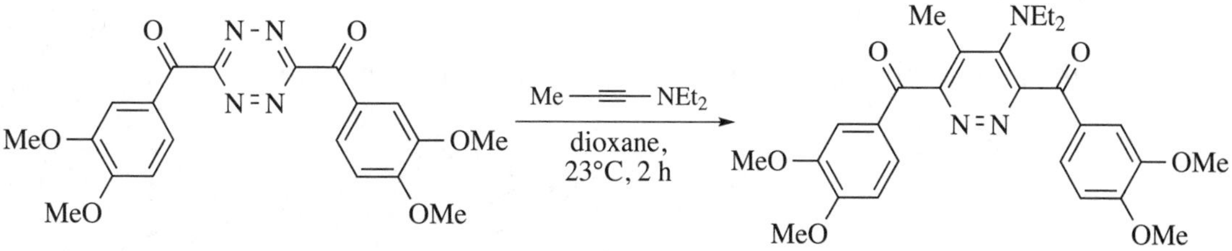

4. Give a mechanistic rationale for the transformation that accounts for one product being formed preferentially and also accounts for the stereochemistry of both products.

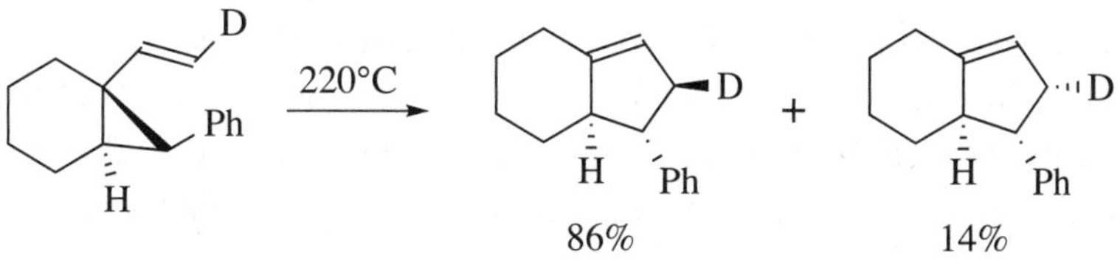

5. Explain the different stereochemical outcomes of these reactions:

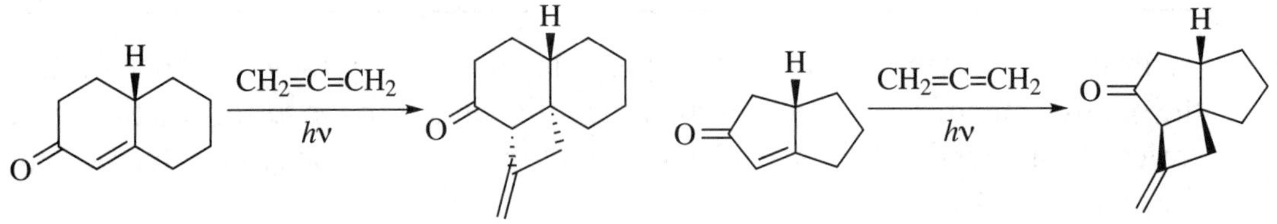

6. Explain the following transformation:

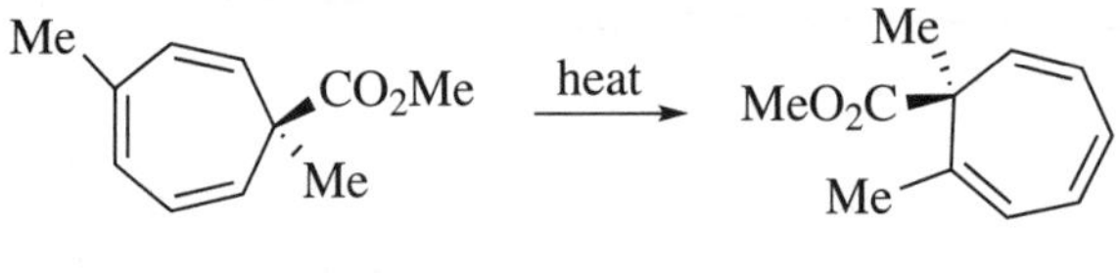

7. Explain the following transformation:

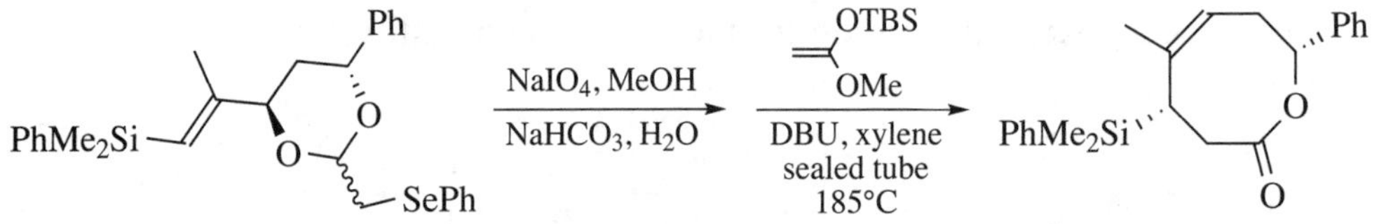

8. Draw a reasonable transition state for the reaction (sec 11.12.E.ii.) that converted **651** to **652**.

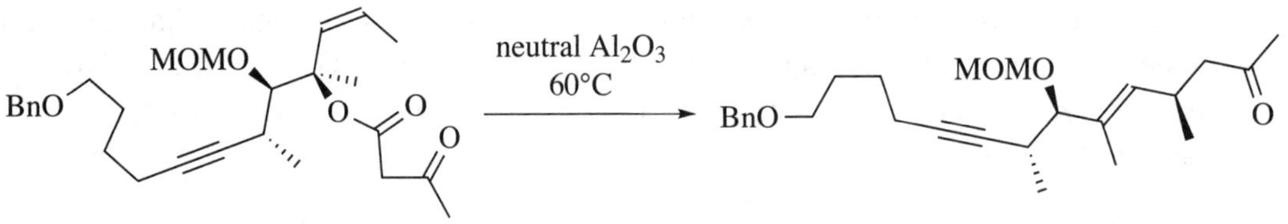

9. Give the product for this reaction, and rationalize its formation.

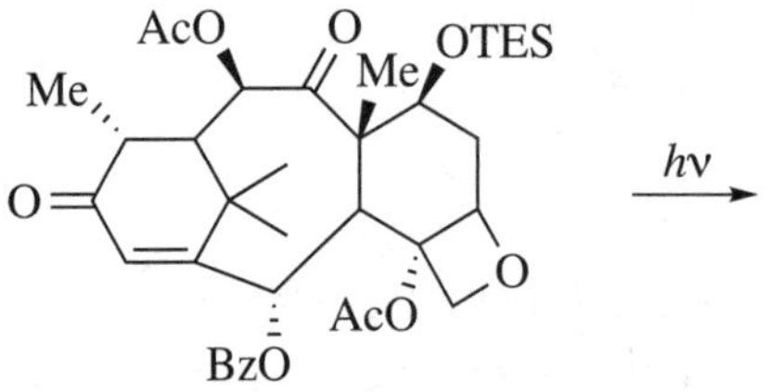

10. Reaction of *N*-tosylimines and allylic sulfonium salts under basic conditions leads to aziridines in this particular work. When the imine shown was treated with the sulfonium salt given in the reaction, a mixture of aziridine and a dihydroazepine were formed. Draw both products and explain formation of the dihydroazepine.

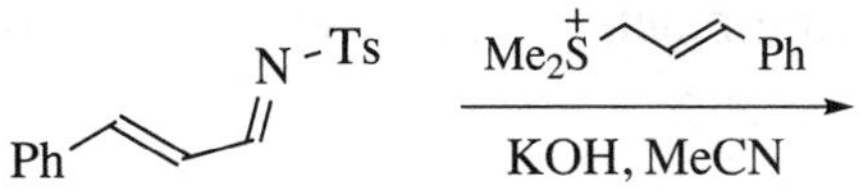

$Ph\diagdown\diagup\diagdown\diagup N\text{-}Ts$ $Me_2\overset{+}{S}\diagup\diagdown Ph$

$\xrightarrow{\quad KOH,\ MeCN \quad}$

11. Use the tables in this chapter (where possible) to <u>estimate</u> the correct regiochemistry for the products in the following reactions.

    (a) $C_5H_{11}-C\equiv\overset{+}{N}-O^-$ + $H-C\equiv C-(CH_2)_{12}Me$ ⟶

    (b) 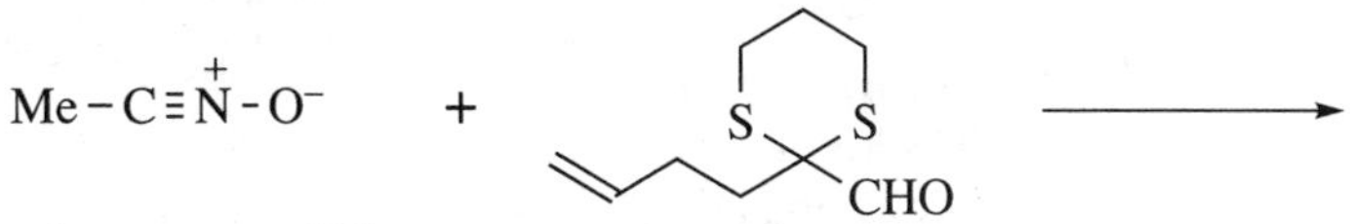 $Me-C\equiv\overset{+}{N}-O^-$ + [dithiane with butenyl and CHO substituents] ⟶

    (c)   [N-methyl pyridinium with ^-O and OMe substituents] + [$CH_2=CH-CO_2Me$] ⟶

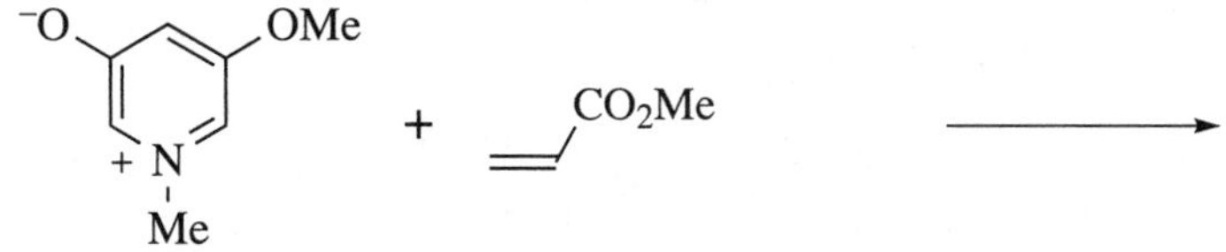

12. The macrocyclic compound shown, generated *in situ*, spontaneously reacted to give the indicated product in 63% yield. Explain the formation of this product.

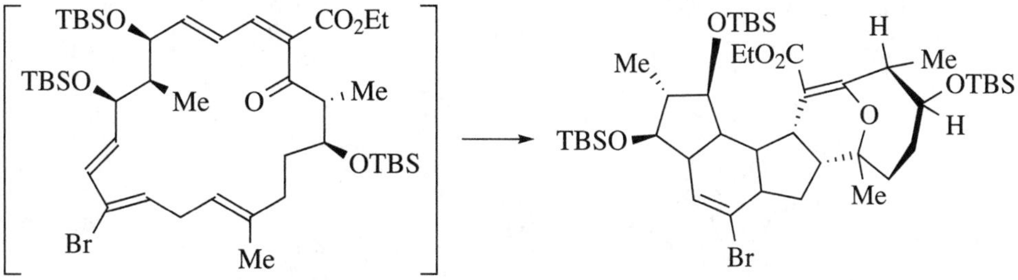

13. In a synthesis of merrilactone A (see *J. Am. Chem. Soc.*, **2002**, *124*, 2080), Danishefsky reports the Diels-Alder reaction gives the cycloadduct shown. This product results from an *exo* approach of the anhydride to the diene, assuming a disrotatory motion of the OTBS group. Although this result is not discussed in the cited reference, offer a suggestion why the Alder endo rule does not predict the major diastereomer.

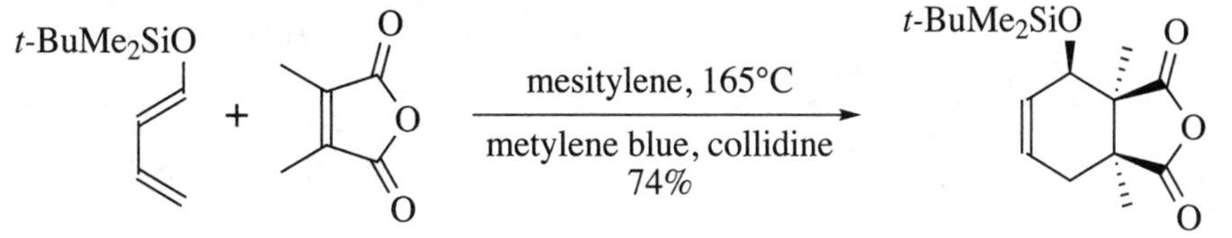

$t\text{-BuMe}_2\text{SiO}$ [diene] + [dimethylmaleic anhydride] $\xrightarrow[\substack{\text{metylene blue, collidine}\\74\%}]{\text{mesitylene, }165°C}$ [cycloadduct with $t\text{-BuMe}_2\text{SiO}$]

14. For each of the following reactions give the major product, with correct regiochemistry and stereo-chemistry:

a)

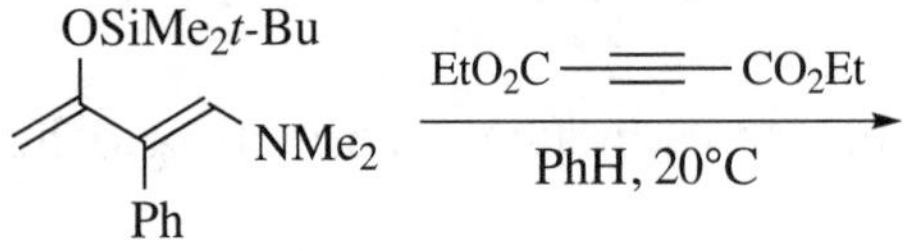

EtO$_2$C —=— CO$_2$Et

PhH, 20°C

b)

O$_2$N —= CO$_2$Et

CHCl$_3$, –20°C, 120 h

c)

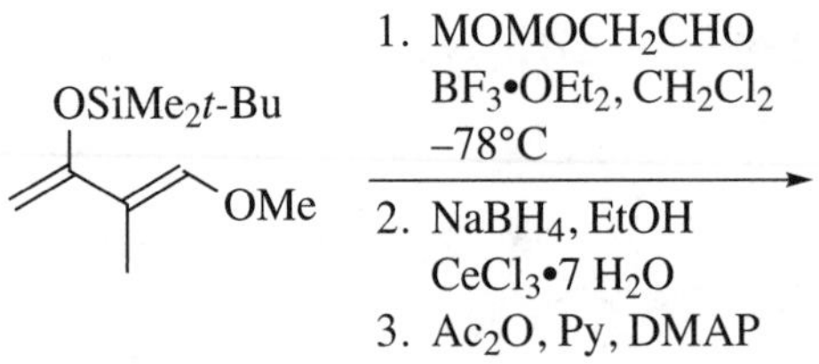

1. MOMOCH$_2$CHO
   BF$_3$•OEt$_2$, CH$_2$Cl$_2$
   –78°C
2. NaBH$_4$, EtOH
   CeCl$_3$•7 H$_2$O
3. Ac$_2$O, Py, DMAP

d)

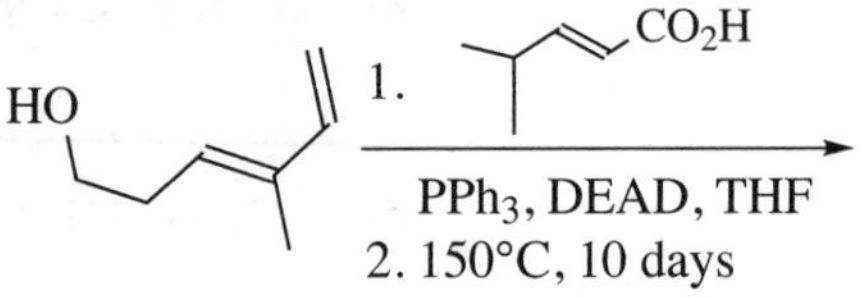

1. (isovaleric acid derivative, CO$_2$H)
   PPh$_3$, DEAD, THF
2. 150°C, 10 days

e)

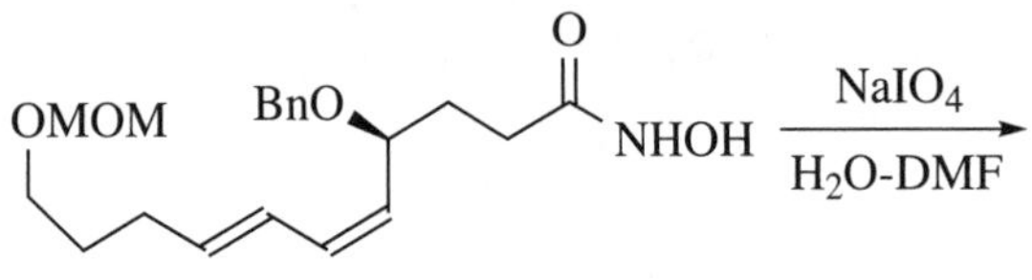

NaIO$_4$

H$_2$O-DMF

f)

PhH, reflux

g)

heat

h)

neat, 50°C

i)

1. toluene
   110°C
2. hν
   THF

j)

300°C, 1 h

k)

5% EtCO$_2$H/EtC(OEt)$_3$

140°C, 3 h

l) BnO —= CO$_2$H

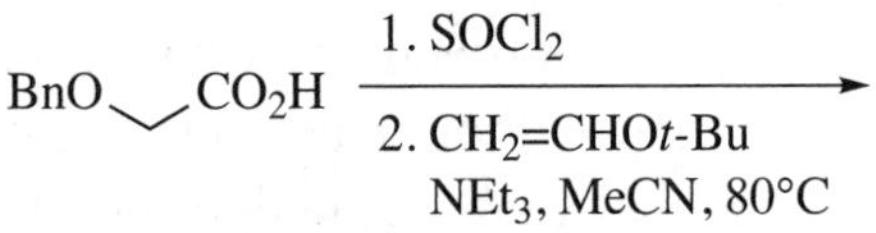

1. SOCl$_2$
2. CH$_2$=CHOt-Bu
   NEt$_3$, MeCN, 80°C

m)

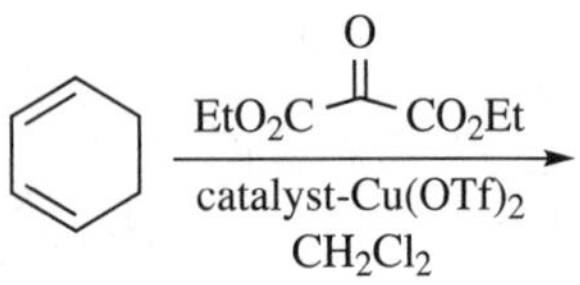

EtO$_2$C —C(=O)— CO$_2$Et

catalyst-Cu(OTf)$_2$
CH$_2$Cl$_2$

n)

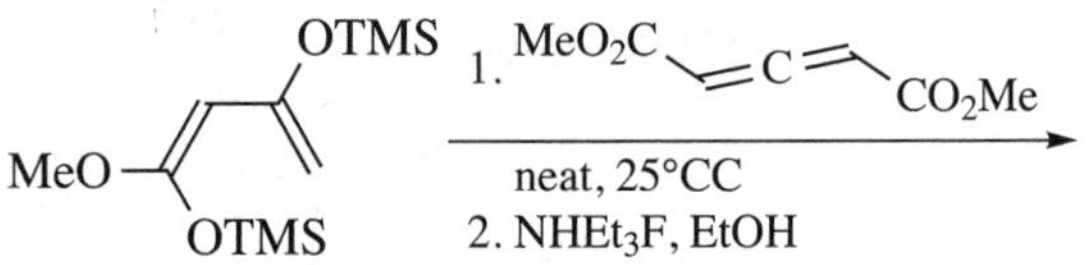

1. MeO$_2$C —= C=—= CO$_2$Me
   neat, 25°CC
2. NHEt$_3$F, EtOH

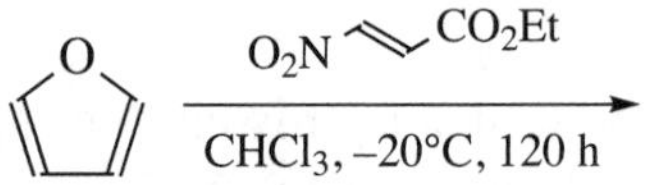

o)

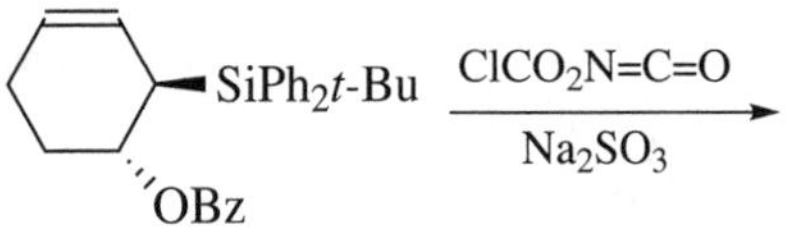

ClCO$_2$N=C=O

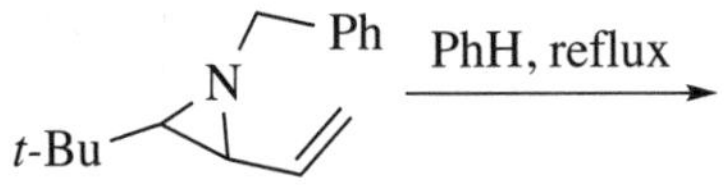

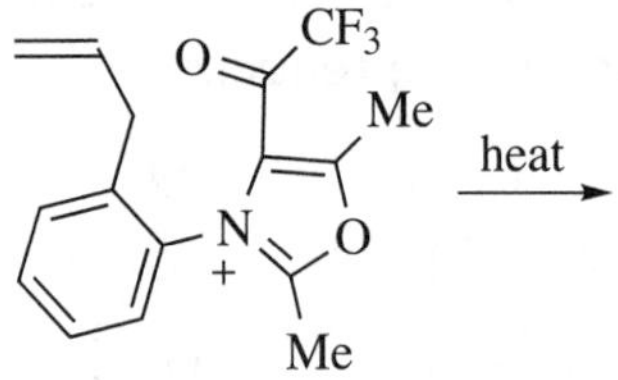

Na$_2$SO$_3$

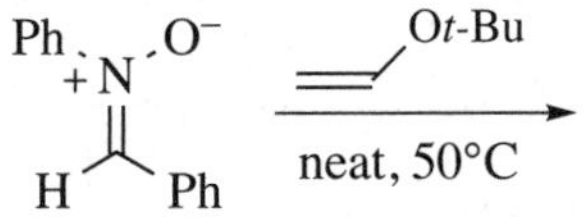

p)

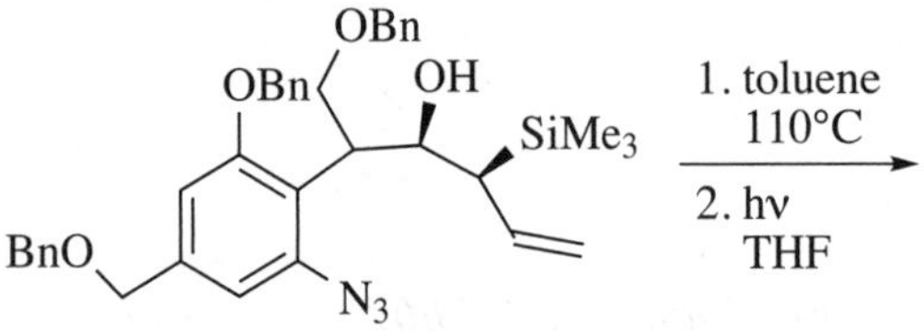

Br —= CO$_2$Et

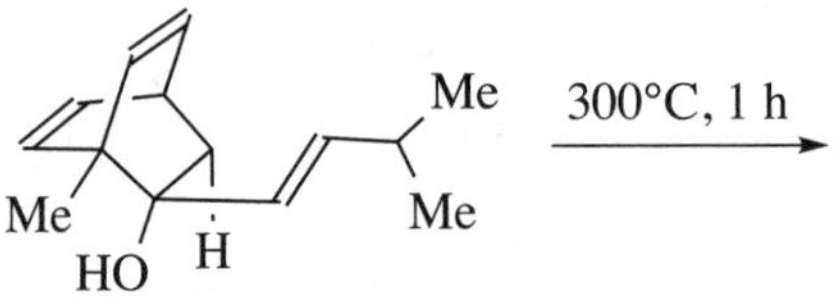

toluene, rt

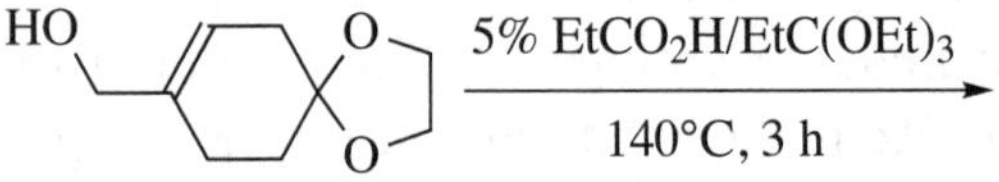

chapter 11   1156

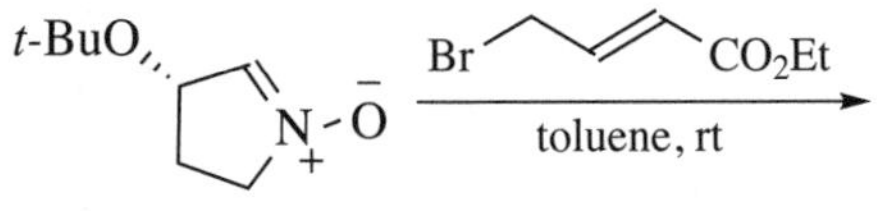

q)

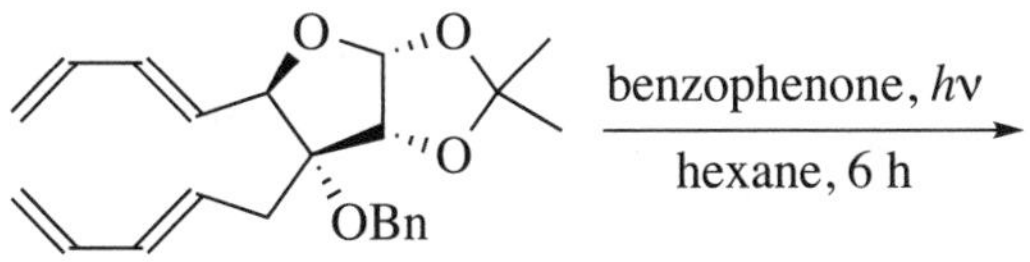

benzophenone, *h*v

hexane, 6 h

r)

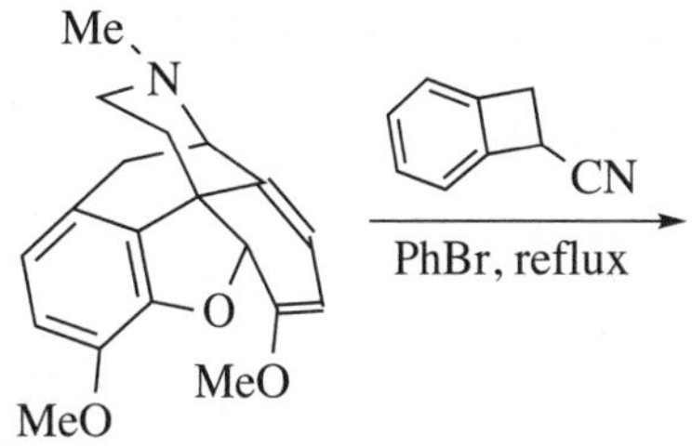

Me–N

CN

PhBr, reflux

MeO   MeO

MeO

s)

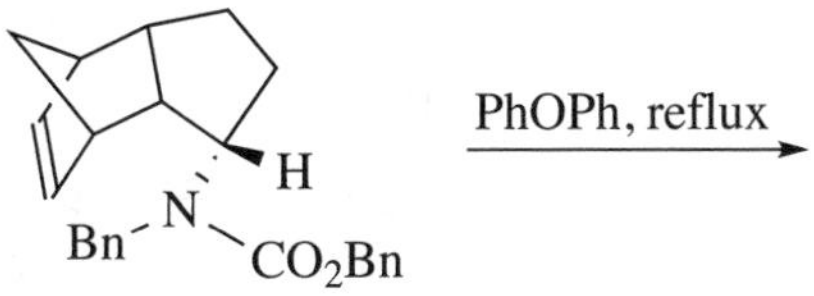

Bn–N   H

CO₂Bn

PhOPh, reflux

t)

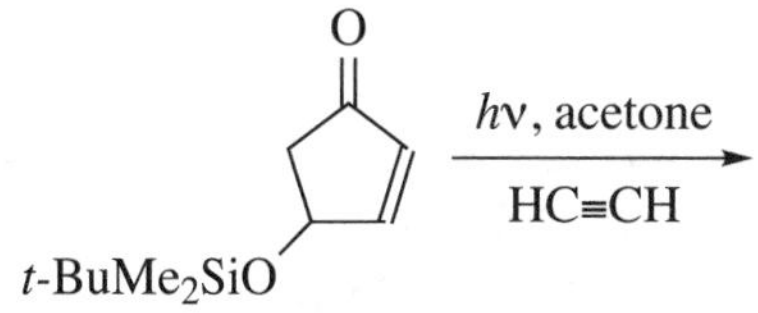

t-BuMe₂SiO

O

*h*v, acetone

HC≡CH

u)

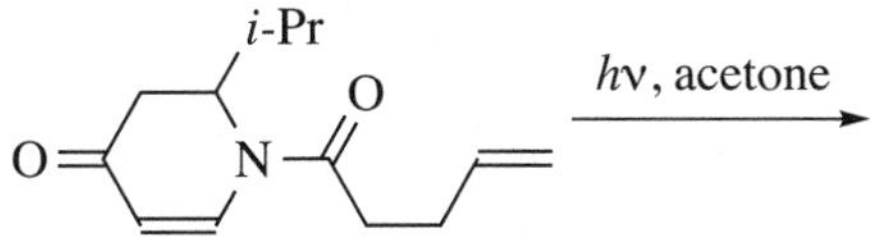

i-Pr

O   N

O

*h*v, acetone

v)

Br

O   O

Me   OMe

1. TMSO—⧸—OMe, NEt₃

OTMS

2. hydrolysis

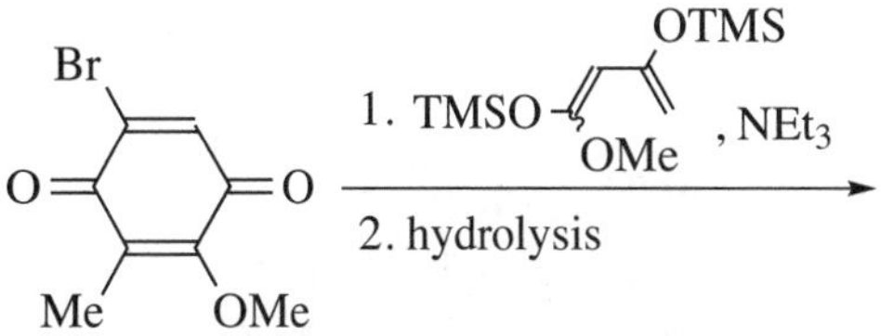

w)

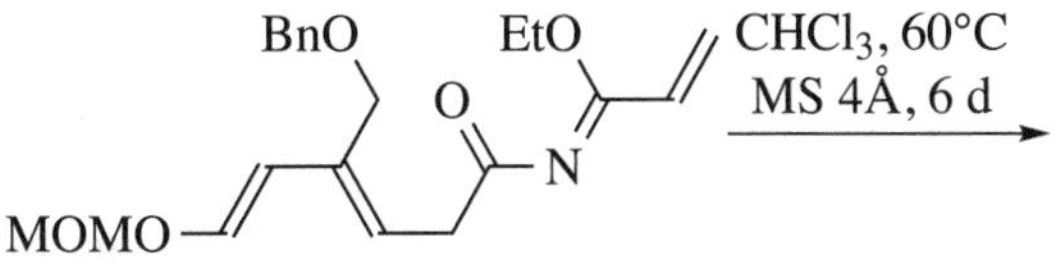

BnO   EtO

O

MOMO   N

CHCl₃, 60°C

MS 4Å, 6 d

x)

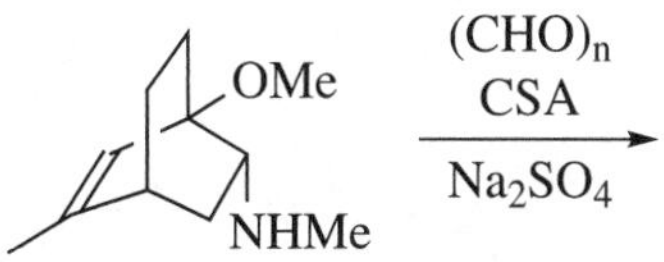

OMe

NHMe

(CHO)ₙ

CSA

Na₂SO₄

y)

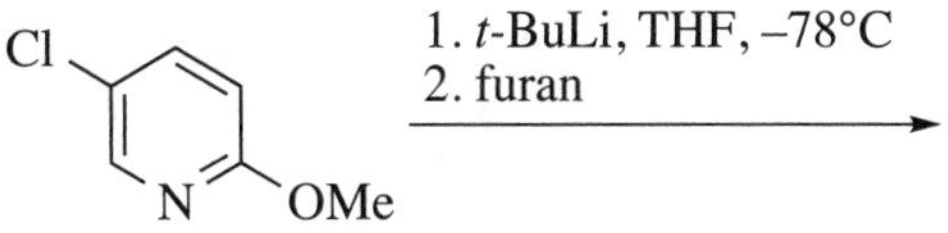

Cl

N   OMe

1. *t*-BuLi, THF, −78°C

2. furan

z)

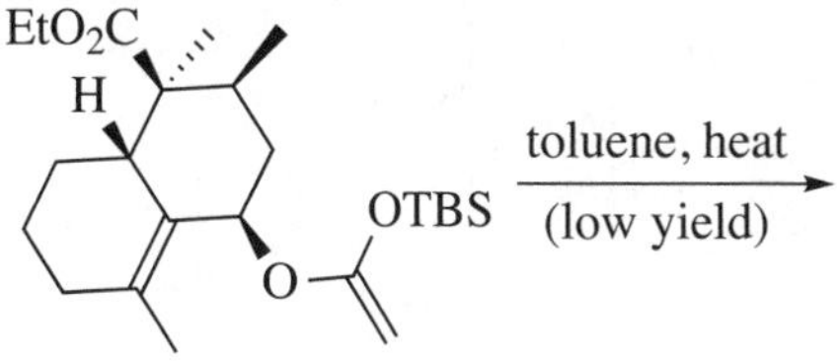

EtO₂C

H   Me

Me

OTBS

O

toluene, heat

(low yield)

aa)

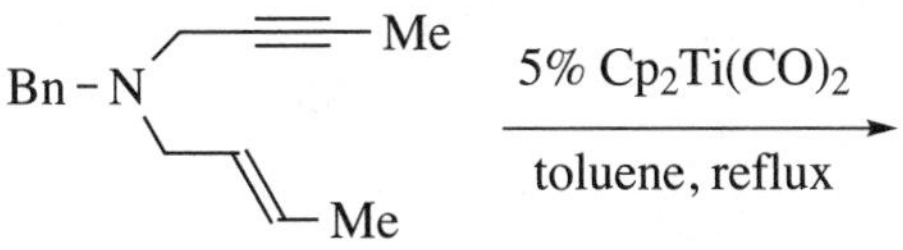

Bn–N   Me

Me

5% Cp₂Ti(CO)₂

toluene, reflux

ab)

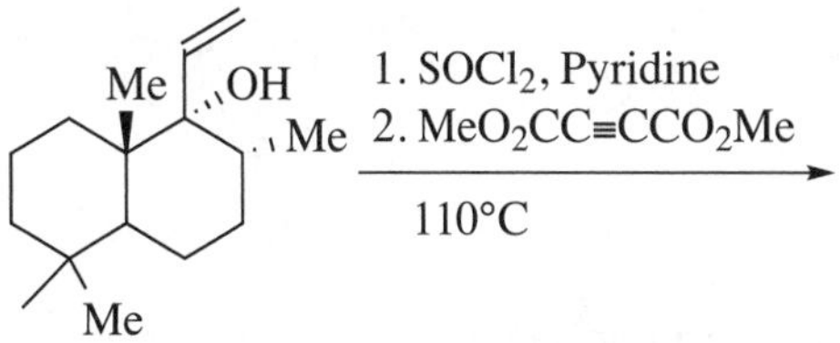

Me   OH

Me

Me

Me

1. SOCl₂, Pyridine

2. MeO₂CC≡CCO₂Me

110°C

ac)

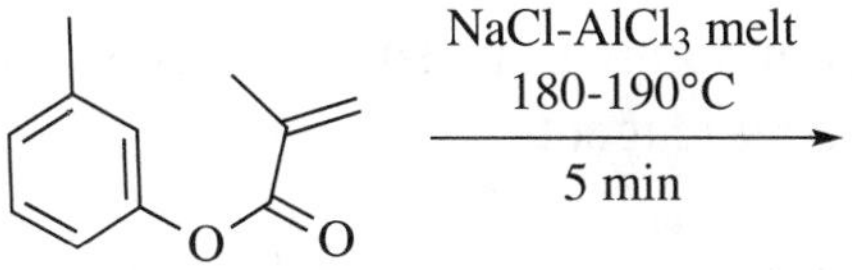

O   O

NaCl-AlCl₃ melt

180-190°C

5 min

ad)

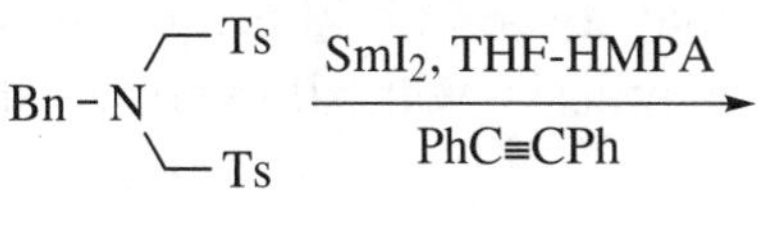

Bn–N   Ts

Ts

SmI₂, THF-HMPA

PhC≡CPh

ae)

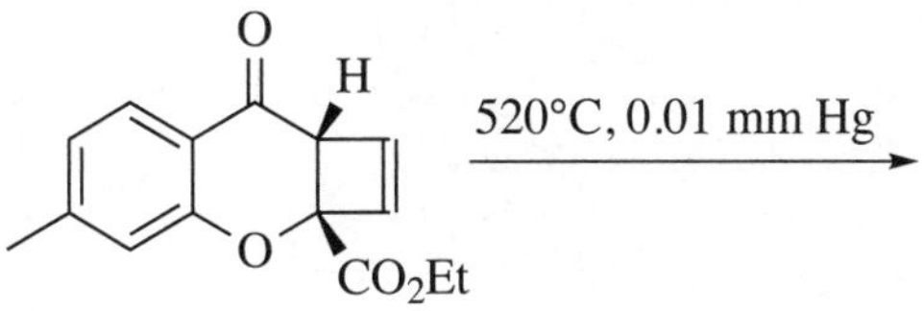

O   H

O   CO₂Et

520°C, 0.01 mm Hg

15. In each case, provide a suitable synthesis. Show all intermediate products and all reagents.

a)

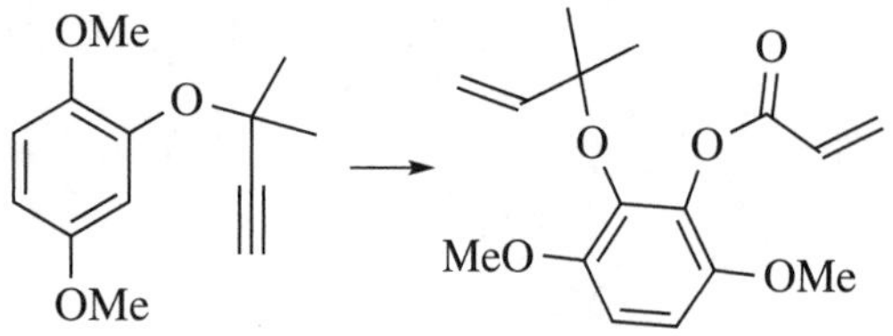

b)

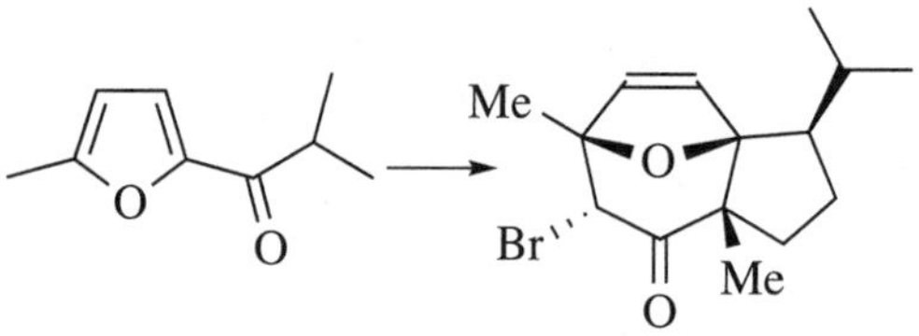

c)

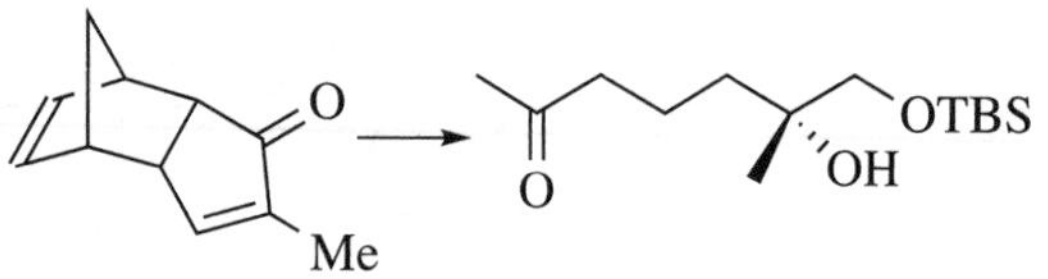

d)

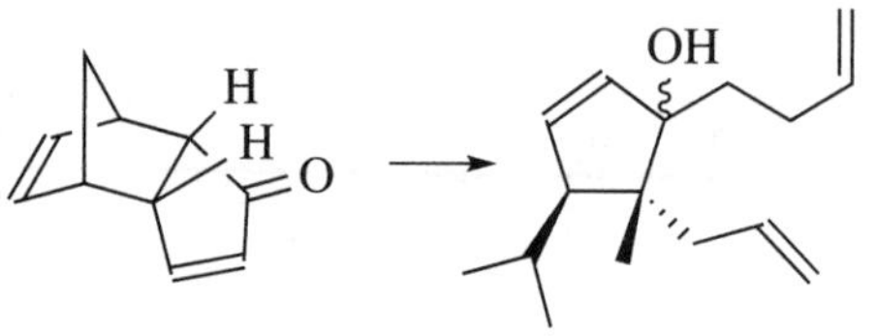

e)

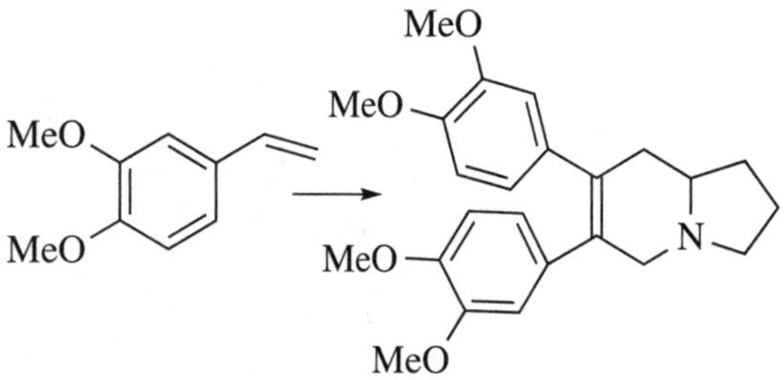

f)

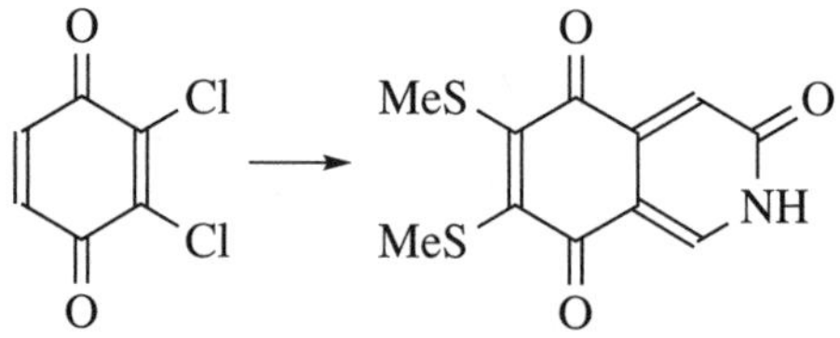

g)

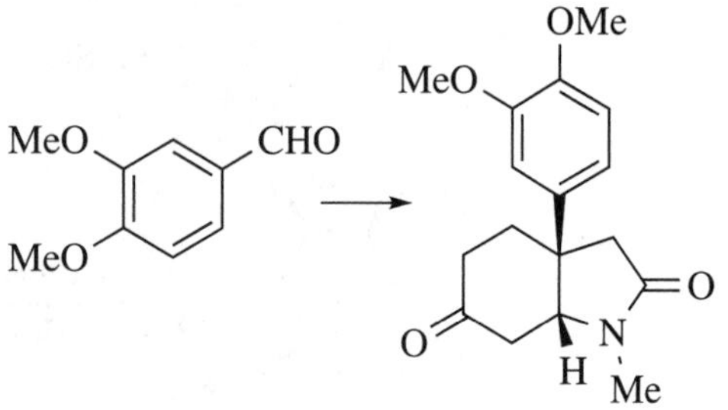

h)

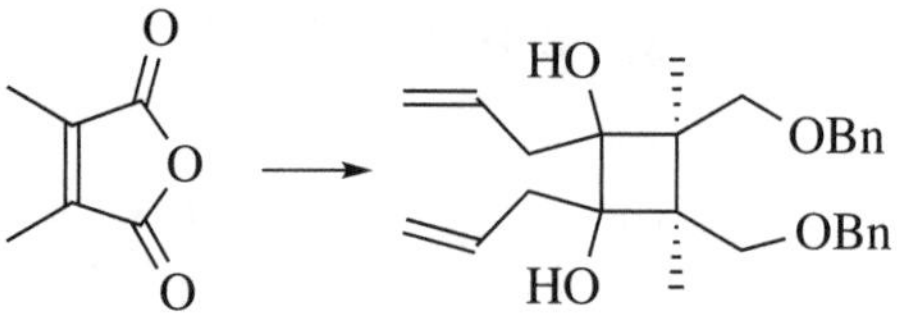

i)

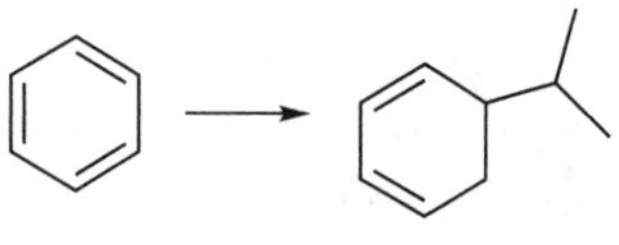

j)

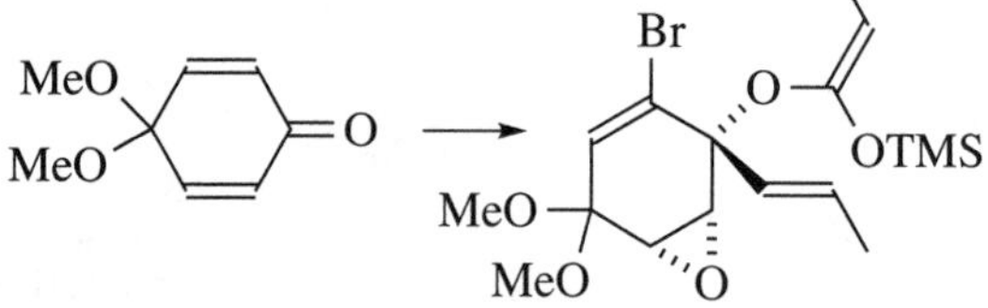

16. In each case, devise a retrosynthetic strategy that will lead to a suitable starting material that costs < $1.00 per g (cite the price and your commercial source). Show the complete synthesis, with all intermediate products and all reagents.

a)

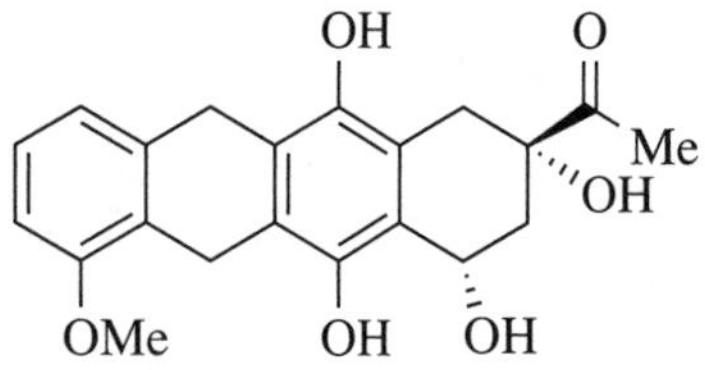

b)

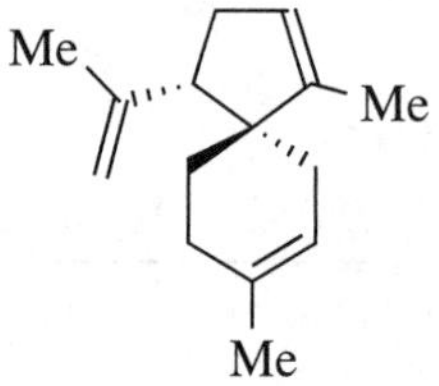

c)

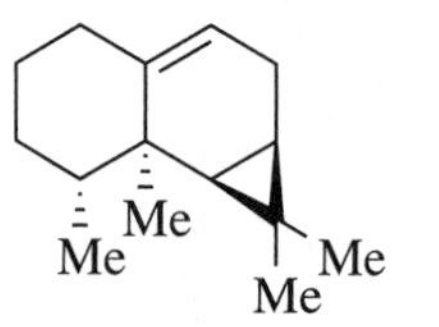

d)

e)

f)

g)

h)

chapter 12

Cᵃ Disconnect Products: Electrophilic Carbon-Carbon Bond-Forming Reactions

12.1. INTRODUCTION

Chapters 8, 9 and 11 introduced many reactions for making carbon-carbon bonds. Molecules possessing a polarized electrophilic carbon (δ^+) were common in previous chapters, as with the reaction of an alkyl halide or a carbonyl compound reacting with nucleophiles and organometallics in Chapter 8, or with nucleophilic enolate anions in Chapter 9. In section 1.2, disconnection of carbon-carbon bonds generated C^d and C^a fragments and Chapters 8 and 9 focused on C^d fragments. This chapter will focus on reagents and reactions that effectively generate equivalents for a C^a fragment, including substrates possessing a $C^{\delta+}$ unit, and organometallic reagents that give reactions similar to a $C^{\delta+}$ unit. Carbocations obviously correspond to a C^a fragment, but they are reactive intermediate rather than reagents. However, if carbocations can be stabilized, they may be used in ways similar to reagents. This concept will be elaborated later in this chapter. Chapter 2 briefly discussed the structure and nature of carbocations in the context of S_N1 reactions (sec. 2.6.B). It was clear that the nucleophilic partners reacting with a carbocation were limited to alkenes, alkynes, aromatic rings or compounds containing heteroatoms such as oxygen, sulfur, halogen or nitrogen. Reactions of carbocations with these species will also be discussed in this chapter, with an emphasis on their synthetic utility for making carbon-carbon bonds. Many organometallic reagents (particularly organopalladium and organonickel reagents) generate species that react with nucleophiles, they can be categorized as a C^a equivalent. Therefore, many organometallic reactions are included in this chapter. This chapter will begin with a discussion of free carbocations and exploitation of their molecular rearrangements to generate carbon-carbon bonds and different structures.

12.2. CARBOCATIONS

12.2.A. Cation Stability

A carbocation is an sp² hybridized carbon bearing three substituents, with an empty p orbital perpendicular to the plane of the other atoms (see **1**). The sp²-hybridization leads to a trigonal planar geometry. This intermediate has been termed a **carbenium ion** by Olah.[1] It is noted that the old term carbonium ions has been applied to **1**, but carbocation is the more useful term

---

1. Olah, G.A. *J. Am. Chem. Soc.* **1972**, *94*, 808.

and will be used in lieu of carbenium ion or carbonium ion.

From the standpoint of reactivity, a carbocation is a transient intermediate that reacts with reagents which are able to provide a pair of electrons to form a new bond to the electrophilic carbon. Formation of this cation is usually the slowest (the rate determining step) in solvolysis or other ionization reactions, and subsequent reactions with a nucleophilic species are fast. The classical S_N1 reaction of *tert*-butyl bromide and aqueous potassium iodide (KI) to give *tert*-butyl iodide is an example where the slowest step (k_1, sec. 2.6.B).is ionization of the bromide to the planar carbocation **1**. The second reaction, trapping the cation, is the fastest step (k_2). There are many different methods for generating carbocations in addition to the classical ionization, and the rate-determining step can vary with each method. One of the most common methods is to treat an alkene with a suitable acid to generate a carbocation such as **2**. In this reaction the alkene π-bond functions as a base in the presence of the acid, H^+. The subsequent reaction with the nucleophilic chloride in is similar to the k_2 sequence shown for carbocation **1**, and the overall reaction constitutes an addition reaction. Addition of an acidic reagent (HX) to an alkyne generates a vinyl carbocation intermediate (sec. 2.10.A).

Formation of **2** raises the issue of regioselectivity in formation of a carbocation. When the terminal alkene shown attacks HCl, the proton can attach to either carbon of the C=C unit to form a primary carbocation or the secondary carbon **2**. Reactions of acids HX with alkenes and alkynes give the more stable cation, where the order of stability is 3° > 2° > 1° (see below). Dipolar effects (inductive and field) largely explain this order, although hyperconjugation (also see sec. 12.2.C) is also invoked. If we focus of dipolar effects, carbocation **1** can be viewed as a carbon atom with an empty p orbital, and redistribution of electron density towards the positive center (inductive effect) occurs, using electrons from the adjacent bonds, diminishing the net charge and stabilizing the cationic center. When R^1, R^2, and R^3 in **1** are alkyl groups, the carbon atoms of the alkyl substituents release electrons to the positive center. The effect increases with increasing numbers of alkyl groups, leading to greater stability. A 3° cation ($R^1 = R^2 = R^3$ = alkyl in **1**) will be more stable than a 2° cation ($R^1 = R^2$ = alkyl, R^3 = H), which is more stable than a 1° cation ($R^1 = R^2$ = H, R^3 = alkyl). The least stable cation in this alkyl series will be the methyl cation ($R^1 = R^2 = R^3$ = H in **3**). The energy difference between 1°, 2° and 3° cations is about 11-15 kcal mol^{-1} (46.1-62.8 kJ mol^{-1}): ΔE 2°↔3° = 11-15 kcal mol^{-1} and ΔE 2°↔1° = 11-15 kcal mol^{-1} with 3° > 2° and 2° > 1°.[2] In the transition state leading to carbocation formation, one pathway will generate a 2° cation and another to a 3° cation. The lower energy

---

2. (a) Lossing, F.D.; Semeluk, G.P. *Can. J. Chem.* **1970**, *48*, 955; (b) Radom, L.; Pople, J.A.; Schleyer, P.v.R. *J. Am. Chem.Soc.* **1972**, *94*, 5935; (c) Arnett, E.M.; Petro, C. *Ibid.* **1978**, *100*, 5408.

pathway to the tertiary cation ensures that it will be formed with high selectivity.

If we focus on stabilization of a carbocation by hyperconjugation,[3] resonance contributors involving the C–H bonds are invoked, as in resonance contributors **4** and **5** that are drawn for

carbocation **3**. *"The hyperconjugation concept arises solely from our model-building procedures.* When we ask whether hyperconjugation is important in a given situation, we are asking only whether the localized model is adequate for that situation at the particular level of precision we wish to use, or whether the model must be corrected by including some delocalization in order to get a good enough description."[4] Using this model, one can draw more resonance contributors for a tertiary carbocation than for a primary or secondary carbocation, leading to greater stability.

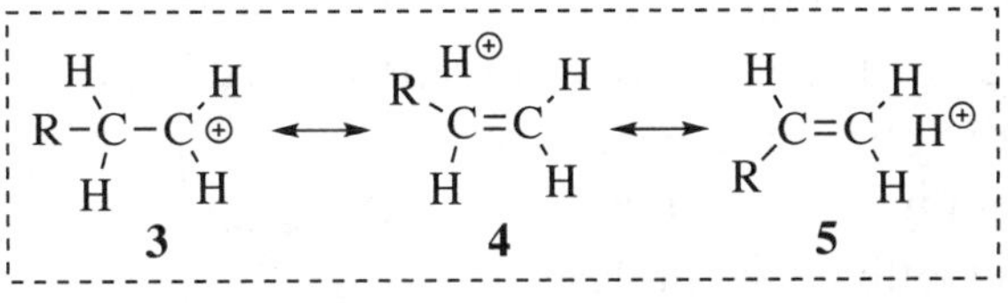

Treatment of a secondary or tertiary alcohol with an acid catalyst gives an onium salt (**6**), which loses water (a good leaving group) to give the cation, and 3° alcohols react faster than 2° alcohols. Heterolytic bond cleavage of a halide or sulfonate ester (C–X, X = Br, Cl, I, OMs, OTs, etc.) in aqueous solvents (solvolysis often requires heating) generates tertiary cations easily, secondary cations with difficulty and primary cations with extreme difficulty. Reaction of amines with nitrous acid (HONO) initially gives a diazoalkane **7**, which can generate carbenes when generated under different conditions (see sec. 13.9.B), but in this case **7** readily decompose to a carbocation.[5]

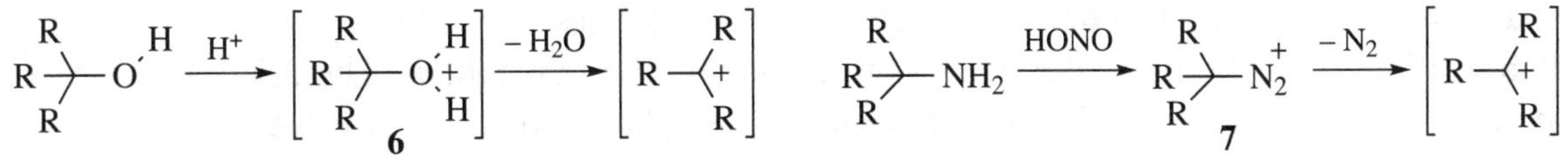

Electron releasing substituents (those with a negative or δ- charge, i.e., high electron density) attached to the positive carbon of a carbocation will stabilize the charge, leading to a more stable ion. Both oxygen stabilized cations (**8**) and sulfur stabilized cations (see **9**) are well known. The reaction of aldehydes and ketones with an

acid catalyst (Brønsted-Lowry or Lewis) gives the oxygen-stabilized cation **10** or **11**.

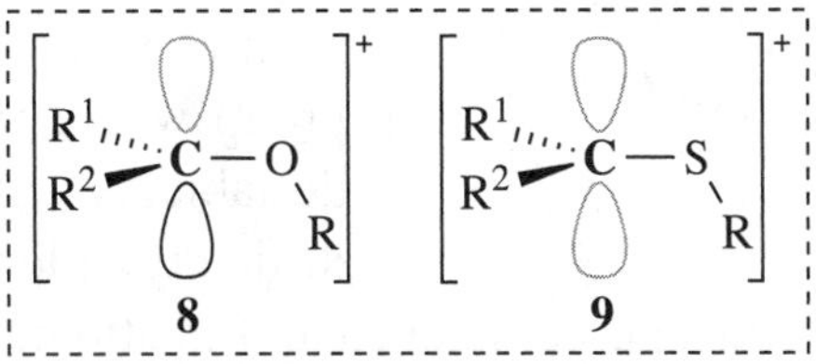

3. See Radom, L.; Poppinger, D.; Haddon, R.C. in Olah, G.A.; Schleyer, P.v.R. *Carbonium Ions, Vol. 5.*; Wiley: NY, *1976*, pp. 2303-2426.
4. Lowry, T.H.; Richardson, K.S. *Mechanism and Theory in Organic Chemistry, 3rd Ed.*, HarperCollins, NY, *1987*, p. 68.
5. (a) Collins, C.J. *Acc. Chem. Res.* *1971*, 4, 315; (b) Friedman, L. in *Carbonium Ions, Vol. II*, Olah, G.A.; Schleyer,P.v.R. (Eds.), Wiley, New York, *1970*, p 655; (c) Kirmse, W. *Angew. Chem. Int. Ed.* *1976*, 15, 251.

A nitrogen atom that is adjacent to a cation gives substantial stabilization (see resonance form **13**), and the actual structure is iminium salt **12**, a discreet and well-characterized species with a formal π bond between the carbon and nitrogen. In terms of chemical reactivity, it is useful to consider iminium salts as amino cations (resonance contributor **13**) since they react with nucleophiles at the electrophilic carbon, but *"amino-cations" such as* **8** *are not usually invoked in a mechanistic pathway.*

The stability provided by an adjacent oxygen atom is apparent in the solvolysis data for substituted primary chloroalkanes.[6] The oxygen atom is a significant contributor to the increase in the rate of solvolysis of $EtOCH_2Cl$ to form $EtOCH_2^+$ when compared to formation of the primary cation derived from 1-chlorobutane (109 versus 1).

On the other hand, moving the EtO group one atom further from the carbon bearing the chlorine has the opposite effect. The electron-withdrawing effect at the α-carbon in $EtOCH_2CH_2Cl$ destabilizes the resulting carbocation and lowers the rate (0.2 versus 1 for 1-chlorobutane). When a cyclopropyl group is conjugated to a cation (cyclopropylcarbinyl cation **14**), that cation is more stable.[7] The cyclopropane bonds have a relatively high degree of **s character** (sec. 2.2) that provides

stabilization to a cationic center only when conjugated to that center. The 3D drawing of **14** indicates the relationship of the cyclopropane bonds relative to the positive center.

In general, a heteroatom attached to the positively charged carbon stabilizes a carbocation. If the cationic center bears an electron withdrawing group where the (+)- and δ+ centers repel, the carbocation (see **15**) is destabilized. Destabilization of a carbocation is usually associated with difficulty in formation of the carbocation. If a π system (not bearing a δ+ center) is conjugated to the cationic center, the positive charge will be dispersed via resonance, diminishing the net charge on the initially formed cationic carbon, and leading to overall stabilization of the carbocation. An allylic cation, for example, disperses the charge over three atoms (see **16**). If the cationic carbon is conjugated to a diene (as in **17A** and **B**) the charge is dispersed over five atoms, and in benzylic cations (**18A** and **B**) the charge is dispersed over seven atoms including the benzene ring.

---

6. (a) Reference 4, see p 391; (b) Streitweiser Jr., A. *Solvolytic Displacement Reactions*, McGraw-Hill, New York, *1962*, pp 102-103.
7. (a) Hart, H.; Law, P.A. J. *Am. Chem. Soc.* *1962*, *84*, 2462; (b) Hart, H.; Sandri, J.M. *Ibid.* *1959*, *81*, 320.

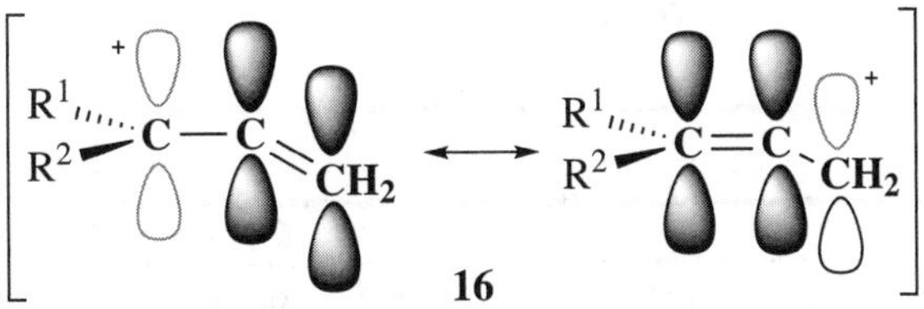

16

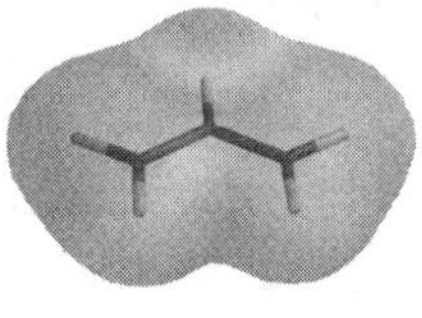

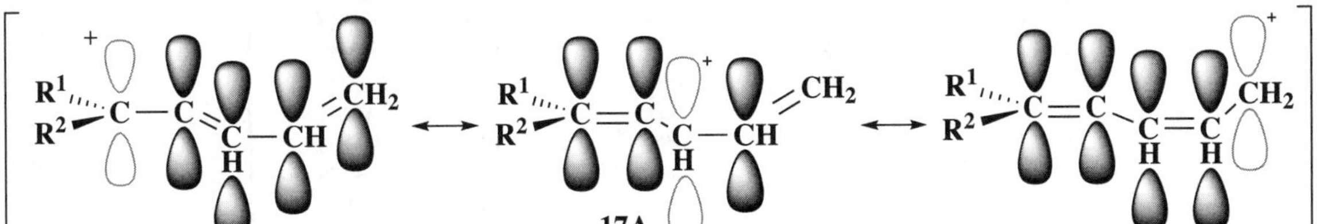

17A

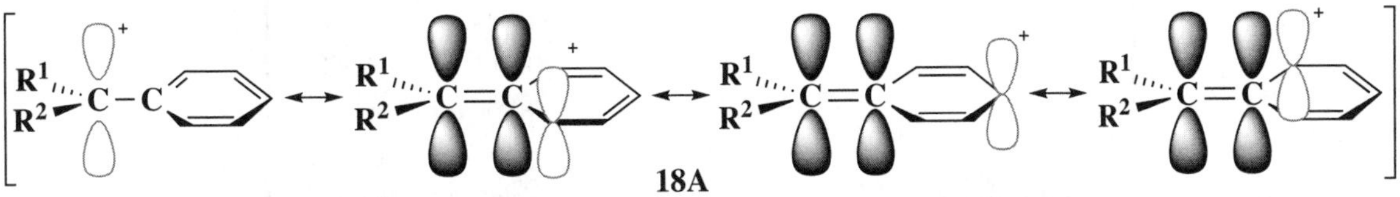

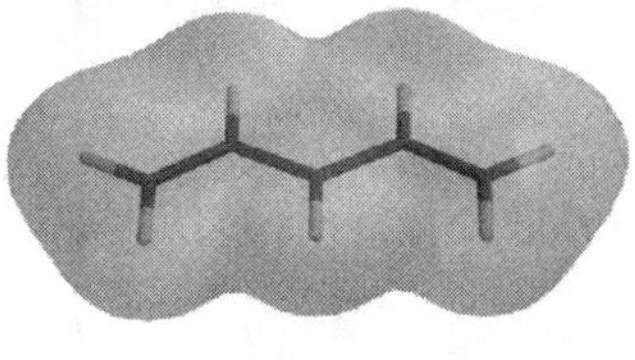

17B

18A

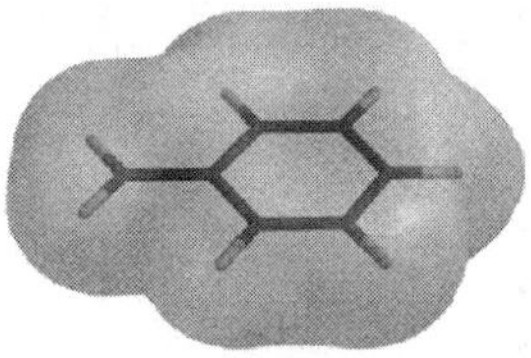

18B

MOLECULAR MODELING: Carbocation Rearrangements

In the presence of acid (silica gel is acidic), the phenylcyclopropane derivative shown loses HCl and yields a substituted indene. The process is believed to involve initial loss of Cl⁻, followed by rearrangement of the carbocation intermediates, **A** to **B** to **C,** and finally loss of H⁺ (see *J. Chem. Soc., Perkin Trans. I 1992*,535, and homework problem 5). Carbocation rearrangements are known to be rapid and reversible, and the indene will only be formed if it is the most stable product.

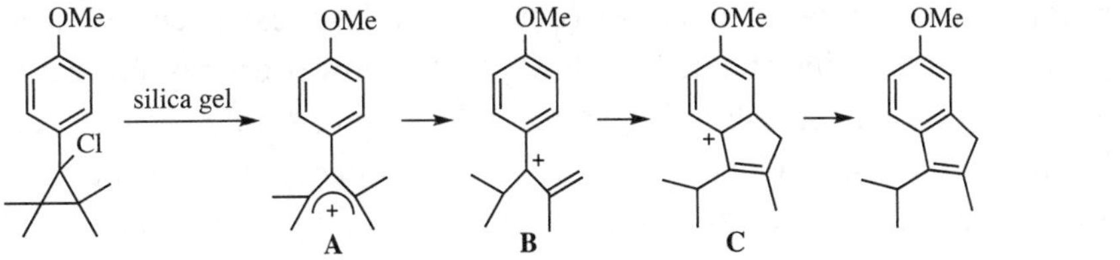

Build the reactant and use the properties icon to obtain its energy. Click on > to the right of the tab at the top of the screen and select *Continue* to bring up a fresh build panel. Construct the indene product and obtain its energy. Repeat the process to obtain the energy for HCl, and calculate the energy of the overall reaction. Is there a large thermodynamic driving force behind the formation of the indene product? Next, obtain energies for the neutral products resulting from deprotonation of both A and B (isomers of the indene product resulting from deprotonation of C). In terms of overall thermochemistry, is either of these competitive with the substituted indene?

Table 12.1 Ionization Potential and $\Delta H^{\ddagger}_{calc}$ for Cations

$$R\bullet + e^- \longrightarrow R^+ + 2e^-$$

R	IP (kcal mol⁻¹)	IP (kJ mol⁻¹)	$\Delta H^{\ddagger}_{calc}$ (kcal mol⁻¹)	$\Delta H^{\ddagger}_{calc}$ (kJ mol⁻¹)
Me	950	3979.6	1080	4524.1
Et	835	3497.8	943	3950.2
n-Pr	835	3497.8	912	3820.4
n-Bu	835	3497.8	885	3707.3
i-Pr	760	3183.6	810	3393.1
t-Bu	715	2995.1	730	3058.0
t-Pentyl	685	2869.5	678	2840.1
Neopentyl	805	3372.2	815	3414.0
cyclopropyl cation (△+)			1000	4189.0
cyclobutyl cation (□+)			890	3728.2
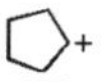 cyclopentyl cation (⬠+)			815	3414.0
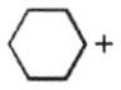 cyclohexyl cation (⬡+)			782	3275.8
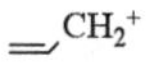 =CH₂⁺			955	4000.5
≡CH₂⁺			1105	4628.9
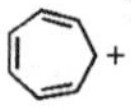 cycloheptatrienyl cation (+)			920	3853.9
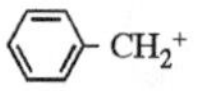 benzyl, C₆H₅–CH₂⁺	748	3133.4		
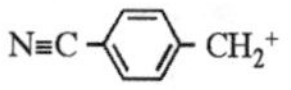 N≡C–C₆H₄–CH₂⁺	806	3376.3		
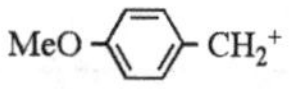 MeO–C₆H₄–CH₂⁺	660	2764.7		
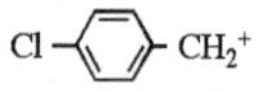 Cl–C₆H₄–CH₂⁺	766	3208.8		

[Reprinted with permission from Isaacs, N.S. *Reactive Intermediates in Organic Chemistry*, Wiley, London/New York, *1974*, p. 151. Copyright © *1974* by John Wiley and Sons, Inc.]

The relative order of carbocation stability is related to the ionization potential of that carbocation, which is experimentally determined by electron bombardment in a mass spectrometer. The ionization potential and relative energies (determined by mass spectrometry) can be used as a measure of the relative order of cation stability.[8] The lower energy ionization (7.42 eV versus 8.64 eV for *tert*-butyl versus *n*-butyl) represents the more stable cation. As expected, tertiary is lower in energy than secondary, which is lower in energy than primary. Table 12.1[9] shows the ionization potential for several cations, as determined by mass spectrometry, along with the calculated energy of cations for a chemical ionization pathway ($\Delta H^{\ddagger}_{calc}$). The two energy values are not related, however, and cannot be correlated. The mass spectral data obviously exclude solvent effects, which can have a profound influence on the relative stability of the cation. For alkyl cations 3° is more stable than 2°, which is more stable than 1°, as seen in Table 12.1,

8. Deno, N.C. in *Progress in Physical Organic Chemistry, Vol. 2*, Cohen, S.G.; Streitwieser Jr., A.; Taft, R.W. (Eds.), Interscience, New York, *1964*, p 135.
9. Isaacs, N.S. *Reactive Intermediates in Organic Chemistry*, Wiley, London/New York, *1974*, p 151.

which also shows several other types of carbocations, and it is clear that the primary benzyl cation is less stable than the *tert*-butyl cation, but more stable than the secondary isopropyl cation. Electron-releasing groups on the aromatic ring make the benzylic cation more stable, and electron-withdrawing groups make it less stable due to the usual inductive effects (sec. 2.2). The $\Delta H^{\ddagger}_{calc}$ data suggest that the primary allyl cation is less stable than a primary alkyl cation but more stable than methyl. Interestingly, the allylic cation is taken to be only slightly less stable than the benzylic cation, and the propargyl cation is significantly less stable than the allylic cation. The stability of secondary cations derived from cyclic molecules is influenced by the size of the ring, with a stability order C6 > C5 > C4 > C3, and the cyclopentyl cation is about equal in energy to the isopropyl cation.

The relative stability of carbocations can also be measured by determining the ΔH_{rxn} for hydrolysis of alkyl halides. Based on a hydrolysis reaction for which ΔH_{rxn} was measured, the relative stability of various carbocations formed as intermediates from the appropriate chloroalkane.[10] An advantage of this measurement is that solvent effects, which are clearly important in most organic transformations, play a significant role. Cyclic 2° and 3° alkyl cations are more stable than their acyclic counterparts. Tertiary benzylic cations are more stable than the tertiary alkyl cations. An important conclusion is that the order 3° > 2° > 1° applies to all cations, and 3° benzylic > 2° benzylic > 1° benzylic. It is not surprising that a 3° benzylic cation is more stable than a 3° alkyl cation since the former is resonance stabilized. The tertiary alkyl cation is more stable than the primary benzylic cation, however, which provides an important benchmark for the following order of carbocation stability.

3° benzylic > 2° benzylic > 3° alkyl > 1° benzylic ≈ 1° allylic ≈ 2° alkyl > 1° alkyl > methyl

Table 12.2 Relative Rate of Solvolysis of Substituted Chloroalkanes[11]

R^1	R^2	Rate	R^1	R^2	Rate
H	Me	1.57×10^{-6}	H	Ph	0.216
Me	Me	0.086	Me	Ph	394
Ph	Me	394	Ph	Ph	19,900

[Reprinted with permission from Brown, H.C.; Rei, M. *J. Am. Chem. Soc* **1964**, *86*, 5008. Copyright © **1964** American Chemical Society.]

The relative rates of solvolysis for several chloroalkanes (**19**) provide another practical measure of carbocation stability. The rate of ionization to carbocation **20** is shown in Table 12.2.[10]

10. Arnett, E.M.; Hofelich, T.C. *J. Am. Chem. Soc.* **1983**, *105*, 2889.
11. (a) Lowry, T.H.; Richardson, K.S. *Mechanism and Theory in Organic Chemistry, 3rd Ed.*, Harper and Row, New York, **1987**, p 385; (b) Brown, H.C.; Rei, M. *J. Am. Chem. Soc.* **1964**, *86*, 5008.

The more stable cations are formed fastest, leading to a faster reaction since this is the rate determining step. The presence of a second phenyl group in chlorodiphenylmethane led to greater conjugation, greater stability and a significantly higher rate of ionization.

Vinyl carbocations can be generated from a variety of sources, but they are generally less stable than alkyl carbocations.[12] Vinyl cations were first detected by Grob and co-workers in solvolysis reactions of α-aryl vinyl halides,[13] but have been observed in solvolysis reactions of other alkenyl halides[14] and in electrophilic reactions of alkynes.[15] As with other carbocations, the stability of this intermediate depends upon the groups attached to the positive center. The X-ray structure of a bis(silyl)-stabilized vinyl cation was reported, for example,[16] indicating the special stability of that cation.

With the data presented thus far on carbocation stability, the goal is to estimate the relative stability of common carbocations in order to make a prediction concerning product distribution when there are competitive processes. In a simple example, when alkene **21** is treated with HX, do we obtain product **22** via carbocation **23**, or product **25** via carbocation **24**? Which is more stable, the primary benzylic cation **24** or the oxygen-stabilized cation **23**? The relative energies for various carbocations has been compiled,[17] and Figure 12.1 shows the order of stability based on $D_{(R+-H-)}$, which is equivalent to the ΔH_f (RH) for the reaction $R–H \rightarrow R^+ + H^-$. The common carbocations H_3C^+, MeH_2C^+, Me_2HC^+ and Me_3C^+ are shown at 313, 271, 252 and 233 kcal mol^{-1} (1308, 1132, 1053, 974 kJ mol^{-1}), respectively.[17] Note that the vinyl cation (287 kcal mol^{-1}, 1199 kJ mol^{-1}) is less stable than the ethyl cation, but more stable than the methyl cation. Replacing a hydrogen atom in $+CH_3$ with an electron-withdrawing group such as cyano ($+CH_2CN$, 305 kcal mol^{-1}, 1274 kJ mol^{-1}) is destabilizing, relative to replacing it with a methyl group (the ethyl cation). Replacing hydrogen with an OH ($+CH_2OH$, 252 kcal mol^{-1}, 1053 kJ mol^{-1}) leads to an ion that is as stable as the secondary isopropyl cation. The primary allylic carbocation (256 kcal mol^{-1}, 1070 kJ mol^{-1}) is almost as stable as a secondary alkyl such as isopropyl cation (252 kcal mol^{-1}, 1053 kJ mol^{-1}). The propargyl cation (271 kcal

12. *Dicoordinated Carbocations*, Rappoport, Z; Stang, P.J. (Eds), Wiley, New York, *1997*.
13. Grob, C.A.; Csapilla, J.; Csch, G. *Helv. Chim. Acta 1964*, *47*, 1590.
14. Hanack, M. *Acc. Chem. Res. 1976*, *9*, 364.
15. Lucchini, V.; Modena, G.; Pasquato, L. in D*icoordinated Carbocations*, Rappoport, Z; Stang, P.J. (Eds), Wiley, New York, *1997*, p 321.
16. Müller, T.; Juhasz, M.; Reed, C.A. *Angew. Chem. Int. Ed. 2004*, *43*, 1543.
17. Lossing, F.P.; Holmes, J.L. *J. Am. Chem. Soc. 1984*, *106*, 6917.

mol⁻¹, 1132 kJ mol⁻¹) is less stable than the primary allylic. The benzyl carbocation (238 kcal mol⁻¹, 994 kJ mol⁻¹) is almost as stable as the tertiary *tert*-butyl cation (233 kcal mol⁻¹, 974 kJ mol⁻¹), and more stable than the primary allylic and slightly more stable than the secondary allylic carbocation (241 kcal mol⁻¹, 1007 kJ mol⁻¹). Placing the positive charge in a five-membered ring (263 kcal mol⁻¹, 1099 kJ mol⁻¹) is less stable than the cyclohexyl cation (243 kcal mol⁻¹, 1015 kJ mol⁻¹), with the acyclic isopropyl cation in between. Acylium ions are quite stable. The formyl cation (257 kcal mol⁻¹, 1074 kJ mol⁻¹) is about as stable as the primary allyl, and the acetyl cation (230 kcal mol⁻¹, 961 kJ mol⁻¹) is more stable than the tertiary alkyl cation. Note that the amino-cation (iminium ion) is very stable, and the tropylium ion shows the greatest stability in this series. Conversely, the alkyne carbocation is extremely unstable (386 kcal mol⁻¹, 1613 kJ mol⁻¹).

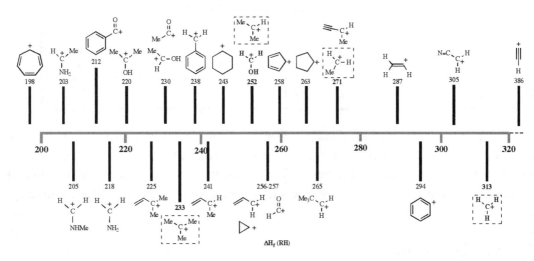

Figure 12.1. Relative Stability of Common Carbocations Based on D$_{(R+-H-)}$, in kcal mol⁻¹.[17] [The data but not the figure is used and reprinted with permission from Lossing, F.P.; Holmes, J.L. *J. Am. Chem. Soc.* **1984**, *106*, 6917. Copyright © 1984 American Chemical Society.]

We can now formulate an answer to the question posed about for **24** versus **23**. A primary benzylic cation (238 kcal mol⁻¹, 994 kJ mol⁻¹) is expected to be less stable than the oxygen-stabilized cation (220 kcal mol⁻¹, 919 kJ mol⁻¹), using Figure 12.1 as a guide. Therefore, **22** is predicted to be the major product. It is likely that a mixture of products will be observed, but the data in Figure 12.1 allows a practical comparison of carbocations that allows one to make reasonable predictions of major and minor products in many cases.

12.2.B. Configurational Instability

As briefly discussed in Section 2.6.B.i and above, carbocations are planar. Therefore, S$_N$1 type reactions are expected to proceed with racemization since there is little or no facial selectivity in the simple cases. Retention or partial retention and even inversion can occur "if the structure contains a configuration-holding group, a nucleophilic group which can interact with the carbonium centre".[18] If a tight ion pair is generated, the S_N1 reaction will show some memory of the departing group since it is in close proximity and partial inversion can occur, which will clearly be solvent dependent. Table 12.3[19] shows the percentage inversion for a series of secondary halides in solvolysis reactions, under a variety of solvolysis conditions. These results indicate that addition of silver ion increases the inversion, presumably via coordination with the halide such that a free cation will not be generated, and the departing bromide is close to the cationic center at the time of the substitution.

Table 12.3 % Inversion in S_N1 Type Solvolysis Reactions

Alkyl Halide	Solvent	% Inversion	Alkyl Halide	Solvent	% Inversion
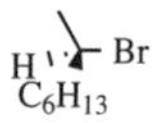 H⋯Br C₆H₁₃	60% aq EtOH	83	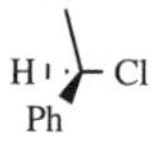 H⋯Cl Ph	H₂O	58
	60% aq EtOH + Ag⁺	97		aq acetone	51
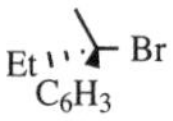 Et⋯Br C₆H₃	80% aq acetone	60	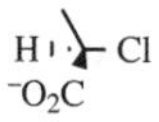 H⋯Cl ⁻O₂C	H₂O	0
				H₂O + Ag⁺	61

[Reprinted with permission from Isaacs, N.S. *Reactive Intermediates in Organic Chemistry*, Wiley, London/New York, *1974*, p. 119. Copyright © *1974* by John Wiley and Sons, Inc.]

As a practical matter, reactions of carbocations do not necessarily proceed with complete inversion or retention, but many proceed with partial inversion. It is reasonable, however, to assume that racemization or partial racemization will be the stereochemical outcome of cationic reactions unless there is experimental evidence or literature precedent to support inversion or retention.

12.2.C. Cationic Rearrangements

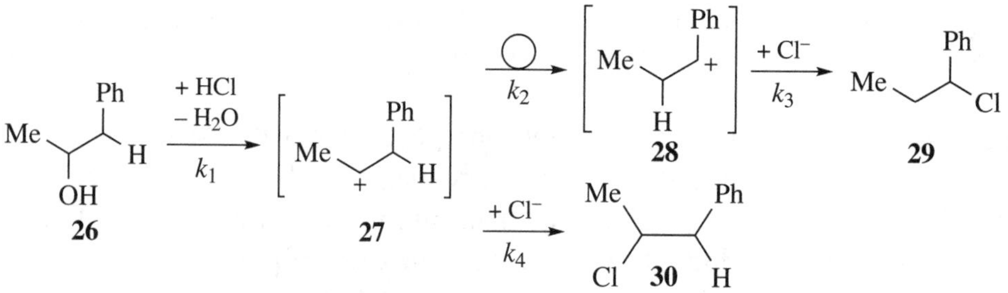

A second property of carbocations that influences their synthetic utility is their propensity to rearrange to a more stable cation. When alcohol **26** was treated with HCl, two products are possible. The first is the direct S_N1 product **30** (see sec. 2.6.B.ii) that results from trapping the initially formed cation (**27**) with chloride ion (rate constant = k_4). The second product is the rearranged benzylic chloride (**29**). Rearrangement via a 1,2-hydride shift can occur (rate

---

18. Reference 9, pp 118-119.
19. Reference 9, p 119.

constant = k_2) because the rearrangement product, benzylic cation **28** is more stable than the initially formed **27** (see Fig. 1), and subsequent trapping with chloride ion (rate constant = k_3) gives **29**. In the limiting cases, if $k_2 \gg k_4$ only **29** will be observed, whereas only **30** will be observed if $k_4 \gg k_2$. In both cases, the initial ionization step (rate constant = k_1) is the slowest (rate-determining) step, and $k_3 \approx k_4$. If the two cation intermediates are close in energy then $k_2 \approx k_4$, and a mixture of products will result or only a modest amount of rearrangement will occur. In many cases, the rearrangement (k_2) proceeds faster than the direct substitution (k_4) and the rearranged product predominates. If the cation and nucleophile are particularly reactive, however, k_4 may be much faster than k_2. A simplifying *assumption* is that $k_2 > k_4$ in most cases, which not always true, but provides a good working model for relatively simple substrates.

This type of rearrangement is understood by viewing the carbocation intermediate as having an electron deficient p orbital (the positive center), which can gain electron density from an adjacent σ bond. When the σ-bond (C–H) adjacent to the p-orbital is parallel (as in **31**), the σ and p-orbitals can overlap and electron density (i.e. the bond) migrates toward the positive center. The transition state for this process is usually represented as a bridged electrophilic species, **32**. The transformation of **31** to **33** via **32** requires that the hydrogen atom is transported as the electrons in the bond move, and this migration is described as a 1,2-hydride shift of the hydrogen atom to the adjacent carbon to give **33**. This rearrangement occurs when the reaction (**31** →

33) is exothermic, and if the bond bearing the migrating group can become parallel with the p orbital of the cationic center. In this case, the more stable benzylic cation is favored over the secondary cation, and the 1,2-hydrogen shift is an exothermic reaction. In general, secondary → tertiary, primary → secondary, and primary → tertiary rearrangements are exothermic reactions, and the rearrangement is facile.

MOLECULAR MODELING: Phenonium Ions

Treatment of the ^{14}C labeled amine with nitrous acid leads to equal amounts of alcohols **A** and **B**, that is, the label is fully scrambled (see homework problem 8).

This fascinating result may be explained by proposing that the initially-formed diazonium salt decomposes to give a secondary carbocation, which either reacts with water to give **A**, or first rearranges to an equivalent carbocation which then reacts with water to give **B**. Connecting the two secondary carbocations is a bridged carbocation (a ***phenonium ion***), a spirocyclic benzenonium

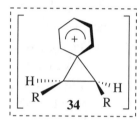

Why did the hydrogen group in **31** migrate and not the phenyl? First, migration of the phenyl would lead to a significantly less stable primary cation and the reaction would be endothermic. Second, migration of the larger phenyl group may require more energy than the smaller hydrogen. Note that the phenyl group can migrate, and a relatively low-energy phenonium ion such as **34** has been invoked in many aryl migrations.[20]

In a transition state such as **32**, considerable positive charge can be present and charge dispersal (as in **34**) can facilitate the rearrangement. A lesson from this observation is that careful analysis of a system is required to determine *a priori* which atom will preferentially migrate.

In the absence of electronic effects, smaller alkyl groups will generally migrate before larger ones: $H > Me > CHMe_2 > CMe_3$

It is difficult to give an absolute scale for migratory aptitude since migratory aptitude is inevitably linked to the stability of the cation being formed and the relative energy of the transition state. An example is cation **35**, where the larger ethyl group can migrate preferentially (to give **37**) rather than the smaller methyl group to give **36**. The *p*-methoxyphenyl group will stabilize cation **37** to a greater extent than the *p*-nitrophenyl group stabilizes **36** however. The electron releasing methoxy diminishes the net charge, but the electron-withdrawing nitro group destabilizes the cationic center in **36**. In other words, the larger ethyl group can migrate rather than the smaller

---

20. (a) Lancelot, C.J.; Cram, D.J.; Schleyer, P.v.R. in *Carbonium Ions. Vol. III*, Olah, G.A.; Schleyer, P.v.R. (Eds.), Wiley, New York, *1972*; (b) Reference 4, p. 367.

methyl due to formation of a more stable cation. The analysis for determining which group migrates must take into account both the stability of the cation being formed (relative to that initially available), and the size of the migrating group. If the rearrangement of either hydrogen or methyl will generate a cation of similar stability, the relative size of the migrating group is usually the determining factor and the hydrogen will migrate preferentially.

The presence of an atom or group that can *assist* migration in some way makes it faster, which is another example of neighboring-group participation noted in several previous chapters.[21] The term σ-**participation** has been used to describe the through-space interaction (sec. 2.2) of an empty p orbital with the electrons in a neighboring bond. In Figure 12.2, **38** represents a transition state for a 1,2-shift that assumes a concerted migration, with ionization of the leaving group (this is effectively a symmetrical bridging transition state).[22] The term σ **conjugation** (usually the same as **hyperconjugation**) "implies sideways interaction of orbitals, as in **39** in Figure 12.2. In Figure 12.2, **39a** represents a hyperconjugative interaction of Z with C_a, whereas **39b** represents the case where there is a small amount of σ participation.[22] Such neighboring-group effects can both accelerate the rate of rearrangement and stabilize the cation.

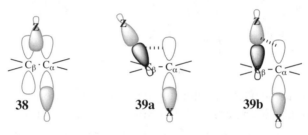

Figure 12.2. σ Conjugation in 1,2-Bond Migrations.

Cationic rearrangements can be highly selective, and several named reactions involve cationic rearrangement. Mundy's monograph is an excellent source that presents many examples of these named reactions.[23] In the Srikrishna et al. synthesis of 6-epijunicedranol, normylataylanone (**43**) was also synthesized,[24] when conjugated ketone **40** was treated with boron trifluoride•etherate. Coordination of the carbonyl and Michael-type addition of the exocyclic alkene unit gave tertiary cation **41**. Despite the fact that this is a tertiary cation, it rearranged via a 1,2-alkyl shift

21. Reference 4, pp 429-439.
22. Reference 4, p 431.
23. Mundy, B.P.; Ellerd, M.G.; Favaloro Jr., F.G. *Name Reactions and Reagents in Organic Synthesis, 2nd Ed.*, Wiley, New York, *2005*.
24. Srikrishna, A.; Yelamaggad, C.V.; Kumar, P.P. *J. Chem. Soc. Perkin Trans. 1 1999*, 2877.

to give spirocyclic carbocation **42**. The migration occurred on the face opposite the methyl group, and the enol unit in **42** attacked the tertiary cation from that same face to give ketone **43** selectively, in 60% yield.

Cyclopropylcarbinyl systems (where the cationic carbon is connected directly to a cyclopropyl ring) are subject to ring-expanding cationic rearrangements. In Section 12.2.A, it was noted that the cyclopropyl group imparts special stability to an adjacent cationic, and migration of the strained cyclopropyl bonds in cyclopropylcarbinyl cations such as **14** leads to the less strained homoallylic cation **46**.[25] Participation of the cyclopropane ring is apparent in the solvolysis of cyclopropylcarbinyl tosylate (**44**), which occurs 10^6 times faster than solvolysis of the tosylate of 2-methyl-1-propanol (**45**).[26] A cyclobutyl cation (**47**) can also form under these conditions via a ring-expansion reaction (see below). The cyclobutylcarbinyl system (**48**) is less strained than **44**, but also undergoes ring expansion via cyclobutylcarbinyl cation, **49**. Solvolysis of tosylate **48** leads to the cyclopentyl cation (**50**), via ring expansion of **49**.

Cationic rearrangement can be used in other ring expansion reactions, such as the transformation of **51** to **54** upon treatment with acid, albeit in only 9% yield.[27] Solvolysis of the tosyl group in **51** produced the unstable primary cation **52**. Rearrangement led to ring expansion to the more stable secondary cation **53**, which lost a proton to give the alkene product (**54**). The overall result was expansion of the eight-membered ring to a nine-membered ring. If bond migration leads to a medium size ring (8-13), the high energy inherent in these rings may inhibit the bond migration to form a larger ring (sec. 6.6.B) and gives poor yields. Despite the fact that a secondary cation is much more stable than a primary cation, the yield of **54** was rather poor. A contributing factor is the ring expansion that generates a nine-membered ring, which is higher in energy than the eight-membered ring precursor.

25. (a) Roberts, J.D.; Mazur, R.H. *J. Am. Chem.Soc.* **1951**, *73*, 2509; (b) Mazur, R.H.; White, W.N.; Semenow, D.A.; Lee, C.C.; Silver, M.S.; Roberts, J.D. *Ibid.* **1959**, *81*, 4390; (c) Brady, S.F.; Ilton, M.A.; Johnson, W.S. *Ibid.* **1968**, *90*, 2882; (d) Julia, M.; Mouzin, G.; Descoins, C. *C.R. Hebd. Seances Acad. Sci.* **1967**, *264*, 330.
26. Roberts, D.D. *J. Org. Chem.* **1964**, *29*, 294; (b) *Idem Ibid.* **1965**, *30*, 23.
27. Huisgen, R.; Seidl, G. *Tetrahedron* **1964**, *20*, 231.

Carbocations are available from precursors other than halides or sulfonate esters. Treatment of an amine with nitrous acid (HONO) generates an unstable diazonium salt, which decomposes to a transient primary cation. The initially formed cation rearranged to the larger ring cation in what is now known as the **Demjanov rearrangement**.[28] In the initially reported reaction, water was present, which trapped the ring-expanded cation as the alcohol.[29,5ac]

Synthetic applications of this ring expansion generally use the milder reaction conditions associated with alcohol or sulfonate ester precursors. In Fitjer and Mandelt's synthesis of laurene,[30] alcohol **55** was heated with *p*-toluenesulfonic acid in benzene to generate cation (**56**), with the positive charge adjacent to a strained four-membered ring. Opening the strained four-membered ring to the lower energy five-membered ring product occurred by a 1,2-alkyl shift to give **57**. Under the reaction conditions, elimination occurred to give a 90% yield of **58** along with 10% of **59**. Note that the stereochemistry of the groups adjacent to the cation can have a significant effect on the migratory aptitude of the groups adjacent to the cationic center.[31]

A relatively simple variation of this reaction is the **Tiffeneau-Demjanov ring expansion**,[32,28c] involving treatment of amino alcohols with nitrous acid to produce ring-expanded ketones. The conversion of **60** to a mixture of isomeric ketones in 81% yield is an example.[33] This example illustrates a common problem in ring-expansion reactions in particular and in rearrangements in general, that of regioselectivity. Treatment of **60** with nitrous acid produced cation **61**. Two ring expansion pathways are possible, paths *a* and *b*. The major pathway is path *a*, first giving ring expansion to cation **62** and then a 55% yield of **63**. This pathway also produced 0.7% of the trans ketone. Path *b* led to **64**, which produced 25.6% of ketone **65**. Incorporation of silicon and tin substituents into the molecule mediates the ring expansion.[34]

28. (a) Demjanov, N.Y.; Lushnikov, M. *J. Russ. Phys. Chem.* **1903**, *35*, 26; (b) *Idem, Chem. Zentr.* **1903**, *I*, 1468; (c) Smith, P.A.S.; Baer, D.R. *Org. React.* **1960**, *11*, 157; (d) The Merck Index, 14th ed., Merck & Co., Inc., Whitehouse Station, New Jersey, **2006**, p ONR-23.
29. Friedman, L. in *Carbonium Ions, Vol. II*, Olah, G.A.; Schleyer, P.v.R. (Eds.), Wiley, New York, **1970**, p 655.
30. Mandelt, K.; Fitjer, L. *Synthesis* **1998**, 1523.
31. Nikishova, N.G.; Bundel, Yu.G.; *Vestn. Mosk. Univ., Ser. 2: Khim* **1985**, *26*, 486 [*Chem. Abstr.* 104: 167800p, **1986**].
32. (a) Tiffeneau, M.; Weill, P.; Tchoubar, B. *Compt. Rend.* **1937**, *205*, 54; (b) The Merck Index, 14th ed., Merck & Co., Inc., Whitehouse Station, New Jersey, **2006**, p ONR-93; (b) Mundy, B.P.; Ellerd, M.G.; Favaloro Jr., F.G. *Name Reactions and Reagents in Organic Synthesis*, 2nd ed., Wiley-Interscience, New Jersey, **2005**, pp. 648-649.
33. Parham, W.E.; Roosevelt, C.S. *J. Org. Chem.* **1972**, *37*, 1975.
34. Chow, L.; McClure, M.; White, J. *Org. Biomol. Chem.* **2004**, *2*, 648.

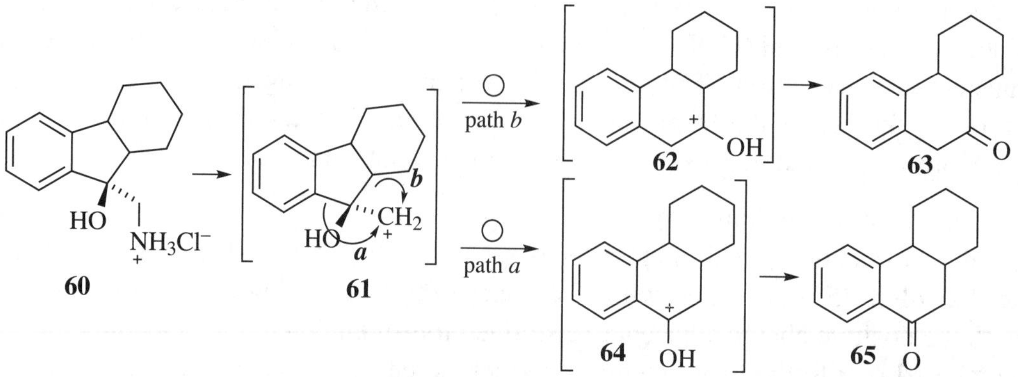

This ring-expansion process is quite useful when applied to polycyclic systems. Corey and Nozie treated **66** with 40% sulfuric acid, and isolated α-caryophyllene alcohol (**69**).[35] The initially formed cyclobutylcarbinyl cation (**67**) rearranged to the cyclopentyl cation (**68**), which was trapped by water to give **70**. Trost and co-workers used ring expansion to generate oxaspiropentane derivatives that were converted to cyclobutane derivatives.[36,37] In section 8.8.B.ii, oxaspiropentanes such as **70** were prepared by the reaction of ketones and sulfur stabilized cyclopropyl ylids. Treatment of **70** with aqueous tetrafluoroboric acid (HBF$_4$) gave cation **71**, which rearranged to **72**. Loss of a proton gave the final product, ketone **73**.[29,28]

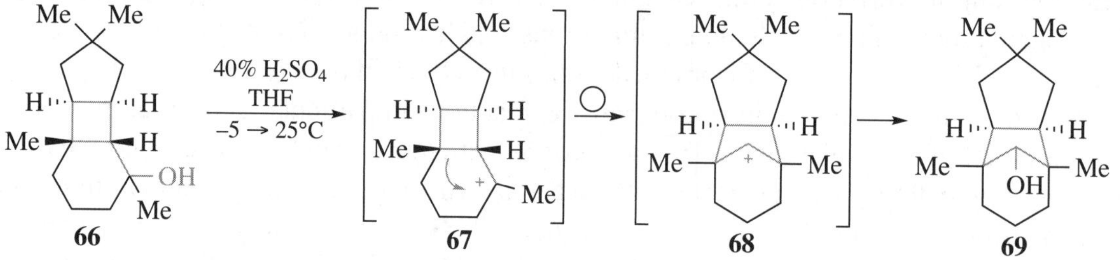

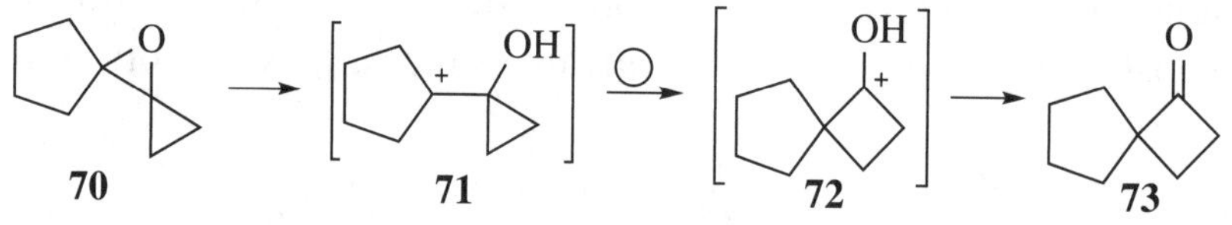

Another variation of this ring expansion used sulfur stabilized cations. When aldehyde **74** was treated with trimethylsilyl methyl sulfide and trimethylsilyl triflate, cation **75** was formed and rearranged to the more stable tertiary cation **76**. Elimination of the silyl group gave a 59% yield of the *exo*-methylene derivative (**77**).[38] The driving force of this reaction is probably formation of the β-silyl cation in **76**. A recent variation treated α-hydroxydithianes with red

35. (a) Corey, E.J.; Nozoe, S. *J. Am. Chem. Soc.* **1964**, *86*, 1652; (b) *Idem Ibid.* **1965**, *87*, 5733.
36. (a) Trost, B.M. *Acc. Chem. Res.* **1974**, *7*, 85; (b) Trost, B.M.; LaRochelle, R.; Bogdanowicz, M.J. *Tetrahedron Lett.* **1970**, 3449.
37. Trost, B.M.; Bogdanowicz, M.J. *J. Am. Chem. Soc.* **1973**, *95*, 5311.
38. Tanino, K.; Sato, K.; Kumajima, I. *Tetrahedron Lett.* **1989**, *30*, 6551.

HgO and HBF$_4$, leading to good yields of ring-expanded 1,2-diketones.[39]

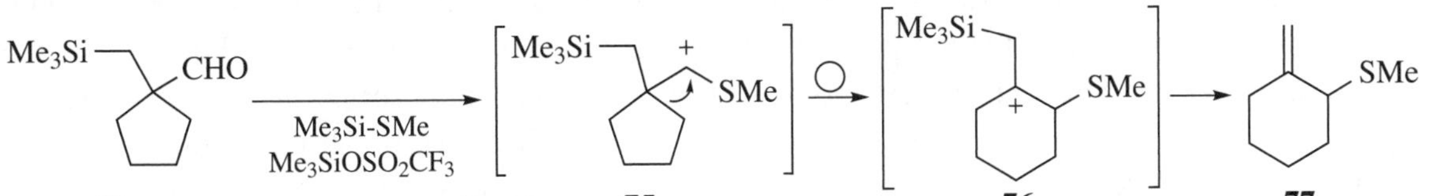

74 **75** **76** **77**

Epoxides react with protonic or Lewis acids to generate cations that are also subject to rearrangement. In Tanino, Kuwajima and co-worker's synthesis of ingenol,[40] treatment of **78** with trimethylaluminum gave **81** in >80% yield. The epoxide unit reacted with Lewis acid to generate oxonium ion **79**, which rearranged by a 1,2-alkyl shift to give oxygen-stabilized cation **80**. Loss of a proton and protonation of the O-Al unit gave ketone **81**.

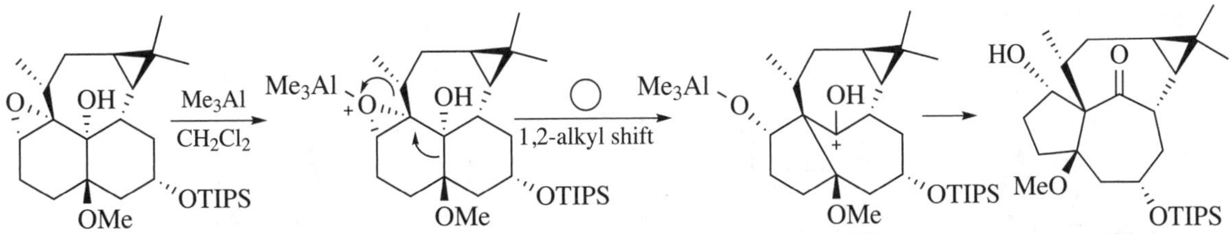

78 **79** **80** **81**

A specialized cation rearrangement first observed with bicyclo[2.2.1] systems was the acid catalyzed reaction of camphene hydrochloride (**82**) to isobornyl chloride (**85**).[41] The reaction, which proceeded by a 1,2-alkyl shift called the **Wagner-Meerwein rearrangement**,[42] initially generated bornyl cation **83**, but rearrangement to **84** occurred prior to trapping of chloride ion. The term Wagner-Meerwein is often applied to the acid-catalyzed rearrangement of [*m.n.o*]-bicyclo and [*m.n.o*]-tricyclo species, although it applies formally to bicyclo[2.2.1]heptane derivatives.

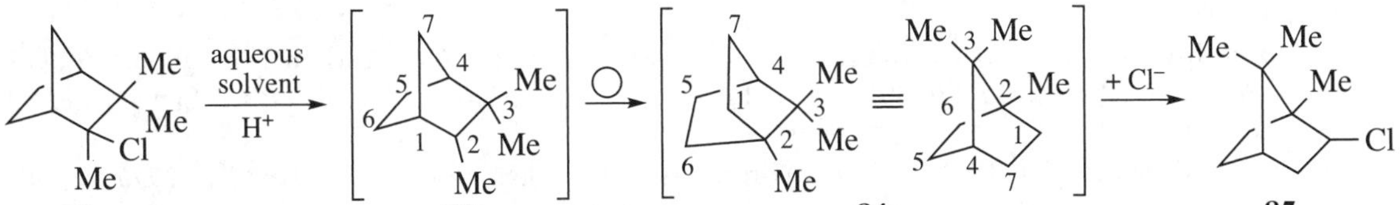

82 **83** **84** **85**

This reaction has been controversial concerning the structure of norbornyl cations such as **83**

39. Ranu, B.C.; Jana, U. *J. Org. Chem.* **1999**, *64*, 6380.
40. Tanino, K.; Onuki, K.; Asano, K.; Miyashita, M.; Nakamura, T.; Takahashi, Y.; Kuwajima, I. *J. Am. Chem. Soc.* **2003**, *125*, 1498.
41. (a) Wagner, G. *J. Russ. Phys. Chem. Soc.* **1899**, *31*, 690; (b) Meerwein, H. *Annalen* **1914**, *405*, 129.
42. (a) The Merck Index, 14th ed., Merck & Co., Inc., Whitehouse Station, New Jersey, **2006**, p ONR-98; (b) Mundy, B.P.; Ellerd, M.G.; Favaloro Jr., F.G. *Name Reactions and Reagents in Organic Synthesis, 2nd ed.*, Wiley-Interscience, New Jersey, **2005**, pp. 678-679.

and **84**.[43,44] Norbornyl cations have been described in terms of the resonance structures shown in **86**, or as the nonclassical ion **87**.[45] This view was questioned, primarily by Brown,[43] who did not believe that the evidence proved the existence of a bridged intermediate.[46] Brown suggested that the facile rearrangements could be explained by a series of fast 1,3-Wagner-

86   **87**

Meerwein shifts,[47] but there is good evidence that the norbornyl cation rearranges with considerable σ-participation.[48] NMR evidence for the nonclassical ion in super acids at low temperatures is strong.[49] Olah is more forceful, stating "it can be concluded that the original views of Winstein[50] on the nonclassical nature of the norbornyl cation, based on kinetic and stereochemical reactivity, have been substantiated by direct spectroscopic studies of the long-lived ion".[51]

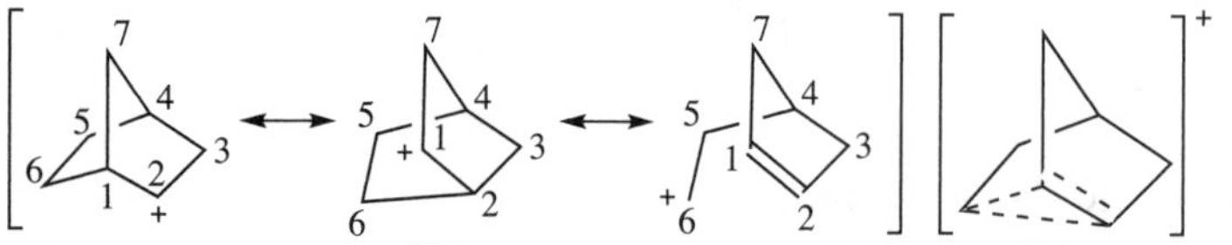

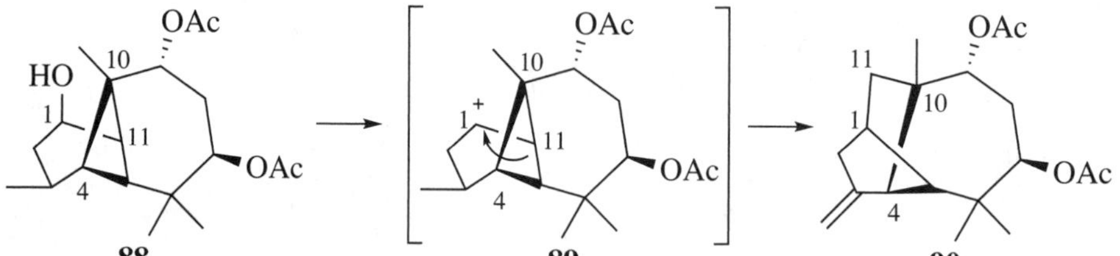

88   **89**   **90**

The mechanistic controversies have not diminished the synthetically interesting transformations possible with controlled cation rearrangements.[52] An example is taken from a synthesis of the longipinane skeleton by Joseph-Nathan and co-workers, in which **88** was treated with sulfuric acid to give a 64% yield of the rearranged product **90**.[53] The reaction involved rearrangement

43. Brown, H.C. *The Non-Classical Ion Problem*, Plenum, New York, *1977*.
44. Story, R.R.; Clark, B.C. in *Carbonium Ions, Vol. III*, Olah, G.A.; Schleyer, P.v.R. (Eds.), Wiley, New York, *1972*, p 1007.
45. (a) Berson, J.A. in *Molecular Rearrangements, Vol. I*, Mayo, P. (Ed.), New York, p 111; (b) Sargent, G.D. *Quart. Rev. Chem. Soc.* *1966*, *20*, 301; (c) Olah, G.A. *Acc. Chem. Res.* *1976*, *9*, 41; (d) Scheppelle, S.E. *Chem. Rev.* *1972*, *72*, 511.
46. (a) Brown, H.C. *Tetrahedron* *1976*, *32*, 179; (b) Brown, H.C.; Kawakami, J.H. *J. Am. Chem. Soc.* *1970*, *92*, 1990.
47. Brown, H.C.; Ravindranathan, M. *J. Am. Chem. Soc.* *1978*, *100*, 1865.
48. (a) Coates, R.M.; Fretz, E.R. *J. Am. Chem. Soc.* *1977*, *99*, 297; (b) Brown, H.C.; Ravindranathan, M. *Ibid.* *1977*, *99*, 299.
49. Olah, G.A. *Carbocations and Electrophilic Reactions*, Verlag Chemie/Wiley, New York, *1974*, pp 80-89.
50. (a) Winstein, S. *Quart. Rev. Chem. Soc.* *1969*, *23*, 141; (b) Winstein, S.; Trifan, D.S. *J. Am. Chem. Soc.* *1949*, *71*, 2953; (c) *Idem, Ibid.* *1952*, *74*, 1154.
51. (a) Reference 49, p 89; (b) Olah, G.A.; White, A.M.; DeMember, J.R.; Commeyras, A.; Lui, C.Y. *J. Am. Chem. Soc.* *1970*, *92*, 4627.
52. Mundy, B.P.; Ellerd, M.G.; Favaloro Jr., F.G. *Name Reactions and Reagents in Organic Synthesis, 2nd Ed.*, Wiley, New York, *2005*, pp 678-679.
53. Cerda-García-Rojas, C.M.; Flores-Sandoval, C.A.; Román, L.U.; Hernández, J.D.; Joseph-Nathan, P. *Tetrahedron* *2002*, *58*, 1061.

of the secondary carbocation **89**, with migration of bond 10-11.

Although they are difficult to generalize, several interesting disconnections are possible utilizing cationic rearrangements. Some general examples are

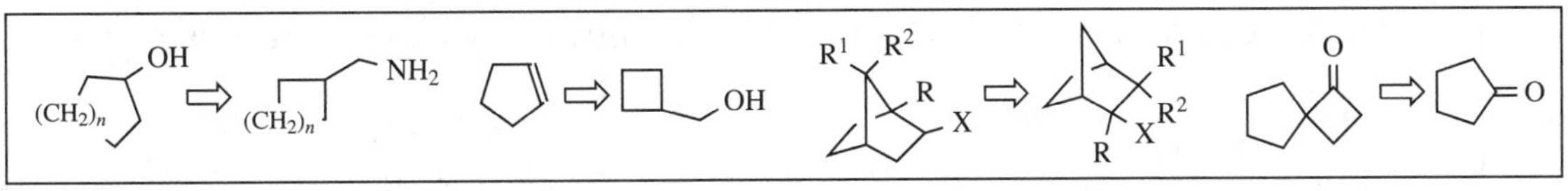

A highly specialized rearrangement is the shift observed during liver microsome-mediated oxidations of aromatic substrates, and during solvolysis of labeled arene oxides,[54] now called the **NIH shift**.[55] "In most enzymatic hydroxylations of aromatic substrates the substituent (^{2}H, ^{3}H, or halogen) present at the position of the entering oxygen migrates to either one of the adjacent ring positions."[56] As observed with an ipso nitration reaction, this shift probably involves an intramolecular hydrogen shift.[57] A specific example involves enzymatic hydroxylation of chlorobenzene to give **91**. The NIH shift produces **92** and deprotonation, with aromatization, gives the substitution product, 4-chlorophenol.[55]

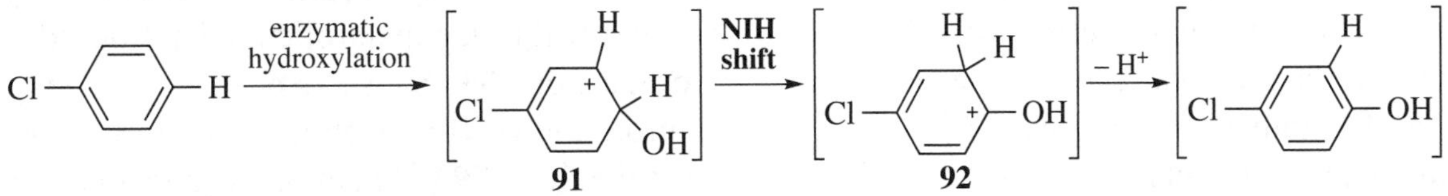

12.2.D. The Pinacol Rearrangement

The **Pinacol rearrangement** is an important cationic rearrangement reaction used for synthesis.[58] Fittig was the first to discover that treatment of 2,3-dimethyl-2,3-butanediol (**93**) with sulfuric acid generated 3,3-dimethyl-2-butanone (pinacolone, **94**).[59,58a] This transformation gave rise to the name of the reaction, and was shown to be applicable to the acid catalyzed rearrangement of most 1,2-diols. The reaction can be generalized to show that 1,2-diols (**95**) are converted first to a β-hydroxy

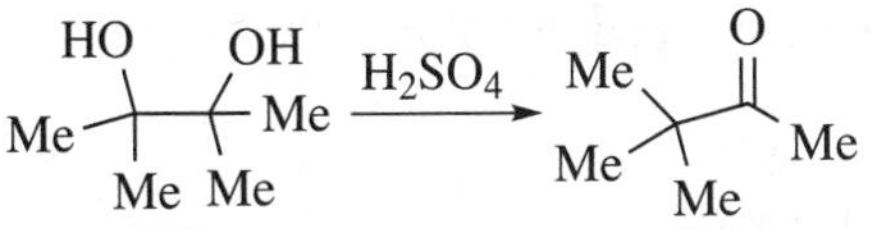

---

54. (a) Daley, J.W.; Hernia, D.M.; Witkop, B. *Experentia* **1972**, *28*, 1129; (b) Bruice, T.C.; Bruice, P.Y. *Acc. Chem. Res.* **1976**, *9*, 378.

55. Guroff, G.; Daly, J.; Jerina, D.; Renson, J.; Witkop, B.; Udenfriend, S. *Science* **1967**, *157*, 1524.

56. Jerina, D.M.; Daly, J.W.; Witkop, B. *J. Am. Chem. Soc.* **1967**, *89*, 5488.

57. Feldman, K.S.; McDermott, A.; Myhre, P.C. *J. Am. Chem. Soc.* **1979**, *101*, 505.

58. (a) Fittig, R. *Annalen* **1860**, *114*, 54; (b) Collins, C.J. *Quart. Rev. Chem. Soc.* **1960**, *14*, 357; (c) Mundy, B.P.; Otzenberger, R.D. *J. Chem. Educ.* **1971**, *48*, 431; (d) The Merck Index, 14th ed., Merck & Co., Inc., Whitehouse Station, New Jersey, **2006**, p ONR-74; (e) Mundy, B.P.; Ellerd, M.G.; Favaloro Jr., F.G. *Name Reactions and Reagents in Organic Synthesis, 2nd ed.*, Wiley-Interscience, New Jersey, **2005**, pp. 514-516.

59. Fittig, R. *Annalen* **1859**, *110*, 17.

carbocation (**96**). Rearrangement of pinacol (2.3-dimethyl-2,3-butanediol, for which the reaction is named) to pinacolone (3,3-dimethyl-2-butanone) is the classical example, but the reaction can be done with a variety of 1,2-diols (cyclic and acyclic), utilizing both protonic and Lewis acid initiators. Rearrangement to the more stable **97** (an oxygen-stabilized cation rather than a tertiary alkyl cation, sec. 12.2.A and Fig. 2.1) is followed by loss of a proton to give **98**.[60]

Two important factors are the acid used to initiate the reaction, and the concentration at which the reaction is done.[61] With increasing dilution of the acid as well as the conjugate base, a higher percentage of elimination products are formed rather than the rearrangement product.[61] When Lewis acids are used rather than protonic acids, differences in the reaction pathway are often observed. Treatment of **99** with aqueous perchloric acid gave an 81% yield of **102**, but reaction with boron trifluoride in dichloromethane gave an 88% yield of **105**.[62] In aqueous media, tertiary cation **100** was formed from **99**. Ring expansion to cycloheptyl cation **101** followed by loss of a proton gave ketone **102**. Reaction with boron trifluoride in aprotic media proceeded by coordination with the cyclohexanol moiety to form cation **103**. Subsequent 1,2-methyl shift (to **104**) rather than ring expansion and loss of a proton gave **105**. Cation formation can be influenced by rather subtle factors. A double bond in the ring of **99**[63] led to ring expansion with no products such as **105**. This reaction is useful for synthesis. An example is the conversion of **106** to **107**,[64] taken from Pettit and co-worker's synthesis of hydroxyphenstatin. Pinacol rearrangements are useful for the preparation of spirocyclic ketones, when cyclic diol precursors are used. Many spirocyclic compounds[65,66] are formed by this type of rearrangement. The pinacol rearrangement has been used to construct stereogenic centers adjacent to heterocycles.[67]

60. For reactions of this type, see Larock, R.C. *Comprehensive Organic Transformations, 2nd ed.*, Wiley-VCH, New York, *1999*, pp 1281-1282.
61. De Lezaeta, M.; Sattar, W.; Svoronos, P.; Karimi, S.; Subramaniam, G. *Tetrahedron Lett. 2002, 43*, 9307.
62. Corey, E.J.; Danheiser, R.L.; Chandrasekaran, S. *J. Org. Chem. 1976, 41*, 260.
63. Bhushan, V.; Chandrasekaran, S. *Chem. Lett. 1982*, 1537.
64. Pettit, G.R.; Lippert III,. J.W.; Herald, D.L. *J. Org. Chem. 2000, 65*, 7438.
65. Mundy, B.P.; Kim, Y.; Warnet, R.J. *Heterocycles 1983, 20*, 1727.
66. Krapcho, A.P. *Synthesis 1976*, 425.
67. Shinohara, T.; Suzuki, K. *Tetrahedron Lett. 2002, 43*, 6937.

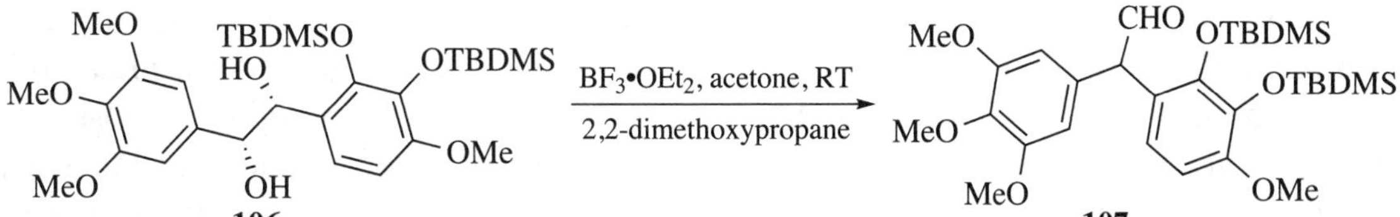

The disconnection for these reactions is

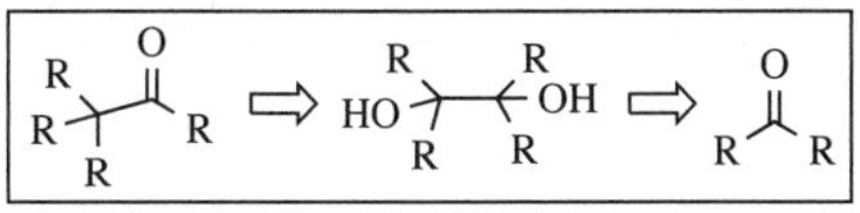

Pinacol-like rearrangements are possible from precursors other than diols, especially oxiranes[68] and allylic alcohols.[69] In principle, any molecule that generates an intermediate with a positive charge on a carbon α to one bearing an OH group can undergo rearrangement. β-Amino alcohols, for example, rearrange on treatment with nitrous acid via the so-called the *semipinacol rearrangement*,[70] as do iodohydrins[71] and allylic alcohols, upon treatment with a strong acid that protonates the double bond. An example of the semipinacol rearrangement is taken from Wood and co-worker's synthesis of welwitindolinone A isonitrile,[72] in which allylic alcohol **108** was treated with sodium hypochlorite in the presence of cerium chloride to give a 78% yield of chloro-ketone **109**. Epoxides rearrange when treated with acidic reagents.[73] The **Meinwald rearrangement**[74] is a related rearrangement induced by the enzyme pig liver esterase.[75]

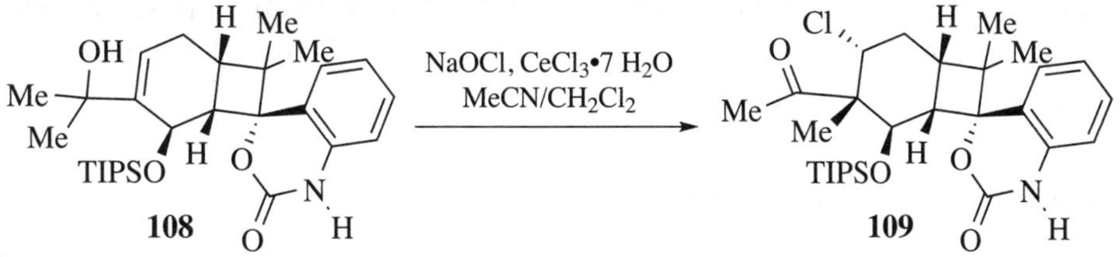

68. Alarcón, P.; Pardo, M.; Sato, J.L. *J. Heterocyclic Chem.* **1985**, *22*, 273.

69. (a) Suzuki, K.; Katayama, E.; Tsuchihashi, G. *Tetrahedron Lett.* **1984**, *25*, 1817; (b) *Idem Ibid.* **1983**, *24*, 4997.

70. (a) Pollack, P.I.; Curtin, D.Y. *J. Amer. Chem. Soc.* **1950**, *72*, 961 (b) Curtin, D.J.; Pollack, P.I. *J. Amer. Chem. Soc.* **1951**, *73*, 992; (c) Curtin, D.Y.; Harris, E.E.; Pollack, P.I. *J. Amer. Chem. Soc.* **1951**, *73*, 3453; (d) Curtin, D.Y.; Meislich, E.K. *J. Amer. Chem. Soc.* **1952**, *74*, 5905; (e) Curtin, D.Y.; Crew, M.C. *J. Amer. Chem. Soc.* **1954**, *76*, 3719. Also see (f) Spivak, C.E.; Harris, F.L. *J. Org. Chem.* **1972**, *37*, 2494.

71. Krief, A.; Laboureur, J.L.; Dumont, W.; Labar, D. *Bull. Soc. Chim. Fr.* **1990**, 681.

72. Reisman, S.E.; Ready, J.M.; Hasuoka, A.; Smith, C.J.; Wood, J.L. *J. Am. Chem. Soc.* **2006**, *128*, 1448.

73. (a) Yandovskii, V.N.; Ershov, B.A. *Russ. Chem. Rev.* **1972**, *41*, 403, 410; (b) Also see Sudha, R.; Narashimhan, K.M.; K.M.; Saraswathy, V.G.; Sankararaman, S. *J. Org. Chem.* **1996**, *61*, 1877; (c) Ranu, B.C.; Jana, U. *J. Org. Chem.* **1998**, *63*, 8212.

74. (a) Meinwald, J.; Labana, S.S.; Chadha, M.S.. *J. Am. Chem. Soc.* **1963**, *85*, 582; (b) Mundy, B.P.; Ellerd, M.G.; Favaloro Jr., F.G. *Name Reactions and Reagents in Organic Synthesis, 2nd ed.*, Wiley-Interscience, New Jersey, **2005**, pp. 420-421.

75. Niwayama, S.; Noguchi, H.; Ohno, M.; Kobayashi, S. *Tetrahedron Lett.* **1993**, *34*, 665.

The term semipinacol rearrangement is applied to the reaction of many substrates, including α-hydroxy aldehydes[76] epoxy alcohols,[77] and epoxy ethers,[78] although the term **pinacol-like rearrangement** is also applied to rearrangement of these substrates. A highly stereoselective rearrangement of 2,3-epoxy alcohols was reported, using the initiator bis(iodozincio)methane.[79] Pinacol-like rearrangements can be useful in synthesis. In a synthesis of (+)-asteltoxin by Cha and co-worker's,[80] epoxide **110** was treated with titanium tetrachloride and gave a 96% yield of aldehyde **111**. In another variation, taken from a synthesis of 2-thiocyanaotoneopupukeanane by Uyehara and co-workers,[81] hydroxy ether **112** was treated with tosic acid to give a 96% yield of ketone **113**.

12.3. CARBON-CARBON BOND FORMING REACTIONS OF CARBOCATIONS

There are several synthetically useful and selective carbocation reactions and some classical named reactions involve cationic intermediates. Variations in functional groups, and the reaction conditions used to generate the cation lead to chemoselective transformations that are often highly stereoselective. The intent of this section is to focus attention on the variety of reaction types, and on synthetic applications.

12.3.A. Carbocation Reactions with Alkene Nucleophiles

Alkenes are nucleophiles in reactions with carbocations, donating two electrons from the π bond to form a new bond to one carbon of the C=C. Note that this is the same definition as that for a Lewis base but applied to a carbocation, making the carbocation a Lewis acid in this reaction. Forming this new bond leaves behind a cationic center at the other carbon of the C=C unit (see **115**). Both intermolecular (as in the formation of **115**) and intramolecular (as

76. For example, see (a) Miller, T.C. *J. Org. Chem.* **1969**, *34*, 3829; (b) Schor, L.; Gros, E.G.; Seldes, A.M. *J. Chem. Soc. Perkin Trans. 1* **1992**, 453; (c) Joshi, A.P.; Nayak, U.R.; Dev, S. *Tetrahedron* **1976**, *32*, 1423; (d) Benjamin, L.J.; Adamson, G.; Mander, L.N. *Heterocycles* **1999**, *50*, 365; (e) Marson, C.M.; Oare, C.A.; McGregor, J.; Walsgrove, T.; Grinter, T.J.; Adams, H. *Tetrahedron Lett.* **2003**, *44*, 141.
77. Marson, C.M.; Khan, A.; Porter, R.A.; Cobb, A.J.A., *Tetrahedron Lett.* **2002**, *43*, 6637.
78. For a synthesis of (+)-grindelic acid using a pinacol ring expansion involving hydroxy ethers, see Paquette, L.A.; Wang, H.-L. *Tetrahedron Lett.* **1995**, *36*, 6005.
79. Matsubara, S.; Yamamoto, H.; Oshima, K. *Angew. Chem. Int. Ed.* **2002**, *41*, 2837.
80. Eom, K.D.; Raman, J.V.; Kim, H.; Cha, J.K. *J. Am. Chem. Soc.* **2003**, *125*, 5415.
81. Uyehara, T.; Onda, K.; Nozaki, N.; Karikomi, M.; Ueno, M.; Sato, T. *Tetrahedron Lett.* **2001**, *42*, 699.

in the formation of **118**) versions are known. The intermolecular process is often plagued by cationic polymerization and/or competitive elimination reactions (E1, sec. 2.9.B), depending of the solvent and relative basicity of the counterion (sec. 2.7). The intramolecular reaction is subject to competing inter- and intramolecular reactions (sec. 6.6.B), but is generally less prone to polymerization although elimination remains a problem. In an intermolecular example, 2-methyl-2-butanol is treated with acid, leading to protonation of the oxygen atom. Loss of

water from this oxonium salt leads to tertiary cation **114**. In the presence of a second molecule (the alkene), the π bond attacks the cationic center in an intermolecular reaction to form a new cation (**115**). In a similar manner, alcohol **116** was the precursor to carbocation **117** when treated with acid. Subsequent intramolecular attack by the alkene moiety on the other side of the same molecule led to carbocation **118**. Formation of **118** is a 6-endo-trig type reaction (see Baldwin's rules, sec. 6.6.A), which can compete with the 5-exo-trig reaction resulting from attack of the cationic center by the other carbon of the alkene moiety. That cyclization process would generate a less stable primary cation so formation of the more stable **118** is favored. Side-reactions are less important in intramolecular cases, so they are used more commonly in synthesis. The remainder of this section will focus on cationic cyclizations as well as controlled rearrangements.

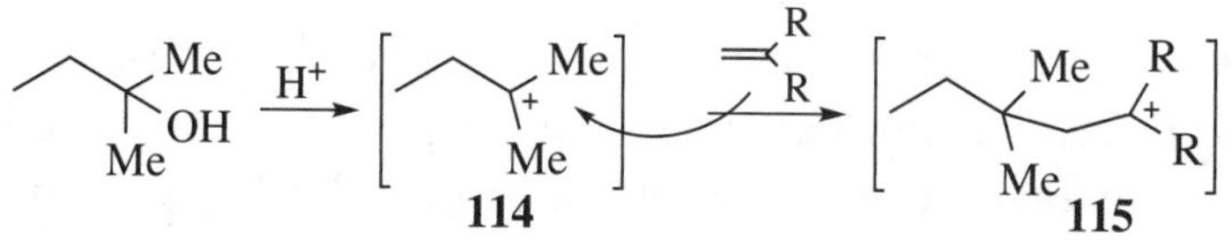

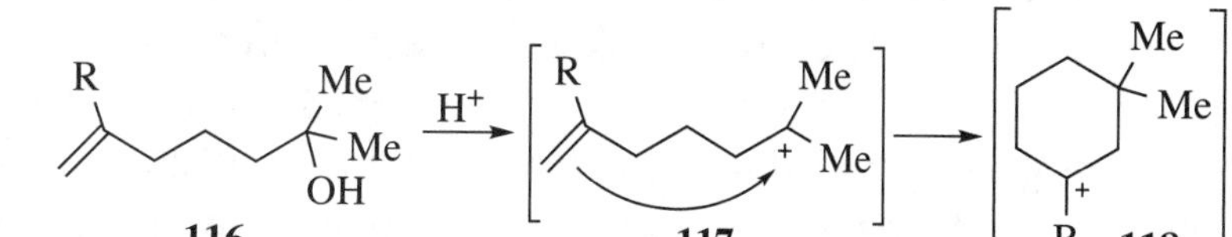

As seen for **116**, alcohols can generate cations, which then react with an alkene unit elsewhere in the molecule. Reaction of alcohol **119** with 90% sulfuric acid gave **120** as a reactive intermediate that was trapped intramolecularly by the alkene to give the chlorine-stabilized cation **121**. Reaction with water led to a chlorohydrin and elimination of HCl gave an enol, which tautomerized to ketone **122** (obtained in 15% yield).[82]

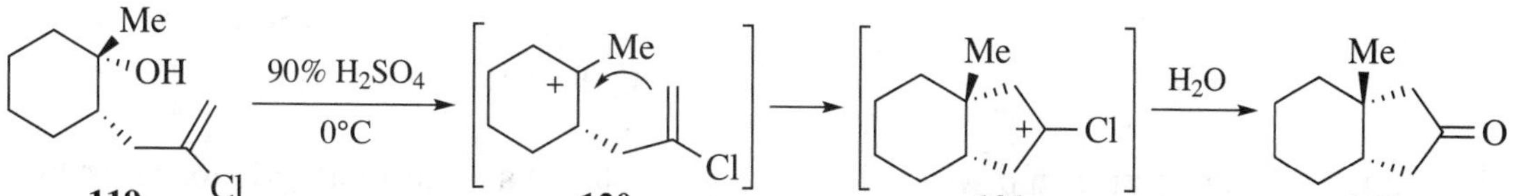

Alkynes are useful partners in cationic cyclization reactions. Initial reaction of the allylic alcohol moiety in **123** with formic acid gave allyl cation **124**. Subsequent attack by the alkyne moiety across the molecule generated vinyl cation **125**. Trapping the formate anion generated a formate enol ester (C=C–OCHO) and hydrolysis liberated the final ketone product **126**, which Lansbury and Serelis converted to damsinic acid.[83]

---

82. Lansbury, P.T.; Nienhouse, E.J. *J. Am. Chem. Soc.* **1966,** *88,* 4290.
83. Lansbury, P.T.; Serelis, A.K. *Tetrahedron Lett.* **1978,** 1909.

123   90% HCO₂H → 124 → 125   H₂O / HCO₂⁻ → 126

The **Johnson polyene cyclization**[84] **(the Stork-Eschenmoser postulate)**, described in Section 10.8.A, is an example of a *cascade cationic cyclization*. Polyenes such as squalene are expected to assume a steroid-like conformation in the lowest energy conformation (sec. 1.5.E), based on the biogenetic preparation of cholesterol from squalene.[85] In practice, treatment of polyenes with acid led to a very low yield of tri- or tetracyclic products, giving instead significant amounts of polymeric material. Diligent work over many years prevailed, however, and Johnson and his co-workers solved the many problems (as described in sec. 10.8.A) to make this reaction an excellent and efficient synthetic route to di-, tri- and tetracyclic molecules. One of the later examples of polyene cyclization uses an allyl silane to quench the cyclization process.[86] Initial reaction of the dioxolane unit in of **127** with tin tetrachloride (SnCl₄) led to a 23% yield of tetracyclic **128a** (2:3 α:β 17-vinyl) and 11% of **128b** (1:1 α:β 17-vinyl). In this case, the Lewis acid partially cleaved the ketal protecting group (sec. 7.3.B.i) during the cation initiation process. The cyclization generated the stereochemical centers at the ring junctures with high selectivity but provided little stereocontrol for the final reaction, which gave a mixture of 17α- and 17β-vinyl groups. Better results were obtained by use of a cyclopentenyl moiety rather than a dioxolane unit to conformationally anchor the cationic cyclization process and incorporating an alkyne to terminate that process (with a vinyl cation that was converted to a ketone).[87]

127   0.2 M SnCl₄ / pentane, 0 → 15°C →

(a) R¹ = -OCH₂CH₂OH, R²=H, 23%
(b) R¹ = H, R² = -OCH₂CH₂OH, 11%

R¹ R² **128**

Cyclization can be accomplished when there are heteroatom substituents. Reaction of amide-diene **129** with paraformaldehyde and formic acid generated iminium salt **130**. Cationic cyclization (the formate anion is the nucleophilic trapping agent) generated the perhydroisoquinoline **131**.[88] A related cyclization procedure that involves *N*-acyl iminium

84. (a) Johnson, W.S. *Acc. Chem. Res.* **1968**, *1*, 1; (b) Mundy, B.P.; Ellerd, M.G.; Favaloro Jr., F.G. *Name Reactions and Reagents in Organic Synthesis, 2nd ed.*, Wiley-Interscience, New Jersey, **2005**, pp. 348-349.
85. Clayton, R.B. *Quart. Rev. Chem. Soc.* **1965**, *19*, 168.
86. Johnson, W.S.; Chen, Y.-Q., Kellogg, M.S. *J. Am. Chem. Soc.* **1983**, *105*, 6653.
87. Gravestock, M.B.; Johnson, W.S.; McCarry, B.E.; Parry, R.J.; Ratcliffe, B.E. *J. Am. Chem. Soc.* **1978**, *100*, 4274.
88. Kano, S.; Yokomatsu, T.; Yuasa, Y.; Shibuya, S. *Tetrahedron Lett.* **1983**, *24*, 1813.

intermediates was developed by Speckamp and Hiemstra,[89] and has been used for the preparation of alkaloids.[90] This route has been used to prepare chiral heterocycles.[91] In this reaction an *N*-alkenyl, ω-hydroxy (or alkoxy) lactam such as **132** is treated with formic acid, generating the acyl iminium salt **133**. Subsequent alkene cyclization led to **134**[92] and trapping with formate (an S_N1 reaction) gave the final indolizidine product **135**. Selective reduction of an imide precursor gave the ethoxy lactam (**132**) required for generating the iminium salt (sec. 4.4.A). The reaction appears to be under kinetic control and proceeds via a chair-like transition state (**133**), where the incoming nucleophile (HCO_2^-) and the carbon being attacked are antiperiplanar. Acyl iminium ions have become useful synthetic intermediates.[93] Speckamp and Wijnberg applied this method to the synthesis of mesembrine.[94] Overman and Robichaud used a variation of this reaction to a synthesis of geissoschizine.[95] Initial reaction of **136** with paraformaldehyde and camphorsulfonic acid generated iminium salt **137** *in situ*. Reaction of the vinyl silane moiety with NaF led to addition of the C=C to the iminium salt (sec. 7.3.A.i). The alkylidene product (**138**) was produced in 80% yield as a 9.1:1 mixture of (*Z/E*) isomers.

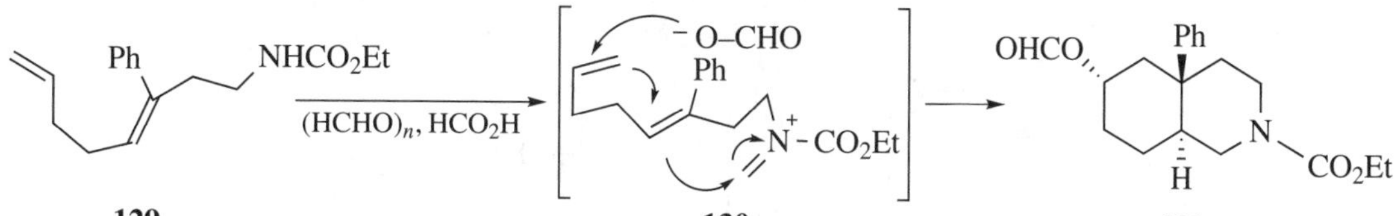

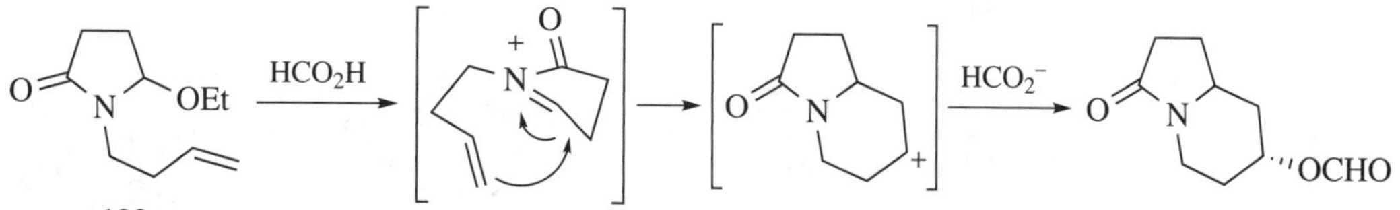

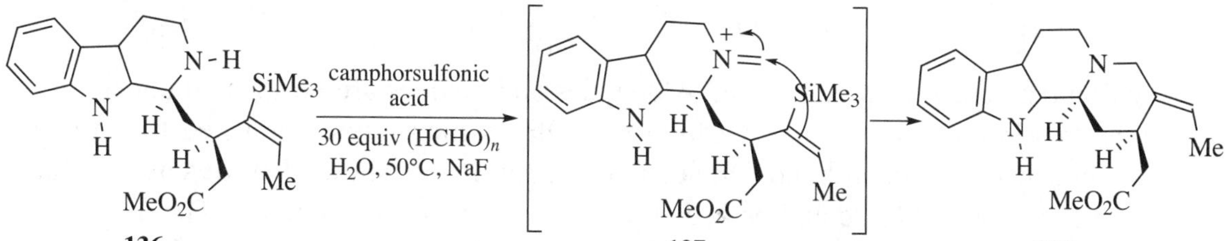

89. Speckamp, W.N.; Hiemstra, H. *Tetrahedron* **1985**, *41*, 4367.
90. (a) Speckamp, W.N.; Hiemstra, H. *Tetrahedron* **1985**, *41*, 4367; (b) Speckamp, W.N.; DeBoer, J.J.J. *J. R. Neth. Chem. Soc.* **1983**, *102*, 405 [*Chem. Abstr. 100*: 51412b, *1984*].
91. Royer, J.; Bonin, M.; Micouin, L. *Chem. Rev.* **2004**, *104*, 2311.
92. Speckamp, W.N. *Rec. Trav. Chim. Pay-Bas* **1981**, *100*, 345.
93. Maryanoff, B.E.; Zhang, H.-C.; Cohen, J.H.; Turchi, I.H.; Maryanoff, C.A. *Chem. Rev.* **2004**, *104*, 1431.
94. Wijnberg, J.B.P.A.; Speckamp, W.N. *Tetrahedron Lett.* **1975**, 3963.
95. Overman, L.E.; Robichaud, A.J. *J. Am. Chem. Soc.* **1989**, *111*, 300.

A similar cyclization procedure uses *C*-acylnitrilium ions for the synthesis of heterocyclic compounds.[96] A typical example treated isonitrile **139** with trimethylacetyl chloride to give **140**.[97] Subsequent treatment with CF_3SO_3Ag (silver trifluoromethanesulfonate) gave an intermediate *C*-acylnitrilium ion (**141**), which cyclized to give dihydroisoquinoline derivative **142** in 82% yield. Livinghouse and co-workers used this strategy to synthesize dendrobine,[98] where a key step was the coupling of **143** with **144** in the presence of molecular sieves. Subsequent treatment with silver tetrafluoroborate gave an 88% yield of **145**. Stevens and Kenney had previously reported a mercury(II) nitrate-mediated Ritter coupling that involved formation of an organomercury compound.[99] Addition to an isonitrile gave an iminium salt that could be reduced. Although this method did not involve an acylnitrilium ion, it is related to this chemistry.

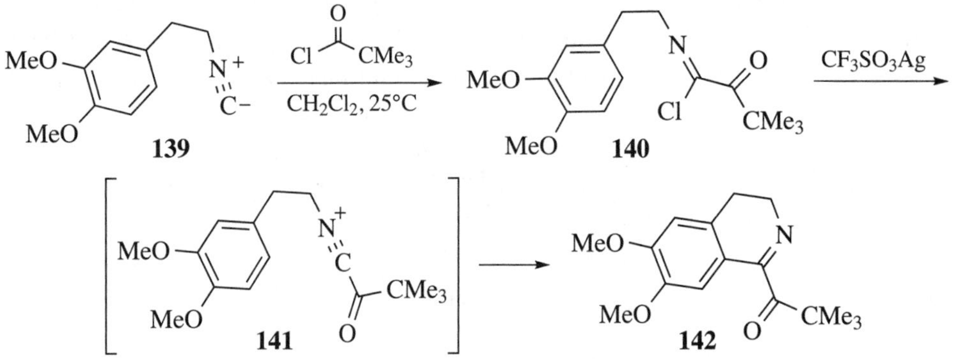

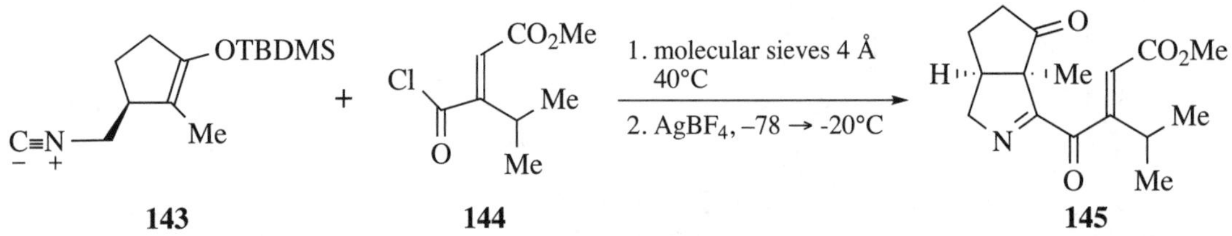

Other cationic precursors are available. Aldehydes and ketones are protonated under various conditions to form an oxo-stabilized cation (^+C-OH), which is then attacked by an alkene to form a new carbon-carbon bond. In section 3.2.B.ii, it was shown that oxidation of acid-sensitive alcohols with PCC could be accompanied by secondary reactions. The oxidation of citronellol (**146**) with PCC not only gave the expected aldehyde, but the aldehyde oxygen was protonated to give **147**.[100] Once formed, this oxo-cation is attacked by the adjacent π-bond of the alkene to form cation **148**. Elimination to an alkene (E1, sec. 2.9.B), and oxidation to the ketone under the reaction conditions leads to the final product, isopulegone (**149**).

96. Livinghouse, T. *Tetrahedron* **1999**, *55*, 9947.
97. Westling, M.; Livinghouse, T. *Tetrahedron Lett.* **1985**, *26*, 5389.
98. Lee, C.H.; Westling, M.; Livinghouse, T.; Williams, A.C. *J. Am. Chem. Soc.* **1992**, *114*, 4089.
99. Stevens, R.V.; Kenney, P.M. *J. Chem. Soc. Chem. Commun.* **1983**, 384.
100. (a) Corey, E.J.; Boger, D.L. *Tetrahedron Lett.* **1978**, 2461; (b) Corey, E.J.; Ensley, H.E.; Suggs, J.W. *J. Org. Chem.* **1976**, *41*, 380.

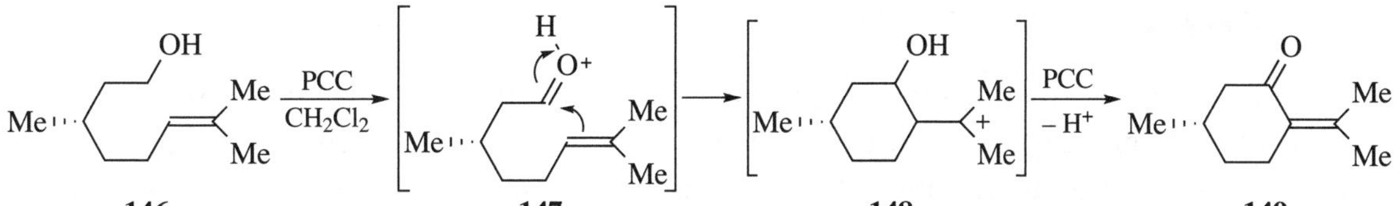

146   **147**   **148**   **149**

A variety of disconnections are possible utilizing intramolecular cationic reactions. A few of the more useful are

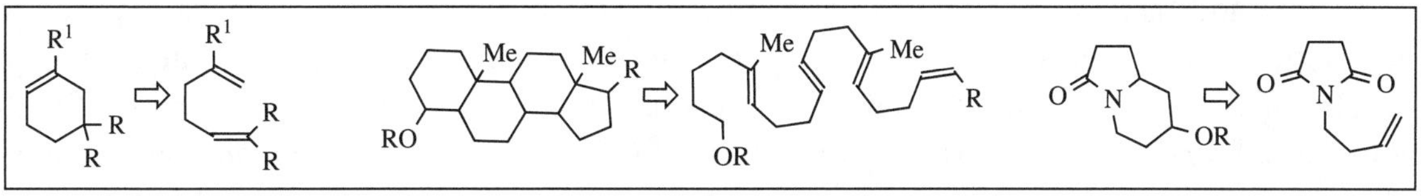

12.3.B. Koch-Haaf Carbonylation[101]

When an alcohol is heated in strong acid, in the presence of carbon monoxide (CO), the initially generated cation traps carbon monoxide to give a carboxylic acid in what is known as the **Koch-Haaf carbonylation**. As with any other carbocation intermediate, rearrangement can occur, usually prior to reaction with carbon monoxide. Reaction of **150** with aqueous sulfuric acid initially gave cation **151**. In the presence of carbon monoxide (generated *in situ* from formic acid and sulfuric acid), acylium ion **152** was formed and reaction with water gave the carboxylic acid (**153**) in 90% yield. Similar treatment of alkenes with these reagents also gave carboxylic acids, but the yields were usually poorer. In most cases, using a sulfuric-formic acid mixture to generate carbon monoxide is more efficient than bubbling in carbon monoxide gas. Lactones are formed when diols are treated with carbon monoxide and acid.[102]

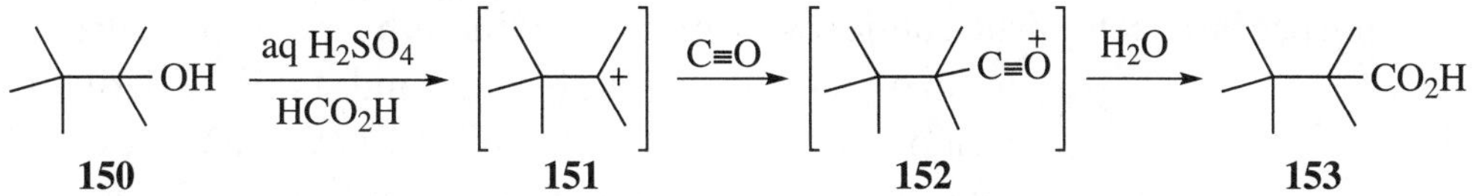

150   **151**   **152**   **153**

The Koch-Haaf disconnection is

$$R_2HC - CO_2H \implies R_2HC - OH$$

101. (a) Koch, H. *Brennst. Chem.* **1955**, *36*, 321 [*Chem. Abstr. 50*:6019g **1956**]; (b) Koch, H.; Haaf, W. *Annalen* **1958**, *618*, 251; (c) Möller, K.E. *Brennst. Chem.* **1966**, *47*, 10 [*Chem. Abstr. 64*:12563b **1966**]; (d) The Merck Index, 14th ed.,Merck & Co., Inc., Whitehouse Station, New Jersey, **2006**, p ONR-52; (e) Mundy, B.P.; Ellerd, M.G.; Favaloro Jr. F.G. *Name Reactions and Reagents in Organic Synthesis, 2nd ed.*, Wiley-Interscience, New Jersey, **2005**, pp. 368-369.244-245 (see p 328).
102. Takahashi, Y.; Yoneda, N.; Nagai, H. *Chem. Lett.* **1982**, 1187.

12.3.C. Nazarov Cyclization[103]

Divinyl ketones are useful precursors for the synthesis of cyclopentenones. In a typical example, reaction of a divinyl alkyne such as **154** with aqueous acid generated a conjugated ketone (**155**), via hydration of an intermediate vinyl cation.[103b] Subsequent treatment with a mixture of phosphoric and formic acids led to a new, oxygen-stabilized carbocation (**155**) that reacted with the adjacent alkene (a cyclization reaction) to form **157**. Elimination led to the conjugated ketone (**158**).[103b] This cationic ring-closing reaction is called the **Nazarov cyclization**.[103,104104] Several methods are available for the preparation of the bis(conjugated) ketones required for this reaction. Most involve coupling a vinyl carbanion equivalent with a conjugated carbonyl. The strongly acidic conditions used in the reaction that gave **158** are not required for the cyclization,[105] and Lewis acids such as tin chloride ($SnCl_4$) or iodotrimethylsilane can be used in a Nazarov cyclization. A synthetic example, taken from Danishefsky and co-worker's synthesis of (–)-scabronine G, involved treatment of **159** with ferric chloride give the cyclopentenone unit in **160**, in 72% yield.[106]

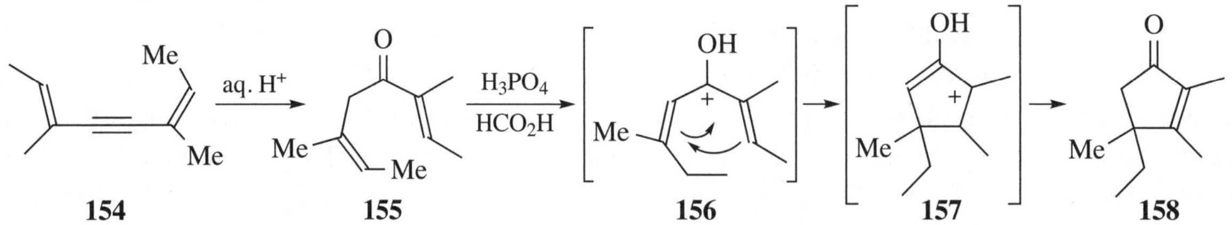

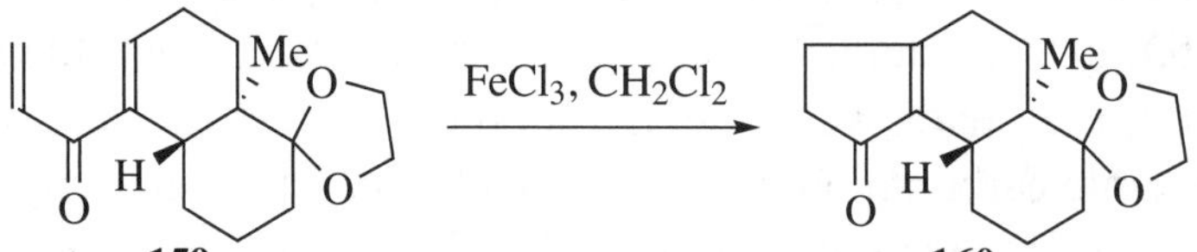

The cyclization can be accomplished under even milder conditions by incorporating a stabilizing group on the alkene. Both silane and stannane derivatives have been prepared. Cyclization of **161** gave **162**, for example, but both the temperature and the nature of the Lewis acid had a significant effect on the reaction.[107] Ferric chloride ($FeCl_3$) gave the best yield of the Nazarov product (**162**) in this study. Vinyl

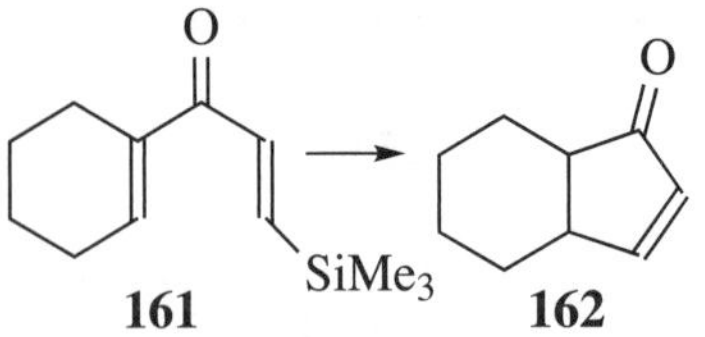

103.   (a) Nazarov, I.N.; Torgov, I.B.; Terekhova, L.N. *Izv. Akad. Nauk. SSSR otd. Khim. Nauk* **1942**, 200; (b) Braude, E.A.; Forbes, W.F. *J. Chem. Soc.* **1953**, 2208.
104.   For reactions of this type, see reference 79, p 1308. Also see (a) The Merck Index, 14th ed., Merck & Co., Inc., Whitehouse Station, New Jersey, **2006**, p ONR-64; (b) Mundy, B.P.; Ellerd, M.G.; Favaloro Jr., F.G. *Name Reactions and Reagents in Organic Synthesis*, 2nd ed., Wiley-Interscience, New Jersey, **2005**, pp. 448-449.
105.   Marino, J.P.; Linderman, R.J. *J. Org. Chem.* **1981**, *46*, 3696.
106.   Waters, S.P.; Tian, Y.; Li, Y.-M.; Danishefsky, S J. *J. Am. Chem. Soc.* **2005**, *127*, 13514.
107.   Denmark, S.E.; Jones, T.K. *J. Am. Chem. Soc.* **1982**, *104*, 2642.

silane cyclization can proceed with modest diastereoselectivity.[108] West and Giese[109] showed that a normal Lewis acid catalyzed Nazarov cyclization in the presence of triethylsilane was accompanied by formation of a reduced product. When **163** was treated with $BF_3 \cdot OEt_2$ and two equivalents of triethylsilane, followed by 1N HCl in a second step, a 71% yield of **164** was obtained along with 14% of **165**.[109] Other variations are known. Vinyl stannanes are good Nazarov precursors, for example, generating cyclopentenone derivatives.[110]

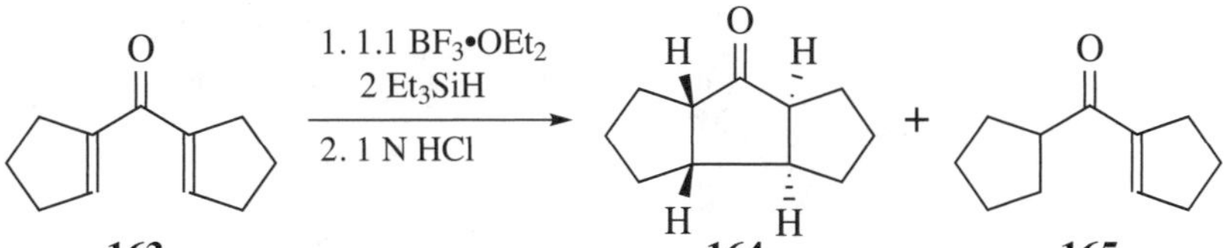

The Nazarov disconnection is

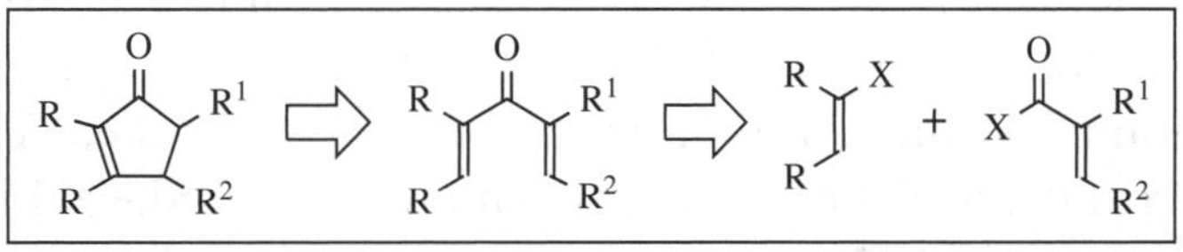

Note that Harmata and Lee reported a retro-Nazarov reaction.[111] When **166** was treated with triethylamine in trifluoroethanol, for example, a 65% yield of **167** was obtained along with 9% of **168**. This reaction appears to be a good route to dienones.

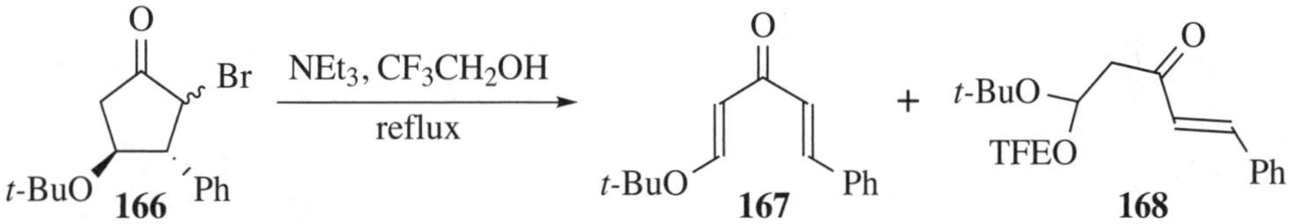

12.3.D. Noyori Annulation

Another synthetically useful carbon bond-forming reaction is related to the Nazarov cyclzation, but involves reaction of diiron nonacarbonyl (see Sect. 12.10) with halocarbonyl compounds. Noyori and co-workers found that α,α'-dibromoketones (**169**) react with diiron nonacarbonyl $[Fe_2(CO)_9]$ to give an iron stabilized alkoxy zwitterion (**170**). The intermediate π-allyl iron species reacts with alkenes in a stepwise manner (initially producing **171**) to give cyclic ketones such as **172**,[112] and the product is equivalent to the product of a [3+2]-cycloaddition with an

108. Jones, T.K.; Denmark, S.E. *Helv. Chim. Acta* **1983**, *66*, 2397.
109. Giese, S.; West, F.G. *Tetrahedron* **2000**, *56*, 10221.
110. Peel, M.R.; Johnson, C.R. *Tetrahedron Lett.* **1986**, *27*, 5947.
111. Harmata, M.; Lee, D.R. *J. Am. Chem. Soc.* **2002**, *124*, 14328.
112. (a) Noyori, R.; Yokoyama, K; Hayakawa, Y. *J. Am. Chem. Soc.* **1973**, *95*, 2722; (b) *Idem Ibid.* **1978**, *100*, 1791; (c) Noyori, R. *Acc. Chem. Res.* **1979**, *12*, 61.

alkene (sec. 11.11). This cyclization method is now known as **Noyori annulation**,[113] but is clearly related to the Nazarov cyclization just discussed. Enamines can react with **169**, but the initially formed enamino ketone product eliminates the amino group to form cyclopentanone derivatives.[114] Intermediates such as **170** may actually exist as cations bound to a metal rather than as the alkoxide-iron structures shown.[112b-d] It is also noted that $Zn/B(OEt)_3$ is also an effective reagent and is a less toxic and less expensive alternative to $Fe_2(CO)_9$.[112e]

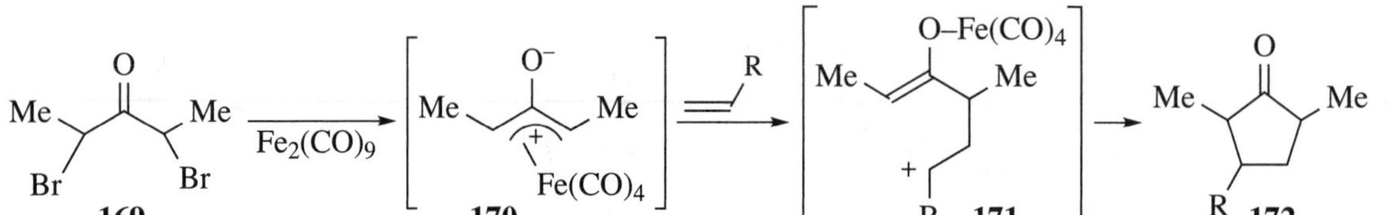

Dienes react to give cycloheptenones, as in the reaction sequence initiated by the reaction of **173** with $Fe_2(CO)_9$ to give **174**. In the presence of 2,3-dimethyl-1,3-butadiene generates the cycloheptenone derivative (**175**) in 55% yield.[115] This variation is the formal equivalent of a [4+3]-cycloaddition, and furan or pyrrole can be used as the diene component.[114] Furan is a useful partner in this annulation, reacting with **176** to produce **177** in a synthesis of showdomycin.[116] Cycloadduct **177** was isolated after reduction of the bromide moieties with Zn/Cu (sec. 4.9.G) in the initially formed adduct. A similar strategy was used for a synthesis of nezukone.[117] *N*-Boc-pyrroles[118] cyclize to 8-azabicyclo[3.2.1]octan-3-one derivatives using diethyl zinc to initialize the Noyori annulation.

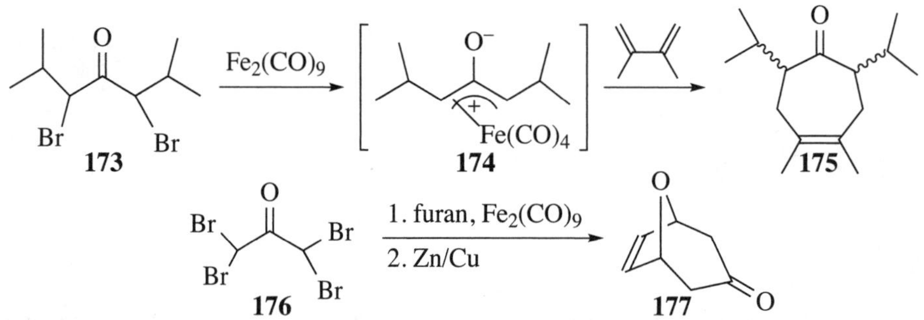

113. Mundy, B.P.; Ellerd, M.G.; Favaloro Jr., F.G. *Name Reactions and Reagents in Organic Synthesis, 2nd ed.*, Wiley-Interscience, New Jersey, **2005**, pp. 464-465.
114. (a) Hayakawa, Y.; Yokoyama, K.; Noyori, R. *J. Am. Chem. Soc.* **1978**,*100*, 1799; (b) see *J. Am. Chem. Soc.* **1972**, *94*, 3201; (c) *J. Am. Chem. Soc.* **1972**, *94*, 1373, 3201, 3940, 6213, 7832; (d) *Angew. Chem. Int. Ed.* **1973**, *12*, 239; (e) *Tetrahedron Lett.* **1975**, 4487.
115. Takaya, M.; Makino, S.; Hayakawa, Y.; Noyori, R. *J. Am. Chem. Soc.* **1978**, *100*, 1765.
116. Sato, T.; Ito, R.; Hayakawa, Y.; Noyori, R. *Tetrahedron Lett.* **1978,** 1829.
117. (a) Hayakawa, Y.; Sakai, M.; Noyori, R. *Chem. Lett.* **1975**, 509; (b) Hayakawa, Y.; Baba, Y.; Makino, S.; Noyori, R. *J. Am. Chem. Soc.* **1978**, *100*, 1786.
118. Paparin, J.-L.; Crévisy, C.; Grée, R. *Tetrahedron Lett.* **2000**, *41*, 2343.

The Noyori annulation disconnections are

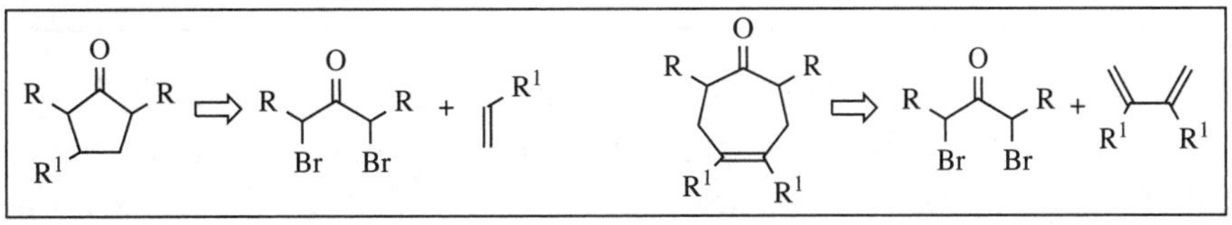

12.3.E. The Prins Reaction[119]

Formaldehyde can be coupled to an alkene in the presence of an acid to give a diol (**178**) or a 1,3-dioxane derivative (**180**) in what is known as the **Prins reaction**.[119] Allylic alcohols such as **179** can also be produced in this reaction. The use of alternative acids, including Lewis acids, may lead to other products. In a synthesis of (–)-centrolobine by Rychnovsky and co-workers,[120] α-acetoxy ether **181** was treated with tin bromide, to give an 84% yield of cyclized product **182** as the bromide. This cyclization proceeded without side-chain exchange or partial racemization by reversible 2-oxonia Cope rearrangement, which are potential problems occur during Prins cyclization. Prins cyclization reactions also occur using cyclic α,β-unsaturated acetals with allylsilane nucleophiles, inside Lewis acidic micelles in water.[121]

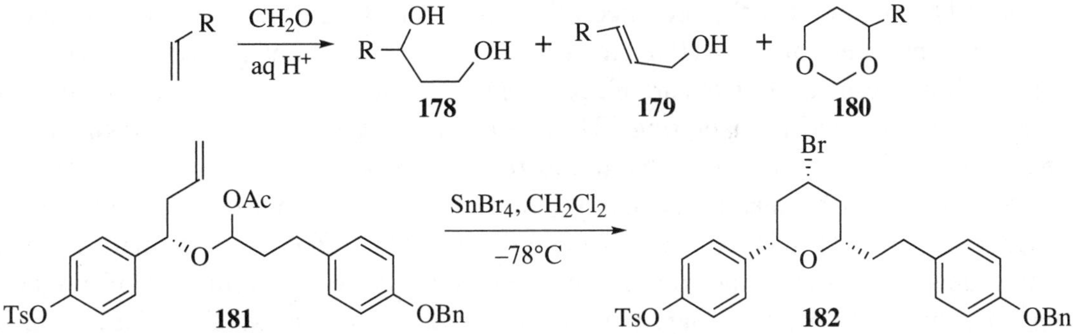

The Prins disconnection is

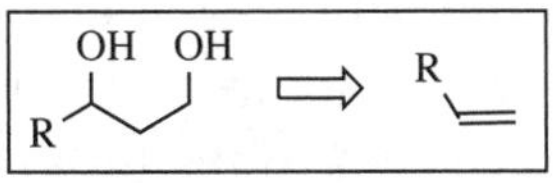

The examples in this section demonstrate that stable cations can be generated and trapped with alkenes, alcohols or carbonyl derivatives. The resulting reaction products contain new carbon-carbon bonds, often produced with controllable and predictable stereo- and regiochemistry.

119. (a) Prins, H.J. *Chem. Weekblad* *1919*, 16, 64, 1072, 1510 [*Chem. Abstr. 13*: 3155, *1919*] and [*Chem. Abstr. 14*: 1119, *1920*]; (b) Arundale, R.; Mikeska, L.A. *Chem. Rev. 1952*, 51, 505; (c) Adams, D. R.; Bhatnagar, S. P. *Synthesis 1977*, 661; (d) The Merck Index, 14th ed., Merck & Co., Inc., Whitehouse Station, New Jersey, *2006*, p ONR-76; (e) Mundy, B.P.; Ellerd, M.G.; Favaloro Jr., F.G. *Name Reactions and Reagents in Organic Synthesis, 2nd ed.*, Wiley-Interscience, New Jersey, *2005*, pp. 532-533.
120. Marumoto, S.; Jaber, J.J.; Vitale, J.P.; Rychnovsky, S.D. *Org. Lett. 2002*, 4, 3919.
121. Aubele, D.L.; Lee, C.A.; Floreancing, P.E. *Org. Lett. 2003*, 5, 4521.

12.4. FRIEDEL-CRAFTS REACTIONS

This section discusses a specialized reaction of carbocations and a nucleophilic species. In this section, however, the nucleophilic species is an aromatic ring and reaction with a carbocation generates a Wheland-type intermediate (sec. 2.11.A) that leads to a substituted aromatic system. Known as Friedel-Crafts reactions, they are among the most important reactions in Organic chemistry.

12.4.A. Electrophilic Aromatic Substitution

In chapter 2, electrophilic reagents such as the bromonium ion [Br^+, generated by the reaction of bromine (Br_2) and aluminum bromide ($AlBr_3$)], were shown to react with benzene to give a resonance-stabilized cation called a **Wheland intermediate** (sec. 2.11). Loss of a proton from the Wheland intermediate and concomitant aromatization gave bromobenzene. In a similar manner, section 2.11 discussed formation of several substituted benzene derivatives.[122] This section will discuss electrophilic substitution reactions of aromatic derivatives that form carbon-carbon bonds, and the Wheland intermediate will be observed again.[122]

In section 2.11, we saw that the presence of electron-releasing groups of the aromatic ring (alkyl, oxygen, nitrogen, sulfur, aryl) led to ortho- and para-substituted derivatives as the major products. All of these substituents are electron rich, and most contain a $\delta-$ dipole or have a (–) charge adjacent to the aromatic ring. These groups are known as **activating substituents** because they increase the rate of substitution relative to benzene.[122] The rate of nitration of anisole was 9.7×10^6 times faster than the rate of nitration of benzene, for example. As discussed in Section 2.11, both ortho and para products can be formed from aromatic molecules containing an electron-releasing group. This selectivity is due to formation of a Wheland intermediate that has a positive charge in the ring, adjacent to a substituent that has a (–) or $\delta-$ center (including an electron releasing alkyl group). Such an intermediate is more stable, which favors formation of the ortho- and para-substituted products. For this reason, activating substituents are commonly called ortho/para directors. Reaction of anisole with nitric acid/ sulfuric acid mixture, for example, gave 44% of the ortho product (2-nitroanisole), 56% of the para product (4-nitroanisole) and < 1% of the meta product 3-nitroanisole.[123] Meta substitution generates an intermediate where the charge is *not* adjacent to the electron-releasing substituent.

Electron-withdrawing groups (NO_2, C=O, SO_2R, SO_3R, NR_4^+) all have a (+) charge or $\delta+$ dipole adjacent to the aromatic ring, and show the opposite effect when compared with activating substituents. They slow the rate of reaction relative to benzene (they are called **deactivating groups**), and are meta directors because the meta product is usually the major product. With a deactivating (electron withdrawing) substituent on an aromatic ring, attack at either the ortho

---

122.   Stock, L.M. *Aromatic Substitution Reactions*, Prentice-Hall, Englewood Cliffs, New Jersey, *1968*.
123.   Reference 122, p 43.

or the para position will generate a Wheland intermediate that has a (+)-charge adjacent to a (+) or a δ+ center. The proximity of the two like charges destabilizes that intermediate and it is more difficult to form, so ortho/para substitution is slower when compared to benzene. Substitution at the meta position generates a Wheland intermediate that does not have the cationic center adjacent to the (+) or δ+ center, and although it is not particularly stabilized it is not strongly destabilized. Therefore, meta substitution is the major product. Selectivity for the meta product is usually good. Nitration of nitrobenzene with a mixture of nitric and sulfuric acid gave 1,3-dinitrobenzene in 94% yield with only 6% of the ortho product and < 1% of the para product.[122] The ortho/para or meta selectivity is illustrated in Table 12.4,[124,122] which shows the relative of ortho, meta and para products observed in the nitration of aromatic derivatives. In accord with the concept of activating and deactivating substituents, the presence of heteroatoms in the aromatic ring have a strong influence on electrophilic aromatic substitution reactions. Electron-rich aromatic rings such as phenols and aryl ethers react very well. Such *activated* heteroaromatic compounds include pyrrole, furan or thiophene. However, pyridine has an electron pair external to the ring, which is deactivating, so it reacts more slowly. This simple analysis is consistent with the order of reactivity shown, with pyrrole being the most reactive and pyridine being rather unreactive to electrophilic aromatic substitution:[125] In part, pyridine and related compounds are bases, and they react with the added Lewis acid (via the one electron pair) to divert or even suppress formation of a Wheland intermediate.

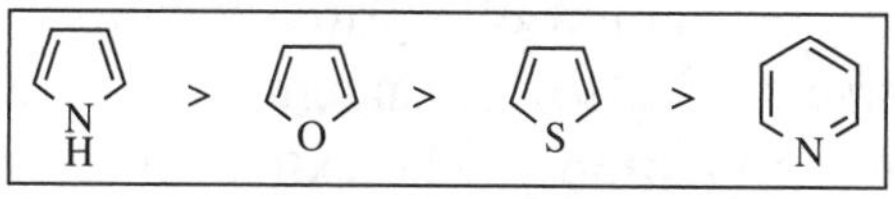

Table 12.4   Regiochemical Preferences for Nitration of Aromatic Substrates

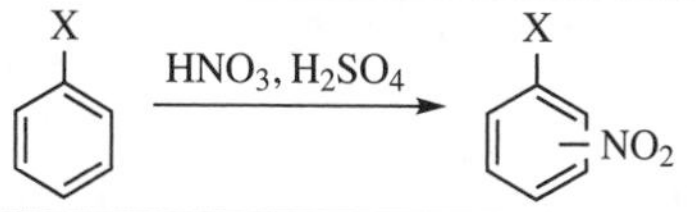

	NITRATION				**NITRATION**		
X	**% ortho**	**% meta**	**% para**	**X**	**% ortho**	**% meta**	**% para**
OMe	44	<1	56	Cl	30	1	69
NHAc	20	<1	80	Br	37	1	62
Ph	69	<1	31	CO_2H	22	76	2
Me	59	4	37	NO_2	6	94	<1
F	9	<1	91	NMe_3^+	<1	89	11

[Leon M. Stock, *AROMATIC SUBSTITUTION REACTIONS*, ©*1968*, p. 44. Reprinted by permission of Prentice Hall, Englewood Cliffs, New Jersey.]

124.   Reference 122, p 44.
125.   Reference 129, p 42.

12.4.B. Friedel-Crafts Alkylation

183 **184**

12.4.B.i. Alkyl Cations and Regioselectivity. This chapter deals with reactions that form carbon-carbon bonds via carbocation intermediates, including coupling reactions with aromatic rings. An aromatic compound will react with a carbocation intermediate (also see section 12.4.A) to form a Wheland intermediate (namely, **183**). Subsequent loss of a proton, with concomitant aromatization, completes the substitution reaction to form arene **184** via electrophilic aromatic substitution. The reaction with an alkyl carbocation is called **Friedel-Crafts alkylation**,[126] first reported by Friedel and Crafts in 1877.[127] The carbocation required for this reaction can be generated from an alkene, an alcohol or an alkyl halide.

In chapter 2, solvolysis reactions of alkyl halides in aqueous or highly protic media were shown to form carbocations that typically lead to S_N1 or E1 products. Such reactions were limited to tertiary and secondary halides, due to the energy demands for the initial ionization. Alkyl halides also react

185

with Lewis acids, but tertiary, secondary or primary halides react to form tertiary, secondary or primary carbocations (**185**). Formation of a primary cation (**185**, $R^1=R^2=H$, $R^3=$ alkyl) is somewhat slower and is, of course, subject to rearrangement. As a result, primary alkyl derivatives are rarely observed under these conditions. The extent of rearrangement from primary and secondary cations depends on the reaction conditions, the strength of the nucleophile, and the

186 **187**

nature of the cation. Reaction of benzene and 1-phenyl-2-chloropropane (**186**), for example, gave 60% of **187**,[128] with little rearrangement to the benzylic cation. This result suggests that the initially formed cation can be trapped very rapidly, possibly via a tight ion pair. In this particular case, the solvent (carbon disulfide, CS_2) played a major role in minimizing solvent separation of the ions.

Alkyl fluorides are the most reactive of the alkyl halides. All the boron trihalides catalyze the reaction of alkyl fluorides and benzene, although they are ineffective catalysts for the analogous

---

126. For reactions of this type, see reference 79, pp 129-133. Also see (a) The Merck Index, 14th ed., Merck & Co., Inc., Whitehouse Station, New Jersey, *2006*, p ONR-34; (b) Mundy, B.P.; Ellerd, M.G.; Favaloro Jr., F.G. *Name Reactions and Reagents in Organic Synthesis, 2nd ed.*, Wiley-Interscience, New Jersey, *2005*, pp. 256-257.

127. (a) Friedel, C.; Crafts, J.M. *Compt. Rend. 1877, 84*, 1392, 1450; (b) Price, C.C. *Org. React. 1946, 3*, 1; (c) Gore, P. *Chem. Rev. 1955, 55*, 229; (d) Baddeley, G. *Quart. Rev. Chem. Soc. 1954, 8*, 355.

128. Masuda, S.; Nakajima, T.; Suga, S. *Bull. Chem. Soc. Jpn. 1983, 56*, 1089.

reaction with alkyl chlorides and alkyl bromides.[129] In a reaction with mixed halides, the C–F bond reacts faster than the C–Cl, C–Br or C–I bonds.[130] This is consistent with the observed order of reactivity for alkyl halides with aluminum chloride ($AlCl_3$):

$$R–F > R–Cl > R–Br > R–I^{131}$$

Table 12.5   The ortho Effect and ortho-para Substitution in Friedel-Crafts Reactions of Anisole and Toluene

R^1	R–X	Catalyst	Solvent	Time (min)	Temp. (°C)	% ortho	% meta	% para
OMe	MeF	BF_3	anisole	30	25	75	<1	25
OMe	MeF	BF_3	nitromethane	15	25	55	<1	45
Me	MeF	BF_3	anisole	30	25	54	18	28
OMe	MeCl	$AlCl_3$	nitromethane	5	25	68	2	30
Me	MeCl	$AlCl_3$	nitromethane	52	17	31		
OMe	MeBr	$AlCl_3$	nitromethane	30	25	67	2	31
Me	MeBr	$AlCl_3$	nitromethane	54	17	29		
OMe	MeI	$AlCl_3$	nitromethane	60	25	67	<1	32
Me	MeI	$AlCl_3$	nitromethane	57	12	31		

[Reprinted with permission from Olah, G.A.; Olah, J.A.; Ohyama, T. *J. Am. Chem. Soc.* **1984**, *106*, 5284. Copyright © **1984** American Chemical Society.]

As with other electrophilic aromatic substitution reactions, Friedel-Crafts alkylation of anisole, toluene or aromatic derivatives bearing an electron-releasing group is faster than benzene, giving primarily ortho- and para substituted products. The reaction is slowed by the presence of electron-withdrawing substituents (such as NO_2 in nitrobenzene and C=O in aryl ketones) on the aromatic ring. Indeed, Friedel-Crafts alkylation is very difficult with deactivated aromatic nuclei. The product distribution varies with the halide, the catalyst, the solvent and the aromatic substrate. As shown in Table 12.5,[132] ortho-substituted products are generally preferred except with highly hindered alkyl halides such as *tert*-butyl. This preference for the ortho product has been called the **ortho effect** (additional stabilization of cation intermediates by an electron-releasing group to stabilize the intermediate generated by ortho-substitution relative to the intermediate generated by para substitution).[132] Inspection of Table 12.5 shows that the ortho effect is more pronounced when the aromatic ring contains the stronger electron-releasing methoxy group (anisole) when compared to a methyl group (toluene). This effect is usually explained by the linear coordination of a heteroatom substituent with the incoming reagent.

129.   (a) Olah, G.A. *Friedel-Crafts Chemistry*, Wiley, New York, **1973**, p 40; (b) Oláh, G.A.; Kuhn, S.; Oláh, J. *J. Chem. Soc.* **1957**, 2174.

130.   Reference 129, p 41 (citation 471 in chapter II).

131.   Calloway, N.O. *J. Am. Chem. Soc.* **1937**, *59*, 1474.

132.   Olah, G.A.; Olah, J.A.; Ohyama, T. *J. Am. Chem. Soc.* **1984**, *106*, 5284.

In the case of oxygen substituents, an *O*-alkyl product can be formed, followed by an O→C migration, where migration to the ortho carbon is faster than migration to the para carbon.[133] Increasing the ability of the activating group to release electrons also correlates with a decrease in the yield of meta product. Separation of the isomeric arene products can be a problem.

When very active catalysts such as aluminum chloride are used for Friedel-Crafts alkylation with alkyl halides, significant amounts of meta products are often observed, even when the aromatic ring contains activating substituents.[129a,b,130] Such products are due to the thermodynamic nature of the reaction with aluminum chloride, which shows a preference for the meta product. Excess catalyst and high reaction temperatures also favor more meta product. The use of BF_3, sulfuric acid, ferric chloride ($FeCl_3$) or zinc chloride ($ZnCl_2$) in lower concentrations and at lower temperatures, are preferred if primarily ortho-para products are the desired targets.[134]

Friedel-Crafts alkylation with alkyl halides proceeds via a carbocation, and chiral alkyl halides are expected to give racemic arene products. The extent of racemization is dependent on the Lewis acid and the reaction conditions. As shown in Table 12.6,[135] Suga et al. found some asymmetric induction during the aromatic substitution, if the reaction temperature was maintained below 0°C and the reaction time was kept to a minimum. Reaction of (*S*)-2-chlorobutane and benzene gave **188** in up to 24% ee (*R*) using short contact times at low temperature. Some asymmetric induction was observed when mild Lewis acids were used, even at ambient temperatures. The use of strong Lewis acids, higher temperatures and longer contact times, however, led to the expected racemization. Presumably, at lower temperatures the C–Cl–M⁺X⁻ species is generated as a tight ion pair, where some asymmetric character is retained in the transition state of the reaction.

Table 12.6   Asymmetric Induction in the Friedel Crafts Reaction of (*S*)-2-Chlorobutane

Lewis Acid	Temp. (°C)	Time (min)	% 188	% ee (*R*)
AlCl₃	–30	0.8	100	24
	–30	1.6	100	14
	–30	2	100	12
	–10	20	100	0
FeCl₃	0	40	70	24
AlCl₃•MeNO₂	30	15	30	13
AlBr₃	–30	1	100	0

133.   Kovacic, P.; Hiller Jr., J.H. *J. Org. Chem.* **1965**, *30*, 1581.
134.   (a) Cullinane, N.M.; Leyshon, D.M. *J. Chem. Soc.* **1954**, 2942; (b) Slanina, S.J.; Sowa, F.J.; Nieuwland, J.A. *J. Am. Chem. Soc.* **1935**, *57*, 1547.
135.   Suga, S.; Segi, M.; Kitano, K.; Masuda, S.; Nakajima, T. *Bull. Chem. Soc. Jpn.* **1981**, *54*, 3611.

The typical Friedel-Crafts alkylation disconnection is

12.4.B.ii. Isomerization, Polyalkylation and Deactivation. Several problems are associated with the Friedel-Crafts alkylation. The first occurs when primary and secondary halides are used, because the cations that are formed are subject to rearrangement. Reaction of benzene with 1-bromopropane and aluminum bromide ($AlBr_3$), for example, generated primary cation **189**. Rearrangement to the more stable secondary cation (**190**) was followed by reaction with benzene to give the final product, isopropylbenzene (cumene).[136] Due to this propensity for rearrangement, preparation of primary arenes (straight chain alkyl benzenes) is very difficult via Friedel-Crafts alkylation. On the other hand, the preparation of tertiary arenes (such as *tert*-butylbenzene), is straightforward because the easily formed tertiary cations usually do not rearrange.

Rearrangement can be accompanied by isomerization of the initially formed product due to the action of the Lewis acid (such as $AlCl_3$) required for the reaction.[137] Isomerization of groups can occur[138] via 1,2-shifts or via dissociation to a cation and readdition. 1,1-Dimethylpropylbenzene (**191**) dissociated in the presence of aluminum chloride to give **192**, for example. Under these conditions, the reaction was reversible and addition of the cation to the aromatic ring generated two new carbocations, **193** (giving **194**) or **195** (giving **196**). The extent to which **196** is formed is related to the length of exposure of **191** to $AlCl_3$. This latter reaction is a thermodynamic process that favors the most stable isomer, and polyalkylation is also possible. Exposure of *sec*-butylbenzene (**197**) to the mixed-acid $HF–BF_3$ led to a mixture of 36.7% of benzene, 10.9% of *n*-butylbenzene, 21.8% of *sec*-butylbenzene and 30.6% of di-*sec*-butylbenzene.[139] Similarly, when *n*-butylbenzene was heated with aluminum chloride (100°C, 3 h), 45.2% of butylbenzene (99.4% *n*-butylbenzene and 0.6% *sec*-butyl-benzene), 15.2% of benzene, and 27.7% of dibutylbenzene (>90% meta) along with 11.9% of polyalkylated benzene products were obtained.[140] Lewis acid induced isomerization of dialkylbenzenes usually leads to an increase in the relative percentage of meta isomer, as seen with *n*-butylbenzene, based on the greater thermodynamic stability of the meta product than that of the ortho and para products.[141]

136. Reference 129, p 68.
137. See (a) Nenitzesco, C.D.; Avram, M.; Sliam, E. *Bull. Chim. Soc. Fr.* **1955**, 1266; (b) Nenitzescu, C.D.; Necsçoiu, I.; Glatz, A.; Zalman, M. *Berichte* **1959**, *92*, 10.
138. Reference 129, p 70.
139. McCaulay, D.A.; Lien, A.P. *J. Am. Chem. Soc.* **1953**, *75*, 2411.
140. Kinney, R.E.; Hamilton, L.A. *J. Am. Chem. Soc.* **1954**, *76*, 786
141. Taylor, W.J.; Wagmen, D.D.; Williams, M.G.; Pitzer, K.S.; Rossini, F.D. *J. Res. Nat. Bur. Stand.* **1946**, *37*, 95.

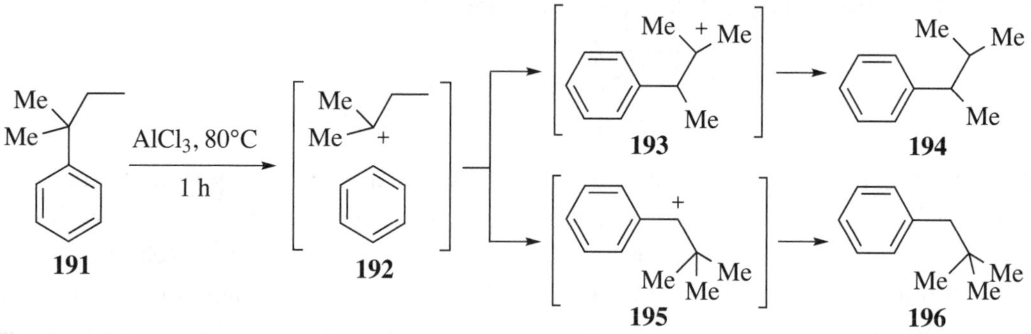

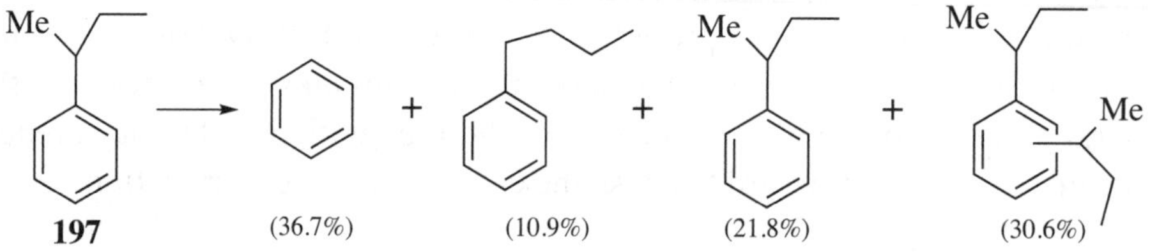

Me Me AlCl₃, 80°C Me Me +
191 1 h **192**

Me + Me
Me
193

Me Me
Me
194

+
Me Me
Me
195

Me Me
Me
196

Me

→ + + +

197 (36.7%) (10.9%) (21.8%)

Me Me
(30.6%)

In Friedel-Crafts alkylation reactions, the alkyl substituents in the products are weakly electron releasing (and therefore activating), making the arene product more reactive than the benzene starting material. This increased reactivity leads to the second observed problem with Friedel-Crafts alkylation, **polyalkylation**. Alkylation of the arene leads to disubstituted products, which are often contaminants or occasionally, the major product in Friedel-Crafts alkylations. If benzene, for example, reacts with 2-chloropropane, the initial product is isopropylbenzene. Since isopropylbenzene is more reactive than benzene, it reacts with the carbocation intermediate in competition with benzene, leading to both *o*-diisopropylbenzene (**198**) and *p*-diisopropylbenzene (**199**) as secondary products. The steric hindrance inherent to formation of the ortho-disubstituted product can lead to a large proportion of the para product, if the steric interaction is severe. In a large excess of benzene, polyalkylation[142,127b] can be suppressed. Francis[143] reported alkyl groups on benzene have only a small effect on the rate of Friedel-Crafts alkylation. Francis showed that "alkylation occurred in a heterogeneous reaction system, specifically in the catalyst layer and that the reason for polysubstitution is the preferential extraction of the early reaction product by this catalyst layer".[144] Using a solvent to solubilize both hydrocarbon and catalyst minimized polyalkylation, as does efficient (high speed) stirring and higher temperatures (which solubilizes the aluminum chloride).[138]

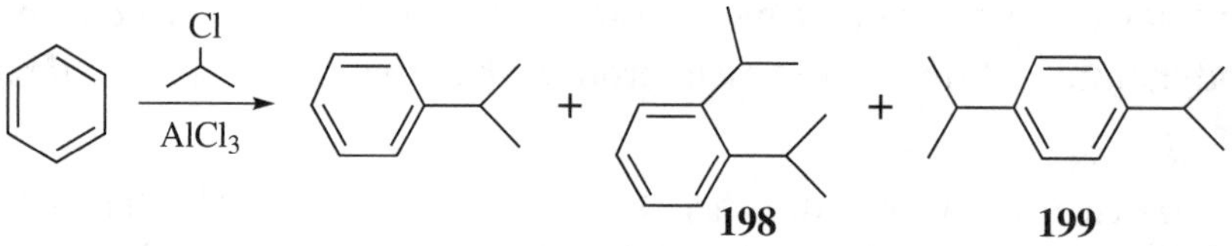

Cl

AlCl₃

+ +

198 **199**

142. (a) Price, C.C. *Chem. Rev. 1941*, 29, 37; (b) Thomas, C.A. *Anhydrous AlCl₃ in Organic Chemistry*, Reinhold, New York, *1941*.
143. Francis, A.W. *Chem. Rev. 1948, 43*, 257.
144. Reference 129, p 39.

When an aromatic ring contains several alkyl groups or complex alkyl groups, treatment with strong acids can lead to formation of isomeric polyalkyl benzene derivatives at relatively low temperatures (<100°C for methyl, room temperature for ethyl, and –80°C for isopropyl and *tert*-butyl).[141] With weaker catalysts, higher temperatures are required. In an example involving simple groups, sulfonation of durene (**200**) led to **203**,[145] presumably via initial sulfonation to **201**, rearrangement to **202** and loss of sulfur trioxide (SO_3) to give **203**. This transformation has come to be known as the **Jacobsen rearrangement**.[146]

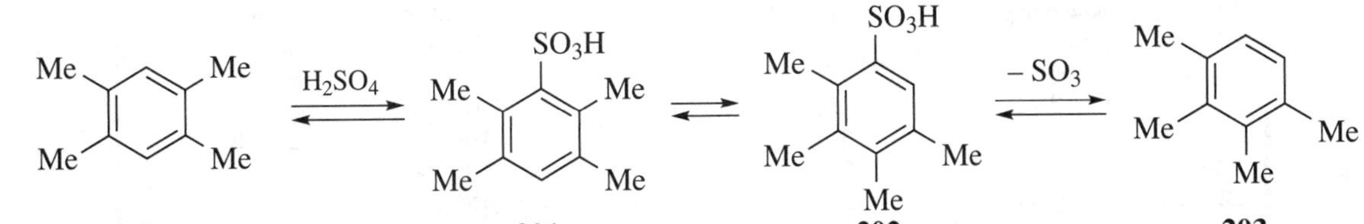

| 200 | 201 | 202 | 203 |

Isomerization is minimized when an electron-withdrawing group is on the benzene ring since the ring is deactivated, and less prone to further reaction. Acetophenone reacted with chloroethane (this reaction is sluggish and required using HF/SbF_5 as a catalyst) to give 3-ethylacetophenone (**204**) in 78% yield.[147]

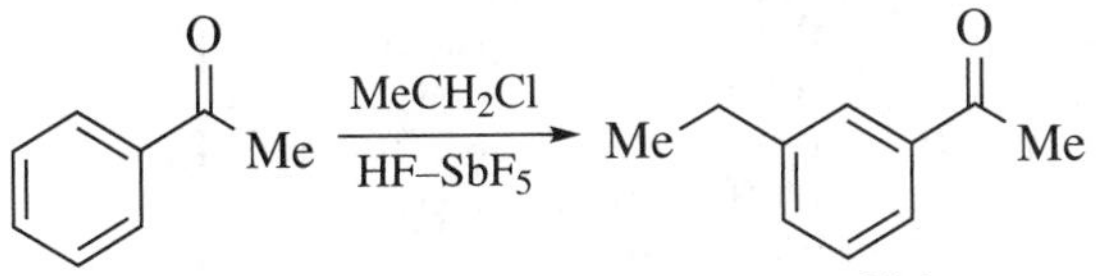

204

A third problem arises when a basic atom is present within the aromatic ring itself, as in pyrrole or pyridine, or is a substituent (as with aniline derivatives). Lewis acids react with the basic nitrogen to give the 'ate' complex (**205**, sec. 2.3), where the nitrogen takes a positive charge. This change in electron distribution deactivates the aromatic ring to substitution. The reaction also removes the catalyst from the reaction and the cation cannot be generated unless a large excess of Lewis acid is used. With aniline derivatives, the basic nitrogen atom can be protected (sec. 7.3). Conversion to an amide (such as *N*-acetylaniline from aniline, called acetanilide) withdraws electron density and diminishes the basicity of the amino group, allowing the Friedel Crafts reaction to give **206**.

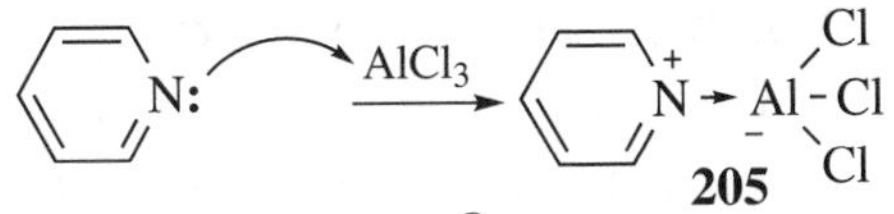

205

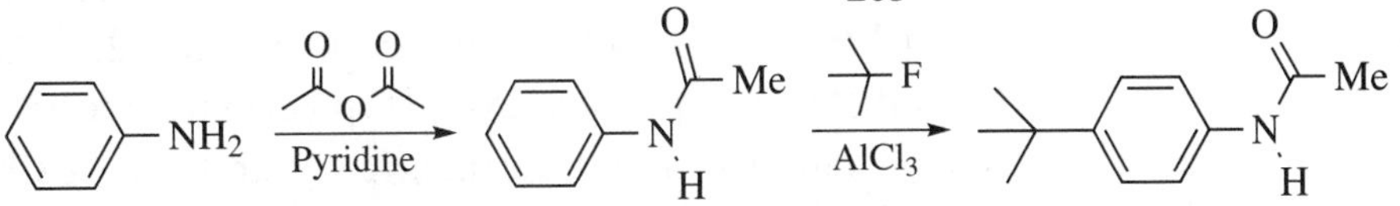

206

145. Hart, H.; Janssen, J.F. *J. Org. Chem.* **1970**, *35*, 3637.
146. (a) Jacobsen, O. *Berichte* **1886**, *19*, 1209; (b) Smith, L.I. *Org. React.* **1942**, *1*, 370; (c) The Merck Index, 14th ed., Merck & Co., Inc., Whitehouse Station, New Jersey, **2006**, p ONR-48.
147. Yoneda, N.; Fukuhara, T.; Takahashi, Y.; Suzuki, A. *Chem. Lett.* **1979**, 1003.

Friedel-Crafts isomerization disconnections are

Table 12.7. Selected Friedel-Crafts Catalysts and The Reactions They Catalyze

Catalyst	Reactions[a]	Catalyst	Reactions[a]	Catalyst	Reactions[a]
$AlBr_3$	general	PBr_3	al, deal	$TiCl_2Et$	al
$AlCl_3$	general	PCl_3	al, deal	$ZnEt_2$	al
AlF_3	poly, iso	PF_3	al, deal	$Al_2Br_2Et_3$	general
AlI_3	general	$SiBr_4$	ac	HBF_4	al, poly
AsF_6	poly, al, ac, nitr	$SiCl_4$	ac	HBr	poly
BBr_3	al, ac, poly	SiF_4	ac, al, nitr	HCl	al, poly
BCl_3	al, ac, poly	$SnCl_2$	ac, poly	HF	al, ac, poly
BF_3	general	$SnCl_4$	ac, al, halo, poly	HI	ac
BI_3	al, ac	$TiCl_3$	ac, al, poly	$HClO_4$	ac, al, poly
$CaCl_2$	al	$TiCl_4$	general	$BF_3 \cdot OH_2$	al
$CaBr_2$	ac	$ZnBr_2$	ac, al, halo	H_2SO_4	general
$CdCl_2$	poly	$ZnCl_2$	general	H_3PO_4	ac, al
$CeCl_3$	acetal formation	ZnI_2	al, ac, poly	PPA	ac, cycl, rearr
$CoCl_2$	for, al dehydr	$Al_2X_2R_n$	al	$ClCH_2CO_2H$	ac, poly
$CuBr_2$	halo, iso	$AlEt_2Br$	al	$(ClCH_2CO)_2O$	ac
$CuCl_2$	iso	$AlEt_3$	al, poly	CCl_3CO_2H	chlor, poly
$FeBr_3$	al, ac, halo, poly, for	$AlMe_3$	al	CF_3CO_2H	ac, chlor, iso
$FeCl_3$	general	$Al(t\text{-}Bu)_3$	al	$(CF_3CO)_2O$	ac, al, poly, cycl
$HgBr_2$	ac, halo, poly	$Al(i\text{-}Bu)_3$	al, poly	RSO_3H	al, poly
$HgCl_2$	ac, al, iso, poly	Br_3	al	$PhSO_3H$	al, poly
$MgBr_2$	al, ac, halo	MgR_2	al, poly	$TolSO_3H$	al, poly
$MgCl_2$	al, iso	$TiCl_3Me$	al		

[a] **PPA** = polyphosphoric acid ; **al** = alkylation; **ac** = acylation; **chlor** = chlorination;   **iso** = isomerization; **halo** = halogenation; **cycl** = cyclization; **poly** = polymerization; **for** = formylation; **dehydr** = dehydration; **nitr** = nitration; **general** = active in most Friedel-Crafts reactions; **deal** = dealkylation; **rearr** = rearrangement.

[Reprinted with permission from Olah, G.A. *Friedel-Crafts Chemistry*, Wiley, New York, *1973*, pp. 298-304. Copyright © *1973* by John Wiley and Sons, Inc.]

12.4.B.iii. Influence of the Catalyst. Lewis acids catalyze several different types of aromatic reactions, including alkylation, polymerization, isomerization, acylation (sec.12.4.D) and halogenation. Table 12.7.[148] shows the dependence of these reactions on various Lewis acids. This table shows a few common Lewis acids and the type of reaction(s) for which each is best suited. The relative strength of Lewis acids was discussed in Section 2.3. Olah gives a comprehensive list of Lewis acid catalysts, and includes several typical synthetic applications.[148] Metal alkyl catalysts are also effective catalysts, but the aromatic substrate is usually converted to a mono-, di-, or trialkyl derivative. Both metal halides and their alkyl derivatives are effective Friedel-Crafts catalysts as shown in Table 12.7,[149] as are common inorganic and organic acids.[150]

148.   Reference 129, pp 298-304.
149.   Reference 129, p 327.
150.   Reference 129, p 342.

Triflate derivatives are actually trifluoromethanesulfonate esters ($-SO_2CF_3$, **OTf**), and they can be prepared with various metal counter ions. Triflates are effective catalysts in Friedel Crafts reactions. Some of the more common catalysts are $B(OTf)_3$, $Al(OTf)_3$, and $Ga(OTf)_3$.[151]

It is clear that many Lewis acids and protonic acids will catalyze various Friedel-Crafts reactions. The choice of halide, alkene, or alcohol substrate will determine which catalyst is used, although, as seen in Table 12.7, the type of reaction that is of interest is also dependent on the choice of catalyst. The following sections provide several examples to illustrate common catalysts used in various reactions, as well as the synthetic utility of the reaction.

12.4.C. Friedel-Crafts Reactions from Alkene and Alcohol Substrates

Friedel-Crafts alkylation occurs between an aromatic substrate and a carbocation. Carbocations can be generated from many precursors, including alkenes, alkynes or alcohols. Treatment of an alkene or an alkyne with a protonic acid catalyst generates a carbocation intermediate. A limiting feature of this process is the fact that a cationic intermediate formed in this manner can also react with the alkene or alkyne starting material, leading to cationic polymerization as a competitive side reaction. The usual problems of polyalkylation and rearrangement associated with aromatic substitution persist. For example, toluene reacts with propene to give isopropylbenzene in 52%,[152] 20% of diisopropyl-benzene, 3% of trisubstituted products and 1.3% of tetrasubstituted products were formed when 0.75 equivalents of propene were used. As shown in Table 12.8,[152] toluene and ethene react to give a wide array of products. The nature of those products is completely dependent on the relative amount of ethene.

Table 12.8. Polyalkylation Products from the Reaction of Toluene and Ethene

Equivalents of Ethene	PhH	PhEt	di-Et	tri-Et	tetra-Et	penta-Et	hexa-Et
0.20	87.9	10.9	1.24				
0.50	80.8	15.5	2.14	0.67	0.15	0.06	0.67
0.56	80.5	15.2	1.67	0.71			1.91
1.14	31.6	35.2	21.1	4	2.4	2.8	2.82
1.68	35.4	23.4	13.6	6.3	4.2	3.6	13.4
3.66	0.2	1.1	10.6	56.2	11.6	1.5	18.7
5.24				7.00	18.8	2.4	7.8
5.74					8.5	0.9	90.6

[Reprinted with permission from OBerry, T.M.; Reid, E.E. *J. Am. Chem. Soc.* **1927**, *49*, 3142. Copyright © **1927** American Chemical Society.]

Hydrogen fluoride (HF) is often used as a catalyst for alkenes since there is less decomposition and fewer polymeric by-products. In addition, the volatile catalyst is easily removed.[153] Both

---

151. Olah, G.A.; Farooq, O.; Farnia, S.M.F.; Olah, J.A. *J. Am. Chem. Soc.* **1988**, *110*, 2560.
152. Berry, T.M.; Reid, E.E. *J. Am. Chem. Soc.* **1927**, *49*, 3142.
153. Reference 129, p 44.

ferric chloride ($FeCl_3$) and BF_3 are common catalysts for coupling alkenes with aromatic rings. Typical alkyne catalysts[154] are aluminum chloride, gallium trichloride ($GaCl_3$), BF_3 or sulfuric acid. Reaction of a mixture of xylenes with a trace amount of BF_3 in liquid HF gave the thermodynamic mixture of 18% ortho, 60% meta and 22% para, but treatment with excess BF_3 gave virtually 100% *m*-xylene.[143,142b,155]

Friedel-Crafts alkylations using alkene substrates are most useful when stable cations are formed and when done intramolecularly, as observed with alkyl halide precursors. Internal Friedel-Crafts cyclization using alkene substrates forms the basis of many natural product syntheses. Majetich and Zhang treated **207** with $TiCl_4$ in a synthesis of perovskone,[156] and obtained a 95% yield of **208**. In this case, a primary cation led to the final product, due to stabilization by extended conjugation with the carbonyl unit. In the Sames and co-worker's synthesis of the core of teleocidin B4,[157] **209** was treated with methanesulfonic acid in dichloromethane. The alkene unit was converted to a carbocation that reacted with the proximal benzene ring by an intramolecular Friedel-Crafts alkylation to give an 83% yield of **210**.

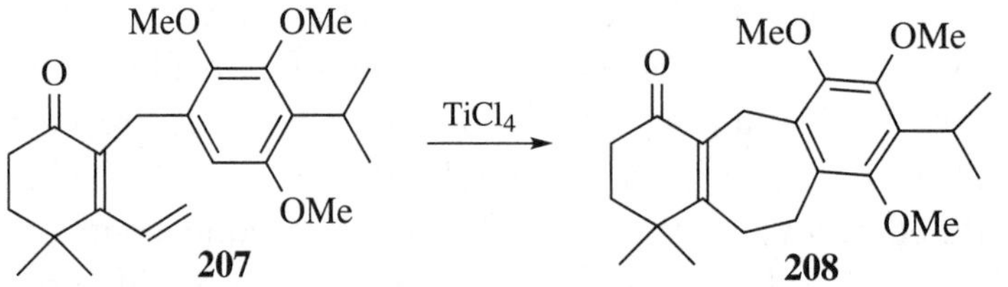

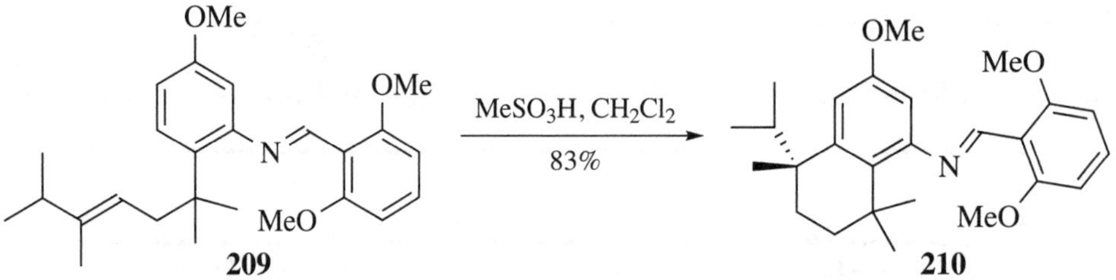

Alcohols are common precursors to carbocations, and they are often used as substrates in Friedel-Crafts alkylation reactions. Sulfuric acid is used as a catalyst with alcohols, forming an alkyl sulfate that reacts with the aromatic substrate.[158] Alcohols are more reactive than halides, although large quantities of a Lewis acid are usually required for Friedel-Crafts reactions.[159] Alcohols form a complex with aluminum chlorides [R(H)O→$AlCl_3$], and a complicating side reaction is loss of HCl from the complex, with formation of an alkyl chloride.[160] A full equivalent

154. Reference 129, ,p. 45.
155. McCaulay, D.A.; Shoemaker, B.H.; Lien, A.P. *Ind. Eng. Chem.* **1950**, *42*, 2103.
156. Majetich, G.; Zhang, Y. *J. Am. Chem. Soc.* **1994**, *116*, 4979.
157. Dangel, B.D.; Godula, K.; Youn, S.W.; Sezen, B.; Sames, D. *J. Am. Chem. Soc.* **2002**, *124*, 11856.
158. (a) Günther, F. *U.S. Patent* 1,670,505, **1928** [*Chem. Abstr., 22*: 2378, **1928**]; (b) Reference 129, p 47.
159. Norris, J.F.; Sturgis, B.M. *J. Am. Chem. Soc.* **1939**, *61*, 1413.
160. Reference 129, p 46.

of the Lewis acid is required to form the complex. In a synthesis of (−)-aphanorphine, Zhai and co-workers[161] treated alcohol **211** with aluminum chloride and obtained an 88% yield of **212**. This is an interesting example because it generated an azabicyclo[3.2.1] system from a tertiary alcohol with good selectivity for the diastereomer shown. Other oxygen-stabilized carbocations add to aromatic rings, including heteroaromatic rings. In Movassaghi and Ondrus' synthesis of tricyclic *Myrmicarin* alkaloids,[162] treatment of the acetal unit in **213** with aqueous acetic acid generated the carbocation and cyclization led to a quantitative yield of **214** with >10:1 regioselectivity after elimination of the initially formed hydroxy compound. Other cation precursors can be used in these reactions. Locher and Peerzada used a benzotriazole unit in a synthesis of tetrahydroisoquinolines.[163] In this reaction, **215** was converted to **216** in 87% yield under very mild conditions.

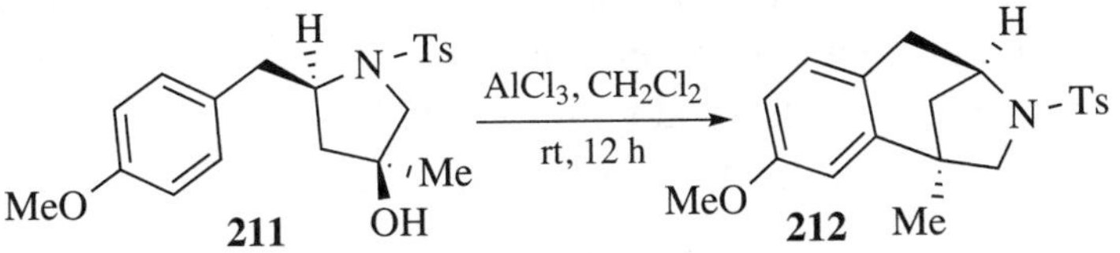

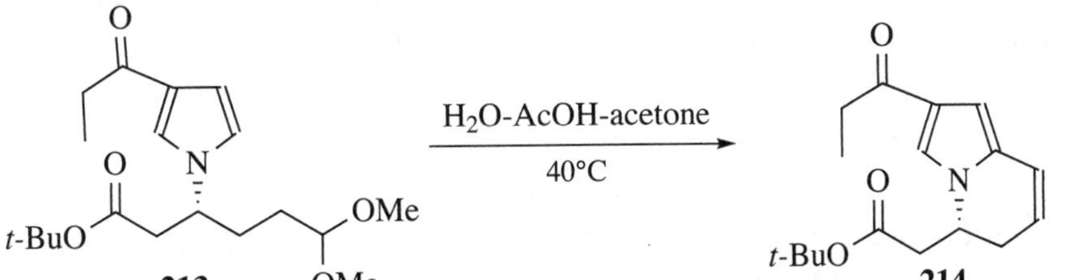

Typical Friedel-Crafts disconnections involving alcohol and alkene precursors are

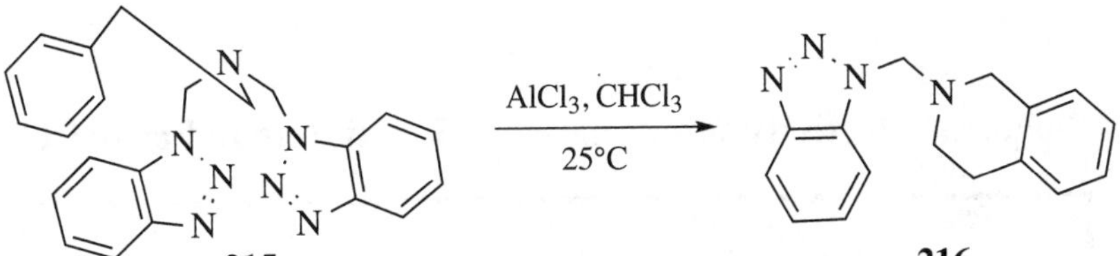

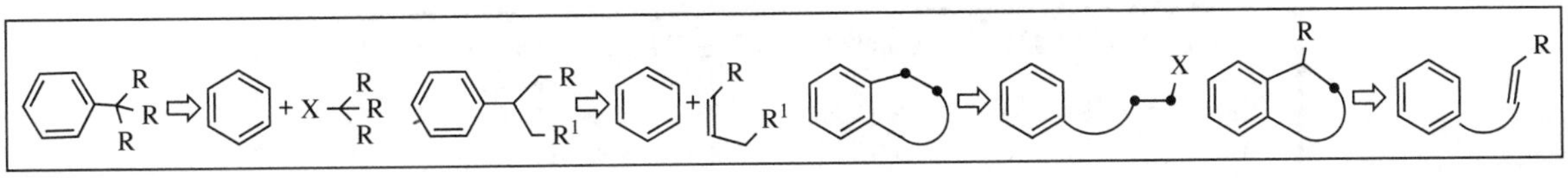

12.4.D. Friedel-Crafts Acylation

The reaction of a Lewis acid reacts and an acyl halide or an anhydride produces a resonance stabilized acylium ion (**217**). The π bond of a benzene ring attacks carbocation **217** to produce the usual Wheland intermediate **218**, and loss of a proton accompanied by aromatization gives

161. Zhai, H.; Luo, S.; Ye, C.; Ma, Y. *J. Org. Chem.* **2003**, *68*, 8268.
162. Movassaghi, M.; Ondrus, A.E. *Org. Lett.* **2005**, *7*, 4423.
163. Locher, C.; Peerzada, N. *J. Chem. Soc. Perkin Trans. 1* **1999**, 179.

an aryl ketone (**219**). This reaction is known as **Friedel-Crafts acylation**.[164,127a] The carbonyl group in the ketone product is polarized $\delta+$ and attached directly to the aromatic ring, so it is deactivated relative to benzene. Since **219** is less reactive than benzene, further reaction to give a polyacylated derivative is not a major problem. In other words, Friedel-Crafts acylation does not lead to polyacylation. The resonance stabilized acylium ion (**217**) is not subject to skeletal rearrangement prior to reaction with the aromatic ring. Isomerization of the ketone product is not a problem.

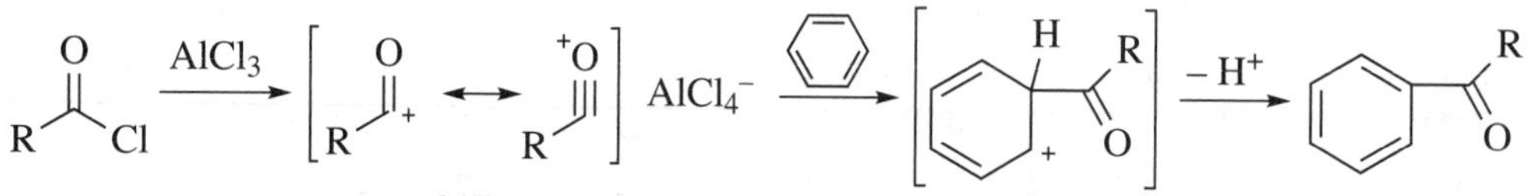

Acylium ions can be produced from acyl bromides and iodides, but the more common and more stable acid chlorides are typically used. Table 12.9[165] shows the efficacy of several catalysts for the reaction of benzoyl chloride and anisole. In all cases, the para product (**221**) predominated over the ortho product (**220**) by about 31:1.[165] Mild Lewis acids can be used to effect Friedel-Crafts cyclization, which is particularly important with synthetic targets bearing other functional groups. The acylation is not limited to benzene, of course, but deactivated aromatic rings either do not undergo Friedel-Crafts reactions at all, or do so with difficulty. Activated aromatic rings react in most cases. In a synthesis of calothrixin A by Chai and Bernardo,[166] acid chloride **222** reacted with indole in the presence of $MeMgCl/ZnCl_2$ to give a 90% yield of **223**.

Table 12.9. ortho/para Selectivity as a Function of Lewis Acid in Friedel-Crafts Acylation

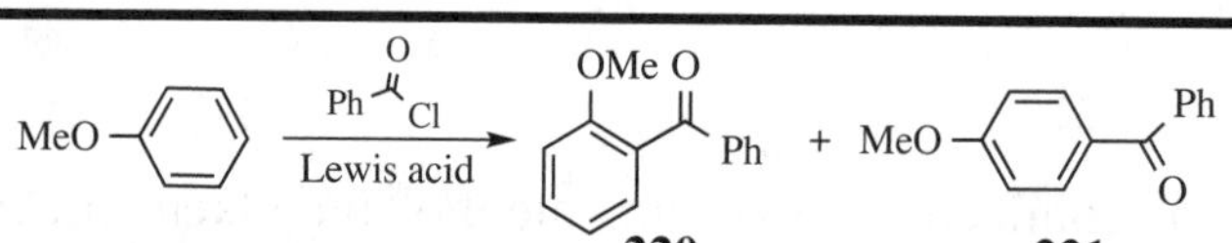

Catalyst	% 220 + 221	Catalyst	% 220 + 221	Catalyst	% 220 + 221
$ZnCl_2$	93	$SnCl_4$	90	$MoCl_5$	80
$FeCl_3$	83	$SbCl_5$	43	$NbCl_5$	54
$AlCl_3$	52	$TaCl_5$	52	$BiCl_3$	49
WCl_6	47	$ZrCl_4$	39	$TiCl_4$	35
$TiCl_2$	32	VCl_3	10		

164. For reactions of this type of reaction, see reference 79, pp 1422-1433. Also see (a) The Merck Index, 14th ed., Merck & Co., Inc., Whitehouse Station, New Jersey, *2006*, p ONR-34; (b) Mundy, B.P.; Ellerd, M.G.; Favaloro Jr., F.G. *Name Reactions and Reagents in Organic Synthesis, 2nd ed.*, Wiley-Interscience, New Jersey, *2005*, pp. 254-255.

165. (a) Tsukervanik, I.P.; Veber, N.V. *Dokl. Akad. Nauk. SSSR* *1968, 180,* 892; (b) Reference 129, p 195.

166. Bernardo, P.H.; Chai, C.L.L. *J. Org. Chem.* *2003, 68,* 8906.

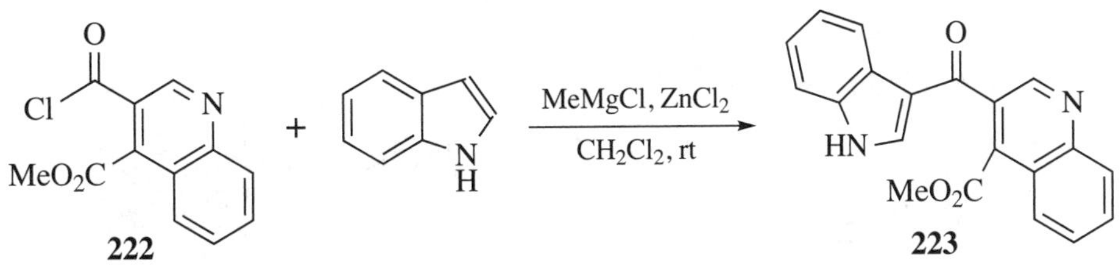

222 + (indole) → **223**

Reagents: MeMgCl, ZnCl₂, CH₂Cl₂, rt

Other carboxylic acid derivatives can be used in Friedel-Crafts acylation reactions, including anhydrides and esters. Chai and co-workers prepared **224** in 85% yield using dibutyl anhydride with perchloric acid, in a synthesis of hybocarpone.[167] Lactones are also effective substrates in this cyclization, generating ketones.[168]

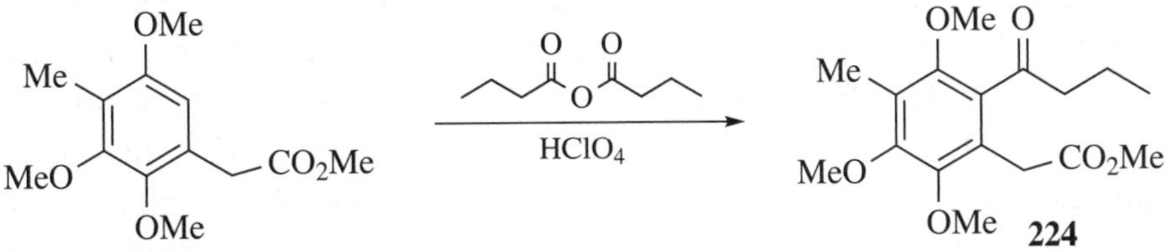

Intramolecular Friedel-Crafts acylations are extremely valuable for the synthesis of polycyclic compounds. The length of the tether directly influences the site of attack on the aromatic nucleus and the facility of the cyclization. The aromatic nucleus can be benzene, another simple aromatic compound, a polynuclear aromatic hydrocarbon or a heteroaromatic compound such as pyrrole, furan or thiophene. The importance to tether length can be illustrated by the reaction of pyrrole-diester **225** with BBr₃, which led to ketone **226** in Vallée and coworker's synthesis of *(R)*-(+)-myrmicarin 217.[169] In intermolecular Friedel-Crafts

acylation reactions, pyrroles usually give the α-acyl derivative with the soft acylium electrophile (sec. 2.4). Using hard electrophiles such as trimethylsilyl triflate (Me₃SiOTf), however, led to the β-derivative.[170] When pyrrole was treated with acetic anhydride (at 150-200°C), a mixture of 2-acetylpyrrole and 2,5-diacetylpyrrole was formed by reaction with the soft acylium ion.[171] Acylation of an *N*-silyl pyrrole led predominantly to the β isomer rather than the α isomer (an 83:17 mixture in 46% yield), when reacted with either anhydrides or acyl chlorides.[172] The bulky -SiMe₃ group presumably blocked the C2 position. A somewhat more traditional example taken from Baati, Misokowski and co-worker's synthesis of rhein and diacerhein,[173] treated acid **227** with trifluoroacetic acid to give ketone **228** in 96% yield.

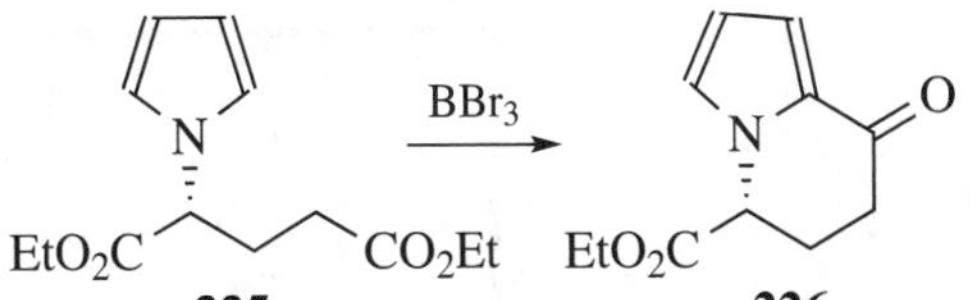

167. Chai, C.L.L.; Elix, J.A.; Moore, F.K.E. *J. Org. Chem* **2006**, *71*, 992.
168. Inouye, Y.; Uchida, Y.; Kakisawa, H. *Chem. Lett.* **1975,** 1317.
169. Sayah, B.; Pelloux-Léon, N.; Vallée, Y. *J. Org. Chem.* **2000**, *65*, 2824.
170. Simchen, G.; Majchrzak, M.W. *Tetrahedron Lett.* **1985**, *26*, 5035.
171. Katritzky, A.R. *Handbook of Heterocyclic Chemistry*, Pergamon Press, Oxford, **1985**, p 254.
172. McCombie, S.W.; Shankar, B.B.; Ganguly, A.K. *Tetrahedron Lett.* **1987**, *28*, 4123.
173. Gonnot, V.; Tisserand, S.; Nicolas, M.; Baati, R.; Mioskowski, C. *Tetrahedron Lett.* **2007**, *48*, 7117.

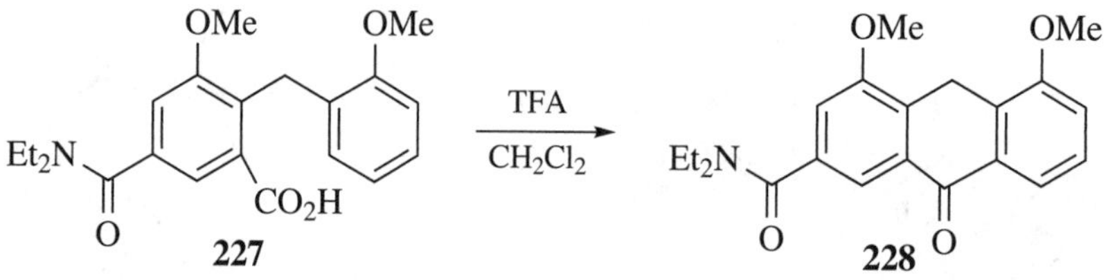

For benzene derivatives, both ortho positions are subject to acylation if they do not bear substituents. As shown in Table 12.10,[174] Friedel-Crafts acylation of **229** gave a mixture of coumarins **230** and **231**. Sulfuric acid gave a mixture favoring **231**, whereas a mixture of phosphorus pentabromide and aluminum chloride (PBr$_5$/AlCl$_3$) was highly regioselective for the formation of **230**. It is clear that the ortho selectivity depends upon the catalyst used, as well as upon the nature of the substrate. A related acylation procedure uses iminium salts rather than acyl halides. The **Vilsmeier-Haack** reaction[175] is a well-known process, illustrated by reaction of pyrrole with the POCl$_3$ complex of N,N-dimethylacetamide (**232**, which can decompose to a chloroiminium salt). The acylation reaction initially gave **233**, which was converted to 2-acetylpyrrole by hydrolysis with aqueous sodium acetate.[176] A synthetic example is taken from a synthesis of lactonamycin by Barrett and co-workers,[177] in which **234** reacted with POCl$_3$ and DMF and then aqueous sodium acetate to give aldehyde **235** in 75% yield.

Table 12.10. Regioselectivity in a Friedel-Crafts Synthesis of Coumarins

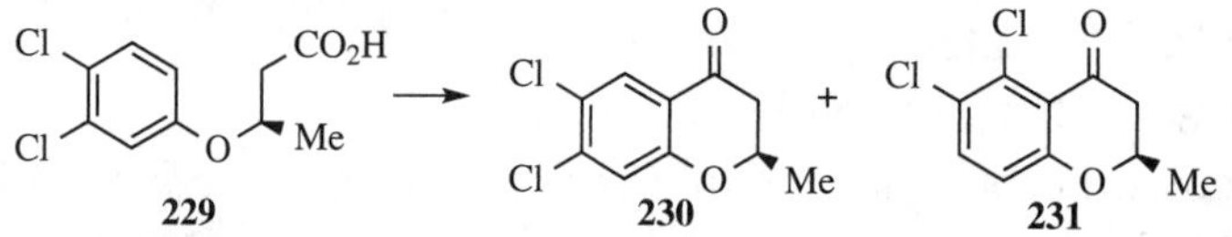

Reagent	% Yield	230 : 231
H$_2$SO$_4$	65	35 : 65
MeSO$_3$H/P$_2$O$_5$	70	50 : 50
PCl$_5$/AlCl$_3$	90	85 : 15
PBr$_3$/AlCl$_3$	87	95 : 5

[Reprinted with permission from Dr. Frank Urban, Pfizer Central Research, Groton, CT.]

174. Provided through the courtesy of Dr. Frank Urban, Pfizer Central Research, Groton, CT, from (a) Lipinski, C.A. *U.S. Patent 4,853,410*, **1989**; EP 230379A2, **1988** [*Chem. Abstr. 108* : 75224h, **1988**]; (b) Moore, B.S. Unpublished results.
175. (a) Vilsmeier, A.; Haack, A. *Berichte* **1927**, *60*, 119; (b) deMaheas, M.R. *Bull. Chim. Soc. Fr.* **1962**, 1989; (c) The Merck Index, 14th ed., Merck & Co., Inc., Whitehouse Station, New Jersey, **2006**, p ONR-96; (d) Mundy, B.P.; Ellerd, M.G.; Favaloro Jr., F.G. *Name Reactions and Reagents in Organic Synthesis, 2nd ed.*, Wiley-Interscience, New Jersey, **2005**, pp. 558-559.
176. Reference 171, p 254.
177. Wehlan, H.; Jezek, E.; Lebrasseur, N.; Pave, G.; Roulland, E.; White, A. J.P.; Burrows, J.N.; Barrett, A.G.M.. *Org. Chem.* **2006**, *71*, 8151.

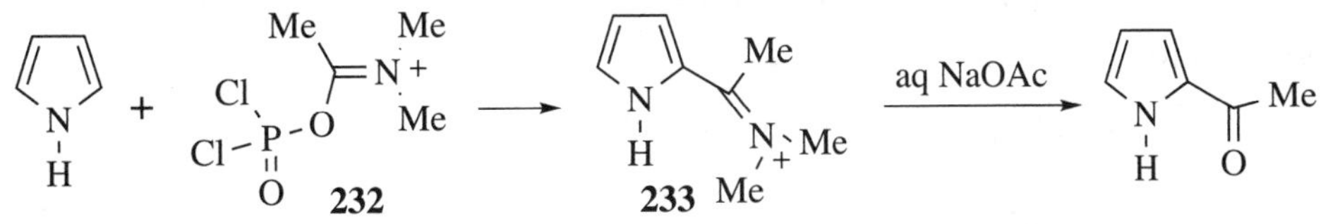

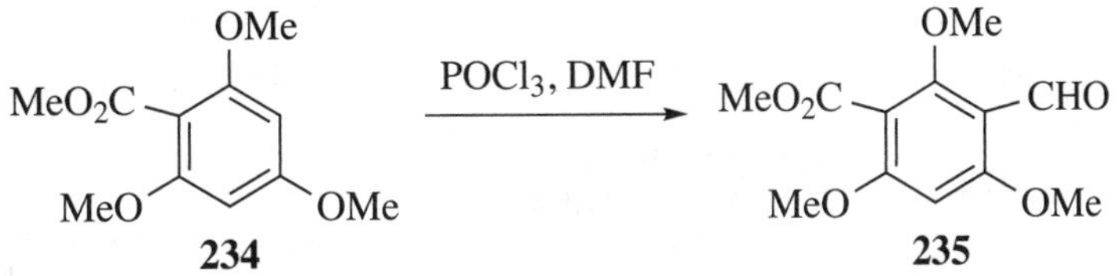

A variation of the Vilsmeier-Haack reaction uses a formamide such as **236** with benzene (or another aromatic compound) to give the aromatic aldehyde and the amine precursor to the formamide (*N*-methylaniline in this case).[175a] A solvent-free Vilsmeier reaction has been reported using microwave irradiation on silica gel.[178]

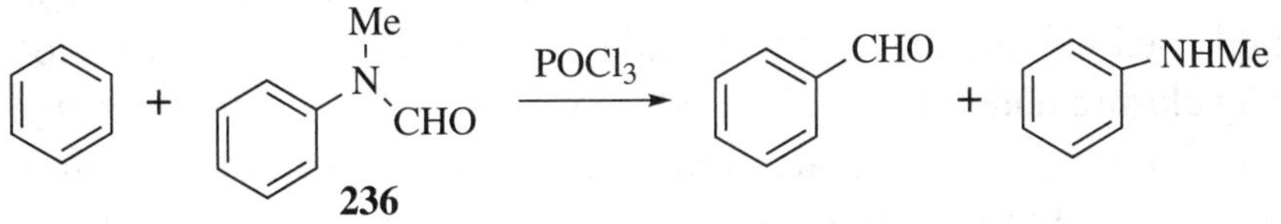

12.4.E. Synthesis of Polycyclic Aromatics That Do Not Contain Nitrogen

A classical reaction that exploits the power inherent to Friedel-Crafts techniques is the **Haworth phenanthrene synthesis**,[179] which uses anhydrides in a Friedel-Crafts reaction with aromatic compounds, to prepare polynuclear aromatic derivatives in a stepwise manner. When 1-methylnaphthalene (**237**) was heated with succinic anhydride in the presence of $AlCl_3$, keto acid **238** was formed. **Clemmensen reduction** (sec. 4.9.G) of the ketone gave **239**, and a second Friedel-Crafts cyclization in the presence of sulfuric acid led to the tricyclic ketone, **240**. Reduction of the carbonyl and aromatization (by heating with selenium led to phenanthrene **241**. Direct cyclization of the acid occurred when a more vigorous catalyst (sulfuric acid) was used, and there was no need to form the acid chloride first. If benzene were used as a starting material in this sequence, a naphthalene derivative would be formed. Intermediate products such as **240** are important for other synthetic work, making this sequence even more useful. In a synthesis of hamigeran B by Clive and Wang,[180] *m*-cresol was converted to **242** by the modified Haworth synthesis shown. Note the use of $Zn/HCl/HgCl_2$ for reduction of the ketone

178. Paul, S.; Gupta, M.; Gupta, R. *Synlett* **2000**, 1115.
179. (a) Haworth, R.D. *J. Chem. Soc.* **1932**, 1125; (b) Haworth, R.D.; Mavin, C.R. *Ibid.* **1932**, 2720; (c) Haworth, R.D. *Ibid.* **1932,** 2717; (d) Haworth, R.D.; Letsky, B.M.; Mavin, C.R., *Ibid.* **1932**, 1784; (e) Haworth, R.D.; Bolam, F.M. *Ibid.* **1932**, 2248; (f) Haworth, R.D.; Mavin, C.R.; Sheldrick, G., *Ibid.* **1934,** 454; (g) The Merck Index, 14th ed., Merck & Co., Inc., Whitehouse Station, New Jersey, **2006**, p ONR-41.
180. Clive, D.L.J.; Wang, J. *J. Org. Chem.* **2004**, *69*, 2773.

after acylation with succinic anhydride, and $POCl_3$ to initiate the Friedel-Crafts acylation with the carboxylic acid.

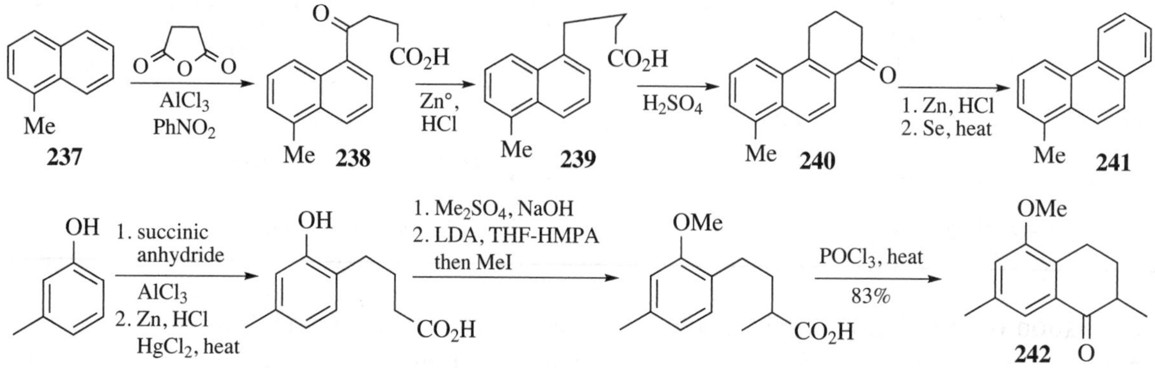

Another useful sequence that produces polynuclear aromatic compounds is the **Bally-Scholl synthesis**.[181] This variation involves addition of a glycerol derivative (a 1,2,3-propanetriol) to an aromatic ring. An example is the reaction of anthraquinone (**243**) with 1,2,3-pentanetriol (**244**, also called α-ethylglycerol) to form an alkenyl diol, **245**. Elimination to diene **246** was followed by ring closure under the acidic conditions to give **247** (1-ethylmesobenzanthrone) in about 5% yield.[182] Poor yields with such multistep processes are common under the strongly acidic conditions used. In many other types of reactions, milder conditions involving milder Lewis acids and structural changes in the substrate led to enhanced yields. Poor yields such as that observed for **247** may be acceptable if the synthesis is relatively straightforward, and there are few alternative synthetic routes.

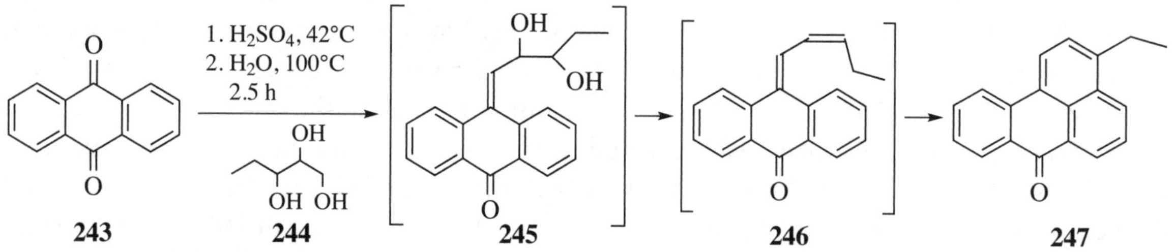

Another Friedel-Crafts route to phenanthrenes is the **Bardhan-Sengupta phenanthrene synthesis**.[183] When phenethylcyclohexanol **248** was cyclized with phosphorus pentoxide at 140°C, the product was **249**.[183] Aromatization was accomplished by heating with selenium to give phenanthrene derivative **250** (1-methyl-7-isopropylphenanthrene, otherwise known as retene).[183] An improvement in the reaction used hydrogen fluoride (HF) to induce cyclization.[184]

181. (a) Bally, O. *Berichte* **1905**, *38*, 194; (b) Bally, O.; Scholl, R. *Ibid.* **1917**, *44*, 1656; (c) Meerwein, H.; Klinz, J. *J. Prakt. Chem.* **1918**, *97*, 235.
182. Baddar, F.G.; Warren, F.L. *J. Chem. Soc.* **1938**, 401.
183. Bardhan, J.C.; Sengupta, S.C. *J. Chem. Soc.* **1932**, 2520, 2798.
184. Renfrow, W.B.; Renfrow, A.; Shoun, E.; Sears, C.A. *J. Am. Chem. Soc.* **1951**, *73*, 317.

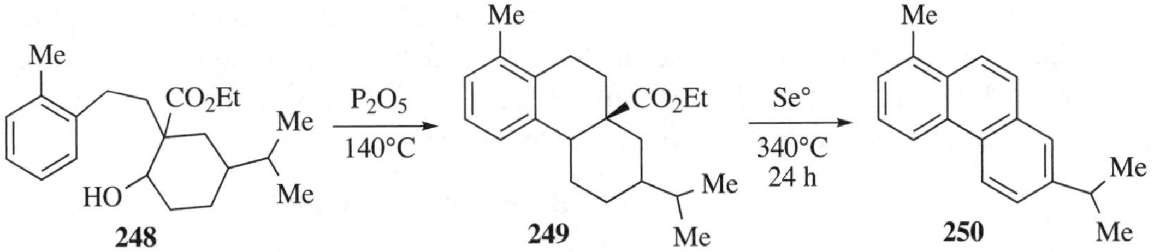

Both phenanthrenes and hydronapthalenes are available by the **Bogert-Cook synthesis**.[185,183] When the Grignard reagent derived from **251** reacted with 2,5-dimethylcyclohexanone, alcohol **252** was produced in 43% yield. Treatment with 85% sulfuric acid effected dehydration of the alcohol to a cyclohexene derivative, which closed on the aromatic ring to give **253** in 81% yield. Dehydrogenation with selenium at 300°C gave retene (**250**), whereas hydrogenation with Raney nickel (sec. 4.8.F) gave abietane (**254**) in 39% yield.[186] The use of 1-bromonaphthalene derivatives as a starting material will lead to larger polycyclic rings.[185b] An alkene unit can also be used to generate a cation and subsequent cyclization will generate a polycyclic aromatic compound. The reaction of **255** with acetaldehyde gave **256**, which was cyclized by treatment with sulfuric acid to give hydronapthalene (**257**). This modification does not incorporate the additional ring of the cyclic ketones used in the Bogert-Cook reaction; it is sometimes called the **Bogert synthesis**.

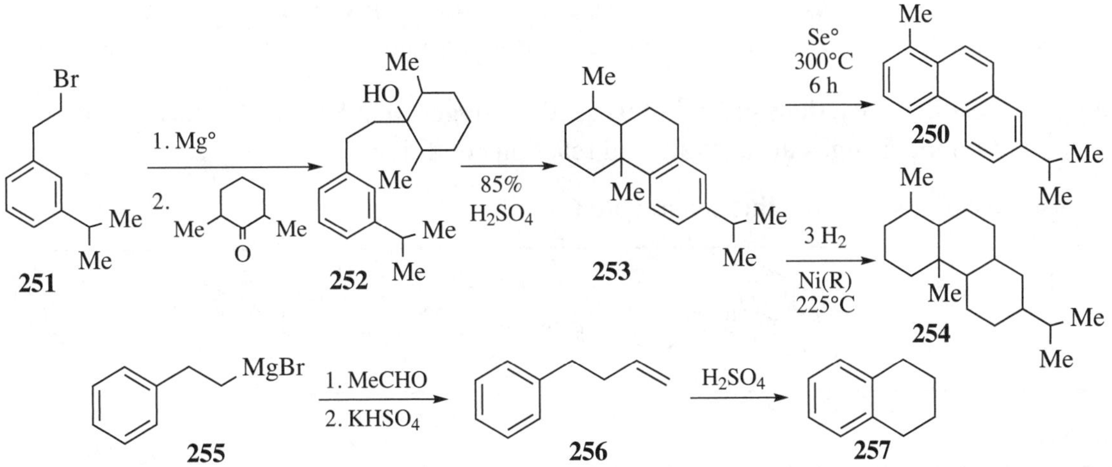

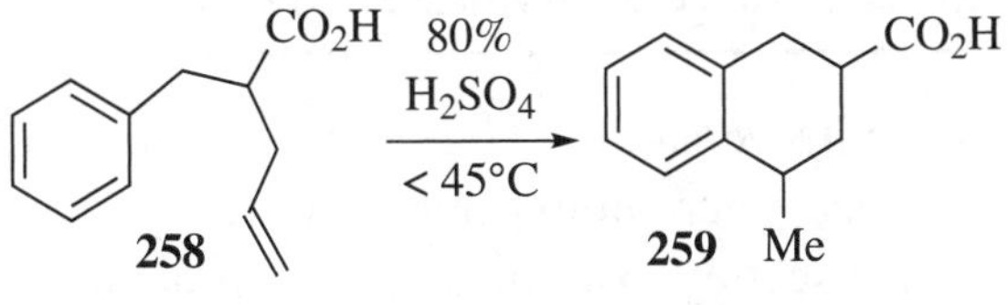

A related modification is called the **Darzens tetralin synthesis**.[187] In this process, the alkenyl group is used directly as the cation precursor. An example is treatment of **258** with

185. (a) Bogert, M.T. *Science* **1933**, *77*, 289; (b) Cook, J.W.; Hewett, C.L. *J. Chem. Soc.* **1933**, 1098.
186. Sterling, E.D.; Bogert, M.T. *J. Org. Chem.* **1939**, *4*, 20.
187. (a) Darzens, G. *Compt. Rend.* **1926**, *183*, 748; (b) *Idem, Ibid.* **1935**, *201*, 730; (c) Darzens, G.; Lévy, A. *Ibid.* **1936**, *202*, 427; (d) *Idem, Ibid.* **1936**, *203*, 669; (e) The Merck Index, 14th ed., Merck & Co., Inc., Whitehouse Station, New Jersey, **2006**, p ONR-22.

concentrated sulfuric acid, followed by gentle heating to give the tetralin (tetrahydronaphthalene, **259**).[188] The difference between this and the Bogert synthesis is the method used to form the alkenyl arene. Tetralin formation from an alkenylbenzene derivative is most often referred to as the Darzens tetralin synthesis rather than the Bogert synthesis.

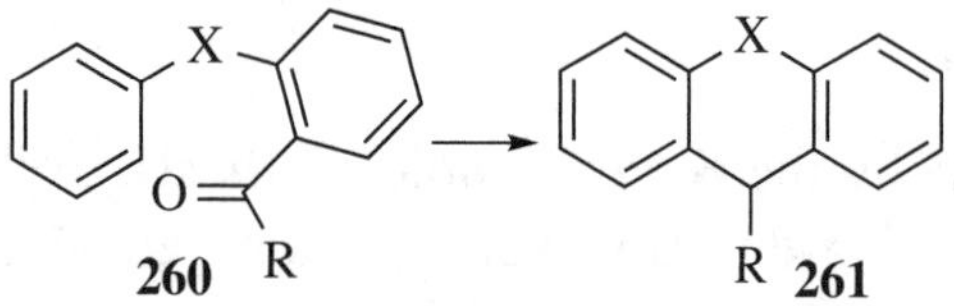

The **Bradsher reaction**[189] is a synthetic route that converts aryl ketones (**260**) to polycyclic hydrocarbons such as **261**. This reaction can also be applied to heterocyclic derivatives of **260** (X = O, S, Se). A common route to ketones such as **260** is shown for the preparation of **266**. Two aryl groups were incorporated by a Grignard reaction of aldehyde **262** with phenylmagnesium bromide, producing **263** in 81% yield. Subsequent treatment with cuprous cyanide (also see sec. 2.11.E) led to nitrile **264**, and reaction with ethylmagnesium bromide gave the ketone (**265**). Subsequent reaction with HBr (reflux for four days) generated ethyl anthracene (**266**) in 69% yield.[189a]

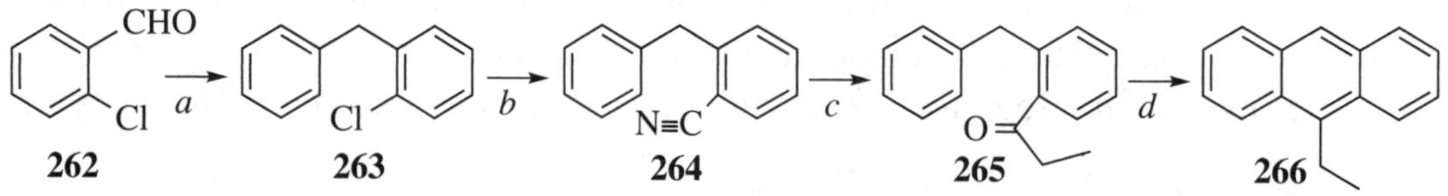

(*a*) i. PhMgBr ii. I₂, P, AcOH, reflux, 29 h (*b*) CuCN, 250°C, 23 h (*c*) i. EtMgBr, PhH, reflux, 20 h ii. H₃O⁺ (*d*) 30% HBr, AcOH, 4 d

Note that the alkene reactions given in this section suggest that Friedel-Crafts cyclizations are sensitive to minor changes in substrate and reaction conditions.

The disconnections found in this section are

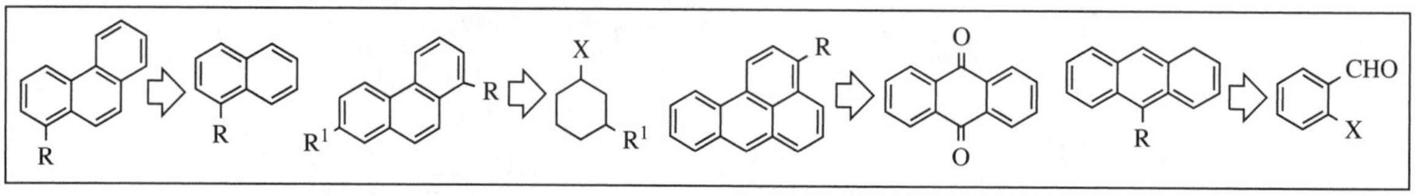

12.5. FRIEDEL-CRAFTS REACTIONS: FORMATION OF HETEROATOM-CONTAINING DERIVATIVES

Friedel-Crafts alkylation and acylation reactions are quite versatile. They form the basis of several named reactions that have been used to synthesize heteroatom-containing compounds such as quinolines, isoquinolines, and many others. This section examines several of the more common transformations that will show the scope and utility of these processes. It will also introduce a modest amount of heterocyclic chemistry into our synthetic discussion.

188.   Darzens, G.; Lévy, A. *Compt. Rend.* **1935**, *200*, 469.
189.   (a) Bradsher, C.K. *J. Am. Chem. Soc.* **1970**, *62*, 486; (b) The Merck Index, 14th ed., Merck & Co., Inc., Whitehouse Station, New Jersey, **2006**, p ONR-14.

12.5.A. Synthesis of Quinoline Derivatives

The quinoline unit is an essential feature of many natural products, and there are several reactions that generate quinolines. One of the older, yet still useful methods, is the **Knorr quinoline synthesis**.[190] This reaction couples an aromatic amine (usually aniline or a derivative) with a β-keto ester. This particular reaction is subject to formation of regioisomeric products, depending on the temperature and reaction conditions. Heating aniline and ethyl acetoacetate to 140°C, for example, gave 50% of **268**. However, if aniline and ethyl acetoacetate were reacted at room temperature for 5 days (or for 3 days with a catalytic amount of HCl), 60% of the isomeric hydroxyquinoline **267** was isolated. A 70% yield of **267** was also obtained when aniline was treated with ethyl acetoacetate in ethanol with a catalytic amount of acetic acid in the presence of Drierite® In an example taken from a synthesis of diplamine, by Heathcock and Szczepankiewicz,[191] initial coupling of aniline derivative **269** with β-keto ester **270** gave keto amide **271** in 83% yield. Subsequent treatment with 90% sulfuric acid at 75°C led to the Knorr quinoline reaction that gave **272** in 82% yield. Further modification was required for the synthesis, and heating this pyridone derivative with $POCl_3$ gave a 94% yield of the 2-chloroquinoline derivative **273**.

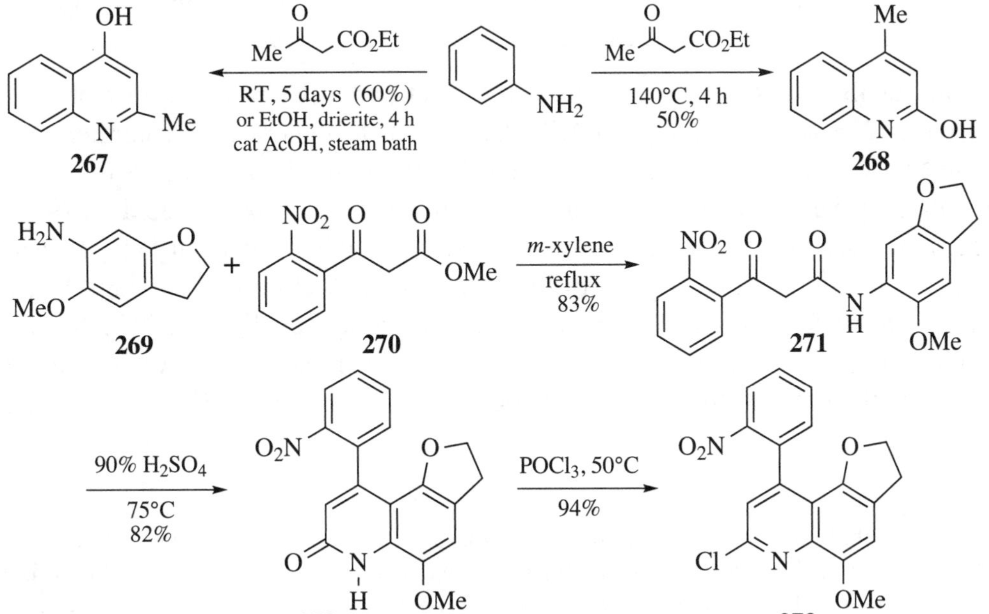

190. (a) Knorr, L. *Annalen* **1886**, *236*, 69; (b) *Idem, Ibid.* **1888**, *245*, 357; (c) Bergström, F.W. *Chem. Rev.* **1944**, *35*, 77 (see p 157). Also see (c) Hodgkinson, A.J.; Staskun, B. *J. Org. Chem.* **1969**, *34*, 1709; (d) Hauser, C.R.; Reynolds, G.A. *J. Am. Chem. Soc.* **1948**, *70*, 2402; (e) The Merck Index, 14th ed., Merck & Co., Inc., Whitehouse Station, New Jersey, **2006**, p ONR-52;.
191. Szczepankiewicz, B.G.; Heathcock, C.H. *J. Org. Chem.* **1994**, *59*, 3512.

The lower temperature variation of this reaction initially forms an imine or an enamine. Friedel Crafts cyclization gave the 4-hydroxyquinoline in what is called the **Conrad-Limpach reaction**.[192] This reaction generally gives the opposite regioisomeric product to that obtained by the Knorr quinoline synthesis. The initially formed product is usually the enamine (as in the formation of **274** from aniline and ethyl acetoacetate).[193] Under acidic conditions the iminium salt was formed and cyclized with the aromatic ring. A more efficient method simply heated **274** to 250°C in mineral oil, giving a 90% yield of **275**. A variety of other functional groups can be tolerated in the molecule when this procedure is used.

The Knorr and Conrad-Limpach disconnections are

The **Gould-Jacobs reaction**[194] is related to the reactions just described, in that an aromatic amine such as **276** is reacted with $EtOCH=C(CO_2Et)_2$ to generate an enamino ester (**277**). Heating to 250°C induced cyclization to give 4-hydroxyquinoline ester **278** in 83% yield.[194a] Saponification gave 4-hydroxy-5,6-benzoquinidine-3,7-dicarboxylic acid, **279**.[195] This sequence can be used with many aromatic substrates, but when secondary amines are used as precursors the reaction takes a slightly different course. 2-Methoxy-*N*-ethylaniline (**280**) was converted to **281** (80% yield) in the same manner. Subsequent cyclization[194a] in the presence of P_2O_5 required only 140°C to give an 85% yield of **282**. The presence of a tertiary amine moiety in this product prevented aromatization, and decarboxylation led to **283** in 41% yield.

192. (a) Conrad, M.; Limpach, L. *Berichte* **1887**, *20*, 944; (b) *Idem Ibid.* **1891**, *24*, 2990; (c) The Merck Index, 14th ed., Merck & Co., Inc., Whitehouse Station, New Jersey, **2006**, p ONR-19; (d) Mundy, B.P.; Ellerd, M.G.; Favaloro Jr., F.G. *Name Reactions and Reagents in Organic Synthesis, 2nd ed.*, Wiley-Interscience, New Jersey, **2005**, pp. 166-167.
193. (a) Manske, R.H. *Chem. Rev.* **1942**, *30*, 113 (see p 121); (b) Reitsema, R.H. *Ibid.* **1948**, *43*, 43 (see p 47).
194. (a) Gould Jr., R.G.; Jacobs, W.A. *J. Am. Chem. Soc.* **1939**, *61*, 2890; (b) Reference 193b, see p 53; (c) The Merck Index, 14th ed., Merck & Co., Inc., Whitehouse Station, New Jersey, **2006**, p ONR-37; (d) Mundy, B.P.; Ellerd, M.G.; Favaloro Jr., F.G. *Name Reactions and Reagents in Organic Synthesis, 2nd ed.*, Wiley-Interscience, New Jersey, **2005**, pp. 278-279.
195. (a) Markees, D.G.; Schwab, L.S. *Helv. Chim. Acta* **1972**, *55*, 1319; (b) Albrecht, R.; Hoyer, G.-A. *Berichte* **1972**, *105*, 3118.

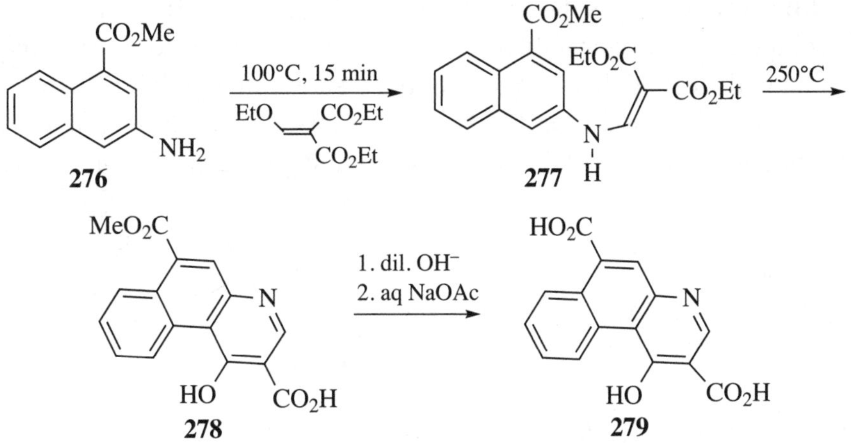

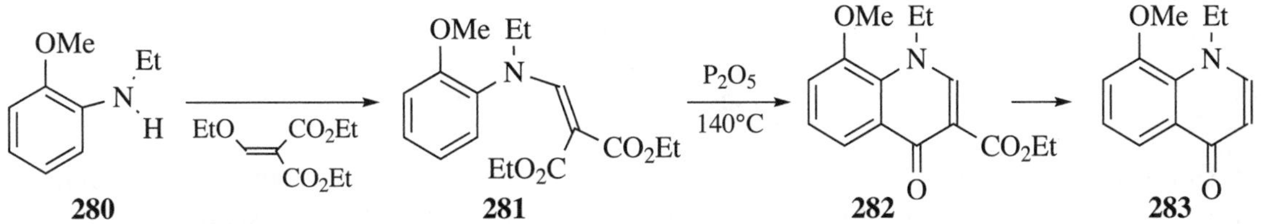

The Gould-Jacobs disconnection is

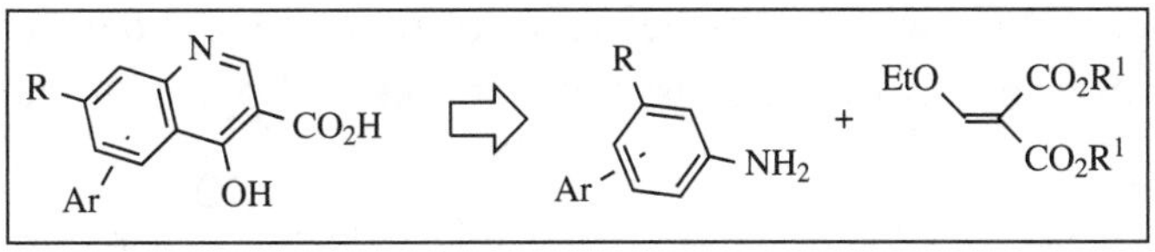

If a β-diketone is used rather than a β-keto ester, the result is a 4-alkylquinoline in what is known as the **Combes quinoline synthesis**.[196] Reaction of aniline with acetyl acetonate (2,5-pentanedione), for example, generated enamine **284** (the tautomer of imine **285**). Enolization to **286** in the presence of HF was followed by cyclization to give quinoline **287** in 96% yield.[197,156a]

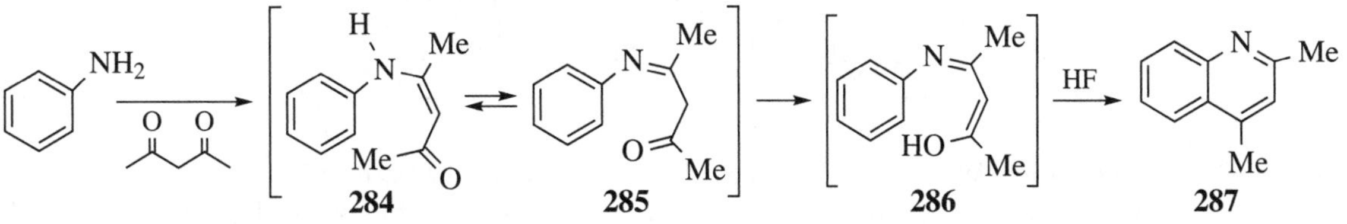

---

196. (a) Combes, A. *Bull. Chim. Soc. Fr.* *1888*, *49*, 89; (b) Reference 190c, see p 156; (c) The Merck Index, 14th ed., Merck & Co., Inc., Whitehouse Station, New Jersey, *2006*, p ONR-19; (d) Mundy, B.P.; Ellerd, M.G.; Favaloro Jr., F.G. *Name Reactions and Reagents in Organic Synthesis, 2nd ed.*, Wiley-Interscience, New Jersey, *2005*, pp. 164-165.

197. Born, J.L. *J. Org. Chem.* *1972*, *37*, 3952.

The **Doebner-Miller reaction**[198] is the reaction of a primary aromatic amine (such as aniline) with a carbonyl compound such as acetaldehyde in the presence of acid to give a 2-alkylquinoline. Reaction of aniline with acetaldehyde gave **288**, but a subsequent aldol condensation with the imine derived from acetaldehyde (sec. 9.4.F.i) led to **289**.[199] Friedel-Crafts cyclization was followed by loss of aniline to give **290**, and aromatization gave 2-methylquinoline (**291**).

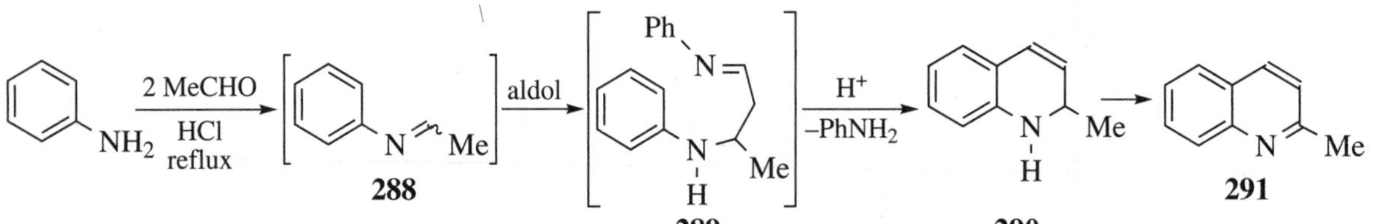

The Combes and Doebner-Miller disconnection is

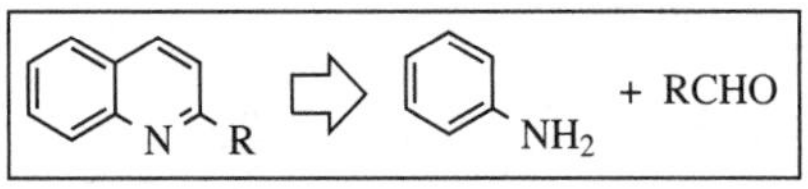

The **Doebner reaction**[200,196b] condensed aromatic amines and aldehydes with pyruvic acid derivatives. Reaction of *p*-toluidine with pyruvic acid (**292**) gave the γ-aminopyruvic acid (**293**) *in situ*, and subsequent cyclization and air oxidation under Friedel-Crafts conditions generated the 4-carboalkoxyquinoline (**294**).[201] Presumably, the reaction proceeds via initial imine formation (via reaction with benzaldehyde) followed by condensation with pyruvate in its enol form.[190cd] Similar cyclization occurs with imines and pyruvate derivatives.[202]

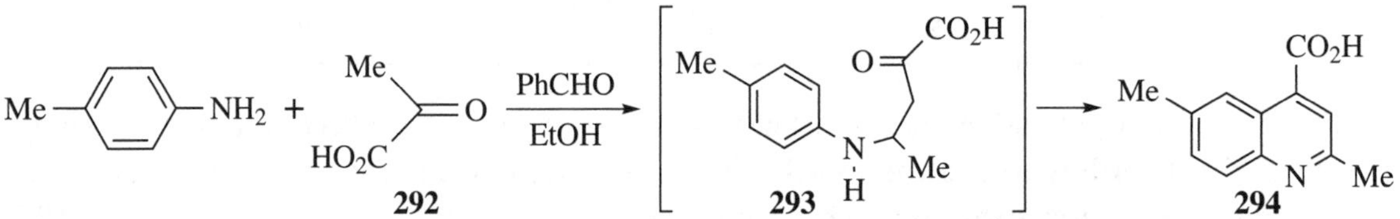

The Doebner disconnection is

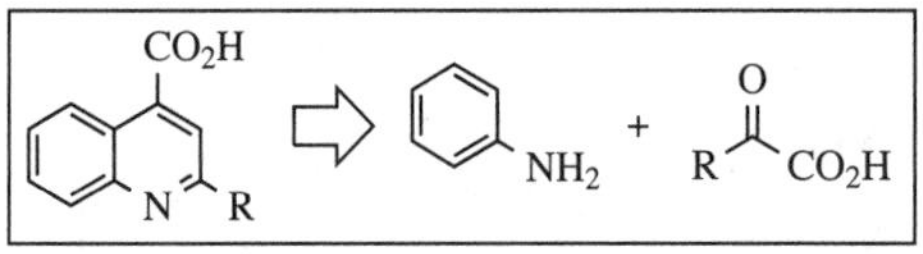

198. (a) Doebner, O.; Miller, W. *Berichte* **1883**, *16*, 2464; (b) Reference 190c, see p 154.
199. (a) Ogata, Y.; Kawasaki, A.; Suyama, S. *J. Chem. Soc. B* **1969,** 805; (b) Forrest, T.P.; Dauphinee, G.A.; Miles, W.F. *Can. J. Chem.* **1969**, *47*, 2121.
200. (a) Döbner, O. *Annalen* **1887**, *242*, 265; (b) *Idem, Berichte* **1887**, *20*, 277; (c) *Idem Ibid.* **1894,** *27*, 352, 2020; (c) The Merck Index, 14th ed., Merck & Co., Inc., Whitehouse Station, New Jersey, **2006**, p ONR-25; (d) Mundy, B.P.; Ellerd, M.G.; Favaloro Jr., F.G. *Name Reactions and Reagents in Organic Synthesis, 2nd ed.*, Wiley-Interscience, New Jersey, **2005**, pp. 212-213.
201. Döbner, O.; Gieseke, M. *Annalen* **1887**, *242*, 290.
202. Cuisa, R.; Musajo, L. *Gazz. Chim. Ital* **1929**, *59*, 796.

The **Skraup reaction**[203] presents another variation in which an aniline derivative reacts with glycerol in the presence of acid, to generate a quinoline derivative. Rapoport reacted 3-aminopyridine (**295**) with glycerol to produce **296** (1,5-naphthyridine) in 31% yield.[204] It is important to note that none of the isomeric 1,7-naphthyridine (**297**) was produced in this reaction. Attack at the carbon adjacent to both nitrogens seems to be the preferred regioselectivity with aminopyridines. The reaction is regioselective[205] rather than regiospecific, however, since Rapoport reported about 4% of 3-methyl-1,5-naphthyridine and 3-ethyl-1,5-naphthyridine. This is obviously related to the **Bally-Scholl synthesis** introduced in sec. 12.4.E.

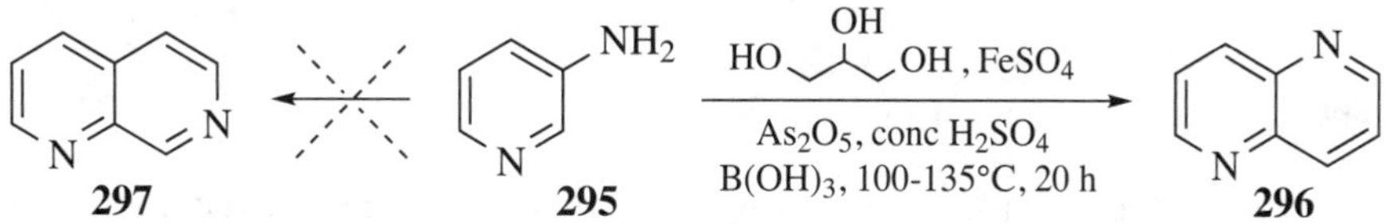

The Skraup disconnection is:

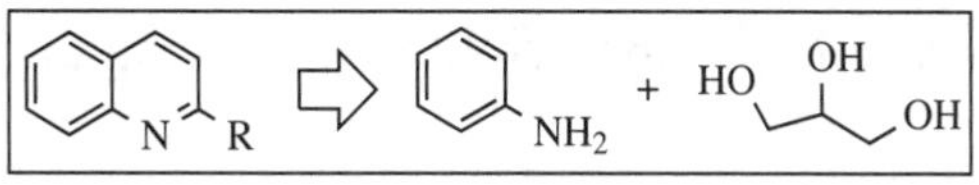

12.5.B. Synthesis of Isoquinoline Derivatives

Isoquinolines are as important as quinolines in the chemistry of natural products, and there are also many syntheses of these compounds. One of the most important Friedel-Crafts routes is the **Bischler-Napieralski reaction**.[206] This process is the reaction of an *N*-acyl amine such as **298** derived from phenethylamine, with a reagent such as PPA or P_2O_5 to give a dehydroisoquinoline derivative (**299**). For example, in a synthesis of stepharine[207] Honda and Shigehisa treated acetamide **300** with $POCl_3$, and obtained **301**, which gave **302** in >60% overall yield after treatment with trifluoroacetic acid and triethylamine.

203. (a) Skraup, Z.H. *Berichte* **1880**, *13*, 2086; (b) Manske, R.H.; Kulka, M. *Org. React.* **1953**, *7*, 59; (c) The Merck Index, 14th ed., Merck & Co., Inc., Whitehouse Station, New Jersey, **2006**, p ONR-87; (d) Mundy, B.P.; Ellerd, M.G.; Favaloro Jr., F.G. *Name Reactions and Reagents in Organic Synthesis, 2nd ed.*, Wiley-Interscience, New Jersey, **2005**, pp. 602-603.
204. Rapoport, H.; Batcho, A.D. *J. Org. Chem.* **1963**, *28*, 1753.
205. (a) Perche, J.-C.; Saint-Ruf, G.; Buu-Hoï, N.P. *J. Chem. Soc. Perkin Trans. 1* **1972**, 260; (b) Buu-Hoï, N.P.; Jacquignon, P.; Thang, D.C.; Bartnik, T. *Ibid.* **1972**, 263.
206. (a) Bischler, A.; Napieralski, B. *Berichte* **1893**, *26*, 1903; (b) Bischler, A. *Ibid.* **1893**, *26*, 189, 1891; (c) Whaley, W.M.; Govindachari, T.R. *Org. React.* **1951**, *6*, 74; (d) Fodor, G.; Gal, J.; Phillips, B.A. *Angew. Chem. Int. Ed.* **1972**, *11*, 919; (e) The Merck Index, 14th ed., Merck & Co., Inc., Whitehouse Station, New Jersey, **2006**, p ONR-11; (f) Mundy, B.P.; Ellerd, M.G.; Favaloro Jr., F.G. *Name Reactions and Reagents in Organic Synthesis, 2nd ed.*, Wiley-Interscience, New Jersey, **2005**, pp. 96-97.
207. Honda, T.; Shigehisa, H. *Org. Lett.* **2006**, *8*, 657.

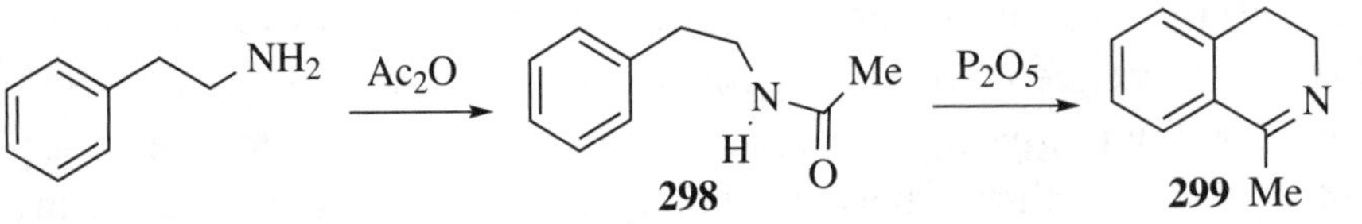

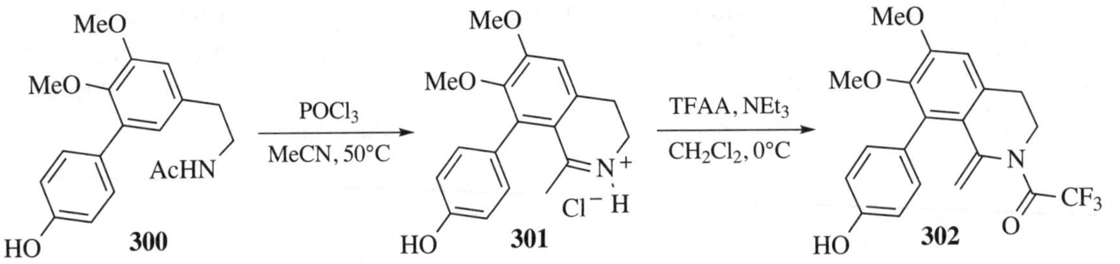

A useful variation first converts the amino group to a formamide derivative, usually by treatment of a ketone (such as **303**) with formamide and formic acid. Treatment of **303** with these reagents led to formamide **304**. When treated with POCl$_3$, cyclization occurred via the N-acyl group to give the isoquinoline, **305**, in 66% yield. Zee-Cheng and Cheng used this sequence in a synthesis of nitidine.[208]

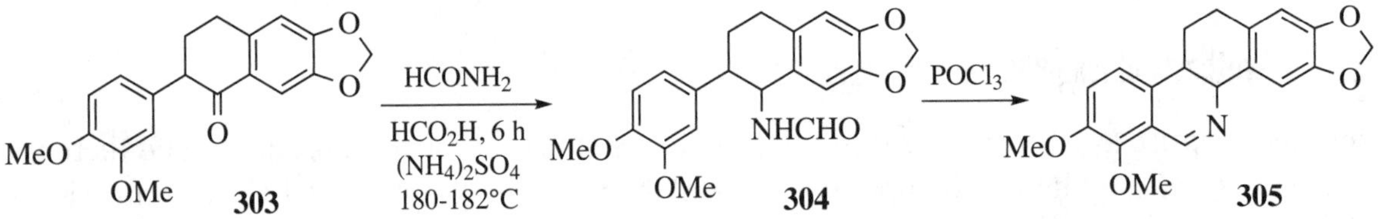

Bischler-Napieralski disconnections are

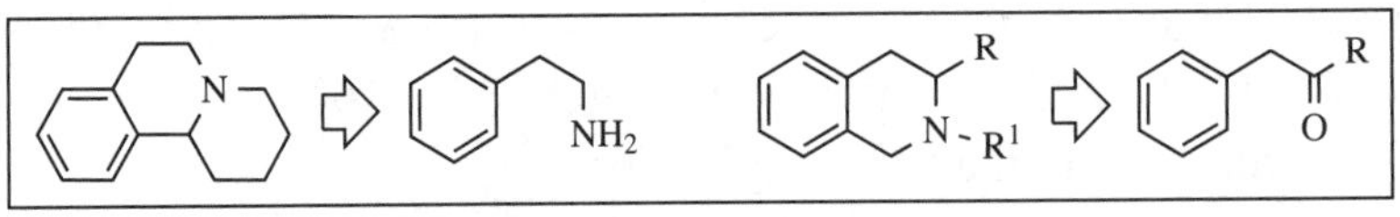

The **Pictet-Spengler isoquinoline synthesis** is another classical approach.[209,206c] This variation generates an iminium salt from an amine and an aldehyde (a Schiff base), which cyclizes with an aromatic ring to complete the reaction.[210] A synthetic example is taken from a synthesis of (−)-fuchsiaefoline by Cook and co-workers.[211] The reaction of amino ester **306** with the indicated aldehyde gave a near quantitative yield of **307**, but as a mixture of cis/trans isomers rather than the single epimer shown. Treatment of this mixture with about five equivalents of

208. ZeeCheng, K.Y.; Cheng, C.C. *J. Heterocyclic Chem.* **1973**, *10*, 85.
209. (a) Pictet, A.; Spengler, T. *Berichte,* **1911**, *44*, 2030; (b) Cox, E.D.; Cook, J.M. *Chem. Rev.* **1995**, *95*, 1797; (c) The Merck Index, 14th ed., Merck & Co., Inc., Whitehouse Station, New Jersey, **2006**, p ONR-73; (d) Mundy, B.P.; Ellerd, M.G.; Favaloro Jr., F.G. *Name Reactions and Reagents in Organic Synthesis, 2nd ed.,* Wiley-Interscience, New Jersey, **2005**, pp. 508-509.
210. Ong, H.H.; May, E.L. *J. Heterocyclic Chem.* **1971**, *8*, 1007.
211. Zhou, H.; Liao, X.; Cook, J.M. *Org. Lett.* **2004**, *6*, 249.

trifluoroacetic acid in dichloromethane and stirring for seven days gave a 92% yield of **307**.

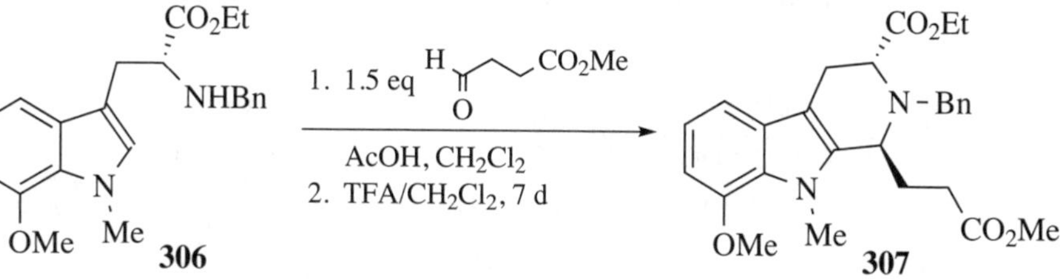

Modifying groups on the aromatic ring can control the regioselectivity of the reaction. A typical Pictet-Spengler reaction with formaldehyde converted **308** to **309** in 22% yield. The silyl directed reaction, however, converted **310** to **311** in 72% yield.[212] Jacobsen and Taylor reported a Pictet-Spengler reaction, catalyzed by a chiral thiourea derivative that gave the product in good yield, with excellent enantioselectivity.[213] A microwave-accelerated Pictet-Spengler reaction has also been used to prepare indole alkaloids.[214]

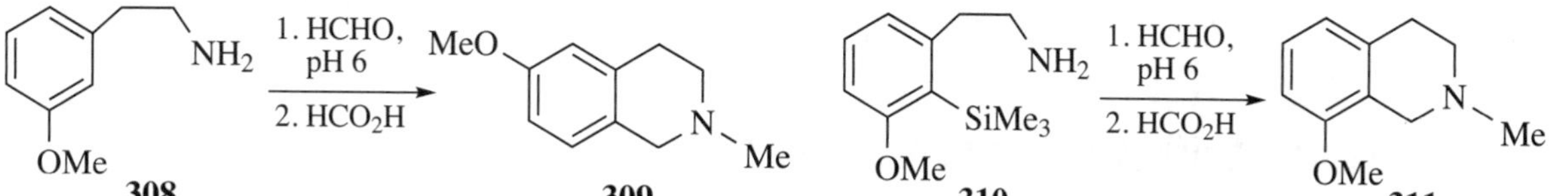

The Pictet-Spengler disconnection is

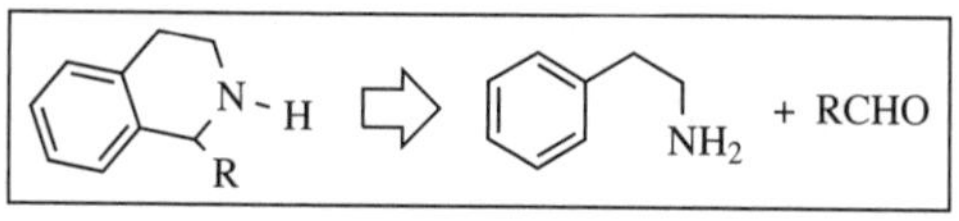

Another method for the synthesis of isoquinoline derivatives is the **Pomeranz-Fritsch reaction**.[215] Simple aniline derivatives can be used, but the reaction is most useful for acid sensitive substrates such as thiophene or pyrrole derivatives. In those cases, it is a good alternative to the Pictet-Spengler reaction. An example of an acid sensitive substrate is **312**, which reacted with glycinal diethyl acetal (**313**) to generate Schiff base **314** in 91% yield. Subsequent treatment with PPA/POCl$_3$ gave a 21% yield of pyrrolo-[1,2a]-pyrazine (**315**).[216] When the reactive partners were reversed to α-methylbenzylamine (**316**) and a glyoxal semiacetal (**317**), a different Schiff base was formed (**318**). Treatment with acid led directly to

212. Miller, R.B.; Tsang, T. *Tetrahedron Lett.* **1988**, *29*, 6715.
213. Taylor, M.S.; Jacobsen, E.N. *J. Am. Chem. Soc.* **2004**, *126*, 10558.
214. Kuo, F.-M.; Tseng, M.-C.; Yen, Y.-H.; Chu, Y.-H. *Tetrahedron* **2004**, *60*, 12075.
215. (a) Pomeranz, C. *Monatsh* **1893**, *14*, 116; (b) Fritsch, P. *Ber.* **1893**, *26*, 419; (c) Gensler, W.J. *Org. React.* **1951**, *6*, 191; (d) Popp, F.D.; McEwen, W.E. *Chem. Rev.* **1958**, *58*, 321 (see p 328); (e) The Merck Index, 14th ed., Merck & Co., Inc., Whitehouse Station, New Jersey, **2006**, p ONR-75; (f) Mundy, B.P.; Ellerd, M.G.; Favaloro Jr., F.G. *Name Reactions and Reagents in Organic Synthesis, 2nd ed.*, Wiley-Interscience, New Jersey, **2005**, pp. 526-527.
216. Herz, W.; Tocker, S. *J. Am. Chem. Soc.* **1955**, *77*, 6355.

isoquinoline **319**.[217] The yield of **319** was only 15% by the Pomeranz-Fritsch method, but 40% with this variation known as the **Schlittler-Müller modification**.[217]

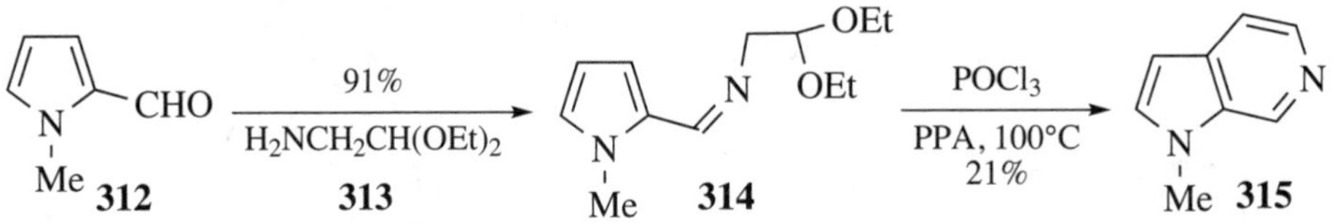

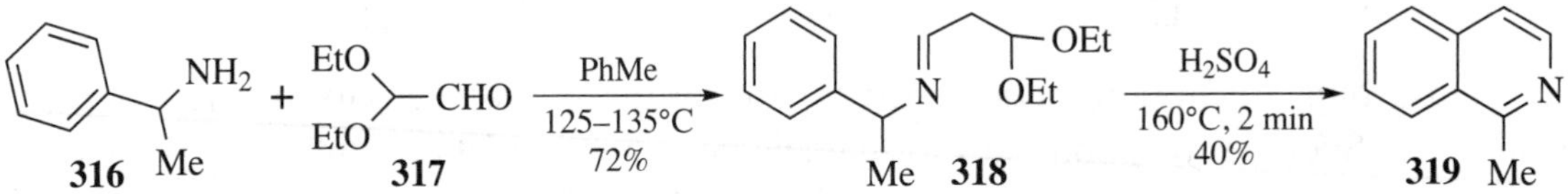

The Pomeranz-Fritsch and Schlittler-Müller disconnections are

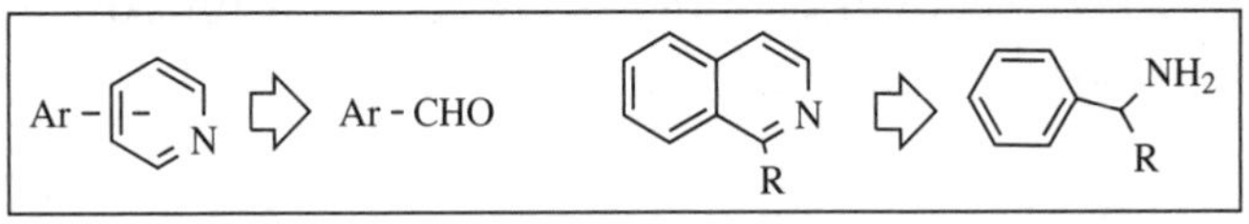

12.5.C. Synthesis of Acridines, Carbazoles, and Phenanthridines

Acridines are useful and medicinally important compounds. They can be produced in a straightforward manner by a Friedel-Crafts reaction the **Bernthsen acridine synthesis**,[218] which couples diaryl amines (such as diphenylamine) with a carboxylic acid (such as benzoic acid) in the presence of a Lewis acid (zinc chloride is a typical reagent). Heating diphenylamine to 260°C (10 h) with benzoic acid and zinc chloride gave a 48% yield of acridine **320**.[218b]

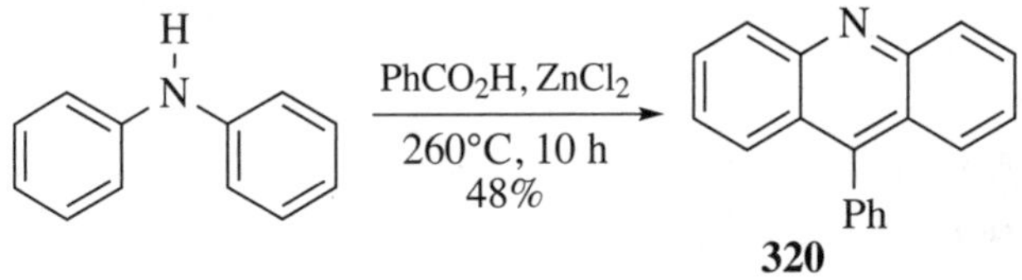

The carbazole nucleus (carbazole itself is **323**) is also an important structural feature of many alkaloids. The **Borsche-Drechel cyclization**[219] is a Friedel-Crafts route to these compounds, related to the Fischer indole synthesis discussed in Section 12.5.E. A typical reaction used phenylhydrazone derivatives of ketones such as cyclohexanone phenylhyrazone (**321**), which was cyclized with sulfuric acid to give **322** in good yield. Aromatization with lead oxide (Pb$_3$O$_4$) gave **323** in low yield. Glacial acetic acid gave cleaner products in the dehydrogenation step,[220]

217. Schlittler, E.; Müller, J. *Helv. Chim. Acta* **1948**, *31*, 914, 1119.
218. (a) Bernthsen, A. *Annalen* **1878**, *192*, 1; (b) *Idem Ibid.* **1884**, *224*, 1; (c) The Merck Index, 14th ed., Merck & Co., Inc., Whitehouse Station, New Jersey, **2006**, p ONR-10.
219. (a) Dreschsel, E. *J. Prakt. Chem.* **1888**, *38*, 69; (b) Borsche, W.; Feise, M. *Berichte* **1907**, *40*, 378; (c) Campbell, N.; Barclay, B.M. *Chem. Rev.* **1947**, *40*, 359 (see p 361); (d) The Merck Index, 14th ed., Merck & Co., Inc., Whitehouse Station, New Jersey, **2006**, p ONR-13; (e) Mundy, B.P.; Ellerd, M.G.; Favaloro Jr., F.G. *Name Reactions and Reagents in Organic Synthesis, 2nd ed.*, Wiley-Interscience, New Jersey, **2005**, pp. 108-109.
220. Perkin Jr., W.H.; Plant, S.G.P. *J. Chem. Soc.* **1921**, *119*, 1825.

but tetrachlorobenzoquinone (also known as chloranil) led to the best yield of **323**.[221]

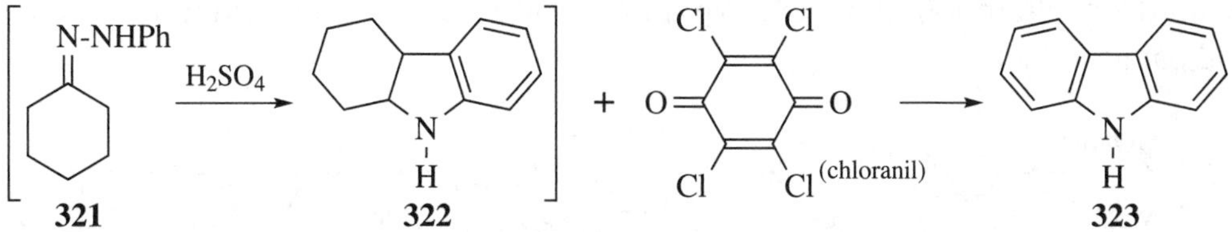

Phenanthridines can be prepared by modification of the Friedel-Crafts reaction. The **Pictet-Hubert reaction**[222] reacted an acyl *o*-aminobiphenyl (**324**) with zinc chloride at 300°C,[223] or more commonly with POCl$_3$ in refluxing nitrobenzene,[224] to give the Friedel-Crafts cyclization

product, phenanthridine **325**. Zinc chloride required long heating and wasteful purification, and gave poor results with reactive substrates. The use of both POCl$_3$ and a high-boiling solvent improved the yields of phenanthridine products.

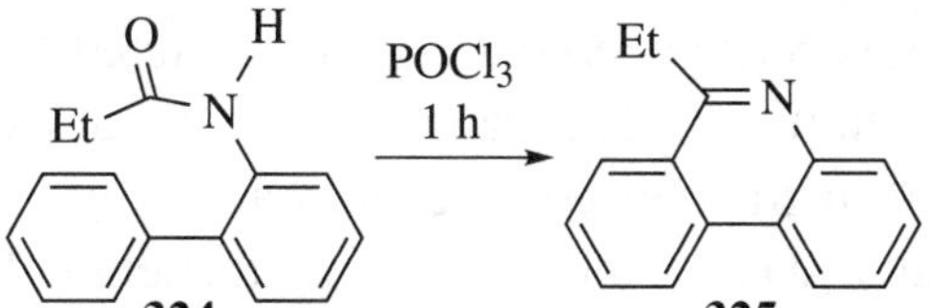

The disconnections discussed in this section are

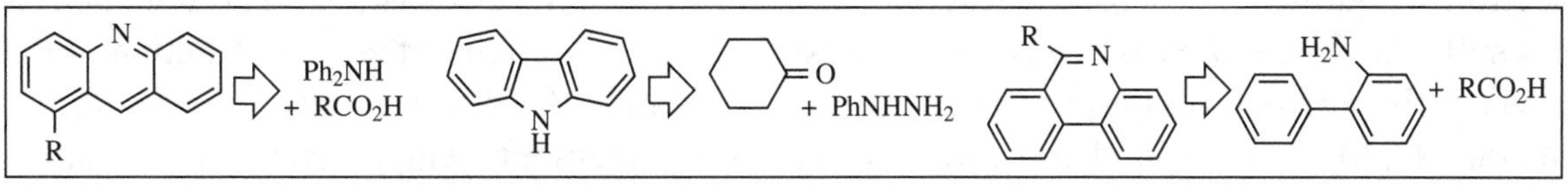

12.5.D. Synthesis of Oxygenated Aromatic Derivatives

The **Bradsher reaction** (see **260** in sec. 12.5.A, where X = O) can generate aromatic derivatives containing an oxygen atom in a polycyclic framework. There are also many phenol derivatives, where the oxygen atom is appended to a simple aromatic ring or to polycyclic aromatic compounds. An interesting route to such compounds exploits the propensity of aryls to rearrange under Friedel-Crafts conditions. When a phenolic ester (such as phenyl acetate, **326**) is heated with a Lewis acid such as AlCl$_3$, a rearrangement occurs to generate an ortho ketone (**330**) and a para ketone (**329**) in what is known as the **Fries rearrangement**.[225] The

221. Barclay, B.M.; Campbell, N. *J. Chem. Soc.* **1945**, 530.
222. (a) Pictet, A.; Hubert, A. *Berichte* **1896**, *29*, 1182; (b) Morgan, C.T.; Walls, L.P. *J. Chem. Soc.* **1931**, 2447; (c) *Idem Ibid.* **1932**, 2225; (d) Eisch, J.; Gilman, H. *Chem. Rev.* **1957**, *57*, 525; (e) The Merck Index, 14th ed., Merck & Co., Inc., Whitehouse Station, New Jersey, **2006**, p ONR-73; (f) Mundy, B.P.; Ellerd, M.G.; Favaloro Jr., F.G. *Name Reactions and Reagents in Organic Synthesis, 2nd ed.*, Wiley-Interscience, New Jersey, **2005**, pp. 506-507.
223. Theobald, R.S.; Schofield, K. *Chem. Rev.* **1950**, *46*, 171 (see p 175).
224. Morgan, G.T.; Walls, L.P. *J. Chem. Soc.* **1931**, 2447.
225. (a) Fries, K.; Finck, G. *Berichte* **1908**, *41*, 4271; (b) Fries, K.; Pfaffendorf, W. *Ibid.* **1910**, *43*, 212; (c) Blatt, A.H. *Org. React.* **1942**, *1*, 342; (d) The Merck Index, 14th ed., Merck & Co., Inc., Whitehouse Station, New Jersey, **2006**, p ONR-35; (e) Mundy, B.P.; Ellerd, M.G.; Favaloro Jr., F.G. *Name Reactions and Reagents in Organic Synthesis, 2nd ed.*, Wiley-Interscience, New Jersey, **2005**, pp. 260-261.

reaction generates an acylium ion (**327**), which fragments to an aluminum alkoxide (**328**) ion-paired with the acylium ion MeC≡O⁺. Migration and Friedel-Crafts acylation generates both ortho and para products.

326   **327**   **328**   **329**   **330**

This reaction is probably used most often to prepare acetate derivatives,[226] as in the conversion of **331** to **332** in quantitative yield in a synthesis of heliannuol D by Macías and co-workers.[227]

331   **332**

The Fries rearrangement is not restricted to acetate unit, however, and other acyl units have been transferred to a aromatic ring.[228] In many cases, there is a regiochemical preference for the para product due to steric effects that hinder reaction at an available ortho position.[229] When the aromatic ring contains electron-withdrawing groups, the reaction does not work well. A thia-Fries rearrangement of aryl sulfonate esters such as phenyl *p*-toluenesulfonate has been reported.[230] Irradiation of this tosylate with microwaves on silica gel, for 10 min, gave an 87% yield of *o*- and *p*-toluenesulfonyl phenol (80:20). There is also a **photo Fries rearrangement**, which proceeds via radical intermediates formed by the initial photolysis (sec. 11.10.B, for a brief overview of photochemical theory). The ketone is formed from the bis(acyl) derivative under photolytic conditions.[231]

The Fries rearrangement disconnection is

Coumarin derivatives (see **336**) can be formed via Friedel-Crafts techniques. The **Pechmann condensation**[232] is illustrated by esterification of phenol with a β-keto-ester (a transesterification reaction), which gave **333**. Subsequent cyclization with AlCl₃ generated coumarin **334**.

226. For an example using BF₃ as a catalyst with hydroquinone diester, see Boyer, J.L.; Krum, J.E.; Myers, M.C.; Fazal, A.N.; Wigal, C.T. *J. Org. Chem.* **2000**, *65*, 4712.
227. Macías, F.A.; Chinchilla, D.; Molinillo, J.M.G.; Marín, D.; Varela, R.M.; Torres, A. *Tetrahedron* **2003**, *59*, 1679.
228. For an example see Cairns, N.; Harwood, L.M.; Astles, D.P. *Tetrahedron Lett.* **1988**, *29*, 1311.
229. Jefferson, A.; Wangchareontrakul, S. *Aust. J. Chem.* **1985**, *38*, 605.
230. Moghaddam, F.M.; Dakamin, M.G. *Tetrahedron Lett.* **2000**, *41*, 3479.
231. Anderson, J.C.; Reese, C.B. *Proc. Chem. Soc.* **1960**, 217.
232. (a) von Pechmann, H.; Duisberg, C. *Berichte* **1883**, *16*, 2119; (b) Sethna, S.; Shah, N.M. *Chem. Rev.* **1945**, *36*, 1 (see p 10); (c) Sethna, S.; Phadke, R. *Org. React.* **1953**, *7*, 1; (d) The Merck Index, 14th ed., Merck & Co., Inc., Whitehouse Station, New Jersey, **2006**, p ONR-70; (e) Mundy, B.P.; Ellerd, M.G.; Favaloro Jr., F.G. *Name Reactions and Reagents in Organic Synthesis, 2nd ed.*, Wiley-Interscience, New Jersey, **2005**, pp. 490-491.

Isolation of esters such as **333** is not always necessary and protonic acids can be used. In a synthesis of (+)-aflatoxin B$_1$ by Trost and Toste,[233] catechol **335** was condensed with diethyl 3-oxo-1,6-hexanedioate in the presence of ethanolic HCl to give a 47% yield of coumarin **336**. The Pechmann condensation is facilitated by the presence of hydroxyl (OH), dimethylamino (NMe$_2$) and alkyl groups (R) meta to the hydroxyl of the phenol.[234,235] Pechmann condensation in an ionic liquid (1-butyl-3-methylimidazolium chloroaluminate, [bmim]Cl•2AlCl$_3$]) using ethyl acetate has also been reported.[236]

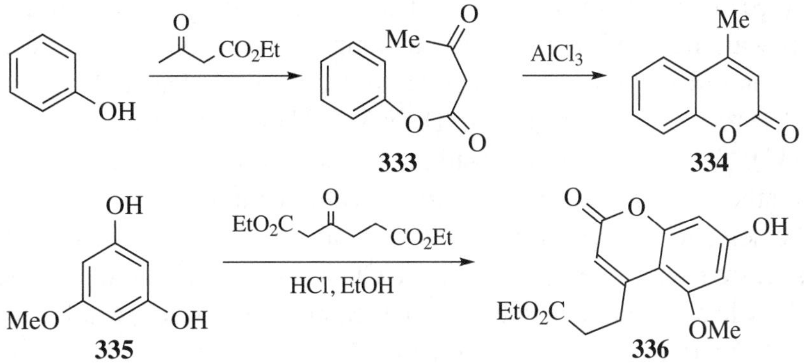

The Pechmann disconnection is:

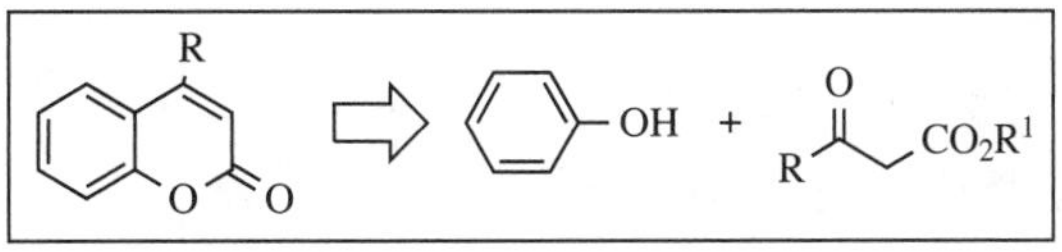

12.5.E. Other Aromatic Substitution Reactions. The Fischer Indole Synthesis[237,240]

This section has been devoted to variations of electrophilic aromatic substitution to make carbon-carbon bonds. There are other alkaloid-forming reactions that do not readily fit into this category, but since they involve acid-catalyzed processes are placed here. The indole nucleus (see **340**) is a component of many natural products and medicinal compounds.[238] In 1883, Fischer reported a straightforward synthesis of indoles that is today considered a powerful reaction (sec. 10.4.B). Although many total syntheses of indole alkaloids begin with indole,

233. Trost, B.M.; Toste, F.D. *J. Am. Chem. Soc.* **2003**, *125*, 3090.
234. Shah, M.M.; Shah, R.C. *Berichte* **1938**, *71*, 2075.
235. Miyano, M.; Dorn, C.R. *J. Org. Chem.* **1972**, *37*, 259.
236. In bmim chloroaluminate, 1-butyl-3-methylimidazolium chloroaluminate, [bmim]Cl•2AlCl$_3$: Potdar, M.K.; Mohile, S.S.; Salunkhe, M.M. *Tetrahedron Lett.* **2001**, *42*, 9285.
237. (a) Fischer, E.; Jourdan, F. *Berichte* **1883**, *16*, 2241; (b) Fischer, E.; Hess, O. *Ibid.* **1884**, *17*, 559; (c) The Merck Index, 14th ed., Merck & Co., Inc., Whitehouse Station, New Jersey, **2006**, p ONR-32; (d) Mundy, B.P.; Ellerd, M.G.; Favaloro Jr., F.G. *Name Reactions and Reagents in Organic Synthesis, 2nd ed.*, Wiley-Interscience, New Jersey, **2005**, pp. 244-245.
238. For a review of methods for indole ring synthesis, see Gribble, G.W. *J. Chem. Soc. Perkin Trans. 1* **2000**, 1045.

some syntheses prepare the indole nucleus as part of the synthesis and the Fischer indole synthesis remains an important method for the preparation of indoles.[239] A synthesis begins with conversion of an aldehyde or ketone to a phenylhydrazone ($ArNHN=CR_2$), by treatment with phenylhydrazine. Subsequent treatment with a Lewis acid (zinc chloride is a good choice) or a protonic acid induces a reaction cascade that results in indole **340**. The mechanism of this transformation involves formation of a hydrazone that reacts via the *N*-amino enamine form of the hydrazone (see **337**), in what is probably an electrocyclic process analogous to a Claisen rearrangement (sec. 11.12.D) to form **338**. The imine nitrogen of **338** attacks an activated C=N (NH reacts with the Lewis acid to form a complex) with concomitant rearomatization of the benzene ring and formation of the five-membered ring that has a NHX group (a protonic acid converts =NH to $=NH_2$ which gives $-NH_2$ in **339** whereas reaction with $ZnCl_2$ would give $-NHZnCl_2$ in **339**).[240]   Loss of an ammonium species by what is effectively an E1 process completes the synthesis of the indole ring (**340**). An application of the Fischer indole reaction is taken from Roberson and Woerpel's synthesis of peduncularine.[241] In this synthesis, protected acetal **341** was treated with phenylhydrazine•HCl in 4% sulfuric acid. Deprotection of the acetal to the aldehyde allowed the Fischer indole reaction to proceed, catalyzed by the sulfuric acid, to give a 75% yield of the indole product (**342**).

The cyclization proceeds with reasonable regioselectivity. The ketone moiety in **343** has two α-carbons, and two different enamine structures could be formed as well as two bis(imine) intermediates.[242] The final indole products were **344** and **345**. As the concentration of acid catalyst was increased, the yield decreases but the regioselectivity is about the same (2:1 **344**/**345**).[242,243] The stereochemical course of the reaction also depends on the stereochemistry of the hydrazone substrate. In another example, hydrazone **346** (with a cis ring juncture) was treated with HCl•ether,

239. For a list of other named reactions that produce indoles, see Li, J.-J.; Corey, E.J. *Name Reactions in Heterocyclic Chemistry*, **2005**, Wiley Interscience, New Jersey, pp. 99-158.
240. Robinson, B. *Chem. Rev.* **1969**, *69*, 227 and references cited therein.
241. Roberson, C.W.; Woerpel, K.A. *J. Am. Chem. Soc.* **2002**, *124*, 11342.
242. Bryan Reed, G.W.; Cheng, P.T.W.; McLean, S. *Can. J. Chem.* **1982**, *60*, 419.
243. Rashidyan, L.G.; Asratyan, S.N.; Karagezyan, K.S.; Mkrtchyan, A.R.; Sedrakyan, R.O.; Tatevosyan, G.T. *Arm. Khim. Zh.* **1968**, *21*, 793 [*Chem. Abstr. 71:* 21972z, *1969*].

in dichloromethane at room temperature to give a 61% yield of **347**.[244] Reaction of **346** with 85% sulfuric acid, however, gave a 42% yield of the isomeric **348** and the course of the reaction depended on the acid strength of the medium. When the trans derivative (**349**) was treated with ethereal HCl under the same conditions, 38% of a single product (**350**) was isolated. Treatment with 85% sulfuric acid gave no cyclization, and only unreacted starting material. Treatment with more concentrated acid led to decomposition of the starting material but no cyclization.

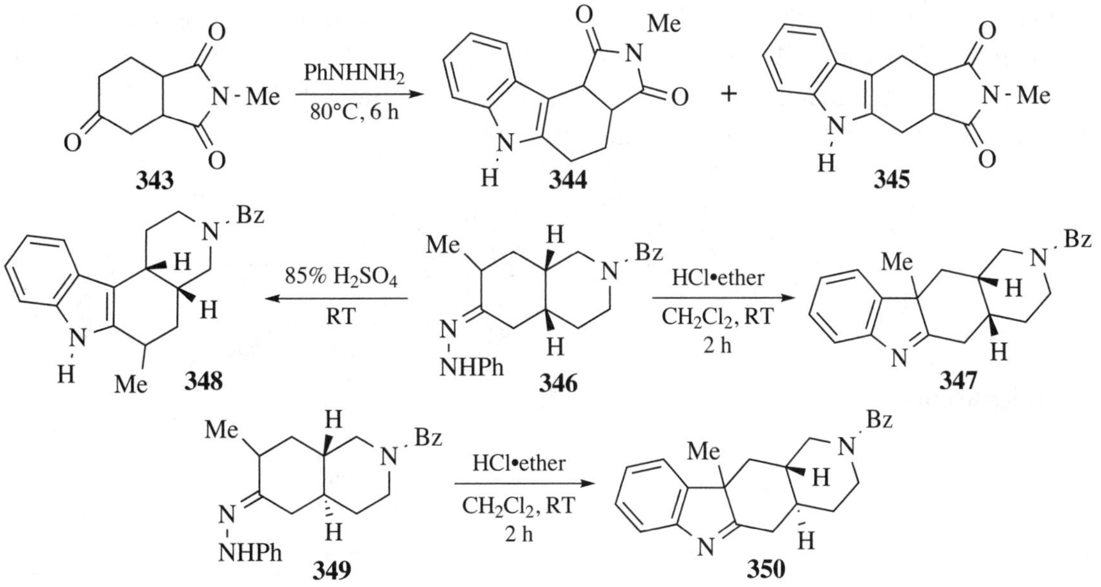

The initially formed imine product may be subject to rearrangement under the strongly acidic reaction conditions. Treatment of the phenylhydrazone of cyclopentane carboxaldehyde (**351**) with acetic acid, for example, led to a 90% yield of **353** via the acid catalyzed rearrangement of imine **352**.[245] Cyclopentane carboxaldehyde derivatives gave only fused ring indoles, but larger ring carboxaldehydes led to spiro derivatives or mixtures of fused-ring and spirocyclic products.[245]

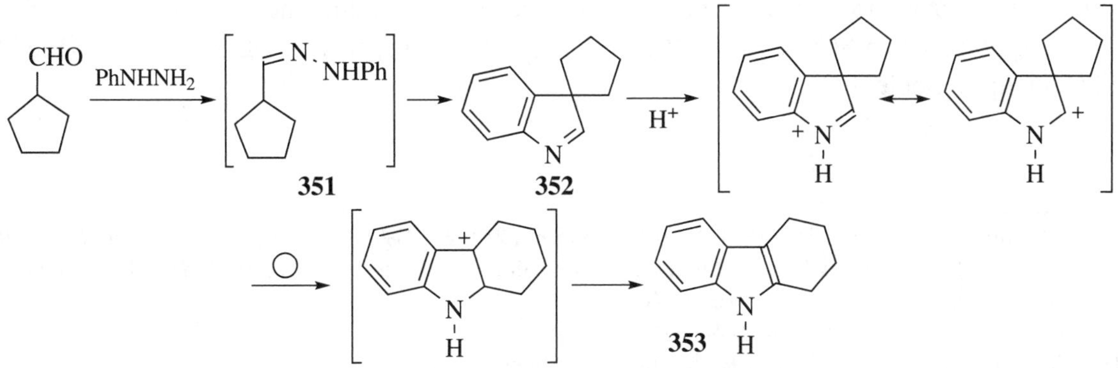

244. Freter, K.; Fuchs, V.; Pitner, T.P. *J. Org. Chem.* **1983**, *48*, 4593.
245. Rodríquez, J.G.; Benito, Y.; Temprano, E. *J. Heterocyclic Chem.* **1985**, *22*, 1207.

The Fisher indole disconnection is

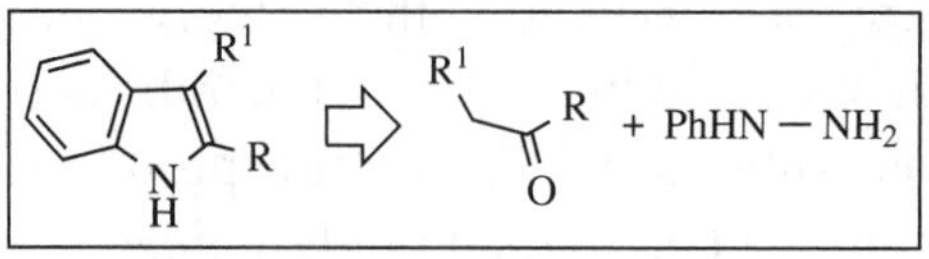

A related route to indolones is the **Stollé synthesis**,[246] where a secondary aryl amine such as diphenylamine reacats with a α-chloro acid chloride or a α-bromo acid bromide, to give a α-bromoamide (such as **355**) or a α-chloroamide. Oxalyl chloride can be used, but a more typical example is the reaction of 2-bromopropanoyl bromide with **354** to give **355**. Friedel-Crafts cyclization with $AlCl_3$ led to the indolone **356**.[247]

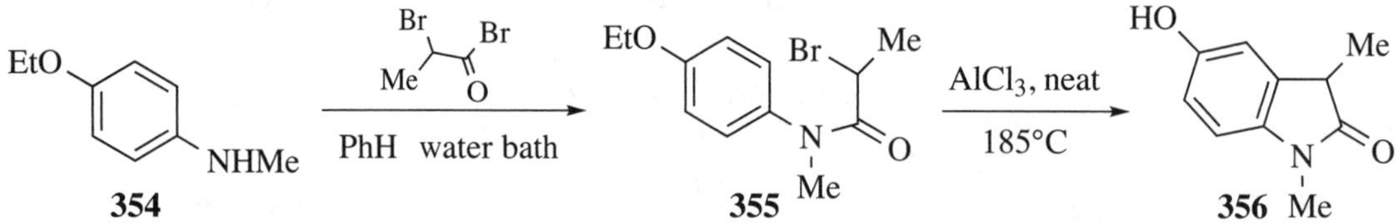

The Stollé disconnection is

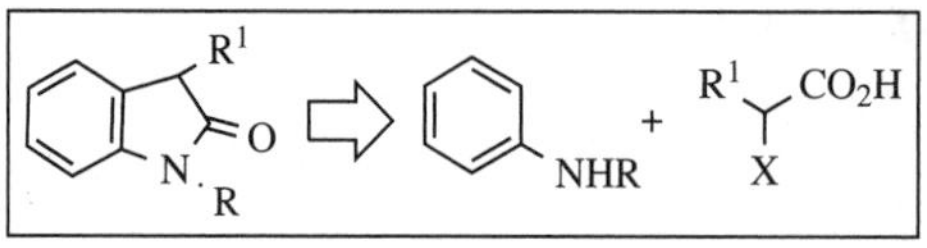

There are, of course, other methods to prepare the indole nucleus.[248,239] A few methods are presented here to illustrate some of the new methodology. In one approach, Barluenga and co-workers treated **357** with *tert*-butyllithium to give **358**.[249] Cyclization and elimination gave lithiated indole **359**, which reacted with electrophiles such as 4-chlorobenzaldehyde to give **360** (71% yield). Another approach used α-diazophosphonates. O'Shea and Coleman reported an organolithium addition strategy.[250] Nakamura and Ukita[251] treated 2-aminobenzophenone with triethyl diazophosphonoacetate, in the presence of the rhodium acetate catalyst (see Sect. 13.9), and obtained an 86% yield of **361**. Subsequent treatment with DBU (sec. 2.9.A) led to a 99% yield of indole **362**.

246.   (a) Stollé, R. *Berichte,* **1913**, *46*, 3915; (b) *Idem Ibid.* **1914**, *47*, 2120; (c) *Idem*, *J. Prakt. Chem.* **1923**, *105*, 137; (d) Stollé, R.; Bergdoll, R.; Luther, M.; Auerhahn, A.; Wacker, W. *Ibid.* **1930**, *128*, 1; (e) Sumpter, W.C. *Chem. Rev.* **1945**, *37*, 443 (see p 446).
247.   Julian, P.L.; Pikl, J. *J. Am. Chem. Soc.* **1935**, *57*, 563.
248.   For a review, see Humphrey, G. R.; Kuethe, J. T. *Chem. Rev.* **2006**, *106*, 2875.
249.   Barluenga, J.; Sanz, R.; Granados, A.; Fañanás, F.J. *J. Am. Chem. Soc.* **1998**, *120*, 4865.
250.   Coleman, C.M.; O'Shea, D.F. *J. Am. Chem. Soc.* **2003**, *125*, 4054.
251.   Nakamura, Y.; Ukita, T. *Org. Lett.* **2002**, *4*, 2317.

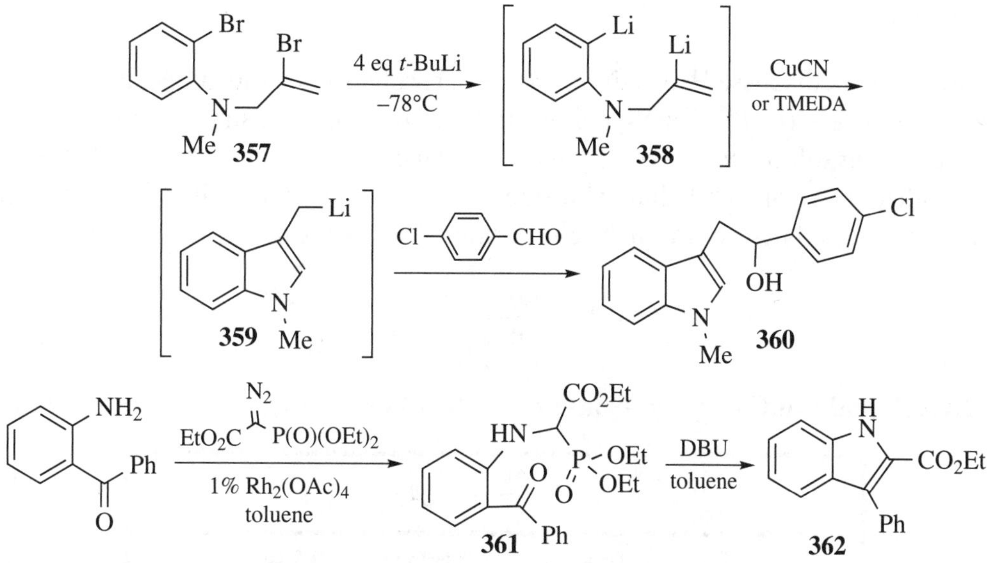

Hiroya and co-workers reported a method for the synthesis of indole-2-carboxylates, based on a palladium catalyzed coupling strategy (see sec. 12.7), in a synthesis of duocarmycin SA.[252a] Coupling methyl propiolate and **363** under what are called **Negishi coupling**[252b] conditions led to a 69% yield of indole **364**. Palladium catalyzed coupling reactions are discussed in sec. 12.7.

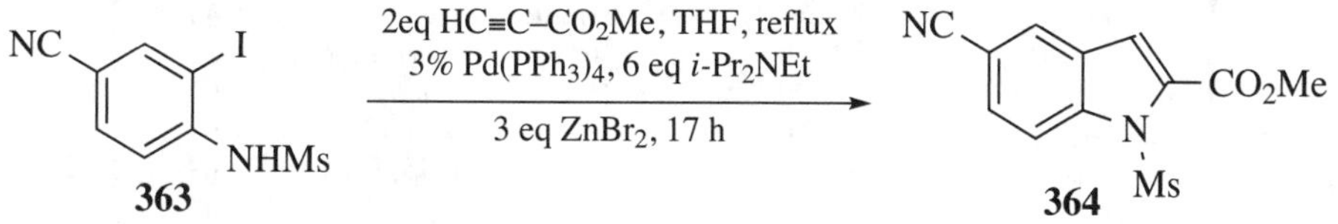

12.6. π-ALLYL PALLADIUM COMPLEXES

There are a number of organometallic compounds that give products from reaction with nucleophiles. In terms of the product, although not necessarily in terms of the formal mechanism, such organometallics effectively function as electrophilic reagents. Palladium salts react with allylic hydrocarbons and with functionalized allylic derivatives to produce π-allyl complexes. These complexes can react with nucleophilic species, including enolate anions, to produce new carbon-carbon bonds.[253] In these reactions, π-allyl palladium complexes function as C^a equivalents; thus they are included in this chapter.

252. (a) Hiroya, K.; Matsumoto, S.; Sakamoto, T. *Org. Lett.* **2004**, *6*, 2953; (b) Negishi, E.; Okukado, N.; King, A. O.; Van Horn, D. E.; Spiegel, B.I. *J. Am. Chem. Soc.* **1978**, *100*, 2254.
253. Poli, G.; Giambastiani, G.; Heumann, A. *Tetrahedron* **2000**, *56*, 5959.

12.6.A. The Wacker Process

The first process to be discussed is a traditional one that probably does not proceed via a π-complex. The **Wacker process** (or the **Wacker oxidation**) is used in industry to convert ethylene to acetaldehyde using soluble palladium catalysts.[254] An example is the reaction of ethylene with cupric chloride ($CuCl_2$) and palladium chloride ($PdCl_2$) to give acetaldehyde. Other alkenes can be oxidized to carbonyls as well, as in the conversion of an alkene to a ketone (see Table 12.11).[255]

$$H_2C = CH_2 \xrightarrow[\text{H}_2\text{O}]{\text{CuCl}_2,\ \text{PdCl}_2} \underset{Me}{\overset{O}{\parallel}} H$$

Table 12.11. Palladium Catalyzed Conversion of Alkenes to Carbonyl Compounds

R	Temp. (°C)	Time (min)	% Ketone
H	20	5	85
Me	20	5	90
n-Pr	20	10	81
C_5H_{11}	50	30	65
C_6H_{13}	50	30	42
C_8H_{17}	70	60	34

Alkene or Diene	Temp. (°C)	Time (min)	Product	% Yield
$CH_2=CCH=CH_2$	80	30	3-Buten-2-one	34
$MeCH=CHCH=CH_2$	20	15	2-Pentenal	91
Cyclopentene	30	30	Cyclopentanone	61
Cyclohexene	30	30	Cyclohexanone	65
Vinyl bromide	10	15	Acrolein	98
1-Bromo-1-pentene	50	15	2-Pentanone	61
1-Bromo-2-phenylethene	40	20	Acetophenone	60
Allyl chloride	50	30	Methyl glyoxal	65
Allyl bromide	50	30	Methyl glyoxal	80

[Reproduced with permission from Smidt, J.; Hafner, W.; Jira, R.; Sedlmeier, J.; Sieber, R.; Rüttinger, R.; Kojer, H. *Angew. Chem* **1959**, *71*, 176. Copyright **1959** VCH Weinheim.]

Both aldehydes and ketones can be formed in moderate to excellent yields. Conjugated dienes such as 1,3-butadiene give conjugated aldehydes under these conditions. Cyclic alkenes give the expected cyclic ketones. Vinyl and allylic halides are converted to the corresponding carbonyl compound.[256,257] Wacker oxidation has been reported using molecular oxygen. A catalytic

254. (a) Smidt, J.; Hafner, W.; Jira, R.; Sieber, R.; Sedlmeier, J.; Sabel, A. *Angew. Chem. Int. Ed.* **1962**, *1*, 80; (b) Jira, R.; Freiesleben W. *Organomet. React.* **1972**, *3*, 1; (c) The Merck Index, 14th ed., Merck & Co., Inc., Whitehouse Station, New Jersey, **2006**, p ONR-98; (d) Mundy, B.P.; Ellerd, M.G.; Favaloro Jr., F.G. *Name Reactions and Reagents in Organic Synthesis, 2nd ed.*, Wiley-Interscience, New Jersey, **2005**, pp. 676-677.
255. Smidt, J.; Hafner, W.; Jira, R.; Sedlmeier, J.; Sieber, R.; Rüttinger, R.; Kojer, H. *Angew. Chem.* **1959**, *71*, 176.
256. Byrom, N.T.; Grigg, R.; Kongkathip, B. *J. Chem. Soc. Chem. Commun.* **1976**, 216.
257. Clement, W.H.; Selwitz, C.M. *J. Org. Chem.* **1964**, *29*, 241.

amount of palladium(II) acetate, pyridine, and oxygen converted dodec-1-ene to 2-dodecanone, in up to 70% yield.[258] Similarly, cyclopentene was oxidized to cyclopentanone under an oxygen atmosphere using palladium acetate and molybdovanadophosphate on activated carbon.[259]

The Wacker oxidation can be used with more complex substrates than those shown, or to construct relatively simple fragments as part of a longer synthesis. An example is taken from Simpkins and co-worker's synthesis of (+)-demethoxyerythratidinone,[260] in which the terminal alkene unit in **366** was converted to methyl ketone **367** in 84% yield.

12.6.B. Formation of π-Allyl Palladium Complexes

The Wacker process involved alkenes, but when the alkene has an allylic position, a new type of organometallic can be formed. With palladium reagents, 1-propene reacts to form a π-allyl metal complex such as **368**, stabilized by back-donation from the metal atom. Substitution reaction of this π-allyl complex with a suitable nucleophile will generate the allylic species, **369**, where X is the nucleophile. Early work by Hüttel and Christ,[261] and also by Volger[262] established that π-allyl complexes could be prepared, but often in very poor yield.[261c] This low yield was probably due to preferential formation and attempted oxidation of a hydropalladium species such as **370** (derived from cyclohexene), which readily decomposed. Heck showed that arylpalladium chlorides added to conjugated dienes.[263] It was later discovered that oxidation of **370** converts it to the more stable halo-bridged dimer (**371**).[264] This observation allowed π-allyl palladium complexes to be examined for their synthetic utility. Trost reacted allylic compounds with disodium tetrachloropalladate (Na_2PdCl_4), in the presence of sodium acetate (NaOAc) and acetic acid (AcOH), and found that halogen dimers were formed. The alkene was used in excess, and a ratio of 2:1 alkene/Pd was optimal. A variety of alkenes can be converted to their palladium complex using Trost's conditions, including (Z)-2-pentene, which generated an 87:13 mixture of **372/373** in 83% yield.[264]

258. Nishimura, T.; Kakiuchi, N.; Onoue, T.; Ohe, K.; Uemura, S. *J. Chem. Soc. Perkin Trans. 1* **2000**, 1915.

259. Kishi, A.; Higashino, T.; Sakaguchi, S.; Ishii, Y. *Tetrahedron Lett.* **2000**, *41*, 99.

260. Zhang, F.; Simpkins, N.S.; Wilson, C. *Tetrahedron Lett.* **2007**, *48*, 5942..

261. (a) Hüttel, R.; Christ, H. *Chem. Ber.* **1963**, *96*, 3101; (b) *Idem, Ibid.* **1964**, *97*, 1439; (c) Hüttel, R.; Dietl, H.; Christ, H. *Ibid.* **1964**, *97*, 2037; (d) Hüttel, R.; Christ, H.; Herzig, K. *Ibid.* **1964**, *97*, 2710; (e) Hüttel, R.; Dietl, H. *Ibid.* **1965**, *98*, 1753; (f) Hüttel, R.; Schmid, H. *Ibid.* **1968**, *101*, 252.

262. Volger, H.C. *Recl. Trav. Chim. Pays-Bas*, **1969**, *88*, 225.

263. Heck, R.F. *J. Am. Chem. Soc.* **1968**, *90*, 5542.

264. Trost, B.M.; Strege, P.E.; Weber, L.; Fullerton, T.J.; Dietsche, T.J. *J. Am. Chem. Soc.* **1978**, *100*, 3407.

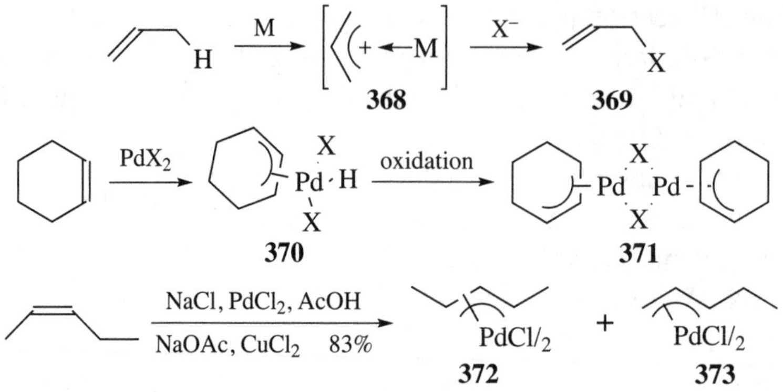

The reaction follows a Markovnikov orientation (see secs. 2.10.A, 6.2.A, 6.3.A), with the removed hydrogen being allylic to the most substituted end (as in the removal of H_a in the conversion **374** → **375**). Addition of cupric chloride (CuCl$_2$) influences the ratio of the Markovnikov-like to the anti-Markovnikov-like product.[264] Palladation of 2-methyl-2-butene (**376**) without CuCl$_2$ removed

H_a to give **378** (71% **378** + 29% **377**). Addition of CuCl$_2$ led to removal of H_b to give 74% of **377** and 26% of **378**.[265] The general order[264] for removal of allylic protons appears to be CH$_3$ > CH$_2$ >> CH. Conformational effects can alter this order or reactivity, however. The syn-complex usually predominates, as in the reaction of 4-methyl-3-heptene to give **375**. Both (*E*)- and (*Z*)-isomers of the alkene led to the syn-isomer **375**, without a trace of the anti-isomer (**379**).

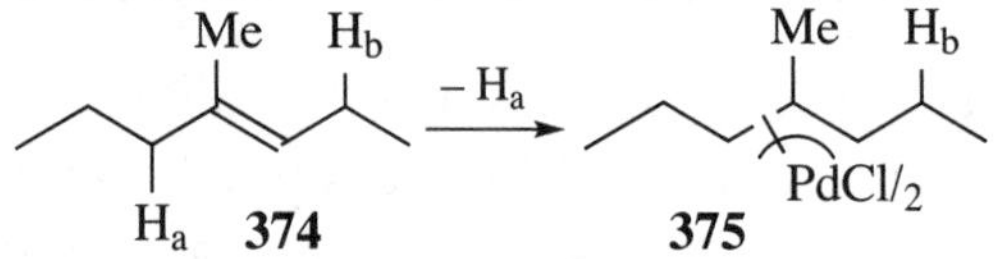

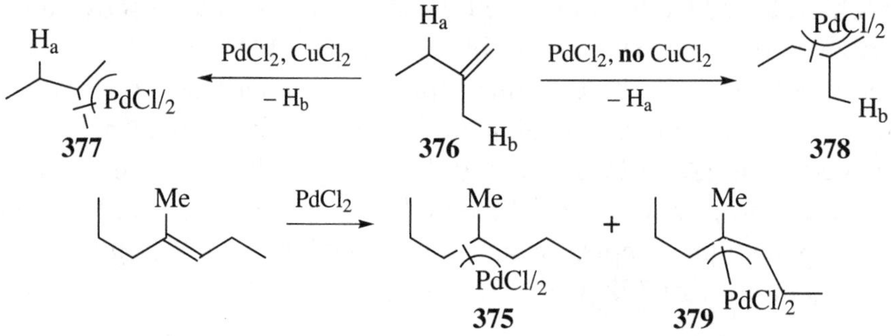

More highly substituted alkenes react faster (trisubstituted > disubstituted > monosubstituted),[265] as illustrated by the reaction of 4,8-dimethylnona-1,7-diene (**380**) to give **381**. Allylic systems conjugated to a carbonyl react slower than unconjugated alkenes. Conjugated alkenes do react if there is no competition.[264,266]

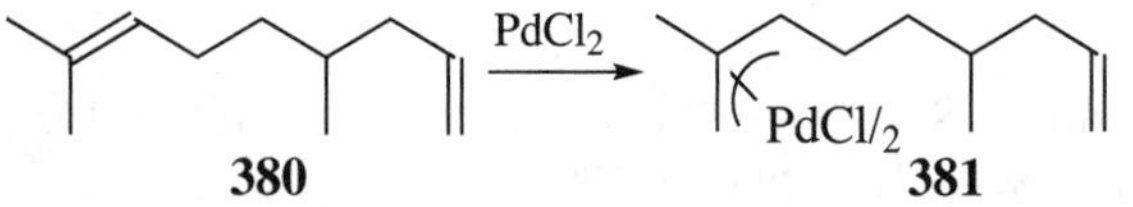

---

265.   Hüttel, R.; McNiff, M. *Chem. Ber. 1973, 106,* 1789.

266.   (a) Jones, D.N.; Knox, S.D. *J. Chem. Soc. Chem. Commun. 1975,* 165; (b) Harrison, I.T.; Kimura, E.; Bohme, E.; Fried, J.H. *Tetrahedron Lett. 1969,* 1589; (c) Susuki, T.; Tsuji, J. *Bull. Chem. Soc. Jpn 1973, 46,* 655.

Trost proposed the following mechanism to explain formation of π-allyl complexes from alkenes.[264] A mechanism for the palladium-mediated arylation of alkenes by arylmercury compounds had been proposed earlier by Heck.[267] Initial reaction with $PdCl_4$ generates a π-alkene complex (**382**), which is in equilibrium with the π-allyl palladium hydride complex **383**. The presence of a base (such as an amine or a phosphine) or $CuCl_2$ led to removal of the hydrogen from Pd-H to give the π-allyl palladium dimer **384**. The $CuCl_2$ oxidative procedure is more efficient than addition of a base.

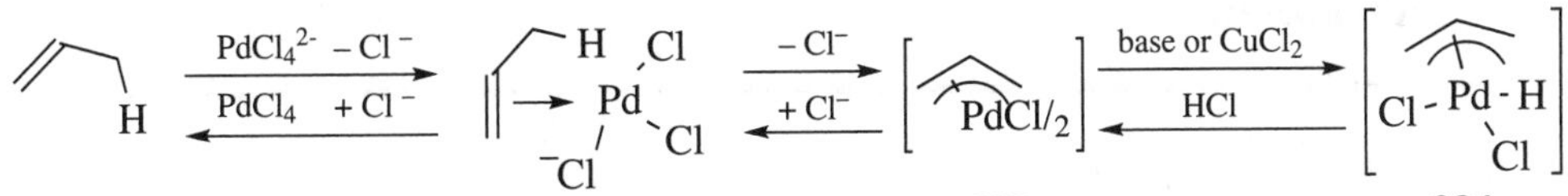

12.6.C. Reactions of π-Allyl Palladium Complexes

Initial reactions with π-allyl palladium complexes involved treatment with nucleophiles. Heck reported the arylation, methylation, and carboxyalkylation of alkenes by organomercury compounds, mediated by palladium, and other transition metals derivatives.[268] Trost found that **385** reacted with diethyl sodiomalonate in the presence of triphenylphosphine (PPh_3). The phosphine acted as a basic ligand, and coordinated with the palladium. The major product from reaction with sodium malonate was **386**, formed in 68% yield.[269] This reaction completes the initially set goal (in sec. 12.1) of forming a new carbon bond by reacting a stable cation with a carbon nucleophile. The distribution of products shows a preference for attack at the less substituted carbon [the ratio (**386** + **387**)/**388** was about 8:1, and the ratio **386**/**387** was 1.6:1]. When **385** reacted with methyl methanesulfonyl sodioacetate (**389**), only one product was observed (**390**) in 80% yield.[269] The extent of attack at the less substituted carbon was dependent on the nature and amount of the added phosphine ligand, as shown in Table 12.12.[269] Many common ligands are shown there. It is seen that **391** was converted to the π-allyl complex **392** using various ligands, which then reacted with methyl methanesulfonyl sodioacetate (**389**) to form a mixture of **393** and **394**. A variety of carbanion nucleophiles can be reacted with π-allylpalladium complexes.[270]

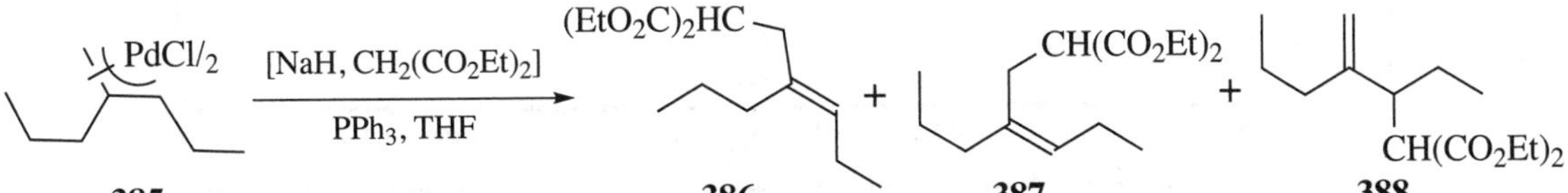

267. Heck, R.F. *J. Am. Chem. Soc.* **1969**, *91*, 6707.
268. Heck, R.F. *J. Am. Chem. Soc.* **1968**, *90*, 5518.
269. Trost, B.M.; Weber, L.; Strege, P.E.; Fullerton, T.J.; Dietsche, T.J. *J. Am. Chem. Soc.* **1978**, *100*, 3416.
270. Trost, B.M.; Weber, L.; Strege, P.E.; Fullerton, T.J.; Dietsche, T.J. *J. Am. Chem. Soc.* **1978**, *100*, 3426.

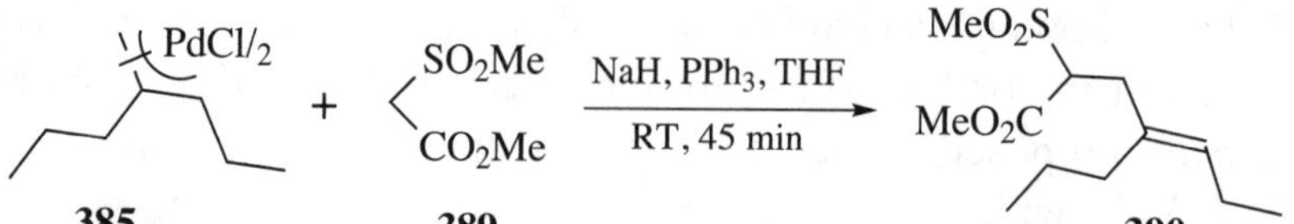

Table 12.12. Dependence of Regioselectivity on Ligands in Alkylation of Di-μ-chlorobis (1,2-tetra-methylene-π-allyl)dipalladium (II) with Methyl Methanesulfonyl Sodioacetate

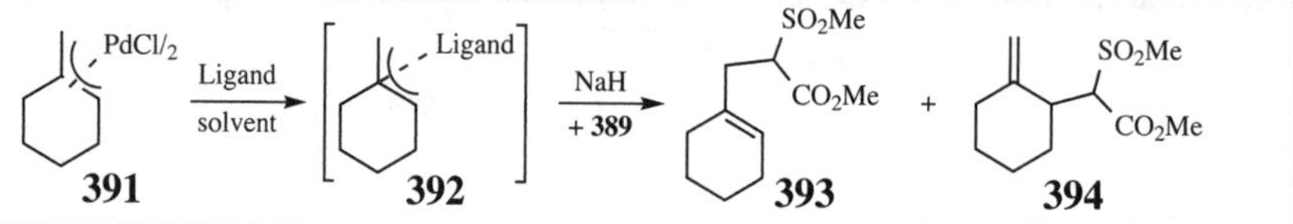

Ligand	Solvent	% Yield	% 393	% 394
Triphenylphosphine	DMSO	75	62	38
	THF	81	89	11
Tributylphosphine	DMSO	100	100	
	THF	57	90	10
dppe [1,2-bis(diphenylphosphino)ethane]	DMSO	64	76	24
	THF	100	75	25
dppp [1,3-bis(diphenylphosphino)propane]	DMSO	100	75	25
	THF	72	78	22
Diphenyl-o-tolylphosphine	DMSO	66	54	46
	THF	53	87	13
TOT [tri-o-tolylphosphine]	DMSO	90	18	82
	DMSO	79	12	18
	THF	0		
Tri-o-anisylphosphine	DMSO	80	70	30
Tri-o-trifluoromethylphosphine	DMSO	79	49	51
HMPT [hexamethylphosphorus triamide]	DMSO	58	100	
	THF	90	100	
Trimethylphosphite	DMSO	91	87	13
	THF	88	91	9
Triphenylphosphite	DMSO	86	93	7
	THF	69	91	9
TMEDA [N,N,N',N'-tetramethylethylenediamine]	DMSO	12	66	34
	THF	17	100	
Triphenylarsine	DMSO	4	62	38
	THF	86	100	
NONE	DMSO	0		
	THF	0		

When TOT was used in place of HMPT, the regioselectivity was reversed[270] as seen in Table 12.12. The coordinating ligand is critically important, but there does not seem to be a strong solvent effect on the stereochemistry of the reaction. Addition of bis(phosphine) ligands such as bis(diphenylphosphinoethane), dppe, generates a complex **394** rather than the dimeric palladium chloride structure observed previously. Reaction of **391** with **389** in the presence of dppe gave a 4:1 mixture of **393** and **394**. A synthetic application is Marchand and co-worker's reaction of **395** and the allylic sulfone anion (from **396**) gave a 52% yield of **397**.[271] Elimination of the sulfone moiety generated vitamin A alcohol **398**.

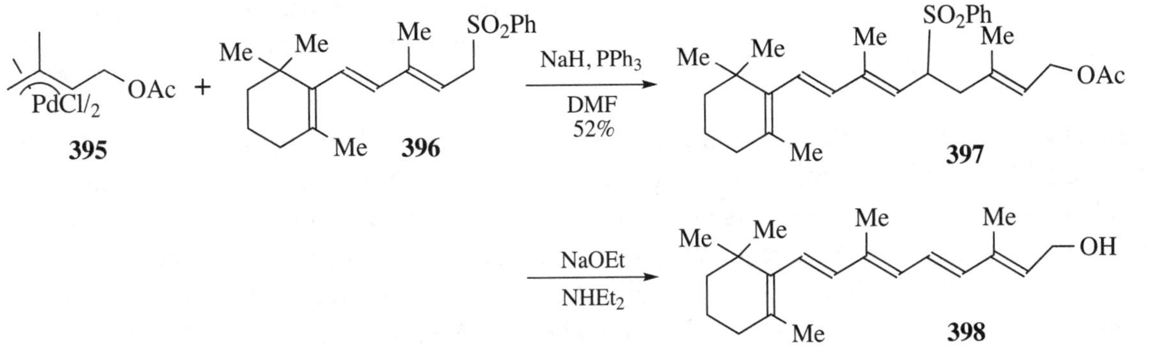

The π-allyl palladium coupling disconnections are

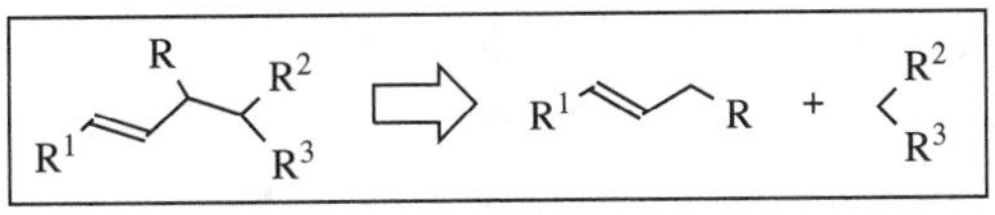

12.6.D. Catalytic π-Allyl Palladium Reactions

The entire allyl palladium alkylation process discussed in Sections 12.6.A and 12.6.B can be made catalytic in palladium. Indeed, Heck showed that the palladium-catalyzed arylation of enol esters gave aryl-substituted alkenes in modest yield.[272] Heck also reported the palladium-catalyzed arylation of allylic alcohols with aryl halides.[273] The basis for this catalytic activity involves the use of an allylic substituted species (**369**, allylic acetate, allylic chloride, etc.) rather than an allylic hydrocarbon. A *tetrakis*-palladium(0) complex (L₄Pd) reacts with the allylic substrate to form a π-allyl palladium complex (**399**). Substitution with a nucleophile generates **400** with liberation of palladium species L₄Pd, which is recycled to react with additional **369**.[274]

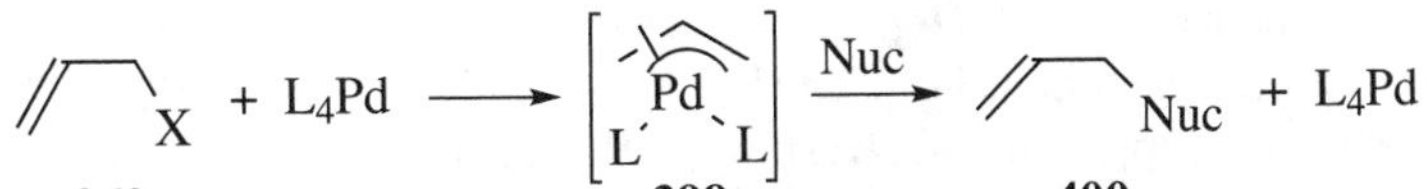

271. Manchand, P.S.; Wong, H.S.; Blount, J.F. *J. Org. Chem.* **1978**, *43*, 4769.
272. Heck, R.F. *J. Am Chem .Soc.* **1968**, *90*, 5535.
273. Melplder, J.B.; Heck, R.F. *J. Org. Chem.* **1976**, *41*, 265.
274. (a) Trost, B.M.; Verhoeven, T.R. *J. Am. Chem. Soc.* **1976**, *98*, 630; (b) *Idem Ibid.* **1978,** *100*, 3435; (c) Takahashi, K.; Miyake, A.; Hata, G. *Bull Chem. Soc. Jpn.* **1970**, *45*, 230,1183; (d) Trost, B.M.; Verhoeven, T.R. *J. Org. Chem.* **1976**, *41*, 3215; (e) *Idem J. Am. Chem.Soc.* **1980**, *102*, 4730.

Two of the common Pd(0) catalysts are the commercially available *tetrakis*-(triphenylphosphino) palladium (**401**) and bis[1,2-bis-(diphenylphosphino)ethane] palladium [**402**, i.e. Pd(dppe)$_2$],

prepared by reduction of palladium acetate [Pd(OAc)$_2$] in the presence of triphenylphosphine or dppe.[274c,d] The use of these ligands was important to the reactions discussed in with Table 12.12.

12.6.D.i. Reactions with Nucleophiles. Previously, a π-allylic palladium complex was generated by reaction of palladium reagents with allylic hydrocarbons prior to reaction with nucleophiles. In the catalytic version of this reaction, an allylic halide or an allylic acetate is used with a palladium(0) reagent. Why use a palladium complex when enolate alkylation is a well-known process (sec. 9.3.A)? A typical enolate coupling reaction is the conversion of 2-methylcyclopentane-1,3-dione (**403**) to the enolate anion by reaction with NaOH, allowing reaction with allyl bromide. Under these conditions only 34% of **404** was obtained. When allyl acetate was used in place of allyl bromide in this reaction and *tetrakis*(triphenylphosphino) palladium was used as a catalyst, a 94% yield of

404 was obtained.[275] In this reaction, formation of the π-allyl palladium complex facilitated coupling with the nucleophilic enolate derived from **403**, which exhibited poor reactivity in the normal enolate alkylation sequence.

This version of the π-allyl palladium reaction uses an allylic acetate or chloride. The use of the acetate is more common, in part because acetate is a weaker nucleophile than chloride. More recently, methyl carbonate groups have been shown to be good leaving groups in this reaction.[276] When it involves a substrate where diastereomeric products can result, the stereochemistry of the nucleophilic displacement is an important issue. The addition of chiral ligands and chiral additives lead to asymmetric induction.[277,278] Palladium assisted alkylation proceeds with net retention of configuration of the acetate or chloride. The displacement can proceed by an

275. (a) Newman, M.S.; Manhart, J.H. *J. Org. Chem.* **1961**, *26*, 2113; (b) unpublished observations of Trost, B.M.; Curran, D. reported in citation 31 of Trost, B.M. *Acc. Chem. Res.* **1980**, *13*, 385.

276. For a –OCO$_2$Ph leaving group, see Ito, K.; Kashiwagi, R.; Hayashi, S.; Uchida, T.; Katsuki, T. *Synlett* **2001**, 284.

277. (a) Kuwano, R.; Kondo, Y.; Matsuyama, Y. *J. Am. Chem. Soc.* **2003**, *125*, 12104; (b) Faller, J.W.; Wilt, J.C. *Tetrahedron Lett.* **2004**, *45*, 7613.

278. (a) Evans, P.A.; Uraguchi, D. *J. Am. Chem. Soc.* **2003**, *125*, 7158; (b) Evans, P.A.; Leahy, D.K.; Slieker, L.M. *Tetrahedron Asymmetry* **2003**, *14*, 3613; (c) Evans, P.A.; Leahy, D.K. *J. Am. Chem. Soc.* **2000**, *122*, 5012; (d) López, F.; Ohmura, T.; Hartwig, J.F. *J. Am. Chem. Soc.* **2003**, *125*, 3426; (e) Matsushima, Y.; Onitsuka, K.; Kondo, T.; Mitsudo, T.-a.; Takahashi, S. *J. Am. Chem. Soc.* **2001**, *123*, 10405; (f) Ohmura, T.; Hartwig, J.F. *J. Am. Chem. Soc.* **2002**, *124*, 15164; (g) Evans, P.A.; Robinson, J.E.; Baum, E.W.; Fazal, A.N. *J. Am. Chem. Soc.* **2002**, *124*, 8782; (h) Bandini, M.; Melloni, A.; Umani-Ronchi, A. *Org. Lett.* **2004**, *6*, 3199.

S_N2'-type displacement with retention of configuration as seen in the conversion of **405** to **406** (99% yield; >50:1 dr), taken from Kündig and co-worker's synthesis of acetoxytubipofuran.[279] A variety of active methylene compounds can be used as nucleophiles,[280] including enolate anions.[281] Intramolecular cyclization is possible when the active methylene compound and an allylic acetate or carbonate is incorporated into the same molecule.[282] This reaction has been called **Trost allylation**, or the **Tsuji-Trost reaction**.[283]

Trost proposed the following mechanism to account for these catalytic transformations. Reaction of the palladium catalyst with **407** generates π-alkene palladium complex **408**. Palladium removes the allylic hydrogen atom, and loss of the acetate moiety gives π-allyl palladium complex **409**. Nucleophilic attack at C_a leads to **410**, with expulsion of the PdL_2 species, whereas attack at C_b leads to **411**. Palladium coordinates on the face of the alkene *distal* to the acetate (distant from the acetate: C_a rather than C_b). Palladium displaces acetate with inversion (**408** → **409**). When the nucleophile displaces the palladium, a second inversion occurs at C_a or C_b, whichever is less sterically hindered, to give a net retention of configuration for the conversion **407** → **410** and/or **411**.

Trost and Genêt provided another example where an alkene and an aromatic ring were coupled using bis(acetonitrile)palladium chloride to convert **412** to desethylibogamine (**415**).[284] Initial reaction of the palladium with the enamine moiety of **412** generated **413**, and cyclization with the allylic amine moiety gave **414**. Sodium borohydride was required to reduce the initially formed palladium chloride species to **415**. Organopalladium catalysts are now common for the

279. Kündig, E.P.; Cannas, R.; Laxmisha, M.; Ronggang, L.; Tchertchian, S. *J. Am. Chem. Soc.* **2003**, *125*, 5642.
280. Kazmaier, U.; Zumpe, F.L. *Angew. Chem. Int. Ed.* **1999**, *38*, 1468.
281. (a) You, S.-L.; Hou, X.-L.; Dai, L.-X.; Zhu, X.-Z. *Org. Lett.* **2001**, *3*, 149; (b) Evans, P.A.; Leahy, D.K. *J. Am. Chem. Soc.* **2003**, *125*, 8974; (c) Evans, P.A.; Lawler, M.J. *J. Am. Chem. Soc.* **2004**, *126*, 8642.
282. (a) Castaño, A.M.; Méndez, M.; Ruano, M.; Echavarren, A.M. *J. Org. Chem.* **2001**, *66*, 589; (b) Zhang, Q.; Lu, X.; Han, X. *J. Org. Chem.* **2001**, *66*, 7676.
283. The Merck Index, 14th ed., Merck & Co., Inc., Whitehouse Station, New Jersey, **2006**, p ONR-94.
284. (a) Trost, B.M.; Genêt, J.P. *J. Am. Chem. Soc.* **1976**, *98*, 8516; (b) Trost, B.M. *Tetrahedron* **1977**, *33*, 2615 and citation 10 therein.

coupling reactions of alkenes, alkynes or aryls.[285]

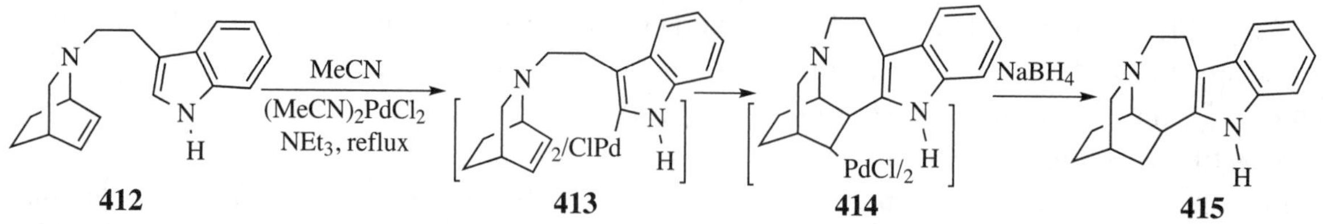

412 **413** **414** **415**

When chiral ligands are used with the palladium catalyst, good enantioselectivity can be achieved in the alkylation reaction.[286] Two common ligands[287] are the bis(phosphine) complex from chlorodiphenylphosphine, and optically active amino alcohols such as **416**. Phosphino-ester binaphthyls such as **417** have also been used.[288] The enantioselectivity can be quite good, as when allylic acetate **418** reacted with diethyl malonate in the presence of **417** and *N,O*-bis(trimethylsilyl)acetamide (BSA) and LiOAc as additives, to give an 89% yield of **419** with 99% ee.[288] The identical reaction with **418**, but using **416** as a ligand gave a 93% yield of **419** but in only 56% ee.[287]

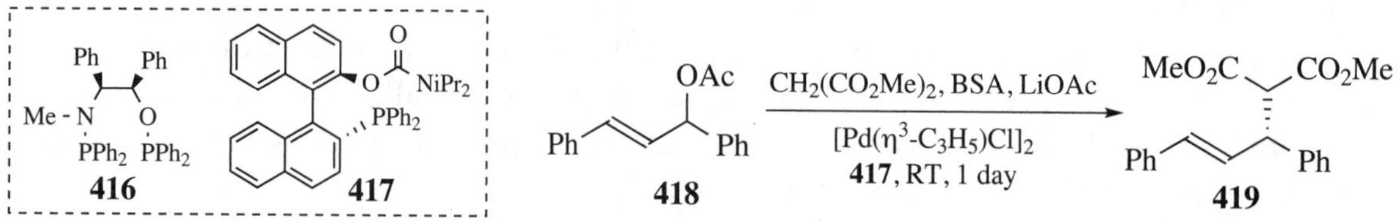

416 **417** **418** **419**

12.6.D.ii. Palladium Mediated Cyclization Reactions. It is possible to use organopalladium chemistry for intramolecular coupling (cyclization) reactions. Treatment of **420** with the palladium catalyst ([dba]$_3$Pd$_2$, where dba = dibenzylidene acetone, PhCH=CHCOCH=CHPh)

420 **421**

led to cyclization, and generated the five-membered ring in **421** in 78% yield.[289] There is a template effect in the π-allyl palladium reactions[290] that can be exploited to form large rings (sec. 6.6.B). Ester **422** was converted to **423** in 70% yield by this procedure, for example, but the yields have been poor in some cases.[290]

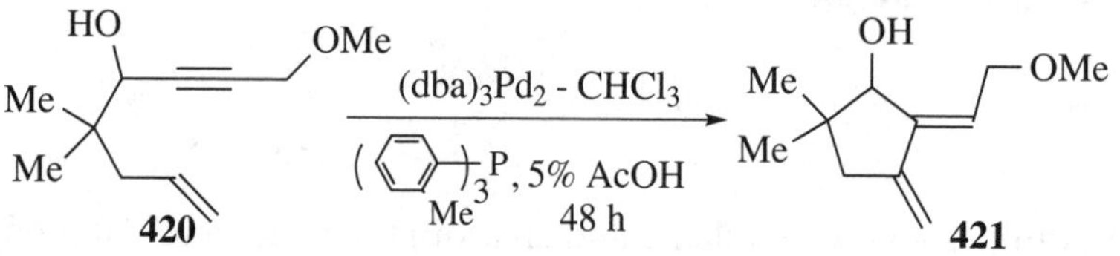

285. (a) Sonogashira, K.; Tohda, Y.; Hagihara, N. *Tetrahedron Lett.* **1975**, 4467; (b) Dieck, H.A.; Heck, R.F. *J. Organomet. Chem.* **1975**, *93*, 259; (c) Edo, K.; Yamanaka, H.; Sakamoto, T. *Heterocycles* **1978**, *9*, 271; (d) Ohsawa, A.; Abe, Y.; Igeta, H. *Chem. Lett.* **1979**, 241.
286. Trost, B.M.; Crawley, M.L. *Chem. Rev.* **2003**, *103*, 2921.
287. Gong, L.; Chen, G.; Mi, A.; Jiang, Y.; Fu, F.; Cui, X.; Chan, A.S.C. *Tetrahedron: Asymmetry* **2000**, *11*, 4297.
288. Kodama, H.; Taiji, T.; Ohta, T.; Furukawa, I. *Tetrahedron: Asymmetry* **2000**, *11*, 4009.
289. Trost, B.M.; Lee, D.C.; Rise, F. *Tetrahedron Lett.* **1989**, *30*, 651.
290. (a) Trost, B.M.; Verhoeven, T.R. *J. Am. Chem. Soc.* **1980**, *102*, 4743; (b) *Idem Ibid.* **1977**, *99*, 3867; (c) *Idem Tetrahedron Lett.* **1978**, 2275.

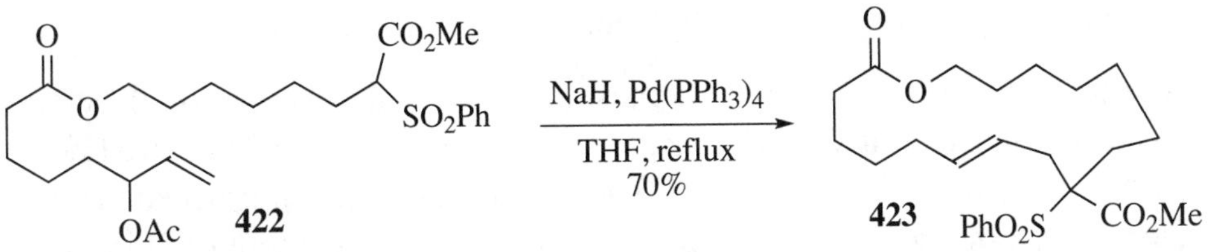

12.6.D.iii. Palladium-Mediated Arylation Reactions. In the presence of palladium catalysts, enolate anions of malonic esters react with aryl halides to give 2-aryl malonates.[291] The reaction of diethyl malonate with sodium hydride and bromobenzene, with the palladium(0) catalyst generated *in situ*, gave an 89% yield of diethyl 2-phenylmalonate (**424**). Similar coupling reactions had been reported earlier, but aryl iodides were the substrates and stoichiometric amounts of copper derivatives were used.[292] Arylation of malonates and cyanoesters was later reported using catalytic amounts of copper derivatives.[293]

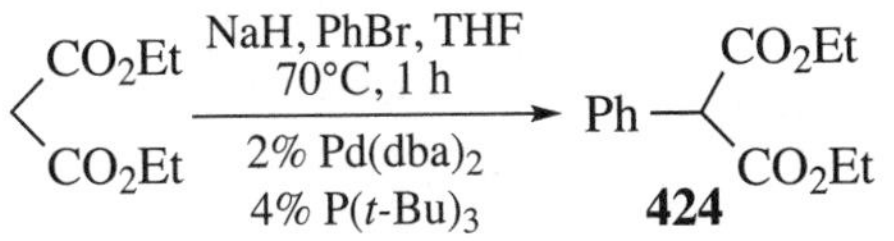

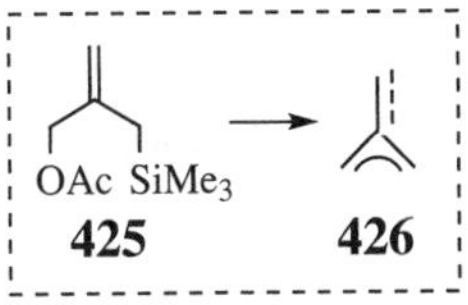

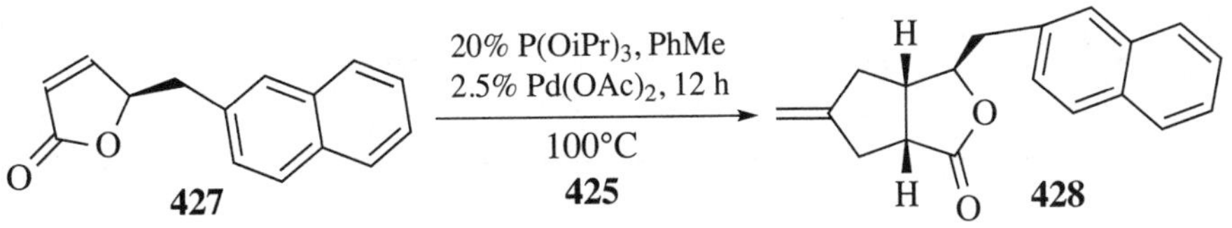

12.6.D.iv. Trimethylenemethane Equivalents. Palladium catalysts can be used to convert trimethylsilyl allylic acetate **425** to a trimethylene methane equivalent (TMM, **426**). A [3+2]-cycloaddition reaction (sec. 11.11) with alkenes generates cyclopentanes in what is called **quinane annulation**.[294] In this reaction, the trimethylsilyl unit is a carbanion equivalent and acetate is a carbocation equivalent. In an example taken from a synthesis of (+)-brefeldin A, Trost and Crawly reacted **425** and **427** in the presence of 2.5 mole % palladium acetate and triisopropyl phosphite [P(Oi-Pr)$_3$] (toluene,100°C) to generate **428** in 93% yield (>93:2 dr).[295] The reaction is characterized by retention of the geometry of the dipolarophile in the cycloadduct, and proceeds with high diastereoselectivity and good regioselectivity. This reaction can also be applied to the formation of larger rings if a diene is used as the dipolarophile.[296] Trost later showed that Z-alkenes show greater selectivity in this cyclization process.[297] Reagents related to **425** react with aldehydes, as well as alkenes, to produce THF derivatives.[298] Alkynes also

291. Beare, N.A.; Hartwig, J.F. *J. Org. Chem.* **2002**, *67*, 541.
292. (a) Osuka, A.; Kobayashi, T.; Suzuki, H. *Synthesis* **1983**, 67; (b) Setsune, J.-I.; Matsukawa, K.; Wakemoto, H.; Kitao, T. *Chem. Lett.* **1981**, 367.
293. Okuro, K.; Furuune, M.; Miura, M.; Nomura, M. *J. Org. Chem.* **1993**, *58*, 7606.
294. Trost, B.M. *Angew. Chem. Int. Ed.* **1986**, *25*, 1.
295. Trost, B.M.; Crawley, M.L. *J. Am. Chem. Soc.* **2002**, *124*, 9328.
296. Trost, B.M.; MacPherson, D.T. *J. Am. Chem. Soc.* **1987**, *109*, 3483.
297. Trost, B.M.; Yang, B.; Miller, M.L. *J. Am. Chem. Soc.* **1989**, *111*, 6482.
298. (a) Trost, B.M.; King, S.A. *J. Am. Chem. Soc.* **1990**, *112*, 408; (b) *Idem Tetrahedron Lett.* **1986**, *27*, 5971.

react with TMM equivalents to form methylenecyclopentene derivatives.[299]

When TMM equivalents are tethered to alkenes, intramolecular cyclization generates bicyclic methylenecyclopentane derivatives. Cyclization of **429**, for example, led to a 51% yield of **430** along with 18% of uncyclized triene **431**.[300] The cyclization reaction proceeded with excellent diastereoselectivity, as shown. Trost has also examined the attachment of chiral auxiliaries to the alkene partner to give asymmetric induction in reactions with TMM equivalents.[296]

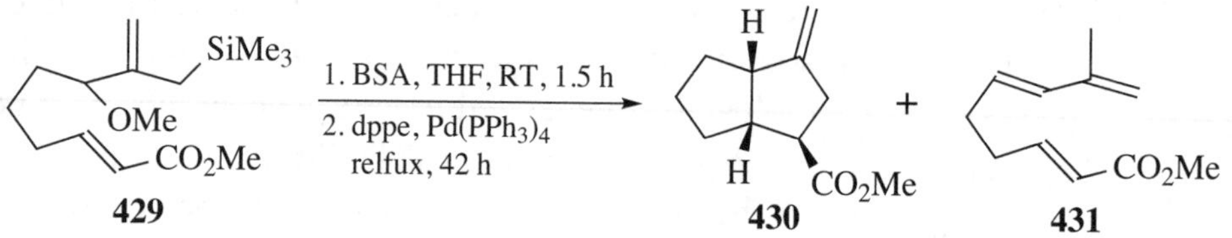

The disconnections for catalytic allyl palladium reactions are essentially the same as for the uncatalyzed version. A few representative examples are

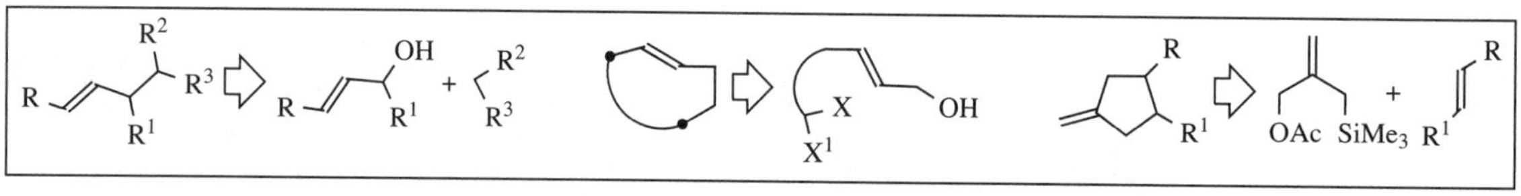

12.7. NAMED PALLADIUM COUPLING REACTIONS

As organometallic chemistry in general, and organopalladium chemistry in particular, have become more popular, many new reactions have been developed. Several of these are so useful that they have joined the list of named reactions. This section will explore several of named reactions that involve palladium catalysts.

12.7.A. The Heck Reaction

There are several variations of the organopalladium reaction, some involving alkenyl palladium and others σ-palladium complexes. One variation generates the palladium species from a heterocyclic alkene. As mentioned in section 12.6.C and 12.6.D, Heck reported the palladium catalyzed coupling of aryl compounds with alkenes. Heck later showed[301] that substituted alkenes could be prepared by the palladium-catalyzed coupling of alkenes and aryl halides. This

299. Trost, B.M.; Balkovec, J.M.; Angle, S.R. *Tetrahedron Lett.* **1986**, *27*, 1445.
300. Trost, B.M.; Grese, T.A.; Chan, D.M.T. *J. Am. Chem. Soc.* **1991**, *113*, 7350.
301. Heck, R.F.; Nolley Jr., J.P. *J. Org. Chem.* **1972**, *37*, 2320.

latter transformation is now known as the **Heck reaction**,[302] and usually involves the coupling of aryl halides or aryl sulfonate esters with alkenes.[303] As Heck reported, the reaction was discovered independently at about the same time by Mizoroki and co-workers.[302d,e] The aryl palladium complexes that are formed via oxidative coupling of aryl halides with palladium(0) can undergo the Heck reaction.[304,305] The reaction proceeds with formation of a palladium species such as **432**, with elimination of palladium to give arylated alkene derivatives.[306] Aryl halides differ greatly in their reactivity, ArI being the most reactive followed by aryl bromides. In general, aryl chlorides are very unreactive in the Heck reaction unless the chlorine is attached to a heteroaromatic ring. Aryl triflates are partners in this reaction. Coupling occurs with many alkenes and aryl derivatives. Styrene reacted with phenylbis(triphenylphosphino) palladium chloride (**433**), in the presence of triethylamine, to give *trans*-stilbene with loss of the palladium complex **434**.[307] The palladium reagent (**433**) was generated *in situ* by reaction of palladium acetate [Pd(OAc)$_2$] and triphenylphosphine in the presence of iodobenzene.

The reaction is thought to proceed by syn addition of the organopalladium, followed by syn elimination of palladium hydride.[307] The use of chiral ligands allows asymmetric induction in some Heck reactions.[308] Conjugate addition is usually preferred when the alkene contains an electron-withdrawing group. Without such a group, addition occurs at the less substituted carbon.[307] In Ruchirawat and Namsa-aid's[309] synthesis of ningalin C, aryl bromide **435** reacted with 3-aryl-1-propene derivative **436** in the presence of Pd(0) to give a 65% yield of **437**. New

302. (a) See Heck, R.F. *Synlett* **2006**, 2855, and the articles in special issue No. 18 of *Synlett, 2006,* 2855-3184; (b) The Merck Index, 14th ed., Merck & Co., Inc., Whitehouse Station, New Jersey, **2006**, p ONR-42; (c) Mundy, B.P.; Ellerd, M.G.; Favaloro Jr., F.G. *Name Reactions and Reagents in Organic Synthesis, 2nd ed.,* Wiley-Interscience, New Jersey, **2005**, pp. 296-297; (d) Mizoroki, T.; Mori, K.; Ozaki, A *Bull. Chem. Soc. Jpn. 1972, 44,* 581; (e) Mizoroki, T.; Mori, K.; Ozaki, A. *Bull. Chem. Soc. Jpn. 1973, 46,* 1505.

303. For other reactions of this type, see reference 79, pp 386-392.

304. Heck, R.F. *Org. React. 1982, 27,* 345.

305. (a) Patel, P.A.; Ziegler, C.B.; Cortese, N.A.; Plevyak, J.E.; Zebovitz, T.C.; Terpko, M.; Heck, R.F. *J. Org. Chem. 1977, 42,* 3903; (b) Cortese, N.A.; Ziegler Jr., C.B.; Hrnjez, B.J.; Heck, R.F. *Ibid. 1978, 43,* 2952; (c) Mori, M.; Kudo, S.; Ban, Y. *J. Chem. Soc. Perkin Trans 1 1979,* 771.

306. For a review of mechanistic considerations for the Heck reaction, see Amatore, C.; Jutland, A. *Acc. Chem. Res. 2000, 33,* 314.

307. Dieck, H.A.; Heck, R.F. *J. Am. Chem. Soc. 1974, 96,* 1133.

308. Deng, W.-P.; Hou, X.-L.; Dai, L.-X.; Dong, X.-W. *Chem. Commun. 2000,* 1483.

309. Namsa-aid, A.; Ruchirawat, S. *Org. Lett. 2002, 4,* 2633.

catalysts have been developed that allow Heck coupling with deactivated aryl chlorides.[310] Beller and co-workers showed that chlorobenzene reacted with styrene, for example, to give *trans*-stilbene in 63% yield in the presence of K_2CO_3 and a palladium catalyst having a di(1-adamantyl)-*n*-butylphosphine ligand.[310] A ligand-free version of the Heck reaction (does not require the use of phosphine ligands) was recently reported, using palladium acetate and potassium phosphate in *N,N*-dimethylacetamide.[311] Aryl chlorides undergo the Heck reaction at ambient temperatures, with high selectivity using a Pd/P(*t*-Bu)$_3$/Cy$_2$NMe catalyst.[312]

The intramolecular version of this reaction is well known[313,314] and quite useful in natural product synthesis, including asymmetric syntheses.[315] An example of this variation is taken from Enders and co-worker's synthesis of the 1-*epi*-aglycon of the cripowellins A and B,[316] where palladium catalyzed reaction of **438** with the gave a 59% yield of **439** with >98% de and ee. Other highly enantioselective intramolecular Heck reactions have been reported using a monodentate ligand.[317]

Heterocyclic substrates can be used in the Heck reaction. For example, 2-bromothiophene and 4-vinylpyridine reacted to a give 57% yield of the 2-thiophenyl-4-pyridyl ethene derivative.[318] In a synthesis of tangutorine by Hsung and co-workers, Heck coupling of bromoindole derivative **440** and methyl acrylate gave an 82% yield of **441**.[319] The Heck reaction can also be applied to heterocyclic vinyl halides[320] and vinyl triflates. Pure cis or trans alkenes can isomerize to a mixture of cis and trans products.[320] In some cases, however, the selectivity is good as can

310. Ehrentraut, A.; Zapf, A.; Beller, M. *Synlett* **2000**, 1589.
311. Yao, Q.; Kinney, E.P.; Yang, Z. *J. Org. Chem.* **2003**, *68*, 7528.
312. Littke, A.F.; Fu, G.C. *J. Am. Chem. Soc.* **2001**, *123*, 6989.
313. Zhang, Y.; Negishi, E. *J. Am. Chem. Soc.* **1989**, *111*, 3454.
314. Abelman, M.M.; Oh, T.; Overman, L.E. *J. Org. Chem.* **1987**, *52*, 4130.
315. Dounay, A. B.; Overman, L. E. *Chem. Rev.* **2003**; *103*, 2945.
316. Enders, D.; Lenzen, A.; Backes, M.; Janeck, C.; Catlin, K.; Lannou, M.-I.; Runsink, J.; Raabe, G. *J. Org. Chem* **2005**, *70*, 10538.
317. Imbos, R.; Minnaard, A.J.; Feringa, B.L. *J. Am. Chem. Soc.* **2002**, *124*, 184.
318. Frank, W.C.; Kim, Y.C.; Heck, R.F. *J. Org. Chem.* **1978**, *43*, 2947.
319. Luo, S.; Zificsak, C.A.; Hsung, R.P. *Org. Lett.* **2003**, *5*, 4709.
320. (a) Dieck, H.A.; Heck, R.F. *J. Org. Chem.* **1975**, *40*, 1083; (b) Patel, B.A.; Heck, R.F. *Ibid.* **1978**, *43*, 3898.

be seen in Kirschning and co-worker's synthesis of the *N*-acetylcysteamine thioester of seco-proansamitocin.[321] In this work, vinyl iodide **443** reacted with the terminal alkene moiety of **442** to give a 72% yield of the coupled product **444**, using the so-called **Jeffery protocol**,[322] with high selectivity for the (*E*)- alkene unit of the diene.

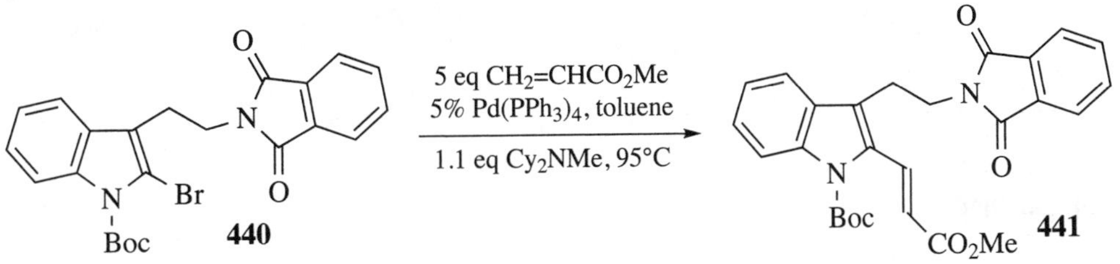

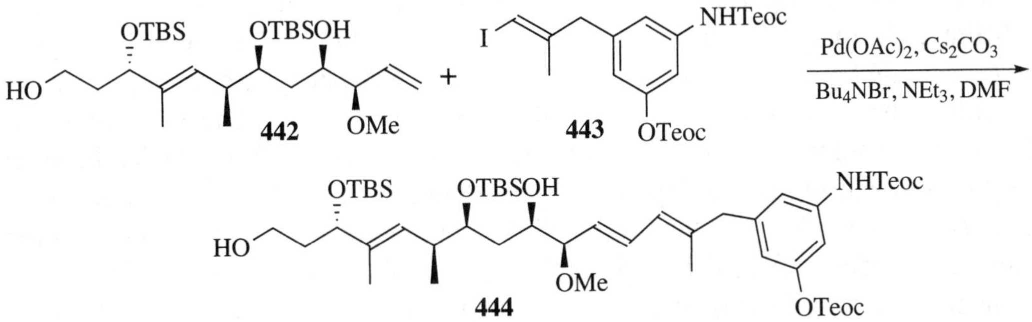

There are several variations that involve heteroatom substituents, including biaryl ethers, amines or ketones, which can be coupled to alkenes using palladium acetate.[323] Medium ring heterocycles can be prepared via an intramolecular Heck reaction.[324] Silyl enol ethers can also be coupled to another alkene upon treatment with palladium acetate to form conjugated ketones.[325] Solvents can change with the reactive partners. Aryldiazonium salts couple to alkenes using palladium acetate in aqueous media.[326a] Carbohydrate-based vinyl bromides have been coupled to methyl acrylate using Pd(dba)$_2$ in aqueous DMF.[327] Alkynes have been coupled to pyrrole using Heck-like conditions.[328] Recyclable Pd (II) catalysts with bis(imidazole) ligands, are used with Heck reactions in ionic liquids.[329] A noncatalytic Heck reaction was reported using

321. Frenzel, T.; Brunjes, M.; Quitschalle, M.; Kirschning, A. *Org. Lett.* **2006**, *8*, 135.

322. Jeffery, T. *Tetrahedron* **1996**, *52*, 10113.

323. (a) Åkermark, B.; Eberson, L.; Jonsson, E.; Pettersson, E. *J. Org. Chem.* **1971**, *40*, 1365; (b) Itahara, T.; Sakakibara, T. *Synthesis* **1978,** 607; (c) Itahara, T. *Ibid.* **1979,** 151.

324. Arnold, L.A.; Luo, W.; Guy, R.K. *Org. Lett.* **2004**, *6*, 3005.

325. Ito, Y.; Aoyama, H.; Hirao, T.; Mochizuki, A.; Saegusa, T. *J. Am. Chem. Soc.* **1979**, *101*, 494.

326. (a) Sengupta, S.; Bhattacharya, S. *J. Chem. Soc. Perkin Trans. 1* **1993**, 1943; (b) For a Heck coupling with an aryl diazonium salt used in a synthesis of (−)-codonopsinine see Severino, E.A.; Correia, C.R.D. *Org. Lett.* **2000**, *2*, 3039; and in a synthesis of the defucogilvocarcin M chromophore, see Patra, A.; Pahari, P.; Ray, S.; Mal, D. *J. Org. Chem.* **2005**, *70*, 9017.

327. Hayashi, M.; Amano, K.; Tsukada, K.; Lamberth, C. *J. Chem. Soc. Perkin Trans. 1* **1999**, 239.

328. Lu, W.; Jia, C.; Kitamura, T.; Fujiwara, Y. *Org. Lett.* **2000**, *2*, 2927.

329. Park, S.B.; Alper, H. *Org. Lett.* **2003**, *5*, 3209.

supercritical water as the medium.[330]

The Heck and related disconnections are

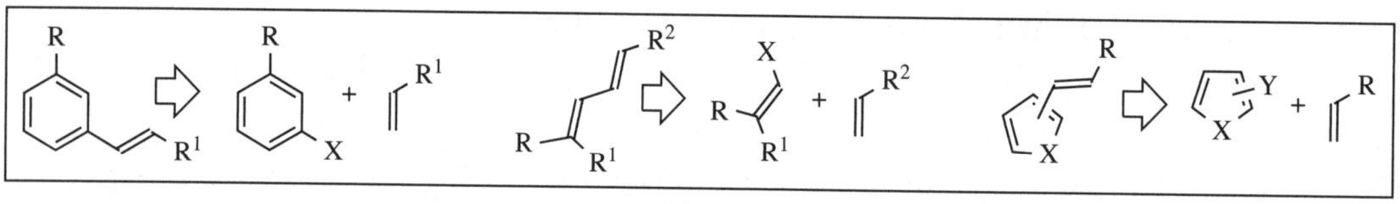

12.7.B. Stille Coupling

Vinyl triflates (C=C–OSO$_2$CF$_3$) react with vinyltin derivatives to give dienes, in what is known as the **Stille coupling**.[331,332] Preparation of vinyl triflate **445** is illustrated by trapping the enolate of 2-methylcyclohexanone (sec. 9.2, 9.3.B) with *N*-phenyl triflimide,[333] and reaction with vinyltin compound **446** in the presence of a Pd(0) reagent gave diene **447** quantitatively as a 95:5 (*E,E*)/(*E,Z*) mixture.[333] Coupling two different alkene units in this way has been referred to as **Stille cross-coupling**, and vinyl iodides are commonly used.[334] The reaction can also be applied to coupling aryl units with vinyl or allylic units. In Phillips and Keaton's synthesis of (–)-7-demethylpiericidin A$_1$, (**450**),[335] pyridyl-tributyltin **448** was coupled to allylic carbonate **449**, in the presence of Pd$_2$(dba)$_3$ to give a 55% yield of **450**. Aryl bromides have been coupled to tributylvinyltin compounds.[336]

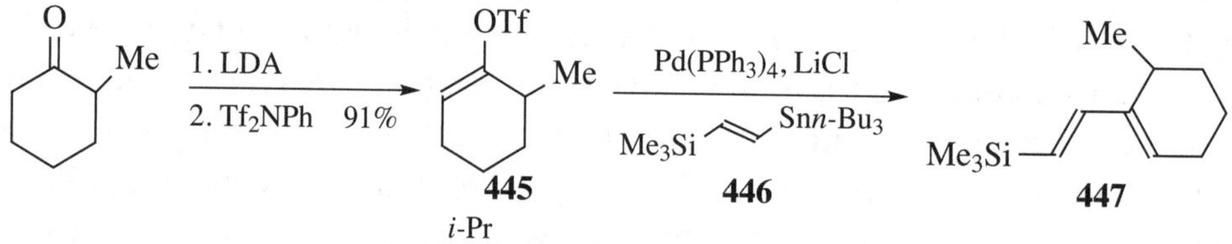

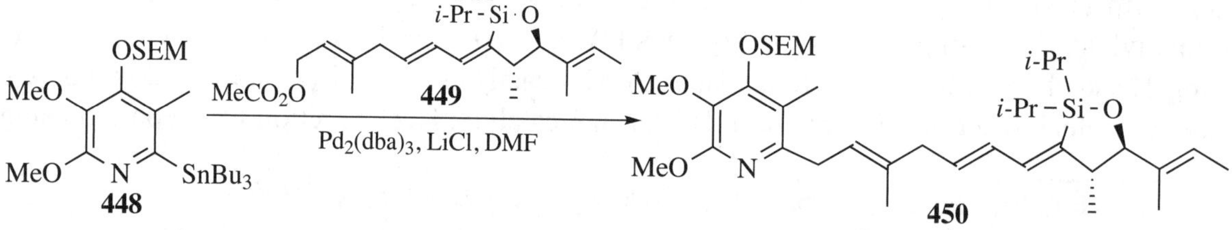

330. Zhang, R.; Zhao, F.; Sato, M.; Ikushima, Y. *Chem. Commun.* **2003**, 1548.
331. (a) Scott, W.J.; Crisp, G.T.; Stille, J.K. *J. Am. Chem. Soc.* **1984**, *106*, 4630; (b) The Merck Index, 14th ed., Merck & Co., Inc., Whitehouse Station, New Jersey, **2006**, p ONR-89; (c) Mundy, B.P.; Ellerd, M.G.; Favaloro Jr., F.G. *Name Reactions and Reagents in Organic Synthesis, 2nd ed.*, Wiley-Interscience, New Jersey, **2005**, pp. 620-621.
332. For reactions involving organotin compounds, see reference 79, pp 434-437.
333. McMurry, J.E.; Scott, W.J. *Tetrahedron Lett.* **1983**, *24*, 979.
334. For the use of a vinyl iodide in a Stille coupling, in a synthesis of elysiapyrones A and B, see Barbarow, J.E.; Miller, A.K.; Trauner, D. *Org. Lett.* **2005**, *7*, 2901.
335. Keaton, K. A.; Phillips, A. J. *J. Am. Chem. Soc.* **2006**, *128*, 408.
336. For an example taken from a synthesis of ageladine A, see Meketa, M.L.; Weinreb, S.M. *Org. Lett.* **2006**, *8*, 1443.

Mechanistic rationales for the Stille reaction have been reviewed,[337] and the Stille coupling has been shown to be catalytic in tin.[338,339] Reaction of alkyne **451** with vinyl bromide **452** in the presence of a palladium catalyst and 6% Me₃SnCl, for example, gave a 90% yield of **453**.[338] This reaction required syringe pump addition of the vinyl bromide over 15 h to the rather complex reaction medium, but the yields were quite good. A method has been developed that allows coupling with aryl chlorides.[340] There has been increasing interest in intramolecular Stille reactions in organic synthesis.[341] An example is taken from Lee and co-workers synthesis of lasonolide A,[342] in which **454** was converted to **455** in 55% yield as a 2:1 E/Z mixture.

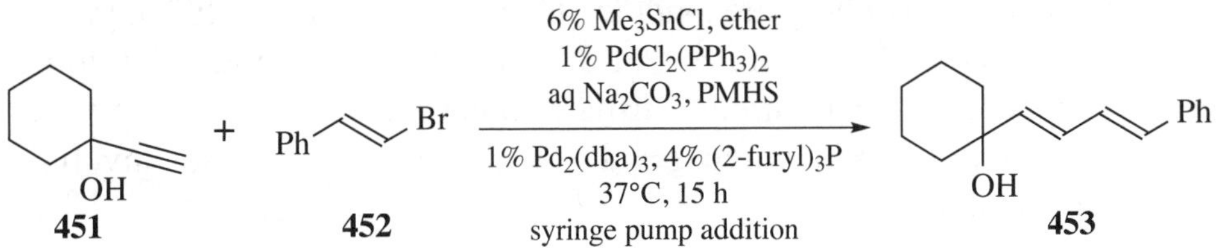

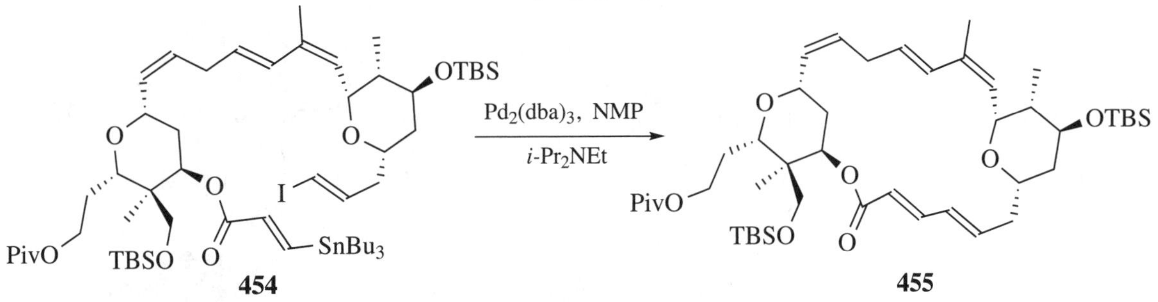

The Stille coupling disconnections are

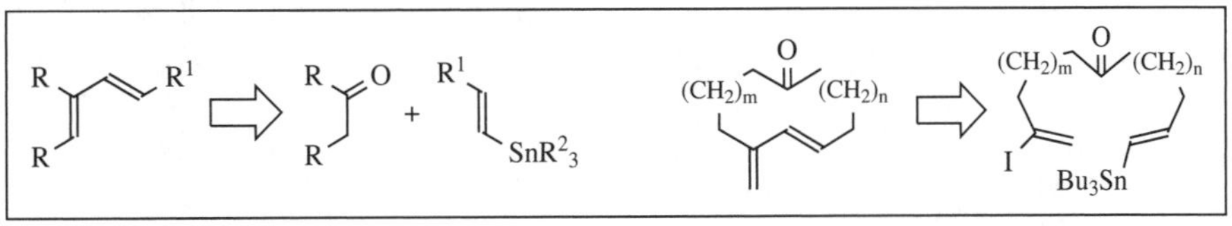

12.7.C. Suzuki-Miyaura Coupling

This reaction involves organoboronic acids [RB(OH)₂], so there is a connection with the chemistry presented in chapter 5 for organoboranes. The chemistry involves organopalladium

337. Espinet, P.; Echavarren, A.M. *Angew. Chem. Int. Ed.* **2004**, *43*, 4704.
338. Maleczka, Jr., R.E.; Gallagher, W.P.; Terstiege, I. *J. Am. Chem. Soc.* **2000**, *122*, 384.
339. The mechanism of the Stille coupling has been studied. See Casado, A.L.; Espinet, P.; Gallego, A.M. *J. Am. Chem. Soc.* **2000**, *122*, 11771.
340. Littke, A.F.; Fu, G.C. *Angew. Chem. Int. Ed.* **1999**, *38*, 2411.
341. For a review, see Duncton, M.A.J.; Pattenden, G. *J. Chem. Soc. Perkin Trans. 1* **1999**, 1235.
342. Song, H.Y.; Joo, J.M.; Kang, J.W.; Kim, D.-S.; Jung, C.-K.; Kwak, H.S.; Park, J.H.; Lee, E.; Hong, C.Y.; Jeong, S.W.; Jian, K.; Park, J.H. *J. Org. Chem.* **2003**, *68*, 8080.

catalysts so it is related to the Heck reaction and Stille coupling and is presented here as a synthetic route to biaryls.[343]

Suzuki and co-workers found that aryl triflates react with arylboronic acids [ArB(OH)$_2$],[344] in the presence of a palladium catalyst, to give biaryls in a reaction that is now known as **Suzuki coupling**[345] or **Suzuki-Miyaura coupling**. In an example taken from a synthesis of vialinin A by Takahashi and co-workers, boronic acid **457** reacted with aryl bis(triflate) **456** in the presence of Pd(PPh$_3$)$_4$ to give **458** in 96% yield.[346] Even hindered boronic acids give good yields of the coupled product.[347] The reaction is not restricted to simple aromatic compounds, as illustrated by the reaction of borylated indole **460** with iodopyrazine **459** to give a 76% yield of **461** in Stoltz and co-worker's synthesis of dragmacidin D.[348] In addition to boronic acid derivatives, organoboranes react with aryl triflates under these conditions to form biaryls.[349] A microwave-promoted Suzuki-type coupling was reported that did not involve the use of transition metal compounds.[350]

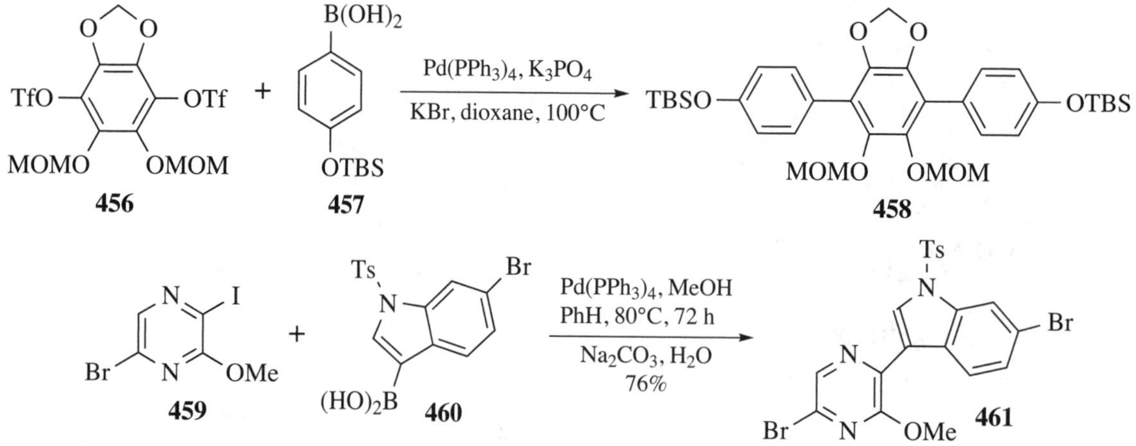

343. For reactions of this type, see reference 79, pp 362-364.
344. (a) Miyaura, N.; Yanagi, T.; Suzuki, A. *Synth. Commun.* **1981**, *11*, 513; (b) Cheng, W.; Snieckus, V. *Tetrahedron Lett.* **1987**, *28*, 5097; (c) Badone, D.; Baroni, M.; Cardomone, R.; Ielmini, A.; Guzzi, U. *J. Org. Chem.* **1997**, *62*, 7170.
345. (a) Miyaura, N.; Suzuki, A. *Chem. Rev.* **1995**, *95*, 2457; (b) Farinola, G.M.; Fiandanese, V.; Mazzone, L.; Naso, F. *J. Chem. Soc. Chem. Commun.* **1995**, 2523; (c) Genêt, J.P.; Linquist, A.; Blart, E.; Mouriès, V.; Savignac, M.; Vaultier, M. *Tetrahedron Lett.* **1995**, *36*, 1443; (d) The Merck Index, 14th ed., Merck & Co., Inc., Whitehouse Station, New Jersey, **2006**, p ONR-92; (e) Mundy, B.P.; Ellerd, M.G.; Favaloro Jr., F.G. *Name Reactions and Reagents in Organic Synthesis*, 2nd ed., Wiley-Interscience, New Jersey, **2005**, pp. 636-637.
346. Ye, Y.Q.; Koshino, H.; Onose, J.-i.; Yoshikawa, K.; Abe, N.; Takahashi, S. *Org. Lett.* **2007**, *9*, 4131.
347. (a) Watanabe, T.; Miyaura, N.; Suzuki, A. *Synlett* **1992**, 207; (b) Chaumeil, H.; Signorella, S.; Le Drian, C. *Tetrahedron* **2000**, *56*, 9655.
348. Garg, N.K.; Sarpong, R.; Stoltz, B.M. *J. Am. Chem. Soc.* **2002**, *124*, 13179.
349. Fürstner, A.; Seidel, G. *Synlett* **1998**, 161.
350. Leadbeaater, N.E.; Marco, M. *J. Org. Chem.* **2003**, *68*, 5660.

The reaction is not restricted to the formation of biaryls, and there are many synthetic applications.[351] Aryl boronic acids can also react with vinyl halides and vinyl triflates in the presence of palladium (0). A synthetic example is taken from Felpin and co-worker's synthesis of the isoflavanone, griffonianone D,[352] in which **462** reacted with the vinyl iodide unit in **463** to give an 73% yield of coupling product **464**. Another modification uses alkynes as coupling partners. In a synthesis of fiscuseptine,[353] Bracher and Daab coupled aryl iodide **465** with propargyl alcohol, in the presence of the palladium catalyst and CuI, to give an 81% yield of **466**.

Thallium (I) ethoxide[354] is known to promote Suzuki-Miyaura cross coupling reactions for vinyl or arylboronic acids with vinyl and aryl halide partners. A mixture of $Pd_2(dba)_3$ and tri-*tert*-butylphosphine has proven to be very effective.[355] Nickel on charcoal,[356] rhodium catalysts,[357] and silver oxide-mediated palladium catalysts[358] have also been used. The latter catalyst was effective for the preparation of 4-substituted 2(5*H*)-furanones.[359] Palladium-on-carbon was shown to be a reusable catalyst in aqueous media,[360] as in the preparation of **464**.

Another variation of this coupling reaction modified the boronic acid moiety. Molander and Rivero have shown the use of potassium trifluoroborates and a palladium catalyst leads to cross coupling reaction with aryl and vinyl halides.[361] In a synthesis of serofendic acids A and B, Toyota, Ihara and Asano coupled potassium trifluorovinylborate with vinyl bromide **467** to give a 51% yield of **468**.[362] Potassium organotrifluoroborates (RBF_3K) are prepared by the

351. Kotha, S.; Lahiri, K.; Kashinath, D. *Tetrahedron* **2002**, *58*, 9633.
352. Felpin, F.-X.; Lory, C.; Sow, H.; Acherar, S. *Tetrahedron* **2007**, *63*, 3010.
353. Bracher, F.; Daab, J. *Eur. J. Org. Chem.* **2002**, 2288.
354. Frank, S.A.; Chen, H.; Kunz, R.K.; Schnaderbeck, M.J.; Roush, W.R. *Org. Lett.* **2000**, *2*, 2691.
355. Littke, A.F.; Dai, C.; Fu, G.C. *J. Am. Chem. Soc.* **2000**, *122*, 4020.
356. Lipshutz, B.H.; Sclafani, J.A.; Blomgren, P.A. *Tetrahedron* **2000**, *56*, 2139.
357. Ueda, M.; Miyaura, N. *J. Org. Chem.* **2000**, *65*, 4450.
358. Chen, H.; Deng, M.-Z. *J. Org. Chem.* **2000**, *65*, 4444.
359. Yao, M.-L.; Deng, M.-Z. *J. Org. Chem.* **2000**, *65*, 5034.
360. Sakurai, H.; Tsukuda, T.; Hirao, T. *J. Org. Chem.* **2002**, *67*, 2721.
361. Molander, G.A.; Rivero, M.R. *Org. Lett.* **2002**, *4*, 107.
362. Toyota, M.; Asano, T.; Ihara, M. *Org. Lett.* **2005**, *7*, 3929.

addition of inexpensive KHF_2 to organoboron intermediates,[363] and they can be used in several of the applications where boronic acids or esters are used.[364]

12.7.D. Sonogashira and Stephens-Castro Coupling

Under certain conditions, alkynes can be coupled to aryl halides.[365] When aryl halides react with copper acetylides to give 1-aryl alkynes such as **469**, the reaction is known as **Castro-Stephens coupling.**[366] Both aliphatic and aromatic substituents can be attached to the alkyne unit, and a variety of aryl iodides have been used. A palladium-

catalyzed variation is also known in which an aryl halide or a vinyl derivative reacts with a terminal alkyne to give derivatives such as **472**. In this variation, called the **Sonogashira coupling,**[367] the reaction is catalyzed by a palladium-copper complex at or near ambient temperatures. In a synthesis of the maduropeptin chromophore aglycon, Inoue, Hirama and co-worker's reacted vinyl triflate **470** and alkyne **471** with palladium(0) in the presence of

363. (a) Vedejs, E.; Chapman, R.W.; Fields, S.C.; Lin, S.; Schrimpf, M.R. *J. Org. Chem.* **1995**, *60*, 3020; (b) Vedejs, E.; Fields, S.C.; Hayashi, R.; Hitchcock, S.R.; Powell, D.R.; Schrimpf, M.R. *J. Am. Chem. Soc.* **1999**, *121*, 2460.

364. (a) Molander, G.A.; Ito, T. *Org. Lett.* **2001**, *3*, 393; (b) Molander, G.A.; Biolatto, B. *Org. Lett.* **2002**, *4*, 1867; (c) Molander, G.A.; Biolatto, B. *J. Org. Chem.* **2003**, *68*, 4302; (d) Molander, G.A.; Katona, B.W.; Machrouhi, F. *J. Org. Chem.* **2002**, *67*, 8416; (e) Molander, G.A.; Yun, C.; Ribagorda, M.; Biolatto, B. *J. Org. Chem.* **2003**, *68*, 5534; (f) Molander, G.A.; Ribagorda, M. *J. Am. Chem. Soc.* **2003**, *125*, 11148.

365. For reactions of this type, see reference 79, pp 596-599.

366. (a) Castro, C.E.; Stephens, R.D. *J. Org. Chem.* **1963**, *28*, 2163; (b) Stephens, R.D.; Castro, C.E. *J. Org. Chem.* **1963**, *28*, 3313; (c) Sladkov, A.M.; Ukhin, L.Yu.; Korshak, V.V. *Bull. Acad. Sci. USSR., Div. Chem. Sci.* **1963**, 2043; (d) Sladkov, A.M.; Gol'ding, I.R. *Russ. Chem. Rev.* **1979**, *48*, 868; (e) Bumagin, N.A.; Kalinovskii, I.O.; Ponomarov, A.B.; Beletskaya, I.P. *Doklad. Chem.* **1982**, *265*, 262; (f) The Merck Index, 14th ed., Merck & Co., Inc., Whitehouse Station, New Jersey, **2006**, p ONR-16; (g) Mundy, B.P.; Ellerd, M.G.; Favaloro Jr., F.G. *Name Reactions and Reagents in Organic Synthesis, 2nd ed.*, Wiley-Interscience, New Jersey, **2005**, pp. 136-137.

367. (a) Sonogashira, K.; Tohda, Y.; Hagihara, N. *Tetrahedron Lett.* **1975**, 4467; (b) Sonogashira, K. in *Comprehensive Organic Synthesis*, Trost, B.M.; Fleming, I. (Eds.), Pergamon Press, New York, **19991**, Vol. 3, chapter 2.4; (c) Rossi, R.; Carpita, A.; Bellina, F. *Org. Prep. Proceed. Int.* **1995**, *27*, 127; (d) Sonogashira, K. in *Metal-Catalyzed Cross-Coupling Reactions*, Diederich, F.; Stang, P.J. (Eds.), Wiley-VCH, New York, **1998**, chapter 5; (e) The Merck Index, 14th ed., Merck & Co., Inc., Whitehouse Station, New Jersey, **2006**, p ONR-88; (f) Mundy, B.P.; Ellerd, M.G.; Favaloro Jr., F.G. *Name Reactions and Reagents in Organic Synthesis, 2nd ed.*, Wiley-Interscience, New Jersey, **2005**, pp. 612-613.

cuprous iodide to give **472**.[368] Several new reaction variations have brought renewed attention to the coupling. A copper-free Sonogashira coupling has been reported in an ionic liquid medium.[369] A combination of Pd(PhCN)$_2$Cl$_2$ and (*t*-Bu)$_3$P is very effective for Sonogashira coupling at room temperature,[370] and the reaction has been done without solvent, on alumina, using microwave irradiation.[371]

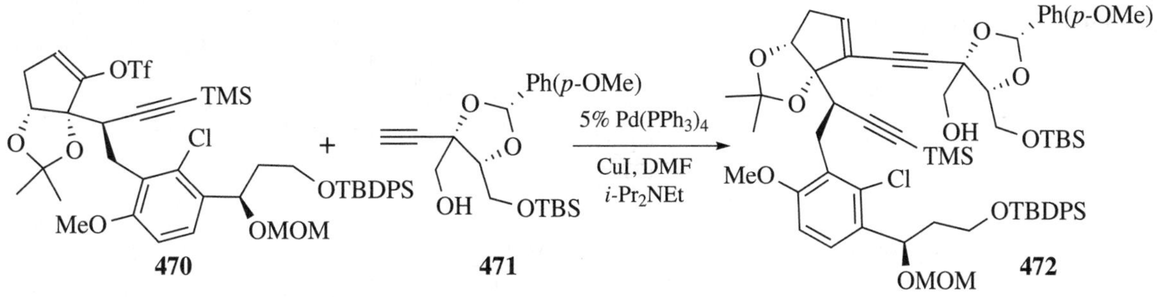

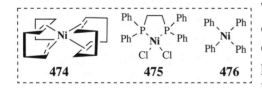

470 + **471** → **472**

12.8. π-ALLYL NICKEL COMPLEXES

Nickel can stabilize allylic cations in a manner similar to palladium.[372] Indeed, the use of organonickel complexes for generating carbon-carbon bonds[373] is synthetically older than the analogous organopalladium chemistry. Nickel chloride (NiCl$_2$) is the nickel analog of PdCl$_2$ and reacts similarly to form zero-valent complexes [Ni(0)], which are the reactive species in π-allyl nickel complexes

such as **473**. A common Ni(0) complex is nickel tetracarbonyl [Ni(CO)$_4$], but this is a volatile, *extremely toxic* and dangerous material.

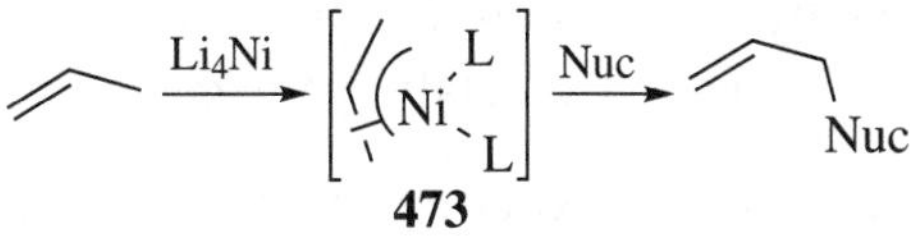

*Nickel tetracarbonyl should be used with **great caution**
and only in a specially ventilated hood.*

474   **475**   **476**

Tetrakis(triphenoxyphosphino)nickel is also a useful organonickel reagent. The *bis*-complex of nickel and cyclooctadiene (cod) **474** [Ni(cod)$_2$], bis(diphenyl-phosphinoethane)nickel chloride (**475**) and tetraphenyl-nickel (**476**) are widely used Ni(0) complexes.

368. Komano, K.; Shimamura, S.; Inoue, M.; Hirama, M. *J. Am. Chem. Soc.* **2007**, *128*, 14184.
369. Fukuyama, T.; Shinmen, M.; Nishitani, S.; Sato, M.; Ryu, I. *Org. Lett.* **2002**, *4*, 1691.
370. Hundertmark, T.; Littke, A.F.; Buchwald, S.L.; Fu, G.C. *Org. Lett.* **2000**, *2*, 1729.
371. Kabalka, G.W.; Wang, L.; Namboodiri, V.; Pagni, R.M. *Tetrahedron Lett.* **2000**, *41*, 5151.
372. For reactions of this type, see reference 79, pp 116-117.
373. For a review, see Montgomery, J. *Acc. Chem. Res.* **2000**, *33*, 467.

A typical reaction involving nickel(0) treats an allylic halide such as 3-chloro-2-methyl-1-propene with Ni(CO)$_4$ to form **477**,[374] a nickel dimer analogous to the palladium chloride dimer discussed in the previous section. This complex reacts with an alkene to form a new complex (**478**), and subsequent reaction with a variety of electrophilic reagents removes nickel. In this example, the chlorine was converted to an acetate ligand (in **479**) and reacted with methanolic carbon monoxide to give the ester (**480**).[374]

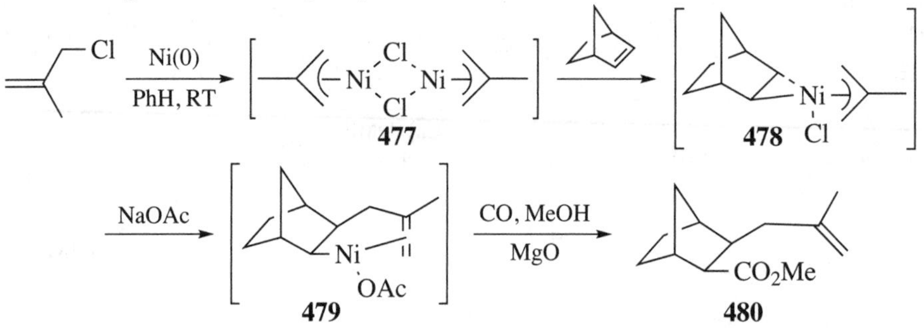

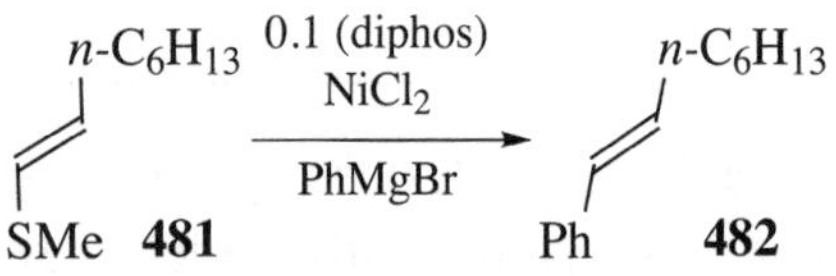

π-Nickel complexes facilitate the coupling of vinyl ethers and vinyl sulfides with Grignard reagents. Wenkert and co-workers showed that **481** was converted to **482** in 80% yield (as a 4:1 trans/cis mixture) by reaction with phenylmagnesium bromide and bis(diphenylphosphinoethane)nickel chloride.[375] Zero-valent nickel can also catalyze the coupling of aryl halides and alkenes.[376,375a,b]

A nickel-catalyzed ene cyclization (sec. 11.13) has been reported that uses Ni(cod)$_2$. The reaction proceeds by initial formation of a π-allylnickel complex, which facilitates the intramolecular ene reaction with an allylic amine unit.[377] π-Allylnickel complexes can be used in coupling reactions with both aryl and alkyl halides. Enolate anions react with nickel(0) reagents to form a complex that subsequently couples to aryl iodides. Semmelhack's final step in the

synthesis of cephalotaxinone (**484**) treated **483** with Ni(cod)$_2$ to produce **484** in 94% yield.[378] The nickel coupling reaction is not limited to aryl halides. Alkyl halides also react, as seen in Marshall and Wuts'

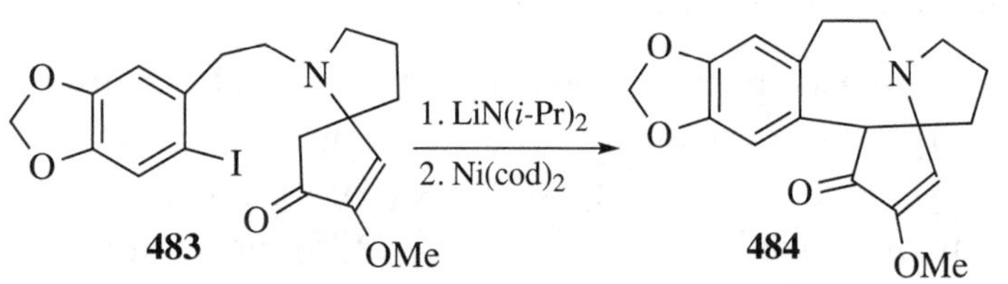

374. (a) Gallazzi, M.C.; Hanlon, T.L.; Vitulli, G.; Porri, L. *J. Organomet. Chem.* **1971**, *33*, C45; (b) Hughes, R.P.; Powell, J. *Ibid.* **1971**, *30*, C45; (c) Guerrieri, F.; Chiusoli, G.P. *Ibid.* **1968**, *15*, 209.
375. (a) Wenkert, E.; Ferreira, T.W.; Michelotti, E.L. *J. Chem. Soc. Chem Commun.* **1979**, 637; (b) Wenkert, E.; Michelotti, E.L.; Swindell, C.S. *J. Am. Chem. Soc.* **1979**, *101*, 2246.
376. Mori, M; Ban, Y. *Tetrahedron Lett.* **1976,** 1803, 1807.
377. Oppolzer, W.; Bedoya-Zurita, M.; Switzer, C.Y. *Tetrahedron Lett.* **1988**, *29*, 6433.
378. Semmelhack, M.F.; Chong, B.P.; Stauffer, R.D.; Rogerson, T.D.; Chong, A.; Jones, L.D. *J. Am. Chem. Soc.* **1975**, *97*, 2507.

coupling of **485** with the π-allyl nickel dimer **486** to give dictyolene **487**.[379]

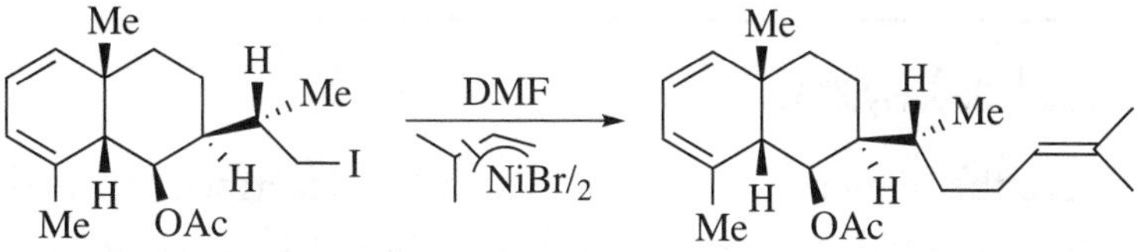

485　　**486**　　**487**

The nickel(0) coupling disconnections are:

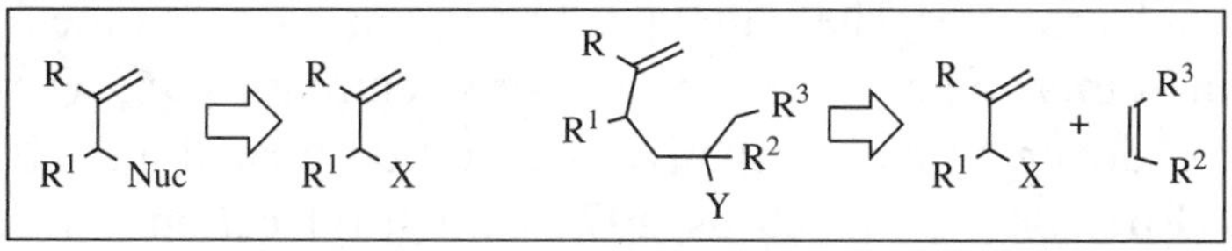

Zero-valent nickel is especially effective for the cyclization of *bis*(allyl) bromides.[380,381] Corey used this technique to cyclize **488** to **489**. Subsequent photolysis in the presence of diphenyl disulfide (PhSSPh) gave humulene, **490**.[380] Wender et al. reported a novel [4+4]-cycloaddition reaction in which two diene moieties are coupled with a nickel(0) catalyst to form eight-membered rings.[382] In a synthesis of asteriscanolide, **491** was converted to **492** in 67% yield using the nickel cyclooctadiene catalyst (**474**), with < 5% of the [4+2]-cycloaddition product.[382] In this particular case, attempts to use a palladium-catalyzed reaction gave poor yields, presumably due to instability of the vinyl iodide intermediates.

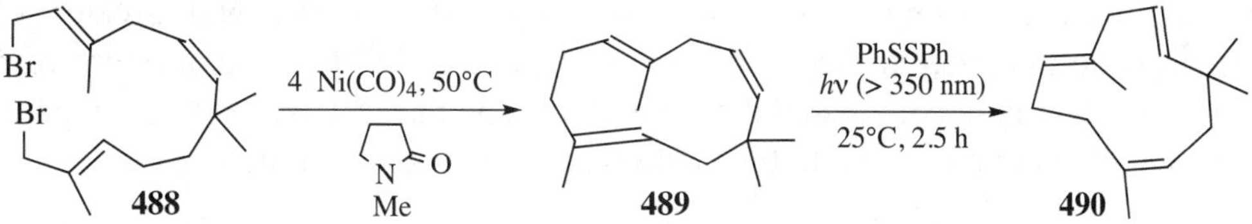

488　　**489**　　**490**

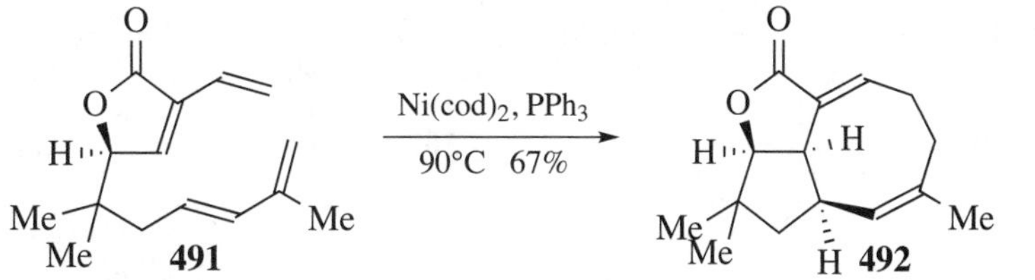

491　　**492**

The nickel catalyzed cyclization disconnections are

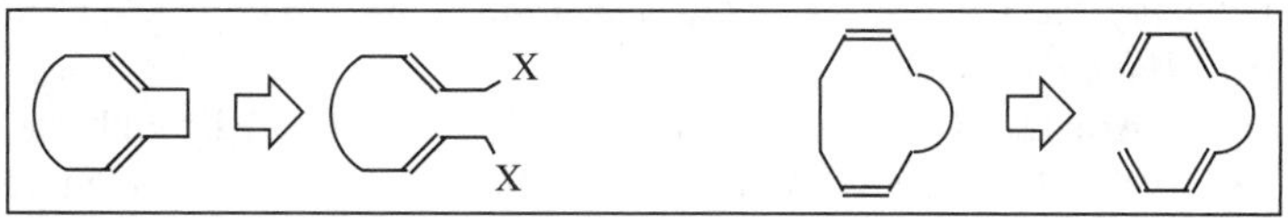

379. Marshall, J.A.; Wuts, P.G.M. *J. Am. Chem.Soc. 1978, 100*, 1627.
380. (a) Corey, E.J.; Hamanaka, E. *J. Am. Chem. Soc. 1967, 89*, 2758; (b) *Idem Ibid. 1964, 86*, 1641; (c) Corey, E.J.; Wat, E.K.W. *Ibid. 1967, 89*, 2757.
381. Corey, E.J.; Kirst, H.A. *J. Am. Chem. Soc. 1972, 94*, 667.
382. Wender, P.A.; Ihle, N.C.; Correia, C.R.D. *J. Am. Chem. Soc. 1988, 110*, 5904.

12.9. ELECTROPHILIC IRON COMPLEXES

12.9.A. Alkene and Diene Iron Complexes

In iron reactions where the reagent was equivalent to C^d, described in Section 8.9.A, the iron moiety was used as an auxiliary. Iron can also stabilize cations, which then react with nucleophiles to generate new carbon-carbon bonds.[383] These cations are formed as iron-alkene complexes, usually by reaction of cyclopentadienyl dicarbonyl ferrate anion (**494**) with an allylic halide such as 3-chloro-2-methyl-1-propene. The initial product in this case was the σ-iron adduct (**495**). Subsequent reaction with an acid generated the π-iron complex **496**, where the double bond was a two-π-electron donor.[384] An alternative preparation treated an alkyl cyclopentadienyl dicarbonyl ferrate (**Fp**) complex (such as **497**) with trityl cation (Ph_3C^+) to form **498**.[385]

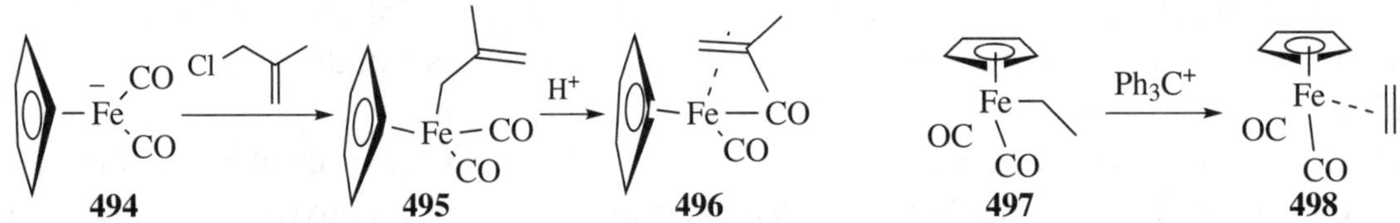

Diene complexes can also be formed by reaction iron pentacarbonyl [$Fe(CO)_5$].[386] The presence and nature of the substituents on the diene will influence the regiochemical outcome of the complex, as does the method of preparation. Treatment of **499** with iron pentacarbonyl under thermal conditions, for example, gave a 4:1 mixture of **500** and **501**, with greater selectivity for reaction at the unhindered face. Photolysis, however, led to a 1:1 mixture of **500** and **501**.[387] Treatment of an unconjugated dienes with iron pentacarbonyl led to conjugated dienyl complexes, which were converted to the isomeric π complexes with trityl cation.[388]

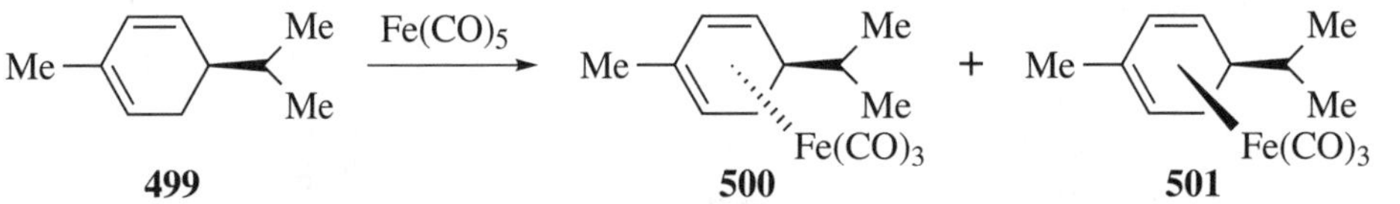

383. For reactions of this type, see reference 79, p 445.
384. Green, M.L.H.; Nagy, P.L.I. *J. Chem. Soc.* **1963**, 189.
385. (a) Green, M.L.H.; Nagy, P.L.I. *J. Organomet. Chem.* **1963**, *1*, 58; (b) Laycock, D.E.; Hartgerink, J.; Baird, M.C. *J. Org. Chem.* **1980**, *45*, 291.
386. (a) Whitesides, T.M.; Arhart, R.W. *Inorg. Chem.* **1975**, *14*, 209; (b) Brookhart, M.; Whitesides, T.M.; Crockett, J.M. *Ibid.* **1976**, *15*, 1550; (c) Whitesides, T.M.; Arhart, R.W.; Slaven, R.W. *J. Am. Chem. Soc.* **1973**, *95*, 5792; (d) Impastato, F.J.; Ihrman, K.G. *Ibid.* **1961**, *83*, 3726.
387. (a) McArdle, P.; Higgins, T. *Inorg. Chim. Acta* **1978**, *30*, L303; (b) Davies, S.G. *Organotransition Metal Chemistry: Applications to Organic Synthesis*, Pergamon Press, Oxford, **1982**, p 54.
388. (a) Birch, A.J.; Chamberlain, K.B. *Org. Synth. Coll. Vol. 6* **1988**, 996; (b) Birch, A.J.; Cross, P.E.; Lewis, J.; White, D.A.; Wild, S.B. *J. Chem. Soc. A* **1968**, 332; (c) Birch, A.J.; Williamson, D.H. *J. Chem. Soc. Perkin Trans. 1* **1973**, 1892.

12.9.B. Reactions of Iron Complexes with Nucleophiles

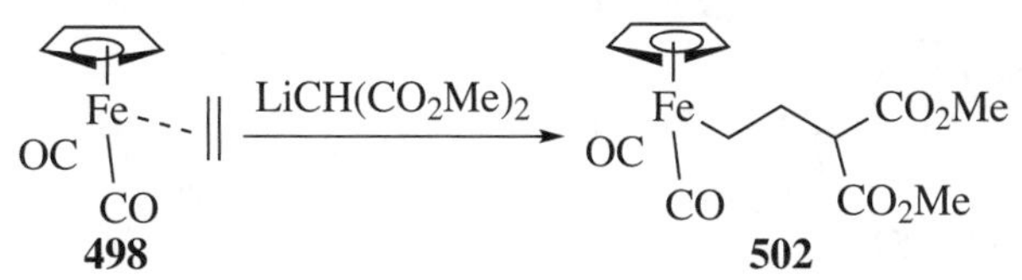

498 **502**

Alkyl iron complexes behave as electrophiles in the presence of nucleophiles. Rosenblum and Rosan showed that the ethene complex of cyclopentadienyl dicarbonyl iron (abbreviated as **Fp**), Fp-ethene (**498**),
reacted with dimethyl lithiomalonate to give **502**.[389] Conjugated iron complexes can also be prepared.[388] These Fp-alkene complexes react with a variety of other nucleophiles,[390] including water, alcohols, amines,[391] phosphines or thiols as well as carbon nucleophiles (enamines, organocuprates, enolates),[392,393] and dialkyl cadmium reagents.[394] Diene complexes such as **503** were converted to the corresponding cationic complexes (**504**) and reaction with malonate gave **505**. The iron complex was removed with trimethylamine N-oxide (Me₃N–O) to give **506**.[395]

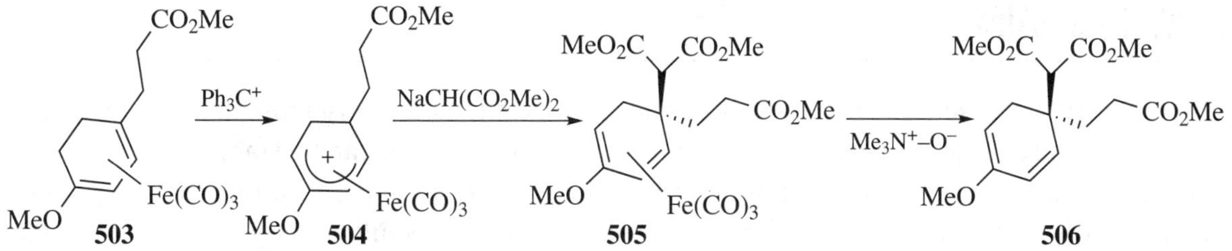

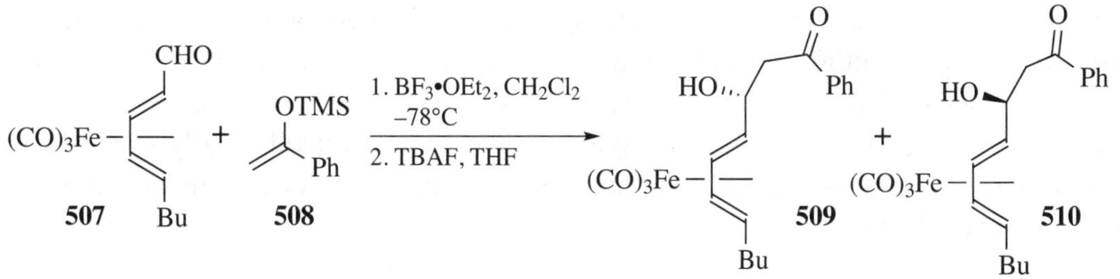

503   **504**   **505**   **506**

507 **508**   **509**   **510**

The iron carbonyl unit is a protecting group for dienes. Iron diene complex **507**,[396] for example, reacted with **508** in a Mukaiyama aldol reaction (sec. 9.4.C) to give an 81% yield of **509** and **510** in an 84:16 ratio. In a synthesis of heptitol derivatives,[397] Pearson and Katiyar protected tropone as the iron tricarbonyl derivative (**511**), and prepared **512** using a multistep sequence

389.   Rosan, A.; Rosenblum, M. *J. Org. Chem.* **1975**, *40*, 3621.
390.   Davies, S.G. *Organotransition Metal Chemistry: Applications to Organic Synthesis*, Pergamon Press, Oxford, **1982**.
391.   Berryhill, S.R.; Rosenblum, M. *J. Org. Chem.* **1980**, *45*, 1984.
392.   (a) Reference 390, p 132; (b) Busetto, L.; Palazzi, A.; Ros, R.; Belluco, U. *J. Organomet. Chem.* **1970**, *25*, 207; (c) Lennon, P.; Madhavarao, M.; Rosan, A.; Rosenblum, M. *Ibid.* **1976**, *108*, 93.
393.   Reference 390, p 142.
394.   Bayoud, R.S.; Biehl, E.R.; Reeves, P.C. *J. Organomet. Chem.* **1979**, *174*, 297.
395.   Pearson, A.J. *J. Chem. Soc. Perkin Trans. 1* **1979**, 1255.
396.   Harvey, D.F.; Selchau, V.B. *J. Org. Chem.* **2000**, *65*, 2282.
397.   Pearson, A.J.; Katiyar, S. *Tetrahedron* **2000**, *56*, 2297.

before deprotecting the diene to give **513**.

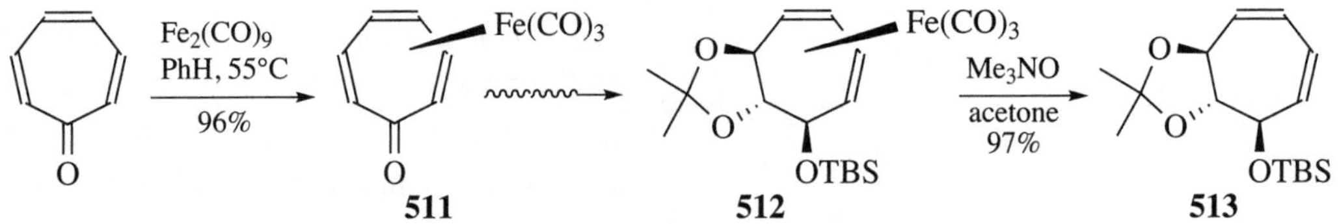

The disconnections for this section are

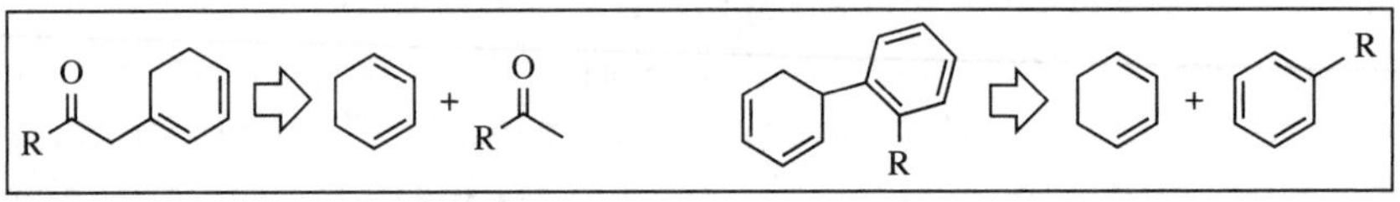

12.10. CONCLUSION

This chapter discussed chemical equivalents for C^a, cations and organometallics. Although carbocations are subject to racemization and rearrangement, they can be stabilized so they can react with carbon nucleophiles ranging from alkenes and alkynes, to aromatic compounds, to Grignard reagents or enolate anions. Cations can be used synthetically for intramolecular coupling and/or if a tertiary or oxygen stabilized cation is generated. The Friedel-Crafts reactions are among the more powerful and useful of all organic reactions for constructing carbon-carbon bonds when aromatic nuclei are involved. The more recent contributions of metal stabilized cations, including π-allylpalladium, π-allylnickel, π-allyliron and other organometallic reagents allow nucleophiles such as Grignard reagents and enolate anions to be coupled with those C^a species.

In Chapter 13, methods for generating carbon-carbon bonds via radical intermediates and carbenes will be discussed, completing the collection of chapters for generating carbon-carbon bonds in organic synthesis.

HOMEWORK

1. Friedel-Crafts alkylation is not a useful synthetic method for the synthesis of primary arenes such as 1-phenylhexane. Explain why not.

2. Briefly explain the regioselectivity for this reaction.

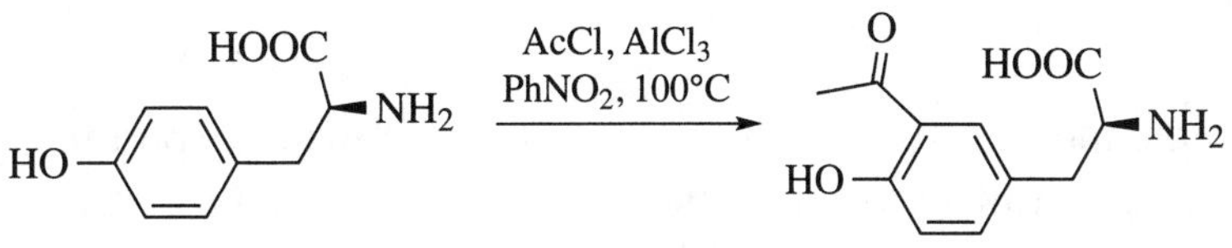

3. Provide a mechanistic rationale for the following transformation:

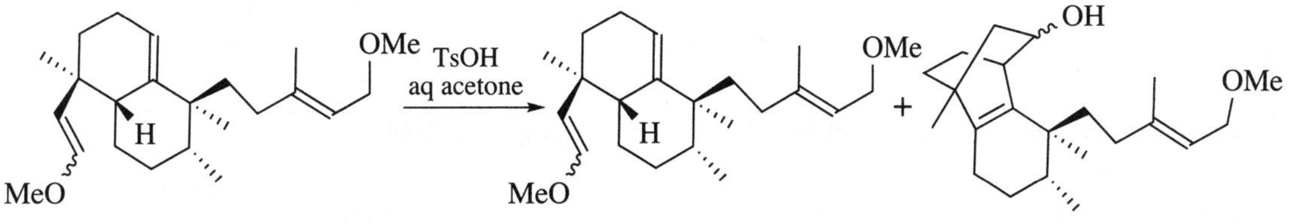

4. Offer a mechanistic rationale for the following transformation:

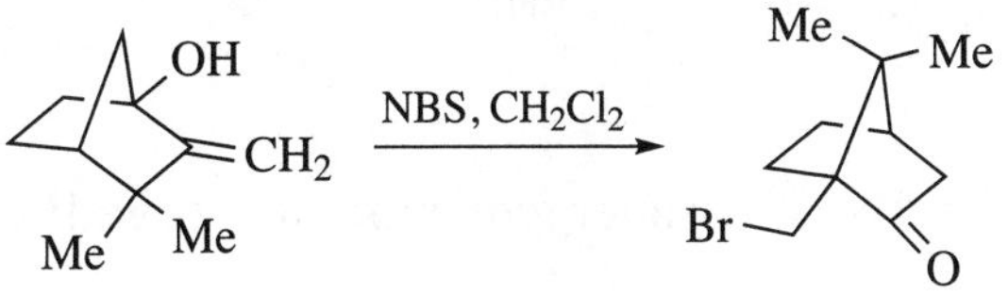

5. Give a mechanistic rationale for the following transformation:

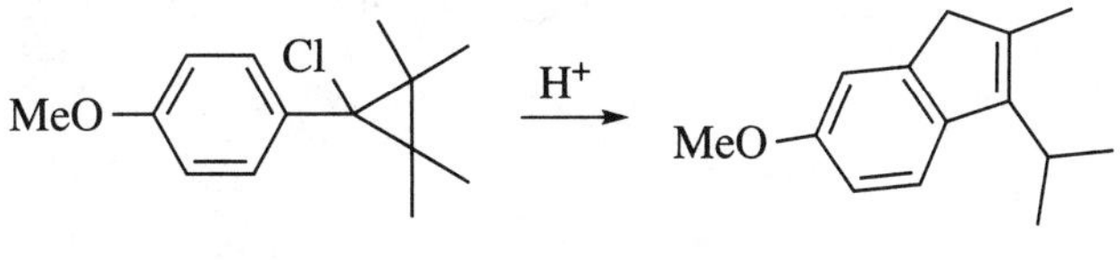

6. Give a mechanistic rationale for this transformation:

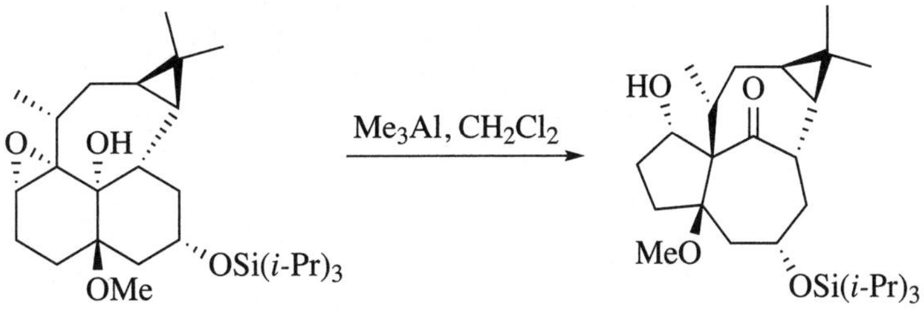

7. Give the complete mechanism for the following transformation.

thebaine $\xrightarrow{\text{dil HCl}}$ thebenine

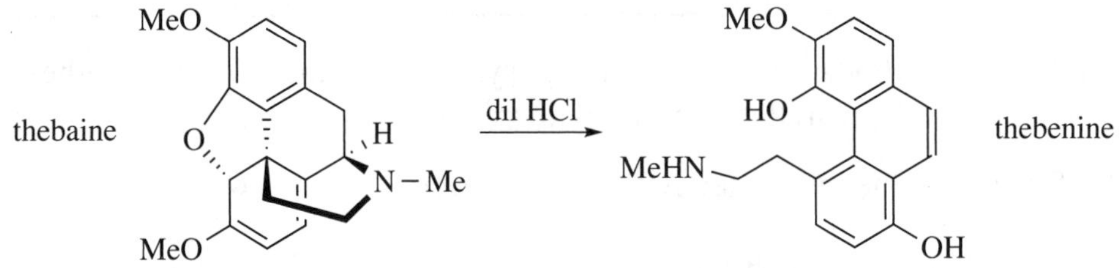

8. What are the mechanistic implications of the following experiment? Of what possible value is this experiment to synthesis if the R–NH$_2$ → R–OH transformation is planned?

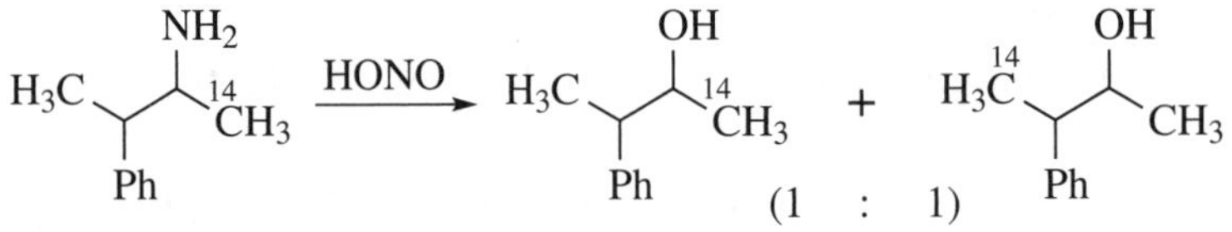

(1 : 1)

9. Give the complete mechanism for the following transformation.

$\xrightarrow[\text{toluene/THF, 2 h}]{\text{HCOOH, 0°C}}$

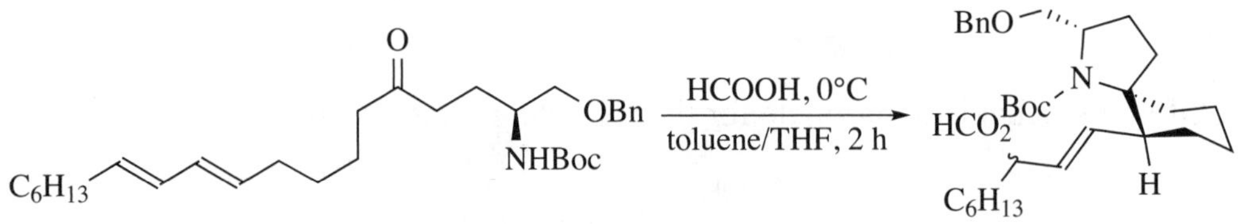

10. Explain the fact that Wolff-Kishner reduction of A gave B but Clemmenson reduction of A gave C.

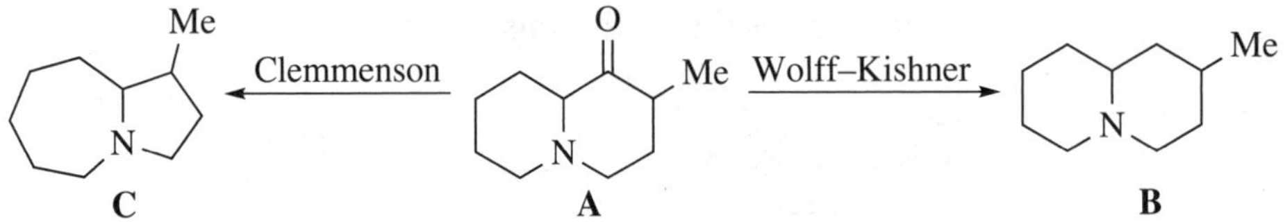

C $\xleftarrow{\text{Clemmenson}}$ A $\xrightarrow{\text{Wolff–Kishner}}$ B

11. Give a reasonable mechanism for the following transformation.

$\xrightarrow[\text{0°C}]{\text{BF}_3\text{•OEt}_2, \text{CH}_2\text{Cl}_2}$

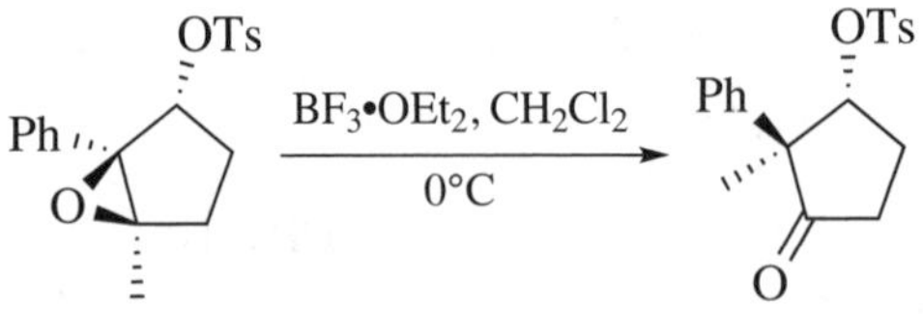

12. Provide a mechanism for the following transformation:

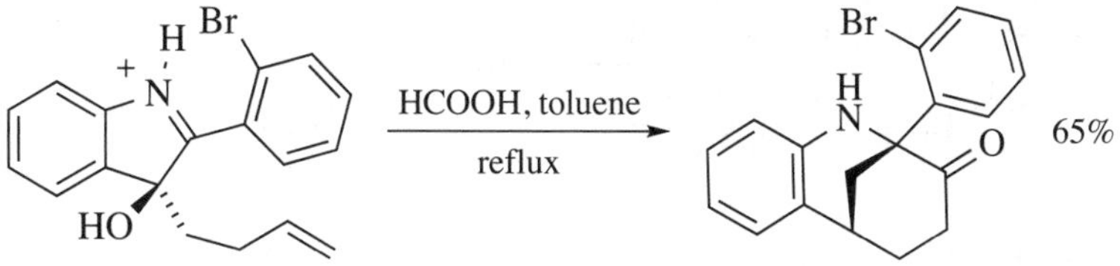

HCOOH, toluene
reflux

65%

13. For each of the following, give the major product. Show the correct stereochemistry where appropriate.

(a)

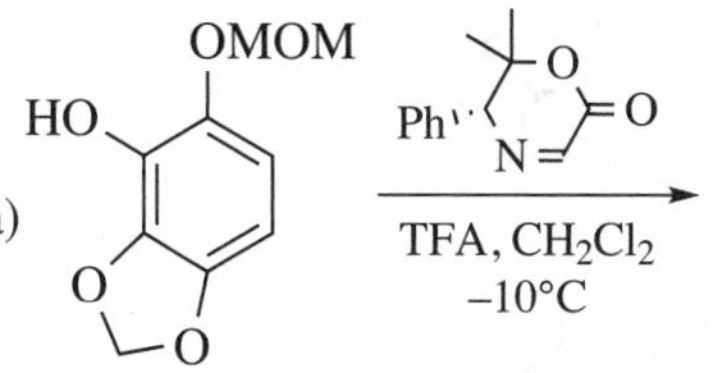

OMOM
HO

Ph'''N=

O

O

TFA, CH₂Cl₂
−10°C

(b)

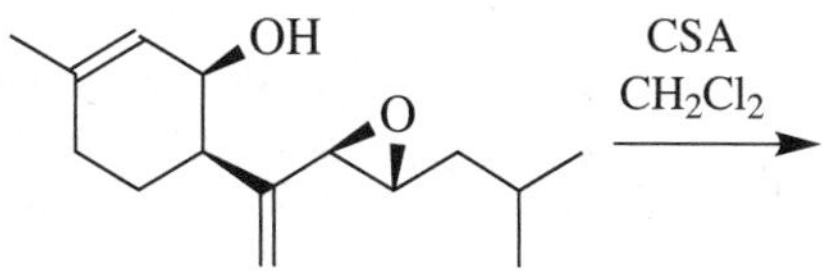

OH

CSA
CH₂Cl₂

(c)

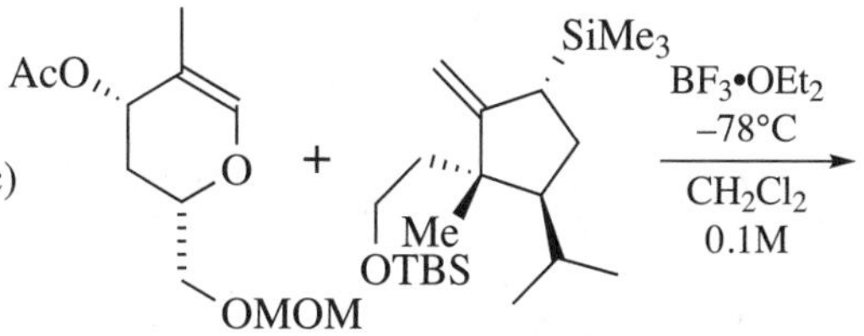

AcO

O

OMOM

+

SiMe₃

Me
OTBS

BF₃•OEt₂
−78°C

CH₂Cl₂
0.1M

(d)

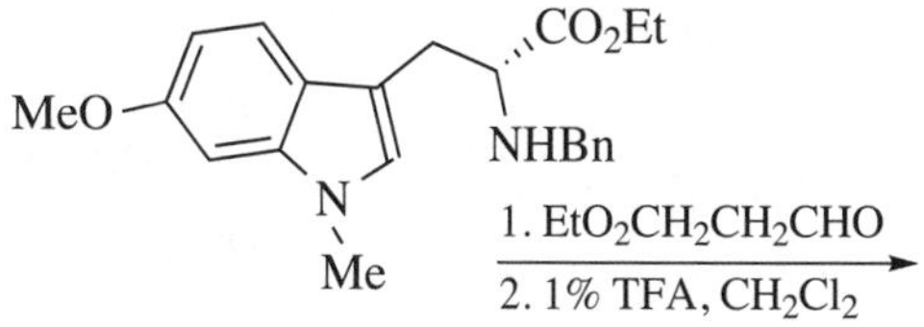

MeO

CO₂Et

NHBn

N
Me

1. EtO₂CH₂CH₂CHO
2. 1% TFA, CH₂Cl₂

(e)

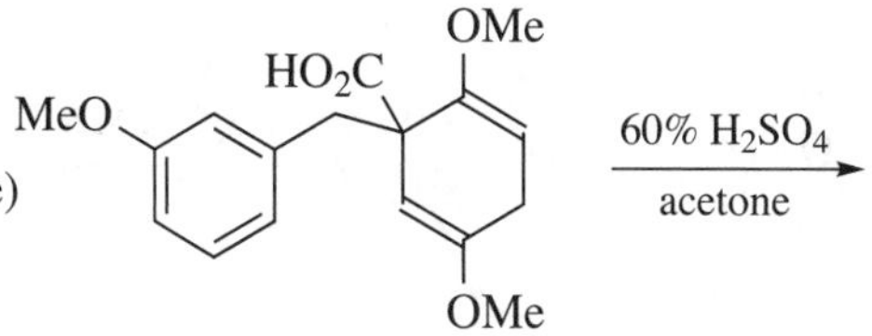

MeO

HO₂C

OMe

OMe

60% H₂SO₄

acetone

(f)

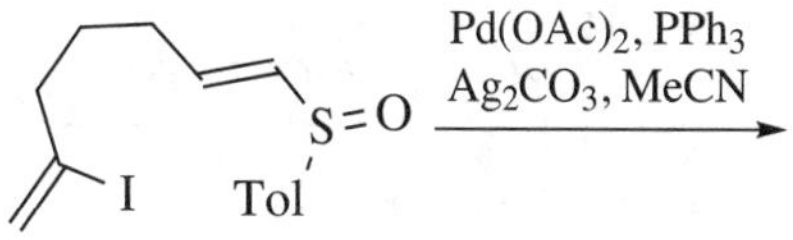

I

S=O

Tol

Pd(OAc)₂, PPh₃
Ag₂CO₃, MeCN

(g)

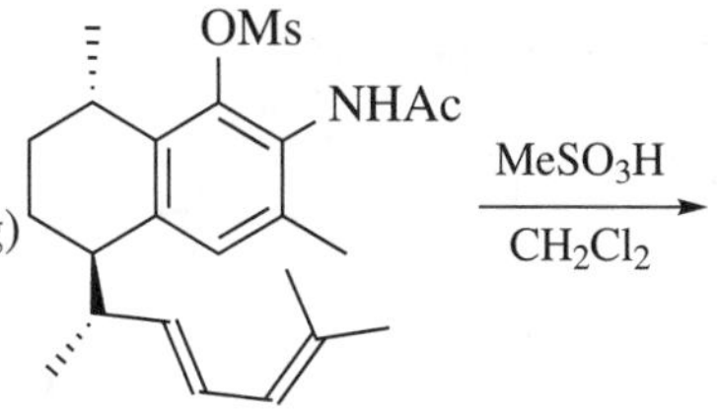

OMs

NHAc

MeSO₃H

CH₂Cl₂

(h)

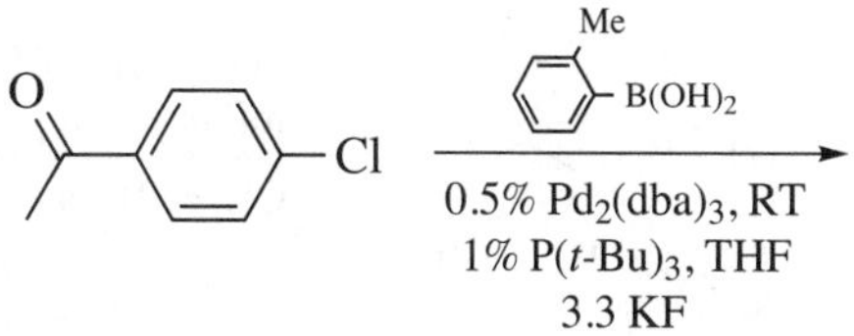

O

Cl

Me

B(OH)₂

0.5% Pd₂(dba)₃, RT
1% P(t-Bu)₃, THF
3.3 KF

(i)

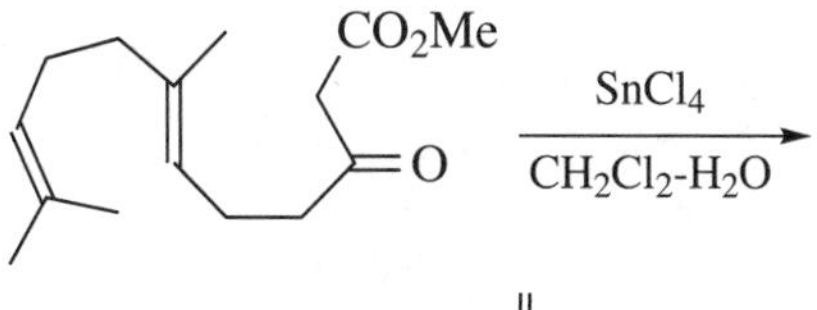

CO₂Me

O

SnCl₄

CH₂Cl₂-H₂O

(j)

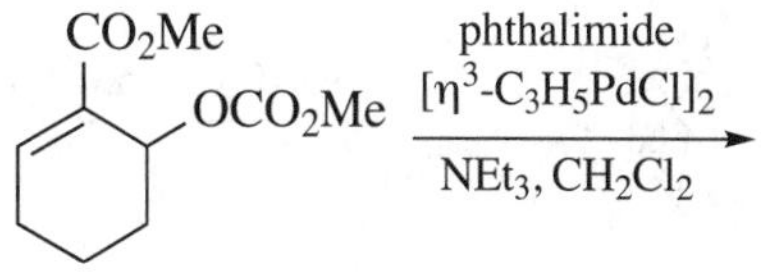

CO₂Me

OCO₂Me

phthalimide
[η³-C₃H₅PdCl]₂

NEt₃, CH₂Cl₂

(k)

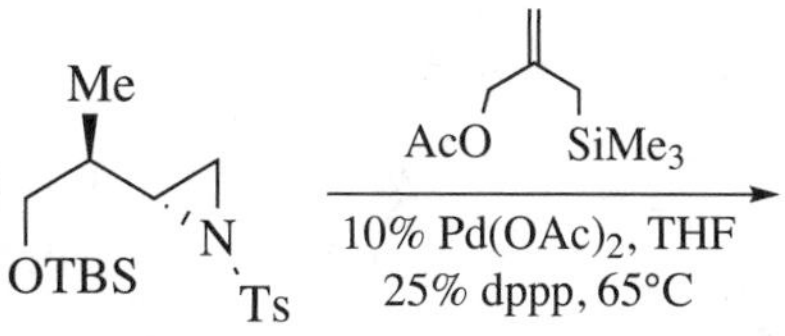

Me

OTBS

N
Ts

AcO   SiMe₃

10% Pd(OAc)₂, THF
25% dppp, 65°C

(l)

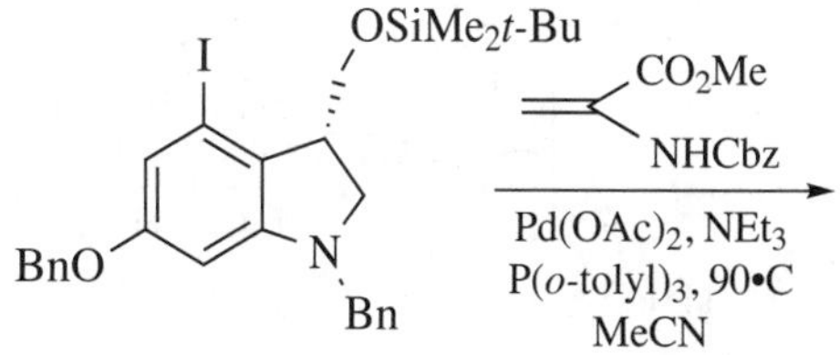

I

OSiMe₂t-Bu

BnO

N
Bn

CO₂Me

NHCbz

Pd(OAc)₂, NEt₃
P(o-tolyl)₃, 90•C
MeCN

(m) 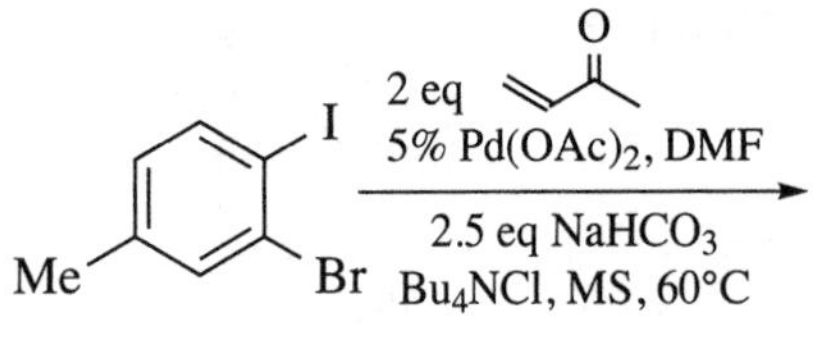 2 eq [methyl vinyl ketone]
5% Pd(OAc)$_2$, DMF
2.5 eq NaHCO$_3$
Bu$_4$NCl, MS, 60°C

(n) 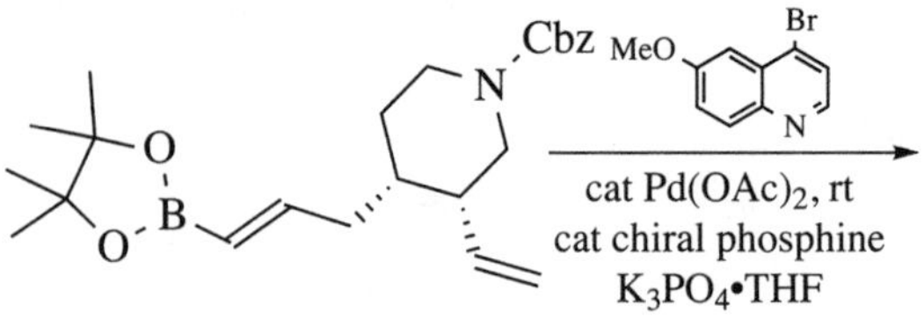 Cbz MeO Br
cat Pd(OAc)$_2$, rt
cat chiral phosphine
K$_3$PO$_4$•THF

(o) 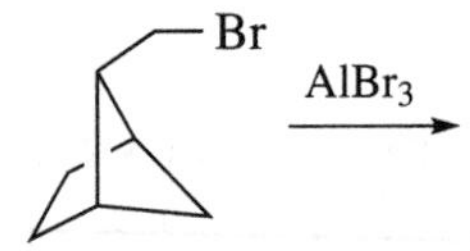 Br
AlBr$_3$

(p) 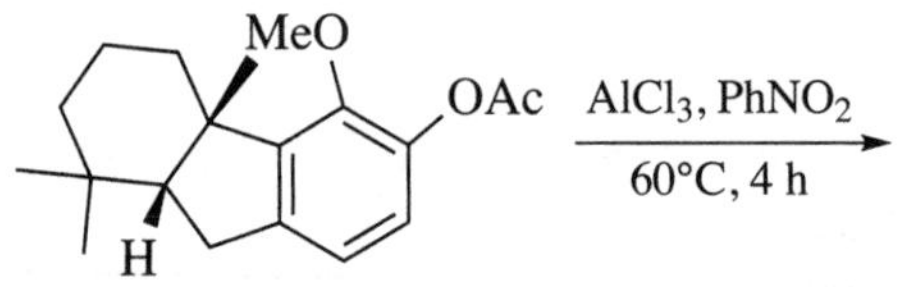 MeO OAc
AlCl$_3$, PhNO$_2$
60°C, 4 h

(q) 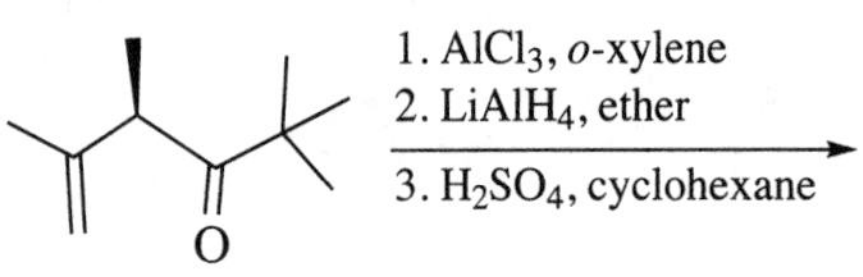 1. AlCl$_3$, o-xylene
2. LiAlH$_4$, ether
3. H$_2$SO$_4$, cyclohexane

(r) 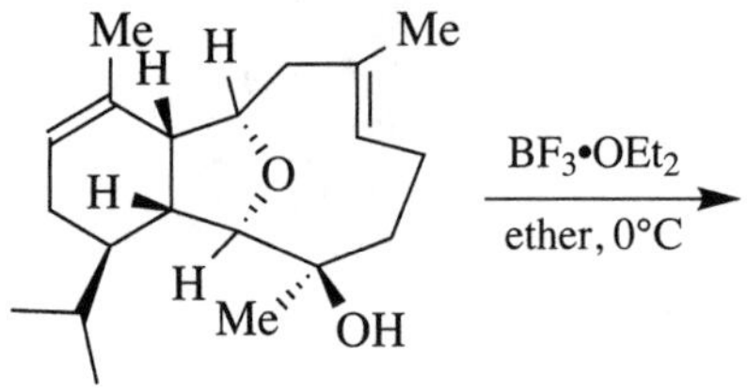 Me H H Me
H
O
H
H Me OH
iPr
BF$_3$•OEt$_2$
ether, 0°C

(s) 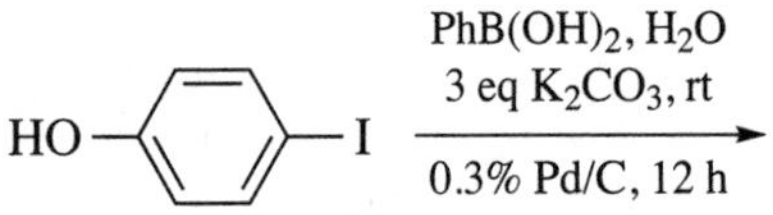 HO—I
PhB(OH)$_2$, H$_2$O
3 eq K$_2$CO$_3$, rt
0.3% Pd/C, 12 h

(t) 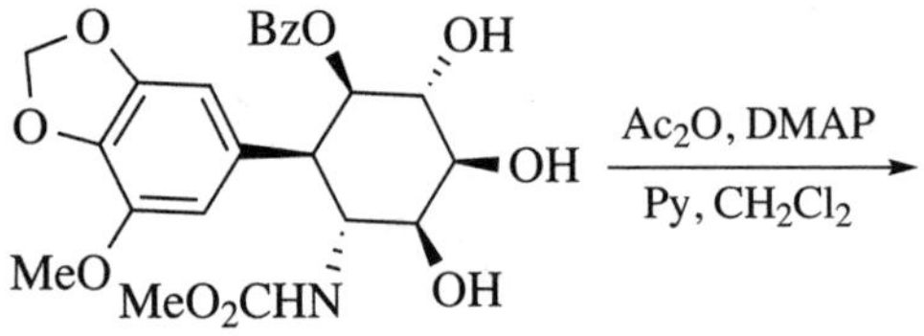 BzO OH
OH
OH
MeO MeO$_2$CHN OH
Ac$_2$O, DMAP
Py, CH$_2$Cl$_2$

(u) 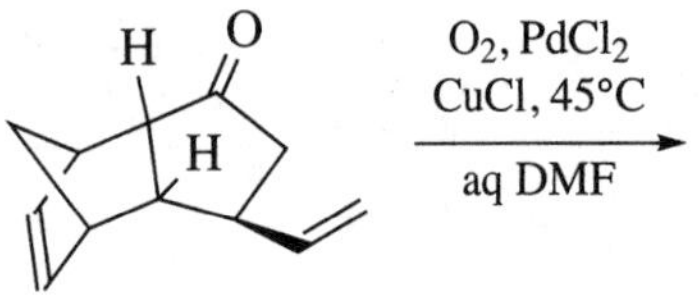 H O
H
O$_2$, PdCl$_2$
CuCl, 45°C
aq DMF

(v) 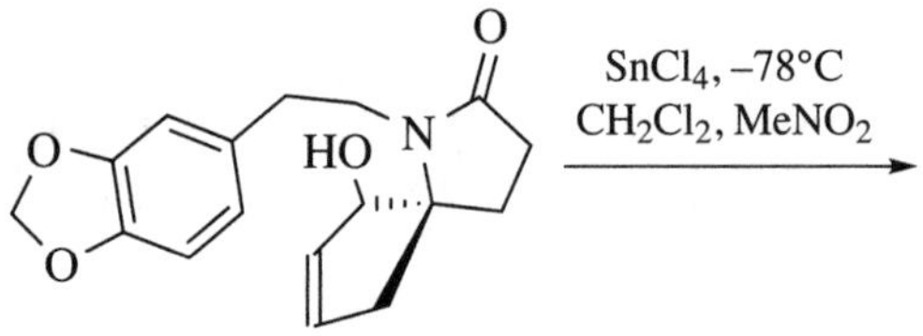 O HO N O
SnCl$_4$, −78°C
CH$_2$Cl$_2$, MeNO$_2$

(w) 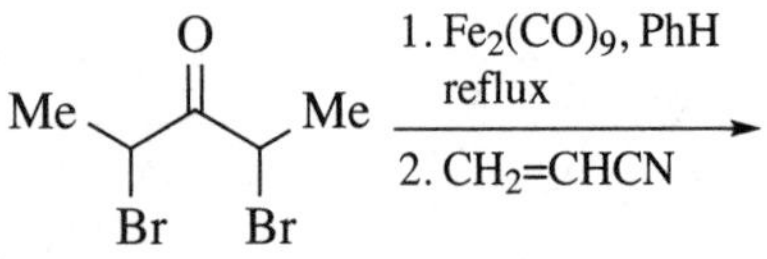 Me O Me
Br Br
1. Fe$_2$(CO)$_9$, PhH
reflux
2. CH$_2$=CHCN

(x) 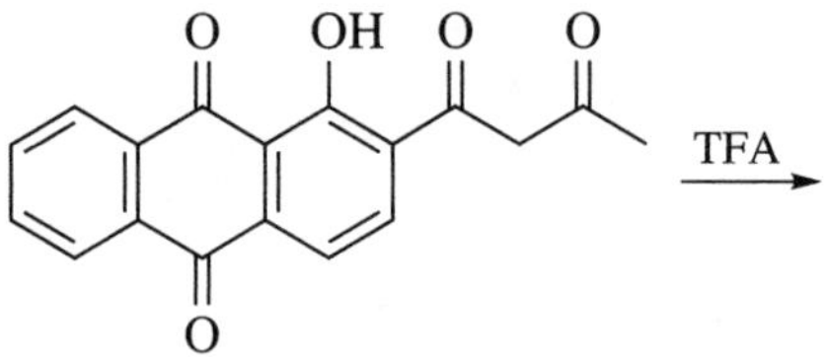 O OH O O
O
TFA

(y) 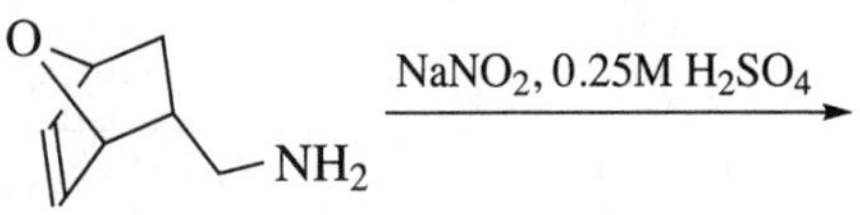 O
NH$_2$
NaNO$_2$, 0.25M H$_2$SO$_4$

(z) 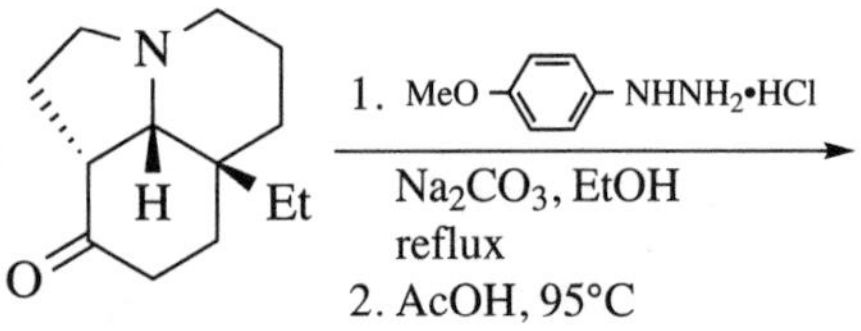 N
H Et
O
1. MeO—NHNH$_2$•HCl
Na$_2$CO$_3$, EtOH
reflux
2. AcOH, 95°C

(aa) 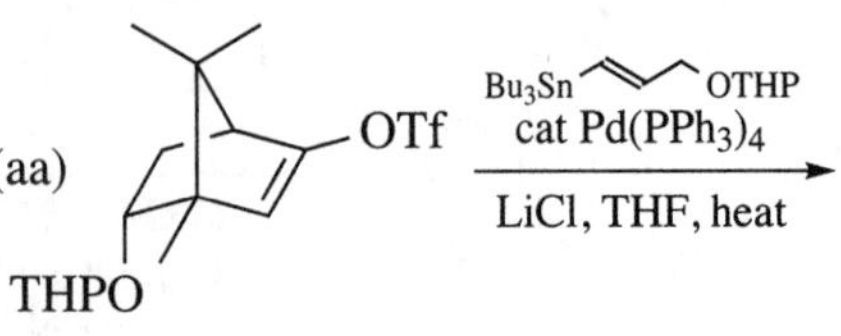 OTf
THPO
Bu$_3$Sn OTHP
cat Pd(PPh$_3$)$_4$
LiCl, THF, heat

(ab) 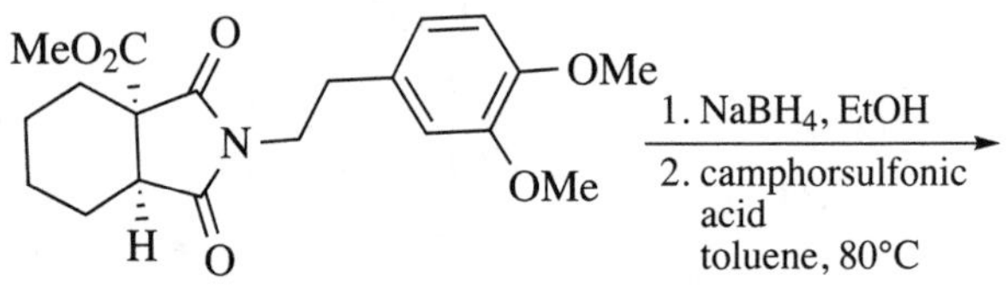 MeO$_2$C O
N
OMe
OMe
O
H
1. NaBH$_4$, EtOH
2. camphorsulfonic acid
toluene, 80°C

(ac) 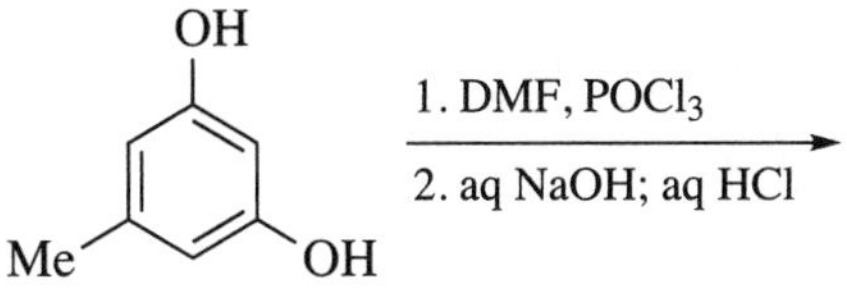 [structure: 5-methylbenzene-1,3-diol with OH, OH, Me]
1. DMF, POCl₃ → 1. DMF, POCl$_3$
2. aq NaOH; aq HCl

(ad) 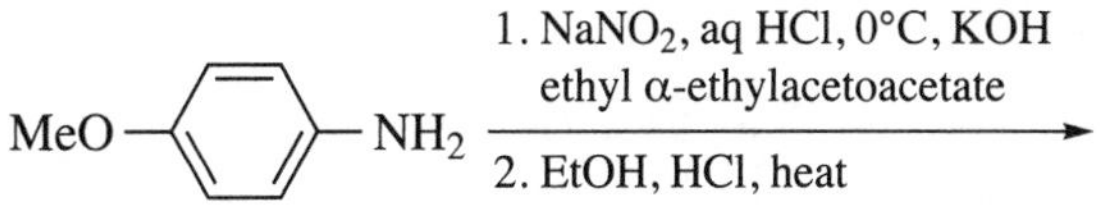 MeO—⟨ ⟩—NH$_2$
1. NaNO$_2$, aq HCl, 0°C, KOH
   ethyl α-ethylacetoacetate
2. EtOH, HCl, heat

(ae) 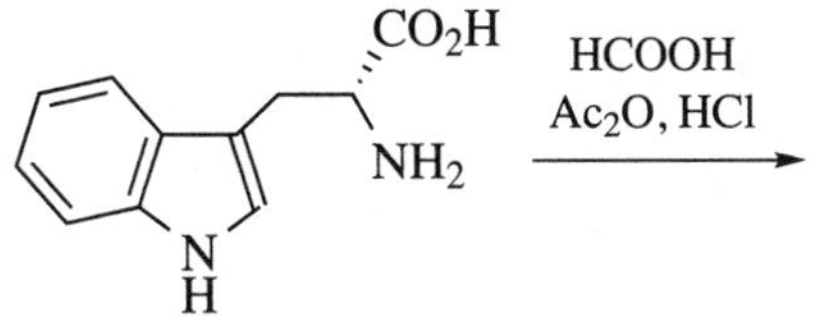 [indole structure with CO$_2$H, NH$_2$]
HCOOH
Ac$_2$O, HCl

(af) 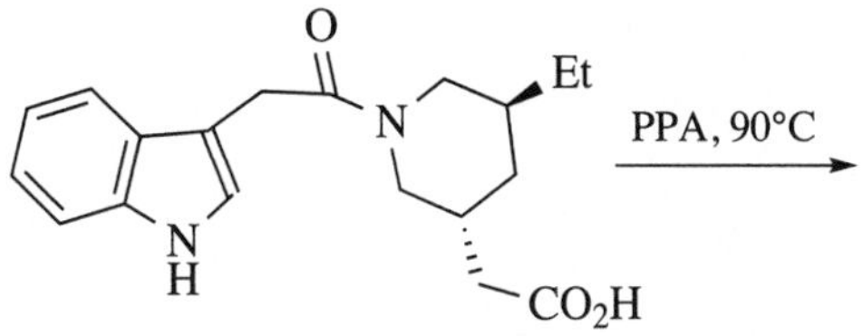 [indole structure with piperidine, Et, CO$_2$H, O]
PPA, 90°C

(ag) 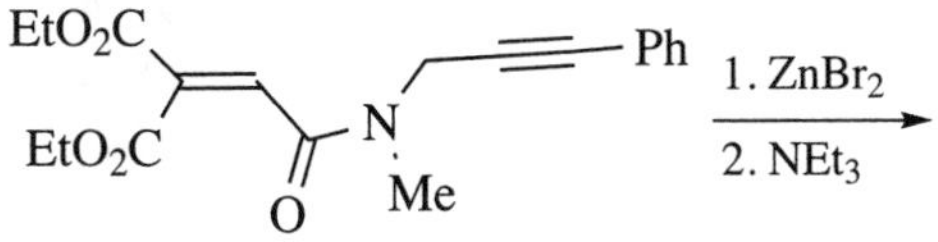 EtO$_2$C / EtO$_2$C [structure with N, Me, O, Ph alkyne]
1. ZnBr$_2$
2. NEt$_3$

14. In each case give a complete synthesis of the target from the designated starting material. Show all intermediate products and give all reagents.

(a) 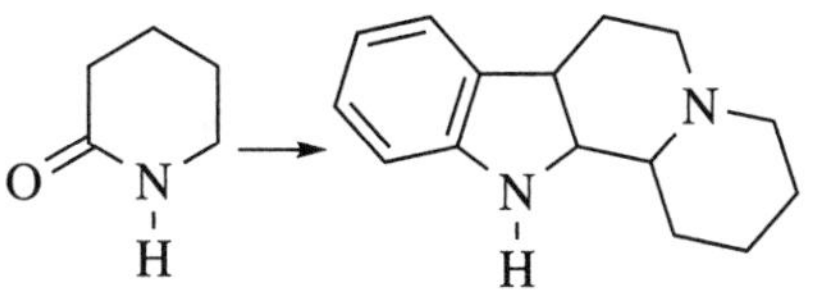 [structure: 2-piperidinone with NH] → [indolizidine/yohimbane-type polycyclic structure with N, N, H]

(b) [indole with CO$_2$Me, NH$_2$] → [polycyclic carbazole-type structure with H, O, CH$_2$Ph, N—Me, H]

(c) 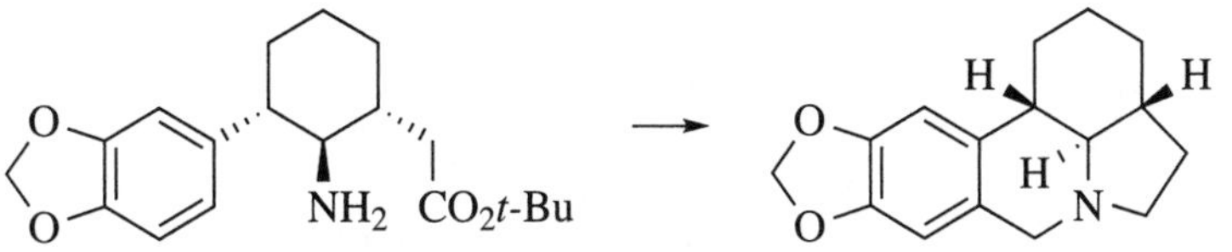 [cyclohexane with methylenedioxyphenyl, NH$_2$, CO$_2$t-Bu] → [polycyclic methylenedioxy structure with H, H, H, N]

(d) 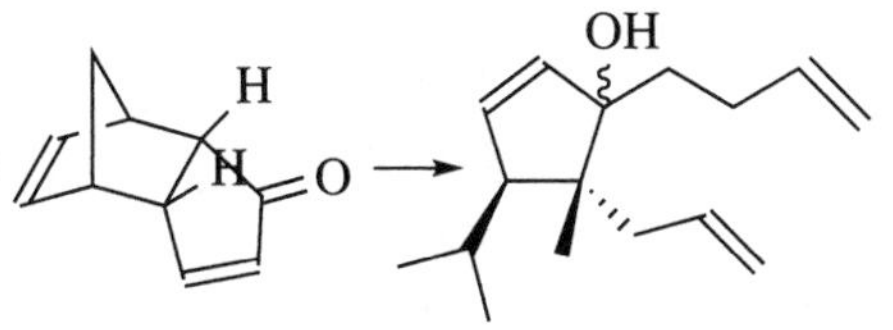 [norbornene structure with H, N, O] → [cyclopentene structure with OH, isopropyl, allyl groups]

(e) [structure: ketone with cyclohexanedione, Me, O, O, OH] → [decalin/steroid-type structure with Me, H, OH, O]

(f) 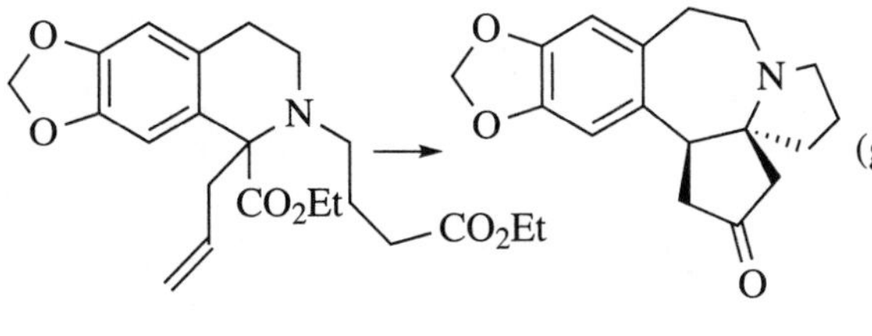 [methylenedioxy isoquinoline structure with N, CO$_2$Et, CO$_2$Et, allyl] → [methylenedioxy polycyclic structure with N, O]

(g) 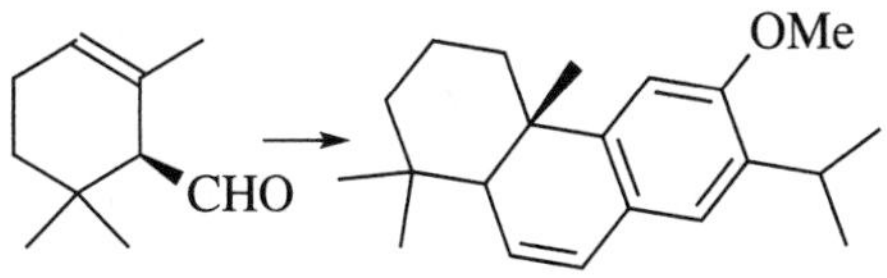 [cyclohexene with CHO and methyl groups] → [polycyclic structure with OMe, Me, isopropyl]

(h) MeO / Ar [styrenyl alcohol structure with OMe, OH] → MeO / Ar [dihydroisoquinoline structure with Me, N, OMe, Me]
Ar = [naphthalene with OMe, OMOM, Me, Me]

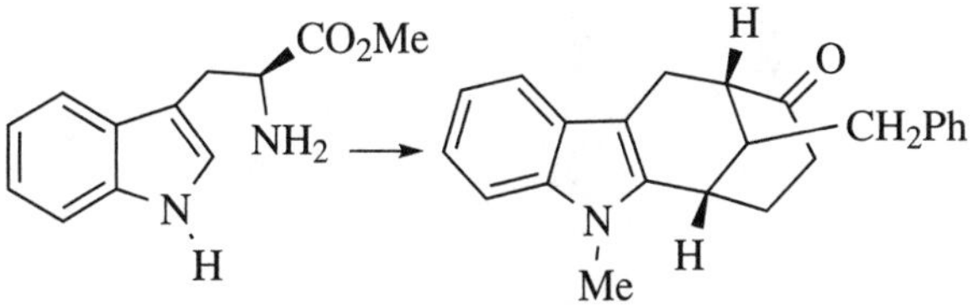

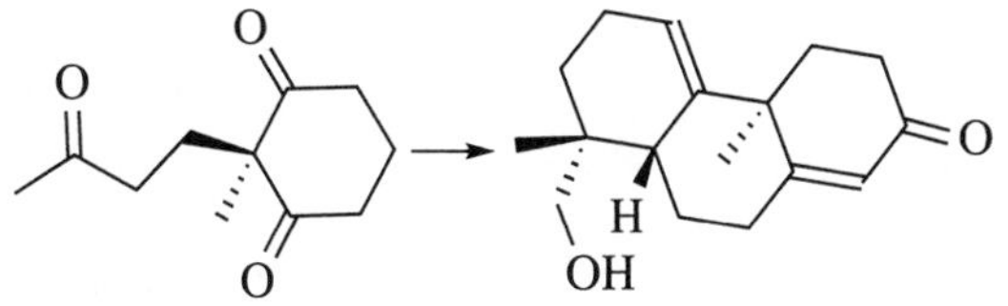

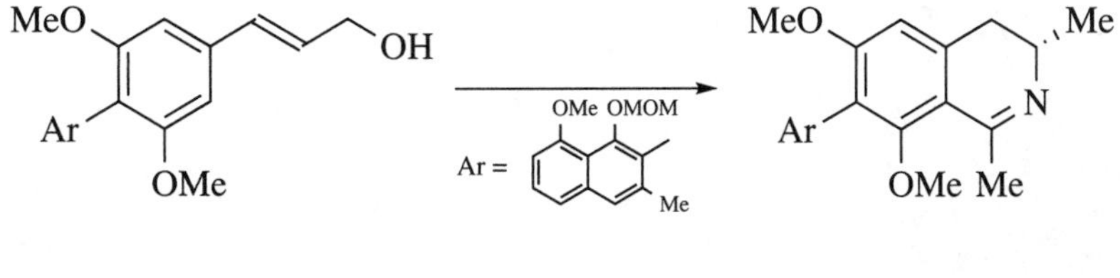

Carbon Radical Disconnect Products: Formation of Carbon-Carbon Bonds via Radicals and Carbenes

Carbanion reagents, electrophilic reagents and molecules that undergo pericyclic reactions are important for the formation of carbon-carbon bonds. Reactions that involve radical intermediates (molecules that possess a single electron in a reactive orbital) are also important. When a C–C bond is disconnected, ionic disconnect products C^a or C^d may occasionally lead to synthetic equivalents that are either impractical, or have side reactions that are intolerable. A useful alternative disconnect product may be a radical that can be represented as $C^{\cdot}$. Radical reactions have become an important and largely controllable synthetic tool in recent years, and should be considered as an important part of synthetic planning. Carbenes are also important for generating carbon-carbon bonds, and they are included in this chapter. This chapter will discuss methods for preparing radicals and carbenes as well as their applications to synthesis.

13.1. INTRODUCTION

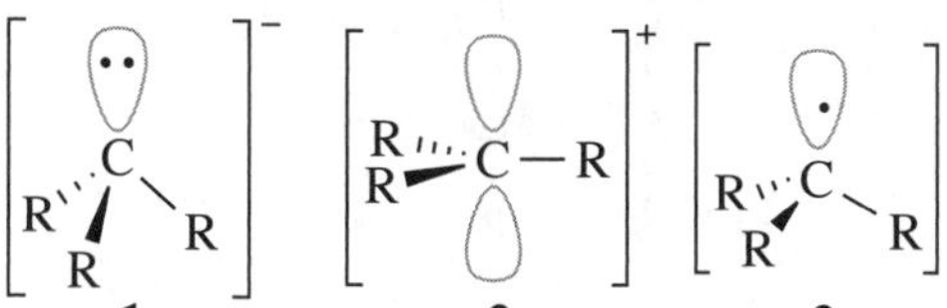

A carbon radical can be viewed as a trivalent species containing a single electron in a p orbital. A carbanion can be viewed as a tetrahedral species containing a pair of electrons in an orbital (**1**). We have viewed a carbocation (carbenium ion) as an sp^2 hybridized, trigonal planar carbon with an empty p orbital (**2**). A radical, which contains one electron in an orbital, can be tetrahedral, planar or 'in between' with properties of both a carbanion and a carbocation. As shown in **3**, a reasonable 'in between' structure is a flattened tetrahedron (the actual structure of radicals will be discussed below). In terms of its reactivity, radical **3** could be considered electron rich or electron poor. In most of its reactions, the electron-deficient characterization is useful for predicting products.

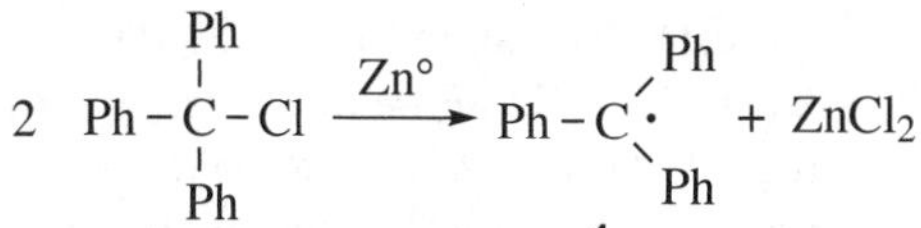

Gomberg was the first to characterize a free radical when, in 1900, he generated triphenylmethyl radical **4** by reacting chlorotriphenylmethane with zinc metal.[1] Triphenylmethyl radical is unusual in that it is quite stable, and its formation is probably the first experimental verification of a free radical. Frankland, however, may have been the first to generate transient methyl and ethyl radicals

---

1. (a) Gomberg, M. *J. Am. Chem. Soc.* **1900**, *22*, 757; (b) *Idem Chem. Ber.* **1900**, *33*, 3150.

in the reaction of iodomethane and iodoethane with zinc in 1849.[2] A good deal of attention is now focused on the reactivity and applications of radicals to organic synthesis. Excellent monographs by Davies and Parrott,[3] Lazár et al.,[4] Hay,[5] Giese,[6] Kochi,[7] Togo,[8] Zard[9] or Renaud and Sibi,[10] describe radical chemistry in great detail, as well as the many applications to synthesis. A review by Curran[11] placed an emphasis on synthetic applications.

13.2. STRUCTURE OF RADICALS

As mentioned in the introduction, there are several possible structures for a radical. The two most common are pyramidal (see **3**) and planar (see **7**). Formation of pyramidal radical **3** "involves a transition state that differs from the original molecules only in the length of the central bond", represented by **5**.[12] Dissociation to a planar radical (**7**) is also possible, but requires transition state **6** where both fragments are flattened.[12] Transition state **6** is generally favored over **5**, where **5** represents an "instantaneous bond dissociation". The lower energy dissociation of the molecule, in the transition state prior to cleavage, is graphically presented in Figure 13.1.[13] For **path 1b**, the C–C bond is changing from C^{sp3}–C^{sp3} to C^p–C^p, which is facile. Formation of the C–R bond is bond strengthening, and energy is released as the radical changes from a pyramidal structure to the planar arrangement. **Path 1a** is the instantaneous bond dissociation, and does not possess as much stabilizing energy. Note that if flattening in **6** is hindered by large steric interactions, then E_{act} increases and in those cases **3** may be the intermediate.

We can use the methyl radical ($H_3C\bullet$) to begin a discussion of the properties and structure of free radicals. Structure **3** (R = H) represents the ground state of the radical, experimentally

2.   Frankland, E. *Annalen 1849, 71*, 171, 213.
3.   Davies, D.I.; Parrott, M.J. *Free Radicals in Organic Synthesis* Springer-Verlag, Berlin, *1978*.
4.   Lazár, M.; Rychlý, J.; Klimo, V.; Pelikán, P.; Valko, L. *Free Radicals in Chemistry and Biology*, CRC Press, *1989*.
5.   Hay, J.M. *Reactive Free Radicals*, Academic Press, London, *1974*.
6.   Giese, B. Radicals in *Organic Synthesis: Formation of Carbon-Carbon Bonds*, Pergamon Press, Oxford, *1986*.
7.   Kochi, J.K. (Ed.) *Free Radicals*, Volumes 1 and 2, Wiley, New York, *1973*.
8.   Togo, H. *Advanced Free Radical Reactions for Organic Synthesis* Elsevier, Amsterdam, *2004*.
9.   Zard, S.Z. *Radical Reactions in Organic Synthesis* , Oxford University Press, Inc., New York, *2003*.
10. Renaud, P.; Sibi, M.P. (Eds.), *Radicals in Organic Synthesi, Volume 1. Basic Principles, Volume 2. Applications*, Wiley-VCH, Weinheim and New York, *2001*.
11. Curran, D.P. *Synthesis 1988*, 417, 489.
12. Reference 5, p 42.
13. Reference 5, p 43.

determined to be planar or nearly so.[14] The inversion energy for the methyl radical has been estimated to be only 4.2 kcal mol⁻¹ (17.6 kJ mol⁻¹).[15] As the temperature increases, the pyramidal model (**7**, R = H) explains the ¹H and ¹³C NMR as well as the electron spin resonance (ESR) spectra.[16] Three factors control the configuration of a radical: "(*a*) the energy of the unpaired electrons, (*b*) the energies of the bond in the radical and (c) the H–H interaction,[17] which is illustrated by a **Walsh diagram**,[18] shown in Figure 13.2 for the methyl radical".[17] The total energy for inversion supports a double barrier for inversion between the two pyramidal forms, proceeding through the planar form (see the total energy curve in Figure 13.2). The energy of the most stable configuration of the methyl radical was calculated to be about 5 kcal mol⁻¹ (20.9 kJ mol⁻¹),[14] and the energy of the pyramidal methyl was calculated to be about 10 kcal mol⁻¹ (41.9 kJ mol⁻¹) higher than the ground state.[17,19]

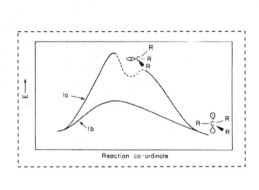

Figure 13.1 Energy Diagram For Dissociation to Planar and Pyramidal Radicals.

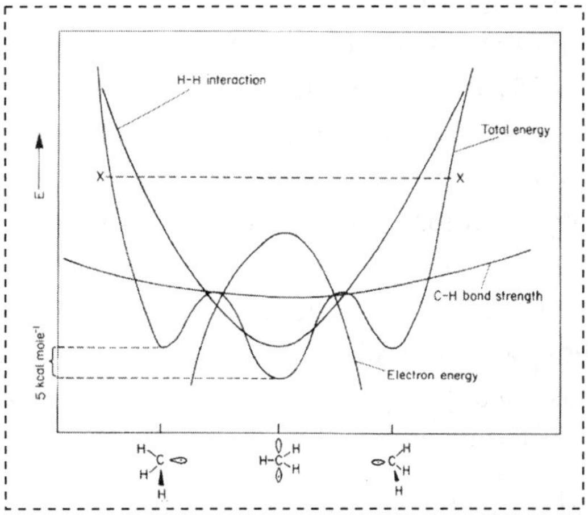

Figure 13.2. Walsh Diagram For Inversion of the Pyramidal Methyl Radical.

13.3. FORMATION OF RADICALS BY THERMOLYSIS

Radicals can be formed in several ways. Many involve dissociative **homolytic cleavage** (one electron is transferred to each adjacent atom from the bond) as a key step, as depicted by X–Y that gives two radical products. Another major route to radical intermediates involves the reaction of a radical (X·) and a neutral molecule (X–Y), producing a new radical (Y·) and a

14. (a) Fessenden, R.W.; Schuler, R.H. *J. Chem. Phys.* **1963**, *39*, 2147; (b) Herzberg, G. *Proc. R. Soc. London, A* **1961**, *262*, 291.
15. Andrews, L.; Pimentel, G.C. *J. Chem. Phys.* **1967**, *47*, 3637.
16. Chang, S.Y.; Davidson, E.R.; Vincow, G. *J. Chem. Phys.* **1970**, *52*, 5596.
17. Reference 5, p 45.
18. Walsh, A.D. *J. Chem. Soc.* **1953**, 2288, 2306, 2325.
19. Kibby, C.L.; Weston Jr., R.E. *J. Am. Chem. Soc.* **1968**, *90*, 1084.

new neutral molecule (X–X). This latter pathway will be discussed under radical reactions in section 3.5. The equilibrium constant for these processes depends on both the relative bond strength of X–Y and also on the relative stabilities of X• and Y•. For homolytic cleavage, raising the temperature of the reaction will generally shift the equilibrium toward a higher concentration of free radicals.[20] This equilibrium

$$X-Y \rightleftharpoons X^{\bullet} + Y\bullet \qquad X^{\bullet} \quad X-Y \rightleftharpoons X-X + Y\bullet$$

makes it convenient to correlate homolytic cleavage with bond dissociation energy. Lazár[21] published a table of bond energies, in ascending order of bond strength,[20,22] that correspond to the instantaneous bond dissociation energies (D). In principle, lower bond dissociation energies should correlate with an increased propensity for homolytic cleavage. Dissociation of C–C (85 kcal mol^{-1}; 357 kJ mol^{-1}),[21] C–O (84 kcal mol^{-1}; 351 kJ mol^{-1}),[18] H–O (111 kcal mol^{-1}; 464 kJ mol^{-1}),[21] and C-halogen bonds (55 kcal mol^{-1} for C–I to 79.8 kcal mol^{-1} for C–Cl; 230 kJ mol^{-1} to 351 kJ mol^{-1})[21] all require high reaction temperatures for homolytic cleavage. Alkali metals (K–K, Na–Na, Li–Li) with bond dissociation energies of 55, 69 and 103 kcal mol^{-1}, respectively (229.9, 188.4 and 430.5 kJ mol^{-1}),[21] undergo homolytic cleavage under very mild reaction conditions. The potent electron-transfer capabilities of these metals for dissolving metal reductions were discussed in Section 4.9. Particularly weak bonds that are expected to give homolytic cleavage are the peroxide bond (O–O, 140 kcal mol^{-1})[21] and those of the halogens, which is reflected in the chemical methods used to generate organic radicals. The methods usually begin with thermolysis of peroxides (H_2O_2, ROOR, etc.), diazo compounds (RN_2), or halogens.

Peroxides are potentially dangerous compounds because many of them can detonate upon exposure to heat or light, or under conditions where there is friction or shock. Peroxides are known to oxidize alkenes to oxiranes (especially in the presence of transition metals, see sec. 3.4.B), but if heated they decompose to hydroxy radicals (HO•) or alkoxy radicals (RO•). Lazár et al. published a table of approximate decomposition temperatures for several common peroxides.[23] Hydrogen peroxide requires the rather high temperature of 380°C for a decomposition half-life of 1 h,[23] and it is not very soluble in organic solvents. Peroxide **8**, however, decomposes at such a low temperature (–16°C)[23] that controlling the reaction is difficult in most instances. Benzoyl peroxide (**9**) cleaves at about 95°C,[23] a temperature that is compatible with many organic reactions, allowing some control of the reaction, and is reasonably soluble in organic solvents. Di-*tert*-butyl peroxide (**10**) decomposes at about 150°C[23] to produce $Me_3C-O•$. Although this peroxide is a useful radical initiator, the higher temperature required for decomposition can occasionally causes problems.

---

20. Reference 4, p 8.
21. Reference 4, p 9.
22. (a) Sanderson, R.T. *Chemical Bonds and Bond Energies*, Academic Press, New York, *1976*; (b) *Idem J. Am. Chem. Soc. 1983, 105*, 2259.
23. Reference 4, p 12.

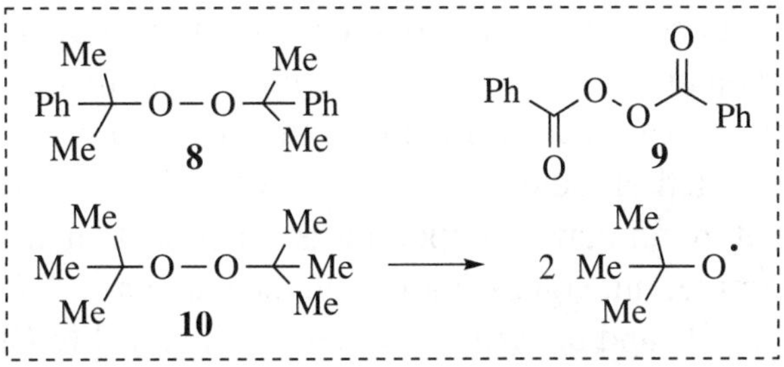

tert-Butyl hydroperoxide (**11**) is another useful peroxide that decomposes to give Me$_3$C–O•, **12**, and water. The presence of a α-hydrogen, as in diisopropyl peroxide (**13**), increases the facility of additional decomposition reactions. Initial homolytic cleavage gives the expected Me$_2$HC–O•, but this reacts with **13** to remove the α-hydrogen to give **14** and 2-propanol. The carbon radical (**14**) fragments to acetone and Me$_2$HC–O•.[24] Acyl peroxides such as the commonly used benzoyl peroxide **9**, react to give phenacyl radicals (**15**) that decarboxylates under the reaction conditions to give an aryl radical (such as **16**). Decomposition of alkyl radical from an alkyl acyl precursor [(RCO$_2$)$_2$] leads to an alkyl radical. Decarboxylation of primary alkyl acyl peroxides such as acetyl peroxide (**17**) is very facile, leading to **18** and the methyl radical •CH$_3$.[24] Decarboxylation of aryl acyl radicals is slower. If a peroxide decomposes, the oxygen radical remains in a "cage" for about 10^{-11} s before diffusing away. It can recombine (dimerize) or react further (see below).

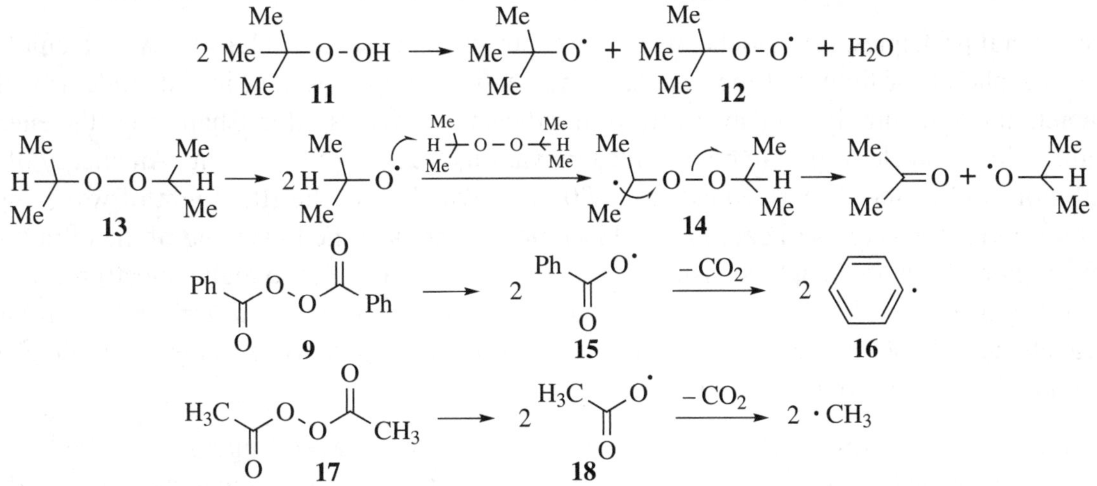

Azo compounds, characterized by a -N=N- bond, are free radical precursors that liberate stable nitrogen gas (N≡N) upon decomposition. One of the most used azo compounds is *azobis*(isobutyronitrile) (AIBN, **19**), which decomposes much faster than dibenzoyl peroxide to give nitrogen and the cyano stabilized radical **20**.[25] Just as with carbanion and carbocation

---

24. Reference 4, p 13.
25. (a) Yoshino, K.; Ohkatsu, J.; Tsuruta, T. *Polym. J.* **1977**, *9*, 275; (b) Hinz, J.; Oberlinner, A.; Rüchardt, C.*Tetrahedron Lett.* **1973**, 1975.

intermediates, the presence of the conjugating π bond in C≡N will delocalize the single electron of the radical via resonance, increasing the stability of that intermediate. Homolytic dissociation of symmetrical diazo compounds may be stepwise,[26] as in the decomposition of **21** to **22** and triphenylmethyl radical (**23**). Azo radical **22** loses nitrogen to form **23**.[27] The decomposition temperature for diazo compounds is dependent on the groups attached to the α-carbon.[28] AIBN, for example, undergoes homolytic cleavage at ~ 67°C,[28] whereas MeN=NMe decomposes at about 275°C and *azobis*(2,2-dimethylethane) (Me₃CN=NCMe₃) decomposes at 172°C.[28] The stability imparted by the electron-withdrawing cyano group leads to greater stability of the radical and more facile decomposition.

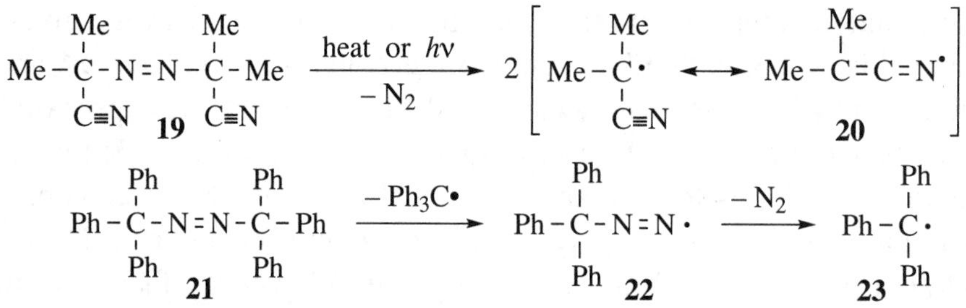

13.4. PHOTOCHEMICAL FORMATION OF RADICALS

The general principles of photochemistry were introduced in section 11.10.B. A molecule can absorb a photon of light and the resultant photoactivated species can, in the context of this chapter, undergo homolytic cleavage to form radicals. As discussed in Chapter 11, the energy required for bond cleavage can be provided by various wavelengths of light. An energy of 10 kcal mol⁻¹ (42 kJ mol⁻¹) corresponds to 3500 cm⁻¹, which is in the IR. Light of wavelength 2000 Å corresponds to 143 kcal mol⁻¹ (598 kJ mol⁻¹), and is at the lower end of the ultraviolet (UV) region. Light in the UV or visible region can excite molecules to higher electronic states. Wavelengths in the UV are usually expressed in nanometers (1 nm = 1 mμ = 1 x 10⁻⁷ cm) and a wavelength of 200-400 nm is used most often for inducing photolysis in the UV region[29] for common functional groups (sec. 11.10.B).

If a molecule were irradiated with a light source generating several intense wavelengths of light, it is possible for several different functional groups to absorb the light and react, including the product we are trying to synthesize. A convenient method for limiting the wavelength of light (and thereby the energy of irradiation) is to filter out higher energy wavelengths (and

26. Dannenberg, J.J.; Rocklin, D. *J. Org. Chem.* **1982**, *47*, 4529.
27. Newman Jr., R.C.; Lockyer Jr., G.D. *J. Am. Chem Soc.* **1983**, *105*, 3982.
28. Reference 4, p 15.
29. Johnson, E. in *The Physiology and Pathophysiology of the Skin, Vol. 8* of *The Photobiology of the Skin: Lasers and the Skin*, Jarrett, A. (Ed.), Academic Press, **1984**, p 2401.

lower energy if necessary). The filter is imposed between the light source and the reactants. Sometimes this is simply the reaction vessel, made of Pyrex or quartz glass,[30,31] but more often the filter material is attached to the light source.

When a molecule absorbs a photon of light, an electron is promoted from the bonding to an antibonding molecular orbital (n → σ*, n → π*, or π → π*) to generate the singly occupied molecular orbital (a **SOMO**) pertinent to radical reactions (sec. 11.10.B).[32] If the excited state is expressed as AB*, there are several pathways available for dissipation of this energy as shown by Wayne (Figure 13.3).[33] The pathway that is pertinent to formation of radicals is dissociation of AB* to A• and B•:

$$AB \xrightarrow{h\nu} [AB]^* \longrightarrow A\bullet + B\bullet$$

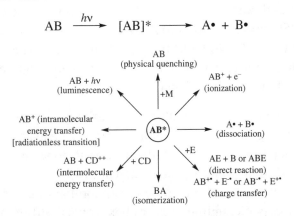

Figure 13.3. Routes Available For Loss of Electronic Excitation.

Photochemical homolysis of the carbon-carbon bond of alkanes is difficult. The carbon-carbon bond of ethane has a bond strength of 83 kcal mol^{-1} (347.4 kJ mol^{-1}), and it fragments to methyl radicals (•CH$_3$) only at temperatures approaching 600°C. Absorption of a photon of light is also difficult since alkanes absorb only in the "vacuum" UV (methane absorbs at 144 nm = 198.6 kcal mol^{-1} = 831.4 kJ mol^{-1}).[34] Loss of molecular hydrogen (H$_2$) usually accompanies conversion of an alkane (**24**) to the alkyl radical (**25**).[35] If a long alkyl chain is present, loss of H• can occur at many points, leading to a large number of different radicals.

30. Murov, S.L. *Handbook of Photochemistry*, Marcel-Dekker, New York, *1973*, pp 107-108.
31. Calvert, J.G.; Pitts Jr., J.N. *Photochemistry*, Wiley, New York, *1966*, pp 474, 490-491.
32. DePuy, C.H.; Chapman, O.L. *Molecular Reactions and Photochemistry*, Prentice-Hall, Englewood Cliffs, New Jersey, *1972*, p 6.
33. Wayne, R.P. *Principles and Applications of Photochemistry*, Oxford University Press, Oxford, *1988*, p 6.
34. Reference 33, pp 47 and 51.
35. Reference 33, p 51.

Alkenes show a $\pi \rightarrow \pi^*$ absorption for the π bond at $\approx$ 180 nm (158.9 kcal mol^{-1}, 665.1 kJ mol^{-1}).[35] Conjugated alkenes show a shift in absorption toward the visible spectrum (lower energy). Isomerization (of ethene to ethyne) and fragmentation are observed.[35] The yield of alkenyl radical decreases as the size of the alkyl portion of 1-alkenes increases.[36] Photolysis of a terminal alkene generates radical **26** in a typical photolysis reaction.

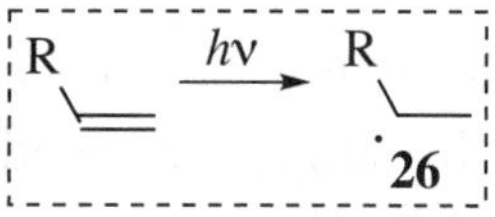

Carbonyl compounds show a symmetry forbidden $n \rightarrow \pi^*$ transition at ~290 nm (98.6 kcal mol^{-1}, 412.8 kJ mol^{-1}) for aliphatic aldehydes, and at ~280 nm (102.1 kcal mol^{-1}, 427.6 kJ mol^{-1}) for ketones. Conjugation to an aromatic ring (as in benzophenone) shifts the absorption to longer wavelength (~340 nm, 84.1 kcal mol^{-1}, 52.1 kJ mol^{-1}). Acid derivatives absorb at a shorter wavelength [<250 nm (114.4 kcal mol^{-1}, 478.9 kJ mol^{-1}) for acids, anhydrides, and so on, and <260 nm (110 kcal mol^{-1}, 460.5 kJ mol^{-1}) for amides].[37] The allowed transitions ($\pi \rightarrow \pi^*$ and $n \rightarrow \sigma^*$) absorb at shorter wavelengths [about 180 and 160 nm (158.9 and 178.8 kcal mol^{-1}, 665.2 and 748.5 kJ mol^{-1})] for aldehydes.[37]

There are two main dissociative fragmentations for carbonyls upon photolysis, **Norrish type I**, and **Norrish type II** fragmentations.[38] Norrish type I cleavage involves loss of carbon monoxide (CO) and generation of an alkyl radical. For acetone, absorption of a photon leads to cleavage to the methyl radical and the acyl radical (**18**). Subsequent fragmentation of **18** leads to another equivalent of methyl radical with concomitant loss of CO. Loss of CO is not always observed, particularly if coupling of the radicals gives a different product, and is a faster process. In a synthesis of the hamigerans, for example, Nicolaou and co-workers observed that irradiation of ketone **27** led to a 93% yield of a 1:3 mixture of **27:29**.[39] This photochemical epimerization likely proceeded by diradical **28**, the Norrish type I process.

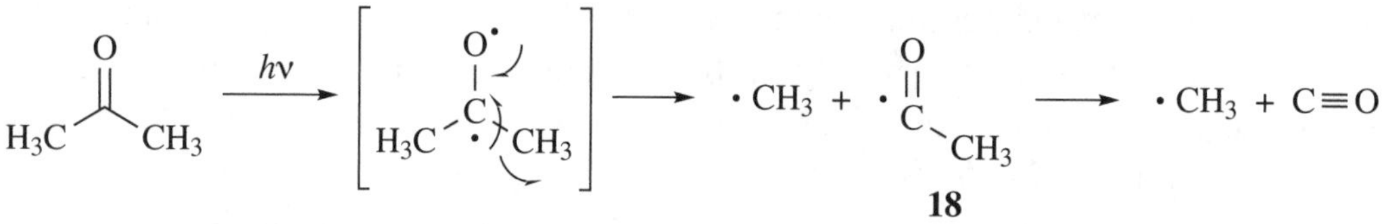

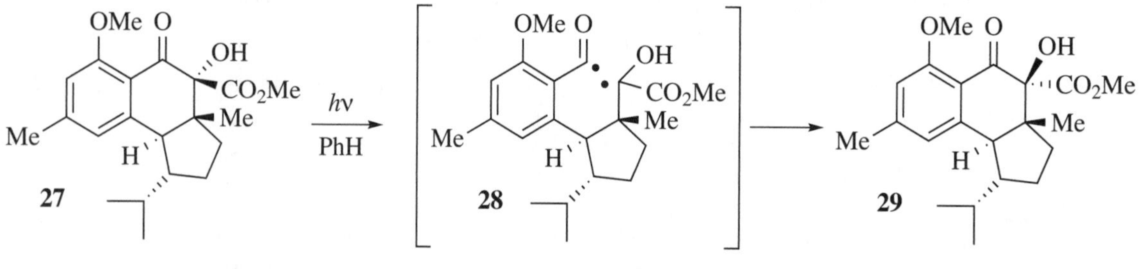

---

36. Shimo, N.; Nakashima, N.; Ikeda, N.; Yoshihara, K. *J. Photochem.* **1986**, *33*, 279.
37. Reference 33, p 52.
38. (a) Norrish, R.G.W.; Bamford, C.H. *Nature (London)* **1937**, *140*, 195; (b) *Idem Ibid.* **1936**, *138*, 1016; (c) The Merck Index, 14th Ed. Merck & Co., Inc., Whitehouse Station, New Jersey, **2006**, p ONR-66.
39. Nicolaou, K.C.; Gray, D.L.F.; Tae, J. *J. Am. Chem. Soc.* **2004**, *126*, 613.

The Norrish type II process is an intramolecular fission of the bond α-β to the carbonyl (as in the fragmentation of **30** to 2,3-dimethyl-2-butene and enol **31**). The enol tautomerizes to a carbonyl compound, here acetone.

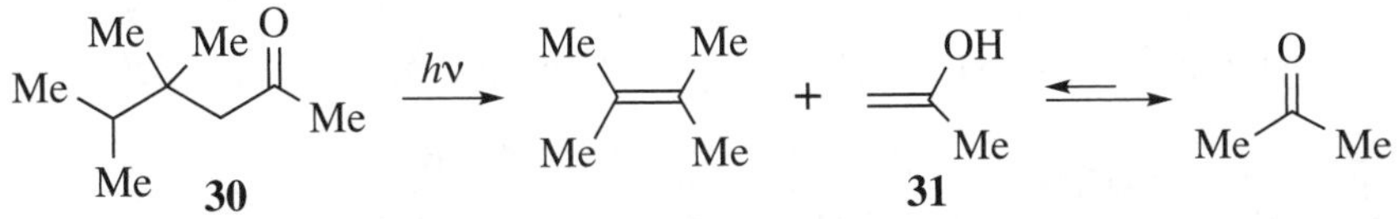

Similarly, photolysis of 2-pentanone undergoes cleavage to ethene and **31**, likely via a six-center transition state such as **32**.[40] This reaction is analogous to a **McLafferty rearrangement**,[41] which is observed in the mass spectrum of ketones having at least three saturated carbon atoms attached to the carbonyl unit. A synthetic example of a Norrish-type II process is the preparation of azepine-fused oxetanols from alkoxy-substituted maleimides by Booker-Milburn and co-workers.[42] Irradiation of maleimide **33** led to **34** via a [5+2] cycloaddition.[43] Further irradiation of **34** under the reaction conditions led to diradical **35**, and a Norrish type II process gave diradical **36**. β-Fission with expulsion of acetone led to **37**, and tautomerization gave **38** in 31% yield. Note that a different pathway formed 47% of other products.

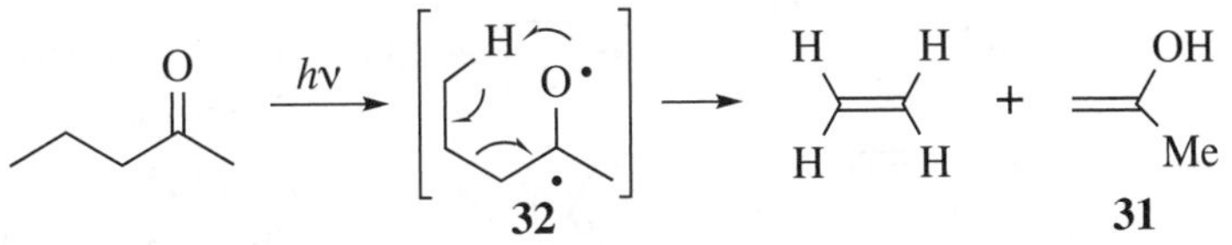

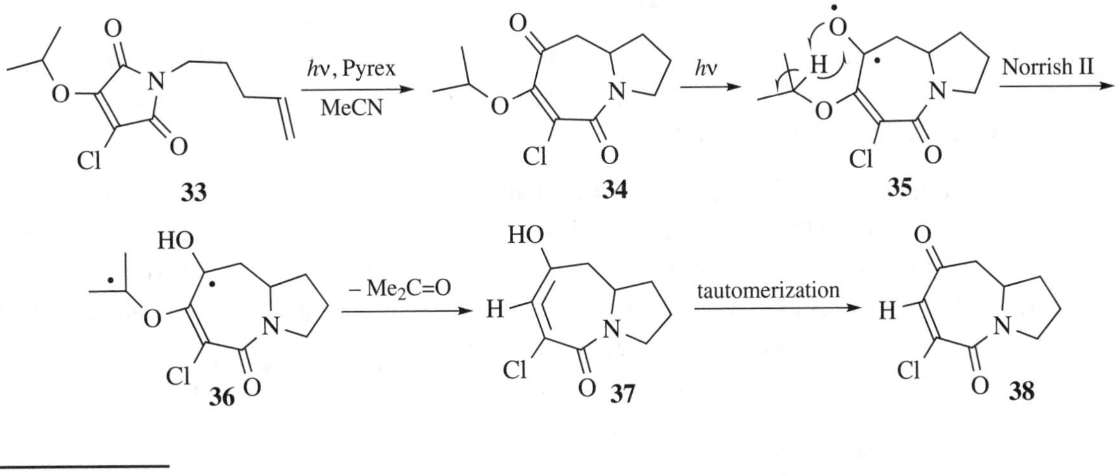

---

40. Reference 33, p 53.
41. (a) McLafferty, F.W. *Anal. Chem.* **1959**, *31*, 82; (b) Kingston, D.G.I.; Bursey, J.T.; Bursey, M.M. *Chem. Rev.* **1974**, *74*, 215; (c) The Merck Index, 14th Ed. Merck & Co., Inc., Whitehouse Station, New Jersey, **2006**, p ONR-58; (d) Mundy, B.P.; Ellerd, M.G.; Favaloro Jr., F.G. *Name Reactions and Reagents in Organic Synthesis, 2nd Ed.* Wiley-Interscience, New Jersey, **2005**, pp. 414-415.
42. Booker-Milburn, K.I.; Baker, J.R.; Bruce, I. *Org. Lett.* **2004**, *6*, 1481.
43. Booker-Milburn, K. I.; Anson, E. A.; Clissold, C.; Costin, N. J.; Dainty, R. F.; Murray, M.; Patel, D.; Sharpe, A. *Eur. J. Org. Chem.* **2001**, 1473.

The Norrish type I process is considered to be the mechanism by which radicals are formed. The regiochemical outcome of Norrish type I cleavage depends on the wavelength of light employed. Photolysis of 2-butanone at 313 nm (91.4 kcal mol^{-1}, 382.5 kJ mol^{-1}) gave a 40:1 mixture of **39** and ethyl radical (CH$_3$CH$_2$•). Irradiation at 253.7 nm (112.7 kcal mol^{-1}, 471.9 kJ mol^{-1}), however, gave a 2.4:1 mixture of methyl radical and acyl radical **18**.[44]

Aldehydes show an elimination reaction (loss of carbon monoxide, CO) that is not possible with ketones. Butanal, for example, photodissociates to propane and carbon monoxide. Apart from the reaction shown for **34**, cyclic ketones such as cyclopentanone dissociate to a diradical (in this case **40**), which then reacts in any of several ways including elimination to ethene or **41** and coupling to cyclobutane. Formation of cyclobutane and ethene is accompanied by expulsion of CO prior to formation of the alkene or the cyclic alkane.[40] The Norrish reactions can occur with the reaction of an alkane and an aromatic ketone such as benzophenone,[45] to give radicals **42** and R•.[46,47] Benzophenone acts as a sensitizer in this reaction. As discussed in Section 11.10.B, a **sensitizer** (S*) is added to facilitate electron transfer between two molecules.[48] The use of a sensitizer involves transfer of excitation energy from an electronically excited molecule (the sensitizer) to the ground state of another molecule, usually providing a triplet excited state.[49,32,33]

Bimolecular halogens (Cl$_2$, Br$_2$, etc.) and alkyl halides also serve as a source of radicals. Gaseous molecular iodine (I$_2$) gave the radical (I•) upon photolysis at about 500 nm (57.2 kcal mol^{-1}, 239.4 kJ mol^{-1}).[50] Other halogens behave similarly, although they absorb at shorter wavelength. Alkyl halides cleave into alkyl radicals and halogen radicals upon absorption of light, presumably via a diradical cage species represented by **43**.[51] Alkyl iodides appear to generate intermediates

$$R-X \xrightarrow{h\nu} [\,R\bullet\ \ \bullet X\,] \longrightarrow R\bullet + X\bullet$$
$$\mathbf{43}$$

---

44. Reference 33, pp 52-53.
45. (a) Murai, H.; Jinguji, H.; Obi, K. *J. Phys. Chem.* **1978**, *82*, 38; (b) Wagner, P.J.; Truman, R.J.; Scaiano, J.C. *J. Am. Chem. Soc.* **1985**, *107*, 7093.
46. Sonawane, H.R.; Nanjundiah, B..; Rajput, S.I.; Udaya Kumar, M. *Tetrahedron Lett.* **1986**, *27*, 6125.
47. Reference 4, p 22.
48. Eaton, D.F. *Pure App. Chem.* **1984**, *56*, 1191.
49. Breslow, R. *Chem. Soc. Rev.* **1972**, *1*, 553.
50. Reference 4, p 19.
51. Kropp, P.J. *Acc. Chem. Res.* **1984**, *17*, 131.

with significant ionic character, whereas alkyl bromide intermediates show much more radical character.

As discussed in Section 11.10.B, the efficiency of the photolysis is measured by the **quantum yield** (or **quantum efficiency**): the number of reacted or formed molecules per absorbed light quantum.[52] Quantum yield is measured for formation of a product (Φ_{form}) as just defined. It can also be defined in terms of the number of molecules of starting material (SM), which disappear per quantum of light (Φ_{dis}):[53]

$$\Phi_{form} = \frac{\text{No. of molecules of product formed}}{\text{No. of quanta absorbed}} \qquad 13.1$$

$$\Phi_{dis} = \frac{\text{No. of molecules of SM destroyed}}{\text{No. of quanta absorbed}} \qquad 13.2$$

13.5. REACTIONS OF FREE RADICALS

Davies and Parrott[54] discuss several types of reactions available to radicals once they have been generated. General aspects of radical chemistry have also been discussed.[55] This section will focus on the reaction types that are pertinent to synthetic applications or must be considered as competing processes in synthetic applications. It is noted that a chiral auxiliary has been developed for reactions involving hydroxyalkyl radicals.[56] Indeed, a body of work has appeared that shifts attention from chemoselectivity and regioselectivity, to control of stereochemistry in radical reactions.[57] Sibi and co-workers have reviewed enantioselective radical processes.[58]

The reactivity of free radicals is linked to their stability. Reactivity is a function of spin density of the atom and the type of orbital occupied by the unpaired electrons. Increasing the number of atoms in a radical generally decreases reactivity, making monatomic radicals such as Li• and F• very reactive. Conjugation decreases the spin density and therefore the reactivity. The resultant delocalization does not necessarily lead to a very stable radical.[59] Steric effects play a significant role in radical reactivity.

44   45

---

52. Reference 4, p 20.
53. Reference 32, p 76.
54. Reference 3, pp 1-14.
55. Alfasi, Z.B. *General Aspects of the Chemistry of Radicals,* John Wiley, Chichester, *1999.*
56. Garner, P.; Anderson, J.T.; Cox, P.B.; Klippenstein, S.J.; Leslie, R.; Scardovi, N. *J. Org. Chem.* *2002, 67,* 6195.
57. (a) Renaud, P.; Gerster, M. *Angew. Chem. Int. Ed.* *1998, 37,* 2563; (b) Smadja, W. *Synlett* *1994,* 1; (c) Porter, N.A.; Giese, B.; Curran, D.P. *Acc. Chem. Res.* *1991, 24,* 296; (d) Giese, B. *Angew. Chem. Int. Ed.* *1989, 28,* 969.
58. Sibi, M.P.; Manyem, S.; Zimmerman, J. *Chem. Rev.* *2003, 103,* 3263.
59. Reference 4, p 147.

"If the bulkiness of the surrounding substituents exceeds about 12-fold that of the central atom containing a radical site", the radical is unreactive and fails to react at all in many cases.[59] Triphenylsilyl radical (**44**) is very reactive, but perchlorobenzyl radical (**45**) is not.

The reactivity of radicals can be explained by molecular orbital theory, in a manner similar to that used for Diels-Alder reactions (secs. 11.2-11.4). Lázar et al. listed four rules that define an allowed reaction of a radical:[60]

1. The HOMO and LUMO of the reactant are decisive for the reaction course. At the mutual approach of reactants, there occurs the stabilization interaction between HOMO and LUMO. This interaction may be understood as a partial flow of electrons from the filled molecular orbitals to the empty ones.

2. The frontier orbitals should be of a proper symmetry to achieve the most efficient overlap. Only molecular orbitals of the same symmetry can be mixed.

3. The energy difference between HOMO and LUMO should be low (6 eV maximum). Presuming that point (*2*) is not fulfilled for some reason, the role of the frontier orbitals may be shifted to other orbitals of proper symmetry having the required energy difference.

4. If both orbitals are bonding, the HOMO should correspond to decaying bonds and LUMO to forming bonds. The opposite is true for an antibonding orbital.

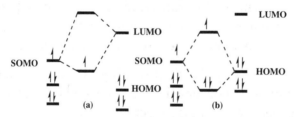

Figure 13.4. The Interaction of a SOMO of a Radical With the Frontier Orbitals of a Molecule: (A) Somo-Lumo Interaction, (B) Somo-Homo Interaction.[61] [Reprinted with permission from Fleming, I. *Frontier Orbitals and Organic Chemical Reactions*, Wiley, London, *1976*, pp. 183. Copyright © *1976* by John Wiley and Sons, Inc.]

A **SOMO**, defined above and in Section 11.10.B, can react with either a HOMO or a LUMO of another molecule as shown in Figure 13.4.[61] The SOMO-LUMO interaction (as in Figure 13.4) occurs with nucleophilic radicals (high energy SOMO) that react best with molecules possessing a low energy LUMO (electron-deficient alkenes, sec. 11.2, Table 11.1). Conversely, the SOMO-HOMO interaction dominates with electrophilic radicals (low-energy SOMO), which react best with molecules possessing a high-energy HOMO (electron rich alkenes). Radical **46** has unpaired electrons closer to the carbonyl and is expected to have a low-energy

60. Reference 4, p 137.
61. Fleming, I. *Frontier Orbitals and Organic Chemical Reactions*, Wiley, London, *1976*, p 183.

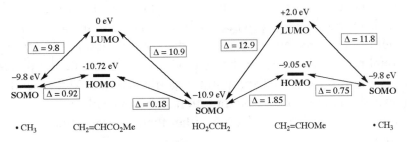

SOMO, making it an electrophilic radical. Radical **47** possesses a radical center closer to the oxygen, has a high energy SOMO, and is a nucleophilic radical. Radical **46** is expected to react best with electron-rich alkenes such as vinyl ethers, and **47** with electron deficient alkenes such as acrylates. The SOMO energy is taken to be the ionization potential (IP, see sec. 11.2) of the radical.

Table 13.1[62] shows the SOMO energies for several free radicals. In Section 11.2, Table 11.1, the LUMO and HOMO energies of a variety of alkenes were listed. The LUMO of methyl acrylate is 0 eV and the HOMO is –10.72 eV. The LUMO of methyl vinyl ether is +2.0 eV and the HOMO is –9.05 eV. Comparison of the SOMO energy of methyl radical ($H_3C\bullet$ in Table 13.1, –9.8 eV) and the acetate radical ($HO_2CCH_2\bullet$ in Table 13.1, –10.9 eV) shows that both reactions are $SOMO_{radical}$-$HOMO_{alkene}$ controlled. The methyl radical will react somewhat faster with methyl vinyl ether than with methyl acrylate, but acetate radical will react significantly faster with methyl acrylate. In general, the reactant with the lower energy SOMO will react faster than the radical with the higher energy SOMO. The greater the $SOMO_{radical}$-$HOMO_{alkene}$ (or $SOMO_{radical}$-$LUMO_{alkene}$) energy difference, the faster the reaction (see Figure 13.5).

Table 13.1   The SOMO Energies (IP) for Common Radical Types

Radical (R•)	SOMO Energy (eV)	Radical (R•)	SOMO Energy (eV)	Radical (R•)	SOMO Energy (eV)
$Me_3C\bullet$	–6.8	$C_{10}H_{23}CH_2\bullet$	–8.7	$t\text{-BuO–O}\bullet$	–11.5
⬡•	–7.8	⬡•	–9.2	$HO_2CCH_2\bullet$	–10.9
$C_5H_{11}CH(Me)\bullet$	–7.8	$H_3C\bullet$	–9.8	$Cl\bullet$	–13.0
$Et_3Si\bullet$	–7.0	$H\bullet$	–13.6	$Br\bullet$	–11.8
$Me_3Si\bullet$	–7.0	$t\text{-BuO}\bullet$	–12.0	$Cl_3C\bullet$	–8.8

[Reprinted with permission from Fleming, I. *Frontier Orbitals and Organic Chemical Reactions*, Wiley, London, *1976*, pp. 186. Copyright © *1976* by John Wiley and Sons, Inc.]

Figure 13.5. The SOMO-HOMO Interactions For Methyl And Acetate Radicals With Methyl Acrylate and Methyl Vinyl Ether.

62. Reference 61, p 186.

There are several major reaction types available to radicals: coupling, addition, substitution, fragmentation, rearrangement and oxidation. However, there are three main reaction types: coupling, addition and atom transfer. The frontier molecular orbital treatment shown here for addition to alkenes can, with modification, be used to predict reactivity in virtually all radical reactions. The following sections will describe the various radical reaction types.

MOLECULAR MODELING: Identifying an Effective Radical Precursor

One reason to expect that the chloride shown below will lead to a "non-reactive" radical is that the unpaired electron is delocalized onto both fluorenyl rings as well as onto the phenyl ring.

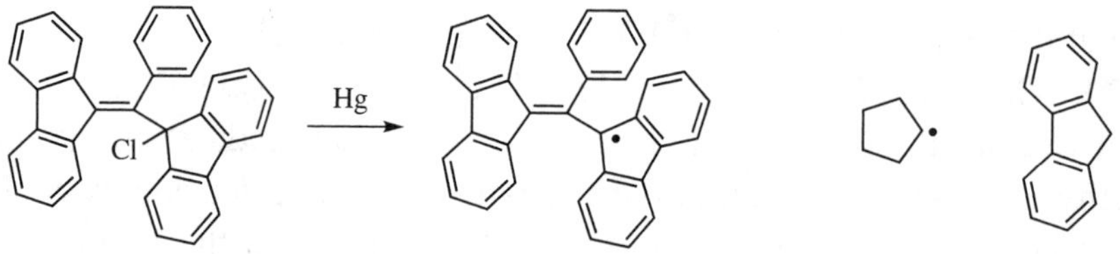

Spin density maps portray the number of unpaired electrons on an electron density surface. By convention, "blue" regions indicate an excess of unpaired electrons. Compare the map for this radical with maps for cyclopentyl and fluorenyl radicals. Which shows least delocalization and which shows greatest delocalization? Which would you expect to be least reactive and which most reactive?

A second reason is that the sp^3 carbon in the radical precursor is likely to be very crowded. *Click* on > and the select **Continue**. Build the precursor chloride and obtain its geometry. Repeat the process for the chloride precursors to the cyclopentyl and fluorenyl radicals. Is the double bond in the bis (fluorenyl) radical unusually long (a typical C=C length is 1.32Å) or is it twisted or puckered or both?

13.5.A. Coupling

Two radicals can combine to form a new carbon bond, as in the coupling of two ethyl radicals (CH$_3$CH$_2$•) to give butane. When one of the radical moieties possesses a β-hydrogen (or a β-leaving group), disproportionation can occur to form an alkene and an alkane. Ethyl radicals, for example, react to give ethane and ethene by disproportionation. Disproportionation can compete with coupling, but depends on the method used to generate the radical. Resonance stabilized radicals are less prone to disproportionation reactions. Methyl and benzyl radicals have no β-carbon and do not undergo disproportionation. In a solvent, a radical is more likely to encounter a solvent molecule or a substrate rather than disproportionate, although coupling is sometimes competitive.

If the radical is particularly stable (such as diphenylmethyl radical, **48**), coupling with a second radical molecule (to give **49**) is preferred to disproportionation, but this is not always the case. Triphenylmethyl radical (**4**), for example, possesses significant steric encumbrance at the radical carbon such that attack at a phenyl ring is preferred (to give **50** rather than Ph_3CCPh_3).[63]

$$CH_2{=}CH_2 + CH_3CH_3 \longleftarrow CH_3CH_2\bullet + \bullet CH_2CH_3 \longrightarrow CH_3CH_2{-}CH_2CH_3$$

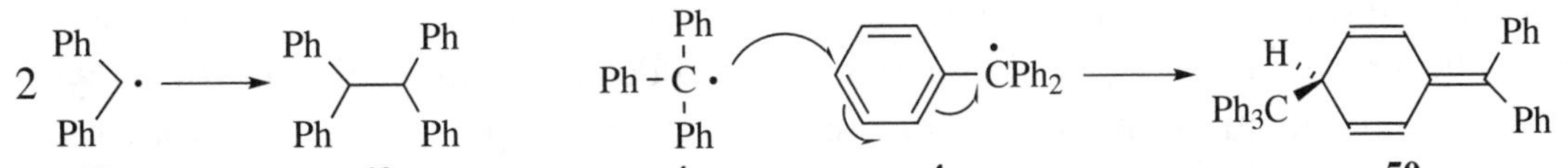

48 **49** **4** **4** **50**

13.5.B. Addition Reactions

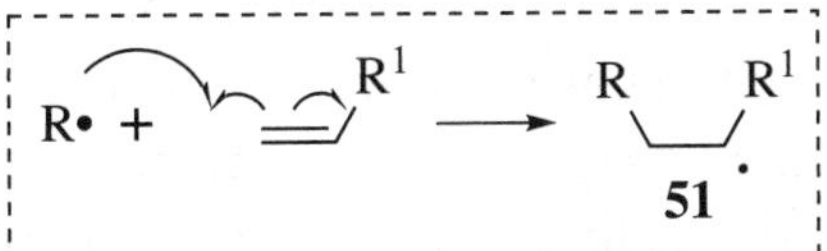

51

Radical addition to π bonds (see above) generates a new alkyl radical species (**51**), which can undergo coupling, disproportionation or other reactions. Since **51** is generated in the presence of the alkene, a chain reaction is possible in which the newly formed radical adds to unreacted alkene to generate another radical. An example is the radical polymerization of an alkene, as shown in Figure 13.6 (cationic polymerization tends to be more efficient in many cases and may be accomplished under milder conditions).

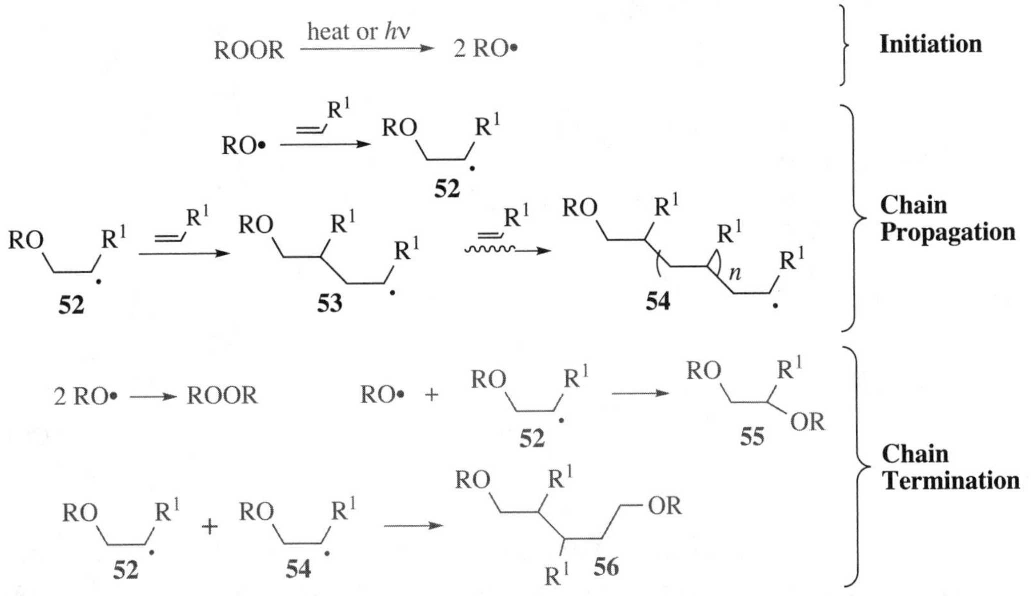

Figure 13.6. Chain Radical Addition Reactions.

An **initiation step** generates the radical, which in this case involves fragmentation of an alkyl peroxide to give RO•. This radical adds to the alkene to give a new reactive species (**52**). Addition of a second molecule of alkene generates **53**, and sequential addition to *n* molecules of alkenes generates **54**. These are called **chain propagation** steps, in which each step

63. Lankamp, H.; Nauta, W.T.; MacLean, C. *Tetrahedron Lett.* ***1968***, 249.

generates a reactive radical intermediate. If the radicals undergo coupling (several examples are shown, including formation of ROOR and **55**, as well as coupling to give **56**) the chain process stops since no radical is available to continue the process. This coupling is called a **chain termination** step because the product is *not* a radical. A new initiation step must occur to begin the chain process again. If a compound is added that will readily transfer a hydrogen atom, or another atom, to the radical this may stop the chain process in a controlled manner. Such a compound is called a hydrogen-transfer agent, or a radical quenching agent. Tributyltin hydride is an excellent hydrogen transfer agent in such reactions (see secs. 13.5.D, 4.9.H).

If the product radicals are stable, and/or if the stoichiometry of the reagents is controlled, a single product can be obtained in high yield. An example is the free radical addition of carbon tetrachloride (CCl_4) to styrene, which gives the benzylic radical **57** as an intermediate. Subsequent addition of **57** to styrene gives **59**. Addition of the benzyl radical to styrene (to produce another benzylic radical) is faster than chlorination[64] with a chlorine atom from carbon tetrachloride (which would give **58**), so polystyrene is the final product. When 1-octene reacted with CCl_4, however, the alkyl radical (**60**) was produced and coupling with a chlorine atom to give **62** was favored over addition to 1-octene (to give **61**, which was not observed). In fact, this reaction gave an 85% yield of **62**.[65]

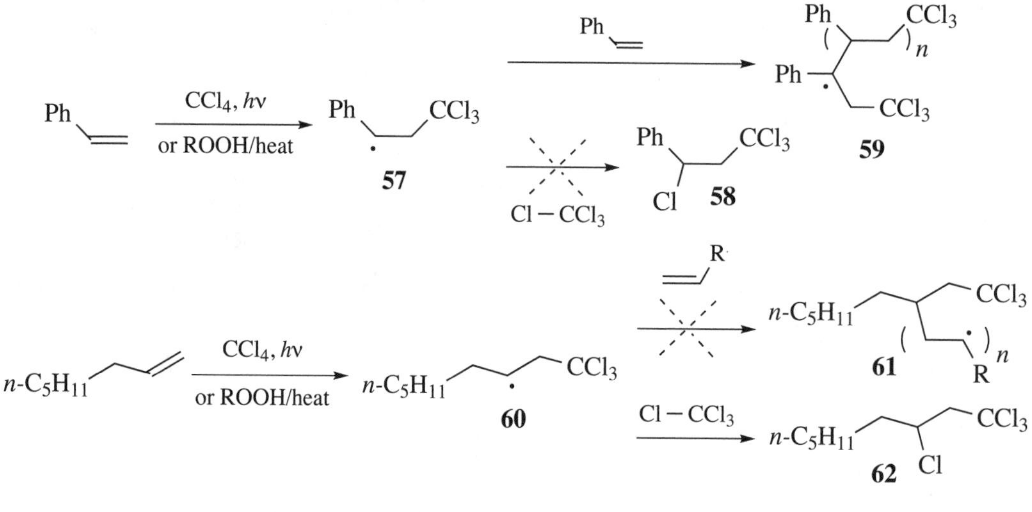

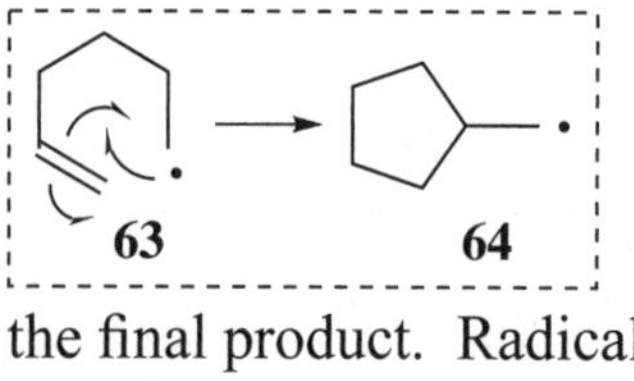

If a radical is generated in a substrate that also has a π bond in the molecule, intramolecular cyclization can occur (as in **63** → **64**). The highly reactive radical center is positioned to react with the π bond of the alkene, as shown, and the product (**64**) can be trapped to give the final product. Radical cyclization reactions will be discussed in Section 13.7.

64. (a) Gregg, R.A.; Mayo, F.R. *J. Am. Chem. Soc.* **1948,** *70,* 2373; (b) Mayo, F.R. *Ibid.* **1948,** *70,* 3689.
65. Kharasch, M.S.; Jensen, E.V.; Urry, W.H. *J. Am. Chem. Soc.* **1947,** *69,* 1100.

13.5.C. Substitution Reactions

A classical radical substitution reaction forms alkyl halides from the reaction of a diatomic halogen with a hydrocarbon such as toluene. Initiation of the reaction of bromine and toluene begins with photolytic cleavage of bromine, to give bromine radicals ($Br_2 + hv \rightarrow 2\ Br\bullet$). Diatomic bromine can also be converted to Br• by adding a peroxide or another radical initiator. Bromine radical abstracts a hydrogen atom from the α-carbon of toluene to give a resonance delocalized benzylic radical (**65**) and HBr. The benzylic radical abstracts a bromine atom from a second molecule of diatomic bromine to give benzyl bromide and Br•, the radical carrier in a chain propagation step.[66] This reaction can be applied to simple alkanes, if a relatively stable tertiary radical can be produced. The relative order of hydrogen abstraction has been shown to be[67]

$$CH_3\text{--}H < CH_3CH_2\text{--}H < CH_3CH_2CH_2\text{--}H < Me_2CHCH_2\text{--}H < Me_2CH\text{--}H < Me_3C\text{--}H$$

There is a great preference for forming tertiary radicals upon reaction of a hydrocarbon with bromine (via hydrogen atom abstraction by Br•). The relative rates for hydrogen atom abstraction by a bromine radical are 1 for primary hydrogen, 82 for secondary hydrogen and 1640 for a tertiary hydrogen, which means that a tertiary hydrogen will be abstracted to form the corresponding radical 1640 times faster than that for a primary hydrogen. When the radical reacts with more bromine to give the alkyl bromide, this selectivity leads to good yields of the tertiary bromide with smaller or trace amounts of isomeric bromides. This selectivity is not shared with the analogous chlorination reaction, however. For chlorination reactions, the relative rate of hydrogen atom abstraction is 1 for primary, 3.9 for secondary, and only 5.2 for tertiary, which means that a tertiary hydrogen atom is removed only 5.2 times faster than a primary, so there is much less selectivity. Halogen radicals will abstract all hydrogen atoms in the molecule, and when the rate of abstraction is close for different hydrogen atoms, as with the chlorine radical, the resulting carbon radicals react to give close to a statistical distribution of the corresponding isomeric chloride products. The significant difference in the rate of hydrogen atom abstraction by the bromine radical leads to good selectivity. The difference in selectivity between chloride and bromide radicals can also be explained by molecular orbital

66. (a) Reference 3, p 7; (b) Kochi, J.K.; Buchanan, D. *J. Am. Chem. Soc.* **1965**, *87*, 853.
67. Fukui, K.; Kato, H.; Yonezawa, T. *Bull. Chem. Soc. Jpn.* **1962**, *35*, 1475.

theory. A tertiary hydrogen (as in 2-methylpropane, HOMO $= -10.4$ eV)[68] is selectively removed by a bromine radical (SOMO $= -11.8$ eV, $\Delta E = 1.4$ eV), but removal of this hydrogen by a chlorine radical (SOMO $= -13.0$ eV, $\Delta E = 3.4$ eV) is less selective than with secondary and primary hydrogens. In other words, the difference in energy for removal of a tertiary versus a secondary hydrogen is greater in reactions with bromine radicals than with chlorine radicals. The bromine reactions are therefore more selective for removal of tertiary hydrogens. Ingold and Roberts discussed free radical substitution reactions in great detail.[69] It is important to note that free radical substitution of hydrogen by bromine is accomplished with N-bromosuccinimide (NBS), and that by chlorine with N-chlorosuccinimide (NCS) if the alkane substrate contains an allylic or benzylic hydrogen (sec. 2.8.B). An example is the conversion of 1-phenylbutane (**66**) to 1-bromo-1-phenylbutane (**67**).

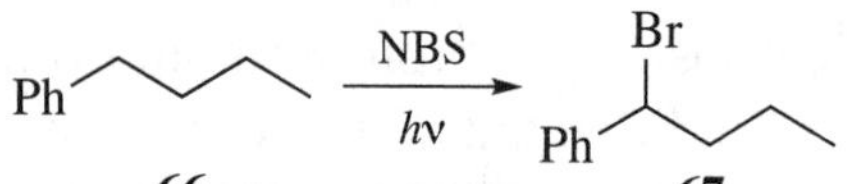

13.5.D. Reduction

When a radical is generated in the presence of a hydrogen transfer agent, it is possible to reduce various functional groups. Since an X group is replaced with H, this is a form of the substitution reaction described in Section 13.5.C, but it is separated into a different section because of it synthetic importance and because it is formally a reduction. The best example may be the one presented in Section 4.9.H in

which an alkyl halide is reduced by treatment with tributyltin hydride, illustrated here by the reduction of iodide **68** in the presence of AIBN and Bu$_3$SnH to give a 98% yield of **69** in an example taken from Davis and co-worker's[70] synthesis of pyrrolidine (–)-197B. AIBN (**19**) is a radical initiator, and leads to loss of iodide to give the radical. The intermediate radical reacted with Bu$_3$SnH via hydrogen transfer to give **69** and Bu$_3$Sn•, which generated Bu$_3$SnI. This reaction is a very effective method for the controlled reduction of halides.

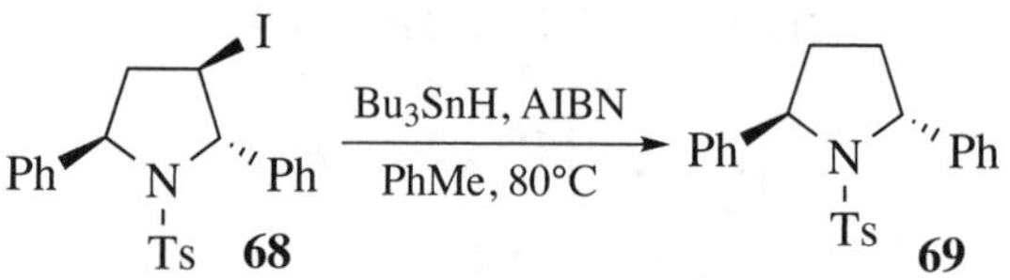

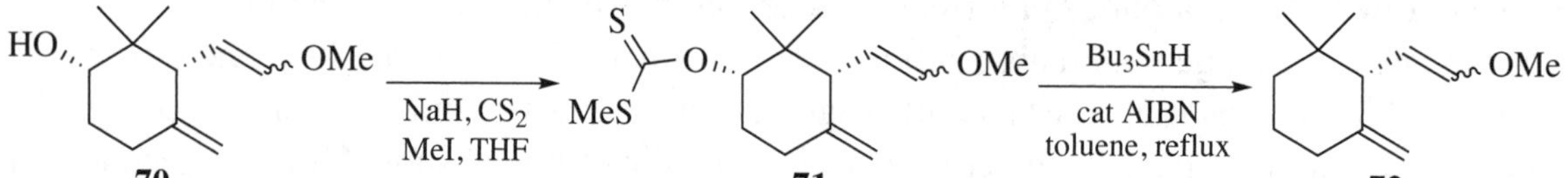

A very useful radical-based reaction has been developed that can be applied to alcohols. As seen in chapter 4, conversion of an alcohol to a thionocarbonate followed by treatment with tributyltin hydride under radical conditions gives cleavage to the C–O bond to give the reduction product. This transformation called the **Barton deoxygenation (Barton-McCombie**

---

68. Fukui, K.; Kato, H.; Yonezawa, T. *Bull. Chem. Soc. Jpn.* **1960**, *33*, 1197.
69. Ingold, K.U.; Roberts, B.P. *Free Radical Substitution Reactions*, Wiley, New York, **1971**.
70. Davis, F.A.; Song, M.; Augustine, A. *J. Org. Chem.* **2006**, *71*, 2779.

reaction).[71] In a synthesis of crispatenine by Commeiras, Parrain and co-workers,[72] treatment of **70** with sodium hydride and carbon disulfide, followed by iodomethane, gave xanthate **71**, and subsequent reduction with tributyltin hydride and AIBN gave a 90% yield of **72**.

13.5.E. Fragmentation

A radical can fragment to form a new radical and a neutral molecule. Typically, the neutral molecule is an alkene, making the overall process a simple elimination.[73] Addition of trichloromethyl radical ($Cl_3C\bullet$), generated by thermolysis of the acyl peroxide **73** to 2,3-dimethyl-2-butene, gave **74**. Abstraction of a hydrogen atom[74] by $\bullet CCl_3$ gave trichloromethane and **76**. Transfer of a hydrogen atom to **74** (from solvent, the alkane, or a hydrogen atom), however, led to **75**.

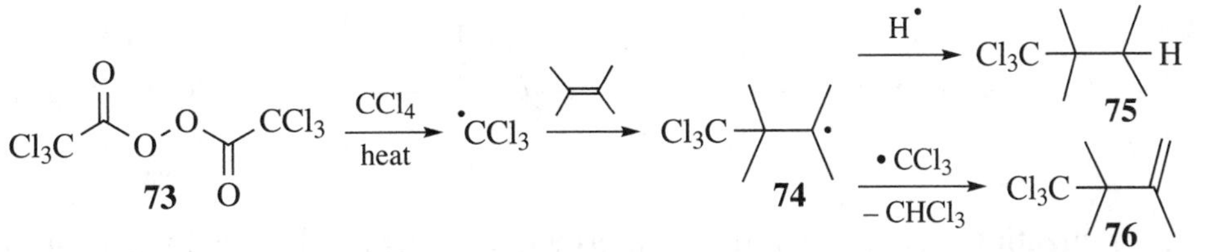

Both processes occurred in this reaction, and are typical of radical additions. Alkoxy radicals such as **78** (formed via loss of a chlorine atom from hypochlorite **77**) often lose a hydrogen atom (to give 2-butanone when R = H) or an alkyl radical (to give **79** when R = Me) via cleavage of the bond β to the radical center.[75] Thermolysis of hypochlorite **80**, for example, gave acetone and chloroethane via coupling of the intermediate ethyl radical with alkoxy radical **81**.[76] Hypochlorites are excellent sources of alkoxy radicals, and can be used to initiate other radical reactions.

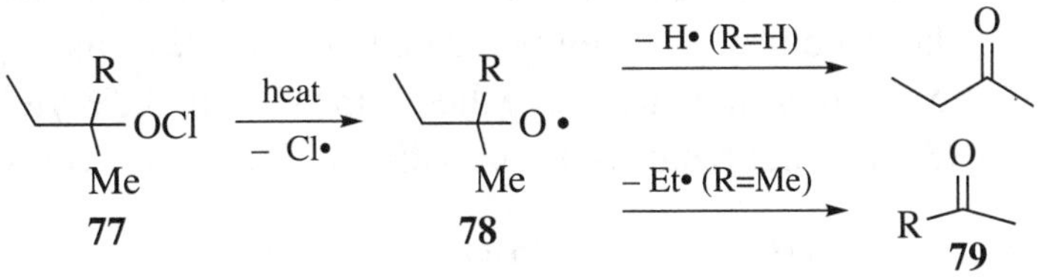

71. (a) Barton, D.H.R.; McCombie, S.W. *J. Chem. Soc. Perkin Trans. 1* **1975**, 1574; (b) Lopez, R.M.; Hays, D.S.; Fu, G.C. *J. Am. Chem. Soc.* **1997**, *119*, 6949. Phenyl thionocarbonates were first used by (c) Robins, M.J.; Wilson, J.S.; Hansske, F. *J. Am. Chem. Soc.* **1983**, *105*, 4059; (d) Robins, M.J.; Wilson, J.S. *J. Am. Chem. Soc.* **1981**, *103*, 932; See (e) The Merck Index, 14th Ed. Merck & Co., Inc., Whitehouse Station, New Jersey, **2006**, p ONR-6; (f) Mundy, B.P.; Ellerd, M.G.; Favaloro Jr., F.G. *Name Reactions and Reagents in Organic Synthesis, 2nd Ed.* Wiley-Interscience, New Jersey, **2005**, pp. 68-69.

72. Bourdron, J.; Commeiras, L.; Audran, G.; Vanthuyne, N.; Hubaud, J.C.; Parrain, J.-L. *J. Org. Chem.* **2007**, *72*, 3770.

73. Wilt, J.W. , Chapter eight in *Free Radicals, Vol 1*, Kochi, J.K. (Ed.), Wiley, New York, **1973**, p 333.

74. Loken, H.Y.; Lawler, R.G.; Ward, H.R. *J. Org. Chem.* **1973**, *38*, 106.

75. Reference 3, p 9.

76. Chattaway, F.D.; Baekeberg, O.G. *J. Chem. Soc.* **1923**, *123*, 2999.

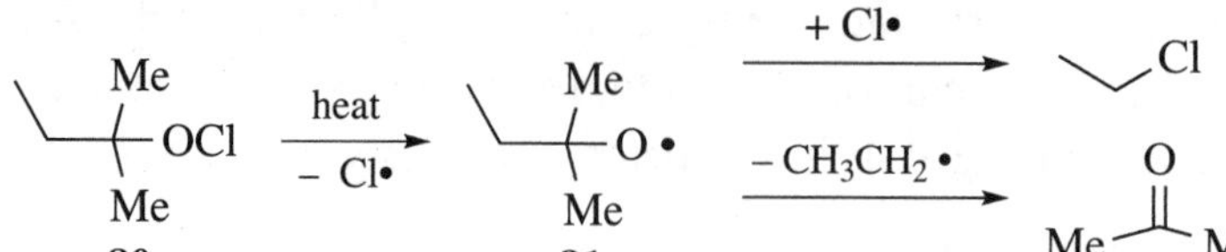

13.5.F. Rearrangement and Hydrogen Abstraction

There are a few rearrangement processes of free radicals.[77,64] When chlorine, phenyl, acetoxy, and acyl groups are in the β-position with respect to the radical center, 1,2-shifts have been observed. An example is the rearrangement of **82** to **84** via **83**.[78]

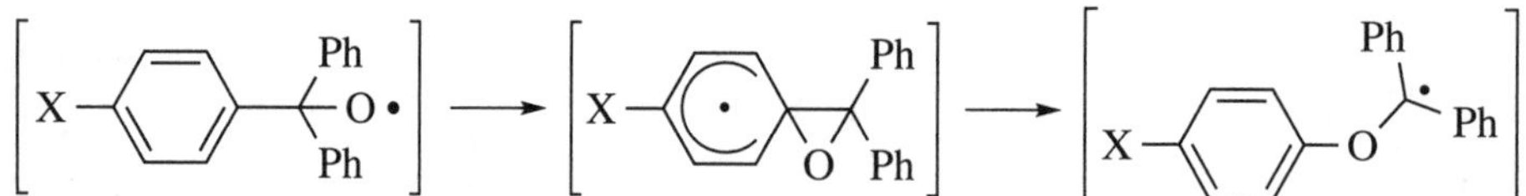

Skeletal reorganization occurs in radical intermediates primarily via hydrogen atom shifts observed that are probably hydrogen abstraction reactions. An example is the **Barton reaction**,[79] in which an alkoxy radical (**86**), generated from the hypochlorite (**85**), abstracts a hydrogen atom via a six-center transition state to give **87**. Subsequently, **87** abstracts a chlorine atom from **85** to give **88** and regenerate reactive radical **87**.[80] In a synthetic example, Barton and co-workers converted 3β-acetoxy-5α-pregnan-20β-ol (**89**) to the nitrite derivative (**90**) via reaction with nitrosyl chloride (NOCl).[79a] Photochemical cleavage (mercury lamp and a Pyrex filter) of **90** generated alkoxy radical **91**, which abstracted the hydrogen atom from the neighboring methyl that was properly positioned for reaction via a six-center transition state by virtue of the rigid steroidal structure. The newly formed radical (**92**) was trapped by the nitrosyl radical to give **93** and treatment with water gave the oxime (**94**) in 34% yield. Hydrolysis with dilute acid gave lactol **95**. The net result is an oxidation (-CH_3 → -CHO; sec. 3.8).

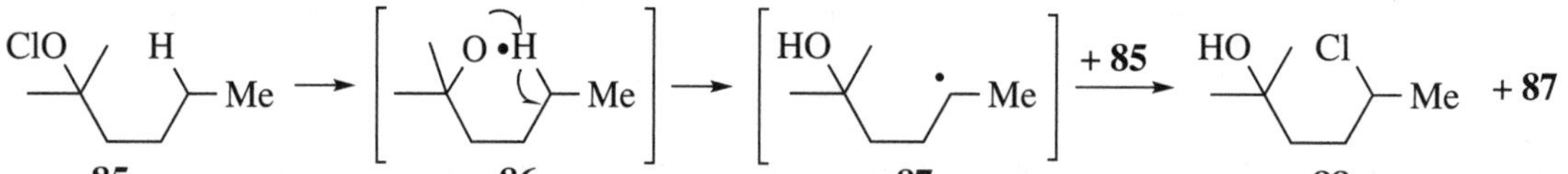

77. (a) Walling, C. *Free Radical Rearrangements* in *Molecular Rearrangements, Vol 1*, chapter 7, de Mayo, P. (Ed.), Wiley, New York, *1963*, p 407; (b) Freidlina, R.Kh. *Adv. Free Radical Chem.* *1965*, *1*, 211.
78. Kharasch, M.S.; Poshkus, A.C.; Fono, A.; Nudenberg, W. *J. Org. Chem.* *1951*, *16*, 1458.
79. (a) Barton, D.H.R.; Beaton, J.M.; Geller, L.E.; Pechet, M.M. *J. Am. Chem. Soc.* *1960*, *82*, 2640; (b) *Idem Ibid.* *1961*, *83*, 4076; (c) Barton, D.H.R. *Pure Appl. Chem.* *1968*, *16*, 1; (d) Hesse, R.H. *Adv. Free Radical Chem.* *1969*, *3*, 83; (e) The Merck Index, 14th Ed. Merck & Co., Inc., Whitehouse Station, New Jersey, *2006*, p ONR-6; (f) Mundy, B.P.; Ellerd, M.G.; Favaloro Jr., F.G. *Name Reactions and Reagents in Organic Synthesis, 2nd Ed.* Wiley-Interscience, New Jersey, *2005*, pp. 66-67.
80. Walling, C.; Padwa, A. *J. Am. Chem. Soc.* *1963*, *85*, 1593, 1597.

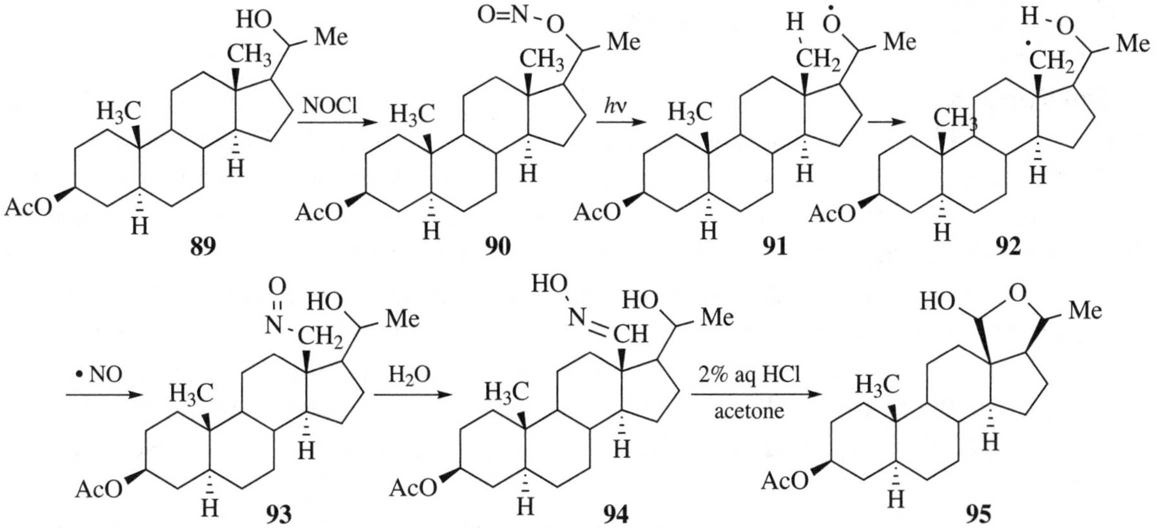

The reaction is not restricted to steroids, but simply requires that the abstracted hydrogen atom is positioned close enough to react with the alkoxy radical. Reaction of alcohol **96** with bromine and mercuric bromide gave the corresponding hypobromite (-OH → -OBr). Subsequent photolysis gave a 43% yield of **97**.[81a,b] Oxidation with chromium trioxide in acetic acid gave the lactone (**98**), which was used by Magnus in a synthesis of grandisol.[81d] Gibson and Erman used similar intermediates and methodology for the synthesis of cis-bergamotenes.[81c]

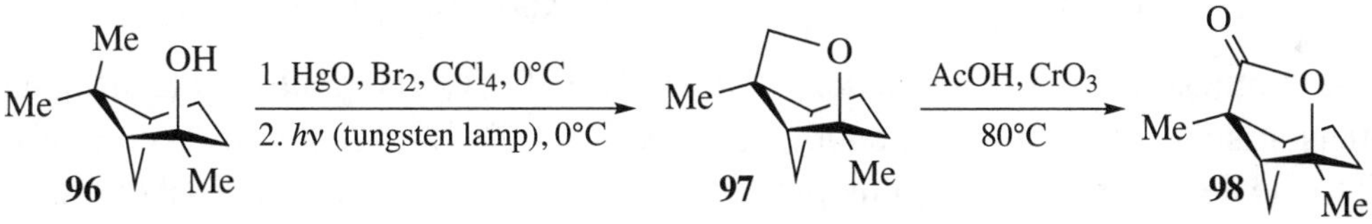

Note that there are potential side reactions. Once the oxygen radical is generated, the hydrogen can be abstracted via a four-center transition state if the six-center transition state is not readily accessible. A Barton-like reaction involved photolysis of **99** to give the oxygen radical **100**.[82] Removal of H_a in **100** via a five-center transition state ($\bullet O–C–C–C–H$) gave the corresponding carbon radical (**101**), which gave two products via either cyclization in the presence of the alkene and trapping by NO (to give **102**, 25% yield) or by direct trapping with NO to give **103** (14% yield).

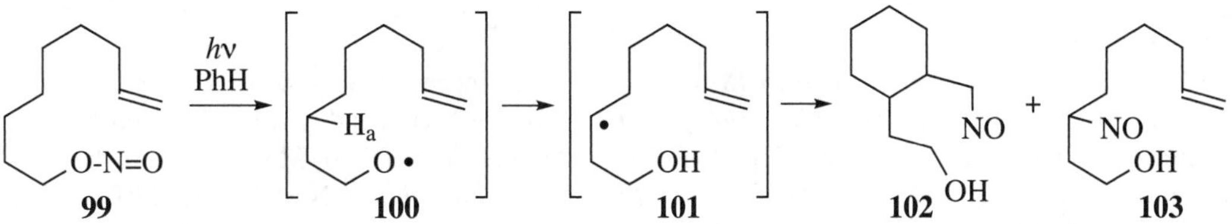

81. (a) Bosworth, N.; Magnus, P.D. *J. Chem. Soc. Perkin Trans. 1* ***1972***, 943; (b) Hortmann, A.G.; Youngstrom, R.E. *J. Org. Chem.* ***1969***, *34*, 3392; (c) Gibson, T.W.; Erman, W.F. *J. Am. Chem. Soc.* ***1969***, *91*, 4771; (d) Hobbs, P.D.; Magnus, P.D. *J. Am. Chem. Soc.* ***1976***, *98*, 4594.
82. Čeković, Ž.; Ilijev, D. *Tetrahedron Lett.* ***1988***, *29*, 1441.

13.6. INTERMOLECULAR RADICAL REACTIONS

In 1849 Kolbe found that electrolysis of potassium acetate generated carbon dioxide (CO_2) and ethane, presumably via acetoxy radical **104** that lost carbon dioxide to give the methyl radical (•CH_3).[83] The methyl radical (•CH_3) coupled with another methyl radical to give ethane. The reaction works very well with long, straight chain alkanoic acids. This reaction is known as the **Kolbe electrolytic synthesis**[83,84] and it was used to convert hexadecanoic acid to *n*-triacontane (n-$C_{30}H_{62}$) in 88% yield.[85] Kolbe electrosynthesis has been used to prepare rigid, rod-shaped hydrocarbons based on oligo-bicyclo[2.2.2]octane derivatives.[86] Much larger and more complex molecules can be synthesized by this method. Stork used the Kolbe electrosynthesis use to dimerize **105** to **106** (in 40% yield) in a synthesis of α-onocerin.[87] As illustrated by the synthesis of **106**, the reaction is very efficient when two identical acids are coupled together.

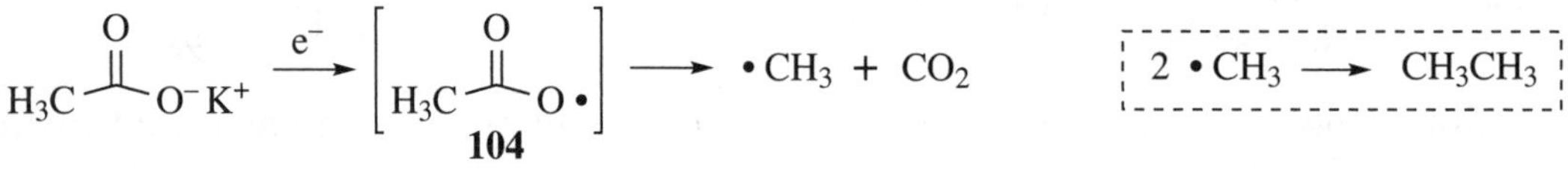

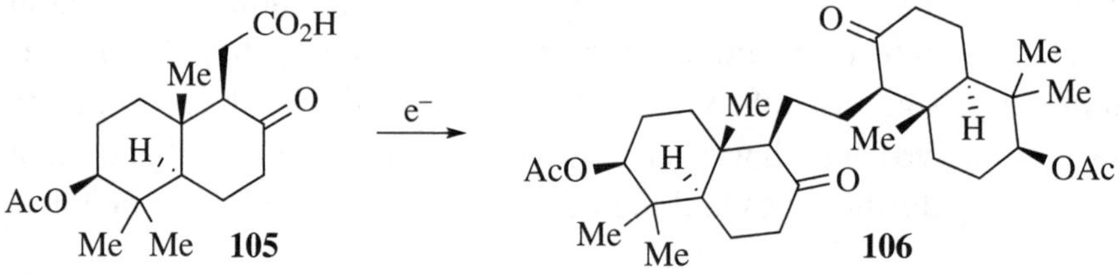

The Kolbe synthesis of two different acids (RCOOH + R'COOH) generally leads to a mixture of products arising from statistical coupling (R-R, R-R', R'-R'). Using an excess of one acid can lead to the desired mixed-coupling product (R-R'), as in the reaction of **107** with an excess of heptanoic acid to give **108** in 80% yield.[88] There are several variations.[88]

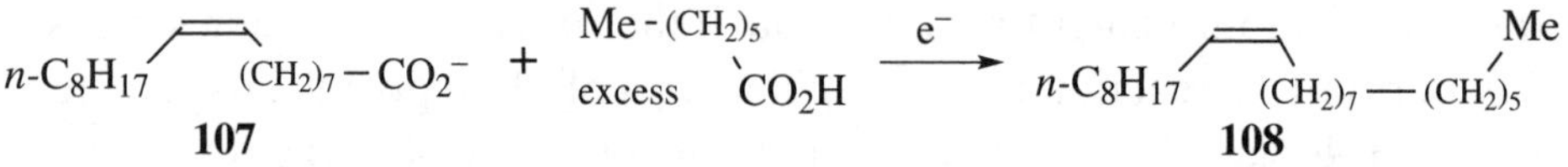

Aldehydes can also be a source of radicals via reaction with transition metal salts such as Mn(III) acetate. Aldehyde-based radical addition to alkenes generates ketones such as **113**

83. Kolbe, H. *Annalen* **1849**, *69*, 257.
84. (a) Crum Brown, A.; Walker, J. *Annalen* **1891**, *261*, 107; (b) Lindsey, A.S.; Jeskey, H. *Chem. Rev.* **1957**, *57*, 583; (c) Vijh, A.K.; Conway, B.E. *Ibid.* **1967**, *67*, 623; (d) The Merck Index, 14th Ed. Merck & Co., Inc., Whitehouse Station, New Jersey, **2006**, p ONR-53; (e) Mundy, B.P.; Ellerd, M.G.; Favaloro Jr., F.G. *Name Reactions and Reagents in Organic Synthesis, 2nd Ed.* Wiley-Interscience, New Jersey, **2005**, pp. 372-373.
85. Petersen, J.B. *Z. Electrochem.* **1906**, *12*, 141.
86. Nuding, G.; Vögtle, F.; Danielmeier, K.; Steckhan, E. *Synthesis* **1996**, 71.
87. Stork, G.; Meisels, A.; Davies, J.E. *J. Am. Chem. Soc.* **1963**, *85*, 3419.
88. (a) Seidel, W.; Knolle, J.; Schäfer, H.J. *Chem. Ber.* **1977**, *110*, 3544; also see (b) Knolle, J.; Schäfer, H.J. *Angew. Chem. Int. Ed.* **1975**, *14*, 758; (c) Klünenberg, H.; Schäfer, H.J. *Ibid.* **1978**, *17*, 47.

in good yield.[89] An acyl radical is initially generated via reaction with the transition metal [Mn(III) and Fe(II) are the most common].[90] Reaction of butanal and 1-heptene in the presence of manganese acetate [Mn(OAc)$_3$] began with addition of the acyl radical (**109**) to both carbons of the π bond, giving regioisomers **110** (48% yield) and **111** (3% yield) as well as a 27% yield of **113** and a 22% yield of **114**. Removal of the hydrogen atom α to the carbonyl led to **112**, and addition to the alkene gave **113**.[91] This reaction also occurs with ketones, but usually requires catalysis (with di-*tert*-butyl peroxide, for example).[92] A simple application is seen in the coupling reaction of cyclopentanone with 1-decene to give 2-decylcyclopentanone in 71% yield.[92b]

Giese describes many synthetic applications of radical coupling with alkenes.[6,93] Reactions with simple alkenes are not very efficient, but reactions with alkenes bearing an electron-withdrawing group are usually quite facile. Perhaps the most common method for generating the radicals for reaction witn alkenes is to heat an alkyl halide with AIBN (**19**), in the presence of tri-*n*-butyltin hydride (*n*-Bu$_3$Sn–H). In a typical reaction, AIBN is thermally decomposed (refluxing benzene) or decomposed by exposure to light, and the resulting radical intermediate reacts with iodocyclohexane to give **115**. This radical reacts with ethyl acrylate to give **116**. Tributyltin hydride is an efficient intermolecular hydrogen donor, reacting with **116** to form the product (**117**), and generate a chain-carrying radical Bu$_3$Sn•, which can react with additional halide to produce **115**. Giese and Gerth used this technique to couple **118** with methyl methacrylate, giving **119** in 70% yield.[94] Removal of the benzylidene acetal (sec. 7.3.B.i) by hydrogenation and subsequent lactonization gave malyngolide (**120**).[95] Table 13.2[95] shows the

89. Reference 3, p 69.
90. (a) Sosnovsky, G. *Free Radical Reactions in Preparative Organic Chemistry*, MacMillan, New York, *1964*; (b) inogradov, M.G.; Nikishin, G.I. *Usp. Khim.* *1971*, *40*, 1960; (c) Nikishin, G.I.; Vinogradov, M.G.; Il'ina, G.P. *Synthesis* *1972*, 376.
91. Nikishin, G.I.; Vinogradov, M.G.; Verenchikov, S.P.; Kostyukov, I.N.; Kereselidze, R.V. *J. Org. Chem. USSR* *1972*, *8*, 539 (Engl. p 544).
92. (a) Nikishin, G.I.; Somov, G.V.; Petrov, A.D. *Dokl. Akad. Nauk.* *1961*, *136*, 1099; (b) *Idem Izvest. Akad. Nauk.* *1961*, 2065 (Engl. p 1924); (c) Glukhovtsev, V.G.; Spektor, S.S.; Golubev, I.N.; Nikishin, G.I. *J. Org. Chem. USSR* *1973*, *9*, 316 (Engl. p 317).
93. (a) Giese, B. *Angew. Chem. Int. Ed.* *1983*, *22*, 753; (b) *Idem Ibid.* *1985*, *24*, 553.
94. Gerth, D.B.; Giese, B. *J. Org. Chem.* *1986*, *51*, 3726.
95. Reference 6, p 58.

radical coupling of representative alkyl halides and selected alkenes (to give **121**), under both photochemical and thermal conditions.

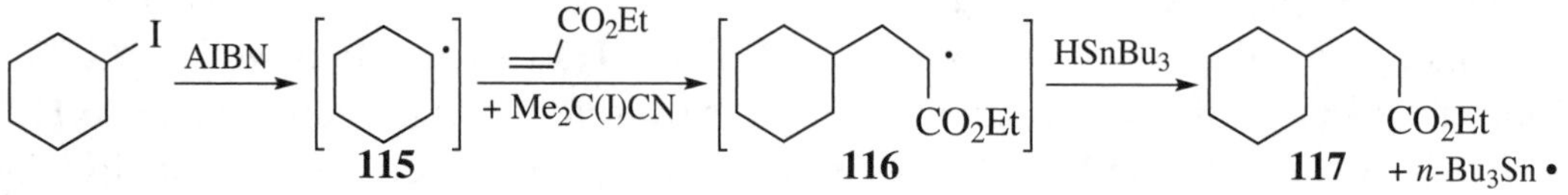

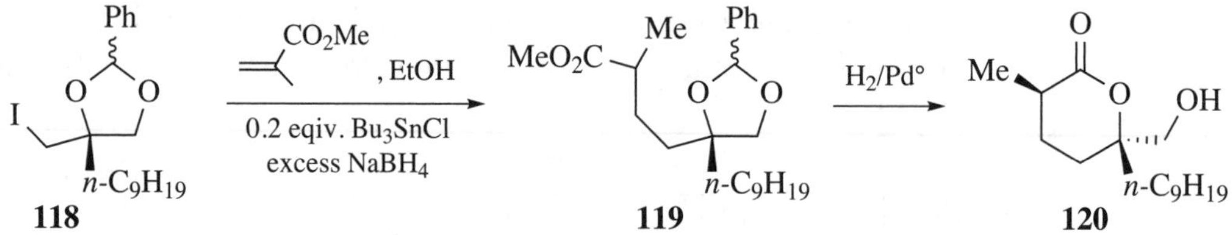

Table 13.2   Intermolecular Radical Coupling of Alkyl Halides and Alkenes

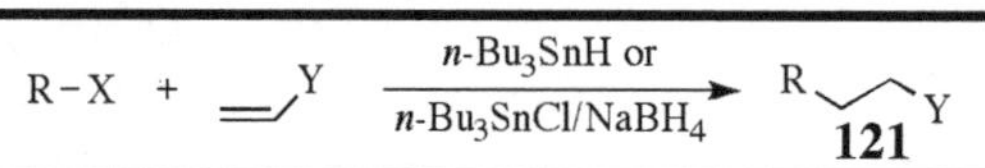

		% 121				% 121	
Alkyl Halide	**Alkene**	**hv**	**Δ (AIBN)**	**Alkyl Halide**	**Alkene**	**hv**	**Δ (AIBN)**
n-C$_6$H$_{13}$I	CH$_2$=CHCN	80		c-C$_6$H$_{11}$I	Methyl vinyl ketone	85	
c-C$_6$H$_{11}$I		95			Acrolein		90
t-C$_4$H$_9$I		87	98		Ethyl acrylate	85	
n-C$_6$H$_{13}$Br			68		Styrene	83	
c-C$_6$H$_{11}$Br			80		1,1-Dichloroethene	87	
t-C$_4$H$_9$Br		62			(E)-Dicyanoethene		72
					(E)-1-Cyano-1-propene	86	

[Reprinted with permission from Giese, B. *Radicals in Organic Synthesis: Formation of Carbon-Carbon Bonds*, Pergamon Press, Oxford, *1986*, p. 58, Copyright *1986*, with permission from Elsevier Science.]

A problem with radical coupling in the presence of a hydrogen donating species such as tin hydride is competing hydrogen transfer to directly reduce the alkyl halide (-CH$_2$I → -CH$_2$H), as described in Section 13.5.D. If the initially formed radical is long lived and/or if the radical coupling reaction is slow, that radical can be reduced by hydrogen abstraction from the tin hydride. In some cases, the reduction process can be the major reaction, but about 20-25% yields are typical. α-Halo esters add to alkenes to form either lactones or substituted ester derivatives.[96,97]

An alternative route has been used for radical coupling, involving alkyl halides and allyl tin derivatives (such as allyltributyltin, n-Bu$_3$SnCH$_2$CH=CH$_2$).[98,101] Migata[99] and Peryne,[100]

96. Kharasch, M.S.; Skell, P.S.; Fisher, P. *J. Am. Chem. Soc. 1948, 70*, 1055.

97. (a) Reference 11, p 501; (b) Giese, B.; Horler, H.; Leising, M. *Chem. Ber. 1986, 119*, 444.

98. Keck, G.E.; Yates, J.B. *J. Am. Chem. Soc. 1982, 104*, 5829.

99. (a) Kosugi, M.; Kurino, K.; Takayama, K.; Migata, T. *J. Organomet. Chem. 1973, 56*, C11; (b) Migata, T.; Nagai, K.; Kosugi, M. *Bull. Chem. Soc. Jpn. 1983, 56*, 2480.

100. (a) Grignon, J.; Pereyne, M. *J. Organomet. Chem. 1973, 61*, C33; (b) Grignon, J.; Servens, C.; Pereyre, M. *Ibid. 1975, 96*, 225.

introduced allyltin compounds used in this context, but Keck and co-workers contributed significantly to the synthetic development of these reagents.[101,98] Allyl tin compounds react with alkyl halides to give the corresponding coupling product (see sec. 12.9.A). An electron-withdrawing substituent at C2 of the allyl stannane increased the reactivity, illustrated by the coupling of bromoadamantane (**122**) and **123** to give **124** in 70% yield.[102] It is not usually possible to use allyl stannanes bearing substituents at C1 or C3,[103] but there are exceptions.[104]

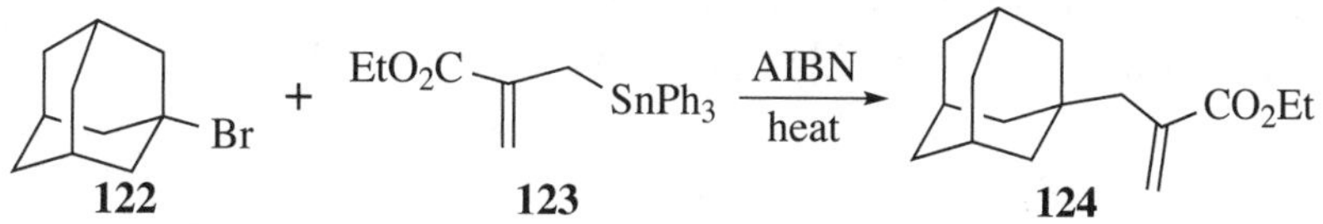

When there is a substituent at the C3 position, addition of the radical is slow. When C1 is substituted the stannane can rearrange, which can be competitive with addition.[105] This latter problem was solved, at least for prenylation, by the use of an allyl sulfide precursor such as **125**. Keck and Byers showed that **126** reacted with **125** in the presence of tributyltin dimer (Bu_3Sn-$SnBu_3$) to give **127** in 76% yield.[106] Photolysis of tributyltin dimer generates two equivalents of $Bu_3Sn\bullet$, which initiates the reaction. Good diastereoselectivity may be achieved in radical addition reactions. When **128** was treated with *tert*-butyl iodide, Bu_3SnH and $Sc(OTf)_3$, a 58% yield of **129** was obtained as a 1:99 anti/syn mixture.[107]

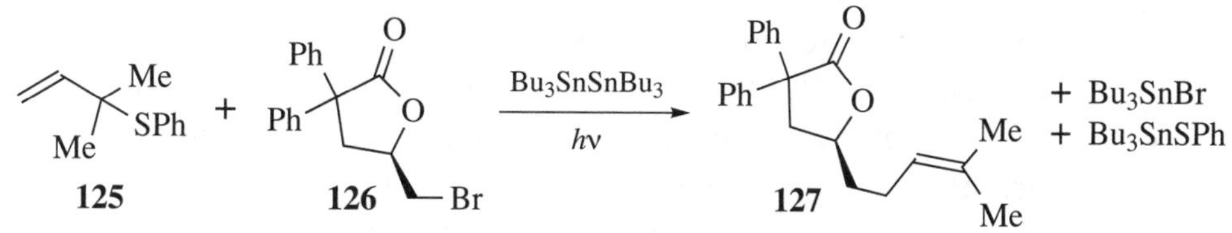

When heteroatom containing substrates react with peroxides or other radical initiators, hydrogen atom transfer can occur as in the transfer of hydrogen from an acetal to the radical to give an alkane and the α-alkoxy radical **130**. The presence of the heteroatom α to the carbon

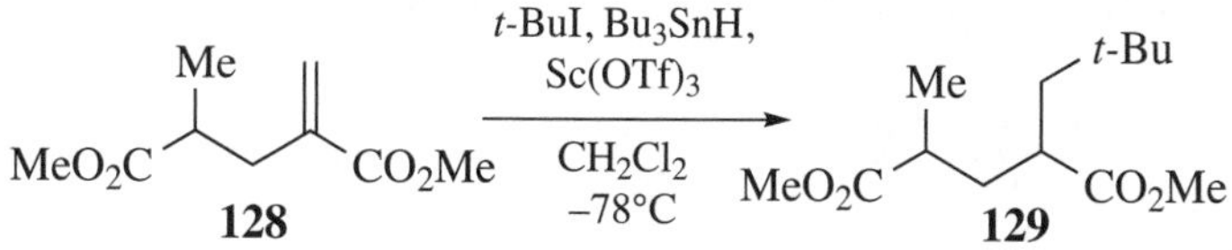

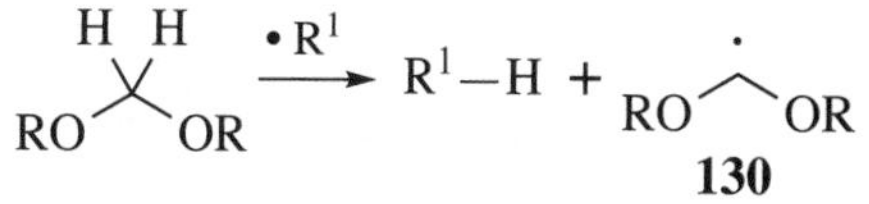

101. Keck, G.E.; Enholm, E.J.; Yates, J.B.; Wiley, M.R. *Tetrahedron* **1985**, *41*, 4079.
102. Baldwin, J.E.; Adlington, R.M.; Birch, D.J.; Crawford, J.A.; Sweeney, J.B. *J. Chem. Soc. Chem. Commun.* **1986**, 1339.
103. Reference 11, p 491.
104. Fliri, H.; Mak, C.-P. *J. Org. Chem.* **1985**, *50*, 3438.
105. (a) Keck, G.E.; Yates, J.B. *J. Organomet. Chem.* **1983**, *248*, C21; (b) Baldwin, J.E.; Adlington, R.M.; Basak, A. *J. Chem. Soc. Chem. Commun.* **1984**, 1284.
106. Keck, G.E.; Byers, J.H. *J. Org. Chem.* **1985**, *50*, 5442.
107. (a) Hayen, A.; Koch, R.; Saak, W.; Haase, D.; Metzger, J.O. *J. Am. Chem. Soc.* **2000**, *122*, 12458. Also see (b) Hayen, A.; Koch, R.; Metzger, J.O. *Angew. Chem. Int. Ed.* **2000**, *39*, 2758.

bearing the radical center leads to enhanced stability. Such radicals add to alkenes, usually with anti-Markovnikov orientation, as in the radical induced addition of HBr to alkenes (sec. 2.10.A). The reaction usually involves slow addition of a large excess (5-10 equivalents) of the heteroatom containing substrate. Peroxides and peroxyesters are the usual radical initiators. The hydrogen atom transfer reaction is often very slow, and an excess of the hydrogen atom donor (the heteroatom substrate) and higher reaction temperatures are essential.[108] Curran showed several general examples of this type of reaction involving alcohols (to give **131**), esters (to give **132**), amines (to give **133**) or aldehydes (to give **134**).[109]

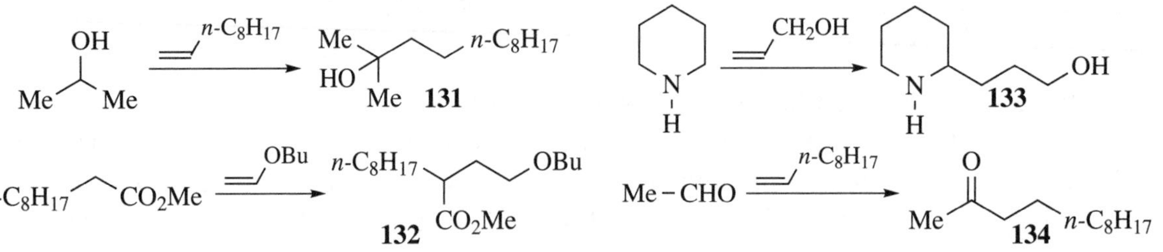

Alkynes are generally less reactive than alkenes in radical coupling reactions since the LUMO of the alkyne is energetically higher than the LUMO of an alkene (sec. 11.2.A, Table 11.1). Formation of the requisite SOMO is therefore more difficult.[110] The intermediate vinyl radical (**135**) undergoes coupling to give a mixture of cis and trans isomers, as shown but when R is bulky the cis compound (**136**) is favored via trans-attack. An example is formation of **115** from cyclohexyl mercuric acetate, which added to 1-phenylethyne to give a 70:30 mixture of **137/138** in 8% yield.[110] Addition to methyl propiolate showed better reactivity, but poorer selectivity.[110] As the size of R increases, syn approach of the hydrogen donor is increasingly hindered.[110] Nonradical nucleophiles usually react faster with alkynes than with alkenes, however.[111]

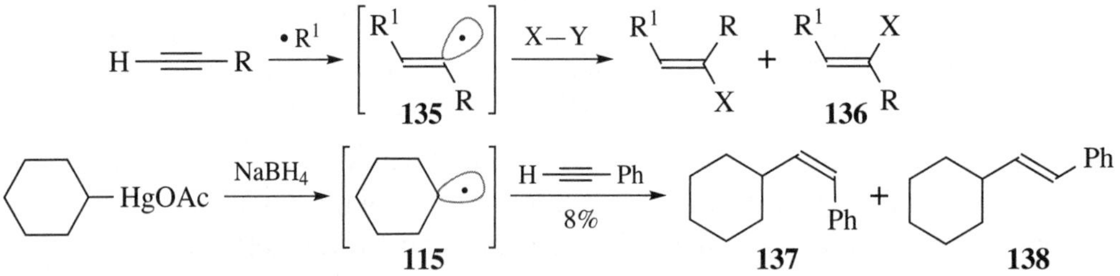

The disconnections that are pertinent to intermolecular radical coupling reactions include

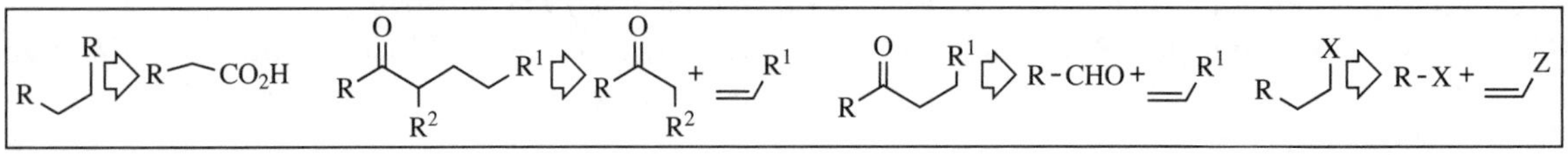

108. Reference 11, p 497.
109. Reference 11, p 498.
110. (a) Giese, B.; Lachhein, S. *Angew. Chem. Int. Ed.* **1982**, *21*, 768; (b) Giese, B.; Meixner, J. *Ibid.* **1979**, *18*, 154.
111. Dickstein, J.I.; Miller, G.I. in *The Chemistry of Carbon Carbon Triple Bonds, Vol. 2*, Patai, S. (Ed.), Wiley, New York, **1978**.

13.7. INTRAMOLECULAR RADICAL REACTIONS (RADICAL CYCLIZATION)

13.7.A. Carbocyclic Ring Systems

There are many examples of radical reactions that produce rings.[11,112] The fundamentals of the reaction can be illustrated by the AIBN induced reaction of tributyltin hydride and **139** (X = I or Br), which generates radical **63**. Once formed, at least three pathways are possible under these conditions.[113] The first is **path a**, a 5-exo-trig-cyclization to give **64**. **Path b** is a 6-endo-trig-cyclization to give **115**. These descriptors are based on **Baldwin's rules** (sec. 6.6.A) and both pathways are favored. The last pathway is the usual hydrogen-transfer process where tin hydride reacts with **63** to give **140** (**path c**), as discussed in Section 13.5.D. In general, the 5-exo-trig-pathway is lower in energy than **path b** and faster than **path c**, leading to **64** as the major product. This outcome is subject to steric effects or electronic effects that arise from interaction of substituents on the alkenyl moiety. At high concentrations (> 5 M), hydrogen transfer to give **140** usually predominates so radical cyclization reactions are typically done at low concentrations (< 0.05 M).[11] In cases where path c is faster than path a or b, photolysis in the presence of n-Bu$_3$Sn-Snn-Bu$_3$ and a halogen transfer agent such as an iodoalkane usually gives cyclized products such as **64** or **115** rather than **140** (see below). Dilauroyl peroxide has also been used to induce radical cyclization with alkyl iodides, and constitutes a tin-free procedure for radical iodine atom transfer reactions.[114]

Formation of the five-membered ring via a 5-exo-trig reaction is a kinetic process. With the possibility of a larger or a smaller ring, radical cyclization generally gives the smaller ring. There is a better SOMO-LUMO overlap ($\approx 109°$ angle of attack) than is possible in the 6-endo-trig transition state, which favors the 5-exo-trig preference. Beckwith[115] calculated the transition state strain energies for radical cyclization of ω-alkenyl radicals H$_2$C=CH(CH$_2$)$_n$CH$_2$•, and found that the energy increased as the size of the ring being formed increased (i.e., the value of

---

112. For a review of the synthesis of medium-sized rings using free radical conditions, see Yet, L. *Tetrahedron* **1999**, *55*, 9349.
113. Reference 11, p 419.
114. Ollivier, C.; Bark, T.; Renaud, P. *Synthesis* **2000**, 1598.
115. Beckwith, A.L.J.; Ingold, K.U. in Vol 1 of *Rearrangements in Ground States and Excited States*, de Mayo, P. (Ed.), Academic Press, New York, *1980*, pp 162-283.

n increased).[116] The ΔE_s for the exo transition state is lower in energy than the corresponding endo transition state, except for octenyl radical and 2-methyl-1-hexenyl radical. The former reflects the difficulty in forming eight-membered rings, and the latter reflects the increase in steric strain due to the methyl group. Beckwith used **141** as the model for exo-trig ring closure, and **142** as the model for endo-trig ring closure.[116]

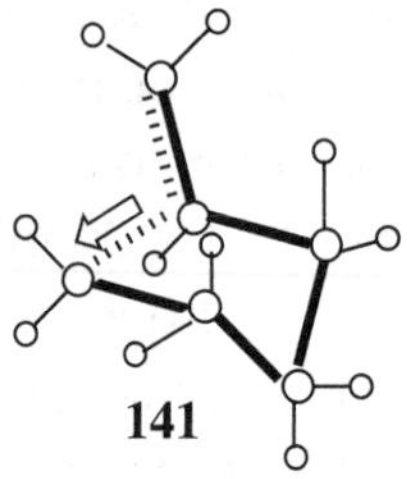

141 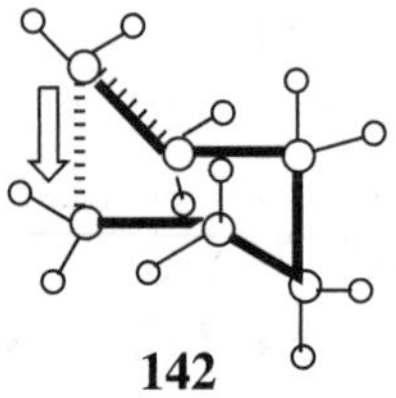   **142**

[Reprinted with permission from Beckwith, A.L.J.; Schiesser, C.H. *Tetrahedron Lett.* **1985**, *26*, 373, Copyright 1985, with permission from Elsevier Science.]

MOLECULAR MODELING: Thermodynamic vs. Kinetic Control of Radical Cyclization

In the presence of AIBN and Bu_3SnH, 6-bromo-1-hexene reacts with AIBN to give the hex-5-enyl radical. This may either abstract hydrogen from Bu_3SnH, leading to the unrearranged reduction product, 1-hexene, or first cyclize either to cyclopentylmethyl radical or to cyclohexyl radical before abstracting hydrogen, leading to methylcyclopentane and cyclohexane, respectively. Experimentally, the ratio of 1-hexene:methylcyclopentane:cyclohexane is 17:81:2.

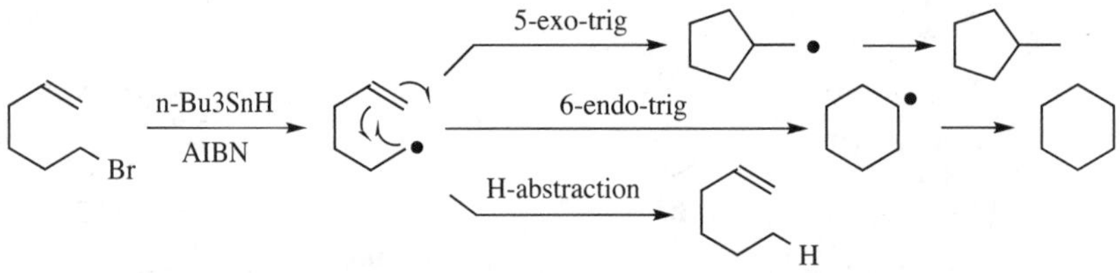

Calculations show that the barrier for closure to cyclohexyl radical is 10 kJ/mol larger than that for closure to cyclopentylmethyl radical. Assuming that there is no unrearranged product, this corresponds to a ratio of cyclopentylmethyl to cyclohexyl radicals of 95:5. Does this support the notion that the ring closure reaction is kinetically controlled? Examine the transition states that are provided on screen for the ring closure reactions leading to the two radicals. Are there are structural differences that might explain the noted preference?

Obtain energies for both cyclohexyl to cyclopentylmethyl radicals. For each in turn, *click* on > to the right of the tab at the top of the screen and select ***Continue***. A radical may be built from its corresponding products by deleting the appropriate hydrogen atom. Which radical is more stable? Why? Were the reaction under thermodynamic control, what would the ratio of the products be at room temperature (assume that there is no unrearranged product)? Does your result support the notion that the ring closure reaction is thermodynamically controlled?

---

116.   (a) Beckwith, A.L.J.; Schiesser, C.H. *Tetrahedron Lett.* **1985**, *26*, 373; (b) Beckwith, A.L.J. *Tetrahedron* **1981**, *37*, 3073.

This preference is seen for the cyclization reaction of **143**, where the five-membered ring radical product (**144**) predominated with a variety of substituents R rather than the cyclohexyl radical (**145**).[117] When R is hydrogen or an alkyl group, the ratio of **144/145** is >95:5. That ratio is 93:7 when R is CO_2R^1, and about 75:25 when R is COR or COPh.[117] Another potential problem with this reaction is formation of cis or trans isomers. In Table 13.3,[117] iodoalkene **146** gives essentially a 1:1 mixture of cis (**147**) and trans (**148**) products under a variety of conditions, but the 6-endo-trig product (**149**) is formed as a minor constituent. In the absence of special controlling factors (steric, electronic, neighboring group effects, etc.) the cyclization proceeds with poor cis-trans selectivity.

Table 13.3   *cis-trans* Selectivity in the Radical Cyclization of 146

Conditions	% Yield	147 : 148 : 149
10% Bu₃SnH, AIBN, 5h	48	48 : 45 : 7
10% AIBN, 24h	78	49 : 43 : 8
Me₃Sn-SnMe₃, heat, 16h	83	53 : 40 : 7
Bu₃Sn-SnBu₃, hv, 24h	42	50 : 40 : 10

[Reprinted with permission from Curran, D.P.; Chang, C.-T. *J. Org. Chem.* **1989**, *54*, 3140. Copyright © **1989** American Chemical Society.]

In the formation of polycyclic systems, the cyclization process can lead to more than one regioisomer or stereoisomer. As noted in Table 13.3, the ratio of (*E*) and (*Z*) isomers is often ~1:1. In most cyclizations, cis-ring fusion predominates when the final product is a bicyclo[3.3.0]- or bicyclo[4.3.0]-system, but formation of a bicyclo[5.3.0]-system usually leads to significant amounts of trans-fused products.[118] In the cyclization of **150**, a mixture of (*E*) and (*Z*) isomers is possible, and there are two different alkene moieties in **150** that lead to different regioisomeric products.[118] Abstraction of the bromide in **150** with Bu₃Sn• generated a radical that cyclized, and was quenched with tin hydride via hydrogen transfer to give a 1:1 mixture of **151/152** in 89% yield. Note that there are diastereomers resulting from the presence of OMe and the ratio of α/β-methoxyl isomers in **152** was 3:1.[119] All products were cis fused. The reaction was controlled by frontier molecular orbitals (SOMO-LUMO, sec. 13.4) rather than stabilities of radical intermediates.

117.   Curran, D.P.; Chang, C.-T. *J. Org. Chem.* **1989**, *54*, 3140.
118.   (a) Reference 11, p.421; (b) Clive, D.L.J.; Cheshire, D.R.; Set, L. *J. Chem. Soc. Chem. Commun.* **1987**, 353.
119.   Beckwith, A.L.J.; O'Shea, D.M.; Roberts, D.H. *J. Chem. Soc. Chem. Commun.* **1983**, 1445.

There are many synthetic examples that use radical cyclization as a key step, and the radical precursor is not limited to unfunctionalized alkyl iodides or bromides. In Lee and co-worker's synthesis of (+)-IKD-8344,[120] the primary bromide unit on the silyloxy tether of **153** was cyclized to give **154** in 91% yield. Note the use of triethylborane as the radical initiator. Cyclization via aryl radicals is also possible. In a synthesis of (+)-dinapsoline by Sit and co-workers[121] aryl bromide **155** was cyclized to **156** in good yield under standard conditions in degassed benzene. Radical cascade reactions can be used for the synthesis of polycyclic ring systems. In these reactions, polyenes are subjected to radical cyclization, generating tricyclic or even tetracyclic ring systems.[122] Chiral auxiliaries have been used effectively in radical cyclization reactions.[123]

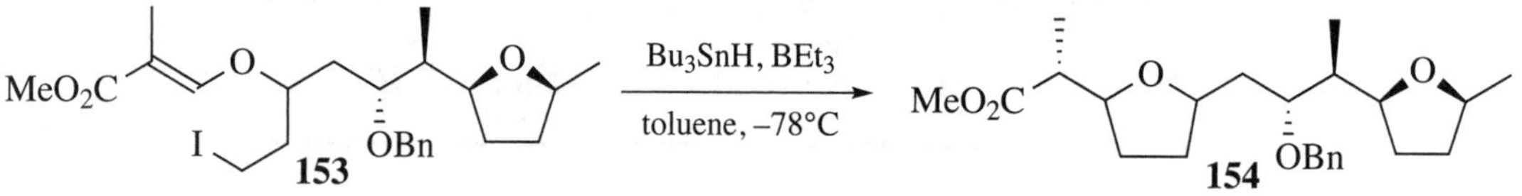

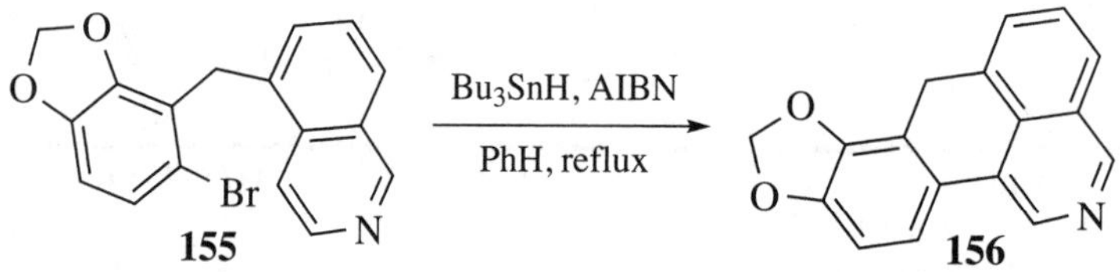

Alkynes are common precursors to *exo*-methylene compounds, as in Sha and co-worker's synthesis of (−)-bakkenolide.[124] Treatment of iodide **157** with tributyltin dimer and then AIBN/Bu₃SnH led to the radical intermediate and cyclization. The second step was required to reduce the iodide formed in the first reaction of the radical by tributyltin hydride to give a 93% yield of **158**. The trimethylsilyl group attached to the alkyne unit is important to sterically inhibit a 6-endo-dig reaction that can complete with the 5-exo-dig process. The silyl group in **158** was removed by treatment with trifluoroacetic acid in dichloromethane, giving **159** in 80% yield along with 10% of the conjugated ketone in which the C=C unit migrated to the endocyclic position.

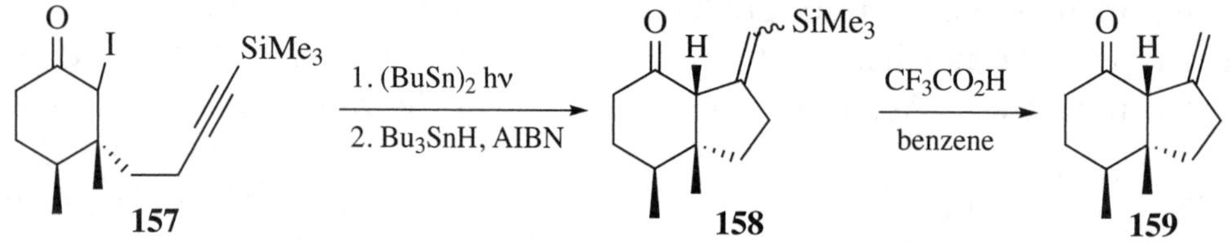

120. Kim, W.H.; Hong, S.K.; Lim, S.M.; Ju, M.-A.; Jung, S.K.; Kim, Y.W.; Jung, J.H.; Kwon, M.S.; Lee, E. *Tetrahedron* **2007**, *63*, 9784.
121. Sit, S.-Y.; Xie, K.; Jacutin-Porte, S.; Taber, M.T.; Gulwadi, A.G.; Korpinen, C.D.; Burris, K.D.; Molski, T.F.; Ryan, E.; Xu, C.; Wong, H.; Zhu, J.; Krishnananthan, S.; Gao, Q.; Verdoorn, T.; Johnson, G. *J. Med. Chem.* **2002**, *45*, 3660.
122. For examples, see (a) Boehm, H.M.; Handa, S.; Pattenden, G.; Roberts, L.; Blake, A.J.; Li, W.-S. *J. Chem. Soc. Perkin Trans. 1* **2000**, 3522; (b) Fensterbank, L.; Mainetti, E.; Devin, P.; Malacria, M. *Synlett*, **2000**, 1342.
123. For an example taken from a synthesis of (+)-triptocallol, see Yang, D.; Xu, M.; Bian, M.-Y. *Org. Lett.* **2001**, *3*, 111.
124. Jiang, C.-H.; Bhatacharyya, A.; Sha, C.-K. *Org. Lett.* **2007**, *9*, 3241.

Vinyl radicals are useful in radical coupling and addition reactions.[125] Stork used the cyclization of vinyl radical **161** (derived from vinyl bromide **160**) to complete a synthesis of norseychellanone (**162**).[126] The cyclization step proceeded in 70% yield. Coupling one aromatic ring to another is also possible using radical techniques. In a synthesis of goniopedaline,[127] Couture and co-workers refluxed **163** with AIBN and Bu_3SnH in benzene and obtained a 75% yield of **164**.

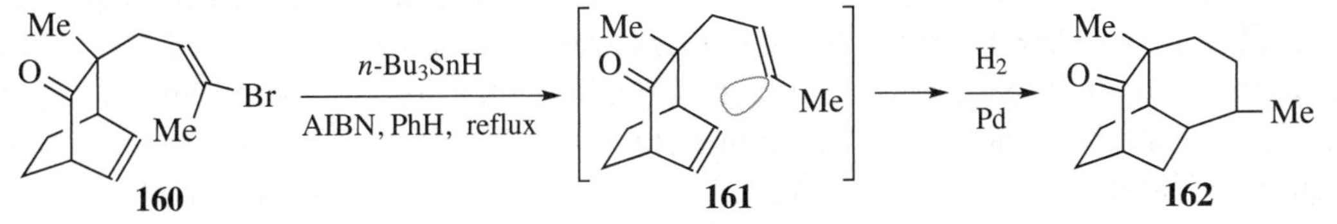

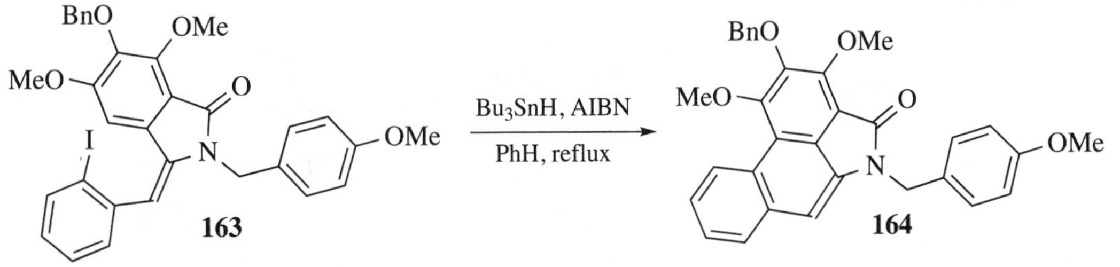

The radical cyclization disconnections are

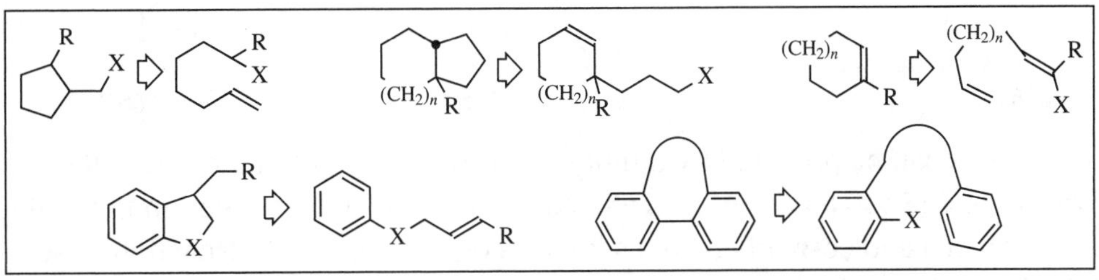

13.7.B. Heteroatom Ring Systems

Radical cyclization can be used to prepare many heterocyclic rings.[128] The standard AIBN and tributyltin hydride in refluxing benzene conditions are common, but other solvents can also be used. For instance, radical cyclization of **165** in aqueous ethanol using a combination of phosphoric acid and a base led to **166** in 83% yield.[129] In this example, it is clear that

125. (a) Stork, G.; Baine, N.H.. *Am. Chem. Soc.* **1982**, *104*, 2321; (b) Stork, G.; Mook, Jr., R. *Ibid.* **1983**, *105*, 3720; (c) Nozaki, K.; Oshima, K.; Utimoto, K. *Ibid.* **1987**, *109*, 2547; (d) Stork, G.; Mook, Jr., R. *Ibid.* **1987**, *109*, 2829; (e) Beckwith, A.L.J.; O'Shea, D.M. *Tetrahedron Lett.* **1986**, *27*, 4525; (f) Stork, G.; Mook, Jr., R. *Ibid.* **1986**, *27*, 4529.

126. Stork, G.; Baine, N.H. *Tetrahedron Lett.* **1985**, *26*, 5927.

127. Rys, V.; Couture, A.; Deniau, E.; Grandclaudon, P. *Eur. J. Org. Chem.* **2003**, 1231.

128. (a) Majumdar, K.C.; Basu, P.K.; Mukhopadhyay, P.P. *Tetrahedron*, **2004**, *60*, 6239; (b) Bowman, W.R.; Fletcher, A.J.; Potts, G.B.S. *J. Chem. Soc. Perkin Trans. 1* **2002**, 2747.

129. Yorimitsu, H.; Shinokubo, H.; Oshima, K. *Chem. Lett.* **2000**, 104.

heteroatoms can be part of the chain that links the radical precursor and the alkene. Stork prepared protected lactones via radical cyclization of alkenyl ethers.[130] White and coworkers used this reaction in a synthesis of rhizoxin D.[131] Cyclization of **167** gave **168A** and **168B**, but the nature of R^1 and R^2 played a key role in the cyclization. Cyclization of **167** (R^1= H, R^2 = CO_2Me) gave essentially a 1:1 mixture of anti (**168A**) and syn (**168B**) diastereomers in 92% yield. When **167** (R^1 = CO_2Me, R^2 = H) was similarly cyclized, however, a 15:1 mixture of anti:syn (**168A:168B**) diastereomers was obtained in 93% yield. The electron withdrawing carbomethoxy group facilitated formation of the six-membered ring. The position of the CO_2Me unit influenced the transition state of cyclization, where the transition state for cyclization of Z-ester showed a marked preference for a less hindered conformation that led to high diastereoselectivity.[132]

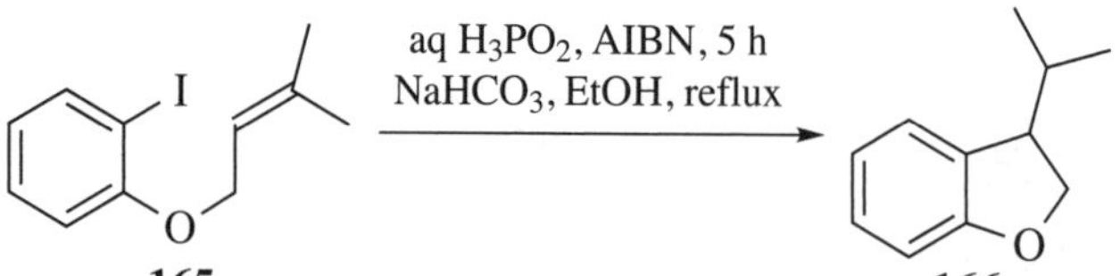

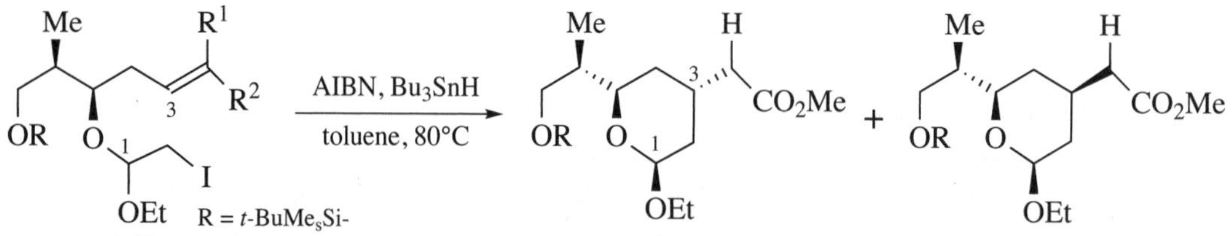

Heterocyclic rings can be prepared by simply incorporating a heteroatom into the chain, as in the cyclization of **165** to **166** or **167** to **168**. An example with a nitrogen in the molecule was reported by Koreeda and co-workers in a synthesis of (−)-sibirine.[133] Reaction of selenide **169** with AIBN/Bu₃SnH in refluxing toluene led to a close to 1:1 mixture of the 5-exo-trig product **170** and the 7-endo-trig product, **171**. Based on Baldwin's rules (sec. 6.6.A), it is clear that there is little difference in the transition state energies leading to the two products.

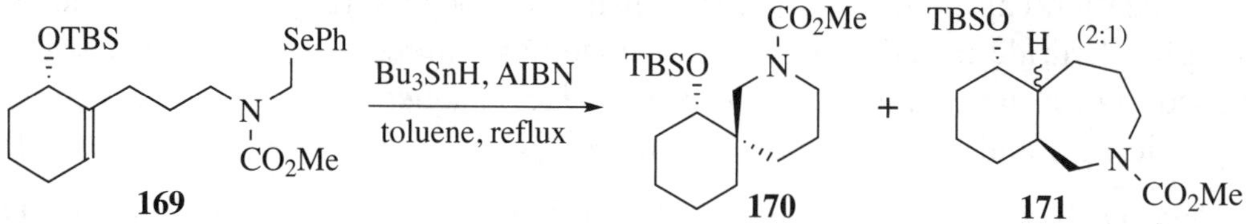

130. (a) Stork, G.; La Clair, J.J.; Spargo, P.; Nargund, R.P.; Totah, N. *J. Am. Chem. Soc.* **1996**, *118*, 5304. Also see (b) Stork, G.; Mook Jr., R.; Biller, S.A.; Rychnovsky, S.D. *J. Am. Chem. Soc.* **1983**, *105*, 3741.
131. White, J.D.; Blakemore, .R.; Green, N.J.; Hauser, E.B.; Holoboski, M.A.; Keown, L.A.; Kolz, C.S.N.; Phillips, B.W. *J. Org. Chem.* **2002**, *67*, 7750.
132. Ishibashi, H.; Matsukida, H.; Toyao, A.; Tamura, O.; Takeda, Y. *Synlett* **2000**, 1497.
133. Koreeda, M.; Wang, Y.; Zhang, L. *Org. Lett.* **2002**, *4*, 3329.

Both phenylthio (-SPh) and phenylseleno (-SePh) groups can be used as leaving groups for radical cyclization (see above), where sulfur or selenium atom transfer leads to radical formation. Hart and Tsai used this technique to cyclize **172** to the racemic diastereomer **173** in a synthesis of isoretronecanol.[134c] Smith and co-workers, however, used a chiral template approach and prepared alkenyl iodide **174** from the chiral precursor pyroglutamate.[135] Radical cyclization gave pyrrolizidinone **175** in a synthesis of pseudoheliotridane, with high asymmetric induction.[135] Knapp et al. also used pyroglutamate derivatives for asymmetric radical cyclizations.[136] Other heterocycles can be used in radical cyclizations.[137] Treatment of pyrrole derivative **176** with AIBN and tributyltin hydride gave a 60% yield of **177**. The sulfonyl group on the pyrrole ring was required for good yields of cyclized product.[138]

Radical cyclization with high diastereoselectivity and high asymmetric induction is possible when chiral precursors are used (as with **174** → **175** from pyroglutamate). Livinghouse and Jolly[139] reported a chiral synthesis of (–)-trachelanthamidine via cyclization of **178**, prepared

from prolinal, which gave a 30:1 mixture of **179** and **180** in 55% yield. In this reaction, initiation with AIBN in the presence of tri-*n*-butyltin hydride led to significant amounts of the hydrogen transfer product (2-ethenyl-*N*-acetyl-2-pyrrolidinone). Hydrogen transfer is a common problem, and it was solved in this work by changing the method by which the radical is generated. Radical cyclization in the presence of tributyltin dimer (Bu₃SnSnBu₃) gave the desired products (as shown), but required photochemical initiation with a sunlamp.

134. (a) Choi, J.-K.; Hart, D.J.; Tsai, Y.-M. *Tetrahedron Lett.* **1982**, *23*, 4765; (b) Burnett, D.A.; Choi, J.-K.; Hart, D.J.; Tsai, Y.-M. *J. Am. Chem. Soc.* **1984**, *106*, 8201; (c) Hart, D.J.; Tsai, Y.-M. *Ibid.* **1982**, *104*, 1430; (d) *Idem Ibid.* **1984**, *106*, 8209; (e) Choi, J.-K.; Hart, D.J. *Tetrahedron* **1985**, *41*, 3959.

135. (a) Keusenkothen, P.F.; Smith, M.B. *J. Chem. Soc. Perkin Trans. 1* **1994**, 2485; (b) Keusenkothen, P.F.; Smith, M.B. *Tetrahedron Lett.* **1989**, *30*, 3369.

136. Knapp, S.; Gibson, F.S.; Choe, Y.H. *Tetrahedron Lett.* **1990**, *31*, 5397.

137. Bowman, W.R.; Bridge, C.F.; Brookes, P. *J. Chem. Soc. Perkin Trans. 1* **2000**, 1.

138. Antonio, Y.; de la Cruz, E.; Galeazzi, E.; Guzman, A.; Bray, B.L.; Greenhouse, R.; Kurz, L.J.; Lustig, D.A.; Maddox, M.L.; Muchowski, J.M. *Can. J. Chem.* **1994**, *72*, 15.

139. Jolly, R.S.; Livinghouse, T. *J. Am. Chem. Soc.* **1988**, *110*, 7536.

MOLECULAR MODELING: Experiments and Calculations are both Similar and Different

Experiments and calculations are both able to provide information about structures, energies and properties. Experiments "report" chemistry, whereas calculations simulate (model) the fundamental processes underlying chemistry. Neither experiments nor calculations tell "why" something happens, for example, why one product is favored over another. What chemists might explain in terms of simple language ("this is more crowded than that") may in fact be much more complicated.

Radical cyclization of the *E*-conjugated ester (**A**) gives a 1:1 mixture of *syn* and *anti* diastereomers in >90% yield. Cyclization of the corresponding *Z*-conjugated ester (**B**) under the same conditions gives the same products, but in a 15:1 mixture of *anti:syn* diastereomers.

Transition state calculations reproduce these results. The *syn* and *anti* transition states for the reaction of **A** are separated by less than 1 kJ/mol, while the *syn* transition state for the reaction of **B** is 13/kJ below the *anti* transition state. Examination of the transition states for both reactions (**A**→*syn*, top left; **A**→*anti*, top right; **B**→*syn*, bottom left; **B**→*anti*, bottom right), however, does *not* reveal the cause of the individual preferences or for the change in preference from *E* to *Z* alkenes.

You can expect modeling to help you assign major and minor products, determine whether the reaction is under kinetic or thermodynamic control, estimate reactivity as well as regio and stereoselectivity, and compare products and rates of one reaction with another. Don't expect the models to tell you why. If a synthetic chemist needs to modify product distribution, an understanding of mechanism is essential. Thoughtful experiments will usually be necessary, often in combination with thoughtful modeling.

The use of dibutyltin dimer under photochemical methods is an atom-transfer method that is most useful when the cyclization step is slow (see above). When cyclization is slow relative to hydrogen transfer, the dominant reaction is usually hydrogen atom transfer to give reduced products. Photolysis of the tin dimer generates the radical intermediate without a hydrogen atom transfer agent present (other than the substrate and the solvent). Another example illustrating the utility of this reagent in cyclization reactions is Curran and Chang's attempt to cyclize **181**, which gave only **182** on treatment with tin hydride/AIBN and no **183**.[140] Cyclization with dibutyltin dimer and photochemical induction gave a radical intermediate that was quenched by reaction with the iodine radical byproduct to give **183**. The iodine could be reduced by reaction with tin hydride (see sec. 13.5D).[140]

---

140.   Curran, D.P.; Chang, C.-T. *Tetrahedron Lett.* **1987**, *28*, 2477.

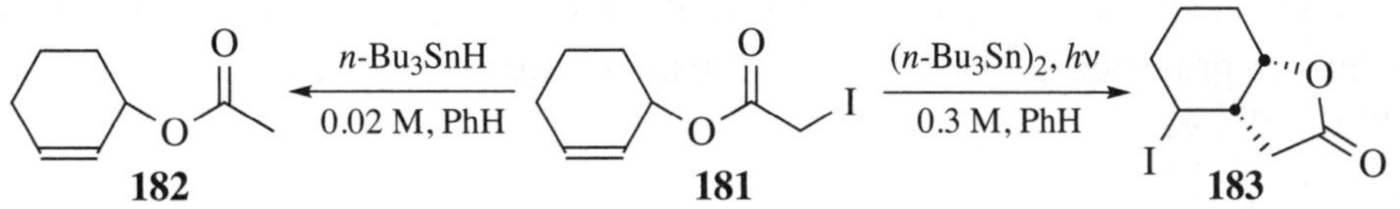

MOLECULAR MODELING: Thermodynamic vs. Kinetic Control Revisited

Reaction of **A** with AIBN gives rise to the radical **B**, which then cyclizes either to a cyclohexyl radical (**C**) via pathway *a*, or to a tetrahydropyranyl radical (**D**) via pathway *b*. The products are **E** and **F**, respectively. Experimentally, the ratio of **E:F** is 1:3 (pathway *b* is favored).

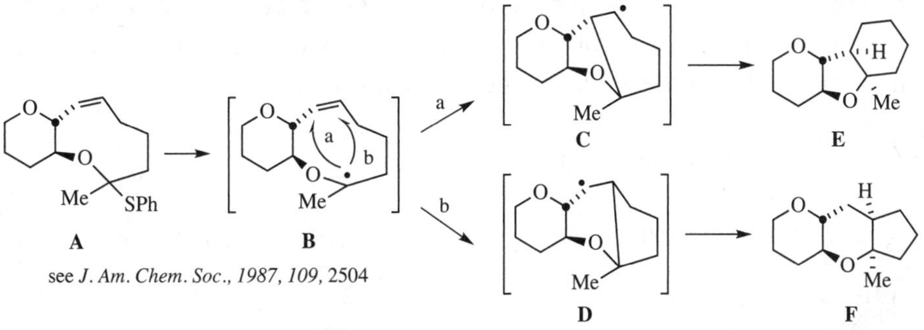

see *J. Am. Chem. Soc.*, *1987, 109,* 2504

Calculations show that the barrier to **D** is 3 kJ/mol less than that to **C**. This does not necessarily mean that this reaction is under kinetic control, as **D** may also be the thermodynamic product. Obtain energies for **C** and **D**. Start with **E** and **F**, and delete the appropriate hydrogen atom. Is **D** the thermodynamic product? Is the cyclization necessarily kinetically controlled? Elaborate.

Where cyclization involves formation of a six-membered ring, allylic hydrogen abstraction may be preferred. Attempted cyclization of **184** gave only **187** via hydrogen abstraction (**179** to **180**), and no cyclization was observed.[141] A six-center transition state for removal of the allylic hydrogen is possible in **185**, and this is preferred over the 6-exo-trig transition state for ring formation. The methyl groups in **185** prevent close approach of hydrogen to the radical center. If an activating group is attached to the alkene receptor, formation of the six-membered ring is accelerated relative to hydrogen transfer.

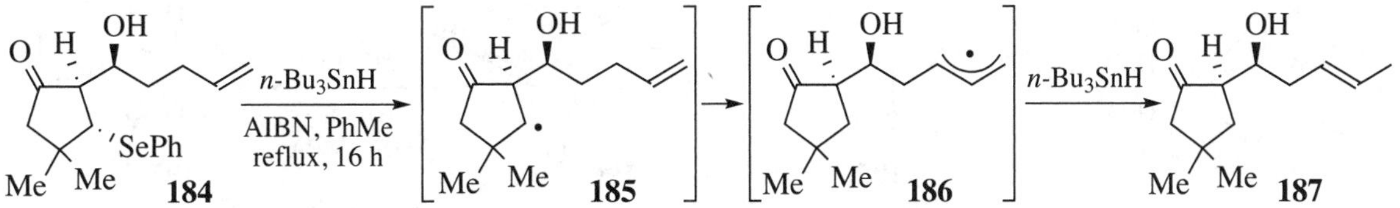

Bach and co-workers reported an asymmetric radical cyclization,[142] in which alkenyl-iodide **188** cyclized under radical conditions (initiated by BEt₃) in the presence of the chiral complexing

141. Leonard, W.R.; Livinghouse, T. *Tetrahedron Lett.* *1985, 26,* 6431.
142. Aechtner, T.; Dressel, M.; Bach, T. *Angew. Chem. Int. Ed.* *2004, 43,* 5849.

agent **189**. The selectivity is believed to rise from hydrogen-bond mediation of the radical cyclization in the presence of **189**. Under these conditions, cyclization of **188** gave a 71% yield of **190** in 79% ee.

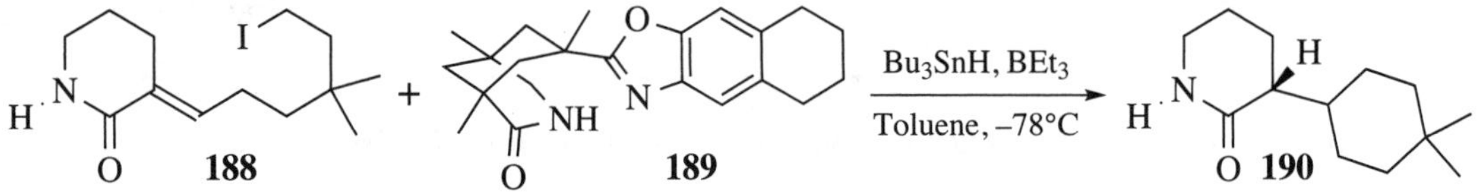

Radical conjugate additions of this type are very facile, and many internal cyclizations have been reported.[143] Relatively large rings can be prepared, as illustrated by the cyclization of **191** to **192** and **193** (63% and 22% yield, respectively).[144] The cyclization process is faster than the hydrogen transfer under high dilution conditions. The two remote reactive fragments of the long chain must be in close proximity for the cyclization.

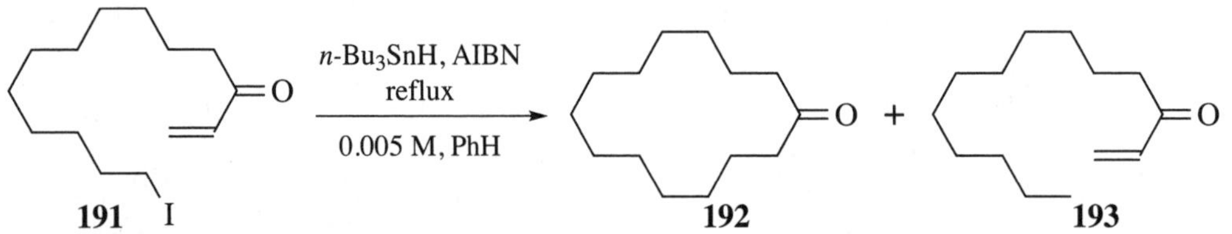

The alkenyl and alkynyl radical disconnections are

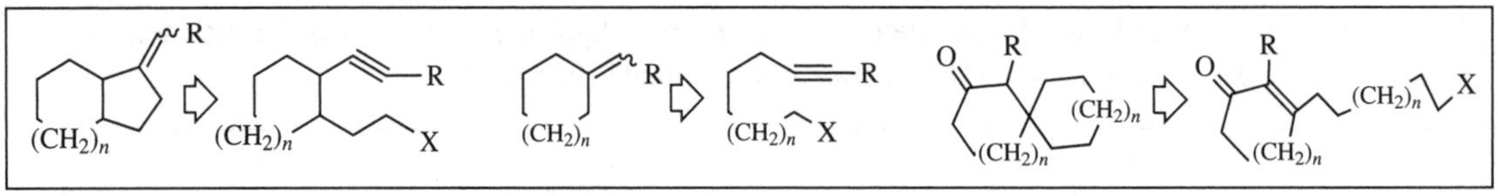

Samarium (II) iodide mediates radical-type cyclizations, and has been used in the synthesis of natural products.[145] Two examples illustrate the utility, and similarity to the transformations described above. Curran and co-workers shown that aryl ether **194** reacted with SmI_2 to give the cyclized product, and quenching with D_2O provide benzofuran **195** in 80% yield.[146] Ohta and co-workers used this type of strategy for a synthesis of pyrrolam A.[147] Alkynyl lactam **196** was treated with excess SmI_2, leading to a 90% yield of cyclized product, indolizidinone **197**.

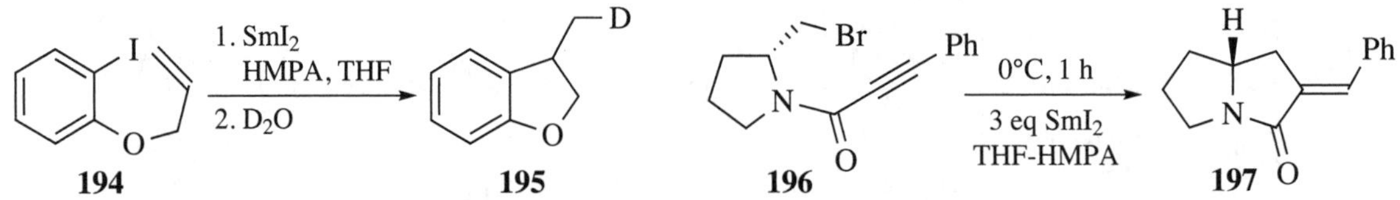

---

143. For example, see Winkler, J.D.; Sridar, V. *Tetrahedron Lett.* **1988**, *29*, 6219.
144. Porter, N.A.; Chang, V.H.-T. *J. Am. Chem. Soc.* **1987**, *109*, 4976.
145. Edmonds, D.J.; Johnston, D.; Procter, D.J. *Chem. Rev.* **2004**, *104*, 3371.
146. Curran, D.P.; Fevig, T.L.; Totleben, M.J. *Synlett* **1990**, 773.
147. Aoyagi, Y.; Manabe, T.; Ohta, A.; Kurihara, T.; Pang, G.-L.; Yuhara, T. *Tetrahedron* **1996**, *52*, 869.

13.7.C Cyclization to C=X and Cyclization of Heteroatom Radicals

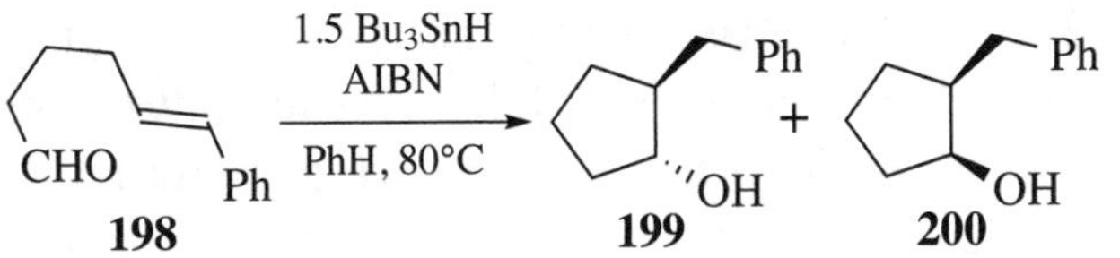

Radical cyclization is not limited to addition to alkenes and alkynes. Carbonyls and imines can also be radical receptors. Enholm and co-workers showed that when **198** was treated with AIBN and tributyltin hydride, the major product was a mixture of alcohols **199** and **200**.[148] Fu and Hays showed that this could be made catalytic in tributyltin hydride.[149] When **198** was treated with 0.1 Bu₃SnH, 0.5 PhSiH₃, two equivalents of EtOH and AIBN, an 85% yield was obtained as a 1.1:1 **199/200** mixture.

In Keck and co-worker's synthesis of 7-deoxypancratistatin, radical precursor **201** was cyclized to the oximine ether unit under radical conditions to give a 3:1 mixture of **202** and **203** in 90% yield.[150] Another variation by Weinreb and co-workers involving generation, cyclization, and trapping of amidyl radicals.[151] When hydroxamic acid **204** was cyclized with TEMPO in the presence of diphenyl diselenide, lactam **205** was obtained in 64% yield.

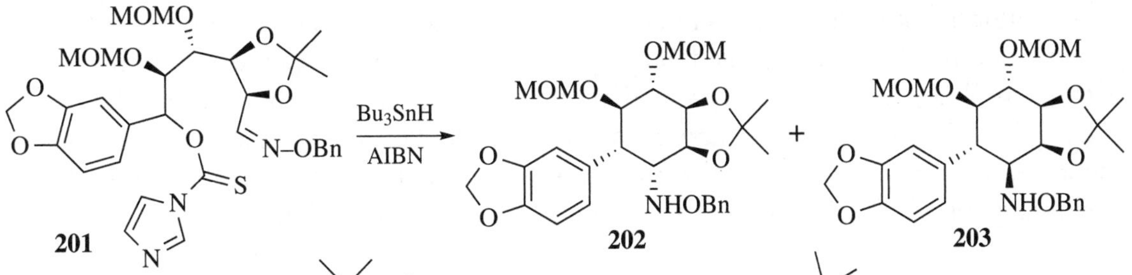

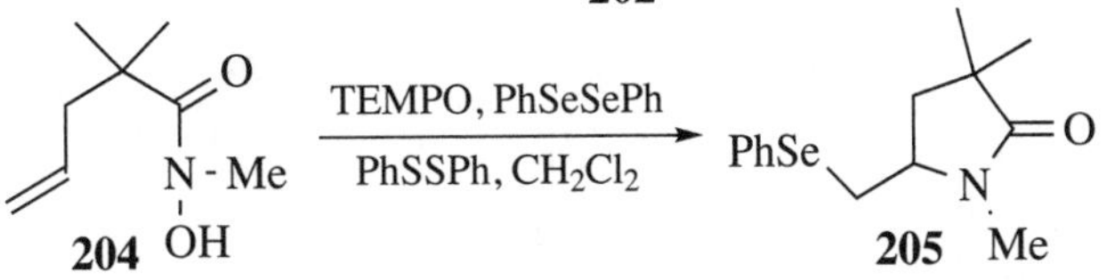

The C=X radical cyclization disconnection is

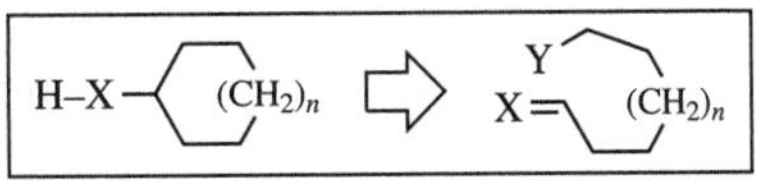

A classical named reaction provides another example of a hydrogen atom transfer reaction that produces a heterocyclic compound. When N-halo-amines[152] such as **206** are treated with

148. (a) Enholm, E.J.; Prasad, G. *Tetrahedron Lett.* **1989**, *30*, 4939; (b) Enholm, E.J.; Burroff, J.A. *Tetrahedron Lett.* **1992**, *33*, 1835; (c) Enholm, E.J.; Burroff, J.A. *Tetrahedron*, **1997**, *53*, 13583; (d) Enholm, E.J.; Kinter, K.S. *J. Org. Chem.* **1995**, *60*, 4850.
149. Hays, D.S.; Fu, G.C. *Tetrahedron* **1999**, *55*, 8815.
150. Keck, G.E.; McHardy, S.F.; Murry, J.A. *J. Org. Chem.* **1999**, *64*, 4465.
151. Lin, X.; Stien, D.; Weinreb, S.M. *Tetrahedron Lett.* **2000**, *41*, 2333.
152. Kovacic, P.; Lowery, M.K.; Field, K.W. *Chem. Rev.* **1970**, *70*, 639.

concentrated sulfuric acid, the amino radical cation **207** is produced. Hydrogen atom transfer via a six-center transition state generates the ammonium radical (**208**), which reacts with the protonated form of **206** (ammonium salt **209**) to give **210**. When **210** is neutralized, the amine displaces the primary halide to give a hexahydroindolizine (**211**).[153] The conversion of *N*-haloamines to cyclic amines is called the **Hofmann-Löffler-Freytag reaction**.[154]

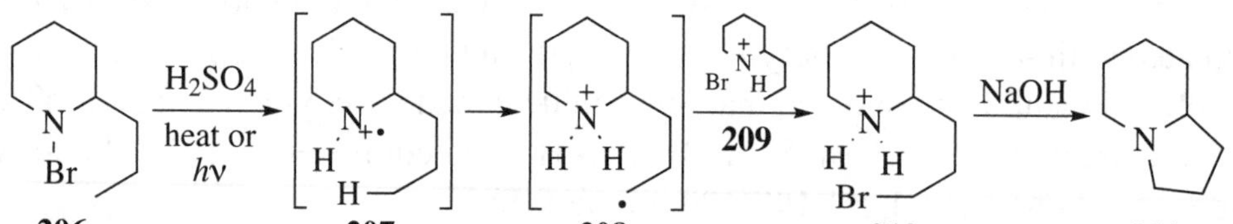

In its original form, the reaction generated pyrrolidine derivatives and other cyclic amines as in the conversion of **212** to **213**. Treatment with a base generated the free amine, and subsequent nucleophilic displacement of chloride (cyclization) led to **214** as the final product. This acid mediated reaction can be initiated or photochemically.[155] as in the conversion of **215** to **216** in 99 x 53% overall yield, taken from Sanz-Cervera and Williams' synthesis of (–)-VM55599.[156] Initial *N*-chlorination of the primary amine (to give a 99% yield of the chloroamine) was followed by irradiation in sulfuric acid. The final step was basification to generate the pyrrolidine (in this case a proline) derivative, and protection of the nitrogen as a *tert*-butylcarbamate (*N*-Boc, see sec. 7.3.C.iii). The high temperatures required in older versions of this conversion can lead to poor stereoselectivity. One solution to this problem is to use a mixture of cuprous chloride and cupric chloride (CuCl–CuCl$_2$) in acidic media to effect cyclization.[157]

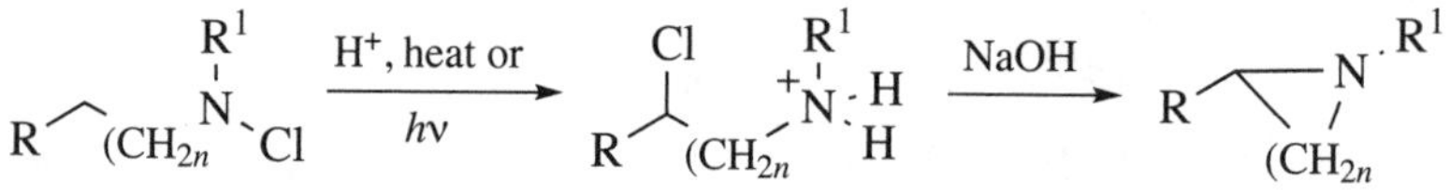

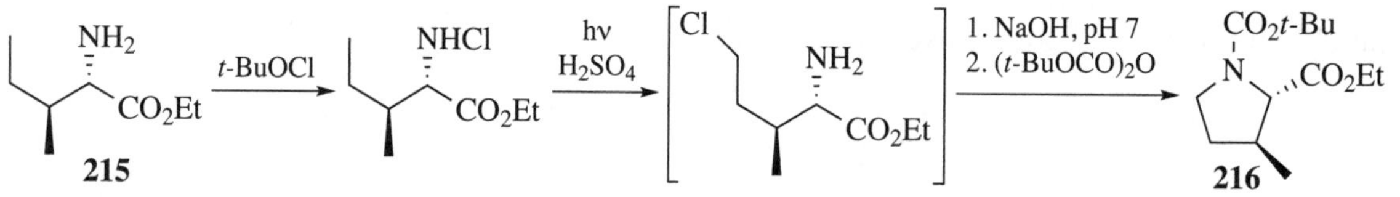

153. Corey, E.J.; Hertler, W.R. *J. Am. Chem.Soc.* **1960**, *82*, 1657.
154. (a) Hofmann, A.W. *Berichte* **1883**, *16*, 558; (b) *Idem Ibid.* **1885**, *18*, 5; (c) Löffler, K.; Freytag, C., *Ibid.* **1909**, *42*, 3427; (d) Wolff, M.E. *Chem. Rev.* **1963**, *63*, 55; (e) The Merck Index, 14th Ed. Merck & Co., Inc., Whitehouse Station, New Jersey, **2006**, p ONR-45; (f) Mundy, B.P.; Ellerd, M.G.; Favaloro Jr., F.G. *Name Reactions and Reagents in Organic Synthesis, 2nd Ed.* Wiley-Interscience, New Jersey, **2005**, pp. 328-329.
155. Furstoss, R.; Teissier, P.; Waegell, B. *Tetrahedron Lett.* **1970**, 1263.
156. Sanz-Cervera, J.; Williams, R.M. *J. Am. Chem. Soc.* **2002**, *124*, 2556.
157. Broka, C.A.; Eng, K.K. *J. Org. Chem.* **1986**, *51*, 5043.

The Hofmann-Löffler-Freytag disconnection is

13.7.D Bergman Cyclization

Ene-diynes undergo a cycloaromatization reaction that is formally a thermally allowed electrocyclic reaction. In principle, this reaction belongs in Chapter 11, but it is known that the reaction proceeds by a diradical. For that reason and since a ring is formed it is placed in this

section, although it is not formally a radical cyclization. In a generic example, ene-diyne **217** cyclizes to give diradical **218**, which loses two hydrogen atoms via aromatization to give the benzene ring in **219**. This conversion of an ene-diyne to a benzene derivative is called the **Bergman cyclization**.[158] The activation barrier for this process can be greatly influenced by electronic substituent effects.[159] A metal-accelerated Bergman cycloaromatization has been reported, using an arene complex, $[(\eta^5\text{-}C_5Me_5)Ru(\eta^6\text{-}3,4\text{-benzocyclodec-3-ene-1,5-diyne}][OTf]$.[160] This reaction has taken on great importance with the discovery that the biological action of several anticancer antibiotics [calicheamicin, dynemicin, esperamicin, and kedarcidin] is linked to Bergman cyclization of an ene-diyne core structure accompanying cleavage of double-stranded DNA.[161]

From a purely synthetic viewpoint, the radicals generated by the Bergman cyclization have been used to initiate further cyclization, including the preparation of aromatic compounds. In one example, Jones and Plourde prepared a series of chlorinated ene-diynes and used the Bergman cyclization to convert them to chlorinated aromatic compounds.[162] Heating **220**, for example, gave a 40% yield of **221**.[162] Interestingly, heating **220** with tetramethylpiperidinyloxy free radical (TEMPO) led to a 30% yield of the chlorinated benzoquinone **222**. The additional chorine atoms in **221** are attributed to atom transfer chemistry from carbon tetrachloride to the diradical intermediate. This reaction is not limited to carbocyclic compounds, and an aza-

158. (a) Jones, R.R.; Bergman, R.G. *J. Am. Chem. Soc. 1972, 94,* 660; (b) Bergman, R.G. *Acc. Chem. Res. 1973, 6,* 25;(c) Darby, N.; Kim, C.U.; Shelton, K.W.; Takada, S.; Masamune, S. *J. Chem. Soc. (D) 1971, 23,* 1516; (d) The Merck Index, 14th Ed. Merck & Co., Inc., Whitehouse Station, New Jersey, *2006,* p ONR-9; (e) Mundy, B.P.; Ellerd, M.G.; Favaloro Jr., F.G. *Name Reactions and Reagents in Organic Synthesis, 2nd Ed.* Wiley-Interscience, New Jersey, *2005,* pp. 90-91.

159. Schmittel, M.; Kiau, S. *Chem. Lett. 1995,* 953.

160. O'Connor, J.M.; Lee, L.I.; Gantzel, P. *J. Am. Chem. Soc. 2000, 122,* 12057.

161. (a) Nicolaou, K.C.; Stabila, P.; Esmaeli-Azad, B.; Wrasidlo, W.; Hiatt, A. *Proc. Natl. Acad. Sci., USA 1993, 90,* 3142; (b) Zein, N.; Solomon, W.; Casazza, A.M.; Kadow, J.F.; Krishnan, B.S.; Tun, M.M.; Vyas, D.M.; Doyle, T.W. *Bioorg. Med. Chem. Lett. 1993, 3,* 1351.

162. Jones, G.B.; Plourde, II, G.W. *Org. Lett. 2000, 2,* 1757.

Bergman rearrangement reaction has been reported.[163]

13.8   METAL-INDUCED RADICAL REACTIONS

13.8.A   General Principles

Metal-induced radical reactions can take at least two forms: (*1*) electron transfer of a metal to a peroxide, leading to reactive free radicals, and (*2*) direct electron transfer from a metal to an alkene, carbonyl or aromatic substrate. An example of the first type of reaction is oxidation with **Fenton's reagent** (sec. 3.3.A),[164] which initiates oxidation of aryls to phenols, although the yields are usually only 5-20%. In this reaction, ferrous ion (Fe^{2+}) is oxidized to ferric ion (Fe^{3+}) by hydrogen peroxide which gives hydroxide and hydroxyl radicals (HO•).[164] The presence of the radicals initiates a chain process, which can be terminated by reaction of HO• with Fe^{2+}. Oxidation of benzene with Fenton's reagent gave only a small amount of phenol, but large amounts of biphenyl, arising from coupling of the aryl radicals. Another metal system is Ti^{3+}-H_2O_2 (at pH 2), which generates HO• and HOO• radicals.[165] Transition metals such as titanium, cobalt or manganese are often used in conjunction with hydroperoxides, and decompose under these conditions to alkoxy radicals (RO•) and alkyl peroxyls (ROO•).[164a,166] In all cases, reaction with the metal generates a radical, which then reacts with the carbon fragment to induce the desired radical reaction. The latter metal-peroxide reaction was discussed previously in section 3.4.B in connection with the transition metal-peroxide oxidation of alkenes to oxiranes.[167]

13.8.B   Phenolic Oxidative Coupling

A classical oxidation reaction that involves radicals is the **phenolic oxidative coupling**,[168] which is introduced in Section 10.5.A.i. In this reaction, electron transfer from a metal salt to a bis(phenol) leads to intramolecular coupling and a quinone product. In early experiments, yields were poor. For example, Barton and Kirby reacted **223** with potassium ferricyanide [$K_3(Fe(CN)_6$]

163.   Feng, L.; Kumar, D.; Kerwin, S.M. *J. Org. Chem.* **2003**, *68*, 2234.
164.   (a) Haines, A.H. *Methods for the Oxidation of Organic Compounds*, Academic Press, London, **1985**, p 173; (b) Smith, J.R.L.; Norman, R.O.C. *J. Chem. Soc.* **1963**, 2897.
165.   (a) Reference 4, p 31; (b) Irvine, M.J.; Wilson, I.R. *Aust. J. Chem.* **1979**, *32*, 2131, 2283.
166.   Black, J.F. *J. Am. Chem. Soc.* **1978**, *100*, 527.
167.   Hiatt, R. in *Oxidation, Vol. 2*, Augustine, R.L.; Trecher, D.J. (Eds.), Marcel Dekker, New York, **1971**, pp 117-124.
168.   (a) Barton, D.H.R.; Deflorin, A.M.; Edward, O.E. *J. Chem. Soc.* **1956**, 530; (b) Scott, A.I. *Quart. Rev.* **1965**, *19*, 1.

and the initial product was aryl radical **224**. This radical reacted with the second phenolic ring in an intramolecular process that gave **225**. Loss of a hydrogen atom led to the quinone narwedine (**226**), as part of a synthesis of galanthamine.[169] The yield of **226**, however, was only 1.4% under optimal conditions. Kametani and co-workers improved the yield of the oxidation only slightly (5% yield) by using an amide derivative of **223**, in a synthesis of galanthamine.[170]

A similar oxidation procedure was used by Gervay et al. to generate erythrinadienone in 35% yield,[171] which constituted a great improvement in the yield. The yield remained relatively low, however, due to the shortness of the tether between the two aryl moieties or from the steric hindrance provided by the hydroxyl and alkoxyl groups.[172] Indeed, success for phenolic oxidative coupling is dependent on the nature of the substituents on the aromatic rings, as well as the length of the tether.[173] Other oxidative methods can be used, including anodic oxidation under electrolysis conditions. Good yields can be obtained in such systems, as in the anodic oxidation of tetramethoxy derivative **227** to give **228** in 80% yield, where **228** was a key intermediate in Tobinaga and co-worker's synthesis of colchicine.[172] In another variation, Stevenson and co-workers used VOF_3 to convert **229** to deoxyschizandrin (**230**) in 54% yield.[174] Phenyliodine (III)bis(trifluoroacetate) has also been used to effect phenolic oxidative coupling,[175] as have other hypervalent iodine (II) reagents.[176] In a synthesis of (+)-puupehenone, Quideau and co-workers[177] treated **231** with [bis(trifuoroacetoxyiodo]benzene and obtained a 67% yield of **232**.

169. Barton, D.H.R.; Kirby, G.W. *J. Chem. Soc. 1962,* 806.
170. (a) Kametani, T.; Shishido, K.; Hayashi, E.; Seino, C.; Kohno, T.; Shibuya, S.; Fukumoto, K. *J. Org. Chem. 1971, 36,* 1295; (b) Kametani, T.; Seino, C.; Yamaki, K.; Shibuya, S.; Fukumoto, K.; Kigasawa, K.; Satoh, F.; Hiiragi, M.; Hayasaka, T. *J. Chem. Soc. C 1971,* 1043; (c) Kametani, T.; Yamaki, K.; Yagi, H.; Fukumoto, K. *Ibid. 1969,* 2602.
171. (a) Gervay, J.E.; McCapra, F.; Money, T.; Sharma, G.M.; Scott, A.I. *J. Chem. Soc. Chem. Commun. 1966,* 142; (b) Mondon, A.; Ehrhardt, M. *Tetrahedron Lett. 1966,* 2557.
172. (a) Kotani, E.; Miyazaki, F.; Tobinaga, S. *J. Chem. Soc. Chem. Commun. 1974,* 300; (b) Tobinaga, S.; Kotani, E. *J. Am. Chem. Soc. 1972, 94,* 309.
173. See for example, Krauss, A.S.; Taylor, W.C. *Aust. J. Chem. 1992, 45,* 925, 935.
174. Biftu, T.; Hazra, B.G.; Stevenson, R. *J. Chem. Soc. Perkin Trans. 1 1979,* 2276.
175. Kita, Y.; Gyoten, M.; Ohtsubo, M.; Tohma, H.; Takada, T. *Chem. Commun. 1996,* 1481.
176. (a) Quideau, S.; Pouységu, L. *Org. Prep. Proc. Int. 1999, 31,* 617; (b) Quideau, S.; Pouységu, L.; Oxoby, M.; Looney, M.A. *Tetrahedron 2001, 57,* 319.
177. Quideau, S.; Lebon, M.; Lamidey, A.-M. *Org. Lett. 2002, 4,* 3975.

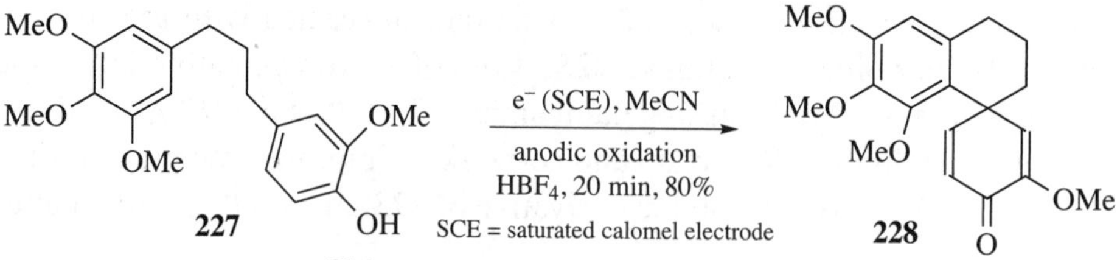

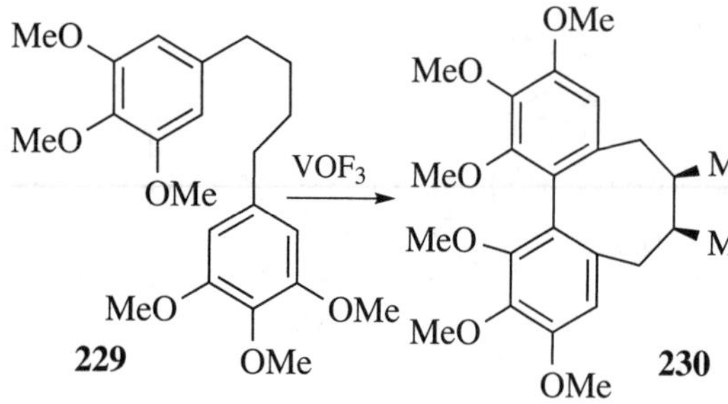

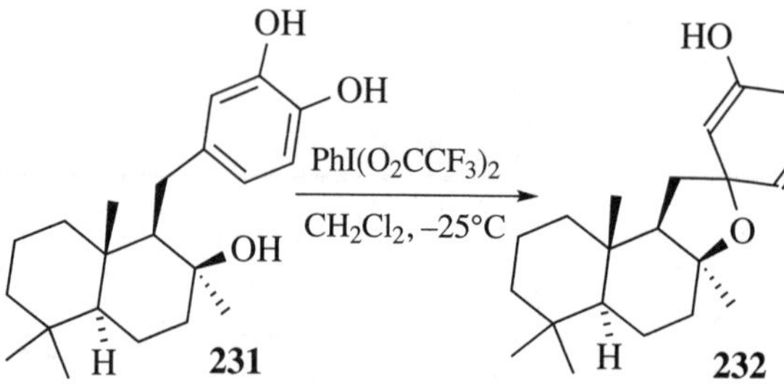

Note that many phenolic coupling reactions may proceed via radical cation intermediates and/ or phenoxonium ions, especially when the coupling is induced by electrochemical methods. In such cases, it is not always clear that the mechanism involves a radical-radical coupling reaction.

The phenolic oxidative coupling disconnection is

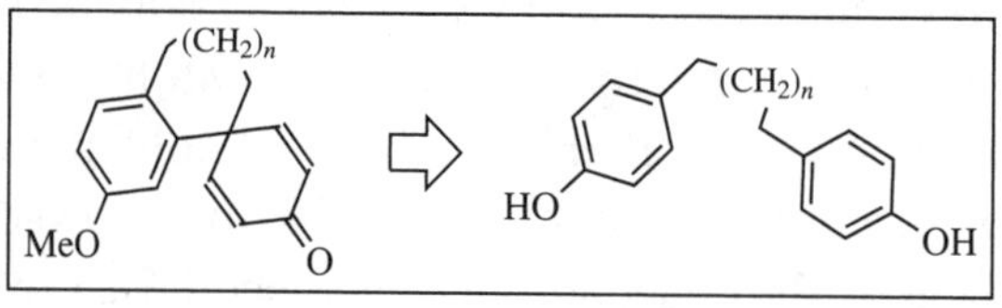

13.8.C Copper-Catalyzed Coupling Reactions

Many methods have been developed that form an aryl-aryl bond,[178] including the metal-catalyzed radical reaction known as **Meerwein arylation**.[179] This reaction combines an aryl diazonium salt (such as phenyldiazonium chloride, **233**) with an electron-deficient alkene (containing an electron-withdrawing group such as carboethoxy in ethyl acrylate) to give the phenyl-substituted acrylate derivative (**235**) with a cupric chloride (CuCl₂) catalyst. Under these conditions, CuCl₂ induces loss of nitrogen (N₂) from **233** to give the aryl radical (**16**), and formation of this radical is followed by addition to the alkene to give **234**. Subsequent reaction of this reactive intermediate with CuCl₂ generated the alkene (**235**) via loss of HCl, also generating CuCl. This reaction is obviously similar to the Heck reaction discussed in

---

178. Hassan, J.; Savignon, M.; Gozzi, C.; Schulz, E.; Lemaire, M. *Chem. Rev.* **2002**, *102*, 1359.

179. (a) Meerwein, H.; Büchner, E.; van Emster, K. *J. Prakt. Chem.* **1939**, *152*, 237; (b) Rondestvedt, Jr., C.S. *Org. React.* **1960**, *11*, 189; (c) *Idem Ibid.* **1976**, *24*, 225; (d) The Merck Index, 14th Ed. Merck & Co., Inc., Whitehouse Station, New Jersey, **2006**, p ONR-59.

section 12.7.A. Titanium trichloride ($TiCl_3$) has been used to initiate 1,4 addition to conjugated carbonyl derivatives.[180]

$$\text{233} \xrightarrow[\text{CuCl}_2]{} [\text{16} + N_2 + CuCl + Cl_2] \xrightarrow{\text{CO}_2\text{Et}} [\text{234}] \xrightarrow{\text{CuCl}_2} \text{235}$$

There are many variations, and a wide range of aromatic diazonium salts can be used. One method for generating the diazonium salt is treatment of an aromatic amine with nitrous acid (HONO via reaction of HNO_3/HCl or $NaNO_2$/2 HCl.[181] The coupling of diazonium salts to other aromatic compounds[182] in the presence of base is commonly called the **Gomberg-Bachmann reaction**. Treatment of 4-bromoaniline with nitrous acid, for example, gave the diazonium salt (**236**), which reacted with benzene to give the biphenyl derivative (**237**) in 46% yield.[183] The aromatic amine precursors are usually obtained by hydrogenation of nitrobenzene derivatives (secs. 4.8.D, 4.8.F).[184]

$$\text{Br}\text{—}\text{NH}_2 \xrightarrow[\text{NaNO}_2]{\text{2 equiv HCl}} \text{Br}\text{—}\text{N}_2^+\text{Cl}^- \xrightarrow[\text{40\% aq NaOH}]{} \text{237}$$

A related process that involves diazonium salts is the **Pschorr reaction**,[185] which also couples aryl diazonium compounds to other aromatic rings. This diazonium salt coupling can be done under acidic conditions, but addition of copper powder usually promotes the radical process. Aryl amines generate aryl diazonium salts upon treatment with nitrous acid.[186] An example is the reaction of **238** to give an aryl diazonium salt, which cyclized in the presence of copper to give thaliporphine (**239**) in 43% yield. Kupchan called this transformation an improved Pschorr reaction.[187]

180. Citterio, A.; Cominelli, A.; Bonavoglia, F. *Synthesis* **1986**, 308.
181. (a) Ridd, J.H. *Quart. Rev. Chem. Soc.* **1961**, *15*, 418; (b) Hegarty, A.F. *The Chemistry of Diazonium and Diazo Groups, Part 2*, Patai, S. (Ed.), Wiley, New York, **1978**, pp 511-591.
182. (a) Bachmann, W.E.; Hoffman, R.A. *Org. React.* **1944**, *2*, 224 (see pp 225-226); (b) The Merck Index, 14th Ed. Merck & Co., Inc., Whitehouse Station, New Jersey, **2006**, p ONR-37.
183. Gomberg, M.; Bachmann, W.E. *J. Am. Chem. Soc.* **1924**, *46*, 2339.
184. Rylander, P.N. *Catalytic Hydrogenation Over Platinum Metal*, Academic Press, New York, **1967**, pp 168-202.
185. (a) Pschorr, R. *Berichte* **1896**, *29*, 496; (b) Leake, P.H. *Chem. Rev.* **1956**, *56*, 27; (c) DeTar, D.F. *Org. React.* **1957**, *9*, 409 (see pp 411-416); (d) The Merck Index, 14th Ed. Merck & Co., Inc., Whitehouse Station, New Jersey, **2006**, p ONR-76; (e) Mundy, B.P.; Ellerd, M.G.; Favaloro Jr., F.G. *Name Reactions and Reagents in Organic Synthesis, 2nd Ed.* Wiley-Interscience, New Jersey, **2005**, pp. 534-535.
186. (a) Kametani, T.; Fukumoto, K.; Satoh, F.; Yagi, H. *J. Chem. Soc. Chem. Commun.* **1968**, 1398; (b) *Idem. J. Chem. Soc. C.* **1969**, 520; (c) Kametani, T.; Ihara, M.; Fukumoto, K.; Yagi, H. *Ibid.* **1969**, 2030.
187. Kupchan, S.M.; Kameswaran, V.; Findlay, J.W.A. *J. Org. Chem.* **1973**, *38*, 405.

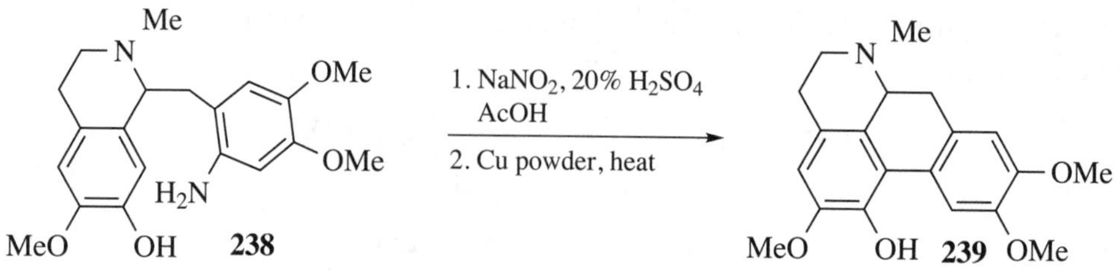

The **Ullmann reaction**,[188,189] is very similar in that aryl halides are coupled to form biaryls (such as biphenyl) by heating with copper. Once again, the copper generates an aryl radical (in this case **16**), which reacts with Cu^+ to form an arylcopper(I) species (**240**). Subsequent reaction of the aryl copper with iodobenzene leads to a coupling reaction that gives biphenyl, which is related to the organocuprate coupling reactions discussed in Section 8.7.A.ii. The reaction is placed here because of the radical intermediate. This reaction is similar to the Ni(0) coupling reactions discussed in Section 12.8.[190,191]

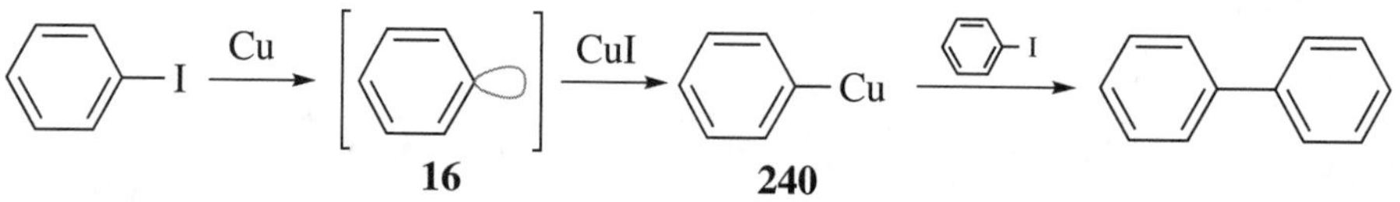

A halide reactivity depends on the halide: I > Br > Cl.[189] An electronegative group in the ortho position activates the reaction. Nitro groups on the aromatic ring are activating in this type of coupling:[192] ortho > para > meta. 2-Iodonitrobenzene is very reactive, but the 3-iodo and 4-iodo derivatives are about as reactive as iodobenzene. Aromatic rings containing substituents are readily coupled using copper or copper salts.[189] In a synthesis of (+)-isokotanin A,[193] Bringmann and co-workers coupled aryl bromide **241** with copper in DMF to give symmetrical biaryl **242** in 89% yield. Functionalized aryls such as ferrocene can be coupled using copper salts,[194]

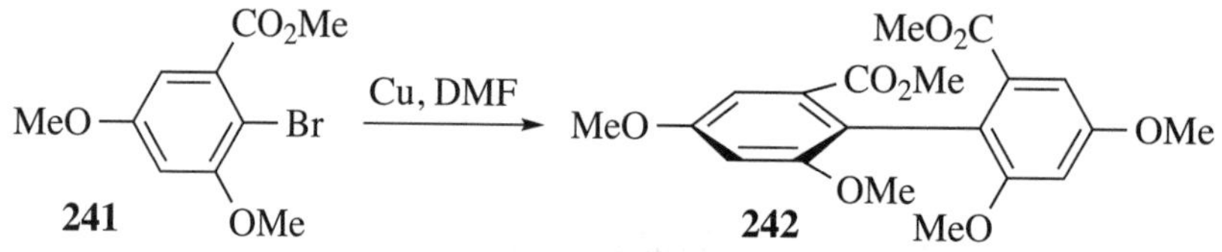

188. (a) Ullmann, F.; Meyer, G.M.; Loewenthal, O.; Gilli, E. *Annalen* **1904**, *332*, 38; (b) Ullmann, F.; Sponagel, P. *Berichte* **1905**, *38*, 2211; (c) Fanta, P.E. *Chem. Rev.* **1946**, *38*, 139; (d) The Merck Index, 14th Ed. Merck & Co., Inc., Whitehouse Station, New Jersey, **2006**, p ONR-96; (e) Mundy, B.P.; Ellerd, M.G.; Favaloro Jr., F.G. *Name Reactions and Reagents in Organic Synthesis, 2nd Ed.* Wiley-Interscience, New Jersey, **2005**, pp. 662-663.
189. Fanta, P.E. *Chem. Rev.* **1964**, *64*, 613.
190. Semmelhack, M.; Helquist, P.; Jones, L.D.; Keller, L.; Mendelson, L.; Ryono, L.S.; Smith, J.G.; Stauffer, R.D. *J. Am. Chem. Soc.* **1981**, *103*, 6460.
191. See Tsou, T.T.; Kochi, J.K. *J. Am. Chem. Soc.* **1979**, *101*, 7547.
192. Davey, W.; Latter, R.W. *J. Chem. Soc.* **1948**, 264.
193. Bringmann, G.; Hinrichs, J.; Henschel, P.; Kraus, J.; Peters, K.; Peters, E.-M. *Eur. J. Org. Chem.* **2002**, 1096.
194. Rausch, M.D. *J. Org. Chem.* **1961**, *26*, 1802.

For unsymmetrical coupling (two different aryl halides) the best yields are observed when one aryl halide is very reactive and the other is relatively unreactive. Reactive aryl halides have electronegative groups, such as nitro or carbomethoxy, ortho to the halogen. Aryl bromides and chlorides are best for unsymmetrical coupling, while iodo compounds tend to give symmetrical coupling (Ar–Ar rather than Ar–Ar').[195] Unreactive halides lack an ortho electronegative group and are usually iodo derivatives. When unsymmetrical coupling is attempted with two unreactive aryls, three biaryls are usually produced in approximately equal amounts.[189] Ultrasound is effective for the synthesis of diaryl ethers via the Ullmann reaction.[196]

The disconnections for copper promoted coupling are

Alkynes can be coupled under a variety of conditions,[197] including reactions that involve alkynyl copper derivatives. There are two classical alkyne coupling reactions. The first is the **Glaser reaction**[198] in which an alkyne such as phenylacetylene reacts with basic cupric chloride (CuCl$_2$) to give a diyne (1,4-diphenyl-1,3-butadiyne) in 90% yield upon oxidation with air. The mechanism of this reaction has been studied, and Cu (II) has been proposed as an oxidant in the reaction.[199] It has been shown that molecular oxygen forms adducts with Cu(I) supported by tertiary amines[200] which might be the intermediates, and mechanistic considerations[201] for this variation have also been reported. Note the similarity to the Sonogashira coupling using a palladium catalyst in section 12.7.D. The coupling can be done both inter- and intramolecularly.[202] An example of the latter is the coupling of the two terminal alkyne units in **243** to give diyne **244** (65% yield), in Myers and co-worker's[203] synthesis of kedarcidin. This

195. Forrest, J. *J. Chem. Soc.* **1960**, 566, 574, 581, 589, 592, 594.

196. Smith, K.; Jones, D. *J. Chem. Soc. Perkin Trans. 1* **1992**, 407.

197. Cadiot, P.; Chodkiewicz, W. in *Chemistry of Acetylenes*, Viehe, H.G. (Ed.), Marcel-Dekker, New York, **1969**, pp 597-648.

198. (a) Glaser, C. *Berichte* **1869**, *2*, 422; (b) *Idem Annalen* **1870**, *154*, 137. For a recent review, see (c) Siemsen, P.; Livingston, R.C.; Diederich, F. *Angew. Chem. Int. Ed.* **2000**, *39*, 2632; (d) The Merck Index, 14th Ed. Merck & Co., Inc., Whitehouse Station, New Jersey, **2006**, p ONR-37; (e) Mundy, B.P.; Ellerd, M.G.; Favaloro Jr., F.G. *Name Reactions and Reagents in Organic Synthesis, 2nd Ed.* Wiley-Interscience, New Jersey, **2005**, pp. 276-277.

199. Bohlmann, F.; Schoenowsky, H.; Inhoffen, E.; Grau, G. *Chem. Ber.* **1964**, *97*, 794.

200. Wieghardt, K.; Chaudhury, P. *Prog. Inorg. Chem.* **1988**, *35*, 329.

201. Fomina, L.; Vazquez, B.; Tkatchouk, E.; Fomine, S. *Tetrahedron* **2002**, *58*, 6741.

202. For an example taken from a synthesis of the kedarcidin core structure, see Myers, A.G.; Goldberg, S.D. *Angew. Chem. Int. Ed.* **2000**, *39*, 2732.

203. Myers, A.G.; Hogan, P.C.; Hurd, A.R.; Goldberg, S.D. *Angew. Chem. Int. Ed*, **2002**, *41*, 1062.

example clearly shows that the Glaser reaction is compatible with molecules bearing a vast array of functionality and stereochemistry. The second synthetic route is the **Cadiot-Chodkeiwicz coupling**,[204] in which a bromoalkyne reacts with a terminal alkyne in the presence of copper derivatives and an amine to give a diyne. An example taken from Danishefsky and Yun's synthesis of panaxytriol,[205] in which bromoalkyne **245** reacted with **246**, in the presence of CuCl, ethylamine, and hydroxylamine hydrochloride. Coupling gave diyne **247** (panaxytriol) in 63% yield. Glaser type coupling of alkynes can be accomplished with other copper derivatives, and with bases other than hydroxide. The mixture of cupric acetate and pyridine is an effective reagent for this coupling.[206] It is also possible to couple alkynyl copper derivatives with aryl halides[207,208] or other aromatic substrates.[209] This methodology can also be applied in an Ullmann-type reaction, analogous to that described for aryl halides where a vinyl iodide is coupled in the presence of copper.[210] Alkynylsilanes can also be coupled using Cu(I) salts.[211]

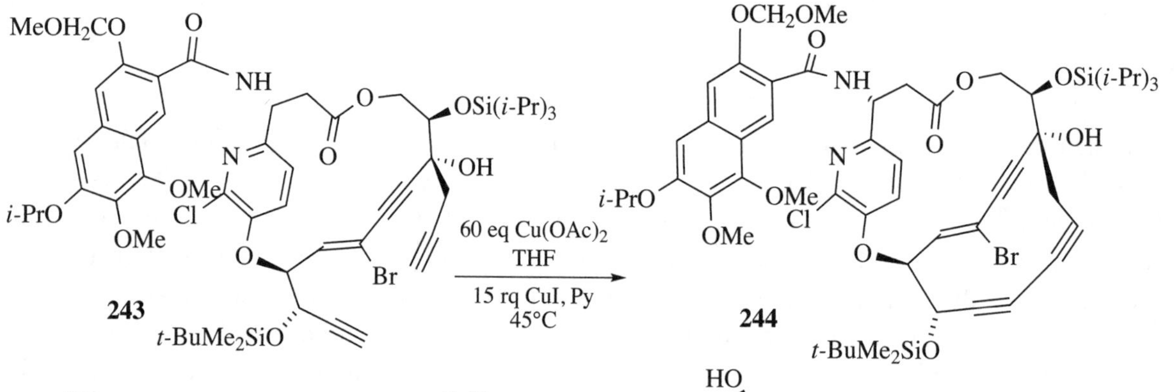

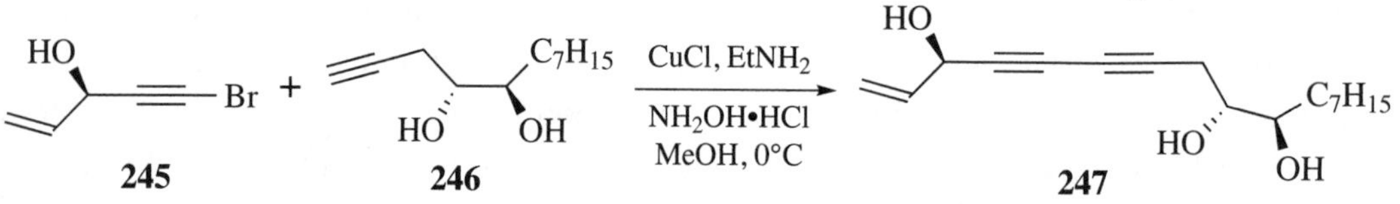

The disconnections for copper promoted coupling of alkynes and alkenes are

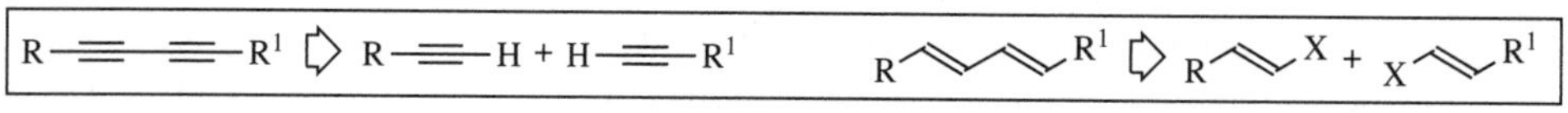

204. (a) Chodkiewicz, W.; Cadiot, P. *Compt. Rend.* **1955**, *241*, 1055; (b) Chodkiewicz, W. *Ann. Chim.* **1957**, *2*, 819; (c) Mundy, B.P.; Ellerd, M.G.; Favaloro Jr., F.G. *Name Reactions and Reagents in Organic Synthesis, 2nd Ed.* Wiley-Interscience, New Jersey, **2005**, pp. 128-129.
205. Yun, H.; Danishefsky, S.J. *J. Org. Chem.* **2003**, *68*, 4519.
206. Brown, A.B.; Whitlock, Jr., H.W. *J. Am. Chem. Soc.* **1989**, *111*, 3640.
207. (a) Posner, G.H. *An Introduction to Synthesis Using Organocopper Reagents*, Wiley, New York, **1980**, p 41; (b) House, H.O.; Umen, M.J. *J. Org. Chem.* **1973**, *38*, 3893.
208. (a) Hatchard, W.R. *J. Org. Chem.* **1963**, *28*, 2163; (b) Stevens, R.D.; Castro, C.E. *Ibid.* **1963**, *28*, 3313.
209. (a) Bohlmann, F.; Kleine, K.-M.; Arndt, C. *Berichte* **1966**, *99*, 1642; (b) Bohlmann, F.; Bonnet, P.-H.; Hofmeister, H. *Ibid.* **1967**, *100*, 1200; (c) Bohlmann, F.; Zdero, C.; Gordon, W. *Ibid.* **1967**, *100*, 1193.
210. Cohen, T.; Poeth, T. *J. Am. Chem. Soc.* **1972**, *94*, 4363.
211. Nishihara, Y.; Ikegashira, K.; Hirabayashi, K.; Ando, J.-i.; Mori, A.; Hiyama, T. *J. Org. Chem.* **2000**, *65*, 1780.

13.8.D Pinacol Coupling

A classical metal transfer reaction is the **pinacol reaction** or **pinacol coupling**,[212] in which alkali metals react with a ketone such as 3-pentanone to produce a radical anion (ketyl **248**) via electron transfer (secs. 4.9.A, 4.9.B).[213] Other reagents developed that give pinacol coupling,[214] include samarium iodide,[215] titanium(IV) iodide[216] or $TiCl_4$-Bu_4NI[217] and lanthanide metals in the presence of chlorotrimethylsilane.[218] If a monovalent metal such as sodium or potassium is used in the presence of an alcohol (most commonly, ethanol), the carbonyl is reduced to the alcohol (sec. 4.9.B). If a bivalent metal such as magnesium is used, however, the ketyl is stabilized, and reduction is slow. Here, the ketyl behaves more like a radical, leading to the coupling product (**249**), hydrolysis of which leads to the diol **250**.[212a,219] This reaction is a preferred method for the synthesis of 1,2-diols.[219,220]

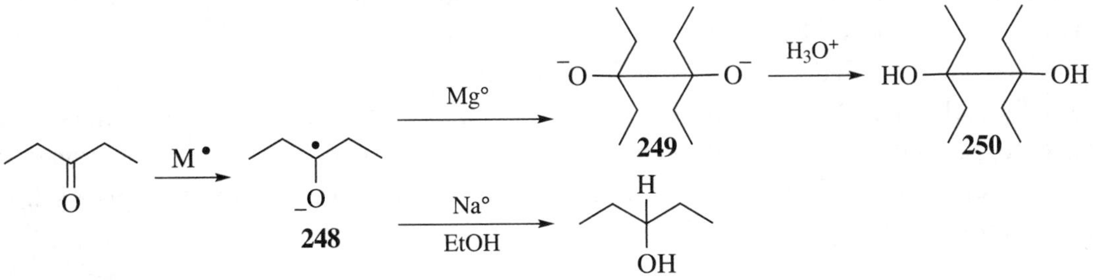

MOLECULAR MODELING: Radical Coupling

Radical coupling typically proceeds with little if any activation barrier, which means that is is under thermodynamic control. Consider recombination of α-(*tert*-butyl)benzyl radicals leading either to *meso* or *dl* forms of 2,2,5,5-tetramethyl-3,4-diphenylhexane.

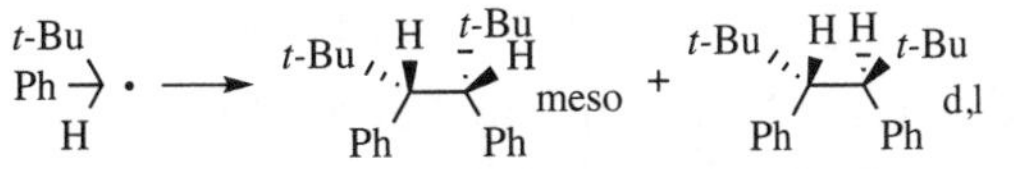

Build the *meso* product as drawn, with the two phenyl groups in a *syn* planar arrangement. This will ensure that you get the *meso* form and not the *d,l* form. Obtain the energy. Then rotate the central bond

---

212. (a) Schreibman, A.A.P. *Tetrahedron Lett.* **1970**, 4271; (b) The Merck Index, 14th Ed. Merck & Co., Inc., Whitehouse Station, New Jersey, **2006**, p ONR-74; (c) Mundy, B.P.; Ellerd, M.G.; Favaloro Jr., F.G. *Name Reactions and Reagents in Organic Synthesis, 2nd Ed.* Wiley-Interscience, New Jersey, **2005**, pp. 512-513.
213. For reactions of this type, see Larock, R.C., *Comprehensive Organic Transformations*, Wiley-VCH, New York, **1999**, pp. 1111-1114.
214. Wirth, T. *Angew. Chem. Int. Ed.* **1996**, *35*, 61.
215. Ghatak, A.; Becker, F.F.; Banik, B.K. *Tetrahedron Lett.* **2000**, *41*, 3793.
216. Hayakawa, R.; Shimizu, M. *Chem. Lett.* **2000**, 724.
217. Tsuritani, T.; Ito, S.; Shinokubo, H.; Oshima, K. *J. Org. Chem.* **2000**, *65*, 5066.
218. Ogawa, A.; Takeuchi, H.; Hirao, T. *Tetrahedron Lett.* **1999**, *40*, 7113.
219. Popp, F.D.; Schultz, H.P. *Chem. Rev.* **1962**, *62*, 19 (see pp 27-30).
220. Weber, J.E.; Boggs, A.D. *J. Chem. Educ.* **1952**, *29*, 363.

MOLECULAR MODELING: Radical Coupling, con't.

to generate the rotamer with the two phenyl groups *anti* planar. To rotate about a bond, first *click* on it and then move the mouse up and down while simultaneously holding down on the left button and the **Alt** key (**Option** key on Macintosh). Obtain its energy, and choose the better (lower-energy) conformer of the *meso* product. Next, build the *d,l* product, again starting with the two phenyl groups in a *syn* planar arrangement, and obtain its energy. Rotate into a conformation where the two phenyl groups are *anti*. Obtain the energy of this conformation and select the better (lower-energy) *d,l* product. Is the *meso* or *dl* form more stable? Assuming thermodynamic control, which diastereomer is predicted to be the major product?

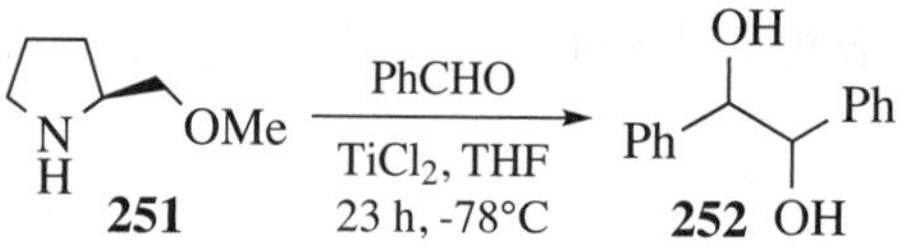

An asymmetric pinacol coupling of aromatic aldehydes was reported by Enders and Ullrich, using $TiCl_2$ and an enantiopure amine or hydrazine reagent.[221] When benzaldehyde was treated with $TiCl_2$ and chiral amine **251**, a 31% yield of **252** was obtained as an 81:19 mixture of dl/meso diastereomers, and the *dl*-diastereomer was obtained in 65% ee (*S,S*).[220] When the reaction time was increased to 48 hours, the yield improved to near quantitative but the diastereomeric ratio was 56:44 dl/meso and the enantioselectivity was only 37% ee.

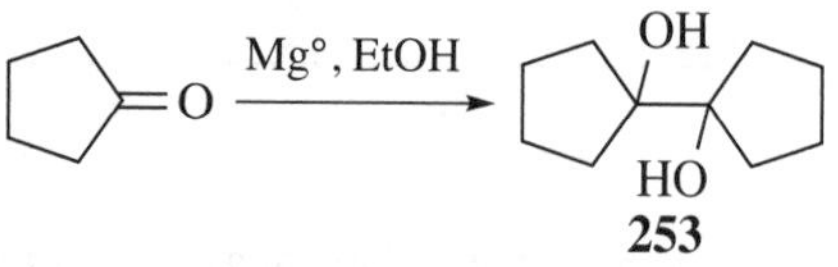

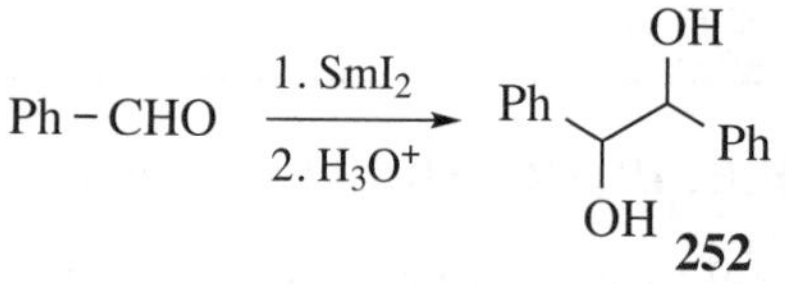

The preferred method for reductive coupling uses a mixture of magnesium and magnesium iodide, although the amalgam (Mg/Hg) can also be used.[222] A classical example is reductive condensation of cyclopentanone to the 1,2-glycol **253**.[212a] 2,3-Dimethyl-2,3-butanediol (pinacol) is formed by reduction of acetone and gives the reaction its name.[212a,220,219] Many other reductants can be used. Samarium iodide (SmI_2), mentioned above, was used for the coupling of benzaldehyde to **252** in 95% yield. Ytterbium as the free metal, has been used for the reduction of ketones but extensive coupling was observed. The extent of coupling depended on the ratio of ytterbium to ketone.[223] A related ketyl coupling reaction is Molander and co-worker's conversion of **254** to the samarium stabilized ketyl **255**. Conjugate radical addition to the acrylate moiety gave **256**, and hydrolysis led to an 88% yield of **257** (200:1 diastereoselectivity).[224] A catalytic and enantioselective pinacol coupling was reported that used chiral titanium (III) complexes, where the chiral ligand was a salen compound.[225]

221. Enders, D.; Ullrich, E.C. *Tetrahedron: Asymmetry* ***2000***, *11*, 3861.
222. Adams, R.; Adams, E.W. *Org. Synth. Coll. Vol. 1* ***1941***, 459.
223. Hou, Z.; Takamine, K.; Aoki, D.; Shiraishi, H.; Fujiwara, U.; Taniguchi, H. *J. Org. Chem.* ***1988***, *53*, 6077.
224. (a) Molander, G.A.; Kenny, C. *J. Am. Chem. Soc.* ***1989***, *111*, 8236; (b) Molander, G.A.; Etter, J.B.; Zinke, P.W. *Ibid.* ***1987***, *109*, 453.
225. Chatterjee, A.; Bennur, T.H.; Joshi, N.N. *J. Org. Chem.* ***2003***, *68*, 5668.

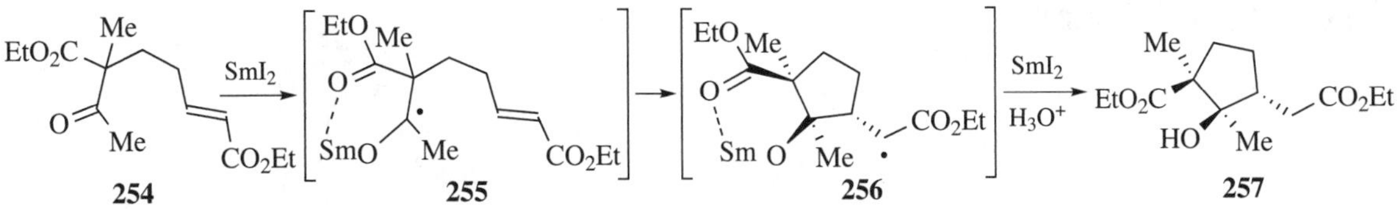

254   255   256   257

The pinacol coupling reaction can be applied to most ketones and aldehydes when two identical molecules are being coupled. An improvement in the reaction involves trapping the initial coupling product as the bis(trimethylsilyl) derivative. Coupling of 4'-methylacetophenone (**258**) with zinc[226] in the presence of chlorotrimethylsilane gave 76% of the bis(*O*-silyl)ether **259**. Treatment with tetrabutylammonium fluoride removed the silyl protecting group (sec. 7.3.A.i) to give the pinacol product **260**. Coupling of two different ketones may lead to several different products in relative proportions close to those expected from statistical coupling. One product may predominate but the yields can be low. When cyclopentanone was coupled to cycloheptanone in the presence of aluminum and mercuric chloride (HgCl$_2$), for example, only 23% of the cross-coupled diol (**261**) was formed.[227]

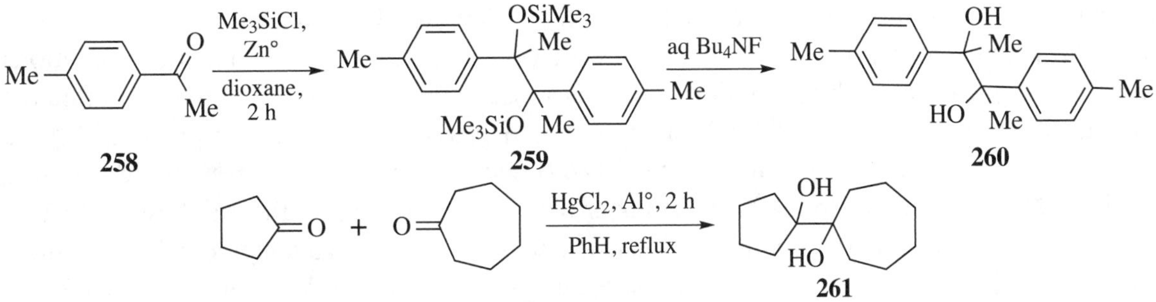

258   259   260

261

The pinacol disconnection is:

$$R_2C(OH)\text{–}C(OH)R_2 \implies R_2C=O$$

Photoreductive coupling of aldimines leads to the synthesis of C_2-symmetrical diamines. An example is the photocoupling of aldimine **262** to a 52:48 mixture of **263**/**264** in 85% yield.[228]

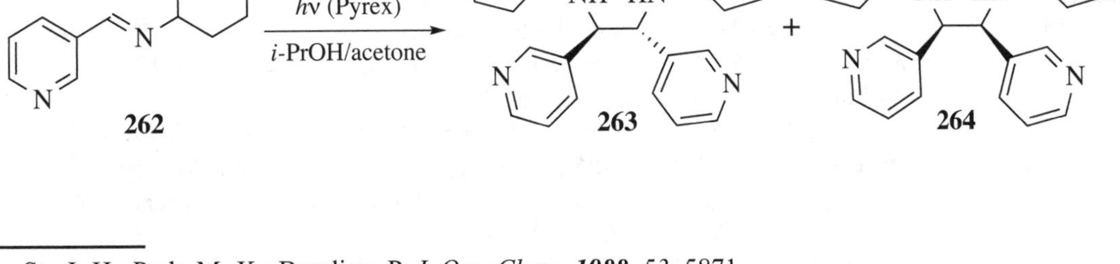

262   263   264

226. So, J.-H.; Park, M.-K.; Boudjou, P. *J. Org. Chem.* **1988**, *53*, 5871.
227. (a) Sands, R.D. *Tetrahedron,* **1965**, *21*, 887; (b) Sands, R.D.; Botteron, D.G. *J. Org. Chem.* **1963**, *28*, 2690.
228. Campos, P.J.; Arranz, J.; Rodríguez, M.A. *Tetrahedron* **2000**, *56*, 7285.

13.8.E. The Acyloin Condensation

A variation of the pinacol reaction has been applied to esters for the preparation of α-hydroxy ketones (**acyloins**). A simple example is the reaction of **265** with sodium metal (Na°) in refluxing xylene,[229] which gave a good yield of 2-hydroxycycloheptanone (**266**) after hydrolysis. Bouveault and Loquin reported the condensation of a α,ω-diester to a α-hydroxy ketone,[230] the **acyloin condensation**,[231,229] in 1905.[232] The reaction works well with relatively high molecular weight esters in an intermolecular reaction. Methyl laurate (**267**), for example, was treated with Na° (in xylene at 110°C) to give a 90% yield of lauroin (**268**).[233] The reaction conditions shown are typical, where high dilution techniques (sec. 6.6.B) were not required but effective stirring (high speed stirrer at 2000-2500 rpm) was necessary. The reaction proceeded[234] via initial transfer of an electron to form a ketyl (**269**), which is in equilibrium with the coupling product (**270**) and the diketone (**271**). Transfer of a second electron to this mixture led to **272**, and protonation generated a dihydroxy intermediate (**273**), which gave acyloin **273**. Radical coupling of the ketyl intermediate is most efficient when the ketyl is bound to, or closely associated with the surface of the metal. The high-speed stirrer generates a fine "sodium sand", which maximizes the surface area for effective electron transfer and radical coupling of the ketyl. Under these conditions, both intermolecular coupling and intramolecular cyclization reactions are possible as shown in Table 13.4 for the conversion of diesters **275** to cyclic acyloins **276**. The preparation of **269** is one example of an intermolecular acyloin condensation.

229. (a) McElvain, S.M. *Org. React* **1948**, *4*, 256; (b) Finley, K.T. *Chem. Rev.* **1964**, *64*, 573; (c) House, H.O. *Modern Synthetic Reactions, 2nd Ed.* W.A. Benjamin, Menlo Park, CA., **1972**, pp 169-172.
230. For transformations of this type, see reference 213, pp 1313-1315.
231. (a) Bloomfield, J.J.; Owsley, D.C.; Nelke, J.M. *Org. React.* **1976**, *23*, 259; (b) The Merck Index, 14th Ed. Merck & Co., Inc., Whitehouse Station, New Jersey, **2006**, p ONR-1; (c) Mundy, B.P.; Ellerd, M.G.; Favaloro Jr., F.G. *Name Reactions and Reagents in Organic Synthesis, 2nd Ed.* Wiley-Interscience, New Jersey, **2005**, pp. 4-5.
232. Bouveault, L.; Loquin, R. *Compt. Rend.* **1905**, *140*, 1593.
233. (a) Hansley, V.L. *J. Am. Chem. Soc.* **1935**, *57*, 2303; (b) Ruzicka, L.; Plattner, Pl.A.; Widmer, W. *Helv. Chim. Acta* **1942**, *25*, 604, 1086.
234. Reference 229a, pp 257 and 259.

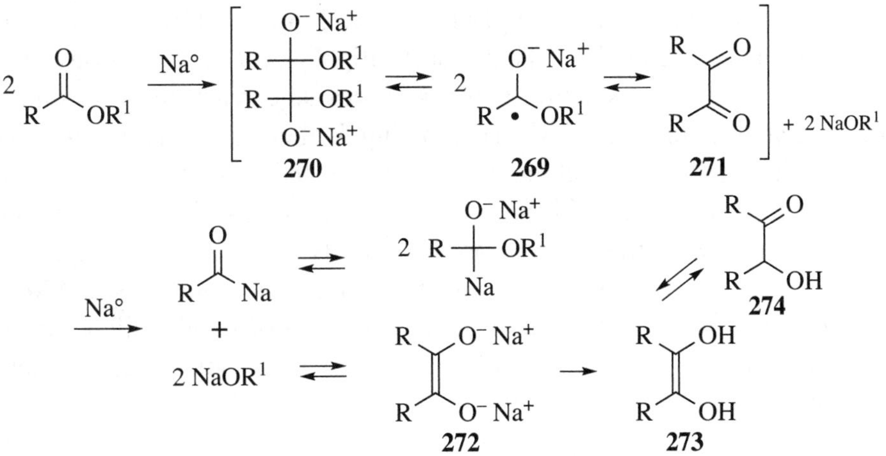

[Reprinted with permission from McElvain, S.M. *Org. React.* **1948**, *4*, 256. Copyright © **1948** by John Wiley and Sons, Inc.]

Table 13.4 Intramolecular Acyloin Condensations of 275

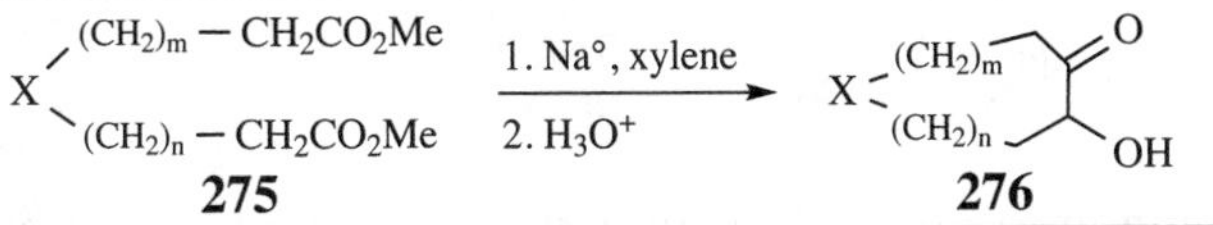

X	m	n	Ring Size	% 276	X	m	n	Ring Size	% 276
CH₂	2	1	8	47[c]	cis- CH=CH	2	2	10	80[c]
	2	2	9	9[b]	trans- CH=CH	2	2	10	51[c]
	2	3	10	45[b]	Ph–N	1	1	7	10[c]
	3	4	12	76[a,b]		2	2	9	3[c]
	4	4[a]	13	72[c]	Me–N	2	2	9	75[c]
	4	5[a]	14	79[b]		3	3	11	64[c]
	5	5[a]	15	77[b]	Et–N	3	3	11	85[c]
	5	6[a]	16	84[b]	C₆H₁₁-N	2	2	9	50[c]
	6	6[a]	17	85[b]	p-Me-Ph–N	2	2	9	38[c]
	6	7[a]	18	96[b]	1,4-C₆H₄	6	6	20	0[c]
	7	8[a]	20	96[b]		5	5	18	70[c]
	13	14[a]	34	20[c]		4	4	16	76[c]
MeCH	2	2	9	60[c]		3	4	15	36[c]
C≡C	3	3	12	73[c]		2	5	15	0[c]
	3	4	11	0[c]		1	6	15	75[c]

[a] diethyl ester, [b] Ref. 229a, [c] Ref. 229b.

[Reprinted with permission from Finley, K.T. *Chem. Rev.* **1964**, *64*, 573. Copyright © **1964** American Chemical Society.]

A modification of the acyloin condensation adds chlorotrimethylsilane to trap the alkoxide intermediate as a bis(silyl enol ether). This modification has become the standard version of the acyloin condensation. Rühlmann and co-workers showed that ethyl butanoate reacted with excess sodium to generate the expected bis(alkoxide) **277**. Addition of chlorotrimethylsilane led to a 92% yield of the bis(trimethylsilyl)derivative, **278**,[235] and hydrolysis with 1N HCl in THF gave 96% of the acyloin (**279**). This modification serves two purposes. First, trapping the enolate product allows its isolation and minimizes side reactions. Second, the *O*-TMS group protects the acyloin (see sec. 7.3.A.i), allowing further manipulation prior to unmasking the relatively sensitive α-hydroxyketone moiety. When there are no enolizable hydrogens in the *O*-TMS derivative, treatment with bromine generates a 1,2-diketone.[236] When enolizable hydrogens are present the normal acyloin product is observed. A synthetic example is taken from Sih and co-worker's synthesis of castanospermine,[237] in which diester **280** derived from hydroxy-proline was refluxed with sodium metal and chlorotrimethylsilane in toluene. This reaction led to formation of **281**, which could be isolated by chromatography, and converted to the corresponding acyloin by mild acid hydrolysis.

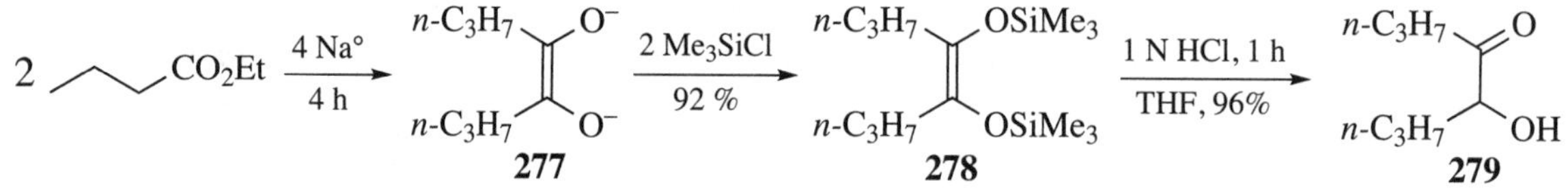

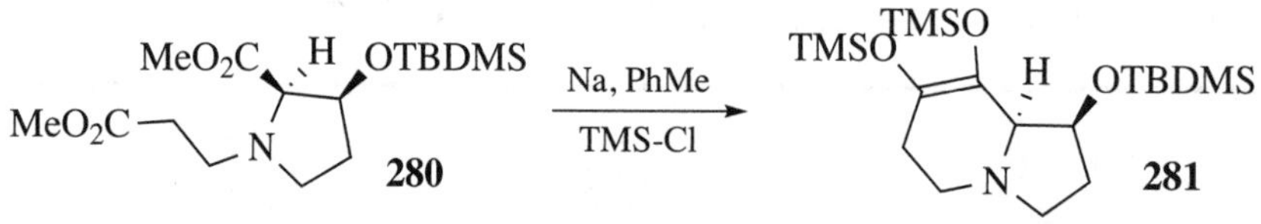

McMurry and Rico generated an acyloin-like reaction using a titanium trichloride (TiCl₃)–zinc-copper couple. α,ω-Dialdehydes are coupled with this reagent to give 1,2-diols.[238] This procedure generates mixtures of cis and trans diols with the cis predominating for smaller rings and the trans predominating for larger rings.[239] Cyclization of 1,6-hexanedial [OHC–(CH₂)₄–CHO] under these conditions gave only *cis*-1,2-cyclohexanediol in 85% yield, but cyclization of 1,12-dodecanedial [OHC–(CH₂)₁₀–CHO] gave a 25:75 cis/trans mixture of 1,2-cyclododecanediol[239] in 75% yield. As with the acyloin condensation, good yields are realized even with medium size (8-13 membered) rings. An example, taken from Williams and Heidebrecht's synthesis of (+)-4,5-deoxyneodolabelline,[239] cyclized **282** with TiCl₃ and Zn/Cu (see Sect. 13.8.F) to give an 8:2:1:1 (*R,R;S,R;R,S;S,S*) mixture of **283** in 85% yield.

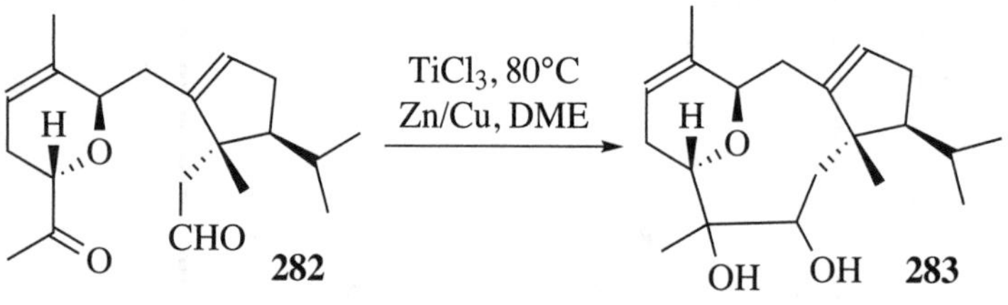

235. (a) Schräpler, U.; Rühlmann, K. *Chem. Ber.* **1963**, *96*, 2780; (b) Rühlmann, K.; Poredda, S. *J.Prakt. Chem.* **1960**, *12*, 18.
236. Strating, J.; Reiffers, S.; Wynberg, H. *Synthesis* **1971**, 209.
237. Bhide, R.; Mortezaei, R.; Scilimati, A.; Sih, C.J. *Tetrahedron Lett.* **1990**, *31*, 4827.
238. McMurry, J.E.; Rico, J.G. *Tetrahedron Lett.* **1989**, *30*, 1169.
239. Williams, D.R.; Heidebrecht Jr., R.W. *J. Am. Chem. Soc.* **2003**, *125*, 1843.

The acyloin disconnections are

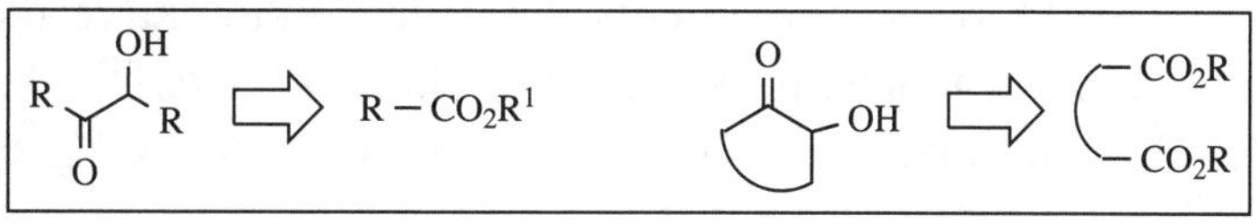

13.8.F McMurry Olefination

A useful relative of the acyloin condensation is the **McMurry olefination reaction**.[240] In this reaction, ketones or aldehydes are treated with Ti(0) (TiCl$_3$ + LiAlH$_4$; TiCl$_3$ + K; TiCl$_3$ + Li)[241] to give alkenes.[242] Reductive coupling is possible with a wide variety of ketones and aldehydes.[241a,243] Retinal (**284**), for example, was coupled with this reagent to give β-carotene (**285**) in 85% yield.[243]

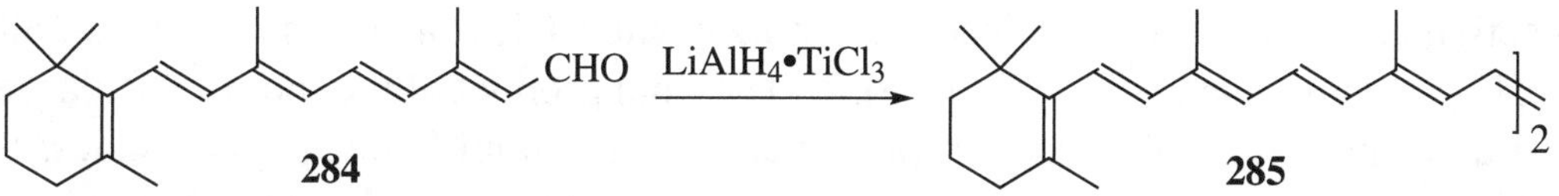

The TiCl$_3$/K mixture is a commonly used reagent, and a variety of carbonyl compounds can be coupled to give the corresponding alkene.[241a,244] Decanal was coupled with TiCl$_3$ and K to give a 60% yield of the C-20 alkene (10-icosene).[244] Cross-coupling is also possible as illustrated by the treatment of a mixture of cycloheptanone and acetone with TiCl$_3$/Li to give a 50% yield of **286** along with 26% of **287**.[244]

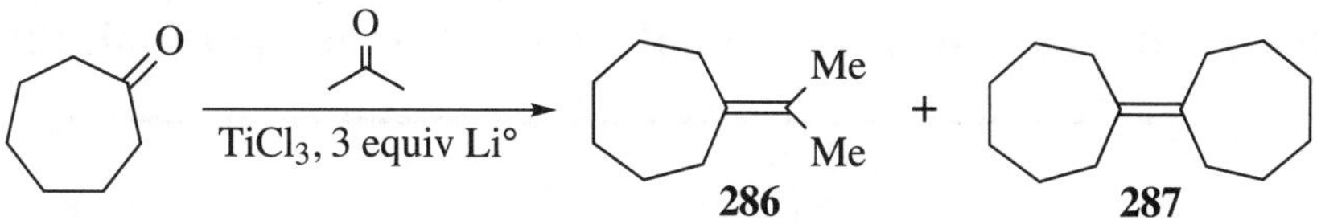

The mechanistic studies of this coupling suggest it occurs "in a heterogeneous process on the surface of an active titanium particle", as outlined in Figure 13.7.[242] The byproduct is titanium dioxide (TiO$_2$) and electron transfer to the carbonyl generates the dianion (**288**), presumably via a ketyl-like intermediate. It has been suggested that the metallopinacol is not the only precursor, and that carbenoid intermediates and/or nucleophilic intermediates can be important.[244] In the commonly accepted mechanism, the initially formed dianion binds to the

240. (a) McMurry, J.E. *Acc. Chem. Res.* ***1974****, 7*, 281; (b) The Merck Index, 14th Ed. Merck & Co., Inc., Whitehouse Station, New Jersey, ***2006***, p ONR-58; (c) Mundy, B.P.; Ellerd, M.G.; Favaloro Jr., F.G. *Name Reactions and Reagents in Organic Synthesis, 2nd Ed.* Wiley-Interscience, New Jersey, ***2005***, pp. 416-417.
241. McMurry, J.E.; Fleming, M.P.; Kees, K.L.; Krepski, L.R. *J. Org. Chem.* ***1978****, 43*, 3255.
242. For transformations of this type, see reference 213, pp 305-308.
243. McMurry, J.E.; Fleming, M.P. *J. Am. Chem. Soc.* ***1974****, 96*, 4708.
244. For discussions of the mechanism of this reaction, see (a) Stahl, M.; Pidun, U.; Frenking, G. *Angew. Chem. Int. Ed.* ***1997****, 36*, 2234; (b) Villiers, C.; Ephritikhine, M. *Angew. Chem. Int. Ed.* ***1997****, 36*, 2380.

titanium particle as shown, and sequential homolytic cleavage (via **289**) liberates the alkene and titanium dioxide. McMurry's experiments discredited alternative mechanisms such as formation of a discrete cyclic (or acyclic) titanium dialkoxy species.[241a] This reagent also led to elimination of 1,2-diols to alkenes, as in the conversion of *trans*-1,2-cycloheptanediol to cycloheptene in 55%.[245,241a]

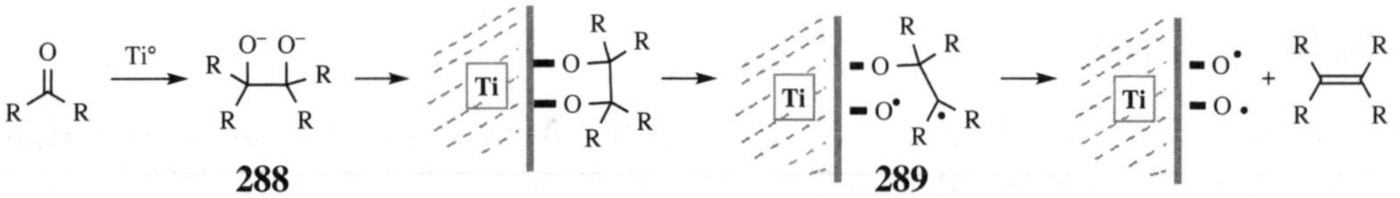

Figure 13.7. Mechanism of the McMurry Coupling Reaction. [Reprinted with permission from McMurry, J.E.; Fleming, M.P.; Kees, K.L.; Krepski, L.R. *J. Org. Chem.* **1978**, *43*, 3255. Copyright © 1978 American Chemical Society.]

This reaction can be done intramolecularly and Zn–Cu/TiCl$_3$ is a common catalyst, although Zn and TiCl$_4$ are also used (see Table 13.5).[242] The final product of the intramolecular reaction is a cyclic alkene. Since the reaction proceeds via a titanium bound ketyl species, the two reactive ends are in close proximity, allowing efficient coupling of even medium and large rings analogous to the acyloin condensation.[246] Table 13.5[242] shows the power of this reaction for the preparation of medium- and large-ring alkenes (**291**) from both diketones and dialdehydes (**290**). In a synthetic example, Corey and Hu converted diketone **292** to δ-araneosene (**293**) in 90% yield using McMurry coupling.[247] It is noted that the McMurry coupling of esters leads to enol ethers.

Table 13.5 Intramolecular McMurry Coupling of Diketones and Dialdehydes (290)

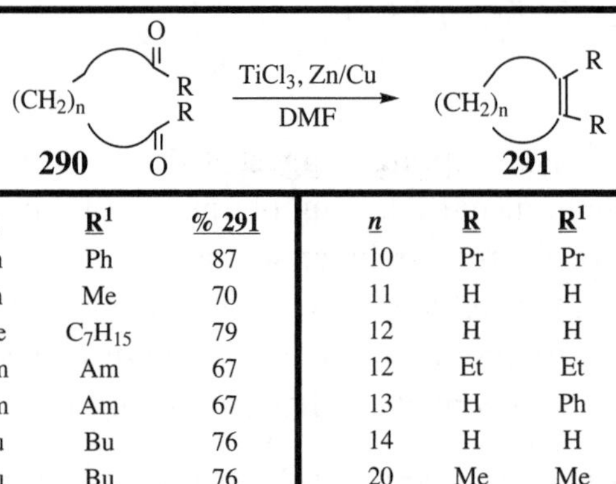

n	R	R^1	% 291	n	R	R^1	% 291
2	Ph	Ph	87	10	Pr	Pr	71
3	Ph	Me	70	11	H	H	52
4	Me	C$_7$H$_{15}$	79	12	H	H	71
6	Am	Am	67	12	Et	Et	75
7	Am	Am	67	13	H	Ph	80
8	Bu	Bu	76	14	H	H	85
9	Bu	Bu	76	20	Me	Me	83
10	H	H	76				

[Reprinted with permission from McMurry, J.E.; Fleming, M.P.; Kees, K.L.; Krepski, L.R. *J. Org. Chem.* **1978**, *43*, 3255. Copyright © *1978* American Chemical Society.]

245. McMurry, J.E.; Fleming, M.P. *J. Org. Chem.* **1976**, *41*, 896.
246. McMurry, J.E.; Kees, K.L *J. Org. Chem.* **1977**, *42*, 2655.
247. Hu, T.; Corey, E.J. *Org.Lett.* **2002**, *4*, 2441.

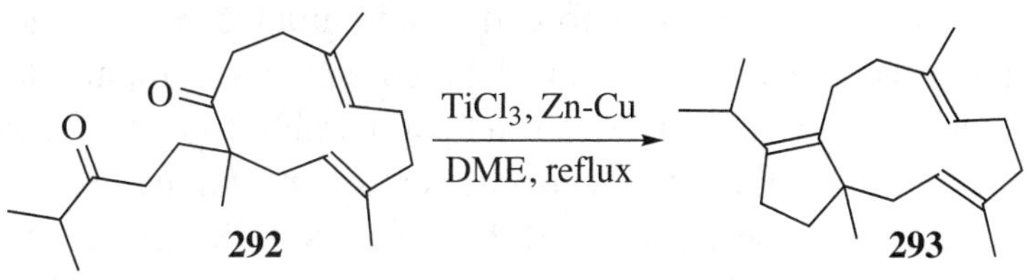

The disconnections for McMurry coupling are

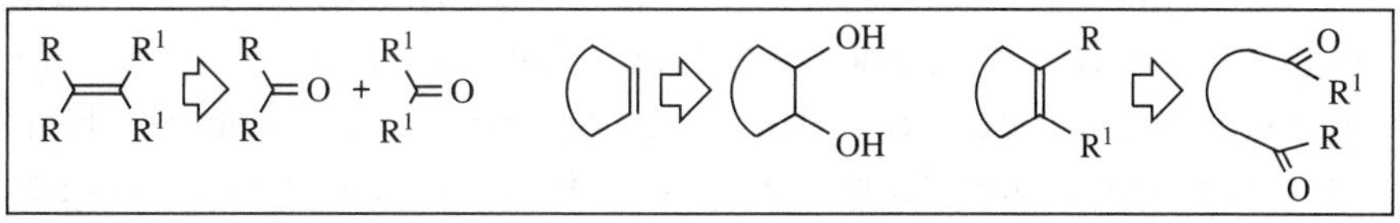

13.9 CARBENES AND CARBENOIDS

There is another class of molecules called carbenes, that in many ways can be considered to be *radical-like*. Carbenes have been known since the work of Curtius[248] in the late 19th century and Staudinger[249] in the early 20th century, and earlier work[250] is known. Carbenes are important in several synthetic methods and are growing in importance, especially the intramolecular versions. *N*-Heterocyclic carbenes are important as universal ligands in organometallic and inorganic coordination chemistry.[251] "*N*-Heterocyclic carbenes both stabilize and activate metal centers in quite different key catalytic steps of organic syntheses".[252] The remaining sections of this chapter discuss the preparation and reactions of carbenes and related compounds.

13.9.A Definition of a Carbene

A carbene is a divalent carbon species linked to two adjacent groups by covalent bonds,[252] possessing two nonbonded electrons and six valence electrons (as in **294**).[253] The two nonbonded electrons can have anti-parallel spins in a single orbital (see **295**, a **singlet carbene**), or parallel spins in different orbitals (see **296**, a **triplet carbene**). Substituted carbenes (**297**)

are generated in the singlet state from most precursors, but "intersystem crossing to the triplet

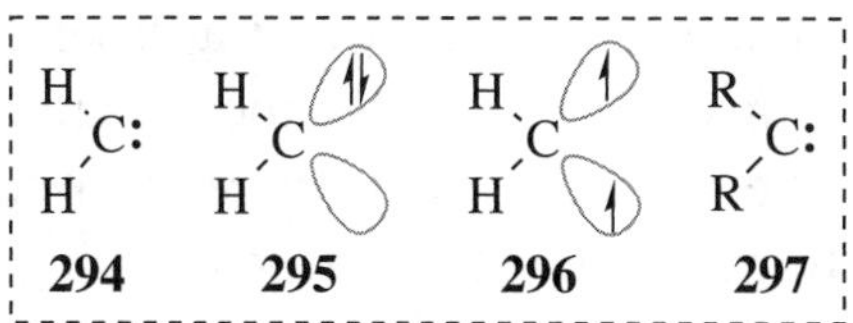

248. Buchner, E.; Curtius, T. *Ber. Dtsch. Chem. Ges.* **1885**, *8*, 2377.
249. Staudinger, H.; Kupfer, O. *Ber. Dtsch. Chem. Ges.* **1912**, *45*, 501.
250. Arduengo III, A.J.; Krafczyk, R. *Chem. Unserer Zeit.* **1998**, *32*, 6.
251. Herrmann, W.A. *Angew. Chem. Int. Ed.* **2002**, *41*, 1291.
252. For a discussion of stable carbenes, see Arduengo, III, A.J. *Acc. Chem. Res.* **1999**, *32*, 913.
253. (a) Kirmse, W. *Carbene Chemistry*, Academic Press, New York, **1971**, pp 5-7; (b) Chinoporos, E. *Chem. Rev.* **1963**, *63*, 235; (c) Closs, G.L. *Top. Stereochem.* **1968**, *3*, 193; (d) Parham, W.E.; Schweizer, E.E. *Org. React.* **1964**, *13*, 55; (e) Brinker, U.H. (Ed.), *Advances in Carbene Chemistry, Vol 1 and Vol. 2*, Jai Press, Greenwich and Stamford, **1994** and **1998**; (f) Regitz, M. *Angew. Chem. Int. Ed.* **1996**, *35*, 725.

(sec. 11.10.B) may or may not occur before the individual carbene reacts with a suitable precursor".[254] "Singlet carbenes are electron deficient species comparable to carbocations," but "they possess a nonbonding pair of electrons comparable to that of carbanions."[254] If an adjacent group on the carbene withdraws electrons, the carbene will have electrophilic character. If that group supplies electrons the carbene will have nucleophilic character. "Triplet carbenes may be considered as diradicals, although the interaction of two unpaired electrons in orbitals on the same carbon atom gives rise to some peculiarities."[255] "The term carbenoid has been suggested" by Closs and Moss "for the description of intermediates which exhibit reactions qualitatively similar to those of carbenes without necessarily being free divalent carbon species".[255] Although generally considered to reactive intermediates, stable carbenes are known, particularly persistent triplet diarylcarbenes and heteroatom-substituted singlet carbenes.[256]

13.9.B Preparation of Carbenes

There are several methods for the preparation of carbenes. A simple approach is to photolyze ketenes (**298**) at 270-370 nm (100.5-76.5 kcal mol^{-1}, 438.8-320.2 kJ mol^{-1}) in the presence of oxygen.[257] Irradiation of **298** at 270 nm gave carbon monoxide (CO, with a quantum yield = 2) and the yield of ethene was independent of the pressure of oxygen. At 370 nm, the quantum yield of CO varies with the pressure of applied oxygen and also with the temperature. The presence of additional oxygen diminished the yield of ethene by $\approx$ ⅔.[226] The reaction proceeded via formation of excited ketene (**299**), which fragments to methylene (**294**) and CO. The reactive carbene reacts with additional ketene to give ethene and CO. It was noted that a possible intermediate to **299** is an excited cyclopropanone species.[258]

Photolysis of alkyl and aryl ketenes generates alkyl and aryl carbenes. Methylketene and dimethylketene led to methylcarbene[259] and dimethylcarbene,[260] respectively, upon irradiation. The quantum yields were close to unity at 270 nm.[261] Methylcarbene coupled with itself to form an activated ethene, which gave both ethene and ethyne.[260] 2-Butene was also formed by

254.   Reference 253a, p 5
255.   Reference 253a, p 6.
256.   Bourissou, D.; Guerret, O.; Gabbai, F.P.; Bertrand, G. *Chem. Rev. **2000**, 100*, 39.
257.   Reference 253a, p 9.
258.   (a) Knox, K.; Norrish, R.G.W.; Porter, G. *J. Chem. Soc. **1952**,* 1477; (b) Kistiakowsky, G.B.; Sauer, K. *J. Am. Chem. Soc. **1956**, 78*, 5699; (c) *Idem Ibid.* ***1958**, 80*, 1066.
259.   (a) Kistiakowsky, G.B.; Mahan, B.H. *J. Am. Chem. Soc. **1957**, 79*, 2412; (b) Chong, D.P.; Kistiakowsky, G.B. *J. Phys. Chem. **1964**, 68*, 1793.
260.   Halroyd, R.A.; Blacet, F.E. *J. Am. Chem. Soc. **1957**, 79*, 4830.
261.   Reference 253a, p 14.

reaction with methyl carbene and the starting ketene. Similarly, photolysis of dimethylketene led to cyclopropanone (via dimethylcarbene and the ketene), which lost CO under the photolytic conditions to give 2,3-dimethyl-2-butene.[261]

13.9.B.i. Diazomethane. Important precursors for the preparation of carbenes and related molecules are diazoalkanes. Although diazomethane is not used in carbene preparations, it has other synthetic applications, and is presented here for comparative purposes. There are four convenient preparations of diazomethane, which is a highly reactive gas. *Since diazomethane can detonate on contact with ground glass, its preparation requires the use of specialized glassware. Diazomethane is also toxic.*

All preparative methods for diazomethane (CH_2N_2) involve *N*-nitroso compounds. Treatment of commercially available *N*-methyl-*N*-nitroso-*p*-toluenesulfonamide (Diazald®, **300**) with hydroxide liberates diazomethane and sulfonate ester **301**.[262] Similar treatment of *N,N'*-dimethyl-*N,N'*-dinitrosoterephthalamide (**302**),[263] *N*-methyl-*N'*-nitro-*N*-nitrosoguanidine (MNNG, **303**),[264] or *N*-Methylnitrosourea (**304**)[265] give diazomethane. It is noted that *MNNG is allergenic and is potentially mutagenic.* These reagents and synthetic uses of diazomethane are outlined in an excellent review by Hopps.[266] It is noted that at least two other structural isomers of diazomethane are known, diazirine (**307**)[267] and isodiazomethane (**308**).[268]

13.9.B.ii. Diazoalkanes. One of the most convenient and useful sources of reactive carbenes are diazoalkanes. Diazoalkanes (RN_2) have been known and used synthetically for over 50 years.[269] The simplest diazoalkane is diazomethane, represented by several resonance hybrid structures (**309**), which are zwitterionic, although it can also be written as a nitrene. It is noted

262. (a) de Boer, T.J.; Backer, H.J. *Org. Synth., Coll. Vol. 4* **1963,** 250; (b) *Idem Rec. Trav. Chim* **1954,** *73*, 229.
263. Moore, J.A.; Reed, D.E. *Org. Synth. Coll. Vol. 5* **1973,** 351.
264. (a) McKay, A.F. *J. Am. Chem. Soc.* **1948,** *70*, 1974; (b) McKay, A.F.; Ott, W.L.; Taylor, G.W.; Buchanan, M.N.; Crooker, J.F. *Can. J. Res.* **1950,** *28B*, 683.
265. Arndt, F. *Org. Synth. Coll. Vol. 2* **1943,** 165.
266. Hopps,. H.B. *Aldrichimica Acta* **1970,** *3*, 9.
267. (a) Schmitz, E. *Angew. Chem. Int. Ed.* **1964,** *3*, 333; (b) *Idem Adv. Heterocyclic Chem.* **1963,** *2*, 83 (see p 122); (c) Schmitz, E.; Ohme, R. *Tetrahedron Lett.* **1961,** 612; (d) Paulsen, S.R. *Angew. Chem.* **1960,** *72*, 781.
268. (a) Anselme, J.P. *J. Chem. Educ.* **1966,** *43*, 596; (b) Müller, E.; Ludsteck, D. *Chem. Ber.* **1954,** *87*, 1887.
269. (a)Huisgen,R.*Angew. Chem.***1955,***67*,439;(b)Cowell,G.W.;Ledwith,A.*Quart. Rev. Chem. Soc.***1970,***24*,119.

that the chemistry of nitrenes is quite useful, particularly for generating aziridines, although we will not discuss it here.[270] The preparation and reactions of other alkyl diazoalkanes are described in a review by Cowell.[270b,271] Hydrazones (such as that from acetophenone, **310**) are convenient sources of carbenes via their reaction with *p*-toluenesulfonyl azide in THF to give the corresponding diazoalkane (**311**) along with *p*-toluenesulfonamide.[272]

$$\left[H_2C = \overset{+}{N} = \overset{-}{N} \quad \longleftrightarrow \quad H_2\overset{+}{C} - N = \overset{-}{N} \quad \longleftrightarrow \quad H_2\overset{-}{C} - N \equiv N \quad \longleftrightarrow \quad H_2\overset{-}{C} - N = \overset{+}{N} \right]$$

309

The first aliphatic diazo compound known was ethyl diazoacetate (**313**), prepared by Curtius in 1883.[273] Indeed, several types of amines can be converted to diazo compounds by treatment with potassium nitrite or nitrous acid. Treatment of ethyl glycinate hydrochloride (**312**) with potassium nitrite (KNO$_2$) gave **313**. This technique is most useful when an electron-withdrawing group is adjacent to the amino group.

$$EtO_2CH_2NH_2 \bullet HCl \xrightarrow{KNO_2} EtO_2CCH = N_2 \;+\; KCl \;+\; 2\,H_2O$$

$$\textbf{312} \qquad\qquad\qquad\qquad \textbf{313}$$

Photolysis of diazo compounds (**314**) generates nitrogen and the carbene **297**. Photolysis of diazomethane (**314**, R = H), for example, generated methylene (H$_2$C:).[274] Benzophenone is often added as a sensitizer. Under these conditions, triplet methylene is formed via intersystem crossing $(S_1 \rightarrow T_1)$[275] (sec. 11.10.B). Energy transfer from triplet benzophenone to triplet diazomethane followed, and triplet diazomethane decomposed to triplet methylene.[276] A similar thermal reaction gives a carbene that reacts normally.[277]

Diazoalkanes (**314**) can undergo valence isomerization to diazirines (**315**) upon photolysis, but reversal to the diazoalkane occurs both photochemically and thermally.[278,268cd] Diazirines

270. Müller, P.; Fruit, C. *Chem. Rev.* **2003**, *103*, 2905.
271. (a) Jones, W.M.; Muck, D.L. *J. Am. Chem. Soc.* **1966**, *88*, 3798; (b) Jones, W.M.; Muck, D.L.; Tandy Jr., T.K. *Ibid.* **1966**, *88*, 68; (c) Applequist, D.E.; McGreer, D.E. *Ibid.* **1960**, *82*, 1965.
272. Fischer, W.; Anselme, J.P. *Tetrahedron Lett.* **1968**, 877.
273. Curtius, T. *Chem. Ber.* **1883**, *16*, 2230.
274. Herzberg, G.; Shoosmith, J. *Nature (London)* **1959**, *183*, 1801.
275. (a) Bäckström, H.L.J.; Sandros, K. *Acta Chem. Scand.* **1958**, *12*, 823; (b) *Idem Ibid.* **1960**, *14*, 48; (c) Hammond, G.S.; Moore, W.M. *J. Am. Chem. Soc.* **1959**, *81*, 6334.
276. (a) Kopecky, K.R.; Hammond, G.S.; Leermakers, P.A. *J. Am. Chem. Soc.* **1961**, *83*, 2397; (b) *Idem Ibid.* **1962**, *84*, 1015.
277. Staudinger, H.; Kupfer, O. *Berichte* **1912**, *45*, 501.
278. (a) Schmitz, E.; Oehme, R. *Angew. Chem.* **1961**, *73*, 115; (b) *Idem Chem. Ber.* **1961**, *94*, 2166.

can be photolyzed back to the diazoalkane or directly to the carbene.[279]

The diazoalkane disconnections are

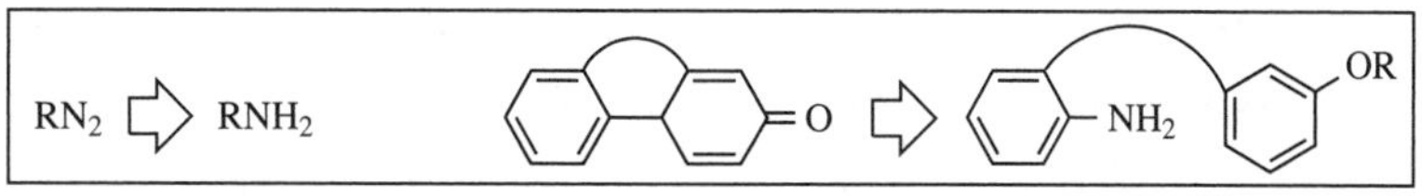

13.9.B.iii. α-Diazocarbonyl Compounds.

Another reaction that generates carbenes is a rearrangement of α-diazocarbonyl compounds (called the **Wolff rearrangement**),[280] in which a diazocarbonyl compound (**316**) loses nitrogen (N$_2$) to give an acyl carbene (**317**). This reaction occurs under both thermal and photochemical conditions, and the resulting carbene rearranges to a ketene (**318**) (completing the Wolff rearrangement of diazoketone to ketene). In this case, the Wolff rearrangement to ketene **318** was followed by an internal [2+2]-cycloaddition (secs. 11.10.A, 11.10.C) to give **319** in 79% yield.[281] It is noted that hydrolysis of a ketene will generate an acid (or an ester in the presence of an alcohol).[282] Meinwald and Gassman used this rearrangement to "shrink" bicyclic systems. Irradiation of **320** in water gave a 68% yield of **321** via initial formation of a carbene (see **322**), and rearrangement to the ketene, **323**.[283] Hydrolysis gave the carboxylic acid.

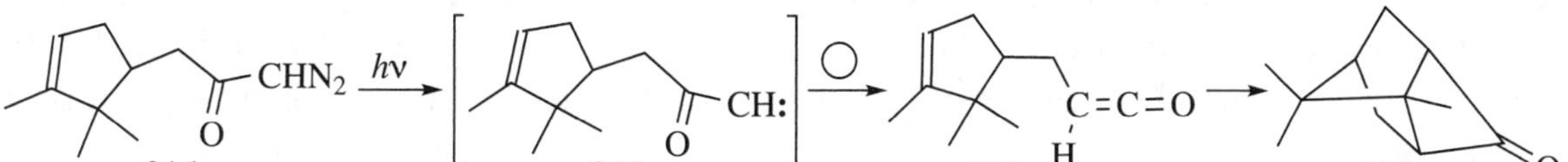

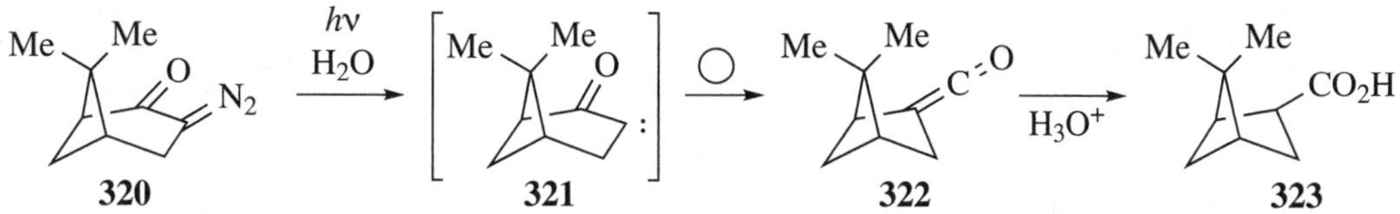

Diazoketones such as **316** can be formed from acid chlorides by reaction with diazomethane. Subsequent treatment with aqueous Ag$_2$O leads to the Wolff rearrangement and formation of a carboxylic acid of one carbon more than the starting acid chloride. This sequence is known

279. (a) Frey, H.M. *Pure Appl. Chem.* **1964**, *9*, 527; (b) *Idem. Advan. Photochem.* **1966**, *4*, 225; (b) Overberger, C.G.; Anselme, J.P. *Tetrahedron Lett.* **1963**, 1405.
280. (a) Wolff, L. *Annalen* **1912**, *394*, 23; (b) Reference 253a, pp 475-492; (c) The Merck Index, 14th Ed. Merck & Co., Inc., Whitehouse Station, New Jersey, **2006**, p ONR-103; (d) Mundy, B.P.; Ellerd, M.G.; Favaloro Jr., F.G. *Name Reactions and Reagents in Organic Synthesis, 2nd Ed.* Wiley-Interscience, New Jersey, **2005**, pp. 702-703.
281. Yates, P.; Fallis, A.G. *Tetrahedron Lett.* **1968**, 2493.
282. Bergmann, E.D.; Hoffmann, E. *J. Org. Chem.* **1961**, *26*, 3555.
283. Meinwald, J.; Gassman, P.G. *J. Am. Chem. Soc.* **1960**, *82*, 2857.

as the **Arndt-Eistert synthesis**[284] and the ketene can be trapped as an ester or an amide. A synthetic example is taken from Panek and Zhang's synthesis of herboxidiene/GEX 1A,[285] in which ester **324** was saponified and then reacted with (*1*) oxalyl chloride, (*2*) CH$_2$N$_2$ to give a 97% yield of diazoketone **325**. Subsequent reaction with silver benzoate and pseudoephedrine gave an 80% yield of **326** via reaction with the ketone formed by a Wolff rearrangement.

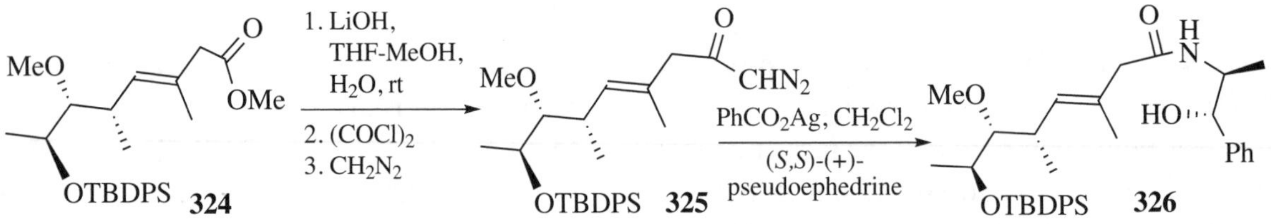

The disconnections pertinent to these α-diazocarbonyl compounds are

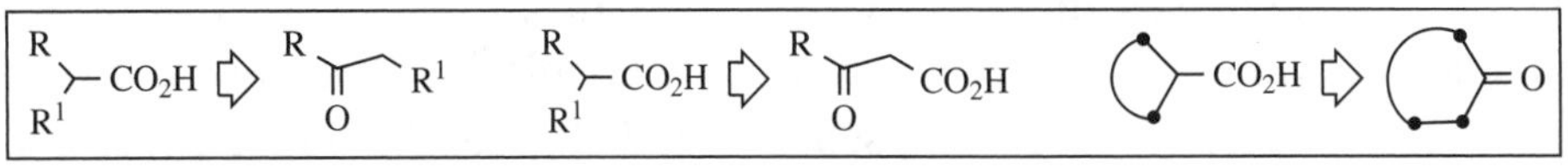

13.9.B.iv. Photolysis of Cyclopropanes.

Photolysis of cyclopropane derivatives[286] will generate a carbene via elimination of an alkene. Photolysis of 1,1-dimethylcyclopropane in the gas phase gave dimethylmethylene (Me$_2$C:) and ethene,[287] and irradiation of 9,10-dihydro-9,10-methanophenanthrene (**327**) gave carbene and phenanthrene.[288] Oxiranes can be photolyzed to carbenes as well.[289]

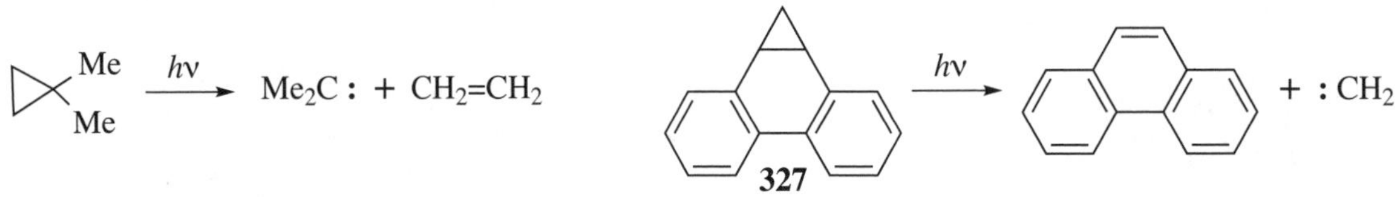

The cyclopropane-alkene disconnection is

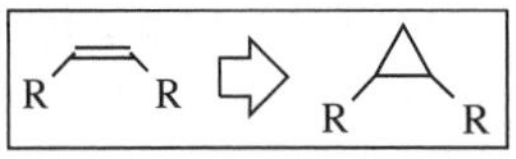

284. (a) Arndt, F.; Eistert, B. *Berichte* **1935**, *68*, 200; (b) Bachmann, W.E.; Struve, W.S. *Org. React.* **1942**, *1*, 38; (c) The Merck Index, 14th Ed. Merck & Co., Inc., Whitehouse Station, New Jersey, **2006**, p ONR-3; (d) Mundy, B.P.; Ellerd, M.G.; Favaloro Jr., F.G. *Name Reactions and Reagents in Organic Synthesis, 2nd Ed.* Wiley-Interscience, New Jersey, **2005**, pp. 44-45.
285. Zhang, Y.; Panek, J.S. *Org. Lett.* **2007**, *9*, 3141.
286. Griffen, G.W.; Bertoniere, N.R. in *Carbenes, Vol 1*, Jones Jr., M.; Moss, R.A. (Eds.), Wiley, New York, **1977**, pp 305-317.
287. Dhingra, A.K.; Koob, R.D. *J. Phys. Chem.* **1970**, *74*, 4490.
288. Richardson, D.B.; Durrett, L.H.; Martin Jr., J.M.; Putnam, W.E.; Slaymaker, S.C.; Dvoretzky, I. *J. Am. Chem. Soc.* **1965**, *87*, 2763.
289. Reference 287, pp 318-331.

13.9.B.v. The Bamford-Stevens Reaction. Aldehydes and ketones such as **328** can be converted to a tosylhydrazone (**329**) via reaction with tosylhydrazine. Subsequent treatment with base followed by heating results in the loss of *p*-toluenesulfinate to give the corresponding diazo compound (**330**),. which generates a carbene by loss of nitrogen and forms an alkene. The formation of alkenes from tosyl hydrazones when they are treated with base is called the **Bamford-Stevens reaction**.[290,291a] The carbene nature of the intermediate was detected by examination of camphor tosylhydrazone,[291] which on heating with hydroxide give a mixture of camphene and tricyclene. When the reaction was done in acidic alcohols such as ethylene glycol, camphene was the main product (via rearrangement). In aprotic solvents, the carbene insertion products observed (sec. 13.9.C.iii). A synthetic example of the Bamford-Stevens reaction is taken from a synthesis of (+)-lasonolide A by Shishido and co-workers in which ketone **331** was converted to the tosylhydrazone (**332**) in 97% yield, and subsequent treatment with base gave alkene **333** in 86% yield.[292]

Although this is *not* a carbene mechanism and involves extrusion of nitrogen via a carbanion, it is included here for comparison with the reactions just cited. Grieco et al. used this method to convert **334** to diene **335** (>98% yield) in a synthesis of (±)-ivangulin.[293] When an organolithium is used as the base, the alkene-forming reaction is sometimes called the **Shapiro reaction**.[294,297b] Two equivalents of organolithium are required, and the reaction generates the *less substituted* alkene. In Sha and co-worker's synthesis of (−)-pinguisenol, tosylhydrazone **336** (generated from the ketone by reaction with tosylhydrazone in 91% yield) was treated with butyllithium

290. (a) Bamford, W.R.; Stevens, T.S. *J. Chem. Soc.* **1952**, 4735; (b) The Merck Index, 14th Ed. Merck & Co., Inc., Whitehouse Station, New Jersey, **2006**, p ONR-4; (c) Mundy, B.P.; Ellerd, M.G.; Favaloro Jr., F.G. *Name Reactions and Reagents in Organic Synthesis, 2nd Ed.* Wiley-Interscience, New Jersey, **2005**, pp. 56-57.
291. (a) Bayless, J.H.; Friedman, L.; Cook, F.B.; Schechter, H. *J. Am. Chem. Soc.* **1968**, *90*, 531; (b) Davies, H.W.; Schwarz, M. *J. Org. Chem.* **1965**, *30*, 1242; (d) Powell, J.W.; Whiting, M.C. *Tetrahedron* **1959**, *7*, 305; (e) Clarke, P.; Whiting, M.C.; Papenmeier, G.; Reusch, W. *J. Org. Chem.* **1962**, *27*, 3356.
292. Yoshimura, T.; Yakushiji, F.; Kondo, S.; Wu, X.; Shindo, M.; Shishido, K. *Org. Lett.* **2006**, *8*, 475.
293. Grieco, P.A.; Oguri, T.; Wang, C.-L, J.; Williams, E. *J. Org. Chem.* **1977**, *42*, 4113.
294. (a) Shapiro, R.H. *Tetrahedron Lett.* **1968**, 345; (b) Mundy, B.P.; Ellerd, M.G.; Favaloro Jr., F.G. *Name Reactions and Reagents in Organic Synthesis 2nd Ed.* Wiley, New York, **2005**, pp 590-591.

and TMEDA to give a 66% yield of **337**.[295] The Bamford-Stevens reaction generates the *more substituted* alkene.

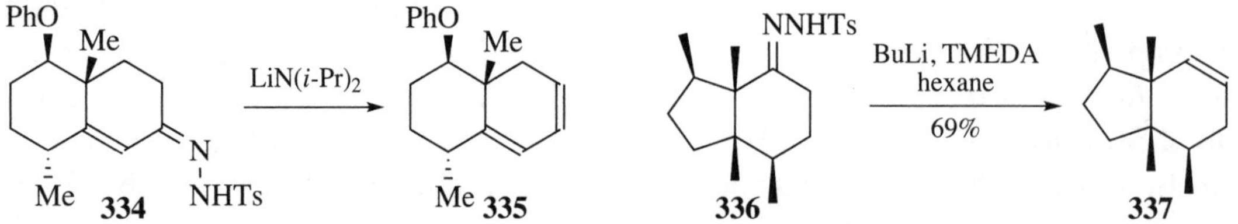

The relative amount of base has a significant effect on the product distribution. When the tosyl hydrazone of pinacolone was treated with > 1.8 equivalents of *n*-butyllithium, 3,3-dimethyl-1-butene was formed. When only 1 equivalent of *n*-butyllithium was used, however, a 57% yield of 3,3-dimethyl-1-butene, 40% of 1,1,2-trimethylcyclopropane, and 3% of the rearranged alkene (2,3-dimethyl-1-butene) were isolated.[296] The type of base employed is also important. The product distribution from the hydrazone of cyclopentanone, for example, gave a mixture of cyclopentene/(*E*)-2-pentene/(*Z*)-2-pentene.[297] When sodium methoxide was used, a 2:84:14 mixture of these products was obtained whereas sodium hydride led to a 83:14:3 mixture and sodium amide a 60:35:3 mixture.[298]

The Bamford-Stevens disconnection is

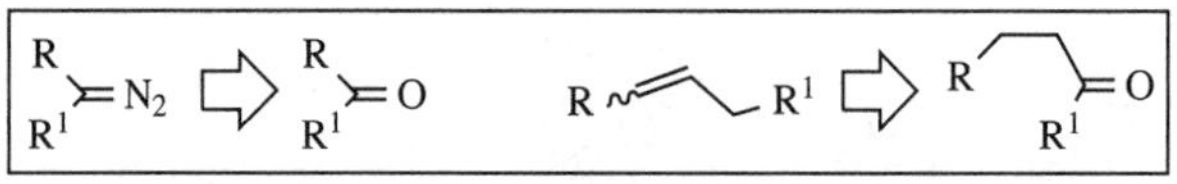

13.9.B.vi. Photolysis of Carbonyls. Under certain conditions, carbonyl compounds can be photolyzed to diradicals, and these can rearrange to carbenes.[298] Photolysis of cyclobutanone **328** generated diradical **329**, and in one pathway it eliminated to form an alkene (**341**), thereby generating ketene **340**. It decarbonylated to form cyclopropane **342** via an alternative pathway. Yet another pathway involved ring expansion to give carbene **343**, which reacted with methanol to give **344**.[299] When R = R¹ = H, 48% of **340** and **341** was observed, with 8% of **344** and a trace of **342**.[299a] When R = R¹ = Me, however, 68% of **344** was obtained, along with 13% of **340 + 341** and 11% of **342**.[299a]

295. Sha, C.-K.; Liao, H.-W.; Cheng, P.-C.; Yen, S.-C. *J. Org. Chem.* **2003**, *68*, 8704.
296. (a) Kaufman, G.; Cook, F.; Shechter, H.; Bayless, J.; Friedman, L. *J. Am. Chem. Soc.* **1967**, *89*, 5736; (b) Shapiro, R.H.; Heath, M.J. *Ibid.* **1967**, *89*, 5734.
297. Kirmse, W.; von Bülow, B.G.; Schepp, H. *J. L. Ann. Chem.* **1966**, *691*, 41.
298. Reference 253a, pp 47-51.
299. (a) Turro, N.J.; Southam, R.M. *Tetrahedron Lett.* **1967**, 545; (b) Morton, D.R.; Lee-Ruff, E.; Southam, R.M.; Turro, N.J. *J. Am. Chem. Soc.* **1970**, *92*, 4349.

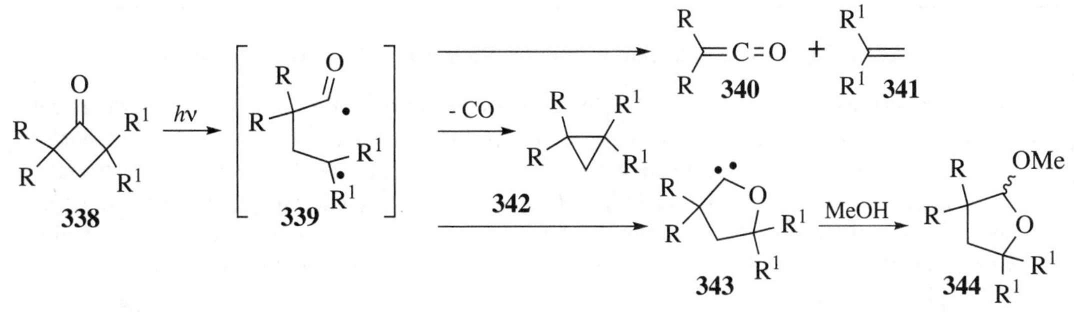

The carbonyl disconnection is:

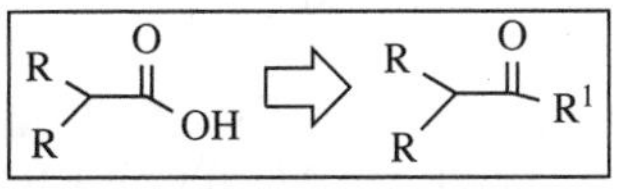

13.9.B.vii. Halocarbenes. Polyhalomethanes can be converted to halocarbenes via photolysis, thermolysis, or treatment with base. When cyclohexene was treated with a mixture of chloromethane and phenylsodium, methylene ($:CH_2$) was formed, and it added

to the alkene gave norcarane (**345**), but in only 3.2% yield.[300] Obviously, chloromethane is not the best source of carbenes. Polyhalomethanes give much better results, and the most common method for generating halocarbenes is treatment of dihalomethanes and trihalomethanes with sodium metal or with another base.[301]

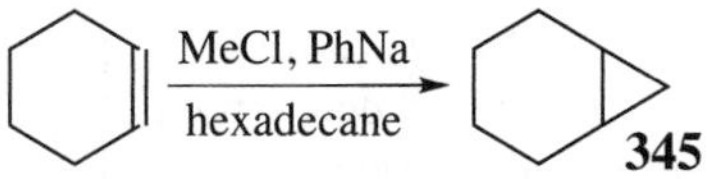

Chloroform is a particularly useful carbene precursor. Reaction of chloroform with hydroxide begins with an acid-base reaction to generate the anion ($Cl_3C:^-$) and water. Loss of chloride ion generates dichlorocarbene ($Cl_2C:$). Geuther suggested this type of carbene in 1862.[302] Potassium *tert*-butoxide is the base most commonly used to generate dichlorocarbene, after its initial introduction by von E. Doering and Hoffmann.[303] A typical reaction is that of cyclohexene and chloroform with potassium *tert*-butoxide to give **346**. In general, reactions with

alkenes give dichlorocyclopropanes [dibromocyclopropanes with bromoform ($CHBr_3$) and diiodocyclopropanes with iodoform (CHI_3)]. Tri-*n*-butyltin hydride (Bu_3SnH) can be used to reduce the product dihalide to a monohalide (sec. 4.9.H).

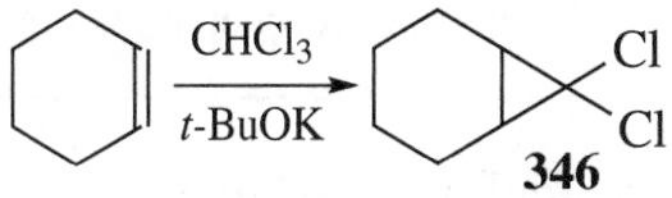

300. Friedman, L.; Berger, J.G. *J. Am. Chem. Soc.* **1960**, *82*, 5758.
301. Reference 253a, pp 129-150.
302. Geuther, A. *Annalen* **1862**, *123*, 121.
303. von E. Doering, W.; Hoffmann, A.K. *J. Am. Chem. Soc.* **1954**, *76*, 6162.

MOLECULAR MODELING: Addition of Singlet Dichlorocarbene to Ethylene

The term *singlet* is used to describe a molecule in which all electrons are paired, while the term ***triplet*** is used to describe a molecule in which two of the electrons are unpaired. While the vast majority of organic molecules are singlets, divalent carbon compounds (carbenes) may also exist as triplets. Most notable, the parent compound methylene, CH_2, is a triplet. The chemistry of singlet and triplet carbenes is quite different, allowing them to be easily distinguished. Triplet carbenes generally behave as radicals, whereas singlet carbenes are typically electrophiles in that they add to carbon-carbon double bonds. For example, $:CCl_2$, adds to ethylene to give 1,1-dichlorocyclopropane.

$$:CCl_2 \;+\; H_2C=CH_2 \longrightarrow$$

The transition state for this reaction is shown, as well as the energy profile for the addition. Examine the energy profile and identify the transition state as the structure at the top of the profile. "Walk" forward and back through the individual structures using the **Step** keys or animate the sequence using the **Play** key (all are found at the bottom left of the screen). What happens to the two reactants as the product (1,2-dichlorocyclopropane) is approached?

Rationalize the geometry of the transition state and the motion of the two reactants by examining the HOMO and LUMO for singlet dichlorocarbene. *Click* on > and then select ***Continue*** (not ***New***) from the screen to bring up the builder panel. Start with methylene chloride and delete both hydrogen atoms to get dichlorocarbene. Are the highest-energy pair of electrons associated with the σ or π system of dichlorocarbene? Will the next pair of electrons be added to σ or π system? Next, examine an electrostatic potential map for $:CCl_2$. Recall that this colors an electron density surface which gives to overall molecular size and shape (where a reagent can approach) with the value of the electrostatic potential (the energy of a point positive charge with the nuclei and electrons of a molecule). By convention, electron-rich regions on are "red", electron-deficient regions are "blue" and neutral regions are "green". Where are the most "available" electrons located? Where are additional electrons most easily accommodated? Are your conclusions based on the electrostatic potential map consistent with those based on the HOMO and LUMO? What would you expect is the best geometry for approach of the two reagents? What would be the final geometry?

Typical halocarbene disconnections are

13.9.C. Reactions of Diazomethane

As with other diazoalkanes, diazomethane reacts with alkenes to form cyclopropane derivatives (sec. 13.9.C.i).[304] Reaction with aromatic derivatives leads to ring expansion to cycloheptatriene derivatives.[305] Both of these reactions (addition to an alkene or arene insertion) involve generation of an intermediate carbene and addition to a π bond.

304. Muck, D.L.; Wilson, E.R. *J. Org. Chem.* **1968**, *33*, 419.
305. Müller, E.; Kessler, H.; Fricke, H.; Kiedaisch, W. *Annalen* **1964**, *675*, 63.

Both reactions will be discussed below. Many of the reactions of diazomethane tend to be ionic in nature and are, therefore, set aside from the other diazoalkane chemistry in this section. One of the most prevalent uses of diazomethane itself is for esterification of small quantities of acids, especially acids that are precious for one reason or another. The reaction proceeds in near quantitative yield, producing a single product, as in Vidari and co-worker's conversion of **347** to methyl ester **348** in 90% isolated yield[306] in a synthesis of J_2 isoprostane. Note that the esterification reaction is faster than reaction with the C=C unit.

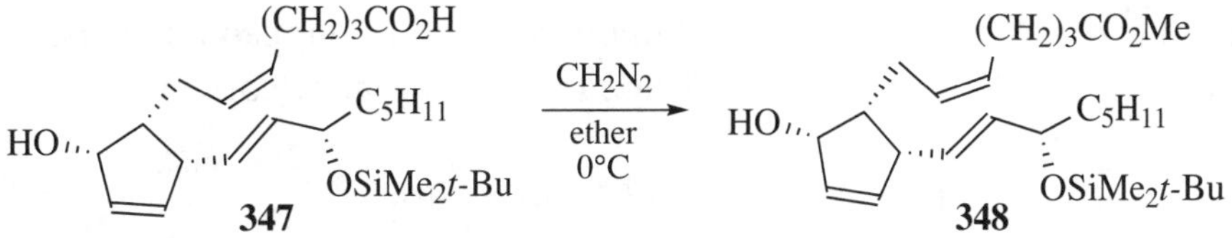

Diazoalkanes undergo [3+2]-cycloadditions in the presence of alkenes to give pyrazolines, as discussed in Section 11.11.E. Diazomethane reacts with α-methylacrylonitrile, for example, to give pyrazoline **349** in 76% yield.[307] In another ionic reaction, diazomethane reacts with aldehydes to give methyl ketones,[308] as in the conversion of benzaldehyde to acetophenone (97% yield), which is a useful reaction for extending carbon chains. Diazomethane also reacts with cyclic ketones to give ring-expanded ketones, along with epoxides as minor products. This method was used to prepare cycloheptanone from cyclohexanone in 65% yield by reaction with diazomethane, along with 15% of oxaspirooctane.[309,263a]

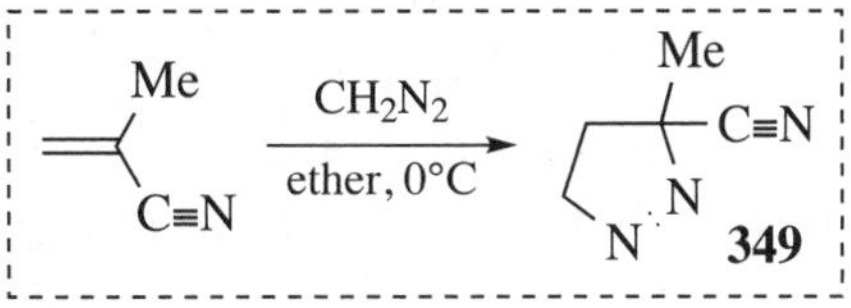

Some of these specialized diazoalkane disconnections are

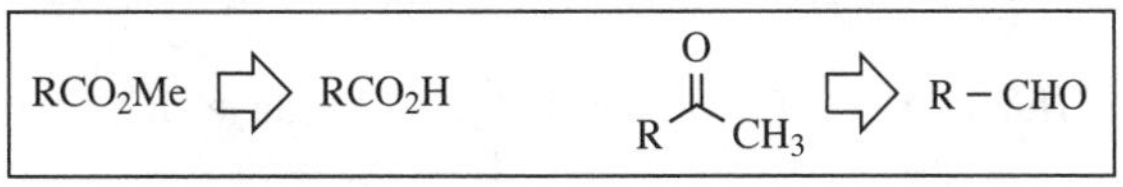

13.9.D Carbene Reactions of Diazoalkanes

Carbenes are relatively easy to generate and they have several synthetic uses. Diazoalkanes are converted to the corresponding carbene photolytically, and in the presence of copper derivatives

306. Zanoni, G.; Porta, A.; Castronovo, F.; Vidari, G. *J. Org. Chem.* **2003**, *68*, 6005.
307. Gotkis, D.; Cloke, J.B. *J. Am. Chem. Soc.* **1934**, *56*, 2710.
308. (a) Schlotterbeck, F. *Chem. Ber.* **1907**, *40*, 479; (b) Gutsche, C.D. *Org. React.* **1954**, *8*, 364.
309. (a) House, H.O.; Grubbs, E.J.; Cannon, W.F. *J. Am. Chem. Soc.* **1960**, *82*, 4099; (b) Cram, D.J.; Helgeson, R.C. *Ibid.* **1966**, *88*, 3515.

such as Cu(acac)$_2$. Dirhodium tetraacetate [Rh$_2$(OAc)$_4$] was found to be a better catalyst in some cases, initiating reaction with electron deficient alkenes.[310] Metal carbenes are important in synthesis.[311] Carbenes can undergo 1,2-hydrogen shifts to generate alkene products. When diazine **350** was photolyzed (355 nm, 103.7 kcal mol^{-1}, 433.9 kJ mol^{-1}), chlorocarbene **351** was formed. In the presence of 2,3-dimethyl-2-butene, a 73% yield of the cyclopropane product (**353**) was produced.[312] In addition to **353**, alkene **352** was formed by a 1,2-hydrogen shift during the course of the reaction and was competitive with the addition. The energy barrier for this rearrangement was calculated to be about 6.4 ± 2 kcal mol^{-1} (26.79 ± 8.37 kJ mol^{-1}) and proceeded at a rate of 4.9-6.7 x 10^7 s^{-1}.[312a] Although the addition reaction is usually faster, the 1,2-shift should be considered an important side reaction in many carbene reactions.

13.9.D.i. Addition to Alkenes. Perhaps the most common reaction of carbenes is their addition to alkenes to generate cyclopropanes,[313] as shown in the formation of **353** above. A generalized example is the preparation of **355** from carbene **297** and alkene **354**.[314] The carbenes can be generated by any of the methods discussed in Section 13.9. The addition is usually exothermic[315] (a typical value is 91 kcal mol^{-1}, 380.9 kJ mol^{-1}), with an activation energy of ≈ 64 kcal mol^{-1} (267.9 kJ mol^{-1}). Reaction of *cis*-and *trans*-2-butene with diazomethane

(under photolytic conditions) showed that the yield of cyclopropanes increased with increasing pressure. Insertion (see sec. 13.9.D.iv) into allylic and vinyl bonds was competitive in many cases. The reaction is stereoselective for retention of the geometry of the starting alkene, when diazomethane is used.[316] For other methylene precursors, *trans*-1,2-dimethylcyclopropane is

310. Aggarwal, V.K.; Smith, H.W.; Hynd, G.; Jones, R.V.; Fieldhouse, R.; Spey, S.E. *J. Chem. Soc. Perkin Trans. 1* **2000**, 3267.
311. Dörwald, F.Z. *Metal Carbenes in Organic Synthesis,* Wiley-VCH, New York, **1999**.
312. (a) Jackson, J.E.; Soundararajan, N.; White, W.; Liu, M.T.H.; Bonneau, R.; Platz, M.S. *J. Am. Chem. Soc.* **1989**, *111*, 6874; (b) LaVilla, J.A.; Goodman, J.L. *Ibid.* **1989**, *111*, 6877.
313. For transformations of this type, see reference 213, pp 135-152. Also see Lebel, H.; Marcoux, J.-F.; Molinaro, C.; Charette, A.B. *Chem. Rev.* **2003**, *103*, 977.
314. (a) Reference 253a, chapter 8, pp 267-362; (b) Moss, R.A. in *Carbenes, Vol. 1*, Jones, Jr., M.; Moss, R.A. (Eds.), Wiley, New York, **1973**, pp 153-304.
315. Reference 253a, p 268.
316. (a) Reference 253a, p 271; (b) McKnight, C.; Lee, P.S.T.; Rowland, F.S. *J. Am. Chem. Soc.* **1967**, *89*, 6802; (c) Eder, T.W.; Carr, Jr., R.W. *J. Phys. Chem.* **1969**, *73*, 2074; (d) Ring, D.F.; Rabinovitch, B.S. *Ibid.* **1968**, *72*, 191; (e) Duncan, F.J.; Cvetanovic, R.J. *J. Am Chem. Soc.* **1962**, *84*, 3593; (f) Montague, D.C.; Rowland, F.S. *J. Phys. Chem.* **1968**, *72*, 3705; (g) Frey, H.M. *J. Am. Chem. Soc.* **1960**, *82*, 5947.

favored for both *cis*- and *trans*-2-butene.[317] The ratio of cis to trans isomers is, however, dependent upon the wavelength of light used, and on the additive. The reaction was more stereoselective at higher pressures, and when long wavelengths of light were used.[317] Molecular oxygen appears to be the best additive, and irradiation at 313 nm (91.4 kcal mol^{-1}, 382.6 kJ mol^{-1}) is optimal for methylene generation from ketenes. The yield of cyclopropanes from 2-butene was 87% at 366 nm (78.1 kcal mol^{-1} = 326.9 kJ mol^{-1}), but only 29% at 313 nm.[317c] Oxygen scavenges triplet methylene, which gives diminished stereoselectivity relative to singlet methylene, although isomerization of the active intermediates is also a factor.[318]

Table 13.6   cis-trans Selectivity of 2-Butene with Various Diazo Compounds

Alkene Geometry	Diazo Compound	hν	356 : 357	Alkene Geometry	Diazo Compound	hν	356 : 357
cis-	CH$_2$N$_2$	direct	0 : 100	trans-	Ph acyl N$_2$ (O)	direct	73 : 27
		sensitized	34 : 66			sensitized	74 : 26
	Ph Me C=N$_2$	direct	4 : 96		cyclohexanone diazo (O)	sensitized	86 : 14
		sensitized	15 : 85				
	Ph acyl N$_2$ (O)	direct	50 : 50		MeO$_2$C / MeO$_2$C C=N$_2$	direct	90 : 10
		sensitized	55 : 45			sensitized	86 : 14
	Me acyl N$_2$ (O)	sensitized	56 : 44				
	cyclohexanone diazo (O)	sensitized	86 : 14				
	MeO$_2$C / MeO$_2$C C=N$_2$	sensitized	90 : 10				

When the reaction occurs in the liquid phase, which is more common, the reaction is stereoselective unless the carbene "has a readily accessible triplet state".[319] Photolysis of diazomethane in the presence of *cis*-2-butene, for example, gave *cis*-1,2-dimethylcyclopropane (**357**) with complete retention of configuration.[320] Other diazoalkanes led to a mixture of **356** and **357**. A decrease in stereoselectivity was observed when triplet methylene was produced (by photolysis of diazomethane) in the presence of a large excess of perfluoropropane (CF$_3$CF$_2$CF$_3$).[317d] Reaction of triplet methylene with a cis alkene (**358**) proceeds by a two-step mechanism. The initially generated diradical (**359**) is in rotational equilibrium with **360**. Both

317. (a) Frey, H.M. *Proc. Roy. Soc. Ser. A* **1959**, *250*, 409; (b) *Idem Ibid.* **1959**, *251*, 575.
318. Reference 253a, p 272.
319. Reference 253a, p 275.
320. (a) Skell, P.S.; Woodworth, R.C. *J. Am. Chem. Soc.* **1956,** *78*, 4496, 6427; (b) *Idem Ibid.* **1959**, *81*, 3383; (c) von E. Doering, W.; LaFlamme, P. *Ibid.* **1956**, *78*, 5447.

triplet diradicals are slowly converted to singlet diradical **361** and **362**, which give the cis-cyclopropane **363** and the trans-cyclopropane **364**.[321,321]

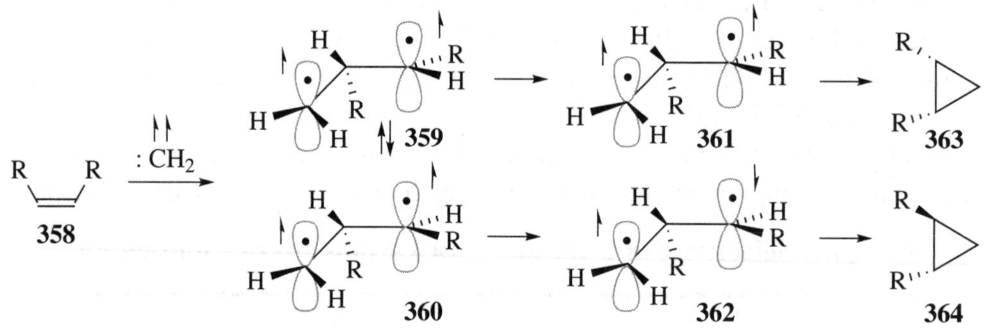

Singlet carbenes are produced from singlet diradicals, and close with high selectivity. Bond rotation in the triplet diradical is faster than $T_1 \rightarrow S_1$ conversion or ring closure. Ring closure of the singlet radical is faster than bond rotation (an equilibrium between **361** and **362** and unfavorable relative to conversion to **359** or **364**). Table 13.6[322,277] summarizes the reaction of various diazo compounds (and the carbenes they produce) with 2-butene. There was a decrease in stereoselectivity for the sensitized reaction in comparison with direct photolysis. The rate of carbene addition also varies with the type of alkene, as shown in Table 13.7.[323]

Table 13.7   Rate of Carbene Addition as a Function of Both Alkene and Carbene

Alkene	Carbene Generation (Relative Rate)						
	A	**B**	**C**	**D**	**E**	**F**	**G**
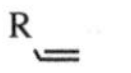 R	0.89	0.66	0.56	0.90	0.011	0.045	
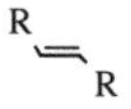 R, R	0.79		1.10		0.15		0.25
 R R	0.78		1.01		0.23		0.53
 ⬡		1.00		1.00	0.13	0.27	
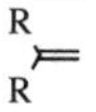 R, R	1.0		1.0		1.0	1.0	1.0
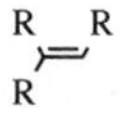 R R, R	1.17	3.32	1.9	1.62	2.8	2.0	2.5
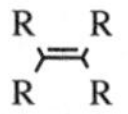 R R, R R	1.13	3.46		1.6	6.5	1.85	4.4

A. $CH_2=C=O$, 260 nm. B. CH_2N_2, Cu(II). C. $PhCHN_2$/hv. D. $EtO_2C\text{-}CHN_2$/CuSO$_4$.

E. $CHCl_3$/KOt-Bu.   F. $CHBr_3$/KOt-Bu.   G. 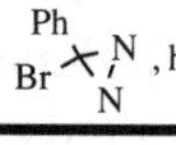 Br⤫N, hv.

---

321.   Reference 253a, p 283.
322.   (a) Reference 253a, p 279; (b) Moritani, I.; Yamamoto, Y.; Murahashi, S. *Tetrahedron Lett.* **1968**, 5697; (c) Cowan, D.O.; Couch, M.M.; Kopecky, K.R.; Hammond, G.S. *J. Org. Chem.* **1964**, *29*, 1922; (d) Jones, Jr., M.; Ando, W. *J. Am. Chem. Soc.* **1968**, *90*, 2200; (e) Jones, Jr., M.; Ando, W.; Kulczycki Jr., A. *Tetrahedron Lett.* **1967**, 1391.
323.   Reference 253a, pp 294-297.

When more than one alkene moiety is present, there is the potential for formation of isomers via addition to one or both of the double bonds. In general, carbenes add to the more electron-rich alkene. Reaction of **365** with dichlorocarbene (Cl₂C:) gave only **366**, but **367** gave an 80% yield of **368**, with 20% of **369**.[324] Where it is appropriate, there seems to be a preference for trans double bonds, as seen in the reaction of methylene with *cis,trans,trans*-1,5,9-cyclodecatriene (**370**), which gave a 97:3 mixture of **371/372**.[325] It is noted that the trans double bond may be more sterically accessible, and this can account for the stereochemical preference. Similar reaction with dichlorocarbene and also with the Simmons-Smith reagent (sec. 13.9.D) showed the same preference for reaction with the (*E*)-alkene (this preference ranged from 2:1 to 9:1).

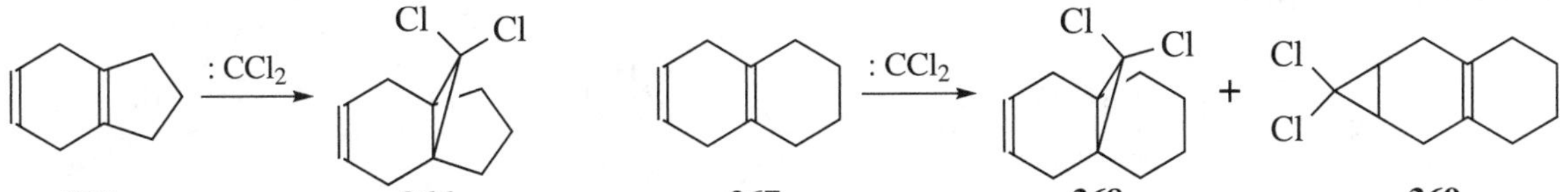

A wide range of cyclopropane derivatives can be prepared by this carbene addition to alkenes.[326] There is a general preference for the less congested exo product in reactions with cyclic alkenes. Initial addition of the carbene can be accompanied by rearrangement of the product, as discussed in the previous section (see above).[327] In many cases, however, the cyclopropanation proceeds without rearrangement. Diazoalkanes add to alkenes give cyclopropane derivatives in the presence of transition metal reagents. An example is the reaction of diazoketone **373** in the presence of copper triflate to give **374** (in 95% yield and 71% ee with the correct chiral ligand), taken from Nakada and co-workers' synthesis of (−)-methyl jasmonate.[328]

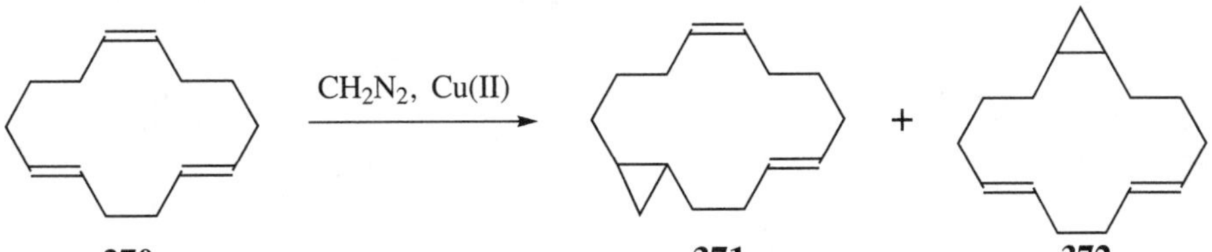

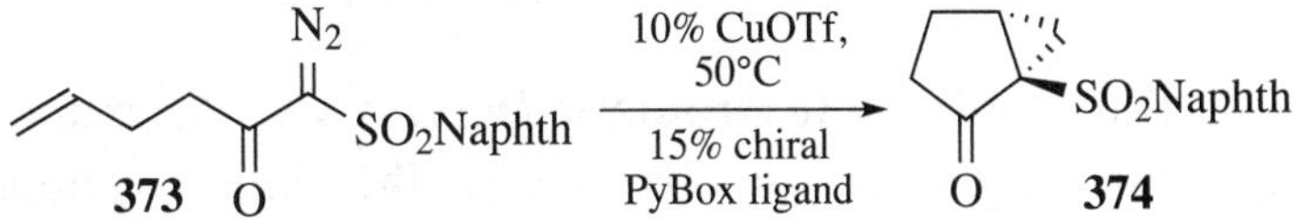

324. Sims, J.J.; Honwad, V.K. *J. Org. Chem.* **1969**, *34*, 496.
325. (a) Nozaki, H.; Kawanisi, M.; Noyori, R. *J. Org. Chem.* **1965**, *30*, 2216; (b) Nozaki, H.; Katô, S.; Noyori, R. *Can. J. Chem.* **1966**, *44*, 1021; (c) Locke, J.M.; Duck, E.W. *Chem. Ind. (London)* **1965**, 1727; (d) Mühlstädt, M.; Graefe, J. *Chem. Ber.* **1966**, *99*, 1192; (e) *Idem Z. Chem.* **1969**, *9*, 303 [*Chem. Abstr. 71*:90918y **1969**].
326. Reference 253a, pp 304-320.
327. Reference 253a, pp 321-328.
328. Takeda, H.; Watanabe, H.; Nakada, M. *Tetrahedron* **2006**, *62*, 8054. For a synthesis of communesin F which employed an intramolecular cyclopropanation of an indole derivative with a diazolactone see Yang, J.; Wu, H.; Shen, L.; Qin, Y. *J. Am. Chem. Soc.* **2007**, *129*, 13794.

Carbenes generated the more traditional way react similarly. In Trost and Oslob's synthesis of (−)-anatoxin,[329] the alkene unit in **375** reacted with dibromocarbene, generated from bromoform and base, to give an 85% yield of **376** and **377** as a 3.5:1 mixture of exo and endo isomers. Carbenes react with enamines in the normal way, as seen in the reaction of **373** to give **374**. However, in some cases the reaction is accompanied by a ring expansion that generates ketones via rearrangement of the initially formed aminocyclopropane. The cyclopropane product can be very stable and is usually isolated as the major product. Dichlorocarbene reacted with 1-piperidinocyclopentene (**378**) (derived from piperidine and cyclopentanone, see sec. 9.6.A) to give **379**. Loss of chloride ion generated the conjugated iminium chloride (**380**), and hydrolysis gave **381**.[330]

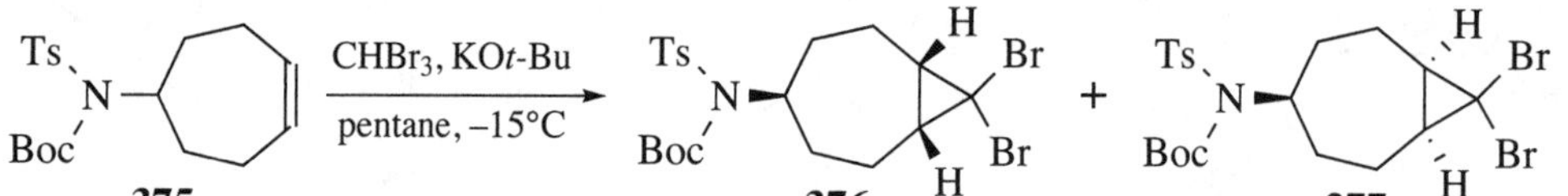

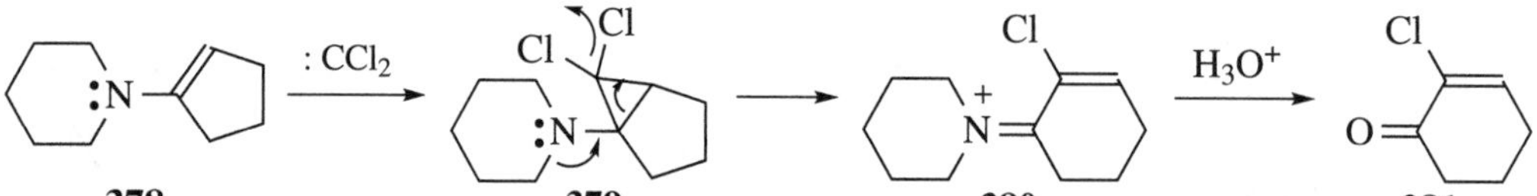

The disconnections available from carbene additions are

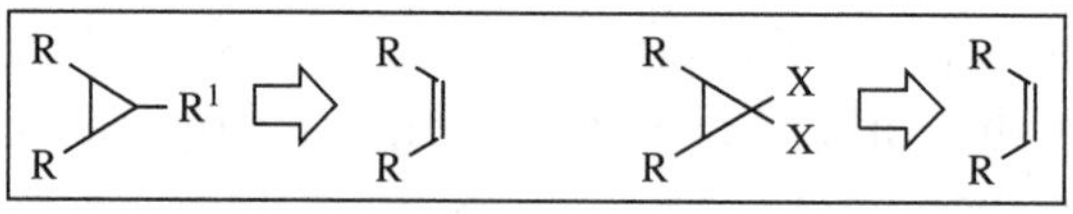

13.9.D.ii. Addition to Aromatic Derivatives. Aromatic compounds also react with carbenes, but ring expansion usually follows the initial cyclopropanation. In a typical example, 2-methoxynaphthalene (**382**) reacted with dichlorocarbene to give **383** and subsequent ring expansion gave **384**.[331] This is a general reaction of enol ethers to give either unsaturated acetals or unsaturated carbonyl compounds.[332]

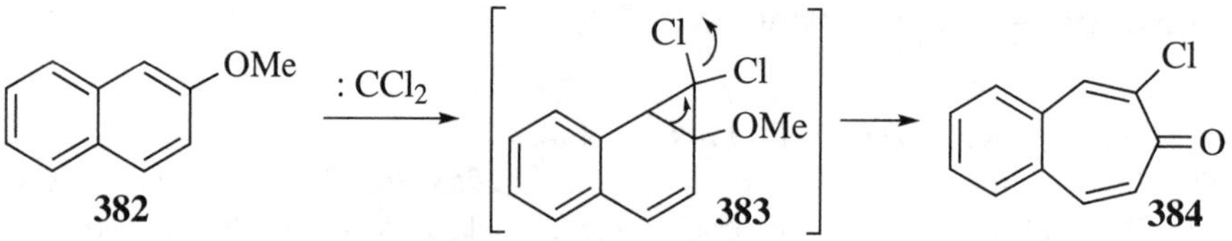

329. Trost, B.M.; Oslob, J.D. *J. Am. Chem. Soc.* **1999,** *121,* 3057.
330. (a) Wolinsky, J.; Chan, D.; Novak, R. *Chem. Ind. (London)* **1965,** 720; (b) Ohno, M. *Tetrahedron Lett.* **1963,** 1753.
331. Parham, W.E.; Bolon, D.A.; Schweizer, E.E. *J. Am. Chem. Soc.* **1961,** *83,* 603.
332. (a) Buddrus, J.; Nerdel, F.; Hentschel, P.; Klamann, D. *Tetrahedron Lett.* **1966,** 5379; (b) Nerdel, F.; Buddrus, J.; Brodowski, W.; Hentschell, P.; Klamann, D.; Weyerstahl, P. *Annalen* **1968,** *710,* 36.

Rearrangement or ring opening of the cyclopropane adducts derived from aromatic compounds are possible, depending on the carbene. Addition of carbene to benzene gave norcarane **385** (R = H) and cycloheptatriene (**386**, R = H) in equilibrium. With functionalized norcarane products, ring opening usually occurs to give the substituted benzene, **387** (R = R^1 = CN, for example). Product **386** is favored when **385** is generated by reaction of benzene with diazomethane or ester carbenes such as :CHCO$_2$R.[333] Photolysis of diazomethane in benzene, for example, gave a 4.8:1 mixture of cycloheptatriene and toluene.[334] This reaction is not restricted to simple benzene derivatives, and other aromatic rings can participate. In synthetic studies directed toward the guanacastepenes by Trauner and co-workers,[335] for example, diazo derivative **388** was treated with the rhodium catalyst to give **389** by cyclopropanation of the aromatic furan ring. Ring opening led to a 50% yield of **390** by ring expansion and formation of an enol that tautomerized to the aldehyde unit.

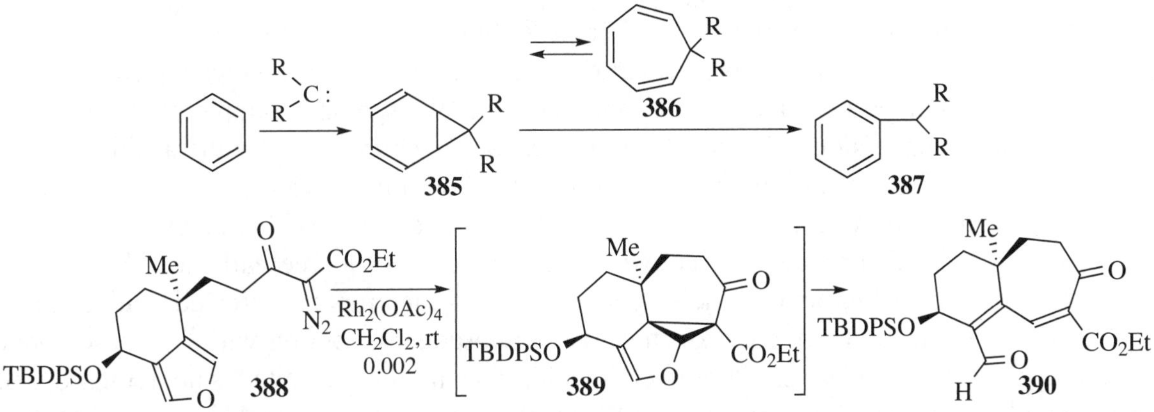

Other carbenes react similarly, and the reaction can be applied to much more sophisticated substrates. In a synthetic example taken from Feldman and coworker's synthesis of pareitropone, alkynyl-tin biaryl **391** was first converted to hypervalent iodine derivative **392**.[336] Subsequent treatment with base led to a cyclization reaction that formed the six-membered nitrogen-containing ring, and a vinyl carbene (**393**). Carbene insertion to the phenyl ring and ring expansion in the expected manner led to formation of the seven-membered ring found in pareitropone, **394**, in 64% overall yield.[337] The course of the reaction is dictated by the inherent stability of the final product.[337]

333.   Reference 253a, p 381.
334.   (a) von E. Doering, W.; Knox, L.H. *J. Am. Chem. Soc.* **1950**, *72*, 2305; (b) Meerwein, H.; Disselnkötter, H.; Rappen, F.; von Rintelen, H.; van de Vloed, H. *Annalen*, **1957**, *604*, 151; (c) Lemmon, R.M.; Strohmeier, W. *J. Am. Chem. Soc.* **1959**, *81*, 106; (d) Russell, G.A.; Hendry, D.G. *J. Org. Chem.* **1963**, *28*, 1933.
335.   Hughes, C.C.; Kennedy-Smith, J.J.; Truaner, D. *Org. Lett.* **2003**, *5*, 4113.
336.   Feldman, K.S.; Cutarelli, T.D.; Di Florio, R. *J. Org. Chem.* **2002**, *67*, 8528.
337.   (a) Sonnenberg, J.; Winstein, S. *J. Org. Chem.* **1962**, *27*, 748; (b) Skell, P.S.; Sandler, S.R. *J. Am. Chem. Soc.* **1958**, *80*, 2024; (c) Gatlin, L.; Glick, R.E.; Skell, P.S. *Tetrahedron* **1965**, *21*, 1315.

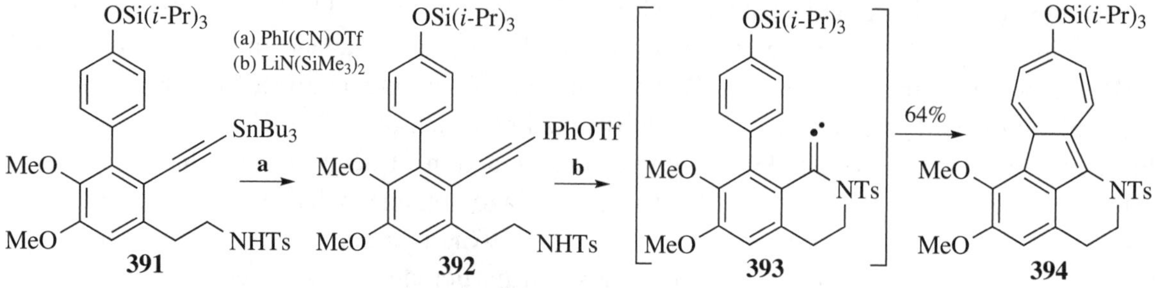

The scheme shows structures **391**, **392**, **393**, **394** with reagents (a) PhI(CN)OTf, (b) LiN(SiMe$_3$)$_2$; arrows **a** and **b**; substituents OSi(i-Pr)$_3$, SnBu$_3$, IPhOTf, NHTs, NTs, MeO; yield 64%.

The carbene ring expansion disconnection is

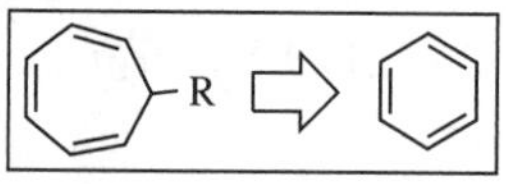

$$\text{(cycloheptatriene)}-R \Rightarrow \text{(benzene)}$$

13.9.D.iii. Cyclization Reactions of α-Diazo Ketones.

An attractive application of the cyclopropanation reaction is intramolecular trapping of carbenes to form bicyclic compounds, with a three-membered ring appended to another ring. We saw an example of this reaction previously in the conversion of **373** to **374**. When diazoalkanes are treated with transition metals in the presence of an alkene, particularly copper or rhodium derivatives, cyclopropanation occurs although the reactive intermediate may not be a free carbene. With rhodium compounds, in particular, a C=Rh unit is the reactive species rather than a free carbene. A synthetic example from Srikrishna and Dethe's synthesis of (–)-cucumin H[338] treated acid chloride **395** with diazomethane to give diazoketone **396**. Subsequent reaction with Cu-CuSO$_4$ gave cyclopropane derivative **397** in >70% overall yield. The mixture just cited is not required, and heating with cupric sulfate or copper metal[339] can also be used to initiate this cyclization, as in von E. Doering's synthesis of barbaralone (**398**) by heating **399** with copper.[340]

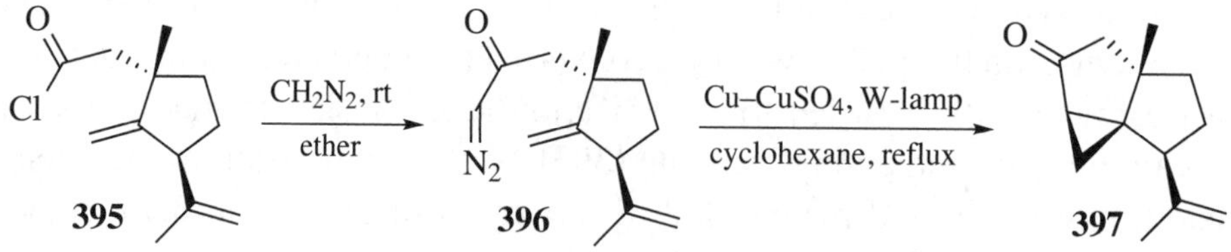

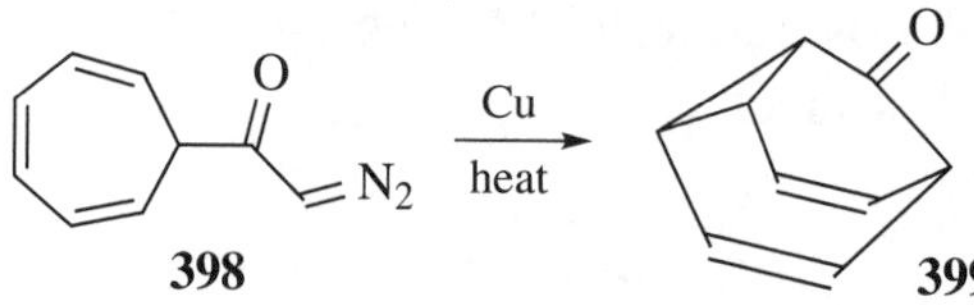

Scheme: **395** $\xrightarrow[\text{ether}]{\text{CH}_2\text{N}_2,\ \text{rt}}$ **396** $\xrightarrow[\text{cyclohexane, reflux}]{\text{Cu–CuSO}_4,\ \text{W-lamp}}$ **397**

398 $\xrightarrow[\text{heat}]{\text{Cu}}$ **399**

This carbene insertion reaction has been used in a variety of syntheses and coupled with other synthetic techniques. Taber et al. used carbene cyclopropanation in several syntheses,

338. Srikrishna, A.; Dethe, D.H. *Org. Lett.* **2003**, *5*, 2295.
339. Julia, S.; Linstrumelle, G. *Bull. Soc. Chim. Fr.* **1966**, 3490.
340. von E. Doering, W.; Ferrier, B.M.; Fossel, E.T.; Hartenstein, J.H.; Jones Jr., M.; Klumpp, G.; Rubin, R.M.; Saunders, M. *Tetrahedron* **1967**, *23*, 3943.

including a copper-catalyzed synthesis of (+)-isoneonepatelactone.[341] Other diazo-carbonyl compounds can be used. In a synthesis of (–)-spirotryprostatin B,[342] Carreira and Meyers reacted diazolactam **400** (prepared from isatin)[343] with 1,3-pentadiene, in the presence of the rhodium catalyst, to give a 71% yield of cyclopropyl derivative **401**. In another example, the reaction of diazoketone **402** and rhodium octanoate [$Rh_2(oct)_4$] produced a 52% yield of **403**, along with 28% of **404**.[344] These compounds were used in further synthetic transformations in Taber and co-worker's synthesis of the ethyl ester of the major urinary metabolite of prostaglandin E_2.

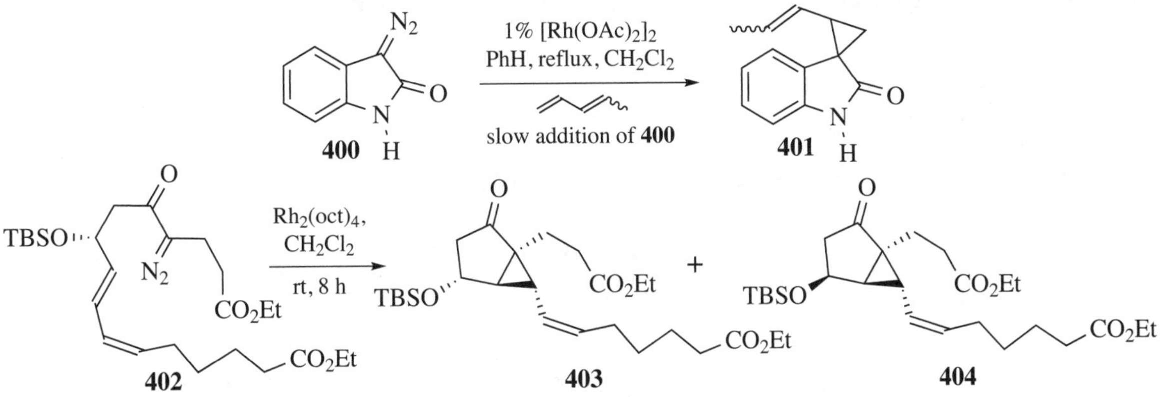

The intramolecular carbene addition disconnections are

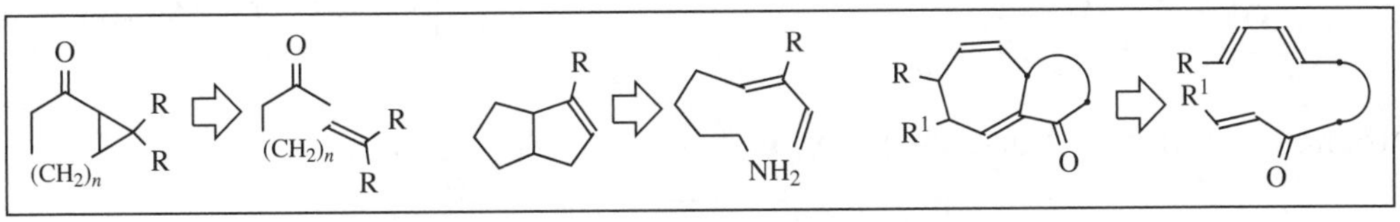

13.9.D.iv. C-H Insertion Reactions of α-Diazo Ketones.

Taber and co-workers showed that diazocarbonyls undergo C-H insertion reactions in the presence of rhodium acetate [$Rh_2(OAc)_4$],[345b,346] as illustrated in syntheses of (±)-pentalenolactone E methyl ester[347] and α-cuparenone.[348] Other rhodium catalysts can be used. Fukuyama and co-workers applied this reaction to a synthesis of (–)-ephedradine A (also known as orantine).[349] When **405** was treated

341. Taber, D.F.; Amedio Jr., J.C.; Raman, K. *J. Org. Chem.* **1988**, *53*, 2984.
342. Meyers, C.; Carreira, E.M. *Angew. Chem. Int. Ed.* **2003**, *42*, 694.
343. Cava, M.P.; Little, R.L.; Napier, D.R. *J. Am. Chem. Soc.* **1957**, *80*, 2257.
344. Taber, D.F.; Teng, D. *J. Org. Chem.,* **2002**, *67*, 1607.
345. (a) Wenkert, E.; Davis, L.L.; Mylari, B.L.; Solomon, M.F.; da Silva, R.R.; Shulman, S.; Warnet, R.J.; Ceccherelli, P.; Curini, M.; Pellicciari, R. *J. Org. Chem.* **1982**, *47*, 3242; (b) Taber, D.F.; Petty, E.H. *Ibid.* **1982**, *47*, 4808.
346. Taber, D.F.; Ruckle Jr., R.E. *J. Am. Chem. Soc.* **1986**, *108*, 7686.
347. Taber, D.F.; Schuchardt, J.L. *J. Am. Chem. Soc.* **1985**, *107*, 5289.
348. Taber, D.F.; Petty, E.H.; Raman, K. *J. Am. Chem. Soc.* **1985**, *107*, 196.
349. Kurosawa, W.; Kan, T.; Fukuyama, T. *J. Am. Chem. Soc.* **2003**, *125*, 8112.

with 0.3% of a chiral rhodium catalyst known as the **Davies catalyst**[350] [Rh$_2$(*S*-DOSP)$_2$, see the structure of the chiral DOSP ligand in the reaction scheme], the *trans*-dihydrobenzofuran derivative **406** was obtained in 63% yield (the yield is for two steps: conversion of the ester precursor to the diazo compound followed by cyclization by reaction with the rhodium complex). Note that Doyle and co-workers[351] developed several important rhodium catalysts, including the **Doyle catalysts**,[352] that are widely used in cyclopropanation and C–H insertion reactions. In a different example with a different catalyst, taken from a synthesis of (−)-tetrodotoxin,[353] Du Bois and Hinman used Rh$_2$(HNCOCPh$_3$)$_4$ to convert **407** to **408**.

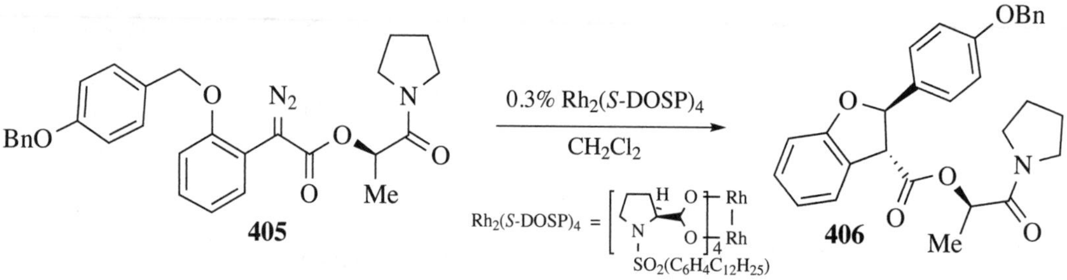

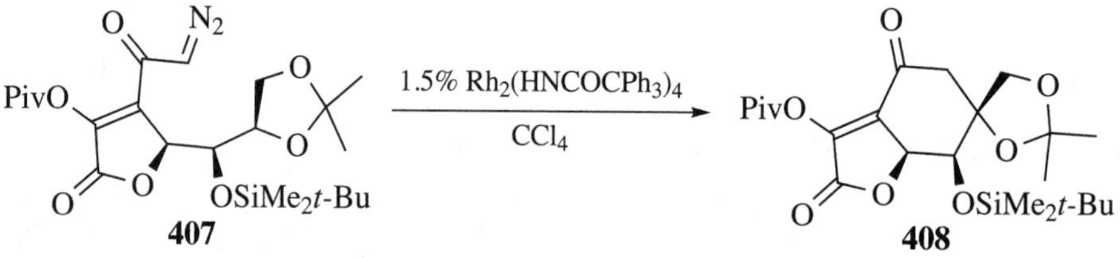

The carbene C–H insertion disconnection is

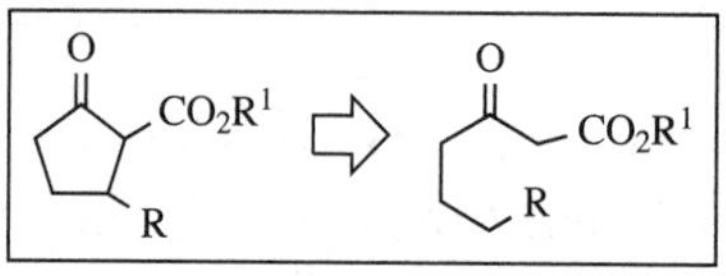

There is an interesting variation of this reaction that involves intramolecular N–H insertion of α-diazocarbonyls. When diazoketone **409** was treated with cupric bis(acetylacetonate), for example, a 74% yield of azetidinone (**410**) was obtained.[354] It was shown that competitive

350. (a) Davies, H.M.L.; Hansen, T.; Churchill, M.R. *J. Am. Chem. Soc.* **2000**, *122*, 3063; (b) Davies, H.M.L.; Hansen, T. *J. Am. Chem. Soc.* **1997**, *119*, 9075. Also see (c) Davies, H.M.L. *Aldrichimica Acta* **1997**, *30*, 107.

351. (a) Doyle, M.P. *Pure & Appl. Chem.* **1998**, *70* 1123 (b) Doyle, M.P.; Protopopova, M.N. *Tetrahedron,* **1998**, *54* 7919 (c) Martin, S.F.; Spaller, M.R.; Liras, L.; Hartman, B. *J. Am. Chem. Soc.* **1994**, *116* 4493 (d) Doyle, M.P.; Kalinin, A.V. *J. Org. Chem.* **1996**, *61* 2179 (e) Doyle, M.P.; Dyatkin, A.B.; Roos, G.H.P.; Cañas, F., Pierson, D.A.; van Basten, A.; Mueller, P.; Polleux, P. *J. Am. Chem. Soc.* **1994**, *116* 4507. Also see Doyle, M.P.; Hu, W.; Valenzuela, M.V. *J. Org. Chem.* **2002**, *67*, 2954 for a synthesis of (S)-(+)-imperanene using these catalysts.

352. Colacot, T.J. *Proc. Indian Acad. Sci. (Chem. Sci.)* **2000**, *112*, 197.

353. Hinman, A.; Du Bois, J. *J. Am. Chem. Soc.* **2003**, *125*, 11510.

354. Wang, J.; Hou, Y.; Wu, P. *J. Chem. Soc. Perkin Trans. 1* **1999**, 2277.

intramolecular C–H insertion by the carbenoid was not a problem, illustrated by the presence of ten equivalents of styrene in the reaction. The styrene actually seemed to enhance the N–H insertion reaction.

13.9.E Carbenoids

This section involves a class of compounds called carbenoids, which are reactive intermediates that react similarly to a carbene, but carbenes are not formed. The most commonly used carbenoid

$CH_2I_2 \longrightarrow$

411 ZnI   **412** ZnI   **345**

is generated by reaction of diiodomethane and a Zn/Cu couple. When the carbenoid formed in this manner adds to alkenes, it is called the **Simmons-Smith reaction**.[355]

A simple example is the conversion of cyclohexene to bicyclo[4.1.0]heptane (norcarane, **345**). Initial reaction of diiodomethane with zinc gave an iodozinc compound (**411**, see sec. 9.8.B for a brief discussion of organozinc compounds), which reacted with the alkene to give **412**. Loss of zinc iodide (ZnI_2) gave the cyclopropane derivative (**345** in this case).[356] A one-step mechanism has also been proposed that involves **413**,[356a,357] and generates the cyclopropane derivative. The iodomethylene zinc intermediate is represented as **414**,[358] based in part on the structure of known organozinc compounds.[359]

413

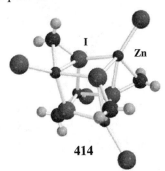

414

[Reprinted with permission from Simmons, H.E.; Cairns, T.L.; Vladuchick, S.A.; Hoiness, C.M. *Org. React.* **1973**, *20*, 1. Copyright © **1973** by John Wiley and Sons, Inc.]

355. (a) Simmons, H.E.; Smith, R.D. *J. Am. Chem. Soc.* **1958**, *80*, 5323; (b) *Idem. Ibid.* **1959**, *81*, 4256; (c) Denis, J.M.; Girard, J.M.; Conia, J.M. *Synthesis* **1972**, 549; (d) The Merck Index, 14th Ed. Merck & Co., Inc., Whitehouse Station, New Jersey, **2006**, p ONR-87; (e) Mundy, B.P.; Ellerd, M.G.; Favaloro Jr., F.G. *Name Reactions and Reagents in Organic Synthesis, 2nd Ed.* Wiley-Interscience, New Jersey, **2005**, pp. 600-601.
356. Sawada, S.; Inouye, Y. *Bull. Chem. Soc. Jpn.* **1969**, *42*, 2669.
357. (a) Wittig, G.; Schwarzenbach, K. *Annalen* **1961**, *650*, 1; (b) Simmons, H.E.; Blanchard, E.P.; Smith, R.D. *J. Am. Chem. Soc*, **1964**, *86*, 1347; (c) Wittig, G.; Wingler, F. *Chem. Ber.* **1964**, *97*, 2146; (d) Blanchard, E.-P.; Simmons, H.E. *J. Am. Chem. Soc.* **1964**, *86*, 1337.
358. Simmons, H.E.; Cairns, T.L.; Vladuchick, S.A.; Hoiness, C.M. *Org. React.* **1973**, *20*, 1.
359. Shearer, H.M.M.; Spencer, C.B. *J. Chem. Soc. Chem. Commun.* **1966**, 194.

As with carbenes, more highly substituted alkenes react somewhat faster than less substituted alkenes.[357d] Even highly substituted alkenes such as **415** gave an 87% yield of the Simmons-Smith product (**416**).[360] It is known that 1-substituted cyclic alkenes react faster than unsubstituted alkenes.[361]

$$\text{415} \xrightarrow[\text{Zn-Cu}]{\text{CH}_2\text{I}_2} \text{416}$$

415   **416**

1-Methylcyclohexene (relative rate 21.4) and methylenecyclohexane (relative rate 3.84) both react faster than cyclohexene (relative rate 1.0). 1,2-Dimethylcyclohexene, however, reacts slightly slower, with relative rate of 0.94. Cyclohexene (relative rate 1.0) reacts slower than cyclopentene (relative rate 1.60) or norbornene (relative rate 1.70).[362] In addition, 1-methylcyclopentene (relative rate 5.14) reacts faster than cyclopentene (relative rate 1.60).[362]

Preparation of the zinc is critical to good yields in the Simmons-Smith reaction. Both Zn–Cu and Zn-Ag couples have been used with both iodine and lithium.[362] The zinc must be activated or the yield of cyclopropane product is poor. A useful development is the use of ultrasound to activate the zinc. Treatment of

$$\text{417} \xrightarrow[\underset{\text{ether}\quad 67\%}{\text{Zn (ultrasound), CH}_2\text{I}_2}]{\overset{\text{Zn-Cu, CH}_2\text{I}_2\text{, ether}}{12\text{ h}\quad 12\%}} \text{418}$$

417   **418**

(−)-α-pinene (**417**) with the usual Simmons-Smith reagent gave only 12% of **418**.[363] When the zinc was activated by ultrasound, however, a 67% yield of **418** was obtained.[364]

An example of the Simmons-smith cyclopropanation is taken from a synthesis of cascarillic acid by Piva and Salim in which **419** was converted to a 90:10 mixture of cyclopropanes **420**:**421** in 72% yield.[365a] Electron-rich alkenes such as vinyl ethers and silyl enol ethers usually react faster than other alkenes.[365b-d]

$$\text{419} \xrightarrow[\text{ether, 1 d}]{\text{CH}_2\text{I}_2\text{, Zn/Cu}} \text{420} \text{—CO}_2\text{H} + \text{421} \text{—CO}_2\text{H}$$

419   **420**   **421**

There is a pronounced neighboring-group effect when a hydrogen-bonding oxygen is a substituent. The oxygen will "coordinate with the reagent, increasing the rate and control the stereochemistry of the addition."[366,361a,b] Reaction of **423**, for example, gave **425** via coordination of the oxygen in **424**. In a synthetic example, Casares and Maldonado used this

360. Koch, S.D.; Kliss, R.M.; Lopiekes, D.V.; Wineman, R.J. *J. Org. Chem.* **1961**, *26*, 3122.
361. (a) Reference 253a, p 299; (b) Rickborn, B.; Chan, J.H.H. *J. Org Chem.* **1967**, *32*, 3576; (c) Balenkova, E.S.; Kochnova, G.P.; Khromov, S.I. *Neftekhimiya* **1969**, *9*, 29 [*Chem. Abstr. 70*: 106025x, **1969**].
362. Rieke, R.D.; Li, P.T-J., Burns, T.P.; Uhm, S.T. *J. Org. Chem.* **1981**, *46*, 4323.
363. Filliatre, C.; Guéraud, C. *C.R. Acad. Sci. Paris, Ser. C* **1971**, *273*, 1186.
364. Repič, O.; Vogt, S. *Tetrahedron Lett.* **1982**, *23*, 2729.
365. (a) Salim, H.; Piva, O. *Tetrahedron Lett.* **2007**, *48*, 2059; (b) Ragauskas, A.J.; Stothers, J.B. *Can. J. Chem.* **1985**, *63*, 2969; (c) Hoyano, Y.; Patel, V.; Stothers, J.B. *Ibid.* **1980**, *58*, 2730; (d) Jurlina, J.L.; Patel, H.A.; Stothers, J.B. *Ibid.* **1984**, *62*, 1159.
366. (a) Reference 253a, p 95; (b) Winstein, S.; Sonnenberg, J. *J. Am. Chem. Soc.* **1961**, *83*, 3235; (c) Dauben, W.G.; Berezin, G.H. *Ibid.* **1963**, *85*, 468; (d) Chan, J.H.; Rickborn, B. *Ibid.* **1968**, *90*, 6406.

neighboring group effect to convert **426** to **427** in a synthesis of β-cuparenone.[367]

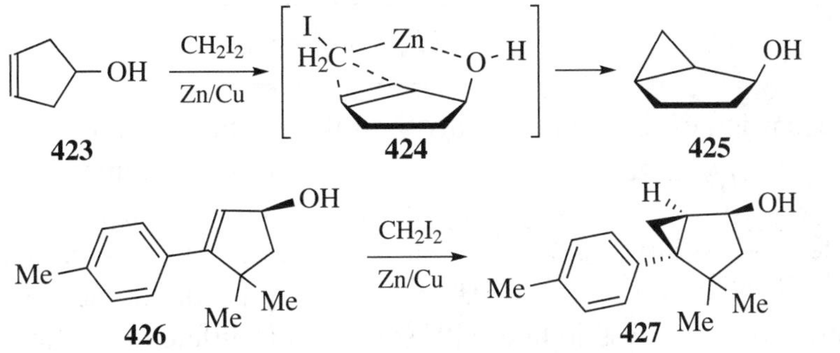

An alternative preparative method avoids the use of the Zn/Cu couple, using diethyl zinc (Et$_2$Zn).[368] An example is taken from the synthesis of solandelactones E and F, by White and co-workers.[369] In this work, **428** was treated with diethylzinc and diiodomethane to give a 97% yield of **429**.

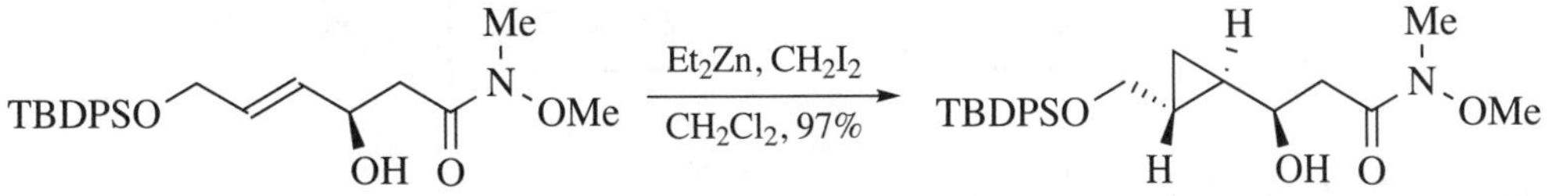

Asymmetric induction is possible in the Simmons-Smith reaction if a suitable auxiliary is used, or if the chirality is built in via a chiral template precursor (sec. 10.9). When **430** was prepared from diethyl tartrate, subsequent treatment with diethylzinc gave a 99% yield of **431**.[370] Subsequent hydrolysis gave **432** (74% yield and 90% ee), which allowed Yamamoto to complete an asymmetric synthesis of 5,6-methanoleukotriene A$_4$. Imai et al. has also reported a chiral boronic acid auxiliary that leads to good enantioselectivity in the Simmons-Smith reaction.[371] The ability to produce the highly useful cyclopropane derivatives with good-to-excellent enantiomeric purity greatly enhances the utility of an already classical reaction.

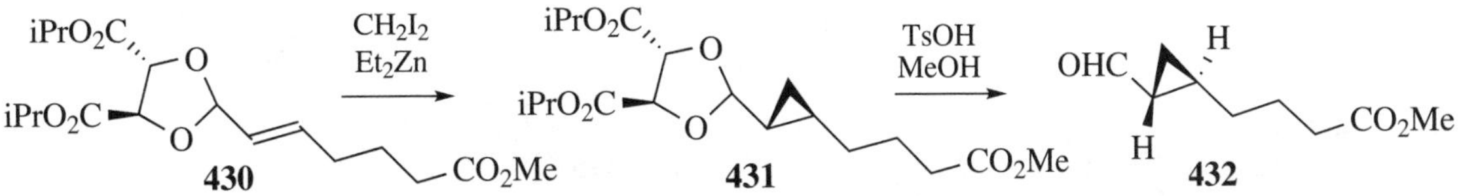

367.   (a) Casares, A.; Maldonado, L.A. *Synth. Commun.* **1976**, *6*, 11; also see (b) Ando, M.; Sayama, S.; Takase, K. *Chem. Lett.* **1979**, 191.
368.   Furukawa, J.; Kawabata, N.; Nishimura, J. *Tetrahedron* **1968**, *24*, 53.
369.   White, J.D.; Martin, W. H.C.; Lincoln, C.; Yang, J. *Org. Lett.* **2007**, *9*, 3481.
370.   (a) Arai, I.; Mori, A.; Yamamoto, H. *J. Am. Chem. Soc.* **1985**, *107*, 8254; (b) Mori, A.; Arai, I.; Yamamoto, H. *Tetrahedron,* **1986**, *42*, 6447; (c) Mash, E.A.; Nelson, K.A. *Ibid.* **1987**, *43*, 679; (d) Mash, E.A.; Nelson, K.A. *Tetrahedron Lett.* **1986**, *27*, 1441.
371.   Imai, T.; Mineta, H.; Nishida, S. *J. Org. Chem.* **1990**, *55*, 4986.

The Simmons-Smith disconnection is

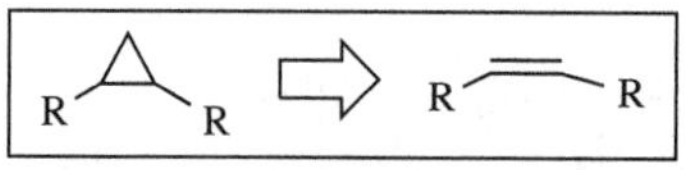

An interesting extension of reactivity that involves the Simmons-Smith reagent chain extends carbonyl compounds, as reported by Zercher. In a simple example, keto-ester **433** was treated with diethylzinc and diiodomethane followed by

iodine.[372] Destruction of excess iodine with sodium thiosulfate, and elimination with DBU (Sec 2.9.A) gave an 86% yield of **434**. This sequence is a carbene-mediated chain extension.

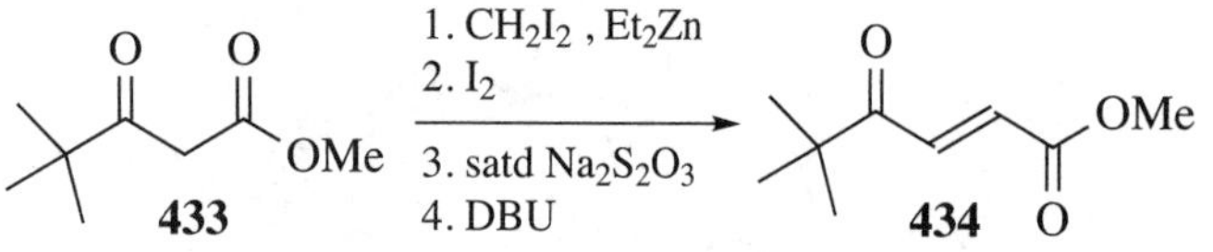

The carbene-mediated chain extension disconnection is

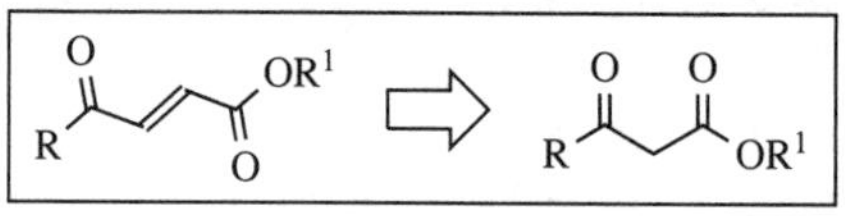

13.10 METATHESIS REACTIONS

In the presence of transition metal catalysts, two alkenes such as 2,3-dimethyl-2-butene and 2,3-diethyl-3-hexene react to form the original alkenes along with 3-ethyl-2-methyl-2-pentene. This reaction is an example of **metathesis** or **alkene metathesis** (or **olefin metathesis**),[373] and an alkene[374] is produced from an alkene or diene precursor. As implied in the reaction shown, a mixture of alkenes can be formed, but the reaction can be controlled in many cases, as will be seen below. The mechanism of this reaction is probably a chain mechanism that involves initial formation of a metal-carbene complex.[375] The metal-carbene reacts with each alkene to form

372. Ronsheim, M.D. Zercher, C.K. *J. Org. Chem.* **2003**, *68*, 4535.
373. (a) Drăguțn, V.; Balaban, A.T.; Dimonie, M. *Olefin Metathesis and Ring-Opening Polymerization of Cycloolefins*, Wiley, NY, **1985**; (b) Ivin, K.J. *Olefin Metathesis*, Academic Press, NY, **1983**; (c) Schrock, R.R. *J. Organomet. Chem.* **1986**, *300*, 249; (d) Grubbs, R.H. in Wilkinson *Comprehensive Organometallic Chemistry*, Vol. 8, Pergamon, Elmsford, NY, **1982**, pp 499; (e) Grubbs, R.H. *Prog. Inorg. Chem.* **1978**, *24*, 1; (f) Calderon, N. *Acc. Chem. Res.* **1972**, *5*, 127; (g) Haines, R.J.; Leigh, G.J. *Chem. Soc. Rev.* **1975**, *4*, 155; (h) Bailey, G.C. *Catal. Rev.* **1969**, *3*, 37; (i) Grubbs, R.H. *Tetrahedron* **2004**, *60*, 7117; (j) The Merck Index, 14th Ed. Merck & Co., Inc., Whitehouse Station, New Jersey, **2006**, p ONR-67.
374. For transformations of this type, see reference 213, p 401.
375. (a) Collman, J.C.; Hegedus, L.S.; Norton, J.R.; Finke, R.G. *Principles and Applications of Organotransition Metal Chemistry*, 2nd Ed. University Science Books, Mill Valley, CA, **1987**, pp 459-520; (b) Lindner, E. *Adv. Heterocyclic Chem.* **1986**, *39*, 237.

a four-membered ring containing a metal (**437** and **438** from **436**; **439** and **440** from **435**).[376] Decomposition of the four-membered ring generates the two possible alkenes (2,3-dimethyl-2-butene and 3-ethyl-2-methyl-2-pentene from **440**, for example), and regenerates a metal carbene complex. The net result is formation of new alkenes. A single unsymmetrical alkene gives a mixture of itself and two other alkenes, and a mixture of two alkenes gives different alkenes, but the number depends on the symmetry of the reactants. The reaction can be done intermolecularly between two alkenes, or intramolecularly using dienes. Both intermolecular and intramolecular reactions of dienes are known.[377] Simple alkenes usually give a statistical mixture of products,[378] but thermodynamic stability can influence the relative proportion of products. The equilibrium can be shifted if one of the products is volatile, such as ethylene. Terminal and internal alkenes have been used and the facility of reaction changes with branching according to the order $CH_2=>RCH_2CH=>R_2CHCH=>R_2C=$.[379]

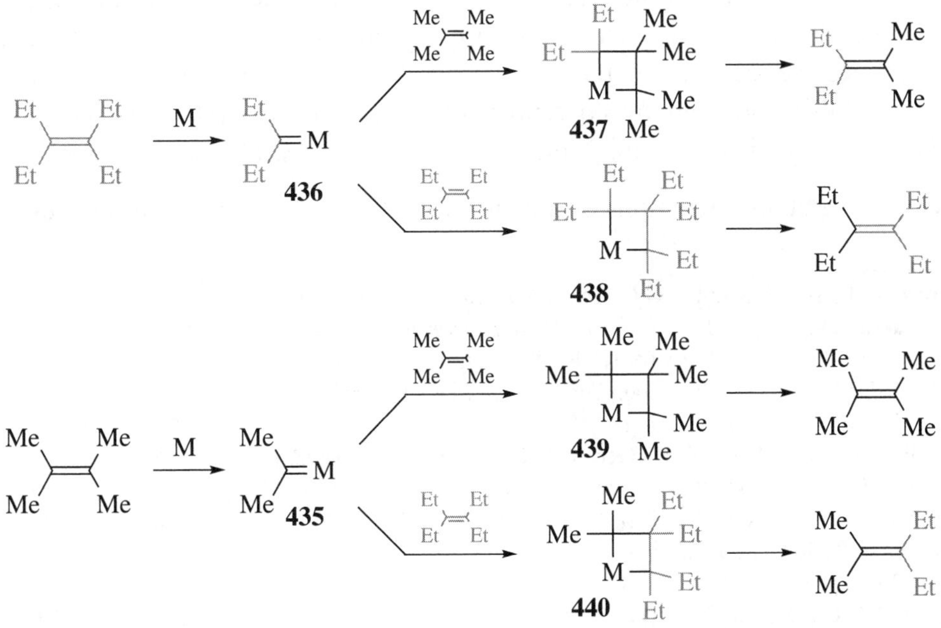

376. For reviews of the mechanism, see (a) Grubbs, R.H. *Prog. Inorg. Chem.* **1978**, *24*, 1; (b) Calderon, N.; Ofstead, E.A.; Judy, W.A. *Angew. Chem. Int. Ed.* **1976**, *15*, 401; (c) McLain, S.J.; Wood, C.D.; Schrock, R.R. *J. Am. Chem. Soc.* **1977**, *99*, 3519; (d) Casey, C.P.; Polichnowski, S.W. *J. Am. Chem. Soc.* **1977**, *99*, 6097; (e) Mango, F.D. *J. Am. Chem. Soc.* **1977**, *99*, 6117; (f) Stevens, A.E.; Beauchamp, J.L. *J. Am. Chem. Soc.* **1979**, *101*, 6449; (g) Lee, J.B.; Ott, K.C.; Grubbs, R.H. *J. Am. Chem. Soc.* **1982**, *104*, 7491; (h) Kress, J.; Osborn, J.A.; Greene, R.M.E.; Ivin, K.J.; Rooney, J.J. *J. Am. Chem. Soc.* **1987**, *109*, 899; (g) Feldman, J.; Davis, W.M.; Schrock, R.R. *Organometallics* **1989**, *8*, 2266.

377. (a) Kroll, W.R.; Doyle, G. *Chem. Commun.* **1971**, 839; (b) Grubbs, R.H.; Miller, S.J.; Fu, G.C. *Acc. Chem. Res.* **1995**, *28*, 446.

378. Calderon, N.; Ofstead, E.A.; Ward, J.P.; Judy, W.A.; Scott, K.W. *J. Am. Chem. Soc.* **1968**, *90*, 4133.

379. (a) McGinnis, J.; Katz, T.J.; Hurwitz, S. *J. Am. Chem. Soc.* **1976**, *98*, 605; (b) Casey, C.J.; Tuinstra, H.E.; Saeman, M.C. *J. Am. Chem. Soc.* **1976**, *98*, 608.

Metathesis reactions[380] are well known, using both homogeneous[381] and heterogeneous[382] catalysts. Although other catalysts have been used,[383] ruthenium complexes[384] are the most common. With the development of the **Grubbs I catalyst (441)**[385] and the **Schrock catalyst (442)**,[386] the reaction has taken on increased synthetic utility. The most common application is formation of cyclic alkenes [ring-closing metathesis (RCM) reactions].

The Grubbs catalyst is stable to Lewis acid conditions.[387] Catalysts have been developed that allow the reaction to take place in aqueous and alcohol solution,[388] and polymer-bound catalysts have been used.[389] Microwave-assisted ring-closing metastasis reactions have also been reported.[390] Note that there are some problems in handling the Grubbs' catalyst, in that prolonged exposure to air and moisture deactivates the complex. Therefore, the catalyst is usually stored under inert atmosphere. Attaching the active ruthenium complex to a polymer support improves the stability of the catalyst while retaining its reactivity in ring-closing metathesis.[391] Taber and Frankowski reported that Grubbs catalyst dispersed in paraffin is easily handled, and retains its activity indefinitely with no special storage precautions.[392]

Intramolecular metathesis reactions dominate synthetic applications, and alkenes containing

380. Grubbs, R.H.; Miller, S.J.; Fu, G.C. *Acc. Chem. Res.* **1995**, *28*, 446.
381. (a) Calderon, N.; Chen, H.Y.; Scott, K.W. *Tetrahedron Lett.* **1967**, 3327; (b) Hughes, W.B. *Organomet. Chem. Synth.* **1972**, *1*, 362; (c) Toreki, R.; Schrock, R.R. *J. Am. Chem. Soc.* **1990**, *112*, 2448.
382. Banks, R.L.; Bailey, G.C. *Ind. Eng. Chem. Prod. Res. Dev.* **1964**, *3*, 170.
383. Banks, R.L. *Fortschr. Chem. Forsch.* **1972**, *25*, 41.
384. (a) Gilbertson, S.R.; Hoge, G.S.; Genov, D.G. *J. Org. Chem.* **1998**, *63*, 10077; (b) Maier, M.E.; Bugl, M. *Synlett*, **1998**, 1390; (c) Stefinovic, M.; Snieckus, V. *J. Org. Chem.* **1998**, *63*, 2808.
385. (a) Schwab, P.; Grubbs, R.H.; Ziller, J.W. *J. Am. Chem. Soc.* **1996**, *118*, 100; (b) Mundy, B.P.; Ellerd, M.G.; Favaloro Jr., F.G. *Name Reactions and Reagents in Organic Synthesis, 2nd Ed.* Wiley-Interscience, New Jersey, **2005**, p. 783.
386. (a) Bazan, G.C.; Oskam, J.H.; Cho, H.-N.; Park, L.Y.; Schrock, R.R. *J. Am. Chem. Soc.* **1991**, *113*, 6899, and references cited therein; (b) Mundy, B.P.; Ellerd, M.G.; Favaloro Jr., F.G. *Name Reactions and Reagents in Organic Synthesis, 2nd Ed.* Wiley-Interscience, New Jersey, **2005**, p. 845.
387. Bentz, D.; Laschat, S. *Synthesis* **2000**, 1766.
388. (a) Kirkland, T.A.; Lynn, D.M.; Grubbs, R.H. *J. Org. Chem.* **1998**, *63*, 9904; (b) Lynn, D.M.; Mohr, B.; Grubbs, R.H.; Henling, L.M.; (c) Day, M.W. *J. Am. Chem. Soc.* **2000**, *122*, 6601.
389. (a) Yao, Q. *Angew. Chem. Int. Ed.* **2000**, *39*, 3896; (b) Schürer, S.C.; Gessler, S.; Buschmann, N.; Blechert, S. *Angew. Chem. Int. Ed.* **2000**, *39*, 3898.
390. (a) Mayo, K.G.; Nearhoof, E.H.; Kiddle, J.J. *Org. Lett.* **2002**, *4*, 1567; (b) Garbacia, S.; Desai, B.; Lavastre, O.; Kappe, C.O. *J. Org. Chem.* **2003**, *68*, 9136.
391. (a) Dowden, J.; Savovic, J. *Chem. Commun.* **2001**, 37; (b) Jafarpour, L.; Nolan, S.P. *Org. Lett.* **2000**, *2*, 4075; (c) Ahmed, M.; Barrett, A.G.M.; Braddock, D.C.; Cramp, S.M.; Procopiou, P.A. *Tetrahedron Lett.* **1999**, *40*, 8657. (d) Ahmed, M.; Arnauld, T.; Barrett, A.M.G.; Braddock, D.C.; Procopiou, P.A. *Synlett* **2000**, 1007.
392. Taber, D.F.; Frankowski, K.J. *J. Org. Chem.* **2003**, *68*, 6047.

a $=CH_2$ unit are commonly used so that one product is ethylene. Ethylene escapes from the reaction medium and drives the equilibrium toward the cyclized product. A representative example is taken from the Kerr and Young synthesis[393] of (+)-nakadomarin A, in which diene **443** was treated with 0.4 equivalents of **441** to give a 66% yield of the cyclized product **444** as a 5:3 mixture of *E:Z* isomers, along with the volatile ethylene.

With some of the early catalysts, alkenes containing functional groups[394] did not work very well. This limitation prompted the development of alternative metathesis catalysts. In Leighton and co-worker's synthesis of dolabelide D,[395] for example, the highly functionalized substrate **445** gave a 31% yield of **447** (dolabelide D) using **446** (the so-called **second generation Grubbs catalyst**, or **Grubbs II catalyst**, where Mes = mesitoate and Cy = cyclohexyl). Compounds with nitrogen in the ring being formed have also been prepared.[396] An example is taken from Nishida and co-worker's synthesis of nakadomarin A,[397] where the two vinyl units **448** were cyclized to form the eight-membered lactam ring in **449** using **446**[398] Catalyst **446** is more stable and more active than **441**. There are now many synthetic examples that employ **446** rather than **441**. One of the more attractive features of this methodology is the ability to form macrocyclic compounds, as seen in the synthesis of **447**. The Schrock catalyst (**442**) also widely used. An example is taken from Nelson and co-worker's synthesis of (−)-laulimalide,[399] in which **450** was converted to **451** in excellent yield. Other ruthenium catalysts are known, and in a variety of applications the metathesis reaction has been shown to proceed with good stereoselectivity.[400]

393. Young, I.S.; Kerr, M.A. *J. Am. Chem. Soc. 2007, 129,* 1465.
394. (a) Mol, J.C. *Chemtech 1983,* 250-255; (b) Bosma, R.H.A.; van den Aardweg, G.C.N.; Mol, J.C. *J. Organomet. Chem. 1983, 255,* 159; (c) *Idem Ibid. 1985, 280,* 115;(d) Xiaoding, X.; Mol, J.C. *J. Chem. Soc. Chem. Commun. 1985,* 631; (e) Crisp, C.T.; Collis, M.P. *Aust. J. Chem. 1988, 41,* 935.
395. Park, P.K.; O'Malley, S.J.; Schmidt, D.R.; Leighton, J.L. *J. Am. Chem. Soc. 2006, 128,* 2796.
396. Phillips, A.J.; Abell, A.D. *Aldrichimica Acta 1999, 32,* 75.
397. Nagata, T.; Nakagawa, M.; Nishida, A. *J. Am. Chem. Soc. 2003, 125,* 7484.
398. (a) Scholl, M.; Ding, S.; Lee, C.W.; Grubbs, R.H. *Org. Lett. 1999, 1,* 953. Also see (b) Love, J.A.; Sanford, M.S.; Day, M.W.; Grubbs, R.H. *J. Am. Chem. Soc. 2003, 125,* 10103.
399. Nelson, S.G.; Cheung, W.S.; Kassick, A.J.; Hilfiker, M.A. *J. Am. Chem. Soc. 2002, 124,* 13654.
400. Lee, C.W.; Grubbs, R.H. *Org. Lett. 2000, 2,* 2145.

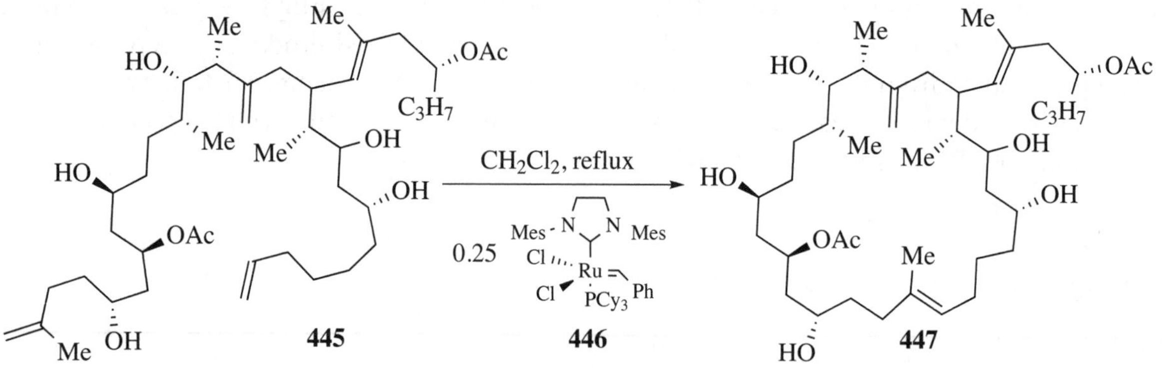

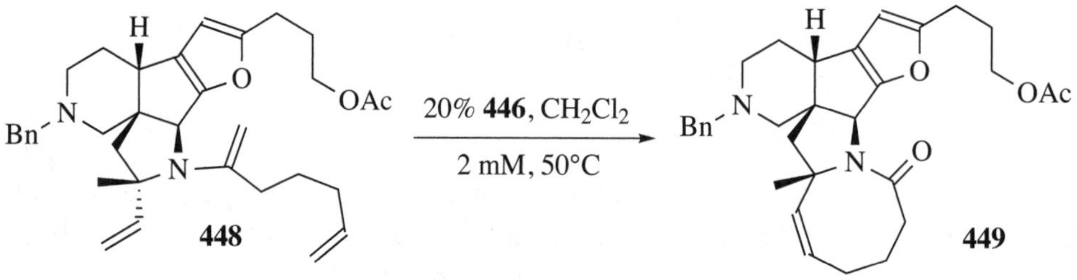

As a synthetic method, cross metathesis reactions[401] have been limited by the lack of predictability of product selectively and stereoselectivity. Grubbs and co-workers have advanced a general model that is useful for predicting selectivity.[402] As stated by Grubbs, a general ranking of alkene reactivity is achieved by categorizing alkenes by their relative ability to undergo homodimerization via cross metathesis, and the susceptibility of their homodimers toward secondary metathesis reactions.[402] Suppressing the rate of homodimerization of one component and controlling the rate of secondary metathesis on the desired cross product gives product selectivity.[402] Grubbs provided a table (see Table 13.8) to assist ranking alkenes as to their reactivity.[402]

401.   Connon, S.J.; Blechert, S. *Angew. Chem. Int. Ed.* **2003**, *42*, 1900.
402.   Chatterjee, A.K.; Choi, T.-L.; Sanders, D.P.; Grubbs, R.H. *J. Am. Chem. Soc.* **2003**, *125*, 11360.

Table 13.8 Alkene categories for selective cross metathesis.[402]

Metathesis catalyst ⟹ Alkene Type (⟱)	446	441	442
Type I (fast homodimerization)	terminal alkenes, 1° allylic alcohols, esters, allyl boronate esters, allyl halides, styrenes (no large ortho substitutents), allyl silanes, allyl phosphontes, allyl sulfides, allyl phosphine oxides, protected allyl amines	terminal alkenes, allyl silanes, 1° allylic alcohols, ethers, esters, allyl boronate esters, allyl halides	terminal alkenes, allyl silanes
Type II (slow homodimerization)	styrenes (large ortho substituents), acylates, acrylamides, acrylic acid, acrolein, vinyl ketones, unprotected, 3° allylic alcohols, vinyl epoxides, 2° allylic alcohols, perfluorinated alkanes, alkenes	styrene, 2° allylic alcohols, vinyl dioxolanes, vinyl boronates	styrene, allyl stannanes
Type III (no homodimerization)	1,1-disubstituted alkenes, non-bulky trisubstituted alkenes, vinyl phosphonates, phenyl vinyl sulfone, 4° allylic carbons (all alkyl substitutents, 3° allylic alcohols (protected)	vinyl siloxanes	3° allyl amines, acrylonitrile
Type IV (spectators to cross metathesis)	vinyl nitro alkenes trisubstituted allylic alcohols (protected)	1,1-disubstituted alkenes, disubstituted α,β-unsaturated carbonyls, 4° allylic carbon alkenes, perfluorinated alkanes, alkenes, 3° allyl amines (protected)	1,1-disubstituted alkenes

[Reprinted with permission from Chatterjee, A.K.; Choi, T.-L.; Sanders, D.P.; Grubbs, R.H. *J. Am. Chem. Soc.* *2003*, *125*, 11360. Copyright © *2003* American Chemical Society.]

Metathesis with alkynes is also quite useful in synthesis,[403] particularly for internal alkynes[404] although terminal alkynes are not good partners in this reaction.[405] Internal metathesis reactions with alkynes are known,[406] including the conversion of **452** to **453** (in 73% yield) in Fürstner's synthesis of prostaglandin E$_2$-1,15-lactone.[407] Note the use of a molybdenum metathesis catalyst for this reaction. Diynes also react with alkynes in an intermolecular reaction to form aromatic rings. An example is the conversion of **454** to a 6:1 mixture of **455/456** in 82% yield.[408a] A similar, palladium-catalyzed cycloaromatization is also known.[409] Shmidt and co-

403. Bunz, U.H.F.; Kloppenburg, L. *Angew. Chem. Int. Ed.* *1999*, *38*, 478.
404. (a) Pennella, F.; Banks, R.L.; Bailey, G.C. *Chem. Commun.* *1968*, 1548; (b) Villemin, D.; Cadiot, P. *Tetrahedron Lett.* *1982*, *23*, 5139; (c) McCullough, L.G.; Schrock, R.R. *J. Am. Chem. Soc.* *1984*, *106*, 4067; (d) Tamao, K.; Kobayashi, K.; Ito, Y. *Synlett* *1992*, 539.
405. McCullough, L.G.; Listemann, M.L.; Schrock, R.R.; Churchill, M.R.; Ziller, J.W. *J. Am. Chem. Soc.* *1983*, *105*, 6729.
406. (a) Trost, B.M.; Trost, M.K. *J. Am. Chem. Soc.* *1991*, *113*, 1850; (b) Gilbertson, S.R.; Hoge, G.S. *Tetrahedron Lett.* *1998*, *39*, 2075.
407. Fürstner, A.; Grela, K. *Angew. Chem. Int. Ed.* *2000*, *39*, 1234.
408. (a) Witulski, B.; Stengel, T.; Fernández-Hernández, J.M. *Chem. Commun.* *2000*, 1965. Also see (b) Yamamoto, Y.; Ogawa, R.; Itoh, K. *Chem. Commun.* *2000*, 549.
409. Yamamoto, Y.; Nagata, A.; Itoh, K. *Tetrahedron Lett.* *1999*, *40*, 5035.

workers reported cautionary note,[410] in that difficult-to-cyclize substrates can have a radical cyclization process (see sec. 13.7) compete effectively. The first and second-generation (see below) catalysts exhibit different activities when competing with radical cyclization.

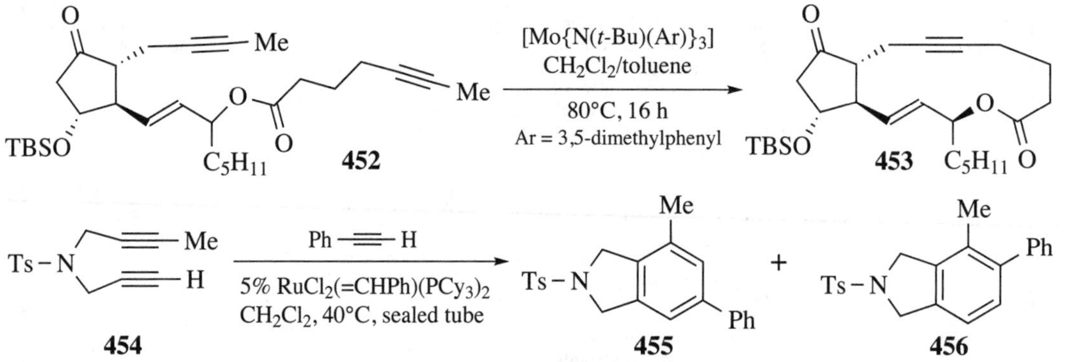

Grubbs and co-workers have reported another useful variation of the metathesis reaction, for the synthesis of macrocyclic compounds. When bis(conjugated) ketone **457** reacted with cyclooctene[411] in the presence of **446**, metathesis led to the ring expanded diketone **458** in 45% yield.

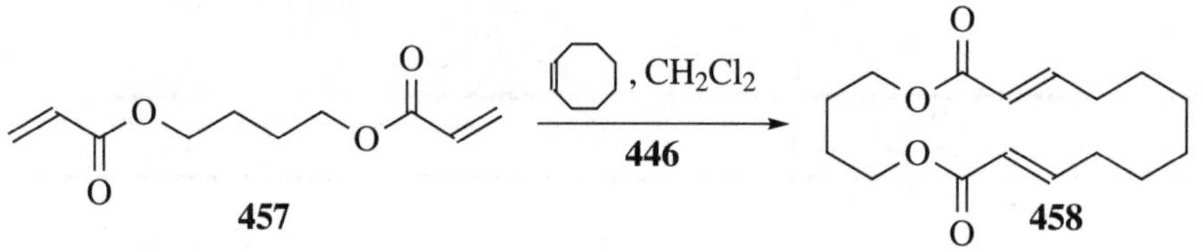

The ring-closing metathesis disconnections are

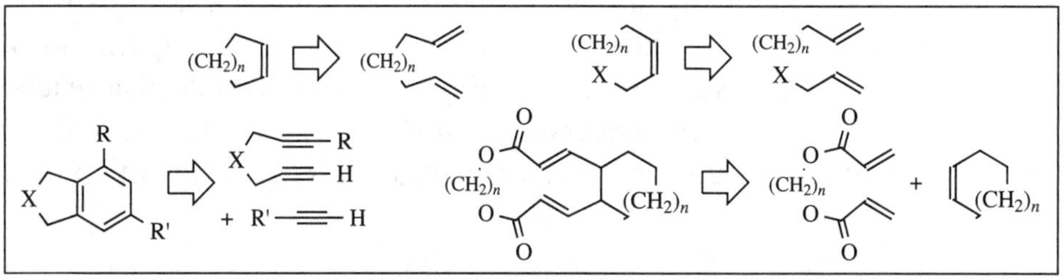

Another important application of metathesis reactions is to break C=C bond to form dienes, in what is called **ring-opening metathesis** (ROM).[412] Both the first and second-generation Grubbs' catalysts have been used, as well as other catalysts. This methodology has been

410. Schmidt, B.; Pohler, M.; Costisella, B. *J. Org. Chem.* **2004**, *69*, 1421.
411. Lee, C.W.; Choi, T.-L.; Grubbs, R.H. *J. Am. Chem. Soc.* **2002**, *124*, 3224.
412. (a) Bespalova, N.B.; Bovina, M.A.; Sergeeva, M.B.; Oppengeim, V.D.; Zaikin, V.G. *J. Mol. Catal.* **1994**, *90*, 21; (b) Zuercher, W.J.; Scholl, M.; Grubbs, R.H. *J. Org. Chem.* **1998**, *63*, 4291; (c) Randl, S.; Connon, S.J.; Blechert, S. *Chem. Commun.* **2001**, 1796; (d) Morgan, J.P.; Morrill, C.; Grubbs, R.H. *Org. Lett.* **2002**, *4*, 67.

used extensively in polymerization reactions called ring-opening metathesis polymerization (ROMP),[413] but only recently in total synthesis. An example of ring opening metathesis in this latter context is taken from a synthesis of F2-isoprostanes, Snapper and co-worker reacted **459** with **446** to give a 67% yield of **460**,[414] as a 1.4:1 mixture of *Z:E* isomers.

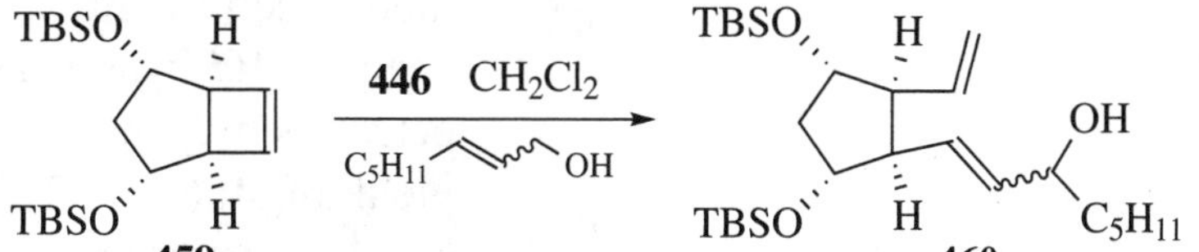

The ring-opening metathesis disconnection is:

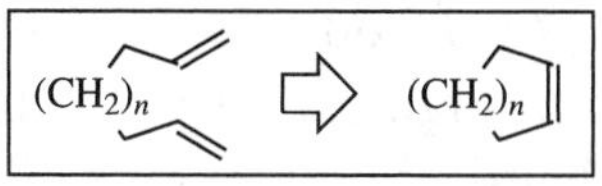

The synthetic importance of ring-closing and ring-opening metathesis reactions has led to the development of several new catalysts.[415] Air stability, reactivity, robust character and improved reactivity are the goals of such new catalysts. A sample of newer catalysts includes **461** by Hoveyda,[416] **462** by Grela,[417] **463** by Blechert,[418] and **464** by Moore.[419] The latter catalyst (**464**) was particularly useful for the metathesis homocoupling of alkynes. Hoveyda, Schrock and co-workers developed a polymer-supported, recyclable molybdenum catalyst.[420]

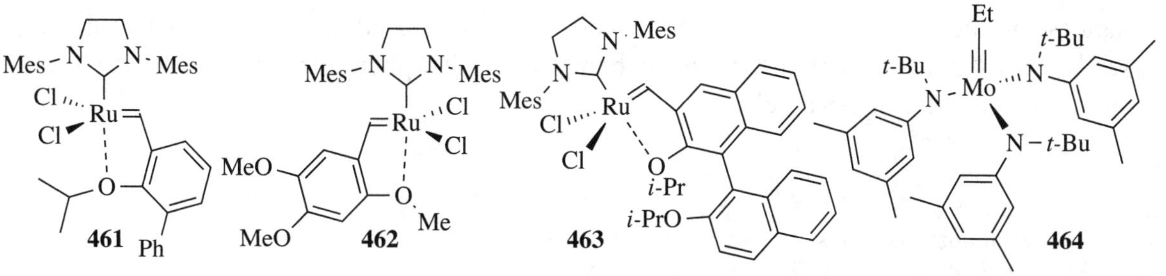

413. (a) Novak, B.M.; Grubbs, R.H. *Encycl. Polym. Sci. Eng. 1990, Suppl. Vol.* 420; (b) Amass, A.J. *New Methods Polym. Synth. 1991,* 76; Grubbs, R.H.; Khosravi, E. *Materials Science and Technology 1999, 20 (Synthesis of Polymers)* 65.
414. Schrader, T.O.; Snapper, M.L. *J. Am. Chem. Soc. 2002, 124,* 10998.
415. Schrock, R.R.; Hoveyda, A.H. *Angew. Chem. Int. Ed. 2003. 42,* 4592.
416. (a) Kingsbury, J.S.; Harrity, J.P.A.; Bonitatebus, P.J.; Hoveyda, A.H. *J. Am. Chem. Soc. 1999, 121,* 791; (b) Garber, S.B.; Kingsbury, J.S.; Gray, B.L.; Hoveyda, A.H. *Ibid, 2000, 122,* 8168.
417. (a) Grela, K.; Kim, M. *Eur. J. Org. Chem. 2003,* 963; (b) Grela, K.; Harutyunyan, S.; Michrowska, A. *Angew. Chem. Int. Ed. 2002, 41,* 4038.
418. (a) Wakamatsu, H.; Blechert, S. *Angew. Chem. Int. Ed. 2002, 41,* 794; (b) Conon, S.J.; Dunne, A.M.; Blechert, S. *Angew. Chem. Int. Ed. 2002, 41,* 3835.
419. Zhang, W.; Kraft, S.; Moore, J.S. *J. Am. Chem. Soc. 2004, 126,* 329.
420. Hutlzsch, K.C.; Jernelius, J.A.; Hoveyda, A.H.; Schrock, R.R. *Angew. Chem. Int. Ed. 2002, 41,* 589.

13.11 PAUSON-KHAND REACTION

An interesting cyclization reaction was reported that involved the reaction of dienes, diynes or ene-ynes with transition metals to form cyclopentenone derivatives in the presence of carbon monoxide.[421]   In a simple example, ene-yne **465** was heated with dicobalt octacarbonyl and CO to give a 68% yield of **466**.[422]   This transformation has become an important synthetic tool known as the **Pauson-Khand reaction**.[423]   The mechanism probably involves insertion of the alkene (or alkyne) into the transition metal bond, which is why it is presented in this section.   Formally, it is a [2+2+1]-cycloaddition, but the accepted mechanism is the one proposed by Magnus,[423] and shown in Figure 13.8.[424]   It has been stated that further study is required to firmly establish the mechanism[425] although Krafft's work supports the one reported by Magnus.[425]   Recently, Gimbert and co-workers showed that CO is lost from the Pauson-Khand complex prior to alkene coordination and insertion.[426]   Milet, Gimbert and co-worker's performed calculations that concluded the LUMO of the coordinated alkene plays a crucial role in alkene reactivity, by largely determining the degree of back-donation in the complex.[427]   Formation of the C–C bond is strongly influenced by the effectiveness of the LUMO overlap with the HOMO of the $Co_2(CO)_x$-complex, and such overlap is impacted by the degree of back-donation.[428]   Other metal complexes have been used,[428] and the addition of primary amines enhances the rate of the reaction.[429]   Aldehydes have been used as a source of CO in some cases.[430]   Gibson has reported the use of heptacarbonyl(triphenylphosphine)cobalt (0) as a robust, and stable Pauson-Khand catalyst.[431]

421. For transformations of this type, see reference 213, pp 1369, 1371-1372.
422. Magnus, P.; Principe, L.M. *Tetrahedron Lett.* **1985**, *26*, 4851.
423. (a) Khand, I.U.; Knox, G.R.; Pauson, P.L.; Watts, W.E.; Foreman, M.I. *J. Chem. Soc. Perkin Trans. 1* **1973**, 977; (b) Khand, I.U.; Pauson, P.L.; Habib, M.J. *J. Chem. Res. (S)* **1978**, 348; (c) Khand, I.U; Pauson, P.L. *J. Chem. Soc. Perkin Trans. 1* **1976**, 30; (d) The Merck Index, 14th Ed. Merck & Co., Inc., Whitehouse Station, New Jersey, **2006**, p ONR-69; (e) Mundy, B.P.; Ellerd, M.G.; Favaloro Jr., F.G. *Name Reactions and Reagents in Organic Synthesis, 2nd Ed.* Wiley-Interscience, New Jersey, **2005**, pp. 486-487.
424. For a review, see Brummond, K.M.; Kent, J.L. *Tetrahedron* **2000**, *56*, 3263.
425. Krafft, M.E. *Tetrahedron Lett.* **1988**, *29*, 999.
426. Gimbert, Y.; Lesage, D.; Milet, A.; fournier, F.; Greene, A.E.; Tabet, J.-C. *Org. Lett.* **2003**, *5*, 4073.
427. de Bruin, T.J.M.; Milet, A.; Greene, A.E.; Gimbert, Y. *J. Org. Chem.* **2004**, *69*, 1075.
428. (a) Koga, Y.; Kobayashi, T.; Narasaka, K. *Chem. Lett.* **1998**, 249; (b) Hoye, T.R.; Suriano, J.A. *J. Am. Chem. Soc.* **1993**, *115*, 1154.
429. Sugihara, T.; Yamada, M.; Ban, H.; Yamaguchi, M.; Kaneko, C. *Angew. Chem. Int. Ed.* **1997**, *36*, 2801.
430. Shibata, T.; Toshida, N.; Takagi, K. *Org. Lett.* **2002**, *4*, 1619.
431. Gibson, S.E.; Johnston, C.; Stevenazzi, A. *Tetrahedron* **2002**, *58*, 4937.

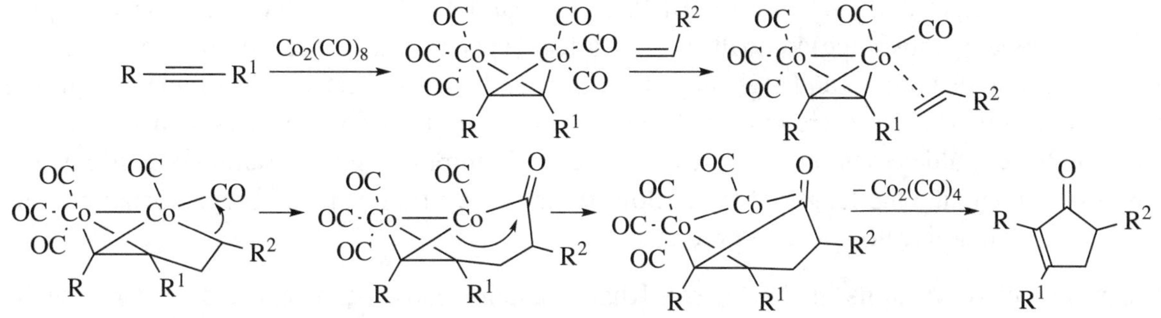

Figure 13.8. Magnus Mechanism for the Pauson-Khand Reaction. [Reprinted with the permission from Magnus, P.; Principe, L.M. *Tetrahedron Lett.* **1985**, *26*, 4851, Copyright 1985, with permission from Elsevier Science.]

This reaction has found its way into organic synthesis,[425] and is attractive because of its generality and selectivity. In a synthetic study aimed at nitiol that prepared two 1,22-dihydroxynitianes,[432] Dake and co-workers used a Pauson-Khand reaction of the alkene-alkyne **467** to give an 86% yield of **468** and **469** as a 6.2:1 mixture of diastereomers. Very sophisticated ring systems can be produced by this technique. An example is Cook and co-worker's synthesis of [5.5.5.5]-tetracyclic systems (pentalenes) in which **470** was treated with $Mo(CO)_6$ to give an 89% yield of **471**.[433]

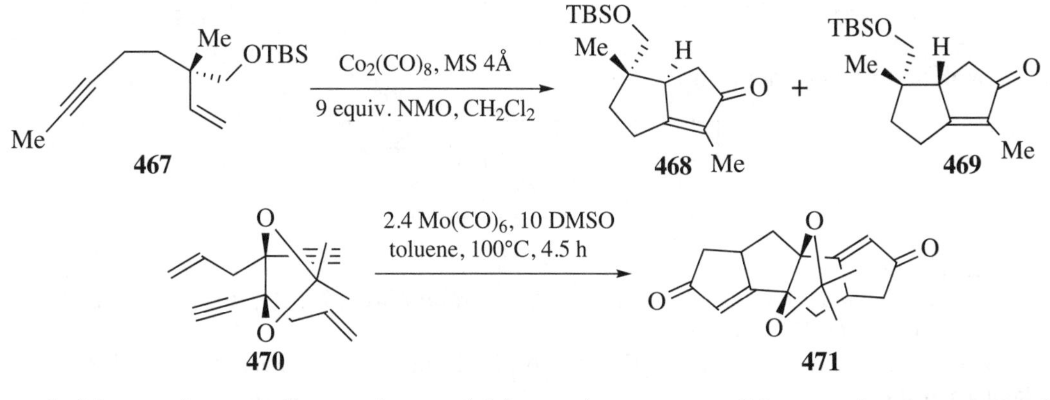

Although this reaction usually requires stoichiometric amounts of the metal carbonyl, Kraft and co-workers have shown that substoichiometric amounts can be used[434] (35-50% of $Co_2(CO)_8$, nitrogen atmosphere, in dimethoxyethane with three equivalents of cyclohexylamine. The use of amines to promote the reaction was mentioned above. Kraft and Boñaga also developed dodecacarbonyltetracobalt as a viable catalyst,[435] and catalytic Pauson-Khand reactions are

432.   Wilson, M.S.; Woo, J.C.S.; Dake, G.R. *J. Org. Chem.* **2006**, *71*, 4237.
433.   Cao, H.; Van Ornum, S.G.; Cook, J.M. *Tetrahedron Lett.* **2000**, *41*, 5313.
434.   Krafft, M.E.; Bañaga, L.V.R. *Synlett,* **2000**, 959.
435.   Krafft, M.E.; Boñaga, L.V.R. *Angew. Chem. Int. Ed.* **2000**, *39*, 3676.

now well established.[436]   A more robust Pauson-Khand catalyst has been developed, heptac arbonyl(triphenylphosphine)dicobalt(0).[437]   Molecular sieves have been used to promote the Pauson-Khand reaction,[437] and high intensity ultrasound is also effective.[438] Polymer-supported promoters of this reaction are known.[439]   In the presence of rhodium catalysts under solvent-free conditions, aldehydes serve as a CO source.[440]   Manipulating the reaction conditions, the catalyst, solvent and the length of time allowed for the reaction can give bicyclooctanones, as well as the usual bicyclooctenone products.[441]

Enantioselective versions of the Pauson-Khand reaction have been reported.[442]   Brucine-*N*-oxide promotes an asymmetric Pauson-Khand reaction, for example.[443]   An enantiospecific variation[444] has been reported by using a menthol chiral auxiliary and by replacing one of the cobalt centers with an isoelectronic molybdenum fragment. Chiral phosphine derivatives have also proven to be effective,[445] as have catalytic amounts of chiral aryl biphosphites.[446]   An aza-Pauson-Khand reaction was used in a synthesis of physostigmine by Mukai and co-workers,[447] the reaction of alkyne-diimide **472** and the cobalt catalyst gave a 55% yield of **473** with a pyrrolo[2,3-*b*]-indol-2-one unit.

The Pauson-Khand disconnection is

436.   (a) Gibson, S.E. ; Johnstone, C. ; Stevenazzi, A. *Tetrahedron* **2002**, *58*, 4937; (b) Gibson, S.E.; Stevenazzi, A., *Angew. Chem. Int. Ed.* **2003**, *42*, 1800.

437.   Pérez-Serrano, L.; Blanco-Urgoiti, J.; Casarrubios, L.; Domínguez, G.; Pérez-Castells, J. *J. Org. Chem.* **2000**, *65*, 3513.

438.   Ford, J.G.; Kerr, W.J.; Kirk, G.G.; Lindsay, D.M.; Middlemiss, D. *Synlett* **2000**, 1415.

439.   (a) Kerr, W.J.; Lindsay, D.M.; McLaughlin, M.; Pauson, P.L. *Chem. Commun.* **2000**, 1467; (b) Brown, D.S.; Campbell, E.; Kerr, W.J.; Lindsay, D.M.; Morrison, A.J.; Pike, K.G.; Watson, S.P. *Synlett*, **2000**, 1573.

440.   Shibata, T.; Tshida, N.; Takagi, K. *J. Org. Chem.* **2002**, *67*, 7446.

441.   Krafft, M.E.; Boňaga, L.V.R.; Wright, J.A.; Hirosawa, C. *J. Org. Chem.* **2002**, *67*, 1233.

442.   Ingate, S.T.; Marco-Contelles, J. *Org. Prep. Proceed. Int.* **1998**, *30*, 121.

443.   Kerr, W.J.; Lindsay, D.M.; Rankin, E.M.; Scott, J.S.; Watson, S.P. *Tetrahedron Lett.* **2000**, *41*, 3229.

444.   Fletcher, A.J.; Rutherford, D.T.; Christie, S.D.R. *Synlett* **2000**, 1040.

445.   Verdaguer, X.; Moyano, A.; Pericàs, M.A.; Riera, A.; Maestro, M.A.; Mahía, J. *J. Am. Chem. Soc.* **2000**, *122*, 10242.

446.   Sturla, S.J.; Buchwald, S.L. *J. Org. Chem.* **2002**, *67*, 3398.

447.   Mukai, C.; Yoshida, T.; Sorimachi, M.; Odani, A. *Org. Lett.* **2006**, *8*, 83.

13.12 CONCLUSION

This chapter has shown the great control and selectivity of modern radical reactions, especially intramolecular radical coupling reactions. Carbenes, especially those derived from diazoalkenes, have also played a prominent role in organic synthesis. The chemical reactions illustrated in this chapter are excellent additions to the list of nucleophilic, pericyclic, and electrophilic reactions presented previously. This chapter concludes the methodology for generating carbon-carbon bonds. With the functional group reactions in the first part of the book, all the tools for pursuing a synthesis are at hand.

The last chapter will present several actual syntheses done by students, not their teachers. Examination of these syntheses will show common mistakes made by beginning synthetic chemists. They will also show some of the correct approaches to presenting a synthesis.

1. Predict the major product of the following reaction and briefly discuss why it is formed and other possible products are not (or are minor products).

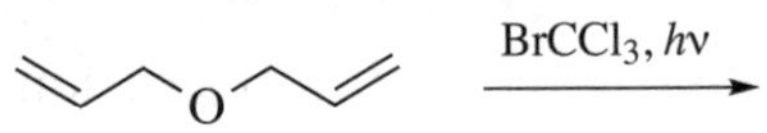

BrCCl$_3$, $h\nu$

2. Treatment of this molecule with mercury produced a radical. This radical product literally sat on the shelf for 25 years before the work was published. Give the structure of the radical and discuss its stability.

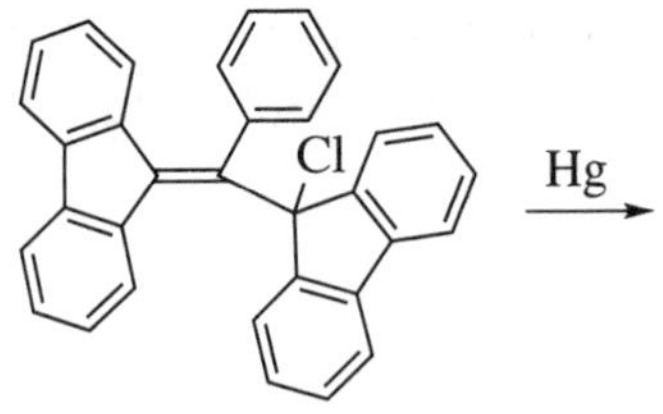

Hg

3. The structure of Fremy's salt is given. It is a rather stable free radical. Discuss the stability of this salt and also discuss its synthetic utility.

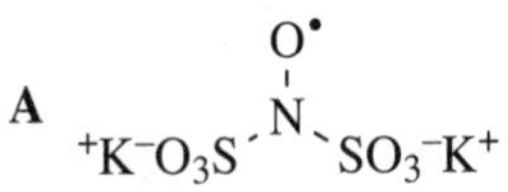

A

$^+K^-O_3S$ ^{N} $SO_3^-K^+$

Chem. Ind., **1953**, 244
Ann. Chim. Phys., **1845**, 15, 459

4. The two products of this reaction show stereochemical variation in the five-membered ring. Explain why the relative stereochemistry of the methyl group and two ring-juncture hydrogen atoms are the same in both products.

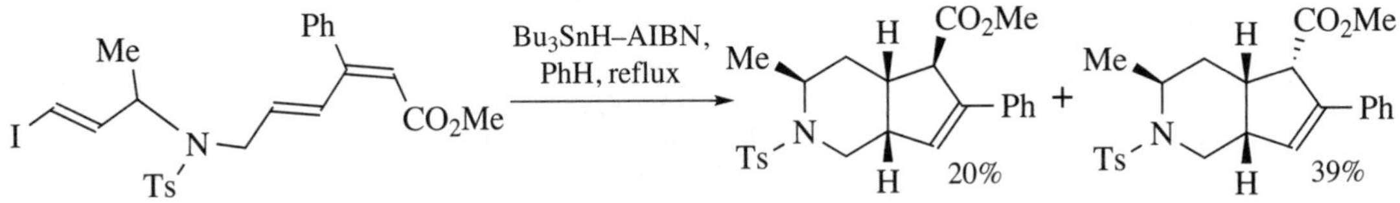

Bu$_3$SnH–AIBN,
PhH, reflux

20% + 39%

5. Explain the following results.

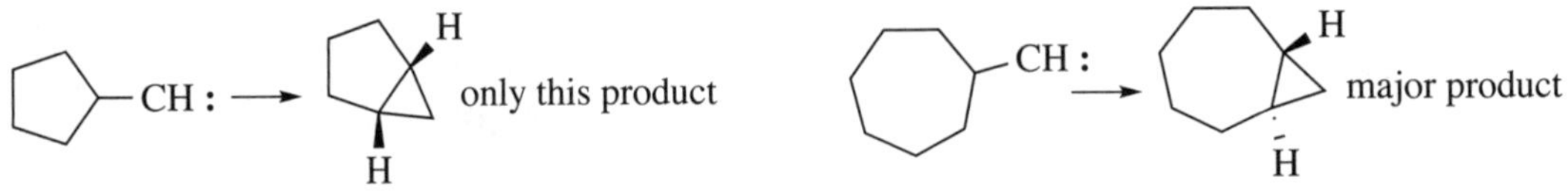

—CH : → only this product

—CH : → major product

6.   Explain the following rate data for the reaction shown.

R (relative rate): Ph- (17.0); *n*-BuO- (3.9); MeO$_2$C- (2.0); *n*-C$_6$H$_{13}$- (1.0); ClCH$_2$- (0.7).

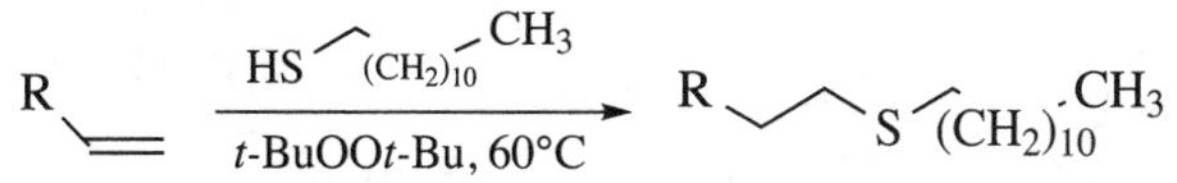

7.   Why is benzophenone added to the following reaction? Give a mechanistic rationale for this transformation.

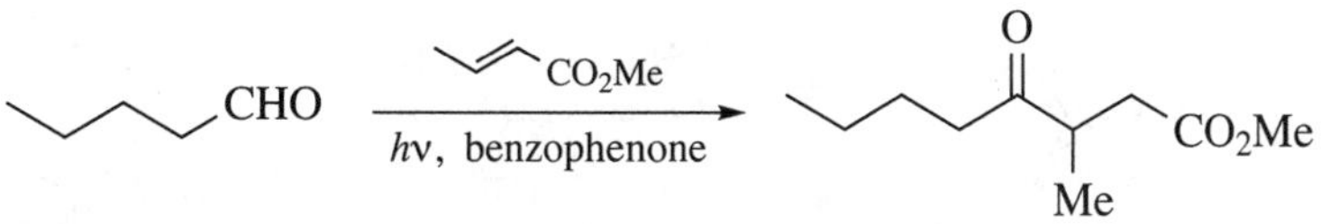

8.   Give the mechanism for this transformation.

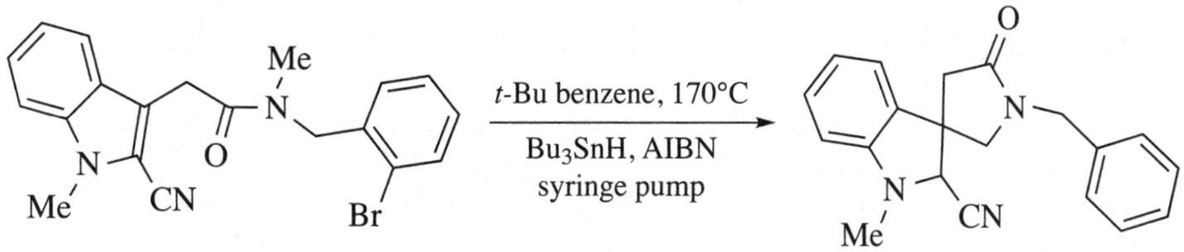

9.   Explain the formation of the indicated product from β-pinene and carbon tetrachloride.

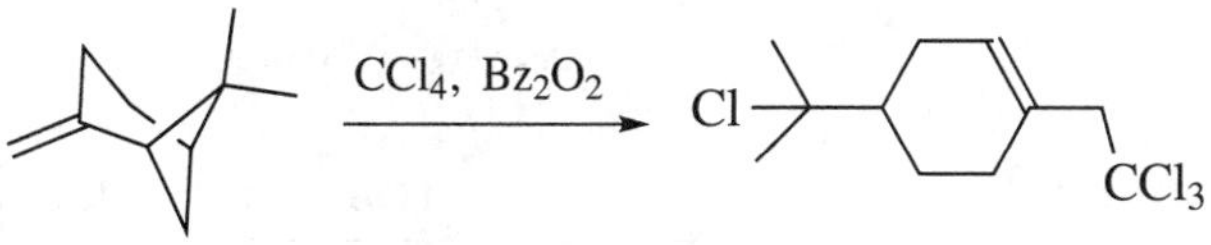

10.   Explain the following transformation:

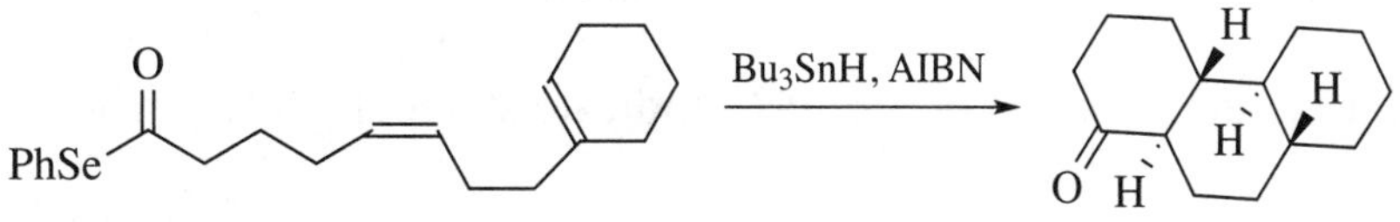

11.   Explain the following transformation:

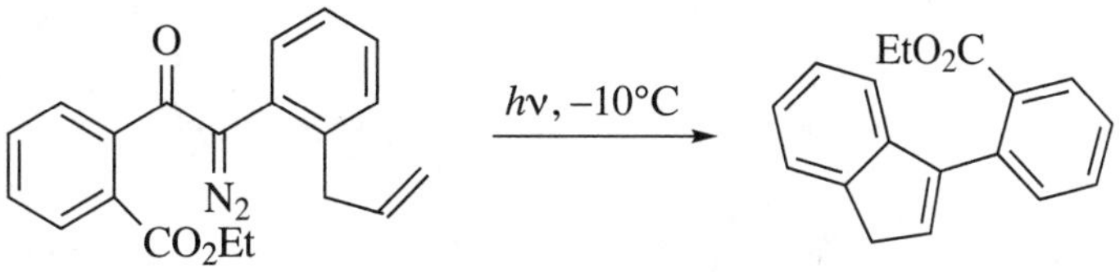

12. Two major cyclized products are produced in this reaction. Draw their structures and discuss the mechanism of their formation.

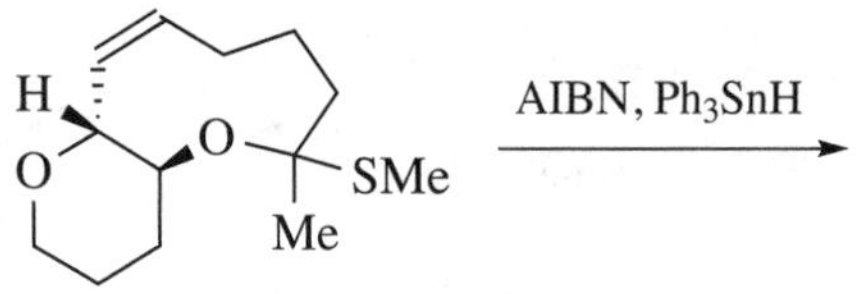
H
O
O
SMe
Me
AIBN, Ph3SnH

13. Explain the following transformation:

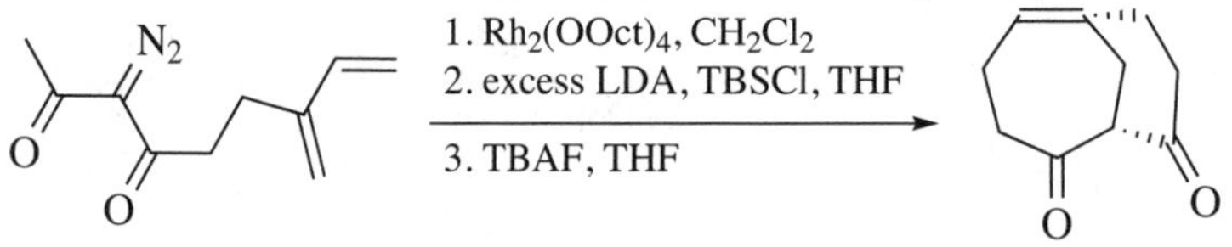
N2
O
O
1. Rh2(OOct)4, CH2Cl2
2. excess LDA, TBSCl, THF
3. TBAF, THF
O
O

14. Three products are formed in this reaction, two minor and one major. Show all three products and indicate which one is the major product. Explain your choice.

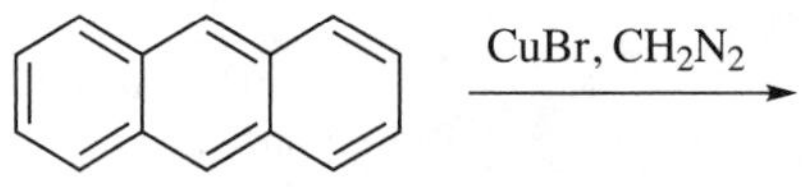
CuBr, CH2N2

15. Explain the following transformation.

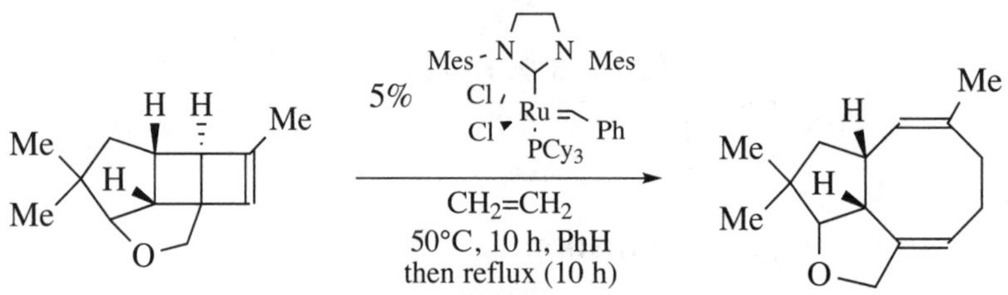
H H
Me
Me
H
Me
O
5%
Mes
N
N
Mes
Cl
Ru
Ph
Cl
PCy3
CH2=CH2
50°C, 10 h, PhH
then reflux (10 h)
Me
H
Me
H
Me
O

16. In each case show the major product. Indicate stereochemistry where appropriate.

(a) 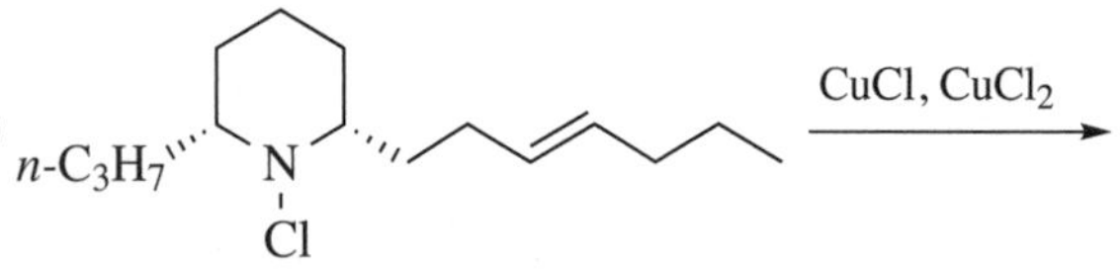
n-C3H7
N
Cl
CuCl, CuCl2

(b) 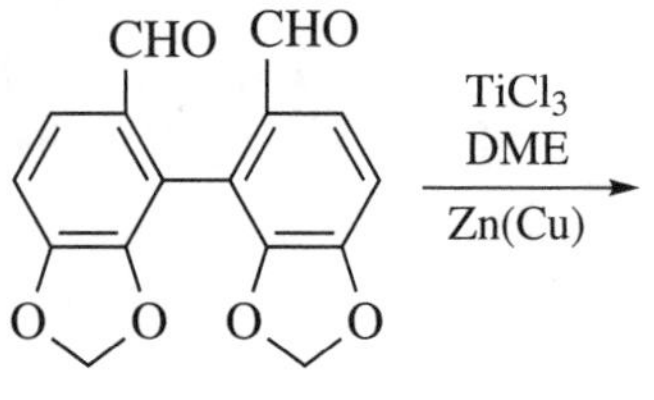
CHO CHO
O O O O
TiCl3
DME
Zn(Cu)

(c) 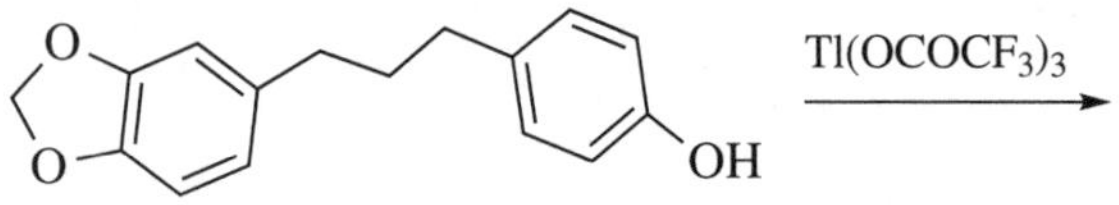
O
O
OH
Tl(OCOCF3)3

(d) 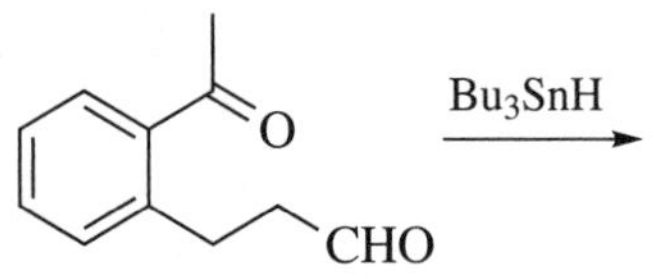
O
CHO
Bu3SnH

(e) 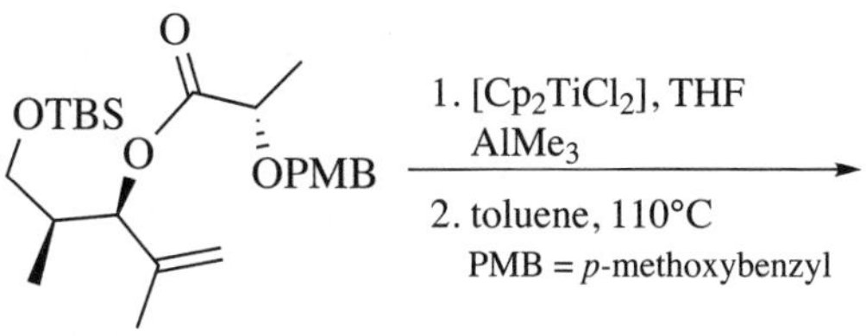 OTBS / O / OPMB →
1. [Cp₂TiCl₂], THF AlMe₃
2. toluene, 110°C
PMB = *p*-methoxybenzyl
→ (f) 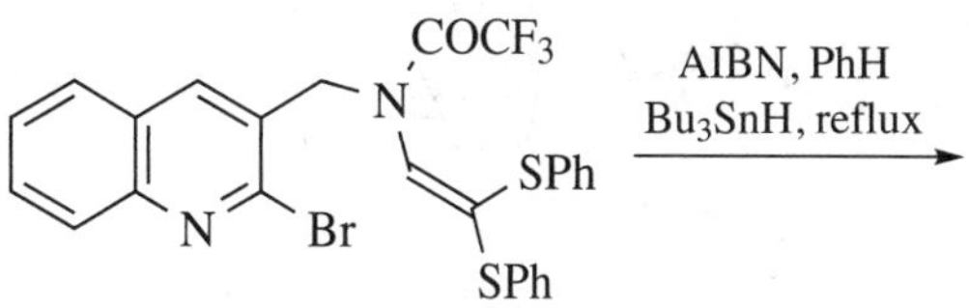 ... COCF₃ / N / SPh / SPh / Br
AIBN, PhH
Bu₃SnH, reflux →

(g) 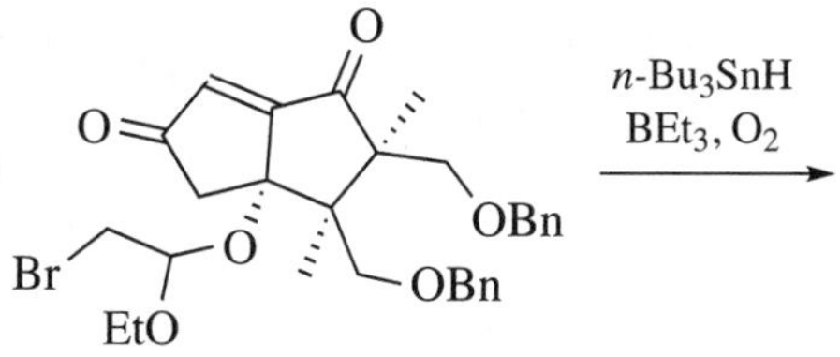 O / O / Br / EtO / OBn / OBn
→ *n*-Bu₃SnH / BEt₃, O₂ →
(h) 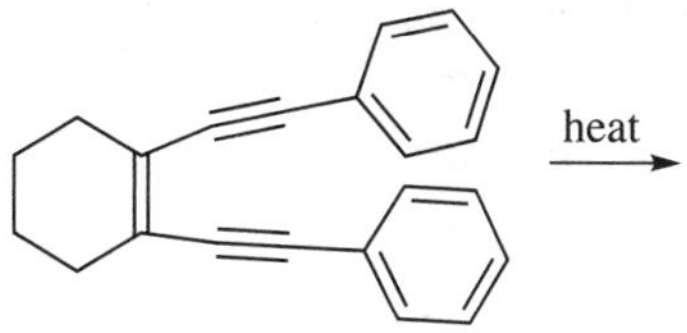 (cyclohexene with two PhC≡C groups)
heat →

(i) 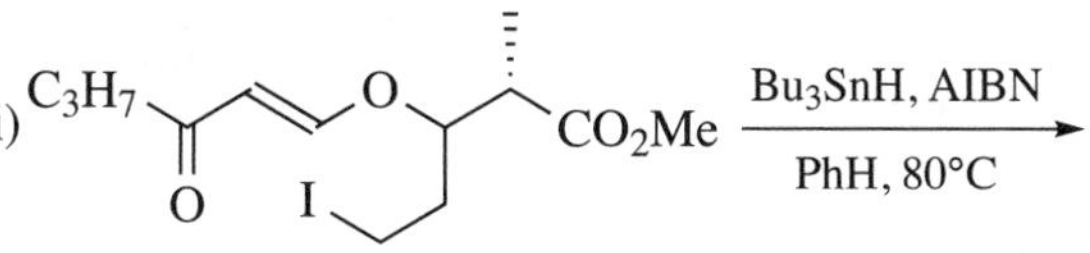 C₃H₇ / O / O / I / CO₂Me
→ Bu₃SnH, AIBN / PhH, 80°C →
(j)  2 Ph—≡—SiMe₃
CuCl, DMF / 60°C →

(k) 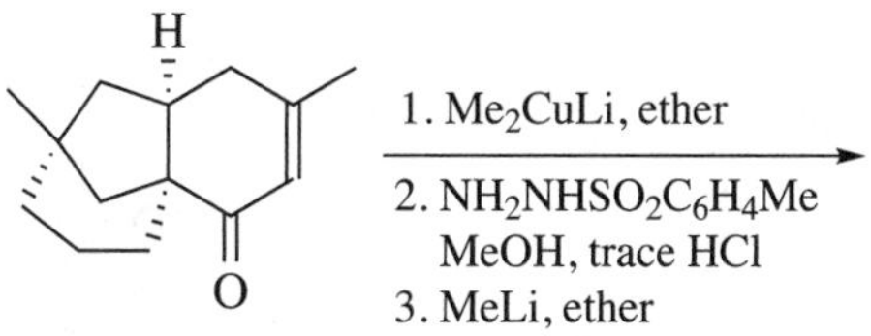 H / Me structure / O
1. Me₂CuLi, ether
2. NH₂NHSO₂C₆H₄Me MeOH, trace HCl
3. MeLi, ether
→ (l) 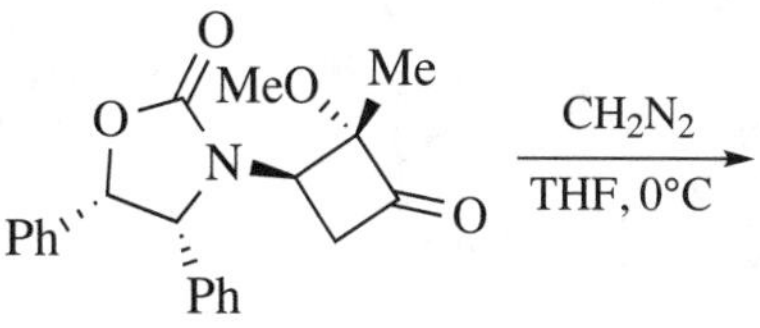 O / MeO, Me / N / Ph / Ph / O
CH₂N₂ / THF, 0°C →

(m) 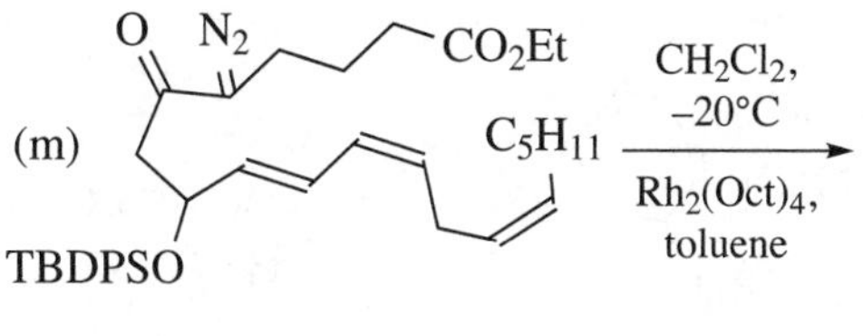 O / N₂ / CO₂Et / C₅H₁₁ / TBDPSO
CH₂Cl₂, −20°C / Rh₂(Oct)₄, toluene →
(n) 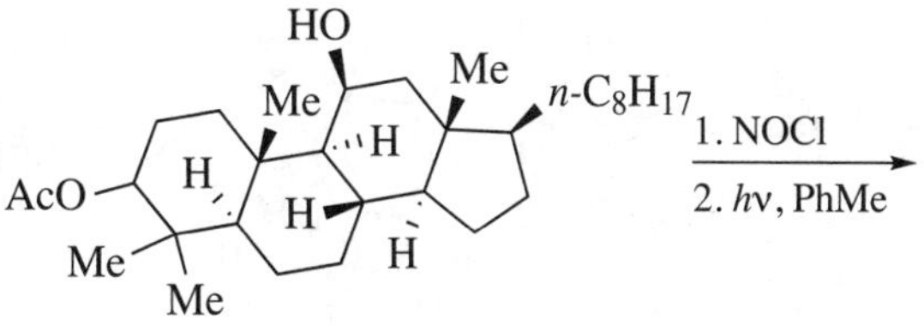 HO / Me / Me / H / H / *n*-C₈H₁₇ / AcO / H / H / Me / Me
1. NOCl
2. *h*ν, PhMe →

(o) 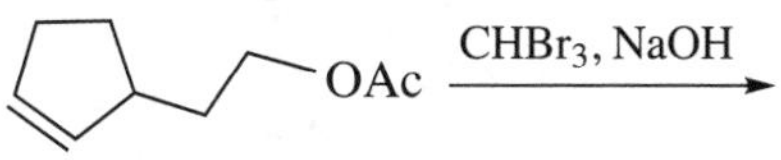 (cyclopentene)—OAc
CHBr₃, NaOH →
(p) 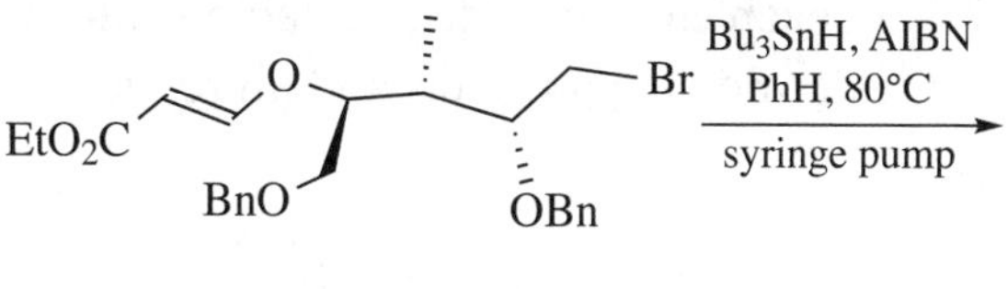 EtO₂C / O / Br / BnO / OBn
Bu₃SnH, AIBN / PhH, 80°C / syringe pump →

(q) 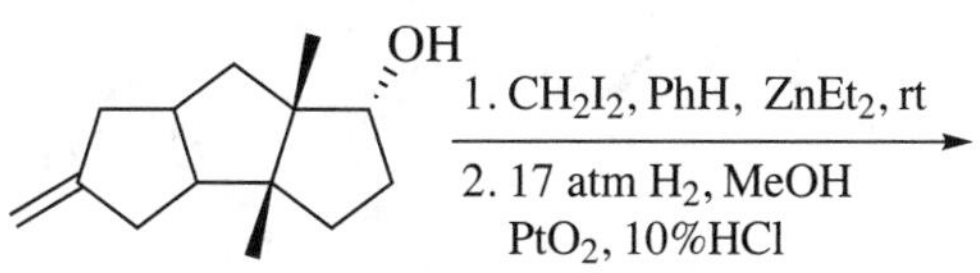 OH structure
1. CH₂I₂, PhH, ZnEt₂, rt
2. 17 atm H₂, MeOH PtO₂, 10%HCl →
(r) 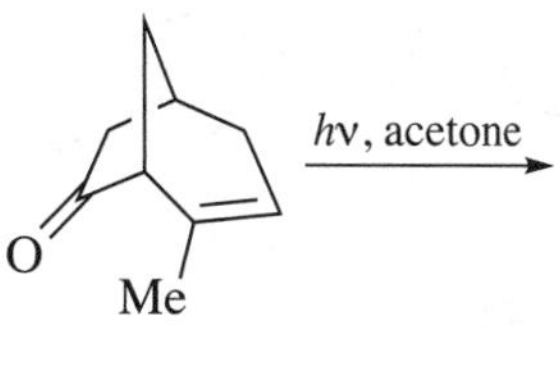 O / Me (bicyclic)
*h*ν, acetone →

(s) 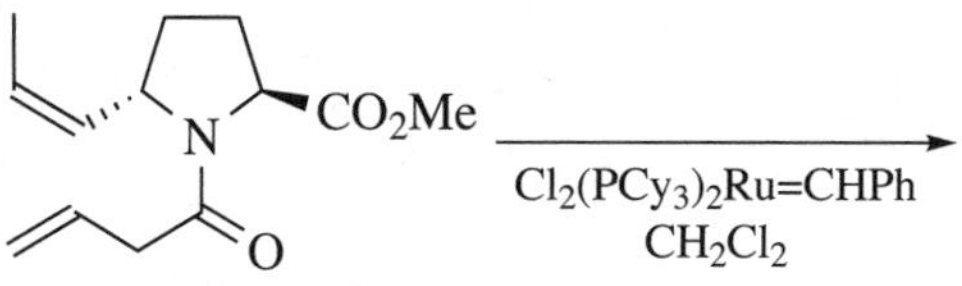 N / CO₂Me / O (pyrrolidine)
Cl₂(PCy₃)₂Ru=CHPh / CH₂Cl₂ →
(t) 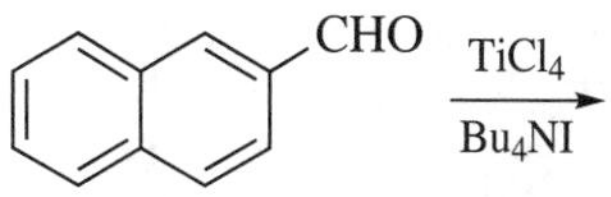 (naphthalene)—CHO
TiCl₄ / Bu₄NI →

(u) 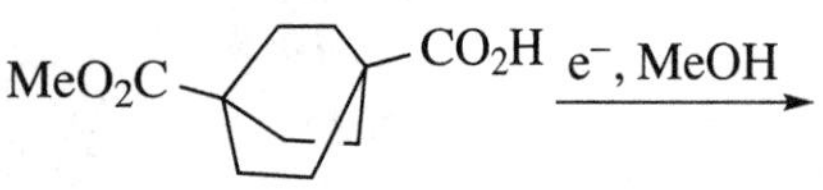 MeO$_2$C —[bicyclic]— CO$_2$H $\xrightarrow{\text{e}^-,\ \text{MeOH}}$

(v) 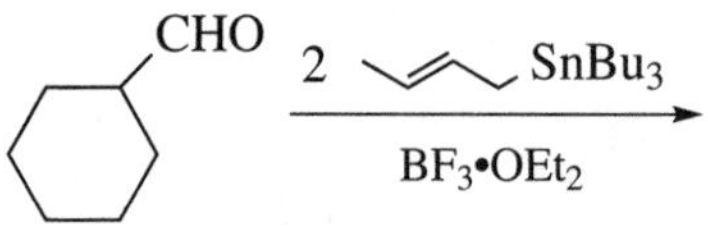 [cyclohexyl]CHO $\xrightarrow[\text{BF}_3\cdot\text{OEt}_2]{2\ \diagup\!\!\diagdown\ \text{SnBu}_3}$

(w) 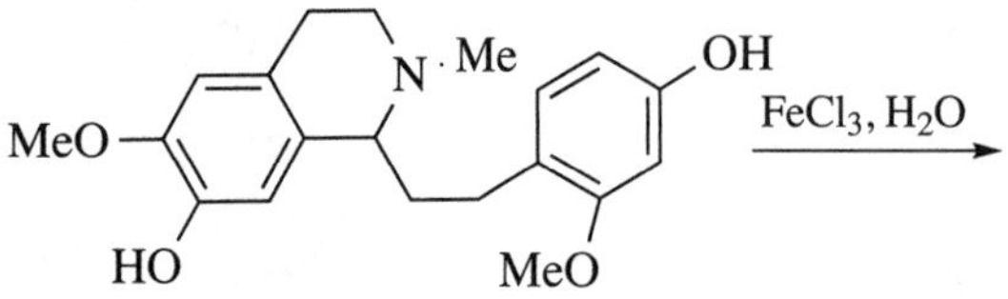 [tetrahydroisoquinoline structure, MeO, HO, N·Me, OH, MeO] $\xrightarrow{\text{FeCl}_3,\ \text{H}_2\text{O}}$

(x) 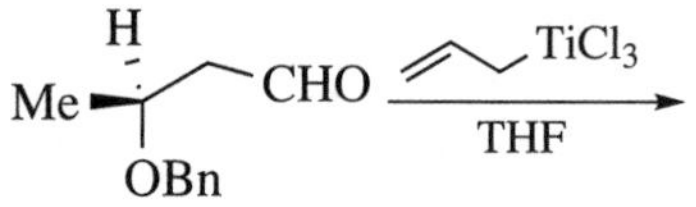 Me—[OBn]—CHO $\xrightarrow[\text{THF}]{\diagup\!\!\diagdown\ \text{TiCl}_3}$

(y) 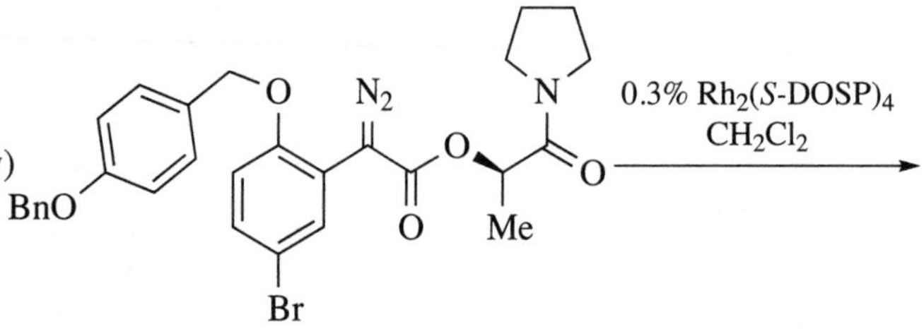 [diazo ester structure, BnO, O, N$_2$, pyrrolidine amide, Me, Br] $\xrightarrow[\text{CH}_2\text{Cl}_2]{0.3\%\ \text{Rh}_2(S\text{-DOSP})_4}$

(z) 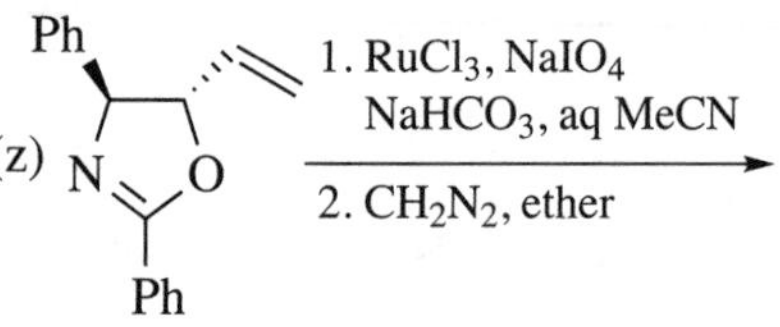 [oxazoline structure, Ph, vinyl, N, O, Ph] $\xrightarrow[\text{2. CH}_2\text{N}_2,\ \text{ether}]{\text{1. RuCl}_3,\ \text{NaIO}_4 \quad \text{NaHCO}_3,\ \text{aq MeCN}}$

(aa) 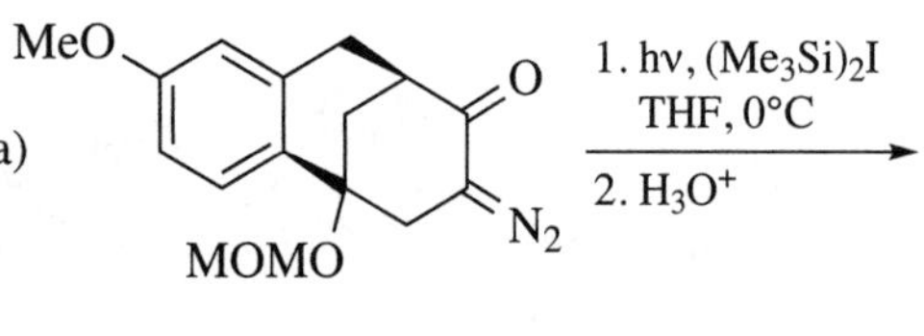 [tricyclic structure, MeO, O, N$_2$, MOMO] $\xrightarrow[\text{2. H}_3\text{O}^+]{\text{1. hv, (Me}_3\text{Si)}_2\text{I} \quad \text{THF, 0°C}}$

(ab) 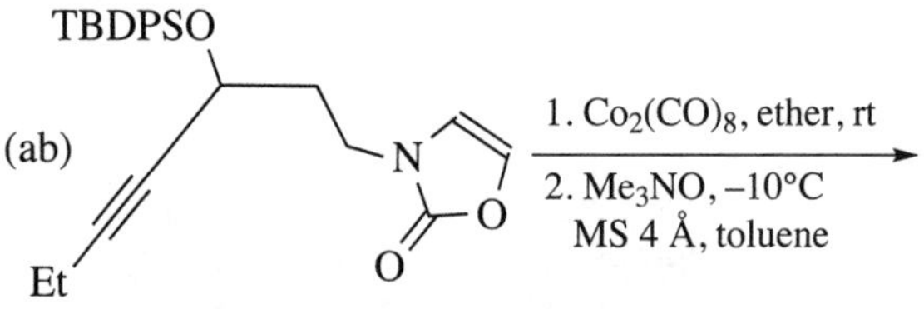 [TBDPSO, alkyne, oxazolidinone N, O, Et] $\xrightarrow[\text{MS 4 Å, toluene}]{\text{1. Co}_2\text{(CO)}_8,\ \text{ether, rt} \quad \text{2. Me}_3\text{NO, }-10°\text{C}}$

(ac) 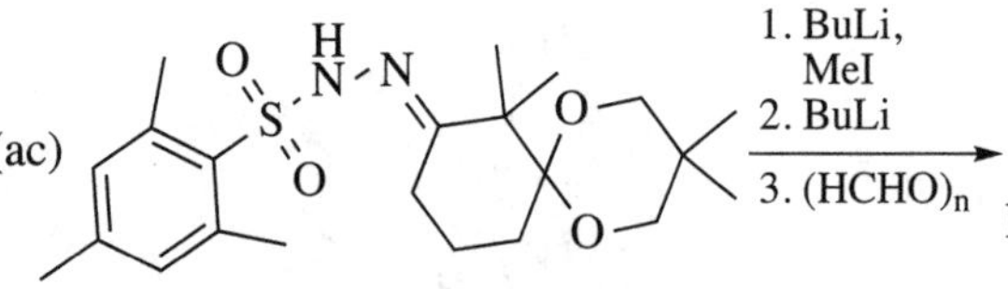 [sulfonyl hydrazone, dioxolane structure, mesityl] $\xrightarrow[\text{3. (HCHO)}_n]{\text{1. BuLi, MeI} \quad \text{2. BuLi}}$

(ad) 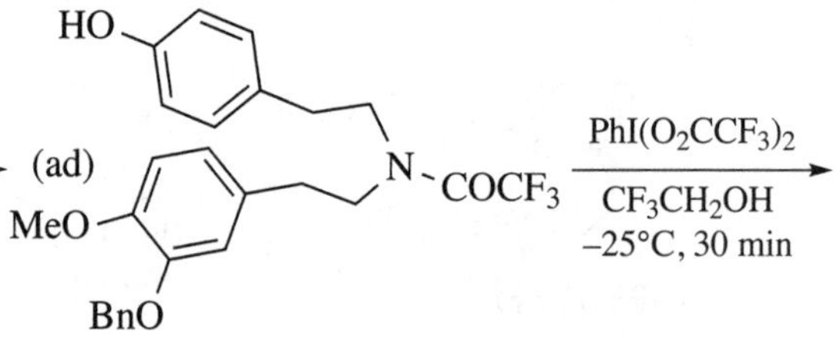 [HO-phenethyl, MeO, BnO, N-COCF$_3$] $\xrightarrow[\text{−25°C, 30 min}]{\text{PhI(O}_2\text{CCF}_3)_2 \quad \text{CF}_3\text{CH}_2\text{OH}}$

17. In each case provide a suitable synthesis from the designated starting material. Show all intermediate products and all reagents.

(a) 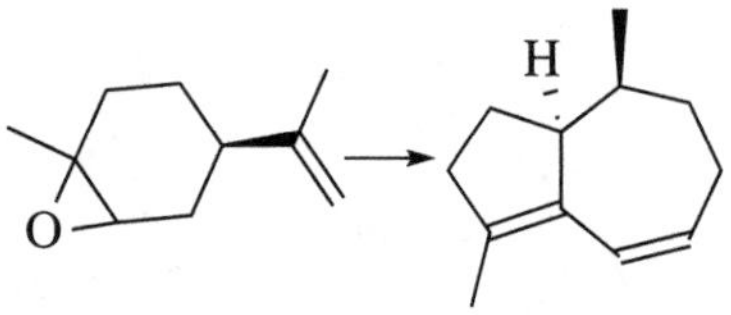 [epoxide terpene structure] → [bicyclic azulene structure]

(b) 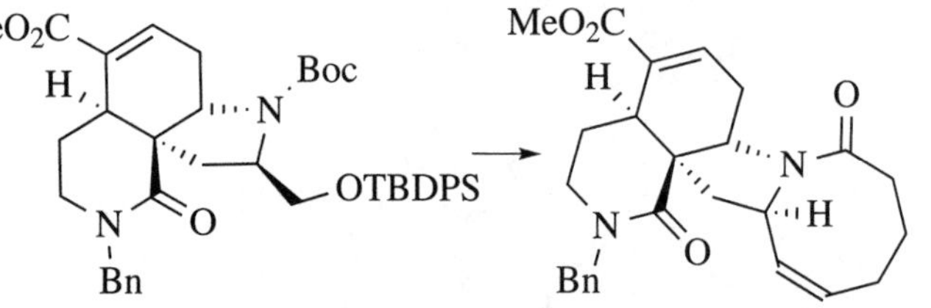 [MeO$_2$C, H, N·Boc, N, O, Bn, OTBDPS structure] → [MeO$_2$C, H, N, O, Bn structure]

(c) MeO—[OSiMe$_3$ diene] → [bicyclic pyran structure, H, H, OAc, O, O, OH]

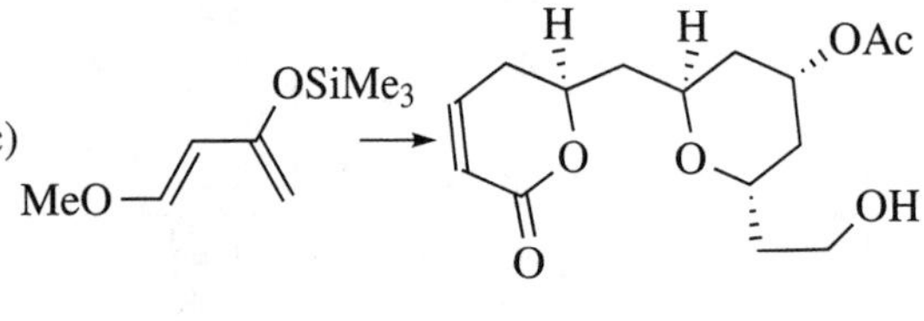

(d) 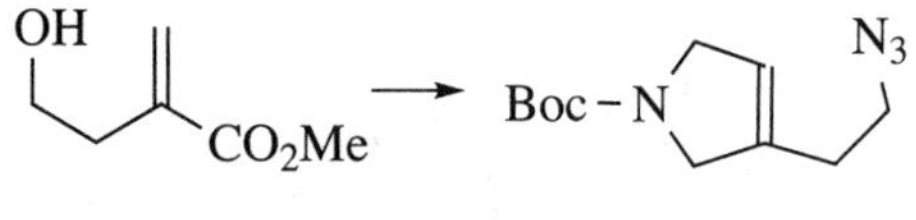 [OH, CO$_2$Me structure] → [Boc—N pyrroline, N$_3$]

(e)
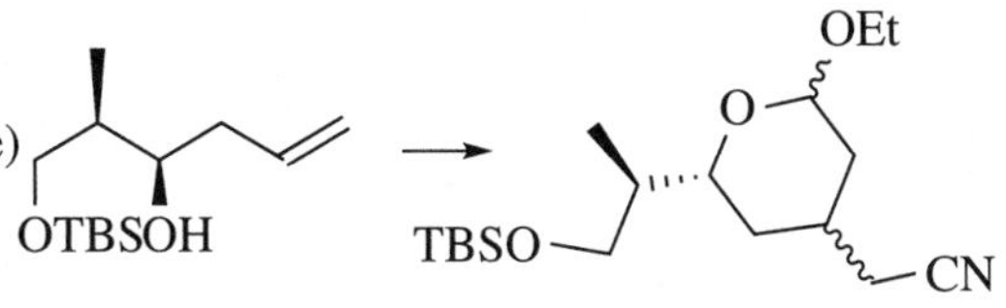
OTBSOH
TBSO
OEt
CN

(f)
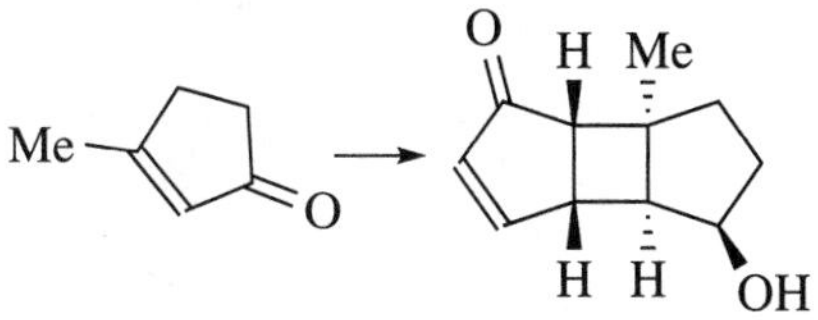
Me
O
O
H
Me
H
H
OH

(g)
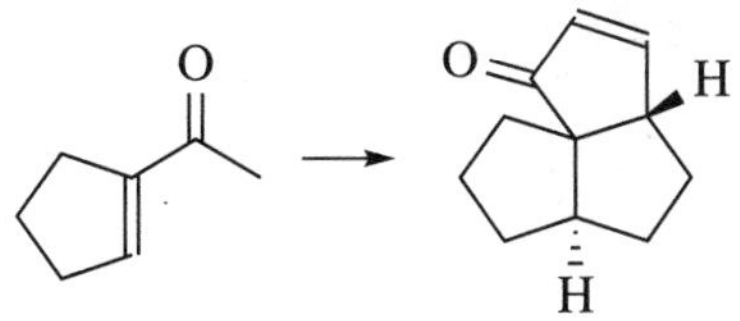
O
O
H
H

(h)
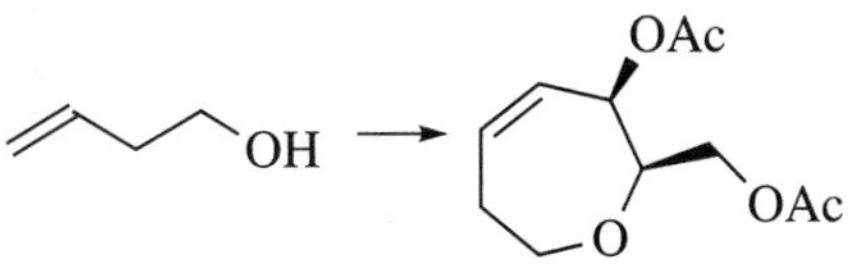
OH
OAc
OAc
O

(i)
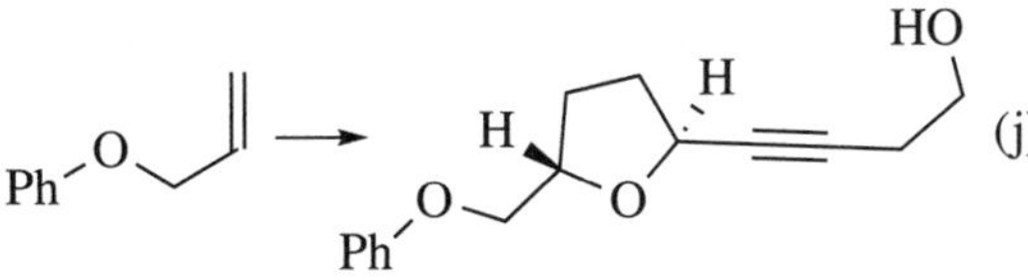
Ph
O
HO
H
H
Ph
O
O

(j)
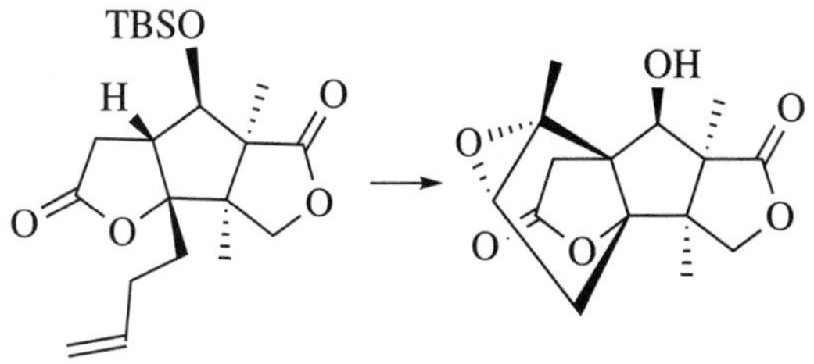
TBSO
H
O
O
O
O
OH
O
O
O
O
O

(k)
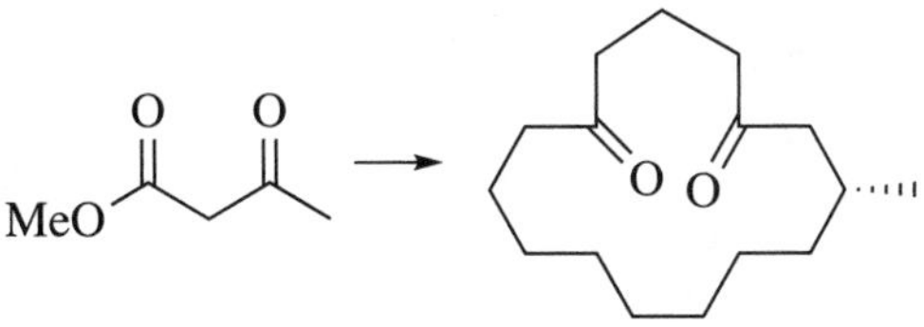
O
O
MeO
O
O

chapter ⬤14

Student Synthesis: The First Synthetic Problem

14.1   INTRODUCTION

It is one thing to learn all the reactions and processes described in this book. It is another to apply them to a complex molecule. A total synthesis first demands a review and understanding of the various disconnections and retrosynthesis strategies. Once a strategy is chosen, functional group exchange reactions and carbon bond forming reactions must be reviewed in order to choose those reactions best suited to a particular strategy. Reactions pertinent to the problem are then studied, which should force a thorough literature search (in the library or using one of the various literature searching techniques available for computers). Every step forces a complete review of chemical principles, stereochemical and conformational relationships in order to justify the use of each reagent that is not obvious. This collection of steps must then be examined for continuity, and the continuum of functional groups must all be compatible. In other words, groups introduced in the first synthetic step must not react with any succeeding steps. If they do, protection of those groups or redesign of the strategy is required. This discussion clearly suggests that a total synthesis focuses attention on organic chemistry in a way few, if any, other area of organic chemistry is able to do. For these reasons, a synthesis and reactions course often includes a proposal by the student for the synthesis of a natural product.

In most of the papers where a total synthesis is presented, the strategy and rationale for the synthesis is given. Most of the paper is devoted to explaining how and why the reactions worked, the stereochemical and conformational importance of various steps and, in some cases, a detailed explanation of things that did not work well in the initial planning. If a detailed study of published syntheses is of interest, and it should be to anyone proposing to synthesize a molecule, there is an excellent series that will provide information about how and why a synthesis was done. Lindberg's excellent series "Strategies and Tactics in Organic Synthesis", Volumes 1–3 [QD 262.s85] (1984, 1989, and 1991), Academic Press, San Diego is *strongly* suggested for anyone who is using this book in a synthesis course.

The course that spawned this book asked the students to prepare a synthesis proposal. There are many formats for the write-up, including those used by the National Institutes of Health (NIH) and the National Science Foundation (NSF). Each have somewhat different formats and page requirements. It is convenient to write a synthesis proposal following one of these formats, although any arbitrary format is acceptable. Proposals are submitted electronically to both organizations, but adherence to the older format will suffice for our purposes.

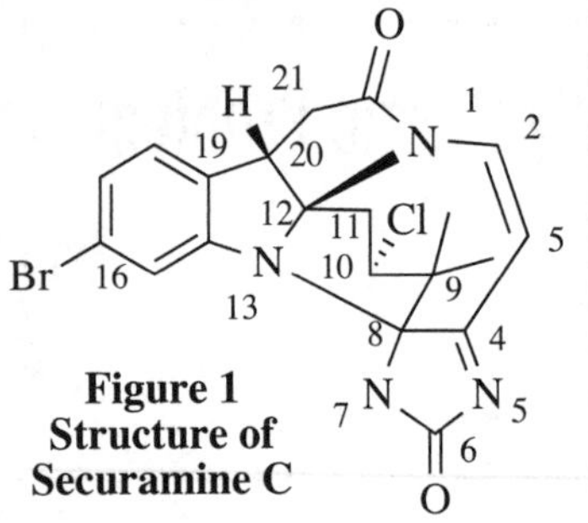

Figure 1
Structure of
Securamine C

This chapter is intended to show the first synthesis submitted by students who have had a course based on the material in this book. Scanning the literature will provide many examples of finished and polished total syntheses. Many have been used as source material to illustrate various reactions and those literature citations will lead to a broader explanation of that particular synthesis. The work behind such publications, however, is sometimes lost in the highly focused final paper. Various attempts to solve problems are sometimes, but not always fully discussed. A study of known syntheses is very useful and provides a basis for the thoughts and strategies of the investigator (refer to Linderg's work, *vide supra*). As a teaching tool for the beginner, however, many points are glossed over or are not obvious. The next two sections will discuss first-time proposals submitted by students in a reactions/synthesis course that targeted securamine C (sec. 14.2) and variecolol (sec. 14.3).

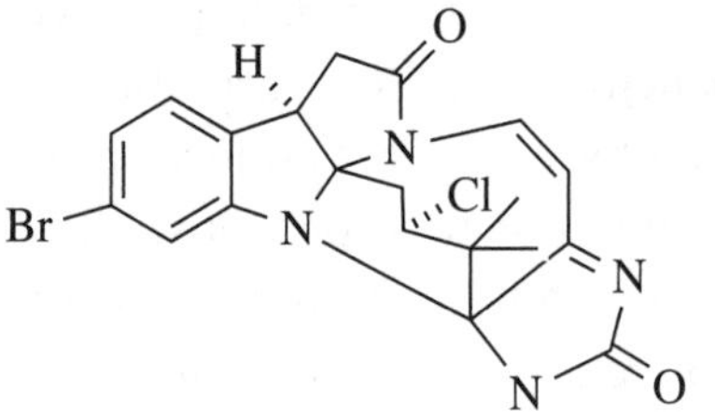

securamine C

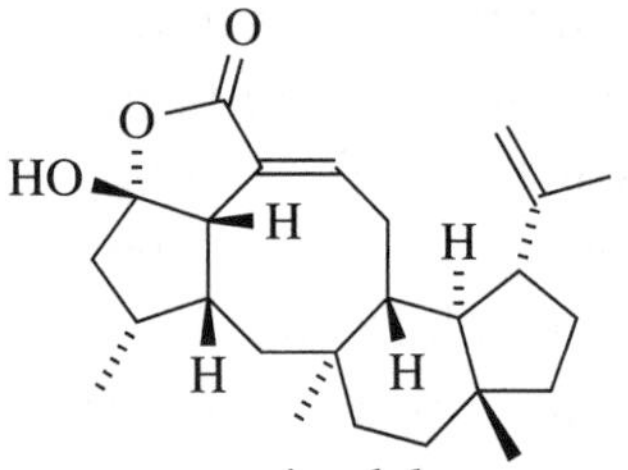

variecolol

The intent of this chapter is <u>not</u> to simply criticize the proposal, but rather to show common mistakes that creep into that first proposal. In each section there will be three parts. The first part is the proposal by the student, <u>as submitted</u>. The second part is a list of all reaction types proposed by the student, and the third is the instructor's critique. The comments in the critique are often aimed at specific chemical problems related to that target and will serve as only a brief review of some chemistry. Others, however, are more general and will hopefully illustrate things to avoid while planning a synthesis. The proposals are reproduced exactly as submitted to show the importance that a presentation can make to the visual and conceptual interpretation of the proposal.

There are many ways for the reader to approach this chapter. The reader could construct a synthetic proposal of before reading further. The reader could act as the reviewer and construct their own critique. Of course, one could simply read my comments about each proposal. It is my belief that an active role in reviewing each proposal will be far superior to a passive role. Compare my criticisms with your own. Agree or disagree! That is a good first step to a good proposal.

14.2 TOTAL SYNTHESIS OF SECURAMINE C

14.2.A The Proposal

This synthesis is reasonable, but there are some problems with the presentation. In several places there is the potential for side reactions that have not been addressed. In one or two places, the proposed reaction maybe difficult. In general, however, the proposal has a good literature basis.

I. Introduction

II. Retrosynthetic Analysis

Securamine C from the marine bryozoan *Securiflustra Securifrons* farnily has recently been isolated and characterized by researchers at the University of Copenhagen[1]. This interesting hexacyclic structure is an indole-imidazole alkaloid from a family of alkaloids that has yielded important biological activity[2]. This proposal outllines a total synthesis of **1**.

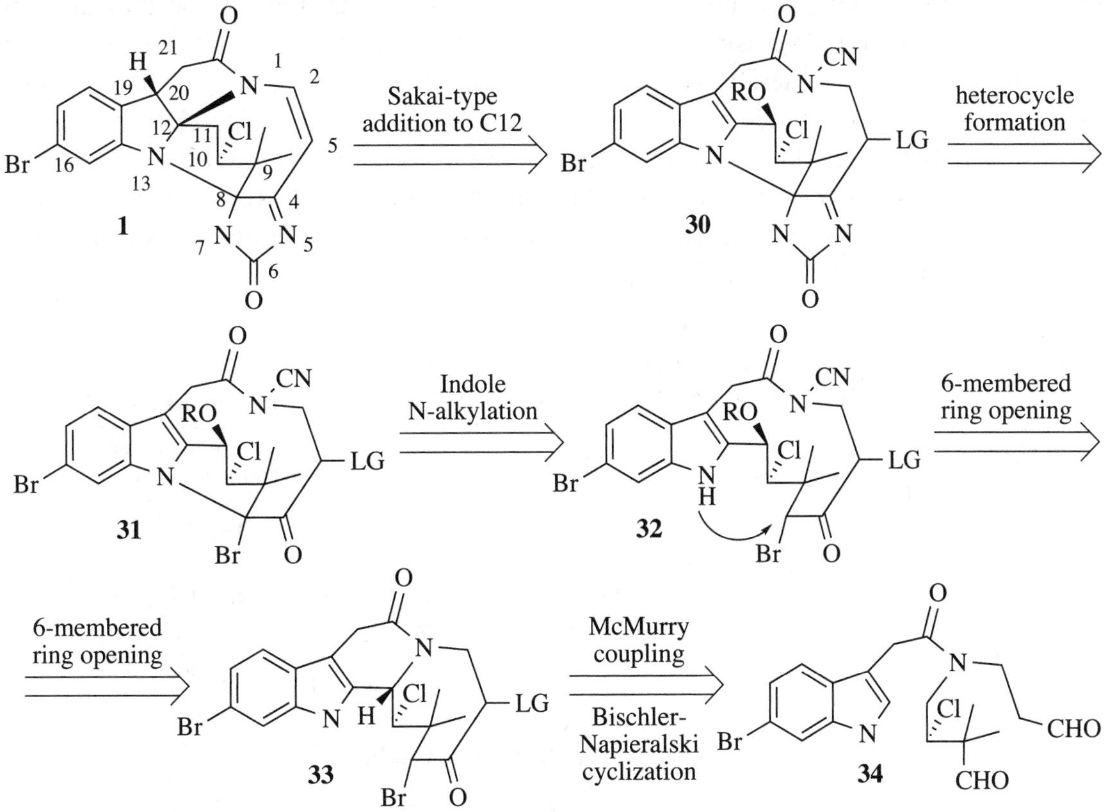

Scheme 1
Retrosynthetic Analysis

Our approach to the synthesis of **1** is shown in **Scheme 1**. The crucial N_1-C_{12} bond making reaction is based on a serendipitous discovery of a Japanese group working on the related Picraline alkaloid system[3]. Using an intermediate closely related to **32**, they were hoping to form a bond between C_8 (bearing the leaving group) and C_{20} (the 3-position of the indole); instead the bond was formed between the indole nitrogen and C_8. The formation of this new six-membered ring resulted in an isopleiocarpamine-type skeleton similar to **31**. From there, under acetic acid/ammonium acetate conditions, C_{20} was protonated then N_1 attacked the resulting iminium function at C_{12} (the 2-position of the indole) to give the unusual skeletal structure analogous to **1**.

① **32** in turn can be made by opening of the yohimbane-type ring structure pictured in **33** with the use of cyanogen bromide.[3] The C–D rings of yohimbane-type alkaloids are usually constructed by way of a Bischler-Napieralski reaction of an amide intermediate such as **34**.[4] Not shown but critically important in the transformation from **34** to **33** is the construction of the 8-membered ring. It is envisioned that this difficult task can be accomplished using a radical McMurry coupling[5]. **34** can be made by subsequent alkylation and acylation of N1 on a tryptamine-type of precursor.

The synthesis begins with the known 6-bromoindole **2**,[6] which has recently been made in high yield from the 6-aminoindole using a Sandmeyer reaction (**Scheme 2**). After protection of the indole nitrogen as the Boc derivative,[7] the electron rich three position of the indole moity can be alkylated with an alkyl halide using iron sulfate with hydrogen peroxide in DMSO to give **3**.[8] The nitrile can then be hydrolyzed enzymatically to the acetamide using the nitrile hydratase taken from the *Rhodococcus rhodochrous J1* organism.[9] Intermediate **4b** can then be alkylated with the commercially available propionaldehyde **4b** in EtOH with sodium carbonate as base[10] to give **5**. ② This alkylation attaches C_2–C_4.

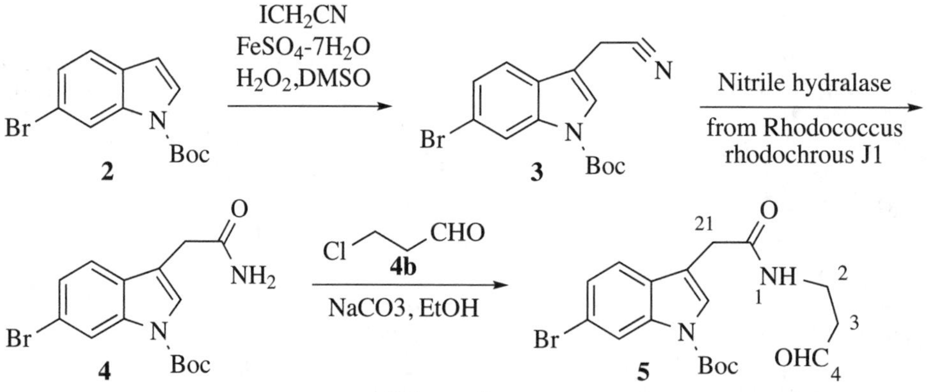

Scheme 2
Construction of the C_{21}-C_4 Side Chain

From our retrosynthetic scheme we now need to attach C_8–C_{11} to N_1 in order to set up the construction of the medium sized ring *via* McMurry coupling. This C_8–C_{11} fragment (intermediate **6**, **Scheme 3**) can be attached using a dicyclohexylcarbodiirnide (DCC) coupling of the carboxylic acid of **6** to N_1 in the presence of hydroxy-benzotriazole in DMF as the solvent.[10,11] The synthesis of fragment **6** begins with known **6a**.[12] This interesting compound can undergo ozonolysis ③ with a reductive workup to give aldehyde **6b**,[13,14] which can be hydrolyzed with potassium t-butoxide and water to give the carboxylic acid **6c**.[15] Resolution of racemic **6c** will follow the protocol of Daniel Comins; the procedure calls for esterification with a chiral alcohol ⑤ to produce a pair of diastereomers which can be separated with a chiral HPLC column.[16] The particular chiral alcohol to be used here is *(+)-trans-2-(alpha-cumyl)*cyclohexanol((+)-TCC). ④ The desired diastereomer will then be hydrolyzed to the carboxylic acid **6** using the *t*-butoxide/water conditions mentioned above. This hydrolysis would also regenerate the expensive chiral alcohol.

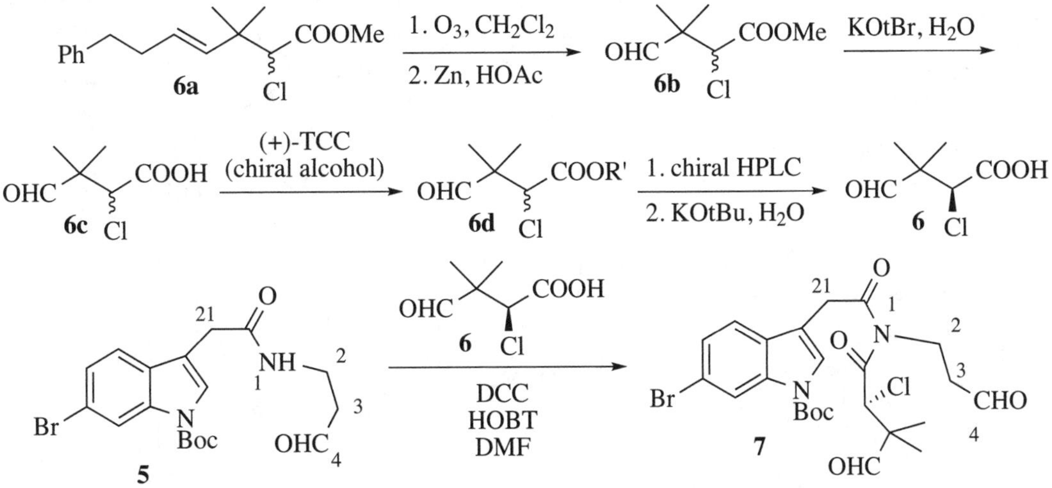

Scheme 3
Synthesis and attachment of the C_8 - C_{11} Side Chain

Dialdehyde was prepared with a view towards making the 8-membered ring which includes N_1 to C_4 and C_8 to C_{11}. These medium sized rings are notoriously difficult to make because of the transannular strain generated. The one called for here would be particularly difficult since the dimethyl group at C_9 would add an even greater dimension of steric hindrance. The method of choice to accomplish this task would be the use of a radical coupling, which has been successfully employed in the formation of a key 8-membered ring in the important taxol structure.[17] Radical cyclizations are less susceptible to the steric effects which cause other methods to fail.[5] For the present synthesis, **7** can be cyclized using a titanium induced reductive coupling of the two carbonyl units in the presence of a zinc-copper couple ⑥ using dimethoxyethane as solvent. The zinc-copper mixture is generated by adding zinc dust to a solution of copper sulfate in deoxygenated water.[5]

Once the C_4,C_8 diol is in hand, ⑦ the two hydroxy bearing carbons need to be differentially oxidized. The same C_9 methyl groups that would make the ring closure difficult could become an asset in differentiating C_8 from C_4. In particular, it may well be possible to oxidize the C_4 hydroxy group to the ketone and give **13** directly from **8**, without the need for the protection/deprotection method shown in **Scheme 4**. The Swern oxidation,[18] for example, uses the bulky base TEA; this could lead to selective oxidation of C_4, since C_8 should be sterically less accessible. Failing that, the method shown in **Scheme 4** could be employed. C_4 would be protected with a silyl group.[19] Then the C_8 hydroxyl could be acetylated with acetic anhydride and pyridine.[20] Deprotection at C_4 with fluoride, followed by oxidation[18] to the ketone with Swern conditions(oxalyl chloride and DMSO in methyene chloride followed by the addition of TEA), and finally deprotection of the acetate with potassium carbonate in methanol would give intermediate **13**.

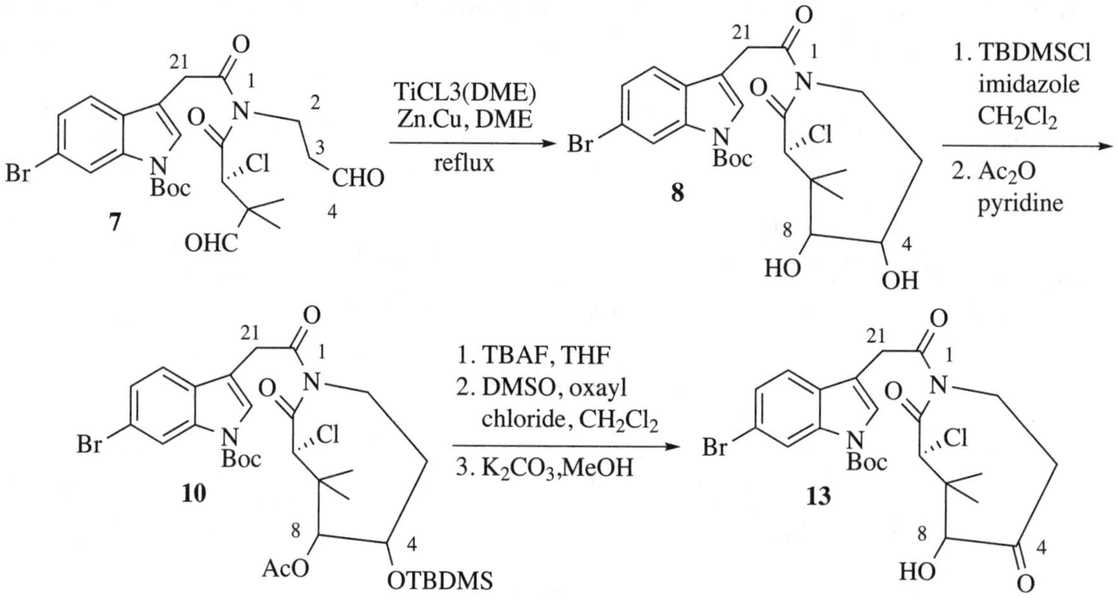

Scheme 4
Synthesis of hydroxy-keto 8-membered ring

The final target securamine contains a dou le bond between C_2 and C_3. We would like to introduce that double bond after th important Sakai-type addition of N_1 to C_{12} since an sp^2 center at C_1 may adversely affedt that ring closure. The strategy employed here is to introduce a functional group at C_3 which can be easily transformed into the C_2–C_3 double bond at a later stage. For this reason, **13** ⑨ will be converted to the selenide *via* addition of phenyl selenium chloride to the enolate generated at C_3 with LDA,[21] to give **14**. ⑩ The selenide will be oxidized later and will eliminate thermally at mild temperature to give the olefin.

The hydroxy group at C_8 needs to be converted to a leaving group for eventual bond formation with the indole nitrogen. This can be done by conversion to the bromide **15** ⑪ with thionyl bromide and pyridine in methylene chloride at ambient temperature[22]. The leaving group could just as easily have been tosylate by converting the hydroxyl group with tosyl chloride and base in methylene chloride.

The closure of the new 6-membered lactam ring to give **16** from **15** is based on extensive literature precedent from synthetic work in the yohimbane-type indole alkaloids.[23] The mechanism of this Bischler-Napieralski reaction involves conversion of the amide function to an imidoyl chloride, ⑫ ⑬ which makes C_{11} electrophilic enough to add to the two position of the indole.[24]

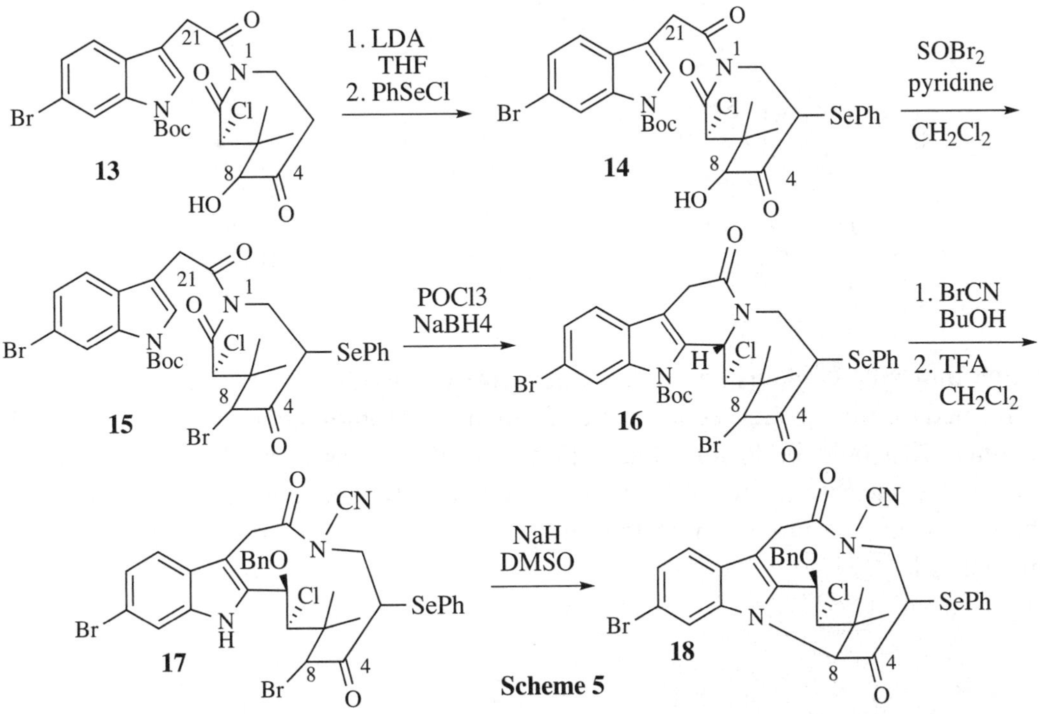

Scheme 5

In preparation for the important N_{13}(indole N)–C_8 bond forming reaction(**17** to **18**), the newly formed 6-membered ring of **16** must now be reopened with cyanogen bromide to give, after deprotection of the Boc group, intermediate **17**. ⑭ The stereochemistry at C_{11} is critical because if the OR(in this case OBn) group is *alpha*(behind the plane of the page) then you will not get Sakai-type addition of N_1 to C_{12} but rather N_1 will simply add back into C_{11} to form the 6-membered ring again. With the OBn group at C_{11} *beta,* as would be expected from the literature precedent, intermediate **17** is now poised for the immediate N_3–C_8 bond formation and subsequent Sakai ring closure.[3] In the original paper the cyanogen bromide reaction was done in MeOH which gave an OMe group at C_{11}. In the current target, C_{11} needs to ultimately become a methylene group. Towards that end, the cyanogen brotnide reaction would be done in benzyl alcohol to give the benzyl ether at C_{11}, which can be removed(hydrogenolysis) and reduced(triethyl silane reduction of the ketone) more easily than the methyl ether.

⑮ When **17** is subjected to sodium hydride in DSMO to generate the indole anion, the isopleiocarpamine-type skeleton **18** should be formed by the formation of the new sixmembered ring.[3] In our case there is the danger that N_{13} could displace the C10 chlorine to give a new four-membered ring; modeling studies would need to be done to evaluate that possibility.

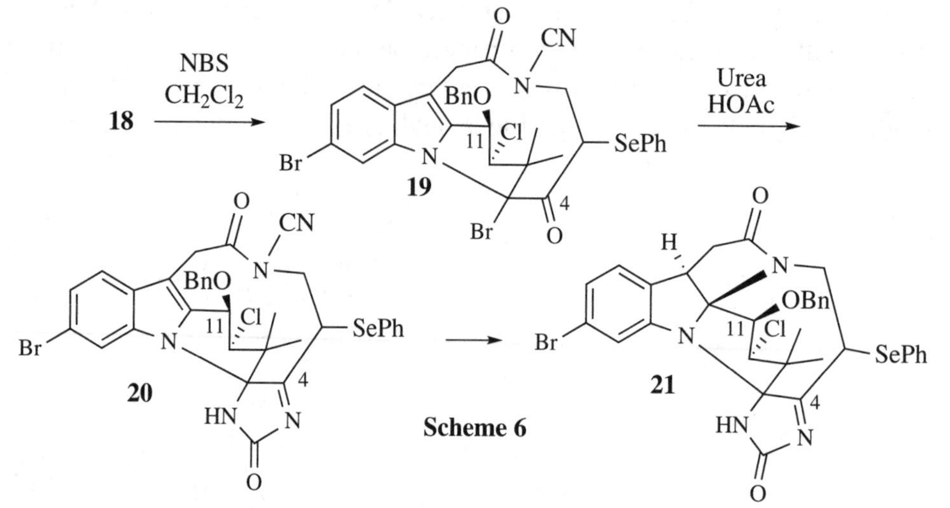

Scheme 6

The target molecule contains an imidazolinone off of the C_4, C_8 positions. This heterocycle can be formed by adding a molecule of urea to an alpha-bromoketone. ⑯ ⑰ ⑱ Accordingly, intermediate **18** would be brominated with a radical process using N-bromosuccinimide in methylene chloride,[25] then urea ⑲ would be added under acetic acid conditions to give the new heterocycle.[26] Ammonium acetate in acetic acid would then complete the hexacyclic intermediate **21**.[3] ㉓

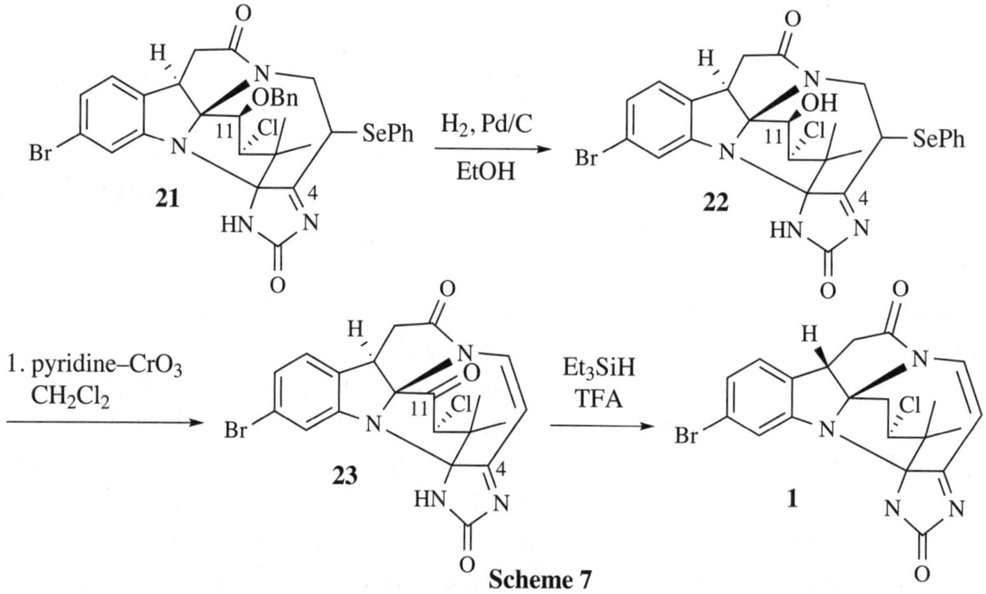

Scheme 7

What is left to complete the synthesis of Securamine C is essentially the ㉔ reduction of C_{11} to a methylene. Removal of the benzyl ether can be done by hydrogenolysis using palladium on carbon in ethanol.[27] Literature precedent shows that the oxidation of the resulting alcohol to the ketone using the Collins reagent (pyridine-chromium trioxide in methylene chloride) ㉔ ⑳ would also oxidize the C_3 selenide to the selenoxide which would immediately eliminate under the reaction conditions.[28] Reduction of the C_{11} ketone with triethyl silane would complete the synthesis of **1**. ㉑

References

1. Rahbaek, Lisa; Christophersen, Carsten; *Journal of Natural Products*, **1997**, *60*, 175.

2. Rahbaek, L; et al; *Journal of Organic Chemistry*, **1996**, *61*, 887. ㉒

3. Sakai, Shin-ichiro; et al; *Chemical and Pharmaceutical Bulletin*, **1991**, *39(7)*, 1677.

4. Wender, Paul A; et al; *Journal of Organic Chemistry*, **1996**, *6]*, 842.

5. McMurry, J.E.; et al; *Journal of Organic Chemistry*, **1978**, *43*, 3255; Furstner, A; Langemann, K; *Journal of Organic Chemistry*, **1996**, *61*, 8747; Furstner, A; et al; *Journal of the Chemical Society Perkin I*; **1988**, 1729.

6. Toste, F; et al; *Organic Preparations and Procedures*, **1995**, *27(5)*, 576.

7. Greene, T; Wuts, P; *Protective Groups in Organic Synthesis*, **1991**, John Wiley & Sons, p.327.

8. Baciocchi, E; et al; *Journal of Organic Chemistry*, **1992**, *57*, 6817.

9. Manger, J; et al; *Tetrahedron*, **1989**, 1347.

10. Wender, Paul A; et al; *Journal of the American Chemical Society*, **1995**, *117*, 1843.

11. March, J; *Advanced Organic Chemistry*, 1992, *John Wiley & Sons*, p.420.

12. Elworthy, Todd R; et al; *Tetrahedron Letters*, **1994**, 4951.

13. Marshall, James A; Greene, A.E; *Journal of Organic Chemistry*, **1972**, *37*, 982.

14. Smith, Michael B; *Organic Synthesis*, **1994**, McGraw-Hill, p.309.

15. Ref. 7, p.230.

16. Comins Daniel L; Salvador, James M; *Tetrahedron Letters*, **1993**, *34(5)*, 801; Comins, Daniel L; et al; *Journal of the American Society*, **1994**, *116*, 4719.

17. Kende, J; et al; *Journal of the American Society*, **1986**, 3513.

18. Omura, K; Swern, D; *Tetrahedron*, **1978**, 1651.

19. Ref. 7, p.77.

20. Ref. 7, p.88.

21. Ref. 14, pp.169, 323.

22. Ref. 14, p.288.

23. Ref. 4; Aube, Jeffrey; et al; *Journal of the American Society*, **1994**, *116*, 9009.

24. Fodor, G; Nagubandi, S; *Tetrahedron*, **1980**, 1279.

25. Zinnes, H; et al; *Journal of Organic Chemistry*, **1965**, 2241.

14.2.B The Critique

The reaction types used in this proposal are as follows.

1. Bischler–Napieralski reaction (sec. 12.5.B).
2. McMurry coupling (sec. 13.8.F).
3. Sandmeyer reaction (sec. 2.11).
4. DCC coupling (not discussed in this book).
5. Ozonolysis (sec. 3.7.B).
6. Radical cyclization (sec. 13.7).
7. Birch reduction (sec. 4.9.E).
8. Swern oxidation (sec. 3.2.C.i).
9. Sakai coupling (not discussed in this book).
10. Selenoxide elimination (sec. 2.9.C.vi).
11. Alcohol to bromide with thionyl bromide (sec. 2.8.A).
12. N-Boc protection (sec. 7.3.C.iii).
13. NBS allylic bromination (sec. 2.8.B).
14. Alcohol protection - benzyl ether (sec. 7.3.A.i).
15. Silane reduction of ketones (sec. 4.10.C).

The number of each critique appears in the body of the proposal as a circled number. The critique of this proposal is:

1. Do not begin a sentence with a number.
2. For the conversion of **4** to **5**, the CHO group may not be compatible with this reaction? Does it have to be protected?
3. In the conversion of **6a** to **6b**, there may be a problem because zinc and acetic acid can reduce halides. There is a question as to whether or not the Cl will be reduced.
4. How will TCC be used? It would be appropriate to give some reaction conditions.
5. In the conversion of **5** to **7** there maybe a problem putting the second C=O on the amide. This can be difficult. The chloro-acid is rather hindered and there is a question whether this will have an effect on the reaction with DCC.
6. In the conversion to **8** the Cl adjacent to the carbonyl may be reduced?
7. What is the stereochemistry of the diol?
8. Since one secondary alcohol is adjacent to a gem-dimethyl unit, it might be possible to selectively oxidize the less hindered OH to the acyloin (compound **13**) without all those steps?

9. In the conversion of **13** to **14**, the OH group probably requires protection.

10. Once the enolate is formed, is displacement of the chlorine across the ring possible?

11. In the conversion of **14** to **15**, does thionyl bromide react with seleno ethers?

12. In the conversion of **15** to **16**, this is actually 2 steps over the arrow, not one as shown

13. Will the chlorine survive this step?

14. In the conversion of **16** to **17** - is seems unlikely that the SePh unit, the Cl unit, the Br unit will survive without some secondary reactions, particularly when TFA is being used.

15. In the conversion of **17** to **18** - **17** NaH may form the enolate at the O=C–C–SePh carbon. Although it may be less favored, it formation of a three-membered ring possible (see Favorskii rearrangement)? Can the enolate anion displace the Cl across the ring?

16. In the conversion of **18** to **19** bromine radicals are generated. Will the presence of the chlorine pose a problem?

17. Under these conditions, SePh can be lost to form a radical. Will this cause any problems?

18. It is also possible to brominate at the C bearing the SePh unit. To form the targeted bond requires displacement of bromide from a neopentyl carbon, which is very difficult

19. In the conversion of **19** to **20**, the new bond generating the heterocycle occurs at a quaternary carbon. Can this occur as planned?

20. The Collins oxidation is applied to a hindered substrate. Will this cause problems?

21. In the conversion of **23** to **1** - will Cl survive treatment with silane and TFA?

22. In the references, do not use et al - cite all authors for each paper.

23. For the transformation of **18** to **1**, there is minimal discussion. This section should be expanded.

24. Note the typographical error.

14.3   TOTAL SYNTHESIS OF VARIECOLOL

This proposal is reasonable considering it is a difficult target. The main problems have to do with stereo-chemistry of proposed products and, in some cases, choice of reagents. There is at least one reaction, the last one, that is questionable, but in general it is a reasonable approach.

14.3.A   The Proposal

I.   Introduction

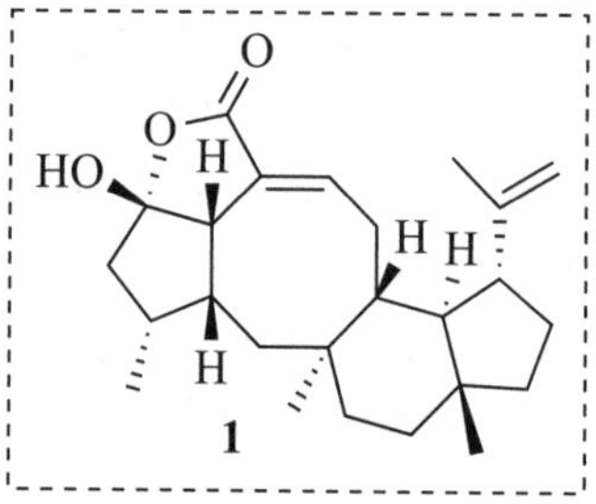

Variecolol (**1**), a new sesterterpene was recently isolated from the mycellium of *Emericella purpurea*.[1]   Presently, a study of variecolol's ability as an agent for preventing and treating hypertension, cardiovascular and cerebrovascular diseases and renal disease is taking place.[2] However, variecolol can only be obtained by extraction from the mycellium of *Emericella purpurea*. This is due to the fact of variecoloactone's highly functionalized complex fused ring structure making its total synthesis rather difficult.

Herein is described a synthetic strategy to obtain the basic ring structure of variecolactone which can be applied to other natural products with a similar ring structure.[1]

II.   Retrosynthetic Analysis

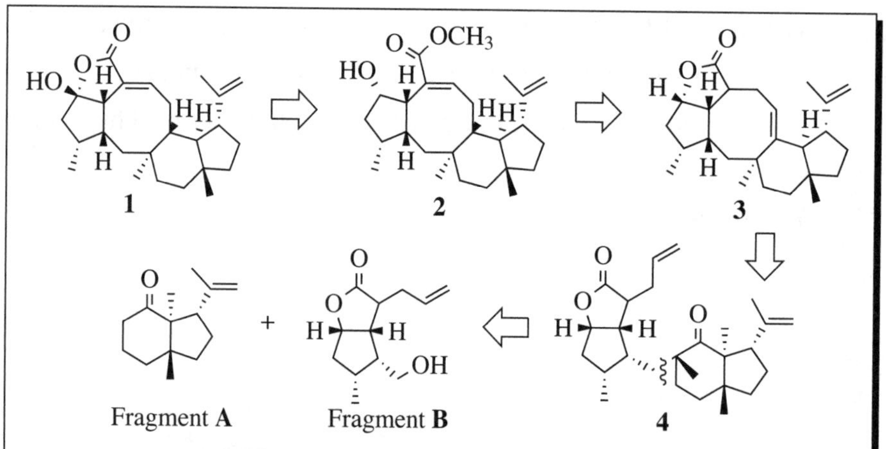

Scheme 1: Retrosynthetic Analysis of Variecolol (**1**)

The key features of the synthesis of variecolol are outlined retrosynthetically in Scheme 1. Working one step back from (**1**) gives the _-hydroxy-carboxylic acid (**2**). Manipulation of the double bond contained within the eight membered ring gives (**3**) followed by a ring closing metathesis reaction utilizing Grubb's catalyst leads to (**4**). Cleavage of the indicated bond leads to Fragments **A** and **B** which can be synthesized using slightly modified literature procedures.

A. Synthesis of Fragment A

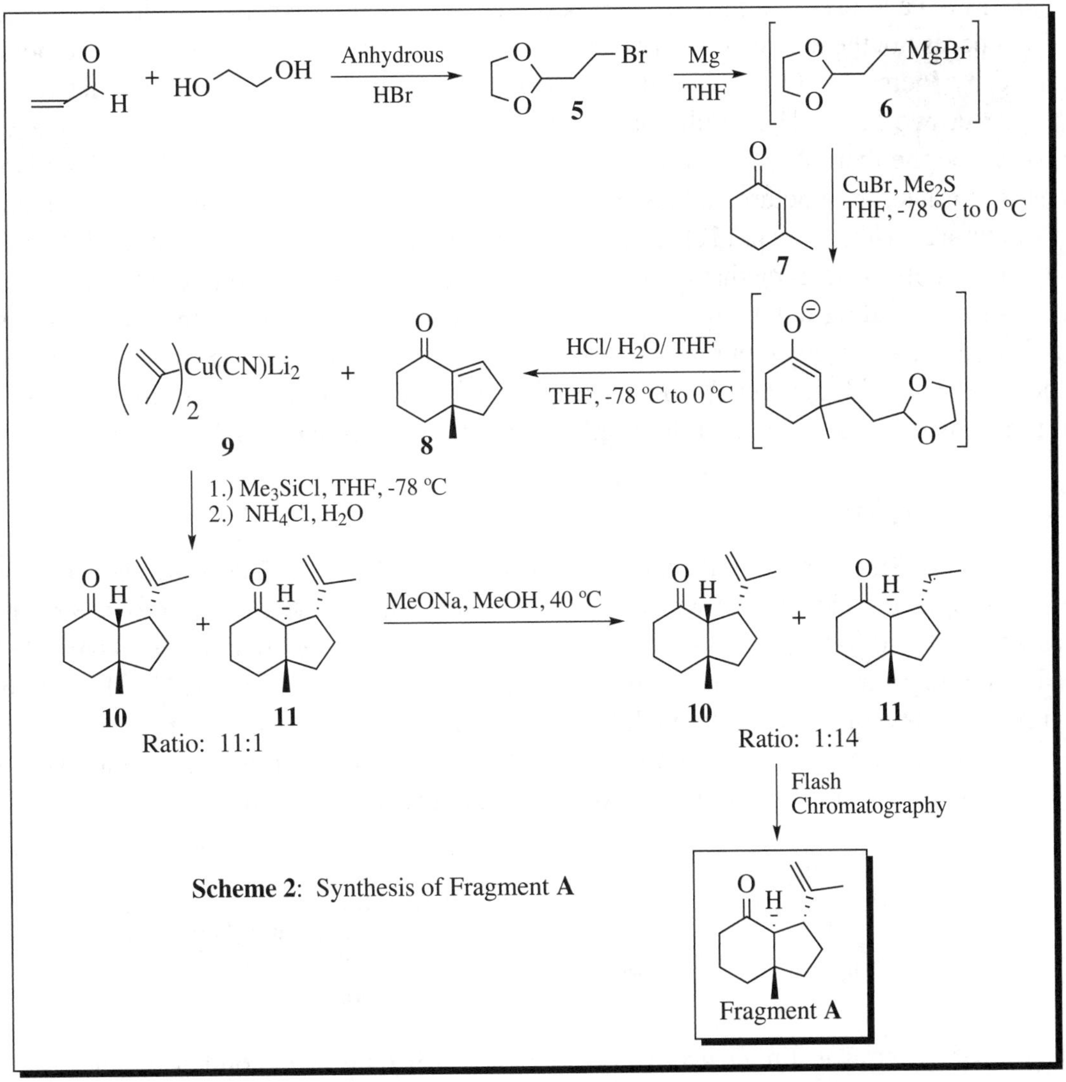

Scheme 2: Synthesis of Fragment A

The complete synthesis of the known[3] Fragment A will follow literature procedures and is outlined in Scheme 2. Bromoacetal (**5**) can be prepared following the procedure by Buchi and West.[4] Condensation of acrolein with ethylene glycol in the presence of anhydrous hydrogen bromide affords the bromoacetal (**5**) in good yield. The bromoacetal (**5**) is then converted to the corresponding grignard[5] (**6**) by treatment with magnesium in tetrahydrofuran at room temperature. Treatment of (**6**) with a mixture of the α,β-unsaturated ketone (**7**), cuprous

III. Synthesis of Variecolol

bromide and dimethyl sulfide in tetrahydrofuran yields the 1,4-addition product (**8**) after aqueous acidic workup in moderate yield. Treatment of (**8**) with the previously prepared[6] lithium dialkyl cyano cuperate (**9**) adds stereoselectively in a Michael fashion to provide a mixture of (**10**) and (**11**) in a ratio of approximately 11:1 and in excellent yield, however, the major isomer is not the desired one. This problem can be circumvented using the methodology described by Piers[7] and Oballa and the conformational studies of 4-hydrindanones done by Dana[8] and co workers. The results of the work done by Dana and co workers suggest that (**11**) would be the thermodynamically more stable isomer and the rationale is logical: isomer (**11**) has the trans ring juncture and less gauche interactions when compared to (**10**) which has the cis juncture. Thus, isomer (**11**) can be obtained preferentially by treating the 11:1 mixture with sodium methoxide in methanol at 40 °C causing epimerization at the α-carbon and hence approximately yielding a 1:14 mixture of (**10**) and (**11**). Separation of the two epimers by flash chromatography gives pure (**11**) in modest yield. In summary, a literature precedent for the synthesis of Fragment **A** will be used and is quite convenient, setting the absolute configurations in three of the nine stereogenic centers in Variecolactone

B. Synthesis of Fragment B

The synthesis of fragment **B** will be done according to Scheme 3. The known hexacarbonyl-dicobalt complexed alkyne (**12**) will be prepared according to the work done by Krafft and co workers.[9] First, 3-butyn-1-ol (**13**) will be converted to its *tert*-butyldimethylsilyl (TBS) ether by treating a mixture of (**13**) with TBS–Cl in triethylamine and DMAP. Treatment of the protected alkynol with *sec*-butyl lithium in THF followed by low temperature quench with ethyl chloroformate yields the alkynoate (**14**). The alkynoate is then complexed with dicobaltoctacarbonyl in petroleum ether to yield (**12**) in excellent yield.

R' > R"
R''' > R'''' in steric bulk

Major   Minor   **Fig 1**

Production of the tetrasubstituted cyclopentenone (**15a, 15b**)) involves condensing the protected vinyl ether (**16**) with (**12**) via a Pauson-Khand reaction.[10,11] ① Studies on the Pauson-Khand reaction[10,11] have shown that upon condensing an unsymmetrical alkene with an unsymmetrical alkyne in the presence of a cobalt carbonyl compound give rise to mixtures of diastereomers (Fig 1) with the bulkier substituents of the alkene and alkyne being α to the carbonyl of the cyclopentenone. However, the stereochemistry α to the carbonyl is unimportant as it will be lost in later steps. ④ The stereochemistry at the β position is important, and in both products gives the correct absolute configuration, thus, no separation is necessary.

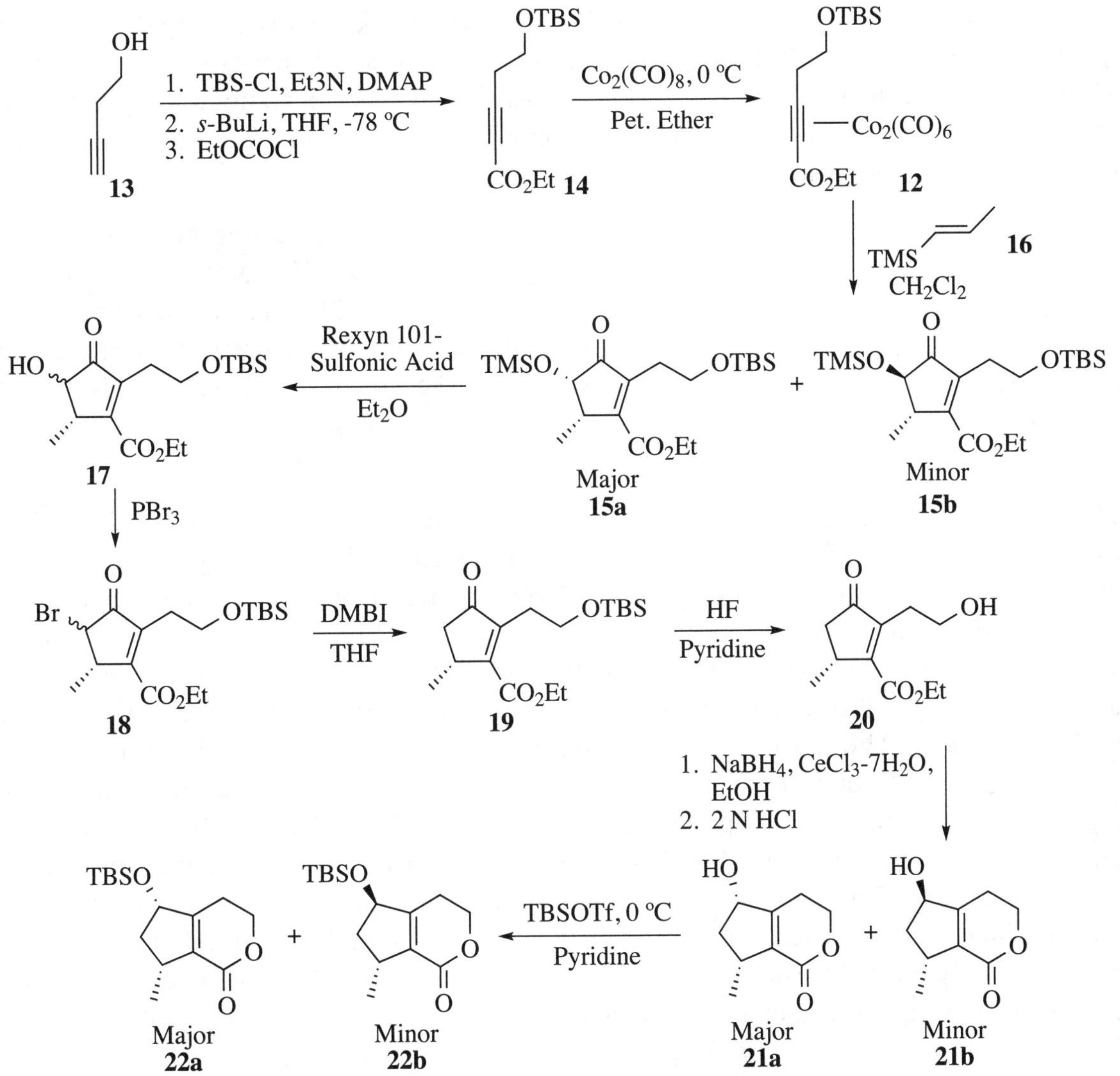

Scheme 3: Synthesis of Fragment **B**

The next step involves selectively deprotecting the trimethylsilyl ether to produce (**17**) by a method described by Bunce and Hertzler.[12] Their method involves treating TMS ethers with sulfonic acid type exchange resins which will remove the TMS group and not the TBS group.[13]

The two preceding steps serve to eliminate the hydroxyl group of (**16**) by first converting the alcohol to a bromine and then replacing the bromine with a hydrogen. The bromo analog will be made using a method described by Sacks and Aston[14] for converting α-hydroxyketones to α-bromoketones. Thus, treatment of (**17**) with phosphorus tribromide yields the α-bromoketone

(**18**). Subsequent removal of the bromine and replacement with hydrogen to produce (**19**) will follow work done by Chikashita and Itoh.[15] Their work involves using 1,3-dimethyl-2-phenylbenzimidazoline (DMBI)② as the reducing agent and is believed to go through a radical reaction mechanism. Through their studies, DMBI was shown not to react with other functioanlities contained within the α-bromoketones such as esters, α,-systems, and ethers making this methodology attractive for this synthesis.

Removal of the TBS protecting group to produce (**20**), cyclization (**21a, 21b**), protection (**22a, 22b**), hydrogenation (**23a, 23b**), and rearrangement (**24a, 24b**) will follow the procedure done by Krafft and coworkers.[9] The structure (**25**) they did this series of transformations on to produce (**26**) is very similar to (**20**) and we believe it to work without incident. ㉒

Thus treating a pyridine/acetonitrile solution of (**19**) with hydrofluorous acid will lead ⑥ to (**20**) in excellent yield. Subsequent cyclization to (**21a, 21b**)⑦ by treating (**20**) first with an ethanolic solution of sodium borohydride and cerium (III) heptahydrate, and second a two normal solution of hydrochloric acid. This reaction is predicted to be highly diasteroselective. Because of the methyl's orientation, hydride transfer will occur preferentially on the opposite face of the methyl. Protecting the alcohol again with the TBS group using TBS-triflate in pyridine produces (**22a, 22b**). Hydrogenation of this mixture is also predicted to be highly diastereoselective. With the bulky TBS group present, hydrogen should add across the double bond on the opposite face thus producing a mixture of (**23a, 23b**).⑤ ⑧   The three step rearrangement process to form (**24a, 24b**) is as follows. Upon reducing the lactone moiety in the mixture from the previous step with diisobutyl aluminum hydride in toluene, gives a lactol which is then treated with ten equivalents of methyltriphenylphosphonium bromide and *n*-butyl lithium in THF to give the corresponding alkene. The large excess of ylide is required to avoid epimerization of the stereocenter adjacent to the aldehyde prior to the Wittig olefination. Oxidation of the primary alcohol using Jone's Reagent[9] to the carboxylic acid, followed by refluxing in two normal hydrochloric acid will yield, by and Sn2 reaction, (**24a, 24b**). ⑨   Separation of the mixture of diastereomers by flash chromatography is predicted to yield pure (**24a**) in moderate yield. Ozonolysis of (**24a**) to produce (**25**) will follow the pioneering work done by Pappas and co workers.[16] A mixture of (**24a**) in methanol is treated with ozonized oxygen followed by reductive workup with dimethyl sulfide to produce (**25**). The attractive methodology developed by Pappas and co workers has shown that carbonyl moieties present in olefinic compounds do not interfere with the oxidation by ozone. Reduction of the aldehyde moiety of (**25**) with sodium borohydride[17] to produce the corresponding alcohol is a well known reaction in which the lactone group is known to be left unmolested by the reducing agent, will complete the synthesis of Fragment **B**.

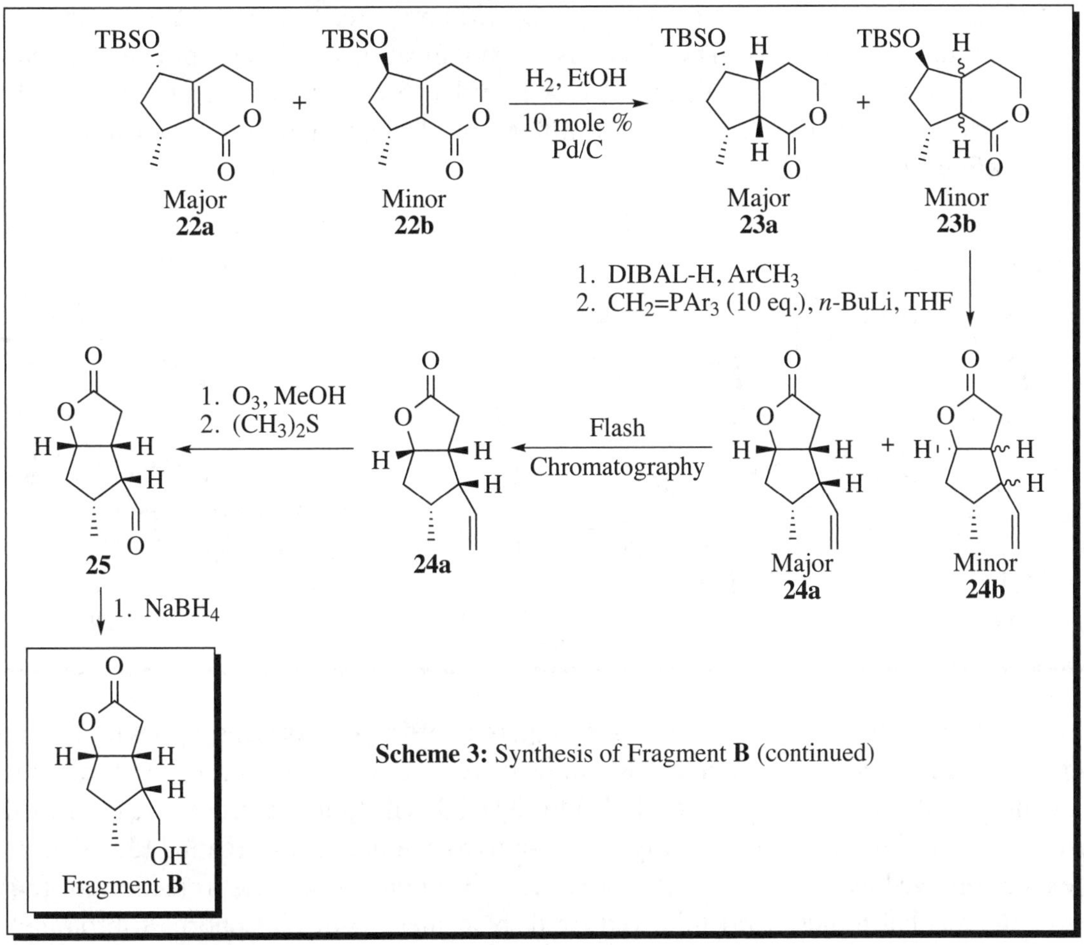

Scheme 3: Synthesis of Fragment **B** (continued)

C. D. Modification of Fragment A Before Condensation with Modified

Fragment B

The modifications of Fragment **A** which are required before it's condensation with Fragment **B** are outlined in Scheme 4. Conversion of Fragment **A** to (**26**) will follow a procedure described by Bailey and Khanolkhar.[18] Their work ⑩ describes a procedure using potassium hexamethyldisilizane as a non-nucleophilic base to produce an enolate and subsequently substituting a methyl group α to the carbonyl of a phenyl ketone. Although the hydrogens at each α position of Fragment **A** are available for abstraction, it is predicted that the kinetic hydrogen will be removed due to the steric bulk of the base employed. Thus, treatment of Fragment **A** with KHMDS followed by methyl iodide will yield (**26**). Reduction of (**26**) with sodium borohydride[17] in THF leads to the corresponding alcohol mixture (**27**). ⑪ Protection of this alcohol mixture with TMS-Cl[19] to (**28**) prepares the fragment for subsequent ozonolysis[16] to produce the ketone (**29**). Protection of this ketone with ethylene glycol[20] to make (**30**) followed by deprotection[9,17] (**31**) ⑫ and reoxidation[9,17] ⑬ (**32**) prepares Fragment **A** to be condensed with the modified Fragment **B**. The above list of modifications indirectly protects the double bond of the isopropenyl group of Fragment **A** in order to conduct a ring closing metathesis utilizing Grubb's catalyst in later steps. **Modification of Fragment B before Condensation with Modified Fragment A**

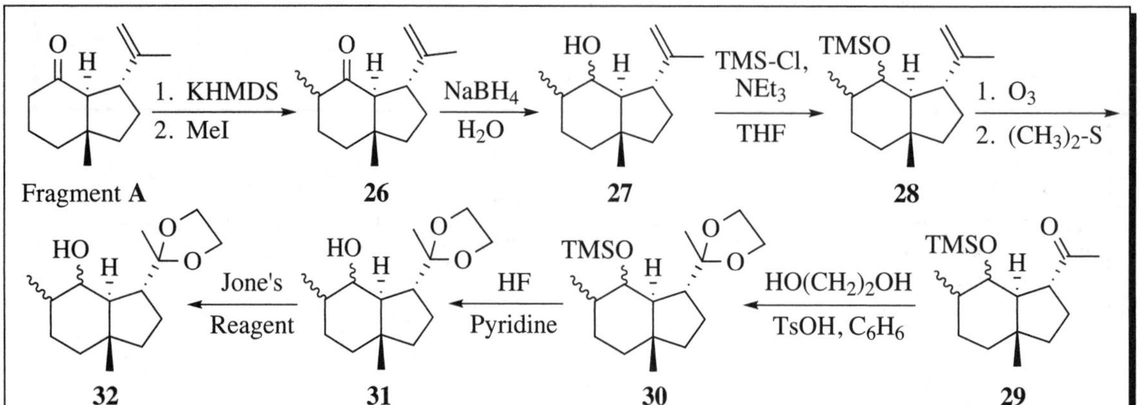

Scheme 4

The modifications of Fragment **B** which are required before it's condensation with Fragment **A** are outlined in Scheme 5. Fragment **B** will be converted to its TMS protected form (**33**) by the method cited above.[19] Allylation of (**33**) to (**34**) ⑭ will follow a procedure described by Walton and Reid[21] in their work of allylating γ-butyrolactone. Thus, treating (**33**) with LDA in hexane followed by HMPA and allyl bromide will produce (**34**). Deprotection[9] of (**34**) to recover the alcohol moiety (**35**) followed by displacement with phosphorus tribromide[17] to produce (**36**) prepares Fragment **B** to be condensed with the modified Fragment **A**.

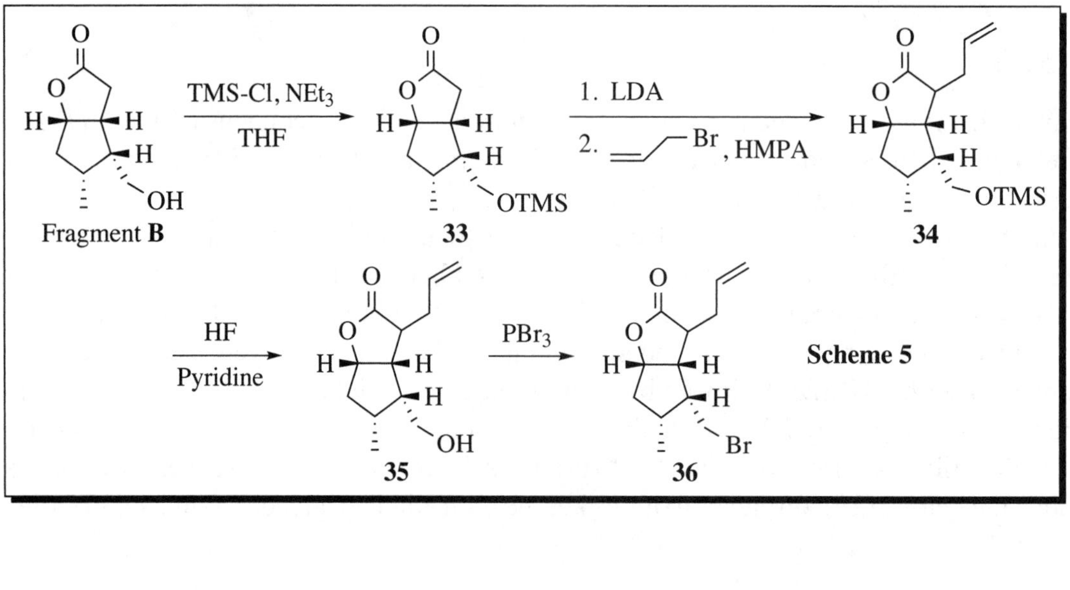

Scheme 5

E. Completion of the Synthesis of Variecolactone

The completion of the synthesis of variecolactone is outlined in Scheme 6. First, treatment of (32) with KHMDS[18] is predicted to remove the _-hydrogen adjacent to the methyl group due to the fact of the nature of the bulky base employed and the steric hinderance of the 1,3-dioxalane group adjacent to the other α-hydrogen. Once the enolate has formed it is predicted that the incoming electrophile will preferentially come in on the same face as the angular methyl ⑮ thus producing (37). If this prediction turns out to be wrong, the problem can be alleviated by first condensing Fragment A with (36) followed by methylation. Subsequent Wittig olefination[22] ⑰ of (37) to produce (38) under basic conditions will not molest the ketal or ester moiety. Cyclization of (38) using Grubb's catalyst will follow a procedure done by Prasad and Hoppe.[23] In their work, several 8-membered rings were synthesized containing ester and cyclic ether linkages using Grubb's catalyst. It is believed that their methodology can be applied in converting (38) to (39). The stereochemical outcome ⑯ concerning the hydrogen α to the ester is unimportant as it will be transformed to a double bond in a later step. Hydrogenation of (39) to (40) using Wilkison's catalyst[24] is predcited to occur on the same face as the angular methyl due to the bulky 1,3-dioxalane group present α to the prochiral center. Hydrogenation with Wilkinson's catalyst has also been show not to interfere with ester functionalities.

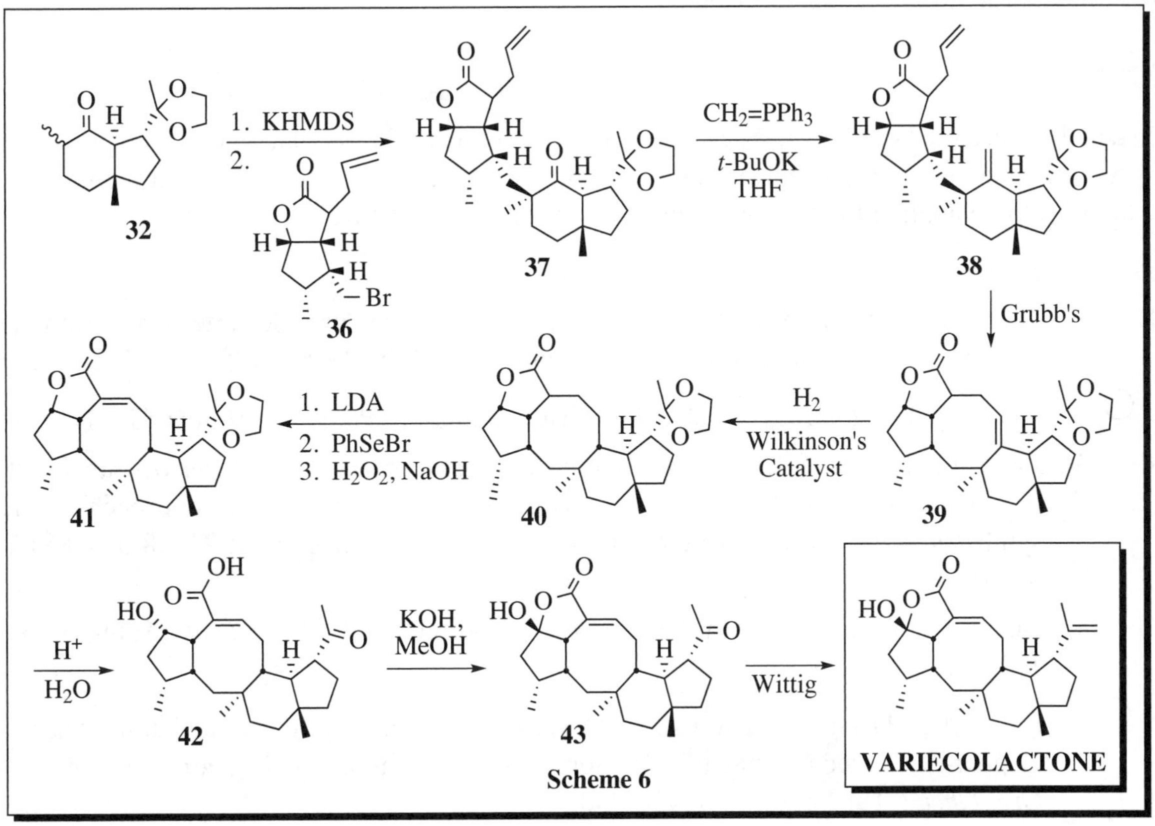

Scheme 6

Incorporation of the double bond to produce (41) will follow work done by Bella and Pirgo[25] as well as Fallis and co workers.[26] Their work involves incorporating a double bond into saturated γ-lactones containing a ketal group using phenyl selenium bromide followed by hydrogen peroxide and sodium hydroxide to execute a syn elimination to produce the double bond in conjugation with ester carbonyl. It is predicted that when (41) is enolized with LDA, the phenyl selenium bromide will preferentially approach from the bottom face due to the three angular hydrogens present α and β to the enolized carbon. Once treated with peroxide a and sodium hydroxide, the only hydrogen which can be abstracted is from the bottom face thus producing (41). Treatment of (41) with aqueous acid[17] will serve two purposes, to hydrolyze the ester ⑱ and ketal moiety to produce(42). The next step will follow work done by Dutler and co workers in their work involving steroids. One key step of their synthesis was converting a α-hydroxy carboxylic acid to a α-hydroxy lactone by using a dilute solution of potassium hydroxide in methanol. It was also noted that the stereochemistry of the hydroxyl group was observed to invert upon completion of this transformation as well as the ketone moiety being left unmolested. It is predicted that this methodology will work on our system to produce (43) ⑲ with one possible problem and that is the stereochemistry of the hydroxyl group. If it turns out to be the wrong stereochemistry , Mitsunobu[28] conditions can be utilized to invert the hydroxyl group before the cyclization. Another Wittig olefination[22] reaction on (43) shall complete the synthesis of Variecolactone.

Summary

In conclusion, a methodology has been presented to synthesize Variecolactone. More generally, a synthesis has been proposed to attain the complex fused ring structure of variecolactone which could be applied to other products with this similar structure.

References, ⑳

D 1. Takahashi, H.; Hosoe, T.; Nozawa, K.; Kawai, K. "Two New Sesterterpenes from the Ascomyycetous Fungus *Emericella purpurea*" *J. Nat. Prod.* **1999**, *62*, 1712-1713

E ③ 2. Aiya, S.; Morishita, T.; Takeshi, H. No Title Available; Patent# JP10306087, 11-17-98

F 3. Piers, E.; Boulet, S.L. "Total Syntheses of Diterpenoids (+/-)-Verrucosan-2_-ol, (+/-)-Neoverrucosan-5_-ol, and (+/-)-Homoverrucosan-5_-ol. An approach to the synthesis of the Sesterterpenoid Variecolin" *Tetrahedron Lett.* **1997**, 38(*51*), 8815-8818

G 4. Buchi, G.; Wuest, H. "Synthesis of (+/-)-Nuciferal" *J. Org. Chem.* **1969**, 34(*4*), 1122-1123

H ③ 5. Bal, S.A.; Marfat, A.; Helquist, P. "Cyclopentene and Cyclohexene Annulation via Copper-Catalyzed Conjugate Additiion of Acetal-Containing Grignard Reagents" *J. Org. Chem.* **1982**, 47(*26*), 5045-5050

I   6. Lipshutz, B.; Sengupta, S. "Organocopper Reagents: Substitution, Conjugate Addition, Carbo/Metalcupration, and Other reactions" *Org. React.* **1992**, *41*, 135-590

J③ 7. Piers, E.; Oballa, R.M. "Iterative Annulations Leading to Functionalized Tricyclo[6.4.0.0^{1,5}]Dodecanes and Tricyclo[5.3.2.0^{4,11}]Dodecanes" *Tetrahedron Lett.* **1995**, 36(*33*), 5857-5860

K   8. Dana, G.; Weisbuch, F.; Cicero, B.L. "Conformational Analysis and Stability of Substituted 4-Hydrindanones. A Thermodynamic and Magnetic Resonance (^{1}H and ^{13}C) Study" *J. Org. Chem.* **1981**, *46*, 914-919

L ㉑ 9. Kraffit, M.E.; Cheung, Y.Y.; Capucao, A.J. "Synthesis of the First "Inside-Outside" Eight-Membered Ring via Ring-Closing Metathesis: A Total synthesis of (+/-)-Asteriscanolide" *Synthesis* **2000**, *7*, 1020-1026

M   10. Krafft, M.E. "Regiocontrol in the Intermolecular Cobalt-Catalyzed Olefin-Acetylene Cycloaddition" *J. Am. Chem. Soc.* **1988**, *110*, 968-970

N   11. Krafft, M.E.; Julianoa, C.; Scott, I.; Wright, C.; McEachin, M. "The Directed Pauson-Khand Reaction" *J. Am. Chem. Soc.* **1991**, *113*, 1693-1703

O   12. Bunce, R.A.; Hertzler, D.V. "A Selective Method for Oxygen Deprotection in Bistrimethylsilylated Terminal Alkynols" *J. Org. Chem.* **1986**, *51*3451-3453

P   13. Greene, T. W.; Wuts, P.G.M. "Protective Groups in Organic Synthesis" John Wiley & Sons, Inc., New York, 1991

Q   14. Sacks, A.A.; Aston, J.G. "_-Halo Ketones. V. The Preparation, Metathesis and Rearrangement of Certain _-Bromo Ketones" *J. Am. Chem. Soc.* **1951**, 3902-3906

R   15. Cikashita, H.; Itoh, K. "1,3-Dimethyl-2-Phenylbenzimidazoline as a Novel and Efficient Reagent for Mild reductive Dehalogenation of _-Halo Carbonyl Compounds and Acid Chlorides" *J. Org. Chem.* **1986**, *51*, 5400-5405

S③ 16. Pappas, J.J.; Keaveney, W.P.; Gancher, E.; Berger, M. "A New and Convenient Method for Converting Olefins to Aldehydes" *Tetrahedron Lett.* **1966**, *36*, 4273-4278

T   17. March, J. "Advanced Organic Chemistry. Reactions, Mechanisms, and Structure" John Wiley & Sons, New York, 1992

U   18. Bailey, W.F.; Khanolkar, A.D. "Construction of a Sterically Congested Carbon Framework via 5-Hexenyllithium Cyclization. Synthesis of (+/-)-Cuparene" *Tetrahedron Lett.* **1990**, 47(*37*), 7727-7738

V ③ 19. Corey, E.J.; Snider, B.B. "A Total Synthesis of (+/-)-Fumagillin" *J. Am. Chem. Soc.* **1978**, *94*2549-2550

W   20. Fiesser, L.F.; Stevenson, R. "Cholesterol and Companions IX. Oxidation of _^{5}-Cholestene-3-one with Lead Tetraacetate" **1954**, *76*, 1728-1733

X   21. Walton, R.; Reid, B.F. "Studies on the Itramolecular Competitive Addition of Carbon Radicals to Aldehydo and Alkenyl Groups" **1991**, *113*, 5791-5799

Y   22. Maryanoff. B.E.; Reitz, A.B. "The Wittig Olefination Reaction and Modifications Involving Phosphoryl-Stabilized Carbanions. Stereochemistry, Mechanism, and Selected Synthetic Aspects" *Chem Rev.***1989**, *89*, 863-927

@A   23. Prasad, K.R.; Hoppe, D. "Applications of Homoallylic Alcohols Derived from (-)-Sparteine-Mediated Asymmetric Homaldolization:Entry to 5- to 8-Membered Unsaturated Oxacycles through Metathesis Reaction" *Synlett* **2000**, *7*, 1067-1069

AA   24. Mohrig, J.R.; Dabora, S.L.; Foster, T.F.; Schultz, S.C. "Stereospecific Synthesis of 2-Deuterio-3-hydroxybutanoate Esters. Regiochemistry and Stereochemistry of Homogeneous Hydrogenation with Wilkinsons' Catalyst" *J. Org. Chem.* **1984**, *49*, 5179-5182

BA   25. Piancatelli, G.; Bella, M.; Pigro, C. "3-Phenylselanyl-2(5H)-one: a Versatile Building Block in the Synthesis of Lignans. A New Approach Towards 3,4-Dibenzyl _-Butyrolactones" *Tetrahedron* **1999**, *55*, 12387-12398

CA   26. Wilson, P.D.; Wong, T.; Romero, M.A.; Fallis, A. "Synthesis of Functionalized Ring A Building Blocks for Taxoids" *Tetrahedron Lett.* **1995**, 36(*34*), 6039-6042

DA   27. Dutler, H.; Gantar, H.; Utzinger, E.C.; Weinberg, K. "Photochemische Reaktionen" *Helv. Chima. Acta.* **1962**, 2346-2381

EA   28. Mitsunobu, O. "The Use of Diethyl Azodicarboxylate and Triphenylphosphine in Synthesis and Transformation of Natural Products" *Synthesis* **1981**, 1-2

14.3.B The Critique

The reactions used in this proposal are as follows:

1.   Grignard reactions (sec. 8.4).

2.   Magnesium cuprate conjugate additions (sec. 8.4.D).

3.   Higher order cuprates, conjugate additions (sec. 8.7.B).

4.   Pauson-Khand reaction (sec. 13.11).

5.   Alcohol protection - trimethylsilyl ether (ec. 7.3.A.i).

6.   DMBI reduction (not covered in this book).

7.   Luche reduction (sec. 4.4.B).

8.   Dibal reduction of lactones (sec. 4.6.C).

9.   Wittig olefination (sec. 8.8.A).

10.   Ozonolysis (sec. 3.7.B).

11.   Sodium borohydride reduction of aldehydes (sec. 4.4.A).

12. Enolate alkylation of ketones (sec. 9.3).

13. Ketone protection - dioxolane (sec. 7.3.B.i).

14. Jones oxidation (sec. 3.2.A.iii).

15. Enolate alkylation of lactones (sec. 9.3).

16. Conversion of alcohol to bromide with PBr_3 (sec. 2.8.A).

17. Olefin metathesis (sec. 13.9.E).

18. Selenoxide elimination (sec. 2.9.C.vi).

The Critique. The number for each critique appears in the body of the proposal.

1. There is a discussion about the stereochemistry of the Pauson–Khand, saying it give the correct stereochemistry. Is this speculation, or based on the two references in a similar or identical system? This needs to be specific.

2. Is DMBI compatible with conjugated ketones also?

3. Note typographical error.

4. For the cyclization step–it is not clear why the reaction is diastereoselective n this system, or what the selectivity will be. Use models to explain.

5. Why is this hydrogenation diastereoselective? Is it subject to equilibration or isomerization?

6. In the conversion of **19** to **20**, will the use of HF cause problems with the ester?

7. In the cyclization, use models and discussion to show that **21a** is the major product.

8. Prove the stereochemistry of **23a** and discuss the selectivity.

9. Is there a step missing in the conversion of **23** to **24?**

10. While it is OK to cite a recent reference, this transformation was reported much earlier and the original reference should be cited –see House's book, for example.

11. Although the stereochemistry of **27** is not important for the final product, it is important for isolation and handling and there should be a comment.

12. Is the dioxolane unit stable to treatment with HF?

13. The conditions used with Jone's reagent include strong acid, which might cause problems if the contact time is long. There are many other oxidizing agents where this is not potentially a problem. Is one of them better?

14. What is the stereochemistry of the allylic unit in **34?**

15. The statement concerning stereochemistry of **37** requires proof - use models?

16. Prove the stereochemistry of **39** to **40** using models?

17. Compound **37** contains a sterically hindered carbonyl. The Wittig reaction can be slow

in such systems and/or give poor yields. Is this a problem here?

18.	This compound is a lactone, not an ester. When the lactone is treated with aqueous acid, it can open. However, the equilibrium of most five-membered ring lactones lies towards the lactone. Why should this behave differently, and can the equilibrium be manipulated to obtain the target?

19.	It is not at all obvious that the transformation of **42** to **43** will work. It may have precedent, but there should be more discussion of this reaction, particularly regarding the stereochemistry of the system and the relative distance of the two reactive centers.

20.	Why are the letters included with the references?

21.	Reference 9. Krafft is misspelled.

22.	This scheme indicates the sequence will follow the work of Krafft and that it is similar. More detail should be provided as to the similarities and particularly differences between the proposed work and the literature work. Specifically, the differences must be highlighted since they are a potential source of problems.

disconnection index

A

acridines, from diarylamines, 1219
acyl cyclobutanes
  from alkenes, 1097
  from conjugated ketones, 1100
acyloins, from esters, 1309
alcohols, allylic
  from aldehydes, 0678
  from alkenes, 0328, 0702
  from conjugated carbonyls, 0369, 0476
  from ketones, 0678
  from propargylic alcohols, 0363
  from vinyl halides, 0678
alcohols, bicyclic, from bicyclic alcohols, 1179
alcohols, cyclic, from amines, 1179
alcohols, from acid chlorides, 0357
alcohols, from acyl halides, 0648
alcohols, from aldehydes, 0357, 0373, 0456, 0468, 0476, 0527, 0646, 0668, 0680, 0694
alcohols, from alkenes, 0183, 0363, 0519, 0533, 0533, 0655, 0718
alcohols, from alkyl halides, 0680, 0684, 0688, 0718
alcohols, from ketones, 0476
alcohols, from alkylsilanes, 0331
alcohols, from anhydrides, 0357
alcohols, from aryls, 0694
alcohols, from carboxylic acids, 0357, 0648, 0684
alcohols, from epoxides, 0363, 0655, 0688, 0718
alcohols, from esters, 0357, 0648
alcohols, from ketones, 0357, 0373, 0456, 0468, 0527, 0646, 0668, 0680, 0694
alcohols, homoallylic
  from aldehydes, 0768, 0770
  from alkenes, 0770
  from ketones, 0770
  from vinyltin compounds, 0768
aldehyde-ketones, from ketones, 0710
aldehydes, conjugated
  from acids, 0846
  from aldehydes, 0702
  from alkenes, 0328

from allylic alcohols, 0255
from benzylic alcohols, 0255
from hydroxy-aldehydes, 0822
aldehydes, cyclopropyl, from alkyl halides, 0748
aldehydes, cyclopropyl, from conjugated aldehydes, 0748
aldehydes, from acids, 0846
aldehydes, from alcohols, 0239, 0255, 0258
aldehydes, from aldehydes, 0805, 0846, 1282
aldehydes, from alkenes, 0317, 0322, 0333, 0530, 0533, 1282
aldehydes, from alkyl halides, 0723, 0805
aldehydes, from alkynes, 0519
aldehydes, from conjugated aldehydes, 0369, 0476, 0533, 0723
aldehydes, from diols, 0324
aldehydes, from glycidic esters, 0835
aldehydes, from halo esters, 0835
aldehydes, from hydroxy ketones, 0317
aldehydes, from ketones, 0846
aldehydes, from vinyl ethers, 1142
aldehdyes, from vinyl sulfides, 1142
aldehydes-ketones, from aldehydes, 0710
alkanes, cyclic
  from alkene-halides, 1287
  from alkenes, 0514
  from dienes, 0514
  functionalized, from alkenes, 1293
alkanes, from active methylene compounds, 1236
alkanes, from alkenes, 0514
alkenes, bicyclic,
  from conjugated carbonyls, 0935
  from cyclopentadiene, 0935
alkenes, from vinyl halides, 0718
alkene-aldehydes,
  from allyl vinyl ethers, 1131
  from allyl vinyl sulfides, 1131
  from allylic alcohols, 1121
  from allylic halides, 0713
  from allylic sulfides, 0713
  from vinyl sulfides, 0713
alkene-conjugated ketones, from acyl cyclobutanes, 1097

alkene-imines, from allyl enamines, 1131
alkenes, bicyclic,
  from alkenes, 1036
  from cyclopentadienes, 1036
  from fulvalenes, 1036
alkenes, cyclic,
  from alcohols, 1179
  from alkene-vinyl halides, 1287
  from alkyne-halides, 1287
  from allylic alcohols, 1236
  from dialdehydes, 1311
  from dienes, 1329, 1340
  from diketones, 1311
  heteroatom substituted, from aldehydes or ketones, 1287
alkenes, from aldehydes, 0744, 1311
alkenes, from alkenes, 0694, 0768, 1231, 1247
alkenes, from alkyl halides, 0158, 0643, 0718, 0744, 1292
alkenes, from alkynes, 0456, 0507, 0509, 0763, 1292
alkenes, from allylic alcohols, 1236
alkenes, from allylic halides, 1247
alkenes, from allylic silanes, 0771
alkenes, from amine oxides, 0170
alkenes, from ammonium salts, 0170
alkenes, from aryl halides, 0768
alkenes, from carboxylic acids, 1282
alkenes, from cyclopropanes, 1316
alkenes, from dienes, 0456
alkenes, from diols, 1311
alkenes, from esters, 0170
alkenes, from homoallylic alcohols, 0177
alkenes, from ketones, 0718
alkenes, from ketones, 0744, 1311, 1318
alkenes, from nucleophilic carbon, 1231
alkenes, from selenoxides, 0170
alkenes, from sulfones, 0702
alkenes, from sulfoxides, 0170
alkenes, from vinyl halides, 0643
alkenes, from xanthate esters, 0170
alkyl halides, from alkenes, 0180
alkyls, from ketones, 0465
alkylidene compounds,
  from alkyl halides, 1292
  from alkynes, 1292

alkylidene derivatives, from fulvalenes, 0886
alkyls, from alcohols, 0363
alkyls, from aldehydes, 0465, 0474, 0476
alkyls, from alkenes, 0476, 0507, 0509, 0514
alkyls, from alkyl halides, 0363, 0369, 0465, 0643, 0677, 0702, 0718
alkyls, from ketones, 0474, 0476
alkyls, from sulfonate esters, 0363, 0369
alkyls, from sulfones, 0468
alkyls, from sulfoxides, 0468
alkyne-alcohols, from aldehydes, 0635
alkyne-alcohols, from alkynes, 0519, 0635
alkyne-alcohols, from ketones, 0635
alkynes, from aldehydes, 0160
alkynes, from alkyl halides, 0635
alkynes, from alkynes, 0635
alkynes, from vinyl halides, 0160
amide-esters, from aldehydes, 0627
amide-esters, from carboxylic acids, 0627
amide-esters, from isonitriles, 0627
amide-esters, from ketones, 0627
amides, bicyclic,
    from alkenes, 1053
    from dihydropyridines, 1053
amides, from amides, 0765
amides, from ketones, 0193
amine-ketones, from amines, 0754
amine-ketones, from ammonium ketones, 0754
amine-ketones, from esters, 0754
amines, cyclic,
    from haloamines, 1287
    from N-haloamines, 1295
amines, from aldehydes, 0373
amines, from alkenes, 0537
amines, from alkyl halides, 0363
amines, from amides, 0357
amines, from ammonium salts, 0754
amines, from azides, 0363
amines, from carboxylic acids, 0193
amines, from enamines, 0373
amines, from ketones, 0373
amines, from lactams, 0357
amines, from nitriles, 0363
amines, from sulfonamides, 0369
amino acids, from alkenes, 0537
amino nitriles, from aldehydes, 0629
amino nitriles, from ketones, 0629

anilines, from aldehydes, 1218
anthracenes, from aryl aldehydes, 1210
arenes, from alkyl halides, 0765
aryls, from alcohols, 1203
aryls, from alkenes, 1100, 1203, 1240, 1301
aryls, from alkyl halides, 0677, 0694, 0765, 1197
aryls, from alkynes, 1340
aryls, from anilines, 1301
aryls, from aryl halides, 0643, 0677, 1240, 1287
aryls, from aryls, 0687, 0694, 1200, 1203, 1210
aryls, from benzene, 0765
aryls, from benzocyclobutanes, 1100
aryls, from diynes, 1340
aryls, from lactones, 0886, 1036
aryls, from pyrones, 0886
aryls, from quinines, 1210
azadienes, from alkynes, 1058
azides, from alkyl halides, 0363
aziridines, from alkenes, 1115
aziridines, from alkyl azides, 1115
aziridines, from triazoles, 1115
azomethine imines, from aldehydes, 1113
azomethine imines, from hydrazines, 1113

B

biaryls, from anilines, 1301
biaryls, from aryl halides, 1287, 1301
biaryls, from aryls, 1301
bicyclohexanes, from alkenes, 1097
bromohydrins, from alkenes, 0188

C

carbazoles, from phenylhyrazones, 1219
carboxylic acids, conjugated,
    from aldehydes, 0827, 0831
    from anhydrides, 0827
carboxylic acids, dienyl,
    from aryl carboxylic acids, 0460
carboxylic acids, from alcohols, 1187
carboxylic acids, from aldehydes, 0260
carboxylic acids, from alkenes, 0317, 0322
carboxylic acids, from alkyl halides, 0651, 0685, 0762, 0809

carboxylic acids, from alkynes, 0519
carboxylic acids, from aryls, 0694
carboxylic acids, from carbon dioxide, 0651, 0685
carboxylic acids, from carboxylic acids, 0317, 0762
carboxylic acids, from cyclic ketones, 1316
carboxylic acids, from keto acids, 1316
carboxylic acids, from ketones, 1316, 1319
carboxylic acids, from malonates, 0809
carboxylic acids, from malonic acids, 0177
carboxylic acids, from methyl ketones, 0885
carboxylic acids, from primary alcohols, 0260
carboxylic aids, conjugated, from malonates, 0831
chlorohydrins, from alkenes, 0188
coumarins, from keto esters, 1221
coumarins, from phenols, 1221
cyclic alkanes, from dienes, 0514
cyclic alkenes, from alkenes, 0514
cyclic ketones, conjugated, from alkenes, 1051
cyclic ketones, conjugated, from dienes, 1051
cyclobutanes, from alkenes, 1094
cyclobutanes, from conjugated carbonyls, 1094
cyclobutanes, from dienes, 1094
cyclobutanones, from aldehydes, 0750
cyclobutanones, from alkenes, 1081
cyclobutanones, from cyclopropylcarbinyl halides, 0750
cyclobutanones, from ketenes, 1081
cyclobutanones, from ketones, 0750
cycloheptatrienes, from aryls, 1328
cycloheptenones, from dibromo ketones, 1191
cyclohexenes, from alkenes, 0921
cyclohexenes, from dienes and alkenes, 1017
cyclohexenes, from dienes, 0921, 1187
cyclopentanones, from dibromo ketones, 1191
cyclopentanones, from ketones, 0750
cyclopentenes, from vinylcyclopropanes, 1121
cyclopentenones, from alkenes, 1191

cyclopentenones, from dienes, 1191
cyclopentenones, from divinyl ketones, 1189
cyclopropanes, from alkenes, 1111, 1320, 1326, 1329, 1334
cyclopropanes, from pyrazolines, 1111

D

diazoalkanes, from amines, 1315
diazoalkenes, from ketones, 1318
dicarboxylic acids, from 1,2-diols, 0260
dienes, cyclic, from aryls, 0460
dienes, cyclic, from bicyclohexanes, 1097
dienes, cyclic, from tetraenes, 1247
dienes, from cyclobutenes, 1100
dienes, from alkenes, 1240
dienes, from alkyl halides, 0768
dienes, from alkynes, 0513
dienes, from aryls, 1250
dienes, from bis(allylic halides), 1247
dienes, from cyclic alkenes, 1340, 1341
dienes, from cyclobutenes, 1100
dienes, from dienes, 0768, 1121, 1131, 1250
dienes, from ketones, 1241, 1250
dienes, from vinyl halides, 1240, 1302
dienes, from vinyltin, 1241
dihydrofurans, bicyclic,
   from alkenes, 1036
   from cyclohexenones, 0935
   from furans, 0935, 1036
dihydroisoquinolines, from aldehydes, 1217
dihydroisoquinolines, from phenethylamines, 1216, 1217
dihydronapthalens, from alkenes, 1100
dihydronapthlanes, from benzocyclobutanes, 1100
dihydropyrans, from aldehydes, 1055
dihydropyrans, from alkenes, 1057
dihydropyrans, from conjugated aldehydes, 1035, 1057
dihydropyrans, from conjugated ketones, 1057
dihydropyrans, from dienes, 1055
dihydropyrans, from ketones, 1055
dihydropyrans, from vinyl ethers, 1035
diketones, from alkyl halides, 0805
diketones, from ketone-acids, 0805
dinitriles, from dihalides, 0833

diols, from aldehydes, 1305
diols, from alkenes, 0303, 1191
diols, from aryls, 0306
diols, from ketones, 1305
diols, from lactones, 0357
diynes, from alkynes, 1302
diynes, from haloalkynes, 1302

E

enamines, from aldehydes, 0373
enamines, from conjugated imines, 1035
enamines, from ketones, 0373
epoxides, from aldehydes, 0748
epoxides, from alkenes, 0269, 0277, 0284, 0363, 0655, 0688, 0718, 0813
epoxides, from alkyl halides, 0748
epoxides, from ketones, 0748
epoxy-alcohols, from allylic alcohols, 0284
epoxy-aldehydes,
   from conjugated aldehydes, 0277
epoxy-ketones, from conjugated ketones, 0277
esters, conjugated,
   from aldehydes, 0846
   from acids, 0846
   from conjugated aldehydes, 0255
   from ketones, 0846
esters, from aldehydes, 0313
esters, from alkenes, 0534
esters, from carboxylic acids, 1321
esters, from hydroxy ketones, 0317
esters, from ketones, 0313
esters, sulfonate, from alcohols, 0363
esters-amides, from aldehydes, 0627
esters-amides, from carboxylic acids, 0627
esters-amides, from isonitriles, 0627
esters-amides, from ketones, 0627
ethers, from alkenes, 0183
ethers, vinyl, from phenolic compounds, 0460
exo-methylene compounds, from alkenes, 1236

G

Grignard reagents, from alkyl halides, 0646
Grignard reagents, from aryl halides, 0646
halides, alkyl, from ketones, 0885

halides, alkenyl, from ketones, 0718
halides, alkyl,
   from alcohols, 0153
   from alkenes, 1282
   from alkyl halides, 1282
halides, aryl,
   from alkyl halides, 0718
   from aryl halides, 0718
halides, benzylic, from aryls, 0765
halides, bicyclic, from bicyclic halides, 1179
halides, vinyl,
   from aldehydes, 0643
   from alkyl halides, 0643
   from ketones, 0643, 0678
heterocycles, from alkenes, 0933, 1240
heterocycles, from heteroatom dienes, 0933
heterocycles, from heterocyclic halides, 1240
hydroxy acids, from aldehydes, 0629, 0762
hydroxy acids, from carboxylic acids, 0762
hydroxy acids, from ketones, 0629, 0762
hydroxy esters, from aldehydes, 0806
hydroxy esters, from esters, 0806
hydroxy esters, from ketones, 0806
hydroxy ketones, cyclic, from dieters, 1309
hydroxy ketones, from aldehydes or ketones, 0822
hydroxy ketones, from esters, 1309
hydroxy-ketones, cyclic, conjugated,
   from diketones, 0823
   from keto-aldehydes, 0823
hydroxy-ketones, from acid derivatives, 0813
hydroxy-ketones, from alkenes, 0813
hydroxyketones, from epoxides, 0813
hydroxy-ketones, from ketones or aldehydes, 0822
hydroxy-ketones, from ketones, 0712

I

imidazolidinones, from alkenes, 1110
imidazolidinones, from nitrones, 1110
imines, azomethine, from aldehydes, 1113
imines, azomethine, from hydrazines, 1113

imines, from alkenes, 1058
imines, from azadienes, 1058
imines, from enamines, 1121, 1142
indoles, from aldehydes, 1224
indoles, from anilines, 1224
indoles, from carboxylic acids, 1224
indoles, from ketones, 1224
indoles, from phenylhydrazines, 1224
indolones, from halo-acids, 1224
indolones, from anlines, 1224
iodohydrins, from alkenes, 0188
isocyanates, from alkyl halides, 0627
isocyanates, from isonitriles, 0627
isonitriles, from alkyl halides, 0627
isoquinolines, from aldehydes, 1217, 1218
isoquinolines, from phenethylamines, 1216-1218
isoxazolines, from alkenes, 1107
isoxazolines, from nitrile oxides, 1107

K

ketenes, from carboxylic acids, 1081
keto acids, from alkyl halides, 0805
keto acids, from carboxylic acids, 0827
keto acids, from keto acids, 0805
keto esters, conjugated, from keto esters, 1334
keto esters, from alkyl halides, 0805
keto esters, from carboxylic acids, 0827
keto esters, from keto esters, 0805
keto-acids, from 1,3-diketones, 0177
keto-esters, cyclic, from dicarboxylic acids, 0828
ketone-aldehydes, from ketones, 0710
ketones from homoallylic alcohols, 0177
ketones, acetoxy,
    from ketones, 0324
    from vinyl esters, 0324
ketones, aryl,
    from aryl halides, 0884
    from ketones, 0884
ketones, chloro, from alkenes, 0333
ketones, conjugated,
    cyclic, from conjugated ketones, 0881
    cyclic, from diketones, 0823
    cyclic, from ene-ynes, 1344
    cyclic, from hydroxy ketones, 0823
    cyclic, from keto-aldehydes, 0823
    cyclic, from ketones, 0881

from alkyl halides, 0712
from allylic alcohols, 0255, 0328
from benzylic alcohols, 0255
from conjugated aldehydes, 0712
from hydroxy-ketones, 0822
from ketones, 0702
spirocyclic from bis(phenols), 1298
ketones, cyclic,
    from anisole derivatives, 0461
    from dicarboxylic acids, 0828
    from dienyl-conjugated ketones, 1329
    from dinitriles, 0833
    from keto esters, 1330
    from ketones, 0750
ketones, cyclopropyl,
    from alkene-ketones, 1329
    from alkyl halides, 0748
    from conjugated ketones, 0748
ketones, divinyl,
    from alkenes, 1189
    from conjugated ketones, 1189
    from conjugated acyl halides, 1189
ketones, from acids, 0846
ketones, from acyl halides, 0684
ketones, from acyloins, 0465
ketones, from alcohols, 0239, 0255, 0258
ketones, from aldehydes, 0705, 0709, 0835, 0846, 1282, 1321
ketones, from alkenes, 0317, 0322, 0333, 0530, 0533, 0655, 1282
ketones, from alkyl halides, 0723, 0651, 0684, 0685, 0705, 0709, 0712, 0719, 0761, 0805, 0809
ketones, from alkynes, 0183, 0519
ketones, from amides, 0684
ketones, from carbon dioxide, 0685
ketones, from carboxylic acids, 0684, 0719
ketones, from conjugated ketones, 0369, 0456, 0465, 0476, 0533, 0723, 1292
ketones, from diols, 0324, 1181
ketones, from esters, 0684
ketones, from halo-ketones, 0643
ketones, from hydroxyketones, 0465
ketones, from keto-acids, 0177, 0809, 0846
ketones, from ketones, 0643, 0712, 0805, 1181, 1282
ketones, from nitriles, 0651, 0685
ketones, from vinyl ethers, 1142
ketones, from vinyl sulfides, 1142
ketones, halo, from ketones, 0884

ketones, spirocyclic,
    from cyclic ketones, 1179
    from phenols, 0772
ketones-aldehydes, from aldehydes, 0710
keto-nitriles, from nitriles, 0833

L

lactams from amines, 0771
β-lactams, from alkenes, 1081
β-lactams, from isocyanates, 1081
lactams, from cyclic ketones, 0193
lactams, from imides, 0373, 1187
lactols, from lactones, 0357
lactones, bicyclic,
    from alkenes, 1036
    from lactones, 1036
lactones, from aldehydes, 1035
lactones, from anhydrides, 0373
lactones, from conjugated ketones, 1035
lactones, from cyclic ketones, 0313
lactones, from diols, 0258
lactones, from pyrones, 0886

N

nitriles, allylic from allylic halides, 0126
nitriles, aryl, from aryl halides, 0629
nitriles, from alkyl halides, 0363, 0625, 0685, 0833

O

oxabicycloheptenes, from alkenes, 1051
oxabicylcoheptenes, from furans, 1051
oxetanes, from aldehydes, 1096
oxetanes, from alkenes, 1096
oxetanes, from ketones, 1096

P

phenanthrenes, from alkyl halides, 1210
phenanthrenes, from naphthalenes, 1210
phenanthridines, from ketones, 1219
phenathridines, from aminobiphenyls, 1219
phenolic ketones, from carboxylic acids, 1220
phenolic ketones, from phenolic esters, 1220

phenols, from aryls, 0268
phenols, from formate esters, 0313
phenols, from phenols, 1220
piperidines, from pyridine, 0373
pyrazolines, from alkenes, 1111
pyrazolines, from diazoalkanes, 1111
pyrazolizidines, from alkenes, 1113
pyrazolizidines, from azomethine
    imines, 1113
pyridines, from azadienes, 1058
pyrroles, from alkenes, 1240
pyrrolidines, from alkenes, 1112
pyrrolidines, from azomethine ylids,
    1112
pyrrolines, from alkenes, 1106
pyrrolines, from nitrile ylids, 1106

Q

quinolines, from aldehydes, 1213
quinolines, from anilines, 1212-1215
quinolines, from diesters, 1213
quinolines, from keto esters, 1212
quinolines, from pyruvic acids, 1214
quinolines, from triols, 1215
quinolones, from anilines, 1212
quinolones, from keto esters, 1212
quinones, from cathechols, 0266
quinones, from phenols, 0266
quinones, from resorcinols, 0266

S

spirocycles, from conjugated ketones,
    1292
spirocycles, from alkene-halides, 1287
steroids, from polyenes, 1187
succinate esters, alkylidene, from
    alcohols, 0834
succinate esters, alkylidene, from
    ketones, 0834
succinic esters, alkylidene, from
    succinates, 0834
sulfides, allylic, from allylic halides,
    0713
sulfides, allylic, from vinyl sulfides,
    0713
sulfides, from alkenes, 0537

T

tetrahydropyridines, from alkenes, 1058

tetrahydropyridines, from azadienes,
    1058
tetrahydropyridines, from dienes, 1056
tetrahydropyridines, from imines, 1056
triazoles, from alkenes, 1115
triazoles, from alkyl azides, 1115

index

A

A^{1,3}-strain, 0055, 0056
abelsonite, 0932
abietane, 1209
abrine esters, as chiral catalysts, 1072
absolute configuration,
  and computers, 0023
absorption bands, for alkenes, 1264
absorption maxima,
  of functional groups, 1084
absorption spectrum,
  and photochemistry, 1088
absorption, and photochemistry, 1087
abstraction, of hydrogen
  and the **Barton reaction**, 1276
  by radicals, 1276
Ac. *See* acetate
acceptor sites, 0007
acetals,
  acid catalysts, 0609
  aldehyde protecting group, 1333
  and hydrogenation, 1279
  and hydrogenolysis, 0607
  and the Fischer indole synthesis, 1222
  and the **Mukaiyama aldol
    reaction**, 0839
  and the **Pomeranz-Fritsch
    reaction**, 1217
  as aldehydes protecting groups, 0608
  by the Paternò-Büchi reaction, 1095
  from aldehydes, 0108, 0608
  from diols, 0607
  reaction with Lewis acids, 0610
  reaction with ozone, 0609
  with silyl enol ethers, 0839
  with tosic acid, 1333
acetals, diethyl, protecting group
  for aldehydes, 0609
acetals, dimethyl, protecting group
  for aldehydes, 0609
acetals, dithioketene, 0709
acetals, ethylenedioxy, 0610
acetals, hemi. *See* hemi-acetals
acetals, propylenedioxy, 0610
acetamides
  amine protecting groups, 0615
  with LiAlH$_4$, 0615
acetate ion, and resonance, 0086

acetate radicals, SOMO, 1269
acetate
  alcohol protecting group, 0605
  loss of carbon dioxide, 1278
  potassium, buffer in peroxyacid
    epoxidation, 0278
  silver, and the **Prévost reaction**,
    0302
  and the **Baeyer-Villiger reaction**,
    0308
  buffer in peroxyacid epoxidation,
    0278
acetates, enol. *See* enol acetates
acetates
  iodo, 0303
  reduction with LiAlH$_4$, 0353
  with **Grignard reagents**, 0648
acetic acid
  with Cr(VI), 0232
  and hydrogenation, 0444
  with alcohols, 0605
  with zinc, 0464
acetic anhydride. *See* anhydride
acetic anhydride
  and DMSO, 0245
  and SeO$_2$, 0325
  with alcohols, 0605
  with amines, 0592
  with aniline, 1199
  with ketones, 0815
acetoacetic ester synthesis, 0808
acetone
  and **Jones reagent**, 0233
  and **Oppenauer oxidation**, 0250
  triisopropoxyaluminum, 0467
acetonides, diol protecting groups,
  0606
acetonitrile, solvent effects in acidity,
  0088
acetophenone, as a sensitizer, 1090
acetoxy ketones, from enols, 0323
acetoxy-5β-pregnan-20α-ol, 1276
acetoxyheptanoic acid, 0314
acetoxytubipofuran, 1233
acetyl chloride, with alcohols, 0605
acetyl chromate ion, 0232
acetyl coenzyme A, 0948, 0949
acetylene, with borane, 0492
acetylides. *See* alkyne anions

acid catalysts
  and chromium(VI) oxidation
    of alcohols, 0228
  and hydrogenation, 0435
  and TEMPO, 0248
  with alcohols, 1163
acid chloride, with enolate anions,
  0826
acid chlorides, *See* chlorides
acid chlorides
  and acyl substitution, 0109
  reduction with LiAlH$_4$, 0353
  with amines, 0613
  with aryls, 0007
  with enamines, 0875
acid derivatives, reactivity
  to hydrogenation, 0440
acids
  and alkenes, 0514
  and Boc-amines, 0617
  with zinc, for reduction, 0463
acid-base equilibrium ketones
  and aldehydes, 0782
acid-catalyzed aldol condensation
  0820
acidity
  and the Periodic table, 0083
  and +I and –I effects, 0085
  and dithianes hydrolysis, 0706
  and electronegativity, 0084
  and electrostatic potential maps,
    0785
  and field effects, 0085
  and functional groups, 0692, 0784,
    0785
  and hydrogen bonding, 0088
  and hyperconjugation, 0085
  and ion aggregation, 0088
  and ionic size, 0084
  and ion-pairing, 0088
  and PCC, 0236
  and peroxyacids, 0277
  and resonance, 0086
  and steric effects, 0085
  and the conjugate acid-base pair,
    0084
  and the field effect model, 0084
  and the inductive model, 0084
  and the **Kirkwood-Westheimer
    model**, 0085

and through-bond effects, 0085
common carbon acids, 0633
inductive effect of halogens, 0085
of alkanes, 0639
of alkynes, and dissolving metal
   reduction, 0455
of ethers, 0673
solvent effects, 0087
acids and bases
  and enolate anions, 0797
  reactions, 0080
acids, amino. *See* amino acids
acids
  alkynes, 0632
  and hydrogenation, 0447
  and hydrogenolysis, 0440
  and hydrosilylation, 0475
  and permanganate, 0224
  and sulfide oxidation, 0335
  Brønsted-Lowry, 0081-0088
  Brønsted-Lowry, definition, 0081
  carbon acids, acidity, 0633
  carboxylic. *See* carboxylic
  formic. *See* formic
  hard. *See* hard acids
  Lewis. *See* Lewis
  reaction with alcohols, 0082
  reaction with alkenes, 0177-0188
  reactions with diols, 1180
  soft. *See* soft acids
  strong acid definition, 0083
  table of pK_a, 0693
  with aldehydes or ketones, 1163
  with dienones, 1188
  with epoxides, 1181
  with **Fremy's salt**, 0262
  with polyenes, 0950
  with zinc, 0464
Acinetobacter NCIB 9871,
  and the **Baeyer-Villiger reaction**,
  0312
Acinetobacter sp. NCIB 9872,
  and the **Baeyer-Villiger reaction**,
  0312
acorenone, 0765
acridines
  and **Bernthsen acridine synthesis**,
  1218
  from diarylamines, 1218
acrolein
  in the **Diels-Alder reaction** 1027,
  1038, 1056
  with alkenes, 1056

with BF_3, 1039
acrylate, methyl, with
  dimethylaminoisobutene, 1076
acrylates, chiral, and facial
  selectivity, 1070
activating groups, electrophilic
  aromatic substitution, 0202
activating substituents, electrophilic
  aromatic substitution, 1192
activation barrier, and
  Cope rearrangement, 1127
activation energy
  and **Arrhenius equation**, 1018
  and hydroboration, 0517
  and pinacol coupling, 1303
  carbenes with alkenes, 1322
  Diels-Alder reaction, 1021-1023,
  1026
activation of reaction, and **Lewis
  acids**, 1038
activation volume,
  and high pressure, 1045
  and rate of the **Diels-Alder
  reaction**, 1045
  and the **Diels-Alder reaction**,
  1045, 1048
active methylene compounds, 0829
  with allyl palladium, 1233
  with palladium(0), 1233
active sites, in catalytic
  hydrogenation, 0427
acutiphycin, 0990
acyl addition. *See* addition, acyl
acyl addition, 0108
  of sulfoxide anions, 0698
acyl anion equivalents, 0009, 0704
acyl azides. *See* azide
acyl halides. *See* halides
acyl hydrazines, with nitrous acid,
  0192
acyl imines, in the **Diels-Alder
  reactions**, 1055
acyl peroxides, thermolysis, 1261
acyl radicals, 1261
acyl substitution. *See* substitution
acylation
  Friedel-Crafts acylation, 0007,
  0929, 0932, 1203-1210
  of enolate anions, 0811
acylium ions, 1203
  and **Friedel-Crafts acylation**, 1204
  and **Fries rearrangement**, 1220
  and phenols, 1220

and resonance, 1204
and the **Fries rearrangement**, 1220
acylnitrilium salts, cyclization, 1186
acylnitroso compounds, 1056
acyloin condensation, 1306
  and chlorotrimethylsilane, 1307
  and sodium sand, 1307
  intramolecular, 1306, 1307
  powerful in synthesis, 0929
acyloins. *See* hydroxy ketones
acyloins
  and $TiCl_3$-Zn-Cu, 1308
  cyclic, 1307
  from diesters, 1307
  from esters, 1306
  with zinc and acetic acid, 0464
acyloxyborohydrides, 0377
  reductive alkylation, 0378
acylureas
  by the **Curtius rearrangement**,
  0192
  by the **Hofmann rearrangement**
  0193
adamantanes, bromo, 1281
Adams' catalyst, 0438
  and hydrogenation, 0439
  with ferric chloride, 0439
addition reactions, 0177-0188
  and radicals, 1270
  and radicals, 1271
  chain radical,1271
  conjugate, 0107, 0111
addition reactions, acyl, 0106-0109
  and alkyne anions, 0107, 0108
  and amines, 0108
  and base strength, 0786
  and electrophiles, 0107
  and enolate anons, 0816
  and HSAB, 0104
  and nucleophiles, 0107
  and vinylogy, 0106
  order of nucleophilic strength, 0107
  to carbonyls, 0108
  with alkyne anions, 0635
  with **Grignard reagents**, 0644
addition, conjugate
  and enamines, 0875
  and enolate anions, 0112
  and resonance, 0112
  and **Robinson annulation**, 0112
  and vinylogy, 0106
  conjugate, silyl cuprates, 0331
  Michael addition, 0112, 0877

adipic acid, 0313
AD-mix-α, 0300, 0306
AD-mix-β, 0300
adociasulfate 1, 0716
adsorption
  and **Fétizon's reagent**, 0257
  and hydrogenation, 0434, 0435, 0436
aeruginosin 298-A, 0460
aflatoxin B$_1$, 1221
agalacto fragments,0169
agalastatin A, 0990, 0128, 0342
aggregates, of dialkylamides, 0789
aggregation state
  and nitrile enolate anions, 0832
  dialkylamide bases, solvent effects,
    0794
  dialkylamides, 0789
  enolate anions, 0796
  enolate anions, E/Z ratio, 0795
  HMPT and enolate anions, 0791
  isopropyllithium, 0671
  LDA, 0788
  LDA, and structure, 0793
  of enolate anions, 0791
  of **Grignard reagents**, 0637
  of organolithium reagents, 0671,
    0672, 0679, 0687
  of organolithium reagents,
    and solvent effects, 0672
AIBN [*azobis*(isobutyronitrile)]
  and alkenes, 1279, 1280
  and **Baldwin's rules**, 0565
  and **Chugaev elimination**, 0167
  and HBr, 0180
  and heteroatom radical cyclization,
    1293
  and radical cyclization, 1284, 1285,
    1289, 1290, 1293
  and radical reduction, 1274
  and triphenyltin hydride, 0478
  N-bromosuccinimide and
    halogenation, 0151
  radical initiator, 1274
  thermolysis, 1261
  with tributyltin hydride, 0467
aigialospirol, 0687
air oxidation, and **Doebner reaction**,
  1214
air, and metathesis catalysts, 1336
alanates, vinyl, 0762
alantrypinone, 1058, 1105
AlCl$_3$. *See* aluminum chloride

alcohol dehydrogenase, 0483
  from *Geotrichum candidum*, 0479
alcohols
  2-(trimethylsilyl)ethoxymethyl
    ether protecting group, 0598
  acetate protecting groups, 0604
alcohols, allylic
  and metal catalyzed epoxidation,
    0273
  and metal catalyzed epoxidation,
    mechanism, 0273, 0274
  and **Mislow-Evans rearrangement**,
    0699
  and **Sharpless Asymmetric
    Epoxidation**, 0284
  by **Nozaki-Hiyama-Kishi reaction**,
    0849
  by rearrangement of allylic
    sulfoxides, 0699
  by reduction of conjugated
    carbonyls, 0352, 0366
  carbocation rearrangement, 1181
  Claisen rearrangement, 1135
  epoxidation, 0560
  from conjugated carbonyls, 0370
  from propargyl alcohols, 0362
  halogenation, 0149
  model for **Sharpless Asymmetric
    Epoxidation**, 0285, 0287
  neighboring group effects, 0560
  oxidation with MnO$_2$, 0253, 0254
  S$_N$2' and halogenation, 0149
  with ortho esters, 1136
  with PCC, 0236
  with PDC, 0237
  with peroxyacids, 0560
alcohols, amino. *See* amino alcohols
alcohols
  and alkoxysulfoxonium salts, 0241
  and **Bogert-Cook synthesis**, 1209
  and carboxylic acids, 0110
  and chromate esters, 0228-0230
  and **Collins oxidation**, 0593
  and DEAD, 0545
  and dimethyl sulfide, 0217
  and dissolving metal reduction, 0452
  and esters, 0110
  and **Friedel-Crafts alkylation**, 1201
  and ketones, 0108
  and oxidation number, 0221
  and phase transfer catalysts, 0594
  and pyridine, 0145
  and S$_N$i reactions, 0145

  and **Swern Oxidation**, 0241, 0242
  and the **Claisen condensation**, 0824
  and the **Cram cyclic model**, 0406,
    0407
  and the **Darzens' procedure**, 0146
  and the E1 reaction, 0161
  and the **Grob fragmentation**, 0176
  and the Mitsunobu reaction, 0123
  and the **Mitsunobu reaction**, 0545
  and the **Williamson ether synthesis**
    0594
  and triphenylphosphine, 0545
  as a base, 0082
  as hydrogen transfer agents, 0467
  benzoate protecting groups, 0605
  benzyl ether protecting group, 0595
alcohols, benzylic
  hydrogenation, 0446
  hydrogenolysis, 0368, 0431
  oxidation with MnO$_2$, 0253, 0254
  reduction to alkyls, 0368
alcohols
  by addition of boranes to aldehydes,
    0527
  by dissolving metal reduction of
    epoxides, 0458
  by electrolytic reduction of amides,
    0472
  by electrolytic reduction of
    carboxylic acids, 0472
  by electrolytic reduction of epoxides,
    0471
  by electrolytic reduction of esters,
    0472
  by enzyme reduction of ketones,
    0480
  by **Grignard** addition to aldehydes,
    0644
  by **Grignard** addition to ketones,
    0547, 0644
  by **Grignard** reduction of aldehydes
    or ketones, 0667
  by hydroboration of alkenes, 0495,
    0551
  by hydrogenation of epoxides, 0443
  by hydrolysis of
    tetrahydroaluminates, 0350
  by LiAlH$_4$ reduction of acetates, 0353
  by LiAlH$_4$ reduction of sulfonate
    esters, 0361
  by **Luche reduction**, 0545
  by **Meerwein-Ponndorf-Verley
    reduction**, 0467

by photoreduction of ketones, 0562
by reduction of acid chlorides, 0353, 0373
by reduction of aldehydes, 0373, 0467
by reduction of anhydrides, 0353
by reduction of carboxylic acids, 0353, 0386
by reduction of epoxides, 0367, 0373, 0375, 0386
by reduction of esters, 0353, 0373
by reduction of ketones, 0373, 0467
by reduction of ozonides, 0371
by reduction of sulfonamides, 0361
by reduction of sulfonate esters, 0368
by the **Demjanov rearrangement**, 1175
by the **Tiffeneau-Demjanov ring expansion**, 1175
alcohols
chelation with peroxyacids, 0549
alcohols, chiral
by hydroboration, 0520
in the Diels-Alder reaction, 1069
alcohols
chloromethyl methyl sulfide protecting group, 0597
Collins oxidation, 0235, 0236
complex with Lewis acids, 1202
conformations, 0043
conversion to esters, 0604
Cr(VI) oxidation, rate determining step, 0228
Cr(VI) oxidation, rate of reaction, 0229
Cr(VI) oxidation, solvent effects, 0229, 0230
alcohols, cyclic
by reduction of epoxides, 0358
by reduction of ozonides, 0358
alcohols, cyclopropylcarbinyl
with copper triflate and Hünig's base, 1098
with HBr/ZnBr$_2$, 1098
alcohols
dissolving metal reduction and hydrogenolysis, 0456
DMSO and oxalyl chloride, 0241
elimination with sulfurane, 0591
ester protection groups, 0604
ethoxyethyl ether protecting group, 0600
formation of carbocations, 1163

alcohols
from acid chlorides and organolithium reactions, 0683
from aldehydes and allyltin, 0766
from aldehydes and organolithium reagents, 0677
from aldehydes, 0350, 0668, 0669, 0687
from alkenes, 0181-0183, 0514
from alkenes, enantioselectivity, 0526
from alkylboranes, 0514, 0529
from amines, 1175
from amino alcohols, 1175
from benzylic ethers, 0457
from carboxylic acids, 0684
from epoxides and organocuprates, 0718
from epoxides, 0128
from esters with organolithium reagents, 0683
from esters, 0648
from **Grignard reagents** and epoxides, 0654
from **Grignard reagents** and esters, 0110
from **Grignard reagents**, 0639
from ketones and allyltin, 0766
from ketones, 0347, 0350, 0668, 0669
from silanes, 0330
from silyl ethers, 1065
alcohols, halogenation, 0148
stereochemistry, 0145
alcohols, homoallenyl,
Saucy-Marbet reaction, 1135
alcohols, homoallylic
elimination, 0177
from alcohols, 0767
from aldehydes, 0767
from oxetanes, 0563
isomerization with **Moffatt oxidation**, 0245
alcohols
in alkoxymercuration, 0183
inversion of configuration, 0124
ionization with acid, 0130
Jones oxidation, 0233
Ley oxidation, 0249
methoxyethoxymethyl ether protecting group, 0598
methoxymethyl ether protecting group, 0596
methyl ester protecting groups, 0594
Moffatt oxidation, 0243

neighboring groups in epoxidation, 0560
Oppenauer oxidation, 0250
alcohols, oxidation, 0226-0261
and chromate esters, 0228
rate of reaction, 0228, 0230
with Ce reagents , 0261
with chromium trioxide, 0067, 1360
with Cr(VI) and manganous ion, 0231
with Cr(VI) in water, 0229, 0230
with Cr(VI) reagents that are structurally modified, 0238
with Cr(VI), mechanism, 0229
with Dess-Martin peridiodinane, 0247
with DMSO and acetic anhydride, 0246
with DMSO and DCC, 0243
with DMSO and phosphorus pentoxide, 0246
with DMSO, 0239-0246
with **Fétizon's reagent**, 0256
with **Jones reagent**, 0233, 1368
with manganese dioxide, 0036, 0252
with manganese dioxide, mechanism, 0253
with Mo reagents, 0261
with oxammonium salts, 0248
with PCC, 0236
with PDC in DMF, 0238
with pyridine-sulfur trioxide, DMSO, 0246
with silver carbonate, 0255
with sulfoxonium salts, 0239
with TEMPO, 0248
with TPAP, 0249
with **Bobbitt's reagent**, 0248
alcohols
pK$_a$, 0082, 0091
primary, with silver(II) oxide, 0260
protected as benzyl ether, 0444
protecting groups, 0594-0608
rate of oxidation with MnO$_2$, 0253, 0254
alcohols, reaction
with acids, 0082, 0144
with Cr(VI), 0228
with HBr, 0130
alcohols
Sarett oxidation, 0234, 0235
silyl protected, 0316
silyl protecting groups, 0269, 0600
solvent for enolate anions, 0800
solvents in ozonolysis, 0318

tert-butyl ether protecting group, 0596
tert-butyldimethylsilyl ether
   protecting groups, 0603
tert-butyldiphenylsilyl ether
   protecting groups, 0604
tetrahydropyranyl ether protecting
   group, 0599
triethylsilyl ether protecting groups,
   0602
triisopropylsilyl ether protecting
   groups, 0602
trimethylsilyl ether protecting groups,
   0601
alcohols
   with 2-(trimethylsilyl)ethoxy-methyl
      chloride, 0598
   with acetic acid, 0605
   with acetic anhydride, 0605
   with acetyl chloride, 0605
   with acid catalysts, 1163
   with bases, 0786
   with benzoic acid, 0605
   with benzoyl chloride, 0605
   with bromine and PPh$_3$, 0147
   with chloromethyl methyl ether, 0596
   with chloromethyl methyl sulfide, 0597
   with chlorotriethylsilane, 0602
   with chlorotriisopropylsilane, 0602
   with chlorotrimethylsilane, 0601
   with chromium trioxide-pyridine, 0234
   with Cr(VI) in acetic acid, 0232
   with cyclopentadienyl dicarbonyl
      and alkenes, 1249
   with diborane, 0499
   with dihydropyran, 0590
   with dihydropyran, 0599
   with dimethyl sulfate, 0594
   with ethyl vinyl ether, 0600
   with formic acids, 1187
   with **Grignard reagents**, 0639
   with HCl or HBr, 0144
   with isobutylene and acid, 0596
   with LHMDS, 0176
   with LiAlH$_4$, 0364
   with manganese dioxide, 0252
   with **Meerwein's reagent**, 0594
   with mesitoic acid, 0605
   with methoxyethoxy chloromethyl
      ether, 0598
   with NaBH$_4$, 0370
   with NaH, 0594
   with P and I$_2$, 0148
   with PBr$_3$, 0148, 0542

with PCC, 1186
with PDC, 0237
with phosphorus halides, 0146
with phosphorus oxychloride, 0146
with phosphorus pentabromide, 0146
with phosphorus pentachloride, 0146
with phosphorus tribromide, 0146
with phosphorus trichloride, 0146
with phosphoryl chloride, 0146
with pivaloyl chloride, 0605
with PPh$_3$ and iodine, 0148
with pyridine and POCl$_3$, 0148
with sulfuric acid, 1187
with sulfuric or perchloric acid, 0161
with *tert*-butyldimethylsilyl chloride,
   0603, 1366
with *tert*-butyldiphenylsilyl chloride,
   0604
with thionyl chloride and amines,
   0544
with thionyl chloride, 0145, 0543,
   0544
with thionyl chloride-pyridine, 0552
with TPAP and NMO, 0555
with trimethyloxonium
   tetrafluoroborate, 0594
alcohols-amines, 0306
alcyonin, 0509
aldehydes, 0438, 0645
aldehydes
   acetal formation, 0608
   acid-base equilibria, 0782
   addition of alkylboranes, 0527
   adsorption bands, 1264
aldehydes
   and acyl addition, 0108
   and betaines, 0730
   and bis(dimethylamino)naphthalene,
      0788
   and catalysts reactivity, 0430
   and enols, 0518
   and **Horner-Wadsworth-Emmons**
      olefination, 0593
   and kinetic control, 0800
   and **Knoevenagel condensation**, 0829
   and **Lewis acids**, in the **Diels-Alder**
      reaction, 1054
   and **McMurry olefination**, 1309
   and oxidation number, 0221
   and Proton Sponge, 0788
   and radicals , 1303
   and **reductive amination**, 0371
   and selenium dioxide, 0327, 0328

and **Stephen reduction**, 0465
and the **aldol condensation**, 0816
and the **Baeyer-Villiger reaction**,
   0307-0313
and the **Barbier reaction**, 0637
and the **Baylis-Hillman reaction**,
   0878, 0879
and the **Bürgi-Dunitz trajectory**,
   0795
and the **Corey-Fuchs procedure**,
   0159, 0630, 0733
and the **Cram chelation model**, 0550
and the **Grignard reaction**, 0636
and the **Henry reaction**, 0830
and the **Kamlet reaction**, 0830
and the **Knoevenagel reaction**, 0751
and the **Lemieux-Johnson reagent**,
   0314
and the **Noyori Open-Chain Model**,
   0858
and the **Passerini reaction**, 0627
and the **Perkin reaction**, 0827
and the **Petasis reagent**, 0756
and the **Reformatsky reaction**, 0885
and the **Strecker synthesis**, 0628
and the **Takai reaction**, 0757
and the **Tebbe reagent**, 0754
and the **Ugi reaction**, 0627
and the **Wacker process**, 0333
and the **Wittig reaction**, 0730, 0732
and tris(diethylamino)sulfonium, 0858
and umpolung, 0704
and **Wolff-Kishner reduction**, 0472
pinacol coupling, 1303
aldehydes, aromatic, and the
   Claisen-Schmidt reaction, 0817
aldehydes
   by carbonylation of alkylboranes, 0532
   by chain extension of aldehydes, 0533
   by **Darzens' glycidic ester**
      condensation, 0834
   by hydrolysis of hydrooxazolines,
      0845
   by hydrolysis of imines, 0466
   by hydrolysis of imines, 0847
   by hydrolysis of nitrones, 0907
   by hydrolysis of oxazolidines, 0844,
      0845
   by hydrolysis of oximes, 1277
   by hydrolysis of silyl enol ethers, 0837
   by hydrolysis of vinyl ethers, 0733
   by **Oppenauer oxidation**, 0250

by reduction of a **Weinreb amide**, 0389

by reduction of acid chlorides, 0365

by reduction of esters, 0366, 0390

by reduction of nitriles, 0390, 0465, 0466

by **Swern Oxidation**, 0241

by the **Barton reaction**, 1277

chain extension reactions, 0835

aldehydes, chiral

  enantioselectivity in aldol condensation, 0862

aldehydes, complexation

  with Lewis acids, 1038

aldehydes, condensation

  with amides enolates, 0824

  with anhydrides enolates, 0824

  with ester enolates, 0823

  with lactam enolates, 0824

  with lactone enolates, 0824

aldehydes, conjugated

  acyl addition, 0107

  and epoxidation, 0275

  and hydroperoxide anions, 0275

  and $LiAlH_4$ reduction, 0352

  and $NaBH_4$, 0370

  by **Claisen-Schmidt reaction**, 0817

  by rearrangement, 0699

  by **Wacker oxidation**, 1226

  conjugate addition, 0107

  from ketones, 0844, 0845

  from dienes, 1226

  hydrogenation, 0438, 0439

  in the **Diels-Alder reaction**, 1056

  Mukaiyama aldol reaction, 0840

  with alkenes, 1056

  with **Grignard reagents**, 0652

  with hydroperoxide anions, 0275

  with organocuprates, 0714, 0720

aldehydes,

  cyclization, 0558

  diastereoface selection, 0862

aldehydes, diastereoselectivity

  of **Grignard** addition, 0659

  of organolithium addition, 0681, 0682

aldehydes

  diethyl acetals, 0609

  dimerization upon electrolysis, 0469

  dimethyl acetals, 0609

  dioxane protecting groups, 0610

  dioxolane protecting groups, 0610

  dissolving metal reduction, 0453

  dithianes protecting groups, 0611

dithioacetals protecting groups, 0611

dithioacetals protecting groups, 0611

dithiolanes protecting groups, 0611

double stereodifferentiation, 0862

electrolytic reduction, 0469

elimination of CO, 1265, 1266

enantioselectivity of **Grignard** addition, 0659

ene reactions, 1149

enol content, and pK_a, 0783

enolate anions from **Grignard** reagents, 0669, 0670

enolate anions, 0802

formation of acetals, 0108

formation of dithianes, 0704

formation of radicals, 1278

aldehydes

  from acetals, 1333

  from alcohols, 0233-0255, 0256

  from aldehydes, 0835

  from alkenes, 0319, 0324, 0333, 1226

  from alkylboranes, 0528, 0531

  from alkynes, 0518

  from amides, 0364

  from diols, 0322

  from DMF and **Grignard** reagents, 0684

  from esters, 0825

  from glycidic ester, 0834

  from nitriles, 0364

  from nitro compounds, 0831

  from oxazines or oxazolines, 0843

  from reduction of amides, 0355

  from the **Nef reaction**, 0831

  from thioaldehydes, 1141

aldehydes

  hydrogen acceptors in **Oppenauer oxidation**, 0251

  hydrosilylation, 0475

  in the Diels-Alder reactions, 1053

  in the **Pauson-Khand reaction**, 1342

  intramolecular aldol condensation, 0822

  keto. *See* glyoxals

Kornblum aldehyde synthesis, 0240

mechanism of $LiAlH_4$ reduction, 0350

Meerwein-Ponndorf-Verley reduction, 0467

non-enolizable, 0803

oxidation with Cr(VI), 0231

oxidation with silver(I) oxide, 0260

oxidation with silver(II) oxide, 0259

photodissociation, 1265, 1266

photolysis, 1264

pK_a, 0782

protecting groups, 0608

radical formiation with Cr(VI), 0232

rate of reduction versus ketones, 0351

reaction with acids, 1163

reaction with chiral sulfoxide anions, 0698

reagents for oxidation, 0259

aldehydes, reduction

  with borane, 0386

  with chiral complexes, 0397

  with Dibal, 0391

  with **Grignard** reagents, 0667

  with tin, 0465

aldehydes

  self-condensation, 0817

  simple diastereoselection, 0862

  substituent effects and pK_a, 0782

  UV, 1264

  via ozonolysis, 0319

aldehydes

  with alkoxyaluminum hydrides, 0365

  with alkylboranes, 0533

  with alkyne anions, 0635

  with allylsilanes, 0769

  with allyltin, 0766

  with amines, 0108, 0371, 0873, 0874

  with amino alcohols, 0845

  with anhydride enolate anions, 0827

  with aryllithium reagents, 0687

  with BF_3, 1164

  with boranes, 0395, 0397

  with boron enolates, 0842

  with carboxylic acid dianions, 0835

  with CBr_4 and PPh_3, 0159

  with chromyl chloride, 0849

  with cyanide and MnO_2, 0255

  with cyanide, 0628

  with dialkylamide bases, 0794

  with diols, 0611

  with dithioalkyl phosphonate esters, 0709

  with dithiols, 0444, 0611

  with DMF and organolithium reagents, 0684

  with enolate anions, 0007, 0816, 0817

  with esters, and the **Claisen reaction**, 0826

  with **Grignard** reagents, 0589, 0590, 0644, 1209, 1210

  with halo ester enolate anions, 0834

  with HCN, 0628

with hydrazine, 0473
with hydroxylamine, 0189
with hydroxylamines, 1107, 1109
with imine carbanions, 0846
with keto esters, 0832
with LDA, 0787
with LiAlH$_4$, 0350
with lithiated dithianes, 0705
with lithiated enol ethers, 0707
with lithium borohydride, 0373
with lithium triphenylmethide, 0689
with magnesium, 1304
with malonate enolate anions, 0829
with manganese acetate, 1278
with N$_2$O, 0831
with NaBH$_4$, 0370, 1368
with nitrile enolates, 0832
with non-nucleophilic bases, 0787
with **Nysted reagent**, 0757
with organocerium reagents, 0668
with organocuprates, 0719
with organoiron compounds, 0761
with organolithium reagents, 0677, 0679, 1224
with phenylhydrazine, 1222
with phosphine oxide carbanion, 0741, 0742
with phosphonothioates, 0743
with potassium, 1303
with Red-Al, 0367
with samarium iodide, 1304
with Selectride, 0381
with silica gel, 0558
with silyl enol ethers, 0839
with silylketenes, 1080
with SnCl$_4$, 0558
with sodium, 1303
with succinic ester enolates, 0833
with sulfur ylids, 0746
with titanium(0), 1309
with TosMIC, 0624
with tosylhydrazine, 1317
with trimethylsilylmethyllithium, 0709
with vinyl triflates, 0849
with ylids, 0007, 0733
Alder endo rule, 1031, 1155
and FMO, 1030
Diels-Alder reaction, 1030
secondary orbital interactions, 1030
Alder ene reaction. *See* ene reaction
aldimines. *See* imines
aldol condensation, 0001, 0025, 0518,

0816, 0933, 0953, 0957
acid catalyzed, 0820
acid catalyzed, diastereoselectivity, 0821
acid-catalyzed, thermodynamic control, 0820, 0821
aldol transfer, 0821
aldol condensation
and a Dean-Stark trap, 0822
and **Baldwin's rule**, 0571
and boat transition state, 0857
and boron enolate, 0765
and boron enolates, 0856
and chelation control, 0868
and chiral auxiliaries, 0864
and chiral complexes, 0866
and chiral templates, 0866
and Cram's rule, 0863, 0864
and double stereodifferentiation, 0867
and enolate anions, 0790, 0802
and **Evans' auxiliaries**, 0865, 0869
and hydroboration, 0518
and macrolides, 0953
and nitromethane, 0830
and **Oppenauer oxidation**, 0252
and **Oppolzer's sultam**, 0864
and proline, 0866
and self-condensation, 0817
and stereodifferentiation, 0868
and sulfonamides, 0864
and the **Baylis-Hillman reaction**, 0879
and the **Evans' Model**, 0857
and the **Knoevenagel reaction**, 0832
and the **Zimmerman-Traxler Model**, 0852
and TMEDA, 0866
and tris(diethylamino)sulfonium, 0858
and ultrasound, 0817
aldol condensation
chair transition state, 0853, 0854, 0855
diastereoface selectivity, 0861, 0862
diastereoselectivity, 0851, 0858, 0859
double stereodifferentiation, 0862
enantioselectivity and **Cram's rule**, 0863, 0864
equilibration of syn/anti isomers, 0859
intramolecular, 0570, 0822, 0941, 0951

intramolecular, and ring size, 0823
isomerization, influence of base, 0859
matched-mismatched, 0867
aldol condensation, mixed, 0817, 0818, 0819
kinetic and thermodynamic conditions, 0818, 0819
aldol condensation
powerful in synthesis, 0929, 0931
proline as a chiral base, 0820
retro-aldol, 0821, 0860
reverse aldolization, 0860
simple diastereoselection, 0861, 0862
steric effects, 0861
syn/anti isomers, 0851, 0852
transition state, 0852
with aluminum compounds, 0821
with chiral substrates, 0861-0868
with Lewis acids, 0820
aldol, nitro, 0830
aldol reaction
and aldol boronates, 0841
and combinatorial chemistry, 0987
Mukaiyama aldol reaction, 0837
aldol transfer reaction, 0821
and equilibration, 0860
aldol. *See* Mukaiyama aldol
aldolates, 0818, 0819
aldolates, boron, 0843
with hydrogen peroxide, 0843
with MoOPh, 0843
aldolates, syn/anti, 0857
aldosterone, 0110
alkaloid 223A, 0473
alkaloids 0022
and photoreduction, 0478
and the Diels-Alder reaction, 1058
tetrahydroisoquinoline, 0976, 0977
alkanes
and extended conformation, 0041
chlorination, 1272
cyclic, by hydroboration of allylic halides, 0511
from alkenes, 0513
from alkylboranes, 0511
from radicals, 1263
HOMO, 1274
Markovnikov addition of water, 0514
pK$_a$, 0639
via coupling of boranes, 0513
with borane-dimethyl sulfide, 0516
with OsO$_4$, 0548

alkene categories, and metathesis, 1339
alkene metathesis. *See* metathesis
alkene reactivity
  and electronic effects, 1016
  and substituent effects, 1016
alkene-aldehydes, from allyl
  vinyl ethers, 1132
alkene-alkynes. *See* en-ynes
alkenes
  absorption bands, 1264
  alkylboranes, and prochiral centers,
    0497
  allowed and forbidden reactions,
    1010
  allylic, oxidation with SeO$_2$, 0325
alkenes
  and [2+2]-cycloaddition, 1076
  and [2+2]-cycloreversion, 1096
  and AIBN, 1279, 1280
  and alkynes, radical coupling, 1282
  and aminohydroxylation, 0306
  and anti-Markovnikov addition,
    0180, 0181
  and asymmetric epoxidation, 0291
  and asymmetric hydrogenation,
    0448-0451
  and **Bogert-Cook synthesis**, 1209
  and borane isomerization, 0501
  and **Bredt's rule**, 0553, 0554
  and carbocation formation, 0179, 0180
  and catalysts reactivity, 0430
  and chloramines, 0306
  and chromium trioxide, 0317
  and diisopinocampheylborane, 0519,
    0520
  and dioxiranes, 0290
  and dipolarophile, 1236
  and dissolving metal reduction, 0452
  and **Friedel-Crafts alkylation**, 1201
  and **Friedel-Crafts cyclization**, 1209
  and halonium ions, 0184
  and haptophilicity, 0434, 0436
  and **Hofmann elimination**, 0162
  and iodolactonization, 0186
  and **Jacobsen hydrolytic kinetic
    resolution**, 0290
  and **Johnson polyene cyclization**,
    1184
  and **Markovnikov addition**, 0179
  and **Meerwein arylation**, 1299
  and metalocycle formation, 0295
  and *N*-methylmorpholine *N*-oxide,
    0548

  and osmium tetraoxide, 0294-0301
  and oxidation number, 0222
  and oxymercuration, 0181-0183
  and **Pauson-Khand reaction**, 1342
  and **Payne epoxidation**, 0271
  and permanganate, 0292
  and peroxides, 0180
  and POCl$_3$ with pyridine, 0701
  and polyalkylation, 1201
  and potassium peroxomonosulfate,
    0291
  and radical coupling, 1279
  and radical formation, 0180
  and salen catalysts, 0288
  and **Saytzeff (Zaitsev) elimination**,
    0155, 0160
  and **Sharpless asymmetric
    aminohydroxylation**, 0306
  and **Sharpless asymmetric
    dihydroxylation**, 0298-0301
  and **Simmons-Smith reaction**, 1331
  and sodium hypochlorite, 0289
  and sodium periodate, 0324
  and substituents, HOMO and
    LUMO energy, 1002, 1003, 1004
  and the **ene reaction**, 1143
  and the **Étard reaction**, 0329
  and the **Hammond postulate**, 0155
  and the **Jacobsen catalyst**, 0289
  and the **Lemieux-Johnson reagent**,
    0314
  and the **Prévost reaction**, 0301, 0302
  and the **Prilezhaev (Prileschajew)
    reaction**, 0277
  and the **Prins reaction**, 1191
  and the S$_N$1 reaction, 0178
  and the **Woodward modification**,
    0303
  and transition metal reagents 0754-
    0758
  and tributyltin hydride, 1283
  and **Wacker oxidation**, 0333, 1226
alkenes
  as Brønsted-Lowry bases, 0177-0183
  as Lewis bases, 0181-0188
  bromination, 0158
alkenes
  by dissolving metal reduction of
  alkynes, 0456
  by electrolysis of carboxylic acid
    salts, 1278
  by electrolytic reduction of alkynes,
    0470

  by elimination of alkyl halides, 0077
  by elimination, 0153
  by fragmentation of radical, 1275
  by Grignard reduction of
    aldehydes or ketones, 0667
  by **Grob fragmentation**, 0174
  by **Horner-Wadsworth-Emmons
    olefination**, 0593, 0739
  by hydrogenation of alkynes, 0390,
    0436, 0546
  by **McMurry olefination**, 1309, 1310
  by metathesis, 1334-1341
  by photolysis of ketones, 1318
  by reduction of alkynes, 0636
  by reduction of conjugated alkynes,
    0465
  by the **Bamford-Stevens reaction**,
    1317
  by the **Chugaev elimination**, 0167
  by the **Ramberg-Bäcklund
    reaction**, 0701
  by the **Shapiro reaction**, 1317
  by the **Wittig reaction**, 0730, 0732
  by the **Wittig reaction**, stereocontrol,
    0738
  by thermolysis of selenoxides, 0815
  by thermolysis of sulfoxides, 0814
alkenes
  catalysts and hydroperoxide
    epoxidation, 0271
  catalysts and peroxide epoxidation,
    0270
  chromophoric, 0102
  cis, enantioselectivity
    in hydroboration, 0520
  cis/trans isomers, 0029
  cis/trans, by hydrogenation
    of alkynes, 0546
  cleavage with ozone, 0318-0322
  conformations, 0044
alkenes
alkenes, conjugated
  and dihydroxylation, 0297
  and epoxidation, 0282, 0283
  versus nonconjugated,
    hydrogenation, 0431
  with PdCl$_2$, 1228
alkenes, coupling
  cross metathesis, 1337
  with alkyl halides, 1280
  with aryl halides, 1246
alkenes, cyclic
  and **Birch reduction**, 0459

by elimination, 0551
from dialdehydes, 1310
from diketones, 1310
alkenes
  diastereomers, 0033
  diastereoselectivity of hydroboration, 0516
  dihydroxylation, 0291-0306, 0548
  electron effects, 1016
  enantioselectivity, reaction with carbenes, 1325
alkenes, epoxidation
  and density ionization map, 0278
  and ionization potential maps, 0281
  and stereoselectivity, 0284
  and substituent effects, 0282
  with mcpba, 0558, 0655
alkenes, exocyclic, 0171, 0177
alkenes, formation of diols, 0291-0306
alkenes
  from alcohols, 0148
  from aldehydes, 0730, 0742
  from alkyl halides, 0154
  from alkynes, 0455, 0509
  from alkynes, with the **Lindlar catalyst**, 0437
  from alkynes, with the **Rosenmund catalyst**, 0437
  from amine oxides, 0165
  from carbocations, 1186
  from cyclobutanes, 1096
  from diazoalkanes, 1317
  from dienes, 0438
  from dihalides, 0158
  from en-ynes, 0438
  from episulfones, 0701
  from esters, 0166
  from halo-sulfones, 0701
  from hydroxylamines, 0165
  from ketones, 0730
  from organolithium reagents, 1318
  from phosphonothioates, 0743
  from selenoxides, 0169
  from sulfoxides, 0552
  from the **Burgess reagent**, 0170
  from tosylhydrazones, 1317
  from xanthate esters, 0167
alkenes, halo
  hydrogenolysis, 0437
  with **Grignard reagents**, 0665
  with organolithium reagents, 0688
alkenes
  Henbest rule, 0283

HOMO and LUMO, 1001-1004
hydroboration, 0543, 0561
hydroboration, regioselectivity, 0524
alkenes, hydrogenation, 0428, 0429, 0546
  and bond migration, 0432
  and cis/trans isomers, 0433
  and rearrangement, 0432
  and steric hindrance, 0436
alkenes
  hydrosilylation, 0475
  iron complexes, 1248
  isomerization during hydrogenation, 0432
  Jacobsen-Katsuki Asymmetric Epoxidation, 0288-0290
  long chain, conformations, 0041
  LUMO, 1342
  LUMO, and the **Diels-Alder reaction**, 1013
  Markovnikov addition of alkylboranes, 0493
  Markovnikov addition, 0551
alkenes, mechanism
  metal catalyzed epoxidation of allylic alcohols, 0273, 0274
alkenes, model
  for **Sharpless Asymmetric Epoxidation**, 0285, 0287
alkenes
  monosubstituted, and epoxidation, 0281
  neighboring group effects, 0434, 0436
  orbital coefficients, 1001
  order of stability, 0155
  oxidation with mercuric salts, 0333
  oxymercuration, 0514
  ozonolysis, 0563
  palladium oxidation, 1226
  partial cleavage, 0321
  peroxyacids epoxidation, and buffers, 0277
  photochemistry, 1091
  photocycloaddition, 1092
  photolysis, 1264
  prochiral, 0519
  radical coupling with aryl diazonium salts, 1299
  rate of addition of carbenes, 1324
  rate of reaction with carbenes, 1324
  rate of Simmons-Smith reaction, 1332

alkenes, reaction
  with carbenes, 1325
  with carbocations, 1182
  with HCl, 0132
  with hydrogen peroxide, 0271
alkenes, reactivity
  with diazoalkanes, 1110
  with hydrogenation, 0432
  with peroxides, 0269
alkenes
  reduction with diimide, 0474
  reduction, 0474
  regioselectivity of hydroboration, 0516
  sacrificial, and alkylboranes, 0501
  sacrificial, in hydroboration, 0501
  Sharpless Asymmetric Epoxidation, 0284
  silyl, with LDA, 1123
  stability, and **Saytzeff (Zaitsev) elimination**, 0155
  stereoisomers, 0029
  stereoselectivity in the E2 reaction, 0154
  steric hindrance in addition reactions, 0184
  sterically hindered, and ozonolysis, 0320
alkenes, stucture
  and dihydroxylation, 0296, 0297, 0299
  and epoxidation, 0272
  and metathesis, 1335
  and **Sharpless asymmetric dihydroxylation**, 0299
alkenes
  substituent effects, 1016
  substitution, and hydrogenation reactivity, 0432
  substitution, and peroxyacid epoxidation, 0280
  suprafacial and antarafacial addition, 1011
  trans, enantioselectivity in hydroboration, 0520
  transition state for borane addition, 0492
  UV, 1264
alkenes
  via **Cope elimination**, 0165
  via **Lombardo's reagent**, 0757
  via the **Nysted reagent**, 0757
  via the **Takai reaction**, 0757

alkenes, visible spectrum, 1264
alkenes
  diiodomethane, 1332
  with acids, 1183
  with acrolein, 1056
  with alkenes, 1143
  with alkyl halides, 1279, 1280
  with alkyl hydroperoxides, 0269, 0271
  with aqueous acid, 0514
  with arylmercury, 1229
  with azomethine imines, 1111
  with borane, 0491
  with borane, anti-Markovnikov addition, 0496
  with borane, electronic effects, 0494
  with borane, steric hindrance, 0492
  with boranes, 0493, 0513
  with boranes, transition state, 0494, 0497
  with carbenes, 1319, 1322
  with carbenes, activation energy, 1322
  with carbenes, stereoselectivity, 1323
  with carbenes, transition state modeling, 1320
  with CCl_4, 1272
  with chromyl chloride, 0332
  with conjugated carbonyls, 1056
  with diazomethane, 1320
  with diborane, 0515
  with dichlorocarbenes, 1319
  with dienes, 0007, 1013
  with dihalogens, 0183-0188
  with dihalogens, solvent effects, 0184
  with diisopinocampheylborane, 0522, 0523
  with disiamylborane, 0496, 0516
  with diynes, metathesis, 1339
  with formaldehyde, 1191
  with **Grubbs I catalyst**, 1336
  with **Grubbs II catalyst**, 1337
  with HBr or HCl, 0178
  with HBr, 0063, 0542
  with HBr, and AIBN, 0180
  with HBr, and peroxides, and radicals, 0542
  with HBr, and radicals, 1282
  with hydrogen peroxide, 0269
  with hydroperoxides, 0272
  with hydroxamic acids, 1293
  with hypohalous acids, 0187

  with iodozinc compounds, 1331
  with lead tetraacetate, 0333
  with mercuric acetate, 0182
  with methylborane, 0531
  with monoisopinocampheylborane, 0519, 0520
  with Ni(0), 1246
  with nitrile ylids, 1104, 1105
  with nitriles, 0976
  with nitrones, 1108
  with organolithium reagents, 0688
  with palladium chloride, 0333
  with palladium compounds, 1227
  with permanganate and periodic acid, 0313
  with permanganate, 0293
  with peroxides, 0269
  with peroxyacids, 0277-0284
  with radicals, 1269, 1270, 1277, 1279, 1283, 1298, 1299
  with *tetrakis*-palladium(0), 1231
  with thallium nitrate, 0333
  with the **Lemieux-von Rudloff reagent**, 0313
  with thexylborane, 0529, 0531
alkenyl **Grignard** reagents. *See* Grignard
alkenyl organolithium. *See* organolithium
alkenylborinates, 0503
alkoxides
  and alcohol conjugate acids, 0800
  and basicity, 0091
  and enolate anions, 0800
  and $LiAlH_4$ reduction of carbonyls, 0350
  and organolithium reagents, 0675
  and the **Claisen condensation**, 0824
  and the **Grob fragmentation**, 0175
  and thermodynamic control, 0800
alkoxides
  as bases, 0786
  bis, and **acyloin condensation**, 1307
  bond lengths, 0856
  by hydride reduction of aldehydes and ketones, 0350
  enantioselectivity with organocuprates, 0725
  silyl, 1308
  solubility, 0589
  with alkyl halides, 0545
  with chlorotrimethylsilane, 0749
alkoxy radicals, 1260, 1261, 1277

  and the **Barton reaction**, 1277
alkoxyaluminum hydrides, *See* hydrides
alkoxyboranes, 0513
alkoxyborinate, 0500
alkoxymercuration. *See* oxymercuration
alkoxysulfoxonium salts, and alcohols 0241
alkyl anions
  with alkyl halides, 0634
  with alkyl sulfonate esters, 0634
alkyl azides. *See* azides
alkyl compounds
  by reduction of alkyl halides, 0360
  by reduction of sulfonate esters, 0361
alkyl fragments
  and chromium trioxide, 0328
  oxidation with permanganate, 0314
  oxidation, 0324
alkyl halides. *See* halides
alkyl shifts, and carbocations, 0135
alkylation,
  and **Julia olefination**, 0699
  and lithium hexamethyldisilazide, 0804
  and Pd(0), 1232
  and the **Birch reduction**, 0461
  and the thio-Claisen rearrangement 0713
  enolate anion, diastereoselectivity 0851
  Friedel Crafts. *See* Friedel-Crafts
  O versus C, for enolate anions, 0810, 0811
  O-, with enolate anions, 0812
  of enamines, 0873
  of sulfides, 0698
  of sulfones, 0699
  sulfoxide-sultone anions, 0707
  sultone carbanions, 0701
  with enolate anions, 0803
  with Pd(0), inversion of configuration, 1233
  with sulfur carbanions, 0697
alkylboranes. *See* boranes, alkyl
alkylborinate, 0505
alkylidene cyclohexanes, and chirality 0028
alkylidene enolates, 0870
alkylidene malonates. *See* malonates
alkyls
  and **Kuhn-Roth oxidation**, 0329
  and ruthenium compounds, 0330

by hydrogenation of alkenes, 0429
by hydrogenation of alkynes, 0390
by hydrogenolysis of cyclopropane, 0457
by hydrogenolysis of dithianes, 0464
by hydrogenolysis of sulfides, 0457
by hydrogenolysis of sulfones, 0457
by hydrogenolysis of sulfoxides, 0457
by reduction of aldehydes or ketones, 0472
by reduction of alkyl halides, 0368,0383
by reduction of benzylic alcohols, 0368
by reduction of benzylic amines, 0368
by reduction of mesylates, 0379
by reduction of sulfonate esters, 0368, 0383
by reduction of sulfones, 0468
by reduction of sulfoxides, 0468
by **Wolff-Kishner reduction**, 0472
alkyls
from alkyl halides, 0641
from alkynes, 0438
from dienes, 0438, 0514
from Grignard reagents, 0641
alkyltin. *See* organotin
alkyne-alcohols, from ketones, 0635
alkyne anions, 0632
and acyl addition, 0107, 0108
and **Castro-Stephens coupling**, 1244
and **Sonogashira coupling**, 1244
as carbanions, 0629
formation of, 0629
from terminal alkynes, 0632
with aldehydes, 0635
with alkyl halides, 0007, 0629
with D$_2$O, 0633
with epoxides, 0635
with ketones, 0635
with sulfonate esters, 0629
alkyne silanes, coupling with aryl halides, 1302
alkyne-alcohols,
and hydrogenolysis, 0362
and LiAlH$_4$, 0362
and the **Whiting reaction**, 0362
from aldehydes, 0635
from epoxides, 0635
alkyne-ketones, 0564

alkynes
acidity, and conjugation, 0633
acidity, and substituent effects, 0633
alkynes
and alkenes, radical coupling, 1282
and **Baldwin's rules**, 0568
and **Castro-Stephens coupling**, 1244
and ene reactions, 1146
and homogeneous hydrogenation catalysts, 0438
and metathesis, 1339
and radical coupling, 1282
and radical cyclization, 1286, 1292
and s character, 0632
and SOMO, 1282
and **Sonogashira coupling**, 1244
and **Stille coupling**, 1241
and the **Cadiot-Chodkiewicz coupling**, 0631
and the **Fritsch-Buttenberg-Wiechell rearrangement**, 0700
and the **Glaser reaction**, 0631, 1301
and the **Heck reaction**, 1238
and the **Lindlar catalyst**, 0437
and the **Pauson-Khand reaction**, 1342
and the **Rosenmund catalyst**, 0437
and trimethylene methane equivalents, 1236
alkynes
anti-Markovnikov cross coupling, 0631
aryl, by **Sonogashira coupling**, 1244
by hydrogenation of alkynes, 0436
by the **Corey-Fuchs procedure**, 0630
cis hydrogenation, 0433
conjugated, with zinc and KCN, 0464
coupling reactions, 0631, 1301
dianions, 0633
dissolving metal reduction, 0455
dissolving metal reduction, and amines, 0456
electrolytic reduction, 0470
formation of radical anions, 0455
formation of vinyl carbocations, 1168
alkynes
from aldehydes, 0733, 0734
from alkyl halides, 0629, 0634
from alkyl sulfonate esters, 0634
from alkynes, 0629, 0634
from aryl halides, 1303
from dihalides, 0158

from **Grignard reagents**, 0700
from organolithium reagents, 0700
from sulfonate esters, 0629
from vinyl dihalides, 0159, 0733, 0734
from vinyl halides, 0158, 0630
alkynes, halo,
and **Cadiot-Chodkiewicz coupling**, 0631
with alkynes, 1302
alkynes
HOMO and LUMO, 1003, 1004
homo-coupling, 1341
hydroboration, 0511, 0512
hydroboration, and enol formation, 0518
alkynes, hydrogenation, 0390, 0436, 0546
catalysts reactivity, 0436
formation of cis/trans isomers, 0436
overreduction, 0436
with the **Lindlar catalyst**, 0546
alkynes
in **Diels-Alder reactions**, 1058
lithio, with boranes, 0513
LUMO, 1282
pK$_a$, 0088, 0632, 0633
radical coupling, 1301
reaction with alkynes, 1183
reaction with bases, 0632, 0633
reduction, 0465
regioselectivity with boranes, 0504, 0505
silane protecting groups, 1286
silyl, 0518
stereoselectivity of dissolving metal reduction, 0456
terminal, acidity of, 0632
terminal, and dissolving metal reduction, 0455
terminal, isomerization, 0456
alkynes
with 9-BBN, 0503, 0509
with alkylboranes, 0503, 0509, 0528
with alkylboranes, stereoselectivity, 0509
with aryl halides, 1244, 1302
with butyllithiums, 0632, 0633, 1366
with catecholborane, 0504
with copper compounds, 1301
with Dibal, 0762
with disiamylborane, 0503
with enamines, 1076

with Grignard reagents, 0632
with haloalkynes, 1302
with indium, 0631
with monochloroborane, 0503, 0504
with nitrones, 1108
with organoaluminum compounds,
    0762, 0763
with organolithium reagents, 0632
with trimethylaluminum, 0763
with vinyl triflates, 1244
alkynes-alkynes (diynes), 0564
alkynyl alcohols, 0518
alkynyl copper, and organocuprates,
    0717, 0718
allamandin, 0749
allene-ketene cycloaddition, 1080
allenes
    and chirality, 0027
    from alkynes, 0634
    from propargyl halides, 0634
    HOMO and LUMO, 1003, 1004,
        1005
    sigmatropic rearrangement, 1116
    thermal rearrangement, 1116
    with ketenes, intramolecular, 1080
Alloc, 0111
allokainic acid, 1152
allowed reactions, 1010
allyl bromide, with Pd(0), 1232
allyl nickel. See nickel
allyl palladium. See palladium
allyl silanes. See silanes
allyl stannanes. See tin
allyl(diisopinocampheyl)borane, 0526
allylic acetate, with Pd(0), 1232
allylic carbocations. See carbocations
allylic compounds
    and SeO$_2$, 0325
    and the **Étard reaction**, 0329
    with Pd compounds, 1227
allylic Grignard reagents. See Grignard
allylic sulfides. See sulfides
allylic systems, and sigmatropic
    rearrangement, 1118
allylic thiols. See thiols
allylsilanes, 0769
    chelation control in coupling, 0769
    with SnCl$_4$, 0769
    with TMSOTf, 0769
allyltin compounds
    and prenylation, 1281
    and radical coupling, 1280, 1281
    and ZnCl$_2$, 0768

conjugate addition, 0767
    enantioselectivity, 0766
allyltin. See tin
allyltributyl stannane, 0767
aloperine, 0331, 0456
Alpine borane, 0395
alstonerine, 0476
altromycin aglycone, 0301
alumina
    and hydrogenation, 0438
    and PCC, 0237
    as a hydrogenation support, 0426
aluminum amalgam
    and hydrogenolysis, 0468
    and sulfoxides, 0725
    reduction of sulfur compounds, 0698
aluminum borohydride, 0374
aluminum catalysts, for
    the **ene reaction**, 1151
aluminum chloride
    and **Friedel-Crafts alkylation**, 1202
    and LiH, 0348
    and PBr$_5$, 1206
    complex with alcohols, 1202
    with alkyl halides, 1194
    with phenolic esters, 1219
aluminum compounds
    acid-catalyzed aldol, 0821
    and **Meerwein-Ponndorf-Verley**
        reduction, 0467
    and reduction, 0467
    and the **Tebbe reagent**, 0755
aluminum hydride. See hydride
aluminum alkoxides
    triisopropoxide, and
        Oppenauer oxidation, 0250, 0251
    tri-*n*-propoxide, and
        Oppenauer oxidation, 0250, 2051
    tri-*tert*-butoxide, and
        Oppenauer oxidation, 0251
aluminum, and **pinacol coupling**, 1305
AM6898A, 0600
AM6898D, 0832
Amalgamal
    aluminum, reduction of sulfur
        compounds, 0698
    and Fe(CO)$_5$, 0760
    with sulfones, 0700
Amaryllidaceae alkaloids, 0022
ambergris, 0951
Amberlite, and enzyme reduction,
    0481
Amberlyst A21, hydrolysis of esters,

0545
ambrox, 0951
ambruticin S, 1122
amide bases
    and enolization, 0789
    and isomerization in aldol
        condensation, 0859
    colligative properties, 0794
    dialkyl, aggregation state, 0792-0795
    dialkyl, and TMEDA, 0794
    dialkyl, E/Z isomers, 0792-0795
    dialkyl, solvent and
        aggregation state, 0794
    dialkyl, structure and
        aggregation state, 0792-0795
    dialkyl, with aldehydes, 0794
    dialkyl, with esters, 0794
    dialkyl, with ketones, 0794
amide dimethylacetals
    and the **Claisen rearrangement**,
        1137
    and the **Eschenmoser-Claisen**
        rearrangement, 1137
amide esters
    by the **Passerini reaction**, 0627
    from isonitriles, 0627
amide-alkenes, iodolactamization,
    0187
amides
    and acyl substitution, 0109
    and aluminum borohydride, 0374
    and basicity, 0091
    and Friedel-Crafts alkylation, 1199
    and glycosylation, 0615
    and hypobromites, 0193
    and ketenes, 1316
    and phosphorus oxychloride, 0372
    and the **Haller-Bauer reaction**, 0803
    and the **Hofmann rearrangement**,
        0193
    and the **Vilsmeier reaction**, 0372
amides
    benzamides, from amines, 0617
    by the **Arndt-Eistert synthesis**, 1316
    by the **Passerini reaction**, 0627
    by the **Ritter reaction**, 0976
    by the **Ugi reaction**, 0627
    dialkyl, aggregates, 0789
    dialkyl, aggregation state, 0789
    dialkyl, as non-nucleophilic bases,
        0789
    electrolytic reduction, 0472
    elimination in LiAlH$_4$ reduction,

0355

amides, enolate anions, 0802
  alkylation, 0805
  E/Z isomers, 0792
  enolate geometry, 0792
  with aldehydes, 0824
  with ketones, 0824
amides
  from acids chlorides, 0613
  from amines and acid chlorides, 0359
  from amines, 0613, 0615
  from ketones, 0191, 0803
  from oximes, 0189
amides
  magnesium, and HSAB, 0103
  mechanism, of LiAlH$_4$ reduction,
    0355
  potassium, and aryl halides, 0207
  protecting groups for amines, 0615
  reaction with alkylboranes, 0527
  reduction and steric hindrance, 0355
  reduction and enolate condensation,
    0357
  reduction with LiAlH$_4$, 0355
  steric hindrance and reduction, 0364
  trifluroaetamide, from amines, 0616
  Weinreb. *See* Weinreb amide
amides
  with alkoxyaluminum hydrides, 0364
  with **Grignard reagents**, 0649
  with LiAlH$_4$, 0615
  with lithiated dithianes, 0705
  with **Meerwein's reagent**, 0615
  with organolithium reagents, 0683
  with triethyloxonium
    tetrafluoroborate, 0615
amidinium ion, catalysts of Diels-
  Alder, 1039
amido-amides, by the **Ugi reaction**,
  0627
amido-dienes, Diels-Alder reactions,
  1051
amidofurans, 1059
amidyl radicals, 1293
amination, of alkylboranes, 0534, 0535
amine oxides
  and elimination, 0165
  and intramolecular **Diels-Alder
    reactions**, 1065
  and OsO$_4$, 0296
  oxidation of boronates, 1065
amine-alcohols, 0306
  and the **Tiffeneau-Demjanov**

ring expansion, 1175
  by dihydroxylation of enamines, 0561
  by hydroboration of enamines, 0561
  by reduction of amino acids, 0386
  by reduction of cyanohydrins, 0368
  with aldehydes, 0845
  with **Grignard reagents**, 0662
  with nitrous acid, 1181
amine-dienes, **Diels-Alder reactions**,
  1051
amine-esters, polymer bound, 0979
amine-ketones
  and **Mannich bases**, 0881
  and the **Mannich reaction**, 0881
  from amines, 0881
  from conjugated ketones, 0881
  from iminium salts, 0649
  from lactams, 0649
amines
  acetamides, 0615
  allyl, and **aza-Claisen
    rearrangement**, 1141, 1142
  amide protecting groups, 0615
amines
  and basicity, 0090
  and **Benkeser reduction**, 0462
  and dissolving metal reduction
    of alkynes, 0456
  and dissolving metal reduction, 0453
  and E2, 0158
  and fluxional inversion, 0020
  and **Grignard reagents**, 0092, 0638
  and **Hofmann elimination**, 0162
  and HSAB, 0100
  and **Michael addition**, 0877
  and palladium complexes, 1229
  and phosphorus oxychloride, 0372
  and polyalkylation, 0122
  and S$_N$2 reactions, 0012, 0122
  and S$_N$2' reactions, 0127
  and the **Demjanov rearrangement**,
    1175
  and the **Gabriel synthesis**, 0125
  and the Pauson-Khand reaction, 1343
  and the **Schlenk equilibrium**, 0638
  and the **Vilsmeier-Haack complex**,
    0372
amines
  aromatic, from cyclohexene
    enamines, 0876
  aromatic, with nitrous acid, 0208
  as a poison for hydrogenation, 0426
  as acids, 0786

as leaving groups, 0162
  basicity and ammoniums alts, 0090
  benzyl protecting groups, 0613
amines, benzylic
  dissolving metal reduction
    and hydrogenolysis, 0456
  hydrogenolysis, 0368, 0431
amines
  bicyclic, by **Hofmann-Löffler-
    Freytag reaction**, 1294
  Boc protecting groups, 0617
  Boc, with acid, 0617
  Boc, with trimethylsilyl triflate, 0617
amines
  by electrolytic reduction of azides,
    0472
  by electrolytic reduction of nitriles,
    0472
  by hydrogenation of aniline
    derivatives, 0446
  by hydrogenation of anilines, 0446
  by hydrogenation of aryl nitro
    compounds, 0360
  by hydrogenation of azides, 0442
  by hydrogenation of aziridines, 0443
  by hydrogenation of nitriles, 0442
  by hydrogenation of nitro
    compounds, 0442
  by reduction of amides, 0355
  by reduction of aryl nitriles, 0368
  by reduction of azides, 0359, 0477
  by reduction of carbamates, 0356
  by reduction of enamines, 0371
  by reduction of imines, 0371
  by reduction of iminium salts, 0371
  by reduction of lactams, 0367, 0388
  by reduction of nitriles, 0358
  by reduction of nitro compounds,
    0359
  by reduction of pyridinium salts,
    0372
  by **reductive amination**, 0371
  by the **Curtius rearrangement**,
    0192
  by the **Vilsmeier reaction**, 0372
amines
  carbamate protecting groups, 0617
  Cbz protecting groups, 0618
  Cbz, catalytic hydrogenation, 0618
  Cbz, hydrogenolysis, 0618
amines, chiral
  and enamines, 0876, 0877
  and organocuprates, 0724

and **pinacol coupling**, 1304
complex with organocuprates, 0724
with arylchromium compounds, 0765
amines
conformations, 0043
conversion to amides, 0615
conversion to iminium salts, 1079
conversion to ketones, 1176
amines, cyclic
by the **Hofmann-Löffler-Freytag reaction**, 1294
by radical cyclization, 1288
by reduction of lactams, 0356, 0357, 0386
from lactams, 0649
amines
dialkyl, pK_a, 0788
diaryl, and **Bernthsen acridine synthesis**, 1218
diaryl, with carboxylic acids, 1218
dibenzyl, 0614
fluxional inversion, 0020
formation of carbocations, 1163
formylation, 0626
amines
from acyl azides, 0192
from acyl halides, 0193
from alkenes, 0534
from alkenes, enantioselectivity, 0526
from alkylboranes, 0534
from amides, 0355, 0372
from ammonium salts, 0752
from azides, 0359, 0536
from benzylic amine, 0457
from carboxylic acids, 0191
from imines, 0446
from iminium salts, 0685
from isocyanates, 0191
from lactams, 0356, 0357
from lactams, 0527, 0649
from nitriles, 0347, 0359, 0685
from nitro compounds, 0359
amines, halo
and **Hofmann-Löffler-Freytag reaction**, 1294
and radial cyclization, 1294
amines
hydrolysis, 0685
methylation of, 0355, 0384, 0476
N-halo, photolysis, 1294
order of basicity, 0090
oxidation, 0165, 0339

ozonolysis, 0339
photochemical coupling, 1305
pK_a, 0088, 0788
protecting groups, 0613-0618
protection as benzyl ether, 0444
solvation and basicity, 0090
structure, and **Hofmann elimination**, 0163, 0164
trimethylsilyl, protecting groups, 0614
amines
with acetic anhydride, 0592
with acid chlorides, 0613, 1078
with aldehydes or ketones, 0108
with aldehydes, 0371
with benzoyl chloride, 0617
with benzyl chloroformate, 0618
with benzyl halide, 0613
with calcium, 0462
with carboxylic acids, 1218
with chlorotrimethylsilane, 0615
with cyclopentadienyl dicarbonyl and alkenes, 1249
with di-*tert*-butyldicarbonate, 0617
with nitrous acid, 1314
with α-bromoketones, 1189
with formaldehyde, 1217
with ketones, 0873
with $NaNO_2$ and KBr, 0545
with nitrous acid, 1163, 1175, 1299, 1299
with peroxides, 0339
with peroxyacids, 0339
with potassium nitrite, 1314
with TFAA, 0616
amino acids
3-letter codes, 0979
and carbamate protecting groups, 0617
and synthesis, 0979, 0980
by the Merrifield synthesis, 0979
by the **Strecker synthesis**, 0628
chiral additives in hydrogenation, 0449
chiral, 0629
from aldehydes, 0628
from cyanides or HCN, 0628
from ketones, 0628
poly, and **Juliá-Colonna epoxidation**, 0276
reduction to amino alcohols, 0386
reduction, 0386
via oxidation of cyclic amino acids, 0330

with nitrous alcohols, 1175
amino radicals, 1294
amino. *See* amine
amino-(methoxymethyl)pyrrolidine. *See* RAMP or SAMP
aminoalanine, 0974
amino-alcohols. *See* amine-alcohols
aminobiphenyls, with $ZnCl_2$, 1219
amino-cyclitols, 0305
aminodienes, and **Diels-Alder reactions**, 1051
aminohydroxylation, and chloramines, 0306
aminomethylation, 0355, 0356
ammonia
and dissolving metal reduction, 0452
pK_a, 0788
with benzyl amines, 0613
with conjugated dienes, 0452
with lithium, 0457
with metals, hydrogenolysis, 0457
with sodium, 0595
ammonium carbanions, 0750, 0751
ammonium radical, 1294
ammonium salts
and basicity of amines, 0090
and Bredt's rule, 0554
and dihydroxylation, 0297
and **Hofmann elimination**, 0162
and inductive effects, 0085
and S_N2 reactions, 0122
and **Sommelet-Hauser rearrangement**, 0752
and **Stevens' rearrangement**, 0752
and the **Burgess reagent**, 0170
by S_N2 reaction, 0163
chiral, catalyst for [2+2]-cycloaddition, 1080
with organolithium reagents, 0752
amphidinol 3, 0248
amphidinolide P, 0575
amphidinolide T3, 0374, 0516, 0704
amphidinolide W, 0576
amphorogynine A, 1078
anaferine, 0431
analog studies, and synthesis, 0904
anatoxin, 1326
androstan-6a-ol, 0257
angiogenesis inhibitor, 0840
angle of attack, and **Baldwin's rules**, 0564, 0565
anhydride, acetic. *See* acetic anhydride
anhydride, benzoic. *See* benzoic

anhydride

anhydride, trifluoroacetic, and
Swern Oxidation, 0241

anhydrides
and acyl substitution, 0109
and **Friedel-Crafts acylation**, 1205
and the **Perkin reaction**, 0827
enolate anions, with aldehydes, 0827
mixed, and macrolactonization, 0576
O-alkylation with enolate anions,
0813
reaction with alkylboranes, 0527
reduction with LiAlH$_4$, 0353
reduction with NaBH$_4$, 0373
reduction, 0373
with amines, 0616
with enolate anions, 0812

anhydro-D-mannitol, as a chiral
template, 0958

aniline
and **Friedel-Crafts alkylation**, 1199
and the **Knorr quinoline synthesis**,
1211
derivatives from aryl nitro
compounds, 0442
derivatives, hydrogenation, 0446
from aryl nitro compounds, 0339
regioselectivity in **Friedel-Crafts
reactio**ns, 1211, 1212
with acetic anhydride, 1199
with diketones, 1213

anion, acyl, 0704

anion, dithioacetals, 0710

anionic accelerated Claisen
condensation, 0791

anionic **oxy-Cope rearrangement**,
0932, 1130
and diradicals, 1130
and phosphazene bases, 1130
and **Wender's macroexpansion
reaction**, 1130

anions
alkyne, 0632
dithiolane, 0706, 0707
ketene dithioacetal, and resonance,
0710, 0711

anisole
and **Birch reduction**, 0460
reaction energy, 0702

anisomycin, 0956, 0991

annelation. *See* annulation

annelation, definition, 0880

annonacin, 0738

annulation
and **Baldwin's rules**, 0564
definition, 0880

anodic oxidation, of phenols, 1298

anomeric effect, 0064, 0422

antarafacial shifts, 1118

antarafacial, cycloadditions, 1011

antarafacial, sigmatropic
rearrangement, 1116, 1117

anthraquinone, 0321

anti and syn, definition, 0025

anti conformations, 0039

anti/syn isomers,
aldol condensation, 0851, 0852
aldolates, 0857
and dihydroxylation, 0548
and ester enolate , 0777
and radical coupling, 1281
and the **Reformatsky reaction**,
0886, 0887
and the **Zimmerman-Traxler
Model**, 0852
with **Grignard reagents**, 0659

antibodies, and [3+2]-cycloaddition,
1107

antibonding orbitals, 1000
and perturbation theory, 1015

anti-Cram selectivity, 0406, 0407

anti-**Markovnikov** addition, 0541
boranes and alkenes, 0496
hydroboration, 0497
cross coupling of alkynes, 0631

aphanorphine, 1203

aphidicolin, 0746, 0918, 0919, 1139

apicul020 A, 0685

apicularen, 0549

Aplysia kurodai, 0065

aplysiapyranoids, 0065

apoptolidinone, 0500

aprotic solvents, and substitution, 0138

aqueous media
and the **Claisen rearrangement**,
1139
rate enhancement of the
Diels-Alder reaction, 1041
rate enhancement, 1139

aqueous Wittig reactions, 0732

Ar$_1$-5 mechanism, 0772

araneosene 0932, 1310

arborone, 0236

Arbuzov reaction, and S$_N$2, 0121

archazolid B, 0396

arene oxides, 1179

arenes
from aryl diazonium salts, 0210
from **Grignard reagents**, 0643

aristolochene, 0169

Arndt-Eistert synthesis, and the
Wolff rearrangement, 1316

arnottin I, 1018

aromatic compounds
and electrophilic substitution, 0195-
0205
and Lewis acids, 0095
and nucleophilic substitution, 0194
and the **Bradsher reaction**, 1219
nitration, 1192

aromatic rings
and strategic bonds, 0936
bromination, 0147
with POBr$_3$, 0147
substitution. *See* substitution

aromaticity, 0901
and basicity, 0785
and phenylene, 0901
and pKa, 0692

aromatization, 0936
and azadienes, 1058
and **Bergman cyclization**, 1295
with lead oxide, 1218
with selenium, 1208

Arrhenius equation, 1018
and the **Diels-Alder reaction**, 1021

aryl ally ethers, **Claisen
rearrangement**, 1133

aryl biphosphites, chiral
and the **Pauson-Khand reaction**,
1344

aryl compounds
and calcium, 0462
and the **Benkeser reduction**, 0462
Birch reduction and electronic
effects, 0459
Birch reduction, 0459
by electrolytic reduction of
aryl halides, 0472
catalyst reactivity in hydrogenation,
0445
electronic effects in hydrogenation,
0446
hydrogenation and coupling, 0447
hydrogenation, 0445
partial reduction, 0459, 0460

aryl diazonium salts. *See* diazonium

aryl halides. *See* halides

aryl ketones, 0643

aryl migration, carbocations, 1172
aryl radicals, 1297-1299
aryl shifts, and carbocations, 0135
aryl triflates, with boronic acids, 1242
arylation
 of bromoesters, 0534
 palladium catalyzed, 1231, 1234
arylchromium carbanions
 with iodine, 0764
 with TFA, 0764
arylchromium compounds, 0763
 nucleophilic aromatic substitution,
  0764
 with chiral amines, 0765
 with lithiocarbanions, 0764
 with organolithium reagents, 0764
 with organolithium reagents, 0765
arylchromium. *See* chromium
arylcopper reagents
 with aryl halides, 1300
 and **Ullmann reaction**, 1300
aryldiazonium salts
 and **Gomberg-Bachmann reaction**,
  1299
 and the **Heck reaction**, 1238
aryllithium reagents. *See*
 organolithium reagents
arylmercury, with alkenes, 1229
aryls
 and carbocations, 1201, 1202
 and catalyst reactivity, 0430
 and decarboxylation, 0173
 and dissolving metal reduction, 0453
 and **Elbs persulfate oxidation**, 0267
 and **Fenton's reagent**, 0266
 and peroxyacetic acids, 0266
 and *Pseudomonas putida*, 0304-0306
 and **Udenfriend's reagent**, 0266
 by **Bally-Scholl synthesis**, 1208
 by **Bardhan-Sengupta
  phenanthrene synthesis**, 1208
 by **Bogert-Cook synthesis**, 1209
 by **Bradsher reaction**, 1210
 by **Darzens tetralin synthesis**, 1210
 by **Haworth phenanthrene
  synthesis**, 1207, 1208
 by reduction of aryl halides, 0368
 by the **Heck reaction**, 1237
 dihydroxylation, 0303
 from acyl halides, 1204
 from arylchromium compounds,0764
 from diols, 1208
 from metathesis of alkynes, 1340

functionalized, 1300
hydrogenation, 1209
liver microsome-mediated oxidation,
 1179
oxidation and Lewis acids, 0266
oxidation to phenols, 0266
oxidation with **Fenton's reagent**,
 1296
oxidation with lead tetraacetate, 0268
partial reduction, powerful in
 synthesis, 0928
with acyl halides, 0007
with alkyl halides, 0007
with carbenes, 1326
with diazomethane, 1320
with POBr$_3$, 0147
ascochlorin, 1081
ascorbic acid
 and ferrous ion, 0266
 as a chiral template, 1075
 template and the **Diels-Alder
  reaction**, 1075
asmarine A, 0996
Aspergillus niger, and reduction
 0481
Aspergillus ochraceus WC76466,
 0902
aspidophytine, 0626, 0627
aspidospermidine, 0192, 0466, 0931
asteltoxin, 1182
asteriscanolide, 1247
A-strain
 and E2, 0156
 cyclohexane, exo-methylene, 0063
astrogorgin, 0732
asymmetric
 Baylis-Hillman reaction, 0879
 enamines, 0876, 0877
 epoxidation, 0291
 epoxidation. *See* epoxidation
 hydroboration, 0519
 hydrogenation, 0448-0451
 induction, reduction with
  Grignard reagents, 0667
 synthesis. *See* synthesis
asynchronicity, transition state, 1001
asynchronous **Diels-Alder** transition
 state, 1051
ate complex, 0093
 amination of alkylboranes, 0535
 and hydroboration, 0508
 and oxidation of alkylboranes, 0514
 borane, with iodine, 0513

cyano, 0528
 in hydroboration, 0512
 in **Oppenauer oxidation**, 0250
 with halogens, 0762
atom priorities, IUPAC, 0019
atomic number, and priority rules,
 0015
atropic acid, 0449
austrodoral, 0157
auxiliaries
 benzopyranoisoxazolidine, 0741
 camphor based, 1071
 chiral, and radical cyclization, 1286
 chiral, **Evans' auxiliary**, 0865
 chiral, oxazolidines, 0865
 Evan's auxiliary, 1070, 1071
 imides, 1072
 in the **Baylis-Hillman reaction**, 0879
 lactones, 1071
 oxazolidines, 0869
 transient, 0866
auxiliary, chiral. *See* chiral
 and lithium borohydride, 0374
 and **Oppolzer's sultam**, 0361, 0864
 and sulfonamides, 0864
 organoiron compounds, 0761
avermectin, 0605
Avogadro's number, 1083
axial and equatorial
 hydrogen atoms, 0049
 and E2, 0156
 conformational stability, 0558
axis, chiral, 0026-0029
aza-Bergman rearrangement, 1295-
 1296
azabicyclo[3.2.1]octan-3-ones, 1190
azabicyclooctanes, and
 dihydroxylation, 0299
aza-Claisen rearrangement, 1141
aza-Cope rearrangement, 1131
azadienes
 in **Diels-Alder reactions**, 1051, 1053,
  1057, 1058
 in **Diels-Alder reactions**,
  enantioselectivity/ tereoselectivity,
  1052
 inverse electron demand
  Diels-Alder reactions, 1033
 isomerization, 1057, 1058
azaoxindoles, 0150
aza-**Pauson-Khand reaction**, 1344
azaspiracid-1, 0367
azaspirene, 0840

aza-Wittig Rearrangement, 1123
azepine-fused oxetanol, 1265
azetidinone, 1330
azide ion, with alkyl halides, 0359
azide, sodium
  with acid, 0191
  with alkyl mesylates, 1114
  with sulfonate esters, 0544
azide-alcohols, from epoxides, 0128
azides acyl
  and the **Curtius rearrangement**,
    0192
  and the **Schmidt reaction**, 0192
azides
  acyl, and the **Curtius
    rearrangement**, 0192
  acyl, and the **Schmidt reaction**, 0190
  acyl, from acyl hydrazines, 0192
  alkyl, [3+2]-cycloaddition, 1113
  and S$_N$2 reaction with alkyl halides,
    0359
  and S$_N$2, 0117
  and the Mitsunobu reaction, 0125
  diphenyl phosphoryl, 0192
  electrolytic reduction, 0472
  from alcohols, 0544
  from alkyl halides, 0359, 0545
  hydrogenation, 0442
  mesyl, and alkyl azides, 1115
  reaction with epoxides, 0128
  reduction with LiAlH$_4$, 0359
  thermolysis, 0933
  tosyl, and alkyl azides, 1115
  trimethylsilyl, with ketones, 0192
  with alkylboranes, 0536
  with ammonium formate and Pd/C,
    0477
  with conjugated ketones, 1114
aziridines
  from alkyl azides, 1114
  from allylic sulfonium salts, 1154
  from azides, 0933
  from azomethine ylids, 0751
  from tosylimines, 1154
  from triazoles, 1113
  from triazolines, 1114
  hydrogenation, 0443
  ring opening, 0947
  vinyl, and vinyl cyclopropane
    rearrangement, 1114
  vinyl, rearrangement, 1114
  vinyl, thermolysis, 0933

azo compounds
  and radicals, 1261
  by electrolytic reduction of
    nitro compounds, 0472
  by LiAlH$_4$ reduction of aryl
    nitro compounds, 0360
  mechanism of decomposition, 1262
  thermolysis, 1261
azo radicals, 1262
azobis(2,2-dimethylethane), 1262
azobis(isobutyronitrile). *See* AIBN
azo-Cope rearrangement, 1131
azodicarboxylate, potassium,
  and diimide, 0474
azomethine imines
  [3+2]-cycloaddition, 1111
  and intramolecular [3+2]-
    cycloaddition, 1113
  from aldehydes or ketones, 1112
  from hydrazones, 1112
  with alkenes, 1111
azomethine ylids. *See* ylids
2-azonia rearrangement, 1142
2-azonia rearrangement. *See* aza-
  Claisen
azonia salts, inverse electron
  demand **Diels-Alder**, 1032
azoxy compounds, by electrolytic
  reduction of nitro compounds, 0472

B

back donation
  and oxymercuration, 0182
  by mercury, 0182
  in the **Pauson-Khand reaction**, 1342
back strain
  and alkyl halides, 0134
  and carbocations, 0134
Baeyer strain
  and chair conformations, 0049
  and macrocyclization, 0573
  definition, 0047
Baeyer-Villiger reaction, 0307-0313,
  0926
  and *Acinetobacter* NCIB 9871, 0312
  and *Acinetobacter sp.* NCIB 9872,
    0312
  and buffers, 0308
  and Candida *anatarctica* lipase, 0312
  and epoxidation, 0279
  and peroxyacids, 0307
  and transesterification, 0308

catalysis, 0307
  electronic effects, 0309
  enzymatic, 0311
  migratory aptitudes, 0309
  reactivity of ketones, 0310
  reactivity of peroxyacids, 0310
bafilomycin, 0528
baker's yeast (*Saccharomyces
  cerevisiae*), 0035
  and reduction, 0035, 0479
bakkenolide A, 0463
bakkenolide, 1286
balanol, 0386
Baldwin's rules, 0564-0572, 0582
  and carbocations, 1183
  and enolate alkylation, 0810
  and enolate anions, 0570
  and epoxides, 0569
  and radical cyclization, 0565, 1283,
    1284, 1286, 1288
  and S$_N$2 reactions, 0570
  and the aldol condensation, 0571
  and thermodynamic control, 0570
  exceptions, 0568
  for enolate anions, 0570
Bally-Scholl synthesis, 1208
  and the **Skraup reaction**, 1215
Bamford-Stevens reaction, 1317
  and **Shapiro reaction**, 1317
  regioselectivity, 1318
Bao Gong Teng A, 0308
barbaralone, 1328
Barbier reaction, 0637, 0913
  and the **Grignard** reaction, 0637
  retro-**Barbier** fragmentation, 0637
Barbier-Wieland degradation, 0317,
  0648
  and **Miescher degradation**, 0317
**Bardhan-Sengupta phenanthrene
  synthesis**, 1208
  and HF, 1209
barium sulfate
  and the **Lindlar catalyst**, 0437
  and the **Rosenmund catalyst**, 0437
  as a hydrogenation support, 0426
  hydrogenation, 0438
Barton deoxygenation. *See also*
  Barton-McCombie reaction
Barton deoxygenation, 0467
Barton reaction, 1276
Barton-McCombie reaction, 0168,
  0467, 1274
base and acid reactions, 0080

base strength
  and acyl addition, 0786
  and conjugate acids, 0088
  and enolate anions, 0799
  and molecular modeling, 0785
  and nucleophilic strength, 0786
  and proton affinity, 0786
bases
  and acids, and enolate anions, 0797
  and aryl halides, 0207
  and permanganate, 0224
  and the E2 reaction, 0153, 0154
  Brønsted-Lowry, 0088-0092
  Brønsted-Lowry, definition, 0081
  hard. *See* hard bases
  HSAB, classification, 0099
  Lewis. *See* Lewis
  non-nucleophilic, 0786
  non-nucleophilic, and the E2
    reaction, 0157
  reaction with alcohols, 0786
  reaction with amines, 0786
  soft. *See* soft bases
  tethered, and elimination, 0164
  used with alkynes, 0632
basicity
  +I and –I effects, 0089
  +R and –R effects, 0089
  amines and ammonium salts, 0090
  amines and solvation, 0090
  and alkoxides, 0091
  and amide bases, 0091
  and amines, 0090
  and aromaticity, 0785
  and conjugate acids, 0088
  and DABCO, DBU, DMAP, 0090
  and electronic effects, 0089
  and field effects, 0089
  and **Hofmann elimination**, 0164
  and hydroxide, 0091
  and inductive effects, 0089
  and mesomeric effects, 0089
  and organolithium reagents, 0092
  and pK$_a$, 0088
  and primary steric effect, 0089
  and proton sponge, 0090
  and quinuclidine, 0090
  and secondary steric effect 0089
  and solvent effects, 0090, 0091
  and the conjugate acid, 0089
  of **Grignard reagents**, 0639
  of organolithium reagents, 0689
  order, with amines, 0090

  versus nucleophilicity, 0141
batzelladine F, 0902
Baylis-Hillman reaction, 0878, 0879,
  0987
  amine additives, 0879
  and aldol condensation, 0879
  and chiral auxiliaries, 0879
  asymmetric, 0879
  intramolecular, 0879
  Oppolzer's sultam, 0879
bazeispirol A, 0994
9-BBN, 0496, 0515
  and boron enolates, 0842
  and reduction, 0387
  complex with pinene, 0395
  iodo, 0509
  with alkenes, 0505, 0514, 0516
  with alkynes, 0503, 0509
  with heteroatom alkenes, 0502
Beauveria sulfurescens, reduction,
  0482
Beckmann rearrangement, 0188-
  mechanism, 0189
  migratory aptitude, 0189
belactosin C, 0444
Benkeser reduction, 0461
  and aryl compounds, 0462
  and LiAlH$_4$, 0462
benzalacetone, 0352
benzamides, with Dibal, 0617
benzene
  Diels-Alder reaction with furan,
    1018
  with carbenes, 1326
  with chromium hexacarbonyl, 0764
benzenesulfenyl chloride, with
  enolate anions, 0814
benzhydryltrimethylammonium
  bromide, 0752
benzo[*c*]fluorene, nitration, 0198
benzoate, alcohol protecting group,
  0605
benzoates, and the **Prévost reaction**,
  0301, 0302
benzocyclobutanes
  and the **Diels-Alder reactions**, 1099
benzocyclobutenes, ring opening, 1099
benzodiazepines, 0986, 0987
benzofurans, 0118
  by radical cyclization, 1292
benzoic acid derivatives, and
  decarboxylation, 0173
benzoic acid, with alcohols, 0605

benzoic anhydride, with alcohols, 0605
benzophenone, and
  Oppenauer oxidation, 0251
benzophenone, as a sensitizer, 1089,
  1091, 1266, 1314
benzopyranoisoxazolidine,
  auxiliary with ylids, 0741
benzoquinone, **Diels-Alder reaction**,
  1017, 1025, 1026
benzoquinones, and cyclopentadiene,
  1013
benzothiadiazine-7-sulfonamide,
  6-chloro-1,1-dioxide, 0904
benzotriazol-1-
  yloxytripyrrolidinophosphonium
  hexafluorophosphate (PyBOP), 0581
benzoxepin, 0126
benzoyl chloride, with alcohols, 0605
benzoyl peroxide. *See* peroxide
benzyl alcohols, hydrogenolysis, 0431
benzyl amines, hydrogenolysis, 0431
benzyl chloroformate, with amines,
  0618
benzyl ethers. *See* ether
benzyl groups, and transfer
  hydrogenation, 0613
benzyl protecting groups, amines, 0613
benzyldimethylamine, 0752
benzylic carbocation. *See* carbocation
benzylic groups, hydrogenolysis, 0440
benzyloxycarbonyl. *See* Cbz
benzyloxymethyl cyclopentadiene
  and the **Diels-Alder reaction**, 1068
benzylpenicillin methyl ester, 0336
benzyltriethylammonium chloride
  and dihydroxylation, 0294
benzyltrimethylammonium
  hydroxide. *See* Triton B
benzyne, 0194
  and aryl halides, 0207
  and intramolecular **Diels-Alder**, 1065
  and organolithium reagents, 0207
  and substitution, nucleophilic
    aromatic, 0207
  and the **Diels-Alder reaction**, 0208,
    1018, 1065
  from organolithium reagents, 1065
  LUMO map, 0207
berbine methiodide, 0369
bergamotenes, 1277
Bergman cyclization, 1295
Bergman rearrangement, aza-, 1295-
  1296

Bernthsen acridine synthesis, 1218

betaines, and phosphorus ylids, 0730
  questionable intermediates, 0737
  stereochemistry, 0735

BF_3. *See* boron trifluoride

BHT (2,6-di-*tert*-butyl-4-
  phenylphenol)), 1048
  and the **Diels-Alder reaction**, 1036,
    1048
  and intramolecular **Diels-Alder**, 1064

biaryls
  and chirality, 0028
  by **Meerwein arylation**, 1298, 1299
  by **Suzuki-Miyaura coupling**, 1242
  by the **Gomberg-Bachmann
    reaction**, 1299
  by the **Pschorr reaction**, 1299
  by the **Ullmann reaction**, 1300
  chiral, and **Dess-Martin
    periodinane**, 0247
  from anilines, 1299
  from aryl diazonium salts, 0210
  from aryl **Grignard reagents**, 0642
  from aryl halides, 1300
  from aryl triflates, 1242
  from aryldiazonium salts, 1299
  from boronic acids, 1242
  heterocyclic, 1242
  symmetrical, 1300

biatractyolide, 0459

bicyclic compounds
  and **Wagner-Meerwein
    rearrangement**, 1177

bicyclic hexanes, conformations, 0067

bicyclic ketones. *See* ketones

bidentate nucleophiles, cyanide, 0625

Biellmann alkylation, 0697

bielschowskysin A, 1094

bigelovin, 0813

bilosespene A, 0932

BINAL, 0028

binapthols, with allyltin, 0767

BINOL
  and [3+2]-cycloaddition, 1109
  and **Michael addition**, 0878
  chiral complex, 1075
  in the **Diels-Alder reactions**, 1075
  with allyltin, 0767

biogenetic synthesis. *See* biomimetic

biological activity, and synthesis, 0903

biomimetic (biosynthetic) synthesis,
  0948-0954
  deoxyerythronolide B, 0953, 0954

sparteine, 0952

biotin, 0651

biphenyls, amino, with $ZnCl_2$, 1219

biphosphites, aryl, chiral, and
  Pauson-Khand reaction, 1344

bipinnatin J, 0850

bipyridyl (bpy), as a redox mediator,
  0470

biradicals, and photochemistry, 1094

Birch reduction, 0459, 0936, 1049
  ammonia free, 0460
  and alkylation, 0461
  and electronic effects, 0459
  and heterocycles, 0461
  and the **Benkeser reduction**, 0461
  anisole derivatives, 0460
  electrolytic, 0470
  electronic effects, 0459
  mechanism, 0460
  powerful in synthesis, 0928, 0931
  resonance contributors, 0459

bis(acetonitrile)palladium chloride,
  1233

bis(bipyridylcopper) permanganate,
  0260

bis(diisopropoxytitanium) dibromide,
  and chiral ene reactions, 1152

bis(dimethylamino)naphthalene,
  and aldehydes or ketones, 0788

1,8-bis(dimethylamino) naphthalene.
  See also bis[1,2-bis
  (diphenylphosphino) ethane
  palladium, 1231, 1232
  Proton Sponge

bis(diphenylphosphinoethane)nickel
  chloride, 1245

bis(imidazole), ligand in the
  Heck reaction, 1238

bis(imines), and **Fischer Indole
  Synthesis**, 1222

bis(iminoaluminate), 0359

bis(iodozincio)methane,
  carbocation initiation, 1182

bis(methanesulfonate), and
  spirocyclization, 0772

bis(oxazolidine), and **Mukaiyama
  aldol reaction**, 0839

bis(oxazoline)-Fe(III) complex,
  chiral catalysts, 1073

bis(phosphines). *See* phosphines

bis(salicylidene)ethylenediamine. *See*
  salen

[bis(trifuoroacetoxyiodo)]benzene,
  1297

bis(ylids), 0734

bisabalol, 1028

bisabolol, 0358

bisbolene, 0265

Bischler-Napieralski reaction, 1215,
  1356, 1359

bisorbicillinoids, 1119

bisulfite in aq ethanol, with
  osmium tetraoxide, 0295

blastmycinone, 0437

blazeisirol A, 0994

bleomycin A_2, 0973, 0974

bleomycin, 0442

bmim(butyl-3-methylimidazolium
  chloroaluminate), 1037, 1221
  and the **Pechmann condensation**,
    1221

boat and chair, energy difference, 0055

boat conformations, 0048, 0061, 0062

boat transition state, 0060
  aldol condensation, 0857
  Claisen rearrangement, 1132
  Cope rearrangement, 1124

Boat versus chair transition state,
  Claisen rearrangement, 1132

boat-boat conformation, 0050

Bobbitt's reagent, 0248

Boc (tert-butoxycarbonyl), with
  amines, 0617

Boc, amine protecting groups, 0617

Boc-anhydride, and
  macrolactonization, 0576

Boc-On, 0617

Bogert synthesis. *See also*, **Bogert-
  Cook synthesis**, 1209

Boltzmann distribution, 0040, 0065,
  0421, 0523, 0558
  and the ene reactions, 1147
  arene nitration, 0198
  hydroboration, 0517

bond construction, and SYNGEN,
  0969, 0970

bond disconnection, 0908

bond dissociation energy
  and functional groups, 1084, 1085
  and photochemistry, 1085
  and radicals, 1260

bond distance, and S_N2 transition
  states, 0115

bond energy, and syn elimination, 0170

bond length, B-O, 0856

bond length, O-Li, 0856
bond migration
    and decarboxylation, 0173
    and hydrogenation of alkenes, 0432
    and photoreduction, 0478
bond opposition strain, definition, 0048
bond polarization
    and disconnection, 0011
    and functional groups, 0011
    and syn elimination, 0170
    and the E2 reaction, 0153
    in **Grignard reagents**, 0636
bonding orbitals, 1000
    and perturbation theory, 1015
bonding, hydrogen. *See* hydrogen
    bonding
bonds, C-Li, covalent character, 0671
bonds, Fe-C, 0758
bonds, functional groups and
    transforms, 0080
bonds, π-bonds
    and electron donation, 0080
    and transforms, 0080
    and leaving groups, 0080
    and transforms, 0080
bondset, and SYNGEN, 0963
bondsets, for estrone, 0963, 0964
9-borabicyclo-[3.3.1]nonane.
    See 9-BBN
borane, Alpine. *See* Alpine
borane
    and diborane, 0386
    chlorothexylborane, 0510, 0512
    difficulty in handling, 0387
    dimethyl sulfide complex, 0510
    dimethyl sulfide, 0516
    flammability, 0387
    from NaBH₄ and BCl₃ or BF₃, 0492
    in ether solvents, 0491
    preparation, 0385
    reduction of aldehydes, 0386
    reduction of carboxylic acids, 0386
    reduction of ketones, 0386
    reductions, 0385-0387
    transition state in reaction with
        alkenes, 0492
borane
    with acetylene, 0492
    with alkenes, 0491
    with alkenes, electronic effects, 0494
    with alkenes, steric hindrance, 0492
    with amino acids, 0386
    with boron trifluoride etherate, 0524

with carene, 0526
with catechol, 0500
with cyclododecatriene, 0531
with dienes, 0495
with dimethyl sulfide, 0386
with enamines, 0561
with epoxides, 0386
with lactams, 0386
with **Lewis acids**, 0491
with longifolene, 0525
with NaBH₄, 0386
with pinene, 0495
borane-dimethyl sulfide, 0494
    with alkenes, 0516
boranes, alkenyl, 0492, 0503, 0509,
    0518
    cis/trans isomers, 0509
    protonolysis, 0509
    stereochemistry, 0503
    trans/cis isomers, 0509
    with BBr₃, 0510
    with methylcopper, 0510
boranes, alkenylborinates, 0503
boranes, alkyl,
    9-BBN, 0496
    alkoxyborinate, 0500
    alkyl, from alkenes, 0491
    alkyl, with alkynes, 0509
    alkyl, with alkynes, stereoselectivity
        0509
    alkylborinate, 0505
    allyl(diisopinocampheyl)borane,
        0526
    amination, 0534, 0535
    and aldol condensation, 0518
    and **Boltzmann distribution**, 0523
    and borinates, 0514
    and conjugated dienes, 0502
    and contrathermodynamic
        isomerization, 0501
    and cyanogen bromide, 0510
    and dienes, 0531
    and dihydrofuran, 0523
    and dihydropyran, 0523
    and dihydropyrrole, 0523
    and dihydrothiophene, 0523
    and ether solvents, 0491
    and galvinoxyl, 0533
    and heterocycles, 0523
    and lithium trimethoxyaluminum
        hydride, 0532
    and potassium triisopropoxy-
        borohydride, 0505, 0509

and prochiral centers, 0497
and pyridines, 0502
and sacrificial alkenes, 0501
and thiophenes, 0502
and trienes, 0531
and trifuoroacetic anhydride, 0528
boranes, alkyl
    anti-**Markovnikov** selectivity, 0497
    Boltzmann distribution, 0517
    borinanes, 0496, 0518
    bromoborane, 0506
    carbonylation, 0530, 0531
    catecholborane, 0500
    chain extension of aldehydes, 0533
    chain extension, 0518
    chloroalkylboranes, 0500
    chlorothexylborane, 0509
    cis and trans, 0498
    complex with dimethyl sulfide, 0509
    conjugate addition, 0533
    coupling, 0511, 0512
    crotyldiisopinocampheylborane, 0527
    cyclic, 0518
    cyclopropylboranes, 0511
    dialkoxyborates, 0499
    dialkyl, 0495
    dibromoborane, 0503, 0504, 0505,
        0510
    dichloroborane, 0500
    dicyclohexylborane, 0509
    dihydroboration, 0503
    diisopinocampheylborane, 0495
    dilongifolylborane, 0525
    dimethoxyborane, 0499
    disiamylborane, 0495
    from carene, 0526
    halogenation, 0507
    heteroatom, 0500
    isomerization, 0501
    Markovnikov addition to alkenes,
        0493
    mechanism of protonolysis, 0506
    mixed, 0509, 0510
    monochloroborane, 0500
    monoisopinocampheylborane, 0524
    oxidation to aldehydes, 0528
    oxidation to ketones, 0528
    oxidation, 0513, 0518
    palladium-catalyzed coupling, 0511
    protonolysis, 0505, 0509
    reaction with NaOH and H₂O₂, 0513
    thexylborane, 0495
    trans and cis, 0498

transition state modeling, 0517
transition state modeling, 0521
trialkyl, 0495
vinyl. *See* alkylboranes, alkenyl
vinylborinates, 0503
boranes, alkyl
  with aldehydes, 0533
  with alkenes, 0543, 0561
  with alkenes, regioselectivity, 0551
  with alkynes, 0503
  with azides, 0536
  with bromine, 0507, 0533
  with bromoesters, 0533
  with bromoketones, 0534
  with bromosulfoxides, 0534
  with carboxylic acids, 0506
  with chloramines, 0534
  with chloronitriles, 0534
  with chromic acid, 0528
  with dienes, 0502
  with furans, 0502
  with galvinoxyl, 0536
  with Grignard reagents, 0535
  with HBr, 0506
  with hydroxylamine *O*-sulfonic acid,
    0534
  with iodine, 0507
  with NaCN, 0528
  with NaOH and peroxide, 0495
  with oxygen, 0505
  with PCC, 0528
  with propargylic halide, 0511
  with propionic acid, 0506
  with rhodium catalysts, 0519
  with silver nitrate, 0511
  with sulfides, 0537
boranes, allyl, 0526
  enol formation, 0518
  oxidation, 0528
  reaction with aldehydes, 0527
  reaction with amides, 0527
  reaction with anhydrides, 0527
  reaction with carboxylic acids, 0527
  reaction with esters, 0527
  reaction with lactams, 0527
  transition state modeling, 0523
boranes
  and galvinoxyl, 0536
  and HSAB, 0106
  and hydride reducing agents, 0498
  and **Lewis acids**, 0535
  and methylcopper, 0511
  and oxazaborolidines, 0396

and radial cyclization, 1291, 1292
and radical scavengers, 0536
anti-Markovnikov addition, 0496
as radical initiator, 1286, 1291, 1292
asymmetric hydroboration, 0519
boronic acids, 1242
bromine and NaOMe, 0508
cis addition to alkenes, 0492
conversion to alkyl borohydrides,
  0381
cyano, 0529
cyclic, from dienes, 0530
dicyclohexylborane, 0528
halo, with iodine, 0512
halo, with NaOMe, 0513
methylborane, 0531
oxidation, 0514
protonolysis, 0510, 0512
radical chain mechanism, 0533
rearrangement, 0515, 0530
reduction of aldehydes, 0395, 0397
reduction of ketones, 0395, 0397
stereoselectivity in oxidation, 0515
substituted, reduction, 0387
trialkyl, with conjugated ketones,
  0841
trialkyl, with diazoketones, 0841
tricyclohexylborane, 0528
triethyl, with LiH, 0379
triethyl, with tributyltin hydride, 0466
trinorbornylborane, 0534
boranes, vinyl. *See also* boranes,
  alkenyl
boranes, vinyl, 0841
  halogenation, 0508
  protonolysis, 0509
  rearrangement, 0510-0512
boranes
  with alkenes, 0493, 0513
  with alkenes, transition state, 0497
  with allylic halides, 0511, 0514
  with carbon monoxide, 0530, 0532
  with cyanogen bromide, 0511
  with lithioalkynes, 0513
  with longifolene, 0519, 0520
  with NaOH and H_2O_2, 0514
  with NaOMe, 0510, 0512
  with O-sulfonylhydroxylamine, 0522
  with palladium catalyst, 0514
  with pinene, 0519, 0520
  with silver nitrate and NaOH, 0513
borane-THF, 0524
boranyl alkoxyaluminate, 0532

borate, trimethyl, and NaH, 0348
borates, with KH or NaH, 0377
borinanes, 0496, 0518
  alkyl, and iodine, 0518
  oxidation, 0518
  with lithium dichloromethane, 0536
borinates, and oxidation
  of alkylboranes, 0514
borneol, by **Grignard** reduction
  of camphor, 0667
borneol, oxidation with Cr(VI), 0230
bornyloxyaluminum, chiral catalyst,
  1073
Borodin reaction, 0150
borohydride, acyloxy. *See*
  acyloxyborohydrides
borohydride
  alkyl, 0381
  aluminum, 0374
  cerium, 0376
  cerium, and 1,2-reduction, 0417
  cerium, the **Luche reagent**, 0376
  cerium, with conjugated carbonyls,
    0376
  cyano. *See* cyanoborohydride
  lithium solubility, 0374
  lithium triethyl (Super hydride), 0378
  lithium, preparation, 0373
  lithium, reduction of chiral auxiliary,
    0374
  lithium, with acid chlorides, 0373
  lithium, with aldehydes, 0373
  lithium, with epoxides, 0373
  lithium, with esters, 0373
  lithium, with ketones, 0373
  lithium, with lactones, 0373, 0374
  magnesium, 0374
  metal, 0376
  potassium, with carboxylic acids,
    0378
borohydride, sodium, 1368
  preparation, 0348
  and enamines, 0561
  and oxymercuration, 0182
  and ozonolysis, 0319
  and the Manhattan project, 0369
  and water, 0370
  as a reducing agent, 0369-0373
  preparation, 0370
  reactivity, 0370
  reduction of cyclohexanones, 0414-
    0416
  reduction of imines, 0843

reduction of ketones, 0742
reduction, 0104
solubility, 0370
solvents, 0370
with acid derivatives, 0370
with aldehydes or ketones, 0370
with anhydrides, 0373
with BCl_3, 0492
with BF_3, 0385, 0492
with borane, 0386
with conjugated carbonyls, 0370
with enamines, 0371
with HCN, 0382
with hydrooxazolines, 0845
with imines, 0371, 0685
with iminium salts, 0371, 0562
with ketones, diastereoselectivity,
   0555
with lithium bromide, 0373
with metal salts, 0376
with oxazolidines, 0844, 0845
with ozonides, 0371
with pyridinium salts, 0372
with trifluoroacetamides, 0616
with trimethyltin chloride, 0466
reduction of ketones, 0036
borohydride, tetramethylammonium
   triacetoxy, 0377
borohydride, zinc, 0035, 0043, 0375
   and chelation control, 0550
   and conjugated carbonyls, 0375
   and **Cram chelation model**, 0550
   and reduction, 0004
   and silver tetrafluoroborate, 0375
   and stereoselectivity, 0375
   on silica gel, 0375
   with epoxides, 0375
   with ketones, 0560
borohydrides, and chiral additives, 0397
boron aldolates. *See* aldol
boron annulation, 0530
boron enolates. *See* enolates
boron tribromide, with
   alkenylalkylboranes, 0510
boron tribromide, with ethers, 0595
boron trichloride, with $NaBH_4$, 0492
boron triflates, and boron enolates,
   0842
boron trifluoride etherate, with borane,
   0524
boron trifluoride
   and acrolein, 1039
   and Friedel-Crafts alkylation, 1202

and the **Mukaiyama aldol** reaction,
   0837, 0838
and organocopper reagents, 0723
and Pinacol rearrangement, 1180
and the **Diels-Alder reaction**
1038, 1039
and the **Nysted reagent** 0757
Diels-Alder reaction and molecular
   modeling, 1039
dithioacetal formation, 0611
dithioketal formation, 0611
organocuprate complex ,0723
with aldehydes or ketones, 1164
with disilanes, 0476
with epoxides, 0635
with ketones, 1173
with LiH, 0386
with $NaBH_4$, 0385
with $NaBH_4$, 0492
boron, and HSAB, 0099
boronates, oxidation with amine
   oxides, 1065
boronic acids
   and potassium trifluoroborates, 1243
   and **Suzuki-Miyaura coupling**,
      0527, 1241
   aryl, 1243
   vinyl, 0511
   with aryl triflates, 1242
   with bromine, 0508
   with iodine 0508
   with sorbic alcohol, 1064
boronic ester *ate* complexes,
   and reduction, 0398
boronolide, 0258
Borsche-Drechel cyclization, 1218
   and **Fisher indole synthesis**, 1218
botryococcene, 0600
C34-botryococcene, 0996
bppm, 0450
bpysee bipyridyl
Bradsher reaction, 1210, 1219
Bredt alkenes, 0554
Bredt's rule, 0553, 0554
   and ring strain, 0554
   and SeO_2, 0326
   and intramolecular **Diels-Alder**
      reactions, 1060
brefeldin A, 0956, 1235
brevicomin, 1056
brevioxime, 0172, 0441
briarellin E, 0680
bridged carbocations, 1171

bridgehead alkenes, epoxidation, 0281
bridgehead atoms, and **Bredt's rule**,
   0553, 0554
bromide, benzyl. *See also* halide
bromide, benzyl, with amines, 0613
bromide, lithium, with $NaBH_4$, 0373
bromide, magnesium, with selectrides,
   0381
bromide, phosphoryl, with aryls, 0147
bromide, potassium, with $NaNO_2$, 0545
bromide, propargyl, with
   Grignard reagents, 0633
bromides, alkyl, with
   sodium cyanoborohydride, 0383
bromination
   and AIBN, 0151
   and the **Wohl-Ziegler reaction**, 0151
   benzylic compounds, 0151
   of alkanes, 0151
   of aromatic rings, 0147
   of ketones, 0152
   of maleic acid and fumaric acid, 0034
bromine
   and electrophilic aromatic
      substitution, 0195
   and NaOH, 0193
   and the **Prévost reaction**, 0302
   boranes and NaOMe, 0508
   in water, 0187, 0268
   oxidation of organoiron compounds,
      0761
   photolysis, 0151
   with alkenes, 0158, 0183-0188
   with alkenes, suprafacial and
      antarafacial addition, 1011
   with alkylboranes, 0507, 0533
   with boronic acids, 0508
   with ketones, 0883, 0884
bromine-lithium exchange, 0512
bromoacids, and macrocyclizations,
   0573
bromobenzene, with
   Pseudomonas putida, 0304
bromobenzenesulfonyl, as a
   leaving group, 0772
bromoborane, 0506
bromoesters, with organoboranes, 0533
bromoethane, reaction with
   nucleophiles, 0120
bromoform, 0884
bromohydrins, from alkenes, 0188
bromoindoles, and the **Heck reaction**,
   1237

bromoketones
  from ketones, 0883, 0884
  with organoboranes, 0534
bromolactam, 0187
bromolactones, 0186
bromolactonization, 0186
bromonium ions, 0184
  and the **Prévost reaction**, 0302
bromophenylsulfonyl, with phenols, 0771
bromoselenides, and **Claisen** rearrangement, 1135
bromosulfoxides, with organoboranes, 0534
bromozinc aldolate, 0886
Brønsted-Lowry acids and bases, 0079, 0081-0088
  and bases, definitions, 0081
Brucine-*N*-oxide, in the **Pauson-Khand reaction**, 1344
brunsvigine, 0125
bryostatin 7, 0467
buffers
  and peroxyacid epoxidation, 0277
  disodium phosphate, 0308
  in the **Baeyer-Villiger reaction**, 0308
Burgess reagent, 0170
Bürgi-Dunitz angle, and LDA, 0787
Bürgi-Dunitz trajectory, 0412
  and aldehydes, 0795
  and **Baldwin's rules**, 0565
  and enolate anions, 0795
  and **Evans' model**, 0857
  and ketones, 0556, 0795
butadiene
  conformations, 0047
  molecular orbital diagram, 1006, 1007
  triplet, 1089
butane
  conformations, 0040
  energy barrier, 0040
  rotamers, 0040
butene, conformations, 0044-0046
butyl permanganate, 0260
butyl-3-methylimidazolium chloroaluminate. *See* bmim
butylhydroperoxide. *See* peroxide
butylimidazolium, Ionic liquids, 1037
butyllithium (tert)-, with vinyl halides, 0512
butyllithium (sec), with alkynes, 1366
n-butyllithium
  and DME, 0672

and sparteine, 0672
and the **Corey-Fuchs procedure**, 0630
and THF, 0672
and TMEDA, 0672
as a base, 0092
reaction with enolate anions, 0802
with alkenyl halides, 0677
with alkyl halides, 0672
with alkynes, 0633
with dithianes, 0705
with terminal alkynes, 0632
with trimethylsilylacetylene, 0634
with vinyl dihalides, 0159
with vinyl halides, 0630
Bz. *See* benzoate
Bz, benzamide protecting group, 0617

C

cacospongionolide F, 0148
cadinene, 0552
Cadiot-Chodkiewicz coupling, 0631, 1302
cadmium chloride, with **Grignard** reagents, 0647
caged radical pair, 0752
caged radicals, 1261
Cahn-Ingold-Prelog selection rules, 0014-0023
calafianin, 1106
calcium sulfate, as a hydrogenation support, 0426
calcium
  and the **Benkeser reduction**, 0462
  with amines, 0462
calculations, and **Cram's rule**, 0402
calicheamicin, 0463, 1295
calicoferol B, 0473
callystatin A, 0390, 0480, 0870
calothrixin A, 1204
calyculin A, 0271
cameroonan-7α-ol, 0809
cameroonanol-7α-ol , 0181
caminoside A, 0996
camphene hydrochloride, rearrangement, 1177
camphene, 1317
  and **Bredt's rule**, 0553, 0554
camphor, 0957
  and chiral auxiliaries, 1071
  and **Oppolzer's sultam**, 0361
  reduction with **Grignard reagents**, 0667

with LiAlH$_4$, 0393
camphorsulfonic acid, 0879
CAN (ceric ammonium nitate)
  oxidation of catechols, 0265
  and dithioacetals, 0612
  and dithioketals, 0612
Candida anatarctica lipase, and the **Baeyer-Villiger reaction**, 0312
cannabisativine, 0170, 0613
capronitrile, 0624
carane, 0433, 0506
carbacephams, 1050
carbamates
  amine protecting groups, 0617
  and Curtius rearrangement, 1051
  and radical cyclization, 1294
  Boc, for amines, 0617, 0618
  by the **Curtius rearrangement**, 0192
  conversion to iminium salts, 1055
  from amines, 0356, 0613
  reduction with LiAlH$_4$, 0356, 0618
carbanions, alkylation, 0461
carbanions
  from sulfides, 0696
  from sulfonate esters, 0696, 0701
  from sulfones, 0696, 0699
  from sulfoxides, 0696
  from sultones, 0701
  phosphonate, 0739
  pyridinium ylids, 0751
  silane, 0770
  sulfur, 0444, 0696
  sultone, 0701
  via Birch reduction, 0461
  with arylchromium reagents, 0764
carbazoles
  by **Borsche-Drechel cyclization**, 1218
  from amides, 1218
carbene insertion, 1328
carbene, dimethyl, 1313
carbene, methyl, 1312
carbenes, 0954, 1311-1331
  1,2-hydrogen shifts, 1322
  acyl, and **Wolff rearrangement**, 1315
  acyl, from diazocarbonyl compounds, 1315
carbenes
  and copper compounds, 1322
  and copper triflate, 1325
  and cyclopropanes, 1120
  and diazoalkanes, 1313, 1321
  and diazocarbonyl compounds, 1315

and diazocarbonyls, 1329
and diazoketones, 1328
and diazomethane, 1313
and dimethylketene, 1312
and diradicals, 1312
and DOSP, 1330
and **Doyle catalysts**, 1330
and excited cyclopropanones, 1312
and hydrazones, 1314
and intersystem crossing, 1311
and metathesis, 1334, 1335
and oxygen, 1323
and quantum yield, 1312
and radical, 1311
and rearrangement, 1325
and rhodium catalysts, 1330
and ring expansion, 1326
and the **Davies catalyst**, 1330
and ylids, 0734, 0750
carbenes
  by photolysis of cyclopropanes, 1316
  by photolysis of diazines, 1322
  by photolysis of ketenes, 1312
  by photolysis of oxiranes, 1316
  C-H insertion, 1329
  definition, 1311
  dichloro, 1325
  dichlorocarbene, 1325
  electrophilic character, 1312
  electrostatic potential map, 1320
  enantioselectivity, addition
    to alkenes, 1325
  ester, 1327
  formation in the presence of oxygen,
    1312
carbenes
  from chloroform, 1319
  from diazirine, 1315
  from diazo compounds, 1314
  from diazoketones, 1328
  from diazomethane, 1325
  from ketenes, 1312, 1313
  from ketones, 1318
carbenes
  halo, formation of cyclopropanes,
    1319
  insertion, 1327
  intramolecular trapping, 1328
  metal, 1334, 1335
  methylene, 1312
  molecular modeling, 1320
  preparation 1312
  pressure and carbene formation, 1322

rate of addition to alkenes, 1324
rearrangement, 1322
regioselectivity of addition
  to alkenes, 1325
ring closure versus bond rotation,
  1324
ring expansion reaction, 1326, 1327
singlet and triplet, 1312
singlet, 1311, 1312
singlet, from singlet diradicals, 1324
titanium, 0756
triplet, 1311
carbenes
  with alkenes, 1319, 1322
  with alkenes, activation energy, 1322
  with alkenes, exo-endo addition, 1325
  with alkenes, rate, 1324
  with alkenes, stereoselectivity, 1323
  with alkenes, transition state
    modeling, 1320
  with aryls, 1326
  with benzene, 1327
  with diazolactams, 1329
  with enol ethers, 1326
  with furans, 1327
  with naphthalene, 1327
carbenium ions. *See* carbocations
carbenium ions, definition, 1161
carbenoids, 1311, 1331-1334
  and diiodomethane, 1331
  and iodozinc compounds, 1331
  and **McMurry olefination**, 1310
  and **Simmons-Smith reaction**, 1331
  and zinc-copper, 1331
  definition, 1312
  from **Grignard** reagents, 0658
carbocations
  1,2-alkyl shift, 1173
  1,2-hydride shift, 1171
  allylic, 1164, 1165
  allylic, and resonance, 1165
  allylic, structure, 0133
carbocations
  and acyl halides, 1204
  and acylnitrilium salts, 1186
  and allyl nickel, 1245
  and anti-**Markovnikov** addition,
    0180, 0181
  and aromatic substitution, 0193
  and aryls, 1201, 1202
  and back strain, 0134
  and **Baldwin's rules**, 1183
  and **Bradsher reaction**, 1210

and cyclopropanes, 1164
and dipolar effects, 1162
and E1, 0160
and enthalpy, 1166
and **Friedel-Crafts alkylation**, 1194,
  1196
and **Friedel-Crafts reactions**, 1192-
  1225
and **Grob fragmentation**, 0174
and halohydrin formation, 0187
and hybridization, 0130, 0133
and iminium ions, 1185
and iminium salts, 1164
and inductive effects, 1162
and ion pairing, 0130, 0137
and **Johnson polyene cyclization**,
  1184
and **Markovnikov addition**, 0179
and **Meinwald rearrangement**, 1181
and molecular modeling, 1171
and **Nazarov cyclization**, 1189
and NMR, 1178
and **Noyori annulation**, 1189
and orbitals, 0135
and oxonium ions, 0136
and reactions of alkenes, 0179, 0180
and resonance, 0132
and **Saytzeff elimination**, 0160
and s-character, 1164
and **semipinacol rearrangement**,
  1181
and S_N1 reactions, 0129, 1162
and the E1 reaction, 0160, 0161, 1186
and the **Fries rearrangement**, 1220
and the **Haworth phenanthrene
  synthesis**, 1207, 1208
and the NIH shift, 1179
and trimethylsilyl triflate, 1176
and **Wagner-Meerwein
  rearrangement**, 1177
and **Wheland intermediate**, 1192
aryl migration transition state, 1172
aryl migrations, 1172
carbocations, benzylic, 1164, 1165
  and resonance, 1165
  structure, 0133
carbocations
  bis(silyl)-stabilized vinyl, 1168
  bridged, 1171
  by ionization of alkyl halides, 0129
  carbon-carbon bond forming
    reaction, 1182
  chlorine-stabilized, 1183

configurational instability, 1169
cyclization, powerful in synthesis, 0928, 0931
cyclobutylcarbinyl, 1174
cyclopropylcarbinyl, 0137, 1163, 1174
definition, 1161
Demjanov rearrangement, 1175
destabilization, 1164
energy of, 1162
carbocations
from alcohols and strong acid, 0161
from alcohols, 0130, 0144, 1163
from alkenes, 0178
from alkyl halides, 1194
from amines, 1163, 1175
from amino alcohols, 1175
from diazonium salts, 1175
from epoxides, 1177
carbocations
halo, 0153
homoallylic, 1174
hydride shift, 0134
hyperconjugation, 1162, 1163
iron stabilized, 1248
labeling, 1171, 1172
mechanism of rearrangement, 1171
mercury stabilized, 0182
nitrogen stabilized, 1164
nonclassical, 1178
norbornyl, 1177
onium salts, 1163
order of stability, 0132
oxocarbenium ions, 1163
oxygen stabilized 0133, 0181, 0656, 1163, 1180, 1186
phenonium ions, 1171
pinacol-like rearrangement, 1182
polyene cyclization, 0860
racemization, 1170
reaction with alkenes, 1182
reaction with alkynes, 1183
reaction with formate anion, 1183
reaction with halide anions, 0130
reaction with vinyl chlorides, 1183
reactivity with halide ions, 0132
carbocations, rearrangement, 0134-0124, 1170
and allylic alcohols, 1181
and amino alcohols, 1181
and cyclopropanes, 0137
and **Grignard reactions**, 0656
and hyperconjugation, 1173

and molecular modeling, 1165
and oxiranes, 1181
and silyl groups, 0136
and σ-conjugation, 1173
energy, 0134
iodohydrins, 1181
migratory aptitude, 1172
powerful in synthesis, 0929
rate of reaction, 0134
s-conjugation, 1173
transition state modeling, 1165
carbocations
relative energy, 1168, 1169
relative stability, 1169
resonance, 1164, 1165
retention versus inversion, 1170
secondary versus primary, 0132, 0134
carbocations, stability, 1162
and ionization potential, 1166
and rate of hydrolysis, 1167
and rate of solvolysis 1167
and reactivity, 0179
and structure, 1168, 1169
and the **Pinacol rearrangement**, 1180
substituent effects, 1162
substituent effects, 1167
carbocations
stereochemistry, 1170
substituent effects, 1162
sulfur stabilized, 0133, 1163, 1176
Tiffeneau-Demjanov ring expansion, 1175
transition state for rearrangement, 1171
trityl, 1248
vinyl, 1168, 1188
vinyl, from alkynes, 1168
vinyl, from vinyl halides, 1168
vinyl, with water, 1183
with CO, 1187
with nucleophiles, 0178
X-ray structure, 1168
β-silyl, 1176
carbocyclic acids
and the **Hunsdiecker reaction**, 0150
from alkyl halides, 0760
carbocyclic rings, strategic bonds, 0934
carbohydrates, 0246
as chiral templates, 0957, 0958
template and **Diels-Alder** reaction, 1075
carboline, 1142

carbon dioxide
and decarboxylation, 0172
and **Kolbe electrolytic synthesis**, 1278
and retro-**Diels-Alder** reactions, 1036
as a poison for hydrogenation, 0426
from carboxylic acid salts 1278
from halo carboxylates, 0177
loss from halo-carboxylates, 0177
supercritical, 0839
supercritical, and [3+2]-cycloaddition, 1107
supercritical, and density, 1049
supercritical, and **Diels-Alder**, reactions, 1049, 1066
supercritical and **Mukaiyama aldol reaction**, 0839
supercritical, and the **Glaser reaction**, 0631
supercritical, enzyme reduction, 0479
with enolate anion, 0800
with **Grignard reagents**, 0651
with organoaluminum compounds, 0762
with organolithium reagents, 0684, 0685
carbon disulfide, and **Friedel-Crafts alkylation**, 1194
carbon monoxide
and boranes, 0530
and **Koch-Haaf carbonylation**, 1187
and **Norrish type I** fragmentation, 1264
and **Pauson-Khand reaction**, 1342
and photolysis of ketenes, 1312
and sodium tetracarbonyl ferrate, 0760
and water in hydroboration, 0531
extrusion and carbene formation, 1312, 1313
extrusion from aldehydes, 1265, 1266
extrusion from ketenes, 1312, 1313
extrusion from organoiron compounds, 0759
insertion, 0760
poison for hydrogenation, 0426
with boranes, 0532
with carbocations, 1187
carbon radicals, 1257
carbon tetrabromide
and PPh₃, 0733
and the **Corey-Fuchs procedure**, 0630
and triphenylphosphine, 0630
and ylids, 0733
carbon tetrachloride
and radicals, 1275

radical addition, 1272
with alkenes, 1272
with HMPT and alcohols, 0147
carbon, as a hydrogenation support, 0426
carbonates, Ba or Ca, and dithiane hydrolysis, 0706
carbon-carbon bond forming reactions, 0007
carbon-carbon bonds, and synthesis, 0006
carbonium ion. *See* carbocations
carbonyl compounds, and adical cyclization, 1293
carbonyl compounds
  ene reaction, 1145
  hydrogenation, 0438
  reduction of, 0350
carbonyl groups, and umpolung, 0703, 0704
carbonyl ylids
  [3+2]-cycloaddition, 1115
  from diazoketones, 1115
carbonylation
  and alkoxyaluminum hydrides, 0532
  of alkylboranes, 0531, 0532
carbonyldiimidazole, and lactonization, 0579
carbonyls, conjugated. *See* conjugated conjugated, with **Grignard** reagents, 0651
  ene reactions, 1149
  protonation, 0133
carboxylate anions, 0732
  with LDA, 0836
  with organolithium reagents, 0836
carboxylate salts
  buffers in peroxyacid epoxidation, 0278
  fusion with cyanogen bromide, 0628
carboxylates, silver, and the **Prévost reaction**, 0301, 0302
carboxylic acid derivatives
  enolate anions, 0823
carboxylic acids
  acidity and resonance, 0086
  and alcohols, 0110
  and aluminum borohydride, 0374
  and **Bernthsen acridine synthesis**, 1218
  and chiral organoiron compounds, 0761
  and **Grignard reagents**, 0648

and **Haworth phenanthrene synthesis**, 1208
and **Kolbe electrolytic synthesis**, 1278
and **Meyers' aldehyde synthesis**, 0844
and peroxides, 0277
and peroxyacids epoxidation, 0279
and the **Barbier-Wieland degradation**, 0317, 0648
and the **Gallagher-Hollander degradation**, 0317
and the **Ivanov condensation**, 0853
and the **Miescher degradation**, 0317
and the **Mitsunobu reaction**, 0545
and the **Ugi reaction**, 0627
carboxylic acids
  by alkyl oxidation, 0329
  by chromium oxidation of alkenes, 0316
  by hydrolysis of carboxylate salts, 0651
  by hydrolysis of esters, 0545
  by hydrolysis of ketenes, 1315
  by **Ireland-Claisen** rearrangement, 1138
  by **Koch-Haaf carbonylation**, 1187
  by **Kuhn-Roth oxidation**, 0329
  by **Lieben iodoform reaction**, 0884
  by **malonic ester synthesis**, 0807
  by oxidation of alcohols, 0259
  by oxidation of aldehydes, 0231
  by saponification of esters, 0725
  by saponification of esters, 0834
  by the **Favorskii** rearrangement, 0887
  by **Wolff** rearrangement, 1315
carboxylic acids
  chain extension, 0648
  chain shortening, 0317
  conjugated, and decarboxylation, 0173
  conjugated, from alkynes, 0762
  conjugated, from organoaluminum compounds, 0762
  dianions, with aldehydes, 0835
  dianions, with ketones, 0835
  electrolytic reduction, 0472
  enolate anions, alkylation, 0805
carboxylic acids
  from alcohols and PCC/DMF, 0238
  from alcohols, 1187
  from alkenes, 0313
  from carbocations, 1187
  from CO_2, 0685
  from diesters, 0808
  from diols, 0313

from directed ortho metalation, 0686
from esters, 0648
from **Grignard reaction** with CO_2, 0651
from halo ketones, 0887
from ketenes, 1315
from methyl ketones, 0884
from organolithium reagents and carbon dioxide, 0684
from organolithium reagents, 0685
from oxidation of aldehydes, 0259
from silyl enol ethers, 1138
carboxylic acids
  halo, decarboxylation, 0177
  halo, with zinc-copper, 1078
  hydroxy, 0067
  oxidation of alkyl fragments, 0314
  pK_a, 0085, 0691
  reaction with alkylboranes, 0527
  reaction with base, 0732
  reduction by borane, 0386
  reduction with $LiAlH_4$, 0353
carboxylic acids
  with alkylboranes, 0506
  with amines, 1218
  with diarylamines, 1218
  with diazomethane, 0110, 1321
  with hydrazoic acid, 0191
  with KBH_4, 0378
  with LDA, 0835
  with organolithium reagents, 0684
  with organolithium reagents, 0835
  with Red-Al, 0367
  with $ZnCl_2$, 1218
carene, 0506, 0507
  with borane, 0526
carnitine, 0479
Carroll reaction, Claisen rearrangement, 1135
Carroll variant. *See* **Carroll reaction**
Carroll-Claisen rearrangement. *See* **Carroll reaction**
carvone, 0457, 0748, 0956, 0957
  as a chiral template, 0959
C-aryl glycosides, 1065
caryophyllene alcohol, 1176
caryophyllene, 0559, 1144
cascade cationic cyclization, 1184
cascade reactions, radical, 1286
cascarillic acid, 1332
castanospermine, 0931, 1308
castoramine, 0021
Castro-Stephens coupling, 1244

catabolites, 0443

catalysis, acid. *See* acid catalysts

Catalysis

  chiral, and **Diels-Alder**, 1029

  Lewis acid, carbonyl ene reaction, 1150, 1151

  of the **Moffatt oxidation**, 0244

catalysts

  (*1R,2S*)-*N*-methylephedrine, 1073

  (*S*)-prolinol, 1073

  abrine esters, 1072

  akylhydroperoxides with sulfides, 0335

  and chemisorption, 0427

  and haptophilicity, 0434

  and hydrogenolysis, 0430

  and polyalkylation, 1197, 1198

  and the **Baeyer-Villiger reaction**, 0307

  bis(oxazoline)-Fe(III) complex, 1073

  bornyloxyaluminum, 1073

  chiral, and the Diels-Alder reaction, 1072

  chiral, pyrrolidines, 1034

  cinchonidine, 1073

  cinchonine, 1073

  cobalt octacarbonyl, 0440

  europium, in the **Diels-Alder** reaction, 1072

  for **Friedel-Crafts alkylation**, 1196

  functional group reactivity, 0425, 0430

  heterogeneous, 0424

  heterogeneous, supported, 0424

  heterogeneous, unsupported, 0424

  homogeneous, 0424

catalysts, hydrogenation, 0422

  active sites, 0427

  and functional groups, 0425

  and hydrogenolysis, 0430

  and isomerization, 0432

  and poisoning, 0426

  and poisons, 0612

  aniline derivatives and coupling, 0447

  geometry, 0429

  of alkenes, and bond migration, 0432

  of alkenes, and rearrangement, 0432

  of alkynes, reactivity, 0436

  order of reactivity, 0425

catalysts

  influence on enantioselectivity, 1072

  influence on **Friedel-Crafts alkylation**, 1200, 1201

  matrix effects, 0427

  menthoxyl, 1073

metal, and isomerization, 0432

metathesis, 1341

organoaluminum reagents, 1074

oxazoborolidines, 1074

palladium and **Suzuki-Miyaura coupling**, 0527

palladium, and boronic acids, 0527

particles sites, 0427

particles, and hydrogenation, 0426

proline esters, 1072

quinidine, 1073

quinine, 1073

rare-earth, and **Diels-Alder**, 1066

reactivity order with aryl compounds, 0445

reactivity with aldehydes, 0430

reactivity with alkenes, 0430

reactivity with aryls, 0430

reactivity with ketones, 0430

rhodium, and hydroboration, 0519

surface area effects, 0427

surface sties, 0427

surface, and hydrogenation, 0426

transition metal, **Claisen rearrangement**, 1140

vanadium, and epoxidation, 0273, 0274

Vaska's catalyst, 0428, 0430

Wilkinson's catalyst, 0428, 0429, 0437

catalytic hydrogenation. *See* hydrogenation

catalytic OsO_4, 0296

catechol

  with borane, 0500

  with keto esters, 1221

catecholborane, 0500, 0519

  with alkynes, 0504

catechols

  oxidation with CAN, 0265

  oxidation, 0265

  oxidation, dimerization, 0265

cathepsin S inhibitors, 0624

cationic polymerization, 1271

cations, bis(benzofurans), 0388

Cbz (benzyloxycarbonyl)

  with amines, 0618

  amine protecting groups, 0618

  and amines, 0356

  and hydrogenolysis, 0444

cedrene, 0772

Celastraceae sesquiterpenoids, 0548

Celite

  and **Fétizon's reagent**, 0256, 0257

and silver carbonate, 0256

6-center transition state, **Diels-Alder**, 1013

centrolobine, 1191

cephalotaxinone, 1246

ceramides, 0564

ceratopicanol, 0340

ceric ammonium nitrate, 0261, 0265

ceric ammonium nitrate. *See* CAN

ceric ammonium sulfate, 0261

ceric sulfate, oxidation of catechols, 0265

cerium borohydride. *See* borohydride

cerium chloride

  and NaOCl, 1181

  and organolithium reagents, 0668, 0669

  with **Grignard** reagents, 0668

cetyltrimethylammonium bromide, 0471

cetyltrimethylammonium tribromide

  and bromination, 0185

C-H insertion, 1329

  and rhodium catalysts, 1330

chain extension reactions, 0835

  and the **Barbier-Wieland degradation**, 0648

  carboxylic acids, 0648

  of aldehydes, 0533

  with alkyl boranes, 0518

chain length, and macrocyclizations, 0573

chain radical addition reactions, 1271

chain termination steps, radicals and addition reactions, 1271, 1272

chair and boat, energy difference, 0055

chair conformations, 0048, 0060

  and NMR, 0049

  and the E2 reaction, 0156

  interconversion, 0049

chair transition state

  Claisen rearrangement, 1132

  Cope rearrangement, 1124

chair versus boat transition state

  of the **Claisen rearrangement**, 1132

charge delocalization, enolate anion, 0799

charge distribution, and hydrogen bonding, 0784

charge transfer complex, and electrophilic aromatic substitution, 0195

charge transfer

  and **Diels-Alder** transition states, 1023

and organocuprates, 0717
charge, and oxidation number, 0221
charge-transfer complex, **Grignard**
  with ethers, 0638, 0639
charomium (VI), and pK$_a$, 0227
cheimonophyllon E, 0273
chelation control
  and allylsilanes, 0769
  and diastereoselectivity, 0868
  and enolate anions, 0868
  and haptophilic effects, 0549
  and LDA, 0868
  and LiAlH$_4$, 0550
  and organocuprates, 0719
  and Red-Al, 0550
  and **Sharpless asymmetric
    epoxidation**, 0549
  and the **Grignard reaction**, 0663
chelation
  and alkoxy ketones, 0406, 0407
  and ion pairing, 0130
  and reduction, 0405, 0406
  and the **Cram chelation model**, 0550
  and the **Cram cyclic model**, 0406
  and zinc borohydride, 0375
  peroxyacids and alcohols, 0549
chelidonine, 1099
chemical degradation, 0975
  and synthesis, 0973
chemical genetic assays, 0987
chemical ionization pathway, 1166
chemical tag, combinatorial chemistry,
  0985
chemiluminescence, and
  photochemistry, 1082
chemisorption, and hydrogenation,
  0427, 0449
chemoselective and chemospecific,
  definitions, 0032
chemoselectivity, 0363
  and NaBH$_4$, 0370
  hydrogenation, 0431
  photocycloaddition, 1093
chiral additions, and methionine, 0449
chiral additives
  and enantioselectivity, 0866
  and hydrogenation, 0449
  and silk fibroin, 0449
  and the Diels-Alder reactions, 1072
  in reduction, 0392
  leucine, 0449
  menthoxy acetate, 0449
  tartaric acid, 0449

chiral alcohols. *See* alcohol
chiral amines, and enamines, 0876, 0877
chiral atom. *See* stereogenic
chiral auxiliaries
  and **Diels-Alder** reactions, 1069
  and facial selectivity, 1067
  and LiAlH$_4$, 1072
  and radical cyclization, 1286
  and radicals, 1267
  lactones 1071
  Oppolzer's sultam, 1071
  removal, 1072
  sultams, 1071
chiral auxiliary. *See* auxiliary
chiral auxiliary
  and **Diels-Alder** reactions, 1066, 1067
  and iodolactamization, 0187
  chiral alcohols, 1069
chiral axis, 0026-0029
chiral **Burgess reagent**, 0170
chiral catalysts, 0957
  and the **Diels-Alder reactions**, 1072
  Mukaiyama aldol reaction, 0839
chiral complex, 0725
  and aldol condensation, 0866
chiral diols, and the **Diels-Alder**
  reaction, 1075
chiral ene reactions, 1151
chiral lactams, and the **Diels-Alder**
  reaction, 1052
chiral ligands
  and Pd(0) alkylation, 1234
  asymmetric hydrogenation, 0451
chiral phosphines, and **Pauson-Khand**
  reactions, 1344
chiral pool, 0956
chiral reagents, 0957
chiral synthon. *See* Chiron
chiral template approach, 0954, 1075
chiral template synthesis, 0947
chiral templates
  and enantioselective **Diels-Alder**
    reactions, 1066
  and radical cyclization, 1289
  and synthesis, 0954-0960
  and the **aldol condensation**, 0866
  and the **Diels-Alder** reaction, 1075
  and the **Simmons-Smith** reaction,
    1333
  ascorbic acid, 1075
  carbohydrates, 0957, 0958
  diethyl tartrate, 1333
  glucose, 0958

hidden, 0956
  in the **Diels-Alder** reaction, 1075
  norvaline, 1075
chirality
  and alkylidene cyclohexanes, 0028
  and allenes, 0027
  and biaryls, 0028
  and chiral axes, 0026-0029
  and mirror images, 0013
  and spirans, 0028
  of sulfoxides, 0336
  transfer, and organocuprates, 0726
  without stereogenic centers, 0026
chiraphos, 0450
chiron analysis
  and synthesis, 0958
  erythronolide A, 0958
  meroquinene, 0960
Chiron approach, 0954-0960
CHIRON, and synthesis, 0960
chirons, and synthesis, 0954-0960
chloral, in the **Diels-Alder** reaction,
  1054
chloramines, and aminohydroxylation,
  0306
chloramines, with alkylboranes, 0534
chloranil (tetrachlorobenzoquinone),
  1219
  and dehydrogenation, 1219
chlorate, potassium, acidic, with
  osmium tetraoxide, 0295
chlorexolone, 0904
chloride, 2-(trimethylsilyl)-
  ethoxymethyl, with alcohols, 0598
chloride, aluminum
  with LiAlH$_4$, 0387
  with NaBH$_4$, 0374
chloride, benzoyl, with amines, 0617
chloride, benzyl. *See* halide
chloride, benzyl
  and **Gilman titration**, 0675
  and organolithium reagents, 0675
  with amines, 0613
chloride, magnesium with NaBH$_4$, 0374
chloride, zinc, with NaBH$_4$, 0375
chlorides, acid. *See also* halides, acyl
chlorides, acid
  reduction to aldehydes, 0365
  with lithium tri-*tert*-butoxy-
    aluminum hydride, 0365
  with organocuprates, 0718
  with organolithium reagents, 0683
chlorides, acyl

and PCl$_5$, 0146
and SOCl$_2$, 0146
with amines, 1078
with dialkylcadmium, 0647
with diazomethane, 1316, 1328
with ester enolates, 0826
with **Grignard reagents**, 0646
with lithium borohydride, 0373
with Zn-Cu, 1078
chlorination
and AIBN, 0151
and radicals, 1272
and the **Wohl-Ziegler reaction**, 0151
benzylic compounds, 0151
of alkanes, 0151, 1272
of ketones, 0152
chlorine
and electrophilic aromatic
substitution, 0195
in water, 0187
photolysis, 0151
with alkenes, 0183-0188
chlorine-stabilized carbocations, 1183
chloroalkylboranes, 0500
chloroaluminum hydride, 0388
chloroamines, and radical cyclization,
1294
chlorobenzene, with *Pseudomonas
putida*, 0304
chloroborane, 0106, 0500
with trimethylaluminum, 0535
chlorochromate, pyridinium. *See* PCC
B-chlorodiisopinocampheylborane,
and boron enolates, 0843
chlorodiphenylphosphine, and
palladium(0), 1234
chloroform
photolysis, 1319
thermolysis, 1319
treatment with base, 1319
chloroformate, benzyl, with amines,
0618
chloroformate, methyl, with
organolithium reagents, 0734
chlorohydrins
and chromyl chloride, 0332
from alkenes, 0187, 0332
chloroiminium salts, and **Vilsmeier-
Haack** reaction, 1206
α-chloroketones, and Grignard
reagents, 0663
chlorolactonization, 0186
chloronitriles, with organoboranes, 0534

chloronitrones, 0906, 0907
Diels-Alder reaction, 0907
chloronium ions, 0184
chloroperoxybenzoic acid (*m-*). *See*
mcpba, and peroxyacid
chlorophyll a, as a sensitizer, 1090
chlorosulfinate esters. *See* esters
chlorosulfite, alkyl
and thionyl chloride, 0145
from alcohols, 0145
chlorosulfonic acid, and electrophilic
aromatic substitution, 0196
chlorothexylborane, 0509, 0510, 0512
chlorothexylborane-dimethyl sulfide,
0513
Chlorothiazide, 0904
chlorotirphenylsilane, with zinc, 1257
chlorotrialkylsilanes, with **Grignard**
reagents, 0770
chlorotriethylsilane, with alcohols,
0602
chlorotriisopropylsilane, with alcohols,
0602
chlorotrimethylsilane
and acyloin condensation, 1307
and conjugated ketones, 0719, 0720
and enolate anions, 1120
and **Mukaiyama aldol reaction**, 0837
and organocuprates, 0722
and pinacol coupling, 1305
with alcohols, 0601
with alkoxides, 0749
with amines, 0615
with cyanocuprates, 0729
with enolate anions, 0837, 0858
with ester enolate anions, 0826
with higher order cuprates, 0729
with organocuprates, 0719
chlorotriphenylphosphonium chloride,
0149
cholestan-3-one, 0656
cholestanone, 0419
cholesterol esterase, 0945
cholesterol, biogenesis, 1184
cholesterol, conformations, 0070
cholesteryl methyl ether, 0435
chrmoic acid, oxidation of sulfides,
0335
chromate esters, and oxidation of
alcohols, 0228-0230
chromate ion in solution, 0226, 0227
chromate, tert-butyl, 0230

chromatography
and the **Horner-Wadsworth-
Emmons** reaction, 0742
of alcohols, 0515
chromic acid
and alkylboranes, 0528
and dichromate, 0227
and HSAB, 0096
and pKa, 0227
functionalized, 0227
chromium complex, chiral, catalyst
for the ene reaction, 1150
chromium compounds
aryl, with carbanions, 0764
aryl, with organolithium reagents,
0763
aryl, with sparteine, 0765
arylchromium, 0763
chromium hexacarbonyl, with benzene,
0764
chromium reagents, and aldehydes,
0849
chromium trioxide
in solution, 0226, 0227
and alkenes, 0317
and alkyl oxidation, 0328
and dimethylpyrazole, 0329
and HCl, 0236
and **Jones reagent**, 0233
and pyridine complex, 0230, 1360
and the **Gallagher-Hollander
degradation**, 0317
in acetone, 0233
in aqueous acid, 0229, 0230
in water, 0229, 0230
oxidation of alcohols, 0067
with ethers, 1277
chromium trioxide-pyridine, 0234
chromium troixde, in aquous acid, 0233
chromium troxide-pyridine complex,
0235
chromium(VI)
alcohols to esters, 0232
and **Collins oxidation**, 0235, 1360
and **Jones oxidation**, 0233
and **Jones reagent**, 0233
and mineral acids, 0227
and oxidative cleavage, 0316
and PCC, 0236
chromyl chloride, 0239
cleavage of carbonyls, 0884
modified reagents, 0238

oxidation alcohols, and steric effects, 0230

oxidation of alcohols, rate of reaction, 0229

oxidation of alcohols, solvent effects, 0229, 0230

oxidation of aldehydes, 0231

oxidation, and the E2 reaction, 0231, 0240

reaction with alcohols, 0228

reduction potential, 0226

with additives, 0234

chromium, aryl. *See* arylchromium

chromium, compounds, aryl, enantioselectivity, 0765

chromophoric, alkenes, 0102

chromyl chloride

and chlorohydrins, 0332

and **Nozaki-Hiyama-Kishi reaction**, 0849

and the **Étard reaction**, 0332

and the **Takai reaction**, 0757

formation of chloro ketones, 0332

oxidation of alcohols, 0239

with aldehydes, 0849

with alkenes, 0332

Chugaev elimination, 0167

and the ene reaction, 0167

chyrsotricine, 0361

Cieplak model, 0414

ciguatoxin, 0023, 0577, 0910, 0913, 0995

cinatrin B0186, 0258

Cinchona alkaloids, 0174

and dihydroxylation, 0299

and **Robinson annulation**, 0882

and **Sharpless asymmetric dihydroxylation**, 0301

in chiral reductions, 0393

cinchonidine, prolinol, chiral catalyst, 1073

cinchonine

and **Diels-Alder**, 1071

chiral catalyst, 1073

with $LiAlH_4$, 0393

cine substitution, 0208

cinnamaldehyde, 0241, 0260

cinnamic acid, 0827

CIP rules. *See* Cahn-Ingold-Prelog

cis principle, **Diels-Alder** reaction, 1024

cis, s. *See* s-cis

cis/trans isomers

alkenes, 0029

alkenylalkylboranes, 0509

alkylboranes, 0498

and diazoalkanes, 1323

and E2 reaction, 0559

and hydrogenation of alkyne, 0546

and radical cyclization, 1285

cyclohexanes, 0062

Diels-Alder reaction, 1024

enantioselectivity in alkene hydroboration, 0520

cisoid. *See* s-cis

citreoviral, 0374

citreoviridin, 0663

citronellol, 0237, 0482, 0956

citrus mealy bug, 0353

cladocoran A, 0699, 0902

Claisen condensation, 0824

and enolate anions, 0791

and LICA 0825

anionic accelerated, 0791

crossed Claisen, 0825

mixed Claisen, 0825

mixed, 0825

Claisen reaction, 0826

Claisen rearrangement, 1045, 1132

allyl alcohols, 1135

and decarboxylation, 1136

and **Fischer Indole Synthesis**, 1222

and HMPA, 1138

and microwaves, 1132

and molecular modeling, 1139

and ortho esters, 1136

and palladium catalyst, 1140

and $PdCl_2$, 1140

and sulfone anions, 1139

aryl allyl ethers, 1133

boat transition state, 1132

bromoselenides, 1135

Carroll reaction, 1135

dimethylacetamide dimethyl acetal, 1137

enolate, rate enhancement, 1139

Eschenmoser-Claisen rearrangement, 1137

in water, 1139

Ireland-Claisen rearrangement, 1138

ketene acetals, 1136

Lewis acids, 1133, 1134

molecular modeling, 1134

powerful in synthesis, 0930, 0932

rate enhancement in aqueous media, 1139

stereochemistry, 1133

syn/anti isomers, 1139

thio-Claisen rearrangement, 1141

thio-Claisen, and umpolung, 0713

thio-Claisen, with organolithium reagents, 0713

transition metal catalysts, 1140

transition state energy, 1139

transition state modeling, 1134

transition state, 1132

variants, 1134

with allyl amines, 1142

Claisen, condensation, **Johnson-ester variant**, 1136

Claisen, rearrangement, **Saucy-Marbet reaction**, 1135

Claisen-Schmidt reaction, 0742

clavepictine A, 0991

clavosolide A, 0360

clavubicyclone, 1128

cleavage

C-N in $LiAlH_4$ reduction of amides, 0355

electrocyclic, 1081

in oxidation of alcohols with Cr(VI), 0229, 0231

of cyclobutanes, 1096

organoiron compounds, 0759

cleavage, oxidative, 0313-0324

and chromium(VI), 0316

and permanganate, 0313

and ruthenium tetraoxide, 0315

and ruthenium trichloride, 0316

and sodium periodate, 0324

and the **Gallagher-Hollander degradation**, 0317

and the **Lemieux-von Rudloff reagent**, 0313

and the **Miescher degradation**, 0317

Barbier-Wieland degradation, 0317

Lemieux-Johnson reagent, 0314

of alkenes, 0313

of diols, 0313, 0322-0324

with lead tetraacetate, 0322

cleavage, photochemical, 1276

cleavage, reductive

and hydrosilylation, 0476

of sulfones, 0700

of sulfoxides, 0698

of sultones, 0701

with aluminum amalgam, 0698

with LiAlH$_4$, 0701
with sodium amalgam, 0700
Clemmensen reduction, 0464, 1207, 1252
  and **Wolff-Kishner reduction**, 0472
clovene, 1080
cobalt chloride
  and Grignard reagents, 0640, 0641
  with alkyl halides, 0640, 0641
cobalt complex
  and HOMO, 1342
  a **Pauson-Khand reaction**, 1342
cobalt octacarbonyl
  hydrogenation catalyst, 0440
  Pauson-Khand reaction, 1342
codeine, 0256, 0327
codeinone, 0256
codes, 3-letter, for amino acids, 0979
coding tags, combinatorial chemistry, 0985
codonopsinine, 1238
coefficients, orbital, alkenes, 1001
colchicines, 1115, 1297
coleophomone B, 0811
colletodiol, 0248
Colligative properties, and
  amide bases, 0794
Collins oxidation, 0235, 0236, 0593
color-coding, combinatorial chemistry, 0986
Combes quinoline synthesis, 1213
combinatorial chemistry, 0978
haloaromatic tags, 0985
  and electrophilic tags, 0984
  and enzymatic cleavage, 0980
  and epitopes, 0986
  and GC, 0984
  and GC/mass spectrometry, 0984
  and genetic tags, 0986
  and HPLC, 0984
  and HPLC/mass spectrometry, 0984
  and linkers, 0984
  and Merrifield synthesis, 0979
  and molecular tags, 0985
  and nucleotide tags, 0986
  and pins, 0981, 0982
  and solid supports, 0982
  chemical tag, 0985
  coding tags, 0985
  color-coding, 0986
  decoding, 0986
  deconvolution, 0983
  disks, 0982

enzymatic cleavage, 0981
identification of library components, 0984
libraries, 0980
photocleavable groups, 0984
pin approach, 0981, 0982
pooling strategy, 0979
protecting groups, 0982
recursive deconvolution, 0983
solution phase, 0987
spacers, 0980
split synthesis, 0982, 0983
tags, 0984
teabag approach, 0981
UV tags, 0986
communesin B, 0355
communesin F, 1325
Compendium of Organic Synthetic Methods, 0073, 0588
complex formation
  and HSAB, 0098
  in the **Grignard** reaction, 0648
complex targets, and synthesis, 0910
complexation
  and cerium borohydride, 0376
  and conjugate reduction, 0376
  Lewis acids, and aldehydes, 1038
Complexes
  and chromic acid, 0227
  arylchromium, 0764
  asymmetric, 0661
  ate, 0093
  ate. *See also* ate complex
  ate, and hydroboration, 0508
  borane-dimethyl sulfide, 0510
  chiral, 0725
  chiral, reduction of carbonyls, 0397
  ethers with **Grignard reagents**, 0638, 0639
  organocuprate-BF$_3$, 0723
  organocuprates and chiral amines, 0724
  oxygen-vanadium in epoxidation, 0274
  π-allyl Pd, 1225-1236
  pinene and 9-BBN, 0395
  pyridie and chromium trioxide, 0235
  pyridinium ylids with LiBr, 0750, 0751
  Sharpless asymmetric epoxidation, 0285
complex-induced proximity effect, 0686

complexing agents, and **Mukaiyama aldol** reaction, 0838
computer
  and LHASA, 0918, 0919
  and SYNGEN, 0961
  automation, combinatorial chemistry, 0981
  CHIRON, 0960
concentration effects
  [2+2]--cycloaddition, 1093
  and substitution, 0142
concentrations, and radical cyclization, 1283
condensation reactions, with
  enolate anions, 0816
condensation, acid derivatives,
  enolate anions, 0823
condensation, intramolecular, 0828
conduritol F, 0304
confertin, 0530
configuration, inversion, *See* inversion
configuration
  of Grignard reagents, 0547
  organolithium reagents, 0680, 0681
  R/S, and computational methods, 0023
configurational instability, carbocations 1169
configurational stability
  for Grignard reagents, 0662
  of Grignard reagents, 0657
conformational energy, 0056
  cycloalkenes, 0061, 0062
  table, 0059
conformational perspective, 0073
conformational rigidity and mobility, 0065
conformations, 0005, 0035-0070
  acyclic molecules, 0037-0042
  alcohols, 0043
  alkenes, 0044
  alkoxy pyrans, 0064, 0065
  amines, 0043
conformations
  and **Boltzmann** distribution, 0040
  and **Dale's systems**, 0053
  and density ionization map, 0278
  and dipoles, 0045
  and energy barriers, 0041
  and enthalpy, 0041, 0057-0059
  and entropy, 0057
  and epoxidation, 0278, 0279
  and free energy, 0057-0059
  and heteroatom substituents, 0042

and heteroatoms, 0052
and **Hofmann elimination**, 0162, 0163
and hydrogen bonding, 0784
and intramolecular **Diels-Alder**, 1060, 1061
and IR, 0040
and **Johnson polyene cyclization**, 1184
and LHASA, 0058
and molecular modeling, 0060, 0062, 0403-0405
and NMR, 0040, 0049, 0065
and polar map analysis, 0053
and pseudorotation, 0051
and pyrolysis of esters, 0166
and quantum mechanical calculations, 0038
and radical cyclization, 1288
and reaction rate, 0060
and reduction 0403-0405
and stereoselectivity, 0278, 0279
and the anomeric effect, 0064, 0065
and the **Boltzmann** distribution, 0558
and the diamond lattice, 0051
and the **Grob fragmentation**, 0174
and wedge notation, 0053
and π character, 0047
conformations
  anti and gauche, 0039
  boat, 0061, 0062
  butadiene, 0047
  butene, 0044-0046
  chair, and the E2 reaction, 0156
  cholesterol, 0070
  cyclic molecules, 0047-0070
  cycloalkenes, 0061, 0062
  cyclodecane, 0050
  cycloheptane, 0050
  cyclohexanes, 0558
  cyclohexanol, 0063
  cyclohexanone, 0063
  cyclohexenes, 0062
  cyclononane, 0052
  cyclooctadecane, 0052
  cyclooctane, 0050-0052
  cyclooctanone, 0052
  cyclooctene, 0052
  cycloundecane, 0051
  decalenes, 0069
  decalines, 0068
  definition, 0035
  dienes, 0046

enamines, 0063
enolate anions, 0063
estimating populations, 0054-0062
estradiol, 0070
ethers, 0043
ethylene glycol, 0043
exo-methylene, molecular modeling, 0064
extended, 0041
gauche interaction, 0056
hirsutene, 0069
hydrazines, 0043
hydrindanes, 0068
hydronaphthalenes, 0069
indanes, 0068
indenes, 0069
lactones, 0053, 0054
long chain alkanes, 0041
macrocycles, 0053
methanol, 0042
methylamine, 0042
of lactones, 0067
oxaphosphatanes, and reactivity, 0738
oxocane, 0052
peroxides, 0044
piperidines, 0065
polycyclic molecules, 0067
propene, 0044-0046
s-cis and s-trans, 1014
steroids, 0070
styrene, 0047
substituted cyclohexanes, 0057
transition state modeling, 0060
zigzag, 0025
conformers, definition, 0038
conjugate acids
  and base strength, 0088
  and basicity, 0088
  and enolate anions, 0798
conjugate addition. *See also* addition
conjugate addition
  allyltin compounds, 0767
  and **Grignard** reactions, 0652
  and Lewis acids in **Grignard** reagents, 0653
  and **Michael addition**, 0877
  copper salts with **Grignard** reagents, 0653
  Grignard reagents, 0651
  of alkylboranes, 0533
  of dithianes anions, 0706
  of **Grignard** reagents, and steric hindrance, 0653

of organocuprates, 0714
of organocuprates, 0720
of radicals, 1304
conjugated aldehydes. *See* aldehydes
conjugated carbonyls
  and enolate anions, 0720
  and **Mannich base**, 0881
  and **Mannich reaction**, 0881
  and the Luche reagent, 0417
  and zinc borohydride, 0375
  electrolytic reduction, 0470
  enzyme reduction, 0482
  hydrogenation, 0439
  hydrosilylation, 0475
  in the **Diels-Alder** reaction, 1053
  $LiAlH_4$ reduction, 0352, 0353
  mechanism of $LiAlH_4$ reduction, 0353
  Michael addition, 0877
  Mukaiyama aldol reaction, 0840
  organocuprates and chlorotrimethylsilane, 0719
  reduction to aldehydes, 0366
  reduction with zinc, 0463
  steric hindrance and reduction, 0352
  with alkenes, 1056
  with aluminum hydride, 0388
  with aryl halides, 1237
  with cerium borohydride, 0376
  with enamines, 0875
  with higher order cuprates, 0728, 0729
  with lithium tri-*tert*-butoxy-aluminum hydride, 0366
  with $NaBH_4$, 0370
  with RedAl, 0367
  with sodium azide, 1114
  with the **Luche reagent**, 0376
conjugated compounds, and dissolving metal reduction, 0454
conjugated esters. *See* esters
  and **Baylis-Hillman reaction**, 0878, 0879
  $LiAlH_4$ reduction, 0354
  with organocuprates, 0722
conjugated ketones. *See* ketones
conjugated ketones
  1,2- versus 1,4-reduction, 0377
  and **Robinson annulation**, 0880
conjugated, alkenes, and dissolving metal reduction, 0452
conjugation
  and alkyne acidity, 0633
  and epoxidation, 0282, 0293

and photolysis, 1264
and pKa, 0690
and UV, 1264
effect on HOMO and LUMO, 1005, 1006
effect on orbital coefficients, 1005, 1007
extended, and Diels-Alder, 1031
σ, carbocations rearrangement, 1173
α-conjugation
  and hyperconjugation, 1173
  carbocation rearrangement, 1173
conjugative effects, and pKa, 0690, 0691
Conrad-Limpach reaction, 1212
conrotatory
  and **Cope rearrangement**, 1126
  electrocyclic reactions, 1098
  in the Diels-Alder reactions, 1025
consecutive versus convergent synthesis, 0912, 0913
consonance, and **Sharpless asymmetric epoxidation**, 0286
consonant stereodifferentiation, 0867
consonant, double stereo-differentiation, 0867
construction sets, 0964
contrathermodynamic isomerization, and alkylboranes, 0501
convergent synthesis, 0920, 0958, 0965, 0974
  and SYNGEN, 0912, 0971
convergent versus consecutive synthesis, 0912, 0913
coordination complex, reduction, 0043
coordination
  alkenes with hydrogenation catalysts, 0433
  and orbital coefficients, 1039
  organolithium reagents, 0682
copaene, 0558
Cope elimination, 0165
Cope elimination, reverse, 0165
Cope rearrangement, 1045, 1110, 1123
  activation barrier, 1127
  and Hammond postulate, 1128
  and molecular modeling, 1126, 1128
  and molecular orbitals, 1124
  and rotamers, 1125
  and strain, 1127
  and tandem reactions, 1129
  and thermodynamic stability, 1126, 1127

conrotatory and disrotatory, 1126
cyclopropane accelerated, 1129
divinyl cyclobutane, 1127
enolate accelerated, 1129
oxonia, 1191
rate acceleration, 1127
stereochemistry, 1124, 1125
stereoselectivity, 1124, 1125
transition state modeling, 1128
transition states, 1124, 1125
copper bromide, dimethyl sulfide complex, 0714
copper catalyzed coupling, 1298
copper chloride
  and **Cadiot-Chodkiewicz** coupling, 0631
  and coupling of alkynes, 0631
  and the **Glaser reaction**, 0631
copper compounds
  and alkyne coupling, 1301
  and alkynes, 1302
  and aryl halides, 1302
  and **Cadiot-Chodkiewicz** coupling, 1302
  and carbenes, 1322
  and diazoketones, 1328
  and Dibal, 0390
  and radical cyclization, 1294
  and the **Pschorr reaction**, 1299
  and Ullmann reaction, 1300
  with alkynes, 1301
  with diazoalkanes, 1321
copper cyanide
  and aryl diazonium salts, 0209
  with aryl halides, 0628
  with organolithium reagents, 0726, 0727
copper free **Sonogashira coupling**, 1245
copper iodide
  and **Sonogashira coupling**, 1245
  and **Suzuki-Miyaura coupling**, 1243
copper salts
  with diazomethane, 1325
  with **Grignard** reagents, 0653
  with organolithium reagents, 0653, 0654
copper triflate
  and carbenes, 1325
  bis(oxazolidine) complex, and [3+2]-cycloaddition, 1108
  copper(I) bromide-dimethyl sulfide, and boranes, 0511

copper(II) salts, and diimide, 0474
copper
  alkynyl, 0717
  and the **Gattermann** reaction, 0209
  and the **Reformatsky** reaction, 0887
  coordinated **Diels-Alder** catalysts, 1073
  with diazoketones, 1328
coprostanone, 0419
cordiachromene, 0466
Corey-Fuchs procedure, 0159, 0630, 0733
Corey-Nicolaou macrocyclization, 0578
Corey-Nicolaou reagent, 0578, 0579, 0584
Corey-Seebach procedure, 0612
coriolin, 0269, 0609
Cornforth model, 0408
  and the Grignard reaction, 0662
coronafacic acid, 1130
cortisone, 0898, 0915, 1025
cosmene, 0362
Coulomb integral, and molecular orbitals, 1016
Coulomb's law, 0116
coumarins
  by **Friedel-Crafts acylation**, 1206
  by Friedel-Crafts reactions, 1220
  by **Pechmann condensation**, 1220
  from catechol, 1221
  from keto esters, 1220
  from phenols, 1220
coupling reactions
  alkyl halides and organocuprates, 0715
  alkyl halides and organocuprates, mechanism, 0714
  alkylboranes, 0513
  alkynes, 1301
  allylsilanes, 0769
  and dissolving metal reduction, 0454
  and electrolytic reduction, 0469
  and hydrogenation of aniline derivatives, 0447
  and ketyls, 1304
  and photoreduction, 0478
  and radical stability, 1270
  and radicals, 1270
  aryl diazonium salts, 0210
  aryl halides and **Grignard** reagents, 0643
  Cadiot-Chodkiewicz coupling, 0631

copper catalyzed, 1298
four component, 0627
Grignard reagents and alkyl halides, 0640-0643
Grignard reagents with metal salts, 0642
multi-component, 0627
of alkyl halides, 0640, 0641
of alkylboranes, 0511, 0512
of alkynes, 0631, 1301
of imines, 1305
organocuprates, 0714, 0715
palladium-catalyzed, with boranes, 0511
phenolic oxidative coupling, 1297
radical, 1270, 1280, 1297
radical, **acyloin condensation**, 1306
radical, and alkenes, 1279
radical, and electrolytic reduction, 0469
radical, and **McMurry olefination**, 1309
reductive, 1308
the **Glaser reaction**, 0631
covalent character, C-Li bond, 0671
Cram's chelation model, 0550
Cram's cyclic model, 0406
and the **Grignard reaction**, 0663
Cram's model, 0037, 0402, 0774
and LUMO maps, 0410
and organolithium diastereoselectivity, 0681, 0682
and the Grignard reaction, 0662
comparison with LUMO maps, 0410
reduction of cyclic ketones, 0412
Cram's open chain model, 0402
Cram's rule, 0402, 0957
and calculations, 0402
and formic acid reduction, 0476
and kinetic reactions, 0402
and **Newman projections**, 0402
and organocuprates, 0719
and steric hindrance, 0402
and the **aldol condensation**, 0863, 0864
and the **Grignard reaction**, 0661, 0662
versus anti-**Cram** selectivity, 0406, 0407
versus anti-**Cram** models, 0402
crambescidin 359, 1108
crinamine, 0932
crinine, 0022

cripowellins A and B, 1237
crispatenine, 0390, 1275
crispine A, 0356
CrO₃. *See* chromium trioxide
cross coupling
    Kolbe electrolytic synthesis, 1278
    McMurry olefination, 1309
    Stille coupling, 1240
    Suzuki-Miyaura coupling, 1243
cross metathesis. *See* metathesis
crossed-**Claisen**. *See* **Claisen**
crotyldiisopinocampheylborane, 0527
crotyltin. *See* tin
crown conformation, 0050
crown ethers, 0313, 0934
    and periodic/periodic acid, 0313
    and phosphonate ester ylids, 0740
cryptates, with allylic thiol anions, 0712
cryptophycin 52, 0397
Crystal structure, organolithium reagents, 0671
C-toxiferin I, 0920
cubebane, 0592
cucumin H, 1137, 1328
culmorin, 0454, 0878
cumyl hydroperoxide. *See* peroxide
cuparene, 0132, 0165
cuparenone, 1329, 1333
cuprate, lithium tetrachloro, 0642, 0643
cuprates, dialkyl. *See* organocuprates
cuprates
    higher order, and inversion of configuration, 0728
    higher order, and kinetics, 0727
    higher order, and Sₙ2 reactions, 0727
    higher order, enantioselectivity, 0727
    lithium, conjugate addition, 0653, 0654
    magnesium, conjugate addition, 0653, 0654
    silyl, 0331
cupric chloride
    and organopalladium complexes, 1228, 1229
    and radical cyclization, 1294
cupric triflate, **Mukaiyama aldol** reaction, 0839
cupric. *See also* copper(II)
cuprous. *See also* copper(I)
cuprous bromide
    and dimethyl sulfide, 0654
    and trimethylsilyl chloride, 0654
    with aryl diazonium salts, 0209
    with **Grignard** reagents, 0642

with lithium trimethoxy-aluminum hydride, 0367
    with Red-Al, 0367
cuprous chloride
    and radical cyclization, 1294
    with aryl diazonium salts, 0209
cuprous cyanide, and coupling, indoles, 1224
cuprous iodide
    with **Grignard** reagents and epoxides, 0655
    with organolithium reagents, 0714
cuprous ion, and HSAB, 0096
cuprous salts
    with **Grignard** reagents and epoxides, 0655
    with **Grignard** reagents, 0642
    with organolithium reagents, 0714
Curtin-Hammett principle, 0058
Curtius rearrangement, 0192, 0945, 1051
    and acyl azides, 0192
Curvularia lunata CECT, 2130, reduction, 0481
cyanide
    and MnO₂, 0255
    and Sₙ2, 0117
    and the **Rosenmund-von Braun reaction**, 0628
    and the **Strecker synthesis**, 0628
    bidentate nucleophiles, 0625
    potassium, with zinc, 0464
    silver, with alkyl halides, 0625, 0626
    and silver(I) oxide) oxide, 0260
    with alkylboranes, 0528
    with boranes and dienes, 0529
    solvent effects with alkyl halides, 0624
    trimethylsilyl, with aldehydes, 0629
    with alkyl halides, 0007, 0145, 0623, 0624
    with aryl halides, 0628
    with epoxides, 0624
    with ketones, 0628
    with sulfonate esters, 0623, 0624
    aldehydes, 0628
cyanides, acyl, and enolate anions, 0811
cyano ate complexes, 0528
cyanobacterium, 0482
cyanoborohydride,
    and alkoxymercuration, 0183
    and formaldehyde, 0384
    and pH, 0382, 0383
    and solvents, 0382

of ketones, 1094
reduction of aldehydes or ketones, 0383
sodium, 0382
with alkyl halides, 0383
with enamides, 0383
with enamines, 0383
with iminium salts, 0384
with sulfonate esters, 0383
cyanocholestane, 0626
cyanocuprates, 0727
with chlorotrimethylsilane, 0729
cyanogen bromide
and alkylboranes, 0510
fusion with carboxylate salts, 0628
with boranes, 0511
cyanohydrins
and protection of aldehydes and ketones, 0612
via the **Strecker synthesis**, 0628
with Red-Al, 0368
cyanoketones, from aldehydes, 0255
cyanthiwigin AC, 0728
cyclcohexanol, substituted, oxidation with Cr(VI), 0230
cyclic amines, by radical cyclization, 1288
cyclic boranes, 0518
from dienes, 0530
cyclic compounds, and stereoselectivity, 0563
cyclic ketones, from dienes, 0530
cyclization
and **Baldwin's rules**, 0564-0572
and entropy, 0573
and **Grignard** reactions, 0666
and organolithium reagents, 0688
and Pd(0) reactions, 1234
and the **Pauson-Khand** reaction, 1342-1344
cascade cationic, 1184
of aldehydes, 0558
of ene-diynes, 1295
of radicals, 1283-1296, 1357
of trienes, 0949
organolithium reagents, 0666
palladium(0) mediated, 1234
radical. *See* radical cyclization
radical, 0946, 1272
radical, powerful in synthesis, 0929, 0931
tetraenes, 0950
thermodynamic, 0949

with sulfur carbanions, 0698
cycloaddition, [2,2]-, 1057, 1076-1097
alkenes and dienes, 1014
allenes with ketenes, 1080
and chemiluminescence, 1082
and diradicals, 1076
and diradicals, 1091
and fluorescence, 1082
and orbital coefficients, 1093
and photochemistry, 1082
and the **ene reaction**, 1077
chiral ammonium salts, 1080
concentration effects, 1093
Danheiser annulation, 1081
dichloroketene, 1078
electronic effects, 1076
enantioselectivity, 1080
formation of cyclobutenes, 1076
formation of oxetanes, 1095
formation of quinines, 1081
intramolecular, 1078, 1079
ketene iminium salts, 1079
ketenes with imines, 1080
Paternò-Büchi reaction, 1094
photochemistry, 1091-1097
quinuclidine catalyst, 1080
solvent effects, 1093
SOMO, 1082
stereochemistry, 1076
thermal, 1076
Types I, II, III, 1079
with alkenes, 1076
with dimethylketene, 1078
with imines, 1080
with ketenes, 1077
cycloaddition, [3+2]-, 0933, 1101-1115, 1189
1,3-dipoles, 1101
and azides, 0442
and BINOL, 1109
and diradicals, 1102
and ethyl diazoacetate, 1111
and imidazolidinones, 1109
and iminium salts, 1111
and isoxazolidines, 1107
and isoxazolines, 1106
and racemization, 1108
and solvent dielectric, 1102
catalyzed by antibodies, 1107
computational study, 1102
concerted thermal mechanism, 1102
dipolarophiles, 1102
Huisgen's mechanism, 1102

in supercritical carbon dioxide, 1107
intramolecular, 1109, 1111, 1113
of azomethine ylids, 1111
of nitrones, 1107
of trimethylene methane with alkenes, 1236
of alkyl azides, 1113
of azomethine imines, 1111
of azomethine ylids, 0751
of carbonyl ylids, 1115
of diazoalkanes, 1110
of nitrile oxides, 1106
of nitrile ylids, 1104
of nitrones, exo selectivity, 1108
of ozone, 1103
of permanganate, 0292
of permanganate, 1103
orbital coefficients, 1103, 1104
stereochemistry and solvent, 1102
with hydrazoic acid, 1113
with ozone and alkenes, 0318
cycloaddition, [4+2]-. *See* **Diels-Alder**
cycloaddition, [4+3]-, and **Noyori annulation**, 1190
cycloaddition, [4+4]-, and nickel(0), 1247
cycloaddition, [5+2]-, 1265
cycloadditions
allowed and forbidden, 1010
and ground state, 1012
and **Woodward-Hoffman rules**, 1011, 1012
antarafacial, 1011
ketene-allene, 1080
suprafacial, 1011
symmetry allowed or forbidden, 1011
cycloalkanes, enthalpy, 0050
cycloalkenes, conformations
and energy, 0061, 0062
cycloaromatization, 1295
with palladium, 1339
cyclobutane, conformation, 0048
cyclobutanes
and [2+2]-cycloreversion, 1096
by [2+2]-cycloaddition, 1076
by photocycloaddition, 1092
divinyl, and **Cope rearrangement**, 1127
from alkenes, 1091, 1092
photochemical cleavage, 1096
cyclobutanones
from ketenes, 1077
from oxaspiropentanes, 0749

cyclobutenes
  by [2+2]-cycloaddition, 1076
  electrocyclic reactions, 1098
  electrocyclic reactions, 1098
  from cyclopropylcarbinyl alcohols, 1097
  preparation, 1098, 1099
  ring opening, 1098
cyclobutanones
  by [2+2]-cycloaddition, 1081
  from alkynes and ketenes, 1081
  photolysis, 1081
cyclobutylcarbinyl carbocations, 1174
cyclocitral, 0259
cyclodecane
  conformations, 0050
  diamond form, 0053
cyclodextrins, and **Diels-Alder** reactions, 1041
cyclododecatriene, with borane, 0531
cyclogeranic acid, 0259
cycloheptaamylose. *See* cyclodextrins
cycloheptane, conformation, 0050
cycloheptatrienes
  and rearrangement, 1118
  from aryls and carbenes, 1326
  from aryls and diazomethane, 1321
cycloheptenones
  by **Noyori annulation**, 1190
  from dibromo ketones, 1190
cyclohexadiene, as a hydrogen donor, 0614
cyclohexadienes
  and electrocyclic reactions, 1100
  from arylchromium reagents, 0764
  from hexatrienes, 1100
cyclohexane carboxylic acid, enolate anions, 0871
cyclohexane
  conformations, 0048
  exo-methylene, A-strain, 0063
  exo-methylene, conformation, 0063
cyclohexanecarboxaldehyde, enolate anions, 0871
cyclohexanes
  and $A^{1,3}$ strain, 0056
  and E2, 0155
  and the S_N2 reaction, 0558
  by hydrogenation of aryls, 0446
  conformations and molecular modeling, 0060, 0062
  conformations, 0558
  substituted, conformations, 0057

substituted, conformer modeling, 0060, 0062
cyclohexanone oxygenase, 0311
cyclohexanone
  and **Oppenauer oxidation**, 0251
  conformation, 0063
  enolate anion conformation, 0063
  enolate anions, 0872
cyclohexanones
  and LUMO maps, 0416
  reduction, 0414-0416
  steric hindrance, and reduction, 0414-0416
cyclohexenes, by **Diels-Alder** reaction, 1013-1075
cyclohexenes
  conformer modeling, 0062
  from alkenes, 1017
  from dienes, 1017
cyclohexenone
  conjugated, by isomerization, 0460
  nonconjugated, by **Birch** reduction, 0460
cyclononane, 0052
cyclooctadecane, conformations, 0052
cyclooctadienes, 0315
cyclooctane, conformations, 0050-0052
cyclooctanol, oxidation, 0067
cyclooctanone
  conformation, 0052
  reduction, 0067
cyclooctene, conformation, 0052
cyclopentadiene
  and benzoquinones, 1013
  and **Diels-Alder reactions**, 1014
  Diels-Alder reaction, 1017, 1021, 1026, 1038
  with menthyl acrylate, 1067, 1068
cyclopentadienes
  and retro-**Diels-Alder** reactions, 1035
  and **Diels-Alder** reactions, 1040, 1073, 1117
  sigmatropic rearrangements, 1117
cyclopentadienyl anions, alkylation, 1117
cyclopentadienyl dicarbonyl and alkenes
  with alcohols, 1249
  with amines, 1249
  with enamines, 1249
  with enolate anions, 1249
  with nucleophiles, 1249
  with organocuprates, 1249

with phosphines, 1249
with thiols, 1249
with water, 1249
cyclopentadienyl dicarbonyl ferrate anion, 1248
cyclopentadienyl zirconium dihydride, 0252
  and **Oppenauer oxidation**, 0252
cyclopentadienylbis(carbonyl)iron, 0759
cyclopentane, conformation, 0048
cyclopentanone, enolate anions, 0872
cyclopentanones
  LUMO map, 0413
  reduction with LiAlH$_4$, 0412
  reduction, 0412
cyclopentenones
  by **Nazarov cyclization**, 1188
  by **Noyori annulation**, 1189
  from divinyl ketones, 1188
  photocyclization, 1094
cyclopropanation
  and **Cope rearrangement,** 1129
  and diazolactone, 1325
  and sulfur ylids, 0748
cyclopropane accelerated **Cope** rearrangement, 1129
cyclopropane
  conformation, 0048
  excited, 1312
  hydrogenation, 0438
  hydrogenolysis, 0457
cyclopropanes
  and carbocation rearrangement, 0137
  and carbocations, 1164
  and sigmatropic rearrangement, 1118
  and the **Simmons-Smith reaction**, 1331
  by [3+2]-cycloaddition, 1110
  cis/trans isomers, 1323
  dichloro, from alkenes and dichlorocarbene, 1325
  enantioselectivity in formation, 1325
  formation of carbenes, 1316
  from alkenes and carbenes, 1322, 1331
  from carbenes, 1120
  from diazoalkanes, 1110, 1120
  from diazoketones, 1328
  from dichlorocarbenes, 1319
  from halocarbenes, 1319
  from ketenes, 1323
  from pyrazolines, 1110

photolysis of, 1316
cyclopropyl Grignard reagents, 0658
cyclopropylcarbinyl carbocations, 1163, 1174
cycloreversion, [2+2]-, 1096
cyclosarkomycin, 1072
cycloundecane, conformations, 0051
cylindricine C 0249
cylindrospermopsin, 0255, 0381, 0902, 1055
cytochalasin, 0886
cytochlasin B, 0956
cytoxazone 0371, 0644

D

d orbitals, and Lewis acids, 0094
D_2O, with alkyne anions, 0633
DABCO (1,4-diazabicyclo[2.2.2 octane)
  and organolithium reagents, 0674
  and basicity, 0090
  and decarboxylation, 0172
  and enolate anions, 0808
  and the **Baylis-Hillman** reaction, 0879
  with allylic thiol anions, 0712
Dakin reaction, 0311
Dale's system, and conformation, 0053
damsinic acid, 1183
Danheiser annulation, 1081
Danishefsky's diene and **Diels-Alder reactions**, 1050, 1054
Darvon alcohol, with LiAlH$_4$, 0394
Darvon alcohols, 0393, 0394
Darvon, 0393
Darzens' glycidic ester synthesis, 0834, 0883
Darzens' **procedure** and alcohols, and thionyl chloride, 0146
Darzens tetralin synthesis, 1209
Daucus carota root, and reduction, 0480
davanone, 0851
Davies catalyst, and carbenes, 1330
Davis reagent (sulfonyloxazidine), 0814
dba see dibenzylideneacetone DBN (1,5-diazabicyclo[4.3.0]-non-5-ene)
  as a base, 0092
  and E2 elimination, 0158
  and elimination, 0140, 0141
  substitution versus elimination, 0157
DBU (1,8-diazabicyclo-[5.4.0]-undec-7-ene), 1224
  and basicity, 0090

and E2 elimination, 0157, 0158
and elimination, 0140, 0141, 1334
and **Horner-Wadsworth-Emmons olefination**, 0739
and the **Baylis-Hillman** reaction, 0879
as a base, 0092, 0118
hydrogen tribromide, and bromination, 0185
with ketones, 0877
DCC (dicyclohexylcarbodiimide), and DMSO, 0243, 1357
DDQ (2,3-dichloro-5,6-dicyano-1,4-benzoquinone)
  oxidation of phenols, 0265
  and electrocyclic reactions, 1100
  and hexatrienes, 1100
  with dioxolanes, 0610
  with methoxybenzoate esters, 0606
  %de, 0035
deacetoxyalcyonin acetate, 0172, 0320
deactivating substituents electrophilic aromatic substitution, 0202, 1192
DEAD (diethylazodicarboxylate)
  and macrolactonization, 0577
  and the **Mitsunobu reaction**, 0124, 0545, 0577
  and triphenylphosphine, 0545
deamido-bleomycin A$_2$, 0442
Dean-Stark trap, and the **aldol** condensation, 0822
decalenes, conformations, 0069
decalines, A-strain and conformations, 0068
decarboxylation, 0171
decarboxylation, 0913
  and 1,3-elimination, 0171
  and bond migration, 0173
  and conjugated acids, 0829
  and DABCO, 0172
  and **Knoevenagel condensation**, 0829
  and the **Carroll reaction**, 1136
  and the **Dieckmann condensation**, 0828
  and the **malonic acid synthesis**, 0172
  Krapcho decarboxylation, 0172
  of aryl acids, 0173
  of keto acids, 0828
decarestrictine C$_2$, 0926
decoding, combinatorial chemistry, 0986
decomposition, of osmate ester, 0295
deconvolution
  combinatorial chemistry, 0983
  recursive, 0983

defucogilvocarcin M chromophore, 1238
degenerate rearrangement, 0121
degradation, chemical, 0975
degree of bridging, in synthesis, 0935
dehydration
  of formamides, 0626
  with sulfuric acid, 0921
dehydroferruginyl methyl ether, 0738
dehydrogenase, horse liver alcohol dehydrogenase, 0479
dehydrogenase, yeast alcohol dehydrogenase, 0479
dehydrogenation
  and **Borsche-Drechel cyclization**, 1219
  and selenium, 1209
dehydrovitamin D$_3$ analogs, 1116
delocalization, and pK$_a$, 0690, 0691
demethoxyerythratidinone, 1227
demethoxypauciflorine 0196
demethylpiericidin A$_1$, 0599, 1240
Demjanov rearrangement, 1175
dendrimers, 0400
Dendrobatid alkaloid 251F, 0720
denopamine, 0831
density functional study, and **Diels-Alder** reactions, 1051
density ionization map
  and dihydroxylation, 0297, 0298
  and epoxidation, 0278
deoxycholic acid, 0454, 0915
deoxyephedrine, 0368
deoxyerythronolide B, 0953
deoxyerythropentose, 0387
deoxymannojirimycin, 0297, 0331
deoxyneodolabelline, 1308
deoxypancratistatin 0942, 0945, 0946, 0994, 1293
deoxyribose, 0590
deoxyschizandrin, 1297
deoxytridachione, 1100
deoxyvinblastine, 0561
deprotection, definition, 0587
desbromohinckdentine A, 0824
deserpidine, 0921, 0923, 0924
desethylibogamine, 1233
desogestrel, 0390
desoxycholic acid, 0317
desoxystemodinone, 1149
Dess-Martin periodinane, 0246, 0259
destabilization, of carbocations, 1164
desulfonylation, dissolving metal reduction, 0457

desulfurization. *See* hydrogenolysis
detergent, and aqueous **Diels-Alder** reaction, 1041
deuterated alkynes, 0633
deuteration, of enolate anions, 0807
deuteride, and super hydride, 0379
deuterium
  and super hydride, 0379
  and the E2 reaction, 0156
deuteroacetic acid, 0507
deuterobutanol, 0395
deuteropropionic acid, 0506
(DHQ)$_2$PHAL, 0300, 0306
(DHQD)$_2$PHAL, 0300
di(1-adamantyl)-*n*-butylphosphine ligands, 1237
diacerhein, 1205
DIAD
  and lactonization, 0581
  and the Mitsunobu reaction, 0124
dialdehydes
  cyclization to diols, 1308
  pK$_a$, 0782
dialkoxyborates, 0499
dialkyl cuprates. *See* organocuprates
dialkyl peroxides, thermolysis, 1260, 1261
dialkylamides. *See* amides
dialkylcadmium
  from **Grignard reagents**, 0647
  with acyl halides, 0647
dialkylimidazolium cations, ionic liquid, 0290, 0298
dialkylmagnesium reagents, 0638
dials, cyclization to diols, 1308
diamide base, with imines, 0846
diamines, from imines, 1305
diamond form, cyclodecane, 0053
diamond lattice
  and conformations, 0051
  and enzyme reductions, 0479
  and **Prelog's rule**, 0479
dianions
  enolate anions, 0802
  of alkynes, 0633
  of esters, 1136
diaryls. *See* biaryls
diastereoface selection
  aldol condensation, 0861
  enolate anions, 0862
diastereomeric excess, 0035
diastereomeric ratio, 0035
diastereomers, 0024-0025

and alkenes, 0033
and meso compounds, 0024
and palladium(0) catalyzed reactions, 1233
and superimposability, 0024
and symmetry, 0024
definition, 0024
meso compound, 0024
diastereoselection, simple, 0850
diastereoselectivity, 0555
  and enzyme reduction, 0481
  and heteroatom radical cyclization, 1288, 1289
  and [2,3]-sigmatropic rearrangement, 1122
  and allyltin reagents, 0767
  and boron enolates, 0841
  and chelation control, 0868
  and **Cram's rule**, 0402
  and enamines, 0876, 0877
  and enolate anions, 0556
  and enolate geometry, 0853, 0854, 0855
  and **Grignard** addition to carbonyls, 0644
  and hydroboration, 0517
  and LUMO maps 0403-0405, 0413, 0418, 0419
  and **Michael addition**, 0878
  and molecular modeling, 0413
  and photochemistry, 1092
  and radical cyclization, 1290
  and rate of reaction, 0416
  and reduction of cyclopentanones, 0412
  and reduction of steroids, 0419
  and Selectride, 0381
  and steric effects, 0850
  and synthesis, 0960
  and the **Cieplak model**, 0414
  and the **Cornforth model**, 0408
  and the **Cram cyclic model**, 0405-0407
  and the **Felkin-Ahn model**, 0409
  and the **Karabatsos model**, 0408
  and the **Nazarov cyclization**, 1189
  and the **Reformatsky reaction**, 0886
  and torsion strain, 0850
  and trimethylene methane equivalent, 1235
  in dihydroxylation, 0549
  in dissolving metal reduction of carbonyls, 0454

in enolate anion alkylation, 0851
in **Grignard** reactions, 0660, 0663
in hydroboration of alkenes, 0516
in photocycloadditions, 1092
in pinacol couplings, 1304
in reductions, 0392
in the acid catalyzed **aldol condensation**, 0820, 0821
in the **aldol condensation**, 0851, 0858, 0859
in the **Dieckmann condensation**, 0829
in the **Diels-Alder reaction**, 1066
in the **ene reaction**, 1146
in the LiAlH$_4$ reduction of cyclopentanones, 0412
in the **Mukaiyama aldol reaction**, 0838
in the NaBH$_4$ reduction of ketones, 0555
in the reduction of ketones, 0401
in the Wittig rearrangement, 1122
simple, definition, 0852
with boron enolates, 0856
with enolate anions and alkyl halides, 0850, 0851
with organocuprates and carbonyls, 0719
with organolithium reagents and carbonyls, 0681
with organolithium reagents and ketones, 0680
diastereospecific reactions
  and diastereoselective, 0034
  of alkenes with dihalogens, 0184
Diatraea saccharalis, 0507
1,4-diazabicyclo[2.2.2]octane. *See* DABCO
1,5-diazabicyclo[4.3.0]-non-5-ene. *See* DBN
1,8-diazabicyclo-[5.4.0]-undec-7-ene. *See* DBU
Diazald, and diazomethane, 1313
diazines
  and diazoalkanes, 1314
  and diazomethane, 1313
  and valence isomerization, 1314
  from tetrazines, 1153
  photolysis, 1314
  photolysis, 1322
diazo compounds
  and **Bamford-Stevens reaction**, 1317
  photochemistry, 1314

diazoalkanes
  [3+2]-cycloaddition, 1110
  and **Bamford-Stevens reaction**, 1317
  and carbenes, 1313, 1321
  and cyclopropanation, 1120
  and diazirines, 1314
  and nitrenes, 1314
  and rhodium compounds, 1322
  and sensitizer, 1314
  and valence isomerization, 1314
  from diazirine, 1315
  from tosylhydrazones, 1317
  photochemistry, 1110, 1322
  with alkenes, reactivity, 1110
  with transition metals, 1324
diazocarbonyl compounds, 1315
  and C-H insertion, 1329
  and rhodium compounds, 1329
  and **Wolff rearrangement**, 1315
  carbenes and diastereoselectivity, 1330
  N-H insertion, 1330
diazoketones, 0959
  and carbenes, 1328
  and C-H insertion, 1329
  and rhodium compounds, 1115
  and the **Gallagher-Hollander degradation**, 0317
  cyclization, 1328
  cyclopropanation, 1328
  from carboxylic acids, 0317
  photolysis, 1328
  rearrangement, 1051
  with copper compounds, 1328
  with copper, 1328
  with rhodium catalysts, 1330
  with rhodium compounds, 1328
  with trialkylboranes, 0841
diazolactams, 1329
  with carbenes, 1329
diazolactone, and cyclopropanation, 1325
diazomethane
  and Diazald, 1313
  and diazirine, 1313
  and isodiazomethane, 1313
  and MNNG, 1313
  and nitroso compounds, 1313
  and silver oxide, 1315
  esterification, 1321
  HOMO and LUMO, 1110
  preparation, 1313
  reactions 1320

  with acid chlorides, 1316, 1328
  with acrylonitriles, 1321
  with acyl halides, 1316, 1328
  with alkenes, 1321
  with aryls, 1320
  with carboxylic acids, 0110, 1321
  with copper salts, 1325
diazonamide A, 0618, 0932
diazonium salts
  and the **Demjanov** rearrangement, 1175
  and the **Tiffeneau-Demjanov** ring expansion, 1175
diazonium salts, aryl
  and **Gomberg-Bachmann** reaction, 1299
  and **Meerwein arylation**, 1298, 1299
  and **Pschorr reaction**, 1299
  and substitution, nucleophilic aromatic, 0208
  and the **Heck reaction**, 1238
  coupling with alkenes, 1299
  coupling with copper compounds, 1298, 1299
  formation of biaryls, 0210
  from aryl amines, 0209
  with copper chloride, 0209
  with copper cyanide, 0209
  with cuprous salts, 0209
  with hypophosphorus acid, 0210
  with potassium iodide, 0209
  with water, 0208
diazonium salts
  and the **Heck reaction**, 1238
  from amines, 1175
  from amino alcohols, 1175
diazophosphonates, and formation of indoles, 1224
diazophosphonoacetate triethyl, 1224
diazosilylalkanes, 0519
diazotization, aryl amines, 0210
Dibal (diisobutylaluminum hydride), 0353
  and HMPA, 0390
  and Triisobutylaluminum, 0389
  with aldehydes, 0391
  with alkynes, 0762
  with benzamides, 0617
  with copper compounds, 0390
  with esters, 0389
  with ketones, 0391
  with lactones, 0389, 0390, 1368
  with nitriles, 0389, 0913

dibenzylamine, 0614
dibenzylidene acetone
  and palladium(0), 1234, 1240
  and Suzuki-Miyaura coupling, 1243
diborane. *See also* borane
  and borane, 0386
  in ether solvents, 0491
  with alcohols, 0499
  with alkenes, 0515
dibromoborane, 0503, 0504, 0505, 0510
  dimethyl sulfide complex, 0505, 0512
dibromomethane, with zinc and $TiCl_4$, 0757
dibutylcuprates. *See* organocuprates
2,6-di-*tert*-butyl-4-phenylphenol). *See* BHT
dibutyltin oxide, and lactonization, 0580
dicaffeoyltartaric acid, 0834
dicarboxylic acids, and decarboxylation, 0172
dicarboxylic acids, from 1,2-diols, 0260
dichloroacetic acid, and
  Moffatt oxidation, 0244
dichloroaluminum hydride, 0388
dichloroborane, 0500
dichlorocarbene, 1325
  with alkenes, 1319
2,3-dichloro-5,6-dicyano-1,4-benzoquinone. *See* DDQ
dichloroketene, [2+2]-cycloaddition, 1078
dichlorolactam, 0187
dichloromethane
  and PCC, 0236
  and PDC, 0237
  lithio, 0536
  lithium with borinanes, 0536
dichroanal B, 0209
dichroanone, 0822
dichromate, and chromic acid, 0227
dicrotaline, 0441
dictyolene, 1247
dictyopterene C, 1127
dictyostatin, 0159, 0510, 0511, 0527
dicyalanil, 0996
dicyclohexyl urea
  and **Moffatt oxidation**, 0244
  and oxalic acid, 0245
dicyclohexyl-18-crown-6. *See* crown ethers
dicyclohexylborane, 0509, 0528
dicyclohexylborinate, 0507
dicyclohexylcarbodiimide. *See* DCC

Dieckmann condensation, 0004, 0827
  and ring size, 0828
  diastereoselectivity, 0829
dielectric constant
  and ionization, 0138, 0139
  and solvents, 0138, 0139
Diels-Alder reaction, 0007, 0305, 0927, 0931, 0943, 0944, 0947, 0966, 0976, 1001, 1013-1075, 1150, 1247
  6-center transition state, 1013
  activation energy and molecular modeling, 1026
  activation energy, 1021, 1022, 1023, 1026
  Alder endo rule, 1030
  amidinium ion catalysis, 1039
  amido dienes, 1051
Diels-Alder reaction
  and activation volume, 1045, 1048
  and acylnitroso compounds, 1056
  and alkene structure, 1015, 1017
  and aminodienes, 1051
  and Arrhenius equation, 1021
  and benzocyclobutanes, 1099
  and benzyne, 0189, 1018, 1065
  and BF$_3$, 1038, 1039
  and BHT, 1036, 1048
  and BINOL, 1075
  and chiral alcohol, 1069
  and chiral templates, 1075
  and **Danishefsky's diene**, 1050
  and density functional study, 1051
  and diene reactivity, 1014
  and dienyl lactams, 1052, 1053
  and electrocyclic reactions, 1099
  and electronic effects, 1030
  and electrostatic potential map, 1023
  and electrostatic potentials, 1029
  and enamines, 1051
  and **Evan's auxiliary**, 1070, 1071
  and extended conjugation, 1031
  and ferric chloride, 1074
  and ferric iodide, 1074
  and FMO, 1029
  and guanidinium perchlorate, 1042
  and iminium salts, 1043, 1055
  and Lewis acids, 0095, 1028, 1038, 1046, 1049, 1117
  and LiCl, 1041
  and maleic anhydride, 1057
  and microwaves, 1042, 1045
  and molecular modeling, 1019
  and molecular orbital theory, 1066,

1268
  and **Oppolzer's sultam**, 1071
  and orbital coefficients, 1038
  and orbital overlap, 1030
  and oxazaborolidine, 1074
  and polarity, 1044
  and pressure, 1029
  and pyrroles, 1053
  and rare earth catalysts, 1066
  and steric hindrance, 1074
  and the **Arrhenius equation**, 1018
  and ultrasound, 1045
  and π-stacking, 1043
Diels-Alder reaction
  azadienes, 1033, 1051
  azonia salts, 1032
  benzoquinone, 1025, 1026
  benzyloxymethyl cyclopentadiene, 1068
  benzynes with furans, 1018
  carbohydrate templates, 1075
  chiral additives, 1072
  chiral auxiliaries, 1069
  chiral auxiliary, 1066, 1067
  chiral catalysts, 1072
  chiral diols, 1075
  chiral lactams, 1052
  cinchonine-catalyzed, 1071
  cis principle, 1024
  cis/trans isomers 1024
  conrotatory versus disrotatory, 1025
  Cu-coordinated catalysts, 1073
Diels-Alder reaction
  cyclopentadienes, 1021, 1026, 1035, 1040, 1073, 1117
  diastereoselectivity, 1066
  dienamines, 1048
  electronic effects, 1015, 1017, 1024
Diels-Alder reaction,
  enantioselectivity, 1029, 1030, 1050, 1066-1075
  and facial selective, 1070
  and ionic liquids, 1037
  chiral templates, 1066
Diels-Alder reaction
  endo rule, 1024, 1030
  endo selectivity, 1029
  equilibrium geometry, 1019
  europium catalysts, 1072
  formation of dihydropyrones, 1074
  heteroatom, 1049
  high pressure, and heterocycles, 1047
  HOMO and LUMO, 1039

HOMO-LUMO interactions, 1013
hydrophobic effect, 1041
in ethylene glycol, 1042
in ionic liquids, 1037
in micelles, 1043
in supercritical CO$_2$, 1066
in water, 1041, 1044
Diels-Alder reaction, intramolecular 1018, 1044, 1048, 1059-1065
  and amine oxides, 1065
  and BHT, 1064
  and **Bredt's rule**, 1060
  and conformation 1060, 1061
  and E/Z dienes, 1061
  and furans, 1060
  and regioisomers, 1061
  and secondary orbital interactions, 1064
  dienyl amides, 1064
  kinetic versus thermodynamic, 1063
  of trienes, 1062, 1063
  regiochemistry, 1064
  stereochemistry, 1062, 1063
  tether length, 1059, 1060
  transition state modeling, 1063
  transition state, 1060, 1061
  transition state, 1061
Diels-Alder reaction, inverse
  electron demand, 1032
  enantioselectivity, 1034
Diels-Alder reaction,
  ketenes, 1032
  kinetic product, 1022
  Lewis acid catalysts, 1035
  Lewis acid catalyzed, rate of reaction, 1039
  Lewis acid catalyzed, regioselectivity, 1039
  lithium perchlorate in ether, 1044
  methyl vinyl ether, 1032
  modeling and regioisomers, 1043
  molecular modeling and rate, 1026
  naphthalene, 1047
  organoaluminum catalyst, 1074
  ortho effect, 1024, 1028
  oxazaborolidinium salt, 1074
  powerful in synthesis, 0928
  quinines, 1049
  quinones, 0928, 1042
  rate enhancement in aqueous media, 1041
  rate enhancement, 1037
  rate of reaction, 1018, 1020, 1021,

1026
reactivity, 1015, 1017
regioselectivity and orbital
   coefficients, 1027
regioselectivity, 1022, 1023, 1026
Diels-Alder reaction, retro, 0914, 0976,
1032, 1035
Diels-Alder reaction
reversibility, 1035
s-cis and s-trans isomers, 1014
secondary orbital interactions, 1028,
   1030
selectivity, 1020, 1021
silyloxybutadienes, 1049
solvent effects, 1044
stereoselectivity, 1025, 1026, 1073
substituent effects, 1015, 1017, 1025
tethered intramolecular, 1064, 1065
transition state modeling and
   regioselectivity, 1020-1023
transition states, 1013, 1028-1031
transition states, asynchronous, 1051
transition states, destabilization, 1031
under high pressure, 1044-1049
Diels-Alder reaction
with acrolein, 1027, 1038, 1056
with acyl imines, 1055
with aldehydes, 1053
with aldehydes, **Lewis acids**, 1054
with alkynes, 1058
with azadienes, 1053, 1057, 1058
with chloral, 1054
with choronitrones, 0907
with conjugated carbonyls, 1053
with conjugated carbonyls, 1056
with cyclodextrins, 1041
with cyclopentadiene, 1014, 1038
with diethyl aluminum chloride, 1054
with dimenthyl fumarate, 1068
with furans, 1018
with glyoxalates, 1054
with imidazoles, 1033
with imines, 1050, 1053, 1055
with iminium salts, 1055
with ketones, 1053
with **Lewis acids**, 1070, 1071
with methacrolein, 1029
with MVK, 1056
with nitroso compounds, 1056
with oxygenated dienes, 1050
with pyrones, 1071
with pyrroline, 1055
with quinones 1044

with sultams, 1071
with tetracyanoethylene, 1014
   with tosylimines, 1055
dienamines, and the **Diels-Alder**
   reaction, 1048
diene reactivity, and electronic effects,
   1016
dienes
1,5, and **Cope rearrangement**, 1123
1,6, and ene reactions, 1145
1,7, and ene reactions, 1145, 1147,
   1148
allowed and forbidden reaction, 1010
dienes
and alkene metathesis, 1335
and **boron annulation**, 0530
and hydrogenation, 0437
and ionization potential maps, 0281
and **Noyori annulation**, 1190
and the **Whiting reaction**, 0362
as dipolarophiles, 1235
azido, 0933
by [2+2]-cycloreversion, 1096
by electrocyclic reactions, 1098
by hydroboration of alkynes, 0510
by **McMurry olefination**, 1310
by retro Diels-Alder reaction, 1035
by **Stille coupling**, 1240
complex, with iron pentacarbonyl, 1248
conformations, 0046
dienes, conjugated
dissolving metal reduction, 0452
from alkynes, 0510
with alkylboranes, 0502
with sodium and ammonia, 0452
dienes
cyclobutane, and **Cope**
   rearrangement, 1127
cyclohexadiene, from arylchromium
   reagents, 0764
cyclopropane, and **Cope**
   rearrangement, 1127
electrocyclization, 1089
electron poor, HOMO and LUMO,
   1006, 1008
electron rich, HOMO and LUMO,
   1006, 1008
dienes
from aldehydes, 0756
from alkene-alkynes, 1234
from alkenes, 1096
from alkyne-alcohols, 0362
from alkynes, 0510-0512, 0763

from boranes, 0511, 0512
from cyclobutenes, 1098
from dienes, 1116
from vinyl halides, 0512
from vinyl triflates, 1240
from vinyltin compounds, 1240
dienes
HOMO and LUMO, 1008, 1009
HOMO, and the **Diels-Alder**
   reaction, 1013
hydrogenation, 0438
in the **Diels-Alder** reaction, 0943,
   0944, 0947
iron complexes, 1248, 1249
nonconjugated, ozonolysis, 0563
orbital coefficients, 1008, 1009
oxygenated, in the **Diels-Alder**
   reaction, 1050
peroxyacid epoxidation, 0281
reactivity in the **Diels-Alder**
   reaction, 1014
reactivity, and substituent effects,
   1016
stereoselectivity in formation, 0510
triplet state, 1089
dienes
with alkenes, 0007, 1013
with alkylboranes, 0502, 0529, 0531
with borane, 0495
with iron pentacarbonyl, 1248
with mcpba, 0280
with PdCl$_2$, 1228
dienones, with Lewis acids, 1188
dienophile, definition, 1013
dienyl amides, intramolecular **Diels-**
   Alder, 1064
dienyl carbamates, and **Curtius**
   rearrangement, 1051
dienyl lactams, **Diels-Alder reactions**,
   1052, 1053
diesters
and **acyloin condensation**, 1306
pK$_a$, 0782, 0808
with sodium, 1307
diethyl aluminum chloride, in the
   Diels-Alder reactions, 1054
diethyl azodicarboxylate. *See* DEAD
diethyl azodicarboxylate, and
   lactonization, 0581
diethyl ether. *See* ether, diethyl
diethyl tartrate. *See* tartrate
diethylaluminum chloride, catalyst for
   the ene reaction, 1151

diethylaluminum-*N*-methylaniline, with oxetane, 0563
diethylamide. *See* lithium, etc.
diethylazodicarboxylate. *See* DEAD
diethylenediamine, and the **Benkeser** reduction, 0462
diethylzinc
  and diiodomethane, 1333
  and **Simmons-Smith reaction**, 1333
dig, and **Baldwin's rules**, 0564, 0567
diglyme
  and boranes, 0506
  and LiAlH₄ 0363
  in hydroboration, 0520
dihalides
  elimination to alkenes, 0158
  elimination, with indium, 0158
  geminal, reduction, 0368
  in E2 reactions, 0630
  malonic ester synthesis, 0809
  vinyl and the **Corey-Fuchs procedure**, 0159, 0733, 0734
  vinyl, from aldehydes, 0734
  vinyl, with butyllithium, 0159
  vinyl, with organolithium reagents, 0734
  with base, 0630
  with tert-butoxide, 0630
dihalo ketones, 0152
dihydroaminoquartynic acid, 0992
dihydroazepines, 1154
dihydroboration, 0502
dihydroceramides, 0642
dihydrochamacantrhin B, 0997
dihydroconfertifolin 0438
dihydrocrinine, 0467
dihydrofuran, and alkyl boranes, 0523
dihydroisoquinolines, 1186
  by **Bischler-Napieralski reaction**, 1215
  electrocyclic reactions, 1101
dihydropinidine, 0375
dihydropyran
  and alkyl boranes, 0523
  with alcohols, 0590
  with alcohols, 0599
dihydropyrans
  by the **Diels-Alder reaction**, 1057
  from conjugate carbonyls, 1057
dihydropyrones, by **Diels-Alder** reactions, 1074
dihydropyrrole, and alkyl boranes, 0523

dihydroquinidines
  OsO₄ complex, X-ray, 0299
  and dihydroxylation, 0299
dihydroquinones, from vinyl quinone imides, 1101
dihydrospiniferin-1, 1100
dihydrosubincanadine B, 0828
dihydrothiophene, and alkyl boranes, 0523
dihydrothiophenes, from thioketones, 0744
dihydroxylation
  and AD-mix-α, 0300
  and AD-mix-β, 0300
  and alkene structure, 0296, 0297, 0299
  and azabicylcooctanes, 0299
  and *Cinchona* alkaloids, 0299
  and density ionization map, 0297, 0298
  and dihydroquinidines, 0299
  and diols, 0291-0306
  and dipolar addition, 0292
  and electronic effects, 0296
  and ionic liquids, 0298
  and osmium tetraoxide, 0294-0301
  and permanganate, 0292
  and pH, 0293
  and phase transfer catalysts, 0294
  and stereoselectivity, 0297, 0298
  and tetraethylammonium acetate, 0297
  and the **Woodward modification**, 0303
  and water, 0301
  enzymatic, 0303
  model for selectivity, 0301
  of alkenes, 0291-0306, 0548
  of aryls, 0303
  of enamines, 0561
  OsO₄, mechanism, 0295
  permanganae, mechanism, 0292
  Sharpless asymmetric dihydroxylation, 0298-0301
  stereochemistry, 0292
  stereoselectivity, 0549
  via the **Prévost reaction**, 0301, 0302
  with *Pseudomonas putida*, 0303-0306
dihydroxynitianes, 1343
dihydroxystearic acid, 0293
dihydroxytyramine (DOPA), 0266
dihydroxyvitamin D₃, 1101
diimide
  and the **Mitsunobu reaction**, 0124

preparation, 0474
  reduction of alkenes, 0474
  with copper(II) salts, 0474
diinositols, 0305
diiodomethane
  and carbenoids, 1331
  and the **Simmons-Smith** reaction, 1331
  and zinc-copper, 1331
  with alkenes, 1332
  with diethylzinc, 1333
diiron nonacarbonyl
  and **Noyori annulation**, 1189
  with dibromo ketones, 1189
diisobutylaluminum hydride. *See* Dibal
diisopinocampheyl chloroborane, 0397
diisopinocampheylborane, 0495, 0519, 0520
  and reduction, 0395
  transition state modeling, 0521
  with alkenes, 0522, 0523
diisopropyl azodicarboxylate. *See* DIAD
diisopropyl tartrate. *See* tartrate
diisopropylamide. *See* LDA
diisopropylamine and kinetic versus thermodynamic control, 0801
diisopropylethylamine (**Hünig's base**), 0598
diketones
  from hydroxydithianes, 1177
  from ketones, 0327, 0328, 0877
  pKₐ, 0782
  reduction, 0406, 0407
  with anilines, 1213
  with peroxyacids, 0311
diketopiperazines, conformation, 0066
dilauroyl peroxide, and radical cyclization, 1283
dilongifolylborane, 0525
dimethyl fumarate, in the **Diels Alder** reaction, 1068
dimerization
  in oxidation of cathechols, 0265
  in oxidation of phenols, 0265
  of aldehydes upon electrolysis, 0470
  of cyclopentadiene, 1035
dimethoxybenzyl alcohol, and organolithium reagents, 0675
dimethoxyborane, 0499
  with Grignard reagents, 0661
dimethoxyethane. *See* DME
dimethyl cuprates. *See* organocuprates
dimethyl sulfate. *See* sulfate

dimethyl sulfide. *See* sulfide
dimethyl sulfoxide. *See* DMSO
dimethylacetamide dimethyl acetal
   and **Claisen rearrangement**, 1137
dimethylaluminum chloride catalyst
   for the **ene reaction**, 1150
dimethylaminoisobutene, with methyl
   acrylate, 1076
dimethylaminopyridine. *See* DMAP
dimethylborinic acid, 0506
dimethylcarbene, 1313
dimethyldioxirane. *See also* dioxiranes
dimethyldioxirane. *See* DMDO
dimethylformamide. *See* DMF
dimethylketene
   [2+2]-cycloaddition, 1078
   and carbenes, 1312
dimethyl-*N*,*N*'-dinitrosotere-
   phthalamide, 1313
dimethylpyrazolium and
   chromium trioxide, 0329
dimethylsulfoxonium methylid, 0746
dimethylsulfoxonium methylid. *See*
also ylids, sulfur
dimsyl sodium
   and enolate anions, 0558
   and phosphonium salts, 0731
dinapsoline, 1286
dinitriles, by electrolytic dimerization
   of nitriles, 0470
dinitromethane
   and **Knoevenagel condensation**, 0830
   enolate anions, 0830
diolides, 0441
   via macrolactonization, 0578, 0579
diols
   1,2-, oxidation with silver(II) oxide,
     0260
   acetonides, 0606
   and lead tetraacetate, 0322
   and osmium tetraoxide, 0295
   and oxocarbenium ions, 1180
   and permanganate, 0292
   and *Pseudomonas putida*, 0303-0306
   and rearrangement, 1180
   and **Sharpless asymmetric**
     dihydroxylation, 0298-0301
   and the **acyloin condensation**, 1308
   and the **Bally-Scholl synthesis**, 1208
   and the **pinacol rearrangement**, 1179
   and the **Woodward modification**,
     0303
   by dihydroxylation, 0548

by **pinacol coupling**, 0454, 1303
by reduction of lactones, 0373, 0374
conversion to acetals or ketals, 0607
from aldehydes, 1304
from alkenes, 0291-0306, 0313, 0548,
   1191
from anhydrides, 0353
from aryls, 0303
from boranes, 0515
from dialdehydes, 1308
from dienes, 0515
from dissolving metal reduction, 0454
from epoxides, 0303, 0548
from formaldehyde, 1191
from **Grignard** addition to lactones,
   0649
from ketones, 1304, 1310
from lactones, 0354
from vinyl, 0849
oxidative cleavage, 0322-0324
protecting groups, 0606
reaction with acid, 1180
unsymmetrical, from ketones, 1305
via the **Prévost reaction**, 0301, 0302
with aldehydes, 0611
with ketones, 0611
with the **Fétizon's reagent**, 0258
diop, 0450
dioxanes
   protecting group for aldehydes or
     ketones, 0610
dioxiranes
   and asymmetric epoxidation, 0291
   and epoxidation, 0290
   and potassium peroxomonosulfate,
     0291
   with alkenes, 0290
dioxolanes
   and diol protecting groups, 0606
   hydrolysis rate versus dioxanes, 0610
   hydrolysis rate versus dioxolanes,
     0610
   hydrolysis, 0951
   protecting group for aldehydes or
     ketones, 0610
   reaction energy, 0702
   with DDQ, 0610
dipamp, 0449, 0450
dipeptide nitriles, 0624
diphenyl phosphoryl azide, 0192
diphenylacetic acid, and organolithium
   reagents, 0675

diphenylacetone tosylhydrazone and
   organolithium reagents, 0675
diphenylcuprate. *See* organocuprates
diphenylcyclopropylsulfonium salts,
   0749
diphenylphosphinomethane, 0450
diphenylphosphoryl azide. *See* DPPA
diphos, 0450
   with Ni(0), 1246
diplamine, 1211
dipolar addition. *See* cycloaddtion,
   [3+2]
1,3-dipolar addition, retro, 0976
dipolar transition states, 1045
dipolarophiles, 1102
   dienes, 1235
dipoles, 1,3-, 1101
dipoles
   and conformations, 0045
   orbital coefficients, 1104
diradical mechanism, [3+2]-
   cycloaddition, 1102
diradicals 1295
   and [2+2]-cycloaddition, 1076, 1095
   and **anionic oxy-Cope**
     rearrangement, 1130
   and **Bergman cyclization**, 1295
   and cages, 1266
   and carbenes, 1312
   and **Norrish** Type I fragmentation,
     1264
   and **Norrish** Type II fragmentation,
     1265
   and photochemistry, 1091
   and the **Paternò-Büchi reaction**,
     1095
   and vinylcyclopropane
     rearrangement, 1119
   by photolysis of ketones, 1318
   from ketones, 1265
   rotational equilibrium, 1323, 1324
   singlet, 1324
   triplet, 1324
directed ortho metalation, 0686, 0687,
   0944
directing effects, of functional groups,
   0283
discodermolide, 0546
disconnect products, 0002
   and synthetic equivalents, 0009
disconnection approach
   and synthesis, 0908
   definition, 0898

disconnection protocol, 0005
disconnections, 0002, 0006, 0908
  and chemical degradation, 0975
  and functional group proximity, 0011
  and synthesis, 0006
  definition, 0909
  Diels-Alder reaction, 0943
  of bicyclic systems, 0935
  of polycyclic systems, 0935
  of strategic bonds, criteria, 0924
  propargylic bonds, 0636
  strategic bonds in rings, 0934
discs, and combinatorial chemistry, 0982
disfavored/favored, and **Baldwin's rules**, 0564, 0567
disiamylborane, 0495, 0515
  and reduction, 0387
  with alkenes, 0496, 0516
  with alkynes, 0503
disodio species, 0456
disodium phosphate. *See* phosphate
disproportionation,
  and LiAlH$_4$ reduction, 0351
  and sodium trimethoxyborohydride, 0377
  lithio-ethers, 0673
  of **Grignard reagents**, 0641
  of radicals, 1270
  organoiron compounds, 0759
disrotatory
  and **Cope rearrangement**, 1126
  electrocyclic reactions, 1098
  in the **Diels-Alder reaction**, 1025, 1026
dissociative fragmentation, 1264
dissolving metal reduction. *See* metals or reduction
dissonance and **Sharpless asymmetric epoxidation**, 0286
dissonant, double stereodifferentiation, 0867
di-tert-butyl peroxide, 1260, 1261
di-*tert*-butyldicarbonate, with amines, 0617
dithianes
  and umpolung, 0704
  anions, conjugate addition, 0706
  from aldehydes or ketones, 0444
  hydrogenolysis, 0444
  hydrolysis rate versus dithiolanes, 0611
  hydrolysis, reagents, 0706

lithiated, reactions, 0705
organotin, 0704
oxidizing reagents, 0640
pK$_a$, 0704
protecting group for aldehydes or ketones, 0611
reaction energy, 0702
reagents for oxidation, 0704
reduction with Raney nickel, 0473
with butyllithium, 0705
with NCS or AgNO$_3$, 0705
with organolithium reagents, 0704
with Raney nickel, 0464
dithioacetals
  anions, 0710
  and BF$_3$, 0611
  and CAN, 0612
  and hydrogenolysis, 0611, 0612
  and mercuric compounds, 0612
  and NBS, 0612
  and the **Corey-Seebach** procedure, 0612
  from aldehydes, 0444
  hydrolysis rate versus dithioketals, 0611
  ketene, 0709
  poisons for catalytic hydrogenation, 0612
  protecting group for aldehydes, 0611
dithioalkyl phosphonate esters. *See* esters
dithiocarboxylic acids, with LDA, 0709
dithioketals
  and BF$_3$, 0611
  and CAN, 0612
  and hydrogenolysis, 0611, 0612
  and mercuric compounds, 0612
  and NBS, 0612
  and the **Corey-Seebach** procedure, 0612
  from ketones, 0444
  hydrolysis rate versus dithioacetals, 0611
  poisons for catalytic hydrogenation, 0612
  protecting group for ketones, 0611
dithioketene acetals, 0709
dithiolanes, 0444
  anions, Michael addition, 0706
  anions, 0706, 0707
  from aldehydes or ketones, 0444
  hydrolysis rate versus dithianes, 0611
  lithio, stability, 0707

protecting group for aldehydes or ketones, 0611
reaction energy, 0702
reduction with Raney nickel, 0473
umpolung, 0706, 0707
with organolithium reagents, 0706, 0707
dithiols, with aldehydes or ketones, 0611
diynes
  by **Cadiot-Chodkiewicz** coupling, 0631, 1302
  by **Castro-Stephens** coupling, 1244
  by **Sonogashira** coupling, 1244
  by the **Glaser** reaction, 0631, 1301
  from alkynes, 0631, 1301, 1302
  from haloalkynes, 1302
  with alkynes, metathesis 1339
DMAP (dimethylaminopyridine)
  and basicity, 0090
  and **Shiina macrolactonization**, 0576
  and the **Yamaguchi protocol**, 0576
  with *tert*-butyldimethylsilyl chloride, 0603
  and 2,4,6-trichlorobenzoyl chloride, 0576
DMDO (3,3-dimethyldioxirane), with silyl enol ethers, 0813
DME (dimethoxyethane), 0672
  and butyllithium, 0672
  and organolithium reagent aggregation, 0672
DMF (dimethylformamide)
  and PDC, 0238
  and phosphorus oxychloride, 0372
  and the **Vilsmeier reaction**, 0372
  solvent effects in acidity, 0088
  with Grignard reagents, 0684
  with organolithium reagents, 0684
DMSO (dimethyl sulfoxide)
  and acetic anhydride, 0245
  and **Cope elimination**, 0165
  and DCC, 0243
  and enolate anion alkylation, 0811
  and epoxides in aqueous acid, 0548
  and HBr, and bromination, 0196
  and lithium-hydrogen exchange, 0697
  and **Moffatt oxidation**, 0243
  and oxalyl chloride, 0241
  and oxidation of alcohols, 0239-0246
  and phosphorus pentoxide, 0246
  and phosphorus ylids, 0731
  and pK$_a$ of sulfur compounds, 0697

and pyridine-sulfur trioxide, 0246
and sulfur ylids, 0747
and **Swern Oxidation**, 0241, 0242
and trifluoroacetic anhydride, 0241
and ylids, 0746
in S_N2 reactions, 0145
oxidation of halo ketones, 0240
solvent effects in acidity, 0088
with alkyl halides, 0240
DNA, cleavage, 1295
dodecacarbonyltetracobalt, 1343
dodecahedrane, 0905
Doebner condensation, 0830, 1214
Doebner-Miller reaction, 1214
dolabelide D, 0333, 0516, 1337
dolabriferol, 0606, 0820
dolastatin 19, 0602
donor sites, 0007
DOPA (dihydroxytyramine), 0266
d-orbitals, and pK_a, 0692
DOSP, and carbenes, 1330
double stereodifferentiation, 0867
    aldol condensation, 0862
    consonant and dissonant, 0867
    enolate anions, 0862
doublet, definition, 1312
Doyle catalysts, and carbenes, 1330
DPPA (diphenylphosphoryl azide) and
    the Mitsunobu reaction, 0125
dppe, 0450
    allyl palladium, 1230, 1231
%dr, 0035
dragmacidin D, 1242
Dreiding Model, 0053
    and reduction, 0397
Drierite, and **Knorr quinoline
    synthesis**, 1211
drug discovery, 0987
dumsin, 1150
duocarmycin SA, 1225
duocarmycin, 0772
durene, 1199
dyads, 0118
dynemicin, 1295
Dysidea etheria, 0900
dysidiolide, 0233. 0900
dysiherbaine, 0555

E

E and Z isomers, 0029
E/Z dienes, and intramolecular **Diels-
    Alder** reactions, 1061

E/Z enolates, and **Zimmerman-
    Traxler** model, 0854, 0855
E/Z isomers
    aggregation state of enolate anions,
        0795
    alkenyl **Grignard reagents**, 0547,
        0642, 0664, 0665
    alkylidene enolates, 0870
    allylic **Grignard reagents**, 0664,
        0665
    amides, enolate anion, 0792
    and [2,3]-sigmatropic rearrangement,
        1122
    and boron enolates, 0842
    and enolate anions, 0547, 0858
    and enolate anions, geometry, 0795
    and metathesis, 1337
    and radical cyclization, 1285
    and **Stille coupling**, 1241
    and the **aldol condensation**, 0858
    and the **Wittig reaction**, 0735
    and the **Zimmerman-Traxler
        Model**, 0852
    and **Wittig rearrangement**, 1122
    crotyl **Grignard** reagents, 0665
    enolate anions, 0790, 0791, 0853,
        0854, 0855
    enolate anions, and the **Ireland
        model**, 0791, 0792
    enolate anions, correlation with base,
        0791
    enolate anions, steric hindrance, 0795
    enolate, and diastereoselectivity, 0853,
        0854, 0855
    esters, enolate anions, 0792
E1 elimination. *See* elimination
E2 reaction. *See* elimination
echnisosporin, 0994
ectocarpene, 1110
EDTA (ethylenediaminetetraacetic acid)
    and oxidation, 0266
    and **Udenfriend's reagent**, 0266
Edward's equation, and HSAB, 0096
%ee, 0035
EE. *See* ether, ethoxyethyl
Elbs persulfate oxidation, 0267
Electrochemistry
    and **Bobbitt's reagent**, 0248
    and radicals, 1298
electrocyclic cleavage, 1081
electrocyclic reactions, 1097-1101
electrocyclic reactions
    and DDQ, 1100

and **Diels-Alder reactions**, 1099
and dihydroisoquinolines, 1101
and stereochemistry, 1001
and symmetry, 1001
benzocyclobutenes, 1099
conrotatory vs. disrotatory, 1098
cyclobutene ring opening, 1098
hexatrienes, 1100
vinyl quinone imides, 1101
electrocyclic ring opening, and
    torquoselectivity, 1099
electrocyclization. *See also* electrocyclic
electrocyclization
    norcaradiene, 1118
    of dienes, 1089
    photochemical 1089
electrode potential, 0223
electrodes, and electrolysis, 0470
electrolysis
    See also reduction, electrolytic
        electrolysis, 0468
    and oxidation, 0222
    and reduction potential, 0222
    and steroids, 0471
    and sulfuric acid, 0471
    of aldehydes or ketones, 0470
    of phenols, 1298
    of potassium acetate, 1278
    of pyridine, 0470
    reduction, 0468
electrolytic reduction. *See also* reduction
electromagnetic spectrum,
    and photochemistry, 1083
electron affinity
    and Milliken electronegativity, 0100
    negative, and LUMO, 1000
electron capture GC, 0986
electron density surface, LUMO. *See*
LUMO map
electron density, and molecular
    orbitals, 0999
electron donating groups, and acidity,
    0084
electron donation, and π-bonds, 0080
electron flow, and **Baldwin's rules**,
    0564
electron pairs, and priority rules, 0021
electron release, and HSAB, 0098
electron releasing groups, and HSAB,
    0098
electron transfer
    and electrolytic reduction, 0468
    and **Grignard** reagents, 0644

and hydrogenolysis of benzylic
ethers, 0457
electron volt, definition, 1001
electron withdrawing groups
and acidity, 0084
and pK$_a$, 0783
electronegativity
and acidity, 0084
and electrophilicity index, 0119
and hardness parameter, 0098
and hardness parameter, 0099, 0101
and HOMO/LUMO, 0101
and HSAB, 0097, 0098
and oxidation, 0219
and rate of reaction, 0416
Milliken, 0090
electronic effects
and [2+2]-cycloaddition, 1076
and acidity, 0084
and basicity, 0089
and **Birch reduction**, 0459
and dihydroxylation, 0296
and electrophilic aromatic
substitution, 0201
and **Friedel-Crafts alkylation**, 1199
and hydrogenation of aryls, 0446
and isomerization, 1199
and ketene dithioacetals, 0711
and phosphorus ylids, 0731
and regioselectivity, 1027
and the **Baeyer-Villiger reaction**,
0309
and the **Heck reaction**, 1237
and ylids, 0731, 0732
borane addition to alkenes, 0494
electrophilic aromatic substitution,
1192
in hydroboration, 0504
in peroxyacid epoxidation, 0280
in the Birch reduction, 0459
in the **Diels-Alder reaction**, 1015,
1017, 1024, 1030
on HOMO and LUMO, 1015, 1017
peroxyacid epoxidation of alkenes,
0280
regioselectivity in the **Diels-Alder**
reaction, 1028
electronic excitation, pathways, 1263
electronic factors, and **Baldwin's**
rules, 0568
electronic interactions, and molecular
orbitals, 0999
electrons, loss of, and oxidation, 0220,
0221

electrophile, definition, 0107
electrophilic aromatic substitution. *See*
substitution
electrophilic
atoms, 0007
radicals, 1268
tag, combinatorial chemistry, 0984
electrophilicity index, 0119
electrostatic potential map
and acidity, 0785
and carbenes, 1320
and enol content, 0784
and the **Diels-Alder reaction**, 1023
and the S$_N$2 reaction, 0115
electrostatic potential
and pK$_a$, 0694, 0702
and the **Diels-Alder reaction**, 1029
of ketones, 0702
of sulfur compounds, 0702
elemane sesquiterpenes, 1097
elemene, 1029, 1069
elenic acid, 0763
elimination and substitution, 0544
elimination reactions, 0153-0160
phosphonate esters, 0716
elimination reactions
1,3-, 0171, 0173, 0934
1,3, decarbonylation, 0172
1,3-, **Grob fragmentation**, 0174
1,3-transition state, 0177
elimination reactions
and **Bredt's rule**, 0553, 0554
and cyclic alkenes, 0551
and decarboxylation, 0171
and **Grob fragmentation**, 0174,
0934
and **Hofmann elimination**, 0753,
0754
and HSAB, 0103
and **Knoevenagel condensation**,
0829
and nitrogen ylids, 0753, 0754
and sulfurane, 0591
and the **Barton-McCombie**
reaction, 0168
and the **Burgess reagent**, 0170
and the **Knoevenagel** condensation,
0829
elimination
bimolecular. *See* elimination, E2
Cope elimination, reverse, 0165
Cope. *See also* Cope
Cope, and microwaves, 0165

E2, versus S$_N$2, 0141
decarboxylation, 0171
elimination, E1, 0160-0161, 0949
and aromatic substitution, 0194
and carbocations, 0160
and ionization, 0160
and racemization, 0160
and rate of reaction, 0160
and **Saytzeff elimination**, 0160
and the E2 reaction, 0161
and the S$_N$1 reaction, 0161
mechanism, 0160
rate of reaction, 0160
elimination, E2, 0033, 0153-0160,
0158
and A-strain, 0156
and axial halogens, 0156
and bond polarization, 0153
and Brønsted-Lowry bases, 0153
and Cr(VI) oxidation, 0231, 0240
and DBN, 0158
and DBU, 0157, 0158
and **Hofmann elimination**, 0162
and leaving groups, 0157
and methoxide, 0544
and molecular modeling, 0156
and **Saytzeff (Zaitsev) elimination**,
0155
and sulfonate esters, 0158
and the E1 reaction, 0161
and the **Hammond postulate**, 0155
and the S$_N$2 reaction, 0153
leaving group position, 0154
LUMO map, 0157
mechanism, 0154, 0156
of cyclohexanes, 0155
reaction, and SeO$_2$, 0328
reactions, 0079
stereoselectivity, 0154, 0559
sulfonate esters leaving groups, 0159
transition state, 0154
versus S$_N$2, 0158
with dihalides, 0630
elimination
from selenides with thiocarbamates,
0170
Grob fragmentation, 0934
Hoffmann, and S$_N$2, 0122
in halogenation of ketones, 0152
in reduction of amides, 0355
intermolecular, 0161
nucleophilicity versus basicity, 0141

elimination
  of amine oxides, 0165
  of ammonium salts, 0164
  of esters, 0166
  of esters, stereochemistry, 0166
  of halo-carboxylates, 0177
  of homoallylic alcohols, 0177
  of selenoxides, 0169, 0815
  of sulfoxides, 0168, 0698, 0814
  of xanthate esters, 0167
elimination, pyrolysis of esters, 0166
elimination, syn, 0161-0171
  and bond energies, 0170
  and bond polarization, 0170
  Hofmann elimination, 0162
  syn, selenoxides, 0815
  sulfoxide, 0814
elimination
  unimolecular. *See also* elimination, E1
  versus substitution, 0140
  with DBU, 1334
  with phosphonate ester ylids, 0742
  with phosphonothioates, 0743
  with vinyl halides, 0630
  β, of alkenes. *See also* elimination, E2
  β, of sulfones, 0158
ELISA (enzyme-linked immunosorbent
  assay), 0981
elisapterosin B, 1044
elutherobin, 0136, 0605
elysiapyrones A and B, 1240
Emericella purpurea, 1364
emim, 1037
emission spectrum, and
  photochemistry, 1088
emission, and photochemistry, 1087
enamides, with sodium
  cyanoborohydride, 0383
enamine-imine, tautomerism, 1213
enamines
  and dienophiles, 1051
  and lead tetraacetate, 0323
  and **Michael addition**, 0877
  asymmetric, 0876, 0877
  by **aza-Claisen rearrangement**,
    1141, 1142
  by **aza-Cope rearrangement**, 1131
  by **Conrad-Limpach reaction**, 1212
  by the **Stork enamine synthesis**, 0873
  conformations, 0063
  cyclohexene, formation of aromatic
    amines, 0876
  diastereoselectivity, 0876, 0877

  dihydroxylation, 0561
  formation, and ketones, 0875
  formation, and steric hindrance, 0873,
    0874
  from aldehydes and ketones, 0108
  from aldehydes, 0873, 0874
  from amines, 0108
  from ketones, 0873, 0874
  hydroboration, 0561
  hydrogenation, 1115
  in **Diels-Alder reactions**, 1051
  nucleophilicity, 0873, 0874
  preparation, 0873
  reactions, 0873, 0874
  tautomerism with imines, 0873
  tautomerization to iminium salts, 0476
  with acyl halides, 0875
  with alkyl halides, 0873, 0874
  with alkynes, 1076
  with borane, 0561
  with conjugated carbonyls, 0875
  with cyclopentadienyl dicarbonyl
    and alkenes, 1249
  with dibromo ketones, 1190
  with $NaBH_4$, 0371, 0561
  with OsO_4, 0561, 0562
  with palladium(0), 1234
  with sodium cyanoborohydride, 0383
  with thallium(III) acetate, 0561, 0562
enamino esters
  by **Gould-Jacobs reaction**, 1212
  via **Firedel-Crafts** reactions, 1212
enantiomeric excess, 0035
enantiomers, 0014-0023
  and chiral axes, 0026
  and stereogenic centers, 0013-0029
  definition, 0013
enantioselective
  Diels-Alder reactions, 1066-1075
  synthesis, 0931
enantioselectivity, 0960
  [2,3]-sigmatropic rearrangement, 1121
  [2+2]-cycloaddition, 1080
  [3+2]-cycloaddition, 1109
  acyl, addition of chiral sulfoxides, 0698
  alcohols from boranes, 0520
  aldol condensation, and **Cram's rule**,
    0863, 0864
  alkylborane addition to aldehydes,
    0527
  allyl(diisopinocampheyl)borane,
    0526
  and abrine esters, 1072

  and allyltin compounds, 0767
  and Alpine borane 0395
  and arylchromium compounds, 0765
  and carene-based hydroboration, 0526
  and chiral additives, 0866
  and chiral auxiliaries, 1067
  and chiral lactams, 1052
  and chiral templates, 1075
  and *Cinchona* alkaloids, 0393
  and **Cram's rule**, 0402
  and **Grignard reagents**, 0658
  and **Jacobsen-Katsuki asymmetric
    epoxidation**, 0289
  and oxazaborolidine, 0400
  and oxazaborolidines, 0396
  and palladium(0) catalyzed reactions,
    1233, 1234
  and **pinacol coupling**, 1304
  and proline esters, 1072
  and radical cyclization, 1289
  and **Robinson annulation**, 0882
  and salen catalysts, 0289
  and **Sharpless asymmetric
    aminohydroxylation**, 0306
  and sparteine, 0689
  and the **Cram cyclic model**, 0405-
    0407
  and the **Pictet-Spengler
    isoquinoline** synthesis, 1216
  and transient auxiliaries, 0866
  chiral alcohols in the **Diels-Alder**
    reactions, 1069
  chiral template approach, 0954
  conjugate addition of **Grignard**
    reagents, 0653
  **Darzens' glycidic ester
    condensation**, 0834, 0835
  diazocarbonyl carbenes, 1330
  in the carbonyl ene reaction, 1151
enantioselectivity, in the **Diels-Alder**
  reaction, 1029, 1030, 1050, 1066-
  1075
  Diels-Alder of azadienes, 1052
  Diels-Alder, and chiral auxiliaries,
    1066, 1067
  Diels-Alder, and facial selectivity,
    1070
  Diels-Alder, influence of catalyst,
    1072
  in ionic liquids, 1037
enantioselectivity
  and higher order cuprates, 0727
  and hydroboration, 0519, 0521

and **Michael addition**, 0878

and organocuprates and alkoxides, 0725

and organocuprates, 0723, 0724

and organoiron compounds, 0761

and organoiron reactions, 0761

and organolithium reagents, with sparteine, 0672

and organolithium reagents, 0680, 0681

and pinacol coupling, 1304

and polyene cyclization, 0951

and proline, in the **aldol condensation**, 0866

and radical cyclization, 1292

and radical reactions, 1267

and reduction of ketones with **Grignard** reagents, 0667

and sparteine, with **Grignard** reagents, 0661

and the **Zimmerman-Traxler model**, 0867

Grignard reagents with amino alcohols, 0662

Grignard reagents, solvent effects, 0661

in dihydroxylation, 0549

in **ene reactions**, 1151

in enzyme reductions, 0481, 0483

in **Friedel-Crafts alkylation**, 1196

in haloalkene-organolithium cyclization, 0689

in hydroboration of cis/trans alkenes, 0520

in palladium(0) reactions, 1234

in reductions, 0392

in **Sharpless asymmetric dihydroxylation**, 0298-0301

in **Sharpless asymmetric epoxidation**, 0286

in the **Heck reaction**, 1237

in the **Mukaiyama aldol reaction**, 0838

in the oxidation of sulfides with oxaziridines, 0337, 0338

in the oxidation of sulfides, 0336

in the **Pauson-Khand reaction**, 1344

in the **Pictet-Spengler** isoquinoline synthesis, 1217

in the **Simmons-Smith reaction**, 1333

in the Simmons-Smith reaction, 1333

in the **Strecker synthesis**, 0629

in the **Wittig rearrangement**, 1121

inverse electron demand **Diels-Alder** reactions, 1034

iodolactonization, 0186

matched and mismatched, 0867

menthyl acrylate with cyclopentadiene, 1067, 1068

with allyltin compounds, 0766

with **Burgess reagent**, 0170

with diisopinocampheylborane, 0519, 0520

enantiospecific and enantioselective, 0035

Enantrane, 0397

Enantride, 0397

encapsulation, and solvation, 0906

endo mode, 1029

endo rule

in the **Diels-Alder** reaction, 1024, 1030

endo selectivity

and secondary orbital interactions, 1030

and the **Diels-Alder reaction**, 1029

endo/exo, and **Baldwin's rules**, 0564, 0567

ene reaction, 0558, 1143-1152

8 versus 6 membered ring formation, 1147

aluminum catalysts, 1151

and [2+2]-cycloaddition, 1077

and alkynes, 1146

and allylnickel, 1246

and **Chugaev elimination**, 0167

and **Lewis acids**, 1147

and the **Hammond postulate**, 1148

and zinc bromide, 1148

carbonyl compound, 1145

carbonyl, **Lewis acid** catalyzed, 1150, 1151

carbonyl, **Lewis acids**, and LUMO, 1150

chiral chromium complex catalysts, 1151

chiral, 1151

chiral, bis(diisopropoxytitanium) dibromide, 1152

diastereoselectivity, 1146

diethyl mesoxalate, 1145

diethylaluminum chloride catalysts, 1151

dimethylaluminum chloride catalysts 1150

enantioselectivity, 1151

ethylaluminum dichloride catalyst, 1150, 1034

hetero-ene, 1150

HOMO and LUMO, 1143

imino, 1151

imino, and **Lewis acids**, 1151

internal, powerful in synthesis, 0929

intramolecular, 1143, 1145, 1147, 1148

ketones, 1149

powerful in synthesis, 0932

SeO_2 reactions, 0325

substituent effects, 1148

titanocene catalyzed, 1147

transition state modeling, 1144, 1147, 1148

transition state, 1143, 1144, 1146

with 1,6-dienes, 1145

with 1,7-dienes, 1145, 1147, 1148

with aldehydes, 1149

with carbonyls, 1149

with imines, 1151

with maleic anhydride, 1144

ene, definition, 1143

ene-diynes, and **Bergman cyclization**, 1295

energy barrier

and conformations, 0041

butane, 0040

carbene rearrangement, 1322

ethane, 0038

s-cis and s-trans, 1014

substituted ethanes, 0040

energy diagrams

and HSAB, 0102

and SOMO, 0102

energy free. *See* free energy

energy

profile, S_N2 reactions and nucleophilic strength, 0116

and light, 1083

transition, photochemistry, 1086

and ring size, 1174

boat versus chair, 0055

bond dissociation, and peroxides, 1260

bonds and hemolytic cleavage, 1260

carbocation rearrangement, 0134

conformational, 0056

conformational, cycloalkenes, 0061, 0062

conformational, table, 0059

free, *See* free energy

free, and enthalpy, 0058

HOMO of alkenes, 1002, 1003, 1004
HOMO-LUMO, and **Diels-Alder**
  reactions, 1013
inversion, of radicals, 1259
LUMO of alkenes, 1002, 1003, 1004
of carbocations, 1168, 1169
radicals, 1258
reaction. *See* reaction energy
SOMO, of radicals, 1269
ene-ynes, and the **Pauson-Khand**
  reaction, 1342, 1343
enol acetates
  from enolate anions, 0820
  from ketones, 0815
  thermodynamic enolates, 0815
  with methyllithium 0815, 0820
  with organolithium reagents, 0815
enol content
  and electrostatic potential map, 0784
  and pK$_a$, 0783
  and substituent effects, 0784
  solvent effects 0783
enol esters
  and **McMurry olefination**, 1310
  from esters, 1310
enol ethers. *See* ethers
enol ethers
  and PDC, 0237
enol ethers, silyl. *See* silyl enol
enol ethers, with carbenes, 1326
enolate accelerated **Cope**
  rearrangement, 1129
enolate alkylation, and **Baldwin's**
  rules, 0810
enolate anions
  acids derivatives, 0823
  acylation, 0811
  additive for alkylation, 0804
  aggregation state of metal ions, 0793
  aggregation state, E/Z ratios, 0795
  aggregation, 0791, 0796
  alkylation and modeling, 0807
  alkylation, 0572, 0803
  alkylation, and HMPA, 0804
  alkylidene, 0870
  alkylidene, with alkyl halides, 0871
  amide, E/Z isomers, 0792
  amide, with aldehydes or ketones,
    0824
enolate anions
  and [3,3]-sigmatropic
    rearrangement, 1137, 1138
  and **acetoacetic ester synthesis**, 0808

and acid-base chemistry, 0797
and acyl addition, 0816
and acyl cyanides, 0811
and alcohol solvents, 0800
and alkoxides, 0800
and alkyl halide reactivity, 0810
and anhydrides, 0812
and anionic **oxy-Cope**
  rearrangement, 1130
and **Baldwin's rules**, 0570, 0571
and chelation control, 0868
and **Claisen condensation**, 0791
and Claisen rearrangement, 1137,
  1138
and conjugate acids, 0798
and conjugate addition, 0112
and DABCO, 0808
and dimsyl sodium, 0558
and dissolving metal reductions, 0458
and **Doebner condensation**, 0830
and enols, 0783
and equilibrium, 0800
and HMPA, 0547, 0808, 1138
and HMPT, 0804
and HSAB, 0105
and ionic character, 0800
and ionization potential maps, 0846,
  0865
and K hexamethyldisilazide, 0812
and keto-esters, 0802
and **Knoevenagel condensation**, 0829
and molecular modeling, 0806
and **Nagata's reagent**, 0811
and organocuprates, 0720
and reaction time, 0800
and silyl triflates, 0840
and stereoselectivity, 0850
and steric hindrance, 0810
and synthetic equivalents, 0009
and the **aldol condensation**, 0790,
  0802, 0816
and the **Bürgi-Dunitz trajectory**,
  0795
and the **Claisen condensation**, 0824
and the **Dieckmann condensation**,
  0828
and the **Felkin-Anh model**, 0860
and the **Haller-Bauer reaction**, 0803
and the **Henry reaction**, 0830
and the **Johnson-ester variant**, 1137
and the **Kamlet reaction**, 0830
and the **malonic ester synthesis**, 0807

and the mixed **aldol condensation**,
  0817, 0818, 0819
and the **Noyori Open-Chain** model,
  0858
and the **Perkin reaction**, 0827
and the **Perkin salicylic synthesis**,
  0808
and the retro-**aldol reaction**, 0790
and the **Zimmerman-Traxler** model,
  0852, 0860
and transesterification, 0809
and vinylogy, 0812
enolate anions
  anhydrides, with aldehydes, 0824,
    0827
  anhydrides, with ketones, 0824
  by Grignard addition to conjugated
    carbonyl, 0651
  carboxylic acids, 0835
  charge delocalization, 0799
  condensation reactions, 0816
  conformations, 0063
  cyclohexane, 0872
  cyclopentanone, 0872
  definition, 0781
  deuteration, 0807
  dianions, 0802
  dianions. *See also* enolate dianions
  dianions, nucleophilicity, 0803
  diastereoface selection, 0862
  diastereoselectivity, 0556
  double stereodifferentiation, 0862
  E/Z isomers, 0790, 0791, 0853, 0854,
    0855
  E/Z isomers, and the **Ireland model**,
    0791, 0792
  E/Z isomers, correlation with base,
    0791, 0792
  E/Z selectivity, 0858
enolate anions, esters, 0791
  and the **Zimmerman-Traxler**
    Model, 0856
  E/Z isomers, 0792
  syn/anti isomers, 0855
  with aldehydes or ketones, 0823
  with TMSCl, 0826
enolate anions, formation, 0781
  of alkylthio ketones, 0814
  of enol acetates, 0820
  gas phase, 0799
  regioselectivity, 0798, 0799
  stereoselectivity, 0859

enolate anions
from amides, 0683
from cleavage of 1,3-diketones, 0176
from enol acetates, 0815
from **Grignard reagents**, 0669, 0670
from ketones, 0558
from lactams, 0683
enolate anions, gegenion and reactivity, 0810, 0811
enolate anions, geometry
and stereochemistry, 0795
and diastereoselectivity 0853, 0854, 0855
and enolate transition state, 0854, 0855
and ketone structure, 0790
of amides, 0792
of esters, 0792
enolate anions
halogenation, 0883
HMPA as a complexing agent, 0805
homoenolates, 0691
influence of metal cations, 0800
intramolecular alkylation, 0801, 0808
ionization potential maps, 0806, 0807
Ireland-Claisen rearrangement, 1138
K_a, 0797
ketones, 0791
kinetic and thermodynamic control, 0795
kinetic, 0571, 0791
kinetic, and **Zimmerman-Traxler** model, 0854, 0855
kinetic, definition, 0796
lactams, with aldehydes or ketones, 0824
lactones, with aldehydes or ketones, 0824, 1370
mechanism of equilibration, 0860
molecular modeling, 0797
nitrile reduction, 0359
nitrile, 0832
nitro compounds, 0830
enolate anions
O versus C alkylation, 0810, 0811
O-alkylation with alkyl halides, 0813
O-alkylation with anhydrides, 0813
O-alkylation, 0812
enolate anions
of acid derivatives, 0802
of aldehydes, 0802
of amides, 0802

of esters, 0802
polyalkylation, 0805
protic versus aprotic solvents, 0798
reaction with alkyl halides, 0809
reaction with butyllithium, 0802
reactivity and steric hindrance, 0846
reactivity, 0826
regioselectivity, 0805
relative energies, 0799
resonance, 0782
retro aldol, 0860
role of base, 0799
simple diastereoselection, 0862
solvent effects, 0798
stability and molecular modeling, 0799
stability, 0790
succinic ester, 0833
syn/anti isomers, 0858
thermodynamic, 0571, 0791
thermodynamic, and **Zimmerman-Traxler** model, 0854, 0855
thermodynamic, definition, 0796
unsymmetrical ketones, 0796
enolate anions
with acid chlorides, 0826
with aldehydes or ketones, 0547
with alkyl halides, 0007, 0801
with alkyl halides, diastereoselectivity, 0850, 0851
with benzenesulfenyl chloride, 0814
with carbon dioxide, 0800
with carbonyl compounds, 0007
with chlorotrialkylsilanes, 0815
with chlorotrimethylsilane, 0837, 0858, 1120
with cyclopentadienyl dicarbonyl and alkenes, 1249
with epoxides, 0813
with iminium salts, 0952
with iodomethane, 0805
with iodomethane, 0851
with MoOPh, 0813
with oxaziridines, 0814
with Pd(0), 1232
with sulfonyloxazidine, 0814
with the **Davis reagent**, 0814
enolate anons, energy, 0797
enolate condensation, and amide reduction, 0357
enolate dianions, 0834
and **Carroll reaction**, 1136
with alkyl halides, 0835
with epoxides, 0813, 0835, 0836

with esters, 0835, 0836
with ketones, 0835, 0836
enolates anions
and **Stobbe condensation**, 0833
and the **Thorpe reaction**, 0832
cyclohexane carboxylic acid, 0871
cyclohexanecarboxaldehyde, 0871
E/Z isomers, 0858
enolates boron, and **Zimmerman-Traxler** model, 0856
enolates. *See* enolate anions
enolates, boron, 0533, 0841, 0841
and 9-BBN, 0842
and the **aldol condensation**, 0841, 0842, 0856
and B-chloro diisopino-campheylborane, 0843
and boron triflates, 0842
diastereoselectivity, 0841, 0856
from ketones, 0842
oxidation, 0841
stereoselectivity, 0842
with aldehydes, 0842
with ketones, 0843
enolates, intramolecular condensation, 0828
enolates, nitrile
alkylation, 0832
intramolecular reaction, 0833
Thorpe reaction, 0832
Thorpe-Ziegler reaction, 0833
with aldehydes or ketones, 0832
with nitriles, 0832
enolates, nitro, with sulfuric acid, 0831
enolendo, and **Baldwin's rules**, 0571
enol-enol tautomerism, 0749
enolexo, and **Baldwin's rules**, 0571
enolization, and amide bases, 0789
enol-keto tautomerism, 0749, 1265, 1327
enols
and dissolving metal reduction, 0454
and enolate anions, 0783
and hydroboration, 0518
and keto-esters, 0826
and lead tetraacetate, 0323
and **Norrish Type II** fragmentation, 1265
and **oxy-Cope rearrangement**, 1130
and the **Claisen reaction**, 0826
by hydroboration of alkynes, 0528
energy, 0797
from aldehydes, 0528

from cyanohydrins, 1183
keto-enol tautomerization, 1130
of aldehydes, 0518
enophile, definition, 1143
enophiles
and the ene reaction, 1143
electronic effects, 1024
enterodiol, 0354
enthalpy
and carbocations, 1166
and conformation, 0041, 0057-0059
and lactones, 0574
cycloalkanes, 0050
magnitude, 0059
entropy
and free energy, 0058
and conformation, 0057
and cyclization, 0573
and lactones, 0574
and macrocyclizations, 0573
ent-Sch 47554, 0766
envelope conformation, 0048
enyl transition states, 0720
en-ynes, hydrogenation, 0438
enzymatic **Baeyer-Villiger reaction**, 0311
enzymatic cleavage,
and combinatorial chemistry, 0980
combinatorial chemistry, 0981
enzymatic resolution, 0945
enzyme reductions, 0479
enzyme reductions. *See also* reduction
enzyme-linked immunosorbent assay (ELISA), 0981
enzymes
and aryl dihydroxylation, 0303
and the **Baeyer-Villiger reaction**, 0311
cholesterol estrase, 0945
cyclohexanone oxygenase, 0311
hydrolysis, 1356
pig liver esterase, 1181
Pseudomonas putida, 0304
reduction, 0398
ephedradine A, 1329
ephedrine, and organocuprates, 0724
epi-12-methyl prostaglandin F_2, 1141
epiaustraline, 0990
epibatidine, 0122
epichlorohydrin, 0546
epicholesteryl methyl ether, 0435
epi-indolizidine 223A, 0720
epijuneol, 1092

epijunicedranol, 1173
epiquinamide, 0995
episulfones, and the **Ramberg-Bäcklund reaction**, 0701
epitopes, and combinatorial chemistry, 0986
epi-zephyranthine, 1018
epothilone, 0900, 0997
epoxidation
and conformation, 0278, 0279
and conjugation, 0282, 0283
and density ionization map, 0278
and dioxiranes, 0290
and epoxy alcohols, 0549
and HSAB, 0104
and hydroperoxide anion, 0275
and ionization potential maps, 0281
and **Jacobsen hydrolytic kinetic resolution**, 0290
and mcpba, 0281
and perphthalic acid, 0283
and salen catalysts, 0288
and sodium hypochlorite, 0289
and stereoselectivity, 0284
and structure of alkenes, 0272
and substituent effects, 0282
and the **Baeyer-Villiger reaction**, 0279
and the **Jacobsen catalyst**, 0289
and the **Prilezhaev (Prileschajew) reaction**, 0277
epoxidation
asymmetric, 0284-0291, 0291
asymmetric, chiral complex, 0285
asymmetric, **Sharpless asymmetric epoxidation,** 0284
in water, 0278
Jacobsen-Katsuki asymmetric epoxidation, 0288-0290
Juliá-Colonna epoxidation, 0276
mcpba and stereoselectivity, 0282
mechanism of vanadium catalyzed, 0274
model for **Sharpless asymmetric epoxidation,** 0285. 0287
neighboring group effects, 0560
epoxidation
of alkenes, catalysts and hydroperoxide, 0272
of alkenes, catalysts and peroxides, 0270
of alkenes, peroxyacid, electronic effects, 0280

of alkenes, peroxyacid, mechanism, 0279, 0280
of alkenes, peroxyacid, steric effects, 0280
of allylic alcohols, 0560
of dienes with mcpba, 0281
of monosubstituted alkenes, 0281
rate of vanadium catalyzed, 0273
Sharpless asymmetric epoxidation, 0284
stereoselectivity of vanadium catalyzed, 0274
stereoselectivity, 0869
with hydroperoxide anion, 0275
with mcpba, 0558
with peroxyacids and buffers, 0277
with peroxyacids, 0277-0284
with peroxyacids, **Henbest rule**, 0283
with peroxyacids, neighboring group effects, 0280
epoxides. *See also* oxiranes
epoxides
acyl, from conjugated carbonyls, 0275
and aluminum borohydride, 0374
and **Baldwin's rules,** 0569
and chelation control, 0549
and hydrogenolysis, 0443
and magnesium halides, 0655
and oxidation number, 0222
and oxygen-stabilized carbocations, 0656
and S_N2 reactions, 0128, 0268
and the **Baeyer-Villiger reaction,** 0307
and the **Benkeser reduction,** 0462
and the **Schlenk equilibrium,** 0655
and **Wilkinson's catalyst,** 0443
and **Williamson ether synthesis,** 0268
by chromium oxidation of alkenes, 0316
by **Darzens' glycidic ester condensation,** 0834
carboalkoxy, rearrangement, 0282
dissolving metal reduction, 0458
electrolytic reduction, 0471
from alkenes, 0268-0291, 0269, 0321, 0549, 0655
from halohydrins, 0268
Grignard reactions, and carbocations, 0656

Grignard reactions, and Lewis acids, 0655
halo, rearrangement, 0282
hydrogenation, 0443
hydrolysis, 0548
hydrolysis, and stereoselectivity, 0548
hydrolysis, to diols, 0303
in S_N1 reactions with Grignard, 0656
reaction with azides, 0128
reduction, and regioselectivity, 0375
reduction, with LiAlH$_4$, 0358
regioselectivity, Grignard reactions, 0654
regioselectivity, reduction, 0386
stereoselectivity of reduction, 0367
steric hindrance and reduction, 0367
via ozonolysis, 0321
with acid and aqueous DMSO, 0548
with alkoxyaluminum hydrides, 0364
with alkyne anions, 0635
with aqueous hydroxide, 0548
with BF$_3$, 0635
with dianions, 0836
with enolate anions, 0813
with enolate dianions, 0813, 0835, 0836
with **Grignard** reagents and copper salts, 0655
with **Grignard** reagents, 0654
with **Lewis** acids, 0129, 1177
with lithium borohydride, 0373
with NaCN or KCN, 0624
with organocuprates, 0718
with organolithium reagents, 0688
with Red-Al, 0367
with sulfur carbanions, 0698
with super hydride, 0379
with trimethylaluminum, 1177
with zinc borohydride, 0375
epoxy alcohols
and epoxidation, 0549
and **Sharpless asymmetric epoxidation**, 0549
epoxy-amides, from nitriles, 0276
epoxy-esters
by **Darzens' glycidic ester condensation**, 0834
epoxy-alcohols, rearrangement, 1182
epoxy-alcohols, to xanthate esters, 0168
epoxycholestane, 0655
epoxy-ethers, rearrangement, 1182
epoxy-ketones, hydrogenation, 0443

epoxy-oxonium salts, rearrangement, 1177
epoxyquinol A, 0186
equatorial and axial hydrogen, 0049
equilenin, 0897
equilibration
and aldol transfer, 0860
of **Grignard reagent**s, 0657
of oxaphosphatanes, 0738
syn/anti **aldol condensation**, 0859
equilibria, in acids-bases reactions, 0081
equilibrium
alkene metathesis, 1335
and electrocyclic reactions, 1097
and enolate anions, 0571, 0800
and **Grignard reagents**, 0637
and homolytic cleavage, 1260
and ylids, 0736
constant, and proton affinity, 0786
geometry, and the **Diels-Alder reaction**, 1019
hydroxy acids and lactones, 0575
in **Grignard reagents**, 0657
potential, and **Faraday's constant**, 0223
potential, and oxidation, 0222
rotational, and diradicals, 1323, 1324
eremantholide, 0806
erythrinadienone, 1297
erythro and threo, definition, 0025
erythromycin B, 0579
erythronolide A, 0597, 0958
erythronolide, 0718
Eschenmoser ring cleavage, 0564
Eschenmoser variant, 1137
Eschenmoser-Claisen rearrangement, 1137
esperamicin, 1295
ester carbenes, 1327
esterification, with diazomethane, 1321
esters
allylic, catalytic palladium(0), 1232
allylic, with Pd(0), 1232
esters
and acyl substitution, 0109
and **acyloin condensation**, 1306
and aluminum borohydride, 0374
and **Friedel-Crafts acylation**, 1205
and ketenes, 1316
and **Knoevenagel condensation**, 0829
and **Krapcho decarboxylation**, 0173
and **McMurry olefination**, 1310
and the **Johnson-ester variant**, 1137

and the **Baeyer-Villiger reaction**, 0307-0313
and the **Barbier-Wieland degradation**, 0648
and the **Claisen condensation**, 0824, 0826, 1136
and the **Claisen rearrangement**, 1137, 1138
and the **Dieckmann condensation**, 0828
and the **Doebner condensation**, 0830
and the **Grignard reaction**, 0317
and the **Ireland-Claisen rearrangement**, 1138
and the **Miescher degradation**, 0317
and the **Mitsunobu reaction**, 0545
and the mixed Claisen condensation, 0825
and the **Petasis reagent**, 0756
and the **Tebbe reagent**, 0755
esters
benzoates, leaving group in S_N2' reactions, 0126
by **Arndt-Eistert synthesis**, 1316
by oxidaton of alcohols, 0232
chlorosulfinate, 0544
complex formation with **Grignard** reagents, 0648
esters, conjugated
and **Baylis-Hillman reaction**, 0878, 0879
and hydroperoxide anions, 0275
formaldehyde, 0879
from conjugated esters, 0879
Grignard addition, enantioselectivity, 0653
HOMO and LUMO, 1002, 1003
LiAlH$_4$ reduction, 0354
with acyloxyborohydrides, 0378
with magnesium, 0454
with nitrones, 1108
with organocuprates, 0722
esters
dithioalkyl phosphonate, 0709
electrolytic reduction, 0472
enol, and **McMurry olefination**, 1310
enol, from esters, 1310
esters, enolate anions, 0791, 0802
alkylation, 0805
and the **Claisen condensation**, 0824
E/Z isomers, 0792
geometry, 0792
self-condensation, 0825

with aldehydes, 0824
with ketones, 0824
esters
  epoxy, rearrangement, 0282
  formate, with peroxyacids, 0311
  formation of ketyls, 1306
esters
  from aldehydes, 0255
  from alkyl halides, 0760
  from carboxylic acids and
    diazomethane, 1321
  from diazomethane and acids, 0110
  from ketones, 0311
  halo, and **Darzens' glycidic ester
    condensation**, 0834
  halo, with zinc, 0885
  hydrolysis with Amberlyst A21, 0545
  hydrolysis, 0310, 0605
  kinetic control and the **Claisen
    condensation**, 0825
  malonic, and decarboxylation, 0172
  menthyl, and **Grignard reagents**,
    0653
  mesitoate, and **Grignard reagents**,
    0648
  non-enolizable, 0803
  ortho, and Claisen rearrangement,
    1136
  ortho, with allylic alcohols, 1136
esters, phosphonate
  and Hünig's base, 0741
  and potassium hexamethyldisilazide,
    0740
  with organolithium reagents, 0740
  ylids, 0739
  ylids, and crown ethers, 0740
  ylids, and metal salts, 0740
  ylids, and Triton B, 0741
  ylids, stereochemistry, 0740
esters, pK_a, 0782
esters
  reaction with alkylboranes, 0527
  reduction to aldehydes, 0366
  reduction with LiAlH₄, 0353, 1068
  saponification, 0310, 0725, 0829,
    0834, 0959
  selenous enol, 0328
  S_N2 reaction with **Grignard**
    reagents, 0648
  sterically hindered, with **Grignard**
    reagents, 0648
  succinic enolate, with aldehydes,
    0833

succinic enoalte, with ketones, 0833
succinic, and **Stobbe condensation**,
  0833
succinic, enolate anions, 0833
esters, sulfonate
  and hydrogenolysis, 0361
  and Red-Al, 0368
  reduction with LiAlH₄, 0361
  solvolysis, 1174
  triflate, with organocuprate reagents,
    0717
  with alkyl anions, 0629
  with NaCN or KCN, 0623, 0624
  with organocuprates, 0716
  with sodium azide, 0544
  with sodium cyanoborohydride, 0383
esters,
  thermodynamic control , 0826
  thio. *See* thioesters
  trichloroacetimidic, 1052
  vinyl, and lead tetraacetate, 0323
  vinyl, epoxidation, 0282
esters,
  with alcohols, 0110
  with aldehydes, and the **Claisen
    reaction**, 0826
  with dialkylamide bases, 0794
  with dianions, 0836
  with Dibal, 0389
  with enolate dianions, 0835, 0836
  with **Grignard reagents**, 0110, 0648
  with ketones, and the **Claisen
    reaction**, 0826
  with LDA, 0825, 0841, 1136
  with LiAlH₄, 0605, 0606
  with lithium borohydride, 0373, 0374
  with lithium tri-*tert*-butoxy-
    aluminum hydride, 0366
  with magnesium borohydride, 0374
  with NaBH₄, 0370
  with organolithium reagents, 0683
  with phosphine oxide carbanions, 0674
  with Red-Al, 0367
  with sodium, 1306
esters, xanthates
  and **Barton deoxygenation**, 0467
  and the **Barton-McCombie reaction**,
    0168, 0467
  from epoxy-alcohols, 0168
  pyrolysis, 0168
estradiol, conformations, 0070
estrone, 0897, 0931, 0961, 1074
  bondsets, 0963, 0964

retrosynthetic analysis, 0963
SYNGEN analysis, 0963
synthesis, 0964
Étard reaction, 0329
  and chromyl chloride, 0332
ethane
  conformations/energy barriers, 0038
  rotamers 0038
  substituted, conformations, 0040
  substituted, energy barrier, 0040
  substituted, rotamers, 0040
ethene, frontier orbitals, 1002
ether, 2-(trimethylsilyl)ethoxymethyl
  (SEM)
  alcohol protecting group, 0598
ether, chloromethyl methyl, with
  alcohols, 0596
ether, diethyl
  and **Grignard** reagents, 0638, 0639
  and organocuprates, 0714
  solvent effect on acidity, 0088
ether, ethoxyethyl, alcohol protecting
  group, 0600
ether, methoxyethoxy chloromethyl
  with alcohols, 0598
ether, methoxyethoxymethyl (MEM)
  alcohol protecting group, 0598
ether, methoxymethyl (MOM) alcohol
  protecting group, 0596
ether, methyl vinyl, **Diels-Alder**, 1032
ether, methylthiomethyl (MTM)
  alcohol protecting group, 0597
ether, tetrahydropyranyl, alcohol
  protecting group, 0599
ethers
  acidity, 0673
  alkenyl, and umpolung, 0708
  alkenyl, with organolithium reagents,
    0708
  allyl vinyl, [3,3]-sigmatropic
    rearrangement, 1132
  allyl vinyl, **Claisen** rearrangement,
    1132
  allylic, and [2,3]-sigmatropic
    rearrangement, 1121
  allylic, and **Wittig** rearrangement, 1121
  and LiAlH₄ solubility, 0349
  and organolithium reagents, 0673
  and phase transfer catalysts, 0594
  and tetrabutylammonium iodide, 0594
  aryl allyl, **Claisen rearrangement**,
    1133
  as a solvent with alkylboranes, 0491

ethers, benzyl
  and hydrogenolysis, 0431, 0595
  dissolving metal reduction and
    hydrogenolysis, 0456
  hydrogenation, 0595
  protecting groups for alcohols, 0595
  sodium and ammonia, 0595
ethers
  by **Ullmann reaction**, 1301
  cleavage with Lewis acids, 0595
  conformations, 0043
  crown. *See* crown
  cyclic, and **Baldwin's rules**, 0569,
    0570
  cyclic, by radical cyclization, 1288
  deprotonation by organolithium
    reagents, 0673
ethers, enol. *See* enol ethers
ethers, enol
  and tert-butyllithium, 0707
  and umpolung, 0707
  with carbenes, 1326
ethers
  ethyl vinyl, with alcohols, 0600
  from alcohols with dimethyl sulfate,
    0594
  from alcohols, 0545
  from alkenes, 0183
  from alkyl halides, 0127
  from **Meerwein's reagent** and
    alcohols, 0594
  lithio, fragmentation, 0673
  methyl, alcohol protecting groups,
    0594
  pK_a, 0082
  reaction with tosic acid, 0136
ethers, silyl enol. *See also* silyl enol
ethers, silyl enol
  and the **Diels-Alder reactions**, 1050
  and the **Ireland-Claisen**
    rearrangement, 1138
  and the **Mukaiyama aldol reaction**,
    0837
  and the **Simmons-Smith reaction**,
    1332
  halogenation, 0883, 0884
  hydrolysis, 0354, 1120
  with fluoride, 1094
  with **Lewis acids**, 0838
ethers
  silyl, with fluoride, 0601
  silylalkyl, 1065

solubility of alkoxyaluminum
  hydrides, 0365
solvent for elimination, 0167
stability with organolithium reagents,
  0673
tert-butyl, protecting groups for
  alcohols, 0596
tert-butyldimethylsilyl, alcohol
  protecting groups, 0603
tert-butyldiphenylsilyl, alcohol
  protecting groups, 0604
tetrahydropyranyl, 0590
triethylsilyl, alcohol protecting
  groups, 0602
triisopropylsilyl, alcohol protecting
  groups, 0602
triisopropylsilyl, with alcohols, 0602
trimethylsilyl, alcohol protecting
  groups, 0601
trimethylsilyl, from alkoxides, 0749
via alkoxymercuration, 0183
via S_N2 reaction, 0143
ethers, vinyl
  from lactones, 0755
  hydrolysis, 0733
  with radicals, 1269
  and lead tetraacetate, 0323
  epoxidation, 0282
  from esters, 0755
  HOMO and LUMO, 1269
  hydrolysis, 0552, 1049
  Simmons-Smith reaction. 1332
  with **Grignard reagents**. 1246
ethers
  with BBr$_3$, 0595
  with **Grignard reagents**, 0638, 0639
  with trimethylsilyl iodide, 0594
ethoxide
  with ketones, 0082, 0822
  with ketones, 0799
ethoxyethyl ethers. *See* ether
ethoxyethyne, and lactonization, 0581
ethyelenedioxy acetals, 0610
ethyelenedioxy ketals, 0610
ethyl chlorooximinoacetate, 1106
ethyl diazoacetate
  [3+2]-cycloaddition, 1111
  with methacrolein, 1111
ethyl radical. *See* radicals
ethylaluminum dichloride, catalyst
  for the ene reaction, 1150, 1034
ethyldiisopropylamine. *See* Hünig's
  base

ethylene glycol
  and **Bamford-Stevens reaction**, 1317
  and **Diels-Alder** reaction, 1042
  conformations, 0043
ethylene
  and metathesis, 1337
  molecular orbitals, 1000
ethylenediaminetetraacetic acid. *See*
  EDTA
ethyllithium, 0670
  crystal structure, X-ray, 0671
  with fluorene, 0671
ethylpyridinium, ionic liquid, 1037
eucannabinolide, 0956, 0957
eudesmanes, 0499
eupenoxide, 0990
European convention, and oxidation,
  0223
europium catalyst in the **Diels-Alder**
  reaction, 1072
eutipoxide B, 1070
Evans' auxiliaries, 0869
  and imine carbanions, 0869
  and the **aldol condensation**, 0865,
    0869
  and the **Diels-Alder** reaction, 1070,
    1071
Evans' Model, 0857
  and the **Zimmerman-Traxler** model,
    0857
  and the **Bürgi-Dunitz trajectory**,
    0857
exchange
  organolithium and alkyl halides, 0676
  organolithium -halide, 0688
exciplex, and photochemistry, 1088
excited states, and photochemistry,
  1082, 1088
exo/endo, and **Baldwin's rules**, 0564,
  0567
exo-tet, and **Baldwin's rules**, 0571
exo-trig, and **Baldwin's rules**, 0571
extended conformation, 0041
extended conformation, and syn/anti,
  0025
extended tetrahedron, 0026, 0028
extrusion
  and triazoles, 1113
  in radical reactions, 1265
  of carbon monoxide, 0759
  of nitrogen from diazonium salts,
    1298, 1299
  of SO$_2$, 0701

of N₂, from triazolines, 1114

F

F2-isoprostanes, 1341
facial selectivity
  and chiral auxiliaries, 1067
  and steric hindrance, 1074
  of chiral acrylates, 1070
Fajans' rules, and Lewis acids, 0084
Faraday's constant
  and equilibrium potential, 0223
  and oxidation, 0223
faranal, 0438
farnesol, 0354
fasicularin, 1046, 1056
favored/disfavored, and **Baldwin's**
  rules, 0564, 0567
Favorskii rearrangement, 0887
fawcettimine, 0363
FD-891, 0288, 0996
Felkin-Ahn model, 0037, 0409, 0411,
  0774
  and allyltin, 0767
  and enolate anions, 0860
  and LUMO maps, 0410
  and the **Grignard reaction**, 0662
  comparison with LUMO maps, 0410
  reduction of cyclic ketones, 0412
Fenton's reagent, 0266, 1296
ferrate anion, cyclopentadienyl
  dicarbonyl, 1248
ferric acetonylacetonate, and **Grignard**
  reagents, 0643
ferric chloride
  and **Friedel-Crafts alkylation**, 1202
  and the **Diels-Alder reaction**, 1074
  with **Adams' catalyst**, 0439
  with dienones, 1188
  with Grignard reagents, 0647
ferric iodide, and the **Diels-Alder**
  reactions, 1073
ferric salts, with **Grignard** reagents,
  0640, 0641
ferricyanide, potassium
  and dihydroxylation, 0300
  and **phenolic oxidative coupling**,
  1296
ferrocene, 0759
  and **Ullmann reaction**, 1300
ferrocenyl derivative, and
  hydrogenation, 0450

ferrous chloride, and
  chlorolactamization, 0187
ferrous ion
  and ascorbic acid, 0266
  and **Fenton's reagent**, 0266, 1296
Fétizon's reagent 0256
  mechanism of alcohol oxidation, 0257
  solvents effects, 0257
  with diols, 0257
field effects
  and acidity, 0085
  and basicity, 0089
  model, and acidity, 0084
filters
  and photochemistry, 1262, 1263
  photochemical, 1085, 1086
Finkelstein reaction, and alkyl halides,
  0149
Fischer Indole Synthesis, 1221
  and imine-enamine tautomerism, 1222
  and named reactions, 1222, 1223
  and rearrangements, 1223
  and the **Borsche-Drechel cyclization**,
  1218
  and the **Claisen rearrangement**, 1222
  formation of isomers, 1223
  powerful in synthesis, 0930, 0932
  regiochemistry and stereochemistry,
  1222, 1223
  with acetals, 1222
Fischer projection, 0015, 0024
  and the E2 reaction, 0154
fiscuseptine, 1243
flagpole hydrogen, 0049
flash vacuum pyrolysis. *See* FVP
flavocommelin, 0990
flavone 8-acetic acid, 0905
fluorene, with ethyllithium, 0671
fluorenyl radicals, 1270
fluorescence spectrum, and
  photochemistry, 1088
fluorescence
  and photochemistry, 1082, 1086
  and radiative deactivation, 1086
fluoride ion
  and silyloxybutadienes, 1049
  with silyl ethers, 0601
fluoride, potassium
  and H_2O_2, **Tamao-Fleming** oxidation,
  0331
fluoride
  with vinyl silanes, 1185
  tetrabutylammonium. *See* TBAF

with allylic silanes, 0770
  with silyl enol ethers, 1094
fluorides, alkyl
  and **Friedel-Crafts alkylation**, 1194
fluorine, pK_a, 0731
fluoroform, supercritical, 0839
fluoroprogestin, 0294
fluorous compounds, and the
  Mitsunobu reaction, 0125
fluxional inversion
  amines, 0020
  energy barriers, 0020
  of phosphines, 0449
  phosphines, 0020
FMO. *See also* molecular orbital
FMO, 1001
  and the **Alder endo rule**, 1030
  and the **Diels-Alder reaction**, 1029
forbidden reactions, 1010
formal charge, and ylids, 0731
formaldehyde
  and cyanoborohydride, 0384
  and the **Prins reaction**, 1191
  with alkenes, 1191
  with amines, 0355, 0384, 1217
  with formic acid, 0355
  with formic acid, and methylation,
  0476
formamide, and formic acid, with
  ketones, 1216
formamides
  dehydration to nitriles, 0625
  dehydration, 0626
formate anions
  and iminium salts, 1184
  and the **Dakin reaction**, 0311
formate, ammonium with Pd/C, 0477
formates, reaction with carbocations,
  1183
formation of radicals, 1259
formic acid
  and formaldehyde, 0355
  and formamide, with ketones, 1216
  and hydride transfer, 0476
  reduction with, 0476
  reduction, and Cram's rule, 0476
  the **Beckmann rearrangement**, 0189
  with alcohols, 1187
  with amines, 0355
  with formaldehyde, and methylation,
  0476
  with iminium salts, 0476, 1185
formylation, of amines, 0626

fostriecin, 0769
four component coupling, 0627
four-center transition state, metal-hydrogen exchange, 0689
FR182877, 0742
FR66979, 0768
FR900848, 0023
FR901483, 0036, 0649, 0820, 1109
fragmentation
 and radicals, 1270
 dissociative, 1264
 of 1,3-diketones, 0176
 of lithio-ethers, 0673
 of radicals, 1275
 of tosylhydrazones, 0564
Franck-Condon principle, and photochemistry, 1087
frangomeric effect, 0175
free energy
 and entropy, 0058
 and conformation, 0057-0059
 and oxidation, 0223
free radicals. *See* radicals
Fremy's salt, 0262, 1346
 and pH, 0262
 and radicals, 0262
 and steric hindrance, 0263
frenolicin, 0765
Friedel-Crafts acylation, 0929, 1203-1210
 and acylium ions, 1203, 1204
 and anhydrides, 1205
 and coumarins, 1206
 and esters, 1205
 and furan, 1205
 and heterocycles, 1205
 and HSAB, 1205
 and PBr_3-$AlCl_3$, 1206
 and $POCl_3$, 1208
 and pyrrole, 1205
 and silanes, 1206
 and succinic anhydride, 1208
 and sulfuric acid, 1206
 and tether length, 1205
 and the **Bally-Scholl synthesis**, 1208
 and the **Bogert-Cook synthesis**, 1209
 and the **Bradsher reaction**, 1210
 and the **Darzens tetralin synthesis**, 1209
 and the **Haworth phenanthrene synthesis**, 1207, 1208
 and the **Pictet-Spengler isoquinoline synthesis**, 1216

and the **Vilsmeier-Haack** reaction, 1206
 and thiophene, 1205
 formation of coumarins, 1206
 intramolecular, 1205
 ortho/para isomers, 1204
 powerful in synthesis, 0932
 regioselectivity, 1204, 1206
 with lactones, 1205
Friedel-Crafts alkylation, 0007, 0203, 1194-1203, 1251
 alkyl halide reactivity, 1195
 and activating aryl substituents, 1195
 and $AlCl_3$, 1202
 and alcohols, 1201
 and alkenes, 1201
 and amides, 1199
 and $BaCl_3$, 1202
 and BF_3, 1202
 and carbocations, 1196
 and deactivated aromatic rings, 1195
 and electronic effects, 1199
 and $FeCl_3$, 1202
 and HF, 1201
 and HF-SbF_5, 1199
 and isomerization, 1197
 and **Jacobsen rearrangement**, 1199
 and Lewis acids, 1200
 and polyalkylation, 1197, 1198
 and pyridine, 1199
 and racemization, 1196
 and rearrangement, 1197
 and sulfonate esters, 1201
 and sulfuric acid, 1202
 and tight ion pairs, 1196
 and triflates, 1201
 catalyst, 1196
 enantioselectivity 1196
 influence of catalyst, 1200, 1201
 intramolecular, 1202
 ortho effect, 1195
 regioselectivity, 0203, 1194
Friedel-Crafts cyclization, 1204
 and HF, 1208
 and $POCl_3$, 1216
 and sulfuric acid, 1209
 powerful in synthesis, 0929
 with alkenes, 1209
 with sulfuric acid, 1218
Friedel-Crafts reactions, 1192-1225
 alkylidene malonates, 1212
 and **Borsche-Drechel cyclization**, 1218

and **Fries rearrangement**, 1219, 1220
 and **Lewis acids**, 0095
 and rearrangements, 1219
 and the **Bardhan-Sengupta phenanthrene** synthesis, 1208
 and the **Bernthsen acridine** synthesis, 1218
 and the **Bischler-Napieralski** reaction, 1215
 and the **Bradsher reaction**, 1219
 and the **Combes quinoline** synthesis, 1213
 and the **Conrad-Limpach** reaction, 1212
 and the **Doebner reaction**, 1214
 and the **Doebner-Miller** reaction, 1214
 and the **Gould-Jacobs** reaction, 1212
 and the **Knorr quinoline** synthesis, 1211
 and the **Pomeranz-Fritsch** reaction, 1217
 and the **Skraup reaction**, 1215
 formation of enamino-esters, 1212
 formation of quinolines, 1211
 heteroatom, 1210-1225
Fries rearrangement, 1219, 1220
 and acylium ions, 1220
 and steric effects, 1220
 photo, 1220
Fritsch-Buttenberg-Wiechell rearrangement, 0700
frondosin C, 0474
frontalin, 0168
frontalin, 0683
frontier molecular orbital theory. *See also* FMO
frontier molecular orbital theory, 0999
 and the **Cope rearrangement**, 1124
 and **Lewis acids**, 1038, 1039
frontier orbitals, 0999
 and radical reactivity, 1268
fruit flies, gland secretion, 0590
fuchsiaefoline, 1216
fucosidase inhibitors, 1055, 1056
fulvalenes, and **Diels-Alder**, 1035
fulvalenes, with maleic anhydride, 1035
fumagillin, 0342, 0361
fumaric acid, bromination, 0034
functional group exchange reaction wheel, 0010
functional group exchanges, 0077

functional group interchange, 0001, 0003

functional groups
absorption maxima, 1084
and acidity, 0692, 0784, 0785
and asymmetric reduction, 0399
and bond dissociation energy, 1084, 1085
and bond polarization, 0011
and dissolving metal reduction, 0452
and electrostatic potential maps, 0785
and hydrogenation, 0425
and pK$_a$, 0784, 0785
and rearrangements, 0188-0193
and synthesis, 0918, 0925
and UV, 1083, 1084
bonds and transforms, 0080
correlation with pK$_a$, 0692, 0693
heteroatom, hydrogenation, 0438
reactivity with organocuprates, 0714

furan
and **Friedel-Crafts acylation**, 1205
and **Noyori annulation**, 1190
electrophilic aromatic substitution, 1193

furanones, 1243

furanose derivatives, 0769

furans
and hydrogenation, 0448
and retro-**Diels-Alder**, 1036
cyclopropanation, 1327
Diels-Alder reaction with benzyne, 1018
in the **Diels-Alder reaction**, 1018
intramolecular **Diels-Alder**, 1060
with borane, 0502
with carbenes, 1327

furansulfonic acid, and electrophilic aromatic substitution, 0196

furyllithium, 0684

FVP (flash vacuum pyrolysis), 0933
a cyclobutene ring opening, 1099
and the retro-**Diels-Alder reaction**, 1036
and vinylcyclopropanes, 1120
vinyl aziridines, 1114

G

G. *candidum*, 0481

Gabriel synthesis, 0125
and the **Ing-Manske modification**, 0125

galactose, as a chiral template, 0958
galanthamine, 0118, 1297
galbonolide B, 1138
Gallagher-Hollander degradation, 0317
gallium trichloride, and **Friedel-Crafts** alkylation, 1202

galvinoxyl
and boranes, 0536
and hydroboration, 0533
with alkylboranes, 0536

gambierol, 0612
Garcinia natural products, 1133
garsubellin A, 0607, 1133
gas phase enolate formation, 0799
Gattermann reaction, 0209

gauche conformations, 0039
and amines, alcohols, ethers, 0043

gauche interaction, 0056

GC
and combinatorial chemistry, 0984
mass spectrometry, and combinatorial chemistry, 0984

gegenion, enolate anions and reactivity, 0810, 0811

gegenions, and sulfonium salts, 0745
geissoschizine, 1185
geldanamycin, 0390, 0684
gelsamine, 0478, 0183

genetic tags, and combinatorial chemistry, 0986

geometry
and intramolecular **Diels-Alder**, 1059, 1060
and permanganate dihydroxylation, 0292
of anions, and stereochemistry, 0795

Geotrichum candidum, 0479
geranial, 0252, 0273, 0482
geraniol, 0238, 0252, 0284
germacradiene sesquiterpenes, 1097
germanium catalysts, and the **Reformatsky reaction**, 0887
gibberellins (GA$_{38}$), 0456
gibberellins, 0757
gibbilimbol A, 0363
gilbertine, 0172, 0932

Gilman reagents, 0714
and one electron transfer, 0714
preparation, 0714

Gilman titration, and organolithium reagents, 0675

gland secretion of fruit flies, 0590

Glaser reaction, 0631, 1301
and **Sonogashira coupling**, 1301

globulol, 0366

glucofuranose
and organocuprates, 0723
with LiAlH$_4$, 0393

glucopyranose, conformation, 0064
glucoride, and reduction, 0397

glucose, 0956, 0995
as chiral template, 0958
tetraacetyl, 0960

glutamic acid, 0956, 0993
glycerol, and the **Skraup reaction**, 1215

glycerols, and **Bally-Scholl synthesis**, 1208

glycidic anions, stereoselectivity of formation, 0869

glycosidase inhibitors, 0461
glycosylation, and amides, 0615
glyoxal semiacetal, 1218

glyoxalates
enzyme reduction, 0481
in the **Diels-Alder** reactions, 1054

glyoxals
by oxidant of halo ketones, 0240
from aldehydes, 0327, 0328

Gomberg-Bachmann reaction, 1299
goniofufurone, 0420
goniopedaline, 1287
goniothalamin, 0826
gossyplure, 0323, 0324
Gould-Jacobs reaction, 1212
grahamimycin A, 0248
grandisol, 0316, 0321, 0365, 0720, 1277
griffonianone D, 1243

Grignard reaction, 0636, 0774, 0944, 0957
1,2- versus 1,4-, 0652
and carbocation rearrangement, 0656
and **Cram's rule**, 0661, 0662
and four-center transition state, 0652
and **Lewis acids**, 0652
and LUMO maps, 0663
and **Meerwein-Ponndorf-Verley reduction**, 0667
and **Oppenauer oxidation**, 0667
and oxygen-stabilized carbocations, 0656
and S$_E$2' reactions, 0665
and S$_N$2' reactions, 0655
and the **Barbier reaction**, 0637
and the **Cornforth model**, 0662

and the **Cram cyclic model**, 0663
and the **Cram model**, 0662
and the **Felkin-Anh mode**l, 0662
and the **Karabatsos model**, 0662
complex formation with esters, 0648
cyclization, 0666
diastereoselectivity, 0660, 0663
mechanism of formation, 0637
rearrangements, 0665
reduction, 0667
stereoselectivity, 0645
with carbonyls, transition state, 0666
with esters, 0317
with ultrasound, 0637
Grignard reagents, 0620, 0636-0670, 1365
acyl substitution, 0646
aggregation state, 0637
α-halo, 0660
alkenyl, and E/Z isomers, 0642
alkenyl, stereochemistry, 0663, 0664
allylic, 0639, 0664
allylic, E/Z isomers, 0664, 0665
allylic, S_E2' rearrangement, 0664, 0665
Grignard reagents,
and allylic silanes, 0770
and amines, 0092, 0638
and asymmetric complexes, 0661
and carbenoids, 0658
and chelation control, 0663
and CO_2, 0003
and cobalt chloride, 0640, 0641
and coupling reactions, 0640
and dialkylmagnesium, 0638
and diethyl ether, 0638, 0639
and enolate anion formation, 0669, 0670
and ferric acetonylacetonate, 0643
and halide structure 0638, 0639
and HSAB, 0105
and hydroperoxide anion, 0639
and **Ivanov condensation**, 0853
and **Kumada-Corriu coupling**, 0657
and **Lewis acids**, 0653, 0655
and LUMO maps, 0660
and magnesium halides, 0653
and NMR, 0657
and organocerium reagents, 0668
and protected aldehydes, 0610
and radicals, 0637
and reduction of ketones, 0667
and single electron transfer, 0637

and single electron transfer, 0644
and synthetic equivalents, 0008, 0011
and tetrahedral intermediate, 0646
and the **Barbier-Wieland degradation**, 0317, 0648
and the **Bogert-Cook synthesis**, 1209
and the Kharasch reaction, 0640, 0679
and the **Miescher degradation**, 0317
and the **Schlenk equilibrium**, 0638
and the **Wurtz reaction**, 0640
and THF, 0638, 0639
and transition metals, 0640, 0641
Grignard reagents
aryl, formation of biaryls, 0642
as bases, 0786
basicity, 0639
basicity, and the halide counterion, 0639
bond polarization, 0636
configurational stability, 0547, 0657, 0662
conjugate addition, 0651
conjugate addition, and steric hindrance, 0653
conjugate addition, enantioselectivity, 0653
coupling with alkyl halides, 0640, 0641
crotyl, 0665
crotyl, E/Z isomers, 0665
diastereoselectivity with aldehydes, 0659
diastereoselectivity with carbonyls, 0644
diastereoselectivity with ketones, 0659
disproportionation, 0641
E/Z isomers, 0664, 0665
enantioselectivity with aldehydes or ketones, 0659
enantioselectivity, 0658
enolate formation versus addition, 0669
equilibration, 0657
equilibrium strictures, 0637
equilibrium, 0657
formation of alkynes, 0700
in dimethoxybutane, 0661
influence of the halide, 0638, 0639
isomerization, 0665
LUMO maps, 0645

mechanism of addition to ketones, 0644
monomer concentration, 0638
polarity inversion, 0636
prochiral centers, 0662
reaction with chiral sulfoxides, 0698
reduction of camphor, 0667
reduction versus addition, 0669
regioselectivity of reaction with epoxides, 0654
S_N1 with epoxides, 0656
S_N2 reaction with esters, 0648
solvent effects, 0638, 0639, 0661
stereochemistry, 0657
structure, 0637
syn/anti isomers, 0659
transition state for addition to carbonyls, 0645
transition state for conjugate addition, 0653
vinyl, 0639
Grignard reagents
with acetates, 0648
with acid derivatives, 0824
with acyl halides, 0646
with alcohols, 0639
with aldehydes, 0589, 0590, 0644, 1209, 1210
with alkenyl sulfoxides, 0700
with alkyl halides, 0640
with alkylboranes, 0535
with alkynes, 0632
with amides, 0649
with amino alcohols, 0662
with aryl halides, 0643
with cadmium chloride, 0647
with carbon dioxide, 0651
with Cbz groups, 0618
with cerium chloride, 0668
with chlorotrialkylsilanes, 0770
with conjugated aldehydes or ketones, 0652
with copper salts, 0653
with cuprous bromide, 0642
with cuprous salts, 0642
with cyclic ketones, 0682
with epoxides and copper salts, 0655
with epoxides, 0654
with esters, 0110, 0648
with ethers, 0638, 0639
with ferric chloride, 0647
with ferric salts, 0640, 0641
with halo-alkenes, 0665

with halo-ketones, 0643
with haloketones, 0883, 0884
with ketones, 0410, 0411, 0547, 0644, 0786
with lactams, 0649
with lactones, 0649
with Li$_2$CuCl$_4$, 0642, 0643
with menthyl esters, 0653
with mesitoate esters, 0648
with nitriles, 0624, 0650
with oxygen, 0639
with phenylisothiocyanate, 0658
with propargyl halides, 0633, 0634
with sparteine, 0661
with sterically hindered ester, 0648
with sulfoxides, 0658
with TMEDA, 0640
with transition metals, 0640, 0641, 0646, 0647
with vinyl ethers, 1246
with vinyl sulfides, 1246
with water, 0639
with Weinreb amides, 0649, 0742
with α-chloroketone, 0663
Grignard reagents
α-halo, 0658
Grob fragmentation, 0174, 0559, 0934
and conformation, 0174
and the frangomeric effect, 0175
radical induced, 0176
rate of reaction, 0175
ground state, definition, 1012
Grubbs catalyst, 1365, 1371
and ROMP, 1340
stability, 1336
Grubbs I catalyst, and metathesis, 1336
Grubbs II catalyst, 1337
G-strain, 0056
guaiane sesquiterpenes, 0748
guaiane, 0182
guanacastepene A, 0237
guanacastepene, 1091
guanacastepenes, 0637, 1327
guanidine bases, and phosphonium salts, 0731
guanidinium perchlorate and the **Diels-Alder reaction**, 1042
gulonolactone, 0946
gulose, 0974
gustastatin, 0124
gymnocin A, 0458
gymnodimine, 1029

H

Hajos diketone, 0882
half-chair transition state, 0060
half-reactions, and SYNGEN, 0967
haliclamine A, 0732
haliclamine, 0372
halide counterion, and basicity of **Grignard reagents**, 0639
halides, acyl. *See also* acid chlorides
halides, acyl
and carbocations, 1204
and diazomethane, 1328
and **Friedel-Crafts acylation**, 1204
and PCl$_5$, 0133
and the **Hell-Volhard-Zelinsky reaction**, 0150
and the **Lossen rearrangement**, 0193
bromo, and **Stollé synthesis**, 1224
reduction to aldehydes, 0365
with amines, 0359, 0875
with aryls, 0007
with dialkylcadmium, 0647
with diazomethane, 1316
with enamines, 0875
with **Grignard reagents**, 0646
with lithiated dithianes, 0705
with lithium tri-*tert*-butoxy-aluminum hydride, 0365
with nitriium salts, 1186
with organocuprates, 0718
with organolithium reagents, 0683
halides, alkenyl. *See also* halides, vinyl
halides, alkenyl,
and **Kumada-Corriu coupling**, 0657
with base, 0630
with organocuprates, 0716
with organolithium reagents, 0630, 0677, 0700
halides, alkyl
amides, enolate anions, 0805
and **acetoacetic ester synthesis**, 0808
and allylbutyltin, 1280, 1281
and back strain, 0134
and **Biellmann alkylation**, 0697
and chromyl chloride, 0757
and E2 reactions, 0079
and **Gilman reagents**, 0714
and halogenation of ketones, 0152
and HSAB, 0100
and ionization, 0129
and **Perkin salicylic synthesis**, 0808

and radical cyclization, 1283
and radicals, 1283
and Red-Al, 0368
and S$_N$1 reactions, 1162
and sodium, 0681
and **Stork enamine synthesis**, 0873
and sulfur carbanions, 0697
and synthetic equivalents, 0008
and the **Barbier reaction**, 0637
and the **Birch reduction**, 0461
and the **Borodin reaction**, 0150
and the E2 reaction, 0157
and the **Finkelstein reaction**, 0149
and the **Grignard reaction**, 0636
and the **Hunsdiecker reaction**, 0150
and the **Kharasch reaction**, 0640
and the **malonic ester synthesis**, 0807
and the **Menschutkin reaction**, 1045
and the **Meyer's** aldehyde synthesis, 0843, 0844
and the S$_N$2 transition state, 0114
and the thio-Claisen rearrangement, 0713
and the **Wittig-Gilman reaction**, 0676
and the **Wurtz reaction**, 0678
and tributyltin hydride, 1283
halides, alkyl
C versus O alkyltin with enolate anions, 0811
carboxylic acid enolate anions, 0805
coupling reactions, 0640, 0641
coupling with alkenes, 1280
coupling with allylnickel, 1246
coupling with **Grignard reagents**, 0640, 0641
coupling with organocuprates, 0714, 0715
coupling with organocuprates, 0715
coupling with organocuprates, mechanism, 0714
cyclohexanes and S$_N$2 reactions, 0558
diastereoselectivity, 0850, 0851
elimination to alkenes, 0077
enolate anions, 0805
enolate anions, and intramolecular alkylation, 0801
exchange with organolithium reagents, 0676
formation of carbocations, 1194
formation of radicals, 1266, 1279

halides, alkyl
  from alcohols and SOCl₂, 0543, 0544
  from alcohols, 0127, 0130, 0144, 0148, 0542
  from alkenes, 0178
  from allylic alcohols, 0149
  from amines, 0545
  from carbocations, 0144
  from CCl₄, alcohols and HMPT, 0147
  from **Grignard reagents**, and esters, 0648
  from LiBr and mesylates, 0147
  from triphenylphosphine and N-chlorosuccinimide, 0146
halides, alkyl
  ω-halo, hydrogenolysis, 0437
  hydrogenolysis, 0431
  hydrogenolysis, with LiAlH₄, 0360
  in water, 0131
  inversion, in S_N1 reactions, 1170
  inversion, with higher order cuprates, 0728
  ionization, 1162
  ionization, and steric strain, 0134
  lactams and lactones, enolate anions, 0805
  leaving groups ability, 0143
  nucleophiles versus bases, 0141
  O-alkylation with enolate anions, 0813
  oxidation with DMSO, 0240
  photolysis, 1266
  photoreduction, 0478
  polyalkylation with enolate anions, 0805
  radical coupling with allyltin compounds, 1280, 1281
  rate with KI, 0115
  rates of hydrolysis and solvolysis, 1167
  reaction with enolate anions, 0809
  reaction with phosphines, 0730
  reactivity and enolate anions, 0810
  reactivity in **Friedel-Crafts alkylation**, 1195
  reactivity order with organocuprates, 0715, 0716
  reduction, with LiAlH₄, 0360
  reduction, with tributyltin hydride, 1319
  regioselectivity with unsymmetrical ketones, 0805
  S_N2 reaction with azide, 0359

S_N2 reactions with higher order cuprates, 0727
S_N2 transition state modeling, 0115
S_N2 with cyanide, 0358
structure and **Grignard** formation, 0638, 0639
structure and substitution, 0140
substitution versus elimination, 0141
halides, alkyl
  with AlCl₃, 1194
  with alkenes, 1279, 1280
  with alkoxides, 0143, 0545
  with alkylidene enolate anions, 0871
  with alkyne anions, 0007, 0629, 0634
  with aryls, 0007
  with azide ion, 0359
  with butyllithium, 0672
  with carboxylic acid dianions, 0835
  with cobalt chloride, 0640, 0641
  with cyanide, 0007, 0623, 0624
  with cyanide, solvent effects, 0624
  with cyclopentadienyl anions, 1117
  with DMSO, 0240
  with enamines, 0873, 0874
  with enolate anions, 0007, 0572, 0801, 0803
  with **Grignard reagents**, 0640
  with higher order cuprates, 0727
  with hydrazone carbanions, 0848
  with KI, 1162
  with KI, in water, 0131
  with lithiated dithianes, 0705
  with lithiated vinyl sulfides, 0708
  with magnesium, 0636
  with metal cyanides, 0625, 0626
  with NaCN, 0145
  with nitrile enolates, 0832
  with organocuprates, 0007, 0714
  with organocuprates, solvent effects, 0716
  with organoiron compounds, 0761
  with organolithium reagents, 0676, 0677
  with oxime anions, 0848, 0849
  with phosphines, 0730
  with silver cyanide, 0625, 0626
  with sodium cyanoborohydride, 0383
  with sodium tetracarbonyl ferrate, 0760
  with sulfide carbanions, 0698
  with sulfides, 0745
  with sulfur carbanions, 0697
  with thiol anions, 0545

  with thiophenol anions, 0127
  with transition metals, 0640, 0641
halides, alkynyl, configuration of **Grignard reagents**, 0547
halides, allylic, 0151
  and triphenyltin hydride, 0679
  catalytic palladium(0), 1232
  formation of Ni(0) complex, 1246
  formation, 0151
  hydroboration, 0511
  S_N2· reactions with organocuprates, 0716
  with boranes, 0514
  with organocuprates, 0716
  with Pd(0), 1232
halides, aryl
  and benzyne, 0207
  and electrophilic substitution, 0194
  and **Gilman reagents**, 0714
  and Pd(0) coupling, 1235
  and Red-Al, 0368
  and substitution, nucleophilic aromatic, 0205, 0207
  and the **Darzens tetralin synthesis**, 1209
  and the **Rosenmund-von Braun reaction**, 0628
  and the **Ullmann reaction**, 1300
  coupling with alkenes, 1246
  coupling with alkyne silanes, 1302
  coupling with allylnickel, 1246
  coupling with malonates, 1235
  electrolytic reduction, 0472
  from aryl amines, 0209
  from aryl diazonium salts, 0209
  halide reactivity in the **Ullmann reaction**, 1300
  palladium(0) catalyzed reactions, 1237
  reduction with LiAlH₄, 0360
  with alkynes, 1244, 1302
  with arylcopper reagents, 1300
  with conjugated carbonyls, 1237
  with CuCN, 0628, 1210
  with cyanide, 0628
  with **Grignard reagents**, 0643
  with organocuprates, 0716
  with organolithium reagents, 0677, 1018
  with potassium amide, 0207
  with potassium trifluoroborates, 1243
halides
  as a poison for hydrogenation, 0426

as leaving groups, 0142, 0143
halides, benzylic, 0151
  formation, 0151
  hydrogenolysis, 0431
  with amines, 0613
  with carbanions, 0461
halides, cyclopropylcarbinyl
  and the S_N2' reaction, 0126, 0127
  with amines, 0127
  with bromine and PPh₃, 0147
  with organocuprates, 0128, 0716
halides
  dihalides. *See* dihalides
  from alkanes, 0151
  from carboxylic acids, 0150
  influence on Grignard reagents,
    0638, 0639
  intramolecular alkylation, 0697
halides, phosphorus
  and alcohols, 0146
  formation of vinyl halides, 0153
  halogenation of ketones, 0152
  halogenation of lactams, 0152
  with alcohols, 0146
halides, propargyl
  rearrangement, 0634
  with **Grignard** reagents, 0633
  hydroboration, 0511
  with hydrazone carbanions, 0848
halides
  sulfenyl, with ylids, 0708
  trialkylsilyl, with enolate anions,
    0815
  trimethylsilyl, and CuBr, 0654
halides, vinyl. *See also* halides, alkenyl
halides, vinyl
  and Stille coupling, 1240
  and the E2 reaction, 0158
  elimination to alkynes, 0158
  elimination, 0158
  epoxidation, 0282
  from aldehydes, 0159
  from ketones, 0153
  protonolysis, 0512
  reaction with base, 0158
  reaction with carbocations, 1183
  solvolysis, 1168
  stereochemistry of coupling with
    organocuprates, 0717
  with alkyls, 1244
  with alkynes, 1244
  with borane ate complexes , 0512
  with magnesium, 0639

with organolithium reagents, 1224
with potassium trifluoroborates, 1243
with tert-butyllithium, 0512
halides-lithium exchange, 0688
halipeptins A and D 0839
Haller-Bauer reaction, 0803
halo amines, and radical cyclization,
  1294
halo ketones
  and the **Cornforth model**, 0408
  and the **Favorskii rearrangement**,
    0887
  oxidant with DMSO, 0240
halo sulfones. *See* sulfones
halo-alkenes
  with Grignard reagents, 0665
  with organolithium reagents, 0688
  and **Cadiot-Chodkiewicz coupling**,
    0631
haloalkynes, with alkynes, 1302
haloamines
  and **Hofmann-Löffler-Freytag
    reaction**, 1294
  and hydrogen transfer, 1294
  with sulfuric acid, 1294
haloaromatic tags, combinatorial
  chemistry, 0985
haloboranes
  with iodine, 0512
  with NaOMe, 0513
halocarbenes, formation of
  cyclopropanes, 1319
halo-carbonyls, 0152
halo-esters, with zinc, 0885
haloform reaction, 0884
haloforms, and the haloform reaction,
  0884
halogen substituents, and ylids, 0733
halogenation
  allylic and benzylic, 0151
  and AIBN, 0151
  of alcohols, 0148
  of alcohols, stereochemistry, 0145
  of alkylboranes, 0507
  of enolate anions, 0883
  of ketones, 0152, 0883
  of lactams, 0152
  of vinylboranes, 0508
  silyl enol ethers, 0883, 0884
  unsymmetrical ketones, 0152
halogens
  formation of radicals, 1266
  photolysis, 1266

substitution with alkenes, 1273
  with ate complexes, 0762
  with isonitriles, 0626
halo-**Grignard reagents**, 0660
α-halo **Grignard reagents**, 0658
halohydrins
  and ion pairing, 0187
  from alkenes, 0269
  with base, 0268
halo-ketones. *See* ketones
halo-ketones, with **Grignard reagents**,
  0643
halonium ions, 0184
  and hydroboration, 0510
  and the **Prévost reaction**, 0302
hamigeran B, 0932, 1207
hamigerans, 1264
Hammett's sigma function, and
  oxidation, 0219
Hammond postulate, 1001
  and **Baldwin's rules**, 0565
  and electrophilic aromatic
    substitution, 0198
  and the **Cope rearrangement**, 1128
  and the **ene reaction**, 1148
  and transition states, 0198
  definition, 0155
haouamine A, 1036
hapalindole Q, 0830
haptophilic effects, and chelation
  control, 0549
haptophilicity, 0434, 0435, 0436
  and neighboring group effects, 0434,
    0436
hard acids and hard bases, definition,
  0097
hard soft acid base theory. *See* HSAB
hardness and HOMO/LUMO, 0102
hardness parameter, and
  electronegativity, 0098, 0099, 0101
hardness, and electrophilicity index,
  0119
hardness, definition, 0102
hasubanonine, 1130
Haworth phenanthrene synthesis,
  1207, 1208
  and carboxylic acids, 1208
HBr
  and AIBN, 0180
  and alkenes, 0180
  and alkylboranes, 0506
  in DMSO, and bromination, 0196
  reaction with alcohols, 0130

with alcohols, 0144
with alkenes and radical, 1282
with alkenes, 0063, 0178, 0542
with peroxide, 0542
HCl
and chromic acid, 0227
and chromium trioxide, 0236
with alcohols, 0144
with alkenes, 0132
with alkenes, 0178
with zinc, 0464
HCN
and the **Strecker synthesis**, 0628
with aldehydes, 0628
with ketones, 0628
with NaBH$_4$, 0382
Heck reaction, 0047, 1236, 1237, 1298
and the **Jeffery protocol**, 1238
electronic effects, 1237
enantioselectivity, 1237
in ionic liquids, 1238
in supercritical water, 1239
intramolecular, 1237
powerful in synthesis, 0930, 0932
with aryldiazonium salts, 1238
with bromoindoles, 1237
with diazonium salts, 1238
with heterocycles, 1237
with alkynes , 1238
heliannuol B, 0683
heliannuol D, 0201, 1220
Hell-Volhard-Zelinsky reaction, 0150
hemiacetals
and oxidation of aldehydes, 0231
oxidation with PCC, 1034
hemibrevetoxin B, 0748, 1054
hemin, 0897
Henbest rule, 0283
directing effects of functional groups,
0283
Henry reaction, 0830
and hydrotalcites, 0831
and microwaves, 0831
and proazaphosphatranes, 0831
and the **Kamlet reaction**, 0830
heptacarbonyl(triphenylphosphine)-
cobalt, 1342
heptacarbonyl(triphenylphosphine)-
dicobalt (0), 1343, 1344
heptatrienes, sigmatropic
rearrangement, 1118
heptitol derivatives, 1249
herbarumin I, 0769

herbertenediol, 0383
herboxidiene/GEX 1A, 0117, 1316
heteroatom alkylboranes, 0500
heteroatom **Diels-Alder**. *See* Diels-
Alder
heteroatom **Friedel-Crafts reactions**,
1210-1225
heteroatom radical cyclization, 1287
and hydrogen transfer, 1289
diastereoselectivity, 1288, 1289
syn/anti isomers, 1289
heteroatom radicals, and cyclization,
1293
heteroatoms
and chelation control, 0550
and conformation, 0042, 0052
heterocycles
and **Birch reduction**, 0461
and dissolving metal reduction, 0453
and electrolytic **Birch reduction**,
0470
and **Friedel-Crafts acylation**, 1205
and hydroboration, 0523
and reductive alkylation, 0448
and the **Diels-Alder reaction**, 1058
by radical cyclization, 1287
electrolytic reduction, 0471
electrophilic aromatic substitution,
0199, 1193
HOMO and LUMO, 1009
hydrogenation, 0440, 0447
in the **Heck reaction**, 1237
oxygenated, and hydrogenolysis,
0448
heterocyclic rings, strategic bonds,
0938, 0940
hetero-ene reaction, 1150
heterogeneous catalysts, 0424
supported, 0424
unsupported, 0424
heuristic, definition, 0918
hexachloroacetone, and PPh$_3$, 0149
hexafluoroantimonate complex,
oxazoborolidines, 1074
hexahydroindolizines, 1294
hexamethyldisilazide
lithium, 0176
potassium, and enolate anions, 0812
potassium, and phosphonate esters,
0740
hexamethylphosphoramide. *See* HMPA
hexamethylphosphorus triamide. *See*
HMPT

hexanenitrile, 0624
hexanes, bicyclic, conformations, 0067
hexatriene-cyclohexene isomerization,
0932
hexatrienes
and DDQ, 1100
disrotatory ring closure, 1100
electrocyclic reactions, 1100
photochemistry, 1100
HF,
and **Bardhan-Sengupta
phenanthrene synthesis**, 1209
and **Friedel-Crafts alkylation**, 1201
and **Friedel-Crafts cyclization**, 1208
with SbF$_5$, and **Friedel-Crafts
alkylation**, 1199
hidden chiral templates, 0956
high dilution method, and
macrocyclizations, 0573
high pressure
and activation volume, 1045
and **Diels-Alder reactions**, 1044-
1049
and rate constants, 1046
higher order cuprates, 0726, 0727
highest occupied molecular orbital.
See HOMO
high-throughput screening, 0988
himandrine, 0296
himbacine, 1063
himbeline, 1063
himgaline, 0741
hindrance, steric. *See* steric
hinesol, 0176
hirsutene, 0994, 1120
hirsutene, conformations, 0069
hirsutic acid, 0163
histidine (hydroxy), 0974
histrionicotoxin, 0925
HIV-1 protease inhibitor, 0900
HMPA (hexamethylphosphoramide)
and **Claisen** rearrangement, 1138
and Dibal, 0390
and enolate alkylation, 0804
and enolate anions, 0547, 0808, 1138
and lithium-hydrogen exchange, 0697
and organocuprates, 0714
complexing agent with enolate
anions, 0805
with allylic thiol anions, 0712
HMPT (hexamethylphosphorus
triamide)
and CCl$_4$ and alcohols, 0147

and allyl palladium, 1230
and enolate anion stability, 0791
and enolate anions, 0804
Hofmann elimination, 0162, 0753, 0754
  and basicity, 0164
  and conformation, 0162, 0163
  and E2, 0162
  and leaving groups, 0164
  and nitrogen ylids, 0162, 0163
  and S_N2, 0122
  and solvent effects, 0162
  and structure of amines, 0163, 0164
  solvent effects, 0164
Hofmann rearrangement, 0193
Hofmann-Löffler-Freytag reaction, 1294
Hojo's reagent, 0374
HOMO energy, 1000
  1,3-butadiene, 1006, 1007
  alkanes, 1274
  alkenes, 1001, 1003, 1004
  alkynes, 1003, 1004
  allenes, 1003, 1004, 1005
  and [2+2]-cycloreversion, 1096
  and conjugating substituents, 1005, 1006
  and **Cope rearrangement**, 1126
  and electronegativity, 0101
  and hardness, 0102
  and HSAB, 0101
  and inverse electron demand **Diels-Alder reactions**, 1033
  and Lewis acids, 1038, 1039
  and molecular modeling, 1019, 1039
  and negative ionization potential, 1000
  and orbital coefficients, 1027
  and perturbation theory, 1015
  and photochemistry, 1082
  and radical reactivity, 1268
  and SOMO, 1268
  and symmetry, 1001
  and the **ene reaction**, 1143
  cobalt complex, 1342
  diazomethane, 1110
  diene, and **Diels-Alder reactions**, 1013
  dienes, 1008, 1009
  electron rich or poor dienes, 1006, 1008
  electronic effects, 1015, 1017
  heterocycles, 1008, 1009

ketenes, 1003, 1004, 1005, 1077
nitrile ylids, 1105
orbitals coefficient, 1002
ozone, 1103, 1104
substituent effects, 1015, 1017
vinyl ethers, 1269
homoconjugative effects, and pK_a, 0691
homocoupling, of alkynes, 1341
homodimerization, and cross metathesis, 1338
homoenolate, 0691
homogeneous catalysts, 0424
homogynolide B, 0366
HOMO-LUMO interactions, 1011
  and the **Diels-Alder reaction**, 1013
homolytic cleavage
  and bond dissociation energy, 1260
  and formation of radicals, 1259
  equilibrium, 1260
HOMO-SOMO interactions, for radicals, 1268, 1269
Horner reaction. *See* **Horner-Wadsworth-Emmons**
Horner-Emmons olefination. *See* **Horner-Wadsworth-Emmons olefination**
Horner-Wadsworth-Emmons olefination, 0121, 0593, 0739
  mechanism, 0740
horse liver alcohol dehydrogenase, 0479
horsfiline, 0323
HPLC
  and combinatorial chemistry, 0984
  mass spectrometry, and combinatorial chemistry, 0984
HSAB (Hard Soft Acid Base) Theory, 0007, 0096-0106
  and acyl addition, 0104
  and alkyl halides, 0100
  and amines, 0100
  and boranes, 0106
  and boron, 0099
  and chromic acid, 0096
  and complex formation, 0098
  and conjugate reduction, 0104
  and cuprous ion, 0096
  and electron releasing groups, 0098
  and electronegativity 0097, 0098
  and elimination reactions, 0103
  and enolate anions, 0105
  and epoxidation, 0104
  and **Friedel-Crafts acylation**, 1205
  and **Grignard reagents**, 0105

and HOMO/LUMO, 0101
and hydroboration, 0106
and iodine, 0100
and K_a, 0096
and mercuric ion, 0097
and molecular orbital theory, 0101
and organocuprates, 0105
and organolithium reagents, 0105
and phosphines, 0100
and polarizability, 0097
and Selectride, 0105
and silver ion, 0100
and softness parameter, 0096
and strength parameter, 0096
and substitution, 0103
and symbiosis, 0099
and the **Edward's equation**, 0096
and the proton, 0098
and π character, 0098
classification of hard and soft acids, 0097, 0098
Huang-Minlon modification, and **Wolff-Kishner reduction**, 0473
Huisgen's mechanism, 1102
humulene, 1247
Hünig's base (diisopropylethylamine), 0598, 0953
  and boron enolates, 0842
  and electrocyclic reactions, 1098
  and phosphonate esters, 0741
  and photochemistry, 1091
Hunsdiecker reaction, 0150
  and microwaves, 0150
  and the **Prévost reaction**, 0302
hybocarpone, 0145, 0235, 1205
hybridization
  and **Baldwin's rules**, 0564
  and carbocation formation, 0133
  and carbocations, 0130
  and electrophilic aromatic substitution, 0204
  and organolithium reagents, 0690
hydrates, from aldhdyes or ketones, 0232
hydratropic acid, 0449
hydrazine derivatives, and diimide formation, 0474
hydrazine
  and imides, 0122
  and phthalimides, 0125
  and S_N2, 0122
  and the **Wolff-Kishner reduction**, 0473

chiral, and pinacol coupling, 1304
tosyl, with aldehydes or ketones, 1317
with aldehydes or ketones, 0473
hydrazines
  acyl, with nitrous acid, 0192
  conformations, 0043
hydrazoic acid
  [3+2]-cycloaddition, 1113
  and the **Schmidt reaction** , 0191
  orbital coefficients, 1113
  with carboxylic acids, 0191
  with ketones, 0191
hydrazone carbanions, 0846
  with alkyl halides, 0848
hydrazones
  and [3+2]-cycloaddition, 1112
  and carbenes, 1314
  and nitrenes, 1314
  and protection of aldehydes and ketones, 0612
  and **Wolff-Kishner reduction**, 0473
  carbanions, 0870
  from aldehydes or ketones, 0473, 1112
  from ketones, 0564
  RAMP and SAMP, 0870
  tosyl. *See* tosylhydrazones
  tosyl, and **Shapiro reaction**, 1317
  tosyl, carbanions, 0848
  with LDA, 0848
hydride shift
  and S$_N$1 reactions, 0134
  carbocation rearrangement, 0134
  in carbocations rearrangement, 1171
hydride transfer
  and Alpine borane, 0397
  and formic acid, 0476
hydride, aluminum
  reduction, 0387
  transition state for reduction, 0388
  with conjugated carbonyl, 0388
  with sultams, 0701
hydride, diisobutylaluminum. *See* Dibal
hydride, lithium aluminum. *See* LiAlH$_4$
hydride, lithium diethoxyaluminum, 0365
hydride, lithium ethoxyaluminum, 0365
hydride, lithium tert-butoxyaluminum, 0365
hydride, lithium triethoxyaluminum, 0364

hydride, lithium trimethoxyaluminum, 0364
hydride, lithium trimethoxy-aluminum, reduction, 0105
hydride, lithium tri-*tert*-butoxy aluminum, 0364
  and epoxides, 0365
hydride, lithium tri-*tert* butoxyaluminum
  with conjugated carbonyls, 0366
  with esters, 0366
hydride, lithium with triethylborane, 0379
hydride, lithium, and AlCl$_3$, 0348
hydride, lithium, with BF$_3$, 0386
hydride, sodium *bis*(2-methoxyethoxy) aluminum, also known as Red-Al
  and aryl halides, 0368
  and hydrogenolysis, 0368
  and sulfonate esters, 0368
  as a base, 0369
  inverse addition, 0366
  preparation, 0366
  solubility, 0366
  with alkyl halides, 0368
  with aryl nitriles, 0368
  with benzylic alcohols, 0368
  with cuprous bromide, 0367
  with cyanohydrins, 0368
  with **Steven's rearrangement**, 0369
hydride, sodium borohydride. *See* NaBH$_4$
hydride, sodium
  and trimethylborate, 0348
  with alcohols, 0594
  with keto esters, 0802
hydride, super hydride. *See* super hydride
hydride, tributyltin
  and alkenes, 1283
  and alkyl halides, 1283
  and **Barton deoxygenation**, 0467
  and **Barton-McCombie reaction**, 0467
  and radical reduction, 1274
  and reduction, 0465
  catalysts for radical cyclization, 1293
  catalytic, and **Barton-McCombie reaction**, 0467
  reduction of halides, 1319
  with AIBN, 0467
  with aldehydes, 0465
  with thionocarbonates, 0467

  with xanthate esters, 0467
hydride, trimethoxyaluminum, 0364
  with cuprous bromide, 0367
hydride, triphenyltin
  and AIBN, 0478
  and allyl halides, 0679
hydride, tri-sec-butyl (Li and K). *See* Selectride
hydrides, alkoxyaluminate, 0363-0369
  and carbonylation of boranes, 0532
hydrides, alkoxyaluminum
  comparison with LiAlH$_4$, 0364
  lithium tert-butoxyaluminum, 0365
  solubility, 0365
  with aldehydes, 0365
  with amides, 0365
  with epoxides, 0364
  with functional groups, 0364
  with ketones, 0365
  with lactones, 0365
  with nitriles, 0364
  with sulfonate esters, 0364
hydrides
  K or Na, with borates, 0377
  metal, and reduction, 0348-0422
  with *Cinchona* alkaloids, 0393
hydrindanes, 0166
  conformations, 0068
hydrindanones, 1047
hydroboration, 0913
hydroboration. *See* alkylboranes
hydroboration
  activation energy, 0517
  and digylme, 0506
  and electronic effects, 0504
  and HSAB, 0106
  and stereoselectivity, 0506
  and steric hindrance, 0504, 0524
  asymmetric, 0519
  ate complex, 0512
  catalysis, 0500
  diastereoselectivity, 0516, 0517
  enantioselectivity, 0521
  exo/endo isomers, 0498
  halogenation, cis/trans isomers, 0508
    of alkynes, 0511, 0512
  oxidation of boranes, 0514
  reactivity as a function of ring size, 0502
  regioselectivity, 0516, 0517
  sacrificial alkenes, 0501
  syn/anti selectivity, 0523
  transition state modeling, 0499, 0523

transition state, 0494
hydrobromic acid. *See* HBr
hydrochloric acid. *See* HCl
hydrofluoric acid. *See* HF
hydrogen abstraction
  and radical cyclization, 1291
  and radicals, 1280
  by radicals, 1276
hydrogen acceptor, and **Oppenauer oxidation**, 0250
hydrogen atom abstractions, by radicals, 1273
hydrogen atom donors, in photoreduction, 0477
hydrogen atom transfer
  and haloamines, 1294
  and radicals, 1294
hydrogen bonding
  and acidity, 0088
  and charge distribution, 0784
  and conformation, 0784
  and molecular modeling, 0783
hydrogen bonding, and pKa, 0783
hydrogen cyanide. *See* HCN
hydrogen donors, cyclohexadiene, 0614
hydrogen peroxide. *See* peroxide
hydrogen peroxide
  thermolysis, 1260
  urea complex, 0276
  with boron enolates, 0841
hydrogen shifts, of carbenes, 1322
hydrogen transfer
  and radical coupling, 1280
  and heteroatom radical cyclization, 1289
  and homogeneous hydrogen ton, 0429
  and radical cyclization, 1283
  and radicals, 1272, 1274
  and sulfide oxidation, 0334
  in the **Meerwein-Ponndorf-Verley reduction**, 0467
hydrogen
  and **Vaska's catalyst**, 0430
  and **Wilkinson's catalyst**, 0428, 0429
hydrogenation, 0441
  active sties, 0427
  Adams' catalyst and ferric chloride, 0439
  alkenes, and isomerization, 0432
  amino acids as chiral additives, 0449
  and acid catalysts, 0435
  and acids, 0447

  and **Adams' catalyst**, 0438, 0439
  and adsorption, 0434, 0435, 0436
  and catalysts surface area, 0427
  and chemisorption, 0427, 0449
  and functional groups, 0425
  and haptophilicity, 0434, 0436
  and hydrogenolysis, 0427, 0439
  and kinetic control, 0433
  and LiAlH$_4$, 0447
  and menthoxy acetate, 0449
  and modified Wilkinson's catalyst, 0449
  and overreduction of alkynes, 0436
  and phosphines, 0449
  and reductive alkylation, 0448
  and ruthenium on carbon, 0440
  and silica gel, 0449
  and silk fibroin, 0449
  and steric hindrance, 0435
  and tartaric acid, 0449
  and the **Lindlar catalyst**, 0437
  and the **Rosenmund catalyst**, 0437
  and thermodynamic control, 0433
  and **Vaska's catalyst**, 0428, 0430
  and **Wilkinson's catalyst**, 0428, 0429, 0437
hydrogenation
  aryl compounds, 0445
  aryl compounds, and coupling, 0447
  aryls, 1209
hydrogenation, asymmetric, 0448-0451
  and phosphine ligands, 0450
  catalytic, mechanisms, 0451
  chiral ligands, 0451
  effect of ligands, 0451
  isomerization, 0450
  ligand effects, 0450
  rhodium complexes, 0451
hydrogenation
  benzylic alcohols, 0446
  carbonyl compounds, 0438
hydrogenation, catalysts, 0422, 0424
  and poisons, 0612
  matrix effects, 0427, 0428
  particle sites, 0427, 0428
  reactivity with aryls, 0445
  surface sites, 0427, 0428
hydrogenation, catalytic 0422-0451, 0536, 0933, 0975
  and MTM ethers, 0597
  with Cbz amines, 0618
hydrogenation
  chemoselectivity, 0431

conjugate carbonyls, 0439
conjugated aldehydes, 0438
conjugated versus nonconjugated alkenes, 0431
electronic effects in aryls, 0446
fixed gel, 0424
functional group reactivity, 0425, 0430
heteroatom functional groups, 0438
heterocycles, 0440
heterogeneous catalysts, supported, 0424
heterogeneous catalysts, unsupported, 0424
homogeneous catalysts, 0424
homogeneous, mechanism, 0430
in acetic acid, 0444
Lindlar catalyst, 1101
mechanism, 0428, 0429
nitro compounds, and palladium-on-carbon, 0442
hydrogenation
  of aldehydes, 0438
  of alkenes, 0428, 0429, 0546
  of alkenes, and bond migration, 0432
  of alkenes, and rearrangement, 0394
  of alkenes, and steric hindrance, 0432
  of alkenes, cis/trans isomers, 0433
  of alkenes, substituent effects, 0432
  of alkynes, 0390, 0436, 0546
  of aniline derivatives, 0446
  of aryl nitriles, 0441
  of aryl nitro compounds, 0360
  of azides, 0442
  of aziridines, 0443
  of benzyl amines, 0613
  of benzyl ethers, 0595
  of benzylidene acetals, 1279
  of Cbz, 0444
  of cyclopropane, 0438
  of dienes, 0437, 0438
  of enamines, 1115
  of en-ynes, 0437
  of epoxides, 0443
  of epoxy-ketones, 0443
  of furans, 0448
  of heterocycles, 0447
  of imines 0446
  of isoquinolines, 0447
  of keto-esters, 0439
  of ketones, 0438
  of nitriles, and rhodium hydroxide, 0441

of nitriles, and rhodium oxide, 0441
of nitro compounds, 0442
of nitrobenzene, 1299
of pyrroles, 0447
hydrogenation
  particle sites on a catalyst, 0426
  phosphine ligands, 0428
  poisoning, 0426
  pressure and temperature, 0440
  rate of reaction, 0426
  reactivity of acid derivative, 0440
  reactivity of catalysts, 0425
  ruthenium oxide, 0440
  slurry, 0424
  solution, 0424
  sulfur as a poison, 0445
  support, 0426
  synthetically useful hydrogenolysis, 0443
  transfer hydrogenation, 0613
hydrogenolysis
  and acidity, 0440
  and alkynes, 0437
  and catalyst reactivity, 0430
  and Cbz, 0444
  and Cbz-amines, 0618
  and dithioacetals or dithioketals, 0444, 0611, 0612
  and epoxides, 0443
  and hydrogenation of aryls, 0446
  and hydrogenation, 0427
  and oxygenated heterocycles, 0448
  and Red-Al, 0368
  and substituent effects, 0360
  and the **Whiting reaction**, 0362
  and **Wilkinson's catalyst**, 0443
hydrogenolysis
  benzylic ethers, mechanism, 0457
  definition, 0431
  dissolving metal reduction, 0456
  dithianes and Raney nickel, 0464
  during hydrogenation, 0439
  in synthesis, 0443
hydrogenolysis
  of acetals, 0607
  of alkyl halides, with $LiAlH_4$, 0360
  of benzyl amines, 0613
  of benzyl ethers, 0595
  of benzylic groups, 0440
  of cyclopropane, 0457
  of dithianes or dithiolanes, 0444, 0473
  of keto-esters, 0439
  of sulfides, 0457

of sulfonate esters, 0361
of sulfones, 0457
of sulfoxides, 0457
with aluminum amalgam, 0468
with metals in ammonia, 0457
hydrolysis
  allylic sulfides, 0713
  and silyl enol ethers, 1120
  enol ethers, 0708
  enzymatic, 1356
  imine-nitriles, 0832
  iminium salts, 0685
  iminium salts, 0873, 0874
  iminium salts, 1079
  iminium salts, 1326
  isonitriles, 0626
  ketene dithioacetals, 0710
hydrolysis
  of acetals and ketals, 0607
  of acid derivatives, 0110
  of carboxylate salts, 0651
  of dioxolanes, 0951
  of dithianes, reagents, 0706
  of epoxides, 0548
  of esters, 0310
  of esters, 0605
  of hydrooxazolines, 0845
  of imines, 0466, 0650
  of iminium salts, 0650
  of ketenes, 1315
  of ketyls, 1306
  of lactones, 0310, 0575
  of lithio imines, 0685
  of lithio-nitriles, 0684
  of nitrones, 0907
  of organoiron compounds, 0759
  of oxazolidines, 0844, 0845
  of oximes, 1276
  of silyl enol ethers, 0722, 0749
  of tetrahydroaluminates, 0350
  of vinyl sulfides, 0708
hydrolysis
  rates of chloroalkenes, 1167
  silyl enol ethers, 0837
  sulfonamides, 0864
  thioaldehydes, 1141
  vinyl ethers, 1049
hydronapthalenes, 1209
  by **Bogert-Cook synthesis**, 1209
  conformations, 0069
hydrooxazolines, with $NaBH_4$, 0845
hydropalladium species, 1227

hydroperoxide anions
  and epoxidation, 0275
  and **Grignard reagents**, 0639
  with alkenes, 0275
hydroperoxides. *See* peroxides, hydro
hydrophobic effect,
  Diels-Alder reaction, 1041
hydrosilylation, 0475
  and BF_3, 0476
  and reductive cleavage, 0476
  conjugated carbonyls, 0475
  in acidic media, 0475
  of aldehydes or ketones, 0475
  of alkenes, 0475
hydrotalcites, and the **Henry reaction**, 0831
hydroxamic acids
  and radicals, 1293
  and the **Lossen rearrangement**, 0193
  with alkenes, 1293
  with base, 1107
  with tetrapropylammonium periodate, 1056
hydroxide, and basicity, 0091
hydroxide
  potassium, and silver(II) nitrate, 0259
  sodium and H_2O_2, with boranes, 0514
  sodium with ω-halo boranes, 0514
    and alkylboranes, 0495
    with silver nitrate, 0511
  with alkyl boranes, 0513
hydroxy acids
  by hydrolysis of lactones, 0575
  cyclization to lactones, 0573, 0574
  from aldehydes or ketones, 0629
  lactone formation, 0067
  via the **Baeyer-Villiger reaction**, 0310
hydroxy ketones. *See also* acyloins
  and osmium tetraoxide, 0297
  and permanganate, 0297
  and the **aldol condensation**, 0816
  and the **Davis reagent**, 0814
  and the mixed aldol condensation, 0818, 0819
  by hydrogenation of epoxy-ketones, 0443
  by **Mukaiyama aldol reaction**, 0838
  enolate anions with MoOPh, 0814
  enolate anions with sulfonyloxazidine, 0814
  from **acyloin condensation**, 1306
  from diesters, 1306
  from hydroxy-ketones, 0821

from ketones, 0247
from silyl enol ethers, 0813
with zinc and acetic acid, 0464
hydroxy radicals, 1260
hydroxy thioesters, lactonization, 0577
hydroxyacids
from aldehydes, 0762
from carboxylic acids, 0836
macrolactonization, 0579
hydroxy-aldehydes
and the **aldol condensation**, 0816
and the mixed aldol condensation, 0818, 0819
rearrangement, 1182
hydroxyazulene, nitration, 0198
hydroxycarvone, 0278
hydroxycholesten-4-en-6-ones, 0279
hydroxydithianes, with red mercury, 1176
hydroxy-esters
by the **Reformatsky reaction**, 0885
from ketones, 0885
hydroxyindane, electrophilic aromatic substitution, 0204
hydroxyjasionone, 1120
hydroxylamine-*O*-sulfonic acid
and the **Lossen rearrangement**, 0193
with alkylboranes, 0534
with ketones, 0189
hydroxylamines
and **Cope Elimination**, 0165
by oxidation of amines, 0340
oxidation, 1108
with aldehydes or ketones, 0189, 1107, 1109
hydroxylation
by **Tamao-Fleming oxidation**, 0330
of tyramine, 0266
hydroxyphenstatin, 1180
hydroxypipecolic acid, 0383
hydroxyquinolines, by **Knorr quinoline synthesis**, 1211
hydroxysolanascone, 0822
hydroxytetralin, electrophilic aromatic substitution, 0204
hydroxytyramines, 0266
hydroxyvincadifformine, 0561
hygrine, 0755
hyperaspine, 0382
hyperbranched dendritic amino acids, 0400
hyperconjugation, 0038
and acidity, 0085

and carbocations, 1162, 1163
and σ-conjugation or σ-participation, 1173
hyperconjugative effects, 0414
and the **Cram model**, 0414
hypervalent iodine reagents, 0247
hypervalent iodine, 1297
and **Dess-Martin periodinane**, 0247
hypobromite, and amides, 0193
hypobromites, and radicals, 1277
hypobromous acid, 0187
with alkenes, 0170
hypochlorite
and CeCl$_3$, 1181
and the **Barton reaction**, 1276
and epoxidation, 0289
hypochlorites, thermolysis, 1275
hypochlorous acid, with alkenes, 0187
hypohalous acids, with alkenes, 0187
hypomanganate, and oxidation, 0223
hypophosphorus acid, with aryl diazonium salts, 0210
hyrtiosal, 0235

I

+I and –I effects, 0084
and acidity, 0085
and basicity, 0089
and electrophilic aromatic substitution, 0200, 0201, 0202
I effects, substitution, nucleophilic aromatic, 0206
ibogamine, 0367, 0927
icosahedral symmetry, 0905
10-icosene, 1309
idarubicin 0993
identification, library components in combinatorial chemistry, 0984
IKD-8344, 1286
imaginary frequency, IR, and transition states, 0499
imidate ylids, 0771
imidazole disulfide, and lactonization, 0578
imidazoles
and oxiranes, 0748
in the **Diels-Alder reaction**, 1033
with PPh$_3$ and I$_2$, 0148
with *tert*-butyldimethylsilyl chloride, 0603
with *tert*-butyldiphenylsilyl chloride, 0604

with triisopropylchlorosilane, 0603
imidazolidines, and protection of aldehydes and ketones, 0612
imidazolidinones, and [3+2]-cycloaddition, 1109
imidazolium salts, and lactonization, 0580
imides
and hydrazine, 0122
and S$_N$2, 0122
as chiral auxiliaries, 1072
by electrolytic reduction of lactams, 0472
photolysis, 1265
reduction to lactams, 1185
imine carbanions, 0846
and Schiff bases, 0847
with aldehydes or ketones, 0846
imine-chloride, by reduction of nitriles, 0465
imine-enamine tautomerism, 1213, 1222
imine-nitriles
by the **Thorpe reaction**, 0832
hydrolysis, 0832
imines and cyanoboranes, 0529
imines carbanions, and Evans' auxiliaries, 0869
imines
[2+2]-cycloaddition, 1080
and **aza-Cope rearrangement**, 1131
and **Doebner-Miller reaction**, 1214
and radical cyclization, 1293
azomethine, [3+2]-cycloaddition, 1111
azomethine, and intramolecular [3+2]-cycloaddition, 1113
azomethine, from aldehydes or ketones, 1112
azomethine, from hydrazines, 1112
by **Bischler-Napieralski reaction**, 1215
by **Conrad-Limpach reaction**, 1212
carbanions, 0847
conjugated, and the **Diels-Alder reaction**, 1057
cyclic, by [3+2]-cycloaddition, 1111
Diels-Alder, 1050
ene reaction, 1151
imines
from acylnitrilium salts, 1186
from aldehydes, 0371, 1216
from allylborane addition to lactams, 0527

from amines, 0108
from **Grignard** addition to nitriles, 0650
from nitriles, 0685
imines
　hydrogenation, 0446
　hydrolysis, 0466, 0685, 0832, 0847
　in the **Diels-Alder reaction**, 1053, 1055
　lithiated, 0847
　photochemical coupling, 1305
　photoreduction, 0478
　radical coupling, 1305
　rearrangements, 1223
　reduction with $NaBH_4$, 0371, 0685
　sterically hindered, hydrolysis, 0650
　with acyloxyborohydrides, 0378
　with LDA, 0846
　with pyruvates, 1214
iminium ions, and carbocations, 1185
iminium salts
　and [3+2]-cycloaddition, 1111
　and **aza-Claisen rearrangement**, 1141, 1142
　and azomethine ylids, 1111
　and carbocations, 1164
　and **Diels-Alder reactions**, 1043, 1055
　and formic acid, 1185
　and **Grob fragmentation**, 0174
　and phosphorus oxychloride, 0372
　and **Pictet-Spengler isoquinoline synthesis**, 1216
　and **Vilsmeier-Haack** reaction, 1206
　by $LiAlH_4$ reduction of amides, 0355
　by reduction of nitriles, 0359
　formate anions, 1184
　from amines, 1079
　from carbamates, 1055
　from enamines, 0562, 0873
　from **Grignard** reactions, 0650
　hydrolysis, 0873, 0874, 1079, 1326
　in the Diels-Alder reactions, 1055
　tautomerism with enamines, 0873
　tautomerization to enamines, 0476
　with enolate anions, 0952
　with formic acid, 0476
　with $LiAlH_4$, 0356
　with $NaBH_4$, 0371, 0562
　with sodium cyanoborohydride, 0384
　with super hydride, 0380
imino ene reactions, 1151
imperanene, 1328
indanes, conformations, 0068

indanomycin, 0956
indenes, conformations, 0069
indinavir, 0900
indium
　and alkyne coupling, 0631
　and elimination of dihalides, 0158
　and the **Barbier reaction**, 0637
　with alkynes, 0631
indole alkaloids, 0151, 1221
indole synthesis, 1221
indoles
　and **Suzuki-Miyaura coupling**, 1242
　by **Fischer Indole Synthesis**, 1221
　by **Negishi coupling**, 1225
　from aldehydes, 1221
　from alkynes, 1225
　from hydrazones, 1221
　from ketones, 1221
　from phenylhydrazine, 1221
　fused ring, 1222, 1223
indolizidine 167B, 0128
indolizidine 223AB, 0290
indolizidines, 1185
　from lactams, 1185
indolizidinones, 1292
indolizine, hexahydro, 1294
indolones
　by the **Stollé synthesis**, 1224
　from bromo acyl halides, 1224
inductive effects
　+I and –I, 0084
　and basicity, 0089
　and carbocations, 1162
　and electrophilic aromatic substitution, 0200, 0201
　and pK_a, 0690, 0691
　halogens and acidity, 0085
inductive model, 0085
　and acidity, 0084
infrared
　and carboxylic acids, 0340
　and transition state modeling, 0499
infrared. *See also* IR
ingenol, 1177
Ing-Manske modification, 0125
　and **Wolff-Kishner reduction**, 0473
inhibitors, radical, and galvinoxyl, 0533
initiation steps, radicals and addition reactions, 1271
inner salts, 0170
insertion
　C-H, 1329
　of carbenes, 1327, 1328

of CO, 0760
of CO, **Pauson-Khand reaction**, 1342
rhodium catalyzed, 0959
interconversion
　chair conformations, 0049
　twist and chair, 0046
intermediates
　closely related, in synthesis, 0926
　cyclic, and lead tetraacetate, 0322
　in aromatic substitution, 0194
　nucleophilic aromatic substitution, 0205
　Sharpless asymmetric epoxidation, 0285
　tetrahedral, 0109, 0110
intermolecular radical reactions, 1277
internal conversions, and photochemistry, 1087
intersystem crossing
　and carbenes, 1311
　and photochemistry, 1087, 1088
　and sensitizers, 1089
intramolecular [2+2]-cycloaddition, 1078, 1079
intramolecular [3+2]-cycloaddition, 1113
intramolecular **acyloin condensation**, 1306, 1307
intramolecular **aldol condensation**, 0822, 0941
intramolecular alkylation
　enolate anions, 0808
　with enolate anions, 0801
intramolecular **Baylis-Hillman reaction**, 0879
intramolecular **Diels-Alder reactions**, 1044, 1048, 1059-1065
　and **Bredt's rule**, 1060
　and E/Z dienes, 1061
　and regioisomers, 1061
　tether length, 1059, 1060
　transition state, 1061
intramolecular **Diels-Alder reaction**. *See also* **Diels-Alder**
intramolecular **ene reaction**, 1143, 1145, 1147, 1148
intramolecular **Friedel-Crafts acylation**, 1205
intramolecular **Friedel-Crafts alkylation**, 1202
intramolecular **Heck reactions**, 1237
intramolecular **McMurry olefination**, 1310

intramolecular metathesis, 1335, 1336
intramolecular **Mukaiyama aldol reaction** , 0840
intramolecular strategy, in synthesis, 0940
Intramolecular **Wittig reactions**, 0733
inverse addition
  of LiAlH₄ and nitriles, 0359
  of Red-Al, 0366
inverse electron demand **Diels-Alder**, 1032
  nantioselectivity, 1034
inversion energy, of radicals, 1259
inversion of configuration, 0541, 0931
  and Pd(0) alkylation, 1233
  and S_N2, 0114, 0123, 0555
  and the **Mitsunobu reaction**, 0124
  and thionyl chloride, 0544
  of **Grignard reagents**, 0657
  with alcohols and SOCl₂, 0543, 0544
inversion versus retention, carbocations, 1170
inversion
  alkyl halides, with higher order cuprates, 0727
  and higher order cuprates, 0728
  and radicals, 1259
  and the S_N2 reaction, 0113
iodide
  potassium, and substitution, 1162
  potassium and ozonolysis, 0321
  potassium rate with alkyl halides, 0114
  potassium with aryl diazonium salts, 0209
  zinc, with selectrides, 0381
iodides, alkyl
  from alcohols, 0148
  with sodium cyanoborohydride, 0383
iodine
  with arylchromium carbanions, 0764
  and HSAB, 0100
  and iodolactamization, 0187
  and iodolactonization, 0186
  and PPh₃, 0148
  hypervalent, 1297
  hypervalent, and **Dess-Martin periodinane**, 0247
  oxidation of organoiron compounds, 0761
  with alkylboranes, 0507
  with borane ate complexes, 0513
  with boronic acid, 0508

  with halo boranes, 0512
  with ketones, 0643
  with phosphorus, 0148
iodo-9-BBN, 0509
iodobenzoic acid, and **Dess-Martin periodinane**, 0246
iodoform reaction. *See* Lieben iodoform
iodoform, 0884
  and the **Lieben iodoform reaction**, 0884
iodohydrins, carbocation rearrangement, 1181
iodolactamization, 0187
iodolactams, 0187
iodolactones, and **Baldwin's rules**, 0569
iodolactonization, 0186
iodomethane
  and conjugate addition of organocuprates, 0722
  and MTM ethers, 0597
  and titanium reagents, 0756
  rate with nucleophiles, 0120
  with enolate anions, 0805, 0851
  with imine carbanions, 0844
iodonium ions
  and hydroboration, 0510
  and the **Prévost reaction**, 0302
iodozinc compounds
  and cyclopropanation, 1331
  and **Simmons-Smith reaction**, 1331
ion aggregation, and acidity, 0088
ion pairing, and carbocations, 0130
ion pairs
  and carbocations, 0137
  and halohydrin formation, 0187
  in the E2 reaction, 0154
  tight, and amines, 0127
  tight, and **Hofmann elimination**, 0164
  tight, and ionization, 0130
  tight, and S_N2' reactions, 0127
ionic character
  and enolate anion, 0800
  organocuprates, 0715
ionic liquids
  and **Diels-Alder reactions**, 1037
  and dihydroxylation, 0298
  and electrophilic aromatic substitution, 0205
  and **Jacobsen-Katsuki asymmetric epoxidation**, 0290
  and the **Heck reaction**, 1238
  bmim, 1221

  dialkylimidazolium cations, 0290
  dialkylimidazolium compounds, 0298
  enantioselective **Diels-Alder reactions**, 1037
  in **Pechmann condensation**, 1221
ionic size, and acidity, 0084
ionization potential maps
  and asymmetric induction, 0846
  and chiral auxiliaries, 0865
  and enolate anions, 0806, 0807, 0846, 0865
  and stereoselectivity, 0806, 0807, 0865
  enolate alkylation, 0806, 0807
  epoxidation, 0281
  local, and aromatic substitution, 0203
ionization potential
  and carbocation stability, 1166
  and mass spectrometry, 1166
  and Milliken electronegativity, 0100
  and oxidation, 0224, 0225
  and reduction potential, 0224, 0225
  negative, and HOMO, 1000
  of common organic compounds, 0225
ionization
  alkyl halides and water, 0131
  and alkyl halides, 1162
  and dielectric constant, 0138, 0139
  and E1, 0160
  and **Grob fragmentation**, 0175
  and ion-pairing, 0130
  and solvent effects, 0131, 0138, 0139
  mass spectral, 0976
  of alcohols, 0130
  of alkyl halides, 0129
ion-pairing, and acidity, 0088
ipalbidine, 1055
IPL576,092, 0330
ipso carbon, and electrophilic aromatic substitution, 0200, 0201
IR
  and conformation, 0040
  and the ene reaction, 1144, 1145
  and transition state modeling, 1144, 1145
Ireland model, E/Z selectivity of enolate anions, 0791, 0792
Ireland variant. *See* Ireland-Claisen
Ireland-Claisen rearrangement, 1138
iridium compounds, **Vaska's catalyst**, 0428, 0430
iridomyrmecin, 1057

irofulven, 0644
iron complexes
    and **Mukaiyama aldol**, 1249
    with alkenes or dienes, 1248
    with dienes, 1249
    with nucleophiles, 1249
    with trimethylamine *N*-oxide, 1249
iron compounds, and **Fenton's reagent**, 1296
iron pentacarbonyl, 0759
    diene complexes, 1248
iron tricarbonyl derivatives, 1249, 1250
iron tricarbonyl, 0760
    protecting group, 1249, 1250
iron(II), and oxidant of phenols, 0266
iron-carbon bonds, 0758
irradiation with light, photoreduction, 0477
irrational drug design, 0978
isabelin, 1097
isatin, 1329
isoborneol, by Grignard reduction of camphor, 0667
isobornyl chloride, and rearrangement, 1177
isobutylene, and acid with alcohols, 0596
isocaryophyllene, 0559
isocomene, 1080
isocyanates
    [2+2]-cycloaddition with alkenes, 1080
    and the **Curtius rearrangement**, 0192
    and the **Lossen rearrangement**, 0193
    and the **Schmidt reaction**, 0190, 0191
    and the **Ugi reaction**, 0627
    from acyl azides, 0190
    from diazoketones, 1051
    from hydroxamic acids, 0193
    from isonitriles, 0626
isocyanides. *See* isonitriles
isocyanocholestane 0626
isodiazomethane, and diazomethane, 1313
isodiplamine, 0996
isodomoic acid G, 0991
isofebrifugine, 0440
isoflavanones, 1243
isokotanin A, 1300
isomer, definition, 0012
isomerization
    aldol condensation, influence of base, 0859
    and asymmetric hydrogenation, 0450
    and electronic effects, 1199
    and Lewis acids, 1197
    and photoreduction, 0478
    and the aldol condensation, 0858
    Friedel-Crafts alkylation, 1197
    hexatriene-cyclohexene, 0932
    in **Moffatt oxidation**, 0245
    of alkenes during hydrogenation, 0432
    of alkylboranes, 0501
    of boranes, mechanism, 0501
    of crotyl **Grignard reagents**, 0665
    of nonconjugated cyclic ketones, 0460
    of terminal alkynes, 0456
    valence, diazirine, 1314
    valence, diazoalkanes, 1314
    with azadienes, 1057, 1058
isomers
    cis/trans, alkylboranes, 0498
    cis/trans, and alkene hydrogenation, 0433
    cis/trans, and halogenation of boranes, 0508
    cis/trans, and hydrogenation of alkynes, 0436
    cis/trans, and hydrogenation, 0546
    cis/trans, in ozonides, 0319
    E/Z, 0029
    exo/endo, and hydroboration, 0498
    in the Fischer indole synthesis, 1223
    syn/anti, LTA oxidation of diols, 0322
    syn/anti, sulfoxides, 0336
    threo/erythro, sulfoxides, 0336
isoneonepatelactone, 1329
isonitriles, 0625, 0626
    and the **Passerini reaction**, 0627
    by dehydration of formamides, 0626
    hydrolysis, 0626
    thermolysis, 0626
    versus nitriles, 0625, 0626
    with halogens, 0626
isopentenylpaxilline, 0615
isopinocamphone, 0291
isopropanol. *See* 2-propanol
isopropylidene (*L*)-ribose γ-lactone, 0305
isopropyllithium
    aggregation, 0671
    X-ray, 0671
isopulegone, 1186
isoquinolines
    and **Pictet-Spengler isoquinoline synthesis**, 1216
    and **Schlittler-Müller modification**, 1218
    by **Bischler-Napieralski reaction**, 1215
    by **Friedel-Crafts reactions**, 1215
    by **Pomeranz-Fritsch reaction**, 1217
    from acetals, 1217
    from **Schiff bases**, 1218
    hydrogenation, 0447
isoretronecanol, 1289
isosafrole, 0742
isoshyobunone, 0823
isotopes, and priority rules, 0015
isoxazolidines
    by [3+2]-cycloaddition, 1107
    from nitrones, 1107
isoxazolines
    by [3+2]-cycloaddition, 1106
    from alkenes, 1106
    from nitrile oxides, 1106
I-strain, 0052
IUPAC, atom priorities, 0019
ivangulin, 1317
Ivanov condensation, 0853
    and the **Zimmerman-Traxler model**, 0853

J

J_2 isoprostane, 1321
Jablonski diagram, 1090
    and photochemistry, 1087
Jacobsen catalyst, and epoxidation, 0289
Jacobsen hydrolytic kinetic resolution, 0290
Jacobsen rearrangement, 1199
Jacobsen-Katsuki asymmetric epoxidation, 0288-0290
jamtine, 0825
jasmonate, methyl, 1325
jasplakinolide, 0124, 0175
Jeffery protocol, and the **Heck** reaction, 1238
jesterone, 0993
jiadifenin, 0814
Johnson polyene cyclization reaction. *See also* polyene cyclization
Johnson polyene cyclization, 1184
Johnson-ester variant, **Claisen** condensation, 1136
Jones oxidation, 0233, 1368
Jones reagent, 0620

Julia olefination, 0699
Juliá-Colonna epoxidation, 0276

K

K, and HSAB, 0096
K_a
  and enolate anions, 0797
  definition, 0082
kainic acid, 0167, 0320, 0437, 0956, 1111, 1123
kalihinol C, 0595, 0626
kallolide A, 1122
Kamlet reaction, 0830
  and the Henry reaction, 0830
Karabatsos model, 0408, 0774
  and the Grignard reaction, 0662
kedarcidin, 1295, 1301
kendomycin, 0756
kessanol, 0592
ketals
  acid catalyst, 0609
  and diol protecting groups, 0606
  and transketalization, 0609
  as ketones protecting groups, 0608
  by the Paternò-Büchi reaction, 1095
  diethyl, protecting group for ketones, 0609
  dimethyl, protecting group for ketones, 0609
  ethylenedioxy, 0610
  formation of, 0108
  from diols, 0607
  from ketones, 0108, 0608
  hydrolysis of, 0108
  propylenedioxy, 0610
  reaction with Lewis acids, 0610
  reaction with ozone, 0609
ketene acetals
  and the Claisen rearrangement, 1136
  from esters, 0841
ketene dithioacetals, 0709
  and umpolung, 0710
  electronic effects, 0711
  hydrolysis, 0710
  vinylogous, and umpolung, 0713
  with organolithium reagents, 0711
ketene iminium salts, [2+2]-cycloaddition, 1079
ketene
  dimethyl, and carbenes, 1312
  orbital coefficients and orbital diagram, 1077

ketene-allene cycloaddition, 1080
ketenes
  [2+2]-cycloaddition, 1077
  and enamine formation, 0875
  and quantum yields, 1312
  and the Arndt-Eistert synthesis, 1316
  and the Wolff rearrangement, 1315
  by photolysis of ketones, 1318
  extrusion of carbon monoxide, 1312
  extrusion of CO, 1312, 1313
  from acid chlorides, 1078
  from amines and acyl halides, 0875, 1078
  from diazocarbonyl compounds, 1315
  from diazoketones, 1315
  from halo carboxylic acids, 1078
  HOMO and LUMO, 1003, 1004, 1005, 1077
  hydrolysis, 1315
  inverse electron demand Diels-Alder, 1032
  photolysis, 1312
  silyl. See silylketene
  vinyl, 1079, 1080
  with allenes, intermolecular, 1080
keto acids
  from ester enolates and acid chlorides, 0826
  and Carroll reaction, 1136
  and decarboxylation, 0172
  and Dieckmann condensation, 0828
  and Knorr quinoline synthesis, 1211
  Claisen rearrangement, 1136
  decarboxylation, 0828
  enzymatic reduction, 0479
  for acid chlorides, 0826
  from diketones, 0176
  reduction with yeasts, 0480
  reduction, 0406, 0407
  with aldehydes, 0832
  with phenols, 1220
keto aldehydes. See glyoxals
keto aldehydes, from aldehydes, 0327, 0328
keto esters
  and Simmons-Smith reagent, 1334
  and hydrogenolysis, 0439
  enol form, 0826
  enolate anion, 0802
  hydrogenation, 0439
  with LDA, 1136
  with peroxyacids, 0311

with sodium hydride, 0802
keto nitriles, by the Thorpe reaction, 0832
ketodicyclopentadiene, 1036
keto-enol tautomerism, 1265, 1327, 1130
keto-etiolithocholic acid, 0915
ketolides, 0190
ketols, by chromium oxidation of alkenes, 0316
ketone, methyl vinyl. See MVK
ketones
  absorption bands, 1264
  acid-base equilibria, 0782
  alkoxy stereoselectivity in reduction, 0392
  alkoxy, and the Cram cyclic model, 0406, 0407
  alkoxy, transition state for reduction, 0406, 0407
  alkylation, and enolate anions, 0814
ketones
  and acyl addition, 0108
  and alcohols, 0108
  and aluminum borohydride, 0374
  and betaines, 0730
  and bis(dimethylamino)naphthalene, 0788
  and catalysts reactivity, 0430
  and chelation, 0405, 0406
  and Cram's rule, 0402
  and Luche reduction, 0545
  and LUMO maps, 0403-0405, 0410, 0411
  and McMurry olefination, 1309
  and Norrish type I fragmentation, 1264
  and Norrish type II fragmentation, 1264
  and pinacol coupling, 1303
  and Proton Sponge, 0788
  and radical anion formation, 0470
  and radicals, 1303
  and reductive amination, 0371
  and selenium dioxide, 0327, 0328
  and sodium tetracarbonyl ferrate, 0760
  and tautomerization, 0749
  and the aldol condensation, 0816
  and the Baeyer-Villiger reaction, 0307-0313
  and the Barbier reaction, 0637
  and the Bradsher reaction, 1210
  and the Bürgi-Dunitz trajectory, 0412, 0556, 0795

and the **Cieplak model**, 0414
and the **Clemmensen reduction**, 0464
and the **Corey-Fuchs procedure**, 0630
and the **Cornforth model**, 0408
and the **Cram chelation model**, 0550
and the **Cram cyclic model**, 0406
and the **Darzens tetralin synthesis**, 1210
and the **Felkin-Ahn model**, 0409, 0411
and the **Grignard reaction**, 0636
and the **Haller-Bauer reaction**, 0803
and the **Karabatsos model**, 0408
and the **Lemieux-Johnson reagent**, 0314
and the **Passerini reaction**, 0627
and the **Petasis reagent**, 0756
and the **Reformatsky reaction**, 0885
and the **Strecker synthesis**, 0628
and the **Takai reaction**, 0757
and the **Tebbe reagent**, 0755
and the **Ugi reaction**, 0627
and the **Wacker process**, 0333
and the **Wittig reaction**, 0730, 0732
and transketalization, 0609
and umpolung, 0704
and **Wolff-Kishner reduction**, 0472
ketones
  aromatic, and the **Claisen-Schmidt reaction**, 0817
  aryl, 0643
  as acids, 0082
  as hydrogen transfer agents, 0467
  bicyclic, exo versus endo reduction, 0418
  bicyclic, LUMO maps, 0418, 0419
  bicyclic, stereoselectivity of reduction, 0418
  α-bromo, with amines, 1189
  α-bromo, with Z n/Cu, 1190
ketones
  by alkyl oxidation, 0328
  by **boron annulation**, 0530
  by carbonylation of alkylboranes, 0530, 0531
  by chromium oxidation of alkenes, 0316
  by coupling of alkylboranes, 0529
  by **Fries rearrangement**, 1220
  by **Haworth phenanthrene synthesis**, 1207, 1208

by hydrolysis of enol ethers, 0708
by hydrolysis of imines, 0650, 0832
by hydrolysis of iminium salts, 0650, 0873, 0874
by hydrolysis of silyl enol ethers, 0354, 0837
by **Kuhn-Roth oxidation**, 0329
by **Oppenauer oxidation**, 0250
by oxidation of oximes, 0237
by oxidation with **Fétizon's reagent**, 0255
by **Pinacol rearrangement**, 1179
by **Swern Oxidation**, 0241
by the **acetoacetic ester synthesis**, 0808
by the **Étard reaction**, 0329
by the **Gallagher-Hollander degradation**, 0317
ketones
  carboalkoxy, from epoxy esters, 0282
  chloro, and chromyl chloride, 0332
  chloro, from alkenes, 0332
  cleavage with Cr(VI), 0884
  condensation with amides enolates, 0824
  condensation with anhydrides enolates, 0824
  condensation with ester enolates, 0823
  condensation with lactams enolates, 0824
  condensation with lactones enolates, 0824
ketones, conjugated, acyl addition, 0107
ketones, conjugated
  addition of alkylboranes, 0533
  conjugate addition, 0107
  from conjugated ketones, 0706
  with alkenes, 1056
  with organocuprates, 0720
  and enolate anions, 0720
  1,2- versus 1,4-reduction, 0377
  and epoxidation, 0275
  and hydroperoxide anions, 0275
  and $LiAlH_4$ reduction, 0352
  and **Michael addition**, 0877
  and $NaBH_4$, 0370
  by **Claisen-Schmidt reaction**, 0817
  by isomerization, 0460
  by **Nazarov cyclization**, 1188
  by **phenolic oxidative coupling**, 1297
  dissolving metal reduction, 0454
  electrolytic reduction, 0470

enzyme reduction, 0482
from phenols, 1297
hydrogenation, 0439
hydrosilylation, 0475
in the **Diels-Alder reaction**, 1056
Juliá-Colonna epoxidation, 0276
LUMO and LUMO+1 maps, 0421
LUMO map, 0417
Mukaiyama aldol reaction, 0840
organocuprates and chlorotrimethylsilane, 0719
reduction with zinc, 0463
with azides, 1114
with dithiane anions, 0706
with **Grignard Reagents**, 0651
with **Grignard reagents**, transition state, 0653
with higher order cuprates, 0728, 0729
with hydroperoxide anion, 0275
with magnesium, 0454
with organocuprates, 0714
with the **Luche reagent**, 0417
with trialkylboranes, 0841
ketones, cyclic
  and the **Baeyer-Villiger** reaction, 0307, 0308
  conjugated, by **Robinson** annulation, 0880
  from anisole derivatives, 0460
  from dienes, 0529
  from the **Dieckmann** condensation, 0828
  rate of reaction for reduction, 0416
  stereoselectivity of reduction, 0411-0422
  with **Grignard reagents**, 0682
  with organolithium reagents, 0682
ketones, diastereoselectivity
  of **Grignard** addition, 0659
  of organolithium addition, 0681, 0682
  of reduction, 0401
  with organolithium reagents, 0680
ketones, diazo. *See* diazoketones
ketones
  dibromo, and **Noyori annulation**, 1189, 1190
  dibromo. cationic cyclization, 1189
  diethyl ketals, 0609
  dihalo, 0152
  dimethyl ketals, 0609
  dioxane protecting groups, 0610
  dioxolane protecting groups, 0610
  dissolving metal reduction, 0453

dithiane protecting groups, 0611
dithioketal protecting groups, 0611
dithiolane protecting groups, 0611
divinyl, and **Nazarov cyclization**,
    1188
divinyl, cationic cyclization, 1188
divinyl, **Nazarov cyclization**, 1188
enantioselectivity of **Grignard**
    addition, 0659
ene reactions, 1149
enol content and pK$_a$, 0783
enolate anions formation, 0558
enolate anions from **Grignard**
    reagents, 0669, 0670
enolate anions, 0791
enolate anon stability, 0799
enolates and dissolving metal
    reduction, 0454
enzyme reduction, 0480
ketones
    formation of diradicals, 1265
    formation of dithianes, 0704
    formation of enol acetates, 0815
    formation of enolate anions, 0547,
        0556
    formation of ketals, 0108
    formation of ketyls, 0453, 0454
    formation of silyl enol ethers, 0815
ketones
    from acid chlorides, 0718
    from acids and organocuprates, 0718
    from acyl cyanides, 0811
    from acyl halides, 0646, 0647, 1204
    from alcohols, 0233-0255
    from aldehydes, 1279
    from alkenes, 0319, 0324, 0329, 1183,
        1226, 1279
    from alkyl halides, 0760
    from alkylboranes, 0528-0530
    from alkynes, 0518
    from amides, 0683
    from amines, 1176
    from carboxylic acids, 0684, 0808
    from diols, 1179, 1180
    from dithianes, 0705
    from epoxides, 1177
    from esters, 0648, 1220
    from **Grignard** addition to nitriles,
        0650
    from halo-ketones, 0643
    from imines, 0650
    from iminium salts, 0649, 0685
    from lactams, 0649

from nitriles, 0650, 0685
from oxaspiropentanes, 1176
from silyl enol ethers, 0722, 0749,
    1094, 1120
from vinyl carbocations, 1183
from vinyl ethers, 0552
from vinyl sulfides, 0733
from Weinreb amides, 0649, 0683
ketones, halo
and the **Favorskii rearrangement**,
    0887
from halo epoxides, 0282
with **Grignard reagents**, 0643, 0883,
    0884
ketones
halogenation of, 0152, 0883
hydrogenation, 0438
hydrosilylation, 0475
hydroxy. *See also* acyloins
hydroxy. *See also* hydroxy-ketones
hydroxy, via dihydroxylation, 0297
in the **Diels-Alder reactions**, 1053
intramolecular aldol condensation,
    0822
ketal formation, 0608
LUMO maps, 0421, 0645
mechanism of LiAlH$_4$ reduction, 0350
Meerwein-Ponndorf-Verley
    reduction, 0467
methyl, and **Lieben iodoform**
    reaction, 0884
migratory aptitudes in the **Baeyer-**
    Villiger reaction, 0309
nonconjugated, by **Birch reduction**,
    0460
non-enolizable, 0803
photolysis, 1264, 1318
photoreduction, 0562
pK$_a$, 0082, 0782, 0799
protecting groups, 0608
rate of reduction versus aldehydes,
    0351
reaction with a base, 0082
reaction with acids, 1163
reaction with chiral sulfoxide anions,
    0698
reaction with **Grignard reagents**,
    0410, 0411
reactivity in the **Baeyer-Villiger**
    reaction, 0310
reduction with borane, 0386
reduction with chiral complexes, 0397

reduction with cyanoborohydride,
    1094
reduction with Dibal, 0391
reduction with **Grignard reagents**,
    0667
reduction with NaBH$_4$, 0742
reduction with tin, 0465
reduction, 0066
self-condensation, 0817
silyl alkenyl, Michael acceptors, 0882
sterically hindered, and **Wolff-**
    Kishner reduction, 0473
structure, and enolate geometry, 0790
substituent effects and pK$_a$, 0782
transition state for addition of
    Grignard reagents, 0645
unsymmetrical, and alkylation, 0805
unsymmetrical, and pK$_a$, 0796
unsymmetrical, enolate anion
    formation, 0796
unsymmetrical, regioselectivity, 0805
UV, 1264
via ozonolysis, 0319
ketones
with acetic anhydride, 0815
with alkoxyaluminum hydrides, 0365
with alkyne anions, 0635
with allyltin, 0766
with amines, 0108, 0873
with BF$_3$•etherate, 1173
with boranes, 0395, 0397
with boron enolates, 0843
with boron sulfides, 0842
with BR$_3$, 1164
with bromine, 0883, 0884
with carboxylic acid dianions, 0835
with cyanide, 0628
with DBU, 0877
with **Dess-Martin periodinane**, 0247
with dialkylamide bases, 0794
with dianions, 0836
with dimsyl sodium, 0558
with diols, 0611
with dithioalkyl phosphonate esters,
    0709
with dithiols, 0444, 0611
with enolate anions 0007, 0816, 0817
with enolate dianions, 0835, 0836
with esters, and the **Claisen reaction**,
    0826
with ethoxide, 0799
with formamide and formic acid, 1216

with **Grignard reagents**, 0547, 0644, 0786

with halo ester enolate anions, 0834

with HCN, 0628

with hydrazine, 0473

with hydrazoic acid, 0191

with hydrazone carbanions, 0848

with hydroxylamine, 0189

with hydroxylamine-*O*-sulfonic acid, 0189

with hydroxylamines, 1109

with imine carbanions, 0844, 0845, 0846

with iodine, 0643

with LDA, 0547, 0556, 0787

with LiAlH$_4$, 0350

with lithiated dithianes, 0705

with lithium borohydride, 0373

with lithium triphenylmethide, 0689

with magnesium, 0454, 1304

with methyllithium, 0552

with Na ethoxide, 0822

with NaBH$_4$, 0370

with NaBH$_4$, diastereoselectivity, 0555

with NBS or NCS, 0152, 0643

with nitrile enolates, 0832

with non-nucleophilic bases, 0787

with **Nysted reagent**, 0757

with organocerium reagents, 0668

with organocuprates, 0719

with organolithium reactions, 0786

with organolithium reagents, 0552

with organolithium reagents, 0679

with PBr$_3$, 1367

with PCl$_5$, 0153

with peroxyacids, 0926

with phenylhydrazine, 1222

with phosphine oxide carbanion, 0741, 0742

with phosphonothioates, 0743

with potassium, 1303

with Selectride, 0381

with silyl enol ethers, 0839

with Red-Al, 0367

with sodium, 1303

with succinic ester enolates, 0833

with sulfur ylids, 0746, 1176

with Ti(0), 1310

with titanium(0), 1309

with TosMIC, 0624

with tosylhydrazine, 0564, 1317

with triflimides, 1240

with trimethylsilylmethyllithium, 0709

with ylids, 0007

with zinc borohydride, 0550, 0560

ketones, Wittig reaction, 0552

ketone-selenides, 0814

ketyls, 1304

acyloin condensation, 1306

and dissolving metal reduction, 0453, 0454

and electrolytic reduction, 0469

and esters, 1306

and **pinacol coupling**, 1303

and radicals, 1303

Kharasch reaction, 0640

and the **Grignard reagent**, 0640

and the **Wurtz reaction**, 0679

KI, with alkyl halides in water, 0131

kifunensine, 0305

Kimel-Cope rearrangement. *See* Carroll reaction

kinesin motor protein inhibitor, 0716

kinetic control

and nucleophilic substitution, 0120

and conjugate reduction, 0376

and enolate anions, 0795

and hydrogenation, 0433

and LDA, 0800

and metal cations of enolate anions, 0800

and radical cyclization, 1284

and the **Claisen condensation**, 0825

and time, 0800

reaction conditions, 0797

kinetic enolate, 0791

Kinetic product, Diels-Alder, 1022

kinetic reaction, and **Cram's rule**, 0402

kinetic resolution

and **Sharpless asymmetric epoxidation**, 0286

Jacobsen hydrolytic kinetic resolution, 0290

kinetic versus thermodynamic control, 0198

and the **Diels-Alder reaction**, 1022-1023

kinetics

and diastereoselectivity, 0416

and the **Diels-Alder reaction**, 1022

and higher order cuprates, 0727

Kirkwood-Westheimer model, 0085

Kloeckera saturnus, and reduction, 0480

KMnO$_4$. *See* permanganate

Knoevenagel condensation, 0829, 0913

and microwaves, 0830

and nitro compounds, 0830

and the **aldol condensation**, 0832

and pyridinium ylids, 0751

Knorr pyrrole synthesis, powerful in synthesis, 0932, 0930

Knorr quinoline synthesis, 1211

Koch-Haaf carbonylation, 1187

kojic acid, 0911

Kolbe electrolytic synthesis, 1278

and cross coping, 1278

Kornblum aldehyde synthesis, 0240

Krapcho decarboxylation, 0172

K-selectrides. *See* Selectride

Kuhn-Roth oxidation, 0329

Kumada-Corriu coupling, 0657

kumausyne, 0310

L

labeling

and aromatic substitution, 0212

carbocation rearrangement, 1171, 1172

labile groups, and synthesis, 0924, 0925

lactams, 0904

alkylation, 0805

and cationic cyclization, 1185

as chiral auxiliaries, 1067

β-, from alkenes, 1080

β-, from alkynes, 1108

β-, by [2+2]-cycloaddition, 1080

β-, from imines, 1080

β-, from isocyanates, 1080

β-, from ketenes with imines, 1080

β-, from nitrones, 1108

by radical cyclization, 1292, 1293

by the **Beckmann rearrangement**, 0190

electrolytic reduction, 0472

from alkene-amides, 0187

from alkenes, 1331

from cyclic ketones, 0189, 0191

from diazocarbonyls, 1331

from diazoketones, 1331

from hydroxamic acids, 1293

from imides, 1185

from ketones, 0190, 0620

from lactones, 0946

from keto-acids, 0845

from oximes, 0189

halogenation, 0152

reaction with alkylboranes, 0527

reduction to amines, 0388
reduction with AlH₃, 0388
reduction with borane, 0386
reduction with LiAlH₄, 0356, 0357
with aldehydes, 0824
with **Grignard reagents**, 0649
with ketones, 0824
with Red-Al, 0367
lactols
  by dissolving metal reduction of a
    lactone, 0458
  oxidation, 0235
  by hydrolysis of oximes, 1276
  by reduction of lactones, 0390
  by the **Barton reaction**, 1276
  from lactones, 0354, 0355, 0365
lactonamycin, 0611, 1206
lactones, 0278, 0909
  and aluminum borohydride, 0374
  and **Baldwin's rules**, 0569
  and Bredt's rule, 0554
  and carbonyldiimidazole, 0579
  and dibutyltin oxide, 0580
  and diethyl azodicarboxylate, 0581
  and diisopropyl azodicarboxylate,
    0581
  and enthalpy, 0574
  and entropy, 0574
  and ethoxyethyne, 0581
  and **Fétizon's reagent**, 0258
  and hydrolysis, 0575
  and imidazolium, 0580
  and iodolactonization, 0185
  and macrocyclizations, 0572
  and mercuric trifluoroacetate, 0577
  and molecular modeling, 0067
  and the **Baeyer-Villiger reaction**,
    0307, 0308
  and the **Corey-Nicolaou reagent**,
    0578
  and the enzymatic **Baeyer-Villiger
    reaction**, 0311
  and the **Mukaiyama** reagent, 0578
  and the **Yamaguchi protocol**, 0576
  and thiopyridone, 0578
  and triphenylphosphine, 0578
lactones
  as chiral auxiliaries, 1071
  β-, from aldehydes and silylketenes,
    1080
lactones
  by cyclization of hydroxy acids, 0573,
    0574

by oxidation of ethers, 1277
by radical cyclization, 1288
by reduction of anhydrides, 0373
by **Shiina macrolactonization**, 0576
by the **Masamune protocol**, 0577
by the **Mitsunobu reaction**, 0577
lactones
  conformations, 0053, 0054
  conformer modeling, 0067
  dissolving metal reductions, 0458
  enolate, alkylation, 0805
  enolate, with aldehydes or ketones,
    0824
  exo-methylene, 0169
  Friedel-Crafts acylation, 1205
lactones
  from aldehydes or ketones, 0833
  from anhydrides, 0353
  from carboxylic acids, 0067
  from cyclic ketones, 0307, 0308
  from diols, 0258
  from hydroxy-acids, 0579
  from iodo-carboxylic acids, 0186
  from ketones, 0311
  from lactols, 0235
lactones
  hydrolysis, 0310
  macrolactonization, 0953
  reactivity, 0050
  reduction with Dibal, 0390
  reduction with LiAlH₄, 0354
  saponification, 0583
  spirocyclic, 1094
lactones
  with acid, 0620
  with alkoxyaluminum hydrides, 0365
  with Dibal, 0389, 1368
  with **Grignard reagents**, 0649
  with LDA, 0872, 1370
  with LiAlH₄, 0975
  with lithium borohydride, 0374
lactonization, of ketones or aldehydes,
  0307-0313
lagerstronolide, 0166
lanosterol 0948, 0949
lanthanides, and **pinacol coupling**, 1303
lanthanoid triisopropoxides,
  and **Robinson annulation**, 0880
lasiodiplodin, 0320
lasonolide A, 1241, 1317
lasubine II, 0457
lateriflorone, 0196
laulimalide, 0375, 0624, 1337

lauraldehyde, 0465
laurencin, 0405
laurene, 1175
lauroin, 1306
lauronitrile, 0465
LDA
  aggregation state, 0788
  and amine conjugate acids, 0800
  and chelation control, 0868
  and chiral organoiron compounds,
    0761
  and cyclopropyl sulfides, 1120
  and enolate alkylation, 0804
  and enolate dianions, 1136
  and intramolecular aldol
    condensation, 0822
  and **Ireland-Claisen
    rearrangement**, 1138
  and kinetic versus thermodynamic
    ontrol, 0800
  and organolithium reagents, 0787
  and steric hindrance, 0787
  and the **Bürgi-Dunitz angle**, 0787
  and the **Carroll reaction**, 1136
  and the **Nef reaction**, 0831
  and THF, 0789
  as a base, 0083
  cyclohexane carboxylic acid, 0871
  cyclohexanecarboxaldehyde, 0872
  solvation, 0789
  structure and aggregation state, 0793
  with carboxylate anions, 0836
  with carboxylic aids, 0835
  with dithiocarboxylic acids, 0709
  with esters, 0825, 0841, 1136
  with hydrazones, 0848
  with imines, 0846
  with ketones, 0066, 0086, 0547, 0556
  with lactones, 0872, 1370
  with nitriles, 0624
  with organoiron compounds, 0761
  with oximes, 0848
  with oxiranes, 0749
  with RAMP or SAMP, 0870
  with sulfur compounds, 0698
  with vinyl silanes, 1123
  X-ray, 0789
lead chloride, and olefination, 0456
lead compounds and the **Lindlar**
  catalyst, 0437
lead compounds,
  and the **Rosenmund** catalyst, 0437
  and aromatization, 1218

lead tetraacetate (LTA)
  and diols, 0322
  and alkenes, 0332
  and enamines, 0323
  and enols, 0323
  and silyl enol ethers, 0323
  and vinyl esters or ethers, 0323
  oxidation of aryls, 0268
  oxidation of phenols, 0265
lead, and the **Barbier reaction**, 0637
leaving group ability, and rate of
    reaction, 0142
leaving groups
  alkyl halides and ethoxide, 0143
  amines, 0162
  and benzyne, 0207
  and **Grob fragmentation**, 0174
  and **Hofmann elimination**, 0164
  and nucleophilic aromatic
      substitution, 0206
  and nucleophilicity, 0143
  and polarized bonds, 0080
  and resonance, 0143
  and second row elements, 0143
  and strain, 0143
  and substitution, 0142
  and the **Baeyer-Villiger reaction**,
      0307
  and the **Mitsunobu reaction**, 0123
  bromobenzenesulfonyl, 0772
  halides, 0143
  in the E2 reaction, 0157
  orientation in the E2 reaction, 0154
  phenylseleno, 1288, 1289
  phenylthio, 1288, 1289
  S_N2, and molecular modeling, 0116
  sulfonate esters, 0142, 0143
  tosylates, 0117, 0143, 1358
  with phenolic carbanions, 0771
lejimalide B, 0576
Lemieux-Johnson reagent, 0314
  and NMO, 0314
Lemieux-von Rudloff reagent, 0313
  and crown ethers, 0313
lemonomycin, 0997
lepadiformine, 0902
leptofuranin D, 0149, 0518
leucascandrolide A 0149, 0183, 0362,
    0560, 0825, 0848
leucine, and asymmetric hydrogenation,
    0449
leucotrine A_4, 0289
leustroducsin B, 0569

Level 1 intramolecular cyclization, 0950
Level 1 Process, in synthesis, 0941
Level 2 Process, in synthesis, 0941
Lewis acidity, and transition metals,
    0093-0095
Lewis acids, 0093-0095
  AlH_3, 0387
  and activation of reaction, 1038
  and allyltin, 0766
  and aromatic compounds, 0095
  and ate complex, 0093
  and d orbitals, 0094
  and electrophilic aromatic
      substitution, 0195
  and **Evan's auxiliary**, 1070, 1071
  and **Fajans' rules**, 0094
  and Friedel-Crafts alkylation, 1196,
      1200
  and **Friedel-Crafts** reactions, 0095
  and frontier molecular orbitals, 1038,
      1039
  and **Grignard reactions**, 0652
  and **Grignard reagents**, 0653
  and isomerization, 1197
  and **Nazarov cyclization**, 1188
  and **Oppenauer oxidation**, 0250
  and orbital coefficients, 1038, 1039
  and oxidation of aryls, 0266
  and peroxy acids, 0266
  and polyalkylation, 1197, 1198
  and pyrrole, 0199
  and rearrangement, 1197
  and stereoselectivity, 0095
  and the **aza-Claisen rearrangement**,
      1141
  and the **Bradsher reaction**, 1219
  and the **Claisen rearrangement**, 1133,
      1134
  and the **Curtius rearrangement**, 0192
  and the **Diels-Alder reaction**, 0095,
      1028, 1035, 1038, 1046, 1049, 1117
  and the **ene reaction**, 1147
  and the **Grubbs catalyst**, 1336
  and the imino **ene reaction**, 1151
  and the **Mukaiyama aldol** reaction,
      0837
  and the periodic table, 0094
  and the **Prins reaction**, 1191
  and the retro-**Diels-Alder** reaction,
      1036
  and the **Schmidt reaction**, 0191
  and the **Wheland intermediate**, 0195
  and thiophene, 0095

  and transition metals, 0094
  and valency, 0093
  and ylids, 0736
Lewis acids
  carbonyl ene reaction, 1150, 1151
  complex with alcohols, 1202
  definition, 0093
  for the aldol condensation, 0820
  in the **Diels-Alder reaction**, 1070,
      1071
  influence of HOMO and LUMO,
      1038, 1039
  rate enhancement, 1038
  rate of **Diels-Alder reaction**, 1039
  rate of **Diels-Alder**, molecular
      modeling, 1039
  reaction with diols, 1180
  reaction with ethers, 0595
  regioselectivity of **Diels-Alder**, 1039
Lewis acids
  with acetals and ketals, 0610
  with borane, 0491
  with boranes, 0535
  with Cbz groups, 0618
  with dienones, 1188
  with epoxides, 0129
  with epoxides, 1177
  with **Grignard reagents** and
      epoxides, 0655
  with methoxyethoxy chloromethyl
      ethers, 0598
  with methylthiomethyl ethers, 0597
  with oxaspiropentanes, 0749
  with phenolic esters, 1219
  with phenylhydrazones, 1222
  with pyridine, 1199
  with silyl enol ethers, 0838
Lewis base, definition, 0093
Lewis bases
  and ate complex, 0093
  and nucleophiles, 0107
  and organolithium reagents, 0674
Ley oxidation. *See* TPAP
LHASA, 0898, 0926, 0962
  and conformations, 0058
  and synthesis, 0908, 0914, 0918, 0919
  powerful reactions, 1023
$LiAlH_4$ (lithium aluminum hydride)
  preparation, 0348
  amides and enolate condensation, 0357
  and **Benkeser reduction**, 0462
  and chelation control, 0550
  and conjugated esters, 0354

and diglyme, 0363
and disproportionation, 0351
and **Hojo's reagent**, 0374
and hydrogenation, 0447
and isolation of alcohol products, 0351
and ketones, 0066
and **Oppolzer's sultam**, 0864
and ozonolysis, 0319
and silanes, 0331
and **Tamao-Fleming oxidation**, 0331
and $TiCl_3$, 1309
and the **Whiting reaction**, 0362
and water, 0349
chiral additives, 0393
comparison with alkoxyaluminum
  hydrides, 0364
comparison with Red-Al, 0367
hydrolysis of alkyl halides, 0360
mechanism for reduction of amides,
  0355
mechanism of carbonyl reduction,
  0350
mechanism of nitrile reduction, 0359
nitriles, and **Thorpe condensation**,
  0359
on silica gel, 0374
$LiAlH_4$
  reduction of acid chlorides, 0353
  reduction of alkyl halides, 0360
  reduction of amides, 0355
  reduction of anhydrides, 0353
  reduction of aryl halides, 0360
  reduction of azides, 0359
  reduction of carbamates, 0356
  reduction of carboxylic acids, 0353
  reduction of conjugated carbonyls,
    0352, 0353
  reduction of cyclohexanones, 0414-
    0416
  reduction of cyclopentanones, 0412
  reduction of epoxides, 0358
  reduction of ester, 1068
  reduction of esters, 0353
  reduction of lactams, 0356, 0357
  reduction of lactones, 0354
  reduction of nitriles, 0358
  reduction of nitro compounds, 0359
  reduction of **Oppolzer's sultam**, 0361
  reduction of ozonides, 0358
  reduction of sulfonamides, 0361
  reduction of sulfonate esters, 0361
  reduction of sultams, 0361

$LiAlH_4$
  removal of chiral auxiliaries, 1072
  solubility, 0349
  versus **Meerwein-Ponndorf-Verley
    reduction**, 0467
$LiAlH_4$
  with $AlCl_3$, 0387
  with alcohols, 0364
  with aldehydes and ketones, 0349
  with amides, 0355, 0615
  with camphor, 0393
  with carbamates, 0618
  with Cbz groups, 0618
  with cinchonine, 0393
  with Darvon alcohol, 0394
  with esters, 0605, 0606
  with glucofuranose, 0393
  with iminium salts, 0356
  with ketones, 0077
  with lactams, 0356
  with lactones, 0975
  with menthol, 0393
  with monosaccharides, 0393
  with quinidine, 0393
  with quinine, 0393
LiBr, complex with pyridinium ylids,
  0751
libraries
  and combinatorial chemistry, 0978,
    0980
  and deconvolution, 0983
  of dipeptides, 0980
LICA (lithium
isopropylcyclohexylmide), 0789
  and the Claisen condensation, 0825
LiCl, and Alder-Alder reactions, 1041
Lieben iodoform reaction, 0884
ligands
  with osmium tetraoxide, 0295
  and catalytic allylic palladium, 1232
  chiral, asymmetric hydrogenation,
    0451
  effect on asymmetric hydrogenation,
    0451
  for iron, 0758
  phosphine, and hydrogenation, 0428
light and energy, 1083
liglularone, 1049
lignans, 0445
linchpin assembly, 0705
Lindlar catalyst, 0437, 0465
  and hydrogenation of alkynes, 0546
Lindlar hydrogenation, 1101

lineramycin A, 0995
linkers, and combinatorial chemistry,
  0984
lipoprotein receptor (LDL-R) promoter,
  0431
liposidomycin diazepanone nucleoside,
  0288
lipstatin, 0994
lithiated dithianes, reactions, 0705
lithio imines, hydrolysis, 0685
lithio sulfides, 0537
lithiocarbanions, with arylchromium
  reagents, 0764
lithioimines, 0847
lithiophosphonic acid bis(amides, 0742
lithiospermic acid, 0734
lithium aluminum hydride. *See* $LiAlH_4$
lithium and ammonia, hydrogenolysis,
  0456
lithium borohydride. *See* borohydride
lithium dichloromethane, 0536
lithium diethylamide, 0787
  as a base, 0092
  solution structure, 0789
lithium diisopropylamide. *See* LDA
lithium hexamethyldisilazide, 0788
  and enolate alkylation, 0804
lithium hydride. *See* hydride
lithium in ammonia, 0455
lithium perchlorate in ethers, and
  Diels-Alder, 1044
lithium pinacolate, X-ray, 0793
lithium tetrachlorocuprate, 0642, 0643
lithium tetramethylpiperidide, 0787
  as a base, 0092
lithium triethylborohydride. *See*
  triethylborohydride
lithium trifluoromethanesulfonate, 1044
lithium trimethoxyaluminum hydride.
  See hydride
lithium triphenylmethide
  with aldehydes, 0689
  with ketones, 0689
lithium tri-*tert*-butoxyaluminum
  hydride. *See* hydride lithium
  and the **Barbier reaction**, 0637
  in ammonia, 0457
  with $TiCl_3$, 1309
lithium-halogen exchange, 0688
lithium-hydrogen exchange
  and ethers, 0673
  arylchromium complexes, 0764
litsenolide C1, 0437

liver microsome-mediated oxidation, 1179

local ionization maps, and aromatic substitution, 0203

local ionization potential maps. *See also* ionization potential maps

loganin, 0522, 1071

Lombardo's reagent, 0757

longifolene, 0383, 0913, 0915, 0916
  with borane, 0525
  with boranes, 0519, 0520

longipinane, 1178

Lossen rearrangement, 0193

louisianin C, 0235

Lowe-Brewster rules, 0027

lowest unoccupied molecular orbital. *See* LUMO

L-Selectride. *See* Selectride

LTA. *See* lead tetraacetate

LTMP (tetramethylpiperidide) lithium, 0790
  with ketones, 0547

Luche reagent, 0376
  and facial selectivity, 0417
  with conjugated carbonyls, 0417

Luche reduction, 0545

luffolide, 0148

LUMO energies, alkenes, table, 1003, 1004

LUMO maps
  and diastereoselectivity, 0418, 0419
  and **Grignard reactions**, 0663
  and **Grignard reagents**, 0645, 0660
  and ketones, 0410, 0411
  and reduction of cyclohexanones, 0416
  and reduction of ketones, 0410, 0411
  and reduction, 0416
  and stereoselectivity of reduction, 0403-0405
  and sulfur compounds, 0702
  and the E2 reaction, 0157
  benzyne, 0207
  bicyclic ketones, 0418, 0419
  comparison with other models, 0410
  conjugated ketones, 0417, 0421
  cyclopentanones, 0413
  for ketones, 0403-0405, 0421
  spiroketals, 0421

LUMO, 1000
  1,3-butadiene, 1006, 1007
  alkene, and Diels-Alder reaction, 1013
  alkenes, 1001
  alkynes, 1003, 1004, 1282

allenes, 1003, 1004, 1005

LUMO
  and [2+2]-cycloreversion, 1096
  and conjugating substituents, 1005, 1006
  and electronegativity, 0101
  and hardness, 0102
  and HSAB, 0101
  and inverse electron demand **Diels-Alder**, 1033
  and Lewis acids, 1038, 1039
  and LUMO maps, 0157, 0402
  and molecular modeling, 1019, 1039
  and negative electron affinity, 1000
  and orbital coefficients, 1002, 1027
  and perturbation theory, 1015
  and photochemistry, 1082
  and radical reactivity, 1268
  and SOMO, 1268
  and symmetry, 1001
  and the **ene reaction**, 1143

LUMO
  diazomethane, 1110
  dienes, table, 1008, 1009
  electron density surface. *See* LUMO map
  electron poor and rich dienes, 1006, 1008
  electronic effects, 1015, 1017
  heterocycles, 1008, 1009
  ketenes, 1003, 1004, 1005, 1077
  Lewis acids and carbonyl ene reaction, 1150
  nitrile ylids, 1105
  of alkenes, 1342
  ozone, 1103, 1104
  substituent effects, 1015, 1017
  vinyl ethers, 1269

LUMO+1 maps, conjugated ketones, 0421

LUMO-HOMO interactions, 1011

LUMO-SOMO interactions, 1285

luotonin A, 0990

lupinine, 0876, 0877

lutidine
  and boron enolates, 0842
  with silyl triflates, 0602

Lutzomyia longipalpis, 0864

lycoperdinoside, 0635

lycopodine, 0938, 0939, 0940

lycorane, 0433

M

+M and –M effects and electrophilic aromatic substitution, 0200, 0201, 0202

M effects, substitution, nucleophilic aromatic, 0206

macquarimicin A, 1062

macquarimicin S, 0997

macrocycles
  and **Hofmann elimination**, 0164
  antibiotics, 1054
  by radical cyclization, 1292
  conformations, 0053

macrocyclic compounds, 0572
  by metathesis, 1340

macrocyclizations, 0572, 0922
  and Baeyer strain, 0573
  and chain length, 0573
  and entropy, 0573
  and high dilution method, 0573
  and Pitzer strain, 0573
  and rate of reaction, 0573, 0574
  and transannular strain, 0573
  of bromoacids, 0573
  transition states, 0573

macroine, 0603

macrolactin A, 0269

macrolactonization, 0579, 0953

macrolide antibiotics, 0953

macrolides, and aldol condensation, 0953

macrolides, and ylids, 0741

macrosphelide B, 0237

macroviracin A, 0604

maduropeptin chromophore aglycone, 1244

magellanine, 0370, 0926

magnesium amalgam
  and **pinacol coupling**, 0454, 1304

magnesium borohydride, 0374

magnesium halides
  and epoxides, 0655
  and Grignard reagents, 0653

magnesium iodide, and pinacol coupling, 1304

magnesium monoperphthalate. *See* monoperphthalate

magnesium
  and **pinacol coupling**, 0454, 1304
  and the **Barbier reaction**, 0637
  and the **Grignard reaction**, 0636
  dialkyl, 0638

in methanol, 0455
with alkyl halides, 0636
with conjugated esters, 0454
with conjugated ketones, 0454
with ketones, 0454
majvinine, 0371, 0991
malayamycin A, 0268, 0555
MALDI (matrix-assisted laser
  desorption ionization), 0985
maleic acid, bromination, 0034
maleic anhydride
  and the **Diels-Alder** reaction, 1025,
    1057
  in the ene reaction, 1144
  with fulvalenes, 1035
maleimide, photolysis, 1265
malic acid, 0956
malonates
  alkylation, 0807
  alkylidene, epoxidation with peroxide
    anions, 0275
  alkylidene, **Friedel-Crafts reactions**,
    1212
  and **Doebner condensation**, 0830
  and **Knoevenagel condensation**,
    0829
  and the **Perkin salicylic synthesis**,
    0808
  and transesterification, 0809
  coupling with aryl halides, 1235
  enolate anions, 0807
  pK$_a$, 0808
  with aldehydes, 0829
  with allyl palladium, 1229
malonic acid synthesis, and
  decarboxylation, 0172
malonic acids
  and decarboxylation, 0171
  with piperidine, 0830
malonic ester synthesis, 0809
malyngolide, 1279
manganate, and oxidation, 0223
manganese acetate
  and radicals, 1279
  with aldehydes, 1278
manganese dioxide
  activated, 0253
  and cyanide, 0255
  and manganese esters 0253, 0254
  and pH, 0252
  from MnSO$_4$ and KMnO$_4$, 0252
  mechanism for oxidation of alcohols,
    0253

oxidation of alcohols, 0036, 0252
oxidation of alcohols, radical
  intermediates, 0253
oxidation of allylic and benzylic
  alcohols, 0253, 0254
oxidation of sulfides, 0335
oxidation, solvent effects, 0253
rate of oxidation of alcohols, 0253,
  0254
reduction potential, 0252
manganese ester, and oxidation of
  alcohols, 0253, 0254
manganese salts, relative oxidizing
  power, 0224
manganese sulfate and potassium
  permanganate, 0252
manganese
  and dihydroxylation, 0292
  and salen, 0289
  and the **Jacobsen catalyst**, 0289
manganous ion, and Cr(VI)
  oxidation of alcohols, 0231
Manhattan project, and NaBH$_4$, 0369
Mannich base, 0881
Mannich reaction 088, 0952
  mannitol, 0546
  alkaline, with osmium tetraoxide, 0295
mannose, carbamoyl, 0974
mannosidase inhibitors, 1055, 1056
manoalide, 0475
manzacidin A, 1111
manzamine A, 1064, 1114
maps, LUMO. *See* LUMO map
Marbet-Saucy variant. *See* **Saucy-
  Marbet reaction**
marcfortine B, 0389
Marchantia polymorpha, and reduction,
  0480
maritidine, 0372
Markovnikov addition, 0541
  alkylboranes with alkenes, 0493
  and alkenes, 0179
  and organopalladium complexes, 1228
  hypohalous acids, 0187
  to alkenes, 0551
  water and alkenes, 0514
MARSEIL/SOS
  and REKEST, 0972
  and synthesis, 0971
  reaction types, 0973
martimol, 0808, 0833
MASA, 0919
Masamune protocol, 0577

mass spectrometry of mass
  spectrometry, 0978
mass spectrometry
  and a synthesis tree, 0977
  and ionization potential, 1166
  and solvent effects, 1166
  and synthesis, 0976
  and the Petasis reagent, 0756
matched, in enantioselectivity, 0867
matched-mismatched
  in **Sharpless asymmetric
    epoxidation**, 0287
  aldol condensation, 0867
matrix effects, in hydrogenation
  catalysts, 0427, 0428
matrix-assisted laser desorption
  ionization (MALDI), 0985
maximum bridging ring, 0935
m-chloroperoxybenzoic acid. *See*
  mcpba
McLafferty rearrangement,
  and **Norrish** Type II fragmentation,
    1265
McMurry olefination, 1309, 1356
  and carbenoids, 1310
  cross-coupling, 1309
  intramolecular, 1310
  mechanism, 1310
mcpba (*m*-chloroperoxybenzoic acid)
  and epoxidation, 0281
  and KHF$_2$, **Tamao-Fleming
    oxidation**, 0331
  and NMO, 0289
  and sulfoxides, 0169
  and the **Baeyer-Villiger reaction**, 0279
  and the **Jacobsen catalyst**, 0289
  epoxidation and stereoselectivity,
    0282
  epoxidation, 0558
  oxidation of amines, 0165
  with alkenes, 0655
  with dienes, 0280
  with sulfoxides, 0337
mechanism
  ab inito, dihydroxylation, 0299
  alcohols with Cr(VI), 0228
  alcohols with HX, 0144
  alkene metathesis, 1335
  alkenes +HX, 0179
  and stereoselectivity, 0548
  Ar$_1$-5 mechanism, 0772
  Beckmann rearrangement, 0189
  Birch reduction, 0460

bromination of alkylboranes, 0508
Burgess reagent, 0170
carbocation rearrangement, 0137
catalytic asymmetric hydrogenation, 0451
chromium(VI) oxidation of alcohols, 0229
continuum, and the Wittig reaction, 0737
Cope rearrangement, 1124
coupling of alkyl halides and organocuprates, 0714
decomposition of azo compounds, 1262
diradical, and [3+2]-cycloaddition, 1102
dissolving metal reduction, alkynes, 0455
dissolving metal reduction of carbonyls, 0453
E1 reaction, 0160
E2 reaction, 0154, 0156
E2 reaction, 0153
electrophilic aromatic substitution, 0200-0203
enolate equilibration, 0860
formation of **Grignard reagents**, 0637
Friedel-Crafts alkylation, 1194
Glaser reaction, 1301
Grignard addition to ketones, 0644
Grob fragmentation, 0175
homogenous catalytic hydrogenation, 0430
Horner-Wadsworth-Emmons olefination, 0740
Huisgen's mechanism, 1102
hydrogenation, 0428, 0429
hydrogenolysis of benzylic ethers, 0457
isomerization of boranes, 0501
mechanism
  Jacobsen-Katsuki asymmetric epoxidation, 0290
$LiAlH_4$ and conjugated carbonyls, 0353
$LiAlH_4$ reduction of amides, 0355
$LiAlH_4$ reduction of nitriles, 0359
$LiAlH_4$ with carbonyls, 0350
lithium-hydrogen exchange, 0690
McMurry olefination, 1310
metal catalyzed epoxidation of allylic alcohols, 0273, 0274
metathesis, 1334, 1335

MnO_2 oxidation of alcohols, 0253
Moffatt oxidation, 0244
nucleophilic aromatic substitution, 0205
of carbocation rearrangement, 1171
organolithium reagents with amides, 0684
OsO_4 dihydroxylation, 0295
oxidation of alcohols with **Fétizon's reagent**, 0232,0233
oxidation of aldehydes with Cr(VI), 0231
oxidation or alkylboranes, 0515
ozonolysis, 0318
mechanism
palladium complexes with alkenes, 1228, 1229
palladium(0) catalyzed reactions, 1233
Pauson-Khand reaction, 1342, 1343
permanganate dihydroxylation, 0292
peroxide decomposition, 1261
peroxyacids epoxidation of alkenes, 0279, 0280
protonolysis of alkylboranes, 0506
radical addition, 1271
radical coupling, 1271
radical formation, 1266
radical polymerization, 1271
radical, and boranes, 0533
rearrangement of cyanoboranes, 0529
Simmons-Smith reaction, 1331
Stille coupling, 1240, 1241
sulfide oxidation, 0338
Swern Oxidation, 0241, 0242
vanadium-catalyzed epoxidation, 0274
vinylcyclopropane rearrangement, 1119
with cyclohexanone oxygenase, 0311
Wittig reaction, 0730, 0737
medium, and enzyme reduction, 0481
Meerwein arylation, 1298
Meerwein-Ponndorf-Verley reduction, 0250, 0467
and **Grignard reactions**, 0667
and **Oppenauer oxidation**, 0467
solvent effects, 0467
versus $LiAlH_4$, 0467
Meerwein's reagent
with alcohols, 0594
with amides, 0615
megaphone, 0086

mehtinine, 0992
Meinwald rearrangement, 1181
melodic ester synthesis, 0807
melon fly pheromone, 0388
MEM. *See* ether, methoxyethoxymethyl
membrenone A, 0993
Menschutkin reaction, 1045
menthol, with $LiAlH_4$, 0393
menthoxy acetate, chiral additive in hydrogenation, 0449
menthoxyl, chiral catalyst, 1073
menthyl acrylate
  Diels-Alder reaction, 1067, 1068
  with cyclopentadiene, 1067, 1068
mercuric acetate
and alkenes, 0182
and **Tamao-Fleming oxidation**, 0331
with silanes, 0331
mercuric chloride, and pinacol coupling, 1305
mercuric compounds
and dithioacetals or dithioketals, 0612
mercuric compounds, and hydrolysis of vinyl sulfides, 0708
mercuric ions, and HSAB, 0097, 0099
mercuric oxide, and dithianes hydrolysis, 0706
mercuric salts
and alkenes, 0181-0183
oxidative rearrnagement, 0333
with alkenes, 0333
mercuric sulfate
and pyridine, 0199
and pyrrole, electrophilic aromatic substitution, 0199
mercuric trifluoroacetate, and lactonization, 0577
mercury (II) bis(trifluoroacetate), and oxymercuration, 0182
mercury lamps, and photochemistry, 1085
mercury salts, back donation, 0182
mercury
and radicals, 1346
as a poison for hydrogenation, 0426
red, with hydroxydithianes, 1176
meroquinene, 0960
Merrifield synthesis, 0979
merrilactone A, 0258, 0291, 0463, 1095, 1155
Mes. *See* mesitoate
mesembrine, 0844, 0845, 1185

mesitoate, alcohol protecting group, 0605

mesitoic acid, with alcohols, 0605

meso compounds, definition, 0024

mesomeric effect. *See* M effects

mesomeric effects, and basicity, 0089

mesoxalate, diethyl, ene reaction, 1145

mesylates
  and LiBr, 0147
  reduction to methyl, 0368
  reduction with LiAlH$_4$, 0361

mesylates, reduction with super hydride, 0379

meta directors, 0202

metal carbenes, 1334, 1335

metal cations
  and kinetic versus thermodynamic control, 0800
  as a poison for hydrogenation, 0426

metal hydrides. *See* hydrides

metal induced radical reactions, 1296

metal ions, and aggregation state of LDA, 0793

metal salts, and selectrides, 0381

metalation
  and organolithium reagents, 0678, 0679
  and the **Wurtz reaction**, 0678, 0679
  directed ortho, 0686, 0687, 0944

metal-halogen exchange
  and organolithium reagents, 0676, 0677

metal-hydrogen exchange, transition state, 0689

metallocycles, and metathesis, 1334, 1335

metallopinacol, 1310

metallocycles
  and osmate esters, 0295
  and OsO$_4$ dihydroxylation, 0295

metals
  in ammonia, hydrogenolysis, 0456
  and **Birch reduction**, 0459
  and **Meerwein-Ponndorf-Verley reduction**, 0467
  as hydrogenation catalysts, 0422, 0424
  as reducing agents, 0347
  dissolving metal reduction, 0451-0467
  transition, and **Grignard reagents**, 0640, 0641
  transition, and Lewis acids, 0093-0095

transition, with silanes, 0475

metathesis catalysts, 1341
  and air, 1336
  and paraffin, 1336
  correlation with alkene type, 1339
  reactivity, 1339
  recyclable, 1341
  stability, 1336

metathesis, 1334-1341
  alkene reaction categories, 1339
  and dienes, 1335
  and E/Z isomers, 1337
  and microwaves, 1336
  and radical cyclization, 1340
  and structure of alkenes, 1335
  cross coupling of alkenes, 1337
  cross, model, 1338
  cross, model, 1338
  diynes and alkynes, 1339
  equilibrium, 1335
  formation of macrocycles, 1340
  Grubbs I catalyst, 1336
  Grubbs II catalyst, 1337
  homodimerization, 1338
  intramolecular, 1335, 1336
  mechanism, 1334, 1335
  ring closing, powerful in synthesis, 0930
  ring expansion, 1340
  ring opening metathesis polymerization, 1340
  ring opening, 1340
  ROMP, and **Grubbs** catalyst, 1340
  Schrock catalyst, 1336
  stereoselectivity, 1337
  types of coupling, 1337
  with alkenes or alkynes, 1339

methacrolein, Diels-Alder, 1029

methanol, conformation, 0042

methanoleukotriene A$_4$, 1333

methanophenanthrenes, 1316

methionine, and asymmetric hydrogenation, 0449

methoxide, sodium
  boranes and bromine, 0508
  with boranes, 0510, 0512
  with haloboranes, 0513

methoxybenzoate, with DDQ, 0606
  alcohol protecting group, 0605

methoxybenzoyl chloride, with alcohols, 0605

methoxyborinane, 0518

methoxyethoxymethyl ether. *See* ether

methoxyhydnocarpin-D, 0931

methoxymethyl ether. *See* ether

methoxymethyl ylids, 0733

methoxytryptamine, 0922

methoxyyohimban-17-ol, 1142

methyl chloroformate. *See* chloroformate

methyl groups, by reduction of sulfonate esters, 0368

methyl radical. *See* radical, methyl

methyl vinyl ketone. *See* MVK

methyl(8-*O*-)tetrangomycin (MM 47755), 0367

methyl-A-homo-β-norcholestan-4aβ-ol, 0655

methylamine, conformation, 0042

methylation
  formic acid and formaldehyde, 0476
  of amines, 0355, 0384

methylazithromycin, 0190

methylborane, with alkenes, 0531

methylborohydride, lithium, 0531

methylcarbapenam TA-949, 0545

methylcarbene, 1312

methylcopper
  and boranes, 0511
  with alkenylalkylboranes, 0510

methylene blue, as a sensitizer, 1090

methylene, carbene, 1312
  from ketenes, 1323

methyleneoxindole, 1058

methylephedrine, and reduction, 0397

methylimidazolium, Ionic liquids, 1037

methyllithium
  and enolate acetates, 0820
  as a base, 0092
  solid-state structure, 0671
  with enol acetates, 0815
  with ketones, 0552
  X-ray crystal structure, 0671

methylmaytansine, 0717

2-methyl-6-nitrobenzoic anhydride, and **Shiina macrolactonization**, 0576

methyl-*N*-nitroso-*p*-toluene-sulfonamide. *See* Diazald

methyloctahydroindolizine, 0448

methylproline, and organocuprates, 0723

methylpropylphenylphosphine, 0449

methyls
  by hydrogenation of cyclopropane, 0438, 0457

methylstearic acid, 0329

methyltetrangomycin, 0603

methylthiomethyl ether. *See* ether
methyltrichlorotitanium, 0768
methyl-α-himachalene, 0864
methynolide, 0652
Meyers' aldehyde synthesis, 0843
Micelles
  and the **Diels-Alder reaction**, 1043
  and the **Prins reaction**, 1191
Michael acceptors
  and umpolung, 0709
  silyl vinyl ketones, 0882
Michael addition. *See also* addition
Michael addition, 0877, 1173
  and **Baldwin's rules**, 0568, 0570
  and BINOL, 0878
  and enamines, 0877
  and **Robinson annulation**, 0880
  and sulfur ylids, 0747
  diastereoselectivity, 0878
  dithiolanes anions, 0706
  enantioselectivity, 0878
  of amines, 0877
  of pyridinium ylids, 0751
  solvent effects, 0878
Michael reaction, 0921, 1173
  and umpolung, 0713
  and vinylogous ketene dithioacetals,
    0713
Michaels-Arbuzov reaction, and S$_N$2,
  0121
micrandilactone A, 1040
microbial degradations, 0311
microsclerodermin E, 0616
microwaves
  and **aza-Cope rearrangement**, 1131
  and **Claisen rearrangement**, 1132
  and **Cope elimination**, 0165
  and **Knoevenagel condensation**, 0830
  and metathesis, 1336
  and **Sonogashira coupling**, 1245
  and **Suzuki-Miyaura coupling**, 1242
  and the **Diels-Alder reaction**, 1042,
    1045
  and the **Henry reaction**, 0831
  and the **Hunsdiecker reaction**, 0150
  and the **Pictet-Spengler
    isoquinoline synthesis**, 1217
  and the **Vilsmeier-Haack** reaction,
    1207, 1208
Miescher degradation, 0317
  and **Barbier-Wieland reaction**, 0317
migrastatin, 0991

migration,
  of double bonds in **Oppenauer
    oxidation**, 0251
migratory aptitude,
  carbocation rearrangement, 1172
  in the **Baeyer-Villiger reaction**, 0309
migratory aptitude, in the
  Beckmann rearrangement, 0189
Milliken electronegativity, 0090
Mills-Nixon effect, 0204
mineral acid
  and chromium(VI), 0227
  and functionalized chromic acid, 0227
  with alkenes, 0178
minwanenone, 0931
mirror images, and superimposability,
  0013
Mislow-Evans rearrangement, 0699,
  1123
mismatched, and enantioselectivity,
  0867
mitochondrial complex inhibitor, 0122
Mitsunobu reaction, 0123-0126,
  0545, 0577, 1372
  and azides, 0125
  and DEAD or DIAD, 0124
  and diimide, 0124
  and fluorous compounds, 0125
  and inversion of configuration, 0124
  and phosphonium salts, 0124
  and sonication, 0124
  and succinimide, 0125
  with alcohols, 0123
mixed **aldol condensation**, 0817,
  0818, 0819
mixed Claisen. *See* Claisen
mixed organocuprates, 0722, 0726, 0727
mixed organocuprates. *See also*
  organocuprates
mixed pinacol coping, 1305
MM 47755, 0603
MNNG, and diazomethane, 1313
MO Theory. *See* molecular orbital
model, selectivity
  in **Sharpless asymmetric
    epoxidation**, 0285, 0287
modeling
  and enolate alkylation, 0807
  and hydrogen bonding, 0783
  molecular. *See* molecular modeling
  molecular, and nucleophilic strength,
    0116
  transition state for hydroboration, 0499

Wheland intermediate, 0198
modhephene, 1146
Moffatt oxidation, 0243
  and dicyclohexyl urea, 0244
  and pyridinium salts, 0244
  catalysis, 0244
  mechanism, 0244
molecular complexity, and synthesis,
  0916
molecular modeling, 0062, 0067, 0403-
  0405, 0499, 0517, 0521, 0523, 0570,
  0575, 0645, 0660, 0694, 0702, 0783,
  0785, 0786, 0797-0799, 0806, 0807,
  0846, 0865, 0925, 1019, 1020-1023,
  1026, 1039, 1043, 1063, 1126, 1128,
  1134, 1139, 1144, 1147-1149, 1165,
  1171, 1290, 1291, 1303, 1320
molecular modeling
  activation energy in **Diels-Alder**
    reactions, 1026
  and base strength, 0785
  and carbenes, 1320
  and carbocation rearrangement, 1165
  and carbocations, 1171
  and conformations, 0064
  and **Cope rearrangement**, 1126
  and diastereoselectivity, 0413
  and **Diels-Alder** regioselectivity, 1022,
    1023
  and enolate anion stability, 0799
  and enolate anions, 0797, 0806, 0807
  and lactones, 0067
  and nucleophilic strength, 0116
  and pinacol coupling, 1303
  and proton affinity, 0786
  and pyrans, 0065
  and radical cyclization, 1290, 1291
  and rate of **Diels-Alder** reactions,
    1022, 1026, 1039
  and reduction, 0413, 0416, 0418,
    0419, 0421
  and the **Claisen rearrangement**, 1139
  and the **Diels-Alder** reaction, 1019
  and the **Diels-Alder** transition state,
    1019, 1021-1023, 1043
  and the **ene reaction**, 1144
  and the **Hammond postulate**, 0198
  and thermodynamic stability, 0067
  and transition states, 0198
  conformations, 0060, 0062
  limitations, 0198
  radical stability, 1270

regioselectivity in **Diels-Alder** reactions, 1040
molecular models, 1284
molecular orbital calculations, and conformations, 0060, 0062
molecular orbital theory
  and **Diels-Alder reactions**, 1268
  and HSAB, 0101
  highest occupied. *See also* HOMO
  lowest unoccupied. *See also* LUMO
  singly occupied. *See also* SOMO
molecular orbitals. *See also* orbitals
molecular orbitals, 1000
  1,3-butadiene, 1006, 1007
  and [2+2]-cycloreversion, 1096
  and conjugating substituents, 1005, 1006
  and **Cope rearrangement**, 1124
  and Coulomb integral, 1016
  and electron density, 0999
  and electronic interactions, 0999
  and FMO, 1001
  and ground state, 1012
  and inverse electron demand **Diels-Alder**, 1033
  and perturbation theory, 1001, 1015
  and sigmatropic rearrangements, 1117
  and SOMO, 1268
  and symmetry, 1268
  and the **Diels-Alder reaction**, 1066
  and **Woodward-Hoffman rules**, 1012
  antarafacial motion, 1011
  ethylene, 1000
  nitrile ylids, 1105
  nodes, 1002
  orbital coefficients, 1003, 1004
  suprafacial motion, 1011
molecular sieves
  and the **Pauson-Khand reaction**, 1343
  and **Shiina macrolactonization**, 0576
molecular tag, and combinatorial chemistry, 0985
molybdenum compounds, **Schrock catalyst**, 1337
molybdenum hexacarbonyl
  and the **Pauson-Khand reaction**, 1343
  with hydroperoxides, 0272
molybdenum oxidizing agents, 0261
molybdenum trioxide, and hydrogen peroxide, 0270
molybdenum, alkyne metathesis, 1340

molybdovanadophosphate, and Pd(OAc)$_2$, 1227
MOM. *See* ether, methoxymethyl
momilactone A, 0171
monochloroborane, with alkynes, 0503, 0504
monoisopinocampheylborane, 0519, 0520, 0524
monoperoxycamphoric acid, oxidation of sulfides, 0337
monoperphthalate, magnesium, and epoxidation, 0278
monosaccharides, with LiAlH$_4$, 0393
Montmorillonite clay
  and [3,3]-sigmatropic rearrangement, 1131
  and **aza-Cope rearrangement**, 1131
MoOPh
  and the **Nef reaction**, 0831
  structure, 0814
  with boron enolates, 0841, 0843
  with enolate anions, 0813
Morita-Baylis-Hillman reaction. *See* **Baylis-Hillman reaction**
mosin B, 0149
motuporin, 0233
MTM. *See* ether, methylthiomethyl
mucocin, 0655, 0767, 0958, 0959
Mukaiyama aldol reaction, 0837
  and cupric triflate, 0839
  and iron complexes, 1249
  and **Noyori Open-Chain Model**, 0858
  and stereodifferentiation, 0868
  chiral catalysts, 0839
  complexing agents, 0838
  diastereoselectivity, 0838
  in supercritical CO$_2$, 0839
  in supercritical fluoroform, 0839
  intramolecular, 0840
  water accelerated, 0838
  with acetals, 0839
  with conjugated carbonyls, 0840
  with Lewis acids, 0837, 0838
Mukaiyama reagent, 0578
multi-component coupling, 0627
multidrug resistance reversal, 0987
multiple bonds, and priority rules, 0018
multistriatin, 0279
muscone, 0438, 0564
muscopyridine, 0430
mutisianthol, 0464
MVK (methyl vinyl ketone)
  and **Robinson annulation**, 0880

and the **Diels-Alder** reaction, 1028, 1056
Diels-Alder reaction, computational study, 1038
mycalamide A, 0718
mycoepoxydiene, 0991
mycolactones A and B, 0635
myoinositol-1,4,5-triphosphate, 0304
myraymycin, 0995
myrcene, 0768, 1091
myristinin A, 0301
myrmicarin 215A, 0353
myrmicarin 217, 0624
Myrmicarin alkaloids, 1203
myrmicarin, 1205
myrocin C, 0813

N

N,O-bis(trimethylsilyl)acetamide (BSA), 1234
N$_2$O
  and the **Nef reaction**, 0831
  with aldehydes, 0831
naamidine A, 0649
NaBH$_4$. *See* borohydride, sodium
Nagata's reagent, 0811
nakadomarin A, 1049, 1337
named reactions, and **Fischer indole** synthesis, 1222
N$_\alpha$-methylsarpagine, 0733
naphthalene, and electrophilic aromatic substitution, 0197
naphthalene
  as a sensitizer, 1090
  in the **Diels-Alder reaction**, 1047
  with carbenes, 1327
naphthyridines, 1215
narciclasine, 0188, 0545
narcotine, 0449
Nazarov cyclization, 1188
  and **Lewis acids**, 1188
  and vinyl silanes, 1188
  diastereoselectivity, 1189
  retro, 1189
NBS (*N*-bromosuccinimide)
  and bromohydrins, 0188
  and bromolactonization, 0187
  and dithioacetals or dithioketals, 0612
  and substitution reaction, 1274
  and the **Miescher degradation**, 0317
  bromination, 0151
  in water, 0268

with ketones, 0152, 0643
NCS (*N*-chlorosuccinimide)
  and chlorohydrins, 0188
  and substitution reactions, 1274
  and triphenylphosphine, 0146
  chlorination, 0151
  in water, 0268
  with dithianes, 0705
  with ketones, 0152, 0643
near-far rule, 0028
Nef reaction, 0831
negamycin, 0238
Negishi coupling, 1225
neighboring group effects, 0560
  and epoxidation, 0560
  and haptophilicity, 0434, 0436
  and peroxyacids epoxidation, 0280
  Henbest rule, 0283
  in the Simons-Smith reaction, 1332
neocembrene, 0126, 0128
neodysiherbaine A, 0632
neomenthyl thiols, and organocuprates, 0723
neopupukean-10-one, 0959, 0960
neosymbioimine, 0832
neplanocin A, 1109
neral, 0482
new reactions, and synthesis, 0906
Newman projections, 0038
  and **Cram's rule**, 0402
nezukone, 1190
N-H insertion, diazocarbonyls, 1330
nickel catalysts, and **Suzuki-Miyaura coupling**, 1243
nickel chloride, and nickel(0), 1245
nickel compounds, and **Kumada-Corriu coupling**, 0657
nickel tetracarbonyl, and nickel(0), 1245
nickel(0) complexes, 1245, 1247
  and diphos, 1246
nickel, allyl, 1245
  and the **ene reaction**, 1246
  bis(diphenylphosphinoethane)-nickel chloride, 1245
  coupling of alkyl halides, 1246
  tetrakis(triphenoxyphosphino)nickel, 1245
  tetraphenylnickel, 1245
nickel, aryl, coupling of aryl halides, 1246
nickel, **Raney**
  and hydrogenation of epoxides, 0443

reduction of dithianes or dithiolanes, 0464, 0473
nickel, zero valent. *See* nickel(0)
nickel; on charcoal, and **Suzuki-Miyaura coupling**, 1243
nicotinamide, 0483
NIH shift, 1179
ningalin C, 1237
niphatoxin B, 0340
nitidine, 1216
nitiol, 1343
nitrate, silver
  with alkylboranes, 0511
  with NaOH, 0511
nitrate, thallium
  and oxidative rearrangment, 0333
  oxidation of alkenes, 0333
  oxidation of pyrans, 0333
nitration
  and electrophilic aromatic substitution, 0196
  aromatic substitution, regiochemistry, 1193
  benzo[*c*]fluorene, 0198
  electrophilic aromatic substitution, 1192
  hydroxyazulene, 0198
  of aromatic compounds, 0359, 1192
nitrenes
  and diazoalkanes, 1314
  and hydrazones, 1314
  from nitrous acid, 1314
nitric acid
  and electrophilic aromatic substitution, 0196
  oxidation of sulfides, 0335
nitrile enolate. *See* enolates
nitrile oxides
  [3+2]-cycloaddition, 1106
  from hydroxamic acids, 1107
  from nitrolic acids, 1107
  from oximes, 1107
nitrile ylids
  [3+2]-cycloaddition, 1104
  HOMO and LUMO, 1105
  molecular orbitals, 1105
  with alkenes, 1104, 1105
nitriles
  acyl, from aldehydes, 0255
  and aluminum borohydride, 0374
  and diisobutylaluminum hydride, 0913
  and **Stephen reduction**, 0465

and the **Thorpe reaction**, 0833
and the **Thorpe-Ziegler reaction**, 0833
aryl, from aryl amines, 0209
aryl, hydrogenation, 0441
aryl, reduction to amines, 0368
aryl, with Red-Al, 0368
by dehydration of formamides, 0625
by thermolysis of isonitriles, 0626
dipeptide, 0624
electrolytic reduction, 0470, 0472
enolate anions, 0832
from aldehydes, 0624
from alkyl halides, 0623, 0624
from amides, 0372, 0683
from aryl halides, 1210
from carboxylate salts, 0628
from cyanogen bromide, 0628
from ketones, 0624
from sulfonate esters, 0623, 0624
hydrogenation, 0441
mechanism of LiAlH$_4$ reduction, 0359
pK$_a$, 0782
reduction and **Thorpe condensation**, 0359
reduction with Dibal, 0390
reduction with LiAlH$_4$, 0358
reduction with tin, 0465
reductive cleavage, 0375
versus isonitriles, 0625, 0626
with alkenes, 0976
with alkoxyaluminum hydrides, 0364
with amides, 0683
with Dibal, 0389
with **Grignard reagents**, 0624, 0650
with hydroperoxide anion, 0276
with LDA, 0624
with lithiated dithianes, 0705
with organolithium reagents, 0624, 0685
nitrite
  potassium, with amines, 1314
  sodium with acid, 0208, 0209, 0191
  with KBr, 0545
nitrites
  from nitrosyl chloride, 1276
  from radicals, 1276
nitro aldol, 0830
nitro compounds
  and the **Henry reaction**, 0830
  and the **Kamlet reaction**, 0830
  and the **Nef reaction**, 0831

aryl, and catalytic hydrogenation, 0360

aryl, LiAlH₄ reduction, 0360

by oxidation of amines, 0339

electrolytic reduction, 0471

enolate anions, 0830

hydrogenation, 0442

reduction with LiAlH₄, 0359

nitro enolates, with sulfuric acid, 0831

nitrobenzene, hydrogenation, 1299

nitrobenzoate esters, and the **Mitsunobu** reaction, 0124

nitrogen stabilized carbocations, 1164

nitrogen ylids, 0750-0754

and elimination, 0753, 0754

and the **aza-Wittig** rearrangement, 1123

and the **Hofmann elimination**, 0753, 0754

nitrolic acids, heating, 1107

nitromethane, enolate anions, 0830

nitrones

[3+2]-cycloaddition, 1102, 1107

[3+2]cycloaddition, enantioselective, 1109

by oxidation of amines, 0340

chloro, 0906, 0907

from aldehydes, 1108

from hydroxylamines, 0340, 1108

intramolecular cycloaddition, 1109

orbital coefficients, 1108

with alkenes, 1108

with alkynes, 1108

with conjugated esters, 1108

nitronium ion

and electrophilic aromatic substitution, 0196

with silver ions, 0906

nitroso compounds

and diazomethane, 1313

by oxidation of amines, 0339

from hydroxamic acids, 1056

in the **Diels-Alder reaction**, 1056

nitrosyl chloride

and the **Barton reaction**, 1276

with radicals, 1276

nitrosyl radicals, 1276

nitrosyl sulfuric acid,
and **Beckmann rearrangement**, 0189

nitrous acid

and the **Demjanov rearrangement**, 1175

and the **Tiffeneau-Demjanov ring expansion**, 1175

with acyl hydrazines, 0192

with amines, 1163, 1175, 1299, 1314

with amino alcohols, 1175, 1181

with aryl amines, 0208

N-methylephedrine chiral catalyst, 1073

N-methylmorpholine *N*-oxide. *See* NMO

N-methylmorpholine *N*-oxide,
and OsO₄, 0548

NMO (*N*-methylmorpholine *N*-oxide)

and epoxidation, 0289

and mcpba, 0289

and OsO₄, 0296

and periodic acid and sodium periodate, 0315

and the **Jacobsen catalyst**, 0289

and the **Lemieux-Johnson reagent**, 0314

with TPAP, 0555

NMR

and aggregation state of dialkylamides, 0789

and alkylation of enolate anion, 0804

and betaines, 0737

and carbocations, 1178

and carboxylic acids, 0340

and conformation, 0040, 0049, 0065

and **Grignard reagents**, 0657

and higher order cuprates, 0727

and nonclassical carbocations, 1178

and oxaphosphatanes, 0737

and s character, 0690

and structure of radicals, 1259

and super acids, 1178

and synthesis, 0902, 0903

and the **Wittig reaction**, 0737

and ylids, 0737

node, definition, 1002

nominine, 0384

nonactic acid, 0448, 0922

nonactin, 0922, 0924

nonanoic acid, 0316

nonclassical carbocations, 1178

non-enolizable carbonyl compounds, 0803

non-nucleophilic bases, 0786

LDA and LICA, 0789

lithium hexamethyldisilazide, 0789, 0790

nor-1,6-germacradien-5-ols, 0835

norbornene, 0553, 0554

norbornyl carbocation, 1177

norcaradiene, electrocyclization, 1118

norcarane, 1327, 1331

norcharaciol, 0601

nordesoxycholic acid, 0317

nor-germacradien-5-ols, 0836

normethyljatrophone, 0840

normylataylanone, 1173

Norrish type I fragmentation, 1264

and radical formation, 1265, 1266

Norrish type II fragmentation, 1264

and **McLafferty rearrangement**, 1265

norseychellanone, 1287

norvaline, as a chiral template, 1075

norzoanthamine, 0931

N-oxides, and **Cope elimination**, 0165

Noyori annulation, 1189, 1190

Noyori Open-Chain Model,
and Mukaiyama aldol, 0858

Nozaki-Hiyama-Kishi reaction, 0849

and the **Takai-Utimoto reaction**, 0849

nucleophile strength, in acyl addition, 0107

nucleophiles, 0077, 0079

and acyl addition, 0107

and conjugate addition, 0107

and halonium ions, 0184

and Lewis bases, 0107

and *tetrakis*-palladium(0), 1231, 1232

and the S_N2 transition state, 0114

bidentate, cyanide, 0625

rate with nucleophiles, 0119

reaction with bromoethane, 0120

strength in S_N2, 0119

with allyl palladium, 1229, 1230

with arenium ions, 0194

with aromatic compounds, 0194

with benzyne, 0194

with carbocations, 0178

nucleophilic aromatic substitution. *See* substitution

nucleophilic atoms, 0007

Nucleophilic radicals, 1268

nucleophilic strength

and base strength, 0786

and molecular modeling, 0116

in S_N2 reactions, 0119

of cyanide, 0625

nucleophilicity, 0119

and **Baldwin's rules**, 0568

and leaving group ability, 0143

and resonance, 0119

and steric hindrance, 0123

and the Periodic table, 0119, 0121
enamines, 0873, 0874
of enolate dianions, 0803
versus basicity, 0141
nucleotide tags, combinatorial
  chemistry, 0986
nucleotides, and genetic tags, 0986
nudenoic acid, 0325
nupharolutine, 0361
Nysted reagent, 0757

O

octahydroquinoline, 0478
octalin, 0461
octylbenzolactam-V9, 0629
okaspirodiol, 0600
olefin metathesis. *See* metathesis
olefination. *See* **Wittig**
olefination. *See* ylids
olefination
  and the **Petasis reagent**, 0756
  and the **Tebbe reagent**, 0755
  with titanium reagents, 0755
  and the **Wittig reaction**. *See* **Wittig**
oligo-bicyclo[2.2.2]octane derivatives,
  1278
oligomers
  and macrocyclizations, 0572
  and organolithium reagents, 0672
oligonucleotides, and tags, 0986
one electron transfer, and **Gilman
  reagents**, 0714
onium salts, carbocations, 1163
onocerin, 1278
opioid antagonists, 0371
Oppenauer oxidation, 0249
  and aldehydes as hydrogen acceptors,
    0251
  and ate complexes, 0250
  and bond migration, 0251
  and Cp$_2$ZrH$_2$, 0252
  and ketones, 0251
  and **Meerwein-Ponndorf-Verley
    reduction**, 0467
  and reduction potential, 0251
  and the **aldol condensation**, 0252
  and the **Grignard reaction**, 0667
  hydrogen acceptor, 0250
Oppolzer's sultam
  and **aldol condensation**, 0864
  chiral auxiliary, 0864
  in the **Baylis-Hillman reaction**, 0879

in the **Diels-Alder reaction**, 1071
LiAlH$_4$ reduction, 0361, 0864
orantine, 1329
orbital coefficients
  and [2+2]-cycloaddition, 1093, 1103,
    1104
  and conjugating substituents, 1005,
    1007
  and coordination, 1039
  and HOMO and LUMO, 1027
  and **Lewis acids**, 1038, 1039
  and photocycloaddition, 1093
  and the **Diels-Alder** reaction, 1038
  definition, 1002
  of alkenes, 1001
  of diazoalkanes, 1110
  of dienes, 1008, 1009
  of dipoles, 1104
  of hydrazoic acid, 1113
  of ketene, 1077
  of nitrones, 1108
  regioselectivity in **Diels-Alder
    reaction**s, 1027
orbital diagram, for ketene, 1077
orbital magnitude, 1002
orbital overlap, and the **Diels-Alder
  reaction**, 1030
orbital symmetry, and sigmatropic
  rearrangements, 1116
orbitals
  and carbocation rearrangements, 1035
  and carbocations, 0135
  and stabilization of the Fe-C bond, 0758
  d-orbitals, and pK$_a$, 0692
  molecular. *See* molecular orbitals
  p, and enolate anions, 0570
organic compounds, ionization
  potentials, 0225
organic synthesis, and radicals, 1258
organoaluminum catalysts, **Diels-
Alder reactions**, 1074
organoaluminum compounds
  with alkynes, 0762, 0763
  with CO$_2$, 0762
  with organolithium reagents, 0762
organoaluminum reagents, chiral
  catalysts, 1074
organocerium reagents, 0668, 0713-
  0729
organocopper, with BF$_3$, 0723
organocuprate. higher order, with
  chlorotrimethylsilane, 0729
organocuprates. *See also* cuprates

organocuprates, 0957
  1,2- versus 1,4-addition, 0721
  alkoxides and enantioselectivity, 0725
  and alkynyl copper, 0717
  and charge transfer, 0717
  and chiral amines, 0724
  and conjugated ketones, 0719, 0720
  and **Cram's rule**, 0719
  and dummy ligands, 0717
  and enolate anions, 0720
  and enyl transition states, 0720
  and ephedrine, 0724
  and glucofuranose, 0723
  and HSAB, 0105
  and methylproline, 0723
  and neomenthyl thiols, 0723
  and organolithium reagents, 0714
  and proline, 0723
  and rate of reaction, 0722
  and silyl enol ethers, 0722
  and S$_N$2' reactions, 0126, 0128, 0716
  and sparteine, 0724
  and steric effects, 0722
  and steric hindrance, 0723
  and sulfoxides, 0725
  and synthetic equivalents, 0008
  and the **Wurtz reaction**, 0715
  BF$_3$ complex, 0723
  chelation control, 0719
  chirality transfer , 0726
  complex with chiral amines, 0724
  conjugate addition and iodomethane,
    0722
  conjugate addition, 0653, 0654, 0714,
    0720
  coupling with alkyl halides, 0714,
    0715
  Cu/RLi ratio, and selectivity, 0721
  cyanocuprates, 0727
  enantioselectivity, 0723, 0724
  functional group reactivity, 0714
  Gilman reagent, 0714
cuprates
  higher order, 0726, 0727
  higher order, alkyl halides,
    stereoselectivity, 0727
  higher order, and conjugated
    carbonyls, 0728, 0729
  higher order, and NMR, 0727
  higher order, with alkyl halides, 0727
  ionic character, 0715
  magnesium, 0128
  mixed, 0717, 0722, 0726, 0727

oxidation potential, 0715
reactivity order with alkyl halides, 0715, 0716
solvent effects, 0714
stereochemistry, of coupling with vinyl halides, 0717
stereoselectivity, 0722
stereoselectivity, of conjugate addition, 0721
thienyl ligands, 0728
organocuprates
  with acyl halides, 0718
  with aldehydes or ketones, 0719
  with alkyl halides, 0007, 0714
  with alkyl halides, solvent effects, 0716
  with allylic or aryl halides, 0716
  with chiral substrates, 0726
  with chlorotrimethylsilane, 0719
  with conjugated aldehydes, 0714
  with conjugated esters, 0722
  with conjugated ketones, 0714
  with cyclopentadienyl dicarbonyl and alkenes, 1249
  with cyclopropylcarbinyl halides, 0128, 0716
  with epoxides, 0718
  with sulfonate esters, 0716
  with vinyl halides, 0716
  with vinyl triflate, 0717
  with **Wittig reagents**, 0744
organoiron compounds, 0758
  as chiral auxiliaries, 0761
  chiral, 0761
  cleavage, 0759
  disproportionation, 0759
  hydrolysis, 0759
  oxidation, 0761
  sodium tetracarbonyl ferrate, 0760
  with aldehydes, 0761
  with iodomethane, 0761
  with LDA, 0761
organolithium reagents, 0552, 0670-0696
  aggregation state, 0671, 0672, 0679, 0687
  alkenyl, stereo defined, 0677
  alkenyl, Wurtz coupling, 0678
organolithium reagents
  and [2,3]-sigmatropic rearrangement, 1122
  and alkoxides, 0675
  and aryl halides, 0677
  and benzyl chloride, 0675

and benzyne, 0207
and carbon dioxide, 0684
and DABCO, 0674
and dialkylamines, 0787
and dimethoxybenzyl alcohol, 0675
and diphenylacetic acid, 0675
and diphenylacetone tosylhydrazone, 0675
and ethers, 0673
and fragmentation of ethers, 0673
and **Gilman reagents**, 0714
and **Gilman titration**, 0675
and HSAB, 0105
and hybridization, 0690
and LDA, 0787
and **Lewis bases**, 0674
and **Meyers' aldehyde synthesis**, 0843
and organocerium reagents, 0668, 0669
and organocuprates, 0714
and phosphonate esters, 0740
and phosphorus ylids, 0731
and pK_a, 0676
and radicals, 0690
and reduction, 0680
and salicylaldehyde phenylhydrazone, 0675
and s-character, 0690
and sodium, 0681
and sparteine, 0672
and substitution, nucleophilic aromatic, 0207
and sulfonium salts, 0749
and tetrahydropyran, 0673, 0674
and the **Shapiro reaction**, 1317
and the **Corey-Fuchs procedure**, 0630
and the triple-ion model, 0686
and the **Wurtz reaction**, 0678
and thio-**Claisen rearrangement**, 0713
and TMEDA, 0674
and Wittig rearrangement, 1122
organolithium reagents
  as bases, 0786
  basicity, 0689
  by the **Wittig-Gilman reaction**, 0676
  commercially available, 0675
  complex-induced proximity effect, 0686
  configurational stability, 0680, 0681
  cyclization, 0666
  diastereoselectivity, 0680

enantioselectivity, 0680, 0681
exchange with alkyl halides, 0676
formation of benzynes, 1065
from alkyl halides, 0676, 0677
from arenes, 0687
from organotin compounds, 0679
Li-H exchange, 0673
metalation, 0678, 0679
metal-halogen exchange, 0676, 0677
metal-hydrogen exchange, 0689
oligomer formation, 0672
preparation, 0670, 0676
solvation, 0679
solvent effects and aggregation state, 0672
stability in solvents, 0673
stability, 0670, 0673, 0674
standardization, 0675
stereoselectivity, 0680, 0681
structure and stability, 0673, 0674
phosphonium salts, 0731
organolithium reagents
  with acid chlorides, 0683
  with aldehydes, 0679, 0687, 1224
  with alkenes, 0688
  with alkenyl halides, 0677, 0700
  with alkenyl sulfoxides, 0700
  with alkyl halides, 0676, 0677
  with allylic reagents, 0712
  with allylic thiols, 0711
  with amides, 0683
  with ammonium salts, 0750, 0752
  with aryl halides, 0677, 1018
  with arylchromium reagents, 0763-0765
  with carboxylic acids, 0684, 0835
  with carboxylate anions, 0836
  with Cbz groups, 0618
  with chloropyridines, 1018
  with CO_2, 0685
  with CuCN, 0726, 0727
  with cuprous iodide, 0714
  with cuprous salts, 0714
  with cyclic ketones, 0682
  with dithianes, 0704
  with dithiolanes, 0706, 0707
  with DMF, 0684
  with enol acetates, 0815
  with epoxides, 0688
  with esters, 0683
  with halo-alkenes, 0688
  with ketene dithioacetals, 0711
  with ketones, 0679, 0786

with methyl chloroformate, 0734
with nitriles, 0624, 0685
with organoaluminum compounds, 0762
with oxazolidines or oxazolines, 0843, 0844
with oximes, 0848, 0849
with phosphine oxides, 0743
with pyridinium salts, 0750
with sparteine, 0689
with sulfonium salts, 0746
with sulfur compounds, 0697
with terminal alkynes, 0632
with TMEDA, 0686, 0687, 0689
with vinyl dihalides, 0734
with vinyl ethers , 0708
with vinyl halides, 1224
with Weinreb amides, 0683
organolithium reagents, X-ray, 0671
organomagnesium. *See also* **Grignard**
organomagnesium compounds, 0636-0670
organometallic transform, 0918
organoosmium, 0295
organopalladium complexes
and CuCl$_2$, 1228, 1229
Markovnikov addition, 1228
organopalladium reagents, 0927
organotin compounds, with organolithium reagents, 0679
organotin reagents, 0766
dithianes anions, 0704
organotitanium compounds, 0768
organotrifluoroborates, 1243
organozinc compounds, and the **Simmons-Smith** reaction, 1331
ornatipolide, 0578
ortho effect, 0203, 1028
and activating substituents, 1195
in **Friedel-Crafts alkylation**, 1195
in **Diels-Alder reactions**, 1024
ortho ester Claisen rearrangement.
See **Johnson-ester variant**
ortho esters. *See* esters
ortho metalation, directed, 0944
ortho quinones, 0262
ortho substitution rearrangement, 0753
ortho/para directors, 0202
ortho/para isomers, and **Friedel-Crafts acylation**, 1204
osmate esters
and metalocycles, 0295

reagents for decomposition, 0295
osmate, potassium, and dihydroxylation, 0300
osmium tetraoxide
additives for decomposition of osmate esters, 0295
and amine oxides, 0296
and dihydroxylation, 0294-0301, 0548
and hydroxy ketones, 0297
and ligands, 0295
and *N*-methylmorpholine *N*-oxide, 0548
and NMO, 0296
and osmate esters, 0295
and **Sharpless asymmetric dihydroxylation**, 0298-0301
and tert-butyl hydroperoxide, 0296
and the **Lemieux-Johnson reagent**, 0314
catalytic , 0296
dihydroquinidine complex, X-ray, 0299
model for stereoselectivity, 0301
reduction potential, 0294
with acidic potassium chlorate, 0295
with alkaline mannitol, 0295
with bisulfite in aq ethanol, 0295
with enamines, 0561, 0562
with pyridine, 0295
with sulfite, 0295
ottelione A, 0456
ouabain, 0560, 0881
ovalicin, 0474
overreduction, of alkynes, 0436
oxalic acid
and dicyclohexyl urea, 0245
hydrolysis of acetals or ketals, 0609
oxalyl chloride
and **Swern Oxidation**, 0241
and the **Stollé synthesis**, 1224
oxammonium salts
and oxidation, 0248
Bobbitt's reagent, 0248
oxaphosphatanes
conformation and reactivity, 0738
equilibration, 0738
stereochemistry and reactivity, 0738
stereochemistry, 0735
oxaspirooctane, 1321
oxaspiropentanes, 1120, 1176
from sulfur ylids, 0749
reaction with Lewis acids, 0749
with HBF$_4$, 1176

oxatitanacycled, and the **Petasis reagent**, 0756
oxazaborolidines
and borane, 0396
and reduction, 0400
oxazaborolidinium salt, **Diels-Alder reactions**, 1074
oxazaborolidinone,
Mukaiyama aldol reaction, 0839
oxazepine-7-ones, 0987
oxazepineones, 0987
oxazines, and the **Meyers' aldehyde synthesis**, 0843
oxaziridines
chiral, oxidation of sulfides, 0337, 0338
with enolate anions, 0814
oxazoborolidines
and **Diels-Alder reactions**, 1074
chiral catalysts, 1074
hexafluoroantimonate complex, 1074
oxazolidines
and protection of aldehydes and ketones, 0612
as an **Evans' auxiliary**, 0865
as chiral auxiliaries, 0869
with NaBH$_4$, 0844, 0845
with organolithium reagents, 0843, 0844
oxazolines
and **Meyers' aldehyde synthesis**, 0843
with organolithium reagents, 0843, 0844
oxazolone, complex, 1029
oxetanes
from aldehydes or ketones, 1095
metal ring opening, 0563
with diethylaluminum-*N*-methylaniline, 0563
oxetanols, azepine-fused, 1265
oxidation number, 0220
and the periodic table, 0220
rules, 0220, 0221
oxidation potential, organocuprates, 0715
oxidation stages, 0220-0222
and reduction, 0347
oxidation state, 0220
oxidation
alcohols with Cr(VI), rate-determining step, 0228

allylic and benzylic alcohols, 0253, 0254

oxidation
  and chromyl chloride, 0332
  and electrode potential, 0223
  and electrolysis, 0222
  and electronegativity, 0219
  and equilibrium potential, 0222
  and **Faraday's constant**, 0223
  and free energy, 0223
  and **Hammett's** sigma function, 0219
  and ionization potential, 0224, 0225
  and **Kuhn-Roth oxidation**, 0329
  and loss of electrons, 0220, 0221
  and oxidation number, 0220, 0221
  and **phenolic oxidative coupling**, 1296
  and radicals, 1270
  and reduction potential, 0222
  and selenium dioxide, 0324
  and strain, 0052
  and **Taft's** sigma function, 0219
  and the **Barton reaction**, 1277
  and the **Étard reaction**, 0329
  and the **European convention**, 0223

oxidation
  anodic, of phenols, 0264
  anodic, of phenols, 1298
  asymmetric, 0337
  Baeyer-Villiger reaction, 0307-0313
  by air, 1301
  cerium reagents, 0261
  Collins oxidation, 0235, 0236
  cyclohexanol, 0063
  definition, 0219
  Jones oxidation, 0233
  Ley oxidation, 0250
  MnO_2, solvent effects, 0253
  molybdenum reagents, 0261

oxidation
  of alcohols, 0226-0261, 1360
  of alcohols, and chromate esters, 0228-0230
  of alcohols, steric effects, 0230
  of alcohols, with Cr(VI), mechanism, 0229
  of aldehydes, reagents for, 0259
  of aldehydes, with Cr(VI), 0231
  of alkenes, 0292
  of alkyl fragments, 0324
  of alkylboranes, 0518, 0528
  of alkyls, with CrO_3, 0328

  of alkyls, with ruthenium compounds, 0330
  of amines, 0339
  of aryls to phenols, 0266
  of boranes, 0514, 0518
  of boron enolates, 0841
  of cyclic amino acids, 0330
  of dithianes, 0704
  of hydroxylamines, 1108
  of nitrogen, 0334
  of organoiron compounds, 0761
  of ozonide, 0320
  of phenols, 0262
  of phenols, reagents, 0264
  of selenides, 0338, 0815
  of selenium compounds, 0334
  of sulfides, 0168, 0698, 0814
  of sulfoxides, 0697
  of sulfur compounds, 0334

oxidation
  Oppenauer oxidation, 0250
  reagents for sulfur compounds, 0704
  sulfides, and hydrogen transfer, 0334
  Swern, 0241-0243, 1358

oxidation
  with **Bobbitt's reagent**, 0248
  with chiral oxaziridines, 0337, 0338
  with chromium trioxide, 1277
  with chromyl chloride, 0239
  with Cr(VI), with manganous ion, 0231
  with DDQ, 0265
  with **Dess-Martin periodinane**, 0247
  with DMSO and acetic anhydride, 0246
  with DMSO and phosphorus pentoxide, 0246
  with **Fenton's reagent**, 1296
  with **Fétizon's reagent**, 0256
  with **Jones reagent**, 0233
  with manganese dioxide, 0252
  with oxammonium salts, 0248
  with PCC, 0236
  with permanganate, 0292
  with pyridine-sulfur trioxide, DMSO, 0246
  with **Sarett oxidation**, 0234, 0235
  with silver carbonate, 0255
  with silver(I silver(I) oxide, and sodium cyanide, 0260
  with silver(II) oxide, 0259
  with TEMPO, 0248
  with TPAP, 0249

oxidative bond cleavage. *See also* cleavage
oxidative bond cleavage, 0313-0324
  with Cr(VI), 0884
oxidizing power, manganese salts, 0224
oxime anions, with alkyl halides, 0848, 0849

oximes
  and protection of aldehydes and ketones, 0612
  and radical cyclization, 1293
  and radicals, 1276
  and the **Beckmann** rearrangement, 0188
  by the **Barton** reaction, 1276
  hydrolysis, 1276
  oxidation, 1107
  oxidation with PCC, 0237
  tosyl, 0190
  with LDA, 0848
  with organolithium reagents, 0848
  with PCl_5, 0188

oxiranes. *See also* epoxides
oxiranes
  and imidazole, 0748
  by the **Paternò-Büchi reaction**, 1095
  carbocation rearrangement, 1181
  from aldehydes or ketones, 0746
  from sulfur ylids, 0746
  photolysis, 1316
  reaction with acid, 1181
  with LDA, 0749
oxo-9H-xanthene-4-acetic acid, 0905
oxocane, conformation, 0052
oxocarbenium ions, 1163
  from diols, 1180
oxochromium(VI)-amine reagents, 0234
Oxone. *See* peroxomonosulfate
oxonia **Cope rearrangement**, 1191
  and the **Prins reaction**, 1191
oxonium ions
  and carbocations, 0136
  and elimination, 0161
  from alcohols, 0144
  from epoxides, 0129
oxonium salts, from alkenes, 1183
oxphosphatanes, decomposition, 0731
oxyamination. *See* aminohydroxylation
oxy-Cope rearrangement, 0934, 1130
  anionic, 0932
  powerful in synthesis, 0930, 0932
oxygen radicals, 1277

oxygen stabilized carbocations, 0133
oxygen
  and carbene formation, 1312
  and carbenes, 1323
  and palladium chloride, 0333
  and trimethylphosphite, 0813
  and **Wacker oxidation**, 1226
  reduction potential, 0270
  with alkylboranes, 0505
  with **Grignard reagents**, 0639
oxygenated aromatic compounds, and
  the **Bradsher reaction**, 1219
oxygen-stabilized carbocations, 1163
oxymercuration, 0181-0183
  and alcohols, 0183
  and alkenes, 0514
  and NaBH$_4$, 0182
oxymercuration-demercuration. *See*
  oxymercuration
ozone
  and [3+2]-cycloaddition, 1103
  cleavage of alkenes, 0318-0322
  HOMO and LUMO, 1103, 1104
  reaction with ketals and acetals, 0609
  reduction potential, 0318
  resonance contributors, 1103
ozonides
  and ozonolysis, 0318
  and trimethylphosphite, 0321
  by [3+2]-cycloaddition, 1103
  cis/trans isomerism, 0319
  decomposition, 0319
  from alkanes, 1103
  oxidation, 0320
  reduction with LiAlH$_4$, 0358
  reduction with NaBH$_4$, 0371
ozonolysis, 0318-0322, 1369
  and dimethyl sulfide, 0320
  and LiAlH$_4$ or NaBH$_4$, 0319
  and ozonide formation, 0318
  and partial cleavage, 0321
  and rearrangement, 0321
  and solvents, 0318
  and the **Schreiber protocol**, 0320
  and trioxolane formation, 0318
  and triphenylphosphine, 0319
  cyclic alkenes, 0563
  decomposition of ozonide, 0319
  mechanism, 0318
  nonconjugated dienes, 0563
  of amines, 0339
  of phosphinimines, 0340
  of sterically hindered alkenes, 0320

P

p character
  and conformation, 0047
  and HSAB, 0098
p orbitals, and enolate anions, 0570
palladation, 1228
palladium (0)
  alkylation, 1232
  with enolate anions, 1232
  with nucleophiles, 1232
palladium acetate
  and molybdovanadophosphate, 1227
  with triphenylphosphine, 1237
palladium allyl, formation, 1227
palladium catalysts
  and boranes, 0514
  and boronic acids, 0527
  and **Claisen rearrangement**, 1140
  and formation of indoles, 1225
  and **Suzuki-Miyaura coupling**, 0527
  coupling reactions, 1225
palladium catalyzed arylation, 1231
palladium chloride
  and conjugated dienes, 1228
  and the **Wacker process**, 0333
  Claisen rearrangement, 1140
  with alkenes, 0333
  with dienes, 1228
palladium complexes
  and amines, 1229
  and CuCl$_2$, 1228, 1229
  and phosphines, 1229
  and sodium malonates, 1229
  from alkenes, 1228
  Markovnikov addition, 1228
  with alkenes, mechanism, 1228, 1229
palladium compounds
  and the **Heck reaction**, 1236
  and the **Lindlar catalyst**, 0437
  and the **Rosenmund catalyst**, 0437
  cycloaromatization, 1339
  Heck coupling, powerful in
    synthesis, 0930
  with alkenes, 1227
  with allylic compounds, 1227
palladium on carbon, with ammonium
  formate, 0477
palladium oxidation, of alkenes, 1226
palladium(0)
  alkylation, and chiral ligands, 1234
  and **Castro-Stephens coupling**, 1244
  and dibenzylidene acetone, 1234

and inversion of configuration, 1233
and quinane annulation, 1235
and **Sonogashira coupling**, 1244
and **Stille coupling**, 1240
and **Suzuki-Miyaura coupling**, 1242
and the **Heck reaction**, 1237
bis(imidazole) ligands, 1238
catalyst for arylation, 1235
reactions, mechanism, 1233
with enamines, 1234
palladium, allyl, 1225
  and dppe, 1230, 1231
  and HMPT, 1230
  and tri(ortho-tolyl)phosphine, 1230
  catalytic reactions, 1231
  reactions, 1229
  regioselectivity of reactions, 1229,
    1230
  with malonates, 1229
  with phosphines, 1229
  with triphenylphosphine, 1229
palladium
  and cis hydrogenation of alkynes, 0433
  and coupling reactions, 0932
  and **Negishi coupling**, 1225
  and **Wacker oxidation**, 1226
  coupling of alkylboranes, 0511
  hydrogenation catalyst,
    hydrogenolysis, 0430
  π-allyl, 1225-1236
  regioselectivity with alkenes, 1228
palladium-on-carbon, hydrogenation
  of nitro compounds, 0442
π-allyl palladium complexes, 1225-
  1236
palmerolide A, 0741
palytoxin, 0849
pamamycin-607, 0579
panaxytriol, 0631, 1302
pancracine, 1111
pancratistatin 0306, 0900, 0903, 0904
  total syntheses, 0942-0947
Pancratium littorale, 0903
panepophenathrin, 0379
paniculidine B, 0913
paraffin, and metathesis catalysts, 1336
paraformaldehyde, with amide-dienes,
  1184
paraherquamide A, 0480, 0595
parallel synthesis, 0978
para-quinones, 0262
pareitropone, 1327
paroxetine, 0356, 0374, 0716

partial cleavage, of alkenes, 0321
partial reduction, of aryl compounds, 0459, 0460
α-participation, and through space interactions, 1173
particles sites, in catalysts, 0427
parviflorin, 0288
Passerini reaction, 0627
patchoulenone, 0306
Paternò-Büchi reaction, 1094
Pauli exchange steric repulsion, 0038
Pauling's electronegativity scale, 0219
Pauson-Khand reaction, 1342-1344, 1366
  and amines, 1343
  and chiral aryl biphosphites, 1344
  and chiral phosphines, 1344
  and cobalt octacarbonyl, 1342
  and molybdenum hexacarbonyl, 1343
  and ultrasound, 1344
  aza, 1344
  enantioselective, 1344
  mechanism, 1342, 1343
  of ene-yens, 1343
  substoichiometric metals, 1343
pavoninin-4, 0329
Paxil, 0716
Payne epoxidation, 0271
PBr₃-AlCl₃, and **Friedel-Crafts acylation**, 1206
PCC (pyridinium chlorochromate), 0620
  and alumina, 0237
  oxidation of alcohols, 1186
  oxidation of hemi-acetals, 1034
  oxidation of oximes, 0237
  rearrangements, 0236
  with alcohols, 0236
  with alkylboranes, 0528
PDC (pyridinium dichromate)
  oxidation of alcohols, 0237
  and DMF, 0238
Pechmann condensation, 1220
  in ionic liquids, 1221
pectenotoxin 2, 0023
peduncularine, 1222
peloruside A, 0526, 0838
penitrem D, 0243
pentalenes, 1343
pentalenolactone E methyl ester, 1329
pentenocin B, 1036
peptides
  and the **Merrifield synthesis**, 0979
  combinatorial analysis, 0984

libraries, 0980
synthesis, 0979, 0980
perbromide, potassium, and **Dess-Martin periodinane**, 0247
perchloric acid
  and chromic acid, 0227
  and hydrogenation, 0435
perfluoropropane, and carbene with alkenes, 1323
perfumes, and ambergris, 0951
perhydrohistrionicotoxin, 0444, 0705
perhydroisoquinolines, 1185
perhydrophenalenol, 0531
perhydrothiophene, 0448
pericyclic reaction types, 1011
periodate, sodium
  and alkenes, 0324
  and NMO, 0315
  and periodic acid, 0314
  and ruthenium trichloride, 0316
  and sulfide oxidation, 0168
  and the **Lemieux-Johnson** reagent, 0314
  with sulfides, 0334, 0335
periodic acid
  and NMO, 0315
  and permanganate, 0313
  and ruthenium tetraoxide, 0315
  and sodium periodate, 0314
  and the **Lemieux-von Rudloff** reagent, 0313
Periodic table
  and acidity, 0083
  and Lewis acids, 0094
  and nucleophilicity, 0119, 0121
  and oxidation number, 0220
periplanone B, 0456, 0757
periplanone C, 0460, 0934
Perkin reaction, 0827
Perkin salicylic synthesis, 0808
permanganate
  alternative oxidants, 0260
  and [3+2]-cycloaddition, 1103
  and benzyltriethylammonium chloride, 0294
  and phase transfer catalysts, 0294
  and reduction potential, 0223
  in acids or bases, 0224
  potassium, and dihydroxylation, 0292
  potassium, and hydroxy ketones, 0297
  potassium, and manganese sulfate, 0252
  potassium, and periodic acids, 0313

potassium, and pyridine, 0234
potassium, and the **Lemieux-von Rudloff reagent**, 0313
potassium, oxidation of alkyl fragments, 0314
potassium, oxidative bond cleavage, 0313
reduction potential, 0292
with alkenes, 0293
perophoramidine, 0732
perovskone, 0668, 1202
peroxide
  acyl, mechanism of decomposition, 1261
  benzoyl, and halogenation, 0151
  benzyl, thermolysis, 1260
  cumyl hydro-, reaction with alkenes, 0272
  dilauroyl, and radical cyclization, 1283
  di-tert-butyl, 1260, 1261
peroxide, hydrogen
  and alkylboranes, 0495
  and **Fenton's reagent**, 0266
  and ferrous ion, with aryls, 0266
  and KF, **Tamao-Fleming oxidation**, 0331
  and ozonide, 0320
  and **Tamao-Fleming oxidation**, 0331
  and the **Baeyer-Villiger reaction**, 0307
  and transition metals, 0270
  and urea, complex, 0276
  oxidation of selenides, 0169
  oxidation of sulfides, 0335
  reaction with alkenes, 0271
  reduction potential, 0269
  solvents for, 0271
  with alkenes, 0269
  with alkyl boranes, 0513
  with base, 0275
  with boron aldolates, 0843
peroxide, tert-butyl hydroperoxide. *See* TBHP
peroxide
  tert-butyl hydroperoxide, and **Sharpless asymmetric epoxidation**, 0272
peroxides. *See also* hydroperoxide
peroxides, acyl, thermolysis, 1261
peroxides, alkyl
  mechanism of decomposition, 1261

model for **Sharpless asymmetric epoxidation**, 0285, 0287
Sharpless asymmetric epoxidation, 0284
  with alkenes, 0269, 0271
peroxides
  and alkenes, 0180
  and carboxylic acids, 0277
  and sulfides, 0334
  benzoyl, thermolysis, 1260
  bond dissociation energy, 1260
  conformation, 0044
  diacyl, and oxidation of phenols, 0267
  diacyl, photochemistry, 0267
  dialkyl, thermolysis, 1260, 1261
  formation of radicals, 1260
peroxides, hydro, tert-butyl
  alkyl, with base, 0275
  and nitriles, 0276
  and transition metals, 0272
  carboalkoxy, from ozonoloysis, 0318
  tert-butyl, and OsO_4, 0296
  tert-butyl, and oxidation of sulfides, 0337
  tert-butyl, and ruthenium chloride, 0330
  tert-butyl, and SeO_2, 0326
  with transition metals, 0272
  with vanadium compounds, 0273
peroxides
  metal, 0273
  radical initiators, 1281
  reactivity in epoxidation reactions, 0270
  reactivity with alkenes, 0269
  with aldehydes and manganese acetate, 1279
  with alkenes, 0269
  with amines, 0339
  with HBr, 0542
peroxomonosulfate, potassium (Oxone), and dioxiranes, 0291
peroxyacarnoate A, 0646
peroxyacetic acids, oxidation of aryls, 0266
peroxyacids
  and acidity, 0277
  and buffers, 0308
  and conjugated carbonyls, 0282, 0293
  and epoxidation, 0277-0284
  and ionization potential maps, 0281
  and **Lewis acids**, 0266
  and neighboring group effects, 0560

and substituent effects, 0282
and **Tamao-Fleming oxidation**, 0331
and the **Baeyer-Villiger reaction**, 0307
and the **Dakin reaction**, 0311
and the **Henbest rule**, 0283
and the **Prilezhaev (Prileschajew) reaction**, 0277
electronic effects of epoxidation, 0280
mechanism of epoxidation, 0279, 0280
neighboring group effects in epoxidation, 0280
oxidation of silanes, 0331
oxidation of sulfides, 0335
reactivity in the **Baeyer-Villiger reaction**, 0310
stereoselectivity, 0284
steric effects of epoxidation, 0280
with alkenes, 0277-0284
with amines, 0339
with diketones, 0311
with formate esters, 0311
with keto-esters, 0311
with ketones, 0926
with quinones, 0311
with silanes, 0331
with sulfides, 0334, 0336
with tosylhydrazones, 0564
peroxybenzimidic acid
  and alkenes, 0271
  and **Payne epoxidation**, 0271
peroxybenzoic acid. *See* peroxyacid
peroxycamphoric acid oxidation of sulfides, 0336
peroxycarboxylic acids, and chelation with alcohols, 0549
peroxyformic acid. *See* peroxyacid
perphthalic acid, and epoxidation, 0283
persulfate, potassium,
  Elbs persulfate oxidation, 0267
  oxidation of aryls, 0267
perturbation theory
  and HOMO and LUMO, 1015
  and molecular orbital theory, 1001
  and secondary orbital interactions, 1031
  and transition states, 1001
PET imaging, 0294
Petasis reagent, 0756
  with aldehydes or ketones, 0756
  with esters, 0756
Peterson olefination, 0709

and the **Wittig reaction**, 0709
petroporphyrins, 0932
pH
  and reduction of aldehydes or ketones, 0383
  and ester hydrolysis, 0605
  and **Fremy's salt**, 0262
  and manganese dioxide, 0252
  and permanganate dihydroxylation, 0293
  and reduction, 0382
  and sodium cyanoborohydride, 0382, 0383
phase transfer catalysts
  and dihydroxylation, 0294
  and ether formation, 0594
phenanthrenes
  and electrophilic aromatic substitution, 0197
  by **Bardhan-Sengupta synthesis**, 1208
  by **Bogert-Cook synthesis**, 1209
  by **Haworth synthesis**, 1207, 1208
  nitration and molecular modeling, 0198
phenanthridine, 1219
phenanthridones, 0945
phenanthro[9,10,3',4']indolizidine, 0388
phenolic esters, with **Lewis acids**, 1219
phenolic oxidative coupling, 0921, 1296
phenols
  and **Baldwin's rules**, 0568
  and electrolysis, 1298
  and **Fenton's reagent**, 0266
  and **Fries rearrangement**, 1220
  and **phenolic oxidative coupling**, 1296
  and pKa, 0086-0087
  and resonance, 0087, 0771
  and spirocyclization, 0772
  and tautomerism, 1081
  and the **Williamson ether synthesis**, 0771
  and **Udenfriend's reagent**, 0266
  anodic oxidation, 0264
  by [2+2]-cycloaddition, 1081
  by **Danheiser annulation**, 1081
  by oxidation of aryls, 0266
  from aryl amines, 0209
  from aryl diazonium salts, 0208
  from aryl halides, 1179

from ketenes and alkenes, 1081
oxidation with **Fremy's salt**, 0262
oxidation, dimerization, 0265
reagents for oxidation, 0264
steric hindrance and oxidation, 0263
substituent effects with Fremy's salt, 0263
with base, 0771
with keto esters, 1220
phenonium ions
   and aryl migration, 1172
   modeling, 1171
phenoxide ion, and resonance, 0087
phenylalanine, 0386, 0536, 0956
   from phenylacetaldehyde, 0628
phenylalinol, 0386
phenylbis(triphenylphosphino)-
   palladium chloride, 1237
phenylene, 0901
phenylglycinol, 0629
phenylhydrazine
   and the **Fischer Indole Synthesis**, 1221
   with aldehydes or ketones, 1222
phenylhydrazones
   from aldehydes or ketones, 1222
   with **Lewis acids**, 1222
   with sulfuric cid, 1218
phenyliodine (III)bis(trifluoroacetate), 1297
phenylisothiocyanate, with
   Grignard reagents, 0658
phenyllithium, as a base, 0092
pheromone gland of a stinkbug, 0609
pheromones, 0956, 0957
   of the citrus mealy bug, *Planococcus citri*, 0353
phomactin A, 0630
phomoidride B, 0807
phomoidrides, 0847
phorboxazole A, 0243, 0381, 0756
phorboxazole, 0183, 0419
phoreactionsholide J, 1135
phosphate, disodium buffer in the
   Baeyer-Villiger reaction, 0308
phosphazene bases,
   and **anionic oxy-Cope rearrangement**, 1130
phosphazines, ylid formation, 0730
phosphine ligands, and rhodium
   catalysts, 0450
phosphine oxides
   β-hydroxy, 0742

β-hydroxy, with silica gel, 0743
carbanion, with a **Weinreb amide**, 0742
carbanion, with carbonyls, 0741, 0742
with organolithium reagents, 0743
ylids, 0739
phosphines
   and ally palladium, 1229
   and HSAB, 0100
   and hydrogenation, 0449
   and palladium complexes, 1229
   and S_N2 reactions, 0121
   and the **Baylis-Hillman reaction**, 0879
   and the **Mitsunobu reaction**, 0545
   and Wilkinson's catalyst, 0449
   and ylids, 0121
   as a poison for hydrogenation, 0426
   chiral ligands, 0449
   chiral, 0449
   chiral, and hydrogenation, 0449, 0450
   chiral, and the **Pauson-Khand reaction**, 1344
   fluxional inversion, 0021
   fluxional inversion, 0449
   ligand effects in asymmetric hydrogenation, 0450
   ligands in hydrogenation, 0428
   S_N2 reaction with alkyl halides, 0730
   with alkyl halides, 0730
   with cyclopentadienyl dicarbonyl and alkenes, 1249
phosphinimines, ozonolysis, 0340
phosphite, trimethyl, and ozonide, 0321
phosphites, alkyl, and S_N2, 0121
phosphodiesterase-4 (PDE-4) inhibitors, 0987
phosphonate carbanions, 0739
phosphonate esters. *See* esters
phosphonate esters, elimination reactions, 0716
phosphonic esters, and S_N2 reactions, 0121
phosphonium salts
   alkenyl, 0744
   and guanidine bases, 0731
   and pK_a, 0731
   and the **Mitsunobu reaction**, 0124
   structural characteristics, 0731
   vinyl, with thioketones, 0744
   water soluble, 0733
   with base, 0730

with dimsyl sodium, 0731
with organolithium reagents, 0731
with pyrrole-carboxaldehydes, 0744
phosphonothioates,
   with aldehydes or ketones, 0743
phosphoranes, and ylids, 0732
phosphorescence
   and photochemistry, 1086
   and radiative deactivation, 1086
phosphoric acid
   and chormic acid, 0227
   and **Moffatt oxidation**, 0244
phosphorous oxybromide
   with aromatic rings, 0147
   and amides, 0372
phosphorous pentabromide, and $AlCl_3$, 1206
phosphorus halides. *See* halides
phosphorus oxychloride
   and amines, 0372
   and **Bischler-Napieralski reaction**, 1215
   and **Borsche-Drechel cyclization**, 1218
   and DMF, 0372
   and iminium salts, 0372
   and the **Vilsmeier reaction**, 0372
   and the **Vilsmeier-Haack complex**, 0372
   and the **Vilsmeier-Haack** reaction, 1206
   with alcohols, 0146
   with aryl amides, 1218
   with pyridine, 0148, 0701
phosphorus pentabromide, with alcohols, 0146
phosphorus pentachloride, with alcohols, 0146
phosphorus pentoxide, and Bardhan-Sengupta phenanthrene synthesis, 1208
phosphorus pentoxide
   and DMSO, 0246
   and **Gould-Jacobs reaction**, 1212
phosphorus tribromide. *See* tribromide
phosphorus tribromide, with alcohols, 0127, 0146
phosphorus trichloride, with alcohols, 0146
phosphorus ylids. *See* ylids
phosphorus ylids, 0729-0745
phosphorus, with iodine, 0148
phosphoryl bromide. *See* bromide

phosphoryl chloride, with alcohols, 0146

photo Fries rearrangement, 1220

photoaddition, 0945

photochemical cleavage, and radicals, 1276

photochemical filters, 1085, 1086

photochemistry
  [2+2]-cycloaddition, 1091-1097
  absorption maxima, 1084

photochemistry
  and [2+2]-cycloaddition, 1082
  and absorption spectra, 1088
  and absorption, 1087
  and biradicals, 1094
  and bond dissociation energy, 1085
  and carbenes, 1312, 1323
  and diazoalkanes, 1110
  and diazocarbonyl compounds, 1315
  and dibutyltin dimer, 1289, 1290
  and diradicals, 1091
  and emission spectra, 1088
  and emission, 1087
  and exciplex, 1088
  and excited states, 1082
  and excited states, 1088
  and filters, 1085, 1086
  and fluorescence spectrum, 1088
  and fluorescence, 1086
  and formation of radicals, 1262
  and functional groups, 1084
  and ground state, 1012
  and **Hofmann-Löffler-Freytag reaction**, 1294
  and **Hünig's base**, 1091
  and imine coupling, 1305
  and internal conversion, 1087
  and intersystem crossing, 1087, 1088
  and **Jablonski Diagrams**, 1087
  and mercury lamps, 1085
  and phosphorescence, 1086
  and **Planck's constant**, 1083
  and Pyrex filters, 1091
  and radiative deactivation, 1086
  and radical cyclization, 1289, 1294
  and radicals, 1276
  and sensitizer, 1088, 1089
  and sodium lamps, 1085
  and SOMO, 1086
  and SOMO, 1263
  and suprafacial addition, 1092
  and the **Franck-Condon principle**, 1087

and the **Barton reaction**, 1276

and the electromagnetic spectrum, 1083

and UV spectra, 1083, 1084

and **Wolff rearrangement**, 1315

photochemistry
  chemiluminescence, 1082
  diacyl peroxides, 0267
  diastereoselectivity, 1092
  diazo compounds, 1314
  electrocyclization, 1089
  energy transition states, 1086
  energy transitions, 1086
  filters, 1262, 1263
  fluorescence, 1082
  formation of radicals, 1262
  general principles, 1082-1090
  HOMO and LUMO, 1082
  of aldimines, 1305
  of alkenes, 1091
  of cyclopropanes, 1316
  of diazirines, 1314
  of diazoalkanes, 1322
  of diazoketones, 1328
  of imines, 1305
  of ketenes, 1312, 1318
  of oxiranes, 1316
  of pyrazolines, 1110
  of triazolines, 1114
  photo Fries rearrangement, 1220
  photocycloaddition, 1092
  Pyrex filters, 1085, 1086
  quantum efficiency, 1090
  quantum yield, 1090
  quartz filters, 1085, 1086
  reduction of ketones, 0562
  remote functionalization, 0562
  tethers in photocycloaddition, 1092

photocleavable groups, combinatorial chemistry, 0984

photocyclization, 1091
  cyclopentenone, 1094
  powerful in synthesis, 0930, 0932
  solvent effects, 1093
  and orbital coefficients, 1093
  Paternò-Büchi reaction, 1095
  tether groups, 1092

photodissociation, of aldehydes, 1265, 1266

photolysis
  of cyclobutenones, 1081
  and conjugation, 1264
  and hydrogen atom donors, 0477

and photoreduction, 0477

in Pyrex, 1265

iron-diene complexes, 1248

of aldehydes, 1264

of alkenes, 1264

of alkyl halides, 1266

of bromine or chlorine, 0151

of chloroform, 1319

of cyclopropanes, 1316

of diazines, 1322

of diazirine, 1314

of diazo compounds, 1314

of diazoketones, 1328

of halogens, 1266

of ketenes, 1312

of ketones, 1264, 1318

of maleimide, 1265

of N-haloamines, 1294

of oxiranes, 1316

photoreduction, 0477
  competing pathways, 0478
  of alkyl halides, 0478
  of imines, 0478
  of ketones, 0562

phthalazines, and **Sharpless asymmetric dihydroxylation**, 0300

phthalhydrazide, and S_N2 reactions, 0122

phthalimide, and S_N2 reactions, 0122

phthalimides, and the **Gabriel synthesis**, 0125

phyllostine, 0275

physostigmine, 0442, 1344

phytuberin, 0274

picraline alkaloid system, 1356

picrotoxinin, 0956

Pictet-Hubert reaction, 1219

Pictet-Spengler isoquinoline synthesis, 1216
  chiral, 1216
  enantioselective, 1217
  with microwaves, 1217

pig liver esterase, 1181

pin approach, combinatorial chemistry, 0981, 0982

pinacol coupling (pinacol reaction), 1303
  and dissolving metal reduction, 0454
  and electrolytic reduction, 0469
  and metals, 1304
  and molecular modeling, 1303
  and silanes, 1305
  diastereoselectivity, 1304

enantioselectivity, 1304
kinetic versus thermodynamic, 1303
mixed, 1305
stereoselectivity, 1304
pinacol rearrangement, 1179
pinacol reduction, and electrolytic
reduction, 0469
pinacol-like rearrangements, 1182
pinacolone, 1318
pinane, and **Bredt's rule**, 0553, 0554
pinene, 0278, 0493, 0497, 0501, 1091,
1144, 1332, 1347
and borane, 0524
complex with 9-BBN, 0395
with borane, 0495
with boranes, 0519, 0520
pinguisenol, 0457, 0680, 1317
pinitol, 0304
pinnatal, 0991
pins, and combinatorial chemistry,
0981, 0982
piperidine
and the Knoevenagel condensation,
0829
with malonic acid, 0830
piperidines
conformation, 0065
via S_N2' reaction, 0126
pironetin, 0452
Pitzer strain
and chair conformations, 0049
and macrocyclizations, 0573
definition, 0048
Piv. *See* pivaloyl
pivaloyl chloride, with alcohols, 0605
pK_a
alcohols, 0091
aldehydes or ketones, 0781, 0782
alkynes, 0088, 0632, 0633
amines, 0088, 0788
ammonia, 0788
and aromaticity, 0692
and basicity, 0088
and chormic acid, 0227
and conjugation, 0690
and conjugative effects, 0690, 0691
and DMSO, 0697
and d-orbitals, 0692
and electron withdrawing groups,
0783
and electrostatic potential, 0694, 0702
and enolate anions, 0690
and functional groups, 0784, 0785

and homoconjugative effects, 0691
and hydrogen bonding, 0783
and inductive effects, 0690, 0691
and organolithium reagents, 0676
and oxidation of sulfides, 0335
and phosphonium salts, 0731
and resonance delocalization, 0690,
0691
and s-character, 0690
and solvation, 0088
and solvent effects, 0091, 0798, 0800
carboxylic acids, 0085, 0691
correlation with functional groups,
0692, 0693
definition, 0082
dialdehydes, 0782
dialkylamines, 0788
diesters, 0782, 0808
diketones, 0782
dithianes, 0704
esters, 0782
fluorine, 0731
ketones, 0799
malonates, 0808
nitriles, 0782
of alcohols, 0082
of alkanes, 0639
of ethers, 0082
of sulfur compounds, 0702
phenols, 0086-0087
solvent effects, 0788
substituent effects, 0784
sulfur compounds in DMSO, 0697
table, 0693
unsymmetrical ketones, 0796
water, 0091
planar radicals, 1258
planarity
and strain, in electrophilic aromatic
substitution, 0203
in aromatic substitution, and the
Mills-Nixon effect, 0204
of radicals, 1270
Planck's constant, 1083
platinum black
and hydrogenation, 0439
and Pd/C, as a hydrogenation
support, 0426
platinum catalysts, **Adams'** catalyst,
0438
platinum oxide. *See* **Adams'** catalyst
platinum oxide, and hydrogenation,
0439

platinum
and cis hydrogenation of alkynes,
0433
hydrogenation catalyst,
hydrogenolysis, 0430
plueromutilin, 0461
plumerinine, 0439
PMB. *See* methoxybenzoyl
$POCl_3$
and **Friedel-Crafts acylation**, 1208
and **Friedel-Crafts cyclization**, 1216
and ketene formation, 1078
point of difference, and priority rules,
0017
poison-frog alkaloid (–)-205B, 0379
poisoning
and hydrogenation, 0426
hydrogenation and cobalt
octacarbonyl, 0440
poisons
and hydrogenation catalysts, 0612
and the **Lindlar catalyst**, 0437
and the **Rosenmund catalyst**, 0437
for hydrogenation, 0426
for hydrogenation, 0445
polar effects, and carbocations, 1162
polar map analysis, 0053
polarity inversion, and **Grignard
reagents**, 0636
polarity, and **Diels-Alder** reactions,
1044
polarizability
and HSAB, 0097
and the E2 reaction, 0153
polarization
and umpolung, 0703
bond. *See* bond
bond, and transforms, 0080
C-Li, 0670
C-Mg, 0636
of bonds, and leaving groups, 0080
of bonds, and transforms, 0080
polarized molecules and synthesis, 0006
poly-alanine, **Juliá-Colonna
epoxidation**, 0276
polyalkylation
and catalysts, 1197, 1198
polyalkylation, and Lewis acids, 1197,
1198
and S_N2, 0122
and solvent effects, 1198
Friedel-Crafts alkylation, 1197,
1198

suppression, 1197, 1198
toluene and alkenes, 1201
with enolate anions, 0805
polycavernoside A, 0315
polycyclic molecules, conformations, 0067
polyene cyclization, 0948
and BINOL, 0951
enantioselectivity, 0951
polyenes
and **Johnson polyene cyclization**, 1184
nonnatural, 0987
with tin tetrachloride, 0951, 1184
polyketides, 0987
poly-leucine, and **Juliá-Colonna epoxidation**, 0276
polymer beads, and synthesis 0979, 0980
polymer bound amino esters, 0979
polymerization
and radicals, 1271
cationic, 1271
ring opening metathesis, 1340, 1341
polymers
and cascade cationic cyclization, 1184
and macrocyclizations, 0572
and synthesis, 0979
polymethylene, 0750
polymethylhydrosiloxane, and the **Barton-McCombie reaction**, 0467
polyperoxide, 0318
polyphosphoric acid, in the **Schmidt reaction**, 0191
Pomeranz-Fritsch reaction, 1217
and **Schlittler-Müller modification**, 1218
formation of **Schiff bases**, 1217
pooling strategy, combinatorial chemistry, 0979
porphyrins, 0932
Porphyromonas gingivalis, 0642
potassium acetate. *See* acetate
potassium amide. *See* amide
potassium azodicarboxylate. *See* azodicarboxylate
potassium ferricyanide. *See* ferricyanide
potassium hydride, and the **Heck reaction**, 1237
potassium hydrogen difluoride, and mcpba, 0331
potassium hydroxide. *See* hydroxide
potassium nitrosodisulphonate. *See*

Fremy's salt
potassium osmate. *See* osmate
potassium peroxomonosulfate. *See* peroxomonosulfate
potassium persulfate. *See* persulfate
potassium trifluoroborates. *See* trifluoroborates
potassium
and **pinacol coupling**, 1303
and titanium trichloride, 1309
with aldehydes or ketones, 1303
potential symmetry, and synthesis, 0920
potential, ionization. *See* ionization
PPA
and **Friedel-Crafts** reactions, 1217
and the **Pomeranz-Fritsch** reaction, 1217
pramanicin, 0331
precapnelladiene, 0315
prelactone C, 0931
preliminary scan, and synthesis, 0920
Prelog's rule
and **Veschambre's rule**, 0482
and enzyme reduction, 0479
prenylation, and allyltin compounds, 1281
pressure
and aromatic substitution, 0206
and carbene formation, 1322
and **Diels-Alder reactions**, 1029
and hydrogenation, 0440
high. *See* high pressure
Prévost reaction, 0301, 0302
and the **Hunsdiecker reaction**, 0302
and the **Woodward modification**, 0303
Prilezhaev (Prileschajew) reaction, 0277
primary rings, and synthesis, 0934, 0936
Prins reaction, 1191
and allylsilanes, 1191
and oxonia **Cope** rearrangement, 1191
in micelles, 1191
priority rules
and atomic number, 0015
and electron pairs, 0021
and first point of difference, 0017
and isotopes, 0015
and multiple bonds, 0018
for atoms, IUPAC, 0019
proazaphosphatranes, and the **Henry reaction**, 0831
prochiral alkenes, 0519

prochiral centers, 0030
and **Grignard reagents**, 0662
and reduction, 0348, 0401
prochiral centers, with alkylboranes, 0497
progesterone receptor, 0294
progesterone, 0950
prolinal, and radical cyclization, 1289
proline esters, as chiral catalysts, 1072
proline
and organocuprates, 0723
and the **Robinson annulation**, 0882
chiral base in the **aldol condensation**, 0820
in the **aldol condensation**, 0866
prolinol, and **Evans' auxiliaries**, 0869
propagation steps, radicals and addition reactions, 1271
propane, conformations, energy, rotamers, 0039
propanenitrile, 0623
2-propanol, and **Meerwein-Ponndorf-Verley reduction**, 0467
propargyl bromide. *See* bromide
propargylic alcohols. *See* alkyne-alcohols
propargylic bonds, disconnection, 0636
propene, conformation, 0044-0046
prophos, 0450
propionic acid, with alkylboranes, 0506
propylenedioxy acetals or ketals, 0610
pro-R and pro-S, 0030
prostaglandin E_2, 0877, 1329
prostaglandin E_2-1,15-lactone, 1339
prostaglandins, 0305, 0510, 1146
prostanoids, 0305
protecting groups, 0951
alcohol, silyl, 0269
alcohols, benzyl, 0444
aldehydes and ketones, 0608
amide protecting groups for amines, 0615
amines, benzyl, 0444
and retrosynthesis, 0910
benzyl ethers, 0595
combinatorial chemistry, 0982
definition, 0587
for alcohols, 0594-0608
for amines, 0613-0618
iron tricarbonyl, 1249, 1250
trimethylsilylamines, 0614
protection, definition, 0587
protein phosphatase, 0900

protic solvents, and substitution, 0138
protodesilylation, 0331
proton affinity
  and base strength, 0786
  and equilibrium constants, 0786
  and molecular modeling, 0786
Proton Sponge [1,8-bis
  (dimethylamino)-naphthalene], 0788,
  1080
  and [2+2]-cycloaddition, 1080
  and aldehydes or ketones, 0788
  and basicity, 0090
proton transfer, and dissolving metal
  reduction, 0452
protonolysis
  alkenylboranes, 0509
  of alkylboranes, 0505, 0509, 0510,
    0512
  of alkylboranes, mechanism, 0506
protons, and HSAB, 0098
Pschorr reaction, 1299
pseudocodeine, 0699
pseudoheliotridane, 1289
Pseudomonas putida, 0944
  and aryl dihydroxylation, 0303-0306
pseudorotation, 0052
pseudorotation, and conformation, 0051
psilocin, 0614
π-stacking, and **Diels-Alder**, 1043
psymberin, 0757
pterocellin A, 0911, 0913
pulegone, 0237, 0889
pumiliotoxin 209F, 0111, 0146
pumiliotoxin 251D, 0111, 0146
pumiliotoxin C, 1075
Pummerer rearrangement, and
  Swern Oxidation, 0241
purine nucleosides, 1033
puupehenone, 1297
PyBOP, and lactonization, 0581
pyramidal radicals, 1258
pyranicin, 0993
pyranoside homologation, 0958
pyrans
  alkoxy, conformations, 0064, 0065
  by the Diels-Alder reaction, 1054
  conformational modeling, 0065
  oxidation with thallium nitrate, 0333
pyrazino[2.1-b]quinazoline-6-one,
  1058
pyrazolines
  by [3+2]-cycloadditions, 1110
  extrusion of nitrogen, 1111

  from acrylonitriles and
    diazomethane, 1321
  from diazoalkanes, 1110
  photochemistry, 1110
pyrazolizidines
  from azomethine imines, 1111
  from alkenes, 1111
pyretic acid, and **Doebner reaction**,
  1214
Pyrex
  and photochemistry, 1091
  and radicals, 1276
  filer in photochemistry, 1085, 1086
  photolysis of imides, 1265
pyridine and pyridines
  and alcohols, 0145
  and alkylboranes, 0502
  and **Bobbitt's reagent**, 0248
  and chromium trioxide, 0234
  and chromium trioxide, complex,
    0230, 1360
  and **Collins oxidation**, 0235
  and **Friedel-Crafts alkylation**, 1199
  and potassium permanganate, 0234
  and sulfuric acid with mercuric
    sulfate, 0199
  chlromium trioxide complex, 0235
  electrophilic aromatic substitution,
    0199, 1193
  lithio, 0976
  with Lewis acids, 1199
  with osmium tetraoxide, 0295
  with POCl$_3$, 0148, 0701
  with TFAA, 0616
  with thionyl chloride, with alcohols,
    0552
pyridinium bromide perbromide, and
  bromination, 0185
pyridinium chlorochromate. *See* PCC
pyridinium dichromate. *See* PDC
pyridinium salts
  and halogenation of alcohols, 0145
  and **Moffatt oxidation**, 0244
  ionic liquids, 1037
  with NaBH$_4$, 0372
pyridinium trifluorroacetate, and
  Moffatt oxidation, 0244
pyridinium ylids. *See also* ylids
pyridinium ylids, 0750
  and **Michael addition**, 0751
pyridones, and high pressure **Diels-
  Alder** reactions, 1047

pyridoxine, 0442
pyridyl-tributyltin, 1240
pyrimidines, by **Diels-Alder** reactions,
  1033
pyrocatechols, 0267
pyrogallol, 0944
pyroglutamate, and radical cyclization,
  1289
pyrolysis
  of amine oxides, 0165
  of esters, 0166
  of selenoxides, 0169
  of sulfoxides, 0168
  of xanthate esters, 0167
pyrones, by the **Diels-Alder** reaction,
  1054, 1071
  Diels-Alder reactions, high pressure,
    1047
  retro-**Diels-Alder**, 1036
pyrrolam A, 1292
pyrrole
  and **Friedel-Crafts acylation**, 1205
  and **Noyori annulation**, 1190
  electrophilic aromatic substitution,
    0199, 1193
  Vilsmeier-Haack reaction, 1207
pyrroles, 0461
  and **Diels-Alder reactions**, 1053
  base strength, 0785
  hydrogenation, 0447
  carboxaldehydes, with vinyl
    phosphonium salts, 0744
pyrrolidine (−)-197B, 1274
pyrrolidines
  and enamine formation, 0873, 0874
  base strength, 0785
  by [3+2]-cycloaddition, 1111
  chiral catalyst, 1034
  from azomethine ylids, 1111
pyrrolidiniminiumbutanoic acid, 0476
pyrrolines
  and **Diels-Alder reactions**, 1055
  by [3+2]-cycloaddition, 1105, 1113
  from alkyl azides, 1113
  from aziridines, 1113
  from azomethine imines, 1113
  from nitrile ylids, 1105
  from triazoles, 1113
  with siloxydienes, 1055
pyrrolizidine alkaloids, 0374, 1113-
  1115
pyrrolizines, from pyrroles-
  carboxaldehydes, 0744

pyrrolo-[1,2a]-pyrazine, 1217
pyrrolo[2,3-b]-indol-2-ones, 1344
pyruvates, with imines, 1214

Q

quantum chemical calculations, 0116
quantum efficiency
  and radical formation, 1267
  photochemistry, 1090
quantum mechanical calculations, and
  conformation, 0038
quantum yield
  and carbenes, 1312
  and radical formation, 1267
  photochemistry, 1090
  and ketenes, 1312
quartromicins, 0169, 1071
Quartz, filter in photochemistry, 1085,
  1086
quassinoids, 1042
quenching agents, for radicals, 1272
quinane annulation, 1235
  with palladium(0), 1235
quinazolinones, and **Diels-Alder**, 1058
quinidine, 0556
  chiral catalyst, 1073
  with LiAlH₄, 0393
quinine, 0174, 0447, 0556, 0898
  chiral catalyst, 1073
  with LiAlH₄, 0393
quinines
  an Diels-Alder reaction, 1042
  and air oxidation, 1214
  and the **Bally-Scholl synthesis**, 1208
  by [2+2]-cycloaddition, 1081
  from glycerol, 1215
  from triols, 1215
quinoids. See quinones
quinoline
  and decarboxylation, 0173
  and the **Lindlar catalyst**, 0437
  and the **Rosenmund catalyst**, 0437
  electrophilic aromatic substitution,
  0199
quinolines, 0611
  by **Combes quinoline synthesis**, 1213
  by **Conrad-Limpach reaction**, 1212
  by **Doebner-Miller reaction**, 1214
  by **Friedel-Crafts reactions**, 1211,
  1212
  by **Gould-Jacobs reaction**, 1212
  by **Knorr quinoline synthesis**, 1211

by the **Doebner reaction**, 1214
by the **Skraup reaction**, 1215
from anilines, 1211
from diketones, 1213
from keto esters, 1211
from pyruvic acids, 1214
quinols, 1118
quinones
  and **Diels-Alder** reactions, 0928,
  1044, 1049
  and **Fremy's salt**, 0262
  by oxidation of catechols, 0265
  by oxidation, 0261-0266
  from resorcinols, 0265
  ortho and para, 0262
  reagents for oxidation of phenols,
  0264
  with peroxyacids, 0311
quinuclidine
  and basicity, 0090
  catalyst for [2+2]-cycloaddition, 1080

R

+R and –R effects, and basicity, 0089
R/S configuration, 0014-0023
  and computational methods, 0023
racemization
  and [3+2]-cycloaddition, 1108
  and E1, 0160
  and **Friedel-Crafts** alkylation, 1196
  and organolithium reagents, 0681
  and S_N1 reactions, 0131
  and solvolysis, 1170
  of carbocations, 1170
radiative deactivation
  and fluorescence and
    phosphorescence, 1086
  and photochemistry, 1086
radical anions
  and dissolving metal reduction, 0453
  and electrolytic reduction, 0468
  dissolving metal reduction of alkynes,
  0455
radical coupling. See coupling
radical cyclization, 0466, 0946, 1277,
  1283-1296, 1357
  activation barrier, 1284
  and AIBN, 1284, 1285, 1289, 1290,
  1293
  and alkene-halides, 1283
  and amidyl radicals, 1293

and **Baldwin's rules**, 0565, 1283,
  1284, 1286, 1288
and boranes, 1291, 1292
and carbamates, 1294
and chiral templates, 1289
and concentration, 1283
and conformations, 1288
and copper compounds, 1294
and cupric chloride, 1294
and cuprous chloride, 1294
and diastereoselectivity, 1290
and dibutyltin dimer, 1289, 1290
and dilauroyl peroxide, 1283
and E/Z isomers, 1285
and ene-diynes, 1295
and heteroatom radicals, 1293
and hydrogen abstraction, 1291
and hydrogen transfer, 1283
and metathesis, 1340
and molecular modeling, 1284, 1290,
  1291
and photochemistry, 1289, 1294
and samarium iodide, 1292
and selenides, 1288, 1289
and sulfides, 1288, 1289
and TEMPO, 1293
and transition state modeling, 1290,
  1291
and tributyltin hydride, 1284, 1285,
  1289, 1290
radical cyclization
  catalytic tributyltin hydrides, 1293
  cis/trans isomers, 1285
  cis-trans selectivity, 1285
  concentration effects, 1283
  enantioselectivity, 1289, 1292
radical cyclization, formation of
  benzofurans, 1292
  cyclic amines, 1288
  cyclic ethers, 1288
  lactams, 1292, 1293
  lactones, 1288
  macrocycles, 1292
radical cyclization, heteroatom, 1287
  and hydrogen transfer, 1289
  diastereoselectivity, 1288, 1289
  syn/anti isomers, 1289
radical cyclization
  kinetic and thermodynamic product,
  1291
  kinetic vs. thermodynamic control,
  1284
  of carbonyl compounds, 1293

of imines, 1293
photolysis of N-haloamines, 1294
powerful in synthesis, 0929, 0931
ring size, 1291
SOMO-LUMO, 1283
stereoselectivity, 1285, 1288
transition state energy, 1288
transition state strain energy, 1283, 1284
transition states, 1283, 1284, 1290, 1291
radical cyclization
  with alkynes, 1286, 1292
  with chiral auxiliaries, 1286
  with haloamines, 1294
  with oximes, 1293
  with vinyl radicals, 1286
radical inhibitors, galvinoxyl, 0533
radicals
  and **Jacobsen-Katsuki asymmetric epoxidation**, 0290
  substitution reactions, 1273
  acetate, SOMO, 1269
  acyl, 1261
  acyl, and manganese acetate, 1279
  addition reactions, 1271
  addition, carbon tetrachloride, 1272
  alkoxy 1260, 1261, 1275-1277
  alkoxy, and the **Barton reaction**, 1277
  alkyl, 1263
  amidyl, and radical cyclization, 1293
  amino, 1294
  ammonium, 1294
radicals
  and [2+2]-cycloaddition, 1076
  and **acyloin condensation**, 1306
  and addition reactions, 1270
  and aldehydes, 1303
  and alkenes with HBr and peroxides, 0542
  and alkyl halides, 1283
  and **anionic oxy-Cope rearrangement**, 1130
  and anti-Markovnikov addition, 0180
  and **Barton-McCombie reaction**, 1274
  and **Bergman cyclization**, 1295
  and bond dissociation energy, 1260
  and carbenes, 1311
  and CCl$_4$, 1275
  and chiral auxiliaries, 1267
  and chlorination, 1272

and copper compounds, 1298
and coupling reactions, 1270
and electrochemistry, 1298
and electrolytic reduction, 0468
and electronic excitation, 1263
and **Fenton's reagent**, 1296
and fragmentation, 1270
and **Fremy's salt**, 0262, 0263
and **Grignard reagents**, 0637
and hydrogen transfer agents, 1272
and hydrogen transfer, 1274
and hydrogenolysis of benzylic ethers, 0457
and hydroxamic acids, 1293
and hypobromites, 1277
and ketones, 1303
and ketyls, 1303, 1304, 1306
and lithium-hydrogen exchange, 0690
and manganese acetate, 1278
and **Meerwein arylation**, 1298
and mercury, 1346
and MnO$_2$ oxidation of alcohols, 0253
and nitrites, 1276
and **Norrish type I** fragmentation, 1264
and Norrish type I reactions, 1266
and **Norrish type II** fragmentation, 1264
and organic synthesis, 1258
and oxidaitn of aldehydes to carboxylic acids, 0232
and oxidation, 1270
and **phenolic oxidative coupling**, 1296
and photochemistry, 1094, 1276
and photolysis of ketones, 1318
and photoreduction, 0478
and **pinacol coupling**, 1303
and planarity, 1270
and polymerization, 1271
and Pyrex filters, 1276
and reactions of alkenes, 0180
and rearrangement, 1270, 1276
and scandium triflate, 1281
and selenides, 1293
and the **Barton reaction**, 1276
radicals
  and six-center transition states, 1276
  and sodium sand , 1306
  and SOMO, 0102
  and spin density, 1270
  and steric hindrance, 1270
  and substitution, 1270

and sulfides, 1281
and TEMPO, 0248, 1295
and the **Barton-McCombie reaction**, 0168
and the **Gomberg-Bachmann reaction**, 1299
and the **Grob fragmentation**, 0176
and the **Paternò-Büchi reaction**, 1095
and the **Pschorr reaction**, 1299
and the **Ullmann reaction**, 1300
and thionocarbonates, 1274
and tributyltin hydride, 1279, 1286
and **Walsh diagram**s, 1259
and xanthates, 1275
and zinc, 1257
radicals
  aryl, 1297-1299
  azo, 1262
  benzylic, 1273
  by **Kolbe electrolytic synthesis**, 1278
  by thermolysis of alkyl peroxides, 1260
  by thermolysis, 1259
  carbon, 1257
  cascade reactions, 1286
  chain propagation steps, 1271
  chain radical addition reactions, 1271
  chain termination steps, 1271,1272
  chlorine, substitution reactions, 1273
  conjugate addition, 1304
radicals, coupling reactions, 1270, 1280
  and hydrogen transfer, 1280
  and alkynes, 1282
  and allylbutyltin, 1280, 1281
  and allyltin compounds, 1280, 1281
  and syn/anti isomers, 1281
radicals
  cyclization, 1272, 1277
  diphenylmethyl, 1270
  diradicals cages, 1266
  disproportionation, 1270
  electrophilic, 1268
  enantioselective reactions, 1267
  energy and stability, 1258, 1284
  energy diagram, 1259
  ethyl, 1257, 1275
  fluorenyl, 1270
  fluorine, 1267
radicals, formation, 1259
  by homolytic cleavage, 1259
  by **Norrish** Type I fragmentation,

1265, 1266
formation of H_2, 1263
formation with UV, 1262
and quantum efficiency, 1267
and quantum yield, 1267
formation, and sensitizers, 1266
radicals
  fragmentation, 1275
  from aldehydes, 1278
  from alkyl halides, 1266, 1279
  from azo compounds, 1261
  from carboxylic acid salts, 1278
  from halogens, 1266
  from hypochlorite, 1275
  from peroxides, 1260, 1261
  from sulfides with alkylboranes, 0537
radicals
  ground state energy, 1259
  halogen, SOMO, 1274
  halogen, substitution reactions, 1273
  HBr and alkenes, 1282
  heteroatom, and cyclization, 1293
  hydrogen abstraction, 1276, 1280
  hydrogen atom abstraction, 1273
  hydrogen atom transfer, 1294
  hydroxide, and **Fenton's reagent**, 1296
  hydroxy, 1260
  in a cage, 1261
  initiation step, 1271
  initiator, AIBN, 1274
  initiator, boranes, 1291, 1292
  initiators, 1273
  initiators, boranes, 1286
  initiators, peroxides, 1281
  intermolecular reactions, 1277
  intramolecular cyclization. *See* radical cyclization
  inversion, 1259
  kinetic and thermodynamic control, 1284
  lithium, 1267
  mechanism of addition, 1271
  mechanism of coupling, 1271
  mechanism of peroxide decomposition, 1261
  mechanism of polymerization, 1271
  metal induce, 1296
  methyl energy diagram, 1259
  methyl, 1257, 1258, 1263, 1264
  methyl, SOMO, 1269
  methyl, **Walsh diagram**, 1259
  nitrosyl, 1276, 1277

nucleophilic, 1268
oxygen, 1277
perchlorobenzyl, 1267, 1268
photochemical formation, 1262
planar, 1258, 1259
pyramidal, 1258, 1259
quenching agents, 1272
rate of hydrogen abstraction, 1273
radicals
  reactions, 1267
  reactivity, 1267
  reactivity, and frontier orbitals, 1268
  reactivity, and HOMO and LUMO, 1268
  reactivity, and SOMO, 1268
  reactivity, and steric effects, 1267
  reactivity, and symmetry, 1268
  reduction, 1274, 1280
  reduction, with tributyltin hydride, 1274
  scavenger, and boranes, 0536
  SOMO energies, 1268, 1269
  SOMO, 1263
  SOMO-HOMO interactions, 1268, 1269
  spin density maps, 1270
  stability, 1267
  stability, and coupling, 1270
  stability, and molecular modeling, 1270
  stability, and spin density, 1267, 1270
  structure, 1258
  structure, and NMR, 1259
  substitution reactions, 1273
  substitution, and NBS or NCS, 1274
  substitution, and resonance, 1273
  thermodynamic product, 1284
  thermodynamic stability, 0478
  transition state strain energy, 1283, 1284
radicals
  tributyltin hydride, 1293
  tributyltin, 1279
  triphenylmethyl, 1257, 1262, 1270
  triphenylsilyl, 1267, 1268
  vinyl, and radical cyclization, 1286
  with alkenes, 1269, 1270, 1277, 1279, 1283, 1298, 1299
  with nitrosyl chloride, 1276
  with silanes, 1293
  with styrene, 1272
  with sulfides and alkylboranes, 0537
  with vinyl ethers, 1269

radulanins, 0126
Ramberg-Bäcklund reaction, 0701
ramoplanin A2, 0306
RAMP, chiral hydrazone, 0870
Raney nickel, 0959
Raney nickel. *See* nickel
rate acceleration, and **Cope rearrangement**, 1127
rate constant, and high pressure, 1046
rate determining step,
  Cr(VI) oxidation of alcohols, 0228
  S_N1, 0131
  substitution, nucleophilic aromatic, 0206
  vanadium-catalyzed epoxidation, 0273
rate enhancement
  and the **Claisen rearrangement**, 1139
  and the **Diels-Alder reaction**, 1037
  Diels-Alder reactions, in aqueous media, 1041
  in aqueous media, 1139
  with **Lewis acids**, 1038
rate of hydrogen abstraction, by radicals, 1273
rate of hydrolysis, alkyl halides, 1167
rate of reaction
  alkyl halides and KI, 0114
  alkyl halides with KI in water, 0131
  and activation volume, 1045
  and **Arrhenius equation**, 1018
  and carbocation rearrangement, 0134
  and conformation, 0060
  and E1 reactions, 0160
  and E1 reactions, 0160
  and electronegativity, 0416
  and electrophilic aromatic substitution, 0203
  and hydrophobic effects, 1041
  and leaving group ability, 0142
  and nucleophilic aromatic substitution, 0206
  and organocuprates, 0722
  and oxidation of alcohols, 0228
  and S_N1 reactions, 0131, 1162
  and S_N2 reactions, 0113
  and the S_N2 transition state, 0114
  Baeyer-Villiger reaction, 0307
  Cr(VI) oxidation of alcohols, 0229
  Diels-Alder reactions, 1018, 1020, 1026
  Grob fragmentation, 0175
  hydrogenation, 0426

kinetic resolution, and **Sharpless asymmetric epoxidation**, 0286
Lewis acid catalyzed **Diels-Alder**, 1039
macrocyclizations, 0573, 0574
MnO_2 oxidation of alcohols, 0253, 0254
nucleophiles with iodomethane, 0120
oxidation of alcohols, 0230
rearrangement versus trapping, 0134
reduction of cyclic ketones, 0416
substitution and concentration, 0142
substitution versus rearrangement, 1171
vanadium catalyzed epoxidation, 0273
rate of reduction, aldehydes vs. ketones 0351
rate of solvolysis, alkyl halides, 1167
R-camp, 0449
RCM. *See* metathesis
re and si, 0031
  and reduction, 0401
  and Alpine borane, 0395
  and **Diels-Alder reactions**, 1068
reaction energy, saturated heterocycles, 0702
reaction rate. *See* rate
reaction rate, and conformation, 0060
reaction type, and reagents, 0079
reactivity
  acyl compounds to acyl substitution, 0111
  and SeO$_2$, 0326
  heterocycles in electrophilic aromatic substitution, 1193
  hydrogenation catalysts, 0425
reactivity
  nucleophiles in S$_N$2, 0119
  of enolate anions, 0826
  of NaBH$_4$, 0370
  peroxides in epoxidation reactions, 0270
reagents
  and reaction type, 0079
  for oxidation of phenols, 0264
rearrangement, [1,5]-sigmatropic and vinyl ketones, 1080
rearrangement, [2,3]-sigmatropic and organolithium reagents, 1122
  and SeO$_2$, 0325
  aza-Wittig Rearrangement, 1123
  diastereoselectivity, 1122
  Mislow-Evans rearrangement, 1123

Sommelet-Hauser rearrangement, 1123
rearrangement, [3,3]-sigmatropic, 0768, 1132
  and enolate anions, 1130
  and enols, 1130
  and palladium catalyst, 1140
  aza-Claisen rearrangement, 1141
  Claisen rearrangement, 1132
  Ireland-Claisen rearrangement, 1138
  Saucy-Marbet reaction, 1135
rearrangement
  1,2-alkyl shift, 1173
  2-azonia, 1142
  2-azonia. *See* aza-Claisen
  alkyl and aryl shifts, 0135
rearrangement
  and alkylboranes, 0501
  and bridged carbocations, 1171
  and carbenes, 1325
  and diols, 1180
  and **Fétizon's reagent**, 0257
  and **Friedel-Crafts reactions**, 1219
  and functional group exchange, 0188-0193
  and **Grignard reactions**, 0665
  and hydrogenation of alkenes, 0432
  and isobornyl chloride, 1177
  and **Lewis acids**, 1197
  and nitrogen ylids, 1123
  and orbitals, 0135
  and oxymercuration, 0182
  and ozonolysis, 0321
  and radicals, 1270, 1276
  and S$_N$1 reactions, 0131
  and super hydride, 0380
rearrangement
  anionic oxy-Cope, 1130
  aza-Claisen, 1131
  aza-Cope, 1131
  aza-Wittig, 1123
  azo-Cope, 1131
  Baeyer-Villiger reaction, 0309
  Beckmann, 0188
  camphene hydrochloride, 1177
  carbenes, 1322
rearrangement, carbocation
  1,2-hydride shift, 1171
  and allylic alcohols, 1181
  and amino alcohols, 1181
  and cyclopropanes, 0137
  and Grignard reactions, 0656

and hyperconjugation, 1173
and iodohydrins, 1181
and oxiranes, 1181
and silyl groups, 0137
and σ-conjugation, 1173
energy, 0134
hydride shift, 0134
mechanism, 1171
Meinwald rearrangement, 1181
migratory aptitude, 1172
pinacol-like rearrangement, 1182
powerful in synthesis, 0929, 0932
rate of reaction, 0134
ring expansion, 1174
semipinacol rearrangement, 1181
transition state, 1171
rearrangement, **Claisen**. *See also* **Claisen**
rearrangement, **Claisen**, powerful in synthesis, 0930
rearrangement, **Claisen**, 1045
rearrangement, **Cope**. *See also* **Cope**
rearrangement, **Cope**, 1045, 1123
  activation barrier, 1127
  and molecular orbitals, 1124
  and rotamers, 1125
  and strain, 1127
  cyclopropane accelerated, 1129
  divinyl cyclobutane, 1127
  transition state, 1124, 1125
rearrangement
  Curtius, 1051
  cyclobutylcarbinyl carbocations, 1174
  cyclopropanes, 1326
  cyclopropylcarbinyl carbocations, 0137, 1174
  degenerate, 0121
  Demjanov, 1175
  epoxy esters, 0282
  epoxy-alcohols, 1182
  epoxy-ethers, 1182
  Eschenmoser-Claisen, 1137
  Friedel-Crafts alkylation, 1197
  hydride shifts, 0134
  hydroxy-aldehydes, 1182
  in oxidation of alkylboranes, 0515
  in the **Fischer indole synthesis**, 1223
  in the S$_N$1 reaction, 1170
  Ireland-Claisen, 1138
  isonitriles to nitriles, 0626
  N → C shift, 0752
  NIH shift, 1179
  nonclassical carbocations, 1178

rearrangement
of allylic selenoxides, 0699
of allylic sulfoxides, 0699
of aziridines, 1113
of boranes, 0530
of carbocations, 0134-0124, 1170
of cycloheptatrienes, 1118
of diazoketones, 1051
of epoxy-oxonium salts, 1177
of imines, 1223
of propargyl halides, 0634
of triazolines, 0933
of tropilidenes, 1118
of vinylboranes, 0510-0512
oxidative, with mercuric salts, 0333
oxonia Cope rearrangement, 1191
oxy-Cope, 1130
oxy-Cope, powerful in synthesis, 0930
Pinacol rearrangement, 1179
rate of reaction, carbocations, 1171
Schmidt reaction, 0190
rearrangement, sigmatropic, 0749, 1116-1142
[1,3], 1118
[1,7], 1118
[2,3], 1121
[2,3], enantioselectivity, 1121
[3,3], 1123
[3,3], **anionic oxy-Cope**
rearrangement, 1130
[3,3], **oxy-Cope** rearrangement, 1130
[m,n], 1004
allylic systems, 1118
and **Cope** rearrangement, 1123
and orbital symmetry, 1116
and seco-alkylation, 1120
and the **walk rearrangement**, 1118
antarafacial shifts, 1118
cyclopentadienes, 1117
cyclopropanes, 1118
heptatrienes, 1118
of 1,5-dienes, 1123
of allenes, 1116
vinylcyclopropane, 1119
Wittig rearrangement, 1121
rearrangement
Sommelet-Hauser, 0752, 1123
Stevens, 0752
thio-Claisen, 1141
Tiffeneau-Demjanov ring expansion, 1175
vinyl aziridines, 1114
vinylcyclopropane, 0749, 0933, 1119,

1129
vinylcyclopropane, concerted and
diradical, 1119
Wagner-Meerwein rearrangement, 1177
with PCC, 0236
Wittig, 1121
Wittig, and organolithium reagents, 1122
Wittig, diastereoselectivity, 1122
recursive deconvolution, 0983
Red-Al
solubility, 0366
and chelation control, 0550
reducing agents, asymmetric, 0399
reductases, 0480
reduction potential
and electrolysis, 0222
and electrolytic reduction, 0469, 0470
and ionization potential, 0224, 0225
and **Oppenauer oxidation**, 0251
and oxidation, 0222
and permanganate, 0223
and reduction, 0347
chromium(VI), 0226
hydrogen peroxide, 0269
manganese dioxide, 0252
of permanganate , 0292
osmium tetraoxide, 0294
oxygen, 0270
ozone, 0318
ruthenium tetraoxide, 0315
silver salts, 0255
reduction
1,2- versus 1,4, 0377, 0417
1,2-, with LiAlH$_4$, 0352
1,4-, with LiAlH$_4$, 0352, 0353
aminomethylation, 0355, 0356
reduction
and aluminum compounds, 0467
and chelation, 0405, 0406
and chiral additives, 0392
and conformation, 0403-0405
and coordination complexes, 0043
and **Cram cyclic model**, 0406
and **Cram's rule**, 0402
and diisopinocampheylborane, 0395
and LUMO maps, 0403-0405, 0416
and molecular modeling, 0413, 0416, 0418, 0419, 0421
and monosaccharides, 0393
and organolithium reagents, 0680
and oxazaborolidine, 0400

and oxidation stage, 0347
and pH, 0382
and prochiral centers, 0348, 0401
and radical anions, 0468
and reduction potential, 0347
and rotamers, 0401
and sodium cyanoborohydride, 0382
and steric hindrance in
cyclohexanones, 0414-0416
and strain, 0052
and the **Bürgi-Dunitz trajectory**, 0412, 0556
and the **Cornforth model**, 0408
and the **Felkin-Ahn model**, 0409, 0411
and the **Karabatsos model**, 0408
and triisopropoxyaluminum, 0467
reduction
aryl diazonium salts, 0210
bicyclic ketone, 0418
conjugate, and HSAB, 0104
cyclopentanones, 0412
diastereoselectivity, 0392
reduction, dissolving metal, 0451-0467
alkynes and stereoselectivity, 0456
alkynes, 0452
alkynes, and amines, 0456
and alkenes, 0452
and amines, 0453
and aryls, 0453
and **Benkeser reduction**, 0461
and **Birch reduction**, 0459
and enolate anions, 0458
and enols, 0454
and epoxides, 0458
and functional groups, 0452
and heterocycles, 0453
and **pinacol coupling**, 0454
and terminal alkynes, 0455
conjugated alkenes, 0452
conjugated compounds, 0454
conjugated dienes, 0452
desulfonylation, 0457
enantioselectivity, 0392
formation of ketyls, 0453, 0454
hydrogenolysis, 0456
isomerization of terminal alkynes, 0456
of aldehydes, 0453
of carbonyls, mechanism, 0453
of ketones, 0453
solvents, 0452
sulfides, 0457
with alkynes, 0455

with calcium, 0462
with conjugated esters, 0455
reduction, electrolytic, 0468
  and ketyl formation, 0469
  and **pinacol coupling**, 0469
  and reduction potential, 0469, 0470
  Birch reduction, 0470
  conjugated carbonyls, 0470
  coupling, 0469
  dimerization of aldehydes, 0470
  halopyridines, 0469
  of alkynes, 0470
  of aryl halides, 0472
  of azides, 0472
  of epoxides, 0471
  of heterocycles, 0471
  of nitriles, 0470
  of nitro compounds, 0472
reduction, enzymatic, 0398, 0479
  and **Prelog's rule**, 0479
  and reaction medium, 0481
  and steric hindrance, 0479
  and the diamond lattice, 0479
  and **Veschambre's rule**, 0482
  conjugated carbonyls, 0482
  keto esters, 0479
  of glyoxalates, 0481
  stereoselectivity, 0481
reduction
  hydrogenation, 0422-0451
  hydrosilylation, 0475
  iminium salts, 0562
  in the **Grignard reaction**, 0667
  ketones, 0066
  Meerwein-Ponndorf-Verley reduction, 0467
  mesylate, 0379
  metal, **Clemmensen reduction**, 0464
  metal, with zinc, 0462
  nonconjugated alkenes, 0474
reduction
  of a **Weinreb amide**, 0389
  of alkynes, 0636
  of anhydrides, 0373
  of aromatic systems, 0936
  of carbonyl compounds, 0350
  of carbonyls, by **Grignard** reagents, 0667
  of cyclohexanones, 0414-0416
  of ozonide, 0320
  of ozonides, with trimethylphosphite, 0321
  of radicals, 1274, 1280

of steroids, 0419
of sultams, 0361, 0701
reduction
  overreduction of aliens, 0436
  partial, of aryl compounds, 0459, 0460
  photoreduction, 0477
  pinacol, 0469
  rate of aldehydes versus ketones, 0351
  reagent versus application, 0385
  remote, 0562
  stereoselectivity, 0391-0422
reduction
  with alkoxyaluminum hydrides, 0367, 0368
  with Alpine borane, 0395
  with aluminum hydride, 0387
  with bakers yeast, 0035, 0480
  with borane, 0385-0387
  with chiral complexes, 0397
  with *Cinchona* alkaloids, 0393
  with Dibal, 0389
  with diimide, 0474
  with formic acid, 0476
  with **Grignard reagents**, models, 0668
  with **Hojo's reagent**, 0374
  with hydrides, stoichiometry, 0350
  with metal borohydride, 0376
  with metal hydrides, 0348-0422
  with metals, 0347
  with NaBH$_4$, 0369
  with Red-Al, 0367
  with selectrides, 0380
  with silanes, 0475
  with substituted boranes, 0387
  with tin compounds, 0465
  with tributyltin hydride, 0466
  with zinc and acetic acid, 0464
  with zinc borohydride, 0004, 0043
reduction
  Wolff-Kishner reduction, 0472
reductive alkylation
  and heterocycles, 0448
  and hydrogenation, 0448
  with acyloxyborohydrides, 0378
reductive amination, 0371, 0384
reductive cleavage
  of nitriles, 0375
  of sulfones, 0700
  of sulfoxides, 0698
  of sultones, 0701
  sulfides, 0457

sulfones, 0457
sulfur compounds, 0473
  with aluminum amalgam, 0698
  with LiAlH$_4$, 0701
  with sodium amalgam, 0700
reductive coupling, 1308
Reformatsky reaction, 0885
  and **Rieke zinc**, 0886
  and SmI$_3$, 0885
  and TiCl$_2$/Cu, 0887
  and ultrasound, 0885
  diastereoselectivity, 0886
  syn/anti isomers, 0886, 0887
regiochemistry
  electrophilic aromatic nitration, 1193
  in the **Fischer indole synthesis**, 1222, 1223
  intramolecular **Diels-Alder**, 1064
regioisomer, definition, 0012
  and azomethine ylid resonance, 1112
  and intramolecular **Diels-Alder reactions**, 1061
  modeling of **Diels-Alder reactions**, 1043
regioselectivity, 0541
  and alkenylalkylboranes, 0505
  and **Bredt's rule**, 0553, 0554
  and **Diels-Alder** transition states, 1022, 1023
  and electronic effects in **Diels-Alder** reactions, 1028
  and electronic effects, 1027
  and enolate anions, 0804
  and formation of cyclic alkenes, 0552
  and **Friedel-Crafts acylation**, 1204, 1206
  and **Friedel-Crafts alkylation**, 1194
  and **Friedel-Crafts reactions** of aniline, 1211, 1212
  and orbital coefficients in **Diels-Alder** reactions, 1027
  and reduction of epoxides, 0375
  and secondary orbital interactions, 1028, 1031
  and the ortho effect in **Diels-Alder** reactions, 1028
  and the **Tiffeneau-Demjanov ring expansion**, 1175
  and transition state modeling in the **Diels-Alder reaction**, 1020, 1021
  and trimethylene methane equivalents, 1235
  dienes with iron pentacarbonyl, 1248

for alkenes with borane, 0494

for alkylboranes with alkenes, 0497, 0498

for aromatic substitution, and local ionization maps, 0203

for boranes with alkynes, 0504, 0505

for carbene addition to alkenes, 1325

for hypohalous acids and alkenes, 0188

for **Lewis acid** catalyzed **Diels-Alder** reactions, 1039

for nitrone [3+2]-cycloaddition, 1108

for palladation, 1228

in allyl palladium reactions, 1229, 1230

in azomethine ylid cycloaddition, 1111

in borane addition to alkenes, 0493

in cyclic systems, 0551

in **Diels-Alder** reactions, 1022, 1023, 1026

in **Diels-Alder** reactions, and molecular modeling, 1022, 1023

in **Diels-Alder** reactions, modeling, 1040

in electrophilic aromatic substitution, 0201

in enolate alkylation, and modeling, 0807

in enolate anion formation, 0798

in **Grignard** addition to epoxides, 0654

in hydroboration of alkenes, 0516, 0524

in hydroboration of cyclic alkenes, 0551

in hydroboration, 0497, 0498, 0517

in photocycloaddition, 1093

in the **Bamford-Stevens reaction**, 1318

with super hydride, 0379

REKEST, and MARSEIL/SOS, 0972

and synthesis, 0972

reaction types, 0973

relaxation energy, 0038

remote functionalization, 0562

remote reduction, 0562

renieramycin O, 0996

reserpine, 0458, 0898, 0921-0924, 1017

resolution, 0956

and synthesis, 0956

enzymatic, 0945

kinetic. *See* kinetic

resonance

and acetate ion, 0086

and acidity of carboxylic acids, 0086

and acidity, 0086

and acylium ions, 1204

and allylic carbocation, 1165

and arenium ions, 0197

and aromatic substitution, 0194

and azomethine ylids, 1111

and benzylic carbocation, 1165

and carbocations, 0132, 1164, 1165

and conjugate addition, 0112

and decarboxylation, 0173

and diazoalkanes, 1314

and diazomethane, 1314

and dissolving metal reduction, 0453

and electrophilic aromatic substitution, 0197, 0200-0202

and enolate anions, 0782

and homoenolates, 0691

and leaving group ability, 0143

and nucleophilic aromatic substitution, 0206

and nucleophilic strength, 0119

and ozone, 0318, 1103

and phenols 0087, 0771

and phenoxide ion, 0087

and phosphorus ylids, 0730

and radical substitution, 1273

and the **Birch reduction**, 0459

delocalization, and pK$_a$, 0690, 0691

ketene dithioacetals anions, 0710, 0711

of ylids, 0735

retene, 1209

retention versus inversion, 0541

and carbocations, 1170

retinal, 0252, 1309

retro aldol, steric effects, 0861

retro **Diels-Alder reaction**, 0914, 0976, 1032, 1035

and FVP, 1036

and high pressure, 1047

and Lewis acids, 1036

loss of carbon dioxide, 1036

of cyclopentadiene, 1035

of furans, 1036

pyrones, 1036

retro mass spectral degradation, and synthesis, 0976

retro Wagner-Meerwein, 0976

retro-1,3-dipolar addition, 0976

retro-**aldol condensation**, 0821, 0860

and enolate anions, 0790

retro-**Barbier fragmentation**, 0637

retro-Nazarov, 1189

retro-**Ritter reaction**, 0976

retrosynthesis, 0003, 0907

and chemical degradation, 0975

and mass spectrometry, 0976

biomimetic, 0948-0954

definition, 0898, 0908

strategies, 0917

retrosynthetic analysis, 0003, 0899

retrosynthetic transforms, 0077

reverse aldolization, 0860

reverse **Cope elimination**, 0165

reversibility

and **Diels-Alder reactions**, 1035

and **Michael addition**, 0881

and the **Robinson annulation**, 0881

of sulfur ylids, 0747

rhamnose, 0993

as a chiral template, 0959

rhein, 1205

rhinocerotinoic acid, 0885

Rhizopus arrhizus, and reduction, 0480

rhizoxin D, 0389, 0865, 1288

rhodium catalysts

and carbenes, 1330

and C-H insertion, 1330

and hydroboration, 0519

and phosphine ligands, 0450

and **Suzuki-Miyaura coupling**, 1243

Davies catalyst, 1330

Doyle catalysts, 1330

with diazoketones, 1330

rhodium complexes, and asymmetric hydrogenation, 0451

rhodium compounds

and carbenoids, 1331

and diazoalkanes, 1322

and diazocarbonyls, 1329

and DOSP, 1330

Wilkinson's catalyst, 0428, 0429

with diazoketones, 1115, 1328

rhodium hydroxide, hydrogenation of nitriles, 0441

rhodium oxide, hydrogenation of nitriles, 0441

rhodium tetraacetate, with diazoketones, 1115

rhodium, and modified **Wilkinson's catalyst**, 0450

Rhodococcus rhodochrous J1, 1356

Rhodococcus ruber DSM 44541, reduction, 0483

rhododaurichromanic acid A, 1092

ricciocarpin A, 0036, 0124
ricinelaidic lactone, 0580
Rieke zinc, and the **Reformatsky reaction**, 0886
ring cavity, and strain, 0052
ring closing metathesis. *See* metathesis
ring expansion
   and carbenes, 1326, 1327
   carbocation, 1174
   Demjanov rearrangement, 1175
   metathesis, 1340
   Tiffeneau-Demjanov ring expansion, 1175
ring opening metathesis. *See* metathesis
ring size
   and **Bredt's rule**, 0553, 0554
   and the intramolecular **aldol condensation**, 0823
ring strain, and Bredt's rule, 0554
ring-forming reactions, 0564
ristocetin aglycone, 0598
Ritter coupling, 1186
Ritter reaction, 0976
   retro, 0976
Robinson annulation, 0880
   and *Cinchona* alkaloids, 0882
   and conjugate addition, 0112
   and lanthanoid triisopropoxides, 0880
   and proline, 0882
   and reversibly of **Michael addition**, 0881
   and thermodynamic conditions, 0881
   enantioselectivity, 0882
   powerful in synthesis, 0928
   selectivity, 0882
   solvent effects, 0883
rolliniastatin 1, 0607
ROMP. *See* metathesis
rose bengal, as a sensitizer, 1090
Rosenmund catalyst, 0437
Rosenmund-von Braun reaction, 0628
rotamers
   and **Cope rearrangement**, 1125
   and pyrolysis of esters, 0166
   and reduction, 0401
   definition, 0035
rotational isomers, and diradicals, 1323, 1324
roxaticin, 0767
rubromycin, 0932
rules for oxidation number, 0220, 0221
ruthenium chloride, and tert-butyl hydroperoxide, 0330

ruthenium compounds
   and metathesis, 1336
   Grubbs I catalyst, 1336
   Grubbs II catalyst, 1337
   oxidation of alkyls, 0330
ruthenium on carbon, hydrogenation, 0440
ruthenium oxide, hydrogenation, 0440
ruthenium tetraoxide
   and periodic acid, 0315
   reduction potential, 0315
ruthenium trichloride
   and sodium periodate, 0316
   in oxidative cleavage, 0316
ruthenium, hydrogenation catalyst, hydrogenolysis, 0430

S

s character
   and carbocations, 1164
   and pK_a, 0632
   of alkynes, 0632
Saccharomyces cerevisiae. *See* bakers yeast
Saccharomyces delbrueckii, and reduction, 0480
sacrificial alkenes, in hydroboration, 0501
saframycin alkaloids, 0881
salen (bis(salicylidene)ethylene-diamine), 0289
   and **Jacobsen hydrolytic kinetic resolution**, 0290
   and manganese, 0289
   and pinacol coupling, 1304
   and the **Jacobsen catalyst**, 0289
   catalysts, and epoxidation, 0288
   catalysts, and sodium hypochlorite, 0289
   catalysts, in sulfide oxidation, 0337
salicylaldehyde phenylhydrazone, and organolithium reagents, 0675
salinosporamide A, 0931
salt free ylids, and the **Wittig reaction**, 0736
salts, ammonium, and **Bredt's rule**, 0554
salvadione-A, 0283
salviasperanol, 0280, 0932
samarium iodide
   and **pinacol coupling**, 1303, 1304
   and radical cyclization, 1292

   and the **Barbier reaction**, 0637
   with aldehydes, 1304
samarium triiodide, and the **Reformatsky** reaction, 0885
SAMP chiral hydrazone, 0870
sand fly, 0864
Sandmeyer reaction, 0209
sanglifehrin A, 0378
santalene, 0421, 0701
saponification, 0959
   and the **Prévost reaction**, 0302
   of esters, 0310, 0725, 0829, 0834
   of lactones, 0583
sarain A, 0118
Sarett oxidation, 0234, 0235
Saucy-Marbet reaction, 1135
saudin, 0812
Saytzeff elimination, 0155, 0160
   and the **Burgess reagent**, 0171
scabronine G, 1188
scandium triflate, 0310
scandium triflate, and radicals, 1281
SCH 351448, 0233, 0386, 0431
Sch 47554, 0767
s-character
   and organolithium reagents, 0690
   and pK_a, 0690
Schiff base
   and imine carbanions, 0847
   via **Pomeranz-Fritsch reaction**, 1217, 1218
Schlenk equilibrium, 0638
   and amines, 0638
   and dialkylmagnesium, 0638
   and epoxides, 0655
Schlittler-Müller modification, 1218
Schmidt reaction, 0190
Schreiber protocol, and ozonolysis, 0320
Schrock catalyst, 1337
   and metathesis, 1336
schweinfurthin, 0351
s-cis and s-trans dienes, energy barrier, 1014
   and **Diels-Alder reactions**, 1014
sclareol, 0166
sclerophytins, 0902
S_E2' rearrangement, allylic
   Grignard reagents, 0664, 0665
seco-acids, 0575, 0579, 0926, 0953
seco-alkylation, 1120
secondary orbital interactions, 1028
   and **Alder endo rule**, 1030

and perturbation theory, 1031
intramolecular **Diels-Alder**, 1064
seco-proansamitocin, 1238
secosyrin 1, 0461
securinine, 0610, 0615, 0812, 0932, 0975
securamine C, 1354, 1355-1363
Securiflustra Securifrons, 1355
selectivity
cis/trans, 0541
cis-trans, and radical cyclization, 1285
Cu/RLi ratio in organocuprates, 0721
definitions, 0032
exo, nitroalkene cycloaddition, 1108
in the **Diels-Alder reaction**, 1020, 1021
in the **Wittig** reaction, 0732
oxidation with SeO_2, 0326
re and si, 0032
reduction and solvent effects, 0410
Robinson annulation, 0882
syn/anti, hydroboration, 0523
Selectrides
and diastereoselectivity, 0381
and HSAB, 0105
and metal salts, 0381
and reduction, 0380
and steric hindrance, 0381
K and L, 0380
KS and LS, 0382
reduction of cyclopentanones, 0412
reduction of ketones, 0420
with aldehydes or ketones, 0381
selenides
and thiocarbamates, 0170
and radical cyclization, 1288, 1289
and radicals, 1293
oxidation with peroxide, 0169
oxidation, 0169
via S_N2 reactions, 0121
selenium dioxide, 0324
and acetic anhydride, 0325
and aldehydes, 0327, 0328
and allylic compound, 0325
and **Bredt's rule**, 0326
and E2 reactions, 0328
and ketones, 0327, 0328
and reactivity, 0326
and solvent effects, 0327
and substituent effects, 0327
and tert-butyl hydroperoxide, 0326
and the **ene reaction**, 0325
oxidation, of sulfides, 0335

oxidation, substituent effects, 0326
selectivity, 0326
selenium
and aromatization, 1208
and dehydrogenation, 1209
selenones, from selenoxides, 0338
selenoxides
allylic, rearrangement, 0699
by oxidation of selenides, 0815
from selenides, 0169, 0338
pyrolysis, 0169
self-condensation, and the **aldol condensation**, 0817
SEM [ether, 2-(trimethylsilyl ethoxymethy]. *See* ether
semiconductors, organic, 0118
semipinacol rearrangement, 1181
semisynthesis, 0315
sensitizer
and acetophenone, 1090
and benzophenone, 1089, 1091, 1266
and chlorophyll a, 1090
and diazoalkanes, 1314
and intersystem crossing, 1089
and methylene blue, 1090
and naphthalene, 1090
and photochemistry, 1088
and radical formation, 1266
and rose bengal, 1090
triplet, 1089
sequestration, of Li, in the **Wittig reaction**, 0738
serofendic acids A and B, 1243
sertraline, 0455
Shapiro reaction, and **Bamford-Stevens reaction**, 1317
Sharpless asymmetric aminohydroxylation, 0306
Sharpless asymmetric dihydroxylation, 0298-0301
and reduction, 0393
model for selectivity, 0301
Sharpless asymmetric epoxidation, 0272, 0284-0288, 0337, 0549, 0867
complex formation, 0285
consonance and dissonance, 0286
enantioselectivity, 0286
kinetic resolution and rate of reaction, 0286
kinetic resolution, 0286
matched-mismatched, 0287
model for selectivity, 0285, 0287
stereoselectivity 0285

shift, sigmatropic. *See* rearrangement
Shiina macrolactonization, 0576
shikonin, 0265
showdomycin, 1190
shyobunone, 1092, 1097
si and re, 0031
and Alpine borane, 0395
and **Diels-Alder reactions**, 1068
and reduction, 0401
sibirine, 1288
siccanin, 0699
sigialomycin D, 0590, 0591
sigmatropic rearrangement. *See* rearrangement
Sigmatropic rearrangement, [3.3]-. *See* **Claisen** rearrangement
Sigmatropic rearrangement, [3.3]-. *See* **Cope** rearrangement
sigmatropic rearrangement, antarafacial, 1116, 1117
sigmatropic rearrangements, 0749, 1116-1142
[1,5], 1116
2,3-, and SeO_2, 0325
and molecular orbitals, 1117
powerful in synthesis, 0930, 0932
suprafacial, 1116, 1117
sigmatropic shift. *See* rearrangement
sila-acyloin condensation, powerful in synthesis, 0929
silanes
alkyl, with TBAF, 0519
alkynyl, coupling with aryl halides, 1302
allylic. *See also* allylsilanes
allylic, 0769
allylic, and the **Prins** reaction, 1191
allylic, with aldehydes, 0769
allylic, with fluoride, 0770
allylic, with TBAF, 0770
and **Friedel-Crafts acylation**, 1206
and **Mukaiyama aldol reaction**, 0837
and **pinacol coupling**, 1305
and the Tamao-Fleming oxidation, 0330
carbanions, 0770
protecting group for alkynes, 1286
reduction, 0475
tethers, 1065
triethyl, and trifuoroacetic acid, 0475
vinyl, and **Nazarov cyclization**, 1188
vinyl, with NaF, 1185
with BF_3, 0476

with radicals, 1293
with TBAF, 1065
with transition metals, 0475
sildenafil, 0900
silica gel
　and **Bobbitt's reagent**, 0248
　and cationic cyclization, 0558
　and hydrogenation, 0449
　and **Juliá-Colonna epoxidation**, 0276
　and oxammonium salts, 0248
　and phosphorus ylids, 0732
　and urea-hydrogen peroxide complex, 0276
　and zinc borohydride, 0375
　with β-hydroxyphosphine oxides, 0743
　with LiAlH$_4$, 0374
silk fibroin, and hydrogenation, 0449
siloxydienes, with pyrrolines, 1055
silphinene, 0994
silyl enol ethers
　and **Diels-Alder reactions**, 1050
　and **Ireland-Claisen** rearrangement, 1138
　and lead tetraacetate, 0323
　and **Mukaiyama aldol reaction**, 0837
　and organocuprates, 0722
　and the **Simmons-Smith reaction**, 1332
　halogenation, 0883, 0884
　hydrolysis to ketones, 0722
　hydrolysis, 0749, 1120
　with azide and ketones, 0192
　with fluoride, 1094
　with **Lewis acids**, 0838
silver carbonate
　and Celite, 0256, 0257
　oxidation of alcohols, 0255
silver cyanide. *See* cyanide
silver fluoride, and silylamines, 1112
silver ions
　and **Grob fragmentation**, 0174
　and HSAB, 0100
　with nitronium ion, 0906
silver nitrate. *See also* nitrate
silver nitrate
　and elimination, 0174
　with dithianes, 0705
　with NaOH and boranes, 0513
silver oxide
　and ammonium salts, 0162
　and diazomethane, 1315

and **Hofmann elimination**, 0162
and **Suzuki-Miyaura coupling**, 1243
and **Wolff rearrangement**, 1315
silver salts,
　and the **Prévost reaction**, 0301, 0302
　reduction potential, 0255
silver tetrafluoroborate, 1186
　and sulfonium salts, 0745
　and zinc borohydride, 0375
silver(I) oxide, oxidation of aldehydes, 0260
silver(II) nitrate, and KOH, 0259
silver(II) oxide
　and oxidant, 0259
　oxidation of primary alcohols, 0260
silver-zinc, and the
　Simmons-Smith reaction, 1332
silyl alkoxides, 1308
silyl enol ethers. *See also* enol ether
silyl enol ethers
　and **Ireland-Claisen** rearrangement, 1138
　and **Mukaiyama aldol reaction**, 0837
　and **Diels-Alder reactions**, 1050
　and **Simmons-Smith reaction**, 1332
　from ketones, 0815
　halogenation, 0883, 0884
　hydrolysis, 1120
　with acetals, 0839
　with aldehydes, 0839
　with fluoride, 1094
　with ketones, 0839
　with **Lewis acids**, 0838
silyl groups, and carbocations
　rearrangement, 0136
silyl protecting groups, 0316
　for alcohols, 0600, 1366
silyl vinyl ketones, as **Michael**
　acceptors, 0882
silyl-acyloin cyclization, powerful in
　synthesis, 0931
silylalkyl ethers, 1065
silylcarbocations, 1176
silylketenes, with aldehydes, 1080
silylketone, 0518
silyloxybutadienes, in **Diels-Alder**
　reactions, 1049
Simmons-Smith reaction, 1331
　and diethylzinc, 1333
　and ultrasound, 1332
　enantioselectivity, 1333
　enantioselectivity, 1333
　intermediate, 1331

mechanism, 1331
neighboring group effects, 1332
with keto-esters, 1334
Simmons-Smith reagent, 1325
simple diastereoselection, 0850
　in the **aldol condensation**, 0861
　enolate anions, 0862
　definition, 0852
single electron transfer, and **Grignard**
　reagents, 0637
singlet carbenes, 1311, 1312
singlet diradicals, 1324
singlet to triplet intersystem crossing,
　1089
singlet, definition, 1312
singly occupied molecular orbitals. *See*
　SOMO
siphonarienedione, 0862
siphonarin B, 0821
six-center transition state, 0949
Skraup reaction, 1215
　and the Bally-Scholl synthesis, 1215
S_N1. *See* substitution
S_N2. *See* substitution
S_Ni. *See* substitution
sodioacetate, with allyl palladium, 1229
sodium acetate. *See* acetate
sodium amalgam, 0457
　and Fe(CO)$_5$, 0760
　and sulfones, 0700
sodium azide. *See* azide
sodium borohydride. *See* borohydride
sodium cyanide. *See* cyanide
sodium cyanoborohydride. *See*
　cyanoborohydride
sodium hydride. *See* hydride
sodium hydroxide. *See* hydroxide
sodium hypochlorite. *See* hypochlorite
sodium lamps, and photochemistry,
　1085
sodium nitrite. *See* nitrite
sodium sand, 1306
　and **acyloin condensation**, 1306
sodium tetracarbonyl ferrate, 0760
　with alkyl halides, 0760
sodium
　and alkyl halides, 0681
　and esters, 1306
　and organolithium reagents, 0681
　and **pinacol coupling**, 1303
　in ammonia, 0455, 0595
　with aldehydes or ketones, 1303
　with ammonia, and benzyl amines,

0613

soft acids and soft bases, definitions, 0097

soft laser desorption time-of-flight mass spectrometry, 0985

softness parameter, and HSAB, 0086

solandelactones E and F, 1333

solavetivone, 0135

solid support, and combinatorial chemistry, 0982

solid-state synthesis, 0978

solubility
    alkoxyaluminum hydrides, 0365
    LiAlH$_4$ in solvents, 0349
    lithium borohydride, 0374
    of alkoxides, 0589
    of NaBH$_4$, 0370

sodium *bis*(2-methoxyethoxy)-aluminum hydride. *See* Red-Al

solution phase combinatorial chemistry, 0987

solvating, LDA, 0789

solvation
    and amine basicity, 0090
    and encapsulation, 0906
    and pK$_a$, 0088
    and S$_N$2 transition state modeling, 0116
    and substitution, 0139
    and transition states, 0140
    definition, 0088
    of organolithium reagents, 0679

solvent effect, and **Diels-Alder** reactions, 1044

solvent effects
    [2+2]--cycloaddition, 1093
    aggregation state of organolithium reagents, 0672
    and acidity, 0087
    and allylic thiol anions, 0712, 0713
    and basicity, 0090, 0091
    and dielectric constants, 0138, 0139
    and **Fétizon's reagent**, 0257
    and **Hofmann elimination**, 0164
    and inversion reactions of **Grignard reagents**, 0657
    and ionization, 0131
    and mass spectrometry, 1166
    and **Michael addition**, 0878
    and oxidaton of sulfides, 0337
    and photocyclization, 1093
    and pK$_a$, 0091, 0788, 0798
    and polyalkylation, 1198

and selectivity in reduction, 0410

and selenium dioxide, 0327

and S$_N$1 reactions, 0131

and S$_N$2 transition state modeling, 0117

and the **Dess-Martin periodinane**, 0247

and the **Hofmann elimination**, 0162

and ylid stereochemistry, 0735, 0736

Cr(VI) oxidation of alcohols, 0229, 0230

cyanide with alkyl halides, 0624

dihalogens with alkenes, 0184

dissolving metal reduction of alkynes, 0455

enol content, 0783

for organocuprates with alkyl halides, 0716

in **Meerwein-Ponndorf-Verley reduction**, 0467

in the **Robinson annulation**, 0882

in the **Schlenk equilibrium**, 0638, 0639

on MnO$_2$ oxidation, 0253

with **Grignard reagent**s, 0638, 0639, 0661

with organocuprates, 0714

solvent free
    Sonogashira coupling, 1245
    Vilsmeier-Haack reaction, 1207

solvents
    and S$_N$1 reactions, 0134
    and ionization of alkyl halides, 0129
    and ionization, 0138, 0139
    and LTA oxidation, 0322
    and NaBH$_4$, 0370
    and nucleophilic substitution, 0138
    and sodium cyanoborohydride, 0382
    and transition states, 0140
    aprotic, 0138, 0139
    aprotic, and enolate anions, 0798
    for hydrogen peroxide, 0271
    in substitution and elimination reactions, 0544
    protic, 0138, 0139
    protic, and enolate anions, 0798
    substitution versus elimination, 0141

solvolysis
    and inversion, 1170
    and racemization, 1170
    of vinyl halides, 1168
    rates of chloroalkenes, 1167
    sulfonate esters, 1174

tosylates, 1174

Sommelet versus **Stevens rearrangement**, 0754

Sommelet-Hauser rearrangement, 0752, 1123
    also known as **Sommelet rearrangement**
    and sulfur ylids, 0745

SOMO energy of radicals, 1268, 1269

SOMO
    acetate radical, 1269
    and [2+2]-cycloaddition, 1082
    and formation of radicals, 1263
    and HOMO or LUMO, 1268
    and HSAB, 0102
    and photochemistry, 1012, 1082, 1086
    and radical reactivity, 1268
    and radicals, 0102
    halogen radicals, 1274
    methyl radical, 1269
    with frontier molecular orbitals, 1268

SOMO-HOMO interactions, for radicals, 1268, 1269

SOMO-LUMO interactions, 1285
    and radical cyclization, 1283

sonication, and the **Mitsunobu** reaction, 0124

Sonogashira coupling, 1244
    and **Glaser reaction**, 1301
    and microwaves, 1245
    copper free and solvent free, 1245

sorangiolides, 0578

sorbic alcohol, with boronic acid, 1064

sorbic anhydride, 0812

sordaricin, 0253, 0467, 1062

spacers, in combinatorial chemistry, 0980

sparagamine A, 1051

Spartan, 0005, 0056, 0060, 0062, 0064, 0065, 0067, 0115-0117, 0156, 0198, 0203, 0207, 0278, 0281, 0297, 0299, 0403-0405, 0410, 0413, 0416, 0418, 0419, 0421, 0499, 0517, 0521, 0523, 0570, 0575, 0645, 0660, 0694, 0702, 0783, 0785, 0786, 0797-0799, 0806, 0807, 0846, 0865, 0923, 1006, 1019-1023, 1026, 1028, 1039, 1043, 1063, 1126, 1128, 1134, 1139, 1144, 1147-1149, 1165, 1171, 1270, 1284, 1290, 1291, 1303, 1320
    and absolute configuration, 0023

sparteine, 0672

and butyllithium, 0672
and organocuprates, 0724
and organolithium reagent
  aggregation, 0672
and tert-butyllithium X-ray, 0672,
  0673
synthesis, 0952
with arylchromium compounds, 0765
with **Grignard reagents**, 0661
with organolithium reagents, 0689
sphinganine, 0359
sphingosines, 0159, 0282
spiculoic acid A, 0596
spin density maps, of radical, 1270
spin density
  and molecular modeling, 1270
  and radical stability, 1267
spinosyn A, 0581
spirans, 0136
  and chirality, 0026
spirastrellolide A, 0843
spirobenzylisoquinolines, 0369
spirocyclization, and phenols, 0772
spirofungin A, 0992
spirolactams, 0678
spirolide B, 0494
spirotryprostatin B, 1329
split synthesis, combinatorial
  chemistry, 0982, 0983
spongidipepsin, 0992
spongistatin 1, 0755
spongistatin, 0705
squalene, 1135
  and biogenesis of cholesterol, 1184
squalene-2,3-epoxide, 0948, 0949
squalene-like polyene, 0948, 0949
squamocin A, 0455
squamocin, 0315, 0316
SS20846A, 0550
stability, configurational,
  of **Grignard reagents**, 0657
  of organolithium reagents, 0680, 0681
stability
  of alkenes, 0155
  of carbocations, 0132
  of carbocations, 1167
  of lithio dithiolanes, 0707
  of metathesis catalysts, 1336
  of organolithium reagents, 0670, 0673,
    0674
  of organolithium reagents, in solvents,
    0673
  of radicals, 1267

of the **Dess-Martin periodinane**,
  0247
of the **Grubbs catalyst**, 1336
order, for alkenes, 0155
relative, of carbocations, 1169
stachybotrylactam, 0628
Stachybotrys spirolactams, 0678
standardization, of organolithium
  reagents, 0675
stannanes. *See* tin
  and **Nazarov cyclization**, 1188
  with aldehydes or ketones, 0767
stannous. *See* tin(II)
starting material, 0002
  definition, 0908
steering wheel model, 0014
stemonamide, 0802
stemonidine, 1108
stenine, 0157
stephacidin A, 0902
Stephen reduction, 0465
stereocenters
  cyclic arcs and synthesis, 0936, 0937
  removal in synthesis, 0924, 0925
stereochemistry, 0012-0030
  and **Cope rearrangement**, 1124
  and dihydroxylation, 0292
  and electrocyclic reactions, 1001
  and enolate anion geometry, 0795
  and halonium ions, 0184
  and prochiral centers, 0030
  and pyrolysis of esters, 0166
  and symmetry, 1001
  and synthesis, 0917, 0937, 0941
  and the S_N1 reaction, 0131
  and the S_N2' **reaction**, 0126
  in the **Baeyer-Villiger reaction**, 0310
  in the Fischer indole synthesis, 1222,
    1223
  intramolecular Diels-Alder, 1062,
    1063
  of [2+2]-cycloadditions, 1076
  of alkenes, 0029
  of alkenyl alkylboranes, 0503
  of alkenyl **Grignard reagents**, 0663,
    0664
  of alkenyl organolithium reagents,
    0677
  of betaines, 0735
  of carbocations, 1170
  of coupling vinyl halides and
    organocuprates, 0717
  of **Grignard reagents**, 0657

of halogenation of alcohols, 0145
of **Horner-Wadsworth-Emmons**
  olefination, 0740
of lithiated sulfoxides, 0698
of oxaphaphatanes, and reactivity,
  0738
of oxaphosphatanes, 0735
of phosphonate ester ylids, 0740
of substitution reactions, 0543
of the **Claisen rearrangement**, 1133
of the **Wittig reaction**,
  oxaphosphatane model, 0737
syn/anti, of LTA oxidation of diols,
  0322
stereodefined alkenyl organolithium
  reagents, 0677
stereodifferentiation, 0867, 0957
  consonant, 0867
stereogenic atom, 0013
stereogenic centers, and S_N2 reactions,
  0114
stereoisomers
  definition, 0013
  and 2^n rule, 0024
  and meso compounds, 0024
  cis and trans, 0029
  E and Z, 0029
stereoselectivity
  and alkene hydrogenation, 0433
  and alkenylboranes, 0503, 0505
  and alkylboranes with alkenes, 0497,
    0498
  and alkylboranes with alkynes, 0509
  and Alpine borane, 0395
  and boron enolates, 0842
  and conformation, 0278, 0279
  and **Cram's rule**, 0402
  and dihydroxylation, 0297, 0298
  and enolate anions, 0850
  and enolate formation, 0859
  and epoxidation, 0869
  and formation of dienes, 0510
  and haptophilicity, 0434, 0435, 0436
  and hydroboration of alkenes, 0561
  and hydroboration, 0497, 0498, 0519
  and hydrogenation of alkynes, 0436
  and hydrogenation, 0433
  and hydrolysis of epoxides, 0548
  and inversion of configuration, 0931
  and ionization potential maps, 0806,
    0807, 0865
  and **Lewis acids**, 0095
  and LiAlH$_4$ reduction, 0352

and LUMO maps, 0403-0405
and mcpba epoxidation, 0282
and neighboring group effects, 0434, 0435, 0436
and olefination with Ti reagents, 0756
and organocuprates, 0722
and organolithium reagents, 0680, 0681
and oxymercuration, 0182
and Pd(0) alkylation, 1233
and photoreduction, 0477
and radical cyclization, 1288
and **Sharpless asymmetric dihydroxylation**, 0298-0301
and super hydride, 0380
and the **Bürgi-Dunitz trajectory**, 0412
and the **Cieplak model**, 0414
and the **Cram cyclic model**, 0405-0407
and the **Felkin-Ahn model**, 0409
and the **Karabatsos model**, 0408
and vanadium-catalyzed epoxidation, 0274
stereoselectivity
 anti-Cram selectivity, 0406, 0407
 cis/trans, with borane reactions, 0522
stereoselectivity
 for organocuprates and conjugate addition, 0721
 for pinacol coupling, 1304
 for reduction of alkoxy ketones, 0392
 for reduction of bicyclic ketones, 0417
 for reduction of cyclic ketones, 0411-0422
 for reduction of cyclopentanones, 0412
 for reduction of epoxides, 0367
 for reduction of ketones, 0561
stereoselectivity
 from cyclic precursors, 0563
 glycidic ester formation, 0869
 in E2 reactions, 0559
 in elimination reactions, 0559
 in epoxidations, 0284
 in hydroboration, 0506
 in metathesis reactions, 1337
 in oxidation of boranes, 0515
 in reductions, 0391-0422
 in **Sharpless asymmetric epoxidation**, 0285- 0287
 in the **Diels-Alder reaction**, 1073

in the **Grignard reaction**, 0645
in the **Schmidt reaction**, 0191
model for dihydroxylation, 0301
stereoselectivity, of alkene hydrogenation
 and **Lindlar** hydrogenation, 0437
 and **Rosenmund** hydrogenation, 0437
stereoselectivity,
 of alkenes with borane, 0494
 of alkyne hydrogenation,
 of alkyne hydrogenation,
 of borane addition to alkenes, 0493
 of carbenes with alkenes, 1323
 of cyclobutene ring opening, 1098
 of **Diels-Alder reactions** of azadienes, 1052
 of **Diels-Alder reactions**, 1024-1026
 of diimide reduction of alkenes, 0474
 of dissolving metal reduction of alkynes, 0456
 of E2 reactions, 0154
 of enolate alkylation and modeling, 0807
 of enolate anions, 0547
 of enzymatic **Baeyer-Villiger reactions**, 0312
 of formic acid reduction, 0476
 of **Grignard reagents**, 0547
 of hydroperoxide anion epoxidation, 0275
 of oxetane ring opening, 0563
 of radical cyclization, 1285
 of sulfide oxidation, 0336
 of the **Cope rearrangement**, 1124, 1125
 of the **Stille coupling**, 1241
 of the **Wittig reaction**, 0735
 of yeast reduction of keto esters, 0480
 of ylids, solvent effects, 0735, 0736
stereoselectivity
 with alkyl halides and organocuprates, 0727
 with zinc borohydride, 0375
stereospecific and stereoselective, 0034
steric effecs
 and acidity, 0085
 and diastereoselectivity, 0850
 and enolate anion stability, 0790
 and organocuprates, 0722
 and oxidation of alcohols, 0230
 and radical reactivity, 1267
 in peroxyacid epoxidation of alkenes, 0280

in the **aldol condensation**, 0861
in the **Fries rearrangement**, 1220
in the retro **aldol condensation**, 0861
primary, and basicity, 0089
secondary, and basicity, 0089
steric hindrance
 and reduction of epoxides, 0367
 and acyl addition of organolithium reagents, 0682
 and borane addition to alkenes, 0492
 and conjugate reduction, 0352
 and **Cram's rule**, 0402
 and electrophilic aromatic substitution, 0203
 and enamine formation, 0873, 0874
 and enolate anion reactivity, 0846
 and enzyme reductions, 0479
 and facial selectivity, 1074
 and **Hofmann elimination**, 0163
 and hydroboration, 0524
 and hydrogenation, 0435
 and hydrogenation, of alkenes, 0432
 and hydrolysis of imines, 0650
 and LDA, 0787
 and $LiAlH_4$ reduction, 0351
 and nucleophilicity, 0123
 and organocuprates, 0723
 and ortho substitution, 0203
 and oxidation of alcohols, with MnO_2, 0253
 and oxidation of phenols, 0263
 and palladium(0) reactions, 1233
 and **Prelog's rule**, 0479
 and radicals, 1270
 and rate of hydride reduction, 0351
 and reduction of amides, 0355, 0364
 and reduction of bicyclic ketones, 0418
 and reduction of cyclohexanones, 0414-0416
 and reduction of ketones, 0413
 and selectrides, 0381
 and **Swern Oxidation**, 0241
 and the **Bürgi-Dunitz trajectory**, 0412
 and the **Diels-Alder reaction**, 1074
 and the S_N2' **reaction**, 0126
 and the S_N2 transition state, 0114
 and the **Zimmerman-Traxler** model, 0854
 and transforms, 0081
 ane enolate anions, 0810
 in chromate esters, 0230

in dihalogen addition to alkenes, 0184

in **Grignard addition** to conjugated
ketones, 0653

in **Grignard addition** to esters, 0648

in hydroboration, 0504

in S_N2 reactions, 0114

steric repulsion, **Pauli** exchange, 0038

steric strain

and **Bredt's rule**, 0553, 0554

and the S_N1 reaction, 0133

steroid NGA0187, 0325

steroids, 0234, 0245, 0250, 0257,
0435, 0948, 0949, 0961, 1276, 1277

and **Johnson polyene cyclization**,
1184

conformations, 0070

electrolysis, 0471

reduction, 0419

sterpurene, 0528

Steven's rearrangement, 0752

and Red-Al, 0369

and **Sommelet rearrangement**, 0752,
0753

versus Sommelet, 0754

Stille coupling, 1240

and alkynes, 1241

cross coupling, 1240

mechanism, 1240, 1241

powerful in synthesis, 0932

syringe pump addition, 1241

with vinyl iodides, 1240

stinkbug pheromone, 0609

stirrer speed, and acyloin condensation,
1306

Stobbe condensation, 0833

stoichiometry, hydride reduction of
carbonyls, 0350

Stollé synthesis, 1224

Stork enamine synthesis, 0873

**Stork-Eschenmoser hypothesis or
postulate.** *See* **Johnson polyene
cyclization**

strain energy, transition state, and
radicals, 1283, 1284

strain, $A^{1,3}$, 0055, 0056

strain

and **Cope rearrangement**, 1127

and electrophilic aromatic
substitution, 0203, 0204

and leaving group ability, 0143

and peroxyacid epoxidation, 0281

and ring cavity, 0052

and the **Mills-Nixon effect**, 0204

and transforms, 0081

and transition states, 0049

back, and alkyl halides, 0134

back, and carbocations, 0134

G-strain, 0056

in oxidation and reduction, 0052

I-strain, 0052

transannular, 0052

U-strain, 0058

s-trans, dienes, and **Diels-Alder**
reactions, 1014

strategic bond analysis, and synthesis,
0920, 0938, 0940

strategic bonds

and aromatic rings, 0936

and synthesis, 0908

carbocyclic rings, 0934

criteria for disconnection, 0924

heterocyclic rings, 0938

in rings, 0934

strategy, for retrosynthesis, 0917

Strecker synthesis, 0628

enantioselective, 0629

strength parameter, and HSAB, 0096

streptazolin, 0283, 0550

streptonigrin, 0147

strigol, 0259

structural challenges, and synthesis,
0905

structure determination, and **Kuhn-
Roth oxidation**, 0329

structure

and phosphonium salts, 0731

and stability, organolithium reagents,
0673, 0674

of enolate anions, 0792-0796

of **Grignard reagents**, 0637

of radicals, and NMR, 1259

strychnine, 0898

strychnofoline, 0324

styrene

conformation, 0047

with radicals, 1272

subersic acid, 0327, 0997

substituent effects

alkyne acidity, 0633

and carbocation stability, 1167

and enol content, 0784

and HOMO and LUMO, 1015, 1017

and hydrogenation of alkenes, 0432

and hydrogenolysis, 0360

and oxidation with Fremy's salt, 0263

and pK_a, 0784

and pK_a, in aldehydes and ketones,
0782

and SeO_2, 0327

and the Henbest rule, 0283

HOMO and LUMO of alkenes, 1003,
1004

in peroxyacid epoxidation of
alkenes, 0282

in **Diels-Alder reactions**, 1015, 1017,
1025

in electrophilic aromatic substitution,
0202, 0203, 1192

in **ene reactions**, 1148

in nucleophilic aromatic substitution,
0206

in SeO_2 oxidation, 0326

on carbocation stability, 1162

on carbocations, 1162

substituents, and electrophilic aromatic
substitution, 0201, 0206

substitution, acyl, 0109-0111

and acid chlorides, 0109

and amides, 0110

and anhydrides, 0110

and **Baldwin's rules**, 0568

and esters, 0110

and **Grignard reagents**, 0646

and the tetrahedral intermediate, 0109,
0646

and **Weinreb amides**, 0111

order of reactivity of carbonyls, 0111

substitution

aliphatic, 0112-0153

and concentration effects, 0142

and elimination, 0544

and HSAB, 0103

and leaving group, 0142

and radicals, 1270

and solvation in the transition state,
0140

and solvation, 0139

and structure of alkyl halides, 0140

aromatic, 0193-0210

aromatic, **Friedel-Crafts** acylation,
0929, 0932

by halogen, 0144-0153

substitution, electrophilic aromatic,
0195-0205, 1192

+I and –I effects, 0200, 0201, 0202

+M and –M effects, 0200, 0201, 0202

activating substituents, 1192

and charge transfer complexes, 0195

and chlorosulfonic acid, 0196

and electronic effects, 1192
and furansulfonic acid, 0196
and halogens, 0196
and inductive effects, 0200, 0201
and Lewis acids, 0195
and local ionization maps, 0203
and naphthalene, 0197
and nitration, 0196
and nitronium ions, 0196
and ortho substituents, 0203
and phenanthrene, 0197
and resonance, 0197, 0200-0203
and strain, 0203
and substituent effects, 1192
and sulfonic acids, 0196
and sulfur trioxide, 0196
and the **Hammond postulate**, 0198
and the **Wheland intermediate**, 0195, 1192
deactivating substituents, 1192
in ionic liquids, 0205
mechanism, 0200-0203
nitration, 1192, 1193
planarity and strain, 0203
rate of reaction, 0204
reactivity of heterocycles, 1193
regioselectivity, 0201
substituent effects, 0202
the ortho effect, 0203
with heterocycles, 0199
with furan, 1193
with heterocycles, 1193
with pyridine, 0199, 1193
with pyrrole, 0199, 1193
with quinoline, 0199
with thiophene, 1193
substitution
halogens and alkenes, 1273
intramolecular, 0772
substitution, nucleophilic aromatic, 0194, 0205-0210, 0913
and aryl halides, 0205
and aryldiazonium salts, 0208
and benzyne, 0207
and **Gattermann reaction**, 0209
and organolithium reagents, 0207
and rate of reaction, 0206
and resonance, 0206
and **Sandmeyer reaction**, 0209
cine substitution, 0208
I and M effects, 0206
intermediate, 0205
mechanism, 0205

substituent effects, 0206
with arylchromium reagents, 0764
substitution, nucleophilic, 0003
and the solvent, 0138
substitution, nucleophilicity vs. basicity, 0141
substitution, radical, 1273
substitution, S_E2', and **Grignard reactions**, 0665
substitution, S_N1, 0129-0137, 0950, 1162
and carbocations, 0129
and epoxides, 0656
and hydride shifts, 0134
and racemization, 0131, 0543
and rearrangement, 0131, 1170
and solvents, 0134
and stereochemistry, 0131
and steric strain, 0133
and the E1 reaction, 0161
and water, 0131
inversion with alkyl halides, 1170
nucleophilic bimolecular, 0079
rate of reaction, 0131, 1162
rate-determining step, 0131
solvent effects, 0131
transition state, 1162
substitution, S_N2 (aliphatic bimolecular), 0003, 0112-0126
alcohols and HX, 0144
alkyl halides and phosphines, 0730
amines and ammonium salts, 0122
and alkyl halides, 0358
and azides, 0117, 0555
and **Baldwin's rules**, 0569, 0570
and cyanide, 0117
and DBN, 0157
and electrophilicity index, 0119
and electrostatic potential maps, 0115
and epoxides, 0128, 0268
and **Grignard** addition to epoxides, 0654
and higher order cuprates, 0727
and inversion of configuration, 0113, 0114, 0123, 0543
and kinetic control, 0120
and molecular modeling, 0115-0117
and nitriles, 0358
and nucleophilicity, 0119
and oxidation of sulfides, 0336
and phthalimide, 0122
and polyalkylation, 0122
and rate of reaction, 0113

and stereogenic centers, 0114
and steric hindrance, 0114
and structure of alkyl halides, 0114
and super hydride, 0379
and the E2 reaction, 0153
and the **Mitsunobu reaction**, 0123-0126
and the pentacoordinate transition state, 0113
and the **Swain-Scott equation**, 0119
and the **Williamson** ether synthesis, 0118, 0594
and transition state modeling, 0115
and Umpolung, 0121
and ylids, 0734
substitution, S_N2
cyanide with alkyl halides, 0623, 0624
cyclization, powerful in synthesis, 0929
cyclohexanes, 0558
definition, 0113
diastereoselectivity, 0555
formation of ammonium salts, 0163
Grignard reagents and esters, 0648
intramolecular, 0569, 0570
inversion of configuration, 0555
leaving group ability, 0143
nucleophilic bimolecular, 0079
nucleophilic strength and modeling, 0116
nucleophilic, 0003
order of nucleophilic strength, 0119
phthalhydrazide, 0122
powerful in reaction, 0931
transition state, 0153
versus E2, 0141, 0158
versus S_N2', 0127, 0214
with amines, 0121
with enolate dianions, 0803
with phosphines, 0121
with selenide or sulfide anions, 0121
substitution, S_N2', 0126-0128, 0947
and cyclopropylcarbinyl halides, 0126
and displacement of esters, 0126
and **Grignard reactions**, 0655
and halogenation of allylic alcohols, 0149
and organocuprates, 0126, 0128, 0716
and stereochemistry, 0126
and steric hindrance, 0126, 0127
formation of piperidines, 0126
with amines, 0127
substitution, S_Ni, 0145

substitution, versus elimination, 0140
subtrees, and SYNGEN, 0962
succinic anhydride, and **Friedel-Crafts** acylation, 1208
succinimide, and the Mitsunobu reaction, 0125
sufloxinium salts, and acetic anhydride, 0245
sugar cane borer pheromone, 0507
sulcatol, 0479
sulfate, dimethyl, with alcohols, 0594
sulfide, chloromethyl methyl, with alcohols, 0597
sulfide, dimethyl
  and **Moffatt oxidation**, 0244
  and oxidant of alcohols, 0240
  CuBr complex, 0714
  and ozonolysis, 0320
  as a leaving group, 0746
  borane complex, 0510
  with borane, 0386, 0516
sulfides
  alkenyl, and umpolung, 0708
  alkenyl, and **Wittig reagents**, 0708
  alkenyl, hydrolysis, 0708
  alkenyl, with mercuric compounds, 0708
  alkenyl, with organolithium reagents, 0700
  allylic, and umpolung, 0711
  allylic, hydrolysis, 0713
  allylic, with organolithium reagents, 0711
  and mcpba, 0169
  and peroxides, 0334
  and radical cyclization, 1288, 1289
  and radicals, 1281
  and tartrate, 0337
  and titanium tetraisopropoxide, 0337
  boron, with ketenes, 0842
  carbanions, 0696
  cyclopropyl, and LDA, 1120
  dimethyl, compels with alkylboranes, 0509
  diphenyl, photolysis, 1247
  from alkyl halides, 0545
  from thiols, 0545
  hydrogenolysis, 0457
  lithio, 0537
  mechanism of oxidation, 0338
  oxidation, 0334, 0335
  oxidation, and pK_a, 0335
  oxidation, and salen catalysts, 0337

oxidation, solvent effects, 0337
oxidation, with oxaziridines, 0337, 0338
oxidizing agents, 0335
reductive cleavage, 0457
via S_N2 reactions, 0121
vinyl, and **thio-Claisen rearrangement**, 1141
vinyl, by **Wittig reactions**, 0733
vinyl, conversion to ketones, 0733
vinyl, with alkyl halides, 0708
vinyl, with **Grignard reagents**, 1246
with alkyl halides, 0745
with alkylboranes, 0537
with $NaIO_4$, 0168
with peroxyacids, 0334, 0336
sulfonates
  and alcohols, 0146
  and thionyl chloride, 0146
sulfinic acid, and pyrrole, electrophilic aromatic substitution, 0199
sulfite, with osmium tetraoxide, 0295
sulfobacin B, 0159
sulfonamides
  and nitrenes, 1314
  from amines, 0613
  hydrolysis, 0864
  in the aldol condensation, 0864
  Oppolzer's sultam, 0361, 0864
  reduction with $LiAlH_4$, 0361
sulfonate esters. *See also* esters
sulfonate esters
  alkyl, with alkyne anions, 0634
  and E2, 0158
  and **Friedel-Crafts alkylation**, 1201
  and the E2 reaction, 0159
  as leaving groups for elimination, 0174
  as leaving groups, 0142, 0143, 0934
  carbanions, 0696, 0701
  vinyl, and **Stille coupling**, 1240
  vinyl, with organocuprates, 0717
  with alkoxyaluminum hydrides, 0364
sulfone anions, and **Claisen rearrangement**, 1139
sulfones, 0445
  alkylation, 0699
  and **Julia olefination**, 0699
  and the **Fritsch-Buttenberg-Wiechell rearrangement**, 0700
  and β-elimination, 0158
  by oxidation of sulfoxides, 0697
  carbanions, 0696, 0699

extrusion of SO_2, 0701
from sulfides, 0334
from sulfoxides, 0334, 0337
halo, and formation of alkenes, 0701
hydrogenolysis, 0457
reaction energy, 0702
reductive cleavage, 0457
with aluminum amalgam, 0468
with sodium amalgam, 0700
sulfonic acids, and electrophilic aromatic substitution, 0196
sulfonium ions, chiral, 0835
sulfonium salts, 0974
  and the gegenion, 0745
  from sulfides, 0745
  with base, 0745
  with organolithium reagents, 0746, 0748
  with silver tetrafluoroborate, 0745
sulfonium ylids, 0745
sulfonylhydroxylamine, with boranes, 0522
sulfonyloxazidine (**Davis reagent**), 0814
  with enolate anions, 0814
sulfoxides, 0445
  alkenyl, formation of alkynes, 0700
  alkenyl, with **Grignard** reagents, 0700
  alkenyl, with organolithium reagents, 0700
  allylic, rearrangement, 0699
  and aluminum amalgam, 0725
  and **Mislow-Evans** rearrangement, 0699
  and organocuprates, 0725
  anions, acyl addition, 0698
  by oxidation of selenoxides, 0815
  by oxidation of sulfides, 0169
  by oxidation of sulfides, 0698
  carbanions, 0696
  chiral, 0336
  chiral, lithio derivatives, 0698
  chiral, reaction with **Grignard** reagents, 0698
  elimination, 0698
  from oxidation of sulfides, 0168
  from sulfides, 0334
  hydrogenolysis, 0457
  oxidation, 0334
  pyrolysis, 0168
  stereoselectivity in oxidation, 0336
  with aluminum amalgam, 0468
  with **Grignard reagents**, 0658

sulfoxide-sulfones, lithiation, 0707
sulfoxonium salts
  and **Moffatt oxidation**, 0244
  and phosphorus pentoxide, 0246
  and oxidation of alcohols, 0239
sulfur carbanions, 0444, 0695
  with alkyl halides, 0697
  with epoxides, 0698
sulfur compounds
  and umpolung, 0703
  as a poison for hydrogenation, 0426
  LUMO maps, 0702
  pK_a, 0697
  reduction with aluminum amalgam, 0698
  table of umpolung reagents, 0703
  with LDA, 0698
  with organolithium reagents, 0697
sulfur dioxide, extrusion, 0701
sulfur stabilized carbocations, 0133, 1163
sulfur trioxide, and electrophilic aromatic substitution, 0196
sulfur ylids, 0745-0750
  with aldehydes or ketones, 0746
sulfur, poison for hydrogenation, 0445
sulfurane
  and elimination, 0591
  with alcohols, 0591
sulfuric acid
  and **Bogert-Cook synthesis**, 1209
  and chormic acid, 0227
  and dehydration, 0921
  and electrophilic aromatic substitution, 0196
  and **Fischer Indole synthesis**, 1222
  and **Friedel-Crafts acylation**, 1206, 1202
  and **Friedel-Crafts cyclization**, 1209, 1218
  and haloamines, 1294
  and hydrogenation, 0435
  and **Jones reagent**, 0233
  and **Knorr quinoline synthesis**, 1211
  and permanganate dihydroxylation, 0293
  and pyridine, 0199
  and the **Schmidt reaction**, 0191
  DMSO, and epoxides, 0548
  in electrolysis, 0471
  the **Beckmann rearrangement**, 0189
  with alcohols, 1187
  with C-toxiferin I, 0920

with nitro enolates, 0831
with phenylhydrazones, 1218
with titanium (III) sulfate, 0472
sulfurous acid, and the **Étard reaction**, 0332
sultams
  as chiral auxiliaries, 1071
  in the **Diels-Alder reaction**, 1071
sultones
  carbanions, 0701
  reductive cleavage, 0701
  with aluminum hydride, 0701
sultone-sulfoxides, lithiation, 0707
super acids, and carbocations, 1178
super hydride, 0378
  with epoxides, 0379
  with iminium salts, 0380
supercritical carbon dioxide. *See* carbon dioxide
supercritical carbon dioxide, 0839
  and [3+2]-cycloaddition, 1107
  and the **Glaser reaction**, 0631
supercritical fluids, and **Oppenauer oxidation**, 0250
supercritical fluoroform, 0839
supercritical water, and the **Heck** reaction, 1239
superimposability, 0013
  and diastereomers, 0024
superpolycyclopentanoids, 0905
superstolide A, 0617
support, hydrogenation, definition, 0426
supported heterogeneous catalysts, 0424
suprafacial addition, and photochemistry, 1092
suprafacial ring opening, 1096
suprafacial
  cycloadditions, 1011
  sigmatropic rearrangement, 1116, 1117
surface area, and catalytic hydrogenation, 0427
surface sites, in catalysts, 0427
(Suzuki coupling) Suzuki-Miyaura coupling, 0511, 0527, 1241
  and microwaves, 1242
  and nickel catalysts, 1243
  and rhodium catalysts, 1243
  and silver oxide, 1243
  cross coupling, 1243
  metal catalyst, 1243
  powerful in synthesis, 0932

with heterocycles, 1242
Swain-Scott equation, 0119
Swern Oxidation, 0241, 0242, 1358
  and steric hindrance, 0241
  and trifluoroacetic anhydride, 0241, 0242
symbiosis, and HSAB, 0099
symmetry
  allowed and forbidden reactions, 1011
  and diastereomers, 0024
  and electrocyclic reactions, 1001
  and molecular orbitals, 1268
  and radical reactivity, 1268
  and stereochemistry, 1001
  and synthesis, 0920
  HOMO and LUMO, 1001
  icosahedral, 0905
syn and anti, definition, 0025
syn elimination. *See* elimination, syn
syn/anti aldolates, 0857
syn/anti isomers
  and **aldol condensation**, 0851, 0852
  and **Claisen rearrangement**, 1139
  and dihydroxylation, 0548
  and ester enolates, 0855
  and heteroatom radical cyclization, 1289
  and radical coupling, 1281
  and the **Reformatsky reaction**, 0886, 0887
  and **Zimmerman-Traxler Model**, 0852
  equilibration in aldol condensation, 0859
  with **Grignard reagents**, 0659
synchronous mechanisms, E2, 0153
Synechococcus sp. PCC 7942, reduction, 0482
SYNGEN, 0912, 0969, 0970
  and bondset, 0963
  and half reactions, 0967
  and subtrees, 0962
  and synthesis, 0961, 0967
  and the synthesis tree, 0961, 0962
  bond construction, 0969, 0970
  characteristics, 0968
  construction sets, 0964
  weighting scheme, 0965, 0966
synthesis tree, 0994
  and mass spectrometry, 0977
  and SYNGEN, 0961, 0962
  and synthesis, 0908, 0909

synthesis
  and amino acids, 0979, 0980
  and analog studies, 0904
  and bicyclic systems, 0935
  and biological activity, 0903
  and carbon-carbon bonds, 0006
  and chemical degradation, 0973, 0975
  and chiral template, 0954-0960
  and CHIRON, 0961
  and chirons, 0954-0960
  and complex targets, 0910
  and disconnections, 0006, 0909
  and functional group strategy, 0918,
    0925
  and LHASA, 0898, 0908, 0914, 0918,
    0919
  and MARSEIL/SOS, 0971
  and MASA, 0919
  and molecular complexity, 0916
  and new reactions, 0906
  and new reagents, 0906
  and NMR, 0902, 0903
  and polarized molecules, 0006
  and polymer beads, 0979
  and potential symmetry, 0920
  and preliminary scan, 0920
  and REKEST, 0972
  and resolution, 0956
  and retrosynthesis, 0908
  and stereochemical strategy, 0917
  and stereochemistry, 0937, 0941
  and stereoselectivity, 0960
  and strategic bonds, 0908
  and structural challenges, 0905
  and symmetry, 0920
  and SYNGEN, 0961, 0967
  and synthons, 0917
  and the synthesis tree, 0908, 0909,
    0956, 0958
  and topology, 0905, 0917
  and transform strategy, 0917
  and X-ray, 0902, 0903
synthesis
  atoms with certain functional groups,
    0933
  biomimetic, 0948-0954
  bonds related by known transforms,
    0934
  bridging goal atoms, 0927
  chiral template approach, 0954
  chiral template, 0947
  chiron analysis, 0958
  closely related intermediates, 0926

  consecutive versus convergent, 0912
  convergent, 0920, 0958, 0965
  criteria for disconnection of strategic
    bonds, 0924
  degree of bridging, 0935
  disconnection approach, 0908
  enantioselective, 0931
  intramolecular strategy, 0940
  irrational, 0978
  labile groups, 0924, 0925
  Level 1 and Level 2 Processes, 0941
  libraries of compounds, 0978
  list of powerful reactions, 0928-0930
  maximum bridging ring, 0935
  Merrifield, 0979
  of deoxyerythronolide B, 0953, 0954
  of pancratistatin, 0942-0947
  of peptides, 0979, 0980
  of sparteine, 0952
  parallel, 0978
  peptides, 0979, 0980
  pooling strategy, 0979
  primary rings, 0934, 0936
  problem like one solved, 0921
  rationale, 0899
  removal of interfering substructures,
    0926
  removal of stereocenters, 0924, 0925
  retro-mass spectral degradation, 0976
  split, combinatorial chemistry, 0982,
    0983
synthesis
  stereocenters and cyclic arcs, 0936,
    0937
  strategic bond analysis, 0920, 0938,
    0940
  strategic bonds and aromatic rings,
    0936
  strategic bonds in carbocyclic rings,
    0934
  strategic bonds in heterocyclic rings,
    0938
  strategic bonds in rings, 0934
  strategies, 0897
  string of available pieces, 0922
  structural verification, 0902
  structure goal strategy, 0917
  symmetric, 0110
  synthetic tree, 0938
  target selection, 0898
synthetic equivalents, 0002, 0008, 0009,
  0589, 0933, 0952
  and disconnect products, 0009

synthetic strategies, 0914
synthetic surrogate, 0002
synthetic tree, 0899, 0956
  and LHASA, 0918
  and synthesis, 0938, 0958
synthon approach, definition, 0898
synthons, definitions, 0917
syringe pumps, and **Stille coupling**,
  1240, 1241

T

2^n rule, 0024
tabersonine, 0561
table, umpolung reagents, 0703
Taft's sigma function, and oxidation,
  0219
Tags
  and oligonucleotides, 0986
  in combinatorial chemistry, 0984
Takai reaction, 0757
Takai-Utimoto reaction, 0849
  and **Nozaki-Hiyama-Kishi** reaction,
    0849
Tamao-Fleming oxidation, 0330
tamoxifen, 0161
tandem reactions, and **Cope
  rearrangement**, 1129
tangutorine, 1237
target structure, 0002
tartaric acid, 0956
  chiral additive in hydrogenation, 0449
tartrate, 0992
  and asymmetric dihydroxylation,
    0549
  and chiral sulfoxides, 0337
  and oxidation of sulfides, 0337
  and S_N2 reactions, 0117
  chiral template, 1333
  model for **Sharpless asymmetric
    epoxidation**, 0285, 0287
  and **Sharpless asymmetric
    epoxidation**, 0284, 0549
TAS. *See* tris(diethylamino) sulfonium
tautomerism
  and phenols, 1081
  imine-enamine, 1213
  keto-enol, 1265, 1327
tautomerization
  and **anionic oxy-Cope
    rearrangement**, 1130
  and hydroboration of alkynes, 0518
  and **oxy-Cope rearrangement**, 1130

enamines-iminium salts, 0476
keto-enol, 0749, 1130
of enols, 0528
of ketones, 0749
taxanes, 0849
taxol, 0900, 0994
TBAF
in the **Barton-McCombie reaction**, 0168
with allylic silanes, 0770
with silanes, 0519, 1065
with silyl protecting groups, 0599
with *tert*-butyldimethylsilyl ethers, 0603
with *tert*-butyldiphenylsilyl ethers, 0604
with triethylsilyl ethers, 0602
with triisopropylsilyl ethers, 0602
with trimethylsilyl ethers, 0601
TBDMS (*tert*-butyldimethylsilyl). *See* ethers
TBDPS (*tert*-butyldiphenylsilyl). *See* ethers
TBHP, reaction with alkenes, 0272
TBS (*tert*-butyldimethylsilyl). *See* ethers
teabag approach, combinatorial chemistry, 0981
Tebbe reagent, 0755
with aldehydes or ketones, 0755
with ester, 0755
tedanolides, 0402
teleocidin B4, 1202
temperature
and aromatic substitution, 0206
and hydrogenation, 0440
templates chiral. *See* chiral templates
TEMPO
and acid catalysis, 0248
and oxidation, 0248
and radical cyclization, 1293
and radicals, 1295
terpenes, 0305
terpenoids, 0176
terpestacin, 0634
terpineol, 0897
tert-butoxide, with dihalides, 0630
tert-butoxycarbonyl. *See* Boc
tert-butyl ether. *See* ether
tert-butyldimethylsilyl chloride
with alcohols, 0603
with dimethylaminopyridine, 0603
with imidazole, 0603

tert-butyldimethylsilyl ethers. *See* ethers
for alcohols, 0603
tert-butyldiphenylsilyl ethers. *See* ethers
tert-butyldiphenylsilyl chloride
with alcohols, 0604
with imidazole, 0604
tert-butyldiphenylsilyl ethers, for alcohols, 0604
tert-butyllithium
and benzyne formation, 1065
and sparteine, X-ray, 0672, 0673
and enol ethers, 0707
TES (triethylsilyl). *See* ethers
TES, 0601
testrone, SYNGEN analysis, 0971
testudinariol A, 0763
tet, and **Baldwin's rules**, 0564, 0567
tether length
and **Friedel-Crafts acylation**, 1205
and **phenolic oxidative coupling**, 1297
intramolecular Diels-Alder, 1059, 1060
tethered bases, and elimination, 0164
tethered intramolecular **Diels-Alder**, 1064, 1065
tethers
boronate esters, 1065
in photocycloaddition, 1092
silane, 1065
tetrabutylammonium fluoride. *See* TBAF
tetrabutylammonium iodide, and ether formation, 0594
tetrabutylammonium triacetoxyborohydride, 0033
tetrabutylammonium tribromide, and bromination, 0185
tetrachlorobenzoquinone. *See* chloranil
tetrachloropalladate, disodium, 1227
tetracyanoethylene, in the **Diels-Alder** reaction, 1014
tetraenes, cyclization, 0950
tetraethylammonium acetate, and dihydroxylation, 0297
tetrafibricin, 0146
tetrafluoroboric acid, with oxaspiropentanes, 1176
tetrahedral intermediate, 0109, 0110
and acyl substitution, 0646
and enolate anions of acid

derivatives, 0823
tetrahedron, extended, 0026
tetrahydro-1-oxpyranobenzopyrans, 0826
tetrahydroaluminates, and LiAlH$_4$ reduction, 0350
tetrahydrofuran. *See* THF
tetrahydroisoquinoline alkaloids, 0976, 0977
tetrahydroisoquinolines, 0976, 1203
tetrahydronaphthalenes, and **Darzens tetralin synthesis**, 1210
tetrahydropyran, and organolithium reagents, 0673, 0674
tetrahydropyranyl ethers. *See* ether
tetrahydropyrazins. *See* pyrazolizidines
tetrahydropyrimidines, by the **Diels-Alder reactions**, 1055
tetrakis(triphenoxyphosphino)nickel, 1245
tetrakis-(triphenylphosphino) palladium, 1231, 1232
tetrakis-palladium(0), 1231
tetralin, 0445
tetramethoxyaluminum, lithium, 0364
tetramethylethylenediamine. *See* TMEDA
tetramethylguanidine, and phosphonium salts, 0732
tetramethylpiperidide. *See* lithium etc.
tetramethylpiperidide. *See* LTMP
tetramethylpiperidine, as a base, 0092
tetraoxone, 0318
tetraphenylnickel, 1245
tetrapropylammonium periodate, with hydroxamic acids, 1056
tetrapropylammonium perruthenate. *See* TPAP
tetrazines, conversion to diazines, 1153
tetrodotoxin, 1330
tetronolide, 0185, 1065
TFA, hydrolysis of acetals or ketals, 0609
TFA
and tert-butyl ethers, 0596
trifluroaetamide protecting group, 0616
TFAA
and alkylboranes, 0528
and DMSO, 0241
and lactonization, 0576
and **Swern Oxidation**, 0241, 0242
thaliporphine, 1299

thallium ethoxide, 1243
and **Suzuki-Miyaura coupling**, 1243
thallium nitrate. *See* nitrate
thallium(II) acetate, with enamines, 0561, 0562
thapsigargin, 0909, 0913
Thermoanaerobacter ethanolicus, reduction, 0483
Thermoanaerobium brockii alcohol dehydrogenase, 0398
Thermoanaerobium brockii, 0483
thermodynamic conditions, and **Robinson annulation**, 0881
thermodynamic control
acid-catalyzed aldol condensation, 0820, 0821
and **Baldwin's rules**, 0570
and enolate anions, 0795
and hydrogenation, 0433
and LDA, 0800
and metal cations of enolate anion, 0800
and radical cyclization, 1284
and the **Claisen condensation**, 0826
and time, 0800
enolate anion, reaction conditions, 0798
thermodynamic enolate, 0791
and enol acetates, 0815
thermodynamic stability
and molecular modeling, 0067
in photoreduction, 0478
thermodynamic versus kinetic control, 0198
thermolysis
and formation of radicals, 1259
of acyl peroxides, 1261
of azides, 0933
of benzoyl peroxide, 1260
of chloroform, 1319
of dialkyl peroxides, 1260, 1261
of di-tert-butyl-peroxide, 1260, 1261
of esters, 0166
of hydrogen peroxide, 1260
of hypochlorites, 1275
of sulfoxides, 0552
of triazolines, 1113
thexylborane, 0495, 0516
thexylborane, with alkenes, 0516, 0529, 0531
THF
and borane, 0524
and butyllithium, 0672

and enolate anions, 0799
and **Grignard reagents**, 0638, 0639
and LDA, 0789
and organocuprates, 0714
and organolithium reagent aggregation, 0672
sequestration of Li in the **Wittig reaction**, 0738
thia-**Fries rearrangement**, 1220
thiazinotrienomycin E, 0342
Thiele's acid, 0941
thienamycin, 0546
thienyl ligands, and organocuprates, 0728
thienyllihtium, 0685
thioaldehydes
by **thio-Claisen rearrangement**, 1141
hydrolysis, 1141
thiocarbamates, and selenides, 0170
thiocarbanions. *See* sulfur carbanions
thio-**Claisen** rearrangement. *See also* **Claisen**
thio-**Claisen rearrangement**, 0713, 1141
and Umpolung, 1141
thiocyanaotoneopupukeanane, 1182
thioesters
and macrolactonization, 0577
and the **Masamune protocol**, 0577
with dialkylboryl triflates, 0842
thioethers. *See* sulfides
thioethers, cyclic, and **Baldwin's rules**, 0570
thioketones, with vinyl phosphonium salts, 0744
thiol
allylic, anions, 0711- 0713
anions, with alkyl halides, 0545
with cyclopentadienyl dicarbonyl and alkenes, 1249
thionocarbonates
and **Barton deoxygenation**, 0467
and the **Barton-McCombie** reaction, 0467, 1274
thionyl chloride
and chlorosulfinate esters, 0544
and pyridine, with alcohols, 0552
and S_Ni, 0145
and the **Darzens' procedure**, 0146
with alcohols, 0145, 0543, 0544
with amines and alcohols, 0544

thiophenes
and alkylboranes, 0502
and **Friedel-Crafts acylation**, 1205
and Lewis acids, 0095
Diels-Alder reactions, high pressure, 1047
electrophilic aromatic substitution, 1193
thiophenol, anion, with alkyl halides, 0127
thiopyridone, and lactonization, 0578
thiosemicarbazides, 1113
thioureas, chiral, and the **Pictet-Spengler isoquinoline synthesis**, 1216, 1217
Thorpe condensation, and $LiAlH_4$ reduction of nitriles, 0359
Thorpe reaction, 0624, 0832
mixed condensation, 0832
Thorpe-Ziegler reaction, 0833
though-bond effects, and acidity, 0085
THP (tetrahydropyranyl). *See* ether
three-carbon extension, 0533
three-component coupling, 0627
three-letter codes, for amino acids, 0979
threitol, 0993
threo and erythro, definition, 0025
threonine, 0974
thromboxane B_2, 1137
through space interactions, and σ-participation, 1173
thyronamines, 0360
Tiffeneau-Demjanov ring expansion, 1175
tight ion pair. *See* ion pair
tight ion pairs, and **Friedel-Crafts** alkylation, 1196
Timaeus, 0905
time, and enolate anions, 0800
tin (II) chloride, with nitriles, 0465
tin compounds, and **Stephen** reduction, 0465
tin tetrachloride
and **Johnson** polyene cyclization, 1184
and **Mukaiyama aldol** reaction, 0837, 0838
with allylsilanes, 0769
with polyenes, 0951, 1184
tin(II) bromide, 0466
tin(II) chloride, 0465
tin(IV) chloride, 0466
and cationic cyclization, 0558

tin, allyl. *See* allyltin
  and prenylation, 1281
  and the Felkin-Anh model, 0767
  binapthols, 0767
  diastereoselectivity, 0767
  enantioselectivity, 0767
  with aldehydes, 0766
  with BINOL, 0767
  with ketones, 0766
  with Lewis acids, 0766
tin, allylbutyl, and radical coupling, 1280, 1281
tin
  and reduction, 0465
  crotyl, with aldehydes, 0767
TIPS (triisopropylsilyl). *See* ethers
titanium (0), and **McMurry olefination**, 1309
titanium carbenes, 0756
titanium compounds, with salen catalysts, 0337
titanium dichloride, and the **Reformatsky** reaction, 0887
titanium dioxide, and **McMurry olefination**, 1310
titanium reagents
  and olefination, 0755
  with iodomethane, 0756
titanium tetrachloride
  bis(tetrahydrofuran) complex, 1039
  and **Mukaiyama aldol reaction**, 0837
  and **pinacol coupling**, 1303
  in the **Schmidt reaction**, 0192
  with Zn and dibromomethane, 0757
titanium tetraisopropoxide, and oxidation of sulfides, 0337
titanium trichloride
  and **acyloin condensation**, 1308
  and acyloins, 1308
  and conjugate addition, 1299
  and potassium, 1309
  with LiAlH$_4$, 1309
  with lithium, 1309
  with potassium, 1309
titanium(0), and zinc, **McMurry olefination**, 1310
titanium(III) sulfate, 0471
  with sulfuric acid, 0472
titanocene, catalyst for ene reaction, 1147
TMEDA
  and aldol condensation, 0866

and butyllithium, 0672
  and dialkylamide bases, 0794
  and lithium hydrogen exchange, 0697
  and organolithium reagent aggregation, 0672
  and organolithium reagents, 0674, 0686, 0687, 0689
  with **Grignard reagents**, 0640
TMS (trimethylsilyl). *See* ethers
tocotrienoloic acid, 0367
toluene
  and polyalkylation, 1201
  with *Pseudomonas putida*, 0304
toluenesulfonic acid. *See* tosic acid
topology, and synthesis, 0905, 0917
topopyrone, 0931
torquoselectivity, 1099
torsion strain, and diastereoselectivity, 0850
tosic acid
  with acetals, 1333
  reaction with ethers, 0136
TosMIC (tosylmethylisocyanide)
  with ketones, 0624
  with aldehydes, 0624
tosylates
  in S$_N$2 reactions, 0117
  leaving groups, 1358
  leaving group for elimination, 0174
  reduction to methyl, 0368
  reduction with LiAlH$_4$, 0361
  solvolysis, 1174
  with sodium cyanoborohydride, 0383
tosylhydrazine
  with aldehydes or ketones, 1317
  with ketones, 0564
tosylhydrazones
  and diimide formation, 0474
  and **Eschenmoser ring cleavage**, 0564
  and the **Bamford-Stevens reaction**, 1317
  and the **Shapiro reaction**, 1317
  carbanions, 0848
  fragmentation, 0564
  with peroxyacids, 0564
tosylimines, in the **Diels-Alder** reactions, 1055
tosylmethylisocyanide. *See* TosMIC
tosyloximes, 0190
TPAP (tetrapropylammonium perruthenate), 0246, 0249

Ley oxidation, 0249
  oxidation of alcohols, 0249
  with NMO, 0555
trachelanthamidine, 1289
trajectory, and **Baldwin's rules**, 0564, 0565
trans, s. *See* s-trans
trans/cis isomers
  alkenes, 0029
  alkenylalkylboranes, 0509
  alkylboranes, 0498
  and diazoalkanes, 1323
  and E2 reactions, 0559
  and hydrogenation of alkynes, 0546
  and radical cyclization, 1285
  and the **Diels-Alder reaction**, 1024
  cyclohexanes, 0062
  enantioselectivity in alkene hydroboration, 0520
transannular interactions, 0054
transannular strain, 0049, 0052
  and macrocyclizations, 0573
transesterification, 0110, 0397
  and enolate anions, 0809
  and the **Baeyer-Villiger reaction**, 0308
transfer hydrogenation, 0613
transforms, 0002, 0078
  and bond polarization, 0080
  and functional groups, 0080
  and LHASA, 0918
  and steric hindrance, 0081
  and strain, 0081
  and synthesis, 0917
  and π-bonds, 0080
  organometallic, 0918
  retrosynthetic, 0078
transition metal reagents, olefination, 0754-0758
transition metals. *See* metals
  and hydroperoxides, 0272
  and metathesis, 1334, 1335
  catalyst for **Claisen** rearrangement, 1140
  with carbenes, 1324
  with **Grignard reagents**, 0640, 0641, 0646, 0647
  with hydroperoxides, 0272
transition state
  1,3-elimination, 0177
  6-centered, **Diels-Alder**, 1013
  aldol condensation, 0852
  AlH$_3$ reductions, 0388

alkenes with borane, 0499
alkenes with boranes, 0497
transition state
  and allowed-forbidden reactions, 1010, 1011
  and asynchronicity, 1001
  and **Baldwin's rules**, 0565
  and molecular modeling, 1019
  and **Noyori Open-Chain Model**, 0858
  and Perturbation theory, 1001
  and radical cyclization, 1288, 1290
  and solvation, 0140
  and strain, 0049
  and the **Hammond postulate**, 0198, 1001, 1128
transition state
  and the **Claisen rearrangement**, 1132
  and the **Cope rearrangement**, 1124, 1125
  aryl migration in carbocations, 1172
  boat, and **aldol condensation**, 0857
  boat, and **Cope rearrangement**, 1124
  borane addition to alkenes, 0492
  boron enolates, 0857
  carbene with alkenes, 1320
  carbocation rearrangement, 1165, 1171
  chair, and **aldol condensation**, 0853, 0854, 0855
  chair, and **Cope rearrangement**, 1124
  chair, and iminium ion cyclization, 1185
transition state, **Diels-Alder reaction**, 0923, 1013, 1014, 1019, 1028, 1030
  and charge transfer, 1023
  and regioselectivity, 1022, 1023
  destabilization, 1031
  modeling and regioselectivity, 1020, 1021
  molecular modeling, 1043
  dipolar, 1045
transition state
  and ester enolates, 0857
  E2 reaction, 0154
  endo, and **Diels-Alder reactions**, 1029, 1030, 1031
  ene reactions, 1143, 1144, 1146
  ene reactions, and modeling, 1144, 1147, 1148
  energy, and radical cyclizatio, 1288
  energy, the **Claisen rearrangement**,

1139
enolate, E/Z geometry, 0854, 0855
en-yl, 0720
Evans' Model, 0857
  for the **Cram cyclic model**, 0406, 0407
transition state
  four-center, and **Grignard reactions**, 0652
  Grignard addition to conjugated ketones, 0653
  Grignard addition to ketones, 0645
  Grignard reaction and selectivity, 0661
  Grignard reactions with carbonyls, 0666
  Grignard reduction of carbonyls, 0667
transition state
  half-chair and boat, 0060
  hydroboration, 0494
  intramolecular **Diels-Alder** reactions, 1060, 1061
  macrocyclizations, 0573
  metal-hydrogen exchange, 0689
transition state, modeling
  Claisen rearrangement, 1134
  Cope rearrangement, 1128
  for radical cyclization, 1290, 1291
  hydroboration, 0499, 0517, 0521, 0523
  intramolecular **Diels-Alder**, 1063
  S_N2 reactions, 0115
  S_N2, bond distance, 0115
transition state
  of decarboxylation, 0173
  Oppenauer oxidation, 0250
  pentacoordinate, 0113
  radical cyclization, 1283, 1284, 1290, 1291
  reactant like, and reduction, 0408
  reduction of alkoxy ketones, 0406, 0407
  six-centered, 0949
  six-centered, and **Norrish Type II** fragmentation, 1265
  six-centered, and radicals, 1276
  six-centered, decarboxylation, 0172
  S_N1 reaction, 1162
  S_N2 reaction, 0153
  S_N2, and alkyl halides, 0114
  S_N2, and rate of reaction, 0114
  S_N2, and steric hindrance, 0114

Zimmerman-Traxler model, 0857
transketalization, 0609
transmetallation, and **Nozaki-Hiyama-Kishi reaction**, 0849
transoid. *See* s-trans
tree, synthesis. *See* synthesis
treprostinil, 0686
tri(ortho-tolyl)phosphine, and allyl palladium, 1230
triallylborane, 0527
triazabicyclo[4.4.0]dec-5-ene, and phosphonium salts, 0731
triazacyclopenta[cd]pentalene, 1113
triazines, by **Diels-Alder reactions**, 1033
triazoles
  by [3+2]-cycloaddition, 1113
  from alkenes, 1113
  from alkyl azides, 1114
  from azomethine imines, 1113
  nitrogen extrusion, 1113
  via [3+2]-cycloaddition, 1114
triazolines
  thermal decomposition, 0933
  by [3+2]-cycloadditions, 0933
  from alkyl azides, 1113
  from azides, 0933
  photochemistry, 1114
  rearrangement, 0933
  thermolysis, 1113
tribromide, phosphorus, with alcohols, 0148
  with ketones, 1367
tributylphosphine, and the **Mitsunobu reaction**, 0545
tributyltin dimer, and radical cyclization, 1289, 1290
tributyltin hydride. *See* hydride
  and alkenes, 1283
  and alkyl halides, 1283
  and **Baldwin's rules**, 0565
  and radical cyclization, 1284-1286, 1289, 1290
  and radical reduction, 1274
  and radicals, 1279, 1293
  reduction of halides, 1319
trichloroacetimidic esters, 1052
trichlorobenzoyl chloride (2,4,6)
  and DMAP, 0576
  and the **Yamaguchi protocol**, 0576
trichlorotitanium compounds, 0768
trichodermamides A and B, 0813
trichodiene, 0453

tricyclene, 1317
tricycloclavulone, 0243
tricyclohexylborane, 0528
trienes
  cyclization, 0949
  intramolecular **Diels-Alder** reactions, 1062, 1063
  with alkylboranes, 0531
triethylamine, with TFAA, 0616
triethylborohydride. *See* borohydride, super hydride
triethyloxonium tetrafluoroborate, with amides, 0615
triethylsilane. *See* silane
triethylsilyl ethers. *See* ethers
triethylsilyl ethers, for alcohols, 0602
triflates, alkenyl, with organolithium reagents, 0717
triflates
  trimethylsilyl, 0602
  trimethylsilyl, with allylsilanes, 0769
  and **Friedel-Crafts alkylation**, 1201
  and the E2 reaction, 0159
  silyl, with enolate anions, 0840
  trialkylsilyl, with lutidine, 0602
  vinyl with alkynes, 1244
  vinyl, and **Stille coupling**, 1240
  vinyl, with aldehydes, 0849
  vinyl, with vinyltin compounds, 1240
triflimides, with ketones, 1240
trifluoroacetamides, with NaBH₄, 0616
trifluoroacetic acid
  and **Dess-Martin periodinane**, 0247
  cleavage of silane tether, 1065
  with arylchromium carbanions, 0764
trifluoroacetic anhydride. *See* TFAA
trifluoroborates, potassium
  and **Suzuki-Miyaura coupling**, 1243
  with aryl halides, 1243
  with vinyl halides, 1243
trifluoroperoxyacetic acid. *See* peroxyacid
trifuoroacetic acid
  and lactonization, 0576
  with triethylsilane, 0475
trig, and **Baldwin's rules**, 0564, 0567
trihydroxybenzene, [1,7]-sigmatropic rearrangement, 1119
triisobutylaluminum, 0389
  and **Meerwein-Ponndorf-Verley reduction**, 0467
  and reduction, 0467

triisopropoxyborohydride, potassium, 0505
  and carbonylation, 0532
  with borane, 0509
  with boranes, 0510, 0512
triisopropoxyborohydride, sodium, 0377
triisopropylchlorosilane, with imidazole, 0603
triisopropylsilyl ethers. *See* ethers
triisopropylsilyl ethers, for alcohols, 0602
trilobolide, 0887
trimethoxyaluminum hydride, 0364
trimethoxyaluminum hydride, lithium
  and boranes, 0532
  and carbonylation, 0532
trimethoxyborohydride, sodium, and disproportionation, 0377
trimethylaluminum, 0932
  with alkynes, 0763
  with chloroborane, 0535
  with epoxides, 1177
trimethylamine *N*-oxide, with iron complexes, 1249
trimethylborate, and NaH, 0348
trimethylcetylammonium permanganate, 0260
trimethylene methane equivalent, 1235
  diastereoselectivity, 1235
  regioselectivity, 1235
  with alkynes, 1236
trimethyloxonium tetrafluoroborate, with alcohols, 0594
trimethylphosphite
  and oxygen, 0813
  and ozonide, 0321
trimethylsilyl allylic acetate, 1235
trimethylsilyl amines, protecting groups, 0614
trimethylsilyl cyanide. *See* cyanide
trimethylsilyl ethers. *See* ethers
trimethylsilyl ethers, for alcohols, 0600
trimethylsilyl iodide, with ethers, 0594
trimethylsilyl triflate, 0602
  and iodolactamization, 0187
  with Boc-amines, 0617
trimethylsilylacetylene, with butyllithium, 0634
2-(trimethylsilyl)ethoxymethyl ether. *See* ether
trimethylsilylmethyllithium, with aldehydes ornes, 1120
trimethyltin chloride, with NaBH₄, 0466

trinorbornylborane, 0534
triols
  and the **Skraup reaction**, 1215
  by reduction of malic acid, 0386
trioxolanes
  by [3+2]-cycloaddition of ozone, 1103
  from alkenes, 1103
  via ozonolysis, 0318
triphenylcarbinol, reduction, 0476
triphenylmethoxyborinane, 0518
triphenylmethyl radicals, 1257, 1270
triphenylphosphine
  and bromine, 0147
  and CBr₄, 0630, 0733
  and DEAD, 0545
  and hexachloroacetone, 0149
  and iodine, 0148
  and macrolactonization, 0578
  and *N*-chlorosuccinimide, 0146
  and ozonolysis, 0319
  and the **Corey-Fuchs procedure**, 0630
  and the **Mitsunobu reaction**, 0545
  complex with bromine, 0448
  with allyl palladium, 1229
  with palladium acetate, 1237
triple-ion model, 0686
triplet
  butadiene, 1089
  carbenes, 1311
  diene, 1089
  diradicals, 1324
  sensitizer, 1089
  definition, 1312
triquinacene, 0941
triquinacenecarboxylic acid, 0941
triquinane derivatives, 1080
1,1,1-tris(acyloxy)-1,1-dihydro-1,2-beniodoxo-3-9(1*H*)-one, 0247
tris(diethylamino) sulfonium (TAS), and **aldol** condensation, 0858
trisaccharides, 0987
trisiamylborane, conversion to alkyl borohydrides, 0381
tri-tert-butylphosphine, in the **Suzuki-Miyaura coupling**, 1243
Triton B (benzyltrimethylammonium hydroxide), 0741
  and phosphonate ester ylids, 0741
trityl cation, 1248
tropilidenes, and rearrangement, 1118
tropinone, 0897
tropone, iron tricarbonyl protected,

1249, 1250

Trost allylation, 1233

tryptamine, 0922, 0927

Tsuji-Trost reaction, 1233

tubelactomicin A, 0879

tuberonic acid, 1137

tuberostemonine, 0246, 0376, 0278

tubifoline, 0358

tungsten trioxide, and hydrogen peroxide, 0270

turmerone, MARSEIL/SOS generated retrosynthesis, 0972

twist conformations, 0048, 0049

twistane, 0941

twistanone, 0941

twist-boat conformations, 0050, 0060

twist-boat conformations. *See* twist

tylosin, 0723

Tynacantha marginata Dallas, 0609

tyramine, hydroxylation, 0266

U

Udenfriend's reagent, and oxidation of aryls, 0266

Ugi reaction, 0627

Ullmann reaction, 1300

and ultrasound, 1301

aryl halide reactivity, 1300

unsymmetrical coupling, 1301

ultrasound

and **Diels-Alder reactions**, 1045

and **Pauson-Khand reaction**, 1344

and **Simmons-Smith reaction**, 1332

and the **aldol condensation**, 0817

and the **Grignard reaction**, 0637

and the **Reformatsky reaction**, 0885

and **Ullmann reaction**, 1301

umpolung equivalents, 0704

umpolung reagents, table, 0703

Umpolung, 0121, 0702

acyl anion equivalent, 0704

and allylic sulfides, 0711

and allylic thiols, 0711

and carbonyl groups, 0703, 0704

and dithiolanes, 0706, 0707

and enol ethers, 0707

and ketene dithioacetals, 0710

and **Michael acceptors**, 0709

and **Michael reactions**, 0713

and sulfoxides-sulfoxides, 0707

and **thio-Claisen** rearrangement, 0713, 1141

and vinyl sulfides, 0708

and vinylogous ketene dithioacetals, 0712

definition, 0703

dithianes, 0704

of sulfur compounds, 0703

vinyl ethers, 0707

unsupported heterogeneous catalysts, 0424

urea-hydrogen peroxide complex, 0276

ureas, by the **Hofmann** rearrangement, 0193

usnic acid, 0920, 0921

ustiloxin D, 0311

U-strain, and conformation, 0058

UV light

and conjugation, 1264

and formation of radicals, 1262

and functional groups, 1083, 1084

and photochemistry, 1083, 1084

and sulfides with alkylboranes, 0537

in the **Schmidt reaction**, 0192

tags, combinatorial chemistry, 0986

vacuum, and radical formation, 1263

wavelength range, 1083

with alkenes, 1264

with carbonyl compounds, 1264

V

vacuum UV, and radical formation, 1263

valance isomerization

diazoalkanes, 1314

and diazirine, 1314

valency, and Lewis acids, 0093

valerolactone, 0310

valienamine, 0245

valinol, and **Evans' auxiliaries**, 0869

valsartan, 0152

vanadium acetylacetonate, and epoxidation, 0273

vanadium pentoxide, and hydrogen peroxide, 0270

vanadium, and allylic alcohols, 0273

vancomycin-resistant bacteria, 0987

vanillin, 0595

variecolol, 1354, 1364-1376

Vaska's catalyst, 0428, 0430

verapliquinone A, 0263

Veschambre's rule

and enzyme reduction, 0482

and **Prelog's rule**, 0482

Viagra, 0900

vibsanin E, 1129

Vilsmeier reaction, 0372

and DMF, 0372

Vilsmeier-Haack complex, 0372

Vilsmeier-Haack reaction, 1206

and chloroiminium salts, 1206

and microwaves, 1207

and $POCl_3$, 1206

solvent free, 1207

with pyrrole, 1207

vinblastine, 0309

vinegrol, 0176, 0659

vinyl alanates, 0762

vinyl alkylboranes. *See* alkylboranes, alkenyl

vinyl boranes, 0841

vinyl carbocations, 1168

vinyl carbocations. *See also* carbocations

vinyl cyclopropane rearrangement, 0749

and vinyl aziridines, 1114

vinyl dihalides. *See* dihalides

Vinyl ethers, HOMO and LUMO, 1002, 1003

vinyl Grignard. *See* Grignard, alkenyl

vinyl halides. *See* halides, alkenyl

vinyl ketenes, 1079, 1080

vinyl quinone imides, electrocyclic reactions, 1101

vinyl radicals, and radical cyclization, 1286

vinyl sulfides. *See* sulfides, alkenyl

vinyl sulfones. *See* sulfones

vinyl sulfoxides. *See* sulfoxides

vinylborinates, 0503, 0504, 0505

vinylboronic acid, 0511

vinylcyclopropane rearrangement, 0933, 1119, 1129

vinylcyclopropanes, flash vacuum pyrolysis, 0749

vinylmagnesium halides, 0639

vinylogous ketene dithioacetals, and umpolung, 0712

vinylogy

and enolate anions, 0812

definition, 0106

vinyltin compounds,

Stille coupling, 1240

with vinyl triflates, 1240

viridin, 1099

visible spectrum, alkenes, 1264

vitamin A alcohol, 1231

vitamin A, 0252
vitamin B-12, 0906
Vitride. *See* Red-Al
VM55599, 1294
volubilide, 0903

W

Wacker oxidation, 1226
  and molecular oxygen, 1226
Wacker process, and alkenes, 0333
Wagner-Meerwein rearrangement,
  1177
  retro, 0976
walk rearrangement, 1118
Wallach degradation, 0887
Walsh diagram
  and methyl radical, 1259
  and radicals, 1259
water
  and **Diels-Alder** reactions, 1041, 1044
  and dihydroxylation, 0301
  and dissolving metal reduction, 0452
  and epoxidation of alkenes, 0278
  and ionization, 0131
  and LiAlH$_4$, 0349
  and NaBH$_4$, 0370
  and the S$_N$1 reaction, 0131
  in hydroboration, 0531
  in the **Prévost reaction**, 0302
  ionization of alkyl halides, 0131
  pK$_a$, 0091
  rate of reaction, KI-alkyl halides, 0131
  supercritical, and the **Heck** reaction,
    1239
  with aryl diazonium salts, 0208
    with cyclopentadienyl dicarbonyl
    and alkenes, 1249
  with **Grignard reagents**, 0639
  with vinyl carbocations, 1183
wavelength and wavenumber, 1083
wedge notation, 0053
weighting scheme, for SYNGEN, 0965,
  0966
Weinreb amides
  and acyl substitution, 0111
  reduction, 0389
  with phosphine oxide carbanions,
    0742
  with **Grignard reagents**, 0649, 0742
  with organolithium reagents, 0683
welwistatin, 0804
welwitindolinone A, 1181

Wender's macroexpansion reaction,
  1130
Wheland intermediate, 0194, 1192
  and **Friedel-Crafts acylation**, 1204
  and the ipso carbon, 0200, 0201
  modeling, 0198
Whiting reaction, 0362
Wieland-Miescher ketone, 0915
Wilkinson's catalyst, 0428, 0429,
  0437, 0500, 1371
  and epoxides, 0443
  modification, 0449
  with phosphines, 0449
Williamson ether synthesis, 0594
  and epoxides, 0268
  and phenols, 0771
  and S$_N$2 reactions, 0118
  intramolecular, 0118, 0268
Witcop-type photocyclization, 0932
Wittig olefination (**Wittig** reaction),
  0007, 0121, 0590, 0730, 0732, 0950
  and NMR, 0737
  and oxaphosphatane conformation,
    0738
  and oxaphosphatanes equilibration,
    0738
  and **Peterson olefination**, 0709
  and synthetic equivalents, 0009
  and the **Corey-Fuchs procedure**,
    0630
  concentration effects, 0738
  continuum of mechanisms, 0737
  formation of E/Z isomers, 0735
  formation of vinyl sulfides, 0733
  in water, 0732
  intramolecular, 0733
  mechanism, 0730, 0737
  oxaphosphatanes model, 0737
  salt free conditions, 0736
  sequestration of lithium, 0738
  stereocontrol, 0738
  stereoselectivity, 0735
  with ketones, 0552
Wittig reagents
  and vinyl sulfides, 0708
  oxaphaphatanes, 0731
  with organocuprates, 0744
Wittig rearrangement, 1121
  diastereoselectivity, 1122
  enantioselectivity, 1121
  organolithium reagents, 1122
Wittig, aza, rearrangement. *See* aza-
  Wittig

Wittig-Gilman reaction, 0676
Witting reaction, selectivity, 0732
Wohl-Ziegler reaction, 0151
Wolff rearrangement, 1315
  Arndt-Eistert synthesis, 1316
Wolff-Kishner reduction, 0472, 0952,
  1252
  and **Clemmensen reduction**, 0472
  and the **Huang-Minlon**
    modification, 0473
  Ing-Manske procedure, 0473
  with sterically hindered ketones, 0473
Woodward modification, to the
  Prévost reaction, 0303
Woodward-Hoffman rules, 1011,
  1012
  and **Cope rearrangement**, 1126
Wurtz coupling, with alkenyl
  organolithium reagents, 0678
Wurtz reaction
  and alkyl halides, 0678
  and **Grignard reagents**, 0640
  and organocuprates, 0715
  and organolithium reagents, 0678
  and the **Kharasch reaction**, 0679

X

xanthate esters. *See* esters
xanthates esters
  and **Chugaev elimination**, 0167
  pyrolysis, 0167
xanthates
  and radical reduction, 1275
  and the **Barton deoxygenation**, 0467
  and the **Barton-McCombie** reaction,
    0168, 0467, 1275
xanthatin, 0159
X-ray crystallography, 0065
  and ethyllithium, 0671
  and methyllithium, 0671
  and phosphorus ylids, 0730
  and synthesis, 0902, 0903
  isopropyllithium , 0671
  LDA, 0789
  lithium pinacolate, 0793
  of carbocations, 1168
  OsO$_4$-dihydroquinidine complex,
    0299
  tert-butyllithium and sparteine, 0672,
    0673
xylopinene, 0976, 0977
xyloside derivatives, 0851

SpartanModel
An Electronic Model Kit

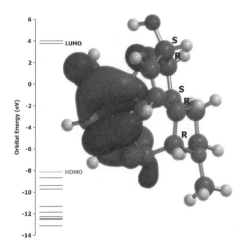

SpartanModel is the 21st century version of the "plastic model kit". While both support students translating 2D chemical drawings into 3D structure portrayals, *SpartanModel* uses quantum mechanics to go far beyond structure. Energies are available for determining the most stable isomer or conformer and for assigning reactions as endothermic or exothermic. Dipole moments, atomic charges, and electrostatic potential maps help distinguish between polar and non-polar molecules. Molecular orbitals and orbital energy diagrams assist prediction of chemical reactivity.

SpartanModel allows you to easily build molecules with dozens of atoms, assign R/S chirality and to compare molecules to emphasize bonding or molecular size and shape. It can seamlessly access and display structures from the ~70,000 experimental biological macromolecules in the on-line Protein Data Bank (PDB).

Getting started with *SpartanModel* is easy. "Hands-on" tutorials can be completed in an hour and build up the experience needed to tackle the set of 68 problems keyed to *Organic Synthesis* as well as an additional set of ~150 organic chemistry problems.

••• Download your copy at *www.wavefun.com/OrganicSynthesisDownload.html* •••

Click on the appropriate link to download *SpartanModel* for your machine (Windows or Macintosh).

Windows

- Choose to save the download on your desktop.

- When the download completes, double click on the executable to begin installation.

- Agree to the license agreement, and follow the installation prompts. When prompted, enter the 21-digit activation code below to activate your *SpartanModel* license.

Macintosh

- Download the disk image, double-click to open, and follow the installation directions.

- Upon opening the program for the first time, agree to the license agreement and enter the 21-digit activation code below to activate your *SpartanModel* license.

If there is no silver foil covering the activation code, it may already have been redeemed, and therefore may no longer be valid. In that case, you can purchase an activation code by e-mailing sales@wavefun.com.